AF746830

saur

Who's Who in the Soviet Union

A biographical encyclopedia of 5,000 leading personalities in the Soviet Union

Edited by
Borys Lewytzkyj

K · G · Saur
München · New York · London · Paris 1984

CIP-Kurztitelaufnahme der Deutschen Bibliothek
Who's who in the Soviet Union : a biograph. encyclopedia of 5000 leading personalities in the Soviet Union / ed. by Borys Lewytzkyj. – München ; New York ; London ; Paris : Saur, 1984.
ISBN 3-598-10467-7
NE: Lewytzkyj, Borys [Hrsg.]

Copyright © 1984 by K · G · Saur Verlag
Printed in the Federal Republic of Germany

All rights reserved. No part of this publication may be reproduced, stored in a retrieval system, or transmitted in any form or by any means, electronic, mechanical, photocopying, recording, or otherwise, without permission in writing from the publisher.

Cover design by Manfred Link, München
Phototype-setting by Druckerei Appl, Wemding
Printed and bound by Druckerei Pustet, Regensburg

ISBN 3-598-10467-7

Contents

Kazakh SSR

Georgian SSR

Azerbaidzhan SSR

Lithuanian SSR

Moldavian SSR

Latvian SSR

Foreword to Who's Who in the Soviet Union

Who's Who in the Soviet Union provides biographical information otherwise generally unavailable in the West about leading Soviet figures in all fields of activity. This more narrowly focussed edition not only updates the first edition of 1978 (which covered leading figures in all the Socialist countries), but also makes the information more accessible through indexes which tell immediately who is what where.

The biographical information given attempts to be as complete and accurate as possible. It was provided by the private archives of the office of B. Lewytzkyj in Munich, which are themselves based on the most reliable Soviet sources available.

All changes in the biographies of Party, State, Komsomol, and Trade Union leaders and functionaries between the end of 1976 and 1982 are fully noted. Less completely noted are changes in the biographies of authors, artists, and other groups since no new reference work dealing with them has appeared since 1976, nor are they referred to in the new *Great Soviet Encyclopedia*. However, as much information as possible has been gathered from other sources.

Especially valuable in this edition is a new feature which arranges names of party, government and military and other leaders according to their areas of activity. Thus one can see immediately, for example, who the members of the Central Committee are, or who commands the Baltic Military District, and can thereby selectively use the biographical data. Such information is, in general, extremely valuable in shedding light on the 'grey' area of knowledge about Soviet leaders and functionaries, and this type of index makes it readily accessible.

Also new are alphabetically arranged name-indexes of authors, artists, scientists, etc., which provide a useful guide to and overview of those active in these areas. Names of people who died, retired, or were dismissed since the publication of the first edition have been similarly listed, and the biographies of their successors incorporated in the main entry section.

It is hoped that the updated information and the new features will be of value to all those engaged in research on the Soviet Union.

Borys Lewytzkyj

Abbreviations and Terms

A

acad.	academy, academic
admin.	administration, administrative
Ag.	August
a.i.	ad interim
alt.	alternate
amb.	ambassador
anniv.	anniversary
Antifa	Antifaschistischer Kampfbund (Anti-Fascist Fighting Union)
Ap.	April
approx.	approximately
asscn.	association
assoc.	associate
ASSR	Autonomous Soviet Socialist Republic
asst.	assistant
AWP	Albanian Workers' Party

B

b.	born
B.A.	Bachelor of Arts
BCP	Bulgarian Communist Party
B.L.	Bachelor of Laws
brig.	brigadier, brigade

C

cand.	candidate
capt.	captain
CCP	Cuban Communist Party
CDU	Christlich-Demokratische Union (Christian Democratic Party, FRG)
cen.	central
CERN	Conseil Européen pour la recherche nucléaire (European Council for Nuclear Research)
ČKD	Českomoravská Kolben Daněk, Engineering Works, CSSR
cl.	class
col.	colonel
comm.	commission
commr.	commissar, commissioner
COMECON	Council for Mutual Economic Aid
Comintern	Communist International
confed.	confederation
CP	Communist Party
CPC	Communist Party of China
CPSU	Communist Party of the Soviet Union
CPY	Communist Party of Yugoslavia
CSR	Czech Socialist Republic
cttee.	committee

D

d.	daughter
D.	December
D.D.	Doctor of Divinity
D.Econ.	Doctor of Economics
decor.	decoration
DEFA	Deutsche Film-Agentur (German Film Agency, GDR)
del.	delegate, delegation
dept.	department, departmental
Dipl.(-)Ing.	Graduate Engineer
Dipl.-Landwirt	Graduate Agricultural Scientist
Dipl.-Volkswirt	Graduate Economist
dir.	director
div.	division
D.Math.	Doctor of Mathematics
DP	Democratic Party, Poland
Dr.Agr.	Doctor of Agriculture
Dr. ès Lettres	Doctor of Letters
Dr.habil.	Doctor with university teaching qualification
Dr.h.c.	Honorary Doctor
Dr.(-)Ing.	Doctor of Engineering
Dr.iur. (jur.)	Doctor of Laws
Dr.med.	Doctor of Medicine
Dr.med.vet.	Doctor of Veterinary Medicine
Dr.oec.	Doctor of Business Management
Dr.of Agric.Sc.	Doctor of Agricultural Sciences
Dr.of Arch.Sc.	Doctor of Archaeological Sciences
Dr.of Biol.Sc.	Doctor of Biological Sciences
Dr.of Chem.Sc.	Doctor of Chemical Sciences
Dr.of Econ.Sc.	Doctor of Economic Sciences
Dr.of Geog.Sc.	Doctor of Geographical Sciences
Dr.of Geol.-Min.Sc.	Doctor of Geological and Mineralogical Sciences
Dr.of Hist.Sc.	Doctor of Historical Sciences
Dr.of Math.Sc.	Doctor of Mathematical Sciences
Dr.of Mil.Sc.	Doctor of Military Sciences
Dr.of Ped.Sc.	Doctor of Pedagogical Sciences
Dr.of Phil.Sc.	Doctor of Philosophical Sciences
Dr.of Philol.Sc.	Doctor of Philological Sciences
Dr.of Phy.Sc.	Doctor of Physical Sciences
Dr.of Phy.-Math.Sc.	Doctor of Physical and Mathematical Sciences
Dr.of Tech.Sc.	Doctor of Technical Sciences
Dr.paed.	Doctor of Pedagogy
Dr.phil.	Doctor of Philosophy
Dr.rer.agr.	Doctor of Agriculture
Dr.rer.nat.	Doctor of Natural Sciences
Dr.rer.oec.	Doctor of Economics
Dr.rer.pol.	Doctor of Political Sciences

Dr.sc.nat.	Doctor of Natural Sciences
Dr.sc.oec.	Doctor of Economic Sciences
Dr.sc.pol.	Doctor of Political Sciences
Dr.techn.	Doctor of Technical Sciences
Dr.theol.	Doctor of Theology
D.Sc.	Doctor of Sciences

E

E.	East
econ.	economic, economics
ed.	editor, edition, edited
educ.	education, educated
EEC	European Economic Community
elem.	elementary
emb.	embassy
Engl.	English
exam.	examination
exec.	executive

F

F.	February
fed.	federation, federal, federative
FRG	Federal Republic of Germany
FZO	Fabrichno-zavodskoe obuchenie (Factory School, USSR)

G

GB	Great Britain
GDR	German Democratic Republic
gen.	general
GmbH	Gesellschaft mit beschränkter Haftung (Limited Liability Company, FRG)
GOSPLAN	Gosudarstvenii planovii komitet (State Planning Commission, USSR)
GOSSTROI	Gosudarstvennii komitet po delam stroitel'stva (State Committee for Construction, USSR)
gouvt.	gouvernement (administrative and territorial district in Russia during Tsarist regime)
govt.	government
grad.	graduation, graduated
gromada	smallest administrative unit in Poland
gymnasium	secondary school, grammar-school

H

HQ	headquarters
HSWP	Hungarian Socialist Workers' Party

I

ind.	industry, industrial
Ing.Arch.	Graduate Architect
Ing.Chem.	Graduate Chemist
insp.	inspector
inst.	institute
int.	international

J

Ja.	January
Je.	June
Jl.	July

K

K.C.M.G.	Knight Commander of St. Michael and St. George (British order)
KGB	Komissariat gosudarstvennoi bezopasnosti (Commissariat for State Security, USSR)
km.	kilometre(s)
kolkhoz	collective farm, USSR
KPD	Kommunistische Partei Deutschlands (Communist Party of Germany)
Krai	administrative unit larger than an Oblast, USSR

L

Landtag	diet, parliament of a Land, FRG,
LC	League of Communists, Yugoslavia
LCY	League of Communists of Yugoslavia
LDP	Liberal-Demokratische Partei (Liberal Democratic Party, GDR)
L. en D.	Licentiate of Laws
Lic.de Philos.	Licentiate of Philosophy
Lic.ès Lettres	Licentiate of Letters
lieut.	lieutenant
LL.B.	Bachelor of Laws
LL.D.	Doctor of Laws
LL.M.	Master of Laws

M

M.A.	Master of Arts
maj.	major
mem.	member
Mgr.Ing.	Master Engineer
mil.	military
min.	minister, ministry
movt.	movement
Mr.	March
MTS	Mashino-traktornaia stantsia (machine and tractor station, USSR)
My.	May

N

N.	North, November
Narodno Sabranie	National Assembly, Bulgaria
nat.	national, nationalist
NDP	Nationaldemokratische Partei Deutschlands (National Democratic Party of Germany)
NLF	National Liberation Front
nr.	near
NSDAP	Nationalsozialistische Deutsche Arbeiterpartei (German National Socialist Party)

O

O. — October
Oberschule — secondary school, FRG, GDR
oblast — basic administrative and territorial unit, USSR
okrug — administrative and territorial subdivision of an Oblast or Autonomous Republic, USSR
ord. — ordinary
org. — organization, organizational

P

PAX — Pro-Communist Catholic group, Poland
perm. — permanent
Ph.D. — Doctor of Philosophy
pol. — policy, politics, political
pres. — president, presidential
prod. — production
prof. — professor, professorial
publ. — published, publication
PUWP — Polish United Workers' Party
PWP — Polish Workers' Party

R

raion — small administrative and territorial division, USSR
R.C. — Roman Catholic
RCP — Rumanian Communist Party
RCP(B) — Russian Communist Party (Bolsheviks)
realgymnasium — type of secondary school in German-speaking countries
reg. — regional
regt. — regiment, regimental
rep. — representative, representation
resp. — respectively
Rs.Dr. — Doctor of Political Sciences
RSFSR — Russian Soviet Federative Socialist Republic
RWP — Rumanian Workers' Party

S

s. — son
S. — September, South
SA — Sturmabteilung (Stormtroopers, Nazi Germany)
SALT — Strategic Arms Limitation Treaty
sec. — secretary
secr. — secretariat
SED — Sozialistische Einheitspartei Deutschlands (Socialist Unity Party, GDR)
Sejm — Parliament, Poland
sen. — senior
soc. — society
sov. — soviet
sovkhoz — state farm, USSR
SPD — Sozialdemokratische Partei Deutschlands (Social Democratic Party of Germany)
SRR — Socialist Republic of Rumania
SS — Schutzstaffel (elite paramilitary corps in Nazi Germany)
SSR — Slovak Socialist Republic, Soviet Socialist Republic
suppl. — supplement
supr. — supreme

T

Tanjug — Yugoslav News Agency
TASS — Soviet News Agency
tech. — technical
temp. — temporary
Th.D. — Doctor of Theology

U

UK — United Kingdom
UN(O) — United Nations (Organization)
UN ECE — United Nations Economic Commission for Europe
UN ECOSOC — United Nations Economic and Social Council, New York
UNESCO — United Nations Educational, Scientific and Cultural Organization
UNIDO — United Nations Industrial Development Organization, Vienna
univ. — university
UPP — United Peasant Party, Poland
US(A) — United States of America
USPD — Unabhängige Sozialdemokratische Partei Deutschlands (Independent Social Democratic Party of Germany)
USSR — Union of Soviet Socialist Republics

V

VEB — Volkseigener Betrieb (state-owned enterprise, GDR)
Voivode — head of a Voivodship, Poland
Voivodship — province, administrative district in Poland
vol. — volume
volost — rural soviet, USSR
VVB — Vereinigung Volkseigener Betriebe (federation of state-owned enterprises, GDR)
VVO — Vaterländischer Verdienstorden (Fatherland Order of Merit, GDR)
VWP — Vietnam Workers' Party

W

W. — West
WHO — World Health Organization
WMO — World Medical Organization

Y

YMCA — Young Men's Christian Association

Z

ZNAK — "Sign", group of Catholic deputies in the Sejm, Poland

Biographies

Aarela (Roos), Reet Iakovlevna
Graphic artist; b. 1912, Gatchino, Petersburg Oblast.
Attended Tallinn School of Arts 1927–33, Inst. of Fine Arts 1945–49, pupil of E. Adamson-Erik, B. Lukats and P. Lukhtein.

Aavik, Priidu Eduardovich
Painter; b. 1905, Novgorod.
Studied with A. Laikmaa 1922–32, student at Pallas School of Arts in Tartu 1934–38; exhibitions 1938–; mem. CPSU 1948–.

Abaev, Vasilii Ivanovich
Philologist; b. D. 22, 1899 (Ja. 3, 1900), Kobi, Georgia.
Osset; grad. Leningrad Univ. 1925; specialist in Iranian languages.
Publs.: "Jazyk kak ideologija i jazyk kak technika" in "Jazyk i myšlenie" vol. 2, Moscow/Leningrad 1934, "Ponjatie ideosemantiki" ibid., vol. 11, Moscow/Leningrad 1948, "Nartovskij ėpos" 1945, "Osetinskij jazyk i fol'klor" 1949, "Istoriko-ėtimologičeskij slovar' osetinskogo jazyka" vol. 1, A-K, 1958, "Skifoevropejskie izoglossy" Moscow 1965, "Le cheval de Troie, P.", Annales 6, 1963.

Abakirov, Dzhumabek Abdylasovich
State official.
Min. of Justice Kirghiz SSR D. 1980–; mem. Auditing Comm. Kirghiz CP 1981–.

Abakirov, Emilbek
Trade union official.
1st City Sec. of Kirghiz CP in Osh 1970–73; mem. Cen. Cttee. 1971–, cand. mem. Bureau Cen. Cttee. Kirghiz CP 1976–; Chair. of Kirghiz Republican Trade Union Council D. 1973–; Dep. to Supr. Sov. Kirghiz SSR 1975–.

Abakumtsev, Nikolai Il'ich
Sculptor; b. 1905, Moscow.
Studied Higher State Studios for Applied Arts with B. D. Korolev, N. P. Krymov, D. A. Shcherbinovskii 1921–24, Inst. of Musicology Moscow 1924–30; exhibitions 1935–.

Abashidze, Grigor Grigor'evich
Poet; b. Ag. 1, 1913, Chiatura, Georgia.
Grad. Tbilisi Univ., Dept. of Philology 1936; mem. CPSU 1944–; cand. mem. 1964–71, mem. Cen. Cttee. Georgian CP 1971–; 1st Sec. and Chair. of Georgian Branch of Union of Sov. Writers 1967–; Sec. of USSR Writers' Union 1970–. Decors.: Stalin Prize, Hero of Socialist Labour 1974, 2 orders and various medals.
Publs.: "Vragi" (Enemies) 1941, "Poedinok tankov" (Tank Duel) 1941, verse cycles "Lenin v Samgori" (Lenin in Samgori), "Na iuzhnoi granitse" (On the Southern Border), 1949, poems "Georgii shestoi" (George the Sixth) 1942, "Legenda o pervykh tbilistsakh" (About the First Men of Tbilisi) 1959, "Dolgaia noch'" (A Long Night) 1963, "Lasharela" (in Russian translation) 1965, and others.

Abashidze, Iraklii Vissarionovich
Poet; b. 1909, Khoni, Georgia.
Grad. State Univ. of Georgia; mem. CPSU 1939–; managing ed. of various periodicals 1930–53; 1st Sec. of Writers' Union of Georgian SSR 1953–64, Chair. 1964–67; mem. Cen. Cttee. Georgian CP 1954–71, and 1981–, cand. mem. 1971–81; mem. Georgian Acad. of Sciences 1960–, Vice-Pres. 1970–76; Dep. to and Chair. of Supr. Sov. Georgian SSR 1971–. Decor.: Order of Lenin.
Publs.: collection "Novye stikhi" (New Poems) 1938, poems "Kapitan Bukhaidze" (Captain Bukhaidze), "Pesnia ranenogo" (Song of a Wounded Man), "Kogda priblizhaeshsia k Brandenburgu" (As You Approach Brandenburg) 1941–45, "Gimn Partii" (Hymn to the Party), "Proshchanie s Tsulukidze" (Parting from Tsulukidze), and others.

Abdrakhimov (Abdurakhimov), Adel Khakimovich
Sculptor; b. 1915, Eniseisk.
No higher education; mem. CPSU 1947–; instructor at Childrens' Art School in Krasnoiarsk 1960–; exhibitions 1944–.

Abdrakhimova, D. E.
Trade union official.
Sec. of Kazakh Republican Trade Union Council D. 1978–.

Abdrakhmanov, S. A.
Komsomol official.
1st Sec. and mem. Bureau of Cen. Cttee. Kazakh Komsomol Mr. 1982–.

Abdrakmanov, Sadyrbek
State official.
Cand. mem. Cen. Cttee. Kirghiz CP 1966–71, mem. 1971–; Min. of Supplies Kirghiz SSR S. 1982–.

Abdukadyrov, Kalmakan Abdukadyrovich
Author; b. 1903, Kazakhstan.
Publs.: "Batrak" (The Farm Hand) 1928, "Tempy" (Tempi) 1932, "Shakhter" (The Miner) 1934, "Collected Verse" 1948, collection "Kliatva" (The Oath) 1945, novelette "Komsomol'skoe zveno" (The Komsomol Team) 1948, children's books "Amantai", "Sladkie dyni" (Sweet Musk-Melons) 1936, and others.

Abdulkhalyk, Abdul Ali ogly
Painter; b. 1920, Baku.
Grad. Azerbaidzhan School of Arts 1939; studied Surikov Art Inst. Moscow 1944–50; teacher at Azerbaidzhan School of Arts Baku 1950–; exhibitions 1951–.

Abdullaev, Abdulakhad
State official.
Chair. of State Cttee. for Cinematography. Uzbek SSR Ja. 1974–; cand. mem. Cen. Cttee. Uzbek CP 1976–81.

Abdullaev, Abdulkhak Aksakovich
Painter; b. 1918, Turkestan.
Studied Tech. School of Art Samarkand with P. P. Benkov, L. A. Bure, Z. M. Kovalevskii 1931–36; studied Inst. of Fine Arts Moscow with V. P. Efanov and A. A. Osmerkin; exhibitions 1934–.

Abdullaev, Agalar Bakhman ogly
Painter; b. 1923, Baku.
Grad. Azerbaidzhan School of Art 1944; studied Surikov Art Inst. Moscow with V. P. Efanov; exhibitions 1945–.

Abdullaev, Gasan Mamed Bagir ogly, Dr. of Phy.-Math. Sc.
Scientist; b. 1918, Azerbaidzhan.
Mem. CPSU 1942–; grad. Azerbaidzhan Pedagogical Inst. 1941; physicist; doctoral degree in physical-mathematical sciences; Prof.; Pres. of Acad. of Sciences Azerbaidzhan SSR and corresponding mem. USSR Acad. of Sciences 1970–; mem. staff. Inst. of Physics at Acad. of Sciences Azerbaidzhan SSR 1949–50, 1954–, Dir. 1957–; doctoral cand. at Physical-Technical Inst. of USSR Acad. of Sciences 1950–53; Dep. to USSR Supr. Sov. 1970–.

Abdullaev, Lutfulla Abdullaevich
Painter and designer; b. 1912, Mirzachul, Tashkent Oblast; brother of S. A. Abdullaev.
Studied Techn. School of Arts Samarkand with P. P. Benkov, L. A. Bure, Z. M. Kovalevskii 1930–35; exhibitions 1934–.

Abdullaev, Mikail Gusein ogly
Painter; b. 1921, Baku.
Studied Azerbaidzhan Tech. School of Arts with S. Bakhlulzade 1935–39 and Surikov Art Inst. Moscow with S. V. Gerasimov, V. A. Favorskii, I. I. Chekmazov 1939–49; mem. CPSU 1954–; corresponding mem. USSR Acad. of Arts 1958–; exhibitions 1940–. Decor.: Order of Lenin.

Abdullaev, Sagdulla Abdullaevich
Painter; b. 1916, Mirzachul, Tashkent Oblast; brother of L. A. Abdullaev.
Grad. Tech. School of Arts Samarkand 1937, student of P. P. Benkov; exhibitions 1937–.

Abdullaev, Sagid Nukhkadievich
Author and philologist; b. 1903, Mugi, Dagestan.
Publs.: poems "la veruiu" (I Believe) 1930, "Zhaloba Mully" (Mulla's Lament) 1925, "Proch, soroka" (Shoo, Magpie!) 1929, "Frontovye ocherki" (Sketches of the Front) 1948, works on the research on the Dagestan language.

Abdullaev, Samyg Faizullaevich
Painter; b. 1917, Tashkent.
Grad. Tashkent School of Arts 1937, pupil of A. N. Volkov and N. G. Karakhan; grad. Inst. of Theatre and Art Tashkent

1960; mem. CPSU 1943–; exhibitions 1944–. Decor.: Hero of the Soviet Union.

Abdullaeva, Geiat Gamdullaevna
Sculptor; b. 1912, Derbent.
Studied Azerbaidzhan School of Arts with P. V. Sabsai, Repin Inst. for Painting, Sculpture and Architecture of USSR Acad. of Arts 1945–50; exhibitions 1948–.

Abdullaeva, Rano Khabibovna, Cand. of Hist. Sc.
State official; b. 1935.
Uzbekian; grad. Higher Party School in Tashkent, mem. CPSU 1955–; intermediate school teacher, 1st sec. of a raion cttee. Uzbek Komsomol, dept. head of a raion cttee. Uzbek CP 1955–61; Sec. and mem. Bureau of Cen. Cttee. Uzbek Komsomol 1961–62; 1st Sec. of Samarkand Oblast Cttee. Uzbek CP 1963; 1st Sec. and mem. Bureau of Cen. Cttee. Uzbek Komsomol 1963–71; mem. Cen. Cttee. Uzbek CP 1966–; Dep. to and mem. Mandate Comm. at Sov. of Nationalities USSR Supr. Sov. 1966–74; cand. mem. Bureau Cen. Cttee. USSR Komsomol 1967–71; Head of Cultural Dept. at Cen. Cttee. Uzbek CP 1971; Vice-Chair. of Council of Mins. Uzbek SSR Jl. 1971–.

Abdullin, Ibragim Akhmetovich
Author and dramatist; b. 1920, Zirikle, Bashkir ASSR; s. of a peasant.
Mem. CPSU 1947–; grad. Gorkii Inst. of Literature Moscow 1957; author and dramatist in Bashkir ASSR; works publ. 1938–.
Publs.: stories: "On byl s Urala" (He Came from the Ural Region) 1946; plays: "Glubokoe dykhanie" (Deep Breathing) 1947, "My ne rasstanemsia" (We Shall Not Part) 1949, "Prekrasnye doliny Ak-Ideli" (The Beautiful Valleys of Ak-Idel) 1950, "General Shaimuratov" 1958, and others.

Abdulov, Boris Nikolaevich
Painter; b. 1926, Tashkent.
No formal art educ.; learned painting in studios of Painters' Asscn. and branch of Painters' Union Kirov; mem. CPSU 1955–; active in Kirov TV studios 1958–.

Abdulzhalilov, Fazil Apasovich
Author and poet; b. 1913, Kangly, Stavropol Krai.
Grad. Cherkess Tech. Pedagogical School 1932; mem. CPSU 1940–.
Publs.: poems "Shumit Kuban'" (Kuban' Surges) 1955, "Asantai" 1956, novel "Burnyi potok" (The Raging Stream) 1959, and others.

Abdumomunov, Toktobolot
Author; b. 1922, Ataika, Kirghiz SSR; s. of an office worker.
No completed higher educ., studied medicine; medical aid at a raion health dept., head of a medical care station 1938–41; proof-reader, Dept. Ed. of Kirghiz State Press, translator on staff of various republican newspapers and periodicals 1941–42; literary activities 1942–; mem. CPSU 1959–; cand. mem. Cen. Cttee. 1960–61, mem. Cen. Cttee. Kirghiz CP 1961–; Dep. to USSR Supr. Sov. 1962–66; Chair. of Kirghiz Branch Novosti Press Agency 1966–; works published 1945–.
Publs.: plays "Sovremenniki" (Contemporaries) 1948, "Peschanoe predgor'e" (Sandy Foothills) 1948, novel "Pod golubym nebom" (Beneath at Azure Sky) 1956, children's books "Zhukesh" 1948, "Neraskrytaia taina" (Unsolved Secret) 1950, and others.

Abdurakhmanov, Ilia Abramovich
Painter; b. 1923, Dushanbe.
Studied Moscow Higher School of Applied Arts 1945–50; mem. CPSU 1949–; teacher at Art School Dushanbe 1950–52; exhibitions 1950–.

Abdurakhmanov, Nadir Gambar ogly
Painter; b. 1925, Lachin, Azerbaidzhan SSR.
Studied Repin Inst. of Painting, Sculpture and Architecture USSR Acad. of Arts 1947–53; teacher at Azerbaidzhan Art School 1954-62; mem. CPSU 1957–; exhibitions 1946–.

Abdurazakov, Bakhodir Abbasovich
State official.
Mem. Auditing Comm. Uzbek CP 1966–71 and 1981–; Head of Dept. of Science and Educ. Institutions of Cen. Cttee. Uzbek CP 1966–70; Sec. of Tashkent Oblast Cttee. Uzbek CP 1970–74; Min. of Foreign Affairs Uzbek SSR D. 1980–.

Abdushukurov, Telman Rustamovich
Party official.
Sec. od Samarkand Oblast Cttee. Uzbek CP 1975–78; Head of Dept. of Science and Educ. Institutions of Cen. Cttee. Uzbek CP 1978–80, of Dept. of Propaganda and Agitation Mr. 1980–; mem. Cen. Cttee. Uzbek CP 1981–.

Abdykalykov, O.
Komsomol official.
1st Sec. and mem. Bureau Cen. Cttee. Kirghiz Komsomol N. 1982–.

Abdykerimov, Annaoraz
State official.
Min. of Posts and Telecommunications Turkmen SSR Jl. 1976–; mem. Auditing Comm. Turkmen CP 1981–.

Abegian, Mger Manukovich (also Meger)
Painter and graphic artist; b. Ja. 13 (26), 1909, Echmiadzin.
People's Artist of the Armenian SSR (1960). Studied at the Yerevan Art and Industry Technicum (1922–27) under S. Agadzhanian and S. Arakelian, at the Moscow State Advanced Arts and Engineering Institute (1927–30), and at the Leningrad Academy of Arts (1930–31). His principal teacher was S. V. Gerasimov. He produced lyrical landscapes and colorful still lifes, as well as portraits and landscape scenes from everyday life. His works include *On Sevan Island* and *Midday* (both 1956), *The Fields of Biurakan* (1957), and *Lullaby* (1959), all at the Armenian Picture Gallery in Yerevan; *Summer* (1959, at the Tret'iakov Gallery); and numerous graphic easel landscapes (drawings, watercolors, etchings, and line engravings) imbued with a subtle mood. Decors.: two orders, medals.

Abelite, Olgert Mintauatovich
Graphic artist; b. 1919, Riga.
Taught in Riga by Ia. K. Liepin 1929–30, studied at Latvian Univ. with S. I. Vidberg, R. Ia. Suta 1930–34.

Abesadze, Dzhabulo Parmenovich
Sculptor; b. 1924, Kursebi, Georgia.
Studied Moscow Inst. for Applied and Decorative Arts 1950–52, and at V. I. Mukhina Higher School of Applied Arts in Leningrad with L. M. P. Pisarevskii and V. L. Simonov 1952–56; mem. CPSU 1954–; teacher at Children's Art School Kutaisi 1956–; exhibitions 1961–.

Abilev, Dikhan
Poet; b. 1907, Baian-Aulskii Raion, Pavlodar Oblast.
Grad. Inst. of Journalism Alma-Ata; active as journalist and teacher; war correspondent 1941–45; works publ. 1938–.
Publs.: poem collection "Energiia" (Energy) 1937, poems "Shalkyma" 1938, "Maidanbek" 1943, "Ot serdtsa" (From the Heart) 1945, "Tulga" 1947, novel "Serdtse Altaia" (The Heart of the Altai) 1953, poem "Ognennye volny" (Waves of Fire) 1956, and others.

Abishev, Alzhappar
Author; b. 1907, Karaganda Oblast, Kazakhstan; s. of a farmer.
Mem. CPSU 1939–. Decor.: Order of Red Banner of Labour.
Publs.: narratives and sketches "Zaval" (Burial) 1936, novellettes "Liudi, osushchetvivshie svoiu mechtu" (People Who Have Realized Their Dreams) 1938, "Za Rodinu" (For the Fatherland) 1939, "Tovarishchi" (Comrades) 1940, "Bditel'nost'" (Alertness) 1941, "Gvardiia chesti" (The Guard of Honour) 1942, play "Druzhba i Liubov'" (Friendship and Love) 1947, narratives "Tulegen Tokhtarov" 1942, "Sarzhan" 1944, plays "Zavist'" (Envy) 1955, "Kto moi otets?" (Who is my Father?) 1957, and others.

Ablesov, Sadykbek Ablesovich
State official.
Mem. Cen. Cttee. Kirghiz CP 1981–; Min. of Construction Kirghiz SSR Je. 1982–.

Aboev, Zaur-Bek Petrovich
Graphic artist; b. 1931, Zaramat, North Ossetia.
Studied Dept. of Painting and Graphic Arts of Pedagogical School in Ordzhonikidze 1947–51, and Faculty of Graphic Arts of Moscow Polygraphic Inst. with A. D. Goncharov and I. I. Chekmazov 1952–57; mem. CPSU 1957–; exhibitions 1952–.

Abol, Oiar Gustavovich, Cand. of Art Sc.
Painter and graphic artist; b. 1922, Riga. Learned painting in studio of R. Ia. Suta 1938–40, studied at Acad. of Arts of Latvian SSR with O. E. Skulme 1946–51; mem. CPSU 1947–; completed lecturer training at Repin Inst. of Painting, Sculpture and Architecture of USSR Acad. of Arts 1955; Cand. of Art Sc. 1956–; exhibitions 1951–.

Abolin (Abolinsh), Mikhail Andreevich
Stage painter; b. 1913, Kurshavka Cossak village, Stavropol Krai.
Studied Kiev Art Inst. with A. G. Petritskii 1933, and at Repin Inst. of Painting, Sculpture and Architecture USSR Acad. of Arts with M. P. Bobyshov 1933–37; stage painter in Riga.

Abov, Iurii Georgevich, Dr. of Phy.-Math. Sc.
Physicist; b. 1922.
Mem. CPSU 1955–; grad. Moscow Univ. 1947; Dr. of Physical-Mathematical Sciences 1969; employed at Inst. of Theoretical and Experimental Physics 1947–, Sen. Research Asst. 1963–, Chief of Laboratory. Decor.: Lenin Prize 1974.

Abraamian, Artashes Saakovich (also Abramian)
Painter; b. 1921, Leninakan.
Mem. CPSU 1944–; studied Art. Inst. Erevan with A. A. Avetisian and E. A. Isabekian 1945–52; exhibitions 1952–.

Abraamian (Abramian), Elena (Egine) Levonovna
Painter; b. 1912, Turkey.
Studied Repin Inst. for Painting, Sculpture and Architecture USSR Acad. of Arts 1934–40; exhibition 1940–.

Abramavichius, Kazis Valdemaras Antano
Graphic artist; b. 1928, Kukautuve, Lithuania. Studied Art Inst. of Lithuanian SSR 1947–54; mem. CPSU 1954–; exhibitions 1956–.

Abramenkov, Iakov Vasil'evich
Painter; b. 1914, Dnepropetrovsk.
Attended Workers' Faculty of Arts in Dnepropetrovsk 1931–33, and courses in Krupskaia Cen. House of Folk Art 1956–59; pupil of M. N. Panin and N. T. Sorokin; teacher in an intermediary school 1956–; exhibitions 1952–.

Abramov, Aleksandr Arushanovich
Sculptor; b. 1911, Shusha.
Studied Art Inst. Kharkov with L. A. Blokh 1929–32; exhibitions 1940–.

Abramov, Aleksei Fedorovich
Painter; b. 1920, Iaroslavl.
Studied Art School Iaroslavl with A. A. Churin and S. F. Shitov 1936–40; exhibitions 1950–.

Abramov, Artur Aleksandrovich
Sculptor; b. 1930, Tashkent; s. of Aleksandr Arushanovich Abramov.
Studied Art School Kharkov with L. A. Tverdianskaia 1947–55, Moscow Pedagogical Oblast Art School and Leningrad Sychev Art School; sculptor in Ashkhabad; exhibitions 1950–.

Abramov, Fedor Aleksandrovich
Author; b. F. 29, 1920, Verkola, Arkhangelsk Oblast.
Russian; mem. CPSU 1945–; grad. Dept. of Philology of Leningrad Univ. 1948; Head of Chair of Sov. Literature at Leningrad State Univ. in the fifties. Decors.: Order of the Patriotic War, various medals.
Publs.: "Bratia i sestry" (Brothers and Sisters) 1958, "Bezottsovshchina" (Fatherless) 1961, "Dve zimy i tri leta" (Two Winters and Three Summers) 1968, "Pelageia" 1969, and others.

Abramov, Georgii (Iurii) Dmitrievich
Painter; b. 1914, Tula.
Studied Workers' Faculty at Inst. of Proletarian Pictorial Art in Leningrad 1930–33; painter in Kaluga; exhibitions 1938–.

Abramov, Kuz'ma Grigor'evich
Author; b. 1914, Starye Naimany, Mordovian ASSR; s. of a peasant.
Works publ. 1935–, amongst others: collection of poems "Stikhi" 1940, plays "Novaia sila" (New Strength) 1951, "Sorevnovanie" (Competition), "Traktoristy" (Tractor Driver), novelette "Rastushchaia sila" (Increasing Power) 1950, novel "Naiman" 1957.

Abramov, Nikolai Alekseevich
Painter; b. 1930, Leningrad.
Studied Leningrad School of Art 1945–50 and Repin Inst. for Sculpture, Painting and Architecture Leningrad 1950–57.

Abramov, Nikolai Nikolaevich
Scientist; b. F. 10 (23), 1901, Moscow.
Mem. CPSU 1945; grad. Engineering and Construction Faculty of Higher Tech. Inst. Moscow 1956; Prof. and Prof. Chair. for Water Supply at Inst. of Civil Engineering Moscow 1942–. Decors.: Order of Lenin and several other decors.
Publs.: "Osnovy proektirovanija i rasčeta setej gorodskich vodoprovodov" Moscow/Leningrad 1940, "Obosnovanija i metody zonirovanija vodoprovodnych setej" Moscow/Leningrad 1949, "Vodosnabženie" Moscow 1967.

Abramov, Vladimir Efremovich
State official.
Head of Dept. of Construction and Communal Econ. with Cen. Cttee. Turkman CP 1966–70; 1st Dep. Chair of GOSPLAN Turkmen SSR 1970–76; Chair. D. 1977–; Dep. Chair of Council of Mins. Turkmen SSR Jl. 1976–. mem. Auditing Comm. 1966–71, cand. mem. Cen. Cttee. Turkmen CP 1971–77, mem. D. 1977–.

Abramova, Sofia Alekseevna
Sculptor; b. 1917, Boguchar, Voronezh Oblast.
Studied Repin Inst. for Sculpture, Painting and Architecture Leningrad 1952–58; pupil of D. P. Schwarz; exhibitions 1957–.

Abrasimov, Petr Andreevich
Diplomat; b. 1912, Belorussia.
Mem. CPSU 1940–; electrical mechanic until 1941; various trade union and local Sov. posts in Vitebsk and Brest; served in Sov. Army 1941–45; Perm. Rep. of Belorussian Council of Mins. to USSR Council of Mins. 1946–48; grad. Belorussian State Univ. 1951; 1st Dep. Chair. of Belorussian Council of Mins. 1948–50, 1952–55; Dep. to Foreign Affairs. Comm. of USSR Supr. Sov. 1950–58, Dep. and mem. 1962–66; Sec. of Cen. Cttee. Belorussian CP 1950–52, 1955–56; Dep. to Belorussian Supr. Sov. 1955–59; mem. Bureau Cen. Cttee. Belorussian CP 1956, mem. Cen. Cttee. Belorussian CP 1956–58; Sov. Amb. to Poland 1957–61; 1st Sec. of Smolensk Oblast CPSU Cttee. 1961–62; mem. Cen. Cttee. CPSU 1961–; Sov. Amb. to the GDR 1962–71, to France 1971–73, to the Republic of Madagascar 1972–73; Head of CPSU Cen. Cttee. Dept. for Liaison with Communist Parties of Socialist Countries Ap. 1973–75; Sov. Amb. to the GDR Mr. 1975–. Decors.: Order of Lenin, Order of Red Banner, Order of Patriotic War 2nd Cl., Order of Red Star, Hero of Socialist Labour.

Abrikosov, Aleksei Alekseevich
Physicist; b. Je. 25, 1928.
Grad. Moscow Univ. 1948; corresponding mem. USSR Acad. of Sciences 1964–; mem. staff. Inst. of Theoretical Physics with USSR Acad. of Sciences 1965–; Prof. at Moscow Univ. 1965–.
Publs.: "Metody kvantovoj teorii polja v statističeskoj fizike" Moscow 1962 (co-author), "O magnitnyh svojstvah sverhprovodnikov vtoroj gruppy" in "Žurnal èksperimental'noj i teoretičeskoj fiziki" (32, 1442) 1957, "Teoriia èlektronnogo ènergitčeskogo spektra metallov rešëtkoj tipa vismuta" (ibid. 43, 1089) 1962 (co-author), "On the Quantum Theory Fields" in "Nuovo cimento" (supplement to 3, series 10, 80) 1956 (co-author).

Abseliamov, Momet
Painter; b. 1911, Simferopol.
Attended N. S. Samokish's Studio in Simferopol 1929–35; exhibitions 1947–.

Abugov, Miron Efimovich
Sculptor; b. 1921, Ashkhabad.
Studied Repin Inst. for Painting, Sculpture and Architecture Leningrad 1948–52, pupil of I. V. Krestovskii.

Abulgasan (Alekper-Zade, Abulgasan Ali Baba ogly)
Author; b. 1906, Azerbaidzhan; s. of a craftsman.
Grad. at a pedagogical inst.; mem. CPSU 1941–; works publ. 1927–.
Publs.: "Mir rushitsia" (The World is Collapsing) 1933, "Synovia i ottsy" (Fathers and Sons) 1944, "Na beregu" (On the Shore) 1943, "Voina" (War) novel, vols. 1–3 1948–55, and others.

Abutkova, Natalia Vsevolodovna
Painter; b. 1916, Verkhodvorie, Viatka Oblast.
Attended Omsk Art School 1935–39 and Kiev Inst. of Art 1945–51; pupil of A. A. Shovkunenko and T. N. Iablonskii.

Abzianidze, Georgii Nikolaevich
Literary historian; b. 1907, Kutaisi; s. of a teacher.
Grad. Philological Faculty of Tbilisi Univ. 1929.
Publs.: "Akakii Tsereteli. Zhizn' i tvorchestvo" (Akakii Tsereteli, Life and Works) 1940, collection of investigations and monographs, "Gruzinskie klassiki" (Georgian Classicists) 1950, "Ocherki po istorii gruzinskoi obshchestvennoi mysli XIX v." (Sketches on the History of Georgian Social Thought in the 19th Century), and others.

Adalian, Ruben Martirosovich
Painter and graphic artist; b. 1929, Simferopol.
Studied Art and Theatre Inst. Erevan 1956; exhibitions 1954–.

Adamanov, O.
Trade union official.
Sec. of Uzbek Republican Trade Union Council Mr. 1977–.

Adamian, Nora (real name: Adamova, Eleonora Grigor'evna)
Author; b. 1910, Baku.
Attended Baku Univ. 1927–30; worked for newspaper "Kommunist" in Erevan 1938–45; literary activity 1939–; first works published 1953.
Publs.: collections "Nachalo zhizni" (The Beginning of Life) 1955, "U sinikh gor" (In the Blue Mountains) 1957, "Devushka iz ministerstva. Povesti i rasskazy" (The Girl from the Ministry. Stories) 1959, story "Pokinutyi dom" (The Deserted House) 1959, and others.

Adamkevich, Viktor Vladislavovich
Painter and graphic artist; b. 1915, Ekaterinoslav.
Studied Art School Dnepropetrovsk 1932–36 and Kiev Art Inst. 1945–51; pupil of K. N. Eleva, K. D. Trokhimenko, I. N. Stilman; exhibitions 1951–.

Adamov, Efim Borisovich
Graphic artist; b. 1924, Moscow.
Mem. CPSU 1945–; studied Moscow Polygraphic Inst. 1950–55, pupil of A. D. Goncharov and P. G. Zakharov; aspirant at same Inst. 1955–59; exhibitions 1957–.

Adamova, Evgenia Mikhailovna
Soviet painter; b. 1913, Livny, Orel Oblast.
Attended Art School in Ashkhabad 1936–39; pupil of S. N. Begliarov and A. I. Khvorostenko; exhibitions 1940.

Adamova, Galina Vasil'evna
Artist in applied art; b. 1919, Elizavetgrad.
Studied Textile Inst. Moscow 1937–44; pupil of P. P. Pashkov and V. M. Shugaev; artist in applied art, designs patterns for furnishing fabrics taken from motifs of Georgian and Usbek ornaments; exhibitions 1945.

Adamovich, Aleksandr Mikhailovich
Author; b. 1927, Koniukhi, Minsk Oblast.
Grad. Minsk Univ.; partisan 1941–45; first article on Belorussian prose publ. 1953.
Publs.: "Belorusskii roman v stikhakh" (A Belorussian Novel in Verse), "Problemy zhanra romana" (Problems in the Genre of the Novel), "Put' do masterstva" (The Path to Mastery) 1958, novelette "Viktoriia" 1965, novel "Partizany" (Partisans) 1965, and others.

Adamovich, Sergej Faddeevich
Graphic artist and painter; b. 1922, Proletarske, Lugansk Oblast.
Mem. CPSU 1963–; attended Kiev Art Inst. 1944–50, pupil of A. A. Shovkunenko; lived in Kiev until 1955 and then in Riga; exhibitions 1949.

Adamovskaia-Andreeva, Busia Markovna
Stage designer; b. 1919, Elizavetgrad.
Attended Leningrad Art School 1941–44 and I. R. Repin Inst. for Painting, Sculpture and Architecture Leningrad 1944–51, pupil of M. P. Bobyshov.

Adashkevich, Sergej Aleksandrovich
Sculptor; b. 1918, Peretsiatki, Minsk Oblast.
Attended Vitebsk Art School 1935–37 and courses for raising the qualification of sculptors in Minsk 1937–39; exhibitions 1949–.

Adleiba, Boris Viktorovich
Party official; b. 1931.
Abkhazian; mem. CPSU 1958–; served in the Soviet Army 1951–55; 2nd, 1st Sec. Ochamchirskii Raykom Georgian Komsomol 1955–62; Head of a Section with Abkhazian ASSR Oblast Cttee. Georgian CP 1962–65; Sec. Sukhumi City Cttee. Georgian CP 1965–69; mem. Apparat Cen. Cttee. Georgian CP 1969–71; 1st Sec. Ochamchirskii Raykom Georgian CP 1971–75; 1st Sec. Tkvarcheli City Cttee. Georgian CP 1975–77; 1st Dep. Chair. Council of Mins. Abkhazian ASSR 1977–78; 1st Sec. Abkhazian ASSR Oblast Cttee. Georgian CP Ap. 1978–; cand. mem. Bureau Cen. Cttee. Georgian CP Ja. 1979–.

Adomonis, Iuozas Antonovich
Ceramic artist; b. 1932, Seirijai, Lithuania.
Studied Art Inst. of the Lithuanian SSR 1951–57, pupil of I. I. Mikenas and T. I. Mikniavichiene; exhibitions 1958–.

Advadze, Vakhtang Nikolaevich
Painter; b. 1919, Akhaltsikhe, Georgia.
Studied Acad. of Art Tbilisi 1950, pupil of S. S. Kobuladze and A. K. Kutateladze; taught at Rostov School of Art 1950–60; exhibitions 1951–. Decor.: Merited Artist of the Georgian SSR 1967.

Adylov, Naufal Ismagilovich
Sculptor; b. 1932, Kazan.
Attended Art School in Kazan 1947–51, pupil of S. S. Akhun, and Art School in Penza 1955–57, pupil of A. A. Fomin; exhibitions 1958–.

Adylov, Vakhid
Painter; b. 1916, Tashkent.
Attended Art School in Tashkent 1933–37, pupil of N. G. Karakhan; exhibitions 1947–.

Adzhiashvili, Iosif Iakovlevich
Artist in applied art; b. 1928, Bandza, Georgia.
Mem. CPSU 1944–; studied Georgian Acad. of Arts 1939 and 1946–52, pupil of Dzhaparidze; artist in applied art (pottery, procelain, crystal); exhibitions 1953–.

Adzhie, Dzhindi (real name: Dzhauari, Adzhie Dzhindi)
Author; b. 1908, Iamanchair.
Raised in orphanage in Aleksandropol, now Leninakan; grad. Pedagogical Tech. School 1929 and Pedagogical Inst. Erevan 1933; writes in Kurdish and Armenian; works published 1930–.
Publs.: "Kurdskie narodnye skazki" (Kurdish Folktales) 1940 and 1952, collection of stories "Novoe utro" (A New Morning) 1947, "Kurdskii folklor" (Kurdish Folklore) 1957, and others.

Adzhiev, Anvar Abdulgamidovich
Poet, author, translator; b. 1914, Kostek, Dagestan; s. of a peasant.
Translated works of Pushkin, Lermontov, Ketagurov, Tsadas and Stal'skii into Kumyk; works publ. 1933–. Decor.: Order of the Red Banner of Labor (twice).
Publs.: collections "Obnovlenie" (The Renewal) 1934, "Nastuplenie" (The Offensive) 1942, "Schastlivye gory" (The Happy Mountains) 1948, collected stories and verse for children.

Aerova, Basia Mendelevna
Artist in applied art; b. 1917, Iaroslavl.
Attended Faculty of Art at Moscow Textile Inst. 1935–40; pupil of M. V. Leblan and P. P. Pashkov; artist in applied art, designs patterns for furnishing fabrics and carpets; exhibitions 1956–.

Afanasev, Georgii Dmitrievich
Geologist and petrographer; b. Mr. 4 (17), 1906, Novorossiisk.
Mem. CPSU 1948–; grad. Leningrad Univ. 1930; Scientific Sec. of Dept. of Geological Sciences, Dep. Scientific Sec. of Presidium USSR Acad. of Sciences 1949–53; sen. mem. of research staff at Inst. of Geology of Ore Deposits, Petrography, Mineralogy and Geochemistry of USSR Acad. of Sciences 1956–. Decors.: Order of Lenin and several other orders and medals.

Afanasev, Sergei Aleksandrovich
State official; b. 1918.
Grad. Bauman Tech. College Moscow 1941; mem. CPSU 1943–; worked for munitions ind. 1946–57; foreman, design engineer, dep. dept. head, dep. chief mechanic at an ind. plant. 1951–56; Dep. Chair. of Leningrad Econ. Council 1957–58, Chair. 1958–61; Dep. to Supr. Sov. RSFSR 1959–67; Dep. Chair. of Council of Mins. RSFSR, Chair. of Econ. Council RSFSR 1961–65; mem. Cen. Cttee. CPSU 1961–; mem. Draft Bills Comm. at Sov. of Nationalities USSR Supr. Sov. 1962–66, Dep. 1962–; USSR Min. of General Engineering Mr. 1965–. Decor.: Order of Lenin.

Afanasev, Viktor Grigorevich, Dr. of Phil. Sc.
Journalist; b. 1922.
Mem. CPSU 1943–; grad. State Pedagogical Inst. N. G. Chernyshevskii Chita 1950; served in Sov. Army, student 1940–53; teacher 1953–54; Dep. Dir. 1954–59, Prof. Chair of Philosophy at State Pedagogical Inst. Chita 1959–60; Prof. Chair for Scientific Communism at Acad. of Sociology Cen. Cttee. CPSU 1960–68; Dep. and 1st Dep. Chief Ed. of "Pravda" 1968–74; Chief Ed. of theoretical party organ "Kommunist" Cen. Cttee. CPSU 1974–76; mem. Cen. Cttee. CPSU 1976–; Chief Editor "Pravda"; Dr. of Phil. Sc. 1964; Prof. 1970; corresponding mem. Acad. of Sciences USSR 1972.

Afanasev, Vladimir Vasil'evich
Sculptor; b. 1928, Liski, Voronezh Oblast.
Attended Leningrad Art School 1948–50; exhibitions 1954–.

Agababaev, Odtekin Iskanderovich
Painter; b. 1928, Dushanbe.
Studied Surikov Art Inst. Moscow 1955; exhibitions 1957–.

Agababaev, Tofik Mamed Ali ogly
Artist in applied art; b. 1928, Baku.
Grad. Azerbaidzhan Art School 1951 and Mukhina High School for Arts and Crafts, Leningrad 1959; mem. CPSU 1963–; artist in applied art (crystal, glass, ceramics).

Agababian, Suren Bardukhovich
Armenian literary scholar and critic; b. Ja. 22, 1922, Karaklis, now Kirovakan. Member of the CPSU since 1949. Doctor of philology. The author of monographs on A. Bakunts 1963; Russian translation (1965), two books of articles of literary criticism – *Responses* (1963) and *Literature and Contemporary Life* (1965) – and many articles on modern Armenian literature. He is also one of the authors of *The History of Soviet Armenian Literature* (vols. 1–2, 1961–65; Russian translation 1966).

Agadzhanian, Norair Gegamovich
Artist in applied art; b. 1927, Erevan.
Grad. Erevan Art and Theatre Inst. 1956, pupil of A. K. Kodzhiian; artist in applied art (ceramics); exhibitions 1955–.

Agaev, Iusuf Ismail ogly
Graphic artist; b. 1915, Shusha.
Mem. CPSU 1948–; grad. Moscow Inst. of Fine Arts 1941.

Agaian, Tsatur Pavlovich
Scholar; b. D. 30, 1911 (Ja. 12, 1912).
Grad. Historical Faculty of Moscow Inst. of Philosophy, Literature and History 1938; historian; Dir. of Armenian Branch Office of Inst. of Marxism-Leninism Cen. Cttee. CPSU 1961–68; Acad. Sec. of Sociology Dept. and mem. Acad. of Sciences Armenian SSR 1968–.

Agakov, Leonid Iakovlevich
Author and dramatist; b. 1910, Kanash, Chuvash ASSR.
Grad. at a pedagogical technical college and Faculty of Theatre Directing at State A. V. Lunacharskii Inst. for Drama in Moscow; mem. CPSU 1943–.
Publs.: stories "Lesnoi otriad" (The Forest Divison), "Posledniaia skazka" (The Last Tale), plays "Obman" (Deception) 1931, "Moshch'" (Power) 1932, novelette "Odnazhdy vesvoi" (Once in Spring) 1939, "Frontovye rebiata" (Front-Line Boys) 1943, "Partizan Murat" (The Partisan Murat) 1943, "Zolotaia tsepochka" (The Golden Chain) 1945, satiric stories "Smeiatsia khochetsia" (One Would Like to Laugh) 1955, "Kozly" (Billy-Goat) 1959, and others.

Agaltsov, Filipp Aleksandrovich
Marshal of aviation; b. Ja. 8 (20), 1900. Member of the CPSU since 1919. He was born into an impoverished peasant family in the village of Soldatskoe, Tula Oblast. He worked in the Obukhov plant from 1915 to 1919, when he joined the Soviet Army to participate in the Civil War. He graduated from the Military Political School in 1925, from the Lenin Military Political Academy in 1932, and from a pilot-training course in 1934. He fought as a volunteer in the Spanish Civil War in 1937–38.
First active in party political work, he moved to military leadership in 1940. At the start of the Great Patriotic War he commanded a bomber regiment; from 1941 to 1943 he directed a military aviation school for pilots, and from 1943 to 1945 he commanded an air attack division and the First Polish Mixed Air Corps. He was first deputy commander of the air force from 1949 to 1956 and for the next two years was air force inspector-general for the Chief Inspectorate of the Ministry of Defense. He was a deputy commander of the air force from 1958 to 1962, at which time he took command of long-range aviation. Decors.: three Orders of Lenin, five Orders of the Red Banner, the Order of Suvorov Second Class, the Order of Kutuzov Second Class, the Order of the Patriotic War First Class, the Order of the Red Star, two Polish orders, and various medals.

Agamalova, Lia Nikitichna
Artist in applied art; b. 1926, Buguruslan, Orenburg Oblast.
Grad. Moscow Inst. of Applied and Decorative Art 1952, pupil of V. S. Kondratev and N. N. Tikhonov; artist in applied art (glass); exhibitions 1957–.

Aganbegian, Abel Gezevich
Economist; b. N. 8, 1932, Tbilisi.
Corresponding member of the Academy of Sciences of the USSR since 1964. Member of the CPSU since 1956.
He was graduated from the Moscow State Institute of Economics in 1955. From 1955 to 1961 he worked for the State Committee of the Council of Ministers of the USSR on problems of labor and wages. In 1961 he became the chief of the laboratory of the Institute of Economics and Organization of Industrial Production of the Siberian Section of the Academy of Sciences of the USSR; he was named its director in 1966. His scientific activity was concerned with problems of labor productivity, wages, and the standard of living and with developing models for optimal future planning. He participated in international economic conferences.
Works: *Zarabotnaia plata v SSSR*. Moscow 1959. (With V. F. Mayer.) *Primenenie*

matematiki i elektronnoi tekhniki v planirovanii. Moscow 1961. (Editor and author of a number of chapters.) *Analiz mezhotraslevykh sviazei v SSSR.* Moscow 1968. (With Granberg.)

Agapov, Igor Aleksandrovich
Painter; b. 1930, Moscow.
Grad. Surikov Inst. of Art Moscow 1957, pupil of P. D. Okarzhevskii.

Agapov, Vitalii Ivanovich
Diplomat.
USSR Amb. to Benin Ap. 1979–.

Agapova, Nina Nikitichna
Graphic artist; b. 1925, Rostov/Don.
Grad. Tashkent Art School 1949; pupil of P. V. Gan.

Agashina, Margarita Konstantinovna
Author and poet; b. 1924, Bor, Iaroslavl Oblast.
Grad. Gorkii Inst. for Literature 1950; first poems publ. 1949.
Publs.: poems "Moe slovo" (My Word) 1951, "Bab'e leto" (Indian Summer) 1956, "Sorok trav" (Forty Blades of Grass) 1959, and others.

Agasian, Grigorii (Grigor) Khorenovich
Painter and graphic artist; b. 1926, Erevan.
Mem. CPSU 1953–; grad. Erevan Art School 1950, pupil of S. V. Gerasimov, N. Kh. Maksimov; exhibitions 1950–.

Ageev, Mikhail Sergeevich
Painter; b. 1918, Ivanovo.
Mem. CPSU 1943–; grad. Ivanovo Art School 1938; exhibitions 1947–.

Ageev, Nikolai Vladimirovich
Chemical and metallurgical scientist; b. Je. 17 (30), 1903, Tbilisi.
Mem. CPSU 1944–; grad. Leningrad Polytech. Inst. 1926; mem. USSR Acad. of Sciences 1968–; mem. staff Leningrad Polytech. Inst. 1926–38; mem. staff Inst. of General and Inorganic Chemistry 1938–51 and Inst. of Metallurgy at USSR Acad. of Sciences 1951–. Decors.: Order of Lenin, 2 other orders and various medals.

Agishev, Rustam Konstantinovich
Author; b. 1913, Staro-Timoshkino, now Ulianovsk Oblast; s. of a worker.
Emigrated to the Far East 1936. Publs.: story "Ded Permiak" (Grandfather Permiak) 1940, "Zelenaia kniga" (The Green Book) 1954, novelette "Ouenga" 1947, play "Severnaia legenda" (North Legend), and others.

Agoshkov, Mikhail Ivanovich
Scientist; b. O. 30 (N. 12), 1905.
Mem. CPSU 1943–; grad. Far East Polytech. Inst. Vladivostok 1931; Mining expert; corresponding mem. USSR Acad. of Sciences. Decors.: USSR State Prize 1951, 4 orders.

Agranovich, Leonid (Lev) Danilovich
Writer, actor and film producer; b. 1915.
Attended school at Theatre of Workers' Youth 1932–34; film producer and director; actor at Sovremennyi Theatre and V. Meierhold Theatre Moscow 1934–37; employed at several theatres in Kirov, Blagoveshchensk, Petropavlovsk-Kamchatskii and at a front theatre 1937–43; teacher at State Union Film Inst. 1943–46; literary activity 1946–; author of numerous dramatic works. Decor.: USSR State Prize 1971.

Agriian, Lazar Samsonovich
Painter and graphic artist; b. 1911, Kagan, Uzbekistan.
Attended Azerbaidzhan Tech. College of Art 1931–32 and Inst. of Painting, Sculpture and Architecture Leningrad 1932–33; exhibitions 1933–.

Aide, Bruno Ansovich
Graphic artist; b. 1913, Riga.
Attended Art School in Riga 1945–48 and Acad. of Art of the Latvian SSR 1948–54, pupil of Ia. Kh. Tilberg; exhibitions 1954–; teacher at School for Applied Art in Riga 1955–.

Aimurzaev, Zholmurza Murzaevich
Author; b. 1910.
Grad. Turtkul Pedagogical Tech. School 1930; completed higher literary courses at Gorkii Inst. of Literature Moscow 1956; mem. CPSU 1946–; founder of Kara-Kalpak literature; translated works of Pushkin, Krylov, Lermontov, Nekrasov, Maiakovskii, Shevchenko, Franko, Shakespeare and Baron into Kara-Kalpak; works publ. 1927–.
Publs.: "Na beregakh Amu-Darii" (On the Banks of the Amu-Daria) 1949, "Prodelki Kamekbaia" (The Escapades of Kamekbay) 1958, play "Berdakh" 1950 (2nd version 1958), and others.

Airapetian, Grant Aramovich
State official.
Min. of Construction and Use of Highways Armenian SSR S. 1970–F. 1977; mem. Auditing Comm. 1971–76, Dep. Chair. 1976–81, mem. Cen. Cttee. Armenian CP 1981; Dep. Chair. Council of Mins. Armenian SSR F. 1977–.

Airapetian, Rafik Armenakovich
Sculptor; b. 1929, Kafan, Armenia.
Grad. Acad. of Art of the Latvian SSR 1958; exhibitions 1959–.

Airian, Aleksandr Arsenovich
State official.
Min. of Timber and Timber-Processing Inds. Azerbaidzhan SSR Ag. 1972–; cand. mem. Cen. Cttee. Azerbaidzhan CP 1976–.

Airiev, Aram Arutiunovich
Sculptor; b. 1915, Geldza, Azerbaidzhan.
Studied Faculty of Architecture and Construction at Azerbaidzhan Industrial Inst. 1935–41, pupil of Ia. I. Keilikhis, and at I. Ie. Repin Inst. for Painting, Sculpture and Architecture Leningrad 1945–51, pupil of V. A. Sinaiskii; exhibitions 1951–.

Aitiev, Gapar Aitievich
Painter and state official; b. S. 15, 1912, Tuleiken, now Osh Oblast.
Mem. CPSU 1939–; grad. Acad. of Fine Arts Moscow 1938; Dep. to 1955–, mem. 1967–71, Dep. Chair of Presidium Supr. Sov. Kirghiz SSR 1971–80. Decors.: People's Artist of the Kirghiz SSR 1954, Order of Lenin, 3 other orders and medals.

Aitmatov, Chingiz
Author; b. 1928.
Sec. of a village soviet, active in a raion financial dept. 1943–53; grad. Kirghiz Agricultural Inst. 1953; Chief Zoo Technician of an experimental enterprise of Kirghiz Scientific Research Inst. for Livestock-Raising 1953–56; attended higher literary courses 1956–58; mem. CPSU 1959–; Ed. of periodical "Literaturnyi Kirgizstan", correspondent of "Pravda" for Kirghiz SSR 1959–65; cand. mem. Cen. Cttee. Kirghiz CP 1966–71, mem. 1971–; 1st Sec. of Union of Film Workers Kirghiz SSR 1964–69, Chair. 1969–; Dep. to 1966–; mem. Comm. for Public Educ., Science and Culture 1966–74, mem. Comm. for Foreign Affairs of Sov. of Nationalities USSR Supr. Sov. 1974–; works publ. 1953–. Decors.: Medal for Valiant Labour, Lenin Prize.
Publs.: "Litsom k litsu" (Face to Face), "Dzhamiila" 1960, "Topolek moi v krasnoi kosynke" (My Little Poplar in Her Red Neckerchief) 1961, and others.

Aitmuratov, Erezhep
Party official; b. 1929.
Uzbek; mem. CPSU 1955–; grad. Central Asian Polytechnical Inst.; teacher at an educational institution, student 1945–1952; electrical engineer, chief energy technician at a power plant Karakalpak ASSR 1952–57; head of a dept. of an Oblast Cttee. CP Uzbekistan, Dep. Chair. of Council of Mins. Karakalpak ASSR, Chair. of Uzbek republican asscn. for agricultural equipment "Uzselkhoztekhnika", Sec. of Party Cttee. of an industrial production admin. 1957–63; Chair. of Council of Mins. Karakalpak ASSR 1963–81; mem. Cen. Cttee. CP Uzbekistan 1966–; Sec. and mem. Bureau of Cen. Cttee. CP Uzbekistan F. 1981–.

Aiusheev, Erkhite Erdynievich
Painter; b. 1925, Kurumkan, Buriat ASSR.
Attended Irkutsk Art School 1951–56; pupil of S. I. Vinogradov; mem. CPSU 1957–; exhibitions 1957–.

Aivazian, Vladimir Tigranovich
Graphic artist; b. 1915, Por, Armenia.
No formal education, studied with S. M. Agadzhanian in Erevan 1935–38;

exhibitions 1937–. Decor.: Merited Artist of the Armenian SSR 1961.

Aizman, Nikolai Spiridonovich
Dramatist; b. 1905.
Former actor; author of 40 plays and comedies.
Publs.: "Derevnia v vetlakh" (The Village in the Meadows) 1929, "Podruga chernookaia" (Black-Eyed Girl Friend) 1950, "Vyidi zamuzh za Ivana" (Marry Ivan) 1956, and others.

Akazeev, Leonard Lvovich
Graphic artist; b. 1927, Ioshkar-Ola.
Attended Moscow Polygraphic Inst. 1948–53, pupil of A. D. Goncharov and G. T. Goroshchenko.

Akgaev, Ata
Party official; b. 1922.
Turkman; mem. CPSU 1945–; grad. Cen. Cttee. Party Higher School, Cand. of Economic Sciences, Hero of Socialist Labor, teacher at a secondary school 1940–42; served in border defense troops 1942–46; held Komsomol and Party posts 1946–58; 1st Raion Cttee. Sec. CP Turkmenistan in Sakar-Chaga, Sec. of Party Cttee. of the Collective and State Farm Production Admin. in Bairam-Ali 1958–65; mem. Auditing Comm. 1960–63, mem. Cen. Cttee. CP Turkmenistan 1963–; 1st Raion Cttee. Sec. CP Turkmenistan in Bairam-Ali 1965–74; Chair. of Exec. Cttee. of Oblast Sov. of Peoples' Deps. in Mary 1974–76; 1st Oblast Cttee. Sec. CP Turkmenistan in Mary Ag. 1976–.

Akhavni (real name: Grigorian, Akhavni Arshakovna)
Author; b. 1911, Kars.
Attended Pedagogical Tech. School at Leninakan; works publ. 1930–.
Publs.: collection "Mantash" 1934, "Voennye pesni" (War Songs) 1942, "Moi pesennik" (My Songbook) 1944, novel "Shirak" 1954, and others.

Akhiezer, Aleksandr Ilich, Dr. of Phy.-Math. Sc.
Scientist; b. O. 31, 1911, Cherikov, now Mogilev Oblast.
Grad. Polytech. Inst. Kiev 1934; theoretical physicist; mem. Acad. of Sciences Ukrainian SSR 1964–; Dr. of Physics and Mathematics 1940–; Prof. 1941–; after graduation employed at Ukrainian Physical-Tech. Inst. (now Physical.-Tech. Inst. of Acad. of Sciences Ukrainian SSR, Kharkov); Prof. Chair for Physics at Kharkov Univ. 1941–; research work on nuclear physics, quantum electrodynamics, plasma theory and theory of magnetic hydrodynamics and of solid bodies. Decors.: Lenin Prize 1967, Badge of Honour and various medals.
Publs.: "Nekotorye voprosy teorii iadra" 1950, "Kvantovaia elektrodinamika" 1959, and others.

Akhiezer, Naum Ilich, Dr. of Phy.-Math. Sc.
Mathematician; b. Mr. 6, 1901, Cherikov, now Mogilev Oblast.
Grad. Mathematics Dept. at Inst. of Public Educ. Kiev 1923; corresponding mem. Acad. of Sciences Ukrainian SSR 1934–; Dr. of Physics and Mathematics 1936; Prof. 1933; Asst. Lecturer and Lecturer at Inst. of Public Educ. Kiev 1933; Prof. Chair of Mathematics at Inst. of Aviation Kharkov 1933–41; Prof. Chair of Mathematics at Kazakh Mining Inst. 1941–43; Prof. at Moscow Energy Inst. 1943–47; Prof. Chair of Mathematical Physics at Kharkov Univ. 1947–61; Chief of Dept. of Low Temperatures at Physical-Tech. Inst. of Acad. of Sciences Ukrainian SSR 1961–; author of more than 100 scientific works incl. 8 monographs. Publs.: "Aerodinamika" 1932, "Elementy teorii ellipticheskikh funktsii" 1948, "Lektsii po teorii approksimatsii" 1965, "Teoriia lineinykh operatorov v gilbertovom prostranstve" 1966.

Akhloian, Vram (Vramshapu) Khorenovich
Sculptor; b. 1930, Krasnodar.
Attended Art and Theatre Inst. Erevan 1951–58; teacher at Pedagogic Inst. Erevan 1963–.

Akhmadulina, Bella (Izabella) Akhatovna
Poet; b. Ap. 10, 1937, Moscow.
Russian; grad. Gorkii Inst. of Literature Moscow 1960; first works publ. 1955.
Publs.: poetry collections "Struna" (The String) 1962 and "Uroki muzyki" (Music Lessons) 1970, poem "Moia rodoslovnaia" (My Family Tree) are among her bestknown works; translated poetry from Georgian and other languages, among them poetry by the Georgian poet Anna Kalandadze.

Akhmatova, Raisa Soltamuradovna
Poet; b. D. 13, 1928, Groznyi; d. of a worker.
Chechen; first works publ. 1957.
Publs.: poetry collection "Respublika rodnaia" (Dear Republic) 1958, "Bei mne, veter, v litso" (Wind, Beat in My Face) 1959, "Idu k tebe" (I Come to You) 1960, "Trudnaja liubov'" (A Difficult Love) 1963, "Otkrovenie" (Openness) 1964, and others.

Akhmedbaev, Akram Guliamovich
State official.
Chair. of State Cttee. for Supervision of Safe Working Practices in Industry and Mines N. 1981–.

Akhmedov, Faik Mikailovich
Painter and graphic artist; b. 1928, Baku.
Attended Azerbaidzhan School of Art 1942–46 and the Repin Inst. for Painting, Sculpture and Architecture at USSR Acad. of Arts 1947–54, pupil of M. P. Bobyshov and A. I. Segal; exhibitions 1954–.

Akhmedov, Fikrat Gamidovich
State official.
Cand. mem. Cen. Cttee. Azerbaidzhan CP 1971–76, mem. 1976–; Head of Dept. of Science and Educ. Institutions with Cen. Cttee. Azerbaidzhan CP 1976–80; Dep. Chair. of Council of Mins. Azerbaidzhan SSR Mr. 1980–.

Akhmedov, Kudratilla Akhmedovich
State official.
Sec. of Tashkent Oblast Cttee. of Uzbek CP 1962–65 (for Agriculture 1963–64); Dep. Chair. of Oblast Exec. Cttee. of Workers' Deputies in Tashkent 1965–69; Chair. of Oblast Exec. Cttee. of Workers' Deputies in Syr-Daria 1969–71; 1st Sec. of Syr-Daria Oblast Cttee. of Uzbek CP 1971–74; Dep. to Supr. Sov. Uzbek SSR 1970–; mem. Cen. Cttee. Uzbek CP 1971–; Dep. Chair. of Council of Mins., Chair. of GOSPLAN Uzbek SSR Mr. 1974–.

Akhmedov, Rakhim
Painter; b. 1924, Tashkent.
Attended Tashkent Art School 1937–41 and the Repin Inst. for Painting, Sculpture and Architecture at USSR Acad. of Arts 1947–53, pupil of Iu. M. Neprintsev, V. M. Oreshnikov, I. A. Serebriannyi; exhibitions 1938; teacher at Theatre and Art Inst. in Tashkent 1954–. Decor.: Artist of the People of the Uzbek SSR 1961.

Akhmetova, Manura Mergalievna
State official.
Dep. Min. of Local Ind. Kazakh SSR 1974–80; Min. of Public Welfare Kazakh SSR S. 1980–; cand. mem. Cen. Cttee. Kazakh CP 1981–.

Akhobadze, Mamiia Apollonovich
Painter; b. 1933, Samtredia, Georgia.
Grad. Acad. of Arts Tbilisi 1960; exhibitions 1958–.

Akhromeev, Sergei Fedorovich
Military official; b. 1923.
Mem. CPSU 1943–, grad. Mil. Acad. of Tank Troops 1952 and Mil. Acad. of Gen. Staff 1967, Army Gen.; entered Sov. Army 1940; command posts at the front 1941–45; chief of staff of a regiment, regiment commander, division commander 1946–64; command posts 1967–72; chief of an admin. in USSR Min. of Defense 1974–79; 1st Dep. Chief of Gen. Staff of USSR Armed Forces 1979–; cand. mem. Cen. Cttee. CPSU 1981–. Decor.: Lenin Prize.

Akhumian, Semen Tigranovich
State official.
Min. of Educ. of Armenian SSR My. 1973–. Mem. Auditing Comm. 1976–81, cand mem. Cen. Cttee. Armenian CP 1981–.

Akhundov, Nadir Safarovich
Painter and graphic artist; b. 1924, Nakhichevan.
Attended Leningrad School of Arts and Crafts 1938–41, pupil of P. V. Ovchinnikov; exhibitions 1955–.

Akhundov, Senan Samandar
State official.
Min. of Local Ind. Azerbaidzhan SSR D. 1982–.

Akhverdov (Akhverdiev), Samed Iunus ogly
Painter; b. 1921, Baku.
Mem. CPSU 1950–; grad. Azerbaidzhan Art School 1944; exhibitions 1946–.

Akhviediani, Vakhtang Grigor'evich
Painter; b. 1924.
Grad. Tbilisi Acad. of Arts 1951.

Akimenko, Vitalii Ivanovich
Painter; b. 1928, Nevinnomyssk, Stavropol Krai.
Attended Rostov Art School 1946–51, pupil of V. G. Len, I. N. Fomin; exhibitions 1954–.

Akimushkin, Nikolai Nikitich
Graphic artist; b. 1925, Ukhor, Moscow Oblast.
Attended Moscow Kalinin High School for Arts and Crafts 1940–45, pupil of B. V. Lange, P. V. Surikov; exhibitions 1956–.

Akimushkina, Vera Mikhailovna
Sculptor; b. 1924, Moscow.
Attended Moscow Kalinin High School for Arts and Crafts 1939–45 and Moscow Higher Inst. for Arts and Crafts 1945–50, pupil of G. I. Motovilov, G. A. Shults.

Akmatov, Dzhusupbek
State official.
Head of Dept. of Admin. organs with Cen. Cttee. Kirghiz CP 1977–80; Student 1978–82; cand. mem. Cen. Cttee. Kirghiz CP 1981–; Min. of Interior Kirghiz SSR Jl. 1982–.

Aknuni, Perch Armenakovich
Painter; b. 1921, Constantinople.
Grad. Art and Theatre Inst. Erevan 1957; exhibition 1954–.

Akopian, Alesha Arshakovich
State official.
Chair. of State Cttee. for Supervision of Safe Working Practices in Industry and Mining Armenian SSR N. 1981–.

Akopian, G.
Komsomol official.
Sec. and mem. Bureau Cen. Cttee. Armenian Komsomol N. 1982–.

Akopian, Grant Stepanovich
Sculptor; b. 1915, Tbilisi.
Studied Acad. of Arts in Tbilisi 1935–41; exhibitions 1953–.

Akopian, Granush Akopovna
Sculptor; b. 1929, Chananai, Nakhichevan ASSR.
Attended Art and Theatre Inst. Erevan 1951–57, pupil of K. M. Metsaturian, A. M. Sarksian; exhibitions 1958–.

Akopian, Greta Surenovna
Trade union official.
Sec. and mem. Bureau Cen. Cttee. Armenian Komsomol 1956; Sec. of Leninakan City Cttee. Armenian CP 1959–61; cand. mem. Cen. Cttee. Armenian CP 1960–61 and 1966–76, mem. 1961–66; Sec. of Erevan City Cttee. Armenian CP 1961–66; Sec. and mem. Presidium of Armenian Republican Trade Union Council Ag. 1973–.

Akopian, Karush Vagramovich (Bagramovich)
Sculptor; b. 1926, Leninakan.
Attended Erevan Art School 1950–51 and Acad. of Arts of the Latvian SSR 1952–58, pupil of K. Ia. Zemdega, E. E. Melderis; lives in Riga; exhibitions 1956–.

Akopian, Naira Surenovna
Painter; b. 1932, Stepanakert.
Grad. Moscow Middle School of Art 1952, attended Surikov Inst. of Art Moscow 1952–58, pupil of M. I. Kurilko; lives in Moscow.

Akopian, Shagen Oganesovich
Stage designer; b. 1920; Constantinople, Turkey.
Attended Art School 1936–41 and Art Inst. 1945–51 in Erevan, pupil of K. A. Minasian, M. S. Sarian; exhibitions 1949–.

Akopian, Shmavon Aramovich
Painter; b. 1918, Aigestan.
Attended Art Inst. Erevan 1946–52, pupil of E. A. Isabekian, A. A. Nalbandian; exhibitions 1956–.

Akritas, Albina Georgievna
Painter and graphic artist; b. 1934, Moscow.
Attended Repin Inst. for Painting, Sculpture and Architecture Moscow 1955–61.

Aksenenok, Georgii Aleksandrovich
Jurist; b. N. 18, 1910, Polotsk.
Mem. CPSU 1930–; lawyer; mem. USSR Acad. of Sciences 1966–; Chief of Dept. of Pol. Science and Law at USSR Acad. of Sciences 1955–. Decors.: 2 orders.
Publs.: "Pravo gosudarstvennoj sobstvennosti na zemlju v SSSR" 1950, "Zemel'nye pravootnošenija v SSSR" 1958, "Pravovoe položenie kolhozov v SSSR" 1963.

Aksenov, Aleksandr Nikiforovich
Party official; b. 1924.
Belorussian; grad. Higher Party School of Cen. Cttee. CPSU; kolkhoz farmer 1941–42; served in Sov. Army 1942–43; dir. of an elem. school 1943–44; sec. of a raion cttee. of Komsomol Chkalov Oblast; mem. CPSU 1945–; sec. of a city and oblast cttee. Komsomol, 1st Sec. of Cen. Cttee. Belorussian Komsomol 1944–57; mem. Cen. Cttee. Belorussian CP 1956–; cand mem. Bureau Cen. Cttee. Belorussian CP 1956–57; Sec. and mem. Bureau of Cen. Cttee. Belorussian Komsomol 1957–60; Min of the Interior Belorussian SSR 1960–62, Min. for the Maintenance of Public Order 1962–65; Dep. to USSR Supr. Sov. 1966–; 1st Sec. Vitebsk Oblast Cttee. Belorussian CP 1966–71; 2nd Sec. 1971–78 and mem. Bureau of Cen. Cttee. Belorussian CP 1971–; mem. Cen. Cttee. CPSU 1976–; Chair. of Council of Mins. Belorussian SSR D. 1978–.

Aksenenok, Georgii Aleksandrovich
Lawyer; b. N. 18, 1910, Polotsk.
Corresponding member of the Academy of Sciences of the USSR (1966); specialist in the field of land and kolkhoz law. Member of the CPSU since 1930.
He is a state councillor of justice, third class, and has been the director of a section at the Institute of State and Law of the Academy of Sciences of the USSR since 1955. His basic works are *The Right of State Ownership of Land in the USSR* (1950), *Legal Land Relationships in the USSR* (1958), and *The Legal Status of the Kolkhozes in the USSR* (1963). Decors.: two orders.

Aksenov, Nikolai Fedorovich
Party official; b. 1928.
Russian; mem. CPSU 1954–; grad. Altai Agricultural Inst. and CPSU Cen. Cttee. Party Higher School; asst. to a combine operator on a state farm, student 1944–53; asst. a Altai Agricultural Inst., chief zoo technician of a machine and tractor station 1953–55; instructor, chief of a dept., Sec., 2nd Sec. of Altai Krai Cttee. CPSU 1955–73; Chair. of Altai Krai Exec. Cttee. of Sov. of Peoples' Deps. 1973–76; Dep. to USSR Supr. Sov. 1974–; 1st Sec. of Altai Krai Cttee. CPSU 1976–; mem. Cen. Cttee. CPSU 1981–.

Aksenov, Nikolai Stepanovich
Painter; b. 1915, Chadaevka, Saratov Oblast.
Attended Omsk Art School 1936–39, pupil of A. I. Tikhonov, and Saratov Art School 1939–40, pupil of I. N. Shcheglov; exhibitions 1941–.

Aksenov, Vasilii Pavlovich
Author; b. 1932, Kazan.
Grad. Leningrad Medical Inst. 1956; doctor in Zapolare 1956–60; doctor in Leningrad Commercial Port and in Karelia; same at Moscow Tubercular Dispensary 1960–; works publ. 1959–.
Publs.: novelette "Kollegi" (Colleagues) 1960, novel "Zvezdnyi bilet" (A Ticket to the Stars) 1961, "Na polputi k lune" (Halfway to the Moon) 1962, "Apelsiny

iz Marokko" (Oranges from Morocco) 1963, "Pora, moi drug, pora" (It's Time, My Friend, It's Time) 1964, "Tovarishch krasivyi Furatskin" (Handsome Comrade Furatskin) 1964, "Zhal', chto vas ne bylo s nami" (What a Pity You Were Not With Us) 1965, and others.

Aksirov, Muaed Alievich
Painter; b. 1931, Psygansu, Kabardino-Balkar ASSR.
Attended Saratov Art School 1947–52, pupil of V.F.Gurov and B.V.Milovidov.

Akulov, Georgii Valentinovich
Graphic artist; b. 1931, Riazan.
Attended Surikov Art. Inst. Moscow 1953–59, pupil of B.A.Dekhterev.

Aladzhadzhian, Stepan Egiaevich
Armenian writer; b. Ja. 3, 1924, Aleppo, Syria.
His collection of poems *Flowers of Sorrow* was published in Aleppo in 1942. In 1946 he was repatriated to Soviet Armenia. He graduated from the philology department of the University of Yerevan in 1951 and began writing prose in Soviet Armenia. His novella *In the Desert* which was published in 1952, shows the growth of consciousness among the soldiers of the Arab countries and their fight against the British colonialists. His novel *Defeat* (1956) is dedicated to the same subject. In 1960 he published the novella *Phoenix* about the repatriation. His novel *The Reeds Have Not Bent* (books 1–2, 1966–67) describes the struggle of the Armenian people against the Turhish yoke.
Works: *In Russian translation: Ploskie krovli: Romany.* Moscow 1961.

Alberg, Valdis Karlovich
Sculptor; b. 1922, Riga.
Attended Acad. of Art of the Latvian SSR, pupil of K.Ia. Zemdega and E.E. Melderis; exhibitions 1950–.

Aldasheva, Shakira Baigazievna
State official.
Min. of Public Services Kirghiz SSR D. 1982–.

Alderbaev, Moldan
State official.
Min. of the Timer and Timber-Processing Ind. Kazakh SSR Ap. 1975–; mem. Auditing Comm. Kazakh CP 1976–.

Aldoshin, Vladimir Vasil'evich
Diplomat.
USSR Amb. to Somalia D. 1978–.

Aleev, Shavkat Makhmudovich
Painter; b. 1928, Staryi Vermaut, Penza Oblast.
Attended Kazakh Theatre and Art School 1945–50 and Acad. of Art of the Latvian SSR 1952–58; lives in Riga; exhibitions 1958–.

Aleeva, Tatiana Aleksandrovna
Artist of monumental and decorative art; b. 1920, Esthonia.
Grad. Acad. of Art of the Latvian SSR; exhibitions 1955–.

Aleinikov, Aleksandr Abramovich
Graphic artist; b. 1915, Petrograd.
Attended Moscow Tech. College of Art 1931–33 and Moscow Inst. for Fine Arts 1933–40, pupil of P.Ia. Pavlinov, M.S. Rodionov.

Alekhin, Nikolai Alekseevich
Painter; b. 1928, Bashkir ASSR.
Attended Sverdlovsk Art School 1946–51, pupil of G.A.Melentev; exhibitions 1952–.

Alekin, Oleg Aleksandrovich
Hydrochemist; b. Ag. 10 (23), 1908.
Corresponding member of the Academy of Sciences of the Soviet Union (1953). Member of the CPSU since 1942. He is the author of works on the chemistry of natural waters, the methods of their chemical analysis, and andrology. He is the rector of the Leningrad Hydrometeorological Institute. Decor.: State Prize of the USSR in 1951.
Works: *Gidrokhimiia rek SSR,* parts 2–3. Leningrad 1948–49. (In *Trudy Gidrologich, instituta,* vols. 10 and 15.) *Osnovy gidrokhimii.* Leningrad 1953. *Khimicheskii analiz vod sushi.* Leningrad 1954. *Khimiia okeana.* Leningrad 1966.

Alekperov, Kamal Dzhamal ogly
Sculptor; b. 1928, Fizuli, Azerbaidzhan.
Grad. Azerbaidzhan Art School 1948; attended Surikov Art Inst. Moscow 1948–54, pupil of M.G.Manizer; exhibitions 1953–.

Aleksandr, Nora Mikhailovna
Painter; b. 1976, Vladikavkaz.
Attended Moscow Inst. of Fine Arts 1937–42, pupil of P.D.Pokarzhevskii; exhibitions 1943–.

Aleksandrin, Mikhail Sergeevich
Stage designer; b. 1925, Sadovka, Saratov Oblast.
Attended Art School, Saratov 1944–49, pupil of N.A.Arkhangelskii and I.V.Sevastianov.

Aleksandrov, Aleksandr Aleksandrovich
Graphic artist; b. 1921, Zaporozhe; s. of A.I.Aleksandrov.
Attended Kiev Middle Art School 1938–40 and Kharkov Tech. College of Art 1940–41.

Aleksandrov, Aleksandr Danilovich
Mathematician; b. Jl. 22 (Ag. 4) 1912, Volyn, Riazan Gouvt.
Mem. CPSU 1951–; grad. Leningrad Univ. 1933; mem. USSR Acad. of Sciences 1964–; Rector of Leningrad Univ. 1952–64; employed in Siberian Branch of USSR Acad. of Sciences 1964–; founder of Sov. school of geometry in the large; set up intrinsic geometry of general surfaces; visited Univ. of California at Berkeley Ap. 1959. Decors.: Int. Lobachevskii Prize 1951, Lenin Order.
Publs.: "Vnutrennjaja geometrija vypuklyh poverchnostej" Moscow-Leningrad 1948, "Vypuklye mnogogranniki" Moscow-Leningrad 1950.

Aleksandrov, Anatolii Petrovich
Scientist; b. 1903, Tarashcha, Ukraine.
Mem. CPSU 1962–; grad. Kiev State Univ. 1930; Dr. of Physics and Mathematics, nuclear physicist; mem. USSR Acad. of Sciences 1953–; electrical fitter 1920–22; taught physics and chemistry at a secondary school 1922–30; researcher, Laboratory Head of Leningrad Inst. of Physics and Engineering 1930–46; Dir. of Inst. of Physical Problems of USSR Acad. of Sciences 1946–55; Dep. Dir. Kurchatov Inst. of Atomic Energy of USSR Acad. of Sciences 1955–60, Dir. 1960–; mem. Cen. Cttee. CPSU 1966–; Pres. of USSR Acad. of Sciences N. 1975–; Chair of Cttee. for Lenin Prizes and State Prizes in Science and Technology of USSR Council of Mins. 1978–.
Decors.: Hero of Socialist Labour (twice), Order of Lenin (six), Lenin Prize, USSR State Price (four).

Aleksandrov, Boris Aleksandrovich
Composer and conductor; b. Jl. 22 (Ag. 4), 1905, Bologoe.
Became People's Artist of the USSR in 1958. He is a colonel and son of A.V.Aleksandrov.
He graduated from the Moscow Conservatory in 1929, where he had studied composition with R.M.Glière. He was conductor and music director of the Central Theater of the Soviet Army. Since 1937, he has been with the Order of the Red Banner Ensemble of Song and Dance of the Soviet Army, at first as a conductor and since 1946 (after his father's death) as head and artistic director of the ensemble. He has written several songs, including *Song of Victory, Long Live Our State, Soldiers on the March,* and the oratoria *The Soldier of October Defends Peace* (1967). He also wrote the operettas *The Girl From Barcelona* (1942) and *My Guizel* (1946) and the ballet *Lefty* (1955). His best-known work is the operetta *Wedding in Malinovka* (1937), which draws on a wealth of folk songs, primarily Ukrainian; a film version of it was made in 1968. He won the State Prize of the USSR in 1950 for his concert performances. Decors.: two Orders of Lenin, the Order of the Red Banner of Labour, and medals.

Aleksandrov, Grigorii Vasil'evich (pseudonym of G.V.Mormonenko)
Film director; b. Ja. 10 (23), 1903, Ekaterinburg (now Sverdlovsk).
People's Artist of the USSR (1948). Member of the CPSU since 1954. Hero of Socialist Labour (1973).

In 1921, he became an actor in Proletkult's first workers' theater in Moscow. He took part in plays staged by S. M. Eisenstein. He began working in the film industry in 1924. In the Films *Battleship Potemkin* (1925), *October* (1927), and *The Old and the New* (1929), directed by Eisenstein, he worked as both an actor and a co-director; in the latter two he was also a co-author of the screenplays.
In 1934 he directed the film *Jolly Boys*, a musical comedy in which he employed eccentric stunts, circus devices, and techniques from the music hall, operetta, and musical sideshow. In further works he continued to develop the genre of eccentric musical comedy: he made the comedy films *The Circus* (1936), *Volga-Volga* (1938), *The Blazing Path* (1940), and *Spring* (1947). These films are filled with buoyant music composed by I. O. Dunaevskii, and they express an optimistic tone. The actors' performances are of a high level (L. P. Orlova, I. V. Il'inskii, and V. S. Volodin). He filmed *Meeting on the Elbe* (1949), which was about the events of the first days after the conclusion of the Great Patriotic War, the biographical film *The Composer Glinka* (1952), and the experimental films *Man a Man ...* (1958) and *Russian Souvenir* (1960). He wrote a number of film scripts.
He has taught in the All-Union State Institute of Cinematography. Decors.: State Prize of the USSR (1941 and 1950), two Orders of Lenin, other orders and medals.
Works: "Oktiabr' (stsenarii)." *Iskusstvo kino,* 1957, no. 10. (Jointly was S. M. Eisenstein.)

Aleksandrov, Iurii Vladimirovich
Sculptor; b. 1930.
Attended Moscow High School for Arts and Crafts 1945–54.

Aleksandrov, Kirill Sergeevich, Dr. of. Phy.-Math. Sc.
Physicist; b. 1931.
Mem. CPSU 1953–; grad. Elektrotechnical Inst. Leningrad 1954; doctoral degree in physical-mathematical sciences 1968; Prof.; corresponding mem. USSR Acad. of Sciences 1972–; Asst. 1954–57, Sen. Research Asst. at Inst. of Crystallography of USSR Acad. of Sciences 1957–58; Inst. of Physics at Siberian Branch of USSR Acad. of Sciences 1958–, Head of Laboratory 1959–68, Dep. Dir. 1968–; Prof. at Krasnoiarsk Univ. 1971–.

Aleksandrov, Kuzma Ivanovich
Sculptor; b. 1918, Verkholomovsk, Rostov Oblast.
Attended Krasnodar Art School 1935–39; pupil of N. D. Sharikov; mem. CPSU 1943–; exhibitions 1948–.

Aleksandrov, Nikolai Pavlovich
Painter and graphic artist; b. 1924, V. Ubinka, Kazakhstan.
Grad. Alma-Ata Art School 1951 and Mukhina High School for Arts and Crafts Leningrad 1957; exhibitions 1960–.

Aleksandrov, Semen Lavrentevich
Painter; b. 1918, Yakut.
Grad. Repin Inst. for Painting, Sculpture and Architecture Moscow 1954; pupil of V. M. Oreshnikov; exhibitions 1955–.

Aleksandrov-Agentov, Andrei Mikhailovich
Party official; b. 1918.
Grad. Leningrad State Univ. 1940; mem. CPSU 1948–; active in Soviet News Agency TASS and in Foreign Min. of USSR 1940–61; Chief Spokesman of Sec. of Presidium USSR Supr. Sov. 1961–63; active in CPSU Cen. Cttee. Je. 1963–; Aide to Gen. Sec. of CPSU Cen. Cttee. L. I. Brezhnev 1964–; mem. CPSU Cen. Auditing Comm. 1971–76; cand. mem. Cen. Cttee. CPSU 1976–81; mem. 1981–. Decors.: Lenin Prize 1980.

Aleksandrov-Sedelnikov, Sergei Konstantinovich
Painter; b. 1929, Moscow.
Attended Moscow Middle School of Art 1948, Tartu Art School 1948–50 and Surikov Art Inst. Moscow 1950–55.

Aleksandrova, Elizaveta Nikolaevna
Graphic artist; b. 1930, Leningrad.
Attended Repin Inst. for Painting, Sculpture and Architecture at the USSR Acad. of Arts 1951–57.

Aleksandrova, Irina Dmitrievna
Sculptor; b. 1928, Moscow.
Attended Moscow High School for Arts and Crafts 1946–54, pupil of G. I. Motovilov and G. A. Shults; exhibitions 1956–.

Aleksandrova, Zinaida Nikolaevna
Poet; b. 1907, St. Petersburg; d. of a teacher.
Worked in textile mill in her youth; works publ. 1926–; has written poems and song lyrics for children and translations of same from other languages of the Sov. peoples 1930–.
Publs.: verse collections "Fabrichnye pesni" (Factory Songs) 1928, children's song "Veter na rechke" (The Wind By The River) 1930, and others.

Aleksandrova-Roslavleva, Valentina Valentinovna
Sculptor; b. 1921, Moscow.
Grad. Moscow Inst. of Applied and Decorative Arts 1947.

Aleksandrova-Sanamiants, Ariadna Arkadevna
Sculptor; b. 1928, Leningrad.
Attended Leningrad Art School 1940–46 and Repin Inst. Moscow 1946–49; exhibitions 1954–.

Aleksanian-Shakhaziz, Nadezhda Nikolaevna
Graphic artist; b. 1914, Kars, Turkey.
Attended Tbilisi Acad. of Arts 1932–38, pupil of I. A. Sharleman; exhibitions 1939–.

Aleksankin, Aleksandr Vasil'evich
State official; b. 1929.
Mem. CPSU 1956–; grad. Belorussian Agricultural Acad. 1953; held posts in Min. of Agriculture apparatus Belorussian SSR 1953–56 and 1959–65; dir. of a machine melioration station in Grodno Oblast 1956–59; Min. of Melioration and Water Econ. Belorussian SSR 1965–71; cand. mem. Cen. Cttee. Belorussian CP 1966–71, mem. 1971–; 1st Dep. USSR Min. of Melioration and Water Econ. 1971–74; Dep. Chair. of Council of Mins. RSFSR 1974–; mem. Cen. Auditing Comm. CPSU 1976–. Decor.: Order of Red Banner of Labor.

Alekseev, Adolf Evgenevich
Painter and graphic artist; b. 1934, Leningrad.
Attended Surikov Art Inst. 1953–59, pupil of V. P. Efanov, F. I. Nevezhin; exhibitions 1957–.

Alekseev, Aleksandr Emelianovich
Electrical engineer; b. N. 15 (27), 1891, Sorokino, now Kalinin Oblast.
Grad. Electrotech. Inst. Leningrad 1925; corresponding mem. USSR Acad. of Sciences 1953–; Prof. at Engineering Inst. for Railway Transport 1936–; additional appointment at Inst. of Electromechanics of USSR Acad. of Sciences 1953–. Decors.: Order of Lenin, Order of the Red Banner of Labour, and other medals.
Publs.: "Tubogeneratory" Leningrad-Moscow 1939 (co-author), "Konstrukcija ėlektričeskih mašin" Leningrad-Moscow 1949, "Tjagovye ėlektričeskie mašiny i preobrazovateli" Leningrad 1967.

Alekseev, Aleksandr Ivanovich
Diplomat; b. 1913.
Active in diplomatic service 1941–; Emb. Counsellor at USSR Emb. in Cuba 1960–62; USSR Amb. to Cuba 1962–68; USSR Amb. to Republic of Madagascar Mr. 1974–Je. 1980.

Alekseev, Anatolii Ivanovich
Painter; b. 1929, Altai.
Attended Irkutsk Art School 1946–50 and Kharkov Art Inst. 1950–56, pupil of S. F. Besedin, A. P. Liubimskii; exhibitions 1956–.

Alekseev, Boris Alekseevich
Actor; b. 1911.
Mem. CPSU 1944–; employed at Chuvash Academic Theatre 1929–. Decor.: People's Artist of the USSR 1969.

Alekseev, Boris Vladimirovich
Painter; b. 1923, Samara, Voronezh Oblast.
Attended Voronezh Art Studio 1937–39 and Art School 1939–41, pupil of A.A. Buchkuri; mem. CPSU 1948–; exhibitions 1950–.

Alekseev, Georgii Leonidovich
Graphic artist; b. 1929, Odessa.
Attended Kiev Art School 1949–55, pupil of V.I. Kasiian; exhibitions 1956–.

Alekseev, Mikhail Iurevich
Graphic artist; b. 1927, Moscow.
Attended Middle School of Art Moscow 1939–46 and Union Inst. for Films 1946–52.

Alekseev, Mikhail Nikolaevich
Author; b. 1918, Monastyrskoe, Saratov Oblast; s. of a peasant.
Served in Sov. Army 1938–55; completed higher literary courses of USSR Writers' Union 1957.
Publs.: novel "Soldaty" (Soldiers) 1951–53, story collection "Nash leitnant" (Our Lieutenant) 1955, "Nasledniki" (Successors) 1957, "Zhili-byli dva tovarishcha" (There Were Once Two Friends) 1958, novelette "Divizionka" (The Girl Soldier in the Division) 1959, and others.

Alekseev, Nikolai Borisovich
Diplomat; b. 1912.
Active in diplomatic service 1943–; USSR Chargé d'Affaires in Uruguay 1949–54; held responsible positions in USSR Min. of Foreign Affairs 1954–59; Counsellor of Soviet Republics in UN 1959; USSR Amb. to Argentina 1959–66; active in USSR Min. of Foreign Affairs 1966–68; USSR Amb. to Chile 1968–71; Head of Latin American Dept. in USSR Min. of Foreign Affairs 1975–80.

Alekseev, Petr Fedorovich
Editor; b. 1913.
Held Komsomol and party positions in Uzbek SSR 1931–32; literary associate, correspondent of various newspapers, acting sec., dep. ed., ed.-in-chief of various reg. and republican newspapers in Tashkent, Samarkand, Novgorod 1932–50; mem. CPSU 1940–; grad. Tashkent Pedagogical Inst. 1948, grad. CPSU Cen. Cttee. Party Higher School; Ed.-in-Chief of "Sovetskoe khlopkovodstvo" 1950–53; active in ed. staff of "Pravda" 1953–59; high positions within org. of CPSU Cen. Cttee. 1959–60; Dep. Ed.-in-Chief 1960–62, Chief Ed. of "Selskaia zhizn'" and "Sovetskaia Rossiia" 1962–76; mem. CPSU Cen. Auditing Comm. 1966–71; cand. mem. 1971–76, mem. Cen. Cttee. CPSU 1976–. Chief Ed. "Izvestiia" 1976–; Dep. USSR Supr. Sov. 1979–. Decors.: Order of Red Banner of Labour (twice).

Alekseev, V.
Historian.
Corresponding mem. USSR Acad. of Sciences 1981–. Researcher at the N.N. Mikukho-Maklay Inst. of Ethnography USSR Acad. of Sciences; expert in the field of historical anthropology; author of more than 300 publications; expert on problems of ethnogeny of the different peoples of the world, kinship between existing and extinct peoples, the relation between the social and biological in the history of man, and the anthropogeny and demography of the native population; founder of the hypothesis of race formation through herd behavior.

Alekseev, Valentin Alekseevich
Stage designer and painter; b. 1923, Russkoe, Rostov/Don Oblast.
Grad. Art School Rostov/Don 1950, attended Repin Inst. for Painting, Sculpture and Architecture Moscow 1950–57, pupil of M.P. Bobyshov; mem. CPSU 1948–.

Alekseev, Vasilii Petrovich
Painter; b. 1923, Baku.
Attended Repin Inst. for Painting, Sculpture, and Architecture 1948–54, pupil of V.M. Oreshnikov; exhibitions 1957–.

Alekseev, Viktor Mikhailovich
Graphic artist; b. 1925, Shatra-Kasy.
Attended Cheboksary Art School 1946–48, pupil of N.K. Sverchkov and M.S. Spiridonov.

Alekseevskii, Nikolai Evgenevich
Physicist; b. My.10 (23), 1912, Petropavlovsk.
Grad. Polytech. Inst. Leningrad 1936; corresponding mem. USSR Acad. of Sciences 1960–; mem. staff Physical-Tech. Inst. Kharkov 1936–41; mem. staff. Inst. for the Problems of Physics at USSR Acad. of Sciences 1942–. Decors.: State Prize of the USSR 1967, 2 orders.

Aleksidze, Nelli Aleksandrovna
Sculptor; b. 1921, Tbilisi.
Attended Acad. of Art Tbilisi 1943–49, pupil of S.Ia. Kakabadze; exhibitions 1950–.

Aleksii (A. M. Ridiger), Cand. of Theol.
Russian Orthodox clergyman; b. F.23, 1929, Tallinn.
Grad. Leningrad Theological Seminary 1949; deacon, priest 1950; completed studies by correspondence course at Leningrad Acad. of Theology, Cand. of Theology 1953; Chair. of Tartu Cathedral; monk, Archimandrite, Bishop of Tallinn and Estonia 1961; worked actively abroad; Vice-Chair. of Dept. of External Affairs of the Moscow Patriarchate 1962–; Archbishop, Admin. Man. of Moscow Patriarchy, permanent mem. of the Holy Synod, Chair. of Teaching Cttee. of the Moscow Patriarchate 1964; Metropolitan of Tallinn and Esthonia 1968–; Vice-Chair. 1971, Chair. of Conference of European Churches 1972–.

Aleksii (V. A. Konoplev), Cand. of Theol.
Russian Orthodox clergyman; b. 1910, Pavlovsk, Voronezh Gouvt.
Grad. Moscow Theological Seminary 1951; priest; grad. Moscow Acad. of Theology 1955; monk, Archimandrite, Bishop of Molotov (Perm) 1956; Bishop of Luga, Substitue for the Bishopric of Leningrad 1957; Bishop of Tula 1961; Archbishop 1964, Archbishop of Riga 1966; Archbishop of Krasnodar and Kuban O. 1966–78, Archbishop of Kalinin and Kashir Ap. 1978–.

Aleksin, Anatolii Georgievich
Author; b. 1924, Moscow.
Grad. Moscow Inst. for Oriental Studies 1950; writes primarily for children; first works publ. 1945.
Publs.: novelette "Tridtsat' odin den'" (Thirty-One Days) 1950, "V odnom pionerskom lagere" (In a Pioneer Camp) 1954, "Zapiski Elviry" (Elvira's Notes) 1956, "O druzhbe serdets" (On the Friendship of Hearts) 1958, "Shkola na novom puti" (The School on the New Path) 1959, and others.

Aleshin, Samuil Iosifovich (real name: Kotliar), Cand. of Tech. Sc.
Dramatist; b. Jl.8 (21), 1913, Zambrów (Poland).
Grad. Mil. Acad. of Motorization and Mechanization 1935; Cand. of Tech. Sc. 1946; active as dramatist. Decor.: Order of Red Banner of Labour.
Publs.: plays "Direktor" (The Director) 1950, "Gogol" 1952, "Strogaia devushka" (A Strict Girl) 1953, "Odna" (Alone) 1956, "Vse ostaetsia liudiam" (Everything Rests With the People) film was made under the same title in 1963, "Tochka opory" (Base) 1960, "Glavnaia rol'" (The Main Role) 1964, "Kazhdomu svoe" (To Each His Own) 1966, "Diplomat" (The Diplomat) 1967, "Drugaia" (The Other Woman) 1968, and others.

Aleskerov, Gasanaga Aga Ragim ogly
Painter and artist in applied art; b. 1929, Baku.
Attended Azerbaidzhan Art School 1947–52 and Leningrad High School for Arts and Crafts 1952–58; exhibitions 1959–.

Alferov, Aleksandr Porfirevich
Painter; b. 1936, Rtishchev, Saratov Oblast.
Attended Saratov Art School 1952–57, pupil of A.I. Borodin; exhibitions 1959–.

Alferov, Slav Petrovich
Graphic artist and painter; b. 1927, Frunze.
Attended Kazakh Art School 1942–46, pupil of S.F. Besedin, A.I. Bortnikov and A.M. Cherkasskii; exhibitions 1954–.

Alferov, Zhores Ivanovich, Dr. of Phy.-Math. Sc.
Physicist; b. 1930.
Mem. CPSU 1965–; grad. Electrotech. Inst. Leningrad 1952; doctoral degree in physical-mathematical sciences 1971; Prof. 1972; corresponding mem. USSR Acad. of Sciences 1972–; mem. staff 1953–. Sen. Research Asst. 1964–67, Section Chief at Physical-Tech. Inst. of USSR Acad. of Sciences 1967–. Decor.: Lenin Prize 1972.

Alga, Aleksandr Egorovich
Poet; b. 1913, Chuvash.
Mem. CPSU 1943–; grad. Pedagogical Inst. Cheboksary; former teacher.
Publs.: verse collection "Solnechnoe utro" (A Sunny Morning) 1940, "Gnev i laska" (Anger and Tenderness), "Ich bylo 28" (They Were 28), "Doroga v plameni" (Street in Flames), "Na Rodine" (At Home) 1951, "Volzhskie volny" (The Waves of the Volga) 1958, and others.

Ali-zade, Abdulkerim Ali ogly
Scholar; b. Ja. 11, 1906, Belgia, former rural district Baku.
Grad. Oriental Inst. Leningrad 1930; orientalist; specialist in the medieval history of Azerbaidzhan and neighbouring countries; mem. Acad. of Sciences Azerbaidzhan SSR 1955–; Dir. of Inst. for the History of the Acad. of Sciences Azerbaidzhan SSR 1944–50; Dir. of Inst. of Oriental Studies Acad. of Sciences Azerbaidzhan SSR 1958–63; has written several books about the history of feudal relations. Decor.: Merited Scientist of the Azerbaidzhan SSR.
Most important publs.: "Social'noėkonomič. i politič. istorija Azerbajdžana XIII–VIV v.v." 1956.

Aliamovskii, Arkadii Dmitrievich
Graphic artist; b. 1924, Kadom, Riazan Oblast.
Attended Leningrad High School for Arts and Crafts 1948–54, pupil of G. V. Pavlovskii, G. Savinov and I. P. Stepashkin; exhibitions 1955–.

Aliev, Akper (pseudonym: Rukhi)
Author and literary scholar; b. 1908, Ashkhabad, Turkmenistan.
Mem. CPSU 1943–; works publ. 1927–. Publs.: poem "Komsomol" 1936, "Annagiul" 1952, novelette "Pis'ma Siulguin" (Siulguin's Letters) 1955, and others.

Aliev, Geldar Ali Rza ogly
Party official; b. 1923, Azerbaidzhan.
Mem. CPSU 1945–; grad. Azerbaidzhan State Univ. 1957; active in State Security and Ministerial Council of Nakhichevan ASSR 1941–49; active in Cttee. for State Security of Azerbaidzhan Council of Mins. 1950–64, Dep. Chair. 1964–67, Chair. 1967–69; Maj.-Gen.; cand. mem. Cen. Cttee. Azerbaidzhan CP 1966–71, mem. 1971–; 1st Sec. and mem. Bureau of Cen. Cttee. Azerbaidzhan CP 1969–82; Deputy 1970–, mem. Comm. for Youth Affairs 1970–74, Dep. Chair. of Sov. of the Union with Supr. Sov. USSR 1974–; mem. Cen. Cttee. CPSU 1971–; cand. mem. Politbureau Cen. Cttee. CPSU Mr. 1976–82, mem. D. 1982–; 1st Dep. Chair. of USSR Council of Mins. D. 1982–. Decors.: 2 Orders of Lenin, Orders of Red Star. Hero of Socialist Labour, medals.

Aliev, Kurban Gasanovich
State official.
Cand. mem. Cen. Cttee. Azerbaidzhan CP 1961–64, mem. 1964–; Dep. to Supr. Sov. Azerbaidzhan SSR 1963–; 1st Sec. of "26 Commissars" Raion Cttee. of Azerbaidzhan CP in Baku 1963; Min. of Higher and Secondary Specialist Educ. Azerbaidzhan SSR Je. 1971–.

Aliev, Makhmud Isag ogly
State official.
Min. of Melioration and Water Econ. Azerbaidzhan SSR Mr. 1980–; mem. Auditing Comm. Azerbaidzhan CP 1981–.

Aliev, Mukhtar
State official.
Min. of Public Health Kazakh SSR Ag. 1982–.

Aliev, Namik Kiamal Mamed ogly
State official.
Min. of Oil-Processing Ind. and Petrochemistry Azerbaidzhan SSR Ja. 1980–.

Aliev, Nuriddin Medzhid ogly
State official.
Mem. Auditing Comm. 1964–66, cand. mem. Cen. Cttee. 1955–76 and 1981–, mem. Cen. Cttee. Azerbaidzhan CP 1976–81; Chair of State Cttee. for Supervision of Safe Working Practices in Industry and for Mine Azerbaidzhan SSR Mr. 1982–.

Aliev, Surkhai Kara ogly
Painter; b. 1930, Shusha.
Attended Azerbaidzhan Art School 1944–49; exhibitions 1957–.

Alieva, Fazu
Avar poet; b. D. 5, 1932, Ginichutl', Khunzakh Raion, Dagestan ASSR.
People's Poet of Dagestan (1969). Member of the CPSU since 1965. Graduated from the M. Gorkii Literary Institute in 1961.
She was first published in 1949. She is the author of the poetry collections *Native Village* (1959), *Spring Wind* (1962), *I Give Out Rainbows* (1963), and *The Moment* (1967); the narrative poems *On the Seashore* (1961) and *In the Heart of Every Person Is Il'ich* (1965); and the novel *Fate* (1964). Her poems – the collections *The Blue Road* (1959), *The Carving on the Stone* (1966), and *The Eighteenth Spring* (1968) – have been translated into Russian. Her novel *The Wind Won't Carry Away the Clod of Earth* (1967) was awarded a prize in the N. Ostrovskii Competition. Decor.: Order of the Badge of Honor.

Alieva, Meredgul
State official.
Min. of Educ. Turkmen SSR Ja. 1979–; mem. Cen. Cttee. Turkmen CP 1981–.

Alieva, Sakina Abbas kyzy
State official.
Mem. Auditing Comm. 1964–66, cand. mem. Cen. Cttee. Azerbaidzhan CP 1966–; Dep. to and Dep. Chair. of Presidium Supr. Sov. Azerbaidzhan SSR 1964–; Chair. of Presidium Supr. Sov. Nakhichevan ASSR 1964–.

Aliger, Margarita Iosifovna
Poet; b. 1915, Odessa.
Attended Gorkii Inst. of Literature 1934–37; mem. CPSU 1942–; works publ. 1933–. Decors.: Order of Red Banner of Labour, Stalin Prize.
Publs.: collected verse "God rozhdeniia" (Year of Birth) 1938, "Zheleznaia doroga" (The Railroad) 1939, "Kamni i travy" (Stones and Grass) 1940, "Pamiati khrabrykh" (In the Memory of the Brave) 1942, poem "Zoia" 1942, "Lirika" (Lyric Verse) 1943, "Tvoia pobeda" (Your Victory) 1945, "Pervyi grom" (First Thunder) 1947, "Pervye primety" (The First Signs) 1948, "Krasivaia Mecha" (Beautiful Mecha) 1951, "Leninskie gory" (Lenin Heights) 1953, play "Skazka o pravde" (A Tale about Truth) 1945, "Neskol'ko shagov" (Several Steps) 1962, 1964, and others.

Alikhanian, Artem Isaakovich, Dr. of Phy.-Math. Sc.
Physicist; b. 1908.
Grad. Univ. Leningrad 1931; doctoral degree in physical-mathematical sciences; corresponding mem. USSR Acad. of Sciences 1946–; Dir. of Physical Inst. at USSR Acad. of Sciences 1950–. Decors.: State Prizes of the USSR 1941, 1948, Lenin Prize as head of a scientific team for the successful development of new means for the study of elementary particles 1970.

Alikhanov, Abram Isaakovich
Nuclear physicist; b. F. 20 (Mr. 4), 1904, Tbilisi.
Grad. Polytech. Inst. Leningrad 1929; mem. USSR Acad. of Sciences 1943–; Dir. of Thermodynamic Laboratory 1945–68 (Inst. for Theoretical and Experimental Physics of USSR Acad. of Sciences 1957–). Decors.: Hero of Socialist Labour, Lenin Order (twice), Order of the Red Banner of Labour.

Alikhanov, Enver Nazarovich, Dr. of Geol.-Min. Sc.
Scientist; b. Ap. 17, 1917, Baku.
Mem. CPSU 1943–; grad. Azerbaidzhan Ind. Inst. 1941; doctoral degree in geological-mineralogical sciences; former

Party and State official; several posts in petroleum ind. and Party positions with Azerbaidzhan SSR; Sec. of Cen. Cttee. Azerbaidzhan CP 1959–61; Chair. of Council of Mins. Azerbaidzhan SSR 1961–70. Decors.: Lenin Order (twice) and other decorations.

Alikov, Sergei Vasil'evich
Graphic artist; b. 1916, Vladikavkaz.
Attended Penza Art School 1933–37; exhibitions 1939–.

Alimarin, Ivan Pavlovich
Chemical scientist; b. Ag. 29 (S. 11), 1903, Moscow.
Expert in analytical chemistry; mem. USSR Acad. of Sciences; Prof. Chair of Analytical Chemistry at Moscow Univ. 1950–; Laboratory Head at Inst. of Geochemistry and Analytical Chemistry of USSR Acad. of Sciences 1949–. Decors.: Lenin Order (twice) and others.
Publs.: approx. 500 scientific works.

Alimov, Aleksandr Nikolaevich
Economist; b. S. 30, 1923, Donetsk.
Mem. CPSU 1943–; grad. in correspondence courses of Union Financial Inst. without interrupting regular employment; specialist in the econ. of the coal industry; corresponding mem. Ukrainian Acad. of Sciences 1966–; served in Sov. Army 1941–47; active in admin. of trust "Donbas-mashbud" 1949–51 and of collective combine "Donetsk-shakhtobud" 1952–53; cand. of Moscow Mining Inst. 1953–57; after graduation Section Chief in Dept. of Mining Econ. at Coal Research Inst. Donetsk, later Chief of same Dept.; Dep. Dir. of Econ. Inst. at Acad. of Sciences Ukrainian SSR Je. 1965–. Decors.: Order of Red Banner (twice), Order of the Fatherland War 1st Cl., Aleksandr Nevskii Order, several medals.
Publs.: "Normirovanie oborotnykh sredstv v ugolnoi promyshlennosti" 1960, "Proizvodstvennye zapasy i oborotnye sredstva v ugolnoi promyshlennosti" 1963, "Voprosy materialno-tekhnicheskogo snabzheniia v ugolnoi promyshlennosti" 1964, and approx. 60 scientific works.

Alimov, Arif Alimovich
State official.
Mem. CPSU 1941; 1st Sec. Samarkand Oblast Cttee. Uzbek CP until 1950; Min. of Cotton Ins. Uzbek SSR 1950–51; 1st Sec. Oblast Cttee. Uzbek CP in Bukhara, Tashkent and Samarkand 1951–59; mem. Cen. Cttee. and mem. Bureau Cen. Cttee. Uzbek CP 1954–61; Chair. Council of Mins. Uzbek SSR 1959–61; Chair. of Agricultural-Industrial Union (working on basis of economic records and calculations) for Production, Acquisition, Industrial Processing and Wholesale Marketing of Potatoes, Vegetables, Melons, Fruit, Grapes – "Uzplodovoshchvinprom" – with Council of Ministers Uzbek SSR Mr. 1975–81; Min. of Fruits and Vegetables Uzbek SSR Ja. 1981.

Alimov, Boris Aleksandrovich
Graphic artist; b. 1932, Moscow.
Grad. Surikov Art Inst. Moscow 1958.

Alimov, Gennadii Vasil'evich
Graphic artist; b. 1922, Orekhovo-Zuevo.
Attended Moscow Polygraphic Inst. 1950–55, pupil of B. A. Dekhterev and P. G. Zakharov.

Alimzhanov, Anuar (Anuarbek)
Author; b. My. 2, 1930, Karlygash, Taldy-Kurgan Oblast.
Kazakh; writes mainly in Russian; mem. CPSU 1953–; grad. Kazakh State Univ. 1954.
Publs.: "Vechnye korni" (Eternal Roots) 1960, "Karavan idet k solntsu" (The Caravan Moves Toward the Sun) 1963, "Sinie gory" (Blue Mountains) 1964, "Suvenir iz Otrara" (Souvenir From Otrar) 1966, "Piat'desiat tysiach mil' po vode i sushe" (Fifty Thousand Miles Over Water and Land) 1962, and others.

Aliverdieva (Allierdova), Sudaba Sadykh kyzy
Sculptor; b. 1927, Agdam, Azerbaidzhan.
Grad. Leningrad High School for Arts and Crafts 1958; exhibitions 1957–.

Alkhimov, Vladimir Sergeevich
State official; b. 1919.
Mem. CPSU 1942–, grad. Leningrad Economic Finance Inst. 1947 and All-Union Acad. of Foreign Trade 1950, Cand. of Econ. Sc. 1951; at the front during World War II: platoon commander, battery commander, topographical service chief of a regiment 1941–45; dept. chief, dep. dir. of a scientific research design inst. of USSR Min. of Foreign Trade 1950–58; trade councillor in USSR Emb. to US 1958–60; chief of an admin. in USSR Min. of Foreign Trade 1961–67; Dep. USSR Min. of Foreign Trade 1967–76; Chairman of the board of USSR State Bank O. 1976–; Dep. to USSR Supr. Sov. 1974–; cand. mem. Cen. Cttee. CPSU 1981–82, mem. N. 1982–. Decor.: Hero of the Sov. Union 1945.

Allamuradov, B. A.
Komsomol official.
1st Sec. and mem. Bureau Cen. Cttee. Uzbek Komsomol Mr. 1982–.

Allendorf, Marina Aleksandrovna
Graphic artist; b. 1915, Moscow.
Attended Surikov Art Inst. Moscow 1943–50, pupil of P. Ia. Pavlinov; exhibitions 1950–.

Almaev, Talgat Mazgutovich
State official.
Head of Main Statistical Admin. Kirghiz SSR Je. 1969–. Chair. Auditing Comm. Kirghiz CP 1976–.

Alovov, Nedirmamed
State official.
Chair. of State Cttee. on Labour Turkmen SSR D. 1976–. Cand. mem. Cen. Cttee. Turkmen CP 1981–.

Alsnik, Petr Ivanovich, Dr. of Agric. Sc.
Scientist; b. F. 14 (27), 1907, Vydreia, Mogilev Gouvt.
Grad. Belorussian Agricultural Acad. 1929; selection potato grower; corresponding mem. V. I. Lenin Acad. of Agricultural Sciences of the USSR (VASKHNIL) 1956–; Dr. of Agric. Sc. 1964; mem. Acad. of Sciences Belorussian SSR 1966–; employed in several agricultural insts. of Belorussian SSR. Decors.: Hero of Socialist Labour 1966, State Prize of the USSR 1953.

Altaev, Ilia Khabich
Painter and graphic artist; b. 1927, Bortoi, Irkutsk Oblast.
Attended Irkutsk Art School 1950–55, pupil of D. I. Barkalov, S. I. Vinogradov and A. P. Krylov; exhibitions 1957–.

Altaev, Serafim Petrovich
Painter; b. 1931, Krasnyi Log, Voronezh Oblast.
Attended Moscow High School for Arts and Crafts 1950–58, pupil of V. A. Vasilev, S. V. Gerasimov.

Altaiskii, Konstantin Nikolaevich
Author; b. 1902, Syzran.
Publs.: various essays and novels, verse collections "Lenin" 1925, "Deti ulitsy" (Street Children) 1927, "Budennovka" (Budenny Helmet) 1928, "Kazakhstan moi" (My Kazakhstan) 1958, and others.

Altshuller, Moisei Solomonovich
Sculptor; b. 1922, Pereiaslav-Khmelnitskii, Kiev Oblast.
Attended Moscow Inst. for Applied and Decorative Art 1946–52; mem. CPSU 1946–; exhibitions 1953–.

Altunin, Aleksandr Terentevich
Military official; b. 1921.
Russian; mem. CPSU 1943; grad. Frunze Mil. Acad. and Acad. of the Gen. Staff; sec. of a village council 1937–39; served in Sov. Army 1939–; train driver, battalion commander, dep. commander of a regiment 1941–45; held command and staff posts 1945–68; Commander of Troops in North Caucasus Mil. District 1968–71; Head of Cadre Admin. in USSR Ministry of Defence 1971–72; Dep. USSR Min. of Defence and Head of USSR Civil Defence 1972–; Army-Gen.; mem. Cen. Cttee. CPSU 1976–.

Alver, Betti (pseudonym of Elizabeth Lepik)
Esthonian writer and translator; b. N. 23, 1906, Jigeva.
Born into a worker's family. She studied at the University of Tartu and published her first novel, *Mistress of the Wind,* in

1927. She also wrote the short story "The Invalids" (1930), the narrative poem *Song About a White Crow* (1931), the poem in prose *The Comedy of Poverty* (1935), and the collection of poems *Dust and Fire* (1936). Her work is characterized by painstaking care with form and flexibility and lightness of language. She translated Pushkin's *Eugene Onegin* and M. Gorky's *Childhood* and *My Universities* into Esthonian.
Works: *Tähetund*. Tallinn 1966. *In Russian translation:* In *Antologiia estonskoi poezii*, vol. 2. Moscow-Leningrad 1959.

Alybaev, Aripbai Alybaevich
State official; b. 1926.
Cand. of Agrarian Sciences; mem. CPSU 1949–, grad. Inst. of Veterinary Med. in Alma-Ata 1950; held council and economic posts 1951–55; Sec., 1st Sec. of Gvardeiskii and Kipalskii Raion Cttees. CP Kazakhstan in Taldy-Kurgan 1955–61; Oblast Cttee. Sec., 2nd Oblast Cttee. Sec. CP Kazakhstan in Alma-Ata, East Kazakhstan, and Pavlodar 1961–68; Chair. of Exec. Cttee. of Oblast Sov. of Peoples' Deps. in Pavlodar 1968–72; cand. mem. 1966–71, mem. Cen. Cttee. CP Kazakhstan 1971–; 1st Oblast Cttee. Sec. CP Kazakhstan in Taldy-Kurgan 1972–78; Min. of the Meat and Dairy Industry Kazakh SSR Ja. 1978–.

Alygulov, Bekten
Designer; b. 1922, Kirghizia.
Attended Art Schools in Frunze and Dnepropetrovsk 1948–53, pupil of G. A. Aitiev and A. N. Mikhalev; exhibitions 1958–.

Amangeldyev, Aman
Painter; b. 1930, Ashkhabad.
Attended Surikov Art Inst. Moscow 1952–58; mem. CPSU 1963–; exhibitions 1957–.

Amangeldyev, Chary
Painter; b. 1934, Ashkhabad.
Grad. Ashkhabad Art School 1951, attended Surikov Art Inst. Moscow 1959–65; exhibitions 1957–.

Amankozhoev, Kubanychbek
Painter; b. 1934, Kairma, Kirghizia.
Attended Art School Frunze 1949–54 and Surikov Inst. Moscow 1954–57; exhibitions 1957–.

Amashukeli, Elgudzha Davidovich
Sculptor; b. 1928, Tbilisi.
Attended Acad. of Art Tbilisi 1947–54; exhibitions 1954–. Decor.: Merited Artist of the Georgian SSR 1965.

Amasian, Pavel Grigorevich
Painter; b. 1922, Przhevalsk, Kirghizia.
Attended Pedagogical School of Art Astrakhan 1940–41; exhibitions 1953–.

Ambardnishvili, Otari Alekseevich
Painter and sculptor; b. 1924, Kakabeti, Georgia.
Grad. High School for Arts and Crafts Leningrad 1954, pupil of A. A. Dejneka; artist in monumental and decorative art; exhibitions 1959–.

Ambartsumian, Rafik (Rafael) Ervandovich
Sculptor; b. 1926, Erevan.
Studied Erevan Art and Theatre Inst., pupil of A. M. Sarkisian.

Ambartsumian, Ruben Sergeevich
Scientist; b. Ag. 28 (S. 10), 1911, Shusha.
Mem. CPSU 1942–, specialist in physical metallurgy and metal corrosion; corresponding mem. USSR Acad. of Sciences 1966–. Decors.: State Prize of the USSR 1949, 1953, Lenin Prize 1960.

Ambartsumian, Viktor Amazaspovich
Astrophysicist; b. S. 5 (18), 1908, Tbilisi.
Mem. CPSU 1940–; mem. USSR Acad. of Sciences 1953–; mem. 1943–, Pres. of Acad. of Sciences Armenian SSR 1947–; Vice-Pres. 1948–55, Pres. of Int. Astronomical Confederation 1961–64; mem. several int. astronomical insts.; Pres. of Int. Council of Scientific Confederations 1968. Decors.: Hero of Socialist Labour, 3 Lenin Orders and other int. decorations.
Publs.: "Teoretičeskaja astrofizika" Moscow-Leningrad 1933, "Evolucija svezd i astrofizika" Erevan 1947, "Naučnye trudy", vols. 1–2, Erevan 1960, "Problemy vnegalaktičeskih issledovanij" in: "Voprosy kosmologii 8" 1962, "Problemy evolucii Vselennoj" Erevan 1968.

Ambitskii, Iurii Ivanovich
Artist in applied art; b. 1927, Prusik, Poland; brother of M. I. Ambitskii.
Studied Lvov Art School 1946–57; woodcarver; exhibitions 1960–.

Ambitskii, Miron Ivanovich
Artist in applied art; b. 1928, Prusik, Poland; brother of Iu. I. Ambitskii.
Studied Lvov School for Applied and Decorative Art 1946–51; lives in Lvov; woodcarver; exhibitions 1957–.

Ambokadze, Eduard Mikhailovich
Graphic artist and painter; b. 1935, Tbilisi.
Attended Tbilisi Art School 1950–55 and Acad. of Art 1955–61, pupil of K. K. Gzelishvili, S. S. Kobuladze; exhibitions 1961–.

Ambraziunas, Alfonsas-Vintsentas Iono
Sculptor; b. 1933, Valdmantishkiai, Lithuania.
Studied Art Inst. of the Lithuanian SSR 1952–60; exhibitions 1959–.

Ambur, Lidia Martovna
Artist in applied art; b. 1919, Riidaku, Esthonia.
Attended Tallinn School of Applied Art 1942–44, pupil of V. Mellik, R. Niman and L. L. Erm; artist in applied art (textiles); teacher at Pedagogic Inst. Tallinn 1963–; exhibitions 1947–.

Amiian, Ivan Aleksandrovich
Painter and graphic artist; b. 1915, Shusha.
Attended Moscow Inst. of Fine Arts 1933–40, pupil of D. S. Moor, V. A. Favorskii, M. F. Shemiakin; exhibitions 1937–.

Amin-zade, Mukhiddin
Poet; b. 1904, Khodzhent, now Leninabad, Tadzhikistan.
Publs.: verse collection "Lug" (The Meadow) 1937, "Vesna Rodiny" (Spring at Home) 1940, "Tovarishcham khlopkorobam" (To the Cottonplanter Comrades) 1954, and others.

Amine, Avdal
Author; b. 1906, Iamanchair; s. of a peasant.
Lost both parents 1919; grew up in orphanages in Kars and Aleksandropol, now Leninakan; moved to Tbilisi 1923; grad. Philological Dept. of Erevan Univ. 1936; writes in Kurdish and Armenian.
Publs.: children's poems "Vesna" (Springtime) 1953, "Kurdy v Soveskoi Armenii" (Kurds in Soviet Armenia) 1953, poem "Tri brata" (Three Brothers) 1957, collection "Kurdskie narodnye skazki" (Kurdish Folktales) in Armenian 1957, "Byt kurdov Zakavkazia" (Way of Life of the Trans-Caucasian Kurds) 1957, and others.

Aminov, Akhad Umarovich
Artist in applied art; b. 1920, Namangan, Uzbekistan.
Attended Samarkand Tech. College of Art 1935–38; pupil of V. K. Razvadovskii; artist in applied art (ceramics).

Amiranashvili, Shalva Iasonovich
Historian of Georgian and Iranian art; b. Mr. 26 (Ap. 7), 1899, Oni, Georgian SSR.
Corresponding member of the Academy of Sciences of the USSR (1943); academician of the Academy of Sciences of the Georgian SSR (1955). Member of the CPSU since 1939.
He studied at the University of Tbilisi (1918–22); he has taught there since 1925 and held a professorship since 1936. Since 1939 he has been director of the Museum of Arts of the Georgian SSR in Tbilisi. His basic works are *Beka Opizari* (1939, expanded edition 1956), *Iranian Paintings of the XVIII–XIX Centuries* (1941), *History of Georgian Art* (1961, 1963), and *Georgian Miniatures* (1966).
Decors.: Order of Lenin, three other orders, and a medal.

Amiri, Gainan Gimazetdinovich
Author; b. 1911, Novyi Artaul, Bashkir ASSR; s. of a peasant.
Grad. Bashkir Pedagogical Inst.; mem. CPSU 1942–; works publ. 1928–.
Publs.: "Shamsutdin i Nuretdin" (Shamsutdin and Nuretdin) 1932, "Serdtse soldata" (A Soldier's Heart) 1943, "Zoloto" (Gold) 1940, "Bol'shoi sad" (The Big Garden) 1947, and others.

Amirkhanova, Marlena Shamirovna
Painter; b. 1914, Tbilisi.
Attended Tbilisi Acad. of Art 1936–42; pupil of D. N. Kakabadze, S. S. Kobuladze, A. G. Tsimakuridze; mem. CPSU 1941–; exhibitions 1945–.

Amirov, Mirsaiaf Masalimovich (pseudonym: Mirsai Amir)
Author and dramatist; b. 1906, Zirgan, Bashkir ASSR.
Grad. Gorkii Inst. of Literature; mem. CPSU 1943–; works publ. 1926–.
Publs.: "Chelovek nashei derevni" (The Man from our Village) 1933, "Agidel" 1936, plays "Minnikamal" 1944, "Pesnia zhizni" (The Song of Life) 1947, novels "Liudi iz Ialantau" (People from Ialantau) 1954, "Chistoe serdtse" (A Pure Heart) 1959, play "Svoboda" (Freedom) 1960, and others.

Amosov, Nikolai Mikhailovich
Medical scientist; b. D. 6 (19), 1913.
Grad. Medical Inst. Arkhangelsk 1939; surgeon; corresponding mem. at USSR Acad. of Medical Sciences; Chief of Clinic for Thorax Surgery at Ukrainian Janovskyi Research Inst. for Tuberculosis and Thorax Surgery 1952–; Prof. Chair of Thorax Surgery at Kiev Inst. for the Further Educ. of Doctors 1954–. Decors.: Merited Scientist of the Ukrainian SSR, 3 Lenin Orders and others.

Amurskii, Sergei Il'ich
Painter; b. 1925, Moscow.
Studied Moscow High School for Arts and Crafts 1948–56; artist in monumental and decorative art.

Ananev, Anatolii Andreevich
Author; b. Jl. 18, 1925, Dzhambul, Kazakhstan.
Russian; mem. CPSU 1950–; grad. Dept. of Philology of Kazakh State Univ. 1957. Decors.: 2 orders, medals.
Publs.: "Vernenskie rasskazy" (Tales of Vernensk) 1958, "Malyi zaslon" (Small Cover) 1959, "Tanki idut rombom" (The Tanks Move in Rhomb Formation) 1963, "Kozyri monakha Grigoriia" (The Trumps of the Monk Grigorii) 1964, "Ten' Isusa" (The Shadow of Jesus) 1961, and others.

Ananev, Boris Gerasimovich
Psychologist; b. Ag. 1 (14), 1907, Ordzhonikidze.
Grad. Pedagogical Inst. Ordzhonikidze 1928, Bekhterev Inst. for Brain Research 1930; mem. USSR Acad. of Pedagogical Sciences 1966–; Dean of Faculty of Psychology at Leningrad Univ. 1967–; author of several works on the study of sensations, of the transition from emotional to intellectual perception and of questions concerning developmental, differential and applied psychology.

Ananev, Iurii Aleksandrovich
Painter; b. 1920, Ogarevka, Saratov Oblast.
Attended Kazan Art School 1937–49, pupil of V. K. Timofeev; exhibitions 1954–.

Ananev, Mikhail Ananevich
Painter; b. 1925, Moscow.
Studied Kalinin School of Arts and Crafts Moscow 1950–55, pupil of Ch. A. Akhmarov, E. A. Lvov; exhibitions 1957–.

Ananian, Vakhtang Stepanovich
Author; b. 1905, Shamakhian, Armenia; s. of a peasant.
Shepherd until 1923; correspondent of krai newspaper "Pakhar", later ed. of same. Decor.: Order of Red Banner of Labour.
Publs.: "V ognennom koltse" (In the Fiery Ring) 1930, "Na okhote" (On the Hunt) 1934, "Rasskazy okhotnika" (From a Hunter's Tales) 1947–53, "Na beregu Sevana" (On the Banks of the Sevan) 1951, "Plenniki Barsova Ushchelia" (The Prisoners of the Barson Ravine) 1956, and others.

Ananikian, Akop Arutiunovich
Painter; b. 1919, Leninakan.
Attended Tech. College of Art and Inst. of Art Erevan, pupil of S. Agadzhanian, G. Giurdzhian; exhibitions 1940–; teacher at Leninakan Art School 1945–; mem. CPSU 1946–.

Ananin, Aleksandr Mikhailovich
Painter and graphic artist; b. 1918, Kolchugino, Siberia; brother of K. M. Ananin.
Attended Vladivostok Art Studio 1945–48; exhibitions 1955–; mem. CPSU 1944–.

Ananin, Iurii Aleksandrovich
Sculptor; b. 1933, Lyseve, Perm Oblast.
Attended Repin Inst. for Painting, Sculpture and Architecture Moscow 1953–59; exhibitions 1953–.

Ananin, Konstantin Mikhailovich
Painter; b. 1918, Leninsk-Kuznetskii; brother of A. M. Ananin.
Attended Vladivostok Art Studio 1949–52, pupil of V. V. Bezrodnyi, V. S. Zdanovich.

Ananko, Vladimir Iosifovich
Sculptor; b. 1929, Minsk.
Attended Belorussian Theatre and Art Inst. 1953–59, pupil of V. V. Volkov; exhibitions 1957–.

Anatolev, Nikolai Anatolevich
Painter; b. 1911, Moscow.
Attended Moscow Oblast Pedagogical Art School 1930–34, pupil of V. A. Favorskii, G. F. Fedorov, V. R. Eiges; exhibitions 1936–.

Anatolii (Evgenii Vlasovich Kuznetsov)
Russian Orthodox clergyman; b. My. 28, 1930, Irkutsk.
Altar server, psalm-reader, subdeacon; mil. service 1951–54; deacon 1954; grad. Moscow Theological Seminary and ordained priest 1956; studied Theological Acad., aspirant; monk 1960; abbot 1963; Archimandrite 1968; teacher at Theological Seminary Moscow; lecturer at Moscow Theological Acad. 1967–; Bishop of Vilnius and Lithuania 1972–74; Bishop of Zvenigorod, Representative of Moscow Patriarchate at seat of the Patriarch of Antioch 1974–80; Bishop of Ufa and Sterlitamak Mr. 1980– .

Anatskii, Semen Denisovich
Sculptor; b. 1924, Ostapovka, Poltava Oblast.
Attended Lugansk Art School 1947–52, pupil of V. I. Agibalov, M. M. Shevchenko; exhibitions 1950–.

Anchupan, Aleksandr Petrovich
Trade union official.
Cand. mem. Cen. Cttee. Latvian CP 1960–63, mem. 1963–; Sec. and mem. Presidium of Latvian Republican Trade Union Council D. 1967–.

Anderson, Edvin Aleksandrovich
Graphic artist; b. 1929, Valnyshi, Latvia.
Attended Acad. of Art of Latvian SSR 1948–56; exhibitions 1955–.

Andreenkov, Vladimir Emilianovich
Graphic artist; b. 1930, Chausy, Mogilev Oblast.
Attended Middle School of Art Moscow 1944–51, pupil of D. N. Domagatskii, and Surikov Art Inst. Moscow 1951–58.

Andreev, Aleksandr (real name: Andreev, Vasilii Dmitrievich)
Author; b. 1915, Krasnoe.
Engaged in literary activities 1942–.
Publs.: "Iasnye dali" (Clear Horizons) vols. 1–2, 1950–55 (collection 1958), "Ochen' khochetsia zhit'" (I So Much Want to Live) 1958, "Grachi prileteli" (Rooks Have Arrived) 1960, and others.

Andreev, Anatolii Evgenevich
State official; b. 1916.
1st Dep. Head of Main Bureau for Transportation with Council of Mins. Belorussian SSR 1960–61, Head 1961–63; Min. of Transportation Belorussian SSR Ap. 1963–; mem. Cen. Cttee. Belorussian CP 1966–.

Andreev, Gennadii Nikolaevich
Party official; b. 1936.
Russian; mem. CPSU 1960–; grad. S. M. Kirov Ural Polytechnical Inst. 1962; technical engineer, Cand. of Econ. Sc.; shift supervisor, sec. of party cttee. of Kuibyshev Chemical Plant in Togliatti 1962–64; 2nd Sec. of Togliatti City Cttee. CPSU 1964–66, 1st Sec. 1966–71; after graduating from Cen. Cttee. Party Higher School Czech. CP, Dep. Chief of Organizational Party Work Dept. of Kuibyshev Oblast Cttee. CPSU, instructor of sector in Organizational Party Work Dept. of Cen. Cttee. CPSU 1971–75, sector chief 1975–79; 2nd Sec. and mem. Bureau of Cen. Cttee. Armenian CP Ap. 1979–; cand. mem. Cen. Cttee. CPSU 1981–.

Andreev, Ivan Ivanovich
Painter; b. 1915, Prakhovo, Smolensk Oblast.
Attended Repin Inst. for Painting, Sculpture and Architecture at the USSR Acad. of Arts 1945–51; pupil of L. A. Ostrova, B. A. Vogel; exhibitions 1954–; mem. CPSU 1942–.

Andreev, Konstantin Grigor'evich
Painter and graphic artist; b. 1914, Totma, Vologda Oblast.
Attended Acad. of Arts of the Latvian SSR 1945–51, pupil of Ia. Kh. Tilberg; mem. CPSU 1945–. Decor.: Merited Artist of the Latvian SSR 1964.

Andreev, Nikolai Andreevich
Painter; b. 1920, Verete, Pskov Oblast.
Attended Leningrad Pedagogic Art School 1936–39.

Andreev, Nikolai Nikolaevich
Acoustical physicist; b. Jl. 16 (28), 1880, Kurman', Poltava Gouvt.
Grad. Basle Univ. 1909; mem. USSR Acad. of Sciences 1953–; founder of Soviet school of acoustics; Univ. Lecturer and Chief of Scientific Research Works at Moscow Univ. 1912–; author of several scientific and popular-scientific books and articles. Decors.: 3 Orders of Lenin, Order of Red Banner of Labour.

Andreev, Viacheslav Mikhailovich, Cand. of Phy.-Math. Sc.
Electronic engineer; b. 1941.
Grad. Electrotech. Inst. Leningrad 1963; Cand. of Phy.-Math. Sc. 1970–; mem. staff and Research Fellow at Physical-Tech. Inst. of USSR Acad. of Sciences 1966–. Decor.: Lenin Prize 1972.

Andreev, Vladimir Aleksandrovich
Painter; b. 1917 Bogchino, Kostroma Oblast.
Attended Art Schools in Perm 1933–38 and Leningrad 1938–39 and the Repin Inst. for Painting, Sculpture and Architecture Moscow 1947–53; pupil of V. M. Oreshnikov; teacher at Leningrad High School for Arts and Crafts 1953–.

Andreev, Vladimir Ivanovich
Diplomat.
USSR Amb. to Colombia Jl. 1971–78, to Suriname 1976–78.

Andreeva (Rumiantseva), Galina Alekseevna
Painter; b. 1927, Sestroretsk, Leningrad Oblast.
Attended Repin Inst. for Painting, Sculpture and Architecture at the USSR Acad. of Arts, pupil of B. V. Ioganson; exhibitions 1958–.

Andreeva, Kapitolina Gerasimovna
Painter; b. 1911, Bolokhov, Orel Oblast.
Attended Simferopol Art School 1943–48; pupil of O. A. Avsiian.

Andreichenko, Semen Semenovich
Sculptor; b. 1919, Sofievka nr. Ekaterinoslav.
Grad. Kiew Art Inst. 1951, pupil of M. G. Lysenko, A. P. Oleinik; exhibitions 1955–; mem. CPSU 1962–.

Andreichuk, Ivan Ignatevich
Sculptor; b. 1919, Kiev.
No formal artistic training; exhibitions 1953–.

Andrianov, Gennadii Ivanovich
Graphic artist; b. 1930, Moscow.
Attended Art Studios at Moscow Pedagogic Inst. 1949–51 and with magazine "Krokodil" 1959–61, pupil of K. S. Eliseev, E. A. Shukaev; exhibitions 1960–.

Andrianov, I.
Trade union official.
Mem. Auditing Comm. Lithuanian CP 1976–81; Sec. of Lithuanian Republican Trade Union Council F. 1982–.

Andrianov, Kuz'ma Andrianovich
Chemist; b. D. 15 (28), 1904.
Academician of the Academy of Sciences of the USSR (1964; corresponding member, 1953). Hero of Socialist Labor (1969). Member of the CPSU since 1949.
Upon graduation from Moscow University in 1930, he worked at the All-Union Electrotechnical Institute, and since 1954 has worked at the Institute of Hetero-organic Compounds of the Academy of Sciences of the USSR. From 1941 to 1959 he taught at the Moscow Power Institute (professor since 1946). Since 1959 he has directed the subdepartment of polymer syntheses at the Moscow Institute of Fine Chemical Engineering. His principal works are in the synthesis and technology of high-molecular-weight compounds, especially organic silicons. He first synthesized polyorganic siloxanes in 1937. Since 1947, he has been working on the basic principles of the synthesis of polymers with chains of inorganic molecules, including polyorganic metallosiloxanes. Under this guidance, work has been conducted in the synthesis of thermostable organosilicon polymers and materials based on them. These materials have found wide application in the insulation of electrical machines and apparatus, in the manufacture of lubricants, plastics, paints and varnishes, and so on. Decors.: State Prize of the USSR in 1943, 1946, 1950, and 1953 and the Lenin Prize in 1962.
Works: *Vysokomolekuliarnye kremniiorganicheskie soedineniia.* Moscow 1949. (With M. V. Sobolevskii.) *Organicheskie dielektriki i ikh primenenie v promyshlennosti szedstv sviazi.* Moscow-Leningrad 1949. (With S. A. Iamanov.) *Kremniiorganicheskie soedineniia.* Moscow 1955. *Teplostoikie kremniiorganicheskie dielektriki.* Moscow-Leningrad 1957. *Polimery s neorganicheskimi glavnymi tsepiami molekul.* Moscow 1962. "Kremnii." (*Metody elementoorganicheskoi khimii,* vol. 5.) Moscow 1968.

Andrianov, Petr Nikolaevich
Painter; b. 1916, Moscow.
Attended Moscow Oblast Pedagogic Art School 1937–39, grad. Surikov Art Inst. Moscow 1948.

Andrianov, Vasilii Ivanovich
Military official; b. Ag. 13, 1920, Ivanisovo, Kalinin Oblast.
Colonel in the Army. Member of the CPSU since 1944. Was graduated from the war aviation school for pilots in 1943. During the Great Patriotic War, he moved up through the ranks from pilot to squadron commander of the 141st Air Assault Regiment, which participated in the battle of Kursk, operated in the Ukraine, Rumania, and Poland, and took part in the Berlin operation. He completed 177 sorties, participated in 37 air battles, and shot down six planes. For his valor and courage he was twice awarded the title Hero of the Soviet Union (Jl. 1, 1944, and Je. 27, 1945). After the war he was graduated from the Air Force Academy and the General Staff Academy. He served in command positions and subsequently engaged in teaching. Decors.: Order of Lenin, three Orders of the Red Banner, Orders of the Patriotic War First Class, of Glory Third Class, of Alexander Nevsky, and of the Red Star, as well as medals.

Andrienko, Vladimir Timofeevich
Painter; b. 1926, Uspenovka, Kharkov Oblast.
Attended Kharkov Art. Inst. 1945–46 and Acad. of Arts of the Latvian SSR 1946–52; exhibitions 1952–.

Andriiashev, Anatolii Petrovich
Ichthyologist and zoogeographer; b. Ag. 6 (19), 1910.
Corresponding member of the Academy of Sciences of the USSR since 1966.
He graduated from Leningrad State University in 1933 and worked there until 1939. He subsequently worked at the Sevastopol' Biological Station and, begin-

ning in 1944, at the Zoological Institute of the Academy of Sciences of the USSR. A member of the first (1955–56) and third (1957–58) Antarctic expeditions, he served as the USSR representative in biology on the International Committee for the Study of the Antarctic.
Works: *Ocherk zoogeografii i proiskhozhdeniia fauny ryb Beringova moria i sopredel'nykh vod.* Leningrad 1939. *Ryby severnykh morei SSSR.* Moscow-Leningrad 1954. "Obzor fauny ryb Antarktiki." In *Issledovaniia fauny morei,* issue 2 (10). Moscow-Leningrad 1964.

Andriiuk, Pavel Stepanovich
Painter and graphic designer; b. 1923, Krasnoselkovo, Zhitomir Oblast.
Attended Omsk Art School 1938–41; pupil of K.P. Belov, T.P. Kozlov; exhibitions 1954–.

Andrikanis, Evgenii Nikolaevich
Cameraman and director; b. D. 14 (27), 1909, Paris.
Honored Art Worker of the RSFSR (1957). Member of the CPSU since 1968. In 1932, he graduated from the cinematography department of the All-Union State Institute of Cinematography. His first film were *Swamp Soldiers* (1938) and *Mashen'ka* (1942). He collaborated with the director S.I. Iutkevich in the creation of a number of color motion pictures – *Przheval'skii* (1952), *Skanderbeg, the Great Warrior of Albania* (1954), *Othello* (based on Shakespeare's tragedy; 1956), and the episode *The Last Autumn* in the film *Stories About Lenin* (1958). In 1958, Andrikanis participated with cameramen V.V. Nikolaev and R. Singh in the creation of the Soviet-Indian film *Journey Beyond the Three Seas.* Since 1960 he has worked as a director. He directed the films *A Northern Story* (1960), based on K.G. Paustovskii's work of that name, and *Executed at Dawn* (1965), which deals with the tragic death of Aleksandr Ul'ianov. He is the author of articles on the art of the cameraman. Decors.: Badge of Honor and medals.
Work: *Zapiski kinooperatora.* Moscow 1956.

Andruishchenko, Timofei Timofeevich
Painter; b. 1915, in Cossack village of Veselaia/Kuban.
Attended Krasnodar Art School 1936–39, pupil of G.A. Avetisian, N.D. Sharikov; exhibitions 1940–.

Andronikashvili, Eteri Nikolaevna
Graphic artist; b. 1920, Tbilisi.
Attended Tbilisi Acad. of Arts 1940–45; pupil of V.D. Grigoliia, I.A. Sharleman.

Andronikashvili, Georgii Aleksandrovich
Party official.
1st Sec. Tbilisi City Cttee. Georgian Komsomol until 1961; 2nd Sec. and mem. Bureau Cen. Cttee. Georgian Komsomol 1961–65; cand. mem. Cen. Cttee. Georgian CP 1961–66, mem. 1981–; 1st Sec. Lenin-Raykom Georgian CP in Tbilisi 1965–70; Sec. Tbilisi City Cttee. Georgian CP 1970–71; Head of Dept. for Building and Municipal Econ. with Cen. Cttee. Georgian CP N. 1976–; Sec. and mem. Bureau Cen. Cttee. Georgian CP Ja. 1981–.

Andronikov, Iraklii Luarsabovich (real name: Andronikashvili)
Author; b. 1908, St. Petersburg.
Mem. CPSU 1949–; grad. Historical-Philological Faculty of Leningrad Univ. 1930; author and literary critic; first work publ. 1929.
Publs.: "Varvara Zakharovna" 1924, "V gostiakh u diadi" (A Visit to Uncle) 1930, "Rasskazy-Razgovory" (Stories-Discussions) 1935, "Zemliak Lermontova" (Lermontov's Compatriot) 1952, "Oshibka Salvini" (Salvini's Mistake) 1959, "Tagilskaia nakhodka" (The Find at Tagil) 1956, and others.

Andronov, Nikolai Ivanovich
Painter; b. 1929, Moscow.
Attended Leningrad Middle School of Art, Repin Inst. for Painting, Sculpture and Architecture at the USSR Acad. of Arts 1948–52 and Surikov Art Inst. Moscow 1952–54, pupil of L.N. Solovev, R.R. Ferents; exhibitions 1951–.

Andropov, Iurii Vladimirovich
Party official; b. Je. 2 (15) 1914, Nagutskaja, Stavropol Krai.
Russian; mem. CPSU 1939–; interrupted univ. educ.; Army-Gen. 1976–; telegraph worker, apprentice cinematographic mechanic, seaman 1930–32; grad. tech. school for waterway transport; Komsomol organizer at a shipyard; Sec., then 1st Sec. of Iaroslavl Oblast Cttee. of All-Union Komsomol 1936–40; 1st Sec. of Cen. Cttee. Karelian Komsomol 1940–44; 1st Sec. of Petrozavodsk City Cttee. All-Union CP (B) 1944–47; 2nd Sec. of Cen. Cttee. Karelian CP (B) 1947–51; Dep. 1950–54, 1962–, mem. Foreign Affairs Comm. at Sov. of the Union USSR Supr. Sov. 1962–67; Inspector 1951–53, Section Chief at Cen. Cttee. CPSU 1951–53; USSR Amb. to Hungary 1953–57; Chief of Cen. Cttee. CPSU Liaison Dept. for Communist and Workers' Parties of Socialist Countries 1957–67; mem. Cen. Cttee. CPSU 1961–; Sec. of Cen. Cttee. CPSU 1962–67; Chair. of State Security Cttee. with Council of Mins. My. 1967–82; cand. mem. 1967–73, full mem. Politbureau Cen. Cttee. CPSU Ap. 1973–; Sec. Cen. Cttee. CPSU My 1982–; mem. Presidium USSR Supr. Sov. D. 1982–83, Chair. Je. 1983–; Chair. of Council of Defence of the USSR Je. 1983–. Decors.: 2 Orders of Lenin, 4 Orders of Red Banner of Labour, various medals.

Andrushkevich, Vladimir Ilich
Graphic artist; b. 1923, Pimenovka in the Urals.
Attended Art Studio for War Invalids Moscow 1947–52 and Surikov Art. Inst. Moscow 1952–58; exhibitions 1958–; mem. CPSU 1950–.

Andzhaparidze, Vera Ivlianovna (also Veriko I. Ivlianovna)
Georgian actress; b. S. 23 (O. 6), 1900, Kutaisi.
People's Artist of the USSR (1950). Studied at the drama studios of S.V. Aidarov in Moscow and G. Dzhabadari in Tbilisi. From 1920 to 1927, she worked at the Rustaveli Georgian Theater in Tbilisi, and in 1928 she became an actress at the Mardzhanishvili Georgian Theater there. She is an actress of brilliant creative individuality and considerable artistic culture; her performances further the romantic traditions of the Georgian theater. Her best roles have been Ginatre (*Guriia Ninoshvili* by Dadiani), Dzhavara (*The Exile* by Vazha Pshavela), Iudif' (*Uriel' Akosta* by Gutskov), Cleopatra (*Antony and Cleopatra* by Shakespeare), Marguerite Gautier (*Camille* by Dumas *son*), Mary Stuart (the play of the same name by Schiller), Grandmother (*Trees Die Standing Up* by Casona), and Mother (*Mother* by Čapek). She first appeared in films in 1923 – for example, she played Rusudan in *Georgii Saakadze* and Otar's Widow in the film of the same name. Decors.: State Prize of the USSR on three occasions – 1943, 1946, and 1952 – and Order of Lenin, three other orders, and medals.

Andzhaparidze, Zurab Ivanovich
Opera singer (lyrical-dramatic tenor); b. Ap. 12, 1928, Kutaisi.
People's Artist of the USSR (1966). Member of the CPSU since 1954.
After graduating from the Tbilisi Conservatory (1952, class of professor D.Ia. Andguladze), he was accepted by the Paliashvili Theater of Opera and Ballet. His parts have included José (Bizet's *Carmen*), Cavaradossi (Puccini's *Tosca*), Malkhaz (Paliashvili's *Daisi*), Radames (Verdi's *Aïda*), and Mindiia (Taktakishvili's *Mindiia*). Since 1959 he has been soloist at the Bolshoi Theater (Moscow). In 1964 he performed the role of Hermann in Tchaikovsky's opera *The Queen of Spades.* He has performed on tour in Czechoslovakia, Bulgaria, Italy, France, Yugoslavia, Canada, Rumania, and elsewhere.

Anfimov, Leonid Vasil'evich
Graphic artist; b. 1928, Tbilisi.
Attended Techn. College of Art 1943–45 and Acad. of Tbilisi 1949–55; pupil of I.A. Sharleman; exhibitions 1955–.

Angervaks, Hugo (real name: Piall, Eduard Nikolaevich)
Author; b. 1903, Esthonia.
Mem. CPSU 1920–; grad. Pedagogical Herzen Inst. Leningrad 1929; Esthonian author and literary critic; Red Army volunteer 1918; teacher in Siberia 1924–26; Govt. and Party official 1940–50; Dir. of Inst. of Languages and Literature at Acad. of Sciences Esthonian SSR 1955–; corresponding mem. Acad. of Sciences Esthonian SSR 1961–; works first publ. in the USSR 1926.
Publs.: "Shestaia lenta" (The Sixth Ribbon) 1931, "Ispytanie vremenem" (Test of Time) 1957, "Utro" (The Morning) 1959, "Starye berezki" (Old Birches) 1960, author of works on linguistics and literary criticism, and others.

Anikeenok, Aleksei Avdeevich
Painter; b. 1925, Detskoe Selo, Leningrad Oblast.
Attended Kazan Art School 1952–57; exhibitions 1957–.

Anikeev, Mikhail Korneevich
Painter and graphic artist; b. 1925, Nizovaia, Tambov Oblast.
Attended Kishinev Art School and Surikov Art Inst. Moscow 1950–56; exhibitions 1954–.

Anikeichik, Anatolii Aleksandrovich
Sculptor; b. 1932, Novo-Borisov, Belorussia.
Attended Belorussian Theatre and Art Inst. 1953–59, pupil of A.O. Bembel, V.V. Volkov; teacher at Theatre and Art Inst. 1959–; exhibitions 1959–.

Anikst, Aleksandr Abramovich
Russian literary and theater critic; b. Jl. 16, 1910, Zurich.
Member of the CPSU since 1942. Doctor of Art Studies (1963).
His works, which were first published in 1930, are devoted to the history of English and American literature and to aesthetics. In the area of dramatic history, he has concentrated on W. Shakespeare, writing works that illustrate the humanism and profundity of the English playwright's artistic images: *The Works of Shakespeare,* 1963; *Shakespeare,* 1964; and *Theater in the Era of Shakespeare,* 1965. He has also written articles on contemporary Soviet and foreign theater and films.
Works: *Istoriia angliiskoi literatury.* Moscow 1956. *B. Shou.* Moscow 1956. "Lev Tolstoi – nisprovergatel' Shekspira." *Teatr,* 1960, no. 11. *Teoriia dramy of Aristotelia do Lessinga.* Moscow 1967. "'Nosorogi' v N'iu-Iorke." *Novyi mir,* 1965, no. 8. "Takoi chelovek nuzhen vsiakomy vremeni." *Iskusstvo kino,* 1968, no. 1.

Anikushin, Mikhail Konstantinovich
Painter; b. 1917.
Mem. CPSU 1944–; grad. Inst. of Painting, Sculpture and Architecture Leningrad 1947; Prof.; served in Sov. Army 1941–45; Asst. 1947–51, Studio Head at Faculty of Sculpture at Inst. of Painting, Sculpture and Architecture Leningrad 1958–62; Sec. of RSFSR Painters' Fed. 1960–; Chair. of Leningrad Branch of RSFSR Painters' Fed. 1962–; mem. USSR Art Acad. 1962–; Sec. of USSR Painters' Fed. 1963–; mem. Cen. Auditing Comm. CPSU 1966–. Decor.: Lenin Prize.

Anisiforova, Nina Nikolaevna
Artist; b. 1915, Ufa.
Grad. Bashkir Tech. College of Art 1937; pupil of A.E. Tiulkin, I.I. Uriadov; exhibitions 1947–.

Anisimkin, Ivan Georgievich
Party official; b. 1909.
Dep. to Supr. Sov. Uzbek SSR 1955–; dir. of an electro-chemical collective combine 1955; cand. mem. 1956–60, mem. Cen. Cttee. Uzbek CP 1960–; Chief of Exec. Office for Chemical Ind. of Cen. Asian Econ. Council 1963; Sec. and mem. Bureau of Cen. Cttee. Uzbek CP Ag. 1964–.

Anisimov, Aleksei Semenovich
Sculptor; b. 1926, Koreevo, Riazan Oblast.
Attended Moscow Oblast Pedagogical Art School 1937–41, and Moscow Oblast Pedagogical Inst. 1946–51; exhibitions 1957–.

Anisimov, Anatolii Vasil'evich
Diplomat; b. 1919.
Mem. Diplomatic Service 1946–; Emb. Councellor at USSR Embassy in Iran 1961–65; mem. Apparat USSR Min. of Foreign Affairs 1965–68 and 1972–77; USSR Amb. to Jordan 1968–72, to Libya S. 1977–.

Anisimov, Leonid Petrovich
Graphic artist; b. 1931, Novosibirsk.
Grad. Kazan Art School 1957, pupil of V.K. Timofeev; exhibitions 1957–.

Anisimov, Pavel Petrovich
State official; b. 1928.
Mem. CPSU 1952–; grad. Aeronautical Inst. Kazan, engineer-mechanic; designing engineer, sec. of Komsomol cttee. of a works in Kuibyshev 1952–53; engineer and designer in a dept. 1953–55, Sec. of Party Cttee. of Zhdanov Works in Leningrad 1955–61; 1st Sec. of Kirov Raion Cttee. CPSU in Leningrad 1961–63; Dept. Chief at Oblast Cttee. CPSU in Leningrad 1963–64; City Sec. of CPSU in Leningrad 1964–68; Dep. Head of Dept. for Party org. Work with Cen. Cttee. CPSU 1968–73; 2nd Sec. and mem. Bureau of Cen. Cttee. Armenian CP Mr. 1973–79; mem. Cen. Cttee. Armenian CP 1976–81; cand. mem. Cen. Cttee. CPSU 1976–. Dep. Chair. USSR Gosplan 1979–. Decor.: Order of October Revolution.

Aniskin, Pavel Sergeevich
Painter; b. 1926, Azarashino, Penza Oblast.
Grad. Penza Art School 1957, pupil of I.S. Goriushkin-Sorokopudov, A.I. Postnov; exhibitions 1954–.

Annakulieva, Annagul'
Turkmen opera singer (soprano); b. D. 31, 1924, Gasan-Kuli Raion, Ashkhabad Oblast.
People's Artist of the USSR (1965). Member of the CPSU since 1952.
She has been a performer with the Turkmen Theater of Opera and Ballet in Ashkhabad since 1941. Her roles include Zokhre in Shaposhnikov's *Zokhre and Takhir,* Shasenem in Shaposhnikov and Ovezov's *Shasenem and Garib,* Ogul'bek in Shaposhnikov and Mukhatov's *The Poet and the Judge,* Tat'iana and Liza in Tchaikovsky's *Eugene Onegin* and *Queen of Spades,* and Natasha in Dargomyzhskii's *The Mermaid.* Decors.: Order of Lenin, the Badge of Honor, and various medals.

Annakurbanov, Chary
State official.
Chair. of State Cttee. for Television and Radio Turkmen SSR D. 1972–; mem. Cen. Cttee. Turkmen CP 1976–.

Annaorazov, Poda
Party official; b. 1922.
Turkmen; mem. CPSU 1946, grad. Party Higher School in Tashkent; served in the Soviet Army 1941–46; Instructor, Sec. of Ashkhabad Oblast Cttee. Turkmen Komsomol 1947–50; Sec. Cen. Cttee. Turkmen Komsomol 1950–53; Dep. Head of Turkmen Republican Admin. for Labour 1954–58; Dep. Head, 1st Dep. Head of a Dept. with Cen. Cttee. Turkmen CP 1962–70; Cand. mem. 1963–71, mem. Cen. Cttee. Turkmen CP 1971–; mem. Bureau for Agriculture at Cen. Cttee. Turkmen CP 1963–64; 1st Sec. Mary City Cttee. Turkmen CP 1970–71; 1st Oblast Sec. of Turkmen CP in Mari F. 1971–76; Dep. to USSR Supr. Sov. 1974–; 1st Sec. of Ashkhabad Oblast Cttee. Turkmen CP 1976–; mem. Bureau Cen. Cttee. Turkmen CP Ag. 1976–. Decor.: Order of Lenin.

Annenkov, Nikolai Aleksandrovich (real surname: Kokin)
Actor; b. S. 8 (20), 1899, village of Inzhavino, Tambov Province.
People's Artist of the USSR (1960). Member of the CPSU since 1942. Instructor at the Shchepkin Theatrical School since 1946; professor since 1961. In 1924, he graduated from the Higher Theatrical Workshops attached to the Malyi Theater in Moscow and joined that theater's troupe. A character and dramatic actor, he was especially associated with the plays of A.N. Ostrovskii and M. Gorky, performing the roles of Milovzorov and Neznamov in Ostrov-

skii's *Guilty Without Guilt*, Andrei Belugin in Ostrovskii and N. Solov'ev's *Belugin's Marriage*, and Nil, Cherkun, and Basov in Gorky's *Petit Bourgeois*, *Barbarians*, and *Summer Residents*. His roles in plays by Soviet dramatists include Ognev in Korneichuk's *The Front*, Maksimov in Lavrenev's *For Those and Sea!*, Shvandia and Koshkin in Trenev's *Liubov' Iarovaia*, Plakun in Zorin's *Eternal Source*, and Novikov in Aleshin's *The Ward*. Decors.: State Prize of the USSR in 1947, 1948, and 1949, Order of Lenin, two other orders, and various medals.

Anni, Albert Avgustovich
Painter; b. 1920, Iadroshino, Moscow Oblast.
Attended Kelinin School of Arts and Crafts Moscow 1937–41 and Tartu Art Inst. 1945–50; teacher at Fine Arts School Tartu 1952–60 and 1964–; exhibitions 1952–.

Anokhin, Iurii Nikolaevich
Painter; b. 1926, Moscow.
Attended Surikov Art. Inst. Moscow 1950–56; exhibitions 1951–.

Anokhin, Petr Kuz'mich
Physiologist; b. Ja. 14 (26), 1898, Tsaritsyn, now Volgograd.
Academician of the Academy of Sciences of the USSR since 1966; member of the Academy of Medical Sciences of the USSR (AMN SSSR) since 1945.
A graduate of the Leningrad Institute of Medical Sciences (1926), he began to work in the Brain Institute under the direction of V. M. Bekhterev in 1921. Between 1922 and 1930 he worked in I. P. Pavlov's laboratory at the Military Medical Academy. From 1930 to 1934 he was a professor in the dept. of physiology at the Gorkii Medical Institute. During 1934–44, he headed a department at the All-Union Institute of Experimental Medicine in Moscow and during 1944–50, at the Institute of Physiology of the AMN SSSR. He became director of the Institute of Physiology in 1946. In 1950, he became head of the Neurophysiological Laboratory of the AMN SSSR and later also head of the department of neurophysiology at the Institute of Normal and Pathological Physiology of the AMN SSSR. He has been teaching in institutions of higher learning in Moscow since 1936 and has been head of a department since 1955.
His principal works have been devoted to the study of the activity of the total organism, especially the brain, based on the functional-system theory which he worked out. The application of this theory to the evolution of function has made it possible for him to formulate the concept of systemogenesis as a general principle of the evolutionary process. Decors.: I. P. Pavlov Gold Medal in 1968, three orders, and also some medals.
Works: *Problema tsentra i periferii v fiziologii nervnoi deiatel'nosti*. Editor-in-chief, P. Anokhin. Gorky 1935. (Collection of works.) "Sistemogenez kak obshchaia zakonomernost' evoliutsionnogo protsessa." *Biulleten' eksperimental'noi biologii i meditsiny*, 1948, vol. 26, no. 8. "Teoriia funktsional'noi sistemy kak predposylka k postroeniiu fiziologicheskoi kibernetiki." In *Biologicheskie aspekty kibernetiki*. Moscow 1962. "Kibernetika i integrativnaia deiatel'nost' mozga." *Voprosy psikhologii*, 1966, no. 3. *Biologiia i neirofiziologiia uslovnogo refleksa*. Moscow 1968.

Anosov, Anatolii Ivanovich
Graphic artist; b. 1936, Ulan-Ude.
Attended Irkutsk Art School 1959–64; pupil of G. V. Bogdanov, M. D. Voronko and L. B. Grimov; exhibitions 1961–.

Anosov, Vasilii Maksimovich
Painter; b. 1921, Ivanovka, Riazan Oblast.
Attended Surikov Art Inst. Moscow 1944–50; exhibitions 1953–.

Ansulis, Voldemar Fritsevich
Artist in applied art; b. 1919, Liepāja, Latvia.
Studied Faculty of Architecture Latvian Univ. 1941–43; artist in applied art (amber, silver); teacher at School for Applied Art Liepāja 1945–; exhibitions 1956–.

Antadze, Isaak (Dodo) Konstantinovich
Stage director; b. 1900.
Georgian; attended G. Dzhabadari Drama Studio Tbilisi 1919; one of the winners of the 2nd Nat. Drama Theatre Kutaisi; Artistic Adviser and Dir. of K. Mardzhanishvili Theatre Tbilisi 1933–38; Dir. of Meskhishvili Theatre Kutaisi 1938–52; Dir. of Russian A. S. Griboedov Theatre 1952–57; Dir. of Sh. Rustaveli Theatre Tbilisi 1957–62; Chair. of Georgian Theatre Soc. 1962–. Decor.: People's Artist of the USSR 1971.

Antanaitis, Iu.
Trade union official.
Cand. mem. Cen. Cttee. Lithuanian CP 1976–; Sec. of Lithuanian Republican Trade Union Council F. 1977–.

Antinite-Povilaitiene, Daiva Roberto
Graphic designer; b. 1935, Kaunas; d. of R. Ia. Antinis.
Attended Art. Inst. of the Lithuanian SSR 1953–59; exhibitions 1959–.

Antipov, Nikolai Grigorevich
Painter; b. 1923, Bogdanovka, Orel Oblast.
Attended Penza Art School 1946–48, pupil of I. S. Goriushkin-Sorokopudov, and Elets Art School 1950–52, pupil of M. I. Lapshin, V. S. Sorokin; teacher at an intermediate school 1952–; exhibitions 1955–.

Antipova, Evgeniia Petrovna
Painter; b. 1917, Toropets, Pskov Oblast.
Attended Leningrad School for Arts and Crafts 1935–39, grad. Repin Inst. for Painting, Sculpture and Architecture at the USSR Acad. of Arts 1950, pupil of S. L. Abugov, A. D. Zaitsev, A. A. Osmerkin; exhibitions 1954–.

Antonchik, Mikhail Vladimirovich
Painter; b. 1921, Kiev.
Attended Sudarin Art School Kiev and Kiev Art Inst. 1944–50; exhibitions 1949–.

Antonenko, Petr Kuzmich
Painter and graphic artist; b. 1921, Buran, E. Kazakhstan.
Attended Samarkand Art School 1939–41, pupil of P. P. Benkov, and Theatre and Art School Alma-Ata 1946–49, pupil of A. M. Cherkasskii; exhibitions 1948–.

Antonenko-Davidovich, Boris Dmitrievich
Author; b. 1899, Romny; s. of a railwayman.
Ukrainian; attended Univ. Kiev; first works publ. 1923.
Publs.: "Rytsari 'absurda'" (Knights of the Absurd) 1924, "Zaporosheni siluety" (Dust-Covered Silhouettes) 1925, "Synia voloshka" (The Blue Cornflower) 1927, "Liudy i vugillia" (People and Coal) 1932, "Za shirmoiu" (Behind the Screen) 1963, and others.

Antonian, Ludmilla Georgievna
Painter; b. 1926, Shalakhovo, Iaroslavl Oblast.
Attended Iaroslavl Art School 1943–48; exhibitions 1951–.

Antonii (O. I. Vakarik), Cand. of Theol.
Russian Orthodox clergyman; b. Jl. 26, Stavchany, Chernovtsy Oblast.
Monk, Admin. Man. at St. John of the Theologians Monastery in Bishopric of Chernovtsy 1947; priest-monk 1947; completed studies at Theological Seminary and Acad. Moscow 1956; aspirant of Moscow Theological Acad. 1964; priest in Moscow 1964–65; Bishop of Smolensk 1965–67, of Simferopol and the Crimea 1967–73 and of Chernigov and Nezhin My. 1973–; Archbishop and temporary mem. of the Holy Synod 1974–.

Antonii (Anatolii Sergeevich Melnikov), Cand. of Theol.
Russian Orthodox clergyman; b. 1924, Moscow.
Grad. Moscow Theological Acad. 1950, priest-monk 1950; lecturer and later supervisor at Odessa Theological Seminary 1950–52; supervisor at Saratov Theological Seminary 1952–56; Archimandrite, Rector of Minsk Theological Seminary and Prior of Zhirovits Uspenskii Monastery 1956–64; Bishop of Belgorod-Dnestrovskii and Substitute Bishop of Odessa

1964–65; Bishop 1965, Archbishop of Minsk and Belorussia O. 1965–78; Chair. of Ed. Board of collections "Theological Works" 1967; temporary mem. of the Holy Synod 1976–78, mem. 1978–; Metropolitan of Leningrad and Novgorod S. 1978–.

Antonii (Aleksandr Mikhailovich Savgorodnii), Cand. of Theol.
Russian Orthodox clergyman; b. S. 12, 1938, Kislovodsk.
Grad. Stavropol Theological Seminary 1960; monk, deacon 1961; grad. Leningrad Theological Acad. 1964; Priest, Head Priest, Sacristan of Nikolai Cathedral Leningrad 1964; Archimandrite 1967, Chair. of Church Representation in Jerusalem until 1970; active in Bishopric of Leningrad 1970–; foreign visits, Delegate of the Russian Church at 4th Gen. Assembly of World Council of Churches in Uppsala; Bishop of Stavropol and Baku 1975–.

Antonii (A. Blum), Dr. med.
Russian Orthodox clergyman; b. Je. 19, 1914, Lausanne.
Studied medicine in Paris; Dr. med. 1942; monk 1943; after World War II attached to Patriarchate of Moscow; monk-deacon, then priest-monk; ecclesiastical leader of the Anglican-Orthodox Soc. of St. Alban and Sergi in England; abbot 1953; Archimandrite, Chair. of the Cathedral of Mary the Deceased in London 1956; Bishop 1957; Archbishop with title "Surozhskii", Head of the newly established Diocese in Great Britain and Ireland 1962; responsible as Exarch of the Moscow Patriarchate in W. Europe 1963–; Metropolitan and Exarch of the Moscow Patriarchate in W. Europe 1963–.

Antonov, Aleksei Konstantinovich
State official; b. 1912.
Mem. CPSU 1940–; grad. Leningrad Polytech. Inst. 1935; locksmith, student, postgraduate studies at the Inst. 1930–35; engineer, dept. chief, dep. chief designer, chief engineer at an industrial plant 1935–37; Chief Engineer, Board Chair. 1957–59, Dep. Chair. 1959–61, Chair. of Leningrad Econ. Council 1961–65; cand. mem. 1961–71, full mem. Cen. Cttee. CPSU 1971–; Dep. to USSR Supr. Sov. 1962–; USSR Min. of the Electrical Engineering Ind. O. 1965–80; Dep. Chair. of USSR Council of Mins. 1980–. Decor.: State Prize.

Antonov, Evgenii Ivanovich
Painter; b. 1930, Moscow.
Attended Repin Inst. for Painting, Sculpture and Architecture at the USSR Acad. of Arts 1955–61; exhibitions 1948–.

Antonov, Evgenii Nikolaevich
Painter and graphic artist; b. 1925, Smolensk.
Attended Penza Art School 1951–55, pupil of M. E. Vakulin, and Moscow Polygraphic Inst. 1955–61, pupil of Iu. N. Burdzhelian, G. Y. Goroshchenko; exhibitions 1948–.

Antonov, Iurii Ivanovich
Graphic artist; b. 1929, Moscow.
Attended Moscow High School for Arts and Crafts 1951–56, pupil of N. Ia. Belianin, V. A. Ermolov; exhibitions 1957–.

Antonov, Nikolai Afanasevich
Party official; b. 1921.
Mem. CPSU 1944–; grad. Leningrad Industrial Inst. 1951; served in Sov. Army 1941–45; dept. chief, chief engineer, dir. of plant in Novgorod Oblast 1945–55; party and sov. posts 1956–60; sec. of Novgorod Oblast Cttee. CPSU 1960–66; Dep. to Supr. Sov. RSFSR 1966–; Chair. of Novgorod Oblast Exec. Cttee. of Sov. of Peoples' Deps. 1966–72; 1st Sec. of Novgorod Oblast Cttee. CPSU My. 1972–; Dep. to USSR Supr. Sov. 1974–; cand. mem. Cen. Cttee. CPSU 1976–.

Antonov, Oleg Konstantinovich, Dr. of Tech. Sc.
Aircraft designer; b. F. 7, 1906, Troitskoe, now Moscow Oblast; s. of a building engineer.
Mem. CPSU 1945–; grad. Polytech. Inst. Leningrad 1930; corresponding mem. Acad. of Sciences Ukrainian SSR 1964–; Dr. of Tech. Sc. 1960; chief designer at a glider factory 1930–62; employed in several design offices; Sen. Designer with USSR Min. of Aviation Ind. 1962–; has written over 50 works on gliders, aircraft construction, economics. Decors.: Lenin Order, Order of the Red Banner of Labour, Order of the Patriotic War 1st Cl., and various medals.
Publs.: "Na kryliakh iz dereva i polotna" 1962, "Dlia vsekh i dlia sebia" 1965.

Antonov, Sergei Fedorovich
State official; b. 1911.
Mem. CPSU 1937–; grad. Leningrad Inst. of Dairy Farming 1937 and CPSU Cen. Cttee. Party College 1941; worked in a creamery, attended master school of dairy farming, taught at a technical college, Dep. Chief Engineer and Dept. Chief at USSR Exec. Office of Dairy Farming 1928–39; attended CPSU Cen. Cttee. Party College 1939–41; served in Sov. Army 1941–46; Chief of Exec. Office, later Dep. USSR. Min. for Meat Production and Dairy Farming, Dept. Chief at USSR Council of Mins. 1946–54; USSR Min. for Meat Production and Dairy Farming 1954–57; USSR Emb. Counsellor in China 1958–60; USSR Amb. to Afghanistan 1960–65; USSR Min. of Meat Production and Dairy Farming O. 1965–; Chair. of USSR Soc. for Friendship and Cultural Relations with Afghanistan 1966–; mem. CPSU Cen. Auditing Comm. 1966–76; mem. Cen. Cttee. CPSU 1976–; Dep. to USSR Supr. Sov. 1966–. Decor.: Order or Lenin.

Antonov, Sergei Georgievich
Painter and graphic artist; b. 1932, Naro-Fominsk, Moscow Oblast.
Attended Moscow School for Arts and Crafts 1948–56; pupil of V. I. Finogeev; exhibitions 1960–.

Antonov, Sergei Petrovich
Author; b. 1915, Petrograd.
Grad. Inst. of Highways Leningrad 1938; civil engineer; Russian author, first work publ. 1944. Decors.: Stalin Prize, Order of the Red Banner of Labour.
Publs.: "Vesna" (Spring) 1947, "Po dorogam idut mashiny" (Cars Drive Along the Street) 1950, "Mirnye liudi" (Peaceful People) 1950, "Pervaia dolzhnost'" (The First Office) 1952, "Delo bylo v Penkove" (It Happened in Penkov) 1956, "Alenka" 1960, and others.

Antonova, Galina Aleksandrovna
Artist in applied art; b. 1926, Moscow.
Grad. Leningrad High School for Arts and Crafts 1953; artist in applied art (glass, crystal); teacher at Moscow High School for Arts and Crafts 1962–; exhibitions 1954–.

Antonovich, Iosif Arkadevich
State official.
Min. for Installation and Special Construction Work Belorussian SSR D. 1970–; mem. Auditing Comm. Belorussian CP 1971–81, cand. mem. Cen. Cttee. Belorussian CP 1981–.

Antonovskii, Valerii Aleksandrovich
Painter; b. 1921, Aleksinskaia, Arkhangelsk Oblast.
Attended Iaroslavl Pedagogic Art School 1937–41, pupil of A. A. Churin, S. F. Shitov; exhibitions 1940–.

Antroshchenko, Vasilii Ivanovich, Dr. of Tech. Sc.
Scientist; b. Jl. 3, 1906, Iuzovka, now Donetsk; s. of a railwayman.
Grad. Chemical-Technological Inst. Odessa 1931; chemical technologist; corresponding mem. Acad. of Sciences Ukrainian SSR 1964–; Dr. of Tech. Sc. 1945; Prof. 1945; Asst. and Cand. at Chemical-Technological Inst. Odessa 1931–35; Asst., Lecturer, Prof., Prof. Chair for Technology of Inorganic Substances, Dep. Dir. of Teaching Dept. at Chemical-Technological Inst. (now Polytech. Inst.) Kharkov 1935–50; Prof. Chair for Technology of Inorganic Substances and Dep. Dir. of Teaching Dept. at Polytech. Inst. Kharkov 1950–; author of over 100 scientific works. Decors.: Merited Scientist of Ukrainian SSR, Lenin Order, Order of the Red Banner of Labour, various medals.
Publs.: "Tekhnologiia azotnoi kisloty" 1949, "Metody raschetov po tekhnologii sviazannogo azota" 1960, "Tekhnologiia sviazannogo azota" 1962, and others.

Anufriev, Evgenii Ivanovich
Painter; b. 1923, Rotanovka, Orel Oblast. Attended Surikov Art Inst. Moscow 1944–50, pupil of P.D. Pokarzhevskii, G.M. Shegal; exhibitions 1953–.

Apanovich, Viktor Antonovich
Painter; b. 1922, Tbilisi.
Attended Repin Inst. for Painting, Sculpture and Architecture at the USSR Acad. of Arts 1946–50; exhibitions 1953–.

Apine-Zhiglevitsa, Benita Karlovna
Painter; b. 1929, Riga.
Attended Acad. of Arts of the Latvian SSR 1951–58, pupil of E.F. Kalnin, O.E. Skulme; exhibitions 1958–.

Apostolova, V.G.
Komsomol official.
Sec. and mem. Bureau Cen. Cttee. Moldavian Komsomol Mr. 1982–.

Appollonov, Ivan Grigor'evich
Artist in applied art; b. 1933, Podol, Poltava Oblast.
Attended Leningrad High School for Arts and Crafts 1954–60, pupil of B.A. Smirnov; artist in applied art (glass); exhibitions 1957–.

April, Aron Isaakovich
Painter and graphic artist; b. 1932, Vilkaviškis, Lithuania.
Grad. Moscow Oblast Pedagogic Art School 1948, attended Surikov Art Inst. Moscow 1954–60, pupil of F.I. Nevezhin; exhibitions 1951–.

Arabidze, Vasilii Valerianovich
Painter; b. 1932, Zestafoni, Georgia.
Attended Tbilisi Acad. of Art 1953–59, pupil of Sh. P. Mamaladze, K.B. Sanadze; exhibitions 1959–.

Araef'ev, Anatolii Vasil'evich
Designer of stage sets; b. N. 24, 1918, Stavropol', Smarskaia Province.
People's Artist of the USSR (1967). Became a member of the CPSU in 1945 and a candidate member of the Central Committee of the Communist Party of Kirghizia in 1967.
He studied at the Art and Pedagogical Technicum in Penza (1936–38). Since 1947 he has worked in the Kirghiz Theater of Opera and Ballet in Frunze. (He has been chief artist there since 1949.) He has designed the sets for more than 70 opera and ballet productions, including *Carmen* by G. Bizet (1949); *Swan Lake* by P.I. Tchaikovsky (1949); *Prince Igor* by A.P. Borodin (1954); *Toktogul* by V.A. Vlasov, A. Maldybaev, and V.G. Fere (1958); *Cholpon* by M.R. Raukhverger (1958); and *The Unknown Soldier* by K.M. Molchanov (1967). Decors.: Order of Lenin, two other orders, and medals.

Arakcheev, Boris Vladimirovich
Painter and graphic artist; b. 1926, Turbanovo, Iaroslavl Oblast.
Attended Belorussian Theatre and Art Inst. 1953–59; exhibitions 1959–.

Arakelian, Arakel Sedrakovich
Painter; b. 1927, Erevan, s. of S.A. Arakelian.
Grad. Erevan Art and Theatre Inst. 1955; exhibitions 1954–.

Arakelian, Babken Nikolaevich
Archaeologist and historian; b. F. 1 (14), 1912, village of Gecherlu, Echmiadzin District.
Corresponding member of the Academy of Sciences of the Armenian SSR (1968). Member of the CPSU since 1940.
He has been director of the Institute of Archaeology and Ethnology of the Academy of Sciences of the Armenian SSR since 1959. He is a specialist on the history and culture of ancient and medieval Armenia. Under his direction, excavations are being conducted at Garni (since 1949) and Armavir (since 1962), which have provided rich material for characterizing the ancient Armenian culture. Decors.: three orders and various medals.
Works: [Arhakhyelyan, B.] *Hahgagan badgyerakhanghagnyere, IV–VII tt.* Yerevan 1949. *Khaghakhnyere g arhyesdnyere hahasdanwolm, IX–XIII tt.*, vols. 1–2. Yerevan 1958–64. *Garni*, vols. 1, 3, and 4. Yerevan 1951–62. (Vol. 4, with G.O. Karakhanian, in Armenian.)

Arakelian, Grair Gavrilovich
Painter; b. 1922, Novorossiisk.
Mem. CPSU 1944–; attended Art Studios in Novorossiisk 1939 and Samarkand 1946, pupil of N.K. Korotkov and P.P. Benkov; exhibitions 1940–.

Arakelian, Gurgen Ashotovich
State official.
Head of Dept. for Culture with Cen. Cttee. Armenian CP 1976–80; cand. mem. Cen. Cttee. Armenian CP 1976–81, mem. 1981–; Min. of Culture Armenian SSR Ja. 1980–.

Arakelian, Sokrat Vagarshakovich
State official; b. 1915.
Armenian; Mem. CPSU 1938–; grad. Polytech. Inst. Erevan; electromechanic in engineering works 1933–39; student; various party posts 1942–60; Dep. to Supr. Sov. Armenian SSR 1955–; mem. Cen. Cttee. Armenian CP 1956–; cand. mem. Bureau Cen. Cttee. Armenian CP 1960–69; Chair. of Armenian Republican Trade Union Council 1960–69; Min. of Food Ind. Armenian SSR Ag. 1969–81; Min. of Fruits and Vegetables Armenian SSR F. 1981–.

Aramian, Gurgen Levonovich
Graphic artist; b. 1919, Tbilisi.
Mem. CPSU 1958–; attended Erevan Art and Theatre Inst. 1952–58, pupil of M.M. Abegian, E.A. Isabekian.

Arapov, Vasilii Mikhailovich
Painter; b. 1934, Putiatino, Riazan Oblast.
Attended Riazan Art School 1949–54 and Repin Inst. for Painting, Sculpture and Architecture at the USSR Acad. of Arts 1954–60; exhibitions 1960–.

Arbatov, Georgii Arkadevich, Dr. of Hist. Sc.
Scientist; b. 1923.
Mem. CPSU 1943–; grad. Moscow Inst. for Int. Relations 1949; doctoral degree in historical sciences; Prof.; served in Sov. Army 1941–44; ed. activity in publishing house for foreign literature and for periodicals "Voprosy filosofii", "Novoe vremia" and "Kommunist" 1949–60; editorial columnist of periodical "Problemy mira i sotsializma" Prague 1960–62; Section Chief at Inst. of World Econ. and Int. Relations of USSR Acad. of Sciences 1962–64; employed in Cen. Cttee. CPSU 1964–67; Dir. of Inst. of US and Canadian Studies of USSR Acad. of Sciences 1967–; mem. Cen. Auditing Comm. CPSU 1971–76; cand. mem. Cen. Cttee. CPSU 1976–81, mem. 1981–.

Arbuzov, Akim Vasil'evich
Painter and graphic artist; b. 1919, Medveditsa, Saratov Oblast.
Mem. CPSU 1953–; attended Penza Art School and Kharkov Art Inst. 1953–59, pupil of A.V. Chernov; exhibitions 1949–.

Arbuzov, Aleksei Nikolaevich
Dramatist; b. My. 26, 1908, Moscow.
Russian; his dramas are part of the permanent repertoire of Sov. theatre; deals especially with problems of Sov. youth; first works publ. 1930.
Publs.: (in part German editions) "Der weite Weg" (The Long Path) 1935, "Tanja" 1938, "Die Stadt in der Morgenröte" (City at Dawn) 1940, "Verschlungene Wege" (Twisted Paths) 1954, "Irkutsker Geschichte" (Irkutsk Story) 1959, "Poterjannyi syn" (The Lost Son) 1961, "Nas gde-to zhdut" (Someone Somewhere is Waiting for Us) 1962, "Moi bednyi Marat" (My Poor Marat) 1965, "Nochnaia ispoved'" (Nocturnal Confession) 1967, "Glückliche Tage eines Unglücklichen" (Happy Days of an Unhappy Man) 1968, and others.

Arbuzov, Boris Aleksandrovich
Organic chemist; b. O. 22 (N. 4), 1903. Grad. Inst. of Agriculture and Forestry Kazan 1926; mem. USSR Acad. of Sciences 1953–; Dir. of Inst. of Organic Chemistry; Arbuzov's research has been in the field of terpenes, diene compounds and application of physical methods in

the study of the composition of organic compounds; discovered the isomerization of α-pinene to allocymene and of α-oxypinene to campholene aldehyde. By the method of diene synthesis he obtained a number of diene products. Decors.: State Prize of the USSR 1951, two Orders of Lenin and other orders and medals.

Arif (Dadash-zade, Mamed Arif Mager-ram ogly)
Literary scholar and critic; b. Je. 12, 1904, Baku.
Mem. CPSU 1941–; mem. Acad. of Sciences Azerbaidzhan SSR 1958–; Dir. of Nizami Inst. for Language and Literature 1939–59. Decors.: Merited Scholar of the Azerbaidzhan SSR, two orders.
Publs.: "Vladimir Maiakovskii" 1940, "Vissarion Belinskii" 1948, 2nd ed. 1954, "Dramaturgia Sameda Vurguna" 1964, and others.

Aristakesova, Stella Leonovna
Graphic artist; b. 1934, Baku.
Attended Moscow Polygraphic Inst. 1952–57; exhibitions 1958–.

Ariste, Paul Aleksandrovich
Linguist; b. F. 3, 1905.
Academician of the Academy of Sciences of the Esthonian SSR (1954).
He graduated from the University of Tartu in 1929. He furthered his studies of the Finno-Ugric languages, especially their phonetics, in Uppsala, Helsinki, and Hamburg. A professor at the University of Tartu, he is an honorary member of international societies in Finland and an honorary member of the Hungarian Academy of Sciences. He is editor-in-chief of the journal *Sovetskoe finno-ugrovedenie*. His major works are devoted primarily to the Esthonian language and other Baltic-Finnish languages and dialects and to problems of general Finno-Ugric language studies, particularly to the experimental and historical study of Esthonian phonetics and phonology. He also studied the Vod', Livian, and Izhora languages, as well as Esperanto.
Works: *Eesti keele foneetika,* 3rd ed. Tartu, 1965. *Vadja keele grammatika.* Tartu 1948. *Sõna sõna korvale.* Tallinn 1965. "K voprosu o razvitii livskogo iazyka." *Tr. In-ta iazykoznaniia AN SSSR,* 1954, vol. 4.

Aristov, Boris Ivanovich
Diplomat; b. 1925.
Mem. CPSU 1945–; grad. Ulianov Electrotech. Inst. Leningrad 1950; served in Sov. Army 1942–45; electromechanic, engineer, Chief Technologist at Leningrad "Svetlana" Works 1947–52; various Party and Sov. posts 1952–69; Dep. Chair. of City Sov. of Leningrad 1969–71; 1st Sec. of City Cttee. CPSU in Leningrad 1971–78, mem. Cen. Cttee. CPSU 1971–; Sov. Amb. to Poland My. 1978–. Dep. to USSR Supr. Sov. 1974–. Decor.: Order of the October Revolution.

Aristov, Ivan Nikolaevich
Painter; b. 1917, Krasnyi Iar, Astrakhan Oblast.
Attended Astrakhan Art School 1934–38, pupil of S. I. Sakharov and A. M. Tokarev; exhibitions 1946–; mem. CPSU 1943–.

Aristova, Irina Fedorovna
Painter; b. 1926, Baku.
Mem. CPSU 1953–; grad. Moscow Middle School of Art 1945 and attended Surikov Art. Inst. 1945–51, pupil of S. V. Gerasimov and V. K. Nechitailo; exhibitions 1952–.

Arkhangelskii, Gennadii Vasil'evich
Party official.
Mem. Cen. Cttee. Uzbek CP 1976–; Head of Dept. of Admin. Organs Cen. Cttee. Uzbek CP 1973–.

Arkharov, Vladimir Ivanovich, Dr. of Tech. Sc.
Physicist; b. F. 14, 1907, Odessa; s. of an army doctor.
Grad. Physical-Tech. Inst. Leningrad 1931; mem. Ukrainian SSR Acad. of Sciences 1965–; doctoral degree in technical sciences 1945; Prof. 1947; Chief Engineer of Physical-Tech. Inst. Leningrad 1931–32; employed in Physical-Tech. Inst. Sverdlovsk (Urals) 1932–34, in Gorkii 1937–47, again in Sverdlovsk 1937–65; Prof. of Solid-State Physics at Uralsk State Univ. 1947–65; Chief of Dept. for Transformation of Crystals at Physical-Tech. Inst. of Ukrainian SSR Acad. of Sciences in Donetsk 1966–; main experimental and theoretical research work on inner adsorption of polycrystalline alloys. Decors.: Order of the Red Star, Badge of Honour, medals.
Publs.: "Gazovoe khromirovanie. Trudy Instituta Fiziki metallov UFAN SSSR" 1945 in 4 vols, "Okislenie metallov pri vysokikh temperaturakh" 1945, "Kristallografia zakalki stali" 1951.

Arkhipets, Nikolai Timofeevich
State official.
Min. of Ind. Engineering Belorussian SSR N. 1969–79; mem. Cen. Cttee. Belorussian CP 1971–81; Dep. Chair. of USSR State Cttee. of Material and Technical Supply Ap. 1979–.

Arkhipov, Aleksei Petrovich
Painter; b. 1925, Baigulovo, Chuvash ASSR.
Attended Kazan Art School 1946–51, pupil of V. S. Podgurskii, S. A. Rotnitskii, Kh. A. Iakupov; exhibitions 1925–.

Arkhipov, Andrei Alekseevich
Sculptor; b. 1931, Moscow.
Attended Leningrad High School for Arts and Crafts 1946–54, pupil of V. K. Rybalko, V. I. Simonov, R. K. Taurit.

Arkhipov, Ivan Vasil'evich
State official; b. 1907.
Mem. CPSU 1928–; grad. Moscow Machine Tools Inst. 1932; lathe operator in various railroad workshops 1921–24; senior foreman, shop superintendant, chief mechanic in various metallurgical plants 1932–38; 1st Sec. of Krivoi Rog City Cttee. Ukrainian CP 1938–39; post in Cen. Cttee. CPSU apparatus 1939–43; Dep. USSR Min. of Nonferrous Metallurgy 1943–57; Dep. Chair. of State Cttee. on Foreign Econ. Relations of USSR Council of Mins. 1958–59, 1st Dep. Chair. 1959–74; Dep. to USSR Supr. Sov. 1974–; Dep. Chair. of USSR Council of Mins. 1974–80, 1st Dep. Chair. 1980–; mem. Cen. Cttee. CPSU 1976–. Decor.: Hero of Socialist Labor 1977.

Arkhipov, Mikhail Vasil'evich
Painter; b. 1928, Streletskie Vyselki, Riazan Oblast.
Attended Moscow School for Arts and Crafts 1943–49; exhibitions 1954–.

Arkhipov, Trofim Arkhipovich
Udmurt writer; b. Jl. 26 (Ag. 8), 1908, village of Novaia Biia, now in the Mozhga Raion, Udmurt ASSR.
People's writer of the Udmurt ASSR (1960). Member of CPSU from 1939. Edited the magazine *Molot* (1956–66). His main theme is the transformation of the Udmurt village. The novella *By the Ludzinka River* (parts 1–2, 1949–57) deals with the everyday working life of a village during the Great Patriotic War and post-war years. The novel *The Beauty of Man* (1963) is devoted to the builders of the Votkinsk hydroelectric plant on the Kama River. The novella *You Have Started on the Road* (1968–69) tells of the Civil War and the establishment of Soviet rule in an Udmurt village. He is the author of sketches, short stories, and critical articles.
Works: *In Russian translation: U reki Ludzinki.* Moscow 1961. *Stremnina.* Izhevsk 1968.

Arkhipov, Vasilii Sergeevich
Military official; b. D. 9 (22), 1906, Tiutniary, now in Cheliabinsk Oblast.
Colonel-general in the tank forces and twice Hero of the Soviet Union. Member of the CPSU since 1931.
He first entered the Soviet Army in 1928. He completed advanced courses for commander (1938 and 1943) and graduated from the Academy of the General Staff in 1950. He was awarded the title of Hero of the Soviet Union on Mr. 21, 1940, for his command of a tank company in the successful penetration of the enemy's defenses during the Soviet-Finnish war, and on S. 23, 1944 for his skillful leadership of the 53rd Tank Brigade during the capture of the town of Peremyshl' and the assault crossing of the Vistula River. After the war, Arkhipov was engaged in command

and leadership duties in the Soviet Army. Decors.: three Orders of Lenin, five Orders of the Red Banner, the Order of Kutuzov Second Class, the Order of the Red Star, two foreign (Polish and Czechoslovak) orders, and medals.

Arkhipova, Irina Konstantinovna
Russian opera singer (mezzosoprano); b. D. 2, 1925, Moscow.
People's Artist of the USSR (1966). Member of the CPSU since 1963.
She was graduated from the Moscow Architectural Institute in 1948 and from the Moscow Conservatory in 1953. She made her debut in the Sverdlovsk Theater of Opera and Ballet and in 1956 was made a soloist with the Bolshoi Theater in Moscow. Her artistry is highly refined; she possesses a powerful voice and is outstanding for her richness of timbre colorations and dramatic talent. She has more than 30 parts in her repertoire. The best of these are her Carmen (in Bizet's work of the same name) and Amneris (in Verdi's *Aida*). She was the first at the Bolshoi Theater to perform the roles of the deacon's wife in Janáček's opera *Her Stepdaughter*, of Hélène in Prokofiev's *War and Peace*, and others. She often appears in concert recitals and has toured many foreign countries. She was a deputy to the Sixth Convocation of the Supreme Soviet of the USSR.

Arkhipova (Surenkova), Iulia Nikolaevna
Sculptor; b. 1930, Leningrad.
Attended Leningrad High School for Arts and Crafts 1946–53, pupil of V. L. Rybalko, V. I. Simonov, R. K. Taurit.

Arkhipova (Arkhi), Lidia Fedorovna
Painter; b. 1924, Moscow.
Attended Surikov Art. Inst. 1943–50, pupil of S. V. Gerasimov, A. A. Osmerkin.

Armalas, Villius Juozo
Graphic artist; b. 1923, Dovidishkiod, Lithuania.
Attended Kaunas Inst. for Applied and Decorative Art 1941–48, pupil of A. A. Kuchas; exhibitions 1954–.

Armen, Mkrtich (real name: Arutiunian, Mkrtich Grigorevich)
Author; b. 1906, Aleksandropol, now Leninakan, Armenia; s. of a craftsman.
Publs.: "Zubeida" 1928, "Pervye pionery" (The First Pioneers) 1935, "Rodnik Egnar" (The Agnar Spring) 1935, "Iasva" 1953, and others.

Arnautov, Aleksei Vasil'evich
Painter; b. 1915, Orsha, Belorussia.
Attended Vitebsk Tech. College of Art 1931–34; exhibitions 1946–.

Arndt, Iurii Vladimirovich
Architect, graphic artist and stage designer; b. 1924, Kharkov.
Attended Kiev School of Art 1940–41, pupil of I. M. Beklemisheva, and Moscow Inst. of Architecture 1947–53, pupil of M. I. Kurilko and V. N. Iakovlev; mem. CPSU 1946–; works as architect; exhibitions 1956–.

Arnshtam, Leo Oskarovich
Film director and screenwriter; b. Ja. 2 (15), 1905, Dnepropetrovsk.
Became People's Artist of the RSFSR in 1969; graduated from Leningrad Conservatory in 1923. He worked in the Meierkhol'd Theater as director of the musical section. In film he began as a sound technician, and then became a director and screenwriter. He has made many historical films about the revolution and historical biographical films; he produced *Girl Friends* (1936), *Friends* (1938), *Zoia* (1944), *Glinka* (1947), and *A Lesson Taught by History* (1957), dedicated to G. Dimitrov, the leader of the Bulgarian people. He produced the film *Five Days, Five Nights* (1961) with East German cinematographers. His film *Sofia Perovskaia*, about the life of the Russian woman revolutionary, was released in 1968. Decors.: State Prize of the USSR (1946, 1947), Order of the Badge of Honor.

Arrak, Erikh Iohannesovich
Painter; b. 1925, Vyrumaa, Esthonia.
Attended Art Inst. of the Esthonian SSR 1945–47, pupil of I. H. Vyerahansu, Iu. Ia. Piutsepp; exhibitions 1956–; mem. CPSU 1964–.

Arrak (until 1958 Soosuu), Melanie Ernstovna
Painter; b. 1923, Tartu, Esthonia.
Attended Art Inst. of the Esthonian SSR 1945–53, pupil of Iu. N. Piutsepp. E. Ia. Einmann; exhibitions 1955–.

Ars, Iurii Lukianovich
Painter; b. 1924, Ussuriisk, Primore Krai. No formal artistic educ.; exhibitions 1946–.

Arsenin, Dmitrii Dmitrievich
Graphic artist; b. 1926, Sloboda, Vladimir Oblast.
Attended Repin Inst. for Painting, Sculpture and Architecture at the USSR Acad. of Arts 1954–60.

Arsentev, Aleksandr Ivanovich, Dr. of Tech. Sc.
Graphic artist; b. 1918, Fedino, Smolensk Oblast.
Grad. Moscow Inst. for Nonferrous Metals and Gold 1941, Dr. of Tech. Sc. 1964; no formal artistic educ.; exhibitions 1948–; mem. CPSU 1962–.

Arshakuni, Zaven Petrosovich
Painter; b. 1932, Leningrad.
Attended Repin Inst. for Painting, Sculpture and Architecture at the USSR Acad. of Arts 1954–61.

Arslan, Timer (real name: Arslanov, Timer Gareevich)
Author; b. 1915, Kakribash, Bashkir ASSR.
Mem. CPSU 1953–; grad. Bashkir Teachers' College; first work publ. 1934.
Publs.: poems "Nasha Iunost" (Our Youth) 1938, "Iz okopov" (From the Trenches) 1944, "Inzhener Izat Islamov" (The Engineer Azat Islamov) 1950, "Doch' gor" (The Daughter of the Mountains) 1949, "Paren' iz Bashkirii" (The Boy from Bashkiria) 1953, and others.

Artamonov, Mikhail Illarinovich, Dr. of Hist. Sc.
Scholar; b. N. 23 (D. 5), 1898, Vygolevo, Tver Gouvt.
Mem. CPSU 1940–; archaeologist and historian; doctoral degree in historical sciences; mem. Polish Acad. of Sciences 1960–; Dir. of Hermitage Leningrad 1951–64. Decors.: Lenin Order, Order of the Red Banner of Labour, various medals.

Artamonov, Viktor Evgenevich
Painter; b. 1919, Armavir.
Grad. Azerbaidzhan Art Inst. 1940, attended Surikov Art Inst. Moscow 1946–52, pupil of F. A. Modorov and V. G. Tsyplakov; teacher at Surikov Art Inst. Moscow 1953–.

Artiukhin, Mikhail Filippovich
Painter; b. 1927, Selivanovo, Tula Oblast.
Attended Surikov Art Inst. Moscow 1947–52, pupil of F. A. Modorov and V. G. Tsyplakov; exhibitions 1952–.

Artiukhov, Dmitrii Fedorovich
Painter; b. 1928, Cherniaevka, Omsk Oblast.
Attended Vladivostok Art School 1946–49, pupil of M. A. Kostin and V. I. Fomin; exhibitions 1954–.

Artmane, Via (real name: Alida Fritsevna Artmane)
Latvian actress; b. Ag. 21, 1929.
People's Artist of the USSR (1969). In 1949, she graduated from the studio of the Rainis Art Theater in Riga, and she was accepted into the theater's company. Among her roles are Ophelia and Juliet in Shakespeare's *Hamlet* and *Romeo and Juliet* and Laimdota in Rainis' *Fire and Night*. She has also acted in motion pictures, where her best roles were Datse (*Beyond the Swanlike Flock of Clouds*, 1957), Sonia (*Native Blood*, 1964), Ona (*Nobody Wanted to Die*, 1966), and Kristina (*Edgar and Kristina*, 1967). As an actress, she creates lyric, soft, feminine characters; her heroines are self-sacrificing, kind, and poetic. Decor.: Badge of Honor.

Artsikhovskii, Artemii Vladimirovich
Archaeologist; b. D. 13 (26), 1902, St. Petersburg.
Mem. CPSU 1952–; corresponding mem.

USSR Acad. of Sciences 1960–; Prof. Chair of Archaeology at Moscow Univ. 1939–; special field of study: Slavic-Russian archaeology; for more than 30 years supervisor of archaeological excavations in Novgorod.

Artsimovich, Lev Andreevich
Nuclear physicist; b. F. 12 (25), 1909, Moscow.
Grad. Belorussian Univ. 1928; mem. USSR Acad. of Sciences 1953–; mem. Presidium USSR Acad. of Sciences 1957–; employed in Inst. for Atomic Energy at USSR Acad. of Sciences 1944–; mem. USA Acad. of Art and Science 1966–; during the war years (1943–46) he was concerned with electron optics and the theory of chromatic aberrations of the electron optical system and carried out theoretical and experimental research in the field of electron optical converters; in 1945, Artsimovich and I. Ya. Pomeranchuk did theoretical research on the role of radiation losses in the betatron; this work permitted the establishment of the maximum energy achieved by this type of electron acceleration; Artsimovich was one of a group who developed an electromagnetic method of isotope separation; from currents then available in the mass spectrometer laboratory (of the order of 10^{-10} amperes), it was necessary to use currents of the order of an ampere; in reaching a solution to this problem, Artsimovich carried out a careful analysis of the problems of the aberrationless focussing of ion beams in axially-symmetric magnetic fields; he designed the optics of the ion source; as a result of this work, the Soviet physicists developed a successful production of separated isotopes; in the beginning of the fifties, Artsimovich began his work on a controlled thermonuclear reaction; the group of physicists under Artsimovich began the study of high current pulse discharge in evacuated deuterium; in the course of these experiments, the group succeeded in obtaining for a short time a highly ionized plasma of a million degrees; in 1952, this group of scientific workers discovered that a powerful discharge in deuterium at low pressure is a source of neutrons and X-ray radiation of short wave length; further studies showed that the gas-discharge plasma, compressed in the presence of the longitudinal magnetic pole possessed paramagnetic properties; it was also shown that neutrons originate, not as a result of the thermonuclear reaction, but as a result of a specific acceleration process; the work of studying the means of obtaining a controlled thermonuclear reaction is now being conducted extensively under Artsimovich's direction. Decors.: Hero of Socialist Labour, State Prize of the USSR 1953, Lenin Prize 1958, four Orders of Lenin, two further orders and various medals.
Publ.: "Upravlenye termoiadernye reakcii" Moscow 1961.

Arustamian, Robert Samvelovich
Painter and graphic artist; b. 1925, Dagraz, Azerbaidzhan.
Attended Erevan Art Inst. 1946–52, pupil of A.A. Avetisian, G.M. Giurdzhian; aspirant at Repin Inst. for Painting, Sculpture and Architecture at the USSR Acad. of Arts 1954–56; exhibitions 1952–; mem. CPSU 1960–.

Arutiunian, Agasi Nazaretovich
Painter, b. 1924, Leninakan.
Grad. Erevan Art Inst. 1951; exhibitions 1953–; mem. CPSU 1953–.

Arutiunian, Aleksandr Grigor'evich
Composer; b. 1920.
Armenian; mem. CPSU 1952–; grad. Erevan Conservatory 1941. Decor.: State Prize of the USSR for "Kantaty o Rodine" 1949.

Arutiunian, Ara Armenovich
Sculptor; b. 1928, Erevan.
Attended Erevan Art and Theatre Inst. 1948–54, pupil of G. P. Agaronian; exhibitions 1950–; teacher at Erevan Art School 1955–; mem. CPSU 1954–. Decor.: Merited Artist of the Armenian SSR.

Arutiunian, Donara Ashkanazovna
Party official.
Mem. Auditing Comm. 1966–71, cand. mem. Cen. Cttee. Armenian CP 1971–81, mem. 1981–; 2nd Sec. of Leninakan City Cttee. Armenian CP 1975–80, 1st Sec. 1980–; cand. mem. Bureau Cen. Cttee. Armenian CP Ja. 1981–.

Arutiunian, Mais (Aramais) Rubenovich
Painter; b. 1928, Goris, Armenia.
Grad. Erevan Art and Theatre Inst. 1957; exhibitions 1957–.

Arutiunian, Meruzhan Gegamovich
Painter; b. 1927, Tsakhkaovit, Armenia.
Attended Erevan Art and Theatre Inst. 1947–53, pupil of M.S. Sarian; exhibitions 1952–.

Arutiunian, Vazgen Frangulovich
Trade union official.
Mem. Cen. Cttee. Armenian CP 1964–76; Sec. and mem. Presidium of Armenian Republican Trade Union Council Ja. 1974–.

Arutiunian, Zhenia (Evgenia) Armenakovna
Painter; b. 1920, Erevan.
Attended Moscow High School for Arts and Crafts 1939–41, pupil of V. N. Iakovlev, and Erevan Art Inst. 1945–50, pupil of A. A. Avetisian, M. S. Sarian; exhibitions 1942–.

Arutiuniants (Arutiunian), Anatolii Sarkisovich
Graphic artist; b. 1921, Omsk.
Attended Kazan Art School 1940–46, pupil of A. M. Cherkasskii; exhibitions 1954–; satirical drawings for the magazine "Perets" 1947–, for the newspapers "Pravda Ukrainy" 1949–57, "Radianska Ukraina" 1950–57, "Vechirnii Kyiv" 1954–.

Arutiunov, Aleksandr Ivanovich
Neurosurgeon; b. D. 21, 1903 (Ja. 3, 1904), Yerevan.
Academician of the Academy of Medical Sciences of the USSR (1967), Member of the CPSU since 1940. Graduated from the medical department of Northern Caucasus University (1929).
Between 1951 and 1964, he was the director of the Kiev Scientific Research Institute of Neurosurgery. Since 1964 he has been director of the N. N. Burdenko Institute of Neurosurgery of the Academy of Medical Sciences of the USSR and simultaneously (since 1965) head of the department of neurosurgery at the Central Advanced Training Institute for Doctors. His main work has been devoted to brain tumors, the neurosurgical pathology of vessels, and damage to the central nervous system. He is vice-president of the World Federation of Neurosurgical Societies (since 1965) and president of the All-Union Society of Neurosurgeons (since 1966). Decors.: Order of Lenin, four other orders, and medals.
Works: *Tuberkulomy golovnogo mozga.* Kiev 1959 (co-author). "O nekotorykh printsipial'nykh voprosakh patologii i khirurgicheskogo lecheniia mezhopozvononchnykh diskov." *Voprosy neirokhirurgii,* 1962, issue 4. "Khirurgicheskoe lechenie ekhinokocka tsentral'noi nervnoi sistemy." *Klinicheskaia khirurgiia,* 1963, no. 5.

Arystanbekova, A. Kh.
Komsomol official.
Sec. and mem. Bureau of Cen. Cttee. Kazakh Komsomol Mr. 1982–.

Arzhanov, Mikhail Nikolaevich
Painter; b. 1924, Sinelnikovka, Saratov Oblast.
Attended Saratov Art School 1941–47 and Moscow Polygraphic Inst. 1948–54, pupil of I. I. Zakharov, I. N. Shcheglov; exhibitions 1955–.

Arzhavkin, S. A.
Trade union official.
Sec. of Belorussian Republican Trade Union Council O. 1970–.

Arzhekaev, Boris Alekseevich
Graphic artist; b. 1922, Saransk.
Attended Moscow Polygraphic Inst. 1949–51, pupil of N. F. Lapin; exhibitions 1954–.

Arzumanian, Movzes Bagdazarovich
Party official.
Mem. Cttee. for Party and State Control at Cen. Cttee. Armenian CP and Council of Mins. Armenian SSR 1963–65; cand. mem. 1964–76, mem. Cen. Cttee. Arme-

nian CP 1976–; Chief of Dept. of Party Org. Affairs at Cen. Cttee. Armenian CP 1975–; cand. mem. Bureau Cen. Cttee. Armenian CP Jl. 1975–.

Arzumanov, Lev Stepanovich
Painter and graphic artist; b. 1932, Tashkent.
Attended Dushanbe Art School 1953–58, pupil of A.T. Bespertsov; exhibitions 1958–.

Asadov, Eduard Arkadevich
Poet; b. 1923 Merv, Turkmenistan.
Grad. Gorkii Inst. of Literature Moscow 1951; front volunteer 1941; became blind after serious injury 1944; first works publ. 1948.
Publs.: poems "Svetlye dorogi" (Bright Streets) 1951, "Snezhnyi vecher" (Snowed Up Evening) 1956, "Soldaty vernulis' voiny" (The Soldiers Returned From the War) 1957, and others.

Asaev, Revaz Nikolaevich
Author and poet; b. 1923, Kakhetia.
Studied Staliniri Pedagogical Tech. College 1938–41; Ossetian writer and translator; served in Sov. Army 1942–43; first works publ. 1938.
Publs.: "Kaplia krovi" (A Drop of Blood) 1952, "Proizvedeniia" (Works) 1954, "Pesnia serdtsa" (Song of the Heart) 1950, "Stikhi i poemy (Verse and Poems) 1955, "Kolkhoz imeni K. Khetagurova" (Khetagurov Kolkhoz) 1952, "Mesiats na Altae" (A Month on Altai) 1955, and others.

Asanov, Dadash Mirza ogly
State official.
Dep. Min. of Rural Construction Azerbaidzhan SSR 1966–73, Min. 1978–82; Min. of Construction and Utilization of Motorways Azerbaidzhan SSR 1973–78; cand. mem. Cen. Cttee. Azerbaidzhan CP 1981–; Dep. Chair. of Council of Mins. Azerbaidzhan SSR O. 1982–.

Asaris, Gunar Konstantinovich
Architect; b. Je. 23, 1934, Riga.
He graduated from the department of architecture at the Riga Polytechnic Institute in 1959. His works include the experimental shop at the Riga Glass Works (1959), a number of individual residences in Riga and Iurmala (1959–69), and a memorial group commemorating the victims of fascist terror in Salaspils (1963–67; with other architects). Decor.: Lenin Prize in 1970.

Asatiani (Astiani), Teimuraz Georgievich
Sculptor; b. 1916.
Grad. Tbilisi Art Acad. 1946; mem. CPSU 1948–. Decor.: Merited Artist of the Georgian SSR 1962.

Asatiani, Tina Levanovna, Dr. of Phy.-Math. Sc.
Physicist; b. 1918.
Grad. Tbilisi Univ. 1940; Dr. of Phy.-Math. Sc. 1970; mem. staff Inst. of Physics in Erevan, Chief of Laboratory 1944–. Decor.: Lenin Prize for contributing to the development of new means for the observation of elementary particles, 1970.

Aseev, Petr Prokhorovich
Designer; b. 1919, Uriupinsk.
Attended Acad. of Art of the Latvian SSR 1946–51, pupil of L.S. Svemp, K.E. Uban; mem. CPSU 1944–; exhibitions 1950–.

Asetskii, Ludvig Petrovich
Graphic artist; b. 1929, Prodvino, Mogilev Oblast.
Attended Belorussian Theatre and Art Inst. 1954–60; exhibitions 1960–; teacher at Belorussian Theatre and Art Inst. 1962–.

Ashimov, Baiken Ashimovich
State official; b. 1917, Kazakhstan.
Mem. CPSU 1940–; grad. Leningrad Inst. of Zoology and Phytopathology 1957 and Party College of Cen. Cttee. CPSU 1969; served in Sov. Army 1938–41, 1942–45; held various sov. and party posts 1945–55; 2nd Sec. of Oblast Cttee. Kazakh CP in Kokchetav 1959–61; mem. Auditing Comm. Kazakh CP 1960–61; mem. Cen. Cttee. Kazakh CP 1961–; Chair. of Oblast Sov. in Karaganda 1961–63; 1st Sec. for Agriculture at Oblast Cttee. Kazakh CP in Karaganda 1963–64; again Chair. of Oblast Sov. in Karaganda 1964–68; 1st Sec. of Oblast Cttee. Kazakh CP in Taldy-Kurgan 1968–70; Chair. of Council of Mins. Kazakh SSR Ap. 1970–; mem. Bureau Cen. Cttee. Kazakh CP Ap. 1970–; mem. Cen. Cttee. CPSU 1971–.

Ashirov, Chary
Poet; b. 1910, Keshi, now Ashkhabad Oblast; s. of a peasant.
Turkmen; mem. CPSU 1944–; studied Inst. of Pedagogics; Ed. of satirical journal "Tokmak".
Publs.: poems "Razluka" (Separation) 1939, "V tylu vraga" (At the Enemy's Rear) 1941, "Oshibka molodogo mel'nika" (A Young Miller's Mistake) 1957, and others.

Ashirov, T.
Trade union official.
Sec. and mem. Presidium of Turkmen Republican Trade Union Council Ja. 1982–.

Ashkenazi, Izrail Mordukhovich (Markovich)
Painter; b. 1917, Novgorod.
Attended Kiev Art Inst. 1951–57, pupil of V.G. Puzyrkov, M.I. Khmelko; mem. CPSU 1943–; exhibitions 1957–.

Ashrafmurad (Muradov), Ashraf Murad ogly
Painter; b. 1925, Baku.
Attended Repin Inst. for Painting, Sculpture and Architecture at the USSR Acad. of Arts 1947–54, pupil of R.R. Frents; exhibitions 1958–.

Ashuraliev, Rashid
Party official.
Min. of Supplies Uzbek SSR Mr. 1975–78; Dep. Chair. of Council of Mins. Uzbek SSR 1978–82; 1st Sec. of Samarkand Oblast Cttee. Uzbek CP F. 1982–.

Ashurov, Irgash
State official; b. 1917, Uzbekistan.
Mem. CPSU 1944–; grad. Pedagogical Inst. Fergana; club manager, lecturer at a teachers' college, dep. manager of a reg. office for public educ. 1935–44; leading Komsomol posts 1944–50; sec. of an oblast cttee. Uzbek CP, Chair. of Oblast Exec. Cttee. of Workers' Deputies in Andizhan, section chief of an oblast cttee. Uzbek CP, 1st Sec. of Andizhan City Cttee. Uzbek CP 1951–61; mem. Cen. Cttee. Uzbek CP 1956–66 and N. 1982–, mem. Auditing Comm. 1976–81, cand. mem. Cen. Cttee. 1981–82; 1st Sec. of Andizhan Oblast Cttee. Uzbek CP 1961–63; Dep. to and mem. Econ. Comm. at Sov. of Nationalities USSR Supr. Sov. 1962–66; Chair. of State Asscn. of Co-operative Socs. for Building Construction in Kolkhozes "Uzkolkhozstroi" with Council of Mins. Uzbek SSR Ap. 1974–.

Asimov, Mukhamed Saifitdinovich
Scholar and state official; b. 1920.
Grad. Physical-Mathematical Faculty of Uzbek State Univ. 1941; philosopher; corresponding mem. USSR Acad. of Sciences 1974–, mem. and Pres. of Tadzhik SSR Acad. of Sciences 1965–; served in Sov. Army 1941–46; teaching and scientific activity 1946–; until 1955 aspirant at Acad. of Sociology Cen. Cttee. CPSU; Rector of Polytech. Inst. Dushanbe 1956–62; Min. of Public Educ. Tadzhik SSR 1962; mem. Presidium and Sec. of Cen. Cttee. Tadzhik CP, Dep. Chair. of Council of Mins. and Chair. of Cttee. for Party and State Control at Cen. Cttee. Tadzhik CP and Council of Mins. Tadzhik SSR 1963–65.

Asipenko, Aleksandr Kharitonovich
Writer; b. S. 7, 1919, Pushkari, Vitebsk Province.
Studied at the Vitebsk Pedagogical Institute. Participated in the partisan movement in Belorussia during the Great Patriotic War.
He writes primarily about life in the kolkhoz village and about young people and their upbringing. He is the author of the novellas and stories "The Latest Version" (1956), "Along a Bumpy Road" (1957), "The Ice Is Melting" (1958), "The Breathing of Flowers and Labor" (1960),

"The Road Into the Distance" (1962), "The Long-occupied Corner" (1964), and "The Grain" (1968) and the novel *The Fiery Azimuth* (1966). He has been the editor-in-chief of the magazine *Youth* since 1966.

Asitashvili (Asiatishvili), Teona Georgievna
Graphic artist; b. 1932, Tbilisi.
Attended Tbilisi Art Acad. 1950–56, pupil of I. A. Sharlemanian; exhibitions 1956–.

Askarov, Asanbai Askarovich
Party official; b. 1922, Kazakhstan.
Mem. CPSU 1944–; grad. Party College of Cen. Cttee. CPSU 1954; dir. of studies at a secondary school 1939–42; served in Sov. Army 1942–46; 1st sec. of a raion cttee., 1st Sec. of Oblast Cttee. Kazakhstan Komsomol in Dzhambul 1946–51; 1st sec. of a raion cttee. Kazakh CP, chair. of an oblast sov. 1954–59; 1st Sec. of Oblast Cttee. Kazakh CP in Dzhambul 1959–65; mem. Cen. Cttee. Kazakh CP 1960–; cand. mem. 1966–71, mem. Bureau Cen. Cttee. Kazakh CP 1971–78; 1st Sec. of Oblast Cttee. Kazakh CP in Alma-Ata Ap. 1965–78, in Chimkent Ap. 1978–; mem. Cen. Cttee. CPSU 1966–; Dep. to USSR Supr. Sov. 1966–.

Askerov, Izmail Nasrulla ogly
Party official; b. 1914.
Mem. CPSU 1939–; grad. Forestry Tech. College Baku 1933, Marx Econ. Inst. and Party High School of Cen. Cttee. CPSU; instructor of a raion cttee. of Komsomol, dep. sec. of a raion cttee. Azerbaidzhan Komsomol 1933–43; Sec. 1943–45, 1st Sec. of Cen. Cttee. Azerbaidzhan Komsomol 1945–47; held party posts 1947–50; Chair. of Azerbaidzhan Republican Trade Union Council 1950–52; Dep. to Supr. Sov. Azerbaidzhan SSR 1951–; 1st Oblast Sec. of Azerbaidzhan CP in Nakhichevan ASSR 1952–56; Dep. Min. for Agriculture Azerbaidzhan SSR 1956–58; mem. Cen. Cttee. Azerbaidzhan CP 1956–; Chief of Dept. for Trade, Finance and Planning Organs at Cen. Cttee. Azerbaidzhan CP 1956–59; Chair. of Party Comm. at Cen. Cttee. Azerbaidzhan CP 1959–76 and Ja. 1981–; cand. mem. Bureau Cen. Cttee. and Chief of Dept. for Party Org. Affairs at Cen. Cttee. Azerbaidzhan CP 1976–81.

Askerov, Mamed Gasan ogly
State official; b. 1918; s. of a farmer.
Azerbaidzhani; mem. CPSU 1945–; grad. Azerbaidzhan Agricultural Acad.; head, chief agronomist of Dept. of Agriculture of Raion Sov., chief agronomist, Dir. of a machine and tractor station 1942–54; Chair. of Exec. Cttee. of Raion Sov. of Peoples' Deps. in Barda 1954–59; Dep. Chair. of Council of Mins. Nakhichevan ASSR 1959, Chair. 1959–65; mem. Cen. Cttee. CP Azerbaidzhan 1960–; 1st Raion Sec. CP Azerbaidzhan in Agdzhabedi 1965–68; head of Admin. for Cotton Cleaning and Procurement, Dep. Min. of Light Ind. Azerbaidzhan SSR 1968–77; Min. of Agriculture Azerbaidzhan SSR D. 1977–.

Aslamazian, Mariam Arshakovna
Painter; b. N. 20 (D. 3), 1907, Bash-Shirak, near Leninakan.
People's Artist of the Armenian SSR (1965). Member of the CPSU since 1946. In 1929, she graduated from the Yerevan Artindustrial Technicum, where she studied with S. Arakelian and S. Agadzhanian. In 1933 she graduated from the Leningrad Academy of Art, where she studied with K. N. Istomin and K. S. Petrov-Vodkin. She is a representative of the Armenian school of decorative, flat still life (*Armenian Still Life*, 1955, and *The Flowering Red Cactus*, 1957– both in the Tret'iakov Gallery). She painted the following thematic paintings: *The Return of the Hero*, 1943; *The Song of the Hero*, 1944; and *The Mother-Heroine*, 1949 – all in the Picture Gallery of Armenia in Yerevan. She also painted portraits.

Aslanian, Vaginak Sirekanovich
Painter; b. 1918, Erevan.
Attended Erevan Art Inst. 1945–50, pupil of E. A. Isabekian, M. S. Sarian; mem. CPSU 1950–; exhibitions 1950–. Decor.: Merited Artist of the Armenian SSR 1963.

Aslanov, Armais Amirovich
State official.
Chair. of Oblast Exec. Cttee. of People's Deputies in Autonomous Oblast of Nagorno-Karabakh Je. 1975–; Dep. to and Dep. Chair. of Presidium Supr. Sov. Azerbaidzhan SSR 1975–; mem. Cen. Cttee. Azerbaidzhan CP 1976–.

Aslanova, Zamina Sardar kyzy
Party official; b. 1940.
Mem. Cen. Auditing Comm. CPSU; mem. CPSU 1962–, grad. Azerbaidzhan Econ. Inst. 1973; village club leader 1960–63; commander of cotton growers' brigade at the 26th Baku Commissars' collective farm 1963–; mem. Cen. Auditing Comm. CPSU 1981–. Decors.: Hero of Socialist Labor 1980, USSR State Prize 1976.

Aslian, Albert Artemovich
Graphic artist; b. 1922, Alaverdi, Armenia.
Grad. Erevan Art Inst. 1952; exhibitions 1952–.

Asmanov, Evgenii Aleksandrovich (Makhmud Akhmedovich)
Graphic artist; b. 1933, Nadezhdinsk, Sverdlovsk Oblast.
Attended Moscow Polygraphic Inst. 1953–58, pupil of A. D. Goncharov and I. I. Chekmazov; works as graphic artist for various book publishers.

Asmanova, Tatiana Ivanovna
Graphic artist; b. 1934, Moscow.
Attended Moscow Polygraphic Inst. 1953–58, pupil of A. D. Goncharov and I. I. Chekmazov; works as graphic artist for various book publishers.

Asmolov, Aleksei Nikitich
Military official; b. Mr. 17 (30), 1906, Saratov Province.
Major-general (1945); member of the CPSU since 1932. Born into the family of a poor peasant.
He joined the Soviet Army in 1928. He graduated from the M. V. Frunze Military Academy (1939) and from the Academy of the General Staff (1952). During the Great Patriotic War he was one of the leaders of the partisan movement, a member of the Military Council, and a representative of the Central Staff of the partisan movement on the Northwest, Southwest, and Third Ukrainian fronts. From 1941 to 1943 he led the partisan struggle in Leningrad Oblast, where large partisan-controlled regions were created. From 1944 to 1945, he was commander of the partisan movement in Czechoslovakia where he was sent at the request of the Central Committee of the Communist Party of Czechoslovakia. After the war he was engaged in important work in the Ministry of Interior Affairs. After 1956 he was in the reserves; he has also participated in the work of the Soviet Committee of War Veterans. Decors.: Order of Lenin, four Orders of the Red Banner, the Orders of Bogdan Khmel'nitskii First Class and of the Red Star, three Czechoslovak orders, and also medals. He is an honorary citizen of the Czechoslovak towns of Baňská Bystrica, Martin, Zvolen, and Sliač.

Asratian, Ezras Asratovich
Physiologist; b. My. 18 (31), 1933, Metsik, Turkish Armenia.
Student of I. P. Pavlov. Corresponding member of the Academy of Sciences of the USSR (1939) and academician of the Academy of Sciences of the Armenian SSR (1947). Member of the CPSU since 1929.
In 1930, he graduated from the medical department of the University of Yerevan. From 1930 to 1941 he worked at the Physiological Institute (Leningrad). From 1941 to 1943 he was professor at the Tashkent Medical Institute. From 1944 on he worked at the Academy of Sciences of the USSR: from 1944 to 1950 he was director of the Laboratory for the Restoration of the Functions of the Nervous System; from 1950 to 1952 he was director of the Institute of Higher Nervous Activity; in 1952 he became head of the Physiological Laboratory and in 1960, director of the Institute of Higher Nervous Activity and Neurophysiology. His theory of the adaptability of the nervous system sheds light on the mechanism of regenerative phenomena in a

damaged organism and the decisive role in this context of the cortex of the cerebral hemispheres. He used Pavlov's concept of the protective role of inhibition as the basis for methods of treating traumatic shock and the effects of certain kinds of organic brain damage. He developed the notions of the morphological and functional multilevels of the unconditioned reflexes; of tonic, or attitudinal conditioned reflexes; of the mechanisms of switching in higher nervous activity; and so on. Decors.: I. P. Pavlov Prize (1951), I. P. Pavlov gold medal (1963), Order of Lenin, two other orders, and a medal.

Astafev, Leonid Eliseevich
Painter; b. 1922, Uman, Kiev Oblast.
Attended Surikov Art Inst. 1953–59; exhibitions 1957–.

Astafev, Viktor Petrovich
Author; b. My. 1, 1924, Ovsianka, Krasnoiarsk Krai.
Russian; was brought up in an orphanage; served in World War II; first works publ. 1951.
Publs.: "Do budushchei vesny" (Until Next Spring) 1953, "Taiut snega" (The Snow Is Melting) 1958, "Zvezdopad" (Shooting Star) 1962, "Krazha. Gdeto gremit voina" (Theft. Somewhere the War is Droning) 1968, "Sinie sumerki" (Blue Dusk) 1968, and others.

Astakhin, Vasilii Iosifovich
Graphic artist; b. 1926, Krutye Verkhi, Kaluga Oblast.
Attended Moscow Oblast Pedagogic Art School 1952–57; exhibitions 1960–.

Astakhov, Evgenii Pavlovich
Painter; b. 1926, Volyn, Riazan Oblast.
Attended Riazan Art School 1945–49, pupil of V. E. Kurakin, G. N. Rakov; exhibitions 1950–.

Astapov, Vasilii Pavlovich
Sculptor; b. 1981, Groznyi.
Attended Odessa Art School 1937–39 and Repin Inst. for Painting, Sculpture and Architecture at the USSR Acad. of Arts 1945–48 and 1959–60; exhibitions 1954–; mem. CPSU 1960–.

Astashev, Eduard Ivanovich
Graphic artist and painter; b. 1932, Georgievsk, Stavropol Krai.
Attended Krasnodar Art School 1948–56; exhibitions 1960–.

Astaulov, Sviatoslav Petrovich
Sculptor; b. 1924, Kiev.
Attended Kiev Art Inst. 1950–56, pupil of M. K. Vronskii, M. I. Gelman; exhibitions 1956–.

Astaurov, Boris Lvovich
Biologist; b. O. 14 (27), 1904, Moscow.
Grad. Moscow Univ. 1927; mem. USSR Acad. of Sciences 1966–; has held several key positions in research institutes; his basic works deal with the theory and practice in the heredity of the mulberry silkworm. He has worked out methods of thermal artificial parthenoses and found a way of obtaining complete experimental androgens. Decors.: 2 orders and several medals.
Publs.: for a list of his works see periodical "Citologija 7" 1965.

Astavin, Sergei Timofeevich
Diplomat; b. 1918.
Joined Diplomatic Service 1954; Chief of Pol. Section of USSR Emb. in the GDR and USSR High Commissioner's Office in Germany 1954–55; Emb. Counsellor 1955–56 and Emb. Counsellor-Envoy at USSR Emb. in the GDR 1956–59; Chief of European Dept. V in USSR Min. of Foreign Affairs 1960–70; USSR Amb. to Iceland 1970–73; USSR Amb. to Cyprus Jl. 1973–.

Astrakhantsev, Aleksandr Aleksandrovich
Graphic artist; b. 1922, Dorogobuzh, Smolensk Oblast.
Attended Moscow Polygraphic Inst. 1947–52, pupil of A. D. Goncharov, P. N. Zakharov.

Astrauskas, Vitautas Stasevich
Party official; b. 1930.
Grad. Party High School in Vilna; mem. Cen. Cttee. 1961–, cand. mem. Bureau Cen. Cttee. Lithuanian CP 1971–81, mem. 1981–; sec. of party cttee. of a kolkhoz-sovkhoz production admin. 1963; Dep. to Supr. Sov. Lithuanian SSR 1963–; Chief of Dept. for Trade and Public Services until 1971, Chief of Dept. for Party Org. Affairs at Cen. Cttee. Lithuanian CP Mr. 1971–81; Sec. Cen. Cttee. Lithuanian CP Ja. 1981–.

Astsatrian, Egishe Tevosovich
State official; b. 1914.
Armenian; mem. CPSU 1944–; grad. Ind. Inst. at Tbilisi; trackwalker, dep. chief engineer in a mine, in a mining trust, employed in USSR Min. of Coal Ind. and in USSR Min. of Non-Ferrous Metallurgy 1938–50; sec. of a raion cttee., district cttee. Armenian CP, Dept. Chief at Cen. Cttee. Armenian CP 1950–60; Dep. to Supr. Sov. Armenian SSR 1955–; mem. Cen. Cttee. Armenian CP 1956–; Chair. of Econ. Council Armenian SSR 1960–62; mem. 1960–66, cand. Bureau Cen. Cttee. Armenian CP 1966–81; Sec. of Cen. Cttee. Armenian CP 1962–66; Dep. Chair. of Council of Mins. and Chair. of Cttee. for Party and State Control at Cen. Cttee. Armenian CP and Council of Mins. Armenian SSR 1962–65; Chair. of Cttee. for People's Control Armenian SSR D. 1965–76; Chair. of State Cttee. for Material-Tech. Supply Armenian SSR Jl. 1976–. Decor.: Order of the Red Banner of Labour.

Ata, Salikh
Turkmen popular poet; b. 1908, Shorden, near the city of Mary.
Member of the CPSU since 1941. Born into a poor peasant family.
At the age of three, he lost his eyesight. His first attempts at poetry date from 1919. In his verse he glorified the Party ("The Bolshevik Is Invincible, Friends!" 1920) and socialist labor ("I will Not Lose a Single Gram" 1936), and he came out against survivals of the past ("Stop!" 1935). During the Great Patriotic War he created many poems and songs imbued with patriotism. His postwar works are devoted to peaceful labor, friendship among peoples, and the denunciation of warmongers (the lyric poem "A Word in All Sincerity" 1947). He is the originator of the fable genre in Turkmen literature ("The Lion and the Fox" 1945; "The Pumpkin's Laughter" 1955; "The Jackal and the Cock" 1955).
Works: *Saylanan ëserler.* Ashkhabad 1960. *In Russian translation: Izbr. stikhi.* Ashkhabad 1954.

Ataev, Bairamburdy
Party official.
Sec. of Oblast Cttee. CP Turkmen in Mary, 1st Sec. of Oblast Cttee. CP Turkmen in Tashauz, Sec. of Party Cttee. of Prod. Admin. in Kerkin 1955–71; cand. mem. Cen. Cttee. 1956, mem. Cen. Cttee. CP Turkmen 1960–; 1st Sec. of Oblast Cttee. CP Turkmen in Tashauz F. 1971–.

Ataian, Armen Arshakovich
Painter; b. 1922, Erevan.
Grad. Erevan Art School 1941 and Art Inst. 1952; exhibitions 1952–.

Atamanov, Lev Konstantinovich
Producer of animated cartoon films; b. F. 8 (21), 1905, Moscow.
Honored Art Worker of the RSFSR (1964).
In 1926, he graduated from the First State School of Cinematography. He is one of the originators of Soviet animated cartoon cinematography. He worked in Armenia where he produced *The Dog and the Cat* (1938) and *The Magic Carpet* (1948). In 1949 he became director of the All Union Animated Cartoon Films Studio. He has received prizes and recognition at international film festivals for such films as *The Yellow Stork* (1950), *The Red Flower* (1952), *The Golden Antelope* (1954), *The Snow Queen* (1957), *The Shepherdess and the Chimney Sweep* (1965), and *The Bench* (1968).

Atanovskii, Leonid Nikolaevich
State official.
Mem. Auditing Comm. 1971–76, mem. Cen. Cttee. CP Moldavia 1976–81; Chair. of the Union republic cooperative state Scientific Production Asscn. for Agro-Chemical Supply to Agriculture Moldavian SSR Mr. 1981–.

Atashev, Kurmanbek Kasievich
State official.
Min. of Trade Kirghiz SSR F. 1976–79; cand. mem. Cen. Cttee. Kirghiz CP 1971–76, mem. 1976–; Dep. Chair. of Council of Mins. Kirghiz SSR My. 1981–.

Atlanov, Iurii Mikhailovich
Graphic artist; b. 1934, Rostov/Don.
Attended Rostov/Don Art School 1949–54, pupil of P.F. Ostashchenko, and Surikov Art Inst. Moscow 1954–60, pupil of M.A. Dobrov; exhibitions 1954–.

Atlantov, Vladimir Nikolaevich
Painter; b. 1924.
Mem. CPSU 1944–; grad. Kiev Art Inst. 1953, pupil of V.I. Kasiian; exhibitions 1953–.

Atlas, Zakharii Veniaminovich
Economist; b. Je. 7, 1903, Astrakhan.
Doctor of economics (1939), professor (1939). Graduated from the economics department of Don State University (Rostov-on-Don, 1924). His main works have been on the circulation of money, finance, credit, and pricing. He was chairman of the sub-department of money circulation and credit from 1935 to 1963, and he has been consultant and professor at the Moscow Finance Institute since 1963. He was an expert on the board of Gosbank (State Bank) and participated in the preparation of the 1947 currency reform. He was also a member of the scientific and economic commission of the presidium of the Academy of Sciences of the USSR.
Works: *Den'gi i kredit pri kapitalizme i v SSSR*. Moscow 1930. *Ocherki iz istorii denezhnogo obrashcheniia v SSSR*. Moscow 1940. *Infliatsiia i valiutnyi krizis v Anglii posle vtoroi mirovoi voiny*. Moscow 1949. (Translated into Polish.) *Khozraschet, rentabel'nost' i kredit*. Moscow 1966. *Denezhnoe obrashchenie i kredit v SSSR*. Moscow 1947. Second edition: Moscow 1957. (By an authors' collective under the leadership of Atlas.) *Sotsialisticheskaia denezhnaia sistema*. Moscow 1969.

Atoian, Rafik Gegamovich
Painter; b. 1931, Leninakan.
Grad. Erevan Art and Theatre Inst. 1959; exhibitions 1958–.

Atroshchenko, Vasilii Ivanovich, Dr. of Tech. Sc.
Chemical technologist; b. 1906.
Mem. CPSU 1925–; grad. Chemotechnological Inst. Odessa 1931; Dr. of Tech. Sc. 1945; Prof.; corresponding mem. USSR Acad. of Sciences 1964–; asst., aspirant at Chemotechnological Inst. Odessa 1931–35; mem. staff Chemotechnological Inst. Kharkov 1935–50, Polytech. Inst. Kharkov 1950–; Prof. Chair of Technology of Inorganic Substances 1962–. Decor.: State Prize of the USSR 1969.

Atsmanchuk, Aleksandr Pavlovich
Painter; b. 1923, Odessa.
Attended Odessa Art School and Repin Inst. for Painting, Sculpture and Architecture of the USSR Acad. of Arts 1947–53, pupil of M.I. Avilov; exhibitions 1953–.

Auelbekov, Erkin Nurzhanovich
Party official; b. 1930, Kazakhstan.
Mem. CPSU 1952–; grad. Timiryazev Acad. of Agricultural Sciences Moscow 1953; chief agronomist, dir. of an agricultural machine centre, dir. of a grain sovkhoz in North Kazakhstan 1953–61; held various sov. state and party posts 1961–63; 2nd Sec. of North Kazakh Oblast Cttee. for Agriculture Kazakh CP, Chair. of North Kazakh Oblast Sov. for Agriculture 1963–65; 1st Dep. Min. of Agriculture Kazakh SSR 1965–67; cand. mem. 1966–68, mem. Cen. Cttee. Kazakh CP 1968–; Min. for Grain Products and Mixed Fodder Ind. Kazakh SSR 1967–68; 1st Sec. of Oblast Cttee. Kazakh CP Kokchetav Mr. 1968–78, Turgaiskii Oblast Ja. 1978–; cand. mem. Cen. Cttee. CPSU 1971–76; mem. 1976–; Dep. to USSR Supr. Sov. 1970–.

Aukhadiev, Kenes Mustakhanovich
Party official; b. 1938.
Kazakh; mem. CPSU 1962–, grad. Kazakh State Agricultural Inst. 1968; leader of a tractor brigade on a collective farm 1960–62; 1st Sec. of Panfilov Raion Komsomol Cttee. Kazakh SSR 1962–66; responsible organizer, dep. chief of a dept. of Cen. Cttee. Kazakh Komsomol, 1st Sec. of Alma-Ata Oblast Komsomol Cttee., Sec. of Cen. Cttee. and mem. Bureau of Cen. Cttee. Kazakh Komsomol 1966–71; 1st Sec. of Sovetskii Raion Cttee. Kazakh CP in Alma-Ata 1971–75; Chair. of Exec. Cttee. of Alma-Ata City Sov. of Peoples' Deps. 1975–78; 1st Sec. of Alma-Ata Oblast Cttee. Kazakh CP Ap. 1978–; mem. Bureau of Cen. Cttee. Kazakh CP D. 1978–; Dep. to USSR Supr. Sov. 1979–; mem. Cen. Cttee. CPSU 1981–.

Aunapuu, Herman
Painter and graphic artist; b. 1915, Kikand.
Attended Pallas Art High School in Tartu 1935–40; exhibitions 1940–.

Aushkap, Erik Ianovich
Party official; b. 1928.
Latvian; mem. CPSU 1953–; grad. Latvian State Univ. and Party Higher School of Cen. Cttee. CPSU; worker of a Sovkhoz in Novosibirsk oblast 1942–44; pupil, student 1944–51; Engineer, Dir. of a factory 1955–63; Head of a Dept. of Cttee. of People's Control Latvian SSR 1963–66; mem. Auditing Comm. Latvian CP 1963–66; mem. Apparat Council of Mins. Latvian SSR 1966–72; 1st Sec. of Moskovskii Raion Cttee. Latvian CP in Riga 1972–73; 1st Sec. of Riga City Cttee. Latvian CP 1973–75; Dep. to USSR Supr. Sov. 1975–; Sec. and mem. Bureau Cen. Cttee. Latvian CP Jl. 1975–; mem. Cen. Cttee. Latvian CP 1976–.

Auza, Lidia Karlovna
Painter; b. 1914, Vitebsk.
Attended Acad. of Art of the Latvian SSR 1937–49; exhibitions 1949–.

Avagimov, Saven Andreevich
Painter; b. 1910, Baku.
Attended Azerbaidzhan Art School 1929–32; exhibitions 1932–; mem. CPSU 1944–.

Avakian, Eduard Pashoevich
State official.
Mem. Cen. Cttee. Armenian CP 1976–; Min. of Ind. Engineering Armenian SSR Ja. 1979–.

Avakian, Gevork Semenovich
Graphic artist; b. 1922, Aleksandropol.
Grad. Erevan Art and Theatre Inst. 1959; exhibitions 1948–.

Avakian, Grant Arshakovich
State official.
Chair. of State Cttee. for Forestry Armenian SSR Je. 1974–.

Avakian, Vardkes Aikovich
Sculptor; b. 1931, Gemur, Azerbaidzhan.
Attended Erevan Art School 1949–54; exhibitions 1957–.

Avakimian, Artsvik Bakshevich
Painter; b. 1920, Piatigorsk.
Mem. CPSU 1943–; attended Erevan Art School 1936–40; exhibitions 1951–.

Avalishvili, Bidzina Georgievich
Sculptor; b. 1922, Tbilisi.
Mem. CPSU 1953–; grad. Tbilisi Art Acad. 1947, pupil of Ia. I. Nikoladze; exhibitions 1953–.

Avanesov, Ruben Ivanovich
Scholar; b. F. 14, 1902.
Linguist; Prof. at Moscow Univ. 1937–; corresponding mem. USSR Acad. of Sciences 1958–.
Publs.: "Očerki russkoj dialektologii" (vol. 1, 1949), "Očerki dialektologii rjazanskoj meščery" in: "Materialy i issledovanija po russkoj dialektologii", vol. 1, Moscow-Leningrad 1949, "Fonetika sovremennogo russkogo literaturnogo jazyka" Moscow 1956, "Atlas russkih narodnyh govorov central'nyh oblastej k vostoku ot Moskvy," vols. 1–2, 1957 (ed.), "Udarenie v sovremennom russkom literaturnom jazyke" Moscow 1958, "Slovari drevnerusskogo jazyka 11–14 vv." 1966 (ed.), "Russkoe literaturnoe proiznošenie", Moscow 1950, Moscow 1968.

Avdeenko, Aleksandr Ostapovich
Author; b. 1908, Makeevka; s. of a miner. Russian; mem. CPSU 1945–; homeless in his early youth, later worked in coal

mining region of Donets Basin and in the construction of the city of Magnitogorsk. Publs.: novels: "Ia liubliu" (I Love) 1933, "Sud'ba" (Fate) 1936, "Trud" (Work) 1951, "Nad Tissoi" (On the Theiss) 1954, "Gornaia vesna" (Mountain Spring) 1955, and others.

Avdeev, Dimitrii Ivanovich
Painter; b. 1915, Safronovo, Ivanovo Oblast.
Grad. Moscow Inst. of Fine and Decorative Art 1952.

Avdeev, Iurii Konstantinovich
Painter; b. 1918, Kiiasov, Moscow Oblast.
Attended Orel and Krasnodar Art Schools 1936–40, pupil of A. A. Buzovkin and N. I. Sharimov; exhibitions 1940–.

Avdeev, Iurii Sergeevich
Graphic artist; b. 1930, Moscow.
Attended Surikov Art Inst. Moscow 1956, pupil of S. S. Glushkov, M. V. Matorin, P. I. Suvorov.

Avdeev, Valentin Nikolaevich
Scientist; b. My. 3 (16), 1915.
Mem. CPSU 1951–; specialist in electronics; corresponding mem. USSR Acad. of Sciences 1958–; Chief of Electronics Laboratory at USSR Acad. of Sciences 1961–.

Avdeev, Viktor Fedorovich
Author; b. 1909, Cossack village of Uriupinskaia/Don.
Grad. Moscow Gorkii Inst. of Literature 1938; first short stories "Karapet" and "Torzhestvo Zhigi" (Zhiga's Triumph) publ. in almanac "Vchera i Segodnia" (Yesterday and Today) 1931.
Publs.: short stories: "Len'ka Okhnar'" 1957, "Konets Gubana" (Guban's End) 1957, "Khleb" (Bread) 1957, "Moia Odisseia" (My Odyssey) 1960, and others.

Avdeeva, Tamara Timofeevna
Sculptor; b. 1923, Astrakhan.
Grad. Repin Inst. for Painting, Scultpure and Architecture Leningrad 1961, pupil of I. V. Krestovskii and V. V. Lishev; exhibitions 1961–.

Avdyshev, Aleksei Ivanovich (also Avedyshev)
Graphic artist; b. 1928, Leningrad.
Grad. Pedagogical Inst. of Art Leningrad 1950; exhibitions 1952–.

Avdysheva, Valentina Mikhailovna (also Avedysheva)
Painter; b. 1931, Leningrad.
Grad. Pedagogical Inst. of Art Leningrad 1952; exhibitions 1955–.

Avedikov, Nikolai Vaganovich
Sculptor; b. 1913, Rostov/Don.
Grad. Rostov Art School, pupil of I. I. Sonkin; exhibitions 1953–.

Averbakh, Aleksandr Mikhailovich
Stage designer; b. 1933, Moscow.
Grad. Theatre School at E. Vakhtangov Theatre 1956, pupil of S. N. Akhvlediani, G. A. Fedorov.

Averbukh, Iakov Abramovich
Graphic artist; b. 1922.
Grad. Kishinev Art School 1950.

Averbukh, Lev Volfovich
Sculptor; b. 1913, Kishinev.
Grad. Kishinev Art School 1948, pupil of K. S. Kobizeva, I. Ia. Khazov; exhibitions 1946–.

Averianov, Boris Iakovlevich
Graphic artist; b. 1931, Dorokhovo, Moscow Oblast.
Grad. Surikov Art Inst. Moscow 1960, pupil of M. V. Matorin, N. A. Ponomarev.

Avetisian, Amaiak Avetisovich
Painter and graphic artist; b. 1912, Mandan, Turkey.
Grad. Leningrad. Inst. for Painting, Sculpture and Architecture 1937, pupil of V. P. Belkin, E. E. Lansere, A. P. Ostroumova-Lebedeva; mem. CPSU 1945–.

Avetisian, Minas Karapetovich
Painter and stage designer; b. 1928, Dzhadzhur, Armenia.
Grad. Repin Inst. for Painting, Sculpture and Architecture Leningrad 1960, pupil of A. D. Zaitsev, L. V. Khudiakov; mem. CPSU 1952–; exhibitions 1956–.

Avetisian, Robert Grigor'evich
Painter; b. 1928, Leninakan.
Studied Erevan Art and Theatre Inst. 1948–54, pupil of G. M. Giurdzhian, O. M. Zardarian; exhibitions 1954–.

Avgustovich, Aleksei Ivanovich
Painter and graphic artist; b. 1914, Turno, Poland.
Grad. Moscow Inst. of Fine Arts 1941, pupil of A. A. Osmerkin.

Avidzba, Khuta Tukanovich
Painter; b. 1932, V. Esheri, Abkhazian.
Grad. Tbilisi Art Acad. 1963, pupil of U. M. Dzhaparidze and V. I. Shukhaev.

Avizhius (Avyžius), Jonas
Author; b. 1922, Lithuania; s. of a peasant.
Publs.: short stories: "Pervye borozdy" (The First Furrows) 1948, short novels: "Nasledstvo" (The Heritage) 1949, "Slava" (The Praise) 1949, drama "Belye ivy" (White Pastures) 1954, short stories and sketches "Osvobozhdenie" (The Liberation) 1951, "Liudi i sobytiia" (People and Events) 1954, "Butkus Mstitel'" (Butkus the Avenger) 1957, and others.

Avramchenko, Oksana Nikitichna
Sculptor; b. 1912, Sevastopol.
Mem. CPSU 1946–; grad. Odessa Art Inst. 1935; exhibitions 1951–.

Avramenko, Ilia Kornilevich
Poet; b. 1907, Volodka Devitsa, Chernigov Oblast.
Russian; mem. CPSU 1940–.
Publs.: collected poems "Shestoi gorizont" (The Sixth Horizon) 1931, "Pokhodnaia tetrad'" (Field Book) 1940, "Noch' nakanune bessmertiia" (The Night Before Immortality) 1943, novel in verse "Dom na Moike" (The House in Moika) 1957, and others.

Avramenko, Stepan Stepanovich
Party official; b. 1918, Ukraine.
Mem. CPSU 1950–; grad. Belaia Tserkov' Agricultural Inst. 1941; veterinarian, chief of a raion cattle procurement agency, chair. of a raion sov. 1941–55; 1st Sec. of CPSU Barabinsk City Cttee. 1955–59; Chair. of Novosibirsk Oblast Sov. 1959–64; 1st Sec. of CPSU Amur Oblast Cttee. 1964–; Dep. 1962–, mem. Credentials Comm. 1966–74, mem. Comm. for Transport, Posts and Telecommunications at Sov. of Nationalities USSR Supr. Sov. 1974–; cand. mem. 1966–71, mem. Cen. Cttee. CPSU 1971–.

Avrorin, Valentin Aleksandrovich
Scholar; b. D. 23, 1907, Tambov.
Grad. Historical-Ethnological Faculty of Leningrad Univ. 1930; philologist, specialist in Tunguso-Manchurian languages.
Publs.: "Očerki po sintaksisu nanajskogo jazyka" Leningrad 1948, "Grammatika nanajskogo jazyka," vols. 1–2, 1959–61, "O klassifikacii tunguso-man'čžurskich jazykov (XXV Meždunarodnyi kongress vostokovedov. Doklady delegacii SSSR)" Moscow 1960, "Jazyki i fol'klor narodov sibirskogo Severa" 1966, "Nanajskij jazyk" in: "Jazyki narodov SSSR. Mongol'skie, tungusoman'čžurskie i paleoaziatskie jazyki," vol. 5, Leningrad 1968.

Avrutis, Khaim Avraamovich
Graphic artist and painter; b. 1928, Baryshevka nr. Kiev.
Grad. Surikov Art Inst. Moscow 1953, pupil of P. I. Kotov, V. G. Tsyplakov; exhibitions 1956–.

Avshalumov, Khizgil Davidovich
Tat writer; b. Ja. 18, 1913, Niugdy, Derbent Raion.
Member of the CPSU since 1940. He writes in Tat and in Russian. He took part in the Great Patriotic War. Until 1965 he was a journalist. In 1939 he published his first novella, *Lovers*.
He gained recognition through his collections of short stories and essays – *Friendship* (1956), *Meeting at the Spring* (1960). *How I Revived* (1961, in Russian), and *The Bride With a Surprise* (1966, in Russian). His novellas include *Sosun's Fate* (1957; Russian translation. 1960, under the title *Under the Plane Trees*), *The Sash of Poverty* (1963), and *Interpreter of the Imam* (1967, in Russian). He also wrote the play *Love in Danger*. He is one of the

first compilers of Tat folklore; in 1940 he published the collection *Tat Folklore.* Decors.: Badge of Honor and medals. Works: *Vstrecha u rodnika.* Moscow 1960. *Pod chinarami.* Moscow 1960.

Avsiian, Osip (Iosif) Abramovich
Graphic artist and painter; b. 1918, Pavlograd, Ukraine.
Grad. Moscow Inst. of Fine Arts 1942, pupil of A. V. Lentulov, N. E. Radlov and A. M. Solovev; exhibitions 1945–.

Avsiuk, Grigorii Aleksandrovich
Scientist; b. D. 16 (29), 1906, St. Petersburg.
Mem. CPSU 1947–; glaciologist; corresponding mem. USSR Acad. of Sciences 1960–.
Publs.: "Gory jugo-vostočnogo Kazahstana" 1945 (co-author), "Temperatura l'da v lednikah" in: "Trudy Instituta geografii AN SSSR", vol. 67, 1956.

Avtokratova, Ksenia Mikhailovna
Artist in applied art; b. 1924, Moscow.
Grad. M. I. Kalinin School for Arts and Crafts Moscow 1949 and Faculty of Applied Art at Textile Inst. Moscow 1964, pupil of N. N. Flerov; artist in applied art, designs patterns for tapestries and furnishing fabrics; exhibitions 1952–.

Avvakumov, Boris Petrovich
Stage designer and graphic artist; b. 1917, Nizhnii Novgorod.
Grad. Perm Art School 1938, pupil of I. E. Shangin; exhibitions 1957–; works in industrial advertising.

Azan, Vladislav Domenikovich
State official.
Mem. Cen. Cttee. Latvian CP 1971–; Min. of Trade Latvian SSR Mr. 1977–.

Azanov, Nikolai Fedorovich
Painter of miniatures; b. 1926, Ovchinnikovo, Perm Oblast.
Attended Pakekh Art School 1948–53, pupil of D. N. Nutorin and I. P. Vakurov; exhibitions 1956–.

Azarov, Vsevolod Borisovich
Poet; b. 1913, Odessa; s. of a doctor.
Russian; grad. Leningrad Univ. 1934; first works publ. 1932.
Publs.: collection "Spat' vospreshcheno" (Sleeping Forbidden) 1933, "Svet maiaka" (The Light of the Lighthouse) 1956, poem "Tovarishch Thälmann" (Comrade Thälmann) 1956, collection "Leningradu" (To Leningrad) 1942, "Baltiiskie ballady" (Baltic Ballades) 1942, poem "Vooruzhennyi Komsomol" (Armed Komsomol) 1932, "Muzhestvo" (Courage) 1932, "Gorod moei iunosti" (The Town of my Youth) 1940, and others.

Azarova, Ludmila Pavlovna
Artist in applied art; b. 1922, Borovsk, Kaluga Oblast.
Attended Moscow Inst. for Applied and Decorative Art 1948–52 and Moscow High School for Arts and Crafts 1952–54; artist in applied art (ceramics, porcelain, clay).

Azeev, Iurii Pavlovich
Sculptor; b. 1926, Zlatoust.
Attended Penza Art School 1946–50, pupil of N. K. Krasnov and M. I. Maliutin; exhibitions 1952–.

Azeroglu (real name: Abizade, Balash)
Poet; b. 1921, Ardebil; s. of a worker.
Iranian Azerbaidzhan; grad. Azerbaidzhan Univ. 1952.
Publs.: poetry: "Stikhi" (Verse) 1948, "Savalan" 1949, "Na putiakh bor'by" (On the Paths of Combat) 1950, "Druzia iunosti" (Friends of Youth) 1950, "Moi pesni" (My Songs) 1952, "Utrenniaia zaria" (Daybreak) 1953, and others.

Azgur, Galina Gavrilovna
Painter; b. 1921, Okrovka, Moscow Oblast; d. of G. N. Gorelov.
Attended Moscow Oblast Pedagogical Inst. of Art 1938–42, pupil of V. N. Baksheev; exhibitions 1947–.

Azgur, Zair Isaakovich
Sculptor; b. 1908.
Mem. CPSU 1943–; studied Acad. of Arts Leningrad and Art Inst. Kiev; mem. USSR Acad. of Arts 1958–. Decors.: People's Artist of the USSR 1973, State Prizes of the USSR 1946, 1948.

Azhgikhin, Boris Stepanovich
Painter and graphic artist; b. 1924, Sergeevo, Perm Oblast.
Attended Sverdlovsk Art School 1950–55; exhibitions 1956–.

Azhiev, Uke
Painter; b. 1924, Kemir-Togan, Kazakhstan.
Grad. Alma-Ata Art School 1959, pupil of N. I. Krutilnikov and A. M. Cherkasskii; exhibitions 1950–.

Azimov, Pigam Azimovich
Linguist; b. Ja. 1, 1915.
Turkmen; mem. CPSU 1939–; mem. 1951–, Pres. of Acad. of Sciences Turkmen SSR 1966–75; Decors.: Order of Lenin and other decors.
Works: *Turkmen dili (söz yasayjï goshülmalar).* Ashkhabad 1950.

Azimov, Sarvar Alimdzhanovich
Diplomat; b. 1923.
Chief of Dept. of Science and Culture at Cen. Cttee. Uzbek CP 1955–56; cand. mem. 1956–58, mem. Cen. Cttee. Uzbek CP 1958–71; Dep. Min. of Culture 1956–57, Min. of Culture Uzbek SSR 1957–59; Dep. Chair. of Council of Mins. and Min. for Foreign Affairs Uzbek SSR 1959–69; USSR Amb. to Lebanon 1969–74; USSR Amb. to Pakistan D. 1974–N. 1980.

Azizbekov, Shamil Abdulragim
Geologist; b. F. 16, 1906.
Mem. CPSU 1942–; Acad. Sec. of Dept. of Geographical Sciences, Chief of Dept. of Petrography at Inst. of Geology Azerbaidzhan SSR Acad. of Sciences; mem. of same Acad. Decors.: several orders and medals.
Publs.: "Magmatizm i metallogenija Azerbajdžana" in "Magmatizm i svjaz' s nim poleznyh iskopaemyh" Moscow 1960 (co-author).

Azizian, Grigor Minasovich
Painter; b. 1923, Armavir.
Attended Erevan Art and Theatre Inst. 1947–53; exhibitions 1953–.

Azizian, Suren Khachaturovich
Painter; b. 1922, Erevan.
Mem. CPSU 1949–; grad. Erevan Art Inst. 1951; exhibitions 1952–.

Azizian, Tsolak Levonovich
Painter; b. 1912, Karbi, Armenia.
Mem. CPSU 1946–; grad. Erevan Tech. College of Art 1930, attended Inst. for Painting, Sculpture and Architecture Leningrad 1930–35; exhibitions 1935–.

Baal Khabyryys (real name: Veshnikov, Gavriil Grigorevich)
Poet; b. 1917.
Yakut; grad. Yakut Pedagogical Inst.; worked as teacher.
Publs.: verse collection "Bessmertie" (Immortality) 1945, "Prekrasnye vechera" (Wonderful Evenings) 1947, "Malen'kim druziam" (For My Little Friends) 1948, "Moe pokolenie" (My Generation) 1951, and others.

Bababev, Akmal Maksumovich
State official.
Chair. of Tadzhik Republican Asscn. for Agricultural Technology "Tadzhikselkhoztekhnika" Ap. 1977–78; Chair. of State Cttee. for the Supply of Production Equipment for Agriculture Tadzhik SSR Ag. 1978–; mem. Cen. Cttee. Tadzhik CP 1981–.

Babadzhan, Ramz (real name: Babadzhanov, Ramz Nasyrovich)
Author; b. 1921.
Uzbek; mem. CPSU 1951–; grad. Pedagogical Inst. Tashkent 1944; Dep. Chair. of Uzbek Writers' Union; Chair. of Uzbek Republican Cttee. on Relations with Asian and African Writers; first works publ. 1935. Decor.: State Prize of the USSR 1972.
Publs.: author of some 20 collections in Uzbek and 5 books in Russian, "Dorogie druzia" (Dear Friends), "Spasibo, rodnoi" (Thank You, My Dear), "Serdtse ne spit nikogda" (The Heart Never Sleeps), "Izbrannaia lirika" (Selected Poetry), "Dvazhdy zhivet poet" (A Poet Lives Twice), poem "Zhivaia voda" (Living Water) was publ. in Uzbek and Russian.

Babadzhanian, Arno Arutiunovich
Composer and pianist; b. 1921.
Armenian; mem. CPSU 1956–; grad. Erevan Conservatory 1947 and Moscow Conservatory 1948. Decors.: State Prize of the USSR 1951, People's Artist of the USSR 1971.
Works: romantic ballad "Geroičeskaja Ballada" for piano and orchestra, and others.

Babadzhanov, A. A.
Trade union official.
Sec. and mem. Presidium of Tadzhik Republican Trade Union Council 1976–.

Babaev, Agadzhan Geldyevich, Dr. of Geog. Sc.
Geographer; b. 1929.
Mem. CPSU 1954–; grad. Dept. of Geography at State Pedagogic Inst. in Ashkhabad 1949; Dr. of Geog. Sc. 1968; Prof. 1959–; mem. Acad. of Sciences Turkmen SSR 1975–; teacher, Lecturer, Head of Chair of Geography at Turkmen State Univ. 1950–60; Dir. of Inst. for Desert Research at Acad. of Sciences Turkmen SSR 1960–75; Pres. of Acad. of Sciences Turkmen SSR 1975–; Chair. of Turkmen Geographical Soc. 1959–; Chair. of "Znanie" Soc. Turkmen SSR 1971–; Chief Ed. of journal "Problemy osvoeniia pustyn'" (Problems in Developing Deserts) 1968–; author of some 110 scientific papers on problems of developing the deserts of the USSR and on physical-geographical division of Turkmen SSR into raions.

Babaev, Aloviddin
State official.
Sec. of Leninakan Oblast Cttee. Tadzhik CP 1974–78; 1st Sec. of Gorno-Badakhshan Oblast Cttee. Tadzhik CP 1978–82; mem. Cen. Cttee. Tadzhik CP 1981–; Min. of Agriculture Tadzhik SSR O. 1982–.

Babaev, Iurii Nikolaevich
Physicist; b. My. 21, 1928, Moscow.
Grad. Moscow Univ. 1950; corresponding mem. USSR Acad. of Sciences 1968–; research work in nuclear physics. Decors.: State Prize of the USSR 1953, Lenin Prize 1959, Hero of Socialist Labour 1962, Order of Lenin.

Babaev, Mamarsul
Author; b. 1911, Katta-Kurgan; s. of an artisan.
Uzbek; grad. Pedagogical Acad. Samarkand and Pedagogical Inst. Tashkent.
Publs.: play "Beloe zoloto" (White Gold) 1933, poem "Ulugbek" 1940, "Sadovnik" (The Gardener) 1948, poetry collection "Iranskaia tetrad'" (Iranian Notebook) 1946–48, and others.

Babaev, Mirzo
Party official; b. 1926.
Mem. CPSU 1951–; mem. Auditing Comm. 1963–66, mem. Cen. Cttee. Tadzhik CP 1963–; 1st Sec. of Voseiskii Raion Cttee. Tadzhik CP until 1973; Sec. and mem. Bureau of Cen. Cttee. Tadzhik CP Mr. 1973–.

Babaev, Nabi Khazri (full name: Babaev, Nabi Alekper ogly)
Poet; b. 1924, Khurdalen nr. Baku; s. of a peasant.
Grad. Gorkii Inst. of Literature Moscow 1952; first works publ. 1944.
Publs.: poetry "Rastsvet mechty" (The Flowering of a Dream) 1950, "Bakinskie kuranty" (Baku Chimes) 1954, "Rodina i Chuzhbina" (Native and Foreign Land) 1958, "Dub na skale" (Oak Tree on a Rock) 1959, and others.

Babaev, Rasim (Rakhim) Ganifa ogly
Graphic artist and painter; b. 1927, Baku.
Grad. Azerbaidzhan Art School 1949, attended Surikov Art Inst. Moscow 1949–56, pupil of P. I. Kotov, P. P. Sokolov-Skal; exhibitions 1958–. Decor.: Merited Artist of the Azerbaidzhan SSR 1946.

Babaev, Salikh
State official.
Dep. to Supr. Sov. Turkmen SSR 1959–; 1st Sec. of Serakhs Raion Cttee. Turkmen CP 1959; cand. mem. Cen. Cttee. Turkmen CP 1960–61; Sec. of Party Cttee. of Production Admin. in Takhta-Bazar 1963; Min. of Supplies Turkmen SSR Ja. 1970–.

Babaevskii, Semen Petrovich
Author; b. 1909, Kune, Kharkov Oblast; s. of a peasant.
Mem. CPSU 1939–; grad. Correspondence Dept. at Gorkii Inst. of Literature 1939; worked for Rostov newspapers, literary journal "Na pod'eme" and others 1929–41; correspondent 1941–45; Sec. of the Board of RSFSR Writers' Union 1959–65; first works publ. 1929. Decors.: Order of Lenin, Order of Red Banner of Labour, 3 Stalin Prizes.
Publs.: short-story collection "Gordost'" (Pride) 1936, novel "Kavaler Zolotoi Zvezdy" (Cavalier of the Gold Star) 1947, collection of sketches about China "Vetvi Starogo Viaza" (Branches of an Old Elm) 1947, "Svet nad Zemlei" (Light Over the Earth) 1949, "Po putiam dorogam" (On the Highways) 1958, "Sestry" (Sisters) 1958, "Chetyre Raisy" (Four Raisas) 1959, "Synovnii bunt" (Rebellion of Sons) 1961, and others.

Babaian, Rafael Surenovich
Graphic artist; b. 1927, Dilizhan, Armenia.
Attended Erevan Art and Theatre Inst. 1949–55; exhibitions 1956–.

Babak, Petr Ivanovich
Painter; b. 1921, Makhnachi, Chernigov Oblast.
Attended Kiev Art Inst. 1946–52, pupil of K. D. Trokhimenko; exhibitions 1951–.

Babakulov, Akhmad
Singer; b. 1931.
Tadzhik; grad. Moscow Conservatory 1954; heroic tenor; soloist with Tadzhik Philharmonic 1948, 1956–66; soloist with Tadzhik Aini Opera and Ballet Theatre 1954–56, 1966; Head of Chair for Singing at Art Faculty of Tadzhik Pedagogic Inst. 1966–. Decor.: People's Artist of the USSR 1971.

Babanova, Mariia Ivanovna
Russian actress; b. O. 29 (N. 11), 1900, Moscow.
People's Artist of the USSR (1954). She graduated from the studio of the Theater of the Artistic and Educational Union of Workers' Organizations. From 1920 to 1927 she worked in the First Theater of the RSFSR (later renamed the Meierkhol'd Theater). Her roles include Stella in Crommelynck's *The Magnanimous Cuckold,* Boi in Tret'iakov's *Roar, China!,* and Mar'ia Antonovna in Gogol's *The Inspector-General.* At the same time she acted in the Theater of the Revolution, where she played the role of Polina in V. E. Meierkhol'd's production of Ostrovsky's *A Profitable Place.* In 1927 she permanently joined the Theater of the Revolution (now the Mayakovsky Moscow Theater). Here she played the roles of Goga in Faiko's *Man With a Briefcase,* Kolokol'chikova and Masha in Pogodin's *My Friend* and *After the Ball,* Juliet in Shakespeare's *Romeo and Juliet,* Diana in Lope de Vega's *Dog in the Manger,* and the title role in Arbuzov's *Tania.* Complex psychological conflicts and the theme of the struggle for human dignity are especially close to her. A refined lyric actress, she creates vivid individualized stage characters. Her work is characterized by high craftsmanship, plasticity of gesture, musicality of speech, and deep emotion. Her roles in Russian and Soviet classical theater include Larisa in Ostrovsky's *The Dowerless Girl* and Sof'ia in Gorky's *Zykovy.* She played the role of Kei in the Japanese playwright Kaoru Morimoto's *A Stolen Life.* Decors.: State Prize of the USSR in 1941, two orders.

Babentsov (Babintsev), Viktor Vladimirovich
Painter and graphic artist; b. 1921, Litvinovka, Ukraine.
Attended Dnepropetrovsk Art School 1938–39 and Kiev Art Inst. 1946–52, pupil of V. N. Kostetskii, K. D. Trokhimenko, exhibitions 1951–; mem. CPSU 1943–.

Babian, Rafael (Rafik) Arturovich
Painter and stage designer; b. 1920, Erevan; s. of A. S. Babian.
Attended Erevan Art School 1937–40 and Art Inst. 1945–50, pupil of Ts. L. Azizian, V. N. Gaifedzhian, A. G. Kodzhoian, M. S. Sarian; exhibitions 1939–; teacher at Kirovakan Art School 1957–; mem. CPSU 1950–.

Babii, V.V.
Komsomol official.
1st Sec. and mem. Bureau Cen. Cttee. Moldavian Komsomol Mr. 1980–; mem. Cen. Cttee. Moldavian CP 1981–.

Babikov, Stanislav Gennadievich
Painter; b. 1934, Ashkhabad.
Attended Leningrad School for Arts and Crafts 1948–53, grad. Repin Inst. for Painting, Sculpture and Architecture at the USSR Acad. of Arts 1959; exhibitions 1954–; mem. CPSU 1965–.

Babinets, Andrei Evtikhievich, Dr. of Geol.-Min. Sc.
Hydrogeologist; b. D.13, 1911, Kiev; s. of a worker.
Mem. CPSU 1939–; grad. Mining and Geological Inst. Kiev; corresponding mem. Acad. of Sciences Ukrainian SSR 1964; Dr. of Geol. Min. Sc. 1961; Prof. 1963; Research Fellow of Inst. of Geology of Acad. of Sciences Ukrainian SSR 1935–41; served in Sov. Army 1941–45; Sen. Research Fellow at Inst. of Geology of Acad. of Sciences Ukrainian SSR 1945–48; Dep. Dir. 1948–49, 1953–, Acting Dir. of Teaching Dept. at Inst. of Geology 1949–53; author of 98 scientific works, among them 3 monographs. Decors.: 2 Orders of the Red Star, Order of the Patriotic War 2nd Cl., Badge of Hon., medals.
Publs.: "Mineralni dzherela Radjanskoho Zakarpattia ta perspektyvy ikh vykorystannia" in "Visnyk Akademii Nauk Ukrainskoi RSR" No. 4/1948, "Podzemnye vody iugozapada Russkoi platformy" 1961, "Issledovanie plotnosti i vlazhnosti gruntov metodami radioaktivnykh izluchenii" 1961, "Deiaki rezultaty vyvchennia vmistu izotopu O^{18} u vodakh tropichnoi Atlantyky" in "Dopovidi Akademii Nauk Ukrainskoi RSR" No. 3/1967.

Babitsin, Viktor Alekseevich
Painter and graphic artist; b. 1926, Leningrad.
Attended Moscow School for Arts and Crafts 1939–45 and Surikov Art Inst. Moscow 1945–52; exhibitions 1942–.

Babkin, Nikolai Aleksandrovich
State official.
Mem. Cen. Cttee. Latvian CP 1961–; Chair. of Latvian Republican Asscn. for Agricultural Technology "Latvselkhoztekhnika" Ap. 1962–78; mem. Bureau of Agriculture at Cen. Cttee. Latvian CP 1962–64; Dep. to Supr. Sov. Latvian SSR 1963–; Chair. of State Cttee. for the Supply of Production Equipment for Agriculture Latvian SSR S. 1978–.

Babko, Anatolii Kirillovich, Dr. of Chem. Sc.
Chemical scientist; b. O. 15, 1905, Sudzhenskoe, Novosibirsk Oblast; s. of a railway worker.
Mem. CPSU 1940–; grad. Polytech. Inst. 1927; specialist in analytical and physical chemistry; mem. Acad. of Sciences Ukrainian SSR 1957–; Dr. of Chem. Sc. 1941; Prof. 1943; Asst. at Prof. Chair of Analytical Chemistry at Polytech. Inst. Kiev 1927–30; Lecturer at Technological Inst. of Foodstuffs Ind. Kiev 1930–33; Lecturer at Prof. Chair of Analytical Chemistry Kiev Univ. 1933–41; Head of Laboratory for Analytical Chemistry at Inst. of Gen. and Inorganic Chemistry of Acad. of Sciences Ukrainian SSR 1941–; additionally mem. staff. Univ. Kiev 1944; his scientific works relate to the chemistry of complex compounds and to the theory and practice of chemical analysis. Decors.: Order of Lenin, Badge of Honour, medals.
Publs.: "Kolorimetricheskii analiz" 1951, "Fiziko-khimicheskii analiz kompleksnykh soedinenii v rastvorakh" 1955, "Kolichestvennyi analiz" 1962, "Fiziko-khimicheskie metody analiza" 1966, "Khemiliuminestsentsnyi analiz" 1966.

Babkov, Sergei Fedorovich
Painter and graphic artist; b. 1920, Zlatoust.
Grad. Leningrad School for Arts and Crafts 1941, attended Repin Inst. for Painting, Sculpture and Architecture at the USSR Acad. of Arts 1946–52, pupil of R.R. Frents; exhibitions 1939–; mem. CPSU 1944–.

Babloian, Grant Tovmasovich
Graphic artist; b. 1917, Iğdir, Turkey.
Attended Erevan Art Inst. 1945–51, pupil of G.M. Giurdzhan, A.G. Kodzhoian; exhibitions 1952–; mem. CPSU 1947–.

Babskii, Evgenii Borisovich, Dr. of Biol. Sc.
Physiologist; b. Ja. 28, 1902, Gerusi, now Goris, Armenian SSR; s. of a doctor.
Mem. CPSU 1932; grad. Moscow Univ. 1925 and Biological-Scientific Timiriazev Research Inst. 1929; mem. Acad. of Sciences Ukrainian SSR 1948–; Dr. of Biol. Sc. 1939; Prof. 1932; Prof. Chair for Physiology at Pedagogical Lenin Inst. Moscow 1930–49; additionally Head of Laboratory for Neurophysiology at All-Union Inst. for Experimental Medicine 1933–37; Head of Laboratory for Physiology at Inst. of Biology and Medical Chemistry of USSR Acad. of Sciences 1945–49; Chief of Dept. for Gen. Physiology at Inst. for Clinical Physiology of Acad. of Sciences Ukrainian SSR 1949–50; mem. staff. of various institutions at USSR Acad. of Medical Sciences 1952–; Head of Laboratory for Clinical Physiology at Inst. of Normal and Pathological Physiology of USSR Acad. of Medical Sciences 1958–; co-ordinates scientific research work in the field of medical electronics and medical cybernetics.
Publs.: "I.P. Pavlov" 1959, "Elektricheskaia stimuliatsiia serdtsa" 1961, "Dinamokardiografiia" 1963, "Fiziologiia, meditsina i tekhnicheskii progress" 1965, "Fiziologiia cheloveka" 1966.

Bachinin, Nikolai Ivanovich
Painter; b. 1921, Barovlianka, Tiumen Oblast.
Attended Omsk Art School 1939–41 and Repin Inst. for Painting, Sculpture and Architecture at the USSR Acad. of Arts 1943–46, pupil of P.P. Belousov, B.V. Iohanson, B.A. Vogel; exhibitions 1947–.

Bachurin, Evgenii Vladimirovich
Graphic artist; b. 1934, Leningrad.
Attended Moscow Polygraphic Inst. 1951–54 and 1957–59 and Repin Inst. for Painting, Sculpture and Architecture at the USSR Acad. of Arts 1954–57.

Badakva, Viktor Pavlovich
Painter; b. 1918, Nikolaev, Saratov Oblast.
Attended Saratov Art School 1937–42; exhibitions 1946–; teacher at Saratov Art School 1948–.

Badalian, Grigorii Babaevich
Sculptor; b. 1922, Rostov/Don.
Attended Erevan Art Inst. 1946–52, pupil of A.A. Avetisian, G.P. Agaronian, S.L. Stepanian; exhibitions 1952–; mem. CPSU 1949–.

Bader, Otto Nikolaevich
Archaeologist; b. Je. 16 (29), 1903, Aleksandrovskoe, Gadiach District, Poltava Province.
Doctor of historical sciences (1964). Studied many relics of the Paleolithic period (the Sungir' camp, paleolithic painting in the Kapova cave, and others) and the Bronze Age (the Balanovo and Turbino burial grounds, and others).
Works: *Ha zare istorii Prikam'ia.* Perm' 1958. (With V.A. Oborin.) *Poseleniia turbinskogo tipa v srednem Prikam'e.* Moscow 1961. *Balanovskii mogil'nik: Iz istorii lesnogo Povolzh'ia v epokhu bronzy.* Moscow 1963. *Kapovaia peshchera: Paleoliticheskaia zhivopis'.* Moscow 1965.

Badova, Rina Grigor'evna
Sculptor; b. 1918, Elizavetgrad.
Attended Surikov Art Inst. Moscow 1938–48; exhibitions 1948–.

Baenskaia, Irina Venediktovna
Artist in applied art; b. 1924, Baku.
Attended Moscow Inst. for Applied and Decorative Art 1945–52, pupil of E.F. Belashova, E.S. Zernova, N.G. Tikhonov, artist in applied art (ceramics); exhibitions 1956–.

Baev, Aleksandr Aleksandrovich, Dr. of Biol. Sc.
Biochemist; b. 1903.
Mem. CPSU 1964–; grad. Kazan Univ. 1927; Dr. of Biol. Sc. 1967; corresponding mem. USSR Acad. of Sciences 1968–; doctor 1927–30; Asst. at Medical Inst. Kazan 1930–35; mem. staff Bach Inst. for

Biochemistry of USSR Acad. of Sciences 1935–59; Head of Laboratory at Inst. of Molecular Biology USSR Acad. of Sciences 1959–. Decor.: State Prize of the USSR 1969.

Bagautdinov, Rem Makhmudovich (Mikhailovich)
Graphic artist; b. 1927, Kazan.
Attended Kiev Art Inst. 1951–57, pupil of E.V. Volobuev, K.D. Trokhimenko, T.N. Iablonskaia; exhibitions 1957–.

Bagdasarian, Grigorii Gaikovich
Painter; b. 1922, Tsur, Azerbaidzhan.
Attended Kharkov Art Inst. 1945–52, pupil of S.F. Besedin, A.A. Kokel, S.M. Prokhorov; exhibitions 1952–.

Bagdasar'ian, Khristofor Stepanovich
Physical chemist; b. N. 18, 1908, Paris.
Corresponding member of the Academy of Sciences of the USSR (1968). He graduated from the Moscow Chemical Engineering Institute in 1931. He began to work in the Karpov Institute of Physics and Chemistry as a senior scientist and laboratory chief in 1936. Much of his research deals with the kinetics and mechanism of the reactions of free radicals and elementary photochemical and radiation chemical reactions. He proposed a general theory based on the concept of an activated complex to account for the effect that molecular structure exerts on the rate of radical reactions. The previously established empirical laws in this field (Polanyi's rule, Alfrey-Price equation, and others) were in essence found to follow this theory. He was the first to demonstrate the existence of two-quanta photochemical reactions – a special class of photochemical reactions in which the chemical act results from the successive absorption of two light quanta, when the second quantum is absorbed by the molecule in the triplet state. He is the author of the monograph *Teoriia radikal'noi polimerizatsii* (Theory of Radical Polymerization [the second revised edition appeared in 1966]). The book has been translated into several foreign languages.
Works: "Dvukhkvantovye fotokhimicheskie protsessy obrazovannia radikalov i kation-radikalov." *Zhurnal Vsesoiuznogo khimicheskogo ob-va im. D.I. Mendeleeva,* 1966, vol. 11, no. 2, p. 216. "Kinetika dvukhkvantovykh fotokhimicheskikh reaktsii: Sistema karbazol-etilovyi spirt." *Kinetika i kataliz,* 1967, vol. 8, no. 2, p. 283. (With Z.A. Sinitsyn.)

Bagdasarian, Levon Israelovich
Painter; b. 1919, Pekhlevi, Iran.
Attended Erevan Art and Theatre Inst. 1947–54, pupil of A.V. Bekarian, M.S. Sarian; exhibitions 1955.

Bagdasarian, Oganes Mnatsakanovich
State official; b. 1914.
Armenian; mem. CPSU 1939–; grad. of a pedagogical inst.; instructor, dept. chief, sen. ed. of a city newspaper 1934–45; Sec. of Leninakan City Cttee. Armenian CP, 1st Sec. of Akhrianskii Raion Cttee. Armenian CP, mem. staff Armenian Min. of Culture 1945–53; Sen. Ed. of newspaper "Sovetakan Aiastan" 1953–57; mem. Cen. Cttee. 1954–, cand. mem. Bureau 1956–60, mem. Bureau Cen. Cttee. Armenian CP 1960–66; Dep. 1955–59, 1967–, Dep. Chair. of Presidium Supr. Sov. Armenian SSR 1967–; 1st Sec. of Erevan City Cttee. Armenian CP 1957–61; Dep. to USSR Supr. Sov. 1958–66; 2nd Sec. of Cen. Cttee. Armenian CP 1961–66; mem. Cen. Auditing Comm. CPSU 1961–66; Min. of Culture Armenian SSR 1966–67.

Bagdasarian, Sergei Ivanovich
Sculptor; b. 1923, Krasnovodsk.
Attended Erevan Art Inst. 1946–52, pupil of G.P. Agaronian, A.M. Sarksian; exhibitions 1942–. Decor.: Merited Artist of the Armenian SSR 1961.

Bagdavadze, Ekaterina Nikolaevna
Painter; b. 1922, Tbilisi.
Attended Tbilisi Acad. of Art 1936–40, pupil of U.M. Dzhaparidze, M.I. Toidze; exhibitions 1946–; mem. CPSU 1941–. Decor.: Merited Artist of the Georgian SSR 1959.

Bagdonas, Anitsetas Ignovich
State official.
Cand. mem. Cen. Cttee. Lithuanian CP 1960–; Min. of Rural Construction Lithuanian SSR O. 1968–.

Bagirov, Kyamran Mamed ogly
Party official; b. 1933; s. of a white-collar worker.
Mem. CPSU 1961–; grad. Azerbaidzhan Polytechnical Inst. and CPSU Cen. Cttee. Party Higher School; mechanic, engineer, dep. head of various building trusts in Azerbaidzhan, dep. head of the Dept. of Construction Engineering of the Council of Mins. Azerbaidzhan SSR until 1971; head of the Dept. of Construction Engineering Cen. Cttee. CP Azerbaidzhan, 1st City Sec. CP Azerbaidzhan in Sumgait 1971–78; mem. Auditing Comm. 1971–76, mem. Cen. Cttee. CP Azerbaidzhan 1976–; mem. Bureau 1978–, Sec. 1978–82, 1st Sec. Cen. Cttee. CP Azerbaidzhan D. 1982–.

Bagirov, Oktai Alizatarovich
Party official; b. 1941.
Azerbaidzhani; mem. CPSU 1963–; mechanic at a cinema 1955–60; served in Sov. army 1960–63; head of milling operative brigade at machine shop in Baku 1963–; mem. Cen. Cttee. and mem. Bureau of Cen. Cttee. CP Azerbaidzhan Ja. 1981–.

Bagirov, Zakir Nariman ogly
State official.
Min. of Culture Azerbaidzhan SSR F. 1971–; cand. mem. Cen. Cttee. Azerbaidzhan CP 1971–76, mem. 1976–; Chair. of Azerbaidzhan Soc. of Book Lovers 1974–.

Baglanova, Roza Tazhibaevna
Singer (soprano); b. Ja. 1, 1922, Kazalinsk, Kzyl-Orda Oblast.
People's Artist of the USSR (1967). Member of the CPSU since 1962.
She studied at the Kzyl-Orda Pedagogical Institute from 1939 to 1941. From 1941 to 1947 she was a soloist with the Song and Dance Ensemble of the Uzbek Philharmonic; from 1947 to 1949, with the Kazakh Theater of Opera and Ballet; from 1949 to 1960, with the Kazakh Philharmonic; and after 1960, with the Kazakh Concert. She performs folk songs (Russian, Kazakh, Uzbek, Tartar, Mongolian, Korean, and others) and Soviet show songs. Beginning in 1949, she toured abroad, visiting Poland, the German Democratic Republic, Belgium, Hungary, Austria, Czechoslovakia, the Chinese People's Republic, the Korean People's Democratic Republic, India, Burma, Canada, and other countries. Decors.: Order of Lenin and medals.

Bagmut, Ivan Andrianovich
Author; b. Babaikovka, Dnepropetrovsk Oblast; s. of a teacher.
Ukrainian; studied Kharkov Agricultural Inst.; first works publ. 1927.
Publs.: "Podorozh do Nebesnykh gir" (Journey to the Heavenly Mountains) 1930, "Gospodari Okhotskykh gir" (The Masters of the Okhotsk Mountains) 1949, "Sluzhu Radianskomu Soiuzu" (I Serve the Soviet Union) 1950, "Blakytne Pleso" (The Blue Pool) 1959, children's stories "Shmatok pyroga" (A Piece of Pie) and "Zlydni" (Poverty) 1957–58, and others.

Bagrov, Genrikh Konstantinovich
Painter and graphic artist; b. 1929, Leningrad.
Attended Leningrad College of Arts and Crafts 1945–53, pupil of A.A. Kazantsev, E.V. Pavlovskii; exhibitions 1955–.

Bagrov, Leonid Vasil'evich
State official.
1st Dep. Min. of Inland Water Transport RSFSR 1974–78, Min. Jl. 1978–.

Baialinov, Kasymaly
Kirghiz writer; b. S. 15, 1902, area of Kok-Moinok, now in Issyk-Kul' Raion, Kirghiz SSR.
He was born into the family of a nomadic herdsman. He began writing in 1923, and his novella *Adzhar* (1928), concerning the tragic lot of women during the Middle Asian uprising, was the first realistic work in Kirghiz prose. From 1929 to 1940 he wrote short stories about the new life in Soviet Kirghizia (for example, "Murad" and "The Lucky Drover"). The cycle of stories entitled *Under Fire* was devoted to the subject of war. In the novella *On the Banks of the Issyk-Kul'* (1947) the life of a kolkhoz village during wartime was

shown. The novella *Kurman Valley* (1958) describes one of the episodes of the 1916 uprising. His novel *Brotherhood* (1962; Russian translation 1965, reissued 1967) is imbued with ideas of the brotherhood of peoples and proletarian internationalism.
Works: *Köl boyunda*. Frunze 1959. *Boordoshtor*, 2nd ed. Frunze 1967. *In Russian translation: Schast'e: Povesti i rasskazy*, Moscow 1948. *Dolina Kurmana: Povesti, rasskazy, ocherki*. Frunze 1958. *Adzhar*. Frunze 1965.

Baialinov, Marklen Kasymovich
State official.
Chair. of State Cttee. on Film Kirghiz SSR O. 1972–. Mem. Auditing Comm. Kirghiz CP 1976–.

Baibakov, Nikolai Konstantinovich, Dr. of Tech. Sc.
State official; b. 1911.
Russian; mem. CPSU 1939–; grad. Azerbaizhan Petroleum Inst. 1932; dir. of an oilfield 1932–37; Chief Engineer, then Dir. of Lenin Petroleum Trust, Head of Eastern Oil Drilling Asscn. in Kuibyshev, Head of Main Eastern Oil Extraction Board 1937–40; USSR Dep. People's Commr. 1940–44, People's Commr. for the Petroleum Ind. 1944–46; USSR Min. of the Petroleum Ind. of the S. and W. Regions 1946–48; Dep. to Supr. Sov. USSR 1946–50; 1954–62 and 1962–; USSR Min. of the Petroleum Ind. 1948–55; mem. Cen. Cttee. CPSU 1952–61 and 1966–; Chair. of USSR GOSPLAN 1955–57; 1st Dep. Chair. of RSFSR Council of Mins., Chair. of RSFSR State Planning Org. 1957–58; Chair. of Krasnodar Econ. Council, later Chair. of N. Caucasian Econ. Council 1959–63; Dep. to RSFSR Supr. Sov. 1963–67; Chair. of USSR GOSPLAN State Cttee. for Chemistry, then Chair. of State Cttee. for Chemical and Petroleum Ind. with rank of Min. of USSR 1963–64; Chair. of USSR GOSPLAN State Cttee. for Petroleum Ind. 1964–65; Dep. Chair. of USSR Council of Mins. and Chair. of USSR GOSPLAN O. 1965–; Chair. of COMECON Cttee. for Cooperation on Planning Sector. Decors.: Lenin Prize, 2 Orders of Red Banner of Labour, 3 Orders of Lenin, various medals.

Baidashvili, Madlena Alekseevna
Artist in applied art; b. 1928, Tbilisi.
Attended Tbilisi Acad. of Art. 1948–54, pupil of D. N. Tsitsishvili, artist in applied art (ceramics, porcelain); exhibitions 1955–.

Baidukov, Georgii Filippovich
Military official; b. My. 13 (26), 1907, at the Taryshta station of the Omsk railroad. Colonel-general of the air force (1961); Hero of the Soviet Union (Je. 24, 1936); member of the CPSU since 1936. He was born into the family of a railroad worker. He began working as a laborer. In Mr. 1926 he joined the Soviet Army as a volunteer. He graduated from the military pilot school (1928) and became a test pilot. In 1936, together with V. P. Chkalov and A. V. Beliakov, he completed a nonstop flight from Moscow to Udd Island and in 1937, across the North Pole to the USA. He fought in the Soviet-Finnish War of 1939–40. From N. 1941, during the Great Patriotic War, he was deputy commanding officer and then commander of the 31st and later of the 211th Combined Air division, and commander of the Air Force 4th Shock Army. He was commander of the 212th (later the 4th Guards) Attack Air Division and from D. 1943, commander of the 4th Attack Air Corps. After the war he had command responsibilities. In 1947–49 he was chief of the main directorate of the Civilian Air Force. He graduated from the Military Academy of the General Staff (1951). After 1952 he held a responsible position in the Defense Ministry of the USSR. He is the author of the books *Our Flight to America* (1937), *Notes of a Pilot* (1938), and *About Chkalov* (1939). Decors.: two Orders of Lenin, four Orders of the Red Banner, two Orders of Suvorov Second Class, Orders of Kutuzov First and Second Class, Order of the Patriotic War Second Class, four Orders of the Red Star, one Polish order, and various medals.

Baimakhanova, Annabibi
State official.
Min. of Local Ind. Turkmen SSR Mr. 1977–79; Min. of Light Ind. Turkmen SSR O. 1979–; mem. Cen. Cttee. Turkmen CP 1981–.

Baimatov, Abdukhakim
State official.
Cand. mem. 1966–77, mem. Cen. Cttee. Tadzhik CP Ap. 1977–; Min. of Local Ind. Tadzhik SSR Ap. 1976–.

Baimirov, Tukhtamysh
Party official; b. 1935.
Mem. CPSU 1961; grad. Tashkent Inst. of Railroad Engineering 1957; Engineer 1957–70, Head Admin. "Golodnostepstroi" 1970–78; cand. mem. Cen. Cttee. Uzbek CP 1971–76, mem. 1976–; 1st Sec. of Dzhizak Oblast Cttee. Uzbek CP F. 1978–.

Bairamov, Bairam Salman ogly
Author; b. 1918, Alimadatly.
Azerbaidzhani; grad. Azerbaidzhan Univ. 1950; first works publ. 1950.
Publs.: "Odinokii chelovek" (A Lonely Person) 1954, "Sviataia obiazannost'" (The Holy Duty) 1955, "V Raionnom Gorodke" (In a Small Raion Town) 1956, "Listia" (Leaves) 1958, and others.

Bairamsakhatov, Nursakhat
State official.
1st Dep. Chair. of Turkmen Assoc. "Znanie" 1968; Sec. of Mary Oblast Cttee. Turkmen CP 1971–75; cand. mem. Cen. Cttee. Turkmen CP 1971–76, mem. 1976–; Head of Dept. for Prop. and Agitation with Cen. Cttee. Turkmen CP 1975–79; Chair. of State Cttee. of Vocational Training Turkmen SSR N. 1979–.

Bairamukova, Khalimat (pseudonym of Khalimat Bashchievna Kubanova)
Karachai poet; b. Ag. 15, 1917, *aul* (village) of Kurzuk.
Member of the CPSU since 1963.
She wrote the verse narrative *Zalikhat* (1963) about the struggle of the Soviet people against German fascism, the collection of poems *I Love Life* (in Russian, 1957), the novel *Years and Mountains* (1964), and other works in which the new life of the Karachai-Cherkess Autonomous Oblast is vividly portrayed.
Works: [Bayramukova, Khalimat.] *Jïrladedile taularïm*. Stavropol' 1969. *In Russian translation: Vesennii polden': Stikhi*. Cherkessk 1966.

Baiseitov, Rymbek Smakovich
State official; b. 1922.
Mem. CPSU 1943–, grad. Union Inst. of Finance 1956; served in Sov. army 1941–45; held posts in Min. of Finance system 1946–51; held posts in Min. of Finance apparatus Kazakh SSR 1951–61; 1st Dep. Min. of Finance Kazakh SSR 1961–73; Chair. of State Cttee. on Prices Kazakh SSR 1973–74; Min. of Finance Kazakh SSR 1974–; mem. Cen. Cttee. CP Kazakhstan 1976–.

Baishev, Saktagan Baishevich
Economist and historian; b. S. 18 (O. 1), 1909, Aktiubinsk Oblast.
Mem. CPSU 1931–; grad. Inst. of Marxism-Leninism Alma-Ata 1937; mem. 1956–, mem. Presidium Acad. of Sciences Kazakh SSR 1967–; Dir. of Kazakh Branch Office of Inst. of Marxism-Leninism with Cen. Cttee. CPSU, Dir. Inst. of History Kazakh CP 1946–56; Dir. of Econ. Inst. at Acad. of Sciences Kazakh SSR 1956–68.

Baitala, Vasilii Demyanovich
State official; b. 1922.
Mem. CPSU 1947; grad. Inst. of Forestry in Voronezh; forest ranger, dir. of a forest range in Khmel'nitsky Oblast, Dir. of Forestry School in Lubny in Poltava Oblast, Head of Admin. of Forest Cultures of Min. of Agriculture Ukrainian SSR, Head of Admin. of Reforestry under Council of Mins. Ukrainian SSR until 1966; 1st Dep. Min. 1966–80, Min. of Forestry Ukrainian SSR Ja. 1980–.

Baitemirov, Nasirdin
Poet and author; b. 1916, Kegeti, now Frunze Oblast; s. of a peasant.
Studied Gorkii Inst. of Literature Moscow 1950–55; mem. Board of Kirghiz Writers' Union 1954–61; first works publ. 1934. Decor.: Order of Red Banner of Labour.

Publs.: "Skazki" (Tales) 1939, "Azamat" 1948, "Torzhestvo" (Celebration) 1949, "Molodoe pokolenie" (The Younger Generation) 1951, "Zhyldyzgan" 1958, "Moia liubov" (My Love) 1957, and others.

Baitsur, Gleb Grigorevich
State official.
USSR Dep. Min. of Communications 1975–79; Min. of Communications RSFSR Ag. 1979–.

Baizhanov, Sabit Mukanovich
State official.
Min. of Posts and Telecommunications Kazakh SSR Ja. 1981–; cand. mem. Cen. Cttee. Kazakh CP 1981–.

Bakaev, Sergei Ivanovich
Painter; b. 1922, Semipalatinsk.
Attended Simferopol Art School 1947–51; exhibitions 1954–.

Bakhchinian, Manik Margarovna
State official.
2nd Sec. of Erevan City Cttee. Armenian CP 1966–71; Chief of Dept. for Light and Food Ind. at Cen. Cttee. Armenian CP 1971–75; Dep. 1967–, Sec. of Presidium Supr. Sov. Armenian SSR 1975–; mem. Cen. Cttee. Armenian CP 1976–.

Bakhirev, Viacheslav Vasil'evich
State official; b. 1916.
Russian; mem. CPSU 1951–; grad. Moscow State Univ. 1941; engineer-constructor, dep. head, then head construction bureau, chief constructor, chief engineer of a plant 1941–60; dir. of a machine-building factory in Vladimir Oblast 1960–65; Dep. to USSR Supr. Sov. 1962–; USSR 1st Dep. Min. of the Defence Ind. 1965–68; USSR Min. of Machine-Building, F. 1968–; mem. Cen. Cttee. CPSU 1971–.

Bakhlulzade, Sattar Bakhlul ogly
Painter; b. D. 2 (15), 1909, Amiradzhany, now within the city limits of Baku. People's Artist of the Azerbaidzhan SSR (1964).
He studied at the Moscow Art Institute (1933–40) under V. A. Favorskii and G. M. Shegal'. He paints large lyrical landscapes, celebrating nature in his native region. His works *Gudialchai Valley* (1953), *Evening Above the Caspian* (1959), and *Spring of My Homeland* (1967) are all in the Azerbaidzhan Museum of Art (Baku). Decors.: two Orders of the Red Banner of Labor, medals.

Bakhshaliev, Bakhshali Gasan ogly
State official.
Min. of Finance Azerbaidzhan SSR Ag. 1970–; mem. Cen. Cttee. Azerbaidzhan CP 1971–.

Bakhtin, Iurii Alekseevich
Artist in applied art; b. 1921, Baku.
Attended Tbilisi Art Acad. 1945–52, pupil of G. D. Sesiashvili, D. N. Tsitsishvili, B. G. Shebuev, artist in applied art (ceramics); exhibitions 1955–.

Bakin, Boris Vladimirovich
State official, b. 1913.
Mem. CPSU 1941–; grad. All-Union Energy Inst. 1959; electrician, foreman, work superintendent, section chief, assembly admin. chief, manager of a trust in Cheliabinsk, Sverdlovsk, and Moscow Oblasts 1930–67; Dep. USSR Min. of Installation and Special Construction Work 1967–75, Min. My. 1975–; mem. Cen. Cttee. CPSU 1976–; Decors.: Hero of Socialist Labor 1975, USSR State Prize (twice) 1951, 1967.

Baklanov, Grigorii Iakovlevich
Author; b. 1923, Voronezh.
Russian; mem. CPSU 1942–; grad. Gorkii Inst. of Literature Moscow 1951; first works publ. 1950.
Publs.: "V Snegiriakh" (In Snegiri) 1954, "Deviat' dnei" (Nine Days) 1958, "Mertvye sramu ne imut" (The Dead are not Ashamed) 1961, and others.

Bakradze, Demna Konstantinovich
Painter; b. 1929, Parakheti, Georgia.
Attended Tbilisi Art Acad., pupil of M. I. Toidze; exhibitions 1958–.

Bakuradze, Karlo Nikolaevich
Sculptor and artist in monumental and decorative art; b. 1927, Telavi, Georgia.
Attended Tbilisi Art Acad. 1945–52, pupil of S. I. Kakabadze, N. P. Kandelaki; exhibitions 1955–.

Balabanov, Petr Nikolaevich
Painter; b. 1923, Borskoe, Samarkand Oblast.
Attended Art Schools in Alma-Ata 1941, Samarkand 1947–49, and Tashkent 1951–52; exhibitions 1954–.

Balabin, Aleksandr Fedorovich
Painter; b. 1922, Lichadeevo, Nizhnii Novgorod Oblast.
Attended Penza Art School 1938–41 and 1946–48, pupil of I. S. Goriushkin-Sorokopudov, and Kharkov Art Inst. 1948–54, pupil of S. I. Dudnik, P. I. Kotov; exhibitions 1954–; teacher at Kursk Pedagogic Inst. 1960–.

Balabin, Boris Nikolaevich
Painter; b. 1935, Labinsk, Krasnodar Krai.
Attended Kiev Middle School of Art 1952–56, pupil of A. I. Lopukhov; exhibitions 1957–.

Balaian, N. M.
Trade union official.
Sec. and mem. Presidium of Armenian Republican Trade Union Council Ja. 1982–.

Balakai, Vladimir Leonidovich
Painter; b. 1928, Uspenovlia, Donetsk Oblast.
Attended Kiev Middle School of Art 1946–47; exhibitions 1956–.

Balakhmetov, Kazhakhmet Balakhmetovich
State official.
Min. of Educ. Kazakh SSR Ag. 1974–; mem. Cen. Cttee. Kazakh CP 1976–.

Balanchivadze, Andrei Melitonovich
Composer; b. My. 19 (Je. 1), 1906, St. Petersburg.
People's Artist of the USSR (1968). Son of the composer M. A. Balanchivadze.
He studied at the Tbilisi Conservatory with M. M. Ippolitov-Ivanov and at the Leningrad Conservatory with A. M. Zhitomirskii. He wrote the first Georgian ballet, *Heart of the Hills* (1936, Georgian Theater of Opera and Ballet, Tbilisi) and the ballets *The Pages of Life* (1961, Bolshoi Theater, Moscow), and *Mtsyri* (1964, Georgian Theater of Opera and Ballet). Of his two symphonies, the first, composed in 1944, earned him the State Prize of the USSR in 1946. He has also composed four concerti for piano and orchestra, instrumental chamber music, operas, and music for the theater and movies. His Second Piano Concerto earned him the State Prize of the USSR in 1947. Balanchivadze started to teach at the Tbilisi Conservatory in 1935, becoming a professor there in 1942. He has been chairman (1953–56) and first secretary of the board of the Composers' Union of the Georgian SSR (1956–61; and since 1968) as well as honorable chairman of the Georgian Musical and Choreographic Society (since 1957). Decors.: Order of Lenin, two other orders, and various medals.

Balandin, Anatolii Nikiforovich
Party official; b. 1927.
Russian; mem. CPSU 1954–, grad. Orenburg Agricultural Inst. 1952; chief agronomist, dir. of a machine and tractor station in Orenburg Oblast 1952–57; instructor in Agriculture Dept. of Orenburg Oblast Cttee. CPSU, 1st Sec. of Chkalovsk Raion Cttee. CPSU 1957–62; Chief of Collective and State Farm Prod. Admin. in Sorochinsk Raion, Chief of Agriculture Dept. of Orenburg Oblast Cttee. CPSU, Sec. of Orenburg Oblast Cttee. CPSU 1962–66; Chair. of Exec. Cttee. of Orenburg Oblast Sov. of Peoples' Deps. 1966–80; Dep. to USSR Supr. Sov. 1966–; 1st Sec. of Orenburg Oblast Cttee. CPSU D. 1980–; mem. Cen. Cttee. CPSU 1981–.

Balandin, Iurii Nikolaevich
Party official; b. 1925.
Mem. CPSU 1944–; grad. K. A. Timiriazev Agricultural Acad. in Moscow 1951; Cand. of Econ. Sciences 1955; served in Sov. Army 1943–46; chief agronomist of

a machine and tractor station 1955–58; post in Moscow City Cttee. CPSU 1959–61; post in Cen. Cttee. CPSU apparatus 1961–71; Dep. to USSR Supr. Sov. 1970–; 1st Sec. of Kostroma Oblast Cttee. CPSU O. 1971–; mem. Cen. Cttee. CPSU 1976–.

Balandiuk, Mikhail Vasil'evich
Graphic artist; b. 1924, Novo-Olchedaev, Vinnitsa Oblast.
Attended Kiev Art Inst. 1952–58; exhibitions 1957–.

Balanova, Klavdia Erofeevna
Painter; b. 1915, Taganrog.
Attended Repin Inst. for Painting, Sculpture and Architecture at the USSR Acad. of Arts 1950–56, pupil of B. V. Iohanson, L. V. Khudiakov; exhibitions 1946–; mem. CPSU 1953–.

Balasanian (Balasaniants), Sergei Artem'evich
Composer, teacher, musical and public figure; b. Ag. 13 (26), 1902, Askhabad (Ashkhabad).
People's Artist of the Tadzhik SSR (1957). Honored Art Worker of the RSFSR (1963). Member of the CPSU (since 1944).
He studied in the history and theory department of the Moscow Conservatory. He worked in Tadzhikistan (1936–43). He was a teacher at the Moscow Conservatory from 1948 to 1951 and again from 1955, becoming a professor in 1965. The development of Soviet multinational music is connected with the work of him. He is one of the founders of the Tadzhik musical theater.
His music is noted for its originality of tunes and inventiveness of rhythmic thought, which are associated with his use of elements of the folklore of Eastern peoples. He wrote the first Tadzhik opera, *The Uprising at Vosse* (staged in 1939 by the Tadzhik Theater of Opera and Ballet in Dushanbe); the musical show *Lola* (staged in 1939 in Dushanbe); the operas *The Blacksmith Kova* (staged in 1941 in Dushanbe) and *Bachtior and Nisso* (staged in 1954 in Dushanbe); the ballets *Leili and Medzhnun* (staged in 1947 in Dushanbe); and *Shakuntala* (staged in 1963 by the Latvian Theater of Opera and Ballet in Riga); symphonic works (including *Seven Armenian Songs*, 1955, and *Choral Symphony*, 1967); instrumental chamber works; romances; and music for dramatic performances and films. Decors.: State Prize of the USSR (1949) and five orders.

Balashka, Iazep Iazepovich
Painter; b. 1929, Krivany, Latvia.
Attended Art Acad. of the Latvian SSR 1951–57; exhibitions 1957–.

Balavadze, Iago Il'ich
Painter; b. 1929, Tbilisi.
Attended Tbilisi Art Acad. 1948–54, pupil of A. K. Kutateladze; exhibitions 1954–.

Balchunas, Vitautas-Kazimeras Mechislovo
Architect; b. 1924.
Grad. Faculty of Architecture Kaunas Univ. 1954; Lecturer at Inst. of Engineering in Vilnius 1953–. Decors.: Merited Architect of the Lithuanian SSR 1974, Lenin Prize 1974.

Baldano, Namzhil Garmaevich
Buriat playwright, actor, director; b. O. 20, 1907, *ulus* (village) of Orongoi, Ivolginsk aimak.
People's Artist of the Buriat ASSR (1938). Member of the CPSU since 1940. In his plays, he portrayed the vicious class struggle in the countryside and the victory of collectivization and the kolkhoz structure: *Breakthrough* (1932), *One of Many* (1937), *Baikal Fishermen* (1942), *Baiangol* (1947), and *Flame* (1953). He is the author of the librettos of the operas and ballets *Light Over the Valley* (1950), *The Beautiful Angara* (1957), and *Adopted Brothers* (1958). He compiled the composite text of the epos *Gesar* (1959 edition). Decors.: two orders.
Works: *Sugluulagdamal zokhëolnuud*, vols. 1–2. Ulan-Ude 1967. *P'esënüüd*. Ulan-Ude 1957. *In Russian translation: Izbrannoe: P'esy*. Moscow 1959.

Baldina, Irina Mikhailovna
Painter; b. 1922, Moscow.
Attended Moscow Inst. for Applied and Decorative Art 1940–41 and Repin Inst. for Painting, Sculpture and Architecture at the USSR Acad. of Arts 1946–52, pupil of M. P. Bobyshov; exhibitions 1952–.

Baliasnaia, Riva Naumovna
Poet; b. 1910, Radomysl, Ukraine.
Jewish; grew up in orphanage; worked in shoe factory; mem. CPSU 1929–; first works publ. 1934.
Publs.: poetry collection "Pereklichka" (Summons) 1934, "Junost' moia" (My Youth) in Ukrainian, 1961, "Sneg i tishina" (Snow and Stillness) in Russian, 1951, and others.

Balikov, Iurii Evdokimovich
Painter; b. 1924, Pliskovo, Podolia.
Attended Art Inst. 1952–58, pupil of V. G. Puzyrkov; exhibitions 1960–.

Balmont, Boris Vladimirovich
State official; b. 1927.
Mem. CPSU 1956–, grad. N. E. Bauman Moscow Higher Tech. Univ. 1952, Asst. Prof. 1965; senior shop engineer, shop superintendent, chief engineer of a plant, dir. of a plant 1952–65; chief of a main admin. of USSR Min. of Gen. Machine Building 1965–73; 1st Dep. USSR Min. of the Machine Tool and Tool-Building Ind. 1976–81, Min. F. 1981–; cand. mem. Cen. Cttee. CPSU 1981–. Decor.: Hero of Socialist Labor 1978.

Balodis, Andrei Martynovich
Poet; b. 1908.
Latvian; worked as a farm-hand and loader; joined Komsomol 1925; active in underground movt.; confined in Latvian prison for revolutionary activities 1928–39; after Sov. occupation of Latvia Ed. of journal "Brivais zemnieks" 1940; fought in World War II in 1st Latvian Fighter Battalion of Sov. Army; Ed. of journal "Karogs" 1947; first works publ. 1924. Decor.: Red Banner of Labour.
Publs.: poetry collection: "Ty liubish Rodinu" (You Love Your Fatherland) 1943, "V luchakh bor'by i pobed" (In the Light of Battles and Victories) 1945, "Seiateli zhizni" (Creators of Life) 1948, "Ot vsego serdtsa" (From the Bottom of the Heart) 1950, "Pust pushki molchat'" (Let the Cannons be Silent) 1951, "Krylatye gody" (The Winged Years) 1958, and others.

Balsy, Eduardas
Composer and public figure; b. D. 20, 1919, Nikolaev.
People's Artist of the Lithuanian SSR (1965). In 1950, he graduated from the conservatory in Kaunas (class of A. Račiunas), and in 1953 he completed graduate studies at the Leningrad Conservatory (class of V. V. Voloshinov). He wrote the ballet *Ėglė, Queen of the Snakes* (1960, Lithuanian Theater of Opera and Ballet, Vilnius), works for symphony orchestras, violin concertos, *Dramatic Frescoes* for violin, piano, and orchestra (1965), chamber music, songs and film scores. Since 1953, he has taught composition at the Lithuanian Conservatory (since 1959 he has been head of a subdepartment). He became chairman of the board of the Union of Composers of Lithuania in 1962. He was a deputy to the sixth and seventh convocations of the Supreme Soviet of the Lithuanian SSR. Decor.: Badge of Honor.

Baltakis, Algimantas
Poet; b. 1930; s. of a peasant.
Lithuanian; grad. Dept. of History and Philology of Vilnius Univ.; mem. CPSU 1957–; first works publ. 1952.
Publs.: "Kogda morosit dozhdik" (When it Drizzles) 1955, "Chertov most" (The Devil's Bridge) 1957, "Chetyre struny" (Four Strings) 1959, and others.

Baltin, Gunnar Andreevich
State official.
Head of Main Statistical Admin. Latvian SSR S. 1970–; mem. Auditing Cttee. Latvian CP 1971–.

Baltrunas, Aleksas
Writer; b. 1925, Kupishkis; s. of a worker.
Lithuanian; studied Art Inst. Vilnius and Kaunas Univ.; active as teacher and journalist; first works publ. 1939.

Publs.: children's stories "Malchik s Goluboi Dachi" (The Boy from the Blue Dacha) 1951, "Malchiki nashego goroda" (The Boys From Our Town) 1953, "Linas i Valentinas" 1958, collection "Vesennii potok" (Spring Stream) 1953, "Dozhd'" (Rain) 1960, "Dal'nie dorogi'" (Distant Streets) 1958, and others.

Baltrunas, Iurii Fedorovich
Painter; b. 1923, Petrograd.
Attended Leningrad Middle School of Art 1935–41, Art Acad. of the Latvian SSR 1949–53 and Art Inst. of the Lithuanian SSR 1953–55; exhibitions 1943–; mem. CPSU 1946–.

Baltrunas, Valerionas Stesevich
Party official.
Mem. Bureau 1966–82, 1st Sec. of Cen. Cttee. Lithuanian Komsomol F. 1974–82; mem. Cen. Cttee. and cand. mem. Bureau Cen. Cttee. Lithuanian CP 1976–.

Baltrushaitis (Baltrušaitis), R.
Trade union official.
Sec. and mem. Bureau of Cen. Cttee. Lithuanian Komsomol D. 1972–78; Sec. of Lithuanian Republican Trade Union Council 1982–.

Baltushis, Iuozas (real name: Iuozenas, Albertas)
Author; b. 1909, Riga; s. of a worker.
Mem. CPSU 1943–; Dep. to Supr. Sov. Lithuanian SSR 1947–63; Exec. Ed. of journal "Pergale" 1947–65; cand. mem. Cen. Cttee. Lithuanian CP 1958–64; first works publ. 1932.
Publs.: collected stories "Nedelia nachinaetsia khorosho" (The Week Begins Well) 1940, "Belyi klever" (White Clover) 1942, "Noch' pered bol'shoi pakhotoi" (The Night Before the Great Ploughing) 1952, "Prodannye gody" (Bartered Years) 1957, "O chem v pesne ne poetsia" (What the Song does not Tell) 1959, and others.

Baluev, Veniamin Georgievich
State official.
Chair. of Cttee. for State Security (KGB) Belorussian SSR 1980–; mem. Cen. Cttee. and mem. Bureau Cen. Cttee. Belorussian CP Ja. 1981–.

Balykin, Aleksei Pavlovich
Painter and graphic artist; b. 1924, Arzamas.
Attended Gorkii Art School 1938–41 and 1948–50, pupil of A. V. Samsonov, L. N. Khnygin; exhibitions 1953–.

Balzhak (Bolzhak), Rem Ivanovich
Graphic artist; b. 1924, Berdichev, Ukraine.
Attended Moscow High School for Arts and Crafts 1948–51, pupil of S. P. Markelov; exhibitions 1951–.

Bandrovskii, Genrikh Iosifovich
Party official.
Mem. CPSU 1953–; various posts in Lvov Oblast, 1st Sec. of Lvov City Cttee. Ukrainian CP 1960–80; cand. mem. Cen. Cttee. Ukrainian CP 1966–71; mem. Auditing Comm. Ukrainian CP 1971–76; mem. Cen. Cttee. Ukrainian CP 1976–; 1st Sec. of Transcarpathian Oblast Cttee. Ukrainian CP D. 1980–.

Bandzeladze, Aleksandr Matveevich
Painter and graphic artist; b. 1927, Tulgun.
Attended Tbilisi Art Acad. 1947–49; painter and graphic artist.

Banis, Talivaldis Kristapovich
Painter and graphic artist; b. 1926, Sloka, Latvia.
Grad. Art Acad. of the Latvian SSR 1955; exhibitions 1955–.

Bannikov, Nikolai Vasil'evich
Party official; b. 1914.
Russian; mem. CPSU 1940–; grad. Kuibyshev Ind. Inst. 1937, then worked in ind. plant as dept. mechanic, dept. head, chief works mechanic, dep. dir. and party organizer 1937–45; dep. sec. of an oblast cttee., 1st sec. of a raion party cttee., Dept. Head in Kuibyshev Oblast Cttee. of CPSU 1945–55; 1st Sec. of Kuibyshev City Cttee. of CPSU 1955–59; 2nd Sec. of Karaganda Oblast Cttee. of Kazakh CP 1959–63, 1st Sec. 1963–68 (for Industrial Affairs 1963–64); Dep. to Supr. Sov. RSFSR 1959–63; cand. mem. 1960–63, mem. Cen. Cttee. Kazakh CP 1963–71; Dep. to Supr. Sov. Kazakh SSR 1963–67; cand. mem. Cen. Cttee. CPSU 1966–68, mem. 1968–; Dep. to Supr. Sov. USSR 1966–; 1st Sec. of Irkutsk Oblast Cttee. of CPSU F. 1968–. Decor.: Order of Lenin.

Bannikov, Sergei Grigor'evich
Legal official.
Chair. of Cttee. for State Security with Council of Mins. Turkmen SSR 1956; Dep. Chair. of USSR Supr. Court O. 1967–.

Banykin, Viktor Ivanovich
Author; b. 1916, Stavropol.
Russian; first works publ. 1937.
Publs.: "Rasskazy o Chapaeve" (Stories about Chapaev) 1939, "Fomichevy" (The Fomichevs) 1952, "Andrei Snezhkov uchitsia zhit" (Andrei Snezhkov Learns to Live) 1959, and others.

Barabanova, L. M.
Trade union official.
Cand. mem. Cen. Cttee. Belorussian CP 1966–76; Sec. of Mogilev Oblast Cttee. Belorussian CP 1971–72; Sec. of Belorussian Republican Trade Union Council Mr. 1972–.

Baraisov, Mazhit Shagievich
Painter and graphic artist; b. 1918, Nizhnii Baskunchak, Astrakhan Oblast.
Attended Tashkent Art School 1935–39, pupil of P. V. Gan, N. G. Karakhan, exhibitions 1949–.

Baramidze, Aleksandr Georgievich
Literary critic; b. Mr. 27, 1902, Lanchkhuti.
Academician of the Academy of Sciences of the Georgian SSR (1961).
He graduated from the department of philology of the University of Tbilisi in 1925. He was first published in 1927. He has written works in Georgian and Russian about the history of ancient Georgian literature and has published scholarly-critical texts of ancient Georgian authors. His most significant works are *Essays on the History of Georgian Literature* (vols. 1–4, 1945–64); the monographs *The Narrative Poems of Shota Rustaveli* (1958), *Sulkhan-Saba Orbeliani* (1959), and *David Guramishvili* (1955); the book *Ancient Georgian Literature (Fifth to 18th Centuries)* (1954, with Academician K.S. Kekelidze); and *On the Question of Literary Interrelationships in Ancient Georgia* (1960). He is the director of the Sh. Rustaveli Institute of Georgian Literature of the Academy of Sciences of the Georgian SSR (1966). Decors.: Order of Lenin, the Badge of Honor, and medals.
Works: *Narkkvevebi kharthuli litteratturis isttoriidan*, vols. 1–4. Tbilisi 1945–64. *Sulhan-Saba Orbeliani*. Tbilisi 1959. *Kharthuli litteratturis isttoria*, vol. 1. Tbilisi 1954. (With K. S. Kekelidze.) *Shotha Rusthaveli da misi ppoema*. Tbilisi 1966. *In Russian translation: Shota Rustaveli*. Moscow 1966. *Istoriia gruzinskoi literatury: Kratkii ocherk*. Tbilisi 1958. (With Sh. Radiani and B. Zhgenti.)

Barancheev, Mikhail Sergeevich
Painter and graphic artist; b. 1920, Kalinovo, Moscow Oblast.
Attended Surkov Art Inst. Moscow 1947–53, pupil of F.A. Modorov, D.K. Mochalskii, exhibitions 1946–; mem. CPSU 1964–.

Barannikov, Nikolai Aleksandrovich
Graphic artist; b. 1931, Gubashevo, Samara Oblast.
Grad. Penza Art School 1958, pupil of I. S. Goriushkin-Sorokopudov, A. I. Postnov; exhibitions 1958–.

Baranov, Aleksei Pavlovich
Painter and graphic artist; b. 1923, Gnezdovo, Vladimir Oblast.
Attended Penza Art School 1938–41, pupil of M. S. Pyrin, M. N. Troitskii; exhibitions 1940–; mem. CPSU 1950–.

Baranov, Georgii Nikolaevich
Sculptor; b. 1917, Berezovo, Kemerovo Oblast.
Grad. Alma-Ata Pedagogic Inst. 1942, no formal art training, exhibitions 1950–.

Baranov, Nikolai Varfolomeevich
Architect; b. N. 14 (27), 1909, at the station of Starosel'tsa in the former Grodno Province (today in Poland).
Member of the CPSU since 1938. He graduated from the Leningrad Institute of Civil Engineers (1931) and the Academy of Fine Arts (1934). He studied with L. A. Il'in, V. A. Shchuko, and V. G. Gel'freikh. From 1938 to 1950 he was the chief architect of Leningrad and from 1956 to 1961 the vice-president and secretary-academician of the Academy of Construction and Architecture of the USSR; in 1963 he became deputy president of the Committee on Civil Construction and Architecture of the Gosstroi (State Committee on Construction) of the USSR. He was a professor at the Moscow Architecture Institute (1956). His main works (with co-authors) include blueprints for the general development of Leningrad (1938–40; 1945–48), the architectural ensemble of Lenin Square and the new building of the Finland Station (1943–60), the building on Primorskii Prospect (1948–56), the Primorskii zone of Vasil'evskii Island (1943–66), a series of large government structures in Leningrad (including a multipurpose spectator sports hall seating 25000; since 1967), and plans for the restoration and development of Pskov (begun in 1946). He is the author of a series of works on city planning and architecture. Decors.: Order of Lenin, two other orders, and medals.

Baranov, Sergei Nikitovich, Dr. of Chem. Sc.
Chemist; b. N. 14, 1918, Kustanai; s. of a junior doctor.
Mem. CPSU 1945; grad. Leningrad Univ. 1941; corresponding mem. Acad. of Sciences Ukrainian SSR 1965–; Dr. of Chem. Sc. 1963; Prof. 1963; served in Sov. Army 1941–46; Asst. at Chair of Inorganic Chemistry of Medical Inst. Lvov 1946–53; Prof. Chair of Organic Chemistry at same Inst. 1953–66; Chief of Dept. of Low Temperature Chemistry and Special Methods of Synthesis in Dept. of Physical-Organic Chemistry at Inst. of Physical Chemistry Acad. of Sciences Ukrainian SSR in Donetsk 1966–; author of more than 50 scientific works. Decor.: Order of Red Star.
Publs.: "Issledovanie reaktsionnoi sposobnosti vodorodnykh atomov metilenovoi gruppy nekotorykh azolidinov" in "Zhurnal obshchei khimii" vol. 21/1962.

Baranovskaia, Olga Petrovna
Sculptor; b. 1922, Moscow.
Attended Moscow Inst. for Applied and Decorative Art 1939–46, pupil of B. N. Lange.

Baratashvili, Maria Gervasievna
Poet; b. 1908, Chiatura; d. of a teacher.
Georgian; grad. Dept. of Philology of Tbilisi Univ.; first works publ. 1935–.
Publs.: "Nanuli" 1941, "Sapovnella" 1958, "Marine" 1953, and others.

Baratov, Rauf Baratovich
Scientist.
Vice-Pres. of Acad. of Sciences Tadzhik SSR D. 1976–.

Barchenkov, Nikolai Ivanovich
Painter; b. 1918, Zagorsk.
Attended Moscow Oblast Pedagogical Art School 1934–39, pupil of F. N. Gorelov, S. I. Frolov, A. P. Chirkov; exhibitions 1945–.

Bardadze, Tariel Davidovich
Painter; b. 1921, Akhaltsikhe, Georgia.
Attended Tbilisi Art Acad. 1937–46, pupil of V. V. Sidamon-Eristavi, M. I. Toidze; exhibitions 1947–; teacher at Tbilisi Art School 1946–.

Bardash, Viacheslav Nikolaevich
State official.
Cand. mem. Cen. Cttee. Moldavian CP 1963–65 and 1981–, mem. 1966–71; Min. of Fruits and Vegetables Moldavian SSR My. 1981–.

Bardina (Shpirko), Raisa Nikolaevna
Graphic artist; b. 1930, Kuibyshevka-Vostochnaia, Amur Oblast.
Attended Irkutsk Oblast Art School 1949–54, pupil of S. I. Vinogradow, A. P. Krylov; exhibitions 1954–.

Barilkis, Gidonas Leibovich
Painter; b. 1931, Kaunas.
Attended Art Inst. of the Lithuanian SSR 1950–57, pupil of V. S. Matskevichius; exhibitions 1954–.

Barinova, Maria Nikolaevna
Sculptor; b. 1925, Andreevka, Zhitomir Oblast.
Attended Kiev Art Inst. 1946–52, pupil of M. G. Lysenko; exhibitions 1951–.

Barisaite, Laima Iozo
Graphic artist; b. 1930, Penevezhis.
Attended Art Inst. of the Lithuanian SSR 1948–54, pupil of I. M. Kuzminskis, V. M. Iurkunas; exhibitions 1954–.

Barkauskas, Antanas Stasevich, Cand. of Econ. Sc.
State official; b. 1917, Paparchai nr. Kaunas; s. of a peasant.
Lithuanian; mem. CPSU 1942–; grad. Acad. of Social Sciences of Cen. Cttee. CPSU 1959; joined Komsomol 1940; Sov. posts in Kaunas 1940–41; served in Sov. Army 1941; mem. 16th Lithuanian Div. of Sov. Army 1942–44; Sov. and party posts in Kaunas 1944–50; Sec. of Vilnius and Kaunas Oblast Cttee. 1950–53; teacher, then Lecturer at Polytech. Inst. Kaunas 1953–59; Head of Dept. of Propaganda and Agitation at Cen. Cttee. Lithuanian CP 1959–61; mem. 1960–, Sec. of Cen. Cttee. 1961–D. 1975, mem. Bureau Cen. Cttee. Lithuanian CP 1961–; Dep. to 1963–, Chair. of Supr. Sov. Lithuanian SSR 1963–D. 1975; Dep. to USSR Supr. Sov. 1974–; Chair. of Presidium Supr. Sov. Lithuanian SSR D. 1975–; mem. Cen. Auditing Comm. CPSU 1976–81, cand. mem. Cen. Cttee. CPSU 1981–; Decor.: Order of Lenin.

Barkovskii, Anatolii Aleksandrovich
Diplomat; b. 1921.
Mem. Diplomatic Service 1951–; USSR Emb. Counsellor in United Arab Republic 1959–61; USSR Consul Gen. in Damascus 1961; USSR Amb. to Syria 1961–68; mem. staff USSR Min. of Foreign Affairs 1968–71; USSR Amb. to Cyprus 1971–73, to Iraq 1973–Mr. 1982.

Barkovskii, Mikhail Aleksandrovich
Painter; b. 1923, Iudino, Tartar ASSR.
Attended Kazan Art School 1947–52; exhibitions 1953–; mem. CPSU 1959–.

Barmin, Vladimir Pavlovich
Scientist; b. Mr. 4 (17), 1909, Moscow.
Mem. CPSU 1944–; grad. Moscow Tech. College 1930; specialist in mechanics and engineering; mem. USSR Acad. of Sciences 1966–; teacher at Moscow Tech. College.

Barpal, Tamara Viktorovna
Artist in applied art; b. 1927, Dnestr, Odessa Oblast.
Grad. Tashkent Art School 1957; artist in applied art (ceramics); exhibitions 1953–.

Barsanov, Georgii Pavlovich
Scientist; b. D. 29, 1907, Saratov.
Mineralogist. Doctor of geological and mineralogical sciences (1948); professor and head of the subdepartment of mineralogy of the department of geology at Moscow State University since 1953. Director of the A. E. Fersman Mineralogical Museum of the Academy of Sciences of the USSR since 1953.
His main works are devoted to theoretical mineralogy – the classification of minerals, the physics of minerals, and genetic mineralogy – and to the mineralogy of rare elements – niobium and tantalum. Decors.: three orders. The mineral barsanovite was named after him.

Barshch, Galina Aleksandrovna
Painter; b. 1926, Moscow; d. of A. O. Barshch.
Attended Moscow Middle School of Art 1941–45, and Repin Inst. for Painting, Sculpture and Architecture at the USSR Acad. of Arts 1945–51, pupil of A. M. Gritsai, V. N. Iakovlev; exhibitions 1952–.

Barshch, Mikhail Osipovich
Architect; b. Ja. 16 (29), 1904, Moscow.
He graduated from the Moscow Higher Art and Technical Institute (1926), where he studied under I. V. Zholtovskii and A. A. Vesnin. In 1935 he began teaching at

the Moscow Architectural Institute, since 1947 as a professor. From 1925 to 1932 he was a member of the Society of Modern Architects and was an adherent of constructivism (designed the Moscow Planetarium in 1928 in collaboration with M. I. Siniavskii). During the years 1950–58 he took part in the planning and construction of the Lenin Prospekt, the Iakub Kolas Square, and apartment houses in Minsk (State Prize of the Belorussian SSR 1968). He directed the planning of a series of model commercial and municipal utilities buildings (1956). He is one of the designers of the monument *In Commemoration of the Outstanding Achievements of the Soviet People in the Conquest of Outer Space* in Moscow (1964) and of the monument to K. E. Tsiolkovskii in Kaluga (1958). Decors.: Order of the Badge of Honor and various medals.

Barskii, Lev Mironovich
Author; b. 1909, Staraia Kul'naia, now Odessa Oblast.
Moldavian; grad. Gorkii Inst. of Literature 1940; mem. CPSU 1941–; first works publ. 1931.
Publs.: "K novoi zhizni" (Toward a New Life) 1933, "Povorot" (Turning Point) 1935, "Na beregu Dnestra" (On the Banks of the Dniestr) 1939, "Vo Florenakh" (In Florenay) 1952, "Buzhoriane" 1954, collections of novelettes, stories and plays.

Barskii, Vilen Isaakovich
Painter; b. 1930, Kiev.
Attended Kiev Art Inst. 1951–57, pupil of E.V. Volobuev, S.A. Grigorev, T.N. Iablonskii; exhibitions 1957–.

Bartkevich, Leonard Leopoldovich
State official; b. 1932.
Latvian; mem. CPSU 1952–; grad. Latvian State Univ.; Komsomol posts 1950–51; served in the Soviet Army 1951–60; Sec. of Riga City Cttee. Latvian Komsomol 1960–61; Sec. 1961–63, 1st Sec. Cen. Cttee. Latvian Komsomol 1963–69; mem. Cen. Cttee. Latvian CP 1963–; Head of Dept. of Party Org. Work of Cen. Cttee. Latvian CP 1971–74; Chair. of State Cttee. on Radio and Television Latvian SSR My. 1976–.

Barto, Agnila L'vovna
Russian poet; b. F. 4 (17), 1906, Moscow. He began publishing in 1925. She writes for children about life in Soviet schools, families, and pioneer clubs. Among the collections of verse she has written are *Little Brothers* (1928), *The Back-to-front Little Boy* (1934), *The Toys* (1936), *The Bullfinch* (1939), *Verses for Children* (1949); State Prize of the USSR (1950), *I Am Growing* (1968), and others. She has also written scenarios for the motion pictures *The Foundling* (1939), *The Elephant and the Rope* (1946), *Alesha Ptitsyn Develops His Character* (1958), and others. His poems, marked by a gentle humor and clarity of language, have gained great popularity among young readers as well as adults; they have been translated into the languages of the peoples of the USSR and many foreign languages. In her prose book for adults *To Find a Person* (1969), the author tells about her attempts to find the families of children who were lost during the Great Patriotic War. Decors.: three orders, medals.
Works: *Stikhi detiam,* vols. 1–2. Moscow 1966.

Bartulis, Pranas Kazio
Sculptor; b. 1928, Budliai, Lithuania.
Attended Art Inst. of the Lithuanian SSR 1949–55, pupil of R. Iakhimavichius; exhibitions 1957–.

Baruzdin, Sergei Alekseevich
Author; b. 1926, Moscow.
Russian; mem. CPSU 1949–; grad. Gorkii Inst. of Literature 1958, writes primarily for children; first works publ. 1946.
Publ.: poetry collection "Flazhok" (The Little Flag) 1950, "Kto postroil etot dom?" (Who Built This House?) 1950, stories "Pro Svetlanu" (About Svetlana) 1951, poetry "Kto segodnia uchitsia?" (Who Learns Today?) 1955, "Lastochkin mladshii i Lastochkin starchii" (Lastochkin Junior and Lastochkin Senior) 1957, and others.

Barvenko, Viktor Petrovich
Painter and artist in monumental and decorative art; b. 1933, Rostov/Don.
Attended Moscow High School for Industrial Art 1950–58, pupil of S. V. Gerasimov, G. M. Korzhev; exhibitions 1958–.

Baryshnikov, Revold Vladimirovich
Painter and graphic artist; b. 1924, Kharkov.
Attended Kharkov Art Inst. 1948–58, pupil of S. F. Besedin and G. A. Tomenko; exhibitions 1950–; mem. CPSU 1962–.

Basariia, Revaz Vladimirovich
State official.
Head of Main Statistical Admin. Georgian SSR Jl. 1976–; mem. Auditing Comm. Georgian CP 1981–.

Bash, Iakov Vasil'evich (real name: Bashmak)
Author; b. 1908, Milovo, Kherson Oblast; s. of a peasant.
Ukrainian; studied Kharkov and Kiev Univs. 1932–37; mem. CPSU 1938–.
Publs.: "Doba goryt'" (Time is Pressing) 1931, "Dni nastupu" (The Days of the Offensive) 1933, "Syla (Strength) 1934, "Na beregakh Slavuty" (On the Banks of the Slavuta) 1940, "Goriachi pochuttia" (Cordial Sentiments) 1947, "Professor Buiko" 1946, "Nadiia" (Hope) 1959, and others.

Bashilov, Sergei Vasil'evich
State official; b. 1923.
Mem. CPSU 1947–, grad. Moscow Inst. for Rail Transport 1950; work superintendent, chief engineer at "Siblestransstroi" Trust 1950–54; manager of "Irtyshlestransstroi" Trust 1954–58; chief of an admin. of Tiumen Sovnarkhoz 1958–63; dep. chief, chief of territorial Main Admin. of "Glavsreyralstroi" 1963–76; chief of a dept. of USSR GOSPLAN 1976–79; USSR Min. of Construction for the Far East and Transbaikal Area D. 1979–; Dep. to USSR Supr. Sov. 1979–; cand. mem. Cen. Cttee. CPSU 1981–.

Bashirov, Gumer Bashirovich
Author; b. 1901, Ianasala, Tartar ASSR; s. of a peasant.
Tartar; grad. Inst. of Marxism-Leninism Kazan; mem. CPSU 1928–; former teacher; during Civil War mem. Red Army; served as People's Judge; contributor to Tartar newspapers and journals 1925–; Dep. to Supr. Sov. USSR 1954–62; Chair. of Tartar Writers' Union 1956–59; first works publ. 1931.
Publs.: story "Krov' Khashima" (Kashim's Blood) 1931, novelette "Sivash" 1937, collected stories "Mest'" (Revenge) 1942, "Garmonist" (The Accordion Player) 1943, novel "Chest" (Honour) 1947, and others.

Bashkirov, Andrei Nikolaevich
Organic chemist and technological engineer; b. D. 9 (22), 1903, Nizhny Novgorod.
Corresponding member of the Academy of Sciences of the USSR (1958).
He graduated from the D. I. Mendeleev Moscow Chemical-Technological Institute (1929). He directs the laboratories of the Institute of Petrochemical Synthesis of the Academy of Sciences of the USSR (since 1946) and the institute's Department of Fine Chemical Technology (since 1943). His principal works are devoted to studies in the area of the chemistry and technology of fuel and petrochemical synthesis, particularly the oxidation of carbohydrates and synthesis on a base of CO and H_2. Decors.: two orders, medals.
Works: "Sintez vysshikh spirtov zhirnogo riada metodom priamogo okisleniia parafinovykh uglevodorodov." *Khimicheskaia nauka i promyshlennost',* 1956, vol. 1, no. 3, p. 273. "Proizvodstvo vysshikh zhirnykh spirtov okisleniem zhidkikh parafinov." *Khimicheskaia nauka i promyshlennost',* 1959, vol. 4, no. 5, p. 607. (With V. V. Kamzolkin.) "Sintez vysshikh zhirnykh spirtov iz CO i H_2." *Doklady AN SSSR,* 1956, vol. 109, no. 525. (With Iu. B. Kagan and L. I. Zvezdkin.)

Bashmakov, Evgenii Fedorovich
Party official; b. 1932.
Mem. CPSU 1960–, grad. Penza Inst. of Ind. 1955; held technical engineering posts 1955–64; Sec. of Party Cttee. of the

Kazmetallurgstroi trust assoc. 1964–67; head of the dept. of construction engineering of the Oblast Cttee. CP Kazakhstan in Karaganda 1968–72; Sec. of Turgaiskii Oblast Cttee. CP Kazakhstan 1972–75; Dep. Chair. of Council of Mins. Kazakh SSR 1975–80; mem. Cen. Cttee. 1976–, Sec. and mem. Bureau Cen. Cttee. CP Kazakhstan Mr. 1976–.

Bashtaniuk, Gennadii Sergeevich
Laborer; b. 1949.
Mem. CPSU 1977–, finished secondary school; fitter in a plant 1967–68; served in Sov. Army 1968–70; fitter in "Karbolit" Plant 1970–72; leader of fitter brigade at foundry in Naberezhnye Chelny (Tartar ASSR) 1972–; cand. mem. Cen. Cttee. CPSU 1981–.

Basiev, Oleg Aleksandrovich
State official; b. 1919.
Ossetian; mem. CPSU 1947–; grad. Zooveterinarian Inst. in Novocherkassk; served in Sov. Army 1941–48; Asst. at Chair of Horse Breeding, Sec. of Party Bureau at Zooveterinarian Inst. Novocherkassk 1949–54; 2nd Sec. of Tselinskii, 1st Sec. of Dubovskii Raion CPSU Cttee. in Rostov Oblast, instructor in a CPSU Cen. Cttee. Dept. 1954–62; Chair. of Council of Mins. North Ossetian ASSR 1962–; Dep. to USSR Supr. Sov. 1962–; Dep. and Dep. Chair. of Presidium Supr. Sov. RSFSR 1975–.

Baskakov, Nikolai Vladimirovich
Painter; b. 1928, Kadnikov, Vologda Oblast.
Attended Surkov Art Inst. Moscow 1950–56; exhibitions 1958–.

Basov, Aleksandr Vasilevich, Cand. of Agric. Sc.
Diplomat; b. 1912.
Russian; grad. Vologda Agric. Inst. 1938; Asst. Prof.; veterinary specialist 1930–42; mem. CPSU 1945–; researcher at an inst. of microbiology, chief veterinary specialist, dir. of a sovkhoz, researcher at a cattle breeding station, dept. head, dean, then dir. of a veterinary inst. 1942–54; leading party and govt. posts 1954–; Dep. 1958–66, mem. Credentials Comm. at Sov. of the Union USSR Supr. Sov. 1962–66; 1st Sec. of Rostov Oblast CPSU Cttee. 1960–62; mem. Cen. Cttee. CPSU 1961–76; Min. of Agriculture RSFSR Mr.–D. 1965; Sov. Amb. to Rumania 1965–71, to Chile 1971–73; to Australia Ja. 1975–O. 1979 and concurrently to Fiji Islands My. 1975–Je. 1980. Decors.: 2 Orders of Lenin.

Basov, Izrail Matveevich (Mordukhovich)
Painter; b. 1918, Mstislav, Belorussia.
Attended Minsk Art School 1947–51, pupil of I.O. Akhremchik; exhibitions 1947–.

Basov, Lev Stanislavovich
Sculptor; b. 1928, Vladivostok.
Attended Moscow High School for Arts and Crafts 1945–54; exhibitions 1957–.

Basov, Nikolai Gennadievich, Dr. of Phy.-Math. Sc.
Physicist; b. D. 14, 1922, Voronezh.
Mem. CPSU 1951–; grad. Inst. of Engineering and Physics Moscow 1950; radio physicist; mem. staff Inst. of Physics at USSR Acad. of Sciences 1950–; Prof. at Inst. of Engineering and Physics Moscow 1963–; Chief Ed. of periodical "Priroda" 1967–; has written numerous works on quantum radio physics. Decors.: Lenin Prize 1959, Nobel Prize 1964, Order of Lenin.
Publs.: "Kvantovaja radiofizika" in "Vestnik AN SSSR" 1960, No. 4 (co-author), "Poluprovodnikovye kvantovye regeneratory" in "Uspechi fizičeskich nauk" 85, vol. 4, 1965.

Bassus, Oleg Sergeevich
Sculptor; b. 1931, Samarkand.
Attended Art Inst. of the Lithuanian SSR 1948–54, pupil of Iu. I. Mikenas; exhibitions 1956–.

Batalov, Aleksei Vladimirovich
Film actor and director; b. N. 20, 1928, Vladimir.
People's Artist of the RSFSR (1969).
In 1950, after having graduated from the Moscow Art Academic Theater Studio, he joined the troupe of the Central Theater of the Soviet Army. From 1953 to 1956, he was an actor in the Moscow Art Academic Theater. In 1954, Batalov made his film debut in the role of the worker Aleksei Zhurbin *(A Large Family)*. He then played the role of the chauffeur Sasha Rumiantsev (*The Rumiantsev Case,* 1956). The role of the young contemporary occupies a major place in his work. His best roles include Boris (in the film *The Cranes Are Flying,* 1957), the young physics scholar Gusev (in the film *Nine Days in One Year,* 1962), and the doctor Ustimenko (in the film *My Dear Man,* 1958). In film versions of works of classical literature, Batalov played the roles of Pavel Vlasov (*Mother,* after M. Gorky) and Gurov (*The Lady With the Dog,* after A. P. Chekhov). In 1960 he made his debut as a film director (in the film *The Overcoat*). In 1966 he directed the film *Three Fat Men* in which he played the role of the ropewalker Tibul. Decor.: Order of Lenin.

Batechko, Ivan Vasil'evich
Graphic artist; b. 1926, Rudnia, Kiev Oblast.
Attended Kiev Art Inst. 1956–62, pupil of I. M. Selivanov, L. I. Chichkan; exhibitions 1960–; teacher at Kiev Art Inst. 1962–.

Batitskii, Pavel Fedorovich
Military official; b. 1910.
Ukrainian; served in Sov. Army 1924–; mem. CPSU 1938–; grad. Frunze Mil. Acad. 1938, Gen. Staff Acad. 1948; during World War II commanded artillery div., later artillery corps; various commands in Sov. Army 1946–; Chief of Staff of Sov. Anti-Aircraft Defence Forces, then Dep. Commander 1950–53; Commander of Moscow Anti-Aircraft Defence System, 1st Dep. Chief of Staff of Sov. Army and Navy 1954–66; Dep. to Supr. Sov. RSFSR 1955–63; cand. mem. 1961–66, mem. Cen. Cttee. CPSU 1966–; Dep. to Supr. Sov. USSR 1962–; Commander of Sov. Anti-Aircraft Defence System, USSR Dep. Min. of Defence 1966–78; Inspector General Group, USSR Ministry of Defense 1978–. Marshal of the Sov. Union. Decors.: 4 Orders of Lenin, "Golden Star" medal, Order of the October Revolution, 5 Orders of Red Banner of Labour, Kutuzov Order 1st Cl., Suvorov Order 2nd Cl., Order "For Service Rendered to the Fatherland in the USSR Armen Forces" 3rd Cl., foreign orders and medals.

Batov, Pavel Ivanovich
Military official; b. 1897.
Russian; served in Sov. Army 1918–; grad. Frunze Mil. Acad. 1928; mem. CPSU 1929–; command positions in Sov. Army 1928–35; Dep. to Supr. Sov. USSR 1937–66; during Sov.-Finnish War Division Commander at the front 1939–40; Dep. Commander of a front, Commander of an army 1941–45; Commander of an army of Sov. Forces Group in Germany 1945–49; Commander of a Group of Sov. Forces in Kaliningrad Oblast 1949–50; command positions in Sov. Army and mem. staff USSR Min. of Defence 1950–55; Commander of Transcarpathian District 1955–58; Commander of Baltic Mil. District 1959–60; Commander of Southern Forces Group (Hungary) 1961–62; Chief-of-Staff of Warsaw Pact Forces 1962–65; Chair. of Sov. War Veterans' Cttee. 1970–. Decors.: 5 Orders of Lenin, 3 Orders of Red Banner, Order of October Revolution, twice Hero of the Soviet Union, other orders and medals.

Batsanov, Boris Terent'evich
State official; b. 1927.
Mem. CPSU 1951–; grad. Inst. of Foreign Trade 1949 and Higher Diplomatic School of USSR Min. of Foreign Affairs 1961; 2nd Sec. in USSR Emb. in Fed. Repub. Germany 1955–58; post in USSR Min. of Foreign Affairs apparatus 1967–74; head of Secretariat of Chair. of USSR Council of Mins. 1974–; mem. Cen. Auditing Comm. CPSU 1976–.

Battal, Salikh (pseudonym of Salikh Vazykhovich Batalov)
Tartar writer; b. 1905, Bol'shoi Tiganali in the Middle Volga Region; s. of a peasant.

Became a pilot. He began publishing in 1924. His first collection of poems, *Song of the Years*, appeared in 1929; later collections were *Verses* (1940) and *Motherland* (1944). His other works include the novel *Pilots* (1935), the plays *The Hand-Painted Kerchief* (1933) and *The Ordeal* (1941), the novella in verse *On the High Road* (1950), the short stories "Notes of a Pilot" (1951) and "Soviet People" (1962), and the novel in verse *On the Cheremshan* (1967). He has also translated the works of Pushkin and Mayakovsky into the Tartar language.
Works: *Shigïr'lär häm poämalar.* Kazan 1959. *Komediyälär.* Kazan 1959. *In Russian translation: Tem, kto pomog ...: Izbr. stikhi i poemy.* Kazan 1956.

Batyev, Salikh Gilimkhanovich
State official; b. 1911, Novoe Diumeevo, Bashkir ASSR.
Tartar; mem. CPSU 1932–; grad. Inst. of Econ. Finance Kazan and Party High School of Cen. Cttee. CPSU; Dep. Chief of Polit Dept. for Komsomol Matters with Aktashskaia Machine and Tractor Station, Sec. of Arskii Raion Cttee. of Komsomol, Head of Dept. for Komsomol Organs with USSR Oblast Komsomol Cttee. Tartar ASSR 1933–37; teacher of history at Workers' Faculty for Chemistry Kazan 1937–39; Lecturer at Tartar ASSR Oblast CPSU Cttee., Dep. Head of Polit Section in People's Commissariat of Agriculture Tartar ASSR, Dep. Chair. of Council of People's Commrs. Tartar ASSR, Instructor at Dept. of Party Organs Cen. Cttee. CPSU 1939–51; Sec. of Tartar ASSR Oblast CPSU Cttee. 1951–60; Chair. of Presidium Supr. Sov. Tartar ASSR 1960–; Dep. to and Dep. Chair. Presidium Supr. Sov. RSFSR 1967–.

Batyrkanov, Kalkash Mambetovich
State official.
Min. of Melioration and Water Econ. Kirghiz SSR Jl. 1979–; cand. mem. Cen. Cttee. Kirghiz CP 1981–.

Baukov, Ivan Petrovich
Poet; b. 1909, Starikovo, Riazan Oblast.
Publs.: "Vtoraia vesna, Stikhi. 1941–45" (The Second Spring. Poems. 1941–45) 1946, "Iunost'moia" (My Youth) 1951, "V rodnom kraiu" (At Home) 1955, "Govori mne o Rossii" (Tell Me About Russia) 1959, and others.

Bauma, Dzidra Karlovna
Painter and graphic artist; b. 1930, Elgava, Latvia.
Attended Riga Art School 1950–51, pupil of H. P. Veldre, A. Ia. Muceniek; exhibitions 1959–; mem. CPSU 1961–.

Bauman (Baumanis), Karl Ekabovich
Sculptor and graphic artist; b. 1916, Ronnenburg, Latvia.
Grad. Acad. of Art of the Latvian SSR 1947; exhibitions 1945–.

Baumane, Biruta Fritsevna
Painter; b. 1922, Tervete, Latvia.
Attended Art Acad. of the Latvian SSR 1941–48, pupil of Ia. K. Liepinsh; exhibitions 1956–.

Bazanov, Vasilii Grigor'evich
Literary scholar; b. O. 14 (27), 1911, Vozherovo, Kostroma Province.
Corresponding member of the Academy of Sciences of the USSR (AN SSSR) since 1962. Member of the CPSU since 1946.
He graduated from the University of Gorky in 1931. He was director of the Department of Literature at the Karelo-Finnish branch of the AN SSSR from 1948 to 1952. Since 1965, he has been director of the Institute of Russian Literature at the AN SSSR (Pushkin House). His works are devoted to 19th-century Russian literature, to the works of the Decembrists and the revolutionary populists, and to folklore. From 1958 to 1968, he was editor in-chief-of the magazine *Russkaia literature.*
Works: *Ocherki dekabristskoi literatury (Publitsistika, Proza, Kritika).* Moscow 1953. *Kareliia v russkoi literature i fol'kloristike 19 v. Petrozavodsk,* 1955. *Ocherki dekabristskoi literatury: Poeziia.* Moscow-Leningrad 1961. *Uchenaia respublika.* Moscow-Leningrad 1964. "Problema esteticheskogo otnosheniia fol'klora k deistvitel'nosti u N. G. Chernyshevskogo." *Russkaia literatura,* 1958, no. 1. "Dobroliubov i narodoznanie." *Russkaia literatura,* 1962, no. 2. "Zametki fol'klorista." *Russkaia literatura,* 1966, no. 2. "*Kapital* Karla Marksa v gody 'khozhdeniia v narod' v Rossii (1872–1875)." *Russkaia literatura,* 1968, no. 3.

Bazarkulov, Mukash
State official.
Mem. Cen. Cttee. Kirghiz CP 1976–; Min. of Educ. Kirghiz SSR My. 1979–.

Bazarova, Roza Atamuradovna
State official.
Dep. Chair. of Council of Mins. Turkmen SSR D. 1975–; mem. Cen. Cttee. Turkmen CP 1976–.

Bazelian, Ivan Fedorovich
Painter; b. 1929, Antonovo, Vinnitsa Oblast.
Attended Lugansk Art School 1945–51, pupil of M. L. Volshtein, A. A. Filbert; exhibitions 1958–.

Bazelian, Nadezhda Mikhailovna
Painter; b. 1930, Krasnodon, Lugansk Oblast.
Attended Lugansk Art School 1946–51, pupil of V. S. Kravchenko; exhibitions 1958–.

Bazhan, Mykola Platonovich
Author and poet; b. S. 26 (O. 9), 1904, Kamenets-Podolski.
Ukrainian; mem. CPSU 1940–; after grad. from secondary school in Uman' worked in Museum of Literary History and in Theatre Studio in Uman'; studied at an inst. in Kiev, broke off studies, devoted himself to literary work 1921–; first works publ. 1923; served in Sov. Army 1941–43; Dep. Chair. of Council of People's Commrs. resp. Council of Mins. Ukrainian SSR 1943–49; Dep. to Supr. Sov. USSR 1946–62; mem. Acad. of Sciences Ukrainian SSR 1951–; mem. Cen. Cttee. Ukrainian CP 1954–; Sec. of USSR Writers' Union 1967–; Chief Ed. of "Ukrainian Soviet Encyclopaedia" 1967–. Decors.: Rustaveli Prize 1937; State Prize of the USSR 1946 and 1949, Shevchenko Prize 1965, 3 Orders of Lenin, 2 other orders and various medals.
Publs.: "Bessmertie" (Immortality) 1937, "Ottsy i synovia" (Fathers and Sons) 1938, "Kliatva" (The Oath) 1941, "Daniil Galitskii" 1942, "Stalingradskaia tetrad'" (Stalingrad Diary) 1943, "Angliiskie vpechatleniia" (English Impressions) 1949, "U Spasskoi bashni" (At the Spassky Tower) 1952, "Polet skvoz' buriu" (Flight Through a Thunderstorm) 1964, "Chetyre rasskaza o nadezhde" (Four Stories About Hope) 1967, and others.

Bazhbeuk-Melikian, Lavinia Aleksandrovna
Painter; b. 1922, Tbilisi; d. of A. A. Bazhbeuk-Melikian.
Grad. Repin Inst. for Painting, Sculpture and Architecture at the USSR Acad. of Arts 1951; exhibitions 1951–.

Bazhenov, Grigorii Innokentevich
Painter; b. 1927, Verkhnii Mangertui, Bashkir ASSR.
Attended Kiev Art Inst. 1952–58, pupil of V. N. Kostetskii, V. G. Puzyrkov; exhibitions 1958–.

Bazhenova, Elizaveta Ivanovna
Painter; b. 1922, Ananino, Kostroma Oblast.
Attended Vladivostok Art School 1947–50; exhibitions 1951–; mem. CPSU 1959–.

Bazilev, Nikolai Ivanovich
Painter and graphic artist; b. 1914, Iuzovka, Don Basin.
Attended Kiev Art School 1931–40, pupil of F. G. Krichevskii, K. D. Trokhimenko; exhibitions 1946–.

Bazilevich, Anatolii Dmitrievich
Graphic artist; b. 1926, Zhmerinka, Vinnitsa Oblast.
Grad. Kharkov Art. Inst. 1953, pupil of G. A. Bondarenko, I. A. Daits, E. P. Svetlichnyi; exhibitions 1949–.

Bazilevskii, Ivan Ivanovich
Painter; b. 1935, Lugovoe, Crimea Oblast.
Attended Simferopol Art School 1951–56, pupil of K. A. Prokhorov, M. M. Shcheglov; exhibitions 1957–.

Bazorkin, Idris Murtuzovich
Author; b. 1911, Vladikavkaz; s. of a regular soldier.
Teacher by profession; first works publ. 1928.
Publs.: play "Tamara" (Tamara), "Rozhdenie nenavisti" (The Birth of Hate), "Kapitan Ibragimov" (Captain Ibragimov), "V eti dni" (In These Days), "Operatsiia" (Operation), and others.

Bazovskii, Vladimir Nikolaevich
Diplomat; b. 1917.
Russian; mem. CPSU 1942–; locksmith, design technician in various match factories 1932–41; served in Sov. Army 1941–46; grad. Leningrad Inst. of Refrigeration Engineering 1947; instructor for a raion party cttee., 1st Sec. of a raion Komsomol cttee., 1st Sec. of Frunze Raion Cttee. CPSU in Leningrad, Dept. Head in Leningrad City CPSU Cttee., Inspector and Section Head at Cen. Cttee. CPSU 1946–61; 1st Sec. of Novgorod Oblast CPSU Cttee. 1961–72; cand. mem. 1961–76, mem. Cen. Cttee. CPSU 1976–; Dep. to Supr. Sov. USSR 1962–74; Sov. Amb. to Bulgaria My. 1972–82, to Hungary Ja. 1982–.

Beban, Maksim Afanasevich
Poet; b. 1913, Keretino, Mordovian ASSR.
Mordovian; grad. Mordovian Pedagogical Inst. 1949; first works publ. 1930.
Publs.: "Vesne posveshchaiu" (I Dedicate it to Spring) 1933, "Vtoraia pesnia" (The Second Song) 1936, "Ogni v doline" (Lights in the Valley) 1946, "Lirika i Satira" (Lyric and Satire) 1955, novel "Vesennie ptitsy" (Spring Birds) 1959, poem "Po priamoi doroge i po obochine" (On the Straight Road and by the Roadside) 1959, and others.

Beda, Dmitrii Mikhailovich
Painter; b. 1921, Kievka, Stavropol Krai.
Attended Art Acad. of the Latvian SSR 1948–54, pupil of E. F. Kalninsh and O. E. Skulme; mem. CPSU 1945–; exhibitions 1955–.

Bednoshei, Daniil Panteleevich
Painter; b. 1924, Dmitrovichi, Kiev Oblast.
Attended Kiev Art Inst. 1947–53; exhibitions 1951–.

Bednow, Viktor Andreevich
Painter; b. 1938, Kochelaev, Mordovian ASSR.
Attended Penza Art School 1952–57, pupil of M. E. Valukin, A. I. Postnov; exhibitions 1957–.

Bednyi, Boris Vasil'evich
Author; b. 1916, Cossack village of Iaroslavskaia, now Krasnodar Krai; s. of a teacher.
Grad. Gorkii Inst. of Literature 1952; engineer-forestry technician by profession; was active in Komi ASSR and in the Far East; first works publ. 1951.
Publs.: stories "Komary" (The Gnats), "Starshii vozrast" (The Older Generation), novelette "Devchata" (The Girls) 1961, and others.

Bedzyk, Dmytro Ivanovich
Author; b. N. 1, 1898, Vilkhivtsi nr. Sanok, Galicia.
Grad. Dept. of Philology of Inst. of Popular Educ. in Kharkov 1926; moved to the Ukraine 1915, attended an agricultural school in Dnepropetrovsk Oblast and worked as agronomist and teacher; first works publ. 1924; after World War II correspondent for newspaper "Radianska Ukraina".
Publs.: "Ostannii vals" (The Last Waltz) 1959, "Krov' za krov'" (Blood for Blood) 1941, "Korsun-Shevchenkivske poboishche" (The Battle of Korsun-Shevchenkivskyi) 1944, "Olena Khobta" 1947, "Opovidannia pro Oleha Koshovoho" (Stories About Oleh Koshovyi), 1950, "Sertse moho druha" (The Heart of My Friend) 1964, "Naperedodni velykoho shturmu" (On the Eve of the Great Assault) 1967, "Ukradeni hory" (Stolen Mountains) 1969, and others.

Bedzyk, Iurii Dmytrovych
Author; b. N. 25, 1925, Kharkov.
Ukrainian; served in World War II; grad. Dept. of Int. Relations and concluded postgraduate studies at Dept. of Int. Public Law of Kiev Univ.; literary and editorial work 1953–; mem. ed. staff of "Radianskyi pismennyk" publishing firm, Dep. Exec. Ed. of "Literaturna hazeta"; Chief Ed. at O. P. Dovzhenko Film Studio in Kiev 1970.
Publs.: "Poruch z toboiu" (Next to You) 1956, "Divchyno moia khorosha" (My Good Girl) 1959, "Liudyna bez sertsia" (A Man Without Heart) 1957, "Chest' meni dorozhcha" (Honour Is Dearer to Me) 1967, "Sylnyi pomsty ne zhadae" (The Strong Needs no Revenge) 1969, and others.

Begaliev, Sopubek
State official.
Min. of Local Ind. Kirghiz SSR 1962–63; Dep. to Supr. Sov. Kirghiz 1963–; Min. of Local Econ. Kirghiz SSR 1963–64; mem. Cen. Cttee. Kirghiz CP 1966–; Dep. Chair. of GOSPLAN Kirghiz SSR 1967–68; Dep. Chair. of Council of Mins. and Chair. of GOSPLAN Kirghiz SSR Ag. 1968–.

Begel'dinov, Talgat Iakubekovich
Military official; b. Ag. 5, 1922, Frunze.
Twice Hero of the Soviet Union (O. 26, 1944 and Je. 27, 1945); lieutenant-colonel. CPSU member since 1943. A Kazakh by nationality.
He volunteered for the Soviet Army in 1940 and graduated from the Chkalov Military Aviation School in 1942. In the Great Patriotic War he rose from pilot to commander of the 144th Guard Attack Aviation Regiment and made more than 300 combat flights. The title of Hero of the Soviet Union was twice conferred on him for courage and during in attacks on enemy airfields, railroad junctions, strongpoints, fortifications, troops, and matériel. In 1950, Begel'dinov graduated from the Air Force Academy and was deputy commander and chief of staff of an aviation regiment. He served as deputy to the second and third convocations of the Supreme Soviet of the USSR. He retired in 1956 for reasons of health. Decors.: Order of Lenin, two Orders of the Red Banner, Orders of the Patriotic War First and Second Class, the Order of Alexander Nevsky, the Order of the Red Star, the Order of Glory Third Class, and also medals.

Bei-Bienko, Grigorii Iakovlevich
Entomologist; b. F. 7, 1903, Belopole, Kharkov govt.
Grad. Agricultural Inst. Omsk 1925; corresponding mem. USSR Acad. of Sciences 1953–; Prof. at Leningrad Agricultural Inst. 1938–; Laboratory Chief at Inst. of Zoology of USSR Acad. of Sciences 1947; his numerous scientific investigations deal with theoretical and applied entomology and ecology; Pres. of All-Union Entomological Soc. 1966–. Decors.: Merited Scientist of the RSFSR 1966, 2 orders.
Publs.: author of text books, amongst others "Obščaja entomologija"; monographs in the series "Fauna SSSR", and others.

Beibutov, Rashid Madzhid ogly
Azerbaidzhani singer (lyric tenor); b. D. 1 (14), 1915, Tbilisi.
People's Artist of the USSR (1959). Son of a *khanende* (folk singer).
He began his stage activity as a performer of Azerbeidzhani folk songs (Tbilisi, 1933). From 1933 to 1944 he was a soloist with the Yerevan Philharmonic Orchestra and from 1944 to 1956 with the Baku Philharmonic Orchestra. During the period 1957–59, he was the artistic director of the Azerbaidzhani State Concert Ensemble. He is the founder and director of the Azerbaidzhani State Song Society (1966). He is famous as a singer on the *estrada* (the variety stage), performing songs of various peoples of the world (including Russian, Iranian, Turkish, Arabic, and Indian). He appears periodically in the stage productions of the Akhundov Theater of Opera and Ballet; he has also performed the principal roles in the opera *Sevil'* by Amirov and the musical comedy *Arshin mal alan* by Uz. Gadzhibekov. He played in the film version of *Arshin mal alan* (State Prize of the USSR 1946). He has made several tours abroad. He was a deputy to the seventh and eighth convocations of the Supreme Soviet of the USSR. Decor.: Order of the Red Banner of Labor.

Beiko, E. M.
Komsomol official.
Mem. Bureau 1978–, Sec. Cen. Cttee. Moldavian Komsomol Mr. 1980–.

Bekbaev, Dzhekshenaly Bekbaevich
State official.
Chair. of State Cttee. on Forestry Kirghiz SSR Mr. 1971–.

Bekbulatov, Shamil Khairullovich
State official.
Min. of Highways Kazakh SSR Ja. 1981–; cand. mem. Cen. Cttee. Kazakh CP 1981–.

Bekhtereva, Nataliia Petrovna, Dr. of Med. Sc.
Medical scientist; b. 1924.
Mem. CPSU 1959–; grad. Medical Inst. Leningrad 1947; physiologist; Dr. of Med. Sc. 1960; Prof. 1965; corresponding mem. USSR Acad. of Medical Sciences 1963–; Dir. of Inst. of Experimental Medicine at USSR Acad. of Medical Sciences 1970–.

Bekkhozhin, Khalizhan Nurgozhaevich
Poet; b. 1913, Pavlodar.
Kazakh; first works publ. in mid-thirties. Publs.: poem "Lesnaia devushka" (Forest Girl) 1939, "Partizan Baltabai" (Partisan Baltabai), poetry collection "Pokhod" (March) 1944, "Syr-Daria" 1947, "Vesna" (Spring) 1948, poem "Mariia, doch' Egora" (Maria, Egor's Daughter) 1949–54, "Golos druzhby" (The Voice of Friendship) 1954, "Akan Aktaev", "Stepnoi komissar" (Steppe Commissar) 1958–60, play "Esli serdtsa verny" (When Hearts are True) 1961, and others.

Bekoev, Elioz Davidovich
Author; b. 1909, Dzher, Southern Ossetia.
Ossetian; mem. CPSU 1939–; grad. Pedagogical Inst. Staliniri 1949.
Publs.: story collection "Rodnik" (Well), "Voiny" (Warrior), trilogy "Fatima", 3 vols., 1947–59, translated into Ossetian works by Chekhov, A. Kazbegi, M. Sholokhov, A. Tolstoi and others.

Bektemisov, Anet Imanakyshevich
State official.
Sec. of Alma-Ata City Cttee. Kazakh CP 1966–74; mem. Auditing Comm. Kazakh CP 1971–76; Chair. of State Cttee. on Construction Kazakh SSR Jl. 1977–; cand. mem. Cen. Cttee. Kazakh CP 1981–.

Bekturganov, Khasan Shaiakhmetovich
Party official; b. 1922.
Kazakh; mem. CPSU 1942–; grad. Party Higher School of Cen. Cttee. CPSU 1962; in Sov. Army 1941–46; held party posts 1946–; Sec. 1955–58, 2nd Sec. of Kokhetav Oblast Cttee. Kazakh CP 1958–59; cand. mem. 1956–60, mem. Cen. Cttee. Kazakh CP 1960–; 1st Sec. of Aktiubinsk Oblast Cttee. Kazakh CP 1959–64; mem. Bureau for West Kazakhstan of Cen. Cttee. Kazakh CP 1962–64; Chair. of Aktiubinks Oblast Sov. 1964–66; 1st Sec. of Ksyl-Orda Oblast Cttee. Kazakh CP 1966–72; Dep. to Supr. Sov. USSR 1970–; cand. mem. 1971–76, mem. Cen. Cttee. CPSU 1976–; 1st Sec. of Dzhambul Oblast Cttee. Kazakh CP Je. 1972–.

Beletskaia, Irina Petrovna, Dr. of Chem. Sc.
Chemist; b. 1933.
Grad. Univ. Moscow 1955; Dr. of Chem. Sc. 1964; corresponding mem. USSR Acad. of Sciences 1974–; Sen. Laboratory Asst., Jun. Research Fellow, Asst., Sen. Research Fellow at Moscow Univ. 1955–; Prof. Chair 1971–.

Beletskii, Viktor Nikolaevich
Diplomat.
USSR Amb. to the Netherlands Je. 1982–.

Belevich, Anton Petrovich
Belorussian poet; b. My. 1 (14), 1914, Dunbrovka, now in Uzda Raion, Minsk Oblast, Belorussian SSR.
His works have been published since 1937. He fought in the Great Patriotic War of 1941–45. He writes lyrical verse and narrative poems, sketches on kolkhoz village life, and children's verse. His verse is suffused with love for his native region and is close to Belorussian folklore in spirit. Decors.: Badge of Honor and several medals.
Works: *Vybranni tvory.* vols. 1–2. Minsk 1969. *In Russian translation: Chevlovek iz dubravy: Poema i stikhi.* Moscow 1945. *Poemy.* Leningrad 1949. *Dobryi den': Stikhi.* Moscow 1953. *Solov'inaia roshcha: Stikhi.* Moscow-Leningrad 1966.

Belevtsev, Iakov Nikolaevich, Dr. of Geol.-Min. Sc.
Geologist; b. Ap. 7, 1912, Orechovoe, now Kursk Oblast.
Mem. CPSU 1930–; grad. Mining Inst. Donetsk 1937; corresponding mem. Acad. of Sciences Ukrainian SSR 1951–; Dr. of Geol.-Min. Sc. 1952; Prof. 1955–; miner in Donetsk Basin 1928–31; student 1931–37; Chief Geologist of Dzerzhinskii Mine, Chief of Dept. for Geological Photography with Geological Trust in Krivoi Rog 1937–41; employed in geological depts., Kemerovo Oblast 1941–44; Chief Geologist with Geological Trust in Krivoi Rog, mem. staff USSR Min. of Geology 1944–46; Prof. at Kiev Univ., Dept. Chief at Inst. of Geology with Acad. of Sciences Ukrainian SSR 1955–; author of about 110 works, among them 7 monographs. Decors.: Order of Lenin, Order of Red Banner of Labour, 2 Badges of Honour, State Prize 1st Cl. 1951, medals. Publs.: "Klassifikatsiia i uslovnye oboznacheniia porod Krivogo Roga" 1945, "Geneticheskaia klassifikatsiia zheleznykh rud Krivogo Roga" in "Mineralogicheskii sbornik Lvovskogo geologicheskogo obshchestva" No. 6/1952, "Ginezis zheleznykh rud Krivorozhskogo basseina" 1959, "Rudnichnaia geologiia na zheleznorudnykh mestorozhdeniiakh" 1962, "Problemy teorii i eksperimenta v rudoobrazovanii" 1966, and others.

Beliaev, Dmitrii Konstantinovich
Biologist; b. Jl. 4 (17), 1917, Protasovo, Kostroma Oblast.
Grad. Agricultural Inst. Ivanovo 1938; corresponding mem. USSR Acad. of Sciences 1964–; Dir. of Inst. of Cytology and Genetics at Siberian branch of USSR Acad. of Sciences 1959–.

Beliaev, Evgenii Mikhailovich
Russian lyric tenor; b. S. 11, 1926, Klintsy, Briansk Oblast.
People's Artist of the USSR (1967). Member of the CPSU since 1952.
He graduated from the Gnesiny Moscow Music and Pedagogical Institute. In 1947 he became a soloist with the Song and Dance Ensemble of the Carpathian Military District, and in 1955 he joined the Aleksandrov Soviet Army Song and Dance Ensemble as a soloist. He performs Russian and Ukrainian folk songs and works by Soviet composers.

Beliaev, German Vasil'evich
Sculptor; b. 1927, Leningrad.
Attended Leningrad High School for Arts and Crafts 1943–47, pupil of A. I. Bolshakov; exhibitions 1958–.

Beliaev, Sergei Stepanovich
Graphic artist and painter; b. 1922, Kiovo, Iaroslavl Oblast.
Attended Iaroslavl Art School 1937–41; exhibitions 1945–.

Beliaev, Spartak Timofeevich
Physicist; b. O. 27, 1923, Moscow.
Mem. CPSU 1943–. Moscow Univ. 1952; mem. USSR Acad. of Sciences 1968–; mem. staff Kurchatov Inst. of Anatomic Energy 1952–62; mem. staff Inst. of Nuclear Physics at Siberian Branch Office of USSR Acad. of Sciences 1962–; Rector of Novosibirsk Univ. 1965–. Decors.: Order of Lenin, Order of Red Star, several medals.

Beliaev, Vladimir Pavlovich
Writer; b. 1907 in Kamenets-Podolsk, Ukraine.
Russian; metal founder in Berdiansk and Leningrad 1926–; edited factory newspapers; writes in Russian and Ukrainian languages.
Publs.: story "Podrostki" (Adolescents) 1936, collection "Ukrainskie nochi" (Ukrainian Nights), 1939, trilogy "Staraia krepost" (The Old Fortress) 1937–51, film scripts "Golos Tarasa" (The Voice of Taras) 1941, "Chas rasplaty" (The Hour of Revenge) 1941, and others.

Beliak, Konstantin Nikitovich
State official; b. 1916.
Russian; mem. CPSU 1942–; grad. Polytech. Inst. Tomsk; locksmith apprentice, locksmith, mechanic's asst., student, master, technologist, head of a workshop, dept. head in a plant 1930–48; dir. of various factories in Kuibyshev, Komsomolsk-na-Amure, Voronezh 1948–57; Chair. of Voronezh Econ. Council 1957–63; Dep. to Supr. Sov. USSR 1962–66; Chair. of Econ. Council of Cen. Black Earth Zone 1963–64; USSR 1st Dep. Min. of Tractor and Agricultural Machine Construction 1966–73; USSR Min. of Machine Construction for Cattle Ind. and Fodder Prod. O. 1973–; cand. mem. Cen. Cttee. CPSU 1976; mem. O. 1976–.

Beliakov, Aleksandr Vasil'evich
Military official; b. D. 9 (21), 1897, Bezzubovo, Moscow Province; s. of a teacher.
Lieutenant general of the air force (1943); Hero of the Soviet Union (Jl. 24, 1936). Specialist in aerial navigation; doctor of geographic sciences; professor. Member of the CPSU since 1936.
He was in the Red Army from February 1919 and fought in V.I. Chapaev's 25th Rifle Division. He graduated from aviation school in 1921. From 1930 to 1935 he was an instructor and head of a subdepartment of the Zhukovskii Air Force Academy. He was crew navigator with V.P. Chkalov on nonstop flights from Moscow to Lake Udd (1936) and over the North Pole to the USA (1937). From 1936 to 1940 he was chief flight navigator of the air force, as well as of the Workers' and Peasants' Red Army. From 1940 to 1945 he was director of the Riazan' Higher Aviation School for Navigators. As a chief navigator in the air force, he took part in the Berlin Operation of 1945. From 1945 to 1960 he was head of a department at the Air Force Academy. He has been in retirement since 1960. He is the author and editor of many scientific studies in aerial navigation. He was a deputy at the first convocation of the Supreme Soviet. Decors.: 2 Orders of Lenin, three Orders of the Red Banner, the Order of the Patriotic War First Class, two Orders of the Red Star and various medals.

Beliakov, Rostislav Appolonovich, Dr. of Tech. Sc.
Scientist; b. 1919.
Mem. CPSU 1944–; grad. Aviation Inst. Moscow 1941; Dr. of Tech. Sc. 1973; specialist in engineering; corresponding mem. USSR Acad. of Sciences 1974–; leading positions as engineer and designer 1941–. Decors.: Lenin Prize 1972, State Prize of the USSR 1952.

Beliauskas, Alfonsas
Writer; b. 1923, Neverenis nr. Kaunas; s. of a worker.
Lithuanian; mem. CPSU 1944–; grad. Historical-Philological Dept. of Vilnius Univ. 1951; served in Sov. Army 1941–45.
Publs.: "Tak rozhdaetsia schastie" (Thus Happiness is Born) 1949, "Moguchie ruki" (Strong Hands) 1950, "Liudi u stankov" (People at the Workbenches) 1955, "Ulitsa truda" (The Road of Work) 1956, novel "Tsvetut alye rozy" (Red Roses are in Bloom) 1959, and others.

Belinis, Ionas Leonovich
State official.
Head of Dept. of Science and Educ. Establishments with Cen. Cttee. Lithuanian CP 1973–76; Min. of Culture Lithuanian SSR Ja. 1976–; mem. Cen. Cttee. Lithuanian CP 1976–.

Belinnik, Petr Sergeevich
Ukrainian singer (lyric tenor); b. O. 1 (14), 1906, Akhtyrka, Kharkov Province. People's Artist of the USSR (1954). From 1933 to 1936, he studied at the Kharkov Conservatory (R.S. Vain's singing class). Between 1936 and 1940 he was a soloist at the Bolshoi Theater of the USSR (Moscow); from 1941 to 1942 he was a soloist for the Kharkov Theater and beginning in 1942, a soloist at the Kiev Theater of Opera and Ballet. His roles include Lenskii (in Tchaikovsky's *Eugene Onegin*), the idiot (in Mussorgsky's *Boris Godunov*), Toropka (in Verstovskii's *Askol'd's Grave*), Alfredo (in Verdi's *La Traviata*), Petro (in Lysenko's *Natalka-Poltavka*), and Anatole (in Prokofiev's *War and Peace*). He is also well known as a singer of Ukrainian and Russian folk songs. Decors.: Order of Lenin and the Order of the Red Banner of Labor.

Belinskii, Iakov Lvovich
Poet; b. 1909, Krolevets, Sumy Oblast.
Russian; grad. Architectural Inst. Moscow.
Publs.: poetry collection "Vziatye goroda" (Conquered Cities) 1947, "Boi i liubov'" (Combat and Love) 1957, and others.

Belitser, Vladimir Aleksandrovich, Dr. of Biol. Sc.
Biochemist; b. S. 30, Riazan; s. of a veterinarian.
Grad. Faculty of Biology at Moscow Univ. 1930; mem. Acad. of Sciences Ukrainian SSR 1957–; Dr. of Biol. Sc. 1941; Prof. 1946; Scientific Asst. at All-Union Inst. for Experimental Medicine with USSR Council of People's Commrs. 1930–44; Laboratory Chief for Ferments 1944–45, Chief of Dept. for Structure and Function of Protein at Inst. of Biochemistry Acad. of Sciences Ukrainian SSR 1945–; main research in typical transformations and denaturation of protein.
Publs.: "O mekhanizme fosforilirovaniia, sopriazhennogo s dykhaniem" in "Biokhimiia" vol. 4, 1939, "Khimichekie prevrashcheniia v myshtse" 1940, "O strukture belkov" 1960, "Issledovaniia N-kontsevoi posledovatelnosti aminokislot v syvorotochnykh albuminakh razlichnykh zhivotnykh" 1964, "Polimerizatsiia fibrinmonomeru ta ii zalezhnist' vid RN" in "Ukrainskyi biokhimichnyi zhurnal" No. 5/1965.

Belopolskii, Iakov Borisovich
Architect; b. Ap. 23 (My. 6), 1916, Kiev.
Mem. CPSU 1943–; grad. Moscow Inst. of Architecture 1932–37; mem. staff Moscow Inst. of Architecture 1955–. Decors.: Merited Architect of the RSFSR 1967, Order of Lenin, other orders and medals.

Belostotskii, Anatolii Efimovich
Sculptor; b. 1921, Odessa; s. of E.I. Belostotskii.
Attended Kiev Art Inst. 1946–52, pupil of M.G. Lysenko; exhibitions 1951–; mem. CPSU 1943–. Decor.: Merited Artist of the Ukrainian SSR 1960.

Belotserkovskii, Oleg Mikhailovich, Dr. of Phy.-Math. Sc.
Mathematician; b. 1925.
Mem. CPSU 1960–; grad. Moscow Univ. 1952, Dr. of Phy.-Math. Sc. 1965; Prof. 1966, corresponding mem. USSR Acad. of Sciences 1972–; Cand. 1952–55, Research Asst. 1955–60, Lecturer, Rector of Physical-Tech. Inst. Moscow 1962–; in addition mem. scientific staff at Inst. of Mathematics of USSR Acad. of Sciences 1952–55; Sen. Research Asst. at Electronic Data Processing Centre of USSR Acad. of Sciences 1962–. Decor.: Lenin Prize 1966.

Belous, Dmytro Grigor'evich
Poet and satirist; b. 1920, Kurmany, Sumy Oblast; s. of a peasant.
Ukrainian; grad. Kiev Univ. 1945, completed post-graduate studies 1948; first works publ. 1943.
Publs.: satirical works "Oskolochnym" (With Splinter Shells) 1948, "Dobrym liudam na zdorovia-vorogam na bezgolovia" (Here's to the Good People-Off with the Heads of Our Enemies) 1951, "Dekomu na zgadku" (A Reminder to Someone) 1953, "Likarnia v zooparku" (Hospital in the Zoo) 1962, "Tarasovi zharty" (The Jokes of Taras) 1964, translations primarily from Bulgarian.

Belousov, Vladimir Vladimirovich
Geologist; b. O. 17 (30), 1907, Moscow.
Mem. USSR Acad. of Sciences 1953–; Chief of Dept. of Geodynamics at Inst. of Terrestrial Geophysics of USSR Acad. of Sciences 1944–; Prof. at Inst. of Geological Prospecting Moscow 1943–; mem. staff. Univ. Moscow 1953–; Pres. of Int. Union of Geodesy and Geophysics 1960–63; Chair. of Sov. Geophysical Cttee. 1961–; worked in the field of tectonics and tectono-physics; has develop-

ed new methods of studying history of geological development in the Greater Caucasus and the Russian Platform. Decors.: 3 orders, medals.
Publs.: "Osnovnye voprosy geotektoniki" Moscow 1962, "Strukturnaja geologija" Moscow 1961, "Zemnaja kora i verhnjaja mantija materikov" Moscow 1966, "Zemnaja kora i verhnjaja mantija okeanov" Moscow 1968.

Belov, Aleksandr Fedorovich, Dr. of Tech. Sc.
Metallurgist; b. Mr. 15 (28), 1906.
Mem. CPSU 1932–; grad. Moscow Mining Acad. 1929; specializes in light and special alloys; mem. USSR Acad. of Sciences 1972–; Dr. of Tech. Sc. 1963; Prof. 1970–; sen. engineering posts 1929–52; head of an admin. of USSR Min. of the Aircraft Ind. 1952–63; Dir. of Scientific Union Research Inst. of Light and Special Alloys Technology 1961–. Decors.: Hero of Socialist Labour 1966, State Prize of the USSR 1943, 1949, 1950, Lenin Prize 1964.

Belov, Andrei Ivanovich
Military official; b. Ag. 6 (19), 1917, Pskov.
Mem. CPSU 1941–, grad. Electrotechnical Mil. Acad. 1940, Marshal of Signal Troops 1973; entered Sov. Army 1938; engineer for technical intelligence equipment of a tank brigade 1940–42; at the Stalingrad, Southern, 1st Ukrainian, 3rd Belorussian, and 2nd Baltic fronts 1942–45; chief of signal troops at 1st Far Eastern front Ag.–S. 1945; chief of signal troops of a unit, teacher, prof. and chief of a faculty at Mil. Acad. for Intelligence 1945–57; chief of signal troops in Turkestan Mil. District 1957–60, of strategic missile troops 1960–68; 1st Dep. Chief of Gen. Staff for Signal Troops of USSR Min. of Defense 1968–70, Chief 1970–. Decors.: Order of Lenin, Order of Red Banner, Order of Suvorov 3rd Class, Order of Red Star (three), Order "In Service to Homeland in USSR Armed Forces" 3rd Class, medals.

Belov, Iurii Vladimirovich
Painter; b. 1929, Leningrad.
Attended Leningrad Middle School of Art 1941–49 and Repin Inst. for Painting, Sculpture and Architecture at the USSR Acad. of Arts 1949–55; exhibitions 1954–.

Belov, Vasilii Ivanovich
Author; b. O. 23, 1932, Timonikha, Vologda Oblast.
Russian; mem. CPSU 1956–; grad. Gorkii Inst. of Literature 1964; first works publ. 1956.
Publs.: "Dereven'ka moia lesnaia" (My Little Forest Village) 1961, "Znoinoe leto" (An Oppressive Summer) 1963, "Privychnoe delo" (Accustomed Work) 1966, "Tisha da Grisha" (Stillness and Grisha) 1966, "Za tremia volokami" (Beyond the Three Tow-Paths) 1968, and others.

Belskaia, Vera Nikolaevna
Sculptor; b. 1926, Leningrad.
Attended Leningrad High School for Arts and Crafts 1943–52, pupil of V. I. Ingal, V. L. Simonov; exhibitions 1953–.

Beltsov, Georgii Ivanovich
Painter; b. 1920, Veselinovo Nikolaev Oblast.
Attended Odessa Art School 1946–52, pupil of L. E. Muchnik, A. N. Stelianudi, V. V. Tokarev; exhibitions 1952–.

Belyi, Boris Mikhailovich
Painter; b. 1936, Mysovsk, Buriat ASSR.
Attended Krasnoiarsk Art School 1951–54, pupil of A. P. Lekarenko, Iu. I. Khudonogov; exhibitions 1957–.

Belykh, Aleksei Pavlovich
Painter; b. 1923, Krutets, Orel Oblast.
Attended Surikov Art Inst. Moscow 1952–58, pupil of I. I. Orlov, P. D. Pokarzhevskii; exhibitions 1953–; mem. CPSU 1944–. Decor.: Merited Artist of the RSFSR 1965.

Belza, Igor' Fedorovich
Musicologist; b. Ja. 26 (F. 8), 1904, Kielce, Poland.
Doctor of art studies (1954).
In 1925, he graduated from the Kiev Conservatory in B. N. Liatoshinskii's class (composition). He moved to Moscow in 1941; he was a professor at the conservatory (1942–49) and senior scholar at the Institute of Art History (1954–61). In 1961 he became a senior scholar at the Institute of Slavic Studies of the Academy of Sciences of the USSR. Since 1965 he has also headed the Commission for the Study of Dante of the Academy of Sciences of the USSR. His principal works are on the musical culture of Poland and Czechoslovakia and on the Russian classics. He has also written musical works. He became an honorary doctor of philosophy at Charles University in Prague in 1967. Decors.: Order of the Red Banner of Labor, various medals, and the Order of the Polish Renaissance.
Works: *Ocherki razvitiia cheshskoi muzykal'noi klassiki.* Moscow-Leningrad 1951. *Istoriia pol'skoi muzykal'noi kul'tury,* parts 1–3. Moscow-Leningrad 1954–72. *Istoriia cheshskoi muzykal'noi kul'tury,* vol. 1. Moscow 1959. *Shopen.* Moscow 1968. *O slavianskoi muzyke: Izbr. raboty.* Moscow 1963. (With a bibliographical index.)

Bembel', Andrei Onufrievich
Sculptor; b. O. 17 (30), 1905, Velizh Vitebsk Province.
People's Artist of the Belorussian SSR (1955). Member of the CPSU since 1952. He studied at the Vitebsk Art Technicum (1924–27) and Leningrad Academy of Arts (1927–31). From 1950 to 1954, Bembel' was chairman of the Board of the Union of Artists of the Belorussian SSR. Since 1953 he has been teaching in the Belorussian Theatrical Art Institute. (He has been a professor there since 1966.) The works of his include bas reliefs in the Belorussian SSR House of Government in Minsk (plaster, 1932–36, in cooperation with G. K. Izmailov and V. N. Ritter), a bust of N. F. Gastello (bronze, 1943, Art Museum of the Belorussian SSR, Minsk), the monumental statue *Socialist Belorussia* for the Belorussian pavilion at the All-Union Agriculture Exhibition (cement and smalt, 1950–51), and a bust of D. G. Guliaev (plaster, 1967). He is one of the creators of the memorial complex *Burial Mound of Glory* (at the 21st km of the highway from Minsk to Moscow; 1969). Decors.: three orders, medals.

Bendinger, Vladimir Adolfovich
Graphic artist; b. 1924, Krasnoiarsk.
Attended Leningrad Pegagogical Art School 1948–53, pupil of M. V. Sudakov, G. A. Shakh; exhibitions 1948–.

Bendryshev, Vladimir Nikolaevich
Diplomat; b. 1920.
Mem. Diplomatic Service 1946–; USSR Adviser 1962–66 and Perm. USSR Rep. at UN and other int. orgs. in Geneva 1966–68; mem. staff USSR Min. of Foreign Affairs 1968–73; USSR Amb. to Malaysia D. 1973–Ap. 1978.

Berdyev, Redzhep Nuryevich
State official.
Min. of Interior Turkmen SSR F. 1978–; mem. Cen. Cttee. Turkmen CP 1981–.

Beregovoi, Georgii Timofeevich
Astronaut; b. Ap. 15, 1921, Fedorovka, Poltava Oblast.
Mem. CPSU 1943–; grad. Mil. Aviation School Lugansk 1941 and Mil. Flying Acad.; pilot; astronaut 1964–; Maj.-Gen. in Sov. Air Force 1968–; earth orbit in spaceship "Soiuz 3" O. 26–30, 1968, several manoeuvres, among them 2 rendezvous manoeuvres with "Soiuz 2". Decors.: 2 Orders of Lenin, 2 Orders of Red Banner, Alexander Nevskii Order, Bogdan Khmelnitskii Order 3rd Cl., Order of Patriotic War 1st Cl., 2 Orders of Red Star, medals.

Berestov, Valentin Dmitrievich
Poet; b. 1928, Meshchovsk, Kaluga Oblast.
Russian; grad. Dept. of History at Moscow Univ. 1951; completed post-graduate studies at Inst. of Ethnography with USSR Acad. of Sciences 1955; first works publ. 1946.
Publs.: poem collection "Otplytie" (Leaving Port), "V Golodnoi Stepi" (In the Hunger Steppe) 1957, children's book "Skazka pro vykhodnoi den'" (A Tale About a Free Day) 1960, and others.

Berezanskii, Iurii Makarovich, Dr. of Phy.-Math. Sc.
Mathematician; b. My. 8, 1925, Kiev; s. of an agronomist.

Grad. Univ. Kiev 1948; grad. Inst. of Mathematics at Acad. of Sciences Ukrainian SSR 1951; corresponding mem. Acad. of Sciences Ukrainian SSR 1964–; Dr. of Phy.-Math. Sc. 1956; Prof. 1961–; Research Asst. 1951–53, Sen. Research Asst. 1953–60, Chief of Dept. for Mathematical Analysis at Inst. of Mathematics Acad. of Sciences Ukrainian SSR 1960–; Lecturer at colleges and Univ. Kiev 1954–.
Publs.: "Razlozhenie po sobstvennym funktsiiam uravnenii v chastnykh raznostiakh vtorogo poriadka" in "Trudy Moskovskogo matematicheskogo obshchestva" vol. 5/1956, "O teoreme edinstvennosti v obratnoi zadache spektralnogo analiza dlia uravneniia Shredingera" in ibid. vol. 7/1958, "Razlozhenie po sobstvennym funktsiiam samosopriazhennykh operatorov" 1965.

Berezhnoi, Anatolii Semenovich, Dr. of Tech. Sc.
Chemical technologist; b. S. 19, 1910, Ostape, Poltava Oblast; s. of a junior surgeon.
Mem. CPSU 1954–; grad. Chemical-Technological Inst. in Kharkov 1932; corresponding mem. Acad. of Sciences Ukrainian SSR 1948–; Dr. of Tech. Sc. 1947; Prof. 1947–; mem. staff Ukrainian Inst. for Fireproof Materials (Institut Vohnetryviv) 1932–41; Dep. Chief of Dept. for Tech. Control at "Magnezit" Works (Ural) 1941–44; Laboratory Chief 1944–55, Dep. Dir. and Dir. of Ukrainian Inst. for Fireproof Materials 1955–. Decors.: Order of the Banner of Red Labour, Badge of Honour.
Publs.: "Kremnii i ego binornye sistemy" 1958, "Izgotovlenie osnovnykh ogneuporov iz rapnoi okisi magniia" in "Kompleksnoe ispolzovanie soljanykh resursov Sivasha i Perekopskikh ozer" 1958, "Fizikokhimicheskie osnovy tekhnologii forseritovykh ogneuporov" in "Sbornik trudov Vsesoiuznogo nauchno-issledovatelskogo instituta ogneuporov" 1958.

Berezin, Anatolii Ivanovich
Party official; b. 1931.
Grad. Mordovian Pedagogical Inst. 1953 and Mordovian State Univ. 1967; held Komsomol and party posts 1953–; 2nd Sec., then 1st Sec. of Saransk City Komsomol Cttee. 1953–57; 1st Sec. of Mordovian Oblast Cttee. of USSR Komsomol 1957–61; 1st Sec. of Saransk City Cttee. CPSU 1969–70; 2nd Sec. of Mordovian Oblast Cttee. CPSU 1970–71; mem. Auditing Comm. Cen. Cttee. CPSU 1971–76; mem. Cen. Cttee. CPSU 1976–; Chair. of Council of Mins. Mordovian SSR F.–Ap. 1971; 1st Sec. Mordovian Oblast Cttee. CPSU Ap. 1971–.

Berezin, Ilia Vasilevich, Dr. of Tech. Sc.
Chemical scientist; b. 1923.
Mem. CPSU 1943–; grad. Moscow Univ. 1950; Dr. of Tech. Sc. 1963; Prof. 1966–; corresponding mem. USSR Acad. of Sciences 1970–; Dean of Faculty of Chemistry at Moscow Univ. 1969–.

Berezin, Vladimir Semenovich
Sculptor; b. 1925, Dnepropetrovsk.
Attended Moscow High School for Arts and Crafts 1950–57, pupil of G. I. Motovilov, G. A. Shults; exhibitions 1957–; teacher at Moscow School for Arts and Crafts 1957–.

Berezko, Georgii Sergeevich
Author; b. 1905; s. of a teacher.
Russian; served in Sov. Army 1919; grad. Dept. of Ethnology at Moscow Univ. 1927; active as film dir.; served in Sov. Army 1941–45; first poems publ. 1925, first stories 1942. Decor.: Order of Red Banner of Labour.
Publs.: "Krasnaia raketa" (The Red Rocket) 1943, "Komandir divizii" (The Division Commander) 1944, "Noch' polkovodtsa" (The General's Night) 1946, "Mirnyi gorod" (A Peaceful Town) 1951–54, "Sil'nee atoma" (More Powerful than the Atom) 1959, "Vot ia idu" (Here I Go) 1960, and others.

Bergelson, Lev Davidovich, Dr. of Chem. Sc.
Chemical scientist; b. Ag. 8, 1918, Gaisin, Vinnitsa Oblast.
Mem. CPSU 1967–; grad. Moscow Univ. 1941; organic chemist and biochemist; corresponding mem. USSR Acad. of Sciences 1968–; Prof. at Moscow Univ. 1964–. Decors.: Order of Red Star, several medals.

Beriev, Georgii Mikhailovich (also G. M. Beriashvili)
Aircraft designer; b. Ja. 31 (F. 13), 1903, Tbilisi.
Doctor of technical sciences (1961); major-general of the engineering and technical service (1951). Member of the CPSU since 1929.
In 1930, he graduated from the Kalinin Leningrad Polytechnical Institute. From 1934 to 1968 he headed the experimental and construction bureau. The MBR-2, MP-1, MDR-5, MBR-7, and Be-6 hydroplanes; the R-1 and B-10 jet-powered flying boats; the Be-8 and B-12 amphibious airplanes; the Be-2 and Be-4 ship-based catapult airplanes; and the Be-30 overland passenger aircraft were created under his direction. Decors.: State Prize of the USSR in 1947 and 1968, two Orders of Lenin, two other orders, and medals.

Bernatavichius, Iuozas Ionovich
State official; b. 1929.
Lithuanian; grad. Vilna State Univ. and the Lithuanian Acad. of Agriculture; held Komsomol and party posts until 1971; mem. Cen. Cttee. Lithuanian CP 1971–; Head of Dept. of Agriculture with Cen. Cttee. Lithuanian CP 1971–80; Dep. Chair. of Council of Mins. Lithuanian SSR N. 1980–.

Bernotas, Iulius Mikolovich
State official.
Chair. of State Cttee. for Supervision of Safe Working Practices in Industry and for Mine Lithuanian SSR S. 1981–.

Bernov, Iurii Vladimirovich
Diplomat.
USSR Amb. to Ghana My. 1974–79.

Bershadskii, Rudolf Iulevich
Author; b. 1909, Elizavetgrad.
Russian; as a boy in Red Army during Civil War; mem. CPSU 1938–.
Publs.: collections "Start" 1930, "Eto i est voina" (That's War) 1933, "Liniia pritsela" (Line of Sight) 1933, "Rasskazy o vyruchke v boiu" (Stories of Aid in Combat) 1940, film script "V tylu vraga" (Behind Enemy Lines) 1941, "Rasskazy o voine" (Stories of the War) 1942, "Put' k podvigu" (The Path to Valiant Deeds) 1956, "Znakomstvo s Chernovym" (Acquaintance with Chernov) 1960, essays on Vietnam „V dvukh shagakh ot ekvatora" (Two Steps from the Equator) 1961, and others.

Bertse, Vizbul Avgustovich
Author; b. 1916; s. of Latvian revolutionary and poet A. Arais-Berce.
Latvian; first works publ. 1930.
Publs.: stories "Sila sil'nykh" (The Might of the Strong) 1948, "Pervye odinnadtsat'" (The First Eleven) 1949, "Budushchee nachinaetsia Segodnia" (The Future Begins Today) 1951, "Vyshli my vse iz naroda" (We Are All Offspring of the People) 1956, and others.

Berulava, Khuta Mikhailovich
Poet; b. 1924, Zugdidi.
Georgian; mem. CPSU 1944–; studied Pedagogical Inst. Sukhumi and Gorkii Inst. of Literature Moscow; first collection of poetry publ. 1944.
Publs.: "Russkoe serdtse" (The Russian Heart) 1951, "Pesnia o zvezdakh" (The Song of the Stars), and others.

Berzegov, Nukh Aslancherievich
Party official; b. 1925.
Adjar; mem. CPSU 1944–; grad. Pedagogical Inst. in Krasnodar; served in Sov. Army 1943–46; student, dir. of a 7-class school, head of pedagogical dept. of a secondary school, head of an Oblast Dept. of Popular Educ. 1946–57; Dept. Chair of Adjar Oblast Exec. Cttee. of Workers' Deputies 1957–58; Sec. of Adjar Oblast Cttee. CPSU 1958–60, 1st Sec. 1960–; Dep. to and mem. Law Drafting Comm. at Sov. of Nationalities of USSR Supr. Sov. 1962–. Decor.: Order of Red Banner of Labour.

Berzin, Boris Avgustovich
Painter; b. 1930, Riga.
Attended Riga School for applied art 1947–52 and Art Acad. of the Latvian SSR 1952–59; exhibitions 1957–.

Berzin, Ivan Petrovich
State official.
Head of Dept. of Ind. and Transportation at Cen. Cttee. Latvian CP 1962–69; mem. Bureau of Ind. at Cen. Cttee. Latvian CP 1962–64; mem. Cen. Cttee. Latvian CP 1963–; Dep. to Supr. Sov. Latvian SSR 1963–; Min. of Communal Econ. Latvian SSR F. 1969–.

Berzins, Lilita (Liliia Davydovna Priede-Berzinia)
Latvian actress; b. Jl. 4 (17), 1903, Riga. People's Artist of the USSR (1956). Member of the CPSU since 1949.
Since 1922, she has worked in the Reinis Art Theater (Riga). Her work is noted for accuracy in her psychological characterizations, spareness of expressive devices, and nobility of stage form. Among her roles are Asnate, Spidola, and Mother (in Reinis's *Iosif and his Brothers, Fire and Night,* and *Breathe, Wind!*), Liubov' Iarovaia (in Trenev's play of that title), Anna Karenina (from L. Tolstoy's novel), Maria Stuart (in Schiller's tragedy), Filomena Marturano (in De Filippo's play), and the Colonel's Widow (in Smuul's play of that title).
She was a deputy to the second through fourth convocations of the Supreme Soviet of the Latvian SSR. Decors.: State Prize of the USSR (1947), Order of Lenin, the Order of the Red Banner of Labor, and medals.

Besaev, Tazret Urusbievich
Author-poet; b. 1907, Mastinok, Northern Ossetia.
Grad. Inst. of Philosophy, Literature and History in Moscow 1937.
Publs.: poems "Obshchestvennyi pevets" (The Public Singer) 1931, story "Kto kogo? ili Poslednee Pokhozhdenie mully Magometa" (Who Whom? or The Last Adventure of Mulla Mahomed) 1933, "Dolg" (Duty), play "Pesnia ob otvazhnykh" (The Song of the Brave) 1959, and others.

Beschastnov, Ivan Vasil'evich
Sculptor; b. 1925, Panfilovo, Vladimir Oblast.
Attended Ivanovo Art School 1945–47 and the Repin Inst. for Painting, Sculpture and Architecture at the USSR Acad. of Arts 1947–53, pupil of I. V. Krestovskii, A. P. Timchenko; exhibitions 1955–.

Besdetko, Andrei Pavlovich
Oblast soviet official; b. 1919.
Ukrainian; mem. CPSU 1944–, grad. Kharkov Agricultural Inst.; served in Sov. Army 1941–46; agronomist of a machine and tractor station, chief of agriculture dept., Sec. of Balakleia Raion Cttee. Ukrainian CP, 1st Sec. of Chuguev Raion Cttee. Ukrainian CP, chief of an oblast state farm admin. 1946–62; Dep. Chair. of Exec. Cttee. of Kharkov Oblast Sov. of People's Deps. 1962–68, Chair. 1968–; mem. Cen. Cttee. Ukrainian CP 1971–.

Beskakotov, Veniamin Vasil'evich
Graphic artist; b. 1929, Perekopnoe, Saratov Oblast.
Attended Tashkent Art School 1945–50 and Kharkov Art Inst. 1953–59, pupil of A. A. Kokel, V. F. Mironenko; exhibitions 1959–.

Bespalov, Ivan Petrovich
Party official; b. 1915.
Russian; mem. CPSU 1944–; grad. Omsk Inst. for Machine Building 1947–; electrical installation worker in Belovo, assessor in Topki (Kemerovo Oblast) 1931–35; master, sen. master, dept. head in a factory in Omsk 1941–51; chief engineer of a plant in Kirov 1952–61; held party posts 1961–64; 1st Sec. of Kirov City Cttee. CPSU 1964–68; 2nd Sec. 1968–71, 1st Sec. of Kirov Oblast Cttee. CPSU Mr. 1971–; cand. mem. 1971–76, mem. Cen. Cttee. CPSU 1976–; Dep. to Supr. Sov. USSR 1974–. Decor.: Order of Lenin.

Bessarabova, Natal'ia Ivanovna
Ceramic artist; b. Ag. 19 (31), 1895, Voronezh.
She studied in the Voronezh State Higher Arts and Technical Studios (1919–22). From 1944 to 1955 she worked in the ceramic studios of the Institute of Industrial Art in Moscow. She helped bring about a revival of Gzhel' and Skopin ceramic art. By studying the traditions of these crafts she created samples of household articles and of sculptures from porcelain, majolica, and terra-cotta (kept in the Museum of Folk Art and the Museum of History in Moscow and in the Russian Museum in Leningrad). She also worked as a theater set designer, designing F. Z. Iarullin's ballet *Shurale* for the Tartar Theater of Opera and Ballet in Kazan in 1945.

Bevzenko, Ivan Pavlovich
Painter and graphic artist; b. 1920, Elizavetgrad.
Attended Odessa Art School until 1952 and Kiev Art. Inst. 1952–58, pupil of V. N. Kostetskii; exhibitions 1958–.

Beziuk, Lev Leonidovich
Sculptor; b. 1922, Arkhangelsk.
Attended Repin Inst. for Painting, Sculpture and Architecture at the USSR Acad. of Arts 1950–56, pupil of I. V. Krestovskii, V. V. Lishev; teacher at Rovno Art School 1958–; exhibitions 1959–; mem. CPSU 1945–.

Bezklubenko, Sergei Danilovich
State official; b. 1932, Ladyzhynka, Cherkassy Oblast.
Grad. Kiev Univ. and Acad. of Sociology at Cen. Cttee. CPSU; Cand. of Art Sc.; worked for communications media (press and television) after grad.; head of a sector, then Dep. Head of Cultural Dept. of Cen. Cttee. Ukrainian CP 1970–77; Min. of Culture Ukrainian SSR Je. 1977–; mem. Cen. Cttee. Ukrainian CP 1981–.

Bezliudov, Artur Iosifovich
State official.
Mem. Auditing Comm. Belorussian CP 1966–71; Min. of Housing and Communal Econ. Belorussian SSR My. 1973–; cand. mem. Cen. Cttee. Belorussian CP 1981–.

Bezrukavnikov, Boris Vasil'evich
Diplomat; b. 1923.
Mem. Diplomatic Service 1954–; USSR Emb. Counsellor in Japan 1963–67; held sen. positions in USSR Min. of Foreign Affairs 1967–71; USSR Amb. to Singapore Mr. 1971–My. 1975.

Bezrukov, Panteleimon Leonidovich
Geologist; b. F. 2 (15), 1909, Moscow.
Corresponding mem. USSR Acad. of Sciences 1968–; Chief of Oceanogeological Dept. at Shirshov Inst. of Oceanology with USSR Acad. of Sciences 1946–; organized several sea expeditions of the "Vitiaz". Decors.: State Prizes of the USSR 1946, 1951, Badge of Honour.

Bezsmertnyi, Ilia Semenovich
State official.
Min. of Building Materials Ind. Kirghiz SSR Jl. 1975–; mem. Auditing Comm. Kirghiz CP 1976–.

Bgazhba, Khukhut Solomonovich
Author and linguist; b. 1914, Gupy, Abkhazian ASSR; s. of a peasant.
Grad. Moscow Lenin-Pedagogical Inst.; first works publ. 1932.
Publs.: "Ocherki ob abkhazskoi literature" (Essays on Abkhazian Literature) 1940, two fairy-tale books in Russian language "Syn olenia" (Son of a Stag) 1959 and "Abkhazskie skazki" (Abkhazian Tales) 1959, "Ob abkhazskoi literature. Kriticheskie ocherki" (On Abkhazian Literature. Critical Essays) 1960, and others.

Bialik, Boris Aronovich
Literary historian; b. O. 5 (18), 1911, Kiev.
Mem. CPSU 1941–; his main works deal with Maxim Gorkii's works.
Publs.: "Gorkii v bor'be s teatral'noi reaktsiei" (Gorkii and His Fight Against Regression in the Theatre) 1938, "Dramaturgiia M. Gorkogo sovetskoi epokhi" (M. Gorkii's Dramas of the Soviet Era) 1952, "Razvitie traditsii russkoi klassicheskoi literatury v tvorchestve M. Gorkogo" (Development of Tradition of Russian Classical Literature in the Works of M. Gorkii) 1958, "M. Gorkii – Literaturnyi kritik" (M. Gorkii – Literary Critic) 1960, and others.

Bibikov, Sergei Nikolaevich, Dr. of Hist. Sc.
Archaeologist; b. S. 14, 1908, Sevastopol; s. of a clerk.
Mem. CPSU 1940–; grad. Ethnographic Dept. of Faculty of Geography at Leningrad Univ. 1931; corresponding mem. Acad. of Sciences Ukrainian SSR 1958–;

Dr. of Hist. Sc. 1953; mem. staff State Acad. for the History of Material Culture in Leningrad (since 1937 Inst. for the History of Material Culture at USSR Acad. of Sciences) 1934–55; Dir. of Inst. of Archaeology at Acad. of Sciences Ukrainian SSR 1955–; in addition Dep. Chief of Dept. of Sociology at Acad. of Sciences Ukrainian SSR 1956–61; author of more than 60 scientific works. Decors.: 3 Badges of Honour, medals.
Publs.: "Pozdneishii paleolit Kryma" in "Materialy po chetvertichnomu periodu SSSR" 1950, "Iz istorii kamennykh serpov na iugovostoke Evropy" in "Sovetskaia arkheologiia" No. 3/1962, "Khoziaistvennoekonomicheskii kompleks razvitogo Tripolia" ibid. No. 1/1965.

Bieshu, Mariia Lukianovna
Singer; b. 1934.
Moldavian; grad. Kishiniov Conservatory 1961; soprano singer; concerts with Moldavian Folk Instruments Orchestra "Flu Erash" 1958–60; mem. "Scala" Opera House Milan 1965–67; 1st Prize in Int. Competition for the best performance of Puccini's "Madame Butterfly" in Tokyo 1967; soloist with Moldavian Theatre for Opera and Ballet 1961–. Decor.: People's Artist of the USSR 1970.

Biisheva, Zeinab Abdullovna
Author; b. 1908 in Tuimbetovo, Bashkir ASSR.
Mem. CPSU 1941–; grad. Bashkir Pedagogical Tech. College in Orenburg.
Publs.: "Mal'chik partizan" (The Boy Partisan) 1942, "Kiunkhylyu" 1949, "Druzhba" (Friendship) 1950, "Guliamal" 1953, "Budem druziami" (We Shall Be Friends) 1954, "Unizhennye" (The Humble) 1959, and others.

Bikbai, Baiazit Gaiazovich
Author; b. 1909 in Kaltai, Bashkir ASSR; s. of a village teacher.
Grad. Pedagogical Tech. College.
Publs.: "Tekushchie dni" (Daily Life) 1932, "Za lesom" (Behind the Forest) 1934, "Solnechnyi den'" (A Sunny Day) 1934, "Pochemu morskaia voda solenaia" (Why is Seawater Salty?) 1939, "Platochek" (The Small Kerchief) 1942, "Ognennye stroki" (Flaming Lines) 1943, plays "Karlugas" 1938, "Rodina zovet" (The Native Land Calls) 1941, "Deti odnoi sem'i" (The Children of One Family) 1942, "Salavat" 1940–54, "Kakhym Turia" 1957, and others.

Bikchentaev, Anver Gadeevich
Author; b. 1913, Ufa; s. of an office worker.
Bashkir; mem. CPSU 1941–; completed Advanced Courses in Literature of USSR Writers' Union 1956; writes in Bashkir and Russian; first works publ. 1933.
Publs.: collection "Krasnye maki" (Red Poppy) 1944, "Novelly o russkom soldate" (Novelettes about the Russian Soldier) 1946, "Pravo na bessmertie" (The Right to Immortality) 1950, novel "Lebedi ostaiutsia na Urale" (Swans Remain in the Ural) 1956, "Bolshoi orkestr" (The Big Orchestra) 1957, and others.

Bikkulov, Sharif Sagadatovich
Poet; b. 1924, Karlaman; s. of a farmer.
Bashkir; mem. CPSU 1944–; grad. Pedagogical Inst. 1949.
Poems: "Stremlenie" (Striving) 1947, "Moemu sovremenniku" (To my Contemporaries) 1951, "Tian Ge Chan" 1954, "Cherez porogi" (Across the Thresholds) 1957, and others.

Bilai, Vera Iosifovna, Dr. of Biol. Sc.
Botanist; b. Jl. 12, 1908, Nikopol; d. of a clerk.
Mem. CPSU 1942–; grad. Agricultural Inst. Krivoi Rog 1929; corresponding mem. Acad. of Sciences Ukrainian SSR 1961–; Dr. of Biol. Sc. 1955; Prof. 1960–; agronomist 1929–30; Asst. at Inst. of Sugar Ind. Kiev 1933–34; Research Asst. at Inst. of Microbiology at Acad. of Sciences Ukrainian SSR (now Inst. of Microbiology and Virology) 1935–; Chief Dept. of Physiology of Fungi; author of approx. 120 scientific works. Decors.: State Prize of the USSR 1952, Order of Red Banner of Labour, Badge of Honour, medals.
Publs.: "Toksychni hryby na zerni khlibnykh zlakiv" 1946 "Sontse, zhyttia i khlorofil" 1949, "Biologicheski aktivnye veshchestva mikroskopicheskikh gribov i ikh primenenie" 1965, "Zhyttia, viddane liudiam" 1966.

Bildiukevich, Viktor Leontevich, Cand. of Tech. Sc.
State official; b. 1930.
Mem. CPSU 1959–; grad. Belorussian Polytech. Inst. 1952; Dir. of Collective Combine for Building Materials in Minsk 1958–67; Dep. Min. of Building Materials Ind. Belorussian SSR 1967–80, Min. 1980–; cand. mem. Cen. Cttee. Belorussian CP 1981–. Decor.: State Prize of the USSR 1972.

Bilyi, Mikhail Ulianovich, Dr. of Phy.-Math. Sc.
Scientist; b. 1922, Moskali, Chernigov Oblast; s. of a peasant.
Mem. CPSU 1951–; grad. Kiev State Univ. 1948; corresponding mem. Ukrainian Acad. of Sciences; Dr. of Physics and Mathematics, Prof.; served in Soviet Army 1940–43, badly injured; teacher, lecturer, Head of Chair of Experimental Physics 1951, also Dean of Faculty of Physics at Kiev State Univ. 1962–; Rector of Kiev State Univ. 1970–; Chair. of Ukrainian Branch of USSR-Canada Soc. 1971–; cand. mem. Cen. Cttee. Ukrainian CP 1971–76, mem. 1976–; Dep. 1971–, Dep. Chair. 1971–72, Chair. of Supr. Sov. Ukrainian SSR 1972–80.

Biriukova, Aleksandra Pavlovna
Trade union official; b. 1929.
Mem. CPSU 1956–; grad. Moscow Textile Inst. 1952; forewoman, dept. head of a Moscow factory 1952–59; head specialist of tech. dept., head of an admin. dept. for textile and knitwear ind. of Moscow City Econ. Council 1959–63; Chief Engineer of Moscow collective combine "Trekhgornaia manufaktura" 1963–68; Sec. and mem. Presidium of USSR Cen. Council of Trade Unions 1968–; cand. mem. Cen. Cttee. CPSU 1971–76, mem. 1976–.

Biriukova, M. M.
Komsomol official.
Sec. and mem. Bureau Cen. Cttee. Belorussian Komsomol Ja. 1981–.

Birkenfeld, Vsevolod Ianovich
State official.
Mem. Cen. Cttee. Latvian CP 1971–; Min. of Wood Processing Ind. Latvian SSR F. 1975–.

Birze, Miervaldis Ianovich (real name: Berzin)
Author; b. 1921, Ruiena.
Grad. Latvian Univ. 1949, Latvian author; entered Komsomol 1940, arrested several times and interned in concentration camps; returned to Riga, worked as doctor 1945. Decor.: State Prize.
Publs.: "Pervye tsvety" (First Flowers) 1956, short story "Odno iabloko" (An Apple) 1956, "I podo ldom reka techet" (The River Flows Under the Ice Too) 1957, humorous stories "Neschastnaia sobaka" (An Unlucky Dog) 1959, and others.

Bitsadze, Andrei Vasil'evich
Mathematician; b. My. 9 (22), 1916, Tskhrukveti, Georgia.
Mem. CPSU 1947–; grad. Tbilisi Univ. 1940; corresponding mem. USSR Acad. of Sciences 1958–; mem. Acad. of Sciences of the Georgian SSR 1969–; mem. staff Inst. of Mathematics at Siberian Branch Office of the USSR Acad. of Sciences and Novosibirsk Univ. 1959–. Decors.: Order of the Red Banner of Labour, several medals.

Biushgens, Georgii Sergeevich
Technologist; b. S. 3 (16), 1916, Moscow.
Mem. CPSU 1963–; grad. Moscow Inst. of Aviation 1940; specializes in mechanics; corresponding mem. USSR Acad. of Sciences 1966–; mem. staff Cen. Inst. of Aerohydrodynamics 1940–. Decors.: Lenin Prize 1961, Order of Red Banner of Labour, several medals.

Blanter, Matvei Isaakovich
Composer; b. Ja. 28 (F. 10), 1903, Pochep, Chernigov Province.
People's Artist of the RSFSR (1965).
He studied in the music colleges of Kursk and Moscow and took lessons in composition from G. E. Konius. From 1921 to

1933 he worked in various theaters as a composer and director of their music divisions. One of the masters of the Soviet popular song, he wrote "Katiusha," which achieved worldwide popularity, and the songs "The Partisan Zhelezniak," "Song of Shchors," "Good-bye, Cities and Huts," "In the Frontline Forest," "My Loved One," "Under the Balkan Stars," "Nothing Better Than Apple Blossom," and "The Birds of Passage Are Flying," among others. Blanter's songs have been written to the words of M. Golodnyi, V. Lebedev-Kumach, K. Simonov, A. Surkov, M. Svetlov, and other poets. More than 20 songs were written in collaboration with the poet M. Isakovskii. Blanter composed several operettas, including *On the Shores of the Amur* (1938; produced in 1939 at the Moscow Operetta Theater), as well as music for shows, films, and radio productions. Decors.: State Prize of the USSR in 1946, the Order of the Badge of Honor, and various medals.

Blatov, Anatolii Ivanovich
Party official; b. 1914.
Russian; mem. CPSU 1940–, grad. Dnepropetrovsk Engineering Inst. of Rail Transport 1940 and CPSU Cen. Cttee. Party Higher School 1945; technician, foreman of Zaporozhe Locomotive Repair Plant 1932–40; employed by railroad in Primore Krai, controller for Primore Krai in USSR Min. of State Control 1940–45; counsellor in USSR Emb. in GDR, dir. of pol. dept. of USSR Emb. apparatus in GDR and of High Commissioner in Germany, posts in USSR Min. of Foreign Affairs apparatus, Chief of 3rd European Dept. of USSR Min. of Foreign Affairs 1945–68; posts in Cen. Cttee. apparatus CPSU: Dep. Chief of Dept. for Liaison with Communist and Workers' Parties of Socialist Countries of Cen. Cttee. CPSU 1968–72; Asst. to Gen.-Sec. of Cen. Cttee. CPSU 1972–; mem. Cen. Auditing Comm. CPSU 1976–81; cand. mem. Cen. Cttee. CPSU 1981–; Dep. to USSR Supr. Sov. 1979–. Decor.: USSR State Prize 1979.

Blekte, Ianis Oskarovich
State official.
Min. of Finance Latvian SSR Jl. 1981–.

Blinov, Vladimir Mikhailovich
State official.
Public Prosecutor of the RSFSR 1963–70; Min. of Justice RSFSR S. 1970–.

Blium, Viktor Adolfovich
State official; b. 1923.
Latvian; mem. CPSU 1946–; grad. Party High School Cen. Cttee. CPSU; served in Sov. Army 1941–45; organized party matters in a rural district 1945–48; Dep. Chief of Latvian Republican Trade Union Council, chair. of a raion soviet of People's Deps., 1st sec. of a raion cttee. Latvian CP 1948–59; Dep. to 1955–, Dep. Chair. of Presidium Supr. Sov. Latvian SSR 1973–; mem. Cen. Cttee. Latvian CP 1958–; Chair. of Latvian Republican Trade Union Council 1959–73; cand. mem. Bureau Cen. Cttee. Latvian CP 1960–76.

Blinova, Ekaterina Nikitichna
Scholar; b. N. 24 (D. 7), 1906, Kamensk-Shakhtinskii, Rostov Oblast.
Specialist in the area of dynamic meteorology. Corresponding member of the Academy of Sciences of the USSR (1953).
In 1928, she graduated from North Caucasus University (Rostov-on-Don). From 1935 to 1943 she worked at the Main Geophysical Observatory; from 1943 to 1958, at the Central Institute of Forecasting; and from 1958 to 1961, at the Institute of Applied Geophysics. Since 1961 she has headed a department of the Meteorological Computer Center.
In 1936 she studied the conditions of stability of an atmospheric front, developed a theory of radiant equilibrium in the atmosphere, and quantitatively explained the existence of so-called centers of atmospheric activity. She proposed methods of long-term weather forecasting in 1943 and 1961. Blinova's work *A Hydrodynamic Theory of Pressure Waves, Temperature Waves, and Centers of Atmospheric Activity* (1943) laid the foundations for hydrodynamic long-term weather forecasting. Decors.: Order of the Badge of Honor and medals.
Works: "Obshchaia tsirkuliatsiia atmosfery i gidrodinamicheskii dolgosrochnyi prognoz pogody." In the collection *Meteorologiia i gidrologiia za 50 let sovetskoi vlasti.* Leningrad 1967.

Blokhin, Anatolii Prokopevich
State official.
Chair. of RSFSR State Cttee. on Labour Je. 1982–.

Blokhin, Nikolai Nikolaevich
Medical scientist; b. Ap. 21 (My. 4), 1912, Lukoianov, Nizhnii Novgorod Gouvt.
Mem. CPSU 1948–; grad. Medical Inst. Gorkii 1934; surgeon and oncologist; Prof. at Medical Inst. Gorkii 1947–; Dir. of Inst. for Experimental and Clinical Oncology at USSR Acad. of Medical Sciences 1952–; mem. 1960–, Pres. of USSR Acad. of Medical Sciences 1960–68; mem. Int. Federation of Cancer Research, mem. Polish Acad. of Sciences, mem. New York Acad. of Sciences, Pres. of Inst. of Sov.-American Relations 1960–. Decors.: 2 Orders of Lenin, Order of Red Star, several medals.

Blokhintsev, Dmitrii Ivanovich
Physicist; b. D. 29, 1907 (Ja. 11, 1908), Moscow.
Corresponding member of the Academy of Sciences of the USSR (1958) and Hero of Socialist Labor (1956). Member of the CPSU since 1943.
After graduating from Moscow State University in 1930, he began teaching there, becoming a professor in 1936. He directed the design and construction of the first atomic power plant in the world. From 1956 to 1965 he was director of the Joint Institute of Nuclear Research in Dubna and has been director of the institute's Laboratory of Theoretical Physics since 1965. His first scientific studies were devoted to the quantum theory of solids, phenomena in semiconductors, acoustics of inhomogeneous moving media, and quantum mechanics. He was worked, since 1947, in the field of nuclear physics and technology (the theory of chain reactions and atomic reactors and problems of the theory of elementary particles). He is the author of the first handbook on quantum mechanics in the USSR. He has been a member of the Soviet Committee for the Protection of Peace since 1963, president of the International Union of Pure and Applied Physics (IUPAP) since 1966, and an advisor to the Scientific Council of the Secretary General of the United Nations since 1967. Decors.: State Prize of the USSR (1952) and the Lenin Prize (1957), four Orders of Lenin, the Order of the Red Banner of Labor, and various medals.

Bobosadykova, Guldzhikhon
Party official; b. 1937.
Tadzhik; mem. CPSU 1960–, grad. Tadzhik State Univ.; mem. Cen. Cttee. 1961–, cand. mem. Bureau Cen. Cttee. Tadzhik CP 1961–76, mem. Ja. 1976–; 1st Sec. and mem. Bureau Cen. Cttee. Tadzhik Komsomol 1961–72; 1st Sec. Tsentralnyi Raion Cttee. Tadzhik CP 1972–75; 1st Sec. Dushanbe City Cttee. Tadzhik CP 1975–78; Sec. Cen. Cttee. Tadzhik CP D. 1978–.

Bobovikov, Ratmir Stepanovich
Oblast Soviet official; b. 1927.
Russian; mem. CPSU 1947–; grad. Ulianov Electrotechnical Inst. in Leningrad 1951; Cand. of Tech. Sc. 1954; machinist in a power station 1944–52; posts in a scientific research inst. for electrophysics equipment 1952–62; posts in Leningrad City and Oblast cttees. CPSU 1962–72; Sec. of Leningrad Oblast Cttee. CPSU 1972–78, 2nd Sec. 1978–80; mem. Cen. Auditing Comm. CPSU 1976–81; cand. mem. Cen. Cttee. CPSU 1981–; Dep. to USSR Supr. Sov. 1979–; Chair. of Exec. Cttee. of Leningrad Oblast Sov. of Peoples' Deps. N. 1980–.

Bobrov, Vsevolod Mikhailovich
Athlete (soccer and hockey); b. D. 1, 1922, Morshansk.
Became Honored Master of Sports in 1948 and Honored Trainer of the USSR in 1967. Officer in the Soviet Army, Member of the CPSU since 1953.
He was several times a champion of the USSR (1945–46, ice hockey; 1946–48, soccer; 1948–49, 1951–52, and 1955–57,

ice hockey). His team won the world championship (1954, 1956), European championship (1954–56), and the gold medal at the 1956 Winter Olympic Games (ice hockey). He was named as the best forward in the World Ice-Hockey-Championship of 1954. He is the only Soviet olympic athlete in sports history to be captain of the national teams in both soccer and hockey. Decors.: Order of Lenin and medals.
Work: *Samyi interesnyi match.* [Moscow] 1963.

Bobrovnikov, Andrei Andreevich
State official.
Min. of Finance RSFSR Mr. 1973–.

Bobylev, Sergei Andreevich
Military official; b. 1922, Lenino, Kalinin Oblast.
Mem. CPSU 1941–; Lieut.-Gen.; until 1941 sec. of a raion cttee. USSR Komsomol; served in Sov. Army 1941–45; pol. activity in Sov. Army, Dep. Chief of Polit Admin. at Main Party HQ, Chief of Polit Dept. in Mil. Acad., Chief of Org. and Instruction Dept. in Pol. Admin. of Kiev Mil. District, 1st Dep. Chief of Pol. Admin. of Leningrad Mil. District and of Sov. Army in Germany 1945–71; mem. Mil. Council, Chief of Pol. Admin. of Leningrad Mil. District 1971–75; Dep. to Supr. Sov. RSFSR 1975–; Dep. Chief of Pol. HQ of Sov. Army and Navy F. 1975–77; Chief of Pol. Admin., USSR Air Defence Forces 1977–.

Bodiul, Ivan Ivanovich
State official; b. 1918, Moldavia.
Mem. CPSU 1940–; grad. Mil. Veterinary Acad. 1942 and Party Higher School of Cen. Cttee. CPSU 1958; agronomist at a machine and tractor station, attended Mil. Veterinary Acad. 1937–42; served in Soviet Army 1942–46; worked for Moldavian Council of Mins. 1946–51; 1st sec. of various raion cttees. of Moldavian CP, attended Party Higher School 1951–58; cand. mem. 1956–58, mem. 1958–80, mem. Bureau 1959–80, 2nd Sec. 1959–61, 1st Sec. of Cen. Cttee. Moldavian CP 1961–80; Dep. 1960–, mem. Credentials Comm. 1962–66, mem. Foreign Affairs Comm. at Sov. of the Union USSR Supr. Sov. 1966–; Instructor for Dept. of CPSU Cen. Cttee. 1958–59; mem. CPSU Cen. Cttee. 1961–; Dep. to and presidium mem. Supr. Sov. Moldavian SSR 1961–81; Dep. Chair. of USSR Council of Mins. D. 1980–. Decor.: Order of Lenin.

Bogatikov, V. F.
Trade union official.
Sec. of All-Union Cen. Council of Trade Unions of the USSR Mr. 1968–.

Bogatyrev, Anatolii Vasil'evich
Composer; b. Jl. 31 (Ag. 13), 1913, Vitebsk.
People's Artist of the Belorussian SSR (1968). Member of the CPSU since 1954. He graduated from the Belorussian Conservatory in 1937 in V. A. Zolotarev's class. He has taught there since 1948 and has been a professor since 1960. Among his students are G. M. Vagner, E. A. Glebov, Iu. V. Semeniako, and E. M. Tyrmand. He has composed the operas *In the Woods of Poles'e* (performed in 1939 in Minsk) and *Nadezhda Durova* (performed in 1956 in Minsk); the cantatas *Belarus'* (1949) and *Belorussian Songs* (1967); two symphonies; concerti for cello and contrabass; music for dramatic presentations and films; and choral works and arrangements of Belorussian folk songs. He was a deputy to the first through the fourth convocation of the Supreme Soviet of the Belorussian SSR. Decors.: State Prize of the USSR (1941), the Order of Lenin, four other orders, and medals.

Bogatyrev, Mikhail Grigor'evich
Painter; b. 1924, Niskuchai, Kalinin Oblast.
Attended Moscow Art Studio for War Disabled 1946–51, pupil of A. I. Ivanov, G. G. Rizhskii; exhibitions 1955–.

Bogdan, Aleksandr Stepanovich
State official.
Can. mem. Cen. Cttee. Belorussian CP 1966–71; 1st Sec. Molodechno City Cttee. Belorussian CP, Dep. Chair. of Cttee. of People's Control Belorussian SSR 1967–76; Chair. of State Cttee. for Labor Safety in Ind. and in the Mines Belorussian SSR F. 1976–.

Bogdanas, Konstantinas Aleksandro, Cand. of Art Sc.
Sculptor; b. 1926, Zhiaimiai, Lithuania. Attended Kaunas Inst. for Applied and Decorative Art 1945–51, pupil of Iu. I. Mikenas and V. F. Palis, aspirant at Repin Inst. for Painting, Sculpture and Architecture at the USSR Acad. of Arts 1953–56, pupil of I. V. Krestovskii and V. V. Lishev; exhibitions 1951–; teacher at Art Acad. of the Lithuanian SSR 1951–; Cand. of Art Sc. 1958. Decor.: Merited Artist of the Lithuanian SSR 1963.

Bogdanov, Anatolii Ivanovich
Sports marksman; b. Ja. 1, 1932, Leningrad.
Honored Master of Sport of the USSR (1952). Officer in the Soviet Army. Member of the CPSU since 1955.
He has been Champion of the USSR several times and champion at the XV Olympic Games (Helsinki 1952) and the XVI Olympic Games (Melbourne 1956). He has been Champion of the World (1954) and Europe (1955, 1959) several times. He is the first Soviet marksman to become an Olympic champion. Decors.: Order of Lenin and medals.

Bogdanov, Gleb Vasil'evich
Painter; b. 1926, Tambov.
Attended Kharkov Art Inst. 1950–56, pupil of S. I. Dudnik, L. I. Chernov; exhibitions 1956–; mem. CPSU 1962–.

Bogdanov, Mikhail Aleksandrovich
Film artist; b. N. 4 (17), 1914, Vyborg. Became People's Artist of the RSFSR in 1969 and Honored Artist of the RSFSR in 1959. Member of the CPSU since 1950. He graduated from the department of art of the All-Union State Institute of Cinematography (VGIK) in 1943. He was a set designer for the Soiuzdetfil'm Studio, and since 1944, for the Mosfil'm Studio. He works jointly with the artist G. A. Miasnikov. They were set designers for the following color films: *The Stone Flower* (1946), *Michurin* (1949), *The Heroes of Shipka* (1955, in collaboration with the Bulgarian artist G. Popov), *The Communist* (1958), *Journey Beyond the Three Seas* (1958, in collaboration with Indian artists), and *Hussar Ballad* (1962). One of the most important works by him and G. A. Miasnikov is the film *War and Peace*, based on the work of the same name by L. N. Tolstoi (1965–67, parts 1–4). He began teaching at VGIK in 1944 and became a professor in 1965. Decors.: Badge of Honor and medals.

Bogdanov, Nikolai Vladimirovich
Author; b. 1906, Kandoma, now Riazan Oblast; s. of a country doctor.
Publs.: First works publ. 1923, "Pervaia devushka" (The First Girl) 1928, "Vyzov" (Challenge) 1931, "Plenum druzei" (The Plenum of Friends) 1934, for children: "O smelykh i umelykh" (For the Brave and Bright) 1950, and others.

Bogdesko, Il'ia Trofimovich
Graphic artist; b. Ap. 20, 1923, Botushany near the city of Rybnitsa, Padol'sk Province, Ukrainian SSR.
People's Artist of the USSR (1963). Member of the CPSU since 1968.
He graduated from the I. E. Repin Institute of Painting, Sculpture, and Architecture in Leningrad in 1951 and has been working in the Moldavian SSR since 1953. He has created lyrical, tensely emotional illustrations for N. V. Gogol's *Sorochintsy Fair* (India ink, pen, and brush; 1951), A. S. Pushkin's *Gypsies* (black watercolor; 1956), and the Moldavian folk ballad *Miorítsa* (watercolor and tempera; 1966) and a series of color engravings, *My Homeland* (1961–63). Decors.: State Prize of the Moldavian SSR in 1966, Order of the Red Banner of Labor.

Bogoliubov, Klavdii Mikhailovich, Cand. of Hist. Sc.
Party official; b. 1909.
Mem. CPSU 1938–; grad. Herzen Pedagogical Inst. Leningrad 1939 and Party Higher School of CPSU Cen. Cttee. 1944; Cand. of Historical Sciences; teacher, head of teaching dept., head of

raion dept. for educ., dir. of an intermediate school in Vologda Oblast 1930–39; served in Soviet Army 1939–41; held Party posts in Vologda Oblast 1941–42; active in admin. of Cen. Cttee. CPSU 1944–63 and 1965–; Dep. Chair. of State Cttee. of USSR Council of Mins. for Printing and Publishing 1963–65; 1st Dep. Head of Gen. Dept. with Cen. Cttee. CPSU Je. 1968–; mem. CPSU Cen. Auditing Comm. 1971–81, mem. Cen. Cttee. CPSU 1981–.

Bogoliubov, Mikhail Nikolaevich
Philologist; b. Ja. 11 (24), 1918, Leningrad. Grad. Leningrad Univ. 1941; specialist in Iranian languages; corresponding mem. USSR Acad. of Sciences 1966–; Prof. at Leningrad Univ. 1959–.

Bogoliubov, Nikolai Ivanovich
Russian actor; b. O. 22, 1899, Ivanovskoe, now in Lipetsk Oblast.
People's Artist of the RSFSR (1945). Member of the CPSU since 1950.
He studied in the studio attached to the Riazan' Theater and then became an actor in that theater. In 1926 he graduated from one of the state experimental theatrical studios – the studio of the First Theater of the RSFSR (which subsequently became the Meierkhol'd Theater); he began working at the Meierkhol'd Theater in 1923. From 1938 to 1958 he was with the Moscow Art Theater. Heroic characters have been especially dear to his creative work. His art organically combines a high degree of emotion with simplicity, and his heroes are profoundly humane and courageous. His best stage roles include Chub, the leader of a partisan detachment in I. L. Sel'vinskii's *Army Commander 2,* and Master Sergeant Bushuev in Vs. Vishnevskii's *The Last and Determined* at the Meierkhol'd Theater; Rybakov in N. F. Pogodin's *The Kremlin Chimes;* General Murav'ev in B. F. Chirskov's *The Victors;* and Kondrat'ev in A. A. Kron's *Officer of the Fleet,* at the Moscow Art Theater. He has been acting in motion pictures since 1931. His most important achievement was the role of Shakhov in the film *A Great Citizen* (parts 1 and 2, 1938–39). His other film roles include Nikolai Kadkin in *The Outskirts,* Nikolai Mironovich in *The Peasants,* and Nikolai Letnikov in *The Seven Bold Ones.* Decors.: State Prize of the USSR in 1941, 1942, 1946, 1947, 1949, and 1950, two Orders of Lenin, the Badge of Honor, and various medals.

Bogoliubov, Nikolai Nikolaevich, Dr. of Math. Sc.
Scientist; b. S. 21, 1909, Nizhnii Novgorod, now Gorkii.
Mathematician, mechanic, physicist; mem. USSR Acad. of Sciences 1953–; mem. Presidium USSR Acad. of Sciences 1964–; Acad. Sec. of Mathematics Dept. at USSR Acad. of Sciences 1964–; mem. Acad. of Sciences Ukrainian SSR 1948–; mem. Presidium Acad. of Sciences Ukrainian SSR 1962–; Dr. of Math. Sc. 1930; Prof. 1936–; mem. several foreign acads. of sciences and scientific institutions; active in Krylov Seminar 1923; mem. staff Acad. of Sciences Ukrainian SSR 1928–; Prof. and Prof. Chair for Mathematical Physics at Kiev Univ. 1936–50; Dean of Mechanical-Mathematical Faculty of Kiev Univ. 1946–49; Dept. Head at Steklov Inst. of Mathematics of USSR Acad. of Sciences 1949–56; Laboratory Head of Joint Inst. for Nuclear Research Dubno 1956–65; Dir. of same Inst. 1965–; Dir. of Inst. of Theoretical Physics at Acad. of Sciences Ukrainian SSR 1965–; scientific work comprises several fields of mathematical analysis, function theory, oscillation theory, stability theory and static physics amongst others. Decors.: 3 Orders of Lenin, Order of Red Banner of Labour, Badge of Honour, medals.
Publs.: "Novi metody v variatsiinomu chyslenni" 1932 "Vvedenie v nelineinuiu mekhaniku" 1937, "O nekotorykh statisticheskikh metodakh v matematicheskoi fizike" 1945, "Problemy dinamicheskoi teorii v statisticheskoi fizike" 1946, "Lektsii z kvantovoii statystyky" 1949, "Novyi metod v teorii sverkhprovodimosti" 1958.

Bogomiakov, Gennadii Pavlovich
Party official; b. 1930.
Mem. CPSU 1959–; grad. Tomsk Polytechnical Inst. 1952, completed postgraduate work there 1957; Cand. of Geological and Mineralogical Sciences; engineer with a hydrogeological expedition in Kuznets Basin 1952–53; posts in geological insts. in Novosibirsk 1957–60; dir. of a branch of Siberian Scientific Research Inst. of Geology, Geophysics, and Mineral Raw Materials 1960–64; Dep. Dir. of West Siberian Scientific Research Inst. of Geological Surveying 1964–67; party posts 1967–69; 1st Sec. of Tiumen City Cttee. CPSU 1969, 2nd Sec. 1969–73; 1st Sec. of Tiumen Oblast Cttee. CPSU 1973–; mem. Cen. Cttee. CPSU 1976–; Dep. to USSR Supr. Sov. 1974–.

Bogomolets, Oleg Aleksandrovich, Dr. of Med. Sc.
Pathophysiologist; b. F. 26, 1911, Odessa; s. of A. Bogomolets.
Mem. CPSU 1944–; grad. Medical Inst. Kiev 1934; corresponding mem. Acad. of Sciences Ukrainian SSR; Dr. of Med. Sc. 1944; Prof. 1944–; cand. at Inst. for Experimental Biology and Pathology in Min. of Health Ukrainian SSR 1934–38; Asst. at Surgery Clinic of Inst. of Orthopaedics and Traumatology of Min. of Health Ukrainian SSR 1935–39; doctoral cand. 1938–41, Sen. Research Asst. at Inst. of Clinical Physiology at Acad. of Sciences Ukrainian SSR 1941–52; in addition Sen. Research Asst. 1942–, Laboratory Chief 1943–, Dir. of Inst. for Experimental Biology and Pathology of Min. of Health Ukrainian SSR 1946–53; in addition Prof. Chair for Gen. Surgery at Kiev Medical Inst. 1944–46; Scientific Dir. of Dept. of Pathophysiology at Ukrainian Scientific Sanitary-Chemical Research Inst. of Min. of Health Ukrainian SSR 1953– (Kiev Scientific Research Inst. for Pharmacology and Toxicology 1965–); in addition Prof. Chair for Pathophysiology at Kiev Inst. for Continuing Educ. of Doctors 1955–65; author of more than 60 scientific works. Decors.: Order of Red Banner of Labour, 2 Badges of Honour, medals.
Publs.: "Vliianie antiretikuliarnoi tsitotoksicheskoi syrovatki na zazhivlenie perelomov" 1944, "K voprosu o mekhanizme deistviia perelivaniia krovi" 1950, "Nekotorye rezultaty eksperimentalnoi terapii luchevoi bolezni" in "Meditsinskaia radiologiia" 1960.

Bogomolov, A. I.
Komsomol official.
2nd Sec. and mem. Bureau Cen. Cttee. Latvian Komsomol Mr. 1982–.

Bogomolov, Aleksei Fedorovich
Scientist; b. My. 20 (Je. 2), 1913, Sitskoe, now Smolensk Oblast.
Mem. CPSU 1944–; grad. Energy Inst. Moscow 1937; radiation technologist; corresponding mem. USSR Acad. of Sciences 1966–; Prof. at Energy Inst. Moscow 1958–; important works on radiation technology and radiation physics, the construction of large-scale radio telescopes for the reception of satellite transmissions. Decors.: Hero of Socialist Labour 1957, 3 Orders of Lenin, 2 other orders and several medals.

Bogomolov, Oleg Timofeevich
Economist; mem. USSR Acad. of Sciences 1981–.
Dir. of Inst. of Economics of World Socialist Systems USSR Acad. of Sciences; specialist on problems of economic devt. of socialism and the socialist world system; author of roughly 200 works; Chair. of comm. on the problems of multilateral economic cooperation of the Acads. of Sciences of the socialist countries "The Socialist World System"; mem. Exec. Cttee. of the International Assoc. of Economists; Chief editor of the periodical *Izvestiya AN SSSR. Seriya ekonomicheskaya* (Proceedings of the USSR Acad. of Sciences: Economic Series); mem. editorial advisory board of the periodical *Gesellschaftswissenschaften* (Social Sciences).

Bogomolov, Sergei Aleksandrovich
Diplomat; b. 1926.
Mem. Diplomatic Service 1948–; held posts at USSR embassies in the USA and France, was Sec.-Gen. of USSR Del. at Geneva Disarmament Conference; USSR Amb. to Spain Ap. 1977–78; Chief Admin. for General International Problems, USSR Min. of Foreign Affairs 1978–.

Bogorodskii, Vitalii Vasil'evich, Dr. of Phy.-Math. Sc.
Scientist; b. 1919.
Mem. CPSU 1942–; grad. Electrotechnical Inst. Leningrad 1952; oceanographer; Dr. of Phy.-Math. Sc. 1956; Prof. 1967; corresponding mem. USSR Acad. of Sciences 1970–; served in Sov. Army 1939–46; junior research scholar 1952–55; mem. staff Arctic and Antarctic Research Inst. 1955–; in addition Prof. at Electrotechnical Inst. Leningrad 1967–.

Bogorov, Veniamin Grigor'evich
Scientist; b. D. 11 (24), 1904, Moscow.
Grad. Biological Faculty of Moscow Univ. 1926; oceanographer, hydrobiologist; corresponding mem. USSR Acad. of Sciences 1958–; Dep. Dir. 1941–61, Chief of Plankton Laboratory at Inst. of Oceanography at USSR Acad. of Sciences 1961–; in addition Prof. at Geographical Faculty 1946–, Prof. Chair for Hydrobiology at Moscow Univ. 1964–; organized several expeditions in the Pacific and the Indian Ocean aboard "Vitjaz". Decors.: State Prize of the USSR 1951, Order of Lenin, three other orders and several medals.

Bogush, Lev Konstantinovich
Medical scientist; b. F. 18 (Mr. 3), 1905, Moscow.
Grad. Medical Faculty of Nizhnii Novgorod Univ. 1928; phthisist; mem. USSR Acad. of Medical Sciences 1963–; Prof. at Cen. Inst. for Continuing Education of Doctors and Chief of Surgery Clinic with Cen. Tuberculosis Inst. at USSR Acad. of Medical Sciences 1951–. Decors.: Lenin Prize 1961, three orders, several medals.

Boiarskii, Aron Iakovlevich
Economist and statistician; b. O. 2 (15), 1906, Liepāja.
Demographer and professor (1934). Doctor of economics (1940).
Honored Scientist of the RSFSR (1967). Member of the CPSU since 1931.
He graduated from the Statistical Division of the department of social sciences of Moscow State University (1926). He is director of the Scientific Research Institute of the Central Statistical Administration of the USSR and head of the subdepartment of statistics in the department of economics at Moscow State University. He is a member of the International Statistical Institute (1966).
His principal works on mathematical statistics, demography, and the use of mathematics in economic research are *The Theory of Mathematical Statistics* (1930, with V. N. Starovskii, V. I. Khotimskii, and B. S. Iastremskii); *A Course in Demographic Statistics* (1945); *Demographic Statistics* (1951), with P. P. Shusherin); and *Mathematical Economic Essays* (1962). Decors.: Order of the Red Banner of Labor, medals.

Boiko, Aleksei Sergeevich
State official; b. 1925.
Russian; mem. CPSU 1946–; grad. Chechen-Ingush Pedagogical Inst.; Maj.-Gen.; operator in a plant 1942–43; held posts in the State Security apparatus 1943–63; Dep. Chair. of State Security Cttee. Chechen-Ingush ASSR; held posts on USSR State Security Cttee. 1978–79; Chair. of State Security Cttee. Turkmen SSR 1979–; mem. Cen. Cttee. and mem. Bureau of Cen. Cttee. CP Turkmenistan 1981–.

Boiko, Anatolii Grigor'evich
State official.
Chair. of State Cttee. for the Supply of Production Equipment for Agriculture Kirghiz SSR Ag. 1978–; cand. mem. Cen. Cttee. Kirghiz CP 1981–.

Boiko, Viktor Grigor'evich
Oblast sov. official; b. 1931.
Ukrainian; mem. CPSU 1954–; grad. Krivoi Rog Metallurgical Inst.; served in Sov. Army 1951–54; work superintendent in "Krivorozhaglostroi" Trust 1960–61; instructor, dept. chief of Krivoi Rog City Cttee. Ukrainian CP 1961–64; dep. chief of a dept. of Dnepropetrovsk Oblast Cttee. Ukrainian CP 1966–70; Chair. of Exec. Cttee. of Dnepropetrovsk City Sov. of Peoples' Deps. 1970–74; 1st Sec. of Dnepropetrovsk City Cttee. Ukrainian CP 1974–76; mem. Cen. Cttee. Ukrainian CP 1976–; 2nd Sec. of Dnepropetrovsk Oblast Ukrainian CP 1976–78; Chair. of Exec. Cttee. of Dnepropetrovsk Oblast Sov. of Peoples' Deps. D. 1978–.

Boiko, Vladimir Eliseevich
Sculptor; b. 1925, Erkovtsy, Kiev Oblast. Attended Lemberg Inst. for Applied and Decorative Art 1948–54, pupil of I. V. Sever; mem. CPSU 1954–.

Boitsov, Vasilii Vasil'evich
State official; b. 1908.
Mem. CPSU 1939–; grad. N. E. Bauman Higher Technical School in Moscow 1937, Prof. 1955, Dr. of Tech. Sc. 1960; engineer, shop superintendent, production chief, chief engineer, dir. of various plants 1937–47; Dep. USSR Min. of Aircraft Industry, dir. of a scientic research inst. 1947–63; Chair. USSR State Cttee. for Standards 1963–; Dep. to USSR Supr. Sov. 1967; Cand. mem. Cen. Cttee. CPSU 1976–. Decor.: Order of Lenin.

Bokii, Georgii Borisovich
Scientist; b. S. 26 (O. 9), 1909, Leningrad. Mem. CPSU 1944–; grad. Mining Inst. Leningrad 1930; crystallographer, chemist; corresponding mem. USSR Acad. of Sciences 1958–; Prof. at Moscow Univ. 1944–63; mem. staff Inst. of Radioelectronics at USSR Acad. of Sciences 1963–.

Bokov, Khazhbikar Khakiashevich
State official.
Chair. of Presidium Supr. Sov. Chechen-Ingush ASSR; Dep. to and Dep. Chair. of Presidium Supr. Sov. RSFSR 1975–.

Bokov, Viktor Fedorovich
Poet; b. 1914, Iazvitsy, now Moscow Oblast; s. of a peasant.
Grad. Gorkii Inst. of Literature 1938; worked as turner and zoo technician; first works publ. 1934.
Publs.: poems "Iar-Khmel" (Khmel-Slope) 1958, "Nad rechkoi Istermoi, Zapiski poeta" (On the Insterma River. Sketches of the Poet) 1960, and others.

Bolbat, Ivan Semenovich
State official.
Min. for Building and Use of Motorways Moldavian SSR S. 1972–; cand. mem. Cen. Cttee. Moldavian CP 1976–81, mem. 1981–.

Bolduman, Mikhail Panteleimonovich
Russian actor; b. Jl. 11 (23), 1898, Zhmerinka.
People's Artist of the USSR (1965).
He began his stage career in 1923. From 1924 to 1931 he worked in the Kiev Theater of Russian Drama, and from 1931 to 1933 he worked in the Moscow Dramatic Theater (formerly the Korsh Theater). Since 1933 he has been an actor in the Moscow Art Academic Theater.
He created a series of images of positive soviet heroes: Platon Krechet (in Korneichuk's play of that name), Maiorov (in Kron's *Deep Reconaissance*), and Gorbunov (in Kron's *Officer of the Fleet*). His roles in the classic reportoire included Sintsov (in Gorky's *The Enemies*), Boris Godunov (in A. K. Tolstoy's *Tsar Fedor Ioannovich*), Vershinin (in Chekhov's *The Three Sisters*), Shamraev (in Chekhov's *The Seagull*), and King Leontes (in Shakespeare's *The Winter's Tale*). Decors.: State Prize of the USSR twice in 1946 and once in 1950, the Order of Lenin, the Order of the Badge of Honor, and various medals.

Boldyrev, Vil Konstantinovich
Diplomat; b. 1924.
Held responsible posts in USSR Min. of Foreign Affairs apparatus and abroad 1956–69; embassy councillor and envoy to Sov. embassy in India 1969–75; dep. chief of South Asia Dept. 1975–78 and chief of Middle East Dept. USSR Min. of Foreign Affairs 1978–82; Sov. amb. to Iran Je. 1982–.

Bolotin, Vladimir Vasilevich, Dr. of Tech. Sc.
Scientist; b. 1926.
Mem. CPSU 1954–; grad. Inst. of Railway Transport Engineering Moscow 1948; mechanic; Dr. of Tech. Sc. 1952; Prof. 1954–; corresponding mem. USSR Acad. of Sciences 1974–; aspirant 1948–58, Asst. Lecturer, Prof. Chair at Energy Inst. Moscow 1958–.

Bolovinov, Konstantin Alekseevich
State official; b. 1911.
Chair. RSFSR State Cttee. for Material and Technical Supply Jl. 1979–. Decor.: Order of October Revolution.

Bol'shakov, Kirill Andreevich
Inorganic chemist; b. D.11 (24), 1906, Riazhsk, now in Riazan' Oblast.
Corresponding member of the Academy of Sciences of the USSR (1958). Specialist in the chemistry and technology of rare trace elements and of the minor metals. One of the founders and organizers of the Russian rare element industry.
He graduated from the University of Kazan. From 1930 to 1947 he directed a laboratory at the State Scientific Research Institute for Rare Metals. In his teaching activity he has been closely associated (since 1933) with the M.V. Lomonosov Moscow Institute of Fine Chemical Technology where he successively occupied the posts of lecturer, assistant professor, professor, head of the subdepartment of the chemistry and technology of rare elements, and head of the laboratory founded by him in the same field; since 1958 he has been rector of the institute. His major research is devoted to the study of the physical-chemical bases of technological processes. He is the author of many scientific works and has many inventions to his credit. He edited the textbook for institutions of higher learning entitled *The Chemistry and Technology of Rare and Trace Elements* (vols. 1–2, 1965–69). Decors.: State Prize of the USSR (1941, 1953), two Orders of Lenin and three other orders.

Bolshev, Login Nikolaevich, Dr. of Phy.-Math. Sc.
Mathematician; b. 1922.
Grad. Moscow Univ. 1951; Dr. of Phy.-Math. Sc. 1966; corresponding mem. USSR Acad. of Sciences 1974–; served in Sov. Army 1940–46; postgraduate studies 1951–54, Asst., Sen. Research Asst. 1955–59, Prof. at Moscow Univ. 1969–; Jun. Research Asst. 1951–60, Dept. Chief at Inst. of Mathematics at USSR Acad. of Sciences 1964–.

Bondaletov, Ivan Vasil'evich
State official.
Dep. Chair. of Bureau for Ind. at Cen. Cttee. Latvian CP 1962–64; Dep. to Supr. Sov. Latvian SSR 1963–; mem. Cen. Cttee. Latvian CP 1963–; Dep. Chair. of Council of Mins. Latvian SSR Ja. 1964–.

Bondar, Aleksei Georgievich
Jurist; b. 1908, Kostiashi, Minsk Oblast. Mem. CPSU 1930–; grad. Minsk Law Inst. 1935; until 1929 worked for Uzda Timber Ind. Board, served in Sov. Army; Asst. 1935–39, then Public Prosecutor in Minsk; Public Prosecutor in Minsk Oblast 1939–41; during World War II organized underground Communist and partisan activities behind German lines 1941–44; Chair. of Belorussian Supr. Court 1944–45; Sec. of Presidium Belorussian Supr. Sov. 1945–49; Belorussian Public Prosecutor 1949–67; mem. Cen. Cttee. Belorussian CP 1956–71; Chair. of Belorussian Supr. Court Ag. 1971–; mem. Auditing Comm. Belorussian CP 1971–. Decors.: Order of Lenin, Order of Red Star, Order of Red Banner, Order of Red Banner of Labour, medals.

Bondarchuk, Vladimir Gavrilovich, Dr. of Geol.-Min. Sc.
Geologist; b. Jl. 29, 1905, Denishi, now Zhitomir Oblast; s. of a peasant.
Mem. CPSU 1945–; grad. Inst. of Public Educ. Volhynia 1926; mem. Acad. of Sciences Ukrainian SSR 1951–; Dr. of Geol.-Min. Sc. 1941; Prof. 1945–; postgraduate studies at Inst. of Geological Sciences at Acad. of Sciences Ukrainian SSR 1926–30; Lecturer at Inst. of Mining and Geology Kiev; Chief Geologist at Dept. of Geological Photography in Ukrainian Geological Admin. 1926–38; in addition mem. staff Kiev Univ. 1935–38; Prof. Chair for Geology at Kiev Univ. 1938–41; Chief of Dept. of Ferrous Metals at Inst. of Geology Acad. of Sciences Uzbek SSR 1941–44; Rector of Kiev Univ. 1944–51; Dep. Chair. of Council of Mins. Ukrainian SSR 1951–53; Dir. of Inst. of Geology at Acad. of Sciences Ukrainian SSR 1953–63; has written more than 175 scientific works on general geology, geomorphology and geotechnics. Decors.: Order of Red Banner of Labour, medals.
Publs.: "Geologichna struktura URSR" 1946, "Tektoorogeniia" 1946, "Geologichna budova URSR" 1947, "Kurs zagalnoi geologii" 1947, "Geomorphologiia URSR" 1949, "Osnovy geomorphologii" 1949, "Radjanski Karpaty" 1957, "Geologiia Ukrainy" 1959, "Tektonika Karpat" 1962, "Geologichna budova URSR" 1963, "Geologiia rodovyshch korysnykh kopalyn Ukrainy" 1966.

Bondarenko, Aleksandr Pavlovich
Diplomat; b. 1922.
Held diplomatic posts 1949–; senior posts in USSR Min. of Foreign Affairs 1962–67; Emb. Counsellor-Envoy at USSR Emb. to the FRG 1967; Head of 3rd European Dept. in USSR Min. of Foreign Affairs 1975–, mem. Collegium of USSR Min. of Foreign Affairs.

Bondarenko, Ivan Afanasevich
Party official; b. 1926, Ukraine.
Mem. CPSU 1950–; grad. Azovo-Chernomorskii Agricultural Inst. 1956; bookkeeper in a kolkhoz 1942–43; served in Sov. Army 1943–45; bookkeeper of a kolkhoz brigade, attended agricultural tech. college, section agronomist of a machine and tractor station, chief agronomist of an erosion protection centre, dir. of an experimental teaching farm, Sec. of Party Cttee., Asst. of Prof. at Azovo-Chernomorskii Agricultural Inst. 1945–59; 1st Sec. of Raion Cttee. CPSU, Dept. Head and later Sec. of Rostov Oblast Cttee. CPSU 1959–64; Chair. of Rostov Oblast Sov. 1964–66; Dep. to USSR Supr. Sov. 1966–; 1st Sec. of Rostov Oblast Cttee. CPSU N. 1966–; mem. Cen. Cttee. CPSU 1971–.

Bondarenko, Mikhail Vasil'evich
State official.
Sec. and mem. bureau of Cen. Cttee. Moldavian Komsomol 1961–63; cand. mem. 1966–68, mem. Cen. Cttee. Moldavian CP N. 1968–; Chair. of Moldavian Republican Union for Agricultural Technology "Moldselkhoztekhnika" Mr. 1973–78; Chair. of State Cttee. for the Supply of Production Equipment for Agriculture Moldavian SSR S. 1978–.

Bondarev, Iurii Vasil'evich
Author and state official; b. 1924, Orsk. Russian; mem. CPSU 1944–; grad. Gorkii Inst. of Literature Moscow 1951; Sec. of USSR Writers' Union; 1st Dep. Chair. of RSFSR Writers' Union; served in Sov. Army 1941–45; Dep. to and Dep. Chair. of Supr. Sov. RSFSR 1975–80; first works publ. 1949.
Publs.: "Na bolshoi reke" (On the Big River) 1953; "Iunost komandirov" (Young Commanders) 1956, "Bataliony prosiat ognia" (Fire for the Battalions) 1957, "Poslednie zalpy" (Last Salute) 1959, and others.

Bondarin, Sergei Aleksandrovich
Russian writer; b. Ja. 14 (27), 1903, Odessa.
He graduated from the Institute of the National Economy in 1929. His works first appeared in print in 1922. He has written much about children and the forming of character in youth – for example, in *Dyndyp From Durgun-Khotok Valley* (1931), a novella of the childhood of a Mongolian Pioneer, and in the autobiographical book *A Novella for My Son* (1935). He was a participant in the Great Patriotic War and dealt with the life of sailors in combat in the novellas *We Shall Make a Landing* (1955) and *Vania Zolushkin* (1957) and in short stories. He wrote a book of reminiscences about writers entitled *A Grape Cluster* (1964).
Works: *5 let.* Moscow 1935. *Liricheskie rasskazy.* Moscow 1957. *Volny Dunaia: Povesti i rasskazy.* Moscow 1961. *Povest' dlia syna: Povesti vospominaniia, rasskazy.* Moscow 1967.

Bongard-Levin, G.
Orientalist.
Corresponding mem. USSR Acad. of Sciences 1981–. Sector head at the Inst. of Oriental Studies USSR Acad. of Sciences; author of more than 100 publications on the history and culture of India, Central and South Asia, the ethnogeny of the peoples of India, the origin of the Indian civilization and the history of Buddhism; Vice-pres. International As-

soc. for Sanskrit Studies; honorary mem. Indian Archaeological Society.

Boreskov, Georgii Konstantinovich
Physical chemist; b. Ap. 7 (20), 1907, Omsk.
Grad. Odessa Polytech. Inst.; mem. USSR Acad. of Sciences 1966–; Dir. of Inst. of Catalyses at Siberian Branch of USSR Acad. of Sciences. Decors.: Hero of Socialist Labour, State Prize of the USSR 1942 and 1953, Order of Lenin, 2 other orders and various medals.

Borisenko, Vasilii Vasil'evich
Literary historian; b. 1904, Novo-Borisov, Belorussia; s. of a worker.
Mem. CPSU 1928–; grad. Belorussian State Univ. 1929; literary historian, specialist in Russian literature; corresponding mem. 1957–69, mem. Acad. of Sciences Belorussian SSR 1969–; Dir. of Inst. for Literature at Acad. of Sciences Belorussian SSR 1937–41, 1946–50 and 1952–. Decors.: 4 orders, medals.
Publ.: monograph "Frantiszek Boguszewicz and the Problem of Realism in Belorussian Literature in the XIX Century" 1957.

Borisenkov, Vasilii Mikhailovich
Party official; b. 1927.
Russian; mem. CPSU 1946–; grad. Moscow State Univ. 1950; Cand. of Hist. Sc. 1965; railroad laborer in Briansk Oblast 1941–42; partisan 1942–44; served in Sov. Army 1944–45; Komsomol posts in Briansk Oblast, posts in USSR Komsomol Cen. Cttee. apparatus 1945–59; posts in Moscow Oblast Cttee. CPSU and in Cen. Cttee. apparatus CPSU 1959–72; Sec. of Moscow Oblast Cttee. CPSU 1972–74, 2nd Sec. 1974–; cand. mem. Cen. Cttee. CPSU 1976–81, mem. 1981–; Dep. to USSR Supr. Sov. 1981–.

Borisevich, Nikolai Aleksandrovich, Dr. of Phy.-Math. Sc.
Scientist; b. S. 21, 1923, Luchny nr. Minsk, Belorussia.
Mem. CPSU 1945–; grad. Belorussian Lenin Univ. Prof.; partisan, served in Sov. Army 1941–45; student, postgraduate studies, scientific asst. at an inst. 1945–55; Dep. Dir. of Inst. for Physics at Acad. of Sciences Belorussian SSR 1955–69; mem. 1966–, Pres. of Acad. of Sciences Belorussian SSR My. 1969–; Dep. 1969–, mem. Comm. for Educ. Science and Culture at Sov. of the Union USSR Supr. Sov. 1970–; mem. Cen. Cttee. Belorussian CP 1971–. Decors.: Order of the October Revolution and other orders and medals.

Boriskovskii, Pavel Iosifovich, Dr. of Hist. Sc.
Archaeologist; b. My. 14 (27), 1911, St. Petersburg.
Mem. CPSU 1941–; specialist in archaeology of the Palaeolithic age; Dr. of Hist. Sc. 1952; Prof. at Leningrad Univ. 1959–. Decors.: 2 orders and medals; decorations from various foreign countries.
Publs.: "Ljudyna kam'janogo viku na Ukraini" Kiev 1940, "Paleolit Ukrainy" Moscow-Leningrad 1953, "Drevnejšee prošloe čeolovečestva" Moscow-Leningrad 1957, "Očerki po paleolitu bassejna Dona" Moscow-Leningrad 1963, "Pervobytnoe prošloe V'etnama" Moscow-Leningrad 1966.

Borisov, Aleksandr Fedorovich
Russian actor; b. Ap. 18 (My. 1), 1905, St. Petersburg.
Became People's Artist of the USSR in 1951.
In 1927, he graduated from the studio of the Leningrad Academic Theater of Drama (as a student of Iu. M. Iur'ev) and was accepted first into the theater's studio, and in 1928, into the theater's company (now the Leningrad Pushkin Theater of Drama). His first major role was Boris Volgin (in Afinogenov's *The Eccentric*). Deep charm, emotional purity, and the modesty of everyday heroism distinguished his heroes – Pavel Korchagin (*How the Steel Was Tempered,* after N. Ostrovskii's novel, 1947) and the fighting man Stepan (Chirskov's *The Victors,* 1946). He later played the roles of Semen Kotko (*A Soldier Was Returning From the Front* by Kataev), the Pretender (Pushkin's *Boris Godunov*), Meluzov, Gavrila, and Arkashka Schastlivtsev (in A. Ostrovskii's *Talents and Admirers, A Passionate Heart,* and *The Forest,* respectively), Rekalo (*For Those at Sea* by Lavrenev), Kisel'nikov (A. Ostrovskii's *The Abyss*), Krivokhatskii (Gorbatov's *One Night*), and Kuzovkin (Turgenev's *The Parasite*). An actor with great dramatic strength, he also plays comic roles and creates brilliant character parts. Since 1937 he has appeared in movies – for example, as Pavlov *(Academician Ivan Pavlov)* and the composer Mussorgsky *(Mussorgsky).* Decors.: State Prize of the USSR in 1947 and 1950, and twice in 1951, two orders and medals.
Work: *Iz tvorcheskogo opyta.* Moscow, 1954.

Borisova, Iuliia Konstantinovna
Russian actress; b. Mr. 17, 1925, Moscow.
Became People's Artist of the USSR in 1969.
In 1949, she graduated from the Shchukin Theatrical Institute and entered the company of the Vakhtangov Theater. The actress's life-asserting art combines a brilliant comical temperament with intense dramatism and penetrating, gentle lyricism. Her repertoire includes the contemporary roles Pavlina (*The Cook,* and *The Cook Married* by Sofronov), Varia (*Alone* by Aleshin), and Val'ka (*An Irkutsk Story* by Arbuzov). She has played Lika Mizinova (*Luck Laughs at Me* by Maliugin), Helen (*Warsaw Melody* by Zorin), Princess Turandot (Gozzi's play of that name), Virineia (the play of that name by Seifullina and Pravdukhin), and others. She played Nastas'ia Filippovna in the play *The Idiot* and in the film of that title (from F. M. Dostoevsky's novel). In 1970 she played Kol'tsova in the film *Ambassador of the Soviet Union.*
She was deputy of the Supreme Soviet of the RSFSR at its sixth convocation. Decors.: State Prize of the RSFSR in 1966, Order of the Red Banner of Labor and a medal.

Borisovskii, Vladimir Zakharovich
State official; b. 1933.
Mem. CPSU 1961–, electromechanical engineer; foreman, shop superintendent, chief engineer in an electrical installation admin. of Min. of Installation and Special Construction Work Ukrainian SSR 1956–63; party posts 1963–77; Dep. Chair. of Exec. Cttee. of Dnepropetrovsk Oblast Sov. of Peoples' Deps. 1977–79; 1st Dep. Min. of Industrial Construction Ukrainian SSR 1979–81; Min. of Installation and Special Construction Work Ukrainian SSR Mr. 1981–; cand. mem. Cen. Cttee. Ukrainian CP 1981–.

Borkovskii, Viktor Ivanovich
Linguist; b. Ja. 6 (18), 1900, Minsk.
Active as linguist and Russian philologist; mem. USSR Acad. of Sciences 1972–; foreign mem. German Acad. of Sciences Berlin, mem. Int. Cttee. on Slavic Studies. Decor.: State Prize of the USSR 1970.

Borodai, Vasilii Zakharovich
Sculptor; b. 1971, Ekaterinoslav.
Mem. CPSU 1945–; sculptor and painter; Dep. Chair. 1962–63, Sec. 1963–68, Chair. of Painters' Union of Ukrainian SSR 1968–73. Well-known works: sculpture "Ivan Bogun" 1954, composition "Zemle moia" (My Country) 1957, portraits of P. Tychyna and L. Revutskii 1963; series of sculptures "Po Egyptu" (Through Egypt) 1961, and others.

Borodin, Leonid Aleksandrovich
Party official; b. 1923.
Russian; mem. CPSU 1948–; grad. Volgograd Agricultural Inst. 1955; served in Sov. Army 1941–46; Prod. Man. of Board "Ogneuporlom" in Volgograd 1946–49; doctor/epizootologist of Macheshanskii Raion Dept. for Agriculture, Instructor, Dept. of Macheshanskii Raion Cttee. CPSU, Instructor, Dep. Head of CPSU Oblast Cttee. in Volgograd, 1st Sec. of Novoanninskii Raion Cttee. CPSU in Volgograd 1949–59; Sec. of Chuvash Oblast Cttee. CPSU 1959–62; Inspector of Agricultural Dept. at CPSU Cen. Cttee. for RSFSR 1962–63; 2nd Sec. of Bashkir Oblast Cttee. CPSU 1963–67; Dep. to USSR Supr. Sov. 1966–; 1st Sec. of Astrakhan Oblast Cttee. CPSU 1967–; cand. mem. Cen. Cttee. CPSU 1971–79, mem. 1979–. Decor.: Order of Lenin.

Borodin, Pavel Dimitrievich
State official; b. 1911.
Mem. CPSU 1932–; grad. Inst. of Roadbuilding Moscow 1939; party work at Moscow Inst. of Roadbuilding 1938–39; active with Cen. Cttee. All-Union CP (B) 1939–47; USSR Min. of Automobile and Tractor Ind. 1947–50; leading econ. posts 1950–54; staff mem. USSR Min. of Automobile, Tractor and Agricultural Machine Building 1954–55; USSR Dep. Min. of Tractor and Agricultural Machine Building 1955–57; 1st Dep. Chair. of Moscow City Econ. Council 1957–59; Dep. Chair. of GOSPLAN's State Cttee. for Automation and Machine Building 1959–63; Dir. of Likhachev Automobile Plant in Moscow 1963–71; mem. Cen. Cttee. CPSU 1966–; Dir. Gen. of Moscow Asscn. for Automobile Production Jl. 1971–. Decor.: Hero of Socialist Labour 1966.

Borodulin, Grigorii Ivanovich
Poet; b. 1935, Gorodok village, Vitebsk Oblast.
Belorussian; grad. Minsk Univ.; first work publ. 1954.
Publs.: poems "Mesiats nad stepiu" (Moon Over the Steppe) 1959, "Runets, krasavats, nalivatstsa" (Mystify, Beautify, Fill Up) 1961, and others.

Borovik-Romanov, Viktor-Andrei Stanislavovich
Physicist and astronomer; b. Mr. 18, 1920, Petrograd.
Grad. Moscow State Univ. 1947; mem. USSR Acad. of Sciences 1972–; mem. staff Inst. of Measures and Measuring Instruments Moscow 1948–55; mem. staff Inst. of Problems in Physics 1956–.

Borovikov, Grigorii Fedorovich
Author; b. 1905, Beloglazovo, now Kirov Oblast.
Mem. CPSU 1945–.
Publs.: children's books "V Kaspiiskikh dzhungliakh" (In the Caspian Jungle) 1946, and others.

Borovkov, Aleksandr Alekseevich
Mathematician; b. Mr. 6, 1931, Moscow.
Grad. Moscow Univ. 1964; corresponding mem. USSR Acad. of Sciences 1966–; Prof. at Novosibirsk Univ. 1965–.

Borshchagovskii, Aleksandr Mikhailovich
Author; b. 1913, Belaia Tserkov.
Mem. CPSU 1940–; Russian author, critic, dramatist; worked as author 1933–.
Publs.: "A. M. Buchma" 1947, "Dramaturgiia Tobilevicha" (Tobilevich's Drama) 1948, novel "Russkii flag" (Russian Flag) 1953, "Ostrov vsekh nadezhd" (The Island of all Hopes) 1960, and others.

Borshosh-Kumiatskii, Iulii Vasil'evich
Poet; b. 1905, Bol'shye Komiaty, Transcarpathians; s. of a peasant.
Ukrainian; grad. Uzhgorod Teaching Seminar.
Publs.: poems for children "Vesniany kvity" (Spring Flowers) 1928, anthology "Z mogo Kraiu" (From My Country) 1929, "Kraina dyv" (A Land of Miracles) 1934, "Dvi doli" (Two Fates) 1948, and others.

Borzov, V. F.
Komsomol official.
Sec. and mem. Bureau Cen. Cttee. Ukrainian Komsomol F. 1980–.

Bosenko, Nikolai Vasil'evich
State official; b. 1918.
2nd Sec. 1961–62, 1st Sec. 1963–64 and 2nd Sec. of Stavropol Krai CPSU Cttee. 1964–73 (for Ind. 1963–64); Dep. 1963, mem. Comm. for Ind., Building and Transport 1963–71, mem. Presidium Supr. of Sov. RSFSR 1971–73; Chair. All-Russian Union for Agricultural Technology "Rosselkhoztekhnika" F. 1973–78; Chair. of Ukrainian SSR State Cttee. for the Supply of Production Equipment for Agriculture S. 1978–. Decor.: Order of Red Banner of Labour.

Bostorin, Oleg Vladimirovich
Diplomat; b. 1931.
USSR Amb. to Kampuchea My. 1979–.

Botsu, Pavel Petrovich
Author and poet; b. Je. 14, 1933, Priozernoe, Izmail Raion, Odessa Oblast.
Moldavian; mem. CPSU 1961–; grad. I. Kriange Pedagogic Inst. Kishinev 1956; Chair. of Moldavian Writers' Union 1965–; Sec. of USSR Writers' Union 1967–; Dep. to USSR Supr. Sov. 1966–70; Dep. to 1971, mem. Presidium 1971–80, Chair. of Supr. Sov. Moldavian SSR Ap. 1980–. Decor.: 1 order.
Publs.: "Zemlia ottsov" (Land of the Fathers) 1959, "Materiki" (Continents) 1967, "Kruglyi stol zemli" (The Round Table of the Earth) 1967, and others.

Botvin, Aleksandr Platonovich
Diplomat; b. 1918, Ukraine.
Mem. CPSU 1943–; grad. Kharkov Aviation Inst. 1941; foreman, dept. head, chief metallurgist, CPSU Cen. Cttee. party organizer in a plant in Khabarovsk 1941–55; 1st Sec. of Raion Cttee. Ukrainian CP in Kharkov, Inspector of Cen. Cttee. Ukrainian CP, 2nd Sec. of Kiev Oblast Cttee. Ukrainian CP 1955–62; 1st Sec. of Kiev City Cttee. Ukrainian CP 1962–63; Dep. to Supr. Sov. Ukrainian SSR 1963–; 2nd Sec. of Kiev Oblast Cttee. Ukrainian CP 1963–65; mem. Cen. Cttee. Ukrainian CP 1965–81; 1st Sec. of Kiev City Cttee. Ukrainian CP Mr. 1965–80; mem. CPSU Cen. Auditing Comm. 1966–71; Dep. to USSR Supr. Sov. 1966–; cand. mem. CPSU Cen. Cttee. 1971–76, mem. 1976–; mem. Politbureau Cen. Cttee. Ukrainian CP 1976–80; USSR Amb. to Czechoslovakia F. 1980–.

Bovin, Aleksandr Evgenevich
Journalist; b. 1930.
Mem. CPSU 1952–; grad. Rostov Univ. 1953; Cand. of Philos. Sc. 1969; party work in Krasnodar Krai 1954–56; served on editorial staff of "Kommunist" 1959–63; held posts in Cen. Cttee. apparatus CPSU 1963–72; polit. observer for "Izvestiia" Ap. 1972–; mem. Cen. Auditing Comm. CPSU 1981–.

Botvinnik, Mikhail Moiseevich
Chess player; b. Ag. 4 (17), 1911, Repino, formerly Kuokkala.
Doctor of technical sciences (1951). A member of the CPSU since 1940. Grand master of the USSR (1935), international grand master (1950). Honored Master of Sport (1945). World champion (1948–57, 1958–60, 1961–63).
He is a strategic chess player and an innovator in the area of chess theory, particularly theory of opening moves. He is the author of many works on chess. Decors.: Order of Lenin, three other orders, and medals.
Works: *Izbrannye partii (1926–1946),* 2nd ed. Moscow 1951. *Sovetskaia shakhmatnaia shkola.* Moscow 1951. *Algoritm igry v shakhmaty.* Moscow 1968.

Bozhenov, Petr Ivanovich, Dr. of Tech. Sc.
Scientist; b. Je. 29 (Jl. 12), 1904, Karachev, Orel Govt.
Mem. CPSU 1948–; grad. Polytech. Inst. Leningrad 1929; specialist in technology of building materials; Dr. of Tech. Sc. 1953. Decors.: Merited Scientist and Technologist of the RSFSR, Lenin Prize 1962, Order of the Red Banner of Labour, several other orders and medals.

Bozhko, Nina Grigor'evna
Graphic artist; b. 1928, Kiev.
Attended Kiev Art Inst. 1950–56, pupil of V. I. Kasian, I. I. Pleshchinskii; exhibitions 1956–.

Bradulov, Nikolai Mikhailovich
State official.
Dep. Public Prosecutor of the Moldavian SSR until 1961; Min. of the Interior 1961–62, Min. of the Protection of Public Order 1962–68, again Min. of the Interior Moldavian SSR 1968–; mem. Cen. Cttee. Moldavian CP 1961–.

Bragin, Aleksandr Konstantinovich
State official.
1st Sec. of Gvardeiskii Raion Cttee. Kazakh CP 1967–73; Chair. of State Cttee. for the Supply of Petroleum Products Kazakh SSR Jl. 1973–.

Bragin, Mikhail Grigor'evich
Author; b. 1906, Krasnopole, Belorussia.
Mem. CPSU 1931–, grad. Mil. Acad. 1934.
Publs.: "Polkovodets Kutuzov" (General Kutuzov) 1941, "Velikoe srazhenie pod Stalingradom" (The Great Battle of Stalingrad) 1943, "Ot Moskvy do Berlina"

(From Moscow to Berlin) 1947, "Put' Generala" (The Way of the General) 1953, "Put' Leitenanta" (The Way of the Lieutenant) 1957, and others.

Braginskii, Iosif Samuilovich
Scholar; b. Je. 5 (18), 1905, Baku.
Mem. CPSU 1925–; orientalist; corresponding mem. Acad. of Sciences Tadzhik SSR 1951–; Ed.-in-Chief of periodical "Narody Azii i Afriki" 1957–.

Braichevskyi, Mykhailo
Historian and archaeologist; b. 1924.
Ukrainian; former mem. staff of Inst. of History at Acad. of Sciences Ukrainian SSR; author of numerous treatises on Russian-Ukrainian relations; dismissed for his civil-rights activities; his works are no longer published.

Brandys, Anatolii Iakovlevich, Cand. of Mil. Sc.
Military scientist; b. Ag. 12, 1923, Nizhnedneprovsk.
Mem. CPSU 1944–; grad. Air Force Acad. 1950 and Acad. of Gen. Staff 1959; Cand. of Mil. Sc.; Col.; served in Sov. Army 1941–; frontline soldier 1941–45; grad. War Flying School 1943; after the war in Command Staff, later teacher at mil. acads. Decors.: Order of Lenin, 4 Orders of Red Banner, Order of Patriotic War 1st Cl., Aleksandr Nevskii Order, Order of Red Star, Hero of the Sov. Union (twice), medals.

Bratchenko, Boris Fedorovich
State official; b. 1912.
Russian; grad. Moscow Mining Inst. 1935; worked in coal mining 1935–43; pit manager, chief engineer of a trust and later of a collective combine 1943–53; USSR Dep. Min. of Coal Ind., Dept. Head with USSR GOSPLAN, Chair. of Karaganda Econ. Council 1953–61; mem. Cen. Cttee. Kazakh CP 1960–66; Dep. Chair. of Kazakh Council of Mins. and Chair. of Kazakh GOSPLAN 1961–65; Dep. 1962–, mem. Budget Comm. of Sov. of the Union USSR Supr. Sov. 1962–66; Dep. to Supr. Sov. Kazakh SSR 1962–67; USSR Min. of Coal Ind. O. 1965–; cand. mem. 1966–71, mem. Cen. Cttee. CPSU 1971–. Decor.: State Prize.

Bratukhin, Ivan Pavlovich
Designer and scientist in the field of helicopter construction; b. F. 12 (25), 1903, Iashchera, now Suna Raion, Kirov Oblast.
Professor (1954); doctor of technology (1962). Honored Scientist and Engineer of the RSFSR (1964). Member of the CPSU since 1920.
He graduated from the Bauman Moscow Technical School in 1930. His principal works deal with the design and construction of helicopters. He has created a number of helicopter designs and a new type of airplane-helicopter (1946). Decors.: State Prize of the USSR in 1946, Order of Lenin, the Order of the Red Banner of Labour, and medals.
Work: *Proektirovanie i konstruktsii vertoletov.* Moscow 1955.

Bratun, Rostyslav Andrievich
Author; b. Ja. 7, 1927, Liuboml, Volhynia; s. of a teacher.
Ukrainian; grad. Dept. of Philology of Lvov Univ. 1950; Chief Ed. of journal "Zhovten" 1965–66; Chair. of Lvov chapter of Ukrainian Writers' Union 1970; first works publ. 1946.
Publs.: "Ia – syn Ukrainy" (I Am a Son of the Ukraine) 1958, "Pora liubovi" (The Time of Love) 1960, "Grudka zemli" (A Lump of Earth) 1962, and others.

Bratus', Sergei Nikitich
Scholar and jurist; b. O. 6 (19), 1904, Konotop.
Doctor of law (1943), professor, and Honored Scholar of the RSFSR (1960). Member of the CPSU since 1954. Senior research worker of the All-Union Research Institute of Soviet Legislation (director from 1963 to 1969) and chairman of the Foreign Trade Arbitration Commission.
The major works of him are in the fields of the general theory of law and Soviet civil law. These works include *Juridical Persons in Soviet Civil Law* (1947), *Subjects of Civil Law* (1950), *Questions of Economic Contract* (1954, co-author), *Subject Matter and System of Soviet Civil Law* (1963), and *General Theory of Soviet Law* (1966, co-author). He has also been the coauthor of textbooks in these disciplines.

Braude, Semen Iakovlevich, Dr. of Tech. Sc.
Physicist; b. Ja. 28, 1911, Poltava.
Grad. Kharkov Inst. for Physics, Chemistry and Mathematics 1932; corresponding mem. Acad. of Sciences Ukrainian SSR 1958–; Dr. of Tech. Sc. 1943; Prof. 1944; mem. staff Ukrainian Physical-Technical Inst. (now Physical-Technical Inst. at Acad. of Sciences Ukrainian SSR) 1933–55; mem. staff Inst. of Radiophysics and Electronics at Acad. of Sciences Ukrainian SSR 1955–; in addition mem. staff Polytech. Inst. Kharkov; author of approx. 120 scientific works, among them 2 monographs. Decors.: State Prize of the USSR 1952, Order of Red Banner of Labour.
Publs.: "K Voprosu o vozdeistvii vneshnei elektrodvizhushchei sily na magnetron s razrezannym anodom" in "Zhurnal tekhnicheskoi fiziki" No. 23/1940, "Issledovanie amplitudnykh, fazovykh i chastotnykh kharakteristik signalov korotkikh i srednikh radiovoln, rasseiannykh morskoi poverkhnostiu" in "Radiookeanograficheskie issledovaniia morskogo volneniia" 1962, "Gustyna potokiv radiovyprominniuvannia deiakykh dyskretnykh dzherel v diapazoni 20–40 Mgz" in "Dopovidi AN URSR" No. 11/1964.

Braun, Nikolai Leopoldovich
Author; b. 1902, Parakhino nr. Tula; s. of a teacher.
Grad. Literary Dept. of Leningrad Pedagogic Inst. 1929; Russian poet; first works publ. 1924.
Publs.: poems "Mir i master" (The World and the Master) 1926, "Novyi krug" (The New Circle) 1928, "Vylazka v budushchee" (Excursion into the Future) 1931, "Vernost'" (Faithfulness) 1936, "Doliny Rodiny moei" (Valleys of My Country) 1947, "Zemlia v tsvetu" (The Earth in Blossom) 1955, "München" (Munich) 1933, "Molodost'" (Youth) 1960, and others.

Braunshtein, Aleksandr Evseevich
Biochemist; b. My. 13 (26), 1902, Kharkov.
Grad. Kharkov Inst. of Medicine 1925; mem. USSR Acad. of Sciences 1964– and USSR Acad. of Medical Sciences 1945–; Sen. Mem. Scientific Staff Inst. of Biochemistry of USSR People's Commissariat for Public Health and Cen. State Inst. of Occupational Diseases 1928–35; Head of Laboratory of Union Inst. of Experimental Medicine and Inst. of Biological and Medical Chemistry 1936–60; Head of Laboratory of Inst. of Molecular Biology at USSR Acad. of Sciences 1960–; author of more than 180 scientific papers. Decors.: State Prize of the USSR 1941, Order of Red Banner of Labour.

Brazaitis, Algimantas Ionas
State official.
2nd Sec. of Kaunas City Cttee. Lithuanian CP 1976–79; Chair. of State Cttee. on Film Lithuanian SSR Je. 1979–; mem. Auditing Comm. Lithuanian CP 1981–.

Brazauskas, Algirdas-Mikolas Kazevich
State official.
Min. of the Construction Materials Ind. Lithuanian SSR 1965–66; 1st Dep. Chair. of GOSPLAN Lithuanian SSR 1966–77; cand. mem. Cen. Cttee. Lithuanian CP 1966–76, mem. 1976–; Sec. and mem. Bureau Cen. Cttee. Lithuanian CP My. 1977–.

Brekhov, Konstantin Ivanovich
State official; b. 1907.
Russian; mem. CPSU 1931–; grad. Kharkov Inst. of Machine Building and Mechanics 1936; asst. engine-driver until 1931, then engine-driver; engineer, chief engineer, dir. of a machine building plant 1931–54; USSR Dep. Min. of Construction and Road Machine Building, Head of the Machine Building Board, Dep. Chair. of Moscow Oblast Econ. Council 1954–59 and Chair. 1959–64; cand. mem. Cen. Cttee. CPSU 1961–71, mem. Cen. Cttee. CPSU 1971–; Dep. to USSR Supr. Sov. 1962–; Chair. of USSR GOSPLAN's State Cttee. for Chemical and Petroleum

Machine Building with the rank of USSR Min. 1964–65; USSR Min. of Chemical and Petroleum Machine Building O. 1965–. Decors.: Hero of Socialist Labour, Order of Lenin and Hammer and Sickle Medal in Gold 1977.

Brekhovskikh, Leonid Maksimovich
Physicist; b. Ap. 23 (My. 6), 1917, Strunkino, Arkhangelsk Oblast.
Mem. CPSU 1950–; grad. Perm Univ. 1939; mem. USSR Acad. of Sciences 1968–; Prof. at Moscow State Univ. 1953; Dir. of Inst. of Acoustics 1954–64, head of a dept. at same inst. 1964–; Acad. Sec. of Dept. of Oceanology, Physics of the Atmosphere and Geography of USSR Acad. of Sciences 1969–. Decors.: State Prize of the USSR 1951, Order of Red Banner of Labour, various medals.

Brenchev, S. A.
Komsomol official.
Mem. Auditing Comm. Latvian CP 1981–; Sec. and mem. Bureau Cen. Cttee. Latvian Komsomol F. 1982–.

Brezhnev, Dmitrii Danilovich
Grower and breeder of plants; b. O. 25 (N. 7), 1905, Pisklovo, Troitskoe *volost* (small rural district), Kursk District, Kursk Province.
Member of the V. I. Lenin All-Union Academy of Agricultural Sciences (1956) and its first vice-president (1956–61 and from 1968). Member of the CPSU since 1926.
He graduated from the Voronezh Agricultural Institute in 1933. He was director of the Akhtubinsk Zone Experimental Station (1933–34) and the Gribovo Selection Experimental Station (1934–36). From 1937 to 1941 and from 1946 to 1950 he was head of the department of vegetable crops at the All-Union Scientific Research Institute of Plant Growing and has been its director since 1966. He participated in the creation of several valuable varieties of tomato. His principal works are on questions of the biology of vegetable crops and the problem of heterosis.
He was a member of the Central Committee of the CPSU in 1956–61. He was a deputy to the fourth convocation of the Supreme Soviet of the RSFSR. Decors.: Order of Lenin, six other orders, and various medals, received the State Prize of the USSR in 1952, member of the Agricultural Academy of the German Democratic Republic (1958).
Works: *Novoe v selektsii rastenii.* Moscow-Leningrad 1954. *Svezhie ovoshchi letom i zimoi.* Moscow 1963. *Tomaty,* 1st ed., Leningrad 1955; 2nd ed., Leningrad 1964. *Ovoshchevodstvo v SShA.* Moscow 1961. *Geterozis v ovoshchevodstve.* Moscow 1968 (editor).

Brezhnev, Iurii Leonidovich
State official; b. 1933.
Son of L. I. Brezhnev, Russian; mem. CPSU 1957–, grad. Dnepropetrovsk Metallurgical Inst. 1955 and All-Union Acad. of Foreign Trade 1960; asst. foreman, superintendent of pipe-welding shop of K. Liebknecht Plant in Dnepropetrovsk 1955–57; senior engineer, chief of a dept. in USSR Trade Mission in Sweden 1960–66; Dep. USSR Trade Representative to Sweden 1966–68, Trade Representative 1968–70; Chair. of All-Union Foreign Trade Asscn. for import and export of pig iron, ferro-alloys, steel wire, and metal products 1970–76; Dep. USSR Min. of Foreign Trade 1976–79, 1st Dep. Min. Mr. 1979–; cand. mem. Cen. Cttee. CPSU 1981–.

Briede, Aleksandra Ianovna
Sculptor; b. Jl. 12 (25), 1901, Liepāja.
People's Artist of the Latvian SSR (1955) and corresponding member of the Academy of Arts of the USSR (1958).
She studied with K. Ronchevskii at the Riga Academy of Arts from 1923 to 1931. Her work includes statues, tombstones, portraits, genre scenes, and small porcelain objects. She has done *The Young Sculptor* (bronze, 1947), *Peace* (marble, 1960), and *Drizzle* (granite, 1969), all of which are in the Art Museum of the Latvian SSR in Riga. Her other works include *May There Always Be Sunshine* (granite, 1963), *Ancite the Kolkhoz Worker* (bronze, 1965), *My Land* (aluminium, 1967), and *Lenin* (granite, 1969). Decors.: Order of Lenin and the Order of the Badge of Honor.

Bril, Anton Piusovich
State official.
Cand. mem. Cen. Cttee. Latvian CP 1971–76, mem. 1976–; Head of Dept. for Agriculture with Cen. Cttee. Latvian CP 1976–80; Min. of Melioration and Water Econ. Latvian SSR Ja. 1980–.

Briukhov, A. O.
Trade union official.
Sec. of Kirghiz Republican Trade Union Council F. 1977–.

Brodele, Anna Iulevna
Author; b. 1910, Tauragė, Lithuania; d. of a forester.
Studied Gorkii Inst. of Literature Moscow; condemned to 4 years imprisonment for participation in revolutionary movt. 1932; worked as editor in various places in Latvia after establishment of Sov. power; first works publ. 1927.
Publs.: plays "Dolg" (Duty) 1942, "Ne daite uiti" (Do Not Let Them Go) 1943, "Neosushchestvivshaiasia ideia" (The Unrealized Idea) 1944, "Vesna v sele Rechnom" (Spring in the Rechnoe Village) 1948, "Uchitel' Straume" (Straume the Teacher) 1948, "Zolotaia niva" (The Golden Cornfield) 1949, "Goriachie serdtsa" (Burning Hearts) 1951, novels "Kroviu serdtsa" (With the Heart's Blood) 1955, "Vernost'" (Loyalty) 1960, and others.

Brodelis, Ian Ianovich
State official.
Head of Main Admin. 1962–66, Chair. of State Cttee. for Vocational Training Latvian SSR Ag. 1966–; mem. Auditing Comm. Latvian CP 1971–81, cand. mem. Cen. Cttee. Latvian CP 1981–.

Brodova, Anna Ivanovna
State official.
Dep. to Supr. Sov. Uzbek SSR 1955–; Sec. of Oblast Cttee., 1st Sec. of Raion Cttee. Uzbek CP in Tashkent 1955–63; mem. Cen. Cttee. Uzbeck CP 1956–; City Sec., Dep. Chair. of Oblast Exec. Cttee. of Workers' Deputies in Tashkent 1963–67; Min. of Public Services Uzbek SSR 1967–.

Bromlei, Iulian Vladimirovich
Historian; b. F. 21, 1921, Moscow.
Specialist on the medieval history of Croatia. Corresponding member of the Academy of Sciences of the USSR (1966). Member of the CPSU since 1944. Graduated from the history department of Moscow University (1950). Doctor of history (1965).
In 1952–58 he was learned secretary on coordination of the History Division of the Academy of Sciences of the USSR and in 1958–1966, learned secretary of the History Division of the Academy of Sciences of the USSR. He has been director of the Ethnography Institute of the Academy of Sciences of the USSR since 1966 and deputy academician-secretary of the History Division of the Academy of Sciences of the USSR since 1968. He is the author of the works *The 1573 Peasant Uprising in Croatia* (1959) and *The Development of Feudalism in Croatia* (1964).

Brovikov, Vladimir Ignatevich
Party official; b. 1931, Vetka, Gomel Oblast.
Belorussian; mem. CPSU 1958–; grad. Belorussian State Univ. 1955 and CPSU Cen. Cttee. Acad. of Social Sciences 1969, Cand. of Philosophical Sc. 1969; posts in press organs 1955–63; party posts 1963–70; Sec. of Vitebsk Oblast Cttee. Belorussian CP 1970–72; posts in Cen. Cttee. apparatus CPSU, Dep. Chief of Organizational Party Work Dept. of Cen. Cttee. CPSU 1972–78; mem. Bureau of Cen. Cttee. and 2nd Sec. of Cen. Cttee. Belorussian CP D. 1978–; Dep. to USSR Supr. Sov. 1979–; mem. Cen. Cttee. CPSU 1981–.

Brovman, Grigorii Abramovich
Literary critic; b. Ag. 1 (14), 1907, Kalarash, Bessarabia.
Mem. CPSU 1942–; grad. Dept. of Philology of Moscow State Univ. 1931; first works publ. 1931.
Publs.: "Pisatel' i zhizn'" (The Writer and

Life) 1936, "Literaturnye zametki o sovremennoi teme" (Literary Notes on a Contemporary Subject) 1938, "Knigi, rozhdennye zhizniu" (Books Born of Life) 1958, "Zhiznennaia pozitsiia geroia" (The Hero's Place in Life) 1960, "V. V. Veresaev. Zhizn' i tvorchestvo" (V. V. Veresaev. His Life and Work) 1959, and others.

Brumel', Valerii Nikolaevich
Field athlete; b. My. 14, 1942, Razvedki, Chita Oblast.
Honored Master of Sport of the USSR (1961). Member of the CPSU since 1964. Graduated from the Central Institute of Physical Culture (Moscow) in 1967.
He has repeatedly set records for the running high jump in the USSR, Europe, and the world. In 1963 he attained a world record of 228 cm. He was champion at the XVIII Olympic Games (1964, Tokyo – 218 cm). In 1965 he sustained a serious injury to his leg in an automobile crash; after effective treatment he resumed training in 1969 (June 1969 – 206 cm). He was three times acknowledged best athlete of the year in the world (1961, 1962, and 1963). In 1962, Brumel' was awarded the highest honors in the world of sports – the Helms World Trophy and the Golden Caravel of Columbus. Decors.: Order of the Red Banner of Labor and a medal.

Brusilovskii, Evgenii Grigor'evich
Composer; b. O. 30 (N. 12), 1905, Rostov-on-Don.
People's Artist of the Kazakh SSR (1936). One of the originators of Kazakh professional arts.
From 1922 to 1924, he studied piano and composition at the Moscow Conservatory, and in 1931 he graduated from M. O. Shteinberg's class in composition at the Leningrad Conservatory. In 1933 he was commissioned to Alma-Ata by the Leningrad Union of Composers. He was music director of the Kazakh Music Theater from 1934 to 1938, an instructor beginning in 1944, art director of the Philharmonic from 1949 to 1951, and professor at the Alma-Ata Conservatory from 1955, heading the subdepartment of composition. His pupils have included B. Baikadamov, M. Tulebaev, B. G. Erzakovich, K. Kuzham'iarov, E. Rakhmadiev, and S. Mukhamedzhanov. He was chairman of the Organization Committee from 1939 to 1948 and chairman of the board of the Union of Composers of the Kazakh SSR from 1948 to 1953.
He wrote down more than 250 Kazakh folk songs and *kiuis* (Kazakh instrumental folk songs) and created the first nine Kazakh operas (*Kyz-Zhibek*, 1934; *Zhalbyr*, 1935; *Er-Targyn*, 1936; and others; all performed at the Kazakh Theater of Opera and Ballet) and seven symphonies. He wrote many chorales and cantata oratorios; chamber, vocal, and instrumental pieces; and film music. The national anthem of the Kazakh SSR was written by him, in collaboration with M. Tulebaev and L. Khamidi (words by A. Mukhametkhanov, G. Musrepov, and A. Tazhibaev; 1945). Decors.: State Prize of the USSR in 1948 and the Kurmangaza State Prize of the Kazakh SSR in 1967, Order of Lenin, three other orders, and medals.

Brusilovskii, Mikhail Shaevich
Graphic artist; b. 1931, Kiev.
Attended Kiev Middle School of Art 1947–52 and the Repin Inst. for Painting, Sculpture and Architecture 1953–59, pupil of V. M. Zvontsov, A. F. Pakhomov; exhibitions 1955–.

Brutane, Valiia Ianovna
Author; b. 1911.
Latvian; translates Russian Sov. classical literature into Latvian; first works publ. 1945.
Publs.: "Sady rastsvetaiut" (The Gardens are in Bloom) 1950, "Shumite, belye berezy moi!" (Rustle, My White Birch Trees!) 1956, and others.

Bryl, Ianka (Ivan Antonovich)
Author; b. 1917, Odessa; s. of a railwayman.
Belorussian; family returned to Zagore, W. Belorussia (now Grodno Oblast) 1922; mil. service in Polish Army 1939; after destruction of Gdańsk (Danzig) harbour imprisoned by Germans 1939, escaped 1941, active participant in Belorussian partisan movt.; first work publ. 1938.
Publs.: novelette "V Zabolot'e svetaet" (Dawn Over Zabolot'e) 1950, "Galia" 1953, "Moi zemliak" (My Countryman) 1946, "Devochka chernobrova" (The Girl with Black Eyebrows) 1949, "Mat'" (Mother), "Ty – moi najlepshi drug" (You Are My Best Friend) 1952, "Serdtse kommunista" (The Heart of a Communist) 1957, and others.

Bryzga, Lidiia Dmitrievna
Party official; b. 1943.
Mem. CPSU 1975–, finished secondary school; collective farmer in Brest Oblast 1959–60; milkmaid at Zhdanov Collective Farm in Brest Oblast 1960–; mem. Cen. Auditing Comm. CPSU 1981–. Decor.: Heroine of Socialist Labor.

Buadze, Aleksandr Il'ich
State official.
Cand. mem. Cen. Cttee. Georgian CP 1971–76, mem. 1976–; Chair. of State Cttee. for Material and Technical Supply Georgian SSR Jl. 1976–.

Bubennov, Mikhail Semenovich
Author; b. 1909, Vtoroe Polomoshnevo, Altai; s. of a peasant.
Russian; mem. CPSU 1951–; teacher in Siberia and Tartar 1927–; first work publ. 1927, Decors.: Order of Red Banner of Labour, Stalin Prize.
Publs.: "Gremiashchii god" (The Thundering Year) 1932, collected stories "V polovodie" (At Flood Time) 1940, novelette "Bessmertie" (Immortality) 1940, novels "Belaia bereza" (The White Birch) 1947, vol. 2, 1952, and others.

Bublev, Valentin Petrovich
Sculptor; b. 1928, Baku.
Attended Azerbaidzhan Art School 1942–45 and Repin Inst. for Painting, Sculpture and Architecture at the USSR Acad. of Arts 1945–51, pupil of A. T. Matveev, V. A. Sinaiskii; exhibitions 1952–.

Bubnov, Igor Dmitrievich
Diplomat.
USSR Amb. to Suriname Jl. 1982–.

Budagov, Ruben Aleksandrovich
Philologist and linguist; b. Ag. 23 (S. 5), 1910, Rostov/Don.
Grad. Pedagogical Inst. Rostov 1933; specialist in general and Romance linguistics and philology; corresponding mem. USSR Acad. of Sciences 1970–; Prof. at Leningrad Univ. 1947–52; mem. staff Moscow Univ. 1952–. Decors.: Order of Red Banner of Labour and several medals.

Budyko, Mikhail Ivanovich
Geophysicist; b. Ja. 20, 1920, Gomel'.
Corresponding member of the Academy of Sciences of the USSR (1964). Member of the CPSU since 1956.
Upon graduating from the Leningrad Polytechnic Institute in 1942, he worked at the A. I. Voeikov Main Geophysical Observatory, becoming its director in 1954. His major works are in the fields of physical climatology, bioclimatology, and actinometry. Together with Academician A. A. Grigor'ev, he formulated the periodic law of geographical zonality. Decors.: Lenin Prize in 1958 for his work on the heat balance of the earth's surface, two orders and various medals.
Works: *Isparenie v estestvennykh usloviiakh*. Leningrad 1948. *Atlas teplovogo balansa*. Leningrad 1955 (Editor). *Teplovoi balans zemnoi poverkhnosti*. Leningrad 1956. *Klimat i zhizn'*. Leningrad 1971.

Bugaev, Boris Pavlovich
State official; b. 1923, Ukraine.
Mem. CPSU 1946–; grad. Higher Flying School for Civil Aviation 1966; Chief Marshal of the Air Force; pilot/instructor at Aktiubinsk School for Civil Aviation 1942–43; pilot, commander of an aviation dept. at Ukrainian Civil Aviation Admin. 1943–46; pilot, capt., commander of a pilots' dept. of civil aviation 1947–66; Dep., 1st Dep. Min. 1966–70 and Min. of USSR Civil Aviation 1970–; mem. Cen. Cttee. CPSU 1971–. Decors.: State Prize of the USSR for Participating in the Construction of the TU-134 1972.

Bugrimova, Irina Nikolaevna
Circus artist and animal trainer; b. Mr. 13, 1910, Kharkov.
People's Artist of the USSR (1969). Member of the CPSU since 1951.
She is the first female trainer of wild animals in the USSR. She began her performing career in 1929 in an act entitled "Sleigh Ride Under the Big Top" (created in collaboration with A. N. Buslaev). Since 1939 she has worked as an animal trainer in circus attractions with trained lions. Decors.: Order of the Badge of Honor and a medal.
Work: *Moia professiia*. Moscow 1957.

Bugrov, B. I.
Party official.
Head of Dept. of Trade and Public Services of Cen. Cttee. Uzbek CP 1979–; mem. Auditing Comm. Uzbek CP 1981–.

Bukeeva, Khadisha
Kazakh actress; b. F. 21, 1917, Kaztalovka, Ural'sk Oblast.
People's Artist of the USSR (1964).
In 1938, she graduated from the Leningrad Theatrical Institute and joined the company of the Chimkent Theater. In 1942 she began to work in the Auezov Theater of Drama (Alma-Ata). Among her roles are Saule (Abishev's *Friendship and Love*), Karlyga (Auezov's *Koblandy*), the narrator (*Abai*, based on Auezov's novel of the same name), Mekhmene Banu (Nazim Hikmet's *Legend About Love*), Katherine (Shakespeare's *The Taming of the Shrew*), Saule (Akhtanov's work of the same name), and Tkhan' (Kuprianov's *Star of Vietnam*). Her best stage roles are marked by deep realism and theatrical expressiveness. Since 1965 she has carried on pedagogical activity (since 1968 as assistant professor at the theatrical department of the Kurmangazy Institute of the Arts. Decors.: State Prize of the USSR (1952), Order of Lenin, the Order of the Badge of Honor, and medals.

Bukin, Sergei Alekseevich
Diplomat.
USSR Amb. to Uganda Jl. 1979–.

Buklis, Marionas Vladovich
State official.
Min. for Meat and Dairy Ind. Lithuanian SSR O. 1965–; cand. mem. Cen. Cttee. Lithuanian CP 1966–76, mem. 1976–.

Bukov, Emilian Nesterovich
Author; b. 1909, Novaia Kiliia, Bessarabia; s. of a gardener.
Moldavian; mem. CPSU 1950–; attended B. P. Khadzheu Lyceum and Faculty of Literature and Philosophy of Bucharest Univ. Decors.: Order of Lenin, Order of Red Banner of Labour.
Publs.: "Trud kipit" (Work in Full Swing) 1932, "Rech' solntsa" (The Speech of the Sun) 1937, "Kitai" (China) 1938, "Ia vizhu tebia, Moldaviia" (I See You, Moldavia) 1942, "Vesna na Dnestre" (Spring on the Dniestr) 1944, "Strana moia" (My Country) 1947, "Rastut etazhi" (Floor Levels are on the Increase) 1952, "Stikhi" (Poems) 1954, "Gorod Reut" (The Town of Reut) 1956, drama "Dunai – bespokoinye vody" (The Danube and its Turbulent Waters) 1956, and others.

Bukovskii, Lev Vladimirovich
Sculptor; b. 1910.
Educ. Latvian Univ. 1932–35, Art Academy Florence 1938–39. Decor.: Lenin Prize 1970.

Bulanzhe, Iurii Dmitrievich
Geophysicist; b. Jl. 28 (Ag. 10), 1911, Moscow.
Corresponding member of the Academy of Sciences of the USSR (1966).
Since his graduation from the Moscow Geodetic Institute in 1934, he has worked at the Institute of Physics of the Earth of the Academy of Sciences of the USSR (department head since 1956). His major works concern experimental studies in gravimetrics (the development of methods and apparatus for highly accurate measurement of the force of gravity) and the study of deformations in the earth's crust. He has been vice-presicent of the Soviet Geodetic Society (since 1957) and of the International Association of Geodesy. Decors.: two orders.
Works: "Formuly dlia vychisleniia oshibok gravimetricheskoi sviazi dvukh punktov pri mnogokratnykh izmereniiakh, vypolnennykh gruppoi gravimetrov." *Izv. AN SSSR: Ser. geofizicheskaia,* 1956, no. 7. "Po povodu izucheniia sovremennykh dvizhenii zemnoi kory na statsionarnykh poligonakh." In *Sovremennoe dvizhenie zemnoi kory,* no. 2. Tartu 1965.

Bulashevich, Iurii Petrovich, Dr. of Phy.-Math. Sc.
Geophysicist; b. 1911.
Grad. Kazan Univ. 1935; Dr. of Phy.-Math. Sc. 1951; Prof. 1952–; corresponding mem. USSR Acad. of Sciences 1970–; mem. staff Urals Branch of USSR Acad. of Sciences 1935–; Dir. of Inst. for Geophysics 1958–, Dir. of Inst. for Geophysics with Urals Research Centre at USSR Acad. of Sciences 1971–.

Bulgakov, Aleksandr Alesandrovich
State official; b. 1907, Ukraine.
Mem. CPSU 1937–; grad. Kharkov Electrical Engineering Inst. 1939; messenger boy, apprentice statistician 1919–29; served in Sov. Army 1929–31; after graduation engineer, chief mechanic, chief design engineer 1939–42; Sec. of CP Cttee., Cen. Cttee. CPSU organizer in various plants 1942–50; Sec. of Kharkov City Cttee. Ukrainian CP, Chair. of Kharkov City Sov., 2nd Sec. of Kharkov Oblast Cttee. Ukrainian CP 1950–59; cand. mem. Cen. Cttee. Ukrainian CP 1954–58; Dep. to Supr. Sov. Ukrainian SSR 1955–63; Sec. of All-Union Cen. Trade Union Council 1959–64; cand. mem. 1961–71, mem. Cen. Cttee. CPSU 1971–; Dep. to USSR Supr. Sov. 1962–; Chair. of USSR State Cttee. for Tech. Training 1964–. Decors.: 2 Orders of Lenin.

Buniatian, Mnatsakan Rubenovich
Mathematician; b. 1932.
Armenian; mem. CPSU 1965–; grad. Erevan Univ. 1954; Inspector at Dept. of Public Educ. in Kamo, Armenia 1954–56; mem. staff Computing Centre at Acad. of Sciences Armenian SSR 1957–59; mem. staff. Scientific Research Inst. for Computing Machinery in Erevan 1959; Laboratory Chief 1963–. Decor.: State Prize of the USSR for contribution to the development and elaboration of computer generation "Nairi" 1971.

Bunin, Andrei Vladimirovich, Dr. of Art. Sc.
Art historian; b. Mr. 15 (28), 1905, Voronezh.
Mem. CPSU 1958–; specialist in the history of architecture; Dr. of Art Sc. 1958–; Prof. at Inst. of Architecture Moscow 1942–. Decors.: 2 orders and several medals.

Bunin, Konstantin Petrovich, Dr. of Tech. Sc.
Scientist; b. N. 2, 1910, Ekaterinograd, now Krasnodar; s. of a worker.
Mem. CPSU 1940; grad. Inst. of Metallurgy Dnepropetrovsk 1932; specialist in physical metallurgy; corresponding mem. Acad. of Sciences Ukrainian SSR 1948–; Dr. of Tech. Sc. 1941; Prof. 1938–; Lecturer, Prof. Chair for Physical Metallurgy with Inst. of Metallurgy Dnepropetrovsk 1932–45; mem. staff. Inst. of Nonferrous Metallurgy at Acad. of Sciences Ukrainian SSR (now Research Inst. for Nonferrous Metallurgy of USSR Min. of Nonferrous Metallurgy) 1948–; author of more than 160 scientific works. Decors.: Order of Red Banner of Labour, 2 Badges of Honour, medals.
Publs.: "Otbelennyi chugun" 1947, "Zhelezouglerodistye splavy" 1949, "Struktura chuguna" 1952, "Rost chuguna i stali pri termotsiklirovanii" 1967.

Bunkin, Boris Vasil'evich
Physicist; b. Jl. 16, 1922, Aksin'ino, Moscow Oblast.
Corresponding member of the Academy of Sciences of the USSR since 1968; Hero of Socialist Labor (1967). Member of the CPSU since 1963.
He worked at the Moscow Aviation Institute after graduating from it in 1947. He has been with the Moscow Engineering-Physics Institute since 1950. His major works are on radio engineering and electronics. Decor.: Order of Lenin.

Burakova, Nina Federovna
Party official; b. 1934.
Mem. CPSU 1976–; railroad worker 1952–53; factory worker in Krasnoiarsk 1953–56; worker in a synthetic fiber plant in Krasnoiarsk 1961–; mem. Cen. Auditing Comm. CPSU 1981–. Decor.: Heroine of Socialist Labor 1977.

Buravkin, Gennadii Nikolaevich
State official.
Chair. State Cttee. for Television and Radio Belorussian SSR S. 1978–; mem. Cen. Cttee. Belorussian CP 1981–.

Burenkov, Sergei Petrovich
State official; b. 1923.
Mem. CPSU 1945–, grad. Naval Medical Acad. 1947, Cand. of Medical Sc. 1970; participated in World War II; physician, chief of a section, chief physician of a tuberculosis hospital 1947–61; Sec., 1st Sec. of Sestroretsk Raion Cttee. CPSU in Leningrad City 1961–66; chief of city health dept. in Leningrad 1966–71; 1st Dep. USSR Min. of Health 1971–80, Min. D. 1980–; cand. mem. Cen. Cttee. CPSU 1981–.

Buriak, Boris Spiridonovich
Author; b. 1913, Rubezhnoe, now Lugansk Oblast; s. of a railwayman.
Ukrainian author and literary historian; mem. CPSU 1945–; grad. Kiev Univ. 1939.
Publs.: "Taras Zhurba" 1951, and others.

Burkhanov, Asli Badriddinovich
Tadzhik actor and director; b. Ja. 8, 1915, Bukhara.
People's Artist of the USSR (1965). Member of the CPSU since 1945.
He began his stage career in 1932 at the Bukhara Music and Drama Theater. In 1933 he began performing at the Lakhuti Theater in Dushanbe, where he created roles of strong, courageous people – fighters for the people's freedom – and played a number of character roles. As an actor, he combines stormy temperament and expressive stage mannerisms with gentle lyricism and rich comic sence. His roles include Khushvakt in Saidmuradov and Ismailov's *Slander,* Salim in Ulagzade's *The Red Stick Carriers,* Alisher Navoi in Uigun and Sultanov's play of the same name. Kent and Iago in Shakespeare's *King Lear* and *Othello,* and Khlestakov in Gogol's *The Inspector-General.* He was the first man to play V. I. Lenin in the Tadzhik theater – in Kapler and Zlatogorova's *Man With a Gun,* Vishnevskii's *The Unforgettable Year of 1919,* and Abdullo and Kiiamov's *Hurricane.* Among the plays produced by him are Mirshakar's *The Teacher of Love* (1945), Saidmuradov and Rabiev's *Saodat* (1956), Mamedkulizade's *Corpses* (1963), and Shakespeare's *The Two Gentlemen of Verona* (1966). He has also appeared in films. Decors.: Order of Lenin and three other orders, a medal.

Burkhanov, Mutal' Muzainovich
Composer and public figure; b. Ap. 22 (My. 5), 1916, Bukhara.
Honored Art Worker of the Uzbek SSR (1964).
In 1949, he graduated from the Moscow Conservatory, where he studied composition with S. N. Vasilenko. He composed the first Uzbek polyphonic choral works, including six a cappella choral pieces based on folk melodies (1954), as well as "Odes to Lenin's Party" (1959), symphonic works, songs and art songs, and music for plays and movies. Decors.: Khamza State Prize of the Uzbek SSR (1959 and 1970), the Order of Lenin, and various medals.

Burkhanov, Shukur
Uzbek actor; b. S. 15, 1910, Tashkent.
People's Artist of the USSR (1959). Member of the CPSU since 1943.
In 1928, he joined the company of the State Uzbek Drama Theater (now the Khamza Theater, Tashkent). The actor's work is diverse: romantic, vivid, spirited characters; delicate, psychologically complex images; deep dramatism; and sharply satirical characterization are congenial to him. His roles include Guliam (Iashen's *Honor and Love*), Gafur (Khamza's *The Landowner and the Hired Man*), General Rakhimov (Iashen's play of that name), Fakhad (*Legend About Love* by Khikment), the mayor (Gogol's *The Inspector General*), the hired man Iulchi (*Holy Blood,* after Aibek), Hamlet and Brutus (Shakespeare's *Hamlet* and *Julius Caesar*), Voinitskii (*Uncle Vanya* by Chekhov), Ulugbek (*Mirzo Ulugbek* by Sheikhzade), and Oedipus (Sophocle's *Oedipus Rex*). He portrayed V. I. Lenin in Pogodin's *Kremlin Chimes* (1947) and Iashen's *Guiding Star* (1957). His film roles include Karabatyr (*Takhir and Zukhra*), director of the kolkhoz *(Bitty-Bird),* Ulugbek *(Ulugbek's Star),* and Ialantush *(Storm Over Asia).* Decors.: Khamza State Prize of the Uzbek SSR (1964), 5 orders, as well as medals.

Burmistrova, Nataliia Mikhailovna
Actress; b. 1918.
Engagement in Tomsk, Petropavlovsk, Orekhovo-Zuevo, Elets, Gorkii among others 1936–48; mem. Griboedov Theatre Tbilisi 1948–. Decors.: People's Artist of the Georgian SSR 1961, People's Artist of the USSR 1972.

Bursov, Boris Ivanovich, Dr. of Phil. Sc.
Literary historian; b. N. 1 (14), 1905, Novoselovka, Voronezh Oblast.
Mem. CPSU 1942–; Dr. of Phil. Sc. 1951; Prof. at Leningrad Univ. 1948–66; Prof. at Herzen Pedagogic Inst. in Leningrad 1966–; his main works deal with Russian revolutionary democrats, G. V. Plekhanov, L. N. Tolstoi, F. M. Dostoevsky and M. Gorkii and with methodological problems in classical Russian and Sov. literature. Decors.: Order of Patriotic War 2nd Cl., medals.
Publs.: "Voprosy realizma v estetike revolutsionnykh demokratov" (Questions of Realism in the Aesthetics of the Revolutionary Democrats) 1953, "Lev Tolstoi" 1960, "Roman M. Gorkogo 'Mat'" (Gorkiis' Novel 'The Mother') 1962, "Lev Tolstoi i russkii roman" (Leo Tolstoi and the Russian Novel) 1963, "Natsional'noe svoeobrazie russkoi literatury" (National Characteristics of Russian Literature) 1967, "Realizm vsegda i segodnia" (Realism For Now and Ever) 1967, "Lichnost' Dostoevskogo" (Dostoievsky's Personality) 1969, and others.

Busakov, Dzhamalleddin
Uigur writer; b. 1918, Dzharkent.
He served as a teacher, a school principal, and a chairman of a kolkhoz. He is the author of the novel *Whirlpool* (books 1–2, 1964–66; Russian translation, 1966), which describes the formation of a Uigur kolkhoz village, and the novellas *Folk Singer* (1959) and *Morning Breeze* (1969), which are devoted to the way of life of his contemporaries.

Bushmin, Aleksei Sergeevich
Literary historian; b. 1910, Levaja Rossosh, Voronezh Oblast.
Mem. CPSU 1940–; grad. Inst. for Zoo Technicians and Veterinary Surgeons Voronezh 1932 and Faculty of Russian Language and Literature at Voronezh Pedagogic Inst. 1939; literary historian; Scientific Collaborator 1946–; Dir. of Inst. for Russian Literature at USSR Acad. of Sciences 1955–.
Publs.: articles "A S. Serafimovich" 1954, "O rannei sovetskoi proze" (About Early Russian Prose) 1953, and others.

Buslaev, Iurii Aleksandrovich
Chemical scientist; b. N. 22, 1929, Syzran.
Mem. CPSU 1959–; grad. Moscow Chemical-Technological Mendeleev Inst. 1952; specialist in organic chemistry; corresponding mem. USSR Acad. of Sciences 1968–; mem. staff Kurnakov Inst. for Gen. and Inorganic Chemistry at USSR Acad. of Sciences 1952–.

Busygin, Mikhail Ivanovich
State official.
USSR Dep. Min. of the Pulp and Paper Ind. 1974–80; USSR 1st Dep. Min. of the Timber, Pulp and Paper, and Wood Processing Industries 1980–82, Min. Mr. 1982–.

But-Gusaim, Ivan Mikhailovich
Party official; b. 1925.
Mem. Cen. Cttee. Belorussian CP 1966–71 and 1976–; Sec., 2nd Sec. Brest Oblast Cttee. Belorussian CP 1967–80; Chair. of Auditing Comm. Belorussian CP N. 1980–.

Butel, R.
Trade union official.
Head of Dept. for Ind. and Transport with Cen. Cttee. Esthonian CP 1971–80;

mem. Cen. Cttee. Esthonian CP 1976–; Sec. and mem. Presidium Esthonian Republican Trade Union Council F. 1982–.

Butenko, Raisa Georgievna, Dr. of Biol. Sc.
Biologist; b. 1920.
Grad. K. A. Timiriazev Agricultural Acad. 1943; Dr. of Biol. Sc. 1963; Prof. 1971–; corresponding mem. USSR Acad. of Sciences 1974–; postgraduate studies K. A. Temiriazev Agricultural Acad. 1943–47, mem. staff of same until 1956; mem. staff Inst. for Plant Physiology at USSR Acad. of Sciences 1956–; Laboratory Chief of same Inst. 1968–.

Butiai, Vladimir Danilovich
Sculptor; b. 1925, Repki, Chernigov Oblast.
Attended Kharkov Art School 1948–52, pupil of L. A. Tverdianskaia; exhibitions 1958–.

Butusov, Sergei Mikhailovich
State official; b. 1911.
Mem. CPSU 1941–; Sec. of Moscow City CPSU Cttee. 1957–60; 1st Dep. Chair. of Exec. Cttee. of Moscow City Workers' Deputies 1960–61; Min. of Local Econ. 1961–71; Min. of Housing and Local Econ. RSFSR Jl. 1971–. Decor.: Order of Lenin.

Bychko, Valentin Vasil'evich
Author; b. 1912, Kharkov; s. of a fitter.
Ukrainian; mem. CPSU 1943–; grad. Kharkov Inst. of Educ. 1932; writes predominantly for children and young people; first works publ. 1929.
Publs.: "Solntse nad golovami" (The Sun Above Our Heads) 1935, "Perekop" (1937, "Veselka" (The Rainbow) 1946, "Siisia, rodysia zerno" (Be Sown and Thrive, Grain) 1959, and others.

Bychkov, Stanislav Aleksandrovich
Painter and graphic artist; b. 1928, Moscow.
Attended Moscow Polygraphic Inst. 1948–53, pupil of A. D. Goncharov.

Bykhanov, Anatolii Georgievich
State official.
Dep. Min. of Domestic Construction RSFSR until 1980, Min. Ja. 1980–.

Bykhovskii, Bernard Emmanulovich
Philospher; b. Ag. 31 (S. 12), 1898, Bobruisk.
Doctor of philosophy (1941); professor (1929). Member of the CPSU since 1920. Since 1923, he has conducted pedagogical and scholarly work in the field of dialectic materialism and the history of Western European philosophy and contemporary foreign philosophy. He is the author of one of the first Soviet textbooks of dialectic materialism (*Essay on the Philosophy of Dialectic Materialism,* 1930). Decor.: State Prize of the USSR (1944) for his participation in the creation of the *History of Philosophy* (1940–43).
Works: *Filosofiia Dekarta.* Moscow-Leningrad 1940. *Metod i sistema Gegelia.* Moscow 1941. *Osnovnye techeniia sovremennoi idealisticheskoi filosofii.* Moscow 1957. "Neotomizm – sovremennaiia filosofiia katolitsizma." In: *Nauka i religiia.* Moscow 1957. *Filosofiia neo-pragmatizma.* Moscow 1959. *Feierbakh.* Moscow 1967.

Bykov, Vasilii Vladimirovich
Author; b. 1924, Cherenovshchina, Vitebsk Oblast, Belorussia.
Decors.: Lenin Prize 1974, Ia. Kolas Literature Prize 1964, Order of the Red Star, several medals.
Publs.: "Smert' cheloveka" (The Death of a Person) 1957, "Dvadtsatyi" (The Twentieth) 1958, "Zhuravlinyi krik" (Cry of the Cranes) 1960, "Tretiaia raketa" (The Third Rocket) 1962, "Alpiiskaia ballada" (The Alps Ballad) 1964, "Odna noch'" (One Night) 1965, "Pro vremia i pro sebia" (About Time and About One-Self) 1966, "Obelisk", "Dozhit' do rassveta" (To Survive until the Dawn), and others.

Bykova, Elizaveta Ivanovna
Chess player; b. N. 1 (14), 1913, Bogoliubovo, in present-day Vladimir Oblast.
Honored Master of Sport of the USSR (1953); International Master (1959).
She has been the women's chess champion of the USSR (1947, 1948, 1950), and of the world (1953–56, 1958–62). Decor.: Badge of Honor.
Works: *Sorevnovaniia na pervenstvo mira po shakhmatam sredi zhenshchin.* Moscow 1955. *Sovetskie shakhmatistki,* 2nd ed. Moscow 1957. *Vera Menchik.* Moscow 1957.

Bykova, Olga Vasil'evna
State official.
Sec. of Cen. Cttee. Moldavian Komsomol 1956; Min. of Social Welfare Moldavian SSR N. 1962–; mem. Cen. Cttee. Moldavian CP 1963–.

Bykovskii, V. I.
Trade union official.
Mem. Cen. Cttee. Belorussian CP 1971–76; Sec. of Belorussian Republican Trade Union Council Je. 1975–.

Bykovskii, Valerii Fedorovich
Cosmonaut; b. Ag. 2, 1934, Pavlovskii Posad, Moscow Oblast.
Mem. CPSU 1963–; grad. Aeroclub School Moscow, Mil. Aviation College Kachinsk 1955 and Zhukovskii Mil. Aviation and Engineering Acad. Moscow; pilot, cosmonaut; made 81 earth orbits in 119 hours Je. 14–19, 1963, 3.3 million km, from Je. 16 simultaneously with "Vostok 6" (pilot V. V. Nikolaeva-Tereshkova); further orbit flights: with "Soiuz 22" S. 1976 and "Soiuz 31" Ag.–S. 1978. Decors.: Order of Lenin, Order of Red Banner, medals.

Bystritskaia, Elina Avraamovna
Russian actress; b. Ap. 4, 1928, Kiev.
Became People's Artist of the RSFSR in 1966. Member of the CPSU since 1970.
In 1953, she graduated from the department of acting at the Kiev Institute of Dramatic Arts. She worked in the Vilnius Russian Theater. Since 1958 she has been in the company of the Malyi Theater. Her roles include Margaret (Wilde's *Lady Windermere's Fan*), Baronness Shtral' (Lermontov's *Masquerade*), and Iuliia Filippovna (Gorkii's *Summer Folk*). She began making films in 1951. Her roles include Elizaveta Maksimovna (*An Unfinished Tale,* 1955), Aksin'ia (Quiet Don, after Sholokhov, 1957–58), Lelia (*Volunteers,* 1958), and Kseniia Rumiantseva (*Everything Is Left to People,* 1963). Decors.: two orders, as well as medals.

Chagin, Boris Aleksandrovich
Philosopher; b. Mr. 10 (22), 1899.
Corresponding member of the Academy of Sciences of the USSR (1960). Member of the CPSU since 1920.
Graduated from the Institute of the Red Professors in 1931. Began his teaching career in 1922, and in 1963 he became a senior research worker in the Leningrad subdepartment of philosophy of the Academy of Sciences of the USSR. His principal works deal with the history of Marxist-Leninist philosophy and with historical materialism. Chagin has been awarded the Order of Lenin, the Order of the October Revolution, four other orders, and various medals.
Works: *Filosofskie i sotsiologicheskie vozzreniia F. Meringa.* Moscow-Leningrad 1934. *Bor'ba marksizma-leninizma protiv filosofskogo revizionizma v kontse 19–nach. 20 vv.* Leningrad 1959. *Leninskii etap v razvitii marksistskoi filosofii.* Leningrad 1960. *G. V. Plekhanov i ego rol' v razvitii marksistskoi filosofii.* Moscow-Leningrad 1963. *Sub"ektivnyi faktor: Struktura i zakonomernosti.* Moscow 1968. *Ocherk istorii sotsiologicheskoi mysli v SSR (1917–1969).* Leningrad 1971. *Razrabotka G. V. Plekhanovym obshchesotsiologicheskoi teorii marksizma.* Leningrad 1977.

Chagina, Lizaveta Petrovna
State official.
Sec. of Trade Unions in Molodechno Oblast 1959; Dep. 1959–, Dep. Chair. of Supr. Sov. 1959–63, Sec. of Presidium Supr. Sov. Belorussian SSR Je. 1970–.

Chagovets, Rostislav Vsevolodovich, Dr. of Biol. Sc.
Biochemist; b. S. 21, 1904, Kiev.
Mem. CPSU 1945–; grad. Agrobiological Faculty at Inst. of Public Educ. Kiev 1925; corresponding mem. Acad. of Sciences Ukrainian SSR 1957–; Dr. of Biol. Sc. 1954; Prof. 1956–; teacher at various schools 1924–31; Asst., Lecturer at Kiev Medical Inst. 1932–50; in addition mem. staff Inst. of Biochemistry at Acad. of Sciences Ukrainian SSR

1933–48; Chief of Dept. of Biochemistry of Vitamins at same Inst. 1948–; mem. Presidium and Acad. Sec. of Dept. of Biochemistry, Biophysics and Physiology at Acad. of Sciences Ukrainian SSR 1966–; author of more than 90 works on biochemistry of muscles, application of vitamins in animal husbandry etc. Decors.: Badge of Honour, medals.
Publs.: "Tkanevyi obmen veshchestv" 1962, "Razvitie ucheniia o vitaminakh i vitaminoterapiia in "Sovetskaia meditsina" No. 6/1958, "Kharchuvannia sportsmeniv" 1959, "Nekotorye voprosy teorii i praktiki vitaminoterapii" in "Trudy VII sezda terapevtov Ukrainskoi SSR (Je. 3-8, 1957)" 1962.

Chagylov, A.
Komsomol official.
Sec. and mem. Bureau Cen. Cttee. Turkmen Komsomol Ag. 1982–.

Chaikovskaia, Elena Anatol'evna
Figure skater; b. D. 30, 1939, Moscow.
Master of Sport (1957) and Honored Coach of the USSR (1970). Skating choreographer. Champion of the USSR (1957).
Graduated from the A. V. Lunacharskii State Institute of Theatrical Arts in 1964 and began coaching the same year. Among her students have been such repeated champions and prizewinners of international, European, and Soviet competitions as L. A. Pakhomova and A. G. Gorshkov, V. N. Kovalev, and N. V. Linichuk and G. M. Karponosov. Chaikovskaia has been awarded the Order of the Red Banner of Labour.
Works: *Uzory russkogo tantsa*. Moscow 1972. *Figurnoe katanie*. Moscow 1975.

Chaikovskii, Boris Aleksandrovich
Composer; b. S. 10, 1925, Moscow.
Honored Art Worker of the RSFSR (1969).
In 1949, graduated from the Moscow Conservatory, where he had studied composition with N. Ia. Miaskovskii; previously he had studied with V. Ia. Shebalin and D. D. Shostakovich. In the late 1950's and early 1960's emerged as a leading composer of Soviet symphonic and chamber music. His principal works include the Symphony No. 2 (1962; State Prize of the USSR, 1969), *Chamber Symphony* (1967), and the orchestral work *Theme and Eight Variations* (1974). Other notable works are a piano quintet; piano, violin, and cello concerti; six quartets; and *The Lyric Poetry of Pushkin* (1972), a cyclic vocal composition. Chaikovskii has also composed *The Signs of the Zodiac* (1974), based on Russian poems, and music for motion pictures and radio.

Chaikovskii, Vasilii Fedorovich
State official.
Chair. of State Cttee. for the Supply of Petroleum Products Belorussian SSR F. 1980–.

Chailakhian, Mikhail Khristoforovich, Dr. of Biol. Sc.
Plant physiologist; b. Mr. 8 (21), 1902.
Grad. Erevan Univ. 1926; mem. USSR Acad. of Sciences 1968–; Prof. 1943–; Instructor for Agricultural Admin. of People's Commissariat of Agriculture Armenian SSR 1928–29; Research Asst. at Chair of Botany of Transcaucasian Inst. of Veterinary Zoo Technology in Erevan 1928–34; mem. staff Timiriazev Inst. of Plant Physiology of USSR Acad. of Sciences 1934–; Head of its Laboratory 1953–.

Chakabaev, Saken Erzhanovich
State official.
Dep. Min. of Geology Kazakh SSR 1972–78, Min. Ap. 1978–.

Chakovskii, Aleksandr Borisovich
Author; b. 1913; s. of a clerk.
Jew; mem. CPSU 1941–; grad. Gorkii Inst. of Literature Moscow 1938; electrical fitter, Dep. Head of Econ. Planning Dept. of Moscow Bulb Factory 1930–36; Literary Sec., Dept. Head, Dep. Ed. of newspaper "Oktiabr'" 1936–41; mem. staff film studio "Mosfilm" 1941–42; newspaper correspondent at the front 1942–45; Acting Sec. for Prose Section with USSR Writers' Union, mem. ed. board of newspaper "Znamia" 1945–54; Chief Ed. of newspaper "Inostrannaia literatura" 1954–62; Chief Ed. of "Literaturnaia gazeta" 1962–; Sec. of USSR Writers' Union 1963–; dep. mem. and mem. Comm. on External Affairs at Sov. of Nationalities USSR Supr. Sov. 1966–; cand. mem. Cen. Cttee. CPSU 1971–. Decors.: Stalin Prize, Order of Red Banner of Labour, 2 Orders of Lenin, Hero of Socialist Labour.
Publs.: trilogy "Eto bylo v Leningrade" (It Was in Leningrad) (vol. I "Voennyi korrespondent" (War Correspondent) 1944, vol. II "Lida" (Lidia) 1945, vol. III "Mirnye dni" (Peaceful Days) 1947), novels "U nas uzhe utro" (It is Already Morning Here) 1949, "Dorogi, kotorye my vybiraem" (The Paths We Choose) 1960, "Blokada" (Blockade) 1969–71, and others.

Chalaganidze, Shota Ivanovich, Cand. of Tech. Sc.
State official; b. 1933.
Mem. CPSU 1954–, grad. Georgian Agricultural Inst. (1956) and CPSU Cen. Cttee. Party Higher School 1971; Dep. Sec., Sec. of Komsomol, lab. asst. at Agricultural Inst. in Tbilisi 1953–56; lab. asst., researcher at Scientific Research Inst. for Mechanization and Electrification of Agriculture 1956–62; sector head, dep. head of Dept. of Agriculture Cen. Cttee. CP Georgia 1962–73; 1st Raion Cttee. Sec. CP Georgia in Borzhomi 1973–76; mem. Cen. Cttee. CP Georgia 1976–81, cand. mem. 1981–; Chair. of Georgian State Cttee. on forestry O. 1976–.

Chalmaev, Viktor Andreevich
Literary historian and critic; b. My. 9, 1932, Gus'-Khrustal'nyi, Vladimir Oblast.
Grad. Dept. of Philology of Moscow State Univ. 1955; first works publ. 1955.
Publs.: "Samye nasushchnye zaboty" (The Most Pressing Problems) 1962, "Puti razvitiia literatur narodov SSSR" (Ways of Development of the Literature of the Peoples of the USSR) 1962, "Literatura sud'by narodnoi" (Literature of the Fate of the People) 1966, "Viacheslav Shishkov" 1969, "Khram Afrodity. Tvorcheskii put' i masterstvo E. Nosova" (The Temple of Aphrodite. The Creative Life and Mastery of E. Nosov) 1972, "Ogon' v odezhde slova. O narodnosti, grazhdanstvennosti, problemakh masterstva sovremennoi prozy" (Fire in the Guise of the Word. On Nationality, Citizenship, Problems of Mastery in Contemporary Prose) 1973, and others.

Chanukvadze, Shota Illarionovich
State official.
Mem. Cen. Cttee. Georgian CP 1954–76; 1st Raion Sec. of Georgian CP in Makharadze 1954–58; Min. of Agriculture Georgian SSR 1959–61; Sec. and mem. Bureau of Cen. Cttee. Georgian CP 1962–73; Min. of Supplies Georgian SSR F. 1973–.

Chaplin, Boris Nikolaevich
Diplomat; b. 1931.
Mem. CPSU 1961–; grad. Moscow Mining Inst. 1955, Cand. of Tech. Sc. 1963, Asst. Prof. 1964; miner in Kemerovo Oblast 1955–57; party posts 1961–74; 2nd Sec. of Oktiabr Raion Cttee. CPSU in Moscow 1965–68; 1st Sec. of Cheremyshki Raion Cttee. CPSU in Moscow 1968–74; USSR Amb. to Vietnamese Democratic Republic 1974–; cand. mem. Cen. Cttee. CPSU 1976–. Decor.: USSR State Prize 1967.

Charyev, Mergen Ataevich
Trade union official.
Chair. of Turkmen Republican Trade Union Council Ja. 1976–79; mem. Cen. Cttee. 1976–, cand. mem. Bureau Cen. Cttee. Turkmen CP 1976–79, mem. My. 1979–; Sec. Cen. Cttee. Turkmen CP My. 1979–.

Charyeva, Zh. K.
Komsomol official.
Sec. 1974–79, 1st Sec. 1979–, mem. Bureau of Cen. Cttee. Turkmen Komsomol F. 1974–, mem. Cen. Cttee. Turkmen CP 1981.

Chavdar, Elizaveta Ivanovna
Ukrainian coloratura soprano and teacher; b. F. 23, 1925, Odessa.
People's Artist of the USSR (1952). Member of the CPSU since 1953.
Graduated from the Odessa Conservatory in 1948 and in the same year became a soloist at the Ukrainian Theater of Opera

and Ballet. In 1968 she joined the teaching staff of the Kiev Conservatory; she has been a professor there since 1979.
Chavdar's singing is melodious, and her articulation has great purity and expressiveness. Her voice is strong and beautiful in timbre. She is a virtuoso with vivid individuality and refined taste.
Among Chavdar's best roles are Antonida in Glinka's *Ivan Susanin,* Liudmila in Glinka's *Ruslan and Liudmila,* Gilda in Verdi's *Rigoletto,* Violetta in Verdi's *La Traviata,* Lakmé in Delibes's *Lakmé,* Iolan in G. Maiboroda's *Milana,* Oksana in Maiboroda's *Taras Shevchenko,* and Iarina in Maiboroda's *Arsenal.* She was the first Ukrainian singer to perform R. M. Glière's Concerto for Voice and Orchestra.
Chavdar is a laureate of the international competitions in Budapest (1949, second prize) and Berlin (1951, first prize). She has toured in the Soviet Union and abroad. Chavdar has been awarded the Order of Lenin, the Order of the October Revolution, and the Order of the Red Banner of Labor.
References: Kyrychenko, T. *Ielyzaveta Ivanivna Chavdar.* Kiev 1958. Kaufman, L. *Elizaveta Ivanovna Chavdar – narodnaia artistika SSSR.* Kiev 1960.

Chazov, Evgenii Ivanovich, Dr. of Med. Sc.
Medical scientist; b. 1929.
Russian; mem. CPSU 1962–; grad. Medical Inst. Kiev 1953; cardiologist; Dr. of Med. Sc. 1963; Prof. 1965–; corresponding mem. USSR Acad. of Medical Sciences 1967–; mem. staff Clinical Consultation Office at 1st Medical Inst. Moscow 1953–58; mem. staff Inst. of Cardiology at USSR Acad. of Medical Sciences 1958–67; Chief of Dept. of Intensive Care at Inst. of Cardiology of USSR Acad. of Medical Sciences 1967–; Chief Therapist at USSR Min. of Health 1965–67; Dep. USSR Min. of Health 1968–; Dep. to USSR Supr. Sov. 1974–; Director-General, All-Union Cardiological Scientific Center, USSR Academy of Medical Sciences 1976–; cand. mem. Cen. Cttee. CPSU 1981–82, mem. 1982–. Decor.: State Prize of the USSR 1969.

Cheban, Tamara Savel'evna
Moldavian singer (soprano); b. O. 9 (22), 1914, Berezlozhi, in what is now Orgeev Raion, Moldavian SSR.
Leading figure in public life and in the world of music. People's Artist of the USSR (1960). Member of the CPSU since 1953.
Graduated from the Kishinev Conservatory in 1946. She was a soloist with Moldavian Radio from 1944 to 1951. She sang with the Fluerash Folk Orchestra and performed with the orchestra on foreign tours, began teaching voice classes at the Institute of Arts in Kishinev in 1973, becoming a docent in 1976. Since 1973 she has been honorary chairman of the Music and Choir Society of the Moldavian SSR. She is an outstanding performer of Moldavian folk songs and songs by Moldavian composers. Decors.: State Prize of the USSR (1950), the Order of Lenin, and two other orders.

Chebrikov, Viktor Mikhailovich
State official; b. 1923.
Mem. CPSU 1944; grad. Dnepropetrovsk Metallurgical Inst. 1950, Maj.-Gen.; served in Sov. Army 1941–46; held party posts 1951–67; mem. Auditing Comm. Ukrainian CP 1956–61; 2nd Sec. 1958–59; 1st Sec. of Dnepropetrovsk City Cttee. Ukrainian CP 1961–63; cand. mem. Cen. Cttee. Ukrainian CP 1961–71; Sec. 1963–65; 2nd Sec. of Dnepropetrovsk Oblast Cttee. Ukrainian CP 1965–67 (for Ind. 1963–64); Head of Cadre Admin. 1967–68; Dep. Chair. USSR Cttee. on State Security (KGB) S. 1968–82, Chair. D. 1982–; cand. mem. Cen. Cttee. CPSU 1971–81, mem. 1981–.

Chechulin, Dmitrii Nikolaevich
Architect; b. Ag. 9 (22), 1901, Shostka, in present-day Sumy Oblast, Ukrainian SSR.
People's Architect of the USSR (1971); Hero of Socialist Labor (1976). Member of the CPSU since 1945.
Graduated from the architecture school of the Moscow VKhutein (Higher Art and Technical Institute) in 1929. Studied with A. V. Shchusev. As chief architect of the city of Moscow from 1945 to 1949, made major contributions to the reconstruction of the city. His designs for the Komsomol'skaia (first line, 1935), Kievskaia (second line, 1937–38), and Dynamo (ground-level vestibule, 1938) subway stations were awarded the State Prize of the USSR in 1941. Chechulin also designed the Tchaikovsky Concert Hall (1940, with K. K. Orlov), the new Moscow Council building (1945, with others), a high-rise apartment house on Kotel'nicheskaia Esplanade (1948–52, with A. K. Rostkovskii; State Prize of the USSR, 1953), the Hotel Rossiia and the Central Concert Hall (1967–74, with P. P. Shteller and others), the Library of Foreign Literature (1967, with V. A. Sitnov and others), and the Hall of Councils of the RSFSR (begun in 1968, with P. P. Shteller and others). Decors.: Order of Lenin, four other orders, and various medals.

Chekanauskas, Vitautas Al'girdo (Vytautas Algirdas Čekanauskas)
Architect; b. My. 13, 1930, Šiauliai.
People's Architect of the USSR (1975).
Graduated from the Art Institute of the Lithuanian SSR in Vilnius in 1955, has designed an art exhibition pavilion in Vilnius (1967) and collaborated on plans for the residential district of Lazdynai, also in Vilnius (1967–early 1970's; Lenin Prize 1974).

Chekaniuk, Andrej Terentevich
Historian and party official; b. O. 29, 1906, Kamenka, Vinnitsa Oblast.
Mem. CPSU 1928–; grad. Sov. Party College Vinnitsa 1930 and Inst. of Communism Kiev 1937; historian; corresponding mem. Acad. of Sciences Ukrainian SSR 1964–; Prof. 1964–; Komsomol and trade union posts 1925–28; Section Chief 1937–38, Managing Ed. of newspaper "Kommunist" (from 1943 "Radianska Ukraina") 1938–43; Dep. Chief of Propaganda Dept. with Cen. Cttee. Ukrainian CP 1943–44; Dep. Chief of Office for Propaganda and Agitation and Chief of Press Section with Cen. Cttee. Ukrainian CP 1945–46; Cen. Cttee. Ukrainian CP activities in Transcarpathians 1944–46; Dir. 1946–48, Rector of Party College of Cen. Cttee. Ukrainian CP 1948–. Decors.: 2 Orders of Lenin, Order of Red Banner of Labour, Order of Patriotic War 2nd Cl., Order of Red Banner, medals.
Publs.: "Kommunistychna partiia Ukrainy virnyi boiovyi zagin KRPS" in "Pid praporom marksyzmu-lininizmu" 1959, "Programa rozgornutogo budivnytstva kommunizmu" in "Suchasnist i deiaky ekonomichni pytannia budivnytstva kommunizmu v SSSR" 1962.

Chekati, Grigorii Iakovlevich
State official.
Chair. of State Cttee. for Pricing Moldavian SSR Ja. 1981–; cand. mem. Cen. Cttee. Moldavian CP 1981–.

Chekharin, Evgenii Mikhailovich
Jurist; b. N. 22, 1924, Gubino, Kozel'sk Raion, Kaluga Oblast.
Corresponding member of the Academy of Sciences of the USSR (1976). Member of the CPSU since 1950.
From 1954 to 1959, taught at Moscow State University, has had a responsible party position since 1959. From 1972 to 1978 he was rector of the Higher Party School of the Central Committee of the CPSU.
Works: *Sotsialisticheskaia demokratiia i sovremennaia ideologicheskaia bor'ba.* Moscow 1970. (With D. A. Kerimov.) *Sovetskaia politicheskaia sistema v usloviiakh rezvitogo sotsializma.* Moscow 1975. *Razvitie politicheskoi sistemy sovetskogo obshchestva na sovremennom etape.* Moscow 1977.

Chekoi, Arkhip Il'ich
State official.
Mem. Cen. Cttee. Moldavian CP 1971–; Min. of the Food Ind. Moldavian SSR Mr. 1977–.

Chelidze, Otar Silovanovich
Author and poet; b. S. 8, 1929, Tbilisi; s. of a worker.
Georgian; mem. CPSU 1958–; grad. Dept. of Philosophy of Tbilisi Univ. 1950; studied Gorkii Inst. of Literature 1950–54; first works publ. 1943.
Publs.: "Smert' kuznetsa" (The Death of

a Smith), "Ballada o tigre" (Ballad of the Tiger), "Vtoraia ruka" (The Second Hand), "Koren'" (The Root), "Kitesa", "Moe magnitnoe pole" (My Magnetic Field); translated "Eugene Onegin" by Pushkin and "Vasilii Terkin" by Tvardovsky into Georgian.

Chelomei, Vladimir Nikolaevich
Mechanical engineer; b. Je. 17 (30), 1914.
Mem. CPSU 1941–; grad. Kiev Aviation Inst. 1937; specialist in the field of mechanics; mem. USSR Acad. of Sciences 1962–; Prof. at Moscow Inst. of Technology 1952–.

Chelyshev, A.
Linguist; corresponding mem. USSR Acad. of Sciences 1981–.
Head of the Dept. of Literature of the People's of Asia at the Institute of Asian Studies USSR Acad. of Sciences; author of roughly 200 publications; deals mainly with the historical and typological study of Asian literature as it relates to world literary movements, the study of laws of development in Indian literature, and the works of leading Indian novelists.

Chemm, Vitalii Aleksandrovich
Party official; b. 1923.
Cand. mem. 1963–66, mem. Cen. Cttee. Latvian CP 1966–71 and 1976–; Head of Dept. of Agriculture at Cen. Cttee. Latvian CP 1965–71; USSR Dep. Min. of Supplies 1971–74; Sec. and mem. Bureau of Cen. Cttee. Latvian CP Ag. 1974–.

Chepurov, Anatolii Nikolaevich
Author and poet; b. Je. 16, 1922, Dukhovshchina, Smolensk Oblast.
Russian; mem. CPSU 1947–; first works publ. 1941.
Publs.: "Svetlyi den'" (A Bright Day) 1951, "Molodost' moia" (My Youth) 1956, "Veter s Nevy" (Wind From the Neva) 1961, "Vesennie fonariki" (Little Spring Lanterns) 1965, "Zolotoi pozhar" (Golden Fire) 1972, "Belaia dorozhka" (The White Path) 1973, and others.

Cherdintsev, Vasilii Makarovich
Laborer; b. 1927.
Mem. CPSU 1956–; combine operator at a machine and tractor station in Orenburg Oblast 1943–58; combine operator of "Rassvet" Collective Farm in Orenburg Oblast 1958–; mem. Cen. Auditing Comm. CPSU 1976–81; cand. mem. Cen. Cttee. CPSU 1981–. Decors.: Hero of Socialist Labor 1966, USSR State Prize 1976.

Cherednik, Vladimir Ivanovich
Diplomat.
USSR Amb. to Ghana 1971–74, to Zambia Jl. 1981–.

Cherenkov, Pavel Alekseevich
Physicist; b. Jl. 15 (28), 1904.
Mem. CPSU 1946–; grad. Voronezh Univ. 1928; mem. USSR Acad. of Sciences 1970–; teacher in Michurinsk 1928–30; mem. staff Physico-Mathematical Inst. 1930–35, mem. staff Inst. of Physics of USSR Acad. of Sciences 1936–. Decors.: State Prize of the USSR 1946, Nobel Prize 1958.

Cherep, Valerii Ivanovich
State official; b. 1940.
Mem. CPSU 1962–; master, foreman, chief engineer, head of a construction admin., head of a trust, Dep. Chair. Exec. Cttee. of Oblast Sov. of People's Deps. in Dnepropetrovsk 1957–76; 1st Dep. Chair. 1976–78, Chair. Ukrainian Inter-Kolkhoz Construction Assoc. 1978–82; cand. mem. Cen. Cttee. CP Ukraine 1981–; Min. of Rural Construction Ukrainian SSR My. 1982–.

Cherevashko, Eduard Ivanovich
State official.
Chair. of Auditing Comm. Cen. Cttee. Esthonian Komsomol 1966–68; 1st Sec. of Narva City Cttee. Esthonian CP 1968–82; Chair. of State Cttee. for Vocational Training Esthonian SSR Ag. 1982–. Mem. Cen. Cttee. Esthonian CP 1981–.

Cherkashina, Valentina Nikolaevna
Laborer; b. 1942.
Mem. CPSU 1972–, finished secondary school; worker in a glass plant 1958–60; weaver at Kamyshin Cotton Combine in Volgograd Oblast 1961–; mem. Cen. Auditing Comm. CPSU 1981–. Decor.: Heroine of Socialist Labor.

Cherkeziia, Otari Evtikhievich
State official; b. 1933, Georgian.
Mem. CPSU 1955–; grad. Georgian Polytech. Kirov Inst.; Dept. Head at Cen. Cttee. Georgian Komsomol, dep. head of a dept. with Cen. Cttee. USSR Komsomol 1957–61; mem. Cen. Cttee. Georgian CP 1961–; can. mem. Bureau Cen. Cttee. Georgian CP 1961–68 and 1971–79, mem. 1979–; 1st Sec. and mem. Bureau of Cen. Cttee. Georgian Komsomol 1961–67; Dep. to Supr. Sov. Georgian SSR 1963–; Dep. to Supr. Sov. USSR 1966–; 1st Sec. of Ordzhonikidze Raion Cttee. Georgian CP in Tbilisi 1968–70; Head of Dept. of Science and Educ. and of Dept. of Org. Party Work at Cen. Cttee. Georgian CP 1971–73; Dep. Chair. of Council of Mins. Georgian SSR 1973–; Chair. of Georgian Soc. for Conservation of Monuments 1974–.

Chernavin, Vladimir Nikolaevich
Military official; b. 1928.
Mem. CPSU 1949–; grad. Naval Acad. 1965 and Mil. Acad. of Gen. Staff 1969, Admiral 1978; held command posts in Sov. Navy 1947–; Commander of Northern Fleet Ja. 1977–; Dep. to USSR Supr. Sov. 1979–; cand. mem. Cen. Cttee. CPSU 1981–. Decor.: Hero of the Sov. Union 1981.

Chernenko, Konstantin Ustinovich
Party official; b. 1911; s. of a peasant.
Russian; mem. CPSU 1931–; grad. Higher School for Party Organizers at Cen. Cttee. CPSU 1945 and Kishinev Teachers' Training Inst. 1953; dept. head of a raion Komsomol cttee., sec. of a border unit's party org., dept. head of a raion all-union CP (B) cttee. 1929–41; Dep. Dept. Head and later Sec. of Krasnoiarsk Krai Cttee. CPSU 1941–43; Sec. of Penza Oblast Cttee. CPSU 1945–48; Head of Agitation and Propaganda Dept. with Cen. Cttee. Moldavian CP 1948–56; Dep. to Moldavian Supr. Sov. 1955–59; mem. Cen. Cttee. Moldavian CP 1956; Head of Propaganda und Agitation Dept. with CPSU Cen. Cttee. 1956–60; Head of Presidium Secr. USSR Supr. Sov. 1960–65; Head of Gen. Dept. at CPSU Cen. Cttee. 1965–; cand. mem. 1966–71, mem. Cen. Cttee. CPSU 1971–; Dep. to USSR Supr. Sov. 1966–; Sec. of Cen. Cttee. CPSU Mr. 1976–; cand. mem. 1977–78, mem. Politbureau Cen. Cttee. CPSU N. 1978–; Secretary-General of Cen. Cttee. CPSU F. 1984–, Pres. of State of USSR Ap. 1984–. Decor.: Order of the Red Banner of Labour.

Cherniaev, Anatolii Sergeevich
Party official; b. 1921.
Mem. CPSU 1942–; grad. Moscow State Univ. 1947, Cand. of hist. sciences 1950; served in Sov. Army 1941–45; taught at Moscow State Univ. 1950–53; posts in Cen. Cttee. CPSU apparatus 1953–58; mem. of editorial staff of journal "Problemy mira i sotsializma" 1958–61; posts in Cen. Cttee. CPSU apparatus 1961–70; Dep. Chief of Int. Dept. of Cen. Cttee. CPSU O. 1970–; mem. Cen. Auditing Comm. CPSU 1976–81; cand. mem. Cen. Cttee. CPSU 1981–.

Cherniakov, Iurii Nikolaevich
Diplomat; b. 1918.
Diplomatic service 1948–; mem. staff USSR Min. of Foreign Affairs 1955–57; Emb. Counsellor at USSR Emb. in Hungary 1957–59; sen. posts in USSR Min. of Foreign Affairs 1959–65; Emb. Counsellor-Envoy at USSR Emb. to USA 1965–67 and 1970; Head of Press Dept. 1971–73 and Sec.-Gen. and mem. Collegium of USSR Min. of Foreign Affairs 1973–77; USSR Amb. to Syria 1977–79; Head, Press Dept., USSR Min. of Foreign Affairs 1979–.

Chernigovskii, Vladimir Nikolaevich
Physiologist; b. F. 16 (Mr. 1), 1907.
Grad. Medical Dept. of Perm Univ. 1930; mem. USSR Acad. of Sciences 1960–; mem. USSR Acad. of Medical Sciences 1950–; Research Asst. at Veterinary Inst. in Orenburg and Medical Inst. in Sverdlovsk 1930–37; mem. staff Dept. of Gen. Physiology at Union Inst. of Experimental Med. in Leningrad 1937–41; mem. staff Naval-Medical Acad. in Leningrad 1941–53; Prof. 1944–; concurrently work

at various institutes of USSR Acad. of Medical Sciences and USSR Acad. of Sciences; Vice-Pres. of USSR Acad. of Medical Sciences 1953–57; Dir. of Inst. of Physiology (after reorganization renamed Inst. of Gen. Pathological Physiology) of USSR Acad. of Medical Sciences 1953–.

Chernikov, Ivan Semenovich
State official.
Head of Dept. of Motor Transport and Highways with Cen. Cttee. Lithuanian CP 1969–76; mem. Cen. Cttee. Lithuanian CP 1971–; Min. of Motor Transport and Highways Lithuanian SSR Ag. 1976–.

Chernyi, Aleksei Klementevich
Party official; b. 1921.
Ukrainian; mem. CPSU 1946–; grad. Moscow Chemical Machine Building Inst. 1942; foreman, tech. dept. head, dep. dir. of a factory in Komsomolsk-na-Amure 1942–49; dept. head of a raion cttee., a city cttee., 2nd Sec. of Komsomolsk-na-Amure City Cttee. CPSU, 1st Sec. of a Raion Cttee. CPSU 1949–56; Sec., 2nd Sec. of Khabarovsk Krai Cttee. CPSU, 1st Sec. of Oblast Cttee. CPSU, Jewish Autonomous Oblast 1956–62; Chair. of Khabarovsk Krai Exec. Cttee. of Workers' Deps. 1962–70; Dep. to USSR Supr. Sov. 1962–; 1st Sec. of Khabarovsk Krai Cttee. CPSU Jl. 1970–; mem. Cen. Cttee. CPSU 1971–. Decor.: Medal for Valiant Labour.

Chernyi, Gorimir Gorimirovich
Scientist in mechanics; b. Ja. 22, 1923, Kamenets-Podol'skii.
Corresponding member of the Academy of Sciences of the USSR (1962). Member of the CPSU since 1954.
He served in the Soviet Army from 1941 to 1945. He graduated from Moscow State University in 1949. From 1949 to 1958, he was on the staff of the Central Institute of Aircraft Engine Construction. He became a professor at Moscow State University in 1958 and was made the director of the university's Research Institute of Mechanics in 1960.
His main works deal with aerodynamics. Chernyi proposed a new method of studying and analyzing gas flows behind strong shock waves, particularly in the case of hypersonic flows past bodies. He has written works on the theory of detonation and combustion, boundary-layer theory, and the theory of gas-fired machines. Decors.: State Prize of the USSR and five orders and various medals.
Works: *Techeniia gaza s bol'shoi sverkhzvukovoi skorost'iu.* Moscow 1959. "Teoriia sverkhzvukovykh techenii gaza." In the collection *Mekhanika v SSSR za 50 let,* vol. 2. Moscow 1970.

Chernyshev, Aleksandr Ivanovich
State official.
Chair. of Petropavlovsk City Exec. Cttee. of Workers' Deps., Sec. of Tselinograd Oblast Cttee. Kazakh CP 1963–69; cand. mem. Cen. Cttee. Kazakh CP 1966–; Min. of Local Econ. Kazakh SSR O. 1969–78; Min. of Housing and Communal Econ. Kazakh SSR N. 1978–.

Chernyshev, Arkadii Ivanovich
Athlete; b. Mr. 3 (16), 1914, Nizhnii Novgorod (present-day Gorkii).
One of the founders of the Soviet school of ice hockey. Honored Master of Sport (1948); Honored Coach of the USSR (1956) Colonel.
He graduated from the Higher Coaches' School in 1940. He was a champion of the USSR in soccer (1937, 1940) and ice hockey (1947). From 1937 to 1940 he played on the Moscow Dynamo bandy team, helping win the USSR Cup four times. From 1948 to 1972, he and A. V. Tarasov coached the national ice-hockey team of the USSR. His pupils included more than 30 Olympic, world, and European champions. Decors.: four orders, as well as medals.

Chernyshev, Vladimir Ivanovich
Diplomat; b. 1929.
Mem. Diplomatic Service 1953–; 3rd Sec., USSR Embassy in Mexico 1958–61; 1st Sec., USSR Embassy in Washington 1964–69; mem. Apparat USSR Min. of Foreign Affairs 1969–79; USSR Amb. to Costa Rica 1979–81, to Brazil S. 1981–.

Chernyshev, Vladimir Vasil'evich
State official.
Mem. Cen. Cttee. Esthonian CP 1961–64 and 1966–; dir. of an engineering works in Moscow 1963; Min. of Timber, Cellulose, Paper and Timber Processing Ind. 1965–68, Min. of Timber and Timber Processing Ind. Esthonian SSR 1968–.

Cherskii, Nikolai Vasil'evich
Scientist; b. Ja. 20 (F. 2), 1905, settlement of Ol'ga, in what is now Primor'e Krai.
Specialist in mechanics and the development of petroleum and gas deposits. Corresponding member of the Academy of Sciences of the USSR (1968); Hero of Socialist Labor (1975). Member of the CPSU since 1942.
He graduated from the Institute of Water Transport Engineers in Vladivostok in 1931 and from the Academy of the Petroleum Industry in 1951. He became chairman of the Presidium of the Yakutsk Branch of the Academy of Sciences of the USSR in 1964 and has also served as the director of the Institute of Physical and Technical Problems of the North since 1973. His chief works deal with the designs of gas wells, the development of methods for calculating the reserves of natural gas deposits and for exploiting gas hydrate deposits, and container-pipeline transportation. He was the co-discoverer of the property of natural gas forming deposits in the crust in the form of a solid gas hydrate. He was a deputy to the seventh through ninth convocations of the Supreme Soviet of the USSR. Decors.: two Orders of Lenin, five other orders, and various medals.

Chervonenko, Stepan Vasilevich, Cand. of Phil. Sc.
Diplomat; b. 1915, Ukrainian.
Mem. CPSU 1940–; grad. Kiev State Univ. 1936; taught in Kiev Oblast 1937–40; completed courses for Marxist-Leninist teachers run by Ukrainian CP (B) Cen. Cttee. 1940; served in Sov. Army 1941–44; taught dialectical and historical materialism, Head of Dept. for Marxist-Leninist Principles at Cherkassy Teachers' Training Inst. 1944–48; postgraduate studies at Acad. of Social Sciences at All-Union CP (B) Cen. Cttee. 1948–49; Head of lecturing staff at Ukrainian CP (B) Cen. Cttee. 1949–51; Head of Science and Culture Dept. with Ukrainian CP Cen. Cttee. 1951–56; cand. mem. 1954–56, mem. Cen. Cttee. Ukrainian CP 1956–60; Dep. and Chair. of Foreign Affairs Comm. with Ukrainian Supr. Sov. 1955–63; Sec. 1956–59, cand. mem. Presidium Cen. Cttee. Ukrainian CP 1957–59; Dep. to 1958–62, Chair. of Credentials Comm. of Sov. of Nationalities USSR Supr. Sov. 1958–60; Sov. Amb. to China O. 1959–Ap. 1965; mem. Cen. Cttee. CPSU 1961–; Sov. Amb. to CSSR Ap. 1965–73, to France My. 1973– and also to Republic of Malagasy 1973–74. Decors.: 2 Orders of Lenin, Order of Red Banner of Labour.

Chervoniuk, Evgenii Ivanovich
Singer (bass); b. Ap. 21, 1924, Brusilov, in what is now Zhitomir Oblast.
People's Artist of the USSR (1967). Member of the CPSU since 1960.
In 1950, he graduated from the Kiev Conservatory, where he studied under D. G. Evtushenko and I. S. Patorzhinskii. He was a soloist at the Ukrainian Theater of Opera and Ballet from 1950 to 1952 and has been a soloist at the Kharkov Theater of Opera and Ballet since 1952. His wide repertoire includes the title roles in Glinka's *Ivan Susanin* and Mussorgsky's *Boris Godunov,* Mephistopheles in Gounod's *Faust,* Basilio in Rossini's *The Barber of Seville,* and Taras and the Elected Official in Lysenko's *Taras Bulba* and *Natalka Poltavka.* He also gives concerts. Since 1960 he has taught at the Kharkov Institute of the Arts.
He was a deputy to the sixth through eighth convocations of the Supreme Soviet of the Ukrainian SSR. He has performed abroad. Decor.: Order of Lenin.

Chesnavichius, Aleksandras Iuozovich
State official.
Mem. Cen. Cttee. Lithuanian CP 1960–66 and 1971–; cand. mem. Bureau Cen. Cttee. Lithuanian CP 1960–64,

mem. 1964–66; 1st Sec. and mem. Bureau Cen. Cttee. Lithuanian Komsomol 1960–66; Sec. and mem. Bureau Cen. Cttee. USSR Komsomol 1966–69; Head of Dept. of Admin. Organs with Cen. Cttee. Lithuanian CP 1969–74; Student 1974–76; Dep. Chair. of Council of Mins. Lithuanian SSR Jl. 1976–.

Chevachin, Valerii Nikolaevich
State official.
Dep. Head of Dept. of Ind. and Transport with Cen. Cttee. Latvian CP 1974; mem. Cen. Cttee. Latvian CP 1976–; Chair. of State Cttee. of Labour Latvian SSR D. 1981–.

Chibisov, Konstantin Vladimirovich
Physical chemist; b. F. 17 (Mr. 1), 1897, Moscow.
Corresponding member of the Academy of Sciences of the USSR (1946).
He graduated from Moscow State University in 1922. He worked at the Air Force Scientific and Testing Institute from 1918 to 1930, the All-Union Scientific Research Institute of Motion Pictures and Photography from 1930 to 1966, and the State Scientific Research and Design Institute of the Chemico-photographic Industry from 1966 to 1977. He gained a professorship at Moscow State University in 1950 and became chairman of the Commission on the Chemistry of Photographic Processes of the Academy of Sciences of the USSR in 1948. He was made honorary president of the International Committee on Scientific Photography in 1974.
His major work is devoted to the theory of the light sensitivity of photographic emulsions and the chemistry of the photographic process. He received the USSR State Prize in 1950. Decors.: 4 Orders of Lenin, six other orders, and several medals.
Works: *Teoriia fotograficheskikh protsessov*, vol. 1. *Kolichestvennyi fotograficheskii metod*. Moscow 1935. *Khimiia fotograficheskikh emul'sii*. Moscow 1975.

Chicherov, Vladimir Stepanovich
Laborer; b. 1933.
Mem. CPSU 1964–; finished evening school for working youth 1977; fitter, work brigade leader in a metal plant in Leningrad 1950–75; work brigade leader at "Leningradskii metallicheskii zavod"; Prod. Asscn. for Turbine Construction 1975–; mem. Cen. Cttee. CPSU 1981–. Decor.: Hero of Socialist Labor.

Chichikanov, V.
Economist.
Corresponding mem. USSR Acad. of Sciences 1981–. Dir. of the Inst. of Economic Research of the Far Eastern Scientific Center USSR Acad. of Sciences in Khabarovsk; expert in the fields of economics, economic management and planning, organization of work and production; took part in a series of extensive research projects on prospects for development of production capacity in the Urals and the Far East; author of more than 100 publications.

Chigrinov, Ivan Gavrilovich
Belorussian writer; b. D. 21, 1934, Velikii Bor, Kostiukovichi Raion, Mogilev Oblast.
Member of the CPSU since 1973.
He graduated from the philology department of Belorussian University in 1957. He was first published in 1961. He is the author of the short story collections *The Birds Fly to Freedom* (1965), *The Happiest Man* (1967), and *A Man Went to War* (1973), which deal with the life and labor of Soviet people and with the heroism of the past war and its consequences. His pair of novels *The Quail's Cry* (1972) and *Blood Acquittal* (1977) depict the dramatic nature of events and the fates of individuals at the beginning of the Great Patriotic War of 1941–45. Chigrinov is a co-author of a television screenplay about the Minsk party underground entitled *The Ruins Are Firing* (1973; State Prize of the Belorussian SSR 1974).

Chikobava, Arnol'd Stephanovich
Linguist; b. Mr. 14 (26), 1898, Sachikobao, in what is now Tskhakaia Raion.
Academician of the Academy of Sciences of the Georgian SSR (1941).
He graduated in 1922 from the University of Tbilisi, where he became a professor in 1933. His major works are devoted to general linguistics, the history and structure of the Caucasian languages, and ergative sentence construction. He compiled the *Chan-Mingrelian-Georgian Dictionary* (1938) and edited the *Defining Dictionary of the Georgian Language* (vols. 1–8, 1950–64) and numerous publications devoted to Caucasian studies. Decors.: three Orders of Lenin and two other orders.
Works: *Chanuris gramatikuli analizi tek'stebit'urt'*. Tbilisi 1936. *Saxelis p'uzis uzvelesi agebuleba k'art'velur enebshi*. Tbilisi 1942. *Ergatiuli konstruk'c'iis problema iberiul-kavkasiur enebshi*, vols. 1–2. Tbilisi 1948–61. *Martivi cinadadebis problema k'art'ulshi*, 2nd ed., [vol.] 1. Tbilisi 1968. *In Russian translation: Problema iazyka kak predmeta iazykoznaniia*. Moscow 1959. *Vvedenie v iazykoznanie*, part 1, 2nd ed. Moscow 1953.

Chikovani, Grigol Samsonovich
Georgian writer; b. Mr. 2 (15), 1910, Khobi, in what is now Khobi Raion, Georgian SSR.
The son of a civil servant, he graduated from the department of law at the University of Tbilisi in 1936 and from the screenwriting department of the Moscow Institute of Cinematography in 1939. He was first published in 1932. During the Great Patriotic War of 1941–45, he wrote stories about the exploits of Soviet servicemen and workers. His most important postwar works are contained in the cycle *Odishi Stories*, in which critical social conflicts unfold against a background of the colorful Western Georgian landscape and sketches of peasant life. He proves himself a master of social satire in such works as the short story "Staircase." In his historical novel about revolution entitled *February Is Here* (1971; Russian translation, 1973) he describes the revolutionary struggle of Georgian workers under the leadership of the Bolsheviks. The novel is noteworthy for the profound historical knowledge it displays of the period. He has also written plays and screenplays.
Works: [Ch'ik'ovani, G.] *T'xzulebani or tomad*, vols. 1–2. Tbilisi 1971–73. *In Russian translation: Pod nebom Odishi*. Moscow 1964. *Lestnitsa: Rasskazy*. Moscow 1966. *Vadilai-Vadila: Rasskazy i povest'*. Moscow 1972. *Sobr. soch.*, vols. 1–2. Tbilisi 1976.

Chiladze, Tamaz Ivanovich
Georgian writer; b. Mr. 5, 1931, Signakhi.
Member of the CPSU since 1967.
He graduated from the department of philology of the University of Tbilisi in 1954. He was first published in 1951. His works deal with contemporary problems, and his hero is usually a young man of the 1950's or 1960's seeking a place in life. He became editor of the journal *Sabchota khelovnebi* (Soviet Art) in 1973.
Works: [Chilaze, T'.] *Lek'sebi*. Tbilisi 1956. *Mot'xrobebi*. Tbilisi 1972. *Mkvleloba. Davicqebuli ambavi. Kiesebi*. Tbilisi 1976. *In Russian translation: Seti zvezd*. Tbilisi 1961. *Progulka na poni*. Tbilisi 1963. *Pervyi den'*. Tbilisi 1965. *Kto zhivet na zvezdakh*. [Foreword by E. Mal'tsev.] Moscow 1970. *Belyi dym: Povesti roman*. Tbilisi 1973.

Chilaia, Sergei Evstaf'evich
Georgian writer, critic, and literary scholar; b. O. 29 (N. 11), 1912, Sergieti, in what is now Gegechkori Raion.
Honored Scientist of the Georgian SSR (1971). Member of the CPSU since 1939.
The son of a peasant, he was first published in 1932. He was secretary of the Writers' Union of Georgia from 1951 to 1974. His works deal with Georgian writers of the 19th and 20th centuries and the development of Soviet Georgian literature. He is the author of the chronicle-like novel *Ekaterina Chavchavadze* (parts 1–2, 1966–67; Russian translation, 1969). Decors.: two orders and various medals.
Works: *Literaturuli cerilebi*. Tbilisi 1946–49. *Meoc'e saukunis k'art'uli mcerloba*, parts 1–3, Tbilisi 1956–62. *Uaxlesi k'art'uli mcerloba*. Tbilisi 1975. *In Russian translation: Galaktion Tabidze*. Moscow 1958. *Gruzinskie pisateli-demokraty kontsa XIX–nach. XX vv*. Tbilisi 1965. *Potomki*. Moscow 1976.

Chilebaev, Toktogul Bekbolotovich
State official.
Head of Dept. of Trade, Planing and Finances with Cen. Cttee. Kirghiz CP 1976–79; mem. Auditing Comm. Kirghiz CP 1976–; Min. of Trade Kirghiz SSR Mr. 1979–.

Chirkov, Boris Petrovich
Russian actor; b. Jl. 31 (Ag. 13), 1901, Lozovaia-Pavlovka, now Dnepropetrovsk Oblast.
People's Artist of the USSR (1950); Hero of Socialist Labor (1975). Member of the CPSU since 1945.
He after graduated from the Leningrad Institute of Stage Arts in 1926, performed with the Leningrad Theater for Young Audiences and the New Theater for Young Audiences. He later appeared in several Moscow theaters: the Central Film Actors' Theater Studio, the Pushkin Theater (1950–65), and the Gogol Theater (since 1966).
He has appeared in motion pictures since 1928. He became especially well known as the title character in the Maksim trilogy: *The Youth of Maksim* (1935), *The Return of Maksim* (1937), and *The Vyborg Side* (1939). In these films he created a romantic and yet deeply realistic portrait of a hero of the revolutionary era, the young worker Maksim, who learns profound lessons from his experiences in the class struggle and becomes a prominent Bolshevik revolutionary.
His other motion-picture roles include Stepan Lautin in *The Teacher* (1939), Makhno in *Aleksandr Parkhomenko* (1942), the title role in *Glinka* (1947), and Chizhov in *True Friends* (1954). In the theater he has played the roles of Kuzovkin in Turgenev's *The Boarder*, Lebedev in Chekhov's *Ivanov*, and Rasputin in *The Empress's Plot* by Shchegolev and A. N. Tolstoy.
He was a deputy to the third convocation of the Supreme Soviet of the USSR. Decors.: State Prize of the USSR in 1941, 1947, 1949, and 1952, three Orders of Lenin, two other orders, and several medals.
Works: *Pro nas, pro akterov*. Moscow 1957. *Rasskazy o tvorcheskom puti*. Moscow 1965.

Chistiakov, Mikhail Nikolaevich
Military leader; b. N. 6 (18), 1896, St. Petersburg.
Marshal of the artillery (1944). Member of the CPSU since 1944.
The son of a railroad employee, he fought in World War I from 1915 as an artillery scout and was twice awarded the Cross of St. George. In 1917 he was promoted to ensign and made a platoon commander. He joined the Red Army in September 1918. During the Civil War of 1918–20, Chistiakov commanded a platoon, a battery, and a battalion. He graduated from courses for senior command personnel in 1926 and from advanced courses for higher command personnel at the M. V. Frunze Military Academy in 1930.
During the Great Patriotic War of 1941–45, he was commander of artillery on the Western Front and later commanded the Thirty-fourth Army on the Northwestern Front (1941). He was named head of the Directorate of Combat Training of the Main Artillery Directorate in December 1941 and deputy artillery commander of the Red Army in June 1943. As a representative of the commander of the artillery, he helped prepare and execute several operations on the Stalingrad, Don, Southern, Voronezh, Steppe, Third and Fourth Ukrainian, Third Belorussian, and First, Second, and Third Baltic fronts. As acting commander of the artillery under the commander in chief of Soviet forces in the Far East, he participated in the defeat of the Japanese forces in the Far East in 1945.
In April 1946, he was named deputy artillery commander of the armed forces of the USSR. He became inspector-general of the Main Inspectorate of the Ministry of Defense in November 1957 and military inspector-adviser of the Group of Inspectors-General of the Ministry of Defense in December 1964. Decors.: three Orders of Lenin, the Order of the October Revolution, three Orders of the Red Banner, the Orders of Suvorov First and Second Class, two Orders of Kutuzov First Class, the Order of Bogdan Khmel'nitskii First Class, the Order of the Patriotic War First Class, the Order of the Red Star, the Order for Service to the Motherland in the Armed Forces of the USSR Third Class, and various medals, also foreign orders and medals.

Chistopliasov, Stepan Ivanovich
State official; b. 1930.
Min. of Food Ind. RSFSR Mr. 1979–. Decor.: Order of Red Banner of Labour.

Chistov, K.
Ethnographist.
Corresponding mem. USSR Acad. of Sciences 1981–. Sector head at the N. N. Mikukho-Maklay Inst. of Ethnography USSR Acad. of Sciences; examines questions of the methodology of historical research, the modern theory of intellectual culture, the development of ethnography and folklore studies, social psychology and sociolinguistics, and contemporary ethnography; author of over 200 publications; chief editor of the periodical *Sovetskaya etnografiya* (Soviet Ethnography). Decor.: USSR State Prize for his collaboration on the collective work "Current Ethnic Trends in the USSR."

Chitanava, Nodar Amvrosievich
State official; b. 1935, Chkadiashi, Sugdidi Raion.
Georgian; mem. CPSU 1958–; construction engineer, Cand. of Econ. Sciences; collective farmer 1954–58; instructor, 2nd, 1st Raion Cttee. Sec. Georgian Komsomol in Sugdidi 1959–62; Raion Cttee. Sec. CP Georgia in Sugdidi, Dep. Sec. of Ind. Production Cttee. CP Georgia in Sugdidi 1962–64; adviser to Republican House for Pol. Ed. Cen. Cttee. CP Georgia 1964–66; attended CPSU Cen. Cttee. Party Higher School 1966–68; instructor at Agriculture Dept. Cen. Cttee. CP Georgia 1968–70; 1st Sec. Tskhakaevskii Raion Cttee. CP Georgia 1970–73; 2nd Oblast Cttee. Sec. CP Georgia Adzhar ASSR; cand. mem. 1971–76, mem. Cen. Cttee. CP Georgia 1976–; Min. of Agriculture Georgian SSR 1974–79; mem. Bureau Cen. Cttee. CP Georgia 1979–; 1st Dep. Chair. of Council of Mins. Georgian SSR Ja. 1979–.

Chivilikhin, Vladimir Alekseevich
Author; b. Mr. 7, 1928, Mariinsk, Kemerovo Oblast; s. of a worker.
Mem. CPSU 1952–; grad. Dept. of Journalism of Moscow State Univ. 1954.
Publs.: "Zhivaia sila. Dokumental'naia povest'" (Living Strength. Documentary Stories) 1957, "Zdravstvuite, mama!" (Greetings, Mother!) 1958, "Pro Klavu Ivanovu" (About Klava Ivanova) 1964, "Nad urovnem moria" (Above Sea-Level) 1967, "Pestryi kamen'" (Coloured Stone) 1969, "Goroda v teni derevev" (Towns in the Shadow of Trees) 1973, "Liubit li ona tebia?" (Does She Love You?) 1968, "Zemlia v bede" (Land in Trouble) 1970, and others.

Chkheidze, Revaz Davidovich
Motion-picture director; b. D. 8, 1926, Kutaisi.
People's Artist of the Georgian SSR (1966). Member of the CPSU since 1958. He studied in the directing department of the Tbilisi Shota Rustaveli Theatrical Institute from 1943 to 1946 and graduated in 1953 from the All-Union State Institute of Cinematography, where he studied in the workshop of M. I. Romm and S. I. Iutkevich. He is a director at the Gruziia-fil'm studio, where, together with the director T. E. Abuladze, he has produced several documentary films, as well as the feature film *Lurdzha Magdany* (1956; prize-winner at the Ninth International Film Festival in Cannes). His most outstanding works as a director have been *The Father of a Soldier* (1965), the leading actor of which, S. A. Zakariadze, was awarded the Lenin Prize in 1966, and *The Seedlings* (1972). His other films have included *Our Yard* (1957), *Maia From Tskhneti* (1960), and *Those Young People!* (1969).
He was a deputy to the ninth convocation of the Supreme Soviet of the USSR. Decors.: Order of Lenin, two other orders, and several medals.

Chkheidze, Zurab Amvrosievich
State official.
Cand. mem. 1966–71, mem. Cen. Cttee. Georgian CP 1971–; Min. of Posts and Telecommunications Georgian SSR D. 1972–75; mem. Bureau of Cen. Cttee. Georgian CP D. 1975–; Sec. of Cen. Cttee. Georgian CP 1975–82; Dep. Chair. of Council of Mins. and Chair. of GOSPLAN Georgian SSR. Jl. 1982–.

Chkhenkeli, Ivan Noevich
Architect; b. S.16 (29), 1910, Batumi.
Honored Artist of the Georgian SSR (1961). Member of the CPSU since 1945.
He studied at the Civil Engineering Institute in Tbilisi from 1928 to 1933. He has taught at the Academy of Arts in Tbilisi since 1944. From 1959 to 1972 he was chief architect of Tbilisi, and since 1962 he has been chairman of the board of the Architects' Union of Georgia. Most of his major works are in Tbilisi. They include apartment houses on Plekhanov Prospect (1950), the renovated Communications Building (1954), and the Concert Hall of the Georgian State Philharmonic (1971). He won the Rustaveli State Prize of the Georgian SSR in 1973. Decors.: two orders and various medals.

Chkhikvadze, Viktor Mikhailovich
Jurist; b. D.19, 1911 (Ja.1, 1912), Zorveti, in what is now Zestafoni Raion, Georgian SSR.
Corresponding member of the Academy of Sciences of the USSR (1964). Member of the CPSU since 1939.
He was director of the Institute of State and Law of the Academy of Sciences of the USSR from 1964 to 1973. He was president of the International Association of Legal Science from 1968 to 1972 and vice-president of the International Political Science Association from 1968 to 1970. He was namend a foreign member of the Bulgarian Academy of Sciences in 1970. Decors.: four orders and the Joliot-Curie Gold Peace Medal of the World Peace Council.
Works: *Voprosy sotsialisticheskogo prava i zakonnosti v trudakh V. I. Lenina*. Moscow 1960. *Gosurdastvo, demokratiia, zakonnost': Leninskie idei i sovremennost'*. Moscow 1967. *Bor'ba za mir – neodolimoe dvizhenie sovremennosti*. Moscow 1969.

Chkhikvishvili, David Iosifovich
State official; b. 1921, Geguti.
Doctor of Historical Sciences, mem. CPSU 1940–; grad. Cen. Cttee. Party Higher School CP Georgia and Pushkin State Pedagogical Inst. in Tbilisi; served in Sov. Army 1939–46; student, in Georgian Society for Friendship and Cultural Relations with Foreign countries and in apparatus of Presidium of Acad. of Sciences Georgian SSR 1946–54; 1st Sec. Kalinin Raion Cttee. CP Georgia in Tbilisi 1954–55; head of Dept. for Science and Culture Cen. Cttee. CP Georgia 1955–56; mem. Cen. Cttee. CP Georgia 1956–61 and 1971–; 1st Sec. of Ordzhonikidze Raion Cttee. CP Georgia in Tbilisi 1956–57; Min. of Culture Georgian SSR 1957–61; Rector of Pushkin State Pedagogical Inst. in Tbilisi 1961–69; head of Dept. of Culture of Central Cttee. CP Georgia 1969–72; Rector of Tbilisi State Univ. 1972–80; Min. of Higher and Secondary Specialized Education Georgian SSR Ja. 1980–. Decors.: Order of Lenin, Order of the Oct. Rev., Order of the Red Banner of Labour, two Orders of the Great Patriotic War, Order of the Red Star, medals.

Chkoniia, Lamara Grigor'evna
Georgian singer (lyric-coloratura soprano); b. D.27, 1930, Batumi.
People's Artist of the USSR (1976). Member of the CPSU since 1974.
She graduated from the Tbilisi Conservatory in 1956, where she studied voice under V.S. Kashakashvili and G.M. Gogichadze. From 1956 to 1960 and since 1968 she has been a soloist at the Georgian Theater of Opera and Ballet; from 1960 to 1968 she was a soloist at the Ukrainian Theater of Opera and Ballet.
She possesses a voice of delicate silvery timbre, as well as extraordinary musicality and impeccable technique. Her roles have included Marfa in Rimsky-Korsakov's *The Tsar's Bride,* Tat'iana and Iolanthe in Tchaikovsky's *Eugene Onegin* and *Iolanthe,* the title roles in Puccini's *Madame Butterfly* and Massenet's *Manon*. Violetta in Verdi's *La Traviata,* Rosina in Rossini's *The Barber of Seville,* and Maro and Eteri in Z. Paliashvili's *Daisi* and *Abesalom and Eteri.*
She was a prize-winner at the Young Opera Singers' Competition in Sofia in 1963 and at the Glinka All-Union Vocalists Competition in Moscow in 1960. She gives recitals in addition to her theater performances. Decors.: Order of the Badge of Honor and several medals.

Cholak, Mikhail Isaevich
State official.
Mem. Cen. Cttee. Moldavian CP 1966–; Head of Dept. for Trade and Public Services at Cen. Cttee. Moldavian CP 1966–70; Chair. for Production, Ind. Processing and Sale of Fruit and Vegetables at Agrarian-Ind. Union with Council of Mins. Moldavian SSR 1972–73; Min. of Trade Moldavian SSR F. 1973–.

Chonoev, Abdyiamat
State official.
Chair. of State Cttee. on Pricing Kirghiz SSR Je. 1977–.

Chudakov, Aleksandr Evgen'evich
Physicist; b. Je. 16, 1921, Moscow; s. of E.A. Chudakov.
Corresponding member of the Academy of Sciences of the USSR (1966).
He graduated from Moscow State University in 1947. From 1946 to 1971, he was on the staff of the Institute of Physics of the Academy of Sciences of the USSR. In 1971 he joined the staff of the Institute of Nuclear Research of the Academy of Sciences of the USSR.
His main works deal with the nature and properties of cosmic rays. He developed a technique for measuring the energy of a particle that caused a cosmic-ray shower. He used the technique to search for local sources of high-energy gamma quarta. Together with S.N. Vernov and others, he discovered and investigated the earth's outer radiation belt. Decors.: Lenin Prize in 1960. Three orders and various medals.

Chudin, Vitalii Ivanovich
State official; b. 1929.
Mem. CPSU 1954; grad. Altai Inst. of Agricultural Mechanical Engineering 1953; shop technologist, shop superintendent, sec. of a party cttee., dir. of a plant 1952–67; Dep. Chair. of Council of Mins. Kirgiz SSR 1967–76; Dep. USSR Min. of Construction, Road, and Municipal Machine Building 1976–80, Min. D. 1980–; mem. Cen. Auditing Comm. CPSU 1981–.

Chukarin, Viktor Ivanovich (Victor Tchoukarine)
Gymnast; b. N.9, 1921, Krasnoarmeiskoe, Donetsk Oblast.
Honored Master of Sport (1951); Honored Coach of the Ukrainian SSR (1972). Docent (1963); head of the subdepartment of gymnastics of the L'vov Institute of Physical Culture (since 1971). Member of the CPSU since 1951.
He was overall men's champion of the Olympics in 1952 and 1956, of the world in 1954, and of the USSR from 1949 to 1951 and in 1953 and 1955. He was Olympic champion in various events five times in 1952 and 1956, world champion in 1954, and champion of the USSR 13 times from 1948 to 1956. Decors.: Order of Lenin and several medals.
Work: *Put' k vershinam*. [Moscow] 1955.

Chukhanov, Zinovii Fedorovich
Scientist; b. O.8 (21), 1912, Novinki, in what is now Krasnogorsk Raion, Moscow Oblast.
Specialist in thermal power engineering and power production technology. Corresponding member of the Academy of Sciences of the USSR (1939). Member of the CPSU since 1944.
He studied at the Moscow Institute of Chemical Engineering. In 1938 he became the head of a laboratory at the Institute of Power Engineering of the Academy of Sciences of the USSR.
His main works deal with problems of the combustion, gasification, and pyrolysis of solid fuels. He developed a theory of the high-speed pyrolysis of fuels and proposed new methods for the energy-efficient use of fuels in electric power plants. A number of his works are devoted to heat-transfer theory, diffusion, and

systems analysis as applied to the evaluation of the efficiency of new equipment. Decors.: three orders and various medals. Works: *Protsess gazifikatsii koksa i problema podzemnoi gazifikatsii topliv.* Moscow 1957. *Nekotorye problemy topliva i energetiki.* Moscow 1961.

Chukhrai, Grigorii Naumovich
Motion-picture director; b. My. 23, 1921, Melitopol'.
People's Artist of the RSFSR (1969). Member of the CPSU since 1944.
In 1953, he graduated from the department of directing of the All-Union State Institute of Cinematography, where he studied under S.I. Iutekevich and M.I. Romm. One of the leading directors of the postwar cinema, he has made a number of significant films that have gained popularity throughout the Soviet Union and the world; several of his films have won prizes at international film festivals. His most outstanding works have been *The Forty-first* (1956, based on the novella by B.A. Lavrenev), *Clear Sky* (1961), and *Memory* (1970). His *Ballad of a Soldier* (1959) was awarded the Lenin Prize in 1961. From 1966 to 1971, he worked at the All-Union State Institute of Cinematography. Decors.: five orders and several medals.
Work: *Ballada o soldate.* Moscow 1960. (With V.I. Ezhov.)

Chukhrov, Feodor Vasil'evich
Mineralogist and geochemist; b. Jl. 2 (15), 1908, Egor'evsk, Moscow Oblast.
Academician of the Academy of Sciences of the USSR (1970; corresponding member, 1953). Member of the CPSU since 1953.
He graduated from the Moscow Geological Prospecting Institute in 1932. From 1936 to 1956 he worked at the Institute of Geological Sciences of the Academy of Sciences of the USSR, where he was appointed deputy director in 1950. In 1956 he became director of the Institute of the Geology of Ore Deposits, Petrography, Mineralogy, and Geochemistry of the Academy of Sciences of the USSR.
His main works deal with the mineralogy of ore deposits and the geochemistry of ore-forming processes. He established the science of natural colloids and developed the principles of the theory of the typomorphism of minerals. He contributed to the development of the mineralogy and geochemistry of the supergene zone: he conducted research on the mineralogy of the oxidation zone of sulfide deposits and studied clay minerals, iron oxides, and sulfur isotopes. He is the editor of the multivolume reference work *Minerals,* which has been published since 1960 using the crystallochemical classification of minerals that he developed. Decors.: State Prize of the USSR in 1951, Order of Lenin, three other orders, and several medals. Chukhrovite, a sulfate-fluoride of calcium, aluminum, and rare-earth elements, was named in honor of him.
Works: *Zona okisleniia sul'fidynkh mestorozhdenii stepnoi chasti Kazakhstana.* Moscow 1950. *Kolloidy v zemnoi kore* [2nd ed.]. Moscow 1955. *Mineralogiia i zonal'nost' Vostochnogo Kounrada.* Moscow 1960.

Chulanov, B. G.
Trade union official.
Sec. of Kazakh Republican Trade Union Council N. 1980–.

Chumakov, Mikhail Petrovich
Microbiologist and virologist; b. N. 1 (14), 1909, Ivanovka, in what is now Volyn' Raion, Orel Oblast.
Academician of the Academy of Sciences of the USSR (1960; corresponding member, 1948). Member of the CPSU since 1940.
He graduated from the First Moscow Medical Institute in 1931. He was director of the D.I. Ivanovskii Institute of Virology of the Academy of Medical Sciences of the USSR from 1950 to 1954. He was director of the academy's Institute of Poliomyelitis and Viral Encephalitides (founded on his initiative) from 1955 to 1972; he served as scientific director of the institute from 1972 to 1976. His work has mainly been devoted to the etiology, immunology, laboratory diagnosis, epidemiology, prophylaxis, and treatment of human viral diseases. Among the diseases he has studied are Russian spring-summer and mosquito-borne encephalitides, poliomyelitis, enteroviral diseases, measles influenza, and trachoma. He discovered several viruses that are pathogenic for man and developed a theory of hemorrhagic fevers. He assisted in the development of vaccines for poliomyelitis, measles, Russian spring-summer and Japanese encephalitides, and other diseases. He also helped organize mass immunization campaigns against poliomyelitis, which resulted in the rapid eradication of poliomyelitis epidemics in the USSR and many other countries. Decors.: three orders, several medals, the Lenin Prize (1963), the State Prize of the USSR (1941), and the D.I. Ivanovskii Prize of the Academy of Medical Sciences of the USSR (1953). He became a member of the Leopoldina German Academy of Naturalists in 1969.

Chunikhin, Sergei Antonovich (Siargei Antonavich), Dr. of Phy.-Math. Sc.
Mathematician; b. S. 8 (21), 1905, Kharkov.
Grad. Moscow Univ. 1929; Prof. 1934–; Dr. of Phy.-Math. Sc. 1936; mem. Acad. of Sciences Belorussian SSR 1966–; mem. staff of various scientific institutions in Moscow, Tomsk and Gomel; Prof. Chair of Algebra at Pedagogical Inst. Gomel 1960– (Univ. 1968–). Decors.: Kuibyshev Prize 1952, Badge of Honour.

Chura, Leonid Maksimovich
State official.
Min. of Rural Construction Belorussian SSR 1980–; mem. Auditing Comm. Belorussian CP 1981–.

Churbanov, Iurii Mikhailovich
State official; b. 1936.
Mem. CPSU 1960–; grad. Moscow State Univ. 1964; Lt.-Gen.; posts in Komsomol 1957–70; Dep. Chief of Pol. Dept. in USSR Min. of Internal Affairs 1970–71; Dep. Chief of Pol. Admin. of Security Police Force (MVD Troops) 1971–75, Chief 1975–77; mem. Cen. Auditing Comm. CPSU 1976–81; cand. mem. Cen. Cttee. 1981–; Dep. USSR Min. of Internal Affairs 1977–80, 1st Dep. Min. 1980–. Decor.: USSR State Prize 1980.

Chutkikh, Aleksandr Stepanovich
Innovator in light industry; b. Ag. 2 (15), 1903, Pavelets, in what is now Skopin Raion, Riazan' Oblast.
Member of the CPSU since 1950.
He spent most of his career at the Krasnokholmskii Worsted Combine in Moscow – from 1922 to 1958. Between 1927 and 1930, without leaving work, he completed evening courses at a general-education school. He served as deputy foreman in a textile shop from 1937 to 1950 and as technical foreman from 1950 to 1958. He has received a personal pension since 1958.
In 1949, he initiated socialist competition of work brigades for high-quality production. As a total master of his work specialty, he serviced 34 (instead of the usual 24) wide looms. He was a deputy to the third and fourth convocations of the Supreme Soviet of the USSR. Decor.: State Prize of the USSR in 1949.

Dadabaev, Abdurakhman
Party official.
Mem. Auditing Comm. 1966–71, mem. Cen. Cttee. Tadzhik CP 1976–; Head of Dept. of Ind. and Transport at Cen. Cttee. Tadzhik CP 1971–73; Sec. and mem. Bureau Cen. Cttee. Tadzhik CP N. 1973–.

Dadabaev, Khasanzhan
State official.
Mem. Cen. Cttee. Kirghiz CP 1963–; Min. of Public Service Kirghiz SSR Ag. 1974–82; Chair. of State Cttee. of Material and Techn. Supplies Kirghiz SSR S. 1982–.

Dadaboev, Rauf
State official.
Mem. Bureau for Ind. and Construction at Cen. Cttee. Tadzhik SSR 1962–63; Dep. to Supr. Sov. Tadzhik SSR 1963–; mem. Cen. Cttee. Tadzhik CP 1963–; Head of Depts. of Science and Culture, Science and Educ., and Propaganda and Agitation at Cen. Cttee. Tadzhik CP 1964–74; Min. of Educ. Tadzhik SSR Je. 1974–.

Dadenkov, Iurii Nikolaevich, Dr. of Tech. Sc.
Scientist; b. O. 28, 1911, Lubny, Poltava Oblast; s. of a teacher.
Mem. CPSU 1939–; grad. Kharkov Inst. of Motor Transport and Highway Engineering 1933; Dr. of Tech. Sc. 1951; Prof. 1962–; corresponding mem. Acad. of Sciences Ukrainian SSR 1961–; Research Asst., Lecturer at Kharkov Inst. of Motor Transport and Highway Engineering 1934–41; served in Sov. Army 1941–44; Dir. of Kiev Inst. of Motor Traffic and Highway Engineering 1944–59; Dep. Min. 1959–60, Min. of Higher and Secondary Specialist Educ. Ukrainian SSR 1960–73; cand. mem. Cen. Cttee. 1960–71, mem. Cen. Cttee. Ukrainian CP 1971–76; author of more than 60 scientific works.
Publ.: "Gidravlicheskii raschet kanalov" 1951 (co-author).

Daian, Kadyr (real name: Kadyr Khakimovich Daianov)
Author; b. 1910, Kunashak; s. of a peasant.
Mem. CPSU 1947–; grad. Bashkir Pedagogical Inst.; first works publs. 1929. Decor.: Order of the Red Star.
Publs.: "Rashida" 1934; children's books "Ptitsy letiat" (The Birds are Flying) 1952, "Khitraia lisa" (The Wily Fox) 1957, and others.

Dallakian, Karlen Levonovich
Party official.
Chair. of Council for Affairs of the Armenian Church with Council of Mins. Armenian SSR 1963–70; mem. Cen. Cttee. Armenian CP 1971–; Head of Dept. of Propaganda and Agitation at Cen. Cttee. Armenian CP 1971–76; mem. Bureau and Sec. of Cen. Cttee. Armenian CP 1976–.

Dal'skii, Vladimir Mikhailovich (real surname: Nesterenko)
Ukrainian actor; b. Ap. 22 (My. 5), 1912, Nikopol'.
People's Artist of the USSR (1960).
He studied at the Kiev Music and Drama Institute from 1930 to 1932. He worked in the Kharkov Young People's Theater, in the Kharkov T. G. Shevchenko Dramatic Theater, and in the dramatic theaters of the Soviet Army (Kiev, Odessa, and L'vov). In 1957 he joined the company of the I. Ia. Franko Ukrainian Dramatic Theater in Kiev. His best roles include Shel'menko in Kvitka-Osnov'ianenko's *Shel'menko, the Orderly;* Morzh, Groza, and Kirill Sergeevich in Korneichuk's *Why the Stars Smiled, Diary Pages,* and *The Heart's Remembrance;* Minutka in Chirskov's *The Victors;* Krutitskii in Ostrovskii's *Not a Penny, and Suddenly an Altyn;* Siplyi in Vishnevskii's *An Optimistic Tragedy;* and the title role in Karpenko-Karyi's *Martyn Borulia.* He possesses a vivid gift for comedy. In his stage characters he combines keen external expressiveness with psychologically profound interpretation.

Damaskin (A. I. Bodryj), Cand. of Theol.
Russian Orthodox clergyman; b. Ap. 20, 1937, Beliny, Velikie Luki Oblast.
Grad. Theological Seminary Leningrad 1966; monk, deacon 1968; completed studies at Theological Acad. Leningrad, priest-monk 1970; abbot 1972; Bishop of Tambov and Michurinsk 1972–74; Bishop of Vologda and Velikii Ustiug 1974–79; Bishop of Poltava and Kremenchug N. 1979–.

Damdinov, Nikolai Garmaevich
Author; b. 1932, Buriat.
Mem. CPSU 1962–; grad. Moscow Gorkii Inst. of Literature.
Publs.: "Barguzin" 1955, "Gudiashchie sosny" (Whistling Pines) 1958; "Udacha v dorogakh" (Success on the Roads) 1962, "Imia ottsa" (Father's Name) 1962, "Bitva za solntse" (Battle for the Sun) 1963, and others.

Damian (D. G. Marchuk)
Russian Orthodox clergyman; b. F. 15, 1899, Korostianin, Volhynia Gouvt.
Priest, Chair. of St. Nicholas Church in Vinnitsa 1921; Chair. of village parish of Knizhkovtsy, Bishopric of Kamenets-Podolskii 1945; Bishop of Chernovtsy, Bishop of Lvov 1961–64; Archbishop of Volhynia and Rovno O. 1964–.

Danchenko, Aleksandr Grigorevich
Graphic artist; b. 1926, Novoalekseevka, Nikolaev Oblast.
Attended Dnepropetrovsk Art School 1944–47 and Kiev Art Inst. 1948–54, pupil of V. Kasian, I. Pleshchinskii, A. Sereda. Decor.: Merited Artist of the Ukrainian SSR 1969.

Danchenko, Aleksandra Stepanovna, Cand. of Arts
Art historian; b. 1930, Kiev.
Attended Kiev Univ. 1948–54; Cand. of Arts 1969.

Daneliia, Georgii Nikolaevich
Film director; b. Ag. 25, 1930, Tbilisi.
Honored Art Worker of the RSFSR (1965) and People's Artist of the RSFSR (1974).
He graduated from the Moscow Institute of Architecture in 1955 and worked as an architect. In 1958 he enrolled in the Advanced Directing Courses. Together with I. V. Talankin he directed the film *Serezha* (1960) based on V. F. Panova's work of the same name. It won several awards, including the Crystal Globe Grand Prize at the 12th International Film Festival at Karlovy Vary. He also directed the films *Road to the Mooring* (1962) and *Thirty-three* (1965). His work is marked by gentle humor, lyricism, and a poetic style. Among his works is the film *I Walk Through Moscow* (1964). In 1969 he directed the musical comedy *Don't Be Sad,* which was filmed in color.

Danenov, N. Zh.
Komsomol official.
Sec. and mem. Bureau of Cen. Cttee. Kazahk Komsomol S. 1981–.

Danielian, Albert Airapetovich
State official.
Min. of Food Ind. Armenian SSR Je. 1981–.

Danilchenko, V. I.
Trade union official.
Mem. Cen. Cttee. Tadzhik CP 1971–81; Head of Dept. of Light and Food Ind. with Cen. Cttee. Tadzhik CP 1971–77; Sec. and mem. Presidium of Tadzhik Republican Trade Union Council F. 1977–.

Danilenko, Viktor Dmitrievich
State official; b. 1936.
Belorussian; mem. CPSU 1962; civil engineer; until 1971 1st Dep. Chair. of State Cttee. for Construction Belorussian SSR; Dep. Chair. of Mogilev Oblast Exec. Cttee. of People's Deps. 1971–75; Min. of Rural Construction Belorussian SSR 1976–80; USSR Dep. Min., 1st Dep. Min. of Rural Construction 1980–82, Min. D. 1982–.

Daniliuk, Ivan Il'ich, Dr. of Phy.-Math. Sc.
Mathematician; b. D. 3, 1931, Reshkov, now Ivano-Frankovsk Oblast; s. of a peasant.
Grad. Lvov Univ. 1955; corresponding mem. Acad. of Sciences Ukrainian SSR 1965–; Dr. of Phy.-Math. Sc. 1963; Prof. 1966–; until 1965 mem. staff Inst. of Hydrodynamics at Siberian Branch of USSR Acad. of Sciences; at the same time Lecturer, Prof. at Novosibirsk Univ.; Dir. of Computing Centre in Donetsk at Acad. of Sciences Ukrainian SSR, Prof. at Donetsk Univ. 1965–; author of more than 40 scientific works.
Publs.: "O zadache s naklonnoi proizvodnoi" in "Sibirskii matematicheskii zhurnal" No. 3/1962, "Ob odnoi nelineinoi zadache so svobodnoi granitsei" in "Doklady AN SSSR" No. 5/1965.

Danilov, Dimorfant Anatolevich
State official; b. 1928.
Vice-Chair. of Oblast Exec. Cttee. of Workers' Deps. in Vitebsk 1973–75; Min. of Local Ind. Belorussian SSR Jl. 1975–77; Dep. Chair. Council of Mins. Belorussian SSR Ag. 1977–; cand. mem. Cen. Cttee. Belorussian CP 1976–81, mem. 1981–.

Danilov, Sofron Petrovich
Author; b. 1922, Yakut ASSR.
Studied Pedagogic Inst., Yakut ASSR; first works publ. 1939.
Publs.: "Geroi Sovetskogo Soiuza Fedor Kuzmich Popov" (Hero of the Soviet Union Fedor Kuzmich Popov) 1944, "Vstrecha" (Meeting) 1945, "Tvoi druzia" (Your Friends) 1948–49, "Vasilii Manchary" 1957, "Ot imeni iakutov" (In

the Name of the Yakuts) 1961, and others.

Dan'kevich, Konstantin Fedorovich
Composer, conductor, teacher, public figure in the field of music; b. D. 11 (22), 1905, Odessa.
People's Artist of the USSR (1954). Member of the CPSU since 1946.
He graduated from the Odessa Institute of Music and Drama in 1929 in the composition class of V.A. Zolotarev and P.I. Molchanov and the piano class of M.I. Rybitskaia. In 1929 he became a teacher and in 1948 a professor of composition at the Odessa Conservatory (director, 1944–51). In 1953 he became a professor at the Kiev Conservatory. He was president of the board of the Composers' Union of the Ukraine during 1956–67.
His music is closely associated with Ukrainian and Russian folklore. His compositions are marked by heroic enthusiasm and dramatic intensity. His major compositions are the operas *Tragic Night* (1934), *Bogdan Khmel'nitskii* (1st version, 1951; 2nd version, 1953), and *Nazar Stodolia* (1959); the ballet *Lileia* (after T.G. Shevchenko, 1939); the oratorio *Oktiabr'* (1957); two symphonies (1937, 1945); four symphonic poems, including *Taras Shevchenko* (1939); chamber and choral works; and music for dramatic productions and films.
He was deputy to the Supreme Soviet of the Ukrainian SSR for the fourth through the sixth convocations. Decors.: Order of Lenin and the Order of the Red Banner of Labor.

Dar, David Iakovlevich
Author; b. 1910, Leningrad.
Russian; first works publ. 1940.
Publs.: "Gospodin Gorillius" (Mr. Gorillius) 1941, "Utro" (The Morning) 1948, "Kamenshchik" (The Builder) 1950, "Olga Nechaeva" 1947, "10,000 kilometrov na velosipede" (10,000 Kilometres on a Bicycle) 1960, "Tam za povorotom" (There, after the Curve) 1962, and others.

Darchiev, David Gavrilovich
Author; b. 1911, Alagir, Ossetia.
Grad. Higher Literary Courses of USSR Writers' Union 1958; first works publ. 1932.
Publs.: "Esli by ty okazalsia moim bratom" (If You Were My Brother) 1938, "Vesna zhizni" (The Spring of Life) 1952; "Zemlia predkov" (The Country of Forefathers) 1963; translated the works of Pushkin, Gorkii, Chekhov etc. into Ossetian, and others.

Darienko, Domnikiia Timofeevna
Moldavian actress; b. Ja. 15, 1919, Dolinskoe, in present-day Anav'ev Raion, Odessa Oblast.
People's Artist of the Moldavian SSR (1957).
She graduated from the Odessa School of Dramatic Arts in 1937. That same year she joined the Moldavian Theater in Tiraspol' (now the A.S. Pushkin Theater of Music and Drama in Kishinev). Her best roles include Kruchinina (*Guilty Without Guilt* by A.N. Ostrovskii), Mavra (*It was Happiness the Girl Was After,* based on Kobylianskaia), Tat'iana (*Enemies* by Gorkii), Liubov' Iarovaia (*Liubov Iarovaia* by Trenev), Corinne (*Ovid* by Alecsandri), Elmira (*Tartuffe* by Molière), Afteniia (*Mother-in-law and Three Daughters-in-law* by Kriange), Vasilutsa (*Casa Mare* by Drutse), and Tankabike (*The Night of the Lunar Eclipse* by Karim). Decors.: Order of the Red Banner of Labor and a medal.

Dauksha, A.
Party official.
Mem. Cen. Cttee. Lithuanian CP 1964–66 and 1981–, cand. mem. 1971–81; Sec. of Lithuanian Republican Trade Union Council 1967–73; Head of Dept. for Construction and Municipal Econ. with Cen. Cttee. Lithuanian CP N. 1973–.

Daulanov, B.D.
Komsomol official.
Sec. and mem. Bureau Cen. Cttee. Uzbek Komsomol S. 1979–.

Davidov, Aleksandr Sergeevich, Dr. of Phy.-Math. Sc.
Physicist; b. D. 26, 1912, Evpatoriia; s. of a worker.
Grad. Dept. of Physics Moscow Univ. 1939 and postgraduate studies 1941; theoretical physicist; mem. Acad. of Sciences Ukrainian SSR 1964–; Dr. of Phy.-Math. Sc. 1950; Prof. 1951–; until 1945 chief of dept. for physical research methods at a machine construction works; Sen. Research Fellow, Dep. Dir. of Instruction Dept. at Inst. of Physics Acad. of Sciences Ukrainian SSR 1945–53; Prof., Prof. Chair of Electrodynamics and Quantum Theory Moscow Univ. and mem. staff Inst. of Physics at USSR Acad. of Sciences 1953–64; Chief of Dept. for Theory of Atomic Nucleus at Inst. of Physics Acad. of Sciences Ukrainian SSR 1964–66; Chief of Dept. for Theory of Atomic Nucleus at Inst. of Theoretical Physics Acad. of Sciences Ukrainian SSR 1966–; author of approx. 90 scientific works on the theory of atomic nucleus absorption and dispersion of light in molecular crystals, and on the theory of luminescence. Decor.: Lenin Prize 1966.
Publs.: "Teoriia pogloshcheniia sveta v molekuliarnykh kristallakh" 1951, "Teoriia atomnogo iadra" 1958, "Kvantovaia mekhanika" 1963, "Vozbuzhdennoe sostoianie atomnykh iader" 1967.

Davitaia, Feofan Ferneevich
Agrometeorologist and climatologist; b. S. 15, 1911, Eki, in present-day Tskhakaia Raion, Georgian SSR.
Doctor of agricultural sciences (1951), academician of the Academy of Sciences of the Georgian SSR (1960), Honored Scientist of the Georgian SSR (1966). Became a member of the CPSU in 1939. Academician-secretary of the Division of Earth Sciences, director of the Vakhushti Institute of Geography of the Academy of Sciences of the Georgian SSR. Became president of the Geographical Society of the Georgian SSR in 1970.
His basic works are devoted to questions of the agroclimatic division of the USSR, the evaluation, of climatic resources in the various natural zones, and the substantiation of forecasting the availability to plants of heat and moisture in regions where these plants have been assimilated.
Decors.: five orders, as well as medals.
Works: *Klimaticheskie zony vinograda v SSSR,* 2nd ed. Moscow 1948. *Prognoz obespechennosti teplom i nekotorye problemy sezonnogo razvitiia prirody.* Moscow 1964. *Klimaticheskie resursy Kuby i ikh ispol'zovanie v narodnom khoziaistve.* Tbilisi 1966. (Together with I.I. Trusov.) *Problema prognoza ispariaemosti i orositel'nykh norm.* Leningrad 1970. (Co-author.)

Davitashvili, Teimuraz Nikolaevich
State official; b. 1919.
Dep. to Supr. Sov. Georgian SSR 1955–; Min. of Motor Transport and Highways 1955–67, Min. of Motor Transport Georgian SSR 1967–; cand. mem. 1956 and 1971–, mem. Cen. Cttee. Georgian CP 1960–66.

Davliatkadamov, Khushkadam, Cand. of Phil. Sc.
State official; b. 1924.
Tadzhik; mem. CPSU 1946–; grad. Party High School Tashkent and Acad. of Social Sciences with Cen. Cttee. CPSU; teacher, head of pedagogical dept. of a school 1943–47; propagandist of a raion cttee. Tadzhik CP, 1st sec. of a raion Komsomol cttee. 1947–51; Lecturer at Gorno-Badakhshan Oblast Cttee. Tadzhik CP 1951–55; Sec. of Rushanskii Raion Cttee. Tadzhik CP 1955–57; head of an oblast centre for pol. educ., postgraduate studies at Acad. of Social Sciences of Cen. Cttee. CPSU 1960–64; Sec. of Gorno-Badakhshan Oblast Cttee. Tadzhik CP, Chair. of Exec. Cttee. Gorno-Badakhshan Oblast Soviet of Workers' Deps. 1964–70; 1st Sec. of Gorno-Badakhshan Oblast Cttee. Tadzhik CP Ap. 1970–78; Dep. to USSR Supr. Sov. 1970–; mem. Cen. Cttee. Tadzhik CP 1971–; Min. of Higher and Secondary Specialist Educ. Tadzhik SSR D. 1978–.

Davtian, Vaagn Armenakovich
Author; b. 1922, Arabkir, Turkey; s. of an artisan.
Armenian; took part in World War II.
Publs.: poems "Pervaia liubov'" (First Love) 1947, "Utro mira" (A Morning of Peace) 1950, "Doroga cherez serdtse" (A

Way Through the Heart) 1952, "Rassvet v gorakh" (Dawn in the Mountains) 1957, and others.

Davydov, Iurii Vladimirovich
Author; b. 1924, Moscow.
Russian; mem. CPSU 1947–; served in the Sov. Navy 1942–49; first works publ. 1945.
Publs.: "V moriakh i stranstviiakh" (On the Seas and on Journeys) 1949, "Iuzhnyi krest" (The Southern Cross) 1957, "Kapitany ishchup put'" (The Captains are Looking for the Way) 1959, "O druziakh tvoikh, Afrika" (About Your Friends, Africa) 1962, and others.

Davydov, N. G.
Party official.
2nd Sec. Chimkent Oblast Cttee. Kazakh CP 1978–82; cand. mem. Cen. Cttee. Kazakh CP 1981–; 1st Sec. Dzhezgazgan Oblast Cttee. Kazakh CP N. 1982–.

Davydov, Vitalii Semenovich
Ice hockey player; b. Ap. 3, 1939, Moscow.
Honored Master of Sport (1963), teacher. Member of the CPSU from 1970 on.
He was a member of the teams that won the world championship several times between 1963 and 1971 and the European championship from 1963 to 1970. He was on the championship team of the Winter Olympic Games of 1964 and 1968 and on the team that was a prize-winner of the USSR national championship several times (1957–60, 1962–65, 1967–70). Through the 1960's to the present (1972) he has been one of the leading defensemen of the Dynamo team of Moscow and of the combined team of the USSR. He was recognized as the best defenseman at the world championship of 1967. Decors.: Badge of Honor and medals.

Davydova, Natalia (real name: Davydova, Maiia Maksimovna)
Author; b. 1925, Leningrad.
Russian; grad. Dept. of Philology at Leningrad State Univ. 1948.
Publs.: "Budni i prazdniki" (Working Days and Holidays) 1953, "Liubov' inzhenera Izotova" (Engineer Izotov's Love) 1960, "Tol'ko odna udacha" (Only One Success) 1958, and others.

Davydova, Vera Aleksandrovna (married name Mchedlidze)
Russian mezzo-soprano and teacher; b. S. 17 (30), 1906, Nizhnii Novgorod, now Gorkii.
People's Artist of the RSFSR (1951) and honored artist of the Georgian SSR (1944). Member of the CPSU since 1951. She graduated from the Leningrad Conservatory from the class of E. V. Devos-Soboleva in 1930. She made her debut in 1929 at the Leningrad Theater of Opera and Ballet. From 1932 to 1956 she was a soloist at the Bolshoi Theater of the USSR. In 1959 she began teaching at the Tbilisi Conservatory, becoming a professor in 1964.
She is a brilliant representative of the Soviet vocal school and has created realistic and truthful characters. Her roles include Liubava in Rimsky-Korsakov's *Sadko* (State Prize of the USSR, 1950), Marfa in Mussorgsky's *Khovanshchina* (State Prize of the USSR, 1951), the title role in Bizet's *Carmen*, Amneris in Verdi's *Aida*, Aksin'ia in Dzerzhinskii's *The Quiet Don*, Nilovna in Zhelobinskii's *The Mother*, and Grunia in Chishko's *The Battleship Potemkin*. Davydova has performed abroad. A deputy to the second and third convocations of the Supreme Soviet of the RSFSR, she has received the State Prize of the USSR for concert performances (1946). Decors.: two orders and various medals.

Davydova-Medene, Lea Matisovna
Portrait sculptor; b. Mr. 31, 1921, Novadnieki, in present-day Saldus Raion, Latvian SSR.
Honored Art Worker of the Latvian SSR (1966).
From 1941 to 1949, she studied at the Academy of Arts in Riga under T. Zaļkalns. The diverse search for plastic means of depicting artistic images reflects the striving of the artist to express the individual uniqueness of the physical and spiritual temper of man, to emphasize his active, creative basis. Her works include portrait sculptures of P. Upītis (granite, 1960), M. Kempe (granite, 1961), and A. Upītis (granite, 1964–67), which are all in the Museum of Arts of the Latvian SSR in Riga. Among her other works are a monumental portrait sculpture of I. Sudmalis (bronze, 1960, placed in Communards Square in Riga) and a portrait sculpture of K. Zemdega (granite, 1962–65, Russian Museum, Leningrad). Decor.: State Prize of the Latvian SSR (1965).

Davydychev, Lev Ivanovich
Author; b. 1927, Solikamsk.
Russian; mem. CPSU 1957–.
Publs.: "Volshebnik dachnogo poselka" (The Magician of a Dacha Community) 1952, "Druzia moi, priiateli" (My Comrades, My Friends) 1957, "Pochemu plakala devochka?" (Why Has the Girl Cried?) 1959, play "Allochka" 1961, "Goriachie serdtsa" (Hot Hearts) 1953, "Trudnaia liubov'" (A difficult Love) 1955, and others.

Dedov, Ivan Fedorovich
Party official.
Cand. mem. 1960–61, mem. Cen. Cttee. Tadzhik CP 1961–; 1st Sec. of Gissar Raion Cttee. Tadzhik CP 1961; mem. Bureau for Agriculture Tadzhik CP 1962–63; Sec. of Party Cttee. of Kolkhoz-Sovkhoz Prod. Admin. in Gissar 1963; Head of Dept. of Party Org. Work at Cen. Cttee. Tadzhik CP 1968–; cand. mem. 1971–76, mem. Bureau Cen. Cttee. Tadzhik CP 1976–.

Degtiarev, Vladimir Ivanovich
State official; b. 1920.
Russian; mem. CPSU 1945–; grad. Moscow Mining Inst. 1942; head of a mine section, chief engineer, pit manager 1942–53; Head of Chistiakov Anthracite Trust 1953–57; Sec. of Donetsk Oblast Cttee. Ukrainian CP 1957–62; Dep. to USSR Supr. Sov. 1962–; Chair. of Donetsk Econ. Council 1962–63; 2nd Sec. Ja.–Jl. 1963; 1st Sec. of Donetsk Oblast Cttee. Ukrainian CP 1963–D. 1975 (for Ind. 1963–64); mem. Cen. Cttee. CPSU 1966–76; mem. Cen. Cttee. 1966–76, cand. mem. Politbureau 1966–71, mem. Politbureau Cen. Cttee. Ukrainian CP 1971–76; Chair. of State Cttee. for Labour Safety in Ind. and Mines. Ukrainian SSR. Decors.: Hero of Socialist Labour, Order of Lenin.

Delets, Mikhail Ivanovich
State official.
Cand. mem. 1960–71 and 1976–, mem. Cen. Cttee. Belorussian CP 1971–76; Sec. of CPSU Cttee. of Kolkhoz-Sovkhoz Prod. Admin. in Baranovichi 1963; Chief Ed. of newspaper "Zviazda", 1st Dep. Chair. of Journalists' Union of Belorussian SSR 1971; Chair. of State Cttee. for Publishing Houses, Printing and the Book Trade. Belorussian SSR F. 1973–.

Delimarskii, Iurii Konstantinovich, Dr. of Chem. Sc.
Chemical scientist; b. My. 23, 1904, Krasnopoltsi, now Vinnitsa Oblast; s. of a teacher.
Mem. CPSU 1940–; grad. Inst. for Public Educ. Kiev 1928 and postgraduate studies at Kiev Polytech. Inst. 1932; specialist in electrochemistry; mem. Acad. of Sciences Ukrainian SSR 1957–; Dr. of Chem. Sc. 1950; Prof. 1950; Lecturer, Prof. Chair of Inorganic Chemistry at Polytech. Inst. Kiev 1932–34; Research Asst., Laboratory Chief, Dep. Dir. 1934–60, Dir. of Inst. of Gen. and Inorganic Chemistry with Acad. of Sciences Ukrainian SSR 1960–; Laboratory Chief at Inst. of Chemistry of Uzbek Branch of USSR Acad. of Sciences 1941–44; mem. Presidium and Chief of Dept. of Chemical and Geological Sciences Acad. of Sciences Ukrainian SSR 1957–60. Decors.: Order of Red Banner of Labour, medals.
Publs.: "Elektrodnye protsessy i metody issledovaniia v poliarografii" 1960, "Elektrokhimiia rasplavlennykh solei" 1960, and others.

Demchenko, Ivan Ivanovich
State official.
Min. of Building Materials Ind. Moldavian SSR Ja. 1980–; mem. Cen. Cttee. Moldavian CP 1981–.

Demchenko, Vladimir Akimovich
State official; b. 1920.
Russian; mem. CPSU 1946–; grad. Bauman Tech. College 1944 and CPSU Cen.

Cttee. Party Higher School 1958; design engineer at a machine construction plant in Moscow 1944–45; Instructor, Dep. Dept. Head with Moscow City Komsomol Cttee., 1st Sec. of Kuibyshev Raion Komsomol Cttee. Moscow 1945–52; Dep. Dept. Head, later 1st Sec. of Stupino City CPSU Cttee. Moscow Oblast, Dept. Head, Sec. of Moscow Oblast CPSU Cttee. 1952–63; 1st Sec. of Moscow Oblast CPSU Cttee. in charge of Ind. 1963–64, 2nd Sec. of Moscow Oblast CPSU Cttee. 1964–67; Dep. and mem. Presidium RSFSR Supr. Sov. 1963–67; cand. mem. CPSU Cen. Cttee. 1966–; Dep. to and mem. Planning and Budgeting Comm. of Sov. of the Union USSR Supr. Sov. 1966–70; Dep. Chair. of RSFSR Council of Mins. Ap. 1967–. Decor.: Order of Lenin.

Dementei, Nikolai Ivanovich
Party official.
Cand. mem. Cen. Cttee. Belorussian CP 1971–76, mem. 1981–; Sec. Vitebsk Oblast Cttee. Belorussian CP 1974–77; Head of Dept. for Agriculture with Cen. Cttee. Belorussian CP 1977–79; Sec. and mem. Bureau Cen. Cttee. Belorussian CP Ap. 1979–.

Dementev, Nikolai Stepanovich
Author; b. 1927, Leningrad.
Russian; grad. Leningrad Inst. of Water Transport Engineering; worked as engineer in Siberia; first works publ. 1952. Publs.: "Sibirskie ogni" (Siberian Lights) 1952, "Pervoe pismo" (The First Letter) 1954, "Sestry" (The Sisters) 1955, "Moi dorogi" (My Ways) 1958, "Idu v zhizn'" (I'm Stepping into Life) 1958, "Prekrasna zima v Sibiri" (The Winter is Wonderful in Siberia) 1960, "Chuzhie bliznetsy" (Strange Twins) 1962, "Zamuzhestvo Tatiany Belovoi" (Tatiana Belova's Marriage) 1963, and others.

Dementeva, Raisa Fedorovna
Party official; b. 1925.
Mem. CPSU 1948–; grad. Moscow Inst. of Econ. Statistics 1957; Komsomol posts 1947–52; CPSU posts 1952–; Sec. of Moscow City Cttee. CPSU 1960–80, 2nd Sec. 1980–; mem. Cen. Auditing Comm. CPSU 1961–66; cand. mem. Cen. Cttee. CPSU 1966–76, mem. 1976–; Dep. to USSR Supr. Sov. 1979–. Decor.: Order of Red Banner of Labor.

Demesinov, Khatip Khabievich
State official.
Sec., 2nd Sec. Pavlodar Oblast Cttee. Kazakh CP 1976–81; Dep. to 1980–, Sec. of Presidium Supr. Sov. Kazakh SSR Je. 1981–; cand. mem. Cen. Cttee. Kazakh CP 1981–.

Demichev, Petr Nilovich
Party and state official; b. Ja. 3, 1918, Pesochnaia, Kaluga Oblast.
Russian; mem. CPSU 1939; grad. Mendeleev Inst. of Chemical Tech. Moscow 1944 and CPSU Cen. Cttee. Higher Party School 1953; Sec. of a raion Komsomol cttee. 1937; served in Sov. Army 1947–44; Asst. Prof. at Mendeleev Inst. Moscow 1944–45; Dept. Head, Sec. of Sov. Raion All-Union CP (B) Cttee. in Moscow, Dep. Dept. Head of Moscow City CPSU Cttee., mem. staff Cen. Cttee. CPSU 1945–56; Sec. of Moscow Oblast CPSU Cttee. 1956–58; sen. post with USSR Council of Mins. 1958–59; Dep. to 1958–, mem. Presidium USSR Supr. Sov. 1962–66; 1st Sec. of Moscow Oblast Cttee. 1959–60 and Moscow City CPSU Cttee. 1960–62; mem. Bureau Cen. Cttee. CPSU for RSFSR 1959–61; mem. Cen. Cttee. CPSU 1961–; Sec. 1961–74, cand. mem. Politbureau Cen. Cttee. CPSU 1964–; Chair. of Bureau for Chemical and Light Ind. with Cen. Cttee. CPSU 1962–64; USSR Min. of Culture N. 1974–. Decors.: 3 orders of Lenin, Order of Red Banner of Labour, various medals, Order of October Revolution.

Demidenko, Vasilii Petrovich
Party official; b. 1930, Ukraine.
Mem. CPSU 1955–; grad. Novosibirsk Agricultural Inst. 1950; agronomist in a sovkhoz, student 1947–50; chief agronomist in a grain sovkhoz and later in a raion dept. for agriculture 1950–54; chief agronomist, dir. of various grain sovkhozes in Kazakhstan 1954–58; 1st sec. of a raion cttee. Kazakh CP, Kazakh Dep. Min. of Sovkhozes 1958–60; Sec. of Virgin Lands Krai Cttee. Kazakh CP 1960–65 (for Agriculture 1964–65); 1st Sec. of N. Kazakhstan Oblast Cttee. Kazakh CP 1965–81; mem. Cen. Cttee. Kazakh CP 1966–; Dep. to USSR Supr. Sov. 1966–; cand. mem. Cen. Cttee. CPSU 1971–77, mem. 1977–; 1st Sec. of Kustanai Oblast Cttee. Kazakh CP 1981–. Decor.: Order of Lenin.

Demidenko, Viacheslav Sergeevich
Cand. mem. Cen. Cttee. CPSU; b. 1935.
Mem. CPSU 1971–, finished secondary school 1954; served in Sov. Army 1954–58; metal worker at "Rakitnoe" State Farm in Poltava Oblast 1958–59; fitter at construction sites of hydroelectric power plants in Kremenchug, Bratsk, Krasnoiarsk, and Kapchagai 1959–73; leader of fitter brigade in Saianogorsk "Spetsgidroenergomontazh" Trust 1973–; cand. mem. Cen. Cttee. CPSU 1981–.

Demidov, Anatolii Mikhailovich, Cand. of Phy.-Math. Sc.
Physicist; b. 1926.
Grad. Moscow Univ. 1951; Cand. of Phy.-Math. Sc. 1959–; mem. staff Kurchatov Inst. of Atomic Energy 1952–. Decor.: State Prize of the USSR 1969.

Demidova, Ekaterina Iakovlevna
Laborer; b. 1940.
Mem. CPSU 1967–; grad. Leningrad Technicum for Light Industry 1976; weaver at "Rabochii" Spinnung and Weaving Works in Leningrad 1959–; mem. Cen. Auditing Comm. CPSU 1976–. Decor.: Heroine of Socialist Labor 1973.

Demin, Lev Stepanovich, Cand. of Tech. Sc.
Cosmonaut; b. 1926.
Mem. CPSU 1956–; grad. Mil. Aviation Acad. for Telecommunications 1945; student 1949, later dep. head of a dept. at Zhukovskii Air Force Engineering Acad.; Cand. of Tech. Sc. 1963; cosmonaut 1963–; Col.-Engineer, flight engineer in spaceship "Soiuz 15"; undertook with G. V. Sarafanov an orbital flight and carried out scientific experiments under various flight conditions relating to manoeuvres and rendezvous manoeuvres with "Saliut 3". Decor.: Hero of the Sov. Union 1974.

Demin, Mikhail (real name: Trifonov, Georgii Evgenevich)
Author; b. 1926, Finland.
Russian; first works publ. 1953.
Publs.: "Pod nezakatnym solntsem" (Beneath the Everlasting Sun) 1956, "Litsom k Vostoku" (Looking Eastwards) 1958, "Paralleli i meridiany" (Parallels and Meridians) 1962, and others.

Demin, Veniamin Alekseevich
Architect; b. 1908.
Grad. Art Tech. School, Dept. of Sculpture 1929 and Inst. of Architecture Moscow 1936. Decor.: Lenin Prize (for his part in designing monument for the Heroes of Stalingrad in Volgograd) 1970.

Demin, Vladimir Grigorevich, Dr. of Phy.-Math. Sc.
Scientist; b. 1929.
Grad. Moscow Univ. 1951; specialist in mechanics of celestial bodies; cand. 1961–69, Dr. of Phy.-Math. Sc. 1969; employed as engineer 1951–52; teacher at Gorkii Univ. 1952–54, Pedagogical Inst. Novozybkov 1954–56; Research Asst. at P. K. Sternberg State Inst. of Astronomy 1956–62; Lecturer at Lumumba Univ. for Friendship Among Nations 1962–65; mem. staff Moscow Univ. 1965–, Prof. Chair of Theoretical Mechanics 1971–. Decor.: State Prize of the USSR 1971.

Demirchian, Karen Seropovich
Party official; b. 1932, Erevan.
Mem. CPSU 1932–; grad. Polytech. Inst. Erevan and Party High School of Cen. Cttee. CPSU 1961; chief engineer-designer, div. head of a scientific research inst. in Leningrad, foreman, Dept. Head with Electrotech. Works in Erevan, Sec. of Party Cttee. of same works 1954–58; Chief Engineer, Dir. of Electrotech. Works in Erevan 1961–66; City Sec., then 2nd City Sec. of Armenian CP in Erevan 1966–72; mem. Cen. Cttee. Armenian CP 1966–; mem. Bureau 1972–, Sec. 1972–74, 1st Sec. of Cen. Cttee. Armeni-

an CP N. 1974–; mem. Cen. Cttee. CPSU 1976–.

Denisiuk, Iurii Nikolaevich
Scientist; b. 1927.
Grad. Shipbuilding Tech. School Leningrad 1947 and Inst. of Precision Mechanics and Optics Leningrad 1954; specialist in physical optics; mem. staff S. I. Vavilov State Inst. of Optics 1954–. Decor.: Lenin Prize for outstanding achievements in the field of holography 1970.

Denisov, Andrei Ivanovich, Dr. of Legal Sc.
Jurist; b. O. 13, 1906, Ozerki, now Tambov Oblast.
Mem. CPSU 1926–; lawyer, political scientist; Prof. 1945–; lectured at various colleges 1935–. Decor.: Merited Scientist of the RSFSR.

Deregus, Marina Mikhailovna
Painter; b. 1929, Kharkov; d. of M. Deregus.
Grad. Kharkov Art School 1950.

Deregus, Mikhail Gordeevich
Painter and graphic artist; b. 1904, Vesele, Kharkov Oblast.
Mem. CPSU 1951–; attended Kharkov Art Inst. 1923–30, pupil of M. Sharonov, S. Prokhorov, P. Kokel. Decors.: People's Artist of the USSR 1963, Shevchenko Prize 1965.

Deregus, Natalia Mikhailovna
Sculptor; b. 1931, Kharkov; d. of M. Deregus.
Grad. Kiev Art Inst. 1957, pupil of M. Lysenko.

Derevianko, A. P., Dr. of Hist. Sc.
Party official.
Cand. mem. 1974–76, Sec. and mem. Bureau of Cen. Cttee. USSR Komsomol Mr. 1976–79; Sec. Novosibirsk Oblast Cttee. CPSU N. 1979–.

Dergausov, Iu. A.
Komsomol official.
Sec. Rostov/Don City Cttee. All-Union Komsomol until 1979; Sec. and mem. Bureau of Cen. Cttee. All-Union Komsomol N. 1979–.

Deriagin, Boris Vladimirovich
Chemical scientist; b. Jl. 27 (Ag. 9), 1902, Moscow.
Grad. Moscow Univ. 1922; physical chemist, molecular physicist; corresponding mem. USSR Acad. of Sciences 1946–; Chief of Laboratory of Surface Forces at Inst. of Physical Chemistry at USSR Acad. of Sciences 1935–. Decors.: M. V. Lomonosov Prize of USSR Acad. of Sciences 1958, Order of Red Banner of Labour and various medals.
Publs.: "Adgezija" Moscow-Leningrad 1949 (co-author), "Fizikohimija nanesenija tonkih sloëv na dvižuščujusja podložku", Moscow 1959 (co-author), "Čto takoe trenie?" Moscow 1963.

Desnitskaia, Agnia Vasilevna
Linguist; b. Ag. 10 (23), 1912, Sednev, now Chernigov Oblast.
Mem. CPSU 1943–; grad. Herzen Pedagogical Inst. Leningrad 1931; corresponding mem. USSR Acad. of Sciences 1964–; Prof. at Leningrad Univ. 1949–; Dep. Dir. of Inst. of Linguistics at USSR Acad. of Sciences 1963–.

Deviatkov, Nikolai Dmitrievich
Electrical engineer; b. Mr. 29 (Ap. 11), 1907, Vologda.
Grad. Polytech. Inst. Leningrad 1931; mem. USSR Acad. of Sciences 1968–; mem. staff Physical-Tech. A. F. Joffe Inst. and various other research insts. 1925–54; Scientific Man. in Dept. of Ultra-High Frequency Electronics at Inst. of Radio Engineering and Electronics at USSR Acad. of Sciences 1954–; Prof. Chair at Physical-Tech. Inst. Moscow 1960–. Decors.: Hero of Socialist Labour, State Prize of the USSR 1949, Lenin Prize 1965, two Orders of Lenin, two other orders and several medals.
Publ.: "Razvitie ėlektronnyh priborov sverchvysokih častot" in "Izv. AN SSSR" 1958, No. 2.

Deviatykh, Grigorii Grigor'evich
Chemist; b. D. 1, 1918, Baranovshchina, Viatka Province.
Corresponding member of the Academy of Sciences of the USSR (1968). Member of the CPSU since 1947.
He graduated from the chemistry department of the University of Gorkii in 1941. He is head of the department of inorganic chemistry at the University of Gorkii as well as head of a laboratory of the Institute of Chemistry of the Academy of Sciences of the USSR (in the city of Gorkii). His principal works deal with the separation of stable isotopes and the preparation and analysis of extremely pure substances. Decors.: Order of the Red Banner of Labor and various medals.

Devin, Ilia Maksimovich
Author; b. 1922, Staraia Terizmorga.
Mordovian; first works publ. 1938.
Publs.: "Utrenniaia zaria" (Dawn) 1945, "Moim tovarishcham po schastiu" (My Friends in Happiness) 1953, "Pis'ma inzhenera" (The Engineer's Letters) 1954, "Poema o syne" (The Son's Poem) 1957, and others.

Diachenko, S. A.
Komsomol official.
2nd Sec. and mem. Bureau and Cen. Cttee. Kazakh Komsomol D. 1980–; cand. mem. Cen. Cttee. Kazakh CP 1981–.

Diachkov, Evgenii Petrovich
State official.
Chair. of State Cttee. for the Supply of Petroleum Products Latvian SSR Ag. 1982–.

Diad'kin, Ivan Afanas'evich (partisan name: Ian Bos)
Partisan; participant in the resistance movement. Lieutenant; b. Jl. 1, 1920, Krasnosel'e, Volgograd Oblast.
Member of the CPSU since 1957.
He graduated from the First Moscow Artillery School in 1940. He served in the Great Patriotic War of 1941–1945 as a commander of a reconnaissance platoon of the 311th Artillery Regiment. He was wounded and captured on the Berezina River in July 1941, kept in camps in Belgium, and forced to work in mines. On Je. 2, 1943, he escaped and by the end of 1943 had organized a partisan detachment in the forests of Limburg Province consisting of over 80 Soviet citizens who had escaped from captivity. In June 1944 the detachment merged with Lieutenant Colonel K. D. Shukshin's detachment into the Russian partisan brigade, Za Rodinu (For the Motherland), in which he was at first deputy commander and then commander. The brigade was active in the enemy's rear until October 1944, inflicting significant losses. He returned to the motherland in 1945.

Diagilev, Vladimir Iakovlevich
Author; b. 1919, Petukhove, Urals.
Russian; mem. CPSU 1945–; grad. War Faculty of 2nd Medical Inst. Moscow 1943; first works publ. 1947.
Publs.: "Gvardeitsy" (Guardsmen) 1952, "Doctor Golubev" 1956, "Ruki i serdtse" (Hands and Heart) 1957, "Khirurg Kuprianov" (Surgeon Kuprianov) 1961, "Da zdravstvuet zhizn" (Long Live Life) 1960, "Piat' kolets druzhby" (Five Rings of Friendship) 1961, "Maiskii zhuk" (May Bugs) 1962, "Volshebnik v belom khalate" (Magician in the White Coat) 1963, and others.

Diakonov, Igor Mikhailovich, Dr. of Hist. Sc.
Scholar; b. D. 30, 1914 (Ja. 12, 1915), Petrograd.
Grad. Leningrad Univ. 1938; historian, orientalist; Dr. of Hist. Sc. 1960; mem. scientific staff of the Hermitage 1937–59; mem. staff Leningrad Branch of Inst. of Orientalism at USSR Acad. of Sciences 1954–; mem. foreign scientific institutions; main works on socioeconomic history and languages of countries of ancient south-western Asia.

D'iakonov, Nikolai Mikhailovich
Komi playwright, director; b. Mr. 28 (Ap. 10), 1911, Ust'-Vym', now in the Komi ASSR.
People's Artist of the Komi ASSR since 1945; Honored Artist of the RSFSR since 1951. Member of the CPSU since 1941.
In the prewar years he collaborated with S. I. Ermolin in writing the plays *Deep Boom, Domna Kalikova, Ravens,* and *Courage*. He won fame with his play *Wedding With a Dowry* (staged 1949; Russian version in collaboration with

A. Glebov, 1950; State Prize of the USSR, 1951); the sequel is *Kurochkin's Quartet* (1962). Decors.: two Orders and several medals.
Work: *In Russian translation: Svad'ba s pridanym.* Moscow 1950.

Diasamidze, David Dursunovich
State official.
Sec. of Adjar ASSR Oblast Cttee. Georgian CP 1963–69; Chair. of Presidium Supr. Sov. Adjar ASSR Mr. 1969–; Dep. to 1969–, Dep. Chair. of Presidium Supr. Sov. Georgian SSR 1971–; cand. mem. Cen. Cttee. 1971–76, mem. Auditing Comm. Georgian CP 1976–.

Dikusarov, Vladimir Grigorevich
Party official; b. 1927.
Ukrainian; mem. 1949–; grad. CPSU Cen. Cttee. Party Higher School and Lvov Veterinary Inst.; chief of a veterinary office 1945–48; sec. of a raion cttee. Ukrainian Komsomol 1948–51; instructor in a dept. of an oblast cttee. Ukrainian CP 1951–54; sec., 2nd sec., 1st sec. of a raion cttee. Ukrainian CP 1954–62; sec. of party cttee. of a collective and state farm prod. admin., 1st sec. of a raion cttee. Ukrainian CP 1962–66; 2nd Sec. of Transcarpathian Oblast Cttee. Ukrainian CP 1966–72; 1st Sec. of Chernovtsy Oblast Cttee. Ukrainian CP Je. 1972–; Dep. to USSR Supr. Sov. 1974–; mem. Cen. Cttee. Ukrainian CP 1976–.

Dimitriadi, Odissei Akhillesovich
Conductor; b. Je. 24 (Jl. 7), 1908, Batumi. People's Artist of the USSR (1958).
He graduated from the Tbilisi Conservatory (class of composition) in 1930 and from the Leningrad Conservatory (class of conducting) in 1936. In 1937 he became a conductor at the Georgian Theater of Opera and Ballet and was chief conductor there from 1952 to 1965. He was also chief conductor of the symphony orchestra of the Georgian SSR from 1947 to 1952. He became a conductor at the Bolshoi Theater in Moscow in 1965. He participated in the production of various operas at the Georgian Theater of Opera and Ballet, including Rimsky-Korsakov's *The Tsar's Bride* (1938), Paliashvili's *Abesalom and Eteri* (1953), Toradze's *Bride of the North* (1958), and Prokofiev's *Semen Kotko* (1964). He also took part in the staging of many ballets, including Balanchivadze's *Heart of the Mountains* (1940) and Machavariani's *Othello* (1958). Decors.: two orders and various medals.

Dinere, Tsetsilia Robertovna
Author; b. 1919, Kraslava.
Latvian; began literary career as radio operator in Sov. Army 1943.
Publs.: poems "Sverkaiushchie krylia" (Shining Wings) 1948, "Pesnia serzhanta Sarmy" (The Song of Sergeant Sarma) 1947, "Fritsis Gailis" 1948, "Nam dorog mir" (Peace is Dear to Us) 1951, "Malen'kii Rizhanin" (Little Riga Boy) 1951, "Katiny igrushki" (Katia's Toy) 1955, and others.

Dinkov, Vasilii Aleksandrovich
State official.
USSR 1st Dep. Min. of Gas Ind. 1979–81, Min. My. 1981–.

Diordiev, Evgenii Iakovlevich
Actor; b. 1912.
Attended Drama School in Odessa; after 1930 mem. Ivanov Theatre Odessa; engagements at several theatres 1935–40; actor and Stage Dir. at Republican Russian Lermontov Drama Theatre Alma-Ata 1946–. Decors.: People's Artist of the USSR 1970, People's Artist of the Kazakh SSR 1958.

Dirzhinskaite-Piliushenko, Leokardia Iuozovna
State official; b. 1921.
Lithuanian; mem. CPSU 1950–; grad. Cen. Cttee. CPSU High School; Sec. of Exec. Cttee. of a volost' sov. of workers' deps. 1940–41; posts in Komsomol 1941–51; Instructor, head of a section at an oblast cttee. Lithuanian CP 1951–53; Sec., 1st Sec. of Shauliai City Cttee. Lithuanian CP 1953–60; Dep. to 1959–, Dep. Chair. of Supr. Sov. Lithuanian SSR 1959–60; Dep. Chair. of Council of Mins. Lithuanian SSR Ja. 1960–75; mem. Cen. Cttee. 1960–, cand. mem. Bureau Cen. Cttee. Lithuanian CP 1960–64; Min. of Foreign Affairs Lithuanian SSR Ja. 1967–75; Dep. Chair. of Presidium Supr. Sov. Lithuanian SSR Je. 1975–. Decors.: 2 Orders of Red Banner of Labour.

Dmiterko, Liubomir Dmitrievich
Ukrainian writer; b. Mr. 5 (18), 1911, Vinniki, present-day L'vov Oblast, Ukrainian SSR.
Member of the CPSU since 1943.
He was the son of a schoolteacher. He was first published in 1928. His first book of poems, *I Am Coming,* was published in 1930. Folk heroism, socialist labor, the struggle for peace, and the battle against imperialism are the main themes of his work. The play *General Vatutin* (1947) and the historical drama *Together Forever* (1949–51) have become widely known. The plays *Human Ways* (1954), *In a Golden Frame* (1958), *A Maiden's Fate* (1959), and *Judgment of the Heart* (1969) are devoted to the problems of morality in modern Soviet society. The plays of him are full of drama and are written in an elevated romantic style. He is also the author of the novels *Parting* (1957), *The Bridge Over the Abyss* (1966), and *Through Days and Nights* (1968), as well as a number of sketches. His works have been translated into many languages spoken in the Soviet Union and other foreign languages. Since 1962, he has been the editor of the magazine *Vitchyzna.*
Works: *Vybrane,* vols. 1–2. Kiev 1963. *Stikhotvoreniia.* Moscow 1951. *Dramy.* Moscow 1955. *Dobrye sosedi.* Moscow 1956. *Pod Iuzhnym Krestom.* Moscow 1958. *Kievskie kruchi.* Moscow 1966. *Opalennye grozami.* Moscow 1969.

Dmitrevskii, Vladimir Ivanovich
Author; b. 1908, Voskresenskoe, Tula Oblast.
Russian; first works publ. 1927.
Publs.: "My mirnye liudi" (We Are Peaceful People) (with B. Cherverikov) 1960, "Pis'ma s 'Putilovtsa'" (Letters from on Board the "Putilovets") 1930, "Nash Kuzbass" (Our Kuznets Basin) 1947, "Dorogi k zvezdam" (Ways to the Stars) 1961, "Cherez gory vremeni" (Over the Mountains of Time) (with E. Brandis) 1963, and others.

Dmitriev, Igor Fedorovich
Party official; b. 1909.
Mem. CPSU 1925–; grad. Leningrad Mechanical Mil. Inst. 1934; designer at a scientific research inst. in Leningrad 1932–38; held posts in USSR Min. of Armaments 1941–44; sec. of party cttee. of USSR Min. of Armaments 1944–46; dept. chief in a comm. of USSR Council of mins. 1958–65; 1st Dep. Chief of Defense Ind. Dept. Cen. Cttee. CPSU 1965–81; Chief Mr. 1981–; cand. mem. Cen. Cttee. CPSU 1981–.

Dmitriev, Iurii Arsen'evich
Student of the theater; b. D. 3 (16), 1911, Moscow.
Historian of the circus and of *estrada* (the variety stage); doctor of art studies (1951). Member of the CPSU since 1940. He graduated from the State Institute of Theatrical Arts (GITIS) in 1935. His literary activity began in 1929. From 1937 to 1967 he taught at GITIS (appointed professor in 1952); he has worked at the Institute for the History of the Arts since 1943 (since 1959, as chief of the Russian theater section). His main works have been devoted to questions of the Russian theater and the history and theory of the Russian and Soviet circus and *estrada*.
Works: *Brat'ia Durovy.* Moscow-Leningrad 1945. *Vitalii Lazarenko.* Moscow-Leningrad 1946. *P. S. Mochalov.* Moscow 1949. *Russkii tsirk.* Moscow 1953. *Ermolova.* Moscow 1954. *Mochalov akter-romantik.* Moscow 1961. *Iskusstvo sovetskoi estrady.* Moscow 1962. *Sovetskoe akterskoe masterstvo.* Moscow 1963. *Sovetskii tsirk.* Moscow 1963. *Sovetskaia estrada.* Moscow 1968. *Iskusstvo tsirka.* Moscow 1964. *Sovetskii tsirk segodnia.* Moscow 1968. *Estrada i tsirk glazami vliublennogo.* Moscow 1971.

Dmitriev, Ivan Nikolaevich
Party official; b. 1920.
Mem. CPSU 1945–; grad. Chernorechenskii Chemical Tech. College 1940 and Chkalov Inst. for Engineering Gorkii 1948; worker, head of the shift, mechanic, Dept. Head of Chernorechenskii chemical Kalinin Factory 1940–43; foreman,

leader, chief engineer, head of admin. during construction of Gorkii Hydro-Power Plant 1948–58; Head of Building and Construction Trust in Dzerzhinsk 1958–64; Sec. of Gorkii Oblast CPSU Cttee. 1964–69; Head of Construction Dept. at Cen. Cttee. CPSU 1969–; Dep. to USSR Supr. Sov. 1970–; mem. CPSU Cen. Auditing Comm. 1971–81, cand. mem. Cen. Cttee. CPSU 1981–. Decor.: State Prize of the USSR 1970.

Dmitriev, P. I.
Trade union official.
Sec. and mem. Presidium of Moldavian Republican Trade Union Council S. 1967–; cand. mem. Cen. Cttee. Moldavian CP 1981–.

Dmitriev, Valentin Ivanovich
Party official; b. 1927.
Mem. CPSU 1947–; grad. Odessa Hydrotechnical Inst. 1955 and CPSU Cen. Cttee. Party Higher Correspondence School 1971; served in Sov. Army 1943–50; chief engineer of construction admin. of "Iuzhuralspetsstroi" Trust 1956–59; Sec. of Kirov Raion Cttee. CPSU in Magnitogorsk City 1959–60; 1st Sec. of Magnitogorsk City Cttee. CPSU 1962–70; 1st Sec. of Cheliabinsk City Cttee. CPSU 1970–79; held posts in Cen. Cttee. apparatus CPSU 1979–80; 2nd Sec. of Cen. Cttee. Latvian CP and mem. Bureau Cen. Cttee. Latvian CP O. 1980–; cand. mem. Cen. Cttee. CPSU 1981–.

Dmitrochenko, Aleksandr Petrovich
Agronomist; b. Ag. 19 (S. 1), 1900.
Grad. at an agricultural inst. 1922; mem. USSR V. I. Lenin Union Acad. of Agricultural Sciences 1967–; Prof. Chair of Feeding of Domestic Cattle at Zoological Inst. Leningrad 1930–35; teaching activity at several insts. 1930–; has written several works on cattle feeding. Decors.: Hero of Socialist Labour, 2 Orders of Lenin, several medals.

Dneprov, Anatolii Petrovich (real name: Mitskevich)
Author; b. 1919, Dnepropetrovsk.
Russian; mem. CPSU 1949–; grad. Dept. of Physics at Moscow State Univ. 1941; first works publ. 1958.
Publs.: science-fiction "Mir, v kotorom ia ischez" (The World I Disappeared in) 1961, "Formula bessmertiia" (The Key to Immortality) 1962, and others.

Dobin, Grigorii Izrailevich
Author; b. 1905, Zhlobin.
Jewish; took part in Belorussian partisan movt.; first works publ. 1929.
Publs.: "U melnitsy" (Near the Mill) 1931, "U Amura" (On the Amur) 1935, "Na paseke" (Bee-keeping) 1941, "Na belorusskoi zemle" (On Belorussian Soil) 1947, "Rasskazy" (Stories) 1956, and others.

Dobrik, Viktor Fedorovich
Party official; b. 1927, Ukraine.
Mem. CPSU 1954–; grad. Dnepropetrovsk Inst. of Railway Engineers 1950; foreman, works manager, dept. head with a construction and assembly train 1950–51; chief engineer, dep. dept. head, chief engineer, dept. head on various construction projects 1951–57; Dep. Chief Engineer, Dep. Sec., Sec. of CPSU Cttee. at construction of Dneprodzerzhinsk Hydroelectric Plant 1957–61; 2nd Sec. 1961, 1st Sec. of Dneprodzerzhinsk City Cttee. Ukrainian CP 1963–69; Chair. of Exec. Cttee. of Dneprodzerzhinsk City Sov. of Workers' Deps. 1961–63; Dep. to 1963–, mem. 1963–64, Chair. of Comm. on Transport, Posts and Telecommunications with Supr. Sov. Ukrainian SSR 1964–; cand. mem. 1966–71, mem. Cen. Cttee. CPSU 1971–; cand. mem. 1966–69, mem. Cen. Cttee. Ukrainian CP 1969–; 1st Sec. of Ivano-Frankovsk Oblast Cttee. Ukrainian CP 1969–73; 1st Sec. of Lvov (Lemberg) Oblast Cttee. Ukrainian CP 1973–; cand. mem. Politbureau Cen. Cttee. Ukrainian CP 1976–.

Dobrovol'skii, Anatolii Vladimirovich
Architect; b. My. 6 (19), 1910, Buki, in present-day Malin Raion, Zhitomir Oblast.
Honored Builder of the Ukrainian SSR (1966). Member of the CPSU from 1950. He, a graduate of the Kiev Civil Engineering Institute (1934), was chief architect of Kiev from 1950 to 1955 and a professor at the Kiev Institute of Art from 1965. His projects in Kiev (done in collaboration with others) include a residential house for the workers of the Bolshevik Plant (1949), a construction plan for Kreshchatik (1945–47, construction began in 1947; USSR State Prize for introducing decorative ceramics into construction, 1950), a dormitory for a school of mining (1950; State Prize of the USSR, 1951), subway stations of the Kreshchatik (with a restaurant; 1960–61), the Bolshevik Plant (1963), the Hotel Moskva (1961–62), and the Ukraine Movie Theater (1963). He also built the airport in Borispol (1964–66, in collaboration with others). Decors.: Order of the Red Banner of Labor and medals.

Dobrovolskii, Vladimir Anatolevich
Author; b. 1918, Kharkov.
Russian; mem. CPSU 1948–; grad. Dept. of Philology of Kharkov Univ. 1941; first works publ. 1938. Decors.: Order of Red Banner, Stalin Prize.
Publs.: novels "Troe v serykh shineliakh" (Three Men in Grey Coats) 1948, "Zhenia Maslova" 1950, "Na vsiu zhizn'" (For the Whole Life) 1957, "Dom v tupike" (The House in the Cul-de-sac) 1959, and others.

Dobrynin, Anatolii Fedorovich, Cand. of Hist. Sc.
Diplomat; b. 1919.
Mem. CPSU 1945–; grad. Moscow Aviation Inst. 1942; entered Diplomatic Service 1946; USSR Asst. Dep. Min. of Foreign Affairs 1949–52; Emb. Counsellor, then Min. in USSR Emb. in USA 1952–55; USSR Asst. Dep. Min. of Foreign Affairs 1955–57; Dep. Sec.-Gen. at UN 1957–60; Board mem. and Head of American Dept. in USSR Min. of Foreign Affairs 1960–62; USSR Amb. to the USA 1962–; cand. mem. 1966–71, mem. Cen. Cttee. CPSU 1971–. Decor.: Order of Red Banner of Labour.

Dobrzhanskaia, Liubov' Ivanovna
Russian actress; b. D. 24, 1908, Kiev.
People's Artist of the USSR (1965).
After her graduation from the studio of the Kiev Russian Theater of Drama in 1924, she joined the theater's company. In 1934 she began working in the Central Theater of the Soviet Army. Her best roles: Katherine (Shakespeare's *The Taming of the Shrew*), Shura Azarova (Gladkov's *A Long Time Ago*), Elena Stigliano (De Filippo's *My Family*), Emilia Marti (Čapek's *The Macropoulos Secret*), Vasiluce (Druce's *Casa Mare*), and Mrs. Warren (Shaw's *Mrs. Warren's Profession*). Her work is distinguished by keen characterizations, versatility, elegance, and vitality. Her portrayals of uncommon, freedom-loving women are especially effective. She is also effective in portraying the grotesque, for example, she successfully created the role of the crotchety, stupid, self-satisfied, hypocritical tyrant Karolina in Chernyshevskii's *The Expert Troublemaker*. In 1964, she played the role of Tolgonai in Ch. Aitmotov's *The Maternal Field* at the Moscow K. S. Stanislavsky Drama Theater. Decors.: State Prize of the USSR in 1950, Order of Lenin and the Order of the Red Banner of Labor.

Dolenko, Grigorii Nazarovich, Dr. of Geol.-Min. Sc.
Geologist; b. F. 24, 1917, Gaivtsi, Kirovograd Oblast.
Mem. CPSU 1953–; grad. Geological-Geographical Faculty of Kharkov Univ. 1940; corresponding mem. Acad. of Sciences Ukrainian SSR 1964–; Dr. of Geol.-Min. Sc. 1961; Prof. 1963–; employed in various geological institutions 1940–45; mem. staff Sov. Petroleum Admin. in Austria 1945–50; Chief Geologist of "Ukrnaftorozvidka" Trust 1950–52; mem. staff Inst. of Geology and Geochemistry of Fuels at Acad. of Sciences Ukrainian SSR 1952–; Dep. Dir. 1959–63, Dir. of same Inst. 1963–.
Publs.: "Geologiia naftovykh rodovyshch Ukrainy" 1959, "Geologiia nefti i gaza Karpat" 1962, "Problema proiskhozhdeniia nefti i gaza i obrazovanie ikh promyshlennykh zalezhei" 1966.

Dolgikh, Vladimir Ivanovich
Party official; b. D. 5, 1924.
Russian; mem. CPSU 1942–; grad. Irkutsk Mining Inst. 1949; served in Sov. Army 1941–43; shift foreman, chief engineer of various works in Krasnoiarsk 1949–58; Chief Engineer 1958–62 and Dir. of metallurgical Zaveniagin Combine in Norilsk 1962–69; Dep. to USSR Supr. Sov. 1966–; 1st Sec. of Krasnoiarsk Krai CPSU Cttee. 1969–72; mem. Cen. Cttee. CPSU 1971–; Sec. of Cen. Cttee. CPSU D. 1972–; Chief, Heavy Industry Dep., Cen. Cttee. CPSU 1976–; cand. mem. Politbureau Cen. Cttee. CPSU My. 1982–. Decors.: Order of Lenin, Hero of Socialist Labor, Order of Patriotic War 1st Class.

Dolgoplosk, Boris Aleksandrovich
Chemical scientist; b. N. 12 (25), 1905, Lukoml, now Vitebsk Oblast.
Mem. CPSU 1945–; grad. Dept. of Chemistry of Moscow State Univ. 1931; mem. USSR Acad. of Sciences 1964–; Scientific Head of Cen. Scientific Research Laboratory at SK-1 Factory in Iaroslavl 1932–46 and of Union Inst. of Synthetic Rubber 1946–61; Head of Laboratory at Inst. of Chemical Physics 1961–63 and Inst. of Petrochemical Synthesis of USSR Acad. of Sciences 1963–. Decors.: Hero of Socialist Labour 1963, State Prize of the USSR 1941 and 1949, Lebedev Prize of the Presidium of USSR Acad. of Sciences 1947 and 1963, 2 Orders of Lenin, 2 other orders and various medals.

Dolgushkin, N. K.
Komsomol official.
1st Sec. Orenburg Oblast Cttee. All-Union Komsomol until 1981; Sec. and mem. Bureau of Cen. Cttee. All-Union Komsomol Mr. 1981–.

Dolidze, Semen Vissarionovich
Film director and screenwriter; b. Ja. 24 (F. 6), 1903, Makharadze.
People's Artist of the USSR (1965). Member of the CPSU since 1941.
He graduated from the department of history and philology at the University of Tbilisi in 1925. One of his first important works was *The Last Crusaders* (1934). His films include *Dariko* (1937), *The Shield of Dzhurgai* (1944, in collaboration with D. E. Rondeli), *The Dragonfly* (1954, based on the play *Marine* by M. G. Baratashvili), *Eteri's Song* (1957), *Fatima* (1959, based on the poem by K. Khetagurov), *Last Day, First Day* (1960), *Paliastomi* (1963), *A Meeting With the Past* (1967), and *The City Awakens Early* (1968). He was a deputy to the fourth convocation of the Supreme Soviet of the Georgian SSR. Decors.: State Prize of the USSR (1950) three orders, medals.

Dollezhal, Nikolai Antonovich
Engineer; b. O. 15 (27), 1899, Omelnik, Zaporozhe Oblast.
Grad. Moscow Inst. of Technology 1923; specialist in mechanics and thermotechnology; mem. USSR Acad. of Sciences 1962–; Tech. Dir. of Inst. of Azor Machine Construction in Leningrad 1932–34; Chief Engineer of "Bolshevik" Plant in Kiev 1935–38; Dir. of Scientific Research Inst. of Chemical Machine Construction in Moscow 1942–53; dir. of a scientific research inst. in Moscow 1953–; chief designer of reactor of first nuclear power station in the world. Decors.: State Prize of the USSR 1949, 1952, 1953, 1970, Lenin Prize 1957, 4 Orders of Lenin, 2 other orders and various medals.

Dolmatov, Vladimir Konstantinovich
Party official.
Cand. mem. Cen. Cttee. Kirghiz CP 1976–79, mem. 1979–; Head of Dept. of Org. Party Work with Cen. Cttee. Kirghiz CP D. 1979–; cand. mem. Bureau Cen. Cttee. Kirghiz CP Ja. 1981–.

Dolmatovskii, Evgenii Aronovich
Author; b. 1915, Moscow; s. of a lawyer. Mem. CPSU 1941–; grad. Gorkii Inst. of Literature Moscow 1937; first works publ. 1934.
Publs.: "Lirika" (Lyric) 1934, "Den'" (The Day) 1935, "Pesnia o Dnepre" (Dnepr Song) 1942, "Stepnaia tetrad'" (Notebook of the Steppes) 1943, "Vera v pobedu" (Belief in Victory) 1944, "Stikhi izdaleda" (Poems from Afar) 1945, "Propal bez vesti" (Missed) (First part of the trilogy "Odna sud'ba" – A Fate – 1942–46), "Dobrovoltsy" (The Volunteers) 1956; has written lyrics of many popular songs, incl.: "Moia liubimaia" (My Love), "Esli by parni vsei zemli" (If the Boys of the Whole World), "Marsh brigad kommunisticheskogo truda" (March of the Brigades of Communist Workers), and others. Decor.: Order of Red Banner of Labour.

Donenbaeva, Kashmat Baigazinovna
Laborer; b. 1943.
Mem. CPSU 1973–; grad. Kazakh State Farm Technicum in Kustanai Oblast 1980; laborer at "Kamensk-Uralskii" State Farm 1961–68; tractor driver at "Kharkovskii" State Farm in Kazakh SSR 1968–; Dep. to USSR Supr. Sov. 1974–; mem. Cen. Auditing Comm. CPSU 1981–. Decors.: Heroine of Socialist Labor 1975; USSR State Prize 1976.

Donskoi, Mark Semenovich
Film director and screenwriter; b. F. 21 (Mr. 6), 1901, Odessa.
People's Artist of the USSR (1966) and Hero of Socialist Labor (1971). Member of the CPSU since 1945.
He graduated from the division of law of the University of Simferopol'. He began his film career in 1926. His first films, made with M. A. Averbakh, were *In the Big City* (1928) and *A Man's Worth* (1929). The film *The Song of Happiness*, made with V. G. Legoshin in 1934, brought him renown. His most important work of the 1930's was the motion picture trilogy on the childhood and youth of M. Gorkii – *The Childhood of Gorkii* (1938, scenario by I. Gruzdev and Donskoi), *In the World* (1939; State Prize of the USSR, 1941), and *My Universities* (1940, scenario by Donskoi). The development of the writer's personality was depicted against the broad background of provincial life in prerevolutionary Russia. The film presents original characterizations and organically combines a universal poetic principle with colorful scenes of everyday life.
During the Great Patriotic War, he filmed *How the Steel Was Tempered* (1942, his scenario based on N. A. Ostrovskii's novel), *The Rainbow* (1944), after V. L. Vasilevskaia's novella; State Prize of the USSR, 1946), and *The Unvanquished* (1945, after B. L. Gorbatov's novella; scenario by him and Gorbatov). A significant achievement was the film *The Rural Schoolmistress* (1947; State Prize of the USSR, 1948; starring V. P. Maretskaia in the leading role), devoted to the beauty of selfless creative work. He made the film versions of Gorkii's *Mother* (1956, scenario by him and N. A. Kovarskii) and *Foma Gordeev* (1959, scenario by B. G. Bialik and him). In the 1960's he turned to the subject of Lenin, making the films *A Mother's Heart* (1966; State Prize of the USSR, 1968) and *A Mother's Fidelity* (1967). Decors.: two Orders of Lenin, the Order of the Red Banner of Labor, and various medals.
Work: "Kak ia stal rezhisserom." In *Kak Ia Stal Rezhisserom.* Moscow 1946, pp. 95–104.

Donzhashvili, Tina Georgievna
Author; b. 1916, Tbilisi.
Georgian; mem. CPSU 1949–; grad. Moscow Medical Inst. and Kirov Acad. of War Medicine Leningrad; first works publ. 1947.
Publs.: "Ia ne pokinu tebia" (I Shall Not Leave You) 1947, "Novyi Kartli" (The New Kartli), "Ulybki pobeditelei" (Smile of the Victors), "Moi drug" (My Friend) 1963, and others.

Dordzhiev, Basang Biuriunovich
Author; b. 1918.
Kalmuk; first works publ. 1934.
Publs.: "Vo imia Rodiny" (In the Name of the Motherland) 1941, "Boskha", has translated works by Pushkin, Lermontov, Maiakovskii, Marshak, Mikhalkov and others into Kalmuk.

Dorfman, Iakov Grigor'evich
Physicist; b. Mr. 26 (Ap. 7), 1898, St. Petersburg.
Doctor of physical and mathematical sciences (1934).

He graduated from the Leningrad Polytechnic Institute in 1925. From 1921 to 1931 he worked at the Leningrad Physicotechnical Institute. In 1958 he began working at the All-Union Institute of Scientific and Technical Information. In 1964 he became head of the physics section of the Institute of the History of the Natural Sciences and Technology of the Academy of Sciences of the USSR. His principal works are on the physics of magnetic phenomena. In 1923, Dorfman expressed the possibility of the resonance absorption of electromagnetic energy in paramagnets. In 1927 he experimentally proved that the molecular field in ferromagnets is not of magnetic origin. In 1930, together with Ia. I. Frenkel', he theoretically substantiated the domain structure of ferromagnets. In 1951, he theoretically established the existence of cyclotron resonance. He is also the author of works on the history of science.
Works: "Einige Bemerkungen zur Kenntnis des Mechanismus magnetischer Erscheinungen." *Zeitschrift für Physik,* 1923, vol. 17, issue 2, pp. 98–111. *Magnitnye svoistva atomnogo iadra.* Moscow-Leningrad 1948. *Lavuaz'e.* Moscow-Leningrad 1948. "Paramagnitnyi i diamagnitnyi rezomans elektronov provodimosti." *Dokl. AN SSSR,* 1951, vol. 81, no. 5, pp. 765–66. *Diamagnetizm i khimicheskaia sviaz'.* Moscow 1961.

Dorizo, Nikolai Konstantinovich
Author; b. 1923, Krasnodar.
Russian; grad. Higher Literary Courses of USSR Writers' Union; first works publ. 1938.
Publs.: "Na rodnykh beregakh" (On Homeland Banks) 1948, "My – mirnye liudi" (We are Peaceful People) 1950, "Stikhi" (Poems) 1952, "Vstrecha s toboi" (Encountering You) 1955, "Lirika" (Lyric) 1958, "Veriu, liubliu, poiu ...!" (I Believe, Love, Sing ...) 1959, "Imia moe – chelovek" (My Name is Man) 1961, and others.

Dorodnitsyn, Anatolii Alekseevich
Mathematician, geophysicist and specialist in mechanics; b. N. 19 (D. 2), 1910, Bashino, Tula Oblast.
Grad. Petroleum Inst. in Groznyi 1931; mem. USSR Acad. of Sciences 1953–; mem. staff Cen. Aero-Hydrodynamic Inst. 1941–55; mem. staff Inst. of Mathematics of USSR Acad. of Sciences 1945–55; Dir. of Computing Centre of USSR Acad. of Sciences 1955–. Decors.: Hero of Socialist Labour, State Prize of the USSR 1946, 1947 and 1951, 3 Orders of Lenin, 3 other orders and various medals.

Dorofeev, Nikolai Ivanovich
State official.
Cand. mem. Cen. Cttee. Latvian CP 1976–; Head of Dept. of Construction and Municipal Econ. with Cen. Cttee. Latvian CP 1976–77; Min. of Building Materials Ind. Latvian SSR Ag. 1977–.

Dorokhov, Aleksei Alekseevich
Author; b. 1902, St. Petersburg.
Russian; first children's books publ. 1940.
Publs.: "Strana stroit" (The Land Builds) 1952, "Sto poslushnykh ruk" (A Hundred Obedient Hands) 1954, "Iz vorot vykhodit gruzovik" (A Lorry Comes Out of the Gate) 1956, "Eto ne melochi!" (Those Are Not Trivialities!) 1961, "Serdtse na ladoni" (Heart on Hand) 1962, and others.

Doronina, Tat'iana Vasil'evna
Russian actress; b. S. 12, 1933, Leningrad. People's Artist of the RSFSR (1969).
She graduated from the Studio-School of the Moscow Art Theater in 1956. From 1956 to 1959 she worked at the Lenin Komsomol Leningrad Theater, from 1959 to 1966 at the M. Gorkii Leningrad Bol'shoi Dramatic Theater, from 1966 to 1971 at the Moscow Art Theater, and from 1971 at the Vladimir Mayakovsky Moscow Theater. Her most important roles include Nadezhda Monakhova in Gorkii's *The Barbarians,* Sof'ia in Griboedov's *Woe From Wit,* Lushka in *Virgin Soil Upturned* (after Sholokhov), and Masha in Chekhov's *The Three Sisters.* She also appears in films. Her best screen roles were created by her on the stage: Nadia in Volodin's *My Elder Sister* (film title, *Elder Sister*) and Natasha in Radzinskii's *Once Again About Love* (film of the same title). She played the role of Niura in the film *Three Poplars on Plushchikha Street.*

Doroshko, Petr Onufrievich
Author; b. 1910, Tupichev, Chernigov Oblast.
Ukrainian; mem. CPSU 1941–; first works publ. 1928. Decors.: Order of the Patriotic War, Order of Red Banner, medals.
Publs.: "Pobediteli stepi" (Victors over the Steppe) 1931, "Bronza" (Bronge) 1937, "Dekady rosta" (Decades of Growth) 1932, "Predgrozie" (Before the Thunderstorm) 1935, "Rodnaia storona" (Native Land) 1941, "Sady Krasnograda" (Krasnograd Gardens) 1943, "Dalnie zareva" (Distant Glow), "Nasushchnyi khleb" (Daily Bread) 1946, "Dumy moi" (My Thoughts) 1963, and others.

Dostian, Richi Mikhailovna
Author; b. 1914, Warsaw.
Russian; first works publ. 1938.
Publs.: "Dva cheloveka" (Two People) 1956, "Kto idet?" (Who is There?) 1958, "Nezhdannyi drug" (Unexpected Friend) 1959, and others.

Dotsenko, Nadezhda Petrovna
Actress; b. 1914.
Ukrainian; mem. CPSU 1952–; grad. Kiev Drama Inst. 1936. Decor.: People's Artist of the USSR 1972.

Dovboshinskii, Daniil Danilovich
Painter; b. 1924, Kolki, Volhynia.
Grad. Lvov Inst. for Applied and Decorative Art 1953, pupil of S. Selskii, I. Bokshai; teacher at Lvov Inst. for Applied and Decorative Art 1953–.

Dovgan, Boris Stepanovich
Sculptor; b. 1928, Kiev.
Grad. Kiev Art Inst. 1956, pupil of M. Lysenko and I. Makogon.

Dovidaitis, Ionas
Author; b. 1914, Vilnius, Lithuania.
Publs.: "Druzia do samoi smerti" (Friends Until Death) 1947, "Rasskazy" (Stories) 1947, "Bespokoinyi chelovek" (A Restless Person) 1952, "Opasnyi put'" (A Dangerous Way) 1956, "Odnazhdy v letniuiu noch' ..." (On a Summer's Night ...) 1961, "Podvig" (A Heroic Deed) 1956, "Bol'shie sobytiia v Naujamiestyje" (Important Events in Naujamiestyje) 1951, and others.

Dovodov, Kh. I.
Komsomol official.
Sec. and mem. Bureau Cen. Cttee. Turkmen Komsomol 1981–.

Drach, Ivan Fedorovich
Author and poet; b. O. 17, 1936, Telizhentsy, Kiev Oblast.
Ukrainian; mem. CPSU 1959–; studied Dept. of Philology of Kiev Univ., concluded Advanced Courses in Theatre Arts in Moscow 1964.
Publs.: "Podsolnechnik" (Sun Flower) 1962, "Ballady Budnei" (Ballads of Everyday Life) 1967, "Idu do tebe" (I Come to You) 1970, and others.

Draganiuk, Kirill Alekseevich
State official.
Min. of Health Moldavian SSR Je. 1974–; mem. Auditing Comm. 1976–81, mem. Cen. Cttee. Moldavian CP 1981–.

Dragunskii, David Abramovich
Military official; b. 1910.
Mem. CPSU 1931; grad. Frunze Mil. Acad. 1941 and Acad. of Gen. Staff 1950, Col.-Gen. of Armoured Troops; served in Sov. Army 1933–; commander of an armoured div. on W. Front, commanding officer of an armoured brigade on 1st Ukrainian Front 1941–45; command posts in Sov. Army, Head of Erevan and Tbilisi Garrisons 1945–69; mem. Cen. Cttee. and Bureau Cen. Cttee. Armenian CP 1961–65; Head of Sen. Officers' Courses "Vystrel" My. 1969–; mem. CPSU Cen. Auditing Comm. 1971–. Decors.: Hero of the Sov. Union (twice), Order of Lenin, 3 Orders of Red Banner, and other orders and medals.

Drampian, Khazhak Andranikovich
State official.
Mem. Cen. Cttee. 1961–66, mem. Auditing Comm. 1966–71 and 1976–81, cand. mem. Cen. Cttee. Armenian CP 1981–;

Party Organizer with Cen. Cttee. Armenian CP in Idzhevan Kolkhoz-Sovkhoz Prod. Admin. 1962; Dep. to Supr. Sov. Armenian SSR 1963–; Head of Cen. Admin. with Council of Mins. Armenian SSR for Gasification Je. 1975–78; Min. of Motor Transport Armenian SSR My. 1978–.

Driba, Dzintars Ianovich
Architect; b. 1928.
Latvian; grad. Latvian Univ. 1953; teacher at Polytech. Inst. Riga 1958–. Decor.: State Prize of the USSR 1972.

Driuchin, Aleksandr Pavlovich
State official.
Head of Cen. Statistical Admin. RSFSR O. 1970–.

Drizul, Aleksandr Arvidovich
State official; b. Je. 29, 1920, Pskov.
Mem. CPSU 1950–; grad. Moscow Historical Archival Inst. 1942; historian; mem. Acad. of Sciences Latvian SSR 1963–; Dep. Dir. 1949–63, Dir. of Inst. of History and Acad. Sec. of Dept. of Social Sciences at Acad. of Sciences Latvian SSR 1963–70; cand. mem. 1961–66, mem. Cen. Cttee. Latvian CP 1966–; Sec. and mem. Bureau of Cen. Cttee. Latvian CP 1970–75; Dep. to 1967–, Dep. Chair. of Presidium Supr. Sov. Latvian SSR 1967–71, 1975–; Vice-Pres. of Acad. of Sciences Latvian SSR 1975. Decors.: State Prize of the Latvian SSR 1967, 2 orders.

Drobnis, Aleksandras Antanovich
State official; b. 1912, Lithuania.
Mem. CPSU 1944–; grad. Vilna State Univ.; bank employee until 1940; Dir. of Finance Dept., Head of Republican Bureau of USSR State Bank. Rep. of Council of People's Commrs. Lithuanian SSR 1940–42; Dep. Commr., Commr. Min. of Finance Lithuanian SSR 1942–57; Dep. to Supr. Sov. Lithuanian SSR 1955–; mem. Cen. Cttee. Lithuanian CP 1956–; Dep. Chair. Council of Mins. Lithuanian SSR 1957–58; Chair. of GOSPLAN Lithuanian SSR 1958–; dep. to and mem. Budget Comm. at Sov. of Nationalities USSR Supr. Sov. 1962–66; again Dep. Chair. of Council of Mins. Lithuanian CP D. 1965–. Decor.: Order of Lenin.

Drozd, Mikhail Filippovich
State official.
Mem. Cen. Cttee. Latvian CP 1971–; 1st Sec. of a Raion Cttee. Latvian CP in Elgava until 1978; Min. of Interior Latvian SSR F. 1978–.

Drozdov, Oleg Alekseevich, Dr. of Geog. Sc.
Climatologist; b. O. 5 (18), 1909, Bucharest.
Grad. Physical-Mathematical Faculty of Kazan Univ. 1930; Dr. of Geog. Sc. 1948; Prof. 1950–; Prof. Chair of Meteorology and Climatology at Geographical Faculty of Leningrad Univ. 1954–; numerous inventions. Decors.: 3 orders, several medals.

Drunina, Iulia Vladimirovna
Author; b. 1924, Moscow.
Russian; grad. Gorkii Inst. of Literature Moscow 1952; medical orderly at the Front 1941–45; first works publ. 1948. Publs.: "V soldatskoi shineli" (In a Soldier's Coat) 1948, "Veter s fronta" (The Wind from the Front) 1958, "Sovremenniki" (Contemporaries) 1960, "Trevoga" (Alarm) 1963, and others.

Druskin, Mikhail Semenovich
Musicologist and teacher; b. Ja. 1 (14), 1905, Kiev.
Honored Art Worker of the RSFSR (1962). Doctor of the Arts (1946).
He graduated as a musicologist from the Institute of the History of the Arts in Leningrad, where he studied under B. V. Asaf'ev. He also completed piano study under O. K. Kalantarova at the Leningrad Conservatory. From 1915 to 1935, he gave concerts. Since 1935 he has taught music history at the Leningrad Conservatory; he has been a professor there since 1947. His research has dealth with a wide range of problems, including questions concerning western musical classics, contemporary music, the history of piano music, pianism, operatic dramaturgy, and the history of revolutionary song.
Works: *Gans Eisler i rabochee musykal'noe dvizhenie v Germanii.* Moscow 1934. *Ocherki po ustorii tantseval'noi muzyki.* Leningrad 1936. *Voprosy muzykal'noi dramaturgiiopery.* Leningrad 1952 *Russkaia revolutsionnaia pesnia.* Moscow 1954. *Brams,* 2nd ed., Moscow 1970. *Klavirnaia muzyka Ispanii, Anglii, Niderlandov, Frantsii, Italii, Germanii XVI–XVIII vekov.* Leningrad 1960. *Vagner,* 2nd ed., Moscow 1963. *Istoriia zarubezhnoi musyki vtoroi poloviny 19 v.,* 2nd ed., Moscow 1967.

Druzhinin, Mikhail Ivanovich
Military official; b. 1920.
Mem. CPSU 1943–; grad. Higher Mil. Pedagogical Inst. 1955 and Mil. Acad. of Gen. Staff 1962, Col.-Gen.; entered Sov. Army 1939; komsomol organizer of a battalion 1944–45; dep. chief, chief of polit. dept. of a division 1955–60; polit. party work in armed forces 1962–63; 1st dep. chief of polit. directorate of a mil. district 1969–73; holds leadership position within context of party polit. work in armed forces 1973–; cand. mem. Cen. Cttee. CPSU 1981–. Decor.: Hero of the Sov. Union 1945.

Druzhinina, Y.
Historian.
Corresponding mem. USSR Acad. of Sciences 1981–. Researcher at the Inst. of History USSR Acad. of Sciences; main area of research: Russian history in the 18th–19th cent.; author of more than 100 publications; shows in her work the positive repercussions of Russia's access to the southern seas, the history of this region's economic devt., the social division of labor between remote areas and central districts of the country.

Drygin, Anatolii Semenovich
Party official; b. 1914.
Russian; mem. CPSU 1940–; grad. Michurinsk Inst. of Vegetable Crops 1935; researcher, dept. head at a fruit and soft-fruit experimental station 1935–41; served in Sov. Army 1941–46; dir. of an experimental base for the All-Union Inst. of Plant Breeding, dir. of a sovkhoz, chair. of a raion sov., sec. of a raion CPSU cttee., 1st Dep. Chair. of Exec. Cttee. Leningrad Oblast Sov. of Workers' Deputies, Sec. of Leningrad Oblast CPSU Cttee. 1946–61; dep. to and mem. Budget Comm. RSFSR Supr. Sov. 1959–63; cand. mem. 1961–71, mem. Cen. Cttee. CPSU 1971–; 1st Sec. of Vologda Oblast CPSU Cttee. S. 1961–; dep. mem. 1962–, mem. Comm. for Educ., Science and Culture at Sov. of the Union USSR Supr. Sov. 1966–74.

Dubinin, Iurii Vladimirovich
Diplomat; b. 1930.
Mem. Diplomatic Service 1955–; mem. Apparat USSR Min. of Foreign Affairs 1960–63 and 1968–78; 1st Sec., Emb. Councellor, USSR Embassy in France 1963–68; USSR Amb. to Spain O. 1978–.

Dubinin, Mikhail Mikhailovich
Physical chemist; b. D. 19, 1900 (Ja. 1, 1901), Moscow.
Grad. Dept. of Chemistry of Moscow Inst. of Technology 1921; mem. USSR Acad. of Sciences 1943–; Head of Dept. of Sorption Processes at Inst. of Physical Chemistry of USSR Acad. of Sciences 1946–. Decors.: Hero of Socialist Labour 1969, State Prize of the USSR 1942, 1950, 2 Orders of Lenin, 6 other orders and various medals.

Dubinin, Nikolai Petrovich
Geneticist; b. D. 22, 1906 (Ja. 4, 1907), Kronstadt.
Mem. CPSU 1969–; grad. Moscow State Univ. 1928; mem. USSR Acad. of Sciences 1966–; worked in various research laboratories 1932–66; Dir. of Inst. of Gen. Genetics of USSR Acad. of Sciences 1966–; his works deal with problems in general and evolutionary genetics and the relationship between genetics and agriculture; mem. of various foreign acads. of sciences and of British Soc. of Geneticists. Decor.: Lenin Prize 1966.

Dubinovskii, Lazar' Isaakovich
Sculptor; b. Ap. 18 (My. 1), 1910, Faleshty.
People's Artist of the Moldavian SSR (1963); corresponding member of the

Academy of Arts of the USSR (1954). Member of the CPSU from 1952.
He studied at the Academy of Art in Bucharest from 1925 to 1930 (in 1929 he studied at the studio of E. A. Bourdelle in Paris). His works include portraits of the Rumanian artist C. Baba (bronze, 1960, Russian Museum, Leningrad). Ho Chi Minh (wood, 1968), and the academician K. I. Skriabin (plaster of Paris, 1961, Tret'iatov Gallery). He created a monument to G. I. Kotovskii (bronze, granite, 1953; in cooperation with K. D. Kitaika, I. G. Pershudchev, and A. I. Posiado) and the *Heroic Komsomol* monument (bronze, granite, 1958–59); both monuments are in Kishinev. His compositions include the trilogy *Fathers and Sons* (wood, 1957) and *The Seamstresses of Tiraspol'* (aluminum, 1962). These works, which are both in the Art Museum of the Moldavian SSR in Kishinev, depict very few figures. Sharp contrasts between light and shadow, created by the dynamic gestures, poses, and turns and by the rough surface texture of the sculptures, determine the intensely emotional nature of art his. Decors.: State Prize of the Moldavian SSR in 1970, Order of Lenin, the Order of the Red Banner of Labor, and various medals.

Dubov, Nikolai Ivanovich
Author; b. 1910, Omsk; s. of a worker.
Russian; lives in the Ukraine; first works publ. 1929. Decor.: State Prize of the USSR.
Publs.: play "U poroga" (On the Threshold) 1948, "Nastupaet utro" (Morning is Coming) 1950, "Na kraiu Zemli" (At the End of the World) 1951, "Ogni na reke" (Lights on the River) 1952, "Sirota" (Orphan) 1955, "Zhestkaia proba" (Difficult Test) 1960 (the latter two were publ. as a novel under the title "Gore odnomu" (The Misfortune of Being Alone) 1967), "Beglets" (Refugee) 1966, and others.

Dubovik, Nikolai Ivanovich
State official.
Chair. of State Cttee. for Gasification Belorussian SSR Ap. 1982–.

Dubovskoi, Boris Vladimirovich
Astronomer; b. 1912.
Grad. Moscow Univ. 1937; specialist in geodesy; mem. staff. Cen. Research Inst. for Geodesy, Aerial Photography and Cartography 1937–42; served in Sov. Army 1942–47; mem. staff Main Admin. of Northern Seaway at USSR Min. of Maritime Navigation ("Glavsevmorput") 1947–53; expedition head; Section Head of State Project and Research Inst. for Sea Transport 1953–. Decor.: State Prize 1971.

Dudenkov, Ivan Grigor'evich
State official; b. 1929.
Min. of Public. Services RSFSR O. 1974–. Decor.: Order of Red Banner of Labour.

Dudin, Mikhail Aleksandrovich
Author and poet; b. N. 7 (20), 1916, Klevnevo, Ivanovo Oblast.
Russian; mem. CPSU 1951–; first works publ. 1934; served in Sov. Army 1939–45; war correspondent for various front newspapers during World War II (1941–45). Decors.: 3 orders, medals.
Publs.: "Fliaga" (Water Bottle) 1943, "Doroga gvardii" (The Way of the Guards) 1944, "Schitaite menia kommunistom" (Regard Me as a Communist) 1950, "Mosty. Stikhi iz Evropy" (Bridges. Poems From Europe) 1958, "Do vostrebovaniia" (Poste restante) 1963, "Vremia. Stikhotvoreniia. 1964–1967" (Time. Poetry 1964–1967) 1969, "Stikhotvoreniia. Poemy. 1935–1969" (Verses. Poems. 1935–1969) 1970, and others.

Dudinskaia, Natal'ia Mikhailovna
Ballerina; b. Ag. 8 (21), 1912, Kharkov. People's Artist of the USSR (1957).
She studied at the Leningrad Choreographic School from 1923 to 1931 with A. Ia. Vaganova. From 1932 to 1962 she was prima ballerina of the S. M. Kirov Theater of Opera and Ballet in Leningrad. She danced the leading roles in Tchaikovsky's *Swan Lake* and *The Sleeping Beauty*. Prokofiev's *Cinderella*, Glazunov's *Raymonda*, and Adam's *Giselle*. A virtuosa in the technique of classical dance, she has created stage roles characterized by psychological depth and emotional expressiveness. Beginning in 1951, she has conducted a class for the dancers of the Kirov Theater. In 1964 she began teaching at the Vaganova Choreographic School in Leningrad. She is a follower of the Vaganova method. Decors.: State Prize of the USSR in 1941, 1947, 1949, and 1951; the Order of the Red Banner of Labor and various medals.

Dudintsev, Vladimir Dmitrievich
Author; b. 1918, Kupiansk, Kharkov Oblast.
Russian; grad. Law Inst. Moscow 1940; served in Sov. Army 1940–42, wounded nr. Leningrad; worked in a mil. court in Siberia 1942–45; story writer for "Komsomolskaia pravda" 1946–51; first works publ. 1933.
Publs.: "U semi bogatyrei" (With the Seven Knights) 1952, novel "Ne khlebom edinym" (Not by Bread Alone) 1956, "Novogodniaia skazka" (New Year's Fairytale) 1960, "Rasskazy" (Stories) 1963, and others.

Duisheev, Arstanbek
State official; b. 1932; s. of a farmer.
Kirghiz; mem. CPSU 1957–; grad. Kirghiz K. I. Skryabin Inst. of Agr. 1954, Cand. of Veterinary Sciences 1970; veterinarian on various state farms, Dir. of Dzhumgal state farm 1954–60; held Party posts 1960–62; 1st Sec. of Raion Cttee. CP Kirghizia in Uzgen 1962–68; Sec. of Cen. Cttee. CP Kirghizia 1968–71; mem. Bureau Cen. Cttee. CP Kirghizia 1968–71 and 1978–; mem. Cen. Cttee. CP Kirghizia 1963; 1st Oblast Cttee. Sec. CP Kirghizia in Issyk-Kul 1971–79; Chair. of Presidium Kirghiz Supr. Sov., Dep. Chair. of Presidium of USSR Supr. Sov. 1979–81; Chair. of Council of Mins. Kirghiz SSR Ja. 1981–; cand. mem. Cen. Cttee. CPSU 1981–. Decors.: Order of the October Revolution, 2 Orders of the Red Banner of Labor, medals.

Dulskas, Stanislovas Albinovich
State official.
Min. of Food Ind. Lithuanian SSR O. 1970–; mem. Auditing Comm. Lithuanian CP 1971–76; cand. mem. Cen. Cttee. Lithuanian CP 1976–.

Dumachev, Anatolii Panteleevich
Party official; b. 1932.
Mem. CPSU 1957–; grad. Leningrad Mil. Mechanical Inst. 1957; Sec. of Lenin Raion Komsomol Cttee. in Leningrad City 1956–58; party posts 1960–64; dep. chief, chief of a dept. of Leningrad Oblast Cttee. CPSU 1964–72; Sec. of Leningrad City Cttee. CPSU 1972–75; Sec. of Leningrad Oblast Cttee. CPSU Jl. 1975–; mem. Cen. Auditing Comm. CPSU 1981. Decor.: USSR State Prize 1973.

Dumbadze, Nina Iakovlevna
Track athlete; b. My. 23, 1919, Odessa. Honored Master of Sport (1943); member of the CPSU since 1946; teacher.
She was USSR discus champion in 1939, from 1943 to 1944, and from 1946 to 1950 and held the USSR record from 1937 to 1960. She was European discus champion in 1946 and 1950, won a medal at the Olympic games in 1952, and held the world discus record from 1948 to 1960. She set her personal record in 1952 with a throw of 57 m 4 cm. Decors.: two orders and various medals.

Dumbadze, Nodar Vladimirovich
Author; b. Jl. 14, 1928, Tbilisi.
Georgian; mem. CPSU 1964–; grad. Dept. of Econ. of Tbilisi Univ. Decor.: Prize of Cen. Cttee. USSR Komsomol.
Publs.: "Ia, babushka, Iliko i Illarion" (Me, Grandmother, Iliko and Illarion) 1960, Russian translation publ. 1967, "Ia vizhu solntse" (I See the Sun) 1960, "Solnechnaia noch'" (A Night of Sun) 1967, "Ne boisia, mama!" (Don't be Afraid, Mama!) 1971, and others.

Durdyev, Khaldurdy
Author; b. 1909, Kongur, Turkmenistan. Mem. CPSU 1951–; translator and journalist; worked for Turkmen newspapers and magazines 1930–.
Publs.: "Moe oruzhie" (My Weapon) 1947. "Sovetskii soldat" (Soviet Soldier), "Dnepr", "Beglets" (Refugee), and others.

Durova, Natalia Iurevna
Author; b. 1934, Moscow; d. of a circus artist.
Russian; mem. CPSU 1964–; grad. Gorkii Inst. of Literature Moscow 1956; appeared in circus from age of 4, first works publs. 1951.
Publs.: children's books "Kotka" 1959, "Pro Nikitu i belku" (Nikita and the Squirrel) 1960, "Bolshoi teater 'Malyshka'" (The Large Theatre 'Malyshka') 1962, "Arena" 1962, "Vash nomer!" (Your Number!) 1964, and others.

Dvalishvili, Akakii Archilovich
State official.
Chair. of State Cttee. on Film Georgian SSR Ag. 1973–; cand. mem. Cen. Cttee. Georgian CP 1976–81; mem. Auditing Comm. Georgian CP 1981–.

Dvalishvili, Demur Noevich
State official; b. 1938.
Georgian, mem. CPSU 1964–, grad. Leningrad Inst. of Finance Economics 1961; economics specialist in Finance Dept., head of Budget Dept., dep. head and head of Finance Dept. of Exec. Cttee. Kutaisi City Sov. of Peoples' Deps. 1961–75; Dep. Chair., Chair. of Exec. Cttee. Kutaisi City Sov. of Peoples' Deps. 1975–77; cand. mem. Cen. Cttee. CP Georgia 1976–81, mem. 1981–; dep. head of Light and Food Ind. Dept. Cen. Cttee. CP Georgia 1977; 1st Sec. of Raion Cttee. CP Georgia in Makharadze Ag. 1977–81; Min. of Finance Georgian SSR My. 1981–.

Dveirin, Anatolii Borisovich
Graphic artist; b. 1935, Dnepropetrovsk.
Grad. Dnepropetrovsk Art School 1955, attended Leningrad High School for Arts and Crafts 1955–56 and Kharkov Art Inst. 1956–62, pupil of O. Khmelnitskii, O. Liubimskii.

Dvoretskii, Ignatii (Izrail) Moiseevich
Author; b. 1919, Sliudianka, Irkutsk Oblast; s. of a clerk.
Russian; lives in Irkutsk 1947–; first works publ. 1948.
Publs.: "Taiga vesenniaia" (Taiga in Spring) 1952, "Polnovodie" (High Water) 1954, "Komandirovka" (Official Journey) 1957, "Vzryv" (Explosion) 1960, "Bol'shoe volnenie" (Great Excitement) 1961, and others.

Dybenko, Nikolai Kirillovich
Party official; b. 1928.
Russian; mem. CPSU 1951–; grad. Siberian Metallurgical Mining Inst. 1952 and CPSU Cen. Cttee. Party Higher School 1968, Cand. of Econ. Sc. 1978; technology engineer, dept. chief, dep. shop superintendent in various plants 1952–55; chief of a dept., sec. of a raion cttee., chair. of an exec. cttee. of a raion sov. of peoples' deps., 1st sec. of a raion cttee. CPSU in Novosibirsk City, 1st Sec. of Novosibirsk City Cttee. CPSU 1955–66; Sec. of Novosibirsk Oblast Cttee. CPSU 1966–73; posts in Cen. Cttee. apparatus CPSU 1973–78; 2nd Sec. of Cen. Cttee. and mem. Bureau of Cen. Cttee. Lithuanian CP D. 1978–; Dep. to USSR Supr. Sov. 1979–; cand. mem. Cen. Cttee. CPSU 1981–.

Dygai, Gleb Grigor'evich
Party official; b. 1922; s. of a blue-collar worker.
Russian; mem. CPSU 1942; higher education; served in Sov. army 1941–47; Komsomol organizer at a trade school, instructor of Tyrnovskii Raion Cttee. CP Moldavia 1947–51; head of Dept. for Party, Trade Union and Komsomol Organs Raion Cttee. CP Moldavia in Bendery 1951–55; Raion Cttee. Sec. CP Moldavia in Kalarashy 1955–60; held various posts in Council of Mins. Moldavian SSR and Cen. Cttee. CP Moldavia 1960–61; mem. Cen. Cttee. CP Moldavia 1961–; 1st Raion Cttee. Sec. CP Moldavia in Edintsy, Party organizer Cen. Cttee. CP Moldavia in the territorial collective and state farm production asscn. 1961–62; head of Dept. for Party-Organizational Work Cen. Cttee. CP Moldavia 1962–; cand. mem. Bureau of Cen. Cttee. CP Moldavia 1971–.

Dyikanbaev, K. D.
Trade union official.
Sec. and mem. Presidium of Khirghiz Republican Trade Union Council Ja. 1968–.

Dymshits, Veniamin Emmanuilovich
State official; b. 1910.
Jew; mem. CPSU 1937–; grad. Bauman Tech. College in Moscow 1945; worked on construction of blast furnaces 1931–50; Head of Main Board for Construction of Lead Ind. Enterprises, USSR Dep. Min. for the Construction of Metallurgical and Chemical Ind. Enterprises 1950–57; chief engineer at construction of a metallurgical plant in Bhilai (India) 1957–59; mem. and Head of Capital Construction Dept. of USSR GOSPLAN with rank of USSR Min. 1959–61; 1st Dep. Chair. of USSR GOSPLAN 1961–62; mem. Cen. Cttee. CPSU 1961–; Dep. to USSR Supr. Sov. 1962–; Dep. Chair. of USSR Council of Mins. 1962–; Dep. Chair. of USSR GOSPLAN Jl.–N. 1962; Chair. of USSR Econ. Council N. 1962–O. 1965; Chair. of State Cttee. for Material and Tech. Supply with USSR Council of Mins. O. 1965–76. Decors.: State Prize, 4 Orders of Lenin, 2 Orders of Red Banner of Labour and various medals.

Dzesov, Kudzag Gabrelovich
Ossetian writer; b. Jl. 1, 1905, Edisa, present-day Dzhava Raion, Iuzhnaia Osetiia Autonomous Oblast.
He, the son of a peasant, first published his works in 1925. A graduate from the Moscow Institute of Journalism in 1927, he was a delegate to the first congress of writers of the USSR in 1934. He has published the collections of stories *For Bread* (1959) and *Strong Mothers* (1962), the children's books *The Police Officer* (1932) and *The Wonderful Doll* (1963), the novellas *Mistress of the Bird Kingdom* (1967) and *Uneasiness* (1964), and the plays *The Unwilling Actress* and *The Bashful Bride* (1957). He vividly depicts field workers and intellectuals; he portrays the life and character of the people in a variety of ways.
Works: Dzestï Gabrelï firt Kudzoeg. *Xorxössäg. Radzïrdtä.* Stalinir 1959. *Alagirag satana.* Ordzhonikidze 1967. *Fïyyauï sagäs.* Ordzhonikidze 1970.

Dzhabadari, Tengiz Konstantinovich
State official; b. 1928.
Georgian; mem. CPSU 1955–; grad. Georgian Polytechnical Inst. 1952, mechanical engineer; construction engineer at a plant in Tbilisi 1952–55; instructor of Dept. for Ind. and Transport of Stalin Raion Cttee. and Tbilisi City Cttee. CP Georgia 1959–63; dep. head of Dept. for Ind., Transport and Communications Cttee. for Peoples' Control Georgian SSR 1963–73, head 1973; 1st Sec. of Talavskii Raion Cttee. CP Georgia 1973–75; 1st Dep. Min. of the Food Ind. Georgian SSR 1975–81; cand. mem. Cen. Cttee. CP Georgia 1976–81; Min. of Fruits and Vegetables Georgian SSR F. 1981–.

Dzhamanova, Roza Umbretovna
Kazakh soprano; b. Ap. 16, 1928, Aktiubinsk.
People's Artist of the USSR (1959). Member of the CPSU from 1963.
She graduated from the voice class of A. M. Kurganov at the Institute of Art in Alma-Ata in 1954. In 1953 she became a soloist at the Kazakh Theater of Opera and Ballet, where she has portrayed several musical stage characters including Sara (Tulebaev's *Birzhan and Sara*), Azhar (Zhubanovs' and Khamidi's *Abai*), Nazugum (Kuzham'iarov's *Nazugum*), Kamar (Rakhmadiev's *Kamar sulu*), Tat'iana (Tchaikovsky's *Eugene Onegin*), and Marguerita (Gounod's *Faust*). She is also a concert singer and has toured abroad (Poland in 1959, India in 1963, Canada in 1967, and Italy in 1968). She was a deputy to the eighth convocation of the Supreme Soviet of the USSR and the seventh convocation of the Supreme Soviet of the Kazakh SSR.

Dzhamgerchinov, Begimaly Dzhamgerchinovich
Historian; b. O. 2 (15), 1914, Kiuntu, the present-day S. M. Kirov Sovkhoz, Kirghiz SSR.
He has been a member of the Academy of Sciences of the Kirghiz SSR since 1954 and of the Communist Party since 1941. He became vice-president of the Kirghiz Academy of Sciences in 1965. From 1951 to 1954 he was rector of Kirghiz State

University. From 1952 to 1956 he was a member of the Central Committee of the Communist Party of Kirghizia. The principal theme of his works is 19th- and 20th-century Kirghiz history. His works include *Essays on the Political History of Kirghizia in the 19th Century* (1966), *History of Kirghizia – From Ancient Times to the Twenty-third Congress of the CPSU* (1967, jointly with V. P. Sherstobitov and A. K. Kanimetov), and *The Marxist-Leninist Solution of the Nationality Question* (1969). Decor.: State Prize of the Kirghiz SSR in 1970.

Dzhamil', Akhmed (pseudonym of Akhmed Sattar ogly Dzhemil'zade)
Azerbaidzhan lyric poet; b. O. 1913, Yerevan.
The son of a craftsman, he graduated from the Azerbaidzhani Pedagogical Institute. He began to publish his work in 1928. His first collection of verse, *Memory* (1941), is devoted to the theme of the socialist motherland. During the Great Patriotic War (1941–45), he wrote of the heroism and fortitude of the Soviet people. After the war he published poems about the labor of industrial and village workers (for example, "Stars of Mingechaur," 1958). His verses are characterized by carefully polished form and by laconic, expressive language. He has translated the poetry of M. Iu. Lermontov, N. A. Nekrasov, T. G. Shevchenko, I. Ia. Franko, A. T. Tvardovskii, M. V. Isakovskii, S. Ia. Marshak, G. N. Leonidze and A. Isaakian into Azerbaidzhani. He has also translated Goethe's *Faust*. Decor.: Order of the Red Banner of Labor.
Works: Jämil, Åhmäd. *She'rlär,* Baku 1961. *Sechilmish äsärläri.* Baku 1967. *In Russian translation: Stikhi.* Moscow 1947. *Izbrannoe.* Baku 1959.

Dzhandosov, Sandzhar Urasovich
State official.
Dep. Chair. of State Cttee. for Television and Radio Kazakh SSR 1967–71; Sec. of Alma-Ata City Cttee. Kazakh CP 1971–74; Head of Dept. of Science and Educ. Institutions with Cen. Cttee. Kazakh CP 1976–79; mem. Cen. Cttee. Kazakh CP 1976–; Chair. of State Cttee. on Vocational Training Kazakh SSR D. 1979–.

Dzhanelidze, Dmitrii Sevast'evich
Theatrical scholar; b. Ap. 26 (My. 9), 1906, Kutaisi.
Honored Art Worker of the Georgian SSR (1958), doctor of the Arts (1965).
In 1931, he began conducting research and teaching at various research institutes and acting studios. He also worked at Sh. Rustaveli Theatrical Institute in Tbilisi, becoming a professor there in 1966. He is the author of *The First Soviet Stage Performance* (1936), *The Kote Mardzhanishvili Georgian State Drama Theater* (1958, written in collaboration with E. Gugushvili), *The Georgian Theater From Ancient Times to the Second Half of the 19th Century* (1959), *History of the Georgian Theater* (vol. 1, 1966), *Sakhioba* (vols. 1–2, 1958–72), and other works devoted to Georgian playwrights, actors, and stage designers. Decors.: Order of the Badge of Honor and various medals.

Dzhanibekian, Gurgen Dzhanibekovich (pseudonym of G. D. Ter-Khachatrian)
Armenian actor; b. My. 6 (18), 1897, Yerevan.
People's Artist of the USSR (1967). Member of the CPSU since 1944.
He began his stage career in 1918 in Yerevan. He proceeded to work in Armenian theaters in Leninakan and Tbilisi. In 1924 he became an actor at the G. Sundukian Armenian Theater in Yerevan. His acting is notable for its national distinctiveness and an inner dramatic intensity, at times attaining tragic heights. He is also successful in comedy roles. His best roles are Seiran in Shirvanzade's *Namus,* Pepel in Gorkii's *The Lower Depths,* Figaro in Beaumarchais's *The Marriage of Figaro,* Othello in Shakespeare's play, Protasov in L. Tolstoy's *The Living Corpse,* Gaev in Chekhov's *The Cherry Orchard,* Barravin in Sukhovo-Kobylin's *The Case,* Baregam Pakhlavuni in Demirchian's *Napoleon Korkotian,* Markar in Papaian's *Yes, the World Has Gone Topsy-turvy!,* Grigoraga in Papazian's *The Cliff,* and Solomon in A. Miller's *The Price.* He has also acted in films, worked as a director, and written several plays. Decors.: State Prize of the USSR (1952), Order of Lenin, the Order of the Red Banner of Labor, and several medals.

Dzhanoian, Dzhanik Artashesovich
State official.
Min. of Finance Armenian SSR Jl. 1975–; cand. mem. Cen. Cttee. 1976–81, mem. Auditing Comm. Armenian CP 1981–.

Dzhaparidze, Revaz Andreevich
Author; b. 1923.
Georgian; mem. CPSU 1945–; grad. Dept. of Philosophy of Tbilisi Univ.; served in Sov. Army 1941–45; first works publ. 1949.
Publs.: "Khevskaia nevesta" (The Bride from Khevis) 1951, "Vdova soldata" (War Widow) 1956–60, "Tainstvennyi golos" (Secret Voices) 1963, "Belye nochi" (White Nights) 1963, and others.

Dzhaparidze, Rusudan Aleksandrovich
State official.
Chair. of State Cttee. on Labour Georgian SSR 1976–; cand. mem. Georgian CP 1981–.

Dzhaparidze, Ucha Malakievich
Painter and graphic artist; b. Ag. 4 (17), 1906, Gari, present-day Oni Raion, Georgian SSR.
People's Artist of the USSR (1963), member of the Academy of Arts of the USSR (1958). Member of the CPSU from 1944.
He studied at the Mose Toidze People's Art Studio (1922–24) and the Tbilisi Academy of Arts (1924–25 and 1928–31). He began teaching at the Tbilisi Academy of Arts in 1936; he became a professor there in 1944. He was the director of the academy from 1942 to 1948. In his historical paintings, he strives for heroic and romantic representation (for example, *May Day Demonstration in Tbilisi in 1901,* tempera, 1939–41, at the Georgian branch of the Institute of Marxism-Leninism, Tbilisi; State Prize of the USSR in 1942). He also paints book illustrations, genre pictures (for example, *Friends of Youth,* 1938 and *A Mother's Thoughts,* 1945, both at the Museum of Arts of the Georgian SSR, Tbilisi) and portraits (for example, *U. Chkheidze,* pastel, 1943). Another example of his work is *Parting Words,* which is in a private collection in the United States. Decors.: Order of Lenin, four other orders, and various medals.

Dzhasme, Dzhalil Oseovich
Kurd poet; b. O. 7 (20), 1908, Kizil-Kula, Kars Province, Turkey.
Member of the CPSU since 1930.
After his parents perished in a slaughter organized by the Turkish sultanate, he was raised in an orphanage in Aleksandropol' (now Leninakan). He graduated from the Armenian Pedagogical Institute in Yerevan in 1941. He was first published in 1930. He is the author of the collections of lyrics and narrative poems *Alagiaz* (1954, in Armenian), *Kurd Narrative Poems* (1955, in Armenian), *Native Spring* (1957, in Armenian), *My Days* (1960, in Kurd), and *My Hearth* (1966, in Armenian); the poem "Lenin" was published in Russian translation in *Pravda* on Ap. 15, 1956. His verses and narrative poems were included in the book *Soviet Kurd Poets* (1956). He has translated into Kurd the verses of A. S. Pushkin, O. Tumanian, A. Isaakian, and E. Charents.
Works: *In Russian translation: Dorogami vremen: [Stikhi i poemy],* Moscow 1968.

Dzhatiev, Totyrbek Ismailovich
Author; b. 1910, S. Ossetia; s. of a peasant.
Mem. CPSU 1939–; grad. Gorkii Inst. of Literature Moscow 1941; first works publ. 1931. Decor.: Order of Red Banner of Labour.
Publs.: "Geroi revoliutsii" (Hero of the Revolution) 1932, "Dobrovoltsy" (Volunteers) 1940–41, "Chest' osetina" (The Honour of an Ossetian) 1945, "Dorogoi zhizni" (On the Road of Life) 1956, "U gory Zilga Khokh" (At Zilga Khokh Mountain) 1954, "Gornaia zvezda" (Mountain Star) 1962, "Na beregakh Tereka" (On the Banks of the Terek) 1952, and others.

Dzhelepov, Boris Sergeevich
Physicist; b. N. 29 (D. 12), 1910, Odessa.
Corresponding member of the Academy of Sciences of the USSR (1953).

He graduated from Leningrad University in 1931. From 1931 to 1943 he worked in the Physicotechnical Institute of the Academy of Sciences of the USSR. Beginning in 1945 he was with the Radium Institute of the Academy of Sciences of the USSR. From 1939 to 1941 and from 1946 he worked in the All-Union Institute of Metrology and from 1935 to 1941 and from 1944, at Leningrad University. His principal works were in nuclear physics, mainly in nuclear spectroscopy. He introduced new methods of γ-spectroscopy and new devices and equipment for γ- and β-spectroscopy. He investigated the decay of α-radioactive isotopes and the influence of the electric field of the nucleus on β-decay. He also discovered a number of radioactive isotopes. Decors.: Order of the Badge of Honor and various medals.
Works: *Vliianie elektricheskogo polia atoma na beta-raspad.* Moscow-Leningrad 1956. (With L. N. Zyrianova.) *Schemy raspada radioaktivnykh iader $A > 100$.* Moscow 1963. (With others.)

Dzhelepov, Venedict Petrovich
Physicist; b. Mr. 30 (Ap. 12), 1913, Moscow.
Corresponding member of the Academy of Sciences of the USSR (1966). Member of the CPSU since 1949.
He graduated from the Leningrad Polytechnic Institute in 1937. From 1941 to 1943 he worked in the Physicotechnical Institute of the Academy of Sciences of the USSR and from 1943 to 1948 in the Institute of Atomic Energy. Between 1948 and 1956 he was deputy director of the Institute of Nuclear Problems of the Academy of Sciences of the USSR. Since 1956 he has been director of a laboratory in the Joint Institute for Nuclear Research in Dubna. His works deal with atomic nuclei and elementary particles. Since 1965 he has been working on applications of nuclear physics to medicine. In the USSR he was the first to develop a proton beam for medical uses. This beam, with an energy as high as 200 MeV, is used in the treatment of oncological diseases. Decors.: State Prize of the USSR (1961, 1963), Order of Lenin, the Order of the Red Banner of Labor, and various medals.

Dzherembaev, Erzhepai Ishmetovich
State official.
1st Dep. Min. of Food Ind. Kazakh SSR 1975–81; Min. of Fruits and Vegetables Kazakh SSR Ja. 1981–; cand. mem. Cen. Cttee. Kazakh CP 1981–.

Dzhienbaev, Sultan Suleimenovich
State official.
Mem. Cen. Cttee. Kazakh CP 1961–; Min. of Trade Kazakh SSR 1962–70. Dep. Chair. of Council of Mins. Kazakh SSR 1970–, cand. mem. Bureau Cen. Cttee. Kazakh CP Mr. 1980–.

Dzhikurashvili, I.
Komsomol official.
Mem. Apparat Cen. Cttee. All-Union Komsomol until 1982; Sec. and mem. Bureau Cen. Cttee. Georgian Komsomol Ag. 1982–.

Dzhimbinov, Bem Okunovich
Author; b. 1914, Kerdata.
Kalmuk; mem. CPSU 1941–; studied Moscow Inst. of Journalism, grad. Acad. of Social Sciences of Cen. Cttee. CPSU 1961; first works publ. 1934.
Publs.: "Iunye stroiteli sotsializma" (The Young Founders of Socialism) 1932, "Netlennaia zvezda" (Everlasting Star) 1962, "Drug stepei" (Friend of The Steppes) 1959, "350 let vmeste s Rossiei" (350 Years with Russia) 1959, and others.

Dzhioev, David Antonovich
Author; b. 1913, Bolshie Sikhiat, S. Ossetia.
Osset; mem. CPSU 1939–.
Publs.: "Nasha Krasnaia Armiia" (Our Red Army), "Front", novel "Iliko" 1951, "Nepobedimaia sila" (Unconquerable Power) 1954–61, "Zaria" (Dawn), and others.

Dzhonua, Aleksei Nestorovich
Author and poet; b. 1920, Chlou, Abkhazia; s. of a peasant.
Mem. CPSU 1947–; first works publ. 1937.
Publs.: poems "Utrenniaia zaria" (The Dawn) 1949, "Vo imia Rodinay" (In the Name of the Motherland) 1953, "Prisiaga" (The Oath) 1956, "Potok" (The River) 1958, "Rasskazy" (Stories) 1961, "U partizanskogo kostra" (By the Partisans' Camp Fire), and others.

Dzhonua, Chichiko Mikhailovich
Poet; b. 1915, Chlou, Abkhazia; s. of a peasant.
Mem. CPSU 1952–; first works publ. 1936.
Publs.: poems "Strana tsvetov" (Flower Country) 1950, "Poemy" (Poems) 1956, "Stikhi i poemy" (Poetry and poems) 1957, "Belaia skala" (The White Cliff) 1961, "Gudisa Shlarba" 1955, and others.

Dzhumaev, Iazy Kurbanovich
State official.
Min. of Food Ind. Turkmen SSR Ja. 1981–; cand. mem. Cen. Cttee. Turkmen CP 1981–.

Dzhumagulov, Apaz
Party official.
Head of Dept. of Ind. and Transport with Cen. Cttee. Kirghiz CP 1973–79; mem. Cen. Cttee. 1976, Sec. and mem. Bureau Cen. Cttee. Kirghiz CP F. 1979–.

Dzhurabekov, Ismail Khakimovich
State official.
Min. of Melioration and Water Supply Uzbek SSR Ja. 1977–; mem. Cen. Cttee. Uzbek CP 1981–.

Dzhusoev, Nafi Grigorevich
Poet and literary historian; b. 1924, S. Ossetia.
Publs.: poems "Serdtse soldata" (The Heart of a Soldier) 1949, "Lirika" (Lyric) 1955, "Tema Kavkaza v russkoi literature i v tvorchestve Kosta Khetagurova" (The Theme of the Caucasus in Russian Literature and in Kosta Khetagurov's Works) 1955, "Kosta Khetagurov" 1958, tragedy "Ezau i Taimuraz" (Ezau and Taimuraz), and others.

Dzhusoity, Nafi Grigor'evich
Ossetian writer and literary scholar; b. F. 27, 1925, Nogkau, in present-day Dzhava Raion, Iuzhnaia Ostiia Autonomous Oblast.
He is the author of the verse collections *Heart of a Soldier* (1949), *Word of the Heart* (1953), *Lyrics* (1955), *Thoughts of Ossetia* (1959), *I Was Born in the Mountains* (1960), *At a Difficult Age* (1962), and *Nameless Book* (1967); the novella *The Sun's Vortex* (1965); and the novels *Blood of the Forefathers* (1965) and *Twelve Wounds Are Like One* ... (1970). He has published the collection of articles *For the Art of the Word* (1966), as well as monographs on the classical Ossetian writers K. Khetagurov, S. Gadiev, E. Britaev, A. Kotsoev, and Ts. Gadiev. He has also translated Russian, Ukrainian, and Georgian classics into Ossetian.
Works: [Jusoytï, Nafi Grigorï Firt] *Äz rayguïrdtän xoxï.* [Tskhinvali] 1960. *Zin karï.* Tskhinvali 1962. *In Russian translation: Mot gornyi krai: Stikhi.* Moscow 1963. *Dobrota: Stikhi.* Moscow 1969.

Dzhusupov, Bekaidar
State official.
Chair. of Supr. Court Kazakh SSR 1962–70; mem. Cen. Cttee. Kazakh CP 1966–; Min. of Justice Kazakh SSR O. 1970–.

Dzialoshinskii, Igor Ekhielevich, Dr. of Phy.-Math. Sc.
Physicist; b. 1931.
Grad. Moscow Univ. 1953; Dr. of Phy.-Math. Sc. 1964; Prof. 1969–; corresponding mem. USSR Acad. of Sciences 1974–; cand. mem. 1954–57, Jun. Research Asst. at Inst. of Physical Problems of USSR Acad. of Sciences 1957–65; Section Chief at Inst. of Theoretical Physics of USSR Acad. of Sciences 1965–; in addition Lecturer and Prof. at Physical-Tech. Inst. Moscow 1965–72; Prof. at Moscow Univ. 1972–.

Dzigan, Efim L'vovich
Film director; b. D. 2 (14), 1898, Moscow.
People's Artist of the USSR (1969). Member of the CPSU from 1943.
He began working in motion pictures in 1924. In 1926 he graduated from B. V. Chaikovskii Motion Picture School. He made his directorial debut with the film *First Cornet Streshnev* (1928, with M. E. Chiaureli). A participant in the

Civil War, he in this film introduced the theme that was to dominate his future work: the development of the consciousness of the masses under the influence of the crucial events of history. His greatest work is the historical revolutionary epic *We Are From Kronstadt* (1936, screenplay by Vs. Vishnevskii), a work innovative in form and content. The plot and ideological content of the film are centered around the organizing influence of the party on sailors disposed to the revolution. He also made the films *Fatali-Khan* (released in 1959) and *Dzhambul* (1953). Subsequently his best films likewise affirmed the genre of the historical revolutionary epic. In 1956, he made one of the first Soviet wide-screen films, *Prologue*, devoted to the Revolution of 1905. The heroic and epic foundation of his work was evident in the film *The Iron Flood* (1967, after the novel by A.S. Serafimovich). In 1937 he began teaching at the All-Union State Institute of Cinematography, where he became a professor in 1965. Decors.: State Prize of the USSR (1941), Order of Lenin, the Order of the October Revolution, the Order of the Red Banner of Labor, and a medal.
Works: *My iz Kronshtadta: Printsipy rezhisserskogo postroeniia fil'ma.* Moscow 1937. "Avtor i rezhisser." In the collection *Voprosy kinodramaturgii,* fasc. 2. Moscow 1956. *O rezhisserskom stsenarii.* Moscow 1961.

Dziuba, Ivan Mikhailovich
Literary critic; b. Jl. 26, 1931, Mykolaivka, Donetsk Oblast.
Grad. Dept. of Philology of Donetsk Pedagogic Inst.; concluded post-graduate studies at Shevchenko Inst. of Literature at Acad. of Sciences Ukrainian SSR; Ed. with Ukrainian State Publishing House, Head of Reviews Dept. of journal "Vitchyzna", consultant for "Molod'" Publishing House, Lit. Ed. of journal "Ukrainskyi biokhimichnye Zhurnal", mem. ed. staff of journal "Dnipro"; first works publ. 1950.
Publs.: collection "Zvychaina liudyna chy mishchanyn?" (An Ordinary Man or a Petit Bourgeois?) 1959, "Internationalism or Russification" London 1968, and others.

Dziuban, Ivan Feodosevich
Graphic artist; b. 1923, Konstantinovka, Donetsk Oblast.
Attended Kiev Art Inst. 1950–55, pupil of V. Kasian; mem. CPSU 1961–.

Dzugaev, Georgii Khasakoevich
Poet; b. 1911, Dzugatykau.
Osset; mem. CPSU 1939–.
Publs.: poems "Luch solntsa" (A Sunbeam) 1938, poem "Mech'" (The Sword) 1937–39, "Komsomolets Soltan" (Soltan, Member of Komsomol) 1938, "Khamat" 1936–37, "Kolkhoz Kosta" (Kosta Kolkhoz), "Liubov" (Love) 1956, poem "Natlia" 1954–55, "Druzia" (Friends) 1956, "Detiam" (To the Children) 1956, "Dobryi gost'" (A Good Guest) 1962, and others.

Efendiev, Gadzhibaba Sharif ogly
State official.
Dep. to Supr. Sov. Azerbaidzhan SSR 1955–; Sec. of Cen. Cttee. Azerbaidzhan CP 1955–58; mem. Cen. Cttee. 1956–61, mem. Auditing Comm. 1961–64, cand. mem. Cen. Cttee. 1964–71, mem. Cen. Cttee. Azerbaidzhan CP 1971–; Chair. of Comm. on Sov. Control/State Control with Council of Mins. Azerbaidzhan SSR 1959–63; Dep. Chair. of Cttee. for Party and State Control with Cen. Cttee. Azerbaidzhan CP and Council of Mins. Azerbaidzhan SSR 1963–65; Chair. of Cttee. for People's Control Azerbaidzhan SSR N. 1970–; cand. mem. Bureau Cen. Cttee. Azerbaidzhan CP 1971–.

Efimenko, Georgii Grigorevich
State official.
Dep. Min. 1955–73; Min. of Higher and Secondary Specialist Educ. Ukrainian SSR 1973–; mem. Cen. Cttee. Ukrainian CP 1976–.

Efimenko, O. V.
Trade union official.
1st Sec. of Voroshilovgrad Oblast Cttee. Ukrainian Komsomol 1972; mem. Bureau 1972–, Sec. of Cen. Cttee. Ukrainian Komsomol Mr. 1973–82; Sec. and mem. Presidium of Ukrainian Republican Trade Union Council F. 1982–.

Efimov, Aleksandr Nikolaevich
Military official; b. 1923.
Mem. CPSU 1943–; grad. Air Force Acad. 1951 and Acad. of Gen. Staff 1951; served in Sov. Army 1941–; at the front 1942–45; command posts in Air Force 1945–; 1st Dep. Commander-in-Chief of USSR Air Force 1969–; Marshal of the Air Force Ap. 1975–. Decors.: two Orders of Lenin, four Orders of the Red Banner, Order of Alexander Nevskii, Order of Patriotic War 1st Cl., Order of the Red Star, a Polish order, various medals.

Efimov, Anatolii Nikolaevich
Economist; b. Jl. 3 (16) 1908, Troitsk, now Cheliabinsk Oblast.
Mem. CPSU 1948–; grad. Ural Polytech. Inst. 1937; mem. USSR Acad. of Sciences 1970–; teaching activity at Ural Polytech. Inst. 1937–48; Chief of Dept. of Econ. Research at Ural Branch of USSR Acad. of Sciences 1948–55; Dir. of Econ. Inst. of GOSPLAN USSR 1955–; Chief Ed. of "Ėkonomičeskaja ėnciklopedija. Promyšlennost' i stroitel'stvo" 1961–65. Decors.: State Prize 1968, Order of Red Banner of Labour, several medals.
Publs.: "Organizacija potoka v mašinostroenii" Sverdlovsk-Moscow 1944, "Voprosy techničeskogo progressa v SSSR" Moscow 1951, "Perspektivy razvitija promyšlennosti SSSR" Moscow 1959, "Metody planirovanija mežotraslevyh proporcij" Moscow 1965 (CH. 1–2), "Sovetskaja industrija (Proizvodstvennyj apparat, upravlenie i planirovanie)" Moscow 1967, "Ėkonomičeskoe planirovanie v SSSR" 1967 (co-author), "Ėkonomika i planirovanie sovetskoj promyšlennosti" Moscow 1970.

Efimov, Boris Efimovich
Graphic artist; b. S. 15 (28), 1900, Kiev.
Corresponding mem. USSR Acad. of Arts 1954–; worked for several newspapers 1922–; designs cartoons and posters dealing with international topics. Decors.: People's Artist of the USSR, State Prizes of the USSR 1950, 1951, 4 orders, several medals.

Efimov, Moisei Dmitrievich
Yakut poet; b. Ag. 6, 1929, Dulgalakh *nasleg* (district), Verkjojanskij Raion, Yakut ASSR.
Member of the CPSU since 1960. Graduated from the Yakut Pedagogical Institute in 1953.
He began to publish in 1948. He is the author of several collections of lyric and narrative poetry, notably *In the Land Where the Sun Does Not Set* (1955), *Man of Stone* (1960), *Holiday of Flowers* (1965; Komsomol of Yakutiia Prize, 1968), and *Beyond the Rainbow* (1968). The poet draws his themes from the natural beauty and people of the Far North and from their new socialist way of life. Decor.: Order of the Red Banner of Labor.
Works: *In Russian translation: V kraiu, gde solntse ne zakhodit.* Moscow 1965. *Zvezdnyi dozhd'.* Moscow 1970.

Efremov, Leonid Nikolaevich
State official; b. 1912.
Russian; mem. CPSU 1941–; grad. Voronezh Inst. for Modernization of Agriculture 1933; tech. head of a tractor repair station 1933–37; engineer, chief mechanic of a works 1937–44; CPSU posts 1944–46; 2nd Sec. of Kuibyshev City CPSU Cttee. 1946–49 and 2nd Sec. of Kuibyshev Oblast CPSU Cttee. 1949–51; Dep. to USSR Supr. Sov. 1950–66; Chair. of Kuibyshev Oblast Exec. Cttee. of Workers' Deps. 1951–52; 1st Sec. of Kursk Oblast CPSU Cttee. 1952–58; mem. Cen. Cttee. CPSU 1952–71; 1st Sec. of Gorkii Oblast CPSU Cttee. 1958–62; mem. 1959–62 and 1st Dep. Chair. of Bureau Cen. Cttee. CPSU for RSFSR 1962–65; cand. mem. Presidium Cen. Cttee. CPSU 1962–66; 1st Sec. of Stavropol Krai CPSU Cttee. 1964–70; 1st Dep. Chair. of State Cttee. on Science and Tech. of USSR Council of Mins. Ap. 1970 mem. CPSU Cen. Auditing Comm. 1971–, 1st Dep. Chair. USSR Main Cttee. for Exhibition of Achievements of the National Economy of the USSR 1979–. Decors.: 3 Orders of Lenin, Order of Red Banner of Labour, Order of Red Star, and medals.

Efremov, Mikhail Timofeevich
Diplomat; b. 1911.
Russian; mem. CPSU 1932–; grad. Kuibyshev Ind. Inst. 1941; sec. of a Komsomol volost' cttee., attended a tech. school, head of an electrical tech. inspection brigade, power technician at various enterprises 1927–42; leading CPSU posts 1942–; 1st Sec. of Kuibyshev Oblast CPSU Cttee. 1952–59; dep. to 1954–, mem. Credentials Comm. of Sov. of the Union USSR Supr. Sov. 1962–66; mem. Cen. Cttee. CPSU 1956–76; Head of Dept. for Party Organs in CPSU Cen. Cttee. Bureau for the RSFSR 1959–61; mem. CPSU Cen. Cttee. Bureau for RSFSR 1959–61; 1st Sec. Cheliabinsk Oblast CPSU Cttee. 1961–62; 1st Sec. of Gorkii Oblast CPSU Cttee. 1963–65 (for Ind. 1963–64); Dep. Chair. of Council of Mins. 1965–71; Sov. Amb. to the GDR 1971–75 and to Austria Mr. 1975–. Decors.: 2 Orders of Lenin.

Efremov, Oleg Nikolaevich
Actor and stage director; b. 1927.
Mem. CPSU 1955–; grad. Drama School of Moscow Acad. Gorkii Art Theatre of the USSR ("MHAT") 1949; Stage Dir. of "Sovremennik" Theatre 1956–70; Sen. Stage Dir. of Moscow Acad. Gorkii Art Theatre of the USSR 1970–. Decors.: State Prize of the USSR 1969, People's Artist of the RSFSR 1969.

Efremov, Vasilii Sergeevich
Military official; b. Ja. 1 (14), 1915, Tsaritsyn, now Volgograd.
Twice Hero of the Soviet Union (My. 1, 1943, and Ag. 24, 1943), lieutenant-colonel (1948). Became a member of the CPSU in 1943.
He joined the Soviet Army in 1934, graduated from Military Aviation Pilot's School in 1937, and took part in the liberation of the Western Ukraine and in the Soviet-Finnish War of 1939–40. During the Great Patriotic War of 1941–45 he took part in the fighting around Voronezh and Orel, in the battle of Stalingrad, and in fighting in the Ukraine, the Crimea, and Belorussia. He held the posts of flight commander, deputy squadron commander, and then squadron commander in the 10th Guards Bomber Regiment. He carried out 340 combat missions, destroying 32 enemy aircraft on the ground and four in aerial battle. In 1949 he graduated from the Military Air Force Academy and then served as deputy regimental commander, instructor, and test pilot. In 1960 he retired. Decors.: two Orders of Lenin, the Order of the Red Banner, the Order of the Patriotic War First Class, two Orders of the Red Star, and medals.

Eftakova, S. Kh.
Komsomol official.
Sec. and mem. Bureau Cen. Cttee. Tadzhik Komsomol Je. 1980–.

Egizbaev, Kosai Alekulovich
State official; b. 1928.
Kazakh; mem. CPSU 1953–; grad. Inst. of Railway Engineering in Tashkent; master, chief engineer with railways 1949–53; Komsomol and party functions 1953–60; Sec. of Alma-Ata Oblast Cttee. Kazakh CP 1960–61; 1st Sec. of Cen. Cttee. Kazakh Komsomol 1961–62; mem. Cen. Cttee. 1961–66 and 1971–81; cand. mem. Bureau Cen. Cttee. Kazakh CP 1961–62 and 1971–77; 1st Sec. of Gurev Oblast Cttee. Kazakh CP 1963; Sec., then 2nd Sec. of E. Kazakhstan Oblast Cttee. 1964–70; Chair. of Kazakh Republican Trade Union Council 1970–F. 1977; cand. mem. Bureau Cen. Cttee. Kazakh CP 1971–Mr. 1977; Dep. Min. of Rural Construction Kazakh SSR F. 1977–.

Egorov, Aleksandr Mikhailovich
State official.
Cand. mem. Cen. Cttee. Kazakh CP 1971–81, mem. 1981–; Chair. of Kazakh Asscn. for Agricultural Technology "Kazselkhoztekhnika" Je. 1974–78; Chair. of State Cttee. for the Supply of Production Equipment for Agriculture Kazakh SSR Ag. 1978–.

Egorov, Anatolii Grigor'evich, Dr. of Phil. Sc.
Party official; b. O. 25, 1920, Skopin, Riazan Oblast.
Russian; mem. CPSU 1944–; grad. Liebknecht Teachers' Training Inst. Moscow 1941; corresponding mem. USSR Acad. of Sciences 1962–; served in Sov. Army 1942–46; mem. staff Teachers' Training Inst. Vladivostok 1946–48; postgraduate studies 1948–51; Consultant, Dep. Ed., Ed. for a section of journal "Kommunist" 1951–56; Chief Ed. of journal "V pomoshch politicheskomu samoobrazovaniiu," renamed "Politicheskoe samoobrazovanie" 1956–61; mem. staff Cen. Cttee. CPSU 1961–65; mem. CPSU Cen. Auditing Comm. 1961–66; Chief Ed. of "Kommunist" 1966–74; cand. mem. 1966–76, mem. Cen. Cttee. CPSU 1976–; Dir. of Inst. for Marxism-Leninism with Cen. Cttee. CPSU 1974–. Decors.: 2 orders and various medals.

Egorov, Boris Andrianovich
Author; b. 1924, Moscow.
Russian; mem. CPSU 1944–; grad. Faculty of Philology of Moscow State Univ. 1956; first works publ. 1945.
Publs.: "Ne prokhodite mimo" (Don't Go By) (together with Ia. Polishchuk and B. Privalov) 1956, "Kaplia vnimaniia" (A Drop of Truth, together with Ia. Polishchuk and B. Privalov) 1959, "Otkroite fortochku!" (Open the Trap Window!) 1960, "Vzdokhi na skameike" (Sigh on the Bank) 1962, "Vchera, segodnia, zavtra" (Yesterday, Today, Tomorrow) 1958, "Romantika nashego vremeni" (Romance of our Time) 1959, and others.

Egorov, Boris Borisovich
Cosmonaut; b. 1937.
Grad. 1st Medical Inst. Moscow 1961 and Scientific Research Insts. of Aviation and Space Medicine; pilot, cosmonaut, Lieut. in Medical Service; space flight as flight physician in spaceship "Voskhod" with V. M. Komarov and K. P. Feoktistov O. 12–13, 1964; 16 earth orbits.

Egorov, Evgenii Pavlovich
Graphic artist and painter; b. 1917. Krasnyi Lyman, Donetsk Oblast.
Mem. CPSU 1949–; attended Kharkov Art School 1936–39 and Art Inst. 1945–49, pupil of O. Kokel, D. Shavikin.

Egorov, Georgii Mikhailovich
Military official; b. 1918.
Admiral of the Fleet 1973–; mem. CPSU 1942–; Dep. Commander-in-Chief, USSR Naval Forces 1967–72; Commander-in-Chief on Northern Fleet 1972–77; cand. mem. Cen. Cttee. CPSU 1976–81; Chief of Staff, USSR Naval Forces 1977–.

Egorov, Iurii Pavlovich
Motion picture director; b. My. 25, 1920, Sochi.
Honored Art Worker of the RSFSR (1965).
In 1950, he graduated from the directing department of the All-Union State Institute of Cinematography (S. A. Gerasimov's studio). He collaborated with G. S. Pobedonostsev in writing the play *Three Soldiers* and staged it at the Film Actors' Theater-Studio. In 1954 he and Pobedonostsev produced the film *Incident in the Taiga,* and in 1955 he made *The Frozen Sea.* He has directed and written the screenplays for a number of films dealing with Soviet young people: *The Volunteers* (1958, based on the narrative poem by E. Dolmatovskii), *A Business Trip* (1962), *If You're Right* (1964), and *Not the Luckiest of Days* (1967). The director's most important films are *They Were the First* (1956) and *A Simple Story* (1960). From 1964 to 1970 he headed the central board for feature film production of the Committee on Cinematography of the Council of Ministers of the USSR. In 1972, he collaborated with a Swedish film company in producing *A Man From the Other Side.* Since 1962, he has been teaching at the All-Union State Institute of Cinematography.

Egorov, Mikhail Vasil'evich
State official; b. 1907.
Russian; mem. CPSU 1938–; grad. Dzerzhinskii Naval Engineering College 1930; engineer in a scientific research inst. in Leningrad 1930–38; posts in USSR Mins. of the Ship Building Ind., Transport, and Heavy Machine Building 1939–55; Dep. USSR Min. of the Ship Building Ind. 1955–58, 1st Dep. Min. 1958–76, Min. Jl. 1976–; Dep. to USSR

Supr. Sov. 1974–; mem. Cen. Cttee. CPSU 1981–. Decor.: Hero of Socialist Labor 1963.

Egorychev, Nikolai Grigor'evich
Diplomat; b. 1920.
Mem. CPSU 1942–; grad. Bauman Tech. School Moscow; student 1938–41; served in Sov. Army 1941–46; Sec. of Komsomol Org. and Sec. of CPSU Cttee. of Bauman Tech. School Moscow, Sec., 1st Sec. of Bauman Raion Cttee. CPSU in Moscow 1947–60; mem. staff Cen. Cttee. CPSU 1960–61; 1st Sec. of CPSU city Cttee. in Moscow 1962–67; mem. Cen. Cttee. CPSU 1961–71; mem. Bureau for RSFSR with Cen. Cttee. CPSU 1962–66; Dep. to 1962–70, mem. Foreign Affairs Comm. of Sov. of the Union 1962–66; mem. Presidium USSR Supr. Sov. 1966–67; Sov. Amb. to Denmark Ap. 1970–.

Eibek (real name: Tashmukhamedov, Musa)
Author; b. 1905, Tashkent, Uzbekistan.
Mem. CPSU 1947–; grad. Faculty of Econ. at Cen. Asian Univ. 1930; mem. Acad. of Sciences Uzbek SSR 1943–; Dep. to USSR Supr. Sov. 1958–66; first works publ. 1922. Decors.: Stalin Prize, Order of Lenin, Order of Red Banner of Labour.
Publs.: collected verse "Chuvstva" (Feelings) 1926, "Fleity serdtsa" (Flute of the Heart) 1929, "Fakel" (The Torch) 1932, novels "Sviashchennaia krov'" (Sacred Blood) 1943, "Navoi" (The Navoys) 1945, "Veter zolotoi doliny" (Wind of the Golden Valley) 1950, "Solntse ne pomerknet" (The Sun Will Not Grow Dim) 1958, and others.

Einoris, Vitautas Nikodemovich
State official.
Min. of Fruits and Vegetables Lithuanian SSR Ja. 1981–; mem. Auditing Comm. Lithuanian CP 1981–.

Eks, Valter Petrovich
State official.
Head of Dept. for Construction and Municipal Econ. with Cen. Cttee. Esthonian CP 1969–79; cand. mem. Cen. Cttee. Esthonian CP 1976–; Min. of Housing and Communal Econ. Esthonian SSR N. 1979–.

Elchenko, Iurii Nikiforovich
Party official; b. 1929.
Mem. CPSU 1953–; grad. Kiev Polytechnical Inst. 1952; foreman at a machine plant in Moscow Oblast 1952–53; held Komsomol posts 1953–68; Sec., 2nd Sec. of Kiev City Cttee. Komsomol 1957–58; Sec., 2nd Sec. of Cen. Cttee. Ukrainian Komsomol 1958–60, 1st Sec. 1960–69; 2nd Sec. of Lvov Oblast Cttee. Ukrainian CP 1969–71; Min. of Culture Ukrainian SSR 1971–73; Dep. Chair. of Propaganda and Agitation Dept. Cen. Cttee. Ukrainian CP 1973–80; 1st Sec. of Kiev City Cttee. Ukrainian CP Ja. 1980–; cand. mem. Politbureau Cen. Cttee. Ukrainian CP Mr. 1981–82, mem. O. 1982–; mem. Cen. Cttee. CPSU 1981–.

Eleukenov, Sheriazdan Rustemovich
State official.
Chair. of State Cttee. for Publishing, Polygraphy and the Book Trade Je. 1971–; mem. Auditing Comm. Kazakh CP 1976–81; cand. mam. Cen. Cttee. Kazakh CP 1981–.

Eliakov, Georgii Borisovich
Chemist and biochemist; b. S. 13, 1929, Kostroma.
Corresponding member of the Academy of Sciences of the USSR (1970). Member of the CPSU since 1965.
He graduated from the department of chemistry of Moscow State University in 1952. From 1955 to 1959 he worked in Moscow. Since 1959 he has been in Vladivostok, where in 1962 he organized and headed the Chemical Laboratory of Natural Physiologically Active Compounds of the Far Eastern Branch of the Siberian Division of the Academy of Sciences of the USSR. Since 1964 he has been the director of the Institute of Biologically Active Substances. His major work has been the synthesis of triterpenoid glycosides, triterpenoids, sterols, and glycosides from the roots of ginseng and other plants of the family Araliaceae. Decors.: Order of the Badge of Honor and medals.

Eliseev, Aleksei Stanislavovich, Cand. of Tech. Sc.
Cosmonaut; b. 1934.
Mem. CPSU 1967–; grad. Bauman Higher Tech. College Moscow ("MVTU") 1966; Cand. of Tech. Sc. 1967–; cosmonaut 1968–; spaceflight with B.V. Volynov and E.V. Khrunov in spaceship "Soiuz 5". Ja. 15, 1969, conducted extra-vehicular activity and docking operations with transfer to spaceship "Soiuz 4" (pilot V.A. Shatalov) with Khrunov, landing Ja. 17, 1969; flight engineer of spaceship "Soiuz 8", participated with V.A. Shatalov in group flight of three spaceships O. 13–18, 1969; flight engineer of "Soiuz 10", orbital flight with docking manoeuvre to scientific space station "Salut" Ap. 23–25, 1971. Decors.: Hero of the Sov. Union Ja. 22 and O. 22, 1969.

Eliseev, N. I.
Komsomol official.
Sec. and mem. Bureau of Cen. Cttee. Kazakh Komsomol Mr. 1978–.

Eliseeva, Natalia Grigor'evna
State official; b. 1927.
Mem. CPSU 1954–; grad. Kirov Textile Inst. Leningrad 1950; Forewoman, Dept. Manager, Chair. of Factory Cttee. at "Rabochii" Factory in Leningrad 1950–60; held various party posts 1960–63; Sec. 1963–68, 2nd Sec. and 1st Sec. of Nevskii Raion Cttee. CPSU in Leningrad 1968–76; mem. CPSU Cen. Auditing Comm. 1971–76, cand. mem. 1976–81; Dep. Chair. of Exec. Cttee. of Leningrad City Sov. of People's Deps. 1976–.

Eliutin, Viacheslav Petrovich, Dr. of Tech. Sc.
State official; b. 1907; s. of a clerk.
Russian; mem. CPSU 1929–; grad. Moscow Steel Inst. 1930; Prof.; corresponding mem. USSR Acad. of Sciences; worked in Komsomol, student of Moscow Steel Inst. 1924–35; Dean of Dept., Dep. Dir. for Research and Instruction at All-Union Ind. Acad. 1935–41; served in Sov. Army 1941–43; Lecturer, Dir. of Moscow Steel Inst. 1943–51; USSR 1st Dep. Min. 1951–54, Min. of Higher Educ. 1954–59; Dep. to Supr. Sov. RSFSR 1955–63; cand. mem. 1956–61, mem. Cen. Cttee. CPSU 1961–; USSR Min. of Higher and Secondary Specialist Educ. 1959–; Dep. Chair. of Lenin Prize Cttee. for Science and Technology 1960–; Dep. to USSR Supr. Sov. 1962–; Chair. of USSR-Algerian Friendship Soc. 1967–. Decors.: State Prize, 2 Orders of Lenin, Order of Red Banner of Labour, various medals.

Elizarov, Viktor Dmitrevich
Architect; b. 1911.
Mem. CPSU 1944–; grad. Inst. of Architecture Moscow 1937; Dep. Chair., 1st Dep. Chair. of State Cttee. on Construction (GOSSTROI) Ukrainian SSR 1964–. Decor.: People's Architect of the USSR 1973.

Elizavetin, Aleksei Ivanovich
Diplomat; b. 1915.
Entered Diplomatic Service 1956; USSR Consul Gen. in Shenyang and Shanghai 1956–60; sen. posts in USSR Min. of Foreign Affairs 1960–61; Counsellor at USSR Emb. in China 1961–64; various posts in USSR Min. of Foreign Affairs 1964–68; Counsellor-Envoy at USSR Emb. in China 1968–70; USSR Amb. to Burma My. 1971–, also to S. Vietnam Je. 1973–D. 1976.

Eltsin, Boris Nikolaevich
Party official; b. 1931.
Russian; mem. CPSU 1961–; grad. Ural Polytechnical Inst. 1955; foreman, work superintendent, chief engineer, chief of a construction office of "Iuzhgorstroi" Trust 1955–63; chief engineer, chief of House Construction Combine in Sverdlovsk 1963–68; chief of a dept., Sec. of Sverdlovsk Oblast Cttee. CPSU 1968–76; 1st Sec. N. 1976–; Dep. to USSR Supr. Sov. 1974–; mem. Cen. Cttee. CPSU 1981–.

Emanuel, Nikolai Markovich
Physical chemist; b. S. 18 (O. 1), 1915.
Mem. CPSU 1948–; grad. Leningrad Ind. Inst. 1938; mem. USSR Acad. of Sciences 1966–; mem. staff Inst. of Chemical

Physics of USSR Acad. of Sciences 1938–; concurrently Lecturer 1944–, Prof. at Moscow State Univ. 1950–; Dep. Chair. of Section Chemo-Technological and Biological Sciences of Presidium of USSR Acad. of Sciences 1963–. Decor.: Lenin Prize 1958.

Emelianov, Ivan Iakovlevich, Dr. of Tech. Sc.
Scientist; b. 1913.
Mem. CPSU 1949–; grad. Engineering Inst. for Posts and Telecommunications Moscow 1940; specializes in energetics; Dr. of Tech. Sc. 1961; Prof. 1964–; corresponding mem. USSR Acad. of Sciences 1974–; served in Sov. Army 1941–46; leading posts as engineer and designer 1946–. Decors.: State Prize of the USSR 1949, 1953, Lenin Prize 1961.

Emelianov, Stanislav Vasil'evich
Scientist; b. My. 18, 1929, Voronezh.
Mem. CPSU 1962–; grad. Aviation Inst. Moscow 1952; specialist in automatic control systems; corresponding mem. USSR Acad. of Sciences 1970–; important inventions in aerotechnology. Decor.: Lenin Prize 1972.
Publ.: "Teorija sistem s peremennoj strukturoj" Moscow 1970.

Emelianov, Vasilii Semenovich
Metallurgist; b. Ja. 30 (F. 12), 1901, Khvalynsk, now Saratov Oblast.
Mem. CPSU 1919; grad. Mining Inst. Moscow 1928; corresponding mem. USSR Acad. of Sciences 1953–; dir. of a factory in Cheliabinsk 1935–37; mem. staff People's Commissariat for Defence Ind. 1937–40; Dep. Chair. 1940, Chair. of Cttee. for Standards 1942–45; Min. for Tank Prod. 1941–45; leading positions in atomic ind. 1945–46; Prof. Chair at Inst. of Physical Engineering Moscow, Chair. of Comm. on Scientific Disarmament Problems with USSR Acad. of Sciences 1946–; mem. UN Scientific Consultation Cttee. 1955–65; one of the Chiefs with Int. Atomic Energy Council UN in Vienna 1957–65; mem. American Acad. of Arts and Sciences 1965–. Decors.: Hero of Socialist Labour, State Prizes of the USSR 1942, 1951, 4 Orders of Lenin, Order of October Revolution, other orders and several medals. Publs.: "Metallurgija jadernogo gorjučego" Moscow 1968, "O vremeni, o tovariščah, o sebe" Moscow 1968, "Na poroge vojny" Moscow 1971.

Emin, Gevork (real name: Muradian, Karlen Grigorevich)
Author and poet; b. O. 30, 1919, Ashtarak, Armenia; s. of a teacher.
Mem. CPSU 1953–; grad. Dept. of Hydrotechnology at Erevan Polytech. Inst. 1940; Chief Ed. of periodical "Literaturnaia Armeniia" 1969–72; first works publ. 1935.
Publs.: poems "Vesna chelovchestva" (Spring of Mankind), "Vesenniaia pesnia" (Spring Song), "Oktiabriu" (To October), "Stikhi o Lenine" (Poems About Lenin), "Ia k vam obrashchaius', liudi mira" (I Turn to You, People of the World) and others; "Dve dorogi" (Two Roads) 1962, "V etom vozraste" (At this Age) 1968 (Russian transl. 1972), "Dvadtsatyi vek" (The Twentieth Century) 1970, "Sem pesen ob Armenii" (Seven Songs About Armenia) 1973, "Vek. Zemlia. Liubov'" (Century. The Earth. Love) 1974, and others.

Endeladze, Nugzar Shalvovich
Party official.
1st Sec. of Lenin Raion Cttee. Georgian CP in Tbilisi 1975–76; Dep. to Supr. Sov. Georgian SSR 1975–; 1st Sec. of Tbilisi City Cttee. Georgian CP 1976–82; can. mem. Cen. Cttee. Georgian CP 1976–; Head of Dept. of Propaganda and Agitation at Cen. Cttee. Georgian CP 1982–.

Endovitskii, Mikhail Vasil'evich
State official.
Cand. mem. Cen. Cttee. Kirghiz CP 1961–66, mem. 1966–81; Chair. of State Cttee. on Labour Kirghiz SSR S. 1976–.

Enikeev, Amirkhan Nigmetzianovich (pseudonym: Eniki, Amirkhan)
Author; b. 1909, Novo-Kargali, Bashkir; s. of a peasant.
Tartar; first works publ. 1926.
Publs.: "Bolotnyi tsvetok" (The Moorland Flower) 1955, "Spasibo, tovarishchi!" (Thanks, Comrades!) 1952, "Tainy serdtsa" (Secrets of the Heart) 1958, "Solnechnoe utro" (A Sunny Morning) 1950, and others.

Enikolopov (Enikolopian), Nikolai Sergeevich
Physical chemist; b. Mr. 13, 1924, Stepanakert, Azerbaidzhan.
Grad. Polytech. Inst. Erevan 1945; corresponding mem. USSR Acad. of Sciences 1966–; mem. staff Inst. of Chemical Physics at USSR Acad. of Sciences 1946–.

Enukidze, Guram Nikolaevich
State official.
Sec. of Tbilisi City Cttee. Georgian CP 1973–75; cand. mem. Cen. Cttee. Georgian CP 1976–78, mem. 1978–; Chair. of State Cttee. on Radio and Television Georgian SSR Mr. 1976–78; Sec. and mem. Bureau Cen. Cttee. Georgian CP D. 1978–.

Epishev, Aleksei Alekseevich
Military official; b. 1908, Astrakhan; s. of a worker.
Russian; mem. CPSU 1929–; grad. Mil. Acad. for Mechanization and Motorization of Sov. Army 1938; Army Gen.; worked in workshops of a fishery trust 1923–27; instructor, dept. head of a raion Komsomol cttee. 1927–30; served in Sov. Army 1930–38; Dep. to 1937–46, 1950–58 and 1962–, mem. Foreign Affairs Comm. of Sov. of the Union USSR Supr. Sov. 1966–; leading CPSU posts 1938–40; 1st Sec. of Kharkov Oblast and City Cttees. Ukrainian CP, Plenipotentiary of Stalingrad Front Mil. Council, mem. staff Cen. Cttee. CPSU, USSR Dep. People's Commr. for Medium Machine Building 1940–43; served in Sov. Army 1943–46; Sec. of Cen. Cttee. Ukrainian CP, 1st Sec. of Odessa Oblast Cttee. Ukrainian CP 1946–51; Dep. to Supr. Sov. Ukrainian SSR 1947–51; USSR Dep. Min. of State Security 1951–53; cand. mem. 1952–64, mem. Cen. Cttee. CPSU N. 1964–; 1st Sec. of Odessa Oblast Cttee. Ukrainian CP 1953–55; Sov. Amb. to Rumania 1955–61, to Yugoslavia 1961–62; Head of Main Pol. Board of Sov. Army and Navy My. 1962–. Decors.: 2 Orders of Lenin, Order of Red Banner of Labour; medals.

Eremei, Grigorii Isidorovich
Trade union official; b. 1935.
Moldavian; mem. CPSU 1957; grad. Kishinev Agricultural Inst.; served in the Soviet Army 1954–56; held Komsomol posts 1956–66; mem. Auditing Comm. 1966–71, mem. Cen. Cttee. Moldavian CP 1971–; 1st Sec. of Kotovskii Raion Cttee. Moldavian CP 1966–70; Dep. Chair. 1970–73, 1st Dep. Chair. of Council of Mins. Moldavian SSR 1973–80; cand. mem. 1974–76, mem. Bureau Cen. Cttee. Moldavian CP F. 1976–; Chair. of Moldavian Republican Trade Union Council 1980–.

Eremenko, Valentin Nikiforovich, Dr. of Chem. Sc.
Physicist and chemist; b. Ag. 12, 1911, Kremennaia, now Voroshilovgrad Oblast; s. of a clerk.
Grad. Kharkov Univ. 1936; corresponding mem. Acad. of Sciences Ukrainian SSR 1964; Dr. of Chem. Sc. 1961; Prof. 1961–; Sen. Research Fellow at Kharkov Univ. 1939–41; mem. staff Inst. of Nonferrous Metallurgy at Acad. of Sciences Ukrainian SSR 1940–53; Sen. Research Fellow Inst. of Metal Ceramics and Special Alloys (Inst. for Problems of Materials 1964–) Acad. of Sciences Ukrainian SSR, Chief of Dept. for Physical and Chemical Methods in Research, Dep. Dir. of same Inst. 1953–; teaching activity at Ukrainian SSR colleges 1935– and Kiev Univ. 1944–.
Publs.: "Titan i ego splavy" 1960, "Mnogokomponentnye splavy titania" 1962, "Karbid titana i zharostoikie materialy na ego osnove" 1954, and others.

Eremian, Suren Tigranovich
Historian; b. Mr. 28 (Ap. 10), 1908, Tbilisi.
Grad. Erevan Univ. 1955; mem. Armenian Acad. 1963–; Prof. 1955–; mem. staff Inst. of Orientalism at USSR Acad. of Sciences Leningrad 1935–41; mem. staff Inst. of History at Acad. of Sciences Armenian SSR 1941–; specialist in the his-

tory of ancient and medieval Armenia and the entire Transcaucasia. Decor.: Merited Scientist of the Armenian SSR.

Ergaliev, Khamid
Author; b. 1916, Novobogatenka, Gurev Oblast, Kazakhstan.
Mem. CPSU 1944–.
Publs.: "Ispoved' ottsa" (The Father's Confessions) 1947, "Devushka iz nashego aula" (The Girl From Our Aul) 1950, "Na bolshom puti" (On the Great Road) 1949, "Tvoia reka" (Your River) 1953, and others.

Ergashev, Kudrat
State official.
Min. of the Interior Uzbek SSR Jl. 1979–; mem. Cen. Cttee. Uzbek CP 1981–.

Erkimbekov, Zheksembek
State official.
Dep. Min. of Culture 1969–76, Min. of Culture Kazakh SSR Ag. 1976–; mem. Cen. Cttee. Kazakh CP 1981–.

Ermash, Filipp Timofeevich
State official; b. 1923.
Russian; mem. CPSU 1945–; grad. Ural State Univ. 1951; served in Sov. Army 1941–46; 2nd Sec., 1st Sec. of Sverdlovsk City Cttee. Komsomol USSR 1951–56; dep. chief, chief of a dept. of Sverdlovsk Oblast Cttee. CPSU 1956–62; chief of a sector, dep. chief of a dept. of Cen. Cttee. CPSU 1962–72; Dep. to USSR Supr. Sov. 1970–; Chair. of USSR State Cttee. for Cinematography Ag. 1972–; cand. mem. Cen. Cttee. CPSU 1981–. Decor.: Order of the October Revolution.

Ermin, Lev Borisovich
Party official; b. 1923, s. of a worker.
Russian; mem. CPSU 1943–; grad. Azov-Black Sea Agricultural Inst. 1952; served in Sov. Army 1941–47; student, chief agronomist at a machinery and tractor station 1947–52; Sec. of Veselovskii Raion Cttee., then 1st Sec. of Razdorskii Raion CPSU Cttee. in Rostov Oblast 1952–59; instructor for a CPSU Cen. Cttee. Dept. 1959–61; 1st Sec. of Penza Oblast CPSU Cttee. Ag. 1967–79 (for Agriculture 1963–64); cand. mem. 1961–71, mem. Cen. Cttee. CPSU 1971–; dep. to 1972–, mem. Comm. on Educ., Science and Culture with Sov. of the Union USSR Supr. Sov. 1966–; 1st Dep. Chair. of RSFSR Council of Mins. 1979–. Decor.: Order of the October Revolution.

Erofeev, Vladimir Iakovlevich
Diplomat; b. 1909.
Entered Diplomatic Service 1939; Dep. Head, Head of Consular Dept. of People's Commissariat for USSR Foreign Affairs 1939–40; Counsellor at USSR Emb. in Turkey 1940–42; Dep. Head of 2nd European Dept. 1942–48, Head of Dept. for Latin America in USSR Min. of Foreign Affairs 1948–49; Counsellor at USSR Emb. in UK 1949–52; mem. staff USSR Min. of Foreign Affairs 1953–54; Counsellor-Envoy at USSR Emb. in France 1954–55; Head of 2nd European Dept. 1955–58 and Head of Middle East Dept. USSR Min. of Foreign Affairs 1958–59; USSR Amb. to United Arab Republic and Yemen 1959–62; USSR Amb. to United Arab Republic 1962–65; leading posts in USSR Min. of Foreign Affairs 1965–68; USSR Amb. to Iran Ja. 1968–77; Head of Dept. for Cultural Relations with Foreign Countries, USSR Min. of Foreign Affairs 1977–.

Erofeev, Vladimir Ivanovich
Diplomat; b. 1920, Leningrad.
Mem. Diplomatic Service 1942–; in sen. positions at USSR Min. of Foreign Affairs and USSR Embs. abroad 1942–55; Emb. Counsellor at USSR Emb. in France 1955–59; Dep. Head of 1st European Dept. in Min. of Foreign Affairs 1959–62; USSR Amb. to Senegal 1962–66; and concurrently to Gambia 1965–66; mem. staff USSR Min. of Foreign Affairs 1966–70; Dep. Dir.-Gen. of UNESCO 1970–O. 1975; Perm. USSR Rep. at int. orgs. in Vienna O. 1975–79.

Eroshin, Valerii Konstantinovich, Cand. of Biol. Sc.
Biologist; b. 1932.
Mem. CPSU 1967–; grad. Moscow Univ. 1955; Cand. of Biol. Sc. 1963; postgraduate studies 1962; mem. staff Chemical-Pharmaceutical Union Research Inst. 1955–69; mem. staff Inst. of Microbiology at USSR Acad. of Sciences 1963–65; mem. staff Inst. of Biology and Physiology of Micro-organisms at USSR Acad. of Sciences 1965–, Sen. Research Asst., Laboratory Chief 1967–, Dep. Dir. 1968–. Decor.: State Prize of the USSR 1971 for participating in developing scientific principles for the microbiological extraction of proteins from petroleum hydrocarbons.

Erpilov, Petr Ivanovich
Party official; b. 1926.
Russian; mem. CPSU 1953–; grad. Dnepropetrovsk Mining Inst.; served in the Soviet Army 1943–45; engineering and party posts 1950–64; 1st Sec. of Stepnogorsk City Cttee. Kazakh CP 1964–66; 1st Sec. City Cttee., Sec. Oblast Cttee. Kazakh CP in Tselinograd 1966–74; 1st Sec. of Alma-Ata City Cttee. Kazakh CP 1974–79; mem. Cen. Cttee. Kazakh CP 1976–; 2nd Sec. of Kokchetav Oblast Cttee. Kazakh CP 1979–82; 1st Sec. of Pavlodar Oblast Cttee. Kazakh CP Ja. 1982–.

Ershov, Andrei Petrovich
Mathematician; b. Ap. 19, 1931, Moscow.
Corresponding member of the Academy of Sciences of the USSR (1970).
He graduated from Moscow University in 1954. In 1969 he became a professor at the University of Novosibirsk. He is recognized as a specialist in the field of the theory and automation of programming. He headed a number of projects dealing with the mathematical foundations of electronic computers. He is a foreign member of the Association for Computing Machinery (USA, since 1965). Decor.: Order of the Red Banner of Labour.
Work: *Programmiruiushchaia programma dlia bystrodeistvuiushchei elektronnoi schetnoi mashiny.* Moscow 1958.

Ershov, Iurii Leonidovich
Mathematician; b. My. 1, 1940, Novosibirsk.
Grad. Novosibirsk Univ. 1963; corresponding mem. USSR Acad. of Sciences 1970; Prof. at Novosibirsk Univ. 1967–; main works on theory of algorithms, theory of models and number theory; mem. Int. Asscn. of Symbolic Logic 1968–.

Ershov, Vladimir Stepanovich
Parasitologist; b. Jl. 15 (28), 1904, Suslovo, now in Brasovo Raion, Briansk Oblast.
Academician of VAS-KhNIL (the V.I. Lenin All-Union Academy of Agricultural Sciences). Member of the CPSU since 1941.
He graduated from the Leningrad Veterinary Institute in 1925. From 1925 to 1929 he worked as a veterinarian. Since 1929 he has done scientific research in helminthology at various institutions. In 1954–55 he was director of the Moscow Institute of Down and Fur. Since 1957, he has been director of the K. I. Skriabin All-Union Institute of Helminthology. His main works are on the study of immunity to helminths. Decors.: four orders and a number of medals.
Works: *Gel'mintozy loshadi.* Moscow-Leningrad 1933. (In collaboration with K. I. Skriabin.) *Gel'mintozy svinei.* Moscow 1963. (Co-author.) *Parazitologiia i invazionnye bolezni sel'skokhoziaistvennykh zhivotnykh,* 3rd ed. Moscow 1964. (Co-author.)

Ershova, Neli Mikhailovna
Laborer; b. 1939.
Mem. CPSU 1973–; grad. secondary school 1957; lathe operator at F.E. Dzerzhinskii Machine-Building Plant in Perm 1964–71; adjuster at industrial asscn. "Permskii mashinostroitel'nyi zavod im F.E. Dzerzhinskogo" in Perm 1971–; cand. mem. Cen. Cttee. CPSU 1981–.

Erzhanov, Boris Mikhailovich
State official.
Min. of Installation and Special Construction Work Kazakh SSR Je. 1972–; mem. Cen. Cttee. Kazakh CP 1976–.

Esenbaev, Ziiat Islamovich
State official.
Chief Ed. of newspaper "Kizil Uzbekistan" 1959–63; mem. Cen. Cttee. Uzbek CP 1961–; Chair. of Uzbek Journalists'

Union 1962–; mem. Cttee. on Party and State Control of Cen. Cttee. Uzbek CP and Council of Mins. Uzbek SSR 1963–65; Chair. of State Cttee. Publishing, Polygraphy and the Book Trade. Uzbek SSR 1963–. Decor.: Order of Lenin.

Esenov, Shakhmardan Esenovich, Cand. of Geol.-Min. Sc.
Scientist; b. 1927, Kazakhstan.
Mem. CPSU 1956–; grad. Kazakh Mining Inst. 1949; mem. Acad. of Sciences Kazakh SSR 1967–; engineer and geologist, Head Geologist, Head Engineer of geological Dzhezgazgan Group Expedition 1949–60; Dep. Min. 1960–61, Min. of Geology and Mineral Resources Kazakh SSR 1961–64; mem. Cen. Cttee. Kazakh CP 1961–76, mem. Auditing Comm. Kazakh CP 1976–81; Dep. to Supr. Sov. Kazakh SSR 1962–; Chair. of State Prod. Cttee. for Geology Kazakh SSR 1964–65; Dep. Chair. of Council of Mins. Kazakh SSR 1965–67; Pres. of Acad. of Sciences Kazakh SSR 1967–74; Dep. to and mem. Comm. on Educ., Science and Culture with Sov. of the Union USSR Supr. Sov. 1970–74.; Min. of Geology Kazakh SSR 1974–78.

Esetov, Takei
Party official; b. 1923.
Kazakh; mem. CPSU 1946–; grad. All-Union Finances school; served in the Soviet Army 1942–43; mem. Apparat USSR People's Commr. for State Security 1943–45; Sec., Head of Finance Dept. of a Raion Exec. Cttee. of People's Deps. 1945–57; Sec., 2nd Sec. of a Raion Cttee. Kazakh CP, Dept. Head of Kzyl-Orda Oblast Cttee. Kazakh CP, 1st Sec. of Aralsk Raion Cttee. Kazakh CP 1957–78; 1st Sec. of Kzyl-Orda Oblast Cttee. Kazakh CP D. 1978–; mem. Cen. Cttee. Kazakh CP 1981–.

Esin, Vasilii Pavlovich
Party official.
2nd Sec. Fergana Oblast Cttee. Uzbek CP 1974–79; mem. Cen. Cttee. Uzbek CP 1976–; Dep. Chair. Council of Mins. Uzbek SSR 1979–82; 1st Sec. Navoi Oblast Cttee. Uzbek CP Je. 1982–.

Esipenko, Pavel Evmenovich
State official.
Min. of Agriculture Ukrainian SSR O. 1973–Ap. 1975; Dep. Chair. of Council of Mins. Ukrainian SSR Ap. 1975–; mem. Cen. Cttee. Ukrainian CP 1976–.

Essenson, Albert Rudolfovich
State official.
Min. of Meat and Dairy Ind. Esthonian SSR D. 1971–; cand. mem. Cen. Cttee. Esthonian CP 1976–.

Evdokimov, Nikolai Semenovich
Author; b. 1922, Bobr, Minsk Oblast.
Russian; grad. Gorkii Inst. of Literature Moscow 1943; first works publ. 1946. Publs.: "Rybaki" (Fishermen) 1948, "Vysokaia dolzhnost" (High Post) 1950, "Greshnitsa" (Sinner) 1960, filmed 1962, and others.

Evseeva, Svetlana Georgievna
Author; b. 1932, Tashkent.
Russian; grad. Gorkii Inst. of Literature Moscow 1962; first works publ. 1954.
Publs.: "Zhenshchina pod iablonei" (A Woman Under the Appletree) 1962, and others.

Evsiukov, Petr Nikitovich
Diplomat.
USSR Amb. to Mozambique N. 1975–80, to São Tomé and Principe Ag. 1981–.

Evstigneev, Kirill Alekseevich
Military official; b. F. 4 (17), 1917, Khokhly, now in Shumikha Raion, Kurgan Oblast.
Twice Hero of the Soviet Union (Ag. 2, 1944, and F. 23, 1945), major-general of aviation (1966). Member of the CPSU since 1943.
He was the son of a poor peasant. He joined the Soviet Army in 1938. In the Great Patriotic War of 1941–45 he fought in combat, beginning in March 1943, on the Voronezh. Steppe, and Second Ukrainian fronts as a pilot, flight and squadron commander, and deputy commander of the 24th Fighter Air Regiment. He made about 300 combat sorties, fought in more than 120 air battles, and shot down 53 enemy aircraft personally and three aircraft with others in his group. After the war he graduated from the Air Force Academy (1955) and from the Military Academy of the General Staff (1960) and served in high command and staff positions. Decors.: Order of Lenin, four Orders of the Red Banner, the Order of Suvorov Third Class, the Order of the Patriotic War Second Class, the Order of the Red Star, two foreign orders, and medals.

Evstratov, Vasilii Fedorovich
Organic chemist; b. Ja. 30 (F. 12), 1908, Dubrovka, in present-day Zhizdra Raion, Kaluga Oblast.
Corresponding member of the Academy of Sciences of the USSR (1970).
He graduated from Leningrad University in 1930. He became deputy director of the Tire Industry Institute in 1941 and a professor at the Moscow Institute of Fine Chemical Engineering in 1966. His principal work was on the relationship between the structure and properties of synthetic rubbers and the performance characteristics of their vulcanizates. He was awarded the State Prize of the USSR in 1951. In 1967 he won the Lenin Prize for developing a Soviet synthetic-rubber radial tire for trucks and the associated manufacturing technology. Decors.: Order of Lenin, three other orders, and various medals.

Evtukh, Vladimir Gavrilovich
State official.
Mem. Cen. Cttee. Belorussian CP 1971–76 and 1981–; Head of Dept. for Building and Municipal Econ. with Cen. Cttee. Belorussian CP 1971–76; Chair. State Cttee. for Construction Belorussian SSR S. 1979–.

Evtushenko, Evgenii Aleksandrovich
Author; b. 1933, Zima, Irkutsk Oblast.
Russian; student at Gorkii Inst. of Literature Moscow 1951–54; first works publ. 1949. Decor.: Badge of Honour.
Publs.: "Razvedchiki griadushchego" (Scouts of the Future) 1952, "Tretii sneg" (The Third Snow) 1955, "Shosse Entuziastov" (The Highway of Enthusiasts) 1956, "Obeshchanie" (The Promise) 1957, "Khotiat li russkie voiny?" (Do the Russians Want War?) 1962, "Vzmakh ruki" (Wave of the Hand) 1962, "Nezhnost'" (Tenderness) 1962, "Bratskaia GES" (Bratsk Hydroelectric Station) 1965, "Idut belye snegi" (White Snow is Falling) 1969, "Kazanskii universitet" (Kazan University) 1970, and others.

Evtykh, Asker Kaderbechevich
Author; b. 1915, Khatukai, Adyg Autonomous Oblast; s. of a farmer.
Publs.: "Nashi dni" (Our Days) 1939, "Bol'shaia sila" (The Great Power) 1939, "Prevoskhodnaia dolzhnost" (A Wonderful Post) 1948, "U nas v aule" (In Our Aul) 1953, "Devushka iz aula" (A Girl From the Aul) 1963, and others.

Ezhevskii, Aleksandr Aleksandrovich
State official; b. 1915, Tulun, Irkutsk Oblast.
Russian; mem. CPSU 1945–; grad. Irkutsk Agricultural Inst. 1939; fitter, foreman at a machine building plant in Irkutsk, attended Workers' Faculty, then Faculty for the Mechanization of Agriculture at Irkutsk Agricultural Inst. 1929–39; taught and conducted research at same 1939–42; head of various repair shops, chief engineer and later dir. of a plant in Irkutsk, dir. of a tractor plant in Altai, Dir. of "Rosselmash" Factory 1942–54; USSR Dep. Min. of Tractors and Agricultural Machine Building, 1st Dep. Min. of same, Head of USSR GOSPLAN's Automobile, Tractors and Agricultural Machine Building Dept. 1954–62; Chair. of "Soiuzselkhoztekhnika" Agricultural Machinery Trust D. 1962–78; Chair. USSR State Cttee. for the Supply of Production Equipment for Agriculture 1978–80; USSR Min. of Tractors and Agricultural Machine Building O. 1980–. Dep. to USSR Supr. Sov. 1966–; cand. mem. 1966–71, mem. Cen. Cttee. CPSU 1971–. Decor.: Order of Lenin.

Ezhov, Valentin Ivanovich
Author; b. Ja. 21, 1921, Samara, now Kuibyshev.
Russian; mem. CPSU 1951–; grad.

Union Inst. of Film-making Moscow 1951; author and scriptwriter. Decor.: Lenin Prize.
Scripts: "Chempion mira" (World Champion) 1954, "Liana" 1955, "Chelovek s planety Zemlia" (Man from Planet Earth) 1958, "Ballada o soldate" (Ballad of a Soldier) 1959, "Svoia golova na plechakh" (Your Head is on your Shoulders) 1960, "Bud' schastliva, Ani" (Be Happy, Ani) 1960, "Techet Volga" (The Volga Flows) 1963, "Dvorianskoe gnezdo" (Nest of the Nobility) 1969, "Beloe solntse pustyni" (White Sun of the Desert) 1971, and others.

Fabrikant, Lev Borisovich
Engineer; b. 1915.
Mem. CPSU 1955–; grad. Inst. of Civil Engineering Kharkov 1949; civil engineer; Chief of Branch Office for Construction and Architecture in Ulianovsk Oblast 1971–. Decor.: Lenin Prize 1972.

Faddeev, Dmitrii Konstantinovich
Mathematician; b. Je. 17 (30), 1907, Iukhnov, Kaluga Oblast.
Corresponding member of the Academy of Sciences of the USSR (1964).
Graduated from Leningrad University in 1928 and was made a professor at the university in 1944. In 1940 he became affiliated with the Leningrad Division of the V. A. Steklov Institute of Mathematics of the Academy of Sciences of the USSR. Faddeev's principal works deal with number theory, algebra, the theory of homology groups, and computer mathematics. He has been awarded the Order of Lenin, the Order of the Red Banner of Labor, and various medals.
Works: "Teoriia gomologii v gruppakh," parts 1–2. *Vestnik LGU:* 1956, no. 7 (*Ser. matematiki, mekhaniki i astronomii,* issue 2), pp. 3–39; 1959, no. 7 (*Ser. matematiki, mekhaniki i astronomii,* issue 2), pp. 72–87. (With Z. I. Borevich.) *Tablitsy osnovnykh unitarnykh predstavlenii fedorovskikh grupp.* Moscow-Leningrad 1961. *Vychislitel'nye metody lineinoi algebry.* 2nd ed. Moscow-Leningrad 1963. (With V. N. Faddeeva.)

Faddeev, Ludvig Dmietrievich, Dr. of Phy.-Math. Sc.
Mathematician and physicist; b. 1934.
Grad. Leningrad Univ. 1956; cand. 1956–59; Dr. of Phy.-Math. Sc. 1964; mem. staff Inst. of Mathematics with Leningrad Branch of USSR Acad. of Sciences 1959–; Sen. Research Fellow of same Inst. 1965–; in addition mem. staff Leningrad Univ. 1967–, Prof. at Mathematical-Mechanical Faculty 1969–; specialist in quantum mechanics. Decor.: State Prize of the USSR 1971.

Faddeev, N. V.
State official.
Sec. of Council for Mutual Econ. Aid (COMECON).
Publ.: "Sovet Ekonomicheskoi Vzaimopomoshchi" (Council for Mutual Econ. Aid) 1974.

Fadeechev, Nikolai Borisovich
Ballet dancer; b. Ja. 27, 1933, Moscow.
People's Artist of the USSR (1976).
In 1952, he graduated from the Moscow Choreographic School and joined the company of the Bol'shoi Theater in Moscow. His roles have included Siegfried in Tchaikovsky's *Swan Lake.* Albrecht in Adam's *Giselle,* Don José in the *Carmen Suite,* to Bizet's music arranged by Shchedrin and Karenin in Shchedrin's *Anna Karenina.* His dancing is noted for its polished technique, lightness, and romanticism. Fadeechev was a frequent partner of G. S. Ulanova and M. M. Plisetskaia. He has been teaching since 1974. Decor.: Order of the Red Banner of Labor.

Faintsimmer, Aleksander Mikhailovich (Alexander Feinzimmer)
Film director; b. D. 31, 1905 (Ja. 13, 1906), Ekaterinoslav (Dnepropetrovsk).
Honored Art Worker of the Belorussian SSR (1935) and the Lithuanian SSR (1954).
He graduated from the department of film directing of the State Technicum of Cinematography in 1928. He won fame with his satiric comedy *Lieutenant Kije* in 1934. The Soviet Navy, which occupies an important place in his work, is the subject of *People From the Baltic* (1938), *The Tanker Derbent* (1941), and *For Those in the Sea!* (1948). A number of his best films deal with the Civil War, the Great Patriotic War, and the underground movement, for example, *Kotovskii* (1943), *Konstantin Zaslonov* (1949, in collaboration with V. V. Korsh-Sablin), and *They Have a Homeland* (1950). Faintsimmer collaborated with Lithuanian film-makers on *Dawn Over the Neman* in 1953. In 1955 he adapted Voynich's novel *The Gadfly* for the screen. He also directed musical comedies, such as *Girl With a Guitar* (1958) and *The Sleeping Lion* (1965). Decors.: State Prize of the USSR (1950, 1951), Order of the Badge of Honor, and various medals.

Faiziev, Latif Abidovich
Film director; b. Ja. 2, 1929, Tashkent.
Honored Art Worker of the Uzbek SSR (1964).
Studied and acted in the Tashkent Theater for Young Audiences from 1941 to 1945. In 1951 he graduated from the All-Union State Institute of Cinematography and began working at the Uzbekfil'm studio. His works include *The Fall of the Emirate* (1955, in collaboration with V. P. Basov), *On Lenin's Permit* (1958), *Tiny Bird* (1961), *Star of Ulug Beg* (1965), *Born in a Storm* (1966), and *Sons of the Fatherland* (1968). He also directed the ballet films *Legend of the Old Fortress* (1970) and *Miniatures Brought to Life* (1971). Faiziev directed the film *Sunrise Over the Ganges* in collaboration with Indian cinematographers at the Tadzhikfil'm studio in 1975. He served as first secretary of the Cinematographers' Union of the Uzbek SSR from 1958 to 1967. Decors.: Order of the Badge of Honor and various medals.

Falin, Valentin Mikhailovich
Party official; b. 1926, Leningrad.
Mem. CPSU 1953–; grad. Moscow Inst. of Int. Relations 1950; apprentice lathe operator, lathe operator in "Krasnyi Proletarii" Plant in Moscow 1942–45; entered Diplomatic Service 1950; posts in USSR Min. of Foreign Affairs apparatus 1952–58 and 1959–70: counsellor, Dep. Chief of 3rd European Dept., Chief of 2nd European Dept., mem. of Collegium, and Chief of 3rd European Dept.; post in Cen. Cttee. CPSU 1958–59; USSR Amb. to FRG 1971–78; mem. Cen. Auditing Comm. CPSU 1976–; 1st Dep. Chief of Int. Information Dept. of Cen. Cttee. CPSU 1978–82; Political observer, Izvestiia D. 1982–.

Farafonov, Georgil Nikolaevich
Diplomat; b. 1919.
Entered Diplomatic Service 1946; Emb. Counsellor at USSR Emb. in Sweden 1961–63; mem. staff USSR Min. of Foreign Affairs 1963–66; Counsellor-Envoy at USSR Emb. in Finland 1966–72; leading posts in USSR Min. of Foreign Affairs 1972–75; USSR Amb. to Iceland Mr. 1975–79; Head, Skandinavian Countries Dept. USSR Min. of Foreign Affairs 1979–.

Farberov, Naum Pavlovich
Scholar and jurist; b. N. 21, 1909, Sherstin, Gomel' Oblast.
Doctor of the science of law (1949). Professor (1950). Honored Scientist of the RSFSR (1970). Member of the CPSU since 1940.
His most important works deal with the theory of the state and law, state law, and the constitutional development of socialist countries. Farberov was a subdepartment head at the Military Law Academy from 1946 to 1952 and at the University of Irkutsk from 1953 to 1956. In 1964 he became head of the section of the theory of the socialist state and law in the Institute of State and Law of the Academy of Sciences of the USSR.
Works: *Gosudarstvennoe pravo zarubezhnykh sotsialisticheskikh stran.* Moscow 1957. *Gosudarstvo i demokratiia v period stroitel'stva kommunizma.* Moscow 1968.

Fataliev, N. S.
Trade union official.
Sec. and mem. Presidium Azerbaidzhan Republican Trade Union Council Mr. 1977–.

Fazylov, Malik Sabirovich
Diplomat.
Sec. of Kokchetav Oblast Cttee. Kazakh CP 1963–68 (for Agriculture 1963–64); cand. mem. Cen. Cttee. Kazakh CP 1966–76, mem. Auditing Comm. Kazakh CP 1976–81; Dep. to Supr. Sov. Kazakh SSR 1967–; Head of Dept. of Science and Educ. with Cen. Cttee. Kazakh CP 1968–73; Min. of Foreign Affairs Kazakh SSR N. 1973–76; USSR Amb. to Mali Jl. 1976–.

Fazylova, Tufa Fazylovna
Tadzhik actress and singer (soprano); b. Ap. 25, 1917, Kanibadam.
People's Artist of the USSR (1957). Member of the CPSU since 1941.
In 1930, she became a singer in the Kanibadam Musical Drama Theater. She later worked in Dushanbe, appearing at the Tadzhik Musical Theater (1934–40), the Aini Tadzhik Theater of Opera and Ballet (1940–48), and the Lakhuti Drama Theater (from 1948). Her opera roles included Gulizor and Nushofarin in Balasanian's *The Uprising at Vosse* and *The Blacksmith Kova,* Tat'iana in Tchaikovsky's *Eugene Onegin,* Zukhra in Lenskii's *Takhir and Zukhra,* and Marfa in Rimsky-Korsakov's *The Tsar's Bride.* Her major roles as an actress included Guli in Uigun and Sultanov's *Alisher Navoi,* Gul'nor in Ikrami's *Dokhunda,* and Goneril in Shakespeare's *King Lear.* She still sings on the variety stage. She has also appeared in films. Decors.: Order of Lenin, two other orders, and various medals.

Fedirko, Pavel Stefanovich
Party official; b. 1932.
Mem. CPSU 1957–, grad. Rostov Engineering Inst. of Rail Transport 1954 and CPSU Cen. Cttee. Party Higher School 1971; Komsomol posts 1954–58; chief engineer, dir. of a machine repair workshop 1958–60; city sov. and party posts 1960–62; 1st Sec. of Igarka City Cttee. CPSU 1962–65; posts in Krasnoiarsk Krai Cttee. Apparatus CPSU 1967–70; 1st Sec. of Krasnoiarsk City Cttee. CPSU 1970–71, 2nd Sec. 1971–72, 1st Sec. of Krasnoiarsk Krai Cttee. CPSU D. 1972–; Dep. to USSR Supr. Sov. 1974–; mem. Cen. Cttee. CPSU 1976–. Decor.: Order of Lenin.

Fedorchenko, Ivan Mikhailovich, Dr. of Tech. Sc.
Scientist; b. O. 31, 1909, Taganrog; s. of a worker.
Mem. CPSU 1944–; grad. evening classes in metallurgy Kemenskoe (now Dneprodzerzhinsk) 1930; specialist in powder metallurgy and metallurgy; mem. Acad. of Sciences Ukrainian SSR 1961–; Dr. of Tech. Sc. 1953; Prof. 1954–; employed in metallurgical works in Dneprodzerzhinsk and Taganrog 1930–35; mem. staff Research Inst. for Agricultural Machine Construction in Moscow 1935–52; mem. staff of Laboratory for Special Alloys at Acad. of Sciences Ukrainian SSR (later Inst. of Metalloceramics and Special Alloys, now Inst. for Problems of Materials) 1952–57; concurrently teacher at Polytech. Inst. Kiev 1953–55; Chief Acad. Sec. of Presidium Acad. of Sciences Ukrainian SSR 1957–62; mem. staff Inst. for Problems of Materials Acad. of Sciences Ukrainian SSR 1962–; mem. Presidium Acad. of Sciences Ukrainian SSR, Acad. Sec. of Dept. for Physical-Tech. Problems of Materials 1963–; author of more than 150 scientific works. Decors.: Order of Red Banner of Labour, 2 Badges of Honour, medals.
Publs.: "Osnovy poroshkovoi metallurgii" 1961, "Metalokeramichni vyroby v narodnomu hospodarstvi" 1962.

Fedorchuk, Vitalii Vasil'evich
State (KGB) official; b. 1918.
Ukrainian; mem. CPSU 1940; grad. Higher School of KGB of USSR Council of Mins., Army General; employed by various raion newspapers in Zhitomir and Kiev oblasts 1934–36; student at a mil. school 1936–39; posts state security organs, participated in World War II 1939–70; Chair. of Cttee. for State Security (KGB) Ukrainian SSR Jl. 1970–82; Dep. to USSR Supr. Sov. 1970–; mem. Cen. Cttee. Ukrainian CP 1971–; cand. mem. Politbureau of Cen. Cttee. Ukrainian CP 1973–76, mem. 1976–82; Chair. USSR Cttee. for State Security (KGB) My.–D. 1982; USSR Min. of the Interior D. 1982–.

Fedorenko, Nikolai Prokofevich
Specialist in chemical technology; b. Ap. 15 (28), 1917.
Mem. CPSU 1942; concluded 4 terms at Moscow Inst. of Chemical Precision Technology 1941 and Dept. of Engineering at Mil. Chemical Acad. 1942; worked in Mil.-Chemical Main Admin. of the Sov. Army 1942–45; student at Party High School 1945–46; mem. staff Lomonosov Inst. of Chemical Precision Technology in Moscow 1946–62; Dep. Acad. Sec. of Econ. Dept. of USSR Acad. of Sciences 1962–63; Dir. of Cen. Econ.-Mathematical Inst. of USSR Acad. of Sciences 1963–. Decor.: State Prize of the USSR 1970.

Fedorenko, Nikolai Trofimovich
Russian philologist and orientalist; b. O. 27 (N. 9), 1912, Piatigorsk.
Professor (1953); corresponding member of the Academy of Sciences of the USSR (1958). Member of the CPSU from 1943. The son of a worker graduated from the Moscow Institute of Oriental Studies in 1937. In 1954 he became an ambassador extraordinary and plenipotentiary of the USSR. He also served as deputy minister of foreign affairs (1955–58), ambassador to Japan (1958–62), permanent representative of the USSR by the United Nations, and representative of the USSR on the UN Security Council (1963–68). In 1970, Fedorenko became editor-in-chief of the journal *Inostrannaia literatura* (Foreign Literature). He is the author of works on the history of Chinese and Japanese culture. Chinese classical and modern literature, and the theory of literature and art. Fedorenko became secretary of the administrative board of the Writers' Union of the USSR in 1971. He was a member of the Central Auditing Commission of the CPSU from 1966 to 1971. Fedorenko is an honorary member of the China Research Institute in Tokyo (1961) and an honorary academician of the Florgance Academy of Fine Arts (1975). Decors.: two Orders of Lenin, four other orders, and several medals.
Works: *Kalskaia literatura.* Moscow 1956. *Kraiskie zapisi.* Moscow 1958. *Skiszin"i ego mesto v kitaiskoi literature.* Moscow 1958. *Zovelia i legendy Kitaia.* Moscow 1961. *Deponskie zapisi,* Moscow 1966; 2nd ed., Moscow 1974. *Diplomaticheskie zapisi.* Moscow 1972. *Kraski vremeni.* Moscow 1972. *Problemy issledovaniia kitaiskoi literatury.* Moscow 1974. *Metkost' slova.* Moscow 1975.

Fedorov, Aleksandr Aleksandrovich
Botanist; b. N. 24 (D. 7), 1906, Tver', now Kalinin.
Corresponding member of the Academy of Sciences of the USSR (1964). Member of the CPSU since 1942. Brother of Andrei Aleksandrovich Fedorov.
He graduated from the Tver' Pedagogical Institute in 1929. From 1929 to 1934 he worked at the Sukhumi and Lenkoran divisions of the All-Union Institute of Plant Growing. Beginning in 1934 he worked at the V. L. Komarov Botanical Institute of the Academy of Sciences of the USSR. He was appointed assistant director of the institute in 1947 and served as director from 1962 to 1976.
His main works are devoted to plant morphology (including teratology), botanical resource management, and the taxonomy of flowering plants. He is co-author of the *Atlas of Descriptive Morphology of Higher Plants: The Leaf* (1956), *The Stem and Root* (1962), and *The Flower* (1975). He is editor-in-chief of the six-volume *Life of Plants* (vol. 1, 1974–). Decors.: four Orders of the Red Banner of Labor and various medals, V. L. Komarov Prize of the Academy of Sciences of the USSR in 1957.
Works: *Teratologiia i formoobrazovanie u rastenii.* Moscow-Leningrad 1958. "Botanicheskoe resursovedenie kak *Teratologiia i formoobrazovanie u rastenii.* Moscow-Leningrad 1958. "Botanicheskoe resursovedenie kak nauka i ego polozhenie v sisteme nauchnykh znanii." *Rastitel'nye resursy.* 1966, vol. 2, fasc. 2.

Fedorov, Andrei Aleksandrovich
Botanist; b. O. 17 (30), 1908, Tver', now Kalinin.
Corresponding member of the Academy

of Sciences of the USSR (1970). Member of the CPSU since 1952. Brother of Aleksandr Aleksandrovich Fedorov.
Graduated from Tver' Pedagogical Institute in 1929. Worked at the Sukhumi division of the All-Union Institute of Plant Growing from 1929 to 1935, at the Armenian branch of the Academy of Sciences of the USSR from 1935 to 1943, and at the Academy of Sciences of the Armenian SSR from 1943 to 1945. Since 1945 he has worked at the V. L. Komarov Botanical Institute of the Academy of Sciences of the USSR in Leningrad; he has been director of the institute's laboratory since 1963.
Principal works are on the systematics of flowering plants, the flora of the USSR and other countries, and the history of flora. He is editor and co-author of *Flora of the European USSR* (vols. 1–, 1974–) and coauthor and USSR consultant of *Flora of Europe* (vols. 1–4, 1964–76). Decors.: two orders of the Red Banner of Labor and various medals.
Works: "Istoriia vysokogornoi flory Kavkaza v chetvertichnoe vremia." In *Materialy po chetvertichnomy periodu SSSR,* vol. 3, Moscow, 1952. "O floristicheskikh sviaziakh Vostochnoi Azii s Kavkazom." In *Materialy po istorii flory i rastitel' nosti SSSR,* vol. 3, Moscow-Leningrad 1958. "Dipterokarpovyi ekvatorial'nyi vlazhnotropicheskii les Tseilona." *Trudy Moskovskogo obshchestva ispytatelei prirody: Otdel biologicheskii,* 1960, vol. 3.

Fedorov, Evgenii Pavlovich, Dr. of Phy.-Math. Sc.
Astronomer; b. Je. 26, 1909, Irkutsk; s. of a doctor.
Grad. Physical-Mathematical Faculty of Irkutsk Univ. 1937; corresponding mem. Acad. of Sciences Ukrainian SSR 1961; Dr. of Phy.-Math. Sc. 1959; Asst., Lecturer, Dir. of Observatory at Irkutsk Univ. 1937–41; served in Sov. Army 1941–44; postgraduate studies at Astronomical Main Observatory of USSR Acad. of Sciences 1944–47; Research Asst., Scientific Sec. at Observatory of Acad. of Sciences Ukrainian SSR in Poltava 1947–59; Dir. of Astronomical Main Observatory of Acad. of Sciences Ukrainian SSR 1959–; author of more than 70 scientific works, among them one monograph. Publ.: "Nutatsiia i vynuzhdennoe dvizhenie poliusov Zemli po dannym shirotnykh nabliudenii" 1958.

Fedorov, Evgenii Petrovich
Military official; b. D. 14 (28), 1911, Strel'na, now under the jurisdiction of the Leningrad city soviet.
Twice Hero of the Soviet Union (Ap. 7, 1940, and Je. 29, 1945). Major-general of aviation (1957). Member of the CPSU since 1932.
He joined the Soviet Army in 1930. In 1933 he graduated from a military pilots' school in Orenburg, and in 1948, from the Air Force Academy. During the Soviet-Finnish War of 1939–40, he was a squadron commander in bomber aviation and flew 24 combat missions. During the Great Patriotic War of 1941–45, he first served as a squadron commander in the 98th Long-Range Bomber Regiment and the 750th (later 3rd Guards) Aviation Regiment. In 1943 he was appointed deputy commander of the 2nd Long-Range Bomber Aviation Division. Fedorov took part in the battles of Moscow, Leningrad, and Stalingrad, in the liberation of Poland and Hungary, and in the capture of Berlin, flying a total of 178 combat missions. After the war he held command positions at higher military educational institutions, and in the standing army. He entered the reserves in 1958.
Decors.: three Orders of Lenin, two Orders of the Red Banner, the Order of Alexander Nevsky, two Orders of the Red Star, and various medals. A monument including a bust of Fedorov has been erected in Leningrad.

Fedorov, Fedor Ivanovich, Dr. of Phy.-Math. Sc.
Physicist; b. Je. 6 (19), 1911, Turets, Minsk Gouvt.
Belorussian; grad. Belorussian State Univ. 1931; Dr. of Phy.-Math. Sc. 1956; Prof. 1957–; mem. Acad. of Sciences Belorussian SSR 1966–; mem. staff Belorussian Univ. for several years; in addition Research Asst. at Physical-Tech. Inst. of Acad. of Sciences Belorussian SSR 1953–, mem. staff Gen. Laboratory of Theoretical Physics with Inst. of Physics of Acad. of Sciences Belorussian SSR 1955–, Acad. Sec. at Dept. of Physical-Mathematical Sciences of same Acad. 1963–. Decors.: State Prize of the Belorussian SSR 1972, Order of Lenin, 2 other orders, several medals.

Fedorov, Gennadii Aleksandrovich
Komi writer; b. Ag. 5 (18), 1909, Troitsko-Pechorsk, in what is now the Komi ASSR.
Member of the CPSU from 1946.
Graduated from the M. Gorkii Institute of Literature in 1940. He took part in the Great Patriotic War of 1941–45. He was the author of the novella *A Country Morning* (1932), which deals with the beginning of collectivization. His later novellas were *In the Days of the War* (1952) and *Mariika* (1960). In his novel *At Dawn* (parts 1–3, 1959–62; Russian translation, *When Daybreak Comes,* 1966), he recreated the character of Domna Kalikova, a heroine of the Civil War. Decors.: the Order of the Badge of Honor and various medals.
Works: *Povest'ias da rasskaz"ias*. Syktyvkar 1955.

Fedorov, Vasilii Dmitrievich
Russian poet; b. F. 23, 1918, Kemerovo. Member of the CPSU since 1945.
He graduated from the M. Gorkii Institute of Literature in 1950. He first published his works in 1944. Fedorov is the author of the poetry collection *Lyrical Trilogy* (1947). *Forest Springs* (1955). *The Dawning Cock-a-Doodle-Doo* (1966), and *Seventh Heaven* (1968), the last two of which were awarded the M. Gorkii State Prize of the RSFSR in 1968. His later collections include *Wings to Noonday* (1971) and *Book of Love and Faith* (1974). Fedorov also wrote the narrative poems *Lenin's Gift* (1934) and *Venus Sold* (1958), the drama in verse *The Golden Velm* (1958), the novellas *Maturity* (1953) and *The Volunteers* (1955), and the book *Such Is Our Time: Of Poetry and Poets* (1973). In addition to lyric poetry on nature and love, his works include many poems in praise of Soviet man and his achievements in work and on the field of battle. Fedorov also publishes articles on poetry. Decors.: two orders.
Works: *Sobr. soch.,* vols. 1–3. Moscow 1975–76.

Fedorov, Viktor Stepanovich, Cand. of Tech. Sc.
State official; b. 1912.
Russian; mem. CPSU 1939–; grad. Groznyi Petroleum Inst. 1932; head of a dept., later dir. of an inst., head of a trust, Dir. of Groznyi Petroleum Combine, Dir. of "Grozneftzavody" Petroleum Asscn. 1937–48; USSR-Dep., then 1st Dep. Min. of Petroleum Ind. 1948–57; Chair. of Bashkir Econ. Council 1957–58; Dep. to USSR Supr. Sov. 1958–; Chair. of State Cttee. on Chemistry with USSR Council of Mins., 1st Dep. Chair., then Chair. of State Cttee. on Petroleum Refining Ind. and Petrochemistry with USSR GOSPLAN 1958–65; cand. mem. Cen. Cttee. CPSU 1961–76, mem. 1976–; USSR Min. of Petroleum Refining Ind. and Petrochemistry O. 1965–. Decors.: Hero of Socialist Labour, State Prize, Order of Red Banner of Labour, 2 Orders of Lenin.

Fedorovich, Boris Aleksandrovich
Physical geographer and geomorphologist; b. Jl. 17 (30), 1902, Warsaw.
Doctor of geographical sciences (1953); professor (1956). Member of the CPSU since 1944.
He graduated from Tavrida University (now University of Simferopol') and worked there as an assistant professor worker for the Commission on Natural Productive Forces of the Academy of Sciences of the USSR. In 1933 he began working at the Institute of Geography of the Academy of Sciences of the USSR. His principal works deal with the relief formation and paleogeography of the plains and mountains of Middle and Central Asia and Kazakhstan, geomorphological cartography, desert landscapes, and the classification of eolian relief formations. Decors.: the State Prize of the USSR in 1952, the Order of Lenin, the Order of the Red Star, and various medals.
Works: *Lik pustyni,* 3rd ed. Moscow

1954. *Kazakhstan*. Moscow 1969. (Editor and co-author.) "Ravniny Srednei Azii." In *Ravniny i gory Srednei Azii i Kazakhstana*. Moscow 1975. *Geomorfologicheskaia karta SSSR v masshtabe 1: 4 000 000*. Moscow 1960. (Editor and co-author.)

Fedoseev, Petr Nikolaevich, Dr. of Phil. Sc.
Scientist; b. 1908.
Russian; mem. CPSU 1939–; grad. Gorkii Teachers' Training Inst. 1930; postgraduate course at Moscow Inst. of History, Philosophy and Literature 1936; Prof.; lecturer at an inst. 1930–36; researcher at Inst. of Philosophy of USSR Acad. of Sciences 1936–41; mem. staff All-Union Cen. Cttee. CP (B) 1941–46; corresponding mem. 1946–60, full mem. USSR Acad. of Sciences 1960–; Chief Ed. of journal "Bolshevik", Dept. Head of CPSU Cen. Cttee. at Acad. of Social Sciences 1946–50; Inspector at CPSU Cen. Cttee. 1950–54; Chief Ed. of journal "Partiinaia zhizn'" 1954–55; Dir. of Inst. of Philosophy at USSR Acad. of Sciences 1955–62; Chair. of Sov.-Hungarian Friendship Soc. 1958–; mem. Cen. Cttee. CPSU 1961–; Vice-Pres. of USSR Acad. of Sciences 1962–67 and My. 1971–; dep. mem. 1962, mem. Foreign Affairs Comm. 1962–70, Chair. of Comm. for Educ., Sciences and Culture with Sov. of Nationalities USSR Supr. Sov. 1970–; Dir. of Inst. for Marxism-Leninism Cen. Cttee. CPSU 1967–71; Decors.: 2 Orders of Red Banner of Labour, Order of Patriotic War 1st Cl., Order of Lenin, medal.

Fedotov, Feliks Nikolaevich
Diplomat; b. 1929.
Entered Diplomatic Service 1952; Emb. Counsellor 1968–69, Counsellor-Envoy at USSR Emb. in Iraq 1969–72; leading posts in USSR Min. of Foreign Affairs 1972; USSR Amb. to Sudan D. 1972–78, to People's Democratic Republic Yemen 1978–82.

Fedotov, Sergei Aleksandrovich
Geophysicist; b. Mr. 19, 1931, Leningrad. Corresponding member of the Academy of Sciences of the USSR (1970).
He graduated from Moscow State University in 1953. From 1957 to 1971 he worked at the O. Iu. Shmidt Institute of Earth Physics of the Academy of Sciences of the USSR. In 1971 he became director of the Institute of Volcanology of the Far East Scientific Center of the Academy of Sciences of the USSR. In 1975 he was elected vice-president of the International Association of Volcanology and Chemistry of the Earth's Interior. His principal works deal with the internal structure and seismicity of the earth, properties of the earth's mantle, the mechanism of deep magmatic and volcanic activity, and long-term earthquake predictions. Decor.: the Order of the October Revolution.
Works: "O seismicheskom tsikle, vozmozhnosti kolichestvennogo seismicheskogo raionirovaniia i dolgosrochnom seismicheskom prognoze." In *Seismicheskoe raionirovanie SSSR*. Moscow 1968. "O sviazi vulkanov s tikhookeanskim fokal'nym sloem, mekhanizme pod"ema magm i vozmozhnom polozhenii mantiinykh oblastei pitaniia vulkanov." In *Geodinamika, magmoobrazovanie i vulkanizm*. Petropavlovsk-Kamchatskii 1974.

Fedulova, Alevtina Vasil'evna
Komsomol official; b. 1940.
Mem. CPSU 1963–; grad. Moscow Oblast Pedagogical Inst. 1963; attended Young Pioneer leader seven-year school 1958–61; taugnt biology and chemistry 1961–67; held komsomol posts 1967–68; 1st Sec. of Elektrostal City Cttee. All-Union Komsomol in Moscow Oblast 1968–70; Sec. of Moscow Oblast Cttee. All-Union Komsomol 1970–71; Chair. of Cen. Council of All-Union Young Pioneer Org. 1971–; Sec. of Cen. Cttee. All-Union Komsomol Ja. 1977–; mem. Cen. Auditing Comm. CPSU 1981–.

Feinberg, Iakov Borisovich, Dr. of Phy.-Math. Sc.
Physicist; b. S. 7, 1918, Zolotonosha, Cherkassy Oblast.
Grad. Kharkov Univ. 1940; corresponding mem. Acad. of Sciences Ukrainian SSR 1964–; Dr. of Phy.-Math. Sc. 1962; Prof. 1963–; postgraduate studies 1940–41; served in Sov. Army 1941–45; mem. staff Physical-Tech. Inst. at Acad. of Sciences Ukrainian SSR 1946–; mem. staff Kharkov Univ. 1949–; author of more than 120 scientific works.
Publs.: "Uskorenie chastits v plazme" in "Atomnaia energiia" No. 6/1959, "Kvazilineinaia teoriia slaboturbulentnoi plazmy s ucheton korreliatsii elektricheskikh polei" in "Zhurnal eksperimentalnoi i teoreticheskoi fiziki" No. 49/1965 (co-author), "Dreifovaia neustoichivost v plazme, pomeshchennoi v vysokochastotnoe elektricheskoe pole" in "Zhurnal eksperimentalnoi i teoreticheskoi fiziki" No. 52/1967, and others.

Feodosii (I. I. Protsiuk), Cand. of Theol.
Russian Orthodox clergyman; b. Ja. 7, 1927, Topulno, Volhynia Oblast.
Grad. Leningrad Theological Acad. 1956, Cand. of Theol. 1959; Bishop of Chernigov 1962; Bishop of Poltava 1964; Bishop of Chernovtsy 1967; Bishop of Smolensk and Viazma 1972–.

Feodosii (M. N. Dikun), Cand. of Theol.
Russian Orthodox clergyman; b. F. 23, 1926, Chernitsa, Rovno Oblast.
Attended Moscow Theological Seminary 1947; grad. Moscow Theological Acad. 1955, Cand. of Theol.; monk, deacon, priest, teacher and Asst. to Supervisor of Volhynia Theological Seminary; Sec. of same Seminary 1960–64; abbot, teacher and Asst. to Supervisor of Odessa Theological Seminary 1964; Archimandrite, Rector of same Seminary 1966; Bishop of Pereiaslav-Khmelnitskii, Substitute for the Bishopric of Kiev 1967; Bishop of Poltava and Kremenchug O. 1967–79; Archbishop of Astrakhan and Enotaevka D. 1979–.

Feofilov, Petr Petrovich
Physicist; b. Mr. 31 (Apr. 13), 1915, Solvychegodsk, now in Arkhangel'sk Oblast. Corresponding member of the Academy of Sciences of the USSR (1964).
He graduated from the Leningrad Polytechnic Institute in 1939. Since then, he has been on the staff of the State Optical Institute. His main works deal with luminescence, spectroscopy, and quantum optics. He studied the luminescence of crystals and of solutions of complex organic compounds, developed an oscillograph method for the study of high-speed physical processes, discovered the latent anisotropy of the luminescent centers in cubic crystals and carried out research in the optics of synthetic crystals, such as fluorite. Together with V. V. Ovsiankin, Feofilov discovered cooperative sensitized luminescence. He also developed crystals for lasers. Decors.: the State Prize of the USSR in 1949 and 1975, two Orders of Lenin, the Order of the Red Banner of Labor, and various medals.
Works: *Poliarizovannaia liuminestsentsiia atomov, molekul i kristallov*. Moscow 1959. "Nekotorye voprosy spektroskopii redkozemel'nykh ionov v kristallakh." In the collection *Spektroskopiia kristallov*. Moscow 1966. "Kooperativnye protsessy v liuminestsiruiushchikh sistemakh." *Izv. AN SSSR: Ser. fizicheskaia*, 1973, vol. 37, no. 2. (With V. V. Ovsiankin.)

Feoktistov, Konstantin Petrovich, Cand. of Tech. Sc.
Cosmonaut; b. 1926.
Grad. N. E. Bauman Higher Tech. College Moscow ("MVTU") 1963; pilot, cosmonaut; Cand. of Tech. Sc. 1955; 16 earth orbits in spaceship "Voskhod" with V. V. Komarov and B. B. Egorov O. 12–13, 1964. Decor.: Order of Red Banner of Labour.

Ferensas, Algirdas Antanovich
Trade union official; b. 1928.
Lithuanian; mem. CPSU 1952–; grad. Polytech. Inst. Kaunas; Sec., 1st Sec. of Kaunas City Cttee. Lithuanian Komsomol 1952–56; Sec. and mem. Bureau Cen. Cttee. Lithuanian Komsomol 1956–60; mem. Cen. Cttee. Lithuanian CP 1960–61 and 1967–; 1st Sec. Shauliai City Cttee. Lithuanian CP 1960–61; mem. apparat Cen. Cttee. CPSU 1961–67; Sec. Cen. Cttee. Lithuanian CP 1967–77, mem. Bureau Ap. 1967–; Chair. Lithuanian Republican Trade Union Council Ap. 1977–; Dep. USSR Supr. Sov. 1979–.

Filaret (Kirill Varfolomeevich Vakhromeev), Cand. of Theol.
Russian Orthodox clergyman; b. Mr. 21, 1935, Moscow.
Grad. Theological Seminary 1957 and Theological Acad. Moscow 1961, Cand. of Theol.; monk, deacon 1959–; Lecturer in Homiletics at Moscow Theological Acad., priest-monk 1961; Sen. Asst. to the Supervisor 1962, Supervisor of Acad. 1963; Abbot, Archimandrite, Head of Aspirants at Moscow Theological Acad. 1963; Bishop of Tikhvin, Substitute of Bishopric of Leningrad 1965; Bishop of Dmitrov, Substitute of Bishopric of Moscow, Rector of Moscow theological schools 1966; hon. mem. Theological Acad. Leningrad, mem. ed. board of "Theological Treatises" 1970–; Archbishop 1971; Metropolitan of Berlin and Cen. Europe, Exarch of Moscow Patriarchate for Cen. Europe 1973–78; Metropolitan of Minsk and Belorussia 1978–81; mem. Comm. of Holy Synod for Matters Concerning Christian Unity and Inter-Church Relations. Chief of Dept. of External Affairs, Moscow Patriarchate, mem. of Holy Synod Ap. 1981–.

Filaret (Mikhail Antonovich Denisenko), Cand. of Theol.
Russian Orthodox clergyman; b. Ja. 28, 1929, Blagodatnoe, Amvrosievka/Donets Basin district.
Studied Moscow Theological Acad.; monk, deacon 1950, priest Je. 1951; Cand. of Theol. 1952; Lecturer, promoted to Sen. Asst. to Supervisor of Moscow Theological Acad. 1953–54, Supervisor of Saratov Theological Seminary 1956, moved to Kiev Theological Seminary, worked in the position of Archimandrite as Supervisor and was promoted Rector of same Seminary 1957; Head of Chancery of Ukrainian Exarchate and Chief Clergyman of Vladimir Cathedral 1960; Chief Clergyman of Alexandr Nevskii Church of the Moscow Patriarchate in Alexandria, United Arab Republic 1961–62; appointed Bishop of Luga and Substitute Bishop of Leningrad Diocese, in addition Administrator of Riga Diocese 1962; Exarch in Cen. Europe Je.–O. 1962; episcopal work in Vienna and the rest of Austria O. 1962–64; Rector of Moscow Theological Inst., mem. Holy Synod D. 1964–; Metropolitan of Kiev and Galicia, Exarch of the Patriarchate in the Ukraine My. 1966–.

Filatov, Alexandr Pavlovich
Party official; b. 1922.
Russian; mem. CPSU 1947–; grad. Novosibirsk Inst. of Railroad Engineers 1947 and CPSU Cen. Cttee. Party Higher School 1958; foreman, engineer of a construction admin. of "Sibstroiput" Trust in Novosibirsk 1947–48; CPSU posts in Novosibirsk 1948–61; Sec. 1961–63, 1st Sec. of Novosibirsk City CPSU Cttee. 1966–73; held various Sov. posts 1963–66; cand. mem. Cen. Cttee. CPSU 1971–81, mem. 1981–; 2nd Sec. 1973–78, 1st Sec. of Novosibirsk Oblast CPSU Cttee. 1978–. Decor.: Order of Lenin.

Filatov, Vladimir Gerasimovich
Diplomat.
USSR Amb. to Gabon Ag. 1974–78.

Filchakov, Pavel Feodosevich, Dr. of Phy.-Math. Sc.
Mathematician; b. S. 24, 1916, Petrograd; s. of an engineer and designer.
Grad. Kiev Univ. 1940; corresponding mem. Acad. of Sciences Ukrainian SSR 1964; Dr. of Phy.-Math. Sc. 1953; Prof. 1958–; mem. staff Inst. of Mathematics of Acad. of Sciences Ukrainian SSR 1945–; concurrently Lecturer at Chair of Higher Mathematics at Inst. for Water Econ. Kiev 1956–59; author of approx. 100 scientific works, among them 7 monographs.
Publs.: "Teoriia filtratsii pod gidrotekhnicheskimi sooruzheniiami" (vols. 1–2) 1959–60, "Priblizhennye metody konformnykh otobrazhenii" 1964.

Filin, Fedot Petrovich
Linguist; b. F. 23 (Mr. 7), 1908, Selino, now Dubna Raion, Tula Oblast.
Corresponding member of the Academy of Sciences of the USSR (1962). Member of the CPSU since 1939.
Graduated from the Second Moscow State University in 1931. From 1964 to 1968 he was director of the Institute of Linguistics of the Academy of Sciences of the USSR and in 1968 was named director of the academy's Institute of the Russian Language. Filin's principal works deal with the history of the Russian language, lexicology, and lexicography. He has done work on the ethnogeny of the Slavs, Russian dialectology and linguistic geography, and problems of general linguistics and sociolinguistics. He has also devoted attention to the development, structures, functions, and stylistic variants of literary languages, taking a comparative and historical approach. Filin was chairman of the editorial board and a compiler of the 17-volume *Dictionary of the Contemporary Russian Literary Language* (1950–65; Lenin Prize, 1970) and he oversaw the compilation of the *Dictionary of Contemporary Russian Folk Dialects* (fascs. 1–11, 1965–76; publication in progress). In 1971, Filin became editor in chief of the journal *Voprosy iazykoznaniia* (Problems of Linguistics). Decors.: four orders and various medals. In 1975 he received an honorary doctorate from the University of Kraków (Jagiellonian University).
Works: *Ocherk istorii russkogo iazyka do XIV stoletiia.* Leningrad 1940. *Leksika russkogo literaturnogo iazyka drevnekievskoi epokhi.* Leningrad 1949. *Obrazovanie iazyka vostochnykh slavian.* Moscow-Leningrad 1962. *Proiskhozhdenie russkogo, ukrainskogo i belorusskogo iazykov, Istoriko-dialektologicheskii ocherk.* Leningrad 1972.

Filipchenko, Anatolii Vasil'evich
Cosmonaut; b. 1928.
Mem. CPSU 1952–; grad. Mil. Aviation College in Chugnev 1950 and Air Force Acad. 1961; cosmonaut 1961–; served in Sov. Army 1950–61; commander of spaceship "Soiuz 7", in which he participated in a group flight of three spaceships together with V. N. Volkov and V. V. Gorbatko O. 12–17, 1969; commander of spaceship "Soiuz 16", in which he undertook an orbital flight D. 2–8, 1974 together with N. N. Rukavishnikov to test the modernized docking system developed for the Sov.-American space rendezvous. Decors.: two Orders of Lenin, medals.

Filippov, Aleksandr Gavrilovich
State official; b. 1914.
Dep. Min. of Culture RSFSR 1958–60; Chair. of State Cttee. 1963–65, Chair. of Cttee. 1965–72, Chair. of State Cttee. on Cinematography. RSFSR S. 1972–. Decor.: Order of the October Revolution.

Filippov, D. N.
Komsomol official.
Sec. and mem. Bureau of Cen. Cttee. USSR Komsomol Ap. 1974–.

Filippov, Genrikh Alekseevich
State official.
Chair. of Belorussian Asscn. for Sale of Agricultural Equipment and Supplies "Belselkhoztekhnika" 1961–62; cand. mem. Cen. Cttee. Belorussian CP 1961–71; Dep. Chair. of Agricultural Bureau at Cen. Cttee. Belorussian CP 1963–64; Head of Agricultural Dept. at Cen. Cttee. Belorussian CP 1964–65; 1st Dep. Min. of Agriculture Belorussian SSR 1965–72; Min. of Peat Ind. Belorussian SSR Je. 1972–77; Min. of Fuel Ind. Belorussian SSR Ap. 1977–; mem. Auditing Comm. Belorussian CP 1981–.

Filonenko, Viktor Lazarevich
State official; b. 1933.
Mem. CPSU 1959–; grad. Ukrainian Acad. of Agricultural Sciences; mechanical engineer; engineer, dir. of a maintenance engineering station (RTS), 2nd Sec., 1st Sec. of Chernigov Oblast Cttee. Ukrainian Komsomol, chair. of exec. cttee. of a raion sov. of peoples' deps., 1st sec. of a raion cttee. Ukrainian CP, Chair. of "Selkhoztekhnika" oblast asscn. to 1970; Sec. of Chernigov Oblast Cttee. Ukrainian CP 1970–73; Chair. of Exec. Cttee. of Chernigov Oblast Sov. of Peoples' Deps. 1973–81; mem. of Auditing Comm. Ukrainian CP 1976–81; cand. mem. Cen. Cttee. Ukrainian CP 1981–; Chair. of State Cttee. for Agricultural Material and Technical Supply Ukrainian SSR Mr. 1981–.

Finogenov, Pavel Vasil'evich
State official; b. 1919.
Mem. CPSU 1948–, grad. Leningrad Mechanical Mil. Inst. 1953; dep. shop superintendent, shop superintendent, chief en-

gineer, dir. of a plant 1941–60; Dep. Chair. of Vladimir Sovnarkhoz 1960–63; Dep. USSR Min. of the Defense Ind. 1965–73. 1st Dep. Min. 1973–79, Min. Ja. 1979–; Dep. to USSR Supr. Sov. 1958–62 and 1979–; mem. Cen. Cttee. CPSU 1981–.

Firisanov, Leonid Semenovich
Party official.
Sec., 2nd Sec. Vitebsk Oblast Cttee. Belorussian CP 1968–78; mem. Cen. Cttee., Sec. and mem. Bureau Cen. Cttee. Belorussian CP S. 1978–.

Firiubin, Nikolai Pavlovich
Diplomat; b. 1908.
Mem. CPSU 1929–; entered Diplomatic Service 1953; Counsellor at USSR Emb. in CSSR 1953–54; USSR Amb. to CSSR 1954–55 and to Yugoslavia 1955–57; cand. mem. Cen. Cttee. CPSU 1956–66; USSR Dep. Min. of Foreign Affairs 1957–. Decors.: Order of Lenin, Order of Red Banner of Labour.

Firsov, Oleg Borisovich
Physicist; b. 1915.
Grad. Leningrad Univ. 1938; mem. staff Inst. for the Construction of Physical Instruments 1938–45; mem. staff Physical-Tech. Inst. of USSR Acad. of Sciences 1945–56, Sen. Research Asst. 1951–; mem. staff I. V. Kurchatov Inst. of Atomic Energy 1956–. Decor.: Lenin Prize 1972.

Firsov, Vladimir Ivanovich
Author, poet and translator; b. Ap. 26, 1937, Zabolote, Smolensk Oblast.
Russian; mem. CPSU 1970–; grad. Gorkii Inst. of Literature 1960; first works publ. 1955. Decor.: Prize of the Lenin Komsomol 1968.
Publs.: "Berezovyi rassvet" (Birch Dawn) 1959, "Vdali ot tebia" (Far Away From You) 1961, "Zelenoe ekho" (Green Echo) 1963, "Predannost" (Devotion) 1964, "Goritsvet" (Lamp Flower) 1965, "Solnechnye kolodtsy" (Sun Fountain) 1969, "Chuvstvo Rodiny" (The Feeling of the Native Country) 1971, "Respublika bessmertiia" (The Republic of Immortality) 1968; translates works of Chuvash and Ukrainian authors into Russian.

Flerov, Georgii Nikolaevich
Physicist; b. F. 17 (Mr. 2), 1913.
Mem. CPSU 1955–; grad. Leningrad Ind. Inst. 1938; mem. USSR Acad. of Sciences 1968–; mem. staff Physico-Tech. Inst. of USSR Acad. of Sciences in Leningrad 1938–44; worked in various institutions of USSR Acad. of Sciences 1944–60; Dir. of Laboratory of United Inst. of Nuclear Research 1960–. Decors.: State Prize of the USSR 1946, Lenin Prize 1967.

Florensov, Nikolai Aleksandrovich
Geologist; b. Ja. 15 (28), 1909, Kiev.
Corresponding member of the Academy of Sciences of the USSR (1960).
Graduated from the University of Irkutsk in 1936. His main works deal with the geology, geomorphology, seismology, and minerals of Eastern Siberia. Florensov was the first to substantiate the possible coal reserves of the Gusinoozersk coal deposit in Transbaikalia and determined the prospects of the Botogol'skii graphite deposit in the Vostochnyi Saian. Florensov contributed to the development of neotectonics by establishing the rift nature of the system of depressions of the Lake Baikal type in Southeastern Siberia and Northern Mongolia. He also proposed a paleoseismogeological approach to evaluation of the present-day level of seismic activity and developed new concepts in geomorphology, such as the structure, formation, and age of topography. Florensov has been awarded three orders and various medals.
Works: *Mezozoiskie i kainozoiskie vpadiny Pribaikal'ia.* Moscow-Leningrad 1960. *(Tr. Vost.-Sib. filiala AN SSSR: Ser. geologicheskaia, issue 19.)* "O nekotorykh obshchikh poniatiiakh v geomorfologii." *Geologika i geofizika.* 1964, no 10. "O neotektonike i seismichnosti Mongolo-Baikal'skoi gornol oblasti." *Geologiia i geofizika,* 1960, no 1.

Florent'ev, Leonid Iakovlevich, Cand. of Econ. Sc.
State official; b. 1911.
Russian; mem. CPSU 1939–; grad. Gorkii Agricultural Inst. 1931; Assoc. Prof.; agronomist at a machine and tractor station, teacher, Dept. Head at CPSU Agricultural High School 1931–39; Dep. People's Commr. for Agriculture of Mari ASSR, then Commr. for Agriculture, Sec. of an Oblast CPSU Cttee., Dep. Chair., then Chair. of an Oblast Sov. 1939–49; Sen. Researcher at All-Union Inst. of Fertilizers, Agrotechnology and Pedology 1949–54; Head of Altai Krai Agricultural Board, Dep. Chair. of Exec. Cttee. Altai Krai of Sov. of Workers' Deps., Sec. of Altai CPSU Cttee. 1954–56; Dep. to Supr. Sov. RSFSR 1955–59; cand. mem. 1956–66, mem. Cen. Cttee. CPSU 1966–; 1st Sec. of Kostroma Oblast CPSU Cttee. 1956–65 (for Agriculture 1963–64); Dep. to 1958–, mem. Agricultural Comm. of Sov. of Nationalities USSR Supr. Sov. 1966–; Min. of Agriculture RSFSR D. 1965–; Chair. of RSFSR Council of Kolkhozes 1970–. Decors.: Order of Red Banner of Labour, Order of the October Revolution.

Fokina, Olga Aleksandrovna
Poet; b. N. 2, 1937, Artemevskaia, Arkhangelsk Oblast; d. of a kolkhoz worker.
Russian; grad. M. Gorkii Inst. of Literature 1962; first works publ. 1955.
Publs.: "A za lesom – chto?" (And What is Beyond the Forest?) 1965, "Alenushka" 1967, "Ostrovok" (Little Island) 1969, "Samyi svetlyi dom" (The Brightest House) 1971, "Izbrannaia lirika" (Selected Lyric) 1971, and others.

Fomenko, Lidia Nikolaevna
Author and critic; b. Mr. 3 (16), 1909, Rostov on Don.
Russian; mem. CPSU 1946–; grad. North Caucasian Univ. 1930; first works publ. 1948.
Publs.: "Obraz sovetskogo voinav khudozhestvennoi literature" (The Soviet Soldier in Literature) 1963, "Cherty vremeni. Idei i kharaktery v proze 60-kh godov" (Trends of the Time. Ideas and Characters in the Prose of the Sixties) 1970, "Odin iz semi" (One Out of Seven) 1968, "Maria Prilezhaeva" 1971, "Iurii Iakovlev. Ocherk tvorchestva" (Iurii Iakovlev. Outline of his Works) 1974, and others.

Fomenko, Mikhail Vladimirovich
State official; b. 1934.
Grad. Pedagogical Inst. in Zaporozhe; teacher, head of a faculty dept., dir. of a school, head of a raion, then city Dept. of People's Education, Dep. Dir. Dept. of Education of Exec. Cttee. of Oblast Sov. of People's Deps. in Donetsk until 1968; Dep. Head of Dept. of Science and Educational Institutions of Oblast Cttee. CP Ukraine in Donetsk, sector head of Dep. of Science and Educational Institutions of Cen. Cttee. CP Ukraine 1968–74; 1st Dep. Min. 1974–79, Min. of Education Ukrainian SSR Jl. 1979–; mem. Auditing Comm. CP Ukraine 1981–.

Fomenko, Vladimir Dmitrievich
Writer; b. S. 16 (29), 1911, Chernigov.
Member of the CPSU since 1949.
In 1941, graduated from the department of language and literature of the Rostov Pedagogical Institute. He took part in the Great Patriotic War of 1941–45. His works, first published in 1946, include the collections of essays *An Affair of Honor* (1946), *Ordinary People* (1947), and *Man in the Steppe* (1949) and the short-story collections *One Line* (1951). *Short Stories* (1952), and *Hunter's Vein* (1953). His novel *Memory of the Earth* (books 1–2, 1961–70) records the great changes brought about in the lives of the Don villagers by the construction of the Volga-Don Ship Canal. It also depicts the struggle to eradicate vestiges of private ownership and explores the role and methods of party leadership. The novel has been translated into foreign languages. Decors.: the Order of the Badge of Honor and various medals.

Fomichenko, Konstantin Efimovich
Diplomat; b. 1927.
Mem. CPSU 1948–; grad. CPSU Cen. Cttee. Party Higher Correspondence School 1958; Cand. of Hist. Sc. 1971; Komsomol organizer in a handicrafts school in Ust-Katav for Cen. Cttee. All-Union Komsomol 1945–46; 1st Sec. of Katav Raion Cttee. All-Union Komsomol in Cheliabinsk Oblast 1950–52; Sec. of Cheliabinsk Oblast Cttee. All-Union Komsomol 1953–54; posts in Chelia-

binsk Oblast Cttee. apparatus CPSU 1955–70; Sec. of Cheliabinsk Oblast Cttee. CPSU 1970–73; posts in Cen. Cttee. apparatus CPSU 1973–77; 2nd Sec. of Cen. Cttee. Kirghiz CP 1977–81; USSR Amb. to Ethiopia 1981–; Dep. to USSR Supr. Sov. 1979–; cand. mem. Cen. Cttee. CPSU 1981–.

Fomin, Andrei Andronovich
Diplomat; b. 1918.
Grad. Nikolaev Ship-Building Inst., engineer; entered Diplomatic Service 1946–; Adviser 1955–57, Chief Adviser to Soviet Rep. at UN 1957–59; Adviser to American Dept. in USSR Min. of Foreign Affairs 1959–61; Counsellor-Envoy at USSR Emb. in Brazil 1961–62; USSR Amb. to Brazil 1962–65; Dep. Sec. Gen. of USSR Min. of Foreign Affairs 1965–67; Head of Dept. for S. Asia in USSR Min. of Foreign Affairs 1967–73; USSR Amb. to Bangladesh F. 1973–76.

Fomin, Gennadii Ivanovich
Laborer; b. 1936.
Mem. CPSU 1963; completed intermediate school 1956; lathe operator 1956–59 and fitter at 3rd State Ball-Bearing Plant in Saratov 1963–; served in Sov. Army 1959–63; cand. mem. Cen. Cttee. CPSU 1971–81, mem. 1981–.

Fomin, Igor' Ivanovich
Architect; b. Ja. 21 (F. 3), 1904, Moscow. People's Architect of the USSR (1971). Son of I. A. Fomin.
In 1926, graduated from the Academy of Arts in Leningrad, where he had studied under A. I. Dmitriev and V. A. Shchuko. He taught in Leningrad at the Academy of Arts and at the I. E. Repin Institute of Painting, Sculpture, and Architecture (from 1935; professor from 1946). Fomin designed the Moscow (1930–35) and Nevsky (1937–40) district soviet buildings, an apartment house on Karpovka Esplanade (1935), Shchemilovka district (including Ivanovskaia Street (1937–48) and an apartment house on Dvortsovaia Esplanade (1946–51), and the Pioshchad' Vosstaniia subway station (1951–55) in Leningrad. He also designed an apartment house on Gagarin Street (1940–47) in Moscow and the branch of the A. A. Zhdanov Leningrad University in Petrodvorets (design, 1969). Decors.: the Order of Lenin, two other orders, and various medals.

Fomin, V. A.
Trade union official.
Sec. of Kazakh Republican Trade Union Council Ja. 1978–.

Fomin, Vasilii Mikheevich
State official.
2nd Sec. of Cen. Cttee. Moldavian Komsomol 1957–59; mem. staff Cen. Cttee. Moldavian CP 1959–66; cand. mem. Cen. Cttee. Moldavian CP 1960–76, mem. 1976–; City Sec. 1966–68, 2nd Sec. of Kishinev City Cttee. Moldavian CP 1968–71; Min. of Motor Transport Moldavian SSR Jl. 1971–.

Fomin, Vladimir Vladimirovich
Chemist; b. Ja. 12 (25), 1909, Damuraty, now in Poland.
Corresponding member of the Academy of Sciences of the USSR (1964). Member of the CPSU since 1947.
After graduating from the Institute of Social Upbringing in Nezhin in 1931, taught in secondary schools until 1938. From 1943 to 1948 he taught at Moscow State University. In 1948 he began working at the Scientific Research Institute of Inorganic Materials. Fomin's principal works deal with the formation of complexes, extraction (particularly extraction equilibria), kinetics, and the chemistry of radioactive elements. Decors.: five orders and a number of medals.
Work: *Khimiia ekstraktsionnykh protsessov,* Moscow 1960.

Fradkin, Efim Samilovich
Physicist; b. F. 24, 1924, Shchedrin, in what is now Gomel' Oblast, Belorussian SSR.
Corresponding member of the Academy of Sciences of the USSR (1970). Member of the CPSU since 1945.
From 1942 to 1947 he served in the Soviet Army. In 1948 he graduated from the University of L'vov. That same year he began working at the Institute of Physics of the Academy of Sciences of the USSR. His major works are devoted to quantum field theory, quantum statistics, and hydrodynamics. Fradkin has used the methods of functional analysis in the theory of relativistic fields. He developed a modified perturbation theory for the theory of interacting particles. He obtained renormalized field theory equations and studied their solution. He also introduced functional methods into quantum statistics, in which he developed the diagram technique. Decors.: the State Prize of the USSR in 1953, three orders and several medals.
Works: "Funktsii Grina dlia vzaimodeistviia nuklonov's mezonami." *Dokl. AN SSSR: Novaia Seriia,* 1954, vol. 98, no. 1. "Ob asimptotike funktsii Grina v kvantovoi elektrodinamike," *Zhurnal eksperimental'noi i teoreticheskoi fiziki,* 1955, vol. 28, no. 6.

Frank, Ilia Mikhailovich
Physicist; b. O. 10 (23), 1908.
Grad. Moscow State Univ. 1930; mem. USSR Acad. of Sciences 1968–; mem. staff State Inst. of Optics 1930–34; mem. staff Inst. of Physics 1934–; Prof. at Moscow State Univ. 1944–. Decors.: Nobel Prize for Physics 1958, 2 State Prizes of the USSR.

Frantsevich, Ivan Nikitovich, Dr. of Chem. Sc.
Scientist; b. S. 3, 1905, Poltava; s. of a worker.
Mem. CPSU 1940–; grad. Inst. of Public Educ. Kharkov 1929; specialist in physical chemistry and materials; mem. Acad. of Sciences Ukrainian SSR 1961–; Dr. of Chem. Sc. 1961; Prof. 1934–; mem. staff Inst. of Public Educ. Kharkov 1929–39, Prof. Chair of Physical Chemistry 1936–39; Dep. Dir. of Inst. of Nonferrous Metallurgy at Acad. of Sciences Ukrainian SSR (now USSR Min. of Nonferrous Metallurgy) 1939–52; Lecturer at Kiev Univ. 1945–57; Chief of Laboratory for Special Alloys at Acad. of Sciences Ukrainian SSR 1952–55; Dir. of Inst. of Metalloceramics and Special Alloys at Acad. of Sciences Ukrainian SSR (now Inst. of Materials) 1955–; author of more than 200 scientific works. Decors.: Merited Scientist of the Ukrainian SSR 1965, 2 Orders of Red Banner of Labour, medals.
Publs.: "Deformatsii stalei pri sovmestnoi teplovoi i mekhanicheskoi obrabotke i fiziko-mekhanicheskie svoistva ikh pri vysokikh temperaturakh" 1933, "Katodnaia zashchita magistralnykh gazoprovodov" 1949, "Poroshkovaia metallurgiia" 1958, "Issledovanie uslovii polucheniia i ekspluatatsii monolitnogo polikristallicheskogo karbida kremniia" in "Karbid kremniia" 1966, and others.

Freindlikh, Bruno Arturovich
Actor; b. 1909.
Attended Tech. Inst. of Theatre Arts Leningrad; contributed in organizing Leningrad branch of Theatre for Workers' Youth ("TRAM") for kolkhoz peasants 1931; mem. Youth Theatre Leningrad ("TIUZ") 1941–43; actor at Great Gorkii Drama Theatre Leningrad 1946–. Decors.: People's Artist of the USSR 1974, State Prize of the USSR 1951.

Frolov, Ivan Timofeevich
Philosopher; b. S. 1, 1929, Dobroe, Dobroe Raion, now in Lipetsk Oblast.
Corresponding member of the Academy of Sciences of the USSR (1976). Member of the CPSU since 1960.
Graduated from the department of philosophy of the M. V. Lomonosov Moscow State University in 1953. From 1952 to 1965 he worked as an editor, first in Moscow and then in Prague. Between 1965 and 1968 he was engaged in party work. From 1968 to 1977 he was editor-in-chief of the journal *Voprosy filosofii.* His work is in the area of dialectical materialism, philosophical problems in contemporary natural science, and the social and philosophical problems of the scientific and technological revolution. Frolov has been awarded two orders as well as various medals.
Works: *O prichinnosti i tselesoobraznosti v zhivoi prirode (filosofskii ocherk).* Moscow 1961. *Ocherki metodologii biologicheskogo*

issledovaniia (sistema metodov biologii). Moscow 1965. *Genetika i dialektika.* Moscow 1968. *Mendel', mendelizm i dialektika.* Moscow 1972. (With S. A. Pastushnyi.) *Progress nauki i budushchee cheloveka.* Moscow 1975. *Mendelizm i filosofskie problemy sovremennoi genetiki.* Moscow 1976. (With S. A. Pastushnyi.)

Frolov, Konstantin Ivanovich
State official.
Chair. of RSFSR State Cttee. on Pricing F. 1970–.

Frolov, Konstantin Vasil'evich
Authority on machine vibrations; b. Jl. 22, 1932, Kirov, Kaluga Oblast.
Corresponding member of the Academy of Sciences of the USSR (1976). Member of the CPSU since 1965.
After graduating from the Briansk Institute of Transportation Machine Building in 1956, joined the staff of the Institute of Machine Science (now the Academician A. A. Blagonravov State Scientific Research Institute of Machine Science) in Moscow became director of the institute in 1975, founded the branch of biomechanics dealing with the effects of exposure to machine vibrations and formulated the scientific principles underlying the theory of systems used to protect human operators from such vibrations.
Work: *Kolebaniia elementov aksial'no-porshnevykh gidromashin.* Moscow 1973. (Co-author.)

Frolov, Nikolai Andrianovich
Komi poet and mathematician; b. Ap. 1 (14), 1909, Tentiukovo (present-day Syktyvkar).
From 1966 to 1972, professor at the Moscow Power Engineering Institute. In 1972 he became head of the subdepartment of higher mathematics at the University of Syktyvkar.
He published works in 1929. His first poetry collection was *In Springtime* (1941). He is also the author of *Domna* (1936), a narrative poem about Domna Kalikova, a heroine of the Civil War in Komi Krai, and *Deep in the Taiga* (1941), a drama in verse based on folklore and history. He has also published a number of textbooks on higher mathematics. Decors.: the Order of the Red Banner of Labor and various medals.
Work: *Ezhva dorin.* Syktyvkar 1958.

Frolov, Vasilii Semenovich
Party official; b. 1914.
Russian; mem. CPSU 1944–; grad. Bauman Inst. of Mechanical Engineering and Machine Building Moscow 1941; attended factory school, steel-worker at Ordzhonikidze Plant in Kolchugino 1930–41; technologist, head of tech. bureau, dept. head at a plant 1941–49; Cen. Cttee. CPSU organizer at a plant 1949–51; 2nd Sec. 1951–54, then 1st Sec. of Nizhnii Tagil City CPSU Cttee. 1951–54; Dep. Head 1955–58, Head of Dept. of Machine Building at Cen. Cttee. CPSU 1958–; cand. mem. Cen. Cttee. CPSU 1961–; dep. mem. 1966–, mem. Comm. for Ind., Transport and Telecommunications 1966–70 and Comm. for Ind. at Sov. of Nationalities USSR Supr. Sov. 1970–. Decor.: Order of Lenin.

Furnike, Ivan Dmitrievich
Ballet dancer; b. 1931.
Soloist with "Zhok" ensemble of the Moldavian Philharmonic 1949–. Decors.: People's Artist of the Moldavian SSR 1960, State Prize of the USSR 1972.

Gabashvili, David Aleksandrovich
Graphic artist, painter and sculptor; b. 1914, Tbilisi.
Grad. Tbilisi Tech. College of Architecture and Construction 1933; attended Tbilisi Art Acad., pupil of I. A. Sharleman; postgraduate studies at Tbilisi Art Acad. 1939–44; exhibitions 1939–; teacher 1944–65 and Prof. 1965– at Tbilisi Art Acad. Decor.: Merited Artist of the Georgian SSR 1958–.

Gabdrakhmanov, Rais Bikmukhametovich
Author; b. 1918, Kucherbaevo, Bashkir ASSR.
Mem. CPSU 1948–; worked as teacher and for various newspapers and magazines; first works publ. 1938.
Publs.: Childrens' books "Novye cherty" (New Lines) 1951, "Samyi blizkii drug" (The Closest Friend) 1954, "Derevenskii mal'chik" (A Boy from the Village) 1956, "Doroga prikliuchenii" (Adventure Way) 1958, "Rasskaz shofera" (The Story of a Driver) 1960, "Liubov' zhivet v serdtse" (Love Lives in the Heart) 1961, and others.

Gabdullin, Malik
Author and literary historian; b. 1915, Kazakhstan.
Decor.: Hero of the Sov. Union.
Publs.: "Moi frontovye druzia" (My Comrades at the Front) 1947, "Zolotaia zvezda" (The Golden Star) 1948, "Frontovye ocherki" (Notes from the Front) 1959, and others.

Gabe, Mira Rufinovna
Graphic artist; b. 1914, St. Petersburg; d. of R. M. Gabe.
Grad. Repin Inst. for Painting, Sculpture and Architecture at the USSR Acad. of Arts 1947; exhibitions 1951–.

Gabitashvili, David Dimitrievich
Painter; b. 1920, Tbilisi.
Attended Tbilisi Art Acad. 1946–52, pupil of V. G. Shebuev; exhibitions 1950–; mem. CPSU 1944. Decor.: Merited Painter of the Georgian SSR 1961.

Gabrielian, Emil Samsonovich
State official.
Min. of Health Armenian SSR Je. 1975–; mem. Auditing Comm. 1976–81, cand. mem. Cen. Cttee. Armenian CP 1981–.

Gabrielian, Vazrik Tserunovich
Graphic artist; b. 1924, Leninakan.
Attended Moscow Intermediate School of Art 1939–41, pupil of N. I. Andriaki, N. A. Korrenberg; exhibitions 1958–.

Gabrilovich, Evgenii Iosifovich
Author and scriptwriter; b. S. 17 (29), 1899, Voronezh.
Russian. Decors.: State Prize of the USSR 1943 and 1967, Order of Lenin, 3 other orders, medals.
Publs.: film scripts "Mashenka" 1942, "Dva boitsa" (Two Soldiers) 1943, "Ubiistvo na ulitse Dante" (Murder in Dante Street) 1956, "Kommunist" 1958, "Lenin v Pol'she" (Lenin in Poland) 1965, "Tvoi sovremennik" (Your Contemporary) 1968, "Nachalo" (The Beginning) 1970; author of essays on the theory of film, especially film directing.

Gabuniia, Guram Dmitrievich
Party official.
Until 1972 held posts in Komsomol; 1st Sec. of 26-Kommissars-Raikom Georgian CP in Tbilisi 1972–74; Sec. of Tbilisi City Cttee. Georgian CP 1974–79, 1st Sec. D. 1982–; Chair. of Tbilisi City Exec. Cttee. of Peoples Deps. 1979–82; cand. mem. Cen. Cttee. Georgian CP 1976–81, mem. 1981–, mem. Bureau D. 1982–.

Gabyshev, Lev Mikhailovich
Painter; b. 1923, Iakutsk.
Attended Moscow Oblast Pedagogic Art School 1939–41, pupil of V. N. Baksheev, P. I. Petrovichev, and the Surikov Art Inst. Moscow 1944–49, pupil of P. D. Pokarzhevskii; exhibitions 1939–.

Gabyshev, Nikolai Alekseevich
Author; b. 1922, Yakut ASSR.
Publs.: "V dalekom Amychane" (In Distant Amychan) 1951, "Deti moi liubimye" (My Dear Children) 1956, "Moi znakomye" (My Acquaintances) 1958, and others.

Gadaskina, Delia Georgievna
Artist in applied art; b. 1927, Baku.
Attended Leningrad High School for Industrial Art 1945–50, works with metal, plastics, glass; exhibitions 1957–.

Gadeliia, Givi Bagratovich
State official; b. 1923, Tbilisi.
Grad. Tbilisi Polytech. Inst. 1959; postgraduate studies at Scientific Inst. for Research on Energetics and Hydrotechnology in Tbilisi 1966; employed in one of the depts. of Min. of Rural Construction Georgian SSR 1966–74; Inspector of Cttee. on People's Control of Georgian SSR 1974–75; Instructor, Dep. Head of Agricultural Dept. of Cen. Cttee. Georgian CP 1975–76; 1st Dep. Min. 1976–77, Min. of Melioration and Water Supply Georgian SSR Ap. 1977–.

Gadzhiev, Akhmed Dzhevdet Ismail ogly
Composer; b. Je. 5 (18), 1917, Nukha, now Sheki.
People's Artist of the Azerbaidzhan SSR (1960). Member of the CPSU since 1951. Between 1935 and 1938, he studied at the Azerbaidzhan Conservatory (Baku) in the class of L.M.Rudol'f. In 1947 he graduated from D.D.Shostakovich's class at the Moscow Conservatory and began teaching at the Azerbaidzhan Conservatory (from 1957 to 1969, rector; since 1963, professor). With K.Karaev, he composed the patriotic opera *Veten* (The Motherland, staged in 1945; State Prize of the USSR, 1946). He has devoted particular attention to symphonic works and has written five symphonies (including *In Memory of Lenin,* 1956), the symphonic poem *For Peace!* (1951; State Prize of the USSR, 1952), an Azerbaidzhani suite, and a small symphony. His compositions also include oratorios and works for string quartet, piano, and chorus. Decors.: two orders as well as medals.

Gadzhieva, Begim Aziz Bala kyzy
Graphic artist; b. 1927, Baku.
Attended Leningrad High School of Arts and Crafts 1950–57, pupil of B.A.Smirnov; exhibitions 1961–.

Gadzhieva, S.R.
Komsomol official.
Sec. and mem. Bureau Cen. Cttee. Azerbaidzhan Komsomol Ap. 1982–.

Gaev, Gennadii Petrovich
Painter; b. 1918, Ekaterinburg.
Attended Sverdlovsk Art School 1935–40, pupil of G.A.Melentev; exhibitions 1947–. Decor.: Merited Painter of the RSFSR 1970.

Gaevskii, Iurii Fedorovich
State official.
Head of Dept. of Food and Light Ind. at Cen. Cttee. Ukrainian CP 1968; Min. of Local Ind. Ukrainian SSR O. 1970–; mem. Auditing Comm. Ukrainian CP 1971–78, Chair. D. 1978–.

Gafurov, Rakhimdzhon
State official.
Min. of Rural Construction Tadzhik SSR Je. 1975–80. Cand. mem. Cen. Cttee. Tadzhik CP 1976–81; Dep. Chair. of Council of Mins. Tadzhik SSR N. 1980–.

Gagarin, Iakov Petrovich
Painter; b. 1922, Udmurt ASSR.
Attended Art Schools in Perm 1938–41 and Nizhnii Tagil 1948–50; exhibitions 1952–.

Gaidaenko, Ivan Petrovich
Author; b. 1914, Semiozerskoe, Kustanai Oblast, Kazakhstan.
Mem. CPSU 1945–; grad. Navigation School Odessa 1932; former sailor, helmsman; served in Sov. Navy 1941–45; first works publ. 1939.
Publs.: Sketches and stories "Zarubezhnye vstrechi" (Encounters Abroad) 1950, "Dva berega" (Two Shores) 1952, "Na morskikh dorogakh" (On the Sea Routes) 1953, "V chuzhikh gavaniakh" (In Foreign Harbours) 1953, play "Poslednii signal" (The Last Signal) 1958, and others. Also writes in Ukrainian.

Gaidai, Ivan Iakovlevich
Artist in monumental and decorative art; b. 1922, Cherepanovo, Tomsk Oblast.
Attended Lvov Inst. for Applied and Decorative Art 1948–54, pupil of I.F. Gutorov, R.E.Selskoi.

Gaiduk, Viktor Grigorevich
Graphic artist; b. 1922, Harbin, China.
Attended Lvov Inst. for Applied and Decorative Art 1950–56, pupil of G.A. Leonov, I.N. Shcheglov; exhibitions 1957–.

Gaigalas, Balis Pranovich
State official.
Chair. State Cttee. of Labour. Lithuanian SSR My. 1967–; mem. Auditing Comm. Lithuanian CP 1976–81.

Gainullin, Mukhamed Khairullovich, Dr. of Philol. Sc.
Philologist; b. Ag. 18 (31), 1903, Kysyltan, now Tartar ASSR.
Mem. CPSU 1941–; grad. East Pedagogical Inst. Kazan 1931; literary scholar; Prof.; Dir. of Inst. of Language, Literature and History at Kazan Branch of USSR Acad. of Sciences 1944–54, 1959–61. Decor.: Merited Scientist of the Tartar ASSR.

Gaipov, Ruzmet
Party official; b. 1922.
Uzbekian; mem. CPSU 1949–; grad. Tashkent Agricultural Inst.; worked at an oblast experimental cotton and lucerne farm, agronomist at a machine and tractor station 1938–42; served in Sov. Army 1942–46; agronomist, head agronomist at a machine and tractor station, head agronomist and head of a raion dept. for cotton growing 1946–51; dir. of a machine and tractor station 1951–57; 1st Sec. of Urgench Raion Cttee. Uzbek CP, Sec. of CPSU Cttee. of Kolkhoz Prod. Admin. in Khorezm Oblast 1957–64; Sec. of Samarkand Oblast Cttee. Uzbek CP 1964–68; 1st Sec. of Kashka-Daria Oblast Cttee. Uzbek CP Ja. 1968–; Dep. to and mem. Comm. for Ind. with Sov. of the Union USSR Supr. Sov. 1970–; mem. Cen. Cttee. Uzbek CP 1971–.

Gairbekova, Mashidat Gadzhievna
Author; b. 1927, Karata, Dagestan.
Avar; grad. Gorkii Inst. of Literature Moscow. Decors.: Order of Red Banner of Labour, Badge of Honour.
Publs.: "Slovo gorianki" (The Word of a Mountain Dweller) 1962, "V puti na vershinu" (On the Way to the Top) 1958, "Nedopisannoe pis' mo" (An Unfinished Letter) 1955, "Dalekaia sestra" (The Distant Sister) 1954, "Za schastie nado borotsia" (One Must Fight for Happiness) 1958, and others. *The Bride's White Kerchief* (1965), *Punishment for a Crime* (1967), and *The Red Partisan's Daughter* (1968). Decors.: two orders.

Gaivanenko, Anna Platonovna
Painter; b. 1916, Rostov/Don.
Attended Rostov/Don Art School 1935–40, pupil of S.I.Zhovmir, G.N.Poliakov exhibitions 1938–.

Gakhov, Fedor Dmitrievich, Dr. of Phy.-Math. Sc.
Mathematician; b. F. 6 (19), 1906, Batalpashinskaia Stanitsa/Kuban.
Mem. CPSU 1964–; grad. Kazan Univ. 1930; Dr. of Phy.-Math. Sc. 1943; Prof.; mem. Acad. of Sciences Belorussian SSR 1966–; posts in various institutions in Belorussian SSR and other republics; Prof. Chair of Mathematical Analysis at Belorussian State Univ. 1961–.

Galaganov, Ivan Alekseevich
Painter; b. 1915, Saratov Oblast.
Attended Kiev Art Inst. 1944–50, pupil of V.N.Kostetskii; exhibitions 1949–.

Galakhov, Nikolai Nikolaevich
Painter; b. 1928, Leningrad.
Attended Kazan Art School 1942–47 and Repin Inst. for Painting, Sculpture and Architecture at the USSR Acad. of Arts 1947–53; exhibitions 1949–.

Galazii, Grigorii Ivanovich
Botanist and hydrobiologist; b. Mr. 5, 1922, Mechebilovo, in present-day Barvenkovo Raion, Kharkov Oblast.
Corresponding member of the Academy of Sciences of the USSR since 1970.
He graduated from the University of Irkutsk in 1942. In 1954 he became director of the Baikal Limnology Station of the Academy of Sciences of the USSR (AN SSSR) and in 1961 of the Limnology Institute of the Siberian Division of the AN SSSR, which had been organized on the basis of the station. His major research has been on the habitat of the ligneous vegetation along the shores of Lake Baikal and the adjoining mountain ranges, as a basis for reconstructing the climate, water level, and topography of the area in the postglacial period, as well as on the protection and rational utilization of the natural resources of Lake Baikal.

Galeev, Gabdrakhman Galeevich (pseudonym: Musa Gali)
Poet; b. 1923, Staromusino, Bashkir ASSR; s. of a peasant.
Mem. CPSU 1947–; worked formerly as teacher; first works publ. 1940.
Publs.: "Svetlyi den'" (A Bright Day) 1953, "Pesnia Ak-Ideli" (Ak-Ideli's Song) 1957, "Nachalo puti" (Beginning of the Journey) 1954, and others.

Galeev, Garif Zakirovich
Author; b. 1903, Urmanovo, Tartar ASSR. Mem. CPSU 1946–; worked as teacher, served in Sov. Army; worked for the press at the front 1941–45.
Publs.: "Rasskazy o Krasnoi Armii" (Stories of the Red Army) 1927, "Tatarskie rasskazy" (Tatar Stories) 1951, one-act play "Nachalos' s pis'ma" (It Began With a Letter), "Drug za drugom" (One After Another) 1952, and others.

Galents, Armine Sarkisovna
Painter; b. 1920, Damascus, Syria.
Attended studios of K. Mishle 1938 and A.T. Galents 1939, Beirut; lived in the Lebanon until 1946. Decor.: Merited Artist of the Armenian SSR 1967.

Galin, Lev Aleksandrovich
Scientist in the field of mechanics; b. S. 15 (28), 1912, Bogorodsk, now in Gorkii Oblast.
Corresponding member of the Academy of Sciences of the USSR (1953). Member of the CPSU since 1951.
Upon graduating from the Moscow Technological Institute of Light Industry in 1939, he began working at the Institute of Mechanics of the Academy of Sciences of the USSR. He has been a professor at Moscow University since 1956. His main works are on the theory of elasticity, on elastoplastic problems, and on questions of the transient filtration of liquids.
Work: *Kontaktnye zadachi teorii uprugosti.* Moscow 1953.

Galitskii, Rostislav Nikolaevich
Painter; b. 1920, Rostov, Iaroslavl Oblast. Attended Surikov Art Inst. Moscow 1950–56, pupil of D. K. Mochalskii, V. G. Tsyplakov; exhibitions 1952–.

Galkin, Aleksandr Aleksandrovich, Dr. of Phy.-Math. Sc.
Physicist; b. Jl. 4, 1914, Berdiansk, now Zaporozhe Oblast.
Grad. Kharkov Univ. 1939; mem. Acad. of Sciences Ukrainian SSR 1965–; Dr. of Phy.-Math. Sc. 1955; Prof. 1956–; mem. staff Ukrainian Physical-Tech. Inst. (after 1939 Physical-Tech. Inst. of Acad. of Sciences Ukrainian SSR Kharkov) while still a student 1937–41; front-line soldier 1941–45; mem. staff Physical-Tech. Inst. of Acad. of Sciences Ukrainian SSR 1945–60; Dep. Dir. of Physical-Tech. Inst. for Low Temperatures of Acad. of Sciences Ukrainian SSR 1960–65; concurrently mem. staff Inst. for Radiophysics and Electronics of Acad. of Sciences Ukrainian SSR in Kharkov 1946–60; Dir. of Physical-Tech. Inst. of Acad. of Sciences Ukrainian SSR in Donetsk, Prof. at Donetsk Univ. 1965–; mem. Presidium Acad. of Sciences Ukrainian SSR 1966–.
Publs.: "Ob izmenenii sverkhprovodimosti olova pod vliianiem neodnorodnoi deformatsii" in "Zhurnal eksperimentalnoi i teoreticheskoi fiziki" vol. 14/1944, "O nekotorykh osobennostiakh perekhoda v sverkhprovodiashchee sostoianie" in "Zhurnal eksperimentalnoi i teoreticheskoj fiziki" vol. 20/1950, "Issledovanie paramagnitnogo rezonansa v kamennykh ugliakh Donetskogo basseina" in "Khimiia i tekhnologiia topliv i masel" No. 7/1958, "O novom tipe ostsiliatsii koeffitsienta pogloshcheniia ultrazvuka v metallakh v magnitnom pole" in "Doklady Adademii Nauk SSR" No. 1/1960, and others.

Galkin, Dmitrii Prokhorovich
State official; b. 1926.
Russian; mem. CPSU 1949–; grad. Magnitogorsk Metallurgical Mining Inst. 1950; electrician, shift supervisor, shop superintendent, chief engineer, sec. of party cttee. at Magnitogorsk Mining and Processing Combine 1943–73; Dep. to USSR Supr. Sov. 1970–; Dir. of Magnitogorsk Metallurgical Combine 1973–79; mem. Cen. Cttee. CPSU 1976–81; mem. Cen. Auditing Comm. 1981–; 1st USSR Dep. Min. of Ferrous Metallurgy 1979–81; Min. of the Metallurgical Ind. Ukrainian SSR N. 1981–. Decors.: Hero of Socialist Labor 1976, USSR State Prize 1974.

Galkov, Rudolf Alekseevich
Artist in monumental and decorative art; b. 1935, Navoloki, Ivanovo Oblast.
Completed Ivanovo Art School 1957 and attended Repin Inst. for Painting, Sculpture and Architecture at the USSR Acad. of Arts 1957–63; exhibitions 1957–.

Galkus, Iuozas Iuozo
Graphic artist; b. 1932, Shimkunai, Lithuania.
Attended Art Inst. of the Lithuanian SSR 1952–58, pupil of I. M. Kuzminskis; exhibitions 1957–; teacher at Art Inst. of the Lithuanian SSR 1960–.

Galochkin, Valentin Andreevich
Sculptor; b. 1928, Dnepropetrovsk.
Attended Dnepropetrovsk Art School 1944–49 and Kiev Art Inst. 1949–55, pupil of A. I. Zhiradkov, M. N. Panin; exhibitions 1956–.

Galperin, Iulii Arkadevich
Graphic artist; b. 1928, Odessa.
Attended Moscow Polygraphic Inst. 1959–65, pupil of G. K. Burdzhelian, V. S. Zhitenev; exhibitions 1954–.

Galsanov, Tseden Galsanovich
Poet; b. 1917, Buriat ASSR; s. of a cattle breeder.
Mem. CPSU 1961–; studied Gorkii Inst. of Literature Moscow; first works publ. 1931.
Publs.: Poems "Chetyre dnia i tri nochi" (Four Days and Three Nights) 1939, "Sorevnovanie i mudrost" (Contest and Wisdom) 1940, "Ekho na Chudskom ozere" (The Echo on Lake Chudskoe) 1943, "Pavel Baltakhinov" 1945, "Poema o piatiletke" (Poem to the Five-Year-Plan) 1947, "Moi veselye druzia" (My Jolly Friends) 1962, and others.

Galstian, Ruben Stepanovich
State official.
Mem. Cen. Cttee. Armenian CP 1961–; Min. of Public Welfare Armenian SSR F. 1974–.

Galumian, Vladimir Bagratovich
Party official.
Cand. mem. 1966–71; mem. Cen. Cttee. Armenian CP 1971–; Head of Agricultural Dept. at Cen. Cttee. Armenian CP 1971–72; mem. Bureau and Sec. of Cen. Cttee. Armenian CP 1972–.

Gambarian, Gaik Ogandzhanovich
State official.
Head of Cen. Admin. for Installation and Special Construction Work Armenian SSR My. 1974–.

Gambarian, Karlen Amaiakovich
Party official.
Dep. to Supr. Sov. Armenian SSR 1959–; 1st Sec. of Akhta Raion Cttee. Armenian CP 1950–61; mem. Cen. Cttee. Armenian CP 1960–; Head of Dept. of Ind. and Transport at Cen. Cttee. Armenian CP 1961–75; Dep. Chair. of Council of Mins. Armenian SSR Jl. 1975–81; Sec. and mem. Bureau Cen. Cttee. Armenian CP Ja. 1981–; Head of Dept. for Construction and Municipal Econ. with Cen. Cttee. Armenian CP Ja. 1981–.

Gamrelidze, Givi Galaktionovich
State official; b. 1929.
Georgian; mem. CPSU 1958–; grad. Georgian Polytechnical Inst. 1953, electrical engineer; engineer at "Gruzenergo" electrical distributing system in Tbilisi 1953–57; chief engineer of Trust No. 2 of Min. of Construction Georgian SSR 1957–61; chief engineer, head of admin. for mechanization of Trust No. 2, head of the section for special assembly, head of the specialized building admin. of Trust No. 9 of Min. of Construction Georgian SSR 1961–71; chief engineer of "Spetsstroi" trust 1971–73; head of "Gruzselstroimontazh" trust 1973–76; 1st dep. head of Cen. Admin. for assembly and specialized construction work Georgian SSR 1976–79, head My. 1979–.

Gamtselidze, Georgii Pavlovich
State official.
Dep. Min. 1971–72, Min. of Light Ind. Georgian SSR Ja. 1972–; mem. Cen. Cttee. Georgian CP 1976–.

Gamzatov, Rasul Gamzatovich
Poet; b. 1923, Tsada, Dagestan; s. of poet Gamzat Tsadary.
Avar; mem. CPSU 1944–; studied at Gorkii Inst. of Literature Moscow 1945–50; mem. staff of Republican newspaper "Bolshevik Gor" 1941–51; radio

ed. of programmes in Avar language broadcasting station in Dagestan; Chair. of Writers' Asscn. of the Dagestan ASSR 1951–; Dep. to and mem. Presidium USSR Supr. Sov. 1962–; first works publ. 1937. Decors.: Order of Lenin, Order of Red Banner of Labour, Stalin Prize, Lenin Prize, Hero of Socialist Labour 1974, and various medals.
Publs.: "Plamennaia liubov' i zhguchaia nenavist" (Hot Love and Burning Hate) 1943, "Zemlia moia" (My Country) 1947, "Nashi gory" (Our Mountains) 1947, "Rodina gortsa" (The Homeland of a Miner) 1950, "Slovo o starshem brate" (Tales of My Elder Brother) 1952, "Dagestanskaia vesna" (Spring in Dagestan) 1955, "God moego rozhdeniia" (The Year of My Birth) 1950, "V gorakh moe serdtse" (My Heart is in the Mountains) 1959, "Vysokie zvezdy" (Stars on High) 1962, "Zarema" 1963, "I zvezda s zvezdoi govorit" (A Star Also Speaks With Another Star) 1964, "Mulatka" (Mulatto Girl) 1966, "Moi Dagestan" (My Dagestan) 1968, "Sobranie sochinenii" (Collected Works), vols. 1–3, 1968–69, and others.

Ganiev, Akhmed Azizovich
Sculptor; b. 1927, Denau, Uzbekistan.
Attended Art Inst. of the Esthonian SSR 1952–58, pupil of A. A. Kaazik, aspirant at Repin Inst. for Painting, Sculpture and Architecture at the USSR Acad. of Arts 1958–61, pupil of V. V. Lishev; exhibitions 1955–.

Ganina, Maiia Anatolevna
Author; b. 1927, Moscow.
Russian; grad. Gorkii Inst. of Literature Moscow 1954; first works publ. 1947.
Publs.: "Pervye ispytaniia" (First Exams) 1955, "Razgovor o schastie" (A Conversation About Happiness) 1959, "Nastiny deti" (Nastia's Children) 1957, "Matvei i Shurochka" 1962, "Ia ishchu tebia, chelovek" (I'm Seeking You, Man) 1963, and others.

Gaponenko, Taras Gur'evich
Painter; b. F. 20 (Mr. 5), 1906, Staraia Zavoron', in present-day Smolensk Oblast. People's Artist of the RSFSR (1968). Corresponding member of the Academy of Arts of the USSR (1954).
He studied in Moscow at the State Higher Arts and Technical Studio and the State Higher Institute of Art and Technology (1924–30). He was a member of the Youth League of the Association of Artists of Revolutionary Russia (1928–29), the Association of Artists of Revolutionary Russia (1929–31), and the Russian Association of Proletarian Artists (1931–32). His works are devoted chiefly to life in the kolkhoz village. They include *The Kolkhoz Women Going Out to Work* (1933, Tret'iakov Gallery) and *In the Little Poultry Town* (1959–60, Russian Museum, Leningrad). The painting *Sucklings on the Way to Their Mothers* (1935, Tret'iakov Gallery) played an important role in the development of Soviet genre painting. He also painted the dramatic picture *After Driving Out the Fascist Occupiers* (1943–46, Tret'iakov Gallery; State Prize of the USSR, 1947).

Gaponov-Grekov, Andrei Viktorovich
Physicist; b. Je. 7, 1926, Moscow.
Grad. Gorkii Univ. 1949; mem. USSR Acad. of Sciences 1968–; mem. staff Gorkii Polytech. Inst. 1952–55; mem. staff Scientific Research Inst. of Radiophysics at Gorkii Univ. 1955–. Decor.: State Prize of the USSR 1967.

Gapurov, Mukhamednazar
Party official; b. 1922.
Turkmenian; mem. CPSU 1944–; grad. Chardzhou Teachers' Training Inst. 1954; secondary-school teacher 1941–42; served in Sov. Army 1942–43; dept. head, then school principal 1943–44; instructor, dept. head, sec. of a raion cttee. Turkmen CP, 1st sec. of an oblast Turkmen Komsomol cttee., Sec. of Cen. Cttee. Turkmen Komsomol, Dept. Head, Sec., 1st Sec. of Chardzhou Oblast Cttee. Turkmen CP 1944–62; mem. Cen. Cttee. Turkmen CP 1958–, Dep. to and mem. Presidium Supr. Sov. Turkmen SSR 1959–63 and 1969–; Sec. Ja. 1962–Mr. 1963, mem. Bureau Cen. Cttee. Turkmen CP Ja. 1962–; Dep. to USSR Supr. Sov. 1962–; mem. Planning and Budget Comm. with Sov. of the Union USSR Supr. Sov. 1974–; Chair. of Turkmen Council of Mins. and Turkmen Min. of Foreign Affairs 1963–69; cand. mem. 1966–71, mem. Cen. Cttee. CPSU 1971–; 1st Sec. of Cen. Cttee. Turkmen CP D. 1969–. Decors.: 2 Orders of Lenin.

Garanko, Elfa Valerianovna
Sculptor; b. 1930, Novorossiisk.
Attended Repin Inst. for Painting, Sculpture and Architecture at the USSR Acad. of Arts 1952–58; exhibitions 1959–.

Garbuzov, Vasilii Fedorovich
State official; b. 1911.
Russian; mem. CPSU 1939–; grad. Kharkov Inst. of Financial Econ. 1933; Assoc. Prof.; former carpenter at a sawmill 1925–30; student and postgraduate student 1930–36; Lecturer Kharkov Inst. of Financial Econ. 1936–42; Chief Inspector, Dept. Head of a Board with Kirghiz People's Commissariat of Finance, 1942–43; Dep. Dir. of Pedagogical and Research Work, Dir. of Kiev Inst. of Financial Econ. 1943–50; Chair. of Ukrainian GOSPLAN 1950–52; USSR Dep. Min., 1st Dep. Min. 1952–60, Min. of Finance My. 1960–; Dep. to USSR Supr. Sov. 1958–; mem. Cen. Cttee. CPSU 1961–. Decors.: 4 Orders of Lenin; Hero of Socialist Labour 1981, medals.

Gardabkhadze, Karl Konstantinovich
State official.
1st Sec. of Ordzhonikidze Raion Cttee. Georgian CP in Tbilisi 1959; mem. Cen. Cttee. 1960–66 and 1976–81, mem. Auditing Comm. Georgian CP 1971–76 and 1981–; Chair. of Georgian SSR State Cttee. for Television and Radio 1962–76; Dep. Chair. of Georgian Journalists' Union 1971–; Min. of Public Welfare Georgian SSR F. 1976–.

Gareev, Iusuf Ibragimovich
Author; b. 1904; s. of a peasant.
Bashkir; first works publ. 1930.
Publs.: "Timer" (poem) 1932, children's books "Po shchuchemu veleniiu" (At a Pike's Command) 1937, "Kak shchenok nauchilsia laiat'" (How a Puppy Learned to Bark) 1937, "Delaem tsvety" (We're Making Flowers) 1939, "Skazki" (Fairytales) 1959, and others.

Gareev, Musa Gaisinovich
Military official; b. Je. 9, 1922, Iliakshidy, Bashkir ASSR.
Twice Hero of the Soviet Union (F. 23, 1945, and Ap. 19, 1945); colonel. Member of the CPSU since 1944. Born into the family of a Bashkir peasant.
He has been in the Soviet Army since 1940 and graduated from a military aviation school in 1942. During the Great Patriotic War, as a pilot and the commander of a flight and squadron of the 76th Guards Attack Air Regiment, he participated in battles near Stalingrad and in the Donbas, the Crimea, Belorussia, Poland, and Germany, and made about 250 successful combat flights. After the war he has held high command positions and graduated from the Frunze Military Academy in 1951 and from the Military Academy of the General Staff in 1959. He was a deputy to the second, third, and fourth convocations of the Supreme Soviet of the USSR. Decors.: Order of Lenin, three Orders of the Red Banner, the Order of Alexander Nevsky, the Order of Bogdan Khmel'nitskii Third Class, the Order of the Patriotic War First Class, three Orders of the Red Star, and various medals. He has been in retirement since 1964.

Garibzhanian, Ludvig Papikovich, Cand. of Hist. Sc.
State official; b. 1922.
Armenian; mem. CPSU 1943–; grad. Erevan State Univ.; sec. of Komsomol cttee. of a medical inst. 1944–47; 1st Sec. of Kirov Raion Cttee., 1st Sec. of Erevan City Cttee. Armenian Komsomol 1947–51; Sec. of CPSU Cttee. Erevan State Univ., Sec. of Erevan City Cttee. Armenian CP, Instructor, Exec. Organizer of Cen. Cttee. Armenian CP, Sec. of a Raion Cttee. Armenian CP 1951–61; Dep. to Supr. Sov. Armenian SSR 1955–; cand. mem. 1956–60, mem. Cen. Cttee. Armenian CP 1960–; 2nd Sec. 1961–66, 1st Sec. of Erevan City Cttee. Armenian

CP 1966–75; mem. Bureau Cen. Cttee. Armenian CP 1966–76; Min. of Higher and Secondary Specialist Educ. Armenian SSR My. 1975–.

Garin, Erast Pavlovich
Actor and stage manager; b. O. 28 (N. 10), 1902, Riazan.
Grad. advanced experimental courses in theatre management 1956; mem. V. Meierhold Theatre Moscow 1922–; actor and stage manager at Comedy Theatre Leningrad 1936–; mem. Drama Studio for Film Actors 1950–. Decors.: People's Artist of the RSFSR 1964, State Prize of the USSR 1941, Badge of Honour and medals.

Garkusha, Nikolai Andreevich
State official; b. 1914.
Sec. of Kiev Oblast Cttee. Ukrainian CP 1955–62; Chair. of State Prod. Cttee. for Agricultural Land Irrigation and Water Econ. Ukrainian SSR 1963–65; Min. of Melioration and Water Econ. Ukrainian SSR O. 1965–; mem. Cen. Cttee. Ukrainian CP 1966–; mem. Council of Kolkhozes Ukrainian SSR 1970–.

Garokalns, Visvaldis Ernestovich
Painter; b. 1933 nr. Kuldiga, Latvia.
Attended Liepaia School for Applied Art 1955–60; exhibitions 1960–; teacher at Liepaia School for Applied Art 1961–.

Garusova, Vladilena Lvovna
Sculptor; b. 1925, Rostov/Don.
Attended Surikov Art Inst. Moscow 1946–52, pupil of R. R. Iodko, D. P. Shvarts; exhibitions 1955–; teacher at a children's art studio 1963–.

Gasanov, Gabib Ragim ogly
State official.
Chair. of State Cttee. on Forestry Azerbaidzhan SSR 1980–81; Min. of Forestry Azerbaidzhan SSR Ja. 1981–.

Gasanov, Gasan Azis ogly
Party official; b. 1940.
Azerbaidzhani; mem. CPSU 1963–; grad. Azerbaidzhan Polytechnic Inst.; Komsomol and Party posts 1962–75; 1st Sec. of a Raion Cttee., 1st Sec. of Sumgait City Cttee. Azerbaidzhan CP 1975–79; cand. mem. Cen. Cttee. Azerbaidzhan CP 1976–80, mem. 1980–; 1st Sec. of Kirovabad City Cttee. Azerbaidzhan CP 1979–81; Sec. and mem. Bureau Cen. Cttee. Azerbaidzhan CP Ja. 1981–.

Gasanova, Zuleikha Magerram kyzy
State official.
Mem. Auditing Comm. Azerbaidzhan CP 1966–76, mem. Cen. Cttee. 1976–81; Dep. to Supr. Sov. Azerbaidzhan SSR 1967–; Chair. of Mingechaur City Exec. Cttee. of Workers' Deps. 1967; Min. of Public Services Azerbaidzhan SSR Jl. 1974–.

Gashinskii, Arkadii Evgenevich
Actor; b. Jl. 29, 1920, Melitopol.
Ukrainian; grad. A. V. Lunacharskii Theatre Inst. Moscow ("GITIS") 1945; mem. of a theatre brigade on 1st Ukrainian Front 1941–45; actor with Ukrainian Dramatic Theatre Kiev 1945–48; mem. I. Franko Ukrainian Drama Theatre Kiev 1948–; teaching activity at same theatre. Decor.: People's Artist of the USSR 1971.

Gashokov, Khusin Khanakhovich
Poet; b. 1913.
Cherkess; mem. CPSU 1939–; first works publ. 1934.
Publs.: "Stikhi i pesni" (Poems and Songs) 1940, "Stikhi i poemy" (Poems and Poetry) 1953, "Golos serdtsa" (The Voice of the Heart) 1955, "Vesna" (Spring) 1958, poem "Otets i syn" (Father and Son), vol. 1, 1959, "Svet v gorakh" (Light in the Mountains) 1952, "Solntse nad aulami" (Sun Over the Village) 1957, "Radostiu polnitsia serdtse" (The Heart Grows Large With Joy) 1960, and others.

Gasparian, Goar Mikaelovna (maiden name: Khachatrian)
Armenian soprano; b. D. 14, 1924, Cairo. People's Artist of the USSR (1956).
She received her musical education in Cairo under the guidance of E. Feldman and V. Carrot. In 1948 she moved to the USSR and in 1949 became a soloist with the Armenian Theater of Opera and Ballet in Yerevan. Her roles include Anush and Shushan in Tigrian's operas *Anush* and *David-bek,* Goar in Stepanian's *The Heroine* (State Prize of the USSR, 1951), Olimpiia in Chukhadzhian's *Arshak II,* Lucia in Donizetti's *Lucia di Lammermoor,* Lakmé in Delibes' *Lakmé,* Violetta and Desdemona in Verdi's *La Traviata* and *Otello,* and Rosina in Rossini's *The Barber of Seville.* Her performances are appealing because of her natural elegance and thoughtful and subtle interpretation of diverse characters. She also performs as a concert singer and has toured Poland, Hungary, England, France, Japan, the USA, Canada, Brazil, Mexico, and other countries. A deputy to the seventh and eighth convocations of the Supreme Soviet of the USSR. Decor.: State Prize of the Armenian SSR in 1964.

Gatin, Sergei Zakharovich
Painter; b. 1923, Rybinsk, Iaroslavl Oblast.
Attended Ivanovo Art School 1939–45, pupil of N. G. Burov, A. M. Kuznetsov; exhibitions 1940–.

Gatov, Aleksandr Borisovich
Russian poet, translator, and literary scholar; b. N. 14 (26), 1899, Kharkov.
He is the author of several collections of lyric verse and of the book *Eugène Pottier: Life and Works* (1933). He is a prolific translator; his translations include the collections *Revolutionary Poetry of the West in the Nineteenth Century* (1930), *Poets of Armenia* (1934), *Poets of the Paris Barricades* (2nd ed., 1935), *Eugène Pottier: Songs, Verse, and Narrative Poems* (1966), and *Bulgarian Poets* (1969). Decors.: Bulgarian Order of Cyril and Methodius and various medals.

Gatovskii, Lev Markovich
Economist; b. Jl. 13 (26), 1903, Minsk.
Mem. CPSU 1927–; grad. Plekhanov Inst. of Econ. Moscow 1924; corresponding mem. USSR Acad. of Sciences 1960–; mem. staff Econ. Inst. of USSR Acad. of Sciences 1939–; Dir. of same Inst. 1965–.

Gatserelia, Akakii Konstantinovich
Georgian writer, literary critic; b. Ap. 4 (17), 1909, Poti.
Honored Scholar of the Georgian SSR. He graduated from the philology department of the University of Tbilisi in 1932. He began to publish his works in 1927. He wrote such short stories as "Chir-Iurt" (1942), "The Death of Baratashvili" (1944), "Mouravi" (1945), and "The Artist's Story" (1958). He is the author of a work on the theory and history of Georgian prosody, *Georgian Classical Poetics During the Seventh to 18th Centuries* (1953; 3rd ed., 1963), *Sketches on Georgian Poetics* (1938), and articles on Georgian and Russian writers, including F. M. Dostoevsky and Iu. N. Tynianov. He edited an academic edition of the works of G. Orbeliani (1959).
Works: Gatserelia, A. *Rtcheuli natserebi,* [vols.] 1–2. Tbilisi 1962–65. *In Russian translation: Nikoloz Baratashvili: Kratkii ocherk zhizni i tvorchestva.* Tbilisi 1968.

Gaumigs, Talivald Iakovlevich
Sculptor; b. 1930, Valka, Latvia.
Grad. Riga School for Applied Art 1951, attended Art Acad. of the Latvian SSR 1951–57; exhibitions 1957–; teacher at Acad. of Arts of the Latvian SSR 1961–.

Gavdzinskii, Albin Stanislavovich
Painter; b. 1923, Bobrik, Odessa Oblast.
Grad. Odessa Art School 1950, pupil of N. A. Pavliuk; exhibitions 1951–.

Gavrilenko, Grigorii Ivanovich
Graphic artist; b. 1927, Kholopkovo, Ukraine.
Attended Kiev Middle School of Art 1945–49 and Kiev Art Inst. 1949–55, pupil of G. K. Titov; exhibitions 1955–.

Gavrilenko, Nikolai Mefodievich
State official; b. 1937.
Cand. of Geological and Mineralogical Sciences; mem. CPSU 1966–; higher education, mining engineer; drilling foreman, chief engineer and head of a geological prospecting expedition, lecturer at Inst. of Mining in Dnepropetrovsk, chief engineer and head of a trust, Gen. Dir. of "Voroshilovgrad" Geological Production Assoc. 1959–80; 1st Dep. Min. 1980–82, Min. Geology Ukrainian SSR Jl. 1982–.

Gavriliuk, Nikolai (Mykola) Stepanovich
Painter; b. 1927, Mikhnov, Ukraine.
(Attended Simferopol Art School 1946–51, pupil of M.P. Kroshitskii, M.M. Shcheglov; exhibitions 1954–.

Gavrillov, Aleksandr Fadeevich
State official.
Dep. to Supr. Sov. Uzbek SSR 1963–; 2nd Sec. of Fergana Oblast Uzbek CP 1963–64 and 1965–73 (for Ind. 1963–64); 1st Dep. Chair. of Fergana Oblast Exec. Cttee. of Workers' Deps. 1964–65; mem. Cen. Cttee. Uzbek CP 1966–; Min. of Motor Transport Uzbek SSR S. 1973–.

Gavrilov, Evgenii Georgievich
Graphic artist; b. 1914, Kharkov.
Attended Kharkov Technical School of Art 1929–32, pupil of M.M. Shcheglov; exhibitions 1948–.

Gavrilov, Mikhail Aleksandrovich
Scientist; b. N. 11 (24), 1903, Moscow.
Grad. Higher Tech. College Moscow 1925; specialist in the theory of automatic control systems; corresponding mem. USSR Acad. of Sciences 1964–; mem. staff Inst. for Automatic Systems and Telemechanics 1937–; Chair. of Scientific Council for Tech. Cybernetics at USSR Acad. of Sciences 1965–; main works in telemechanics. Decors.: P.N. Jablochkov Prize USSR Acad. of Sciences 1958, Order of Red Banner of Labour, several medals.
Publs.: "Telemechanizacija disperčeskogo upravlenija ėnergetičeskimi sistemami" Moscow-Leningrad 1938, "Teorija relejnokontaktnyh schem" Moscow-Leningrad 1950, "Strukturnaja teorija relejnyh ustrojstv" 1–4, Moscow 1960–64, and others.

Gavrilova, Tatiana Andreevna
State official.
Chair. of "Zdobutok Zhovtnia" Kolkhoz in Cherkassy Oblast; Dep. to and Dep. Chair. of Supr. Sov. Ukrainian SSR 1975–.

Gavrusev, Stepan Zakharovich
Poet; b. 1931, Novo-Aleksandrovka, Mogilev Oblast; s. of a peasant.
Belorussian; first works publ. 1948.
Publs.: "Pokhodnye kostry" (Camp Fire) 1955, "Na grebniakh voln" (On the Crests of Waves) 1959, "Shchedrost'" (Generosity) 1962, and others.

Gazenko, Oleg Georgievich
Physiologist; b. D. 12, 1918, Nikolaevka, Stavropol' Krai.
Corresponding member of the Academy of Sciences of the USSR (1966). Member of the CPSU since 1953.
He graduated from the Second Moscow Medical Institute (1941). Since 1969 he has been director of the Institute of Medical-Biological Problems at the Ministry of Public Health of the USSR. Since 1956 he has taken part in the organization of work on space biology and medicine that helped make manned space flights and their equipping possible. His principal works deal with space physiology, particularly in the influence of weightlessness on the organism. Decors.: Order of Lenin and three other orders, as well as various medals.
Works: "Nekotorye problemy kosmicheskoi biologii." *Vestnik AN SSSR 1962, no. 1. "Kosmicheskaia biologiia." In Razvitie biologii v SSSR.* Moscow 1967. "Fiziologicheskie effekty gravitatsii." In the cellection *Problemy kosmicheskoi biologii,* vol. 6. Moscow 1967. (With A.A. Giurdzhian.)

Gazhenko-Zlotnik, Anna Trofimovna
Sculptor; b. 1927, Odessa.
Attended Odessa Art School 1946–50 and Leningrad High School for Industrial Art 1951–57, pupil of L.M. Pisarevskii; exhibitions 1958–; teacher at Minsk Art School 1957–.

Gazizova, Munavara Gabdulbarovna
Painter; b. 1918, Orenburg Oblast.
Grad. Moscow Inst. of Fine Arts 1941, pupil of S.V. Gerasimov, K.N. Istomin; exhibitions 1949–.

Gedeon (Aleksandr Nikolaevich Dokukin), Cand. of Theol.
Russian Orthodox clergyman; b. D. 18, 1929, Novo-Pokrovskaia, Krasnodar Krai.
Grad. Stavropol Theological Inst. 1952; psalm-reader and choirmaster at Uspenskii Church of Makhach-Kala Dagestan ASSR; deacon and priest 1953; correspondence course at Leningrad Theological Acad., Cand. of Theol. 1955–60; archpriest; Chief Clergyman of Panteleimon Church in Kislovodsk, Stavropol Diocese 1961; Chief Clergyman of Polrovskii Church of Mineralnye Vody, Stavropol Diocese 1962; Clergyman of Catherine Church and then Cathedral of the Elevation of the Cross in Petrozavdsk 1965; Chief Clergyman and Provost of Olonets Diocese 1966; Monk, Archimandrite at Nikolo Bogoiavlenski Cathedral in Leningrad 1966; Bishop of Novosibirsk and Barnaul 1972–.

Gedmantaite-Galkuvene, Vanaga Antano
Graphic artist; b. 1935, Kaunas.
Attended Art Inst. of the Lithuanian SSR 1954–60, pupil of I.M. Kuzminskis, A.A. Kuchas; exhibitions 1960–.

Geld, Pavel Vladimirovich, Dr. of Tech. Sc.
Physical chemist; b. 1911.
Mem. CPSU 1944–; grad. Ural Polytech. Inst. 1938; Dr. of Tech. Sc. 1950; Prof. 1952–; corresponding mem. USSR Acad. of Sciences 1970–; served in Sov. Army 1941–42; mem. staff Ural Polytech. Inst. 1943–, Prof. Chair 1952–; in addition Sen. Research Fellow of Inst. of Chemistry 1954–, also of Urals Scientific Centre at USSR Acad. of Sciences 1971–.

Gelfand, Izrail Moiseevich
Mathematician; b. Ag. 20 (S. 2), 1913, Okna, now Odessa Oblast.
Corresponding mem. USSR Acad. of Sciences 1953–; Prof. at Moscow Univ. 1943–; Pres. of Moscow Mathematical Soc. 1966–70; hon. mem. Boston Acad. of Arts and Sciences London Mathematical Soc., Irish Acad. of Sciences, Nat. Acad. of USA. Decor.: State Prize of the USSR 1951.

Genere, Mara Germanovna
Artist in applied art; b. 1943, Tsesis, Latvia.
Attended Acad. of Art of the Latvian SSR 1963–69, pupil of S. Ia. Ozolin, artist in applied art (ceramics, porcelain); exhibitions 1963–.

Georgiev, Georgii Pavlovich, Dr. of Biol. Sc.
Biologist; b. 1933.
Grad. 1st Medical Inst. Moscow 1956; Dr. of Biol. Sc. 1963; Prof. 1968–; corresponding mem. USSR Acad. of Sciences 1970–; Research Asst. at Inst. of Animal Morphology at USSR Acad. of Sciences 1956–63; Laboratory Chief at Inst. of Molecular Biology 1963–.

Georgievskaia, Anastasia Pavlovna
Actress; b. O. 25 (N. 7), 1914, Orel.
Mem. CPSU 1940–; grad. A.V. Lunacharskii State Inst. for Theatre Arts Moscow ("GITIS") 1935; mem. Moscow Acad. M. Gorkii Artists' Theatre of the Sov. Union ("MHAT") 1936–. Decors.: State Prize of the USSR 1951, People's Artist of the USSR 1968, Order of Red Banner of Labour, medals.

Georgievskaia, Susanna Mikhailovna
Russian writer; b. My. 10 (23), 1916, Odessa.
Graduated from Leningrad University in 1935.
She published her first works in 1939. She took part in the Great Patriotic War. She has written children's tales and stories, including *Galia's Mother* (1947), *Grandma's Sea* (1949), *Little Ivanov* (1950), *Youth* (1953), and *Tarasik* (1959). Such stories as "The Golden Word" (1955), drawn from the life of contemporary Tuva, "Young People" (1961), and "Twice Two Makes Four" (1965) are addressed to the adult reader. Decors.: Order of the Patriotic War Second Class and various medals.
Works: *Povesti i rasskazy.* Moscow 1954. *Tri povesti.* Moscow 1957. *Svetlye goroda.* Moscow 1965. *Portnoi osobogo plat'ia.* Moscow 1966. *Povesti.* Moscow 1967. *Rasskazy i skazki.* Moscow 1968.

Gerasimov, Evgenii Nikolaevich
Author; b. 1903, Grodno.
Russian; mem. CPSU 1944–; grad. Leningrad Univ. 1926; first works publ. 1931. Publs.: "General Dovator" 1943, "Tovarishchi iz odnoi shkoly" (Friends From a

School) 1951, "Zapiski Ani Churilinoi" (The Notes of Ania Churilina) 1954, "Igor Petukhov. Povest iz zhizni pochti vzroslykh ludei" (Igor Petukhov. From the Lives of Almost Grown-up People) 1961, "Shelkovyi gorod" (The Silk City) 1962, and others.

Gerasimov, Iakov Ivanovich
Chemist; b. S. 10 (23), 1903, Valdai.
Mem. CPSU 1952–; grad. Moscow Univ. 1925; physical chemist; corresponding mem. USSR Acad. of Sciences 1953–; main works in thermodynamics, nonferrous and rare metals and metal alloys; Chief Ed. of "Žurnal fizičeskoj himii"; Chair. of Scientific Council for Chemical Thermodynamics and Thermochemistry of USSR Acad. of Sciences. Decors.: Order of Lenin, several medals.

Gerasimov, Innokentii Petrovich
Scientist; b. D. 9 (22), 1905, Kostroma.
Mem. CPSU 1945–; grad. Leningrad Univ. 1926; geographer, geomorphologist and soil scientist; mem. USSR Acad. of Sciences 1953–; Chief of Dept. of Geography and Soil Cartography at V. V. Dokuchaev Inst. for Soil Science 1936–56; mem. staff Geographical Inst. at USSR Acad. of Sciences 1945–, Dir. of same Inst. 1951–; organized and took part in various expeditions in the USSR and foreign countries; Chief Ed. of "Physical-Geographical Atlas of the World" 1964; Chair. of Nat. Cttee. of Sov. Geographers 1957–; Pres. of Union Society of Soil Scientists 1963–; Vice-Pres. of Int. Geographical Federation 1960–68; Vice-Pres. of Int. Soc. of Soil Scientists 1968–; mem. numerous foreign acads. Decors.: Order of Lenin, 3 other orders, several medals. Publs.: "Osnovnye čerty razvitija sovremennoj poverchnosti Turana" Moscow-Leningrad 1937, "Lednikovyj period na territorii SSSR" Moscow-Leningrad 1939 (co-author), "Strukturnye čerty rel'efa zemnoj poverchnosti na territorii SSSR i ich proizchoždenie," Moscow 1959, "Očerki po fizičeskoj geografii zarubežnyh stran" Moscow 1959, "Osnovy počvovedenija i geografija počv" Moscow 1960 (co-author), "Počvy Central'noj Evropy i svjazannye s nimi voprosy fizičeskoj geografii" Moscow 1960, "Preobrazovanie prirody i razvitie geografičeskoj nauki v SSSR" Moscow 1967, and others.

Gerasimov, Ivan Aleksandrovich
Military official; b. 1921.
Russian; mem. CPSU 1942–; grad. Mil. Acad. of Tank Troops 1955 and Mil. Acad. of Gen. Staff 1966, Army Gen. 1977; entered Sov. Army 1938; at the front 1941–45; chief of a regiment staff 1945–50; regiment commander, dep. commander, commander of a division 1955–64; command posts, Commander of 1st Tank Army Guard in the GDR; 1st Dep. Commander of Carpathian Mil. District, Commander of Group of Troops North (Poland) 1964–75; Dep. to USSR Supr. Sov. 1970–; Commander of Troops of Kiev Mil. District 1975–; cand. mem. Cen. Cttee. CPSU 1976–; mem. Cen. Cttee. Ukrainian CP 1976–; cand. mem. Politbureau Cen. Cttee. Ukrainian CP 1977–80, mem. Ap. 1980–.

Gerasimov, Pavel Ivanovich
Diplomat; b. 1915.
Entered Diplomatic Service 1945–; Dep. Head of Near and Middle East Dept. at USSR Min. of Foreign Affairs 1953–54; Counsellor at USSR Emb. in Egypt 1954–59; USSR Amb. to Guinea 1959–60; mem. staff USSR Min. of Foreign Affairs 1960–61; USSR Amb. to Luxembourg 1961–62, to Belgium 1962–67; mem. staff USSR Min. of Foreign Affairs 1967–70; mem. Collegium, Head of Cadre Admin. at USSR Min. of Foreign Affairs 1970–73; USSR Amb. to Switzerland 1973–O. 1977.

Gerasimov, Sergei Apollinarievich
Film director, actor, screenwriter, and theorist; b. My. 21 (Je. 3), 1906, Ekaterinburg, now Sverdlovsk.
People's Artist of the USSR (1948). Doctor of the arts (1967). Member of the CPSU since 1943.
He graduated from the Leningrad Institute of Stage Arts in 1928. He began his work in films in 1924 as an actor in the studio of FEKS (Slapstick Actors' Workshop). He first clearly showed his directorial skill in the films *The Courageous Seven* (1936), *Komsomol'sk* (1938), and *The Teacher* (1939), in which he concentrated on an artistic examination of the spiritual world of Soviet youth raised in socialist society. He depicted the genuine heroism, the exalted beauty, and the poetry of the day-to-day creative and constructive labor of the Soviet people. Starting from the experience of prose and making use of its methods, he has been especially attracted to the adaptation of the novel to film. Even in his first films, he emerged as a consistent director-teacher, devoting great attention to his work with actors.
In 1941, on the occasion of the 100th anniversary of the death of M. Iu. Lermontov, he played the role of the Unknown in his production of the film *Masquerade* (based on the play of the same name). Arguing in this film against the traditional melodramatic interpretation of the play and emphasizing rather its social significance, the director showed with great emotional force the tragedy awaiting an individual of uncommon personality in the autocratic Russia of the 19th century. During the Great Patriotic War he was the Committee of Cinematography's deputy chairman for the filming of military newsreels, and director of the Central Studio of Documentary Films. He produced the films *The Invincible* in 1943 (in collaboration with M. K. Kalatozov) and *The Home Front* in 1944. In the film *The Young Guard* (1948, based on the novel by A. A. Fadeev), the director combined epic scope with a deeply penetrating view of individual psychology. Its heroic ardor and brilliant portraits of the Young Guards, reproducing the patriotic atmosphere of the struggle against fascism, have made the film one of the outstanding works of cinematic art. The film *The Quiet Don* (1957–58, based on the novel by M. A. Sholokhov) proved a major event in the development of Soviet film. The portraits of the heroes of this distinguished Soviet novel were revealed by the director with all the complexity and richness of their social and psychological nature. In 1962, he produced the film *Man and Beast,* in which he also played the role of L'vov-Shcherbatskii. He next produced *The Journalist* (1967) and *By The Lake* (1970). These three films are united by their attempt to reflect contemporary life and to deal with the problems that have arisen in the postwar world. The films show the growth of ever higher moral standards in their heroes. He has written the scripts for almost all of his films.
He has taught in the All-Union State Institute of Cinematography (VGIK) since 1931 (as a professor since 1946), where he directs a joint actor-director workshop. He was a deputy of the Supreme Soviet of the USSR at its third and fourth convocations and a deputy at the seventh and eighth convocations of the Supreme Soviet of the RSFSR. He is a member of the presidium of the Soviet Committee for the Defense of Peace. Decors.: State Prize of the USSR in 1941, 1949, and 1951, two Orders of Lenin, three other orders, and various medals.
Works: *Komsomol'sk.* In *Kniga stsenariev.* Moscow 1938. *Uchitel'.* In *Uchitel'.* Leningrad 1940. *Litso sovetskogo kinoaktera.* Moscow 1935. *Zhizn', fil'my, spory.* Moscow 1971.

Geraskina, Liia Borisvna
Dramatist; b. 1910, Novorossiisk.
Russian; mem. CPSU 1944–; first works publ. 1938.
Publs.: plays "Khrustald i Katrinka" 1947, "Attestat zrelosti" (School-leaving Certificate) 1949, "Ee budushchee" (Her Future) 1954, "Sosedi po kvartire" (Apartment Neighbours) 1956, "Nikolai Ivanovich" 1958, and others.

German (Gennadii Evgenevich Timofeev), Cand. of Theol.
Russian Orthodox clergyman; b. N. 11, 1937, Tashkent.
Educ. Theological Seminary; monk, deacon 1965; Cand. of Theol. at Leningrad Theological Acad., priest-monk 1966; postgraduate Byzantine studies at Leningrad Theological Acad., teacher of liturgics and Lecturer in dogmatics 1966–67; abbot; Sen. Asst. to the Supervisor, then Supervisor of Leningrad Theological Acad. and Seminary, Archimandrite 1967; lecturer in ecclesiastical law

until 1968; Provisional Rector of Theological Acad. and Seminary 1968; travelled abroad; Rector of Leningrad Theological Acad. and Seminary, Bishop of Tikhvin 1968; Bishop of Vienna and Austria 1970–74; Bishop of Vilnius and Lithuania 1974–78; Bishop of Tula and Belev Ap. 1978–.

Gernet, Nina Vladimirovna
Author; b. 1904, Odessa.
Russian; first works publ. 1926.
Publs.: "Trinadtsatyi v mire" (The Thirteenth in the World) 1931, "Tri palatki" (Three Tents) 1933, "Knizhka bez kartinok" (A Book Without Pictures) 1942, "Lida i bukvy" (Lida and the Letters) 1946, plays "Lesnye artisty" (Forest Artists) 1951, "Belye krylia" (White Wings) 1952, "Sestrenka" (Little Sister) 1960, "Volshebnaia lampa Aladina" (Aladin's Magic Lamp) 1955, "Katiia i krokodil" (Katia and the Crocodile) 1959, and others.

Gershuni, Grigorii Viktorovich
Medical scientist; b. Jl. 21 (Ag. 3), 1905, Minsk.
Grad. 1st Medical Inst. Leningrad 1927; human and veterinarian physiologist; corresponding mem. USSR Acad. of Sciences 1965–; Prof. Chair of Physiology at 1st Medical Inst. Leningrad 1925–; mem. staff Medical Mil. Kirov Acad. 1931–36; mem. staff Pavlov Inst. of Physiology at USSR Acad. of Sciences 1936–; Chief of Laboratory for Physiology of Hearing 1951–; important scientific discoveries, had major part in developing diagnostic methods in otology. Decors.: I. P. Pavlov Prize of the USSR Acad. of Sciences 1949, Order of Red Banner of Labour.

Getman, Andrei Lavrent'evich
Military official; b. S. 22 (O. 5), 1903, Klepaly, Sumy Oblast.
Mem. CPSU 1927–; grad. Mil. Acad. of Armoured Troops 1937; Gen. of the Army; railway and sugar refinery worker 1920–24; served in Sov. Army 1924–; during the war commanded an armoured div., then armoured corps, dep. commander of an armoured force; after the war command posts in armoured and mechanized forces; commanded Transcarpathian Mil. District 1958–64; Dep. to USSR Supr. Sov. 1958–74; mem. Cen. Cttee. Ukrainian CP 1960–66; cand. mem. Cen. Cttee. CPSU 1961–76; Chair. of Cen. Cttee. All-Union Soc. for the Promotion of the Sov. Army, Air Force and Navy (DOSAAF) 1964–71; Mil. Adviser and mem. Inspectors-Gen. Group in USSR Min. of Defence 1971–. Decors.: Hero of the Sov. Union 1965, 5 Orders of Lenin, Order of Red Banner of Labour, Order of the October Revolution, and other decorations.

Gevorkian, Artavazd Ashotovich
State official.
Chair. of Supr. Court Armenian SSR 1967–70; Dep. to Supr. Sov. Armenian SSR 1967–; Min. of Justice Armenian SSR N. 1970–.

Gevorkian, Artem Ashotovich
State official.
Mem. Cen. Cttee. 1960–64 and 1971–, cand. mem. Cen. Cttee. Armenian CP 1964–71; Dep. to Supr. Sov. Armenian SSR 1963–; 1st Dep. Chair. of GOSPLAN Armenian SSR 1963; mem. Bureau for Ind. at Cen. Cttee. Armenian CP 1964; Min. of Light Ind. Armenian SSR. Mr. 1970–.

Giatsintova, Sof'ia Vladimirovna
Actress and director; b. Jl. 23 (Ag. 4), 1895, Moscow.
People's Artist of the USSR (1955). Member of the CPSU since 1951.
She was a member of the Moscow Art Theater from 1910 to 1924 and participated in the work of its first studio. From 1924 to 1936 she was an actress in the second Moscow Art Academic Theater, and from 1936 to 1938 she worked in the Moscow Oblast Council of Trade Unions Theater. From 1938 she was an actress and director in the Lenin Komsomol Moscow Theater. The subtle, graceful, and psychologically profound art of her, who was a pupil and follower of K. S. Stanislavsky, is distinguished by its broad range and lofty dramatic quality. Among her best roles were Maria in Shakespeare's *Twelfth Night,* Nelly in *The Insulted and Injured,* based on Dostoevsky, Nora in Ibsen's *A Doll's House,* Leonarda in Lope de Vega's *The Widow From Valencia,* Maria Aleksandrovna Ul'ianova in Popov's *The Family,* and Aunt Tasia in Arbuzov's *Years of Wandering.* She directed many plays, including *A Doll's House* (1939, in cooperation with I. N. Bersenev), Turgenev's *A Month in the Country* (1944), *The Family* (1949), and *The Cherry Orchard* (1954, in cooperation with A. A. Pelevin). Her film roles included Maria Aleksandrovna Ul'ianova in *The Ul'ianov Family* (1957) and Pani Maria in *Lost Without a Trace* (1957). Decors.: State Prize of the USSR in 1947, the Order of Lenin, the Order of the Red Banner of Labor, and various medals.

Gikhman, Iosif Ilich, Dr. of Phy.-Math. Sc.
Mathematician; b. My. 26, 1918, Uman; s. of a worker.
Mem. CPSU 1945; grad. Kiev Univ. 1939; corresponding mem. Acad. of Sciences Ukrainian SSR 1965–; Dr. of Phy.-Math. Sc. 1956; Prof. 1959–; postgraduate studies at Kiev Univ. until 1941; served in Sov. Army 1941–45; Lecturer at Kiev Inst. of Highway Engineering 1946–48; Lecturer, Prof., Prof. Chair of Probability Theory and Mathematical Statistics at Kiev Univ. 1948–65; Chief of Dept. of Probability Theory and Mathematical Statistics at Computing Centre Donetsk and Prof. and Prof. Chair at Donetsk Univ. 1966–.
Publs.: "O nekotorykh differentsialnykh uravneniiakh so sluchainymi funktsiiami" in "Ukrainskii matematicheskii zhurnal" No. 3/1950, "Po povodu odnoi teoremy N. N. Bogoliubova" in "Ukrainskii matematicheskii zhurnal" No. 4/1952, "Vvedenie v teoriiu sluchainykh protsessov", 1965.

Gilashvili, Pavel Georgievich
State official; b. 1918.
Georgian; mem. CPSU 1939–; grad. Party High School at Cen. Cttee. CPSU; locksmith 1934–38; 1st sec. of a Komsomol raion cttee. 1938–39; served in Sov. Army 1939–45; head of a dept. of Cen. Cttee. Georgian Komsomol, 1st sec. of a raion cttee. Georgian CP, Sec. of Tbilisi, Tskhinvali and Rustavi City Cttee. Georgian CP 1945–58; mem. staff Georgian CP 1958–67; mem. Cen. Cttee. Georgian CP 1956–58 and 1971–; Chair. of Council of Mins. Abkhazian ASSR 1967–72; 1st Sec. of Tbilisi City Cttee. Georgian CP 1972–Ja. 1976; mem. Bureau Cen. Cttee. Georgian CP N. 1972–; Chair. of Presidium Supr. Sov. Georgian SSR Ja. 1976–; mem. Cen. Auditing Comm. CPSU 1976–.

Gilels, Emil Grigorevich
Pianist; b. O. 6 (19), 1916, Odessa.
Mem. CPSU 1942–; grad. Odessa Conservatory 1935 (class of B. M. Reingbals) and Moscow Conservatory (master class of G. G. Neuhaus) 1938; first concerts 1929; teacher 1936–, Prof. Moscow Conservatory 1952–; numerous performances throughout the world. Decors.: 2nd Prize Int. Piano Contest Vienna 1936, 1st Prize Int. Piano Contest Brussels 1938; People's Artist of the USSR 1954, State Prize of the USSR 1946, Lenin Prize 1962, 2 Orders of Lenin, 2 other orders and various medals.

Giliarov, Merkurii Sergeevich
Zoologist; b. F. 22 (Mr. 6), 1912, Kiev.
Grad. Kiev Univ. 1933; corresponding mem. USSR Acad. of Sciences 1966–; mem. staff Inst. of Morphology and Ecology of Evolution at USSR Acad. of Sciences 1944–55; in addition Prof. at Pedagogical Lenin Inst. Moscow 1949–; main works in research on soil pests, founded a new branch of biology, soil zoology; Chair. of Nat. Cttee. of Sov. Biologists 1959–; Vice-Pres. of Int. Acad. of Zoology (India) 1960–; Vice-Pres. of Union Soc. of Entomology 1967–. Decors.: 2 orders, several medals.

Giliazhev, Lukman-Khakim Gabdrakhmanovich
Poet; b. 1923, Kanly-Turkeevo, Bashkir ASSR; s. of a peasant.
Mem. CPSU 1947; first works publ. 1943. Decor.: Order of the Patriotic War 1st Cl.

Publs.: "Ogni molnii" (Flashes of Lightning) 1947, novel "Soldaty bez pogon" (Soldiers Without Shoulder-Pieces) 1964, children's poems "Aizar" 1957, and others.

Giniunas, Kazis Pranovich
State official.
Chair. of State Cttee. on Nature Conservation Lithuanian SSR 1980–.

Gintsburg, Vitalii Lazarevich
Physicist; b. S. 21 (O. 4), 1916, Moscow. Mem. CPSU 1944–; grad. Moscow Univ. 1938; mem. USSR Acad. of Sciences 1966–; mem. Inst. of Physics of USSR Acad. of Sciences 1940–; concurrently Prof. at Gorkii Univ. 1945–. Decors.: State Prize of the USSR 1953, Lenin Prize 1966, Order of Lenin, 2 other orders and various medals.

Ginzburg, Lev Vladimirovich
Russian writer and translator; b. O. 24, 1922, Moscow.
Member of the CPSU since 1945. Participant in World War II.
In *The Ratcatcher's Fife* (1960) and *The Price of Asher* (1961), books of sketches and essays, he exposes the fascists and neo-Nazis in the German Democratic Republic and describes the construction of a new life-style in the GDR. His journalistic book *The Abyss*, which has been translated into several languages, is based on materials from the Krassnodar trial of the fascist criminals. *Otherworld Encounters*, a book of sketches, deals with the personal responsibility of the surviving Nazi leaders for their crimes. He has translated German folk ballads, 17th-century German poets, the early lyric poems of Schiller, and the poets of the GDR into Russian. Decor.: J. Becher Literature Prize by the GDR (1969).

Gippius, Evgenii Vladimirovich
Music critic and musical ethnographer; b. Je. 24 (Jl. 7), 1903, Tsarskoe Selo, now Pushkin, Leningrad Oblast.
Doctor of art criticism (1958).
He studied in the department of social sciences of the University of Petrograd. He graduated from the Leningrad Institute of Art History in 1924 and from the scientific composition department of the Leningrad Conservatory in 1928. He studied composition with M. O. Shteinberg, conducting with N. A. Mal'ko, and music history with B. V. Asaf'ev. He specialized in gathering and studying folk music and poetry. He took part in expeditions to the Russian North (Pinega, Mezen', Pechora), Central Russia, Belorussian Poles'e, Armenia, and Uzbekistan. He studied the folklore of the peoples of Siberia, the Finno-Ugric and Turkic peoples, and gypsies. He also studied the songs of the Russian workers' and the international revolutionary movements. In 1927 he founded and until 1943 was the scientific director of the phonogram archive of the Academy of Sciences of the USSR. He was chairman of the department of folk music at the Leningrad Conservatory from 1939 to 1941. He was a professor and chairman of the department of musical folklore at the Moscow Conservatory from 1944 to 1949. He was a senior scholar at the Institute of Ethnography from 1931 to 1952 and at the Moscow Institute of Art History from 1959 to 1963.
Works: "Krest'ianskaia lirika." In *Russkii fol'klor*. Leningrad 1935. *Pesni Pinezh'ia*, book 2. Moscow 1937. (With Z. V. Eval'd.) "Intonatsionnye elementy russkoi chastushki." *Sovetskii fol'klor*, 1936, nos. 4–5. "Ei ukhnem" and "Dubinushka." *Istoriia pesen*. Moscow 1962. "Krasnoe znamia." *Iz istorii pesni trekh russkikh revoliutsii*. Moskow 1969. (With P. G. Shiriaeva.)

Girenko, Andrei Nikolaevich
Party official; b. 1936.
Mem. CPSU 1936–; grad. Krivoi Rog Metallurgical Inst. 1958 and CPSU Cen. Cttee. Correspondence Party Higher School 1971; engineer, senior engineer in a mining admin. in Krivoi Rog 1958–62; 1st Sec. of Krivoi Rog City Cttee. Ukrainian Komsomol 1963–67; 1st Sec. of Dnepropetrovsk Oblast Cttee. Ukrainian Komsomol 1967–70; mem. Auditing Comm. Ukrainian CP 1971–76; cand. mem. Cen. Cttee. Ukrainian CP 1976–80, mem. 1980–; 1st Sec. of Cen. Cttee. Ukrainian Komsomol 1972–75; 2nd Sec. of Kherson Oblast Cttee. Ukrainian CP 1975–80, 1st Sec. Mr. 1980–; cand. mem. Cen. Cttee. CPSU 1981–.

Girgenson, Ziegfried Viktorovich
State official.
Cand. mem. 1961–66, mem. Cen. Cttee. Latvian CP 1966–; Dep. to Supr. Sov. Latvian SSR 1963–; head of a dept. at Cen. Cttee. Latvian CP 1963; sec. of a raion cttee. Latvian CP in Riga 1969; Min. of Supplies Latvian SSR N. 1969–.

Gladkii, Viktor Ivanovich
State official.
Min. for Building Materials Ind. Turkmen SSR O. 1979–; mem. Cen. Cttee. Turkmen CP 1981–.

Gladkyi, Vladimir Il'ich
Painter; b. 1929, Dnepropetrovsk.
Grad. Odessa Art School 1952.

Gladkyi, Vladimir Ilich
Trade union official.
Sec. and mem. Presidium of Ukrainian Republican Trade Union Council D. 1970–82; Sec. of All-Union Cen. Council of Trade Unions of the USSR Mr. 1982–.

Gladush, Ivan Dmitrievich
State official.
Min. of Interior Ukrainian SSR Je. 1982–.

Glazkov, Ilia Mikhailovich, Cand. of Tech. Sc.
State official; b. 1922, Genichask, Kherson Oblast; s. of a worker.
Mem. CPSU 1943–; grad. Belorussian Polytech. Inst. 1954; Cand. of Tech. Sc. 1969; Komsomol, econ. and CPSU posts 1944–51; dept. head, sec. of a raion cttee. Belorussian CP in Minsk 1951–55; Sec. of CPSU Cttee., Dep. Chief Designer of Automobile Works in Minsk 1955–61; Dept. Head, Chief Designer, Head of Design and Construction Bureau 1961–71, Dir.-Gen. of a Scientific Prod. Asscn. in Minsk 1971–74; Dep. Chair. of Council of Mins. Belorussian SSR F. 1974–; mem. Cen. Cttee. Belorussian CP 1976–. Decors.: State Prize, Order of Lenin, Order of the Patriotic War, various medals.

Glazov, Grigorii Solomonovich
Author and Poet; b. S. 24, 1925, Artemovsk, Donetsk Oblast; s. of a clerk.
Ukrainian (writes in Russian); grad. Dept. of Philology of Lvov Univ. 1951. Decors.: medals.
Publs.: "My mirnye liudi" (We Are Peaceful People) 1951, "Zhivu na zemle" (I Live on the Earth) 1963, "Chastnoe poruchenie" (Private Mission) 1957, "Eto sluchilos' vchera" (This Happened Yesterday) 1961, "Posle nas" (After Us) 1968, and others.

Gleb (real name: Smirnov, Ivan Ivanovich)
Russian Orthodox clergyman; b. Ag. 23, 1913, Orekhovo-Zuevo.
Grad. secondary school in Riazan, vocational school in Moscow and course for tech. cadres in Moscow 1935; worked in admin.-technological posts in various enterprises in Riazan until 1953; deacon 1953, priest in Riazan, Mikhailov and Letovo 1953–65; Head of Riazan Cathedral 1973–76; took vows as monk 1976; Archimandrite My. 1976–; Bishop of Orel and Briansk.

Glerzerman, Grigorii Erukhimovich, Dr. of Phil. Sc.
Philosopher; b. D. 31, 1906 (Ja. 13, 1907), Tver, now Kalinin.
Mem. CPSU 1940–; attended Econ. Faculty of Plekhanov Inst. of Econ. 1926–30; Dr. of Phil. Sc. 1950; Prof. 1951–; Prof. Chair of Dialectical and Historical Materialism 1955–, Dep. Rector of Acad. of Social Sciences at Cen. Cttee. CPSU 1967–. Decors.: Merited Scientist of the RSFSR, State Prize of the USSR 1951, 2 orders, several medals.

Glikman, Gennadii Naumovich
Graphic artist and sculptor; b. 1921, Berdichev, Zhitomir Oblast.
Grad. Kiev Art Inst. 1952, pupil of I. Pleshchinskii and V. Kasian.

Glushchuk, Fedor Timofeevich
Graphic artist; b. 1925, Starostintsi, Vinnitsa Oblast.
Attended Lvov Inst. for Applied and Decorative Art 1948–50.

Glushko, Valentin Petrovich, Dr. of Tech. Sc.
Scientist; b. Ag. 20 (S. 2), 1908, Odessa; s. of a clerk.
Ukrainian; mem. CPSU 1956–; grad. Leningrad State Univ. 1929; mem. USSR Acad. of Sciences 1958–; scientist working in field of rocket technology; engineer, head of various scientific research insts. 1929–; Dep. to USSR Supr. Sov. 1966–; mem. Cen. Cttee. CPSU 1976–; all Sov. launchers which were launched between 1949 and 1976 were equipped with rocket motors developed under his supervision. Decors.: Lenin Prize 1957, State Prize of the USSR 1967, 3 Orders of Lenin, Order of October Revolution, Order of Red Banner of Labour, medals. Hero of Socialist Labour 1956 and 1961.

Glushkov, Nikolai Timofeevich
State official; b. 1918.
Russian; mem. CPSU 1945–; grad. Sarapul Finance and Economic Technicum 1936 and All-Union Correspondence Law Inst. 1956, Cand. of Econ. Sc. 1966; bookkeeper in an artel 1932–36; chief bookkeeper of Consumers' Cooperatives System 1936–41; chief bookkeeper, chief of finance dept. of Norilsk Metallurgical Combine 1941–49; chief of finance dept. of "Eniseistroi" Main Admin., dir. of admin. for metallurgy in USSR Min. of Nonferrous Metallurgy 1949–57; dept. chief, Chief of Economic and Finance Dept. of Krasnoiarsk Sovnarkhoz 1957–60, Dep. Chair., Chair. 1960–65; Chief of Main Admin. for Economic Planning in USSR Min. of Nonferrous Metallurgy 1965–73; Chair. USSR State Cttee. for Pricing Ag. 1975–; mem. Cen. Auditing Comm. CPSU 1976–; Dep. to USSR Supr. Sov. 1979–. Decor.: Order of Lenin.

Gnatiuk, Dmitrii Mikhailovich
Ukrainian singer (baritone); b. Mr. 28, 1925, Starosel'e, in present-day Kitman' Raion, Chernovtsy Oblast.
People's Artist of the USSR (1960). Member of the CPSU since 1962.
He graduated in 1951 from I. S. Patorzhinskii's class at the Kiev Conservatory and became a soloist in the Kiev Opera and Ballet Theater. His roles include Ostap and Mykola (Lysenko's *Taras Bulba* and *Natalka-Poltavka*), Petrucchio (Shebalin's *Taming of the Shrew*), Mazeppa and Onegin (Tchaikovsky's *Mazeppa* and *Eugene Onegin*), the Demon (Rubinstein's *The Demon*), Figaro (Rossini's *The Barber of Seville*), and Rigoletto (Verdi's *Rigoletto*). He is also a concert singer and has performed abroad. Decors.: Lenin Komsomol Prize (1967), the Order of Lenin, and various medals.

Gnatiuk, Ostap Romanovich
Graphic artist; b. 1933, Kozina, Ivano-Frankovsk Oblast.
Mem. CPSU 1961–, grad. Kosiv School of Applied Art 1962.

Gnedenko, Boris Vladimirovich, Dr. of Phy.-Math. Sc.
Mathematician; b. Ja. 1, 1912, Simbirsk, now Ulianovsk.
Grad. Saratov Univ. 1930; mem. Acad. of Sciences Ukrainian SSR 1948–; Dr. of Phy.-Math. Sc. 1942; Prof. 1942–; employed at Textile Inst. Ivanovo 1930–34; until 1937 cand. Moscow Univ.; mem. staff Moscow Univ. 1938–45 (Prof. 1942–); Chief of Dept. of Probability Theory and Mathematical Sciences at Acad. of Sciences Ukrainian SSR 1945–60, Dir. of same Inst. 1955–58; Chair. of Bureau Dept. of Physics and Mathematics at Acad. of Sciences Ukrainian SSR 1956–58; Prof. at Lvov Univ. 1945–50, Prof. and Prof. Chair at Kiev Univ. 1950–58; Prof. 1960–65, Prof. Chair at Moscow Univ. 1965–; main research work on probability theory, mathematical statistics, mathematical analysis and history of mathematics. Decors.: Order of Red Banner of Labour, medals.
Publs.: "Ocherki po istorii matematiki v Rossii" 1946, "Predelnye raspredeleniia dlia summ nezavisimykh sluchainykh velichin" 1949, "Mikhail Vasilevich Ostrogradskii" 1963, "Elementarnoe vvedenie v teoriiu veroiatnostei" 1964, "Kurs teorii veroiatnostei" 1965, "Vvedenie v teoriiu massovogo obsluzhivaniia" 1967.

Gnedykh, Vladimir Vasilevich
Diplomat; b. 1921.
Mem. Diplomatic Service 1943–; Emb. Counsellor at USSR Emb. in Morocco 1958–62 and in Algeria 1964–65; various posts in USSR Min. of Foreign Affairs 1962–64 and 1965–68; USSR Amb. to Mauretania 1968–72; mem. staff USSR Min. of Foreign Affairs 1972–75; USSR Amb. to Equatorial Guinea O. 1975–77.

Godenko, Mikhail Matveevich
Author; b. 1919, Osipenko, Zaporozhe Oblast.
Russian; mem. CPSU 1942–; grad. Gorkii Inst. of Literature Moscow 1951.
Publs.: "Poslednii" (The Last One) 1956, "Lastochka" (Swallow) 1957, "Minnoe pole" (A Mined Field) 1964, "Tsvetet akatsiia" (The Acacia is Blooming) 1959, and others.

Gogitidze, Iraklii Konstantinovich
State official.
Mem. Cen. Cttee. Georgian CP 1976–81; Head of Main Admin. for Geology Georgian SSR Mr. 1980–; mem. Auditing Comm. Georgian CP 1981–.

Gogoladze, Vladimir Trofimovich
State official; b. 1930.
Georgian; mem. CPSU 1959–; grad. Georgian Polytechnical Inst. (1953); master, foreman of "Transspetsstroi" trust of USSR Min. of Construction in Mezhinsk, Bryansk Oblast 1953–54; engineer, head of a section of a trust in Ivanovo 1954–56; chief engineer, head of an admin. of Trust No. 148 of USSR Min. of Construction in Klin and Kalinin 1956–58; head of Trust No. 6 "Kalininspetsstroi" 1958–63; head of Trust No. 3 of Georgian Min. of Construction, 1st Dep. Min. of Construction Georgian SSR 1973–80; Min. of Construction Georgian SSR Ja. 1980–; mem. Cen. Cttee. CP Georgia 1981–.

Gogolev, Ivan Mikhailovich
Poet; b. 1930, Viliuisk; s. of a teacher.
Yakut; mem. CPSU 1960–; grad. Gorkii Inst. of Literature Moscow 1953; first works publ. 1948.
Publs.: "V dalekii put'" (On the Long Journey) 1956, novel in verse "Solnechnaia gora" (Sun Mountain) 1962, "Na tikhom okeane" (On the Pacific) 1956, "Samoe dorogoe" (The Dearest) 1958, play "Pod maiskim nebom" (Under May Skies), and others.

Gogoleva, Elena Nikolaevna
Russian actress; b. Mr. 25 (Ap. 7), 1900, Moscow.
People's Artist of the USSR (1949). Member of the CPSU since 1948.
In 1918, she was accepted into the troupe of the Malyi Theater from the second course of the dramatic division of the Music-Drama School of the Moscow Philharmonic Society. She is an actress of heroic stature who creates images of strong, willful, energetic women. She is particularly attuned to the portrayal of dramatic and tragic characters. Her roles include Sofia (in Griboedov's *Work From Wit*), Princess Eboli and Lady Milford (in Schiller's *Don Carlos* and *Cabal and Love*), Judith (in Gutzkow's *Uried Acosta*), Nadezhda Monakhova (in Gorkii's *Barbarians*) and Lady Macbeth (in Shakespeare's *Macbeth*). Her portrayals of Panova (in Trenev's *Liubov Iarovaia*), the Comtess Marlborough (in Scribe's *The Glass of Water*), and Mrs. Crawley (in *Vanity Fair*, after Thackeray's novel) were marked by satirical exposé. Soviet dramaturgy occupies a great place in the actress' work; her principal roles were She (in *To Those at Sea!* by Lavrenev), Polozova (in *A Moscow Character* by Sofronov), and the Abbess Melan'ia (in *Dostigaev and Others* by Gorkii). She has acted in films (*Gobsek* and *You Can't Forget About That*). Decors.: State Prize of the USSR (1947, 1948, 1949), the Order of Lenin, two other orders, and various medals.

Gogua, Aleksei Nochevich
Author; b. 1932, Gutsy, Abkhazian ASSR; s. of a peasant.
Grad. Gorkii Inst. of Literature Moscow 1960; first works publ. 1948.
Publs.: "Reka speshit k moriu" (The River Rushes to the Sea) 1957, "Otchego rastet trava?" (Why Does the Grass Grow?) 1960, "Sneg i molniia" (The Snow and the Lightning) 1961, and others.

Goian, Georg Iosifovich (pseudonym of G. I. Ter-Nikogosian)
Theatrical scholar; b. N. 10 (23), 1901, Tbilisi.
Honored Art Worker of the Armenian SSR (1961) and doctor of the arts (1949). Member of the CPSU since 1930.
He graduated from the State Institute of Theatrical Arts (1935) and the Institute of the Red Professoriat (1938). He began his literary career in 1921. In 1936 he began teaching at the State Institute of Theatrical Arts, becoming a professor in 1949, and in 1958 he became head of the institute's subdepartment of the theater of the peoples of the USSR. His works are devoted to the theatrical art of the peoples of the USSR – Russian (*The Development of Russian Theater,* 1939, and *Glikeriia Fedotova,* 1940) and Tadzhik (*A Theater Born in October: A History of Tadzhik Dramatic Art,* 1943). His major work is *2,000 Years of Armenian Theater* (vols. 1–2, 1952). His articles have been combined into an anthology, *What Is National Theater?* (1970).

Goldanskii, Vitalii Iosifovich
Physical chemist; b. Je. 18, 1923, Vitebsk.
Mem. CPSU 1950–; grad. Moscow Univ. 1944; corresponding mem. USSR Acad. of Sciences 1962–; mem. staff Inst. of Chemical Physics 1942–52, 1961–; mem. staff Physics Inst. at USSR Acad. of Sciences 1952–61; Lecturer 1951–, Prof. at Inst. of Physical Engineering Moscow 1956–; main works in nuclear chemistry.

Goldin, Nikolai Vasil'evich
State official; b. 1910.
Russian; mem. CPSU 1929–; grad. Kharkov Electrotech. Inst. 1937; Works Manager, Engineer, Head of Bureau of Bureau of "Elektromontazh" Trust in Kharkov 1937–39; Chief Engineer, Head of Trust of People's Commissariat for Construction USSR 1939–46; various posts in Min. for the Construction of Heavy Ind. Enterprises 1946–50; USSR Dep. Min. 1950–51, 1st Dep. Min. for the Construction of Heavy Ind. Enterprises 1951–53; USSR Dep. Min. of Construction 1953–54; USSR Dep. Min. for the Construction of Metallurgical and Chemical Ind. Enterprises 1954–57; Chair. of Kemerovo Econ. Council 1957–58; Chief Engineer at construction of metallurgical works in Bhilai, India 1958–61; Sov. Amb. to Cuba 1962–63; USSR Dep. Chair. of State Prod. Cttee. for Installation and Special Construction Work 1963–65; USSR Dep. Min. of Installation and Special Construction Work 1965–67; Dep. to USSR Supr. Sov. 1967–; USSR Min. for the Construction of Heavy Ind. Enterprises F. 1967–; mem. Cen. Cttee. CPSU 1971–. Decor.: Order of Lenin.

Goliavkin, Viktor Vladimirovich
Author; b. 1929; s. of a teacher.
Grad. Inst. of Sculpture and Architecture.
Publs.: children's books "Tetradki pod dozhdem" (Notebooks in the Rain) 1959, "Nashi s Vovkoi razgovory" (Our Conversations With Vovka) 1960, "Komu chto udivitel'no?" (Whom does it Surprise?) 1961, "My igraem v Antarktidu" (We're, Playing Antarctis) 1961, "Parovozik v nebe" (A Little Locomotive in the Sky) 1962, "Gorod v more" (A Town in the Sea) 1962, "Moi dobryi papa" (My Good Daddy) 1962, "Privet vam, ptitsy!" (Welcome, Birds!) 1962, and others.

Golikov, Viktor Andreevich
Party official; b. 1914.
Russian; mem. CPSU 1930–; grad. Rostov Pedagogical Inst. 1935 and CPSU Cen. Cttee. Party School 1945, Cand. of Hist. Sc. 1954; librarian in a secondary school 1930–35; history teacher, dir. of a secondary school in Novorossiisk 1935–40; Sec. of Krasnodar Krai Cttee. USSR Komsomol, instructor, chief of a dept. of Oblast Cttee. CPSU in Iakut ASSR 1940–45; chief of a dept., Sec. of Uezd Cttee. Moldavian CP, chief of a sector, Asst. to 1st Sec. Cen. Cttee. Kazakh CP 1945–55; asst. to a sec. of Cen. Cttee. CPSU 1956–60; Asst. to Chair. of Presidium of USSR Supr. Sov. 1960–63; Asst. to Gen.-Sec. of Cen. Cttee. for Agricultural Questions 1963–; Dep. to USSR Supr. Sov. 1979–; cand. mem. Cen. Cttee. CPSU 1981–.

Golobutskii, Vladimir Alekseevich
Historian; b. Jl. 28 (Ag. 10), 1903, Velikii Bor, now in Surazk Raion, Briansk Oblast.
Doctor of historical sciences (1948), professor (1949). Member of the CPSU since 1950.
He graduated from the university of Rostov on Don in 1930. He is the author of many works on the socioeconomic and political history of the Ukraine of the 16th to 19th centuries and on problems of the disintegration of the feudal-serf system and the origins of capitalism. A number of his works are devoted to the war of liberation waged by the Ukrainian people from 1648 to 1654 and to the history of the Zaporozh'e and Black Sea cossacks. Decor.: Order of the Red Banner of Labor.
Works: *Osvoboditel'naia voina ukrainskogo naroda pod rukovodstvom Khmel'nitskogo (1648–1654 gg.).* Moscow 1954. *Chernomorskoe kazachestvo.* Kiev 1957. *Diplomaticheskaia istoriia osvoboditel'noi voiny ukrainskogo naroda: 1648–1654.* Kiev 1962.

Golovachev, Pavel Iakovlevich
Military official; b. D. 2 (15), 1917, Koshelevo, now Gomel' Oblast.
Twice Hero of the Soviet Union (N. 1, 1943, and Je. 29, 1945). Air force major general (1957). Member of the CPSU since 1943.
He was born into the family of a poor peasant. He enlisted in the Soviet Army in 1938 and graduated from a military school for aviators in 1940. During the Great Patriotic War he took part in battles on the Southwestern, Stalingrad, Southern, Fourth Ukrainian, First Belorussian, and Third Belorussian fronts as a pilot and flight and squadron commander in the 9th Guards Fighter Aviation Regiment. He flew more than 450 combat missions, fought in 125 air battles, and personally shot down 26 enemy aircraft. He was wounded four times. After the war he graduated from the Air Force Academy in 1951 and from the Military Academy of the General Staff in 1959.
Decors.: two Orders of Lenin, six Orders of the Red Banner, the Order of the Patriotic War First Class, two Orders of the Red Star, and various medals.

Golovanivskii, Savva Evseevich
Poet and dramatist; b. 1910, Elizavetgradka, Kirovgrad Oblast; s. of a clerk.
Ukrainian; mem. CPSU 1942–; studied Agricultural Insts. in Odessa and Kharkov; first works publ. 1927. Decors.: Order of Red Banner of Labour, Order of the Patriotic War 2nd Cl.
Publs.: "Zheleznymi loshadmi" (With Iron Horses) 1927, "Smert' Ledi Grei" (The Death of Lady Grey) 1934, "Sud'ba poeta" (The Fate of the Poet) 1939, "Solnechnaia storona" (The Sunny Side) 1952, "Pervyi grom" (The First Thunder) 1957, and others.

Golovchenko, Fedor Petrovich
State official; b. 1918, Ukraine.
Mem. CPSU 1949–; grad. Tech. Aviation Inst. Moscow; after grad. head of tech. bureau, design engineer of various repair workshops 1944–47; Head of Tech. Bureau, head of various depts. 1947–53 and Organizer of Cen. Cttee. CPSU in Kharkov Ordzhonikidze Tractor Factory 1953–60; Head of Dept. of Machine Construction at Cen. Cttee. Ukrainian CP 1960–65; cand. mem. 1961–66, mem. 1966–71, cand. mem. Cen. Cttee. Ukrainian CP 1971–; 2nd Sec. 1965–66, 1st Sec. of Kiev Oblast Cttee. Ukrainian CP 1966–70; cand. mem. Cen. Cttee. CPSU 1966–71; Dep. to USSR Supr. Sov. 1966–70; Min. of Motor Transport Ukrainian SSR Ap. 1970–. Decor.: Order of Red Banner of Labour.

Golovkina, Sofia Nikolaevna
Ballet dancer; b. 1915.
Mem. CPSU 1942–; grad. Choreographic College Moscow 1933; with Bolshoi Theatre 1933–; Dir. of Choreographic College Moscow 1960–. Decors.: People's Artist of the USSR 1973, State Prize of the USSR 1947.

Golozubov, Vladimir Vasil'evich
Graphic artist; b. 1925, Kiev.
Grad. Kiev Art Inst. 1958, pupil of I. Pleshchinskii.

Golubev, Vasilii Nikolaevich
Journalist; b. 1913.
Mem. CPSU 1940–; grad. theoretical course at Ordzhonikidze Inst. of Econ. Engineering 1941 and Party High School at Cen. Cttee. CPSU 1948; dep. bookkeeper, planner, sec. of Komsomol cttee. in various factories in Ivanovo and Kostroma 1930–35; served in Sov. Army 1935–37 and 1941–45; CPSU posts 1946–55; Ed. of magazine "Moskovskii propagandist" 1955–57; Ed. of newspaper "Leninskoe znamia" 1958–68; Chief Ed. of magazine "Journalist" 1968–69; Chief Ed. of newspaper "Sotsialisticheskaia industriia" My. 1969–; mem. Cen. Auditing Comm. CPSU 1971–.

Golubeva, Mariia Arkhipovna
Milkmaid; b. 1945.
Mem. CPSU 1968–, finished 8th class at Verkh-Katun Secondary School 1963; milkmaid in "Katun" Breeding Plant in Altai Krai 1963–; cand. mem. Cen. Cttee. CPSU 1981–.

Golubeva, Valentina Nikolaevna
Weaver; b. 1949.
Mem. CPSU 1977–; grad. textile technicum 1978; weaver at Ivanovo Worsted Goods Combine 1967–; mem. Cen. Cttee. CPSU 1981–. Decors.: Heroine of Socialist Labor 1977, USSR State Prize 1979.

Golubtsova, Tamara Vasil'evna
State official; b. 1928.
Mem. CPSU 1954–; grad. Moscow State Univ. 1951; research associate at M. I. Kalinin Museum 1951–53; posts in Komsomol 1956–62; party posts 1962–75; Dep. USSR Min. of Culture 1975–; mem. Cen. Auditing Comm. CPSU 1976–.

Gomiashvili, Aleksandr Konstantinovich
Poet; b. 1911, Kazbegi, Georgia; s. of a peasant.
Georgian; mem. CPSU 1939–; grad. Tbilisi Univ. 1933; served in Sov. Army 1941–45; Acting Sec. of Georgian Writers' Union 1946; first works publ. 1929.
Publs.: "Gornaia ballada" (Mountain Ballad) 1929, poem "Opiat' zovut Germaniiu" (Germany Is Called Again) 1933, "Stikhi" (Poems) 1937, "Darial" 1940, "Ballady" (Ballads) 1946, "Stikhi i ballady" (Poems and Ballads) 1960, "Izbrannoe" (Selected Works) 1961, and others.

Gonchar, Aleksandr Terent'evich
Author; b. 1918, Sukha, Poltava Oblast; s. of a peasant.
Ukrainian; mem. CPSU 1946–; grad. Dnepropetrovsk State Univ., literary contributor to a raion and an oblast newspaper 1933–41; served in Sov. Army 1941–45; postgraduate studies at Inst. of Literature at Acad. of Sciences Ukrainian SSR 1946; Dept. Head with magazine "Dnipro" in Kiev 1947–52; Dep. Chair. 1952–59, Chair. of Ukrainian Writers' Union 1959–71; Board Sec., USSR Union of Writers 1971–; mem. Cen. Cttee. Ukrainian CP 1961–66 and 1971–; Dep. to USSR Supr. Sov. 1962–; cand. mem. Cen. Cttee. CPSU 1976–.
Publs.: novelettes "Vesna za Moraviu" (Spring at the Morava), "Alpy" (Alps), novels "Golubyi Dunai" (the Blue Danube), "Zlata Praga" (Golden Prague), "Tavriia" (Tauria), "Perekop", "Liudyna i zbroia" (Man and the Weapon), poem "Mykyta Bratus'", and others. Decors.: Hero of Socialist Labour 1978, Lenin Prize 1964, State Prize of the USSR 1947 and 1948.

Gonchar, Andrei Aleksandrovich, Dr. of Phy.-Math. Sc.
Mathematician; b. 1931.
Grad. Moscow Univ. 1954; Dr. of Phy.-Math. Sc. 1964; corresponding mem. USSR Acad. of Sciences 1974–; cand. 1954–57, Research Asst., then Lecturer 1957–, Prof. at Moscow Univ. 1967–; in addition Sen. Research Fellow 1965–72, Dept. Chief at Inst. of Mathematics USSR Acad. of Sciences 1972–.

Gonchar, Oles' (Aleksandr Terent'evich Gonchar)
Ukrainian writer; b. Ap. 3, 1918, Sukha na Poltavshchine.
Member of the CPSU since 1946. Born into a peasant family.
He began to appear in print in 1938. In June 1941 he went to the front as a volunteer. He graduated from the University of Dnepropetrovsk in 1946.
Even his early works – "The Cherries Are Blooming," "Healing Waters," and "Ivan Mostovoi" (1938) – attempt to depict strong, whole characters. In 1946 he published *The Alps,* the first part of a trilogy known as the *Standard-bearers*. The second part, *The Blue Danube* (1947), was awarded the State Prize of the USSR in 1948, and the third part, *Golden Prague* (1948), received the same award in 1949. *The Standard-bearers* is one of the best works of Soviet creature on the Great Patriotic War of 1941–45, depicting the nobility, courage, and humanity of the Soviet troops, their desire for peace, and their keen sense of justice. His story "The Earth Roars" shows the heroic struggle of the members of the underground Poltava Komsomol against the fascist occupiers. He also wrote several collections of short stories, the tale *That Light May Shine* (1954), and the screenplay *The Partisan Spark* (1956). In novels devoted to the life of laborers in the southern Ukraine, he drew pictures of forced labor and the awakening of the class consciousness of the prerevolutionary Ukrainian peasantry (*Tavriia,* 1952) and showed the heroism of the people who revolted in 1919–20 in defense of the victories of the socialist revolution (*Retrenching,* 1957). The novel *Man and Arms* (1960), on the fate of men at war, received the T. Shevchenko Prize in 1962. The heroes of the novel *Tronka* (1963; Lenin Prize, 1964) are our contemporaries-shepherds and workers, the transformers of the land. The writer told of people who are generous both in work and in their private lives; he created deep and complex characters. His novel *The Cathedral* was published in 1968. His novel *The Cyclone,* on the continuity of the heroic traditions of the Soviet people, appeared in 1970.
His work is marked by a combination of lyricism and romantic inspiration with deep realism in the depiction of people and events. His books have been translated into many languages of the peoples of the USSR and foreign languages.
He was a deputy to the sixth through eighth convocations of the Supreme Soviet of the USSR, president of the governing body of the Ukrainian Writers' Union (1959–71), and one of the secretaries of the governing body of the Writers' Union of the USSR. Decors.: two Orders of Lenin, four other orders, and medals.
Works: *Tvory,* vols. 1–2. Kiev 1954. *Tvory,* vols. 1–4, Kiev 1959–60. *Tvory,* vols. 1–5. Kiev 1966–67. *In Russian translation: Izbrannye proizvedeniia,* vols. 1–2. Kiev 1951. *Izbrannoe.* Moscow 1969.

Goncharenko, Boris Trofimovich
Party official; b. 1927.
Ukrainian; mem. CPSU 1948–; grad. Kharkov Polytechnical Inst. 1954; engineer, senior engineer, dep. dept. chief of a turbine plant in Kharkov 1948–55; party posts 1955–67; Chief of Machine Building Dept. of Cen. Cttee. Ukrainian CP 1967–73; cand. mem. Cen. Cttee. Ukrainian CP 1971–74, mem. 1974–; 1st Sec. of Voroshilovgrad Oblast Cttee. Ukrainian CP 1973–; Dep. to USSR Supr. Sov. 1974–; mem. Cen. Cttee. CPSU 1976–.

Goncharenko, Ivan Ivanovich
Poet; b. 1908, Iablonovo, Poltava Oblast; s. of a peasant.
Ukrainian; mem. CPSU 1931–; grad. Inst. of Red Professors Kharkov 1934; first works publ. 1926.
Publs.: "Druzia" (Friends) 1930, "Odessa, Sevastopol" 1944, "Pomolvka" (Engagement) 1949, "Selo Tarasa" (The Village of Taras) 1951, "Matros Gaidai" (The Sailor Gaidai) 1956, "Vybrane" (Selected Works) 1959, "Na klych zhyttia" (To the Call of Life) 1960, "Idu rodnoi zemlei" (I'm Walking on My Native Earth) 1956, and others.

Goncharenico, Oleg Georgievich
Athlete, coach; b. Ag. 18, 1931, Kharkov.
Honored Master of Sport (1953). Member of the CPSU since 1953.
He was the first Soviet athlete to win the

title of world champion in speed skating (1953, Helsinki). He was champion of the USSR (1956, 1958), of Europe (1957–58), and of the world (1953, 1956, and 1958). Decor.: Order of Lenin.

Goncharov, Aleksandr Arkhipovich
State official.
Min. of Meat and Dairy Produce Belorussian SSR Ja. 1981–; mem. Auditing Comm. Belorussian CP 1981–.

Goncharov, Andrei Dmitrievich
Graphic artist, primarily for books; b. Jl. 9 (22), 1903, Moscow.
Honored Art Worker of the RSFSR (1969).
He studied in Moscow with private teachers and in the free art studios of Vkhutein (State Higher Institute of Art and Technology; 1918–27) with I.I. Mashkov, A.V. Shevchenko, and V.A. Favorskii. He was one of the founders of the Society of Easel Painters (1925). In 1943–45 he worked in the magazine *Frontovoi iumor* (Frontline Humor). He taught at the Moscow Polygraphic Institute (from 1930 to 1934 and since 1948, as a professor since 1958). His graphic works (mainly wood engravings) are characterized by the close relationship of all the elements of the book (illustrations, headpiece, tailpiece), tense emotional content, a profound harmony of the visual elements with the ideological and imaginal structure of the literary work, and rich black and white tones (illustrations for *The Adventures of Peregrine Pickle* by Smollett, published in 1934–35 and 1955; Shakespeare's *Works,* published in 1953–64; and *The Twelve* by Blok, published in 1966). He is also an easel painter (mainly a portraitist), a monumentalist, and a stage designer.
Works: *Ob iskusstve grafiki.* Moscow 1960. *Khudozhnik i Kniga.* Moscow 1964.

Goncharov, Iurii Danilovich
Author; b. 1923, Voronezh.
Russian; first works publ. 1944.
Publs.: "Rasskazy" (Stories) 1948, "Po lesam budushchego" (In the Woods of the Future) 1950, "Liubimaia nasha zemlia" (Our Beloved Country) 1952, "Povest' o rovesnike" (Story of a Contemporary) 1956, "Dezertir" (Deserter) 1962, and others.

Gontar, Avram Iutkovich
Author; b. 1908, Berdichev.
Mem. CPSU 1940–; Jewish writer; first works publ. at the end of the twenties.
Publs.: "Iasnyi den'" (A Bright Day) 1938, "Prostor" (The Width) 1940, "Poslednii raz vspomnil" (The Last Time that I Remembered) 1936; novel "V zabroshennom uglu" (In a Crumbling Corner) 1937, "Pokhod" (The March) 1942, "Bol'shaia semia" (Large Family) 1948, "Serebriannye niti" (Silver Threads) 1957, "Gorod mechty" (Dream City) 1963, and others.

Gorbachev, Mikhail Sergeevich
Party official; b. 1931.
Russian; mem. CPSU 1952–; grad. Moscow State Univ. 1955 and Stavropol Agricultural Inst. 1967; asst. to combine head at a machine and tractor station in Stavropol Krai 1946–50; posts in Komsomol 1955–56; 1st Sec. of Stavropol City Cttee. Komsomol 1956–58; 2nd, then 1st Sec. of Stavropol Krai Cttee. Komsomol 1958–62; 1st Sec. of Stavropol City CPSU Cttee. 1966–68; 2nd Sec. 1968–70, 1st Sec. of Stavropol Krai CPSU Cttee. Ap. 1970–78; Dep. to USSR Supr. Sov. 1970–, mem. Cen. Cttee. CPSU 1971–; Sec. Cen. Cttee. CPSU 1978–; cand. mem. Politbureau Cen. Cttee. CPSU 1979–80, mem. O. 1980–.

Gorban', Grigorii Iakovlevich
Mem. Cen. Cttee. CPSU, steel founder; b. 1932.
Mem. CPSU 1959–; grad. Zhdanov Metallurgical Tech. Inst. 1960; fitter at "Azovstal'" Works 1949–52; served in Sov. Army 1952–54; steel founder's asst. 1955–62, steel founder at "Azovstal'" Works 1962–; mem. Cen. Cttee. Ukrainian CP 1966–; mem. Cen. Cttee. CPSU 1971–.

Gorbatko, Viktor Vasil'evich
Cosmonaut; b. 1934.
Mem. CPSU 1959–; grad. Mil. Aviation College Bataisk 1956 and Zhukovski Mil. Aviation Engineering Acad. 1968; air force officer 1956–61; cosmonaut 1961–; took part in group flight of three spaceships as research engineer in spaceship "Soiuz 7" together with A.V. Filipchenko and V.N. Volkov O. 12–17, 1969, "Soiuz 24" and "Saliut 5" F. 1977. Decors.: Hero of the Sov. Union 1969, Order of Lenin, Order of the Red Star.

Gorbov, Evgenii Konstantinovich
Author; b. 1906, Teplik.
Russian; first works publ. 1934.
Publs.: "Kurinaia slepota" (Hen's Blindness) 1941, "Mirnye zhiteli" (Peaceful Inhabitants) 1943, "Fenia" 1954, novel "Dom pod topoliami" (The House Under the Poplars) 1957, "Bratia Bezfamilnye" (Brothers Without Names) 1961, and others.

Gorbovskii, Gleb Iakovlevich
Poet; b. 1931, Leningrad.
Russian; first works publs. 1954.
Publs.: poems "Poiski tepla" (Search for Warmth) 1960, and others.

Gorbunov, Grigorii Ivanovich
Geologist; b. 1918.
Mem. CPSU 1944–; grad. Inst. of Geological Prospecting Moscow 1941; corresponding mem. USSR Acad. of Sciences 1972–; geologist and chief of various geological prospecting projects commissioned by USSR Min. of Nonferrous Metallurgy 1941–48; Scientific Sec. 1949–52, Dep. Chair. 1952–59, Chair. of Presidium Kola Branch of USSR Acad. of Sciences 1971–; Sen. Research Fellow 1959–64, Laboratory Chief of Geological Inst. at Kola Branch of USSR Acad. of Sciences 1964–65; Dept. Chief with USSR Min. of Geology 1965–71.

Gorchakov, Petr Andreevich
Military official; b. 1917.
Russian; mem. CPSU 1939–; grad. Mil. Pol. Lenin Acad. 1954; Col.-Gen.; served in Sov. Army 1938–; Sec. of party bureau of a regiment, war commr., commander of a regiment for pol. affairs, head of pol. dept. of a rifle div. on Briansk, Cen. and 4th Ukrainian Front 1941–45; head of pol. dept. of various asscns. 1st dep. head of pol. admin. of a mil. district, mem. Mil. Council, Head of Pol. Admin. of Mil. District of Baltic Provinces 1945–70; mem. Cen. Cttee. Latvian CP 1966–71; mem. Mil. Council, Head of Pol. Admin. of Strategic Missile Units 1970–; Dep. to USSR Supr. Sov. 1970–; cand. mem. Cen. Cttee. CPSU 1971–. Decor.: Hero of the Sov. Union.

Gordanov, Viacheslav Viacheslavovich
Cameraman; b. Jl. 15 (28), 1902, Polotsk.
Honored Worker in the Arts of the RSFSR (1950).
He graduated from the Leningrad Cinematic Technicum and began to work at the Lenfilm Studio in 1927. (He was a cameraman on the front from 1942 to 1944.) He shot the films *Lenin's Address* (1929), *Fritz Bauer* (1930), *The Weir,* and *The Fugitive* (both in 1932). His major works are *Thunderstorm* (1934), *Peter I* (1937, part 1, with V.T. Iakovlev), *Masquerade* (1941), *Academician Ivan Pavlov* (1949, with M.Sh. Magid and L.E. Sokol'skii). He did experimental work in the area of color film (the films *Colored Revue* and *Autumn,* both in 1940). He taught at the Leningrad Institute for Film Engineers (1933–35) and directed the camera workshop of the Lenfilm Studio (1935–38). Decors.: State Prize of the USSR in 1950, Order of the Red Banner of Labor and medals.

Gordeichev, Vladimir Grigor'evich
Poet; b. 1930, Kastornoe, Kursk Oblast.
Russian; mem. CPSU 1957–; grad. Gorkii Inst. of Literature Moscow 1957; first works publ. 1950.
Publs.: Poems "Nikitiny kamenia" (Nikita's Stones) 1957, "Zemnaia tiaga" (Force of Gravity) 1959, "Bespokoistvo" (Unrest) 1961, "Zrelost" (Maturity) 1962, and others.

Gordienko, Aleksei Fedorovich
Party official; b. 1929.
Mem. CPSU 1957–; grad. Dnepropetrovsk Technicum for Rail Transport 1949 and Kharkov Polytechnical Inst. 1960; railroad electrician 1940–50; served in Sov. Army 1950–53; electrician, work brigade leader, foreman, shop superintendent of a metal products plant in

Dnepropetrovsk 1953–60; party posts 1960–73; 2nd Sec. of Dnepropetrovsk City Cttee. Ukrainian CP 1973–75; 1st Sec. Dneprodzerzhinsk City Cttee. Ukrainian CP 1975–; mem. Cen. Auditing Comm. CPSU 1976–81; cand. mem. Cen. Cttee. CPSU 1981–.

Gordienko, Iurii Petrovich
Poet and translator; b. 1922, Altaiskoe, Altai Krai.
Russian; grad. Higher Literary Courses in Moscow 1956; first works publ. in various newspapers on the Front.
Publs.: "Zvezdy na kaskakh" (Stars on the Helmets) 1948, "Vchera i segodnia" (Yesterday and Today) 1950, "Stikhi o Vostoke" (Poems about the East) 1951, "Vblizi okeana" (Near the Ocean) 1962, "Vesnoi v listopad" (In Spring When Leaves Fall) 1963, and others.

Gordon, Il'ia Zinov'evich
Jewish writer; b. D. 2 (15), 1907, Zelenopol', in present-day Zaporozh'e Oblast.
Member of the CPSU since 1944. Born into a peasant family.
He treats rural themes, rarely found in prerevolutionary Jewish literature, in such works as *Weeds* (1928). He is the author of the collection of stories *In the Green Cornfield* (1930), the novel *Ingul the Boyar* (1935), the tale *Three Brothers* (1938; Russian translation, 1962), and the book *In Our Parts* (1969).
Works: *Vildgroz.* Kharkov 1930. *In eigene kantn.* Moscow 1969. *In Russian translation: Povesti i rasskazy.* Moscow 1957. *Vnachale ikh bylo dvoe.* Moscow 1965.

Gordon, Samuil Vulfovich
Author; b. 1909, Kaunas; s. of a worker.
Jew; grew up in an orphanage; grad. 2nd Moscow Univ. 1934.
Publs.: "Elka ubogii" (Poor Elka) 1930, "Mezhdu Azovskim i Chernym" (Between the Azov and the Black Sea) 1934, "Patrioty" (Patriots) 1936, "Birobidzhanskie deti" (The Children of Birobidzhan) 1937, "V dni voiny" (In War Days) 1946, "Birobidzhanskie starozhily" (Old Inhabitants of Birobidzhan) 1947, "Posle zhatvy" (After the Harvest) 1948, "V puti" (On the Way) 1957, and others.

Gorizontov, Petr Dmitrievich
Pathophysiologist; b. S. 3 (16), 1902, Petropavlovsk.
Academician of the Academy of Medical Sciences of the USSR (1962).
He graduated from the Omsk Medical Institute in 1927 and took his further training under A. A. Bogomolets. He has been a professor since 1939. He became head of the subdepartment of pathological physiology at the First Moscow Medical Institute in 1941. From 1953 to 1960 he was head of the subdepartment of pathophysiology at the Central Advanced Training Institute for Doctors, from 1962 to 1968 he was director fo the Institute of Biophysics of the Ministry of Public Health of the USSR; he had been head of the institute's laboratory since 1952. His principal works have been the study of metabolic processes in the brain, the pathogenesis of radiation diseases, and the role of the adrenocortical steroid and pituitary hormones in pathology. He was the first to show the role of the neuroglia in cholesterol metabolism, he worked out the principles for the treatment of acute radiation disease, and he determined the changes in the blood during the various stages of stress reaction. He was chairman of the board of directors of the All-Union Scientific Medical Society of Pathophysiologists from 1956 to 1961. Decors.: Lenin Prize in 1963, two Orders of Lenin, two other orders, and medals.
Works: *Znachenie golovnogo mozga v kholesterinovom obmene.* Moscow 1940. Patologicheskaia fiziologiia luchevykh porazhenii. In *Radiatskonnaia meditsina,* 4th ed. Edited by A. I. Burnazian. Moscow 1968. *Rol' AKTG i kortikosteroidov v patologii.* Moscow 1968. (With P. N. Protasov.)

Gorkin, Aleksandr Fedorovich
Former govt. official; b. 1897, Rameshki nr. Tver', now Kalinin Oblast.
Russian; mem. CP 1916–; did not complete high school educ.; Sec., Chair. of Tver' Exec. Cttee. of Gubernia Sov. of Workers' Deps. 1917–18; mem. Kursk Collegium Gubernia-Cheka 1919; served in Sov. Army 1919–21; Sec., mem. Bureau Gubernia Cttee. RCP(B) in Tver' 1921–22; mem. staff Kirghiz Oblast Cttee. and Gubernia Cttee. RCP(B) in Orenburg 1922–23; various posts in "Selkhozsoiuz" and "Pititsevodsoiuz" 1923–26; mem. staff Cen. Cttee. CPSU 1926–29 and 1932–33; 2nd Sec. of Cen. Volga Krai CPSU Cttee. 1933–34; 1st Sec. of Orenburg CPSU (B) Cttee. 1934–37; Sec. of Presidium Cen. Exec. Cttee. (ZIK) 1937–38; Sec., Dep. Sec., Sec. of Presidium USSR Supr. Sov. 1938–57; cand. mem. Cen. Cttee. 1939–52; mem. Cen. Auditing Comm. CPSU 1952–59 and 1961–; Dep. to USSR Supr. Sov. 1946–74; Chair. of USSR Supr. Court 1957–72; Chair. of Cen. Auditing Comm. CPSU 1959–61. Decors.: Hero of Socialist Labour, 2 Orders of Lenin.

Gorkov, Lev Petrovich
Physicist; b. Je. 14, 1929, Moscow.
Grad. Moscow Physical-Tech. Inst. 1953; corresponding mem. USSR Acad. of Sciences 1966–; mem. staff Inst. of Physical Problems and Inst. of Physical Chemistry at USSR Acad. of Sciences 1955–66; Dept. Chief at Inst. of Theoretical Physics USSR Acad. of Sciences 1966–. Decor.: Lenin Prize 1966.

Gorshkov, Georgii Stepanovich
Volcanologist; b. F. 15, 1921, Irkutsk.
Corresponding member of the Academy of Sciences of the USSR (1966).
He was the director of the Institute of Volcanology of the Academy of Sciences of the USSR from 1966 until 1970 and became the president of the International Association of Volcanology and Chemistry of the Earth's Interior in 1971. His main works concern the present volcanism found in the Kuril Islands and on Kamchatka. He proposed a new classification of explosive eruptions and a method for the seismic location of magma chambers. He is the author of the hypothesis according to which the volcanism generated within the earth's upper mantle is virtually independent of the composition of the earth's crust.
Works: "Nekotorye voprosy teorii vulkanologii." *Izv. AN SSSR: Seriia geologicheskaia,* 1958, no. 11. *Vulkan Bezymiannyi i osobennosti ego poslednego izverzheniia (1955–1963).* Moscow 1965. (With G. E. Bogoiavlenskaia.) *Vulkanizm Kuril'skoi ostrovnoi dugi.* Moskow 1967.

Gorshkov, Leonid Aleksandrovich
Party official; b. 1930.
Mem. CPSU 1952–; grad. Siberian Metallurgical Mining Inst. 1952; asst. section chief, section chief, asst. chief engineer, 1st dep. chief of a mine 1952–54; sec. of party organization of a mine 1953–57; 2nd Sec. of Kemerovo City Cttee. CPSU 1957–60, 1st Sec. 1963–64; chief of a dept., Sec., 2nd Sec. of Kemerovo Oblast Cttee. CPSU 1964–74, 1st Sec. S. 1974–; Dep. to USSR Supr. Sov. 1974–; mem. Cen. Cttee. CPSU 1976–.

Gorshkov, Sergei Georgievich
Military official; b. F. 13 (26), 1910, Kamenets-Podolsk; s. of a teacher.
Russian; mem. CPSU 1942–; grad. Frunze Naval College 1931; Admiral of the Fleet; served in Navy 1927–; commanded various naval units, completed Torpedo Boat Commanders' Course 1937 and Advanced Officers' Training Course at Voroshilov Mil. Acad. 1941; during war commanded various naval units of Black Sea Fleet, Commander of Azov and Danube Flotillas; commanded a squadron 1944–48; Chief-of-Staff, Commander of Black Sea Fleet 1948–55; USSR Dep. Min. of Defence and Commander of Sov. Navy 1956–; Dep. to USSR Supr. Sov. 1954–; cand. mem. 1956–61, mem. Cen. Cttee. CPSU 1961–. Decors.: Hero of the Sov. Union, 4 Orders of Lenin, 4 Orders of Red Banner, 5 other orders and various medals. State Prize of the USSR 1980.

Gorshman, Shira
Author; b. 1906, Krok, Kaunas Oblast, Lithuania.
Jew; emigrated to Israel 1923, returned to USSR 1929; first works publ. 1939.
Publs.: stories "Sila zhizni" (The Power

of Life) 1948, "Tridtsat' tri novelly" (Thirty-three Novelettes) 1961, "Tretie pokolenie" (The Third Generation) 1963, and others.

Goryshin, Gleb Aleksandrovich
Author; b. 1931, Leningrad.
Russian; first works publ. 1952.
Publs.: "Khleb i sol'" (Bread and Salt) 1958, "V tridtsat' let" (At Thirty) 1961, and others.

Gostev, Boris Ivanovich
Party official; b. 1927.
Mem. CPSU 1954–; grad. Moscow Tech. Inst. for Light Ind. 1951; engineer, Chief Head of Service at "Burevestnik" Shoe Factory in Moscow 1951–53; in USSR Min. of Light Ind., Kalinin Econ. Council, State Econ. Council and USSR GOSPLAN 1953–63; various posts in Cen. Cttee. CPSU 1963–66; 1st Dep. Head of Dept. of Trade and Finance, Dept. of Planning and Finance at Cen. Cttee. CPSU 1966–75, Head 1975–; mem. CPSU Cen. Auditing Comm. 1971–76; cand. mem. Cen. Cttee. CPSU 1976–81, mem. 1981–.

Gotua, Levan Partenovich
Georgian writer; b. Mr. 10, 1905, Tbilisi. Graduated from the department of economics of the University of Tbilisi.
He is the author of the historical novellas *The Krtsanisi Satrow* (1956) and *The Wayward Caravan* (1957) and the historical novel *The Destiny of Heroes* (vols. 1–4, 1958–62) drawn from the life of 17th-century Georgia, which was torn by internecine feudal strife. The first book of the historical trilogy *Mithridates* (1968) is a narrative about ancient Colchis defending itself against the incursions of the Romans. His novel *The Lion Cub* (1966) and books of short stories and sketches deal with contemporary themes. During the Great Patriotic War of 1941–45 his historical dramas *King Erekle* (1942) and *David the Builder* (1945) and his play *The Invincible Ones* (1943), about the feats of Soviet sailors, played successfully in Georgian theaters.
Works: [Klot'ua, L.] *Gmirt'a varami: Romani*, vols. 1–4. Tbilisi 1965–67. *In Russian translation: Kamennyi fazan: Rasskazy.* Tbilisi 1962. *Rasskazy.* Tbilisi 1967.

Govorov, Vladimir Leonidovich
Military official; b. 1924.
Mem. CPSU 1946–; grad. M.V. Frunze Mil. Acad. 1949 and Mil. Acad. of Gen. Staff 1963, Sov. Army Gen. 1977; served in Sov. Army 1943–45; dep. commander, commander of a regiment, chief of staff of a division 1946–58; command posts 1958–69; 1st Dep. Commander of Sov. Forces Group in Germany, Commander of Troops in Baltic Mil. District 1969–71; Commander of Troops in Moscow Mil. District 1971–; cand. mem. Cen. Cttee. CPSU 1976–81, mem. 1981–.

Govorun, Nikolai Nikolaevich, Dr. of Phy.-Math. Sc.
Mathematician; b. 1930.
Mem. CPSU 1961–; grad. Moscow Univ. 1953; Dr. of Phy.-Math. Sc. 1969; Prof. 1970–; corresponding mem. USSR Acad. of Sciences 1972–; mem. staff Joint Inst. for Nuclear Research 1958–, Dept. Chief and Dep. Laboratory Dir. 1966–.

Grabin, Vasilii Gavrilovich
Designer of artillery scapons; b. D. 28, 1899 (Ja. 9, 1900), Ekaterinodar; s. of a worker.
Colonel-general of the engineering forces (1945), Hero of Socialist Labor (1940), doctor of engineering sciences (1941), professor (1951). Became a member of the CPSU in 1921.
He began his working life at the age of 13. In 1923 he graduated from the artillery school in Petrograd and in 1930 completed the Artillery Academy. In 1930 he began design work and in 1934 became chief plant designer. From 1942 to 1946 he was head of the Central Artillery Design Bureau and in 1946 became director and chief designer of a scientific research institute. From 1960 he has also been engaged in teaching. He developed and applied methods for highspeed designing of artillery systems with simultaneous designing of the production process. This made it possible to organize mass production of new models of guns in a short time to supply the Soviet Army during the Great Patriotic War of 1941–45. He was a deputy to the second and third convocations of the Supreme Soviet of the USSR. Decors.: State Prize of the USSR (1941, 1943, 1946, 1950), four Orders of Lenin, two Orders of the Red Banner, the Orders of Suvorov First and Second Class, the Order of the Red Banner of Labor, the Order of the Red Star, and medals.

Grachev, Aleksandr Matveevich
Author; b. 1912.
Russian; mem. CPSU 1940–; first works publ. 1935.
Publs.: "Taina Krasnogo Ozera" (The Secret of the Red Lake) 1948, "Pervaia proseka" (The First Lane) 1959, and others.

Grachov, Oleg Valerianovich
Painter; b. 1923, Cheremoshki, Tula Oblast.
Grad. Simferopol Art School 1950, pupil of S. Vladimirov.

Gradova, Nina Borisovna, Cand. of Biol. Sc.
Scientist; b. 1933.
Mem. CPSU 1956–; grad. K.A. Timiriazev Agricultural Acad. 1957; soil scientist, agro-chemist; Cand. of Biol. Sc. 1964; mem. staff K.A. Timiriazev Agricultural Acad. 1957–64 (cand. 1960–63); mem. staff Union Research Inst. for Biosynthesis of Proteins 1964–, Laboratory Chief 1966–. Decor.: State Prize of the USSR 1971.

Grafov, Sergei Sergeevich
State official.
Dep. Chair. of Kishinev City Exec. Cttee. of Workers' Deps. 1973–75; Chair. of State Cttee. for Construction Moldavian SSR Jl. 1975–; mem. Cen. Cttee. Moldavian CP 1976–.

Granatkin, Valentin Aleksandrovich
Athlete, public figure; b. Jl. 3 (16), 1908, Ryblovo, present-day Rybolovo, in Ramenskii Raion, Moscow Oblast.
Honored Master of Sports (1942), Honored Cultural Worker of the RSFSR (1968), engineer.
He became a member of the CPSU in 1939. He graduated from the Higher Party School of the Central Committee of the CPSU. He is vice-president of the International Federation of Association Football (FIFA, from 1946) and chairman of the USSR Football Federation (1950–64 and from 1968). Decors.: two orders, medals.

Granin, Daniil Aleksandrovich (real name: German)
Author; b. 1918, Volyn, Kursk Oblast; s. of a forester.
Russian; mem. CPSU 1942–; grad. Leningrad Polytech. Inst. 1940; Chief Engineer of Energy Laboratory and Construction Bureau at Kirov Works in Leningrad 1940–49; engineer at a scientific research inst. 1949–50; first works publ. 1949.
Publs.: "Pobeda Inzhenera Korsakova" (The Victory of Engineer Korsakov) 1949, "Variant vtoroi" (The Second Version) 1949, "Iskateli" (The Searchers) 1954, "Posle svad'by" (After the Wedding) 1958, "Idu na grozu" (I Attack the Storm) 1962, "Iaroslav Dombrovskii" 1951, "Sobstvennoe mnenie" (Own Opinion) 1956, and others.

Grebenev, Naum Isaevich
Poet and translator; b. 1921, Kharbin.
Russian; mem. CPSU 1945–; grad. Gorkii Inst. of Literature Moscow 1949; first works publ. 1944.
Publs.: "Pravoe delo" (The Right Thing) 1948, "Balkarskaia narodnaia lirika" (The Balkar Folk Lyric) 1959, and others.

Grebeniuk, Vasilii Andreevich
State official; b. 1924.
Russian; mem. CPSU 1954–; grad. Kazakh Inst. of Mining Metallurgy 1951; Cand. of Tech. Sciences 1967; collective farmer 1942; served in Sov. Army 1942–45; foreman, dept. chief, dep. chief engineer, chief engineer, mine supervisor at a lead combine in Zyriansk 1951–63; dir. of a polymetals combine in Leninogorsk 1963–73; 1st Dep. Min. of Nonferrous Metallurgy Kazakh SSR 1973–74, Min. 1974–79; mem. Cen. Cttee. Kazakh CP 1976–; mem. Bureau Cen. Cttee. Ka-

zakh CP 1979–; 1st Dep. Chair. Council of Mins. Kazakh SSR Ja. 1979–; Dep. to USSR Supr. Sov. 1979–; mem. Cen. Auditing Comm. CPSU 1981–. Decor.: Hero of Socialist Labor 1971.

Grebennik, Vitalil Sergeevich
Painter; b. 1928, Krolevets, Sumy Oblast. Grad. Kiev Art Inst. 1956.

Grechanyk, Viktor Vasil'evich
Sculptor; b. 1929, Gorodeshchina, Kiev Oblast.
Grad. Kiev Art Inst. 1955, pupil of M. Lysenko, M. Vronskii.

Grechkina, Elza Robertovna
State official.
Head of Dept. of Science and Educ. Institutions with Cen. Cttee. Esthonian CP 1974–80; Min. of Educ. Esthonian SSR Jl. 1980–; mem. Cen. Cttee. Esthonian CP 1976–.

Greisha-Lisovskii, Viktor Stanislavovich
State official.
Chair. of State Cttee. of Gasification Latvian SSR Mr. 1982–.

Grekov, Leonid Ivanovich
Party official; b. 1928.
Mem. CPSU 1949–; grad. Kharkov Aviation Inst. 1954; served in Sov. Army 1943–44; engineer, Dep. Sec., Sec. of CPSU Cttee. of Cen. Inst. of Aircraft Engine Construction in Moscow 1954–63; Sec. 1963–66, 1st Sec. of Kalinsinskii Raion CPSU Cttee. in Moscow 1966–71; 2nd Sec. of Moscow City CPSU Cttee. Mr. 1971–76; mem. Cen. Cttee. CPSU 1971–; 2nd Sec. and mem. Bureau Cen. Cttee. Uzbekistan 1976–.

Greku, Mikhail Grigor'evich
Painter; b. N. 9 (22), 1916, Faraonovka, in present-day Odessa Oblast.
He studied at the Academy of Arts in Bucharest (1937–40) and at the I. E. Repin Republic Art School in Kishinev (1945–47). He created a series of pictures on themes of the history of the Moldavian people, marked by deep dramatic effect and by heightened emotional expressiveness of color (*The Tatarbunar Uprising,* 1956–57). With the early 1960's a yearning for a poetic understanding of the life of the contemporary Moldavian village and its socially significant phenomena predominates in his works (genre compositions, landscapes). His *Chadyrlung Girls,* 1960; *The Sendoff,* 1964–65; and the triptych *The Story of a Life,* 1966, are all in the Art Museum of the Moldavian SSR in Kishinev. Decor.: Order of the Badge of Honor.

Gren, Arnold Karlovich
State official.
Grad. Party High School of Cen. Cttee. CPSU; served in Sov. Army 1941–45; Vice-Chair. of Council of Mins. Esthonian SSR 1953–58; mem. Cen. Cttee. Esthonian CP 1954–; Min. of Educ. Esthonian SSR 1958–60; again Vice-Chair. of Council of Mins. 1960– and also Min. of Foreign Affairs Esthonian SSR 1962–. Decor.: Order of the Red Star.

Grib, Alla Lavrentevna
State official.
1st Sec. of Central Raikom Belorussian CP in Minsk until 1977; cand. mem. Cen. Cttee. Belorussian CP 1976–81, mem. 1981–; Min. of Public Services Belorussian SSR D. 1977–.

Gribachev, Nikolai Matveevich
Author; b. 1910, village of Lopush, Briansk Oblast; s. of a peasant.
Mem. CPSU 1943–; writer, hydraulic engineer by profession; Chief Ed. of journal "Sovetskii Soiuz" 1950–54 and 1956–; Sec. and mem. Presidium of USSR Writers' Union 1953–54 and 1959–; cand. mem. CPSU 1961–; Chair. RSFSR Supr. Sov. 1980–. Decors.: State Prize of the USSR 1948, 1949, Lenin Prize 1960, Hero of Socialist Labour 1974, Order of the October Revolution, 6 other orders and various medals.
Publs.: "Kolkhoz-'Bol'shevik'" (The 'Bolshevik' Kolkhoz) 1947, "Vesna v 'Pobede'" (Spring at 'Pobeda') 1948, "Avgustovskie zvezdy" (August Stars) 1958, "Nochnaia groza" (Night Thunderstorm) 1964, "Liubov' moia shal'naia" (My Wild Love) 1967, "Belyi angel v pole" (White Angel in the Field) 1967, "Stal' i Mol'" (Steel and Mole) 1963, "Liubov', i trevoga, i boi" (Love, Alarm and Battle) 1966, "A ona vertitsia ..." (And Yet She Turns ...) 1966, "I veter s chetyrekh storon ..." (And the Wind from Four Directions ...) 1968, "Vybor veka" (Election of the Century) 1969, and others.

Gribanov, Iurii Evgenevich
Engineer; b. 1927.
Mem. CPSU 1966–; grad. Shipbuilding Inst. Nikolaev 1950; engineer, shipbuilder; Chief Specialist with Main Admin. of USSR Min. of Shipbuilding 1965–. Decor.: State Prize of the USSR 1972.

Gribanov, Mikhail Grigor'evich
Diplomat; b. 1906.
Entered Diplomatic Service 1941; Adviser to Sov. Mission in Switzerland 1946–47; Dep. Pol. Adviser of Sov. Mil. Admin. in Germany 1947–49; Dep. Head, Head of 3rd European Dept. in USSR Min. of Foreign Affairs 1949–53; leading posts in USSR Min. of Foreign Affairs 1953–55; Head of Dept. of Scandinavian Affairs at USSR Min. of Foreign Affairs 1955–56; USSR Amb. to Norway 1956–62; leading posts in USSR Min. of Foreign Affairs, Head to Admin. of General Int. Problems at USSR Min. of Foreign Affairs 1962–78.

Gribkov, Anatolii Ivanovich
Military official; b. 1919; son of a peasant.
Mem. CPSU 1941–; grad. Frunze Acad. 1942 and Mil. Acad. of Gen. Staff 1951, Army Gen.; entered Sov. Army 1938; represented Gen. Staff on staff of various units 1942–44; served on Gen. Staff of Sov. Armed Forces 1945–51; served on staff of various mil. districts, on Gen. Staff, commander of an army, 1st Dep. Commander, Commander of Troops in Leningrad Mil. District 1951–76; mem. Cen. Cttee. and mem. Bureau Cen. Cttee. Armenian CP 1966–70; Dep. to USSR Supr. Sov. 1974–; cand. mem. Cen. Cttee. CPSU 1976–81, mem. 1981–; 1st Dep. Chief of Gen. Staff of Sov. Army and Navy Fleet, Chief of Gen. Staff of Troops of Warsaw Pact 1976–.

Gribov, Vladimir Naumovich, Dr. of Phy.-Math. Sc.
Physicist; b. 1930.
Grad. Leningrad Univ. 1952; Dr. of Phy.-Math. Sc. 1964; Prof. 1968–; corresponding mem. USSR Acad. of Sciences 1972–; Sen. Laboratory Technician; Jun. Research Asst., Sen. Research Asst., Section Chief at Physical-Tech. Inst. of USSR Acad. of Sciences 1954–71; Dept. Chief at Inst. of Nuclear Physics of USSR Acad. of Sciences in Leningrad 1971–.

Gridasov, Dmitrii Matveevich
Oblast sov. official; b. 1916.
Russian; mem. CPSU 1940–; grad. Cen. Cttee. Party Higher School Ukrainian CP and Kiev Econ. Inst; laborer, work brigade leader, asst. shop superintendent in a plant 1934–39; Komsomol and party posts 1939–47; 1st sec. of a raion cttee., of a city cttee. Ukrainian CP 1950–56; 1st Dep. Chair. of Exec. Cttee. of Donetsk Oblast Sov. of Peoples' Deps. 1956–59, Chair. 1959–; mem. Cen. Cttee. Ukrainian CP 1960–.

Grigaliunas, Medardas Iulevich
State official.
Mem. Cen. Cttee. Lithuanian CP 1961–; mem. Bureau of Agriculture at Cen. Cttee. Lithuanian CP 1962–64; Dep. Chair., 1st Dep. Chair. of Council of Mins. and Min. of Production and Procurement of Agricultural Products Lithuanian SSR 1963–65; Min. of Agriculture Lithuanian SSR Ap. 1965–; mem. USSR Council of Kolkhozes 1969–; Chair. of Council of Kolkhozes Lithuanian SSR 1970–.

Grigor'ev, Sergei Alekseevich
Painter; b. Je. 22 (Jl. 5), 1910, Lugansk, present-day Voroshilovgrad.
People's Artist of the Ukrainian SSR (1951). Member of the Academy of Arts of the USSR (1958). Member of the CPSU since 1941.
He studied at the Kiev Art Institute (1928–32) with F. G. Krichevskii. Later,

he taught there (1934–60), becoming a professor in 1948. His paintings, which are carefully executed, with accurate, lifelike details, are concerned with problems of Soviet morality, the upbringing of young people, and the world of children. His works have had an appreciable influence on the development of Soviet genre painting. His paintings include *At a Meeting* (1947, Kharkov Art Museum) and *Admission to the Komsomol* (1949, Museum of Ukrainian Fine Arts of the Ukrainian SSR, Kiev) an *The Goalie* (1949, Tret'iakov Gallery), both of which received the State Prize of the USSR in 1950. He also painted *Discussion of a "Poor" Grade* (1950, State Prize of the USSR, 1951) and *He Has Returned* (1954), both of which are in the Tret'iakov Gallery. Decors.: three orders as well as medals.

Grigorev, V. V.
Komsomol official.
Mem. Bureau 1970–74, Sec. 1970–72, 1st Sec of Cen. Cttee. Belorussian Komsomol 1973–74; Sec. and mem. Bureau of Cen. Cttee. USSR Komsomol Ap. 1974–80; 2nd Sec. Brest oblast Cttee. Belorussian CP D. 1980–; mem. Cen. Cttee. Belorussian CP 1981–.

Grigorianis, Artem Nikolaevich
Scientist; b. 1916.
Mem. CPSU 1944–; grad. Moscow Univ. 1941; specialist in thermo-energetics. Decors.: State Prize of the USSR for his part in the construction of the I. V. Kurchatov Atomic Energy Plant in Beloiarsk.

Grigorovich, Iurii Nikolaevich
Ballet dancer and choreographer; b. Ja. 7, 1927.
Grad. Choreographic College Leningrad 1946 and Moscow State A. V. Lunacharskii Inst. for Theatre Arts ("GITIS") 1965; ballet dancer and master at Leningrad Kirov Opera and Ballet Theatre 1946–64; ballet master at Bolshoi Theatre Moscow 1964–. Decors.: People's Artist of the RSFSR 1966, Lenin Prize for his part in the choreography of "Spartak" ballet at Bolshoi Theatre 1970.

Grigulis, Arvid Petrovich
Author and literary historian; b. 1906.
Latvian; mem. CPSU 1954–; grad. Law Faculty of Latvian Univ. 1937; first works publ. 1927.
Publs.: poems "Zapiski reportera" (A Reporter's Notes) 1929, "Imitatsiia i serdtse" (Imitation and Heart) 1931, "V dome ustalykh" (In the House of the Weary) 1934, "Liudi v sadu" (People in the Garden) 1939, "V buriu" (In the Storm) 1946, "Skvozt' ogon' i vodu" (Through Fire and Water) 1945; plays "V kakuiu gavan'?" (In Which Harbour?) 1945, "Kak v Garpeterakh istoriiu delali" (How History Was Made in Garpeteri) 1946, "Soldatskaia shinel'" (Soldier's Coat) 1955, "Baltiiskoe more shumit" (The Baltic Rages) 1957, comedy "I tak byvaet na belom svete" (Life Is Like That Sometimes) 1959, and others.

Grin, Elmar (real name: Iakimov, Aleksandr Vasilevich)
Author; b. Je. 2 (15), 1909, Kivennapa, now Pervomaisk, Leningrad Oblast; s. of a peasant.
Russian; mem. CPSU 1943–; first works publ. 1926. Decor.: State Prize of the USSR 1946.
Publs.: "Veter s Iuga" (Wind From the South) 1946, "Drugoi put'" (Another Way) vol. 1, 1956, vol. 2 ("V strane Ivana" – In the Land of Ivan) 1962–67, "Proidennye bolota" (Traversed Moors) 1966, and others.

Grinberg, Georgii Abramovich
Physicist; b. Je. 3 (16), 1900, St. Petersburg.
Grad. Polytech. Inst. Petrograd 1923; corresponding mem. USSR Acad. of Sciences 1946–; mem. staff State Roentgenologic and Radiologic Inst. Leningrad 1919–30; Lecturer at Polytech. Inst. Leningrad 1924–55, Prof. 1930–; mem. staff Physical-Tech. Inst. of USSR Acad. of Sciences 1941–. Decors.: State Prize 1949, Order of Lenin, Order of Red Banner of Labour.

Grinberg, Iosif Lvovich
Author and critic; b. Ap. 26 (My. 9), 1906, Aleksandrovsk, now Zaporozhe.
Grad. Leningrad Univ. 1930; author of numerous works on the Sov. writers E. Bagritskii, N. Tikhonov, V. Inber and others as well as essays on contemporary Sov. literature.
Publs.: "Eduard Bagritskii. Tvorcheskii put' poeta" (Eduard Bagritskii. The Creative Development of a Poet) 1940, "Poeziia Alekseia Surkova" (The Poetry of Aleksei Surkov) 1958, "Tvorchestvo Nikolaia Tikhonova" (The Work of Nikolai Tikhonov), enlarged and revised ed. publ. 1958, "Vera Inber" 1961, "V bor'be za idealy kommunizma" (In the Fight for the Ideals of Communism) 1963, and others.

Grintsov, Ivan Grigor'evich
Party official; b. 1935.
Ukrainian, mem. CPSU 1960–, grad. Lugansk Agricultural Inst. 1957; agronomist at "Kirov" and "Michurin" collective farms in Donetsk Oblast 1957–61; Chair. of "Krasnaia Armiia" Collective Farm in Donetsk Oblast 1961–65; chair. of a collective farm in Marinka Raion in Donetsk Oblast 1965–67; 1st Sec. of Velikaia Novoselka Raion Cttee. Ukrainian CP in Donetsk Oblast 1967–71; Sec. of Donetsk Oblast Cttee. Ukrainian CP 1971–75; Dep. to USSR Supr. Sov. 1974–; 1st Sec. of Sumy Oblast Cttee. Ukrainian CP Ap. 1975–; mem. Cen. Cttee. Ukrainian CP 1976–; cand. mem. Cen. Cttee. CPSU 1976–.

Grishin, Evgenii Romanovich
Ice skater; b. Mr. 23, 1931, Tula.
Honored Master of Sport (1952). Officer of the Soviet Army. Member of the CPSU since 1962.
In the 1950's and 1960's, he was repeatedly a Soviet and world record holder and champion. At the Seventh and Eighth Winter Olympics (Cortina d'Ampezzo, Italy, 1956; Squaw Valley, California, 1960), he won the 500-m and 1,500-m speed skating races. His best times were 39.5 sec for 500 m (1963, Alma-Ata) and 2 min 8.2 sec for 1,500 m (1960, Squaw Valley). He was overall champion of Europe in 1956 at Helsinki. In the early 1950's he was one of the best Soviet bicycle racers (on the track). Decors.: Order of Lenin, Order of the Red Banner.

Grishin, Viktor Vasil'evich
Party official; b. S. 18, 1914, Serpukhov; s. of a railway worker.
Russian; mem. CPSU 1939–; grad. Moscow Tech. College of Geodesy 1932 and Tech. Coll. of Railway Engine Econ. 1937; corresponding student at Party High School of Cen. Cttee. CPSU 1937; land consolidation technician of Raion Dept. of Agriculture in Serpukhov, Moscow Oblast 1932–33; Dep. Head of Engine Dept. in Serpukhov 1937–38 and 1940–41; served in Sov. Army 1938–40; Sec. of CPSU Cttee. at Railway Junction in Serpukhov 1941–42; Sec., 2nd Sec., 1st Sec. of Serpukhov City CPSU Cttee. 1942–50; Head of Dept. of Machine Building at Moscow Oblast CPSU Cttee. 1950–52; mem. Cen. Cttee. CPSU 1952–; Dep. to USSR Supr. Sov. 1950–; 2nd Sec. of Moscow Oblast CPSU Cttee. 1952–56; Chair. of Cen. Council of Trade Unions of USSR 1956–67; cand. mem. 1961–71, mem. Politbureau Cen. Cttee. CPSU Ap. 1971–; 1st Sec. of Moscow City CPSU Cttee. Je. 1967–; mem. Presidium USSR Supr. Sov. 1967–. Decors.: 3 Orders of Lenin, decorations and various medals.

Grishina, Raisa Mikhailovna
State official.
Min. of Trade Tadzhik SSR O. 1974–81; mem. Cen. Cttee. Tadzhik CP 1976–; Dep. Chair. of Council of Mins. Tadzhik SSR My. 1981–.

Grishkiavichius, Piatras Piatrovich
Party official; b. 1924.
Lithuanian; mem. CPSU 1945–; grad. Party High School of Cen. Cttee. CPSU; served in World War II, partisan, after the war worked as journalist and party official 1941–55; Sec. of Vilnius City Cttee. Lithuanian CP 1955–63; mem. Cen. Cttee. Lithuanian CP 1960–; mem. Bureau of Agriculture Cen. Cttee. Lithuanian CP 1962–63; 2nd Sec. of Vilnius City Cttee. Lithuanian CP 1963–64; Head of Dept. of Party Org. Work at Cen. Cttee. Lithuanian CP 1964–71; Dep. to Supr. Sov. Lithuanian SSR 1965–; cand. mem. 1966–71, mem. Bureau Cen. Cttee.

Lithuanian CP 1971–; 1st Sec. of Vilnius City Cttee. Lithuanian CP 1971–74; 1st Sec. of Cen. Cttee. Lithuanian CP F. 1974–; mem. Cen. Cttee. CPSU 1976–. Decor.: Order of Lenin.

Gritsai, Aleksei Mikhailovich
Painter; b. F. 22 (Mr. 7), 1914, St. Petersburg.
People's Artist of the RSFSR (1963). Member of the Academy of Arts of the USSR (1964).
From 1932 to 1939, he studied at the Academy of Arts of Leningrad under I. I. Brodskii and V. N. Iakovlev. He has taught at the V. I. Surikov Moscow Art Institute from 1948 to 1952 and from 1964 to the present. He has been a professor there since 1966. In his landscapes he follows the tradition of Russian landscape painting of the second half of the 19th century. This is especially evident in his *In the Zhiguli: A Stormy Day* (1948–50, Tret'iakov Gallery), which won the State Prize of the USSR in 1951, and in his *Black Water* (1967, painter's private collection). He has painted portraits, such as the group portrait *A Session of the Presidium of the Academy of Sciences of the USSR* (in collaboration with V. P. Efanov and others, 1951, the Russian Museum in Leningrad), which won the State Prize of the USSR in 1952, and has also painted works on industrial themes, such as *The Electric Furnaces* (1947, Tret'iakov Gallery). Decors.: Order of the Patriotic War Second Class, medals.

Gritsenko, Nikolai Olimpievich
Russian actor; b. Jl. 11 (24), 1912, Iasinovataia, in present-day Donetsk Oblast.
People's Artist of the USSR (1964).
In 1940, he graduated from the school of the E. Vakhtangov Theater and joined the company of this theater. Inherent in his art is the free-and-easy mastery of many genres, a disposition to completely transform himself internally and externally, and the ability to make use of techniques of the grotesque as well as exaggeration. Some of his most important comedy roles are Molokov in *Down the Gold Mine* (based on Mamin-Sibiriak's work), Tartaglia in Gozzi's *Princess Turandot*, Mamaev in Ostrovskii's *Even a Wise Man Stumbles*, and Kazanets in Sofronov's *The Cook* and *The Cook Married*. The comical palette of Gritsenko is rich with colors: from ironic, improvised theatrical acting to sharp and sweeping satirical generalizations.
An important part of his creativity is his work in the dramatic roles of the Russian classics: Fedor Protasov in L. N. Tolstoy's *The Living Corpse* and Prince Myshkin in *The Idiot* (based on Dostoevsky's novel). In these roles the actor conveys spiritual goodness, freethinking, and revolt against hypocrisy. He has played the role of Tsvetukhin in *First Joys* and *Kirill Izvekov* (based on Fedin's novels *First Joys* and *An Unusual Summer*). He has also appeared on the screen as Artamashev in *The Bearer of the Golden Star*, as Roshchin in the trilogy *Road to Calvary* (based on A. N. Tolstoy's *Sisters. The Year 1918*, and *A Dull Morning*), and as Karenin in *Anna Karenina*. Decors.: State Prize of the USSR in 1952, and the K. S. Stanislavsky State Prize of the RSFSR in 1970. Order of the Red Banner of Labor and medals.

Gritsuk, Ivan Mikhailovich
State official.
Min. of Communications Belorussian SSR Ja. 1980–; mem. Auditing Comm. Belorussian CP 1981–.

Grizodubova, Valentina Stepanovna
Flyer; b. Ja. 18 (31), 1910, Kharkov.
Hero of the Soviet Union (N. 2, 1938), and colonel (1943). Became a member of the CPSU in 1941. Daughter of the airman and aircraft designer S. V. Grizodubov.
She completed flying-club training in 1929 and worked in the civil air fleet. On S. 24–25, 1938, she made a nonstop flight with P. D. Osipenko and M. M. Raskova in an ANT-37 from Moscow to the Far East, setting a women's world distance record. In the Great Patriotic War (1941–45) the assumed command in March 1942 of the 101st Long-Range Air Group (later the 31st Guards Bomber Group), which lent support to partisan detachments. Decors.: Order of Lenin, the Order of the Red Banner of Labor, the Order of the Patriotic War First Class, Order of the Red Star, medals.

Gromov, Mikhail Mikhailovich
Military official; b. F. 12 (24), 1899, Tver', now Kalinin; s. of a doctor.
Colonel-general of aviation (1944). Hero of the Soviet Union (S. 28, 1934). Honored Pilot of the USSR, professor (1937). Became a member of the CPSU in 1941.
He graduated from the Zhukovskii Theoretical Courses at the Higher Technical School in 1917. In 1918 he joined the Soviet Army and graduated from pilot school. He then took part in the Civil War on the Eastern Front. After the war he was an aviation instructor and test pilot. In 1934 he was proclaimed a Hero of the Soviet Union for his world record nonstop 75-hour flight over a closed course covering more than 12,000 km. In 1937, with A. B. Iumashev and S. A. Danilin, he completed a flight from Moscow to the North Pole and then on to the United States. In 1940–41 he was head of the Institute of Flight Research. Beginning in December 1941 he took part in the Great Patriotic War as commander of an air force division (until February 1942), as commander of the air force on the Kalinin Front (from February to May 1942), and as commander of the Third (until May 1943) and First (until June 1944) air forces. From July 1944 he was chief of the Main Directorate for Combat Training of Aviation at the Front, and from 1946 he was deputy commander of long-range aviation.
He was a deputy to the first and second convocations of the Supreme Soviet of the USSR. Decors.: three Orders of Lenin, four Orders of the Red Banner, the Orders of Suvorov Second Class and of the Patriotic War First Class, three Orders of the Red Star, and various medals. He was the first Soviet pilot to receive an award of the International Aeronautical Federation – the De Lavaud Medal (1937).

Gromova, Mariia Sergeevna
Milkmaid; b. 1929.
Russian; mem. CPSU 1973–; finished 7th class in secondary school 1945; collective farmer 1945–62; milkmaid, foreman of mechanized milking at "Kommunarka" State Breeding Plant in Lenin Raion (Moscow Oblast) 1962–; Dep. to USSR Supr. Sov. 1979–; mem. Cen. Auditing Comm. CPSU 1981–. Decors.: Heroine of Socialist Labor 1975, USSR State Prize 1975.

Gromovich, Ivan Ivanovich
Author; b. 1918, Zakruzhka nr. Minsk; s. of a peasant.
Belorussian; mem. CPSU 1943–.
Publs.: "Rasskazy" (Stories) 1947, "Shirokie ozera" (Vast Lakes) 1951, "Rina-Marina" 1958, "Zhavoronok" (The Lark) 1961, and others.

Gromyko, Anatolii Andreevich, Dr. of Hist. Sc., Cand. of Legal Sc.
Political scientist; b. 1932; s. of the USSR Min. of Foreign Affairs.
1st Sec. at USSR Emb. in London 1961–65; head of a section at African Inst. of USSR Acad. of Sciences 1966–69; Head of Section for General Trends in US Foreign Policy (renamed Inst. of the United States of America and Canada in 1974) at USSR Acad. of Sciences 1969–76; Dir. of African Inst. of USSR Acad. of Sciences D. 1976–; corresponding mem. USSR Acad. of Sciences 1981–; scholarly interests focused on two main areas: interdisciplinary study of the socioeconomic, political, and international problems of the African countries and the central problems of modern imperialism – the analysis of the foreign and domestic plicy of the USA; author of more than 150 publications; organized a series of collective studies on the evolution of the national economy and developmental strategy of the African countries in the 1970's and 1980's, on the social changes in the independent African countries and on their foreign policy; Chair. of the Scientific Council on the Problems of Africa USSR Acad. of Sciences, mem. editorial advisory council of the periodical *Gesellschaftswissenschaften* (Social Sciences).

Gromyko, Andrei Andreevich, Dr. of Econ. Sc.
State official; b. Jl. 5 (18) 1909, Starye Gromyki, Gomel Oblast; s. of a peasant.
Russian; mem. CPSU 1931–; grad. from an econ. inst. 1932 and an agricultural techn. school in Minsk 1936; postgraduate student; sen. researcher at Inst. of Econ. of USSR Acad. of Sciences 1936–39; Diplomatic Service 1939–; Head of American Dept. in USSR People's Commissariat for Foreign Affairs, Min. at USSR Emb. in Washington 1939–43; Sov. Amb. to USA and Sov. Min. in Cuba 1943–46; USSR Perm. Rep. at UN 1946–48; Dep. to USSR Supr. Sov. 1946–50 and 1958–; USSR Dep. 1947–49, 1st Dep. Min. of Foreign Affairs 1949–52; Sov. Amb. to UK 1952–53; cand. mem. 1952–56, full mem. Cen. Cttee. CPSU 1956–; USSR 1st Dep. 1953–57 and Min. of Foreign Affairs F. 1957–; mem. Politbureau Cen. Cttee. CPSU Ap. 1973–. Decors.: 5 Orders of Lenin and other orders and medals. Hero of Socialist Labour 1969, 1979.

Gross, Villem
Author; b. 1922, Tartu; s. of a worker.
Esthonian; mem. CPSU 1946–.
Publs.: "Leto adnogo mal'chika" (A Boy's Summer) 1956, novel "Prodaetsia nedostroennyi individual'nyi dom" (A Halffinished Private House For Sale) 1961, and others.

Grossu, Semen Kuzmich
Party official; b. 1934.
Moldavian; mem. CPSU 1961–; grad. Kishinev Agricultural Inst. 1959, Cand. of Econ. Sc. 1968; chief agronomist, chair. of a collective farm, chief of raion collective and state farm admins. in Moldavian SSR 1959–65; mem. Cen. Cttee. Moldavian CP 1963–; Chief of Suvorov Raion Prod. Admin. for Agriculture 1965–67; Dep. to USSR Supr. Sov. 1966–; 1st Sec. of Kriuliany Raion Cttee. Moldavian CP 1967–70; mem. Bureau Cen. Cttee. Moldavian CP 1970–; Sec. of Cen. Cttee. Moldavian CP 1970–76, 1st Sec. D. 1980–; Chair. of Council of Mins. and Min. of Foreign Affairs Moldavian SSR 1976–D. 1980; mem. Cen. Cttee. CPSU 1981–.

Grubian, Matvei Mikhailovich
Author and poet; b. 1919, Sokolovka, now Kiev Oblast.
Jew; grad. Minsk Pedagogical Inst. 1938; first works publ. 1930.
Publs.: "Iz podvala – k solntsu" (Out of the Cellar – Into the Sun) 1935; "Lirika" (Lyric) 1940, "Pesnia o muzhestve" (The Song of Bravery) 1947, "Po sledam voiny" (On the Tracks of War) 1947, "Ia zval tebia, zhizn'" (I Called You, Life) 1958, "Kliuchi" (The Keys) 1962, and others.

Grunianskii, Ivan Ivanovich
State official.
Min. of Timber, Cellulose, Paper and Timber Processing Ind. Ukrainian SSR 1965–68; Min. of Timber and Timber Processing Ind. Ukrainian SSR 1968–; Chair. of Ukrainian Branch of Sov.-Hungarian Friendship Soc. 1974–; mem. Auditing Comm. Ukrainian CP 1976–81.

Gruodis, Mikolas Alfonsovich
State official.
Chair. of State Cttee. for Pricing Lithuanian SSR Ap. 1970–; mem. Auditing Comm. Lithuanian CP 1976–.

Grushin, Petr Dmitrievich, Dr. of Tech. Sc.
Engineer; b. Ja. 2 (15), 1906, Volsk, Saratov Oblast.
Mem. CPSU 1931–; grad. Moscow Aviation Inst. 1932; specializes in aviation technology; mem. USSR Acad. of Sciences 1966–; sen. engineering-tech. posts 1932–48; dean of a dept. at Moscow Aviation Inst. 1948–51; head of a plant 1953–; mem. Cen. Cttee. CPSU 1966–. Decors.: Hero of Socialist Labour 1958, Lenin Prize 1963, 5 Orders of Lenin, Order of the October Revolution, medals.

Grutso, Aleksei Petrovich, Dr. of Philol. Sc.
Philologist; b. 1924.
Belorussian; grad. Pedagogical Inst. Minsk 1951; Cand. 1958–69, Dr. of Philol. Sc. 1969; served in Sov. Army 1945–47; school teacher in Minsk 1951–58; mem. staff Inst. of Philology at Acad. of Sciences Belorussian SSR 1958–69; Prof. at Pedagogical Inst. Minsk 1969–. Decor.: State Prize of the USSR 1971.

Gruzberg, Semen Borisovich
Graphic artist; b. 1918; Kamianets-Podolskii, Khmelnitskii Oblast.
Attended Kiev Tech. School of Art 1933–35 and Kharkov Art Inst. 1936–40, pupil of S. Besedin, V. Kasian.

Gruzinov, Sergei Sergeevich
Diplomat; b. 1920.
Entered Diplomatic Service 1956–; sen. posts in USSR Min. of Foreign Affairs 1956–63; 1st Raion Sec. of CPSU Frunze Raion in Moscow 1963–70; USSR Amb. to Algeria 1970–Ap. 1975, and to Burma 1976–80.

Grynko, Nikolai Konstantinovich
State official; b. 1928, Mylkovichi, Grodno Oblast.
Belorussian; mem. CPSU 1954–; grad. Leningrad Mining Inst. 1952, Cand. of Tech. Sc.; chief of a section, chief engineer, dir. of a mine, chief engineer of a trust, of a combine, Dep. Min. of the Coal Ind. Ukrainian SSR 1952–70; Chief of Technical Admin. of USSR Min. of the Coal Ind. 1970–78; Dep. USSR Min., 1st Dep. USSR Min. of the Coal Ind. F.–Ap. 1978; 1st Dep. Min. of the Coal Ind. Ukrainian SSR My.–D. 1978, Min. D. 1978–; mem. Cen. Cttee. Ukrainian CP 1981–.

Gubai, Garif (real name: Gabaidullin, Garif Shakirovich)
Author; b. 1907, Kazan; s. of a craftsman.
Tartar; mem. CPSU 1939–; first works publ. 1934.
Publs.: children's books "Kalancha" (Watchtower) 1934, "Deti nashego vremeni" (The Children of Our Time) 1949, "Sud'ba Marata" (Marat's Fate) 1956, "Pervyi dozhdik" (First Rain) 1960, and others.

Gubarevich, Konstantin Leontevich
Dramatist; b. 1906, Raduchi, Mogilev Oblast; s. of a peasant.
Belorussian; first works publ. 1926. Decor.: Order of Red Banner of Labour.
Publs.: poems and stories "Ajda!" (Go On!) 1930, "Zhivye vinty" (Living Screws) 1930, plays "Tsentral'nyi khod" (A Central Course) 1947, "Alazanskaia dolina" (The Alazansk Valley) 1948, "Brestskaia krepost'" (The Brest Fortress) 1950, "Prostaia devushka" (A Simple Girl) 1953, "Na krutom povorote" (On a Sharp Turn) 1955, drama "Glavnaia stavka" (The Main Stake) 1957, filmscript "Devochka ishchet ottsa" (A Little Girl in Search of Her Father) 1959, and others.

Gubnitskii, Alesksandr Matveevich
Author; b. 1912, Monastyrishche, Ukraine; s. of a craftsman.
Jew; mem. CPSU 1943–; first works publ. 1932.
Publs.: "Shofery" (Chauffeurs) 1959, "Put' na Stalingrad" (The Road to Stalingrad) 1963, "Den' v Bratske" (A Day in Bratsk) 1963, "Khoziain moria" (The Lord of the Sea) 1963, and others.

Gudiashvili, Lado Davidovich (also Vladimir Gudiashvili)
Painter and graphic artist; b. Mr. 18 (30), 1896, Tbilisi.
People's Artist of the Georgian SSR (1957).
He studied in Tbilisi at the School of Painting an Sculpture (1910–14). From 1919 to 1926 he lived in Paris, where he studied at a private studio. His work, influenced by medieval Georgian art and Western European developments of the 1920's, is replete with folklore motifs and mythological and fantastic mages – for example, *Serafita's Outing*, a portrait of the actress Tamara Tsitsishvili (both 1940, Museum of Art of the Georgian SSR, Tbilisi), and a series of antifascist drawings (india ink, 1942). His other works include the illustrations for *The Wisdom of Invention* by Sulkhan Saba Orbeliani (sepia, 1939 and 1960) and illustrations

for *Georgian Folksongs* by G. Chkhikvadze (sepia, 1946). Decors.: Sh. Rustaveli State Prize of the Georgian SSR (1965). Order of Lenin, two other orders, and a medal.

Gudkov, Aleksandr Fedorovich
Party official; b. 1930.
Russian; mem. CPSU 1958–; grad. Voronezh Agricultural Inst. 1955 and CPSU Cen. Cttee. Party Higher School 1967; chief agronomist dir. of a machine and tractor station in Kursk Oblast 1955–58; Sec. 1968–70, 1st Sec. of Kursk Oblast CPSU Cttee. Ap. 1970–; Dep. to USSR Supr. Sov. 1970–; cand. mem. 1971–76, mem. CPSU Cen. Cttee. 1976–. Decor.: Order of Lenin.

Gukasov, Erik Khristoforovich
State official.
Cand. mem. Cen. Cttee. Kazakh CP 1976–79, mem. Je. 1979–; Dep. Chair. Council of Mins. Kazakh SSR O. 1978–.

Gukasov, Sergei Bagratovich
State official.
Min. of Meat and Dairy Ind. Turkmen SSR O. 1965–; cand. mem. Cen. Cttee. 1966–71, mem. Auditing Comm. 1971–76, mem. Cen. Cttee. Turkmen CP 1976–.

Gulaev, Nikolai Dmitrievich
Military official; b. F. 26, 1918, in the *stanitsa* (large cossack village) of Aksaiskaia, present-day Rostov Oblast.
Twice Hero of the Soviet Union (S. 28, 1943, and Jl. 1, 1944). Lieutenant-general of aviation (1968). Member of the CPSU since 1943.
He was born into a worker's family. He entered the Soviet Army in 1938 and completed military aviation school in 1940. During the Great Patriotic War (1941–45) he participated in the battles at Stalingrad, Voronezh, and on the First and Second Ukrainian fronts as flight commander, deputy squadron commander, squadron commander, and navigator of the 27th (later the 129th Guards) Fighter Air Regiment. He performed 240 combat flight missions, took part in 69 aerial battles, and shot down 52 enemy aircraft personally and four more in group fighting.
After the war he held command positions. In 1950 he graduated from the Zhukovskii Military Air Academy, and in 1960 he graduated from the Military Academy of the General Staff. He now serves in the National Air Defense Forces. Decors.: two Orders of Lenin, three Orders of the Red Banner, the Order of the Patriotic War First Class, and two Orders of the Red Star, as well as medals.

Gulia, Georgii Dmitrievich
Author; b. 1913, Sukhumi; s. of D.I. Guilia.
Russian; mem. CPSU 1940–; first works publ. 1930.
Publs.: "Vesna v Sakene" (Spring in Saken) 1948, "Dobryi gorod" (A Good Town) 1949, "Kama" 1951, "Belaia noch'" (The White Night) 1958, and others.

Guliaev, Iurii Aleksandrovich
Ukrainian singer (lyric baritone); b. S. 9, 1930, Tiumen'.
People's Artist of the USSR (1968). Member of the CPSU since 1964.
He graduated from the Urals Conservatory (Sverdlovsk) in the class of F. I. Obraztsovskaia in 1954. During the years 1955–60 he sang in the opera theaters of Sverdlovsk and Donetsk. In 1960 he became a soloist with the Ukrainian Theater of Opera and Ballet (Kiev). His best roles include Shchors (*Shchors* by Liatoshinskii), Listnitskii (*The Quiet Don* by Dzerzhinskii), Onegin (*Eugene Onegin* by Tchaikovsky), Figaro (*The Barber of Seville* by Rossini), Count di Luna (*Il Trovatore* by Verdi), Renato (*Un Ballo in Maschera* by Verdi), Escamillo (*Carmen* by Bizet), Robert (*Iolanta* by Tchaikovsky), and Papageno (*The Magic Flute* by Mozart). He has appeared as a concert singer in many cities of the Soviet Union and abroad. His repertory includes many works by Soviet composers.

Gulian, Eduard Khachaturovich
State official.
Head of Admin. of Geology Armenian SSR D. 1965–.

Gulii, Maksim Fedorovich, Dr. of Biol. Sc.
Biochemist; b. Mr. 3, 1905, Nova Basan, now Chernigov Oblast; s. of a peasant.
Mem. CPSU 1940–; grad. inst. of Veterinary Medicine Kiev 1929; Dr. of Biol. Sc. 1942; Prof. 1945–; mem. Acad. of Sciences Ukrainian SSR 1957–; Research Asst. at Inst. of Biochemistry at Acad. of Sciences Ukrainian SSR 1932–; Prof. Chair of Biochemistry at Ukrainian Agricultural Acad. 1944–; mem. Presidium 1957–62, Vice-Pres. of Acad. of Sciences Ukrainian SSR 1958–62; author of more than 170 scientific works on biochemistry of muscle activity, methods of secretion, purification and crystallization of tissue protein etc. Decors.: Order of Red Banner of Labour, medals.
Publs.: "Khimizm biologichnoho okyslennia i syntezu zhyriv ta problema pidvyshchennia zhyrnomolochnosty" 1959, "Biokhimiia zhirovogo obmena" 1961, "Biosintez belka" 1963, "Ferment gliukozooksidaza i ego primenenie" 1964, and others.

Gulmanov, Berdy
State official.
Mem. Auditing Comm. Turkmen CP 1971–76 and 1981–; Min. of Rural Construction Turkmen SSR Ap. 1978–.

Gurbangeldyev, Khodzhamurad
State official.
Min. of Motor Transport Turkmen SSR Ja. 1974–; cand. mem. Cen. Cttee. Turkmen CP 1976–81, mem. 1981–.

Gurbanov, Gaily Gurbanovich
State official.
Chair. of Tashauz Oblast Exec. Cttee. of People's Deps. 1979–80; Min. of Agriculture Turkmen SSR N. 1980–; mem. Cen. Cttee. Turkmen CP 1981–.

Gureev, Aleksei Ivanovich
Author; b. 1913.
Ukrainian; mem. CPSU 1948–; grad. Gorkii Pedagocical Inst. Kiev 1940; first works publ. 1930.
Publs.: poems "Entuziasty shakht" (Mining Enthusiasts) 1932, "Na boevykh dorogakh" (On the Ways of Battle) 1944, "Nasha molodost'" (Our Youth) 1949, "Zhizn' idet" (Life Goes On) 1954, "Druzia ne izmeniaiut" (Friends Don't Betray) 1963, and others.

Gurenko, Stanislav Ivanovich
State official.
Sec. of Donetsk Oblast Cttee. Ukrainian CP 1976–80; Dep. Chair. of Council of Mins. Ukrainian SSR Mr. 1980–; mem. Cen. Cttee. Ukrainian CP 1981–.

Guretskas, Ionas Ionovich
State official.
1st Sec. of Klaipeda City Cttee. Lithuanian CP 1975; Dep. of Supr. Sov. Lithuanian SSR 1975–; mem. Cen. Cttee. Lithuanian CP 1971–; Sec. of Presidium Supr. Sov. Lithuanian SSR S. 1982–.

Gurevich, Isai Izrailevich
Physicist; b. Je. 30 (Jl. 13), 1912, Riga.
Corresponding member of the Academy of Sciences of the USSR (1968). Member of the CPSU since 1951.
He graduated from Leningrad University in 1934. From 1934 to 1945 he worked at the Radium Institute of the Academy of Sciences of the USSR, and since 1945 at the I. V. Kurchatov Institute of Atomic Energy. His main work has been in neutron physics, reactor theory, and the properties of muonium. He discovered the phenomenon of the two-frequency precession of muonium in a magnetic field.

Gurfinkel, Viktor Semenovich, Dr. of Med. Sc.
Biophysicist; b. 1922.
Mem. CPSU 1947; grad. Kirghiz Medical Inst.; Dr. of Med. Sc. 1962; Prof. 1967–; army doctor 1944–46; mem. staff Cen. Research Inst. for Prosthetics and Prosthesis Construction 1946–58; mem. staff Inst. of Experimental Biology and Medicine at Siberian Branch of USSR Acad. of Medical Sciences 1958–60; mem. staff Inst. of Biological Physics 1960–67, Laboratory Chief at Inst. for Problems of Information Transmission of USSR Acad.

of Sciences 1967–. Decor.: State Prize of the USSR (for his part in the construction of a forearm prosthesis with bioelectric control) 1970.

Gurgenidze, Nelli Vladimirovna
Party official.
Sec. of Tbilisi City Cttee. Georgian CP Je. 1975–; Dep. to and Dep. Chair. of Supr. Sov. Georgian SSR 1975–; mem. Cen. Cttee. Georgian CP 1976–.

Gurin, V. V.
Komsomol official.
Sec. and mem. Bureau Cen. Cttee. Belorussian Komsomol Je. 1979–.

Gurinovich, Anatolii Emelianovich
State official; b. 1924, Slatvin, Minsk Oblast; s. of a peasant.
Mem. CPSU 1948–; grad. Belorussian Inst. of Econ. 1949 and Higher Diplomatic School of USSR Min. of Foreign Affairs 1951; worked in Underground City Cttee. Belorussian CP in Minsk, partisan 1941–44; sov. and econ. posts 1944–45; mem. Min. of Foreign Affairs Belorussian SSR 1951–57; USSR and Belorussian SSR Rep. at UN, Head of Dept. of Int. Econ. Orgs. at Min. of Foreign Affairs Belorussian SSR 1957–61; Dep. Min. of Foreign Affairs 1961–66, Min. of Foreign Affairs Belorussian SSR My. 1966–; mem. Cen. Cttee. Belorussian CP 1971–. Decors.: Order of Red Banner of Labour, medals.

Gurtuev, Bert Izmailovich
Author; b. 1910, Ak-Su; s. of a cattle breeder.
Balkar; mem. CPSU 1939–; first works publ. 1928.
Publs. "Na puti k novoi zhizni" (On the Way to the New Life) 1929, "Makhai-efendi" 1931, "Krasnye golosa" (Red Voices) 1935, "Iasnoe utro" (A Clear Morning) 1958, "Svetlye dali" (Light Expanses) 1960, "Nalchik", "XXI s'ezdu partii" (To the XXI Party Convention), "Chetyre veka" (Four Centuries), and others.

Gusakovskii, Iosif Iraklievich
Military official; b. D.12 (25), 1904, Vorod'kovo, Mogilev Province.
General of the army (1968). Twice Hero of the Soviet Union (S.23, 1944, and Ap.6, 1945). Member of the CPSU since 1931.
He was born into a Belorussian peasant family. He entered the Soviet Army in 1928. He graduated from cavalry school (1931), armor courses (1932), staff courses (1934), and the Military Academy of the General Staff (1948). During the Great Patriotic War of 1941–45 he was chief of staff of a tank brigade, and from September 1943 the comander of the 112th Tank Brigade (renamed the 44th Guards Tank Brigade in October 1943). He participated in the battles of Moscow and Kursk, in the liberation of the Ukraine and Poland, and in the fighting for Berlin. For skillful control of his brigade during the L'vov-Sandomierz operation (destruction of the Brody grouping) and the Vistula-Oder operation (crossing the Pilica River and taking the city of Łowicz), he was twice awarded the title Hero of the Soviet Union. After the war he commanded a division and the armored and mechanized troops of a military district, and he served as deputy commander of the troops of a military district. From 1959 to 1963 he was troop commander of the Baltic Military District, and 1963–70 chief of the Main Directorate of Personnel of the Ministry of Defense of the USSR. He was a deputy to the sixth and seventh convocations of the Supreme Soviet of the USSR. Decors.: three Orders of Lenin, four Orders of the Red Banner, the Order of the Patriotic War First Class, two Orders of the Red Star, two Mongolian Orders, and also medals.

Guseinov, Dzhavanshir Tamraz ogly
State official.
Chair. of State Cttee. for Construction Azerbaidzhan SSR N. 1978–.

Guseinov, Isa Mustafa ogly
Author; b. 1928, Muganly, Azerbaidzhan. Grad. Gorkii Inst. of Literature Moscow 1952; first works publ. 1949.
Publs.: "Nashi devushki" (Our Girls) 1953, "Utrenniaia zvezda" (The Morning Star) 1953, "Plamennoe serdtse" (A Burning Heart) 1957–58, "Svoi i chuzhie" (One's Own and Strangers) 1962, and others.

Guseinov, Ragim Ali Gusein ogly
State official.
Head of Cen. Admin. for Material and Tech. Supply with Council of Mins. Azerbaidzhan SSR Ap. 1974–78; mem. Auditing Comm. Azerbaidzhan CP 1976–; Chair. of State Cttee. for Material and Tech. Supply Azerbaidzhan SSR O. 1978–.

Guseinov, V. A.
Party offical.
1st Sec. and mem. Bureau of Cen. Cttee. Azerbaidzhan Komsomol F. 1974–78; cand. mem. Bureau Cen. Cttee. Azerbaidzhan CP 1976–78, mem. Ja. 1981–; mem. Cen. Cttee. Azerbaidzhan CP 1976–; Sec. and mem. Bureau Cen. Cttee. USSR Komsomol 1978–80; 1st Sec. Baku City Cttee. Azerbaidzhan CP D. 1980–.

Guseinzade, Gusein Dzhamal ogly
Author; b. 1924, Kazakh; s. of a peasant. Grad. Azerbaidzhan Univ.; first works publ. 1946.
Publs.: "Na putiakh novoi zhizni" (On the Way to a New Life) 1949, "Mileila", "Most Ibragima" (Ibragim's Bridge), "Poema o liubvi" (Love Poem) "V puti" (On the Way), and others.

Gusev, Vladimir Kuzmich
Party official; b. 1932.
Russian; mem. CPSU 1963–; grad. Saratov State Univ. 1957 and Saratov Econ. Inst. 1969, Cand. of Tech. Sc. 1972; laboratory asst. at a faculty at Saratov State Univ. 1957–58; shift supervisor, dispatcher dep. chief laboratory asst., chief technologist, chief engineer, dir. of a synthetic fiber combine 1958–75; 1st Sec. of Engels City Cttee. CPSU 1975–76, 2nd Sec. 1976, 1st Sec. Saratov Oblast Cttee. CPSU N. 1976–; Dep. to USSR Supr. Sov. 1966–70 and 1979; mem. Cen. Cttee. CPSU 1981–.

Gustov, Ivan Stepanovich
Party official; b. 1911; s. of a peasant.
Russian; mem. CPSU 1932–; grad. Velikie Luki Agricultural Inst. 1965 and CPSU Cen. Cttee. Higher Party School 1949; kolkhoz chair., dep. chair. of a raion kolkhoz asscn., dir of a machine and tractor station 1930–36; head of a raion agricultural dept., chair. exec. cttee. of a raion sov. of workers' deps., 1st sec. of a raion CPSU cttee. Novosibirsk Oblast 1936–49; Head of Agricultural Dept., later Sec. of Velikie Luki Oblast CPSU Cttee., 1st Dep. Chair., then Chair. of Exec. Cttee. of Velikie Luki Oblast Sov. of Workers' Depts. 1949–57; 2nd Sec. 1957–61, 1st Sec. of Pskov Oblast CPSU Cttee. 1961–71; Dep. to and mem. Agricultural Comm. RSFSR Supr. Sov. 1959–63; cand. mem CPSU Cen. Cttee. 1961–81, mem. 1981–; Dep. to 1962–, mem. Comm. for Trade, Public Services and Communal Econ. 1966–74, mem. Draft Bills Comm. at Sov. of the Union USSR Supr. Sov. 1974–; Dep. Chair. of Cttee. for Party Control with Cen. Cttee. CPSU N. 1971–.

Gutman, Grigorii Petrovich
Sculptor; b. 1916, Kiev.
Attended Kiev Art Inst. 1937–45, pupil of M. Gelman, L. Shervud, L. Matveev.

Gutsu, Ivan Timofeevich
Party official.
Mem. Bureau 1973–80, Sec. 1973–74, 1st Sec. of Cen. Cttee. Moldavian Komsomol Jl. 1974–80; mem. Cen. Cttee. Moldavian CP 1976–81, cand. mem. 1981–; 1st Sec. Oktiabrskii Raion Cttee. Moldavian CP 1980–82; Head of Dept. of Ind. with Cen. Cttee. Moldavian CP Je. 1982–.

Gutyria, Viktor Stepanovich, Dr. of Chem. Sc.
Scientist; b. S. 11, 1910, Siniavskoe, now Rostov Oblast; s. of a railwayman.
Mem. CPSU 1940; grad. Azerbaidzhan Petroleum Inst. 1932; specialist in petroleum and gas technology; corresponding mem. USSR Acad. of Sciences 1953–; mem. Acad. of Sciences Ukrainian SSR 1961–; mem. Acad. of Sciences Azerbaidzhan SSR 1949–; Dr. of Chem. Sc., Prof. 1944; mem. staff Azerbaidzhan Pe-

troleum Research Inst. 1932–54; Chief of Tech. Admin., Chair. of Tech. Council of Min. of Petroleum Ind. Azerbaidzhan SSR 1954–55; mem. Nat. Scientific-Tech. Cttee. with Council of Mins. Azerbaidzhan SSR 1958–59; Chief of Dept. of Petrochemistry and Gas at Inst. for Chemistry of High Molecular Compounds at Acad. of Sciences Ukrainian SSR and Dep. Dir. of Teaching Dept. of same 1959–; author of more than 100 scientific works. Decors.: State Prize of the USSR 1942, Order of Lenin, Order of Red Banner of Labour, 2 Badges of Honour, medals.
Publs.: "Kreking i reforming nefteproduktov v prisutstvii aliumosilikatnykh katalizatorov" 1944, "Nefti Azerbaidzhana" 1945, "Kataliticheskaia ochistka destilatov termicheskogo reforminga" 1946, "Krivye razgonok bakinskikh neftei" 1947, and others.

Guzhel, Anatolii Moiseevich
Author; b. 1922, Iaşi, Rumania.
Moldavian; mem. CPSU 1949–; grad. Kishinev Pedagogic Inst.; first works publ. 1945.
Publs.: poems "Moe pokolenie" (My Generation) 1951, "Dnestr, na tvoem beregu ..." (Dnestr, On Your Bank ...) 1956, "Aprel'skii veter" (April Wind) 1959, "Zelenye ostrova" (Green Islands) 1962, "Okno s tremia geraniami" (The Window With Three Geraniums) 1956, and others.

Guzhenko, Timofei Borisovich
State official; b. 1918.
Russian; mem. CPSU 1941–; grad Odessa Inst. of Water Transport Engineering 1942 and Leningrad Marine Acad. 1951; mechanic. dept. head, sec. of CPSU bureau, chief engineer in commerical port of Murmansk 1942–49; head of various commercial ports, Head of Shipping Co. in Sakhalin Oblast 1951–60; worked in USSR Min. of Maritime Navigation 1960–62; responsible posts in Admin. of Cen. Cttee. CPSU 1962–66; 1st Dep. Min. 1966–70, USSR Min. of Maritime Navigation Ja. 1970–; Dep. to USSR Supr. Sov. 1970–; cand. mem. Cen. Cttee. CPSU 1971–76, mem. 1976–; Chair. of USSR-Japan Soc. 1973–.

Gvelesiani, Shota Florentievich
Painter; b. 1925, Shrosha, Georgia.
Attended Tbilisi Art Acad. 1945–52, pupil of A. K. Kutateladze, M. I. Toidze; exhibitions 1953–.

Gverdtsiteli, Iraklii Georgievich
State official.
Chair. of State Cttee. for Science and Technology Georgian SSR Mr. 1980–; mem. Cen. Cttee. Georgian CP 1981–.

Gvetadze, Guram Ivanovich
State official; b. 1931.
Georgian; mem. CPSU 1957–; grad. Lenin Polytechnical Inst. in Georgia, mechanical engineer, Major-Gen. in the police, construction engineer, Sec. of Komsomol Cttee. of the Kirov machine-tool shop in Tbilisi 1955–56; instructor of Georgian Komsomol Cen. Cttee. 1956–57; engineer in an admin. of the mechanical engineering ind. of Economic Council Georgian SSR 1957–60; instructor of Dept. for Ind. and Transport of Tbilisi City Cttee. CP Georgia 1960–63; Sec. of Lenin Raion Cttee. CP Georgia in Tbilisi 1963–65; dep. head of police admin., head of an admin. of Min. of the Interior Georgian SSR 1965–72; Dep. Min. of the Interior Georgian SSR 1972–76; dep. head of Administrative Organs Dept. Cen. Cttee. CP Georgia 1976–77, head 1977–79; Min. of the Interior Georgian SSR My. 1979–; mem. Cen. Cttee. CP Georgia 1981–.

Gvishiani, Dzhermen Mikhailovich.
Philosopher and sociologist; b. D. 24, 1928, Akhaltsikhe, Georgian SSR.
Corresponding member (1970) of the Academy of Sciences of the USSR (AN SSSR). Member of the CPSU since 1951. He graduated from the Moscow Institute of International Relations in 1951. He served in the navy between 1951 and 1955. In 1955 he began work with the State Committee for New Technology of the Council of Ministers of the USSR, and in 1965 he became vice-chairman of the State Committee for Science and Technology of the Council of Ministers of the USSR. During 1960–68 he taught in the philosophy department at Moscow State University. In 1969, he was appointed head of the Laboratory for Research Into Complex Problems of Management of the Institute for Concrete Social Research of the AN SSSR. He conducts scientific work on the problems of management and social organization. He wrote the works *The Sociology of Business* (1962), *The Social Role of Science and Scientific Policy* (1968), and *Organization and Management* (1970). Decors.: Order of the Red Banner of Labor and various medals.

Gvozdetskii, Nikolai Andreevich
Physical geographer; b. D. 2 (15), 1913, St. Petersburg.
Doctor of geographical sciences (1949). Member of the CPSU since 1971. Professor (since 1951) and director of the subdepartment of physical geography of the USSR in the geography department of Moscow State University (since 1959).
He has done research on karst, the physical geographic zones of the USSR, the landscape and mapping of mountain regions (in particular the Caucasus and Middle Asia), and the history of Soviet geography. Decor.: D. N. Anuchin Prize of Moscow State University in 1969.
Works: *Karst,* 2nd ed. Moscow 1954. *Fizicheskaia geografiia Kavkaza,* issues 1–2. Moscow 1954–58. *Kavkaz.* Moscow 1963. *Fizicheskaia geografiia SSSR,* vol. 1, 3rd ed. Moscow 1969; vol. 2, 2nd ed. Moscow 1970. (Co-author.) *Fiziko-geograficheskoe raionirovanie SSSR.* Moscow 1968. (Co-author.) *Sovetskie geograficheskie issledovaniia i otkrytiia.* Moscow 1967. *Istoriia otkrytiia i issledovaniia Sovetskoi Azii.* Moscow 1969. (Co-author.)

Gvozdev, Viktor Aleksandrovich
State official; b. 1931.
Russian; mem. CPSU 1956–; grad. Belorussian Agricultural Acad.; chief engineer of a machine and tractor station, dir. of same, 1st sec. of a raion cttee. Belorussian CP 1954–63; mem. staff Cen. Cttee. Belorussian CP, Head of "Glavpolesevodstroi" 1963–69; 1st Sec. of Gomel Oblast Cttee. Belorussian CP 1969–; Dep. to and mem. Comm. for Construction and Building Materials Ind. with Sov. of the Union USSR Supr. Sov. 1970–; mem. Cen. Cttee. Belorussian CP 1971–; mem. Buro Cen. Cttee. Belorussian CP 1978–82; Dep. Chair. of Council of Mins. Belorussian SSR 1978–79, 1st Dep. Chair. 1979–82; Chair. of GOSPLAN Belorussian SSR 1978–82; Dep. Chair. USSR GOSPLAN Ap. 1982–.

Gzhitskii, Vladimir Zenonovich
Ukrainian author; b. O. 3 (15), 1895, Ostrovets, in present-day Terebovas Raion, Ternopol' Oblast.
Born into the family of a teacher. During World War I, 1914–18, a soldier in the Austrian army.
In 1926, he graduated from the Forestry Institute in Kharkov. He began publishing his work in 1924. The thema of his novel *Black Lake* (1929; Russian translation, 1930) is the socialist transformation of Altai Krai. The work was criticized for inaccurately portraying the nationality problem in the Altai region. In 1957, he published a new edition of *Black Lake.* In 1962 he wrote the historical novel Oprishli about the western Ukrainian popular hero Oleksa Dovbush. His trilogy of novels, *Great Hopes,* is about historical events in the Ukraine from 1914 to 1941. He writes short stories and tales for children and young people – for example, "The Boxwood Grove," "The Deserted Shore," and "Petia's Storks." His novel *Word of Honor* appeared in 1968.
Works: *Trembityni tony: Poezii.* Kharkov 1924. *Nastup.* Kharkov 1932. *Chorne ozero.* Kiev 1957. *In Russian translation: Chernoe ozero.* Moscow 1960. *Bol'shie nadezhdy.* Moscow 1966.

Hiob (Dmitrii Iakovlevich Tyvoniuk), Cand. of Theol.
Russian Orthodox clergyman; b. N. 6, 1938, Pochaev, Ternopol Oblast.
Novice in Holy Ghost Monastery of Pochaev Lavra, later in Holy Trinity Monastery, Odessa Bishopric; studied Theological Seminary Kiev, interrupted by mil. service 1957–65, then studied at Theological Acad. Moscow; entered St. Sergius

Holy Trinity Monastery (Troitse-Sergieva-Lavra); monk-deacon; Cand. of Theol., priestmonk, Chief Clergyman at Church of the Resurrection in Chita, Irkutsk Diocese 1969; aspirant at Moscow Theological Acad. 1970; Adviser and Head of Chancery in Dept. of External Affairs, Moscow Patriarchate 1971–; abbot; Bishop of Zaraisk, Substitute Bishop of Moscow 1974–; Administrator of Russian Churches in Canada and USA 1975–.

Iablonskii, Eduard Ianovich
State official.
Min. of Light Ind. Latvian SSR Jl. 1972–; cand. mem. Cen. Cttee. Latvian CP 1976–.

Iagodkin, Vladimir Nikolaevich, Cand. of Econ. Sc.
State official; b. 1928.
Mem. CPSU 1950–; grad. Moscow State Univ. 1951; Komsomol posts 1951–53; Dep. Dean, Acting Dean, Lecturer at Faculty of Econ. of Moscow State Univ. 1954–67; party posts 1967–71; Sec. of Moscow City CPSU Cttee. Mr. 1971–76; cand. mem. Cen. Cttee. CPSU 1971–76; USSR Dep. Min. of Educ. 1976–.

Iakhiaev, Khaudar Khalikovich
State official.
Head of Admin. Cttee. on State Security at Council of Mins. Uzbek SSR for Surkhan-Daria Oblast 1963; Min. of Maintenance of Public Order Uzbek SSR 1964–68; mem. Cen. Cttee. Uzbek CP 1966–; Min. of the Interior Uzbek SSR 1968–79; 1st Dep. Chair. of Uzbek Cttee. for People's Control 1979–.

Iakimenko, Aleksandr Nikiforovich
Legal official.
Dep. Chair. 1957–63, Chair. of Supr. Court Ukrainian SSR O. 1970–; mem. Auditing Comm. Ukrainian CP 1976–81.

Iakov (Akkersdaik)
Russian Orthodox clergyman; b. Ag. 13, 1914, Holland.
Studied theology with the Benedictines; conversion to Orthodoxy; ordained priest in the Russian Church Abroad 1954; organized groups of Orthodox laymen and monks in Dutch towns; Bishop 1965, translated liturgical texts into Dutch; attached to Moscow Patriarchate 1972; Bishop of the Hague and the Netherlands 1972–.

Iakovlev, Aleksandr Nikolaevich, Dr. of Hist. Sc.
Diplomat; b. 1923.
Mem. CPSU 1944–; grad. Iaroslavl Pedagogic Inst. 1946, Acad. of Social Sciences at CPSU Cen. Cttee. 1960; Prof.; served in Sov. Army 1941–43; mem. staff Iaroslavl Oblast CPSU Cttee. 1946–48 and 1950–53. Dept. Head with Oblast newspaper "Severnyi rabochii", Lecturer at Iaroslavl Oblast newspaper "Severnyi rabochii", Lecturer at Iaroslavl Oblast Party School 1948–50; mem. staff Cen. Cttee. CPSU 1953–56 and 1960–65; 1st Dep. Head of Dept. of Propaganda at Cen. Cttee. CPSU 1965–73; mem. Cen. Auditing Comm. CPSU 1971–76; USSR Amb. to Canada Je. 1973–.

Iakovlev, Aleksandr Sergeevich, Dr. of Tech. Sc.
Aircraft designer; b. 1906.
Russian; mem. CPSU 1938–; grad. N. E. Zhukovskii Air Force Acad. 1931; corresponding mem. USSR Acad. of Sciences 1943–; Col.-Gen. in Engineering-Tech. Services; worker in workshops of Air Force Acad. 1924–27; student 1927–31; engineer in an aircraft plant, dir. of a factory, head of a main admin., Dep. Min. of the USSR Aircraft Ind. 1931–57; Dep. to Supr. Sov. 1946–; Chief Designer of USSR Min. of the Aircraft Ind. 1957–. Decors.: State Prize of the USSR 1941, 1942, 1943, 1946, 1947, 1948, 8 Orders of Lenin, 1 Order of Red Banner, 6 other orders and various medals.

Iakovlev, Boris Pavlovich
Party official; b. 1908.
Mem. CPSU 1931–, grad. CPSU Cen. Cttee. Party Higher School 1955; metal worker at a plant in Kalchugino 1928–33; 1st Sec. of Kalchugino Raion Cttee. CPSU 1937–39; Sec. of Oblast Cttee. Daghestan ASSR CPSU 1940–43; 2nd Sec. of Frunze Oblast Cttee. Kirghizian CP 1945–46; Sec. of Cen. Cttee. Kirghizian CP 1946–52; 1st Sec. of Osh Oblast Cttee. Kirghizian CP 1955–58; Chair. of Comm. of Sov. Control of Council of Mins. Kirghizian SSR 1958–61; Min. of Procurement Kirghizian SSR 1961–62; Sec. of Vladimir Oblast Cttee. CPSU and Dep. Chair. of Exec. Cttee. of Vladimir Oblast Sov. of Peoples' Deps. 1962–66; posts in Cen. Cttee. apparatus CPSU 1966–79; Chief of Letters Dept. of Cen. Cttee. CPSU 1979–; mem. Cen. Auditing Comm. CPSU 1981–.

Iakubov, Narimon Kholovich
State official.
Head of Dept. of Ind. and Transport with Cen. Cttee. Tadzhik CP 1974–77; Min. of Motor Transport Tadzhik SSR D. 1977–; cand. mem. Cen. Cttee. Tadzhik CP 1976–.

Iakubov, Valerii Azizovich
State official.
Mem. Auditing Comm. 1976–81, cand. mem. Cen. Cttee. Tadzhik CP 1981–; Chair. of State Cttee. of Construction Tadzhik SSR Jl. 1981–.

Iakubovskii, Petr Ivanovich
State official.
Mem. Cen. Cttee. Moldavian CP 1971–; 1st Sec. of Beltsy Municipal Cttee. Moldavian CP 1971–73; Head of Dept. of Ind. and Transport with Cen. Cttee. Moldavian CP 1973–77; Chair. of State Cttee. on Labour Moldavian SSR Ja. 1977–.

Iakushev, Ivan Fedorovich
Party official.
Sec. of Gomel Oblast Cttee. Belorussian CP 1971– 72; Head of Dept. of Party Org. Work with Cen. Cttee. Belorussian CP N. 1972–; mem. Cen. Cttee. Belorussian CP 1976–; cand. mem. Bureau Cen. Cttee. Belorussian CP 1976–77, mem. 1977–.

Iakushin, Ivan Nesterovich
Diplomat; b. 1913.
Grad. Moscow Aviation Inst.; mem. Diplomatic Service 1946–; USSR Chargé d'Affaires in Iraq 1953–55; USSR Envoy, Amb. to Thailand 1955–60; Dep. Head of Middle East Dept. at USSR Min. of Foreign Affairs 1960–63; USSR Amb. to Sudan 1963–68; mem. staff USSR Min. of Foreign Affairs 1968–70; USSR Amb. to Libya F. 1970–S. 1977.

Iakushev, Nikolai Stefanovich
State official.
Sec. Minsk Oblast Cttee. Belorussian CP 1975–81; Min. of Fruits and Vegetables Belorussian SSR 1981–; cand. mem Cen. Cttee. Belorussian CP 1981–.

Ianenko, Nikolai Nikolaevich, Dr. of Phy.-Math. Sc.
Mechanics scientist; b. My. 22, 1921.
Mem. CPSU 1952–; grad. Tomsk Univ. 1942; Dr. of Phy.-Math. Sc. 1955; Prof. 1960–; served in Sov. Army 1942–45; postgraduate studies at Moscow Univ. 1946–49; mem. staff Inst. of Geophysics of USSR Acad. of Sciences 1948–53; concurrently Lecturer at Moscow Univ. 1949–56; Head of Laboratory of Computing Centre at Siberian branch of USSR Acad. of Sciences 1963–; concurrently Prof. and Head of Chair at Novosibirsk Univ. 1964–; mem. USSR Acad. of Sciences 1970–.

Ianshin, Aleksandr Leonidovich
Geologist; b. Mr. 15 (28), 1911.
Mem. USSR Acad. of Sciences 1958–, Vice-Pres. S. 1982–.

Ianuitis, Ionas Ionovich
State official.
Cand. mem 1960–64, mem. Cen. Cttee. Lithuanian CP 1964–; Chair. of State Cttee. for Television and Radio. Lithuanian SSR Jl. 1962–; Dep. to Supr. Sov. Lithuanian SSR 1963–.

Iarkovoi, Ivan Mefodievich
Party official; b. 1927.
Ukrainian; mem. CPSU 1957–; grad. Kharkov Zootech. Inst.; kolkhoz farmer 1946–53; chief zoo tech. of a raion dept. of agriculture, chief zoo tech. of a machine and tractor station, head of an agricultural raion inspectorate 1953–61; 1st Sec. of Zborov Raion Cttee. Ukrainian CP 1961–63; Sec. 1963–68, 1st Sec. of Ternopol Oblast Cttee. Ukrainian CP 1968–; Dep. to and mem. Comm. for Posts and Telecommunications at Sov. of

Nationalities USSR Supr. Sov. 1970–; mem. Cen. Cttee. Ukrainian CP 1971–.

Iarutin, Vladimir Kuzmich
State official.
Min. of Rural Construction Moldavian SSR D. 1976–; mem. Auditing Comm. Moldavian CP 1981–.

Iashen, Kamil (real name: Nugmanov)
Author; b. D. 18 (31), 1909, Andizhan.
Uzbek; mem. CPSU 1943–; studied at Inst. of Forestry Leningrad 1925–28; literature teacher at a school 1928–30; Head of Dept. of Literature at Uzbek State Theatre Museum 1930–36; Head of Artistic Affairs Admin. with Council of Mins. Uzbek SSR 1946–49; Chair. respectively 1st Sec. of Uzbek Writers' Union 1958–; Sec. of USSR Writers' Union 1959–; mem. Cen. Cttee. Uzbek CP 1960–; mem. Presidium Sov.-Afro-Asian Solidarity Cttee. 1967–; mem. Uzbek Acad. of Sciences 1969–; Chair. of Sov. Cttee. on Relations with Asian and African Writers 1969–; first works publ. 1930. Decors.: Merited Artist of the Uzbek SSR 1939, People's Writer of the Uzbek SSR 1969, State Prize of the USSR 1951, State Prize of the Uzbek SSR.
Publs.: "Gulsara" 1934, "Smert' okkupantam" (Death to the Occupants) 1942, "General Rakhimov" 1949, "Putevodnaia zvezda" (Guiding Star) 1958, "Zaria Revoliutsii" (Dawn of the Revolution) 1972, "P'esy" (Theatre Plays) 1971, and others.

Iashin, Aleksei Ivanovich
State official; b. 1919.
Russian; mem. CPSU 1941–; grad. Moscow Engineering Inst. of Rail Transport 1942 and CPSU Cen. Cttee. Party Higher School 1948. Cand. of Econ. Sc. 1952, Asst. Prof. 1960; served in Sov. Army 1941–42; engineer for reconstruction of Tula and Serpukhov 1942–43; party posts 1943–54; Dep. USSR Min. of Construction 1954–57; Dep. Chair. Belorussian Sovnarkhoz 1957–63; mem. Cen. Cttee. Belorussian CP 1960–61; Dep. USSR Min. of Installation and Special Construction Work 1963–73; 1st Dep. Chair. of Exec. Cttee. of Moscow City Sov. of Peoples' Deps. 1973–79; Dep. to USSR Supr. Sov. 1979–; USSR Min. of the Construction Materials Ind. Ja. 1979–; cand. mem. Cen. Cttee. CPSU 1981–.

Iasiunas, Stanislovas Piatrovich
State official.
Cand. mem. 1961–64, mem. Cen. Cttee. Lithuanian CP 1964–; Head of Dept. of Ind. and Transport at Cen. Cttee. Lithuanian CP 1961–66; Dep. Chair. of Bureau for Ind. at Cen. Cttee. Lithuanian CP 1962–64; Dep. to Supr. Sov. Lithuanian SSR 1963–; Min. of Building Materials Ind. Lithuanian SSR D. 1966–.

Iasnov, Mikhail Alekseevich
State official; b. 1906.
Russian; mem. CPSU 1925–; secondary educ.; worked at leather factory in Moscow 1918–28; completed Workers' Evening Faculty Moscow Univ. 1925; served in Sov. Army 1928–30; head of a construction office, head of a road building trust, attended Inst. of Advanced Training for Engineering and Tech. Personnel 1930–38; Dep. Chair. 1938–39, Chair. of Moscow City Sov. 1950–56; USSR Dep. Min. of Urban Construction 1949–50; Dep. to 1950–, Chair. of Budget, then Planning Comm. with Sov. of Nationalities USSR Supr. Sov. 1962–67; mem. Cen. Cttee. CPSU 1952–; Dep. to RSFSR Supr. Sov. 1955–; Chair. 1956–57, 1st Dep. Chair. of RSFSR Council of Mins. 1957–D. 1966; mem. Cen. Cttee. CPSU Bureau for RSFSR 1957–66; Chair. of Presidium RSFSR Supr. Sov. D. 1966–; Dep. Chair. of Presidium USSR Supr. Sov. O. 1967–. Decors.: 4 Orders of Lenin, 2 Orders of Red Banner of Labour, Order of Red Star, various medals.

Iastrebov, Ivan Pavlovich
Party official; b. 1911.
Mem. CPSU 1941–; grad. Uralsk Polytech. Inst. 1936; foreman, head of a group, Head of Tech. Dept. of Lysva Metallurgical Works in Perm Oblast 1936–46; CPSU posts 1946–50; 1st Sec. of Lysva City CPSU Cttee. 1950–51; Dept. Head at Perm Oblast CPSU Cttee. 1951–53; 1st Sec. of Perm City CPSU Cttee. 1953–54; Dep. Head 1954–62, 1st Dep. Head of Dept. of Heavy Ind. at Cen. Cttee. CPSU 1962–; mem. Cen. Auditing Comm. CPSU 1971–81; cand. mem. Cen. Cttee. CPSU 1981–.

Iatsimirskii, Konstantin Borisovich, Dr. of Chem. Sc.
Chemical scientist; b. Mr. 22, 1916, Pologi, now Vinnitsa Oblast; s. of a teacher.
Mem. CPSU 1944–; grad. Cen. Asian Univ. 1941; specialist in analytical chemistry and chemistry of complex compounds; mem. Acad. of Sciences Ukrainian SSR 1964; Dr. of Chem. Sc., Prof. 1949; Asst. Prof. at Chair of Physical Chemistry of Cen. Asian Univ. 1941; served in Sov. Army 1941–45; Research Asst., Lecturer at Chair of Inorganic Chemistry, Head of Chair of Analytical Chemistry, Dep. Dir. of Teaching Dept. at Chemical-Technological Inst. Ivanovo 1946–62; Chief of Laboratory for Structure of Complex Compounds at Inst. of Gen. and Inorganic Chemistry at Acad. of Sciences Ukrainian SSR and Prof. at Kiev Univ. 1962–63; mem. Presidium Acad. of Sciences Ukrainian SSR, Acad. Sec. of Dept. of Chemistry and Chemical Technology 1963–; author of more than 260 scientific works including 5 monographs. Decors.: Badge of Honour, medals.
Publs.: "Termokhimiia kompleksnykh soedinenii" 1951, "Konstanty nestoikosti kompleksnykh soedinenii" 1959, "Kineticheskie metody analiza" 1963, and others.

Iaunushan, Alfred Ignatovich
Actor and film director; b. 1919.
Mem. CPSU 1965–; grad. Drama Studio of Latvian Drama Theatre 1948; actor with Latvian Drama Theatre 1947–; teaching activity 1954–. Decor.: People's Artist of the USSR 1973.

Iazkuliev, Bally Iaskulievich
State official; b. 1930.
Mem. CPSU 1953–; grad. Teachers' Inst. in Tashauz 1949, Pedagogical Inst. in Khardzhou 1952, CPSU Cen. Cttee. Party Higher School 1962; teacher 1949–52; held Komsomol posts 1952–60; held state, Party, trade union, and econ. posts 1962–71; mem. Auditing Comm. CP Turkmenistan 1966–71; Chair. of Oblast Exec. Cttee. of People's Deps. in Tashauz 1971–73; Dep. to Supr. Sov. Turkmen SSR 1971–; mem. Cen. Cttee. 1971–, cand. mem. Bureau Cen. Cttee. Ap.–D. 1975, mem. Bureau Cen. Cttee. CP Turkmenistan D. 1975–; Chair. of Turkmen Republican Trade Union Council 1973–75; Chair. of Council of Mins. and Foreign Min. Turkmen SSR 1975–78; mem. Cen. Auditing Comm. CPSU 1976– 81; Chair. of Presidium of Supr. Sov. Turkmen SSR D. 1978–; Dep. Chair. of Presidium of USSR Supr. Sov. Ap. 1979–; cand. mem. Cen. Cttee. CPSU 1981–.

Iazov, Dmitrii Timofeevich
Military official; b. 1923.
Russian; mem. CPSU 1944–; grad. M. V. Frunze Mil. Acad. 1956 and Mil. Acad. of Gen. Staff 1967, Col.-Gen.; entered Sov. Army 1941; served at front 1941–45; command posts 1945–76; 1st Dep. Commander of Troops in Far Eastern Mil. District 1976–79; Commander of Central Group Forces (Czechoslovakia) 1979–80; Dep. to USSR Supr. Sov. 1979–; Commander of Troops in Central Asian Mil. District 1980–; mem. Cen. Cttee. and mem. Bureau of Cen. Cttee. Kazakh CP 1981–; cand. mem. Cen. Cttee. CPSU 1981–.

Ibragimov, Gadzhi Aga Khalil ogly
State official; b. 1923.
Azerbaidzhani; mem. CPSU 1948–; grad. Azerbaidzhan State Univ.; served in Sov. Army 1941–46; Komsomol official 1946–52; party official 1952–; 1st Oblast Sec. in Nakhichevan ASSR Azerbaidzhan CP 1961–70; mem. Cen. Cttee. Azerbaidzhan CP 1961– 81; Sec. and mem. Bureau of Cen. Cttee. Azerbaidzhan CP Mr. 1970–81; Chair. of State Cttee. on Nature Conservation Azerbaidzhan SSR Ja. 1981–.

Ibragimov, Ildar Abdullovich, Dr. of Phy.-Math. Sc.
Mathematician; b. 1932.
Grad. Leningrad Univ. 1956, Dr. of Phy.-Math. Sc. 1967; Prof. 1969; Asst., Lecturer, Prof. Chair of Probability Theory at Leningrad Univ. 1956–. Decor.: Lenin Prize 1970 for his part in a series of works which clarified important regularities of coincidental mass phenomena and processes, which are mathematically expressed in various so-called limit theorems.

Ibragimov, Mirza Adzhar ogly, Cand. of Philol. Sc.
Author; b. 1911, Azerbaidzhan.
Mem. CPSU 1930–; grad. Inst. of Oriental Studies at USSR Acad. of Sciences; mem. Acad. of Sciences Azerbaidzhan SSR; farm-hand, student 1919–33; ed. of a newspaper, head of admin. for artistic affairs 1933–42; Min. of Educ., Dep. Chair. of Council of Mins. Azerbaidzhan SSR, Chair. of Union of Writers Azerbaidzhan SSR 1942–54; Chair. of Presidium Supr. Sov. Azerbaidzhan SSR 1954–58; literary acitvity 1958–; 1st Sec. of Azerbaidzhan Writers' Union SSR 1965–. Decors.: 2 Orders of Lenin, Order of Red Banner of Labour, Stalin Prize, various medals.
Publs.: dramas "Khaiat" 1935, "Madrid" 1938, "Makhabbet" 1942, "Derevenskaia devushka" (A Girl From the Country) 1962, "Khoroshii chelovek" (A Good Man) 1965, novels "Nastupit den'" (The Day is Coming) 1948, "Velikaia opora" (Great Supports) 1957, and others.

Ibragimov, Nasim Sadykhovich
State official.
1st Sec. of Oktiabrskii Raion Cttee. Azerbaidzhan CP 1975–80; mem. Auditing Comm. Azerbaidzhan CP 1976–; Chair. of State Cttee. for Publishing, Printing and the Book Trade Azerbaidzhan SSR S. 1980–.

Ibragimov, Sokhrab Mirkazanfar ogly
State official.
Min. of Light Ind. Azerbaidzhan SSR Je. 1982–.

Ibragimov, Ubaidulla Iakubovich
State official.
Acting Chair. 1957–59, Chair. of Uzbek SSR State Cttee. für Television and Radio 1959–; mem. Auditing Comm. Uzbek CP 1960–76, cand. mem Cen. Cttee. 1976–81, mem. 1981–.

Ibragimov, Vagiz Galimovich
State official.
Min. of Light Ind. Kazakh SSR O. 1965–; mem. Cen. Cttee. Kazakh CP 1966–.

Ieviev, Vitalii Mikhailovich
Physicist; b. My. 15, 1926, Briansk.
Mem. CPSU 1955–; grad. Aviation Inst. Moscow 1948; specialist in thermal physics and mechanics; corresponding mem. USSR Acad. of Sciences 1964–. Decors.: 3 orders, N. E. Zhukovski Prize 1959 for work on aviation theory.

Igamberdyev, Mavlian
State official.
Chair. of State Cttee. for the Supply of Production Equipment for Agriculture Uzbek SSR D. 1981–.

Igemberdiev, Samibek Abazovich
State official.
Head of Admin. of Geology Kirghiz SSR My. 1979–; mem. Cen. Cttee. Kirghiz CP 1981–.

Ignatov, Vadim Nikolaevich
Party official; b. 1931.
Mem. CPSU 1953–; grad. Leningrad Agricultural Inst. 1957; Chief Agronomist of Pedagogical Econ. at same inst., chief agronomist of an oblast admin. of state farms 1952–61; party posts 1961–62; chief of a dept. of Leningrad Oblast Cttee. CPSU 1962–68; Sec., 2nd Sec. of Leningrad Oblast Cttee. CPSU 1968–75; Dep. to USSR Supr. Sov. 1974–; 1st Sec. of Voronezh Oblast Cttee. CPSU 1975–; mem. Cen. Cttee. CPSU 1976–.

Ignatovich, Boris Vsevolodovich
Photography expert; b. Mr. 22 (Ap. 3) 1899, Slutsk, present-day Belorussian SSR.
Member of the CPSU since 1925.
Since 1927, he has worked as a press photographer for *Pravda, Komsomol'skaia pravda, Trud* and *Bednota* (published until 1931). He was the first chairman of the Association of Moscow Press Photographers (1929) and was a member of the Oktiabr' association. His best-known works are those that show the socialist transformations of the 1920's and 1930's and record the leading Soviet cultural figures. In his work the search for photographic means of expression is closely related to his artistic mastery of a new reality.

Ignotas, P.-V. P.
Komsomol official.
Mem. Bureau 1978–, Sec. 1978–82, 1st Sec. Cen. Cttee. Lithuanian Komsomol Mr. 1982–; cand. mem. Cen. Cttee. Lithuanian CP 1981–.

Ikaunieks, Albert Eduardovich
Worker.
Dep. Supr. Sov. Latvian SSR 1980–; Worker in a factory in Riga 1980; mem. Cen. Cttee. and mem. Bureau Cen. Cttee. Latvian CP Ja. 1981–.

Ikhsanova, Liabiba Faizovna
Author; b. 1923, Srednie Shuni, Kirov Oblast; d. of a village teacher.
Tatar; mem. CPSU 1945–; grad. Faculty of Geography at Kazan Univ.
Publs.: "Na reke Serebrianka" (On the Serebrianka River) 1948, "Sem dnei pod zemlei" (Seven Days Under the Earth) 1949, "Niiaz" 1959, "Niiaz v shkole" (Niiaz in School) 1961. "Ulitsa kosmonavtov" (Street of the Cosmonauts) 1963, and others.

Ikrami, Dzhalal
Author; b. 1909, Bukhara; s. of a judge.
Tadzhik; mem. CPSU 1945–; first works publ. 1927. Decor.: Order of Red Banner of Labour.
Publs.: "Zmeia" (The Snake) 1934, "Vrag" (The Enemy) 1933, "Semena liubvi" (Seeds of Love) 1937, "Priznaiu sebia vinovnym" (I Admit I'm Guilty) 1957, "Shodi" 1949, "Neschastnaia devushka" (Unhappy Girl) 1950, "Dokhunda" 1954, and others.

Ikramov, A. S.
Party official.
Head of Dept. of Light and Food Ind. of Cen. Cttee. Uzbek CP 1980–; mem. Auditing Comm. Uzbek CP 1981–.

Iksanov, Mustakhim Belialovich
Party official; b. 1926.
Mem. CPSU 1951–; grad. Kazakh Agricultural Inst. 1952; technician of a section of Kazakh railway, student 1945–52; foreman, dept. head, chief engineer, head of construction admin. "Dzhetysaistroi", head of a trust 1952–61; 1st Sec. of a raion cttee., head of a krai admin. for agriculture 1961–63; 1st Sec. of Kzyl-Orda Oblast Cttee. Kazakh CP 1963–66 (for Agriculture 1963–64); Dep. Chair. of Council of Mins. Kazakh SSR 1966–70; mem. Cen. Cttee. Kazakh CP 1966–; Dep. to USSR Supr. Sov. 1966–; 1st Sec. of Dzhambul Oblast Cttee. Kazakh CP 1970–71; Sec. and mem. Bureau of Cen. Cttee. Kazakh CP 1971– 75; mem. Cen. Auditing Comm. 1971–76, cand. mem. Cen. Cttee. CPSU 1976–; 1st Sec. of Uralsk Oblast Cttee. Kazakh CP Ap. 1975–.

Il'chenko, Aleksandr Eliseevich
Ukrainian writer; b. My. 22 (Je. 4), 1909, Kharkov.
His work first appeared in print in 1929. His collection of essays *Dneprel'stan* about the builders of Dneproges (Dnieper hydroelectric power plant) was published in 1932. His novella *The Heart Waits* (1939; second version entitled *Petersburg Autumn,* 1941; Russian translation 1961), which is devoted to the life of T. G. Shevchenko, was produced as a play in 1954. The novella *The Ordinary Urchin* (1947) is about V. P. Chkalov. A wealth of Ukrainian folklore appears in his novel *No End to Cossack Stock, or Mamai and the Fiery Young Wife* (books 1–2, 1958; Russian translation 1965). This novel is about the Ukraine during the second half of the 17th century and the cossack Mamai, a character in Ukrainian folk graphics. Decors.: three orders and various medals.
Works: [Il'chenko, Oleksandr Ielyseio-

vich.] *Vybrane*. Kiev 1948. *In Russian translation: Bagrianyi list: Povesti i rasskazy*. Moscow 1969.

Ilichev, Boris Ivanovich
Diplomat.
USSR Amb. to Thailand Mr. 1974–Je. 1978.

Ilichev, Leonid Fedorovich
Diplomat; b. Mr. 2 (15), 1906, Krasnodar. Russian; mem. CPSU 1924; grad. N. Caucasian Univ. Rostov 1930 and Inst. of Red Professors for Philosophy 1937; apprentice, master asst. at machine building works "Kubanol" in Krasnodar 1918–24; Komsomol posts, student, scientific and teaching activity, student, CPSU work, journalist, Ed. of Phil. Section of "Malaia Sovetskaja Entsiklopediia", mem. Ed. Board and Acting Sec. of CPSU (B) organ "Bolshevik", Chief Ed. of "Pravda" and "Izvestiia" 1924–37; Prof. at Moscow State Univ. Lecturer at Party High School and Acad. of Social Sciences at Cen. Cttee. CPSU 1940–52; cand. mem. Cen. Cttee. 1952–56, mem. Cen. Auditing Comm. 1956–61, mem. Cen. Cttee. CPSU 1961–66; Head of a Dept. and mem. Collegium USSR Min. of Foreign Affairs 1953–58; Head of Dept. of Propaganda and Agitation at Cen. Cttee. CPSU for the Union Republics 1958–61; mem. Ed. Board of journal "Kommunist" 1959–62; Sec. of Cen. Cttee. CPSU 1961–65; Dep. to USSR Supr. Sov. 1962–66; USSR Dep. Min. of Foreign Affairs Mr. 1965–; mem. Cen. Auditing Comm. CPSU 1976–81, cand. mem. Cen. Cttee. CPSU 1981–. Decors.: Order of Lenin, 3 Orders of Red Banner of Labour, Order of Patriotic War 1st Cl., Badge of Honour, medals, Lenin Prize.

Il'in, Boris Fedorovich
Russian actor; b. Ap. 8 (21), 1901, Saratov.
People's Artist of the USSR (1956).
In 1919–20, he studied at the theatrical school of M. I. Velizarii and later in the Higher Theater Workshops in Saratov. From 1921 to 1925 he acted with the Saratov City Theater. He subsequently worked in the theaters of Odessa, Kuibyshev, Baku, Kharkov, and other cities. In 1936 he joined the troupe of the Sverdlovsk Dramatic Theater. His first important roles, which included Uriel Acosta in Gutzkow's play of the same name and Chatskii in Griboedov's *Woe From Wit*, determined the heroic and romantic direction of his art. This was vividly revealed in Soviet dramatic works, in which he created the characters of active, purposeful, and talented people. His best roles have been Platon and Ognev in Korneichuk's *Platon Krechet* and *The Front*, Godun in Lavrenev's *The Break*, and Murav'ev in Chirskov's *The Victors*. He created the characters of V. I. Lenin in Pogodin's play *The Kremlin Chimes* and of Ia. M. Sverdlov in the play *The Bolshevik* by Del'. His important roles in Russian and foreign classical dramas have included Voinitskii in Chekhov's *Uncle Vanya*, Protasov in L. N. Tolstoy's *The Living Corpse*, Prince Volkovskii in *The Insulted and Injured* after Dostoevsky, Don César de Bazan in Hugo's *Ruy Blas*, the title role in Rostand's *Cyrano de Bergerac*, and Iaropegov in Gorkii's *Somov and Others*.

Ilina, Lidiia Aleksandrovna
Graphic artist; b. 1915.
Grad. Art Inst. Moscow 1939; employed in Kirghiz SSR 1939–; numerous book illustrations. Decors.: People's Artist of the Kirghiz SSR 1963, State Prize of the USSR 1971.

Ilina, Natalia Iosifovna
Author; b. 1914, St. Petersburg.
Russian; grad. Gorkii Inst. of Literature Moscow 1953; emigrated to Harbin with parents in 1920, lived in Shanghai 1936–47, returned to the Sov. Union 1947; worked for TASS and pro-Soviet newspaper "Novaia zhizn'" in China.
Publs.: "Inymi glazami. Ocherki shanghaiskoi zhizni" (With Other Eyes. Sketches from Shanghai Life) 1946, novel "Vozvrashchenie" (The Return) 1957, "Vnimanie: opasnost'!" (Attention: Danger!) 1960, "Ne nado ovatsii!" (No ovations!) 1964, and others.

Il'inskii, Igor' Vladimirovich
Russian actor; b. Jl. 11 (24), 1901, Moscow.
People's Artist of the USSR (1949); Hero of Socialist Labor (1974). Member of the CPSU from 1960.
He studied in the theater school of F. F. Komissarzhevskii in 1917. He made his debut at the V. F. Komissarzhevskaia Theater in Moscow in 1918, subsequently performing in Moscow in musical comedies and at the cabaret The Bat, the Terevsat (Theater of Revolutionary Satire), and the First Studio of the Moscow Art Theater. In 1920 his stage career became closely linked with the activities of V. E. Meyerhold; he appeared in Meyerhold's productions at the First Theater of the RSFSR and the Meyerhold Theater. He portrayed the Compromiser and Prisypkin in Mayakovsky's *Mystery Bouffe* and *The Bedbug*, Bruno in Crommelynck's *The Magnificent Cuckold*, Arkashka Schastlivtsev in Ostrovskii's *The Forest*, and Raspliuev in Sukhovo-Kobylin's *Krechinskii's Wedding*.
At the same time he became well known as a film actor, particularly for his performances in comic roles in the films *Aelita, The Cigarette Salesgirl From Mossel'prom, The Garment Cutter From Torzhok, The Case of the Three Million, Miss Mend*, and *The Doll With Millions*. He played the part of Byvalov in the film comedy *Volga-Volga* (1938) with great success.
Since 1938, he has been performing at the Malyi Theater, where his first role was Khlestakov in Gogol's *The Inspector-General*. He later portrayed Arkashka Schastlivtsev and Shmaga in Ostrovskii's *The Forest* and *Guilty Though Guiltless*, Zagoretskii in Griboedov's *Woe From Wit*, and Chesnok in Korneichuk's *On the Steppes of the Ukraine*. He has demonstrated the ability to perform a wide variety of roles ranging from outspoken buffoons to individuals requiring subtle and deeply psychological portrayals. He is able to express gentle, sympathetic humor and also irate, castigating satire.
His roles during the 1940's and 1950's included Iusov in Ostrovskii's *A Profitable Post* (1946) and Foma Opiskin in *The Hamlet of Stepanchikovo*, an adaptation of Dostoevsky's work. Among his screen roles was Ogurtsov in *Carnival Night*. A new stage in the development of his talent was reached in his interpretation of Akim in L. N. Tolstoy's *The Power of Darkness*. He portrayed the highly dramatic image of the peasant with epic simplicity and power. A masterful speaker, he has performed on radio and on the music-hall stage.
Since 1958 he has been involved in the production of plays. At the Malyi Theater he staged an adaptation of Thackeray's *Vanity Fair* (1958, together with V. N. Tsygankov), Trenev's *Liubov' Iarovaia* (1960, together with Tsygankov), an adaptation of Flaubert's *Madame Bovary* (1963), and Gogol's *The Inspector-General* (1965, he played the Governor). Decors.: State Prize of the USSR in 1941, 1942, and 1951. Order of Lenin, three other orders, and various medals.
Works: *Sam o sebe*. Moscow 1961. *So zritelem naedine*. Moscow 1964.

Iliukhin, Aleksandr Merkulovich
State official.
Mem. Auditing Comm. 1960–66, mem. Cen. Cttee. Uzbek CP 1966–; Sec. of Tashkent Oblast Cttee. Uzbek CP 1960–63; Dep. Chair. of Tashkent Oblast Exec. Cttee. of Workers' Deps. (for Ind.) 1963–64; 2nd Sec. of Kashka-Daria Oblast Cttee. Uzbek CP 1964–68 and Andizhan Oblast 1968–71; Min. of Housing and Communal Econ. Uzbek SSR D. 1971–.

Iliukhin, Avan Alekseevich
Diplomat; b. 1916.
Mem. Diplomatic Service 1956–; mem. Apparat USSR Min. of Foreign Affairs 1961–63 and 1967–78; Councellor-Envoy, USSR Emb. in Rumania 1963–67; USSR Amb. to Togo, S. 1978–.

Iliushin, Aleksei Antonovich
Scholar in continuum mechanics; b. Ja. 7 (20), 1911, Kazan.
Corresponding member of the Academy of Sciences of the USSR (1943). Member of the CPSU since 1940.
Graduated from Moscow University in 1934 and became a professor there in

1938. From 1936 to 1960 he worked at the Institute of Mechanics of the Academy of Sciences of the USSR (he became the director of the institute in 1953). He was rector of Leningrad University from 1950 to 1952. His main works were on the theory of elasticity and plasticity. He developed the theory of viscoplastic flows (1934–40), the theory of small elasticoplastic deformations (1942–48), the theory of the stability of plates and shells beyond the elastic limit (State Prize of the USSR, 1948), the theory of simulation in processes of pressure working of metals (1951–52), and the theory of linear and nonlinear thermoviscoelasticity. He also proposed a method of simulation during supersonic flow around a body, as well as solutions to a number of problems of gas dynamics. He was a deputy to the third and fourth convocations of the Supreme Soviet of the RSFSR. Decors.: six orders, as well as several medals.

Ilko, Ivan Ivanovich
Painter; b. 1938, Dulovo, Transcarpathians.
Grad. Uzhgorod School of Applied Art 1957, pupil of A. Erdeli, I. Bokshai and F. Manailo.

Il'skaia, Fatima Salikhovna
Tartar actress; b. D. 31, 1902 (Ja. 13, 1903), Orenburg.
People's Artist of the RSFSR (1967).
In 1915, she began her dramatic career in the Shirkat Tartar Dramatic Company in Orenburg. In 1916 she joined the Saiar Company. She performed in the Tartar theater in Tashkent from 1921 to 1923 and since 1923 has worked at the G. Kamal Tartar Theater in Kazan. One of the founders of Tatar national dramatic arts, she established realism as a tradition in Tartar theater.
She has played Dilbar in Tinchurin's *Without Sails*, Zuleikha in Kulakhmetov's *Young Life*, Fatima in Amir's *Song of Life*, Katerina in Ostrovskii's *Thunderstorm*, Tat'iana and Sof'ia in Gorkii's *Enemies* and *The Last Ones*, and the title role in Arbuzov's *Tania*. She has also performed in Shakespearean tragedies and in the comedies of Lope de Vega and Beaumarchais. Decors.: four orders and several medals.

Iltner, Edgar Karlovich (Edgars Iltners)
Painter; b. Mr. 3, 1925, in the *khutor* (privately owned homestead) of Buriki, present-day Liepāja Raion, Latvian SSR.
Honored Art Worker of the Latvian SSR (1963).
He studied at the Academy of Arts of the Latvian SSR in Riga under O. Skulme and E. Kalniņš from 1949 to 1956, becoming a professor there in 1960. In his works he strives for a generalized philosophical interpretation of contemporary Soviet themes. His pictures are marked by a clear rhythmic composition and a strict color scheme. Among his works are *The Husbands Return* (1957), *Immortality* (1968), *Masters of Earth* (1960), and *Calloused Hands* (1962), all in the Art Museum of the Latvian SSR in Riga. Decors.: State Prize of the Latvian SSR in 1965. Chairman of the Union of Artists of the Latvian SSR since 1968.

Imashev, Sattar Nurmashevich, Cand. of Hist. Sc.
Party official; b. 1925.
Kazakh; mem. CPSU 1945–; grad. Kazakh State Univ.; teacher at a middle school 1942–43; served in Sov. Army 1943–46; instructor, head of a sector at Aktiubinsk Oblast Cttee. Kazakh CP 1946–50; scientific activity 1950–58; dep. head of a dept. at Cen. Cttee. Kazakh CP 1958–60; Sec. of Virgin Lands Krai Cttee. Kazakh CP 1960–65; cand. mem. 1961–66, mem. Cen. Cttee. Kazakh CP 1966–; Dep. to 1963–. Chair. of Supr. Sov. Kazakh SSR Jl. 1975–79; Sec. 1965–79, and mem. Bureau of Cen. Cttee. Kazakh CP 1965–; mem. Cen. Auditing Comm. CPSU 1976–81, mem. Cen. Cttee. CPSU 1981–; Chair. of Presidium Supr. Sov. Kazakh SSR D. 1979–; Dep. Chair. of Presidium USSR Supr. Sov. Je. 1980–; mem. Cen. Cttee. CPSU 1981–.

Imermanis, Anatol Adolfovich
Author; b. 1914, Moscow; s. of a clerk.
Latvian; moved to Latvia 1919; first works publ. 1930.
Publs.: "U Daugavy" (On the Dvina) 1947, "Gorod u moria" (The Town at the Sea) 1951, "V etu noch' spokoiny tol'ko zvezdy" (Only the Stars Are Peaceful in This Night) 1961, "Gde vstretiatsia vse korabli" (Where All Ships Will Meet) 1953, "Kvartira bez nomera" (Flat Without Number) 1952, "Tovarishch Mauzer" (Comrade Mauzer) 1960, "Tobago meniaet kurs" (Tobago Changes Course) 1961, and others.

Imshenetskii, Aleksandr Aleksandrovich
Microbiologist; b. D. 26, 1904 (Ja. 8, 1905), Kiev.
Grad. Voronezh Univ. 1926; mem. USSR Acad. of Sciences 1962–; mem. staff Inst. of Microbiology of USSR Acad. of Sciences 1930–, Dir. of same 1949–. Decors.: Order of Lenin, 2 other orders and various medals.

Inauri, Aleksei Nikolaevich
State official; b. 1908.
Georgian; mem. CPSU 1932–; grad. Acad. of Gen. Staff of Sov. Army and Fleet; railway worker until 1927; served in Sov. Army 1927–53; regimental, div. and corps commander 1941–45; Min. of the Interior Georgian SSR 1953–54; Chair. of Cttee. for State Security (KGB) Georgian SSR 1954–; Dep. to Supr. Sov. USSR 1954–; mem. Cen. Cttee. 1954–, mem. Bureau Cen. Cttee. Georgian CP 1954–62; cand. mem. Bureau 1964–71, mem. Bureau Cen. Cttee. Georgian CP 1971–. Decors.: Order of Lenin, 2 Orders of Red Banner, Order of Red Star, medals.

Indzhiev, Lidzhi Ochirovich
Author; b. 1913, Abganery, Kalmuk ASSR.
Mem. CPSU 1938–; first works publs. 1934.
Publs.: poems "Radost'" (Joy) 1940, "Rodnik" (The Source) 1941, "Blagodarnost'" (Gratefulness) 1958, novel "Doch' Oldi" (Oldi's Daughter) 1963, and others.

Infantev, Vadim Nikolaevich
Author; b. 1921; Titovskoe, Tomsk Oblast.
Russian; mem. CPSU 1943–; grad. Naval Acad. 1955; took part in World War II.
Publs.: "Shestero s ostrova" (The Six From the Island) 1957, "Oni shli riadom" (They Went Side by Side) 1958, "Kogda chelovek liubit" (When a Person Loves) 1963, "Funkciia vremeni. Malenkaia povest'" (Function of Time. Short Story) 1964, and others.

Inzhievskii, Aleksei Alekseevich
Party official.
Head of Dept. of Agriculture, Stavropol Krai Cttee. CPSU until 1976, 1st Sec. Stavropol City Cttee. CPSU 1976–78; 1st Sec. of Karachay-Cirkassian Autonomous Oblast Cttee. CPSU D. 1978–.

Ioann (Dmitrii Aleksandrovich Razumov)
Russian Orthodox clergyman; b. O. 28, 1898, Kolomna, Moscow Oblast.
Novice at Smolensk Zosima Hermitage, St. Sergius Holy Trinity Monastery 1916; monk, deacon 1924; Econ. Man. of Moscow Patriarchate Centre 1936–; worked in Ulianovsk at Kazan Cathedral 1941–1943; priest 1943; Abbot; Archimandrite; Prior at St. Sergius Holy Trinity Monastery 1946–; Bishop of Kostroma and Galich 1953–54; Bishop 1954–59, Archbishop 1962–72 and Metropolitan of Pskov and Porkhov and Head of Pskov Monastery 1972–; Bishop of Berlin and Germany 1959–62.

Ioann (Ivan Matveevich Snychev), Cand. of Theol.
Russian Orthodox clergyman; b. O. 9, 1927, Novo-Maiachka, Nikolaev Oblast.
Subdeacon 1945; monk, deacon 1946; priest-monk 1948; grad. Theological Seminary Saratov 1951; completed studies at Theological Acad. Leningrad, Cand. of Theol. 1955; postgraduate studies, Lecturer at Chair of Sectarianism; final monastic vows, Lecturer at Theological Seminary Minsk 1956; priest in Cathedral of Archbishop Manuel of Cheboksary and Chuvashia and his Sec. 1957; Dep. Supervisor and Lecturer at Saratov Theological Seminary 1959–60; Custodian of Our Lady the Protectress Cathedral in Kuibyshev 1960–65; abbot 1961; Archimandrite 1964; Bishop of Syzran, Substitute Bishop of Kuibyshev

Diocese 1965–69; Bishop of Kuibyshev and Syzran Mr. 1969–.

Ioann (K. N. Vendland), Cand. of Geol.-Min. Sc., Cand. of Theol.
Russian Orthodox clergyman; b. 1909, Petersburg.
Grad. Leningrad Mining Inst. as geological engineer 1930; monk, priest 1936; decorated for bravery during World War II; Sec. of Archbishop of Tashkent 1946; Archimandrite 1950; grad. Leningrad Theological Acad. (correspondence course), Cand. of Theol. 1956; Rector of Theological Acad. Saratov 1957–58; Rep. of the Moscow Patriarchate at the Patriarchate of Antioch in Damascus 1958–; Bishop of Berlin and Cen. Europe, Exarch of the Moscow Patriarchate in Cen. Europe 1960; Archbishop of the Aleutian Islands and N. America, Exarch of the Moscow Patriarchate in N. and S. America 1962–; Metropolitan 1963; Metropolitan of Iaroslavl and Rostov 1967–.

Ionafan (Ivan Mikhailovich Kopalovich), Dr. of Theol.
Russian Orthodox clergyman; b. Je. 8, 1912, Oleshnik, Transcarpathians.
Educ. in Uzhgorod 1923–27; studied Serbian Theological Seminary in Sremski Karlovci, Yugoslavia 1927–32; deacon 1933; worked in Mukachevo Diocesan Admin. until 1936; priest 1936; sec. of same 1939–; religious instructor at Khust High School 1941–44; sec. again; responsible for Russian and Hungarian Orthodox communities in Hungary 1949–54; archpriest; high decorations; Chief Clergyman of Mukachevo Cathedral and Sec. for Diocesan Admin. 1954–55; mem. staff Czechoslovak Metropolitan Council 1955; until 1964 Vicar-Gen. of Orthodox Diocese of Prešov and studies at Faculty of Orthodox Theology at same; Dr. of Theol. 1963; asst. in Church Dept. of External Affairs, Moscow Patriarchate, teacher at Moscow Theological Acad. 1965; lectures on Orthodoxy and the History of the Foreign Mission of the Russian Orthodox Church; monk, Archimandrite 1965; Bishop 1972–73 and Archbishop of Kishinev and Moldavia 1973–.

Ioannisian, Abgar Rubenovich
Historian; b. My. 30 (Je. 12), 1908, Tbilisi.
Mem. CPSU 1944–; mem. Acad. of Sciences Armenian SSR 1947–; Dep. Rector of Erevan Univ. 1938–46; Dir. of Inst. of History at Acad. of Sciences Armenian SSR 1947–53; Acad. Sec. of Dept. of Sociology at Acad. of Sciences Armenian SSR 1949–60; Vice-Pres. of Acad. of Sciences Armenian SSR 1960–63, 1967–. Decors.: Order of Red Banner of Labour, several medals.

Ioasaf (Vasilii Simenovich Ovsiannikov), Cand. of Theol.
Russian Orthodox clergyman; b. Ja. 1, 1904, Belgorod.
Studied railway engineering in Kharkov; monk 1946; deacon, then priest, appointed 2nd priest in Cathedral of the Transfiguration Kharkov 1943; entry into brotherhood of Pochaev Monastery 1944; Main Clergyman at Bogoiavlenskii Monastery in Kremenets 1950; abbot 1951; correspondence course at Leningrad Theological Seminary; 2nd priest at Poltava Cathedral 1953; priest at Cathedral of the Resurrection in Frunze, Tashkent Bishopric 1955; priest at Uspenskii Cathedral in Tashkent 1959; custodian of same 1961; Archimandrite 1964; grad. Leningrad Theological Acad., Cand. of Theol. 1966; Bishop of Rostov and Novocherkassk My. 1973–.

Iokubonis, Gediminas Al'bino
Sculptor; b. Mr. 8, 1927, Kupishkis.
Honored Art Worker of the Lithuanian SSR (1965). Corresponding member of the Academy of Arts of the USSR (1967). He studied at the Kaunas Institute of Applied and Decorative Art (1946–51) and at the Art Institute of the Lithuanian SSR in Vilnius (1952) under I. I. Mikenas. An audacious generalization of forms and an energetic manifestation of the plastic structure of the volume typify his monuments and sculptures, which display national-romantic spirit. His works include *The Chairman of the Kolkhoz* (bronze, 1957), and a portrait of I. Ianulenis (granite, 1962), both in the Art Museum of the Lithuanian SSR, Vilnius; a monument to the victims of fascism in the village of Pirchiupis (granite, 1960; Lenin Prize, 1963; architect V. P. Gabriunas); a portrait of I. Venozhinskis (granite, 1964); and a monument to Lenin in Il'ich square in Moscow (bronze and granite, 1967; architect V. A. Chekanauskas).

Iordanov, Ivan Efremovich
State official.
Dep. to Supr. Sov. Moldavian SSR 1963–; Sec. of CPSU Cttee. of Kolkhoz-Sovkhoz Prod. Admin. in Orgeev 1963; mem. Cen. Cttee. 1963–71, Chair. of Auditing Comm. Moldavian CP 1971–; Chair. of State Cttee. for Film Moldavian SSR Ja. 1970–.

Iosif (I. M. Savrash)
Russian Orthodox clergyman; b. Ap. 27, 1909, Cherniev nr. Stanislav.
Studied in Innsbruck; Greek Catholic priest, Chair. of Pokrovski Cathedral in Stanislav 1932; conversion to Orthodox Church 1948; monk, Archimandrite, Bishop of Stanislav and Administrator of Bishopric of Mukachevo 1957; after the renaming of Stanislav to Ivano-Frankovsk. Bishop of Ivano-Frankovsk and Kolomyia, Archbishop of same 1965–.

Iovchuk, Mikhail Trifonovich, Dr. of Phil. Sc.
Academician; b. N. 6 (19) 1908, Zanzhove, Brest Oblast.
Mem. CPSU 1926–; grad. Krupskaia Acad. of Communist Educ. 1931; corresponding mem. USSR Acad. of Sciences 1946–; dep. head, head of pol. dept. of various sovkhozes in Belorussian SSR 1933–36; mem. staff "Sotsekgiz" and Head of Chair for Dialectical Materialism at Mendeleev Inst. of Chemical Tech. Moscow 1936–38; leading posts in Komintern Exec. Cttee. 1939–41; mem. staff Cen. Cttee. CPSU 1941–47; Sec. of Cen. Cttee. Belorussian CP and teacher 1941–47; employed in learned insts. (Acad. of Social Sciences at Cen. Cttee. CPSU, Uralsk State Univ., Moscow State Univ.) 1949–57; Head of Sector with Inst. of Philosophy of USSR Acad. of Sciences, Chief Ed. of "Filosofskie nauki" 1957–70; Rector of Acad. of Social Sciences at Cen. Cttee. CPSU 1970–; cand. mem. Cen. Cttee. 1971–. Decors.: Order of Lenin, 4 other orders and various medals.

Iovits, V. D.
Trade union official.
Sec. of Moldavian Republican Trade Union Council Ja. 1982–.

Ird, Kaarel Kirillovich
Director, theater and public figure; b. Ag. 14 (27), 1909, Riga.
People's Artist of the USSR (1970). Member of the CPSU since 1940.
From 1934 to 1936, he studied at the studio of dramatic art in Tartu, and from 1937 to 1939 he directed and acted at the Workers' Theater in Tartu. Since 1940 (with hiatuses), he has headed the Vanemuine Theater (Tartu), which presents all the genres of stage art: opera, drama, operetta, and ballet. He directed the opera productions *The Fires of Vengeance* by E. Kapp (1945), *The Shore of Storms* by Ernesaks (1958), *Betrothal in the Monastery* by Prokofiev (1962), *Katerina Izmailova* by Shostakovich (1966), *The Swan's Flight* by Tormis (1966), and *Johnny From Guiana* by Bush (1969). Among his dramatic productions are Gorkii's *The Petite Bourgeoisie* (1946), and *Egor Bulychov and Others* (1950), Shakespeare's *The Merchant of Venice* (1958) and *Coriolanus* (1964), Brecht's *Galileo* (1961) and *Schweik in the Second World War* (1967), and Smuul's *The Wild Captain* (1965).
An experimentalist, he stages productions characterized by social focus, profound stage treatment, and original resolutions for mass scenes. He organized and teaches at a theatrical studio associated with his theater and is the author of articles on theatrical issues. Decors.: State Prize of the Esthonian SSR (1947), the State Prize of the USSR (1967), two orders, and various medals.
Works: *Postaraemsia poimat' chudo: Stat'i o teatre.* Leningrad 1967. *Ceterum censeo ehk jahedate suvede jutte.* Tallinn 1965. *Tartu Töölisteater, 1937–1939.* Tallinn 1966. *Semper idem ehk mõtterände ja rännumõtteid.* Tallinn 1970.

Irenei
Russian Orthodox clergyman.
Bishop of Serpukhov until 1982; Archbishop of Alma-Ata and Kazakhstan Jl. 1982–.

Irenei (Igor Vladimirovich Zuzemil)
Russian Orthodox clergyman; b. 1919, Chernigov.
Educ. in Berlin, studied medicine; completed pastoral courses in German Bishopric of Russian Church Abroad 1942; deacon, priest in the churches at Constance and Melbourne; conversion to Russian Patriarchate Church 1957; worked in The Hague, Munich, Berlin; monk; Bishop 1966; Bishop 1971–72 and Archbishop of Baden and Bavaria 1972–; responsible for Vienna and Austria concurrently 1975–.

Irenei (Ivan Petrovich Seredni), Cand. of Theol.
Russian Orthodox clergyman; b. My. 10, 1939, Stolpin, Ukraine; s. of a peasant.
Studied Leningrad Theological Seminary and Leningrad Theological Acad.; served in Sov. Army 1963–66; Cand. of Theol. at Theological Acad. Leningrad 1968; deacon My. 21, 1968, ordained as priest at St. Nicholas Epiphanias Cathedral My. 22, 1968; Research Asst. at Chair of Canon Law at Theological Acad. Leningrad 1969–70; priest at Holy Trinity Cathedral of Alexandr Nevskii Lavra in Leningrad 1969–70; Lecturer in Canon Law and Council Sec. and Inspector of Acad. and Seminary Leningrad 1970–71; monastic vows under the name Irenei S. 1, 1970; clergyman with Tokyo Mission 1971–75; promoted to rank of Archimandrite 1971; Bishop of Ufa and Sterlitamak Jl. 1975–80.

Iroshnikova, Irina Ivanovna
Author; b. 1911, Iaroslavl.
Russian; mem. CPSU 1943–.
Publs.: "Gde-to v Sibiri" (Somewhere in Siberia) 1946, "Nadezhda Egorova" 1948, "Sashen'ka" 1955, "Chudesnaia vysota" (Wonderful Height) 1958, "Trudnoe leto" (A Difficult Summer) 1964, sketches "Eto bylo v Odesse" (It Was in Odessa) 1957, "Parashiutisty" (Parachutist) 1962, "Ulitsa Saliama Adilia" (Saliam Adil Street) 1964, "Oni zhivut na sosednei ulitse" (They Live in the Neighbouring Street) 1964, and others.

Isabekov, Abdimir
State official.
Dep. Chair. until 1974, 1st Dep. Chair. 1974–77, Chair. of State Cttee. on Vocational Training. Kazakh SSR Ja. 1977–79; Dep. Min. of Forestry Kazakh SSR D. 1979–.

Isachenko, Anatolii Grigorevich, Dr. of Geol. Sc.
Geographer; b. My. 28, 1922, Gomel.
Physical geographer and cartographer; Prof. at Faculty of Geography of Leningrad Univ. 1964–.

Isaev, Egor (Georgii) Aleksandrovich
Author; b. 1926, Korshevo, Voronezh Oblast.
Russian; grad. Gorkii Inst. of Literature Moscow 1955; first works publ. 1945.
Publs.: "Nad volnami Dunaia" (Over the Waves of the Danube) 1953, "Sud pamiati" (The Court of Memory) 1962, and others.

Isaev, F. A.
Komsomol official.
Sec. and mem. Bureau Cen. Cttee. Azerbaidzhan Komsomol Ap. 1982–.

Isaev, Georg Georgievich
State official.
Min. of Building Materials Ind. Uzbek SSR D. 1981–.

Isaev, Vasilii Iakovlevich
State official; b. 1917; s. of a peasant.
Russian; mem. CPSU 1939–; grad. Leningrad Inst. of Civil Engineering 1960; dir. of works of a construction trust, then at various construction projects in Far East 1937–49; during war dept. head of an oblast all-union CP(B) cttee., all-union CP(B) cen. cttee. party organizer in a construction trust; head of a trust's construction and assembly board, dir. of a trust, Dir. of "Glavleningradstroi" Trust 1949–61; Dep. Chair. 1961–62, Chair. of Exec. Cttee. of Leningrad City Sov. of Workers' Deps. 1962–66; Dep. to USSR Supr. Sov. 1962–; cand. mem. Cen. Cttee. CPSU 1966–81, mem. 1981–. 1st Dep. Chair. of USSR GOSPLAN, responsible for Capital Construction Jl. 1966–. Decor.: Order of Red Banner of Labour.

Isakova, Mariia Grigor'evna
Sportswoman; b. Jl. 5, 1918, Kirov.
Honored Master of Sport (1946). Member of the CPSU since 1941.
She is a trainer and teacher (the Dynamo Society). She was all-around champion of the USSR from 1945 to 1949 and in 1951 and the first Soviet world champion (1948–50) in speed skating. She ist the holder of many USSR and world records in the all-around combined tournament and in races of various distances. She has won 84 medals, including eight gold medals in world championships and 24 in USSR competitions. Decors.: Order of Lenin and various medals.

Isakova, Nina Sergeevna
Russian mezzo-soprano; b. O. 8, 1928, Sebezh, present-day Pskov Oblast.
People's Artist of the RSFSR (1969).
She was a partisan during the Great Patriotic War of 1941–45. She later was a student in F. S. Petrova's class at the Moscow Conservatory. Upon graduation in 1958, she was accepted into the company of the K. S. Stanislavsky and V. I. Nemirovich-Danchenko Music Theater. Her voice is rich and has a deep timbre. It is particularly beautiful in the lower contralto register. She displays great vocal mastery and dramatic talent. She has been most successful in the roles of ambitious, strong-spirited women. Her portrayals of comic roles are characterized by spontaneity and lightness on stage. Her roles have included Ustin'ia in Koval's *Emel'ian Pugachev,* the Commissar in Kholminov's *Optimistic Tragedy,* Sonetka in Shostakovich's *Katerina Izmailova,* Dorabella in Mozart's *Così fan tutte,* and the title role in Offenbach's *La Perichole.* Isakova is also well-known as a chamber singer. Her concert and opera repertoires primarily consist of works by contemporary composers, such as Z. A. Levina, K. V. Molchanov, T. N. Khrennikov, and D. D. Shostakovich. She won second prize at the International Competition of Musical Performance in Geneva in 1958. Decors.: Order of the Badge of Honor and various medals.

Isakovich, Garol'd Grigor'evich
Architect; b. N. 7, 1931, Moscow.
He studied under B. S. Mezentsev at the Moscow Institute of Architecture from 1950 to 1956. His works include housing projects in the towns of Balashikha, Voskresensk, and Dzerzhinsk (1956–59); laboratory buildings for chemical plants in Moscow and Kazan (1958–59); administrative buildings on the highway through the Hindu Kush in Afghanistan (1963–64); the design for a central square in Tula (under construction since 1970); and the Lenin Memorial in Ul'ianovsk (1970, jointly with B. S. Mezentsev, M. P. Konstantinov, and other architects; Lenin Prize, 1972).

Isanbet, Naki Sirazievich (also N. S. Isanbetov)
Poet, dramatist, and folklorist; b. N. 27 (D. 9), 1899, Maloiaz, in present-day Salavat Raion, Bashkir ASSR.
Honored Art Worker of the Tartar ASSR (1957) and the RSFSR (1960).
He began publishing his works in 1914. He is the author of more than 90 poems, ballads, verses, and songs and of 30 plays. His most popular plays are the comedy *Khodzha Nasretdin* (1940), which ran for more than 1,000 performances; the historical drama *Mullanur Vakhitov* (1947; Russian translation, 1957); *Raikhan* (1953); and the tragedy *Musa Dzhalil'* (1955).
He translated A. S. Pushkin's *Poltava* and *The Bakhchisarai Fountain* and Shakespeare's *Hamlet* and *King Lear* from Russian into Tartar. He has written a number of articles of literary criticism. He also compiled the collections *Children's Folklore of the Tartar People* (1941), *Proverbs of the Tartar People* (vols. 1–3, 1959–67; G. Tukai Republic Prize, 1968), and *Riddles of the Tartar People* (1970).
Decor.: Order of Lenin.
Works: [Isänbät, Näkiy.] *Saylanma äsärlär,* vols. 1–3. Kazan 1960–68.

Isangulov, Farid Akhmadullovich
Author; b. 1928, Bashkir ASSR.
Mem. CPSU 1954–; grad. Higher Literary Courses at USSR Writers' Union 1960; first works publ. 1950.
Publs.: children's stories "Pervoe ispytanie" (The First Exam) 1953, "Lestnitsa" (The Ladder) 1954, "Most Khamita" (Khamit's Bridge) 1956, "Sozvezdie" (The Stars) 1959, and others.

Isanin, Nikolai Nikitich
Engineer; b. Ap. 22 (My. 5), 1905, Moscow.
Mem. CPSU 1926–; grad Inst. of Shipbuilding Leningrad 1935; specializes in shipbuilding; mem. USSR Acad. of Sciences 1970–. Decors.: Hero of Socialist Labour 1963, State Prize of the USSR 1942, Lenin Prize 1959, Order of Lenin, Order of Red Banner of Labour, various medals.

Ishankulieva, Ogulnabat Imamkulievna
Trade union official.
1st Sec. and mem. Bureau Cen. Cttee. Turkmen Komsomol 1976–79; mem. Cen. Cttee. Turkmen CP 1976–; cand. mem. Bureau Cen. Cttee. Turkmen CP 1979–; Chair. Turkmen Republican Trade Union Council 1979–.

Ishanturaeva, Sara Abdurakhmanovna
Uzbek actress; b. O. 26 (N. 8), 1911, Beshbulak, Iangi-Kurgan Raion, Namangan Oblast.
People's Artist of the USSR (1951). Member of the CPSU since 1942.
In 1927, she graduated from the Uzbek drama studio affiliated with the Moscow House of Culture. The studio formed the nucleus of the Khamza Uzbek Drama Theater in Tashkent. Among her best roles in plays by Uzbek playwrights are Dil'bar and Onakhon in Iashen's *Two Communists* and *Honor and Love* and Dzhamilia in Khamza's *Landowner and Farmhand.* Her outstanding roles in classical drama and contemporary Soviet and foreign plays include Katerina in *Thunderstorm* by Ostrovskii, Ophelia and Desdemona in *Hamlet* and *Othello* by Shakespeare, Jocasta in *Oedipus Rex* by Sophocles, Guli in *Alisher Navoi* by Uigun and Sultanov, Aini in *Algeria, My Homeland!* based on the novel by M. Dib, and Hurriiat in *Flight* by Uigun.
She is one of the most important representatives of the dramatic art of Soviet Uzbekistan. Her acting is characterized by warmth, and a striving for verisimilitude and dramatically profound character portrayal. She was a deputy to the second, third, and fourth convocations of the Supreme Soviet of the USSR. Decors.: State Prize of the USSR in 1949, two Orders of Lenin, five other orders, and several medals.

Ishchanov, Rakhim
State official.
1st Raion Sec. of Uzbek CP in Kosh-Kupyr, Sec. of Khorezm Oblast Cttee. Uzbek CP 1959–68; Chair. of Khorezm Oblast Exec. Cttee. of People's Deps. 1968–; mem. Cen. Cttee. Uzbek CP 1971–81.

Ishlinskii, Aleksandr Iulevich, Dr. of Phy.-Math. Sc.
Scientist; b. S. 6, 1913, Moscow.
Mem. CPSU 1940–; grad. Moscow Univ. 1935; specialist in mechanics; mem. USSR Acad. of Sciences 1960–; mem. Acad. of Sciences Ukrainian SSR 1948–; Dr. of Phy.-Math. Sc. 1944; Prof. 1945–; Lecturer, Prof. Chair of Elasticity Theory at Moscow Univ. 1935–48; Dir. of Inst. of Mathematics at Acad. of Sciences Ukrainian SSR and Prof. at Mechanic-Mathematical Faculty at Kiev Univ. 1948–55; Dir. of Inst. of Mechanics at Moscow Univ. and Prof. at Moscow Univ. 1955–64; Dir. of Inst. of Mechanics at USSR Acad. of Sciences 1964–; main subjects in his scientific works are theory of gyroscopes, theory of elasticity and plasticity of materials, general mechanics, oscillation theory. Decors.: Lenin Prize, Order of Lenin, Order of Red Banner of Labour, Badge of Honour, medals.
Publs.: "Trenie kacheniia" in "Prikladnaia matematika i mekhanika" vol. 2/1938, "Osesimmetricheskaia zadacha plastichnosti i proba Brinellia" in "Prikladnaia matematika i mekhanika" vol. 8/1944, "Obshchaia teoriia plastichnosti s lineinym uprocheniem" in "Ukrainskii matematicheskii zhurnal" vol. 3/1954, "Mekhanika giroskopicheskikh sistem" 1963.

Ishmurat, Riza Fakhrutdinovich (also Ishmuratov)
Tartar playwright and public figure; b. N. 1, 1903, Badriash, present-day Ianaul Raion, Bashkir ASSR; s. of a peasant.
Honored Art Worker of the Tartar ASSR (1953). Member of the CPSU since 1927. Fought in the Great Patriotic War (1941–45).
After graduating from a dramatic technicum in 1926, he was a director in the Tartar Academic Theater. In 1926 he wrote his first comedies, *In the Realm of Sleep* and *Good-for-Nothing Philosopher.* Notable among his many dramatic works are the play *To Meet the Storm* (1952) about Lenin's student days and the tragedy *Immortal Song* about Musa Dzhalil' (1956). He also translated D. I. Fonvizin's *The Minor* into Tartar. Decors.: four orders.
Works: *Saylanma äsärlär,* vols. 1–2. Kazan 1966–67.

Isinaliev, Mikhail Ivanovich
State official; b. 1928.
Sec., 2nd Sec. Alma-Ata City Cttee. Kazakh CP 1966–71; cand. mem. Cen. Cttee. Kazakh CP 1971–76, mem. 1976–; Head of Cultural Dept. with Cen. Cttee. Kazakh CP 1971–81; Min. of Foreign Affairs Kazakh SSR Jl. 1981–.

Iskander, Fazil Abdulovich
Author; b. Mr. 6, 1929, Sukhumi.
Russian; grad. M. Gorkii Inst. of Literature 1954.
Publs.: "Zelenyi dozhd" (Green Rain) 1960, "Molodost' moria" (The Youth of the Sea) 1964, "Zapretnyi plod" (Forbidden Fruit) 1966, "Vremia schastlivykh nakhodok" (The Time of Lucky Finds) 1970, "Derevo detstva" (The Tree of Childhood) 1970, and others.

Iskenderov, Adil'rza ogly
Director and actor; b. Mr. 10, 1912, Giandzha, now Kirovabad.
People's Artist of the USSR (1959). Member of the CPSU since 1940.
He graduated from the State Institute of Theatrical Arts in 1936. From 1937 to 1960 he was director and manager of the M. Azizbekov Azerbaidzhan Theater (Baku). He directed Korneichuk's *Platon Krechet* (1936), Vurgun's *Vagif* (1938) and *Farkhad and Shirin* (1941), Shirvanzade's *Honor* (1940), Mamedkhanly's *Morning of the East* (1948; State Prize of the USSR), G. Dzhavid's *Sheikh Senan* (1949), Dzhabarly's *In the Year 1905* (1954), Medzhnunbekov's *The Bay of Il'ich* (1958), and Shakespeare's *Othello* (1959).
Since 1961, he has worked in films both as an actor and director. He appeared in the films *This One If Not the Other One; Romeo, My Neighbor; Tale of Love; Twenty-Six Baku Commissars;* and *The Last Crossing.* He has been teaching at the M. A. Aliev Azerbaidzhan Institute of Arts since 1937. (He became a professor there in 1956.) He was a deputy to the fourth and fifth convocations of the Supreme Soviet of the Azerbaidzhan SSR. Decors.: Order of Lenin, two other orders, and various medals.

Iskhak, Akhmet Abdullovich (also A. A. Iskhakov)
Tartar poet; b. Ap. 18 (My. 1), 1905, Kazan.
Member of the CPSU since 1945.
He worked as a journalist from 1925 to 1939. He fought in the Great Patriotic War (1941–45). He began publishing his works in 1923. His first collection, *Songs of Stone Streets,* appeared in 1928. His main poetic themes are the greatness of the Soviet motherland, the friendship of peoples, and the struggle for peace. His verses and narrative poems of the war years ("Vow" and "Song of the Tank Commander and Hero Petr Novikov") are deeply courageous and patriotic. The poetic cycle *Dawn Over Asia* (1953) praises the struggle of peoples against colonialist oppression. He is also author of the satirical collection *A Questionnaire for the Woman in Love* (1966, in Russian). He has translated works by A. S. Pushkin, M. Iu. Lermontov, V. V. Mayakovsky,

T. G. Shevchenko, and Navoi into Tartar. Decor.: Order of the Red Banner of Labor.
Works: *Saylanma äsärlär.* Kazan 1965. *In Russian translation: Stikhi.* Kazan 1956. *Vstrecha v pesne.* Moscow 1960.

Islambekov, Radzhab Kaplanovich
Endocrinologist; b. D. 30, 1925, Tashkent.
Corresponding member of the Academy of Medical Sciences of the USSR (1969), Honored Scientist of the Uzbek SSR (1964). Member of the CPSU since 1968. He graduated from Tashkent Medical Institute in 1947. From 1967 to 1971 he was director in the Institute of Regional Medicine in Tashkent. His principal works have been devoted to the clinical practice, morphology, and histology of the thyroid gland in its normal and pathological states and to the problems of diagnosis and treatment of thyrotoxicoses with the use of radioactive iodine (in particular, to the results of radiotherapy in diffuse and nodular toxic goiter). Decors.: Lenin Prize in 1964, Order of the Red Banner of Labor.
Works: *Zlokachestvennyi zob.* Tashkent 1967. (With I. K. Kadyrov.) *Kliniko-morfologicheskoe issledovanie endemicheskogo zoba s primeneniem radioaktivnogo ioda.* Tashkent 1961. *Uzlovoi toksicheskii zob.* Tashkent 1965. (With M. M. Baltabaev.) *Vospalitel'nye zabolevaniia shchitovidnoi zhelezy.* Tashkent 1968. (With L. B. Platonova.)

Islamov, Erkabai
State official.
Mem. Auditing Comm. 1966–71, mem. Cen. Cttee. Uzbek CP 1971–; Chair. of Dzhizak Oblast Exec. Cttee. of People's Deps. 1974–; Dep. to Supr. Sov. Uzbek SSR Ag. 1974–.

Isliukov, Semen Matveevich
State official; b. 1915.
Chuvash; mem. CPSU 1939–; grad. Law Inst. Kazan and Acad. of Social Sciences at Cen. Cttee. CPSU; teacher, headmaster 1934–41; sec. of a city and oblast CPSU cttee., Chair. of Council of Mins. Chuvash ASSR 1941–55; 1st Sec. of Chuvash ASSR CPSU Oblast Cttee. 1955–68; Dep. to USSR Supr. Sov. 1958–70; cand. mem. Cen. Cttee. CPSU 1956–71; Chair. of Presidium Supr. Sov. Chuvash ASSR 1968–; Dep. to and Dep. Chair. of Presidium RSFSR Supr. Sov. 1971–. Decor.: Order of Lenin.

Ismailov, O. D.
Komsomol official.
Sec. and mem. Bureau Cen. Cttee. Uzbek Komsomol 1981–.

Israelian, Rafael Sergeevich (Sarkisovich)
Architect; b. S. 4 (17), 1908, Tbilisi.
People's Architect of the USSR (1970).
He studied at the Leningrad Academy of Arts from 1929 to 1934. He began working in Yerevan in 1936, becoming an instructor at a polytechnical institute there in 1940. Israelian's buildings are noted for simplicity and clearness of form and silhouette. Contemporary ideological, functional, and constructive features are organically combined with traditional national elements. The simple decoration of large flat surfaces consists of motifs from medieval Armenian architecture. His works in Yerevan include the wine cellars of the Ararat Trust (1937–63, the first stage of which was designed in collaboration with G. B. Kochar and others), the aqueduct across the Razdan River (1949), the pedestal of the monument *To Victory* (1949; State Prize of the USSR, 1950), and the monument *Vaagn* (1965, sculptor A. A. Arutiunian). He designed a complex (two sculptures, stelae, and an obelisk; 1968) in the Oktemberian Raion in memory of the battle of Sardarapat of 1918. He also created a new type of memorial – small fountains in memory of the heroes of the Great Patriotic War.

Istru, Bogdan Spiridonovich
Moldavian author; b. Mr. 31 (Ap. 13), 1914, in the present-day village of Pistrueny, Teleneshty Raion, Moldavian SSR. Member of the CPSU since 1947.
He graduated from the Higher Pedagogical School in Iaşi in 1931. His first work was published in 1936, and the verse collections *Incantation* in 1937 and *Death of an Eagle* in 1938. The liberation of Bessarabia and the formation of the Moldavian SSR in 1940 served as a stimulus in his creative work. The socialist reforms in Moldavia, the achievements of the Soviet people in the struggle with fascist aggressors, and the heroic images of our contemporaries found artistic realization in the verse collections *Voice of the Motherland* (1946), *In the Vanguard* (1951), *Avenue of Peace* (1953), *From Shore to Shore* (1958), and *Selected Poems* (1964) and in the narrative poems *Pokhoarnele* (1947), *Spring in the Carpathians* (1955), and *Pain of the Shadows* (1968).
His poems have been translated into the languages of the peoples of the USSR and foreign languages. Decors.: two orders and various medals.
Works: *Versur'.* Kishinev 1966. *In Russian translation: Rodnye berega.* Moscow 1954. *Poemy.* Kishinev 1958. *V dobryi chas.* Moscow 1959. *Stikhi.* Moscow 1964.

Iuhanson, Nikolai Ottovich
Trade union official; b. 1927.
Esthonian; mem. CPSU 1947–; grad. Higher Party School at Cen. Cttee. CPSU; held Komsomol and Party posts, active in Cttee. for State Security of Esthonian SSR 1951–71; mem. Auditing Comm. 1961–66, cand. mem. Cen. Cttee. 1966–71, mem. Cen. Cttee. Esthonian CP 1971–; 1st Sec. of Tallinn City Cttee. Esthonian CP 1971–82; mem. Bureau Cen. Cttee. Esthonian CP F. 1971–; Chair. of Esthonian Republican Trade Union Council F. 1982–.

Iukalov, Iurii Alekseevich
Diplomat.
Mem. Apparat, USSR Min. of Foreign Affairs, Emb. Councellor, USSR Embassy in Tanzania until 1980; USSR Amb. to Tanzania Ap. 1980–.

Iukhin, Vladimir Ivanovich
Diplomat; b. 1918.
Mem. Diplomatic Service 1946–; Councellor-Envoy, USSR Emb. in Algeria 1964–68; mem. Apparat, USSR Min. of Foreign Affairs 1968–70 and 1974–79; USSR Amb. to Mali 1970–74, to Syria My. 1979–.

Iuldashev, Muminzhon
State official.
Dep. Chair. of Dzhizak Oblast Exec. Cttee. of People's Deps. 1974–75; Head of Dept. of Light and Food Ind. with Cen. Cttee. Uzbek CP 1975–79; Min. of Food Ind. Uzbek SSR D. 1979–.

Iunak, Ivan Kharitonovich
Party official; b. 1918; s. of a peasant.
Ukrainian; mem. CPSU 1944–; grad. Glukhov Agricultural Inst. 1941; bookkeeper on a kolkhoz, student 1933–42; served in Sov. Army 1942–46; agronomist, Chief Agronomist of Talnoe Raion Agricultural Dept. in Kiev Oblast 1946–49; dept. head, then sec. of a raion cttee. Ukrainian CP 1949–54; Chair. of Exec. Cttee. of Dnepropetrovsk Oblast Sov. of Workers' Deps. 1954–61; Dep. to Supr. Sov. Ukrainian SSR 1955–63; cand. mem. Cen. Cttee. Ukrainian CP 1960–61; 1st Sec. of Tula Oblast CPSU Cttee. Jl. 1964– (for Agriculture 1963–64); mem. Cen. Cttee. CPSU 1961–; Dep. to USSR Supr. Sov. 1962–. Decors.: 3 Orders of Lenin.

Iusif-Zade, Ziia Mamedia ogly
State official; b. 1929.
Major-Gen.; cand. mem. Cen. Cttee. Azerbaidzhan CP 1971–80, mem. 1980–; Dep. Chair. of Cttee. for State Security Azerbaidzhan SSR 1975–77, 1st Dep. Chair. 1977–80, Chair. Je. 1980–; cand. mem. Bureau Cen. Cttee. Azerbaidzhan CP 1980–81, mem. Ja. 1981–.

Iusifov, Nariman Gabibovich
State official.
Min. of Justice Azerbaidzhan SSR Ja. 1980–; can. mem. Cen. Cttee. Azerbaidzhan CP 1981–.

Iusufbekov, Rustambek, Cand. of Econ. Sc.
State official.
Head of Dept. of Propaganda and Agitation, Dept. of High Schools and Schools, Dept. of Science, Schools and High Schools at Cen. Cttee. Tadzhik CP 1959–63; cand. mem. 1960–61, mem. Cen. Cttee. Tadzhik CP 1961–; Min. of

Educ. Tadzhik SSR 1963–74; Dep. to Supr. Sov. Tadzhik SSR 1963–; Dep. Chair. of Council of Mins. Tadzhik SSR Ap. 1974–. Min. of Foreign Affairs Tadzhik SSR D. 1981–.

Iusupov, Pulat Iunusovich
State official.
Chair. of Tashkent Oblast Exec. Cttee. of Peoples' Deps. 1975–78; Student 1978–80; mem. Cen. Cttee. Uzbek CP 1976–81, cand. mem. 1981–; Min. of Forestry Uzbek SSR 1980–.

Iuvenalii (Spiridon Alekseevich Tarasov), Dr. of Theol.
Russian Orthodox clergyman; b. Ap. 29, 1929 Rostov/Don Oblast.
Grad. Saratov Theological Seminary 1953 and Moscow Theological Acad. 1969; deacon and priest 1953; priest in Shakhty, Novoshakhtinsk and Rostov 1953–67; monk 1968; Bishop of Voronezh and Lipetsk 1975–82; Bishop of Voronezh and Chita Jl. 1982–.

Iuvenalii (Vladimir Kirillovich Poiarkov)
Russian Orthodox clergyman; b. S.22, 1935, Iaroslavl.
Grad. Leningrad Theological Acad. 1961; monk 1959; priest-monk 1960; teacher at Moscow Theological Seminary 1961–62; Ed. of journal "Golos Pravoslaviia" (Berlin) 1962–63; Dep. Head of Dept. of External Affairs, Moscow Patriarchate 1964–65; extremely active abroad, took part in various conferences, accompanied a group of Russian pilgrims to Jordan, Syria, the Lebanon, Greece, Mount Athos and Israel; Bishop of Saraisk and Substitute Bishop of Moscow Diocese 1965; Moscow Patriarchate observer at Congress for Lay Apostolate in Rome 1967; Bishop of Tula and Belev 1969; Archbishop 1971; Metropolitan 1972–; Chair. of Church Dept. of External Affairs, Moscow Patriarchate, and perm. mem. of Holy Synod My. 1972–77; Metropolitan of Krutitsy and Kolomna and Head of Moscow Eparchy Jl. 1977–.

Iuzbashian, Marius Aramovich
State official; b. 1924.
Armenian; mem. CPSU 1948–; grad. Uzhgorod State Univ.; Major-Gen.; served in the Sov. Army 1941–43; div. posts in the Organs of State Security (KGB) 1943–78; Chair. Cttee. of State Security Armenian SSR Je. 1978–; Dep. USSR Supr. Sov. 1979–; mem. Cen. Cttee. and cand. mem. Bureau Cen. Cttee. Armenian CP Ja. 1981–.

Ivakhnenko, Aleksei Grigorevich, Dr. of Tech. Sc.
Scientist; b. Mr. 30, 1912, Kobeliaki, Poltava Oblast; s. of a teacher.
Grad. Electrotech. Inst. Leningrad 1938; specialist in automation and technical cybernetics; corresponding mem. Acad. of Sciences Ukrainian SSR 1961–; Dr. of Tech. Sc. 1954; Prof. 1956–; mem. staff Electrotech. Union Inst. Moscow 1938–43; cand. at Inst. of Physics of Moscow State Univ. 1943; mem. staff Inst. of Building Mechanics 1944–47 and Inst. of Electrical Engineering at Acad. of Sciences Ukrainian SSR 1947–62; Chief of Dept. for Combined Systems of Automatic Control at Inst. of Cybernetics Acad. of Sciences Ukrainian SSR 1962–; in addition teacher at Kiev Polytech. Inst. 1944–; author of more than 120 works.
Publs.: "Kombinirovanye kiberneticheskie sistemy" 1963, "Kiberneticheskie predskazyvaiushchie ustroistva" 1965 (co-author), "Kiberneticheskie sistemy s kombinirovannym upravleniem" 1966, "Mashyny pochynaiut myslyty" 1967 (co-author).

Ivanenko, Boris Vasil'evich
Party official; b. 1933, Kanev, Cherkassy Oblast.
Grad. Kiev State Univ. and Ukrainian CP Cen. Cttee. Party Higher School; journalist 1956–; instructor, chief of a sector of Propaganda and Agitation Dept. of Cen. Cttee. Ukrainian CP 1966–76; 1st Dep. Chair. of State Cttee. for Television and Radio Broadcasting Ukrainian SSR 1976–78, Chair. 1978–79; Chief of Culture Dept. of Cen. Cttee. Ukrainian CP 1979–; cand. mem. Cen. Cttee. Ukrainian CP 1981–.

Ivanenko, Dmitrii Dmitrevich, Dr. of Phy.-Math. Sc.
Physicist; b. Jl. 16 (29), 1904, Poltava.
Grad. Leningrad Univ. 1927; Prof. at Moscow Univ. 1943–. Decor.: State Prize of the USSR 1950.

Ivanenko, Oksana Dmitrievna
Author; b. 1906, Poltava; d. of a teacher.
Ukrainian; mem. CPSU 1940–; worked in Children's Colony under A. S. Makarenko; first works publ. 1925.
Publs.: novel "Puti Tarasa" (Taras' Ways) 1939–61, "Rodnye deti" (Dear Children) 1951, and others.

Ivanov, Aleksandr Gavrilovich
Film director and screenwriter; b. Jl. 31 (Ag. 12), 1898, Davydovo, in present-day Borovichi Raion, Novgorod Oblast.
People's Artist of the USSR (1964). Member of the CPSU since 1918. Participant in the Civil War of 1918–20.
He graduated form a cinematography school in Leningrad in 1924 and finished his graduate studies at the State Scientific Research Institute of Art in 1935 (earning the title of docent). In 1925 he began working at Leningradkino (present-day Lenfil'm Studios). His films include *The Moon on the Left Side* (1929, based on the play by V. N. Bill'-Belotserkovskii), *Fire Transport* (1930), *Aye, Aye, Captain!* (1930), *On the Border* (1938), *Submarine T-9* (1943), *The Sons* (1946), *Victorious Return* (1948), *The Star* (1953, based on the novella by E. G. Kazakevich), and *The Soldiers* (1957, based on V. P. Nekrasov's tale *In the Dugouts of Stalingrad*). He has an affinity for war themes, combat exploits, and everyday life in the army, and his work expresses the harsh romanticism of life at home and at the front in severe, restrained colors. He directed the film trilogy *Virgin Land Upturned* (1960–61; based on the novel by M. A. Sholokhov), *Should a Comrade Call* (1963), and *The First Russians* (1968). He wrote the screenplays for a number of his films. Decors.: Order of Lenin, four other orders, and medals.

Ivanov, Aleksei Petrovich
Russian singer (baritone); b. S. 9 (22), 1904, Chizhovo, in present-day Kalinin Oblast.
People's Artist of the USSR (1951).
He graduated in 1932 from the Leningrad Conservatory, where he had studied with G. A. Bosse. During 1932–36 he was a soloist at the Leningrad Malyi Opera Theater and from 1938 to 1967, at the Bol'shoi Theater of the USSR. He originated the parts of the Commissar in Kabalevskii's *Under Fire* (1943), Pomazov in Muradeli's *The Great Friendship* (1947), and Ryleev in Shaporin's *Decembrists* (1953). His performance of the title roles in Rubinstein's *Demon* and Verdi's *Rigoletto* and of the Devil in Tchaikovsky's *Cherevichki* earned him the State Prize of the USSR in 1946. His other roles included Prince Igor (Borodin's *Prince Igor*), Peter (Serov's *The Power of Evil;* State Prize of the USSR, 1948), and Mazepa (Tchaikovsky's *Mazepa;* State Prize of the USSR, 1950). He began touring abroad in 1945. He is the author of *On the Art of Singing* (1963) and *The Vocal Image* (1968). Decors.: Order of the Red Banner of Labor and various medals.

Ivanov, Anatolii Stepanovich
Author; b. 1928, Shemonaikha, Kazakhstan; s. of a peasant.
Russian; mem. CPSU 1952–; grad. Faculty of Journalism of Kazakh Univ. 1950; first works publ. 1948.
Publs.: "Alkiny pesni" (Alkas' Songs) 1956, "Teni ischezaiut v polden'" (Shadows Disappear at Noon) 1963, and others.

Ivanov, Artemii Vasil'evich
Zoologist; b. My. 5 (18), 1906, Molodechno.
Grad. Leningrad Univ. 1930; Asst., Lecturer 1935–50, Prof. at Leningrad Univ. 1950–; in addition Chief of Laboratory for Evolutionary Morphology at Zoological Inst. of USSR Acad. of Sciences 1965–. Decor.: Lenin Prize 1961.
Publs.: "Morfologičeskie adaptacii k parazitičeskomu obrazu žizni" in "Už. zap. LGU. Ser. biologičeskih nauk" 1937 vol. 3 No. 4, "Stroenie Udonella caligorum Johnston i položenie Udonellidae v sisteme ploskih červej" in "Parazitologičeskij sbornik" 1952 No. 14, "Beskišečnye turbelljarii južnogo poberež'ja Sahalina" in "Trudy Zoologičeskogo institu-

ta" 1952 vol. 12, "Promyslovye vodnye bespozvonočnye" Moscow 1955, "Pogonofory" in "Fauna SSSR" Moscow-Leningrad 1960, "Embranchement des Pogonophores" in "Traité de Zoologie" 1960 t. 5 fasc. 2, "Proizhoždenie mnogokletočnyh životnyh" Leningrad 1968, "Pogonophora" New York 1963, and others.

Ivanov, Boris Petrovich
State official.
Dep. Min. of Energy Econ. and Electrification Kazakh SSR 1978–80, Min. O. 1980–; cand. mem. Cen. Cttee. Kazakh CP 1981–.

Ivanov, Ianis
Composer; b. S. 26 (O. 9), 1906, Preiļi.
People's Artist of the USSR (1965).
He graduated in 1931 from the Latvian Conservatory, where he studied composition with J. Vītols, piano with N. Dauge, and conducting with G. Šnēfogts. He worked in the radio service from 1931 to 1961 and as artistic director of music broadcasting of the Latvian SSR since 1945. In the 1930's he wrote program works, mainly about his native land, including the suite *Latgale Lakes* (1935) and the symphonic pictures *Mountain Under the Skies* (1938) and *The Rainbow* (1938). His best symphonic works are *Atlantida* (1941) and *Latgales* (1949; State Prize of the USSR, 1950). Among his other works are 14 symphonies (1933–71), a symphonic suite (1935), a symphonic poem (1957), concerti, quartets, and piano pieces. The musical language of his symphonic works is lyrical, dramatically expressive, and melodious, and it often relies on the natural modes of the Latvian folk song. His music has been evolving toward greater harmonic complexity. He began teaching composition at the Latvian Conservatory in 1944 (professor since 1955). He was a deputy of the third and fourth convocations of the Supreme Soviet of the USSR. Decors.: State Prize of the Latvian SSR (1959 and 1970) and two orders.

Ivanov, Konstantin Konstantinovich
Conductor; b. My. 8 (21), 1907, Efremov, now Tula Oblast.
Grad. Conducting Class of Moscow Conservatory 1937; Dir. of Great Symphony Orchestra with Union Broadcasting Cttee. 1941; Conductor of USSR State Symphony Orchestra 1946–65. Decors.: People's Artist of the USSR, State Prize of the USSR 1949, orders, medals.

Ivanov, Leonid Ivanovich
Author; b. 1914, Filatikha, Kalinin Oblast; s. of a peasant.
Russian; studied agronomy.
Publs.: "Sibiriaki" (Siberians) 1953–55, "Sibirskie vstrechi" (Siberian Encounters) 1956, "Desiataia protsenta" (One Tenth of a Percent) 1958, "Kak ukhodit i vozvrashchaetsia slava" (How Fame Fades and Returns) 1958, "Bol'shie peremeny" (Great Changes) 1958, "Glubokaia borozda" (Deep Rut) 1960, "Terentii Maltsev" 1962, "Doverie" (Confidence) 1963, and others.

Ivanov, Mikhail Stepanovich
State official, b. 1917.
Mem. CPSU 1944–, grad. Moscow Inst. of Cooperative Trade 1942, cand. of Econ. Sciences 1977; held posts in trade system 1942–47; chief of Trade Dept. of Oblast Exec. Cttee. of People's Deps. in Pavlodar and City Exec. Cttee. of People's Deps. in Alma-Ata 1947–54; Dep., 1st Dep. Min. of Trade 1954–70, Min. of Trade Kazakh SSR Je. 1970–; cand. mem. 1971–76, mem. Cen. Cttee. CP Kazakhstan 1976–.

Ivanov, Semen Pavlovich
Military official; b. Ag. 18 (31), 1907, Porechno, in present-day Smolensk Raion, Smolensk Oblast; s. of a peasant.
General of the army (1968), Hero of the Soviet Union (S. 8, 1945). Became a member of the CPSU in 1929.
He volunteered for the Red Army in 1926, graduated from the Moscow Infantry School in 1929 and from the M. V. Frunze Military Academy in 1939, and was appointed chief of staff of an infantry corps in January 1940, taking part in the Soviet-Finnish War. In the Great Patriotic War (1941–45), Ivanov was chief of the operations department of the Thirteenth Army (June to December 1941); chief of staff of the Thirty-eighth, First Tank, and First Guard armies of the Southwestern and Don fronts (December 1941 to October 1942); and chief of the operations department and chief of staff of the Southwestern Front (October 1942 to May 1943), the Voronezh and First Ukrainian fronts (May to November 1943), the Transcaucasian Front (November 1943 to October 1944), and the Third Ukrainian Front (October 1944 to May 1945). In 1945 he served in the war against Japan and was chief of staff of the Far Eastern Forces. After the war, he was chief of staff of several military districts and of the Soviet Troop Group in Germany and deputy chief of the General Staff of the Armed Forces of the USSR. He was commander of the troops of the Siberian Military District from 1964 to 1968 and chief of the K. E. Voroshilov Military Academy of the General Staff from April 1968 to 1973. He became an inspector of the Inspectors' General Group in the Ministry of Defence of the USSR in February 1973.
A deputy to the seventh convocation of the Supreme Soviet of the USSR. Decors.: three Orders of Lenin, six Orders of the Red Banner, the Order of Suvorov First Class, the Order of Kutuzov First Class, the Order of the Red Banner of Labor, the Order of the Red Star, six foreign orders, and medals.

Ivanov, Sviatoslav Nesterovich
Geologist; b. F. 3 (16), 1911, Nizhny Novgorod, now Gorkii.
Corresponding member of the Academy of Sciences of the USSR (1970). Became a member of the CPSU in 1970.
After graduating in 1932 from the Sverdlovsk Institute of Geological Prospecting, he worked on the exploration and prospecting of copper deposits in the Urals. From 1940 he directed the laboratory of physical geochemistry at the A. N. Zavaritskii Institute of Geology and Geochemistry of the Urals Branch of the Academy of Sciences of the USSR; since 1966 he has been director of the institute and deputy chairman of the Urals Scientific Center of the Academy of Sciences of the USSR.
His main works deal with the regularities of the location of pyrite-type cupreous deposits. He showed the similarity of the origin of these deposits and ancient submarine volcanic structures. He established the concept of the period and reasons for greenstone metamorphism, which in his opinion occurred before the uplifting of the sea floor and before the formation, in the volcanic region, of mountain structures connected with the intrusion of granite batholiths. Decors.: State Prize of the USSR in 1949, Order of the Red Banner of Labor and medals.

Ivanov, Vadim Mikhailovich
Architect; b. O. 16 (29), 1906, Baku.
Mem. CPSU 1948–; grad. Azerbaidzhan Polytech. Inst. 1930; Chief Architect of City of Baku 1937–67; teacher at Azerbaidzhan Ind./Polytech. Inst. Baku 1931–. Decors.: order, medals.

Ivanov, Valentin Konstantinovich
Mathematician; b. S. 18 (O. 1), 1908, St. Petersburg.
Corresponding member of the Academy of Sciences of the USSR (1970).
He graduated from Leningrad University in 1938. In 1956 he became a professor at Ural (Sverdlovsk) University. His first works dealt with the theory of matrices. Subsequently he worked in various branches of mathematics: generalized functions, the theory of approximations, the inverse problem of potential theory with applications to geophysics, harmonic functions, and entire functions of many complex variables. He devoted a series of works to incorrect problems (the concept of quasisolution, methods of solution). Decor.: Lenin Prize in 1966.
Works: "Obratnaia zadacha potentsiala dlia tela, blizkogo k dannomu." *Izv. AN SSSR: Ser. matematicheskaia,* 1956, vol. 20, no. 6. "O lineinykh nekorrektnykh zadachakh." *Dokl. AN SSSR: Novaia seriia,* 1962, vol. 145, no. 2.

Ivanov, Viacheslav Nikolaevich
Athlete, oarsman of single shells; b. Jl. 30, 1938, Moscow.
Honored Master of Sports (1956). Officer of the Soviet Army.

He won the USSR championship on a one-man shell 11 times in succession (1956–66) and was European champion (1956, 1959, 1961, and 1964) and world champion (1962). He is the only oarsman in the history of rowing who was three times champion of the Olympic Games (1956, 1960, and 1964) in one-man shell races. Decors.: three orders.

Ivanov, Viktor Evgen'evich
Physicist; b. N.9 (22), 1908, Staraia Maina, now in Ul'ianovsk Oblast.
Specialist in the physical chemistry of inorganic substances. Corresponding member of the Academy of Sciences of the USSR (1964) and academician of the Academy of Sciences of the Ukrainian SSR (1967). Member of the CPSU since 1929.
He graduated from the Ashkhabad Pedagogical Institute in 1942 and the Odessa Hydrometeorological Institute in 1945. In 1947 he began working at the Kharkov Physicotechnical Institute, becoming its director in 1965. Also a professor at the University of Kharkov (since 1962), he is one of the founders and directors of the physicotechnical department.
He worked out a method of producing high-purity metals by vacuum distillation and condensation on heated surfaces, which has found industrial application. With his students he devised original equipment for hot rolling and pressing of metals in a high vacuum. His studies of the properties of reactor materials are highly important, as is his work on compositions having special mechanical and physical properties. Decors.: Order of Lenin, the Order of the Red Banner of Labor, and several medals.
Works: *Chistye i sverkhchistye metally.* Moscow 1965. *Zashchita vol'frama ot okisleniia pri vysokikh temperaturakh.* Moscow 1968.

Ivanov-Vano, Ivan Petrovich
Film director and artist; b. Ja. 27 (F. 8), 1900, Moscow.
People's Artist of the RSFSR (1969). Member of the CPSU since 1951.
He graduated from the State Higher Arts and Technical Studios in 1923. He became involved in animated films in 1924. His screen versions of Russian folk tales and classics, such as *The Little Humpedback Horse* (1948), *Wild Geese* (1950), *The Tale of the Dead Tsarevna and the Seven Bogatyrs* (1951), *The Snow Maiden* (1952), *Lefty* (1964), *Seasons of the Year* (1969), and *The Battle at Kerzhenets* (1971), are among his best works. A number of his films have received prizes at international and all-Union film festivals. He began teaching at the All-Union State Institute of Cinematography in 1939, becoming a professor there in 1952. Decors.: State Prize of the RSFSR in 1970, three orders and various medals.
Work: *Ocherk istorii razvitiia mul'tiplikatsii (Do vtoroi mirovoi voiny).* Moscow 1967.

Ivanova, Tatiana Grigor'evna
Party official; b. 1940.
Mem. CPSU 1968–, grad. Moscow State Univ. 1966; junior ed. of "Molodaia gvardiia" Publishers, senior ed. of journal "Novyi Mir", ed. of "Nauka" Publishers 1960–72; party posts 1972–77; 2nd Sec. Kalinin Raion Cttee. CPSU in Moscow City 1977–79, 1st Sec. 1979–; cand. mem. Cen. Cttee. 1981–.

Ivanovskii, Evgenii Filippovich
Military official; b. 1918.
Belorussian; mem. CPSU 1941–; grad. Mil. Acad. for Motorization and Mechanization of the Sov. Army 1941 and Acad. of Gen. Staff 1958; Gen.; served in Sov. Army 1936–; chief of staff, commander of a tank battalion, head of a reconnaissance tank corps, commander of a tank regiment on Western, Stalingrad, South-Western, 1st and 2nd Belorussian Fronts 1941–45; dept. head of a group, dep. commander of tank and mechanized troops of a mil. district, chief of staff, commander of a div., 1st dep. chief of staff of a mil. district, 1st Dep. Commander of Moscow Mil. District 1946–68; Commander of Moscow Mil. District 1968–72; Dep. to USSR Supr. Sov. 1970–; mem. Cen. Cttee. CPSU 1971–; Commander-in-Chief of Sov. Forces in Germany My. 1973–80; Commander of Belorussian Mil. District 1980–; mem. Bureau Cen. Cttee. Belorussian CP 1981–.

Ivantsov, Anatolii Ivanovich
Diplomat; b. 1922, Slavianskaia, Krai Krasnodar.
Grad. Moscow Inst. for Rail Transport and Higher Diplomatic School of the USSR Min. of Foreign Affairs; mem. Diplomatic Service 1944–; Sec. of USSR Mission in New Zealand 1955–59; 1st Sec. USSR Embassy in Canada 1961–66; Adviser, 2nd Europe an Dept., USSR Min. of Foreign Affairs 1966–69; USSR Envoy in New Zealand 1969–73, Amb. 1973–74; mem. Apparat USSR Min. of Foreign Affairs 1974–79; USSR Amb. to Ghana S. 1979–.

Ivashutin, Petr Ivanovich
Military official; b. S. 5 (18), 1909, Brest.
Russian; mem. CPSU 1930–; did not complete higher education; Army Gen. 1971; laborer, metal worker, work brigade leader at a weaving mill 1926–31; entered Sov. Army 1931; command posts at the front 1941–45; posts in USSR Min. of Defense apparatus 1945–63; Dep. to USSR Supr. Sov. 1950–54 and 1966–; Dep. Chief of Gen. Staff of USSR Armed Forces, Chief of Main Intelligence Directorate 1963–. Decors.: Order of Lenin, Order of the October Revolution, Order of Red Banner (four), Order of Bogdan Khmelnitskii 1st Class, Order of Kutuzov 2nd Class, Order of the Great Fatherland War 1st Class, Order of Red Banner of Labor, Order of Red Star (three), medals, foreign orders.

Ivert, Ilmar Ianovich
State official.
Chief ed. of journal "Tsina" (organ of Cen. Cttee. Latvian CP) 1960–72; mem. Cen. Cttee. Latvian CP 1960–; cand. mem. Bureau Cen. Cttee. Latvian CP 1960–62, Sec. 1961–62, 77–82; Chair. of Latvian Journalists' Union 1962–77; Dep. to Supr. Sov. Latvian SSR 1963–; Chair. of State Cttee. for Television and Radio Latvian SSR Je. 1972–76; Chair. of State Cttee. for Film Latvian SSR Ap. 1982–. Decor.: Order of Lenin.

Iziumskii, Boris Vasil'evich
Author; b. 1915, Tsaritsyn.
Russian; mem. CPSU 1940–; first works publ. 1936.
Publs.: "Rannee utro" (Early Morning) 1946, trilogy "Alye pogony" (Red Shoulder Pieces) vols. 1–3, 1948–54, "Prizvanie" (Vocation) 1952–58, "Timofei Novgorodets" 1956, "More dlia smelykh" (The Sea is for the Brave) 1962, and others.

Iziurov, Ivan Vasil'evich
Author; b. 1910, Poztykeros; s. of a peasant.
Komi; first works publ. 1930.
Publs.: "Timkina brigada" (Timka's Brigade) 1932, "Prestuplenie Domentiia" (Domenti's Crime) 1936, "Schastie" (Happiness) 1935, "V verkhoviakh Vychegdy" (On the Upper Vychegda) 1959, "Devushki nashego sela" (Girls From Our Village) 1958, and others.

Izmailova, Galiia Baiazitovna
Ballerina; b. F. 12, 1923, Tomsk.
People's Artist of the USSR (1962).
She studied in the Tashkent ballet school from 1935 to 1941. Upon graduating in 1941, she joined the A. Navoi Theater of Opera and Ballet in Tashkent. Her roles included Maria in *The Fountain of Bakhchisarai* by Asaf'ev, Kitri in *Don Quixote* by Minkus, Scheherazade in the ballet of the same name to the music of Rimsky-Korsakov, and Carmen in *Bolero* to music by Ravel. In 1958 she graduated from the directorial department of the Tashkent Institute of Theatrical Arts. She staged Tchaikovsky's ballet *Swan Lake* and the dances for Ashrafi's opera *Diliaram,* among other works. She toured abroad (the Chinese People's Republic, the Socialist Republic of Rumania, France, Great Britain, India, the United States), performing national Uzbek dances as well as Chinese, Arabic, and Indian dances. Decors.: State Prize of the USSR (1950), Order of Lenin, the Order of the Badge of Honor, and various medals.

Izrael, Iurii Antonievich
State official; b. 1930.
Russian; mem. CPSU 1955–; grad. Central Asian State Univ. 1953, Dr. of Math. and Phys. Sc. 1969, Prof. 1973, Corresponding Mem. of USSR Acad. of

Sciences 1974; engineer, senior engineer, junior research associate of Geophysics Inst. and Inst. of Applied Geophysics of USSR Acad. of Sciences 1953–63; Dep. Dir., Dir. Inst. of Applied Geophysics of Main Admin. of Hydrometeorology and Environmental Control of USSR Council of Mins. 1963–70; 1st Dep. Chief of Main Admin. of Hydrometeorology and Environmental Control of USSR Council of Mins. 1970–74, Chief 1974–78; Chair. of USSR State Cttee. for Hydrometeorology and Environmental Control Mr. 1978–; Dep. to USSR Supr. Sov. 1979–; mem. Cen. Auditing Comm. CPSU 1981–.

Kaal, Aira (real name: Houn, Aira)
Author; b. 1911, Saaremaa; d. of a peasant.
Esthonian; mem. CPSU 1940–.
Publs.: "Raketa pobedy" (Rocket of Victory) 1948, "Ne otdam oruzhiia" (I Won't Give the Weapon Away) 1945, "Svobodnaia Rodina" (Free Fatherland) 1950, "Morskie vetry" (Sea Winds) 1958, "Izbrannye stikhotvoreniia" (Selected Poems) 1961, and others.

Kaalep, Ain
Author and translator; b. 1926, Tartu.
Esthonian; first works publ. 1953.
Publs.: "Samarkandskaia tetrad'" (Samarkand Diary) 1962, "Predrassvetnye peizazhi" (Landscapes at Dawn) 1962, and others; translated works by Shevchenko, Schiller and Heine into Esthonian.

Kabachnik, Martin Izrailevich
Organic chemist; b. Ag. 27 (S. 9), 1908, Ekaterinburg, now Sverdlovsk.
Mem. CPSU 1957–; grad. Moscow Tech.-Technological Inst. 1931; mem. USSR Acad. of Sciences 1958–; Head of Laboratory for Phosphor-Organic Compounds of Inst. of Elemental-Organic Compounds at USSR Acad. of Sciences 1954–. Decors.: State Prize of the USSR 1946, Order of Lenin, 2 other orders and various medals.

Kabaev, Leonid Nikolaevich
Geologist and geophysicist; b. 1935.
Mem. CPSU 1965–; grad. Kiev Univ. 1958; mem. Salekhardsk geological prospecting expedition 1959–68; Chief Engineer 1966–; mem. Tazovsk geophysical expedition 1968–. Decor.: Lenin Prize (for his part in discovery of large petroleum resources on Central Ob) 1970.

Kabalevskii, Dmitrii Borisovich
Composer, teacher, and public figure; b. D. 17 (30), 1904, St. Petersburg.
People's Artist of the USSR since 1963, doctor of the arts since 1965, and full member of the Academy of Pedagogical Sciences of the USSR since 1971. Member of the CPSU since 1940.
He completed N. Ia. Miaskovskii's course in composition (1929) and A. B. Goldenweiser's course in piano (1930) at the Moscow Conservatory. He has been a professor there since 1939. Between 1940 and 1946 he was editor-in-chief of the journal *Sovetskaia Muzyka*. Since 1954 he has been a member of the collegium of the Ministry of Culture of the USSR. In 1952 he became secretary of the Union of Soviet Composers, in which he also heads the commission on the musical and aesthetic education of children and young people. Since 1969 he has headed the Council on Aesthetic Education under the Presidium of the Academy of Pedagogical Sciences of the USSR. He has been a member of the Council of Directors of the International Society for Music Education since 1961 and its honorary president since 1972. Since 1953 he has been a member of the Soviet Committee for the Defense of Peace. In performances of his own works, he appears both as a pianist and as a conductor.
Most of his work draws upon contemporary socially significant themes. He frequently turns for inspiration to Soviet youth and children, as in three of his concerti – the Concerto for Violin (1948; State Prize of the USSR, 1949), the Concerto for Cello No. 1 (1949), and the Concerto for Piano Nr. 3 (1952), as well as cantatas, children's songs, and piano pieces. His work is strongly lyrical, with cheerful tones, a clear, songlike quality, joyous vitality, fervor, and humor. At the same time he captured in music the tragic events of the Great Patriotic War (1941–45) in such works as the opera *Taras' Family* (1950; State Prize of the USSR, 1951), based on B. Gorbatov's novella *The Unvanquished* and depicting the struggle of the Soviet people against the fascist invaders. In the opera *Colas Breugnon* (1937; revised version 1968, Lenin Prize, 1972), based on R. Rolland's novella, a lyrical theme blends with an unfolding of social conflict. He also composed the operas *In the Fire* (formerly *Near Moscow*, 1943), *Nikita Vershinin* (1955), and *The Sisters* (1969).
He has done much work in symphonic and cantataoratorio music. He has written four symphonies (1932, 1933, 1934, 1956), a number of concerti – three for piano (1929, 1936, 1952), one for violin (1948), and two for cello (1949, 1964) – the suite for chorus and orchestra *People's Avengers* (1942), the cantata *The Leninists* (1959), and the *Requiem* for soloists, two choruses, and orchestra, with words by R. Rozhdestvenskii (1963; Glinka State Prize of the RSFSR, 1966). His chamber works include two string quartets (No. 2, 1945; State Prize of the USSR, 1946) and 24 preludes for piano. He has also written music for a number of films and plays. He was a deputy to the seventh and eighth convocations of the Supreme Soviet of the USSR. He is a corresponding member of the Academy of Arts of the German Democratic Republic (1970), honorary professor at the Conservatory in Mexico City (1959), and vice-president of the British Workers' Choral Society (1950). Decors.: two Orders of Lenin, two other orders and medals.
Works: *Izbrannye stat'i o muzyke*. Moscow 1963. *Pro trekh kitov i mnogoe drugoe: Knizhka o muzyke*. Moscow 1970.

Kabanov, Mikhail Fedorovich
State official.
Head of Dept. of Party Org. Work with Cen. Cttee. Kirghiz CP 1969–79; mem. Cen. Cttee. 1971–, cand. mem. Bureau Cen. Cttee. Kirghiz CP 1976–81; Dep. to Supr. Sov. Kirghiz SSR 1975–. Dep. Chair. of Presidium Supr. Sov. Kirghiz SSR Jl. 1979–.

Kabanov, Viktor Aleksandrovich
Chemical scientist; b. Ja. 15, 1934, Moscow.
Grad. Moscow Univ. 1956; corresponding mem. USSR Acad. of Sciences 1968–; Professorial Chair of High Molecular Compounds at Moscow Univ. 1970–.

Kabkov, Iakov Ivanovich
Party official; b. 1908.
Mem. CPSU 1930–; grad. Mechanical-Tech. Inst. of Tinning Ind. in Odessa 1937; worked in food ind. admin. 1938–40; various posts in Apparat of Cen. Cttee. Ukrainian CP 1940–41; served in Sov. Army 1941–43; mem. staff Cen. Cttee. Ukrainian CP 1943–50; 1st Dep. Min. of Trade 1950–55, Min. of Fisheries Ukrainian SSR 1955–57; cand. mem. Cen. Cttee. Ukrainian CP 1956–61; mem. staff Cen. Cttee. Ukrainian CP 1957–58; Head of Dept. of Trade, Finance and Planning at Cen. Cttee. CPSU 1958–62; mem. CPSU Cen. Auditing Comm. 1961–71; Head of a Dept. of Cttee. for Party and State Control with Cen. Cttee. CPSU and USSR Council of Mins. 1962–65; Head of Dept. of Trade and Public Services with Cen. Cttee. CPSU 1965–; cand. mem. Cen. Cttee. CPSU 1971–. Decors.: Order of Red Banner of Labour, Order of Lenin.

Kaboshkin, Vladimir Fedorovich
Diplomat.
USSR Amb. to Democratic Republic of Yemen Ag. 1975–78, in Greece S. 1979–.

Kachalovskii, Evgenii Viktorovich
Party official; b. 1926.
Ukrainian; mem. CPSU 1947; grad. Dnepropetrovsk Inst. of Rail Transport 1948; asst. locomotive driver, locomotive driver, shop brigade leader, dep. chief of a locomotive depot 1948–54; chief engineer, chief of a locomotive depot 1956–62; Dir. of Dnepropetrovsk Electric Locomotive Plant 1962–67; Chair. of Exec. Cttee. of Dnepropetrovsk City Sov. of Peoples' Deps.; 1st Sec. of Dnepropetrovsk City Cttee. Ukrainian CP 1967–74; mem. Cen. Cttee. Ukrainian CP 1971–; cand. mem. Politbureau of Cen. Cttee. Ukrainian CP 1980–; 2nd Sec. of Dnepropetrovsk Ob-

last Cttee. Ukrainian CP 1974–76, 1st Sec. Je. 1976–; Dep. to USSR Supr. Sov. 1979–; mem. Cen. Cttee. CPSU 1981–.

Kacharava, Viliam Iakovlevich, Cand. of Hist. Sc.
State official; b. 1927, Tbilisi.
Komsomol posts 1950–59; 2nd Sec. of Ordzhonikidze-Raion Cttee. Georgian CP in Tbilisi 1959–64; Head of Republican "House of Pol. Educ." 1964–73; 1st Dep. Head of Dept. of Propaganda and Agitation with Cen. Cttee. Georgian CP 1973; Head of Cen. Archives Admin. with Council of Mins. Georgian SSR 1973–77; Chair. of State Cttee. on Nature Conservation. Georgian SSR F. 1977–; mem. Auditing Comm. Georgian CP 1981–.

Kachin, Dmitrii Ivanovich
Party official; b. 1929.
Mem. CPSU 1953–; grad. Tech. Inst. of Fishing Ind. and Econ. Moscow 1964 and Cen. Cttee. CPSU Party High School 1961; sailor on various fishing ships 1945–47; trawler foreman, engineer of admin. "Kamchatrybprom" 1954–59; CPSU and state posts 1961–63; 2nd Sec. 1963–65, 1st Sec. of Petropavlovsk-Kamchatskii, City Cttee. CPSU 1965–68; Sec. of Kamchatka Oblast Cttee. CPSU 1968–69; Chair. of Kamchatka Oblast Exec. Cttee. 1969–71; 1st Sec. of Kamchatka Oblast Cttee. CPSU 1971–; mem. Cen. Auditing Comm. CPSU 1971–76; cand. mem. Cen. Cttee. CPSU 1976–81, mem. 1981–. Decor.: Order of Lenin.

Kachura, Boris Vasil'evich
Party official; b. 1930.
Ukrainian; mem. CPSU 1957–; grad. Kharkov Polytechnical Inst. 1954; senior engineer, shop superintendent, dep. chief energetics specialist in various plants in Donetsk Oblast 1954–63; 2nd Sec. of Zhdanov City Cttee. Ukrainian CP 1963–67, 1st Sec. 1968–74; cand. mem. Cen. Cttee. Ukrainian CP 1971–76, mem. 1976–; 2nd Sec. of Donetsk Oblast Cttee. Ukrainian CP 1974–76, 1st Sec. Ja. 1976–82; cand. mem. Politbureau of Cen. Cttee. Ukrainian CP 1976–80, mem. Ap. 1980–; mem. Cen. Cttee. CPSU 1976–; Dep. to USSR Supr. Sov. 1979–; Sec. Cen. Cttee. Ukrainian CP O. 1982–.

Kachurovskii, A. P.
Trade union official.
Sec. and mem. Presidium of Latvian Republican Trade Union Council D. 1967–.

Kadatsky, Evgenii Mikhailovich
State official; b. 1930.
Ukrainian; mem. CPSU 1954–; cand. of Economic Sciences; grad. Aviation Inst. in Kharkov; after completion of aviation studies in Kharkov technician, student, dep. dir. of a section of aircraft plant in Kharkov, Komsomol Sec. and Party Cttee. Sec. at this plant, instructor in Oblast Cttee. CP Ukraine in Kharkov 1950–67; Head of Oblast Admin. of Professional-Technical Education in Kharkov 1967–76; 1st Dep. Chair. 1976–81, Chair. State Cttee. Professional-Technical Education Ukrainian SSR O. 1981–.

Kadochnikov, Pavel Petrovich
Russian film actor; b. Jl. 29, 1915, Petrograd.
People's Artist of the RSFSR (1965).
He graduated from the Leningrad Theatrical Institute in 1935 and worked until 1944 in the New Young People's Theater of Leningrad. He made his film debut in 1935 as Mikhas' in *Coming of Age*. He played the role of M. Gorkii in the films *Iakov Sverdlov* (where he also played Lenia Sukhov, 1940), *A Teacher's Epic* (*Pedagogical Poem,* 1955), and *Prologue* (1956). He has also successfully appeared in lyrical comedies. These roles include Mukhin in *Anton Ivanovich Is Angry* (1941) and Rybal'chenko in *Honeymoon* (1956). His best roles were Major Fedotov in *A Reconnaissance Scout's Feat* (1947), Aleksei Meres'ev in *The Story of a Real Man* (1948), and Vladimir Staritskii in S. M. Eisenstein's film *Ivan the Terrible* (parts 1 and 2, 1945 and 1958). In this same film he also played Khaldei and the Tsar's confessor, Evstafii.
His directorial debut was the 1965 film *The Musicians of a Regiment* (together with G. S. Kazanskii), in which he also played the role of Chulkovskii. In 1970 he directed the film version of A. N. Ostrovskii's play *The Snow Maiden,* in which he also played Tsar Berendei and for which he wrote the scenario. Decors.: State Prize of the USSR in 1948, 1949, and 1951, Order of the Red Banner of Labor and a medal.

Kadomtsev, Boris Borisovich
Physicist; b. N. 9, 1928, Penza.
Academician of the Academy of Sciences of the USSR (1970; corresponding member, 1962).
Upon graduating from Moscow University (1951), he worked at the Institute of Physics and Energetics in Obninsk. In 1956, he began working at the Institute of Atomic Energy. His principal works are devoted to plasma physics and to the problem of controlled thermonuclear fusion. He has predicted some types of transient plasma and laid the foundations for the theory of transport phenomena (diffusion and thermal conduction) in turbulent plasma. He has provided a quantitative explanation of the phenomenon of anomalous behavior of plasma in a magnetic field. A number of his studies have been devoted to the problem of the thermal insulation of plasma in toroidal chambers. In 1966 he discovered the instability of plasma involving the so-called trapped particles. Decor.: State Prize of the USSR in 1970.
Works: "Gidromagnitnaia ustoichivost plazmy." In *Voprosy teorii plasmy,* fasc. 2. Moscow 1963. "Turbulentnost plasmy." *Ibid.,* fasc. 4. Moscow 1964. "Turbulentnye protsessy v toroidal'nykh sistemakh." *Ibid.,* fasc. 5. Moscow 1967. (With O. P. Pogutse.)

Kadyrbaev, Vladimir Kasymovich
State official.
Min. of Motor Transport Kazakh SSR My. 1978–; mem. Cen. Cttee. Kazakh CP 1981–.

Kadyrov, Gairat Khamidullaevich
Party official.
Cand. mem. Cen. Cttee. Uzbek CP 1976–81, mem. 1981–; Head of Dept. of Heavy Ind. and Machine Building of Cen. Cttee. Uzbek CP F. 1979–; Dep. of Supr. Sov. Uzbek SSR 1980–.

Kadyrov, Gaptrakhman Faizurakhmanovich
Athlete, motor racer; b. Ja. 27, 1941, Shatura, Moscow Oblast.
Honored Master of Sport (1965), and trainer. Champion of the USSR (1964, 1969–70), Europe (1964), and the world (1966, 1968–69, 1971–72) in motor races on an ice track. Decor.: Order of the Badge of Honor.

Kafarov, Viktor Viacheslavovich
Chemical engineer; b. Je. 18, 1914, Shiauliai, now in the Lithuanian SSR.
Corresponding member of the Academy of Sciences of the USSR (1966). Member of the CPSU since 1952.
He graduated from the Kirov Institute of Chemical Engineering in Kazan in 1938. In 1944 he began working at the Mendeleev Institute of Chemical Engineering in Moscow, becoming head of a subdepartment in 1960. His research is devoted to the processes of and equipment used in chemical engineering as well as to the cybernetics of chemical engineering processes. Decors.: Order of the Red Banner of Labor twice, and several medals.
Works: *Osnovy massoperedachi.* Moscow 1962. *Ravnovesie mezhdu zhidkost'iu i parom,* books 1–2. Moscow 1966. (With others.) *Metody kibernetiki v khimii i khimicheskoi tekhnologii.* Moscow 1969. *Metody optimizatsii v khimicheskoi tekhnologii.* Moscow 1969. (With A. I. Boiarinov.) *Spravochnik po rastvorimosti* vols. 1–3. Moscow 1961–70. (With others.) *Programmirovanie i vychislitel'nye metody v khimii khimicheskoi tekhnologii.* Moscow 1972. (With others.)

Kafarova, Elmira Mikail kyzy
State official.
Mem. Bureau 1962–70, Sec. 1962–66, 1st Sec. Cen. Cttee. Azerbaidzhan Komsomol 1966–70; Sec. Baku City Cttee. Azerbaidzhan CP 1974–80; Min. of Education Azerbaidzhan SSR Mr. 1980–; cand. mem. Cen. Cttee. Azerbaidzhan CP 1981–.

Kaftailov, Nikolai Ivanovich
State official.
Min. of Construction Latvian SSR D. 1981–.

Kafurov, Oraz Muhamed
State official.
Min. of Local Ind. Turkmen SSR O. 1979–; cand. mem. Cen. Cttee. Turkmen CP 1981–.

Kagan, Gitta Iakovlevna, Dr. of Med. Sc.
Microbiologist; b. 1916.
Mem. CPSU 1968–; grad. Medical Inst. Kiev 1938; Dr. of Med. Sc. 1964; Prof. 1966; completed postgraduate studies at Mil. Medical Acad. 1941, mem. staff of same 1941–44; mem. staff Inst. of Epidemiology and Microbiology at USSR Acad. of Medical Sciences 1944–, Head of Laboratory of same 1964–. Decor.: Lenin Prize 1974.

Kagan, Iurii Moiseevich
Physicist; b. Jl. 6, 1928, Moscow.
Mem. CPSU 1955–; grad. Inst. of Physical Engineering Moscow 1950; corresponding mem. USSR Acad. of Sciences 1970–; mem. staff Inst. of Atomic Energy 1956–; Prof. at Inst. of Physical Engineering Moscow 1962–. Decor.: Order of Red Banner of Labour.

Kaik, Lembit Iurevich
State official.
Cand. mem. 1964–66, mem. Cen. Cttee. Esthonian CP 1966–; Chair. of State Cttee. of Council of Mins. Esthonian SSR for Printing and Publishing 1964–72; Chair. of Esthonian Branch of Sov.-German Frienship Soc. 1965–; Chair. of State Cttee. for Publishing, Printing and the Book Trade. Esthonian SSR Ag. 1972–.

Kairialis, Alfonsas Konstantinovich
State official.
Sec. of Klaipeda City Cttee. Lithuanian CP, Head of Dept. of administrative organs Lithuanian CP 1955–69; mem. Cen. Cttee. Lithuanian CP 1956–; Public procurator of Lithuanian SSR 1969–82; Chair. of Cttee. on People's Control Lithuanian SSR D. 1982–; mem. Bureau Cen. Cttee. Lithuanian CP D. 1982–.

Kairis, Ksaveras Kostovich
State official; b. 1909.
Lithuanian; mem. CPSU 1946–; grad. of a trade inst.; worker, bookkeeper in various enterprises 1929–42; served in Sov. Army 1942–43; engineer and economist of People's Commissariat of USSR Meat and Dairy Ind., Dept. Head, Dep. Min., Min. of Meat and Dairy Ind. Lithuanian SSR 1943–57; Dep. Chair., Chair. of Econ. Council Lithuanian SSR 1957–61; Dep. to Supr. Sov. Lithuanian SSR 1955–; mem. Cen. Cttee. Lithuanian CP 1956–; mem. Bureau 1960–66 and 1967–, cand. mem. Bureau Cen. Cttee. Lithuanian CP 1966–67; 1st Dep. Chair. 1961–62 and My. 1965–, Dep. Chair. of Council of Mins. Lithuanian SSR 1962–65; Dep. to USSR Supr. Sov. 1962–79; mem. Bureau for Ind. with Cen. Cttee. Lithuanian CP 1962–64. Decors.: Order of Lenin, Order of Red Banner of Labour.

Kaitukov, Georgii Kharitonovich
Author; b. 1911, Khod, N. Ossetia; s. of a peasant.
Osset; mem. CPSU 1940–; grad. Inst. of Philosophy, Literature and History Moscow 1936; first works publ. 1927. Decor.: Order of Red Banner of Labour.
Publs.: "K bor'be" (To the Fight) 1931, "V dni voiny" (In Wartime) 1944, "Oruzhie mira" (Weapon of Peace) 1951, "Vershiny" (The Heights) 1952, "Vesennie pesni" (Spring Songs) 1958, "Spasibo, luidi!" (Thank You, People!) 1959, "Khorosho imet' druga" (It's Good to Have a Friend) 1960, and others.

Kaiumov, Abdukhamid Kaiumovich
State official.
Min. of Motor Transport and Highways 1968–69, Min. of Construction and Utilization of Highways Uzbek SSR N. 1969–; cand. mem. Cen. Cttee. Uzbek CP 1971–81, mem. 1981–.

Kaiumov, Malik Kaiumovich
Cameraman and director of documentary films; b. Ap. 22, 1912, Tashkent.
People's Artist of the USSR (1967). Became a member of the CPSU in 1960.
He began his career in the Uzbek film industry in 1931 as an actor and later became a cameraman. During the Great Patriotic War he was a cameraman with film groups on the front. He began directing in the 1950's and made a number of films dealing with socialist construction in Uzbekistan both before and after the war, with Uzbekistan monuments, and with the relations between the Soviet Middle Asian republics and foreign countries. His films include *The Mighty Stream* (1940), *Soviet Tadzhikistan* (1946), *Socialist Uzbekistan* (1950), *The Dawn of the New India* (1956), *Five Hands of Mankind* (1958), *Vietnam – My Country* (1960), *Registan* (1965), *Tashkent, the Earthquake* (1968), and *Children of Tashkent* and *Samarkand After 2,500 Years* (both 1969). Decors.: Khamza State Prize of the Uzbek SSR in 1965, seven orders, and medals.

Kaiumov, Pulat
State official.
Mem. Cen. Cttee. Uzbek CP 1961–76, mem. Auditing Comm. Uzbek CP 1976–; Dep. to Supr. Sov. Uzbek SSR Je. 1962–; Chair. of Fergana Oblast Exec. Cttee. of Workers' Deps. 1962–63, in Andishan Oblast (for Ind. 1963–64) 1963–73; Chair. of State Cttee. for Vocational Training. Uzbek SSR S. 1973–.

Kalachinskii, Mikhas (Mikhail) Ivanovich
Author and poet; b. 1917, Krupki, Minsk Oblast.
Belorussian; mem. CPSU 1942–; first works publ. 1932. Decors.: Order of Patriotic War 2nd Cl., Order of Red Star, medals.
Publs.: children's poems "Kostia-chekist" 1938, "Paket" (The Package) 1939, "Solntse v sineve" (Sun in the Blue) 1949, "Navstrechu zhizni" (Facing Life) 1951, "V velikom pokhode" (On the Great March) 1952, "Na lesnom polustanke" (At the Small Forestry Station) 1955, "Grozd' riabiny" (A Cluster of Ashberries) 1964, and others.

Kaladze, Karlo Razhdenovich
Poet and playwright; b. Mr. 16 (29), 1904, Kutaisi.
Georgian; member of the CPSU since 1939.
His first book of poems was published in 1926. He created poetic panoramas of the new Georgia ("Challenge of the East" 1926; "Dar'ial" 1926; and "Song of the Horseman" 1927). Cycles of poems such as *Dawns in Khertvis* reflect the construction of socialism in the 1930's. The narrative poem Uchardioni (1933) is dedicated to the collectivization of Georgian agriculture. Heroic themes of the Great Patriotic War (1941–45) are the basis of the cycle of poems *Meetings on the Battlefield* (1942).
A prominent theme in his poetry is the friendship of peoples ("The Muscovites," "Songs on the Shores of the Dnieper," and "We Want Peace"). He is the author of the plays *The Way It Was* (1929), *The House on the Bank of the Kura* (1931), *Khatidzhe* (1933), and *One Night's Comedy* (produced in 1945), which have played a significant role in the growth of Georgian drama and theater. Decors.: three orders, as well as medals.
Works: [Kalaze, K.] *Nat'argmni lek'sebi da poemebi.* Tbilisi 1961. *Reh'euli lek'sebi da poemebi,* vols. 1–2. Tbilisi 1968. *In Russian translation: Stikhi, pesni, ballady.* Moscow 1957. *Stikhotvoreniia.* Moscow 1963. *Molodost' Gruzii: Stikhi i poemy.* Tbilisi 1968. *Stikhotvoreniia i poemy.* Tbilisi 1970.

Kalandarov, Gaib
State official.
Chair. of Tadzhik Journalists' Union 1973–; Chief Ed. of daily "Todzhikistoni Sovieti" 1973–77; mem. Cen. Cttee. Tadzhik CP 1976–; Chair. of State Cttee. on Television and Radio Tadzhik SSR F. 1977–.

Kalandarov, Sh.
State official.
Min. of Light Ind. Tadzhik SSR Jl. 1982–.

Kalashnikov, Aleksei Maksimovich
State official; b. 1914.
Russian; mem. CPSU 1939–; grad. Moscow Inst. of Engineers of Communal

Construction Engineering 1938, CPSU Cen. Cttee. Party Higher School 1957; served in Sov. Army 1938–40 and 1941–45; held state and Party posts 1945–61; Dep. Chair., 1st Dep. Chair. of Oblast Exec. Cttee. of People's Deps. in Moscow 1961–64; CPSU City Sec. in Moscow 1964–71; Dep. Chair. of Council of Mins. RSFSR My. 1971–; Dep. to USSR Supr. Sov. 1974–; cand. mem. Cen. Cttee. CPSU 1976–. Decor.: Order of Lenin.

Kalashnikov, Mikhail Timofeevich, Dr. of Tech. Sc.
Engineer; b. N. 10, 1919, Kuria, now Altai Krai.
Mem. CPSU 1953–; designer of mil. weapons. Decors.: Hero of Socialist Labour, State Prize of the USSR 1949, 2 Orders of Lenin, 2 other orders, several medals.

Kalashnikov, Vladimir Il'ich
State official.
Min. of Melioration and Water Econ. RSFSR S. 1982–.

Kaldma, Arvo Ukuvich
State official.
Min. of Posts and Telecommunications Esthonian SSR Jl. 1975–; cand. mem. Cen. Cttee. Esthonian CP 1976–.

Kalenik, Evgenii Petrovich
Party official; b. 1937; s. of a farmer.
Moldavian; cand. of Agrarian Sciences; mem. CPSU 1961–; grad. Agriculture Inst. in Kishinev; agronomist at "Fruktovyi Donbass" collective farm in Debossary Raion 1959–61; 1st Raion Sec. Moldavian Komsomol in Debossary 1961–62; grad. student at the Agriculture Inst. in Kishinev, held posts in Cuba 1962–66; Dep. Sec. of Party Cttee. of Agriculture Inst. in Kishinev, teacher at this Inst. 1966–68; head of a sector, dep. head of Agriculture Dept. Cen. Cttee. CP Moldavia 1968–70; 1st Raion Sec. CP Moldavia in Kotovsk 1970–76; mem. Cen. Cttee. CP Moldavia 1971–; head of the Dept. for Agriculture and the Food Ind. Cen. Cttee. CP Moldavia 1976–77; Sec. and mem. Bureau of Cen. Cttee. CP Moldavia N. 1976–.

Kalgan, Aleksandr Dmitrievich
Author; b. 1912, Novye Mertli; s. of a peasant.
Mem. CPSU 1943–; grad. Chuvash Pedagogical Tech. Inst. Kazan 1930 and Higher Courses at USSR Writers' Union; Chuvash poet and dramatist; first works publ. 1928.
Publs.: "Blizkie zvezdy" (Adjacent Stars) 1950, "Anissa" 1936, "Alim" 1940, "Volna Oktiabria" (October Wave) 1957, and others.

Kaliaev, Sandzhi Kaliaevich
Kalmyk writer; b. Ja. 2, 1905, Tsagan-Nur, in present-day Priozernyi Raion.
People's Poet of Kalmykiia (1965). Member of the CPSU since 1928.
He is one of the founders of Soviet Kalmyk literature. His first work was published in 1924. He is the author of several collections of poetry, a cycle of poems about V. I. Lenin entitled *Only by the Precepts of Il'ich* (separate edition, 1969), and the narrative poems *Brigadier* (1935) and *Tamara* (1963), the last about a young partisan named Tamara Khakhlynova. He is also an expert on Kalmyk folklore. He is a literary critic and a translator of the plays of Molière.
Works: *In Russian translation: S toboiu, Rossiia: Stikhi i poemy.* Elista 1959. *Pod neob"iatnym nebom: Poemy.* Moscow 1961. *Kogda est' liubov': Poemy.* Elista 1966.

Kalin, Ivan Petrovich
State official; b. 1935; of a collective farmer.
Moldavian; mem. CPSU 1955–; grad. Frunze Agricultural Inst. in Kishinev and CPSU Cen. Cttee. Party Higher School; agronomist, dep. chair. and sec. of party cttee. of "Viatsa Noue" Collective Farm in Orgeev Raion 1960–63; instructor of Party Organs Dept. of Cen. Cttee. Moldavian CP, sec. of party cttee. for agriculture in production admin. in Kalarash, 1st Sec. of Kalarash Raion Cttee. Moldavian CP 1963–67; mem. Cen. Cttee. Moldavian CP 1963–; auditor CPSU Cen. Cttee. Party Higher School 1967–69; 1st Sec. of Kalarash Raion Cttee. Moldavian CP 1969–71; Chief of Agriculture Dept. 1971–74, Chief of Agriculture and Food Ind. Dept. of Cen. Cttee. Moldavian CP 1974–76; mem. Bureau of Cen. Cttee. Moldavian CP 1976–; Sec. of Cen. Cttee. Moldavian CP 1976–80; Chair. of Presidium of Supr. Sov. Moldavian SSR Ap. 1980–; Dep. and mem. of Presidium of USSR Supr. Sov. 1980–; cand. mem. Cen. Cttee. CPSU 1981–.

Kalinin, Anatolii Veniaminovich
Author; b. 1916, Kamenskaia (later renamed Kamensk-Shakhtinskii); s. of a teacher.
Russian; mem. CPSU 1946–. Decors.: Order of Patriotic War 1st Cl., Order of Red Star.
Publs.: "Kurgany" (The Tumuli) 1941, "Na iuge" (In the South) 1944, "Tovarishchi" (Comrades) 1945, "Krasnoe znamia" (Red Banner) 1951, "Neumiraiushchie korni" (Undying Roots) 1947, "Na srednem urovne" (At Average Level) 1954, "Lunnye nochi" (Moonlight Nights) 1955, "Surovoe pole" (Grim Field) 1958, "Ekho voiny" (Echo of War) 1963, "Zapretnaia zona" (Restricted Area) 1962, and others.

Kalinin, Arnold Ivanovich
Diplomat; b. 1929.
Mem. Diplomatic Service 1969–; Emb. Counsellor-Envoy at USSR Emb. in Cuba 1969–74; USSR Amb. to Portugal Jl. 1974–D. 1982.

Kalinovskaia, Valentina Fedorovna
Ballerina; b. Jl. 21, 1938, Kiev.
People's Artist of the USSR 1968.
A graduate of the Kiev Choreographic School, she has been dancing at the T. G. Shevchenko Opera and Ballet Theater since 1957. Technically perfect, her dancing is characterized by light jumps, a lyrical adagio, and a brilliant allegro. She is best suited for dramatic roles. Her best roles include Odette-Odile in Tchaikovsky's *Swan Lake,* Aegina in Khachaturian's *Spartacus,* Kitri in Minkus' *Don Quixote,* Zarema in Asaf'ev's *The Fountain of Bakhchisarai,* Mekhmene Banu in Melikov's *The Legend of Love,* and Dona Anna in Gubarenko's *The Stone Ruler.* Decor.: Order of the Red Banner of Labor.

Kalitko, Aleksandr Iakovlevich
State official.
Cand. mem. Cen. Cttee. 1961–71, mem. Cen. Cttee. Belorussian CP 1971–76; Head of Kolkhoz-Sovkhoz Production Admin. in Pukhovichi 1963; 2nd Sec. of Vitebsk Oblast Cttee. Belorussian CP 1968–74; Min. of Supplies Belorussian SSR F. 1974–.

Kalnroze, Valdis (pseudonym of V. Rozenberg)
Landscape painter; b. Ja. 5 (17), 1894, Kuldiga.
Honored Artist of the Latvian SSR (1963).
He studied in Riga at the Latvian Academy of Arts (1927–32) under V. Purvītis. His landscapes (primarily of the rivers and the Baltic seacoast of Latvia) convey spatial depth and are characterized by a distinctive, decorative painting technique; they contain large areas of silvery gray colors. Among his painting are *Daugava* (1935), *Autumn Landscape* (1942), *Sea* (1956), and *In a River Cove* (1957) – all located in the Art Museum of the Latvian SSR in Riga. He also painted *Cloudy Day* (1962, Directorate of the Art Museums and Exhibitions of the Latvian SSR) and *Landscape With River* (1966).

Kalnyn', Eduard Fridrikhovich
Painter; b. O. 12 (25), 1904, Riga. People's Artist of the USSR (1975). Member of the Academy of Arts of the USSR (1970).
He studied under V. Purvītis at the Latvian Academy of Arts in Riga from 1923 to 1932. He became an instructor at the Academy of Arts of the Latvian SSR in 1945; receiving a professorship there in 1955. His art played a large role in the development of Latvian genre painting. His landscapes and genre scenes are characterized by large-scale compositions and

large, rhythmic planes of color. They are primarily devoted to the difficult work of fishermen and the magnificent beauty of the elements of the sea. His works include *The Raftsmen* (1935), *New Sails* (1945), *The Seventh Baltic Regatta* (1952), *Latvian Fishermen in the Atlantic* (1957), and *In the Distance* (1967) – all except the last painting are in the Museum of Art of the Latvian SSR in Riga.

Kaluga, Hanna Maksimovna
Ceramist; b. 1924, Lubny, Poltava Oblast.
Grad. Kiev School of Applied Art 1948, artist in applied and decorative art (ceramics).

Kaluga, Vasilii Ivanovich
Soviet graphic artist; b. 1923, Novozybkov, Briansk Oblast.
Attended School of Applied Art 1951–52 and Art Inst. 1952–58 in Kiev, pupil of D. Shavykin, O. Pashchenko, V. Kasian and K. Trokhimenko.

Kamaev, Gerontii Leontevich
State official; b. 1919.
Chair. of Tula Oblast Exec. Cttee. of Peoples Deps. (for Ind. 1963–64) 1963–70; Chairs. of RSFSR State Cttee. for Vocational Training S. 1970–. Decor.: Order of Red Banner of Labour.

Kamalidenov, Zakash Kamalidenovich
State official; b. 1936.
Mem. CPSU 1960–; grad. Gubkin Petroleum Inst. in Moscow 1958; CPSU Cen. Cttee. Party Higher School 1970; held technical engineering posts 1958–59; Sec. of Zhilokosinskii Raion Cttee. Komsomol in Gurev region, 1st Sec. of Oblast Cttee. Komsomol in Gurev 1959–64; 2nd City Sec. CP Kazakhstan in Shevchenko 1964–68; Chair. of Regional Trade Union Council in Gurev 1970; 1st Sec. of Cen. Cttee. Kazakh Komsomol 1970–78; mem. Cen. Cttee. CP Kazakhstan 1971–; 2nd Oblast Cttee. Sec. CP Kazakhstan in Tselinograd 1978–79; USSR KGB supervisor, Dep. Chair. of KGB Kazakh SSR 1979–80; mem. 1980–, Sec. 1980–82, mem. of Bureau Cen. Cttee. CP Kazakhstan Mr. 1980–; Chair. of KGB Kazakh SSR F. 1982–.

Kamalov, Kallibek
Party official; b. 1926, Kara-Kalpak ASSR.
Mem. CPSU 1946–; grad. Kara-Kalpak Teachers' Training Inst. 1947 and Kara-Kalpak Pedagogical Inst. 1955; middle school teacher, 2nd sec. of a Komsomol oblast cttee. 1942–51; Min. of Communal Econ. 1953–56 and Min. of Motor Transport and Highways Kara-Kalpak ASSR 1956–58; 1st Sec. of Kuibyshev Raion Cttee. Uzbek CP 1958–59; Chair. of Council of Mins. Kara-Kalpak ASSR 1959–63; mem. Cen. Cttee. Uzbek CP 1960–; Dep. to USSR Supr. Sov. 1962–; 1st Oblast Sec. of Uzbek CP in Kara-Kalpak ASSR 1963–; cand. mem. Cen. Cttee. CPSU 1971–; cand. mem. Bureau Cen. Cttee. Uzbek CP 1971–.

Kamalov, Makhkam
Party official; b. 1925.
Uzbek; mem. CPSU 1948–; grad. Tashkent Pedagogical Inst., Tashkent Agricultural Inst., CPSU Cen. Cttee. Party Higher School; mechanic in clothes factory 1942–43; served in Sov. Army 1943–46; studied at republic's Party School 1946–48; instructor at Cen. Cttee. Uzbek Komsomol, Komsomol Oblast Cttee. Sec. in Tashkent, Chair. of Raion Exec. Cttee. of People's Deps. 1948–62; 1st Sec. of Raion Cttee. CP Uzbekistan 1962–73; mem. Auditing Comm. 1971–76, mem. Cen. Cttee. CP Uzbekistan 1976–; Oblast Cttee. Sec. CP Uzbekistan in Tashkent 1973–76; 1st Oblast Cttee. Sec. CP Uzbekistan in Namangan Ja. 1976–.

Kamanin, Nikolai Petrovich
Military official; b. O. 5 (18), 1908, Melenki, in present-day Vladimir Oblast; s. of a shoemaker.
Colonel-general of aviation (1967), Hero of the Soviet Union (Ap. 20, 1934). Became a member of the CPSU in 1932.
He graduated from the Leningrad Military-Theoretical Flight School in 1928 and then volunteered for the Red Army. In April 1934 he was a pilot in the rescue of the expedition on the steamship *Cheliuskin,* which was sinking in the Arctic Ocean. In 1938 he graduated from the Higher Air Force Academy and took command of an aviation brigade. He fought in the Soviet-Finnish War of 1939–40. During the Great Patriotic War (1941–45), he became commander in May 1942 of the V Assault Aviation Corps, which was active in the liberation of the Ukraine, Poland, Rumania, Hungary, and Czechoslovakia. In 1950 he graduated from the Academy of the General Staff. Between 1966 and 1972 he worked at the main headquarters of the air force; he was in charge of training cosmonauts. He retired in 1972. Decors.: three Orders of Lenin, the Order of the Red Banner, two Orders of Suvorov Second Class, the Order of Kutuzov Second Class, the Order of the Red Star, several foreign orders, and medals.
Work: *Letchiki i kosmonavty.* Moscow 1972.

Kamentsev, Vladimir Mikhailovich
State official; b. 1928.
Mem. CPSU 1954–; grad. Moscow Technical Inst. of the Fish Ind. and Econ. 1950; stoker, lathe operator, motor mechanic of Moscow Steamship Line 1942–44; posts in admin. of Murmansk Basin Trawler Fleet 1950–57; chief engineer, Dep. Chair. of Murmansk Sovnarkhoz 1957–62; Dep. Chair., 1st Dep. Chair. of USSR State Cttee. for the Fish Ind. 1962–65; 1st Dep. USSR Min. of the Fish Ind. 1965–79, Min. F. 1979–; cand. mem. Cen. Cttee. CPSU 1981–.

Kamianskii, Vasilii Kirillovich
Author; b. 1912, Cossack village of Ekaterinovskaia, Krasnodar Oblast.
Russian; mem. CPSU 1940–; lives in Kalinin; first works publ. 1933.
Publs.: "Den' Ivana Groznogo" (A Day in the Life of Ivan the Terrible) 1947, "U sten Kremlia" (At the Kremlin Walls) 1948, "Dunai techet na Vostok" (The Danube Flows East) 1955, "Zvezdy ne merknut" (Stars Don't Go Out) 1958, and others.

Kaminnoi, Nikolai Ivanovich
Graphic artist; b. 1924, Novo-Ustinovska, Kirovograd Oblast.
Attended Kharkov Art Inst. 1951–57, pupil of V. Mironenko, G. Bondarenko, I. Daits; teacher at Kharkov Inst. for Arts and Crafts 1963–.

Kamozin, Pavel Mikhailovich
Military official; b. Jl. 3 (16), 1917, Bezhitsa.
Twice Hero of the Soviet Union (My. 1, 1943, and Jl. 1, 1944); captain (1944). Member of the CPSU since 1943.
His father was a worker; he himself was a locksmith before joining the Soviet Army in 1938. That year he graduated from the Borisoglebsk Military Aviation School. During the Great Patriotic War of 1941–45 he fought on the Southern, Transcaucasian, and Northern Caucasian fronts as flight commander, deputy commander, and commander of a squadron of a fighter aviation regiment. He shot down 35 enemy aircrafts in individual combat and 13 in group combat. Decors.: Order of Lenin, two Orders of the Red Banner, the Order of Alexander Nevsky, the Order of the Patriotic War First Class, and various medals.

Kandelaki, Vladimir Arkad'evich
Russian singer (bass-baritone) and director; b. Mr. 16 (29), 1908, Tbilisi.
People's Artist of the USSR (1971). Member of the CPSU since 1952.
He graduated from the Tbilisi Conservatory in 1928 and from the State Institute of Theatrical Art (Moscow) in 1933. Since 1929 he has been a soloist at the V. I. Nemirovich-Danchenko Music Theater (since 1941 called the K. S. Stanislavsky and V. I. Nemirovich-Danchenko Moscow Music Theater from 1954 to 1964, he was also chief director and actor for the Moscow Musical Comedy Theater.
Performing both comic and dramatic roles, he has created expressive stage portraits, including Stefan in *The Gypsy Baron* by Offenbach, Ollendorf in *The Poor Student* by Millöcker, Sultanbek in *Arshin mal alan* by Gadzhibekov, Salieri in *Mozart and Salieri* by Rimsky-Korsakov, Taras in *The Family of Taras* by Kabalevskii (State Prize of the USSR, 1952), and Magara in *Virineia* by Slonimskii. He has worked as a director since 1943 and has appeared in films.

Kandrenkov, Andrei Andreevich
Party official; b. 1915.
Russian; mem. CPSU 1939–; grad. Veterinary Acad. Moscow 1950; agronomist with a raion agricultural dept., dep. dir. of a machine and tractor station, 1st Sec. of Moscow City Komsomol Cttee., head of a sovkhoz pol. dept. 1936–43; served in Sov. Army 1943–45; Sec. of Kashir City Cttee. CPSU, instructor of a Cen. Cttee. CPSU dept., 2nd Sec. of Kaluga City Cttee. CPSU, Chair. of Kaluga Oblast Exec. Cttee. of Workers' Deps. 1945–61; Dep. to Supr. Sov. RSFSR 1955–63; Dep. to USSR Supr. Sov. 1962–; 1st Sec. of Kaluga Oblast Cttee. CPSU (for Agriculture 1963–64) Ag. 1961–; cand. mem. 1961–76, mem. Cen. Cttee. CPSU 1976–. Decors.: Order of Red Banner of Labour, Order of Lenin.

Kanep, Vilgelm Vilgelmovich
State official.
Min. of Health Latvian SSR D. 1962–; Dep. to Supr. Sov. Latvian SSR 1963–; cand. mem. 1963–66, mem. Cen. Cttee. Latvian CP 1966–.

Kantorovich, Leonid Vitalevich
Mathematician and economist; b. Ja. 6 (19), 1912, St. Petersburg.
Grad. Leningrad Univ. 1930; mem. USSR Acad. of Sciences 1964–; Lecturer 1932–34, Prof. at Leningrad Univ. 1934–60; mem. staff Siberian Branch of USSR Acad. of Sciences 1958–71; mem. staff Inst. of Econ. Planning of State Cttee. on Science and Technology of USSR Council of Mins. 1971–; honorary doctorate from numerous foreign univs.; mem. Budapest and Boston Acads. of Sciences. Decors.: State Prize of the USSR 1949, Lenin Prize 1965, Order of Lenin, 3 other orders and various medals.
Works: *Matematicheskie metody organizatsii i planirovaniia proizvodstva.* Leningrad 1939. *Funktsional'nyi analiz v poluuporiadochennykh prostranstvakh.* Moscow-Leningrad 1950. (Co-author.) *Ekonomicheskii raschet nailuchshego ispol'zovaniia resursov.* Moscow 1959. *Funktsional'nyi analiz v normirovannykh prostranstvakh.* Moscow 1959. (With G.P. Akilov.) *Priblizhennye metody vysshego analiza,* 5th ed. Moscow-Leningrad 1962. (With V.I. Krylov.)

Kantorovich, Vladimir Iakovlevich
Author; b. Ap. 2 (15), 1901, St. Petersburg.
Russian; grad. Dept. of Social Sciences of Moscow State Univ. 1924; suffered reprisals 1937, could not resume literary career until 1956.
Publs.: "Bol'shoi shans" (The Great Chance) 1934, "Rasskazy provodnika Karatsupy" (Tales of the Conductor Karatsupa) 1937, "Istochnik sily" (Source of Power) 1958, "Istoriia inzhenera Ganshina" (The History of Engineer Ganshin) 1957, "V molodom gorode" (In a Young Town) 1960, "'Ty' i 'Vy'" ('Thou' and 'You') 1960, "Zametki pisatelia o sovremennom ocherke" (Notes of a Writer on the Contemporary Essay) 1962, and others.

Kapitan, Aleksei Alekseevich
Graphic artist; b. 1927, Kiev.
Grad. Kiev Art Inst., pupil of V. Kasian.

Kapitan, Leonid Alekseevich
Graphic Artist; b. 1918, Kiev.
Grad. Kiev Art Inst. 1941, pupil of F. Krichevskii, S. Grigorev, V. Kostetskii.

Kapitonov, Ivan Vasil'evich
Party official; b. F. 10 (23), 1915, Serovskoe, Riazan Oblast; s. of a peasant.
Russian; mem. CPSU 1939–; grad. Inst. for the Construction of Public Utilities Moscow 1938; construction engineer in Riazan, Kiev and Stryi 1939–40; Head of Planning and Production Dept. 1940–41, Sec. of Party Org. Work at Krasnopresnenskii Tram Depot in Moscow 1941–42; Dep. Head of Personnel Dept. at Krasnopresnenskii Raion All-Union CP(B) Cttee., Instructor of Personnel Dept. at Moscow City All-Union CP(B) Cttee. 1942–43; Sec. of Krasnopresnenskii Raion All-Union CP(B) Cttee. 1943–47; Chair. of Exec. Cttee. Krasnopresnenskii Raion Sov. 1947–48; Dept. Head with Moscow City All-Union CP(B) Cttee. 1948–51; Dep. to USSR Supr. Sov. 1950–; Sec. of Moscow Oblast and City All-Union CP(B) Cttee. 1951–54; mem. Cen. Cttee. CPSU 1952–; 1st Sec. of Moscow Oblast Cttee. CPSU 1954–59; mem. Presidium 1954–62, mem. Budgetary Comm. at Sov. of the Union, USSR Supr. Sov. 1962–66; 1st Sec. of Ivanovo Oblast Cttee. CPSU 1959–64 (1963–64 for Ind.); Head of RSFSR Party Organs Dept. with Cen. Cttee. CPSU 1964–65; mem. Bureau for the RSFSR with Cen. Cttee. CPSU 1964–66; Head of Dept. of Party Org. Work with Cen. Cttee. CPSU 1965–; Sec. of Cen. Cttee. CPSU D. 1965–. Decors.: 2 Orders of Lenin, Order of Red Banner of Labour, Order of Patriotic War 2nd Cl., Order of Red Star, various medals.

Kapitonov, Viktor Arsen'evich
Athlete; b. O. 25, 1933, Kalinin.
Honored Master of Sport (1958) and Honored Coach of the USSR (1970). Officer in the Soviet Army. Member of the CPSU since 1969.
He was the first Soviet athlete to become an Olympic champion in bicycling (1960, Rome, road racing). He was a seven-time champion of the USSR (in 1956–64, in the marathon road race) and winner of the world bicycle race (as a team member) in 1958 and 1961–62. Decors.: Order of Lenin, the Order of the Badge of Honor, and medals.

Kapitsa, Andrei Petrovich
Geographer and geomorphologist; b. Jl. 9, 1931, Cambridge, England.
Corresponding member of the Academy of Sciences of the USSR (1970). Son of P. L. Kapitsa. Member of the CPSU since 1962. Chairman of the Presidium of the Far Eastern Scientific Center of the Academy of Sciences of the USSR (since 1970); member of the Presidium of the Academy of Sciences of the USSR (since 1971). Director of the Pacific Institute of Geography of the Far Eastern Scientific Center of the Academy of Sciences of the USSR.
He worked at the department of geography of Moscow State University from his graduation there in 1953 until 1970. He became a professor in 1966 and was dean of the department from 1966 to 1970. He participated in four Soviet Antarctic expeditions and in the trans-Antarctic crossings from Mirny Station to Pionerskaia (1956), Mirnyi to the south pole (1959–60), and from Vostok station through the Pole of Inaccessibility to Molodezhnaia Station (1963–64). In 1967–69 he led the Soviet Complex Geophysical Expedition of the Academy of Sciences of the USSR to East Africa. His major works have dealt with the dynamics and morphology of the Eastern Antarctic ice cap. Decors.: State Prize of the USSR in 1971 for his participation in the creation of the *Atlas of Antarctica,* Order of the Red Banner of Labor.
Works: *Dinamika i morfologiia lednikovogo pokrova tsentral'nogo sektora Vostochnoi Antarktidy.* Leningrad 1961. (*Tr. Sovetskoi Antarkticheskoi ekspeditsii,* vol. 19). *Podlednyi rel'ef Antarktiki.* Moscow 1968.

Kapitsa, Mikhail Stepanovich, Dr. of Hist. Sc.
Diplomat; b. 1921.
Prof.; mem. Diplomatic Service 1943–; specialist and consultant in Far East Dept. 1954–56 and Dep. Head of Far East Dept. of USSR Min. of Foreign Affairs 1956–60; USSR Amb. to Pakistan 1960–61; sen. posts in USSR Min. of Foreign Affairs 1961–66; Head of South-East Asia Dept., Head of 1st Far East Dept. 1970–D. 1982; mem. of Collegium of Min. of Foreign Affairs 1970–; USSR Dep. Min. of Foreign Affairs D. 1982–.

Kapitsa, Piotr Iosifovich
Author; b. 1909, Suvalki.
Russian; mem. CPSU 1928–; grad. Inst. of Econ. Leningrad 1934; first works publ. 1931.
Publs.: "Pravila vesny" (Spring Rain) 1931, "Nastoiashchaia rabota" (Real Work) 1932, "Boksery" (Boxers) 1938, "Kogda ischezaet strakh" (When Fear Disappears) 1962, "Muzhestvo" (Bravery) 1942, "Geroi Sovetskogo Soiuza krasnoflotets A. Poskonkin" (Hero of the Sov. Union, Seaman of the Red Fleet A. Poskonkin) 1941, "Dvazhdy geroi So-

vetskogo Soiuza gvardii polkovnik Boris Feoktistovich Safonov" (Two-time Hero of the Sov. Union, Major of the Guards Boris Feoktistovich Safonov) 1943, "Dmitrii Glukhov" 1945, "V otkrytom more" (On the Open Sea) 1946, "V dni razluki" (In the Days of Separation) 1955, "Oni shturmovali Zimnii" (They Stormed the Winter Palace) 1957, and others.

Kaplan, Samuil Solomonovich
Graphic artist; b. 1928, Baku.
Grad. Kiev Art Inst. 1962, pupil of O. Pashchenko, I. Selivanov; mem. CPSU 1955–.

Kaplanian, Rachia Nikitovich
Theatre director; b. 1923.
Mem. CPSU 1947–; grad. Art Theatre Inst. Erevan 1950; actor 1937–, Asst. to the Dir. 1946–52, Principal Dir. of Youth Theatre Erevan 1953–56 and 1959–62; Dir. of Armenian Sundukian Theatre 1956–59 and Principal Dir. of Armenian Spendiazov Opera and Ballet Theatre 1962–65; Principal Dir. of Erevan Theatre founded by him 1968–; Chair. of Presidium Armenian Theatre Soc. 1966–. Decor.: People's Artist of the USSR 1971.

Kapler, Aleksei Iakovlevich
Actor and film script writer; b. S. 23 (O. 11) 1904, Kiev.
Actor 1919; together with S. I. Iutkevich and G. M. Kozintsev founded "Arlekin" Theatre in Kiev 1920; film work 1926–; wrote scripts for documentaries 1929–30. Decors.: State Prize of the USSR 1941, Order of Lenin, one order, medals, Merited Artist of the RSFSR 1969.
Film scripts: "Tri tovarishcha" (Three Friends) 1935, "Lenin v Oktiabre" 1937, "Lenin v 1918 godu" (Lenin in the Year 1918) 1939, "Ona zashchishchaet Rodinu" (She Defends the Fatherland) 1943, "Kotovskii" 1943, "Neobyknovennoe leto" (An Unusual Summer) 1957, "Dve zhizni" (Two Lives) 1962, "Chelovek amfibia" (Amphibious Man) 1962, "Vera, Nadezhda, Liubov'" 1972, and others.

Kapp, Eugen
Composer and public figure; b. My. 13 (26), 1908, Astrakhan.
People's Artist of the USSR (1956). Member of the CPSU since 1947.
He studied piano under P. Ramul and A. Lemba during 1928–31 and composition under his father, A. I. Kapp, after 1926 at the Tallinn Conservatory. He became an assistant in the music theory course at the conservatory in 1935, a lecturer in 1944 (professor in 1947), and the director in 1952 (until 1965). He was chairman of the Composers' Union of the Esthonian SSR during 1944–46.
He contributed to the establishment and development of a national music culture. He composed the first Soviet Esthonian opera, *Fire of Vengeance* (1945); the first ballet, *Kalevipoeg* (Son of Kalev; 1947), based on themes from epic folk literature; and the first Esthonian children's opera, *Winter Fairy Tale* (1958). His other compositions include the opera *Freedom Singer* (1950), the ballet *The Gold Spinners* (1956), symphonies, choral and chamber works, and piano pieces.
He was a deputy to the fourth and fifth convocations of the Supreme Soviet of the USSR. During 1951–61 he was a member of the Central Committee of the Communist Party of Esthonia. Decors.: State Prize of the USSR in 1946, 1949, and 1952, Order of Lenin, three other orders, and medals.

Kapto, Aleksandr Semenovich
Party official; b. 1933.
Ukrainian; mem. CPSU 1955–; grad. Dnepropetrovsk State Univ., Cand. of Philosophical Sc.; Komsomol posts 1957–63; ed. of Ukrainian Komsomol newspaper "Komsomolskoe znamia" 1963–66; 2nd Sec. of Cen. Cttee. Ukrainian Komsomol 1966–68, 1st Sec. 1968–72; mem. Cen. Cttee. Ukrainian CP 1971–; Sec. of Kiev City, than Kiev Oblast Cttee. Ukrainian CP 1972–78; Chief of Culture Dept. of Cen. Cttee. Ukrainian CP 1978–79; Sec. of Cen. Cttee. and mem. Politbureau of Cen. Cttee. Ukrainian CP Ap. 1979–; cand. mem. Cen. Cttee. CPSU 1981–.

Kapustin, Erimei Davidovich
State official.
Mem. Cen. Cttee. Tadzhik CP 1963–; Head of Dept. of Trade and Public Service with Cen. Cttee. Tadzhik CP 1965–75; Head of Main Statistical Admin. Tadzhik SSR N. 1975–.

Kaputikian, Silva (Sirvard) Barunakovna
Author and poet; b. 1919, Erevan.
Armenian; mem. CPSU 1945–; grad. Dept. of Philology of Erevan Univ. 1941; first works publ. 1933. Decor.: Stalin Prize.
Publs.: "V eti dni" (These Days) 1945, "Stikhi" (Poems) 1947, "Na beregu Zangi" (On the Banks of the Zanga) 1947, "Eto moia strana" (This Is My Country) 1949, "Moi rodyne" (My Loved Ones) 1951, "Otkrovennaia beseda" (An Open Conversation) 1955, "Liricheskie stikhi" (Lyric Poems) 1955, "Rozovye kamni" (Pink Stones) 1959, "Rasdumia na polputi" (Halfway Thoughts) 1960, "Pesnia liubvi" (Love Song), "Ia i ona" (She and I), "Vernost'" (Faithfulness), and others.

Karadurdyev, Chary
State official.
Chair. of State Cttee. for the Supply of Petroleum Products Turkmen SSR Ja. 1981–.

Karaffa-Korbut, Sofia Petrovna
Graphic artist; b. 1924, Lvov.
Grad. Lvov Inst. for Applied and Decorative Art 1953, pupil of I. Gutorov, V. Monastyrskii, R. Selskii.

Karaganov, Aleksandr Vasil'evich
Literary and theatrical historian and film critic; b. 1915, Martynovo, now Volgograd Oblast.
Russian; mem. CPSU 1956–; grad. Inst. of Philosophy, Literature and History Moscow 1939; Managing Ed. of English-language edition of journal "Sovetskaia literatura" 1945–48; Head of "Iskusstvo" publishing house 1960–64; Sec. of USSR Union of Filmmakers 1965–; first works publ. 1931.
Publs.: "Chernyshevski i Dobroliubov o realizme" (Chernyshevski and Dobroliubov on Realism) 1955, "Kharaktery i obstoiatel'stva" (Characters and Conditions), article on theatre directing, 1959, "Zhizn' dramaturga. Tvorchestkii put' Aleksandra Afinogenova" (The Live of a Dramatic Adviser. The Creative Life of Aleksandr Afinogenov) 1964, "Vysokoe iskusstvo dramy" (The High Art of Drama) 1952, "Zhizn' i method" (Life and Method) 1955, "Mezhdu pravdoi i lozhiu" (Between Truth and Lie) 1963, "O sovremennike i sovremennosti" (On Contemporary Man and the Present) 1964, and others.

Karakeev, Kurman-Gali
Historian; b. N. 7, 1913, Kurmenty, present-day Tiup Raion, Issyk-Kul' Oblast, Kirghiz SSR; s. of a peasant.
Academician (1960) and president (from 1960) of the Academy of Sciences of the Kirghiz SSR.
Corresponding member of the Academy of Sciences of the USSR (1968). Member of the CPSU from 1938.
He graduated from the Higher Party School (1946) and the Academy of Social Sciences (1959) under the Central Committee of the CPSU. From 1939 to 1959 he was the secretary of the Tien-Shan and Issyk-Kul' oblast committees of the party, an editor of the republic newspaper *Kyzyl Kyrgystan* (Red Kirghizia), the head of the propaganda and agitation section of the Central Committee of the Communist Party of Kirghizia, and the secretary (from 1947) of the part's Central Committee. His primary fields of research were cultural construction, the history of the CPSU, and the history of the USSR. He translated the classic Marxist-Leninist works into Kirghiz; he also edited other translations of these works. He was the co-author and editor of several publications, including *Studies of the Communist Party of Kirghizia* (1966), *History of the Kirghiz SSR* (3rd ed., 1967), *History of the Communists Organizations of Middle Asia* (1967), and the fundamental work. *The Victory of Soviet Power in Middle Asia and Kazakhstan* (1967).
He was a deputy to the sixth, seventh, and eighth convocations of the Supreme Soviet of the USSR. He was also a delegate to the Twenty-second (1961) and Twenty-third (1966) Congresses of the CPSU. Decors.: Order of Lenin, four other orders, and various medals.

Karamian, Artashes Ivanovich
Neurophysiologist; b. Mr. 2 (15), 1908, Kafan, now Armenian SSR.
Mem. CPSU 1940–; grad. Medical Inst. Erevan 1940; corresponding mem. USSR Acad. of Sciences 1970–; corresponding mem. Armenian Acad. of Sciences 1963–; mem. staff Physiological Inst. at USSR Acad. of Sciences 1945–50, Inst. of Experimental Medicine at USSR Acad. of Medical Sciences 1950–59, I. M. Sechenov Inst. of Evolutionary Physiology and Biochemistry at USSR Acad. of Sciences 1959–. Decors.: I. P. Pavlov Prize of USSR Acad. of Sciences 1957, Order of Patriotic War 2nd Cl., several medals.

Karamian, Erazm Aleksandrovich (real surname: Melik-Karamian)
Motion-picture director; b. Mr. 2 (15), 1912, Paris.
Poeple's Artist of the Armenian SSR (1966). Member of the CPSU since 1945. He graduated from the School of Drama in Kharkov in 1935. He began working in Armenian cinema in 1936. In 1947–54 he directed newsreels and in 1954 began making feature films. His most important works (with S. A. Kevorkov) are *The Path of Thunder* (1956), *Known Personally* (1958), *Mission Extraordinary* (1965), and *Explosion Past Midnight* (1966). He also directed *The Ghosts Leave the Summits* (1955), *Nasreddin in Khodzhent, or the Charmed Prince* (1960, with A. I. Bek-Nazarov), and *The Twelve Fellow-Travelers* (1962). He took part in writing the screenplays for several of his films. Decors.: State Prize of the Armenian SSR (1967), two orders, and also medals.

Karandash (pseudonym of Mikhail Nikolaevich Rumiantsev)
Circus artist; clown; b. N. 27 (D. 10), 1901, St. Petersburg.
People's Artist of the USSR (1969).
He graduated from a technicum of circus art in 1930 and made his debut as a floor clown. Initially he appeared as a Charlie Chaplin type, but in 1934–35, while working in the Leningrad Circus, he created a new character, adopting the pseudonym of the French caricature artist Caran d'Ache. His character is an adult who retains the spontaneity and *joie de vivre* of the child; this helps to create comic situations and to lend credence to the stunts performed by the artist. He continually appears during the acts of acrobats, gymnasts, jugglers, and animal trainers. In 1946, he began heading clown troupes. His students and partners included Iu. V. Nikulin and M. I. Shuidin. He has appeared in motion pictures *(Karandash on the Ice)* and has toured abroad on a number of occasions.
Work: *Na arene sovetskogo tsirka*. Moscow 1954.

Karas, Iosip Illich
Painter; b. 1918, Khorol, Poltava Oblast. Mem. CPSU 1944–; attended Kharkov Art School 1934–38 and Art Inst. 1938–42 and 1946–48, pupil of O. Kokel, S. Prokhorov.

Karataev, Mukhamedzhan
Author, critic and translator; b. 1910, Kzyl Orda, Kazakhstan; s. of a nomad Kazakh cattle breeder.
Kazakh; first works publ. 1933.
Publs.: "Dumy o rodnoi literature" (Thoughts on Native Literature) 1958, "Rozhdennaia Oktiabrem" (Born Through October) 1958, "Ilias Dzhansugurov" 1959, "Puteshestvie za pesniami" (Journey in Search of Songs) with A. Bragin, 1958, "Kazakhskaia literatura" (Kazakh Literature) 1960, "Gudok v stepi" (Sirens in the Steppe) with K. N. Altaiskii, 1964, "Saken Seifullin" 1964, "Stanovlenie sotsialisticheskogo realizma v kazakhskoi proze" (The Devolpment of Socialist Realism in Kazakh Prose) 1965, and others.

Karatsiupa, Nikita Fedorovich
Military official; b. Ap. 12 (25), 1911, Alekseevka, in present-day Kuibyshev Raion, Dnepropetrovsk Oblast; s. of a peasant.
Colonel; Hero of the Soviet Union (Je. 21, 1965). Became a member of the CPSU in 1939.
He entered the frontier service in 1932. In 1933 he completed the school for junior staff officers handling service dogs; and in 1937 he graduated from the Central School for Staff Officers Handling Service Dogs in the Border Guard. Between 1933 and 1937, he was a trainer and instructor with service dogs at frontier posts. After 1939 he held various posts in the headquarters of border detachments and frontier military districts. From 1957 to 1961 he served in the headquarters of the border guard of the USSR. In 1961 he entered the reserves.
A skillful tracker, he participated in 120 military encounters with smugglers and diversionary forces; he apprehended 467 persons violating the state frontiers. Decors.: Order of Lenin, two Orders of the Red Banner, the Order of the Red Star, and a number of medals.

Karaulov, Y.
Linguist.
Corresponding mem. USSR Acad. of Sciences 1981–. Academic sec. of the Literature and Language Dept. USSR Acad. of Sciences; expert on general, Russian, and Finno-Ugric linguistics; research on semantics, theory of lexicography, the question of automation of linguistic investigations; originator of the theory of ideography, which laid the foundations of a new concept for the study of the content of language that was named "Thesaurus Semantics"; author of roughly 60 publications.

Karavaev, Georgii Arkadevich
State official; b. 1913.
Russian; mem. CPSU 1940–; grad. Leningrad Inst. of Water Transport Engineers 1935; served in Sov. Army 1935–36; leading engineering and tech. posts 1936–51; head of admin. for construction of Palace of Culture and Sciences Warsaw, USSR Dep. Min. of Construction for Heavy Ind. 1951–54; 1st USSR Dep. Min. of Construction for Metallurgy and Chemical Ind. 1954–57; Dep. Chair. of Sverdlovsk Econ. Council 1957–59; leading posts in USSR State Cttee. of Construction 1959–60; 1st Dep. Chair. of USSR State Cttee. on Construction 1960–61 and 1965–67; Chair. of Union Bank for Financing USSR Capital Investments, USSR Min. 1961–63; 1st Dep. Chair. of USSR State Cttee. of Construction with USSR Council of Mins., USSR Min. 1963–65; Dep. to USSR Supr. Sov. 1966–; USSR Min. of Construction F. 1967–; cand. mem. 1971–1976, mem. Cen. Cttee. CPSU 1976–. Decor.: Order of Red Banner of Labour.

Karavaeva, Anna Aleksandrovna
Author; b. D. 15 (27), 1893, Perm.
Russian; mem. CPSU 1926–; grad. Berstuzhev Advanced Courses for Women 1916; first works publ. 1922; Chief Ed. of journal "Molodaia Gvardiia" 1931–38; correspondent for "Pravda" 1941–43. Decors.: State Prize of the USSR 1951, Order of Lenin, 5 other orders and various medals.
Publs.: "Pervoe pokolenie" (The First Generation) 1959, "Grani zhizni" (Limits of Life) 1963, trilogy "Rodina" (Fatherland), novels "Ogni" (Fires) 1943, "Razbeg" (Running Jump) 1948, "Rodnoi dom" (The Family Home) 1950, "Zvezdnaia stolitsa. Zapiski i vospominaniia sovremennika" (Star Capital. Notes and Reminiscences of a Contemporary) 1968, "Dvor. Povesti i rasskazy" (The Courtyard. Stories) 1969, and others.

Kardin, V. (real name: Kardin, Emil Vladimirovich)
Literary and theatre critic; b. Ag. 3, 1921, Moscow.
Mem. CPSU 1943–; author of numerous articles on contemporary literature, drama, theatre and film.
Publs.: collection "Vernost' vremeni" (Loyalty of Time) 1962, "Sudja po imeni vremia. O geroe sovetskoi dramy" (A Judge Named Time. On the Hero in Soviet Drama) 1964, "Poeziia podviga i podvig poezii" (Poetry of the Heroic Deed and the Heroic Deed of Poetry) 1965, "Oba luchshe" (Both Together is Better) 1965, and others.

Karelin, Lazar Viktorovich
Author; b. 1920, Moscow.
Russian; mem. CPSU 1947–; grad. State Union Film Inst. 1943.
Publs.: "Pis'mo" (The Letter) 1946 (with

L. Gurunts), "Na tikhoi ulitse" (In a Quiet Street) 1954, "V aprele" (In April) 1956, "Obshchezhitie" (Community Lodgings) 1957, "Otkrytyi dom" (An Open House) 1961, "Devochka s kraskami" (A Girl With Colours) 1963, and others.

Karetskas, Liudas Ionovich
State official.
Chair. of State Cttee. for Grain and Mixed Feed at Council of Mins. Lithuanian SSR 1965–69; cand. mem. Cen. Cttee. Lithuanian CP 1966–; Min. of Supplies Lithuanian SSR D. 1969–.

Kargapolov, Mikhail Ivanovich
Scientist; b. N. 9, 1928, Rusakova, Kurgan Oblast.
Mem. CPSU 1965–; grad. Uralsk Univ. Sverdlovsk 1951; corresponding mem. USSR Acad. of Sciences 1966–; Prof. at Novosibirsk Univ. 1964–. Decors.: Order of Red Banner of Labour, several medals.

Karger, Mikhail Konstantinovich, Dr. of Art Sc.
Archaeologist and art historian; b. My. 17 (30), 1903, Kazan.
Grad. Petrograd Univ. 1923; specialist in Slavic-Russian Archaeology and history of Old Russian culture and art; Prof. at Leningrad Univ. 1949; Head of Leningrad Branch of Inst. of Archaeology at USSR Acad. of Sciences 1964–. Decors.: State Prize of the USSR 1952, Order of Lenin.

Karieva, Bernara Rakhimovna
Ballet dancer; b. 1937.
Grad. Choreographic School Moscow 1955; engagements at A. Navoi Opera and Ballet Theatre Tashkent 1955–. Decors.: People's Artist of the USSR 1973, People's Artist of the Uzbek SSR 1964.

Karim, Mustai (real name: Karimov, Mustafa Safich)
Author and poet; b. 1919, Kliashevo, Bashkir ASSR.
Bashkir; mem. CPSU 1944–; grad. Bashkir Pedagogical Inst. 1940; served in Sov. Army 1941–45; Chair. of Bashkir Writers' Union 1951–62; Secr. of RSFSR Writers' Union 1962–; first works publ. 1935–. Decors.: Order of Lenin, State Prize of the USSR 1972, 5 other orders and various medals.
Publs.: "Otriad tronulsia" (The Detachment Moves Off) 1938, "Vesennie golosa" (Spring Voices) 1941, "Dekabrskaia pesnia" (December Song) 1942, "Ulmasbai" 1942–44, "Devushki nashego kolkhoza" (The Girls from Our Kolkhoz) 1948, "Sabantui" 1953, "Evropa-Aziia" (Europe-Asia) 1951–54, "Stikhi o Vietname" (Poems on Vietnam) 1956–67, "Svad'ba prodolzhaetsia" (The Wedding Feast Goes On) 1947, "Odinokaia bereza" (Lonely Birch) 1950, "Pokhishchenie devushki" (Abduction of a Girl) 1959, "Nespetaia pesnia" (A Song Unsung) 1961, "V noch' zatmeniia luny" (On the Night of the Lunar Eclipse) 1964, "Vietnam nedalek" (Vietnam Is Not Far) 1958, "Strana Aigul'" (The Land Aigul) 1969, "Izbrannye proizvedeniia" (Selected Works) vols. 1–2, 1969, "Godam vo sled" (Year by Year) 1971, "Salavat" 1973, and others.

Karimov, Abdukhalik
Party official; b. 1923.
Uzbek; mem. CPSU 1946–; grad. Agricultural Inst. in Samarkand; teacher 1941–42; served in Sov. Army 1942–46; teacher, 1st Sec. of Komsomol Raion Cttee., Chair. of Exec. Cttee. of Sov. of People's Deps. 1946–54; 1st Sec. of Raion Cttee. CP Uzbekistan, Dir. of oblast trust for state farms, instructor of Oblast Cttee. CP Uzbekistan, dir. of a state farm, head of a raion admin. for agriculture 1954–65; 1st Sec. of Raion Cttee. CP Uzbekistan 1965–67; cand. mem. Cen. Cttee. CP Uzbekistan 1966–71; 1st Dep. Min. of Agriculture Uzbek SSR 1967–68; Oblast Cttee. Sec. CP Uzbekistan in Surkhan-Dar'ya F. 1977–; mem. Cen. Cttee. CP Uzbekistan 1981–.

Karimov, Abduvakhid Karimovich
Party official; b. 1932.
Uzbek; mem. CPSU 1959–; grad. Inst. of Engineers of Irrigation and the Mechanization of Agriculture, Cand. of Technical Sciences; hydrologist of raion water resources admin. 1946–56; chief engineer, dept. chief, chief of regional admin. for irrigation systems 1956–65; Dep. Chair. of Oblast Exec. Cttee. of People's Deps., dep. chief of a dept. Cen. Cttee. CP Uzbekistan, 1st Sec. of Raion Cttee. CP Uzbekistan 1965–70; Chair. of Oblast Exec. Cttee. of People's Deps. in Surkhan-Daria 1970–74; mem. Cen. Cttee. CP Uzbekistan 1971 –; Dep. to USSR Supr. Sov. 1974–; 1st Oblast Cttee. Sec. CP Uzbekistan in Surkhan-Daria 1974–77, 1st Oblast Cttee. Sec. in Bukhara F. 1977–.

Karimov, Khanif Karimovich
Author and poet; b. 1910, Verkhnie Kigi, Bashkir ASSR.
Mem. CPSU 1932–; grad. Bashkir Pedagogic Inst. 1938; first works publ. 1930.
Publs.: "Pesnia chasovogo" (The Sentry's Song) 1934, "Devushka" (The Girl) 1935, "Aul" 1940; "Nasha zvezda" (Our Star) 1935, "Odno serdtse" (A Heart) 1936, "Dlia liubvi net smerti" (There is no Death for Love) 1943, "Dorogi voiny" (Streets of War) 1944, "Svetlaia zvezda" (A Light Star) 1949, "Zolotoi kolos" (A Golden Ear) 1950–51, "Liubov Tankhylyu" (Tankhylyu's Love) 1955–58, and others.

Karitskii, Konstantin Dionisievich
Military official; b. S. 13 (26), 1913, Zheltaia Reka, the present-day city of Zheltye Vody, Dnepropetrovsk Oblast.
One of the organizers of the partisan movement in Leningrad Oblast during the Great Patriotic War of 1941–45; a colonel. Hero of the Soviet Union (Ap. 2, 1944). Member of the CPSU since 1932. He served with the border troops between 1932 and 1940. When the Great Patriotic War began he became commander of the 55th Fighter Battalion. From December 1941 through March 1944, he led a partisan detachment and the Fifth Partisan Brigade at the enemy's rear in Leningrad Oblast. In the course of combat operations his brigade killed more than 2,000 German soldiers and officers, disabled 15 tanks and armored vehicles, 18 locomotives, and 160 boxcars and flatcars, and blew up 29 bridges. During the Leningrad-Novgorod offensive of 1944, his brigade captured important sections of the Nikolaevo-Gorodets and Utorgosh-Nikolaevo highways and held them until Soviet troops arrived. Since the war, he has worked in state security agencies and, subsequently, as deputy director of the Museum of the Defense of Leningrad. Decors.: Order of Lenin, the Order of the Red Banner, two Orders of the Patriotic War First Class, and various medals.

Karkarashvili, Shalva Vasil'evich
Party official; b. 1922.
Sec. of Tbilisi City Cttee. Georgian CP 1968–69; Chief Ed. of newspaper "Zarja Vostoka" 1969–73; mem. Cen. Cttee. 1971–, cand. mem. Bureau Cen. Cttee. and Head of Dept. of Party Org. Work with Cen. Cttee. Georgian CP 1973–75; Chair. of Party Comm. with Cen. Cttee. Georgian CP D. 1975–.

Karklin, Mirdza Iakovlevna
State official.
Sec. of Cen. Cttee. Latvian Komsomol 1955–59; Chief Ed. of youth newspaper "Padomiu iaunatne" and mem. Bureau Cen. Cttee. Latvian Komsomol 1960–63; Dep. Min. of Culture Latvian SSR 1964; Head of Dept. of Science and Culture with Cen. Cttee. Latvian CP 1964–69; mem. Cen. Cttee. Latvian CP 1966–; Min. of Educ. Latvian SSR My. 1969–.

Karklinsh, Nikolai Nikolaevich
Trade union official.
Chair. of State Film Cttee., with Council of Mins. Latvian SSR N. 1963–76; mem. Auditing Comm. Latvian CP 1963–76 and 1981; Sec. and mem. Presidium of Latvian Republican Trade Union Council Ja. 1982–.

Karlov, Vladimir Alekseevich
Party official; b. 1914.
Russian; mem. CPSU 1940–; grad. Voronezh Zootech. Inst. 1933 and Higher School of Cen. Cttee. CPSU 1948; zoo technologist, head of stock-raising dept. of an oblast agricultural board, sec. of a raion Komsomol cttee. 1933–41; Instructor, Dep. Dept. Head with Volgograd Ob-

last All-Union Cttee. CP(B) 1941–46; worked for All-Union Cen. Cttee. CP(B) 1948–49; Sec. 1949–51, 2nd Sec. of Volgograd Oblast Cttee. CPSU 1951–53; 2nd Sec. of Balashov Oblast Cttee. CPSU 1954; worked for Cen. Cttee. CPSU 1954–59; 1st Sec. of Kalinin Oblast Cttee. CPSU 1959–60; Dep. to and Chair. of Agricultural Comm. of RSFSR Supr. Sov. 1959–63; mem. staff Cen. Cttee. CPSU 1960–62; cand. mem. Cen. Cttee. CPSU 1961–76, mem. O. 1976–; Dep. to 1962–, Chair. of Agricultural Comm. at Sov. of Nationalities USSR Supr. Sov. 1966–; 2nd Sec. and mem. Bureau of Cen. Cttee. Uzbek CP 1962–65; mem. RSFSR Bureau at Cen. Cttee. CPSU 1965–Ap. 1966; 1st Dep. Head of Agricultural Dept. at Cen. Cttee. CPSU My. 1966–76, Head My. 1976–; Decors.: Order of Lenin, Order of Red Banner of Labour, various medals.

Karm, Kaarel Iukhanovich (pseudonym of Kaarel Länts)
Esthonian actor; b. O. 5 (18), 1906, Narva.
People's Artist of the USSR (1956).
From 1925 to 1949, he was an actor in the Esthonia Theater. He joined the V. Kingissepp Esthonian Theater of Drama in Tallinn in 1949. Among his best parts in the classical repertoire have been Hamlet, Othello, and Antony in Shakespeare's *Hamlet, Othello,* and *Antony and Cleopatra,* respectively; Mephistopheles in Goethe's *Faust;* Arbenin in Lermontov's *Masquerade;* Protasov in L. N. Tolstoy's *The Living Corpse;* Juhani in Kivi's *Seven Brothers;* and Jan in Kitzberg's *Whirlwinds.* Among the parts he has played in Esthonian dramas are Ralf in *Life in the Citadel* and Allan in *The Jackals,* both by Jakobson; Kustas Lokk in *Conscience* and Mart Tuisk in *Prodigal Son,* both by Rannet; Jönn in Schmuul's *Wild Captain;* and Clever Ants in Tammsaare's *New Satan From Põrgupõhja.* In plays by other Soviet writers, he has performed as Borovskii in Lavrenev' *To Those At Sea!* and as Stryzhen' in Korneichuk's *The Destruction of the Squadron.* He is a master of character portrayal. He has a forceful stage presence and good stage delivery. He has also appearad in movies. Decors.: State Prize of the Esthonian SSR in 1949 and 1959.

Karmaliuk, Pavel Petrovich
Ukrainian singer (baritone); b. D. 24, 1907 (Ja. 6, 1908), Osovtsy, in present-day Korostyshev Raion, Zhitomir Oblast.
People's Artist of the USSR (1960). Member of the CPSU since 1950.
He graduated from the Kiev Conservatory in 1941. He became a soloist at the L'vov Theater of Opera and Ballet in 1944. Since 1950 he has taught a course in singing at the L'vov Conservatory, becoming a professor in 1970. Among his best parts have been Griaznoi in Rimsky-Korsakov's *The Tsar's Bride.* Onegin in Tchaikovsky's *Eugene Onegin,* the Demon in Rubinstein's *Demon,* Prince Igor in Borodin's *Prince Igor,* Rigoletto and Germont in Verdi's *Rigoletto* and *La Traviata,* respectively, and Bogdan in Dan'kevich's *Bogdan Khmel'nitskii.* He also performs in concerts. Decors.: Order of the Badge of Honor and a number of medals.

Karmanov, Ivan Vasilevich
State official.
Chair. of State Cttee. for Supervision of Safe Working Practices in Industry and for Mine Kirghiz SSR D. 1981–.

Karmyshev, Lakai
State official.
Mem. Cen. Cttee. Tadzhik CP 1971–; Chair. of Cttee. for People's Control Tadzhik SSR N. 1972–; mem. Bureau Cen. Cttee. Tadzhik CP 1976–.

Karpachev, Sergei Vasil'evich, Dr. of Chem. Sc.
Chemical scientist; b. 1906.
Mem. CPSU 1944–; grad. Uralsk Polytechn. Inst. 1930; Dr. of Chem. Sc. 1940; Prof. 1941; corresponding mem. USSR Acad. of Sciences 1970–; Dir. of Inst. of Electrochemistry at Uralsk Branch of USSR Acad. of Sciences (since 1971 Inst. of Electrochemistry of Uralsk Research Centre) 1963–.

Karpeko, Vladimir Kirillovich
Author and poet; b. 1922, Kazatin.
Russian; participated in Sov.-Finnish War and World War II; first works pub. 1947.
Publs.: poems "Ot serdtska k serdtsu" (From Heart to Heart) 1956, "Litsom k ogniu" (With the Face to the Fire) 1959, "Shirokaia doroga" (The Broad Way) 1950, "Pro mal'chika Petiu" (About a Boy Called Petia) 1954, and others.

Karpiuk, Aleksei Nikoforovich
Author; b. 1920, Strashevo, now Gomel Oblast (W. Belorussia); s. of a peasant.
Belorussian; participated in revolutionary movt. in Poland, sentenced to prison twice (1934 and 1936), partisan in World War II.
Publs.: "V odnom institute" (In an Institute) 1953, "Dve sosny" (Two Pines) 1958; "Danuta" 1959; "Na lesnykh stezhkakh" (On Forest Paths) 1964, and others.

Karpov, Vladimir Borisovich
Author; b. 1912, Khvalynsk, Saratov Oblast.
Belorussian; has lived in Belorussia since 1921, worked in a sawmill and as teacher; fought as partisan in Belorussia. Decors.: Order of Red Star, medals.
Publs.: "Bez neitral' noi polosy" (Without a Neutral Zone) 1950, "Vesennie livni" (Spring Reins) 1961, "Nemigi krovavye berega" (The Bloody Banks of the Nemiga) 1962, and others.

Karpova, Evdokiia Fedorovna
State official; b. 1923.
Russian; mem. CPSU 1952–; grad. Moscow Textile Inst. 1949; forewoman of weaving mill of a combine in Egorevsk 1949–56; 1st Sec. of Egorevsk City CPSU Cttee., in Moscow Oblast CPSU Cttee. and Cen. Cttee. CPSU, USSR Dep. Min. of Light Ind. 1956–66; Dep. Chair. of Council of Mins. RSFSR 1966–; cand. mem. Cen. Cttee. CPSU 1966–76, mem. 1976–. Decor.: Order of October Revolution.

Karryev, Akhmed
State official.
Head of Admin. for Geology Turkmen SSR O. 1978–; cand. mem. Cen. Cttee. Turkmen CP 1981–.

Karryev, Baimukhamed Atalievich
Literary scholar; b. D. 22, 1914, Geokcha, in present-day Ashkhabad Oblast, Turkmen SSR.
Academician of the Academy of Sciences of the Turkmen SSR (1965). Member of the CPSU since 1949.
He has been a professor in the subdepartment of Turkmen literature at the M. Gorkii Turkmen University since 1969. His doctoral dissertation was entitled *Makhtumkuli and His Patriotism* (1943). He is the author of the book *Zelili and His Times* (1943), and has also written pamphlets, articles devoted to Turkmen literature, and textbooks for Turkmen schools. He takes part in the publication of texts of Turkmen classical literature and folklore. Decors.: Order of the Red Banner of Labor, medals.
Work: *Epicheskie skazaniia o Ker-ogly u tiurkoiazychnykh narodov.* Moscow 1968.

Karryev, Chary Soiunovich
State official; b. 1932.
Turkmen; mem. CPSU 1958–; grad. Moscow Gubkin Petroleum Inst. 1955; operator, shop engineer, installation chief, shop superintendent of Krasnovodsk Oil-Refining Plant 1955–65; Chief of Petroleum and Chemical Ind. Dept. of Cen. Cttee. Turkmen CP 1965–74; mem. Auditing Comm. Turkmen CP 1966–71; cand. mem. Cen. Cttee. Turkmen CP 1971–76, mem. 1976–; 1st Sec. of Ashkhabad Oblast Cttee. Turkmen CP 1974–76; mem. Bureau of Cen. Cttee. 1976–, Sec. of Cen. Cttee. Turkmen CP 1976–78; Min. of Foreign Affairs Turkmen SSR 1978–79; Chair. of Council of Mins. Turkmen SSR D. 1978–; mem. Cen. Auditing Comm. CPSU 1981–.

Karryeva, Roza Mamedovna
State official.
Dir. of Inst. of Party History at Cen. Cttee. Turkmen CP (a branch of Inst. of Marxism-Leninism at Cen. Cttee. CPSU); Dep. to Chair. of Presidium Supr. Sov. Turkmen SSR 1975–.

Kartvelishvili, Dmitrii Levanovich
State official; b. 1927.
Georgian; mem. CPSU 1956–; grad. Georgian Polytechnical Inst. 1951; construction engineer, chief construction engineer, chief engineer at Ordzhonikidze Automobile Factory in Kutaisi 1951–71; Chair. of City Exec. Cttee. of People's Deps. in Kutaisi 1971–75; chief of Dept. of Ind., Transport, and Communications of Georgian Council of Mins. 1975–76; 1st Dep. Chair. of GOSPLAN Georgian SSR 1976–78; Chief of Planning and Finance Dept. Cen. Cttee. CP Georgia 1978–79; Dep. Chair. of Council of Mins. and Chair. of GOSPLAN Georgian SSR 1979–82; mem. Cen. Cttee. 1981–, mem. Bureau Cen. Cttee. CP Georgia Jl. 1982–; Chair. of Council of Mins. Georgian SSR Jl. 1982–.

Kasatonov, Vladimir Afanas'evich
Military official; b. Jl. 8 (21), 1910, Peterhof, now Petrodvorets, Leningrad Oblast; s. of a worker.
Admiral of the fleet (1965), Hero of the Soviet Union (N. 25, 1966). Member of the CPSU since 1939.
He joined the navy in 1927. He graduated from the M. V. Frunze Naval School in 1931, from the course preparing commanders at the S. M. Kirov Submarine Training Center in 1932, and from the Naval Academy in 1941. After serving as deputy commander and commander of a submarine from 1933 to 1938 and as commander of a submarine division of the Pacific Fleet from 1938 to 1940, he worked on the Chief Naval Staff from 1941 to 1945 and held staff positions in the Baltic Fleet and on the General Staff from 1945 to 1949. He has held the posts of first deputy commander and chief of staff of the Pacific Fleet (from 1949), commander of the Red Banner Baltic Fleet (from 1954), commander of the Black Sea Fleet (from 1955), commander of the Northern Fleet (from 1962), and first deputy commander in chief of the Navy (from Je. 1964–74). He was elected a deputy of the fifth through eighth convocations of the Supreme Soviet of the USSR. Decors.: three Orders of Lenin, the Order of the Red Banner, the Order of the Patriotic War First Class, the Order of Nakhimov Second Class, two Orders of the Red Banner of Labor, two Orders of the Red Star, and medals.

Kasendeev, Ilias
State official.
Mem. Cen. Cttee. Kirghiz CP 1976–; Chair. of State Cttee. on Vocational Training Kirghiz SSR Mr. 1977–.

Kasianenko, Oleg Iakovlevich
State official.
Min. of Light Ind. Ukrainian SSR My. 1972–; mem. Auditing Comm. Ukrainian CP 1976–81; cand. mem. Cen. Cttee. Ukrainian CP 1981–.

Kasianenko, Vladimir Grigorevich, Dr. of Biol. Sc.
Veterinary anatomist and morphologist; b. N. 16, 1901, Velykii Ianisol, now Velyka Novosilka, Donetsk Oblast; s. of a teacher.
Mem. CPSU 1945–; grad. Veterinary-Zoological Inst. Kiev 1926; Dr. of Biol. Sc. and Prof. 1938; mem. staff Veterinary-Zoological Inst. Kiev 1926–50; concurrently Head of Dept. of Evolutionary Morphology at Inst. of Zoology of Ukrainian Acad. of Sciences 1947–50; Dir. of same inst. 1950–62; Scientific Consultant at Dept. of Evolutionary Morphology at Inst. of Zoology of Ukrainian Acad. of Sciences 1964–; Decors.: Order of Lenin, Order of Red Star, Order "Honorary Badge", medals.
Publs.: author of 64 scientific works on questions of comparative and evolutionary morphology of the musculoskeletal system of vertebrates and functional anatomy of organs and systems of animals; "Apparat dvizheniia i opory loshadi" 1947, "Zakonomernosti prisposobitelnykh preobrazovanii sustavov konechnostei mlekopitaiaushchikhsia" in "Zoologicheskii zhurnal" No. 3/1954, "Einige Gedanken zum Inhalt der funktionalen Morphologie" in "Naturwissenschaftliche Beiträge" No. 9/1957, "A forming influence of the type of support and function of the limbs of mammals" in "XVI International Congress of Zoology", vol. 2, Washington 1963.

Kasimova, Valentina Aleksandrovna
Laborer; b. 1944.
Mem. CPSU 1972–; grad. secondary school 1961; fitter at Tula Plant for Electroelements 1961–; mem. Cen. Auditing Comm. CPSU 1981 –.

Kask, E. B.
State official.
Chair. of State Cttee. for Supervision of Safe Working Practices in Industry and for Mine Esthonian SSR Mr. 1982–.

Kassian (S. N. Iaroslaviskii)
Russian Orthodox clergyman; b. Ag. 22, 1899, Zolotoruche, Iaroslavl Oblast; s. of a priest.
Completed studies at a theological educ. inst. 1915, then 3 classes at Kashin Theological Seminary; studied Law Faculty of Iaroslavl Univ.; mil. service 1919–20; attended theological course in Uglich 1920–22, then preacher; ordained priest 1925; monk 1943; abbot 1949; finished 4th class of Leningrad Theological Seminary by correspondence course 1953; until 1956 main clergyman of various churches in Uglich Oblast, Substitute Bishop of Iaroslavl Diocese; Archbishop of Novosibirsk 1963–64 and of Kostroma and Galich My. 1964–.

Kasumov, Imran Ashum ogly
Author; b. 1918, Baku.
Azerbaidzhani. Decor.: Lenin Prize.
Publ.: filmscripts "Podzemny gul" (Underground Rumble) 1937, "Mechta" (A Dream) 1938, "Novyi gorizont" (New Horizon) 1940, "Na dal'nikh beregakh" (On Distant Shores) 1954, "More liubit otvazhnykh" (The Sea Loves the Brave) 1954, "Prokhodiat gody" (The Years Go By) 1955, "Povest' o neftianikakh Kaspiia" (The Caspian Oil Workers' Story) 1961, "Pokoriteli moria" (Victor of the Sea) 1961, and others.

Kasumov, Taliat Ali ogly
State official.
Min. of Health Azerbaidzhan SSR Je. 1979–; mem. Cen. Cttee. Azerbaidzhan CP 1981–.

Kasumova, Svetlana Chingiz kyzy
State official.
Min. of Trade Azerbaidzhan SSR Je. 1979–; mem. Cen. Cttee. Azerbaidzhan CP 1981–.

Kasybaev, Kakimzhan
Party official.
Dep. Chair. State Cttee. for Printing and Publishing Kazakh SSR 1972–74; Dir. of Information Agency with Council of Mins. Kazakh SSR 1974–82; mem. Auditing Comm. Kazakh CP 1976–81; cand. mem. Cen. Cttee. Kazakh CP 1981–82, Sec. and mem. Bureau Mr. 1982–.

Kasymkanov, Aubakir Kasymkanovich
State official.
Dep. to Supr. Sov. Kazakh SSR 1959–; chair. of exec. cttee. of a raion sov. 1959; 2nd Sec. of Kustanai Oblast Cttee. Kazakh CP 1961–67 (for Agriculture 1963–64); cand. mem. Cen. Cttee. Kazakh CP 1966–81; 1st Dep. Chair. 1967–75, Chair. of State Cttee. for Utilization of Labour Reserves with Council of Mins. Kazakh SSR 1975–76; Chair. of State Cttee. for Labour Kazakh SSR S. 1976–.

Kasymova, Adolat Timurovna
State official; b. 1938.
Tadzhik; grad. Textile Inst. Moscow; Master at Repairs Dept. 1960–61 and Shift Master at Cocoon-Unravelling Dept. in Silk Combine in Dushanbe 1961–64; Dep. to USSR Supr. Sov. 1962–; Dir. of clothes factory "Guliston" in Dushanbe 1964–74; mem. Cen. Cttee. Tadzhik CP 1966–; Min. of Public Services Tadzhik SSR Ag. 1974–.

Kataev, Opokhon Kataevich
State official.
Min. of Trade Tadzhik SSR Jl. 1981–.

Kataev, Semen Isidorovich, Dr. of Tech. Sc.
Television scientist; b. Ja. 27 (F. 9), 1904, in the *posad* (suburb) of Elionka, in what is now Starodub Raion, Briansk Oblast.

Doctor of Technical Sciences (1951), professor (1952), Honored Scientist and Engineer of the RSFSR (1968). Member of the CPSU since 1964.
He graduated from the N. E. Bauman Moscow Higher Technical School in 1929. He developed the first high-vacuum receiving television tube in the USSR (1931–32) and invented a cathode-ray transmitting tube having charge storage – the prototype of the modern iconoscope (1931). He proposed and was the first to produce under laboratory conditions a system for transmitting a high-definition television picture on a narrow-band channel – the slow-scan television, or transmission with a frame of prolonged duration (1934–38). He developed a method (1965–70) of transmitting the sound portion of television programs within the frequency band of the video signal. He studied the problems of converting television standards (1964–70). Decors.: Order of the Red Banner of Labor and medals.
Works: *Elektronno-luchevye televizionnye trubki.* Moscow 1936. *Osnovy televideniia.* Moscow 1940. (Editor.) *Generatory impul'sov televizionnoi razvertki.* Moscow-Leningrad 1951.

Kataev, Turgabek Kataevich
State official; b. 1919.
Mem. CPSU 1942–; grad. CPSU Cen. Cttee. Party Higher School 1957, CPSU Cen. Cttee. Acad. of Social Sciences 1963; Cand. of Hist. Sciences 1963; served in Sov. Army 1941–44; instructor, chief of a dept. of Raion Cttee. CP Kazakhstan 1944–50; instructor, branch chief of Dept. of Propaganda and Agitation Oblast Cttee. CP Kazakhstan in South Kazakhstan and Pavlodar 1950–60; dep. chief of Dept. of Propaganda and Agitation Cen. Cttee. CP Kazakhstan, Sec. Oblast Cttee. CP Kazakhstan in Alma-Ata, chief of Dept. of Science and Educational Insts. Cen. Cttee. CP Kazakhstan 1963–74; Min. for Higher and Secondary Specialized Ed. Kazakh SSR D. 1974–; mem. Cen. Cttee. CP Kazakhstan 1976–.

Kataev, Valentin Petrovich
Author; b. Ja. 16 (28), 1897, Odessa; s. of a teacher, brother of the Russian writer E. P. Petrov.
Mem. CPSU 1958–; first works publ. 1910; served at the front 1915–17; fought against the Whites during Civil War and October Revolution, worked for Russian Wire Service; contributor to journal "Krokodil" and newspaper "Gudok" 1923–; Chief Ed. of journal "Iunost". Decors.: State Prize of the USSR 1946, 2 Orders of Lenin, 2 other orders, various medals; Hero of Socialist Labour 1974.
Publs.: "Syn polka" (Son of the Regiment) 1945, "Beleet parus odinokii" (The White Shimmer of a Lonesome Sail) 1936, "Malen'kaia zheleznaia dver' v stene" (A Small Iron Door in the Wall) 1964, "Sviatoi kolodets" (The Holy Well) 1967, "Kubik" (Little Cubic) 1969, "Sobranie sochinenii" (Collected Works) vols. 1–9, 1968–72, and others.

Katanian, Vasilii Abgarovich
Literary historian; b. Ap. 15 (28), 1902, Moscow.
Studied Tbilisi Polytech. Inst.
Publ.: "Maiakovskii. Literaturnaia khronika" (Maiakovskii. Literary Chronicle) 1945, 2nd ed. 1948, 3rd ed. 1956, 4th ed. 1961, play "Oni znali Maiakovskogo" (They Knew Maiakovskii) 1954, script for the film "Anna Karenina" (based on the novel by Tolstoi) 1963, "Iz vospominanii o V. V. Maiakovskom" (From Recollections About Maiakovskii) 1963, "O sochinenii memuarov" (On the Writing of Memoirs) 1964, and others.

Katrushenko, Ivan Gordeevich
Graphic artist; b. 1925, Nizhnee, Voroshilovgrad Oblast.
Mem. CPSU 1949–; grad. Lvov Inst. for Applied and Decorative Art 1953, pupil of R. Selskii.

Katsiev, Khabu Khadzhikurmanovich (pseudonym: Khabib)
Balkar writer; b. My. 13, 1916, in the *aul* (village) of Gundelen, present-day Baksan Raion, Kabarda-Balkar ASSR.
Member of the CPSU since 1939.
He graduated from the Higher Party School. He is the author of lyrical poems (the collection *Heartfelt Joy,* 1936) and stories (*Stars of the Earth,* 1940), depicting the struggle of the working people of the mountains to achieve Soviet power and to build a new life. In his collection of humorous and satirical short stories *What's New* (1961), he effectively employs elements of the grotesque and popular speech. He has also written the collection of novellas and stories *Muhammad* (1964) and the novel *Tamata* (1971).
Works: *In Russian translation: V gornom aule.* Nal'chik 1966. *Nasmeshnik Omar.* Moscow 1969.

Katsnelson, Abram Isaakovich
Author and poet; b. 1914, Gorodnia, Chernigov Oblast; s. of a teacher.
Ukrainian; mem. CPSU 1945–; grad. Kiev Univ. 1939; first works publ. 1935. Decors.: Order of Patriotic War 2nd Cl., Order of Red Star, medals.
Publs.: "Kapli solntsa" (Drops of Sun) 1935, "Sozrevaiut plody" (Fruits Are Ripening) 1938, "Perednii krai" (The Main Line) 1947, "Vo imia zhizni" (In Life's Name) 1953, "Pogozhie dni" (Serene Days) 1956, "Riadom s toboi" (By Your Side) 1959, "Rakety i solovi" (Rockets and Nightingales) 1960, "Stredi buri" (In the Midst of a Storm) 1963, and others.

Katulin, Aleksei Zakharovich
Athlete, coach, and teacher; b. F. 6 (19), 1906, Moscow.
Honored Master of Sports (1936), Honored Coach of the USSR (1956), and Honored Cultural Worker (1972). Became a member of the CPSU in 1931.
From 1933 to 1935, he was the USSR champion in Greco-Roman wrestling. From 1937 to 1939 and from 1958 to 1972 he was the head of the department of wrestling at the State Central Institute for Physical Education. In 1947, he became the chairman of the Wrestling Federation of the USSR. From 1956 to 1970 he served as the vice-president of the International Amateur Wrestling Federation (FILA); he has been an honorary vice-president of FILA since 1970. Decors.: three orders and various medals.
Works: *Bor'ba klassicheskaia i vol'naia.* Moscow, 1952. (In collaboration with N. M. Galkovskii and N. G. Chionov.) *Klassicheskaia bor'ba.* Edited by A. Z. Katulin, Moscow 1962. *Sportivnaia bor'ba.* Edited by N. M. Galkovskii and A. Z. Katulin. Moscow 1968.

Katushev, Konstantin Fedorovich
Diplomat; b. O. 1, 1927, Boldino, Gorkii Oblast.
Mem. CPSU 1952–; grad. Gorkii Polytech. Inst. 1951; designer, Sec. of CPSU Bureau of a dept. at Gorkii Automobile Plant 1951–59; 2nd Sec. of Gorkii Avtozavod Raion Cttee. CPSU, 1st Sec. of Gorkii City Cttee. CPSU 1959–65; 1st Sec. of Gorkii Oblast Cttee. CPSU D. 1965–Ap. 1968; mem. Cen. Cttee. CPSU 1966–; Dep. to 1966–, mem. 1966–67, Chair. of Comm. for Construction and Building Materials 1967–70, mem. Foreign Affairs Comm. at Sov. of the Union USSR Supr. Sov. 1970–74; Sec. of Cen. Cttee. CPSU Ap. 1968–77; Dep. Chair. of USSR Council of Min. 1977–82; Permanent USSR Representative at COMECON 1977–82; USSR Amb. to Cuba Jl. 1982–. Decor.: Order of Lenin.

Kaupuzh, Vladimir Il'ich
State official.
Min. of Culture Latvian SSR Ja. 1962–; Dep. to Supr. Sov. Latvian SSR 1963–; cand. mem. 1963–66, mem. Cen. Cttee. Latvian CP 1966–.

Kavalerov, Kirill Vladimirovich
Mining engineer and geophysicist; b. 1930.
Mem. CPSU 1967–; grad. Petroleum Inst. Moscow 1953; Chief Engineer of Trust "Iamalneftegazgeofizika" 1968–. Decor.: Lenin Prize for participation in exploration of large natural gas resources in Northern Siberian Raions 1970.

Kaverin, Veniamin Aleksandrovich
Author; b. Ap. 6 (19), 1902, Pskov.
Russian; grad. Inst. of Oriental Languages 1923 and Dept. of History and Philology of Leningrad Univ. 1924. Decors.: 3 orders and various medals.
Publs.: "Ispolnenie zhelanii" (Fulfilment of Wishes) vols. 1–2, 1934–36, "Dva kapitana" (Two Captains) vols. 1–2, 1938–44

(42 reprints publ.), "Otkrytaia kniga" (An Open Book) 1949–56, "Zdravstvui, brat, pisat' ochen' trudno" (Greetings, Brother, Writing Is Hard For Me) 1965, "Dvoinoi portret" (Double Portrait) 1966, "Shkolnyi spektakl'" (A School Performance) 1968, "Pered zerkalom" (In Front of a Mirror) 1971, "V starom dome" (In the Old House) 1971, and others.

Kavun, Vasilii Mikhailovich
State official; b. 1928; s. of a peasant.
Ukrainian; mem. CPSU 1954–; grad. Uman Agricultural Inst. 1953; agronomist at the "Fortieth Anniversary of October" Kolkhoz, Chair. of "Kuibyshev" Kolkhoz, Dzhulinskii Raion; Chair. of "22nd Party Congress" Kolkhoz in Berzhadskii Raion, Vinnitsa Oblast 1958–70; mem. Cen. Cttee. CPSU 1961–; Dep. to 1962–, mem. Presidium USSR Supr. Sov. 1962–74; mem. Cen. Cttee. Ukrainian CP 1971–; Chair. of Vinnitsa Oblast Exec. Cttee. My. 1970–78; 1st Sec. of Zhitomir Oblast Cttee. Ukrainian CP 1978–. Decor.: Order of Lenin.

Kazakevich, V. V.
Komsomol official.
Sec. and mem. Bureau Cen. Cttee. Belorussian Komsomol Mr. 1982–.

Kazakov, Iurii Pavlovich
Author; b. 1927, Moscow.
Russian; grad. Gorkii Inst. of Literature Moscow 1958; first works publ. 1952.
Publs.: "Man'ka" 1958, "Po doroge" (On the Road) 1961, "Na polustanke" (At the Stop) 1956, "V gorod" (To Town) 1961, "Goluboe i zelen" (Blue and Green) 1963, "Legkaia zhizn'" (Easy Life) 1962, "Adam i Eva" 1962, and others.

Kazakov, Konstantin Petrovich
Military official; b. N. 5 (18), 1902, Tula; s. of a worker.
Marshal of artillery (1962). Became a member of the CPSU in 1920.
He was a worker before he joined the Soviet Army in 1921. He graduated from the VTsIK Combined Military School in 1923, from the M. V. Frunze Military Academy in 1936, and from advanced academic courses of the Military Academy of the General Staff in 1948. In the Great Patriotic War (1941–45), he began as a commander of a howitzer artillery regiment; in June 1941 he became chief of operations of the artillery department of the Southwestern Front, and in April 1942 he moved to the staff of the chief of the artillery of the Red Army and in April 1943 to the staff for especially important assignments of the commander of the artillery of the Red Army. From April 1944 to May 1945 he commanded the artillery of the Second Shock Army. He participated in the liberation of the Baltic area and in battles in East Prussia and was an army artillery commander in the war with imperialist Japan.
Holding high positions after the war, he was commander of the rocket troops and of the artillery of the ground forces from 1963 to 1969 and was appointed military inspector-adviser in July 1969. Decors.: three Orders of Lenin, the Order of the October Revolution, four Orders of the Red Banner, two Orders of Suvorov Second Class, Orders of Kutuzov First and Second Class, the Order of Bogdan Khmel'nitskii First Class, three foreign orders, and medals.

Kazakov, Leonid Davydovich
Laborer; b. 1951.
Mem. CPSU 1974–; grad. secondary school 1968, studied at Irkutsk Inst. of Rail Transport; laborer at "Angarstroi" 1968–69; served in Sov. Army 1969–72; installator and concrete worker, geodesy technician, sec. of a Komsomol cttee. USSR 1972–74; brigade leader of installation and concrete workers of the "Angarstroi" construction and assembly admin. 1974–, of the "LenaBAMstroi" Trust 1980–; mem. Cen. Auditing Comm. CPSU 1976–81; mem. Cen. Cttee. CPSU 1981–. Decor.: Hero of Socialist Labor 1981.

Kazakov, Mikhail Il'ich
Military official; b. S. 26 (O. 9), 1901, Velikusha, in present-day Kichmengskii Gorodok Raion, Vologda Oblast; s. of a peasant.
General of the army (1955). Became a member of the CPSU in 1919.
He joined the Soviet Army in 1920 and served in the Civil War as a political personnel soldier on the Southern Front. He graduated from an advanced cavalry training school for commanders in 1927, from the M. V. Frunze Military Academy in 1931, and from the Military Academy of the General Staff in 1937. He served in command and staff positions and was chief of staff of the Middle Asian Military District from 1938 to 1941. In the Great Patriotic War (1941–45), he was chief of staff of the Briansk Front (January to July 1942) and of the Voronezh Front (July 1942 to February 1943) and commander of the Sixty-ninth Army (February to March 1943). He then became assistant commander of the troops of the Reserve and Steppe fronts and deputy commander of the troops of the Briansk and Second Baltic fronts (April to December 1943), after which he commanded the Tenth Guards Army from January 1944 until the end of the war. He fought on the Don and in the battle of Kursk and participated in the liberation of the Ukraine and the Baltic.
Holding command and staff positions after the war, he was commander of the troops of the Ural Military District (1953–56), deputy commander-in-chief of the ground forces (1956), commander of the Southern Force Group (December 1956 to 1960), commander of the troops of the Leningrad Military District (1960–65), chief of staff of the Joint Armed Forces of the countries of the Warsaw Treaty Organization (1965–68), and military inspector-adviser (since 1968). He was a deputy to the second and to the fourth through seventh convocations of the Supreme Soviet of the USSR and a candidate member of the Central Committee of the CPSU from 1961 to 1971. He wrote *Over the Maps of Past Battles* (1965). Decors.: two Orders of Lenin, the Order of the October Revolution, four Orders of the Red Banner, Orders of Suvorov First and Second Class, the Order of Kutuzov First Class, two Orders of the Red Star, two foreign orders, and medals.

Kazakov, Nikolai (Miklai) Ivanovich
Poet; b. 1918, Kiutiuk-Kiner, Mari ASSR.
Mem. CPSU 1943–; grad. Gorkii Inst. of Literature Moscow 1955; first works publ. 1934.
Publs.: "Poeziia – liubimaia podruga" (Poetry – Beloved Friend) 1950, and others.

Kazakov, Vasilii Ivanovich
State official; b. 1927.
Mem. CPSU 1947–; grad. Tech. College of Machine-Tool Construction in Egorevsk 1944 and Union Inst. of Machine Building 1953; master, constructor, dept. head at Sverdlov Machine-tool Construction Factory in Leningrad 1944–54; party official 1954–; 2nd Sec. of Leningrad Oblast CPSU Cttee. 1970–73; mem. Cen. Cttee. CPSU 1971–81, cand. mem. 1981–; Chair. of Leningrad City Sov. 1973–76; Dep. to USSR Supr. Sov. 1974–; Dep. Chair. RSFSR Council of Mins. My. 1976–. Decor.: Order of Red Banner of Labour.

Kazakova, Rimma
Author and poet; b. 1932, Sevastopol.
Russian; grad. Faculty of History of Leningrad Univ. 1954; worked as reader and editor at Khabarovsk Studio for Film Chronicles; first works publ. 1955.
Publs.: "Vstretimsia na Vostoke" (We Shall Meet in the East) 1958, "Tam, gde ty" (Where You Are) 1960, "Stikhi" (Poems) 1962, and others.

Kazanavichius, Vilius-Antanas Antanovich
State official.
Mem. Auditing Comm. 1976–81, mem. Cen. Cttee. Lithuanian CP 1981–; Dep. Chair. Council of Mins. Lithuanian SSR F. 1980–.

Kazanets, Ivan Pavlovich
State official; b. 1918, Lotsmano-Kamianka, Dnepropetrovsk Oblast; s. of a peasant.
Ukrainian; mem. CPSU 1944–; grad. Novokuznetsk Metallurgical Inst. 1944; after completing ind.-tech. training in Dnepropetrovsk electrician, foreman, engineer, Dept. Head at Kuznetsk Steel Combine 1937–44; shift foreman, sec. of

CPSU Cttee., Dept. Head at Enakievo Steel Plant 1944–50; organizer of Cen. Cttee. CPSU at same plant 1950–52; 1st Sec. of Enakievo City Cttee. Ukrainian CP 1952; mem. Auditing Comm. Ukrainian CP 1952–56; 1st Sec. of Donetsk Oblast Cttee. Ukrainian CP 1953–60; Dep. to 1954–, mem. Credentials Comm. with Sov. of the Union USSR Supr. Sov. 1962–66; cand. mem. 1956–61; mem. Cen. Cttee. CPSU 1961–; mem. Cen. Cttee. Ukrainian CP 1956–66; Dep. to 1959–65; mem. Presidium Supr. Sov. Ukrainian SSR Ap.–Jl. 1963; 2nd Sec. 1960–63, mem. Presidium Cen. Cttee. Ukrainian CP 1960–66; Chair. of Council of Mins. Ukrainian SSR 1963–65; USSR Min. of Ferrous Metallurgy O. 1965–; Chair. of Permanent Comm. for Ferrous Metallurgy with COMECON. Decors.: 2 orders of Lenin, Order of the October Revolution.

Kazanskii, Vladimir Borisovich, Dr. of Chem. Sc.
Physical chemist; b. 1913.
Mem. CPSU 1962–; grad. Moscow Univ. 1954; Dr. of Chem. Sc. 1967; Prof. 1972; corresponding mem. USSR Acad. of Sciences 1974–; postgraduate studies at Moscow Univ. 1954–57; Research Asst. at Inst. of Chemical Physics at USSR Acad. of Sciences 1957–67; Dep. Dir. and Head of Laboratory at Inst. of Organic Chemistry at USSR Acad. of Sciences 1967–; concurrently Lecturer 1960–73, Prof. at Physical Techn. Inst. Moscow 1973–.

Kazantsev, Aleksandr Petrovich
Author; b. Ag. 20 (S. 2), 1906, Akmolinsk, now Tselinograd.
Russian; mem. CPSU 1954–; grad. Tomsk Technological Inst. 1930; first works publ. 1939.
Publs.: "Lunnaia doroga" (Moon Street) 1960, "Planeta bur" (The Planet of Storms) 1959, "Gost' iz kosmosa" (Guest From Outer Space) 1958, "Stupeni griadushchego" (Steps of the Future) 1962, "L'dy vozvrashchaiutsia" (The Ice Returns) 1964, and others.

Kazarian, Litvin Eligumovich
State official.
Cand. mem. Cen. Cttee. Armenian CP 1964–71, mem. Auditing Comm. 1976–81, mem. Cen. Cttee. 1981–; Min. of Construction Materials Ind. Armenian SSR Ja. 1980–.

Kazarinova, Nina Nikolaevna
Russian actress; b. O. 27 (N. 9) 1907, Perm'.
People's Artist of the RSFSR (1957). Member of the CPSU since 1944. One of the organizers of, and an actress in, the Young People's Theater in Perm' (1926–28).
She graduated from the Leningrad Technicum of Theatrical Arts in 1931. In 1930 she began to act with the Leningrad Young People's Theater, playing children's parts. Her heroes are impassioned youths with great dreams. Her best roles include Sema in *The Marshal's Childhood* by Vsevolozhskii, Gavrik and Vania Solntsev in *The Lone White Sail* and *Son of the Regiment* by Kataev, Pashka in *Pashka* by Makar'ev, and Shura in *The Red Necktie* by Mikhalkov. She later played the roles of old women (Feklusha in Ostrovskii's *Thunderstorm*, for example).

Kaziev, Mamed Iakubovich
State official.
Min. of Social Welfare Azerbaidzhan SSR N. 1974–; mem. Auditing Comm. Azerbaidzhan CP 1976–81, Chair. 1981–.

Kazimirov, Vladimir Nikolaevich
Diplomat.
USSR Amb. to Costa Rica 1971–75, to Venezuela and concurrently to Trinidad and Tobago Ap. 1975–80; Head, First Latin American Countries Dept., USSR Min. of Foreign Affairs 1980–.

Kazimirova, Ekaterina Grigorevna
Actress; b. D. 8, 1921, Fedorovka.
Moldavian; grad. Drama School Odessa 1939. Decors.: People's Artist of the Moldavian SSR, 2 orders, several medals.

Kazin, Vasilii Vasil'evich
Russian poet; b. Jl. 25 (Ag. 6), 1898, Moscow.
He was among the first to incorporate the theme of labor into Soviet lyric poetry. He was a founder of the Smithy, a literary group of proletarian writers. He first published his work in 1914. His collection of verses *Workers' May* (1922) celebrated free and joyous labor. The simplicity and concreteness of Kazin's verse ("The Mason" and "The Plane") distinguish it favorably from the cosmic and abstract qualities of other Smithy poets.
His major narrative poems are *The Fox Coat and Love* (1926), in which he denounced petit bourgeois morality; *The White Sea Poem* (1937); which depicts the new Soviet man's upbringing in labor; and *The Great Beginning* (1954), which is about the first Communist Subbotniks. Decors.: two orders and various medals.
Works: *Stikhotvoreniia i poemy. [Avtobiografiia: Slovo o sebe.]* Moscow 1957. *Stikhotvoreniia i poemy.* Foreword by B. Solov'ev. Moscow 1964.

Kaziulin, V. A.
Komsomol official.
2nd Sec. and mem. Bureau Moldavian Komsomol Mr. 1980–.

Kebin, Ivan Gustavovich
State official; b. 1905; s. of a peasant.
Esthonian; mem. CPSU 1927–; did not complete higher educ.; seasonal worker in Gachino Raion (Leningrad Oblast) 1923–25; attended Sov. Party School in Leningrad 1925–26; Chair. of Susanino Rural Sov. 1927–35; Instructor of Gachino Raion All-Union CP(B) Cttee., Sec. of "Kommunar" Factory Party Cttee., sec. of party cttee. at a machine-building plant in Leningrad 1931–35; after completing Marxism-Leninism courses in Leningrad, Head of Dept. of Propaganda and Agitation in Tavrichevskii Raion (Omsk Oblast) All-Union CP(B) Cttee. 1935–36; studied Inst. of Red Professors 1936–38; Lecturer on the Principles of Marxism-Leninism at Gubkin Petroleum Inst. Moscow 1938–41; sen. posts with Dept. of Propaganda and Agitation of Esthonian CP(B), Dir. of Inst. of Party History (a branch of Inst. of Marxism-Leninism at Esthonian CP(B) Cen. Cttee.) 1941–48; responsible for the training of Esthonian national cadres 1941–45; Sec. 1948–50, 1st Sec. 1950–78, mem. Bureau of Cen. Cttee. Esthonian CP 1950–; Dep. 1950–, Dep. Chair. of Presidium USSR Supr. Sov. N. 1978–; mem. Cen. Cttee. CPSU 1952–; Dep. to Esthonian Supr. Sov. 1955–, mem. Presidium 1955–78, Chair. of Presidium Jl. 1978–. Decors.: 4 Orders of Lenin, Order of Red Banner of Labour, various medals.

Kėdainis, Juozas Justino
Sculptor; b. D. 18 (31), 1915, Karaliuniškis, in present-day Kupiškis Raion.
People's Artist of the Lithuanian SSR (1965).
From 1933 to 1939, he studied under J. Mikėnas and J. Zikaras at the Kaunas School of Art. From 1946 to 1951 he taught at the Institute of Applied and Decorative Art in Vilnius. In 1951, he became an instructor at the Art Institute of the Lithuanian SSR, which is also located in Vilnius.
Combining the romantic generalization of images and forms with soft picturesque modeling, he imparts a lyrical and intimate quality to his sculpture. His works include *The Team-Leader* (bronze, 1949, Art Museum of the Lithuanian SSR, Vilnius), *The Kolkhoz Stableman* (bronze, 1957, Kaunas Museum of Art; State Prize of the Lithuanian SSR, 1958), *At the Seashore* (plaster of Paris, 1959, Neringa Café, Vilnius), *Repose* (cast iron, in a park in Druskininkai; plaster of Paris, 1959), and the decorative stele *Labor* (granite, 1970, Kaunas).

Kedishvili, Iurii Ivanovich
State official; b. 1927, Tbilisi.
Osset; mem. CPSU 1953–; grad. Inst. of Railway Engineering 1949; various posts in apparat of Cen. Cttee. Georgian CP 1959–66; Head of Dept. of Construction and Communal Econ. with Cen. Cttee. Georgian CP 1966–76; mem. Cen. Cttee. Georgian CP 1966–; Dep. Chair. of Council of Mins. Georgian SSR S. 1976–.

Kedrina, Zoia Sergeevna
Literary critic; b. Ag. 18 (31), 1904, Moscow.
Grad. M. Gorkii Inst. of Literature 1938; her articles deal mainly with the history of Sov. Kazakh literature, the works of M. Auezov, S. Mukanov, G. Musrepov, G. Mustafin, S. Ragimov, D. Gulia, B. Kerbabaev, Ch. Ajtmatov and M. Ibragimov as well as translations from national languages.

Kedrov, Bonifatii Mikhailovich
Philosopher; b. N. 27 (D. 12), 1903, Iaroslavl.
Mem. CPSU 1918–; grad. Dept. of Chemistry of Moscow Univ. 1930; science historian; mem. USSR Acad. of Sciences 1966–; party functions, later research work 1935–41; served in Sov. Army 1941–45; mem. staff. Inst. of Philosophy at USSR Acad. of Sciences 1945–49, 1958–62; Chief Ed. of "Voprosy filosofii" 1947–49; Prof. at Acad. of Social Sciences of Cen. Cttee. CPSU 1946–58 and 1971–; Dir. of Inst. for the History of Natural Sciences and Technology at USSR Acad. of Sciences 1962–73; Dir. of Inst. of Philosophy at USSR Acad. of Sciences 1973–; mem. of several foreign acads. Decors.: 3 orders, several medals.

Kekilov, Aman
Author, poet and literary historian; b. 1912, Keshi nr. Ashkhabad; s. of a peasant.
Turkmen; mem. CPSU 1951–; grad. Inst. of History, Philosophy and Literature Moscow 1939; mem. Acad. of Sciences Turkmen SSR 1960–; wrote text of Turkmen SSR anthem.
Publs.: "Vpered" (Forward) 1935, "Posledniaia vstrecha" (The Last Meeting) 1940, "Liubov'" (Love) vols. 1 and 2, 1947–57, "Turkmenskaia sovetskaia literatura na puti razvitiia" (Turkmen Soviet Literature in the Process of Development) 1943, "Teoriia literatury" (Theory of Literature) 1961, and others. Decors.: 2 orders, several medals.

Keldysh, Iurii (Georgii) Vsevolodovich, Dr. of Art Sc.
Musicologist; b. Ag. 16 (29), 1907, St. Petersburg.
Mem. CPSU 1947–; grad. Dept. of Composition 1930, mem. staff 1930–, Prof. at Moscow Conservatory 1948–; Head of Music Section at Inst. for the History of the Arts Moscow 1961–. Decors.: Merited Artist of the RSFSR, Order of Red Banner of Labour, several medals.

Keldysh, Leonid Veniaminovich
Physicist; b. Ap. 7, 1931, Moscow.
Grad. Moscow Univ. 1954; corresponding mem. USSR Acad. of Sciences 1968–; mem. staff Inst. of Physics at USSR Acad. of Sciences 1954–; Prof. at Moscow Univ. 1969–.

Kemashvili, Vera Baturovna
Laborer; b. 1929.
Mem. CPSU 1955–; grad. Tbilisi Textile Technicum 1949; technician in control shop, asst. foreman for prod. of worsted goods in Tbilisi Worsted Cloth Factory 1949–54; asst. foreman in "Sovetskaia Gruziia" Worsted Cloth Combine in Tbilisi 1954–; mem. Cen. Auditing Comm. CPSU 1981–. Decor.: Heroine of Socialist Labor 1966.

Kemenov, Vladimir Semenovich
Art historian and critic; b. My. 20 (Je. 2), 1908, Ekaterinoslav, present-day Dnepropetrovsk.
Honored Art Worker of the RSFSR (1968), doctor of the arts (1958), member of the Academy of Arts of the USSR (1954; vice-president, since 1966). Became a member of the CPSU in 1939.
He studied at the first Moscow State University from 1928 to 1930. He taught at several educational institutions in Moscow, including the State Institute of Theatrical Arts (1933–38). Between 1938 and 1940, he was the director of the Tret'iakov Gallery. From 1940 to 1948 he served as the chairman of the Board of the All-Union Society for Cultural Relations with Foreign Countries. He was the deputy minister of culture of the USSR from 1954 to 1956 and the permanent representative of the USSR to UNESCO from 1956 to 1958.
His principal works, which include numerous articles in Soviet and foreign publications, are devoted to the conflicts among various trends in contemporary art and aesthetics, the affirmation of socialist realism in the fine arts, and questions concerning classical Russian and foreign art, as well as Soviet art.
Works: *Stat' i ob iskusstve.* Moscow 1956. *Istoricheskaia zhivopis' Surikova: 1870–1880-e gg.* Moscow 1963. *Protiv abstraktsionizma. V sporakh o realizme: Sb. st.*, 2nd ed. Moscow 1969. *Kartiny Velaskesa.* Moscow 1969.

Kerashev, Tembot Magometovich
Adygeian writer; b. Ag. 7 (20), 1902, in the *aul* (village) of Koshekhabl', in present-day Adygei Autonomous Oblast. Member of the CPSU since 1928.
He graduated from the Institute of Industrial Economics in Moscow (1929). His work was first published in 1925. His articles, essay, and short stories of the 1920's and early 1930's depiet the revolutionary changes in the life, work, and psychology of the Adygeian peasantry. He was the first in the literature of the peoples of the Northern Caucasus to write in the genre of the social novel. An example of this genre is his novel *The Road to Happiness* (1940; Russian translation, 1947; State Prize of the USSR, 1948), which deals with the construction of socialist economy and culture in Adygeia. His novel *Competing With a Dream* (book 1, 1955, in Russian) is devoted to the postwar life of an Adygeian *aul*. He is the author of such novellas as *The Daughter of the Shapsugs* (1951, in Russian), *Abrek* (or *Rebel Mountaineers*, 1957; Russian translation, 1959), and *The Revenge of the Horse Herder.* He raised the problem of the upbringing of youth in the novella *The Daughter of a Wise Mother* (1963) and the novel *Kuko* (1968). His works are characterized by realistic descriptions of characters and landscapes, psychological penetration of characters, and a polemically passionate style. Decors.: Order of Lenin, the Order of the October Revolution, and the Order of the Red Banner of Labor as well as medals.
Works: *Abdzekhe shekiozh'yr.* Maikop 1969. *Nasypym ig"ogu,* 3rd ed. Maikop 1970. *In Russian translation: Izbr. proizv.*, vols. 1–2. Krasnodar 1964–65.

Kerbel', Lev Efimovich
Sculptor; b. O. 25 (N. 7), 1917, Semenovka, in present-day Sumy Oblast, Ukrainian SSR.
People's Artist of the RSFSR (1967); corresponding member of the Academy of Arts of the USSR (1962). Became a member of the CPSU in 1963.
He studied at the Moscow Art Institute under L. V. Shervud from 1937 to 1942. He became an instructor there in 1962, receiving a professorship in 1963. He creates monumental sculptures and portraits, which are marked by great civic content. His works in the Tret'iakov Gallery include *The Labor Reserves* (bronze, 1947), the reliefs *The Victory Parade* and *Listening to Lenin* (plaster of Paris, 1948–49; State Prize of the USSR, 1950), and the portraits *V. S. Petrov – Twice a Hero of the Soviet Union* (marble, 1951), *K. S. Mukhtarova – Hero of Socialist Labor* (marble, 1954), and *J. Endicott* (bronze, 1957). He also sculptured monuments to V. I. Lenin in the Vladimir Il'ich Kolkhoz in Gorki Leninskie (granite, 1959) and in Sofia, Bulgaria (granite, 1971). He executed monuments to Karl Marx in Moscow (granite, unveiled in 1961; Lenin Prize, 1962) and in Karl-Marx-Stadt, the German Democratic Republic (bronze, 1971). He also created the monument to the victims of fascism in the city of Rudnia in Smolensk Oblast (granite, 1965). Decors.: two orders and various medals.

Kerdivarenko, Vasilii Aleksandrovich
State official.
Min. of Educ. Moldavian SSR D. 1971–78; Chair. of Moldavian branch of Sov.-Polish Friendship Soc. 1972–; mem. Cen. Cttee. Moldavian CP 1976–; Min. for Higher and Secondary Specialist Educ. Moldavian SSR 1978–.

Kerimov, Ali Gabib ogly
State official; b. 1919.
Azerbaidzhani; mem. CPSU 1943–; grad. Azerbaidzhan Ind. Inst.; Supervisor of Tech. Control Board, 1st sec. of a

Komsomol raion cttee., 1st Sec. of Baku City Cttee. Azerbaidzhan Komsomol, 1st Sec. of Cen. Cttee. Azerbaidzhan Komsomol 1942–52; mem. Apparat Cen. Cttee. Azerbaidzhan CP 1952–56; mem. Cen. Cttee. Azerbaidzhan CP 1953–81; Min. of the Interior 1956–60, Dep. Chair. of Council of Mins. Azerbaidzhan SSR 1960–65; concurrently Chair. of Cttee. of Party and State Control with Cen. Cttee. Azerbaidzhan CP and of Council of Mins. Azerbaidzhan SSR 1962–65; mem. 1962–66 and 1971–81; cand. mem. Bureau Cen. Cttee. Azerbaidzhan CP 1966–71; Sec. of Cen. Cttee. Azerbaidzhan CP 1962–65; Chair. of Cttee. on People's Control 1965–70; 1st Sec. of Baku City Cttee. Azerbaidzhan CP My. 1970–80; Chair. of State Cttee. on Labour Azerbaidzhan SSR D. 1980–.

Kerimov, Dzhangir Ali-Abasovich
Jurist; b. Jl. 18, 1923, Baku.
Azerbaidzhani; mem. CPSU 1944–; corresponding mem. USSR Acad. of Sciences 1966–; mem. Azerbaidzhan Acad. of Sciences 1967–; Prof. Chair of Theory and History of Pol. Science and Law at M. I. Kalinin Inst. of Law and Leningrad Univ. 1951–57; Prof. at German Ulbricht Acad. of Pol. Science and Law and E. Berlin Univ. 1957–59; Prof. Chair 1959–65, Dep. Rector of Leningrad Univ. 1965–68; Sen. Research Staff Mem. at Inst. of Marxism-Leninism of Cen. Cttee. CPSU 1969–71; Prof. Chair of Theory of Pol. Science and Law at Acad. of Social Sciences of Cen. Cttee. CPSU 1971–.

Kerimov, Khalid Gadzhibaba ogly
State official; b. 1930.
Cand. mem. Cen. Cttee. Azerbaidzhan CP 1976–; Chair. of State Cttee. for Supply of Production Equipment for Agriculture Azerbaidzhan SSR Ap. 1979–.

Kerimov, Shakir Kerim ogly
State official.
Min. of Motor Transport Azerbaidzhan SSR Ap. 1979–.

Kesamaa, Manivald Augustovich
Author and poet; b. 1921, Narva; s. of a worker.
Esthonian; mem. CPSU 1947; first poems publ. in a front newspaper.
Publs.: "S goriachim serdtsem" (With a Burning Heart) 1950, "Posle meteli" (After the Snowstorm) 1956, "Doma est' doma" (Home is Home) 1958, "Materi i synovia" (Mothers and Sons) 1961, "Zvezdy nad zemlei" (Stars Over the Earth) 1964, and others.

Keshokov, Alim Pshemakhovich
Author and poet; b. 1914, Shalushka, Kabardino-Balkar ASSR; s. of a peasant.
Mem. CPSU 1941–; grad. N. Caucasian Pedagogical Inst., completed post-graduate studies at Inst. of Languages and Philosophy Moscow and Acad. of Social Sciences of Cen. Cttee. CPSU; first works publ. 1934. Decors.: 2 Orders of Red Banner of Labour.
Publs.: "U podnozhia gor" (In the Foothills of the Mountains) 1941, "Put' vsadnika" (A Rider's Way) 1946, "Zemlia molodosti" (The Country of Youth) 1948, "Teplyi kamen" (A Warm Stone) 1964, and others.

Kevorkov, Boris Sarkisovich
Party official; b. 1932.
Armenian; mem. CPSU 1954–; grad. Azerbaidzhan State Univ.; Komsomol posts 1954–61; mem. Apparat Cen. Cttee. Azerbaidzhan CP 1961–70; 1st Sec. of Kirov Raion Cttee. Azerbaidzhan CP in Baku 1970–73; 1st Oblast Sec. of Azerbaidzhan CP in Nagorno-Karabakh Autonomous Oblast O. 1973–; mem. Cen. Cttee. and cand. mem. Bureau Cen. Cttee. Azerbaidzhan CP Ja. 1976–.

Kevorkov, Stepan Agabekovich
Film director; b. 1903.
Grad. Dept. of Drama State College of Cinematography Moscow 1930; Dir. of film studio "Armenfilm" 1936–; 1st Sec. of Armenian Fed. of Film Workers 1962–66. Decor.: People's Artist of the USSR 1970.

Kezhun, Bronislav Adolfovich
Author and poet; b. 1914, Petrograd; s. of a clerk.
Russian; mem. CPSU 1944–.
Publs.: "Rodina" (Homeland) 1935, "Dorogi v mir" (Roads Into the World) 1939, "Pesnia o dome. Stikhi 1944–46" (The Song of Home. Poems 1944–46) 1947, "Baltiiskie zori" (Baltic Stars) 1959, and others.

Khabeishvili, Soliko Evtikhevich
Party official.
2nd Sec. of Tbilisi City Cttee. Georgian Komsomol 1961–65; mem. Bureau 1964–68, 2nd Sec. of Cen. Cttee. Georgian Komsomol 1965–67; cand. mem. 1966–71, mem. Cen. Cttee. Georgian CP 1976–; cand. mem. Bureau Cen. Cttee. Georgian CP 1975–79, mem. 1979–; Head of Dept. of Party Org. Work at Cen. Cttee. Georgian CP D. 1975–82; Sec. Cen. Cttee. Georgian CP Jl. 1982–.

Khabibullaev, Zukhur Nurdzhanovich
Artist; b. Ja. 4, 1932, Dushanbe.
Honored Artist of the Tadzhik SSR (1967). Member of the CPSU since 1962. From 1953 to 1959, studied at the V. I. Mukhina Leningrad Higher School of the Art Industry. Since 1964 he has taught at the Republic Art School in Dushanbe (he taught there for a short time previously – from 1959 to 1962). His work in many was influenced the stylistic qualities of Soviet Tadzhik painting of the 1960's. His emotional, energetically executed landscapes, portraits, and thematic works are noted for their expressive images, sharp color contrasts, and dynamic lines (*Nurek Hotel,* 1962; *Pamir Still Life,* 1964; *Bakhor,* 1967; *Thirst,* 1972; all in the Bekhzad Republic Museum of History, Local Lore, and Fine Arts in Dushanbe).

Khachaturov, Tigran Sergeevich
Economist; b. S. 23 (O. 6), 1906.
Mem. CPSU 1945–; grad. Moscow Univ. 1926; mem. USSR Acad. of Sciences 1966–; mem. staff. Cen. Statistical Admin. of the USSR 1926–29; mem. staff various scientific railway transport research insts. 1929–45; Ed. of journal "Tekhnika zheleznykh dorog" (Railway Technology) 1945–49; head of a chair at Acad. of Railway Transport 1947–54; head of a chair at Moscow Inst. of Railway Transport Engineering 1949–52; Dir. of Inst. of Complex Transport Problems of USSR Acad. of Sciences 1953–59; head of a sector at Econ. Inst. of USSR Acad. of Sciences 1959–; concurrently Prof. at Acad. of Social Sciences of Cen. Cttee. CPSU 1964–, and Chief Ed. of journal "Voprosy ekonomiki" 1966–.

Khachidze, M. I.
Komsomol official.
Sec. and mem. Bureau Cen. Cttee. Georgian Komsomol Ap. 1982–.

Khaeev, Izatullo
State official.
Chair. of Exec. Cttee. of Workers' Deps. in Gorno-Badakhshan Autonomous Oblast 1973–78; Dep. to 1975– and Dep. Chair. of Presidium Supr. Sov. Tadzhik SSR 1975–78; mem. Cen. Cttee. Tadzhik CP 1976–; Min. of Meat and Dairy Ind. Tadzhik SSR Mr. 1978–.

Khaidarov, Aiub
Trade union official; b. 1922.
Tadzhik; mem. CPSU 1945–; grad. Party High School of Cen. Cttee. CPSU; teacher, head of pedagogical dept. of a school, inspector educ. dept. of a raion sov. 1941–46; studied Republican Party School 1946–48; sec. of a raion cttee. Tadzhik CP and dep. chair. of an oblast exec. cttee. of workers' deputies, 1st sec. of a city party cttee. Tadzhik CP 1948–63; mem. Cen. Cttee. Tadzhik CP 1960–; Chair. of Tadzhik Republican Trade Union Council 1963–; mem. Bureau for Ind. and Construction at Cen. Cttee. Tadzhik CP 1963–64; cand. mem. Bureau Cen. Cttee. Tadzhik CP 1976–81, mem. Ja. 1981–.

Khaidurov, Viktor Alekseevich
Party official; b. 1931.
Mem. CPSU 1960; Russian, grad. Tashkent Agricultural Inst.; agrarian specialist, chief agrarian specialist, dept. chief, chief agrarian specialist, Dir. of a state farm 1949–66; chief of a dept. of Oblast Cttee. CP Uzbekistan, Dep. Chair. of Syr-Daria Oblast Exec. Cttee. of People's Deps., Dep. Min. for state farms Uzbek SSR 1966–71; Chair. of Syr-Dazia Oblast

Exec. Cttee. of People's Deps. 1971 74; mem. Cen. Cttee. Uzbek CP 1971 –; 1st Sec. of Syr-Dazia Oblast Cttee. Uzbek CP 1974 –.

Khaitov, Iusup
State official.
Min. of Justice Turkmen SSR N. 1979–; mem. Cen. Cttee. Turkmen SSR 1981–.

Khakim, Sibgat Tazievich (pen name of Sibgat Tazievich Khakimov)
Tartar poet and public figure; b. D. 4 (17), 1911, Kulle-Kimi, in what is now Arsk Raion, Tartar ASSR.
Member of the CPSU since 1943.
Graduated from the Kazan Pedagogical Institute in 1937 and served in the Great Patriotic War (1941–45). He first published his works in 1931. His first poetry collection, *First Songs* (1938), is concerned with life in the new village. Khakim won fame with his narrative poems about G. Tukai: *A Pair of Bay Horses* (1939) and *A Poet's Childhood* (1940).
During the Great Patriotic War and in the years that followed, Khakim wrote about the heroism of Soviet soldiers and the valor of the Soviet people in battle and labor, for example, in the narrative poems *Gardeners* (1952), *Across the Steep Slopes* (1964–66) and *The Salient* (1968). The theme of Lenin is of special importance in the collections *At Lenin's Call* (1958; G. Tukai Prize of the Republic, 1960) and *My Heart Speaks With Lenin* (1970). Khakim has also written articles on Tartar literature.
Khakim was a member of the Presidium of the Supreme Soviet of the Tartar ASSR from 1963 to 1967. He served as deputy chairman of the administrative board of the Writers' Union of the Tartar ASSR from 1956 to 1960 and secretary of the administrative board of the Writers' Union of the RSFSR from 1965 to 1975. Decors.: A. M. Gorkii State Prize of the RSFSR in 1970. He has been awarded the Order of Lenin, two other orders, and various medals.
Works: *Saylanma shigir'lär häm poëmalar.* [Foreword by Kh. Khäyri.] Kazan 1961. *Torïgïz, musalar! Shigir'lär, poëmalar.* Kazan 1970. *Sibelä chächäk.* Kazan 1973. *Tukay turïnda poëmalar.* Kazan 1974. *Poëmalar,* Kazan 1977. *In Russian translation: Izbrannoe.* Kazan 1952. *Stikhi.* Moscow 1964. *Nebo v glazakh.* Kazan 1966. *Vechnye vetvi.* Moscow 1971. *Stikhi poslednikh let.* Moscow 1973. *Vrata vremeni.* Moscow 1977.

Khakimov, M.
Trade union official.
Sec. and mem. Presidium of Tadzhik Republican Trade Union Council F. 1982–.

Khalafov, Rafik Rasul ogly
State official.
Min. of Construction and Utilization of Motorways Azerbaidzhan SSR Jl. 1979–; cand. mem. Cen. Cttee. Azerbaidzhan CP 1981–.

Khalatnikov, Isaak Markovich, Dr. of Phy.-Math. Sc.
Physicist; b. 1919.
Mem. CPSU 1944–; grad. Dnepropetrovsk Univ. 1941; Dr. of Phy.-Math. Sc. 1953; Prof. 1956; corresponding mem. USSR Acad. of Sciences 1972–; served in Sov. Army 1941–45; postgraduate studies 1946–50, Sen. Research Staff Mem. 1950–54, Head of Laboratory 1954–55, Sen. Research Staff Mem. 1955–62, Dept. Head at Inst. of Problems of Physics at USSR Acad. of Sciences 1962–65; Dir. of Inst. of Theoretical Physics at USSR Acad. of Sciences 1965–; concurrently Lecturer, Prof. at Physical-Tech. Inst. Moscow 1954–. Decor.: State Prize of the USSR.

Khaldeev, Mikhail Ivanovich
Journalist; b. 1921.
Mem. CPSU 1942–; grad. from a commercial college 1939 and Party High School of Cen. Cttee. CPSU 1948; served in Sov. Army 1939–45; Komsomol work 1945–50; Moscow City Sec. 1950–51, Moscow Oblast Sec. 1951–52, 1st Moscow Sec. 1952–53 and 1st Moscow Oblast Sec. of USSR Komsomol 1953–57; Chief Ed. of journal "Molodoi Kommunist" 1957–59; cand. mem. Bureau Cen. Cttee. USSR Komsomol 1958–59; 1st Sec. of Timiriazev Raion Cttee. CPSU in Moscow 1959–63; Head of RSFSR Ideological Dept., Head of RSFSR Dept. of Propaganda and Agitation with Cen. Cttee. CPSU 1963–66; Chief Ed. of journal "Partiinaia zhizn'" Mr. 1966–; mem. Cen. Auditing Comm. CPSU 1966–.

Khalilov, Kurban Ali ogly
State official; b. 1906.
Azerbaidzhani; mem. CPSU 1926–; grad. Azerbaidzhan Ind. Inst. 1933; lathe turner, student 1923–33; engineer, dept. head 1933–37 and dir. of various plants 1937–42; Sec. of Baku City Cttee; Azerbaidzhan CP 1942–45; econ. and state posts 1945–54; Dep. Min. of Building Materials Ind. Azerbaidzhan SSR, Dep. Chair. of Exec. Cttee. of Baku City Workers' Deps., Min. of Local Ind. Azerbaidzhan SSR 1954–58; Min. of Finance Azerbaidzhan SSR 1958–69; Dep. to Supr. Sov. Azerbaidzhan SSR 1959–; mem. Cen. Cttee. 1960–, mem. Bureau Cen. Cttee. Azerbaidzhan CP 1969–; Chair. of Presidium Supr. Sov. Azerbaidzhan SSR 1969–; Dep. to and Dep. Chair. of Presidium USSR Supr. Sov. 1970–; mem. Cen. Auditing Comm. CPSU 1971–.

Khalimov, Abdukhakim
State official.
Min. of Rural Construction Tadzhik SSR F. 1981–.

Khalitov, Fuad Ibragimovich
Tartar actor; b. F. 18, 1909, Kazan.
People's Artist of the USSR (1975).
Member of the CPSU since 1946.
In 1930, graduated from a drama studio and joined the company of the Kazan Workers' Theater. In 1937 he joined the Kamal Tartar Academic Theater. He gave original interpretations of the historical roles of Kaium Nasyri, the Tartar enlightenment figure, in *Kaium Nasyri* by Gali and Urazikov, and Musa Dzhalil', the poet and Hero of the Soviet Union, in *Musa,* by Isanbet.

Khalmiagi, Valter Albertovich
State official.
Mem. 1966–71, Chair. of Auditing Comm. Esthonian CP 1971–76; Dep. to Supr. Sov. Esthonian SSR 1967–75; Vice-Head of Dept. of Trade and Public Services at Cen. Cttee. Esthonian CP 1967; Min. of Public Services Esthonian SSR Mr. 1975–; mem. Cen. Cttee. Esthonian CP 1976–.

Khamidov, Azis Khakimovich
State official.
Min. of Energy and Electrification Uzbek SSR N. 1962–; Dep. to Supr. Sov. Uzbek SSR 1963–; mem. Cen. Cttee. Uzbek CP 1966–.

Khamidov, Kh. Kh.
Komsomol official.
Sec. and mem. Bureau Cen. Cttee. Uzbek Komsomol Mr. 1979–.

Khamidov, Vali Abduvakhidovich
State official.
Head of Dept. for Heavy Ind. and Machine Construction with Cen. Cttee. Uzbek CP 1974–79; mem. Cen. Cttee. Uzbek CP 1976–81; Chair. of State Cttee. for the Supply of Petroleum Products Uzbek SSR My. 1981–.

Khandogin, A. P.
Komsomol official.
Sec. and mem. Bureau Cen. Cttee. Belorussian Komsomol Mr. 1982–.

Khanov, Aleksandr Aleksandrovich
Actor; b. 1904.
Training as actor at Revolution Theatre Moscow 1923; engagements at Proletkult Theatre 1924–34; engagements at Revolution Theatre (now V. Maiakovskii Theatre) 1934–. Decors.: People's Artist of the USSR 1973, State Prize of the USSR 1941.

Khanzadian, Sero Nikolaevich
Armenian writer; b. N. 20 (D. 3), 1915, Goris, in what is now the Armenian SSR.
Graduated from the pedagogical technicum in Goris in 1934 and became a village schoolteacher. He fought in the Great Patriotic War of 1941–45.
In 1950, published his first novel, *The Men of Our Regiment* (Russian translation, 1956), which was devoted to the defenders of Leningrad. His novel *The Land* (books 1–2, 1954–55; Russian translation, books 1–2, 1956–57) deals with the working people of the countryside in the postwar period. The historical

novel *General Mkhitar* (1961; Russian translation, 1971) depicts the Armenian liberation struggle of the 18th century. The subject of the novel *Kadzharan* (1965; Russian translation, 1974) is the formation of the working class in Armenia.
Other novels are *Dawn Over Sevan* (1974) and *Speak, Hills of Armenia* (1976). He also produced the short-story collections *Red Lilies* (1958), *Lost Paths* (1964), *After the Rain* (1969) and the memoirs *A Book About the Past* (1966) and *Three Years and 291 Days* (1972). Decors.: four orders.
Works: [Khanzadyan, S.] *Erkeri zhoghovatsol,* vols. 1–5. Foreword by G. Hovsep'yani. Yerevan 1967–70. *Akandatoln: Hek'iat'ner, aghvesagirk', dzrolyts'ner.* Yerevan 1976. *Khosekh', Hayastani lerhner* (novel). Yerevan 1976. *In Russian translation: Zhazhdu-daite vody. Povesti. Frontovoi dnevnik.* Moscow 1974.

Khappalaev, Iusup Ramazanovich
Lak poet; b. Je. 22, 1916, in the *aul* (village) of Khurukra, in what is now Lak Raion.
People's Poet of Dagestan (1966).
First published in 1932. His poetry reflects the changes that have taken place in Dagestan and provides insight into the spiritual world of contemporary man. Khappalaev has introduced new rhythms and rhymes. His collections of lyric verse include *The Expanses of Love* (1941), *Stars of Happiness* (Russian translation, 1950), and *I Am a Grain of Sand From the Hill* (Russian translation, 1969). He has translated works by such writers as M. Iu. Lermontov, K. Khetagurov, V. V. Mayakovsky, S. Stal'skii, and G. Tsadasa, and many of his poems have been translated into various languages of the USSR. He served as executive secretary of the Writers' Union of Dagestan from 1948 to 1955 and again from 1957 to 1967. Khappalaev has been awarded three orders.
Works: *Zid va chchavu.* Makhachkala 1944. *Zunttal balairdu.* Makhachkala 1953. *Iazimii.* Makhachkala 1968. *Asardal sinttu.* Makhachkala 1972. *Kiiurkhnil diukhlu.* Makhachkala 1974. *In Russian translation: Goriachee serdtse druga* [Foreword by M. Amaev.] Makhachkala 1959. *Kremen' i fialka.* Moscow 1970.

Kharatishvili, Iosif Alekseevich
State official.
1st Dep. Min. 1968–71, Min. of Rural Construction Georgian SSR F. 1971–; cand. mem. Cen. Cttee. Georgian CP 1976–81, mem. 1981–.

Kharazov, Valerii Innokentevich
State official; b. 1918.
Mem. CPSU 1944–; grad. Aviation Inst. Moscow 1945; constructor-designer, technologist, foreman in various plants in Kazan and Moscow 1940–46; Komsomol posts 1946–54; Sec. of Alma-Ata City Cttee. Kazakh CP 1954–55; Sec. of Gurev Oblast Cttee. 1955–56 and Pavlodar Oblast Cttee. Kazakh CP 1956–61; mem. Apparat Cen. Cttee. CPSU 1961–67; 2nd Sec. and mem. Bureau of Cen. Cttee. Lithuanian CP Ap. 1967–78; mem. Cen. Cttee. Lithuanian CP 1971–78; mem. Auditing Comm. CPSU 1971–76; cand. mem. Cen. Cttee. CPSU 1976–81; 1st Dep. Chair. RSFSR Cttee. of People's control N. 1978–. Decor.: Order of Red Banner of Labour.

Kharbediia, L. S.
Trade union official.
Mem. Auditing Comm. 1964–66, mem. Cen. Cttee. Georgian CP 1966–71; Sec. of Georgian Republic Trade Union Council Ja. 1971–.

Kharchev, Konstantin Mikhailovich
Diplomat.
USSR Amb. to Guyana Mr. 1981–.

Khariton, Iulii Borisovich
Physical chemist; b. F. 14 (27), 1904.
Jew; mem. CPSU 1956–; grad. Leningrad Polytech. Inst. 1925; Prof.; mem. USSR Acad. of Sciences 1953–; messenger in a library, then apprenticeship as mechanic 1917–21; mem. scientific staff Leningrad Physico-Tech. Inst., then USSR Acad. of Sciences 1921–; Dep. to USSR Supr. Sov. 1950–74. Decors.: three times Hero of Socialist Labour, Lenin Prize, State Prize of the USSR.

Kharkov, Vladimir Alekseevich
Party official.
Head Dept. of Ind. and Transport with Oblast Cttee. Turkmen CP in Chardzhou until 1973; Sec. Chardzhou Oblast Cttee. Turkmen CP 1973–78; mem. Auditing Comm. Turkmen CP 1976–81, mem. Cen. Cttee. Turkmen CP 1981–; Head Dept. for Org. Party Work with Cen. Cttee. Turkmen CP 1979–; cand. mem. Bureau Cen. Cttee. Turkmen CP 1981–.

Kharlamov, Aleksandr Pavlovich, Cand. of Phil. Sc.
Journalist; b. 1929.
Grad. State Univ. Leningrad and Acad. of Social Sciences of Cen. Cttee. CPSU; Reader and Head of Readers' Group with Orenburg Oblast Cttee. CPSU 1954–70; Ed. of Oblast newspaper "Iuzhnyi Ural" in Orenburg 1959–64; studied Acad. of Social Sciences of Cen. Cttee. CPSU 1964–67; mem. Apparat Cen. Cttee. CPSU 1967–70; Dep. Chief Ed., 1st Dep. Chief Ed. of newspaper "Selskaia Zhizn'" 1976–; mem. Cen. Auditing Comm. CPSU 1976–.

Kharlamov, Mikhail Averkievich
State official; b. 1913.
Russian; mem. CPSU 1940–; grad. Inst. of History, Philosophy and Literature Moscow and Party High School of Cen. Cttee. CPSU; machine fitter 1929–31; Acting Sec. of Radio Cttee. of Council of People's Commissars Belorussian SSR 1931–35; teacher 1935–42; mem. Apparat Cen. Cttee. CPSU 1942–48; on ed. staff of various newspapers and journals 1948–53; mem. Apparat USSR Min. of Foreign Affairs 1953–58; Head of Press Dept. 1958–62 and mem. Collegium of USSR Min. of Foreign Affairs 1961–62; Chair. of State Radio Cttee. of USSR Council of Mins. 1962–64; Dep. to USSR Supr. Sov. 1962–66; Sec. of USSR Journalists' Union 1964–65; Dep. Chief Ed. of "Politizdat'" 1966–68; Head of Historical and Diplomatic Admin. of USSR Min. of Foreign Affairs 1968–75; Dep. Chair. USSR State Cttee. for Publishing, Polygraphy and the Book Trade O. 1980–. Decors.: Order of Lenin, Lenin Prize.

Kharlamov, Nikolai Mikhailovich
Military official; b. D. 6 (19), 1905, Zhukovka, in what is now Briansk Oblast.
Admiral (1949). Member of the CPSU since 1925.
Kharlamov joined the navy in 1922. He graduated from the M. V. Frunze Naval School in 1928, completed advanced training courses for higher command personnel at the Naval Academy in 1941, and graduated from the K. E. Voroshilov Higher Military Academy in 1956.
Served in the Black Sea Fleet as a destroyer commander, a cruiser commander, and, from 1938 to 1941, fleet chief of staff. During the Great Patriotic War of 1941–45, he was a directorate chief in the central headquarters of the Soviet Navy. He became acting head of the Soviet military mission in Great Britain in August 1941 and head of the mission in 1943. In November 1944 he became a directorate chief and then deputy chief of the Main Naval Staff.
After the war, served as deputy chief for the navy on the General Staff of the Armed Forces from 1946 to 1950, commander of the Eighth Fleet in the Baltic from 1950 to 1954, and commander of the Baltic Fleet from 1956 to 1959. From 1961 to 1971 he performed vital duties in the central headquarters of the Soviet Navy, retired in August 1971. Decors.: Order of Lenin, the Order of the October Revolution, three Orders of the Red Banner, the Orders of Ushakov First Class and Nakhimov First Class, the Order of the Red Banner of Labor, two Orders of the Red Star, and various medals.

Kharlamov, Pavel Vasil'evich, Dr. of Phy.-Math. Sc.
Scientist; b. Je. 25, 1924, Gakhov, Kursk Oblast.
Grad. Moscow Univ. 1952; specialist in mechanics; corresponding mem. Ukrainian Acad. of Sciences 1965–; Dr. of Phy.-Math. Sc. 1965; Research Asst., Lecturer, Prof. Chair of Theoretical Mechanics at Ind. Inst. Donetsk 1952–59; Sen. Research Staff Mem. at Inst. of Hydrodynamics at Siberian Branch of USSR

Acad. of Sciences Novosibirsk 1959–65; Head of Dept. of Applied Mechanics in Computing Centre of Ukrainian Acad. of Sciences Donetsk and Prof. Chair of Applied Mathematics at Donetsk Univ. 1965–.
Publs.: "O dvizhenii v zhidkosti tela, ogranichennogo mnogosviaznoi poverkhnostiu" in "Zhurnal prikladnoi mekhaniki i tekhnicheskoi fiziki" No. 4, 1963, "Lektsii po dinamike tverdogo tela" 1965.

Kharlamov, Valerii Borisovich
Ice hockey player; b. Ja. 14, 1948, Moscow.
Honored Master of Sport (1969). Officer in the Soviet Army.
Kharlamov, who plays on the same line with Honored Masters of Sport V. V. Petrov and B. P. Mikhailov, has won championships of the USSR (1970–73, 1975, 1977–80), Europe (1969–70, 1973, 1975, 1978–79), the world (1969–71, 1973, 1975, 1978–79), and the Olympic Games (1972, 1976). He plays on the hockey team of the Central Sports Club of the Army, which won the European Champion Clubs Cup every year from 1969 through 1975. Decors.: two orders and several medals.

Kharlamova, Valentina Borisovna
Actress; b. F. 23 (Mr. 8), 1911, in the *stanitsa* (large Cossack village) of Uriupinskaia (now the city of Uriupinsk, Volgograd Oblast).
People's Artist of the USSR (1959).
Began her stage career in 1933. Appeared with the Gorkii Theater, the Ioshkar-Ola Theater (Mari ASSR), and the Murom Theater. In 1937 joined the Alma-Ata Russian Theater. Her roles have included Ophelia in Shakespeare's *Hamlet,* Larisa in A. Ostrovskii's *The Girl Without a Dowry,* Ranevskaia in Chekhov's *The Cherry Orchard,* the title roles in Trenev's *Liubov' Iarovaia* and De Filippo's *Filumena Marturano,* and Zhustailok in Auezov's *Summer Lightning.*

Khasanov, Kagor Gafurovich
State official.
Mem. Cen. Cttee. Tadzhik CP 1960–; Dep. to Supr. Sov. Tadzhik SSR 1962–; 1st Sec. of Leninabad City Cttee. Tadzhik CP 1963–65; Chair. of State Cttee. for Vocational Training Tadzhik SSR D. 1965–.

Khasanov, Madamindzhan Khasanovich
State official.
Dep. Min. of Agriculture Uzbek SSR 1966–79; cand. mem. Cen. Cttee. Uzbek CP 1971–81, mem. 1981–; Min. of Trade Uzbek SSR Mr. 1979–.

Khasaradze, I. G.
Trade union official.
Grad. Georgian Polytech. Inst.; worker, then master and dep. head of a section in an automobile factory 1945–75; mem. Cen. Cttee. Georgian CP 1971–76; Sec. and mem. Presidium of Georgian Republican Trade Union Council Je. 1975–; Chair. Auditing Comm. Georgian CP 1976–.

Khasenov, Khamit
State official.
Chair. Kazakh SSR State Cttee. for Television and Radio Mr. 1975–; cand. mem. Cen. Cttee. Kazakh CP 1976–.

Khechinashvili, Simon Nikolaevich
Otorhinolaryngologist; b. N. 26, 1919, Kutaisi.
Corresponding member of the Academy of Medical Sciences of the USSR (1963). Member of the CPSU since 1947.
Graduation from the Tbilisi Medical Institute in 1941, worked under the supervision of I. S. Beritashvili. In 1952 he was appointed to the chair of ear, nose, and throat diseases at the Tbilisi Institute for the Advanced Training of Physicians. His principal works deal with the subcortical and cortical vestibular mechanisms, surgical treatment of tumors of the throat, and microsurgery of the ear. He proposed a new method of studying the function of the tonsils. Khechinashvili received a Lenin Prize in 1964 for perfecting and popularizing operations to improve hearing in cases of otosclerosis. Decors.: Order of Lenin, the Order of the Red Banner of Labor, and various medals.
Works: *Vestibuliarnaia funktsiia. (Voprosy fiziologii i metodiki issledovaniia.)* Tbilisi 1958. *Voprosy teorii i praktiki slukhovosstanovitel'noi khirurgii.* Tbilisi 1963.

Kheifits, Josif Efimovich (also Josef Heifitz)
Film director and screenwriter; b. D. 4 (17), 1905, Moscow.
People's Artist of the USSR (1964). Hero of Socialist Labor (1975). Member of the CPSU since 1945.
Graduated from the Leningrad Technicum of Cinema Art in 1927 and from the motion-picture department of the Institute for the History of Arts in 1928. Began his film career in 1928. Until 1950 he worked in collaboration with A. G. Zarkhi. In 1929, Kheifits and Zarkhi organized the first Komsomol production group, which released films similar to the plays staged at the Theater of Working Youth (TRAM), such as *Against the Wind* (1930) and *Midday* (1931). They later produced the lyric comedy *Hectic Days* (1935) and directed the films *The Baltic Deputy* (1937) and *A Member of the Government* (1940). During the Great Patriotic War (1941–45) they made the films *They Call Him Sukhe-Bator* (1942), *Malakhov Mound* (1944), and the documentary *The Defeat of Japan* (1946).
Directed *The Big Family* (1954), *The Rumiantsev Case* (1956), *My Dear Man* (1958), *The Lady With the Dog* (1960), *Salute, Maria!* (1970), and *The One and Only* (1976).
Decors.: State Prize of the USSR (1941, 1946), two Orders of Lenin, two Orders of the Red Banner of Labor, and various medals.

Khesin-Lur'e, Roman Beniaminovich
Biologist; b. Mr. 24, 1922, Moscow.
Corresponding member of the Academy of Sciences of the USSR (1974).
In 1945, graduated from Moscow State University, where he had been a pupil of A. S. Serebrovskii, worked at the university from 1945 to 1948 and at the Institute of Biological and Medical Chemistry of the Academy of Medical Sciences of the USSR from 1949 to 1953. From 1954 to 1956 headed the subdepartment of organic and biological chemistry at the Kaunas Medical Institute, and from 1956 to 1959 a group of biochemical geneticists at the Institute of Biophysics of the Academy of Sciences of the USSR in Moscow. Since 1959 he has been supervisor of a section of the biology department at the I. V. Kurchatov Institute of Atomic Energy in Moscow.
His principal works deal with genetics, biochemistry, and molecular biology. He showed the influence of the cytoplasm of the egg cell (the "maternal effect") on the manifestation of the majority of genes of the zygote in the embryonic development of *Drosophila* (1945–48).
Works: *Biokhimiia tsitoplazmy.* Moscow 1960. "Rol' belkov v reguliatsii biologicheskoi aktivnosti DNK." *Uspekhi sovremennoi biologii,* 1965, vol. 59. "Sostoianie voprosa o mekhanizmakh reguliatsii sinteza RNK u nizshikh i vysshikh organizmov," *Uspekhi sovremennoi biologii,* 1972, vol. 74. "Struktura khromosom, gistony i aktivnost' genov u drozofily." *Molekuliarnaia biologiia,* 1976, vol. 10, no. 1. (With B. A. Leibovich.)

Khimich, Georgii Lukich
Specialist in heavy machine building; b. O. 22 (N. 4), 1908, Ulanovo, in what is now Glukhov Raion, Sumy Oblast.
Corresponding member of the Academy of Sciences of the USSR (1968). Hero of Socialist Labor (1958). Member of the CPSU since 1952.
After graduating from the S. M. Kirov Urals Industrial Institute in 1936, began working at the S. Ordzhonikidze Urals Heavy Machine-building Plant. In 1956 he became chief designer of the plant's rolling equipment. Under his supervision a number of unique rolling mills were developed as well as continuous steel-casting installations. Decors.: State Prize of the USSR in 1950, Order of Lenin, three other orders, and various medals.

Khisamutdinov, Akhmedzhan
Party official; b. 1932.
Uzbekian; mem. CPSU 1953–; grad. Tadzhik Agricult. Inst. and Party Higher School at Cen. Cttee. CPSU; served in Sov. Army 1952–55; Instructor of a Raion Cttee. Tadzhik CP, Sec. of Party Cttee. in a Kolkhoz 1955–56; held Komsomol

posts 1956–64; Head of a Raion Agricult. Prod. Admin. 1964–68; Inspector Cen. Cttee. Tadzhik CP 1968–69; 1st Sec. of Sovetskii and Piandzh Raion Cttee. Tadzhik CP 1969–74; cand. mem. Cen. Cttee. Tadzhik CP 1971–76, mem. 1976–; 1st Sec. of Kuliab Oblast Cttee. Tadzhik CP 1974–; Dep. USSR Supr. Sov. 1974–.

Khitarov, Nikolai Ivanovich
Geochemist; b. S.20 (O.3), 1903, Piatigorsk.
Corresponding member of the Academy of Sciences of the USSR (1964). Member of the CPSU since 1949.
Graduated from the Leningrad Polytechnic Institute in 1930. From 1929 to 1953 worked at the All-Union Scientific Research Geological Institute. Between 1941 and 1944 he was head of the department of metals of the All-Union Commission on Reserves of Mineral Raw Materials. In 1953 he became head of the laboratory of the origin of magmatic processes, which he had organized at the Vernadskii Institute of Geochemistry and Analytic Chemistry of the Academy of Sciences of the USSR.
His principal works deal with experimental studies of various processes of interaction between the earth's crust and upper mantle. Decors.: Vernadskii Prize (in 1960), A. E. Fersman Prize in 1970, the Order of Lenin, two other orders, and various medals.
Works: *Tseolity, ikh sintez i usloviia obrazovaniia v prirode.* Moscow 1970. (With E. E. Senderov.) *Voda v magmaticheskikh rasplavakh.* Moscow 1971. (With A. A. Kadik and E. B. Lebedev.) *Fiziko-khimicheskie osobennosti glubinnykh protsessov i voprosy global'noi tektoniki.* Moscow 1976. *(16-e chtenie im. V. I. Vernadskogo AN SSSR.)*

Khitrun, Leonid Ivanovich
State official; b. 1930.
Mem. CPSU 1955–; grad. Belorussian Agricultural Acad. 1953; tractor operator, driver of machine and tractor station in Lida, Grodno Oblast 1946–48; chief engineer of a machine and tractor station, chief of an oblast admin. for agriculture 1953–60; Chair. of Grodno Oblast asscn. "Selkhoztekhnika" 1961–62; Chair. of Belorussian Asscn. for Agricultural Technology "Belselkhoztekhnika" 1962–71; Dep. Chair. of Council of Mins. Belorussian SSR 1971–72 and 1976–79; 1st Dep. USSR Min. of Agriculture 1972–76; mem. Cen. Cttee. Belorussian CP 1966–72; dep. chief of a dept. of Cen. Cttee. CPSU 1979–80; Chair. of USSR State Cttee. for Supply of Production Equipment for Agriculture O. 1980–; mem. Cen. Auditing Comm. CPSU 1981–.

Khmara, Irina Grigorevna
Party official; b. 1938.
Mem. CPSU 1965–, grad. Kuban Agricultural Inst. 1962 and CPSU Cen. Cttee. Party Higher Correspondence School 1977; collective farmer, work brigade leader on "Kuban" Collective Farm in Krasnodar Krai 1955–62; work brigade leader, agronomist in fruit nursery section of Novotitarovskaia State Farm 1962–66; chief agronomist at Novotitarovskaia Soft Fruit State Farm in Krasnodar Krai 1966–70, dir. 1970–73; Sec. of Starominskaia Raion Cttee. CPSU in Krasnodar Krai 1973–77; Chair. of Exec. Cttee. of Starominskaia Raion Sov. of People's Deps. 1977–80; 1st Sec. of Starominskaia Raion Cttee. CPSU D. 1980–; cand. mem. Cen. Cttee. CPSU 1981–.

Khodorovich, Tatiana Sergeevna
Scientist; b. 1923, Moscow.
Former mem. scientific staff at Inst. of Russian Language at USSR Acad. of Sciences; mem. of Initiative Group for the Defence of Human Rights in the Sov. Union.

Khodos, Piotr Mikhailovich
State official.
Mem. Cen. Cttee. Kirghiz CP 1960–; Head of Dept. of Agriculture with Cen. Cttee. Kirghiz CP 1962–63; 2nd Sec. of Osh Oblast Cttee. Kirghiz CP 1963–73 (for Agriculture 1963–64); 1st Dep. Chair. of Council of Mins. Kirghiz SSR Je. 1973–; mem. Bureau Cen. Cttee. Kirghiz CP S. 1973–.

Khodulov, Dmitrii Fedorovich
Yakut actor; b. Ag. 4 (17), 1912, Megino-Kanchalasskii *ulus* (district).
People's Artist of the USSR (1958).
In 1934 graduated from the drama studio at the Yakut Music and Drama Theater (since 1971, the P. A. Oiunskii Yakut Drama Theater) and was accepted into the theater's company, has played such varied contemporary roles as the legendary youth Lookut in Smetanin's *Lookut and Niurgusun;* the fighter for the people's happiness Manchary in Protod'iakonov's *Manchary;* the first Yakut Communists Dzhur, Chorkin, and Ivan in Sivtsev's *The Blacksmith Kiukiur* and Mordinov's *The Human* and *The Tear in the Spider's Web;* and the Yakut revolutionary and scholar Neustroev in Kliorn's *The First Sparks.* His parts in the classical repertoire have included the title role in Shakespeare's *Othello,* Egor Bulychov and Satin in Gorkii's *Egor Bulychov and the Others* and *The Lower Depths,* and Neschastlivtsev in Ostrovskii's *The Forest.* Khodulov played V. I. Lenin in Danilov's *On Behalf of the Yakuts* in 1961. He has also staged a number of plays. Decors.: Order of the Red Banner of Labor, the Order of the October Revolution, and several medals.

Khodzhaev, Akram Rustamovich
State official.
Cand. mem. 1960–61, mem. Cen. Cttee. Uzbek CP 1961–; Sec. of Andizhan Oblast. Cttee. Uzbek CP 1961–62; Chair. of Econ. Council Uzbek SSR 1962–63; 1st Sec. of Andizhan Oblast Cttee. for Ind. Uzbek CP 1963–64; Chair. of Andizhan Oblast Exec. Cttee. of Workers' Deps. 1964; Chair. of Econ. Council Uzbek SSR My.–O. 1965; Dep. Chair. of Council of Mins. Uzbek SSR O. 1965–.

Khodzhaev, Asadilla Ashrapovich
Party official; b. 1920.
Uzbek; mem. CPSU 1945–; grad. Inst. of Railway Engineering Tashkent; foreman, sen. foreman, engineer, chief engineer, head of a railway section in Cen. Asia 1942–51; instructor of a dept. at Cen. Cttee. Uzbek CP 1951–54; Dep. Min., 1st Dep. Min. of Motor Transport and Highways Uzbek SSR 1954–63; 1st Sec. of Samarkand Oblast Cttee. for Ind. Uzbek CP 1963–64; Chair. of Samarkand Oblast Exec. Cttee. of Workers' Deps. 1964–67; mem. Cen. Cttee. Uzbek CP 1966–; Chair. of Org. Bureau for Namangan Oblast Cen. Cttee. Uzbek CP 1967–68; 1st Sec. of Namangan Oblast Cttee. Uzbek CP 1968–73; Dep. to 1970–, Sec. 1970–74, mem. Comm. on Youth Affairs at Sov. of Nationalities USSR Supr. Sov. 1974–; 1st Sec. of Tashkent City Cttee. Uzbek CP 1973–78; cand. mem. Je. 1973–76, mem. Bureau Cen. Cttee. Uzbek CP 1976–; Sec. Cen. Cttee. Uzbek CP 1978–. Decor.: Order of Red Banner of Labour.

Khodzhaev, Emud
State official.
Head of Dept. of Propaganda and Agitation at Cen. Cttee. Turkmen CP 1961–63; cand. mem. Cen. Cttee. 1963–66, mem. Cen. Cttee. Turkmen CP 1966–; Chair. of State Cttee. for Radio and Television with Council of Mins. Turkmen SSR 1963–67; Head of Dept. of Propaganda and Agitation, Head of Dept. of Science and Educ. Insts. at Cen. Cttee. Turkmen CP 1967–73; Chair. of State Cttee. for Vocational Training Turkmen SSR F. 1973–79; Chair. of State Cttee. for Printing, Polygraphy and the Book Trade Turkmen SSR N. 1979–.

Khodzhamamedov, Kakadzhan
State official.
Mem. Cen. Cttee. Turkmen CP 1971–; Min. of Public Services Turkmen SSR Ja. 1974–.

Khodzhamuradov, Annamurad
State official.
Min. of Melioration and Water Econ. Turkmen SSR D. 1979–; mem. Cen. Cttee. Turkmen CP 1981–.

Khodzher, Grigorii Gibivich
Nanai writer who writes in Russian; b. Ap. 5, 1929, Verkhnii Nergen, Nanai Raion, Khabarovsk Krai.
Member of the CPSU since 1954.
Graduated from the history department of the Leningrad A. I. Herzen Pedagogical Institute in 1956. Works were first

published in 1953, author of the novellas *Gulls Over the Sea* (1958), *Lake Emoron* (1960), and *The Greatgrandson of Dersu Uzala* (1962). He has also written short-story collections and the trilogy *The Broad Amur* (1970–71; A. M. Gorkii State Prize of the RSFSR, 1973), which depicts the past and present of the Nanai people. Khodzher has also written children's books. His works have been translated into the languages of the peoples of the USSR and into foreign languages. Decors.: Order of the Red Banner of Labor and several medals.

Khodzhiev, Rifat
Party official; b. 1925.
Tadzhik; mem. CPSU 1948–; grad. Pedagogical Inst. Leninabad; served in Sov. Army 1943–45; factory worker, asst. to the dir. of a factory school (FZO) 1945–51; instructor of an oblast cttee. Tadzhik CP, Dept. Head of Leninabad City Cttee. Tadzhik CP, dir. of a hydroelectrical plant, head of a regional board for local ind., sec. of party cttee. of a kolkhoz-sovkhoz prod. admin. 1951–66; mem. Cen. Cttee. Tadzhik CP 1963–; 1st Sec. of Leninabad City Cttee. 1966–70, 1st Sec. of Leninabad Oblast Cttee. Tadzhik CP F. 1971–; Dep. to USSR Supr. Sov. 1970–.

Khokhlachev, Mikhail Egorovich
State official.
Sec. of Frunze City Cttee. Kirghiz CP 1959; mem. Cen. Cttee. Kirghiz CP 1963–71 and 1976–; Min. of Rural Construction Kirghiz SSR 1965–69; Chair. of State Cttee. for Construction Kirghiz SSR Mr. 1973–.

Khokhlov, Aleksandr Stepanovich
Chemist; b. Je. 23 (Jl. 6), 1916, Moscow.
Corresponding member of the Academy of Sciences of the USSR (1964). Member of the CPSU since 1961.
Graduated from the Moscow Institute of Fine Chemical Technology in 1941. From 1948 to 1952 he was a research worker at the Institute of Biological and Medical Chemistry of the Academy of Medical Sciences of the USSR. From 1952 to 1954 he was head of a laboratory at the Institute of Experimental Cancer Pathology and Therapy (now the Institute of Experimental and Clinical Oncology). Between 1954 and 1959 he was head of a section at the All-Union Antibiotics Institute. In 1959 he became deputy director of the M. M. Shemiakin Institute of Bioorganic Chemistry and head of one of the institute's laboratories.
His major works deal with the study of antibiotics, antineoplastic agents, microbial bioregulators, and other natural substances. Decors.: Order of the October Revolution, the Order of the Red Banner of Labor, and several medals.
Works: *Khimiia antibiotikov,* 3rd ed., vol. 1. Moscow 1961. (With others.) *Khimicheskie reguliatory biologicheskikh protsessov.* Moscow 1969. (With Iu. A. Ovchinnikov.) Bumazhnaia khromatografiia antibiotikov. Moscow 1970. (With N. O. Blinov.)

Khokhriakov, Viktor Ivanovich
Actor and film director; b. 1913.
Mem. CPSU 1939–; grad. Drama College Leningrad 1933; engagement at Theatre for Russian Drama Kharkov 1933–40, Transport Theatre Moscow 1940–53, Mali Theatre Moscow 1953–. Decors.: People's Artist of the USSR 1973, State Prize of the USSR 1949 and 1951.

Kholov, Makhmadullo
State official; b. 1920.
Tadzhik; mem. CPSU 1947–; grad. Party School of Cen. Cttee. Tadzhik CP 1954; teacher 1940; served in Sov. Army 1940–42 and 1944–47; Sec. of Bolozhar Village Sov. 1942–44; Sec. of Shulmak Raion Cttee. Tadzhik Komsomol 1947–51; student, sec. of a raion cttee. Tadzhik CP 1951–54; 1st Sec. of Obil-Garm Raion Cttee. Tadzhik CP, Chair. of Kuibyshevskii Raion Cttee., 1st Sec. of Moskovskii Raion Cttee. Tadzhik CP 1954–63; mem. 1960–, mem. Bureau Cen. Cttee. Tadzhik CP 1963–; Dep. to Supr. Sov. Tadzhik SSR 1959–; Chair. of Presidium Supr. Sov. Tadzhik SSR Mr. 1963–, Dep. to and Dep. Chair. of Presidium USSR Supr. Sov. 1964–; mem. Cen. Auditing Comm. 1966–71; cand. mem. Cen. Cttee. CPSU 1971–. Decors.: Order of Red Banner of Labour, Order of Lenin, various medals.

Khomenko, Nikolai Grigor'evich
State official.
2nd Sec. of Sumy Oblast Cttee. Ukrainian CP 1978–82; Dep. of Supr. Sov. Ukrainian SSR 1980–; mem. Auditing Comm. Ukrainian CP 1981–; Sec. of Presidium Supr. Sov. Ukrainian SSR D. 1982–.

Khomentovskii, Aleksandr Stepanovich
Geologist; b. Mr. 11 (24), 1908, Leningrad.
Corresponding member of the Academy of Sciences of the USSR (1960). Member of the CPSU since 1941.
Graduated from the Siberian Technological Institute (now the Tomsk Polytechnic Institute) in 1930. From 1930 to 1937 he was head of a geological group in Siberia. From 1938 to 1941 he worked at the Krasnoiarsk Pedagogical Institute, and from 1943 to 1954, at the Iuzhuraluglerazvedka trust in Orenburg. In 1955 he became a professor at the University of Saratov, where he was also head of a subdepartment from 1955 to 1957. In the periods 1958–60 and 1971–72 he was a professor at the Perm' Polytechnic Institute. In 1973 has became a professor at the Orenburg Polytechnic Institute.
His principal works deal with the classification, tectonics, and regularities of formation of coal deposits in Siberia and the Urals. Decors.: State Prize of the USSR in 1950, Order of the Badge of Honor and various medals.

Khomiakov, Aleksandr Aleksandrovich
Party official; b. 1932.
Mem. CPSU 1958–; grad. Novocherkassk Polytechnical Inst. 1955, Cand. of Econ. Sc. 1974; foreman, dep. shop superintendent at an electrical measuring instruments plant in Krasnodar 1955–58; held komsomol, sov. and party posts 1958–65; 2nd Sec. of Krasnodar City Cttee. CPSU 1965–69; Sec. of Krasnodar Krai Cttee. CPSU 1969–71, 2nd Sec. 1971–78; 1st Sec. of Tambov Oblast Cttee. CPSU Mr. 1978–; Dep. to USSR Supr. Sov. 1979–; mem. Cen. Cttee. CPSU 1981–.

Khomulo, M. G.
Military official.
Col.-Gen.; served at the front 1941–45; 1st Dep. Commander-in-Chief of Sov. Forces Group in Germany until 1969; Commander of Siberian Mil. District 1969–79.

Khoroshilov, Ivan Ivanovich
Agronomist; b. 1907.
Mem. CPSU 1938–; grad. Agricultural Inst. Voroshilovgrad 1926; Sen. Specialist, Dep. Chief at Experimental Station Artemovsk 1926–31; dep. dir. of a cereals sovkhoz in Rostov Oblast 1931–39; dir. of an experimental station 1939–43; Head of Agricultural Regional Admin. in Rostov 1943–49; Head of Main Admin. of USSR Min. of Agriculture 1949–53, 1957–58 and 1965–; Consultant to USSR Embassies in Rumania and Canada 1953–57, 1961–65; mem. Apparat Cen. Cttee. CPSU 1958–61. Decor.: Lenin Prize (for active participation in taking preventive measures in the Western Siberian prairies to protect Kazakhstan against wind erosion).

Khorunzhii, Mikhail Vasil'evich
State official; b. 1920.
Mem. CPSU 1943–; cand. mem. 1960–76, mem. Cen. Cttee. Ukrainian CP 1976–; Chair. of Oblast Exec. Cttee. of Workers' Deps. in Odessa 1958–70 (for Agriculture 1963–64) and in Zaporozhe 1970–76; Dep. to 1963–, mem. Presidium 1963–75, Dep. Chair. of Supr. Sov. Ukrainian SSR 1975–76; Min. of Agriculture Ukrainian SSR F. 1976–.

Khovrin, Nikolai Ivanovich
Military official; b. 1922.
Russian; mem. CPSU 1943–; grad. Pacific Ocean Naval College and Naval Acad., Admiral; active war duty in Sov. Navy 1941–45; command posts, commander of various units of Pacific Fleet 1945–70; 1st Dep. Commander of Northern Fleet 1970–74; commander of Black Sea Fleet 1974–; mem. Cen. Cttee. Ukrainian CP 1976–; Dep. to USSR Supr. Sov. 1979–.

Khrapchenko, Mikhail Borisovich
Literary historian; b. 1904.
Mem. CPSU 1928–; grad. Smolensk Univ. 1924; mem. USSR Acad. of Sciences 1966–; Dep. Chair. 1938–39, Chair. of Art Cttee. 1939–48; Sen. Research Staff Mem. at Gorkii Inst. of World Literature 1948–; concurrently Chief Ed. of periodical "Oktiabr" 1954–57; Dep. Acad. Sec. 1957–62, Acting Sec. 1962–66, Acad. Sec. of Dept. of Language and Literature at USSR Acad. of Sciences 1966–. Decor.: Lenin Prize 1974.

Khrennikov, Tikhon Nikolaevich
Composer; b. 1913, Elets.
Russian; mem. CPSU 1947–; grad. Moscow Conservatory 1936; Sec. of Elets Village Sov., attended Music Tech. Coll. Moscow 1929–32; Head of Music Dept. at Cen. Theatre of Sov. Army 1941–54; Sec. Gen. 1948–57, 1st Sec. of USSR Composers' Union 1957–; mem. Cen. Auditing Comm. CPSU 1961–76; Dep. to USSR Supr. Sov. 1962–; cand. mem. Cen. Cttee. CPSU 1976–; Dep. Chair. of USSR Parliamentary Group. Decors.: 3 Stalin Prizes, 2 Orders of Lenin, Order of Red Banner of Labour, State Prize of the USSR 1942, 1946, 1952, 1966, Lenin Prize 1974.
Works: operas "V buriu" (In the Storm) 1936–39, "Odinochestvo" (Solitude), "Frol Skobeev" 1946–49, "Mat'" (Mother) 1957, 2 symphonies (1935, 1940– 43), operetta "Sto chertei i odna devushka" (A Hundred Devils and One Girl) 1963, and others.

Khrenov, Arkadii Fedorovich
Military official; b. Ja.24 (F.5), 1900, Ocherskii zavod, now the city of Ocher, Perm' Oblast; s. of a worker.
Colonel-general of the engineer troops (1944). Hero of the Soviet Union (Mr. 21, 1940). Member of the CPSU since 1931. Joined the Red Army in 1918 and served in the Civil War of 1918–20. Graduated from advanced engineering courses for officers in 1929 and the Higher Academic Courses at the Military Academy of the General Staff in 1949. Served in the Soviet-Finnish War of 1939–40 as chief of the engineer troops of the Seventh Army, became chief of the Main Military Engineering Directorate of the Red Army in July 1940 and chief of the engineering directorate of the Moscow Military District in May 1941. During the Great Patriotic War of 1941–45, he was chief of the engineering directorate of the Southern Front and deputy commander in charge of engineering in the defense of Odessa and Sevastopol' in 1941 and 1942. He was chief of the engineer troops of the Crimean Front from April to May 1942, the Leningrad and Volkhov fronts from 1942 to 1944, the Karelian Front in 1944 and 1945, and the First Far Eastern Front in 1945.
After the war, served as chief of the engineer troops of the Primor'e Military District. He was commander of the engineer troops of the Far East from December 1945 to May 1949 and inspector general of the engineer troops of the Main Inspectorate of the Ministry of Defense of the USSR from 1949 to 1960. He retired in September 1960. Decors.: three Orders of Lenin, three Orders of the Red Banner, two Orders of Kutuzov First Class, the Order of Kutuzov Second Class, the Order of Suvorov Second Class, and various medals. Foreign orders and medals.

Khristianovich, Sergei Alekseevich
Mathematician and mechanics scientist; b. O. 27 (N.9), 1908.
Mem. CPSU 1949–; grad. Leningrad Univ. 1930; mem. USSR Acad. of Sciences 1943–; mem. staff State Hydrological Inst. Leningrad 1930–37; mem. staff Cen. Aero-Hydrodynamic Inst. 1937–53; mem. Presidium USSR Acad. of Sciences 1946–56; mem. staff Inst. of Chemical Physics of USSR Acad. of Sciences 1956–; Dep. Chair. of Presidium of Siberian Branch of USSR Acad. of Sciences 1959–. Decors.: State Prize of the USSR 1942, 1946 and 1952, 3 Orders of Lenin, other orders and various medals.

Khristoradnov, Iurii Nikolaevich
Party official; b. 1929.
Mem. 1951–; grad. All-Union Correspondence Course Inst. of Econ. and Finance 1959; foreman, asst. shop superintendent, shop superintendent of automobile plant in Gorkii 1949–62; party posts 1962–66; 2nd Sec. of Gorkii City Cttee. CPSU 1966–68, 1st Sec. 1968–74; Dep. to USSR Supr. Sov. 1974–; 1st Sec. of Gorkii Oblast Cttee. CPSU My. 1974–; mem. Cen. Cttee. CPSU 1976–. Decor.: Order of Lenin.

Khrizostom (Georgii Fedorovich Martishkin), Cand. of Theol.
Russian Orthodox clergyman; b. My. 3, 1934, Kazinka, Riazan Oblast.
Curator and restorer of monuments 1951–61; grad. Moscow Theological Seminary 1961; grad. Moscow Theological Acad. 1970 as Cand. of Theology; deacon 1964; monk, then priest-monk 1966; pastoral work 1968–71; Archimandrite 1970; mem. Russian Orthodox Church Del. at the 4th Christian Peace Conference in Prague 1971; Head of the Chancery at Church Dept. of External Affairs in Moscow Patriarchate 1971; Bishop of Zaraisk Ap. 1972–74; Substitute Bishop of Diocese of Moscow; Bishop of Kursk and Belgorod 1974–78, of Kirov and Slobodskoi Ap. 1978–.

Khromov, Pavel Alekseevich, Dr. of Econ. Sc.
Economist; b. O. 15, 1907, Koverin, now Vladimir Oblast; s. of a worker.
Mem. CPSU 1929–; completed postgraduate studies at Inst. of Econ. Research of GOSPLAN USSR 1933; specialist in ind. econ. and theory and history of econ.; corresponding mem. Ukrainian Acad. of Sciences 1939–; Dr. of Econ. Sc. 1939; Prof. 1944; active in planning offices in Vladimir and Ivanovo 1929–30; scientific and teaching work 1935–; Dir. of Inst. of Econ. at USSR Acad. of Sciences 1942–48; Prof. at Lomonosov Univ. Moscow 1944–57; mem. staff Acad. of Social Sciences of Cen. Cttee. CPSU 1946–. Decors.: Order of Red Banner of Labour, Order "Honorary Badge".
Publs.: author of some 100 works, "Ekonomicheskoe razvitie Rossii v XIX–XX vekakh" 1950, "Ocherki ekonomiki Rossii perioda monopolisticheskogo kapitalizma" 1960, "Ekonomicheskoe razvitie Rossii" 1967.

Khropotinskii, Vasilii Petrovich
State official.
Mem. Cen. Cttee. Moldavian CP 1966–; Chair. of State Cttee. for Printing and Publishing 1971–72, Chair. of State Cttee. for Publishing, Printing and the Book Trade Moldavian SSR Ag. 1972–.

Khrunov, Evgenii Vasil'evich
Cosmonaut; b. 1933.
Mem. CPSU 1959–; grad. Mil. Aviation College Bataisk 1956 and correspondence courses at Zhukovskii Mil. Aviation Engineering Acad. 1968; Air Force service 1956–60; cosmonaut 1960–; space flight in spaceship "Soiuz 5" together with B.V. Volynov and A.S. Eliseev Ja. 15, 1969; transferred to spaceship "Soiuz 4" (pilot V.A. Shatalov) and landed Ja. 17, 1969. Decors.: Hero of the Sov. Union 1969, Order of Lenin, Order of the Red Star, medals.

Khubiev, Osman Akh'iaevich
Karachai writer; b. F. 17, 1918, in the *aul* (village) of Verkhniaia Teberda, in what is now the Karachai-Cherkess Autonomous Oblast.
Member of the CPSU since 1956.
Served in the Great Patriotic War (1941–45), is author of the poetry collection *Komsomol Songs* (1936), which includes poetry about his native land, the party, friendship among peoples, and the Komsomol. He has also written sketches about workers and the novella *The Abrek* (1938), which depicts the struggle for Soviet power. *The Hostage* (vols. 1–3, 1959–65) is a trilogy based on documentary material concerning the struggle of the Karachai people against fascist aggression. The novella *People* (1966) poses problems concerning the moral development of young people.
Works: *Ant.* Cherkessk 1963. *Murat.* Cherkessk 1968. *Juqüsiz kechele.* Cherkessk 1969. *Dert.* Cherkessk 1971. *Jilla.* Cherkessk 1974. *In Russian translation: Vremia: Stikhi.* Cherkessk 1957.

Khudaiberdyev, Narmankhonmadi Dzhuraevich
State official; b. 1928.
Uzbek; mem. CPSU 1948–; grad. Uzbek Agricultural Inst. 1949; dept. head, sec. of a raion Uzbek Komsomol cttee., postgraduate studies after completing inst., Lecturer and Assoc. Prof. at Agricultural Inst. Samarkand 1943–54; leading CPSU and govt. posts 1954–; Sec. of Bukhara Oblast Cttee. Uzbek CP, Head of Agricultural Dept. at Cen. Cttee. Uzbek CP, 2nd Sec. of Bukhara Oblast Cttee. Uzbek CP 1956–60; Dep. to Supr. Sov. Uzbek SSR 1959–63 and 1967–; mem. Cen. Cttee. Uzbek CP 1960–; Dep. Chair. of Council of Mins. Uzbek SSR 1960–61; 1st Sec. of Surkhan-Daria Oblast Cttee. Uzbek CP 1961–D. 1962; cand. mem. 1961–66, mem. Cen. Cttee. CPSU 1971–; Dep. to 1962–66 and 1974–, mem. Foreign Affairs Comm. at Sov. of the Union USSR Supr. Sov. 1962–66; Sec. and mem. Presidium Cen. Cttee. Uzbek CP D. 1962–65; Chair. of Agricultural Bureau Cen. Cttee. Uzbek CP 1962–64; Chair. of Council of Mins. Uzbek SSR 1971–; mem. Bureau Cen. Cttee. Uzbek CP Mr. 1971–. Decor.: Order of Red Banner of Labour.

Khudaiberdyev, Rozykuly
Party official; b. 1936.
Turkmen; mem. CPSU 1961; grad. Turkmen Agricultural Inst. Dep. Chair. of a Kolkhoz 1959–62; Engineer, Chef of a Raion Assoc. "Turkmenselkhoztekhnika" 1962–68; 2nd Sec. 1968–71, 1st Sec. of a Raion Cttee. Turkmen CP in Chardzhou 1971–77; mem. Cen. Cttee. Turkmen CP 1971–; 1st Sec. Oblast Cttee. Turkmen CP in Chardzhou 1977–.

Khudaibergenov, Abdulla Matkarimovich
State official.
Min. of Health Uzbek SSR Je. 1981–.

Khudaibergenov, Madiar
Party official; b. 1924.
Uzbek; mem. CPSU 1949–; grad. Urgench Teaching Inst. and Party High School of Cen. Cttee. CPSU; teacher 1941–43; served in Sov. Army 1943–45; head of teaching dept., dir. of a school, head of a raion dept. for educ. 1945–51; sec. of a raion cttee. Uzbek CP, Dept. Head of a dept. at Khorezm Oblast Cttee. Uzbek CP, 1st sec. of various raion cttees. Uzbek CP 1951–65; Chair. of Khorezm Oblast Exec. Cttee. of Workers' Deps. 1965–68; mem. Cen. Cttee. Uzbek CP 1966–; 1st Sec. of Khorezm Oblast Cttee. Uzbek CP 1968–; Dep. to Supr. Sov. 1970–.

Khudaikuliev, Meretmamet
State official.
Min. for Forestry Turkmen SSR Ja. 1980–; cand. mem. Cen. Cttee. Turkmen CP 1981–.

Khuramshin, Talgat Zakirovich
State official.
Chair. RSFSR State Cttee. for the Supply of Petroleum Products 1979–81; Chair. USSR State Cttee. for the Supply of Petroleum Products Ag. 1981–.

Khusainov, Iurii Minivalich
State official; b. 1929.
Russian; mem. CPSU 1953–; grad. Inst. of Agricultural Machinery Building in Zaporodzhe, Cand. of Econ. Sciences; foreman of Technical Control Dept., technical engineer at Gomelmash factory 1952–53; chief engineer of a machine tractor station and of regional agriculture admin. in Gomel, 2nd City Sec. CP Belorussia in Mozyr 1953–62; 1st Dep. Chair., Dep. Chair. of Oblast Exec. Cttee. of People's Deps. in Gomel 1962–67; Dep. Min. of Agriculture Belorussian SSR 1967–74; Chair. of Oblast Exec. Cttee. of People's Deps. in Gomel 1974–78; mem. Cen. Cttee. CP Belorussia 1976–; 1st Oblast Cttee. Sec. CP Belorussia in Gomel 1978–82; 1st Dep. Chair. of Council of Mins. Belorussian SSR Jl. 1982–; mem. Bureau Cen. Cttee. Belorussian CP D. 1982–.

Khushvakhtov, Khushbakht Davliatovich
Painter; b. D.9, 1926, Rushan, Rushan Raion, Gorno-Badakhshan Autonomous Oblast.
People's Artist of the Tadzhik SSR (1974). Member of the CPSU since 1962. Studied at the V.I. Surikov Moscow Art Institute from 1953 to 1960. Has served as chairman of the board of the Artists' Union of Tadzhikistan since 1961 taught at the Republic Art School in Dushanbe in 1960–61. Khushvakhtov's landscapes and thematic pictures depicting the Pamir's majestic scenery and its inhabitants are characterized by a restrained palette, vividness, and traditionalist artistic solutions. Examples are *Khoroq* (1961), *The Alai Valley* (1962), *In Years Gone By* (1967), *Native Craftsmen* (1970) and *Yaks* (1970); all are at the Bekhzad Republic Museum of History, Local Lore, and Fine Arts in Dushanbe.

Khusnutdinov, Ibragim Kamalovich
Party official.
Mem. Cen. Cttee. Uzbek CP 1960–61 and 1971–, cand. mem. 1961–71; Head of Dept. of Planning and Finance of Cen. Cttee. Uzbek CP D. 1974–.

Khutsiev, Marlen Martynovich
Motion-picture director; b. O.4, 1925, Tbilisi.
People's Artist of the RSFSR (1977).
Entered the motion-picture industry in 1944 as an assistant to the art director. In 1950 graduated from the directing department of the All-Union State Institute of Cinematography and made his debut as a director. Major themes of Khutsiev's work are the problems and relationships of contemporary youth. Detailed, unhurried unfolding of the plot line, distinguished by documentary accuracy, deep emotional drama, and subtle lyricism, characterizes the films *Spring on Zarechnaia Street* (1956, in collaboration with F.K. Mironer), *Two Fedors* (1959), *I'm Twenty* (1965, awarded a prize at the Twenty-sixth International Film Festival in Venice), *July Rain* (1967), and *It Was May* (1970, made for television). Khutsiev also directed the television documentary *The Red Flag of Paris* (1971). Decor.: Order of the Badge of Honor.

Khynku, Mikhail Sergeevich
State official.
Chair. of the Agrarian-Ind. Asscn. for Production of Volatile Oils, Perfumes and Cosmetics Moldavian SSR Jl. 1981–.

Kiao, Vladimir Aleksandrovich
State official.
Mem. Cen. Cttee. Esthonian CP 1960–61 and 1966–; Vice-Chair. of Econ. Council Esthonian SSR 1963–65; Min. of Local Ind. Esthonian SSR 1965–75; Vice-Chair. of Council of Mins. Esthonian SSR M. 1975–78; mem. Bureau and Sec. Cen. Cttee. Esthonian CP 1978–.

Kiasimov, Telman Mamed ogly
State official.
Head Main Admin. of Installation and Special Construction Projects with Council of Mins. Azerbaidzhan SSR Je. 1976–82; Min. of Installation and Special Construction Azerbaidzhan SSR Ap. 1982–.

Kibal'nikov, Aleksandr Pavlovich
Sculptor; b. Ag.9 (22), 1912, Orekhovo, in present-day Volgograd Oblast.
People's Artist of the USSR (1963); member of the Academy of Arts of the USSR (1954). Became a member of the CPSU in 1963.
He was the chairman of the governing board of the Moscow section of the Artists' Union of the RSFSR from 1963 to 1966. His works, which include monuments and portraits, are distinguished by psychological expressiveness. He has done portraits of N.G. Chernyshevskii (bronze, 1948; State Prize of the USSR, 1949), J.V. Stalin (plaster of Paris, 1950; State Prize of the USSR, 1951); V.V. Mayakovsky (bronze, 1954); and P.M. Tret'iakov (marble, 1962) – all are in the Tret'iakov Gallery. He created a monument to N.G. Chernyshevskii in Saratov (bronze and granite, unveiled in 1953), a monument to V.V. Mayakovsky in Moscow (bronze and granite, unveiled in 1958; Lenin Prize, 1959), and the memorial complex *Brest Hero-fortress* in Brest (with the sculptor A.O. Bembel' and others, the architect V.A. Korol' and others; concrete, stone, and metal; unveiled in 1971). Decors.: Order of the Red Banner of Labor and several medals.

Kiiak, Grigorii Stepanovich, Dr. of Agric. Sc.
Scientist; b. Ja. 3, 1910, Vinniki, Lvov Oblast.
Grad. Dept. of Agronomy of Polytech. Inst. Lvov 1934; specialist in plant breeding; corresponding mem. Ukrainian Acad. of Sciences 1951–; Dr. of Agric. Sc. 1956; Prof. 1949–; Asst. Lecturer 1935–39, Lecturer at Polytech. Inst. Lvov 1939–44; Prof. Chair. of Plant Breeding at Agricultural Inst. Lvov 1944–; concurrently Head of Dept. of Plant Breeding at Inst. of Agricultural Biology of Ukrainian Acad. of Sciences 1951–60. Decors.: Order of Lenin, Order of Red Banner of Labour, medals.
Publs.: Author of more than 90 scientific and popular-science works, "Dosvid vyroshchuvannia kukurudzy (v zakhidnykh raionakh URSR)" 1959; "Agrotekhnika i nasinnytstvo kormovoho liupynu v zakhidnykh raionakh Ukrainy" 1962 (co-author), "Roslynnytstvo" 1964.

Kiiatkhanov, Berdy Ashirovich
State official.
Min. of Health Turkmen SSR Mr. 1980–; mem. Auditing Comm. Turkmen CP 1981–.

Kiikov, Aleksei Iakovlevich
State official; b. 1921, Koltyk, Altai.
Mem. CPSU 1944–; grad. Leningrad Inst. for Raising Qualification Standards with USSR Min. of Wood Ind. 1951 and Party High School of Cen. Cttee. CPSU 1960; various posts in Krasnoiarsk area 1936–41; served in Sov. Army 1941–42; foreman, dept. head, chief engineer in various firms in wood ind. 1942–49; Dir. of "Komintern" Wood Combine, Chair. of Borisov City Exec. Cttee. of Workers' Deps., 1st Sec. of Borisov City Cttee. Belorussian CP 1951–60; Dep. Head of Dept. of Ind. and Transport at Cen. Cttee. Belorussian CP 1960–62; cand. mem. Cen. Cttee. Belorussian CP 1961–81, mem. 1981–; 2nd Sec. of Mogilev Oblast Cttee. for Ind. Belorussian CP 1962–64; Head of Admin. for Wood, Paper and Wood-processing Ind. of Econ. Council Belor ussian SSR 1964–65; Min. of Forestry and Wood-Processing Ind. Belorussian SSR 1965–. Decors.: Order of the October Revolution, 2 other orders and various medals.

Kikoin, Isaak Konstantinovich
Physicist; b. Mr. 15 (28), 1908, Zhagory.
Grad. Polytech. Inst. Leningrad 1930; mem. staff Physical-Tech. Inst. Leningrad 1930–36, Inst. of Physics of Metals Sverdlovsk 1936– and Inst. of Atomic Energy 1943–; mem. USSR Acad. of Sciences 1953–; Prof. at Inst. of Physical Engineering Moscow 1944–; Prof. at Moscow Univ. 1955–; Chair. of Comm. on Physics Curricula; main research work on nuclear physics and nuclear engineering and physics of solid bodies. Decors.: Hero of Socialist Labour, Lenin Prize 1959, State Prize of the USSR 1942, 1949, 1951, 1953, 1967, 5 Orders of Lenin, 3 other orders, several medals.

Kiktenko, Vladimir Konstantinovich
Party official; b. 1932.
Moldavian; mem. CPSU 1959–; grad. Kiev Technological Inst. of Light Ind. and Cen. Cttee. Party Higher School CPSU; technician, engineer 1952–59; engineer, dep. head of a dept., dispatcher, Dir. of a clothing factory, chief engineer of a clothing asscn. 1959–64; dep. head of an admin., Dep. Min. of Light Ind. Moldavian SSR 1964–67; Cen. Cttee. inspector 1967–70; Min. of Domestic Services Moldavian SSR 1970–78; mem. Auditing Comm. CP Moldavia 1971–81, mem. Cen. Cttee. CP Moldavia 1981–; 1st City Sec. CP Moldavia in Kishinev Jl. 1978–.

Kilchevskii, Nikolai Aleksandrovich, Dr. of Phy.-Math. Sc.
Scientist; b. Je. 15, 1909, Kamianets Podolskii; s. of a teacher.
Mem. CPSU 1961–; grad. Kiev Inst. of Physics, Chemistry and Mathematics 1933; scientific work in mechanics and mathematics; Dr. of Phy.-Math. Sc. and Prof. 1941; Research Asst., Lecturer, Prof. Chair at Kiev Aviation and Ind. Inst. 1933–41; Prof. Chair at Kiev Polytech. Inst. 1944–61; concurrently mem. staff Kiev Univ. 1945–58; mem. staff Inst. of Mechanics at Ukrainian Acad. of Sciences 1961–; corresponding mem. Ukrainian Acad. of Sciences 1961–. Decors.: Order of Red Banner of Labour, Order "Honorary Badge", medals.
Publs.: author of more than 100 scientific works, among them 3 monographs; "Elementy tenzornogo ischisleniia i ego prilozhenie k mekhanike" 1954, "Kurs teoretychnoi mekhaniki" vols. 1–2 1955–57, "Osnovy analiticheskoi mekhaniki obolochek" 1963.

Kilchichakov, Mikhail Eremeevich
Poet; b. 1919.
Chakass; grad. Gorkii Inst. of Literature Moscow 1953.
Publs.: "Somnenie" (Doubt) 1947, "Otchego zhelteiut listia?" (Why Do the Leaves Turn Yellow?) 1961, and others.

Kim, Maksim Pavlovich
Historian; b. My. 12 (25), 1908, Putsilovka, now Primore Krai.
Mem. CPSU 1929–; grad. Inst. of History, Philosophy and Literature Moscow 1934; teaching activity at Moscow colleges until 1951; mem. staff Inst. of USSR History at USSR Acad. of Sciences 1951–; Prof. Chair at Acad. of Social Sciences of Cen. Cttee. CPSU 1965–71; Chief Ed. of periodical "Istorija SSSR", Dep. Ed. of multivolume "Istorija SSSR s drevnejših vremen do naših dnej" 1956–60. Decors.: 3 orders, several medals.

Kimantaite, Kazimira Martynovna
Actress and film director; b. Je. 29 (Jl. 12), 1909, Kurshenai, Lithuania.
Lithuanian; mem. CPSU 1945–. Decors.: People's Artist of the Lithuanian SSR, 3 orders, several medals.

Kimask, Georgii Karlovich
State official.
Head of Cen. Statistical Admin. Esthonian SSR N. 1968–; mem. Auditing Comm. Esthonian CP 1971– 81, Chair. 1981–.

Kimiagarov, Bension Arievich
Film director; b. S. 30, 1920, Samarkand.
Mem. CPSU 1944–. Decors.: People's Artist of the Tadzhik SSR, 4 orders, several medals.

Kinkladze, Otar Danilovich
State official; b. 1922.
Georgian; mem. CPSU 1945–; grad. Dept. of Philosophy of Tbilisi State Univ.; Chair. of State Cttee. on Films with Council of Mins. Georgian SSR 1963–65; scientific and teaching work at Ia. Godebashvili Inst. of Pedagogics and at Tbilisi State Univ. 1965–73; Dr. of Pedag. Sc. 1969; Prof. 1971–; Ed. of newspaper "Zaria Vostoka" 1973–76; mem. Cen. Cttee. Georgian CP 1976–; Min. of Educ. Georgian SSR Ag. 1976–.

Kipshakbaev, Nariman
State official.
Dep. Min. of Melioration and Water Econ. Kazakh SSR 1974–81, Min. Jl. 1981–.

Kirakosian, Aleksan Matevosovich
State official; b. 1918.
Dep. to Supr. Sov. Armenian SSR 1955–; 1st Sec. of Gukasian Raion Cttee. Armenian CP 1955; mem. Cen. Cttee. Armenian CP 1956–; 1st Sec. of Oktemberian Raion Cttee. Armenian CP 1959; Head of Dept. for Party Organs, Head of Dept. of Agriculture at Cen. Cttee. Armenian CP 1961–65; Dep. Chair. of Bureau of Agriculture at Cen. Cttee. Armenian CP 1962–64; Min. of Meat and Dairy Ind. Armenian SSR 1965–70; Dep. Chair. of Council of Mins. Armenian SSR D. 1970–77, 1st Dep. Chair. F. 1977–.

Kirakosian, Dzhon Saakovich, Dr. of Hist. Sc.
State official.
Head of Sub.-Dept. of Propaganda and Agitation, Dep. Head of Ideology Dept. at Cen. Cttee. Armenian CP 1963; mem. Bureau for Ind. at Cen. Cttee. Armenian CP 1963–64; mem. Cen. Cttee. Armenian CP 1966–; Chair. of State Cttee. for Radio and Television with Council of Mins. Armenian SSR 1966–69; Head of Dept. of Science and Educ. Insts. at Cen. Cttee. Armenian CP 1969–75; Min. of Foreign Affairs Armenian SSR Jl. 1975–.

Kireenko, Kastus' (Konstantin) Tikhonovich
Author and poet; b. 1918, Gaishin, Mogilev Oblast.
Belorussian; mem. CPSU 1943; first works publ. 1939.
Publs.: "Utro idet" (Morning is Coming) 1945, "Posle grozy" (After the Thunderstorm) 1947, "Moia respublika" (My Republic) 1949, "Liubov' i druzhba" (Love and Friendship) 1955, "Zhivye idut vpered" (The Living Go Forward) 1964, and others.

Kireev, Akhniaf Nureevich (pseudonym: Kirei Mergen)
Author and literary historian; b. 1912, Kigazy, Bashkir ASSR; s. of a teacher.
Mem. CPSU 1941–; grad. Acad. of Social Sciences of Cen. Cttee. CPSU 1954; first works publ. 1930.
Publs.: "Suetlivyi chelovek" (A Fussy Person) 1940, "Bashkiry" (The Bashkirs) 1943, "Dzhigity" (The Djigits) 1944, "Na sklonakh Narysh-Tau" (On the Slopes of Narysh-Tau) 1951, "Puteshestvie v derevniu" (Journey to a Village) 1940, "Moia semia" (My Family) 1942, "Voina" (The War) 1944, "Utro goroda" (The Morning of a Town) 1958, "Bashkirskii narodnyi epos" (The Bashkir Popular Epos) 1961, "Zerkalo dushi narodnoi" (Mirror of the Peoples' Soul) 1960, "Pesennaia tetrad" (Song Notebook) 1964, and others.

Kirichenko, Fedor Grigorevich
Scientist; b. F.17 (Mr.1), 1904, Vladislavka, Kiev Oblast.
Mem. CPSU 1929–; grad. K.A.Timiriazev Inst. for Selection and Seed Cultivation in Maslovka (Kiev Oblast) 1928; specialist in selection and plant breeding; mem. USSR V.I.Lenin Acad. of Agricultural Sciences 1956–; mem. staff 1932–, Dir. of Union Inst. for Selection and Genetics in Odessa 1954–59. Decors.: Hero of Socialist Labour, State Prize of the USSR 1949, Lenin Prize 1959, 2 Orders of Lenin, Order of the October Revolution, 3 other orders, several medals.

Kirichenko, Iurii Alekseevich
Diplomat.
USSR Amb. to Iceland 1973–75 and to Norway Ap. 1975– F. 1982.

Kirichenko, Nikolai Karpovich
Party official; b. 1923, Ukraine; s. of a peasant.
Mem. CPSU 1944–; grad. Kharkov Teachers' Inst. 1950 and Kharkov Pedagogical Inst. 1955; served in Sov. Army 1941–45; various CPSU, Sov. and Komsomol posts 1946–47; 1st Sec. of Kupriansk City Komsomol Cttee. 1947–49; 2nd, 1st Sec. of Kharkov Oblast Komsomol Cttee. 1949–55; 2nd Sec. and mem. Bureau of Cen. Cttee. Ukrainian Komsomol 1955–60; Sec., 2nd Sec. of Poltava Oblast Cttee. Ukrainian CP 1960–62; Chair. of Poltava Oblast Exec. Cttee. of Workers' Deps. 1962–63; 1st Sec. of Kirovograd Oblast Rural Cttee. Ukrainian CP 1963–64; Chair. of Kirovograd Oblast Exec. Cttee. of Workers' Deps. 1964–65; 1st Sec. of Kirovograd Oblast Cttee. Ukrainian CP 1965–67; mem. Cen. Cttee. Ukrainian CP 1966–; Dep. to USSR Supr. Sov. 1966–; 1st Sec. of Crimean Oblast Cttee. Ukrainian CP Ap. 1967–77, in Odessa 1977–; cand. mem. Cen. Cttee. CPSU 1971–77, mem. 1977–. Decor.: Order of October Revolution.

Kirichenko, Olena Stepanovna
Painter; b. 1937, Kiev.
Grad. Kiev Art Inst. 1963, pupil of O.Shovkunenko.

Kirichenko, Stepan Andreevich
Painter; b. 1911, Chopovichi, Zhitomir Oblast.
Mem. CPSU 1959–; grad. Kiev Art Inst. 1941, pupil of F.Krichevskii, A.Taran, M.Rokitskii. Decor.: Merited Artist of the Ukrainian SSR 1963.

Kiriiak, Nellia Pavlovna
State official; b. 1935; d. of a white-collar worker.
Moldavian; mem. CPSU 1961–; grad. Moscow Technological Inst. of Light Ind. and Cen. Cttee. Party Higher School CPSU; master, head of a shift, head of a shop at an enterprise for imitation leather and rubber products in Kishinev 1958–63; instructor in the Dept. for Ind. and Transport Cen. Cttee. CP Moldavia 1963–64; attended Cen. Cttee. Party Higher School CPSU, dep. head of Dept. for Ind. and Transport Cen. Cttee. CP Moldavia 1964–69; Dep. Min., 1st Dep. Min. of Local Ind. Moldavian SSR 1969–72; 1st Sec. of Lenin Raion Cttee. CP Moldavia in Kishinev 1972–74; head of Planning and Finance Organs Dept. Cen. Cttee. CP Moldavia 1974–76; Dep. to USSR Supr. Sov. 1975–; head of Dept. for Light Ind., Trade and Domestic Services Cen. Cttee. CP Moldavia 1976–78; mem. Cen. Cttee. CP Moldavia 1976–; studied 1978–80; Dep. Chair. of Council of Mins. Moldavian SSR Jl. 1980–.

Kiriushchenko, Iakov Vasil'evich
State official.
Chair. State Cttee. of Material and Tech. Supplies. Lithuanian SSR F. 1971–; mem. Auditing Comm. Lithuanian CP 1971–76; mem. Cen. Cttee. Lithuanian CP 1976–.

Kirnasovskii, Boris Efremovich
Diplomat; b. 1918.
Higher tech. and classical educ.; mem. Diplomatic Service 1950–; various posts in USSR Min. of Foreign Affairs and abroad 1950–58; Emb. Counsellor at USSR Amb. in Indonesia 1958–62; sen. posts in USSR Min. of Foreign Affairs 1962–64 and 1968–70; USSR Amb. to Laos 1964–68, Nepal 1970–My. 1975, Ethiopia 1978–81, Sri Lanka and Maldives D. 1981–.

Kirpotin, Valerii Iakovlevich
Literary scholar and critic; b. O. 29 (N.10), 1898, Kaunas.
Mem. CPSU 1918–; grad. Inst. of Red Professors 1925; worked in party Apparat 1932–36; Prof. at M.Gorkii Inst. of Literature 1956–. Decors.: Merited Scientist of the RSFSR, Order of Red Banner of Labour, several medals.

Kirsanov, Aleksandr Vasil'evich, Dr. of Chem. Sc.
Organic chemist; b. N.20, 1902, Moscow; s. of an engineer.
Mem. CPSU 1962–; grad. Tech. Acad. Moscow 1924; Dr. of Chem. Sc. 1939; Prof. 1941; Research Asst., Lecturer at Tech. Acad. Moscow 1924–30; Tech. Chief of Chemical Laboratory, Consultant at Scientific Research Inst. for Mechanical Ore-Processing in Sverdlovsk 1932–39; Prof. Chair of Organic Chemistry at Medical Inst. Sverdlovsk 1940–44; Prof. Chair of Organic Chemistry at Inst. of Metallurgy Dnepropetrovsk 1944–56; Head of Laboratory, Dep. Dir. 1956–, Dir. of Inst. of Organic Chemistry at Ukrainian Acad. of Sciences 1960–; mem. Ukrainian Acad. of Sciences 1961–; main research on chemistry of phosphor and sulphur-organic compounds. Decors.: Merited Scientist of the Ukrainian SSR 1962, Mendeleev Gold Medal 1965.
Publ.: "Fosfatsoedineniia" 1965.

Kiselev, Ivan Ivanovich
Engineer; b. 1917.
Mem. CPSU 1944–; grad. Automobile Mechanical Inst. 1937 and Gorkii Evening Polytech. Inst. 1945; technologist, Head of Standards Bureau, Head of Tech. Supervision, Dept. Head, Dep. Head of Motors Block 1937–54, Chief Engineer 1954–58, Dir. of Gorkii Automobile Plant 1958–; Dep. to RSFSR Supr. Sov. 1959–; mem. Cen. Cttee. CPSU 1961–.

Kiselev, Vladimir Leontevich
Author; b. My.3, 1922, Kiev; s. of a teacher.
Ukrainian; studied Kiev Theatre Inst.; served in World War II; seriously wounded, after release from service worked on ed. staff of newspapers "Na budovi gazoprovodu" and "Radianska Ukraina"; for many years correspondent for "Literaturnaia Gazeta" in the Ukraine; Chief Ed. of Yearbook of the Acad. of Sciences Ukrainian SSR of "Znanie" Soc. of the Ukrainian SSR and of "Nauka i kultura" (Science and Culture); writes in Russian. Decors.: several govt. decorations.

Kiselitsia, Odarka Alekseevna
Painter; b. 1912, Beregomozh, Chernigov Oblast.
Attended Prague Acad. of Plastic Arts 1930–39.

Kishkin, Sergei Timofeevich
Metallurgist; b. My. 17 (30), 1906, Lugansk, now Voroshilovgrad.
Mem. CPSU 1939–; grad. Tech. Bauman College 1931; mem. USSR Acad. of Sciences 1966–; Prof. Chair at Aviation Inst. Moscow 1948–60. Decors.: State Prize of the USSR 1942, 1949 and 1968, Order of Lenin, 4 other orders, several medals.

Kishniakov, Ilia Prokofevich
Author; b. 1922, Nagornoe-Aleksovo, Mordovian ASSR.
Mordovian; mem. CPSU 1952–; worked in the Far East, pilot 1939–56; Acting Sec. of journal "Moksha" 1956–63; Acting Sec. of Mordovian Writers' Union 1963–; first works publ. 1950.
Publs.: "Nashi budni" (Our Weekdays) 1954, "Pervaia liubov'" (First Love) 1956, "Vesna" (Spring) 1955, "Skol'zkii put'" (The Slippery Path) 1954, "Bud'te zdorovy!" (God Bless You!) 1956, "Devushka iz nashego sela" (A Girl From Our Village) 1960, "Samolet popal v tuchi" (An Airplane Got Caught in the Storm Clouds) 1961, and others.

Kislik, Naum Zinovevich
Author and poet; b. 1925, Moscow.
Russian; grad. Belorussian Univ. 1950; first works publ. 1943.
Publs.: "Razgovor s druziami" (A Conversation With Friends) 1956, "Poisk" (The Search) 1959, "Pozyvnye zemli" (Call of the Earth) 1961, and others.

Kisun'ko, Grigorii Vasil'evich
Scientist in radioelectronics; b. Jl. 20, 1918, Bel'manka, in what is now Kuibyshev Raion Zaporozh'e Oblast.
Corresponding member of the Academy of Sciences of the USSR (1958) and Hero of Socialist Labor (1956). Member of the CPSU since 1944.
He graduated from the Voroshilovgrad Pedagogical Institute in 1938. His principal works have been in radio engineering and in electronics. Deputy to the seventh and eighth convocations of the Supreme Soviet of the USSR. Decors.: two Orders of Lenin, the Order of the Red Star, and various medals.

Kitaev, Vladimir Sergeevich
Diplomat.
USSR Amb. to Guinea Mr. 1982–.

Kizichenko, Vsevolod Leonidovich
Diplomat.
USSR Amb. to Tunisia Mr. 1981–.

Kizimov, Ivan Mikhailovich
Equestrian sportsman; b. My. 28, 1928, Novocherkassk.
Honored Master of Sports (1968).
He was champion of the USSR (1967–70) and of the world (1970). He took firsts at the 19th Olympic Games in individual dressage and at the 20th in team dressage. He won medals at the 18th and 19th Olympic Games. Decors.: two orders.

Klauson, Valter Ivanovich
State official; b. 1914; s. of a worker.
Esthonian; mem. CPSU 1943–; grad. Party Higher School of Cen. Cttee. CPSU 1965; secondary tech. training; labourer, asst. engine driver 1928–33; tech. dir., group dir. in construction of bridges, roads and railroads 1933–41; served in Sov. Army 1941–44; Chief Mechanic, Head of Roads Board, Min. of Motor Transport and Highways Esthonian SSR 1944–54; 1st Dep. Chair. of Council of Mins. Esthonian SSR 1954–61; Dep. to Supr. Sov. Esthonian SSR 1955–59 and 1963–; mem. and mem. Bureau Cen. Cttee. Esthonian CP 1956–; Dep. to USSR Supr. Sov. 1958–; cand. mem. Cen. Cttee. CPSU 1961–; Chair. of Council of Mins. Esthonian SSR O. 1961–. Decors.: 3 Orders of Lenin, Order of Patriotic War 2nd Cl., Order of Red Banner of Labour, Order of Red Star.

Klauson, Vladimir Rudolfovich
State official.
Min. of Building Materials Ind. Esthonian SSR Ja. 1980–; cand. mem. Cen. Cttee. Esthonian CP 1981–.

Kldiashvili, Sergo Davidovich
Georgian writer; b. O. 6 (18), 1893, Simoneti in present-day Terzhola Reion; son of D. S. Kldiashvili.
He graduated from the department of law of Moscow University in 1917. He was a member of the Blue Horns, a literary group of Georgian symbolists. He wrote the realistic novella *Moon of the Provinces* and the novel *Ashes* (1932; Russian translation, 1934), which told of life in the Georgian countryside during World War I (1914–18). In the second part of this novel (1962) he portrayed in the Georgian countryside during the period of collectivization. In the novel *Peaceful Abode* (1958), he tells of the downfall of the old world, the establishment of Soviet power in Georgia, and the changes in the psychology of the people. The cycle *Short Stories of Svan* (1935) is notable for its high artistic merits. He wrote the plays *Generation of Heroes* (1937), *Deer Gorge* (1944), and *The Return* (1952).
Works: *T'xzulebani,* vols. 1–2. Tbilisi 1966. *T'xzulebat'a krebuli 4 tomad,* vols. 1–. Tbilisi 1971. *Axlobeli da shoreuli.* [*T'xz. kreb.*]. Tbilisi 1964. *In Russian translation: Povesti i rasskazy.* Moscow 1957. *Irina Tukhareli.* Tbilisi 1958. *Pepelishche: Povesti i rasskazy.* Tbilisi 1965. *Dalekie zarnitsy: Rasskazy.* Moscow 1969. (Introduction by G. Natroshvili.)

Klepikov, Mikhail Ivanovich
Mechanic; b. 1927.
Russian; mem. CPSU 1956–; uncompleted secondary educ.; tractor driver, asst. head, head of a tractor brigade at Ust'-Labinskaia Machine and Tractor Station, mechanic at "Kuban" Kolkhoz 1943–60; Brigade Leader at "Kuban" Kolkhoz in Ust'-Labinskii Raion, Krasnodar Krai 1960–; Dep. to USSR Supr. Sov. 1966–; cand. mem. 1966–71, mem. Cen. Cttee. CPSU 1971–.

Kleshchenko, Anatolii Dmitrievich
Author; b. 1921, Porokoi, Iaroslavl Oblast.
Russian; suffered reprisals 1937, rehabilitated 1957; first works publ. 1937.
Publs.: "Gusi letiat na Sever" (The Geese Are Flying North) 1957, "Dobraia zavist'" (Good Envy) 1958, "Rasputitsa konchaetsia v aprele" (The Mud Season Ends in April) 1962, "Kogda raskhoditsia tuman" (When the Fog Lifts) 1962, "Bez vystrela" (Without a Shot) 1963, and others.

Kletskov, Leonid Gerasimovich
Party official; b. 1918.
Belorussian; mem. CPSU 1940–; grad. CPSU Cen. Cttee. Party Higher School 1955; held Komsomol, sov. and party posts 1935–41; served in Sov. Army 1941–47; 1st Sec. of Molodechno Oblast Cttee. Belorussian Komsomol 1947–52; held party and sov. posts 1952–66; Chief of Administrative Organs Dept. and Organizational Party Work Dept. of Cen. Cttee. Belorussian CP 1966–72; mem. Cen. Cttee. Belorussian CP 1966–; mem. Bureau Cen. Cttee. Belorussian CP 1971–76; 1st Sec. of Grodno Oblast Cttee. Belorussian CP 1972–; Dep. to USSR Supr. Sov. 1974–; mem. Cen. Auditing Comm. CPSU 1976–81; cand. mem. Cen. Cttee. CPSU 1981–. Decor.: Hero of Socialist Labor 1976.

Klevtsov, Nikolai Nikitovich
State official.
1st Dep. Min. of Supplies Kazakh SSR 1971–82, Min. S. 1982–.

Klibik, Valentina Sergeevna
Party official.
Cand. mem. Cen. Cttee. Latvian CP 1966–81, mem. 1981–; Sec. of Riga City Cttee. Latvian CP 1970–80; Dep. to 1971–, Chair. of Supr. Sov. Latvian SSR Je. 1975–; Head of Dept. of Science and Educ. Inst. with Cen. Cttee. Latvian CP 1981–.

Klimchenko, Evgenii Ivanovich
Laborer; b. 1924.
Mem. CPSU 1944–; attended factory and plant apprenticeship school 1940–41, metal worker and toolmaker; served in Sov. Army 1941–49; metal worker and toolmaker at Minsk Tractor Plant 1949–75; metal worker and toolmaker at Minsk Tractor Building and Prod. Asscn. 1975–; mem. Cen. Cttee. CPSU 1976–. Decor.: Hero of Socialist Labor 1966.

Klimenko, Ivan Efimovich, Cand. of Econ. Sc.
Party official; b. 1921.
Russian; mem. CPSU 1945–; grad. Rostov Inst. of Railway Transport Engineers 1942, Party Higher School of Cen. Cttee.

CPSU 1954 and Acad. of Social Sciences of Cen. Cttee. CPSU 1958; asst. engine driver, engineer, teacher, head of teaching dept. of an engine-drivers' school, Dep. Head of Cadre Admin. with Northern Railway 1942–49; CPSU and state posts 1949–52; 2nd Sec. of Iaroslavl City Cttee. CPSU 1952–53; Chair. of Iaroslavl Oblast Exec. Cttee. of Workers' Deps. 1953–54 and 1962–64; teacher 1959–60; in admin. of Iaroslavl Oblast Cttee. CPSU 1960–61; 2nd Sec. 1961–62, Sec. of Iaroslavl Oblast Cttee. CPSU 1964–65; 2nd Sec. 1965–69, 1st Sec. of Smolensk Oblast Cttee. CPSU D. 1969–; Dep. to RSFSR Supr. Sov. 1967–; Dep. to USSR Supr. Sov. 1970–; cand. mem. 1971–76, mem. Cen. Cttee. CPSU 1976–.

Kliment (Andrei Adamovich Perestiuk)
Russian Orthodox clergyman; b. O. 15, 1904, Astrakhanka, Primore Krai.
Novice 1917–25; psalm-reader 1925; subdeacon in Khabarovsk 1929–31; monk, deacon and sacristan 1931–; priest-monk and provost at the Church and Brotherhood of the Monastery of Mercy at the Kamchatka branch 1933; grad. Theological Seminary Kharbin 1934; studied Theological Faculty of St. Vladimir Inst.; mem. Brotherhood of Kazano-Bogoroditskii Monastery in Diocese of Kharbin 1937; reader 1941–45; priest-monk/abbot in parishes in Diocese of Kharbin 1937–51; Archimandrite with function of Main Clergyman in Kazano-Bogoroditskii Monastery in Kharbin 1951; appointed priest by Archbishop Palladii of Irkutsk and Chita 1955, and Custodian of Znamenskii Cathedral 1956, Main Clergyman of Pokrovskii Church in Ussurilsk, Primore Krai, and Provost of Churches in this area (Khabarovsk Diocese) 1957; Bishop of Sverdlovsk and Kurgan 1966–80.

Klimov, Ivan Dmitrievich
Military official; b. Je. 11 (24), 1904, Volosovo-Zevegino, in present-day Kozel'sk Raion, Kaluga Oblast; s. of a railroad worker.
Colonel-general of aviation (1944). Became a member of the CPSU in 1939.
He joined the Soviet Army in 1926. A graduate of the technical war school of the air force (1927), of the military pilots' school (1928), and of the Military Academy of the General Staff (1948), he commanded an air division on the eve of the Great Patriotic War (1941–45). In the Great Patriotic War he participated in the defense of Moscow and in combat in the Crimea and on the Voronezh Front. Commander of a corps of the Air Defense Forces, deputy commander of the National Air Defense Forces, and commander of an air group and of a fighter air division in 1941 and 1942, he commanded a fighter air corps from February to September 1943 and the fighter aviation arm of the Air Defense Forces from 1943 to January 1947. In February 1949 he became inspector-general and deputy chief inspector of the Ministry of Defense.
He has been in the reserve since February 1959. Deputy to the first convocation of the Supreme Soviet of the USSR. Decors.: two Orders of Lenin, three Orders of the Red Banner, the Order of Kutuzov First Class, the Order of Savorov Second Class, two Orders of the Red Star, and medals.

Klimov, Ivan Frolovich
Party official; b. 1903.
Belorussian; mem. CPSU 1925–; grad. Communist Lenin Univ. and Party High School of Cen. Cttee. CPSU; railway worker, chair. of a workers' union cttee., sec. of a raion Komsomol cttee., dept. head of a raion cttee. Belorussian CP 1923–31; in admin. of Cen. Cttee. Belorussian CP 1931–34; teacher, Dir. of Party Courses at Cen. Cttee. Belorussian CP 1934–40; sec. of an oblast cttee. Belorussian CP 1940–53; 1st Dep. Chair. 1953–62, Dep. Chair. of Council of Mins. Belorussian SSR 1962–68; Dep. to USSR Supr. Sov. 1954–66; Dep. to Supr. Sov. Belorussian SSR 1955–; mem. Cen. Cttee. 1956–76, Dep. Chair. of Auditing Comm. Belorussian CP 1976–; mem. Bureau Cen. Cttee. Belorussian CP 1956–62; Min. of Supplies Belorussian SSR 1961; Dep. Chair. of Presidium Supr. Sov. Belorussian SSR Ja. 1968–75. Decors.: 2 Orders of Lenin, Order of Red Banner of Labour, Order of Red Banner, Order of Patriotic War 1st Cl., medals.

Klimova, Aleksandra Ivanovna
Actress; b. O. 1, 1921, Zatobolovka, Kustanai Raion, Kustanai Oblast.
People's Artist of the USSR (1969).
She graduated from the M.S. Shchepkin Theater School in 1949. She worked in the Magnitogorsk Theater from 1942 to 1945, the Odessa Theater from 1949 to 1951, the Lesia Ukrainka Kiev Theater from 1951 to 1953, and the Karkov Russian Theater from 1953 to 1956. She has been with the M. Gorkii Russian Theater of the Belorussian SSR in Minsk since 1956.
She has played both character parts and lyrical heroines. Her best early roles included Katia Redozubova in Gorkii's *The Barbarians* and Ania in Chekhov's *The Cherry Orchard.* She went on to mainly dramatic portrayals, including Vera Pavlovna in *Extraordinary People* (after N.G. Chernyshevskii's novel *What Is to Be Done?*), Vera in *The Precipice* (after I.A. Goncharov's novel of the same name), the Commissar in Vishnevskii's *Optimistic Tragedy,* the title role in Trenev's *Liubov' Iarovaia,* Nila Snizhko in Salynskii's *The Drummer Girl,* Rakitina in Virta's *Vast Distances,* and Cleopatra in Shakespeare's *Antony and Cleopatra.* Decors.: Order of the Red Banner of Labor and various medals.

Klimuk, Petr Il'ich
Cosmonaut; b. 1942.
Mem. CPSU 1963–; grad. Higher Mil. Aviation College Chernigov 1964; pilot; served in Sov. Air Force 1964–65; cosmonaut 1965–; Commander of spaceship "Soiuz 13" in which he performed a number of scientific experiments during orbital flight D. 18–26, 1973; Commander of spaceship "Soiuz 18" during orbital flight My. 24–Jl. 26, 1975, when – together with V.I. Sevastianov – he performed scientific, tech. and medico-biological experiments in orbital station "Saliut 4" which was linked to the spaceship, "Soiuz 30" Je.–Jl. 1978. Decors.: Hero of the Sov. Union 1973 and 1975.

Kliuev, Vladimir Grigor'evich
Party official; b. 1924.
Mem. CPSU 1949–; grad. Ivanovo Textile Inst. 1949; worked in textile ind. enterprises in Ivanovo Oblast: foreman, chief of prod., dir. of a cotton combine 1949–61; chief of a dept. of Ivanovo Oblast Cttee. CPSU 1961–64; 1st Sec. of Ivanovo City Cttee. CPSU 1964–66; 2nd Sec. of Ivanovo Oblast Cttee. CPSU 1966–72, 1st Sec. Jl. 1972–; Dep. to USSR Supr. Sov. 1970–; mem. Cen. Cttee. CPSU 1976–.

Klokov, Viacheslav Mikhailovich
Sculptor; b. 1928, Kharkov.
Attended Kiev Art Inst. 1947–53, pupil of M. Gelman. Decor.: Merited Artist of the Ukrainian SSR.

Klova, Vitautas Juliono
Composer; b. Ja. 31, 1926, Tirkšliai, Mažeikiai Raion.
Honored Artist of the Lithuanian SSR (1959).
He graduated from A.I. Račiūnas' course in composition at the conservatory of the Lithuanian SSR in 1951 (until 1948 he had studied with J.M. Gruodis). He is the author of the operas *Pilėnai* (1956; State Prize of the Lithuanian SSR, 1957), *Vaiva* (1958), *The Daughter* (1960; State Prize of the Lithuanian SSR, 1960), *Two Swords* (1966), and *An American Tragedy* (1969). He has also written symphonic works (*Pictures of Vilnius,* 1962; orchestral suite), concerti (for violin and orchestra, 1950; for piano and orchestra, 1953; for cello and orchestra, 1963), and other vocal and instrumental works. He is a docent at the Conservatory of the Lithuanian SSR.

Klychev, Izzad Nazarovich
Painter; b. 1923.
Grad. I.E. Repin Inst. for Painting, Sculpture and Architecture Leningrad 1953; corresponding mem. USSR Acad. of Arts 1970–. Decors.: People's Artist of the USSR 1973, People's Artist of the Turkmen SSR 1964, State Prize of the USSR 1967.

Klychev, Niiaz-Durdy
State official.
Chair. of State Cttee. for the Supervision of Safety Precautions for Workers in Ind. and Mining. Turkmen SSR Mr. 1977–.

Knebel', Mariia Osipovna
Director and teacher; b. My. 6 (18), 1898, Moscow.
People's Artist of the RSFSR since 1958. She began studying theater in 1918 at the M. A. Chekhov Studio and continued her education at the Second Studio of the Moscow Art Theater. She joined the troupe of the Moscow Art Theater in 1924, concentrating mainly on poignant single-faceted parts (for example, Charlotta in Chekhov's *The Cherry Orchard* and Karpukhina in *Uncle's Dream,* after Dostoevsky).
She directed her first play in 1935 with the Ermolova Moscow Theater. Productions followed on Arbuzov's *A Long Road* (1936), Gorkii's *The Last Ones* (1937) and *Children of the Sun* (1937; jointly with N. P. Khmelev), and Shakespeare's *As You Like It* (1940). Her Moscow Art Theater productions included *The Kremlin Chimes* by Pogodin (1942; jointly with V. I. Nemirovich-Danchenko and L. M. Leonidov) and *Ivan the Terrible* (*The Difficult Years*) by A. N. Tolstoy (1946; jointly with A. D. Popov). She became a director of the Central Children's Theater in 1950 and was its head director between 1955 and 1960. Her productions there included *A Page From Life* by Rozov (1952).
She has a strong affinity for the plays of Chekhov, and she staged *Ivanov* in 1955 at the Pushkin Moscow Theater and *The Cherry Orchard* in 1965 at the Central Theater of the Soviet Army and the Abbey Theatre in Dublin. She staged in Ostrovskii's *Talents and Admirers* at the Mayakovsky Moscow Theater in 1969.
She began teaching in 1932. After 1948 she taught in the department of directing of the A. V. Lunacharskii State Institute of Theatrical Art. She became a professor there in 1960. Decors.: two orders and various medals.
Works: *Slovo v tvorchestve aktera.* Moscow 1954. *O deistvennom analize p'esy i roli.* Moscow 1961. *Vsia zhizn'. [Vospominaniia aktrisy].* Moscow 1967.

Kniazev, Filipp Kirillovich
Party official; b. 1916.
Russian; mem. CPSU 1940–; grad. Voronezh Teachers' Training Inst. 1939; teacher, head of a dept., inspector of a raion educ. dept., sec. of a raion Komsomol Cttee. 1934–41; served in Sov. Army 1941–42; sec. of an oblast Komsomol Cttee., 1st sec. of various raion CPSU Cttees. in Tambov Oblast 1942–55; sec. of Kurgan Oblast Cttee. CPSU 1955–59; Chair. of Kurgan Oblast Sov. 1959–66; Dep. to 1960–62 and 1966–, mem. Planning and Budget Comm. at Sov. of the Union USSR Supr. Sov. 1966–; Dep. to and mem. Draft Bills Comm. RSFSR Supr. Sov. 1963–67; cand. mem. 1966–76, mem. Cen. Cttee. CPSU 1976–; 1st Sec. of Kurgan Oblast Cttee. CPSU Ap. 1966–.

Kniazeva, Lidiia Nikolaevna
Actress; b. 1925.
Mem. CPSU 1954–; studied State Lunacharskii Inst. of Theatre Arts (GITIS) Moscow 1943–45; mem. Youth Theatre (TIUZ) Moscow 1948–. Decor.: People's Artist of the USSR 1970.

Kniazik, Aleksandr Beniaminovich
Sculptor; b. 1932, Dnepropetrovsk.
Mem. CPSU 1959–; attended Kiev Art Inst. 1954–59, pupil of M. Gelman, M. Vronskii.

Knipovich, Evgeniia Fedorovna
Literary scholar and critic; b. S. 14 (26), 1898, St. Petersburg.
Her work was first published in 1919. She wrote the books *Heinrich Heine* (1931), *Heine as a Political Lyricist* (1932), and *A. Fadeev's "The Rout" and "The Young Guard"* (1964). She is also the author of the collections of articles of literary criticism *In Defense of Life* (1958), *The Power of Truth* (1965), and *The Artist and History* (1968) and numerous articles devoted to Soviet as well as foreign literature (H. Mann, I. Feuchtwanger, A. Zweig, A. Seghers). Decors.: two orders and various medals.

Knorozov, Jurii Valentinovich
Ethnologist and linguist; b. N. 19, 1922, Kharkov.
Doctor of historical sciences (1955).
He graduated from Moscow State University in 1948. His research has been devoted to the ethnology of Central America and the study of ancient writing systems, especially that of the Mayan Indians, and also to the problems of semiotics and comparative linguistics.
Works and translations: *Pis'mennost' indeitsev maiia.* Moscow-Leningrad 1963. Introductory article, notes, and translation in D. Landa, *Soobshcheniia o delakh v Iukatane.* Moscow-Leningrad 1955.

Knuniants, Ivan Ludvigovich
Organic chemist; b. My. 22 (Je. 4), 1906, Shusha.
Mem. CPSU 1941–; grad. Moscow Inst. of Technology 1928; mem. USSR Acad. of Sciences 1953–; Maj.-Gen./Engineer; Head of Laboratory of Inst. of Elemental-Organic Compounds at USSR Acad. of Sciences. Decors.: Hero of Socialist Labour 1966, Lenin Prize 1972, State Prize of the USSR 1943, 1948, 1950, Order of Lenin, 3 other orders and various medals.

Kobakhidze, Vladimir Il'ich
State official.
Min. of Posts and Telecommunications Georgian SSR D. 1975–; mem. Cen. Cttee. Georgian CP 1976–.

Kobilov, Umirzak
State official.
Dep. Min. of Light Ind. Tadzhik SSR until 1971; mem. Cen. Cttee. Tadzhik CP 1971–81; Head of Admin. 1977–79, Min. of Cotton-Cleaning Ind. Tadzhik SSR O. 1979–.

Kobizeva, Klavdiia Semenovna
Sculptor; b. Mr. 1 (14), 1905, Kishinev.
People's Artist of the Moldavian SSR (1965).
She studied under A. M. Plamadiala at the Kishinev School of Art from 1926 to 1931, at the Academy of Arts in Brussels from 1931 to 1934, and at the workshop of C. Medrea in Bucharest from 1934 to 1936. Beginning in 1936 she worked in Moldavia. Kobizeva portrayed the women of Soviet Moldavia, glorified working people, and appealed for peace. Her works include *The Head of a Moldavian Woman* (wood, 1947), *The Woodcutter* (plaster of Paris, 1962), *Let There Be Peace Forever!* (plaster of Paris, 1965), and the relief *On Liberated Land* (plaster of Paris, 1967; State Prize of the Moldavian SSR, 1968) – all are in the Art Museum of the Moldavian SSR in Kishinev. She also created the sculptural group *The Gifts of Moldavia* (bronze, 1959), which is in Tiraspol'. Decors.: two orders.

Koblianidze, Guram Levanovich, Cand. of Tech. Sc.
State official; b. 1935.
Georgian; mem. CPSU 1966–; grad. Engels Trade Inst. in Leningrad 1959; responsible for market research, Dir. of a warehouse in Syvtykar, Komi ASSR 1959–63; held various posts in trade orgs. in Gori, asst. in dept. of market research at Tbilisi State Univ. 1963–67; lecturer in this dept. 1971–74; head of Trade, Planning and Finance Organs Dept. Tbilisi City Cttee. CP Georgia 1974–75; 1st Dep. Min. of Trade Georgian SSR 1975–80, Min. My. 1980–; mem. Cen. Cttee. CP Georgia 1981–.

Kobzarev, Iurii Borisovich, Dr. of Tech. Sc.
Radio physicist; b. 1905.
Grad. Inst. of Public Educ. Kharkov 1926; Dr. of Tech. Sc. and Prof. 1949; Sen. Research Asst. at Physical-Tech. Inst. of USSR Acad. of Sciences 1926–43; Prof. Chair at Inst. of Energetics Moscow 1944–55; mem. staff 1955–, Dept. Head at Inst. of Radio Technology and Electronics of USSR Acad. of Sciences 1968–; mem. USSR Acad. of Sciences 1970–. Decors.: State Prize of the USSR 1941 and 1949.

Kobzev, Igor Ivanovich
Author and poet; b. 1924, Rostov/Don; s. of a railwayman.
Russian; first poems publ. in a front newspaper 1943.
Publs.: "Pervaia liubov'" (First Love), "Legenda" (Legends), "Spetskorre-

spondent" (Special Correspondent), "Ia v goru shel" (I Went Into the Mountain), "Moi znakomye" (My Acquaintances) 1956, "Da zdravstvuet romantika!" (Long Live Romanticism!) 1959, "Litsom k solntsu" (With the Face to the Sun) 1962, "Lebedi v Moskve" (The Swans in Moscow) 1964, and others.

Kocharian, Suren Akimovich
Variety performer; b. Ag. 23 (S. 5), 1904, Tbilisi.
Professional reciter. People's Artist of the Armenian SSR (1945); People's Artist of the RSFSR (1967).
He studied at the Moscow Armenian School for Drama (1921–25). From 1925 to 1932 he acted with the G. Sundikian Armenian Theater in Yerevan, playing nearly 40 roles. In 1939 he joined the Moscow Philharmonic Society. He has arranged more than 30 literary readings, including the cycles *Treasures of World Literature, Armenian and Russian Classics,* and *Works by Soviet Writers.* His readings from literary epics (*Vladimir Il'ich Lenin* by Mayakovsky, *The Man in Panther's Skin* by Rustaveli, *The Odyssey* by Homer, and *Shakhrazad,* based on *A Thousand and One Nights*) were important contributions to the development of the genre. Decors.: State Prize of the USSR (1951), the Order of the Badge of Honor, and medals.
Works: K'ocharyan, S.A. *Kenyani Khost'i olobtnebolm.* Yerevan 1963. *V poiskakh zhivogo slova.* Moscow 1960.

Kochemasov, Viacheslav Ivanovich
State official; b. 1918.
Mem. CPSU 1942–; completed 5 terms at Gorkii Inst. of Hydraulic Engineering; leading Komsomol posts in Gorkii Oblast 1942–48; 2nd Sec. 1943–46, 1st Sec. of Gorkii Oblast Komsomol Cttee. 1946–48; Dep. Chair. 1948–49, Chair. of Anti-Fascist Sov. Youth Cttee. 1949–55; Sec. of Cen. Cttee. All-Union Komsomol 1949–55; Adviser 1955–58, Min. at USSR Emb. in GDR 1958–60; Dep. Dept. Head at USSR Min. of Foreign Affairs 1960–61; Dep. Chair., 1st Dep. Chair. of State Cttee. for Cultural Relations with Foreign Countries with USSR Council of Mins. 1961–62; Dep. Chair. of RSFSR Council of Mins. Jl. 1962–; cand. mem. Cen. Cttee. CPSU 1966–; Chair. RSFSR Comm. on Questions Concerning Minors 1980–. Decors.: Order of Red Banner of Labour, Order of Lenin.

Kochetkov, Nikolai Konstantinovich
Organic chemist; b. My. 5 (18), 1915, Moscow.
Grad. M. V. Lomonosov Inst. of Chemical Technology Moscow 1939; corresponding mem. USSR Acad. of Medical Sciences 1957 and USSR Acad. of Sciences 1960–; Dir. of Inst. of Organic Chemistry at USSR Acad. of Sciences 1966–; specializes in organic synthesis, chemistry of pharmaceutical drugs, chemistry of matter and chemistry of natural compounds. Decors.: Order of Lenin, medals.

Kochin, Nikolai Ivanovich
Russian writer; b. Jl. 2 (15), 1902, Gremiachaia Poliana, now in Dal'ne Konstantinovo Raion, Gorkii Oblast; s. of a peasant.
Member of the CPSU since 1940.
In the early postrevolutionary years, he worked for the Committee of the Poor and was a village correspondent for the newspaper *Bednota* ("The Poor"). He graduated from the Nizhny Novogorod Pedagogic Institute in 1924 and became a teacher. His first works appeared in print in 1925. His novel *Village Girls* (books 1–2, 1928–31) was a major work in Soviet literature about the life of the peasantry on the eve of collectivization. The novel *Fellows* (1934), the novella *Notes of a Village Correspondent* (1929; separate edition, 1930), and the autobiographical tale *Youth* unfold the life of the Soviet country side at various periods in history. He is also the author of *Kulibin* (revised, 1940) and the novel *The Nizhny Novgorod Otkos* (1970), which describes student life right after the Revolution. His collections of publicistic articles include *The Initiative of the Village of Pochiaki* (1931) and *The Countryside in the Days of the War* (1942). Decors.: two orders and various medals.

Kochina (Polubarinova-Kochina), Pelageia Iakovlevna
Hydrodynamics scientist; b. My. 1 (13), 1899, Astrakhan.
Grad. Petrograd Univ. 1921; mem. USSR Acad. of Sciences 1958–; mem. staff Geophysical Main Observatory in Petrograd 1919–; Lecturer at Inst. of Traffic Engineering 1925–31, at Inst. of Civil Aviation Engineering 1931–35; Prof. at Leningrad Univ. 1934; mem. staff Inst. of Mathematics 1935–39, Inst. of Mechanics at USSR Acad. of Sciences 1939–59, Inst. of Hydrodynamics at Siberian Branch of USSR Acad. of Sciences 1959–71, Inst. of Problems in Mechanics at USSR Acad. of Sciences. Decors.: Hero of Socialist Labour, State Prize of the USSR 1946, 4 Orders of Lenin, Order of Red Banner, medals.

Kochnev, Mikhail Kharlampievich
Author; b. 1914, Shegodskoe, Vladimir Oblast; s. of a peasant.
Russian; mem. CPSU 1941–; grad. Dept. of Philology of Moscow State Univ. 1939; first works publ. 1938.
Publs.: "Obiatiia zemli" (Embraces of the Earth) 1943, "Serebriannaia priazha" (Silver Thread), "Raspisnoi uzor" (Painted Pattern) 1948, "Sorok vereten" (Forty Spindles) 1948, "Delo chelovekom slavitsia" (A Matter Is Made Famous by Man) 1950, "Zhivoi rodnik" (A Live Spring) 1951, "Tsvet zari" (The Colour of Dawn) 1955, film script "Ilia Muromets" 1959, novel "Potriasenie" (The Tremor) 1964, and others.

Kogan, Leonid Borisovich
Violinist; b. N. 14, 1924, Ekaterinoslav, now Dnepropetrovsk.
Mem. CPSU 1954–; grad. Moscow Conservatory 1948; concluded post-graduate studies 1953; soloist with Moscow Philharmonic Orchestra 1944–; teacher 1952–, Prof. at Moscow Conservatory 1963–; numerous performances throughout the world. Decors.: People's Artist of the USSR 1966, Lenin Prize 1965, Order of Red Banner of Labour, medals.

Kokhanskii, Vasilii Ivanovich
Trade union official.
Sec. of Cen. Cttee. Moldavian Komsomol 1945; mem. Cen. Cttee. 1956–66 and 1971–76, mem. Auditing Comm. Moldavian CP 1966–71; 1st Sec. of Atakskii Raion Cttee. Moldavian CP 1959; Head of Dept. of Party Organs at Cen. Cttee. Moldavian CP 1960; mem. Bureau 1960–61, cand. mem. Bureau Cen. Cttee. Moldavian CP 1961–62; 1st Sec. of Kishinev City Cttee. Moldavian CP 1961–62; Sec. and mem. Presidium of Moldavian Republican Trade Union Council 1965–.

Kokin, Mikhail Aleksandrovich
Painter; b. 1921, Dnepropetrovsk.
Mem. CPSU 1943–; attended Dnepropetrovsk Art School 1937–41, grad. Kiev Art Inst. 1954, pupil of S. Grigorev, V. Kostetskii.

Kokkinaki, Vladimir Konstantinovich
Pilot; b. Je. 12 (25), 1904, Novorossiisk; s. of a railroad official.
Honored Test Pilot of the USSR (1959), major-general of aviation (1943), twice Hero of the Soviet Union (July 17, 1938, and Sept. 17, 1957), Honored Master of Sport (1959). Became a member of the CPSU in 1938.
He joined the Red Army in 1925 and graduated from the Borisoglebsk Flight School in 1930. He was a test pilot from 1931 to 1965, specializing in high altitude flights from 1932. On N. 21, 1935, he climbed to 14,575 m using a single-seat plane, and on Jl. 17, 1936, he climbed to 11,458 m with a cargo of 500 kg, surpassing world records. The high speed flights he made in 1936 included the flights Moscow–Sverdlovsk–Moscow, and Moscow–Sevastopol'–Sverdlovsk–Moscow.
On Je. 27–28, 1938, he made a nonstop flight from Moscow to Vladivostok in 24 hours 36 minutes, and on Ap. 28, 1939, he made a nonstop flight from Moscow to Miscou Island (Canada) in 22 hours and 56 minutes. He set a total of 22 world records. During the Great Patriotic War (1941–45) he worked as a test pilot while serving as chief inspector of the aviation industry and a leader of the industry's flight-testing service. He has been a test

pilot and methods specialist since 1965. Decors.: Lenin Prize (1960), five Orders of Lenin, three Orders of the Red Banner, Orders of the Patriotic War First and Second Class, four Orders of the Red Star, and medals, gold medal of the International Aeronautical Federation (FAI) and the Wind Rose diamond necklace. Vice-president (1961) and president of the FAI, he has been honorary president of the FAI since December 1968.

Kolbin, Gennadii Vasil'evich
Party official; b. 1927.
Mem. CPSU 1954–; grad. Ural Polytechnical Inst. 1955; technological designer, chief of a technical office, shop superintendent, dep. chief engineer in a plant in Nizhnii Tagil 1947–59; party posts 1959–62; 2nd Sec. of Nizhnii Tagil City Cttee. CPSU 1962–64, 1st Sec. 1964–70; Sec. of Sverdlovsk Oblast Cttee. CPSU 1970–71, 2nd Sec. 1971–75; 2nd Sec. of Cen. Cttee. and mem. Bureau of Cen. Cttee. Georgian CP Ap. 1975–; cand. mem. Cen. Cttee. CPSU 1976–81, mem. 1981–.

Kolchin, Pavel Konstantinovich
Athlete; b. Ja. 1, 1930, Iaroslavl.
Honored Master of Sports (1956), Honored Coach of the USSR (1970). Became a member of the CPSU in 1962. Olympic champion (ski relay, 1956) and medal winner of winter Olympic games (15-km and 30-km Nordic combined events, 1956; relay, 1964). Eight-time champion of the USSR (individual races and relays, 1953–64). Decors.: two orders and several medals.

Kolchina, Alevtina Pavlovna
Woman athlete; b. F. 11, 1930, Pavlovsk, in present-day Ocher Raion, Perm' Oblast.
Honored Master of Sport (1958), teacher. Became a member of the CPSU in 1962. Olympic champion (ski relay, 1964) and medal winner of winter Olympic games (5-km race, relays, 1956, 1964, 1968). Champion of the world (seven times from 1958 to 1966) and the USSR (eight times from 1956 to 1967) in individual races and relays. Decors.: Order of the Badge of Honor and medals.

Kolchina, Olga Pavlovna
State official; b. 1918; d. of a teacher.
Russian; mem. CPSU 1946; grad. Voronezh Econ. Planning Inst. 1939; teacher, econ. planner, section head of an oblast planning cttee. 1939–43; organizer for the All-Union Komsomol Cen. Cttee., teacher at a tech. college 1943–47; instructor, dept. head, Sec., 1st Sec. of Krasnaia Gorka City CPSU Cttee. 1947–60; cand. mem. Cen. Cttee. CPSU 1961–; Dep. to and Dep. Chair. of Draft Bills Comm. at Sov. of the Union USSR Supr. Sov. 1962–66; Sec. of Moscow Oblast CPSU Cttee. 1964–Ap. 1967; Dep. to RSFSR Supr. Sov. 1967–; Dep. Chair. of Presidium RSFSR Supr. Sov. Ap. 1967–. Decor.: Order of Red Banner of Labour.

Koldunov, Aleksandr Ivanovich
Military official; b. 1923.
Mem. CPSU 1944–; grad. Air Force Acad. 1952 and Acad. of Gen. Staff 1960, Marshal of Aviation; entered Sov. Army 1941; served at front 1943–45; commander of a fighter aviation regiment, of a fighter aviation division, Dep. Commander, Commander of Aviation of Baku Air Defense District (PVO), 1st Dep. Commander of Baku Air Defense District (PVO) 1952–70; Commander of Moscow Air Defense District (PVO) 1970–75; cand. mem. Cen. Cttee. CPSU 1971–76, mem. 1981–; 1st Dep. Commander-in-Chief of USSR Air Defense Forces (PVO) 1975–78, Commander-in-Chief 1978–; Dep. USSR Min. of Defense 1978–. Decors.: Hero of the Soviet Union 1944 and 1948.

Kolesnik, Piotr Iosifovich
Author and literary historian; b. 1904, Baryshevka, Kiev Oblast; s. of a peasant. Ukrainian; mem. CPSU 1962–; grad. Kiev Inst. of Educ. 1928; first works publs. 1928.
Publs.: "Na fronte priozoshli peremeny" (Changes Have Occurred at the Front) 1935, "Tern na puti" (Blackthorn on the Road) 1959, "Syn naroda" (Son of the People) 1957, "Kotsiubinskii – khudozhnik slova" (Kotsiubinskii – Literary Artist) 1964, and others.

Kolesnikov, Aleksandr Iakovievich
Worker; b. 1930.
Russian; mem. CPSU 1966–; grad. Mining College Krasnodon 1967; miner in Voroshilovgrad Oblast mines 1947–; cand. mem. Cen. Cttee. CPSU 1971–81, mem. 1981–.

Kolesnikov, Arkadii Georgievich, Dr. of Phy.-Math. Sc.
Scientist; b. 1907.
Mem. CPSU 1952–; grad. Higher Tech. N. E. Bauman College (MVTU) Moscow 1930; specializes in marine physics; Dr. of Phy.-Math. Sc. 1944; Prof. 1947; mem. Ukrainian Acad. of Sciences 1967–; Sen. Mem. Research Staff and Dept. Dir. of Hydrophysical Marine Laboratory of USSR Acad. of Sciences 1942–48; Head of Laboratory 1948–60; Prof. (Chair) at Moscow Univ. 1944–62; Dir. of Hydrophysical Marine Inst. of Ukrainian Acad. of Sciences 1962–. Decor.: State Prize of the USSR 1970.

Kolesnikov, Boris Pavlovich, Dr. of Biol. Sc.
Biologist; b. 1909.
Mem. CPSU 1945–; grad. Far Eastern Forestry Inst. 1931; Dr. of Biol. Sc. and Prof. 1953; postgraduate studies 1931–33; mem. research staff at several institutions of USSR Acad. of Sciences in the Far East and extreme North 1934–49; Dep. Chair., later Acting Chair. of Far Eastern branch of USSR Acad. of Sciences 1951–55; Head of Laboratory at Inst. of Biology of Uralsk Branch of USSR Acad. of Sciences 1955–64; Rector 1963–68, Prof. Chair of Geobotany and Geology at Uralsk Univ. 1968–; corresponding mem. USSR Acad. of Sciences 1970–.

Kolesnikov, Ivan Stepanovich
Medical scientist; b. D. 2 (15), 1901, Podosinovka, Voronezh Oblast.
Mem. CPSU 1928–; grad. S. M. Kirov Mil. Medical Acad. 1931; surgeon; mem. USSR Acad. of Medical Sciences 1971–; Maj.-Gen. in Medical Corps; army and front surgeon 1941–45; Head of Surgical Hospital of S. M. Kirov Mil. Medical Acad. 1953–. Decors.: Merited Scientist of the RSFSR, Order of Lenin, 12 other orders, several medals.

Koliakin, A. N.
Komsomol official.
1st Sec. Leningrad Oblast Cttee. All-Union Komsomol until 1981; Sec. and mem. Bureau, Cen. Cttee. All-Union Komsomol Mr. 1981–.

Kolmanovskii, Eduard Savel'evich
Composer; b. Ja. 9, 1923, Mogilev.
Honored Art Worker of the Buriat ASSR (1963).
In 1945, he completed his training at the Moscow Conservatory, where he had studied composition under V. Ia. Shebalin. He wrote popular songs with the themes of civic spirit and patriotism, such as "Live, I Love You" (1958), "Do the Russians Want War" (1961), and "The Communards Will Not Be Slaves" (1967). He also composed other types of songs, including the lyrical songs "We Will Wait for You" (1963), "Biriusinka" (1963), "A Waltz About a Waltz" (1964), and "The Young Crane" (1965). He was an eminent composer of music for the theater. He wrote the music for the fairy tale *Snow White* (1966, Moscow Children's Musical Theater), Shakespeare's comedy *Twelfth Night* (1955, Moscow Art Theater), and several other productions. His works include a concerto for alto and orchestra (1945), four suites for symphony orchestra (1945, 1948, 1950, 1956), scores for films, and art songs.

Kolmogorov, Andrei Nikolaevich
Mathematician; b. Ap. 12 (25), 1903, Tambov.
Academician of the Academy of Sciences of the USSR (1939); Hero of Socialist Labor (1963).
He graduated from Moscow University in 1925 and became a professor there in 1931. He began his scientific work on the theory of functions of a real variable, where he is credited with fundamental work on trigonometric series, measure theory, set theory, integration theory, and the theory of approximation of functions.

He also made substantial contributions to the development of constructive logic, topology (cohomology theory), mechanics (theory of turbulence), the theory of differential equations, and functional analysis.
His work in probability theory is of fundamental importance. In this field, together with A. Ia. Khinchin, he began (1925) to apply the methods of the theory of functions of a real variable. This enabled him to solve a number of difficult problems and to construct the widely known axiomatic basis of probability theory (1933) and to lay the foundations of the theory of continuous-time random Markov processes. He later developed (in conjunction with the studies of A. Ia. Khinchin) the theory of stationary random processes, processes with stationary increments, and branching processes. He has made important contributions to information theory. He has done research on the theory of firing, statistical inspection methods for mass production, applications of mathematical methods in biology, and mathematical linguistics.
He was the editor of the mathematical section of the first edition of the *Great Soviet Encyclopedia* and a member of the Main Editorial Board of the second edition. He is active in working out problems of mathematical education in secondary schools and universities. He founded a large school of probability theory and the theory of functions. Among his students are Academicians of the Academy of Sciences of the USSR A. I. Mal'tsev, M. D. Millionshchikov, S. M. Nikol'skii, and Iu. V. Prokhorov; corresponding members of the Academy of Sciences of the USSR I. M. Gel'fand and A. S. Monin; Academician of the Academy of Sciences of the Ukrainian SSR B. V. Gnedenko; Academician of the Academy of Sciences of the Uzbek SSR S. Kh. Sirazhdinov; and Lenin Prize laureates I. V. Arnol'd and Iu. A. Rozanov.
He is a foreign member of the Paris Academy of Sciences, the Royal Society of London, and a number of other foreign academies (the Netherlands, Poland, Rumania, the USA), scientific institutions and societies. Decors.: Bolzano International Prize (1963), the State Prize of the USSR (1941), and the Lenin Prize (1965), six Orders of Lenin, the Order of the Red Banner of Labor, and various medals.
Works: *Osnavnye poniatiia teorii veroiatnostei.* Moscow-Leningrad 1936. *Vvedenie v theoriiu funktsii deistvitel'nogo peremennogo,* 3rd ed. Moscow-Leningrad 1938. (With P. S. Aleksandrov.) *Predel'nye raspredelniia dlia summ nezavisimykh sluchainykh velichin.* Moscow-Leningrad 1949. (With B. V. Gnedenko.) *Elementy teorii funktsii i funktsional'nogo analiza.* 3rd ed. Moscow 1972. (With S. V. Fomin.)

Kolokolov, Boris Leonidovich
Diplomat; b. 1924.
Served in Diplomatic Service 1956–; in USSR Min. of Foreign Affairs 1965–69; Head of Protocol Dept. at USSR Min. of Foreign Affairs 1969–73; USSR Amb. to Tunisia N. 1973–81; RSFSR Dep. Min. of Foreign Affairs Ap. 1981–.

Kolokolov, Iurii Borisovich
Party official.
Head of Dept. of Building and Municipal Econ. at Cen. Cttee. Belorussian CP 1965–70 and 1978–; cand. mem 1966–71, mem. Cen. Cttee. Belorussian CP 1971–; Dep. Chair. of Council of Mins. Belorussian SSR My. 1970–78; Sec. and mem. Bureau Cen. Cttee. Belorussian CP 1978–.

Kolomasov, Vasilii Maksimovich
Author; b. 1909, Mordovskoi Naimany, Mordovian ASSR; s. of a peasant.
Mordovian; grad. Mordovian Pedagogical Inst.; chair. of a kolkhoz and teacher of literature and language in an intermediate school; first works publ. 1931.
Publs.: "Prokopych" 1935, novelette "Iakhim Lavginov" 1936, novel "Lavginov" 1941–56, and others.

Kolomiets, Iurii Afanasevich
State official; b. 1925.
Ukrainian; mem. CPSU 1953–; grad. Uman Agricultural Inst. 1954; served in Sov. Army 1944–48; agronomist on a collective farm, chief agronomist at a machine and tractor station, chief of a raion inspection for agriculture, 2nd sec., 1st sec. of a raion cttee. Ukrainian CP, chair. of exec. cttee. of a raion sov. of peoples' deps., chief of a collective and state farm prod., admin., dept. chief, Sec. of Cherkassy Oblast Cttee. Ukrainian CP 1948–70; Dep. Min., Min. of Agriculture Ukrainian SSR 1970–77; Min. of State Farms Ukrainian SSR 1977–79; 1st Dep. Chair. of GOSPLAN Ukrainian SSR 1979–80; cand. mem. Politbureau Cen. Cttee. Ukrainian CP My. 1980–; 1st Dep. Chair. of Council of Mins. Ukrainian SSR My. 1980–; cand. mem. Cen. Cttee. CPSU 1981–. Decors.: Order of Lenin, Order of the October Revolution, Order of Red Banner of Labor (three), medals.

Kolomiichenko, Aleksei Isidorovich
Otolaryngologist; b. Mr. 5 (17), 1898, Shpola, Cherkassy Oblast.
Grad. Medical Inst. Kiev 1924; Prof. Chair of Continuing Educ. for Physicians at Kiev Univ. 1944–; Dir. of Inst. of Otolaryngology Kiev 1960–; Chief Ed. of periodical "Žurnal ušnyh, nosovyh i gorlovyh boleznej" 1958–; Chair. of Managing Board of Ukrainian Otolaryngological Soc. 1964–; corresponding mem. USSR Acad. of Sciences 1967–. Decor.: Lenin Prize 1964.

Kolosov, Mark Borisovich
Author; b. 1904, Mogilev; s. of a clerk.
Russian; mem. CPSU 1928–; grad. Dept. of Philology of Moscow Univ. 1930; first works publ. 1923. Decor.: Order of Red Banner of Labour.
Publs.: "Trinadtsat'" (Thirteen) 1923, "Stengaz" (Wall Newspaper), "Baraban" (The Drum), "Buza" (A Row), "Liudi i podvigi" (People and Heroic Deeds) 1942, "Tovarishch general" (Comrade General) 1955, "Pisma s fabriki" (Letters From the Factory) 1938–61, "Slozhnaia zhizn'" (Complicated Life) 1961, and others.

Kolosov, Mikhail Nikolaevich
Organic chemist; b. My. 11, 1927, Kursk.
Mem. CPSU 1963–; grad. M. I. Lomonosov Inst. of Chemical Technology Moscow 1948; mem. staff Inst. of Biological and Medical Chemistry at USSR Acad. of Medical Sciences 1951–59; mem. staff Inst. of Chemistry of Natural Compounds at USSR Acad. of Sciences 1959–; corresponding mem. USSR Acad. of Sciences 1966–.
Publ.: co-author of "Khimiia antibiotikov", vols. 1–2, Moscow 1961.

Kolotyrkin, Iakov Mikhailovich
Physical chemist; b. N. 1 (14), 1910, Zanino, Smolensk Oblast.
Mem. CPSU 1940–; grad. Dept. of Chemistry of Moscow Univ. 1937; mem. USSR Acad. of Sciences 1970–; Dir. of Karpov Inst. of Physical-Chemical Research 1948–51 and 1957–. Decors.: Order of Lenin, 2 other orders and various medals.

Kolpakova, Irina Aleksandrovna
Ballerina; b. My. 22, 1933.
People's Artist of the USSR (1965).
Upon graduating from the Leningrad Choreographic School from A. Ia. Vaganova's class in 1951, she joined the ballet troupe of the S. M. Kirov Leningrad Theater of Opera and Ballet. Delicate lyricism, musicality, and subtle artistry distinguish her art as a talented classical ballerina. Her repertoire includes the leading roles in such ballets as Tchaikovsky's *The Sleeping Beauty* and *The Nutcracker,* Adam's *Giselle,* and Prokofiev's *Romeo and Juliet* and *Cinderella.* Kolpakova created the leading roles in productions by Iu. N. Grigorovich, I. D. Bel'skii, G. D. Aleksidze, N. D. Kasatkina, and V. Iu. Vasilev: Katerina in Prokofiev's *The Stone Flower,* Shirin in Melikov's *The Legend of Love,* His Beloved in Petrov's *Coast of Hope,* Ala in Prokofiev's *Scythian Suite,* and Eve in Petrov's *Creation of the World.* She received the Gold Star in 1965 at the International Festival of Dance in Paris. Decors.: Order of the Red Banner of Labor and various medals.

Koluntsev, Fedor (real name: Barkhudarian, Tadeos Avisovich)
Author; b. 1923, Moscow.
Russian; first works publ. 1947.
Publs.: "Dorogi zovut" (The Roads Are Calling) 1955, "Pervaia liubov'" (First Love), "Nasten'ka", "Ogni" (Lights), novel "U Nikitskikh vorot" (At the Nikitskii Gate) 1962, and others.

Komar, Anton Panteleimonovich, Dr. of Phy.-Math. Sc.
Physicist; b. Ja. 30, 1904, Berezan, now Kiev Oblast.
Mem. CPSU 1944–; grad. Polytech. Inst. Kiev 1930; Dr. of Phy.-Math. Sc. 1943; mem. staff Physical-Tech. Inst. Leningrad 1930–36 and Sverdlovsk (Urals) 1936–47; mem. Ukrainian Acad. of Sciences 1948–; various posts at P.M. Lebedev Inst. of Physics at USSR Acad. of Sciences Moscow 1948–50; concurrently Prof. Chair of Physics at Uralsk Univ. 1937–47, at Uralsk Ind. Inst. 1943–47 and at Polytech. Inst. Leningrad 1952–67; Dir. of Physical-Tech. Inst. Leningrad 1950–57, mem. staff of same 1957–. Decors.: 2 Orders of Red Banner of Labour, 2 Orders "Honorary Badge", medals
Publs.: author of more than 200 scientific works on nuclear physics; "Geometria plasticheskoi deformatsii kristallov po rentgenogrammam Laue" in "Zhurnal eksperimentalnoi i teoreticheskoi fiziki" vol. 6, 1936, "Galvanomagnitnye svoistva splava AiSi v uporiadochennom i neuporiadochennom sostoianii" in "Izvestiia sektora fiz.-khim. analiza Akademii Nauk SSSR", vol. 16, 1943, "Opyty s elektronnymi i ionnymi proektorom" in "Izvestiia Akademii Nauk SSSR. Seriia fizicheskaia" vol. 20/1956.

Komar, Igor' Valeri'ianovich
Economic geographer; b. Ag. 20 (S. 2), 1907, Moscow.
Doctor of geographic sciences (1960).
He graduated from the Moscow Planning Institute. From 1931 to 1943 he worked for planning and economic bodies of the Urals on the development and location of the region's power and industry. He headed an interdisciplinary expedition to the Upper Kama region in 1944–45. He became a senior research worker at the Institute of Geography of the Academy of Sciences of the USSR in 1945. His basic works are on the economic geography of the Urals and the integrated development of its natural resources, as well as on general theoretical and regional questions of economic geography and the problems of the rational use of natural resources. He has elaborated a concept of resource cycles.
Works: *Komi-Permiatskii natsional'nyi okrug*. Moscow-Leningrad 1948 (co-author). *Ural: Ekonomiko-geograficheskaia kharakteristika*. Moscow 1959. *Geografiia khoziaistva Urala*. Moscow 1964. *Geograficheskie problemy razvitiia krupnykh ekonomicheskikh raionov SSSR*. Moscow 1964 (Co-author). *Ural i Priural'e*. Moscow 1968 (Co-author). "Problemy povysheniia effektivnosti obmena veshchestv mezhdu obshchestvom i priorodoi." In *Prirodnye resursy i ekonomicheskaia geografiia SSSR*. Moscow 1971.

Komarov, Nikolai Dmitrievich
State official; b. 1918.
Mem. CPSU 1945–; grad. Moscow Aviation Inst. 1944 and All-Union Acad. of Foreign Trade 1949; worked in aviation ind. 1937–46; admin. chief 1962–65; Dep. USSR Min. of Foreign Trade 1965–80, 1st Dep. Min. O. 1980–; cand. mem. Cen. Cttee. CPSU 1981–.

Komarova, Domna Pavlovna
State official; b. 1920.
Russian; mem. CPSU 1940–; grad. Party High School at Cen. Cttee. CPSU 1955; worked in a raion library and then a raion Komsomol cttee. 1936–42; CPSU work 1942–48; 1st Sec. of Briansk Raion CPSU Cttee. 1948–60; Chair. of Briansk Oblast Exec. Cttee. of Workers' Deps. 1960–67; Dep. to USSR Supr. Sov. 1958–70; mem. Cen. Auditing Comm. CPSU 1966–; Min. of Public Welfare RSFSR 1967–.

Komendant, Petr Vasil'evich
State official.
Mem. Cen. Cttee. Moldavian CP 1971–76; Min. of Foreign Affairs Moldavian SSR D. 1981–.

Komisarenko, Vasilii Pavlovich, Dr. med.
Pathophysiologist and endocrinologist; b. Ja. 14, 1907, Cherniakhov, now Kiev Oblast.
Mem. CPSU 1939–; grad. Medical Inst. Kharkov 1932; Dr. med. 1941; Prof. 1942; mem. staff 1932–37, Dir. of Research Inst. for Experimental Endocrinology Kharkov (present name: Kharkov Research Inst. for Endocrinology and Chemistry of Hormones) 1937–40; Dep. People's Commissar for Public Health Ukrainian SSR 1940–43; Head of Laboratory for Experimental Endocrinology at Inst. for Experimental Biology and Pathology at Ukrainian Min. of Health 1944–; mem. Ukrainian Acad. of Sciences 1951–; after its reorganizations in 1953 Head of Bogomolets Inst. of Physiology at Ukrainian Acad. of Sciences; concurrently Prof. Chair of Pathological Physiology at Medical Inst. Kiev 1950–54; Dir. of Research Inst. for Endocrinology and Metabolic Diseases at Ukrainian Min. of Health 1964–. Decors.: Merited Scientist of the Ukrainian SSR 1960, Order of Lenin, 2 Orders of Red Banner of Labour, Order "Honorary Badge", medals.
Publs.: author of 120 scientific works on questions of pathophysiology of endocrine glands, author of more than 100 popular-science works; "Do pytannia pro mekhanizm dii insulinu na tsentralnu nervovu sistemu" in "Medichnyi zhurnal" No. 3/1940, "Vvedenie v kliniku zabolevannii zhelez vnutrennei sekretsii" 1950, "Mekhanizm i deistviia gormonov" 1959, "Gipofizkora nadpochechnikov" 1964, and others.

Konarev, Nikolai Semenovich
State official; b. 1927.
Russian; mem. CPSU 1952–; Engineer of Transport, Cand. of Technical Sciences; employed in railway transport over 30 years as engineer, dep. head of a railway station, of a railway section, head of the southern railway; Dep. Min. 1976–77, 1st Dep. Min. 1977–82, USSR Min. of Transport N. 1982–.

Konarygin, Vasilii Semenovich
State official.
Min. of Meat and Dairy Ind. RSFSR S. 1970–.

Kondrashin, Kirill Petrovich
Conductor; b. F. 21 (Mr. 6), 1914, Moscow.
Mem. CPSU 1941–; grad. Moscow Conservatory 1936; conductor at several theatres, at Bolshoi Theatre 1943–; Principal Conductor and Musical Dir. of Moscow Philharmonic 1960–. Decors.: State Prize of the USSR 1948, 1949, State Prize of the Russian SFSR 1969, Order of Red Banner of Labour, several medals.

Kondrashov, Ivan Petrovich
State official.
Dep. Chair. of Osh Oblast Exec. Cttee. of People's Deps. 1971–76; cand. mem. Cen. Cttee. Kirghiz CP 1976–81, mem. 1981–; Head of Dept. of Building and Municipal Econ. with Cen. Cttee. Kirghiz CP 1976–79; Min. of Rural Construction Kirghiz SSR Jl. 1979–.

Kondratev, Kirill Iakovlevich
Geophysicist; b. Je. 14, 1920, Rybinsk.
Mem. CPSU 1943–; grad. Leningrad Univ. 1946; mem. staff 1946–58, Prof. 1958–, Rector of Leningrad Univ. 1964–70; concurrently on staff of Cen. Geophysical Observatory 1961–; corresponding mem. USSR Acad. of Sciences 1968–; mem. Int. Astronautical Acad. Decors.: Order of Lenin, Order of Red Banner of Labour, several medals.

Kondratev, Nikolai Dmitrievich
Author; b. 1914, Shelopugino nr. Leningrad.
Russian; mem. CPSU 1944–; mil. journalist 1941–54; first works publ. 1935.
Publs.: "Klim Matuzov" 1943, "Byvalyi soldat Ivan Konyshev" (Veteran Soldier Ivan Konyshev) 1943, "Syny naroda" (Sons of the People) 1956, "Ian Fabritsius" 1954, "Tovarishch Peterson" (Comrade Peterson) 1959, "Skvoz' revol'vernyi lai ..." (Through the Burst of Revolvers ...) 1962, "Gori, moe serdtse" (Be Aflame, My Heart) 1961, "Nadezhnyi tovarishch" (A Reliable Comrade) 1963, and others.

Kondratkov, Evgenii Fedorovich
State official.
Dep. Min. of Light Ind. RSFSR until 1974, Min. D. 1974–.

Konen, Valentina Dzhozefovna
Musicologist; b. Jl. 29 (Ag. 11), 1909, Baku.
Doctor of art studies (1946).
From 1921 to 1931, she lieved in the USA, where she studied at the Juilliard School in New York (1924–29). In 1938 she graduated from the Moscow Conservatory (history and theory department). She taught the history of foreign music at the Moscow Conservatory (1938–41 and 1944–49; docent from 1945) and simultaneously at the Gnesin Music Teachers College (from 1944). She was a professor at the Urals Conservatory (1949–51). In 1960, she became a senior research worker at the Institute of the History of the Arts, under the Ministry of Culture of the USSR. Her scholarly interest centers chiefly on problems in Western European and American music.
Works: Shubert, 2nd ed. Moscow 1959. Puti amerikanskoi muzyki, 2nd ed. Moscow 1965. *Etiudy o zarubezhnoi muzyke.* Moscow 1968. *Klaudio Monteverdi.* Moscow 1971.

Konetskii, Viktor Viktorovich
Autor; b. 1929, Leningrad.
Russian; mem. CPSU 1953–; grad. Higher Navy School 1952; served as mate aboard emergency rescue ships in the Arctic; first works publ. 1956.
Publs.: "Skvozniak" (Draft) 1957, "Kamni pod vodoi"(Rocks Under Water) 1959, "Luna dnem. Povest' i rasskazy raznykh let" (The Moon by Day. Novelette and Stories of Various Years) 1963, "Ogni na merzlykh skalakh" (Lights on Frozen Cliffs) 1964, co-author of film scripts "Svoimi rukami" (With One's Own Hands) 1957, "Opora" (The Support) 1958, "Polosatyi reis" (The Striped Course) 1961, "Esli pozovet tovarishch" (If a Comrade Should Call) 1963, and others.

Konnov, Veniamin Fedorovich
State official; b. 1921.
Mem. CPSU 1952–; grad. Aviation Inst. in Kuibyshev 1946; served in Sov. Army 1939–40; engineer, dep. chief of Technological Office, dep. chief construction engineer, chief construction engineer, Sec. of Party Cttee. in Kuibyshev plant 1946–59; held posts in Cen. Cttee. apparatus CPSU 1960–67; 2nd Sec. of Oblast Cttee. CPSU in Kuibyshev, Chair. of Oblast Exec. Cttee. of People's Deps. 1971–75; Dep. to USSR Supr. Sov. 1974–; Chair. of RSFSR People's Control Cttee. D. 1975–; mem. Cen. Auditing Comm. CPSU 1976–.

Kononenko, Viktor Olimpanovich, Dr. of Tech. Sc.
Mechanical engineer; b. S. 11, 1918, Koroch, now Belgorod Oblast; s. of a teacher.
Mem. CPSU 1946–; grad. Inst. of Railway Engineering Kharkov 1942; Dr. of Tech. Sc. 1954; Prof. 1958; worked in institutions of USSR People's Commissariat for Highways 1942–46; postgraduate studies, sen. research staff. mem. Inst. of Construction Mechanics at Ukrainian Acad. of Sciences 1946–52; mem. staff Laboratory for Measuring Instruments at USSR Acad. of Sciences 1952–56; mem. staff Inst. of Machine Construction 1956–62, Inst. of Mechanical Engineering of Moscow Univ. 1962–64; mem. Ukrainian Acad. of Sciences 1964–; mem. staff Inst. of Mechanics at Ukrainian Acad. of Sciences 1964–, Dr. of same 1965–. Decor.: 1 medal.
Publs.: author of more than 50 scientific works; "O parametricheskom rezonanse drobnogo poriadka" in "Izvestiia Akademii Nauk SSSR, Otdelenie technicheskikh nauk. Mekhanika i Mashinostroenie" No. 8, 1958, "Kolebatelnye sistemy s orgranichennym vozbuzhdeniem" 1964.

Kononov, Andrei Nikolaevich
Orientalist; b. O. 14 (27), 1906, St. Petersburg.
Mem. CPSU 1944–; grad. Inst. of Oriental Studies Leningrad 1930; specialist in Turkish languages and civilization; Prof. at Leningrad Univ. 1950–; corresponding mem. USSR Acad. of Sciences 1958–; Dir. of Leningrad Branch of Inst. of Oriental Studies at USSR Acad. of Sciences 1961–63. Decors.: 3 orders.

Konoplev, Boris Vsevolodovich
Party official; b. 1919; s. of a clerk.
Russian; mem. CPSU 1945–; intermediate vocational training; technologist, foreman, sen. foreman, dept. head in a works 1939–47; dep. sec. of a CPSU cttee. at a works, sec. of CPSU raion cttee., chair. of exec. cttee. of a raion sov., 1st sec. of a raion cttee. CPSU 1947–52; organizer for Cen. Cttee. CPSU during construction of hydroelectric power station on the Kama, 1st Sec. of Perm City Cttee. CPSU, 2nd Sec. of Perm Oblast Cttee. CPSU 1952–62; Chair. of Perm Oblast Exec. Cttee. of Workers' Deps. 1962–72; Dep. 1966–, mem. Comm. of Trade and Public Services/Comm. of Trade, Public Services and Econ. at Sov. of the Union USSR Supr. Sov. 1966–74; 1st Sec. of Perm Oblast Cttee. CPSU N. 1972–; mem. Cen. Cttee. CPSU 1976–. Decor.: Order of Lenin.

Konotop, Vasilii Ivanovich
Party official; b. 1916, s. of a worker.
Ukrainian; mem. CPSU 1944–; grad. Kharkov Inst. of Mechanics and Machine-Building 1940; chemical technologist at a research inst., then at an experimental plant 1933–40; design engineer 1940–49; sec. of CPSU cttee. and organizer for Cen. Cttee. CPSU at a locomotive works, 1st Sec. of Kolomna Raion Cttee. CPSU, Sec. of Moscow Oblast Cttee. CPSU 1949–59; Dep. 1954–, mem. Budget Comm. at Sov. of the Union 1962–66, mem. Presidium USSR Supr. Sov. 1966–; Chair. of Exec. Cttee. Moscow Oblast Sov. of Workers' Deps. 1959–63; cand. mem. 1961–64, mem. Cen. Cttee. CPSU 1964–; 1st Sec. of Moscow Oblast Cttee. CPSU 1963– (for Agriculture 1963–64); Chair. of Sov.-Rumanian Friendship Soc. Ja. 1967–. Decors.: Order of Lenin, Order of the October Revolution.

Konovalenko, Viktor Sergeevich
Hockey player; b. Mr. 11 1938, Gorkii.
Honored Master of Sport (1963), coach. From 1961 to 1971, he was the goalkeeper of the USSR national ice hockey team. His team was Olympic champion 1964 and 1968, champion of the world and Europe from 1963 to 1968 and in 1970, and world champion in 1971. Decors.: two orders.

Konovalov, Evmenii Grigorevich, Dr. of Tech. Sc.
Engineer; b. S. 6 (19), 1914, Chornorechka, Mogilev Gouvt.
Belorussian; grad. Mechanical-Tech. Inst. Leningrad 1936; specialist in metal processing; Dr. of Tech. Sc. 1961; mem. Belorussian Acad. of Sciences 1969–.

Konovalov, Grigorii Ivanovich
Author; b. 1908, Bogoliubovka, Orenburg Oblast; s. of a peasant.
Russian; mem. CPSU 1930–; grad. Perm Pedagogical Inst. 1935; first works publ. 1927.
Publs.: "Plotnik" (The Carpenter) 1934, "Kalmytskii brod" (Kalmyk Ford) 1935, "Tetka Matrena" (Aunt Matrena) 1940, "Universitet" (The University) 1947, "Stepnoi maiak" (Steppe Beacon) 1949, "Istoki" (Sources) 1959, "Pod odnoi kryshei" (Under One Roof) 1962, and others.

Konovalov, Nikolai Semenovich
Party official; b. 1907.
Russian; mem. CPSU 1929–; grad. Party Higher School of Cen. Cttee. CPSU 1948; sec. of a village sov., head of a raion educ. dept., dept. head of a raion CPSU cttee., chair. of a raion sov. of workers' deps. 1925–36; 1st sec. of a raion CPSU cttee., 1st Dep. People's Commr. for Agriculture Udmurt ASSR, dept. head with an oblast CPSU cttee. 1936–43; Exec. Organizer, Inspector for Cen. Cttee. CPSU, 1st Sec. of Kalinin Oblast Cttee. CPSU 1943–51; Dep. to 1950–54 and 1962–, mem. Credentials Comm. 1962–66, mem. Comm. for Construction and Construction Materials Ind. at Sov. of the Union USSR Supr. Sov. 1966–74; worked for Cen. Cttee. CPSU 1951–55; 2nd Sec. 1956–61, 1st Sec. of Kaliningrad Oblast Cttee. CPSU Je. 1961–; Dep. to RSFSR

Supr. Sov. 1959–63; mem. Cen. Cttee. CPSU 1961–. Decor.: Order of Lenin.

Konstantinov, Anatolii Ustinovich
Military official; b. 1923.
Mem. CPSU 1943–; grad. Mil. Acad. of Gen. Staff 1964; Col.-Gen. of Air Force; entered Sov. Army 1940; pilot, squadron commander 1941–45; Sectional regiment commander 1952–56; command posts 1956–73; Commander of Air Force Defense District (PVO) Baku 1973–80, Moscow 1980–; Dep. to USSR Supr. Sov. 1974–; mem. Cen. Cttee. and Bureau of Cen. Cttee. Azerbaidzhan CP 1976–80; mem. Cen. Auditing Comm. CPSU 1981–. Decor.: Hero of the Sov. Union 1946.

Konstantinov, Anton Sidorovich
State official.
Dep. to Supr. Sov. Moldavian SSR 1955–; 1st Sec. of Cen. Cttee. Moldavian Komsomol 1955–56; mem. Cen. Cttee. 1956–60, 1963–66 and 1971–, cand. mem. Cen. Cttee. Moldavian CP 1960–63; cand. mem. Bureau Cen. Cttee. Moldavian CP 1956; Sec. of Republican Trade Union Council Moldavian SSR 1960–61; 1st Sec. of Kishinev Lenin Raion Cttee. Moldavian CP 1961–62; 1st Sec. of Kishinev City Cttee. Moldavian CP 1962–66; mem. Bureau for Ind. at Cen. Cttee. Moldavian CP 1962–64; Head of Dept. of Propaganda and Agitation at Cen. Cttee. Moldavian CP 1967–73; Min. of Culture Moldavian SSR Jl. 1973–. Decor.: Order of Red Banner of Labour.

Konstantinov, Fedor Vasil'evich
Philosopher; b. F. 8 (21), 1901, Novoselki, Gorkii Oblast.
Mem. CPSU 1918–; grad. Inst. of Communist Ideology 1932; pedagogical and party work 1932–45; mem. staff Inst. of Philosophy at USSR Acad. of Sciences 1945–51; Chief Ed. of periodical "Voprosy filosofii" 1952–54; Rector of Acad. of Social Sciences at Cen. Cttee. CPSU 1954–55; cand. mem. Cen. Cttee. CPSU 1956–61; Head of Agitprop Dept. for Union Republics at Cen. Cttee. CPSU 1955–58; Chief Ed. of party publ. "Kommunist" 1958–62; mem. USSR Acad. of Sciences 1964–; Dir. of Inst. of Philosophy at USSR Acad. of Sciences 1962–67; Acad. Sec. of Dept. of Philosophy and Law at USSR Acad. of Sciences 1967–; Chief Ed. of "Filosofskaja enciklopedija" 1960–70; Pres. of USSR Philosophical Soc. 1971–. Decors.: Order of Lenin, Order of the October Revolution, 4 other orders, medals.
Publs.: "Za bol'ševizaciju raboty na filosofskom fronte" in: "Za povorot na filosofskom fronte" Moscow-Leningrad 1931 (co-author), "Značenie ličnyh sposobnosteij i truda pri socializme" Moscow 1938, "Rol'idej v obščestvennom razvitii" Moscow 1940, "Cto takoe marksistsko-leninskaja filosofija" Moscow 1941, "Materialističeskoe i idealističeskoe ponimanie istorii" Moscow 1946, "Istoričeskij materializm kak nauka" Moscow 1949, "O dvižuščih silah razvitija socialističeskogo obščestva" Moscow 1951, "Formy obščestvennogo soznanija" Moscow 1951, "B.I. Lenin. Biografija" Moscow 1972 (co-author).

Konstantinov, Mikhail Panteleimonovich
Architect; b. S. 27, 1924, Moscow.
He studied under B.S. Mezentsev at the Moscow Institute of Architecture from 1944 to 1950. His works in Moscow include residential buildings on Osipenko Street (1952–54) and on Vernadskii Boulevard (1965), as well as the Krasnopresnenskaia subway station (1953, in collaboration with others). He also designed the river terminal in Kazan (1956–59, together with the architect I.G. Gainutdinov), the city communication center in Kislovodsk (1970), the Lenin Memorial in Ul'ianovsk (1970, together with B.S. Mezentsev, G.G. Isakovich, and others). Decor.: Lenin Prize 1972.

Konstantinov, Vasilii Nikiforovich
State official.
Mem. 1964–66, cand. mem. Cen. Cttee. Esthonian CP 1966–; Sec. 1965, 2nd Sec. of Tallinn City Cttee. Esthonian CP 1967–76; Chair. of State Cttee. on Labour Esthonian SSR S. 1976–.

Kontrimavichus, Vitautas Leonovich, Dr. of Biol. Sc.
Biologist; b. 1930.
Grad. Veterinary Inst. Leningrad 1952; Dr. of Biol. Sc. 1969; corresponding mem. USSR Acad. of Sciences 1970–; Dir. of Inst. of Biology at Far Eastern Research Centre of USSR Acad. of Sciences 1970–.

Kontselidze, Rezan Khasanovich
State official.
2nd Sec. and mem. Bureau Cen. Cttee. Komsomol Georgian SSR 1967–69; Head of Dept. of Food and Light Ind. at Cen. Cttee. Georgian SSR 1973–74; Min. of Food Ind. Georgian SSR Je. 1974–; mem. Cen. Cttee. Georgian CP 1976–81.

Konurbaev, Marat Omurzakovich
State official.
Min. of Local Ind. Kirghiz SSR Ja. 1973–, mem. Auditing Comm. 1976–81, cand. mem. Cen. Cttee. Kirghiz CP 1981–.

Koptelov, Afanasii Lazarevich
Russian writer; b. O. 24 (N. 6), 1903, Shatunovo, in present-day Zalesovo Raion, Altai Krai. Member of the CPSU since 1944.
In his early works (the novella *Confusion*, 1927, and the novel *The New Fields*, 1929), he depicted the class struggle in a village of Kerzhaks (Old Believers). His novel *The Great Nomad Encampment* (1934–35) deals with the Altais' national revival under Soviet power. Many of his works, including the novel *The Garden* (1955) and the essays *The Transformation of Siberia* (1959), focus on man's struggle to transform nature. He has also written novel dealing with V.I. Lenin – *The Great Beginning* (1963) and *The Flame Will Flare Up* (parts 1–2, 1965–68). Decors.: four orders as well as medals.
Works: *Forposty sotsializma: Ocherki.* Moscow 1931. *Svetlaia krov': Roman.* Novosibirsk 1933. *Na-gora: Roman.* Novosibirsk 1947. *Po putiam-dorogam Evropy.* Novosibirsk 1959. *Snezhnyi pik.* Novosibirsk 1964. *Minuvshee i blizkoe.* Novosibirsk 1972.

Koptiaeva, Antonina Dmitrievna
Russian writer; b. O. 25 (N. 7), 1909, at the Iuzhnyi Mine, now in Zeia Raion, Amur Oblast.
She began to publish in 1935 and graduated from the Gorkii Literary Institute in 1947. She is the author of the novella *Kolyma Gold* (1936, under the pen name A. Zeite), the collection of essays *The Tale of Aldan* (1937), and the novels *Luck* (1940), *Comrade Anna* (1946), and *Gift of the Earth* (1963). She also wrote the trilogy *Ivan Ivanovich* (1949; State Prize of the USSR, 1950), *Friendship* (1954), and *Daring* (1958), the central character of which is a talented doctor and humanist. Her books deal largely with questions of love, family, marriage, and man's social duty. Her books have been translated into many languages of the peoples of the USSR and into foreign languages. Decors.: two orders and various medals.
Works: *Na Urale-reke,* book 1. Introduction by S. Vasil'ev. Moscow 1971. *Po sledam Ermaka: Iz dnevnika pisatelia.* Sverdlovsk 1972.

Koptiug, Valentin Afanasevich
Scientist; b. 1931.
Mem. CPSU 1961–; grad. Moscow Inst. of Chemical Technology 1954, Dr. of Chemical Sc. 1966, Prof. 1968, mem. USSR Acad. of Sciences 1979; engineer at an inst. of chemical technology 1957–58; junior research associate, senior research associate, dep. laboratory asst. in Siberian Dept. of USSR Acad. of Sciences 1959–78; Rector of Novosibirsk Univ. 1978–80; Vice-Pres. of USSR Acad. of Sciences and Chair. of Siberian Dept. of USSR Acad. of Sciences Mr. 1980–; mem. Cen. Auditing Comm. CPSU 1981–. Decor.: Order of Lenin.

Korban, Vladimir Ivanovich
Author; b. 1910, Baran, Vitebsk Oblast; s. of a worker.
Belorussian; mem. CPSU 1932–.
Publs.: "My ikh znaem" (We Know Them) 1950, "Basni" (Fables) 1953, "Tochnye primety" (Accurate Signs) 1958, "Gde eta ulitsa?" (Where is This Street?) 1961, "Brigadnyi kum" (Brigade Gossip) 1962, "K teshche na bliny" (To Mother-in-Law's for Pancakes) 1963, "Siniak" (Black and Blue) 1964, and others.

Koretskii, Viktor Borisovich
Graphic and poster artist; b. My. 5 (18), 1909, Kiev.
Honored Artist of the RSFSR (1964).
From 1922 to 1929, he studied at the Moscow Secondary School of Fine Arts. In his works he combines photographs of nature with pencil and gouache drawings. His most outstanding posters, which were executed primarily during the Great Patriotic War of 1941–45, are distinguished by their clarity of design and by their intensely emotional depiction of the dramatic experiences of the Soviet people. His works include the posters *Greetings to the Great Union of SSR!* (1937), *Warrior of the Red Army, Save Us!* (1942), *Peace Will Be Victorious!* (1950), *The Immortality of Lenin – In Our Achievements* (1964), and *Be Damned!* (1965). Decors.: State Prize of the USSR in 1946 and 1949.
Works: *Zametki plakatista*. Moscow 1958.

Koretskii, Vladimir Mikhailovich
Lawyer; b. F. 5 (17), 1890, Ekaterinoslav, present-day Dnepropetrovsk.
Specialist in international law and the history of the state and law. Academician of the Academy of Sciences of the Ukrainian SSR (1948).
From 1919 to 1941 and again from 1944 to 1949, he headed the Sector of State and Law of the Academy of Sciences of the Ukrainian SSR. (In 1970 the sector was reorganized as the Institute of State and Law.) He was a member of the Soviet delegation to the UN General Assembly in 1946, 1947, and 1949 and participated in the work of various international conferences and commissions of the UN. In 1957 he was appointed to the Permanent Court of Arbitration, and from 1961 to 1970 he was a member of the UN International Court of Justice, serving as its vice-president from 1967 to 1970. He joined the Mexican Academy of International Law in 1969. Decors.: Order of Lenin and the Order of the Red Banner of Labor.
Works: *Lektsii po istorii gosudarstva i prava*. Moscow 1947. *Ocherki anglo-amerikanskoi doktriny i praktiki mezhdunarodnogo chastsogo prava*. Moscow 1948. *"Obshchie printsipy prava" v mezhdunarodnom prave*. Kiev 1957.

Korkiia, Otar Mikhailovich
Sportsman; b. My. 10, 1923, Kutaisi.
Honored Master of Sport (1951) and Honored Coach (1967). Became a member of the CPSU in 1953.
In the 1940's and 1950's, he was one of the leading players on the USSR national basketball team. He was champion of Europe in 1947, 1951, and 1953 and of the USSR in 1950–51 and 1953, as well as an Olympic prize winner (silver medal, Helsinki, 1952). Decors.: Order of Lenin and various medals.

Korkin, Aleksandr Gavrilovich
Party official; b. 1927.
Russian; mem. CPSU 1949–; grad. Magnitogorsk Metallurgical Mining Inst. 1949; work superintendent, section chief of admin. for construction in State Electrical Power Plant in Karaganda Raion 1949–51; chief of a dept., 1st Sec. of Temir-Tau City Cttee. Kazakh CP, Sec. of Party cttee. of construction site of State Electrical Power Plant in Karaganda Raion, instructor in Cen. Cttee. Kazakh CP 1951–61; manager of "Kazmetallurgstroi" Trust 1961–72; mem. Cen. Cttee. Kazakh CP 1961–; Min. of Construction of Heavy Ind. Enterprises Kazakh SSR 1971–75; mem. of Bureau of Cen. Cttee. Kazakh CP 1975–79; Sec. of Cen. Cttee. Kazakh CP 1975–76, 2nd Sec. 1976–79; 1st Sec. of Karaganda Oblast Cttee. Kazakh CP 1979–; Dep. to USSR Supr. Sov. 1979–; cand. mem. Cen. Cttee. USSR 1981–.

Kormashov, Nikolai Ivanovich
Painter; b. Ag. 28, 1929, Turgenevo, Melenki Raion, Vladimir Oblast.
He, who studied in Tallinn at the Art Institute of the Esthonian SSR from 1951 to 1957, works in Esthonia. He paints thematic compositions and industrial landscapes, represented by *Salt of the Earth* (1965, Art Museum of the Esthonian SSR, Tallinn), *Reinforced Concrete* (1965, Art Museum of the Esthonian SSR) and *The Dinner Table* (1968, Art Museum of the Esthonian SSR), the triptych *The Young Builders* (1967, in the All-Union Production Art Combine of the Ministry of Culture of the USSR, Moscow), and the triptych *To the Victims of War* (1969, Tartu Art Museum). He also paints portraits (*G. Reindorff*, 1965, Art Museum of the Esthonian SSR). His works are distinguished by broad generalization, intense emotional feeling, and concise expressive devices.

Kornev, Vasilii Ivanovich
Diplomat; b. 1915.
Mem. Diplomatic Service 1942–; Emb. Counsellor at USSR Emb. in Turkey 1951–56; Dep. Head of Middle East Dept. at USSR Min. of Foreign Affairs 1956–58; Consul-Gen. in Damascus 1958–61; USSR Amb. to Lebanon 1961–62; sen. posts in USSR Min. of Foreign Affairs 1962–72; USSR Amb. to Arab Republic of Yemen My. 1972–80.

Kornienko, Anatolii
Komsomol official.
2nd Sec. of Kiev City Cttee. Ukrainian Komsomol 1962–72; mem. Bureau 1972–, 2nd Sec. 1972–75, 1st Sec. of Cen. Cttee. Ukrainian Komsomol Ja. 1975–; mem. Cen. Cttee. Ukrainian CP 1976–.

Kornienko, Georgii Markovich
Diplomat; b. 1925.
Ukrainian; mem. CPSU 1947–; grad. Moscow Law Inst. 1953; entered Diplomatic Service 1949; dep. chief of a dept. in USSR Min. of Foreign Affairs 1959–60; Emb. Counsellor in USSR Emb. in the USA 1960–63, Emb. Counsellor-Envoy 1963–65; Dep. Chief, Chief of American Dept. in USSR Min. of Foreign Affairs 1965–78; Dep. USSR Min. of Foreign Affairs 1975–77, 1st Dep. Min. 1977–; Dep. to USSR Supr. Sov. 1979–; mem. Cen. Cttee. CPSU 1981–.

Kornilov, Vladimir Nikolaevich
Author and poet; b. 1928, Dnepropetrovsk; s. of an engineer.
Russian; studied Gorkii Inst. of Literature 1945–50; first works publ. 1953.
Publs.: "Shofer" (Chauffeur) 1961, "Pristan'" (Harbour) 1964, and others.

Koroid, Aleksei Stepanovich
Economist; b. Jl. 10, 1911, Sliudianka, Irkutsk Oblast; s. of a railwayman.
Mem. CPSU 1939–; grad. Dept. of Econ. of Kharkov Univ. 1933; teaching activity at Chemical-Tech. Inst. Kharkov 1933–42; party assignments 1942–49; Dep. USSR Min. of Higher and Secondary Specialist Educ. 1952–53; Dep. Dir. of Inst. of Econ. at Ukrainian Acad. of Sciences 1955–65; Rector of Econ. Kiev 1965–.
Publs.: has written some 30 scientific works, co-author of monographs "Radianska Ukraina" 1957, "Ukrainska RSR", 2 vols., 1957–58, "Ekonomichni pytannia budivnytstva kommunizmu v SRSR" 1960, "Nauka i kommunizm" 1962.

Korol', Vladimir Adamovich
Architect; b. D. 14 (27), 1912, Igumen, present-day Cherven', in Minsk Oblast.
People's Architect of the USSR (1970). Became a member of the CPSU in 1940.
From 1934 to 1941, he studied at the Leningrad Academy of Arts. He has worked in the Belorussian SSR, primarily in Minsk, since 1945; in 1955 he became the chairman of the State Committee of the Council of Ministers of the Belorussian SSR for Affairs of Construction and Architecture. He has been an instructor at the Belorussian Polytechnic Institute since 1947. His works in Minsk, which were designed in collaboration with others, include the city's general plan (1948–69), Lenin Prospect (1950's; State Prize of the Belorussian SSR, 1968), Central Square (1948–55), the Main Post Office (1950–53), the Central Telegraph Office (1954–56), and the monument at Victory Square (1954). He also participated in the design of the memorial complex *Brest Hero-Fortress* (1966–71). He directed the design of a number of plans for raions and for many villages in the Belorussian SSR. Decors.: Order of Lenin, the Order of the October Revolution, two other orders, and various medals.

Korolev, Anatolii Maksimovich
Laborer; b. 1936.
Mem. CPSU 1967–; finished evening secondary school in Sverdlovsk 1979; served in Sov. Army 1955–58; lathe operator at Ural Plant for Heavy Machine Building 1960–; mem. Cen. Cttee. CPSU 1981–.

Korolev, Nikolai Fedorovich
Boxer; b. Mr. 14, 1917, Moscow.
Honored Master of Sport (1942), and trainer. Became a member of the CPSU in 1950.
He was heavyweight champion of the USSR nine times between 1936 and 1953 and overall champion of the USSR in 1936–37 and 1944–45. He was also a victor at the World Workers' Olympiad in Antwerp in 1937 and at international tournaments in Helsinki and Prague (1946). In the Great Patriotic War he served in the partisan detachment of Hero of the Soviet Union D. N. Medvedev. Decors.: two orders and also medals.

Koroliuk, Vladimir Dorofeevich
Historian; b. F. 28, 1921, Kutaisi.
Doctor of historical sciences (1965).
Koroliuk graduated from Moscow State University in 1943. In 1965 he was appointed head of a section at the Institute of Slavic and Balkan Studies of the Academy of Sciences of the USSR and assistant editor-in-chief of the journal *Sovetskoe Slavianovedenie* (Soviet Slavic Studies). His principal works are devoted to the emergence of the Slavic and eastern Romance peoples, to the development of feudal relations and states among the eastern and western Slavs, to the history of international relations in the early feudal period in Central and Eastern Europe, to the history of Russo-Polish relations from the 16th to the 18th centuries, to the history of Slavic-German relations, and to the history of Slavic studies in the USSR. He wrote a number of chapters for the collaborative *History of Poland* (vols. 1–3, 1954–56) and *History of Czechoslovakia* (vols. 1–2, 1956–59) and served on the editorial board of these publications.
Works: *Livonskia voina*. Moscow 1954. *Drevnepol'skoe gosudarstvo*. Moscow 1957. *Zapadnye slaviane i Kievskaia Rus' v X–XI vv.* Moscow 1964. *Polska i Rosija a wojna północna*. Warsaw 1954.

Korostovtsev, Mikhail Aleksandrovich, Dr. of Hist. Sc.
Historian; b. Ap. 10 (23), 1900, Popovka, Dnepropetrovsk Oblast.
Mem. CPSU 1929–; grad. Azerbaidzhan Univ. 1934; Egyptologist; Prof. 1942; Research Asst. and Scientific Sec. of Leningrad Branch of Inst. of History at USSR Acad. of Sciences 1935–41; research staff mem. 1943–65, Head of Dept. of Ancient Oriental Studies at Inst. of Oriental Studies USSR Acad. of Sciences 1965–; worked in Egypt-Sudan 1943–47; mem. French-Egyptian Soc. 1958. Decor.: Dr. h.c. from Egypt Inst. of Charles Univ. Prague 1965.

Korotkevich, Vladimir Semenovich
Author; b. 1930, Orsha, Vitebsk Oblast.
Belorussian; grad. Dept. of Philology of Kiev Univ. and Higher Courses in Literature Moscow 1961; first works publ. 1955.
Publs.: "Dusha materi" (A Mother's Soul) 1958, "Vechernie parusa" (Evening Sailing) 1960, and others.

Korotych, Vitalii Alekseevich
Author and poet; b. My. 26, 1936, Kiev.
Ukrainian; grad. Kiev Medical Inst.; physician 1959–66; Exec. Ed. of journal "Ranok" 1966–; Sec. of Ukrainian Writers' Union 1966–69.
Publs.: "Zoloti ruky" (Golden Hands) 1961, "Zapakh neba" (The Smell of Heaven) 1962, "Vulytsia voloshok" (Cornflower Street) 1963, "Poezii" (Poetry) 1967, "O Kanado!" (Ah Canada!) 1966; translated works of English and Slavic poets into Ukrainian.

Korsakas, Kostas Pranovich
Author, poet and literary historian; b. 1909, Pašvitinys, Lithuania; s. of a peasant.
Lithuanian; grad. Kaunas Univ.; Ed. of monthly journal "Kultura" 1933–40; lived in Penza and Moscow 1941–45; Dir. of Inst. of Lithuanian Language and Literature at Lithuanian Acad. of Sciences, Prof. at Vilnius Univ. 1946–; mem. Lithuanian Acad. of Sciences 1949–; Dep. 1955–63 and 1975–, Chair. of Supr. Sov. Lithuanian SSR 1959–63 and 1975–80; first works publ. 1925. Decors.: 2 Orders of Red Banner of Labour, Order of Patriotic War 2nd Cl., medals.
Publs.: "Stat'i o literature" (Articles on Literature) 1932, "Kritika" (Critique) 1936, "Pisateli i knigi" (Writers and Books) 1940, "Literatura i kritika" (Literature and Critique) 1949, "Roman A. Venuolise 'Puodžiunkiemis'" (A. Vienuoli's Novel 'Puodzhiunkiemis') 1953, "Ptitsy vozvrashchaiutsia" (The Birds Return) 1945, "Izbrannoe" (Selected Poetry) 1950, "V razluke" (In Parting) 1944, and others.

Korsakiene, Galina Petrovna
Author; b. 1910, d. of a doctor.
Lithuanian; grad. Dept. of History and Philology of Kaunas Univ.; returned to Lithuania with parents 1921.
Publs.: "V rodnom gorode" (In the Hometown) 1949, "Podarok" (Present) 1950, "Chest' klassa" (Class Honour) 1952, "Pervyi god" (The First Year) 1955, "Vremennaia stolitsa" (Provisional Capital) 1962, and others.

Korshak, Vasilii Vladimirovich
Chemical scientist; b. D. 27, 1908 (Ja. 9, 1909), Vysokoe, Chernigov Oblast.
Mem. CPSU 1940–; grad. Mendeleev Chemical-Technologist Inst. Moscow 1931; corresponding mem. USSR Acad. of Sciences 1953–; Chief Ed. of Periodical "Vysokomolekuliarnye soedineniia". Decors.: State Prize of the USSR 1949, 1951, 4 orders, several medals.
Publs.: author of more than 16 monographs on the chemistry of polymers, "Himija vysokomolekuljarnyh soedinenij" Moscow-Leningrad, co-author of "Ravnovesnaja polikondensacija" Moscow 1968, "Termostojkie polimery" Moscow 1969, co-author of "Neravnovesnaja polikondensačija" Moscow 1972.

Kortelainen, Karl Efremovich
State official.
Chair. Cttee. of State Security Esthonian SSR Je. 1982–.

Korzh, Nikolai Afanasevich
State official; b. 1921.
Ukrainian; mem. CPSU 1943–; grad. Uman Agricultural Inst. and CPSU Cen. Cttee. Party Higher School; served in Sov. Army 1939–46; party posts 1947–67; Chair. of Exec. Cttee. of Poltava Oblast Sov. of Peoples' Deps. 1967–73; cand. mem. Cen. Cttee. Ukrainian CP 1971–76, mem. 1976–; 1st Sec. of Volyn Oblast Cttee. Ukrainian CP 1974–79; Dep. to USSR Supr. Sov. 1974–; Min. of State Farms Ukrainian SSR Mr. 1979–.

Korzhavin, N. (real name: Mandel, Naum Moiseevich)
Author and poet; b. 1925, Kiev.
Grad. Karaganda Mining Inst. and Gorkii Inst. of Literature Moscow 1959; first works publ. 1941.
Publs.: "Gody" (The Years) 1963, "Gde vy, gde vy ...?" (Where Are You, Where Are You ...?), "Khleb" (Bread), "Deti v Osventsime" (Children in Auschwitz), "Osen' v Karagande" (Autumn in Karaganda), and others.

Korzhev, Gelii Mikhailovich (also G. M. Korzhev-Chuvelev)
Painter; b. Jl. 7, 1925, Moscow.
People's Artist of the RSFSR (1972). Member of the Academy of Arts of the USSR (1970).
He studied under S. V. Gerasimov at the V. I. Surikov Moscow Art Institute from 1944 to 1950. A teacher at the Moscow Higher Industrial Arts School (1951–58 and from 1964), he received a professorship there in 1966. In 1968, Korzhev became the chairman of the administrative board of the Artists' Union of the RSFSR.
His works are marked by clear composition, restrained colors, a large-scaled plan, and carefully modeled figures. He focuses on important civic themes and depicts Soviet people in dramatic and heroic situations. His paintings in the

Russian Museum in Leningrad include the triptych *The Communists* (1957–60; I. E. Repin State Prize of the RSFSR, 1966), *People in Love* (1959), and *The Send-off* (1967; from the series *Singed by the Fire of War,* 1962–67). His painting *The Artist* (1960–61) is in the Tretiakov Gallery.

Korzhinskii, Dmitrii Sergeevich
Geologist, petrographer and mineralogist; b. S. 1 (13), 1899, St. Petersburg.
Grad. Leningrad Mining Inst. 1926; mem. USSR Acad. of Sciences 1953–; mem. Geological Cttee. 1926–29, then mem. staff Scientific Union Research Inst. of Geology (VSEGEI); teacher, then Prof. at Leningrad Mining Inst. 1929–40; mem. staff USSR Acad. of Sciences 1937–, Dir. of its Inst. of Experimental Mineralogy 1969–; mem. numerous foreign socs. and acads.; Vice-Pres. of Union Soc. of Mineralogists. Decors.: State Prize of the USSR 1946, Lenin Prize 1958, A. P. Karpinskii Prize 1949, Vernadskii Medal in Gold 1972, Order of Lenin, Order of Red Banner of Labour, medals.

Kosarev, Evgenii Aleksandrovich
Diplomat; b. 1919.
Mem. Diplomatic Service 1947–; sen. posts in Sov. Control Comm. in Germany 1947–54; mem. Apparat USSR Min. of Foreign Affairs 1954–57; USSR Consul in Karl-Marx-Stadt 1957–61; sen. posts in USSR Min. of Foreign Affairs 1961–65; Emb. Counsellor at USSR Emb. in the FRG 1965–69; USSR Amb. to Luxemburg Ag. 1969–S. 1979.

Kosarik, Dmitrii Mikhailovich (real name: Kovalenko)
Author and literary historian; b. 1904, Fidrovka, Poltava Oblast; s. of a peasant.
Ukrainian; first works publ. 1923.
Publs.: "Krasnaia kupel'" (The Red Font) 1925, "Partyzanskii kolodiaz'" (The Partisan Well) 1932, "Pushkin na Ukraine" (Pushkin in the Ukraine) 1937, "David Guramishvili" 1950, "Zhyttia i diial'nist' Shevchenka" (Shevchenko's Life and Works) 1955, "D. Guramishvili na Ukraine" (D. Guramishvili in the Ukraine) 1948, and others.

Koshlakov, Georgii Vladimovich
State official.
Head of Admin. of Geology with Council of Mins. Tadzhik SSR Jl. 1975–78; mem. Auditing Comm. 1976–81, mem. Cen. Cttee. Tadzhik CP 1981–; Dep. Chair. of Council of Mins. Tadzhik SSR O. 1978–.

Koshoev, Temirbek Khudaibergenovich
State official; b. 1931.
Kirghiz; mem. CPSU 1952–; grad. Kirghiz Agricultural Inst. 1957; forester, student 1950–57; instructor in Cen. Cttee. Kirghiz CP, 2nd sec. of a raion cttee. Kirghiz CP, chair. of exec. cttee. of a raion sov. of peoples' deps., 1st sec. of a raion cttee. Kirghiz CP 1957–63; mem. Auditing Comm. Kirghiz CP 1960–61; mem. Cen. Cttee. Kirghiz CP 1963–; 1st Dep. Chair. of Exec. Comm. of Osh Oblast Sov. of Peoples' Deps. 1963–66, Chair. 1966–78; 1st Sec. of Osh Oblast Cttee. Kirghiz CP 1978–81; mem. Bureau of Cen. Cttee. Kirghiz CP D. 1978–; Dep. to USSR Supr. Sov. 1979–; Chair. of Presidium of Supr. Sov. Kirghiz SSR, Dep. Chair. of Presidium of USSR Supr. Sov. 1981–; mem. Cen. Auditing Comm. CPSU 1981–.

Kosmodamianskii, Aleksandr Sergeevich, Dr. of Tech. Sc.
Mechanical engineer; b. Mr. 24, 1923, Revni, now Briansk Oblast.
Mem. CPSU 1943–; grad. Saratov Univ. 1949; Dr. of Tech. Sc. 1964; Prof. 1965; completed postgraduate studies at Saratov Univ. 1952; mem. staff Rostov/Don Univ. 1952–53 and Saratov Univ. 1953–65; Dep. Rector of Donetsk Univ. and mem. staff Computing Centre of Ukrainian Acad. of Sciences Donetsk 1965–; corresponding mem. Ukrainian Acad. of Sciences 1965–. Decors.: 1 order, medals.
Publs.: author of approx. 50 scientific works, "Deistvie sosredotochennykh sil v mnogosviaznykh oblastiakh" in "Kontsentratsiia napriazhenii" 1965, "O kvazireguliarnosti beskonechnykh sistem v zadache o kontsentratsii napriazhenii vozle krivolineinykh otverstii" in "Prikladnaia mekhanika" No. 1/1965.

Kosolapov, Rikhard Ivanovich
Journalist; b. 1930.
Mem. CPSU 1957–; grad. Moscow State Univ. 1955, Dr. of Philosophical Sc. 1971; posts in Komsomol 1955–58; teacher and researcher at Moscow State Univ. and USSR Acad. of Sciences 1958–66; posts in Cen. Cttee. CPSU apparatus 1966–74; 1st Dep. Chief Ed. of "Pravda" 1974–76; Chief Ed. of theoretical organ of CPSU "Kommunist" 1976–; cand. mem. Cen. Cttee. CPSU 1976–81, mem. 1981–. Decor.: Order of the October Revolution.

Kostandov, Leonid Arkad'evich
State official; b. 1915, Kerki, Turkmen SSR.
Armenian; mem. CPSU 1942–; grad. Inst. of Chemical Machine-Building Moscow 1949; machine operator in a cotton mill in Chardzou, brigade leader at a silk spinning mill 1930–49; completed a workers' faculty 1934; Dir. of Chirchik Electro-Chemical Combine 1949–53; Head of Main Nitrade Ind. Board with USSR Min. of the Chemical Ind., 1st Dep. Chair. of State Cttee. for Chemistry with USSR Council of Mins. 1953–63; Chair. of USSR GOSPLAN State Cttee. for Chemical and Petroleum Machine-Building with rank of USSR Min. 1963–64; Chair. of USSR GOSPLAN State Cttee. for Chemistry 1964–65; USSR Min. of Chemical Ind. O. 1965–80; Dep. to USSR Supr. Sov. 1966–; cand. mem. 1966–71, mem. Cen. Cttee. CPSU 1971–; Dep. Chair. USSR Council of Minsk. 1980–. Decor.: Stalin Prize, Lenin Prize, Order of Lenin.

Kostarev, Iurii Borisovich
State official.
Chair. of State Cttee. of Gasification Tadzhik SSR Mr. 1980–.

Kostenko, Lina Vasilevna
Author and poet; b. 1930, Rzhishchev, Kiev Oblast.
Ukrainian; grad. Gorkii Inst. of Literature 1956.
Publs.: "Luchi zemli" (The Earth's Rays) 1957, "Parusa" (Sails) 1958, "Stranstviia serdtsa" (Journeys of the Heart) 1961, and others.

Kostenko, Mikhail Vladimirovich
Scientist; b. D. 15 (28), 1912, Nikolaev.
Mem. CPSU 1945–; grad. Polytech. Inst. Leningrad 1938; specialist in energetics; Lecturer 1938–55, Prof. Chair of Technology of High Voltage at Polytech. Inst. Leningrad 1955–; corresponding mem. USSR Acad. of Sciences 1962–. Decors.: Order of Lenin, medals.

Kostenko, Nikolai Fedorovich
Moldavian writer; b. D. 21, 1913, Kishinev.
He studied at the University of Iaşi. He made his literary debut in 1934. His first collection, *Poems,* was published in 1937, followed by the collections *Hours* (1939) and *Pagan Elegies* (1940). After the liberation of Bessarabia in 1940, he worked on the magazine *Nistrul* (October). His other collections include *New Poems* (1960) and *Poems* (1961) in Moldavian and *Constancy* (1959) and *Moldavian Songs* (1961) in Russian. He has also written sketches and stories from the life of Moldavian villagers. His novel *Severograd* (1962) deals with the workers of the North. He translated the works of A.S. Pushkin, M. Iu. Lermontov, T.G. Shevchenko, S.A. Esenin, and A.A. Blok into Moldavian.
Works: *Poezii alese.* Kishinev 1957. *Versur'.* Kishinev 1966. *In Russian translation: Severograd: Roman [i rasskazy].* Kishinev 1967.

Kostin, Vitalii Semonovich
Laborer; b. 1938.
Mem. CPSU 1961–, grad. Prokopev Mining Technicum 1958; served in Sov. Army 1958–61; traffic foreman, mine foreman, asst. section chief, tunneller at "Ziminka" Mine in Kemerovo Oblast 1961–73; stope miner in "Ziminka" Mine N. 1973–; cand. mem. Cen. Cttee. CPSU 1981–82, mem. My. 1982–.

Kostiuk, Grigorii Silovich
Psychologist; b. N. 23 (D. 5), 1899, Mogil'no, in present-day Gaivoron Raion, Kirovograd Oblast.
Active member of the Academy of Pedagogical Sciences of the USSR (1967).
He became a professor at the Kiev Pedagogical Institute in 1935 and director of the Ukrainian Psychology Research Institute in 1945. His principal works deal with the psychology of thought; the learning of concepts; the relations between schooling, upbringing, the development of the child's personality, and the child's abilities; and the history of psychology. Decors.: Order of Lenin, two other orders, and several medals.
Works: "Pro henezys poniattia chysla u ditei." *Naukovi zapysky Yn-tu-psykhologii,* 1949, vol. 1. "Narysy z istorii vitchyznianoi psykhologii (XVII–XVII st.)." *Zbirnik st.* Kiev 1952. "Voprosy psikhologii myshleniia." In the collection *Psikhologicheskaia nauka v SSSR,* vol. 1. Moscow 1959. "Printsip razvitiia v psikhologii." In *Metodologicheskie i teoreticheskie problemy psikhologii.* Moscow 1969.

Kostiuk, Platon Grigorevich, Dr. of Biol. Sc.
Physiologist; b. Ag. 2, 1924, Kiev; s. of a teacher.
Mem. CPSU 1947–; grad. Dept. of Biology of Kiev Univ. 1946; Dr. of Biol. Sc. 1956; Prof. 1960; Head of Dept. of General Physiology at Inst. of Veterinary Physiology of Kiev Univ. 1946–58; Head of Laboratory for General Physiology at Bogomolets Inst. of Physiology at Ukrainian Acad. of Sciences 1958–66, Dir. of same inst. 1966–; corresponding mem. Ukrainian Acad. of Sciences 1964–, of USSR Acad. of Sciences 1966–. Decor.: Order of Red Banner of Labour.
Publs.: author of more than 100 works on electrophysiology and neurophysiology; "Dvukhneironnaia reflektornaia duga" 1959, "Mikroelektrodnaia tekhnika" 1960, "Mekhanizm pesymalnoho halmuvannia skeletnoho m'iaza" in "Fiziologichnyi zhurnal" No. 11/1965.

Kostiukov, Ivan Ivanovich
Trade union official; b. 1926.
Mem. CPSU 1951–; grad. Dnepropetrovsk Mining Inst. 1957; locksmith 1941–43, electrical welder, technician in works in Kamensk-Uralskii 1947–53; engineering and tech. posts at Lenin Works in Dnepropetrovsk 1953–59; Dep. Sec., Sec. of CPSU Cttee. at same works 1959–62; Chair. of Dnepropetrovsk Oblast Cttee. of Union of Metallurgical Ind. Workers 1962–64; Chair. of Cen. Cttee. of Union of Metallurgical Ind. Workers F. 1964–; mem. CPSU Cen. Auditing Comm. 1971–.

Kostiukovskii, Boris Aleksandrovich
Author; b. 1914, Kansk, Krasnoiarsk Oblast.
Russian; mem. CPSU 1938–; first works publ. 1940.
Publs.: "Sibiriaki" (Siberians) 1947, "I snova vesna ..." (And It Is Spring Again ...) 1948, "V gorakh Akatuia" (In the Mountains of Akatui) 1952, "Utro Andreia Shilina" (Andrei Shilin's Morning) 1954, "Sud'ba druga" (The Fate of a Friend) 1963, "Doroga k solntsu" (The Way to the Sun) 1963, "Priznanie v liubvi" (Declaration of Love) 1965, and others.

Kostiukovskii, Iakov Aronovich
Author; b. 1921, Zolotonosha.
Russian; mem. CPSU 1944–; studied Moscow Inst. of Philosophy, Literature and History 1939–41; worked for various Army newspapers 1941–45, writes for journal "Krokodil" and others in satiric genre 1945–.
Publs.: co-author of satiric verses "Mozhete zhalovatsia" (You May Make a Complaint) 1951, "Zanimaite vashi mesta" (Take Your Seats) 1954, "Tarapunka i Shtepsel" (Tarapunka and Shtepsel) 1957, "Kniga bez basen" (Book Without Fables) 1960, and others; co-author of film scripts: "Shtrafnoi udar" (Penalty Kick) 1963, "Operatsiia 'Y' i drugie prikliucheniia Shurika" (Operation 'Y' and Shurik's Other Adventures) 1965.

Kostrov, Vladimir Andreevich
Author and poet; b. 1935, Vlasikha, Kostroma Oblast.
Russian; grad. Dept. of Chemistry of Moscow Univ. 1958; worked as engineer; first works publ. 1957.
Publs.: "Pervyi sneg" (The First Snow) 1963, and others.

Kosygin, Iurii Aleksandrovich
Geologist; b. Ja. 9 (22), 1911, St. Petersburg.
Grad. Moscow Petroleum Inst. 1931; mem. USSR Acad. of Sciences 1970–; worked in "Turkmenneft'" Trust 1931–35; mem. staff USSR Acad. of Sciences 1935–; Dir. of Complex Scientific Research Inst. in Khabarovsk 1970–; Dir. of Inst. of Tectonics and Geophysics 1971–. Decors.: Order of Lenin, 3 other orders, medals.

Kotelnikov, Vladimir Aleksandrovich
Radio and electronics engineer; b. 1908.
Mem. CPSU 1948–; grad. Molotov Inst. of Power Engineering Moscow 1931; Prof.; instructor and Dean of Dept. of Radio Engineering at Inst. of Power Engineering Moscow 1931–37; Prof. and Head of Dept. of Radio Engineering Principles at Inst. of Power Engineering Moscow 1947–; mem. USSR Acad. of Sciences 1953; Dir. of Inst. of Radio Engineering and Electronics with USSR Acad. of Sciences 1954–; Vice-Pres. of USSR Acad. of Sciences Mr. 1970; Dep. to 1971–, Chair. of Supr. Sov. RSFSR 1973–80. Decors.: Lenin Prize, 2 Stalin Prizes, Order of Lenin, medals.

Kotenev, Vladimir Vladimirovich
Diplomat.
USSR Amb. to Guyana F. 1976–Mr. 1981.

Kotenko, Nikolai Stepanovich
State official; b. 1928.
Mem. CPSU 1957–; grad. Gubkin Inst. of the Petrochemical and Gas Ind. in Moscow; after discharge from the Sov. Army employed in system of oil marketing organizations 1957–63; Dir. of Oil Base in Smela, Head of the deliveries dept. of the Main Admin. for Material Supply to Enterprises of the Oil Ind. in Cherkassy; Dep. Dir. of the Freight Transport Admin. in Crimea and in Kiev 1963–78; 1st Dep. Dir. of Main Admin. for Material Supply to Enterprises of the Oil Ind. Ukrainian SSR 1978–80; Chair. of State Cttee. on Supply of Oil Products Ukrainian SSR Ja. 1980–.

Kotiatsy, Ivan Aleksandrovich
State official.
Dep. to Supr. Sov. Moldavian SSR 1963–; Sec. of Edintsy CPSU Cttee. of Kolkhoz-Sovkhoz Prod. Admin. 1963; mem. Cen. Cttee. 1963–71, mem. Auditing Comm. Moldavian CP 1971–; Chair. of State Cttee. for Nature Conservation Moldavian SSR O. 1968–.

Koton, Mikhail Mikhailovich
Organic Chemist; b. D. 16 (29), 1908, Novyi Petergof station, now Petrodvorets, Leningrad Oblast.
Corresponding member of the Academy of Sciences of the USSR (1960).
He graduated from Leningrad State University in 1935. In 1952 he became head of a laboratory at the Leningrad Institute of Macromolecular Compounds of the Academy of Sciences of the USSR; in 1960 he became the institute's director. His main works are devoted to organic and organometallic chemistry and to macromolecular compounds. He studied the reactivity of organic derivatives of mercury, lead, tin, bismuth, arsenic, antimony, and silicon and discovered a new reaction in the mercuration of aromatic compounds. He also established the regularities of the formation of many new polymers of the aromatic and heterocyclic series. He ist the author of the monograph *The Chemistry of Organic Dielectrics* (1957).

Kotov, Iurii Borisovich
State official.
Min. of Rural Construction Ukrainian SSR Ap. 1975–82; mem. Auditing Comm. Ukrainian CP 1976–81; Dep. USSR Min. of Rural Construction Ap. 1982–.

Kotov, Vasilii Timofeevich
Epizootologist; b. Mr. 20 (Ap. 1), 1899, Berezovo, Pavlovsk District, Voronezh Province.
Doctor of Veterinary Sciences (1952);

corresponding member of the Lenin All-Union Academy of Agricultural Sciences (1956). Member of the CPSU since 1945. He graduated from the Novocherkassk Veterinary Institute in 1928. In 1932 he became director of the subdepartment of epizootology of the Voronezh Agricultural Institute. His principal work has been devoted to the study of diseases of pigs (plague, erysipelas, paratyphoid, pox, Aujeszky's disease). He obtained a live antierysipelas vaccine and suggested that it be used in medical practice. Decors.: three orders and several medals.
Works: *Bolezni svinei, protekaiushchie s priznakom porazheniia nervnoi sistemy.* Voronezh 1950. "Puti napravlennoi izmenchivosti bakterii rozhi svinei i opyt polucheniia vaktsinnykh shtammov." *Trudy Voronezhskoi oblastnoi nauchnoissledovatel'skoi veterinarnoi opytnoi stantsii,* 1955, issue 4.

Kotov, Vladimir Petrovich
Author and poet; b. 1928, Moscow.
Russian; mem. CPSU 1957–; grad. Dept. of Literature of Moscow Pedagogical Inst.1950.
Publs.: "Letnie dni" (Summer Days) 1955, "Byla u devushki kosa" (A Girl had a Plait) 1959, "Zaitsy na pliazhe" (Hares on the Beach) 1959, "Klenovyi list" (A Maple Leaf) 1961, "Dve ulybki" (Two Smiles) 1962, and others.

Kovalchenko, Ivan Dmitrievich
Historian; b. N. 26, 1923, Novenkoe, Briansk Oblast.
Mem. CPSU 1944–; grad. Dept. of History of Moscow Univ. 1952; Lecturer 1955–, Prof. at Moscow Univ. 1967–; Chief Ed. of periodical "Istorija SSSR" 1969–; Chair. of Comm. for Applied Mathematics and Electronic Data-Processing in Historical Research at Dept. of History USSR Acad. of Sciences 1966–.

Kovaldzhi, Kirill Vladimirovich
Author and poet; b. 1930, Tashlyk, Bessarabia.
Russian; grad. Gorkii Inst. of Literature Moscow 1954; worked as journalist; first works publ. 1947.
Publs.: "Ispytanie" (Test) 1955, "Lirika" (Lyric Poems) 1956, "Na rassvete" (At Dawn) 1958, "Razgovor s liubimoi" (Conversation with the Loved One) 1959, "Chelovek moego pokoleniia" (A Person of my Generation) 1961, "Stikhi" (Poems) 1963, "Piat' tochek na karte" (Five Points on the Map) 1963, and others.

Kovalenko, Aleksandr Vlasovich
State official; b. 1909; s. of a peasant.
Ukrainian; mem. CPSU 1931–; grad. Communist Agricultural Univ. Kharkov 1937; shepherd 1921–28; sovkhoz worker, student, dep. dir., dir. of a machine and tractor station, chair. of exec. cttee. of a raion sov. of workers' deps., 1st sec. of a raion CPSU cttee. 1928–41; partisan during World War II; 1st sec. of various raion CPSU cttees. Chair. of Exec. Cttee. Belgorod Oblast Sov. of Workers' Deps. 1943–60; 1st Sec. of Belgorod Oblast Cttee. CPSU 1960–64 (for Agriculture 1963–64); mem. Cen. Cttee. CPSU 1961–; Dep. to 1962–, mem. Agricultural Comm. at Sov. of the Union USSR Supr. Sov. 1966–; 1st Sec. of Orenburg Oblast Cttee. CPSU Ap. 1964–80 (for Agriculture Ap.–D. 1964); Chair. USSR State Cttee. for Material Reserves 1980–. Decors.: 2 Orders of Lenin, Order of Red Banner of Labour, Hero of Socialist Labour, Hammer and Sickle Medal in Gold 1976.

Kovalenko, Anatolii Dmitrievich, Dr. of Tech. Sc.
Scientist; b. Ja. 16, 1905, Kiev; s. of a worker.
Mem. CPSU 1942–; grad. Polytech. Inst. Kiev 1929; Dr. of Tech. Sc. 1947; Prof. 1949; employed in Kosiakov Works Kiev 1929–36; mem. staff 1936–59, Dir. of Inst. of Construction Mechanics at Ukrainian Acad. of Sciences (now Inst. of Mechanics) 1959–65; mem. Ukrainian Acad. of Sciences 1961–; Head of Dept. of Thermoelasticity and Thermoplasticity of same inst. 1965–; concurrently Prof. at Kiev Univ. 1949–. Decor.: Merited Scientist of the Ukrainian SSR 1964.
Publs.: works dealing with theory of elasticity and thermoelasticity; "Plastiny i obolochki v rotorakh turbomashin" 1955, "Kruglye plastinki peremennoi tolshchiny" 1959, "Vvedenie v termouprugost'" 1965, "Raschet konicheskikh obolochek pri antisimmetrichnykh nagruzkakh" 1966.

Kovalenko, Iakov Romanovich
Veterinary microbiologist; b. Ap. 29 (My. 12), 1906, Ranino, now Vitebsk Oblast.
Mem. CPSU 1929–; grad. Inst. of Veterinary Medicine Moscow 1931; epicytologist; Dir. of Union Inst. of Experimental Veterinary Medicine 1955–; mem. USSR V. I. Lenin Acad. of Agricultural Sciences 1964–. Decors.: 5 orders, several medals.

Kovalenko, Viktor Karpovich
Painter; b. 1930, Tuapse, Krasnodar Krai. Grad. Dnepropetrovsk Art School 1956 and Kharkov Art Inst. 1962, pupil of O. Liubimskii.

Kovalenkov, Aleksandr Aleksandrovich
Author; b. 1911, Novgorod.
Russian; mem. CPSU 1961–; grad. Film Inst. Moscow 1932; first works publ. 1935.
Publs.: "Zelenyi bereg" (The Green Shore) 1935, "Ee Kareliei zovut ..." (It is Called Karelia ...) 1942, "Daleko na Severe" (Far up North) 1943, "Na vysokoi volne" (On a High Wave) 1943, "Lirika" (Poetry) 1954, "Iasnyi den'" (A Clear Day) 1946, "Stikhi" (Poems) 1960, lyrics for songs "Pesnia o Lenine" (Lenin Song), "Studencheskaia" (Student Song), "Alenushka", short stories "Vyshe liubvi" (Higher than Love) 1964, and others.

Kovalev, Aleksandr Aleksandrovich
Sculptor; b. 1915, Rogachov, Belorussia. Attended Kiev Art Inst. 1935–42, pupil of B. Kratko, L. Shervud, O. Matveev. Decors.: State Prize of the USSR 1950, People's Artist of the Ukrainian SSR 1963.

Kovalev, Anatolii Gavrilovich
Diplomat; b. 1923.
Mem. Diplomatic Service 1948–; mem. Apparat USSR Min. of Foreign Affairs 1957–65; Head of 1st European Dept. at USSR Min. of Foreign Affairs 1965–71; USSR Dep. Min. of Foreign Affairs and Head of Admin. for Planning Foreign Policy Measures at USSR Min. of Foreign Affairs 1971–.

Kovalev, Dmitrii Mikhailovich
Author and poet; b. 1915, Vetka, now Belorussian SSR.
Russian; mem. CPSU 1945–; grad. Higher Courses in Literature at Gorkii Inst. of Literature Moscow 1957; first works publ. 1938.
Publs. "Riabinovye nochi" (Ashberry Nights) 1958, "Tishina" (Silence) 1958, "Tikhaia molniia" (Quiet Lightning) 1961, "Rozh'" (Rye) 1963, "Solnechnaia noch'" (A Sunny Night) 1963, "Molchanie groz" (The Silence of Thunderstorms) 1964, and others.

Kovalev, Fedor Lukich
Process engineer; b. Ap. 22 (My. 5), 1909, Glushkovo, in present-day Kursk Oblast. One of the initiators of the mass introduction of advanced working methods into production; candidate of technical sciences (1954). Became a member of the CPSU in 1939.
In 1948, as the chief engineer at the Proletarskaia Pobeda Fine Cloth Mill in Moscow Oblast, he proposed a method for studying and disseminating the experience of production innovators. The method involved the selection of the most advanced work methods and adapting them for mass incorporation. His method has been widely used at enterprises of the USSR and the other socialist countries. In 1956, he became the director of the Central Scientific Research Institute of the Wool Industry. His work has been in the area of production organization and the creation of flow lines and automated lines in the wool industry. Decor.: State Prize of the USSR in 1951.

Kovalev, Feliks Nikolaevich
Diplomat; b. 1927.
Mem. Diplomatic Service 1953–; 1st Sec., Counsellor, USSR Embassy in Argentina 1963–67; Adviser 1967–71, Dep. Head of Treaty and Logal Dept. USSR Min. of Foreign Affairs 1971–80; USSR Amb. to Ecuador Mr. 1980–.

Kovalev, Mikhail Vasil'evich
State official.
Dep. Min. of Construction Belorussian SSR 1966–67; Dep. Min. of Ind. En-

gineering Belorussian SSR 1967–68; Chair. Minsk City Exec. Cttee. of People's Deps. 1968–78; mem. Cen. Cttee. Belorussian CP 1971–; Dep. Chair. of Council of Mins. Belorussian SSR O. 1978–.

Kovalev, Nikolai Ivanovich
Trade union official; b. 1922.
Mem. CPSU 1945–, grad. Dnepropetrovsk Inst. of Rail Transport 1951; served in Sov. Army 1940–45; 2nd Sec. of Krasnogvardeiskii Raion Cttee. Ukrainian CP 1955–58; Chair. of Dnepropetrovsk Oblast Sov. of Trade Unions 1958–70; Chair. of Cen. Cttee. of the Rail Transport Union Ja. 1970–; mem. Cen. Auditing Comm. CPSU 1981–.

Kovalev, Nikolai Nikolaevich
Scientist; b. F. 9 (22), 1908, Potava.
Mem. CPSU 1942–; grad. Inst. of Technology Leningrad 1930; research on mechanics and hydroturbine construction; corresponding mem. USSR Academy of Sciences 1953–; work as teacher and designer. Decors.: Hero of Socialist Labour, State Prizes of the USSR 1946, 1950, Lenin Prize 1959, Order of Lenin, 3 other orders, several medals.

Kovalev, Pavel Nikiforovich
Author; b. 1912, Sklimin, Mogilev Oblast.
Belorussian; mem. CPSU 1940–; critic in the thirties; with cen. staff of partisan movt. 1941–45.
Publs.: "Zhazhda boia" (Eager for Battle) 1945, "Poteriannyi dnevnik" (Lost Diary) 1954, "Andreika" 1948, "Lenka Grom" 1961, and others.

Kovalev, Sergei Adamovich
Biologist and geneticist; b. 1932.
Grad. Moscow Univ.; specialized in mathematical biology and genetics; Sen. Research Asst. at Moscow Univ. until 1969, then dismissed for his public pol. activities; worked in applied biology until his arrest on D. 27, 1974; mem. USSR branch of Amnesty International and of Initiative Group for the Defence of Human Rights in the USSR; signed statements and appeals in defence of Vladimir Bukovsky, Mustafa Dzhemilev, Petr Grigorenko, Viktor Khaustov, Viktor Nekipelov, Leonid Pliushch, Iurii Shikhanovich, Aleksandr Solzhenitsyn and Gabriel Superfin; he is currently in prison in Vilnius, Lithuania.

Kovalev, Sergei Ivanovich
Diplomat.
USSR Amb. to Bolivia Ap. 1981–.

Kovalskii, Aleksandr Alekseevich
Physical chemist; b. Ag. 28 (S. 10), 1906, Orenburg.
Mem. CPSU 1949–; grad. Polytech. Inst. Leningrad 1930; Dir. of Inst. of Chemical Kinetics at Siberian Branch of USSR Acad. of Sciences 1957–; corresponding mem. USSR Acad. of Sciences 1958–. Decors.: Order of Lenin, 3 other orders, several medals.

Kovanov, Vladimir Vasil'evich
Medical scientist; b. F. 28 (Mr. 12), 1909, St. Petersburg.
Mem. CPSU 1939–; grad. Dept. of Medicine of Moscow Univ. 1931; surgeon; Prof. Chair of Surgery and Topographical Anatomy at 1st Medical Inst. Moscow 1947–; Head of same inst. and Head of Laboratory for Organ and Tissue Transplantation at USSR Acad. of Medical Sciences 1956–66; mem. 1963–, Vice-Pres. of USSR Acad. of Medical Sciences 1966–. Decors.: Merited Scientist of the RSFSR, 2 Orders of Lenin, 6 other orders, several medals.

Kovda, Viktor Abramovich
Soil scientist; b. D. 16 (29), 1904, Vladikavkaz, now Ordzhonikidze.
Corresponding member of the Academy of Sciences of the USSR (1953). Became a member of the CPSU in 1927.
He graduated from the Agricultural Institute in Krasnodar in 1927. In 1931 he organized a laboratory of saline soils in the V. V. Dokuchaev Soil Institute and directed it until 1959. He was a professor at Moscow State University from 1939 to 1941, and in 1941–42 he directed the Institute of Botany and Soil Science of the Uzbek branch of the Academy of Sciences of the USSR. He was professor and chairman of the subdepartment of soil science of the biology and soil department of Moscow State University (from 1953). From 1958 to 1965 he directed the Department of Exact and Natural Sciences of UNESCO (Paris), and since 1970 he has directed the Institute of Agricultural Chemistry and Soil Science of the Academy of Sciences of the USSR. His main studies deal with the soils of the USSR, China, and Egypt. Initiator and director (1960–65) of the Soil Map of the World, an international project of the FAO and UNESCO, he became president of the International Society of Soil Science in 1968.
His works have practical implications for the development of new lands, including construction of irrigation systems and reclamation of solonetses and saline soils. Decors.: State Prize of the USSR (1951, 1953), the V. V. Dokuchaev Gold Medal (1967), and the Silver Medal of the French Association of Soil Scientists (1971), three orders.
Works: *Solonchaki i solontsy.* Moscow-Leningrad 1937. *Proiskhozhdenie i rezhim zasolennykh pochv.* vols. 1–2. Moscow-Leningrad 1946–47. *Geokhimiia pustyn' SSSR* (Speeches to the Seventh International Congress of Soil Scientists). Moscow 1954. "Zakonomernosti protsessov solenoakopleniia v pustyniakh Aralo-Kaspiiskoi nizmennosti." *Tr. Pochvennogo in-ta AN SSSR.* 1954, vol. 44. *Ocherki prirody i pochv Kitaia.* Moscow 1959.

Kovtunov, Aleksandr Vasil'evich
Military official; Lieutenant-Gen.
Dep. Supr. Sov. Azerbaidzhan SSR 1980–; mem. Cen. Cttee. and mem. Bureau Cen. Cttee. Azerbaidzhan CP Ja. 1981–.

Kozachenko, Vasilii Pavlovich
Author; b. 1913, Novo-Arkhangelskoe, Kirovograd Oblast.
Ukrainian; mem. CPSU 1952; grad. Dept. of Philology of Kiev Univ. 1938; Ed. of journal "Radianska literatura" 1938–41; served in Sov. Army 1941–42; partisan 1942–44; sen. posts on ed. staff of various journals 1944–48; literary work 1948–; Sec. 1966–71, 1st Sec. of Ukrainian Writers' Union 1971–; cand. mem. 1966–71, mem. Cen. Cttee. Ukrainian CP 1971–. Decors.: 2 Orders of Lenin, Order of Red Banner of Labour, medals "Partisan of the Great Patriotic War" 1st Cl., "For Valiant Labour", "For the Defence of Kiev", and others.
Publs.: collected stories "Zolota gramota" (The Golden Scroll) 1939, "Danylo Skorobogatko", "Pershy vzvod" (The First Platoon), "Tsina zhyttia" (The Prize of Life), "Sertse Materi" (The Heart of a Mother), "Zori nazustrich" (Towards a Star), essays "V boiakh gartovani" (Tempered in Battle), "Niu Iork zblyz'ka" (New York at Close Quarters), and others.

Kozelkova, Elena Georgievna
Actress; b. 1937.
Grad. Dept. of Geology of Moscow Univ. 1960 and drama studio at Mossoviet Theatre 1964; actress at Sovremennik Theatre Moscow 1964–71; mem. Sov. Army Theatre Moscow 1971–. Decor.: State Prize of the USSR 1971.

Kozeruk, Vasilii Petrovich
State official; b. 1927.
Grad. Odessa Credit and Economic Inst. 1961; held posts in budget orgs. and financial organs 1961–66; Chief of Kharkov Oblast Finance Dept. 1966–73; 1st Dep. Min. of Finance Ukrainian SSR 1973–79, Min. Ag. 1979–; mem. Auditing Comm. Ukrainian CP 1976–81; mem. Cen. Cttee. Ukrainian CP 1981–.

Kozhamkulov, Seraly (also Serke Kozhamkulov)
Kazakh actor; b. My. 5, 1896, in *aul* (village) no. 13, in present-day Karabalyk Raion, Kustanai Oblast.
People's Artist of the Kazakh SSR (1936). One of the first professional dramatic actors of the Kazakh SSR.
In 1920, he began performing as an amateur actor. In 1925 he was invited to work as an actor and director to the newly established Kazakh Drama Theater (now the M. O. Auezov Kazakh Drama Theater in Alma-Ata). His best roles include Espembet in Auezov's *Enlik and Kebek,* Karabai and Konkai in Musrepov's *Kozy-Korpesh and Baian-Slu* and *Akhan-*

Sere and Aktokty, Maibasar in an adaption of Auezov's *Abai* (State Prize of the USSR, 1952), and Sly in Shakespeare's *The Taming of the Shrew.* Decors.: K. Baiseitova State Prize of the Kazakh SSR in 1970, two Orders of Lenin, three other orders, and various medals.

Kozhedub, Ivan Nikitovich
Military official; b. Je. 8, 1920, Obrazheevka, Sumy Oblast.
Mem. CPSU 1943–; grad. Chemical-Technological Inst. 1940, Sov. Air Force Acad. 1949 and Gen. Staff Acad. 1956; combat pilot, Air Force Gen.; during World War II 330 bombing missions, 120 air attacks, 62 enemy aircraft shot down leading posts in Sov. Air Force 1971–. Decors.: 3 times Hero of the Soviet Union, Order of Lenin, 7 Orders of Red Banner, Alexander Nevskii Order, 2 Orders of Red Star, several medals.

Kozhevnikov, Aleksei Venediktovich
Russian writer; b. Mr. 6 (18), 1891, Khabazy, in present-day Kirov Oblast; s. of a peasant.
He fought in the Civil War of 1918–20. He worked as a village teacher and then studied at the V. Ia. Briusov Higher Literary and Artistic Institute (1922–25). His works began to appear in print in 1924. They include the novellas *Man-Song* (1927) and *Tansyk* (1931), the novels *The Golden Poor* (1927), *Hello, Road!* (1936; first published as *Highway* in 1934), *The Ocean's Brother* (1939), and *Living Water* (1950; State Prize of the USSR, 1951), and stories and novellas for children and young people. The transformation of Siberia and the creative labor of the Soviet people are the basic themes in his works. Works: *Za kraem sveta.* Moscow 1959. *Dobrye vskhody.* Moscow 1960. *Brat okeana: Roman, Povesti, Rasskazy.* Introductory article by Iu. Lukin. Moscow 1970.

Kozhevnikov, Fedor Ivanovich, Dr. of Legal Sc.
Jurist; b. Je. 2 (15), 1903, Smolensk.
Specialist in int. law; teaching and scientific work 1927–; Prof. at Moscow Univ. 1930–60; Prof. at Inst. of Int. Relations Moscow 1960–; Head of Dept. of Int. Law 1969–; mem. UN Comm. on Int. Law 1952–53; mem. UN Int. Court of Justice 1953–61; permanent mem. Int. Court of Justice in The Hague 1966–. Decors.: Merited Scientist of the RSFSR, 3 orders and several medals.

Kozhevnikov, Vadim Mikhailovich
Author; b. 1909, Narym.
Russian; mem. CPSU 1943–; grad. Moscow Univ. 1933; moved from Siberia to Moscow in 1925; Sec. of USSR Writers' Union 1949–; Chief Ed. of journal "Znamia" 1959–65; Sec. of RSFSR Writers' Union 1970–; first works publ. 1939. Decors.: Order of Lenin, Order of Red Banner of Labour, 2 Orders of Patriotic War 1st Cl., Order of Red Star, State Prize of the USSR 1971, Hero of Socialist Labour 1974, various medals.
Publs.: "Nochnoi razgovor" (Nocturnal Conversation) 1939, "Velikii prizyv" (The Great Summons) 1940, "Groznoe oruzhie" (The Formidable Weapon) 1941, "Rasskazy o voine" (War Stories) 1942, "Mart-aprel'" (March-April) 1942, "Liubimye tovarishchi" (Beloved Friends) 1943, "Dorogami voiny" (On War Roads) 1955, "V Velikom Narodnom Kitae" (In the Great People's Republic of China) 1952, "Zhivoi most" (Living Bridge) 1954, "Zare navstrechu" (To Meet the Dawn) 1956–57, "Shchit i mech" (The Shield and the Sword) 1965, play "Sud'ba Redzhinalda Devisa" (The Fate of Reginald Davis) 1947, film script "Veter s Vostoka" (Wind From the East) 1958, "Znakomets', Baluev" (Make Yourself Known, Baluev) 1960, "Petr Riabinkin" 1968, "V polden' na solnechnoi storone" (At Noon on the Sunny Side) 1973, and others.

Kozhukhov, Anatolii Dmitrievich
Graphic artist; b. 1927, Derevetske, Zaporozhe Oblast.
Mem. CPSU 1963–; attended Lvov Inst. for Applied and Decorative Art 1947–51, pupil of S. Gebus-Baranetska; grad. Kiev Art. Inst. 1953, pupil of V. Kasian and F. Samusev.

Kozlov, Genrikh Abramovich
Economist; b. F. 4, 1901, Vilnius.
Corresponding member of the Academy of Sciences of the USSR (1968). Became a member of the CPSU in 1920.
Since graduation from the economic division of the Institute of the Red Professoriat (1926), he has been engaged in scholarly work and teaching. In 1946 he became head (chief) of the subdepartment of political economy of the Higher Party School under the Central Committee of the CPSU. At the same time, he worked for the State Planning Commission of the USSR in 1928–29, was a member of the board of Gosbank (State Bank) from 1931 to 1934, and was engaged in party work from 1934 to 1937.
His main works are on the theory of commodity circulation, the law of value and currency circulation, and the general methodology of political economy and Lenin's contributions to that methodology. He has helped prepare a number of textbooks and lecture courses on political economy. Decors.: six orders and several medals.
Works: *Sovetskie den'gi.* Moscow-Leningrad 1939. *Khoziaistvennyi raschet, sebestoimost', tsena i pribyl' v sotsialisticheskom khoziaistve.* Moscow 1945. *Teoriia deneg i denezhnogo obrashcheniia* [ch. 1]. Moscow 1946. *Tovarnoe proizvodstvo i zakon stoimosti: Lektsii.* Moscow 1956. *Pervye stupeni v razvitii tovarnogo proizvodstva.* Moscow 1957. *Deistvie zakona stoimosti v usloviiakh sovremennogo kapitalizma.* Moscow 1964. *Predmet i metod politicheskoi ekonomii.* Moscow 1968.

Kozlov, Mikhail Mikhailovich
Military official; b. 1917.
Mem. CPSU 1949; grad. Frunze Military Acad. 1949 and Acad. of the General Staff 1955; Col.-Gen.; entered the Army 1941; Dep. Chief 1973–74, 1st Dep. Chief of Gen. Staff of Sov. Army and Navy Fleet 1974–; cand. mem. Cen. Cttee. CPSU 1976–81.

Kozlov, Nikolai Semenovich, Dr. of Chem. Sc.
Organic chemist; b. My. 4 (17), 1907, Tver.
Mem. CPSU 1929–; Dr. of Chem. Sc. 1935; Prof. 1933; mem. Belorussian Acad. of Sciences 1966–; mem. staff of several scientific insts. in Belorussia and other republics; Dir. of Inst. of Physical-Organic Chemistry at Belorussian Acad. of Sciences 1967–. Decors.: Merited Scientist and Technician of the RSFSR, 2 orders.

Kozlov, Nikolai Timofeevich
State official; b. 1925; s. of a peasant.
Russian; mem. CPSU 1946–; grad. Timiriazev Acad. of Agriculture Moscow 1952; served in Sov. Army 1943–47; student 1947–52; dep. dir. of a machine and tractor station, 2nd Sec. of Kimovskii Raion Cttee. in Moscow Oblast, senior posts with Moscow Oblast Cttee. CPSU 1952–58; Head of Moscow Oblast Statistical Board, Head of Agricultural Dept., then Sec. of Moscow Oblast Cttee. CPSU 1958–63; Dep. to and mem. Pres. RSFSR Supr. Sov. 1963–67; Chair. of Exec. Cttee. Moscow Oblast Sov. of Workers' Deps. 1963–80; cand. mem. 1966–71, mem. Cen. Cttee. CPSU 1971–; Dep. USSR Supr. Sov. 1966–. USSR Min of Fruits and Vegetables 1980–. Decor.: Order of Red Banner of Labour.

Kozlov, Sergei Vasil'evich
Trade union official; b. 1923.
Russian; mem. CPSU 1947–; grad. Timiriazev Agricultural Acad. Moscow 1948; served in Sov. Army 1941–42; mem. scientific staff of a selection and an experimental station in Frunze and Vladimir Oblast 1948–51; party and sov. posts 1951–56; Sec. 1956–60, 2nd Sec. of Vladimir Oblast Cttee. CPSU 1960–62; mem. Apparat Cen. Cttee. CPSU 1963–68; mem. Cen. Cttee., 2nd Sec. and mem. Bureau of Cen. Cttee. Azerbaidzhan CP 1968–77; Dep. to USSR Supr. Sov. Jl. 1969–79; cand. mem. Cen. Cttee. CPSU 1971–. Sec. of All-Union Cen. Council of Trade Unions 1977–. Decor.: Red Banner of Labour.

Kozlov, Viktor Alekseevich
State official.
1st Sec. of Slutsk Raion Cttee. Belorussian CP 1959; cand. mem. Cen. Cttee. Belorussian CP 1960–61; Dep. Chair. of

Minsk Oblast Exec. Cttee. of Workers' Deps. 1964–69; Min. of Agriculture Belorussian SSR Ag. 1972–79; Chair. of State Cttee. on Nature Conservation Belorussian SSR Mr. 1979–; mem. Cen. Cttee. Belorussian CP 1976–81, mem. Auditing Comm. Belorussian CP 1981–.

Kozlov, Vladimir Iakovievich
Mathematician; b. Je. 15 (28), 1914, Moscow.
Mem. CPSU 1941–; grad. Moscow Univ. 1937; corresponding mem. USSR Acad. of Sciences 1966–; Prof. at Moscow Univ. 1952–.

Kozlovskii, Evgenii Aleksandrovich
State official; b. 1929.
Russian; mem. CPSU 1955–; grad. Moscow Geological Survey Inst. 1953, Dr. of Tech. Sc. 1973; served in Sov. Army 1945–48; senior foreman, chief engineer, head of a geological expedition to Amur Oblast and Khabarovsk Krai 1953–65; admin. chief in USSR Min. of Geology 1965–73; dir. of a scientific research inst. 1973–74; Dep. USSR Min. of Geology 1974–75, Min. D. 1975–; cand. mem. Cen. Cttee. CPSU 1976–; Dep. to USSR Supr. Sov. 1979–. Decor.: Order of Lenin.

Kozlovskii, Iakov Abramovich
Author, poet and translator; b. 1921, Istra, Moscow Oblast.
Russian; mem. CPSU 1945–; grad. Gorkii Inst. of Literature 1949; first works publ. 1941.
Publs.: "Moskva 1941 goda" (Moscow 1941), 1947, "Soldatskie stikhi" (Soldiers' Poems) 1951, "Ofitser sviazi" (Signal Corps Officer) 1961, and others; translates from Avar and other languages.

Kozlovskii, Ivan Semenovich
Russian lyric tenor; b. Mr. 11 (24), 1900, Mar'ianovka, near Belaia Tserkov', in present-day Kiev Oblast; s. of a peasant. People's Artist of the USSR (1940).
He studied under E. A. Murav'eva at the Kiev Institute of Music and Drama from 1917 to 1919. He served in the Red Army from 1919 to 1924, participating in the productions of the Poltava Traveling Theater of Music and Drama. It was in this theater that he sang his first leading part – the title role in Gounod's *Faust*. In 1924 he was a soloist at the Kharkov Opera Theater; in 1925, at the Sverdlovsk Opera Theater; and from 1926 to 1954, at the Bolshoi Theater of the USSR.
He is an outstanding representative of the Soviet school of vocal music. His clear silvery smooth voice, with a beautiful mellifluous timbre and a free upper register, his free natural breathing, his fine ear for music, his great artistic taste, and his acting abilities enabled him to create vocal and dramatic images that have become a part of the history of Soviet opera. Paying particular attention to vocal expression, he brings out the musicality of words and emphasizes the importance of musical declamation and phrasing.
His best roles have included the Simpleton in Mussorgsky's *Boris Godunov* (State Prize of the USSR, 1949), Lenskii in Tchaikovsky's *Eugene Onegin*, and the title role in Wagner's *Lohengrin*. He was also very effective in the roles of Berendei in Rimsky-Korsakov's *The Snow Maiden*, Vladimir Dubrovskii in Napravnik's *Dubrovskii*, the Prince in Prokofiev's *The Love for Three Oranges*, and Alfredo in Verdi's *La Traviata*.
From 1938 to 1941, he was the organizer and artistic director of the concert Opera Ensemble. He staged a number of operas, in which he sang the principal parts (for example, Massenet's *Werther* and Leoncavallo's *I Pagliacci*). Seeking to produce operatic concerts with tense stage action, he trained the members of his ensemble to synthesize singing and dramatic acting. Concertizing occupies an important place in his career (he has given concerts since 1919).
His performances of art songs by M. I. Glinka, A. S. Dargomyzhskii, P. I. Tchaikovsky, and S. V. Rachmaninoff are marked by an elegiac quality, a wide range of emotions, and varied moods. He is also a fine interpreter of works by Beethoven, Schubert, Schumann, and Liszt. His broad repertoire includes old art songs and Russian and Ukrainian folk songs. Decors.: State Prize of the USSR in 1941, three Orders of Lenin, the Order of the Badge of Honor, and various medals.

Kozub, Konstantin Ivanovich
State official.
Chair. of State Cttee. on Films with Council of Mins. Moldavian SSR 1963–66; mem. 1963–66 and 1981–, cand. mem. Cen. Cttee. Moldavian CP 1966–71 and 1976–81; Head of Main Statistical Admin. Moldavian SSR S. 1976–.

Kozyrev, Boris Mikhailovich
Physicist; b. Ap. 21 (My. 4), 1905, Ashkhabad.
Grad. Kazan Univ. 1930; mem. staff Kazan Univ. 1930;47; mem. staff Physical-Tech. Inst. of USSR Acad. of Sciences in Kazan 1947–; corresponding mem. USSR Acad. of Sciences 1968–. Decors.: Order of Red Banner of Labour, several medals.
Publ.: "Ėlektronnyj paramagnitnyi rezonans v židkih rastvorah solej" in "Doklady AN SSSR" 1955, vol. 105, p. 53, coauthor of "Elektronnyj paramagnitnyj rezonans" Moscow 1961.

Kozyrev, Semen Pavlovich
Diplomat; b. 1907.
Mem. Diplomatic Service 1939–; Asst. to Sec.-Gen. of USSR People's Commissariat for Foreign Affairs 1939; Asst. to USSR People's Commr. for Foreign Affairs 1939–43; Sec.-Gen. and mem. Board of USSR People's Commissariat for Foreign Affairs 1943; Emb. Counsellor of USSR Emb. at Allied Govts. in London; Emb. Counsellor at French Liberation Cttee. in Algeria and at USSR Emb. in France 1943–45; Head of 1st European Dept. at USSR Min. of Foreign Affairs 1945–49; USSR Amb. to Egypt 1950–53; Head of Cadre Admin. and mem. Board USSR Min. of Foreign Affairs 1953–55; mem. Board USSR Min. of Foreign Affairs 1955–57; USSR Amb. to Italy 1957–66; USSR Dep. Min. of Foreign Affairs 1966–.

Kraft, Iuri Andreasovich
State official.
Min. of Light Ind. Esthonian SSR D. 1977–; mem. Cen. Cttee. Esthonian CP 1981–.

Krainiev, Vladimir
Pianist.
Winner of Tchaikovsky Competition Moscow 1970.

Kramov, Isaak Naumovich
Author and literary critic; b. 1919, Zhitomir.
Russian; mem. CPSU 1953–; grad. Dept. of Philology of Moscow Univ. 1948; first works publ. 1948.
Publs.: "Iakov Osipov" 1955, "Larisa Reisner", "John Read", "Mate Zalka" and others 1957–60, "Effendi Kapiev" 1964, "Aleksandr Malyshkin" 1965, and others.

Krapiva, Kondrat (real name: Atrakhovich, Kondrat Kondratevich), Dr. of Phil. Sc.
Author and playwright; b. 1896, Nizok, Belorussia.
Belorussian; mem. CPSU 1941–; grad. Dept. of Philology of Belorussian Univ. 1930; primary school teacher 1913–15; Army service on Romanian front 1915–18; primary school teacher 1918–20; served in Sov. Army 1920–23; Instructor of Cen. Bureau of Ethnography at Inst. of Belorussian Culture Minsk 1925–32; Desk Chief of journal "Polymia revoliutsii" (Flame of the Revolution) 1932–36; served in Sov. Army 1939–40; Assoc. of Inst. of Literature and Language at Belorussian Acad. of Sciences 1940; on ed. staff of newspaper "Za Sovetskuiu Belorus'" (For Soviet Belorussia) 1941–42; worked for Cen. Cttee. Belorussian CP and as Exec. Ed. of newspaper "Razdavim fashistskuiu gazdinu" (We Shall Crush the Fascist Snake) 1942–45; Ed. of satrical journal "Iozhyk" (The Hedgehog) 1945–47; Head of Linguistics Section at Inst. of Literature and Language at Belorussian Acad. of Sciences 1947–52; Dir. of Inst. of Linguistics at Belorussian Acad. of Sciences 1952–56; Vice-Pres. of Belorussian Acad. of Sciences 1956–; first works publ. 1922. Decors.: 4 Orders of Lenin, Order of Red Banner, Order of Red Banner of Labour, medals.
Publs.: "Krapiva" (Nettles) 1925, plays

"Kanets druzhby" (End of a Friendship) 1933, "Partyzany" (Partisans) 1937, "Liudi i diably" (People and Devils) 1958, "Zbor tvorau" (Collected Works) 1956, scholarly publs. "Dialektychny atlas belorusskai movy" (Dialect Atlas of the Belorussian Language) 1963, "O nekatorykh osobennostiakh belorusskogo literaturnogo iazyka" (On Some Features of the Belorussian Literary Language) 1966, and others.

Krasilnikov, Gennadii Dmitrievich
Author; b. 1928, Alnashi, Udmurt ASSR.
Udmurt; grad. Gorkii Inst. of Literature Moscow 1957; first works publ. in the late forties.
Publs.: "Staryi dom" (The Old House) 1956, "Pustotsvet" (The Dazzler) 1962, "Ostaius s toboi" (I Shall Stay With You) 1960, and others.

Krasinskii, Zinovii Mikhailovich
Poet; b. 1930, place of birth unknown; s. of a family transferred compulsorily from West Ukraine to Siberia.
After attempt to escape arrested and sentenced; returned to Ukraine; univ. educ.; as leading mem. of illegal org. "Ukrains'kyi natsionalnyi front" sentenced to 5 years' imprisonment, 7 years' confinement in correctional labour camp and 5 years' banishment; wrote poetry anthology "Nevolnyčyi plač" (300 poems) in prison; an account of that was transferred to a psychiatric clinic in Sykhovka; some of his poems have been published in the West.

Krasnikov, Boris Anatolevich
Diplomat.
2nd Sec. and mem. Bureau of Cen. Cttee. Uzbek Komsomol 1962–68; cand. mem. Cen. Cttee. Uzbek CP 1966–71; 2nd Sec. of Andizhan Oblast Cttee. Uzbek CP 1971–79; USSR Amb. to Equatorial Guinea S. 1980–.

Krasnov-Asli, Vasilii Ivanovich
Chuvash writer; b. Ja. 6 (18), 1900, Churanovo, in present-day Buin Raion, Tartar ASSR.
Member of the CPSU since 1917.
The son of poor peasants, he took part in the Civil War of 1918–20. He graduated from the Ia. M. Sverdlov Communist University in 1929 and from the Institute of the Red Professoriat in 1936. His novellas *The Fight Against the Whites* (1929), *Giants* (1929–34), and *Roots* (1932–36) make up the autobiographical trilogy *In the Fire* (books 1–3, 1957–61). He wrote the novels *Uphill* (book 1, 1929; rev. ed. 1959; book 2: *The Sun Is the Same for All,* 1964) and *My School* (1963). The main themes in his work are the Revolution, the Civil War, and the collectivization of agriculture. Decors.: Order of Lenin and a number of medals.
Works: *In Russian translation: V ogne,* books 1–3. Afterword by A. Kuz'min. Cheboksary 1962–63.

Krasnova, Zoia Ignatevna
Laborer; b. 1938.
Mem. CPSU 1973–; grad. secondary school for working youth 1976; foreman in a sewing workshop 1957–58; machine operator in carbamic plant of Salavat Petrochemical Combine (Bashkir ASSR) 1961–79; mem. Cen. Auditing Comm. CPSU 1976–81; machine operator in mineral fertilizer plant of "Salavatnefteorgsintez" Asscn. 1979–; cand. mem. Cen. Cttee. CPSU 1981–. Decor.: Heroine of Socialist Labor 1975.

Krasnovskii, Aleksandr Abramovich
Biochemist and biophysicist; b. Ag. 13 (26), 1913, Odessa.
Corresponding member of the Academy of Sciences of the USSR (1962). Graduated from the Moscow Chemical Engineering Institute in 1937. Since 1944 he has worked at the Institute of Biochemistry of the Academy of Sciences of the USSR; since 1959 he has headed the Photobiochemistry Laboratory. He has been a professor in the subdepartment of biophysics at Moscow State University since 1954.
His major works have been on the initial stages of photosynthesis, the photochemistry of chlorophyll, and the state and formation of pigments in organisms. He discovered the reversible photochemical reduction reaction of chlorophyll and its analogues and products (the Krasnovskii reaction). He clarified the mechanism of the photosensitizing action of chlorophyll in the oxidation-reduction conversions of pyridine nucleotides, cytochromes, quinones, and other components of the electron transfer chain. He discovered the monomeric and aggregate forms of chlorophyll, bacteriochlorophyll, and their analogues, which have differing spectral and photochemical properties, and separated the protochlorophyll protein complex from leaves. He proposed a series of models for the chemical evolution of photosynthesis. Decor.: A. N. Bakh Prize 1950 (with A. N. Terenin).
Works: "Fotokhimiia khlorofilla i molekuliarnaia organizatsiia pigmentnoi sistemy organizmov." In the collection *Funktsional'naia biokhimiia kletochnykh struktur.* Moscow 1970.

Krasnyi, Lev Isakovich
Geologist; b. Ap. 4 (17), 1911, St. Petersburg.
Mem. CPSU 1941–; grad. Mining Inst. Leningrad 1936; worked in Far Eastern Geology Admin. 1931–41; served in Sov. Army 1941–45; mem. staff 1946–, Head of Dept. of Geology and Natural Resources of Eastern Regions at Union Geological Research Inst. 1956–; corresponding mem. USSR Acad. of Sciences 1970–; Head of Dept. of Regional Geology at Far Eastern Inst. of Mineral Resources 1971–. Decors.: Lenin Prize 1964, 2 orders, several medals.

Krasovskii, Aleksandr Arkadevich
Scientist; b. Ap. 10, 1921, Kirovgrad, Sverdlovsk Oblast.
Mem. CPSU 1947–; grad. Zhukovskii Air Force Acad. 1945; specialist in automatic control systems; corresponding mem. USSR Acad. of Sciences 1968–; instructor 1945–55, Prof. at Zhukovskii Air Force Acad. 1955–. Decors.: 2 orders, several medals.
Publs.: "Dinamika nepreryvnyh samonastraivajuščihsja sistem" Moscow, 1963, "Statističeskaja teorija perehodnyh processov v sistemah upravlenija" Moscow 1968, "Analitičeskoe konstruirovanie konturov upravlenija letatel'nymi apparatami" Moscow 1969.

Krasovskii, Nikolai Nikolaevich
Scientist; b. S. 7, 1924, Sverdlovsk.
Grad. Uralsk Polytech. Inst. 1949; specialist in mechanics and control processes; mem. USSR Acad. of Sciences 1968–; Prof. at Uralsk Polytech. Inst. 1957–; Prof. at Uralsk Univ. Sverdlovsk 1959–; Dir. of Inst. of Mathematics and Mechanics at USSR Acad. of Sciences 1970–. Decors.: Order of Lenin, Order of Red Banner of Labour.

Krasovskii, Stepan Akimovich
Military official; b. Ag. 8 (20), 1897, Gluchi, Mogilev Oblast.
Mem. CPSU 1918–; completed courses for intensification of specialist knowledge of Air Force Command Staff and Zhukovskii Air Force Acad. 1936; Air Force Marshall 1959–; Prof. 1966–; sen. posts at USSR Min. of Defence 1968–; in leading command positions during Sov.-Finnish War 1939–40 and German-Sov. War 1941–45; Head of Iu. A. Gagarin Air Force Acad. 1956–68. Decors.: Hero of the Sov. Union 1945, 6 Orders of Lenin, 3 Orders of Red Banner, Suvorov Order 1st and 2nd Cl., Kutuzov Order 1st Cl., Bogdan Khmelnitskii Order 1st Cl., Order of Red Star, medals, 4 foreign orders.

Kratis, Kauko Ottovich
Geologist; b. Je. 16, 1914, Sudbury, Ontario, Canada.
Specialist in Precambrian geology and petrography. Corresponding member of the Academy of Sciences of the USSR (1968).
He graduated from the geology department of Leningrad State University in 1939. Between 1939 and 1946 he worked in industrial geological organizations on the Kola Peninsula and in Eastern Siberia. In 1946 he began to work in the system of the Academy of Sciences of the USSR, including the Geological Institute of the Kola Branch of the Academy of Sciences of the USSR. In 1967 he became director of the Institute of Geology and Precambrian Geochronology in Leningrad. His principal work has been on the geology, petrology, and Precambrian geochronology of the Baltic Shield (in-

cluding work on the geology of the Karelian Karelides, 1963).
Works: *Geologiia karelid Karelii.* Moscow-Leningrad 1963.

Kraulin', Karlis Ianovich (also Karlis Kraulinš)
Literary critic; b. Mr. 6 (19), 1904, Koceni Volost (small rural district), present-day Valmiera Raion.
Honored Cultural Worker of the Latvian SSR (1964). Became a member of the CPSU in 1942.
He graduated from the University of Latvia in 1928 and became a teacher. In 1936 he helped run an underground Communist printing house. He participated in the Great Patriotic War of 1941–45. From 1944 to 1948 he was the head secretary on the board of the Union of Writers of Latvia. After 1945 he was an instructor at the Latvian State University, where he has served as professor since 1966. From 1948 to 1951, he directed the Institute of Literature and Language of the Academy of Sciences of the Latvian SSR. Between 1964 and 1967 he edited the journal *Karogs.* He has published many articles on the problems of socialist realism and on the national qualities of Latvian literature. He wrote *The Life and Work of J. Rainis* (1953), *The Creative Work of Vilis Lacis* (1954), *Andrejs Upits: Life and Work* (1963), and *The Contemporary Soviet Novel* (1969). Decors.: Order of the Red Star and medals.

Kravchenko, Ivan Sergeevich
Historian; b. Ag. 29 (S. 11), 1902, Dmitrovka, Voronezh Gouvt.
Belorussian; mem. CPSU 1920–; grad. North Caucasian Communist Inst. 1930, Inst. of Red Professors 1933 and Acad. of Social Sciences of Cen. Cttee. CPSU 1948; scientific activity and party functions; mem. staff 1948–, Dir. of Inst. of History at Belorussian Acad. of Sciences 1958–64; mem. Belorussian Acad. of Sciences 1969–. Decors.: 3 orders, medals.

Kremer, Gidon
Violinist; b. 1947, Riga.
Initial studies in Riga, then attended master class of David Oistrakh at Moscow Conservatory 1965; after successful participation in numerous music contests (e.g. Genova, Brussels and Montreal) started concert career in the Sov. Union and in Western countries; participated in numerous music festivals incl. Vienna, Berlin, Ansbach and Salzburg; numerous recordings (e.g. with Herbert von Karajan and the Berlin Philharmonic Orchestra). Decor.: 1st Prize at Moscow Tchaikovsky competition.

Kreps, Evgenii Mikhailovich
Physiologist and biochemist; b. Ap. 19 (My. 1), 1899, St. Petersburg.
Grad. Mil. Medical Acad. of the Red Army (now S. M. Kirov Mil.-Medical Acad.) 1923; mem. USSR Acad. of Sciences 1966–; head of physiological laboratory of a bio-station in Murmansk 1923–33; Prof. at Leningrad Univ. 1934–37; mem. staff Pavlov Inst. of Physiology of USSR Acad. of Sciences 1935–; Dir. of Sechenov Inst. of Evolutionary Physiology and Biochemistry 1960–; Acad. Sec. of Dept. of Physiology of USSR Acad. of Sciences 1967–. Decors.: Hero of Socialist Labour, 2 Orders of Lenin, 2 other orders and various medals.

Krestiannikov, V. P.
Trade union official.
Sec. of Kokchetav Oblast Cttee. Kazakh CP 1974–82; Sec. of Kazakh Republican Trade Union Council F. 1982–.

Kretovich, Vatslav Leonovich
Biochemist; b. Ja. 14 (27), 1907, Ialta.
Grad. Moscow Univ. 1931; corresponding mem. USSR Acad. of Sciences 1962–; mem. staff Union Research Inst. for Cereals 1931–41; mem. staff 1938–, Head of Laboratory 1944–, Dep. Dir. of A. N. Bakh Inst. of Biochemistry at USSR Acad. of Sciences 1966–; Prof. at Inst. of Foodstuff Ind. Technology Moscow 1946–; Prof. Chair of Biochemistry and Cereal Cultivation at same inst. 1953–. Decors.: Dr. h. c. Humboldt Univ. E. Berlin, A. N. Bakh Prize 1961, State Prize of the USSR 1972, 3 orders, several medals.

Kriuchkov, Nikolai Afanas'evich
Russian actor; b. D. 24, 1910 (Ja. 6, 1911), Moscow.
People's Artist of the USSR (1965). Member of the CPSU since 1953.
He acted in amateur theater groups and joined the troupe of the Moscow Theater of Young Workers in 1928. His teachers included N. P. Khmelev, I. Ia. Sudakov, and I. A. Savchenko. His first motion-picture role was the shoemaker Sen'ka (*The Outskirts,* 1933). In the 1930's and 1940's he created a number of contemporary types who were engaging, vital, and utterly dedicated to their homeland: Andrei Sazonov (*Komsomol'sk,* 1938); Tarasov, the commander of a frontier post (*On the Frontier,* 1938); Klim Iarko (*Tractor Drivers,* 1939); Sergei Lukonin (*One of the Local Boys,* 1942, based on Simonov's play); and Lieutenant Sergei Gorlov (*The Front,* 1943, based on Korneichuk's play).
He portrayed underground revolutionaries, Red Guards, and Soviet partisans in such films as *A Man With the Gun* (1938), *Iakov Sverdlov* (1940), and *The Forty-First* (1956). He displayed outstanding comedic gifts in the roles of Korol'kov (*The Rumiantsev Case,* 1956) and Afanasii Matveevich (*Uncle's Dream,* 1967, based on Dostoevsky's story).
In the 1950's and 1960's he played several dramatic roles that revealed the complex inner life of the characters – for example, the chief of the Criminal Investigation Department in *Cruelty* (1959, based on the novella by Nilin), Teterin in *The Trial* (1962, based on the work by Tendriakov), and the title role in *The Stray Sheep* (1966). He portrayed Colonel Zakharov in *Far in the West* (1969). Decors.: State Prize of the USSR (1941), Order of Lenin, three other orders, and medals.

Kriulin, Gleb Aleksandrovich
State official; b. 1923.
Russian; mem. CPSU 1945–; grad. Party Higher School of Cen. Cttee. CPSU 1956; served in Sov. Army, partisan 1941–44; 1st sec. of a raion cttee., Sec. of Vitebsk City Komsomol Cttee., Sec. of Mogilev Oblast Komsomol Cttee. 1944–54; 2nd Sec., 1st Sec. of Cen. Cttee. Komsomol Belorussian SSR 1954–60; Dep. to and mem. Presidium Supr. Sov. Belorussian SSR 1959–67; Head of Dept. of Admin. Organs at Cen. Cttee. Belorussian CP 1960–62; Head of Dept. of Party Organs at Cen. Cttee. Belorussian CP 1962–63; Chair. of Exec. Cttee. Mogilev Oblast Sov. of Workers' Deps. 1963–64; 1st Sec. of Mogilev Oblast Cttee. Belorussian CP 1964–75; cand. mem. Cen. Cttee. CPSU 1966–; Dep. to USSR Supr. Sov. 1966–79; USSR Amb. to N. Korea Ag. 1974–D. 82; Min. of Social Welfare Belorussian SSR Ja. 1983–.

Krivda, F.
Military official.
Lieut.-Gen.; Commander of Sov. Army Southern Forces Group (Hungary) 1976–.

Kron, Aleksandr Aleksandrovich
Author and dramatist; b. Je. 30 (Jl. 13), 1909, Moscow; s. of a musician.
Mem. CPSU 1939–; grad. Dept. of History and Philosophy of Moscow Univ. 1930.
Publs.: plays "Vintovka Nr. 492116" (Rifle No. 492116) 1929, "Trus" (Coward) 1934, "Ofitser flota" (Navy Officer) 1944, "Kandidat partii" (Party Candidate) 1950, novel "Dom i korabl'" (The House and the Ship), and others.

Kropotkin, Petr Nikolaevich
Geologist; b. N. 11 (24), 1910, Moscow.
Grad. S. Ordzhonikidze Inst. of Geological Prospecting Moscow 1932; Head of Tectonic-Geophysical Laboratory at Inst. of Geology USSR Acad. of Sciences 1959–; corresponding mem. USSR Acad. of Sciences 1966–; main research on regional problems of geology (Kazakhstan and Far East) and on geotectonics. Decors.: Order of Red Star, medals.

Krotov, Viktor Vasil'evich
State official; b. 1912.
Mem. CPSU 1944–; grad. Ural Polytechnical Inst. 1957; dept. chief, mechanic, head of production, chief engineer, Dir. of various defense ind. factories 1936–57; Chief of Admin., Dep. Chair. of Sverdlovsk Econ. Council 1957–58; Dir. of heavy machine construction works in the Urals 1958–63; cand. mem. 1961–66 and

1976–81, mem. Cen. Cttee. CPSU 1981–; Chair. of Econ. Council of Central Urals 1963–65; Chair. of State Cttee. for Heavy, Power and Transport Machine Construction at USSR GOSPLAN, USSR Min. 1965; 1st Dep. Min. of Heavy, Power and Transport Machine Construction of the USSR 1966–75; USSR Min. of Power Machine Construction My. 1975–. Decor.: USSR State Prize 1969.

Krucheniuk, Piotr Aksentevich
Author and poet; b. 1917, Plop, Moldavia.
Moldavian; mem. CPSU 1942–; grad. Gorkii Inst. of Literature Moscow 1957; first works publ. 1939. Decors.: Order of Red Banner of Labour, 2 Orders of Patriotic War 2nd Cl., Order of Red Star, medals.
Publs.: "V luchakh zhizni" (In the Rays of Life) 1939, "Frontovye stikhi" (Front-Line Poems) 1942, "Svezhie borozdy" (Fresh Furrows) 1948, "Postup' vesny" (The Course of Spring) 1951, "U sten Kremlia" (On the Kremlin Wall) 1952, "U nas est' druzia" (We Have Friends) 1960, "Razgovor s liudmi" (A Talk With the People) 1965, and others.

Kruchina, Nikolai Efimovich
Party official; b. 1928.
Russian; mem. CPSU 1949–; grad. Azovo-Chernomorskii Agricultural Inst. 1953; 1st Sec. of Novocherkassk City Cttee. USSR Komsomol 1952–54; 2nd Sec. 1954–55, 1st Sec. of Kamenskoe Oblast Cttee. USSR Komsomol 1955–57; 1st Sec. of Smolensk Oblast Cttee. USSR Komsomol 1957–59; mem. Apparat Cen. Cttee. USSR Komsomol 1959–62 and Cen. Cttee. CPSU 1962–63; Sec. of Tselinograd Krai Cttee. Kazakh CP 1963–65; 1st Sec. of Tselinograd Oblast Cttee. Kazakh CP N. 1965–78; mem. Auditing Comm. CPSU 1966–71; Dep. to USSR Supr. Sov. 1966–; mem. Cen. Cttee. Kazakh CP 1971–78; cand. mem. 1971–76, mem. Cen. Cttee. CPSU 1976–; 1st Dep. Head of Agricultural Dept. at Cen. Cttee. CPSU 1978–.

Kruger, Harald Ottovich
State official.
Mem. Auditing Comm. 1971–76, mem. Cen. Cttee. Esthonian CP 1981–; Min. of Motor Transport and Highways Esthonian SSR O. 1979–.

Krugliakov, Leonid Aleksandrovich
State official.
Head of Raion Admin. for Agriculture in Bairam-Ali 1970–79; Chair. of State Cttee. for the Supply of Production Equipment for Agriculture Turkmen SSR D. 1979–; cand. mem. Cen. Cttee. Turkmen CP 1981–.

Kruglikova, Elena Dmitrievna
Russian soprano; b. Je. 3 (16), 1907, Podol'sk, in present-day Moscow Oblast.
People's Artist of the RSFSR (1947).
She graduated from the Moscow Conservatory in 1933 in the singing class of K. N. Dorliak. She was a soloist with the Bolshoi Theater in Moscow between 1932 and 1956. Kruglikova's lyric and dramatic soprano voice is sonorous and warm, well trained for the stage. She has given sincere renditions of such classical Russian roles as Tatiana and Oksana in Tchaikovsky's *Eugene Onegin* and *Cherevichki* and Fevronia and Vera Sheloga in Rimsky-Korsakov's *Tale of the Invisible City of Kitezh* and *Vera Sheloga*. She sang the first performance of the role of Lushka in Dzerzhinskii's opera *Virgin Soil Upturned* (1937). Kruglikova also performed successfully in operas by Western European composers – for example, as Mathilde in *William Tell* by Rossini (State Prize of the USSR, 1942).
She toured Bulgaria, Rumania, Poland, and Yugoslavia. Kruglikova began teaching at the Moscow Conservatory in 1963. Decors.: two orders and medals.

Kruglov, Georgii Georgievich
Ceramic artist; b. O. 15 (28), 1905, Riga.
Honored Art Worker of the Latvian SSR (1945).
He studied at the Latvian Academy of Arts in Riga from 1921 to 1930 and under A. Hennig at the State School of Ceramics in Bunzlau in Germany (now Bolesławiec, in the People's Republic of Poland) from 1932 to 1933. In Riga, he taught at the School of Artistic Crafts (1934–40) and at the Academy of Arts of the Latvian SSR (from 1944; prorector, 1944–45 and 1947–57; professor, since 1961).
Particularly in his decorative vessels, he strove for sharp plastic resolutions and new forms and means of decoration (texture, overlay, and glazes) that conformed with the properties of the materials and production techniques.

Kruglova, Zinaida Mikhailovna
State official; b. 1923.
Mem. CPSU 1944–; grad. Leningrad Inst. of Aviation Equipment Construction 1951; served in Sov. Army 1942–45; Head of Laboratory, Research Asst. at Leningrad Inst. of Aviation Equipment Construction 1951–54; CPSU posts 1945–50; 1st Sec. of Leningrad-Frunze Raion Cttee. CPSU 1960–63; Sec. of Leningrad City Cttee. 1963–68, Sec. of Leningrad Oblast Cttee. CPSU 1968–74; mem. Cen. Auditing Comm. CPSU 1966–76; USSR Dep. Min. of Culture 1974–S. 1975; mem. Cen. Cttee. CPSU 1976–; Chair. Presidium, Union of Soviet Societies for Friendship and Cultural Relations with Foreign Countries 1975–. Decor.: Order of Friendship Between Nations.

Krukauskas, Cheslovas Stiaponovich
State official.
Chair. State Cttee. for Gasification Lithuanian SSR Ap. 1982–.

Krumin, Viktor Mikhailovich
State official; b. 1923.
Cand. mem. 1956–60, mem. Cen. Cttee. Latvian CP 1960–; Sec. of Riga City Cttee. Latvian CP 1956; Dep. Chair. of Council of Mins. F. 1960–, Min. of Foreign Affairs Latvian SSR Ap. 1967–; Dep. to Supr. Sov. Latvian SSR 1962–.

Krupenina, Lidia Ivanovna
Ballet dancer; b. N. 1, 1928, Moscow.
Grad. Moscow Choreographic College; mem. ballet of Novosibirsk Opera and Ballet Theatre 1947–. Decors.: People's Artist of the USSR 1971, Order of Lenin.

Krushanov, Andrei Ivanovich
Historian; b. Je. 1, 1921, Chet', in present-day Bogotol Raion, in Krasnoiarsk Krai.
Corresponding member of the Academy of Sciences of the USSR (1970). Member of the CPSU (1944).
He graduated from the department of history of the Vladivostok Pedagogical Institute (1949). He is a student of the 19th- and 20th-century history of the Far East. In 1958 he became head of the department of history, archaeology, and ethnography of the peoples of the Far East at the Far Eastern Branch of the Siberian Division of the Academy of Sciences of the USSR, and is deputy chairman of the Far Eastern Scientific Center of the Academy of Sciences of the USSR. Decors.: two orders and a number of medals.
Works: *Bor'ba za vlast' Sovetov na Dal'nem Vostoke i v Zabaikal'e: Ocherki po istorii partiinogo i gosudarstvennogo stroitel'stva (mart 1917–mart 1918).* Vladivostok 1961. *Bor'ba za vlast' Sovetov na Dal'nem Vostoke i v Zabaikal'e (aprel' 1918–mart 1920).* Vladivostok 1962. *Oktiabr' na Dal'nem Vostoke,* parts 1–2. Vladivostok 1968–69.

Krusten, Erni
Esthonian writer; b. Ap. 17 (30), 1900, Muraste, now in Harju Raion, Esthonian SSR.
People's Writer of the Esthonian SSR (1972).
His collection of stories *Heather* was published in 1927 and was followed by novels and novellas that were limited to explorations of psychological problems. His novel *The Book of Peksy* (1946) and the novella collections *Turning Points* (1946) and *In the Name of Peace* (1951) reflect changes in people's social attitudes after Soviet power was restored in Esthonia in 1940.
His major work, the novel *The Hearts of the Young* (books 1–2, 1954–56; Russian translation, 1960), depicts the revolutionary events of 1905 in the Esthonian villages. His later writings include the plays *I Live* (1956) and *The Americans Have Arrived* (1959), the collections of short stories *Funeral of the Knout* (1957) and *In Search of Spring* (1960; Russian transla-

tion, 1962), the novel *Like a Drop in the Sea* (1962; Russian translation, 1964), the collection of verse and miniature sketches *Iuka* (1963), and the short-story collection *Five Stories* (1968). He also wrote several children's books. Decor.: Order of the Badge of Honor.
Works: *Okupatsioon.* Tallinn 1969. *In Russian translation: Gnezdo pod strekhoi.* Moscow 1959. *Bezumnaia lastochka.* Moscow 1970.

Krutikova, Nina Evgenevna
Literary historian; b. Ja. 27, 1913, Łódź, Poland; d. of a teacher.
Mem. CPSU 1944–; grad. Kharkov Inst. of Econ. Planning 1932; literary scholar; completed postgraduate studies at Dept. of Russian Literature of Kiev Univ. 1939; teacher, lecturer at Kiev Univ. 1933–; Sen. Research Staff Mem. (Dep. Dir. 1955–59) of Shevchenko Inst. of Literature at Ukrainian Acad. of Sciences 1941–. Decors.: Order of Red Banner of Labour, medals.
Publs.: author of some 120 scientific works, "Literaturnye sviazi russkogo i ukrainskogo narodov" 1954, "Gogol ta ukrainska literatura" 1957, "Lev Tolstoi i ukrainska literatura" 1958, "Storinky tvorchogo zhyttia (Marko Vovchok v zhytti i pratsi)" 1965.

Krutilin, Sergei Andreevich
Author; b. 1921, Delekhovo, Riazan Oblast.
Russian; mem. CPSU 1945; grad. Dept. of Philosophy of Moscow Univ. 1947; first works publ. 1947.
Publs.: "Rodniki" (Springs) 1953, "Za povorotom" (Around the Bend) 1961, "Gorod na Viatke-reke" (Town on the Viatka River) 1959, "Podsnezhniki" (Snowdrops) 1961, "Lipiagi", 3 vols., 1963–65, and others.

Kruzhilin, Georgii Nikitich
Scientist; b. My. 24 (Je. 6), 1911, Kruzhilinskoe, Rostov Oblast.
Mem. CPSU 1944–; grad. Physical-Mechanical Inst. Leningrad 1934; specialist in thermotechnology and atomic energy; corresponding mem. USSR Acad. of Sciences 1953–; mem. staff Inst. of Atomic Energy 1946–55; Head of Laboratory of G. M. Krzhizhanovskii Inst. of Energetics 1955–. Decors.: Order of Lenin, 2 other orders, several medals.

Kruzhkov, Vladimir Semenovich
Philosopher; b. Je. 3 (16), 1905, St. Petersburg.
Mem. CPSU 1925–; grad. N. K. Krupskaia Acad. of Communist Educ. 1931; assignments in Cen. Cttee. CPSU 1949–55; Prof. at Uralsk Univ. 1955–61; Dir. of Inst. of History of Art at USSR Min. of Culture 1961–73. Decor.: State Prize of the USSR 1952.

Krylov, Aleksandr Petrovich
Scientist; b. Ag. 1 (14), 1904, Tatevo, in present-day Belyi Raion, Kalinin Oblast. Specializing in the exploitation of petroleum deposits; academician of the Academy of Sciences of the USSR (1968; corresponding member, 1953).
He graduated from the Leningrad Institute of Mines in 1926. From 1926 to 1932 he worked in the Baku oil fields and took part in prospecting for coal deposits in the Donbas and on Sakhalin Island. In 1932 he joined the State Petroleum Research Institute; from 1935 to 1956 he was in the subdepartment of petroleum deposit exploitation of the I. M. Gubkin Moscow Petroleum Institute; in 1953 he began working at the All-Union Scientific Research Institute of Petroleum and Natural Gas (director from 1957 to 1971). He originated the theory and proposed designs for various modifications of flow-type and compressor-lift petroleum extraction methods (1938–50). He was one of the creators of new systems for the exploitation of petroleum deposits using perimeter injection of water into producing beds (flooding) and less dense well patterns; he developed exploitation systems using contour flooding. Decors.: State Prize of the USSR in 1949 and the Lenin Prize in 1962, Order of Lenin, four other orders, and medals.
Works: *Nauchnye osnovy razrabotki neftianykh mestorozhdenii.* Moscow-Leningrad 1948. (Co-author.) *Ekspluatatsii neftianykh mestorozhdenii.* Moscow-Leningrad 1949. (With I. M. Murav'ev.) *Proektirovanie razrabotki neftianykh mestorozhdenii.* Moscow 1962. (Co-author.)

Krylov, Aleksei Ivanovich
Party official.
2nd Sec. of Krasnoiarsk Krai Cttee. CPSU until 1971; 1st Oblast Sec. of CPSU Cttee. in Khakass ASSR N. 1971–; Dep. to and mem. Comm. for Consumer Goods at Sov. of Nationalities USSR Supr. Sov. 1974–.

Krylov, Lev Vladislavovich
Diplomat; b. 1920.
Mem. Diplomatic Service 1945–; 2nd, 1st Sec., USSR Embassy in Venezuela 1948–52; Emb. Councellor at USSR Emb. in Kuba 1965–68; Dep. Head, 1st Dep. Head, Press Dept., USSR Min. of Foreign Affairs 1970–80; USSR Amb. to Guinea-Bissau Je. 1980–, to Cape Verde Islands Je. 1980–N. 1981.

Krylovich, Vladimir Nikolaevich
Actor; b. O. 20 (N. 1), 1895, Krylovichi, Minsk Oblast.
Co-founder of Sov. Belorussian Theatre. Decor.: Merited Artist of the Belorussian SSR.

Ktorov, Anatolii Petrovich
Russian actor; b. Ap. 12 (24), 1898, Moscow.
People's Artist of the USSR (1963).
He made his stage debut in 1917 in the V. F. Komissarzhevskaia Theater in Moscow. Between 1920 and 1933, with some interruptions, he worked in the Korsh Theater. His charm, grace, natural artistry, and mastery of the art of transformation made him one of the outstanding masters of character roles. He became famous for his roles in the films *The Three Million Case* (1926, Cascarilla), *Holiday of St. Jorgen* (1930, Mikael Korkis), and *Without Dowry* (1937, Paratov). In 1933 he joined the Moscow Art Theater, where his best roles included Sam Weller in *The Pickwick Club,* based on Dickens' novel; Shervinskii in Bulgakov's *The Days of the Turbins;* Koko in L. N. Tolstoy's *The Fruits of Enlightenment;* Suslov in Gorkii's *The Summer People;* Crossman in Lillian Hellman's *Autumn Garden;* and Nikolai Skrobotov in Gorkii's *The Enemies.* Among his important roles in the 1960's was that of George Bernard Shaw in the play *Dear Liar* by Kilty. His major film roles at this time were those of Prince Bolkonskii in *War and Peace* and the king in *Envoy of the Soviet Union.* Decors.: State Prize of the USSR in 1952, Order of the Red Banner of Labor.

Kuanyshev, Orazbek Sultanovich
Party official; b. 1935.
Kazakh; mem. CPSU 1961–; grad. Kazakh Agricultural Inst. 1958; teacher 1952–58; agronomist, economist on planning comm. of exec. cttee. of a raion sov. of peoples's deps., chief agronomist, dir. of a state farm 1958–65; 2nd Sec. of Astrakhan Raion Cttee. CPSU in Tselinograd City, 1st Sec. of Atbasar Raion Cttee. CPSU in Tselinograd City, 2nd Sec. of Tselinograd Oblast Cttee. CPSU 1965–78; mem. Cen. Cttee. Kazakh CP 1978–; 1st Sec. of Kokchetav Oblast Cttee. Kazakh CP Ja. 1978–; Dep. to USSR Supr. Sov. 1979–; cand. mem. Cen. Cttee. CPSU 1981–.

Kubashev, Sagidolla Kubashevich
Party official; b. 1927.
Mem. CPSU 1953–; grad. Zoological-Veterinary Inst. Alma-Ata 1949; held various leading posts 1949–60; mem. Cen. Cttee. Kazakh CP 1961–; Chair. of Oblast Exec. Cttee. of Peoples Deps. in Gurev 1960–63 and 1965–66, in Aktiubinsk 1966–75, in Kzyl-Orda 1979–82; 1st Sec. of Gurev Oblast Cttee. Kazakh CP (for Agriculture) 1963–64; Dep. Chair. of Council of Mins. Kazakh SSR 1975–79; 1st Sec. of Semipalatinsk Oblast Cttee. Kazakh CP O. 1982–.

Kubasov, Valerii Nikolaevich, Cand. of Tech. Sc.
Cosmonaut; b. Ja. 7, 1935, Viasniki, Vladimir Oblast.
Mem. CPSU 1968–; grad. Aviation Inst. Moscow 1958; cosmonaut 1966–; Cand. of Tech. Sc. 1968–; spaceflight in spaceship "Soiuz-6" as board engineer together with G. S. Shonin, performed some

experimental work, including experimental welding in space O. 11–16, 1969; participated in three-day group flight with spaceships "Soiuz-7" and "Soiuz-8". In Jl. 1975 a spaceflight as part of the EPAS project as a flight engineer (with A. A. Leonov) in the Soyuz 19 spacecraft. During the flight, docking with the American Apollo spacecraft was performed twice, and five scientific and engineering experiments were carried out. Decors.: Order of Lenin, medals.

Kublashvili, Dzhaba Davidovich
State official; b. 1938.
Georgian; grad. Georgian Polytechnical Inst. 1963; foreman, work leader, engineer, chief engineer of technical production dept. at Yuzhstal'konstruktsiya construction and assembly admin. in Rustavi 1963–71; Sov. specialist abroad 1971–74; instructor at Construction and Urban Economy Dept. Cen. Cttee. CP Georgia 1974–77; Dir. of metal constructions factory in Rustavi 1977–81; 1st Dep. Min. 1981–82, Min. of Local Ind. Georgian SSR Jl. 1982–.

Kublitskii, Georgii Ivanovich
Author; b. 1911, Krasnoiarsk.
Russian; first works publ. 1934.
Publs.: "Reis v Evenkiiu" (Passage to the Evenkia) 1939, "Enisei, reka Sibirskaia" (Enisei, Siberian River) 1949, "Bol'shaia Volga" (The Mighty Volga) 1951, "Na velikoi reke" (By the Great River) 1951, "Volga-Volga . . ." 1962, "Fritiof Nansen, ego zhizn i neobyknovennye prikliucheniia" (Fritjof Nansen, his Life and Extraordinary Adventures) 1958, and others.

Kuchas, Antanas Apolinaro (also A. A. Kučas)
Graphic artist; b. Ja. 11 (24), 1909, Deīkiškiai, in present day; Biržai Riaon, Lithuanian SSR.
People's Artist of the Lithuanian SSR (1969).
From 1929 to 1935, he studied at the Kaunas School of Art. Between 1944 and 1951 he was a teacher at the Kaunas Institute of Applied and Decorative Arts. In 1951 he became an instructor at the Art Institute of the Lithuanian SSR in Vilnius, receiving a professorship there in 1970.
He is a book designer and illustrator. He illustrated J. Žemaitė's *The Daughter-in-law* (woodcut, 1963) and P. Cvirka's *Sunsets in the Niki District* (linecut, 1967). He also executed works other than illustrations (*On the Nemunas River,* linecut, 1965). His work is marked by an expression and breadth of rhythms reminiscent of the traditions of the popular print *(lubok)* and folk woodcarving.

Kucher, Roman Vladimirovich, Dr. of Chem. Sc.
Chemical scientist; b. Mr. 12, 1925, Lvov; s. of a clerk.
Grad. Lvov Univ. 1947; Dr. of Chem. Sc. 1964; Prof. 1964; mem. staff Lvov Univ. 1947–62; head of laboratory at a branch research inst. 1962–64; Prof. Chair. of Physical and Colloid Chemistry at Lvov Univ. 1964–65; Head of Dept. of Physical Chemistry at Donetsk Branch of Ukrainian Chemistry at Donetsk Branch of Ukrainian Acad. of Sciences 1966–; corresponding mem. Ukrainian Acad. of Sciences 1965–.
Publs.: author of more than 70 scientific works; "O roli vodnoi fazy v protsesse emulsionnogo okisleniia izopropilbenzola" in "Zhurnal fizicheskoi khimii" No. 33/1959, "Kineticheskie zakonomernosti nakopleniia gidroperekisei pri okislenii p-diizopropilbenzola" in "Neftekhimiia" No. 4/1963.

Kudashev, Saifi Fattakhovich (pseudonym: Saifi Kudash)
Bashkir poet; b. S. 21 (O. 3), 1894, Kliashevo, in present-day Chishmy Raion, Bashkir ASSR.
People's Poet of the Bashkir ASSR (1964). Member of the CPSU since 1939. A teacher by profession, he turned to writing in 1913. His first collections of verses, *Songs of Freedom* (1917) and *Those Unfit for Military Service* (1917), describing the sufferings of the people, were a call to freedom and a protest against war and autocracy. He welcomed the Great October Socialist Revolution, and during the Civil War (1918–20) he appealed to the people to defend the achievements of October. The revolution and the first years of peaceful labor in the Soviet period were the subject of his collections *Plow Songs* (1926) and *Hero of the Struggle* (1928) and of the narrative poems *On the Meadow, October,* and *In the Smithy* (all 1920). His long poem *Letter* (1930) and his novel in verse *Kushkaen* (1936) depict the collectivization of the countryside.
His poetry reflects important stages in the life and struggle of the Bashkir people. Socialist construction is depicted in the narrative poem *The Law of Happiness* (1937) and in the collection of poems *The Republic Sings* (1940). The heroic deeds during the Great Patriotic War (1941–45) are the theme of the narrative poem *Lenin and the Fighting Man* (1943) and the collection *From the Bottom of My Heart* (1944); creative postwar labor and friendship among peoples are the dominant motifs in the collections *My Love* (1956), *Falling Leaves* (1960), and *In My Garden* (1964). Among his popular prose works are his novella *To Meet the Spring* (1954) about the poets G. Tukai and M. Gafuri and his memoirs *Unforgettable Moments* (1957; Russian translation, 1964) and *Treading the Path of Youth* (1964), describing the cultural cooperation of the peoples of Bashkiria, Tataria, and Kazakhstan. His works on literary criticism are also significant. He was a deputy to the Supreme Soviet of the Bashkir ASSR from 1947 to 1963. Decors.: Order of Lenin and four other orders.
Works: *Äthärdhër,* vols. 1–4. Ufa 1946–53. *Haylanma äthärdhër,* vols. 1–3. Ufa 1965–67. *In Russian translation: Izbrannoe.* Moscow 1956. *Izbr. proizv.,* vols. 1–2, Ufa 1970. *Stikhotvoreniia i poemy.* Moscow 1953.

Kudashkin, Vladimir Nikolaevich
Diplomat.
USSR Amb. to Niger Ag. 1978–.

Kudinov, Ivan Pavlovich
Laborer; b. 1922.
Mem. CPSU 1946–; completed 7 classes of middle school 1938 and a craftsmen's school 1956; worked in Moscow Telecommunications Centre 1938–39; milling cutter at Frunze Motor Works in Moscow 1939– and Kuibyshev 1941–; cand. mem. Cen. Cttee. CPSU 1971–81, mem. 1981–.

Kudriashova, Klavdiia Kuzminichna
Singer; b. 1925.
Studied Moscow Conservatory 1946–52; mezzo-soprano; soloist at Belorussian Opera and Ballet Theatre 1952–60; soloist at Opera and Ballet Theatre Perm 1960–. Decor.: People's Artist of the USSR 1970.

Kudriavtsev, Aleksandr Ivanovich
Party official; b. 1928, Vologda Oblast.
Grad. Leningrad Inst. of Mechanics; mechanical engineer; worked in a plant in Ulianovsk 1951–60; Sec. of Party Cttee., head of a dept., Sec. of Oblast Cttee. CPSU in Ulianovsk 1960–66; 1st City Sec. CPSU in Ulianovsk 1966–72; instructor 1972–74, head of the sector of the Dept. for Party-Organizational Work Cen. Cttee. CPSU 1974–82; 2nd Sec. and mem. Bureau of Cen. Cttee. CP Esthonia My. 1982–; Dep. to and mem. of Presidium of Supr. Sov. Esthonian SSR Ap. 1982–. Decors.: two Orders of the Red Banner of Labor, "Znak pocheta" order.

Kudriavtsev, Vladimir Leontevich
State official; b. 1903.
Russian; mem. CPSU 1919–; grad. Inst. of Oriental Studies Leningrad; employed in a cooperative 1918–19; served in Sov. Army 1919–26; sec. of a Party org., instructor, dept. chief, sec. of a raion CPSU cttee. 1926–33; mem. staff East Secretariat of Komintern Exec. Cttee. 1933–34; TASS correspondent in Japan 1934–37; Exec. Ed., Dep. Chief of TASS Foreign Desk 1937–41; served in Sov. Army 1941–45; Dep. Chief of Foreign Desk, Ed. of Foreign Desk "Izvestiia" 1945–61; Pol. Ed. of "Izvestiia" 1961–; Dep. to and mem. Comm. on Foreign Relations at Sov. of Nationalities USSR Supr. Sov. 1966–; Sec. of USSR Parliamentary Group 1966–.

Kugach, Iurii Petrovich
Painter; b. Mr. 8 (21), 1917, Suzdal'.
People's Artist of the RSFSR (1965). Corresponding member of the Academy of Arts of the USSR (1970).
From 1936 to 1942, he studied at the Moscow Art Institute under S. V. Gerasimov and I. E. Grabar'. In his early period he painted portraits (for example, the portraits for the series *Famous People of Moscow,* 1949; State Prize of the USSR, 1950) and scenes dealing with historical and revolutionary subjects. In the early 1960's, he became principally concerned with the new life in the contemporary countryside. He sought to poeticize the traditional elements of Russian folk life (*Before the Dance,* 1961, Tret'iakov Gallery, Moscow; *Haymaking in the Forest,* 1967, Omsk Museum of Fine Arts; *The Farewells,* 1969, the collection of the Board of Directors of Art Funds and the Design of Monuments of the Ministry of Culture of the RSFSR, Moscow). Decor.: I. E. Repin State Prize of the RSFSR in 1969.

Kugul'tinov, David Nikitich
Kalmyk poet; b. Mr. 13, 1922, Gakhan-Avganara, now in Gorodovikovsk Raion, Kalmyk ASSR.
People's Poet of the Kalmyk ASSR (1969). Member of the CPSU since 1943. A participant in the Great Patriotic War of 1941–45.
He made his literary debut in 1936. He graduated from the M. Gorkii Literary Institute in 1960. His narrative poems include *Love and War* (1957), *The Moabite Prisoner* (1958), *The Song of the Miraculous Bird* (1961), *Sar-Gerel* (1963–64), *Memories Awakened by Vietnam* (1966), *Time, the Sovereign* (1967), and *Reason's Revolt* (1965–71). He also wrote the poetry cycles *Life and Reflections* (1963–67) and *All the Years* (1939–67). He received the M. Gorkii State Prize of the RSFSR in 1967 for his collection of poems *I Am Your Coeval* (1966).
He early showed himself to be a romantic poet with a tendency to deal with the philosophical and civic aspects of his themes. The strong and courageous man, the patriot, is the hero of his works. The flashiness of his early poetry was replaced by a severe simplicity that emphasized the weight and profundity of his imagery. Decor.: Order of the Red Banner of Labor.
Works: *In Russian translation: Izbrannye proizvedeniia,* vols. 1–2. Introduction by S. Lipkin. Moscow 1970. *Vozrast.* Moscow 1973.

Kuibakarov, Khushvakt Radzhabovich
State official.
Sec. Kashka-Daria Oblast Cttee. Uzbek CP 1966–71; mem. Cen. Cttee. Uzbek CP 1971–76; Chair. of Kashka-Daria Oblast Exec. Cttee. of People's Deps. 1973–78; Min. of Supplies Uzbek SSR D. 1978–.

Kuiukova, Darkul'
Kirghiz actress; b. My. 15, 1919, Oktiabr'skoe, now in Osh Oblast.
People's Artist of the USSR (1967). Member of the CPSU since 1944.
She joined the Kirghiz Children's Theater (in Frunze) in 1936, mainly in male impersonation roles. She has been one of the leading actresses of the Kirghiz Dramatic Theater (in Frunze) since 1941. She plays heroic, spiritually strong Kirghiz women, for example Akchaim (*Sarynzhi* by Eshmambetov), Aiganysh (*Kurmanbek* by Dzhantoshev), Dzhanyl (*Dzhanyl* by Malikov and Kuttubaev), and Tolgonai (*Mother's Field,* based on the Aitmatov work). Her ability to create incisively satirical portraits and domineering characters was demonstrated in such roles as Panova (*Liubov' Iarovaia* by Trenev), Natal'ia Kovshik (*The Snowball-tree Grove* by Korneichuk), Maisalbiubiu (*The Narrow Gorge* by Abdumomunov), Melan'ia (*Egor Bulychev and the Others* by Gorkii), Kabanikha (*The Thunderstorm* by Ostrovskii), and Goneril (*King Lear* by Shakespeare).
She was a deputy to the seventh convocation of the Supreme Soviet of the Kirghiz SSR. Decors.: Order of the Red Banner of Labor and various medals.

Kukhtenko, Aleksandr Ivanovich, Dr. of Tech. Sc.
Automation specialist; b. Mr. 11, 1914, Gorodnia, Chernigov Oblast; s. of a teacher.
Grad. Ind. Inst. Donetsk 1936; Dr. of Tech. Sc. 1955; Prof. 1956; teacher at Ind. Inst. (now Polytech. Inst.) Donetsk 1937–41; mem. staff Inst. of Mining Mechanics at Ukrainian Acad. of Sciences (now Inst. of Mining Mechanics and Tech. Cybernetics of USSR Min. of Coal Ind.) 1945–55; Lecturer at Kiev Inst. of Civil Aviation 1955–; concurrently Dept. Head at Inst. of Cybernetics of Ukrainian Acad. of Sciences 1963–; corresponding mem. Ukrainian Acad. of Sciences 1964–. Decors.: Order of Red Star, medals.
Publs.: author of more than 60 scientific works on theory of automatic steering and regulating; "Osnovnye problemy avtomatizatsii ugledobyvaiushchikh mashin" 1958, co-author of "Avtomatizatsiia ochistnykh i prokhodcheskikh rabot" 1961, "Problema invariantnosti v avtomatike" 1963.

Kukin, Dmitrii Mikhailovich
Historian; b. My. 10 (23), 1908, Zareche, Orel Oblast.
Mem. CPSU 1929–; studied Pedagogical Inst. Krasnodar and Moscow Univ.; completed postgraduate studies at Belorussian Acad. of Sciences 1936; lecturer, Prof. Chair of History of CPSU, Dep. Dir. of Military-Pol. Lenin Acad. 1938–54; mem. Apparat Cen. Cttee. CPSU 1954–62; Dep. Dir. of Inst. of Marxism-Leninism at Cen. Cttee. CPSU 1962–73; corresponding mem. USSR Acad. of Sciences 1964–; Prof. at Acad. of Social Sciences of Cen. Cttee. CPSU 1973–. Decors.: 7 orders, several medals.

Kuklin, Lev Valerianovich
Poet; b. 1931, Novozybkov, Briansk Oblast.
Russian; mem. CPSU 1960–; grad. Leningrad Tech. Inst. 1954; first works publ. 1954.
Publs.: "Neobkhodimost" (Necessity) 1961, "Rudnik radosti" (A Mine of Joy) 1964, and others.

Kulakova, Galina Alekseevna
Skier; b. Ap. 29, 1942, Logachi, Udmurt ASSR.
Honored Master of Sports (1970). Instructor.
She was a champion of the USSR (1969–71), a world champion (1970), and an Olympic gold-medal winner (three medals, Sapporo, 1972) in individual and relay races. Decors.: Order of Lenin and Order of the Badge of Honor.

Kulakovskii, Aleksei Nikolaevich
Author; b. 1913, Kulaki, Minsk Oblast; s. of a peasant.
Belorussian; mem. CPSU 1948–; company commander 1941–45. Decor.: Order of Red Star.
Publs.: "Sad" (The Garden) 1947, "Zakalka" (Hardening) 1948, "Rastaemsia nenadolgo" (We Are Parting for a Short Time) 1953–62, "Nezabyvaemoe ekho" (Unforgettable Echo) 1956, "K voskhodu solntsa" (Near Sunrise) 1957, "Tut ia zhivu ..." (I Live Here ...) 1960, and others.

Kuleshov, Pavel Nikolaevich
Military official; b. D. 13 (26), 1908, Kailyk, now Uiar Raion, Krasnoiarsk Krai.
Marshal of the artillery (1967). Member of the CPSU since 1939.
In the Soviet Army since 1926, he graduated from an artillery school in 1929, from the F. E. Dzerzhinskii Artillery Academy in 1938, and from the Academy of the General Staff in 1941. In the Great Patriotic War (1941–45), after serving in the Main Directorate of Logistics of the Red Army, he was commander of a task force of multiple rocket launchers on the Northwestern and Volkhov fronts from October 1941, deputy commander of multiple rocket launcher units from June 1943, and deputy chief of staff of the Red Army's artillery from August 1944.
After the war he was head of a department and deputy chief of the Dzerzhinskii Artillery Academy from 1946 to 1952. He held responsible positions in the National Air Defense and in the Ministry of Defense from 1952 to 1957. He was deputy commander in chief of the National Air Defense from 1957 to 1963. He has held responsible positions in the Ministry of Defense since 1963. Decors.: four Orders of Lenin, four Orders of the Red Banner, the Order of Suvorov Second

Class, the Order of Kutuzov Second Class, the Order of the Patriotic War First Class, the Order of the Red Star, and various medals.

Kulibaev, M.
Trade union official.
Sec. of Uzbek Republican Trade Union Council F. 1982–.

Kulibaev, Saliakh Gabitovich
Poet; b. 1910, Tagirovo, Bashkir ASSR; s. of a peasant.
Mem. CPSU 1947–; worked as miner; first works publ. 1927.
Publs.: "Dykhanie chuguna" (The Breath of Cast Iron) 1931, "Schastlivaia zemlia" (A Happy Country) 1939, "Serdtsa trekh" (The Hearts of the Three) 1940, "Dzhigit s Urala" (A Djigit from the Urals) 1943, "Rasskaz starogo shakhtera" (The Story of an Old Miner) 1948, "Posle bitv" (After the Fighting) 1949, "Mastera" (The Masters) 1951, "Veselye rasskazy" (Amusing Stories) 1964, and others.

Kulichenko, Leonid Sergeevich
Party official; b. 1913.
Russian; mem. CPSU 1940–; grad. Inst. of Mechanics Volgograd 1936 and Party Higher School of Cen. Cttee. CPSU 1949; Head of Instruction Dept. of Inst. of Mechanics Volgograd, head of pol. dept. of a machine and tractor station 1936–42; 1st sec. of a raion, dept. head, 1st Sec. of Volgograd City Cttee., 2nd Sec. of Volgograd Oblast Cttee. CPSU 1942–62; Dep. to Supr. Sov. RSFSR 1959–67; Chair. of Volgograd Oblast Cttee. of Workers' Deps. 1962–65; 1st Sec. of Volgograd Oblast Cttee. CPSU N. 1965–; mem. Cen. Cttee. CPSU 1966–; Dep. to 1966–, mem. Comm. for Ind., Transport Posts and Telecommunications 1966–74, mem. Comm. for Ind. at Sov. of the Union USSR Supr. Sov. 1974–. Decors.: Order of Lenin, Order of Red Banner of Labour.

Kulidzhanov, Lev Aleksandrovich
Film producer; b. Mr. 19, 1924, Tbilisi.
Mem. CPSU 1962–; grad. Union Film Inst. 1955; producer at Gorkii Film Studio 1955–; Chair. of Org. Cttee. 1964–65 and 1st Sec. of USSR Union of Film Workers 1965–; mem. Cen. Auditing Comm. 1966–76, cand. mem. Cen. Cttee. CPSU 1976–; Chair. of USSR-Mexico Soc. 1966–; Dep. to USSR Supr. Sov., mem. Comm. on Educ., Science and Culture at Sov. of Nationalities USSR Supr. Sov. 1966–; best-known films: "Otchii dom" (Father's House), 1959, "Poteriannaia fotografiia" (The Lost Photograph) 1959, "Dom v kotorom ia zhivu" (The House in Which I Live) 1957, and others. Decors.: State Prize of the RSFSR 1971, Order of Lenin, 1 medal.

Kuliev, Kaisyn Shuvaevich
Author and poet; b. N. 1, 1917, Verkhnii Chegem, Kabardino-Balkar ASSR.
Balkar; mem. CPSU 1944–; studied A. V. Lunacharskii Inst. of Theatre Arts (GITIS) Moscow 1935–39; served in Sov. Army 1941–45; first works publ. 1937. Decors.: Gorkii State Prize of the RSFSR 1964, Lenin Prize 1974, 3 orders, several medals.
Publs.: "Privet, utro!" (Greetings, Morning!) 1937, "Moi sosedi" (My Neighbours) 1939–45 (publ. 1957), "Gorskaia poema o Lenine" (Mountain Poem in Lenin) 1959–60, "Ranenyi kamen'" (The Wounded Stone) 1964, "Izbrannye proizvedeniia" (Selected Works) 1970, "Kniga zemli" (The Book of the Earth) 1972, and others.

Kulieva, Maia (Mamadzhan)
Turkmen lyric soprano; b. 1920 in the *aul* (mountain village) of Bezmein, Turkmen SSR.
People's Artist of the USSR (1955). Member of the CPSU since 1947.
She studied at the national Turkmen division of the Moscow Conservatory from 1938 to 1941. In 1941 she became a soloist at the Makhtumkuli Turkmen Theater of Opera and Ballet. In the works by composers of the Turkmen SSR and other republics of Middle Asia she has created a number of important portrayals, including Shasenem in Shaposhnikov and Ovezov's opera *Shasenem and Garib*, the title roles in Shaposhnikov and Ovezov's *Aina* and Meitus and Kuliev's *Abadan*, Zokhre in Shaposhnikov and Mukhatov's *Zokhre and Takhir*, and Gunchabi in Shaposhnikov's *The Rose and the Nightingale*. She is the first Turkmen female singer to perform the European opera repertoire, appearing in such roles as Marfa in Rimsky-Korsakov's *The Tsar's Bride*, Tatiana in Tchaikovsky's *Eugene Onegin*, Marguerite in Gounod's *Faust*, and the title role in Puccini's *Madame Butterfly*. Decor.: Order of the Badge of Honor.

Kulik, Boris Trofimovich
Diplomat.
USSR Amb. to Malaysia Ap. 1978–.

Kulikov, Fedor Mikhailovich
Party official; b. 1925.
Mem. CPSU 1949–; grad. Penza Mechanical Engineering Technicum 1949 and Penza Pedagogical Inst. 1959; served in Sov. Army 1943–45; Komsomol posts 1946–55; 1st Sec. Penza Oblast Cttee. USSR Komsomol 1955–58; chief of a dept. of Penza Oblast Cttee. CPSU 1962–73; Sec. of Penza Oblast Cttee. CPSU 1973–79, 1st Sec. Ap. 1979–; mem. Cen. Cttee. CPSU 1981–.

Kulikov, Viktor Georgievich
Military official; b. 1921; s. of a peasant.
Russian; mem. CPSU 1942–; grad. Frunze Mil. Academy 1953 and Voroshilov Gen. Staff Acad. 1959; Marshal of the Soviet Union Ja. 1977–; in Sov. Army 1939–; at the front during the war, chief-of-staff of a tank brigade, after the war commander of a regiment, a division, an army, Commander of Kiev Mil. District 1941–62; Dep. Commander-in-Chief 1962–69, Commander-in-Chief of Sov. Forces Group in Germany 1969–71; Dep. to USSR Supr. Sov. 1970–; mem. Cen. Cttee. CPSU 1971–; Chief of Gen. Staff of Sov. Army and Navy Fleet 1971–77 and USSR 1st Dep. Min. of Defence S. 1971–; Commander-in-Chief of Warsaw Pact Forces Ja. 1977–. Decors.: 2 Orders of Lenin, 3 Orders of the Red Banner, 2 Orders of the Patriotic War 1st Cl., Order of the Red Star, other Ordens and Medals. Hero of Socialist Labour 1977.

Kulinich, Nikolai Fedorovich
State official; b. 1930.
Mem. CPSU 1963–; grad. Kiev Technological Inst. for the Food Ind.; shift supervisor, sec. of party organization of Smela Sugar and Milk Combine, instructor of a dept. of Cherkassy Oblast Cttee. Ukrainian CP, Dir. of Kamenka Sugar Combine, Chief of Cherkassy Oblast Admin. for the Food Ind. to 1972; Dep. Min. of the Food Ind. Ukrainian SSR 1972–73, 1st Dep. Min. 1973–81, Min. F. 1981–; cand mem. Cen. Cttee. Ukrainian CP 1981–.

Kulishev, Oleg Fedorovich
Military official; b. 1928.
Russian; mem. CPSU 1951–; grad. Frunze Mil. Acad. and Mil. Acad. of the Gen. Staff, Senior Gen.; in Sov. Army 1946–; after completion of infantry training in Tambov held command posts in various mil. districts and groups of armed forces 1946–75; troop commander of northern mil. group (Poland) 1975–78; troop commander of Transcaucasian mil. district 1978–; mem. Cen. Cttee. and mem. Bureau of Cen. Cttee. CP Georgia 1978–; Dep. of USSR Supr. Sov. 1979–.

Kullaste, Allan Andreevich
State official.
Mem. Bureau 1965–72, Sec. of Cen. Cttee. Esthonian. Komsomol 1969–72; mem. Auditing Comm. Esthonian CP 1981–; Chair. of State Cttee. for Television and Radio Esthonian SSR Mr. 1982–.

Kulmamedov, Aman
Actor and theatre director; b. D. 31, 1908, Geok-Tepe, Turkmenia.
Mem. CPSU 1940–; Principal Dir. and Managing Dir. of Mollanepes Theatre 1963–. Decors.: People's Artist of the USSR, 2 Orders of Lenin, Order of Red Banner of Labour, medals.

Kulmatov, Kenesh
Party official; b. 1934.
Kirghiz; mem. CPSU 1958–; grad. Kirghiz State Univ.; instructor, Dep. Head of Dept. of Schools and Student Youth at Cen. Cttee. Kirghiz Komsomol 1956–57; 1st Sec. of Chiiskii Komsomol Raion Cttee., Head of Dept. of Propaganda and Agitation and Sec. of Cen. Cttee. Kirghiz Komsomol 1958–59; 1st Sec. of Osh Oblast Cttee. Kirghiz Komsomol 1959–60; 1st Sec. and mem. Bureau of Cen. Cttee. Kirghiz Komsomol 1960–66; cand. mem. 1960–61 and 1966–71, mem. Cen. Cttee. Kirghiz CP 1971–; cand. mem. Bureau 1961–62, mem. Bureau and Sec. of Cen. Cttee. Kirghiz CP 1973–; 1st Sec. of Sverdlov Raion Cttee. Kirghiz CP in Frunze, 1st Sec. of Frunze City Cttee. Kirghiz CP 1970–73.

Kulmatov, Renat Satarovich
Party official.
Head of Dept. of Ind. and Transport with Cen. Cttee. Kirghiz CP 1979–80; Head of Dept. of Transport, Posts and Telecommunications with Cen. Cttee. Kirghiz CP 1980–81; 2nd Sec. Ja.–Je. 1981, 1st Sec. of Osh Oblast Cttee. Kirghiz CP Je. 1981–; mem. Cen. Cttee. Kirghiz CP 1981–.

Kumanchenko, Polina Vladimirovna
Actress; b. O.21, 1910, Muzykovka, Kherson Oblast.
Ukrainian; actress at Ukrainian Shevchenko Theatre Kharkov 1937–61, Ukrainian Ivan-Franko Theatre Kiev 1961–. Decors.: People's Artist of the USSR 1960, State Prize of the Ukrainian SSR 1971, Order of Red Banner of Labour, medals.

Kunaev, Askar Minliakhmedovich, Dr. of Tech. Sc.
Metallurgist; b. 1929.
Mem. CPSU 1971–; grad. Steel and Alloys Inst. Moscow 1951; Prof. 1970–; specializes in ferrous and nonferrous metallurgy; worked in a metallurgical factory in Kazakhstan 1951–53; aspirant, mem. staff and Head of Laboratory at Inst. of Metallurgy and Enrichment at Kazakh Acad. of Sciences 1953–70; Dir. of same Inst. 1970–; mem. and Vice-Pres. 1972–74, Pres. of Kazakh Acad. of Sciences Ap. 1974–; cand. mem. Cen. Cttee. CPSU 1976–; Dep. to USSR Supr. Sov. 1979–. Decor.: Order of Lenin.

Kunaev, Dinmukhamed Akhmedovich, Cand. of Tech. Sc.
Party official; b. D.31, 1911 (Ja.12, 1912), Alma-Ata, Kazakhstan; s. of a clerk.
Mem. CPSU 1939–; grad. Inst. of Non-Ferrous and Fine Metallurgy Moscow 1936; mem. Kazakh Acad. of Sciences; foreman, Dept. Head, Chief Engineer at Balkash Copper Refinery, Dep. Chief Engineer at "Altaipolymetall" Combine, Dir. of Leninogorsk Ore Board 1936–42; Dep. Chair. of Council of Mins. Kazakh SSR, Pres. of Kazakh Acad. of Sciences 1942–55; mem. Cen. Cttee. Kazakh CP 1949–; Dep. to USSR Supr. Sov. 1950–; Chair. of Council of Mins. Kazakh SSR 1955–60; mem. Bureau Cen. Cttee. Kazakh CP 1956–; mem. Cen. Cttee. CPSU 1956–; mem. Presidium USSR Supr. Sov. 1962–; Chair. of Council of Mins. Kazakh SSR 1962–64; Dep. to 1963–, mem. Presidium Supr. Sov. Kazakh SSR 1965–; 1st Sec. of Cen. Cttee. Kazakh CP D. 1964–; cand. mem. 1966–71, mem. Politbureau Cen. Cttee. CPSU 1971–. Decors.: 3 Orders of Lenin, Order of Red Banner of Labour, medals.

Kundzin', Karlis Ernestovich
Historian of literature and theater; b. Jl. 14 (27), 1902, Sasmaka, now Vāldemarpils, Talsi Raion.
Doctor of art studies (1969).
He graduated from the University of Latvia in 1941 and began teaching there, as well as at the Latvian Conservatory. In 1949 he became head of the section for the history of Latvian literature in the Institute of Language and Literature of the Academy of Sciences of the Latvian SSR. His *Latvian Theater* (1963) is the first work in Soviet theater criticism to investigate the development of Latvian theater from its origins to the 1960's. His other works include *The Repertoire of the Latvian Theater Up to 1940* (1953), *History of the Latvian Theater* (vols. 1–2, 1968–72), and chapters in such collective works as *History of Latvian Literature* (vols. 1–6, 1956–63) and *History of Soviet Dramatic Theater* (vols. 1, 4, and 5, 1966–69). Decor.: State Prize of the Latvian SSR in 1972.

Kungurtsev, Evgenii Maksimovich
Military official; b. O.3, 1921, Izhevsk; s. of a worker.
Twice Hero of the Soviet Union (F.3, 1945, and Ap.19, 1945); major-general of aviation (1964). Member of the CPSU since 1943.
He joined the Soviet Army in 1940. He graduated from the Balashov Military Aviation School in 1942, from the Air Force Academy in 1952, and from the K.E. Voroshilov Higher Military Academy in 1957. In the Great Patriotic War (1941–45), he fought on the Leningrad and Third Belorussian fronts as senior pilot, flight commander, and deputy commander and commander of a squadron of the 15th Guards Assault Air Regiment. In March 1945, having been severely wounded, he was captured, but he escaped and returned to his unit. He made 210 successful combat sorties for reconnaissance and attacks on enemy strong points, airfields, and concentrations of troops and equipment. He went into the reserves in 1968. Decors.: Order of Lenin, three Orders of the Red Banner, the Order of Bogdan Khmel'nitskii, the Order of Alexander Nevsky, two Orders of the Red Star, and various medals.

Kuniaev, Stanislav Iurevich
Poet and Author; b. 1932, Kaluga.
Russian; mem. CPSU 1960; grad. Dept. of Philology of Moscow State Univ. 1957; first works publ. 1956.
Publs.: "Kazakhstanskaia tetrad'" (Kazakh Diary) 1961, "Zemleprokhodtsy" (World Travellers) 1960, "Zveno" (Connecting Link) 1962, and others.

Kuperman, Iser Iosifovich
Sportsman (Russian and international checkers); b. Ap.28, 1922, Polesskoe, in present-day Kiev Oblast.
International Grandmaster (1958), Honored Master of Sport and Grandmaster of the USSR (1960); engineer.
He was the Soviet champion in Russian checkers in 1945, 1947, and 1948 and in international checkers in 1954, 1955, 1956, 1962, and 1967. In 1965 he was holder of the Cup of Europe, and from 1958 to 1967 he was five times the world champion in international checkers. Decors.: Order of the Badge of Honor.
Works: *Strategiia v stokletochnykh shashkakh.* Moscow 1964. *Taktika v stokletochnykh shashkakh.* Moscow 1967. *Shkola igry v stokletochnye shashki,* 2nd ed. Kiev 1970. *Novaia teoriia shashechnykh debiutov.* Moscow-Leningrad 1949.

Kuppaev, Tulegen Baigushevich
State official; b. 1937.
Mem. CPSU 1968–; grad. Kurgan Agricultural Inst. 1962; Agronomist, chief agronomist, dir. of a state farm 1962–75; 1st Dep. Head of the Production Admin. for State Farms in Kustanai 1975–76; 1st Dep. Chair. of Kustanai Oblast Exec. Cttee. of Peoples Deps. 1976–77, Chair. 1977–82; mem. Cen. Cttee. Kazakh CP 1981–; Dept. Chair. of Council of Mins. Kazakh SSR N. 1982–.

Kuprianov, Ivan Petrovich
Dramatist; b. 1915, Razdolki, Tula Oblast; s. of a worker.
Russian; mem. CPSU 1943–; grad. Gorkii Inst. of Literature Moscow 1951; began literary career 1938.
Publs.: plays "Snega Finliandii" (Finland Snow) 1941, "Bratia Runkovy" (The Runkov Brothers) 1950, "Nachalo bol'shoi zhizni" (The Beginning of a Great Life) 1957, "Chest' ofitsera" (Officer's Honour) 1955, "Moskvichka" (Moscow Woman) 1957, "Vysokoe napriazhenie" (High Tension) 1959, "Moi brat" (My Brother) 1961, "Galia Rostovtseva" 1964, "Komsorg polka" (Regimental Komsomol Organizer) 1965, and others.

Kurach, A. K.
Komsomol official.
Sec. of Cen. Cttee. Belorussian Komsomol Jl. 1977–81, 2nd Sec. and mem. Bureau Ja. 1981–.

Kurbanov, Ikram Inoiatovich
State official.
Min. of Food Ind. Tadzhik SSR Je. 1975–. Cand. mem. Cen. Cttee. Tadzhik CP 1976–.

Kurbanov, Ismoil
State official.
Dep. of Supr. Sov. Tadzhik SSR 1971–; 1st Sec. Parkhatskii and Kumsangirskii Raion Cttee. Tadzhik CP 1973–80; Min. of Interior Tadzhik SSR Ja. 1980–.

Kurbanov, Makhmut Khasanovich
State official.
Mem. Bureau Cen. Cttee. Uzbek Komsomol 1962–66; Head of Dept. of Food and Light Ind. at Cen. Cttee. Uzbek CP 1974–75; Min. of Light Ind. Uzbek SSR My. 1975–; mem. Cen. Cttee. Uzbek CP 1976–.

Kurbanov, Rakhman Mamadalievich
State official.
Cand. mem. 1961–63, mem. Cen. Cttee. Tadzhik CP 1963–81; Head of Dept. of Construction and Building Materials Ind. at Cen. Cttee. Tadzhik CP until O. 1961; Min. of Construction Tadzhik SSR 1961–63; Dep. to Supr. Sov. Tadzhik SSR 1963–; Head of Territorial Main Admin. of Construction Tadzhik SSR 1963–64; Head of Dept. of Water Econ. and Rural Construction at Cen. Cttee. Tadzhik CP 1966–73; Min. of Communal Econ. Tadzhik SSR S. 1973–; mem. Auditing Comm. Tadzhik CP 1981–.

Kurbanov, Subkhan
State official.
Min. of Social Welfare Turkmen SSR O. 1979–; cand. mem. Cen. Cttee. Turkmen CP 1981–.

Kurbanova, Okuma Abbas kyzy
Azerbaidzhan actress; b. My. 29 (Je. 11), 1913, Baku.
People's Artist of the USSR (1965).
She joined the company of the M. Azizbekov Azerbaidzhani Theater (Baku) in 1938. She brings great artistic skill to dramatic, genre, and comedy roles. Her roles have included Shirin in *Farkhad and Shirin* by Samed Vurgun, Reikhan in *Dzhavanshir* by Mekhti Gusein, Benevshe in *The Country Bumpkin* by Ibragimov, and Cleopatra in *Antony and Cleopatra* by Shakespeare (M.F. Akhundov State Prize of the Azerbaidzhan SSR, 1965). She was a deputy at the seventh and eighth convocations of the Supreme Soviet of the Azerbaidzhan SSR. Decors.: Order of Lenin, Order of the Badge of Honor, and various medals.

Kurbansakhatov, Kurbandurdy
Oblast. Turkmen writer; b. 1919, Karaiab, now in Mary.
He graduated from the Gorkii Pedagogic Institute in Ashkhabad (1941). He published the collection of verses *The Soldier Returned Home* in 1947 and the verse tale *The Foolish Shah* (in Russian) in 1955. Later he turned to prose about contemporary life – for example, the collections of novellas and stories *Surai* (1955), *The Heart Is Not a Stone* (1958), *Forty Coins* (1961; Russian translation, 1964), and *Dear Man* (1965). He wrote a novella about the great Turkmen poet Makhtumkuli (*Invitation,* 1961, in Russian) and a novel about the contemporary Turkmen countryside (*Toily Mergen,* 1970). He also wrote a number of plays, including *Khang* (1966) and *Unshed Blood* (1968). He served as chairman of the board of the Writers' Union of Turkmenia from 1950 to 1955. Decors.: Order of the Red Banner of Labor and a medal.
Works: *Baglaring tëji.* Ashkhabad 1962. *Söyseng.* Ashkhabad 1969.

Kurbatov, Leonid Nikolaevich
Physicist; b. Ja. 2 (15), 1913, Fergana.
Grad. Polytech. Inst. Leningrad 1936; mem. staff Leningrad Univ. 1936–40 and 1956–60; mem. staff Sov. Navy Medical Acad. Leningrad 1941–56; mem. staff State S. I. Vavilov Inst. of Optics 1956–60; dep. dir. of a research inst. 1960–; Prof. at Physical-Tech. Inst. Moscow 1961–; corresponding mem. USSR Acad. of Sciences 1972–. Decors.: State Prize of the USSR 1970, Order of Red Banner of Labour, medals.

Kurdiani, Archil Grigorevich
Architect; b. Mr. 13 (26), 1903, Tbilisi.
Mem. CPSU 1942–; grad. Georgian Polytech. Inst. 1929; Chief Architect of Tbilisi City 1936–44; Chair. of Georgian Federation of Architects 1959–62; lecturer 1948–, Prof. at Acad. of Arts Tbilisi 1960–. Decors.: Merited Artist of the Georgian SSR, Order of Lenin, Order of Red Banner of Labour, medals.

Kurdiumov, Georgii Viacheslavovich, Dr. of Phy.-Math. Sc.
Metal physicist; b. F. 14, 1902, Rylsk, now Kursk Oblast.
Grad. Polytech. Inst. Leningrad 1926; Dr. of Phy.- Math. Sc. 1937; Prof. 1934; mem. staff Physical-Tech. Inst. Leningrad 1925–32; mem. Ukrainian Acad. of Sciences 1939–; various posts at Physical-Tech. Inst. Dnepropetrovsk 1932–44; Dir. of Inst. of Metallography and Physics of Metals at Cen. Metallurgical Inst. Moscow 1944–; concurrently mem. staff Inst. of Metallurgy at Ukrainian Acad. of Sciences 1944–45; Dir. of Laboratory for Physics of Metals at Ukrainian Acad. of Sciences 1946–49; mem. USSR Acad. of Sciences 1953–; Dir. of Inst. of Physics of Solid Bodies at USSR Acad. of Sciences Moscow 1962–. Decors.: 2 Orders of Lenin, 2 Orders of Red Banner of Labour, medals.
Publs.: author of more than 150 scientific works on annealing and tempering of steel and phenomena of phase conversions, hardening and diffusion in metals and alloys; "Fizicheskie osnovy termicheskoi obrabotki stali" 1941, "Primenenie metodov rentgenostrukturnogo analiza dlia issledovaniia otpuska zakalennoi stali" in "Voprosy fiziki metallov i metallovedeniia" No. 2/1950, "Primenenie radio-aktivnykh izotopov v tekhnike, biologii i selskom khoziaistve" 1955, and others.

Kurentsov, Viktor Grigor'evich
Athlete; b. Ap. 15, 1941, Tusinka, Vitebsk Oblast.
Honored Master of Sport (1966). Officer in the Soviet Army. Member of the CPSU since 1970.
He, competing in weight lifting in the middleweight division, won the Gold Medal at the 19th Olympic Games (Mexico, 1968) and the Silver Medal at the 18th (Tokyo, 1964). He is a five-time world champion (1965–70) and a seven-time European and USSR champion (1964–71). Decors.: Order of the Red Banner of Labor and several medals.

Kurilich, Mikhail Vasil'evich
Painter; b. 1921, Zabolotiv, Ivano-Frankovsk Oblast.
Attended Lvov Inst. for Applied and Decorative Art 1947–53; teacher at Lvov Inst. for Applied and Decorative Art 1953–.

Kuris, Pranas Mikolovich
State official.
Cand. mem. Cen. Cttee. Lithuanian CP 1976–81; Min. of Justice Lithuanian SSR O. 1977–.

Kurkotkin, Semen Konstantinovich
Military official; b. 1917.
Russian; mem. CPSU 1940–; grad. Acad. of Tank Troops 1951 and Gen. Staff Acad. 1958; Gen. 1972–; served in Sov. Army 1937–; pol. instructor (Politruk) of a company, commander of a tank battalion, dep. commander of a tank regiment, commander of a tank brigade 1941–58; command posts 1958–; mem. Cen. Cttee. Belorussian CP 1961–66; Dep. Commander-in-Chief of Sov. Forces Group in Germany, Commander of Transcaucasian Mil. District 1966–71; Dep. to USSR Supr. Sov. 1970–; cand. mem. 1971–76, mem. Cen. Cttee. CPSU 1976–; mem. Cen. Cttee. and mem. Bureau Cen. Cttee. Georgian CP Mr.–S. 1971; Commander-in-Chief of Sov. Forces Group in Germany 1971–73; USSR Dep. Min. of Defence, Chief of Rear Services of USSR Armed Forces 1973–.

Kurochikin, Pavel Alekseevich
Military official; b. N. 6 (19), 1900, Gornevo, now Viaz'ma Raion, Smolensk Oblast; s. of a peasant.
General of the army (1959); Hero of the Soviet Union (Je. 29, 1945); professor (1962). Member of the CPSU since 1920. He was a participant in the storming by the Red Guards of the Winter Palace in October 1917. He entered the Soviet Ar-

my in 1918 and commanded a squadron and a cavalry regiment in the Civil War of 1918–20. He graduated from the Petrograd Cavalry Training Courses in 1920 and the M.V. Frunze Military Academy in 1932 and completed postgraduate studies at the academy in 1934. He was a commander of an infantry corps in the liberation of the Western Ukraine and in the Soviet-Finnish War of 1939–40. He commanded an army group of forces, an army, and the forces of the Transbaikal Military District in 1940–41. In the Great Patriotic War (1941–45), he was commander of the troops of the Twentieth and Forty-third armies (July and August 1941); commander and deputy commander of the troops of the Northwestern Front (August 1941–November 1942); commander of the troops of the Eleventh Army (November 1942–March 1943), the Thirty-fourth Army (March–June 1943), and the armies of the Northwestern Front (June–November 1943); deputy commander of the troops of the First Ukrainian Front (November 1943–February 1944); commander of the troops of the Second Belorussian Front (February–April 1944); and commander of the troops of the Sixtieth Army (from April 1944 to the end of the war).
After the war he served as commander of the troops of the Kuban' Military District (from July 1945); deputy commander-in-chief of Soviet Forces in Germany and deputy chief commander of the Soviet Military Administration in Germany (from July 1946); assistant commander-in-chief of the troops of the Far East (from May 1947); deputy chief of the Military Academy of the General Staff (from February 1951); and chief of the M.V. Frunze Military Academy (from May 1954–68). He has held leadership positions in the Ministry of Defense of the USSR since April 1968. He was a deputy to the second convocation of the Supreme Soviet of the USSR. Decors.: M.V. Frunze State Prize (1967), four Orders of Lenin, the Order of the October Revolution, four Orders of the Red Banner, the Order of Suvorov First Class, two Orders of Kutuzov First Class, the Order of the Badge of Honor, and various medals, as well as eight foreign orders.

Kurpnieks, Voldemar Ianovich
State official.
Cand. mem. Cen. Cttee. Latvian CP 1976–81, mem. 1981–; Min. of Fruits and Vegetables Latvian SSR 1981–.

Kursanov, Andrei Lvovich
Physiologist and biochemist; b. O. 26 (N. 8), 1902, Moscow.
Grad. Moscow Univ. 1926; Head of Laboratory of Inst. of Biochemistry at USSR Acad. of Sciences 1935–53; Dir. of K.A. Timiriazev Inst. of Plant Physiology at USSR Acad. of Sciences 1952–; mem. USSR Acad. of Sciences 1953–; founded and Chief Ed. of periodical "Fiziologiia rastenii" 1954–. Decors.: Hero of Socialist Labour, 3 Orders of Lenin, 2 other orders, 1 foreign order, several medals.

Kursanov, Dmitrii Nikolaevich
Organic chemist; b. Ap. 1 (13), 1899, Moscow.
Grad. Moscow Univ. 1924; corresponding mem. USSR Acad. of Sciences 1953–; mem. staff Inst. of Organic Chemistry 1943–54; mem. staff Inst. of Elemental-Organic Compounds 1954–; concurrently mem. staff Inst. of Scientific Information of USSR Acad. of Sciences 1953–. Decors.: Lenin Prize 1963, 2 orders, several medals.

Kusenko, Ol'ga Iakovlevna
Ukrainian actress; b. N. 11, 1919, Kanev, Cherkassy Oblast.
People's Artist of the USSR (1967). Member of the CPSU since 1953. She graduated from the Kiev Institute of Theater Arts in 1941. During the Great Patriotic War of 1941–45 she acted at the front in theatrical brigades. She joined the I. Ia. Franko Ukrainian Theater in 1944. Her roles include Katria (*No Luck* by Staritskii), Olena (*The Predator, or The Spider* by Kropivnitskii), Anna (*Stolen Happiness* by Franko), Andromache (*Cassandra* by Lesia Ukrainka), Ol'ga and Vasilina (*Makar Dubrava* and *Snowball-Tree Grove* by Korneichuk), Melanka and Ingigerda (*Svichka's Wedding* and *Iaroslav the Wise* by Kocherga), Varia (*Cherry Orchard* by Chekhov), Eurydice (*Antigone* by Shakespeare).
She was a deputy to the sixth, seventh, and eighth convocations of the Supreme Soviet of the Ukrainian SSR. Decors.: State Prize of the USSR in 1951, Order of the Red Banner of Labor and various medals.

Kushekov, Unaibai
Party official; b. 1925.
Kazakh; mem. CPSU 1949–; grad. Alma-Ata Party Higher School; worked in a Sovkhoz 1941–43; served in Soviet Army 1943–48; Dep. Chair. of a DOSAAF Oblast Cttee. 1948–51; Instructor, Dept. Head of Gurev City Cttee. Kazakh CP, Instructor of Gurev Oblast Cttee. and East Kazakhstan Krai Cttee. Kazakh CP 1951–63; Dir. of "Baksaiskii" Sovkhoz, Head Agricultural Dept. of Gurev Oblast Cttee. 1st Sec. of a Raion Cttee. Kazakh CP 1963–70; Chair. of Exec. Cttee. Gurev Oblast Sov. of People's Deps. 1970–77; mem. Cen. Cttee. Kazakh CP 1971–; 1st Sec. of Gurev Oblast Cttee. Kazakh CP 1977–.

Kushpel, Vladimir Petrovich
State official.
Chair. of State Cttee. for Construction Turkmen SSR Je. 1982–.

Kuskevich, Ivan Vasil'evich
State official.
Mem. Auditing Comm. Moldavian CP 1961–63; Chair. of Beltsy City Exec. Cttee. of People's Deps. 1963; mem. Cen. Cttee. Moldavian CP 1963–76; Min. of Public Services Moldavian SSR 1966–70 and Ag. 1978–.

Kuskov, Elizar Ilich
Party official; b. 1918.
Mem. CPSU 1939–; grad. Party High School of Cen. Cttee. CPSU 1952; sec. of a village sov., dept. head of a raion paper in Moscow Oblast 1933–36; served in Sov. Army 1938–53; literary contributor, Dept. Head, Acting Sec. of paper "Za prochnyi mir, za narodnuiu demokratiiu" 1953–56; mem. Apparat Cen. Cttee. CPSU 1956–70; 1st Dep. Head of Int. Dept. at Cen. Cttee. CPSU Jl. 1970–; mem. Auditing Comm. CPSU 1971–76.

Kustov, Pavel Petrovich
Poet; b. 1906, Zavalnoe, Lipetsk Oblast; s. of a peasant.
Russian; grad. Dept. of Literature and Art of Moscow State Univ. 1930.
Publs.: "Severo-Vostok" (North-East) 1948, "Na prostorakh Rodiny" (In the Vast Motherland) 1951, "U nas na Severe" (With Us in the North) 1954, "Belomorie" (White Sea Region) 1963, and others.

Kutakhov, Pavel Stepanovich
Military official; b. 1914, Malokirsanovka, Rostov Oblast.
Russian; mem. CPSU 1942–; grad. Gen. Staff Acad. 1957; Chief Marshal of Sov. Air Force 1972–; served in Sov. Army 1935–; dep. commander of a flying squadron, participated in union of Western Ukraine and Belorussia with the Sov. Union, served in Sov.-Finnish War 1939–40; commander of a flying squadron, dep. commander of an airborne regiment, commander of an elite air combat regiment on Karelian Front 1941–45; command posts in Sov. Air Force 1945–67; 1st Dep. Commander-in-Chief 1967–69, Commander-in-Chief of USSR Air Force 1969–; USSR Dep. Min. of Defence 1969–; mem. Cen. Cttee. CPSU 1971–; Dep. to USSR Supr. Sov. 1970–. Decor.: 2 Orders of Lenin, Order of October Revolution, 5 Orders of the Red Banner, Aleksandr Nevskii Order, Order of the Patriotic War 1st Cl., 2 Orders of the Red Star, and others.

Kutateladze, Samson Semenovich
Thermophysicist; b. Jl. 18 (31), 1914, Petrograd.
Mem. CPSU 1943–; grad. College of Thermotechnology Leningrad 1932 and Ind. Inst. for Correspondence Courses Leningrad 1950; mem. staff Cen. Boiler Turbine Inst. 1932–58; served in Sov. Army 1941–45; mem. Siberian Dept. of USSR Acad. of Sciences 1958; co-founder and Dir. of Inst. of Thermophysics

1964–. Decors.: Order of Lenin, 2 other orders, several medals.

Kutateli, Aleksandr Nikolaevich (pseudonym of A. N. Kutateladze)
Georgian writer; b. Ag. 25 (S. 6), 1898, Kutaisi.
He studied in the department of Georgian literature at the University of Tbilisi from 1920 to 1925. His first published work was the drama *Khirse the Snake* (1924). He wrote two books of poetry (1937 and 1941) and a collection of short stories, *The Fighters* (1942). His major work is the novel tetralogy *Face to Face* (books 1–4, 1933–52; new Russian-language version, books 1–4, 1957–58), which depicts the struggle of the Georgian people to overthrow the counter-revolutionary government of the Mensheviks and to bring about the victory of the socialist revolution. Decors.: Order of the Red Banner of Labor and medals.
Works: Kut'at'eli, A. *Leksebi.* Tbilisi 1937. *Leksebi,* Tbilisi 1941. *Abesalom da eteri.* Tbilisi 1959. *Birisbir: Romani,* books 1–4. Tbilisi 1967. *T'xzsulebani 4 tomag.* Vol. 1 –. Tbilisi 1970.

Kutov, Nikolai Nikolaevich
Poet; b. 1917, Velikii Ustiug.
Russian; grad. Cheliabinsk Pedagogical Inst. 1943.
Publs.: "Zolotaia dolina" (The Golden Valley) 1950, "Novyi Krai" (New Country) 1951, "Utro" (Morning) 1952, "Vesennie zori" (Spring Stars) 1957, "Rasstavaniia i vstrechi" (Parting and Reunion) 1962, and others.

Kutsevol, Vasilii Stepanovich
State official; b. 1920; s. of a peasant.
Ukrainian; mem. CPSU 1947–; grad. Nikolaev Shipbuilding Inst. 1942; head of tech. information bureau of a plant, 1st sec. of a raion Komsomol cttee., chief mechanic of a trust, dept. head, chief engineer at various ind. enterprises 1943–52; dept. head with an oblast cttee. Ukrainian CP, Dept. Head with Cen. Cttee. Ukrainian CP, 1st Sec. of Lvov City Cttee. Ukrainian CP 1952–62; Sec. of Lvov Oblast Cttee. Ukrainian CP 1962–73 (for Ind. 1963–64); Dep. to 1962–67, mem. Presidium Supr. Sov. Ukrainian SSR 1963–67; Dep. to USSR Supr. Sov. 1966–; cand. mem. 1966–71, mem. Cen. Cttee. CPSU 1971–76; mem. Cen. Cttee. Ukrainian CP 1966–; Chair. of Cttee. for People's Control Ukrainian SSR N. 1973–. Decors.: Order of Lenin, Order of Red Banner of Labour.

Kutyrkin, Vladislav Georgievich
State official; b. 1931; s. of a white-collar worker.
Russian; mem. CPSU 1950–; grad. Frunze Inst. of Agriculture in Kishinev and Cen. Cttee. Party Higher School CPSU; statistician at a state farm, head of Cadre and Organization Dept., 1st Sec. of Komsomol Raion Cttee. in Kalarashty 1946–53; dept. head of Komsomol District Cttee. in Kagul, dep. head of the Polit. Dept., Sec. of the Party organization of a machine and tractor station, Raion Cttee. Sec. CP Moldavia in Chimishliya 1953–60; Chair. of Karl Marx state farm, 2nd, 1st Raion Cttee. Sec. CP Moldavia in Chimishliya, dep. Party organizer of Cen. Cttee. CP Moldavia in the territorial collective and state farm admin. in Kotovsk, Sec. of Party Cttee. of the Collective and State Farm Production Asscn. in Chimishliya, 1st Raion Cttee. Sec. CP Moldavia in Chimishliya 1960–70; mem. Cen. Cttee. CP Moldavia 1963–; Chair. of State. Cttee. on Prices Moldavian SSR 1970–81; Dep. Chair. of Council of Mins. and Chair. of GOSPLAN Moldavian SSR Ja. 1981–.

Kuusberg, Paul
Author and journalist; b. 1916, Tallinn.
Esthonian; mem. CPSU 1940–.
Publs.: "Steny" (The Walls) 1956, "Vtoroe 'ia' Enna Kalma" (The Second "I" of Enn Kalm) 1960, "Proisshestvie s Andresom Lapeteusom" (Incident With Andres Lapeteus) 1961, "Plamia pod peplom" (Flame Among the Ashes) 1962, and others.

Kuuzik, Tiit (Dietrich Ivanovich)
Opera singer; b. Ag. 29 (S. 9), 1911, Pärnu.
Esthonian; grad. Tallinn Conservatory 1938; baritone; soloist at Esthonian Opera and Ballet Theatre 1940–41 and 1944–; one of the most famous Sov. opera singers; teaching activity 1940–; Prof. at Tallinn Conservatory 1947–. Decors.: People's Artist of the USSR, State Prize of the USSR 1952, Order of Lenin, 2 other orders, several medals.

Kuzin, Aleksandr Mikhailovich
Biochemist and radiobiologist; b. My. 17 (30), 1906, Moscow.
Mem. CPSU 1946–; grad. Moscow Univ. 1929; mem. staff 1st Medical Inst. Moscow 1930–38; Prof. with Chair at 3rd Medical Inst. Moscow 1938–51; Dir., then Head of Dept. of Radiobiology at Inst. of Biophysics of USSR Acad. of Sciences 1952–57; corresponding mem. USSR Acad. of Sciences 1960–; Chief Ed. of periodical "Biofizika" until 1961, of "Radiobiologiia" 1961–; founder of structure-metabolic hypothesis in radiobiology. Decors.: 2 orders, several medals.

Kuzin, Vladimir Semenovich
Athlete; b. Jl. 15, 1930, Lampozhnia, Arkhangel'sk Oblast.
Honored Master of Sports (1954), Honored Coach of the USSR (1970), officer in the Soviet Army, and Candidate of Biological Sciences (1972).
He was the first Soviet athlete to become a world champion in cross country skiing (1954; 30 km and 50 km). He was a champion at the seventh Winter Olympic Games (1956, at Cortina d'Ampezzo, in the cross country relay) and a USSR champion (1953 and 1958; 15 km). Since 1959, he has served as the head ski coach of the armed forces of the USSR. Decors.: Order of Lenin and various medals.

Kuz'kin, Viktor Grigor'evich
Athlete; b. Jl. 6, 1940, Moscow.
Hockey player. Honored Master of Sports (1963); officer in the Soviet Army. Member of the CPSU since 1966.
He has won championships a number of times in the USSR (1961, 1963–66, 1968, 1970–73), in world competitions (1963–69, 1971), in European competitions (1963–69), and in the Olympics (1964, 1968, 1972). He has played on the team of the Central Sports Club of the Army, which won the USSR Cup in 1961 and in 1966–69 and the European Championship in ice hockey in 1969–72. He has been a member of the USSR national team since 1962. Decors.: two orders and several medals.

Kuzmin, Aleksandr Trifonovich
Party official; b. 1918.
Belorussian; mem. CPSU 1942; grad. Belorussian State Univ.; served in the Soviet Army 1937–44; party posts 1944–58; Sec. of Minsk Oblast Cttee. Belorussian CP 1958–62; Head of Ideology Dept. for Ind. and Construction of Bureau Cen. Cttee. Belorussian CP, Head of Ideology Dept. at Cen. Cttee. Belorussian CP 1963–66; mem. Cen. Cttee. Belorussian CP 1966–; Head of Dept. of Propaganda and Agitation at Cen. Cttee. Belorussian CP 1966–71; mem. Bureau and Sec. of Cen. Cttee. Belorussian CP F. 1971–.

Kuzmin, Leonid Filippovich
Diplomat.
USSR Amb. to Peru F. 1975–.

Kuzmin, Viktor Petrovich
Trade union official.
Sec. of Samarkand Oblast Cttee. for Ind., Sec. and 2nd Sec. of Surkhan-Daria Oblast Cttee. Uzbek CP 1963–66; mem. Cen. Cttee. Uzbek CP 1966–76, cand. mem. 1976–; Sec. of Uzbek Republican Trade Union Council Mr. 1972–.

Kuz'mina, Elena Aleksandrovna
Motion-picture actress; b. F. 4 (17), 1909, Tbilisi.
People's Artist of the RSFSR (1950).
She graduated in 1930 from the Factory of the Slapstick Actor, a Leningrad cinema school later reorganized into the Institute of Stage Arts. She made her film debut as the Communard Louise in *The New Babylon* (1929). She next appeared as the schoolteacher in *Alone* (1931). A good character actress, she often created subtle psychological portraits.
Her best roles were in motion pictures directed by M. I. Romm – for example, Anna in *The Dream* (1943), Tania in *Prisoner*

No. 217 (1945), Jessy in *The Russian Question* (1948), and the Soviet intelligence officer Maria-Marta in *Secret Mission* (1950). She played lyrical and comedy roles in the films of other directors. Man'ka in *Outskirts* (1933), Masha in *By the Bluest of Seas* (1936), and Mariia Konstantinovna in an adaptation of Chekhov's *The Duel* (1961). Decors.: State Prize of the USSR in 1946, 1948, and 1951.

Kuzminskis, Ionas Mikolo
Graphic artist; b. D. 1 (14), 1906, Betygala, in present-day Raseiniai Raion, Lithuanian SSR.
People's Artist of the USSR (1965); corresponding member of the Academy of Arts of the USSR (1970). Became a member of the CPSU in 1951.
From 1927 to 1933, he attended the Kaunas School of Art. He became the chairman of the administrative board of the Artists' Union of the Lithuanian SSR in 1958. His work is characterized by a creative use of the traditions of Lithuanian popular prints, by tense tonal relations, and by the diversity and suppleness of expressive brushwork (the series *Vilnius*, engravings on wood, linoleum, and imitation leather, 1942–57; *On the Motifs of Lithuanian Folk and Revolutionary Songs*, engravings on wood, begun in 1960, awarded the State Prize of the Lithuanian SSR in 1967). In 1944, he began teaching at the Art Institute of the Lithuanian SSR in Vilnius, receiving a professorship there in 1957. A member of the Central Committee of the Lithuanian Communist Party (since 1963). Decors.: Order of Lenin, the Order of the October Revolution, two other orders, and a medal.

Kuznetsov, Anatolii Vasil'evich
Author; b. 1929, Kiev.
Russian; mem. CPSU 1955; grad. Gorkii Inst. of Literature Moscow 1960; carpenter and bulldozer operator at Kakhovka Hydroelectric Plant, contributed to factory newspaper, construction worker during construction of Irkutsk Hydroelectric Plant; first works publ. in newspaper "Pionerskaia pravda" 1945.
Publs.: "Prodolzhenie legendy. Zapiski molodogo cheloveka" (Continuation of a Legend. Notes of a Young Man) 1957, "Selenga" 1961, "Bienie zhizni" (The Pulsation of Life) 1961, "U sebia doma" (At Home) 1964, and others.

Kuznetsov, Iurii Alekseevich
Geologist; b. Ap. 19, 1903, Nikolsk, Vologda Oblast.
Grad. Tomsk Univ. 1924; lecturer 1930–, Prof. at Polytech. Inst. Tomsk 1938–; took part in geological prospecting in Western Siberia; mem. staff Inst. of Geology and Geophysics at Siberian branch of USSR Acad. of Sciences 1958–; mem. USSR Acad. of Sciences 1966–; discovered natural resources deposits, among them gold and fire-proof clay. Decors.: 2 Orders of Lenin, 2 other orders, several medals.

Kuznetsov, Iurii Ivanovich
Diplomat.
Mem. Diplomatic Service 1952–; at USSR Emb. in Vietnam and Laos until 1965; Head, South Asian Countries Dept., USSR Min. of Foreign Affairs 1970–78; USSR Amb. to Thailand Je. 1978–.

Kuznetsov, Ivan Maksimovich
Party official; b. 1928.
Russian, mem. CPSU 1956–, grad. Saratov Automobile Road Inst. 1951 and CPSU Cen. Cttee. Party Higher School 1965; senior technological engineer, dep. shop superintendent, shop superintendent, sec. of party cttee. in New State Ball-Bearing Plant in Kuibyshev, 1st Sec. of Kuibyshev Raion Cttee. CPSU 1951–66; dep. chief, chief of a dept. of Kuibyshev Oblast Cttee. CPSU, instructor in a dept. of Cen. Cttee. CPSU 1966–72; 2nd Sec. of Ulianovsk Oblast Cttee. CPSU 1972–77, 1st Sec. Jl. 1977–; Dep. to USSR Supr. Sov. 1974–; mem. Cen. Cttee. CPSU 1981–.

Kuznetsov, Mikhail Artem'evich
Russian actor; b. F. 25, 1918, Noginsk.
People's Artist of the RSFSR (1965).
He graduated in 1941 from the drama section of the K. S. Stanislavsky State Opera and Drama Studio in Moscow. He made his motion-picture debut while still a student. In *Mashen'ka* (1942) he played the chauffeur Aleksei. His best roles were Fedor Basmanov in *Ivan the Terrible* (parts 1 and 2, 1945 and 1958), Doctor Kolesov in *In the Name of Life* (1947), the soldier Skobelev in *Taras Shevchenko* (1951), Chizhik in *Sailor Chizhik* (1956), Political Officer Kovalenko in *An Extraordinary Occurrence* (parts 1–2, 1959), Coach Anton Lutenko in *The Silver Coach* (1963), and Colonel Lartsev in *Game Without Rules* (1965). Decors.: State Prize of the USSR (1952) and the Order of the Badge of Honor.

Kuznetsov, Mikhail Prokop'evich
Russian actor; b. S. 2 (15), 1913, Petrovsk-Zabaikal'skii, now in Chita Oblast.
People's Artist of the USSR (1971).
He began his stage career in amateur productions. He was invited to join the Irkutsk Young People's Theater in 1932. From 1933 to 1945 he worked with the Theater of Working Youth and the V. P. Chkalov Drama Theater in Gorkii. In 1945 he joined the V. P. Chkalov Drama Theater in Tomsk. His main roles were Mitrofan in Fonvizin's *The Minor*, Shvandia in Tenev's *Liubov' Iàrovaia*, Falstaff in Shakespeare's *The Merry Wives of Windsor*, Sergei Tiulenin in an adaptation of Fadeev's *Young Guard*, Khlestakov in Gogol's *The Inspector-General*, Raspliuev in Sukhovo-Kobylin's *Krechinskii's Wedding*, and Luka in Gorkii's *Lower Depths*.
He joined the M. Iu. Lermontov Theater in Stavropol' in 1951. He played Gennadii Dubravin in Romashov's *Bridge in Flames*, Schastlivtsev in Ostrovskii's *The Forest*, Shchukar' in an adaptation of Sholokhov's *Virgin Soil Upturned*, and Skvorets in Shtok's *Leningrad Boulevard*. He portrayed V. I. Lenin in a number of plays: *Youth of the Fathers* by Gorbatov (1944), *The Kremlin Chimes* by Pogodin (1946, 1967), *The Family* by Popov (1950), *In the Name of the Revolution* by Shatrov (1958), *The Third Pathetique* (1959) and *The Flowers Are Alive* (1962) by Pogodin, and *The Sixth of July* by Shatrov (1965). Decors.: Order of Lenin and a number of medals.

Kuznetsov, Mikhail, Vasil'evich
Military official; b. O. 25 (N. 7), 1913, Agarino, in present-day Serpukhov Raion, Moscow Oblast; s. of a worker.
Major general of aviation (1959), twice Hero of the Soviet Union (S. 8, 1943, and Je. 27, 1945). Became a member of the CPSU in 1932.
He entered the Soviet Army in 1933 and graduated from the military school for the naval air force in 1934 and from the Air Force Academy in 1951. In the Great Patriotic War (1941–45), he fought on the Leningrad, Kalinin, Western, Southwestern, and Third and First Ukrainian fronts in the posts of squadron commander, regimental navigator (1941–42), and commander of the 814th Fighter Air Regiment (1942–45). He made 345 combat flights, engaged in 72 air battles, and shot down 28 enemy aircraft, including six assits. He has held high posts in the air force since the war. Decors.: Order of Lenin, four Orders of the Red Banner, the Order of the Red Banner of Labor, the Order of Bogdan Khmel'nitskii Second Class, two Orders of the Red Star, and medals.

Kuznetsov, Nikolai Dmitrievich
Scientist; b. Je. 10 (23), 1911, Aktiubinsk.
Specializing in jet aircraft engines; corresponding member of the Academy of Sciences of the USSR (1968) and Hero of Socialist Labor (1957). Member of the CPSU since 1939.
He graduated from the N. E. Zhukovskii Air Force Academy in 1938. From 1943 to 1946 he was a deputy chief designer and from 1946 to 1956 chief designer. He has been a general designer since 1956. Engines were developed under his leadership for the Tu-114, Tu-154, An-22, and Il-62 aircraft and for the Tu-144 supersonic passenger aircraft. He has been a deputy to the sixth to eighth convocations of the Supreme Soviet of the RSFSR. Decors.: Lenin Prize in 1956 and has been awarded three Orders of Lenin, four other orders, and medals.

Kuznetsov, Pavel Stepanovich
Diplomat; b. 1918.
Mem. Diplomatic Service 1948–; Emb. Counsellor at USSR Emb. in Indonesia 1957–60; leading posts in USSR Min. of Foreign Affairs 1960–63 and 1970–72; Emb. Counsellor 1963–68 and Emb. Counsellor-Envoy at USSR Emb. in Yugoslavia 1968–70; USSR Amb. to Indonesia Je. 1972–Je. 1976.

Kuznetsov, Sergei Ivanovich
Microbiologist; b. N.4 (17), 1900, Poretskoe, Vladimir Oblast.
Grad. Moscow Univ. 1923; Head of Laboratory at Limnological Station in Kosino 1931–41; mem. staff Waste Water Purification Combine in Lubertsy 1941–46; mem. staff Inst. of Microbiology at USSR Acad. of Sciences 1942–. Decors.: Order of Lenin, Order of Red Banner of Labour, medals.

Kuznetsov, Valerii Alekseevich
Geologist and Geographer; b. Ap.12, 1906, Nikolsk, now Vologda Oblast.
Grad. Tomsk Polytech. Inst. 1932; mem. USSR Acad. of Sciences 1970–; worked in various geological institutions in Western Siberia 1932–45; mem. staff Inst. of Mining Geology at Western Siberian Branch of USSR Acad. of Sciences 1945–58; mem. staff Inst. of Geology and Geophysics at Siberian Branch of USSR Acad. of Sciences 1958–. Decors.: 2 orders and various medals.

Kuznetsov, Vasilii Vasilevich, Cand. of Tech. Sc.
State official; b. 1901; s. of a peasant.
Russian; mem. CPSU 1927–; grad. Polytech. Inst. Leningrad 1926; student 1921–26; engineer at Makeevka Metal Plant 1927–31; studied special steels abroad 1931–33; worked for "Elektrostal'" Plant 1934–37; engineer, Chief Engineer of Main Board for Special Steels and Ferro-Alloys with Min. of Ferrous Metallurgy 1937–40; Dep. Chair. of USSR GOSPLAN 1940–44; dep. mem. State Defence Cttee. on Metallurgy 1941–45; Chair. of Ferrous Metallurgy Worker's Union Cen. Cttee., Chair. of All-Union Cen. Trade Union Council 1944–53; Dep. to USSR Supr. Sov. 1946–; mem. 1952–, mem. Presidium Cen. Cttee. CPSU 1952–Mr. 1953; USSR Amb. to China 1953; USSR Dep. Min. 1953–55, 1st Dep. Min. of Foreign Affairs 1955–77; 1st Dep. Chair. Presidium of USSR Supreme Soviet 1977–; Cand. mem. of the Politburo Cen. Cttee. CPSU 1977–. Decors.: 3 Orders of Lenin, Order of Red Banner of Labour, Order of Red Star, Badge of Honour, medals.

Kuznetsov, Viacheslav Nikolaevich
Poet; b. 1932, Dzharkul, Kazakhstan.
Mem. CPSU 1963–; grad. Mozhaiskii Aviation Acad. Leningrad 1956 and Higher Literary Courses at USSR Writers' Union 1960; engineer in Arctic; first works publ. 1949.
Publs.: "Nasledstvo" (The Inheritance) 1955, "Surovyi vozdukh" (Harsh Air) 1957, "Studenyi bereg" (Cold Shore) 1958, "U vysokikh shirot" (At High Altitudes) 1958, "Kedry na skalakh" (Cedars on Rocks) 1961, "Krov' ottsov" (The Blood of Fathers) 1961, "Chuvstvo zemli" (The Sensation of Earth) 1962, "Ia ostaius' romantikom" (I Remain a Romantic) 1963, "Mechta" (A Dream) 1954, "Zvezdy ne gasnut" (The Stars do not Fade) 1960, "Dobrota" (Kindness) 1964, and others.

Kuznetsov, Viktor Ivanovich
Mechanics scientist; b. Ap.4 (17), 1913, Moscow.
Mem. CPSU 1932–; grad. Ind. Inst. Leningrad 1938; specializes in applied mechanics and automatic distribution systems; mem. USSR Acad. of Sciences 1968–. Decors.: State Prize of the USSR 1943, 1946, Order of Lenin, Order of the October Revolution, Order of Red Banner of Labour, medals.

Kuznetsov, Vladimir Nikolaevich
Diplomat; b. 1916.
Mem. Diplomatic Service 1946–; Counselor, USSR Embassy in Netherlands 1956–60, in Indonesia 1961–67; mem. Apparat USSR Min. of Foreign Affairs 1960–61, 1967–68 and 1973–80; USSR Amb. to Malaysia 1968–73, to Burma Ag. 1980–.

Kuznetsov, Vladimir Vasil'evich
Painter; b. 1924, Moscow.
Grad. Kharkov Art Inst. 1954, pupil of F. Fodorovskii, P. Kotov and L. Chernov; teacher at Kharkov Inst. for Arts and Crafts; mem. CPSU 1953–.

Kuznetsova, Irina Davidovna
State official.
Min. of Food Ind. Latvian SSR O. 1965–; mem. Auditing Comm. Latvian CP 1966–76, mem. Cen. Cttee. 1976–; Dep. to Supr. Sov. Latvian SSR 1967–.

Kuznetsova, Svetlana Aleksandrovna
Poet; b. 1934, Irkutsk.
Russian; first works publ. 1956.
Publs.: "Protaliny" (Melted Spots in the Snow) 1962, "Enisei", "Kholodnyi veter" (Cold Wind), and others.

Kvachadze, Zinaida Arsenevna
Trade union official.
Mem. CPSU 1942–; mem. Auditing Comm. 1954–65, mem. Cen. Cttee. Georgian CP 1956–; until 1955 1st Sec. of Kirov Raion Cttee. Georgian CP in Tbilisi; Dep. to Supr. Sov. Georgian SSR 1955–; Sec. of Tbilisi City Cttee. Georgian CP 1955–56; Sec. and mem. Bureau of Cen. Cttee. Georgian CP 1956–57; Min. of Trade 1957–59, Dep. Chair. of Council of Mins. Georgian SSR 1957–62; Sec. of Presidium Supr. Sov. Georgian SSR D. 1962–76; Sec. and mem. Presidium of Georgian Republican Trade Union Council Je. 1976–.

Kvasnikov, Evgenii Ivanovich, Dr. of Biol. Sc.
Microbiologist; b. 1906.
Grad. Inst. of Public Educ. Kiev 1929; Dr. of Biol. Sc. 1960; Prof. 1961; teaching work 1929–41; served in Sov. Army 1941–45; mem. staff Inst. of Botany at Uzbek Acad. of Sciences 1945–60; mem. staff Inst. of Microbiology and Virology at Ukrainian Acad. of Sciences 1961–, Section Chief 1962–; corresponding mem. Ukrainian Acad. of Sciences 1968–. Decor.: State Prize of the USSR 1971.

Kvasnitskii, Aleksei Vladimirovich, Dr. of Biol. Sc.
Animal physiologist; b. F.24, 1900, Lysaia Gora, now Nikolaev Oblast; s. of a peasant.
Mem. CPSU 1941–; grad. of Agricultural Inst. in Kamianets-Podolskii 1925; Dr. of Biol. Sc. 1940; Prof. 1941; mem. staff Agricultural College Vinnitsa 1926–30; mem. research staff 1931–41, Head of Laboratory for Physiology of Animal Reproduction at Inst. for Research on Pig-Breeding Poltava 1944–; Prof. Chair of Physiology of Agricultural Animals at Agricultural Inst. Poltava 1936–; mem. Ukrainian Acad. of Sciences 1951–; mem. former Ukrainian Acad. of Agricultural Sciences 1956–62. Decors.: Merited Scientist of the Ukrainian SSR 1960, Hero of Socialist Labour 1966, 2 Orders of Lenin, 2 Orders "Honorary Badge", medals.
Publs.: author of more than 125 scientific works on the physiology of agricultural animals; "Vyrashchivanie porosiat" 1947, "Nove u fiziologii rozmnozhennia tvaryn" 1950, "Voprosy fiziologii pishchevareniia u svinei" 1951, "Primenenie ucheniia I. P. Pavlova v zhivotnovodstve" 1954, "Iskusstvennoe osemenenie svinei" 1961.

Kvell, Rein Voldemarovich
State official.
Chair. State Cttee. for Construction Esthonian SSR Mr. 1980–; cand. mem. Cen. Cttee. Esthonian CP 1981–.

Kvernadze, Teimuraz Veniaminovich
State official; b. 1923.
Georgian; mem. CPSU 1955–; grad. Georgian Polytechnical Inst. 1948; worked in Donetsk Basin coal mines: assistant to chief engineer of a mine, chief of a strip, Dep. Chief Engineer, Chief Engineer of a mine, of a trust assoc., Dep. Technical Dir. on questions of long-term devt. of mining, Dir. of Capital Construction for the Makeevugol' production assoc. 1948–80; 1st Dep. Chair. 1980–82, Chair. of State Cttee. Georgian SSR for Supervision of Safe Working Practices in Ind. and for Mine Safety Mr. 1982–. Decors.: Order of the Red Banner of Labor, *Znak pocheta* Order.

Kydykeeva, Baken
Actress; b. O.20, 1923, Oktiabr, Kirghizia.
Decors.: People's Artist of the USSR 1970, State Prize of the Kirghiz SSR 1970, Order of Lenin, medals.

Labakhua, Zurab Akakievich
Trade union official.
Sec. and mem. Presidium of Georgian Republican Trade Union Council Ja. 1982–.

Labenok, Leonid Vasil'evich
Painter; b. 1927, Leningrad.
Mem. CPSU 1953–; grad. Leningrad Art High School 1956. Decor.: Merited Artist of the Ukrainian SSR 1973.

Ladynina, Marina Alekseevna
Actress; b. Je.11 (24) 1908, Achinsk, Krasnoiarsk Krai.
Motion picture actress. Decors.: People's Artist of the USSR, State Prize of the USSR 1941, 1942, 1946, 1948, 1951, Order of Red Banner of Labour.

Ladyzhenskaia, Olga Aleksandrovna, Dr. of Phy.-Math. Sc.
Mathematician; b. 1922.
Grad. Moscow Univ. 1947; mem. staff Leningrad Univ. 1949–; Dr. of Phy.-Math. Sc. 1953; Prof. 1955; mem. staff Leningrad Branch of Inst. of Mathematics at USSR Acad. of Sciences 1954–; Head of Laboratory for Mathematical Physics 1962–. Decor.: State Prize of the USSR 1969.

Lafiizov, Dzhanobidin
State official.
Mem. Auditing Comm. Tadzhik CP 1971–; Min. of Finance Tadzhik SSR Ja. 1973–.

Lagin, Lazar' Iosifovich
Russian writer; b. N.21 (D.4), 1903, Vitebsk.
Member of the CPSU from 1920.
His poems and satirical articles were first printed in 1921 in the Komsomol press. He worked for several years for *Pravda* and *Krokodil*. His fairy-tale novella *Old Man Khottabych* (1938) was popular. The novels *Patent "A V"* (1947), *The Island of Disillusionment* (1951), *Ataviia Proksima* (1956), and *The Eroded Archipelago* (1963) manage to combine social pamphleteering with adventure and science-fiction genres. The novel *The Blue Man* (1966) tells about the succession of generations of revolutionaries and about modern and prerevolutionary Moscow. He wrote about the Great Patriotic War (1941–45) in such works as the novella *Battleship "Aniuta"* (1945).
His other works include the lampoons *Major Vell End"iu* (1962) and *The Blond Beast* (1963) and fairy tales for children and adults. His books have been translated into many foreign languages and languages of the peoples of the USSR. Decors.: Order of the Patriotic War Second Class and various medals.

Lagir, Mikhail Ivanovich
State official; b. 1921, Razhava, Minsk Oblast; s. of a peasant.
Mem. CPSU 1942–; grad. Mozyr Pedagogic College 1939 and Party High School of CPSU Cen. Cttee. 1953; teacher 1939–40; served in Sov. Army 1940–46; Sec. of Mozyr City Cttee. Belorussian Komsomol, Sec. of Polesian and Gomel Oblast Cttees. Belorussian Komsomol 1946–52; Inspector at Cen. Cttee. Belorussian CP, 1st Sec. of Lagoisk Raion Cttee. Belorussian CP, Dep. Head, Head of Dept. of Party Organs at Cen. Cttee. Belorussian CP 1952–60; cand. mem. 1956–60, mem. Cen. Cttee. Belorussian CP 1960–; Sec., 2nd Sec. of Minsk Oblast Cttee. Belorussian CP 1960–68 (for Ind. 1963–64); Chair. of Cttee. of People's Control Belorussian SSR O. 1968–; cand. mem. 1971–76, mem. Bureau Cen. Cttee. Belorussian CP 1976–. Decors.: 3 different orders, medals.

Laivin, Vladimir Ivanovich
State official.
Mem. Cen. Cttee. Latvian CP 1966–; Min. of Justice Latvian SSR Mr. 1980–.

Lakshin, Vladimir Iakovlevich
Literary historian; b. 1933, Moscow.
Russian; mem. CPSU 1966–; grad. Faculty of Philology of Moscow Univ. 1955; first works publ. 1954.
Publs.: "Tolstoi i Chekhov" (Tolstoi and Chechov) 1963, article "Ivan Denisovich, ego druzia i nedrugi" (Ivan Denisovich, his Friends and Enemies) 1964, "Pisatel', chitatel', kritik" (Authors, Readers, Critics) 1965, and others.

Lamutskii, Platon (real name: Stepanov, Platon Afanasevich)
Poet; b. 1920, Sebian-Kelia, Yakut ASSR. Event; mem. CPSU 1946–; grad. Teachers' Training Inst.; first works publ. 1958. Publs.: "Stikhi" (Poems) 1961, and others.

Laniauskas, Ionas Iuozovich
State official.
Chair. of State Cttee. for the Supply of Petroleum Products Lithuanian SSR Ag. 1980–; mem. Auditing Comm. Lithuanian CP 1981–.

Lankauskas, Romualdas
Author; b. 1932, Klaipeda, Lithuania.
Attended Lithuanian State Univ.
Publs.: "V dalekom plavanii" (Long Sea Voyage) 1954, "Bluzhdaiushchii pesok" (Wandering Sand) 1960, "Kogda umolkaet truba" (When the Trumpet is Silent) 1961, "Tretia ten'" (The Third Shadow) 1964, "Most v more" (A Bridge in the Sea) 1963, "V seredine bolshogo polia ..." (In the Middle of a Large Field ...) 1962, and others.

Lanskoi, Mark Zosimovich
Author; b. 1909, Nikolaev.
Russian; mem. CPSU 1943–; first works publ. 1928.
Publs.: "Russkii shkolnik iz derevni Komarovo" (A Russian Schoolboy From the Village Komarovo) 1942, "Doroga na Leningrad" (The Way to Leningrad) 1943, "Nezrimyi front" (Invisible Front) 1956, "Prikliucheniia bez puteshestviia" (Adventure Without Travelling) 1958, film scripts "Adygeiskaia Avtonomnaia Oblast" (Adygen Autonomous Oblast) 1951, "Zvezdy sluzhat liudam" (The Stars Serve Men) 1956, and others.

Laos, Ants Iurevich
State official.
Min. of Food Ind. Esthonian SSR D. 1979–; cand. mem. Cen. Cttee. Esthonian CP 1981–.

Lapcheva, Valentina Fedorovna
Geophysicist and meteorologist; b. 1932.
Grad. Odessa Hydrometeorological Inst.; mem. staff 1957–, Head of Laboratory of Geophysical Alpine Inst. 1968–. Decor.: State Prize of the USSR 1969.

Lapin, Konstantin Kirillovich
Author; b. 1914, Moscow.
Russian; first works publ. 1939.
Publs.: "Voennyi korrespondent" (War Correspondent) 1962, "Serdtse serzhanta" (A Sergeant's Heart) 1959, "Na Volge shirokoi" (On the Wide Volga) 1951, "Tri poezdki v Zhiguli" (Three Journeys to Zhiguli) 1955, "Pobeda na Volge" (Victory on the Volga) 1957, "Prostaia istoriia" (A Simple Story) 1958, "Vot ona kakaia, liubov'" (This is What Love is Like) 1960, "Razgovor o liubvi i sem'e" (A Talk About Family and Love) 1957, and others.

Lapin, Sergei Georgievich
State official; b. 1912.
Mem. CPSU 1939–; completed 2 terms at Inst. of History and Linguistics Leningrad 1932 and grad. Party Higher School of Cen. Cttee. CPSU 1942; journalist on ed. board of newspapers in Leningrad and Leningrad Oblast 1932–40; worked for All-Union Cen. Cttee. CP (B) 1942–44; worked for Cttee. on Radio and Broadcasting of USSR Council of Mins. 1944–53; mem. Diplomatic Service 1954–67; Head of 3rd European Dept. in USSR Min. of Foreign Affairs 1955–56; USSR Amb. to Austria 1956–60; RSFSR Min. of Foreign Affairs 1960–62; USSR Dep. Min. of Foreign Affairs 1962–65; USSR Amb. to China 1965–67; mem. Cen. Cttee. CPSU 1966–; Dir.-Gen. of TASS (Sov. news agency) 1967–70; Chair. of USSR State Cttee. for Television and Broadcasting Ap. 1970–. Decor.: Order of Red Banner of Labour.

Lapin, Viktor Ivanovich
Painter and graphic artist; b. 1923, Zolotarevka, Sumy Oblast.
Mem. CPSU 1951–; grad. Kharkov Art Inst. 1954, pupil of G. Tomenko.

Laptev, Konstantin Antonovich
Singer; b. O. 21 (N. 3), 1904, Kiev.
Mem. CPSU 1931–; grad. Kiev Conservatory 1929; baritone, soloist at Opera and Ballet Theatre Odessa 1930–41, Kiev 1941–52 and Leningrad 1952–65. Decors.: People's Artist of the USSR, 2 orders.

Laptsui, Leonid Vasil'evich
Author; b. 1932, Iamalo-Nenets Okrug; s. of a deer breeder.
Nenets; mem. CPSU 1954–; grad. Dept. of Journalism at Higher Courses of Cen. Cttee. USSR Komsomol; worked as shepherd, fisherman, in a fish canning factory and in Komsomol Cttee. in Iamalo-Nenets Okrug.
Publs.: "Tsveti, moi Iamal!" (Blossom, My Iamal!) 1960, "Rasskazy o Iamale" (Stories About Iamal) 1962, "Chelovek rodilsia" (A Man Was Born) 1961, "Kamen' s nadpisiu" (Stone With Inscription) 1963, and others.

Lashchenko, Petr Nikolaevich
Military official; b. D. 6 (19), 1910, Turia, Chernigov Oblast.
Mem. CPSU 1931–; grad. Frunze Mil. Acad. 1940 and Gen. Staff Mil. Acad. 1951; Gen. of the Army; Commander during German-Sov. War; seriously wounded Jl. 1944; 1st Dep. Commander of Kiev and Transcarpathian Mil. District 1959–64; Commander of Transcarpathian Mil. District 1964–68; 1st Dep. Commander-in-Chief of the Army 1968–76. Decors.: 4 Orders of Lenin, 3 Orders of Red Banner, Suvorov and Bogdan Khmelnitskii Order, 2nd Cl., Order of Patriotic War 1st Cl., Order of Red Star, Order of Red Banner of Labour, Hero of the Soviet Union, medals.

Lashkarashvili, Tamara Vasilevna
State official; b. 1916.
Georgian; mem. CPSU 1946; grad. State Univ. Tbilisi; tech. instructor, with railways, student 1932–37; teacher, head of teaching dept., dir. of a middle school 1937–51; sec. of an oblast cttee. Georgian CP, dept. head of an oblast Cttee. Georgian CP, Sec. of Tbilisi City Cttee. Georgian CP, head of school dept. at Cen. Cttee. Georgian CP 1951–60; Dep. to Supr. Sov. Georgian SSR 1955–; mem. Cen. Cttee. Georgian CP 1956–; Min. of Educ. Georgian SSR Ap. 1960–76; Sec. of Presidium Supr. Sov. Georgian SSR Je. 1976–.

Lashkarev, Vadim Evgenevich, Dr. of Phy.-Math. Sc.
Physicist; b. O. 7, 1903, Kiev.
Grad. Kiev. Inst. of Public Educ. 1924–; mem. Ukrainian Acad. of Sciences 1945–; Dr. of Phy.-Math. Sc. and Prof. 1935; mem. staff Kiev Polytech. Inst. 1925–30, Leningrad Physical-Tech. Inst. 1930–35; Lecturer at Arkhangelsk Medical Inst. 1935–39; conducted research on semiconductors at Inst. of Physics at Ukrainian Acad. of Sciences 1939–60; concurrently at Shevchenko State Univ. Kiev 1939–57; Dir. of Inst. for Semiconductors at Ukrainian Acad. of Sciences 1960–. Decors.: Order "Honorary Badge", medals.
Publs.: author of more than 80 scientific works; "Diffuziia nositelei toka v poluprovodnikakh so smeshannoi provodimostiu" in "Izvestiia Akademii Nauk SSR, Seriia fizicheskaia" No. 2, 1952, "Kompleksnoe issledovanie kinetiki protsessov rekombinatsii i infrakrasnogo gasheniia fototoka v sulfide kadmiia" in "Fizika tverdogo tela" vol. 7, 1965.

Laskin, Boris Savelevich
Author; b. 1914, Orsha, Mogilev Oblast.
Russian; grad. State Film Inst. 1935; first works publ. 1938.
Publs.: "Pesni" (Songs) 1943, wrote sketches, variety reviews, film scripts and plays, "Voennyi tovarishch" (War Comrade), "Nebesnoe sozdanie" (Heavenly Being) 1944, "Vremia liubit'" (Time to Love) 1960, "Drug detstva" (Childhood Friend) 1955 "Vesennie golosa" (Spring Voices) 1955, "Karnavalnaia noch'" (Carnival Night) 1956, "Istoriia bolezni" (History of an Illness) 1965, and others.

Laskin, Evgenii Nikolaevich
State official; b. 1922.
Min. of Timber-Processing Ind. Latvian SSR 1972–75; Dep. to Supr. Sov. Latvian SSR 1972–; Chair. of State Cttee. for the Utilization of Labour Reserves with Council of Mins. Latvian SSR F. 1975–76; mem. Auditing Comm. Latvian CP 1976–81; Chair. of State Cttee. for Supervision of Safe Working Practices in Industry and Mines Latvian SSR O. 1981–.

Lavrenko, Evgenii Mikhailovich
Geobotanist and botanic geographer; b. F. 11 (23), 1900, Chuguev, Kharkov Oblast.
Grad. Dept. of Natural Sciences of Kharkov Univ. 1922; mem. USSR Acad. of Sciences 1968–; mem. staff Botanical Gardens Kharkov 1921–28, Inst. of Wildlife and Plant Conservation Kharkov 1926–29, Agricultural Inst. Kharkov 1929–; Prof. 1931–; mem. staff Botanical Inst. of USSR Acad. of Sciences 1934–; Pres. of Botanical Union Soc. 1963–73; main research work on vegetation of prairies and deserts of the USSR, Mongolia and a part of North Korea, vegetation zones of prairies and desert areas of Eurasia and North Africa, cartography of USSR flora and interdisciplinary studies in biological and geographical phenomena. Decors.: 2 V. L. Komarov Prizes (1950 and 1963), 2 Orders of Lenin, Order of Patriotic War 2nd Cl., medals.

Lavrent'ev, Mikhail Mikhailovich
Mathematician; b. Jl. 21, 1932, Moscow. Corresponding member of the Academy of Sciences of the USSR (1968). Son of M. A. Lavrent'ev.
He graduated from Moscow State University in 1954 and has worked in the Siberian Division of the Academy of Sciences of the USSR since 1957. He is one of the creators of the theory of incorrect problems of mathematical physics. He also produced a number of important results on multidimensional inverse problems for differential equations that arise in the study of the structure of the earth on the basis of observed geophysical fields.
Works: "O zadache Koshi dlia uravneniia Lapalsa." *Izv. AN SSSR: Ser. matematicheskaia,* 1956, vol. 20. *O nekotorykh nekorrektnykh zadachakh matematicheskoi fiziki.* Novosibirsk 1962. *Mnogomernye obratnye zadachi dlia differentsial'nykh uravnenii.* Novosibirsk 1969. (With V. G. Romanov and V. G. Vasil'ev.)

Lavrentev, Viktor Vladimirovich
Dramatist; b. F. 14 (27), 1914, Tomsk.
Mem. CPSU 1942–; jurist by profession; first poems publ. in front newspapers; Chief Ed. of journal "Teatr" 1969–72. Decors.: Order of Red Banner of Labour, medals.
Publs.: plays "Kriazhevy" (The Kriazhevs) 1953, "Ivan Budantsev" 1956, "Posledniaia legenda" (The Last Legend) 1958, "Chti ottsa svoego" (Honour Thy Father) 1964, "Chelovek i globus" (Man and the Globe) 1967, "Glava sem'i" (Head of the family) 1970, and others.

Lavrinenkov, Vladimir Dmitrievich
Military official; b. My. 17, 1919, Ptakhino, now in Pochinok Raion, Smolensk Oblast.
Colonel-general of aviation (1971); twice Hero of the Soviet Union (My. 1, 1943, and Jl. 1, 1944). Member of the CPSU since 1942.
He graduated from the Churguev Military Aviation school in 1941, the M. V. Frunze Military Academy in 1948, and the K. E. Voroshilov Higher Military Academy in 1954. In the Great Patriotic War (1941–45) he saw combat on the Western, Voronezh, Stalingrad, Southern, and Third and First Belorussian fronts as flight and squadron commander (1942–44) and commander of the 9th Guards Fighter Air Regiment (1944–45). He made 448 combat sorties, fought in 134 air battles, and personally shot down 35 enemy aircraft and 11 in group combat. In August 1943, while ramming an enemy aircraft, he made a parachute jump and was seized by the Hitlerites, but he escaped to the partisans and rejoined his unit. After the war he held responsible positions in the air force.
Decor: Order of Lenin, six Orders of the Red Banner, the Order of the Red Star, and various medals.

Lavrov, Ilia Mikhailovich
Author; b. 1917, Novosibirsk.
Russian; grad. Novosibirsk Theatre School 1936.
Publs.: "V rodnom kraiu" (In the Native Land) 1953, "Oni letali k oblakam" (They Flew up to the Clouds) 1958, "Devochka i riabina" (The Girl and the Mountain Ash) 1957, "Vstrecha s chudom" (Encounter With a Miracle) 1961, "Pokhozhdeniia Vasi Krivenka i ego brata Kubaria" (Adventures of Vasia Krivenko and His Brother Kubar) 1959, "Ocharovannaia" (The Enchanted One) 1965, and others.

Lavrov, Ivan Mikhailovich
Diplomat; b. 1912.
High school educ.; mem. Diplomatic Service 1941–; various posts in USSR Min. of Foreign Affairs and abroad 1941–57; Emb. Counsellor at USSR Emb. in Austria 1957–60; Dep. Head of 3rd European Dept. in USSR Min. of Foreign Affairs 1960–62; Emb. Counsellor-Envoy at USSR Emb. in Germany (FRG) 1962–65; USSR Amb. to Mauritania 1965–68; sen. posts in USSR Min. of Foreign Affairs 1968–70; USSR Amb. to Zaire D. 1970–S. 1979.

Lavrov, Sviatoslav Sergeevich
Mathematician; b. Mr. 12, 1923, Leningrad.
Mem. CPSU 1950–; grad. Air Force Acad. Leningrad 1944 and Moscow Univ. 1954; specialist in mechanics and applied mathematics; corresponding mem. USSR Acad. of Sciences 1966–; Lecturer at Moscow Univ. 1959–71, Prof. 1965–; Head of Laboratory for Mathematical Systems at Computing Centre of USSR Acad. of Sciences 1966–71; Prof. at Leningrad Univ. 1971–. Decors.: 2 Orders of Lenin, Order of Red Star, medals.

Lavrov, Vladimir Sergeevich
Diplomat; b. 1919.
Mem. Diplomatic Service 1947–; USSR Assist. 1st Dep. Min. of Foreign Affairs 1953–56; Emb. Councellor, USSR Embassy in Washington 1956–59; USSR Chargé d'Affaires in Yemen 1959–60; Dep. Head, Head, 2nd European Dept., USSR Min. of Foreign Affairs 1960–64; USSR Amb. to Kenya 1964–67, to the Netherlands 1967–73, to Switzerland O. 1977–; Mem. Apparat USSR Min. of Foreign Affairs 1973–77.

Lavrovskii, Mikhail Leonidovich
Ballet dancer; b. 1941.
Grad. Moscow Acad. Choreographic College 1961; engagements at Bolshoi Theatre 1961–. Decors.: Lenin Prize 1970, Merited Artist of the RSFSR 1970.

Lazarenko, Evgenii Konstantinovich, Dr. of Geol.-Min. Sc.
Geologist; b. D. 26, 1912, Kharkov; s. of a worker.
Mem. CPSU 1952–; grad. Kharkov Univ. 1934, completed post-graduate studies 1937; Dr. of Geol.-Min. Sc. and Prof. 1948; geologist with several geological institutions 1934–38; Lecturer at Voronezh Univ. 1938–42; Sen. Research Staff Mem. of Ural Branch of USSR Acad. of Sciences 1942–44; Lecturer, Head of Dept. of Geology of Lvov Univ. 1944–48; Prof. Chair. of Mineralogy at Lvov Univ., Dept. Head at Inst. of Geology of Natural Resources at Ukrainian Acad. of Sciences 1948–49; Dep. Rector 1949–51, Rector of Lvov Univ. 1951–63; corresponding mem. Ukrainian Acad. of Sciences 1951–. Decors.: Order "Honorary Badge", medals.
Publs.: "Mineralogiia osadochnykh utvoren Prykarpatia" 1962, co-author of "Mineralogiia Zakarpatia" 1963, "Kurs mineralogii" 1963, "Agronomichni rudy Ukrainy" 1966.

Lazarev, Boris Georgievich, Dr. of Phy.-Math. Sc.
Physicist; b. S. 6, 1906, Miropol, now Sumy Oblast.
Grad. Leningrad Polytech. Inst. 1930; Dr. of Phy.-Math. Sc. 1944; Prof.; mem. staff Physical-Tech. Inst. in Leningrad 1930–36 and in the Urals 1936–37; mem. staff Physical-Tech. Inst. at Ukrainian Acad. of Sciences 1937–; concurrently Lecturer at Kharkov Agricultural Inst. 1944–53; mem. Ukrainian Acad. of Sciences 1951–. Decor.: State Prize 1951.
Publs.: author of more than 100 scientific works; "Ob izmeneniiakh sverkhprovodiashchikh svoistv talliia pod davleniem" in "Doklady Akademii Nauk SSSR" vol. 69, Nr. 2, 1949, "Magnitnye svoistva metallov pri nizkikh temperaturakh" in "Zhurnal teoreticheskoi i eksperimentalnoi fiziki" No. 20, 1950.

Lazarev, L. (real name: Shindel, Lazar Il'ich)
Literary critic; b. 1924, Kharkov.
Mem. CPSU 1951; grad. Dept. of Philology of Moscow Univ. 1950; first works publ. 1949.
Publs.: "Dramaturgii K. Simonova" (The Dramas of K. Simonov) 1952, "Predislovie k zhizni" (Preface to Life) 1960, and others.

Lazarev, Vasilii Grigorevich
Cosmonaut; b. 1928.
Mem. CPSU 1956–; grad. Saratov Medical Inst. 1952 and Mil. Aviation College in Chugnev 1954; pilot instructor, test pilot and medical scientist 1954–66; cosmonaut 1966–; Commander of spaceship "Soiuz 12" in which he tested the improved board systems during an orbital flight (together with O. G. Makarov) S. 27–29, 1973; together with O. G. Makarov went into orbit in "Soiuz" capsule on Ap. 5, 1975; flight was intended as continuation of experiments with orbital station "Saliut 4"; flight had to be discontinued because of deviation from pre-calculated course. Decors.: Hero of the Sov. Union 1973, Order of Lenin, medals.

Lebedev, Aleksandr Aleksandrovich
Literary historian; b. 1928, Moscow.
Russian; mem. CPSU 1956–; first works publ. 1954.
Publs.: "Geroi Chernyshevskogo" (The Heroes of Chernyshevskii) 1962, "Esteticheskie vzgliady A. V. Lunacharskogo" (Aesthetic Views of A. V. Lunacharskii) 1962, "Antonio Gramshi o kulture i iskusstve" (Antonio Gramshi on Culture and Art) 1965, "Chaadaev" 1965, and others.

Lebedev, Evgenii Alekseevich
Actor; b. Ja. 2 (15), 1917, Balakovo, Saratov Oblast.
Russian; grad. Cen. Theatre College Moscow 1940; engagements at Lenin Komsomol Theatre Leningrad 1949–56 and Great Gorkii Drama Theatre 1956–. Decor.: People's Artist of the USSR.

Lebedev, Iurii Vladimirovich
Diplomat; b. 1923.
Mem. Diplomatic Service 1959–; mem. Apparat USSR Min. of Foreign Affairs 1960–61, 1964, 1968–69 and 1975–78; Emb. Councellor at USSR Embassy in Cuba 1961–64, Councellor-Envoy 1964–68; USSR Amb. to Peru 1969–75, to Uruguay Ag. 1978–.

Lebedev, Konstantin Vasil'evich
Author; b. 1921, Sutiazhnoe.
Russian; grad. Kazan Medical Inst.; first works publ. 1946.
Publs.: "Dni ispytaniia" (Days of Examination) 1949, "Mislavskii" 1951, "Nikolai Osipovich Kovalevskii" 1954, and others.

Lebedev, Konstantin Vasil'evich
Party official; b. 1918.
Mem. CPSU 1941–; grad. Party High School of Cen. Cttee. CPSU 1952; inspector of a social-welfare dept., dir. and teacher in a middle school in Western Kazakhstan 1934–41; Sec., 1st Sec. of Oblast Cttee. Kazakh Komsomol in Western Kazakhstan 1941–49; mem. Apparat Cen. Cttee. CPSU 1952–71 and 1982–; mem. Cen. Cttee., mem. Bureau and 2nd Sec. of Cen. Cttee. Esthonian CP F. 1971–82; mem. Cen. Auditing Comm. CPSU 1971–76; cand. mem. Cen. Cttee. CPSU 1976–.

Lebedev, Valentin Vitalevich
Cosmonaut; b. 1942.
Mem. CPSU 1971–; grad. Aviation Inst. Moscow 1966; employment in an engineering office 1966–72; cosmonaut 1972–; board engineer in spaceship "Soiuz 13" in which he performed several scientific-tech. experiments together with P. I. Klymuk D. 18–26, 1973.

Lebedeva, Vera Sergeevna
Laborer; b. 1937.
Mem. CPSU 1969–; grad. Moscow Inst. for the Petrochemical and Gas Industry 1966; operator at Omsk Oil-Refining Combine 1956–61, senior operator

1961–; mem. Cen. Auditing Comm. CPSU 1976–81; cand. mem. Cen. Cttee. CPSU 1981–.

Lein, Voldemar Petrovich
State official; b. 1920; s. of a railwayman. Latvian; mem. CPSU 1946–; grad. Moscow Tech. Inst. of Food Ind. 1945 and Party High School of Cen. Cttee. CPSU 1958; served in Sov. Army 1942–43; head of a prod. technology dept., chief mechanic of a factory 1945–49; instructor of a city cttee. CPSU, chair. of a raion exec. cttee. of workers' deps., Chair. of Liepāja City Exec. Cttee. of Workers' Deps., head of an admin. dept. at Econ. Council Latvian SSR, 2nd Sec. of Riga City Cttee. Latvian CP 1949–61; mem. Cen. Cttee. Latvian CP 1960–71; Sec. and mem. Bureau of Cen. Cttee. Latvian CP 1960–71; Chair. of Bureau of Ind. and Construction at Cen. Cttee. Latvian CP 1962–64; USSR Min. of Food Ind. Ja. 1970–; cand. mem. 1971–76; mem. Cen. Cttee. CPSU 1976–. Decors.: 2 Orders of Lenin.

Lekhtmets, Tynis
Author; b. 1937, Rakvere.
Esthonian; grad. Dept. of Philology of Tartu Univ. 1961.
Publs.: "Pervye stikhi" (First Poems) 1960, "Ulitsy i proselki" (Streets and Country Lanes) 1962, and others.

Lemberanskii, Alish Dzhamilevich
State official.
Dep. to Supr. Sov. Azerbaidzhan SSR 1959–; Chair. of Baku City Exec. Cttee. of Workers' Deps. 1959–66; mem. Cen. Cttee. Azerbaidzhan CP 1960–; 1st Dep. Head of Main Admin. of Microbiological Ind. with USSR Council of Mins. 1966–70; Dep. Chair. of Council of Mins. Azerbaidzhan SSR N. 1970–.

Lengvinas, Kazimeras Kazimerovich
State official.
Mem. Cen. Cttee. Lithuanian CP 1961–; Chair. of City Exec. Cttee. of Workers' Deps., 1st Sec. of Kaunas City Cttee. Lithuanian CP 1963–My. 1975; Head of Main Statistical Admin. Lithuanian SSR My. 1975–.

Lentsman, Leonid Nikolaevich
Party official; b. 1915, Simferopol, Crimea.
Esthonian; mem. CPSU 1942; grad. State Univ. Leningrad 1938; joined Komsomol in 1932; teacher in Murmansk Oblast 1938–41; served in Sov. Army, then pol. work in Esthonia 1941–47; Head of Dept. of Schools 1947–48, Head of Dept. of Propaganda and Agitation at Cen. Cttee. Esthonian CP 1948–50; Min. of Educ. Esthonian SSR 1950–51; Sec. 1951–52 and 1964–71, 2nd Sec. of Cen. Cttee. Esthonian CP 1952–64; mem. Bureau Cen. Cttee. Esthonian CP 1952–; Chair. of Esthonian Republican Trade Union Council F. 1971–82; Chair. of Comm. for Party Control Cen. Cttee. Esthonian CP F. 1982–. Decors.: 2 Orders of Lenin, Order of Great Patriotic War 2nd Cl., Order of Red Star, Order of Friendship between Nations, medals.

Leonid (L. K. Poliakov), Dr. of Theol.
Russian Orthodox clergyman; b. 1913, St. Petersburg.
Doctor; deacon 1949; priest, grad. Leningrad Theological Acad. as Cand. of Theology 1952; monk 1952; Lecturer at Leningrad Theological Acad. 1953–57; Archimandrite 1957; master's degree in theology, Prof. at Theological Acad. 1958; Bishop of Kursk 1959; Bishop of Mozhaisk, Substitute for the Diocese of Moscow, Archbishop, Chair. of Econ. Dept. of Moscow Patriarchate 1962; Archbishop of Iaroslavl 1963; Dr. of Theology 1965; Archbishop of Riga and Latvia 1966–.

Leonov, Aleksei Arkhipovich
Cosmonaut; b. My. 30, 1934, Listianka, Kamerovo Oblast.
Mem. CPSU 1957–; grad. Mil. Aviation College Chuguev 1957 and Zhukovskii Mil. Aviation Engineering Acad. 1968; cosmonaut 1960–; Col.; spaceflight as co-pilot together with P. I. Belaev in spacechip "Voskhod 2" Mr. 18, 1965; first disembarcation into space, space walk of 5 metres and sojourn of 12 minutes in space, "Soiuz 19" Jl. 1975. Decors.: Order of Lenin, Order of Red Star, medals, several foreign decorations.

Leonov, Leonid Maksimovich
Author; b. My. 19 (31), 1899, Moscow.
Russian; no party affiliation, intermediate school educ.; mem. USSR Acad. of Sciences 1972–; Ed. of newspaper of 15th Inzenskaia Div. 1920; Sec. of Ed. Div. of newspaper of Moscow Mil. District "Krasnyi voin" 1921–22; literary work 1922–; Dep. to USSR Supr. Sov. 1946–70. Decors.: Merited Artist of the RSFSR 1949, Hero of Socialist Labour 1967, State Prize of the USSR 1943, Lenin Prize 1957, 4 Orders of Lenin, 4 other orders and various medals.

Leonov, Pavel Artemevich
Party official; b. 1918; s. of a peasant.
Mem. CPSU 1944–; grad. Bauman Tech. College Moscow 1942; technologist, sen. foreman, shop manager in a plant 1942–45; dep. sec., sec. of CPSU cttee. in a plant, Sec. of Moscow Sokolniki Raion Cttee. CPSU, Dept. Head at Moscow Oblast Cttee. CPSU, section head of a Cen. Cttee. CPSU dept. 1945–60; 1st Sec. of Sakhalin Oblast Cttee. CPSU 1960–78; cand. mem. 1961–71; mem. Cen. Cttee. CPSU 1971–; Dep. USSR Supr. Sov. 1961–; 1st Sec. of Kalinin Oblast Cttee. CPSU D. 1978–. Decors.: 2 Orders of Lenin.

Leonov, Piotr Vasil'evich
Ceramist; b. My. 3 (16), 1910, Iiutoe, Orel Oblast.
Mem. CPSU 1947–; grad. Inst. of Architecture and Art Krasnodar 1930; corresponding mem. USSR Academy of the Arts 1973–. Decors.: Merited Artist of the Russian SFSR, Order of Lenin, medals.

Leontev, Aleksei Fedorovich
Mathematician; b. Mr. 14 (27), 1917, Iakovtsevo, Gorkii Oblast.
Grad. Gorkii Univ. 1939; corresponding mem. USSR Acad. of Sciences 1970–; assignments at Gorkii Univ. 1948–54, Mathematical Inst. of USSR Acad. of Sciences and Energetics Inst. Moscow 1954–71; mem. staff Bashkir Branch of USSR Acad. of Sciences 1971–.

Leontev, Oleg Konstantinovich, Dr. of Geog. Sc.
Geomorphologist; F. 19, 1920, Krasnoiarsk Krai.
Mem. CPSU 1944–; Prof. 1960–; Prof. Chair of Geomorphology at Dept. of Geography of Moscow Univ. 1959–; took part in research expeditions (the banks of Lake Baikal, the Caspian Sea, the Sea of Okhotsk, the shores of China, Albania, Cuba and other regions).
Publs.: "Geomorfologiia morskikh beregov i dna" Moscow 1955, "Kratkii kurs morskoi geologii" Moscow 1963, "Dno okeana" Moscow 1968, co-author of "Kaspitiskoe more" Moscow 1969.

Leontii (Ivan Afanasevich Gudimov), Cand. of Theol.
Russian Orthodox clergyman; b. 1928, Sumy Oblast.
Monk 1948; priest 1951; Custodian of Odessa Cathedral 1952; Rector of Theological Seminary Odessa; Bishop of Podolsk, Substitute in Moscow Bishopric, Representative of Moscow Patriarchate at Patriarchate of Antioch 1961; Bishop of Volhynia 1964; Bishop of Simferopol 1965; Bishop of Kharkov 1967; Archbishop 1968; Exarch of the Moscow Patriarchate in Central Europe 1970; Archbishop of Simferopol and the Crimea 1973–.

Leontii (Leonid Fadeevich Bondar)
Russian Orthodox clergyman; b. 1913, Vilnius Oblast.
Grad. Vilnius Theological Seminary 1935; grad. Faculty of Orthodox Theology of Warsaw Univ., master's degree 1939; novice 1940–43; monk and priest 1943; Rector at Shirovitsy Monastery 1946; Inspector and Lecturer at Minsk Theological Seminary 1949; Archimandrite and Prior in Shirovitsky 1953; Bishop 1956; Suffragan Bishop for the Diocese of Minsk; Bishop of Novosibirsk 1961–63; Bishop 1963–71 and Archbishop of Orenburg and Buzuluk S. 1971–.

Leontovich, Mikhail Aleksandrovich
Physicist; b. F. 22 (Mr. 7), 1903, Moscow. Grad. Moscow Univ. 1923; mem. USSR Acad. of Sciences 1946–; mem. Comm. on Research on Kursk Magnet Anomaly; mem. research staff Physics Inst. of Moscow Univ. 1929–; Prof. at Moscow Univ. 1934–45 and 1955–71; mem. staff Physics Inst. of USSR Acad. of Sciences 1934–41 and 1946–52; mem. staff Inst. of Atomic Energy 1951–; specialist in electrodynamics, optics, statistical physics, radiophysics, thermonuclear physics; conducted important research work on the theory of molecular dispersion of light (until 1935), on ultra-acoustic absorption of sound in gases (1936) and in liquids (1936, 1939), on the theory of fluctuations and statistical physics, on the basis of thermodynamic statistics (1932), on gas-kinetic equations derived from the theory of stochastic processes (1935) and on various aspects of radiophysics (thermal fluctuations of the electromagnetic field in solids). Decors.: Lenin Prize 1958, 3 Orders of Lenin, 4 other orders, medals.
Publs.: "Statisticheskaia fizika" Moscow-Leningrad 1944, "Vvedenie v termodinamiku" Moscow-Leningrad 1952.

Lepeshinskaia, Olga Vasil'evna
Ballet dancer; b. S. 15 (28), 1916, Kiev.
Mem. CPSU 1943–; grad. Choreographic College Moscow; ballet teacher; mem. ballet of Bolshoi Theatre 1933–63; teaches ballet also in foreign countries 1963–. Decors.: People's Artist of the USSR, State Prize of the USSR 1941, 1946, 1947, 1950, 3 orders, several medals.

Lepeshkin, Vladimir Aleksandrovich
Party official.
Sec. of Minsk City Cttee. 1966–74, 1st Sec. of Minsk City Cttee. Belorussian CP 1975–; cand. mem. 1971–76, mem. Cen. Cttee. Belorussian CP 1976–.

Leshava, Gela Georgievich
State official; b. 1936.
Mem. CPSU 1965–; grad. State Medical Inst. in Tbilisi 1960, M. D.; researcher at Inst. of Clinical and Experimental Neurology of Georgian Min. of Health, Sec. of Party org. of this inst. 1960–73; head of Dept. of Health and Social Services of Cen. Cttee. CP Georgia 1973–76; Chair. of Exec. Cttee. of Adigenskii Raion Sov. of People's Deps., 1st Raion Cttee. Sec. CP Georgia in Borzhomi 1976–78; Min. of Health Georgian SSR Ja. 1978–; cand. mem. Cen. Cttee. CP Georgia 1981–.

Levada, A. (real name: Kosiak, Aleksandr Stepanovich)
Author; b. 1909, Krivchunka, Cherkassy Oblast; s. of a teacher.
Ukrainian; first works publ. 1925. Decors.: Order of Lenin, other orders, medals.
Publs.: "Nikolai Chernyshevskii" 1939, "Stikhotvoreniia" (Poems) 1941, "Kamo" 1940, "Shliakh na Ukrainu" (The Way to the Ukraine) 1946, "Iugo-Zapad" (South-West) 1948–50, "Poslednaia vstrecha" (The Last Encounter) 1956, "Maria" 1960, and others.

Levakovskaia, Evgeniia Vladimirovna
Author; b. 1914, St. Petersburg.
Russian; studied Gorkii Inst. of Literature Moscow 1939–41; first works publ. 1935.
Publs.: "Kochui schastlivo! ..." (Be a Happy Nomad! ...) 1937, "Step'" (Steppes) 1938, "Veter vstrechaiut grudiu" (One Meets the Wind Head-on) 1943, "U nas v rote" (At Our Company) 1954, "Pervaia zima" (The First Winter) 1952, "Na proshloi nedele" (Last Week) 1961, "Sentiabr – luchshii mesiats" (September – the Best Month) 1964, "Vperedi schastie" (Happiness Lies Before You) 1964, and others.

Levashov, Mstislav Aleksandrovich
Author; b. 1912, Moscow.
Russian; grad. Dept. of Law of Moscow Univ. 1937; first works publ. 1941.
Publs.: "Skazki v stikhakh" (Fairytales in Verse) 1955, "Chestnaia Muska" (Honourable Muska) 1959, "Pod krylom – taiga" (Under the Wing – the Taiga) 1963, "Krylatye serdtsa" (Winged Heart) 1963, "Komendant Berlina" (The Commander of Berlin) 1965, and others.

Levchaev, Piotr Ivanovich
Author; b. 1913, Promzino, Mordovian ASSR.
Worked as teacher and ed.; first works publ. 1929.
Publs.: "Gorkie dni" (Bitter Days) 1935, "Rasskazy" (Stories) 1936, "Zloba" (Malice) 1935, "Les shumit" (The Wood is Rustling) 1959, and others.

Levchenko, Irina Nikolaevna
Author; b. 1924, Kadievka, Lugansk Oblast.
Russian; mem. CPSU 1943–; grad. Tank Acad. and Frunze Mil. Acad.
Publs.: "Povest' o voennykh godakh" (Story of the War Years) 1952, "Doch' komandira" (The Commander's Daughter) 1955, "Nadezhda" 1957, "Bessmertie" (Immortality) 1960, "Chtoby iabloni tsveli ..." (So That the Apple Trees May Blossom ...) 1963, "Schastlivaia" (The Happy One) 1963, "V skorom poezde" (In an Express Train) 1957, "Liudi novoi Germanii" (The People of the New Germany) 1959, "Bez obratnogo bileta" (Without a Return Ticket) 1962, "Liudi, shturm, pobeda" (People, Storm, Victory) 1964, and others.

Levchuk, Timofei Vasil'evich
Director; b. Ja. 6 (19), 1912, Bystrovka, now in Ruzhin Raion, Zhitomir Oblast. People's Artist of the Ukrainian SSR (1969); People's Artist of the USSR (1972). Member of the CPSU since 1942. He graduated from the Kiev Film Institute in 1934 and became a director at the Kiev Feature Film Studio in 1949. He directed *In the Steppes of the Ukraine* (1952), *Snowball Grove* (1953), *The Flame of Rage* (1956), *Ivan Franko* (1956), *The Law of the Antarctic* (1963), *The Cosmic Alloy* (1964), *Two Years Over the Percipice* (1967), and *The Error of Honoré de Balzac* (1968). One of his major motion pictures was *The Woman From Kiev* (parts 1 and 2, 1958 and 1959; part 3, *The Heirs,* 1960). He was awarded the Shevchenko State Prize of the Ukrainian SSR (1971) for *The Kotsiubinskii Family.*
He became secretary of the board of the Cinematographers' Union of the Ukrainian SSR in 1963. He was a deputy to the sixth convocation of the Supreme Soviet of the Ukrainian SSR. Decors.: Order of Lenin, three other orders, and various medals.

Levich, Veniamin Grigor'evich
(Theoretical) physicist; b. Mr. 17 (30), 1917, Kharkov.
Grad. Kharkov Univ. 1937; works in the field of physical chemistry; corresponding mem. USSR Acad. of Sciences 1958–; mem. staff Inst. of Physical Chemistry at USSR Acad. of Sciences 1940–58; mem. staff Inst. of Electrochemistry at USSR Acad. of Sciences 1958–.

Levidova, Valentina Iosifovna
Author; b. 1924, Leningrad.
Russian; grad. Dept. of Law Leningrad Univ.
Publs.: "Pervaia liubov'" (First Love) 1953, "Golos druga" (Voice of a Friend) 1957, "Dalekie druzia" (Distant Friends) 1957, "Neobyknovennoe prikliuchenie Mashi Bulkinoi" (The Unique Adventure of Masha Bulkina) 1964, "Mnogo let spustia" (Many Years Later) 1962, "Fiktivnyi brak" (Simulated Marriage) 1964, "Razgovor s liubimym" (Conversation With the Loved One) 1966, and others.

Levik, Vilgelm Veniaminovich
Literary historian, critic, translator and poet; b. D. 31, 1906 (Ja. 13, 1907), Kiev.
Grad. Painting Class at Higher State Workshops for Arts and Crafts (VKHUTEMAS) 1930; translated works by Shakespeare, Goethe, Schiller, Heine, Lenau and others.
Publs.: "O tochnosti i vernosti" (Exactitude and Faithfulness) 1959, "Vernoe slovo – na vernoe mesto" (The Right Word – in the Right Place) 1964, and others.

Levikov, Valerii Zinovevich
Diplomat.
USSR Amb. to Burundi Ja. 1980–.

Levin, Leonid Mendelevich
Architect; b. 1936.
Grad. Belorussian Polytech. Inst. 1960. Decor.: Lenin Prize for his part in construction of memorial site "Khatyn" 1970.

Levina, Hanna Mironovna
Poet; b. 1900, Dnepropetrovsk.
Jew; first works publ. 1917.
Publs.: "Moi vklad" (My Contribution) 1929, "Svoe" (One's Own) 1941, "Na soniachnii storoni" (On the Sunny Side) 1935, "Vesinni golosa" (Spring Voices), "Vnuchek Tarasik" (The Grandchild Tarasik) 1957, and others.

Levitan, Iurii Borisovich
Announcer for the State Committee on Radio and Television of the Council of Ministers of the USSR; b. S.19 (O.2), 1914, Vladimir.
People's Artist of the RSFSR (1973). Member of the CPSU since 1941.
He became an announcer for the All-Union Radio in 1931. He has a voice of rare timbre and expressiveness. During the Great Patriotic War (1941–45), he, reading reports from Sovinformbiuro, orders of the Supreme Commander, and other news, performed a great agitation and propaganda service with a powerful emotional impact on his listeners. He reads most important political documents, gives reports from Red Square and the Kremlin Palace of Congresses, and does voiceovers for motion pictures. Decors.: two orders and several medals.

Levitanskii, Iurii Davidovich
Poet and translator; b. 1922, Kiev.
Russian; grad. Higher Literary Courses at USSR Writers' Union 1957.
Publs.: "Soldatskaia doroga" (Soldier's Route) 1948, "Vstrecha s Moskvoi" (Seeing Moscow Again) 1949, "Samoe dorogoe" (The Dearest) 1951, "Utro novogo goda" (The Dawning of the New Year) 1952, "Listia letiat" (The Leaves Fly) 1956, "Nashi dni" (Our Days) 1952, and others.

Levitska, Irina Georgievna
Painter; b. 1927, Kiev.
Grad. Lvov Inst. for Applied and Decorative Art 1954, pupil of I. Bokshai and R. Silvestrov.

Lialenkov, Vladimir Dmitrievich
Author; b. 1930, Konstantinovka, Donetsk Oblast.
Russian; grad. Leningrad Polytech. Inst. 1957; construction engineer and hydrotechnologist by profession; first works publ. 1957.
Publs.: "Taina v Podlesakh" (A Secret in Podlesy) 1957, "Boris Kartavin" 1962, "Sestry Strogalevy" (The Strogalev Sisters) 1962, "Marusia" 1964, and others.

Liamin, Mikhail Andreevich
Author; b. 1906, Khomiakovo, Udmurt ASSR.
Mem. CPSU 1942–; grad. Herzen Pedagogical Inst. Leningrad 1931.
Publs.: "Vo imia schastia" (In the Name of Happiness) 1950, "Chetyre goda v shineliakh" (Four Years in a Soldier's Coat) 1965, and others.

Liashchenko, Nikolai Grigorevich
Military official; b. 1908.
Russian; mem. CPSU 1931–; grad. Frunze Mil. Acad. 1941 and Gen. Staff Acad. 1948; Gen. of the Army; groom and smith on a stud farm, chair. of a district cttee. of rural workers, chair. of a trade union cttee. on a stud farm 1919–29; in Sov. Army 1929–; various command posts 1941–48; division, then corps commander 1948–57; Dep. to Supr. Sov. RSFSR 1955–59; 1st Dep. Commander of Turkestan Mil. District 1957–63; Dep. to Kirghiz Supr. Sov. 1959–67; Commander of Volga Mil. District 1963–65; Commander of Turkestan Mil. District 1965–69; mem. and mem. Presidium, then mem. Bureau Cen. Cttee. Uzbek CP 1966–71; cand. mem. 1966–71, mem. Cen. Cttee. CPSU 1971–81; Commander of Cen. Asian Mil. District N. 1969–77; mem. and mem. Bureau Cen. Cttee. Kazakh CP 1971–77. Decors.: 3 Orders of Lenin, 4 Orders of Red Banner, medals.

Liashko, Aleksandr Pavlovich
State official; b. 1915; s. of a worker.
Ukrainian; mem. CPSU 1942–; grad. Donetsk Ind. Inst. 1947; taught a motor mechanics course, mechanic in motor pool of a plant, student 1934–41; served in Sov. Army 1941–45; engineer, Dep. Dept. Head, Dep. Dir., Sec. of Novo-kramatorsk Machine Construction Plant CPSU Cttee. 1945–52; 1st Sec. of Kramatorsk City Cttee. Ukrainian CP 1952–54; 1st Sec. of Donetsk Oblast Cttee. Ukrainian CP 1960–63 (for Ind. Ja.–Jl. 1963); Dep. to 1958–, mem. Draft Bills Comm. 1962–66, Chair. of Comm. for Ind., Transport, Posts and Telecommunications at Sov. of the Union USSR Supr. Sov. 1966–69; Dep. to and mem. Comm. for Ind. and Transport at Supr. Sov. Ukrainian SSR 1959–63; mem. Cen. Cttee. Ukrainian CP 1960–; mem. Presidium, then mem. Politbureau 1963–, Sec. 1963–66, 2nd Sec. of Cen. Cttee. Ukrainian CP 1966–69; mem. Cen. Cttee. CPSU 1961–; Chair. of Bureau for Ind. Cen. Cttee. Ukrainian CP Jl. 1963–N. 1964; Chair. of Presidium Supr. Sov. Ukrainian SSR and Dep. Chair. of Presidium USSR Supr. Sov. 1969–72; Chair. of Council of Mins. Ukrainian SSR Je. 1972–. Decors.: 2 Orders of Lenin.

Lidin, Vladimir Germanovich
Author; b. F.3 (15), 1894, Moscow.
Russian; grad. Dept. of Law of Moscow Univ. 1915; literary work 1915–.
Publs.: "Mogila neizvestnogo soldata" (The Tomb of the Unknown Soldier) 1932, "Velikii ili Tikhii" (The Great or the Quiet One) 1933, "Zima 1941 goda" (The Winter of 1941) 1942, "Dve zhizni" (Two Lives) 1950, "Liudi i vstrechi" (People and Encounters) 1957, 1961, 1965, "Druzia moi – knigi. Zametki knigoliuba" (My Friends – the books. Notes by a Booklover) 1966, "Serdtsa svoego ten'" (The Shadow of One's Own Heart) 1965, and others.

Lifshits, Evgenii Mikhailovich
Physicist; b. N.8 (21), 1915, Kharkov.
Grad. Inst. of Mechanics and Mechanical Engineering Kharkov 1933; mem. staff Physical-Tech. Inst. of Problems of Physics at USSR Acad. of Sciences 1939–; Dep. Chief Ed. of periodical "Žurnal èksperimental'noj teoretičeskoj fiziki" 1946–; corresponding mem. USSR Acad. of Sciences 1966–. Decors.: State Prize of the USSR 1954, Lenin Prize 1962 (for his part as co-author of a basic course in theoretical physics), 2 orders, several medals.

Lifshits, Ilia Mikhailovich
Physicist; b. D.31, 1916 (Ja.13, 1917), Kharkov.
Grad. Kharkov Univ. 1936 and Inst. of Mechanics and Mechanical Engineering Kharkov 1938; mem. USSR Acad. of Sciences 1970–; mem. Ukrainian Acad. of Sciences 1967–; mem. staff Physical-Tech. Inst. of Ukrainian Acad. of Sciences 1967–68; mem. staff Inst. for Problems of Physics at USSR Acad. of Sciences 1968–; Prof. at Kharkov Univ. 1944–, Moscow Univ. 1964–. Decors.: Mandelshtam Memorial Prize of USSR Acad. of Sciences 1952 (for work on dynamic theory of non-ideal crystals), Simon Memorial Prize of London Physical Soc. 1961 (for research on electronic structure of metals), Lenin Prize 1967, Order of Red Banner of Labour, medals.
Publs.: co-author of "Elektronnaia teoriia metallov" Moscow 1971.

Lifshits, Vladimir Aleksandrovich
Author; b. 1913, Kharkov.
Russian; mem. CPSU 1942–; grad. Leningrad Econ. and Finance Inst. 1933; first works publ. 1934.
Publs.: "Aist" (The Stork) 1958, "Leningradskaia storona" (Leningrad Side) 1947, "Sem dnei" (Seven Days) 1948, "Utro pobedy" (The Morning of Victory) 1946, "Respublikantsy" (Republicans) 1938, "Sablia Chapaeva" (Chapaev's Sabre) 1939, "Na beregakh Nevy" (On the Banks of the Neva) 1946, "Kuda khochu, toda lechu!" (I Fly Where I Want!) 1962, "Byla voina" (It Was War) 1965, and others.

Ligachev, Egor Kuzmich
Party official; b. 1920.
Russian; mem. CPSU 1944–; grad. Moscow Aviation Inst. 1943 and Party Higher School of Cen. Cttee. CPSU 1951; tech. engineer, head tech. group of a plant, 1st sec. of a raion Komsomol cttee., Sec., 1st Sec. of Novosibirsk Oblast Komsomol Cttee. 1943–49; Lecturer at Novosibirsk City Cttee. CPSU, Dept. Head with Novosibirsk Oblast Cttee. CPSU, Dep. Chair. of Novosibirsk Oblast Sov., Sec. of Novosibirsk Oblast Cttee. CPSU 1949–61; Dep. Head of Dept. of Propa-

ganda and Agitation, Dep. Head of Dept. of Party Organs (for Ind.) at Cen. Cttee. CPSU for the RSFSR, 1961–65; 1st Sec. of Tomsk Oblast Cttee. CPSU 1965–; cand. mem. 1966–76, mem. Cen. Cttee. CPSU 1976–; Dep. to 1966–, mem. Comm. for Ind., Transport, Posts and Telecommunications 1966–68, mem. Youth Affairs Comm. 1968–74, mem. Planning and Budgets Comm. at Sov. of the Union USSR Supr. Sov. 1974–.

Likhachev, Dmitrii Sergeevich
Literary and cultural historian; b. N. 15 (28), 1906, St. Petersburg.
Grad. Leningrad Univ. 1928; mem. staff Inst. of Russian Literature 1938–, Head of Section for Old Russian Literature 1954–; Prof. at Leningrad Univ. 1946–53; mem. USSR Acad. of Sciences 1970–. Decors.: Order of Red Banner of Labour, medals.

Likhachev, Nikolai Viktorovich
Veterinarian, virologist and immunologist; b. N. 26 (D. 8), 1901, Moscow.
Grad. Zoological Inst. Moscow 1929; mem. USSR V. I. Lenin Acad. of Agricultural Sciences 1956–; Head of Laboratory for Biological Drugs Against Virus Diseases at State Control Inst. for Veterinary Drugs 1937–; main research on virus diseases of animals. Decors.: State Prize of the USSR 1946, Order of Lenin, 4 other orders, medals.

Likhachev, Veniamin Andreevich
Diplomat; b. 1921.
Mem. Diplomatic Service 1946–; Emb. Counsellor at USSR Emb. in Turkey 1958–62; mem. staff USSR Min. of Foreign Affairs 1962–70; USSR Amb. to Iraq 1970–73; Head of 2nd African Dept. in USSR Min. of Foreign Affairs 1973–.

Likhachev, Viktor Ivanovich
Diplomat; b. 1914.
Mem. Diplomatic Service 1942–; USSR Consul in Lantshou (China) 1944–45; Asst. to USSR Dep. Min. of Foreign Affairs 1946–47 and 1949–52; mem. staff USSR Min. of Foreign Affairs 1947–49; Emb. Counsellor at USSR Emb. in China 1952–57; Consultant, Dep. Head of Far East Dept. in USSR Min. of Foreign Affairs 1957–58; Head of South-East Asia Dept. 1958–61, Head of South Asia Dept. 1961–67, Head of Far East Dept. 1967–70, Head of 1st Far East Dept. in USSR Min. of Foreign Affairs 1970; USSR Amb. to Venezuela O. 1970–Ap. 1975.

Likhodeev, Leonid Izrailevich (real name: Lides)
Author; b. 1921, Iuzovka.
Russian; mem. CPSU 1942–; studied Gorkii Inst. of Literature Moscow; first works publ. 1948.
Publs.: "Pokorenie pustyni" (Conquering the Desert) 1953, "Svoimi glazami" (With One's Own Eyes) 1955, "Otkrytoe okno" (Open Window) 1957, "Volga vpadaet v Kaspiiskoe more" (The Volga Flows into the Caspian Sea) 1959, "Istoriia odnoi poezdki" (The Story of a Journey) 1957, "Ja paren' soznatel'nyi" (I am a Reasonable Fellow) 1965, and others.

Lind, Vello Nikolaevich
State official.
Mem. Cen. Cttee. Esthonian CP 1976–; Min. of Agriculture Esthonian SSR Ja. 1981–; Dep. Supr. Sov. Esthonian SSR D. 1981–.

Linde, Edgar Valdemarovich
State official.
Cand. mem. Cen. Cttee. Latvian CP 1971–; Head of Dept. of Science and Educ. Institutions at Cen. Cttee. Latvian CP 1971–75; Min. of Higher and Secondary Specialist Educ. Latvian SSR Je. 1975–.

Line, Velta Martynovna
Actress; b. 1923.
Latvian; studied Drama Studio of People's Theatre in Riga and Drama Theatre of the Latvian SSR. Decor.: People's Artist of the USSR 1973.

Lingis, Iuozas Ionovich (Juozas Lingys)
Ballet dancer and choreographer; b. Ja. 1, 1919, Gyliškiai, Šakiai Raion, Lithuanian SSR.
People's Artist of the USSR (1970).
In 1941, he became a dancer with the Lietuva Honored State Folk Song and Dance Ensemble of the Lithuanian SSR. In 1946 he became choreographer of the ensemble, whose repertoire includes the dances of the fraternal republics and socialist countries. His special contribution lies in gathering, studying. He re-creates native folk rituals and stages little scenes from everyday life. As a choreographer he has succeeded in preserving the national poetic atmosphere and in introducing the dynamics of modern life, which is reflected in the accelerated rhythms, the joyous images, and the distinctive fluidity of the choreographic dance patterns. His best productions include *Sadute, Pakeltkojis, The Rooster, Rugeliai, Džigūņas,* and *Rusnėtis.*
With the ensemble, he has toured the German Democratic Republic, the Polish People's Republic, the Socialist Republic of Rumania, the Czechoslovak Socialist Republic, Belgium, Finland, and the countries of Latin America. Since 1967 he has been doing scholarly work and teaching in the choreographic section of the cultural-educational department of the Vilnius Pedagogical Institute. Decors.: State Prize of the USSR (1950, 1970), two orders and various medals.
Works: *Lietuviu liaudies žaidimai.* Vilnius 1955. *Lietuviu liaudies šokiai,* vol. 3. Vilnius 1959. *Greitpolke.* Vilnius 1967. *Sok mergeles vainikuotos.* Vilnius 1972.

Linkhovoin, Lkhasaran Lodonovich
Buriat opera singer (bass); b. Ja. 1, 1924, Aga-Khangel, now in Aga-Buriat National Okrug.
People's Artist of the USSR (1959).
He graduated from the Leningrad Conservatory in 1954. From 1942 to 1949 he worked in the Buriat Musical Drama Theater in Ulan-Ude. In 1954 he became a soloist with the Buriat Theater of Opera and Ballet, where he was also artistic director from 1962 to 1966. His roles have included Bumal-khan in Frolov's *Enkhe – Bulat-bator,* Bulgan Tabit in Aiusheev's *Blood Brothers,* Lama in Iampilov's *Budamsku,* Khan in Berlinskii's *Bair,* Susanin in Glinka's *Ivan Susanin,* the Miller in Dargomyzhskii's *The Mermaid,* Gremin and Kochubel in Tchaikovsky's *Eugene Onegin* and *Mazepa,* Konchak in Borodin's *Prince Igor,* Mephistopheles in Gounod's *Faust,* and Don Basilio in Rossini's *Barber of Seville.*
He began teaching in 1955 at the school of music in Ulan-Ude. He was a deputy to the sixth convocation of the Supreme Soviet of the USSR. Decors.: Order of the Red Banner of Labor and several medals.
Work: *S pesnei po stranam mira.* Ulan-Ude 1970.

Lipatov, Vil' Vladimirovich
Writer; b. Ap. 10, 1927, Chita.
Member of the CPSU since 1957.
He graduated from the Tomsk Pedagogic Institute in 1952. His first works were published in 1956. He is the author of the novellas *The Six* (1958), *The Captain of the Smelyi* (1959), *Glukhaia Miata* (1960), *The Deep Stream (1961), The Death of Egor Suzun* (1963), *The Village Detective* (1967; screen version, 1969), *Lida Varaksina* (1968), *The Tale of Director Pronchatov* (1969; the television film *Engineer Pronchatov,* 1972), *Gray Mouse* (1970), and *Back Before the War* (1971). He has also written short stories and sketches. He is particularly interested in ethical and moral problems. His usual protagonists are fishermen, river men, woodcutters, and Siberian kolkhoz villagers. Decors.: Order of Red Banner of Labor and a medal.
Works: *Tochka opory: Povesti, rasskazy, ocherki.* Tula 1967. *Povesti.* Moscow 1970. *Zub mudrosti: Povesti.* Moscow 1971. *Dve povesti.* Moscow 1972.

Lipkin, Semen Izrailevich
Russian poet and translator; b. S. 6 (19), 1911, Odessa.
People's Poet of the Kalmyk ASSR (1968).
He graduated from the Moscow Economics Engineering Institute in 1937. He fought in the Great Patriotic War (1941–45). He translated into Russian the Kalmyk epic *Dzangar* (1940), the Kirghiz epic *Manas* (1941), the Kabardinian epic *Narty* (1951), the Buriat epic *Geser* (1968), the epic of the peoples of India

Mahabharata (1969), and the works of classical Tadzhik, Uzbek, Kirghiz, Balkar, and Kalmyk poets. His translations were published as *Kabardinian Epic Poetry* (1956), *Voices of Six Centuries* (1960), and *The Golden Chain: Eastern Poems* (1970). He also has written *Eyewitness* (1967), a book of verses. Decors.: Rudaki State Prize of the Tadzhik SSR (1967), four orders, and a number of medals.

Lishchiner, Mikhail Ivanovich
Soviet painter; b. 1912, Tiraspol.
Grad. Kamianets-Podolskii Technical College of Art 1930 and Kiev Art Inst. 1941, pupil of P. Volokidin.

Lisianskii, Mark Samoilovich
Poet; b. 1913, Odessa.
Russian; mem. CPSU 1938–; grad. Moscow Inst. of Journalism 1934.
Publs.: "Bereg" (The Bank) 1940, "V boi za Rodinu" (Onward in the Battle for the Motherland), "Moia zemlia" (My Country) 1942, "Ot imeni Chernogo moria" (In the Name of the Black Sea) 1947, "Frontovaia vesna" (Front Spring) 1942, "Zolotaia moia Moskva" (My Golden Moscow) 1951, "Druziam i tovarishcham" (To My Friends and Comrades) 1958, "Za vesnoi - vesna" (Spring after Spring) 1959, "Vsegda s nami" (Always With Us) 1955, "Moia Moskva" (My Moscow) 1941, "Zori Moskovskie" (Moscow Stars) 1957, and others.

Lisitsian, Pavel Gerasimovich
Armenian Singer (baritone); b. O. 24 (N. 6), 1911, Vladikavkaz.
People's Artist of the USSR (1956). Member of the CPSU since 1949.
He was a soloist with the Leningrad Malyi Opera Theater (1935–37), the Armenian Theater of Opera and Ballet (1937–40), and the Bolshoi Theater of the USSR (1940–66). A leading Soviet opera singer, he possesses a bright voice with a wide range, beautiful timbre, flawless intonation, and perfect diction.
His numerous roles have included Eletskii, Onegin, Mazepa, and Robert in Tchaikovsky's *Queen of Spades, Eugene Onegin, Mazepa,* and *Iolanthe;* Germont and Amonasro in Verdi's *La Traviata* and *Aida;* Escamillo in Bizet's *Carmen;* Tatul in Spendiarov's *Almast;* the title role in Chukhadzhian's *Arshak II;* and Napoleon in Prokofiev's *War and Peace* (1959; premiere).
He gives recitals. He taught at the Yerevan Conservatory from 1967 to 1973 (department chairman in 1969, professor in 1970). He has toured abroad. Decors.: two Orders of the Red Banner of Labor.

Lisitsyn, Viktor Afanasevich
State official; b. 1930.
Mem. CPSU 1953–; grad. Dnepropetrovsk Party Higher School and Voroshilovgrad Agricultural Inst., economic agronomist; stoker of a locomotive depot, asst. locomotive driver, leader of a locomotive repair brigade, instructor, chief of a dept. of a raion cttee. Ukrainian CP, chair. of a collective farm, dir. of a state farm, 1st Sec. of Kremennaia Raion Ukrainian CP to 1970; 1st Sec. of Voroshilovgrad Oblast Cttee. Ukrainian CP 1970–74; Chair. of Exec. Cttee. of Voroshilovgrad Oblast Sov. of Peoples' Deps. 1974–81; cand. mem. Cen. Cttee. Ukrainian CP 1976–; Min. of Fruits and Vegetables Ukrainian SSR Ja. 1981–.

Lisnianskaia, Inna Lvovna
Author; b. 1928, Baku.
Russian; first works publ. 1949.
Publs.: "Eto bylo so mnoi" (This Happened to Me) 1957, "Vernost'" (Faithfulness) 1958, "Ne prosto - liubov'" (Not Simply - Love) 1963, and others.

Lisnichii, Grigorii Emelianovich
State official.
Head of Cen. Admin. of Oil Processing and Petrochemical Ind. Ukrainian SSR 1971–.

Lisovoi, Timofei Grigor'evich
Party official; b. 1923.
Ukrainian; mem. CPSU 1946–; grad. Kiev Econ. Inst.; teacher at a secondary school 1940–41; served in Sov. Army 1941–44; dir. of a school, chief of a dept., sec. of exec. cttee. of a raion sov. of people's deps. 1944–50; sec. of a raion cttee. Ukrainian CP, chair. of exec. cttee. of a raion sov. of peoples' deps., 1st sec. of a raion cttee. Ukrainian CP 1950–62; sec. of a party cttee. of a collective and state farm prod. asscn., 1st sec. of a raion cttee. Ukrainian CP 1962–70; mem. Cen. Cttee. Ukrainian CP 1966–; inspector of a dept. of Cen. Cttee. Ukrainian CP, Chair. of Exec. Cttee. of Khmelnitskii Oblast Sov. of Peoples' Deps. 1970–72; cand. mem. Cen. Cttee. Ukrainian CP 1971–76, mem. 1976–; 1st Sec. of Khmelnitskii Oblast Cttee. Ukrainian CP Mr. 1972–; Dep. to USSR Supr. Sov. 1974–. Decors.: Order of the October Revolution.

Lisovyi, Vasyl
Scientist; b. 1937.
Former mem. scientific staff Inst. of Philosophy at Acad. of Sciences Ukrainian SSR; together with I. Proniuk author of an open letter to Cen. Cttee. CPSU in which he denounced reprisals against Ukrainian intellectuals; interned in a strict-regime forced labour camp 1973–.

Listov, Konstantin Iakovlevich
Composer; b. S. 19 (O. 7), 1900, Odessa. Honored Art Worker of the RSFSR (1950).
He graduated in 1922 from the Saratov Conservatory, where he studied piano and composition. From 1941 to 1945 he served as musical consultant to the Main Political Directorate of the Navy. His works include the operas *Olesia* (performed in 1959), and *Daughter of Cuba* (performed in 1962); 11 operettas, including *Sevastopol' Waltz* (1962) and *Heart of a Baltic Sailor* (1964); a symphony (1970); and the vocal symphonic poem *Grenada* (words by M. A. Svetlov, 1968). He has also composed numerous popular songs, the best-known of which are "Song About a Machine-gun Cart," "In a Dugout," "We've Been on Cruises," and "Sevastopol' Waltz." Decors.: Order of the Red Star and a number of medals.

Listov, Vladimir Vladimirovich
State official; b. 1931.
Mem. CPSU 1962–; grad. Tomsk Polytechnical Inst. 1955; foreman, shop superintendent, dep. dir. of a plant in Kemerovo 1955–62, dir. 1962–64; 1st Sec. of Kemerovo City Cttee. CPSU 1966–70; Dep. USSR Min. of the Chemical Ind. 1971–77; Chief of the Chemical Ind. Dept. of Cen. Cttee. CPSU 1977–80; Dep. to USSR Supr. Sov. 1979–; USSR Min. of the Chemical Ind. N. 1980–; cand. mem. Cen. Cttee. CPSU 1981–. Decor.: Order of Lenin.

Litvinenko, Leonid Mikhailovich, Dr. of Chem. Sc.
Chemical scientist; b. Ja. 12, 1921, Taganrog; s. of a clerk.
Mem. CPSU 1944–; grad. Kharkov Univ. 1947; completed postgraduate studies 1950; Dr. of Chem. Sc. and Prof. 1962; Research Asst., Lecturer, Prof. Chair of Organic Catalysis and Kinetics at Kharkov Univ. 1947–65; Rector of Donetsk Univ. 1965–; concurrently mem. staff Ukrainian Acad. of Sciences Donetsk and Head of Laboratory of Dept. of Physical-Organic Chemistry at Inst. of Physical Chemistry at Ukrainian Acad. of Sciences in Donetsk 1964–; mem. Ukrainian Acad. of Sciences 1965.
Publs.: author of numerous scientific works on the present state of organic chemistry, "Struktura i reaktsionnaia sposobnost' aromaticheskikh aminov" in "Izvestiia Akademii Nauk SSSR. Otdelenie khimicheskikh nauk" No. 10/1962, "Polifunktsionalnyi kataliz v organicheskoi khimii" in "Ukrainskii khimicheskii zhurnal" No. 4/1964, "Amidnyi kataliz v reaktsiiakh nukleofilnogo zameshcheniia pri karbonilnom atome ugleroda" in "Doklady Akademiia Nauk SSSR" No. 2/1966.

Liubareva, Elena Petrovna
Literary historian and critic; b. 1917, Kiev.
Russian; mem. CPSU 1948–; grad. Inst. of History, Philosophy and Literature Moscow 1939; first works publ. 1948.
Publs.: "Aleksandr Tvardovskii" 1957, "Mikhail Svetlov" 1960, "Eduard Bagritskii" 1964, and others.

Liubeznov, Ivan Aleksandrovich
Russian actor; b. Ap. 19 (My. 2), 1909, Astrakhan'.
People's Artist of the USSR (1970). Member of the CPSU (1940).

In 1931, he graduated from the State Institute of Theatrical Arts and joined the company of the Theatre-Studio of the Malyi Theater; he later worked in the Moscow Satira Theatre. Since 1948 he has been working in the Malyi Theatre. He has appeared in films since 1935. Elements of humor, gentle lyrical feeling, and satire are inherent in his art. He also performs dramatic roles. In the theatre his best roles include Orlov in *Talents* and the title role in *Sashka*, both by Finn, Andrei Belugin in Belugin's *Marriage* by Ostrovskii and Solov'ev, Verba in *Snowball Tree Grove* and Dremliuga in *Wings*, both by Korneichuk, and Dvoetochie in *The Summer People* by Gorkii. In films he played the accountant Kovyn'ko in *The Rich Bride* (1938), Lieutenant Demidov in *At 6 p.m. After the War* (1944), the American sergeant Harry Perebeinoga in *Meeting on the Elbe* (1949), and General Ivolgin in *The Idiot* (1958). He also performs on the variety stage as a reciter. Decors.: State Prize of the USSR (1946) and two orders, as well as a number of medals.
Work: *Zhizn' – luchshaia shkola*. Moscow 1962.

Liubimov, Iurii Petrovich
Actor and theatre director; b. S.17 (30), 1917, Iaroslavl.
Mem. CPSU 1952–; grad. Shchukin Theatre School at V. Vakhtangov Theatre Moscow 1939; served in Sov. Army 1940–46; engagements with E. Vakhtangov Theatre Moscow 1946–; Principal Director of Taganka Drama and Comedy Theatre Moscow 1964–. Decor.: State Prize of the USSR 1952.

Liulka, Arkhip Mikhailovich
Aircraft engine designer; b. Mr.10 (23), 1908, Savarka, Kiev Oblast.
Mem. CPSU 1947–; grad. Kiev Polytech. Inst. 1931; mem. USSR Acad. of Sciences 1968–; worked in aircraft plants in Kharkov, Leningrad and Moscow 1931–33; Chief Designer of Aircraft Engines 1946–. Decors.: State Prize of the USSR 1948, 1951, 3 Orders of Lenin, Order of the October Revolution, Order of Red Banner of Labour, medals.

Liushnin, Grigorii Ivanovich
Poet; b. 1921, Vysokoe, Riazan Oblast.
Russian; grad. Higher Courses in Literature at USSR Writers' Union 1961; served in World War II, wounded 1941; imprisoned in German concentration camp, escaped 1945, again in Sov. Army; before and after the War worker at "Serp i molot" factory Moscow; first works publ. 1952.
Publs.: "Moia pesnia" (My Song) 1952, "Tovarishch Trud" (Comrade Work) 1959, "Poklon tebe, Rossiia" (Reverence to You, Russia) 1964, "Ne vstaiut na koleni soldaty" (Soldiers Do Not Fall on Their Knees) 1961, "Udary serdtsa moego" (My Heartbeats) 1963, and others.

Liusternik, Lazar' Aronovich
Mathematician; b. D. 19 (31), 1899, Zduńska Wola (now in Poland).
Corresponding member of the Academy of Sciences of the USSR (1946).
He graduated from Moscow University in 1922 and was appointed a professor there in 1931. He was the first (1926) to apply the method of finite differences to the solution of the Dirichlet problem. Together with L. G. Shnirel'man, he used topological methods in analysis and proved the theorem of three geodetics. He also worked in functional analysis and computer mathematics. Decors.: State Prize of the USSR (1946), Order of Lenin, the Order of the Badge of Honor, and several medals.

Livanov, Mikhail Nikolaevich
Medical scientist; b. O.7 (20), 1907, Kazan.
Mem. CPSU 1940–; grad. Kazan Univ. 1931; physiologist and neurophysiologist; Head of Laboratory at Inst. of Brain Research Moscow 1933–47; mem. staff Inst. of Pathology and Therapy of Intoxication at USSR Acad. of Medical Sciences 1947–49; mem. staff Inst. of Biophysics at USSR Acad. of Medical Sciences 1949–61; mem. staff Inst. of Neurology and Neurophysiology at USSR Acad. of Sciences 1961–; mem. USSR Acad. of Sciences 1970–. Decors.: 4 orders, several medals.
Publs.: "Ėlektroėntsefaloskopiia" Moscow 1960 (co-author) "Nekotorye problemy deistviia ioniziruiushchei radiatsii na nervnuiu sistemu" Moscow 1962, "Prostranstvennaia organizatsiia protsessov golovnogo mozga" Moscow 1972.

Livanova, Tamara Nikolaevna (married name: Ferman)
Musicologist; b. Ap.5 (18), 1909, Kishinev.
Doctor of art studies (1940). Member of the CPSU since 1948.
She graduated in 1932 from the Moscow Conservatory, where she had studied with V. E. Ferman and M. V. Ivanov-Boretskii. She taught music history at the conservatory from 1932 to 1956 (from 1939 as a professor) and became a senior scientific worker at the Institute of Art History in Moscow in 1944.
She has done extensive research in Russian and Western European music of the 17th–19th centuries and its historical relation to other branches of art. Her main works include *Essays and Materials in the Field of the History of Russian Musical Culture* (1938), *History of Western European Music Before 1789* (1940), *Musical Dramaturgy of J. S. Bach and Its Historical Relations* (1948), *Russian Musical Culture of the 18th Century in Relation to Literature, Theater, and Everyday Life* (vols. 1–2, 1952–53), *Opera Criticism in Russia* (fascs. 1–4, 1966–73; fasc. 1, with V. P. Protopopov), and *Music Bibliography of the Russian Periodical Press of the 19th Century* (fascs. 1–6, 1960–73; fasc. 5 with O. A. Vinogradova). Decors.: two orders and a medal.

Liventsev, Viktor Il'ich
Party official; b. 1914.
Cand. mem. Cen. Cttee. Belorussian CP 1960–81, mem. 1981–; Chair. Cttee. for Physical Culture and Sports with Council of Mins. Belorussian SSR 1968–78; Administrator of Affairs Cen. Cttee. Belorussian CP Jl. 1978–.

Liventsov, Vasilii Andreevich
Party official; b. 1914.
Ukrainian; grad. Inst. of Agricultur Alma-Ata; kolkhoz farmer, agronomist with a machine and tractor station (MTS), Chief Agronomist of Alma-Ata Oblast Dept. of Agriculture 1930–43; dept. head with an oblast cttee. and Dept. Head at Cen. Cttee. Kazakh CP 1943–62, 2nd Sec. of Akmolinsk and Alma-Ata Oblast Cttee. Kazakh CP, Chair. of Exec. Cttee. Dzhambul Oblast Sov. of Workers' Deputies; Dep. to Supr. Sov. Kazakh SSR 1955–67; mem. Cen. Cttee. Kazakh CP 1956–; 1st Sec. of Khimkent Oblast Cttee. Kazakh CP 1962–72; Dep. to Supr. Sov. USSR 1966–; 1st Sec. of Aktiubinsk Oblast Cttee. Kazakh CP Ja. 1972–.

Livshits, Slava Naumovna
Painter; b. 1915, Odessa.
Mem. CPSU 1944–; grad. Kiev Art Inst. 1945, pupil of O. Shovkunenko.

Lobachev, Vladimir Konstantinovich
Diplomat.
USSR Amb. to Kongo Je. 1982–.

Loban, Mikola (Nikolai Pavlovich)
Author; b. 1911, Chaplichi, Minsk; s. of a peasant.
Belorussian; first works publs. 1930.
Publs.: "Irkutianka" (Irkutsk Woman) 1953, "Na poroge budushchego" (On the Threshold of the Future) 1961, "Zemlia moia – sud'ba moia" (My Country – My Fate) 1965, and others.

Lobanok, Vladimir Eliseevich
State official; b. 1907, Ostrov, Minsk Oblast; s. of a peasant.
Belorussian; mem. CPSU 1930–; grad. Belorussian Agricultural Acad. 1931 and Party High School of Cen. Cttee. CPSU 1956; Agronomist of People's Commissariat of Agriculture Belorussian SSR, dir. of various agricultural colleges in Vitebsk Oblast 1932–41; partisan in Belorussian 1941–44; Chair. of Polotsk Oblast Exec. Cttee. of Workers' Deps. 1944–46; Dep. to USSR Supr. Sov. 1946–; 2nd, then 1st Sec. of Polesia Oblast Cttee. Belorussian CP 1948–53; Chair. of Gomel Oblast Exec. Cttee. of Workers' Deps. 1954–56; 1st Sec. of Vitebsk Oblast Cttee. Belorussian CP 1956–62; mem. Cen. Auditing Comm. CPSU 1961–76; 1st Dep. Chair. of Council of Mins. 1962–74 and Min. of Prod. and Procurement of Agri-

cultural Products Belorussian SSR 1962–65; cand. mem. Je.–D. 1962, 1965–66 and 1976– and mem. Bureau Cen. Cttee. Belorussian CP 1966–76; Dep. to 1951–59 and 1963– and Dep. Chair. of Presidium Supr. Sov. Belorussian SSR Ap. 1974–. Decors.: Hero of the Sov. Union, 2 Orders of Lenin, Order of Red Banner, Suvorov Order 1st Cl., Order of Patriotic War 1st Cl., medals.

Lobashev, Vladimir Mikhailovich
Physicist; b. Jl. 29, 1934, Leningrad.
Mem. CPSU 1970–; grad. Leningrad Univ. 1957; corresponding mem. USSR Acad. of Sciences 1970–; mem. staff Physical-Tech. Inst. 1957–72 and Inst. of Nuclear Research at USSR Acad. of Sciences 1972–.

Lobzhanidze, Bakhva Fedorovich
State official; b. 1930.
Georgian; mem. CPSU 1956–; master, foreman, head of a section of a bridge-building crew USSR Min. of Transport Construction 1953–58; engineer, head of a group of Sov. bridge-building specialists in Afghanistan 1958–60; engineer in a dept. of the "Soyuzdorproekt" branch of State Planning Inst. in Tbilisi 1960–62; instructor of Dept. for Construction and Municipal Economy Tbilisi City Cttee. CP Georgia 1962–64; instructor, dep. head of a dept. Cen. Cttee. CP Georgia 1964–73; Chair. of Exec. Cttee. of City Sov. of People's Deps. in Tbilisi 1973–79; mem. Cen. Cttee. CP Georgia 1976–81; Chair. of State Cttee. for Gasification Georgian SSR Ja. 1979–.

Loginov, Vadim Petrovich
Diplomat.
2nd Sec. of Leningrad Oblast Cttee. USSR Komsomol 1956; Sec. and mem. Cen. Cttee. USSR Komsomol 1958–61; 1st Sec. of Vyborg City Cttee. CPSU 1963; Head of 3rd European Dept. in USSR Min. of Foreign Affairs 1975–78; USSR Amb. to Angola Mr. 1978–.

Loginov, Viktor Nikolaevich
Author; b. 1925, Bolshie Veski, Vladimir Oblast.
Russian; grad. Higher Literary Courses at USSR Writers' Union 1959; first works publs. 1945.
Publs.: "Znakomyi marshrut" (Familiar Route) 1959, "Pora landyshei" (Lily of the Valley Time) 1961, "Na to ona i liubov'..." (So is Love...) 1962, "Nash obshchii drug" (Our Mutual Friend) 1962, "Trudnye dni v Beregovoi" (Difficult Days in Beregovaia) 1958, "Dorogi tovarishchei" (Friends' Routes) vols. 1 and 2, 1959–62, "Pered prazdnikom" (Before the Holiday) 1965, and others.

Logunov, Anatolii Alekseevich
Theoretical physicist; b. D. 30, 1926, Obsharovka, Kuibyshev Oblast.
Mem. CPSU 1960–; grad. Moscow Univ. 1951; mem. staff Moscow Univ. 1951–56; Dep. Dir. of Laboratory for Theoretical Physics at Joint Inst. of Nuclear Research Dubna 1956–63; Dir. of Inst. of High Energies Sepukhov 1963–; mem. USSR Acad. of Sciences 1972–; main research on quantum theory of the field and physics of elementary particles. Dep. to USSR Supr. Sov. 1974–; Rector, Moscow State Univ. 1977–; cand. mem. Cen. Cttee. CPSU 1981–. Decors.: Order of Lenin, medals.

Logvin, Ivan Dmitrievich
Painter; b. 1923, Korshevo, Voronezh Oblast.
Mem. CPSU 1951–; grad Penza Art School 1949, and Kharkov Art Inst. 1955, pupil of V. Sizikov and L. Chernov.

Loktev, Konstantin Borisovich
Athlete (ice hockey); b. Je. 16, 1933, Moscow.
Honored Master of Sport (1964), Army officer, coach. Member of the CPSU since 1963. Champion of the USSR (ten times in 1955–66), the world (1964–66), Europe (1958–60, 1964–66), and the Winter Olympics (1964).
From 1957 to 1966, he played with the Central Army Sports Club and USSR combined teams, in the offensive unit with V. V. Aleksandrov and A. D. Al'metov, Honored Masters of Sport and many times Soviet, European, and world, champions and winners of the Winter Olympics in 1964. This offensive unit, with perfect technique and tactical virtuosity, is known in the history of modern hockey under the symbolic name "Grand Master Trio." Decors.: Order of the Badge of Honor and various medals.

Lolashvili, Otar Il'ich
State official; b. 1924.
Georgian; mem. CPSU 1946–; grad. Georgian Polytech. Kirov Inst. and Party High School of Cen. Cttee. CPSU; health instructor with railway 1943–46; instructor, sec. of a raion cttee. Georgian CP, Sec. of CPSU Cttee. of Georgian Econ. Council 1949–60; 1st Sec. of Oktiabrskii Raion Cttee. Georgian CP, Dept. Head of Cen. Cttee. Georgian CP 1960–63; Dep. to Supr. Sov. Georgian SSR 1963–; 2nd Sec. 1963–65, 1st Sec. of Tbilisi City Cttee. Georgian CP 1965–72; mem. Cen. Cttee. Georgian CP 1964–; cand. mem. 1965–71, mem. Bureau Cen. Cttee. Georgian CP 1971–72; Dep. to USSR Supr. Sov. 1966–; Min. of Building Materials Ind. Georgian SSR Jl. 1972–.

Lomakin, Viktor Pavlovich
Party official; b. 1926.
Russian; mem. CPSU 1953; grad. Aviation Inst. Kuibyshev 1949 and Acad. of Aviation Ind. Moscow 1953; technologist, dep. head, 2nd Sec., 1st Sec. of Komsomolsk-na-Amure City Cttee. CPSU 1949–52 and 1953–58; Sec. of Khabarovsk Krai Cttee. CPSU 1961–63; Chair. of Cttee. on Party and State Control at Khabarovsk Krai Cttee. CPSU, Sec. of Khabarovsk Krai Cttee. CPSU and Dep. Chair. of Khabarovsk Krai Sov. 1963–65; Sec. of Khabarovsk Krai Cttee. CPSU 1965–67; mem. Apparat Cen. Cttee. CPSU 1967–69; 1st Sec. of Primore Krai Cttee. CPSU Mr. 1969–; Dep. to and Chair. of Comm. on Construction and Building Materials Ind. at Sov. of the Union USSR Supr. Sov. 1970–; mem. Cen. Cttee. CPSU 1971–.

Lomako, Petr Fadeevich
State official; b. 1904; s. of a peasant.
Russian; mem. CPSU 1925; grad. Inst. of Non-Ferrous Metals and Gold Moscow 1932; Komsomol and trade union work, student 1921–32; brigade leader, foreman, dept. head, dep. chief engineer, dir. of a plant, head of a main board 1933–39; USSR Dep. People's Commiss. 1939–40, People's Commiss. then Min. of Non-Ferrous Metallurgy 1940–48, 1950–53 and 1954–57; Dep. to USSR Supr. Sov. 1946–50 and 1954–; USSR Dep. Min. 1948–50, 1st Dep. Min. of Metallurgical Ind. 1953–54; cand. mem. 1952–61, mem. Cen. Cttee. CPSU 1961–; Chair. of Krasnoiarsk Econ. Council 1957–61; Dep. Chair. of CPSU Cen. Cttee. Bureau for RSFSR 1961–62; Chair. of State Econ. Council of USSR Council of Mins. 1962; Dep. Chair. of USSR Council of Mins. and Chair. of USSR GOSPLAN 1962–65; USSR Min. of Non-Ferrous Metallurgy O. 1965–. Decors.: 5 Orders of Lenin, medals.

Lomidze, Georgii Iosifovich
Literary historian; b. Jl. 28, 1914, Tbilisi.
Russian; mem. CPSU 1942–; grad. Gorkii Inst. of Literature 1947; first works publ. 1934.
Publs.: "K sotsialisticheskomu realizmu v literature" (On Socialist Realism in Literature) 1952, "V poiskakh novogo" (In Search of the New) 1963, "Teoreticheskie problemy razvitija sovetskoi literatury" (Theoretical Problems in the Development of Soviet Literature) 1964, "Gumanizm i internatsionalizm" (Humanism and Internationalism) 1962, "Literatura internatsionalnogo edinstva" (Literature of International Unity) 1964, and others.

Lomonosov, Vladimir Grigor'evich
State official; b. 1928.
Russian; mem. CPSU 1950–; grad. Steel Inst. Moscow 1953; asst. foreman, foreman in a sheet-rolling mill, Sec. of CPSU Cttee. at "Hammer and Sickle" Plant Moscow 1953–58; Sec. of Kalinin Raion Cttee. CPSU Moscow 1958–62; Chair. of Cen. Asia Bureau of Cen. Cttee. CPSU 1962–65; Dep. to 1963–80; mem. Presidium Supr. Sov. Uzbek SSR 1965–67; 2nd Sec. and mem. Presidium, then mem. Bureau Cen. Cttee. Uzbek CP Mr. 1965–76; mem. Cen. Cttee. Uzbek CP 1966–76; mem. Cen. Cttee. CPSU 1966–; USSR Supr. Sov. 1966–; Chair. USSR State Cttee. for Labor and Social Questions Jl. 1976–. Decor.: Order of Lenin.

Lomov, Aleksandr Vasil'evich
State official.
Min. of Construction Projekts of the Heavy Ind. Kazakh SSR O. 1979–; mem. Cen. Cttee. Kazakh CP 1981–.

Lomov, Nikolai Petrovich
State official; b. 1925.
Russian; mem. CPSU 1947–; grad. Party High School in Saratov, Major-Gen.; served in Sov. Army 1942–46; held Komsomol posts, 1st Sec. of a Raion Cttee., Dep. Head of a Dept. of Saratov Oblast Cttee. CPSU 1946–64; with Sov. State Security Service 1964–; Chair. of Cttee. of State Security Kirghiz SSR Ap. 1978–; mem. Cen. Cttee. and mem. Bureau of Cen. Cttee. Kirghiz CP Ja. 1981–.

Lopukhov, Aleksandr Mikhailovich
Painter; b. 1925, Gorodnia, Chernigov Oblast.
Mem. CPSU 1969–; attended Kharkov Art Inst. 1947–53, pupil of O. Shovkunenko. Decor.: People's Artist of the Ukrainian SSR.

Lordkipanidze, Konstantin Aleksandrovich
Author; b. D.25, 1904 (Ja.7, 1905), Didi-Dzhikhaishi, Georgia.
Georgian; grad. Humanistic Tech. College in Kutaisi 1924; Chief Ed. of journal "Literaturnaia Gruziia" 1956–61, of journal "Tsiskari" 1966–. Decors.: 6 orders and various medals, Shota Rustaveli Prize 1971.

Lordkipanidze, V. G.
Komsomol official.
Mem. Bureau Cen. Cttee. Georgian Komsomol D. 1980, Sec. 1980–82, 2nd Sec. Ag. 1982–.

Loriia, Parmen Onisimovich
Georgian writer; b. Jl. 18 (30), 1896, Kvakude, now in Samtredia Raion.
Member of the CPSU from 1940.
The son of peasants, he was self-educated. He served in World War I and began to publish his works in 1920. His work reflects the past and present of Adzhariia. The heroes of the novels *Tea* (parts 1–2, 1937–38) and *Dawn Through a Crack* (1939) are builders of socialist society. The novels *Mevlud* (1941) and *Waves* (1953) were written on historical revolutionary themes. He published the novel *In the Mines,* about the workers of Chiatura, in 1954 and the novel *Sons,* about the builders of a factory, in 1956–60. *Hopes of the Future* (1965) deals with the life of Georgian young people. He has also written for children (*Children's Stories,* 1951). Decors.: Order of the Red Banner of Labor and several medals.
Works: [Loria, B.] *Ert tomeuli (ch'ai, mevludi).* Batumi 1957. *Gzebi da bilikebi.* Tbilisi 1962.

Los', Evdokiia Iakovlevna
Poet; b. 1929, Starina, Vitebsk Oblast.
Belorussian; mem. CPSU 1955–; first works publ. 1948.
Publs.: "Mart" (March) 1958, "Liudi dobrye" (Good People) 1963, and others.

Losev, Sergei Andreevich
State official; b. 1927.
Mem. CPSU 1953–; grad. Moscow Inst. of Int. Relations 1950; posts in Sov. news agency TASS 1950–; correspondent for TASS in Near East 1955–60; chief of Main Editorial Board for African Countries TASS 1960–63; correspondent for TASS in US and chief of its New York bureau 1963–69; Dep. Ed.-in-Chief of Main Editorial Board of Information for Foreign Countries TASS, mem. TASS Collegium and Ed.-in-Chief of Main Board of Information for Foreign Countries TASS 1969–73; Dep. Gen. Dir. of TASS of USSR Council of Mins. 1973–78; 1st Dep. Gen. Dir. 1978–79, Gen. Dir. My. 1979–; mem. Cen. Auditing Comm. CPSU 1981–.

Loshchenkov, Fedor Ivanovich
Party official; b. 1915; s. of a peasant.
Russian; mem. CPSU 1943–; grad. Aviation Inst. Moscow 1943 and Party Higher School of Cen. Cttee. CPSU 1946; technician at a railway car repair works 1932–36; in Sov. Army 1936–38; inspecting engineer at a plant 1943–46; Dep. Dept. Head with Novosibirsk City, then Oblast Cttee. CPSU, 1st sec. of a raion CPSU cttee., 2nd, 1st Sec. of Novosibirsk City Cttee. CPSU, 2nd Sec. of Novosibirsk Oblast Cttee. CPSU 1946–61; Dep. to Supr. Sov. RSFSR 1959–63; 1st Sec. of Iaroslavl Oblast Cttee. CPSU 1961– (for Ind. 1963–64); cand. mem. 1961–76, mem. Cen. Cttee. CPSU 1976–; Dep. to USSR Supr. Sov. 1962–. Decor.: Order of Red Banner of Labour.

Losik, Oleg Aleksandrovich
Military official; b. 1915.
Mem. CPSU 1941–; completed tank school in Saratov 1938 and Gen. Staff Acad. 1950; Marshal of Armoured Troops Ap. 1975–; in Sov. Army 1935–; command posts on front 1941–45; 1st Dep. Commander 1965–67, Commander of Far East Mil. District 1967–69; Head of Malinovskii Armoured Troops Acad. 1969–. Decor.: Hero of the Sov. Union.

Lotsmanova, Galina Pavlovna
Laborer; b. 1944.
Mem. CPSU 1967–; completed school for working youth 1963 and a foreman's school 1966; worker, forewoman 1962–; assembly worker in a plant in Kazan 1966–; mem. Cen. Auditing Comm. CPSU 1971–.

Lott, Iokhannes Adovich
State official.
Mem. Cen. Cttee. Esthonian CP 1971–; 1st Sec. of Tartu City Cttee. Esthonian CP 1971–78; Dep. to 1971–, Chair. of Supr. Sov. Esthonian SSR 1975–78; Min. of Culture Esthonian SSR N. 1978–.

Loveiko, Iosif Ignatevich
Architect; b. F.6 (19), 1906, Prokhory, Primore Krai.
Mem. CPSU 1950–; grad. Inst. of Architecture Moscow 1931; Chair. of Moscow Branch of USSR Federation of Architects 1935–55; Sen. Architect of the City of Moscow 1955–60. Decors.: Merited Master Builder ("stroitel'") of the Russian SFSR, Order of Lenin, Order of the Patriotic War 2nd Cl., medals.

Lozan, Stepan Ivanovich
State official.
Mem. Cen. Cttee. Moldavian CP 1961–; Dep. to Supr. Sov. Moldavian SSR 1963–; Sec. of Lozovskoe CPSU Cttee. for Kolkhoz-Sovkhoz Prod. Admin. 1963; Chair. of State Cttee. for Television and Radio Moldavian SSR 1967–.

Lubenets, Grigorii Kuzmich
State official.
Dep. Min. of Construction of Metallurgical and Chemical Plants Ukrainian SSR 1956–57; Head of Board of Construction of Metallurgical and Chemical Plants in Donetsk (Stalino) Econ. Council 1957–62; Dep. Chair. of Donetsk Econ. Council 1962–63; Min. of Construction 1963–67 and Min. of Construction of Plants for Heavy Ind. Ukrainian SSR 1967–; mem. Cen. Cttee. Ukrainian CP 1966–.

Lubotskii, Lev Davidovich
State official.
Chair. of State Cttee. for Material and Tech. Supplies Latvian SSR Je. 1980–; cand. mem. Cen. Cttee. Latvian CP 1981–.

Luchenko, Evgenii Dmitrievich
Painter; b. 1914, Golinka, Chernigov Oblast.
Grad. Kiev Art Inst. 1946, pupil of F. Krichevskii, K. Trokhimenko.

Luchinskii, Aleksandr Aleksandrovich
Military official; b. Mr. 10 (23), 1900, Kiev.
General of the army (1955); Hero of the Soviet Union (Ap. 19, 1945). Member of the CPSU (1943).
He has been in the Soviet Army since 1919. He fought in the Civil War of 1918–20 as a platoon and squadron commander. He graduated from the M.V. Frunze Military Academy in 1940 and from Higher Academic Training Courses in 1948. In 1937–38, he participated as a volunteer in the Chinese people's struggle against the Japanese imperialists. In the Great Patriotic War (1941–45) he was commander of a rifle division and corps from 1942 to 1944 and commander of the Twenty-eighth and Thirty-sixth armies in 1944–45.

After the war he served as deputy commander in chief of the Group of Soviet Forces in Germany in 1949 and commanded the Leningrad Military District from 1949 to 1953 and the Turkestan Military District from 1953 to 1957. He was deputy chief inspector of the Ministry of Defense of the USSR from 1958 to 1964. He has been military inspector-adviser of the Group of Inspectors General of the Ministry of Defense of the USSR since 1964. He was a deputy to the second through fourth convocations of the Supreme Soviet of the USSR and a candidate member of the Central Committee of the CPSU from 1952 to 1959. Decors.: three Orders of Lenin, four Orders of the Red Banner, three Orders of Suvorov First Class, the Order of Kutuzov First Class, the Order of Suvorov Second Class, the Order of the Red Banner of Labour of the Uzbek SSR, the Order of the Red Star, and various medals.

Luchinskii, P. K.
Party official.
2nd Sec., then 1st Sec. and mem. Bureau of Cen. Cttee. Moldavian Komsomol 1966–71; cand. mem. 1966–67, mem. Cen. Cttee. Moldavian CP 1967–81; mem. Bureau F. 1971–78, Sec. of Cen. Cttee. Moldavian CP 1971–76; 1st Sec. of Kishinev City Cttee. Moldavian CP 1976–78; Dep. Head of Propaganda Dept. of Cen. Cttee. CPSU 1978–.

Luchitskii, Igor Vladimirovich
Geologist; b. Ap. 10 (23), 1912, Warsaw.
Mem. CPSU 1971–; grad. Inst. of Geological Prospecting Moscow 1936; corresponding mem. USSR Acad. of Sciences 1968–; mem. staff Geological Inst. at USSR Acad. of Sciences 1937–49; mem. staff Geological Admin. of "Eniseistroi" and Mining-Geological Inst. at West Siberian Branch of USSR Acad. of Sciences 1949–62; Head of Laboratory for Experimental Tectonics of Inst. of Geology and Geophysics at Siberian Branch of USSR Acad. of Sciences 1962–; Prof. at Novosibirsk Univ. 1964–; main research on paleovolcanology, tectonics, tectonic physics and construction of schemes of tectonic processes. Decors.: 2 orders, several medals.

Luganskii, Sergei Danilovich
Military official; b. O. 1, 1918, Alma-Ata; s. of a peasant.
Major-General of aviation (1957); twice Hero of the Soviet Union (S. 2, 1943, and Jl. 1, 1944). Member of the CPSU since 1942.
He graduated from the Orenburg Pilots School in 1938 and from the Air Force Academy in 1949. In the Great Patriotic War (1941–45) he saw action on the Southern, Voronezh, Steppe, and Second and First Ukrainian fronts as commander of an aviation squadron (1941–44) and commander of a fighter aviation regiment (1944–45). He completed 390 combat flights and personally shot down 37 enemy aircraft and six in group combat. He has been in the reserve since 1964. Decors.: Order of Lenin, two Orders of the Red Banner, the Order of Alexander Nevsky, two Orders of the Red Star, and various medals.

Lukianenko, Aleksandra Mikhailovna
State official; b. 1938 in Dolynska, Kirovgrad Oblast.
Grad. Medical Inst. in Dnepropetrovsk; head of a raion dept. of health, chief physician of City Hospital in Krivoi Rog, head of a dept., then Dep. Head of 4th Main Admin. Min. of Health Ukrainian SSR 1961–73; Dep. Min. 1973–74, 1st Dep. Min. 1974–79, Min. Social Services Ukrainian SSR Mr. 1979–; Dep. Chair. Auditing Comm. CP Ukraine 1979–81.

Lukashevichius, Vitautas-Prantsishkus Vatslovovich
State official.
Mem. Auditing Comm. 1961–64, mem. Cen. Cttee. Lithuanian CP 1964–; Min. of Forestry and Timber Ind. Lithuanian SSR Ag. 1978–.

Lukianov, Anatolii Ivanovich
State official; b. 1930.
Mem. CPSU 1955–; grad. Moscow State Univ. 1953; Dr. of Juridical Sc. 1980; laborer in "Arsenal" Plant 1943; senior Consultant of Legal Comm. of USSR Council of Mins. 1956–61; dep. chief of a dept. of Presidium of USSR Supr. Sov. 1969–76; posts in Cen. Cttee. apparatus CPSU 1976–77; Chief of Secretariat of USSR Supr. Sov. N. 1977–; mem. Cen. Auditing Comm. CPSU 1981–.

Lukianov, Nikolai Nikolaevich
State official.
Mem. Auditing Comm. 1961–66, mem. Cen. Cttee. Moldavian CP 1966–; Chair. of Agro-Ind. Union for Vine-Growing and Wine-Processing with Council of Mins. Moldavian SSR Ap. 1974–.

Lukin, Iurii Borisovich
Critic and film script writer; b. O. 12 (25), 1907, Moscow.
Grad. Dept. of Literature at Inst. of History and Ethnology of Moscow State Univ. 1929; first works publ. 1929; bis main work is devoted to the works of M. A. Sholokhov.
Publs.: "O tvorcheskom puti Mikhaila Sholokhova" (About the Creative Development of Mikhail Sholokhov) 1950, "Mikhail Sholokhov" 1952, 2nd ed. 1962, film script for novels by Sholokhov: "Sud'ba cheloveka" (The Fate of Man) 1959, "Podniataia tselina" (Virgin Soil Upturned) 1960, film script for Leonov's novel "Russkii les" (The Russian Forest) 1963, and others.

Lukovets, Aleksei Illarionovich
Journalist; b. 1921.
Univ. educ.; mem. ed. staff of raion newspaper "Ilanskii rabochii" (Krasnoiarsk Krai) and Republic newspaper "Sovetskaia Moldaviia" 1938–47; correspondent, Dep. Ed., Ed. of Desk of Socialist Countries, mem. Ed. Board, Dep. Chief Ed. of "Pravda" 1947–76; Chief Ed. of "Sovetskaia Rossiia" 1976–; mem. Cen. Auditing Comm. CPSU 1976–. Decors.: Merited Cultural Worker of the RSFSR, Prize of USSR Journalists' Union.

Lunkov, Nikolai Mitrofanovich
Diplomat; b. 1919.
Mem. CPSU 1940–; grad. CPSU Cen. Cttee. Party Higher School 1944; lathe operator, sec. of Komsomol cttee., asst. shop superintendent at a Moscow automobile plant 1937–41; entered diplomatic profession 1943; held posts in USSR Min. of Foreign Affairs apparatus 1943–46, 1949–52 and 1957–62; 1st USSR mission sec., provisional chargé d'affaires in Switz. 1946–49; USSR Amb. to Norway 1962–67; chief of a dept. in USSR Min. of Foreign Affairs 1967–73; USSR Amb. to Great Britain and Malta 1973–80, to Italy S. 1980–; mem. Cen. Auditing Comm. CPSU 1976–81; cand. mem. Cen. Cttee. CPSU 1981–.

Lunts, Lazar' Adol'fovich
Jurist; b. Ja. 17 (29), 1892, Iur'ev.
Doctor of law (1947), professor (1947). Honored Scientific Worker of the RSFSR (1967).
He graduated from the department of law of Moscow University in 1916. From 1918 to 1940 he was a consultant at the People's Commissariat of Finance of the USSR and from 1941, senior associate at the All-Union Institute of Juridical Sciences (from 1963, the All-Union Research Institute of Soviet Legislation). His major works deal with international private law. Decor.: State Prize of the USSR in 1970.
Works: *Den'gi i denezhnye obiazatel'stva.* Moscow 1927. *"Denezhnoe obiazatel'stvo v grazhdanskom i kolizionnom prave kapitalisticheskikh stran."* (In the series *Uchenye trudy VIIuN*, vol. 14). Moscow 1948. *Obshchee uchenie ob obiazatel'stve.* Moscow 1950. (With I. B. Novitskii.) *Kurs mezhdunarodnogo chastnogo pravo.* Vols. 1–4. Moscow 1959–66. *Obshchaia chast',* Moscow 1959; *Osobennaia chast',* Moscow 1966. *Mezhdunarodnoe chastnoe pravo.* Moscow 1970.

Lupan, Andrei Pavlovich
Author; b. 1912, Mukhulen, Moldavia.
Mem. CPSU 1956–; grad. Kishinev Agricultural Inst.; first works publ. 1932. Decors.: Order of Lenin, other orders, medals.
Publs.: "Zabytaia derevnia" (The Forgotten Village) 1940, "Litsom k litsu" (Face to Face) 1945, "Svet" (Light) 1948, "Brat zemli" (Brother of the Land) 1959, and others.

Lupan, Anna Pavlovna
Author and poet; b. 1924, Mikhulen, Moldavia.
Moldavian; grad. Higher Courses in Literature USSR Writers' Union; first works publ. 1952.
Publs.: "Doroga v selo" (The Road to the Village) 1955, "Kolokolchik" (The Little Bell) 1958, "Osennie tsvety" (Autumn Flowers) 1960, "Veter v litso" (With the Face to the Wind) 1958, "Koleso vremeni" (The Wheel of Time) 1959, "Vse v poriadke" (Everything in Fine) 1962, and others.

Lupanov, Oleg Borisovich
Mathematician; b. Je. 2, 1932, Leningrad. Corresponding member of the Academy of Sciences of the USSR (1972).
He graduated from Moscow State University in 1955 and was appointed a professor there in 1964. In 1954 he began working at the Institute of Applied Mathematics of the Academy of Sciences of the USSR, becoming a senior research worker in 1960. He is a specialist on mathematical problems in cybernetics and mathematical logic. His principal studies deal with the investigation of the asymptotic laws of complexity of control systems. Decor.: Lenin Prize in 1966.
Works: "O sinteze nekotorykh klassov upravliaiushchikh sistem." *Problems kibernetiki.* 1963, no. 10. "Ob odnom podkhode k sintezu upravliaiushchikh sistem – printsipe lokal'nogo kodirovaniia." *Ibid.*, 1965, no. 14.

Lupashku, Mikhail Feodosevich
State official.
Min. of Agriculture Moldavian SSR Ap. 1980–; mem. Cen. Cttee. Moldavian CP 1981–.

Lupsiakov, Mikola Rodionovich
Author; b. 1919, Moscow.
Belorussian; grad. Minsk Pedagogical Inst. 1941; first works publ. 1936.
Publs.: "Pervaia ataka" (The First Attack) 1946, "Most" (The Bridge) 1947, "Rasskazy" (Stories) 1947, "Razvedchiki" (The Scouts) 1949, "Krasnyi bereg" (The Red Bank) 1954, "Denprovskaia chaika" (Dnepr Seagull) 1957, "Domik na kolesakh" (The Little House on Wheels) 1957, and others.

Lur'e, Anatolii Isakovich
Scientist in theoretical and applied mechanics; b. Je. 6 (19), 1901, Mogilev.
Corresponding member of the Academy of Sciences of the USSR (1960). Graduated from the Leningrad Polytechnic Institute in 1925. From 1925 to 1941 he worked at the Leningrad Polytechnic Institute (he became a professor in 1935), and in 1944 he became a subdepartment head at the institute. From 1942 to 1944 he was at the Urals Industrial Institute.
His major works concern the theory of elasticity (including nonlinear elasticity), the stability of automatic control systems (the dynamics of space-flight control and the design of optimum control systems), and analytical mechanics. Decors.: Order of the Red Banner of Labor and medals.
Works: *Teoreticheskaia mekhanika,* parts 1–3. Leningrad, 1932–34. With L. G. Loitsianskii.) *Operatsionnoe ischislenie i ego prilozheniia k zadacham mekhaniki,* 2nd ed. Leningrad-Moscow 1950. *Prostranstvennye zadachi teorii uprugosti.* Moscow-Leningrad 1955. *Analiticheskaia mekhanika.* Moscow 1961. *Teoriia uprugosti.* Moscow 1970.

Lur'e, Natan Mikhailovich (literary name: Note Lur'e)
Jewish writer; b. Ja. 2 (15), 1906, Roskoshnoe, Zaporozh'e.
As a boy, he was an agricultural worker. He graduated from the literature department of the Second Moscow State University and served in the Great Patriotic War (1941–45). He worked for the newspaper *Der emes* (The Truth) and began publishing in 1929. The novel *The Steppe Calls* (1932; Russian translation, 1958) describes life in a Jewish village, the sharp class struggle during collectivization, the breakdown of the psychology of private ownership, and the longing of the peasantry for a new life. In the novel *Heaven and Earth* (1963) the author focuses on the events of the Great Patriotic War. His prose is psychological and lyrical. He was a master of the technique of the internal monologue.
Works: *A libe bam iam.* Moscow 1938. *Khiml un erd.* Moscow 1965.

Lushev, Petr Georgievich
Military official; b. 1923.
Russian; mem. CPSU 1951–; grad. Mil. Acad. of the Armored Troop Forces 1954 and Mil. Acad. of Gen. Staff 1966, Army-Gen.; entered Sov. Army 1941; train and company leader at front 1941–45; command posts 1945–73; 1st Dep. Commander-in-Chief of Group of Sov. Troops in Germany 1973–77; Commander of Troops in Central Asian Mil. District 1977–80; Dep. to USSR Supr. Sov. 1979–; Commander of Troops of Moscow Mil. District D. 1980–; mem. Cen. Cttee. CPSU 1981–.

Lutak, Ivan Kondratevich
Party official; b. 1919; s. of a peasant.
Ukrainian; mem. CPSU 1940–; grad. Kazakh Agricultural Inst. 1942; served in Sov. Army, dir. of a sugar refinery 1942–45; chair. of a raion planning cttee., 1st, then 2nd sec. of a raion CPSU cttee., Head of Agricultural Dept. at Kiev Oblast Cttee. Ukrainian CP 1945–54; Chair. of Cherkassy Oblast Exec. Cttee. of Workers' Deps. 1954–61; mem. Auditing Comm. Ukrainian CP 1954–56; Dep. to 1955–63, mem. Foreign Affairs Comm. 1955–57, mem. Agricultural Comm. at Supr. Sov. Ukrainian SSR 1957–63; cand. mem. 1956–61, mem. Cen. Cttee. Ukrainian CP 1961–; 1st Sec. of Crimean Oblast Cttee. Ukrainian CP 1961–67 (for Agriculture 1963–64); cand. mem. 1961–64, mem. Cen. Cttee. CPSU 1964–76; Dep. to 1962–, mem. Trade and Public Services Comm. at Sov. of the Union USSR Supr. Sov. 1970–; mem. Politbureau 1967–76, Sec. 1967–69, 2nd Sec. of Cen. Cttee. Ukrainian CP 1969–76; 1st Sec. of Cherkassy Oblast Cttee. Ukrainian CP Ja. 1976–. Decor.: Order of Lenin.

Lutskevich, Iurii Pavlovich
Painter; b. 1934, Kirovograd.
Mem. CPSU 1959–; grad. Kiev Art Inst. 1959, pupil of M. Melikhov, V. Puzyrkov.

Luzgin, Vasilii Ivanovich
State official.
1st Sec. of Drogichin Raion Cttee. Belorussian CP 1955; mem. Cen. Cttee. Belorussian CP 1956–60; Min. of Social Welfare Belorussian SSR 1973–; mem. Auditing Comm. Belorussian CP 1976–.

Luzhanin, Maksim (pseudonym of Aleksandr Amvros'evich Karatai)
Belorussian author; b. O. 20 (N. 2), 1909, Prusy, in present-day Soligorsk Raion, Minsk Oblast.
Honored Artist of the Belorussian SSR (1969). Member of the CPSU since 1969. He took part in the Great Patriotic War of 1941–45. His poems were first published in 1925. He is the author of the poetry collections *Steps* (1928), *Unanimously* (1931), *Spring Votes for Spring* 931), *City Voices* (1932), *The Broad Battlefield* (1945), *Light of the Homeland* (1952), and *Open Spaces* (1958), all of which reveal the spiritual makeup of contemporary man, his work, and his struggle for peace. His books *With the Eves of Time* (1964; articles and notes on literature), *Kolas Talks About Himself* (1964), and *Twelve Evening Camp Fires* (1968) are well known. He has translated the works of A. S. Griboedov, A. S. Pushkin, N. V. Gogol, A. Mickievicz, and A. A. Fadeev into Belorussian.
He was a deputy to the Seventh and Eighth Convocations of the Supreme Soviet of the Belorussian SSR. Decors.: J. Kolas State Prize of the Belorussian SSR (1965), two orders, and several medals.
Works: *Zbor tvorau.* vols. 1–3. Minsk 1968–70. *In Russian translation: Jakub Kolas rasskazyvaet.* Moscow 1964. *Shumeli berezy: Stikhi, poemy.* Moscow-Leningrad 1971.

Lvov, Mikhail Davydovich
Poet; b. 1917, Nasibash, Bashkir ASSR.
Russian; mem. CPSU 1950–; grad. Gorkii Inst. of Literature 1941.
Publs.: "Vremia" (The Time) 1940, "Eti gody" (These Years) 1963, "U nas v Rossii" (With Us in Russia) 1964, and others.

Lvov, Sergei Lvovich
Author, journalist and literary critic; b. 1922, Moscow.
Russian; mem. CPSU 1943–; grad. Inst.

of History, Philosophy and Literature Moscow 1945; first works publ. 1947.
Publs.: "Poltora mesiatsa v zhizni Tamary" (A Month and a Half in Tamara's Life) 1959, "Spasite nashi dushi" (Save Our Souls) 1960, "Sereznaia muzyka" (Serious Music) 1964, "Serdtse slyshit" (The Heart Hears) 1962, "Gorod ne spit" (The Town Does Not Sleep) 1964, "Otkuda nachinaetsia puteshestvie" (Where Does the Journey Begin?) 1964, and others.

Lvovich, Mark Isaakovich, Dr. of Geol. Sc.
Hydrologist; b. My. 2 (15), 1906, Aleksandrovsk, now Zaporozhe.
Mem. CPSU 1945–; grad. Dept. of Geography of Leningrad Univ.; mem. staff State Hydrological Inst. 1924–52; conducted hydrological research projects in Far East 1930–34; mem. staff Cen. Union Service for Hydrological Prognoses 1937–41; dep. dir. of an inst. 1943–48; played part in hydrological safeguarding of Sov. Army 1941–45; Head of Dept. of Hydrology of Inst. of Geography at USSR Acad. of Sciences 1960–. Decors.: 4 orders, several medals.

Lyiurov, Aleksandr Alekseevich
Author; b. 1923, Zelenets, Komi ASSR; s. of a peasant.
Mem. CPSU 1951–.
Publs.: "Prostye liudi" (Simple People) 1951, "Zolotye ruki" (Golden Hands) 1956, and others.

Lykova, Lidia Pavlovna, Cand. of Econ.
State official; b. 1913.
Mem. CPSU 1938–; grad. Vologda Teachers' Training Inst. 1940 and Acad. of Social Sciences of Cen. Cttee. CPSU 1958; CPSU work in Vologda City and Oblast 1939–47; worked for All-Union CP (B) Cen. Cttee. 1947–49; Sec., 2nd Sec. of Ivanovo Oblast All-Union CP (B) Cttee. 1949–55; cand. mem. 1952–76, mem. Cen. Cttee. CPSU 1976–; Dep. to 1955–, mem. Presidium Supr. Sov. RSFSR 1959–63; 2nd Sec. of Smolensk Oblast Cttee. CPSU 1958–61; Min. of Social Security RSFSR 1961–67; Dep. Chair. of RSFSR Council of Mins. Ap. 1967–; Chair. of Comm. for the Fight Against Alcoholism at Council of Mins. RSFSR 1975–.

Lysenko, Viacheslav Grigor'evich
State official; b. 1931 in Oleksandrivka, Chernigov Oblast.
Mem. CPSU 1955–; Ukrainian; grad. Ukrainian Acad. of Agrarian Sciences 1955; chief engineer at a machine-tractor station, dir. of melioration station in Chernigov Oblast 1955–62; head of a technical production dept. of a specialized republican trust of the State Cttee. on Water Economy Ukrainian SSR 1962–63; held posts in apparat of Cen. Cttee. CP Ukraine 1963–69; Dep. Chair. of "Ukrsel'khoztekhnika" (State Cttee. for Supply of Production Equipment for Agriculture Ukrainian SSR) 1969–70; Dep. Head of Agricultural Dept. of Cen. Cttee. CP Ukraine 1970–72; Oblast Cttee. Sec. CP Ukraine in Crimea 1972–77; Head of Cen. Admin. for Horticulture, Viticulture, and the Wine Ind. Ukrainian SSR S. 1977–.

Maasik, Einar
Author; b. 1929, Narva; s. of a worker.
Esthonian; mem. CPSU 1959–; grad. Tartu Univ. 1956; first works publ. 1954.
Publs.: "Po rodnym preselkam" (On Native Country Lanes) 1959, "Sputnitsa" (Travelling Companion) 1960, "Na dorogakh semi dereven" (On the Streets of the Seven Villages) 1963, "Sem dnei Tuizu Taavi" (Seven Days of Tuiza Taavi) 1966, "Nesnosnyi kharakter" (Intolerable Character) 1962, and others.

Machavariani, Aleksei Davidovich
Composer; b. S. 10 (23), 1913, Gori, Georgia.
Grad. Tbilisi Conservatory 1936; Lecturer 1939–63, Prof. at Tbilisi Conservatory 1963–; famous for his ballet "Othello". Decors.: People's Artist of the USSR, State Prize of the USSR 1951, State Prize of the Georgian SSR 1971, Order of Lenin, several other orders and medals.

Mackevičius, Vytautas Stasjo
Painter; b. Ja. 1 (14), 1911, Šiauliai.
People's Artist of the Lithuanian SSR (1963). Member of the Communist Party of Lithuania (1933–35); member of the CPSU since 1948.
He studied at the Kaunas Art School (1928–35) with J. Vienožinskis, V. Didžiokas, P. Kalpokas, and K. Sklèrius and in Paris (1938–39) under C. Guérine. After becoming a participant in the underground revolutionary movement in 1933, he drew political cartoons for the illegal satirical journal *Šluota* (The Broom). He has taught at the Art Institute of the Lithuanian SSR in Vilnius since 1945 (professor since 1954; rector since 1953).
From the second half of the 1940's to the early 1950's, Mackevičius' work played an important role in the establishment of thematic painting in Lithuanian art. He was awarded the State Prize of the Lithuanian SSR in 1970. His works include *1919 in Lithuania* (1957), *The Rebels of 1863* (1963), and *The Light of Peace* (1969), all in the Art Museum of the Lithuanian SSR, Vilnius; and *By the Oaks of Ablinga* (1973). Decors.: Order of the October Revolution and two other orders.

Maergoiz, Isaak Moiseevich
Economic geographer; b. S. 4 (17), 1908, Ianov, present-day Vinnitsa Oblast.
Doctor of geographical sciences (1965).
He graduated from the geography department of Moscow State University in 1937; he became a professor in that department in 1966. His major research has been on the methodology of economic geography, urban geography, economic cartography, and the economic geography of other European socialist countries. He developed the concept of the industrial regions of foreign countries.
Works: *Ekonomicheskaia geografiia Vengrii.* Moscow 1956. *Chekhoslovatskaia Sotsialisticheskaia Respublika. Ekonomicheskaia geografiia.* Moscow 1964. *Ekonomicheskaia geografiia zarubezhnykh sotsialisticheskikh stran Evropy* (co-author and editor-in-chief). Moscow 1971. *Geograficheskie problemy sotsialisticheskoi ekonomicheskoi integratsii v Evrope* (co-author). Moscow 1971.

Magak'ian, Ivan Georgievich
Geologist; b. Mr. 24 (Ap. 6), 1914, Tbilisi.
Academician of the Academy of Sciences of the Armenian SSR (1948).
He graduated from the Leningrad Institute of Mines in 1935. He was head of a group of the Tadzhik-Pamir expedition organized by the Academy of Sciences of the USSR (1934–40), director of the Institute of Geological Sciences of the Academy of Sciences of the Armenian SSR (1960–63), academician-secretary of the Division of Technical Sciences of the Armenian Academy of Sciences (from 1943), and later academician-secretary of the Division of Earth Sciences of the same body (1963). From 1948 until 1960 he was a professor at the Leningrad Institute of Mines and since 1960 he has been a professor at the University of Yerevan. His principal works are devoted to the geology and geochemistry of ore deposits and to metallogeny. Decors.: State Prize of the USSR 1950, three orders.
Works: *Rudnye mestorozhdeniia,* 2nd ed. Yerevan 1961. *Osnovy metallogenii materikov.* Yerevan 1959. *Tipy rudnykh provintsii i rudnykh formatsii SSSR.* Moscow 1969. *Redkie, rasseiannye i redkozemel'nye elementy.* Yerevan 1971.

Magnitskii, Vladimir Aleksandrovich
Geophysicist; b. My. 30 (Je. 6), 1915, Penza.
Grad. Engineering Inst. for Geodesy, Aerial Photography and Cartography Moscow 1940; corresponding mem. USSR Acad. of Sciences 1964–; lecturer 1940–54, last position Head of Dept. of Geophysics of Moscow Univ.

Magradze, Elgudzha Samuilovich
State official; b. 1927, Khashuri.
Georgian; mem. CPSU 1951–; grad. Union Inst. of Law (correspondence study) 1949 and Tbilisi State Univ. 1952; Sec. of Kalinin Raion Cttee. Komsomol in Tbilisi 1952–53; Dir. of evening univ. courses in Marxism-Leninism for scientists and cultural workers of Tbilisi City Cttee. CP Georgia 1953–54; instructor of Dept. for Science and Culture Cen. Cttee. CP Georgia 1954–56; editor of the periodicals *Niangi* and *Mnatobi,* Dir. of Sabchota Sakartvelo publishing house

1957–78; Chair. of State Cttee. for Publishing Houses, Printing Plants and the Book Trade Georgian SSR N. 1978–; mem. Auditing Comm. CP Georgia 1981–.

Maiboroda, Georgii Illarionovich
Composer; b. N. 18 (D. 1), 1913, Pelekhovshchina, Poltava Oblast.
Grad. Kiev Conservatory 1941; teacher at Kiev Conservatory 1952–58. Decors.: Order of Lenin, Order of Red Banner of Labour.

Maikanova, Sabira
Actress; b. 1914.
Kazakh; mem. CPSU 1941–; studied at and mem. Kazakh Auezov Drama Theatre (major roles). Decors.: People's Artist of the USSR 1970, People's Artist of the Kazakh SSR 1959.

Maiorets, Anatolii Ivanovich
State official; b. 1929.
Mem. CPSU 1957–; grad. Zaporozhe Inst. of Mechanical Engineering 1962; served in Sov. Army 1950–53; electrician, shop superintendent at Zaporozhe Transformer Plant 1953–62, Dir. 1962–65; Dep. USSR Min., 1st Dep. Min. of the Electrical Equipment Ind. 1965–80, Min. D. 1980–; cand. mem. Cen. Cttee. CPSU 1981–. Decor.: USSR State Prize 1978.

Maiorov, Aleksandr Mikhailovich
Military official; b. 1920.
Russian; mem. CPSU 1943–; grad. Frunze Mil. Acad. 1951, Gen. Staff Acad. 1963; Army-Gen.; served in Sov. Army 1940–; company commander, engineer of a division in Voronezh, the Western, 1st and 2nd Ukrainian Front 1941–45; commander of an engineer batallion, in operational dept. of an army staff, 1st Dep. Chief-of-Staff of a mil. district 1945–68; Commander of Czechoslovak Central Forces Group 1968–72; Dep. to USSR Supr. Sov. 1970–; cand. mem. Cen. Cttee. CPSU 1971–81; Commander of Baltic Mil. District 1972–79; mem. Cen. Cttee. and Bureau Cen. Cttee. Latvian CP 1976–81.

Maiorov, Veniamin Stepanovich
State official.
Mem. Auditing Comm. 1964–71, cand. mem. Cen. Cttee. Azerbaidzhan CP 1971–; 1st Dep. Chair. of Baku City Exec. Cttee. of People's Deps. 1969–80; Min. of Housing and Communal Econ. Azerbaidzhan SSR Mr. 1980–.

Maisenberg, Oleg
Pianist; b. 1945, Odessa.
First piano lessons at the age of 5; moved to Kishinev, grad. Cen. Music School in Kishinev 1963; attended master class of Prof. Iokheles at Gnessin Inst. of Music in Moscow; Asst. to Prof. at same inst. 1968–71; soloist with Moscow Philharmonic Orchestra 1971–; accompanist of Sov. violinist Gidon Kremer for the past years; numerous performances in Germany, Austria, Hungary, Yugoslavia, Rumania and Czechoslovakia. Decors.: 2nd Prize at Int. Schubert Contest in Vienna 1967, won prize at "Music of the 20th Century" Contest in Vienna.

Makaed, Nikolai Aleksandrovich
State official.
Min. of Trade Belorussian SSR O. 1981–.

Makarenko, Viacheslav Aleksandrovich
Party official.
2nd Sec. of Osh Oblast Cttee. Kirghiz CP 1973–81, 1st Sec. 1981; mem. Cen. Cttee. Kirghiz CP 1976–; mem. Bureau and 2nd Sec. Cen. Cttee. Kirghiz CP 1981–.

Makarenko, Viktor Sergeevich
Party official; b. 1931.
Ukrainian; mem. CPSU 1960–; grad. Dnepropetrovsk Metallurgical Inst. 1955; asst. foreman, roller, foreman, shift supervisor, senior foreman, engineer of a rolling dept. 1955–62; sec. of a party bureau of a plant shop, dep. sec., sec. of party cttee. of Krivoi Rog Metallurgical Plant 1962–67; 2nd Sec. of Krivoi Rog City Cttee. Ukrainian CP 1967–69; inspector of a dept. of Cen. Cttee. Ukrainian CP 1969–72; 1st Sec. of Sevastopol City Cttee. Ukrainian CP 1972–77; mem. Cen. Cttee. Ukrainian CP 1976–; 1st Sec. of Crimean Oblast Cttee. Ukrainian CP Jl. 1977–; Dep. to USSR Supr. Sov. 1979–; cand. mem. Cen. Cttee. CPSU 1981–82, mem. N. 1982–.

Makarev, Leonid Fedorovich
Actor, stage director and dramatist; b. Ag. 12 (24), 1892, Perm.
Mem. CPSU 1940–; grad. Dept. of History and Philology of Petrograd Univ. 1917 and Higher Pedagogical Courses 1919; Prof. of Drama at Leningrad Inst. of Theatre, Music and Film 1939–. Decors.: People's Artist of the RSFSR, 3 orders and various medals.

Makarevskii, Aleksandr Ivanovich
Aircraft engineer; b. Ap. 3 (16), 1904, Mushkovichi, Smolensk Oblast.
Mem. CPSU 1943–; grad. Bauman Tech. College 1929; mem. USSR Acad. of Sciences 1968–; mem. staff Cen. Aerodynamical Inst. 1927; Dir. of same inst. 1950–60; concurrently Prof. at Moscow Physical-Tech. Inst. 1952–. Decors.: Hero of Socialist Labour, 3 Orders of Lenin, 3 other orders and various medals.

Makarii (Leonid Nikitich Svistun)
Russian Orthodox clergyman; b. S. 14, 1938, Kiev.
Grad. Moscow Theological Acad., studied Bosa Ecumenical Inst. Switzerland 1965; took part in 3rd Christian Peace Conference in Prague 1965; deacon, priest, monk, custodian at Vladimir Cathedral in Kiev 1968; mem. Kiev Branch of Dept. of External Affairs of Moscow Patriarchate; Archimandrite 1970; Bishop of Uman entrusted with leadership of parishes of Moscow Patriarchate in Canada 1970–74; Head of Moscow Patriarchate's Del. at World Council of Churches in Geneva 1974–.

Makarkin, Nikolai Vladimirovich
State official.
Mem. Cen. Cttee. Turkmen CP 1961–; Dep. to Supr. Sov. Turkmen SSR 1963–; 1st Sec. of Chardzhou City Cttee. Turkmen CP 1963; Head of Dept. of Admin. Organs at Cen. Cttee. Turkmen CP 1963–75; Chair. of Cttee. for People's Control Turkmen SSR My. 1975–; mem. Bureau Cen. Cttee. Turkmen CP Jl. 1975–.

Makarov, Oleg Grigor'evich
Cosmonaut; b. 1933.
Mem. CPSU 1961–; grad. Higher Tech. Bauman Inst. Moscow 1957; worked in a bureau for constructional engineering 1957–66; cosmonaut 1964–; flight engineer with spacecraft "Soiuz 12", together with V. G. Lazarev tested improved flight control systems during orbital flight S. 27–29, 1973; together with V. G. Lazarev went into orbit in "Soiuz" capsule Ap. 5, 1975; flight was intended as continuation of experiments with orbital station "Saliut 4" but had to be discontinued because of deviation from pre-calculated course; "Soiuz 27" Ja. 1978. Decor.: Hero of the Sov. Union 1973.

Makarova, Inna Vladimirovna
Actress; b. Jl. 28, 1928, Taiga, now Kamerovo Oblast.
Grad. Union Film Inst. Decors.: People's Artist of the RSFSR, State Prize of the USSR 1949, Order of Red Banner of Labour.

Makarova, Tamara Fedorovna
Russian actress; b. Jl. 31 (Ag. 13), 1907, St. Petersburg.
People's Artist of the USSR (1950). Became a member of the CPSU in 1943.
She completed her training at the Leningrad Institute of Stage Arts in 1930. Her first film roles were the typist Dudkina in *Someone Else's Jacket* (1927) and Greta in *Deserter* (1933). Beginning in 1934, with her role as Natal'ia in the film *Do I Love You,* her career was linked with the work of the director S. A. Gerasimov. She played the roles of Zhenia Okhrimenko in *The Courageous Seven* (1936), Natasha in *Komsomol'sk* (1938), Agrafena in *The Teacher* (1939), Anna Sviridova in *The Home Front* (1944), Elena Koshevaia in *The Young Guard* (1948), Anna Andreevna in *Men and Beasts* (1962), Panina in *The Journalist* (1967), and the architect Petrushkovaia in *To Love a Man* (1972). In these films she created truly Russian characters, distinguished by gentle femininity, naturalness, sincere warmth, and inner sensitivity.
During and after the Great Patriotic War, she portrayed women who had endured

the ordeal of war and, in the process, had become stronger and wiser. In 1941 she played the role of Nina in the film *Masquerade* (adaptation of M. Iu. Lermontov's play). She has also been in films that were not directed by Gerasimov; for example, she portrayed the Mistress of the Copper Mountain in *The Stone Flower* (1946) and Ekaterina Ivanovna in *Memory of the Heart* (1958). In 1944 she became a teacher at the All-Union State Institute of Cinematography, receiving a professorship there in 1968. Decors.: State Prize of the USSR (1941, 1947), two orders, and various medals.

Makaseev, Boris Konstantinovich
Cameraman; b. My. 29 (Je. 6), 1907, Moscow.
Mem. CPSU 1958–; grad. State Tech. Inst. of Cinematography 1927; took part in Spanish Civil War 1936–37; head of camera teams on the front 1941–45. Decors.: Merited Artist of the RSFSR, State Prize of the USSR 1946, 1947, 1948, 1951, 3 orders and various medals.

Makatukha, Vasilii Ivanovich
Painter; b. 1922, Kitaigorod, Dnepropetrovsk.
Mem. CPSU 1959–; grad. Dnepropetrovsk 1947 and Kiev Art Inst. 1954, pupil of O. Shovkunenko, G. Melikhov.

Makeev, Valentin Nikolaevich
State official; b. 1930.
Mem. CPSU 1956–; grad. Moscow Econ. Engineering Inst. 1953, Cand. of Tech. Sc. 1958, Asst. Prof. 1961; employed at Moscow Econ. Engineering Inst. 1953–64; party posts 1964–76; Sec., 2nd Sec. of Moscow City Cttee. CPSU 1976–80; mem. Cen. Cttee. CPSU 1976–; Dep. to USSR Supr. Sov. 1978–; Dep. Chair. of USSR Council of Mins. O. 1980–.

Makeev, Viktor Petrovich, Dr. of Tech. Sc.
Engineer; b. 1924.
Russian; mem. CPSU 1951–; grad. Moscow Aviation Inst. 1948; tech. draughtsman, designer, engineer-designer in aviation works 1939–44 and 1948–50; Komsomol posts 1950–52; employed in works of USSR Min. of Gen. Machine Construction 1952–; Dep. to USSR Supr. Sov. 1966–; cand. mem. 1971–76, mem. Cen. Cttee. CPSU 1976–. Decors.: Hero of Socialist Labour, Lenin Prize.

Makhkamov, Kakhar
State official.
Dep. to Supr. Sov. Tadzhik SSR 1959–; Head of "Tadzhikugol'" Trust 1959; cand. mem. 1960–61, mem. Cen. Cttee. Tadzhik CP 1963–; Chair. of Leninabad City Exec. Cttee. of Workers' Deps. 1963; Chair. of GOSPLAN Jl. 1963–82, Dep. Chair. of Council of Mins. Tadzhik SSR O. 1965–82, Chair. Ap. 1982–; mem. Bureau Cen. Cttee. Tadzhik CP Je. 1982–.

Makhmudov, Abdurakhim
State official.
Chair. of State Comm. on the Fishing Ind. Uzbek SSR F. 1977–.

Makhmudov, Nasyr Makhmudovich
State official; b. 1913.
Uzbek; mem. CPSU 1940–; high-school educ. uncompleted; village teacher, literary contributor to a raion newspaper, instructor and dept. head of a raion sov. 1931–36; served in Sov. Army 1936–38; Chair. of an oblast exec. cttee. of workers' deps. 1938–43; Sec. of Samarkand Oblast Cttee. Uzbek CP 1943–48; Dep. to USSR Supr. Sov. 1946–54 and 1958–70; attended Party High School at Cen. Cttee. CPSU 1948–50; 1st Sec. of Samarkand, Fergana and Tashkent Oblast Cttees. Uzbek CP, dir. of a sovkhoz 1950–55; mem. Cen. Cttee. Uzbek CP 1954–; Chair. of Bukhara Oblast Cttee. of Workers' Deps. 1955–56; 1st Sec. of Oblast Cttee. Uzbek CP in Kara-Kalpak ASSR 1956–63; 1st Sec. of Syr-Daria Oblast Cttee. Uzbek CP 1963–69; Chair. of Cttee. on People's Control Uzbek SSR Jl. 1969–; cand. mem. Bureau Cen. Cttee. Uzbek CP Mr. 1971–81. Decor.: Order of Red Banner of Labour.

Makhmudov, Sharof
State official.
Dep. Min. of Justice Tadzhik SSR 1971–78, Min. O. 1978–; mem. Auditing Comm. Tadzhik CP 1981–.

Makhmudov, Sultan Mamadalievich
State official.
Head of Main Statistical Admin. Uzbek SSR D. 1976–.

Makhmudova, Naima Makhmudovna
Trade union official; b. 1928.
Uzbek; mem. CPSU 1950–; grad. Medical Inst. in Tashkent, M. D.; neurologist, chief doctor of a raion and oblast epidemiology station 1951–63; Dep. Min. of Health Uzbek SSR 1963–71; Sec. of Uzbek republican Council of Trade Unions 1971–77, Chair. My. 1977–; cand. mem. Cen. Cttee. CP Uzbekistan 1976–77, mem. Sept. 1977–; cand. mem. Bureau of Cen. Cttee. CP Uzbekistan F. 1981–.

Makhmudova, Sh. N.
Komsomol official.
Sec. and mem. Bureau Cen. Cttee. Uzbek Komsomol Je. 1976–.

Maksim (Boris Krokha), Cand. of Theol.
Russian Orthodox clergyman; b. D. 20, 1928, Chernigovka, Bashkir ASSR.
Grad. Leningrad Theological Seminary and then Theological Acad. 1955, Cand. of Theology; monk 1949; deacon 1950; priest 1954; after 1970 mem. staff Dept. of External Affairs of Moscow Patriarchate; Bishop of Argentina and South America 1972; Bishop of Omsk and Tiumen 1974–.

Maksimenko, G. D.
Party official.
Sec. and mem. Bureau Cen. Cttee. Ukrainian Komsomol 1970–78; Head of a Section at Dept. of Propaganda and Agitation with Cen. Cttee. Ukrainian CP 1978–79; Sec. of Odessa Oblast Cttee. Ukrainian CP Jl. 1979–.

Maksimov, Iurii Pavlovich
Military official; b. 1924.
Russian; mem. CPSU 1943–; grad. M. V. Frunze Mil. Acad. 1950 and Mil. Acad. of Gen. Staff 1965, Col.-Gen.; entered Sov. Army 1942; served at the front 1943–45; command and staff posts after 1945; Commander of Troops Turkestan Mil. District 1979–; Dep. to USSR Supr. Sov. 1979–; mem. Bureau of Cen. Cttee. Uzbek CP 1981–; cand. mem. Cen. Cttee. CPSU 1981–.

Maksimov, Mark Davydovich (real name: Lipovich)
Poet and author; b. 1918, Snovsk, now Shchors, Chernigov Oblast.
Russian; grad. Dept. of Literature of Gorkii Pedagogic Inst. Kiev 1940; first works publ. 1940.
Publs.: "Nasledstvo" (The Inheritance) 1946, "Rovestniki" (Contemporaries) 1947, "Desiat' let spustia" (Ten Years Later) 1956, "Soldat" 1959, "Navsegda" (Forever) 1950, film script "Lichno izvesten" (Personally Known) 1958, and others.

Maksimov, Vladimir Emelianovich
Author; b. 1932, Leningrad.
Russian; grew up in children's colonies, completed FZO school (works school), trained bricklayer, worked on various building sites; first works publ. 1956.
Publs.: "Pokolenie na chasakh" (Generation on Guard) 1956, "My obzhivaem zemliu" (We Make the Country Inhabitable) 1961, "Zhiv chelovek" (Man Lives) 1962, "Dusia i nas piatero" (Dusia and the Five of Us) 1965, "Iskushenie" (Temptation), "Shag k gorizontu" (Steps Toward the Horizon), "Pozyvnye tvoikh parallelei" (Signals From Your Parallels) 1964, and others.

Maksimova, Ekaterina Sergeevna
Ballerina; b. F. 1, 1939, Moscow.
Grad. Choreographical Inst. Moscow 1958; mem. ballet of Bolshoi Theatre 1958–. Decors.: People's Artist of the USSR, Gold Medal at Int. Ballet Competition Varna 1964, Pavlova Prize of Paris Dance Acad. 1969, Order of Red Banner of Labour.

Maksudov, Latyp Maksudovich
Diplomat.
USSR Amb. to Gabun Jl. 1982–.

Maksumov, Akbar Nustratulaevich
State official.
Mem. Auditing Comm. 1961–66, mem. Cen. Cttee. Tadzhik CP 1966–; Min. of

Agriculture Tadzhik SSR 1965–71; Vice-Pres. of Acad. of Sciences Tadzhik SSR 1971–75; Dep. Chair. of Council of Mins. Tadzhik SSR My. 1975–.

Makukhin, Aleksei Naumovich
State official.
City Sec. of Ukrainian Komsomol Cttee. in Crimea 1958; 2nd Oblast Sec. of Ukrainian CP in Crimea 1964–71; cand. mem. Cen. Cttee. Ukrainian CP 1966–; Min. of Power and Electrification Ukrainian SSR N. 1971–82; 1st Dep. USSR Min. of Power and Electrification Ap. 1982–.

Malbakhov, Timbora Kubatievich
Party official; b. 1917.
Kabardinian; mem. CPSU 1942–; grad. Ordzhonikidze Agricultural Inst. 1940 and CPSU Cen. Cttee. Party Higher School 1953; chief agronomist, dir. of a machine and tractor station 1940–41; served in Sov. Army 1941–46; 1st Sec. of a raion cttee., 2nd Sec. of Kabarda-Balkar Oblast CPSU Cttee. 1946–52; dep. to 1950–, mem. Draft Bills Comm. at Sov. of Nationalities USSR Supr. Sov. 1966–; Chair. of Presidium Kabarda-Balkar Supr. Sov. 1952–56; 1st Sec. of Kabarda-Balkar Oblast CPSU Cttee. 1956–; cand. mem. Cen. Cttee. CPSU 1961–. Decor.: Order of Lenin.

Maldonis, Alfonsas Moteevich
Author and poet; b. Ag. 22, 1929, Naujaplenta, Lithuania.
Lithuanian; mem. CPSU 1961–; grad. Dept. of History and Philology of Vilnius Univ. 1954; Chief Ed. of "Vaga" Publ. House in Vilnius 1962–70; first works publ. 1956. Decor.: State Prize of the Lithuanian SSR 1965.
Publs.: "Seredina leta" (Midsummer) 1958, "Veter tuchu gonit" (The Wind Drives the Cloud) 1960, "Solnechnyi liven" (Sun Rainshower) 1962, "Rastut derevia" (The Trees Grow) 1965, "Eti goda" (These Years) 1966, "Vodianye znaki" (Water Marks) 1969, "Stikhi" (Poems) 1971, and others.

Malikov, Kubanychbek Imanalievich
Poet and dramatist; b. 1911, Uch-Emchek.
Kirghiz; mem. CPSU 1940–; first works publ. 1928.
Publs.: "Stikhi Kubanychbeka" (Kubanychbek's Poems) 1933, "Serdtse betsia" (The Heart Beats) 1957, "Delo Adyla" (Adyla's Business), "My ne te, chto byli" (We Are Not as We Were) 1950, "Duma o zavtrashnem dne" (Thoughts About Tomorrow) 1964, and others.

Maliusov, Vladimir Aleksandrovich
Scientist; b. Je. 21 (Jl. 4), 1913, Moscow.
Mem. CPSU 1948–; grad. D. I. Mendeleev Inst. of Chemical Technology 1940; corresponding mem. USSR Acad. of Sciences 1968–; mem. staff L. Ia. Karpov Inst. of Physical Chemistry 1946–63; Head of Laboratory at N. S. Kurnakov Inst. of General and Inorganic Chemistry of USSR Acad. of Sciences 1963–. Decors.: State Prize 1953, 2 orders, various medals.

Malkin, Aleksandr Stepanovich
Poet; b. 1923, Kishaly, Mordovian ASSR.
First works publ. 1937.
Publs.: "Stikhi" (Poems) 1948, "Slovo tovarishcha" (The Friend's Word) 1952, "Po puti slavy" (On the Way to Fame) 1956, "Slovo druzhby" (The Word of Friendship) 1954, and others.

Malkov, Nikolai Ivanovich
Party official; b. 1932.
Russian; mem. CPSU 1954–; grad. Far Eastern Naval Engineering College in Vladivostok 1955 and CPSU Cen. Cttee. Party Higher Correspondence School 1972; served in Sov. Navy 1955–56; navigator, dep. chief of the nautical school of the Far Eastern Naval Steamship Line in Vladivostok 1956–62; sec. of party cttee., dep. chief of Vladivostok Sea Trade Port, dep. sec., sec. of party cttee. of naval transport in Vladivostok 1962–69; chief of a dept. of Primore Krai Cttee. CPSU, 1st Sec. of Vladivostok City Cttee. CPSU, 2nd Sec. of Primore Krai Cttee. CPSU 1969–78; 1st Sec. of Magadan Oblast Cttee. CPSU D. 1978–; Dep. to USSR Supr. Sov. 1979–; cand. mem. Cen. Cttee. CPSU 1981–.

Malmeister, Aleksandr Kristapovich, Dr. of Tech. Sc.
Scientist; b. 1911.
Mem. CPSU 1942–; grad. Latvian Univ. 1937; Dr. of Tech. Sc. and Prof. 1959–; mem. Latvian Acad. of Sciences 1958–; corresponding mem. USSR Acad. of Sciences 1970–; Instructor, Lecturer, Prof., Head of a Dept. at Latvian Univ. 1945–52; Dir. of Inst. of Civil Engineering and Architecture at Latvian Acad. of Sciences 1953–61; Rector of Polytech. Inst. Riga 1961–63; Dir. of Inst. of Mechanics of Polymers at Latvian Acad. of Sciences 1963–70; Pres. of Latvian Acad. of Sciences 1970–. Decor.: Hero of Socialist Labour 1969.

Malofeev, Anatolii Aleksandrovich
Party official.
1st Sec. of Mozyr Raion Cttee. Belorussian CP until 1975; Sec. of Gomel Oblast Cttee. Belorussian CP 1975–78, 1st Sec. Jl. 1982–; Chair. of Gomel Oblast Exec. Cttee. of Peoples Deps. 1978–82; mem. Cen. Cttee. Belorussian CP 1981–.

Maltinskii, Khaim Izrailevich
Poet; b. 1910, Ponevezhis, Lithuania; s. of a craftsman.
Jew; mem. CPSU 1943; first works publ. 1924.
Publs.: "Bez teni" (Without Shadows) 1936, "Bratishkam i sestrenkam" (For the Little Brothers and Sisters) 1939, "Riadom s synoviami vsemi" (Beside all the Sons) 1963, and others.

Maltsev, Elizar Iurevich (real name: Pupko)
Author; b. 1917, Khonkholoi, Buriat ASSR; s. of a peasant.
Mem. CPSU 1949–; grad. Gorkii Inst. of Literature 1944; first works publ. 1938. Decor.: Stalin Prize.
Publs.: "Goriachie kliuchi" (Hot Springs) 1945, "Ot vsego serdtsa" (With All My Heart) 1948, "Vtoraia liubov'" (Second Love) 1950, "Voidi v kazhdyi dom" (Go Into Every House) 1960, and others.

Maltsev, Nikolai Alekseevich, Cand. of Tech. Sc.
State official; b. 1928.
Russian; mem. CPSU 1953–; grad. Groznyi Petroleum Inst. 1951; shop superintendent, chief engineer of an office of "Almentevneft'" Trust, dep. chief engineer of "Tatneft'" Asscn., chief of an admin. for oil prod. 1951–61; chief of an admin. for oil prod. in Perm Sovnarkhoz, chief of "Permneft'" Prod. Asscn. 1961–72; Dep. to USSR Supr. Sov. 1966–70 and 1974–; 1st Dep. USSR Min. of the Petroleum Ind. 1972–77, Min. Ap. 1977–; cand. mem. Cen. Cttee. CPSU 1981–. Decors.: Hero of Socialist Labour 1971, Order of the October Revolution 1978.

Maltsev, Nikolai Semenovich
State official; b. 1919.
Min. of Rural Construction RSFSR Je. 1973–. Decor.: Order of Red Banner of Labour.

Maltsev, Viktor Fedorovich
Diplomat; b. 1917.
Russian; mem. CPSU 1945–; grad. Inst. of Railway Engineers Novosibirsk 1941 and Moscow Acad. of Railway Transport 1954; engineer, Chief Engineer, Dept. Head with Krasnoiarsk Railway 1941–46; Chief Engineer, Dept. Head, Dep. Head of Transport Admin. of Far-Eastern Railway 1946–51; Dep. Head of Far-Eastern Siberian Railway 1954–61; CPSU and sov. posts 1961–66; Sec. of Irkutsk Oblast CPSU Cttee. 1961–63; Chair. of Irkutsk Oblast Exec. Cttee. of Workers' Deps: 1963–64; Sec. of Irkutsk Oblast CPSU Cttee. 1965; Chair. of Irkutsk Oblast Exec. Cttee. of Workers' Deps. 1966; mem. Diplomatic Service 1966–; USSR Amb. to Sweden 1967–71; USSR Amb. to Finland 1971–73 and to India Ja. 1974–77; cand. mem. 1971–76, mem. Cen. Cttee. CPSU 1976–; USSR 1st Dep. Min. of Foreign Affairs 1977–.

Malygin, Vadim Il'ich
State official.
Chair. of State Cttee. for Supervision of Safe Working Practices in Industry and for Mine Tadzhik SSR O. 1981–.

Malykhin, Vasilii Mikhailovich
Laborer; b. 1935.
Mem. CPSU 1958–; finished 7th class in secondary school 1951; electrician in

Kramatorsk 1952–53; served in Sov. Army 1954–58; electrician in various enterprises in Gorkii 1958–68; adjuster in Volga Automobile Plant in Volga Oblast 1968–72, work brigade leader 1972–; mem. Cen. Auditing Comm. CPSU 1981–. Decors.: Hero of Socialist Labour 1973.

Malyshev, Aleksandr Evgenevich
Diplomat; b. 1919.
Secondary classical educ.; mem. Diplomatic Service 1946–; sen. posts in USSR Min. of Foreign Affairs and abroad 1946–57; Emb. Counsellor at USSR Emb. in Yugoslavia 1957–58; sen. posts in USSR Min. of Foreign Affairs 1958–60; Dep. Head of 2nd African Dept. at USSR Min. of Foreign Affairs 1960–63; Emb. Counsellor at USSR Emb. in CSSR 1963–64; USSR Amb. to Tunisia 1964–69; with USSR Min. of Foreign Affairs 1969–72 and 1978–; USSR Amb. to Cameroon Jl. 1972–78.

Mamadzhanov, Iuldash
State official.
Min. of Meat and Dairy Ind. Uzbek SSR O. 1965–; cand. mem. Cen. Cttee. Uzbek CP 1966–.

Mamarasulov, Salidzhan
Party official; b. 1930.
Uzbek; mem. CPSU 1958–; grad. Tashkent Engineering Inst. for the Irrigation and Mechanization of Agriculture 1953, Cand. of Tech. Sc. 1977; engineer, senior engineer, dep. dir., dir. of "Uzgiprovodkhoz" Inst. 1954–64; Chief of Main Admin. Water Econ. Construction Uzbek SSR 1964–65; Min. of Melioration and Water Econ. Uzbek SSR 1965–77; cand. mem. Cen. Cttee. Uzbek CP 1966–71, mem. 1971–; Dep. Chair. of Council of Mins. Uzbek SSR 1977–78; 1st Sec. of Andizhan Oblast Cttee. Uzbek CP D. 1978–; Dep. to USSR Supr. Sov. 1979–; mem. Cen. Auditing Comm. CPSU 1981–. Decor.: Order of Red Banner of Labour.

Mambeev, Sabur Abdrasulovich
Painter; b. My. 1, 1928 Abai-Bazar, present-day village of Abai, Chardara Raion, Chimkent Oblast.
Honored Art Worker of the Kazakh SSR (1963). Member of the CPSU since 1962. He studied with M. I. Avilov at the I. E. Repin Institute of Painting, Sculpture, and Architecture in Leningrad (1947–53). His genre compositions (*In the Mountains,* 1956–57, Museum of Art of the Eastern Peoples in Moscow; *In My City,* 1960, T. G. Sherchenko Kazakhstan Art Gallery in Alma-Ata) and his portraits (*Girl in Black,* 1964, in the same gallery) are subtly lyrical and characterized by soft harmonies of luminous areas of color. He has been chairman of the Artists' Union of the Kazakh SSR since 1968; he also held that post from 1956 to 1962. Decors.: two orders.

Mamedkhanly, Enver Gafar ogly
Azerbaidzhan writer; b. F. 1913, Geokchai.
Honored Art Worker of the Azerbaidzhan SSR.
He graduated from the All-Union State Institute of Cinematography in 1938. His first novella, *The Whirlpool* (1934), is devoted to the Baku oil workers. He served in the Great Patriotic War of 1941–45 (he was a special correspondent for a frontline newspaper in Tabriz). He has written the stories "The Caravan Stopped" (1944), "On the Main Avenue", and "The Scarlet Buds" (1945). His play *Morning of the East* (1947) describes the establishment of Soviet power in Azerbaidzhan, and his play *In the Fire* (1950) depicts life in southern Azerbaidzhan. His lyrical comedy *A Shirvan Beauty* (1957) was successfully staged. He wrote the film screenplays for *Bakhtiiar* (produced 1942) and *Fatalikhan* (with Mekhti Gusein; produced 1947). He has translated works of A. S. Pushkin, L. N. Tolstoy, M. Gorkii, Shakespeare, G. Flaubert, and T. Dreiser into Azerbaidzhani. Decors.: three orders.
Works: *Sechilmish äsärläri,* 1960. *In Russian translation: Izbrannoe.* Baku 1959. *Alye butony.* Moscow 1964.

Mamedov, Arif Gusein Kuli ogly
State official.
Min. of Meat and Dairy Ind. Azerbaidzhan SSR Jl. 1971–; mem. Auditing Comm. 1976–81, cand. mem. Cen. Cttee. Azerbaidzhan CP 1981–.

Mamedov, Iza Ali ogly
Party official.
Mem. Cen. Cttee. Azerbaidzhan CP 1971–; 1st Sec. of Lenkoran City Cttee. Azerbaidzhan CP until 1981; Sec. and mem. Bureau Cen. Cttee. Azerbaidzhan CP Ja. 1981–.

Mamedov, Kamil Seid ogly
State official.
Min. of Food Ind. Azerbaidzhan SSR Ag. 1972–. Cand. mem. Cen. Cttee. Azerbaidzhan CP 1976–.

Mamedov, Musa Salam ogly
State official.
Chair. of Scientific Production Asscn. for Agro-Chemical Supply to Agriculture Azerbaidzhan SSR 1979–; mem. Auditing Comm. CP Azerbaidzhan 1981–.

Mamiliev, Ashirmurad
State official.
Min. of Culture Turkmen SSR N. 1979–; mem. Auditing Comm. Turkmen CP 1981–.

Mamlin, Gennadii Semenovich
Author; b. 1925, Simferopol.
Russian; writes for children; first works publ. 1950.
Publs.: "Nikita Snegirev" 1956, "A s Aleshkoi my druzia ..." (But We Are Friends With Aleshka ...) 1961, "Chudesa v polden'" (Miracle at Noon) 1964, "Feuerwerk" (Fireworks) 1965, and others.

Managarov, Ivan Mefod'evich
Military official; b. My. 31 (Je. 12), 1898, Enakievo, now in Donetsk Oblast.
Colonel-general (1945); Hero of the Soviet Union (Ap. 28, 1945). Member of the CPSU from 1919.
He entered the Red Army in 1918. He participated in the Civil War of 1918–20 as assistant commander and commander of a regiment. In 1931 he graduated from the Military Political Academy and in 1947 completed the Higher Academic Training Courses. Between 1931 and 1941 he was commander of a cavalry regiment and a cavalry division. During the Great Patriotic War of 1941–45 he commanded a cavalry division and rifle and cavalry corps as part of the Far Eastern Front; then from 1942 he participated in combat actions on the Briansk, Kalinin, Northwestern, Steppe, Second Ukrainian, and Transbaikal fronts as cavalry corps commander (1942) and commander of the Forty-first and Fifty-third armies (1942–45). After the Great Patriotic War he continued in command posts. He entered the reserve in 1953. Decors.: two Orders of Lenin, three Orders of the Red Banner, the Order of Suvorov First Class, the Order of Kutuzov First Class, the Order of Bogdan Khmel'nitskii First Class, the Order of the Red Star, five foreign orders, and various medals.

Manastyrskii, Vitold Antonovich
Painter; b. 1915, Lemberg; s. of A. I. Manastyrskii.
Grad. Warsaw Acad. of Art 1939; teacher at Lemberg Inst. for Applied and Decorative Art 1947–.

Mandelshtam, Nadezhda Iakovlevna, Cand. of Philol. Sc.
Author; b. Ag. 1, 1899, Saratov.
Studied painting in Kiev, was married to poet Osip Emilevich M. who was arrested and liquidated in 1938; lives in retirement in Moscow.
Publs.: "Vospominaniia" (Hope Against Hope. A Memoir) New York 1970, "Vtoraia kniga" Paris 1972 (Engl. translation "Hope Abandoned", Vol. 2 New York 1974).

Maniakin, Sergei Iosifovich
Party official; b. 1923; s. of a worker.
Russian; mem. CPSU 1945–; grad. Stavropol Agricultural Inst. 1948; served in Sov. Army 1941–43; teacher, headmaster 1943–48; agronomist, dir. of a machine and tractor station, kolkhoz chair., Head of Agricultural Dept. with Stavropol Krai Exec. Cttee. of Workers' Deps. 1960–61; 1st Sec. of Omsk Oblast CPSU Cttee. Ag. 1961– (for Agriculture 1963–64). Mem. Cen. Cttee. CPSU 1961–; Dep. to 1962–, mem. Credentials Comm. 1962–66, mem.

Agricultural Comm. at Sov. of the Union USSR Supr. Sov. 1966–74. Decor.: Order of Red Banner of Labour.

Mannapov, Nazrilla
Laborer; b. 1930.
Mem. CPSU 1959–; grad. Tashkent Aviation Technicum 1962; served in Sov. Army 1951–54; fitter at V.P.Chkalov Aviation Prod. Asscn. in Tashkent 1955–; mem. Cen. Auditing Comm. CPSU 1981–. Decors.: Hero of Socialist Labour 1981.

Mannur, Shaikhi Fakhrullovich (also Sh.F.Mannurov)
Tartar poet; b. Ja.2 (15), 1905, Tulbaevo, present-day Mamadysh Raion.
Member of the CPSU (1944).
In 1937, he graduated from a pedagogical institute in Kazan. He fought in the Great Patriotic War of 1941–45. Mannur's works began to appear in print in 1923. His first poetry collections were published in 1928. He was interested in the working class (the poem "Cast Iron Streams", 1930), but he also wrote about the collective farm village (the poems "Grandfather Gaidzhan" and "One of Thousands of Evenings", both written in 1935). During the war years, he wrote patriotic lyric verses and the poem "The Girl From Kazan" (1946). His poem "The Beautiful Girl Is the Daughter of Mother Earth" (1956) is devoted to the oil workers of Tataria.
In 1968, he published the novel *Musa* about M.Dzhahl. He translated *The Tale of Igor's Campaign,* as well as works by A.S.Pushkin, M.Iu. Lermontov, and I.A. Krylov. He has also written books for children. Decors.: two orders, as well as a number of medals.
Works: *Küngel yomgagï. Shigïr'lär häm poëmalar,* vols. 1–2. Kazan 1968.

Manuilova, Ol'ga Maksimilianovna
Sculptor; b. O.17 (29), 1893, Nizhnii Novgorod, present-day Gorkii.
People's Artist of the Kirghiz SSR (1954). From 1913 to 1917, she studied at the Moscow School of Painting, Sculpture, and Architecture. She has worked in Kirghizia since 1939. A master of large-scale decorative sculpture, she also did portraits and small sculptures. Her works include the bas-relief *Soviet Constitution* on the House of the Procurator (cement, 80 sq.m, 1941), a monument to General I.V.Panfilov (granite, cement, 1942, with A.A.Manuilov), decorative groups in the Kirghiz Theater of Opera and Ballet (concrete, 1952–55), a monument to Togolok Moldo (granite, 1963), and a monument to the cosmonaut V.M.Komarov (cement, 1969) – all are in the city of Frunze. Decor.: Order of the Red Banner of Labour.

Marchenko, Vladimir Aleksandrovich, Dr. of Phy.-Math. Sc.
Mathematician; b. Jl.17, 1922, Kharkov; s. of a prof.
Grad. Kharkov Univ. 1948; corresponding mem. Ukrainian Acad. of Sciences 1961–; Dr. of Phy.-Math. Sc. 1952–, Prof. 1953–; mem. scientific staff, Head of Dept. of Mathematical Analysis and Functional Theory at Research Inst. of Mathematics and Mechanics at Kharkov Univ., Lecturer, Prof., Head of Dept. of Mathematics at Kharkov Univ. 1948–61; Dept. Head at Physico-Tech. Inst. of Low Temperatures Kharkov 1961–.
Publs.: co-author of "Razlozhenie po sobstvennym funktsiiam nesamosopriazhennykh singuliarnykh differentialnykh operatorov vtorogo poriadka" in "Matematicheskii sbornik" vol. 52/1960, co-author of "Obratnaia zadacha teorii rasseianiia" 1960.

Marchuk, Gurii Ivanovich
State official; b. Je.8, 1925, Petro-Khersonets, Orenburg Oblast.
Mem. CPSU 1947–; grad. Leningrad Univ. 1949; mem. USSR Acad. of Sciences 1968–; mem. staff Inst. of Physical Energetics Obninsk 1953–62; Inst. of Mathematics at Siberian Branch USSR Acad. of Sciences 1962–64; Dir. of Computer Centre at Siberian Branch USSR Acad. of Sciences 1964–80; Dep. Chair. 1969–75, Chair. of Presidium Siberian Branch USSR Acad. of Sciences 1975–80; Vice-Pres. of USSR Acad. of Sciences 1975–80; cand. mem. Cen. Cttee. CPSU 1976–81; mem. 1981–; Dep. Chair. of USSR Council of Mins., Chair. of USSR State Cttee. for Science and Technology, Chair. Main Cttee. for Exhibition of Achievements of the National Economy of the USSR 1980–; most important works on computer science and applied mathematics. Decors.: Lenin Prize 1961, 2 Orders of Lenin, medals.
Publs.: "Primenenie chislennykh metodov dlia rascheta neitronnykh sechenii" Moscow 1970 (co-author), "Chislennoe reshenie zadach dinamiki atmosfery i okeana na osnove metoda rasshcheplenia" Novosibirsk 1972, "Metody vychislitelnoi matematiki" Novosibirsk 1973.

Marchuk, Ivan Ivanovich
Diplomat; b. 1922, Kozychanka, Oblast Kiev.
Grad. Kiev State University 1944; mem. Diplomatic Service 1947–; div. posts in USSR Min. of Foreign Affairs and USSR embassies abroad 1947–59; Emb. councellor at USSR Embassy in Guinea 1959–62; Dep. Head, Press Dept., USSR Min. of Foreign Affairs 1962–64, 1st Dep. Head 1967–70; USSR Amb. to Burundi 1964–67, to Ecuador 1970–75, in Chad 1977–79, in Zaire S. 1979–; mem. Apparat USSR Min. of Foreign Affairs 1975–77.

Mardzhanishvili, Konstantin Konstantinovich
Mathematician; b. Ag.13 (26), 1903, Moscow.
Mem. CPSU 1939–; grad. Leningrad Univ. 1924; corresponding mem. USSR Acad. of Sciences 1964–; mem. staff V.A.Steklov Inst. of Mathematics of USSR Acad. of Sciences 1934–49 and 1951–. Decors.: Order of Lenin, 2 other orders, various medals.

Mares'ev, Aleksei Petrovich
Pilot; b. My.7 (20), 1916, Kamyshin. major; Hero of the Soviet Union (Ag.24, 1943). Member of the CPSU since 1944. He worked as a lathe operator and took part in the building of Komsomol'sk-na-Amure. He entered the Soviet Army in 1937 and graduated from the Bataisk Military Aviation School in 1940. During the Great Patriotic War of 1941–45 he commanded a flight of a fighter aviation regiment and served as assistant regimental commander and regimental navigator. Early in the war he shot down four enemy aircraft. In March 1942, his plane was hit in combat, and he, gravely wounded, landed in enemy-controlled territory. For 18 days he crawled toward the front line. After his rescue both his legs were amputated at the knee. He then mastered the use of artificial limbs and at his own request was sent to the 63rd Guards Fighter Air Regiment in June 1943. In aerial fighting he shot down seven more enemy aircraft.
He went into retirement in July 1946. He graduated from the Higher Party School in 1952. Since September 1956 he has been executive secretary of the Soviet Committee of War Veterans. His heroic feat is described in B.Polevoi's book *The Story of a Real Man.* Decors.: Order of Lenin, the Order of the October Revolution, the Order of the Red Banner, the Order of the Red Banner of Labour, the Order of the Red Star, and various medals.

Margiani, Revaz Akakievich
Poet; b. 1916, Mulakhi.
Georgian; mem. CPSU 1947–; grad. Dept. of Philology of Tbilisi Univ. 1940; first works publ. 1934.
Publs.: "Gory vysokie" (High Mountains), "Mulakhi – moe selo" (Mulakhi – my Village), and others.

Margiev, Kosta Iosifovich
Poet; b. 1922, Mimiloskhevi; s. of a peasant.
Osset; mem. CPSU 1940–; grad. Higher Literary Courses Moscow 1960; first works publ. 1939.
Publs.: "Ot serdtsa k serdtsu" (From Heart to Heart) 1956, "Na putiakh zhizni" (On Life's Ways) 1964, "Lenin", "Pesni mira" (Songs of Peace), "Les" (The Wood), "Druzia" (Friends), "Kakhetia", and others.

Margulova, Tereza Khristoforovna, Dr. of Tech. Sc.
Scientist; b. 1912.
Mem. CPSU 1942–; grad. Azerbaidzhan Industrial Inst. 1934; specializes in nuclear energy; Dr. of Tech. Sc. 1951, Prof. 1953–; concluded post-graduate studies, mem. staff 1940–, Head of Chair 1956–70, Prof. Chair at Inst. of Energetics Moscow 1970–. Decor.: State Prize of the USSR 1971.

Marisov, Valerii Konstantinovich
Party official; b. 1915.
Russian; mem. CPSU 1940–; grad. Gorkii Ind. Inst. 1940; greaser, asst. machinist on a ship of the Lower Volga Shipping Org. 1931–40; sen. foreman, dep. dept. head, dept. head at a machine-construction plant 1940–47; Dept. Head, Sec., 2nd Sec. of Votkinsk City Cttee. CPSU 1947–51; organizer for Cen. Cttee. CPSU at a machine-construction plant, Dept. Head with Udmurt Oblast CPSU Cttee. 1951–57; Sec. 1957–63, 1st Sec. of Udmurt Oblast CPSU Cttee. 1963–; cand. mem. 1966–71, mem. Cen. Cttee. CPSU 1971–; Dep. to 1966–, mem. Planning and Budget Comm. at Sov. of Nationalities USSR Supr. Sov. 1966–. Decor.: Order of Red Banner of Labour.

Mark, Bishop of Ladoga
Russian Orthodox clergyman.
Ordained Bishop of San Francisco F. 1969.

Markarian, Maro Egishevna
Author; b. 1915, Shulavery, Georgia.
Armenian; grad. Dept. of Philology at Erevan Univ. 1939; first works publ. 1940.
Publs.: "Blizost'" (Nearness) 1940, "Stikhotvoreniia" (Poems) "Golos materi" (Mother's Voice) 1950, "Razdumie" (Thoughts) 1956, and others.

Markevich, Aleksandr Prokofevich
Zoologist; b. Mr. 6 (19), 1905, Ploskoe, Kiev Oblast.
Mem. CPSU 1951–; grad. Kiev Univ. 1930; mem. Ukrainian Acad. of Sciences 1957–; Prof. 1934; mem. staff 1935–70, Dir. of Inst. of Zoology of Ukrainian Acad. of Sciences 1948–50; mem. staff Inst. of Hydrobiology of Ukrainian Acad. of Sciences 1970–; Acad. Sec. of Dept. of General Biology of Ukrainian Acad. of Sciences 1971–72; Chair. of Ukrainian Parasitological Soc. 1945–. Decors.: Merited Scientist of the Ukrainian SSR, 2 orders, various medals.

Markish, Simon Peretsovich
Literary historian and translator; b. 1931, Baku; s. of Perets Markish.
Russian; grad. Dept. of Philology at Moscow Univ. 1954; translated classics such as Plutarch, Lucian, Plato and modern authors such as U. von Hutten, Thomas Mann and others.
Publs.: "Homer i ego poemy" (Homer and his Poems) 1962, "Erazm iz Rotterdama" (Erasmus of Rotterdam) 1966, and others.

Markov, Aleksei Iakovlevich
Poet; b. 1920, Niny, Stavropol Krai.
Russian; grad. Gorkii Inst. of Literature 1951; first works publ. 1940.
Publs.: "Vyshki v more" (Towers in the Sea) 1952, "Nastenka" 1960, "Mikhailo Lomonosov" 1956, "Ilich" 1957, "Ermak" 1962, "Slovo o narode" (The Word of the People) 1958, "Veter v litso" (The Wind in One's Face) 1956, "Osennii grom" (Autumn Thunder) 1961, "Ogon na sebia" (The Fire on Oneself) 1963, "Snova v dorogu" (On the Way again) 1966, and others.

Markov, Andrei Andreevich
Mathematician; b. S. 9 (22), 1903, St. Petersburg.
Mem. CPSU 1953–; grad. Leningrad Univ. 1934; corresponding mem. USSR Acad. of Sciences 1953–; Lecturer 1933–55, Prof. at Leningrad Univ. 1936–; mem. staff Inst. of Mathematics at USSR Acad. of Sciences 1939–72; Prof. Chair of Mathematical Logic at Moscow Univ. 1959–. Decors.: P. L. Chebyshev Prize of USSR Acad. of Sciences 1969, Order of Lenin, various medals.

Markov, Dmitrii Fedorovich
Slavic scholar and literary historian; b. O. 23 (N. 5), 1913, Preslav, Zaporozhe Oblast.
Grad. Dept. of Philology of Kharkov Univ. 1936; corresponding mem. USSR Acad. of Sciences 1966–; Dir. of Inst. of Slavic and Balcan Studies at USSR Acad. of Sciences 1969–; specializes on Bulgaria; mem. Bulgarian Acad. of Sciences. Decors.: Order of Red Banner of Labour, Bulgarian Order of Kirill and Mefodi 1st Cl.

Markov, Georgii Mokeevich
Author; b. 1911, Novo-Kuskovo, Tomsk Oblast; s. of a peasant.
Russian; mem. CPSU 1946–; uncompleted high-school educ.; Komsomol posts 1927–31; exec. ed. for various journals and newspapers in Novosibirsk and Omsk 1931–35; Head of Literary Studio at Irkutsk City Library 1935–41; correspondent of a front newspaper 1941–45; literary activity 1945–; Sec. 1956–71, 1st Sec. of USSR Writers' Union 1971–; mem. Cen. Auditing Comm. 1966–71, mem. Cen. Cttee. CPSU 1971–; Dep. to 1966–, mem. 1966–70, Sec. of Mandate Comm. at Sov. of the Union USSR Supr. Sov. 1970–74; Chair. USSR Cttee. for Lenin Prizes and State Prizes in Literature, Art and Architecture 1979–. First works publ. 1936. Decors.: Stalin Prize, Order of Lenin, Order of Red Banner of Labour, Hero of Socialist Labour 1974, medals.
Publs.: "Strogovy" (The Strogovs) 1939–46, "Soldat pekhoty" (Infantry Soldier) 1948, "Pis'mo v Makeevku" (Letter to Makeevka) 1952, "Sol' zemli" (The Salt of the Earth) 1954–60, "Otets i syn" (Father and Son) 1963–64, "Druzhba narodov" (Friendship Between Nations) 1966, "Sibir'" (Siberia) vols. 1–2, 1969–73, and others.

Markov, Konstantin Konstantinovich
Geographer and geomorphologist; b. My. 7 (20), 1905, Vyborg.
Mem. CPSU 1948–; grad. Leningrad Univ. 1926; mem. USSR Acad. of Siences 1970–; mem. staff Leningrad Univ. until 1937; Head of Dept. of Geomorphology at Inst. of Geography of USSR Acad. of Sciences 1938–; Prof. 1940–, Prof. Chair of Paleogeography at Moscow Univ. 1946–; Head of Dept. of Geography at Moscow Univ. 1945–55; took part in various expeditions, among them to the Pamir, the Tian-Shan and three times to the Antarctic. Decors.: Merited Scientist of the RSFSR, State Prize of the USSR 1971 (for "Atlas Antarktiki"), 2 orders, various medals.

Markov, Moisei Aleksandrovich
Physicist; b. Ap. 30 (My. 13), 1908, Rasskazovo, now Tambov Oblast.
Grad. Moscow Univ. 1930; mem. staff Inst. of Physics at USSR Acad. of Sciences 1934–; mem. USSR Acad. of Sciences 1966–; mem. Presidium and Acad. Sec. at Dept. of Nuclear Physics of USSR Acad. of Sciences 1967–; Chair. of Inter-Authority Comm. for Nuclear Physics 1971–; most important works on quantum mechanics and physics of elemental particles. Decors.: 2 Orders of Lenin, other orders, various medals.

Martianov, Sergei Nikolaevich
Author; b. 1918, Ivanovo.
Russian; mem. CPSU 1940–; lives in Alma-Ata; grad. Iaroslavl Pedagogic Inst. 1940; border guard 1940–55; served in World War II; first works publ. 1948.
Publs.: "Sestry" (Sisters) 1949, "Odnazhdy na granitse" (One Day on the Frontier) 1953, "Piatidesiataia parallel" (The Fiftieth Parallel) 1958, "Veter s chuzhoi storony" (The Wind from the Foreign Side) 1960, "Pervoe zadanie" (The First Task) 1961, "Korotkoe zamykanie" (Short Circuit) 1962, "Pervye zalpy" (First Shots) 1965, film script "Tam, gde tsvetut edelveisy" (Where the Edelweiss bloom) 1965, and others.

Martinson, Sergei Aleksandrovich
Russian actor; b. Ja. 25 (F. 6), 1899, St. Petersburg.
People's Artist of the RSFSR (1964).
In 1923, he graduated from the Institute of Theater Art in Leningrad and appeared in various variety stage troupes. Since 1924 he has performed in Moscow at the Meyerhold Theater (Valerian in Erdman's *The Warrant*) and at the Theater of the Revolution (Mister Pickle in Toller's *Hoppla!,* Karandyshev in Os-

trovskii's *The Dowerless Bride*). He has also appeared at the Moscow Music Hall, where his roles included Skameikina in *Under the Church's Cupola*. In 1945 he joined the troupe of the Film Actors' Theater-Studio (now the Central Film Actors' Studio). Among the roles he played there were Zhivnovskii in Saltykov-Shchedrin's *Death of Pazukhin*, Prince Kostromskoi in Dostoevsky's *Uncle's Dream*, and Harrison in Porter's *Kiss Me Kate*.
Since 1924, he has appeared in motion pictures; his first role was Poincaré in *Adventures of Oktiabrtna*. His best roles have included the barber Sol' in *Marionettes* (1934), the composer Kerosinov in *Anton Ivanovich Is Angry* (1941), the telegraph worker Iat' in *Wedding* (1944), and Prince Kostromskoi in *Uncle's Dream* (1967). In the films *New Adventures of Schweik* (1943) and *The Third Blow* (1948), he created a satirical portrayal of Hitler. His acting, which is mainly satirical, is marked by variety stage and music hall techniques, exaggeration, expressiveness, and some elements of pantomime.

Martirosian, Gerasim Aramaisovich
State official; b. 1926.
1st Dep. Chair. of Council of Mins. Armenian SSR Ag. 1973–Ja. 1977; mem. Auditing Comm. 1971–76, mem. Cen. Cttee. 1976–, mem. Bureau Cen. Cttee. Armenian CP 1976–81, cand. mem. 1981–; Chair. of Cttee. on People's Control of Armenian SSR Ja. 1977–.

Martsinkiavichius, Iustinas Moteiaus
Author; b. 1930, Vazhatkemis, Lithuania. Mem. CPSU 1957–; grad. Dept. of History and Philology of Vilnius Univ. 1954; first works publ. 1953.
Publs.: "Proshu slova" (I Ask to Speak) 1955, "Dvadtsataia vesna" (The Twentieth Spring) 1956, "'Sosna, kotoraia smeialas'" (The Pine that Laughed) 1961, "Krov' i pepel" (Blood and Ashes) 1961, "Ruki, deliashchie khleb" (Hands That Share Out the Bread) 1963, and others.

Martynenko, Vladimir Nikiforovich
State official; b. 1923.
Mem. CPSU 1946–; grad. Faculty of International Relations at Kiev State Univ. and CPSU Cen. Cttee. Acad. of Social Sciences, Cand. of Hist. Sc.; participated in World War II; mem. Ukrainian Society for Friendship and Cultural Relations with Foreign Countries, posts in Cen. Cttee. apparatus Ukrainian CP, 1st Sec. of USSR Emb. in Canada to 1968; Dep. Min. of Foreign Affairs Ukrainian SSR 1968–73; Permanent Rep. of Ukrainian SSR at United Nations 1973–79; Dep. Min. of Foreign Affairs Ukrainian SSR Jn.–N. 1980, Min. N. 1980–; mem. Cen. Cttee. Ukrainian CP 1981–.

Martynov, Aleksandr Konstantinovich
Poet; b. 1913, Sinenkie, Saratov Oblast. Mordovian; mem. CPSU 1942; first works publ. 1931.
Publs.: "K popede" (To Victory) 1942, "Zori pobed" (Stars of Victory) 1944, "Moi tovarishch" (My Comrade) 1956, "Radost'" (Joy) 1964.

Martynov, Leonid Nikolaevich
Author; b. My. 9 (22), 1905, Omsk; s. of a transport technician.
Russian; first works publ. 1921. Decors.: Gorkii State Prize of the RSFSR 1966, Lenin Prize 1974, 2 orders and various medals.
Publs.: "Stikhi" (Poems) 1955, "Pervorodstvo" (Firstborn) 1965, "Golos prirody" (The Voice of Nature) 1966, "Stikhotvoreniia i poemy" (Poetry) 1965, "Giperboly" (Hyperboles) 1972, and others.

Martynov, Nikolai Vasil'evich
State official; b. 1910.
Russian; mem. CPSU 1932–; attended Moscow Energy Inst. four years without graduating to 1933; electrician 1927–29; electrician, dept. chief, shop superintendent, chief plant mechanic, chief engineer, plant dir. in various plants 1929–41; Dep. USSR Min. of Munitions 1941–46; Dep. USSR Min. of Agricultural Machine Building 1946–47, 1st Dep. Min. 1947–53; Chief of a Dept. in USSR Min. of Automobile, Tractors and Agricultural Machine Building 1954–55; Dep. USSR Min. of Tractor and Agricultural Machine Building 1955–57; 1st Dep. Chair. of Tashkent Sovnarkhoz 1957–60; Chair. of Sovnarkhoz Uzbek SSR 1960–62; Sec. of Cen. Cttee. Uzbek CP 1962–64; Dep. Chair. of USSR Sovnarkhoz and USSR Min. 1964–65; 1st Dep. Chair. of USSR State Cttee. for Material and Technical Supply 1965–76, Chair. 1976–; Dep. Chair. of USSR Council of Mins. 1976–; Dep. to USSR Supr. Sov. 1962–; mem. Cen. Auditing Comm. CPSU 1961–66; mem. Cen. Cttee. CPSU 1981–.

Marukas, Kazis (real name: Krasauskas, Marionas Ionovich)
Author; b. 1921, Kaunas; s. of a worker. Lithuanian; mem. underground Komsomol org. in Lithuania 1938; served in Sov. Army during World War II; first Works publ. in front newspapers and almanachs.
Publs.: "Druzhba" (Friendship) 1948, "Obyknovennaia istoriia" (A Common Story) 1958, "Komu vskhodit solntse" (For Whom the Sun Rises) 1965, "Malen'kii soldat" (The Little Soldier) 1956, "Mat' i shofer" (Mother and the Lorry-Driver) 1962, and others.

Marutian, Konstantin Ovanesovich
State official.
Cand. mem. Cen. Cttee. Armenian CP 1961–64, mem. 1964–; Min. of Construction Armenian SSR 1961–65; Chair. of "Armhydroenergostroi" Admin. Armenian SSR 1967; Chair. of State Cttee. for Construction Armenian SSR F. 1978–.

Masaliev, Absamat Masalievich
Party official; b. 1933.
Kirghiz; mem. CPSU 1960–; grad. Mining Inst. in Moscow; Engineer in "Kirghizugol"-Trust 1956–61; Instructor of a Oblast Cttee. Kirghiz CP, Dep. Chair. of a Oblast Cttee. for People's Control 1961–68; Dep. Head of a Dept. with Cen. Cttee. Kirghiz CP, 1st Sec. of Tash-Kumyr City Cttee. Kirghiz CP 1968–71; mem. Cen. Cttee. Kirghiz CP 1971–; Head of Dept. of Ind. and Transport with Cen. Cttee. Kirghiz CP 1971–73; Chair. of Exec. Cttee. Frunze City Sov. of People's Deps. 1973–74; Sec. 1974–79 and mem. Bureau Cen. Cttee. Kirghiz CP 1974–; Dep. to Supr. Sov. Kirghiz SSR 1975–; 1st Sec. Issyk-Kul Oblast Cttee. Kirghiz CP Ja. 1979–.

Masal'skaia, Nina Nikolaevna
Russian actress; b. Ja. 5 (18), 1901, Penza. People's Artist of the USSR (1960).
She graduated from the drama studio directed by M. V. Zenina in Penza. From 1919 to 1924 she performed at the A. V. Lunacharskii Theater in Penza. Later she appeared at the theaters of the People's Commissariat of Railroads in Tambov, Astrakhan, Cheliabinsk, Kharkov, and the Soviet Far East. In 1950, she joined the A. P. Chekhov Kishinev Russian Theater of Drama. She has been particularly successful in portraying strong, determined women. Her first roles were Desdemona in Shakespeare's *Othello* and Judith in Gutzkow's *Uriel Acosta*. Her best roles have included Liubov' Iarovaia in Trenev's *Liubov' Iarovaia*, Negina, Tugina, and Kruchinina in Ostrovskii's *Talents and Admirers, The Last Victim*, and *Guilty Though Guiltless*, Vassa Zheleznova in Gorkii's *Vassa Zheleznova*, Maria Aleksandrovna in Popov's *Family*, and the Mistress of Niskavuori in Wuolijoki's *The Stone Nest*.

Maslakov, Anatolii Vasil'evich
Oblast Sov. official.
Sec. of Mogilev Oblast Cttee. Belorussian CP 1968–74; Chair. of Mogilev Oblast Exec. Cttee. of People's Deps. Ag. 1974–; Dep. to Supr. Sov. Belorussian SSR 1975–; mem. Cen. Cttee. Belorussian CP 1976–.

Maslennikov, Nikolai Ivanovich
State official; b. 1921.
Russian; mem. CPSU 1951; grad. Gorkii Ind. Inst. 1948; served in Sov. Army 1931–43; cultural work in Gorkii Oblast 1944–45; Chief Technologist, Dept. Head, Chief Mechanic, Sec. of CPSU Cttee., Chief Engineer of "Krasnaia Etna" Works in Gorkii 1948–61; 2nd Sec. 1961–66, 1st Sec. of Gorkii City CPSU Cttee. 1966–68; 1st Sec. of Gorkii Oblast CPSU Cttee. 1968–74; Dep. to USSR

Supr. Sov. 1970–; mem. Cen. Cttee. CPSU 1971–; Dep. Chair. of Council of Mins. and Chair. of GOSPLAN of the RSFSR My. 1974–. Decors.: Hero of Socialist Labour, Order of the October Revolution.

Maslov, Vladimir (Vadim) Petrovich
Military official; b. 1925.
Mem. CPSU 1945; grad. Acad. of the General Staff 1966; Admiral 1975; Commander of an atomic submarine detachment of Northern Fleet until 1970; 1st Dep. Commander 1970–74, Commander of Pacific Fleet 1974–79; cand. mem. Cen. Cttee. CPSU 1976–81.

Masol, Vitalii Andreevich
State official; b. 1928.
Ukrainian; mem. CPSU 1956–; grad. Kiev Polytechnical Inst. 1951, Cand. of Tech. Sc. 1972; asst. foreman, foreman, shop superintendent, dep. chief engineer, Dir. of Novo-Kramatorsk Machine Building Plant, Gen. Dir. of Prod. Asscn. of Kramatorsk Heavy Machine Building Plants 1951–72; 1st Dep. Chair. of GOSPLAN Ukrainian SSR 1972–79; mem. Cen. Cttee. Ukrainian CP 1976–; Dep. Chair. of Council of Mins. and Chair. of GOSPLAN Ukrainian SSR Ja. 1979–; Dep. to USSR Supr. Sov. 1979–; mem. Cen. Auditing Comm. 1981–.

Masumi, Nosirdzhon Asadovich
Literary historian; b. Ap. 19 (My. 2), 1915, Karatag, Tadzhikistan.
Mem. CPSU 1945–; grad. 1940, Lecturer at Pedagogical D.G. Shevchenko Inst. Dushanbe 1940–; Dir. of Rudaki Inst. of Linguistics and Literature of Tadzhik Acad. of Sciences 1959–72; corresponding mem. Tadzhik Acad. of Sciences 1966–; Acad. Sec. at Dept. of Social Sciences of Tadzhik Acad. of Sciences 1971–. Decors.: Merited Scientist of the Tadzhik SSR, 2 orders, various medals.

Masyk, Konstantin Ivanovich
State official; b. 1936.
Ukrainian; mem. CPSU 1962–; grad. of Inst. for Water Transport Engineers and of CPSU Cen. Cttee. Party Higher School (correspondence study); construction engineer at a ship-building and repairing plant in Kiev, held Komsomol posts, instructor in Oblast Cttee. CP Ukraine in Kiev 1959–72; 1st Sec. Darnitsa Raion Cttee. CP Ukraine in Kiev 1972–74; inspector Cen. Cttee. CP Ukraine 1974–76; Sec. 1976–77, 2nd Sec. Oblast Cttee. CPSU in Odessa 1977–81; Dep. Chair. Council of Mins. Ukrainian SSR Sep. 1981–; mem. Cen. Cttee. CP Ukraine 1981–.

Matafonov, Mikhail Ivanovich
Party official; b. 1928.
Russian; mem. CPSU 1952–; grad. Chita Pedagogical Inst. 1954; 2nd Sec., 1st sec. of a raion cttee. Komsomol, chief of a dept. of Chita Oblast newspaper 1950–55; Sec. of Chita Oblast Cttee. USSR Komsomol 1955–58; Sec., 1st sec. of a raion cttee. CPSU 1958–62; sec. of party cttee. of state and collective farm prod. admin. 1962–64; Sec., 2nd Sec. of Chita Oblast Cttee. CPSU 1964–73, 1st Sec. 1973–; Dep. to USSR Supr. Sov. 1974–; cand. mem. Cen. Cttee. CPSU 1976–. Decor.: Order of Lenin.

Matevosian, Grant Ignatevich
Armenian author; b. Mr. 3, 1935, Akhnidzor, in present-day Tumanian Raion.
He graduated from the Armenian Pedagogical Institute in 1964 and was employed in a printing house. His first published work appeared in 1959. He is the author of the novella *We and Our Mountains* (1962) and of the short stories "August", "Alkho", "Mesrop" (all in 1967; awarded the prize of the journal *Friendship of the Peoples* in 1967) and "Buffalo Cow" (1968). His works are noted for the attempt to create a "complete" personality. Simplicity of narration is combined with a tension of the moral conflict portrayed.
Works: *In Russian translation: My i nashi gory.* Moscow 1969. *Avgust.* Moscow 1972. "Mat' edet zhenit' syna: Povest'". *Druzhba narodov,* 1973, no. 10.

Matiukhin, V. I.
Komsomol official.
Sec. and mem. Bureau Cen. Cttee. Kirghiz Komsomol S. 1980–.

Matiushevskii, Konstantin Vikentevich
Oblast Sov. official; b. 1917.
Belorussian; mem. CPSU 1940–; grad. Party High School of Cen. Cttee. CPSU; dir. of a children's home, dir. of a school 1937–39; served in Sov. Army 1939–46; sen. party posts 1946–58; Dep. to USSR Supr. Sov. 1954–66; mem. Cen. Cttee. Belorussian CP 1956–; 1st Sec. of Mogilev Oblast Cttee. Belorussian CP 1958–64 (for Agriculture 1963–64); Chair. of Brest Oblast Exec. Cttee. of Workers' Deps. D. 1964–. Decors.: Order of Lenin, 2 Orders of Patriotic War 1st and 2nd Cl., Order of Red Star, medals.

Matora, Ivan Maksimovich, Cand. of Phy.-Math. Sc.
Physicist; b. 1922.
Mem. CPSU 1966–; grad. Leningrad Univ. 1951; served in Sov. Army 1940–46; postgraduate studies 1951–54; Cand. of Phy.-Math. Sc. 1954–, Sen. Scientific Advisor 1967–; mem. staff Research Laboratory for Electrophysical Apparatuses Leningrad 1954–60; mem. staff United Inst. for Nuclear Research Dubna 1962–, head of a section of same 1970–. Decor.: State Prize of the USSR 1971.

Matrosov, Vladimir Evgenevich
Sculptor; b. O. 15, 1928, Leningrad.
Grad. V.I. Surikov Art Inst. Moscow 1952. Decors.: Lenin Prize 1970, Order of Red Banner of Labour.

Matsiavichius, Iuozas
Poet; b. 1928; s. of a worker.
Lithuanian; mem. CPSU 1961–; studied Gorkii Inst. of Literature Moscow.
Publs.: "Shagi vesny" (Steps of Spring) 1949, "Pokolenie stroitelei" (Generation of Builders) 1953, "Osennee plamia" (Autumn Flame) 1957, and others.

Matskevich, Iuzefa Florianovna, Dr. of Phil.
Philologist; b. 1911.
Grad. Pedagogical Tech. Inst. Mogilev 1931 and Pedagogical Inst. Minsk 1938; Cand. 1946–53, Dr. of Phil. 1963–; mem. staff Inst. of Linguistics of Belorussian Acad. of Sciences (interrupted 1955–56 for doctoral thesis at Inst. of Russian Studies at USSR Acad. of Sciences) 1945–, chief of a section of same 1950–. Decor.: State Prize of the USSR 1971.

Matskevichius, Kazimieras Iuzovich
Trade union official; b. 1932.
Lithuanian; mem. CPSU 1952–; grad. Inst. of Maritime Engineering in Odessa; chair. of exec. cttee. of raion sov. of a sports asscn. 1950–56; Dep. Dir. of Klaipeda Maritime Harbour, dep. head of a dept. with Cen. Cttee. Lithuanian CP 1956–59; 1st Sec. of Klaipeda City Cttee. Lithuanian CP 1959–63; mem. Cen. Cttee. Lithuanian CP 1960–77; 1st Sec. of Vilnius City Cttee. Lithuanian CP 1963–71; cand. mem. 1964–71, mem. Bureau Cen. Cttee. Lithuanian CP 1971–My. 1977; Chair. of Lithuanian Republican Trade Union Council Ja. 1972–Ap. 1977; Sec. of Cen. Union Council of USSR Trade Unions Mr. 1977–.

Matulis, Iuozas Iuozovich
Physical chemist; b. Mr. 7 (19), 1899, Tatkonis, Lithuania.
Mem. CPSU 1950–; grad. Univ. Kaunas 1929; mem. 1941–, Pres. of Lithuanian Acad. of Sciences 1946–; corresponding mem. USSR Acad. of Sciences 1946–. Decors.: Hero of Socialist Labour, 4 Orders of Lenin, 3 other orders, various medals.

Matusovskii, Mikhail Lvovich
Poet; b. 1915, Lugansk.
Russian; mem. CPSU 1945–; grad. Gorkii Inst. of Literature 1939, postgraduate studies at Moscow Univ. 1941; first works publ. 1934.
Publs.: "Luganchane" (Lugansk People), "Front" 1942, "Pesnia ob Aidorgy Takhirove i ego druge Andree Savushkene" (The Song of Aidorgy Tashkirov and his Friend Andrei Savushkin) 1943, "Kogda shumit Ilmen-ozero" (When Lake Ilmen Rages) 1944, "Slushaia Moskvu" (Listening to Moscow) 1948, "Ulitsa Mira" (The Street of Peace) 1951, lyrics for the song "Podmoskovnye vechera" (Moscow Nights) and the films "Vernye Druzia" (True Friends), "Ispytanie vernosti" (Test of Faithfulness), "Idiot" and others.

Matveev, Vladimir Vasil'evich
State official.
Dep. Head of Ideological Dept. for Ind., mem. Bureau for Ind. and Construction at Cen. Cttee. Belorussian CP Ja.–N. 1964; mem. 1966–76, cand. mem. Cen. Cttee. Belorussian CP 1976–; Chief Ed. of newspaper "Sel'skaia gazeta" 1965–72; Chair. of State Cttee. for the Film Ind. Belorussian SSR Ag. 1972–.

Matveeva, Novella Nikolaevna
Poet; b. 1934, Pushkin, Leningrad Oblast.
Russian; first works publ. 1958.
Publs.: "Lirika" (Lyric) 1961, "Korablik" (Little Ship) 1963, "Mumiia" (Mummy), "Kraski" (Colours), "Vse skazano na svete . . ." (Everything Has Been Said in the World . . .), and others.

Maulenov, Syrbai
Poet; b. 1922, Kustanai Oblast; s. of a peasant.
Kazakh; mem. CPSU 1944; studied Kzyl-Orda Pedagogic Inst.; first works publ. 1938.
Publs.: "Sbornik stikhov" (Poems – Anthology) 1948, "Golos stepei" (Voice of the Steppes) 1949, "Zelenyi les" (Green Wood) 1950, "Tovarishch" (Comrade) 1954, "Ogni gory Magnitnoi" (Lights of Magnitnyi Mountain) 1952, "Izbrannoe" (Selected Works) 1958, "Chaika" (Sea-Gull) 1962, "Listia goriat" (The Leaves are Burning) 1964, "Sledy borby" (Traces of Battle) 1965, "Polden'" (Midday) 1966, and others.

Mavlianova, Khanifa Mukhiddinovna
Tadzhik singer (soprano); b. 1924, Khodzhent, now Leninabad.
People's Artist of the USSR (1968). Member of the CPSU since 1952.
She graduated from the Leninabad School of Music in 1936. From 1937 to 1939 she performed at the Leninabad Young People's Theater, and between 1939 and 1943 she was a soloist at the musical drama theater. Since 1943 (with an interruption) she has been the lead female singer at the S. Aini Tadzhik Theater of Opera and Ballet. She studied at the Moscow Conservatory from 1952 to 1959. Her voice is remarkable for its power and wealth of timbre nuances. Her best parts have been Gul'ru in Saifiddinov's *Pulat and Gul'ru,* Tatiana in Tchaikovsky's *Eugene Onegin,* Iaroslavna in Borodin's *Prince Igor,* and Leonora in Verdi's *Il Trovatore.* Among the parts she has sung in comic operas are Maisara in Iudakov's *Maisara's Pranks* and Surma in Urbakh's *The Noble Bridegroom.* She also performs as a concert singer and teaches at the Dushanbe Institute of Arts. She was a deputy to the fourth convocation of the Supreme Soviet of the USSR. Decors.: Order of Lenin, two other orders, and medals.

Mazurenko, Aleksei Efimovich
Military official; b. Je. 7 (20), 1917, Leninka, now in Ustinovka Raion, Kirovograd Oblast.
Major-general of aviation (1962); twice Hero of the Soviet Union (O. 23, 1942, and N. 5, 1944). Member of the CPSU since 1942.
He has been in the Soviet Army since 1938. He graduated from the Naval Aviation College in 1940, the Higher Officers' Tactical Flight Courses in 1949, and the Naval Academy in 1952. In the Great Patriotic War (1941–45) he served in the aviation of the Baltic Fleet and commanded the 7th Assault Aviation Regiment in 1944 and 1945. He made about 300 combat flights and personally sank eight enemy ships and 22 in group combat. Since the war he has held command positions in the armed forces and in higher military schools of the navy. Decors.: Order of Lenin, three Orders of the Red Banner, the Order of Ushakov Second Class, the Order of the Patriotic War First Class, the Order of the Red Star, and various medals. A bust of Mazurenko has been erected in the settlement of Ustinovka, Kirovograd Oblast, Ukrainian SSR.

Mdivani, Georgii Davidovich
Georgian playwright and screenwriter; b. S. 13 (26), 1905, Bagdadi, now the settlement of Maiakovskii.
Honored Artist of the Georgian SSR (1961). He writes in Georgian and Russian. Member of the CPSU since 1944.
From 1924 to 1926, he studied at the department of law of the University of Tbilisi. He has lived in Moscow since 1936. He is the author of plays that present sharp conflicts and are imbued with publicistic enthusiasm. Among his plays are *Alcazar,* which was staged in 1936 and deals with the revolutionary struggle in Spain, *Moscow Sky* (1943), *New Times* (1952), *Teresa's Birthday* (1961), which is about revolutionary Cuba, *They Kidnapped the Consul* (1963), *Your Uncle Misha* (1966), and *Big Mama* (1971). Among the films that have been produced using his scenarios are *Private Aleksandr Matrosov* (1948), *Soldier Ivan Brovkin* (1955), and *Ivan Brovkin on the Virgin Lands* (1958). In 1971 the scenario of *Envoys of Eternity* was published. Decors.: two orders as well as medals.
Works: *In Russian translation: P'esy.* Moscow 1955. *Kinostsenarii.* Moscow 1959. *Pereval.* Moscow 1968.

Mednikov, Ivan Semenovich
Military official; b. 1917.
Russian; mem. CPSU 1940–; grad. V. I. Lenin Mil. Polit. Acad. 1949; Col.-Gen.; entered Army 1938; 1st Dep. Chief of Pol. Admin. of Kiev Mil. District, mem. staff Cen. Pol. Admin. of Sov. Army and Navy, mem. Mil. Council and Chief of Pol. Admin. of Baltic Mil. District 1965–71; mem. Cen. Cttee. Lithuanian CP 1971–76; mem. Mil. Council and Chief of Pol. Admin. of Sov. Forces Group in Germany D. 1971–; mem. Cen. Auditing Comm. CPSU 1976–81.

Medunov, Sergei Fedorovich, Cand. of Econ. Sc.
State official; b. 1915.
Russian; mem. CPSU 1942–; grad. Party High School of Cen. Cttee. CPSU; teacher 1931–39; served in Sov. Army 1939–47; sec., then 1st sec. of a CPSU raion cttee. 1947–51; head of a dept. with oblast cttee. Ukrainian CP in Crimea, 1st Sec. of Ialta City Cttee. Ukrainian CP 1951–58; 1st Dep. Chair. of Crimean Oblast Exec. Cttee. of Workers' Deps. 1958–59; 1st Sec. of Sochi City CPSU Cttee. 1959–69; Chair. of Krasnodar Krai Exec. Cttee. of Workers' Deps. 1969–73, 1st Krai Sec. CPSU My. 1973–82; Dep. to 1970–, mem. Comm. for Trade, Public Services and Communal Econ. 1970–74, mem. Draft Bills Comm. at Sov. of the Union USSR Supr. Sov. 1974–, mem. Cen. Cttee. CPSU 1976–; USSR Dep. Min. of Fruits and Vegetables Jl. 1982–. Decor.: Order of Lenin.

Medved', Aleksandr Vasil'evich
Wrestler; b. S. 16, 1937, Belaia Tserkov', Kiev Oblast.
Honored Master of Sports (1964); docent at the Minsk Radio Engineering Institute. Member of the CPSU since 1965.
He has been an Olympic champion (1964, 1968, 1972), many times world champion (1962–63, 1966–67, 1969–71), champion of Europe (1966, 1968, 1971–72), and champion of the USSR (eight times in 1961–70) in free-style middleweight and heavyweight wrestling. Decors.: Order of Lenin and two other orders.

Medvedev, Igor Nikolaevich
Journalist; b. 1914.
Mem. CPSU 1939–; grad. Party High School of Cen. Cttee. CPSU 1941; electro-mechanic, technologist at Likhachev Automobile Works in Moscow 1931–39; party work 1941–52 and 1955–63; state official 1952–55 and 1963–68; Dep. Head of Sov. news agency "Novosti" 1968–. Decor.: State Prize of the USSR 1970 (for documentary film "Čechoslovakija, god ispytanij").

Medvedev, Vadim Andreevich
Party official; b. 1929.
Mem. CPSU 1952–; grad. Leningrad State Univ. 1951, Dr. of Econ. Sc., Prof. 1968; employed at Leningrad State Univ. 1951–56; Asst. Prof. Leningrad Engineering Inst. of Rail Transport 1956–61; head of a faculty at Leningrad Technological Inst. 1961–68; Dep. Chief of Propaganda Dept. of Cen. Cttee. CPSU 1970–78; mem. Cen. Auditing Comm. CPSU 1976–; Rector of CPSU Cen. Cttee. Acad. of Social Sciences 1978–. Decor.: Order of the October Revolution.

Medvedkin, Aleksandr Ivanovich
Film director; b. F.24 (Mr.8), 1900, Penza.
Mem. CPSU 1920–; with Cen. Studio for Documentary Films 1948–. Decors.: Order of Lenin, 3 other orders, various medals.

Medynskii, Grigorii Aleksandrovich (pseudonym of G. A. Pokrovskii)
Russian writer; b. Ja.30 (F.11), 1899, Kozel'sk; s. of a priest.
From 1919 to 1929, he was an educator, working from 1922 to 1928 in the campaign to eliminate child homelessness. He began to publish in 1925 and is the author of the novels *Samstroi* (1930) and *The Ninth "A"* (1939), the novellas *A Tale of Youth* (1954) and *Honor* (1959), the play *The Life and Crime of Anton Shelestov* (1961, written in collaboration with V. Tokarev), and the publicistic books based on documentary material *Without Lowering the Eyes* (1961; 2nd ed., 1963), *Command Good Fortune* (1963, written in collaboration with V. Petrov), and *A Difficult Book* (1964). His best-known work is the novel *Mar'ia* (parts 1–2, 1946–49; State Prize of the USSR, 1950), about the heroic labor of kolkhoz women during the Great Patriotic War of 1941–45.
His interests – fiction and publicism – have been concentrated primarily on acute problems of morality, social pedagogy, and the formation of the human personality. His works have been translated into a number of foreign languages. Decors.: Order of the Red Banner of Labour and various medals.
Work: *Puti i poiski.* Moscow 1970. [With a Foreword by F. Kuznetsov.]

Medzmariashvili, Nodar Akakievich
State official.
1st Dep. Min. Je.–Jl. 1973, Min. of Construction Georgian SSR Jl. 1973–81; mem. Cen. Cttee. Georgian CP 1976–81; 1st Dep. Chair. GOSPLAN Georgian SSR 1981–.

Megrelishvili, Mamiia Amberkovich
State official.
Mem. Cen. Cttee. 1956–64 and 1976–81; cand. mem. Cen. Cttee. Georgian CP 1971–76; Dep. to Supr. Sov. Georgian SSR 1959–; Chair. of Comm. for Sov. Control with Council of Mins. Georgian SSR 1959–61; Min. of Local Econ. 1961–62, Min. of Local Ind. 1962–63, Chair. of State Cttee. for Printing and Publishing Ind. with Council of Mins. Georgian SSR 1963–67; Exec. Sec. of Council of Mins. Georgian SSR 1967–72; Min. of Local Ind. 1972–73, Min. of Trade Georgian SSR 1973–80; Chair. of State Cttee. on Pricing Georgian SSR Mr. 1980–. Decors.: Order of Lenin, medals.

Meilakh, Boris Solomonovich
Literary historian; b. Je.26 (Jl.9), 1909, Lepel, Vitebsk Oblast.
Mem. CPSU 1940–; grad. Dept. of Literature at Moscow State Univ. 1931; scientific work at "Pushkin House" 1935–; Prof. at Leningrad State Univ. 1947–66.
Publs.: "Pushkin i ego epokha" (Pushkin and His Time) 1958, "Talant pisatelia i protsessy tvorchestva" (A Writer's Talent and Processes of Creation) 1969, "Na rubezhe nauki i iskusstva" (On the Border of Science and Art) 1971, and others.

Meisel', Maksim Nikolaevich
Microbiologist and cytologist; b. S.28 (O.11), 1901, St. Petersburg.
Mem. CPSU 1961–; grad. 1st Med. Inst. Leningrad 1926; mem. staff 1934–, Dir. of Sciences 1952–; concurrently Head Laboratory Inst. of Molecular Biology at USSR Acad. of Sciences 1959–; Lecturer 1946–53, Prof. at Moscow Univ. 1953–; corresponding mem. USSR Acad. of Sciences 1960–; one of the founders of Far-Eastern Branch of USSR Acad. of Sciences; Pres. of Union Soc. for Microbiology. Decors.: State Prize of the USSR 1971, 5 orders, various medals.

Meitus, Iulii Sergeevich
Composer; b. Ja.15 (28), 1903, Elizavetgrad, now Kirovgrad.
Mem. CPSU 1954–; grad. Inst. of Music and Drama Kharkov 1931; composer of 12 operas. Decors.: Merited Artist of the Ukrainian SSR, Merited Artist of the Turkmen SSR, State Prize of the USSR 1951, 2 orders, various medals.

Mekhtiev, R. E.
Party official.
Mem. Cen. Cttee. and cand. mem. Bureau Cen. Cttee. Azerbaidzhan CP 1981–; Head of Dept. for Org.-Party Work with Cen. Cttee. Azerbaidzhan CP Ja. 1981–.

Melentev, Iurii Serafimovich
State official.
1st Dep. Chair. of USSR State Cttee. for Publishing, Printing and the Book Trade until 1974; Min. of Culture RSFSR Mr. 1974–.

Melentev, Lev Aleksandrovich
Scientist; b. D.9 (22), 1908, St. Petersburg.
Mem. CPSU 1947–; grad. Polytech. Inst. Leningrad 1930; specializes in energetics; Prof. Chair 1945–60, Dep. Dir. of Inst. of Econ. Engineering Leningrad 1945–51; Sen. Scientific Advisor of Inst. of Energetics at USSR Acad. of Sciences 1942–60; Chair of Presidium E.-Siberian Branch of USSR Acad. of Sciences 1960–65; mem. Presidium Siberian Branch of USSR Acad. of Sciences 1961–66; mem. USSR Acad. of Sciences 1966–; Dir. of Inst. of Energetics at Siberian Branch of USSR Acad. of Sciences 1960–; Head of Dept. of High Temperatures at USSR Acad. of Sciences 1973–; Dep. Acad. Sec. at Dept. of Physical-Technical Problems of Energetics at USSR Acad. of Sciences 1965–. Decors.: 2 Orders of Lenin, 2 other orders, various medals.
Publs.: "Osnovnye voprosy promyšlennoi teploėnergetiki" Moscow 1954, co-author of "Toplivno-ėnergetičeskii balans SSSR" Moscow 1962, "Metody matematičeskogo modelirovaniia v ėnergetike" Irkutsk 1966; Ed. of "Optimizatsiia i upravlenie v bolshih sistemah ėnergetiki" Irkutsk 1970.

Melesh, Ivan Pavlovich
Author; b. 1921, Glinishche, Gomel Oblast; s. of a peasant; died Ag.9, 1976.
Belorussian; mem. CPSU 1940–; served in Sov. Army 1941–45 (severely wounded in 1942); studied Moscow Inst. of History, Philosophy and Literature, grad. Belorussian Univ. 1945; Dep. Chair. of Belorussian Writers' Union 1966–; mem. World Peace Council 1970–; first works publ. 1939. Decors.: People's Writer of the Belorussian SSR 1972, Lenin Prize 1972, 4 orders and various medals.
Publs.: "V meteil'" (In the Snowstorm) 1946, "Goriachii avgust" (Hot August) 1948, "Minskoe napravlenie" (Minsk Direction) 1949–59, "Blizkoe i dalekoe" (Near and Far) 1954, "V gorack dozhdi" (It Rains in the Mountains) 1957, "Poka vy molody" (As Long As You Are Young) 1956, "Dni nashego rozhdeniia" (Our Birthdays) 1958, "Liudi na bolote" (People in the Moorland) 1961, "Dykhanie grozy" (Breath of Thunderstorm) 1965, "V nachale voiny" (At the Beginning of War) 1969, "Zbor tvorau" (Collected Works) vols. 1–6 1969–71, and others.

Melik-Avakian, Grigorii Gevorkovich
Film director; b. Je.21, 1920, Tbilisi.
Grad. Union Film Inst. 1951; with film studio "Armenfilm". Decors.: Order of Red Star, medals.

Melikhov, Georgii Stepanovich
Painter; b. My.11 (24), 1908, Kharkov.
People's Artist of the Ukrainian SSR (1967).
He studied at the Kiev Institute of Art from 1935 to 1941 with P. G. Volokidin and F. G. Krichevskii. From 1945 to 1961 he taught at the institute, where he became a professor in 1960. He was secretary of the board of the Artists' Union of the USSR from 1956 to 1960. In his carefully developed genre scenes, he endeavors to be psychologically convincing. Among his works are *Young Taras Shevchenko in K. P. Briullov's Studio* (1947, Museum of Fine Arts of the Ukrainian SSR, Kiev; State Prize of the USSR, 1948), *Gorky in the Ukraine* (1957, Gorlovka Museum of Art), and *Delegates to the First Congress of the Communist (Bolshevik) Party of the Ukraine in Moscow* (1969, property of the Ministry of Culture of the Ukrainian SSR). Decors.:

Order of the Red Banner of Labour and various medals.

Melikishvili, Georgii Aleksandrovich
Historian; b. D.30, 1918, Tbilisi.
Mem. CPSU 1946–; grad. Tbilisi Univ. 1939; mem. Georgian Acad. of Sciences 1961–; Head of Dept. of Ancient History 1961–, Dir. of I.A. Dzhavakhishvili Inst. of History, Archaeology and Ethnography at Georgian Acad. of Sciences 1965–; most important works on ancient history of the Near East and Transcaucasus, especially Georgia. Decors.: Lenin Prize 1957, orders and medals.

Meliton (Mikhail Dmitrevich Solovev), Cand. of Theol.
Bishop of Tikhvin and Vicar-Bishop of Novgorod; b. O.14, 1897, Svishchevka, Penza Oblast.
Grad. Leningrad Theological Seminary and Theological Acad.; attended religious elem. school in Poime, Cheboksary District 1906–12; served in Tsarist Army 1916–17, in Sov. Army 1917–21; ordained deacon N.24, 1921, and priest Ja.23, 1922; worked in Penza Diocese 1927 and in Moscow Bishopric 1934; secular work 1935–41; officer in Sov. Army 1941–46; teacher 1946; completed pedagogical studies in Kaluga 1948; resumed work as priest, sen. clergyman at Kazan Cathedral in Luga, Leningrad Diocese 1954; Chair. of Trinity Church in Vsevolzhsk, Leningrad Diocese 1966–68; Provost of the Suburban Churches in Leningrad 1968–70; Rector of Leningrad Theological Acad. 1970–D. 1974; Bishop of Tikhvin and Vicar-Bishop of Novgorod 1970–.

Meliukhin, S.
Philosopher, corresponding mem. USSR Acad. of Sciences 1981–.
Holds a professorial chair in the philosophy dept. of Moscow State Univ.; investigates contemporary problems of dialectical materialism, philosophical questions on particle and field physics, methodological problems of the integration of modern scientific achievements; author of more than 100 works.

Melkadze, Otar Valerianovich
State official.
1st Sec. of Batumi City Cttee. Georgian CP until 1975; 2nd Sec. of Oblast Cttee. Georgian CP in Adjar ASSR Ja.–Je. 1975; Chair. of Cttee. of People's Control Georgian SSR Je. 1975–; mem. Cen. Cttee. Georgian CP 1976–.

Melkhisedek (Vasilii Mikhailovich Lebedev)
Russian Orthodox clergyman; b. Ja.26, 1927, Novo-Cherkasovo, Moscow Oblast.
Studied Moscow Theological Seminary 1946–50; deacon, priest, archpriest; main clergyman and provost of churches in the Moscow diocese; began studies at Moscow Theological Acad. 1962; monk at Zagorsk (Troitse Sergieva Lavra) 1963; Bishop of Vologda and Velikii Ustiug 1965; Bishop of Vienna and Austria 1967–70; Bishop of Penza and Saransk Jl. 1970–78; Archbishop of Berlin and Cen. Europe, Exarch of Moscow Patriarchate for Cen. Europe S. 1978–.

Melkumian, Georgii Arshakovich
State official.
Min. of Construction and Utilization of Highways Armenian SSR Ap. 1977–; mem. Auditing Comm. Armenian CP 1981–.

Melkumian, Gurgen Allakhverdovich
Trade union official; b. 1915; s. of a peasant.
Armenian; mem. CPSU 1939–; grad. Party High School of Cen. Cttee. CPSU; teacher, sec., ed. of a raion newspaper 1935–39; served in Sov. Army 1939–45; propagandist, sec., 1st sec. of a raion cttee. Azerbaidzhan CP 1945–61; 2nd Sec. 1961–62, 1st Sec. of Nagorno-Karabakh Autonomous Oblast Cttee. Azerbaidzhan CP 1962–73; mem. Cen. Cttee. 1962–, cand. mem. Bureau Cen. Cttee. Azerbaidzhan CP 1962–73; Sec. of Azerbaidzhan Republican Trade Union Council O. 1973–.

Melkumov, Levon Nikolaevich
State official.
Chair. of Cttee. for State Security Uzbek SSR Mr. 1978–; Mem. Cen. Cttee. and cand. mem. Bureau Cen. Cttee. Uzbek CP F. 1981–.

Melnichenko, Afanasii Kondratevich
State official; b. 1923.
Ukrainian; mem. CPSU 1943–; grad. Odessa Pharmaceutical Inst. 1950, Cand. of Pharmaceutical Sc. 1963; served in Sov. Army 1941–44; manager of an oblast pharmaceutical admin. 1944–52; posts in USSR Min. of Health apparatus 1952–59; Dir. of the Scientific Research Inst. for Pharmaceuticals 1959–64; party, sov., and state posts 1964–68; posts in Moscow City Cttee. apparatus CPSU 1968–70; Dep. Chair. of Exec. Cttee. of Moscow City Sov. of Peoples' Deps. 1970–75; USSR Min. of the Medical Ind. My. 1975–; mem. Cen. Auditing Comm. CPSU 1976–; Dep. to USSR Supr. Sov. 1979–.

Melnichenko, Vladimir Vladimirovich
Painter; b. 1932, Kiev.
Grad. Kiev Art Inst. 1957, pupil of K. Trokhimenko, T. Iablonskaia, L. Chichkan.

Melnichuk, A.
Linguist, corresponding mem. USSR Acad. of Sciences 1981–.
Dept. head at the A.A. Potebin Inst. of Linguists Ukrainian Acad. of Sciences; expert on general and Slavic linguistics; well-known works on the history of the Slavic languages, in particular Ukrainian, and especially in the field of historical lexicography and etymology; directs work on the first multivolume etymological dictionary of the Ukrainian language; author of more than 150 publications.

Melnik, Anna Vasil'evna
State official; b. 1930.
Moldavian; cand. of Biological Sciences; mem. CPSU 1952–; grad. Kishinev State Univ.; 2nd, 1st City Sec. Komsomol in Kishinev 1953–55; instructor of Kishinev City Cttee. CP Moldavia, grad. student, junior researcher at Moldavian affiliate of USSR Acad. of Sciences 1955–62; Sec. of Kishinev City Cttee. CP Moldavia 1962–64; mem. Cen. Cttee. CP Moldavia 1963–; head of Dept. of Science and Educ. Institutions Cen. Cttee. CP Moldavia 1964–76; Dep. to 1971– and Sec. of Presidium Supr. Sov. Moldavian SSR 1975–.

Melnikov, Evgenii Nikolaevich
Diplomat.
USSR Amb. to Central African Republic Jl. 1972–78, to Upper Volta Ag. 1981–.

Mel'nikov, Nikolai Vasil'evich
Mining scientist; b. F.15 (28), 1909, Sarapul, present-day Udmurt ASSR.
Academician of the Academy of Sciences of the USSR (1962; corresponding member, 1953); Honored Worker in Science and Technology of the RSFSR (1960). Member of the CPSU since 1944.
After graduating from the Sverdlovsk Mining Institute in 1933, he worked in various capacities in the mining industry. He has been a professor at the Academy of the Coal Industry (1950–56), deputy director and director of the A.A. Skochinskii Mining Institute (1955–64), chairman of the State Committee on the Fuel Industry of the USSR (1961–65), a member of the Presidium of the Academy of Sciences of the USSR (since 1966), and director of the sector for physicotechnical problems of mining at the Institute of Physics of the Earth of the Academy of Sciences of the USSR (since 1967). His main works deal with the theory of open-pit mining of deposits, control of blast action in mined rock, and methods of engineering-economic analysis and forecasting of developments in mining engineering. He is the originator of new systems for open-pit mining of deposits, methods for engineering calculations in mining and blasting work, and a method for increasing the efficiency of blasting in rock. He was a deputy to the sixth convocation of the Supreme Soviet of the USSR. Decors.: State Prize of the USSR (1946), Order of Lenin, four other orders, and medals.
Works: *Dobycha iskopaemykh otkrytym sposobom.* Moscow-Leningrad 1948. *Razvitie gornoi nauki v oblasti otkrytoi razrabotki mestrorozhdenii v SSSR*, 2nd ed. Moscow 1961. *Energiia vzryva i kon-*

struktsiia zariada. Moscow 1964. (With L. N. Marchenko.)

Melnikov, Oleg Aleksandrovich
Astronomer; b. Mr. 20 (Ap. 2), 1912, Khvalynsk, now Saratov Oblast.
Grad. Kharkov Univ. 1933; Sen. Scientific Adviser at Main Astronomical Observatory of USSR Acad. of Sciences in Pulkovo; corresponding mem. USSR Acad. of Sciences 1960–. Decors.: orders and medals.

Melnikov, Pavel Ivanovich
Geologist; b. Je. 6 (19), 1908, Leningrad.
Mem. CPSU 1929–; grad. Leningrad Mining Inst. 1935; dir. of various stations in permafrost region 1935–56; Dir. of Northeastern Branch of V. A. Obruchev Inst. of Permafrost Geology 1956–60; Dir. of Inst. of Permafrost Geology at Siberian Branch of USSR Acad. of Sciences 1960–; corresponding mem. USSR Acad. of Sciences 1968–. Decors.: 4 orders, medals.

Meniuk, Georgii Nikolaevich
Author; b. 1918, Kishinev; s. of a worker.
Moldavian; studied Bucharest Univ.; first works publ. in the thirties; mem. CPSU 1953–.
Publs.: anthology of verse "Kosmicheskii krug" (Interior Cosmic) 1939, critical work "Obraz v iskusstve" (Imaginea in arta) 1940, poem "Pesnia zari" (Song of the Dawn) 1948, "Ballady i sonety" (Ballads and Sonnets) 1955, "Izbrannye stikhotvoreniia" (Selected Verse) 1958, "Nash Drug – poeziia" (Our Friend – Poetry) 1964, "U soroki na balu" (At the Magpie's Ball) 1956, and others.

Menner, Vladimir Vasil'evich
Geologist and palaeontologist; b. N. 11 (24), 1905, Shatsk, now Riazan Oblast.
Grad. Moscow Univ. 1927; Head of Dept. of Geology and Prof. Chair of Palaeontology at Inst. of Geological Prospecting Moscow 1930–65; Prof. Chair of Palaeontology at Moscow Univ. 1965–; Sen. Scientific Advisor 1934– and Dep. Dir. of Inst. of Geology at USSR Acad. of Sciences 1960–; mem. USSR Acad. of Sciences 1966–; Pres. 1968–72, Vice-Pres. of Stratigraphic Comm. and Pres. of Subcomm. for Stratigraphy of the Palaeocene of Int. Soc. for Geological Sciences 1972–; mem. French and London Geological Soc. Decors.: S. M. Kirov Prize of USSR Acad. of Sciences 1951, 3 orders, various medals.

Menovshchikova, Nina Ivanovna
Ballerina; b. 1934.
Grad. Choreographical Inst. Perm 1953; with Theatre for Opera and Ballet Perm 1953–56; soloist with Theatre for Opera and Ballet Sverdlovsk 1956–. Decor.: People's Artist of the USSR 1973.

Menteshashvili, Tengiz Nikolaevich
State official; b. 1928.
Georgian; mem. CPSU 1952–; grad. Georgian Polytechnical Inst.; construction engineer, chief construction engineer of metallurgical plant in Rustavi 1950–52; Komsomol City Sec. in Rustavi, Sec., 2nd Sec. Cen. Cttee. Georgian Komsomol 1952–61; held posts in Cen. Cttee. apparatus CP Georgia, 1st City Sec. CP Georgia in Rustavi 1961–73; mem. Cen. Cttee. CP Georgia 1964–; 2nd City Sec. 1973–76, 1st City Sec. CP Georgia in Tbilisi Ja. 1976–82; mem. Bureau Cen. Cttee. CP Georgia Ja. 1976–82; Sec. Presidium USSR Supr. Sov. N. 1982–.

Merenishchev, Nikolai Vladimirovich
Party official; b. 1919; s. of an employee.
Russian; mem. CPSU 1945–; grad. Leningrad Polytechnical Inst. 1944; postgrad. student at Leningrad Polytechnical Inst., Komsomol posts 1945–51; senior engineer, group chief, chief of a project at Leningrad Polytechnical Inst. 1951–56; Sec., 2nd Sec. of Petrograd Raion Cttee. CPSU in Leningrad City 1956–62; chief of a dept. of Leningrad Oblast Cttee., Sec., 2nd Sec. of Leningrad City Cttee. CPSU 1962–73; 2nd Sec. of Cen. Cttee. and mem. Bureau Cen. Cttee. Moldavian CP 1973–; Dep. to USSR Supr. Sov. 1974–; cand mem. Cen. Cttee. CPSU 1976–.

Mergelian, Sergei Nikitovich
Mathematician; b. My. 19, 1928, Simferopol'.
Corresponding member of the Academy of Sciences of the USSR (1953) and academician of the Academy of Sciences of the Armenian SSR (1956). Member of the CPSU since 1955.
A graduate of the University of Yerevan (1947), he worked there between 1945 and 1957. From 1954 to 1958 and from 1964 to 1968 he worked at Moscow University. From 1956 to 1960 he was director of the Scientific Research Institute of Mathematical Machines and of the Computing Center of the Academy of Sciences of the Armenian SSR. Between 1961 and 1971 he worked at the Institute of Mathematics of the Academy of Sciences of the USSR. In 1971 he became vice-president of the Academy of Sciences of the Armenian SSR.
His principal works deal with the theory of functions of a complex variable, the theory of approximation, and the theory of the potential and harmonic functions. One of the deputies to the fifth through eighth convocations of the Supreme Soviet of the Armenian SSR. Decors.: State Prize of the USSR in 1952.
Works: "Nekotorye voprosy konstruktivnoi teorii funktsii." *Trudy Matematicheskogo in -ta AN SSSR*, 1951, vol. 37. "Ravnomernye priblizheniia funktsii kompleksnogo premennogo." *Uspekhi matematicheskikh nauk*, 1952, vol. 7, issue 2. "O polnote sistem analiticheskikh funktsii." *Ibid.*, 1953, vol. 8, issue 4. "Vesovye priblizheniia mnogochlenami." *Ibid.*, 1956, vol. 11, issue 5.

Merilaas, Kersti (real name: Kersti Sang)
Author; b. 1913, Narva; d. of a worker.
Esthonian; first works publ. 1935.
Publs.: "Vetry na doroge" (Wind on the Road) 1937, "Beregovaia lastochka" (Bank Swallow) 1962, "Vesennie pastbishcha" (Spring Pastures) 1966, and others.

Merimaa, Otto Ottovich
State official; b. 1920.
Mem. Cen. Cttee. and mem. Bureau Cen. Cttee. Esthonian CP 1954–55 and 1956–; Sec. of Cen. Cttee. Esthonian CP 1954–55 and 1956–65; Dep. to Supr. Sov. Esthonian SSR 1955–; Chair of Cttee. for Party and State Control with Cen. Cttee. Esthonian CP and Council of Mins. Esthonian SSR 1964–65; Vice-Chair. of Council of Mins. Esthonian SSR 1964–65; Chair. of Ctte. for People's Control Esthonian SSR D. 1965–.

Merkulov, Pavel Ivanovich
State official.
Mem. CPSU 1939–; 1st Dep. Chair. of Gorkii Oblast Exec. Cttee. of Workers' Deps. 1959; Dep. Min. of Production and Procurement of Agricultural Products RSFSR 1962; Chair. of State Production Cttee. of Grain and Mixed Feed Ind. RSFSR 1965; Min. of Grain and Mixed Feed Ind. RSFSR 1965–69; Min. of Procurement RSFSR N. 1969–; mem. Council of Kolkhozes RSFSR 1970–.

Merzaev, Bobo Azizovich
State official.
Cand. mem. Cen. Cttee. Tadzhik CP 1966–81, mem. 1981–; Head of Main. Admin. 1971–78, Chair. of State Cttee. for Material and Tech. Supplies Tadzhik SSR Ag. 1978–.

Meshcheriakov, Mikhail Grigor'evich
Physicist; b. S. 4 (17), 1910, Sambek, now Rostov Oblast.
Mem. CPSU 1940–; grad. Leningrad Univ. 1936; mem. staff Radium Inst. of USSR Acad. of Sciences 1936–47; Dep. Dir. of Inst. of Atomic Energy 1947–48; Dir. of Inst. of Nuclear Problems; corresponding mem. USSR Acad. of Sciences 1953–; mem. staff United Inst. of Nuclear Research in Dubna 1956–; Prof. at Moscow Univ. 1954–; has undertaken research on the physics of high-energy particles; he and his associates experimented with and proved the change in interaction of nucleons with nucleons at 460–660 million electron volts; together with D. V. Efremov and A. L. Mints conducted construction of the first Sov. 680-MEV synchrocyclotrone. Decors.: State Prize of the USSR 1951, 1953, 3 Orders of Lenin, 2 other orders, various medals.

Meshkov, Fedor Stepanovich
Party official; b. 1914.
Russian; mem. CPSU 1939–; grad. Borisoglebsk Tech. College of Mechanization of Agriculture 1933 and Party High School of Cen. Cttee. CPSU 1957; mechanic and controller in Kursk Oblast, Head of Teaching Dept., Dir. of Zadonskaia School of Mechanization of Agriculture (Orel Oblast) 1933–43; Sov. and CPSU posts 1943–48; Head, 1st Dep. Head, Head of Orel Oblast Agricultural Admin. 1949–54 and 1957–58; 2nd Sec., Sec. of Orel Oblast CPSU Cttee. 1958–65 (for Agriculture 1962–64); Chair. of Orel Oblast Exec. Cttee. of Workers' Deps. 1965–70; 1st Sec. of Orel Oblast CPSU Cttee. Ap. 1970–; Dep. to USSR Supr. Sov. 1970–; cand. mem. 1971–76, mem. Cen. Cttee. CPSU 1976–.

Meshkov, Leonid Karpovich
Sportsman and swimmer; b. Ja. 1 (14), 1916, Tsaritsyn, present-day Volgograd.
Honored Master of Sports (1940). Honored Coach of the RSFSR (1963). Honored Cultural Worker (*Zasluzhennyi rabotnik kul'tury*) of the RSFSR (1973). Member of the CPSU (1946).
Between 1932 and 1952, he was a champion swimmer of the USSR (42 gold medals), and a record-breaker of the USSR, Europe, and the world (more than 120 records). He and S. P. Boichenko, another outstanding Soviet swimmer, have been leading figures in the rise of swimming in the USSR.
He fought in the Great Patriotic War. He has been a docent at Moscow State University since 1952. Decors.: three orders and a number of medals.

Meshkov, Nikolai Maksimovich
State official.
Dir. of Inst. of Party History at Cen. Cttee. Belorussian CP 1960–62; mem. Cen. Cttee. Belorussian CP 1961–; Head of Dept. of Science, Colleges and Schools at Cen. Cttee. Belorussian CP, Rector of Minsk Party High School 1962–68; Dep. Chair. 1966–72, Chair. of Belorussian Soc. "Znanie" 1972–; Min. of Intermediate and Higher Secondary Specialist Educ. Belorussian SSR Ap. 1968–.

Mesiats, Valentin Karpovich
State official; b. 1928.
Mem. CPSU 1955–; grad. Timiriazev Acad. of Agricultural Sciences 1953–; chief agronomist, dir. of a machine and tractor station in Moscow Oblast 1953–58; Sov., CPSU and state posts 1958–63; 2nd Sec. of Moscow Oblast CPSU Cttee. for Agriculture 1963–64; Sec. of Moscow Oblast CPSU Cttee. 1964–65; various state posts 1965–71; mem. Cen. Cttee. 1971–76, mem. Bureau Cen. Cttee. and 2nd Sec. of Cen. Cttee. Kazakh CP 1971–76; Mem. Cen. Cttee. CPSU 1971–; USSR Min. of Agriculture Mr. 1976–; Chair. of USSR Union Council of Kolkhozes Ap. 1976–. Decors.: Order of Lenin.

Messerer, Asaf Mikhailovich
Ballet dancer, choreographer, and teacher; b. N. 6 (19), 1903, Vilnius.
People's Artist of the USSR (1976) and Honored Art Worker of the Lithuanian SSR (1953). Member of the CPSU since 1944.
Upon graduating from the Moscow Choreographic School in 1921, he was accepted into the ballet company of the Bolshoi Theater where he was a student of A. A. Gorskii and V. D. Tikhomirov. He has contributed much to expanding the technique of male dancing, thus enriching and making more complex the "vocabulary" of ballet dancing. His best roles included Colin in Hertel's *La Fille mal gardée*, Basil in Minkus' *Don Quixote*, Philippe in Asaf'ev's *The Flames of Paris*, the Acrobat in Glière's *The Red Poppy*, and the title role in Stravinsky's *Petrouchka*. His concert *divertissement The Soccer Player* (or *The Football Player*), to the music of A. N. Tsfasman, was particularly popular.
He choreographed his first work in 1926. He began teaching in 1921 and from 1923 to 1960 taught at the Moscow Choreographic School. Since 1946 he has been conducting a *classe de perfectionnement* for ballet dancers of the Bolshoi Theater.
Decors.: State Prize of the USSR (1941, 1947), three orders and various medals.
Work: *Uroki klassicheskogo tantsa*. [Moscow] 1967.

Metreveli, Aleksandr Iraklievich
Tennis player; b. N. 2, 1944, Tbilisi.
Honored Master of Sport (1966). Journalist.
Between 1966 and 1973, he was tennis champion of the USSR 17 times. Between 1967 and 1973 he was champion of Europe nine times in various classes. In 1967–72, he was a repeated winner of open championships in Asia, the Arab Republic of Egypt, India, and in several states of the Commonwealth of Australia.

Mezhelaitis, Eduardas Beniaminovich
Author; b. 1919; s. of a farm labourer.
Lithuanian; mem. CPSU 1943–; uncompleted high-sch. educ.; exec. ed. of Republican Komsomol paper, Instructor, Sec. of Cen. Cttee. Lithuanian Komsomol 1940–46; literary and ed. work 1946–54; Sec. 1954–59, Chair. of Lithuanian Writers' Union, Sec. of USSR Writers' Union 1959–; Dep. to 1955–, Dep. Chair. of Presidium Supr. Sov. Lithuanian SSR 1975–; mem. Cen. Cttee. Lithuanian CP 1960–; first works publ. 1935.
Decors.: Lenin Prize 1962, Order of Lenin, 3 other orders and various medals. Hero of Socialist Labour.
Publs.: "Lirika" (Lyric) 1943, "Veter s Rodiny" (Wind From the Homeland) 1946, "Chuzhie kamni" (Strange Stones) 1957, "Solntse v iantare" (Sun in Amber) 1961, "Luzhnaia panorama" (Southern Panorama) 1963, "Chelovek" (Man) 1961, "Khleb i slovo" (Bread and the Word) 1965, "Nochnye babochki" (Moths) 1966, "Zdes' Litva" (Here Is Lithuania) 1968, "Gorizonty" (Horizons) 1970, "Iantarnaia ptitsa" (Amber Bird) 1972, and others.

Mezhirov, Aleksandr Petrovich
Poet; b. 1923.
Russian; mem. CPSU 1943–; grad. Gorkii Inst. of Literature Moscow 1948; volunteer at front, participant in defence of Leningrad 1941.
Publs.: "Doroga daleka" (The Journey Is Long) 1947, "Vozvrashchenie" (The Return) 1955, "Vospominaniia o pekhote" (Reminiscences of the Infantry), "Liubimaia pesnia" (Favourite Song), "Vetrovoe steklo" (Windscreen) 1961, "Proshchanie so snegom" (Parting From the Snow) 1964, "Ladozhskii led" (The Ice of Ladoga) 1965, and others.

Mgeladze, Guram Davidovich
State official; b. 1932.
Georgian; mem. CPSU 1960–; grad. Georgian Research Inst. of Zootechnical Veterinary and Tbilisi State Univ.; instructor of Raion Cttee. CP Georgia in Samgori 1962–63; Chair. of Cttee. for Party and State Control in Gardabanskii and Mtskhetskii Raions 1963–69; attended the CPSU Cen. Cttee. Party Higher School 1969–71; instructor of the Economics Dept. Cen. Cttee. CP Georgia 1971–73; 1st Sec. of Abasskii Raion Cttee. CP Georgia 1973–82; mem. Cen. Cttee. CP Georgia 1976–; Min. of Agriculture Georgian SSR F. 1982–.

Miandmaa, Endel Iohannesovich
State official.
Min. of Finance Esthonian SSR Jl. 1979–; mem. Cen. Cttee. Esthonian CP 1981–.

Miannik, Kharald Aleksandrovich
State official.
Dep. to Supr. Sov. Esthonian SSR 1959–; Dir. of an RTS 1959; mem. Cen. Cttee. Esthonian CP 1961–71; Vice-Chair. 1962–63, mem. Bureau for Agriculture with Cen. Cttee. Esthonian CP 1963–64; Min. for Agriculture Esthonian SSR 1965–70 and My. 1975–81; Min. of Fruits and Vegetables Ind. Esthonian SSR Ja. 1981–.

Miasnikov, Gennadii Alekseevich
Actor; b. S. 12, 1919, Sosnovka, now Perm Oblast.
Mem. CPSU 1972–; grad. Perm Art Inst. 1938; mem. staff Dept. of Art at State Union Film Inst. 1943–; Lecturer 1960–.
Decors.: People's Artist of the RSFSR, 2 orders.

Miasnikova, Lidiia Vladimirovna
Russian mezzo-soprano; b. S. 8 (21), 1911, Tomsk.
People's Artist of the USSR (1960).
In 1939, she graduated from M. I. Brian's class at the Leningrad Conservatory,

where she was a graduate student (*aspirant*) from 1939 to 1941. In 1944 she became a soloist with the Novosibirsk Theater of Opera and Ballet. She has sung the parts of Marfa in Mussorgsky's *Khovanshchina,* Liubasha in Rimsky-Korsakov's *The Tsar's Bride,* Amneris in Verdi's *Aida,* Azucena in Verdi's *Il Trovatore,* Eboli in Verdi's *Don Carlos,* and Aksin'ia in Dzerzhinskii's *The Quiet Don.* She also performs as a concert singer. Since 1966, she has been an instructor at the Novosibirsk Conservatory. Decor.: Order of the Red Banner of Labor.

Migdal, Arkadii Beinusovich
Physicist; b. F. 26 (Mr. 11), 1911, Lida, now Grodno Oblast.
Grad. Leningrad Univ. 1936; mem. Inst. of Physical Problems of USSR Acad. of Sciences 1943–45; mem. staff Inst. of Atomic Energy 1945–71; mem. USSR Acad. of Sciences 1966–; mem. staff Inst. of Theoretical Physics of USSR Acad. of Sciences 1971–; concurrently Prof. at Inst. of Physical Engineering Moscow 1944–. Decors.: Order of Lenin, Order of Red Banner of Labour, medals.
Publs.: "Teoriia konechnykh fermi-sistem i svoistva atomnykh iader" Moscow 1965, co-author of „Priblizhennye metody kvantovoi mekhaniki" Moscow 1966, "Metod kvazichastits v teorii iadra" Moscow 1967.

Migulin, Vladimir Vasil'evich
Radio physicist; b. Jl. 10 (23), 1911, Sereda, now Ivanovo Oblast.
Mem. CPSU 1945–; grad. Polytech. Inst. Leningrad 1932; mem. staff Inst. of Physics of USSR Acad. of Sciences 1934–41; mem. staff Thermodynamic Laboratory of USSR Acad. of Sciences 1946–51; Dir. of Physical-Tech. Inst. Sukhumi 1951–54; Dir. of Inst. of Earth Magnetism, Ionosphere and Distribution of Radio Waves USSR Acad. of Sciences 1969–; Lecturer 1935–, Prof. at Moscow Univ. 1948–; Dep. Dir.-Gen. of Int. Atomic Energy Agency 1957–59; corresponding mem. USSR Acad. of Sciences 1970–; Vice-Pres. of Int. Radio Fed. 1972–; most important works on theory of oscillation and distribution of radio waves and other problems of radio physics. Decors.: State Prize of the USSR 1946, 1953, Order of Lenin, 3 other orders, various medals.

Mikaelian, Suren Agasievich
State official.
Head of Cen. Admin. 1959–63, Chair. of State Cttee. for Vocational Training Armenian SSR My. 1963–; cand. mem. Cen. Cttee. Armenian CP 1966–; Dep. to Supr. Sov. Armenian SSR 1967–.

Mikalauskas, Iulionas Vladovich
State official.
Mem. Cen. Cttee. Lithuanian CP 1956–; Dep. to Supr. Sov. Lithuanian SSR 1959–; 1st Sec. of Kaunas City Cttee. Lithuanian CP 1959–68; Min. of Maintenance of Public Order S.–D. 1968, Min. of the Interior Lithuanian SSR D. 1968–.

Mikalauskas, Vladas Vladovich
State official.
Chair. of State Produc. Cttee. for Geology Lithuanian SSR 1964–65; Head of Main Admin. for Geology 1980–.

Mikelinskas, Ionas
Author; b. 1922, Lithuania; s. of a peasant.
Grad. Vilnius Univ., Faculty of History and Philology 1952; first works publ. 1957.
Publs.: "Starik pod chasami" (An Old Man Beneath the Clock) 1960, "Teplye ruki" (Warm Hands) 1962, "Uznal li ty ego?" (Did You Recognize Him?) 1963, "Gorst' pesku" (A Handful of Sand) 1965, "Vodonosha" (Waterbearer) 1964, "A chasy idut" (Yet the Hours go by) 1966, and others.

Mikeliunas, V.
Komsomol official.
Sec. and mem. Bureau Cen. Cttee. Lithuanian Komsomol Mr. 1982–.

Mikhail (M. A. Chub), Cand. of Theol.
Russian Orthodox clergyman; b. F. 19, 1912, Pushkino, St. Petersburg Gouvt.
Studied Hydro-Meteorological Inst., external studies at Inst. of Foreign Languages and from 1947 at Leningrad Theological Acad.; teacher at Leningrad Theological Acad., priest 1950–; lecturer at Leningrad Theological Acad., priestmonk 1953, 1953; Bishop of Luga, Substitute for the Bishopric of Smolensk, Bishop of Smolensk 1955–; Bishop of Berlin and Germany 1957; Bishop of Izhevsk 1959, Bishop of Tambov 1961; mem. Ed. Board of "Theological Treatises" 1967–; Archbishop of Voronezh and Lipetsk 1968–; Archbishop of Vologda and Veliki Ustiug 1972–74; Archbishop of Tambov and Michurinsk 1974–.

Mikhail (M. N. Mudiugin), Cand. of Tech. Sc., Cand. of Theol.
Russian Orthodox clergyman; b. My. 12, 1912, St. Petersburg.
Grad. Inst. of Foreign Languages 1933 and Inst. for Metal Ind. and Energetics 1946; Cand. of Tech. Sc. 1953; univ. lecturer in thermo-technology, lecturer at Leningrad Mining Inst. 1954–57; priest 1958; grad. Leningrad Theological Acad. (external course), Cand. of Theology, lecturer at same Acad. 1964; lecturer and deacon of newly founded "Faculta of African Christian Youth", Rector of Leningrad Theological Schools, monk, Archimandrite 1966; Bishop of Tikhvin, Substitute for the Bishopric of Leningrad 1966, Bishop of Astrakhan and Enotaevka; 1968, Archbishop of Vologda and Velikii Ustiug D. 1979–. Master of Arts in Theology 1971.

Mikhailenok, Leonid Lukich
Sculptor; b. 1930.
Grad. Higher Mukhina Art and Ind. Inst. Leningrad 1957. Decor.: State Prize of the USSR 1969.

Mikhailichenko, Ivan Kharlamovich
Military official; b. S. 2, 1920, in the *stanitsa* (large cossack village) of Almaznaia, Kadievka Raion, Voroshilovgrad Oblast.
Twice Hero of the Soviet Union (Jl. 1, 1944 and Je. 27, 1945); lieutenant colonel (1955). Member of the CPSU since 1951. He joined the Soviet Army in 1940 and graduated from the Voroshilovgrad Military Aviation School for Pilots in 1943. In the Great Patriotic War (1941–45) he participated in fighting on the Southwestern, Voronezh, and First and Second Ukrainian fronts as pilot, flight commander, and deputy commander of a squadron of the 667th Assault Air Regiment and later, from 1943, as commander of a squadron of the 141st Guards Assault Air Regiment. He made 179 successful combat flights for bombing and assault strikes on enemy troops and fought 22 air battles. After the war he served in the air force until 1962 and retired that year. Decors.: Order of Lenin, four Orders of the Red Banner, the Order of Alexander Nevsky, the Order of the Patriotic War First Class, the Order of the Red Star, the Order of Glory Second and Third Class, and various medals.

Mikhailin, Vladimir Vasil'evich
Military official; b. Jl. 12 (25), 1915, Moscow.
Admiral (1969). Member of the CPSU since 1941.
He graduated from the M. V. Frunze Higher Naval College in 1941. In the Great Patriotic War (1941–45) he served in the Northern Fleet as commander of communications and a combat unit of navigators and as assistant commander and then commander of a ship; from 1945 to 1948 he served on ships of the Pacific Ocean Fleet. He graduated from the Naval Academy in 1951 and from the Academy of the General Staff in 1960. He was commander of the cruisers *Admiral Kornilov* and *Kuibyshev* from 1955 to 1958 and commander of large ship units of the Black Sea Fleet from 1960 to 1963. He served from 1963 to 1967 as deputy commander and from 1967 to 1975 as commander of the Baltic Fleet. In 1975 he was appointed deputy commander-in-chief of the navy for naval colleges. He was a deputy to the eighth convocation of the Supreme Soviet of the USSR. Decors.: Order of Lenin, three Orders of the Red Banner, the Order of the Patriotic War First Class, the Order of the Red Star, and various medals.

Mikhailov, Aleksandr Alekseevich
Literary critic; b. 1922, Kuia, Arkhangelsk Oblast.
Russian; mem. CPSU 1943–; first works publ. 1952.
Publs.: "Ivan Menshikov" 1959, "Poet i liricheskii geroi" (The Poet and the Lyric Hero) 1960, "Sever v literature" (The North in Literature) 1961, "Lirika serdtsa i razuma" (The Lyric of the Heart and Mind) 1965, and others.

Mikhailov, Boris Mikhailovich
Organic chemist; b. Mr. 21 (Ap. 3), 1906, Znamenka, present-day Kirens Raion, Irkutsk Oblast.
Corresponding member of the Academy of Sciences of the USSR (1963). Graduated from Kazan University in 1929. Student of A. E. Arbuzov. He worked at Kazan and Moscow universities and a number of medical research institutes.
He has been at the Institute of Organic Chemistry of the Academy of Sciences of the USSR since 1943 (he has been a laboratory director since 1954). His main works are devoted to the chemistry of organoboron compounds, polycyclic hydrocarbons, the organic compounds of alkali metals, and heterocyclic compounds. Decors.: two orders and several medals.
Works: "Organicheskie soedineniia bora." *Uspekhi khimii,* 1959, vol. 28, no. 12. "Borazol i ego proizvodnye." *Uspekhi khimii,* 1960, vol. 29, no. 8. *Khimiia borovodorodov.* Moscow 1967. "Serosoderzhashchie organicheskie soedineniia bora." *Uspekhi khimii,* 1968, vol. 37, no. 12.

Mikhailov, Igor Leonidovich
Poet; b. 1913, St. Petersburg.
Russian; grad. Leningrad Univ. 1937; first works publ. 1933.
Publs.: "Matveev kurgan" (Matveev Hill) 1939, "Kochmas, ili vecher vospominanii" (Kochmas, or Evening of Memories) 1940, "Vozvrashchenie" (The Return) 1946, "Poruchik Lermontov" (Lieutenant Lermontov) 1964, "Mir i trud" (Peace and Work) 1953, "Vse, chem zhivem" (Everything We Live For) 1958, and others.

Mikhailov, Nikolai Ivanovich
Physical geographer; b. My. 4 (17), 1915, Rybinsk, now in Iaroslavl Oblast.
Doctor of geographical sciences (1963). Beginning in 1964, a professor in the department of geography at Moscow State University.
His principal works are devoted to the regional physical geography of the USSR, especially Siberia, as well as to the theory of physicogeographical zonation. Decor.: D. N. Anuchin Prize in 1969.
Works: *Sibir'.* 2nd ed. Moscow 1956. *Gory Iuzhnoi Sibiri.* Moscow 1961. *Fizicheskaia geografiia SSSR: Aziatskaia chast'.* 2nd ed. Moscow 1970. (With N. A. Gvozdetskii.) *Fiziko-geograficheskoe raionirovanie (kurs lektsii),* parts 1–3. Moscow 1960–71. *Fiziko-geograficheskoe raionirovanie SSSR.* Moscow 1968. (Coauthor.)

Mikhailov, Nikolai Nikolaevich
Russian writer; b. D. 10 (23), 1905, Moscow.
In 1930, he graduated from the Moscow Institute of Industrial Economics. He traveled a great deal in the USSR and abroad. His first works were published in 1927. His books include *Khan-Tengri* (1933), *The Face of the Country Is Changing* (1937), *The Map of the Motherland* (1947; State Prize of the USSR, 1948), *My Russia* (1964–66; M. Gorkii State Prize of the RSFSR, 1968), *The Amber Republic* (1970), and *Peace Is But a Dream* (1972). He also wrote the sketches *I Walk Along the Meridian* (1957) and the travel novellas (with Z. Kosenko) *The Americans* (1960) and *The Japanese* (1963).
He is one of the recognized masters of Soviet fiction of a scientific nature; he is primarily concerned with the changes in the economic geography of the USSR. Several of his books have been translated into foreign languages. Decors: two orders and a number of medals.

Mikhailov, Viktor Pavlovich
Boxer; b. N. 11 (24) 1907, Moscow.
Honored Master of Sport (1936). Honored Coach of the USSR (1957). Member of the CPSU since 1940.
Between 1927 and 1939, he was light heavyweight boxing champion of the USSR, absolute champion of the USSR (1939), and winner of the World Worker's Olympics in Antwerp (1937). He fought in the Great Patriotic War of 1941–45. From 1946 to 1957 he was coach of the USSR boxing team and from 1946 to 1972 an instructor at the State Central Institute of Physical Culture in Moscow. Decors.: two orders and a number of medals.
Work: *120 vstrech na ringe.* Moscow 1952.

Mikhalkov, Sergei Vladimirovich
Author; b. F. 28 (Mr. 13), 1913, Moscow.
Russian; mem. CPSU 1950–; studied Gorkii Inst. of Literature 1935–37; mem. USSR Acad. of Pedagogic Sciences 1971–; 1st Sec. of Moscow branch of RSFSR Writers' Union 1967–70; Chair. of RSFSR Writers' Union 1970–; Chief Ed. of satirical journal "Fitil" 1962–; first works publ. 1928. Decors.: Merited Artist of the RSFSR 1967, State Prize of the USSR 1941, 1942, 1950, Lenin Prize 1970, 3 Orders of Lenin, 3 other orders and various medals.
Publs.: children's books "Osoboe zadanie" (Special Task) 1945, "Ia khochu domoi" (I Want to Go Home) 1949, "Dorogoi mal'chik" (Dear Boy) 1971, plays "Ilia Golovin" 1950, "Okhotnik" (The Hunter) 1956, "Dikari" (The Savages) 1958, "Raki i krokodil" (The Crayfish and the Crocodile) 1960, books "Sobranie sochinenii" (Collected Works) vols. 1–4 1963–65 and vols. 1–3 1970–71, "Chuvstvo loktia" (The Feeling of Fellowship) 1971, and others.

Mikhailov, Viktor Konstantinovich
State official.
Dep. Chair. Council of Mins. Uzbek SSR Je. 1982–.

Mikhailovskii, Vladimir Nikolaevich
Scientist, specialist in automation, telemechanics and surveying technology; b. Ag. 15, 1914, Konstantinograd, now Krasnograd, Kharkov Oblast.
Mem. CPSU 1945–; grad. Inst. of Energy – Moscow 1939; engineer in various factories 1939–41; served in Sov. Army 1941–45; mem. staff Inst. of Automation and Telemechanics of USSR Acad. of Sciences 1946–51; Head of Dept. of Automation and Telemechanics at Inst. of Mechanical Engineering (now Dept. of Information Selection at Inst. of Physical Mechanics) of Ukrainian Acad. of Sciences 1952–; author of some 130 works.
Publs.: co-author of "Izmerenie krivizny skvashin" 1960, "Elektronnye kommutatory" 1961, and "Teleizmerenie radioaktivnykh izluchenii" 1963.

Mikhnevich, Iurii Mikhailovich
State official; b. 1925, Minsk.
Mem. CPSU 1948–; grad. Belorussian State Univ. 1951; Komsomol posts 1950–55; Dep. Dir. of Minsk Ind. Tech. College, Dep. Head of Cultural Dept. Minsk City Sov., Dep. Head, then Head of Dept. of Propaganda and Agitation at Minsk City Cttee. Belorussian CP, Head of Cultural Admin. at Minsk Oblast Exec. Cttee. of Workers' Deps. 1955–62; Head of Dept. of Propaganda and Agitation at Minsk Oblast Cttee. Belorussian CP 1962–66; Dep. Min. of Culture 1966–71, Min. of Culture Belorussian SSR 1971–; mem. Cen. Cttee. Belorussian CP 1976–. Decors.: Order of Red Star, medals.

Mikirtychev, Eduard Nikolaevich
State official.
Cand. mem. Cen. Cttee. Tadzhik CP 1976–; Min. of Construction Tadzhik SSR Ag. 1982–.

Mikuchiauskas, Vladislovas Kostovich
Party official.
Mem. Cen. Cttee. 1971–; cand. mem. Bureau Cen. Cttee, Lithuanian CP 1976–; Chair. of Exec. Cttee. of Kaunas City Sov. of Workers' Deps. until 1975; 1st Sec. of Kaunas City Cttee. Lithuanian CP My. 1975–.

Mikulich, Vladimir Andreevich
Party official; b. 1920.
Belorussian; mem. CPSU 1943–; grad. Party Higher School of Cen. Cttee. CPSU 1957; worker in a turf factory 1937–40; served in Sov. Army 1940–45; Komsomol

org. in a plant, sec. of a raion cttee. of Belorussian Komsomol, 2nd Sec. of Minsk Oblast Cttee. Belorussian Komsomol 1945–53; sec., 1st sec. of a raion cttee. Belorussian CP in Minsk 1953–62; cand. mem. 1956–60, mem. Cen. Cttee. Belorussian CP 1960–; Dep. to Supr. Sov. Belorussian SSR 1959–; 2nd Sec. 1962–63, 1st Sec. of Brest Oblast Cttee. Belorussian CP 1963–Mr. 1977 (for Agriculture 1963–64); mem. Cen. Auditing Comm. CPSU 1966–76; cand. mem. Cen. Cttee. CPSU 1976–81, mem. 1981–; Dep. to USSR Supr. Sov. 1966–; 1st Sec. of Minsk Oblast Cttee. Belorussian CP and mem. Bureau Cen. Cttee. Belorussian CP Mr. 1977–.

Mikulinskii, Semen Romanovich
Philospher and historian of science; b. Ap. 2, 1919, Kremenchug.
Corresponding member of the Academy of Sciences of the USSR (1968). Member of the CPSU since 1939.
He graduated from the department of philosophy of Leningrad State University in 1949. In 1952 he became a scientific worker at the Institute of the History of the Natural Sciences and Technology of the Academy of Sciences of the USSR; he became the institute's deputy director in 1963 and director in 1974. His main works are on the theory of evolution, general problems of biology and the history of philosophy in Russia in the first half of the 19th century, and philosophical problems of natural science. He is one of the initiators of the development of the science of science in the USSR.
Works: *I. E. Diad'kovskii [Vrach, estestvoispytatel' i filosof-materialist]: Mirovozzrenie i obshchebiologicheskie vzgliady.* Moscow 1951. *K. F. Rul'e i ego uchenie o razvitii organicheskogo mira.* Moscow 1957. *Razvitie obshchikh problem biologii v. Rossii: Pervaia polovina 19 v.* Moscow 1961. *Al'fons Dekandol'.* Moscow 1973. (With L. A. Markova and B. A. Starostin.)

Milashkina, Tamara Andreevna
Singer; b. 1934.
Grad. Moscow Conservatory 1959; lyric-dramatic soprano; soloist with Bolshoi Theatre 1958–; with La Scala, Milan, 1961–62. Decor.: People's Artist of the USSR 1973.

Miliakh, Aleksandr Nikolaevich, Dr. of Tech. Sc.
Engineer; b. Ag. 29, 1906, Simferopol; s. of a railway worker.
Mem. CPSU 1939–; grad. Electrotech. Inst. Kharkov 1931; specializes in electrodynamics; Lecturer at Electrotech. (now Polytech.) Inst. Kharkov 1931–41; mem. staff Inst. of Power Econ. (now Inst. of Electrodynamics and Tech. Thermodynamics) of Ukrainian Acad. of Sciences 1941–47; mem. staff 1947–, Dir. of Inst. of Electrodynamics of Ukrainian Acad. of Sciences 1959–; corresponding mem. Ukrainian Acad. of Sciences 1964–; Dr. of Tech. Sc. 1956–, Prof. 1962–; author of 80 works. Decors.: Badge of Honour, medals.
Publs.: co-author of "Induktivnoemkostnye preobrazovateli istochnikov napriazheniia v istochniki toka" 1964, "Printsip vzaimnosti i obratimost iavlenii v elektrotekhnike" 1967.

Milkin, Anatolii Vasil'evich
State official.
2nd Sec. of Taldy-Kurgan Oblast Cttee. Kazakh CP 1978–81; Chair. of Cttee. of People's Control Kazakh SSR Ag. 1981–; mem. Cen. Cttee. Kazakh CP 1981–.

Mil'kov, Fedor Nikolaevich
Physical geographer; b. F. 17, 1918, Dorovaia, in present-day Vokhma Raion, Kostroma Oblast.
Doctor of geographical sciences (1949), and professor (1949). Honored Scientist of the RSFSR (1970).
He graduated from the Moscow Oblast Pedagogical Institute in 1938. In 1946 he became head of the subdepartment of geography at the Orenburg Pedagogical Institute and in 1950 head of the subdepartment of physical geography at the University of Voronezh. His principal works concern problems of physiographic (landscape) science, physical geographic regionalization, and the physical geography of the USSR. He substantiated the interpretation of the physiographic region (geographic landscape) as a general concept and identified types of provinces as structural units of physical geographic regions. Decors.: P. P. Semenov (Tianshanskii) Gold Medal 1966.
Works: *Lesostep' Russkoi ravniny.* Moscow 1950. *Vozdeistvie rel'efana rastitel'nost' i zhivotnyi mir.* Moscow 1953. *Fiziko-geograficheskii raion i ego soderzhanie.* Moscow 1956. *Landshaftnaia geografiia i voprosy praktiki.* Moscow 1966. *Osnovnye problemy fizicheskoi geografii.* [2nd ed.] Moscow 1967. *Fizicheskaia geografiia SSSR* [3rd ed.], vol. 1. Moscow 1969. (With N. A. Gvozdetskii.) *Landshaftnaia sfera Zemli.* Moscow 1970. *Slovar'-spravochnik po fizicheskoi geografii.* 2nd ed. Moscow 1970. *Chelovek i landshafty: Ocherki antropogennogo landshaftovedeniia.* Moscow 1973.

Miller, Boris Izrailevich (pseudonym: Buzi Miller, D. Riter and others)
Author; b. 1913, Kopai, Vinnitsa Oblast.
Mem. CPSU 1941–; grad. Moscow Pedagogical Lenin Inst. 1936; held state and party posts in Jewish Autonomous Region 1937–46; suffered reprisals 1949–56, rehabilitated 1956.
Publs.: "Smeny zastupaiut" (The Shift Arrives) 1931, "Pod radugoi" (Underneath the Rainbow) 1934, "Birobidschan" 1948, "Na beregu Amura" (On the Bank of the Amur) 1960, "Zemlia Rodnaia. Povesti i rasskazy" (Motherland. Tales and Short Stories) 1965, and others.

Miltinis, Iuozas Ionovich (Juozas Miltinis)
Director and actor; b. S. 3 (16), 1907, Dabikinė, in present-day Akmenė Raion, Lithuanian SSR.
People's Artist of the USSR (1973).
In 1931, he graduated from the Kaunas Theater Studio. From 1932 to 1938 he studied theater and film in France and Great Britain, where he also worked in the theater. In 1939 he became the director of a theater studio. From 1940 to 1954, and since 1959, he has been the principal director of the Panevėžys Drama Theater, which was an outgrowth of his studio.
His best productions include Pogodin's *The Silver Valley* (1941), Ben Jonson's *Valpone* (1941, 1971), Miller's *Death of a Salesman* (1958), Shakespeare's *Macbeth* (1961), *Virgin Soil Upturned,* based on Sholokhov's novel (1964), Borchert's *Outside the Door* (1966), Dürrenmatt's *Frank the Fifth* (1969), and Strindberg's *Dance of Death* (1973).
In 1940, he began teaching in the theater's studio. He has trained many outstanding actors, including D. Banionis and B. Babkauskas. Decors.: State Prize of the Lithuanian SSR (1965), the Order of Lenin, the Order of the Badge of Honor, and a number of medals.

Minachev, Khabib Minachevich
Organic chemist; b. D. 11 (24) 1908, Novye Bikshiki, Chuvash ASSR.
Mem. CPSU 1944–; grad. Moscow Univ. 1939; mem. staff Inst. of Organic Chemistry of USSR Acad. of Sciences 1939–; corresponding mem. USSR Acad. of Sciences 1972–. Decors.: 2 orders, medals.

Minasian, Rafik Armenakovich
State official.
Mem. Cen. Cttee. Armenian CP 1964–; 1st Sec. Kafan Raion Cttee. Armenian CP 1971; Min. of Housing and Communal Econ. Armenian SSR N. 1978–.

Minchkovskii, Arkadii Mironovich (real name: Minchikovskii)
Author; b. 1916, Irkutsk.
Russian; mem. CPSU 1944–; first works publ. 1945.
Publs.: "Shestoi vagon" (The Sixth Waggon) 1957, "Druzia i znakomye" (Friends and Acquaintances) 1959, "Nebo utrom i vecherom" (The Sky in the Morning and Evening) 1962, "Est' na svete liubov'" (There is Love in the World) 1963, "13 raznykh istorii" (13 Different Stories) 1963; "Desiat' dnei odni vtroem" (Ten Days Alone in a Threesome) 1961, "Strannye vzroslye" (Peculiar Adults) 1965, and others.

Minetas, Kazimieras Boleslovovich
State official.
Min. of Furniture and Wood-Processing Ind. Lithuanian SSR N. 1980–; mem. Auditing Comm. Lithuanian CP 1981–.

Minich, Nikolai Grigorevich
State official.
Dep. to Supr. Sov. Kirghiz SSR 1959–; Chair. of Tiup Raion Exec. Cttee. of Worker's Deps. 1959; mem. Cen. Cttee. 1960–, mem. Bureau 1962–66 and 1976–, cand. mem. Bureau Cen. Cttee. Kirghiz CP 1966–76; Sec. of Cen. Cttee. Kirghiz CP 1962–65; Dep. Chair. of Council of Mins. and Chair. of Cttee. of Party and State Control with Cen. Cttee. Kirghiz CP and Council of Mins. Kirghiz SSR 1962–65; Chair. of Cttee. of People's Control Kirghiz SSR D. 1965–.

Minin, Viktor Ivanovich
Diplomat; b. 1926.
Mem. Diplomatic Service 1948–; mem. Apparat USSR Min. of Foreign Affairs 1963–65; Emb. Counsellor at USSR Emb. in Turkey 1965–68; USSR Amb. to Laos 1968–72; Head of Middle East Dept. at USSR Min. of Foreign Affairs 1972–78; USSR Amb. to Guinea 1978–82, to Iraq Mr. 1982–.

Minkevich, Mikhail Gavrilovich
State official; b. 1919, Gulichi, Minsk Oblast; s. of a peasant.
Grad. Belorussian State Univ. 1952; lived in Western Belorussia, arrested and imprisoned there for revolutionary activity 1938–39; pedagogic activity in Baranovichi Oblast 1939–41; served in Sov. Army 1941–45; party and sov. posts, teacher 1945–58; chair. of a raion exec. cttee., 1st sec. of various raion cttees. Belorussian CP 1958–68; Min. of Educ. Belorussian SSR Ag. 1968–; mem. Cen. Cttee. Belorussian CP 1971–. Decors.: 3 orders, medals.

Minko, Vasilii Petrovich
Ukrainian writer; b. Ja. 1 (14), 1902, Minkovka, present-day Valki Raion, Kharkov Oblast.
Member of the CPSU since 1942.
The son of a peasant, he attended the Kharkov Institute of Public Education from 1929 to 1931 and fought in the Great Patriotic War. His first works, plays for amateur theaters, appeared in 1924, and since 1927 he has been publishing essays, short stories, and novellas. Among his best early works are the collection *Local Authorities* (1928) and the novellas *Belladona* (1929) and *Iarina Cherkas* (1936). The novellas *Above the Khorol River* (1949) and *Clear Dawns* (1951) and the short-story collection *A Full Cup* (1950) depict life in the postwar Ukrainian countryside. His satirical comedy *Without Naming Names* (1953) was staged in theaters throughout the country. He also wrote the comedies *At the Farmstead Near Dikan'ka* (1958), *Bridegroom From Argentina* (1960), *A Comedy With Two Heart Attacks* (1966), *Let's Not* (1967), and *Attention, Cockatoo* (1972). His autobiographical novella *My Minkovka,* written between 1962 and 1969, was published in Russian translation in 1973, and his memoirs *Red Parnassus* appeared in 1971. Decors.: three orders, as well as many medals.
Works: *Vybrani tvory.* Kiev 1962. *Komedii.* Kiev 1968. *In Russian translation: Dramy i komedii.* Moscow 1963.

Min'kovskii, Aleksandr Zakhar'evich
Conductor; b. D. 12 (25), 1900, Snezhna, Vinnitsa Oblast.
People's Artist of the USSR (1960).
In 1930, he completed his training in the department of conducting at the N.V. Lysenko Musical Drama Institute. From 1934 to 1941 he was the artistic director and principal conductor of the chorus of the Ukrainian Radio Committee. Since 1946 he has held similar positions with the State Bandurist's Chorus of the Ukrainian SSR. Since 1951, he has taught at the Kiev Conservatory, where he was made chairman of the subdepartment of conducting in 1960 and a professor in 1965. Decors.: T.G. Shevchenko State Prize of the Ukrainian SSR in 1969, Order of Lenin, the Order of the Red Banner of Labor, and various medals.

Minovalova, Nina Fedorovna
Weaver; b. 1936.
Mem. CPSU 1968–, finished 7th class of secondary school 1960; weaver in Rosa Luxemburg Silk Combine in Moscow 1953–; mem. Cen. Auditing Comm. CPSU 1981–.

Mints, Isaak Izrailevich
Historian; b. Ja. 22 (F. 3), 1896, Krinichki, in present-day Dnepropetrovsk Oblast; s. of an office worker.
Member of the Academy of Sciences of the USSR since 1946; corresponding member since 1936. Member of the CPSU since 1917.
He carried out political work in the Red Army during the Civil War of 1918–20. In 1926 he graduated from the Institute of Red Professors in Moscow. He headed subdepartments of history of the USSR at the Moscow Institute of Philosophy, Literature, and History and at Moscow State University (1932–49), and at the Higher Party School under the CPSU Central Committee (1937–49). He was a professor at the Academy of Social Sciences (1947–50) and head of the department of history of the USSR at the Lenin Moscow Institute of Pedagogy (1950–72).
He has been a senior research associate at the Institute of the History of the USSR of the Academy of Sciences of the USSR since 1954. He has been chairman since 1962 of the scientific council for overall study of the history of the Great October Revolution under the Academy of Sciences of the USSR. His major works have been devoted to the study of the history of the CPSU, the October Revolution, the Civil War of 1918–20, the history of Soviet society, and international relations. He was among the authors of *The History of the Communist Party of the Soviet Union* (1959), and a member of the main editorial board of multivolume history of the CPSU. He was a recipient of the State Prize of the USSR (1943 and 1946) for his part in preparation of *The History of the Civil War in the USSR* and *The History of Diplomacy.* He won the Lenin Prize in 1974 for his work on *The History of the Great October* (vols. 1–3, 1967–73). Decors.: Order of Lenin, the Order of the October Revolution, three other orders, and a number of medals.
Work: *Velikaia Oktiabr'skaia sotsialisticheskaia revoliutsiia i progress chelovechestva.* Moscow 1967.

Mirianashvili, Guram Zaalovich
State official; b. 1927, Tbilisi.
Georgian; mem. CPSU 1953–; grad. Georgian Polytechnical Inst. (1949); architect at the "Tbilgorproekt" planning inst. 1949–52; attended faculty foradv. education of architects at Moscow Architectural Inst., teacher there 1952–58; chief architect at "Gruzhydrogorstroi" inst. 1958–64; head of a dept. at "TbilNIIEP" inst. 1964–72; chief architect, Dep. Dir. of "Gruzhydrogorstroi" planning inst., Dir. of this inst. 1972–78; Chair. of State Cttee. for Construction Georgian SSR F. 1978–; cand. mem. Cen. Cttee. CP Georgia 1981–.

Mirkhalikov, Temurbai
State official.
Min. of the Light Ind. Tadzhik SSR Ap. 1976–82; mem. Cen. Cttee. Tadzhik CP 1981–; Dep. Chair. of Council of Mins. and Chair. of GOSPLAN Tadzhik SSR Je. 1982–.

Mirmukhsin (real name: Mirmukhsin Mirsaidov)
Author; b. 1921, Tashkent, Uzbekistan.
Publs.: "Vernost'" (Faithfulness) 1945, "Fergana" 1949, "Gosti" (The Guests) 1954, "Serdtse i filosofiia" (The Heart and Philosophy) 1963, "Usta Giias" (1947), novel "Ziiad i Adiba" 1959, Collected Works "Dzhamilia" 1957, "Rabynia. Belyi mramor" (Slave. White Marble) 1957–64, "Nochnye molnii" (Night Lightning) 1965; also writes for children.

Mirnov, Sergei Vasil'evich
Physicist; b. 1939.
Grad. Physical-Tech. Inst. Moscow 1962; mem. staff I. V. Kurchatov Inst. of Atomic Energy 1962–. Decor.: State Prize of the USSR 1971.

Mironenko, V. I.
Komsomol official.
Mem. Bureau 1981–, Sec. Cen. Cttee. Ukrainian Komsomol 1981–82, 2nd Sec. Ag. 1982–.

Mironov, Vasilii Petrovich
Party official.
Chair. of Donetsk City and Oblast Exec. Cttee. of People's Deps. 1962–82; mem. Cen. Cttee. Ukrainian CP 1976–; 1st Sec. Donetsk Oblast Cttee. Ukrainian CP O. 1982–.

Miroshkin, Oleg Semenovich
Party official; b. 1928.
Russian; mem. CPSU 1959–; grad. Moscow Petroleum Inst. 1951; shop mechanic, shop superintendent of a drilling office, chief mechanic of "Kazakhstanneft'" Asscn., dep. chief of West Kazakh Geological Admin. 1951–70; chief of a dept. of Gurev Oblast Cttee. Kazakh CP 1970–73; 2nd Sec. of Mangyshlak Oblast Cttee. Kazakh CP 1970–73; 2nd Sec. of Mangyshlak Oblast Cttee. (Shevchenko) Kazakh CP 1973–76; Sec. of Cen. Cttee. Kazakh CP 1976–79, 2nd Sec. 1979–; mem. Bureau of Cen. Cttee. Kazakh CP 1976–; Dep. to USSR Supr. Sov. 1979–; mem. Cen. Cttee. CPSU 1981–.

Miroshnichenko, Boris Panteleimonovich, Dr. of Econ. Sc.
Diplomat; b. 1911.
Prof.; mem. Diplomatic Service 1954–; Emb. Counsellor at USSR Emb. in GDR and USSR High Commr. for Econ. Affairs 1954–55; Emb. Counsellor (Econ. Dept.) at USSR Emb. in GDR 1955–57; dep. head of a dept. with Cen. Cttee. CPSU 1957–58; Dep. Head of Dept. of Trade, Finance and Planning Organs, Head of Dept. of Econ. Cooperation with Socialist Countries at Cen. Cttee. CPSU 1961–65; mem. Collegium of USSR Min. of Foreign Affairs, Rector of Moscow Inst. of International Relations at USSR Min. of Foreign Affairs 1965–68; USSR Amb. to Canada 1968–73 and to Kenya N. 1973–.

Miroshnichenko, Evgeniia Semenovna
Singer; b. Je. 12, 1931, Sovetskoe, Kharkov Oblast.
Ukrainian; grad. Kiev Conservatory 1957; lyrical coloratura soprano; soloist with Ukrainian Opera and Ballet Theatre Kiev 1957–. Decors.: T.G. Shevchenko State Prize of the Ukrainian SSR 1972, People's Artist of the USSR, Order of Lenin.

Mirzabekov, Boris Grigor'evich
State official.
2nd Sec. of Andizhan Oblast. Cttee. Uzbek CP 1955; Sec. of Tashkent Oblast Cttee. Uzbek CP 1959; mem. Cen. Cttee. Uzbek CP 1956–61 and 1976–; Min. of Building Materials Ind. Uzbek SSR D. 1972–79; Dep. Chair. of Cttee. of People's Control Uzbek SSR D. 1979–.

Mirzaev, Abdul Taizuevich
Author; b. 1925, Kuli, Dagestan; s. of a railway worker.
Mem. CPSU 1952–; grad. Gorkii Inst. of Literature Moscow 1960; teacher and poet. Publs.: "Vesennie tsvety" (Spring Flowers) 1954, "Zvuki serdtsa" (Notes from the Heart) 1959, "Zvezdy" (Stars) 1962, "Khoroshie liudi" (Good People) 1963, "Deti nashego aula" (The Children from our Aul) 1965, and others.

Mirzagitov, Askhat Masgutovich
Dramatist; b. 1928, Bikmetova, Bashkir ASSR.
Grad. Ufa Theatre School 1948 and Gorkii Inst. of Literature Moscow 1962; worked as actor, producer and director of various kolkhoz and sovkhoz theatres. Publs.: plays "Tikhoe ozero" (Calm Lake) 1953, "Ognennyi vikhr'" (Firestorm) 1957, "Sedye volosy moei materi" (Grey Hairs of My Mother) 1960, "Sosedi" (The Neighbours) 1962, and others.

Mirzo-Zade, Khalik
Literary historian and critic; b. 1911, Ravonak nr. Samarkand.
Tadzhik; mem. CPSU 1946–; grad. Samarkand Univ. 1935; teacher at Samarkand Pedagogical Inst. 1935–; first works publ. 1934.
Publs.: "Shamsiddin Shokhin" 1956, "Abu Abdullo Rudaki" 1958, "Mirsaid Mirshakar" 1962, and others.

Mirzoalieva, Karomat
State official.
Cand. mem. Cen. Cttee. Tadzhik CP 1966–71, mem. 1971–; Sec. of Leninabad Oblast Cttee. Tadzhik CP 1971–78; Min. of Public Welfare Tadzhik SSR O. 1978–.

Mirzoev, Gafar
Author; b. 1929, Dagani Shol, Tadzhikistan.
Mem. CPSU 1954–; grad. Higher Literary Courses at Inst. of Literature of USSR Writers' Union 1956; translator and critic; first works publ. 1952.
Publs.: "Ot vsego serdtsa" (From the Whole Heart) 1952, "Venets schastia" (Crown of Happiness) 1958, "366 granei" (366 Levels) vols. 1 & 2 1963–64, "Zhizn' prodolzhaetsia" (Life Goes On) 1962, "Serdtse na ladoni" (Heart in the Hand) 1966, and others.

Mirzoian, Airiko Bagratovich
State official.
Min. of Posts and Telecommunications Armenian SSR F. 1983–.

Mirzoian, Edvard Mikhailovich
Composer; b. My. 12, 1921, Gori.
Armenian; mem. CPSU 1952–; grad. Erevan Conservatory 1941; mem. Cen. Cttee. Armenian CP 1964–; Prof. at Erevan Conservatory 1965–. Decors.: Order of Lenin, 1 other order, various medals.

Mirzoshoev, Sultan
State official.
Mem. Cen. Cttee. Tadzhik CP 1960–; 1st Sec. and mem. Bureau Cen. Cttee. Tadzhik Komsomol 1960–62; Sec. of Party Cttee. of Kolkhoz-Sovkhoz Prod. Admin. in Voseisk 1962–67; Chair. State Cttee. on Films Tadzhik SSR 1967–73; Min. of Culture Tadzhik SSR D. 1979–.

Mishchenko, Georgii Sergeevich
State official.
Mem. Cen. Cttee. Turkmen CP 1971–; Head of Dept. of Agriculture at Cen. Cttee. Turkmen CP; Dep. 1971–80, Dep. Chair. of Supr. Sov. Turkmen SSR 1975–80; mem. Bureau Cen. Cttee. Turkmen CP N. 1980–; 1st Dep. Chair. Council of Mins. Turkmen SSR N. 1980–.

Mishchenko, Nikolai Vasil'evich
State official.
Worker at "Serp i molot" Engine Factory in Kharkov; Dep. to and Dep. Chair. of Supr. Sov. Ukrainian SSR 1975–.

Mishin, Vasilii Pavlovich
Aircraft engineer; b. Ja. 5 (18), 1917, Orekhovo-Zuevo, Moscow Oblast.
Mem. CPSU 1943–; grad. Moscow Ordzhonikidze Aviation Inst. 1941; specializes in mechanics, control processes and physical-tech. problems in energetics; mem. USSR Acad. of Sciences 1966–; mem. staff of various scientific research and design and engineering organizations 1941–59; Prof. at Moscow Aviation Inst. 1959–. Decors.: Hero of Socialist Labour 1956, Lenin Prize 1957, 3 Orders of Lenin, Order of the October Revolution, Order of Red Star, medals.

Mishin, V. M.
Komsomol official.
Mem. Bureau All-Union Komsomol Cen. Cttee. 1978–, Sec. 1978–82, 1st Sec. D. 1982–.

Mishuk, Aleksandr Stepanovich
State official.
Head of Main Admin. for Viticulture and the Wine-Making Ind. RSFSR Council of Mins. 1975–79; Chair. RSFSR State Cttee. on Viticulture and Wine-Making Ind. Je. 1979–.

Mishustin, Evgenii Nikolaevich
Microbiologist; b. F. 9 (22), 1901, Moscow.
Grad. K.A. Timiriazev Agricultural Acad. 1924; mem. staff various insts. of VASKHNIL (V.I. Lenin Acad. of Agriculture of the USSR) 1924–31; Head of Dept. of Microbiology at USSR Acad. of Sciences 1939; teacher 1930–, Prof. at Moscow colleges 1937–; corresponding mem. USSR Acad. of Sciences 1953–; Prof. Chair of Soil Microbiology at K.A. Timiriazev Agricultural Acad. Moscow 1961–. Decors.: Merited Scientist of the RSFSR, State Prize of the USSR 1951, 1970, 2 Orders of Lenin, Order of Red Banner of Labour, medals.

Mitin, Mark Borisovich
Philosopher; b. Je. 22 (Jl. 5), 1901, Zhitomir.
Mem. CPSU 1919–; grad. Dept. of Phi-

losophy of Inst. of Red Professors 1929; mem. USSR Acad. of Communist Educ., Dep. Dir. of Acad. of Communist Educ., Dep. Dir. of Inst. of Philosophy, Chief Ed. of Journal "Pod znamenem marksizma" 1930–44; Dir. of Inst. of Marxism-Leninism of Cen. Cttee. CPSU 1939–44; Chief Ed. of journal "Za prochnyi mir, za narodnuiu demokratiu" 1950–56; mem. Cen. Cttee. CPSU 1952–61; Chair. of Union Soc. for the Dissemination of Pol. and Scientific Knowledge 1956–60; Chief Ed. of journal "Voprosy filosofii" 1960–67; Chair of Scientific Council on Ideological Trends Abroad with Section of Social Sciences of Presidium of USSR Acad. of Sciences 1967–; author of numerous papers in problems of the Lenin era in the development of Marxist philosophy, partiality in philosophy, on questions of theory of state and socialist democracy. Decors.: State Prize of the USSR 1943, 2 Orders of Lenin, 4 other orders and various medals.

Mitrin, Evgenii Timofeevich
Party official; b. 1931.
Russian; mem. CPSU 1952–; grad. Gubkin Inst. of Petrochemistry Moscow; worked in mineral-oil ind. 1951–53; head dept. of a raion cttee. Turkmen CP, instructor, dep. head of a dept. at Cen. Cttee. Turkmen CP 1953–65; 1st Sec. of Nebit-Dag City Cttee. Turkmen CP 1965–74; mem. Cen. Cttee. Turkmen CP 1966–; Dep. to USSR Supr. Sov. 1970–; 1st Sec. of Krasnovodsk Oblast Cttee. Turkmen CP F. 1974–. Decor.: Order of October Revolution.

Mitropolskii, Iurii Alekseevich, Dr. of Tech. Sc.
Mathematician; b. Ja. 3, 1917, Shishaki, now Poltava Oblast.
Mem. CPSU 1945–; studied Dept. of Mechanical Physics at Kiev Univ., grad. Dept. of Mechanical Physics of Kazakh Univ. 1942; Dr. of Tech. Sc. 1951–; Prof. 1954–; served in Sov. Army 1941 and 1943–45; mem. staff Inst. of Mechanics of Civil Engineering of Ukrainian Acad. of Sciences 1946–50; mem. staff Inst. of Mathematics of Ukrainian Acad. of Sciences 1951–53, Head of Dept. of Mathematical Physics 1953–55, Dep. Dir. 1955–58, Dir. of same Inst. 1958–; concurrently mem. staff Kiev Univ. 1949–; Head of Bureau of Dept. of Physical Mathematics of Ukrainian Acad. of Sciences 1961–63; Acad. Sec. of Dept. of Mathematics, Mechanics and Cybernetics and mem. Presidium Ukrainian Acad. of Sciences 1963–; author of some 100 scientific works. Decors.: Merited Scientist of the Ukrainian SSR 1967–, 2 Orders of Red Star, medals.
Publs.: "Asimptoticheskie metody v teorii nelineinykh kolebanii" 1963, "Doslidzhennia kolyvan v systemakh z rozpodilenymy parametramy" 1961, "Lektsii po metodu usredneniia v nelineinoi mekhanike" 1966.

Mitskevich, Vladimir Fedorovich
State official; b. 1918.
Belorussian; mem. CPSU 1944–; grad. Party High School of Cen. Cttee. CPSU; zootechnologist 1939; served in Sov. Army 1939–46; leading party and state posts 1946–55; Chair. of Minsk Oblast Exec. Cttee. of Workers' Deps. 1955–62; mem. Cen. Cttee. Belorussian CP 1956–; Dep. to USSR Supr. Sov. 1962–; 1st Sec. of Grodno Oblast Cttee. Belorussian CP 1962–68 (for Agriculture 1962–64); Sec. of Cen. Cttee. 1968–74, mem. Bureau Cen. Cttee. Belorussian CP 1968–; 1st Dep. Chair. of Council of Mins. Belorussian SSR Ap. 1974–; mem. Cen. Auditing Comm. CPSU 1976–. Decors.: Order of Lenin, Order of Red Banner of Labour.

Mitskunas, Pranas Pranovich
State official.
Min. of Trade Lithuanian SSR My. 1975–. Cand. mem. Cen. Cttee. Lithuanian CP 1976–81, mem. 1981–.

Mochalin, Fedor Ivanovich
Party official; b. 1920.
Mem. CPSU 1952–; grad. Moscow Textile Inst. 1943; served in Sov. Army 1943–44; master in a spinning-weaving mill, chief engineer of a trust 1944–47; engineer, dep. head, head of a dept. at Min. of Light Ind. Kazakh SSR 1947–57; Dep. Chair. of Econ. Council of South Kazakhstan, head of a dept. at Cen. Cttee. Kazakh CP 1957–64; 2nd Oblast Sec. 1964–65, 1st Sec. of Alma-Ata City Cttee. Kazakh CP 1965–73; mem. Cen. Cttee. Kazakh CP 1966–74; Dep. to USSR Supr. Sov. 1966–; Head of Dept. of Light and Food Ind. with Cen. Cttee. CPSU 1974–; mem. Cen. Auditing Comm. CPSU 1976–81, cand. mem. Cen. Cttee. CPSU 1981–.

Mochal'skii, Dmitrii Konstantinovich
Painter and graphic artist; b. F. 2 (15), 1908, St. Petersburg.
People's Artist of the RSFSR (1969); member of the Academy of Arts of the USSR (1973).
From 1929 to 1936, he studied at the Academy of Arts in Leningrad. In 1937 he became an instructor at the V. I. Surikov Moscow Art Institute, receiving a professorship there in 1950. During the Great Patriotic War of 1941–45 and in the immediate postwar years, he did a series of pictures devoted to Soviet soldiers and partisans (*Victory: Berlin, 1945,* 1947, Tret'iakov Gallery). In 1949 he painted the work *After the Demonstration* (Tret'iakov Gallery). Since 1954 the artist has worked on a series of lyric genre scenes devoted to the labor and life of Soviet youth on virgin lands (the cycles *People of the Virgin Lands* and *Pioneers in the Virgin Lands*). Decor.: Repin State Prize of the RSFSR in 1967.

Modogoev, Andrei Urupkheevich
Party official; b. 1915.
Buriat; mem. CPSU 1940–; grad. Party Higher School of Cen. Cttee. CPSU 1950; teacher; sec. of a city and 1st sec. of an oblast Komsomol cttee. 1932–43; Dept. Head with Buriat Oblast CPSU Cttee., Dept Chair. of Buriat Council of Mins., attended Party Higher School of Cen. Cttee. CPSU, Dept. Instructor with Cen. Cttee. CPSU, Sec. of Buriat Oblast CPSU Cttee. 1943–60; Chair. of Buriat Council of Mins. 1960–62; Dep. to 1962–, mem. Comm. for Health and Social Security at Sov. of the Union USSR Supr. Sov. 1962–66; 1st Sec. of Buriat Oblast CPSU Cttee. Je. 1962–; cand. mem. 1966–71, mem. Cen. Cttee. CPSU 1971–. Decors.: Order of Lenin, Order of Red Banner of Labour.

Moiseenko, Evsei Evseewich
Painter; b. Ag. 15 (28), 1916, Uvarovichi, now Gomel Oblast.
Grad. Leningrad Acad. of the Arts 1947; teacher 1958–63, Prof. at I. E. Repin Inst. of Painting, Sculpture and Architecture 1963–. Decors.: People's Artist of the USSR, Order of Lenin 1974, Order of the Red Banner of Labour, medals.

Moiseenko, Sergei Timofeevich
State official.
Head of Cen. Forestry Admin. with Council of Mins. Belorussian SSR 1960–66; Min. of Forestry Belorussian SSR Ag. 1966–; mem. Auditing Comm. Belorussian CP 1981–.

Moiseev, Igor Aleksandrovich
Ballet dancer and choreographer; b. Ja. 8 (21), 1906, Kiev.
Grad. Inst. of Choreography Moscow 1924; soloist with Bolshoi Theatre ballet until 1939; founded Folkdance Ensemble of the USSR (Ansambl Narodnogo Tantsa SSSR) 1937, Dir. of same to the present. Decors.: People's Artist of the USSR, Lenin Prize 1967, State Prize of the USSR 1942, 1947, 1952, Order of Lenin, 3 other orders, various medals.

Moiseev, Nikita Nikolaevich
Scientist specializing in general mechanics and applied mathematics; b. Ag. 10 (23), 1917, Moscow.
Corresponding member of the Academy of Sciences of the USSR (1966). Member of the CPSU since 1944.
He graduated from Moscow State University in 1941. From 1948 to 1950 he taught at the N. E. Bauman Moscow Higher Technical School, and from 1950 to 1955 he taught at the University of Rostov. He became a professor at the Moscow Physicotechnical Institute in 1956, and in 1961 he was appointed deputy director of the Computer Center of the Academy of Sciences of the USSR.
His chief works deal with the dynamics of a solid containing a liquid, asymptotic and numerical methods in mathematical

physics, the theory of linear and quasi-linear motions, and optimal control theory. Decors.: three orders and several medals.
Works: *Dinamika tela s polostiami, soderzhashchimi zhidkost'.* Moscow 1965. (With V. V. Rumiantsev.) *Asimptoticheskie metody nelineinoi mekhaniki.* Moscow 1969. *Chislennye metody v teorii optimal'nykh sistem.* Moscow 1971.

Moiseev, Sergei Grigorevich
State official.
Mem. Cen. Cttee. Kirghiz CP 1963–; Chair. of GOSPLAN Kirghiz SSR 1964–68; Dep. to Supr. Sov. Kirghiz SSR Jl. 1964–; Dep. Chair. of Council of Mins. Kirghiz SSR N. 1965–.

Molchanov, Aleksandr Alekseevich
Forestry scientist; b. Ag. 19 (S. 1), 1902, Kuzoverskaia, now Arkhangelsk Oblast.
Mem. CPSU 1941–; grad. Inst. of Forestry Science Arkhangelsk 1933; mem. staff Inst. of Forestry Science of USSR Acad. of Sciences 1945–49, Head of Laboratory of Hydrology and Climatology of same Inst. 1952–59; mem. staff Forestry Science Laboratory of USSR Acad. of Sciences and mem. USSR State Cttee. on Forestry with GOSPLAN 1959–66; Dir. of Forestry Science Laboratory of USSR Acad. of Sciences 1966–; corresponding mem. USSR Acad. of Sciences 1968–. Decors.: 2 orders, various medals.

Moldobaev, Karybek
Party official; b. 1932.
Kirghiz; mem. CPSU 1959; grad. Kirghiz State Univ. Head of Dept. of Science and Culture resp. Science and Educ. Institutions with Cen. Cttee. Kirghiz CP 1964–73; cand. mem. Cen. Cttee. 1966–71, mem. Cen. Cttee. Kirghiz CP 1971–; 1st Sec. of Frunze City Cttee. Kirghiz CP 1973–; mem. Bureau Cen. Cttee. Kirghiz CP 1976–.

Moliakov, Nikolai Ivanovich
Diplomat; b. 1910.
Mem. CPSU 1939–; grad. Leningrad Pedagogic Inst. 1936; mem. Diplomatic Service 1944–; 1st Emb. Counsellor at USSR Emb. in the USA 1944–46; 1st Sec. at Sov. Mission at UN 1946–48; Aide and Dep. Head of American Dept. at USSR Min. of Foreign Affairs 1948–53; Emb. Counsellor at USSR Emb. in Norway 1954–57; Dep. Head of Dept. of Int. Orgs. at USSR Min. of Foreign Affairs 1957–61; Head of Southeast Dept. at USSR Min. of Foreign Affairs 1961–62; Permanent USSR Rep. at European Council of UN in Geneva and other int. orgs. in Geneva 1962–66; Head of Consular Admin. at USSR Min. of Foreign Affairs 1966–78.

Mollaeva, Maia
Party official; b. 1934.
Turkmen; mem. CPSU 1956–; grad. Turkmen State Univ.; teacher, Research Asst. at Dept. of Pol. Econ. at Turkmen State Univ. 1957–59; 1st Sec. and mem. Bureau Cen. Cttee. Komsomol Turkmen SSR 1959–64; Dep. to Supr. Sov. Turkmen SSR 1959–; mem. Cen. Cttee. 1960–; mem. 1960–62, cand. mem. Bureau Cen. Cttee. Turkmen CP 1962–63; Dep. to and mem. Econ. Comm. at Sov. of Nationalities USSR Supr. Sov. 1962–66; Dep. Chair. of Council of Mins. Turkmen SSR F. 1971–76; Sec. and mem. Bureau Cen. Cttee. Turkmen CP D. 1976–.

Molokov, Vasilii Sergeevich
Military official; b. F. 1 (13), 1895, Iríninskoe, now Molokovo, Lenino Raion, Moscow Oblast.
Pilot; major-general of aviation (1940); Hero of the Soviet Union (Ap. 20, 1934). Member of the CPSU from 1925.
He entered the army in 1915 and joined the Red Army in January 1918. He served in the Civil War of 1918–20. He graduated from a naval pilots school in 1921 and from refresher courses of the N. E. Zhukovskii Air Force Academy in 1929. He served in the Civilian Air Fleet from 1931. In 1934, he participated with other pilots in the expedition to rescue the crew of the *Cheliuskin.* He completed a flight around the arctic coast of the USSR in 1936 and participated in an expedition to the North Pole in 1937. He was appointed chief of the Main Directorate of the Air Fleet in 1938.
In the Great Patriotic War (1941–45), he commanded air divisions on the Western and Third Belorussian fronts from 1943. He entered the reserves in 1947. He was a Member of the Central Executive Committee of the USSR at the seventh convocation and a deputy to the first convocation of the Supreme Soviet of the USSR. Decors.: three Orders of Lenin, two Orders of the Red Banner, the Order of Suvorov Second Class, the Order of Kutuzov Second Class, the Order of the Patriotic War First Class, the Order of the Red Star, and various medals.

Monakhov, Vladimir Vasil'evich
Cameraman and director; b. S. 30, 1922, Dolgintsevo, Dnepropetrovsk Oblast.
Honored Art Worker of the RSFSR (1964); member of the CPSU since 1943. Participant in the Great Patriotic War.
In 1952, he graduated from the All-Union State Institute of Cinematography. He shot such films as *The Grasshopper* (1955, with F. B. Dobronravov) and *Height* (1957). In the film *The Fate of a Man* (1959), for which he was awarded the Lenin Prize (1960), he organically combines a documentary approach with picturesque contrasts of light and shadow, creating a tense, expressive mood. Originality, monumentality, and a complex and dynamic ordering of scenes eloquently emphasize the epic quality of his film *Optimistic Tragedy* (1963). He directed the films *Unsolicited Love* (1965), *Of Human Miracles* (1968), and *Unexpected Guest* (1972). Decors.: three orders and various medals.

Monin, Andrei Sergeevich
Oceanologist; b. Jl. 2, 1921, Moscow.
Mem. CPSU 1945–; grad. Moscow Univ. 1942; Prof. 1963–; mem. staff Cen. Inst. for Prognosis 1946–51, Inst. of Physics of the Atmosphere at USSR Acad. of Sciences 1951–65; Dir. of P. P. Shirshov Inst. of Oceanology 1965–; corresponding mem. USSR Acad. of Sciences 1972–. Decor.: Order of the October Revolution.

Mordinov, Nikolai Egorovich (pseudonym: Amma Achchygyia)
Yakut writer; b. D. 24, 1905 (Ja. 6, 1906), in the Nizhniaia Amga *nasleg* (village council), Boturussk *ulus* (district), present-day Yakut ASSR.
People's Writer of the Yakut ASSR (1966).
He was the son of a peasant. He graduated from the pedagogical department of the Second Moscow State University in 1931. He was first published in 1927.
His first collection of stories, *The Knife Does Not Cut the Handle,* appeared in 1934. He is the author of the novellas *On Vacation* (1937), *Insult* (1939), *Friends* (1947), and *Calamity* (1961–62). His most important work was the novel *Springtime* (1944; Russian translation, 1958), which depicts the Yakuts' struggle for a new life. His play *Along the Broad Road* (1964) is devoted to those who are developing the country's mineral wealth.
He has translated the works of M. Iu. Lermontov, L. N. Tolstoy, V. G. Korolenki, and other authors into Yakut. He served as a deputy to the seventh convocation of the Supreme Soviet of the USSR. Decors.: four orders.
Works: *Talĭllĭbĭt ayĭmn 'ĭlar,* 4 vols. Yakutsk, 1966–70. *In Russian translation: Povesti.* Moscow 1960. *Beda: Povest'.* Moscow 1970.

Mordiukova, Nonna (Noiabrina) Viktorovna
Russian film actress; b. N. 25, 1925, Konstantinovskaia *stanitsa* (large cossack village), Donetsk Oblast.
People's Artist of the USSR (1974).
In 1950, she graduated from the acting department of the All-Union State Institute of Cinematography (workshop of B. V. Bibikov and O. I. Pyzhova). She works in the Central Film Actor's Studio. While still a student she made her film debut as Ul'iana Gromova in *The Young Guard* (1948; based on a novel by A. A. Fadeev; State Prize of the USSR, 1949). Mordiukova has a flare for depicting wholehearted women with morally strong and generous natures. She has portrayed profoundly national Russian characters. Her best roles include Stesha in *Unrelated Kin* (1956), Sasha Potapova in *A Simple Story* (1960), Donia Trubnikova in *The Chairman* (1965), Fedos'ia

Ugriumova in *The Russian Field* (1972), and Antonina Kashirina in *The Point of No Return* (1974). She also acted the roles of Belotelova in *Bal'zaminov's Marriage* (1965; A. N. Ostrovskii) and Glafira Ogrekhova in *My Little Crane* (1969).
In 1972, for her portrayal of Dar'ia Vasil'eva in the film *The Young* (1970), she was awarded a diploma of the All-Union Festival of Films About Labor and the Working Class held in the city of Gorkii. Decors.: Vasil'ev Brothers State Prize of the USSR (1973) order of the Badge of Honour.

Morgun, Fedor Trofimovich
Party official; b. 1924.
Ukrainian; mem. CPSU 1952–; grad. Dnepropetrovsk Agricultural Inst.; served in Sov. Army 1943–45; asst. head, head of a dept., agronomist, dir. of a sugar-beet sovkhoz, chief agronomist of a sugar-beet trust 1949–54; Dir. of "Tolbukhinskii" Sovkhoz in Kazakh SSR 1954–59; 1st sec. of a raion cttee. Kazakh CP, Head of Oblast Admin. for Procurement, 1st Dep. Chair. of Pavlodar Exec. Cttee. of Workers' Deps., 2nd Sec. of Pavlodar Oblast Cttee. Kazakh CP 1959–65; 1st Dep. Chair. of Exec. Cttee. of Workers Deps. Virgin Lands Krai Council, leading posts with Cen. Cttee. CPSU 1965–69; 1st Dep. Chair. of Council of Mins. Kirghiz SSR 1969–72; mem. Cen. Cttee. and mem. Bureau Cen. Cttee. Kirghiz CP 1971–72; Inspector of Cen. Cttee. Ukrainian CP 1972–73; 1st Sec. of Poltava Oblast Cttee. Ukrainian CP Ja. 1973–; mem. Cen. Cttee. Ukrainian CP 1976–; mem. Cen. Cttee. CPSU 1976–.

Morkunas, Vaclovas Antanasovich
State official.
Sec. 1959–66, 1st Sec. of Cen. Cttee. Lithuanian Komsomol 1966–73; cand. mem. 1964–67, mem. Cen. Cttee. 1967–, cand. mem. Bureau Cen. Cttee. Lithuanian CP 1967–74; Chair. of State Cttee. for Vocational Training Lithuanian SSR Ag. 1973–.

Morozov, Aleksandr Aleksandrovich
Tank designer; b. O. 16 (29), 1904, Bezhitsa, now Briansk Oblast.
Major-general of the technical services (1943); doctor of technical services (1943); doctor of technical sciences (1972); Hero of Socialist Labor (1942). Member of the CPSU from 1943.
He graduated from a machine-building technicum in 1931. He served as chief of a plant's design office group from 1931 to 1938, assistant chief of the design office from 1938, and chief designer from 1940. From 1931 to 1939, he participated in the design of light wheeled and track-laying tanks, including the BT-2, BT-5, and BT-7. Jointly with M. I. Koshkin and N. A. Kucherenko, he was in charge of designing the medium T-34 tank, which was recognized as the best tank of World War II (1939–45). After the war several tanks and other armored weapons were created under his direction. He was a deputy to the fifth convocation of the Supreme Soviet of the USSR. Decors.: State Prize of the USSR (1942, 1946, and 1948) and the Lenin Prize (1967) two Orders of Lenin, the Order of the October Revolution, six other orders, and various medals.

Morozov, Ivan Pavlovich
Party official; b. 1924.
Komi; mem. CPSU 1943–; grad. Party High School of Cen. Cttee. CPSU 1960 and Medical School Syvtyvkar 1942; served in Sov. Army 1942–44; instructor of a raion CPSU cttee., 1st sec. of a raion Komsomol cttee., dept. head of an oblast Komsomol cttee., instructor of a CPSU oblast cttee., dept. head of a CPSU city cttee. 1944–53; Chair. of Vorkuta City Exec. Cttee. of Worker's Deps. 1953–55 and Inta 1957–58; 1st Sec. of Inta City CPSU Cttee. 1958–63; 2nd Sec. 1963–65, 1st Sec. of Komi ASSR Oblast CPSU Cttee. 1965–; mem. Cen. Auditing Comm. 1966–71, cand. mem. Cen. Cttee. CPSU 1971–76, mem. 1976–; Dep. to and mem. Comm. of Health and Social Security at Sov. of Nationalities USSR Supr. Sov. 1966–.

Morozov, Nikolai Efimovich
Party official; b. 1929.
Russian; mem. CPSU 1953–; grad. Party High School in Alma-Ata; master in a factory, master of machine and tractor repair shop at a sovkhoz 1948–50; served in Sov. Army 1950–53; 1st sec. of a Komsomol raion cttee., dep. head of a dept. with Semipalatinsk Oblast Cttee. Kazakh CP 1953–61; 1st Sec. of Urdzharskii Raion Cttee. Kazakh CP 1961–66; cand. mem. 1961–66, mem. Cen. Cttee. Kazakh CP 1966–; Sec. 1966–70, 1st Sec. of Semipalatinsk Oblast Cttee. Kazakh CP Ap. 1970–78; 1st Sec. of Tselinograd Oblast Cttee. Kazakh CP 1978–; Dep. to USSR Supr. Sov. 1970–; mem. Cen. Auditing Comm. CPSU 1976–81, Cand. mem. 1981–. Decor.: Order of Lenin.

Mosashvili, Teimuraz Il'ich
Trade union official; b. 1915.
Georgian; mem. CPSU 1947–; grad. Inst. of Railway Transportation Engineers in Tbilisi; served in Sov. Army 1941–45; held admin. and econ. posts 1945–52; Dep. Chair. of Raion Exec. Cttee. of People's Deps., Sec., 2nd Sec. of Raion Cttee. CP Georgia, Chair. of Raion Exec. Cttee. of People's Deps., 1st Raion Cttee. Sec. CP Georgia in Tbilisi 1962–69; 1st Dep. Chair. of City Exec. Cttee. of People's Deps. in Tbilisi 1969–70; Dep. to USSR Supr. Sov. 1970–; Chair. of Georgian Republican Trade Union Council Jl. 1970–; mem. Cen. Cttee. 1971–, cand. mem. Bureau Cen. Cttee. CP Georgia Mr. 1971–.

Moskalenko, Kirill Semenovich
Military official; b. Ap. 28 (My. 5), 1902, Grishino vi-lage, Donetsk Oblast.
Ukrainian; mem. CPSU 1926–; grad. Dzerzhinski Artillery Acad. 1939; Marshal of the Sov. Union; served in Red Army, attended Courses for Red Commanders, platoon leader, battery commander with 1st Cavalry Army 1920–27; commanded a battery, then a division, chief-of-staff, then commander of an artillery regiment, chief of artillery with a brigade 1927–38; commanded a motorized brigade on Southwestern Front, a rifle, then a cavalry corps, dep. commander of an army, Commander of Southwestern, Stalingrad, Voronezh, 1st and 4th Ukrainian Front 1941–45; Dep. to USSR Supr. Sov. 1946–; commanded Moscow District Anti-Aircraft Defence System 1948–53; commanded Moscow Mil. District 1953–60; mem. Cen. Cttee. CPSU 1956–; Commander-in-Chief of Sov. Rocket Forces 1960–62; USSR Dep. Min. of Defence 1960–; Chief Inspector for USSR Min. of Defence 1962–; Decors.: 6 Orders of Lenin, Order of the October Revolution, 5 Orders of Red Banner, 2 Suvorov Orders 1st Cl., 2 Kutuzov Orders 1st Cl., Bogdan-Khmelnitskii Order 1st Cl., medals, foreign orders and medals, Hero of the Sov. Union 1943, Hero of the CSSR 1969.

Moskovkin, Viktor Flegontovich
Author; b. 1927, Beglitsevo, Iaroslavl Oblast.
Russian; mem. CPSU 1957–; grad. Gorkii Inst. of Literature Moscow.
Publs.: "Valerka i ego druzia" (Valerka and His Friends) 1956, "Ostrov meniaet nazvanie" (An Island Changes Its Name) 1957, "Kak zhizn', Semen?" (How Are You, Semen?) 1958, "Sharik letaet na lunu" (Sharik Flies to the Moon) 1960; "Chelovek khotel dobra ..." (The Man Meant Well ...) 1961, "Potomok sedmoi tysiachi" (The Descendant of the Seventh Thousand) 1964, and others.

Mostovoi, Pavel Ivanovich
State official; b. 1931.
Grad. Ukrainian Polytechnical Correspondence Inst., mechanical engineer; party and sov. posts, Chair. of Exec. Cttee. of Kramatorsk City Sov. of People's Deps., 1st Sec. of Kramatorsk City Cttee. Ukrainian CP, Sec. of Donetsk Oblast Cttee. Ukrainian CP 1964–75; mem. Auditing Comm. Ukrainian CP 1971–81; mem. Cen. Cttee. Ukrainian CP 1981–; 1st Dep. Chief of Main Admin. for Material and Technical Supply of Council of Mins. Ukrainian SSR 1975–77; Chair. of State Cttee. for Material and Technical Supply Ukrainian SSR Ag. 1978–.

Motoriko, Mikhail Georgievich
State official; b. 1920.
Mem. CPSU 1940–; grad. Moscow Zootechnical Inst. of Horse Breeding 1947;

zoo technician, dir. of horse-breeding grounds in Kustanai 1947–58; 1st Sec. of Zabotalskii Raion Cttee. CP Kazakhstan, chief of regional prod. admin., chief of a procurement office, Chair. of Exec. Cttee. of People's Deps. in Kustanai 1958–71; mem. Cen. Cttee. CP Kazakhstan 1966–; Dep. Chair. of Council of Mins. Kazakh SSR 1971; Min. of Agriculture Kazakh SSR D. 1971–.

Motornyi, Dmitrii Konstantinovich
Collective farm chairman; b. 1927.
Mem. CPSU 1951–, grad. Metropol Inst. for the Mechanization of Agriculture 1959; worker at Kherson Ship-Building Plant 1952–53; chief engineer, sec. party cttee. of Kirov Collective Farm in Kherson Oblast 1959–62, Chair. 1963–; mem. Cen. Auditing Comm. CPSU 1981–. Decor: Hero of Socialist Labour 1966.

Motyleva, Tamara Lazarevna
Literary historian and critic; b. My. 16 (29), 1910, Moscow.
Mem. CPSU 1940–; grad. Dept. of Literature of Moscow State Univ. 1930; Head of Chair at Gorkii Inst. of Literature 1956–63; first works publ. 1930.
Publs.: "O mirovom znachenii L. N. Tolstogo" (On the World Significance of L. N. Tolstoi) 1957, "Tvorchestvo Romena Rollana" (The Work of Romain Rolland) 1959, "Zarubezhnyi roman segodnia" (The Foreign Novel Today) 1966, and others.

Movchan, Boris Alekseevich, Dr. of Tech. Sc.
Metallurgist; b. Ja. 9, 1928, Makeevka, Chernigov Oblast; s. of an agronomist.
Mem. CPSU 1964–; grad. Kiev Univ. 1951; specializes in physics of metals; Dr. of Tech. Sc. and Prof. 1962–; mem. staff Paton Inst. of Electrical Welding 1951, Head of Dept. of Metallurgy and Physics of Pure Metals and Alloys of Ukrainian Acad. of Sciences 1964–; author of some 80 scientific works. Publs.: co-author of "Radioaktyvni izotopy v tekhnitsi" 1959, "Mikroskopicheskaia neodnorodnost' v litykh splavakh" 1962, co-author of "Elektronnyi luch v sovremennoi vakuumnoi metallurgii" in "Vestnik AN SSSR" no. 1, 1965.

Movsesian, Aleksandr Organezovich
State official; b. 1936, Tbilisi.
Mem. CPSU 1956–; grad. Lenin Polytechnical Inst. in Tbilisi 1960; specialist in marketing bureau of Georgian Board of Trade 1960–61; technical engineer at factory for sport articles 1961–62; instructor of the Ind. and Transport Dept. of the 26 Commissars Raion Cttee. CP Georgia in Tbilisi 1962–66; Sec. of Party Cttee. of a shoe factory in Tbilisi 1966–71; instructor of the Ind. and Transport Dept. Tbilisi City Cttee. CP Georgia 1971–73; 2nd Sec. of Zavodskoi Raion Cttee. CP Georgia, inspector of Dept. for Party-Organizational Work Cen. Cttee. CP Georgia, 1st Sec. of Akhalkalakskii Raion Cttee. CP Georgia 1973–77; mem. Cen. Cttee. CP Georgia 1976–; Dep. Chair. of Exec. Cttee. of City Sov. of People's Deps. in Tbilisi 1977–78; Min. of the Meat and Dairy Ind. Georgian SSR N. 1978–.

Movsesian, Karapet Melikovich
State official; b. 1919.
Dep. to Supr. Sov. Armenian SSR 1955–; 1st Sec. of Akhurianskii Raion Cttee. Armenian CP 1955–62; mem. Cen. Cttee. Armenian CP 1956–76; Head of Territorial Kolkhoz-Sovkhoz Prod. Admin. in Akhurianskii Raion 1962; Chair. of State Cttee. for Utilization and Protection of Water Supplies Armenian SSR Ja. 1974–.

Movsesian, Vladimir Migranovich
State official.
Head of Dept. for Agriculture with Cen. Cttee. Armenian CP 1976–78; mem. Cen. Ctte. Armenian CP 1976–; Dep. Chair. Council of Mins. Armenian SSR Ag. 1978–.

Movsumov, G. Sh.
Komsomol official.
1st Oblast Sec. Azerbaidzhan Komsomol in Nakhichevan ASSR until 1982; Sec. and mem. Bureau Cen. Cttee. Azerbaidzhan Komsomol D. 1982–.

Mozgov, Ivan Efimovich
Veterinary pharmacologist; b. Ja. 7 (20), 1906, Malygino, Gouvt. Moscow.
Mem. CPSU 1937–; grad. Zoo-Tech. Inst. Moscow 1930; Prof. Chair of Pharmacology at Veterinary Acad. Moscow 1938–; mem. (V. I. Lenin Acad. of Agriculture of the USSR) 1956–; concurrently Dep. Acad. Sec. of Dept. of Animal Husbandry at VASKHNIL 1956–58; Rector of Veterinary Acad. Moscow 58–61; Dep. USSR Min. of Agriculture 1961–62. Decors.: 3 orders, various medals.

Mozgovoi, Ivan Alekseevich
Party official; b. 1927.
Ukrainian; mem. CPSU 1951–; grad. Kherson Agricultural Inst.; agronomist in a sovkhoz, student 1948–53; Sec. of Kherson Oblast Cttee. Ukrainian Komsomol; Sec. of Cen. Cttee. Ukrainian Komsomol 1953–62; 2nd Sec. of Transcarpathian Oblast Cttee. Ukrainian CP 1962–66; 1st Sec. of Rovno Oblast Cttee. Ukrainian CP 1966–72; mem. Cen. Cttee. Ukrainian CP 1966–; Dep. to USSR Supr. Sov. 1966–; 1st Sec. of Kherson Oblast Cttee. Ukrainian CP O. 1972–80; 1st Dep. Chair. of Council of Mins. Ukrainian SSR Mr.–My. 1980; cand. mem. Politbureau Cen. Cttee. Ukrainian CP Ap.–My. 1980, mem. My. 1980–; Sec. Cen. Cttee. Ukrainian CP My. 1980–; mem. Cen. Auditing Comm. CPSU 1976–81, mem. Cen. Cttee. CPSU 1981–.

Mozhaev, Boris Andreevich
Author; b. Je. 1, 1923, Pitelino, Riazan Oblast.
Russian; grad. Higher College of Engineering Leningrad 1948; worked in the Far East for many years; first works publ. 1953.
Publs.: "Udegeiskie skazki" (Fairytales) 1955, "Sania" 1959, "Zemlia zhdet" (The Earth is Waiting) 1961, "V Soldatove, u Nikolaia Lozovogo" (In Soldatov, at Nikolai Lozovoi's) 1961, "Zemlia i ruki" (The Earth and the Hands) 1964, "Eksperimenty na zemle" (Experiments on Earth) 1964, "Iz zhizni Fedora Kuzkina" (Excerpts from Fedor Kuzkin's Life) 1966, "Vlast taigi" (The Power of the Taiga) 1959, "V amurskoi dal'nei storone" (Far Away, on the Amur) 1963, and others.

Mozoliako, Ivan Mikhailovich
State official.
Min. of Ind. Engineering Belorussian SSR Ap. 1979–; mem. Cen. Cttee. Belorussian CP 1981–.

Mravinskii, Evgenii Aleksandrovich
Conductor; b. My. 22 (Je. 4), 1903, St. Petersburg.
Grad, Leningrad Conservatory 1931; Conductor of Kirov Opera and Ballet Theatre 1932–38; Conductor of Leningrad Philharmonic Orchestra 1938–; teacher in Leningrad 1961–63, Prof. at Leningrad Conservatory 1963–. Decors.: People's Artist of the USSR, Hero of Socialist Labour, 1st Prize Union Competition for Conductors 1938; State Prize of the USSR 1946, Lenin Prize 1961, Order of Lenin, 2 other orders, various medals.

Mrevlishvili, Mikhail Nikolaevich
Georgian writer; b. Ag. 6 (19), 1904, Oni, western Georgia.
Honored Art Worker of the Georgian SSR (1961). Member of the CPSU since 1951.
His first work, the novel *Inon,* was published in 1926. He is the author of the novella *Kharateli's Hearth* (1947; his play of the same name was published in 1954), the collection *Tbilisi Stories* (1964), the novel *Dangerous Turn* (1964), and other works dealing with modern Georgia and Soviet young people. He is also a playwriter. His plays include *Nikoloz Baratashvili* (1949) and *Avalanche* (1956), both about the postwar Georgian village, and the historical drama *Fiery Dreamer* (1957), about A. S. Griboedov. Mrevlishvili's works have been translated into Russian and other languages of the peoples of the USSR. Decor.: Order of the Badge of Honor.
Works: *T'xzulebani or tomad,* vols. 1–. Tbilisi 1971. *In Russian translation: P'esy.* Tbilisi 1958. *Staryi tarantas: Rasskazy.* Tbilisi 1960.

Mshvelidze, Shalva Mikhailovich
Composer; b. My. 15 (25), 1904, Tbilisi. People's Artist of the Georgian SSR (1958). Member of the CPSU since 1947. He graduated from M. M. Bagrinovskii's composition class at the Tbilisi Conservatory in 1930 and did graduate work at the Leningrad and Tbilisi conservatories under V. V. Shcherbachev. He wrote the operas *Legend of Tariel'* (1946) and *The Hand of the Great Master* (1961), both presented at the Z. Paliashvili Tbilisi Theater of Opera and Ballet, the oratorios *Kavkasioni* (1949) and *A Legend for the Ages* (1970, dedicated to V. I. Lenin), four symphonies (1943, 1944, 1952, 1968), the symphonic poems *Zviadauri* (1940), *Mindiia* (1950), and *The Youth and the Snow Leopard* (1962), as well as symphonic suites, overtures, song cycles, and music for stage productions and films.
He has taught at the Tbilisi Conservatory (in 1942 he became a professor). From 1941 to 1952 he served as chairman of the Composers' Union of Georgia. He was a deputy to the third convocation of the Supreme Soviet of the Georgian SSR. He received the State Prize of the USSR (1942, 1947) and the Z. Paliashvili State Prize of the Georgian SSR (1971), as well the Jawaharlal Nehru Award (India) for musical works on Indian themes and for a series of essays on Indian folk music (1973). Decors.: Order of Lenin, two other orders, and a number of medals.

Mubariakov, Ar. Arslan Kotlyakhmetovich
Bashkir actor and stage director; b. D. 15, 1908, Asy, what is now Arkhangel'skoe Raion, Bashkir ASSR.
People's Artist of the USSR (1955). Member of the CPSU since 1941.
In 1931 he graduated from the theatrical division of the Technicum of the Arts in Ufa. After 1928, with some interruptions, he worked in the company of the M. Gafuri Bashkir Dramatic Theater in Ufa. From 1937 to 1939 and from 1951 to 1954 he was the theater's principal director.
He has great stage charm and presence; his art is realistic and popular in nature. His best roles have included Shatmurat and Salavat Iulaev in Bikbai's *Karlugas* and *Salavat*, Khudauberdin in Nigmati's *On the Bank of the Belata River*, the title role in Shakespeare's *Othello*, Protasov in Tolstoy's *The Living Corpse*, Voinitskii in Chekhov's *Uncle Vanya*, and Durmet Iarlykapov in Karim's *A Song Unsung*. He created the character of V. I. Lenin in Pogodin's *A Man With a Gun* and *Third Pathétique*. He has staged such plays as Bikbai's *Korlugas* (1938), Akhmetshin's *Tal'ian-Accordion* (1951), and Minko's *Without Naming Surnames* (1953). He is the author of several plays. He has also appeared in films, including *Salavat Iulaev* (1941).
He served as a deputy to the first through fourth convocations of the Supreme Soviet of the Bashkir ASSR. Decors.: Order of Lenin, two other orders, and several medals.

Mugnetsian, Gegam Minoevich
State official.
Mem. Auditing Comm. 1961–71, cand. mem. Cen. Cttee. Armenian CP 1971–76, mem. 1976–; 2nd Sec. of Leninakan City Cttee. Armenian CP until 1975; 1st Sec. of Dilizhan City Cttee. Armenian CP 1975–81; Chair. of State Cttee. for Publishing, Polygraphy and the Book Trade Armenian SSR Jl. 1981–.

Mukaev, Toleubai
State official.
Chair. of State Cttee. for Supervision of Safety in Ind. and Mining Kazakh SSR D. 1981–.

Mukashev, Salamat
Party official; b. 1927.
Kazakh; mem. CPSU 1950–; grad. Trade Union High School at Cen. Union Council of USSR Trade Unions; worked in an oil field 1942–47; after grad. at Gurev Petroleum Tech. College worked in petroleum ind. 1947–51; party and trade union work 1951–61; 1st Sec. of Embinskii Raion Cttee. Kazakh CP 1961–70; 1st Sec. of Gurev Oblast Cttee. Kazakh CP Ap. 1970–77; Dep. to USSR Supr. Sov. 1970–; mem. Cen. Cttee. Kazakh CP 1971–; cand. mem. Bureau Cen. Cttee. Kazakh CP Mr. 1977–81; Chair. of Kazakh Republican Trade Union Council F. 1977–80; 1st Sec. of Mangyshlakskii Oblast Cttee. Kazakh CP D. 1980–.

Mukha, Stepan Nesterovich
State official.
Sec. of Dnepropetrovsk City Cttee. Ukrainian Komsomol 1958; 1st Sec. of Chernigov Oblast Cttee. Ukrainian Komsomol 1959; Dep. to Supr. Sov. Ukrainian SSR 1975–; 1st Dep. Chair. of Cttee. for State Security Ukrainian SSR until 1982, Chair. Je. 1982–; cand. mem. Politbureau Cen. Cttee. Ukrainian CP O. 1982–.

Mukhamed-Rakhimov, Taufik Galeevich
State official.
Sec. N. Kazakhstan Oblast Cttee. Kazakh CP 1971–74; Head of Dept. for Heavy Ind. with Cen. Cttee. Kazakh CP 1974–77; mem. Cen. Cttee. Kazakh CP 1976–; Min. of Local Ind. Kazakh SSR 1977–79; 1st Sec. of Alma-Ata City Cttee. Kazakh CP 1979–81; Dep. Chair. of Council of Mins. and Chair. of GOSPLAN Kazakh SSR Ja. 1981–.

Mukhamedov, Mumtaz
Author; b. 1908, Tashkent; s. of a worker.
Uzbek; mem. CPSU 1943–; grad. Moscow Inst. of Journalism; first works publ. 1931. Publs.: "Les i leopard" (The Wood and the Leopard) 1931, "Malenkii garnizon" (Small Garrison) 1954, "Po sledam geroia" (Following the Hero) 1957, "Etom, chto stanovitsia legendoi" (About Future Legends) 1963, and others.

Mukhamedov, Tair Mukhamedovich
Hydrotechnologist and state official; b. 1912.
Mem. CPSU 1946–; grad. Hydrotech. Inst. Samarkand 1933; sen. positions as engineer in hydrological and cotton-growing orgs. in Tadzhikistan 1933–52; mem. Apparat Council of Mins. Tadzhik SSR 1952–56; Dep. Min. 1956–57, Min. of Water Econ. 1957–66, Min. of Melioration and Water Econ. Tadzhik SSR 1966–72. Decor.: State Prize of the USSR 1971.

Mukhammadiev, Fazliddin
Author; b. 1928, Samarkand.
Tadzhik; mem. CPSU 1951–; first works publ. 1955.
Publ.: "Novyi predsedatel" (The New Chairman) 1956, "Pereselentsy" (The Resettlers) 1959, "Domik na okraine" (The Cottage in the Suburb) 1963, "Vzryv, kotorogo ne bylo" (The Explosion That Didn't Take Place) 1964, and others.

Mukhin, Viktor Ivanovich
Sculptor; b. 1914, Zaitsevo, Donetsk Oblast.
Attended Kharkov Art Inst. 1931–32, pupil of Blokh, grad. Kiev Art Inst. 1938, pupil of Zh. Dindo, L. Shervud. Decors.: Merited Artist of the Ukrainian SSR 1956, Shevchenko Prize 1973.

Mukhitdinov, Nurutdin Akramovich
Diplomat; b. 1917, Tashkent suburb.
Uzbek; mem. CPSU 1942–; grad. Cooperative Union Inst. 1938; served in Sov. Army 1939–46; Lecturer at Cen. Cttee. Uzbek CP 1946–47; Sec. 1947–48, 1st Sec. of Namangan Oblast Cttee. Uzbek CP 1948–50; mem. Cen. Cttee. Uzbek CP 1949–58; Sec. for Propaganda at Cen. Cttee. Uzbek CP 1950; 1st Sec. of Tashkent Oblast Cttee. Uzbek CP 1950–51; Chair. of Council of Mins. Uzbek SSR 1951–55; mem. Bureau Cen. Cttee. Uzbek CP 1952–58; mem. Cen. Cttee. CPSU 1952–66; 1st Sec. of Cen. Cttee. Uzbek CP 1955–57; cand. mem. Presidium 1956–57, mem. Presidium and Sec. of Cen. Cttee. CPSU 1957–61; Dep. Chair. of Cen. Council of USSR Consumer Cooperatives 1962–66; 1st Dep. Chair. of Cttee. of Cultural Relations with Foreign Countries at USSR Council of Mins. 1966–68; USSR Amb. to Syria F. 1968–Mr. 1977. Decors.: 2 Orders of Lenin, medals.

Mukhovatov, Vladimir Semionovich
Physicist; b. 1935.
Grad. Moscow Univ. 1958; mem. staff I. V. Kurchatov Inst. of Atomic Energy 1958–; Sen. Engineer 1968. Decor.: State Prize of the USSR 1971.

Mukhtar, Askad
Author; b. 1921, Fergana.
Uzbek; grad. Dept. of Philology of Cen. Asian Univ.; first works publ. 1939.
Publs.: "Nashi predki" (Our Ancestors)

1939, "Navstrechu budushchemu" (Toward the Future) 1951, "Tam, gde slivaiutsia reki" (Where the Rivers Meet) 1950, "Armiia mira" (Army of Peace) 1960, and others; translated Pushkin, Lermontov, Gorkii, Maiakovskii and other Russian authors into Uzbek.

Muradian, Murad Oganesovich
State official; b. 1930.
Armenian; mem. CPSU 1957–; grad. Pedagogical Inst. in Leninakan and Moscow Inst. of Mechanical Engineering; teacher at a secondary school 1953–58; engineer, master, Dep. Dir. of a grinding machine factory 1958–64; Dir. of the "Mikroelektrodvigatel'" plant in Leninakan 1964–65; head of a dept. of a City Cttee., dep. head of a dept. Cen. Cttee. CP Armenia 1965–67; 2nd Sec. 1967–68, 1st Sec. CP Armenia in Leninakan 1968–75; mem. Cen. Cttee. and cand. mem. Bureau of Cen. Cttee. CP Armenia D. 1969–; Chair. of Exec. Cttee. of City Sov. of People's Deps. in Erevan. My. 1975–.

Muradov, Sakhat Nepesovich
State official.
Head of Dept. for Science and Educ. Institutions with Cen. Cttee. Turkmen CP 1968–71; can. mem. Cen. Cttee. Turkmen CP 1971–81, mem. 1981–; Min. for Higher and Secondary Specialist Educ. Turkmen SSR D. 1979–.

Muradova, Bariiat Soltan Medzhidovna
Kumyk actress; b. Ja. 3 (16), 1914, Makhachkala.
People's Artist of the USSR (1960).
She first performed in amateur groups. Between 1926 and 1930 she was a soloist with Dagestan Radio. In 1930, after graduating from a music and drama school, she joined the A. P. Salavatov Kumyk Music and Drama Theater in Makhachkala.
She acts with great mastery in tragedies, vaudevilles, comedies of everyday life, and incisively satirical dramas. Among her best roles are Susan in Shirvanzade's *Namus,* Kystaman in Salavatov's *Red Partisans,* Dzhumaisat in Kurbanov's *Molla Nasretdin,* Shakerkhan in Rustamov's *Under the Tree,* Dzhaminat in Iakh'iaev's *Irchi Kazak,* and Zhivka in Nušic's *Madame Minister.* She has a beautiful voice and sings folk songs in mountain *auls* (villages).
She was a deputy to the first convocation of the Supreme Soviet of the Dagestan ASSR and the fifth convocation of the Supreme Soviet of the USSR. Decors.: K. S. Stanislavsky State Prize of the RSFSR (1972), Order of Lenin, two other orders, and various medals.

Murakhovskii, Vsevolod Serafimovich
Party official; b. 1926.
Ukrainian; mem. CPSU 1946–; grad. Stavropol Pedagogical Inst. 1954; served in Sov. Army 1944–50; 1st Sec. of Stavropol City Cttee. USSR Komsomol 1954–56; party posts 1956–70; 1st Sec. of Stavropol City Cttee., Stavropol Krai Cttee. CPSU 1970–75; 1st Sec. of Karachay-Cirkassian Autonomous Oblast Cttee. CPSU 1975–78; 1st Sec. of Stavropol Krai Cttee. CPSU D. 1978–; Dep. to USSR Supr. Sov. 1979–; mem. Cen. Cttee. CPSU 1981–. Decor.: Order of the October Revolution 1976.

Murashko, Antineia Grigorevna, Cand. of Philol. Sc.
Philologist; b. 1924.
Belorussian; grad. Belorussian Univ. 1948; Cand. of Philol. Sc. 1958–; mem. Inst. of Linguistics of Belorussian Acad. of Sciences 1950–63; Lecturer at Belorussian Univ. 1963–. Decor.: State Prize of the USSR 1971.

Muratkhodzhaev, Vali Muradovich
State official.
Min. of Finance Uzbek SSR 1957–; cand. mem. Cen. Cttee. 1960–66, mem. Cen. Cttee. Uzbek 1966–. Decor.: Order of Lenin.

Muratov, Igor Leontevich
Author; b. Jl. 15 (28), 1912, Kharkov; s. of a clerk.
Ukrainian; mem. CPSU 1940–; grad. Kharkov Univ. 1939; served in Sov. Army 1941–45, first works publ. 1932.
Publs.: "Zhila na svete vdova" (There Was Once a Widow) 1960, "Posledniia tucha" (The Last Cloud) 1959, "Svizhe povitria dlia materi" (Fresh Air for Mother) 1962, and others. Decor.: State Prize of the USSR 1952.

Muratov, Mikhail Vladimirovich
Geologist; b. F. 29 (Mr. 13), 1908, Moscow.
Corresponding member of the Academy of Sciences of the USSR (1962).
He graduated from Moscow University in 1930. He has served as head of the subdepartment of historical and regional geology of the Moscow Geological Research Institute (since 1959), senior scientific worker of the Geology Institute of the Academy of Sciences of the USSR (since 1957), and chairman of the Interdepartmental Tectonics Committee of the Academy of Sciences of the USSR (since 1965). His principal works deal with theoretical tectonics and, in particular, the tectonic structure of the world's folded zones.
He is one of the compilers and editors of tectonic and geological maps of the USSR, Europe, and the world. In 1951 and 1969 he won the State Prize of the USSR for his *The Tectonics and Developmental History of the Alpine Geosynclinal Area of the Southern Part of the European USSR and Adjacent Countries* (1949) and for participating in compiling *The Tectonic Map of Eurasia* (scale of 1:5,000,000) and the monograph *Tectonic of Eurasia* (1966). Decors.: Order of the Red Banner of Labour and various medals.
Works: "Glavneishie epokhi skladchatosti i megastadii razvitiia zemnoi kory." *Geotektonika,* 1965, no. 1. "Glavneishie strukturnye elementy materikov, ikh vzaimo-otnosheniia vozrast." In *Mezhdunarodnyi geologicheskii kongress, 24 sessiia: Dollady sovetskikh geologov: Problema 3: Tektonika.* Moscow 1972.

Muratov, Valentin Ivanovich
Athlete and gymnast, coach, and instructor; b. Jl. 30, 1928, Kostiukovo, Serpukhov Raion, Moscow Oblast, RSFSR.
Honored Master of Sports (1952). Member of the CPSU since 1958.
At the Olympic Games of 1952 and 1956, he won a total of four gold medals, for both individual events within the combined events and the team championship. In the 1954 and 1958 world championships he earned a total of five gold medals, including one for all-around champion. He was also a champion of the USSR in 1953 and 1956, winning a total of six gold medals, including two for all-around champion. Decor.: Order of Lenin.

Muravev, Evgenii Fedorovich
Party official; b. 1929.
Russian; mem. CPSU 1952–; grad. Karelo-Finnish State Univ. 1952; design engineer, dept. chief, sec. of party cttee. of a plant in Kuibyshev Oblast 1952–59; plant dir. 1959–61 and 1961–62; 1st Sec. of Syrzan City Cttee. CPSU 1962–76; Chair. of Exec. Cttee. of Kuibyshev Oblast Sov. of People's Deps. 1976–79; Dep. to USSR Supr. Sov. 1979–; 1st Sec. of Kuibyshev Oblast Cttee. CPSU Ap. 1979–; mem. Cen. Cttee. CPSU 1981–.

Murtasina, L.
Trade union official.
Sec. and mem. Presidium Esthonian Republican Trade Union Council F. 1982–.

Murzaev, Eduard Makarovich
Physical geographer and toponymist; b. My. 19 (Je. 1), 1908, Simferopol'.
Doctor of geographic sciences (1949); professor (1956). Honored Scientist of the RSFSR (1970).
He graduated from the department of geography of Leningrad State University in 1930. In 1931 he began working at the Geographic Institute of the Academy of Sciences of the USSR. He has taken part in expeditions to Middle Asia, Central Asia, Mongolia, and Vietnam. His principal works deal with the development of the terrain of arid regions, regional geography, the origin and development of geographic names, popular geographic terminology, and the history of geographic exploration in Asia. He has edited and commented on the works of L. S. Berg, N. M. Przheval'skii, and P. K. Kozlov. He is an academician of the German Leopoldine Academy of Researchers in Natural Sciences (1961) and an honorary professor of the Mongolian Academy of

Sciences (1961). He has received the State Prize of the USSR (1951), the V. A. Obruchev Prize of the presidium of the Academy of Sciences of the USSR (1963), the N. M. Przheval'skii Gold Medal (1947) and the A. Humboldt Medal (1959). Decors.: two orders, various medals, and the Mongolian Order of the Polar Star.
Works: *Geograficheskie issledovaniia MNR*. Moscow-Leningrad 1948. *Mongol'skaia Narodnaia Respublika: Fiziko-geograficheskoe opisanie*, 2nd ed. Moscow 1952.
Severo-Vostochnyi Kitai: Fiziko-geograficheskoe opisanie. Moscow 1955. *Zazubeshnaia Aziia: Fizicheskaia geografiia*. Moscow 1956. (Co-author.) *Vladimir Efanas'evich Obruchev*. Moscow 1959. (Co-author.) *Slovar' mestnykh geograficheskikh terminov*. Moscow 1959. (Co-author.) *Sredniaia Aziia*, 3rd ed. Moscow 1961. *Priroda Sin'tsziana i gormirovanie pustyn' Tsentral'noi Azii*. Moscow 1966. *Puteshestvie v zharkuiu zimu*. Moscow 1967. *Istoriia otkrytiia i issledovaniia Sovetskoi Azii*. Moscow 1969. (Co-author.) *Turkmenistan*. Moscow 1969. (Series *Sovetskii Soiuz;* co-author.) *Gody iskanii v Azii*. Moscow 1973. *Toponimika-populiarnaia*. Moscow 1973.

Murzagaliev, Gibadulla Murzagalievich
State official.
Sec. of Alma-Ata City Cttee. Kazakh CP 1974–78, 2nd Sec. 1978–79; 1st Dep. Min. of Local Ind. Kazakh SSR 1979–82, Min. S. 1982–.

Musabaev, Isak Kurbanovich
Specialist in infectious diseases; b. O. 9 (22), 1920, Naukent, now in Fergana Oblast.
Corresponding member of the USSR Academy of Medical Sciences (1961) and academician of the Academy of Sciences of the Uzbek SSR (1966); Honored Scientist of the Uzbek SSR (1955). Member of the CPSU since 1944.
He graduated from the Samarkand Medical Institute in 1935. From 1949 to 1951 he was chairman of the subdepartment of infectious diseases at the Samarkand Medical Institute. In 1951 he became chairman of the subdepartment of infectious diseases at the Tashkent Institute for the Advanced Training of Physicians. He was deputy director of science at the Tashkent Institute of Epidemiology, Microbiology, and Infectious Diseases from 1962 to 1970. His main works deal with Q fever, infectious hepatitis, diphtheria, typhoid-paratyphoid, cholera, and other infectious diseases. He elucidated the nature of the disturbances of the biochemical processes of the body in intestinal infections and influenza. He proposed a classification of the clinical forms of dysentery, cholera, and infectious hepatitis. Decors.: Order of Lenin, four other orders, and medals.
Works: *Likhoradka ku*. Tashkent 1958. (With T. I. Metskan.) *Infeksionnyi gepatit (bolezn' Botkina)*. Tashkent 1961. *Difteriia*. Tashkent, 1967. (With F. Z. Abukakirova.) *Klinika, lechenie i profilaktika kholery*. Tashkent 1967. (With N. N. Zhukov-Verezhnikov and N. K. Zav'ialova.)

Musaeva, R. G.
Komsomol official.
Sec. and mem. Bureau Cen. Cttee. Tadzhik Komsomol Ap. 1982–.

Musakhanov, Mirzamakhmud Mirzarakhmanovich
Party official; b. 1912.
Uzbek; mem. CPSU 1943–; grad. Moscow Textile Inst. 1937; Foreman, Sen. Foreman, Manager of Spinning Mill of "Krasnyi Tekstil'shchik" Factory 1937–40; dept. head, chief engineer of a spinning and weaving mill, Dir. of Tashkent Textile Combine, Dep. Min. of Light. Ind. Uzbek SSR 1940–51; Min. of Light Ind. and Food Ind. 1951–53; Gen. manager for the Uzbek Council of Mins. 1953–55; Sec. of Tashkent Oblast Cttee. Uzbek CP 1955–56 and 1958; Chair. of GOSPLAN with Uzbek SSR Council of Mins. 1956–57; mem. Cen. Cttee. Uzbek CP 1956–58 and 1961–; Dep. Chair. of Council of Mins. Uzbek SSR 1957–58; Chair. of Uzbek Council of Trade Unions 1958–59; Sec. of All-Union. Cen. Trade Union Council 1959–61; Dep. to 1959–, Chair. of Supr. Sov. Uzbek SSR 1961–63; mem. Bureau 1961–, Sec. of Cen. Cttee. Uzbek CP 1961–65; cand. mem. 1961–76, mem. Cen. Cttee. CPSU 1976–; Chair. of Cen. Cttee. Uzbek CP and Council of Mins. Uzbek SSR Cttee. for Party and State Control 1962–65; concurrently Dep. Chair. of Council of Mins. Uzbek SSR; 1st Dep. Chair. of Council of Mins. Uzbek SSR 1965–70; Dep. to 1966–, mem. Planning and Budget Comm. 1966–70, Chair. of Comm. for Trade, Public Services and Econ. at Sov. of Nationalities 1970–74; mem. Draft Bills Comm. Sov. of the Union USSR Supr. Sov. 1974–; 1st Sec. of Tashkent Oblast Cttee. Uzbek CP Ja. 1970–. Decor.: Order of Red Banner of Labour.

Musatov, Leonid Nikolaevich
Diplomat; b. 1921.
Mem. Diplomatic Service 1964–; Counsellor-Envoy at USSR Emb. in Guinea 1964–65; USSR Amb. to Mali 1965–70; mem. Apparat USSR Min. of Foreign Affairs 1970–73; USSR Amb. to Guinea O. 1973–78, concurrently USSR Amb. to Guinea-Bissau 1974–F. 1975, to Madagascar Je. 1980–.

Mushketik, Iurii Mikhailovich
Author; b. Mr. 21, 1929, Verkievka, Chernigov Oblast.
Ukrainian; mem. CPSU 1951–; grad. Kiev Univ. 1953; first works publ 1952.
Publs.: "Vohni sered nochi" (Fires in the Middle of the Night) 1959, "Chernyi khleb" (Black Bread) 1960, "Sertse i kamin'" (The Heart and the Stone) 1961, "Kaplia krovi" (Drop pf Blood) 1964, and others.

Musiiko, Evgenii Vasil'evich
Diplomat.
USSR Amb. to Uganda N. 1975–79.

Musin, Dmitrii Petrovich
Diplomat.
USSR Amb. to Australia 1972–75, to Fiji 1974–75, to Jamaica Mr. 1978–, to Grenada 1980–82.

Musin, Kurgan Nurkhanovich
State official; b. 1926.
Mem. CPSU 1959–; grad. Tashkent Inst. of Melioration and Mechanization of Agriculture 1953; diverse posts 1953–66; 1st Dep. Min. 1967–70, Min. of Agriculture Kazakh SSR Ja. 1970–; mem. Cen. Cttee. Kazakh CP 1971–.

Muslim-Zade, Dzhangir Movsum ogly
Komsomol official.
Mem. Bureau 1974–, 1st Sec. Cen. Cttee. Azerbaidzhan Komsomol My. 1978–; Dep. to Supr. Sov. Azerbaidzhan SSR 1980–; mem. Cen. Cttee. and cand. mem. Bureau Cen. Cttee. Azerbaidzhan CP 1981–.

Muslimov, Marat Nabiullovich (Marat, Muslim)
Author; b. My. 1, 1909, Buraevo, Bashkir ASSR; s. of a peasant.
Bashkir; completed literary courses of USSR Writers' Union 1935; suffered reprisals 1938–57; was rehabilitated 1957; first works publ. 1927.
Publs.: "Stikhi" (Poems) 1936, "Zdravstvuite, druzia, tovarishchi!" (Greeting, Friends, Comrades!) 1958, "Sveti, solntse moe" (Shine, my Sun) 1963, and others.

Musrepov, Gabid Makhmudovich
Author; b. Mr. 22, 1902, Kustanai Oblast, Kazakhstan; s. of a peasant.
Kazakh; mem. CPSU 1926–; grad. Omsk Inst. of Agriculture; mem. Acad. of Sciences Kazakh SSR 1958–; Chair. of Kazakh Writers' Union 1956–62 and 1964–66; Sec. of USSR Writers' Union 1959–; first works publ. 1926. Decors.: 2 Orders of Lenin, Abai Kunanbaev State Prize of the Kazakh SSR 1968, 2 other orders and various medals, Hero of Socialist Labour.
Publs.: "Soldat iz Kazakhstana" (A Soldier From Kazakhstan) 1961, "Probuzhdennyi krai" (Awakening Country) 1958, "Da, eto ego sledy" (Yes, These Are His Tracks) 1962, "Zov zhizni" (The Call of Life) 1963, "Odnazhdy i na vsiu zhizn'" (Once and Forever) 1967, and others.

Mustafaev, Alish Iza ogly
State official.
Chair. of State Cttee. for Supply of Petroleum Products Azerbaidzhan SSR O. 1981–.

Mustafin, Gabiden
Author; b. N. 29, 1902, Karaganda Oblast.
Kazakh; mem. CPSU 1940–; corresponding mem. Kazakh Acad. of Sciences 1958–; first works publ. 1927. Decors.: Order of Lenin, 2 Orders of Red Banner of Labour.
Publs.: "Zhizn' ili smert'" (Life or Death) 1941, "Karaganda" 1952, "Posle buri" (After the Storm) 1959, "Ochevidets" (Eyewitness) 1965, and others.

Mustel', Eval'd Rudol'fovich
Astronomer; b. My. 21 (Je. 3), 1911, Sevastopol'.
Corresponding member of the Academy of Sciences of the USSR (1953). Member of the CPSU since 1964. He graduated in 1935 from Moscow State University, where he began teaching in 1939; from 1944 to 1951 he was a professor. Between 1946 and 1960 he worked at the Crimean Astrophysical Observatory of the Academy of Sciences of the USSR, and since 1957 he has been at the Astronomical Council of the Academy of Sciences of the USSR (appointed chairman in 1963). He developed the theory of radiative equilibrium of stellar atmospheres for frequency-dependent absorption coefficients. He gave a physical picture of processes occurring during nova and supernova outbursts. He studied the causes of corpuscular radiation and physical phenomena in active regions of the sun and the effects of solar corpuscular streams on the earth's atmosphere. Decors.: State Prize of the USSR (1952), he has been awarded the Order of Lenin, the Order of the Red Banner of Labour, and several medals.

Mutafian, Stepan Grigor'evich
State official.
Head of Main Admin. for Gasification Armenian SSR 1978–79; Head of Dept. for Construction and Municipal Econ. with Cen. Cttee. Armenian CP 1979–81; Head of Cen. Statistical Admin. Armenian SSR My. 1981–.

Mutalibov, Aiaz Niiazi ogly
State official.
Min. of Local Ind. Azerbaidzhan SSR 1979–82; mem. Auditing Comm. Azerbaidzhan CP 1981–; Dep. Chair. of Council of Mins. and Chair. of GOSPLAN Azerbaidzhan SSR O. 1982–.

Muzaev, Nurdin Dzhamaldinovich
Chechen poet; b. O. 2 (15), 1913, Belgatoi, now in Shali Raion, Chechen-Ingush ASSR.
Member of the CPSU since 1940.
He served in the Great Patriotic War of 1941–45. He graduated from the history department of the Kirghiz Pedagogical Institute in 1953.
He was first published in 1930, and his early verses reflect the influence of V. V. Mayakovsky. The establishment of kolkhozes in Chechen-Ingushetia is the theme of his play *A Seed Sprouts in Our Epoch* (1933). He is also the author of other plays, as well as of the collections of poems *A Step on the Right Path* (1934), *Collected Verses and Narrative Poems* (1956), *Sweltering Daydreams* (1966), and *The Babble of the Brook* (1968). His novel in verse *Selim Speaks* (1957; Russian translation, 1958) is devoted to the struggle of the Chechen people against the feudal lords and tsarist tyranny. In 1960 he published the novel *The March of the Brave*. The novel *The Strength of a Dream*, which deals with the working class of contemporary Chechen-Ingushetia, appeared in 1971.
He is the author of *Chechen Literature on the Road to Socialist Realism* (1963, in Russian). He has translated works of A. S. Pushkin, T. G. Shevchenko. Sh. Rustaveli, and Shakespeare.
Works: *Mekkhash-Mirza*. Groznyi 1940. *Qonakhiyn bolar*, vols. 1–2. Groznyi 1960. *Dovkha degnash*. Groznyi 1961. *In Russian translation: Gorst' zemli*. Moscow 1960. *Neizbezhnost'*. Moscow 1965. *Prodolzhenie pesni*. Moscow 1967.

Myl'nikov, Andrei Andreevich
Painter; b. F. 22, 1919, Pokrovsk, present-day Engel's, Saratov Oblast.
People's Artist of the RSFSR (1968), member of the Academy of Arts of the USSR (1966).
From 1938 to 1946, he studied at the Institute of Painting, Sculpture, and Architecture in Leningrad under I. E. Grabar' and V. M. Oreshnikov. In 1947 he began teaching at the institute and in 1967 he became a professor there. His works (genre paintings, landscapes, portraits, mosaic panels, and monumental and decorative murals) are distinguished by emotional elevation of style, clarity of composition, and festive brilliance of color.
His major works are *On Peaceful Fields* (1949, Russian Museum, Leningrad; State Prize of the USSR, 1951), *Awakening* (1957 Russian Museum, Leningrad), *Morning* (1972, the Board of Directors of Artistic Depositories and Planning of Monuments of the Ministry of Culture of the RSFSR), and *Lenin* (1961) – a rigid curtain for the Kremlin Palace of Congresses. Other important works by him are mosaics for the Lenin and Young People's Theater (1962). Decors.: Order of the Badge of Honor and several medals.

Mysnichenko, Vladislav Petrovich
Party official; b. 1929.
Mem. CPSU 1955–; grad. Kharkov Aviation Inst. 1952, Cand. of Econ. Sc. 1966; party posts 1957–; 2nd Sec. of Kharkov City Cttee. Ukrainian CP 1964–72; 2nd Sec. of Kharkov Oblast Cttee. Ukrainian CP 1972–80; mem. Cen. Cttee. Ukrainian CP 1976–; 1st Sec. of Kharkov Oblast Cttee. Je. 1980–; mem. Cen. Cttee. CPSU 1981–.

Nabiev, Izzet Akhmed ogly, Cand. of Tech. Sc.
Cybernetic scientist; b. 1923.
Grad. Azerbaidzhan Inst. of Petroleum and Chemistry 1949; specializes in automatics and telemechanics; served in Sov. Army 1941–45; postgrad. studies at Energetics Inst. Moscow 1951–54; Cand. of Tech. Sc. 1954–; lecturer 1956–; mem. staff Azerbaidzhan Inst. of Petroleum and Chemistry 1954–; Lecturer 1956–. Decors.: State Prize of the USSR 1969 for his part in the development and application of systems and means for the automatization of the petroleum ind. in Azerbaidzhan.

Nabiev, Rakhman Nabievich
Party official; b. 1930.
Mem. CPSU 1960–; grad. Tashkent Inst. of Irrigation and Mechanized Agriculture 1954; chief engineer at a machine and tractor station 1954–55 and 1956–58; dir. of a maintenance engineering station 1959–60; held posts in Min. of Agriculture apparatus Tadzhik SSR 1960–61; held posts in Cen. Cttee. apparatus Tadzhik CP 1961–63 and 1964–71; held posts in Cen. Asian Bureau Cen. Cttee. CPSU 1963–64; Min. of Agriculture Tadzhik SSR 1971–73; Chair. of Council of Mins. Tadzhik SSR Jl. 1973–82; Min. of Foreign Affairs Tadzhik SSR 1973–81; mem. Cen. Auditing Comm. CPSU 1976–; mem. Bureau 1973–, 1st Sec. Cen. Cttee. Tadzhik CP Ap. 1982–; Dep. to USSR Supr. Sov. 1974–.

Nabiev, Tuliagan
State official.
Head of Man Construction Admin. of Council of Mins. Uzbek SSR in Tashkent F. 1977–; mem. Cen. Cttee. Uzbek CP 1981–.

Nadezhdina, Nadezhda Sergeevna
Ballet dancer and choreographer; b. My. 21 (Je. 3), 1908, Vilnius.
Grad. 2nd State Ballet School Petrograd 1924; mem. Bolshoi Theatre ballet company 1925–34; ballet mistress at Estrade Theatre Moscow 1943–48; founder of folkdance ensemble "Berezka", expert on Russian folk dance. Decors.: People's Artist of the USSR, Golden Joliot-Curie Peace Medal of the World Peace Council (also for "Berezka" ensemble) 1959, State Prize of the USSR 1950.

Nadiradze, Kolau (Nikolai) Galaktionovich
Georgian poet; b. F. 24 (Mr. 8), 1895, Kutaisi.
Studied at the faculty of law of Moscow University from 1912 to 1916. He was a leading representative of Georgian literary symbolism, and in 1916 his works appeared in the anthology *The Blue Horns*, the organ of the Georgian symbolists. His first book of poems, *Catafalque*, appeared in 1920. In the mid-1920's, he moved away from symbolism and began

depicting Soviet reality in his poetry. He is the author of numerous poetry collections, including *On the Great Road* (1946), *Verses* (1955), and *Verses and a Narrative Poem* (1964). He compiled *Selected Works: Verses and Translations* (1962), which contains Russian poems and translated French poems. Decors.: Order of the Red Banner of Labor and several medals.

Nadzhmi, Nazar (real name: Nazmutdinov, Nazar Nazmutdinovich)
Author; b. N.5, 1918, Minishty, Bashkir ASSR.
Bashkir; grad. Bashkir Pedagogical Inst.; first works publ. 1937.
Publs.: "Volny" (Waves) 1955, "Zemlia i pesnia" (The Earth and the Song) 1962, "Vesenniaia pesnia" (Spring Song) 1951, and others.

Nagibin, Iurii Markovich
Author; b. Ap.3, 1920, Moscow.
Studied Union Film Inst.; served in Sov. Army 1941–43; wounded in action, then war correspondent for "Trud" 1943–45; first works publ. 1940.
Publs.: "Chelovek s fronta" (Man from the Front) 1943, film script "Trudnoe schastie" (Burdensome Happiness) 1958, "Nochnoi gost'" (The Night Guest) 1959, "Neoplachennyi dolg" (The Unpaid Debt) 1959, "Zelenaia ptitsa s krasnoi golovoi" (The Green Bird with the Red Head) 1966, and others.

Nagibovich, Lev Nikolaevich
State official.
Cand. mem. Cen. Cttee. Belorussian CP 1966–76, mem. 1976–; Head of Dept. for chemical and Light Ind. with Cen. Cttee. Belorussian CP 1966–77; Min. of Light Ind. Belorussian SSR Ag. 1977–.

Nagnibeda, Mikola (Nikolai Lvovich)
Author; b. S.7 (20), 1911, Popovka, Zaporozhe Oblast.
Ukrainian; mem. CPSU 1940–; studied Kharkov Inst. of Journalism; first works publ. 1931. Decor.: State Prize of the USSR 1952.
Publs.: "Pravyi bereg" (The Right Shore) 1948, "Step' i more" (The Steppe and the Sea) 1958, "Vechernie kostry" (Evening Camp Fire) 1965, and others.

Nagornyi, Nikolai Nikiforovich
Military official; b. D.5 (18), 1901, Ostashino, now the settlement of Ostashin, Novogrudok Raion, Grodno Oblast, Belorussian SSR.
Colonel-general (1944). Member of the CPSU since 1927.
He entered the Red Army in 1920. He fought in the Civil War of 1918–20. He graduated from the M. V. Frunze Military Academy in 1933 and from courses at the N. E. Zhukovskii Air Force Academy in 1936. He fought as a volunteer in the Civil War in Spain from 1936 to 1938 on the side of the republican government. From 1939 he was head of a subdepartment at the F. E. Dzerzhinskii Artillery Academy and assistant commander of the forces of the Odessa Military District.
In the Great Patriotic War (1941–45), he was chief of staff of the National Air Defense, chief of staff of the Western Front of the National Air Defense, chief of the central staff of the National Air Defense, and deputy commander of the artillery of the Red Army. While holding the last post he was also chief of Main Headquarters of the air defense forces of the Red Army. After the war, he served as chief of Main Headquarters of the National Air Defense, commander of the air defense of a region, commander of the National Air Defense, and assistant commander-in-chief of the National Air Defense. He was placed in the reserves in August 1955.
Decors.: two Orders of Lenin, four Orders of the Red Banner, the Order of Kutuzov First and Second Class, the Order of the Patriotic War First Class, and various medals.

Naiashkov, Ivan Semenovich
State official; b. 1924.
Mem. CPSU 1961–; grad. Moscow Energetics Inst. 1949, Cand. of Tech. Sc. 1954; senior scientific associate, chief of a dept. at Lenin All-Union Electrotechnical Inst. 1956–64, dir. 1964–73; Dep. Chair., 1st Dep. Chair. of USSR State Cttee. for Inventions and Discoveries 1973–79, Chair. Ja. 1979–; mem. Cen. Auditing Comm. CPSU 1981–.

Nakhkur, Arved Iaanovich
State official; b. 1933, Esthonia; s. of a farmer.
Mem. CPSU 1965–; grad. Esthonian Acad. of Agrarian Sciences 1958 and USSR Acad. of Economics 1980; Dep. Min. of Procurement Esthonian SSR 1976–78; studied 1978–80; 1st Dep. Chair. 1980–81, Chair. of State Cttee. for Technical Production Supply to Agriculture My. 1981–.

Nakipov, Shaimerden Kappazovich
State official.
Dep. Chair. 1970–74; Chair. of State Cttee. for Pricing Kazakh SSR S. 1974–.

Nalbandian, Dmitrii Arkadevich
Painter; b. 1906.
Mem. CPSU 1948–; studied Tbilisi Acad. of the Arts 1924–29; mem. USSR Acad. of the Arts 1953–. Decors.: State Prize of the USSR 1946, 1951. Merited Artist of the RSFSR 1951 and of the USSR 1969, Hero of Socialist Labour 1976. Order of Lenin and Hammer and Sickle Medal in Gold 1976.

Nalivkin, Vasilii Dmitrievich
Geologist; b. Ap.17 (30), 1915, Petrograd; s. of D. V. Nalivkin.
Corresponding member of the Academy of Sciences of the USSR (1968).
He graduated from the Leningrad Institute of Mines in 1938. Since 1937 he has worked at the All-Union Research Institute for Geological Oil Prospecting. He is the author of major works on petroleum geology, the tectonics of platform regions, the paleogeography of the USSR, and the geology of the Western Urals. He received the Lenin Prize in 1964 for providing scientific evidence of the oil and gas potentials of the West Siberian Plain and for discovering that province's first Berezovo gas-bearing region. Decors.: two orders and a number of medals.
Works: *Stratigrafia i tektonika Ufimskogo plato i Iurezano-Sylvenskoi depressii.* Leningrad-Moscow 1949. *Fatsii i geologicheskaia istoriia Ufimskogo plato i Iurezano-Sylvenskoi depressii.* Leningrad-Moscow 1950. *Tektonika: Volgo-Ural'skaia neftenosnaia oblast.* Leningrad 1956. (Co-author.) *Analiz vliianiia razlichnykh faktorov na razmeshchenie i formirovanie mestorozhdenii nefti i gaza.* Leningrad, 1971. (Co-author.)

Nametkin, Nikolai Sergeevich
Organic chemist; b. Ag.3 (16) 1916, Moscow; s. of S. S. Nametkin.
Corresponding member of the Academy of Sciences of the USSR (1962), Honored Scientist of the RSFSR (1966). Member of the CPSU since 1949.
He graduated from the University of Kazan in 1942, after which he worked at the Moscow Petroleum Institute. In 1963 he was appointed director of the Institute of Petrochemical Synthesis of the Academy of Sciences of the USSR.
His main works are devoted to the structure and properties of naphthenic acids, alkylation with acetylene hydrocarbons, and the chemistry of naphthenic hydrocarbons. His research on unsaturated and cyclic silicic hydrocarbons led in the formation of a previously unknown class of macromolecular compounds called polysilicic hydrocarbons, which have evaluable technical properties. Many of his inventions have been put to use in industry.
Decors.: State Prize of the USSR (1950), the Order of the October Revolution, and several medals.

Namsinov, Ilia Evgenevich
State official.
Chair. of Presidium Supr. Sov. Kalmuk ASSR; Dep. and Dep. Chair. of Presidium Supr. Sov. RSFSR 1975–.

Narimanbekov, Togrul Farman ogly
Painter and graphic artist; b. Ag.7, 1930, Baku.
People's Artist of the Azerbaidzhan SSR (1967).
He, who attended the Art Institute of the Lithuanian SSR in Vilnius from 1950 to 1955, has produced a number of genre compositions, including *Dawn Over the Caspian* (1957, Azerbaidzhan Museum of Art, Baku), *In the Field Camp* (1967, Tret'iakov Gallery), and *Before the Holi-*

day (1971). His works also include landscapes (*Baku,* 1964, Azerbaidzhan Museum of Art), portraits (*Portrait of S. Bakhlulzade,* 1959, Museum of Art of the Eastern Peoples, Moscow), and still lifes (*Pomegranates and Pears,* 1961, Museum of Art of the Eastern Peoples). He is noted for his energetic application of paint, tension of form, rich coloring, and sometimes primitivist portrayals of his subjects. In 1874 he was awarded the State Prize of the Azerbaidzhan SSR for his work as set designer for the staging of the choreographic work *The Tale of Nasimi* at the M. F. Akhundov Azerbaidzhan Theater of Opera and Ballet.

Narmaev, Morkhadzhi Bambaevich
Author; b. F. 5, 1915, Madzhikiny, Kalmuk ASSR.
Kalmuk; mem. CPSU 1942–; grad. Piatigorsk Inst. of Zoo Technology 1941; served in Sov. Army 1941–45; first works publ. 1935.
Publs.: "Stikhi" (Poems) 1938, "Tovarishchi" (Comrades) 1960, "Tsvetok zhizni" (The Flower of Life) 1963, "Schaste samo ne daetsia" (Happiness Does Not Come by Itself) 1964.

Narochnitskii, Aleksei Leontevich
Historian; b. F. 3 (16), 1907, Chernigov.
Mem. CPSU 1961–; grad. Kiev Univ. 1930; Sen. Scientific Advisor Inst. of History USSR Acad. of Sciences 1946–60; Lecturer and Prof. at Acad. of Social Sciences of Cen. Cttee. CPSU 1955–; Prof. Chair of Modern History at Potemkin Pedagogical Inst. Moscow 1949–60; expert-consultant USSR Min. of Foreign Affairs 1957–; Prof. Chair of Modern and Contemporary History at Lenin Pedagogical Inst. Moscow 1960–; Chief Ed. of Journal "Novaia i noveishaia istoriia" 1962–; mem. Acad. of Pedagogical Sciences 1968–; mem. USSR Acad. of Sciences 1972–; Dir. of Inst. of History of USSR Acad. of Sciences 1974–; specializes in int. relations and colonial policy of the European states and the United States in the Far East in the 19th and 20th century, on history of civil wars and int. relations in Europe in the 18th and 19th century. Decors.: Order of Red Banner of Labour, medals.

Narovchatov, Sergei Sergeevich
Author; b. O. 3, 1919, Khvalynsk, Saratov Oblast.
Russian; mem. CPSU 1943–; grad. Gorkii Inst. of Philosophy, Literature and History 1941; participated in Russo-Finnish War and World War II.
Publs.: "Semen Dezhnev" 1941, "Razgovor nachistotu" (An Open Talk) 1965, "Vasilii Buslaev i Semen Dezhnev" 1967, and others.

Naruzetskii, Mikhail Issakovich
Sculptor; b. 1930, Odessa.
Attended Odessa Art School 1945–50 and Lemberg Inst. for Applied and Decorative Art 1950–56, pupil of A. Overchuk, I. Iakunin.

Narzibekov, Makhmudbek
State official.
Dep. to Supr. Sov. Tadzhik SSR 1959–; 1st Sec. of Moskoviskii Raion Cttee. Tadzhik CP 1959; Chair. of Dushanbe City Exec. Cttee. of Workers' Deps. 1960–67; Mem. Cen. Cttee. Tadzhik CP 1960–63 and 1966–81; Chair. of State Cttee. on Labour Tadzhik SSR Mr. 1967–; mem. Auditing Comm. Tadzhik CP 1981–.

Nasonov, Vsevolod Nikolaevich
Scientist in structural design; b. Jl. 17 (30), 1900, Warsaw; s. of N. V. Nasonov. Doctor of technical sciences (1962). Graduated from the Petrograd Institute of Railroad Engineers (1923).
He has produced a number of designs for large engineering structures and buildings, such as a vehicular bridge over the Iset' River, an overpass in Perm', the assembly shop of the Lugansk Steam Locomotive Plant, a complex of buildings of the M. V. Lomonosov Moscow University, the Palace of Science and Culture in Warsaw, and the V. I. Lenin Central Stadium in Luzhniki (Moscow). His scientific career is associated primarily with the design and development of lightweight building structures, including those using plastics, light alloys, and asbestos cement. From 1957 to 1969, he was director of the V. A. Jucherenko Central Research Institute of Building Construction and the editor-in-chief of the journal *Stroitel'naia mekhanika i raschet sooruzhenii* (Structural Mechanics and Design of Structures). Decors.: State Prize of the USSR (1949), the Lenin Prize (1959), three orders, and medals.

Nasreddinov, Gafur
State official.
Min. of Local Ind. of Uzbek SSR Je. 1976–.

Nasredinov, Khikmatullo
State official.
Head of Dept. of Water Econ. and Rural Construction with Cen. Cttee. Tadzhik CP 1974–80; Min. of Melioration and Water Econ. Tadzhik SSR Ja. 1980–.

Nasretdinova, Zaituna Agzamovna
Ballet dancer; b. Ag. 14, 1923, Ufa.
People's Artist of the USSR (1955). Became a member of the CPSU in 1952.
After her graduation from the Leningrad Choreographic School in 1941, she became a soloist with the Bashkir Theater of Opera and Ballet. Her roles have included Maria and Zarema in Asef'ev's *Fountain of Bakhchisarai,* Tao Khoa in Glière's *The Red Flower,* Zaitungul' in Stepanov's *The Crane's Song.* Zafira in Kliucharev's *Mountain Tale,* and Giselle in Adam's ballet of the same name. She has performed in the German Democratic Republic, Poland, the Mongolian People's Republic, and other countries. Decor.: Order of the Badge of Honor.

Nasyrova, Khalima
Uzbek soprano; b. D. 29, 1913, in the *kishlak* [village] of Taglyk, near Kokand, in present-day Fergana Oblast.
People's Artist of the USSR (1937).
She studied at the Baku Theater Technicum from 1924 to 1927 and at the Uzbek opera studio of the Moscow Conservatory from 1934 to 1937. She made her debut in 1927 as a dramatic actress at the Khamza Theater; later she performed at a theater of music and drama. In 1939, she became the principal soloist at the A. Navoi Uzbek Theater of Opera and Ballet.
She skillfully combines the traditions and methods of the Russian classical vocal school with a distinctive national timbre. Her best roles have included Leili and Giul'sara in Glière's and Sadykov's *Leili i Medzhnun* and *Giul'sara,* Sin Dun-fan in Kozlovskii's *Ulugbek,* Zukhra in Dzhalilov and Brovtsyn's *Takhir i Zukhra,* Maisara in Iudakov's *Maisara's Pranks,* and Carmen in Bizet's opera of the same name.
She was a deputy of the fifth convocation of the Supreme Soviet of the USSR. Decors.: State Prize of the USSR in 1942 and 1951 and the Khamza State Prize of the Uzbek SSR in 1971, two Orders of Lenin, the Order of the October Revolution, four other orders, and several medals.
Work: *Silntse nad Vostokom* (notes of the actress). Moscow 1962.

Natroshvili, Georgii Konstantinovich
Author and literary historian; b. Ja. 10, 1910, Kemo Machkhnai, Georgia.
Georgian; mem. CPSU 1939–; grad. Dept. of Philology of Georgian Univ.; first works publ. 1930.
Publs.: "Na zapadnom fronte" (On the Western Front) 1943, "Smert' bezzabotnogo cheloveka" (The Death of a Careless Man) 1961, and others.

Naumenko, Iurii Andreevich
Military official.
Mem. CPSU 1941–; Lieut.-Gen.; Dep. Commander of Leningrad Mil. District 1967; command posts in the Baltic 1968–71; Commander of Volga Mil. District 1971–75.

Naumenko, Ivan Iakovlevich
Author; b. F. 16, 1925, Vasilevichi, Gomel Oblast; s. of a railway worker.
Belorussian; grad. Belorussian Univ. 1950; first works publ. 1955.
Publs.: "Voina u Titovogo pruda" (The War on Lake Titov) 1959, "Sosna pri doroge" (A Fir on the Roadside) 1962, and others.

Naumov, Aleksei Aleksandrovich
Physicist; b. Ja. 14 (27), 1916, St. Petersburg.
Grad. Moscow Inst. of Postal and Telecommunications Engineering 1942; mem. staff Inst. of Atomic Energy 1945–59; corresponding mem. USSR Acad. of Sciences 1964–; mem. staff Inst. of Nuclear Physics of Siberian Branch of USSR Acad. of Sciences 1959–66; Dep. Dir. of Inst. of High-Energy Physics 1966–. Decors.: State Prize of the USSR 1953, Lenin Prize 1967, Order of Lenin, 2 other orders and various medals.

Naumov, Evgenii Ivanovich
Literary historian; b. Mr. 1 (14), 1909, Smolensk.
Mem. CPSU 1930; grad. Dept. of Philology of Leningrad Univ. 1939; Prof. Chair of Sov. Literature at Dept. of Philology of Leningrad Univ. 1947–; first works publ. 1940.
Publs.: "Maiakovskii v pervye gody sovetskoi vlasti 1917–1922" (Maiakovski in the First Years of Soviet Power 1917–1922) 1950. "D. A. Furmanov" 1951, "Sergei Esenin. Zhizn' i tvorchestvo" (Sergei Esenin. Life and Work) 1960, and others.

Naumov, Petr Ivanovich
Party official; b. 1917.
Head of Dept. of Agriculture with Cen. Cttee. Kirghiz CP until 1971; mem. Cen. Cttee. 1966–, mem. Bureau and Sec. of Cen. Cttee. Kirghiz CP 1971–.

Naumov, Vladimir Ivanovich
State official.
1st Dep. Min. of Agriculture RSFSR 1975–81; Min. of Fruits and Vegetables RSFSR Ja. 1981–.

Naumov, Vladimir Naumovich
Film director; b. D. 6, 1927, Leningrad.
People's Artist of the RSFSR (1974).
In 1951, he graduated from the department of directing of the All-Union State Institute of Cinematography, where he studied in the studio of I. A. Savchenko. He collaborated on all his films with the director A. A. Alov. In 1951 they finished I. A. Savchenko's film *Taras Shevchenko*. Their first independent films were *Uneasy Youth* (1955, based on V. P. Beliaev's *The Old Fortress*), *Pavel Korchagin* (1957, based on N. A. Ostrovskii's novel *How the Steel Was Tempered*), and *Wind* (1959); these films were devoted to the early years of the Komsomol, which was reaching maturity during the Civil War of 1918–20. The film *Peace to Him Who Enters* deals with the last days of the Great Patriotic War (1961; prize at the 22nd International Film Festival in Venice). Another important work was *Flight* (1971, based on M. A. Bulgakov's play). He and Alov have written scenarios for their own films, as well as for the multipart television film *How the Steel Was Tempered* (1974, director N. P. Mashchenko).

Naurzbaev, Khakimzhan Ismakhanovich
Sculptor; b. Ag. 27, 1925, in *aul* (village) No. 1, in present-day Leninskoe Raion, Kustanai Oblast, Kazakh SSR.
People's Artist of the Kazakh SSR (1969). Became a member of the CPSU in 1951. He studied at the Kharkov Art Institute from 1945 to 1951. His works include lifelike portrait busts and statues, marked by thorough modeling. Examples of his sculptures are *Amangel'dy Imanov* (plaster of Paris, 1950), *Young Dzhambul* (bronze, 1958), and *Kurmangazy* (marble, 1958) – all three works are in the T. G. Shevchenko Kazakh Art Gallery in Alma-Ata. The artist also produced monuments to Abai Kunanbaev (bronze and granite, 1960, Alma-Ata), Chokan Valikhanov (bronze and granite, 1969, Alma-Ata; Chokan Valikhanov State Prize of the Kazakh SSR, 1970), and Dzhambul Dzhabaev (cast iron and granite, 1963, in the city of Dzhambul). Decors.: two orders.

Navasarian, Hamlet Gareginovich
Party official.
Chair. of Armenian Republican Asscn. for Agricultural Technology "Armselkhoztekhnika" D. 1975–78; cand. mem. Cen. Cttee. Armenian CP 1976–81, mem. 1981–; Chair. of State Cttee. for the Supply of Production Equipment for Agriculture Armenian SSR 1978–79; Head of Dept. for Agriculture with Cen. Cttee. Armenian CP Ap. 1979–.

Navolochkin, Nikolai Dmitrievich
Author; b. Ja. 5, 1923, Nikolaevka, Jewish Autonomous Oblast.
Mem. CPSU 1946–; grad. Dept. of History of Pedagogical Inst. Khabarovsk 1951; served in Sov. Army 1941–45; First works publ. 1937.
Publs.: "Dorogie moi zemliaki" (My Dear Fellow Countrymen) 1953, "Prikliucheniia Peti Shvyrialkina" (The Adventures of Petia Shvyrialkin) 1959, "Kak lechili Deda Moroza" (How They Cured Father Frost) 1963, and others.

Nazarbaev, Nursultan Abishevich
Party official; b. 1940.
Mem. CPSU 1962–; grad. Higher Tech. Educational Inst. of Karaganda Metallurgical Combine 1967 and CPSU Cen. Cttee. Higher Party Correspondence School 1976; worker in Karaganda Metallurgical Plant 1960–64 and 1965–69; 1st Sec. of Temir-Tau City Cttee. Komsomol 1969–71; 2nd Sec. of Temir-Tau City Cttee. Kazakh CP 1971–73; Sec. of party cttee. of Karaganda Metallurgical Combine 1973–77; Sec., 2nd Sec. of Karaganda Oblast Cttee. Kazakh CP 1977–79; Sec. of Cen. Cttee. and mem. Bureau Cen. Cttee. Kazakh CP D. 1979–; mem. Cen. Auditing Comm. CPSU 1981–.

Nazarov, Ernst Kalandarovich
State official.
Chair. of State Cttee. of Pricing Turkmen SSR Ja. 1980–.

Nazarova, E.
Trade union official.
Sec. and mem. Presidium of Turkmen Republican Trade Union Council Ja. 1982–.

Nazarova, Ogulshirin
State official.
Dep. Chair. of Exec. Cttee. of Sov. of Workers' Deputies in Mari Oblast 1971–75; Dep. to and Sec. of Presidium Supr. Sov. Turkmen SSR 1975–.

Nebylitskii, Boris Rudolfovich
Cinematographer; b. 1909.
Mem. CPSU 1941–; mem. staff Central Studio for Documentary Films 1928–. Decors.: Merited Artist of the RSFSR, State Prize of the USSR 1941, 1951, 1969.

Nechaenko, Aleksandr Vasil'evich
State official.
Mem. Cen. Cttee. Moldavian CP 1960–81, cand. mem. 1981–; Dep. to Supr. Sov. Moldavian SSR 1963–; 1st Sec. of Tiraspol City Cttee. Moldavian CP 1963; Dep. Chair. of Bureau for Ind. with Cen. Cttee. Moldavian CP 1963–64; Head of Dept. of Ind. and Transport at Cen. Cttee. Moldavian CP 1963–65; Min. of Local Ind. Moldavian SSR O. 1965–. Decor.: Order of Red Banner of Labour.

Nechiporenko, Maiia Sergeevna
Painter; b. 1933, Kiev; d. of S. Grigorev. Grad. Kiev Art Inst. 1958, pupil of S. Grigorev, M. Ivanov and M. Khmelko.

Nechiporovich, Vladimir Nikolaevich
State official.
Head of Main Statistical Admin. Belorussian SSR Mr. 1980–.

Nechkina, Militsa Vasil'evna
Historian; b. F. 12 (25), 1901, Nezhin, now Chernigov Oblast.
Grad. Historical-Philological Dept. of Kazan Univ. 1921; mem. staff Inst. of History at Federation of USSR Social Sciences Research Inst. (RANION) 1924–27, concurrently Lecturer at 1st Moscow Univ. 1924–26 and at Workers' Univ. of the East 1925–28; Prof. of History at Moscow Univ. 1934–54; mem. USSR Acad. of Sciences 1958–; mem. RSFSR (since 1966, USSR) Acad. of Pedagogical Sciences 1947–; Prov. 1946, Head of Dept. of History of Sov. Society at Acad. of Social Sciences of Cen. Cttee. CPSU 1946–; mem. staff Inst. of History of USSR Acad. of Sciences 1935–. Decors.: 2 Orders of Lenin, Order of Red Banner of Labour, medals.

Nedbailo, Anatolii Konstantinovich
Military official; b. Ja. 28, 1923, Izium, Kharkov Oblast.
Twice Hero of the Soviet Union (Ap. 19, 1945; Je. 29, 1945); major-general of aviation (1970). Member of the CPSU since 1944.
In the Red Army since 1941, he graduated from the Voroshilovgrad Military Aviation Pilots' School in 1943 and from the Air Force Academy in 1951. In the Great Patriotic War (1941–45), he served on the Southern, Fourth Ukrainian, and Third Belorussian fronts as a pilot, flight commander, and deputy commander and commander of a squadron of the 75th Guards Assault Air Regiment (1943–45). He completed 224 combat flights. He has been teaching since the end of the war. Decors.: Order of Lenin, three Orders of the Red Banner, the Order of the Patriotic War First and Second Class, the Order of Alexander Nevsky, the Order of the Red Star, and various medals.

Nefed, Vladimir Ivanovich
Belorussian theater scholar, educator, and playwright; b. Ja. 14 (27), 1916, Shakhty.
Corresponding member of the Academy of Sciences of the Belorussian SSR (1972). Member of the CPSU since 1945. In 1940, having graduated from the Moscow Institute of Philosophy, Literature, and History, he embarked on a literary career. From 1951 to 1954 he worked in the institutes of the Academy of Sciences of the Belorussian SSR as a research worker in the art section, from 1954 to 1957 as head of the section, and from 1957 to 1970 as head of the theater and cinematography section. In 1970 he became head of the section of monuments of the history and culture of Belorussia. He has made important contributions to the development of Soviet theater studies. In 1947 he began teaching at the Belorussian Institute of Theatrical Arts in Minsk, becoming a professor in 1965. Decors.: State Prize of the Belorussian SSR in 1966, Order of the Badge of Honor and various medals.
Works: *Narodny artyst BSSR Ul. I. Uladamirski.* Minsk 1954. *Odnoaktny p'esy* Minsk 1955. *Narodny artyst SSSR P. S. Malachanau.* Minsk 1958. *Belorusski teatr: Narys historyi.* Minsk 1959. *Stanovlenie belorusskogo sovetskogo teatra.* Minsk 1965. *Malen'kie komedii i dramy.* Minsk 1967. *Belorusski akademichny teatr imia Ianki Kupaly.* Minsk 1970. *Razmyshleniia o dramaticheskom konflikte.* Minsk 1970. "Belorusskii teatr." In *Istoriia sovetskogo dramaticheskogo teatra,* vols. 1–6. Moscow 1966–71.

Negerish, Evgenii Filippovich
State official.
Head of Cen. Admin. 1977–78, Chair. State Cttee. for Material and Tech. supplies Belorussian SSR O.1978–; mem. Cen. Cttee. Belorussian CP 1981–.

Nekrasov, Boris Vladimirovich
Chemical scientist; b. S. 6 (18), 1899, Moscow.
Grad. G. V. Plekhanov Inst. of Econ. 1924; mem. staff various colleges 1924–39; corresponding mem. USSR Acad. of Sciences 1946–; Prof. at Inst. of Nonferrous Metals and Gold Moscow 1924–39; most important works on theoretical problems of chemistry. Decors.: Order of Lenin, Order of Red Banner of Labour, medals.

Nekrasov, Nikolai Nikolaevich
Economist; b. Je. 18 (Jl. 1), 1906, Irkutsk. Mem. CPSU 1950–; grad. Dept. of Econ. of Irkutsk Univ. 1929; Head of Scientific Council on Distribution of Production Resources of the USSR with USSR Acad. of Sciences 1968–; Scientific Dir. of Complex Studies for the Drafting of Long-Range Schemes for the Development and Distribution of Production Resources of the USSR 1968–; Vice-Pres. of Sov.-Japanese Friendship Soc. 1969–; scientific work abroad. Decors.: State Prize of the USSR 1970, 3 orders.

Nekrasov, Viktor Platonovich
Author; b. Je. 4 (17), 1911, Kiev; s. of a doctor.
Russian; mem. CPSU 1944–; grad. Dept. of Architecture of Kiev Inst. of Civil Engineering 1936; studied Theatre Studio of Kiev Theatre of Russian Drama; worked as actor and stage designer at theatres in Kiev, Vladivostok, Kirov and Rostov/Don; served at the front 1941–44. Decor.: State Prize of the USSR 1947.
Publs.: "V okopakh Stalingrada" (In the Trenches of Stalingrad) 1946, "Senka" 1956, "Vtoraia noch'" (The New Man) 1963, "Mesiats vo Frantsii" (One Month in France) 1965, and others.

Nekrasov, Zot Il'ich, Dr. of Tech. Sc.
Metallurgist; b. Ja. 8, 1908, Melitopol; s. of a railway worker.
Mem. CPSU 1939–; grad. Dnepropetrovsk Inst. of Metallurgy 1930; specialist on blast furnaces; Dr. of Tech. Sc. and Prof. 1951–; mem. staff Research Inst. for Metals 1938–41. Dnepropetrovsk Metallurgical Inst. 1943–60 and Magnitogorsk Inst. of Mining and Metallurgy 1941–43; Dep. Dir. 1951–52, Dir. and Head of Dept. of Pig-Iron Metallurgy at Inst. of Ferrous Metals (now Research Inst. for Ferrous Metals of USSR Min. of Ferrous Metals) of Ukrainian Acad. of Sciences 1952–; mem. Ukrainian Acad. of Sciences 1961–; author of some 70 scientific works. Decors.: Order of Lenin, 3 Orders of Red Banner of Labour, Badge of Honour, medals.
Publs.: "Ulushchenie konstruktsii domennykh pechei" in "Stal'", No. 2, 1947, "Isslesdovanie domennogo protsessa na pechakh, rabotaiushchikh s povyshennym davleniem gazov" in "Nauchny trudy Instituta Chernoi Metallurgii Akademii Nauk Ukrainskoi SSR" 1957, "Povedenie myshiaka v protsesse polucheniia neofliusovannogo i ofliusovannogo aglomeratov" in "Trudy Instituta Chernoi Metallurgii" vol. 12, 1960.

Nekroshius, Iuozas Povilovich
State official.
Chair. of State Cttee. for Publishing, Printing and the Book Trade. Lithuanian SSR Mr. 1973–. Mem. Auditing Comm. Lithuanian CP 1976–.

Nenado, Vladimir Spiridonovich
Graphic artist; b. 1935, Kozultsi, Krasnoiarsk Krai.
Grad. Kharkov Art. Inst. 1962, pupil of V. Mironenko, G. Bondarenko.

Nenashev, Mikhail Fedorovich
Journalist; b. 1929.
Mem. CPSU 1952–; grad. Magnitogorsk Pedagogical Inst. 1952, Dr. of Hist. Sc. 1979, Asst. Prof. 1961; engaged in pedagogical sciences in an inst. 1956–63; 2nd Sec. of Magnitogorsk City Cttee. CPSU 1963–67; Sec. of Cheliabinsk Oblast Cttee. CPSU 1968–75; Dep. Chief of Propaganda Dept. Cen. Cttee. CPSU 1975–78; Ed.-in-Chief "Sovetskaia Rossiia" Mr. 1978–; cand. mem. Cen. Cttee. CPSU 1981–.

Neporozhnii, Piotr Stepanovich, Cand. of Tech. Sc.
State official; b. 1910, Iagotin, Poltava Oblast; s. of a peasant.
Ukrainian; mem. CPSU 1940–; grad. Leningrad Inst. of Water Transport Engineering 1933; served in Sov. Army 1933–35; worked for various construction bureaus in Leningrad 1933 and 1935–36; chief engineer at construction of a hydro-electric station in Chirchik 1936–37; Dept. Head with Main Board for the Design of Hydro-Electric Power Stations at USSR People's Commissariat of Heavy Ind. 1937–40; chief engineer at construction of various hydro-electric power stations 1940–54; Dep. Chair. of Council of Mins. Ukrainian SSR 1954–57; cand. mem. Cen. Cttee. Ukrainian CP 1956–58; Chair. of State Cttee. on Construction and Architecture with Council of Mins. Ukrainian SSR 1956–57; Dep. Chair. of Ukrainian GOSPLAN 1957–58; Chair. of State Cttee. for Construction with Council of Mins. Ukrainian SSR 1958–59; 1st USSR Dep. Min. of Power Station Construction 1959–62; USSR Min. of Power Engineering and Electrification 1962–63; Chair. (with rank of USSR Min.) of USSR State Prod. Cttee. for Power Engineering and Electrification 1963–65; USSR Min. of Power Engineering and Electrification O. 1965–; Chair. of Permanent Comm. for Electrical Power with COMECON; cand. mem. 1966–71, mem. Cen. Cttee. CPSU 1971–; Dep. to USSR Supr. Sov. 1966–.

Neprintsev, Iurii Mikhailovich
Painter and graphic artist; b. Ag. 2 (15), 1909, Tbilisi.
Grad. Leningrad Acad. of the Arts 1938; teacher 1948–, Prof. at I. E. Repin Inst. of Painting, Sculpture and Architecture 1954–; mem. USSR Acad. of the Arts 1970–. Decors.: 2 orders, medals.

Neroda, Georgii Vasil'evich
Sculptor; b. Ja. 17 (29), 1895, Chernigov.
People's Artist of the RSFSR (1967); corresponding member of the Academy of Arts of the USSR (1967).
From 1913 to 1917, he studied under S. M. Volnukhin at the Moscow School of Painting, Sculpture, and Architecture. In 1926 he joined the Association of Artists of Revolutionary Russia. His works include portraits of A. V. Lunacharskii (bronze, 1928, Apartment-Museum of A. V. Lunacharskii, Moscow), Ia. M. Sverdlov (plaster of Paris, 1932, Tret'iakov Gallery), and V. I. Lenin (forged copper, 1970, property of the Ministry of Culture of the USSR). He also sculptured the monument to G. K. Ordzhonikidze in Kislovodsk (bronze, 1939, unveiled 1951) and the monument to V. I. Lenin in Perm' (bronze, unveiled 1954). Decor.: Order of the Red Banner of Labor. The artist's son Iurii Georgievich (born 1920) works with him and independently.

Nersesian, Leonid Nersesovich
Party official; b. 1925.
Armenian; mem. CPSU 1945–; grad. Ereven Polytech. Inst.; served in the Sov. Army 1943–46; Student 1946–51; Engineer, Dir. of div. Works in Erevan 1951–65; Chair. of Erevan City raion Exec. Cttee. of People's Deps., 1st Sec. of Lenin Raion Cttee. Armenian CP in Erevan 1965–75; 1st Sec. of Erevan City Cttee. Armenian CP My. 1975–; Dep. to Supr. Sov. Armenian SSR 1975–; mem. Cen. Cttee. and mem. Bureau Cen. Cttee. Armenian CP 1976–.

Nersesov, Evgenii Viktorovich
Diplomat; b. 1921.
Mem. Diplomatic Service 1951–; mem. Apparat USSR Min. of Foreign Affairs 1962–65 and 1974–78; Emb. Councellor at USSR Embassy in Syria 1965–69; USSR Amb. to Chad 1969–74, to Morocco O. 1978–.

Nersisian, Mkrtich Gegamovich
Historian; b. N. 12 (25), 1910, Parbi, now in Ashtarak Raion, Armenian SSR.
Academician of the Academy of Sciences of the Armenian SSR (1950). Honored Scientist of the Armenian SSR (1961). Member of the CPSU since 1938.
He graduated from the department of history and literature of the University of Yerevan in 1931. From 1939 to 1947 he was director of the Institute of History of the Academy of Sciences of the Armenian SSR. From 1947 to 1950 he was head of the subdepartment of history of the Armenian people of the University of Yerevan. Between 1950 and 1960 he was vice-president of the Academy of Sciences of the Armenian SSR; in 1958 he was made editor-in-chief of *Istoriko-filologicheskii zhurnal* (Historical-Philological Journal) of the Academy of Sciences of the Armenian SSR. In 1966 he became rector of the University of Yerevan.
His principal works deal with the history of the Armenian people in the 18th and 19th centuries, Armenian-Russian relations, and Russian-Caucasian relations. He was a deputy to the seventh through ninth convocations of the Supreme Soviet of the USSR. Decors.: Order of Lenin, the Order of the Red Banner of Labor, and various medals.
Works: *In Russian: Iz istorii russko-armianskikh otnoshenii,* books 1–2. Yerevan 1956–61. *Otechestvennaia viona 1912 g. i narody Kavkaza.* Yerevan 1965.

Neshkov, Khrisanf Pavlovich
State official; b. 1917.
1st Sec. of Kalinin Oblast Cttee. for Ind. 1963–64; Dep. to 1963–, mem. Comm. for Ind., Construction and Transport 1963–67, Sec. of Presidium Supr. Sov. RSFSR 1967–. Decor.: Order of the October Revolution.

Nesmelanov, Andrei Nikolaevich
Radiochemist; b. Ja. 15 (28), 1911, Moscow.
Corresponding member of the Academy of Sciences of the USSR (1972). Brother of Aleksandr N. Nesmeianov.
He graduated from Moscow State University in 1934 and was on the staff at the Moscow Aviation Institute from 1934 to 1947 and then at Moscow State University (he became the head of the subdepartment of radiochemistry in 1960). His main works are devoted to the chemistry of atoms formed as a result of nuclear transformations, to the methods of preparation of radioactive isotopes and tracer compounds, and to the use of radioactive isotopes in studying commercially valuable materials. He and his associates investigated the reaction of "hot" atoms with various chemical compounds. He developed an isotope exchange method and a number of other techniques for the use of isotopes in measuring the vapor pressure of relatively nonvolatile substances.
Works: *Poluchenie radioaktivnykh izotopiv.* Moscow 1954. (With A. V. Lapuskii and N. P. Rudenko.) *Davlenie para khimicheskikh elementov.* Moscow 1961. *Rukovodstvo k prakticheskim zaniatiiam po radiokhimii.* Moscow 1961. (With others.) *Rukovodstvo k prakticheskim zaniatiiam po fizicheskim osnovam radiokhimii.* Moscow, 1971. (With others.) *Radiokhimiia.* Moscow 1972.

Nesterenko, Aleksei Efremovich
Diplomat; b. 1915.
Mem. Diplomatic Service 1949–; Adviser to Sov. Rep. at UN 1960–61; USSR Amb. to Pakistan 1961–65; Dep. UN Sec.-Gen. for Pol. Questions and Affairs of the World Security Council 1965–68; Head of Dept. of Int. Econ. Orgs. at USSR Min. of Foreign Affairs 1968–80; USSR Amb. to Ireland Mr. 1980–.

Nesterikhin, Iurii Efremovich
Physicist; b. O. 10, 1930, Ivanovo.
Mem. CPSU 1960–; grad. Moscow Univ. 1953; mem. staff Inst. of Atomic Energy 1954–61, Inst. of Nuclear Physics 1961–67, Dir. of Inst. of Automatics and Electrometrics at Siberian Branch of USSR Acad. of Sciences 1967–; corresponding mem. USSR Acad. of Sciences 1970–; most important works on plasma physics and problems of thermonuclear synthesis; Chief Ed. of journal "Autometriia" of Siberian Branch of USSR Acad. of Sciences. Decor.: one order.

Nesturkh, Mikhail Fedorovich
Anthropologist and primatologist; b. F. 11 (23), 1895, Pskov.
Doctor of biological sciences (1962); professor (1967).
He graduated from the natural science section of the faculty of physics and mathematics of the Novorossiia University in Odessa in 1916. In 1928 he began working at the Scientific Research Institute of Anthropology at Moscow State University and the university's anthropological museum and subdepartment of anthropology. His principal works have been devoted to the ecology, taxonomy, and paleontology of primates, to the origin of man, and to race science. Decors.: Order of the Red Banner of Labor and various medals.
Works: *Chelovek i ego predki.* Moscow 1934. *Chelovecheskie rasy,* 3rd ed. [Moscow 1965.] *Proiskhozhdenie cheloveka,* 2nd ed. Moscow 1970. *Primatologiia i antropogenez.* Moscow 1960.

Netto, Igor' Aleksandrovich
Athlete; b. Ja. 9, 1930, Moscow.
Honored Master of Sports (1954); soccer coach. Member of the CPSU since 1966. He was repeatedly champion of the USSR (five times between 1952 and 1962 as a member of the Moscow Spartak team halfback). From 1952 to 1963 he was a member of the USSR national team and was its captain when it won the 1956 Olympia tournament at Melbourne and the European Cup in 1960 in Paris. Decor.: Order of Lenin.
Works: *Eto-futbol!,* 2nd ed. Moscow 1974.

Neunylov, Boris Aleksandrovich
Soil scientist, agricultural chemist, and specialist in rice growing; b. S. 20 (O. 3), 1908, Dankov, now in Lipetsk Oblast.
Corresponding member of the Academy

of Sciences of the USSR (1970). Academician of the V. I. Lenin All-Union Academy of Agricultural Sciences (1966). Hero of Socialist Labor (1966). Member of the CPSU since 1940.
He graduated from the Far Eastern Agricultural Institute in 1935. He worked at the Far Eastern Rice Experimental Station and on a rice-growing sovkhoz from 1935 to 1963. In 1964 he became acting chairman and between 1966 and 1970 served as chairman of the presidium of the Far Eastern Branch of the Siberian Division of the Academy of Sciences of the USSR. In 1970 he became the first deputy chairman of the presidium of the Far Eastern Science Center of the Academy of Sciences of the USSR, and in 1973 deputy director for science of the All-Union Scientific Research Rice Institute of the V. I. Lenin All-Union Academy of Agricultural Sciences (Belozernoe, Krasnodar Krai).
His principal works deal with the physical chemistry of soils, agricultural chemistry, regulation of oxidation-reduction conditions and acidity of rice paddy soils, and methods of increasing the fertility of rice paddy soils. He proposed new methods of cultivating soils, of introducing organic and mineral fertilizers, and of caring for young crops, all of which promoted the rapid expansion of sown areas and increased the productivity of rice in Primor'e Krai. Decors.: Order of Lenin and various medals.

Nezhevenko, Grigorii Semenovich
Lathe operator and innovator at the V. I. Lenin Machine-tool Plant in Odessa; b. S. 9 (22), 1912, at the Golta railroad station, present-day Pervomaisk, Nikolaev Oblast.
Member of the CPSU since 1943; one of the initiators of high-speed metal cutting as an industrial production process. From 1948 to 1953, he proposed a number of improvements in the design of cutting tools for lathes and developed highly productive methods, of metalworking based on the use of various devices and attachments and on the principle of standardization of technological processes for short-run and unit production. The introduction of advanced labor methods proposed by him made possible a substantial reduction in the time required for machining of parts and an improvement in their quality.
He took part in the organization at the plant of one of the first Institutes for Advanced Labor Methods, which was free and voluntary (late 1958). He was a delegate at the Twenty-second Congress of the CPSU (1961). He has been a member of the Central Committee of the Communist Party of the Ukraine since 1956 and was a deputy of the fourth convocation of the Supreme Soviet of the Ukrainian SSR (1950). Decors.: Order of the October Revolution and the Order of the Red Banner of Labor.
Works: *Moi opyt skorostnoi obrabotki metallov*. Moscow 1950. *Zavodskoi institut peredovogo opyta*. Moscow 1960. *Sovet novatorov*. Odessa 1961.

Niedre, Ianis Ianovich
Latvian writer and literary scholar; b. My. 24, 1909, Tikhvin, now in Leningrad Oblast.
Honored Cultural Worker of the Latvian SSR (1959). Member of the CPSU since 1934.
He was first published in 1927. From 1931 to 1933 he studied at the Riga Commercial Institute. Imprisoned during the fascist regime, he wrote a book of short stories, *In Five Cellbocks* (1941), about his experience. The trilogy *The Village of Pushkany* (vols. 1–3, 1941–61) depicts the revolutionary struggle in pre-Soviet Latgale.
In 1959, he published the novel *Spring Floods* (Russian translation, 1960), and in 1966 the novel *A Burdock Grows Outside My Window*. The novels *To Each His Happiness* (1968; Russian translation, 1970) and *An Island in the Raging Sea* (1970; Russian translation, 1972) are dedicated to the life of the Latvian revolutionary P. I. Stuchka (Stučkas). He is also the author of the works *Latvian Folklore* (1948) and *Latvian Literature* (vols. 1–2, 1952–53). From 1941 to 1957 he was first secretary of the administrative board of the Writers' Union of Latvia. He was a member of the Presidium of the Supreme Soviet of the Latvian SSR from 1940 to 1947. Decors.: two orders, as well as medals.
Works: *Pamošanas. Stasti*. Riga 1963. *Veterans*. Riga 1965.

Nigmanov, Kh. V.
Trade union official.
Mem. Presidium 1967–, Sec. of Uzbek Republican Trade Union Council 1972–.

Niiazi, Fatekh
Author and journalist; b. Ap. 18, 1914, Samarkand.
Tadzhik; mem. CPSU 1941–; grad. Dept. of History at Pedagogical Inst. Tashkent 1949; first works publ. 1928.
Publs.: "Maiak" (Lighthouse) 1933, "Liudi i vstrechi" (People and Encounters) 1962, "Vernost'" (Faithfulness) 1966.

Niiazov, Saparmurad Ataevich
Party official.
Head of Dept. for Ind. and Transport with Cen. Cttee. Turkmen CP 1979–80; 1st Sec. of Ashkhabad City Cttee. Turkmen CP 1980–; mem. Cen. Cttee. and cand. mem. Bureau Cen. Cttee. Turkmen CP 1981–.

Niit, Ellen (real name: Kross)
Author and translator; b. Jl. 13, 1928, Tallinn; d. of a worker.
Esthonian; grad. Tartu Univ. 1952; first works publ. 1952.
Publs.: "Zemlia polna nakhodok" (The Earth is Full of Finds) 1960, "Kak iskali elku" (How They Looked for a Firtree) 1954, and others.

Nikiforov, Dmitrii Semenovich
Diplomat; b. 1917.
Diplomatic Service 1946–; mem. staff USSR Min. of Foreign Affairs 1957–62; USSR Amb. to Lebanon 1962–66; various posts in USSR Min. of Foreign Affairs 1966–68; USSR Amb. to Senegal 1968–73; Head of Protocol Dept. in USSR Min. of Foreign Affairs 1973–; mem. Collegium of USSR Min. of Foreign Affairs.

Nikitin, P. G.
Komsomol official.
Sec. and mem. Bureau of Cen. Cttee. Kazakh Komsomol Mr. 1982–.

Nikitin, Sergei Aleksandrovich, Dr. of Hist. Sc.
Historian; b. Je. 12 (25), 1901, Moscow.
Grad. Dept. of Social Sciences of Moscow Univ. 1922; specializes on history of non-Russian Slavic peoples and history of Russia in the 19th century; Dr. of Hist. Sc. and Prof. 1947–; pedagogical work; Prof. Chair of the History of Southern and Western Slavs at Dept. of History of Moscow Univ. 1947–61; concurrently Head Section of History of non-Russian Slavic Peoples in the Period of Feudalism and Capitalism at Inst. for Slavic Studies of USSR Acad. of Sciences 1957–70; Dep. Chair. 1965–70, Chair. of Int. Comm. on Slavic Research with Int. Cttee. of Historical Sciences 1970–. Decors.: Merited Scientist of the RSFSR, Bulgarian Kirill-Mefodii Order 1st Cl.

Nikitin, Sergei Konstantinovich
Author; b. O. 10, 1926, Kovrov, Vladimir Oblast.
Russian; mem. CPSU 1956–; grad. Gorkii Inst. of Literature 1952; first works publ. 1948, Publs.: "Vozvrashchenie" (The Return) 1952, "Sem Slonov" (Seven Elephants) 1954, "Sobstvennyi dom" (One's Own House) 1962, "Lichnaia zhizn'" (Private Life) 1966, "Vesennim utrom" (On a Spring Morning) 1966, and others.

Nikodim (B. G. Rotov), Cand. of Theol.
Russian Orthodox clergyman; b. 1929, Frolovo, Riazan Oblast.
Studied Theological Faculty of Riazan Pedagogic Inst.; monk 1947; priest-monk 1949; custodian 1952–; Main Clergyman of Iaroslavl Cathedral 1954–; studied Leningrad Theological Seminary and Acad. 1950–55; mem. Ecclesiastical Mission in Jerusalem 1956; Abbot and Head of Mission 1957; Archimandrite; Head of Chancery of Moscow Patriarchate, Vice-Chair. 1959, Chair. of Dept. for External Relations 1960–; Bishop 1960; Archbishop, permanent mem. of Holy Synod 1961–; Archbishop of Minsk and Belorussia 1963; Metropolitan of

Leningrad and Ladoga O. 1963–67; Metropolitan of Leningrad and Novgorod S. 1967; Exarch of Moscow Patriarchate for W. Europe 1974–; activity abroad; Master of Theology 1970; released from duties in Church Dept. of External Relations 1972; Chair. of Cttee. for Christian Unity and Inter-Church Relations.

Nikodim (N. S. Rusnak)
Russian Orthodox clergyman; b. Ap. 4, 1921, Davidovtsy, Chernovtsy Oblast.
Monk 1943; deacon 1945; priest 1946, grad. Moscow Theological Seminary, ordered to serve at Ecclesiastical Mission of Moscow Patriarchate in Jerusalem, Dep. 1958, temporary Head of same Mission 1961–; Bishop of Kostroma 1961; Bishop of Argentina and South America 1964; Archbishop 1968; Archbishop of Kharkov and Bogodukhov 1970–; Exarch 1970, temporary Exarch of Moscow Patriarchate for Cen. and South America 1971–77.

Nikolaev, Adrian Grigor'evich
Cosmonaut; b. S. 5, 1929, Shorshely, Chuvash ASSR.
Mem. CPSU 1957–; grad. Frunze Mil. Aviation College 1954 and Zhukovskii Mil. Aviation Engineering Acad. 1968; fighter pilot; cosmonaut 1960–; 64 earth orbits (a total of 2.6 million km) in spacecraft "Vostok 3" Ag. 11–15, 1962; takeoff of "Vostok 4" Ag. 12, 1962, piloted by P. R. Popovich, group flight lasted 70 hours 44 minutes, contact between the two spaceships was maintained over a distance of 6.5 km; commander of spacecraft "Soiuz 9" Je. 1–19, 1970 (286 earth orbits totalling 11.9 million km in 424 hours 59 minutes) together with V. I. Sevastianov. Decors.: Hero of Socialist Labour, Order of Lenin, Order of Red Star, several foreign decors., mem. Int. Acad. of Astronauts.

Nikolaev, Aleksei Aleksandrovich
State official; b. 1914.
Min. of Construction and Utilization of Highways of RSFSR Jl. 1969–79; Min. of Highways of RSFSR Je. 1979–. Decor.: Order of Red Banner of Labour.

Nikolaev, Georgii Aleksandrovich
Engineer; b. Ja. 4 (17), 1903, Moscow.
Mem. CPSU 1953–; grad. Moscow Engineering Inst. for Railway Transport 1925 and Moscow Univ. 1928; specializes in calculation and design of welded constructions; mem. Scientific-Tech. Cttee. of People's Commissariat for Transport 1925–33; Lecturer 1930–39, Prof. 1939–, Rector of N. E. Bauman Tech. Acad. Moscow 1964–; corresponding mem. USSR Acad. of Sciences 1970–. Decors.: State Prize of the USSR 1972, 3 Orders of Lenin, 4 other orders and various medals.

Nikolaev, Ivan Sergeevich
Architect; b. Jl. 6 (19), 1901, Voronezh.
Mem. CPSU 1952–; grad. Moscow Higher Tech. School 1920; pedagogical work at Higher Tech. School and College of Constructional Engineering at Moscow Inst. of Architecture 1925–, Rector of same 1958–70. Decors.: Order of Lenin, Order of Red Banner of Labour, medals.

Nikolaev, Nikolai Fedorovich
State official.
Chief of Heavy Ind. Dept. of Cen. Cttee. Ukrainian CP to 1980; mem. Cen. Cttee. Ukrainian CP 1971–; Dep. Chair. Council of Mins. Ukrainian SSR Mr. 1980–.

Nikolaev, Sergei Nikolaevich
Mari playwright; b. Mr. 10 (23), 1908, Abdaevo, present-day Mork Raion, Mari ASSR.
People's Writer of the Mari ASSR (1958). Member of the CPSU since 1940.
In 1936, he graduated from the department of literature at the Mari Pedagogical Institute. His first play, *The White Moth,* which dealt with the Civil War, was published in 1930. Nikolaev has written dramas and comedies focusing on the past and on the contemporary experience of the Mari people. His best-known work is the musical comedy *Salika* (1938). In his plays he makes extensive use of Mari folklore and presents uniquely Mari characters.
He has translated L. N. Tolstoy's *The Power of Darkness,* I. Ia. Franko's *Stolen Happiness,* and a number of plays by Soviet playwrights into the Mari language. Decors.: three orders and a number of medals.
Works: *In Russian translation: Fesy.* Ioshkar-Ola 1952. *Bal'shaia volna: Drama.* Ioshkar-Ola 1967. *Komissary: Drama.* Ioshkar-Ola 1972.

Nikolaeva-Tereshkova, Valentina Vladimirovna
Cosmonaut; b. Mr. 6, 1937, Maslennikovo, Iaroslavl Oblast.
Russian; mem. CPSU 1962–; grad. Zhukovskii Aviation Acad. 1969; pilot-cosmonaut; Maj.; worked in a tyre factory and a textile works in Iaroslavl while studying at Textile Tech. College 1955–60; completed course at same college 1960; cosmonaut since 1962; Dep. 1966–, mem. Comm. on Foreign Affairs of Sov. of the Union 1966–74, mem. Presidium USSR Supr. Sov. 1974–; Chair. of Sov. Women's Cttee. 1968–; mem. Cen. Cttee. CPSU 1971–; took part in orbital flight of spacecraft "Vostok 6" Je. 16–19, 1963 (48 earth orbits in 71 hours). Decors.: Order of Lenin, Order of the October Revolution, Hero of the Sov. Union, Hero of Socialist Labour of Czechoslovakia, medals.

Nikolaevskaia, Elena Matveena
Poet and translator; b. Jl. 18, 1923, Moscow.
Grad. Gorkii Inst. of Literature 1946.
Publs.: "Kraski" (Colours) 1961, translated poetry by A. Tsereteli, G. Leonidze, O. Chelidze, S. Kaputikian, R. Gamzatov and others.

Nikolai (Evgenii Nikolaevich Iurik)
Russian Orthodox clergyman; b. D. 6, 1910, Korosten, Lvov Oblast.
Studied medicine and then theology at Lvov Theological Acad. 1930; grad. in Ivano-Frankovsk, priest, pastor in Lvov Diocese and in Ternopol and Lvov 1937; attached to Moscow Patriarchate 1946; co-founder of Leadership and 1st Sec. of Lvov Diocese until 1965; Main Clergyman of Church of the Transfiguration in Lvov 1955–; Bishop, Archbishop 1965–71, Metropolitan of Lvov and Ternopol 1971–.

Nikolai (N. V. Kutepov), Cand. of Theol.
Russian Orthodox clergyman; b. O. 4, 1924, Kutepovo, Tula Oblast.
Subdeacon, Sec. to Archbishop 1944–47; grad. Moscow Theological Acad., deacon 1952; grad. Leningrad Theological Acad. 1958; teacher and Inspector of Kiev Theological Seminary; priest, monk, Supervisor of Saratov Theological Seminary 1959; Bishop of Mukachevo 1961; Bishop of Omsk 1963; Bishop of Rostov 1969; Bishop 1970–72 and Archbishop of Vladimir and Suzdal 1972–75; Archbishop of Kaluga and Borovsk 1975–77; Archbishop of Gorkii and Arzamas Ag. 1977–.

Nikolai (Piotr Sajama)
Russian Orthodox clergyman; b. 1914, Formosa.
Studied Theological Seminary of St. Nicholas Cathedral in Tokyo; ordained as deacon while participating in a course at St. Vladimir Theological Seminary New York 1956; grad. 1959, then spent 18 months on Holy Mount Athos; visited USSR 1961 and 1962; monk 1962; appointed Archimandrite in Moscow 1966; during a visit to Moscow with a group of Japanese Orthodox pilgrims appointed Bishop of Tokyo and Japan 1967; upon recognition of the autonomy of the Japanese Church was released from same office; Bishop of Mozhaisk and Representative of the Moscow Patriarchate with the Archbishop of Tokyo 1970–.

Nikolskii, Boris Petrovich
Physical and radiochemical scientist; b. O. 1 (14), 1900, Menzelinsk, now Tartar ASSR.
Grad. Leningrad Univ. 1925; mem. staff 1925–, Prof. at Leningrad Univ. 1939–; concurrently Head of Laboratory of V. G. Khlopin Radiation Inst. 1946–; mem. USSR Acad. of Sciences 1968–. Decors.: Lenin Prize 1961, State Prize of the USSR 1949, 1973, 2 Orders of Lenin, medals.

Nikolskii, Sergei Mikhailovich
Mathematician; b. Ap. 17 (30), 1905, Talitsa, now Sverdlovsk Oblast.
Mem. CPSU 1952–; grad. Inst. of Popular Educ. in Ekaterinoslav (now Dnepropetrovsk Univ.) 1929; mem. USSR Acad. of Sciences 1972–; mem. Steklov Inst. of Mathematics of USSR Acad. of Sciences 1940–; concurrently Prof. at Physical-Tech. Inst. in Moscow 1974–. Decors.: State Prize of the USSR 1952, Chebyshev Prize 1972, Order of Red Banner of Labour, medals.

Nikon (N. V. Fomichev), Cand. of Theol.
Russian Orthodox clergyman; b. My. 22, 1910, St. Petersburg.
Studied Inst. for Railway Engineering; deacon 1945; priest 1946; grad. Leningrad Theological Acad., Cand. of Theol. 1950; monk, Archimandrite, Bishop of Vyborg, Substitute of Leningrad Diocese, Administrator of Bishopric of Riga 1962; Bishop of Riga and Latvia 1963; Bishop of Arkhangelsk and Kholmogory 1966–77; Bishop of Kaluga and Borovsk 1977–82; Archbishop of Perm and Solikamsk Jl. 1982–.

Nikonov, Viktor Petrovich
State official; b. 1929; s. of a peasant.
Russian; mem. CPSU 1954–; grad. Azovo-Chernomorskii Agricultural Inst. 1950; chief agronomist of a machine and tractor station (MTS); dep. dir. of an agricultural school, dir. of a MTS 1950–58; Dep. Head, Head of Agricultural Dept. of Krasnoiarsk Krai CPSU Cttee. 1958–61; 2nd Sec. of Tartar ASSR Oblast CPSU Cttee. 1961–67; Dep. to USSR Supr. Sov. 1962–; 1st Sec. of Mari ASSR Oblast CPSU Cttee. 1967–79; cand. mem. 1971–76, mem. Cen. Cttee. CPSU 1976–; USSR Dep. Min. of Agriculture, Chair. All-Union Production-Scientific Association for Agrochemical Services of Agriculture 1979–. Decor.: Order of the October Revolution.

Nikulin, Iurii Vladimirovich
Circus artist and carpet clown (clown performing between acts); b. D. 18, 1921, Demidov, Smolensk Oblast.
People's Artist of the USSR (1973). Member of the CPSU since 1943.
In 1950, he graduated from the Conversational Genre Studio of the Moscow Circus. He worked in a clown group headed by the well-known Soviet clown Karadsh. He now performs with M. I. Shuidin. He has created the character of a phlegmatic, outwardly rather simple but actually crafty man – the exact opposite of his energetic and fidgety partner. He has also won acclaim as a film actor. He had played in a series of film comedies *In All Seriousness (The Dog Barbos and an Unusual Cross-country Race,* 1961), *Operation "Y" and Shurik's Other Adventures* (1965), and *A Prisoneress in the Caucasus* (1967). He has also acted in dramatic roles, including Kuz'ma Iordanov in *When the Trees Were Tall* (1962), Glazychev in *Here, Mukhtar!* (1965), and Miachikov in *Old Robbers* (1972). Decors.: (1970) State Prize of the RSFSR for his work in films, Order of the Badge of Honor and a number of medals.

Nikulin, Vladimir Illarionovich
Party official; b. 1928.
Mem. CPSU 1954–; grad. Stavropol Agricultural Inst. 1961 and CPSU Cen. Cttee. Party Higher Correspondence School 1967; driver at Bashanta Machine and Tractor Station 1944–48; served in Sov. Army 1950–54; chief agronomist at "Iuzhnyi" State Farm in Rostov Oblast 1960–63; dir. of "Komsomolets" State Farm in Rostov Oblast 1963–64; 1st Dep. Min. of Agriculture Kalmyk ASSR 1965–66; Min. of Melioration and Water Econ. Kalmyk ASSR 1966–72; 1st Sec. of Gorodovikovsk Raion Cttee. CPSU 1972–75; 2nd Sec. of Oblast Cttee. Kalmyk ASSR 1976–78; 1st Sec. of Oblast Cttee. Kalmyk ASSR D. 1978–; Dep. to USSR Supr. Sov. 1979–; cand. mem. Cen. Cttee. CPSU 1981–.

Nikulishchev, Vladimir Mikhailovich
Oblast soviet official.
1st Sec. of Chernigov City Cttee. Ukrainian CP to 1973; 2nd Sec. of Chernigov Oblast Cttee. Ukrainian CP 1973–81; Chair. of Exec. Cttee. of Chernigov Oblast Sov. of Peoples' Deps. 1981–; cand mem. Cen. Cttee. Ukrainian CP 1981–.

Nikulkov, Anatolii Vasil'evich
Author; b. Ap. 18, 1922, Voronezh.
Mem. CPSU 1944–; grad. Dept. of History of Pedagogical Inst. Novosibirsk 1952. Publs.: "Dostoinye schastia" (Those Worthy of Happiness) 1952, "Gordaia familiia" (The Proud Name) 1959, "Na planete, malo oborudovannoi" (On a Badly Outfitted Planet) 1967, and others.

Nilin, Pavel Filippovich
Author; b. Ja. 3 (16), 1908, Irkutsk.
Russian; mem. CPSU 1944–; worked as locksmith and stoker, then active as journalist for various newspapers; war correspondent for "Pravda" during World War II; first works publ. 1936. Decor.: State Prize of the USSR 1941.
Publs.: "Chelovek idet v goru" (Man Goes Upward) 1936, "Liubimaia devushka" (Beloved Girl) 1936, "Bol'shaia zhizn'" (The Big Life) 1946, "Znakomstvo s Tishkovym" (Acquaintanceship with Tishkov) 1955, "Cherez kladbishche" (Through the Cemetery) 1962, and others.

Nirod, Fedor Fedorovich
Stage designer; b. Mr. 31 (Ap. 13), 1907, St. Petersburg.
People's Artist of the USSR (1965). Member of the CPSU since 1946.
He studied at the Kiev Art Institute from 1926 to 1930. His set designs, which deal with heroic dramatic themes, are characterized by a picturesque treatment of space. He has designed sets for the play *Uriel Acosta* by Gutzkow (1938; the M. Zankovts'a Ukrainian Drama Theater in Zaporozh'e) and the play *Svichka's Marriage* by Kocherga (1959; the I. Franko Academic Ukrainian Dramatic Theater in Kiev). His work at the T. G. Shevchenko Ukrainian Theater of Opera and Ballet in Kiev has included the sets for the operas *The Barber of Seville* by Rossini (1945), *Arsenal* by Maiboroda (1960), and *Abessalom and Eten* by Paliashvili (1972) and the ballet *Spartacus* by Khachaturian (1964). Decors.: Order of Lenin and a number of medals.

Nishanov, Rafik
Diplomat; b. 1926.
Uzbek; mem. CPSU 1949–; grad. Tashkent Pedagogic Inst.; bookkeeper in a kolkhoz, sec. of a village sov., instructor, sec. of a raion Komsomol cttee. 1942–45; served in Sov. Army 1945–50; dept. head of a raion Komsomol cttee., instructor, dept. head, sec., 1st Sec. of Oktiabrskii Raion Cttee. Uzbek CP in Tashkent, Chair. of Tashkent City Exec. Cttee. of Workers' Deps. 1950–63; mem. Cen. Cttee. Uzbek CP 1960–71; Sec. and mem. Bureau Cen. Cttee. Uzbek CP 1963–70; Dep. to USSR Supr. Sov. 1966–70; USSR Amb. to Sri Lanka and Maldive Islands Ag. 1970–78, to Jordan Je. 1978–.

Nishnianidze, Shota Grigor'evich (also S. G. Mamageishvili)
Georgian poet; b. Mr. 18, 1929, Tbilisi.
He graduated from the department of philology at the University of Tbilisi in 1953. His works first appeared in print in 1946. He is the author of the collections of lyrical poetry *You and I* (1956), *The Olive Branch* (1961; Russian translation, 1966), *Autumn in the Village* (1959), and *Sun and Shade* (1963). He has written a narrative poem, *Achilles' Heel and Shield* (1971), and a collection of poetry for children. *Brotseula* (1966). Decor.: Order of the Badge of Honor.

Nisskii, Georghii Grigor'evich
Painter; b. Ja. 8 (21), 1903, Novobelitskaia station, present-day Gomel' Oblast, Belorussian SSR.
People's Artist of the RSFSR (1965). Member of the Academy of Arts of the USSR (1958).
From 1922 to 1930, he studied in Moscow at Vkhutemas (State Higher Arts and Technical Studios) and Vkhutein (Higher Art and Technical Institute). His work, which was influenced by A. A. Deineka, was closely related to that of the Society of Painters (for example, Nisskii's *Autumn-Semphores*. 1932, Tret'iakov Gallery).
Seeking to express the power of a new, socialist reality, as well as the perception of nature by modern man, who has become accustomed to the fast-moving rhythm of life, he invested his works (pre-

dominantly seascapes and industrial landscapes) with a strong dynamism and depicted broad expanses of space. Since the mid-1950's, he has turned increasingly to a decorative and generalized style, developing his compositions with highly rhythmic silhouettes and large planes of intense and, often, local colors. In 1951 the State Prize of the USSR was awarded to his paintings *By the Shores of the Far East* (1950, Moscow State University), *Landscape With a Lighthouse* (1950, L'vov Picture Gallery), and *The Port of Odessa* (1950, the motor ship Rossia). Among his other works are *Winter in the Moscow Region* (1951), Cheliabinsk City Picture Gallery), *Frontline Road Under Moscow* (1957, L'vov Picture Gallery), *The "Zagor'e" Kolkhat* (1959, Russian Museum, Leningrad), and *Above the Snows* (1959–60, Russian Museum, Leningrad). Decors.: Order of the Red Banner of Labour and a number of medals.

Noneshvili, Iosif Eliozovich
Author; b. Ap. 6, 1918, Kardanakhi, Georgia.
Georgian; mem. CPSU 1945–; grad. Dept. of Philology of Tbilisi Univ. 1942; first works publ. 1938.
Publs.: "Gory i serdtse" (Mountains and the Heart) 1960, "Vesna Gruzii" (Georgian Spring) 1964, "Stikhotvoreniia" (Poems) 1967, and others.

Noreika, Virgilius-Kiastutis Leono (Virgilijus-Kestutis Leono Noreiką)
Lithuanian singer (tenor); b. S.22, 1935, Šiauliai.
People's Artist of the USSR (1970). Member of the CPSU since 1971.
He graduated from the Vilnius Conservatory in 1958. In 1957, while a student, he made his debut at the Lithuanian Theater of Opera and Ballet, where he presently performs as one of the leading soloists. In 1965–66 he was a guest performer at La Scala in Italy. He has sung at opera houses in various cities of the USSR, as well as in Berlin, Budapest, and Prague. He has sung many operatic roles, including Lenskii in Tchaikovsky's *Eugene Onegin*, Almaviva in Rossini's *The Barber of Seville*, the Prince in Prokofiev's *Love for Three Oranges*, and Alfredo in Verdi's *La Traviata*. He has also appeared as the Duke of Mantua in Verdi's *Rigoletto*, Don José in Bizet's *Carmen*, Werther in Massenet's opera of the same name, Andriaus in Kova's *The Daughter*, and Tadas in *The Wandering Birds* by Laurušas.
Since 1974, he has been chairman of the Theatrical Society of the Lithuanian SSR. Decors.: State Prize of the Lithuanian SSR in 1960 and 1971.

Nosov, Evgenii Ivanovich
Author; b. Ja.15, 1925, Kursk; s. of a worker.
Russian; grad. Advanced Courses in Literature at Gorkii Inst. of Literature 1962; first works publ. 1947.
Publs.: "Gde prosypaetsia solntse" (Where the Sun Awakes) 1965, "Zatmenie luny" (Lunar Eclipse) 1966, and others.

Novichenko, Leonid Nikolaevich, Dr. of Phil. Sc.
Literary historian and critic; b. Mr.31, 1914, Rusanovka, Sumi Oblast; s. of a village teacher.
Mem. CPSU 1944–; grad. Kiev Univ. 1939; Dr. of Phil. Sc. 1958–; Head of Revies Dept. of "Literaturnaia gazeta" (newspaper) and "Radianska literatura" (periodical) 1939–41; Head of Cultural Dept. of newspaper "Komunist" (now "Radianska Ukraina") 1941–43, concurrently war correspondent for rear frontline during World War II; Managing Ed. of "Literaturnaia gazeta" 1943–46; Head of Dept. of Literature and Art with Propaganda and Agitation Admin. Dept. of Cen. Cttee. Ukrainian CP (B) 1946–47; mem. staff Shevchenko Inst. of Literature at Ukrainian Acad. of Sciences 1945–; concurrently Dep. Managing Ed. of "Vitchysna" 1950–52; corresponding mem. Ukrainian Acad. of Sciences 1958–; author of some 400 articles, reviews and pamphlets, among them 50 works on the history of the Ukrainian pre-October and Sov. literature, on the literature of the peoples of the Sov. Union, on theory of literature, etc. Decors.: 2 Orders of Red Banner of Labour, Order of Red Star, medals.
Publs.: "Pavlo Tychyna" 1941, "Literaturno-kriticheskie ocherki" 1951, "Poeziia i revolutsiia" 1956, "Myroslav Irchan" 1958, "Shevchenko i suchasnost'" 1964, "Ne iliustratsiia – vidkryttia!" 1967.

Novichkov, Viktor Emelianovich
State official; b. 1916.
Russian; mem. CPSU 1941–; grad. Tadzhik State Univ.; locksmith, technician, head of explosion work of a slate combine 1932–41; CPSU posts 1941–42; served in Sov. Army 1942–45; dir. of a works. 1st Dep. Chair. of Dushanbe City Exec. Cttee. of Workers' Deps. 1946–61; 1st Dep. Chair. 1961–62 and 1973–, Dep. Chair. of Council of Mins. Tadzhik SSR 1962–73; mem. Cen. Cttee. 1961–, mem. 1961–63, cand. mem. Bureau Cen. Cttee. Tadzhik CP 1976–; Dep. to and mem. Econ. Comm. at Sov. of Nationalities USSR Supr. Sov. 1962–66; mem. Bureau of Ind. and Construction at Cen. Cttee. Tadzhik CP 1962–64. Decor.: Order of Red Banner of Labour.

Novikov, Anatolii Grigor'evich
Composer; b. O.18 (30), 1896, Skopin, now Riazan Oblast.
Mem. CPSU 1952–; studies Moscow Conservatory 1921–27; conductor of various amateur and soldiers' choirs; composer of "patriotic" songs. Decors.: State Prize of the USSR 1946, 1948, Order of Lenin, Order of the October Revolution, Order of Red Banner of Labour, medals, Hero of Socialist Labour, Order of Lenin, Hammer and Sickle medal in gold 1976.

Novikov, Anatolii Semenovich
Sculptor; b. 1926.
Mem. CPSU 1971–; grad. Moscow Inst. of Applied and Decorative Arts 1952; worked as locksmith 1942–46 and in a ceramic factory 1952–54; sculptor in an arts and crafts combine 1954–. Decor.: Lenin Prize 1970.

Novikov, Ignatii Trofimovich
State official; b. 1906; s. of a worker.
Ukrainian; mem. CPSU 1926–; grad. Dneprodzerzhinsk Metal Inst. 1932; dept. head, dir. of a power plant, chief mechanic of a lead works in Chimkent 1932–37; chief engineer, dir. of a plant in Saratov 1937–41; Sec. of Saratov Oblast CPSU Cttee. 1941–43; mem. Collegium and Head of Main Board for Spare Parts for Power Plants at USSR People's Commissariat, then Min. of Power Plants 1943–50; dir. of works at construction of a hydro-electrical plant in Gorkii 1950–54, in Kremenchug 1954–58; USSR Dep. Min. of Power Plants 1958; USSR Min. for Construction of Electric Power Plants D. 1958–62; mem. Cen. Cttee. CPSU 1961–; Dep. to USSR Supr. Sov. 1962–; USSR Min. of Power and Electrification O.–N. 1962; Dep. Chair. of USSR Council of Mins. and Chair. of State Construction Cttee. N. 1962–. Decor.: Order of Lenin.

Novikov, Igor' Aleksandrovich
Athlete (modern pentathlon); b. O.19, 1929, Drezna, Moscow Oblast.
Honored Master of Sports (1957), Honored Coach of the USSR (1968), and Honored Worker in Physical Education of the Armenian SSR (1969). Member of the CPSU since 1958.
He was an Olympic champion (1956, 1964) and world champion (1957–59; 1961–62) in team competition in the pentathlon. In individual competition he was world champion (1957–59, 1961) and USSR champion (1953, 1956, 1959, and 1964). Decor.: Order of the Red Banner of Labour.

Novikov, Ivan Grigor'evich
Author; b. Ja.26, 1918, Neznan, Mogilev Oblast.
Belorussian; mem. CPSU 1939–.
Publs.: "Riadom druzia" (Friends Close By) 1960, "Dorogi skrestilis v Minske" (The Paths Crossed in Minsk) 1964.

Novikov, Ivan Ivanovich
Physicist; b. Ja.15 (28), 1916, Pobedinka, Riazan Oblast.
Mem. CPSU 1944–; grad. Moscow Univ. 1939; Prof. Chair of Thermophysics at Energetics Inst. Moscow 1948–50; Head of Scientific-Tech. Admin. Dept. at USSR Min. of Medium-Sized Machinery 1950–56; Dir. of Inst. of Physical Engineering Moscow 1956–58; correspond-

ing mem. USSR Acad. of Sciences 1958–; Dir. of Inst. of Thermophysics at Siberian Branch of USSR Acad. of Sciences 1958–64; Prof. Chair of Thermophysics at Novosibirsk Univ. 1961–64; 1st Dep. Chair. of State Cttee. on Norms with USSR Council of Mins. 1964–67; Prof. Chair of Physical-Tech. Measurement at Physical-Tech. Inst. Moscow 1966–70; Head of Laboratory for Electromicroscopy and Phycisal Research Methods at A. A. Baikanov Inst. of Metallurgy 1970–. Decors.: State Prize of the USSR 1951, 1953, Order of Lenin, Order of Red Banner of Labour, medals.

Novikov, Sergei Petrovich
Mathematician; b. Mr. 20, 1938, Gorki; s. of P. S. Novikov.
Grad. Moscow Univ. 1960; mem. staff V. A. Steklov Mathematical Inst. of USSR Acad. of Sciences 1963–; Prof. at Moscow Univ. 1966–; corresponding mem. USSR Acad. of Sciences 1966–. Decors.: Lenin Prize 1967, Prize of Int. Mathematical Soc. 1970.

Novikov, Vladimir Mikhailovich (pen name: Kiunniuk Urastyrov)
Yakut poet; b. Ap. 26 (My. 5), 1907, Emissy *nasleg* (administrative and territorial unit), present-day Amga Raion, Yakut ASSR.
People's Poet of the Yakut ASSR (1967). Member of the CPSU since 1940.
He graduated from the department of language and literature of the Yakut Pedagogical Institute (1942). His works were first published in 1927. The poetry collection *In the Awakened Region* (1932) praises the new Soviet people. The poem "Semen the Communist" (1932) is dedicated to the victory of the kolkhoz.
He is the author of a poetic version of the Yakut epic *The Hero Knight Dzhagaryma on a Chestnut Horse* (1941); the poetry collections *War Days* (1943), *Yakut Khomus* (1946), and *The Bright Stream* (1964); and the opera libretto *Niurgusun and Lookuut* (with D. K. Sivtsev-Omollon). He is also a translator and critic. Decors.: two orders and a number of medals.
Works: *Talilibit ayïmn ïlar,* vols. 1–2, Yakutsk 1967–68. *Taptïbïn subu siri.* Yakutsk 1969. *In Russian translation: Izbrannoe.* Yakutsk 1953. *Severnaia raduga.* Moscow 1957. *Liubimye pereleski. Stikhi.* Poemy. Yakutsk 1971.

Novoselova, Aleksandra Vasil'evna
Chemical scientist; b. Mr. 10 (23), 1900, Verezino, Kalinin Oblast.
Grad. Moscow Univ. 1925; specializes in inorganic chemistry; mem. staff Moscow Univ. 1926–; Head of Dept. of Chemistry and Head of Laboratory at Moscow Univ. 1948–55; mem. USSR Acad. of Sciences 1970–; developed analytical methods for beryllium, and for complexes in system moH_3-H_2O. Decors.: State Prize of the USSR 1948, M. V. Lomonosov Prize of Moscow Univ. 1970, 2 Orders of Lenin, medals.
Publs.: "Ftoristyi berillii ftoroberillaty" in "Uspekhi khimii" 1959, vol. 28, No. 1, co-author of "Analiticheskaia khimiia berillia . . ." Moscow 1966, "Synthese von Silikaten durch chemische Transportreaktionen" in "Kristall und Technik" 1967, vol. 2, No. 4, p. 115, co-author of "Physico-chemical study of germanium, tin, lead chalcogenides" in "Progress in Solid State Chemistry" 1972, vol. 7, p. 85.

Nurberdyev, Pomma
Author; b. 1909, Akiap, Turkmenia; s. of a peasant.
Turkmen; mem. CPSU 1943–; ed. of satirical journal "Tokmak" 1954–58; first works publ. 1932.
Publs.: "Reshaiushchie dni" (Decisive Days) 1943, "Diadia Aba" (Uncle Aba) 1952, "Zhenshchina – geroinia" (Woman – The Hero) 1948, "Budem puteshestvovat' na Lune" (We Shall Travel to the Moon) 1960, "Net, ty ne zabludilsia" (No, You Didn't Lose Your Way) 1961, "Burnoe techenie" (Stormy Current) 1965, and others.

Nurdzhanov, Nizam Khabibullaevich
Tadzhik theater scholar and doctor of the arts (1968); b. D. 18, 1923, Bukhara.
Member of the CPSU since 1948.
In 1947, he graduated from the department of literature of the Pedagogical Institute in Dushanbe. He began his literary career in 1948. In 1952 he became senior research scholar at the Akhmad Donish Institute of History of the Academy of Sciences of the Tadzhik SSR, and in 1953 head of the institute's section of art history.
Since 1963, he has been a teacher. He began teaching in the art department of the Dushanbe T. G. Shevchenko Pedagogical Institute in 1967. His works on the Tadzhik national theater are especially valuable.
Works: *Mukhammedzhan Kasymov.* Moscow 1955. *Tadzhikskii narodnyi teatr.* Moscow 1965. *Starinnye pantomimy tadzhikov.* Moscow 1964. *Istoriia tadzhikskogo sovetskogo teatra (1917–1941 gg.).* Dushanbe 1967. *Tadzhikskii teatr: Ocherk istorii.* Moscow 1968.

Nuretdinov, Zaki Sharafutdinovich (pseudonym: Nuri Zaki)
Author; b. D. 24, 1921, Tatarskie Tiuki, Tartar ASSR; s. of a peasant.
Tartar; mem. CPSU 1948–; concluded Advanced Courses in Literature at Gorkii Inst. of Literature 1958; partisan in Belorussia during World War II; worked as journalist 1949–55; first works publ. 1936. Decor.: Order of Patriotic War 1st Cl.
Publs.: "Stikhi" (Poems) 1946, "Doroga Slavy" (The Road of Fame) 1949, "My zhivem na Volge" (We Live on the Volga) 1951, "I mertvye mstili" (The Dead Revenged Themselves Too) 1962, translated poetry by N. A. Nekrasov, V. V. Maiakovskii, S. P. Shchipachev, V. M. Saianov and others into Tartar.

Nuriev, Ziia Nurievich, Cand. of Econ. Sc.
State official; b. 1915, Verkhnii Lochentau village, Ufa Oblast; s. of a peasant.
Bashkir; mem. CPSU 1939; grad. Birsk Teachers' Training College 1930 and Party Higher School of CPSU Cen. Cttee. 1951; dir. of village school 1933–35; Head of Buraevo Raion Educ. Dept. in Bashkir ASSR 1935–38; served in Sov. Army 1938–40; Head of Buzoviazy Raion Educ. Dept. in Bashkir ASSR 1940–42; Head of Pol. Dept. of Podlubov Machine and Tractor Station (Bashkir ASSR) 1942–43; 2nd Sec. of Buzoviazy Raion CPSU Cttee. 1943–44; 1st Sec. of Kigi Raion CPSU Cttee. 1944–45; Head of Agricultural Dept. of Bashkir Oblast CPSU Cttee. 1945–48; attended Party Higher School of Cen. Cttee. CPSU 1948–51; dept. head with Bashkir Oblast CPSU Cttee. 1951–52; Sec. 1952–57, 1st Sec. of Bashkir Oblast CPSU Cttee. 1957–69; Dep. to 1954–, mem. Presidium USSR Supr. Sov. 1954–69; mem. Cen. Cttee. CPSU 1961–; USSR Min. of Supplies 1969–73; mem. Union Council for Kolkhozes 1969–; Dep. Chair. of USSR Council of Mins. Ap. 1973–. Decors.: 3 Orders of Lenin, Order of Red Banner of Labour, Hero of Socialist Labour, medals.

Nurme, Minni
Author; b. O. 30, 1917, Esthonia.
Esthonian; concluded secondary school in Viliandi.
Publs.: "Stul na kolesakh" (A Chair on Wheels) 1941, "Rodnaia zemlia" (Homeland) 1945, "V more" (On the Lake) 1953, "O chem stuchal diatel" (What the Woodpecker Pecked) 1966, and others.

Nurmukhammedov, Nagim-Bek Dzhemal'-Eddinovich
Painter and art scholar; b. N. 6, 1924, Saitandy, now in Semipalatinsk Oblast.
People's Artist of the Kazakh SSR (1974). He studied at the I. E. Repin Leningrad Institute of Painting, Sculpture, and Architecture under R. R. Frents from 1947 to 1953. Since 1959 he has worked in Alma-Ata. Among the portraits he has painted is one of I. S. Isakov, an admiral of the USSR Navy (1967, Central Naval Museum, Leningrad). He also does landscapes (*The Alatau Foothills,* 1966, T. G. Shevchenko Kazakh Art Gallery, Alma-Ata), genre paintings (*Papa Is Coming,* 1960, T. G. Shevchenko Kazakh Art Gallery, Alma-Ata), and historical paintings (*The Turgai Steppes, 1919;* painted 1971–72). Decors.: two orders and a number of medals.
Works: *Iskusstvo Kazakhstana.* [Moscow 1970.]

Nurpeisov, Abdizhamil Karimovich
Author; b. O.22, 1924, Kugaral (now Kokaral), Kzyl-Orda Oblast.
Kazakh; mem. CPSU 1944–; grad. Gorkii Inst. of Literature 1956; served in Sov. Army 1941–45; first works publ. 1949. Decors.: Lenin Prize 1974, 3 orders.
Publs.: "Kurliandiia" (Kurland) 1950, 2nd ed. under the title "Dolgozhdannyi den'" (The Long-Awaited Day) 1958, "Krov' i pot" (Blood and Sweat) 1973, "Razdumia" (Reflections) 1972, and others.

Nusupova, Dzhumagul Nusupovna
State official.
Sec. of Frunze City Cttee. Kirghiz CP 1961–63; 1st Sec. of Pervomaiskii Raion Cttee. Kirghiz CP in Frunze 1963–69; mem. Cen. Cttee. Kirghiz CP 1963–; Head of Dept. of Culture with Cen. Cttee. Kirghiz CP 1969–80; Min. of Culture Kirghiz SSR Mr. 1980–.

Nuut, Ilmar Iurevich
State official.
Min. of Higher and Secondary Specialist Educ. Esthonian SSR Ja. 1976–; cand. mem. Cen. Cttee. Esthonian CP 1976–.

Obraztsov, Ivan Filippovich, Dr. of Tech. Sc.
State official; b. Jl.28, 1920, Byki, now Kalinin Oblast.
Mem. CPSU 1944–; grad. S. Ordzhonikidze Aviation Inst. Moscow 1944; specialist in construction mechanics and stress in flying devices; corresponding mem. of USSR Acad. of Sciences 1966–; lecturer 1944–57, Prof. 1957–, Rector of S. Ordzhonikidze Aviation Inst. Moscow 1958–72; Min. of Higher and Secondary Specialist Educ. RSFSR 1972–; Chair. of "Znanie" Soc. of the RSFSR; Cand. mem. Cen. Cttee. CPSU 1981–. Decors.: Order of Lenin, 2 other orders and various medals.

Obraztsov, Sergei Vladimirovich
Puppet artist; b. 1901.
Trained at Higher State Workshop for Arts and Crafts (VHUTEMAS) 1918–20; Dir. of famous Cen. State Puppet Theatre Moscow 1931–; Vice-Pres. 1957–; Pres. of Int. Federation of Puppeteers 1958–; corresponding mem. German (GDR) Acad. of the Arts. Decors.: Hero of Socialist Labour 1971, State Prize of the USSR 1946.
Publs.: "Akter s kukloi" (Actor With a Puppet) 1938, "Teatr kitaiskogo naroda" (Theatre of the Chinese People) 1957, and others.

Obreimov, Ivan Vasil'evich
Physicist; b. F.24 (Mr.8), 1894, Annecy, France.
Academician of the Academy of Sciences of the USSR (1958; corresponding member, 1933).
He graduated from Petrograd University in 1915. From 1919 to 1924 and from 1942 to 1944 he worked at the State Optical Institute; from 1924 to 1929, at the Leningrad Physicotechnical Institue, from 1929 to 1941, at the Ukrainian Physicotechnical Institute, from 1944 to 1954, at the Institute of Organic Chemistry of the Academy. Sciences of the USSR; and from 1954 to 1965 at the Institute of Hetero-organic Compounds of the Academy of Sciences of the USSR. Since 1965 he has worked at the Institute of General and Inorganic Chemistry of the Academy of Sciences of the USSR.
His main works deal with crystal physics and molecular spectroscopy. Together with L. V. Shubnikov, he developed a method of growing single crystals from a melt and studied their plastic strain and optical properties. He studied the spectra of organic crystals at low temperatures. He is the author of the monograph *On the Use of Fresnel Diffraction for Physical and Technical Measurements* (1945; State Prize of the USSR, 1946), in which he presented a theory and method he had developed for determining dispersion (the Obreimov method, which has found wide use). Decors.: two orders, and also medals.

Obukhov, Aleksandr Mikhailovich
Geophysicist; b. My. 5, 1918, Saratov.
Grad. Moscow Univ. 1940; mem. staff Inst. of Geophysics at USSR Acad. of Sciences 1940–56; Dir. of Inst. of Physics of the Atmosphere of USSR Acad. of Sciences 1956–; Prof. at Moscow Univ. 1949–; mem. USSR Acad. of Sciences 1970–; main field of work: physics of the atmosphere, turbulence and mathematical statistics. Decors.: orders and medals.

Obukhova, Lidia Alekseevna
Author; b. Ag. 17, 1924, Kutaisi.
Russian; grad. Gorkii Inst. of Literature 1951; first works publ. 1945.
Publs.: "Glubyn-Gorodok" (The Little Town of Glubyn) 1955, "Dobrota" (Goodness) 1962, "Malenkie povesti" (Little Tales) 1963, "Lilit. Fantasticheskaia povest'" (Lilit. A Fantastic Story) 1966.

Obukhovich, A. P.
Trade union official.
Cand. mem. Cen. Cttee. Belorussian CP 1961–; Sec. of Belorussian Trade Union Council 1962–.

Ocheretin, Vadim Kuzmich
Author; b. Je. 6, 1921, Harbin, China.
Russian; mem. CPSU 1943–; grad. Dept. of Philology of Uralsk Univ. 1948 and Advanced Courses in Literature of USSR Writers' Union 1956; lives in Sverdlovsk, worked as porter, draftsman, metalworker in the Urals, contributor to newspaper "Uralskii rabochii"; began his literary career with a factory newspaper.
Publs.: "Ia tvoi, Rodina" (I Am Yours, Motherland) 1950, "Pervoe derzanie" (First Venture) 1953, "Sirena" (The Siren) 1962, and others.

Ochiauri, Iraklii Alekseevich
Sculptor, master of chased metalwork; b. N.20, 1924, Akhiela, Dusheti Raion, Georgian SSR.
Honored Artist of the Georgian SSR (1965).
From 1945 to 1951, he studied in Tbilisi under S. Ia. Kakabadze at the Academy of Arts and in Ia. I. Nikoladze's studio. He revived the ancient art of Georgian chased metalwork, enriching the technique by introducing the freehand scratching of drawings on already chased surfaces and various methods of toning that enhance the color of the metal. His works include the plate *Samaia* (copper, 1962; now in the Picture Gallery of Georgia, Tbilisi), the relief *Love* (brass, 1964; Wedding Palace, Tbilisi), and the panel *The Abduction of the Sun* for the Soviet pavilion at the World's Fair in Osaka (copper, 1970; Ministry of Culture of the USSR, Moscow). Decor.: State Prize of the USSR in 1971.

Odintsov, Arnold Borisovich (real name: Oder)
Author; b. S. 11 (24), 1912, Kharkov.
Russian; mem. CPSU 1943–; grad. Communist Sverdlov Union Univ. 1934; first works publ. 1935.
Publs.: "Iz doma v dom" (From House to House) 1960, "Prikliucheniia v doroge" (Adventure On the Way) 1964, "General Androsov" 1965, "Proryv" (Breakthrough) 1966, and others.

Odintsov, Mikhail Mikhailovich
Geologist; b. O.23 (N.5), 1911, Irkutsk.
Mem. CPSU 1958–; grad. Irkutsk Univ. 1936; mem. staff East Siberian Trust for Geological Prospecting 1931–37; teaching activity at colleges and univ. in Irkutsk 1937–54; Prof. at Irkutsk Univ. 1949–54; Dir. of Inst. for Research on Earth Crust at Siberian Branch of USSR Acad. of Sciences 1954–; corresponding mem. USSR Acad. of Sciences 1964–; had part in discovery of diamond deposits in Siberia. Decors.: 2 Orders of Lenin, medals.

Odintsov, Mikhail Petrovich
Military official; b. N. 18, 1921, Polozovo, Chastye Raion, Perm' Oblast.
Honored Pilot of the USSR (1967); twice Hero of the Soviet Union (F. 4, 1944, and Je. 27, 1945); lieutenant-general of aviation (1968). Member of the CPSU since 1943.
He joined the Soviet Army in 1938 and graduated from the Perm' School for Pilots in 1939, from the V. I. Lenin Military Political Academy in 1952, and from the Military Academy of the General Staff in 1959. In the Great Patriotic War (1941–45), he served on the Southwestern, Briansk, Kalinin, Voronezh, Steppe, and Second and First Ukrainian fronts as a flight commander, deputy commander and commander of a squadron, and deputy commander and navigator of the

155th Guards Assault Aviation Regiment. He made 215 combat flights and personally shot down two enemy aircraft. After the war, he held responsible positions in the air force. He became the commander of aviation of a military district in November 1967. Decors.: two Orders of Lenin, the Order of the October Revolution, four Orders fo the Red Banner, the Order of Alexander Nevsky, the Order of the Patriotic War First and Second Class, the Order of the Red Star, and various medals.

Odintsov, V.E.
Party official.
1st Oblast Sec. of CPSU in North Ossetian ASSR Ja. 1982–.

Oganesian, N.S.
Komsomol official.
Sec. and mem. Bureau Cen. Cttee. Armenian Komsomol Ap. 1982–.

Oganesian, Suren Stepanovich
State official.
Mem. Auditing Comm. 1964–66, mem. Cen. Cttee. Armenian CP 1976–; Chair. of State Cttee. for the Supply of Production Equipment for Agriculture Armenian SSR F. 1979–.

Oganian, Suren Khorenovich
State official.
Mem. Auditing Comm. 1964–66 and 1981–, cand. mem. Cen. Cttee. Armenian CP 1966–71, mem. 1971–76; Min. of the Timber and Wood-Processing Ind. Armenian SSR Ja. 1980–.

Ogarkov, Nikolai Vasil'evich
Military official; b. O. 17 (30), 1917, Molkov, Kalinin Oblast; s. of a peasant.
Russian; mem. CPSU 1945–; grad. Energy Work Faculty 1937, V.V. Kuibyshev Mil. Engineering Acad. 1941, the operations engineering faculty of this acad. 1947, and Mil. Acad. of Gen. Staff 1959, Marshal of the USSR 1977; accountant, sec. of a trade union raion cttee. 1937–38; entered Sov. Army 1938; served in an engineer regiment on Western front, senior engineer for the fortification of an engineer admin. on Karelian front, regiment engineer in 289th Rifle Division, brigade engineer in 61st Marine Rifle Brigade, asst. chief of staff of engineers of 32nd Army, asst. chief of operations dept. of engineer troop forces, staff at the front, division engineer of 122nd Rifle Division at Karelian front and at 2nd and 3rd Ukrainian fronts 1941–45; held important posts on engineer troops forces staff of Carpathian Mil. District and in engineer admin. of Primore Mil. District 1945–48; chief of a dept. of Operations Admin. of Staff of Commander-in-Chief of Far Eastern Troops 1948–53; dep. chief of admin. of Staff of Far Eastern Mil. District 1953–55, chief, Dep. Chief of Staff of Far Eastern Mil. District 1955–59; Commander of Motorized Rifle Division of Group of Soviet Troops in Germany 1959–61; Chief of Staff, Dep. Commander of Troops of Belorussian Mil. District 1961–63, 1st Dep. Commander 1963–65; Commander of Troops of Volga Mil. District 1965–68; cand. mem. Cen. Cttee. CPSU 1966–71, mem. 1971–; 1st Dep. Chief of Gen. Staff of USSR Armed Forces 1968–77, Chief 1977–; Dep. to USSR Supr. Sov. 1966–; Dep. USSR Min. of Defense 1974–77, 1st Dep. Min. 1977–. Decors.: Order of Lenin, Order of Red Banner of Labor, Order of Great Fatherland War 1st and 2nd Class, Order of Red Star (twice), Order "In Service to the Homeland in the Armed Forces of the USSR" 3rd Class, medals, and foreign orders and medals.

Ognetsvet, Edi (real name: Kagan, Edi Semenovna)
Author, poet and translator; b. S.28 (O.11), 1916, Minsk.
Belorussian; mem. CPSU 1940–; grad. Dept. of Literature of Minsk Pedagogical Inst. 1936; first works publ. 1931.
Publs.: "Moe pokolenie" (My Generation) 1935; "Na beregu Volgi" (At the Banks of the Volga) 1940, "Tvoi tovarishchi" (Your Friends) 1957, "Lirika" (Poetry) 1965; translated into Belorussian works by L. Ukrainka, M. Bazhan, A. Barto, Ia. Reinis, Ia. Sudrabkalns and others.

Ognev, Vladimir Fedorovich
Literary historian and critic; b. Jl. 7, 1923, Poltava.
Grad. Gorkii Inst. of Literature 1950; first works publ. 1949; his works deal mainly with theory of poetry, especially contemporary poetry.
Publs.: "Poeziia i sovremennost'" (Poetry and the Present) 1961, "Kniga pro stikhi" (A Book on Poetry) 1963, "Rasul Gamzatov" 1964, and others.

Oistrakh, Igor Davidovich
Violinist; b. Ap. 27, 1931, Odessa; s. of violinist David Oistrakh.
Attended Music School, later Moscow State Conservatory 1949–55; internationally famous violinist; many foreign tours, several concerts with his father, including opening conert of 16th Festival of Theatre and Music East Berlin with David Oistrakh as conductor and Igor Oistrakh as soloist O. 1972. Decors.: 1st Prize at Budapest Violin Competition 1952, 1st Prize at Wieniawski Competition Poznań, Poland, Merited Artist of RSFSR.

Oizerman, Teodor Il'ich
Philosopher; b. My. 1 (14), 1914, Petroverovka, Odessa Oblast.
Mem. CPSU 1941–; grad. Dept. of Philosophy, Literature and History of Moscow Univ. 1938; served in World War II; Prof. of History of non-Russian Philosophy at Dept. of Philosophy of Moscow Univ. 1954–68; corresponding mem. USSR Acad. of Sciences 1966–81, mem. 1981–; Head of Dept. of History of West-European and North-American Philosophy at Inst. of Philosophy of USSR Acad. of Sciences 1971–. Decors.: 4 orders and various medals.
Publs.: mem. ed. board and co-author "Istoriia filosofii" (History of Philosophy) vols. 1–6, 1957–69.

Okhmakevich, Nikolai Fedorovich
State official; b. 1937.
Mem. CPSU 1966–; grad. Kiev State Univ. 1961, journalist; dept. chief, dep. program dir. of Ukrainian television of State Cttee. for Television and Radio Broadcasting 1961–72; instructor in Culture Dept. of Cen. Cttee. Ukrainian CP, Chief of Sector for Radio and Television in Propaganda and Agitation Dept. of Cen. Cttee. Ukrainian CP 1972–78; 1st Dep. Chair. of State Cttee. for Television and Radio Broadcasting Ukrainian SSR 1978–79, Chair. D. 1979–; mem. Auditing Comm. Ukrainian CP 1981–.

Okhotsimskii, Dmitrii Evgen'evich
Scientist in mechanics; b. F. 26, 1921, Moscow.
Corresponding member of the Academy of Sciences of the USSR (1960). Hero of Socialist Labor (1961). Member of the CPSU since 1951.
Between 1945 and 1966, he worked at the Institute of Mathematics of the Academy of Sciences of the USSR; since 1966 he has been at the Institute of Applied Mathematics of the Academy of Sciences of the USSR. He has been a professor in the department of mechanics and mathematics of Moscow State University since 1961. His chief works concern flight dynamics, the control of aircraft flight, and the development of algorithms for the behavior of complex cybernetic locomotive and manipulative systems with elements of artificial intelligence where the systems adapt to an unknown situation. Decors.: Lenin Prize in 1957 and the State Prize of the USSR in 1970, two Orders of Lenin, the Order of the Red Banner of Labour, and various medals.
Works: "Mekhanika kosmicheskogo poleta." In *Mekhanika v SSSR za 50 let*, vol. 1. Moscow 1968. (Co-author.)

Okhromii, D.A.
Komsomol official.
Mem. Bureau Cen. Cttee. Ukrainian Komsomol 1976–78; Sec. and mem. Bureau Cen. Cttee. All-Union Komsomol Ap. 1978–.

Okladnikov, Aleksei Pavlovich
Archaeologist, historian, and ethnologist; b. S. 20 (O. 3), 1908, Konstantinovshchina, present-day Zhigalovo Raion, Irkutsk Oblast.
Academician of the Academy of Sciences of the USSR (1968; corresponding member, 1964). Honored Scientist of the Yakut ASSR (1956), the RSFSR (1957), and the Buriat ASSR (1968). Member of the CPSU since 1946.

From 1938 to 1961, he worked in the Leningrad Division of the Institute of Archaeology of the Academy of Sciences of the USSR (AN SSR). In 1961 he became head of the department of humanities research of the Institute of Economics of the Siberian Division of the AN SSR. In 1966, Okladnikov was appointed director of the Institute of History, Philology, and Philosophy of the Siberian Division. In 1962 he was made a professor and head of the subdepartment of history at the University of Novosibirsk.
He did field work in Siberia, the Far East, Middle Asia, and the Mongolian People's Republic. He discovered and studied the remains of Neanderthal man and his culture in the Teshik-Tash grotto in Uzbekistan; the Paleolithic culture in the Mongolian People's Republic; and primitive petroglyphs along the Lena and Angara rivers, in the Amur region, and in Mongolia. He discovered many local Paleolithic and Neolithic cultures, as well as Bronze and Iron Age cultures in Siberia and the Far East. In 1945, he investigated the remains of a 17th-century expedition of Russian polar seafares on Fadd-Island (off the northeastern coast of the Taimyr Peninsula).
He is the author of general studies on the history of primitive society and culture, Paleolithic and Neolithic art, and the history of Siberia, the Far East, and the Far North from ancient times to the 18th century. Decors.: State Prize of the USSR in 1950 and 1973, Order of Lenin, three other orders, and a number of medals.
Works: *Ocherki iz istorii zapadnykh buriat-mongolov: XVII–XVIII vv.* Leningrad 1937. *Neolit i bronzovyi vek Pribaikal'ia,* parts 1–3. Moscow-Leningrad 1950–55. *Iakutiia do prisoedineniia k Russkomu gosudarstvu* [2nd ed.]. Moscow-Leningrad 1955. *Russkie poliarnye morekhody XVII v. u beregov Taimyra,* 2nd ed. Moscow-Leningrad 1957. *Dalekoe proshloe Primor'ia.* Vladivostok 1959. *Drevnee poselenie na poluostrove Peschanom u Vladivosoka.* Moscow-Leningrad 1963. *Olen' Zolotya Roga.* Moscow-Leningrad 1964. *Petroglify Angary.* Moscow-Leningrad 1966. *Utro iskusstva.* Moscow-Leningrad 1967. *Petroglify Nizhnego Amura.* Leningrad 1971. *Tsentral'noaziatskii ochag pervobytnogo iskusstva.* Novosibirsk 1972.

Okudzhava, Bulat Shalvovich
Poet; b. My. 9, 1924, Moscow; s. of a party official.
Mem. CPSU 1955–; grad. Tbilisi Univ. 1950; served at the front 1942–45; worked as teacher after the war; first works publ. 1953.
Publs.: "Lirika" (Poetry) 1956, "Ostrova" (Islands) 1959, "Veselyi barabanchik" (The Merry Drummer) 1964, "Front prikhodit k nam" (The Front Comes to US) 1965, "Bednyi Avrosimov" (Poor Avrosimov) 1969, and others.

Okun', Lev Borisovich
Theoretical physicist; b. Jl. 7, 1929, Sukhinichi, Kaluga Oblast.
Corresponding member of the Academy of Sciences of the USSR (1966).
He graduated from the Moscow Physics and Engineering Institute in 1953. Since 1954 he has worked at the Institute of Experimental and Theoretical Physics. His main work has been on the theory of elementary particles (theory of weak interactions, composite models of elementary particles, and so on).
Works: *Slaboe vzaimodeistvie elementarnykh chastits.* Moscow 1963. "Nekotorye zamechaniia o medlennykh protsessakh prevrashchenii elementarnykh chastits." *Zhurnal eksperimental'noi i teoreticheskoi fiziki,* 1957, vol. 32, fasc. 6, p. 1587. (With B. M. Pontekorvo.) "Nekotorye zamechaniia o sostavnoi modeli elementarnykh charsts." *Ibid.,* 1958, vol. 34, fasc. 2, p. 469. "Unitarnaia simmetriia i universal'noe slaboe vzaimodeistvie." *Ibid.,* 1962, vol. 42, fasc. 5, p. 1400. (With I. Iu. Kobzarev.)

Okunev, Vasilii Vasil'evich
Military official; b. 1920.
Russian; mem. CPSU 1941–; grad. Dzerzhinski Artillery Acad. 1950 and Gen. Staff Acad. 1961; Col.-Gen.; in Sov. Army 1936–; commander of an anti-aircraft tank squadron, commander of an anti-aircraft div. 1942–44; commander of an artillery regt., a div., dept. head of staff of an anti-aircraft defence district 1944–55; dep. commander 1955–57, commander of anti-aircraft units of various anti-aircraft defence districts 1957–59; commander of various anti-aircraft defence units 1961–66; Commander of Moscow Anti-Aircraft Defence District 1966–70; mem. Cen. Auditing Comm. CPSU 1966–71; Dep. to USSR Supr. Sov. 1970–74; sen. posts in USSR Min. of Defence 1970–74; cand. mem. Cen. Cttee. CPSU 1971–76; 1st Dep. Commander-in-Chief of USSR Anti-Aircraft Defence Troops (PVO) 1974–. Decor.: Order of Red Banner.

Oleinik, Boris Il'ich
Author; b. O. 22, 1935, Zachepilovka, Poltava Oblast.
Ukrainian; mem. CPSU 1961–; grad. Dept. of Journalism at Kiev State Univ. 1958; first works publ. 1950. Decor.: Ostrovskii State Prize of the Ukrainian SSR 1964.
Publs.: "Stal' zakaliaiut kuznetsy" (Steel is Tempered by Smiths) 1962, "Ekho" (Echo) 1970, "Na linii tishiny" (On the Line of Stillness) 1972, "Dvizhenie" (Movement) 1973, and others. Decor.: Ostrovskii State Prize of the USSR 1964.

Oleksich, Vitalii Nikolaevich
State official.
Min. of Melioration and Water Econ. Moldavian SSR Mr. 1977–. Mem. Cen. Cttee. Moldavian CP 1981–.

Oleneko, Iurii Aleksandrovich
State official; b. 1934.
Mem. CPSU 1971–; grad. I. K. Karpenko Theater Inst. in Kiev 1964; electrical welder in a plant in Kiev, served in Sov. Army to 1960; student 1960–64; actor at Ukrainian Theater for Drama in Kharkov, teacher, Dean of Faculty of Cinematography at I. K. Karpenko Theater Inst. in Kiev, Chief of Theater Dept., Dep. Chief of Main Admin. for Theater and Music of Min. of Culture Ukrainian SSR 1964–76; chief of a sector of Culture Dept. of Cen. Cttee. Ukrainian CP 1976–77, Dep. Chief of Culture Dept. 1977–79; Chair. of State Cttee. for Cinematography Ukrainian SSR D. 1979–; Dep. Chair. Auditing Comm. Ukrainian CP 1981–.

Olshanskii, Iosif Grigor'evich
Author and film script writer; b. D. 22, 1917 (Ja. 4, 1918), Kremenchug, Poltava Oblast.
Russian; grad. Inst. of Philosophy, Literature and History 1941.
Publs.: film scripts "Dom v kotorom ia zhivu" (The House in Which I Live) 1957, "Troie vyshli iz lesa" (Three Came From the Wood) 1958, "Pervyi den' mira" (The First Day of Peace) 1959, "A esli eto liubov'?" (And What if it is Love?) 1961, "Ne zabud ... Stantsiia Lugovaia" (Don't Forget – Lugovaia Station) 1965, and others.

Omarov, Seitzhan
Author; b. S. 19, 1907, Keregetas, Tselinograd Oblast.
Kazakh; mem. CPSU 1939–; stud. Kustanai Pedagogic Inst. 1931–32; first works publ. 1932.
Publs.: "Sila vremeni" (The Power of Time) 1949, "Balziia" 1960, "Zolotaia dolina" (The Golden Valley) 1963, and others.

Omel'ianovskii, Mikhail Erazmovich
Philosopher; b. Ja. 19 (F. 1), 1904, Kiev.
Corresponding member of the Academy of Sciences of the USSR (1968) and academician of the Academy of Sciences of the Ukrainian SSR (1948). Member of the CPSU since 1938.
In 1931, he graduated from the Institute of the Red Professors, where he studied philosophy and natural science. From 1931 to 1944 he headed the subdepartment of dialectical materialism (Marxism-Leninism) of the Voronezh Institute of Chemical Engineering. He was senior scientific worker at the Institute of Philosophy of the Academy of Sciences of the USSR (1944–46) and director (1946–52) and head (1952–55) of the department of the philosophy of natural science of the Institute of Philosophy of the Academy of Sciences of the Ukrainian SSR. From 1955 to 1965, he served as deputy director of the department of philosophical problems in natural science of the Institute of Philosophy of

the Academy of Sciences of the USSR; he has been head of the department since 1965. In 1959 he became vice-chairman of the Scientific Council on Philosophical Problems in Contemporary Natural Science under the Presidium of the Academy of Sciences of the USSR.
He studies problems of dialectical materials and philosophical problems in natural science. His research deals primarily with philosophical problems in modern physics. He is one of the authors and the editor of a number of books in the series *Dialectical Materialism and Contemporary Natural Science*. Decors.: Order of Lenin, two other orders, and a number of medals.
Works: *V.I. Lenin i fizika XX veka*. Moscow 1947. "Dialekticheskii materializm-metodologicheskaia osnova sovremennoi fiziki." In *Dialektika i sovremennoe estestvoznanie*. Moscow 1970. "Kommentarii k st. N. Bora." In N. Bohr, *Izbrannye nauchnye trudy*, vol. 2. Moscow 1971. *Dialektika v sovremennoi fizike*. Moscow 1973. "The Problem of Elementary Particles in Quantum Physics." *In Philosophical Problems of Elementary Particle Physics*. Moscow 1968.

Omerov, Server Abdullaevich
State official.
Mem. Uzbek Republican Council of Kolkhozes 1969–; mem. Auditing Comm. 1971–76, mem. Cen. Cttee. Uzbek CP 1976–; Min. of Construction Uzbek SSR Ja. 1974–.

Omurova, A.
Trade union official.
Sec. and mem. Bureau Cen. Cttee. Kirghiz Komsomol 1966–70; Sec. and mem. Presidium Kirghiz Republican Trade Union Council F. 1982–.

Onaitis, Kostas Klemensovich
State official.
Min. of Posts and Telecommunications Lithuanian SSR D. 1968–; mem. Auditing Comm. Lithuanian CP 1971–76; cand. mem. Cen. Cttee. Lithuanian CP 1976–.

Orazmamedov, Saparmamed
State official.
Min. of Communal Econ. Turkmen SSR Jl. 1973–; mem. Auditing Comm. Turkmen CP 1976–.

Orazmukhamedov, Nury
State official.
Cand. mem. Cen. Cttee. Turkmen CP 1966–80, mem. N. 1980–; Chair. of State Cttee. for Construction Turkmen SSR Jl. 1966–80; Dep. Chair. of Council of Mins. Turkmen SSR N. 1980–; Dep. to Supr. Sov. Turkmen SSR 1967–.

Ordzhonikidze, I. N.
Komsomol official.
Mem. Bureau Cen. Cttee. Georgian Komsomol Ag. 1976–, Sec. 1976–78, 2nd Sec. 1978–80, 1st Sec. D. 1980–; mem. Cen. Cttee. Georgian CP 1981–; cand. mem. Bureau Cen. Cttee. Georgian CP Ja. 1981–.

Ordzhonikidze, Shota Markianovich
State official.
Chair. of State Cttee. for Specialized Transport and Systems of its Automation Georgian SSR Ag. 1981–.

Orel, Aleksandr Evstaf'evich
Military official; b. Ag. 12 (25), 1908, St. Petersburg.
Admiral (1964), professor (1970). Member of the CPSU since 1937.
He entered the navy in 1929. He graduated from the M. V. Frunze Naval College in 1932 and from the K. E. Voroshilov Military Academy in 1953. He served on submarines of the Black Sea, Pacific, and Baltic fleets. In the Great Patriotic War (1941–45), he held high positions on the staff of the Red Banner Baltic Fleet until 1943 and then commanded large submarine units of the Baltic Fleet. He held high posts in the Main Headquarters of the Soviet Navy. He was commander of the submarine forces of the Northern Fleet from 1956 to 1959. He was commander of the Baltic Fleet from May 1959 to January 1967 and became chief of the Naval Academy in January 1967. He was a deputy to the sixth and seventh convocations of the Supreme Soviet of the USSR. Decors.: two Orders of Lenin, the Order of the October Revolution, four Orders of the Red Banner, the Order of Ushakov Second Class, the Order of the Patriotic War First Class, the Order of the Red Star, various foreign orders, and a number of medals.

Oreshnikov, Viktor Mikhailovich
Painter; b. Ja. 7 (20), 1904, Perm.
Studied Higher State Inst. for Arts and Crafts (VHUTEIN); postgraduate studies at Acad. of the Arts 1933–36; teacher 1930–, Rector of I. E. Repin Inst. of Painting, Sculpture and Architecture of Acad. of the Arts 1953–. Decors.: State Prize of the USSR 1948 and 1950, 2 orders and various medals.

Oreshnonkova, Anna Vladimirovna, Cand. of Phil. Sc.
Philologist; b. 1928.
Belorussian; grad. Belorussian Univ. 1953; Cand. of Phil. Sc. 1959–; mem. staff Inst. of Linguistics at Acad. of Sciences Belorussian SSR 1956–63 and 1967–; Sen. Mem. Scientific Staff 1969–; Lecturer at Pedagogic Inst. Minsk 1963–67. Decor.: State Prize of the USSR (for her contributions on Belorussian linguistic geography) 1971.

Orlik, Mariia Andreevna
State official; b. Mr. 15, 1930.
Mem. CPSU 1955–; grad. pedagogical inst. 1953; teacher, chief of a teaching dept. of a secondary school 1953–71; Dep. Chair. of Exec. Cttee. Kiev City Cttee. of Peoples' Deps. 1971–75; Chair. of Ukrainian Society for Friendship and Cultural Relations with Foreign Countries 1975–78; cand. mem. Cen. Cttee. Ukrainian CP 1976–81, mem. 1981–; Dep. Chair. of Council of Mins. Ukrainian SSR Ap. 1978–. Decor.: Order of Red Banner of Labour.

Orlov, Aleksandr Kirillovich
Legal official.
1st Dep. Chair. 1962–72, Chair. of RSFSR Supr. Court D. 1972–.

Orlov, Georgii Mikhailovich
State official.
Mem. Bureau for Ind. at Cen. Cttee. Uzbek CP 1963–64; Head of Dept. for Party Organs for Ind. at Bureau Cen. Cttee. Uzbek CP 1963–64; Head of Dept. for Party Org. Work at Cen. Cttee. Uzbek CP 1966–78; mem. Cen. Cttee. Uzbek CP 1966–; cand. mem. Bureau Cen. Cttee. Uzbek CP D. 1971–78; Dep. Chair. Presidium Supr. Soviet Uzbek SSR D. 1978–.

Orlov, Iurii Fedorovich
Physicist; b. 1924.
Corresponding mem. Acad. of Sciences Armenian SSR; mem. USSR branch of Amnesty International, Head of Org. for Supervising Observance of Human Rights in the USSR according to final declaration of Security Conference.

Orlov, Ivan Vasil'evich
Helminthologist; b. Je. 5 (17), 1899, Tambov.
Grad. Kazan Veterinary Inst. 1925; mem. staff Skriabin Union Inst. of Helminthology 1929–34; Prof. Chair. of Parasitology at Voronezh Zoo Tech. Inst. 1934–; Prof. Chair of Parasitology at Technological Inst. of the Meat and Dairy Ind. in Moscow 1945–72; Vice-Pres. of Int. Cttee. on Trichinosis 1969–72. Decors.: 2 orders and various medals.

Orlov, Mikhail Georgievich
Diplomat.
USSR Amb. to Seychelles Ag. 1981–.

Orlov, Pavel Pavlovich
Graphic artist; b. 1924, Kiev.
Mem. CPSU 1952–; attended Lvov Inst. for Applied and Decorative Art 1947–53. Decor.: Merited Artist of the Ukrainian SSR 1970.

Orlov, Vladimir Ivanovich, Cand. of Tech. Sc.
Author; b. O. 31, 1916, Poltava.
Russian; grad. Moscow Power Engineering Inst. 1938; active as writer and journalist; has publ. popular-science features in "Pravda" and other newspapers since 1942. Decor.: Lenin Prize 1960.
Publs.: "Russkoe solntse" (Russian Sun) 1950, co-author of "Litsom k litsu s Amerikoi ..." (Face to Face With America) 1960, "Sto variatsii na temu staroi skazki"

(One Hundred Variations on the Theme of an Old Fairy Tale) 1965, and others.

Orlov, Vladimir Nikolaevich
Literary scholar; b. Je. 22 (Jl. 5), 1908, Valdai, present-day Novgorod Oblast.
He graduated from the division of literature of the Leningrad Institute for the History of Arts (1929). From 1941 to 1945 he was a war correspondent for the Telegraph Agency of the Soviet Union (TASS). From 1956 to 1970 he was editor-in-chief of the series *Biblioteka poeta* (Poet's Library).
He began to publish in 1927. His works deal mainly with the revolutionary and democratic trends in Russian literature in the late 18th and early 19th centuries (*Russian Enlightenment Figures from 1790 to the 1800's*, 1950; State Prize of the USSR, 1951; 2nd ed., 1953) and with such writers as A. N. Radishchev and A. S. Griboedov. He is also concerned with the history and theory of late 19th- and early 20th-century Russian literature. His works on Blok, including *Aleksandr Blok: Essay on His Work* (1956), are of great importance.
He was the editor of *The Diary of W. K. Küchelbecker* (1929) and the anthologies *The Radishchev Poets* (1935, 1952, 1961) and *The Decembrists: Poetry, Drama, Prose, Publicistic Writings, Literary Criticism* (1951). He has also written monographs on M. Tsvetaeva (published in *M. Tsvetaeva: Selected Works*, 1965).
Works: *Puti i sud'by: Literaturnye ocherki.* Leningrad, 1971.

Orlov, Vladimir Pavlovich
State official; b. 1921.
Russian; mem. CPSU 1948–; grad. Ivanovo Textile Inst. 1942; foreman, dept. head at a plant 1943–45; sec. of Komsomol cttee. at a plant, Sec. of Chapaevsk City Komsomol and CPSU Cttees., Chair. of Kuibyshev City Exec. Cttee. of Workers' Deps., Sec. of Kuibyshev City and Oblast CPSU Cttee., Chair. of Kuibyshev Oblast Exec. Cttee. of Workers' Deps. for Ind. 1945–64; Sec. of Kuibyshev Oblast CPSU Cttee. 1964–65; Chair. of Kuibyshev Oblast Exec. Cttee. of Workers' Deps. 1965–67; Dep. to USSR Supr. Sov. 1966–; 1st Sec. of Kuibyshev Oblast CPSU Cttee. 1967–79; 1st Dep. Chair. of RSFSR Council of Mins. 1979–; mem. Cen. Cttee. CPSU 1971–. Decors.: 2 Orders of Lenin.

Orlova, M. L.
Trade union official.
Sec. of Moldavian Republican Trade Union Council Ja. 1982–.

Ormanov, Gali
Kazakh poet; b. Ag. 2 (15), 1907, in present-day Kapal Raion, Taldy-Kurgan Oblast.
Member of the CPSU since 1938.
He graduated from the Kazakh Institute of Public Education in 1929. From 1940 to 1943 he studied at the Kazakh Pedagogical Institute. He worked on the editorial staffs of newspapers and journals of the Kazakh republic. From 1939 to 1946, he was Dzhambul Dzhabaev's secretary. His first works appeared in 1927. His first poetry collection, *The Hoop*, was published in 1934. He is the author of the poetry collections *A Woman's Confession* (1936), *The Oak* (1945), *The Song of Life* (1948), *The Wave* (1956), *The Wings of Thought* (1959), and *The Flaming River* (1965). He also published the short-story collections *Ascent* (1935) and *The Heart* (1963). His works deal principally with contemporary life.
He has translated works by A. S. Pushkin, M. Iu. Lermontov, L. N. Tolstoy, A. P. Chekhov, Nizami Ganjevi, J. Rainis, and S. A. Esenin into Kazakh. He was secretary of the board of the Writers' Union of the Kazakh SSR from 1959 to 1963. Decors.: Order of Lenin, two other orders and a number of medals.
Works: *Sät sparï.* Alma-Ata 1957. *Tangdamalïlar.* Alma-Ata 1961. *Tangdamalï ölengder.* Alma-Ata 1967. *Zher qïzïghï.* Alma-Ata 1972. *Zhan qanatï.* Alma-Ata 1973. *In Russian translation: Glazami mysli.* Alma-Ata 1958. *Golubye dali.* Moscow 1970.

Orovetskii, Pavel Andreevich
Author; b. Mr. 4 (17), 1905, Elizavetovka, Dnepropetrovsk Oblast; s. of a peasant.
Ukrainian; mem. CPSU 1931–; worker, then official for state security organs; served in Sov. Army 1941–45.
Publs.: "Grigorii Mogilchenko" 1951, "V bolshoi sem'e" (In a Large Family) 1952, "Dolina chernogo zolota" (In the Valley of Black Gold) 1960, "Serdtse soldata" (The Heart of a Soldier) 1958, "Vtoraia vstrecha" (The Second Encounter), and others.

Orozaliev, Kydyrma Abyshevich
State official.
Min. of Motor Transport and Highways Kirghiz SSR D. 1979–; cand. mem. Cen. Cttee. Kirghiz CP 1981–.

Orudzhev, Telman Khalil ogly
State official.
Min. of Supplies Azerbaidzhan SSR 1980–81; Min. of Fruits and Vegetables Azerbaidzhan SSR Mr. 1981–.

Osetrov, Evgenii Ivanovich
Author and literary critic; b. Jl. 17, 1923, Kostroma. Russian; mem. CPSU 1946–; grad. Gorkii Inst. of Literature 1953.
Publs.: "Shkatulka Ariny Rodionovny" (The Casket of Arina Rodionovna) 1954, "Vetka Laury" (Laura's Twig) 1960, "Poiski, nakhodki, tainy" (Searches, Finds, Secrets) 1964, "Puteshestvie v stranu chudes" (A Journey to Wonderland) 1964, "Poeziia nashikh dnei" (Poetry of Our Time) 1965, and others.

Osetrov, Timofei Nikolaevich
State official; b. 1920.
Russian; mem. CPSU 1947–; grad. Moscow. Tech. Aviation Inst. 1948; worker in a bread factory 1939–41; technologist in a mechanical plant 1941–47; Komsomol posts 1947–49; Sec., 2nd Sec., 1st Sec. of Cheliabinsk Oblast Cttee. USSR Komsomol 1949–54; mem. staff Cen. Cttee. CPSU 1954–70; 1st Dep. Chair. of Council of Mins. Uzbek SSR Mr. 1970–; Dep. to Supr. Sov. Uzbek SSR 1970–; mem. Cen. Cttee. and mem. Bureau Cen. Cttee. Uzbek CP 1971–; mem. Cen. Auditing Comm. CPSU 1971–.

Oshanin, Lev Ivanovich
Author and lyricist; b. My. 17 (30), 1912, Rybinsk, Iaroslavl Oblast.
Russian; mem. CPSU 1944–; studied Gorkii Inst. of Literature 1936–39. Decor.: State Prize of the USSR 1950.
Publs.: "Etazhi" (Floors) 1930, "Moi drug Boris" (My Friend Boris) 1944, "Deti raznykh narodov" (Children of Various Nations) 1950, "Tebe, moi drug" (For You, My Friend) 1954, "Sto pesen" (One Hundred Songs) 1966, "Prosto ia rabotaiu volshebnikom" (I Simply Work as Magician) 1966, author of numerous lyrics.

Osherovich, Girsh Sholomovich
Author and poet; b. Je. 5 (18), 1908, Lithuania.
Jew; grad. Dept. of Law of Kaunas Univ. 1933; first works publ. 1934.
Publs.: "Bud prokliat, Mengele" (Be Damned, Mengele) 1965, "Preduprezhdenie" (Preliminary Warning) 1961, "Spartak" 1964, and others.

Oshurkov, Mikhail Fedorovich
Cinematographer; b. 1906.
Mem. CPSU 1958–; grad. Union Inst. of Cinematography 1928; active as cameraman; lecturer 1939–; mem. staff Cen. Studio for Documentaries 1926–. Decors.: State Prize of the USSR 1942, 1948, 1949, 1950, 1969.

Osin, Dmitrii Dmitrievich
Author; b. Ag. 24 (S. 6), 1906, Smolensk.
Russian; started literary career 1925; war correspondent for newspaper "Gudok" 1941–45; first works publ. 1933.
Publs.: "Sorevnovanie" (Competition) 1933, "Chernye gody" (Black Years) 1938, "Katia-Katerina" 1941, "U menia na rodine" (In My Native Country) 1946, "Samaia schastlivaia" (The Happiest One) 1966, and others; also translates from the Belorussian and Ukrainian.

Osipov, Afanasii Nikolaevich
Painter; b. F. 28, 1928, Second Irgid *nasleg* (a small administrative unit), now part of Gornyi Raion, Yakut ASSR.
Honored Artist of the RSFSR (1968). Corresponding member of the Academy of Arts of the USSR (1973).
He studied at the V. I. Surikov Moscow

Art Institute from 1949 to 1955. His paintings are devoted to the contemporary life of the peoples of Yakutia. Although they are primarily genre paintings, there is a strong emphasis on elements of portraiture. They tend toward a decorative, two-dimensional style. Examples include *The Kolkhoz Governing Board* (1967), the triptych *The Gray Viliui* (1969), and *Building a House in the Nasleg Mytakh* (1972), all housed at the Board of Art Exhibitions of the Art Fund of the RSFSR, and *Honored Coach of the USSR D. Korkin With his Students at the Churapcha School* (1971), housed at the Board of Art Exhibitions of the Artists' Union of the USSR. Decor.: Order of the Red Banner of Labor.

Osipov, Petr Nikolaevich
Chuvash writer; b. Ja. 15 (27), 1900, Kudemery, present-day Kozlovka Raion, Chuvash ASSR.
One of the founders of Chuvash drama. Honored Art Worker of the Chuvash ASSR (1940). Honored Physician of the RSFSR (1945). Member of the CPSU since 1946.
He graduated from the department of medicine of the University of Kazan in 1924. The life and struggle of the Chuvash people and their efforts to build a new life during the first years of Soviet power are described in the plays *Kuzhar* (staged 1920), *Aidar* (staged 1924; Russian translation, 1964), *Weekday Symphony* (staged 1935; Russian translation, 1937), and *Cherished Dream* (staged 1960). For many years he worked in the K. V. Ivanov Chuvash Theater in Cheboksary as an actor, director, and translator of Russian plays and classic foreign plays.
One of his best works is the novel *The El'geevs* (1973), which is about the October Revolution of 1917 and the events of the first year of Soviet power in the land of the Chuvash. Decors.: 2 orders and a number of medals.
Works: *P'esasem,* vols. 1–2. Cheboksary 1957–60.

Osipov, Vasilii Nikolaevich
Military official; b. D. 17 (30), 1917, Petrograd.
Pilot; major (1944); twice Hero of the Soviet Union (Je. 20, 1942, ad Mar. 13, 1944). Member of the CPSU since 1942. He entered the Soviet Army in 1937 and graduated from the Chkalov Military Aviation School in 1940. In the Great Patriotic War (1941–45) he served from June 1941 as a pilot, flight commander, and deputy commander and commander of a strategic-aviation squadron of the Fifth Guards Bombardment Regiment on the Southern, Southwestern, and other fronts. He made about 400 combat flights. After the war, he continued serving in the air force and graduated in 1949 from the Higher Tactical Flight School for Officers. He entered the reserves in May 1954 for reasons of health. Decors.: Order of Lenin, two Orders of the Red Banner, the Order of Alexander Nevsky, and various medals.

Ossovskii, Petr Pavlovich
Painter; b. My. 18, 1925, Malaia Viska, Kirovograd Oblast, Ukrainian SSR. People's Artist of the RSFSR (1975).
He studied under S. V. Gerasimov at the V. I. Surikov Moscow Art Institute from 1944 to 1950. In 1971 he became first secretary of the board of the Artists' Union of the RSFSR. His works, primarily landscapes and genre paintings, are marked by dynamic composition, severity of color resolution, and a romanticized view of the ordinary phenomena of reality. Examples of his work include *On the City Outskirts* (1957, M. S. Tuganov Severnaia Osetiia Republic Art Museum, Ordzhonikidze), a series of paintings *Cuba* (1961–63), a series *Mexico* (1963–64) and *The Fisherman of Lake Pskov* (1972).
Decors.: Order of the Badge of Honour and a medal.

Ostrianin, Daniil Fomich, Dr. of Phil. Sc.
Philosopher; b. D. 30, 1906, Belsk, now Poltava Oblast; s. of a peasant.
Mem. CPSU 1927–; grad. Inst. of Communist Educ. 1929; corresponding mem. Acad. of Sciences Ukrainian SSR 1957–; Dr. of Phil. Sc. 1953–, Prof. 1954–; Head of Chair of Social Sciences at Kharkov Evening Inst. of Machine Construction 1931–33; Head of Chair of Dialectical Materialism at Agricultural Inst. 1933–35 and at 1st Medical Inst. in Kharkov 1935–45; Head of Chair of Dialectical Materialism at Kharkov Univ. 1945–52; Dir. of Inst. of Philosophy at Acad. of Sciences Ukrainian SSR 1952–62; Head of Chair of Philosophy at Dept. of Humanities of Kiev Shevchenko Univ. 1963–; author of some 169 works (primarily on the history of philosophy and philosophical aspects of natural sciences). Decors.: Order of Red Banner of Labour, Order "Badge of Honour", medals.
Publs.: "Mirovozrenie I. I. Mechnikova" 1948, "I. I. Mechnikov u borotbi za materializm u pryrodoznavstvi" 1954, "V. V. Vorovskii – vydatnyi propagandyst marksystsko-leninskoi filosofii" 1963, and others.

Ostrovoi, Sergei Grigor'evich
Author and poet; b. Ag. 24 (S. 6), 1911, Novosibirsk.
Russian; mem. CPSU 1949–; served in Sov. Army 1941–45; first works publ. 1934.
Publs.: "Ia v Rossii rozheden" (I Was Born in Russia) 1956, "Segodnia ia dumal o vas" (Today I Thought of You) 1964, "Mat" (The Mother) 1963, and others; wrote numerous lyrics for songs.

Otchertsov, V. G.
Party official.
2nd Sec. and mem. Bureau Cen. Cttee. Turkmen Komsomol Je. 1974–78; mem. Auditing Comm. 1976–81, cand. mem. Cen. Cttee. Turkmen CP 1981–; 2nd Sec. of Ashkhabad City Cttee. Turkmen CP D. 1979–.

Otkalenko, Nina Grigor'evna
Athlete, coach, and instructor; b. My. 23, 1928, Kozhlia, Kurchatov Raion, Kursk Oblast.
Honored Master of Sports (1953).
During the 1950's, she broke European and world records 15 times. She broke records in the USSR 22 times and was 17 times declared champion of the USSR in the 400-m and 800-m runs and in relay competitions. She was a six-time winner of the cross-country races sponsored by the newspaper *L'Humanité*. Decor.: Order of the Badge of Honour.

Ovakimian, Onik Tigranovich
State official.
1st Sec. of Kirovakan Raion Cttee. Armenian CP 1959; mem. Auditing Comm. 1960–61, cand. mem. 1961–64, mem. Cen. Cttee. Armenian CP 1964–; Party Organizer for Cen. Cttee. Armenian CP in Artashat Territorial Kolkhoz-Sovkhoz Prod. Asscn. 1962; Min. of Procurement Armenian SSR Ja. 1970–.

Ovalov, Lev Sergeevich (real name: Shapovalov)
Author; b. Ag. 16 (29), 1905, Moscow.
Russian; mem. CPSU 1920–; grad. Moscow Univ. 1929; suffered reprisals 1941–56, was rehabilitated 1956; first works publ. 1925.
Publs.: "Boltovnia" (Talk) 1928, "Zina Demina" 1937, "Utro nachinaetsia v Moskve" (The Morning Begins in Moscow) vols. 1–2 1934–36, "Prikliucheniia majora Pronina" (The Adventures of Major Pronin) 1957, "Sekretnoe oruzhie" (Secret Weapon) 1963, "Istoriia odnoi sud'by" (Story of a Fate) 1963, and others.

Ovanesian, Rachiia Karapetovich
Author and poet; b. D. 8, 1919, Shaab, Armenia; s. of a peasant.
Armenian; mem. CPSU 1948–; studied Dept. of Philology at Erevan Univ.; served in Sov. Army 1941–45; first works publ. 1935.
Publs.: "Pesnia moei zhizni" (The Song of My Life) 1948, "Vtoraia vstrecha" (The Second Encounter) 1951, "Molchanie moria" (The Silence of the Sea) 1964, and others.

Ovanesova, Arsha Ambartsumovna
Film director; b. D. 10 (23), 1906, Baku.
Honored Art Worker of the RSFSR (1947). Member of the CPSU since 1943.
In 1932, she graduated from the State Institute of Cinematography, where she studied directing. From 1931 to 1961 she

worked as a film director at Soiuzkinokhronika (Union Film Current Events), now the Central Documentary Film Studio, Moscow. She helped found the film journal *Pioneriia* in 1931 and was its editor until 1946. Her documentaries include *Youth Festival* (1937), *Young Naturalists* (1939), *Pioneer Truth* (1940), *A Story About Our Children* (1946), *30 Years of the Komsomol* (1948), *Young Builders of Communism* (1949), *World Youth* (1949), and *School Years* (1956). Her film *Unusual Encounters* (1958) traces the lives of the heroes of the early issues of *Pioneriia* for 20 years.
She taught at the All-Union State Institute of Cinematography. Decors.: State Prize of the USSR 1950, the International Peace Prize, two orders and several medals.

Ovanisian, Nar Mikhailovich
Singer; b. Mr. 30 (Ap. 12), 1913, Tbilisi.
Armenian; grad. Tbilisi Conservatory 1941; basso, one of the principal soloists of Armenian Opera and Ballet Theatre 1951–; Sen. Teacher at Erevan Conservatory 1954–63. Decor.: Order of Lenin.

Ovcharenko, Evgenii Ivanovich
Literary historian and critic; b. Ja. 28, 1922.
Mem. CPSU 1948–; grad. Oblast Pedagogic Inst. in Moscow 1942; Prof. at Moscow Univ. 1965–; his works deal mainly with Gorkii and problems of contemporary Soviet and Belorussian literature.

Ovcharenko, Fedor Danilovich, Dr. of Chem. Sc.
Chemical scientist; b. F. 8, 1913, Vasilivshchina, now Sumy Oblast; s. of a peasant.
Ukrainian; mem. CPSU 1939–; grad. Glukhov Pedagogic Inst. 1934; mem. Acad. of Sciences Ukrainian SSR 1961–; Dr. of Chem. Sc. 1956; Prof. 1957–; Research Asst. at Dept. of Chemistry of Glukhov Pedagogic Inst. 1934–37; Research Asst. at Dept. of Inorganic and Analytical Chemistry at Kiev Inst. of Veterinary Medicine 1936–41; served in Sov. Army during World War II 1941–45; Research Asst. 1945–49, Lecturer, Head of Chair of Inorganic and Analytical Chemistry at Kiev Inst. of Veterinary Medicine 1945–49; Sen. Mem. Scientific Staff and Head of Chair of Laboratory for Colloid Chemistry and Dispersion Minerals at Inst. of General and Inorganic Chemistry of Acad. of Sciences Ukrainian SSR 1949–56; Head of Dept. of Science and Culture at Cen. Cttee. Ukrainian CP 1956–58; Head of Laboratory 1958–63, Dept. Head at Inst. of General and Inorganic Chemistry of Acad. of Sciences Ukrainian SSR 1963–68; cand. mem. Politbureau and Sec. of Cen. Cttee. Ukrainian CP 1968–72; scientific work 1972–; author of some 200 scientific papers. Decors.: Order of Lenin, Order of Patriotic War 1st and 2nd Cl., Order of Red Star, Order "Badge of Honour".
Publs.: "Rozvytok kolodnoi khimii na Ukraini" 1955, "Bentonitovye gliny Ukrainy" 1955–60, "Gidrofilnost' glin i glinistykh mineralov" 1961, "Kolloidnaia khimiia palygorskita" 1963, "Issledovanie v oblasti fiziko-khimicheskoi mekhaniki dispersii glinistykh mineralov" 1965, and others.

Ovcharenko, Illia Panteleimonovich
Sculptor; b. 1926, Musiivka, Voroshilovgrad Oblast.
Mem. CPSU 1952–; attended Voroshilovgrad Art School 1944–49, pupil of V. Agibalov and V. Fedchenko. Decor.: Shevchenko Prize 1973.

Ovchinnikov, Iurii Anatolevich
Biochemist; b. Ag. 2, 1934, Moscow.
Mem. CPSU 1962–; grad. Moscow Univ. 1957; mem. 1970–, mem. Presidium 1973–, Vice-Pres. of USSR Acad. of Sciences 1974–; mem. staff Inst. of Chemistry of Natural Compounds (now Inst. of Bio-Organic Chemistry) of USSR Acad. of Sciences 1960–, Dir. of same inst. 1970–; mem. Cen. Auditing Comm. CPSU 1976–81, Cand. mem. Cen. Cttee. CPSU 1981–.
Publs.: "Sovremennye problemy khimii peptidov i belkov" Moscow 1969, co-author of "Membrano-aktivnye kompleksony" Moscow 1974.

Ovchinnikov, Lev Nikolaevich
Geologist; b. S. 26 (O. 9), 1913, Perm.
Mem. CPSU 1944–; grad. Uralsk Univ. 1937; corresponding mem. USSR Acad. of Sciences 1964–; Head of Laboratory for Mineralogy and Geochemistry at Inst. of Mining Geology 1964–; Dir. of Inst. of Geology of Uralsk Centre of USSR Acad. of Sciences 1962–; Dir. of Inst. of Mineralogy, Geochemistry and Cristallography of Rare Elements of USSR Acad. of Sciences and of USSR Min. of Geology 1966–; his most important works deal with geology of the Urals, theory of ore formation and geochemical methods in geological prospecting.
Publs.: "Genezis endogennykh rudnykh mestorozhdenii" Moscow 1968, "Geologicheskie protsessy rudoobrazovaniia (k osnovam sistematiki)" in "Redkometalnye mestorozhdeniia, ikh genezis i metody issledovaniia" Moscow 1972.

Ovezova, M. K.
Komsomol official.
Sec. and mem. Bureau Cen. Cttee. Turkmen Komsomol N. 1979–.

Ovsiannikov, A. A.
Komsomol official.
2nd Sec. and mem. Bureau Cen. Cttee. Turkmen Komsomol Ag. 1982–.

Ovsiannikov, Lev Vasil'evich
Mathematician; b. Ap. 22, 1919, Vasilsursk, Gorkii Oblast.
Grad. Moscow Univ. 1941 and Mil. Acad. of Aviation Engineering in Leningrad 1945; specialist in mechanics; corresponding mem. USSR Acad. of Sciences 1964–; mem. staff Siberian branch of USSR Acad. of Sciences 1959–; Prof. at Novosibirsk Univ. 1963–; Dep. Dir. of Hydrodynamics. Decors.: Lenin Prize 1958, Order of Red Banner of Labour, medals.

Ozerov, Iurii Nikolaevich
Film director; b. Ja. 26, 1921, Moscow; s. of N. N. Ozerov.
Mem. CPSU 1947–; grad. Frunze Mil. Acad. 1944 and Dept. of Film Directing at Union Film Inst. (VGIK) 1951; director with "Mosfilm" 1952–. Decors.: Lenin Prize 1972, 3 Orders of Lenin, various medals.

Ozerov, Lev Adolfovich (real name: Goldberg)
Author, critic and translator; b. Ag. 10 (23), 1914 Kiev.
Russian; grad. Moscow Inst. of Philosophy, Literature and History (MIFLI) 1939; worked for front newspaper "Pobeda za nami" (Victory Will Be Ours) during World War II.
Publs.: poetry "Priznanie v liubvi" (Declaration of Love) 1957, "Dorogi novyi povorot" (A New Turn of the Street) 1965, "Lirika" (Lyric) 1966, essay collection "Rabota poeta" (The Work of the Poet) 1963, and others.

Ozerov, Nikolai Nikolaevich
Sports commentator; b. D. 11, 1922, Moscow.
Honored Master of Sports (1947); People's Artist of the RSFSR (1973).
He, the son of the tenor N. N. Ozerov, graduated from the A. V. Lunacharskii State Institute of Theatrical Arts in 1945. Beginning in 1945 he became an actor in the Moscow Art Theater. He became a sports commentator on the All-Union Radio in 1950. He has captured many USSR tennis championships, winning 45 times, in different categories, between 1939 and 1953. Decors.: two orders and various medals.

Ozerov, Vitalii Mikhailovich, Dr. of Phil. Sc.
Literary historian and critic; b. Mr. 9 (22), 1917, Kislovodsk.
Mem. CPSU 1940–; grad. Moscow Inst. of Philosophy, Literature and History 1940; Chief Ed. of journal "Voprosy literatury" 1959–. Decors.: 4 orders, medals, Bulgarian Kirill i Metodii Order.

Ozolin, Nikolai Georgievich
Athlete and teacher; b. O. 20 (Nov. 2), 1906, Komarowo, present-day Warsaw Wojewodztwo, Polish People's Republic.
Honored Master of Sports (1937).

Honored Coach of the USSR (1957). Professor (1962) and doctor of the pedagogical sciences (1972). Honorary doctor of the Higher School of Physical Education in Leipzig. Member of the CPSU since 1943.
He has been a several-time champion and record holder of the USSR in pole vaulting (1927–50). He has won many international competitions. From 1931 to 1953 and from 1963 he taught at the State Central Institute for Physical Education, and from 1954 to 1963 he was director of the All-Union Scientific Research Institute for Physical Education.
His principal works deal with the theory and methods of sports and problems of athletic training. Decors.: Order of Lenin, two other orders, and a number of medals.
Works: *Trenirovki legkoatleta.* Moscow-Leningrad 1949. *Sovremennaia sistema sportivnoi trenirovki.* Moscow 1970.

Ozols, Egils Martinovich
State official.
Mem. Cen. Cttee. Latvian CP 1976–; Dep. of Supr. Sov. Latvian SSR 1980–; 1st Sec. of Liepaia City Cttee. Latvian CP until 1980; Chair. Cttee. of People's Control Latvian SSR Mr. 1980–; mem. Bureau Cen. Cttee. Latvian CP 1981–.

Paderin, Ivan Grigor'evich
Author; b. Ja. 15, 1918, Bogoslovka, Kemerovo Oblast.
Russian; mem. CPSU 1939–; grad. Mil.-Pol. Lenin Acad. 1945; first works publ. 1936.
Publ.: "V ogne Stalingrada" (In the Fire of Stalingrad) 1958, "Komdiv bessmertnykh" (Division Commander of the Immortal) 1966, novel "Kogda tsvetut kamni" (When the Stones Blossom) 1961, novelettes "Moi druzia" (My Friends) 1965, "Doverie" (Confidence) 1967, and others.

Pagirev, Gleb Valentinovich
Author; b. Ap. 6 (19), 1914, Ekaterinograd.
Russian; mem. CPSU 1949–; first works publ. 1946.
Publs.: "My – mirnye liudi" (We Are Peaceful People) 1951, "Na linii ognia" (On the Firing Line) 1955, "Moi druzia" (My Friends) 1958, "Rasstoianie" (The Distance) 1964, and others.

Pakalnis, Antanas
Author; b. O. 27, 1923, Lithuania; s. of a peasant.
Lithuanian; studied Kaunas and Vilnius Univs.; first works publ. 1947.
Publs.: "Vesna" (Spring) 1950, "Starye znakomye" (Old Acquaintances) 1959, and others.

Pakhmutova, Aleksandra Nikolaevna
Composer; b. N. 9, 1929, Beketovka, now under the jurisdiction of the Volgograd city soviet.
Honored Art Worker of the RSFSR (1971).
In 1953, she graduated from the Moscow Conservatory, where she studied composition with V. Ia. Shebalin, completing graduate study under his guidance in 1956. Since 1968 she has served as secretary of the administrative board of the Composers' Union of the USSR. She has worked in various genres but is especially well known for her songs, among them "The Geologists", "Embracing the Sky", "Tenderness", "Song of Troubled Youth", "The Melody", "Who Will Respond", and the cycle *Gagarin's Constellation.* Other works include a concerto for orchestra (1972; it provided the music for the ballet *Illumination;* premiere 1974, Bolshoi Theater, Moscow), the *Russian Suite* for orchestra (1953), a concerto for trumpet and orchestra (1955), Pioneer cantatas, and music for films. Decors.: two orders.

Pakhomova, Liudmila Alekseevna
Athlete, choreographer; b. D. 31, 1946, Moscow.
Honored Master of Sports (1970).
She graduated from the A. V. Lunacharskii State Institute of Theatrical Arts in 1970. She was national champion in ice dancing nine times between 1964 and 1975, European champion six times between 1970 and 1976, world champion five times between 1970 and 1974, and Olympic champion (1976). From 1964 to 1966 she skated with V. I. Ryzhkin. Since 1967 she has been skating with A. G. Gorshkov, and her coach has been E. A. Chaikovskaia. Decors.: Order of the Red Banner of Labor and the Order of the Badge of Honor.

Pakul', El'frida Ianovna (Elfrīda Pakule)
Coloratura soprano; b. My. 20 (Je. 2), 1912, Riga.
People's Artist of the Latvian SSR (1947). Member of the CPSU since 1952.
She made her debut as a concert singer in 1938. In 1940 and 1941 and from 1944 to 1956 she was a soloist at the Latvian Theater of Opera and Ballet in Riga. Between 1941 and 1944 she worked in Ivanovo and Moscow philharmonic societies, and between 1956 and 1970 in the Latvian Philharmonic Society. Her best roles have included Antonida in Glinka's *Ivan Susanin,* the title role in Delibes's *Lakmé,* Violetta and Gilda in Verdi's *La Traviata,* and *Rigoletto,* and Rosina in Rossini's *Barber of Seville.*
She was a deputy to the second and third convocations of the Supreme Soviet of the USSR. Decors.: 1946 State Prize of the USSR, Order of Lenin, the Order of the Red Banner of Labor, and several medals.

Palazhchenko, Leonid Ivanovich
Party official.
1st Sec. of Chernigovskii Raion Cttee. CP Ukraine until 1978; Oblast Cttee. Sec. 1978–79, 1st Oblast Cttee. Sec. CP Ukraine in Volynia Ap. 1979–; Dep. to Ukrainian Supr. Sov. 1980–; mem. Cen. Cttee. CP Ukraine 1981–.

Pallaev, Gaibnazar
State official.
Mem. Cen. Cttee. Tadzhik CP 1961–; Chair. of Tadzhik Republican Asscn. for Agricultural Tech. "Tadzhikselkhoztekhnika" Mr. 1973–Ap. 1977; 1st Sec. of Kurgan-Tiube Oblast Cttee. Tadzhik CP Ap. 1977–.

Palman, Viacheslav Ivanovich
Author; b. Mr. 13 (26), 1914, Skopin, Riazan Oblast.
Grad. Borovsk Agricultural College 1933; attended Advanced Courses in Literature of USSR Writers' Union; after grad. from Agricultural College 20 years' work as agronomist with various kolkhozes and sovkhozes; first works publ. 1955.
Publs.: "V nashei stanitse" (In Our Cossack Village) 1955, "Razdory na khutore Vishniaki" (Disagreement on Vishniaki Farm) 1956 (2nd ed. publ. under the title "Events on Vishniaki Farm" 1958), "Sud'ba moikh druzei" (The Fate of My Friends) 1958, "Na beregu Kubani" (On the Banks of the Kuban) 1964, "Tvoi sled na zemle" (Your Traces on Earth) 1961, and others.

Palu, Peeter Karlovich
State official; b. 1942, Tallinn.
Esthonian; mem. CPSU 1972–; grad. Polytechnical Inst. in Tallin; (1968); in the Construction Investment Admin. of Exec. Cttee. of City Sov. of Peoples' Deps. in Tallinn 1975–76; Min. of Construction Esthonian SSR 1976–; mem. Cen. Cttee. CP Esthonia 1981–.

Panasenko, Taras Ivanovich
Party official; b. 1930.
Ukrainian; mem. CPSU 1955–; grad. Lvov Party Higher School; zoo technician at a veterinary hospital of a machine and tractor station 1952–54; sec., 1st sec. of a raion cttee. Komsomol 1954–61; instructor, chief of a dept. of Rovno Oblast Cttee. Ukrainian CP 1961–65; 1st Sec. of Rovno Raion Cttee. Ukrainian CP 1965–72; 1st Sec. of Rovno Oblast Cttee. Ukrainian CP O. 1972–; Dep. to USSR Supr. Sov. 1974–; mem. Cen. Cttee. Ukrainian CP 1976–.

Panchenko, Pimen Emelianovich
Author; b. Ag. 23, 1917, Tallinn; s. of a peasant.
Belorussian; grad. Minsk Teachers' Training Inst. 1939; was active as village teacher; first works publ. 1933. Decor.: Order of Friendship Between Nations 1976.
Publs.: "Doroga voiny" (Road of War) 1942, "Dalekie stantsii" (Distant Stations) 1945, spent one year (1944–45) in Iran, wrote "Iranskii dnevnik" (Iran Di-

ary) there; "Kniga puteshestvii i liubvi" (The Book of Travels and Love) 1959, "Pri svete molnii" (In the Light of Lightning) 1966, and others.

Panev, Zosima Vasil'evich
State official; b. 1914; s. of a peasant.
Mem. CPSU 1940–; grad. Party High School of Cen. Cttee. CPSU; teacher, headmaster, inspector of a raion educ. dept. 1933–40; party and sov. posts 1940–50; Chair. of Council of Mins. Komi ASSR 1950–71; Dep. to USSR Supr. Sov. 1954–66; Chair. of Presidium Supr. Sov. Komi ASSR 1971–; Dep. to 1971–, Dep. Chair. of Presidium Supr. Sov. RSFSR 1972–.

Pankin, Boris Dmitrievich
Diplomat.
USSR Amb. to Sweden S. 1982–.

Pankin, Vladimir Efimovich
Cybernetics and computer scientist; b. 1928.
Grad. Leningrad Polytech. Inst. 1952; mem. staff of a design bureau 1953–, dept. head at same 1965–. Decor.: State Prize of the USSR (for his part in the development of electronic control devices and control-computing complexes of the "UM 1-NCH" type and their introduction into the first digital systems in various branches of Sov. economy) 1969.

Pankov, Nikolai Aleksandrovich
Diplomat; b. 1928.
USSR Amb. to Mauritius Jl. 1981–.

Pankov, Viktor Ksenofontovich
Literary historian, author and critic; b. N. 20, 1920, Vichuga, Ivanovo Oblast.
Mem. CPSU 1944–; grad. Dept. of Philology of Moscow Univ. 1945; first works publ. 1949.
Publs.: "Puti razvitiia sovetskoi literatury v poslevoennoe vremia" (doctoral thesis: Development of Soviet Literature in the Post-War Period) 1962, "Sovetskaia deistvitelnost' v izobrazhenii M. Gorkogo" (Soviet Reality in M. Gorkii's Works) 1955, "Vremia i knigi" (Time and Books) 1964, and others.

Pankratov, Iurii Aleksandrovich
State official; b. 1914.
Grad. Moscow Mining Inst.; worked in peat bank 1928–32; after completing studies at inst. Engineer/Service Chief at USSR People's Commissariat for the Fuel Ind., various sen. posts in larger mines and in USSR Min. of Coal Mining Ind. 1932–60; Head of Dept. of Coal, Turf and Shale Ind. at RSFSR GOSPLAN 1960–65; RSFSR Min. of Fuel Ind. O. 1965–. Decor.: Order of Red Banner of Labour.

Pankratov, Iurii Ivanovich
Author; b. S. 12, 1935, Semipalatinsk.
Russian; mem. CPSU 1968–; grad. Gorkii Inst. of Literature 1960; first works publ. 1953.
Publs.: "Volnenie" (Agitation) 1966, "Ne v bure dushevnykh ovatsii. O tvorchestve molodykh poetov" (Not in the Storm of Emotional Ovations. On the Work of Young Poets) 1966, and others.

Panov, Konstantin Nikolaevich
Laborer; b. 1933.
Mem. CPSU 1967–; finished school for working youth in Gorkii 1967; collective farmer 1947–48; worker, work brigade leader at "Krasnoe sormovo" Plant in Gorkii 1950–54 and 1956–; served in Sov. Army 1954–56; cand. mem. Cen. Cttee. CPSU 1981–. Decor.: USSR State Prize 1979.

Panov, Nikolai Nikolaevich (pseudonym: Dir Tumannyi)
Author; b. Ag. 11 (24), 1903, Kozelsk, Kaluga Oblast.
Russian; grad. Moscow Inst. of Journalism 1922 and Dept. of Law of Moscow Univ. 1928; active as war correspondent with Northern Fleet during World War II; first works publ. 1918.
Publs.: "Moskovskaia Amerika" (Moscow America) 1924, "Amerikanskie fashisty" (American Fascists) 1924, "Taina starogo doma" (The Secret of the Old House) 1928, "Komandir tanka" (Tank Commander) 1937, "Goluboe e chernoe" (The Blue and the Black) 1962, and others.

Panteleev, Iurii Aleksandrovich
Military official; b. O. 18 (31), 1901, St. Petersburg.
Admiral (1953), professor (1962). Member of the CPSU since 1940.
He joined the navy in 1918 and saw service in the Civil War. He graduated from the advanced training courses for the command staff in 1925 and from the Naval Academy in 1933. He served in the Black Sea Fleet, in the Directorate of the Naval Forces of the Workers' and Peasants' Red Army (1925–30), and as chief of staff and submarine force commander of the Black Sea Fleet (from 1935). He was chief of staff of the Red Banner Baltic Fleet from 1939 to 1941. In the Great Patriotic War (1941–45), he was commander from October 1941 to April 1942 of the naval defense of Leningrad and the lake region and, at the same time, commander of the Leningrad Naval Base. In 1942–43 (until May) and from December 1943 to July 1944 he was assistant chief of the Naval Staff Headquarters. He took command of the Volga Military Flotilla in 1943 and the White Sea Military Flotilla in July 1944.
After the war, he served as chief of a directorate of the Naval Staff Headquarters. He was deputy chief of the Naval Staff Headquarters in 1947–48, head of the Naval Academy from 1948 to 1951, and commander of the Pacific Ocean Fleet from 1951 to 1956. From 1956 to 1960 he was head of the A. N. Krylov Naval Academy of Shipbuilding and Armament, and from 1960 to 1968, head of the Naval Academy. He retired in 1968.
Decors.: Order of Lenin, four Orders of the Red Banner, the Order of the Red Banner of Labor, the Order of Nakhimov First Class, the Order of the Patriotic War First Class, two Orders of the Red Star, a British order, and various medals.
Work: *Morskoi front*. Moscow 1965.

Panteleev, Nikolai Alekseevich
State official.
Chair. of State Cttee. on Labour Ukrainian SSR O. 1976–.

Panteleimon (Sergii Aleksandrovich Mitrukovskii)
Russian Orthodox clergyman; b. 1912, Vitkinsk, Udmurt ASSR.
Lower orders; deacon 1944, then priest; provost in Perm Diocese 1948–56; in Izhevsk 1960–, Sec. of Diocesan Admin., Provost of Churches in the Diocese, Main Clergyman of Ihzevsk Holy Trinity Cathedral 1967–; Rep. of Clergy in Izhevsk Diocese at Reg. Council of Russian Orthodox Church 1971; Bishop of Kazan and Mariisk 1975–.

Papanin, Ivan Dmitrievich, Dr. of Geog. Sc.
Geographer and arctic explorer; b. N. 14 (26), 1894, Sevastopol.
Mem. CPSU 1919; Dr. of Geog. Sc. 1938; Rear Admiral 1943–; worked for People's Commissariat for Posts and Telecommunications 1923–32; Head of Polar Station in Pacific Bay of Fridtjof Nansen Land 1932–33 and of Polar Station on Cape Cheliuskin 1934–35; Head of first Drift Ice Station "Severnyi polius" (SP-1) 1937–38; Head of Main Admin. Northern Seaways of the Maritime Fleet of the USSR (Glavsevmorput') 1939–46, concurrently Plenipotentiary of State Cttee. on Defence of Transport in the North 1941–45; Dep. Dir. of Expeditions at Inst. of Oceanology of USSR Acad. of Sciences 1948–51; Head of Dept. of Ocean Expedition Projects of USSR Acad. of Sciences 1951–; concurrently Dir. of Inst. of Biology of Inland Waterways at USSR Acad. of Sciences 1952–72; Chair. of Moscow branch of USSR Geographic Soc. 1945–. Decors.: twice Hero of the Soviet Union, 8 Orders of Lenin, Order of the October Revolution, 2 Orders of Red Banner, Nakhimov Order 1st Cl., Order of Red Banner of Labour, Order of Red Star, medals.

Papanov, Anatolii Dmitrevich
Actor; b. 1922.
Grad. A. V. Lunacharskii State Theatre Inst. Moscow (GITIS) 1946; mem. Satirical Theatre Moscow 1949–. Decor.: People's Artist of the USSR 1973.

Papaskiri, Ivan Georgievich
Author; b. D. 12 (25), 1902, Kutol, Abkhazian ASSR.
Abkhaz; mem. CPSU 1938–; studied

Leningrad Inst. of Oriental Languages 1928–29; first works publ. 1932.
Publs.: "Temyr" (Russian transl. "K dolgoi zhizni" – To a Long Life) 1937 resp. 1938, "Zhenskaia chest'" (Woman's Honour) vols. 1–3, 1949, "Rasskazy" (Stories) 1962, and others; his works have been translated into Russian, Polish, Georgian and Kazakh.

Papernyi, Zinovii Samoilovich
Literary historian and critic; b. Ap. 5, 1919, Gaisin, Vinnitsa Oblast.
Grad. Moscow Inst. of Philosophy, Literature and History (MIFLI) 1941; first works publ. 1943.
Publs.: "A. P. Chekhov. Ocherk tvorchestva" (A. P. Chekhov. Outline of his Work) 1954, 2nd ed. 1960, "O masterstve Maiakovskogo" (The Mastery of Maiakovskii) 1953, 2nd ed. 1957, "Chelovek, pokhozhii na samogo sebia" (A Man Who Resembles Himself) 1967; author of numerous essays on Soviet lyric.

Papunidze, Vakhtang Rafaelovich
Party official; b. 1929.
Georgian; mem. CPSU 1952–; grad. Georgian Agricultural Inst., Dr. of Econ. Sciences, Professor; agrarian technician, researcher 1954–57 and held prominent posts in various republican econ. and planning research insts. 1957–72; dep. chief of Econ. Dept. Cen. Cttee. CP Georgia, Oblast Cttee. Sec. of CP Georgia in Adzhar ASSR 1973–75; 1st Oblast Cttee. Sec. CP Georgia in Adzhar ASSR Ja. 1975–; mem. Cen. Cttee. 1976–, cand. mem. Bureau Cen. Cttee. CP Georgia Ja. 1979–; Dep. to USSR Supr. Sov. 1979–.

Paramonov, Albert Mikhailovich
State official; b. 1928.
1st Dep. Min. of Textile Ind. RSFSR until 1977, Min. S. 1977–. Decor.: Order of Red Banner of Labour.

Parasiuk, Ostap Stepanovich, Dr. of Phy.-Math. Sc.
Mathematician; b. D. 20, 1921, Bilzi, now Lvov Oblast; s. of a peasant.
Grad. Lvov Univ. 1947; mem. Acad. of Sciences Ukrainian SSR 1964–; Dr. of Phy.-Math. Sc. 1955; Prof. 1957–; post-graduate studies 1947–49, Sen. Mem. Scientific Staff at Inst. of Mathematics at Acad. of Sciences Ukrainian SSR 1949–51; Dep. Dir. of Inst. of Machine Construction and Automation of Acad. of Sciences Ukrainian SSR in Lvov 1951–52; doctoral work at Inst. of Mathematics at Acad. of Sciences Ukrainian SSR 1952–55, concurrently Research Asst., Lecturer at Lvov Univ. 1947–56; Dep. Dir. of Inst. of Mathematics at Acad. of Sciences Ukrainian SSR 1956–66; Dept. Head at same inst. 1957–; mem. Presidium Acad. of Sciences Ukrainian SSR, Acad. Sec. of Dept. of Physics, Head of Dept. of Mathematical Methods in Theoretical Physics at Inst. of Theoretical Physics of Acad. of Sciences Ukrainian SSR 1966–; mem. staff Kiev Univ. 1956–, Prof. 1957–.
Publs.: "Umnozhennia prichinnykh funktsii pri nesovpadaiushchikh argumentakh" in "Izvestiia Akademii Nauk SSSR – Seriia matematicheskaia" vol. 20, 1956, "Pro matematychnyi aparat teorii elementarnykh chastynok" 1962, and others.

Parfenov, Vitalii Georgievich
State official.
Head of Main Admin. for Material-Tech. Supply with Council of Mins. Moldavian SSR Jl. 1975–78; Chair. of State Cttee. for Material-Tech. Supply Moldavian SSR S. 1978–; cand. mem. Cen. Cttee. Moldavian CP 1981–.

Pariiskii, Nikolai Nikolaevich
Geophysicist and astronomer; b. S. 30, 1900, St. Petersburg.
Corresponding Member of the Academy of Sciences of the USSR (1968).
He graduated from Moscow University in 1924 and taught there from 1924 to 1941 and from 1944 to 1960. In 1935 he joined the Institute of Lithosphere Physics and was appointed head of a department in 1956. He was one of the first to conduct gravimetric research in the USSR. He participated in the first studies and interpretations of the Kursk and Moscow gravitational anomalies and conducted a long series of investigations into the nature of the solar corona and gegenschein. His chief works deal with gravimetry, cosmogony, astrophysics, the earth's tides, and the earth's rotation.
Decors.: Order of Lenin, the Order of the Red Banner of Labor, and several medals.
Works: "Issledovanie vliianiia ucheta parallaksov zvezd i galakticheskogo vrashcheniia na opredelenie postoiannoi lunno-solnechnoi protsessii N'iukoma." *Tr. Gos. astronomicheskogo in-ta im. Shternberg,* 1935, vol. 6, fasc. 1. (With K. F. Ogorodnikov and V. G. Fesenkov.) "Uskorenie sily tiazhesti v osnovnykh opornykh punktakh Soiuza ..." *Izv. Vsesoiuznogo tresta osnovnykh geodezicheskikh i gravimetricheskikh rabot,* 1935, issue 1. *Neravnomernost' vrashcheniia Zemli.* Moscow 1954. "Zemnye prilivy i vnutrennee stroenie Zemli." *Izv. AN SSSR: Ser. geofizicheskaia.* 1963, issue 2.

Parimbetov, Berkimbai
State official.
Dep. Head of Dept. for Construction and Building Materials Ind. with Cen. Cttee. Kazakh CP 1962–65; Min. of the Building Materials Ind. Kazakh SSR 1965–73 and O. 1979–; mem. Auditing Comm. Kazakh CP 1966–71; cand. mem. Cen. Cttee. Kazakh CP 1971–76, mem. 1976–.

Parkhomenko, Mikhail Nikitich
Literary historian; b. S. 9 (22), 1910, Shkuratovka, Sumy Oblast.
Mem. CPSU 1945–; grad. Moscow Oblast Pedagogic Inst. 1935; post-graduate studies at Lenin Pedagogic Inst. Moscow 1938; Prof. Chair of Literary History. Art Appreciation and Journalism at Acad. of Social Sciences of Cen. Cttee. CPSU 1960–; first works publ. 1937; his works deal mainly with the history of Ukrainian literature, the relationship between Ukrainian and Russian literature and theoretical problems of development of multinational Soviet literature.

Parnov, Eremei Iudovich, Cand. of Chem. Sc.
Author; b. O. 20, 1935, Kharkov.
Russian; grad. Moscow Peat Inst. 1958; first works publ. 1935.
Publs.: "Okno v antimir" (Window on the Anti-World – together with E. Glushchenko); co-authorship with N. T. Emtsev 1961–; "Poslednee puteshestvie polkovnika Fosetta" (The Last Journey of Col. Fosett) 1965, "Zelenaia krevetka" (The Green Shrimp) 1966, "Dusha Mira" (The Soul of the World) 1964, "More Diraka" (Dirak's Sea) 1967, and others.

Parsamian, Ruben Ovsepovich
State official.
Mem. Auditing Comm. 1961–64, cand. mem. Cen. Cttee. Armenian CP 1964–81; Dep. to 1971–, Chair. of Supr. Sov. Armenian SSR 1971–72; Min. of Culture Armenian SSR S. 1972–79; Chair. of Council for Affairs of the Armenian Church with Council of Mins. Armenian SSR 1979–.

Parshina, Valentina Romanova
B. 1937.
Mem. CPSU 1965–; grad. Leningrad Agricultural Inst. 1959; leader of vegetable-growing work brigade of "Detskosel'skoe" Specialized Prod. Asscn. of State Farms in Leningrad Oblast 1959–; Dep. to USSR Supr. Sov. 1979–; cand. mem. Cen. Cttee. CPSU 1981–. Decor.: Heroine of Socialist Labor 1978.

Parve, Ralf
Author; b. Je. 25, 1919, Esthonia; s. of an artisan.
Esthonian; mem. CPSU 1947–.
Publs.: "Iz dushi boitsa" (From the Soul of a Warrior) 1945, "O vremeni surovom, prekrasnom i dorogom" (About a Hard, Wonderful and Precious Time) 1946, "Liricheskaia stenogramma" (Lyrical Note in Shorthand) 1964, "V den' semi spiashchikh" (The Day of the Seven Sleepers) 1964, and others.

Pashchenko, Andrei Iakovlevich
State official; b. 1919.
Sec. of Dnepropetrovsk Oblast Cttee. Ukrainian CP 1963–74 (for Ind. 1963–64); Chair. of State Cttee. for Publishing, Printing and Book Trade Ukrainian SSR O. 1974–; cand. mem. Cen. Cttee. Ukrainian CP 1976–.

Pashkov, Anatolii Ignatevich
Economist; b. N.2 (15), 1900, Mazchuki, Lipetsk Oblast.
Mem. CPSU 1919–; grad. Econ. Dept. of Faculty of Social Sciences at Moscow Univ. 1925 and Econ. Inst. of Red Professors 1931; pedagogic and scientific work 1923–. Decors.: 2 orders, medals.

Pasichnik, Mitrofan Vasil'evich, Dr. of Phy.-Math. Sc.
Physicist; b. Je.17, 1912, Zhyrkivtsi, now Poltava Oblast.
Mem. CPSU 1940–; grad. Poltava Inst. of Social Educ. 1931; Dr. of Phy.-Math. Sc. 1956; Prof. 1957–; scientific works at Inst. of Physics of Acad. of Sciences Ukrainian SSR in Kiev, concurrently teaching posts at various colleges in Kiev 1932–40; Head of Chair of Physics at Chernovtsy Univ. 1940–41; served in Sov. Army 1941–45; mem. staff Inst. of Physics at Acad. of Sciences Ukrainian SSR 1946–49; Dir. 1949–65, head of a dept. at same inst. 1965–; concurrently Head of Chair of Nuclear Physics at Kiev Univ. until 1960; author of some 100 scientific papers. Decors.: Merited Scientist of the Ukrainian SSR 1963, Order of Red Banner, Order of Patriotic War 1st and 2nd Cl., Order of Red Star, Order of Red Banner of Labour, medals.
Publ.: "Voprosy neitronnoi fiziki srednikh energii" 1962.

Paskar, Petr Andreevich
State official; b. 1929.
Moldavian; mem. CPSU 1956–; grad. Kishinev Agricultural Inst. 1954; agronomist, chief agronomist of a machine and tractor station (MTS) 1954–58; chief agronomist of a raion inspection cttee. for agriculture. Head of Inspection Cttee. for Agriculture and Seed Growing at Min. of Agriculture Moldavian SSR 1958–59; 1st Sec. of Chadyr-Lungu Raion Cttee. Moldavian CP 1959–62; mem. Cen. Cttee. Moldavian CP 1960–; Dep. to USSR Supr. Sov. 1962–; 1st Dep. Min. of Production and Procurement of Agricultural Products Moldavian SSR 1962; Sec. of Cen. Cttee. Moldavian CP D. 1962–70; mem. Bureau Cen. Cttee. Moldavian CP 1962–76; Chair. of Bureau of Agriculture at Cen. Cttee. Moldavian CP 1962–64; Chair. of Council of Mins. Moldavian SSR Ap. 1970–76; cand. mem. Cen. Cttee. CPSU 1971–; Min. of Foreign Affairs Moldavian SSR Jl. 1971–My. 1976; 1st Dep. Chair. of USSR GOSPLAN My. 1976–. Decor.: Order of Red Banner of Labour.

Passar, Andrei Aleksandrovich
Nanai poet; b. Mr.25, 1925, nomad encampment of Mukha, now in Nanai Raion, Khabarovsk Krai.
Member of the CPSU since 1956.
He began publishing in 1949. His first book of verse was *Sunlight* (1952). In 1960 he graduated from the Advanced Courses in Literature of the Writers' Union of the USSR. He is an expert on the language, folklore, and customs of his native land. Now living on the island of Sakhalin, he celebrates in verse the new life of the Nanai people.
Works: *In Russian translation: Mokona.* Moscow 1958. *Trekhlapyi volk.* Khabarovsk 1967. *Pod stuk bubna: Stikhi.* Moscow 1969.

Pastukhov, Boris Nikolaevich
State official; b. 1933.
Russian; mem. CPSU 1959–; grad. Moscow Bauman College 1958; 2nd Sec. 1958–59, 1st Sec. of Bauman Raion Komsomol Cttee. Moscow 1959–61; 2nd Sec. 1961–62, 1st Sec. of Moscow City Komsomol Cttee. 1962–64; mem. Bureau 1962–, Sec. of All-Union Komsomol Cen. Cttee. 1964–69, 2nd Sec. 1969–77, 1st Sec. 1977–82; cand. mem. Cen. Cttee. CPSU 1966–78, mem. 1978–; Dep. to USSR Supr. Sov. 1966–, mem. Trade and Public Services Comm. 1966–70, mem. Comm. of Youth Affairs at Sov. of Nationalities 1970–77, mem. of the Presidium 1977–; Chair. of USSR State Cttee. on Publishing, Polygraphy and the Book Trade D. 1982–.

Patalov, Evgenii Gurgenovich
State official.
Min. of the Interior Armenian SSR D. 1974–; mem. Cen. Cttee. Armenian CP 1976–.

Patiashvili, Dzhumber Il'ich
Party official.
Mem. Bureau 1969–73, 2nd Sec. 1969–70, 1st Sec. of Cen. Cttee. Georgian Komsomol 1970–73; cand. mem. Bureau Cen. Cttee. USSR Komsomol 1970–73; mem. Cen. Cttee. 1971–, cand. mem. Bureau 1971–73, mem. Bureau Cen. Cttee. Georgian CP N. 1974–; 1st Sec. of Gori Raion Cttee. Georgian CP 1973–74; Sec. of Cen. Cttee. Georgian CP N. 1974–.

Patolichev, Nikolai Semenovich
State official; b. 1908; s. of a peasant.
Russian; mem. CPSU 1928–; grad. Mil. Chemical Acad. 1937; factory worker, sec. of a raion Komsomol cttee., student 1925–38; Dep. to USSR Supr. Sov. 1937–; worked for All-Union CP (B) Cen. Cttee., as Party Organizer for Cen. Cttee. at Iaroslavl Rubber Combine 1938–39; 1st Sec. of Iaroslavl 1939–41, of Cheliabinsk Oblast and City All-Union CP (B) Cttees. 1941–46; cand. mem. 1939–41, mem. Cen. Cttee. CPSU 1941–; Sec. of All-Union CP(B) Cen. Cttee. and Dep. Chair. of Council for Kolkhoz Affairs with USSR Council of People's Commrs. 1946–47; Sec. of Cen. Cttee. Ukrainian CP(B) 1947; 1st Sec. of Rostov Oblast and City All-Union CP(B) Cttees. 1947–50; 1st Sec. of Cen. Cttee. Belorussian CP 1950–56; cand. mem. Presidium Cen. Cttee. CPSU 1952–53; USSR Dep. Min. 1956–57, USSR 1st Dep. Min. of Foreign Affairs 1957–58; USSR Min. of Foreign Trade 1958–; Chair. of Permanent COMECON Comm. for Foreign Trade. Decors.: 6 Orders of Lenin, Order of Red Banner of Labour, medals, Hero of Socialist Labour 1976.

Paton, Boris Evgenevich, Dr. of Tech. Sc.
Scientist; b. N.27, 1918, Kiev; s. of a scientist.
Russian; mem. CPSU 1952–; grad. Kiev Ind. Inst. 1941; engineer at "Krasnoe sormovo" Plant Electrical Laboratory 1941–42; engineer, Research Asst., Head of Electrical Dept., Dep. Dir. 1942–53, Dir. of E. O. Paton Inst. of Electro-Welding at Acad. of Sciences Ukrainian SSR 1953–62; Dep. to 1959–, mem. Comm. for Ind. and Transport 1959–63, mem. Presidium Supr. Sov. Ukrainian SSR 1963–; mem. Cen. Cttee. Ukrainian CP 1960–; cand. mem. 1961–66, mem. Cen. Cttee. CPSU 1966–; Pres. of Acad. of Sciences Ukrainian SSR F. 1962–; Dep. to 1963–; mem. Foreign Affairs Comm. with Sov. of the Union 1962–66, Dep. Chair. of Sov. of the Union USSR Supr. Sov. 1966–; mem. Acad. of Sciences Ukrainian SSR 1958–; his work has been concerned with electro-tech. problems of contact, arc and gas-electric welding of metals; he investigated the basic methods of closed-arc welding 1942–45; developed a theory for controlling automatic electric-arc welding and methods for pipe welding; also had part in the development of new methods of electric-slag welding of massive pieces of metal. Decors.: State Prize of the USSR 1950, Lenin Prize 1957, 2 Orders of Lenin, Order of Red Banner of Labour, medals.

Patsiavichene, Ianina Bronislavovna
State official; b. 1928.
Grad. (by correspondence) Party High School of Cen. Cttee. CPSU; in trade, sov. and CPSU posts until 1961; Chair. of Exec. Cttee. of Novo-Vilianskii Raion Sov. of Workers' Deps. 1961–71; mem. Cen. Cttee. 1964–66 and 1971–, mem. Auditing Comm. Lithuanian CP 1966–71; Head of Dept. of Trade and Public Services at Cen. Cttee. Lithuanian CP 1971–73; Min. of Public Welfare Lithuanian SSR D. 1973–.

Paushkin, Gennadii Aleksandrovich
Author; b. F.28, 1921, Kazan.
Russian; mem. CPSU 1943–; grad. Kazan Univ. 1950; mem. Border Guards 1939–41; first works publ. in front newspapers.
Publs.: "Na dalnei zastave" (At the Distant Border Guard Station) 1952, children's stories "Aleshkin Golub" (Aleshkin's Pigeon) 1953, "Ptitsy uleteli" (The Birds Departed) 1962, "Slushai moi pozyvnye" (Listen to My Call Signal) 1965, and others.

Pavlichko, Dmitrii Vasil'evich
Author and poet; b. S. 28, 1929, Stopchatovo, Ivanovo-Frankovsk Oblast; s. of a tree feller.
Ukrainian; mem. CPSU 1954–; grad. Dept. of Philology of Lvov Univ. 1953; concluded post-graduate studies at same univ. 1957. Decor.: N. Ostrovskii Prize 1961.
Publs.: "Liubov' i nenavist'" (Love and Hatred) 1953, "Moia zemlia" (My Country) 1955, "Na strazhe" (On Guard) 1961, "Govorit Fidel" (Fidel Speaking), "Viktor", "Berezy druzhby" (Birches of Friendship), and others.

Pavliuchuk, Vladimir Iosifovich
State official.
Dep. Min. 1965–71, Min. of Melioration and Water Econ. Belorussian SSR O. 1971–; mem. Cen. Cttee. Belorussian CP 1976–.

Pavlov, Georgii Sergeevich
Party official; b. 1910; s. of a worker.
Russian; mem. CPSU 1939–; grad. Dneprodzershinsk Metal Inst. 1936; foreman, instructor, head of instruction dept. at a factory school 1928–36; dep. dir. of a training combine 1936–40; CPSU posts 1940–41 and 1942–; 2nd Sec., 1st Sec. of Dneprodzerzhinsk City Cttee. Ukrainian CP(B) 1943–47; worked for All-Union CP(B) Cttee. 1947–49; 1st Sec. of Magnitogorsk City All-Union CP(B) 1949–50; 2nd Sec. of Cheliabinsk Oblast All-Union CP(B) Cttee. 1950–51; Dep. Chair. of Kostroma Oblast Exec. Cttee. of Workers' Deps. 1951–54; cand. mem. 1952–56 and 1961–71, mem. Cen. Cttee. CPSU 1971–; Sec. of Kostroma Oblast Cttee. 1954–55 and 1st Sec. of Kostroma City Cttee. CPSU 1955–56; 2nd Sec. of Kostroma Oblast CPSU Cttee. 1956–57; 1st Sec. of Mari Oblast CPSU Cttee. 1957–63; Dep. to 1958–, mem. Budget Comm. at Sov. of Nationalities USSR Supr. Sov. 1962–66; worked for Cttee. for Party and State Control with Cen. Cttee. CPSU and USSR Council of Mins. 1963–65; Dir. of Affairs at Cen. Cttee. CPSU D. 1965–. Decors.: Order of Lenin, Order of Red Banner of Labour, medals.

Pavlov, Grigorii Petrovich
Party official; b. 1913; s. of a peasant.
Russian; mem. CPSU 1940–; grad. Voronezh Agricultural Inst. 1959; apprentice carpenter, carpenter in Moscow, brigade leader, foreman at the construction of Komsomol-na-Amure 1928–36; trade union and Komsomol work, 1st sec. of a raion All-Union CP(B) cttee. in Riazan Oblast 1936–47; 1st sec. of a raion cttee., dept. head with Lipetsk Oblast Cttee., then Sec. of Lipetsk Oblast Cttee. CPSU 1947–56; Dep. Chair., 1st Dep. Chair. 1956–61, Chair. of Lipetsk Oblast Exec. Cttee. of Workers' Deps. 1962–63; Sec. 1961–62, 1st Sec. of Lipetsk Oblast CPSU Cttee. Ja. 1963– (for Agriculture 1963–64); Dep. to 1963–, mem. Comm. for Construction and Building Materials Ind. Sov. of the Union USSR Supr. Sov. 1966–; cand. mem. 1966–76, mem. Cen. Cttee. CPSU 1976–. Decors.: Order of Lenin, Order of Red Banner of Labour.

Pavlov, Iakov Fedotovich
Military official; b. O. 4 (17), 1917, Krestovaia, now in Valdai Raion, Novgorod Oblast.
Hero of the Soviet Union (Je. 27, 1945). Member of the CPSU since 1944.
He joined the Red Army in 1938. He fought in the Great Patriotic War (1941–45) on the Southwestern, Stalingrad, Third Ukrainian, and Second Belorussian fronts as commander of a machine gun squad, a gunner, and a commander of a reconnaissance section with the rank of senior sergeant. On the night of S. 26, 1942, in the battle of Stalingrad of 1942–43, an assault group under his command dislodged the enemy from a four-story apartment house in the center of the city and held the building until the grouping of the fascist German troops in the Stalingrad region was suppressed. The house entered the history of the battle of Stalingrad as the Pavlov house. The fascist German troops furiously attacked the building many times but could not dislodge the Soviet soldiers.
He was demobilized in August 1946 and graduated from the Higher Party School of the Central Committee of the CPSU. He now works in the national economy. Decors.: Order of Lenin, the Order of the October Revolution, two Orders of the Red Star, and various medals.

Pavlov, Igor Mikhailovich
Metallurgist; b. Je. 10 (23), 1900, Sulin, now Krasnyi Sulin, Rostov Oblast.
Grad. Petrograd Polytech. Inst. 1923; corresponding mem. USSR Acad. of Sciences 1946–; worked in metallurgical plants 1923–33; concurrently teacher 1926–39, Prof. at Leningrad Polytech. Inst. 1939–43; Prof. at Steel Inst. (now Steel and Alloys Inst.) Moscow 1943–53; Head of Dept. of Plastic Deformations in Metals at A. A. Baikov Inst. of Metallurgy of USSR Acad. of Sciences 1953–; his main works deal with theory of rolling and pressure processing of metals as well as with general problems in metallurgy and metals. Decors.: Lenin Prize 1966, Order of Lenin, 2 other orders and various medals.

Pavlov, Iurii Ivanovich
Diplomat.
USSR Amb. to Costa Rica D. 1982–.

Pavlov, Sergei Pavlovich
State official; b. Ja. 1, 1929, Rzhev.
Russian; mem. CPSU 1954–; grad. Tech. College of Mechanization and Electrification of Agriculture 1949, completed 5 terms at Moscow Inst. of Physical Culture; head of a library 1944–46; Komsomol work 1952–58; Sec. 1958–59, 1st Sec. and mem. Bureau Cen. Cttee. USSR Komsomol 1959–68; mem. Cen. Cttee. CPSU 1961–71; Dep. to and mem. Foreign Affairs Comm. at Sov. of the Union USSR Supr. Sov. 1962–70; Chair. of Cttee. for Physical Culture and Sports with USSR Council of Mins. 1968–; mem. Cen. Auditing Comm. CPSU 1971–. Decors.: 2 orders and several medals.

Pavlov, Vladimir Iakovlevich
Diplomat; b. O. 26, 1923, Mosalsk, Kaluga Oblast; s. of a clerk.
Russian; mem. CPSU 1948–; grad. Moscow Inst. of Railway Engineering 1950; electrician at construction of railways 1941–44; student 1944–50, Sec. of Komsomol Cttee. at Moscow Inst. of Railway Engineering 1948–50; 1st Sec. of Dzerzhinskii Raion Komsomol Cttee. in Moscow, Sec. of Moscow City Komsomol Cttee. 1950–56; Dep. Dept. Head, Dept. Head at Moscow City CPSU Cttee., 1st Sec. of Dzerzhinskii Raion Cttee. CPSU in Moscow 1956–62; Sec. 1962–65, 2nd Sec. of Moscow City Cttee. CPSU 1965–71; Dep. to 1963–67, mem. Presidium RSFSR Supr. Sov. 1965–67; mem. Cen. Cttee. CPSU 1966–; Dep. to USSR Supr. Sov. 1966–; USSR Amb. to Hungary Mr. 1971–82, to Japan Ja. 1982–. Decors.: Order of Red Banner of Labour, Order of the October Revolution, 2 other orders and various medals.

Pavlova, Dora Georgievna
Author; b. N. 13, 1925, Moscow.
Russian; mem. CPSU 1948–; grad. Moscow Tech. Bauman College 1949.
Publs.: "Sovest'" (Conscience) 1962, "Chastnyi sluchai" (A Personal Case) 1967, and others.

Pavlova, Muza Konstantinova
Author and dramatist; b. D. 28, 1916 (Ja. 10, 1917), Perm.
Russian; first works publ. 1942.
Publs.: "Polosataia smert'" (The Striped Death) 1943, "Den' poezii" (Day of Poetry) 1966, and others.

Pavlovskaia, Aleksandra Georgievna
Graphic artist; b. 1928, Kiev.
Attended Kiev Art Inst. 1947–54, pupil of V. Kasian, I. Pleshchinskii, S. Grigorev.

Pavlovskii, Evgenii Vladimirovich, Dr. of Geol.-Min. Sc.
Geologist; b. Ap. 9 (22), 1901, Moscow.
Grad. Moscow Mining Acad. 1928; Dr. of Geol.-Min. Sc. 1941; mem. staff Geological Inst. of USSR Acad. of Sciences 1932–41 and 1956–; worked in various geological institutions in Eastern Siberia 1941–56; Chair. of Eastern Siberian branch of USSR Acad. of Sciences 1953–56; Prof. at Irkutsk Univ. 1946–56.

Pavlovskii, Ivan Grigor'evich
Military official; b. F.11 (24), 1909, Teremkovtsy, Khmelnitskii Oblast.
Ukrainian; mem. CPSU 1939–; completed first course at Frunze Mil. Acad. 1941; grad. Acad. of Gen. Staff 1948; Gen. of the Army 1967–; agronomist, chief agronomist 1929–31; in Sov. Army 1931–; chief-of-staff, commander of a regiment, a brigade, a div. 1941–45; command posts 1954–58; Dep. to and mem. Presidium Supr. Sov. Armenian SSR 1955–59; 1st Dep. Commander of Transcaucasian Mil. District 1958–61; Commander of Volga Mil. District 1961–63; Dep. to USSR Supr. Sov. 1962–; Commander of Far Eastern Mil. District 1963–67; mem. Cen. Auditing Comm. 1966–71, mem. Cen. Cttee. CPSU 1971–81; USSR Dep. Min. of Defence, Commander-in-Chief of USSR Ground Forces 1967–81; Chief of Sov. Army Delegation to Judia My. 1977. Decors.: 3 Orders of Lenin, 6 Orders of Red Banner, Suvorov Order, Order of Red Star, foreign orders and medals, Hero of the Sov. Union 1969.

Pedaias, August Augustovich
State official.
Mem. Auditing Comm. 1964–66, cand. mem. Cen. Cttee. Esthonian CP 1966–76, mem. 1976–; Min. of Supplies Esthonian SSR Ja. 1970–82; Dep. Chair. of Cttee. of People's Control Esthonian SSR Ag. 1982–.

Pedak, Matti Arnoldovich
Party official.
Dep. of Supr. Sov. Esthonian SSR 1980–; 1st Sec. of Kalinin Raion Cttee. Esthonian CP in Tallinn 1980–82; mem. Cen. Cttee. Esthonian CP 1981–; 1st Sec. of Tallinn City Cttee. Esthonian CP F. 1982–; cand. mem. Bureau Cen. Cttee. Esthonian CP F. 1982–; Chair. of Supr. Sov. Esthonian SSR Jl. 1982–.

Pegov, Nikolai Mikhailovich
Party official; b. 1905.
Mem. CPSU 1930–; uncompleted high-school educ.; worked at "Krasnaia Roza" Factory in Moscow, Dep. Dir. of Gorkii Factory in Moscow, Dep. Dir. of Kuibyshev Factory in Margelan (Uzbekistan) 1919–35; studied at Ind. Acad. 1935–38; Dep. to USSR Supr. Sov. 1937–48; 1st Sec. of Primore Krai Cttee. All-Union CP(B) 1938–47; 1st Sec. of Vladivostok City All-Union CP(B) Cttee. 1940–47; mem. Cen. Cttee. CPSU 1939–; worked for Cen. Cttee. CPSU 1947–52; Sec. and cand. mem. Presidium Cen. Cttee. CPSU 1952–53; Sec. of Presidium USSR Supr. Sov. 1953–56; USSR Amb. to Iran 1956–63, to Algeria 1964–67, to India 1967–71; USSR Dep. Min. of Foreign Affairs Ap. 1973–Ja. 1976; Chief of Cen. Cttee. CPSU Dept. for Cadres Abroad 1975–. Decors.: 3 Orders of Lenin, Order of Red Banner of Labour, Order of Patriotic War 1st Cl., 4 medals.

Peive, Aleksandr Vol'demarovich
Geologist; b. Ja.27 (F.9), 1909, Sementsevo, now Toropets Raion, Kalinin Oblast.
Academician of the Academy of Sciences of the USSR (1964). Member of the CPSU since 1953.
He graduated from the Moscow Geological Prospecting Institute in 1930 and taught there from 1932 to 1935. From 1929 to 1935 he worked at the Scientific Institute for Fertilizers. He began working at the Geological Institute of the Academy of Sciences of the USSR in 1935, becoming the institute's director in 1960.
His principal works are devoted to development of the theory of geosynclines, clarification of the role of horizontal tectonic movements, and study of the oceanic crust of the geological past and the natural laws of formation of various minerals. He founded the doctrine of abyssal fractures (or lineaments) and their relation to magmatism and metallogeny. He also discovered several deposits of bauxites (State Prize of the USSR, 1946), phosphorites, and potassium salts. He received the State Prize of the USSR in 1969 for his work in compiling a tectonic map of Eurasia on a scale of 1:5,000,000. In 1972, he became president of the International Subcommission on the Tectonic Map of the World. Decors.: Order of Lenin, three other orders, and various medals.
Works: "Glubinnye razlomy v geosinklinal'nykh oblastiakh." *Iv. AN SSSR: Seriia geologicheskaia,* 1945, no. 5. *Tektonika Severoural'skogo boksitovogo poiasa.* Moscow 1947. "Obshchaia kharakteristika, klassifikatsiia i prostranstvennoe raspolozhenie glubinnykh razlomov. Glavneishie tipy glubinnykh razlomov." *Izv. AN SSSR: Ser. geologicheskaia,* 1956, no. 1. "Sviaz' osadkonakopleniia, skladchatosti, magmatizma i mineral'nykh mestorozhdenii s glubinnymi razlomami. Glavneishie tipy glubinnykh razlomov." *Ibid.,* no. 3. "Tektonika i magmatizm." *Ibid.,* 1961, no. 3. "Okeanicheskaia kora geologicheskogo proshlogo." *Geotektonika,* 1969, no. 4.

Pekar, Solomon Isakovich, Dr. of Phy.-Math. Sc.
Theoretical physicist; b. Mr. 16, 1917, Kiev.
Grad. Kiev Univ. 1938; mem. Acad. of Sciences Ukrainian SSR 1961–; Dr. of Phy.-Math. Sc. 1941; Prof. 1944–; mem. staff Physics Inst. of Acad. of Sciences Ukrainian SSR 1938–41; head of a laboratory in Ufa 1941–43; again mem. staff Inst. of Physics of Acad. of Sciences Ukrainian SSR 1944–60; Head of Theoretical Dept. of Inst. of Semi-Conductors at Acad. of Sciences Ukrainian SSR 1960–; concurrently Lecturer at Kiev Univ. 1938–41 and 1944–; author of some 100 papers. Decors.: Order "Badge of Honour", medals.
Publs.: "Issledovaniia po elektronnoi teorii kristallov" 1951, "O vliianii deformatsii reshetok elektronami na opticheskie i elektricheskie svoistva kristallov" in "Uspekhi fizicheskikh nauk" 1953, "Nekotorye voprosy teorii poluprovodnikov i dielektrikov i puti dalneishego razvitiia teorii" in "Uspekhi fizicheskikh nauk" vol. 60, 1956.

Pelshe, Arvid Ianovich (Pelše, Arvids)
Party official; b. F. 7, 1899, Mazaia Farm, Bauska District, Kurland, now Latvian SSR; s. of a peasant.
Mem. CPSU 1915–; grad. Inst. of Red Professors 1931; worked in Riga, Vitebsk, Kharkov, Petrograd and as docker in Arkhangelsk 1914–17; Tech. Sec. of Transport Workers' Union 1917–18; mem. Petrograd Sov. after February Revolution 1917; worked for Cheka (state security) in Moscow 1918; official of Latvian govt. 1919; CPSU work in Sov. Army and Navy 1919–29; studied at Inst. of Red Professors, active as teacher 1929–32; CPSU posts in Kazakhstan and Moscow 1933–37; taught in Moscow 1937–40; Sec. of Cen. Cttee. Latvian CP, responsible for Propaganda and Agitation 1941–59; Dep. to USSR Supr. Sov. 1946–; mem. Bureau, then Presidium Cen. Cttee. Latvian CP 1956–66; 1st Sec. of Cen. Cttee. Latvian CP 1959–66; mem. Cen. Cttee. CPSU 1961–; mem. Politbureau and Chair. of Comm. for Party Control at Cen. Cttee. CPSU Ap. 1966–. Decors.: 2 Orders of Lenin, medals.

Penkov, Aleksandr Mikhailovich, Dr. of Tech. Sc.
Mechanics scientist; b. Ap. 11, 1906, Ekaterinoslav, now Dnepropetrovsk; s. of a worker.
Grad. Dnepropetrovsk Mining Inst. 1930; corresponding mem. Acad. of Sciences Ukrainian SSR 1951–; Dr. of Tech. Sc. 1943; Prof. 1943–; mem. staff Inst. of Metallurgy 1930–37, of Dnepropetrovsk State Univ. 1933–41; mem. staff Inst. of Mining Mechanics of Acad. of Sciences Ukrainian SSR (now Fedorov Inst. of Mining Mechanics and Tech. Cybernetics of USSR Min. of the Coal Ind.) 1941–58 (with interruptions) and Inst. of Construction Mechanics of Acad. of Sciences Ukrainian SSR 1948–50; concurrently lecturer at various colleges in Moscow and Kiev; Head of Chair of Strength of Materials and Construction Mechanics of Flying Machines at Kiev Inst. of Civil Aviation Engineering 1954–, concurrently Dep. Rector of same inst. 1954–58; author of some 60 scientific papers. Decors.: Order "Badge of Honour", medals.
Publ.: "Bokovi kolyvannia ta rezonans pidiimalnykh kanativ" in "Dopovidi AN USSR" No. 1, 1947.

Penu, Raimond Aleksandrovich
State official.
Mem. Cen. Cttee. Esthonian CP 1971–; Chair. of State Cttee. of Television and Radio Esthonian SSR S. 1974–82; Chair. of State Cttee. for Film Ind. Esthonian SSR Mr. 1982–.

Peredel'skii, Georgii Efimovich
Military official; b. Mr. 25 (Ap. 7), 1913, Orlovka, now in Chulym Raion, Novosibirsk Oblast.
Marshal of the artillery (1973). Member of the CPSU since 1939.
He joined the Red Army in 1934. He graduated from the Omsk Military School in 1937, from the Higher Officers' Artillery School in 1948, and from the Higher Academic Training Courses of the Military Academy of the General Staff in 1957. By taking correspondence courses, he also graduated from the M. V. Frunze Military Academy in 1965. He saw service in the Soviet-Finnish War of 1939–40 as assistant chief-of-staff of an artillery regiment. In the Great Patriotic War (1941–45), he served on the Karelian Front as assistant chief and chief-of-staff of an artillery regiment. In 1943 he became commander of an artillery regiment.
After the war, he held responsible positions in the artillery headquarters of several military districts. He was appointed chief-of-staff of the artillery of the Northern Military District in 1953 and artillery commander of the district in 1959. He became chief of the missile forces and artillery of the Transcaucasian Military District in 1962. In May 1965 he was appointed deputy commander, and in July 1969 commander, of the missile forces and of ground forces artillery. Decors.: Order of the October Revolution, three Orders of the Red Banner, the Order of Suvorov Third Class, the Order of the Red Star, various medals, and two foreign orders.

Perepelkin, Iurii Iakovlevich
Egyptologist; b. My. 19 (Je. 11), 1903, St. Petersburg.
Doctor of historical sciences (1969).
He graduated from the department of linguistics and material culture of Leningrad University in 1927. In 1948 he became a senior research worker in the Leningrad section of the Institute of Oriental Studies. His main works are devoted to the socioeconomic history of ancient Egypt at the time of the Old Kingdom and to the reforms of Amenhotep IV (Ikhnaton).
Works: *Chastnaia sobstvennost' v predstavlenii egiptianin Starogo tsarstva.* Moscow-Leningrad 1966. *Perevorot Amenkhotpa IV,* part 1. Moscow 1967. *Taina Zolotogo groba.* Moscow 1968.

Peresypkin, Ivan Pavlovich
State official.
Min. of Fish Ind. RSFSR Je. 1982–.

Peresypkin, Oleg Gerasimovich
Diplomat.
Emb. Councellor at USSR Embassy in Arab Republic of Yemen 1971–76; Adviser, Near Eastern Countries Dept., USSR Min. of Foreign Affairs 1976–80; USSR Amb. to Arab Republic Yemen Jl. 1980–.

Perevalov, Nikolai Illarionovich (real name: Kolesnikov)
Author and poet; b. N. 24, 1918, Shabanka, Kirov Oblast; s. of a peasant.
Russian; grad. Gorkii Inst. of Literature 1956; lives in Novosibirsk.
Publs.: "V selskoi shkole" (In a Village School) 1949, "I khleb i sol'" (Bread as well as Salt) 1959, "Ia skoro uedu" (I Shall Soon be Going) 1964, "Natasha" 1960, and others.

Pereverzeva, Nina Vasil'evna
Collective farmer; b. 1929.
Mem. CPSU 1970–; grad. school for working youth 1978; farmer on "Put' Lenina" Collective Farm in Rostov Oblast 1945–66; combine operator on "Put' Lenina" Collective Farm 1966–74; leader of harvest field team on "Put' Lenina" Collective Farm 1974–; cand. mem. Cen. Cttee. CPSU 1981–82, mem. N. 1982–. Decors.: Heroine of Socialist Labor 1973, USSR State Prize 1977.

Permiak, Evgenii Andreevich
Author; b. O. 18 (31), 1902, Perm.
Russian; grad. Dept. of Pedagogics of Perm Univ. 1930; first works publ. 1924. Decors.: 2 orders and various medals.
Publs.: "Dedushkina kopilka" (Grandfather's Savings Box) 1957, "Zamok bez kliucha" (A Lock Without Key) 1962, "Skazka o serom volke" (The Tale of the Grey Wolf) 1960, "Gorbatyi medved" (The Hunchbacked Bear) 1965, "Izbrannye proizvedeniia" (Selected Works) vols. 1–2, 1973, "Iargorod" 1973, and others.

Persianinov, Leonid Semenovich
Gynaecologist and obstetrician; b. Ag. 5 (18), 1908, Staroe Selo, Smolensk Oblast.
Mem. CPSU 1943–; grad. 2nd Medical Inst. Leningrad 1931; mem. USSR Acad. of Medical Sciences 1965–; Head of Chair of Obstetrics and Gynaecology at Minsk Medical Inst. 1951–58, at Moscow 2nd Medical Inst. 1958–67 and at Moscow 1st Medical Inst. 1967–; concurrently Dir. of Scientific Union Inst. for Research on Obstetrics and Gynaecology 1967–; Chair. of Scientific Union Soc. of Obstetricians and Gynaecologists 1960–; permanent mem. Exec. Cttee. 1967–70, Vice-Pres. of Int. Fed. of Obstetricians and Gynaecologists 1967–70; hon. mem. various foreign medical assens. Decors.: Order of Lenin, Order of the October Revolution, Order of Red Star, medals.

Perventsev, Arkadii Alekseevich
Russian writer; b. Ja. 13 (26), 1905, Nagut, now Nagutskoe, Stavropol' Krai.
Member of the CPSU since 1950.
He studied at the Bauman Higher Technical School in Moscow from 1929 to 1933 and began publishing in 1937. During the Great Patriotic War of 1941–45 he was a war correspondent for *Izvestiia.* His first novel, *Kochubei* (books 1–2, 1973; film of the same name, 1958), is a romantic narrative about a hero of the Civil War of 1918–20. The novels *Fiery Land* (1945) and *Honor Since Youth* (1948; State Prize of the USSR, 1949) depict the heroism of the Soviet people in the struggle against fascism.
He has also published the short story collections *From the Same Crew* (1954) and *The Girl from Taman'* (1959) and the novels *Sailors* (1961), *Gamaiun, the Prophetic Bird* (1963), *The Olive Branch* (1965), and *The Secret Fron* (1972). He wrote the screenplays for *The Third Blow* (1948; remade as *Southern Junction*) and *The Shipka Heroes* (1955). Decors.: Order of Lenin, four other orders, and several medals.
Works: *Izbr. proizv.,* vols. 1–2. Simferopol' 1950. *Chernaia buria: Roman.* Moscow 1974.

Perventsev, Evgenii Ivanovich
State official; b. 1926.
Russian; mem. CPSU 1950–; grad. Party Higher School of Cen. Cttee. CPSU, Major-Gen.; served in the Soviet Army 1943–45; Teacher, held Komsomol posts 1945–50; with Sov. State Security Service (KGB) 1950–; Chair. Cttee. of State Security Yakut ASSR 1964–70; mem. Apparat USSR Cttee. of State Security 1970–75; Chair. Cttee. of State Security Tadzhik SSR O. 1975–; mem. Cen. Cttee. and mem. Bureau Cen. Cttee. Tadzhik CP 1976–.

Pervyshin, Erlin Kirikovich
State official; b. 1932.
Russian; mem. CPSU 1959–; grad. Moscow Electrical Engineering Inst. for Communications 1955, Cand. of Tech. Sc. 1971; engineer, assembly section chief, assembly admin. chief, manager of a assembly design trust 1957–69; Dir. Gen. of All-Union Scientific Production Asscn. 1969–70; Dep. USSR Min. of the Radio Ind. 1970–74; Dep. to USSR Supr. Sov. 1974–; USSR Min. of the Communications Equipment Ind. Ap. 1974–; cand. mem. Cen. Cttee. CPSU 1976–. Decors.: USSR State Prize 1978.

Peryshkin, Viktor Aleksandrovich
Diplomat.
USSR Amb. to Djibouti N. 1978–.

Peskov, Vasilii Mikhailovich
Author and journalist; b. Mr. 14, 1930, Orlovo, Voronezh Oblast.
Russian; mem. CPSU 1951–; active as journalist 1953–. Decors.: 2 orders, Lenin Prize 1963.

Publs.: "Zapiski fotoreportera" (Notes of a Press Photographer) 1960, "Shagi po rose" (Steps in the Dew) 1963, "Belye sny" (White Dreams) 1965, "Krai sveta" (The End of the World) 1967, "Po dorogam Ameriki" (On the Roads of America) 1973, and others.

Pestov, Vasilii Serafimovich
Oblast Soviet official; b. 1925.
Mem. CPSU 1956–; grad. Tula Mechanical Inst. 1950; metal worker at Degtiarev Plant in Vladimir Oblast 1942–44; chief of trade union cttee. of Degtiarev Plant 1959–60; Dir. of "Stroiindustriia" Plant 1963–66; Dir. of "Konditsioner" Plant 1966–68; Chair. of Exec. Cttee. of Podolsk Raion Sov. of Peoples' Deps. 1968–70; 1st Sec. of Podolsk City Cttee. CPSU 1970–75; Chair. of Moscow Oblast Cttee. of Peoples' Control 1978–81; Chair. of Exec. Cttee. of Moscow Oblast Sov. of Peoples' Deps. Ja. 1981–; cand. mem. Cen. Cttee. CPSU 1981–.

Peterson, Erik Karlovich
State official.
Cand. mem. 1960–61, mem. Cen. Cttee. Latvian CP 1961–; Head of Dept. of Ind. and Transport until 1960; Head of Dept. of Construction at Cen. Cttee. Latvian CP 1960–63; mem. Bureau for Ind. at Cen. Cttee. Latvian CP 1962–63; Dep. to Supr. Sov. Latvian SSR 1963–; Min. of Construction Latvian SSR 1963–67; Dep. Chair. of Council of Mins. Latvian SSR 1967–70 and Jl. 1975–; Sec. of Cen. Cttee. 1970–75, mem. Bureau Cen. Cttee. Latvian CP 1970–.

Petkiavichius, Iuozas Iuozovich
State official; b. 1924.
Lithuanian; mem. CPSU 1947–; grad. CPSU Cen. Cttee. Party Higher School, Maj.-Gen.; Komsomol posts 1942–52; 1st Sec. of Cen. Cttee. Lithuanian Komsomol 1952–60; cand. mem. Bureau Cen. Cttee. Lithuanian CP 1954–60 and 1976–; mem. Cen. Cttee. Lithuanian CP 1954–60 and 1971–; Dep. to Supr. Sov. Lithuanian SSR 1959–; Dep. Chair. 1960–67, Chair. of Cttee. of State Security. Lithuanian SSR Ja. 1967–. Decor.: Order of Lenin.

Petrenko, Vladimir Vasil'evich
Sculptor; b. 1926, Ivanovka, Donetsk Oblast.
Grad. Dnepropetrovsk Art School 1955 and Kharkov Art Inst. 1961, pupil of M. Riabinin.

Petrianov-Sokolov, Igor Vasil'evich
Physical chemist; b. Je. 5 (18), 1907, Bolshaia Iakshen, now Gorkii Oblast.
Grad. Moscow Univ. 1930; mem. USSR Acad. of Sciences 1966–; mem. staff 1930–, dept. head at Karpov Inst. of Physical Chemistry 1938–; concurrently Prof. at Mendeleev Inst. of Chemical Technology in Moscow 1947–; Chief Ed. of journal "Khimiia i zhizn" 1964–. Decors.: State Prize of the USSR 1941, Lenin Prize 1966, Hero of Socialist Labour 1971, 3 Orders of Lenin, 3 other orders and various medals.

Petrik, Pavel Petrovich
Party official; b. 1925.
Ukrainian; mem. CPSU 1948–; grad. Pedagog. and Polytechn. Inst. in Kishinev and Party Higher School of CPSU Cen. Cttee.; held Komsomol posts 1947–55; Sec., 2nd Sec., 1st Sec. of Floreshty Raion Cttee. Moldavian CP, Partorg at Territorial Kolkhoz-Sovkhoz Admin. in Floreshty 1955–62; mem. Cen. Cttee. Moldavian CP 1961–; Head of Ideological Dept. at Agricultural Bureau of Cen. Cttee. Moldavian CP 1962–64; 1st Sec. of Tiraspol City Cttee. Moldavian CP 1964–72; Chair. of Moldavian Republican Trade Union Council 1972–80; cand. mem. of Bureau Cen. Cttee. Moldavian CP 1972–76, mem. Ja. 1976–; Sec. Cen. Cttee. Moldavian CP My. 1980–.

Petrishchev, Aleksei Georgievich
State official; b. 1924.
Mem. CPSU 1951–; grad. Dnepropetrovsk Inst. of Chemical Technology 1949; shop foreman, dep. shop superintendent, dep. chief mechanic, chief mechanic, sec. of party cttee. of Chernorechensk Plant 1949–62, Dir. 1962–67; Dir. Chernorechensk Combine 1967–70; chief of an all-union asscn. of USSR Min. of the Chemical Ind. 1970–77; Dep. Chair. of State Cttee. for Material and Technical Supply of USSR Council of Mins. 1977–80; USSR Min. for the Prod. of Mineral Fertilizers N. 1980–; Dep. to USSR Supr. Sov. 1966–74; cand. mem. Cen. Cttee. CPSU 1981–. Decor.: Hero of Socialist Labor 1971.

Petrosian, Feliks Gevorkovich
State official.
Head of Dept. for Ind. and Transport resp. of Dept. for Ind. with Cen. Cttee. Armenian CP 1977–81; mem. Cen. Cttee. Armenian CP 1981–; Dep. Chair. Council of Mins. Armenian SSR F. 1981–.

Petrosian, Frunze Amaiakovich
State official.
Mem. Auditing Comm. Armenian CP 1971–; Head of Admin. for Non-Ferrous Metals Armenian SSR Ja. 1975–.

Petrosian, Suren Martirosovich
Party official; b. 1925.
Armenian; mem. CPSU 1951–; incomplete secondary-school education; fitter at the Kirov plant in Erevan 1940–43; served in Sov. Army 1943–48; head of the "Nairit" Scientific Production Asscn. in Erevan 1948–; cand. mem. Cen. Cttee. CP Armenia 1960–64, mem. 1964–; Dep. to USSR Supr. Sov. 1966–; mem. Bureau of Cen. Cttee. CP Armenia Ja. 1981–.

Petrosian, Tigran Vartanovich, Cand. of Phil. Sc.
Chess champion; b. Je. 17, 1929, Tbilisi. USSR and Int. Grand Master 1952; journalist; USSR Champion 1959, 1961, 1969, World Champion 1963–66, 1966–69; mem. of USSR team which gained 1st place at Chess Olympics 1958–74 and of team which participated in European Championship 1957–73; mem. Presidium USSR Fed. of Chess-players 1958–; ed. of weekly supplement "64" of newspaper "Sovetskii sport" 1958–. Decors.: Merited Master of Sports, orders and medals.

Petrosiants, Andronik Melkonovich
State official; b. 1906.
Mem. CPSU 1932–; grad. Uralsk Inst. of Mechanical Engineering 1933; mem. staff Ordzhonikidze Machine Building Plant in the Urals 1933–39; Dep. People's Comm. for Heavy Machine Building and mem. Collegium of Min. of Heavy Machine Building, then USSR 1st Dep. People's Comm. for Machine Tool Building 1939–41; Dep. People's Comm. for the USSR Armoured Vehicles Ind. and mem. staff USSR Min. of Medium Machine Building 1941–55; USSR Dep. Min. of Medium Machine Building 1955–62; Chair. of State Cttee. for the Use of Atomic Energy with USSR Council of Mins. F. 1962–. Decors.: Hero of Socialist Labour, State Prize.

Petrov, Aleksandr Petrovich
Railroad transportation engineer; b. Ag. 19 (S. 1), 1910, Moscow.
Mem. CPSU 1945–; grad. Moscow Inst. of Railroad Transportation 1934; corresponding mem. USSR Acad. of Sciences 1953–; teacher 1936–38, Head of Chair at Inst. of Transportation Engineering in Moscow 1938–41; Head of Cen. Servicing Admin. Depts. of USSR Min. of Transportation 1941–; mem. staff Section for the Scientific Solution of Transportation Problems at USSR Acad. of Sciences 1946–49; Prof. 1950–, Dep. Dir., Head of Scientific Inst. of Railroad Transportation 1959–. Decors.: 2 Orders of Lenin, 3 other orders and various medals.

Petrov, Anatolii Aleksandrovich
Organic chemist; b. Mr. 13 (26), 1913, Livny, Orel Oblast.
Mem. CPSU 1968–; grad. Voronezh Univ. 1934; mem. staff Voronezh Univ. 1934–41; Prof. Voronezh Aviation Inst. 1941–45, then at Inst. of Aviation Instrument Construction in Leningrad; Prof. and Head of Chair at Technological Lensovet Inst. Leningrad 1951–. Decors.: Order of Lenin, 3 other orders and various medals.

Petrov, Georgii Ivanovich
Hydro-aeromechanics and gas dynamics engineer; b. My. 18 (31), 1912, Pinega, now Arkhangelsk Oblast.
Grad. Moscow State Lomonosov Univ.

1935; mem. USSR Acad. of Sciences 1958–; mem. staff of various scientific research institutions 1935–; Dir. 1965–73, head of a dept. at Inst. of Cosmic Research of USSR Acad. of Sciences 1973–; his earlier works are concerned with problems of stability of vortex layers, oscillation in viscous liquids, physical conditions in the destruction of laminar flow; he proved the convergence of Galerkin's method for seeking the characteristic value in a wide range of equations, including non-conservative systems (particularly equations measuring oscillation in viscous liquids). Decors.: 3 Orders of Lenin, 2 other orders and various medals.

Petrov, Ivan Ivanovich (real surname: Krauze)
Russian bass; b. F.29, 1920, Irkutsk.
People's Artist of the USSR (1959).
From 1938 to 1941, he attended the Glazunov School of Music, where he studied in the class of A. K. Mineev. He gave concerts at the Moscow Philharmonic Society and performed in productions of the Opera Ensemble, which was directed by I. S. Kozlovskii. Between 1943 and 1970, Petrov was a soloist at the Bolshoi Theater.
His powerful voice has a beautiful timbre and expressive projection. His roles have included Boris Godunov in Mussorgsky's opera of the same name, Kochubei in Tchaikovsky's *Mazepa* (State Prize of the USSR, 1950), Dositheus in Mussorgsky's *Khovanshchina* (State Prize of the USSR, 1951), and Mephistopheles in Gounod's *Faust*.
He, who has toured abroad, is an honorary member of the Paris Opera. Decors.: Order of Lenin and various medals.

Petrov, Mikhail Nikolaevich
Diplomat.
USSR Amb. to Botswana D. 1977–.

Petrov, Mikhail Platonovich, Dr. of Biol. Sc.
Geographer; b. S.26 (O.10), 1906, Zuevka, Kirov Oblast.
Grad. Dept. of Geography of Leningrad Univ. 1930; active as botanical and physical geographer; Dr. of Biol. Sc. 1940; mem. Acad. of Sciences Turkmen SSR 1951–; Prof. at Leningrad Univ. 1958–; Vice-Pres. of Acad. of Sciences Turkmen SSR 1951–56; Vice-Pres. of USSR Geographical Soc. 1970–. Decors.: Grand Medal in Gold of USSR Geographical Soc., Order of Lenin, 1 other order.

Petrov, Vasilii Ivanovich
Military official; b. 1917.
Russian; mem. CPSU 1944–; grad. M.V. Frunze Mil. Acad. 1948, Army Gen.; teacher of a secondary school 1937–39; entered Sov. Army 1939; command and staff posts 1941–45; commander of a regiment, of a division, chief-of-staff of a mil. district 1945–72; Commander of Troops of Far Eastern Mil. District 1972–76; Dep. to USSR Supr. Sov. 1974–; 1st Dep. Commander-in-Chief of USSR Ground Forces 1976–80, Commander-in-Chief D. 1980–; mem. Cen. Cttee. CPSU 1976–.

Petrov, Vasilii Stepanovich
Military official; b. Je. 22, 1922, Dmitrievka, now in Priazovskoe Raion, Zaporozh'e Oblast.
Twice Hero of the Soviet Union (D.24, 1943, and Je. 27, 1945). Major-general of the artillery (1962). Member of the CPSU since 1945.
He has been in the Red Army since 1939. He graduated from the Sumy Artillery School in 1941. He fought in the Great Patriotic War (1941–45) on the Southwestern, Briansk, Voronezh, Don, and First Ukrainian fronts. He served as a platoon commander, deputy commander of a battery, assistant chiefof-staff of a regiment and deputy commander of the 1850th Tank-destroyer Artillery Regiment. In 1943, he lost both arms in combat near Kremenchug. By his own request he was sent to the army in the field in 1944 and served as deputy commander and then commander of the 248th Tank-destroyer Regiment.
After the war, he continued in the cadres of the armed forces. In 1954 he graduated by correspondence from the history department of the University of L'vov. Decors.: Order of Lenin, the Order of the Red Banner, three Orders of the Red Star, various medals, and two foreign orders.

Petrov, Vladislav Alekseevich
Laborer; b. 1935.
Mem. CPSU 1961–; finished secondary school 1959; lathe operator in a plant on Kharkov 1952–54; served in Sov. Army 1954–57; lathe operator in a plant in Kharkov 1957–; mem. Cen. Cttee. CPSU 1981–. Decor.: Hero of Socialist Labor 1976.

Petrov, V. V.
Komsomol official.
2nd Sec. and mem. Bureau Cen. Cttee. Azerbaidzhan Komsomol Ja. 1980–; cand. mem. Cen. Cttee. Azerbaidzhan CP 1981–.

Petrovichev, Nikolai Aleksandrovich
Party official; b. 1918.
Mem. CPSU 1939–; grad. Moscow Pedagogic Oblast Inst. 1939; apprenticeship as fitter, then fitter, Brigadier, Dep. Sec. of Komsomol Cttee., Dep. Dir. of Factory School (FZU) for Pol. Questions at Leningrad Kirov Works 1932–38; served in Sov. Army 1938–46; 1st Sec. of Krasnaia Gorka (Moscow Oblast) City Cttee. CPSU, Dept. Head at Moscow City Cttee. CPSU 1947–61; mem. staff Cen. Cttee. CPSU 1961–68; 1st Dep. Head of Dept. of Party Org. Work with Cen. Cttee. CPSU 1968–; cand. mem. Cen. Cttee. CPSU 1971–81, mem. 1981–. Decor.: Order of Lenin.

Petushkova, Elena Vladimirovna
Equestrian; b. N.17, 1940, Moscow.
Honored Master of Sports (1970). Candidate in the biological sciences (1967). Research worker at Moscow State University. Member of the CPSU since 1971.
She was champion of the USSR in dressage from 1966 to 1974, world champion in 1970, and Olympic champion at the 1972 Munich games. Decors.: two orders.

Piatibrat, Vasilii Lukich
State official.
Min. of Ind. Construction Azerbaidzhan SSR Ag. 1972–79; mem. Cen. Cttee. Azerbaidzhan CP 1976–79; USSR Dep. Min. of Ind. Construction Je. 1970–.

Piatkov, Viktor Evgenevich
Laborer; b. 1929.
Mem. CPSU 1963–; grad. school for working youth 1961; metal engraver at a plant in Ufa 1946–50; served in Sov. Army 1950–55; petroleum and gas operator at "Oktiabr'skneft'" Admin. in Bashkir ASSR 1956–65, at "Iuganskneft'" Admin. in Tiumen Oblast 1965–; cand. mem. Cen. Cttee. CPSU 1981–. Decor.: Hero of Socialist Labor 1981.

Pichuzhkin, Mikhail Sergeevich
State official; b. 1925.
Russian; mem. CPSU 1945–; grad. Moscow Energetics Inst.; served in Sov. Army 1942–46; shift supervisor, mechanic, shop superintendent, dep. chief mechanic at V. I. Lenin Metallurgical Plant in Krivoi Rog 1953–57, sec. of party cttee. at this plant 1957–62; 2nd Sec. of Krivoi Rog City Cttee. Ukrainian CP 1962–65, 1st Sec. 1965–67; mem. Cen. Cttee. Ukrainian CP 1966–71 and 1980–, cand. mem. 1971–80; 2nd Sec. of Dnepropetrovsk Oblast Cttee. Ukrainian CP 1967–74; 1st Dep. Chair. of Cttee. for People's Control Ukrainian SSR 1974–76; Permanent Rep. of Council of Mins. Ukrainian SSR to USSR Council of Mins. 1976–; Dep. to USSR Supr. Sov. 1979–.

Pidoplichko, Ivan Grigor'evich
Zoologist and palaeontologist; b. Ag.2, 1905, Kozatskoe, now Cherkassy Oblast; s. of a cabinetmaker.
Mem. CPSU 1944–; grad. Advanced Pedagogic Courses in Korsun-Shevchenkovskii 1924 and Inst. of Applied Zoology and Phytopathology in Leningrad 1927; concluded post-graduate studies at Ukrainian Scientific Geology Research Inst. 1933 and at Inst. of Zoology and Biology of Acad. of Sciences Ukrainian SSR 1935; corresponding mem. Acad. of Sciences Ukrainian SSR 1961–; Dr. of Biol. Sc. 1950; Prof. 1952–; mem. staff Inst. of Archaeology of Acad. of Sciences Ukrainian SSR 1934–47; concurrently Head of Dept. of Palaeozoology at Inst. of Zoology and Biology of Acad. of Sciences Ukrainian SSR 1935–39; Lecturer 1939–59, Prof. 1952–, Dep. Rector

of Kiev Univ. 1947–48; served in Sov. Army 1941–45; Dep. Chief Ed. of "Ukrainian Soviet Encyclopedia" 1957–67; Dir. of Inst. of Zoology 1965–, mem. Presidium and Acad. Sec. of Dept. of Gen. Biology of Acad. of Sciences Ukrainian SSR 1966–; author of some 350 scientific papers. Decors.: Order of Red Star, Merited Scientist of the Ukrainian SSR 1965.
Publs.: "Materialy do vyvchennia mynulykh faun USSR" vols. 1–4, 1938–56, "O lednikovom periode" vols. 1–4, 1946–56, "Korotka istoriia zemli" 1958, "Materialy z antropologii Ukrainy" 1960, "Tvarynnyi svit" in "Ukrainska Radianska Entsyklopediia" vol. 17, 1965.

Pidoplichko, Nikolai Makarovich, Dr. of Biol. Sc.
Botanist-mycologist; b. Ap. 3, 1904, Kozatskoe, now Cherkassy Oblast; s. of a peasant.
Grad. Kiev Inst. of Popular Educ. 1929; corresponding mem. Acad. of Sciences Ukrainian SSR 1957–; Dr. of Biol. Sc. 1955; Prof. 1956–; mem. scientific staff Botanical Collection and Herbarium of Acad. of Sciences Ukrainian SSR 1924–28; concurrently teacher at Kiev Inst. of Veterinary Zoo Technology 1926–29; mem. staff Ukrainian Scientific Research Inst. of the Sugar Ind. 1928–32; mem. staff 1931–, Head of Dept. of Mycology of Inst. of Microbiology (now Zabolotnyi Inst. of Microbiology and Virology) at Acad. of Sciences Ukrainian SSR 1933–; author of more than 100 scientific papers. Decors.: State Prize 1952, Order of Lenin, Order "Badge of Honour", medals.
Publs.: "Vyznachnyk grybivshkidnykiv kulturnykh roslyn" 1938, "Gribnaia flora grubykh kormov" 1953, co-author of "Ferment gliukozooksidaza i ego primenenie" 1964.

Pidsukha, Aleksandr Nikolaevich
Author; b. O. 16, 1918, Nezhilovichi, Kiev Oblast; s. of a peasant.
Ukrainian; mem. CPSU 1947–; grad. Kharkov Inst. of Foreign Languages 1939; first works publ. 1939.
Publs.: "Soldaty mira" (Soldiers of Peace) 1948, "Ia khochu mira" (I Want Peace) 1949, "Zhizn' za nas" (Life is For Us) 1950, "Materinskoe zaveshchanie" (A Mother's Legacy) 1964, "Litsom k litsu. Nevymyshlennye istorii" (Face to Face. True Stories) 1962, and others.

Pikhels, Valentina Samuilovna
State official.
Dep. to Supr. Sov. Latvian SSR 1955–; 1st Sec. of Limbazhi Raion Cttee. Latvian CP 1955; Chair. of Valka Raion Exec. Cttee. of Workers' Deps. 1959; cand. mem. 1960–63, mem. Cen. Cttee. Latvian CP 1963–; Min. of Public Welfare Latvian SSR S. 1961–.

Piliar, Iurii Evgenevich
Author; b. O. 14, 1924, Velikii Ustiug; s. of a teacher.
Russian; studied Gorkii Inst. of Literature; volunteer at the front 1941, wounded near Rzhev, taken prisoner and after 3 attempts at escape sent to Mauthausen concentration camp 1943; mem. of the int. anti-fascist underground org. of the camp, took part in uprising of camp inmates 1945; has lived in Moscow 1946–.
Publs.: "Vse eto bylo!" (All This Happened – Mauthausen) 1955, "Pozdnei vesnoi" (In Late Spring) 1960, "Istoriia moege druga" (The Story of My Friend) 1962, "Liudi ostaiutsia liudmi" (Men Remain Men) 1963, and others.

Piliugin, Nikolai Alekseevich
Scientist in automation and remote control; b. My. 5 (18), 1908, Krasnoe Selo, now in Leningrad Oblast.
Academician of the Academy of Sciences of the USSR (1966; corresponding member, 1958). Member of the Presidium of the Academy of Sciences of the USSR since 1967. Twice Hero of Socialist Labor (1956 and 1961). Member of the CPSU since 1940.
He graduated from the N. E. Bauman Moscow Higher Technical School in 1935. He worked at the N. E. Zhukovskii Central Aerohydrodynamics Institute from 1934 to 1941 and was director of a number of scientific research organizations. Since 1969 he has been head of a subdepartment at the Moscow Institute of Radio Engineering, Electronics, and Automation; he has been a professor since 1970. He is a chief designer of control systems and has directed the development of control systems for the first and a number of subsequent booster rockets that placed Soviet artificial earth satellites and many spacecraft in orbit.
He was a deputy to the seventh to ninth convocations of the Supreme Soviet of the USSR. Decors.: Lenin Prize in 1957 and the State Prize of the USSR in 1967, three Orders of Lenin, the Order of the October Revolution, and various medals.

Pilotovich, Stanislav Antonovich
State official; b. S. 25, 1922, Borok, Minsk Oblast.
Belorussian; mem. CPSU 1944–; grad. Minsk Pedagogic Inst. 1956; after completing school for med. personnel head of a med. centre 1940–41; partisan, underground Komsomol work in Minsk Oblast 1941–44; sec. of a raion Komsomol cttee., 1st Sec. of Baranovichi Oblast Komsomol Cttee. Belorussian CP, 1st sec. of a raion cttee. Belorussian CP, 2nd Sec. of Grodno Oblast Cttee. Belorussian CP, head of a dept. at Cen. Cttee. Belorussian CP 1944–62; cand. mem. 1956–60, mem. Cen. Cttee. Belorussian CP 1960–71 and 1981–; cand. mem. Bureau Cen. Cttee. Belorussian CP 1961–62; 1st Sec. of Vitebsk Oblast Cttee. Belorussian CP 1962–65 (for Agriculture 1963–64); Sec. and mem. Bureau Cen. Cttee. Belorussian CP 1965–71; USSR Amb. to Poland Mr. 1971–78; Dep. Chair. Council of Mins. Belorussian SSR My. 1978–; mem. Cen. Cttee. CPSU 1971–81. Decors.: 2 Orders of Lenin, 4 other orders and various medals.

Pimen (Sergei Mikhailovich Izvekov)
Russian Orthodox clergyman; b. Jl. 23, 1910, Bogordsk, Moscow Gouvt.
Ordained monk 1927; monk-deacon 1931; priest-monk 1932; abbot 1946; Prior of Pskov Cave Monastery 1949; Archimandrite 1950; appointed Head of St. Sergius Holy Trinity Monastery in Zagorsk 1954; Bishop of Balta and Substitute for the Diocese of Odessa 1957; moved to Moscow in same year and given title Bishop of Dmitrov; Head of Patriarchate Admin. and Archbishop 1959; Archbishop of Tula and Belev 1961; Metropolitan of Leningrad and Ladoga 1961; Metropolitan of the Bishopric of Krutitsy and Kolomna O. 1963–71; Patriarch of Moscow and all of Russia Je. 1971–; mem. World Peace Council; hon. mem. Acad. Senate of Leningrad and Moscow Theological Acads. Decors.: Order of Sov. Peace Foundation, Gold Medal "Peace Fighter" of Sov. Peace Cttee., decorations from Russian and other Eastern Orthodox Churches, Order of Red Banner of Labour Jl. 1977.

Pimen (P. V. Khmelevskii), Cand. of Theol.
Russian Orthodox clergyman; b. Ag. 26, 1923, Smolensk.
Entered monastery 1943; monk 1944; studied Minsk Theological Seminary; priest-monk 1949; grad. Moscow Theological Acad. 1953; taught at Moscow Theological Acad. 1953–59; abbot, mem. Ecclesiastical Mission of Moscow Patriarchate in Jerusalem 1955; Head of same, Archimandrite 1956; Prior's Asst. at St. Sergius Holy Trinity Monastery 1957; Lecturer at Moscow Theological Acad. 1959; Bishop of Saratov and Volgograd 1965–.

Pinaeva, Liudmila Iosifovna
Athlete (kayaker); b. Ja. 14, 1936, Krasnoe Selo, Leningrad Oblast.
Honored Master of Sport (1964). Teacher.
As a member of various crews, she won many USSR championships (21 between 1960 and 1972), European championships (nine between 1961 and 1971), world championships (five between 1966 and 1971), and Olympic championships (in 1964, 1968, and 1972). Decors.: Order of Lenin, two other orders, and a number of medals.

Pinchuk, Veniamin Borisovich
Sculptor; b. 1908.
Studied Higher State Workshops for Arts and Crafts (VUTEMAS) 1928–30 and at Acad. of the Arts 1930–32; mem. USSR

Acad. of the Arts 1970–. Decor.: People's Artist of the USSR 1969.

Pinegin, Timir Alekseevich
Athlete (yachting); b. Je. 12, 1927, Moscow.
Honored Master of Sports (1960). Coach. Member of the CPSU since 1969.
He was champion of the USSR 16 times between 1953 and 1973. He won a European championship in 1964, and an olympic championship in 1960. He participated in the 16th through 20th Olympic Games. Decors.: Order of the Badge of Honor and various medals.

Piotr (Paul l'Huillier)
Russian Orthodox chlergyman; b. D. 3, 1926, Paris.
Studied Holy Dionisi Inst. and Theological Dept. of School of Advanced Studies at the Sorbonne; diploma from Theological Inst. 1951; Cand. of Theology 1952; teacher and Sec. of Theological Pastoral Courses in West European Exarchate of Moscow Patriarchate 1953; monk, monk-deacon and priest-monk 1954; translated liturgy into French; Archimandrite 1960; has published studies on ecclesiastical history and law, among others; Bishop of Korsun 1968–.

Piotrovskii, Boris Borisovich
Historian, archeologist and oriental scholar; b. F. 1 (14), 1908, St. Petersburg.
Mem. CPSU 1945–; mem. USSR Acad. of Sciences 1970–; mem. Acad. of Sciences Armenian SSR 1968–; Head of Inst. of Archeology at Leningrad branch of USSR Acad. of Sciences 1953–64; Dir. of the Eremitage 1964–, concurrently Head of Chair of History of the Ancient Orient at Leningrad Univ. 1966–; conducted archeological research in the Transcaucasus, the Northern Caucasus and in Cen. Asia 1930–; head of a Sov. archeological expedition to Nubia (Egypt) 1961–63; important excavations of Urartu city and fortification on the Karimir-Blur Hill near Erevan; mem. Bavarian and British Acads. of Sciences. Decors.: Order of Lenin, State Prize of the USSR 1946, 3 other orders and various medals.

Pirkulyev, Araly
State official.
Mem. Cen. Cttee. Turkmen CP 1976–; Head of Main Statistical Admin. Turkmen SSR Mr. 1980–.

Pisarenko, Georgii Stepanovich, Dr. of Tech. Sc.
Specialist in mechanical engineering; b. N. 12, 1910, Poltava; s. of a clerk.
Mem. CPSU 1943–; grad. Gorkii Ind. Inst. 1936; specializes in machine construction mechanics; mem. Acad. of Sciences Ukrainian SSR 1964–; Dr. of Tech. Sc. 1949–; Prof. 1950–; worked in "Krasnoe Sormovo" Plant 1936–40; mem. scientific staff Inst. of Construction Mechanics (now Inst. of Mechanics) of Acad. of Sciences Ukrainian SSR 1939–51; concluded post-graduate studies at Kiev Ind. Inst. 1940; Head of Dept. of High-Temperature Strenght, Head of Section of Strength (of Materials) at Inst. of Metal Ceramics and Special Alloys at Acad. of Sciences Ukrainian SSR 1951–66; Scientific Chief Sec. of Presidium of Acad. of Sciences Ukrainian SSR 1962–66; concurrently assignment at Kiev Polytech. Inst. 1944–; Dir. of Inst. of Problems of Strength at Acad. of Sciences Ukrainian SSR 1966–; author of some 170 scientific papers.
Publs.: "Kolebaniia uprugikh sistem s uchetom rasseianiia energii v lateriale" 1955, "Statystychni teorii mitnosti ta ikh zastosuvannia do metalokeramichnykh materialov" 1961, "Mekhanicheskie kolebaniia i ikh rol' v tekhnike" 1965, "Prochnost' materialov pri vysokikh temperaturakh" 1966, and others.

Pitirim (Konstantin Vladimirovich Nechaev)
Russian Orthodox clergyman; b. 1926, Michurinsk.
Studied Moscow Tech. and Theological Acads.; deacon 1951; priest 1954; monk 1959; Lecturer, then Prof. at Moscow Theological Acad.; Archimandrite; Chair. of Publishing Dept. of Russian Orthodox Church and Chief Ed. of Moscow Patriarchate Journal; Bishop 1963–71, Archbishop of Volokolamsk 1971–, Substitute Bishop of Moscow Diocese 1963–.

Plakhina, A. S.
Trade union official.
Sec. of Kazakh Republican Trade Union Council F. 1982–.

Plataev, Andrei Georgievich
State official.
Sec. East Kazakhstan Oblast Cttee. Kazakh CP 1963–66; cand. mem. Cen. Cttee. Kazakh CP 1971–76, mem. 1976–; Min. of the Interior Kazakh SSR Ja. 1980–.

Plate, Nikolai Al'fredovich
Chemist; b. N. 4, 1934, in Moscow.
Corresponding member of the Academy of Sciences of the USSR (1974). Member of the CPSU since 1972.
He graduated in 1956 from Moscow State University, where he studied under V. A. Kargin. Since 1956 he has been working at Moscow State University, where he was appointed head of the polymer modification laboratory in 1966. At the same time, since 1963, he has been working at the Institute of Petrochemical Synthesis of the Academy of Sciences of the USSR.
He determined the principal behavioral mechanism in graft and block copolymers as compared with the properties of their polymer components and formulated the principles of polymer modification through reactions in macromolecular chains. He developed the statistical theory on the reactivity of links in a polymer chain, taking into account the effect of the adjacent groups. He also studied the structure and physicochemical properties of comb-shaped polymers.

Platon (Vladimir Petrovich Udovenko), Cand. of Theol.
Russian Orthodox clergyman; b. N. 17, 1940, Uspenka, Voroshilovgrad Oblast.
Attended school until 1957; novice in Uspenski Monastery in Odessa, worked at diocese's candlemakers 1958; attended Kiev and Odessa Theological Seminaries 1958–60; mil. service 1961–64; attended Leningrad Theological Acad. 1965–69; Adviser in Church Dept. of External Affairs of Moscow Patriarchate; monk, deacon, priest 1971; Archimandrite 1972; Bishop of Argentina and South America 1972–77; Exarch of Moscow Patriarchate for Cen. and South America 1977–80; Archbishop of Sverdlovsk and Kurgan Ap. 1980–.

Platukis, Ionas Ionovich
State official.
Min. of Health Lithuanian SSR N. 1980–; mem. Auditing Comm. Lithuanian CP 1981–.

Plekhavichius, Kazis Pranovich
State official.
Cand. mem. Cen. Cttee. Lithuanian CP 1964–76, mem. 1976–; Min. of Public Services Lithuanian SSR Je. 1966–.

Pleshakov, Petr Stepanovich
State official; b. 1922.
Russian; mem. CPSU 1944–; grad. Moscow Inst. of Communications Engineering 1944; served in Sov. Army 1944–45; laboratory chief, dept. chief, dir. of a scientific research inst. 1945–64; Dep. Chair. of State Cttee. for Radio Electronics of USSR Council of Mins. 1964–65; Dep. USSR Min. of the Radio Ind. 1965–68, 1st Dep. Min. 1968–74, Min. Ap. 1974–; Dep. to USSR Supr. Sov. 1974–; cand. mem. Cen. Cttee. CPSU 1976–77, mem. 1977–. Decors.: Lenin Prize 1959, USSR State Prize 1968.

Pliev, Gris (Grigorii) Dzambolatovich
Poet and dramatist; b. S. 15, 1913, Ruk, Southern Ossetia.
Osset; mem. CPSU 1938–; grad. Moscow Theatre Inst. 1935; worked as actor and ed. of a journal; served in Sov. Army 1941–45; first works publ. 1930.
Publs.: "V krylatye gody" (During the Inspired Years) 1933, "Soldat" 1948, "Chermen" 1949, "Stikhi" (Poems) 1959, and others.

Pliev, Ilia Konstantinovich
Author; b. Jl. 31 (Ag. 13), 1913, Dzhava, Southern Ossetia; s. of a peasant.
Mem. CPSU 1943–; grad. from a pedagogic inst. 1953; first works publ. 1935.

Publs.: "Rodina" (Native Country) 1951, "Krasnoe znamia" (Red Banner) 1959, "Utro Osetii" (The Morning of Ossetia) 1961, "Stikhi i poemy" (Verse and Poetry) 1964, and others.

Plisetskaia, Maiia Mikhailovna
Ballet dancer; b. N. 20, 1925, Moscow.
Grad. Moscow Choreographic School 1943; mem. ballet of Bolshoi Theatre Moscow 1943–. Decors.: People's Artist of the USSR 1959, Lenin Prize 1964, Order of Lenin, various medals.

Ploshchenko, Vladimir Dmitrievich
State official.
Min. of Communal Econ. Ukrainian SSR 1971–76; cand. mem. Cen. Cttee. Ukrainian CP 1971–76, mem. 1976–; Min. of Housing and Communal Econ. Ukrainian SSR Jl. 1976–.

Plotnikov, Kirill Nikanorovich
Economist; b. My. 11 (24), 1907, Kursk.
Mem. CPSU 1940–; grad. Moscow Inst. of Finance 1930; teaching and scientific work 1931–; Permanent USSR Rep. for Asia and the Far East at UN 1955–59; Dir. of Econ. Inst. of USSR Acad. of Sciences 1959–65; Dep. Acad. Sec. of Econ. Dept. of USSR Acad. of Sciences, head of a sector at Inst. of Econ. of the Socialist World System at USSR Acad. of Sciences 1965–70; head of a chair at Ordzhonikidze Inst. of Econ. Engineering in Moscow 1970–. Decors.: 3 orders and various medals.

Plotnikov, Pavel Artem'evich
Military official; b. Mr. 4, 1920, Gon'ba, now under the Barnaul City Soviet, Altai Krai.
Major-general of aviation (1966), twice Hero of the Soviet Union (Ag. 19, 1944, and Je. 27, 1945), Honored Pilot of the USSR (1966). Member of the CPSU since 1944.
He graduated from the Third Novosibirsk Military Aviation School in 1940, from the Higher Tactical Flight School for Officers in 1945, from the Air Force Academy (now the Iu. A. Gagarin Air Force Academy) in 1951, and from the Military Academy of the General Staff in 1960. In the Great Patriotic War (1941–45) he served on the Southwestern, Southern, Transcaucasian, Voronezh, Steppe, and Second and First Ukrainian fronts. He was a pilot, flight commander, deputy commander of a squadron of the 82nd Guards Bombardment Aviation Regiment, and commander of a squadron of the 81st Guards Bombardment Aviation Regiment. He made 343 combat flights and shot down three enemy aircraft.
Since the war, he has held responsible positions in the air force. Decors.: Order of Lenin, three Orders of the Red Banner, the Order of Alexander Nevsky, the Order of the Patriotic War First Class, the Order of the Red Star, and various medals.

Pluchek, Valentin Nikolaevich
State director; b. Ag. 22 (S. 4), 1909, Moscow.
People's Artist of the USSR (1974).
In 1929, he graduated from the acting department, and in 1932 from the directing department, of the State Experimental Theatrical Studio, headed by V. E. Meyerhold. Beginning in 1929, he acted in the Meyerhold Theater. In 1940 he helped organize the Moscow State Theatrical Studio, which beginning in 1941 gave performances at the war front. In 1940 he directed *City at Dawn,* based on the play by A. N. Arbuzov and members of the studio. From 1942 to 1945 he headed the Theater of the Northern Fleet, and from 1945 to 1950 the Moscow Touring Theater. He became a director at the Moscow Theater of Satire in 1950, and since 1957 has been the theater's chief director.
His most important productions at the Theater of Satire were Mayakovsky's *The Bathhouse* (1953, with N. V. Petrov and S. I. Iutkevich; 1967), *The Bedbug* (1955, with Iutkevich; 1974), and *Mystery-Bouffe* (1957). He also directed Hikmet's *The Sword of Damocles* (1959), Beaumarchais's *The Marriage of Figaro* (1969), Shtein's *A Captive of Time* (1970), Gogol's *The Inspector-General* (1972), and Makaenok's *Pill Under the Tongue* (1973). His productions display a strong interest in current events, are optimistic in character, are intensely satirical, and tend toward hyperbole and the grotesque. He has written the book *On Stage – Mayakovsky* (1962), as well as articles on stagecraft. Decors.: two orders and several medals.

Poberei, Maria Tikhonovna
Party official; b. 1924.
Mem. CPSU 1957–; grad. Volgograd Agricultural Inst. 1958; agronomist at Panfilovskaia Machine and Tractor Station 1941–50; Chief Agronomist at Maiakovskii Machine and Tractor Station 1950–58; Head of Lenin Raion Agricultural Inspection Board, Volgograd Oblast 1958–59; Chair. of Exec. Cttee. Lenin Raion 1959–63 and Srednii Aktiubinsk Raion Cttee. of Workers' Deps. 1963–65; 1st Sec. of Lenin Raion CPSU Cttee. Volgograd Oblast N. 1965–; cand. mem. Cen. Cttee. CPSU 1966–81. Decor.: Hero of Socialist Labour.

Pochivalin, Nikolai Mikhailovich
Author; b. Ap. 7, 1921, Banichi, Chernigov Oblast.
Russian; lives in Penza.
Publs.: "Schastie" (Happiness) 1951; "Prostye geroi" (Simple Heroes) 1951, "Stikhi o moem druge" (Poems About My Friend) 1960, "Iunost" (Youth) 1954, "Vtoraia semia" (The Second Family) 1959, and others.

Podelkov, Sergei Aleksandrovich
Author and poet; b. S. 8, 1912, Pesochnia, Kaluga Oblast; s. of a worker.
Russian; studied Gorkii Inst. of Literature; teacher at Inst. of Literature 1964–; first works publs. 1931.
Publs.: "Stikhi o voine, o slave, o liubvi" (Poems About War, Glory, Love) 1934, "Pole zhizni" (The Field of Life) 1963, "Goriashchie derevia" (Burning Trees) 1963, "Vlast' serdtsa" (The Power of the Heart) 1966, and others.

Podolskii, Mitrofan Grigor'evich
Diplomat.
USSR Amb. to Laos D. 1976–O. 1980.

Pogodin, Radii Petrovich
Author; b. Ag. 16, 1925, Duplevo, Kalinin Oblast.
Russian; volunteered for the front after 8th grade at an intermediate school 1941; worked as pedagogue in a children's sanatorium, fireman, tree feller, locksmith and master in a factory 1946–; writes mainly for children.
Publs.: "Muravinoe maslo" (Ant Oil) 1957, "Rasskazy pro Keshku i ego druzei" (Stories About Keshka and His Friends) 1960, "Rasskazy o veselykh liudiakh i khoroshei pogode" (Stories About Merry People and Fine Weather) 1960, "Otkuda idut tuchi" (Where Do the Clouds Come From) 1966, and others.

Pogorelov, Aleksei Vasil'evich, Dr. of Phy.-Math. Sc.
Mathematician; b. Mr. 3, 1919, Koroch, now Belgorod Oblast; s. of a worker.
Grad. Zhukovskii Air Force Acad. 1945; corresponding mem. USSR Acad. of Sciences 1960–; mem. Acad. of Sciences Ukrainian SSR 1961–; Dr. of Phy.-Math. Sc. 1949; Prof. 1951–; engineer/designer at Zhukovskii Cen. Aerodynamics Inst. 1945–47; Sen. Mem. Scientific Staff, Head of Dept. of Geometry at Research Inst. of Mathematics and Mechanics of Kharkov Univ. 1947–50; head of a chair at Kharkov Univ. 1950–59; Head of Dept. of Geometry of Inst. of Mathematics at Acad. of Sciences Ukrainian SSR in Kiev 1959–60; Head of Dept. of Geometry of Physical-Tech. Inst. of Low Temperatures at Acad. of Sciences Ukrainian SSR in Kharkov and concurrently Prof. at Kharkov Univ. 1960–; author of more than 80 scientific papers. Decors.: State Prize of the USSR 1950, Lenin Prize 1961, Int. Lobachevskii Prize 1959.
Publs.: "Izgibanie vypuklykh poverkhnostei" 1951, "Odnoznachnaia opredelennost' obshchikh vypuklykh poverkhnostei" 1952, "Beskonechno malye izgibaniia obshchikh vypuklykh poverkhnostei" 1959, "Geometricheskaia teoriia ustoichivosti obolochek" 1966, and others.

Pogosian, Stepan Karapetovich
State official.
Sec. of Erevan City Cttee. Armenian Komsomol 1961; mem. Bureau 1961–67, Sec. 1962–63, 1st Sec. of Cen. Cttee. Armenian Komsomol 1963–67; Dep. to Supr. Sov. Armenian SSR 1963–; mem. Cen. Cttee. 1964–, cand. mem. Bureau Cen. Cttee. Armenian CP 1964–68; 1st Sec. of Ararat Raion Cttee. Armenian CP 1968–70; Chair. of State Cttee. for Television and Radio Armenian SSR S. 1970–.

Pogrebniak, Iakov Petrovich
Party official; b. 1928.
Ukrainian; mem. CPSU 1953–; grad. Donetsk Ind. Inst. 1954; Foreman, Sen. Foreman of Mechanics Dept. at Kramatorsk Machine Works 1954–57; Sec., 1st Sec. of Kramatorsk City Cttee. Ukrainian CP, 2nd Sec. of Donetsk Oblast Cttee. Ukrainian CP 1957–63; mem. Auditing Comm. 1961–66. mem. Cen. Cttee. Ukrainian CP 1966–; 1st Sec. for Ind. at Poltava Oblast Cttee. Ukrainian CP 1963–64; 2nd Sec. of Poltava Oblast Cttee. Ukrainian CP 1964–66; 1st Sec. of Ivano-Frankovsk Oblast Cttee. 1966–69 and Nikolaev Oblast Cttee. Ukrainian CP 1969–71; Dep. to USSR Supr. Sov. 1966–; Sec. and cand. mem. Politbureau Cen. Cttee. Ukrainian CP 1971–; cand. mem. Cen. Cttee. CPSU 1971–. Decor.: Order of Red Banner of Labour.

Poida, I. F.
Trade union official.
Sec. of Ukrainian Republican Trade Union Council Ja. 1968–.

Pokrovskii, Boris Aleksandrovich
Operatic director and teacher; b. Ja. 10 (23), 1912, Moscow.
People's Artist of the USSR (1961).
He graduated from the directing department of the State Institute of Theatrical Arts in 1937. From 1937 to 1943 he was a stage director at the Gorkii Theater of Opera and Ballet, and beginning in 1939 the theater's artistic director. He became a stage director at the Bolshoi Theater in 1943 and the theater's principal director in 1968, a post he had also held between 1952 and 1963. Operas staged by him at the Bolshoi include Serov's *The Evil One* (1947; State Prize of the USSR, 1948), Smetana's *The Bartered Bride* (1948; State Prize of the USSR, 1949), Rimsky-Korsakov's *Sadko* (1949; State Prize of the USSR, 1950), Mozart's *The Marriage of Figaro* (1956, with G. P. Ansimov), Prokofiev's *War and Peace* (1959), Dzerzhinskii's *Man's Fate* (1961), Prokofiev's *Semen Kotko* (1970) and *The Gambler* (1974), and Glinka's *Ruslan and Liudmila* (1972). His productions of *War and Peace* at the Leningrad Malyi Opera Theater (1946; State Prize of the USSR, 1947) and at the Sofia People's Opera (1957) are well known.
Since 1972, he has been artistic director of the Moscow Chamber Music Theater, where he has staged Shchedrin's opera *Not Only Love* (1973) and Shostakovich's *The Nose* (1974). He also teaches at the State Institute of Theatrical Arts, where he became a professor in 1954. Decors.: Order of Lenin, the Order of the Red Banner of Labor, and several medals.
Works: "Zametki o rabote rezhissera i aktera v opernom teatre." In *Voprosy muzykal'no-ispolnitel'skogo iskusstva*, collection 2. Moscow 1958. *Ob opernoi rezhissure*. Moscow 1973.

Pokrovskii, Georgii Iosifovich
Physicist and engineer; b. Mr. 31 (Ap. 13), 1901, Kiev.
Mem. CPSU 1950–; grad. Moscow Inst. of Econ. 1923; Maj.-Gen.; teaching posts at various tech. insts. 1925–; Prof. at Mil. Engineering Acad. 1932–; Prof. at N. E. Zhukovskii Acad. of Mil. Aviation Engineering 1944–. Decors.: State Prize of the USSR 1971, 4 orders and various medals.

Pokrovskii, Igor' Aleksandrovich
Architect; b. Mr. 4, 1926, Moscow.
Became a member of the CPSU in 1955. He studied at the Moscow Architectural Institute with B. S. Mezentsev from 1945 to 1950. His urban designs are marked by a compositional diversity that enlivens the urban environment. He collaborated on the following designs in Moscow: ten- to 14-story apartment houses on Semenovskaia Embankment (1952–56), Krasnopresnenskaia subway station (1955), the Palace of Pioneers and Schoolchildren on Lenin Hills (1959–63; State Prize of the USSR, 1967), and the new building of the Moscow Art Theater on Tver' Boulevard (1966–72). He also participated in the planning and construction of Moscow's satellite city – Zelenograd (since 1962; chief architect, since 1964). He helped design the monument honoring the defenders of Moscow on the Leningrad Highway outside Zelenograd (1967; sculptor E. A. Shteiman, painter A. G. Shteiman).

Pokryshkin, Aleksandr Ivanovich
Military official; b. 1913.
Russian; mem. CPSU 1942–; grad. Frunze Mil. Acad. 1948 and Acad. of Gen. Staff 1957; Air Force Marshal 1972–; roofer, machine fitter 1929–32; in Sov. Army 1932–; commander of a squadron, regt., div. 1941–45; command posts in Sov. Army 1945–; Dep. to USSR Supr. Sov. 1946–; mem. Cen. Cttee. Ukrainian CP 1960–71; Chair. of Cen. Cttee. of Volunteer Union Soc. for the Promotion of Army, Navy and Air Force (DOSAAF) D. 1971–81; cand. mem. Cen. Cttee. CPSU 1976–. Decors.: 2 Orders of Lenin, 4 Orders of Red Banner, other orders and medals. 3 times Hero of the Sov. Union.

Pokshishevskii, Vadim Viacheslavovich
Economic geographer; b. Ag. 1 (14), 1905, Moscow.
Doctor of geographical sciences (1952). He graduated from the Azerbaidzhan Polytechnic Institute in 1927. From 1954 to 1968 he was a professor at the V. I. Lenin Moscow Pedagogical Institute. Between 1959 and 1965 he was head of the department of the economic geography of the USSR at the Geographic. Institute of the Academy of Sciences of the USSR. Since 1965, he has been senior research worker at the Institute of Ethnography of the Academy of Sciences of the USSR.
His main works are concerned with the population geography of the USSR and other countries, the economic geography of the USSR, historical geography, and theoretical issues in economic geography. Decors.: Great Gold Medal of the Geographical Society of the USSR.
Works: *Problemy razmeshcheniia promyshlennosti*. Moscow 1932. *Povolzh'e*. Moscow 1951. *Zaselenie Sibiri*. Moscow 1951. *Iakutiia*. Moscow 1957. *Ekonomicheskaia geografiia v SSSR: Istoriia i sovremennoe razvitie*. Moscow 1965. (Co-author and editor.) *Geografia naseleniia SSR*. Moscow 1971. *Geografiia naseleniia zarubezhnykh stran*. Moscow 1971. *Geography of the Soviet Union*. Moscow 1974.

Polevoi, Boris Nikolaevich (real name: Kampov)
Author; b. Mr. 4 (17), 1908, Moscow.
Russian; mem. CPSU 1940–; mem. Sov. Peace Cttee., mem. Bureau World Peace Council, Chief Ed. of journal "Iunost'" 1962–; after grad. Ind.-Tech. College in Tver 1926 worked in Textile Combine "Proletarka" in Kalinin; served in Sov.-Finnish War 1939–40; war correspondent for "Pravda" 1941–45. Decors.: State Prize of the USSR 1947 and 1949, 2 Orders of Lenin, Order of the October Revolution, 6 other orders and various medals, Golden Peace Medal 1968, Hero of Socialist Labour 1974.
Publs.: "Memuary vshivogo cheloveka" (Memoirs of a Lice-Infested Man) 1927, "Ot Belgoroda do Karpat" (From Belgorod to the Carpathians) 1945, "Povest' o nastoiashchem cheloveke" (The Story of a Real Man) 1946, "My – sovetskie liudi" (We Are Soviet People) 1948, "Zoloto" (Gold) 1949–50, "Sovremenniki" (Contemporaries) 1952, "Na dikom brege ..." (On a Wild Shore ...) 1962, "Doktor Vera" 1966, "Izbrannye proizvedeniia" (Selected Works) vols. 1–2, 1969, "Chelovek – cheloveku" (From Man to Man) 1971, "Iz zapisok voennogo korrespondenta" (From the Notes of a War Correspondent) vols. 1–2, 1974, and others.

Poliakov, Anisim Aleksandrovich
Scientist in veterinary sanitation and disinfection; b. F. 15 (25), 1904, Novaia, Iurginskoe Raion, Tiumen' Oblast.
Academician (1966) of the V. I. Lenin All-

Union Academy of Agricultural Sciences (VASKhNIL). Member of the CPSU since 1929.
He graduated from the Siberian Veterinary Institute in the city of Omsk in 1928. From 1934 to 1967 he was director of the All-Union Scientific Research Institute of Veterinary Sanitation (before 1955, laboratory director). From 1969 to 1972 he was academician-secretary of the department of veterinary medition of VASKhNIL. His principal works are on disinfection and insert, rat, and parasite control of animal-husbandry farms, raw-material and food plants, and transport. Decors.: Order of Lenin, the Order of the October Revolution, the Order of the Red Banner of Labor, and various medals.
Works: *Feterinarnaia dezinfektsiia,* 3rd ed. 1964. *Osnovy veterinarnoi sanitarii.* Moscow 1969.

Poliakov, Ilia Mikhailovich, Dr. of Biol. Sc.
Biologist; b. S.16, 1905, Kharkov; s. of a clerk.
Mem. CPSU 1941–; grad. Dept. of Biology Kharkov Inst. of Popular Educ. 1926; corresponding mem. Acad. of Sciences Ukrainian SSR 1948–; Dr. of Biol. Sc. 1941; Prof. 1934–; teacher, Head of Chair of Biology at Kharkov Communist Artem Univ. 1925–34; Head of Chair of Darwinism and Genetics at Kharkov Univ. 1934–41 and 1944–48; Prof. at Tomsk Univ. 1941–44; mem. staff Inst. of Genetics and Selection at Acad. of Sciences Ukrainian SSR 1947–56; Dep. Dir. and Head of Laboratory of Plant Breeding and Insemination at Ukrainian Scientific Research Inst. of Plant Breeding, Selection and Genetics 1956–63; Dir. of same inst. 1963–; author of some 150 scientific papers on problems of experimental botany, theoretical problems in biology and the history of biology. Decors.: Order of Red Star, Order "Badge of Honour", medals.
Publs.: "Kurs Darvinizma" 1941, "Novye puti issledovaniia protsessov oplodotvoreniia vysshikh rastenii pri pomoshchi radioaktivnykh izotopov" in "Zhurnal obshchei biologii" No. 16, 1955, "Oplodotvorenie i geterozis selskokhoziaistvennykh rastenii" 1959, "Protsess oplodotvoreniia i raznokachestvennost' semian" in "Biologicheskie osnovy povysheniia kachestva semian selskokhoziaistvennykh rastenii" 1964, and others.

Poliakov, Ivan Evteevich
State official; b. 1914.
Belorussian; mem. CPSU 1940–; grad. Party Higher School of Cen. Cttee. CPSU 1949; electro-welder 1931–36; served in Sov. Army 1936–38; sec. of a raion Komsomol cttee. 1939–42; Sec. of Gomel Oblast Underground Cttee. Belorussian Komsomol, Commr. of 1st Gomel Partisan Brigade 1942–44; Sec. of Minsk Oblast Cttee. Belorussian Komsomol 1944–49; 2nd, then 1st Sec. of Vitebsk City Cttee. Belorussian CP (B) 1949–53; 1st sec. of a raion cttee. Belorussian CP 1953–56; Chair. of Gomel Oblast Exec. Cttee. of Workers' Deps. 1956–57; Dep. to 1955–59, mem. Presidium Supr. Sov. Belorussian SSR 1957–59; 1st Sec. of Gomel Oblast Cttee. Belorussian CP 1957–64 (for Agriculture 1963–64); Dep. USSR Supr. Sov. 1958–, Dep. Chair. Presidium 1977–; cand. mem. 1961–66, mem. Cen. Cttee. CPSU 1966–; 1st Sec. of Minsk Oblast Cttee. Belorussian CP D. 1964–Mr. 1977; mem. Presidium, then Bureau Cen. Cttee. Belorussian CP Mr. 1965–; Chair. of Presidium Supr. Soviet Belorussian SSR Mr. 1977–. Decors.: 2 Orders of Lenin, medals.

Poliakov, Nikolai Dmitrievich
State official.
Chair. of State Cttee. for Construction Moldavian SSR 1967–74; Dep. to Supr. Sov. Moldavian SSR 1967–; cand. mem. Cen. Cttee. Moldavian CP 1971–77, mem. F. 1977–; Min. of Rural Construction Moldavian SSR Ag. 1974–D. 1976; Dep. Chair. of Council of Mins. Moldavian SSR D. 1976–.

Poliakov, Nikolai Sergeevich
Scientist in the field of mining engineering; b. My. 5 (18), 1903, Dnepropetrovsk. Academician of the Academy of Sciences of the Ukrainian SSR (1967). Member of the CPSU since 1945.
After graduating from the Dnepropetrovsk Mining Institute in 1926, he worked there from 1926 to 1967 on the planning of mining enterprises. In 1967 he was appointed director of the Institute of Mechanics of Geological Engineering of the Academy of Sciences of the Ukrainian SSR. His main works deal with the kinematics and dynamics of mining transport equipment. Decors.: three Orders of Lenin, three other orders, and various medals.
Works: *Kurs rudnichnogo transporta.* Moscow-Leningrad 1944. (With A. O. Spivakovskii.) *Osnovy teorii raschety rudnichnykh transportnykh ustanovok.* Moscow 1962. (With I. G. Shtokman.)

Poliakov, Viktor Nikolaevich
State official; b. 1915.
Russian; mem. CPSU 1944–; grad. Moscow Automobile and Road Inst. 1938; fitter, foreman, dep. shop superintendent in various automobile enterprizes in Moscow 1930–37; served in Sov. Army 1938–46; laboratory chief, shop superintendent, dep. chief designer, chief engineer, Dir. of Small-Displacement Automobile Plant in Moscow 1946–63; 1st Dep. Chair., Chair. of Moscow City Sovnarkhoz 1963–65; Dep. USSR Min. of the Automobile Ind. 1965–75, Min. 1975–; Dep. to USSR Supr. Sov. 1974–; cand. mem. Cen. Cttee. CPSU 1976–81, mem. 1981–. Decor.: Hero of Socialist Labor 1971.

Poliakov, Vladimir Porfirevich, Cand. of Econ. Sc.
Diplomat; b. 1931.
Mem. Diplomatic Service 1956–; Emb. Counsellor at USSR Emb. in Syria 1961–66; mem. staff USSR Min. of Foreign Affairs 1966–67; Emb. Counsellor-USSR Envoy to United Arab Republic 1967–72; USSR Amb. to Sudan F.–Jl. 1972, to People's Republic of Yemen Jl. 1972–74 and to Egypt Ap. 1974–.

Poliakova, Nadezhda Mikhailovna
Author and poet; b. D. 15, 1923, Bosutino, Novgorod Oblast.
Russian; grad. Leningrad Univ. 1949; first works publ. 1940.
Publs.: "Pravo na schastie" (The Right to Happiness) 1955, "Iiul'" (July) 1962, "Priznanie v liubvi" (Declaration of Love) 1963, "Katerina" 1955, "Gosti Indii" (Guests of India) 1959, "My letim vokrug zemli" (We Fly Around the Earth) 1963, and others.

Polianker, Grigorii Isaakovich
Author; b. F. 2 (15), 1911, Uman; s. of an artisan.
Jew; mem. CPSU 1930–; first works publ. 1932.
Publs.: "Pekar iz Kolomen" (A Baker From Kolomeia) 1964, "Serdtse ne kamen'" (The Heart is No Stone) 1957, "Zolotaia dolina" (The Golden Valley) 1958, "Veselyi passazhir" (The Merry Passenger) 1962, and others.

Polianskii, Anatolii Trofimovich
Architect; b. Ja. 29, 1928, Avdeevka, Donetsk Oblast.
Mem. CPSU 1953–; grad. Moscow Inst. of Architecture 1950; Dir. of Cen. Scientific Research Inst. for Experimental Design of Sanatoriums in Spas 1958–; teacher at Moscow Inst. of Architecture 1955–60 and 1973–. Decors.: State Prize of the USSR 1967, 2 orders and various medals.

Polianskii, Dmitrii Stepanovich
Diplomat; b. 1917, Slavianoserbske village, Donets Basin.
Ukrainian; mem. CPSU 1939–; grad. Kharkov Inst. of Agriculture 1939 and Party High School of Cen. Cttee. USSR 1942; joined Komsomol 1931; worked on Dolgii Sovkhoz 1932–35; student 1935–39; Head of Rural Youth Dept. with Kharkov Oblast Cttee. Komsomol, served in Sov. Army 1939–42; CPSU work in the Altai, Head of Pol. Dept. of Khoroshenskaia Machine and Tractor Station, 1st Sec. of Karasuk Raion CPSU Cttee., Exec. Organizer of Cadre Dept. and Inspector of Cen. Cttee. CPSU 1942–49; 2nd Sec. of Crimean Oblast Cttee. Ukrainian CP 1949–52; Chair. of Crimean Oblast Exec. Cttee. of Workers' Deps. 1952–53; 1st Sec. of Crimean Oblast Cttee. Ukrainian CP 1953–55; mem. Cen. Cttee. Ukrainian CP 1954–56; Dep. to USSR Supr. Sov. 1954–; 1st Sec. of

Chkalov (now Orenburg) Oblast CPSU Cttee. 1955–57; mem. Cen. Cttee. CPSU 1956–81; 1st Sec. of Krasnodar Krai CPSU Cttee. 1957–58; Chair. of RSFSR Council of Mins. 1958–62; cand. mem. Presidium 1958–60, mem. Presidium 1960–76, mem. Politbureau Cen. Cttee. CPSU 1960–; mem. Cen. Cttee. CPSU Bureau of RSFSR 1961–66; Dep. Chair. 1962–65, 1st Dep. Chair. of USSR Council of Mins. 1965–73; USSR Min. of Agriculture F. 1973–Mr. 1976; USSR Amb. to Japan Ap. 1976–82, to Norway Ja. 1982–. Decors.: 3 Orders of Lenin.

Polianskii, Iurii (Georgii) Ivanovich
Protozoologist; b. Mr. 2 (15), 1904, St. Petersburg.
Mem. CPSU 1941–; grad. Leningrad Univ. 1924; concluded post-graduate studies 1929; teacher at A. I. Herzen Pedagogic Inst. 1927–41 and 1945–48; served in Sov. Army 1941–45; mem. staff Murmansk Biological Station on Barents Sea 1948–52; Dir. of Inst. of Biology at Karelian branch of USSR Acad. of Sciences 1953–56; concurrently Prof. at Leningrad Univ., Head of Chair of Zoology of Invertebrates 1953–; Head of Laboratory of Inst. of Cytology at USSR Acad. of Sciences 1957–; Pres. of Union Soc. of Protozoologists 1968–. Decors.: Merited Scientist of the RSFSR, Order of Lenin, Order of Red Star, medals.

Polikanov, Sergei Mikhailovich
Atomic physicist; b. S. 14, 1926, Moscow.
Mem. CPSU 1955–; grad. Moscow Inst. of Physical Engineering 1950; mem. staff Inst. of Atomic Energy 1950–57, at United Inst. of Nuclear Research in Dubna 1957–; his most important works are on nuclear physics. Decors.: Lenin Prize 1967, Order of Lenin.

Polkovnikov, Vladimir Ivanovich
Film director; b. My. 28 (Je. 10), 1906, Teriaeva Sloboda, now Teriaevo, Volokolamsk Raion, Moscow Oblast.
Honored Art Worker of the RSFSR (1969).
He graduated from the 1905 Memorial Moscow Art School in 1930. In 1931 he became an animator for Union Film Current Events, and in 1936 a director in the Union Animated Cartoon film studio. From 1943 to 1946 he worked in the Military Film studio.
His films have included *Political Satire Journal No. 1* (1938; with other directors), *Limpopo* (1940), *Barmalei* (1941), *The Peacock's Tail* (1947), *Little Greyneck* (1948), *The Magic Store* (1953; with the director L. A. Amal'rik), *The Enchanted Boy* (1956; with A. G. Blotskaia), *The Braggart Ant* (1961), *The Great Roach* (1963), *Nargis* (1966), *We Are Looking for a Blot* (1969), and *Where Are You Flying, Vitar?* (1971). Many of his films have received prizes on international film festivals. Decors.: Order of the Badge of Honor and several medals.

Polonyk, Vasilii Petrovich
Sculptor; b. 1930, Pryvillia, Donetsk Oblast.
Mem. CPSU 1959; attended Voroshilovgrad Art School 1945–50 and Kiev Art Inst. 1950–56, pupil of M. Lysenko. Decor.: Merited Artist of the Ukrainian SSR 1968.

Polosin, Evgenii Maksimovich
Actor; b. 1912.
Mem. CPSU 1939–; training at State Lunacharski Theatre Inst. (GITIS) in Moscow 1934–39; engagement with Drama Theatre in Gomel; mem. of a cultural brigade at the front during the war; engagements with various theatres in Belorussia 1945–49; mem. Russian Gorkii Drama Theatre of Belorussian SSR 1949–. Decor.: People's Artist of the USSR 1969.

Polozhenko, Nikanor Vladimirovich
State official.
Dep. to Supr. Sov. Moldavian SSR 1955–; 1st Sec. of Leovo Raion Cttee. Moldavian CP 1955; mem. Cen. Cttee. 1956–58 and 1966–81, cand. mem. Cen. Cttee. Moldavian CP 1958–61 and 1981–; Min. of Communal Econ. Moldavian SSR Ap. 1959–78; Min. of Housing and Communal Econ. Moldavian SSR N. 1978–.

Polozov, Nikolai Nikiforovich
Trade union official; b. 1923.
Belorussian; mem. CPSU 1946; grad. Party High School of Cen. Cttee. CPSU; teacher 1939; partisan 1941–45; Komsomol posts, Sec. of Molodechno City Cttee. Belorussian CP, 1st Sec. of Vileika City Cttee. Belorussian CP 1946–60; sen. posts at Cen. Cttee. Belorussian CP, Head of Dept. of Party Organs for Ind. and Construction with Bureau of Cen. Cttee. Belorussian CP, Head of Dept. of Party Organs at Cen. Cttee. Belorussian CP, Head of Dept. of Party Org. Work at Cen. Cttee. Belorussian CP 1960–70; cand. mem. 1961–66, mem. Cen. Cttee. 1966–, cand. mem. Bureau 1968–71, mem. Bureau Cen. Cttee. Belorussian CP F. 1971–; Chair. of Belorussian Republican Trade Union Council Ap. 1970–.

Poltaeva, E. V.
Trade union official.
Sec. and mem. Presidium of Turkmen Republican Trade Union Council Ja. 1972–.

Poltoratskii, Viktor Vasil'evich (real surname: Poltoratskii-Pogostin)
Russian writer; b. Ap. 18 (My. 1), 1907, Ashkhabad.
Member of the CPSU since 1939.
He graduated from the pedagogical institute in Iaroslavl in 1933 and then worked on newspapers in the cities of Ivanov and Moscow. He was editor-in-chief of the newspaper *Literatura i zhizn'* from 1958 to 1961. His first of several poetry collections, *A Word to the Settlement,* was published in 1928. Poltoratskii's impressions of a trip through Europe were the basis of his books of essays *On the Road and at Home* (1950; State Prize of the USSR, 1952). His works depict the natural surroundings of the Meshchera and Vladimir regions and the life and work of the regions' people.
Works: *Struzhan': Izbr. stikhotvoreniia.* Moscow 1970. *Doroga v Suzdal': Kniga o liubimoi zemle.* Moscow 1971. *V deistvuiushchei armii: Iz zapisok voennogo korrespondenta.* Moscow 1973.

Poluboiarov, Pavel Pavlovich
Military official; b. Je. 3 (16), 1901, Tula.
Mem. CPSU 1920–; grad. Armoured-Vehicle School 1926 and Mil. Acad. for the Mechanization and Motorization of the Sov. Army 1938; concluded courses for continuing educ. of higher command cadres at Mil. Acad. of the Gen. Staff 1941; Marshal of Armoured Troops 1962; Mil. Inspector/Adviser of USSR Min. of Defence 1969–; various sen. command posts during World War II and afterwards; Dep. Commander, 1st Dep. Commander of Armoured and Mechanized Troops 1949–54; Commander of Armoured Troops of the Sov. Army 1954–69; Mil. inspector-adviser of the Group of Inspectors General of the USSR Min. of Defense My. 1969–. Decors.: 3 Orders of Lenin, Order of the October Revolution, 5 Orders of Red Banner, 2 Suvorov Orders 2nd Cl., 2 Kutuzov Orders 2nd Cl., 2 Orders of Red Star, 6 foreign orders.

Polukarov, Iurii Ivanovich
Party official; b. 1920.
Russian; mem. CPSU 1951–; grad. N. E. Bauman Higher Technical School in Moscow 1944; section chief, dep. shop superintendent, design office chief, chief engineer, dep. chief engineer of Vladimir Tractor Plant, chief engineer of an admin. of Vladimir Sovnarkhoz 1944–61; 2nd Sec. of Vladimir City Cttee. CPSU 1961–63; 2nd Sec. for Ind. of Vladimir Oblast Cttee. CPSU 1963–64; 1st Sec. of Vladimir City Cttee. CPSU 1964–68; 2nd Sec. of Vladimir Oblast Cttee. CPSU 1968–69; posts in Cen. Cttee. CPSU apparatus 1969–75; Dep. to USSR Supr. Sov. 1974–; mem. Cen. Cttee. Tadzhik CP My. 1975–; 2nd Sec. and Mem. Bureau of Cen. Cttee. Tadzhik CP My. 1975–; cand. mem. Cen. Cttee. CPSU 1976–.

Pomerants, Grigorii Solomonovich
Philosopher and Indian scholar; b. 1918.
Mem. staff Inst. of Scientific Information Moscow and Fundamental Library of USSR Acad. of Sciences (INIBON); mem. staff Asian and African Section of Fundamental Library 1965–; took part in World War II; suffered reprisals under Stalin, was inmate of Stalinist labour camps, released 1953, later rehabilitated; signed the Letter of 170 in protest against

the trail of Alexander Ginzburg and Iurii Galanskov; as a result defence of his thesis at Inst. of the People's of Asia was postponed indefinitely 1968.
Publs.: "Nekotorye techeniia vostochnogo religioznogo nigilizma" (Some Trends in Eastern Religious Nihilism) 1968, "O metode sravnitel'nogo izucheniia vostochnykh i zapadnykh literatur (na primere iaponskoi literatury)" (On the Method of Comparative Study of Eastern and Western Literatures (demonstrated in Japanese Literature)) 1970.

Pomozov, Iurii Fomich
Author; b. O.16, 1926, Leningrad.
Russian; mem. CPSU 1946–.
Publs.: "Nashi tovarishchi" (Our Comrades) 1950, "Est' na Baltike ostrov" (There is an Island in the Baltic Sea) 1960, "Ty po strane idesh ..." (You Go Through the Land ...) 1964, and others.

Ponomar, Fedor Mironovich
Author and poet; b. Mr.20, 1919, Goian, Moldavia.
Moldavian; mem. CPSU 1946–; grad. Tiraspol Pedagogic Inst.; first works publ. 1937.
Publs.: "Druzhba" (Friendship) 1948, "Otvet docheri" (The Daughter's Answer) 1953, "Rodina schastiia" (The Motherland of Happiness) 1954, "Stikhotvoreniia" (Poems) 1961, and others.

Ponomarenko, Evgenii Porfirevich
Actor; b. F.24 (Mr.9), 1909, Kherson.
Ukrainian; mem. CPSU 1944–; at October Revolution Drama Theatre in Odessa 1926–36; concluded courses at Drama Studio at same 1928; at Ukrainian I. Franko Drama Theatre in Kiev 1936–. Decors.: Shevchenko State Prize of the Ukrainian SSR 1971, Order of Lenin, Order of Red Banner of Labour, medals.

Ponomarev, Boris Nikolaevich
Party official; b. Ja.4 (17), 1905, Zaraisk, Moscow Oblast.
Russian; grad. Moscow State Univ. 1926 and Inst. of Red Professors 1932; historian; Prof.; served in Sov. Army 1919; worked for Zaraisk Mil. Revolutionary Cttee. 1919–20; leading Komsomol posts in Riazan Province 1920–22; Sec. of Party Org. in "Krasnyi Vostok" Factory in Moscow Oblast 1922–23; mem. and Dep. Head of Propagandist Group of Cen. Cttee. All-Union CP (B) 1926–28; CPSU work 1928–32; teacher at various high schools 1932–; Dep. Dir. of Inst. of Red Professors 1932–36; sen. posts with Exec. Cttee. of Comintern 1936–43; Dep. Dir. of Marx-Engels-Lenin-Inst. 1943–44; worked for Cen. Cttee. All-Union CP (B) 1944–46; 1st Dep. Head, then Head of Information Bureau at USSR Council of Mins. 1946–49; worked for Cen. Cttee. CPSU 1949–55; cand. 1952–56, mem. Cen. Cttee. CPSU 1956–; dept. head with Cen. Cttee. CPSU 1955–61; Dep. to 1958–, Chair. of Foreign Affairs Comm. with Sov. of Nationalities USSR Supr. Sov. 1966–; Sec. 1961–, cand. mem. Politbureau Cen. Cttee. CPSU My. 1972–; author of various treatises on the history of the CPSU, on USSR foreign policy and on international communist, workers' and national liberation movements; head of authors' collective of textbook "Istoriia KPSS" (1960, 1971–74); mem. of main editorial board of multi-volume 'History of the CPSU'. Decors.: 5 Orders of Lenin, 2 Orders of Red Banner of Labour, Hero of Socialist Labour, medals. Order of the October Rev.

Ponomarev, Mikhail Aleksandrovich
Party official; b. 1918.
Russian; mem. CPSU 1939–; grad. Perm State Univ. 1941; head of a pol. dept. at a machine and tractor station, Dep. Dept. Head with Perm Oblast All-Union CP (B) Cttee. 1941–44; 1st Sec. of Perm Oblast Komsomol Cttee., Cen. Cttee. Inspector for All-Union Komsomol 1944–47; Dept. Head, Sec., 2nd Sec. of Perm Oblast CPSU Cttee. 1947–55; worked for Cen. Cttee. CPSU 1955–59; Dep. to USSR Supr. Sov. 1958–; 1st Sec. of Kalmuk Oblast CPSU Cttee. 1959–61; 1st Sec. of Vladimir Oblast CPSU Cttee. Ag. 1961– (for Ind. 1963–64); cand. mem. 1971–76, mem. Cen. Cttee. CPSU 1976–. Decors.: 2 Orders of Lenin.

Ponomarev, Nikolai Afanasevich
Graphic artist; b. F.4 (17), 1918, Aleksandrovsk-Grushevskii, now Shakhty, Rostov Oblast.
Grad. Moscow V. I. Surikov Inst. 1950; mem. USSR Acad. of the Arts 1973–. Decors.: People's Artist of the RSFSR, State Prize of the USSR 1951, Order of Red Banner of Labour, 1 medal.

Ponomarev, Nikolai Aleksandrovich
State official; b. 1929.
Mem. CPSU 1955–; grad. Kazakh Inst. of Agriculture in Alma-Ata 1951; agronomist at a sovkhoz trust, chief agronomist of a sovkhoz, dir. of various sovkhozes in Kustanai Oblast 1951–63; CPSU posts 1963–65; 1st Sec. of Komsomolskii Raion Cttee. Kazakh CP 1965–71; Chair. of Kustanai Oblast Exec. Cttee. of Workers' Deps. Je. 1971–77; cand. mem. Cen. Cttee. CPSU 1971–76; mem. Cen. Cttee. Kazakh CP 1971–81; Dep. Min. of Agriculture Kazakh SSR O. 1979–.

Ponomarev, Oleg Borisovich
State official.
Dep. Chair. of GOSPLAN Kirghiz SSR 1966–76; Cand. mem. Cen. Cttee. Kirghiz CP 1976–78, mem. Jl. 1978–; Head of Main Admin. for Material-Tech. Supplies with Council of Mins. Kirghiz SSR F.–D. 1976; Dep. Chair. of Council of Mins. Kirghiz SSR D. 1976–.

Pontekorvo, Bruno Maksimovich
Physicist; b. Ag.22, 1913, Pisa, Italy.
Mem. CPSU 1955–; grad. Rome Univ.; mem. USSR Acad. of Sciences 1964–; after grad. mem. staff Rome Univ. and Radium Inst. until 1940; lived in the USA, Canada and Great Britain 1940–50; moved to the Sov. Union 1950; mem. staff Inst. of Nuclear Problems of USSR Acad. of Sciences 1950–56; mem. staff Joint Inst. of Nuclear Research 1956–. Decors.: State Prize of the USSR 1954, Lenin Prize 1963, 2 Orders of Lenin, 3 other orders, medals.

Pontriagin, Lev Semenovich
Mathematician; b. Ag.21 (S.3), 1908, Moscow.
Grad. Moscow Univ. 1929; mem. USSR Acad. of Sciences 1958–; blind since the age of 14; dept. head at V. A. Steklov Inst. of Mathematics of USSR Acad. of Sciences 1939–; concurrently Prof. at Moscow Univ. 1935–; hon. mem. of London Mathematical Soc. 1953–, of Int. "Astronautik" Acad. 1966–; Vice-Pres. of Int. Mathematical Asscn. 1970–74, hon. mem. of the Hungarian Acad. of Sciences 1972–. Decors.: Hero of Socialist Labour 1939, State Prize of the USSR 1941, Lenin Prize 1962, 3 Orders of Lenin, 2 other orders and various medals.

Popkhadze, Nugzar Akakievich
State official; b. 1942.
Georgian; mem. CPSU 1966–; journalist, mechanic at a knitted goods factory in Tbilisi, student, after completion of studies at Tbilisi Univ. dep. head of Dept. for Propaganda and Agitation City Cttee. Georgian Komsomol in Tbilisi 1959–67; instructor, head of cabinet for Komsomol work Cen. Cttee. Georgian Komsomol 1967–68; editor for youth publications of State Cttee. for Television and Radio Broadcasting Georgian SSR, consultant to the House for Political Education Cen. Cttee. CP Georgia 1968–69; head of dept. for students Cen. Cttee. Georgian Komsomol 1970–73; head of section for propaganda and political work in Dept. for Propaganda and Agitation Cen. Cttee. CP Georgia 1973–74; Dep. Min. of Education Georgian SSR 1974–78; 1st dep. head of Dept. for Propaganda and Agitation of Cen. Cttee. CP Georgia 1978–79; of State Cttee. for Television and Radio Broadcasting Georgian SSR Ja. 1979–.

Popkov, Mikhail Danilovich
Military official; b. 1924.
Mem. CPSU 1943–; grad. V. I. Lenin Mil.-Pol. Acad. 1950 and Mil. Acad. of Gen. Staff 1966, Col.-Gen.; entered Sov. Army 1942; Komsomol organizer of an artillery regiment, asst. chief of pol. dept. of brigade, of division 1944–47; chief pol. dept. of Suvorov Mil. School 1959–60; party and pol. posts in armed forces 1960–67; chief of pol. directorate of a mil. district 1973–80; mem. of Mil. Coun-

cil and Chief of Pol. Directorate of USSR Ground Forces D. 1980–; cand. mem. Cen. Cttee. CPSU 1981–.

Popkov, Valerii Ivanovich
Energetics scientist; b. Ja. 21 (F. 3), 1908. Mem. CPSU 1951–; grad. Moscow Energetics Inst. 1930; specialist in electrotechnology; mem. USSR Acad. of Sciences 1966–; after grad. worked in various scientific research establishments; mem. staff Krzhizhanovskii Inst. of Energetics 1943–. Decors.: Iblochkov Prize, Order of Lenin, Order of the October Revolution, Order of Red Banner of Labour, medals.

Popkov, Vitalii Ivanovich
Military official; b. My. 1, 1922, Moscow. Lieutenant-general of aviation (1968), twice Hero of the Soviet Union (S. 8, 1943, and Je. 27, 1945), Honored Pilot of the USSR (1967). Member of the CPSU since 1943.
He joined the Soviet Army in 1940 and graduated from the Chuguev Military Aviation School for Pilots in 1941, from the Bataisk Military Aviation School in 1942, from what is now the Iu. A. Gagarin Air Force Academy in 1951, and from the Military Academy of the General Staff in 1964. In the Great Patriotic War (1941–45) he served from May 1942 on the Kalinin, Southwestern, and Third and First Ukrainian fronts as pilot, flight commander, and deputy commander and commander of a squadron of the Fifth Guards Fighter Aviation Regiment. He made 300 combat flights and shot down 41 aircraft. After the war, he held responsible positions in the air force and has been in the main office of the Ministry of Defense of the USSR since 1964.
Decors.: three Orders of Lenin, two Orders of the Red Banner, the Order of the Patriotic War First and Second Class, the Order of Alexander Nevsky, the Order of the Red Star, and various medals.

Poplavskii, Aleksandr Mitrofanovich
Party official; b. 1914.
Mem. CPSU 1943–; grad. Poltava Thermal Mechanics Technicum 1937 and Moscow Correspondence Inst. of Econ. 1957; technical dir. of a plant in Ulan-Ude 1937–39; chief engineer, dir. of a plant in Chita 1939–44; dir. of a plant in Odessa 1944–50; manager of a trust in Irkutsk 1950–56; dir. of a plant in Volgograd 1956–60; posts in Cen. Cttee. apparatus CPSU 1961–; 1st Dep. Administrator of Affairs CPSU D. 1975–; cand. mem. Cen. Cttee. CPSU 1981–.

Popov, Boris Veniaminovich
Party official; b. 1909.
Russian; mem. CPSU 1931–; grad. Party High School of Cen. Cttee. CPSU 1949; newspaper vendor, machine fitter, chair. of works cttee. of a ship repair works 1927–33; Komsomol work 1933–37; aide to head, Head of Pol. Dept. of Railways 1937–50; 2nd Sec. of Tashkent Oblast Cttee. Uzbek CP 1950–57; mem. staff Cen. Cttee. CPSU 1958–61; mem. Cen. Auditing Comm. 1961–71, cand. mem. 1971–76, mem. Cen. Cttee. CPSU 1976–; 2nd Sec. and mem. Bureau Cen. Cttee. Lithuanian CP 1961–67; Dep. to USSR Supr. Sov. 1962–; 1st Sec. of Arkhangelsk Oblast Cttee. CPSU Mr. 1967–. Decors.: 2 Orders of Lenin.

Popov, Daniil Ivanovich
State official.
Min. of Posts and Telecommunications Tadzhik SSR Je. 1971–; cand. mem. Cen. Cttee. Tadzhik CP 1981–.

Popov, Evgenii Pavlovich
Automation specialist; b. F. 1 (14), 1914, Moscow.
Mem. CPSU 1942–; grad. Bauman Higher Tech. College Moscow 1939; corresponding mem. USSR Acad. of Sciences 1960–; Maj.-Gen./Engineer; mechanic with an Air Force squadron 1939–43; teacher at A. F. Mozhaiskii Acad. of Mil. Aviation Engineering 1943–64, Prof. 1948–; Chair. of Section of Applied Problems at Presidium of USSR Acad. of Sciences 1964–71; head of a chair at Bauman Higher College Moscow (MVTU) 1971–; his most important works are on the theory of automatic controls. Decors.: State Prize of the USSR 1949 and 1972, Order of the October Revolution, 2 other orders and various medals.

Popov, Igor' Aleksandrovich
Painter; b. O. 16, 1927, Kharkov.
Honored Artist of the RSFSR (1970).
He studied at the V. I. Surikov Moscow Art Institute from 1945 to 1951 under S. V. Gerasimov. Since 1974 he has been chairman of the board of the Moscow organization of the Artists' Union of the RSFSR. His paintings – mainly landscapes, genre compositions, and portraits – are poetic renderings of Russian nature and the modern life of small Russian cities and their inhabitants. They are characterized by free painterly modeling; examples include *Marked Day* (1958–60, Iaroslavl' Art Museum), the series of portraits *Fishermen from Galich* (1965, Board of Artistic Funds and the Planning of Monuments of the Ministry of Culture of the RSFSR), *To Your Health, Aleksandr Martynovich!* (1967), and *September: All Is Quiet* (1972; the last two paintings are housed in the Tret'iakov Gallery).

Popov, Konstantin Mikhailovich, Dr. of Econ. Sc.
Geographer and oriental scholar; b. Ap. 15 (28), 1900, Port Arthur.
Grad. Kiev Inst. of Econ. Planning 1923; active as econ. geographer and orientalist; Dr. of Econ. Sc. 1944; Prof. at Moscow Inst. of Oriental Studies 1930–; Expert-Adviser of USSR Min. of Foreign Trade 1943–70; concurrently Head of Dept. of Geography of the Capitalist and Developing Countries of Inst. of Geography at USSR Acad. of Sciences 1950–65; Sen. Mem. Scientific Staff Inst. of Oriental Studies at USSR Acad. of Sciences 1965–.

Popov, Leonid Andreevich
Author; b. Ag. 14, 1919, Yakut ASSR; s. of a teacher.
Yakut; mem. CPSU 1964–; studied Pedagogic Inst.; first works publ. 1937.
Publs.: "Ot vsego serdtsa" (Wholeheartedly) 1950, "Chaiki" (The Seagulls) 1958, "Zhizn', stavshaia pesnei" (Life Which Turned Into a Song) 1948, "Moei respublike" (To My Republic) 1951, "Snegopad" (Snowfall) 1963, and others.

Popov, Nikolai Sergeevich
Designer; b. 1931.
Mem. CPSU 1960–; grad. Kharkov Polytechnical Inst. 1955; design engineer, senior design engineer, sector chief, dept. chief, dep. chief designer of Kirov Plant in Leningrad 1955–73; chief designer of "Kirovskii zavod" Prod. Asscn. in Leningrad 1973–; mem. Cen. Auditing Comm. CPSU 1976–81; mem. Cen. Cttee. CPSU 1981–. Decor.: Hero of Socialist Labour 1975.

Popov, Oleg Konstantinovich
Circus artist and clown; b. Jl. 31, 1930, Vyrubovo, Moscow Oblast.
People's Artist of the USSR (1969).
He graduated from the State School of Circus Art in 1950. Initially he performed as a comic equilibrist on a slack wire. From 1951 to 1953 he worked in a youth company, where he established himself as a carpet clown. A skilled equilibrist, acrobat, and juggler, he combines slapstick, psychological insight, gentle lyric humor, and satire in his acts. He creates the image of the contemporary young man, imparting to him the charm and craftiness of the mischievous heroes of Russian fairy tales. Known as the Sun Clown, he was a prize winner at the Warsaw International Festival of Circus Art (1957) and recipient of the Oscar Prize (1958, Brussels) and other awards. He has performed abroad. Decors.: Order of the Red Banner of Labour and a medal.
Work: "Moi geroi." In the collection *Ogni manezha*. [Moscow] 1961. Pages 77–85.

Popov, Serafim Alekseevich
Poet and author; b. D. 26, 1912 (Ja. 1913), Zheshart, Vologda Oblast.
Komi; mem. CPSU 1942–.
Publs.: "Frontovaia doroga" (Front Street) 1945, "Nash Sever" (Our North) 1948, "Stikhi o druzhbe" (Poems About Friendship) 1955, "Lesnaia skaza" (Forest Fairytale), "Propavshii bez vesti" (Missing Person), and others.

Popov, Viktor Ivanovich
Diplomat; b. 1918.
Mem. CPSU 1947–; grad. Moscow Inst. of Hist., Philos. and Literature 1941 and

Higher Diplomatic School of USSR Min. of Foreign Affairs 1949, Dr. of Hist. Sc., Prof. 1964; factory worker in Astrakhan 1935–36; taught at Higher Diplomatic School 1949–52; Prorector of Higher Diplomatic School 1952–66; councillor at USSR Amb. to Australia 1966–68; Rector of Diplomatic Acad. of USSR Min. of Foreign Affairs 1968–80; USSR Amb. to Malta 1980–81, to Britain 1980–; mem. Cen. Auditing Comm. CPSU 1981–.

Popova, Maria Georgievna
Laborer; b. 1928.
Russian; mem. CPSU 1959–; finished Vladivostok Vocational School 1945; apprentice 1943–45; machine fitter in Suchan and Primore Krai 1945–48; crane operator in Nakhodka Commercial Harbour 1948–; cand. mem. 1966–71, mem. Cen. Cttee. CPSU 1971–.

Popovich, Pavel Romanovich
Cosmonaut; b. O. 5, 1930, Uzin, Kiev Oblast.
Mem. CPSU 1957–; grad. Zhukovskii Acad. of Mil. Aviation Engineering 1960; pilot; cosmonaut 1960–; Col.; group flight as pilot of spaceship "Vostok 4" with "Vostok 3" (A. F. Nikoaev), travelled some 2 million kilometres in 48 earth orbits (flight lasted 70 hours and 44 mins.) Ag. 12–15, 1962; was part of the crew of "Soiuz 14" (together with Iu. P. Artiukhin) which took off Jl. 3, 1974, and docked "Soiuz 14" with scientific space station "Saliut 3" which had been in orbit since Je. 26, 1974 (combined flight lasted 15 days); executed numerous scientific-tech. and medical-biological experiments; Dep. to Supr. Sov. Ukrainian SSR of the 7th and 8th legislative periods. Decors.: 2 Orders of Lenin, Order of Red Banner.

Popovskii, Mark Aleksandrovich
Author; b. Jl. 8, 1922, Odessa.
Russian; grad. Dept. of Philology of Moscow Univ. 1952; first works publ. 1946.
Publs.: "Kogda vrach mechtaet" (When a Doctor Dreams) 1957, "Zhivoe zerno" (Living Grain) 1961, "Sud'ba Doktora Khavkina" (The Fate of Doctor Khavkin) 1963, "Piat' dnei odnoi zhizni" (Five Days of a Life) 1965, and others.

Porai-Koshits, Mikhail Aleksandrovich
Crystallographer; b. Ja. 7, 1918, Viatka, now Kirov; s. of A. E. Porai-Koshits.
Corresponding member of the Academy of Sciences of the USSR (1974).
Upon graduating from the University of Kazan in 1943, he began working at the Institute of General and Inorganic Chemistry of the Academy of Sciences of the USSR, where he was appointed head of the laboratory of crystal chemistry in 1959. Since 1952 he has also been head of the laboratory of crystal chemistry in the subdepartment of chemistry at Moscow State University. His main works are devoted to the crystal chemistry of coordination compounds, the X-ray examination of the structure of theses, and X-ray structural analysis. Many regularities in the stereochemistry of transition metals have been established through the research carried out by him.
Works: *Rentgenostrukturnyi analiz,* 2nd ed., vol. 1. Moscow 1964 (With G. B. Bokii.) *Prakticheskii kurs rentgenostrukturnogo analiza,* vol. 2. Moscow 1960 *Uspekhi kristallokhimii,* parts 1–9. Moscow 1966–74. (With others.)

Porfirev, Boris Aleksandrovich
Author; b. Jl. 24, 1919, Sovetsk, Kirov Oblast.
Russian; studied Leningrad Univ.; served in Sov. Army 1941–45.
Publs.: "Moi tovarishchi" (My Comrades) 1947, "Tsvety poluchaet pobeditel'" (The Winner Gets the Flowers) 1951, "Veter" (Wind) 1953, "Miach i tsvety" (The Ball and the Flowers) 1962, "Koster na l'du" (Camp-Fire on the Ice) 1968, and others.

Porfirev, Vladimir Borisovich, Dr. of Geol.-Min. Sc.
Geologist; b. Jl. 9, 1899, Viatka, now Kirov; s. of a clerk.
Studied Perm Univ. 1919–21; Grad. Leningrad Mining Inst. 1926; mem. Acad. of Sciences Ukrainian SSR 1957–; Dr. of Geol.-Min. Sc. 1938; Prof. 1937–; geologist with Geology Section of Geological Cttee. 1924–29; Head of Cen. Asian Section of Leningrad Inst. of Petroleum Research 1929–37; mem. staff Inst. of Geology of Acad. of Sciences Ukrainian SSR 1938–41; mem. staff Inst. of Geology of Uzbek SSR 1941–44, carried out research work in the oil fields of Cen. Asia; Head of Petroleum Dept. of Geol. Inst. of Acad. of Sciences Ukrainian SSR 1944–51, at Lvov branch of same Inst. 1945–; Dir. of Inst. of Geology Ukrainian SSR 1951–63; Dir. of Inst. of Geology of Acad. of Sciences Ukrainian SSR 1963–; author of some 100 scientific papers on geology of oil and gas deposits. Decors.: Order of Lenin, Order of Red Banner of Labour.
Publs.: "Problema nefteobrazovaniia v svete sovremennykh dannykh" 1941, "Metaforizm iskopaemykh uglei" 1948, "Problema migratsii nefti i formirovaniia neftianykh i gazovykh skoplenii" 1959, and others.

Posebeev, Grigorii Andreevich
Party official; b. 1935.
Mem. CPSU 1959–; grad. Leningrad Agricultural Inst. 1960 and CPSU Cen. Cttee. Party Higher Correspondence School 1972; Komsomol posts 1959–65; Sec. of Leningrad Oblast Cttee. USSR Komsomol 1965–67; 1st Sec. of Kingisepp Raion Cttee. CPSU in Leningrad Oblast 1968–71; chief of a dept. of Leningrad Oblast Cttee. CPSU 1971–75; Sec. of Leningrad Oblast Cttee. CPSU 1975–81; 1st Sec. of Oblast Cttee. CPSU in Mari ASSR Ja. 1981–; cand. mem. Cen. Cttee. CPSU 1981–.

Posokhin, Mikhail Vasil'evich
Architect; b. N. 30 (D. 13), 1910, Tomsk.
Russian; mem. CPSU 1961–; grad. Moscow Inst. of Architecture 1938; Chief Architect of the City of Moscow 1960–; Chair. of Cttee. on Private Housing and Architecture with USSR State Cttee. on Construction (GOSSTROI) and Dep. Chair. of USSR GOSSTROI 1963–67; Dep. to Supr. Sov. 1962–74. Decors.: Lenin Prize 1962, Order of Lenin, 3 other orders and various medals.

Pospelov, Gennadii Nikolaevich
Literary historian; b. Jl. 18 (30), 1899, Kostroma.
Grad. Moscow Univ. 1925; Prof. at Moscow Univ. 1938–; main field of work: methodology problems in literary history, theory of literature, history of Russian literature of the 18th and 19th century.

Pospelov, Germogen Sergeevich
Automation specialist; b. My. 12 (25), 1914, Orekhovo-Zuevo, Moscow Oblast.
Mem. CPSU 1943–; grad. Moscow Energetics Inst. 1940; corresponding mem. USSR Acad. of Sciences 1966–; engineer with an airborne regt. 1941–46; teacher at N. E. Zhukovskii Acad. of Mil. Aviation Engineering 1946–64, Prof. 1957–; Dep. Chair., then Chair. of Section of Applied Problems at Presidium of USSR Acad. of Sciences 1964–73; Dep. Acad. Sec. of Dept. of Mechanics and Control Processes of USSR Acad. of Sciences 1967; head of a chair at Physical-Tech. Inst. in Moscow 1969–; Head of Laboratory at Computer Centre of USSR Acad. of Sciences 1974–; his most important works deal with automatization of flying vehicles, theory of nonlinear systems of automatic control and theory of steering larger systems. Decors.: State Prize of the USSR 1972, 4 orders and various medals.

Postnikov, Stanislav Ivanovich
Military official; b. 1928.
Mem. CPSU 1957–; grad. M. V. Frunze Mil. Acad. 1961 and Mil. Acad. of Gen. Staff 1969, Col.-Gen.; entered Sov. Army 1948; company commander, battalion commander, regimental commander 1951–67; devision commander 1969–73; command posts 1975–79; Commander of Troops of Baltic Mil. District 1979–; mem. Cen. Cttee. and mem. Bureau Cen. Cttee. Latvian CP 1981–; mem. Cen. Auditing Comm. CPSU 1981–.

Postovskii, Isaak Iakovlevich
Organic chemist; b. Mr. 5 (17), 1898, Odessa.
Grad. Munich Inst. of Technology 1925, studied under G. Fischer; Head of Chair of Organic Chemistry at S. M. Kirov Polytech. Inst. in Uralsk 1926–; organized

prod. of Dep. Chief Ed. of "Pravda" 1952–53; Sec. of Cen. Cttee. CPSU, later mem. Bureau for RSFSR of Cen. Cttee. CPSU 1953–60; Dir. of Inst. of Marxism-Leninism of Cen. Cttee. CPSU 1961–67; mem. Presidium USSR Acad. of Sciences 1967–; collaborated on completion of 2nd vol. of "Istoriia grazhdanskoi voiny v SSSR"; Chair. of Ed. Cttee. for multi-vol. works "Istoriia Velikoi Otechestvennoi Voiny Sovetskogo Soiuza" and "Istoriia Kommunisticheskoi Partii Sovetskogo Soiuza", head of ed. collective for a scientific biography of V. I. Lenin. Decors.: Order of Lenin, Order of the October Revolution, 2 other orders and various medals.

Potapenko, Fedor Ivanovich
Diplomat; b. 1927.
Mem. Apparat Cen. Cttee. Belorussian CP 1960–69; Emb. Councellor, Chargé d'Affaires, USSR Emb. in China 1969–72; Councellor, USSR Min. of Foreign Affairs 1972–75; Dep. Head, Far Eastern Dept., USSR Min. of Foreign Affairs 1975–80; USSR Amb. to Singapore Ap. 1980–.

Potapov, Leonid Pavlovich, Dr. of Hist. Sc.
Ethnographer and historian; b. Je. 23 (Jl. 6), 1905, Barnaul.
Mem. CPSU 1946–; specialist in museum science; mem. staff Inst. of Ethnography of USSR Acad. of Sciences 1943–; Dr. of Hist. Sc. 1946; Prof. 1951–. Decors.: Merited Scientist of the RSFSR 1965, 2 orders and various medals.

Potapova, Elena Mikhailovna
Ballet dancer; b. F. 16, 1930, Kuibyshev.
Mem. CPSU 1954–; attended Choreographic Studio of T. G. Shevchenko Opera and Ballet Theatre in Kiev 1946–48; soloist at same theatre 1948–. Decors.: People's Artist of the USSR 1970, Order of Red Banner of Labour.

Povkh, Ivan Lukich, Dr. of Tech. Sc.
Physical-tech. scientist; b. N. 11, 1909, Miropole, Sumi Oblast.
Mem. CPSU 1929–; grad. Leningrad Ind. Inst. (now Polytech. Inst.) 1936; specialist in aero-hydro-gas dynamics; corresponding mem. Acad. of Sciences Ukrainian SSR 1961–; Dr. of Tech. Sc. 1953; Prof. 1954–; Prof. at Physical-Tech. Dept. of Leningrad Ind. resp. Polytech. Inst. 1936–61; Dep. Dir. of Mining Inst. of Acad. of Sciences Ukrainian SSR (now Fedorov Inst. of Mining Mechanics and Tech. Cybernetics of USSR Min. of the Coal Ind.) 1961–63; Head of Physical-Tech. Laboratory of Inst. of Ferrous Metals in Donetsk 1963–; concurrently Head of Chair of Physical Hydrodynamics of Donetsk Univ. 1964–; author of some 80 scientific papers. Decors.: Order of Red Star, medals.
Publs.: "Mykola Egorovych Zhukovskyi – batko rosiiskoi aviatsii" 1952, "Modelirovanie gidravlicheskikh turbin na vozdushnykh potokakh" 1955, "Tekhnicheskaia gidromekhanika" 1964, "Aerodinamicheskii eksperiment v mashinostroenii" 1965, and others.

Pozharskii, Boris Mikhailovich
State official.
Chair. of Belorussian Asscn. for Sale and Supply of Agricultural Equipment "Belselkhoztekhnika" Mr. 1971–78; Chair. of State Cttee. for Agricultural Material and Technical Supply Belorussian SSR D. 1978–; cand. mem. 1971–74, mem. Cen. Cttee. Belorussian CP 1976–.

Poznanskaia, Maria Avvakumovna
Author and poet; b. Jl. 15, 1917, Petrashevka, Kiev Oblast.
Ukrainian; grad. Kiev Pedagogic Inst. 1949.
Publs.: "Budu pionerom" (I'll Be a Pioneer) 1948, "Gde solntse i tsvety" (Where Sun and Flowers Are) 1950, "Chto ia znaiu pro metro" (What I Know About the Metro) 1961, "Valia Kotik" 1959, and others.

Praude, Robert Vladimirovich
State official.
Mem. Auditing Comm. 1961–63; cand. mem. 1963–66, mem. Cen. Cttee. Latvian CP 1966–; Min. of Trade Latvian SSR F. 1963–Mr. 1977; Dep. to Supr. Sov. Latvian SSR 1963–; Min. of Finance Latvian SSR Mr. 1977–81; Dep. Chair. of Council of Mins. Latvian SSR Je. 1981–.

Preobrazhenskii, Nikolai Aleksandrovich
Otorhinolaryngologist; b. N. 27, 1918, Moscow.
Corresponding member of the Academy of Medical Sciences of the USSR (1974). Member of the CPSU since 1959.
He graduated from the First Moscow Medical Institute in 1941. In 1963 he became a professor and in 1972 head of the subdepartment of ear, nose, and throat diseases at that institute. His chief works deal with diseases of the organs of hearing and with angina and chronic tonsillitis. He developed methods of mobilizing the stapes during otosclerosis and otitis and invented protheses for reconstructing the sound-conducting system of the middle ear, as well as a number of medical instruments. He perfected and implemented operations that improved the hearing of patients afflicted with otosclerosis (Lenin Prize, 1964). He has been chairman of the All-Union Scientific Society of Otorhinolaryngologists since 1968, and an honorary member of the Italian Society of Otorhinolaryngologists and Cervicofacial Surgeons since 1973. Decors.: Order of the Patriotic War Second Class and several medals.
Works: *Otoskleroz,* 2nd ed. Moscow 1965. (With K. L. Khilov.) *Stapedektomiia i stapedoplastika pri otoskleroze.* Moscow 1973. (With O. K. Patiakina.)

Press, Irina Natanovna
Track-and-field athlete; b. Mr. 10, 1939, Kharkov.
Honored Master of Sports of the USSR (1960). Member of the CPSU since 1964. In 1962, she graduated from the Leningrad Institute of Railroad Transport Engineering. She received the degree of candidate of pedagogical sciences in 1972.
Between 1959 and 1967, she was women's champion of the USSR 13 times. She was a champion at the Olympic games of 1960 and 1964 and held the world and European records in the hurdles and track-and-field pentathlon events seven times between 1959 and 1965. Since 1967 she has been working in the Central Council of Dynamo. Decors.: two orders.

Press, Tamara Natanovna
Athlete; b. My. 10, 1937, Kharkov.
Honored Master of Sports of the USSR (1960). Member of the CPSU since 1962. She graduated from the Leningrad Construction Engineering Institute in 1962 and from the Higher Party School of the Central Committee of the CPSU in 1967. She received the degree of candidate of pedagogical sciences in 1974.
She was the women's champion in shot put at the Olympic games of 1960 and 1964 and women's champion in discus-throwing at the 1964 games. She was the women's champion of Europe three times in 1958 and 1962 and the women's champion of the USSR 16 times between 1958 and 1966. Since 1967 she has been working for the All-Union Central Council of Trade Unions. Decors.: Order of Lenin and the Order of the Badge of Honor.

Priede, Gunar Reinkholdovich
Dramatist; b. Mr. 17, 1928, Riga.
Latvian; mem. CPSU 1960–; grad. Latvian Univ. 1953; was active as an architect; work in film ind. 1961–; first works (on the aesthetics of architecture) publ. 1949. Decor.: State Prize of the Latvian SSR 1959.
Publs.: plays "Leto mladshego brata" (The Summer of the Younger Brother) 1955, "Pervyi bal Viki" (Vika's First Ball) 1961, "Tvoe dobroe imia" (Your Good Reputation) 1965, and others.

Priezzhev, Nikolai Semenovich
Party official; b. 1919.
Russian; mem. CPSU 1946–; grad. Inst. of Mechanization of Agriculture in Saratov 1944; Engineer, Dep. Dir., Dir. of Saratov Oblast Office of Main Admin. for the Sale of Tractors, Agricultural Machinery and Spare Parts for Lorries and Tractors 1945–49; party and sov. posts 1949–60; mem. staff Cen. Cttee. CPSU 1960–61; 2nd Sec. of Riazan Oblast CPSU Cttee. 1961–62; Chair. of Riazan Oblast Exec. Cttee. of Workers' Deps. 1962–67 (for Agriculture 1962–64); 1st Sec. of Riazan Oblast CPSU Cttee. Ja.

1967–; Dep. to USSR Supr. Sov. 1970–; mem. Cen. Cttee. CPSU 1971–. Decor.: Order of Lenin.

Prigara, Maria Arkadevna
Author; b. F. 7 (20), 1908, Moscow; d. of a physician.
Ukrainian; mem. CPSU 1948–; studied Kiev Inst. of Popular Educ. 1927–30.
Publs.: "Sestry" (The Sisters) 1948, "Iarynka" 1944, "Nashi znakomye" (Our Acquaintances) 1946, "Odna semia" (A Family) 1955, "Kozak Golota" (The Cossack Golota) 1966, and others.

Priimak, Boris Ivanovich
Architect; b. My. 29 (Je. 11), 1909, Novocherkassk.
Mem. CPSU 1952–; grad. Kharkov Inst. of Architecture 1930; Chief Architect of the City of Kiev 1955–72; Prof. at Kiev Art Inst. 1971. Decors.: People's Architect of the USSR 1970, Order of Lenin, 2 other orders and various medals.

Prikhotko, Antonina Fedorovna, Dr. of Phy.-Math. Sc.
Physicist; b. Ap. 26, 1906, Piatigorsk, now Stavropol Krai.
Grad. Leningrad Polytech. Inst. 1930; mem. Acad. of Sciences Ukrainian SSR 1964–; Dr. of Phy.-Math. Sc. 1943; Prof. 1952–; mem. staff Physical-Tech. Roentgen Inst. in Leningrad 1929–30; mem. staff Ukrainian Physical-Tech. Inst. (now Physical-Tech. Inst. of Acad. of Sciences Ukrainian SSR in Kharkov) 1930–41; mem. staff Inst. of Physical Chemistry of Acad. of Sciences Ukrainian SSR 1941–44 (evacuated to Ufa in 1944); mem. staff Physics Inst. of Acad. of Sciences Ukrainian SSR 1944–, Dir. of same inst. 1965–; author of some 100 scientific papers. Decors.: Lenin Prize 1966, Order of Lenin, 1 medal.
Publs.: "Svechenie i tsvet kristallov politsiklicheskikh uglevodorodov" 1952, "Spekty poploshcheniia molekuliarnykh kristallov" 1965, and others.

Prilezhaeva, Maria Pavlovna
Author; b. Je. 9 (22), 1903, Iaroslavl.
Russian; mem. CPSU 1952–; grad. from a pedagogic inst. 1929; worked as teacher and pedagogue in a children's home; first works publ. 1937.
Publs.: "Etod god" (This year) 1941, "Tania Slesareva" 1942, "Iunost' Mashi Strogovoi" (Masha Strogov's Youth) 1948, "Nad Volgoi" (On the Volga) 1952, "Tretia Varia" (The Third Varia) 1963, "Pod severnym nebom" (Under the Northern Sky) 1959, "Povesti o nashei zhizni" (Tales From Our Life) 1966, and others.

Primachenko, Mariia Avksent'evna
Master of decorative graphics; b. D. 30, 1908 (Ja. 12, 1909), Bolotnia, in present-day Ivankov Raion, Kiev Oblast.
Honored Art Worker of the Ukrainian SSR (1970).
From 1935 to 1941, she worked in the experimental studios of the Kiev Museum of Ukrainian Art. She now lives in her native village. She is known primarily for small decorative prints, most of which have folklike floral and animal motifs and are executed in watercolor and gouache on paper or cardboard. In the 1950's she introduced a colored background into her works, achieving a vibrant harmony of bright spots of local color. Decors.: T. G. Shevchenko State Prize of the Ukrainian SSR in 1966 (for works from 1960 to 1965), Order of the Badge of Honor and various medals.

Primakov, Evgenii Maksimovich
Economist and historian; b. O. 29, 1928, Kiev.
Mem. CPSU 1959–; grad. Moscow Inst. of Oriental Studies 1953; corresponding mem. USSR Acad. of Sciences 1974–; worked for State Radio and Television Cttee. 1953–62; Columnist and Dep. Ed. Asian and African Desk of "Pravda" 1962–70; Dep. Dir. of Inst. of World Econ. and Int. Relations of USSR Acad. of Sciences 1970–; specialist on Egypt and other Arab countries; mem. ed. collective "Mezhdunarodnye konflikty" 1972 and "Energeticheskii krizis v kapitalisticheskom mire" 1975. Decor.: Nasser Prize 1975.

Prishchepchik, Vitalii Viktorovich
Party official; b. 1927.
Belorussian; mem. CPSU 1951–; grad. Belorussian Academy of Agricultural Sciences; Partisan 1943–44; Student 1944–51; Komsomol and party posts 1951–59; Sec. of Mogilev Oblast Cttee. 1959–63 and 1964–68, 2nd Sec. of Mogilev Oblast Cttee. Belorussian CP 1963–64 and 1968–74 (for Agriculture 1963–64); mem. Cen. Cttee. Belorussian CP 1971–; Chair. of Mogilev Oblast Exec. Cttee. of Worker's Deps. F.–Ag. 1974; 1st Sec. of Mogilev Oblast Cttee. Belorussian CP Ag. 1974–. Decor.: Order of Red Banner of Labour 1977.

Prisiazhniuk, Andrei Aleksandrovich
Russian actor; b. Jl. 4 (17), 1912, Pogrebishe, now in Vinnitsa Oblast.
People's Artist of the USSR (1973). Member of the CPSU since 1945.
He began his acting career in the Vladivostok Working Youth Theater (1930-32). He then acted in the theaters of Blagoveshchensk (1932), Spassk-Dal'nii (1933-37), and Ussuriisk (1937-38). Since 1938 he has acted in the company of the Vladivostok Dramatic Theater.
A comic actor, he has an infectious gift of humor, charm, masterly movements, and musical talent. At the same time, he has brought some comic roles to a level of tragedy, including Perfishka in *Three*, based on Gorkii's novel, Luzgin in Gorkii's *The Counterfeit Coin*, and Gennaro in De Filippo's *Man and Gentleman*. His other roles have included Schastlivtsev in Ostrovskii's *The Forest*, Perchikhin in Gorkii's *Smug Citizens*, Vas'ka Okorok in Vs. Ivanov's *Armored Train 14–69*, the Watchmaker in Pogodin's *The Kremlin Chimes*, the title role in Kvitka-Osnov'ianenko's *Shel'menko, the Orderly*, Vainonen in Vishnevskii's *An Optimistic Tragedy*, and Grandfather Sliva in Sofronov's *The Cook* and *The Cook Married*. Decors.: Order of the Red Banner of Labour and several medals.

Pristavkin, Anatolii Ignat'evich
Author; b. O. 17, 1931, Liubertsy, Moscow Oblast.
Russian; mem. CPSU 1965–; grad. Gorkii Inst. of Literature 1959; first works publ. 1954.
Publs.: "Malenkie rasskazy" (Little Stories) 1959, "Moi sovremenniki" (My Contemporaries) 1959, "Kostry v taige" (Camp Fire in the Taiga) 1964, "Liricheskaia kniga" (A Lyrical Book) 1969, and others.

Privalov, Boris Avksentevich
Author and poet; b. F. 29, 1924, Moscow.
Russian; grad. Gorkii Inst. of Literature 1944.
Publs.: "Obyknovennaia zhizn'" (An Ordinary Life) 1950, "Taina serzhanta Gorlenko" (The Secret of Sergeant Gorlenko) 1955, "Ne prokhodite mimo" (Don't Pass Me) 1956, "Vsadnik bez borody" (The Horseman Without Beard) 1964, "Skaz pro Ignata - khitrogo soldata" (Stories About Ignat, a Cunning Soldier) 1967, "Zhenikh moei nevesty" (The Bridegroom of My Bride) 1965, "Ulybka Maisary" (Maisara's Smile) 1967, and others.

Prokhorov, Aleksandr Mikhailovich
Radiophysicist; b. Je. 28 (Jl. 11), 1916, Atterton, Australia.
Mem. CPSU 1950–; grad. Leningrad Univ. 1939; mem. USSR Acad. of Sciences 1966–; postgraduate studies at Physics Inst. of USSR Acad. of Sciences 1939–41 and 1944–46; served in Sov. Army 1941–43; mem. staff Physics Inst. of USSR Acad. of Sciences 1946–; Head of Laboratory 1954–, Dep. Dir. 1968–, head of a chair at Moscow Physical-Tech. Inst. 1971–; wrote standard works on laser. Decors.: Hero of Socialist Labour 1969, Lenin Prize 1959, Nobel Prize 1964, mem. American Acad. of Science and Art 1971, 2 Orders of Lenin, medals.

Prokhorov, Aleksei Nikolaevich
Military official; b. Ja. 19, 1923, Rozhdestvenskoe, Voronezh Oblast.
Mem. CPSU 1943–; grad. Mil. Aviation School in Balashov 1942 and Iu. A. Gagarin Air Force Acad. 1950; Col.; teacher at Air Force Mil. School; served in Sov. Air Force 1941–45 (flew 238 missions); after the war various sen. posts in Air Force; teacher 1967–. Decors.: Order of Lenin, 3 Orders of Red Banner, Order of Patriotic War 1st Cl., Aleksandr-Nevskii Order, 2 Orders of Red Star, medals.

Prokhorov, Iurii Vasil'evich
Mathematician; b. D.15, 1929, Moscow. Grad. Moscow State Univ. 1949; mem. USSR Acad. of Sciences 1972–; mem. staff Steklov Inst. of Mathematics of USSR Acad. of Sciences 1949–, Dep. Dir. of same 1969–; Prof. 1957–; mem. Main Ed. Dept. of "Bol'shaia Sovetskaia Entsiklopediia" (3rd ed.). Decor.: Lenin Prize 1970.

Prokhorov, Vasilii Il'ich
Trade union official; b. 1906.
Russian; mem. CPSU 1928–; grad. Bauman Tech. College Moscow 1935; Engineer-Designer at Zhukovskii Cen. Inst. of Aero- and Hydrodynamics 1934–37; Dept. Head, Head of Laboratory, Dep. Chief Engineer of 1st Moscow Car Repair Factory 1937–39 and 1940–42; served in Sov. Army 1939–40; Dept. Head at 1st Moscow Watch Factory 1942–44; CPSU posts 1944–55; Sec. of USSR Cen. Union Council of Trade Unions 1955–77, Dep. Chair. 1977–; Dep. to USSR Supr. Sov. 1966–; mem. Cen. Cttee. CPSU 1971–. Decors.: 2 Orders of Lenin, Order of Red Banner of Labour.

Prokofev, Mikhail Alekseevich, Dr. of Chem. Sc.
State official; b. N.5 (18), 1910, Voskresenskoe, Smolensk Oblast.
Russian; mem. CPSU 1941–; grad. Moscow State Univ. 1935; Prof.; pioneer leader, chemist-presser at Derbenovskii Chemical Works, student, postgraduate student, Dep. Dir. of Inst. for Chemical Research at Moscow State Univ. 1928–41; served in Navy 1941–46; Dep. Dir. and Sen. Scientific Mem. Staff of Inst. of Chemistry at Moscow State Univ. 1948–51; Head of Main Admin. of Univs., USSR Dep. Min. of Higher Educ. 1951–59; USSR 1st Dep. Min. of Higher and Secondary Specialist Educ. 1959–66; Min. of Educ. RSFSR My.–D. 1966; Dep. to USSR Supr. Sov. 1966–; USSR Min. of Educ. D. 1966–; mem. Cen. Cttee. CPSU 1971–. Decors.: Order of Red Banner of Labour, 4 other orders and medals.

Prokopenko, Sergei Kuzmich
State official.
Min. of Supplies Tadzhik SSR O. 1979–; cand. mem. Cen. Cttee. Tadzhik CP 1981–.

Prokopev, Ilia Pavlovich
Party official; b. 1926.
Chuvash; mem. CPSU 1946–; grad. Khabarovsk Pedagogical Inst. 1950, CPSU Cen. Cttee. Party Higher School 1958, and CPSU Cen. Cttee. Acad. of Social Sciences 1963, Cand. of Econ. Sc. 1963; teacher 1943; served in Sov. Army 1943–50; party posts, student 1952–63; Oblast Sec. of Chuvash ASSR Cttee. CPSU 1963–74, 1st Oblast Sec. Ja. 1974–; Dep. to USSR Supreme Sov. 1974–; mem. Cen. Cttee. CPSU 1976–. Decor.: Order of Lenin.

Prokopev, Iurii Nikolaevich
Party official; b. 1932.
1st Sec. of Iakutsk City CPSU Cttee. until 1982; 1st Sec. of Yakut Oblast CPSU Cttee. Je. 1982–. Decor.: Order of the October Revolution.

Prokoshkin, Iurii Dmitrievich
Physicist; b. D.19, 1929, Moscow.
Mem. CPSU 1963–; grad. Moscow Univ. 1952; corresponding mem. USSR Acad. of Sciences 1970–; mem. staff United Inst. of Nuclear Research in Dubna 1953–63; head of a sector at Inst. of High-Energy Physics in Serpukhov 1963–; his most important works are on physics of elemental particles; together with other scientists conducted experiments on discovery of antihelion-3 1968–72. Decor.: Order of Red Banner of Labour.
Publs.: "Nabliudenie antigelija-3" in "Iadernaia fizika", 1970, vol.12, No.2, p. 311–322, "Polnye secheniia vzaimodeistviia π, + k^+-mezonov i protonov s protonami i deitronami v oblasti impulsov 15–60 Gev" in "Iadernaia fizika" 1971, vol.14, No.5, p. 998–1005 (co-author).

Prokushev, Iurii L'vovich, Cand. of Philol. Sc.
Journalist; b. 1920.
Studied Party High School and Acad. of Social Sciences of Cen. Cttee. CPSU 1947–51; Sec. of Moscow Oblast and Moscow City Cttee. USSR Komsomol 1944–47; student 1947–51; Dep. Chief Ed. of Art Programmes of Radio Cttee. of USSR Council of Mins. 1951–54; Lecturer at Chair of Literature and Journalism at Party High School of Cen. Cttee. CPSU 1951–71; mem. USSR Writers' Union; Dir. of "Sovremennik" Publishing House Mr. 1971–.

Promet, Lilly Aleksandrovna
Author; b. F.16, 1922, Petseri, Esthonia; d. of a painter.
Esthonian; studied Tallinn School of Applied Art 1935–40; was active as journalist; first works publ. 1954
Publs.: "Derevnia bez Muzhchin" (A Village Without Men) 1962, "Kto rasprostraniat anekdoty" (Who Is Spreading Anekdotes?) 1969, "Dela zemnye" (Earthly Affairs) 1969, and others.

Promyslov, Vladimir Fedorovich
State official; b. Jl.15, 1908.
Russian; mem. CPSU 1928–; grad. Moscow Inst. of Construction Engineering 1956; fitter at Moscow Construction Trust 1924–29; trade union posts 1930–33; works dir. at a construction project, Dept. Head with Main Board for Planning the Construction of Hydroelectric Plants 1933–38; worked for Moscow City All-Union CP (B) Cttee. 1938–39; leading posts with USSR People's Commissariat for Heavy Machine Construction, worked for USSR Min. of the Tank Ind. and Main Munitions Ind. Construction Board of USSR Council of People's Commrs. 1939–46; worked for Moscow City Cttee. All-Union CP (B) 1946–49; Dep. Chair. of Moscow City Exec. Cttee. of Workers' Deps. 1949–51 and 1953–54; USSR Dep. Min. of Higher Educ. 1951–53; Sec. of Moscow City CPSU Cttee. 1954–55; Head of Main Board for Housing Construction and Civil Engineering with Moscow City Sov., 1st Dep. Chair. of Moscow City Sov. 1955–59; Chair. of State Construction Cttee. at RSFSR Council of Mins. 1959–63; mem. Cen. Auditing Comm. CPSU 1961–66; Dep. to 1962–, mem. Foreign Affairs Comm. with Sov. of the Union USSR Supr. Sov. 1966–; Dep. Chair. of RSFSR Council of Mins. and RSFSR Min. of Construction Ja.–Mr. 1963; Chair. of Moscow City Exec. Cttee. of Workers' Deps. Mr. 1963–; mem. Cen. Cttee. CPSU 1966–. Decors.: 2 Orders of Lenin, Order of the October Revolution.

Proskurin, Petr Lukich
Author; b. Ja.22, 1928, Kosytsy, Briansk Oblast.
Russian; first works publ. 1958.
Publs.: "Taezhnaia pesnia" (Taiga Song) 1960, "Glubokie rany" (Deep Wounds) 1960, "Tsena khleba" (The Price of Bread) 1961, "Liubov' cheloveka" (Human Love) 1965, and others.

Protozanov, Aleksandr Konstantinovich
Party official; b. 1914.
Russian; mem. CPSU 1939–; grad. Moscow Inst. of Non-Ferrous Metals and Gold and Party High School of Cen. Cttee. CPSU; worker 1932–39; sec. of a Komsomol raion cttee. in Moscow, head of a dept. at Altai Krai Cttee. CPSU 1941–44; dep. head of a dept. at Cen. Cttee. Belorussian CP 1944–48; mem. staff Cen. Cttee. CPSU 1948–52; Oblast Sec. of CPSU in Udmurt ASSR, 1st Dep. Chair. of Econ. Council Udmurt ASSR 1952–58; Sec. of Tiumen Oblast Cttee. CPSU (for Ind. 1963–64), Chair. of Tiumen Oblast Exec. Cttee. of Workers' Deps., 1st Sec. of Tiumen Oblast Cttee. CPSU 1959–69; 1st Sec. of East Kazakhstan Oblast Cttee. Kazakh CP S. 1969–; Dep. to USSR Supr. Sov. 1970–; mem. Cen. Cttee. Kazakh CP 1971–; cand. mem. Cen. Cttee. CPSU 1976–.

Protsenko, Dina Iosifovna
State official; b. 1926.
Mem. CPSU 1953–; grad. Kherson Agricultural Inst. 1950; chief zoo technician, dir. of a state farm, instructor, Chief of Agriculture Dept., 1st Dep. Chair. of Kherson Oblast Cttee. Ukrainian CP 1950–69; Chair. of Exec. Cttee. of Kherson Oblast Sov. of Peoples' Deps. 1969–78; mem. Cen. Cttee. Ukrainian CP 1971–; Chair. of State Cttee. for the Protection of Nature Ukrainian SSR Ap. 1978–.

Prozumenshchikova, Galina Nikolaevna (also G. N. Stepanova-Prozumenshchikova)
Athlete; b. N.26, 1948, Sevastopol'.
Honored Master of Sports of the USSR (1964).
She was the first Soviet Olympic champion in swimming (1964). Competing in the 100- and 200-m breaststroke events, she also won two silver and two bronze medals at the 1968 and 1972 Olympic Games and three medals at the 1966 and 1970 European championships. She won Soviet championships 15 times during the years 1963–73. Decor.: Order of the Red Banner of Labour.

Prudkin, Mark Isaakovich
Actor; b. S.1 (13) 1898, Klin.
People's Artist of the USSR (1961). Member of the CPSU since 1941.
He appeared in the productions of the Moscow Art Theater and of the theater's Second Studio beginning in 1911. He graduated from the studio's school in 1920 and joined the theater's company in 1924. He has a special affinity for sharply defined character roles and for satirical and comedy roles. He is a master of characterization and has created a number of dramatic and penetrating psychological portraits. His roles have included Shervinskii in Bulgakov's *Days of the Turbins,* Nezelasov in Vs. Ivanov's *Armored Train 14–69,* Vronskii and Karenin in *Anna Karenina,* adapted from L. N. Tolstoy's novel, Mekhti-Aga in Kron's *Distant Reconnaissance,* Mikhail Skrobotov and Iakov Kolomiitsev in Gorkii's *The Enemies* and *The Last Ones,* and Dul'chin and Krutitskii in Ostrovskii's *The Last Victim* and *Even a Wise Man Stumbles.* He is best known for the role of Fedor Pavlovich Karamazov in the play and film *The Brothers Karamazov,* adapted from Dostoevsky's novel (State Prize of the USSR, 1946, 1947, 1948, State Prize of the RSFSR, 1974). Decors.: Order of Lenin, two other orders, and several medals.

Prut, Iosif Leonidovich
Dramatist; b. N.5 (18), 1900, Taganrog.
Publs.: plays "Doroga na Iug" (Road to the South) 1938, "Dve nochi" (Two Nights) 1940, "Tikhii okean" (The Pacific Ocean) 1947; film scripts "V mirnye dni" (In the Days of Peace) 1951, "Sluchai s efreitorom Kochetkovym" (The Incident with Lance-Corporal Kochetkov) 1955, "P'esy" (Plays) 1963, and others.

Ptashnikov, Ivan Nikolaevich
Author; b. O.7, 1932, Zadrozde, Minsk Oblast.
Belorussian; grad. Dept. of Journalism of Belorussian State Univ. 1957.
Publs.: "Zerna padaiut ne na kamen'" (The Grain Does not Fall on Stone) 1959, "Stepan Zhikar iz Steshits" (Stepan Zhikar From Steshitsy) 1966, and others.

Ptitsyn, Vladimir Nikolaevich
Party official; b. 1925.
Russian; mem. CPSU 1946–; grad. mechanics technicum 1951 and Leningrad Inst. of Precision Mechanics and Optics 1958; collective farmer 1941–43; served in Sov. Army 1943–45; plant security chief, foreman, chief engineer at Leningrad Foundry 1945–56; chief of a dept., 2nd sec. of a raion cttee. CPSU in Leningrad City, Sec. of Leningrad Oblast Cttee., Dep. Chair. of Exec. Cttee. of Murmansk Oblast Sov. of People's Deps., Chair. of Murmansk Oblast Cttee. of People's Control 1956–59; Sec., 2nd Sec. of Murmansk Oblast Cttee. CPSU, Chair. of Exec. Cttee. of Murmansk Oblast Sov. of People's Deps. 1966–71; 1st Sec. of Murmansk Oblast Cttee. CPSU 1971–; Dep. to USSR Supr. Sov. 1974–; cand. mem. Cen. Cttee. CPSU 1976–81, mem. 1981–. Decor.: Order of Lenin.

Puchkov, Nikolai Georgievich
Athlete; b. Ja.30, 1930, Moscow.
Honored Master of Sports (1956) and Honored Coach of the USSR (1971). Officer in the Soviet Army. Member of the CPSU since 1954.
During the period 1952–62, he was goalkeeper on the USSR national ice-hockey team. He was USSR champion eight times during the years 1951–61 as a member of the Central Army Team and the Air Force Team, European champion six times during the period 1954–60, World champion in 1954 and 1956, and Olympic champion in 1956. He was acknowledged to be the best goalkeeper in the 1959 World championship. Decors.: two orders and various medals.

Puchkovskaia, Nadezhda Aleksandrovna
Ophthalmologist; b. My.12 (25), 1908, Smolensk.
Academician of the Academy of Medical Sciences of the USSR (1971). Honored Scientist of the Ukrainian SSR (1968) and Hero of Socialist Labour (1960).
She graduated from the Kiev Medical Institute in 1930. In 1956 she became director of the V. P. Filatov Institute of Eye Disease and Tissue Therapy. Her principal works deal with cornea transplants, the surgical treatment of and immunotherapy for the sequelae of severe eye burns, the use of laser radiation in ophthalmology, and the organization of ophthalmological services for the public. She became editor-in-chief of *Oftal'mologicheskii zhurnal* (Journal of Ophthalmology) in 1956.
She is an honorary member of the International Society of Eye Surgeons and a Bulgarian society of ophthalmologists and is a corresponding member of a society of ophthalmologists in the German Democratic Republic. She was a deputy to the sixth through eighth convocations of the Supreme Soviet of the Ukrainian SSR and a deputy chairman at the seventh and eighth convocations. In 1961 she received the V. P. Filatov Prize of the Academy of Medical Sciences of the USSR. Decors.: Order of Lenin, the Order of the October Revolution, three other orders, and several medals.
Works: *Peresadka rogovoi obolochki pri oslozhnennykh bel'makh.* Kiev 1960. *Wiederherstellende Operationen bei schweren Schädigungen und Erkrankungen der Augen.* Leipzig 1965. *V. P. Filatov.* Moscow 1969. *Osnovy peresadki rogovoi obolochki.* Kiev 1971. (Co-author.) *Patogenez i lechenie ozhogov glaz i ikh posledstvii.* Moscow 1973. (Co-author.)

Pudovik, Arkadii Nikolaevich
Chemist; b. Mr.2 (15), 1916, Tsivil'sk, now the Chuvash ASSR.
Corresponding member of the Academy of Sciences of the USSR (1964). Member of the CPSU since 1953.
He graduated from the University of Kazan in 1938 and completed his graduate study there in 1941. He worked at a plant in Kazan from 1941 until 1945, when he began working at the University of Kazan. In 1946, maintaining his position at the university, he began working at the A. E. Arbuzov Institute of Organic and Physical Chemistry of the Kazan branch of the Academy of Sciences of the USSR, and in 1971 he became the institute's director.
His main works have dealt with the chemistry of the organophosphates. He developed a new method for synthesizing the esters of the phosphonic acids and discovered the phosphonate-phosphate rearrangement as well as the thermal rearrangement of the allyl and propargyl esters of phosphorous acid (the acetylene-allene diene rearrangement). Using electrophilic reagents that do not contain halogen atoms, he has studied numerous reactions of phosphorous acid esters, amidophosphites, and mixed anhydrides. He also obtained many ney types of phosphorus-containing monomers and related polymers. Decors.: Order of the Red Banner of Labour and several medals.
Works: *Reaktsii prisoedineniia fosforsoderzhashchikh soedinenii s podvizhnym atomom vodoroda.* Moscow 1968. (With I. V. Gur'ianova and E. A. Ishmaeva.) *Reaktsii proizvodnykh kislot trekhvalentnogo fosfora s elektrofilnymi reagentami.* Moscow 1973. (With others.)

Pudkov, Ivan Ivanovich
State official; b. 1916.
Russian; mem. CPSU 1945–; grad. Moscow Ordzhonikidze Aviation Inst.; design engineer, head of a design office, sec. of party cttee. of a plant until 1953; prod. head, chief technologist, chief engineer of various plants affiliated with USSR Min. of Aviation 1953–68; 1st USSR Dep. Min. for the Construction of Light Ind. and Foodstuffs Machinery and Household Appliances 1968–77, Min. Mr. 1977–; Dep. to USSR Supr. Sov. 1979–;

mem. Cen. Auditing Comm. CPSU 1981–. Decor.: Lenin Prize.

Pugachev, Iurii Nikolaevich
Party official; b. 1926.
Russian; mem. CPSU 1947–; grad. Tashkent Electromechanical Technicum for Rail Transport 1952 and All-Union Econ. Finance Correspondence Inst. 1964; metal worker, asst. machinist of a locomotive depot at Astrakhan Railroad Station 1942–44; held party posts 1953–62; Sec. of Astrakhan Oblast Cttee. CPSU 1962–69, 2nd Sec. 1969–71; posts in Cen. Cttee. apparatus CPSU 1971–75; Dep. to USSR Supr. Sov. 1974–; 2nd Sec. of Cen. Cttee. and mem. Bureau Cen. Cttee. Kirghiz CP 1975–77; cand. mem. Cen. Cttee. CPSU 1976–; 2nd Sec. of Cen. Cttee. and mem. Bureau Cen. Cttee. Azerbaidzhan CP Ap. 1977–.

Pugachov, Vladimir Semenovich
Engineer; b. Mr. 12 (25), 1911, Riazan.
Mem. CPSU 1943–; grad. N.E. Zhukovskii Air Force Acad. 1931; specialist in mechanics, control processes and mathematics; mem. staff N.E. Zhukovskii Air Force Acad. 1932–72, Prof. at same 1939–; Prof. at Moscow S. Ordzhonikidze Aviation Inst. 1972–; Head of Laboratory at Inst. of Control Problems 1956–; his most important works are on statistical theories of control processes. Decors.: State Prize of the USSR 1948, Order of Lenin, 6 other orders and various medals.

Pugo, Boris Karlovich
State official; b. 1937.
1st Sec. and mem. Bureau Cen. Cttee. Latvian Komsomol 1969–70; Dep. to Supr. Sov. Latvian SSR 1969–70 and 1974–; Sec. and mem. Bureau Cen. Cttee. USSR Komsomol 1970–74; Head of Dept. for Party Org. Work at Cen. Cttee. Latvian CP 1974–75; 1st Sec. of Riga City Cttee. Latvian CP Jl. 1975–O. 1977; mem. Cen. Cttee. 1976–; cand. mem. Bureau Cen. Cttee. Latvian CP 1976–Ap. 1977 and Ja. 1981–. Dep. Chair. of Cttee. of State Security Latvian SSR 1977–80, Chair. N. 1980–.

Pukhov, Georgii Evgenevich, Dr. of Tech. Sc.
Computer scientist; b. Ag. 23, 1916, Sarapul, Udmurt ASSR; s. of a clerk.
Grad. Tomsk Polytech. Inst. 1940; specialist in electro and computer technology; corresponding mem. Acad. of Sciences Ukrainian SSR 1961–; Dr. of Tech. Sc. 1952; Prof. 1954–; served in Sov. Army 1940–42; postgraduate studies at Tomsk Polytech. Inst. 1942–44; Head of Chair of Electrical Machines and Theoretical Principles in Electrotechnology at Lvov Polytech. Inst. 1944–48; Head of Chair of Theoretical Principles in Electrotechnology and Dean of Dept. at Tomsk Polytech. Inst. 1948–53; Head of Chair of Theoretical Principles in Electrotechnology and Electrical Computing Devices at Inst. of Radio Technology in Taganrog 1953–57; Prof., Head of Chair of Gen. Electrotechnology at Kiev Inst. of Civil Aviation Engineering 1957–58; mem. staff and Dep. Dir. of Computing Centre (now Inst. of Cybernetics) of Acad. of Sciences Ukrainian SSR 1959–; author of some 200 scientific papers; made numerous discoveries. Decors.: Order of Red Banner of Labour, medals.
Publs.: "Elektricheskoe modelirovanie sterzhnevykh i tonkostennykh konstruktsii" 1960, "Izbrannye voprosy teorii matematicheskikh mashin" 1964, "Sintez mnogosviaznykh sistem upravleniia po metodu obratnykh operatorov" (co-author) 1966, "Metody analiza i sinteza kvazianalogovykh elektronnykh sistem" 1967, and others.

Pulatov, Salim
State official.
Min. of Higher and Secondary Specialist Educ. Uzbek SSR Jl. 1978–; mem. Cen. Cttee. Uzbek CP 1981–.

Purs, Laimonis (real name: Strazdinsh, Laimonis Ianovich)
Author; b. Je. 3, 1922, Latvia; s. of a worker.
Latvian; studied Chemical Tech. College until 1941; active as journalist 1944–; first works publ. 1945.
Publs.: "Teni ischezaiut" (The Shadows Vanish) 1956, "Skamia podsudimykh" (Bench of the Accused) 1959, "Ideal'nyi chelovek" (An Ideal Man) 1964, and others.

Puzanov, Aleksandr Mikhailovich
Diplomat; b. 1906, Lezhkova, Ivanovo Oblast; s. of a peasant.
Mem. CPSU 1925–; grad. Ples Agricultural Tech. College 1930; chair. of a rural sov., Dep. Chair. of Exec. Cttee. Lezhkova District Sov. 1924–26; attended college, active in collectivization of agriculture, agronomist with Sereda Agricultural Board in Ivanovo Oblast 1926–30; Dep. Dir. of Ivanovo Oblast Suburban Farm Trust 1931–33; worked for Sov. Control Comm. of USSR Council of People's Commrs. 1934–40; Chief Controller 1940–43, USSR Dep. People's Commr. for State Control 1943–44; Chair. of Kuibyshev Oblast Exec. Cttee. of Workers' Deps. 1944–46; 1st Sec. of Kuibyshev City Cttee. 1946–50, of Kuibyshev Oblast CPSU Cttee. 1946–52; Dep. to USSR Supr. Sov. 1946–58; Chair. 1952–56, 1st Dep. Chair. of RSFSR Council of Mins. 1956–57; mem. Cen. Cttee. 1952–76, cand. mem. Presidium 1952–53, mem. Bureau Cen. Cttee. CPSU RSFSR 1956–57; USSR Amb. to Korea 1957–62, to Yugoslavia 1962–67, to Bulgaria 1967–72, to Afghanistan 1972–79. Decors.: 4 Orders of Lenin, 4 medals.

Pyldroo, Aksel Iohannesovich
State official.
Cand. mem. Cen. Cttee. Esthonian CP 1971–76, mem. 1976–81, mem. Auditing Comm. Esthonian CP 1981–; Head Dept. of Agriculture with Cen. Cttee. Esthonian CP 1972–79; Chair. State Cttee. for Melioration and Water Econ. Esthonian SSR S. 1979–.

Pyshnov, Vladimir Sergeevich
Scientist specializing in the aerodynamics of airplanes; b. F. 21 (Mr. 3), 1901, Moscow.
Lieutenant-general of the engineers; Honored Worker in Science and Technology of the RSFSR (1942).
He graduated from the N. E. Zhukovskii Air Force Academy in 1925. He has taught at the academy since 1926; he became a professor in 1939. From 1949 to 1968 he was chairman of the aircraft section of the Scientific and Technical Committee. His main works deal with the theory of spin and the controllability, stability, and maneuverability of aircraft. Decors.: two Orders of Lenin, four other orders, and medals.
Works: *Samovrashchenie i shtopor samoletov.* Moscow 1927. *Aerodinamika samoleta,* parts 1–4. Moscow 1934–38. *Dinamicheskie svoistva samoleta.* Moscow 1951. *Iz istorii letatel'nykh apparatov: Sb. st.* Moscow 1968.

Pysin, Aleksei Vasil'evich
Author and poet; b. Mr. 22, 1920, Vysokii Borov, Mogilev Oblast.
Belorussian; mem. CPSU 1944–; grad. Advanced Courses in Literature 1958; first works publ. 1938. Decor.: State Prize of the Belorussian SSR 1967.
Publs.: "Nash den'" (Our Day) 1951, "Sinee utro" (A Blue Morning) 1967, and others.

Raa, H. H.
Komsomol official.
Sec. and mem. Bureau of Cen. Cttee. Esthonian Komsomol S. 1979–.

Rabotnov, Iurii Nikolaevich
Physicist; b. F. 11 (24), 1914, Nizhni-Novgorod, now Gorkii.
Mem. CPSU 1951–; grad. Moscow Univ. 1935; specializes in mechanical physics; mem. USSR Acad. of Sciences 1958–; teacher at Moscow Energetics Inst. 1935–41 and 1943–47; mem. staff Union Inst. of Electrotechnology 1941–43; mem. staff Inst. of Mechanics of USSR Acad. of Sciences 1947–57; mem. staff Inst. of Hydrodynamics at Siberian branch of USSR Acad. of Sciences 1958–65; Prof. at Dept. of Mechanics and Mathematics of Moscow Univ. 1947–. Decors.: of the October Revolution, 2 other orders and various medals.

Rachkov, Albert Ivanovich
Party official; b. 1927.
Mem. CPSU 1955–; grad. Azovo-Chernomore Inst. for the Mechanization and Electrification of Agriculture 1950 and CPSU Cen. Cttee. Party Higher School 1971; engineer, chief engineer, manager of a machine and tractor station of the Kemerovo Oblast admin. for agriculture 1950–56; 1st Sec. of Chebula Raion Cttee. CPSU in Kemerovo Oblast 1960–62; posts in Cen. Cttee. apparatus CPSU 1965–69 and 1974–80; Sec. of Lipetsk Oblast Cttee. CPSU 1969–74; 2nd Sec. of Cen. Cttee. and mem. Bureau of Cen. Cttee. Turkmen CP D. 1980–; cand. mem. Cen. Cttee. CPSU 1981–.

Radiukevich, Leonid Vladimirovich
Industrial manager; b. 1932.
Mem. CPSU 1959–; grad. Magnitogorsk Metallurgical Mining Inst. 1955; roller, foreman, sheet-rolling shop superintendent 1955–70; chief roller, dep. chief engineer of V.I. Lenin Metallurgical Combine in Magnitogorsk 1970–77; chief of an admin. in USSR Min. of Ferrous Metallurgy 1977–79; Dir. of V.I. Lenin Metallurgical Combine in Magnitogorsk S. 1979–; cand. mem. Cen. Cttee. CPSU 1981–. Decors.: USSR State Prize 1969.

Radzhabov, Nazir Radzhabovich
State official.
Sec. Bukhara Oblast Cttee. Uzbek CP 1975–78; Head of Dept. for Building and Municipal Econ. with Cen. Cttee. Uzbek CP 1978; Min. of Rural Construction Uzbek SSR D. 1978–; mem. Cen. Cttee. Uzbek CP 1981–.

Radzievskii, Aleksei Ivanovich
Military official; b. Jl.31 (Ag.13), 1911, Uman.
Mem. CPSU 1931–; grad. from a cavalry school 1931, M.V. Frunze Mil. Acad. 1938 and Gen. Staff Acad. 1941; Gen. of the Army 1972–; sen. command posts at the front 1941–45; Commander of Northern Forces Group 1950–52; Commander of Turkmen Mil. District 1952–53; Commander of Armoured and Mechanized Troops 1953–54; Commander of Odessa Mil. District 1954–59; 1st Dep. Head of Gen. Staff Mil. Acad. 1959–68; Head of Main Admin. of Mil. Educ. Institutions of USSR Min. of Defence 1968–69; Head of M.V. Frunze Mil. Acad. 1969–78. Decors.: Order of Lenin, 6 Orders of Red Banner, 2 Suvorov Orders 1st Cl., Kutuzov Order 1st Cl., Suvorov Order 2nd Cl., Order of Red Banner of Labour, Order of Red Star, medals, 5 foreign orders.

Radzinskii, Edvard Stanislavovich
Dramatist; b. S.23, 1936, Moscow.
Grad. Inst. of History and Archival Science in Moscow 1959.
Publs.: plays "Mechta moia–Indiia" (My Dream – India) 1960, "Vam 22, stariki!" (You Are 22, Oldsters!) 1962, "104 stranitsy pro liubov'" (104 Pages On Love) 1964, "Obol'stitel' Kolobashkin" (The Seducer Kolobashkin) 1967, and others.

Raevskii, Viktor Ivanovich
State official; b. 1936, Pechory, Pskov Oblast.
Russian; grad. Esthonian Acad. of Agrarian Sciences, mechanical engineer, Cand. of Agrarian Sciences; held posts in Esthonian Scientific Research Inst. of Animal Husbandry and Veterinary Science 1960–75; Dir. of Laeva state farm 1975–79; Dep. Min. of Agriculture Esthonian SSR 1979–82; Min. of Procurements Esthonian SSR Ag. 1982–.

Rafiev, Mirzo Manaevich
State official.
Chair. of State Cttee. on Pricing Tadzhik SSR Mr. 1970–.

Rafikov, Sagid Raufovich
Chemist; b. Ap.6 (19), 1912, Kaishevo, now in Diurtiuli Raion, Bashkir ASSR.
Academician of the Academy of Sciences of the Kazakh SSR (1962) and corresponding member of the Academy of Sciences of the USSR (1970).
After graduating from the Kazan Chemical-Technological Institute in 1937, he worked at various chemical institutes of the academies of sciences of the USSR and the Kazakh SSR. In 1967 he was made chairman of the presidium of the Bashkir branch of the Academy of Sciences of the USSR, and the following year he was also made director of the Institute of Chemistry at the Bashkir branch of the Academy of Sciences of the USSR. His main works deal with the synthesis of high-molecular-weight compounds (polycondensation, polymerization, chemical transformations) and the study of oxidation and oxidative ammonolysis.
He was a deputy to the eight and ninth convocations of the Supreme Soviet of the USSR. Decors.: three medals.
Works: *Vvedenie k izucheniiu vysokomolekuliarnykh soedinenii.* Moscow-Leningrad 1946. (With V.V. Korshak.) *Sintez i issledovanie vysokomolekuliarnykh soedinenii.* Moscow-Leningrad 1949. (With V.V. Korshak.) *Metody opredeleniia molekuliarnykh vesov i polidispersnosti vysokomolekuliarnykh soedinenii.* Moscow 1963. (With S.A. Pavlova and I.I. Tverdokhlebova).

Rafikov, Subbukh Khadievich
Author; b. Jl.25, 1913, Nurkeevo, Tartar ASSR; s. of a peasant.
Tartar; mem. CPSU 1958–; active as writer and journalist; first works publ. 1930. Publs.: "Belye nochi" (White Nights) 1957, "Pervaia vesna" (The First Spring) 1962, "Utro derevni" (Village Morning) 1966, and others.

Ragimov, Kamran Nabi ogly
Party official; b. 1928.
Dr. of Historical Sciences; Azerbaidzhani; mem. CPSU 1953–; grad. Azerbaidzhan Univ.; teacher in dept. of history Azerbaidzhan Univ. 1950–56; head of a dept., dep. editor, editor of the periodical *Azerbaidzhan kommunisti* 1956–75; cand. mem. Cen. Cttee. CP Azerbaidzhan 1966–76, mem. 1976–; 1st Oblast Cttee. Sec. CP Azerbaidzhan in Nakhichevan ASSR 1975–; cand. mem. Bureau of Cen. Cttee. CP Azerbaidzhan 1981–.

Ragimov, Suleiman Gusein ogly
Azerbaidzhan writer; b. Mr.9 (22), 1900, Ain, now in Kubatly Raion, Azerbaidzhan SSR.
People's Writer of the Azerbaidzhan SSR (1960). Member of the CPSU since 1926. In 1931, he graduated from the department of history of the Azerbaidzhan Pedagogical Institute and became a teacher. He began publishing in 1930. His novel *Shamo* (vols. 1–3, 1931–64) deals with the establishment of Soviet power in Azerbaidzhan and the novel *Sachly* (1940–48; Russian translation, 1971), with the achievements of the revolution in the Azerbaidzhan countryside.
During the Great Patriotic War (1941–45) he wrote the novellas *Voice of the Earth* (1941), *Medallion* (1942), and *Common Grave* (1943). The novella *Mekhman* (1944) depicts the fate of a young Soviet lawyer. A strong satirical undercurrent permeates his stories about the prerevolutionary past and vestiges of feudalism; these stories include "The Request for Water" and "The Envious One." The novel *The She-eagle of the Caucasus* (vols. 1–2, 1971–73) depicts the friendship between the peoples of Transcaucasia and Russia.
He was a deputy to the Supreme Soviet of the Azerbaidzhan SSR. Decors.: two Orders of Lenin, two other orders, and several medals.
Works: *Sächilmish äsärläri,* 10 vols., vols. 1–5. Baku 1968–74. *In Russian translation: Shamo,* vols. 1–3. Baku 1950–66. *Izbr. proizv.* vols. 1–2. Moscow 1972.

Raikin, Arkadii Isaakovich
Estrada (variety stage) entertainer; b. O.11 (24), 1911, Riga.
People's Artist of the USSR (1968).
After graduating from the Leningrad Theatrical Technicum in 1935, he performed in the Leningrad Working Youth Theater and the Lenin Komsomol Theater; at the same time he performed on the *estrada.* In 1939 he won a prize at the All-Union Estrada Competition in Moscow. He subsequently performed at the Leningrad Theater of the Estrada and the Miniature, organized in 1939, later becoming the theater's artistic director.
While at the Leningrad Theater of the Estrada and the Miniature, he expanded the range of expressive means and developed

the social and satiric pungency of his art by savagely attacking anything that stood in the way of the new life. As a rule, he has several roles in the same performance or stage miniature and is a master of instantaneous impersonation. He is equally talented as master of ceremonies, pantomimist, and lyrical monologuist. He is also known for his tender, lyrical, and charming roles. His productions include *Around the World in Eighty Days* (1951), *Laughter Is No Sin* (1953), *Seasons of the Year* (1956), *Love and Three Oranges* (1959), *From Two to Fifty* (1961), *Sorcerers Are Living Next Door* (1964), *Traffic Light* (1969), and *Selections '73* (1973).
He has toured abroad. Decors.: three orders and various medals.

Raizman, Iurii Iakovlevich
Film director; b. D. 2 (15), 1903, Moscow. Studied Dept. of Literature and Art of Moscow Univ. Decors.: Hero of Socialist Labour 1973, State Prize of the USSR 1941, 1943, twice 1946, 1950 and 1952, Order of Lenin, 3 other orders and various medals.

Rakhimbabaeva, Zakhra
State official; b. 1923.
Cand. of Historical Sciences; Uzbek; mem. CPSU 1946–; grad. Pedagogical Inst. in Tashkent and Acad. of Social Sciences of the CPSU Cen. Cttee.; instructor, Sec. of Komsomol Oblast Cttee. 1942–46; head of a dept. of Oblast Cttee. Uzbek CP, head of a dept., Sec. of Cen. Cttee. Uzbek CP 1946–51; student 1951–55; 1st Dep. Min. of Culture Uzbek SSR 1955–56; mem. Cen. Cttee. Uzbek CP 1956–71; Sec. and mem. Bureau of Cen. Cttee. Uzbek CP 1956–63; Min. of Culture Uzbek SSR 1963–67 and Ap. 1981–.

Rakhimov, Bektash Rakhimovich
State official; b. 1924.
Uzbek; mem. CPSU 1944–; grad. Cen. Asian State Univ.; teacher 1940–42; served in Sov. Army 1942–45; dir. of a school, inspector of a raion educ. dept., dep. chair., chair. of a raion sov. 1945–50; attended Uzbek Republican Party School, teacher, Sec. of Cttee. of Uzbek CP at same school, Organizer for Cen. Cttee. Uzbek CP, 1st sec. of a raion cttee. Uzbek CP 1950–62; cand. mem. 1961–66, mem. Cen. Cttee. Uzbek CP 1966–; 1st Sec. of Khorezm Oblast Cttee. Uzbek CP 1962–68; Dep. to 1966–, mem. Comm. on Educ., Science and Culture with Sov. of Nationalities 1966–70, mem. Comm. for Nature Conservation with Sov. of the Union USSR Supr. Sov. 1970–; 1st Sec. of Andizhan Oblast Cttee. Uzbek CP 1968–74 and Samarkand Oblast Cttee. N. 1974–82; cand. mem. Cen. Cttee. CPSU 1976–. Dep. Chair. Council of Mins. Uzbek SSR F. 1982–.

Rakhimov, Kh. R.
Party official.
Mem. Auditing Comm. Uzbek CP 1966–71; Head of Dept. of Information and Relations with Foreign Countries of Cen. Cttee. Uzbek CP Mr. 1980–; cand. mem. Cen. Cttee. Uzbek CP 1981–.

Rakhimova, Ibodat
State official; b. 1922.
1st Sec. of Cen. Cttee. Tadzhik Komsomol, Dep. Chair. of Supr. Sov. Tadzhik SSR, 1st Sec. of Oktiabrskii Raion Cttee. Tadzhik CP in Dushanbe, Dep. Head of Dept. of Party Organs with Cen. Cttee. Tadzhik CP 1955–66; mem. Cen. Cttee. 1956–, Sec. and mem. Bureau Cen. Cttee. Tadzhik CP 1966–78; Sec. of Presidium Supr. Sov. Tadzhik SSR D. 1978–.

Rakhlenko, Leonid Gdal'evich (Grigor'evich)
Actor and director; b. Ag. 23 (S. 5), 1907, Terekhova, Gomel Oblast.
Mem. CPSU 1942–; grad. Leningrad Inst. of Theatre Arts 1928; Artistic Dir. of Belorussian Ia. Kupala Theatre in Minsk 1937–43. Decors.: People's Artist of the USSR 1966, State Prize of the Belorussian SSR 1970, Order of the October Revolution, 3 Orders of Red Banner of Labour, medals.

Rakhlin, Natan Grigor'evich
Conductor; b. D. 28, 1905, Snovsk, now Snovskoe, Chernigov Oblast.
Mem. CPSU 1947–; studied Kiev Conservatory; grad. Conducting Dept. of N. V. Lysenko Inst. of Music and Drama in Kiev 1923–27; Head of Ukrainian State Symphony Orchestra 1937–62; Head of USSR State Symphony Orchestra 1941–45; Musical Dir. and Principal Conductor of Symphony Orchestra of the Tartar ASSR 1966–; Prof. at Kazan Conservatory 1967–; one of the most distinguished conductors of the Sov. Union (especially for classical Russian composers such as P. I. Tchaikovsky and A. N. Skriabin and works of Sov. and Ukrainian composers). Decors.: People's Artist of the USSR 1948, 2nd Prize at 1st Union Contest for Conductors 1938, State Prize of the USSR 1952, Order of Lenin, 2 other orders and various medals.

Rakhmanin, Oleg Borisovich
Party official; b. 1924.
Mem. CPSU 1945–; served in Sov. Army 1939–45; mem. staff USSR Min. of Foreign Affairs 1945–63; various posts with Cen. Cttee. CPSU 1963–68; 1st Dep. Head of Dept. of Relations with Communist and Workers' Parties of Socialist Countries at Cen. Cttee. CPSU 1968–; mem. Cen. Auditing Comm. CPSU 1971–76; cand. mem. Cen. Cttee. CPSU 1976–81, mem. 1981–.

Rakhmanov, Leonid Nikolaevich
Author; b. F. 15 (28), 1908, Kotelnich, now Kirov Oblast.
Russian; studied Leningrad Inst. of Electrotechnology 1926–28; first works publ. 1927.
Publs.: film scripts (as co-author) "Deputat Baltiki" (Deputy from the Baltics) 1937, "Kto s mechom voidet ..." (Whoever Comes With the Sword ...) 1953, "Iavlenie Venery" (The Appearance of Venus) 1961, plays "Kamen', kinutyi v tikhii prud" (A Stone Thrown Into a Still Pond) 1964, "Bespokoinaia starost'" (Troubled Age) 1937, "P'esy i stsenarii" (Theatre Plays and Film Scripts) 1956, "Ochen' raznye povesti: pechal'naia, dobraia, starinnaia, voennaia, ozornaia" (Very Diverse Stories: Sad Ones, Kind Ones, Old Ones, War Stories, Naughty Ones) 1965, and others.

Rakitin, Iurii Vladimirovich
Plant physiologist; b. Mr. 23 (Ap. 5), 1911, Dukhovshchina, Smolensk Oblast.
Mem. CPSU 1943–; grad. Gorkii Agricultural Inst. 1932; mem. staff Inst. of Plant Physiology of USSR Acad. of Sciences 1935–; Head of Laboratory 1944–; Chief Ed. of journal "Agrokhimiia".
Publs.: "Stimulatsiia i tormozhenie fiziologicheskikh protsessov u rastenii" in "Istoriia i sovremennoe sostoianie fiziologii rastenii v Akademii nauk" Moscow 1967, p. 1935–46, "Nitrat 2-oksietilrutidefoliant novogo tipa" in "Fiziologiia rastenii" 1974, vol. 21, no. 1, p. 192–204.

Rakitskii, Petr Fomich
Biologist and geneticist; b. Ag. 2 (15), 1903, Kustovnitsa, Gouvernement Minsk.
Belorussian; grad. Moscow Univ. 1927; specialist in plant selection; Head of Chair of Genetics Moscow Univ., then mem. staff of various scientific institutes 1932–38; Pres. of Belorussian Vavilov Soc. of Geneticists and Selectionists 1966–. Decors.: Order "Badge of Honour", medals.

Rakov, Vasilii Ivanovich
Military official; b. Ja. 26 (F. 8), 1909, St. Petersburg; s. of an office worker.
Major-general of aviation (1958). Doctor of naval science (1967); professor (1969). Twice Hero of the Soviet Union (F. 7, 1940, and Jl. 22, 1944). Member of the CPSU from 1932.
He joined the Red Army in 1928. He graduated from the Military Theoretical School for Pilots in 1929, from the First Military School for Pilots, from the Military School for Naval Pilots in 1931, from the Naval Academy in 1942, and from the Military Academy of the General Staff in 1946. He saw service in the Soviet-Finnish War of 1939–40 as a squadron commander in the 57th Bombardment Aviation Regiment. In the Great Patriotic War he served in the Black Sea and Baltic

fleets as commander of a naval aviation brigade, deputy commander of the 3rd Special Mission Sevastopol' Air Group in 1942–43, assistant commander of the 9th Attack Aviation Division in 1944, and commander of the 12th Guards Air Regiment from May 1944 to February 1945. He completed 68 combat flights and participated in the sinking of the German antiaircraft cruiser *Niobe* in the port of Kotka.
After the war, he held responsible positions in the armed forces. He taught at the Naval Academy from 1948 and was head of a department from 1952. He went into the reserve in 1971. Decors.: two Orders of Lenin, three Orders of the Red Banner, the Order of the Red Star, and various medals.

Rakutsev, Iurii Il'ich
State official.
Chair. of State Cttee. for the Supply of Petroleum Products Tadzhik SSR S. 1981–.

Raman, Miervaldis Leonidovich
State official.
Dep. Dir. of Inst. of Power Ind. Latvian SSR 1961; Chair. of State Cttee. for the Coordination of Scientific Research with Council of Mins. Latvian SSR 1962–65; Dep. to Supr. Sov. Latvian SSR 1963–; mem. Cen. Cttee. Latvian CP 1963–; Dep. Chair. of Council of Mins. and Chair. of GOSPLAN Latvian SSR 1965–.

Ramanauskas, Ionas Kazevich
State official; b. 1929.
Grad. Moscow Inst. of Textile Ind.; worker, engineer, Dept. Head, Dep. Dir., Dir. of "Niamunas" Factory until 1963; CPSU posts 1963–65; Dir. of "Niamunas" Factory 1965–71; mem. Auditing Comm. 1964–71, cand. mem. Cen. Cttee. Lithuanian CP 1971–81, mem. 1981–; Head of Dept. of Food and Light Ind. at Cen. Cttee. Lithuanian CP 1971–75; Min. of Light Ind. Lithuanian SSR F. 1975–.

Ramazanov, Gilemdar Zigandarovich, Dr. of Philol. Sc.
Author and poet; b. Je. 16, 1923, Starobalakovo, Bashkir ASSR.
Bashkir; mem. CPSU 1943–; grad. Bashkir Pedagogic Inst. 1949; concluded postgraduate studies at Moscow State Univ. 1953; first works publ. 1939.
Publs.: "Nashe pokolenie" (Our Generation) 1947, "Zvezdy mira" (Stars of Peace) 1951, "Moia zemlia" (My Country) 1959, "Dve reki" (Two Rivers) 1962, "Obraz sovetskogo cheloveka v bashkirskoi poezii" (The Figure of Soviet Man in Bashkir Poetry) 1956, "Stikhi" (Poetry) 1963, and others.

Ramz, Babadzhan (real name: Babadzhanov, Ramz)
Poet and dramatist; b. D. 21, 1921, Tashkent.
Uzbek; mem. CPSU 1951–; grad. Dept. of Literature of Tashkent Pedagogic Inst. 1943; concluded postgraduate studies at same 1955; Sec. of Uzbek Writers' Union 1960–73; Dir. of Belles-Lettres Publishing House 1963–67; first works publ. 1936.
Publs.: "Podarok" (The Present) 1940, "Tainy liubvi" (The Secrets of Love) 1963, "Diadia i plemianniki" (Uncle and the Nephews) 1960, "Tsygana ne obmanesh" (You Don't Deceive a Gipsy) 1964, and others.

Randviir, Tiiu Aleksandrovna (also Tiiu Randviir)
Esthonian ballet dancer; b. Mr. 10, 1938, Tartu.
People's Artist of the USSR (1967).
She graduated from the Tallinn School of Choreography in 1955. She continued her studies at the Moscow Choreographic School and the Bolshoi Theater in 1959 and 1960.
In 1955, she became a soloist with the ballet company of the Esthonia Theater. Her roles have included Meelike in Kapp's *The Gold Spinners*, the Maiden and Joanna in Tamberg's *Ballet Symphony* and *Joanna Possessed*. Odette-Odile in Tchaikovsky's *Swan Lake*, and Carmen in *Carmen Suite* (music by Bizet, arranged by Shchedrin).

Rapoport, Vladimir (Vulf) Abramovich
Cinematographer; b. 1907.
Mem. CPSU 1940–; grad. State College for Cinematography 1929. Decors.: State Prize of the USSR 1942, 1946, 1949, 1951, 1971.

Rashidov, Rashid Medzhinovich
Author and poet; b. My. 1, 1928, Vanashi-Makhi, Dagestan.
Mem. CPSU 1960–; writes primarily for children.
Publs.: "Moe schastie" (My Happiness) 1948, "Moim malenkim druziam" (For My Little Friends) 1954, "Liudi" (People) 1965, "Kodga orly pokidaiut gnezda" (When Eagles Leave Their Nests) 1967, "Sosedi smeiutsia" (The Neighbours Laugh) 1969, and others.

Rashidov, Sharaf Rashidovich
Party official; b. N. 6, 1917; s. of a peasant.
Uzbek; mem. CPSU 1939–; grad. Uzbek State Univ. Samarkand 1941 and Party Higher School of Cen. Cttee. 1948; after completing Teachers' Training College in Dzhizak teacher, student, Exec. Sec., Dep. Ed., Ed. of Samarkand Oblast newspaper "Lenin iuly" 1936–41; pol. work in Sov. Army 1941–42; severely wounded and released from Army 1942; Ed. of "Lenin iuly" 1942–44; Sec. of Samarkand Oblast Cttee. Uzbek CP(B) 1944–47; Ed. of Republican newspaper "Kzyl Uzbekistan" 1947–49; Chair. of Uzbek Writers' Union 1949–50; Chair. of Presidium Supr. Sov. Uzbek SSR 1950–59; Dep. to 1950–, Dep. Chair. 1950–60, mem. Presidium USSR Supr. Sov. 1970–; mem. Bureau Cen. Cttee. Uzbek CP 1954–; cand. mem. 1956–61, mem. Cen. Cttee. CPSU 1961–; 1st Sec. of Cen. Cttee. Uzbek CP 1959–; cand. mem. Politbureau Cen. Cttee. CPSU O. 1961–; mem. Cen. Asia Bureau of Cen. Cttee. CPSU 1962–64. Decors.: 5 Orders of Lenin, Order of Red Banner of Labour, Order of Red Star, medals.

Rasizade, Shamil Alirza ogly
State official.
1st Dep. Min. of Agriculture Azerbaidzhan SSR until 1960; cand. mem. 1960–61, mem. Cen. Cttee. Azerbaidzhan CP 1961–64 and 1971–; Head of Agricultural Dept. of Cen. Cttee. Azerbaidzhan CP 1961–62; 1st Dep. Min. of Prod. and Procurement of Agricultural Products Azerbaidzhan SSR 1962–65; mem. Bureau for Agriculture at Cen. Cttee. Azerbaidzhan CP 1962–64; 1st Dep. Min. of Agriculture Azerbaidzhan SSR 1965–70; Dep. Chair. of Council of Mins. Azerbaidzhan SSR O. 1970–. Decor.: Order of Red Banner of Labour.

Rasputin, Valentin Grigor'evich
Author; b. Mr. 15, 1937, Ust-Uda, Irkutsk Oblast.
Russian; grad. Dept. of History and Philology of Irkutsk Univ. 1959; first works publ. 1961.
Publs.: "Ia zabyl sprosit' u Lioshki" (I Forgot to Ask Lioshka) 1961, "Chelovek s etogo sveta" (A Man of this World) 1965, "Prodaetsia medvezhia shkura" (A Bearskin For Sale) 1966, "Vasilii i Vasilisa" 1967, "Poslednii srok" (The Last Date) 1970, and others.

Rassadin, Stanislav Borisovich
Literary critic; b. Mr. 4, 1935, Moscow.
Grad. Dept. of Philology of Moscow Univ. 1958; writes mainly on contemporary Soviet literature, children's literature and reader psychology.
Publs.: "O stikhakh poslednikh let" (On Poems of the Last Years) 1961, "Nikolai Nosov. Kritiko-biograficheskii ocherk" (Nikolai Nosov. A Critical Biographical Outline) 1961, "Razgovor s chitatelem. Stat'i o literature" (Talk With the Reader. Essay on Literature) 1962, "Kniga pro chitatelia" (Book About the Reader) 1965, "Lipovye allei" (Linden Alley) 1966, and others.

Rasul, Rza (real name: Rzaev, Rasul Ibragim ogly)
Author; b. My. 19, 1910, Geokchai.
Azerbaidzhani; mem. CPSU 1939–; Chair. of Azerbaidzhan Writers' Union 1938–39; Min. of Culture Azerbaidzhan ASSR 1946–49. Decors.: State Prize of the USSR 1951, People's Poet of the Azerbaidzhan ASSR.
Publs.: "Bol'shevistskaia vesna" (Bolshevik Spring) 1931, "Lenin" 1950, "Ballada o negritenke Villi" (Ballad of the Negro

Boy Willy), and others; translated works by Maiakovskii, Pushkin, Shevchenko and Lermontov into Azerbaidzhan.

Rasulbekov, Gusein Dzhumshudovich
State official.
Cand. mem. 1961–71, mem. Cen. Cttee. Azerbaidzhan CP 1971–; Dep. to Supr. Sov. Azerbaidzhan SSR 1967–; Lieut.-Gen.; Min. of Posts and Telecommunications Azerbaidzhan SSR O. 1975–.

Rasulova, Lidia Khudat kyzy
Trade union official.
Mem. Auditing Comm. Azerbaidzhan CP 1976–77, Chair. 1977–81; 1st Sec. Naziminskii Raion Cttee. Azerbaidzhan CP 1980; Dep. Supr. Sov. Azerbaidzhan SSR 1980–; mem. Cen. Cttee. and cand. mem. Bureau Cen. Cttee. Azerbaidzhan CP 1981–; Chair. of Azerbaidzhan Republican Trade Union Council F. 1982–.

Ratanov, Anatolii Petrovich
Diplomat; b. 1921.
Mem. CPSU 1943–; Komsomol posts 1943–54; mem. Ed. Board of journal "Molodoi Kommunist" 1954–58; Permanent Rep. of USSR for Sov. Youth Orgs. at Communist World Fed. of Democratic Youth in Budapest 1958–60; mem. Diplomatic Service 1960–; 1st Emb. Sec. 1960–62, Emb. Counsellor at USSR Emb. in Cambodia 1962–63; USSR Amb. to Cambodia 1965–67; mem. staff USSR Min. of Foreign Affairs 1967–70; USSR Amb. to Guinea 1970–73; USSR Amb. to Ethiopia D. 1973–N. 1978.

Ratner, Evgenii Il'ich
Author; b. F. 5 (18), 1911, Kharkov.
Russian; mem. CPSU 1941–; grad. Kharkov Inst. of Journalism 1934; first works publ. 1930.
Publs.: "Sem' dnei" (Seven Days) 1931, "Na beregakh Baltiki" (On The Shores of the Baltic Sea) 1948, "Avtoritet" (Authority) 1959, "Baltiiskaia simfoniia" (Baltic Symphony) 1961, and others.

Ravenskikh, Boris Ivanovich
Stage director; b. Je. 14 (27), 1914, Moscow.
People's Artist of the USSR (1968). Member of the CPSU since 1954.
In 1935, he graduated from the department of directing at the Leningrad Theatrical Technicum. From 1935 to 1938 he was assistant director of the Meyerhold Theater, he then worked at the Moscow Art Theater. From 1941 to 1950 he was a director at the Drama Studio (later the Moscow Stanislavsky Dramatic Theater). In 1950 he directed D'iakonov's *Wedding With a Dowry* at the Moscow Theater of Satire.
From 1951 to 1960, he was a director at the Malyi Theater. His most important production of this period was L. N. Tolstoy's *The Power of Darkness,* which he treated as a lofty tragedy by developing the theme of man's responsibility. From 1960 to 1970, he headed the Moscow Pushkin Dramatic Theater. He staged modern publicist plays, including Squarzina's *Romagnola* (1963) and Mdivani's *Teresa's Birthday* (1961), as well as the heroic large-scale drama *Virgin Seul Upturned* (1964; adapted from Sholokhov's novel) and the impassioned revolutionary play *Dramatic Song* (1971; adapted by him and Ancharov from N. A. Ostrovskii's novel *How the Steel Was Tempered*).
Since 1970, he has been principal director of the Malyi Theater. In 1972, with Ungurianu, he staged Drube's *Birds of Our Youth,* and in 1973 A. K. Tolstoy's tragedy *Tsar Fedor Ivannovich.* He makes skillful use of music and versatility of expression to reveal the inner meaning of a work. His productions are marked by psychological delineation of character and by vivid and powerful emotional impact.
Decors.: State Prize of the USSR (1951, 1972) and the State Prize of the RSFSR (1967), Order of Lenin, two other orders, and several medals.

Razdukhov, Iurii Ivanovich
Diplomat; b. 1924.
Mem. Diplomatic Service 1958–; Emb. Counsellor 1959–61 and Emb. Counsellor-Envoy at USSR Emb. in China 1966–68; Emb. Councellor-Envoy at USSR Emb. in Bulgaria 1968–75; USSR Amb. to Singapore My. 1975–Ap. 1980.

Razin, Nikolai Vasil'evich
Hydraulic engineer; b. Ap. 26 (My. 9), 1904), Verkhniaia Gora, in present-day Kirillov Raion, Vologda Oblast.
Corresponding member of the Academy of Sciences of the USSR (1968); Hero of Socialist Labor (1958). Became a member of the CPSU in 1951.
He graduated from the Leningrad Polytechnic Institute in 1929. He participated in the design work for connecting the Pechora, Vychegda, and Kama rivers; the Solikamsk hydroengineering complex; and the Shirokovskii Hydroelectric Power Plant on the Kos'va River. In 1949, he became chief construction engineer for the Tsimliansk hydroengineering complex on the Don River; in 1953 he held the same position at the Volga V. I. Lenin Hydroelectric Power Plant. From 1962 to 1967 he was chief engineer of Gidroproekt (full name, the S. Ia. Zhuk All-Union Projekt Surveying and Scientific Research Institute). He later headed the division of regional water problems of the Council for the Study of Productive Forces.
He is the author of works dealing principally with various questions of hydraulic engineering and hydraulic power. Decors.: State Prize of the USSR (1952), two Orders of Lenin, and various medals.
Works: *Tsimlianskii gidrouzel.* Moscow-Leningrad 1954. *Opyt stroitel'stva Volzhskoi gidroelektrostantsii im. V. I. Lenina.* Moscow-Leningrad 1960.

Razumov, Evgenii Zotovich
Party official; b. 1919.
Mem. CPSU 1942–; grad. Mordovian Pedagogical Inst. 1942 and CPSU Cen. Cttee. Party Higher School 1948; Oblast Sec. of All-Union Komsomol Cttee. in Mordovian ASSR 1942–45; party posts 1945–50; ed.-in-chief of oblast newspaper "Kuzbass" 1950–51; Sec. of Kemerovo Oblast Cttee. CPSU 1951–61; posts in Cen. Cttee. CPSU apparatus 1961–67; Dep. Chief of Organizational Party Work Dept. of Cen. Cttee. CPSU My. 1967–; mem. Cen. Auditing Comm. 1976–81; mem. Cen. Cttee. CPSU 1981–.

Razuvaev, Grigorii Alekseevich
Physical chemist; b. Ag. 11 (23), 1895, Moscow.
Grad. Moscow State Univ. 1925; mem. USSR Acad. of Sciences 1966–; after conclusion of studies mem. staff Laboratory of High Pressures of USSR Acad. of Sciences; head of a chair at Leningrad Technological Inst.; head of a chair at Gorkii Univ. 1946–; concurrently Dir. of Scientific Chemistry Research Inst. at same univ. 1956–62; Dir./Organizer of Laboratories for the Stabilization of Polymeres at USSR Acad. of Sciences 1963–; Dir. of Inst. of Chemistry of USSR Acad. of Sciences in Gorkii 1969–.
Decors.: State Prize of the USSR 1971, Lenin Prize 1958, Order of Lenin, medals.
Publs.: "Metalloorganicheskoe soedeniie v elektronike" (co-author) Moscow 1972, "Reactions of organometallic compounds with organic peroxides" (co-author) in "Organic peroxides", vol. 3, New York 1972.

Rebane, Karl Karlovich
Physicist; b. 1926.
Esthonian; mem. CPSU 1948–; Grad. Leningrad Univ. 1952; mem. 1967–, Pres. of Acad. of Sciences Esthonian SSR 1973–; Dep. Dir. of Inst. of Physics and Astronomy of Acad. of Sciences Esthonian SSR 1959–64; Acad. Sec. of Dept. of Physical-Tech. and Mathematical Sciences 1964–68; Vice-Pres. of Acad. of Sciences Esthonian SSR 1968–73; Dir. of Inst. of Physics of Acad. of Sciences Esthonian SSR 1973–; mem. Cen. Cttee. Esthonian CP 1966–71. Decor.: Red Banner of Labour 1976.

Rebrin, Petr Nikolaevich
Author; b. D. 31, 1914 (Ja. 13, 1915), Tiumen.
Russian; grad. Novosibirsk Inst. of Econ. Planning 1938; first works publ. 1939.
Publs.: "Problemy agronomicheskie" (Problems in Agronomy) 1958, "Golovyrino, Golovyrino ..." 1963, "Eto bylo oseniu" (It Was in Autumn) 1964, and others.

Rechkalov, Grigorii Andreevich
Military official; b. F.9, 1918, Khudiakovo, now Zaikovo, Irbit Raion, Sverdlovsk Oblast.
Major-general of aviation (1957), military pilot first class. Twice Hero of the Soviet Union (My.24, 1943, and Jl.1, 1944). Member of the CPSU since 1942.
He graduated from the Perm' Military Aviation School for Pilots in 1939 and from the Air Force Academy in 1951. In the Great Patriotic War of 1941–45 he was a junior pilot in the 55th Fighter Aviation Regiment (1941–42), flight and squadron commander and navigator of the 16th Guards Fighter Aviation Regiment (1943–44), and deputy commander and commander of a regiment (July 1944–February 1945). In February 1945 he became inspector pilot for piloting techniques of the 9th Guards Fighter Aviation Division. He made over 450 combat flights and shot down 56 enemy aircraft alone and five in group fights. He held responsible posts in the army after the war and has been in the reserve since 1959. He is the author of *The Smoke-filled Sky of War* (1968), *Visiting the Youth* (1968), and *In the Sky of Moldavia* (1967). Decors.: Order of Lenin, four Orders of the Red Banner, the Order of Alexander Nevsky, two Orders of the Red Star, and various medals.

Rechkin, N.V.
Trade union official.
2nd Sec. of Kuliab Oblast Cttee. Tadzhik CP 1975–80; mem. Cen. Cttee. 1976–81, mem. Auditing Comm. Tadzhik CP 1981–; Sec. and mem. Presidium of Tadzhik Republican Trade Union Council F. 1982–.

Redlikh, Vera Pavlovna
Stage director; b. Mr.31 (Ap.12), 1894, Nikolaevka, now in Kharkov Oblast.
People's Artist of the RSFSR (1956). Member of the CPSU since 1942.
She studied at the theatrical school of the Moscow Art Theater beginning in 1914 and in the theater's Second Studio. She acted in the Iaroslavl, Vladivostok, and other theaters. In 1932 she became a stage director at the Red Torch theater in Novosibirsk, and from 1943 to 1960 she was principal director. Among her important productions are Chepurin's *Stalingraders* (1944, with N.F. Mikhailov), Gorkii's *The Zykovs* (1944) and *The Last Ones* (1948), Leonov's *An Ordinary Man* (1947), Shakespeare's *Hamlet* and Chekhov's *The Seagull* (both 1952), Shtein's *A Personal Case* (1955), *The Kremlin Chimes* by Pogodin and *The Village of Stepanchikovo,* adapted from Dostoevsky's novella (both 1956), and Salynskii's *The Drummer Girl* (1959).
She worked productively with the Novosibirsk dramatists E. Rogozinskaia (*The Twentieth Anniversary,* 1949) and V.V. Lavrent'ev (*The Kriazhevs,* 1953; *The Radiant One,* 1954). From 1960 to 1964 she headed the Gorkii Russian Theater in Minsk, where her most important production was Shakespeare's *Antony and Cleopatra* (1964).
She began teaching in 1936 and in 1960 joined the staff of the Belorussian Theatrical Institute, where she became a professor in 1966 in the subdepartment of acting. Decors.: three orders and several medals.

Redzhebov, Patyshakuli
State official.
Chair. of State Cttee. for Film Ind. Turkmen SSR My. 1975–.

Reformatskii, Aleksandr Aleksandrovich
Linguist; b. O.16 (29), 1900, Moscow; s. of A.N. Reformatskii.
Doctor of philological sciences (1962).
He graduated from Moscow University in 1923. He became a professor in 1962. Between 1921 and 1954 he taught at Moscow State University, the Second Moscow Pedagogical Institute of Foreign Languages, and other higher educational institutions in Moscow. He headed the section for structural and applied linguistics at the Institute of Linguistics of the Academy of Sciences of the USSR from 1958 to 1970. He was one of the founders of the Moscow school of phonology and formulated the theoretical positions of the school in *The Problem of the Phoneme in American Linguistics* (1941). His chief works concern phonology, morphophonemics, semiotics, applied linguistics, and orthography.
Works: *Vvedenie v iazykovedenie,* 4th ed. Moscow 1967. *Iz istorii otechestvennoi fonologii.* Moscow 1970. *Fonologicheskie etiudy.* Moscow 1975.

Rei, K.
Komsomol official.
Sec. and mem. Bureau of Cen. Cttee. Esthonian Komsomol Mr. 1982–.

Reimane, Irena Albertovna
State official.
Mem. Cen. Cttee. Latvian CP 1976–; Chair. of State Cttee. on Publishing, Polygraphy and the Book Trade Latvian SSR My. 1978–.

Reimers, Fedor Eduardovich
Plant physiologist; b. Jl.12 (25), 1904, Ekaterinoslav, present-day Dnepropetrovsk.
Corresponding member of the Academy of Sciences of the USSR (1970). Member of the CPSU since 1962.
In 1930, he graduated from Odessa Agricultural Institute. He became a professor in 1959. Between 1950 and 1961 he was a department chairman at the Eastern Siberian branch of the Academy of Sciences of the USSR (since 1957, the Siberian division of the Academy of Sciences of the USSR). From 1961 to 1967, he was director of the Biological Institute in Irkutsk; in 1967 he became the director of the Siberian Institute of Plant Physiology and Biochemistry.
His principal works are on the growth and development (mainly the embryogeny) of plants in connection with problems of seed raising in Siberia. He has conducted research on the physiological reasons for low seed germination in Siberia and has worked out methods for increasing seed germination. Decors.: two orders and various medals.
Works: *Osnovnye svedeniia iz zhizni rastenii.* Irkutsk 1951. *Vyrashchivanie ovoshchei v teplitsakh Vostochnoi Sibiri.* Irkutsk 1955. *Fiziologiia rosta i razvitiia repchatogo luka.* Moscow-Leningrad 1959.

Reizen, Mark Osipovich
Russian singer (bass); b. Je.21 (Jl.3), 1895, Zaitsevo, now in Sinel'nikovo Raion, Dnepropetrovsk Oblast.
People's Artist of the USSR (1937).
From 1917 to 1921, he studied at the Kharkov Conservatory. In 1921 he became a soloist at the Kharkov Opera Theater. From 1925 to 1930 he was a soloist at the Leningrad Theater of Opera and Ballet and from 1930 to 1954, at the Boshoi Theater.
As a singer and actor, he created definitive portrayals of characters in Russian operas. He was at his best as Boris Godunov and Dosifei in Mussorgsky's *Boris Godunov* and *Khovanshchina* (State Prize of the USSR, 1949 and 1951). Among his many other parts were Susanin and Ruslan in Glinka's *Ivan Susanin* and *Ruslan and Liudmila,* the Miller in Dargomyzhskii's *The Mermaid,* Mephistopheles in Gounod's *Faust,* and Don Basilio in Rossini's *The Barber of Seville.*
He is a subtle interpreter of chamber music. He has performed on tours abroad. Decors.: State Prize of the USSR (1941), two Orders of Lenin, the Order of the Red Banner of Labor, and a number of medals.

Reizov, Boris Georgievich
Literary critic; b. O.19 (N.1), 1902, Nakhichevan'-na-Donu.
Corresponding member of the Academy of Sciences of the USSR (1970). Honored Scientist of the RSFSR (1969).
He graduated from the Northern Caucasus University at Rostov-on-Don (1926) and became a professor at Leningrad University in 1940. His principal works are devoted to the history of Western European and Russian literature and the theoretical and methodological problems of literary criticism. He has studied the literary process in connection with the social thought of the period, concentrating on its development and historical perspective. He has received an honorary doctorate from the University of Clermont-Ferrand, France (1973). Decor.: State Prize of the USSR (1974).
Works: *Tvorchestvo Bal'zaka.* Leningrad 1939. *Bal'zak: Sb. st.* Leningrad 1960.

Tvorchestvo Flobera. Moscow 1955. (Translated into Polish.) *Frantsuzskaia romanticheskaia istoriografiia*. Leningrad 1956. (Translated into French.) *Frantsuzskii istoricheskii roman v epokhu romantizma*. Leningrad 1958. (Translated into Polish.) *Mezhdu klassitsizmom i romantizmom*. Leningrad 1962. *Tvorchestvo Val'tera Skotta*. Moscow-Leningrad 1965. *Ital'ianskaia literatura XVIII v*. [Leningrad] 1966. *Stendal': Gody ucheniia*. Leningrad 1968. *Frantsuzskii roman 19 v*. Moscow 1969. *Iz istorii evropeiskikh literatur*. [Leningrad] 1970. *Stendal': Filosofiia istorii, Politika, Estetika*. Leningrad 1974.

Rekemchuk, Aleksandr Evseevich
Author; b. D. 25, 1927, Odessa.
Russian; mem. CPSU 1948–; grad. Gorkii Inst. of Literature 1952; was active as journalist in the Komi ASSR; has lived in Moscow 1963–.
Publs.: "Vremia letnikh otpuskov" (The Time of Summer Holidays) 1959, "Tovarishch Hans" (Comrade Hans) 1965, "Mal'chiki" (Boys) 1970, and others.

Rekunkov, Aleksandr Mikhailovich
State official; b. 1920.
Mem. CPSU 1940–; grad. All-Union Law Correspondence Inst. 1952–; served on war front 1941–45; public procurator in various raions in Rostov Oblast 1946–58; public procurator in Briansk and Voronezh Oblasts 1960–71; 1st Dep. Public Procurator RSFSR 1971–76; 1st Dep. Gen. Procurator USSR 1976–81, Gen. Procurator D. 1981–; Dep. to USSR Supr. Sov. 1979–; mem. Cen. Cttee. CPSU 1981–.

Remeslo, Vasilii Nikolaevich
Biologist; b. Ja. 28 (F. 10), 1907, Teplovo, now Teplovka, Poltava Oblast.
Mem. CPSU 1942–; grad. Timiriazev Inst. of Selection and Seed Breeding in Maslovskii (Kiev Oblast) 1928; selection specialist; mem. USSR Acad. of Sciences 1974–; mem. V. I. Lenin Union Acad. of Agricultural Sciences (VASKHNIL) 1964–; Sen. Mem. Scientific Staff of Experimental Selection Station in Novourenskaia (now Ulianovskaia) 1933–37; mem. staff State Selection Station in Severodonetsk 1938–42 and 1947–48; Dir. of same 1941–42; Dep. Dir. 1948–64, Dir. of Mironovskii Scientific Research Inst. for Wheat Selection and Seed Breeding (until 1968 Mironovskaia Selection Station) 1964–; mem. Cen. Cttee. Ukrainian CP 1966–; Dep. to 1963–75, Dep. Chair. of Supr. Sov. Ukrainian SSR 1972–75. Decors.: Lenin Prize 1963, Order of Lenin, Order of the October Revolution, 3 other orders and medals.
Publs.: "Ozimaia pshenitsa Mironovskaia 264 i Mironovskaia 808" 1964, "Mironovskie pshenitsy" (co-author) 1972.

Reshetnikov, Fedor Grigor'evich, Dr. of Tech. Sc.
Physical chemist; b. 1919.
Mem. CPSU 1919–; grad. Moscow Inst. of Non-Ferrous Metals and Gold 1941 and F. E. Dzerzhinskii Artillery Acad. 1944; Dr. of Tech. Sc. 1960; corresponding mem. USSR Acad. of Sciences 1974–; worked in a factory 1944–45; scientific work 1946–; dep. dir. of a scientific research inst. 1966–. Decor.: State Prize of the USSR 1951.

Reshetnikov, Fedor Pavlovich
Painter and graphic artist; b. 1906.
Mem. CPSU 1945–; grad. Higher State Inst. for Arts and Crafts (VHUTEIN) Moscow 1934; mem. 1953–, Vice-Pres. of USSR Acad. of the Arts 1974–; teacher 1953–56, Prof. at Moscow Art Inst. 1954–. Decor.: People's Artist of the USSR 1974.

Reshetnikov, Leonid Vasil'evich
Author; b. Je. 17, 1920, Nolidur, Mari ASSR; s. of a peasant.
Russian; mem. CPSU 1943–; grad. Mil.-Pol. Acad. 1951; first works publ. 1937.
Publs.: "Shla rota s pesnei" (The Company Went Singing) 1959, "Siren' i porokh" (Lilac and Powder) 1960, "Pisateli o sebe" (The Autor About Himself) 1966, and others.

Reshetov, Aleksandr Efimovich
Author and poet; b. O. 21 (N. 3), 1909, Osetki, Pskov Oblast.
Russian; mem. CPSU 1930–; first works publ. 1927.
Publs.: "Tak my zhivem" (This Is How We Live) 1931, "Leningradskaia doblest'" (Leningrad Valour) 1942, "Stikhotvoreniia, Poemy" (Verses, Poems) 1956, "Lirika" (Lyric) 1957, "Ia vernulsia domoi" (I Returned Home) 1967, "Iz moikh desiatiletii" (From My Decades) 1969, and others.

Reutov, Oleg Aleksandrovich
Chemical scientist; b. S. 5, 1920, Makeevka, Donetsk Oblast.
Mem. CPSU 1942–; grad. Moscow State Univ. 1941; mem. USSR Acad. of Sciences 1964–; mem. scientific staff Dept. of Chemistry Moscow State Univ. 1945–57; Head of Laboratory for Problems in Organic Chemistry (founded by him) of Moscow State Univ. and concurrently Head of Laboratory for Application of Isotopes in Organic Chemistry at Inst. of Elemental-Organic Compounds of USSR Acad. of Sciences 1957–; USSR Expert on Questions of Chemical-Bacteriological Warfare with United Nations 1969–; Dep. Chair. of Sov. Peace Cttee. 1974–. Decors.: 4 orders and various medals.
Publs.: "Theoreticheskie problemy organicheskoi khimii" 1956, "Teoreticheskie osnovy organicheskoi khimii" 1964, "Reaction mechanisms of organometallic compounds" (co-author) Amsterdam 1968, "Mekhanizmy reaktsii metalloorganicheskikh soedinenii" (co-author) 1972, and others.

Reva, Konstantin Kuz'mich
Athlete (volleyball); b. Ap. 10, 1921, Sukhodol, now in Glukhov Raion, Sumy Oblast.
Honored Master of Sports (1947). Member of the CPSU since 1943.
He graduated from the Air Force Academy in 1952. From 1963 to 1971 he was a senior instructor at the Lenin Military Political Academy. He was champion of the USSR and as a member of the national volleyball team of the USSR, he was champion eight times from 1940 to 1958, of Europe in 1950 and 1951, and of the world in 1949 and 1952. Decors.: two orders and several medals.

Revich, Aleksandr Mikhailovich
Poet and translator; b. N. 2, 1921, Rostov/Don.
Russian; grad. Gorkii Inst. of Literature 1951; active as poet and translator; first works publ. 1963.
Publs.: "Sled ognia" (The Trace of Fire) 1970, and others.

Riabokliach, Ivan Afanasevich
Author; b. O. 26 (N. 8), 1914, Pivtsy, Kiev Oblast; s. of a peasant.
Ukrainian; studied Rzhev Pedagogic Tech. College; editorial work with newspapers in the Donets Basin 1933–.
Publs.: "Zhavoronki" (The Larks) 1957, "Chaika" (The Seagull) 1960, "Zhivaia voda" (Living Water), "Na belom kone" (On a White Horse) 1963, and others.

Riabov, Iakov Petrovich
State official; b. 1928.
Russian; mem. CPSU 1954–; grad. Uralsk Polytech. Inst. 1952; technician-designer, engineer-designer, dept. head at Sverdlovsk Turbine and Engine Works 1946–58; CPSU posts 1958–63; 1st Sec. of Sverdlovsk City Cttee. CPSU 1963–66; 2nd Sec. 1966–71, 1st Sec. of Sverdlovsk Oblast Cttee. CPSU Ja. 1971–O. 1976; mem. Cen. Cttee. CPSU 1971–, Sec. of Cen. Cttee. CPSU O. 1976–79, 1st. Dep. Chairm. of USSR GOSPLAN F. 1979–; Dep. to USSR Supr. Sov. 1974–. Decor.: Order of Lenin.

Riabushkin, Timon Vasil'evich
Economist; b. D. 30, 1914 (Ja. 12, 1915), Voronezh.
Corresponding member of the Academy of Sciences of the USSR (1966). Member of the CPSU since 1942.
He graduated from the Institute of National Economic Accounting in Voronezh in 1936. Between 1948 and 1953 he was first deputy head and then head of the department of statistical methodology of the Central Statistical Board of the USSR; he was also editor-in-chief at the State Statistical Publishing House. From

1954 to 1961 he was head of the section of economic statistics of the Institute of Economics of the Academy of Sciences of the USSR. From 1961 to 1970 he served as head of the statistics section and deputy director of the Institute of the Economics of the World Socialist System of the Academy of Sciences of the USSR. In 1970 he became head of the department of demography and statistics at the Central Institute of Economics and Mathematics. From 1948 to 1961, Riabushkin was the Soviet representative to the Statistical and Population commissions of the United Nations. He became a member of the International Statistical Institute in 1958; he served as vice-president from 1961 to 1965 and began a new term in 1973.
His chief works are on political economy and statistics. They treat issues of national economic balance, statistical methods and analyzing proportions in the national economy and the interrelations of economic elements, methods of comparing statistical data on an international level, and the regularities in the development of the world socialist economy. He edited the book *V. I. Lenin and Modern Statistics* (vols. 1–3, 1970–73). Decor.: Order of the Red Banner of Labor.
Works: *Ocherki po ekonomicheskoi statistike.* Moscow 1950. *Statisticheskie metody izucheniia narodnogo khoziaistva.* Moscow 1957. *Problemy ekonomicheskoi statistiki.* Moscow 1959. *Mezhdunarodnaia statistika.* Moscow 1965. *Tempy i proportsii razvitiia narodnogo khoziaistva sotsialisticheskikh stran.* Moscow 1966. *Ekonomicheskaia statistika.* Moscow 1966. *V. I. Lenin i statistika.* Moscow 1971.

Riatsep, Vaino Iohannesovich
State official.
Min. of Health Esthonian SSR Jl. 1975–; cand. mem. Cen. Cttee. Esthonian CP 1976–.

Riazanov, Aleksei Konstantinovich
Military official; b. F. 27, 1920, B. Kochetovka, Tambov Oblast.
Mem. CPSU 1942–; grad. V. P. Chkalov Aviation School in Borisoglebsk 1939, M. V. Frunze Mil. Acad. 1950 and Gen. Staff Acad. 1958; Maj.-Gen. of the Air Force 1970; pilot, Wing Commander, Dep. Commander of the 4th Fighter Regiment 1941–45; flew 509 missions, downed 31 enemy aircraft individually and 16 in group combat; sen. posts in Air Force 1959–; mem. Cen. Apparat of Sov. Air Defence (PVO) 1962–. Decors.: 3 Orders of Lenin, 4 Orders of Red Banner, Aleksandr Nevskii Order, Order of Patriotic War 1st Cl., 2 Orders of Red Star, medals.

Riazanskii, Mikhail Sergeevich
Radio engineer; b. Mr. 23 (Ap. 5), 1909, Leningrad.
Corresponding member of the Academy of Sciences of the USSR (1958). Hero of Socialist Labor (1956). Member of the CPSU since 1940.
Since graduating from the Moscow Engineering Institute in 1935, he has worked in various scientific research organizations. Decors.: State Prize of the USSR (1943), four Orders of Lenin, two other orders, and various medals.

Rigvava, Sergei Tarasovich
State official; b. 1927.
Georgian; mem. CPSU 1946–; grad. the Central Komsomol School with the Cen. Cttee. of All-Union Komsomol 1948 and the Tbilisi Pedagogical Pushkin Inst. 1957; 2nd Sec. of Sukhumi City Cttee. Georgian Komsomol 1944–46; Sec. 1948–50, 2nd., 1st Sec. of Abkhazian ASSR Oblast Cttee. Georgian Komsomol 1951–52; Sec., 2nd Sec. Cen. Cttee. Georgian Komsomol 1952–56; diverse Party posts 1956–60; mem. Apparat Council of Mins. Georgian SSR 1960–63; 2nd, 1st Sec. of Kalinin Raion Cttee. Georgian CP in Tbilisi 1963–72; Sec. of Tbilisi City Cttee. Georgian CP 1972–74; 1st Dep. Chair. of Tbilisi City Exec. Cttee. of People's Deps. 1974–79; Chair. of State Cttee. for Vocational Training Georgian SSR Ja. 1979–; mem. Auditing Comm. Georgian CP 1981–.

Rikhter, Sviatoslav Teofilovich
Pianist; b. Mr. 7 (20), 1915, Zhitomir.
Concertmaster at Odessa Opera and Ballet Theatre 1933–37; first concert 1934; studied Moscow Conservatory (with interruptions) 1937–47, pupil of G. G. Neuhaus; concert pianist in the Sov. Union 1940–; performances abroad since the mid-fifties. Decors.: State Prize of the USSR 1950, Lenin Prize 1961.

Rinkule-Zemzare, Dzidra
Author; b. Jl. 8, 1920, Riga.
Latvian; studied Dept. of Philology of Latvian Univ.
Publs.: "Zanoschivyi kuznechik" (The Conceited Grasshopper) 1954, "Put, eshche dalek" (Still Far to Go) 1955, "Chasy kapitana Enrico" (Captain Enrico's Watch) 1967, and others.

Ristlaan, Rein Eduardovich
Party official.
Sec. of Tallinn City Cttee. Esthonian CP 1971–75; cand. mem. Cen. Cttee. Esthonian CP 1971–78, mem. Ag. 1978–; Head Dept. of Trade and Public Services with Cen. Cttee. Esthonian CP 1975–80; Sec. and mem. Bureau Cen. Cttee. Esthonian CP Mr. 1980–.

Riuitel, Arnold Fedorovich
State official; b. 1928 on the island of Saaremaa; s. of a farmer.
Esthonian; mem. CPSU 1964–; grad. Esthonian Acad. of Agrarian Sciences, agronomist, cand. of Agrarian Sciences; agronomist at an agricultural section in Kuressaaresk, served in the navy, teacher at School for the Mechanization of Agriculture in Tartu 1949–57; chief technical zoologist at "RTjachtvere" research station, Dep. Dir. of Esthonian Scientific Research Inst. of Animal Husbandry and Veterinary Science 1957–63; rector at Esthonian Acad. of Agrarian Sciences 1969–77; cand. mem. 1966–71, mem. Cen. Cttee. Esthonian CP 1971–; mem. Bureau of Cen. Cttee. Esthonian CP 1977–; 1st Dep. Chair. of Council of Mins. Esthonian SSR 1979–83.

Rizaev, Kerim Novruz ogly
State official.
Head of Admin. for Non-Ferrous Metallurgy. Azerbaidzhan SSR N. 1965–; mem. Auditing Comm. Azerbaidzhan CP 1981–.

Riznichenko, Iurii Vladimirovich
Seismologist; b. S. 15 (28), 1911, Kiev.
Mem. CPSU 1950–; grad. Kiev Inst. of Mining and Geology 1935; corresponding mem. USSR Acad. of Sciences 1958–; mem. staff O. Iu. Shmidt Inst. of Physics of the Earth of USSR Acad. of Sciences 1938–; Prof. at Moscow Inst. of Geological Prospecting 1945–55; Prof. at Moscow State Univ. 1950–57. Decors.: Order of Red Banner of Labour, various medals.
Publs.: "Issledovanie gornogo davleniia geofizicheskimi metodami" (co-author) 1967, "Seismicheskoe raionirovanie SSSR" (co-author) 1968, and others.

Robitashvili, Georgii Vladimirovich
State official.
Dep. to Supr. Sov. Georgian SSR 1959–; Head of Main Admin. of Highways with Council of Mins. Georgian SSR 1959–68; mem. Auditing Comm. 1961–66, mem. Cen. Cttee. Georgian CP 1966–76; Min. of Highways Georgian SSR D. 1968–. Decor.: Order of Red Banner of Labour.

Robu, I. E.
Komsomol official.
Sec. and mem. Bureau Moldavian Komsomol Je. 1977–.

Rochev, Iakov Mitrofanovich
Author; b. Ja. 10 (23), 1909, Ust-Ukhta, Komi ASSR; s. of a peasant.
Komi; mem. CPSU 1940–; grad. of a pedagogic inst. 1937; first works publ. 1929.
Publs.: "Dva druga" (Two Friends) 1951, "Svetopredstavlenie" (Judgement Day) 1967, and others.

Rodionov, Aleksei Alekseevich
Diplomat; b. 1922.
Mem. CPSU 1942–; sec. of CPSU cttee. in an oil-processing factory in Omsk, Sec., 1st Sec. of Omsk City Cttee., 2nd Sec. of Omsk Oblast Cttee. CPSU 1956–64; mem. Diplomatic Service 1964–; on staff of USSR Min. of Foreign Affairs 1964; Emb. Counsellor-Envoy at USSR Emb. in India 1964–66; USSR

Amb. to Burma 1966–68; mem. Collegium USSR Min. of Foreign Affairs, RSFSR Min. of Foreign Affairs 1968–71; USSR Amb. to Pakistan 1971–74 and to Turkey D. 1974–.

Rodionov, Lev Artemeevich
Sculptor; b. 1929.
Grad. Repin Inst. of Painting, Sculpture and Architecture Leningrad 1957; worked in Master Workshops of USSR Acad. of the Arts until 1960. Decor.: State Prize of the USSR 1969 (for his part in the building of the monument "Legendary Machine-Gun Cart" nr. Kakhovka).

Rodionov, Nikolai Nikolaevich
Diplomat; b. Ap. 17 (30), 1915, Chern, now Tula Oblast; s. of a clerk.
Russian; mem. CPSU 1944–; grad. Moscow Steel Inst. 1941; engineering and tech. posts with Magnitogorsk Metallurgical Combine and at a research inst. in Leningrad 1941–48; CPSU work in Leningrad and Leningrad Oblast 1948–54; 2nd Sec. of Leningrad City Cttee. 1954–56, 2nd Sec. of Leningrad Oblast Cttee. 1956–57, 1st Sec. of Leningrad City Cttee. CPSU 1957–60; Dep. to 1958–79, mem. Draft Bills Comm. 1962–66, Chair. of Credentials Comm. 1966–70, mem. Foreign Affairs Comm. Sov. of the Union USSR Supr. Sov. 1970–79; 2nd Sec. and mem. Bureau Cen. Cttee. Kazakh CP 1960–62; Dep. to and mem. Presidium Supr. Sov. Kazakh SSR 1960–63; cand. mem. 1961–66, mem. Cen. Cttee. CPSU 1966–; Dep. Chair. of Leningrad Econ. Council 1962–65; 1st Sec. of Cheliabinsk Oblast Cttee. CPSU 1965–70; USSR Dep. Min. of Foreign Affairs Jl. 1970–78; USSR Amb. to Yugoslavia My. 1978–. Decors.: 3 Orders of Lenin, Order of the October Revolution, 3 other orders and medals.

Rodnina, Irina Konstantinovna
Figure skater; b. S. 12, 1949, Moscow.
Honored Master of Sport (1969) and teacher. Member of the CPSU since 1975. In 1974, she graduated from the Central Institute of Physical Culture in Moscow. She has won many championships in pairs competition, including the USSR (1970–77) and European and world championships (1969–77). She won a gold medal in the 1972 and 1976 Olympic Games. During the period 1968–1972, she skated with A. N. Ulanov; since 1973 her partner has been A. G. Zaitsev. She has been coached by S. A. Zhuk and, since 1974, by T. A. Tarasova. Decors.: Order of Lenin, the Order of the Red Banner of Labour and various medals.

Roginskii, Iakov Iakovl'evich, Dr. of Biol. Sc.
Anthropologist; b. My. 9 (21), 1895, Mogilev.
Dr. of Biol. Sc. 1949; Prof. 1950–; Head of Chair of Anthropology at Moscow State Univ. 1950–. Decors.: Order "Badge of Honour", medals.
Publs.: "Teorii monotsentrizma i politsentrizma v probleme proiskhozhdeniia sovremennogo cheloveka i ego ras" 1949, "Problemy antropogeneza" 1969, "O formirovanii proportsii tela putem usileniia gradientov rosta (v sviazi s problemoi antropogeneza)" in "Voprosy antropologii" No. 2, 1960, co-author of "Antropologia" and others.

Roitbak, Aleksandr Il'ich
Physiologist; b. F. 17, 1919, Odessa.
Corresponding member of the Academy of Sciences of the USSR (1968).
He graduated from the Kiev Medical Institute in 1941. He was a student of I. S. Beritashvili. In 1944 he joined the staff of the Institute of Physiology of the Academy of Sciences of the Georgian SSR; in 1960 he became head of the Laboratory of General Cerebral-Cortex Physiology. His principal works deal with neurophysiology. He discovered that by rhythmically stimulating the cortex of the large hemispheres with electricity it is possible to create a focus of excitation, or a dominant. He has proposed a hypothesis on the role of neuroglia in the formation of the temporary connections that are the basis of conditioned reflexes.
Works: *Bioelektricheskie iavleniia v kore bol'shikh polusharii,* part 1. Tbilisi 1955. "Neirogliia i obrazovanie novykh nervnykh sviazei v kore mozga." In *Mekhanizmy formirovaniia i tormozheniia uslovnykh refleksov.* Moscow 1973.

Roiter, Vladimir Andreevich, Dr. of Chem. Sc.
Physical chemist; b. Jl. 26, 1903, Nizhnedneprovsk, now part of Dnepropetrovsk; s. of a railway worker.
Grad. Dnepropetrovsk Inst. of Popular Educ. 1926; mem. Acad. of Sciences Ukrainian SSR 1961–; Dr. of Chem. Sc. 1948; Prof. 1934–; mem. scientific staff, Head of Dept. of Catalyses, Dep. Dir. of Pedagogic Dept., Head of Laboratory for Heterogenous Catalyses at Inst. of Physical Chemistry of Acad. of Sciences Ukrainian SSR 1929–; concurrently head of chair of gen. chemistry, physical chemistry and inorganic and physical chemistry at various colleges in Dnepropetrovsk; author of some 80 publs., among them popular-science books and articles on chemistry. Decor.: Merited Scientist of the Ukrainian SSR 1963.
Publs.: "Pro nedootsinku iakisnykh vidminostei v suchasnii fizichnii khimii" in "Filosofski pytannia pryrodoznavstva" 1958, "Razvitie teorii geterogennogo kataliza" in "XX let Instituta fizicheskoi khimii im. L. V. Pisarzhevskogo Akademii Nauk Ukrainskoi SSR" 1950, "Kataliticheskoe okislenie naftalina" 1963, and others.

Romadin, Nikolai Mikhailovich
Painter; b. 1903.
Attended Samara Tech. Art School 1922; studied Higher State Workshops for Arts and Crafts (VHUTEMAS) and Higher State Inst. of Arts and Crafts (VHUTEIN); mem. USSR Acad. of the Arts 1967–. Decor.: People's Artist of the USSR 1971.

Romanenko, Anatolii Efimovich
State official.
Min. of Health Ukrainian SSR Ap. 1975–; mem. Auditing Comm. Ukrainian CP 1976–.

Romanitskii, Boris Vasil'evich
Ukrainian actor and director; b. Mr. 18 (30), 1891, Chernobai, now in Cherkassy Oblast.
People's Artist of the USSR (1944). Member of the CPSU since 1940.
In 1913, he joined P. K. Saksaganskii's troupe while simultaneously studying at the N. V. Lysenko Musical Drama School in Kiev, from which he graduated in 1915. He was a student of M. K. Zan'kovetskaia and Saksaganskii, who both greatly influenced his art. He was one of the founders of the M. K. Zan'kovetskaia Ukrainian Drama Theater in L'vov in 1922 and served as the theater's artistic director until 1948; he also performed at the theater and staged plays.
Vivid characterizations and variety have distinguished his interpretation of his roles in Ukrainian classical plays and Soviet dramas. His roles have included Chalyi in Karpenko-Karyi's *Sava Chalyi,* Mikola Zadorozhnyi in Franko's *Stolen Happiness,* Khrapov in Gorkii's *Vassa Zheleznova,* Gaidai in Korneichuk's *The Destruction of the Squadron,* and Raevich in Bill'-Belotserkovskii's *The Gale.* His best roles in Russian and international classical plays have included Voinitskii in Chekhov's *Uncle Vanya,* Karl in Schiller's *Die Räuber,* and the title role in Shakespeare's *Othello* (performed for the first time on the Ukrainian stage). He staged Korneichuk's *The Destruction of the Squadron* (1933) and Khizhniak's *To a Great Land* (1949; State Prize of the USSR, 1950). He wrote the book *The Ukrainian Theater in the Past and Present* (1950).
Decors.: Shevchenko Prize of the Ukrainian SSR (1974), two Orders of Lenin, two other orders, and various medals.

Romankov, Petr Grigor'evich
Chemical scientist-technologist; b. Ja. 4 (17), 1904, Eisk, now Krasnodar Krai.
Grad. Leningrad Technological Inst. 1929; corresponding mem. USSR Acad. of Sciences 1964–; Research Asst. at Bauman Tech. College 1929–33; teacher at Mendeleev Chemical-Technological Inst. and at Inst. of Chemical Machine Construction in Moscow 1933–34; mem. staff Lensovet Technological Inst. in Leningrad 1934–, Head of Chair of Processes and Apparatuses 1941–; main work in the field of kinetics in processes of chemical technology. Decors.: 5 orders and various medals.

Publs.: "Primery i zadachi po kursu protsessov, i apparatov khimicheskoi tekhnologii" (co-author), 7th ed. 1969, "Gidromekhanicheskie protsessy khimicheskoi tekhnologii" (co-author), 2nd ed. 1974.

Romanov, Aleksandr Iosifovich
Diplomat; b. 1912.
Mem. Diplomatic Service 1946–; sen. posts at USSR Min. of Foreign Affairs 1958–60; Emb. Counsellor-Envoy at USSR Emb. in the United Kingdom 1960–64; USSR Amb. to Nigeria 1964–70; mem. staff USSR Min. of Foreign Affairs 1970–73; USSR Amb. to the Netherlands Ap. 1973– F. 1979.

Romanov, Aleksei Vladimirovich
Journalist; b. 1908; s. of a clerk.
Russian; mem. CPSU 1939–; partial higher educ.; attended higher state courses in literature 1926–29; literary contributor, Dept. Head, Exec. Sec. of newspaper "Gor'kovskaia kommuna", correspondent for "Pravda", Ed. of newspaper "Sovetskaia Belorussiia" 1929–49; correspondence student at Party Higher School of Cen. Cttee. CPSU 1945–49; Exec. Sec. 1949–55 and Dep. Chief Ed. of paper "Za prochnyi mir, za narodnuiu demokratiiu" 1953–55; worked for Cen. Cttee. CPSU, Dep. Head of Dept. of Propaganda and Agitation at Cen. Cttee. CPSU 1955–65; cand. mem. Cen. Cttee. CPSU 1961–; mem. Bureau Cen. Cttee. CPSU for RSFSR 1961–62; Dep. to USSR Supr. Sov. 1962–; Chair. (with rank of USSR Min.) of State Film Cttee. of USSR Council of Mins. 1963–65; Chair. of Film Cttee. of USSR Council of Mins. 1965–72; Chief Ed. of paper "Sovetskaia kul'tura" 1972–. Decors.: Order of Red Banner of Labour, Order of Lenin.

Romanov, Grigorii Vasil'evich
Party official; b. F.7, 1923, Zikhnovo, Novgorod Oblast; s. of a peasant.
Russian; mem. CPSU 1944–; grad. Leningrad Shipbuilding Inst. 1953; attended a tech. school 1938–41; served in Sov. Army 1941–45; Designer, Section Head with Design Bureau of A.A. Zhdanov Works of USSR Min. of Shipbuilding in Leningrad 1946–54; Sec. of CPSU Cttee., party organizer for Cen. Cttee. CPSU at same works 1955–57; Sec. of Kirov Raion Cttee. CPSU in Leningrad 1957–61; Sec. of Leningrad City Cttee. CPSU 1961–62; Sec. 1962–63, 2nd Sec. 1963–70 (1963–64 for Ind.), 1st Sec. of Leningrad Oblast Cttee. CPSU 1970–; mem. Cen. Cttee. CPSU 1966–; Dep. to 1966–, mem. Comm. for Construction and Construction Materials Ind., Sov. of the Union 1966–71; mem. Presidium USSR Supr. Sov. 1971–; cand. mem. 1973–76, mem. Politbureau Cen. Cttee. CPSU Mr. 1976–. Decors.: 3 Orders of Lenin, Order of the October Revolution, 2 other orders and various medals.

Romanov, Leonid Mikhailovich
Diplomat.
USSR Amb. to Suriname 1978–82, to Colombia Ag. 1978–.

Romanov, Moisei Petrovich
State official.
Min. of Interior Moldavian SSR 1959–61; mem. Cen. Cttee. Moldavian CP 1960–61; Chair. of State Cttee. for the Supply of Petroleum Products Moldavian SSR 1965–.

Romanovich, Evgeniia Mikhailovna
Linguist; b. 1922.
Grad. Belorussian Univ. 1948; postgraduate studies at Inst. of Literature and Language of Acad. of Sciences Belorussian SSR 1948–52; mem. staff and Sen. Scientist at Inst. of Linguistics of Acad. of Sciences Belorussian SSR 1952–. Decor.: State Prize of the USSR (for her work in the field of Belorussian linguistic geography) 1971.

Romanovskii, Sergei Kalistratovich
Diplomat; b. 1923.
Russian; mem. CPSU 1943–; grad. Moscow Inst. of Int. Relations; served in Sov. Army 1941–44; student 1944–47; Dep. Chair. of Antifascist Cttee. of Sov. Youth, Chair. of Cttee. of USSR Youth Cttees., Sec. of Cen. Cttee. USSR Komsomol 1947–59; USSR Dep. Min. of Culture, Dep. Chair. of State Cttee. for Cultural Relations with Foreign Countries at USSR Council of Mins. 1959–62; Chair. of State Cttee. and Cttee. for Cultural Relations with Foreign Countries at USSR Council of Mins. 1962–68; Dep. to USSR Supr. Sov. 1966–70; USSR Amb. to Norway 1968–75 and to Belgium Ap. 1975–.

Romishevskii, Igor' Anatol'evich
Athlete and coach; b. Mr.25, 1940, Zhukovskii, Moscow Oblast.
Honored Master of Sports of the USSR (1968). Cand. S. (Tech.) (1974). Member of the CPSU since 1969.
From 1970 to 1974, he was a member of the Central Committee of the All-Union Lenin Communist Youth League of the USSR. He was a champion of the USSR several times during the years 1963–72 as a member of the ice-hockey team of the Central Sports Club of the Army. As a member of the USSR national team he also won the European championship (1968–70), the world championship (1968–71), and the Olympic gold medal (1968, 1972). Since 1974 he has been head of the subdepartment of physical education at the Moscow Physicotechnical Institute. Decors.: two orders and various medals.

Romma, Fedor Dimitrievich
State official; b. 1919.
1st Sec. of Baranovichi City Cttee. Belorussian CP 1959; Chair. 1963–64, Dep. Chair. of Brest Oblast Exec. Cttee. of Workers' Deps. 1964–67 (for Ind. 1963–64); mem. Cen. Cttee. Belorussian CP 1960–61; Chair. of State Cttee. for Utilization of Labour Reserves with Council of Mins. Belorussian SSR Ap. 1967–76; Chair. of State Cttee. on Labour Belorussian SSR O. 1976–.

Ronov, Aleksandr Borisovich
Geologist and geochemist; b. D.3 (16), 1913, Poltava.
Grad. Leningrad Univ. 1938; corresponding mem. USSR Acad. of Sciences 1966–; mem. staff Radium Inst. 1939–43, Inst. of Physics of the Earth of USSR Acad. of Sciences 1944–50; mem. staff V.I. Vernadskii Inst. of Geochemistry and Analytical Chemistry 1950–, Head of Laboratory for Geochemistry and Sediment Rock 1954–. Decors.: Order "Badge of Honour", various medals.
Publs.: co-author of "Atlas litologopaleogeograficheskikh kart Russkoi platformy i ee geosinklinalnogo obramleniia" 1961 and 1962 and "Atlas litologopaleogeograficheskikh kart SSSR" 1967 and 1969.

Roshal', Grigorii L'vovich
Teacher and film director; b. O.8 (20), 1899, Novozybkov, now in Briansk Oblast.
People's Artist of the USSR (1967).
Beginning in 1919, he organized and directed stage productions for children in Kislovodsk and other cities. In 1921 he moved to Moscow, where he studied at the State Higher Directors' Workshop and then was appointed director and artistic advisor of the Pedagogical Theater's workshop. He turned to films in 1925. His first film was *The Skotinins* (1927, adapted from Fonvizin's comedy *The Minor*).
The civic viewpoint characteristic of his work was reflected in the antifascist films *The Salamander* (1928), *The Oppenheim Family* (1939; adapted from L. Feuchtwanger's novel), and *The Madmen's Trial* (1962). His inclination toward sweeping epic productions and heroic subject matter found expression in *Dawn in Paris* (1937) and the trilogy comprising *The Sisters* (1957; adapted from A.N. Tolstoy's novel *Road to Calvary*), *The Year 1918* (1958), and *Gloomy Morning* (1959). Screen adaptations occupy an important place in his work. They include *St. Petersburg Night* (1934; adapted from Dostoevsky's *White Nights* and *Netochka Nezvanova*) and *In Search of Happiness* (1940; adapted from F.I. Panferov's play *Bruski*), both in collaboration with V.P. Stroeva, as well as *The Artamonov Business* (1941; adapted from Gorkii's novel) and *The Freemen* (1956; adapted from F.V. Gladkov's novel).
He also directed a series of historical and biographical films, including *The Songs of Abai* (1946; with E.E. Aron), *Academician Ivan Pavlov* (1949), *Mussorgsky* (1950), *Rimsky-Korsakov* (1953), and *A Year Like a Lifetime* (1966), devoted to the life and work of K. Marx. He wrote a

number of his own screenplays, as well as articles on cinematography. He also taught at the All-Union State Institute of Cinematography. Decors.: State Prize of the USSR in 1950 and 1951, four orders and several medals.
Work: *Kinolenta zhizni.* Moscow 1974.

Roshchin, Anatolii Aleksandrovich
Wrestler; b. Mr. 10, 1932, Gaverdovo, Riazan' Raion, Riazan' Oblast.
Honored Master of Sport of the USSR (1965). Member of the CPSU since 1962. He graduated from the Leningrad Institute of Physical Culture in 1965. He has been an Olympic champion (1972), world champion (1963, 1969–70), champion of Europe (1966), and champion of the USSR (four times in the years 1962–71) in heavyweight Greco-Roman wrestling. Since 1974 he has been an instructor at the Military Institute for Physical Culture and Sports in Leningrad. Decors.: Order of Lenin, the Order of the Red Banner of Labour, and various medals.

Rosliakov, Vasilii Petrovich
Author; b. Mr. 17, 1921, Prikumsk, Stavropol Krai.
Russian; mem. CPSU 1945–; grad. Dept. of Philology of Moscow Univ. 1950.
Publs.: "Odin iz nas" (One of Us) 1962, "Ot vesny do vesny" (From Spring to Spring) 1966, and others.

Rostiashvili, Trifon Vakhtangovich
Party official; b. 1939.
Georgian; mem. CPSU 1962–; secondary school education; served in Sov. army 1958–61; lathe operator at Dimitriev airplane factory in Tbilisi 1963–; mem. Cen. Cttee. 1976–, mem. Bureau of Cen. Cttee. CP Georgia Jan. 1981–.

Rostotskii, Boleslav Norbert Iosifovich
Theater historian and critic; b. Ag. 3 (16), 1912, Moscow. Doctor of art Studies (1967). Member of the CPSU since 1948. He graduated from the State Institute of Theatrical Arts (GITIS) in 1932. He began his literary career in 1931. In 1948 he became a senior researcher in the section for art in the European socialist countries of the Institute for the History of the Arts (Moscow), and in 1962, the head of the section. His chief works deal with the history of the Soviet and Polish theater. He began teaching at GITIS in 1944, becoming a professor there in 1967.
Works: *O. L. Knipper-Chekhova.* Moscow-Leningrad 1946. *Maiakovskii i teatr.* Moscow 1952. *O rezhisserskom tvorchestve V. E. Meierkhol'da.* Moscow 1960. "Teatr Stanislava Vyspianskogo." In S. Wyspiański, *Dramy.* Moscow 1963. (Translated from Polish.) "Teatr." In *Ocherki istorii Narodnoi Pol'shi,* vol. 4. Moscow 1965. "V. E. -Meierkhol'd i ego literaturnoe nasledie." In V. E. Meyerhold, *Stat'i, pis'ma, rechi, besedy,* part 1. Moscow 1968. "Modernizm v teatre." In *Russkaia khudozhestvennaia kul'tura kon. XIX-nach. XX vv.,* book 1. Moscow, 1968. "Pol'skii teatr." In *Istoriia zapadnoevropeiskogo teatra,* vol. 6. Moscow 1974.

Rovnin, Lev Ivanovich
State official.
Min. of Geology RSFSR F. 1970–.

Rozanov, Boris Sergeevich
Surgeon; b. Je. 30 (Jl. 12), 1896, Moscow. Doctor of Medical Sciences (1950). Honored Scientist of the RSFSR (1957).
In 1919, he graduated from the medical department of Moscow University. He began working at the N. V. Sklifosovskii Institute of First Aid under S. S. Iudin in 1926. In 1953 he became a professor and head of the first subdepartment of surgery at the Central Institute for the Advances Training of Physicians and also head surgeon of the S. P. Botkin Hospital. His principal works dealt with surgery of the esophagus, stomach, and gall bladder and with restorative operations performed on nonhepatic bile ducts. He developed original methods for removing foreign bodies from the esophagus and for draining the mediastinum. In 1945 he was one of the first in the USSR to remove successfully the upper section of a cancerous esophagus.
He is an honorary member of the All-Union Society of Surgeons and a member of various international societies of surgeons. Decors.: Order of Lenin, two other orders, and various medals.
Works: *Zheludochnye krovotecheniia i ikh khirurgicheskoe lechenie,* 2nd ed. Moscow 1960. *Inorodnye tela i travmy pishchevoda i sviazannye s nimi oslozhneniia.* Moscow 1961.

Rozen, Aleksandr Germanovich
Author; b. Ja. 6 (19), 1910, St. Petersburg.
Russian; mem. CPSU 1943–; first works publ. 1929.
Publs.: "Shturm" (Storm) 1931, "Zimniaia povest'" (Winter Tale) 1942, "Aleksei Abaturov" 1945, "Polk prodolzhaet put" (The Regiment Continues on its Way) 1947, "Serdechnaia slabost'" (Weakness of Heart) 1949, "Odnopolchane" (Regimental Comrades) 1949, "Vremena i liudi" (Times and People) 1957, "Poslednie dve nedeli" (The Last Two Weeks) 1965, "Pochti vsia zhizn'" (Almost the Whole Life) 1968, "Oskolok v grudi" (A Splinter in the Chest) 1968, and others.

Rozhdestvenskii, Gennadii Nikolaevich
Conductor; b. My. 4, 1931, Moscow.
Grad. Moscow Conservatory 1954; Conductor 1951–60, Principal Conductor of Moscow Bolshoi Theatre 1965–70; concurrently Principal Conductor and Musical Dir. of Great Symphony Orchestra of Sov. Radio and Television 1961–74; Principal Conductor of Stockholm Philharmonic and of Moscow Musical Theatre 1974–; teacher (conducting class) at Moscow Conservatory. Decors.: People's Artist of the RSFSR 1966, Lenin Prize 1970, Order of Red Banner of Labour, other orders, hon. mem. Royal Swedish Acad. 1975.

Rozhdestvenskii, Konstantin Ivanovich
Designer; b. Mr. 19 (Ap. 1), 1906, Tomsk. Honored Artist of the RSFSR (1963); corresponding member of the Academy of Arts of the USSR (1962).
He studied at the Leningrad Institute of Artistic Culture from 1923 to 1927. In his designs for exhibitions, he turns to large-scale, multiplanar spatial solutions, often based on the synthesis of different art forms. He participated in the design of the Soviet pavilions at world's fairs in Paris (1937, awarded Grand Prize) and New York (1939). He was also the chief designer of the Soviet pavilions at the world's fairs in Brussels (1958, awarded Grand Prize) and Osaka (1970) and the Soviet national expositions in New York (1959) and Paris (1961). He also works as a theatrical artist. Decors.: Order of the Red Banner of Labour and various medals.
Work: *Ansambl' i ekspozitsiia.* Leningrad 1970.

Rozhdestvenskii, Robert Ivanovich
Author and poet; b. Je. 20, 1932, Kosikha, Altai Krai.
Russian; grad. Gorkii Inst. of Literature 1956; first works publ. 1950. Decors.: 2 orders and various medals.
Publs.: "Flagi vesny" (Flags of Spring) 1955, "Rovesniku" (To My Contemporary) 1962, "Radius deistviia" (Radius of Action) 1965, "Syn Very" (Vera's Son) 1966, "Posviashchenie" (The Dedication) 1970, "Liniia" (The Line) 1973; "Za dvadtsat let" (In the Twenty Years) 1973, "Pered prazdnikom. Stikhi i poemy" (Before the Holiday. Verses and Poems) 1974, and others; his poetry has been translated into many languages.

Rozhin, Igor' Evgen'evich
Architect; b. S. 17 (30), 1908, St. Petersburg.
Honored Architect of the RSFSR (1969). From 1925 to 1930, he studied at the Leningrad Vkhutein (Higher State Art and Technical Institute) with V. G. Gel'freikh, L. V. Rudnev, I. A. Fomin, and V. A. Shchuko. He taught at the Moscow Architectural Institute from 1935 to 1964 and again in 1972. He was the chief architect of Zelenograd from 1956 to 1963.
His works include the Elektrozavodskaia subway station (1944) and the ground-level lobby of the Novokuznetskaia subway station (1943) in Moscow. These two projects, done in collaboration with Gel'freikh, were awarded the State Prize of the USSR in 1946. He also worked on the USSR Embassy in Warsaw (1954–56, with the architect A. P. Velikanov) and the V. I. Lenin Central Stadium Complex in Luzhniki, Moscow (1955–56, with the architects A. V. Vlasov, A. F. Khriakov, and

others; Lenin Prize, 1959). Decors.: two orders and several medals.

Rozin, Mark Solomonovich
Economic geographer; b. S.29 (O.12), 1898, Vitebsk.
Doctor of Geographic Sciences (1963).
He graduated from the department of social sciences of Moscow State University in 1923 and began working in the People's Commissariat of Foreign Trade in 1924. In 1938 he joined the All-Union Geological Fund of the Ministry of Geology of the USSR; from 1943 to 1953 he was head of the division for records of mineral raw materials in foreign countries. From 1960 to 1974 he headed the editorial board for geography of the Sovetskaia Entsiklopediia Publishing House.
His principal works deal with the mineral resources and geography of the mining industry of foreign countries, the geography of the world economy, and the economic geography of the African countries.
Works: *Geografia poleznykh iskopaemykh kapitalisticheskikh i razvivaiushchikhsia stran*. Moscow 1966. *Geografia mirovogo khoziastva*. Moscow 1971. (Co-author.) *Mineral'nye bogatstva Afriki*. Moscow 1972.

Rozov, Viktor Sergeevich
Dramatist; b. Ag.8 (21), 1913, Iaroslavl.
Grad. Acting Studio at Revolution Theatre and Gorkii Inst. of Literature 1952; worked as actor in Kostroma and Moscow; took part in World War II 1941–45.
Decors.: State Prize of the USSR 1967, 3 orders.
Publs.: plays "Ee druzia" (Her Friends) 1949, "Stranitsy zhizni" (Pages From Life) 1953, "V poiskakh radosti" (In Search of Joy) 1957, "Neravnyi boi" (Disparate Battle) 1960, "Pered uzhinom" (Before Supper) 1962, "Vechno zhivye" (The Immortals – made into film with the title "Letiat zhuravli" – When the Cranes Fly –) 1957, "Traditsionnyi sbor" (Traditional Meeting) 1967, "V den' svad'by" (On the Wedding Day) 1964, "S vechera do poludnia" (From Evening Till Noon) 1970, "Situatsiia" (Situation) 1973, and others.

Rubanenko, Boris Rafailovich, Dr. of Arch. Sc.
Architect; b. Ag.16 (29), 1910, Samara, now Kuibyshev.
Mem. CPSU 1939–; studied Inst. of Civil Engineering 1927–31 and Inst. of Painting, Sculpture and Architecture of All-Russian Acad. of the Arts in Leningrad 1932–34; Dr. of Arch. Sc. 1967; teacher at Moscow Inst. of Architecture 1954–, Prof. at same 1970–; Dir. of Cen. Scientific Research Inst. for Experimental Housing Design 1963–. Decors.: Merited Architect of the RSFSR 1969 and of the Kazakh SSR 1970, Order of the October Revolution, 2 other orders and various medals.
Publ.: "Issledovanie i razrabotka problemnykh voprosov arkhitektury i tipizatsii massovogo zhilishchnogo stroitelstva v SSSR" 1965.

Ruben, Iurii Ianovich
State official; b. 1925.
Russian; mem. CPSU 1953–; grad. Latvian State Univ. 1951 and Party High School of Cen. Cttee. CPSU 1960; served in Sov. Army 1943; engineer-technologist at Riga Compressor Works 1951–52; leading Komsomol posts in Latvia 1952–55; CPSU work 1955–58; 1st Sec. of Liepāja City Cttee. 1960–63 and of Riga City Cttee. Latvian CP 1963–66; mem. Cen. Cttee. Latvian CP 1961–; mem. Bureau for Ind. at Cen. Cttee. Latvian CP 1962–63; cand. mem. 1963–66, mem. Bureau Cen. Cttee. 1966–, Sec. of Cen. Cttee. Latvian CP 1966–70; Chair. of Council of Mins. Latvian SSR My. 1970–; Dep. to USSR Supr. Sov. 1970–; cand. mem. Cen. Cttee. CPSU 1971–.
Decor.: Order of Red Banner of Labour.

Ruben, Vitalii Petrovich
State official; b. F.13 (26), 1914, Moscow; s. of a clerk.
Latvian; mem. CPSU 1939–; grad. Party Higher School of Cen. Cttee. CPSU 1951 and Grass-Seed Production College 1933; agronomist with a raion agricultural dept., teacher in Rybinsk Raion (Iaroslavl Oblast) 1934–41; sen. CPSU, econ. and govt. posts 1941–47; Sec. of Daugavpils District Cttee. Latvian CP (B) 1947–48; Dep. Chair. of Council of Kolkhozes with USSR Council of Mins. Latvian SSR 1952–53; Dep. Plenipotentiary for USSR Min. of Supplies for Latvian SSR 1954–55; dep. head, head of a dept. at Cen. Cttee. Latvian CP 1955–61; mem. Cen. Cttee. Latvian CP 1960–76; Dep. Chair. 1961–62, Chair. of Council of Mins. Latvian SSR 1962–70; mem. Bureau Cen. Cttee. Latvian CP 1962–74; Dep. to USSR Supr. Sov. 1962–; cand. mem. 1966–76, mem. Cen. Cttee. CPSU 1976–; Chair. of Presidium Supr. Sov. Latvian SSR 1970–74; Chair. of Sov. of Nationalities USSR Supr. Sov. Jl. 1974–.

Rubiks, Alfred Petrovich
State official.
Sec. and mem. Bureau Cen. Cttee. Latvian Komsomol until 1969; cand. mem. Cen. Cttee. Latvian CP 1966–71, mem. 1981–; Min. of Local Ind. Latvian SSR Mr. 1982–.

Rubinov, Anatolii Nikolaevich
Physicist; b. 1939.
Mem. CPSU 1972–; grad. Belorussian Univ. 1961; mem. staff Inst. of Physics of Acad. of Sciences Belorussian SSR 1961–, Head of Laboratory 1970–. Decor.: State Prize of the USSR 1972.

Rubins, Ianis Feliksovich
State official.
Mem. Auditing Comm. 1966–71, cand. mem. Cen. Cttee. Latvian CP 1971–; Chair. of State Cttee. for Construction Latvian SSR Ap. 1966–; Dep. to Supr. Sov. Latvian SSR 1967–.

Rudenko, Bella Andreevna
Singer; b. 1933.
Grad. Odessa Conservatory 1956; coloratura-soprano; soloist at Odessa Opera and Ballet Theatre 1955–56; soloist at Kiev Shevchenko Opera and Ballet Theatre 1956. Decor.: State Prize of the USSR 1971.

Rudenko, Larisa Arkhipovna
Ukrainian mezzo-soprano; b. Ja.15 (28), 1918, Makeevka, now in Donetsk Oblast.
People's Artist of the USSR (1960).
In 1940, she graduated from the Kiev Conservatory. From 1939 to 1970 she was a soloist with the Ukrainian Theater of Opera and Ballet in Kiev. A gifted actress, she has a sonorous voice and a fine stage presence. She has sung major roles in many operas by Ukrainian composers, including Solomiia in Dan'kevich's *Bogdan Khmel'nitskii*, Ulia Gromova in Meitus' *The Young Guard* (1947), Var'ka in Zhukovskii's *The First Spring* (1960), and Iarina in Maiboroda's *Taras Shevchenko* (1964). One of her best roles has been that of Liubasha in Rimsky-Korsakov's *The Tsar's Bride*.
She also gives concerts. From 1951 to 1953 she taught at the Kiev Conservatory; she rejoined the staff in 1965, becoming a docent in 1969. From 1966 to 1970 she was head of the subdepartment of musical education at the Karpenko-Karyi Theatrical Institute in Kiev. She tours outside the USSR. She was a deputy to the sixth convocation of the Supreme Soviet of the Ukrainian SSR. Decors.: Order of the Red Banner of Labour and various medals.

Rudenko, Liudmila Vladimirovna
Chess player and economic planner; b. Jl. 14 (27), 1904, Lubny.
Honored Master of Sport of the USSR (1953).
She was USSR women's champion in 1952 and world women's champion from 1950 to 1953. Decor.: Order of the Badge of Honour.

Rudenko, Mikola (Nikolai) Danilovich
Author; b. D.19, 1920, Iurevka, Voroshilovgrad Oblast.
Ukrainian; mem. CPSU 1946–; served in World War II, was seriously wounded; Exec. Ed. of journal "Dnipro" 1947–50; arrested as Sov. mem. of Amnesty International 1975, later released, suffered out-of-court reprisals, was expelled from Ukrainian Writers' Union, expressed his protest in an open letter to Brezhnev.
Publs.: "Ekonomičny monology" (Samisdat publ. in Ukrainian and Russian),

"Leningradtsy" (People From Leningrad) 1948, "Muzhestvo" (Courage) 1952, "Nikolai Krasnoshapka" 1954, "Veter v litso" (Facing the Wind) 1955, "Belaia akatsiia" (White Acacia) 1962, "Sto svityl" (One Hundred Lights) 1971, "Obnovlennia" (The Renewal) 1971, "Vsesvit u Tobi" (The Universe in You) 1974 (was expelled from Party for this work).

Rudenko, Sergei Ignat'evich
Military official; b. O.7 (20), 1904, Korop, Chernigov Oblast; s. of a cobbler.
Mem. CPSU 1928–; grad. 1st Mil. Aviation School 1927, N.E. Zhukovskii Aviation Acad. 1932 and Operative Dept. of same acad. 1936; Air Force Marshal 1955–; Prof. 1972–; in Sov. Army 1923–; commander of an airborne div. at Western Front 1941, Commander of Air Force of 61st Army 1941, Dep. Commander, Commander of Air Force at Kalinin Front Ja.–F. 1942, Dep. Commander of Air Force at Volkhov Front F.–Mr. 1942, Commander of 1st Group and 7th Combat Patrol of Chief HO at Briansk Front Ap.–Je. 1942, Dep. Commander of Air Force at Southwestern Front Je.–S. 1942, Commander of 16th Air Force Army at Stalingrad, Don. Cen. Belorussian and 1st Belorussian Front S. 1942–My. 1945; command posts in Sov. Army 1945–49; Chief of Gen. Staff of Sov. Air Force 1949–50; Commander of Long-Range Air Force, Dep. Chief Commander of Sov. Air Force 1950–53; Chief of Gen. Staff and 1st Dep. Chief Commander of Sov. Air Force 1953–69; cand. mem. Cen. Cttee. CPSU 1961–66; Head of Iu. A. Gagarin Air Force Acad. 1968–73; Mil. Inspector-Adviser of Group of Inspectors-Gen. of USSR Min. of Defence 1973–.
Decors.: Hero of the Sov. Union 1944, 5 Orders of Lenin, Order of the October Revolution, 4 Orders of Red Banner, 2 Suvorov Orders 1st Cl., Suvorov Order 2nd Cl., Kutozov Order 1st Cl., medals and various foreign orders.

Rudenko, Vasilii Timofeevich, Cand. of Tech. Sc.
Engineer and physicist; b. 1930.
Grad. Moscow Inst. of Physical Engineering 1953; Cand. of Tech. Sc. 1971–; mem. staff United Inst. of Nuclear Research Dubna, dept. head at same 1961–.
Decor.: State Prize of the USSR 1971.

Rugoev, Iaakko Vasil'evich
Author; b. Ap. 15, 1918, Suoiarvi, Karelia; s. of a peasant.
Karelian; mem. CPSU 1943–; grad. Teachers' Training Inst. in Petrozavodsk 1939; worked as teacher and journalist; served in Sov.-Finnish War 1939–40 and World War II 1941–45; Chair. of Karelian Writers' Union 1953–57 and 1967–; first works publ. 1934.
Publs.: "Mest'" (The Revenge) 1943, "Skripka i traktor" (The Violin and the Tractor) 1963, "Dvesti piat'desiat' shest' dnei" (Two Hundred And Fifty-Six Days) 1964, and others.

Rukavishnikov, Nikolai Nikolaevich
Soviet cosmonaut; b. 1932.
Mem. CPSU 1970–; grad Moscow Inst. of Physical Engineering 1957; cosmonaut 1967–; test engineer on board of spaceship "Soiuz 10", docking manoeuvre with space lab "Saliut" and tech.-scientific research assignments during orbital flight Ap. 23–25, 1971, "Soiuz 16" D. 1974 and "Soiuz 33" Ap. 1979. Decor.: Hero of the Soviet Union 1971, two Orders of Lenin.

Rukmane, Marite Karlovna
Komsomol official.
Mem. Bureau F. 1978–, Sec. 1978–80, 1st Sec. Cen. Cttee. Latvian Komsomol D. 1980–; Dep. Supr. Sov. Ratvian SSR 1980–, mem. Presidium D. 1982–; mem. Cen. Cttee. Latvian CP 1981–; mem. Bureau Cen. Cttee. USSR Komsomol 1982–.

Rumiantsev, Aleksei Fedorovich
Journalist; b. 1919.
Mem. CPSU 1940–; grad. Leningrad Power Tech. College 1939 and Party High School of Cen. Cttee. CPSU 1948; lecturer; engineering posts in various works in Leningrad, Tomsk, Barnaul and Cheliabinsk 1939–43; mem. staff People's Commissariat for USSR Power Stations 1943–44; Asst. Prof., Prof. at Party High School of Cen. Cttee. CPSU 1948–65; Chief Ed. of "Ekonomicheskaia gazeta" Ag. 1965–; mem. Cen. Auditing Comm. CPSU 1966–. Decor.: Order of Lenin.

Rumiantsev, Aleksei Matveevich, Dr. of Econ. Sc.
Economist; b. 1905, Mintsovo, Kostroma Oblast.
Mem. CPSU 1940–; grad. Kharkov Econ. Inst. 1926; Prof.; labourer 1917; worked for Ukrainian People's Commissariat for Agriculture 1925–27 and 1928–29; served in Red Army 1927–28; worked for People's Commissariat of Justice Ukrainian SSR 1929–30; econ. and teaching posts 1930–43; CPSU work in Kharkov Oblast 1943–49; Dir. of Inst. of Econ. at Acad. of Sciences Ukrainian SSR and Head of Dept. of Social Sciences at same acad. 1950–52; worked for Cen. Cttee. CPSU 1952–55; mem. Cen. Cttee. CPSU 1952–76; Dep. to USSR Supr. Sov. 1954–62; Chief Ed. of periodical "Kommunist" 1955–58; Chief Ed. of periodical "Problemy mira i sotsializma" 1958–64; Chief Ed. of "Pravda" 1964–65; Acad. Sec. of Dept. of Econ. 1965–67, Vice-Pres. of USSR Acad. of Sciences My. 1967–71; mem. Presidium USSR Acad. of Sciences 1971–. Decor.: Order of Lenin.

Rumiantsev, Valentin Vital'evich
Mechanics scientist; b. Jl. 19, 1921, Skatovka, Saratov Oblast.
Mem. CPSU 1948–; grad. Saratov Univ. 1945; specialist in mechanics; corresponding mem. USSR Acad. of Sciences 1970–; mem. staff Inst. of Mechanics of USSR Acad. of Sciences 1945–65; mem. staff Computing Centre USSR Acad. of Sciences 1965–; concurrently Lecturer 1953–, Prof. at Moscow Univ. 1956–.
Decors.: 2 Orders of Red Banner of Labour, various medals.
Publs.: "Dinamika tela s polostiami, soderzhashchimi zhidkost'" (co-author) 1965, "Ob ustoichivosti statsionarnykh dvizhenii sputnikov" 1967.

Rusakov, Konstantin Viktorovich
Party official; b. 1909.
Russian; mem. CPSU 1943–; grad. Kalinin Polytech. Inst. Leningrad 1930; engineering and econ. posts in various construction trusts in Leningrad, Leninakan and Irkutsk 1930–39; on staff of People's Commissariat for the USSR Fish Ind. 1939–46; USSR Dep. Min. of the Fish Ind. of the Western Regions 1946–48; USSR Dep. Min. 1948–50, Min. 1950–52, Dep. Min. 1952–53 and 1955–56, 1st Dep. Min. of the Fish Ind. 1956–57; Emb. Counsellor at USSR Emb. in Poland 1958–60; head of a section 1960–61, dep. head of a dept. at Cen. Cttee. CPSU 1961–62; USSR Amb. to Mongolia 1962–63; mem. staff Cen. Cttee. CPSU 1964–65; 1st Dep. Head 1965–68, Head of Dept. of Relations with Communist and Workers' Parties of Socialist Countries at Cen. Cttee. CPSU 1968–73 and 1977–; mem. Cen. Auditing Comm. 1966–71, mem. Cen. Cttee. CPSU 1971–; Aide to Sec.-Gen. of Cen. Cttee. CPSU 1973–77; Sec. of Cen. Cttee. CPSU My. 1977–. Decor.: Order of Lenin, Hero of Socialist Labour, medals.

Rusakov, Lev Vasil'evich
State official.
Min. of Local Ind. Belorussian SSR Ag. 1977–; cand. mem. Cen. Cttee. Belorussian CP 1981–.

Rusenko, Iurii Lavrent'evich
State official.
Head of Main Statistical Admin. at Council of Mins. Lithuanian SSR 1973–75; Dep. Chair. of Council of Mins. Lithuanian SSR Ap. 1975–, mem. Cen. Cttee. Lithuanian CP 1976–.

Rusetskii, Aleksei Stepanovich (real name: Burdelev)
Author; b. N. 21 (D. 4), 1912, Studenets, Mogilev Oblast.
Belorussian; mem. CPSU 1944–; grad. Moscow Zoo-Veterinary Inst. 1935; worked in various scientific institutions in Moscow and Minsk; served in Sov. Army during World War II; first works publ. 1935.
Publs.: "V zavtrashnii mir" (In the World of Tomorrow) 1951, "Svet okon tvoikh" (The Light in Your Windows) 1959, "Ego Velichestvo" (His Highness) 1965, and others.

Russu, A. A.
Trade union official.
Sec. of Moldavian Republican Trade Union Council Ja. 1982–.

Russu, Vasilii Petrovich
State official.
Min. of Posts and Telecommunications Moldavian SSR Ja. 1966–; mem. Auditing Comm. Moldavian CP 1966–; Dep. to Supr. Sov. Moldavian SSR 1967–.

Rustamova, Kholbuvi
State farmer; b. 1951.
Mem. CPSU 1977–; finished 4th year at Tashkent Agricultural Inst. 1980; nurse in Dzhizak Oblast hospital 1970–75; Komsomol work brigade leader at Ulianov State Farm in Dzhizak Oblast Ap. 1975–; cand. mem. Cen. Cttee. CPSU 1981–.

Rutkevich, Mikhail Nikolaevich
Philosopher; b. O. 2, 1917, Kiev.
Corresponding member of the Academy of Sciences of the USSR (1970). Member of the CPSU since 1944.
He graduated from the department of physics at the University of Kiev in 1939. He began teaching philosophy in 1947. From 1953 to 1966 he was head of the subdepartment of philosophy at Urals University in Sverdlovsk and from 1966 to 1972 served as dean of the university's department of philosophy. From 1972 to 1976 he was director of the Institute of Sociological Research of the Academy of Sciences of the USSR. His works concentrate on dialectical materialism, historical materialism, and sociological research.
Decors.: Order of the Red Star and a number of medals.
Works: *Praktika – osnova poznaniia i kriterii istiny.* Moscow 1952. *Marksizm-leninizm o estestvoznanii i ego roli v zhizni obshchestva.* Sverdlovsk 1952. *Religiia i ee reaktsionnaia rol' v zhizni obshchestva.* Sverdlovsk 1954. *Dvizhenie i razvitie v prirode i obshchestve.* Moscow 1954. *Vek nauki i iskusstva.* [Moscow 1965.] (With A. F. Eremeev.) *Sotsial'nye peremeshcheniia.* Moscow 1970. (With F. R. Filippov.) *Dialekticheskii materializm,* 3rd ed. Moscow 1973.

Rybakov, Aleksei Mironovich
Party official; b. 1925.
Russian; mem. CPSU 1945–; grad. Leningrad Engineering Inst. for Rail Transport 1965; collective farmer 1941–42; served in Sov. Army 1942–44; Komsomol and party posts 1945–62; 2nd Sec. of Velikie Luki City Cttee. CPSU 1962–66, 1st Sec. 1966–71; 1st Sec. of Pskov Oblast Cttee. CPSU N. 1971–; Dep. to USSR Supr. Sov. 1974–; cand. mem. Cen. Cttee. CPSU 1976–.

Rybakov, Anatolii Iakovl'evich
Trade union official; b. 1927.
Mem. CPSU 1961–; grad. Moscow Machine-Tools Evening Technicum 1971; lathe operator in "Frezer" Plant in Moscow 1943–45, in "Krasnyi proletarii" Machine-Tool Plant in Moscow 1945–77 (served in Sov. Army 1945–52); Chair. of Cen. Cttee. of Machine-Building and Instrument-Making Workers Union F. 1977–; mem. Cen. Cttee. CPSU 1976–. Decor.: Hero of Socialist Labor 1976.

Rybakov, Anatolii Naumovich
Russian writer; b. Ja. 1 (14), 1911, Chernigov; s. of an engineer.
He graduated from the Moscow Institute of Railroad Engineers in 1934. He fought in the Great Patriotic War (1941–45).
His first book was an adventure novella for children, *The Dagger* (1948; film of the same title, 1954); its sequel was the novella *The Bronze Bird* (1956). His novels *The Drivers* (1950; State Prize of the USSR, 1951), *Ekaterina Voronina* (1955; film of the same title, 1957), and *Summer in Sosniaki* (1964) depict life in a large enterprise and the complex interactions among the people who work there. These works deal with the problem of the individual's responsibility to himself and society as a whole. He is the author of the novellas *The Adventures of Krosh* (1960; film of the same title, 1962), *Krosh's Vacation* (1966; film entitled *These Innocent Games,* 1968), and *The Unknown Soldier* (1970; film entitled *A Moment of Silence,* 1971; RSFSR Vasil'ev Brothers Prize, 1973). Written from the point of view of an adolescent, the novellas inculcate moral standards; they are marked by lively dialogue, goodnatured irony, and an understanding of contemporary life.
His books have been translated into the languages of the peoples of the USSR and into foreign languages. Decors.: three orders and various medals.
Works: *Povesti.* [Afterword by E. Starikova.] Moscow 1969. *Kortik, Bronzovaia ptitsa, Povesti.* [Foreword by V. Smirnova.] Moscow 1971. *Voditeli, Ekaterina Voronina, Romany, Neizvestnyi soldat, Povest'.* Moscow 1973.

Rybakov, Boris Aleksandrovich
Archaeologist and historian; b. My. 21 (Je. 3), 1908, Moscow.
Mem. CPSU 1951–; grad. Moscow State Univ. 1930; mem. USSR Acad. of Sciences 1958; Prof. at Moscow State Univ. 1943–; Dir. of Inst. of Archaeology of USSR Acad. of Sciences 1956–; Acad. Sec. of Dept. of History of USSR Acad. of Sciences 1973–75; mem. Czechoslovak Acad. of Sciences 1960–, mem. Polish Acad. of Sciences 1970–; mem. Exec. Cttee. of Int. Soc. of Prehistoric and Protohistoric Sciences 1958–; mem. Int. Cttee. of Slavists 1963–; Chair. of USSR-Greece Soc. 1958–. Decors.: 2 Orders of Lenin, 2 other orders, medals, State Prize of the USSR 1949 and 1952.
Publs.: "'Slovo o polku Igoreve' i ego sovremenniki" 1971, "Russkie letopistsy i avtor 'Slova o polku Igoreve'" 1972, "Russkie karty Moskovii XV – nachala XVI vv." 1974.

Rybakov, Vasilii Arsen'evich
Diplomat; b. 1908.
Mem. Diplomatic Service 1939–; Advisor at Sov. Mission in Finland 1945–46; USSR Envoy to Iceland 1946–48, to Ethiopia 1948–54; mem. staff USSR Min. of Foreign Affairs 1954–56; USSR Consul-Gen. in Bombay 1956–58; sen. posts in USSR Min. of Foreign Affairs 1959–.

Rybal'chenko, Stepan Dmitrievich
Military official; b. Je. 28 (Jl. 11), 1903, Novoegorlykskoe, now in Sal'sk Raion, Rostov Oblast; s. of a peasant.
Colonel-general of aviation (1944). Member of the CPSU from 1927.
He fought as a private during the Civil War of 1918–20. He graduated from the M. V. Frunze Military Academy in 1936 and the Higher Academic Training Courses at the K. E. Voroshilov Higher Military Academy in 1951. During the Great Patriotic War he was deputy chief of staff of the directorate of the air forces of the Northern Front in 1941, chief of staff of the directorate of the air forces of the Leningrad Front in 1941–42, commander of the air forces of the Leningrad Front from February to November 1942, and commander of the Thirteenth Air Army from November 1942 until the end of the war.
After the war, he commanded the air forces of the Kiev Military District in 1946–47 and was inspector general and deputy chief inspector of the air force in the Main Inspectorate of the Armed Forces of the USSR from 1947 to 1950. He commanded the air forces of the armies of the Far East and the Far Eastern Military District from 1951 to 1956. He retired in December 1963.
Decors.: two Orders of Lenin, four Orders of the Red Banner, two Orders of Kutuzov First and Second Class, the Order of Suvorov Second Class, the Order of Bogdan Khmel'nitskii First Class, and various medals.

Rybalov, Erim Grigor'evich
State official.
Min. of Trade Turkmen SSR Je. 1973–, mem. Cen. Cttee. Turkmen CP 1976–.

Rybnov, Aleksandr Vasil'evich
Choral conductor; b. Ag. 19 (S. 1), 1906, Moscow.
People's Artist of the USSR (1971). Member of the CPSU since 1944.
He graduated from the choral department of the Moscow Conservatory in 1927 and from the music theory department in 1930. In 1930 he became choirmaster, and in 1958, chief choirmaster of the Bolshoi Theater of the USSR. He has taken part in the Bolshoi's presentations of Mussorgsky's *Boris Godunov* and *Khovanshchina,* Borodin's *Prince Igor,* Rimsky-Korsakov's *Sadko, The Legend of the Invisible City of Kitezh,* and *The Maid of Pskov,* Tchaikovsky's *The Queen of Spades,* Glinka's *Ruslan and Liudmila,*

Verdi's *Aida,* Prokofiev's *War and Peace* and *Semen Kotko,* Shaporin's *The Decembrists,* and Molchanov's *The Dawns Here Are Quiet.* From 1954 to 1967 he was a docent at the Gnesin Institute of Music and Pedagogy. From 1934 to 1950 he directed the amateur choir of the I. A. Likhachev Moscow Automobile Plant. Decors.: State Prize of the USSR (1950) for his work with the chorus in the production of Tchaikovsky's opera *Mazepa,* two orders and various medals.

Rykalin, Nikolai Nikolaevich
Metallurgical scientist; b. S. 14 (27), 1903, Odessa.
Grad. Far Eastern Univ.; specialist in welding and metallurgy; mem. USSR Acad. of Sciences 1968–; mem. staff Far Eastern Univ. 1929–36; mem. staff Bauman Tech. College Moscow 1936–39; mem. staff Section for Scientific Solution of Problems in Electrowelding and Electrothermics of USSR Acad. of Sciences 1939–53; Head of Laboratory of Baikov Inst. of Metallurgy 1953–; Vice-Pres. of Int. Welding Inst., Chair. of Nat. USSR Cttee. for Welding, mem. Serbian Acad. of Sciences and the Arts. Decors.: 4 orders, medals.

Rykov, Gennadii Vasil'evich
Diplomat.
1st Sec. USSR Emb. in Guinea, Emb. Councellor at USSR Embassy in Cameroon until 1978; USSR Amb. to Rwanda My. 1978–.

Rykov, Vasilii Nazarovich
Diplomat; b. 1918.
Russian; mem. CPSU 1943–; grad. Novocherkassk Ind. Inst. 1941; technologist, foreman, dept. head, sec. of CPSU bureau at a plant in Irkutsk 1941–46; chief engineer, sec. of a CPSU bureau at a research inst. 1946–56; 2nd Sec., 1st Sec. of Leningrad Raion CPSU Cttee. in Moscow 1956–61; Inspector for Cen. Cttee. CPSU 1961–63; mem., mem. Bureau and 2nd Sec. of Cen. Cttee. Turkmen CP 1963–Ap. 1975; cand. mem. 1966–71, mem. Cen. Cttee. CPSU 1971–; USSR Amb. to Algeria Ap. 1975–. Decor.: Order of Red Banner of Labour.

Ryl'skil, Iakov Anufrievich
Sabreur; b. O. 25, 1928, Aleksandrovka in Verkhubinka Raion, Vostochayi Kazakhstan Oblast.
Honored Master of Sports (1961). Coach. In the 1950's and 1960's, he was one of the world's best fencers. He was individual world champion in 1958, 1961, and 1963. He was a member of the world championship team in 1965 and a member of the Olympic championship team in 1964. He was individual champion of the USSR five times in the period 1954–58.

Ryndina, Antonina Nikolaevna
State farmer; b. 1942.
Mem. CPSU 1973–; grad. Riazan State Farm Technicum 1977; milkmaid on a state farm 1957–70; cattle farmer and operator at an inter-state-farm terminal point for cattle-feeding 1970–78; work brigade leader at Shilov Inter-State-Farm Enterprise for Cattle-Feeding in Riazan Oblast Ja. 1978; mem. Cen. Auditing Comm. CPSU 1981–.

Rysmendiev, A.
Party official.
Mem. Bureau 1970–72 and 1974–, 1st Sec. of Cen. Cttee. Kirghiz Komsomol F. 1974–78; mem. Cen. Cttee. Kirghiz CP 1976–. Student 1978–80; Head of Dept. of Light and Food Ind. with Cen. Cttee. Kirghiz CP My. 1980–.

Ryspaev, Barpy
State official; b. 1927; s. of a farmer.
Kirghizian; mem. CPSU 1946–; grad. Cen. Cttee. Party Higher School CPSU; Sec. of a People's Court, inspector of a Raion Dept. for People's Education, Dep. Chair. of Exec. Cttee. of a Raion Sov. of People's Deps., 2nd Sec. of a City Cttee. Kirghizian CP 1942–58; 1st Sec. of Cen. Cttee. Kirghizian Komsomol 1958–60; mem. Cen. Cttee. and cand. mem. Bureau of Cen. Cttee. Kirghizian CP 1960–61; head of the Party Organs Dept. Cen. Cttee. Kirghizian CP 1960–61; 1st Raion Cttee. Sec. Kirghizian CP in Issyk-Kul 1961–71; Oblast Cttee. Sec. Kirghizian CP in Osh 1971–81; Min. of Food Ind. Kirghiz SSR Jl. 1981–.

Rythkheau, Iurii Sergeevich
Author; b. Mr. 8, 1930, Uelen, Chukchi National Krai; s. of a hunter.
Chukchi; mem. CPSU 1967–; grad. Leningrad Univ. 1954; first works publ. 1946. Publs.: "Liudi nashego berega" (People From Our Shore) 1953, "Vremia taianiia snegov" (The Time When the Snow Melts) 1958–67, "Samye krasivye korabli" (The Most Beautiful Ships) 1967, "Liudi severnogo siianiia" (People of the Northern Light) 1969, and others.

Rytov, Sergei Mikhailovich
Physicist; b. Je. 20 (Jl. 3), 1908, Kharkov. Corresponding member of the Academy of Sciences of the USSR (1968).
He graduated from Moscow University in 1930. From 1934 to 1958 he worked at the Institute of Physics of the Academy of Sciences of the USSR, and since 1958 he has worked at the academy's Institute of Radio Engineering. Since 1947 he has also been a professor at the Moscow Physicotechnical Institute.
His main work has been devoted to the theory of vibrations and waves and to statistical radio physics and optics. He was one of the discoverers of the theory of thermal fluctuations in electrodynamics. Decors.: Order of the Red Banner of Labour, three other orders, and several medals.
Works: "Modulirovannye kolebaniia i volny." *Tr. fizicheskogo in-ta AN SSSR.* 1940, vol. 2, no. 1. *Teoriia elektricheskikh fluktuatsii i teplovogo izlucheniia.* Moscow 1953. *Vvedenie v statisticheskuiu radiofiziku.* Moscow 1966. *Teoriia ravnovesnykh teplovykh fluktuatsii v elektrodinamike.* Moscow 1967. (With M. L. Levin.)

Ryvina, Elena Izrail'evna
Author; b. Ja. 31 (F. 13), 1910, Mozyr.
Russian; mem. CPSU 1951–; grad. Dept. of Philology of Leningrad Univ. 1934; first works publ. 1931.
Publs.: "Dvadtsat' pesen" (Twenty Songs) 1938, "Nad moei Nevoi" (On My Neva) 1939, "Slovo Leningradki" (The Word of a Leningrad Woman) 1942, "Stikhi o liubvi" (Love Poems) 1957, "Stikhi o Leningrade" (Poems of Leningrad) 1967, and others.

Ryzhikov, Konstantin Minaevich
Helminthologist; b. S. 13 (26), 1912, Smolensk.
Mem. CPSU 1945–; grad. Smolensk Pedagogic Inst. 1937; corresponding mem. USSR Acad. of Sciences 1964–; mem. staff Helminthological Laboratory of USSR Acad. of Sciences 1946–, Dir. of same 1963–. Decors.: 2 orders, medals, foreign orders.
Publs.: "Opredelitel' gelmintov domashnikh kurinykh ptits" 1968, "Gelminty ptits Iakutii i sopredelnykh territorii" (coauthor) vols. 1–2, 1973–74.

Ryzhkov, Nikolai Ivanovich
State official; b. 1929.
Russian; mem. CPSU 1956–; grad. Uralsk Polytechnical Inst. 1959; shift foreman, shop superintendent, dep. dir., chief engineer at S. Ordzhonikidze "Uralmash" Plant 1951–70, Dir. 1970–71; Gen. Dir. of "Uralmash" Prod. Asscn. 1971–75; Dep. to USSR Supr. Sov. 1974–; 1st Dep. USSR Min. of Heavy and Transport Machine Building 1975–79; 1st Dep. Chair. of USSR GOSPLAN Ja. 1979–82; mem. Cen. Cttee. CPSU 1981–; Sec. Cen. Cttee. CPSU N. 1982–. Decors.: USSR State Prize, Order of Red Banner of Labour 1979.

Ryzhov, Nikita Semenovich
Diplomat; b. 1907.
Mem. CPSU 1928–; grad. Moscow Ind. Acad. 1940; weaver, Asst. Foreman, Foreman, Dep. Dir. of Savino Weaving Mill 1924–29; mem. staff People's Commissariat for the Textile Ind. and People's Commissariat for USSR Heavy Ind. 1932–41; Sec. of Tashkent Oblast Cttee. Uzbek CP 1941–42; People's Commr. 1944–46 and Min. of the Textile Ind. RSFSR 1946–49; Min. of Light Ind. resp. Food and Light Ind. 1950–53; 1st USSR Dep. Min. 1953–54 and Min. of Consumer Goods Ind. 1954–55; USSR Min. of the Textile Ind. 1955–56; USSR Amb. to Turkey 1957–66 and to Italy 1966–80; USSR Dep. Min. of Foreign Affairs F.

1980–; mem. Cen. Auditing Comm. CPSU 1966–. Decor.: Order of the Red Banner of Labour.

Ryzhov, Nikolai Ivanovich
Russian actor; b. O.18 (31), 1900, Moscow; s. of V. N. Ryzhova.
People's Artist of the USSR (1971).
He attended the drama courses at the Malyi Theater and the drama school of the Moscow Philharmonic Society. In 1923 he joined the Malyi Theater troupe. A representative of the realistic school of acting, he has created striking and colorful characters, with vivid and sharp details from real life. His best roles have been in the plays of A. N. Ostrovskii: for example, Elesia in *Not a Penny, and Suddenly an Altyn,* Belogubov in *A Profitable Post,* Andrei in *A Hangover at Someone Else's Feast,* Baraboshev in *The Truth Is Good, but Happiness Is Better,* Lyniaev in *Wolves and Sheep,* and Khudobaev in *Shines, but Doesn't Warm.* His other roles have included Joseph Sedley in the stage version of Thackeray's *Vanity Fair,* Motyl'kov in Gusev's *Glory,* Kolosov in Tremer's *Liubov' Iarovaia,* Blokhin in Stepanov and Popov's *Port arthur,* and Lishev in Lavrenev's *To Those in the Sea.* Decors.: Order of the Red Banner of Labor and various medals.

Rzaev, Iunis Kagraman ogly
State official.
Mem. Cen. Cttee. Azerbaidzhan CP 1971–; Chair. of State Cttee. for Viticulture and Wine-Making Azerbaidzhan SSR Je. 1973–.

Rzhanov, Anatolii Vasil'evich
Physicist; b. Ap.9, 1920, Ivanovo.
Mem. CPSU 1943–; grad. Leningrad Polytech. Inst. 1941; corresponding mem. USSR Acad. of Sciences 1962–; mem. staff Inst. of Physics of USSR Acad. of Sciences 1944–62; Dir. of Inst. of Physics of Semiconductors at Siberian branch of USSR Acad. of Sciences 1962–; concurrently Lecturer 1963–, Prof. at Novosibirsk Univ. 1966–. Decors.: 2 orders, various medals.
Publs.: "Elektronnye protsessy na poverkhnosti poluprovodnikov" 1971.

Rzhevskaia, Elena Moiseevna
Author; b. O.27, 1919, Gomel.
Russian; mem. CPSU 1943–; grad. Gorkii Inst. of Literature 1948; first works publ. 1951.
Publs.: "Spustia mnogo let" (Many Years Later) 1958, "Vesna v shineli" (Spring in an Army Coat) 1961, "Zemnoe pritiazhenie" (Attraction of the Earth) 1962, "Ot doma do fronta" (From Home to the Front) 1965, "Berlin, Mai 1945" 1965, and others.

Rzhevskii, Vladimir Vasil'evich
Mining scientist; b. Jl.23, 1919, Vorontsovo-Aleksandrovskoe, now Zelenokumsk, Stavropol' Krai.
Corresponding member of the Academy of Sciences of the USSR since 1966. Member of the CPSU since 1943.
He graduated from the Moscow Mining Institute in 1941, and in 1949 he began teaching at the institute. He was appointed professor in 1958 and rector of the institute in 1962. He initiated the development and introduction of new curricula for mining engineers for the mining industry of the USSR. His main work has been devoted to the theory and practice of opencut mining and the physics of rock and mining processes. He devised a method for geometric analysis of opencut fields and for planning operating conditions. Decors.: honorary doctorates from the Freiberg Mining Academy (German Democratic Republic) and Technical University of Heavy Industry in Miskolc (Hungary), three orders and several medals.
Works: *Osnovy proektirovaniia kar'erov.* Moscow 1958. (With E. F. Sheshko.) *Osnovy fiziki gornykh porod,* 3rd ed. Moscow 1973. (With G. Ia. Novik.) *Protsessy otkrytykh gornykh rabot,* 2nd ed. Moscow 1974.

Saakian, Levon Gurgenovich
Trade union official; b. 1937.
Armenian; mem. CPSU 1961–; grad. Polytech. Inst. in Erevan; held Engineering posts 1960–63; Instr. Cen. Cttee. Armenian CP, Head of a Dept. Erevan City Cttee. Armenian CP 1964–70; 1st Sec. and mem. Bureau Cen. Cttee. Armenian Komsomol 1970–72; mem. Cen. Cttee. 1971–, cand. mem. Bureau Cen. Cttee. Armenian CP 1971–73 and 1976–; 2nd Sec. of Erevan City Cttee. Armenian CP 1972–75; Chair. of Armenian Republican Trade Union Council My. 1975–.

Saarniit, Ia.
Komsomol official.
Sec. and mem. Bureau of Cen. Cttee. Esthonian Komsomol Mr. 1982–.

Sabaneev, Stanislav Nikolaevich
State official.
1st Sec. Rostov City Cttee. CPSU until 1970; Chair. of Rostov Oblast Exec. Cttee. of Peoples' Deps. 1970–79; Chair. of RSFSR State Cttee. on Construction (GOSSTROI) Jl. 1979–.

Sabirova, Malika Abdurakhmanovna
Ballet dancer; b. 1942.
Grad. Leningrad Vaganova Choreographic School; soloist with Tadzhik S. Aini Opera and Ballet Theatre 1961–; numerous guest performances abroad. Decors.: prizewinner in numerous int. competitions (e.g. Silver Medal in Varna 1964, 1st Prize and Gold Medal at 1st Int. Ballet Competition in Moscow 1957), Rudaki State Prize of the Tadzhik SSR 1967, Prize of the Lenin Komsomol 1970, People's Artist of the USSR 1974.

Sachenko, Boris Ivanovich
Author; b. My.15, Velikii Bor, Gomel Oblast.
Belorussian; grad. Dept. of Philology of Belorussian State Univ. 1960; writes in Belorussian (mainly about contemporary topics).
Publs.: "Plen" (Imprisonment), "Poslednie i Pervye" (The Last and the First) 1968, "Oksana" 1968, "Kholostiak" (The Bachelor) 1969, and others.

Sadovskii, Mikhail Aleksandrovich
Physicist; b. O.24 (N.6), 1904, St. Petersburg.
Mem. CPSU 1941–; grad. Leningrad Polytech. Inst. 1930; mem. USSR Acad. of Sciences 1966–; mem. staff Seismological Inst. of USSR Acad. of Sciences 1932–41; mem. staff Dept. of Special Projects of Presidium USSR Acad. of Sciences 1941–44; Dep. Dir. of Inst. of Chemical Physics of USSR Acad. of Sciences 1946–63; Dir. of Inst. of Physics of the Earth of USSR Acad. of Sciences 1963–. Decors.: State Prize of the USSR 1948, 49, 51 and 1953, Lenin Prize 1962, 4 Orders of Lenin, Order of the October Revolution, 2 Orders of Red Banner of Labour, medals.

Sadovskii, Vissarion Dmitrievich
Metallurgist; b. Jl.24 (Ag.6), 1908, Omsk.
Grad. Kazan Univ. 1930; mem. USSR Acad. of Sciences 1970–; mem. staff Precision Tool Works in Zlatoust 1930–35; mem. staff Uralsk branch of USSR Acad. of Sciences in Sverdlovsk 1935–38, at Inst. of Physics of Metals of USSR Acad. of Sciences 1958–; concurrently Lecturer at Kirov Polytech. Inst. in the Urals 1944–59; full Prof. at same 1946–. Decor.: Order of Red Star.
Publ.: "Strukturnaia nasledtvennost' stali" 1973.

Sadri, Mukhamed (real name: Sadretdinov, Mukhamed Khabibullovich)
Author and poet; b. My.7 (20), 1913, Kazan; s. of a worker.
Tartar; mem. CPSU 1938–; studied Lunacharskii Theatre Inst. 1932–36; served in Sov. Army 1941–45; after the war Acting Sec. of Tartar Writers' Union.
Publs.: "Za Sovetskuiu Rodinu!" (For the Soviet Fatherland!) 1939, "Pesnia o Kommunarakh" (Song about the Communards) 1940, "Poi, solovei" (Sing, Nightingale) 1963, "Moi starshie bratiia i sestry" (My Older Brothers and Sisters) 1960, and others.

Sadybekov, Tugelbai
Author; b. My.1 (14), 1912, *aul* (village) Ken-Su, Kirghiz SSR.
Kirghiz; mem. Acad. of Sciences Kirghiz SSR 1954–; first works publ. 1928. Decors.: State Prize of the USSR 1949, Order of Lenin, Order of Red Banner of Labour, various medals.
Publs.: "Poet – solovei" (Poet – Nightingale) 1938, novels "Ken-su" vols. 1–2,

1937–38, "Temïr" 1939–40, "Liudi nashikh dnei" (People of Our Days) 1948, "Deti gor" (Children of the Mountains) 1953, "Sredi gor" (In the Middle of the Mountains) vols. 1–2, 1955–58, "Zhenshchiny" (Women) 1966 (Russian translation 1972).

Sadykov, Abid Sadykovich
Organic chemist; b. N. 2 (15), 1913, Tashkent.
Uzbek; mem. CPSU 1946–; grad. Cen. Asian Univ. in Tashkent 1937; Laboratory Technician at Inst. of Chemistry of Cen. Asian Univ. 1933–37; teacher at Tashkent Textile Inst. 1937–39; teacher at Uzbek Univ. 1939–41; mem. staff 1941–, Head of Chair 1956–, Rector of Cen. Asian Univ. 1958–66; concurrently Dir. of Inst. of Chemistry of Acad. of Sciences Uzbek SSR 1946–50; Head of Dept. of Biochemistry of Acad. of Sciences Uzbek SSR 1973–; cand. mem. 1961–66, mem. Cen. Cttee. Uzbek CP 1966–; Dep. to 1966–, mem. Comm. on Educ., Science and Culture of Sov. of Nationalities 1966–70; mem. Draft Bills Comm. Sov. of the Union 1970–74, mem. Draft Bills Comm. Sov. of Nationalities USSR Supr. Sov. 1974–; Pres. of Acad. of Sciences Uzbek SSR 1966–; mem. USSR Acad. of Sciences 1972–. Decors.: Hero of Socialist Labour 1973, 4 Orders of Lenin, 3 other orders and various medals.

Sadykov, Erk Abdulkarimovich
State official.
Chair. of State Cttee. for the Supply of Petroleum Products Kirghiz SSR D. 1981–.

Sadykov, Ramiz Dzhamil ogly
State official.
Cand. mem. Cen. Cttee. Azerbaidzhan CP 1976–; Head of Dept. for Building and Municipal Econ. with Cen. Cttee. Azerbaidzhan CP 1976–81; Min. of Building Materials Ind. Azerbaidzhan SSR F. 1981–.

Safaraliev, Tofik Shamil ogly
State official.
Min. of Ind. Construction Azerbaidzhan SSR Jl. 1979–; mem. Cen. Cttee. Azerbaidzhan CP 1981–.

Safarian, Stepan Rubenovich
State official.
Min. of Finance Armenian SSR 1969–75; Mem. Cen. Cttee. Armenian CP 1971–; Min. of Trade Armenian SSR Jl. 1975–.

Safarov, Nazir
Author; b. Ja. 9, 1905, Dzhizak; s. of a smith.
Uzbek; mem. CPSU 1929–; grad. Cen. Party School in Samarkand 1929; first works publ. 1926. Decor.: Order of Friendship Between Nations.
Publs.: "Borba za novuiu zhizn'" (Fight For a New Life) 1930, "Teni chernykh dnei" (Shadows of Black Days) 1931, "Shkola Zhizni" (The School of Life) 1954, "Buria nad Aziei" (Storm Over Asia) 1967, and others.

Safronov, Ilia Ivanovich
Diplomat.
USSR Amb. to Singapore 1968–71; mem. Apparat USSR Min. of Foreign Affairs 1971–76; USSR Amb. to Mauritius Ja. 1976–81.

Sagdeev, Roald Zinnurovich
Physicist; b. D. 26, 1932, Moscow.
Grad. Moscow State Univ. 1955; mem. USSR Acad. of Sciences 1968; mem. staff Inst. of Atomic Energy 1956–61, Head of Laboratory of Inst. of Nuclear Physics of Siberian branch of USSR Acad. of Sciences 1961–70; Head of Laboratory of Inst. of Physics of High Temperatures of USSR Acad. of Sciences 1970–73; Dir. of Inst. of Cosmic Exploration 1973–. Decors.: Order of the October Revolution, Order of Red Banner of Labour.
Publs.: "Nelineinaia teoriia plazmy" (publ. in collection "Voprosy teorii plazmy" vol. 7/1972), "Neoklassicheskaia teoriia protsessov perenosa" (ibid.).

Sagdiev, Makhtai Ramazanovich
State official.
Mem. Cen. Cttee. Kazakh CP 1966–76; Min. of the Fish Ind. Kazakh SSR Jl. 1980–.

Sagoian, Grant Samsonovich
State official.
Head of Dept. for Ind. and Transport with Cen. Cttee. Armenian CP 1976–77; mem. Cen. Cttee. Armenian CP 1976–; Dep. Chair. Council of Mins. and Chair. of GOSPLAN Armenian SSR Jl. 1977–.

Saidov, Shody
State official.
Acting Sec. and Chair. of Tadzhik Journalists' Union 1966–73; Chief Ed. of newspaper "Todzhikistoni Soveti" 1968–73; Mem. Cen. Cttee. Tadzhik CP 1971–; Chair. of State Film Cttee. Tadzhik SSR N. 1973–.

Saifiddinov, Sharofiddin Sanginovich
Composer; b. Ag. 24, 1929, Khodzhent, now Leninabad.
Member of the CPSU since 1962.
He graduated in 1952 from V. G. Fere's class in the national division of the Moscow Conservatory, where he was also a student in the department of theory and composition. In 1962 he graduated from B. I. Zeidman's composition class at the Tashkent Conservatory. From 1960 to 1962 he served as director of the Dushanbe School of Music.
He has composed the operas *Pulat and Gul'ru* (1957, S. Aini Tadzhik Theater of Opera and Ballet) and *Rudaki* (1974, S. Aini Tadzhik Theater of Opera and Ballet), the musical comedy *Two Beauties and Two Loves* (1957, Leninabad Musical Dramatic Theater), the cantata for soprano, baritone, mixed chorus, and symphony orchestra *Flourish, Tadzhikistan!* (1954), a cantata for chorus and symphony orchestra (1964), the symphonic poem *The Golden Kishlak* (1962), whose music served as the basis for the ballet film *Mountain Voices* (1965), and a string quartet (1955). He also wrote music for motion pictures, including *The Man Changes His Skin* (1960, together with A. I. Pirumov), *Zumrad* (1962), and *Fire Under the Ashes* (1967), art and popular songs to texts by Tadzhik poets, and adaptations of Tadzhik folk songs.
In 1962, he became first secretary of the directorship of the Composers' Union of Tadzhikistan. In 1970 he received the A. Rudaki State Prize of the Tadzhik SSR for his tone poems for voice and symphony orchestra *The Road to the Mausoleum* and *Caution: Stalingrad!*, as well as for the songs *Ode to the Party* and *My Republic*. Decors.: two orders.

Sakalauskas, Romualdas Mikolovich
State official.
Cand. mem. 1961–64, mem. Cen. Cttee. Lithuanian CP 1964–76; Min. of Construction Lithuanian SSR Ja. 1962–78; 1st Dep. USSR Min. of Construction 1978–82; Chair. State Cttee. of Construction Lithuanian SSR S. 1982–.

Sakalauskas, Vitautas Vladovich
Party official; b. 1933.
Lithuanian; mem. CPSU 1960–; grad. Polytechnical Inst. in Kaunas; senior master, head of a section, chief engineer, Dir. of the "Priekalas" enterprise in Kaunas 1957–65; dep. head, head of a dept. Cen. Cttee. Lithuanian CP 1968–69; Chair. of Exec. Cttee. of Vilnius City Sov. of People's Deps. 1969–74; mem. Cen. Cttee. Lithuanian CP 1971–; 1st City Sec. Lithuanian CP in Vilnius F. 1974–; cand. mem. Bureau of Cen. Cttee. Lithuanian CP 1974–76, mem. Ja. 1976–; Dep. to USSR Supreme Sov. 1974–.

Sakhan, I. Ia.
Komsomol official.
Sec. and mem. Bureau Cen. Cttee. Ukrainian Komsomol Je. 1980–.

Sakharov, Andrei Dimitrievich, Dr. of Phy. Sc.
Physicist; b. My. 21, 1921, Moscow.
Russian; grad. Moscow Univ. 1942; Dr. of Phy. Sc. 1953–; mem. USSR Acad. of Sciences; Prof. mem. staff Lebedev Inst. of Physics of USSR Acad. of Sciences 1945–; as nuclear physicist participated in development of Soviet hydrogen bomb (is known as "Father of the Hydrogen Bomb"); his most important works were classified 'top-secret' and not publ.; articles dealing with the use of nuclear energy for peaceful purposes began appearing in journal "Atomnaja energija" 1958–; an opponent of atomic armament, he demanded a ban on nuclear weapons; because of his convictions he was re-

leased from all his duties and now lives off his pension as mem. of the USSR Acad. of Sciences; publ. his "Thoughts on Progress, Peaceful Coexistence and Intellectual Freedom" in 1968; founded the "Committee on Human Rights in the USSR" in Moscow in 1970; author of numerous appeals, open letters and articles. Decors.: Stalin Prize, 3 Orders of Socialist Labour, Nobel Peace Prize 1975.

Sakh'ianova, Larisa Petrovna
Ballerina; b. F.13, 1930, Kyren, now in Kyren Raion, Buriat ASSR.
People's Artist of the USSR (1963).
She studied at the Moscow Choreographic School from 1937 to 1942 and in 1945 and 1946. In 1946 she became a soloist with the Buriat Theatre of Opera and Ballet in Ulan-Ude. She has danced leading roles in many modern and classical ballets; these roles have included Siuimbike in Iarullin's *Shurale,* Seseg in Riauzov's *Light Above the Valley,* the title role in Adam's *Giselle,* and Odette-Odile in Tchaikovsky's *Swan Lake.* For her performance as Angara in Knipper and Lampilov's *Beautiful Angara,* she was awarded the Glinka State Prize of the RSFSR in 1972.
In 1966, she started teaching and directing the Choreographic School in Ulan-Ude. She has toured many countries. She was a deputy to the seventh convocation of the Supreme Soviet of the USSR. Decor.: Order of the Red Banner of Labour.

Sakhnin, Arkadii Iakovl'evich
Author; b. S.22, 1910, Lugansk, now Voroshilovgrad.
Russian; mem. CPSU 1939–; grad. Inst. of Foreign Trade 1936; first works publ. 1936.
Publs.: "Tuchi na rassvete" (Clouds at Dawn) 1954, "Poedinok" (The Duel) 1959, "Tri minuty do katastrofy" (Three Minutes Before the Catastrophe) 1961, "Chuzhie liudi" (Strange People) 1961, "Sudby liudskie" (Human Fates) 1964, and others.

Sakhniuk, Ivan Ivanovich
Party official; b. 1927.
Russian; mem. CPSU 1960–; grad. Inst. of Motor Transport and Highways in Kharkov 1958; served in the Soviet Army 1944–52; held different posts in the "Serp i Molot" Factory in Kharkov 1956–71; 2nd Sec. Kharkov City Cttee. Ukrainian CP 1972–74, 1st Sec. 1974–76; mem. Cen. Cttee. Ukrainian CP 1976–81; 1st Sec. Kharkov Oblast Cttee. Ukrainian CP 1976–80; mem. Cen. Cttee. CPSU 1976–; Dep. to USSR Supr. Sov. 1979–; Head, Dept. for Agricultural Machine-Building with Cen. Cttee. CPSU 1980–.

Saks, Vladimir Nikolaevich
Geologist and paleontologist; b. Ap.9 (22), 1911, St. Petersburg.
Corresponding member of the Academy of Sciences of the USSR since 1958. Member of the CPSU since 1955.
He graduated from the Leningrad Institute of Mines in 1933. From 1935 to 1940 and from 1944 to 1948 he worked at the Arctic Institute, and from 1940 to 1944 he worked at the Mining Geology Administration of the Main Northern Sea Route Administration. Since 1948 he has been working at the Institute of Geology of the Arctic. In 1958 he became head of the laboratory of Mesozoic and Cenozoic stratigraphy and paleontology of the Institute of Geology and Geophysics of the Siberian Division of the Academy of Sciences of the USSR. He worked out detailed plans of the stratigraphy of the Anthropogenic beds of Belorussia and the Soviet arctic and of the Mesozoic gas- and oil-bearing strata of the Ust'-Enisei and Khatanga depressions. Decors.: three Orders of the Red Banner of Labour, and also various medals.
Works: *Chetvertichnyi period v Sovetskoi Arktike,* 2nd ed. Leningrad-Moscow 1953. (*Trudy In-ta geologii Arktiki,* vol. 77.) *Stratigrafiia uirskoi i melovoi sistem Severa SSSR.* Moscow-Leningrad 1963. (Co-author.)

Sakse, Anna
Author; b. Ja.16, 1905, Latvia; d. of a peasant.
Latvian; studied Dept. of Philology at Latvian Univ.; worked for newspaper "Tsinia" (Cina) in Kirov 1942–44; mem. Sov. Peace Cttee. 1950–58. Decors.: State Prize of the Latvian SSR 1959, State Prize of the USSR 1949, People's Writer of the Latvian SSR.
Publs.: "V goru" (Uphill) 1948, "Iskry v nochi" (Sparks in the Night) vols. 1–2, 1951–52, "Vozvrashchenie v zhizn'" (Return to Life) 1945, "Skazki o tsvetakh" (Tales About Flowers) 1966, and others.

Salakhov, Tair Teimur ogly
Painter and graphic artist; b. N.29, 1928, Baku.
People's Artist of the USSR (1973); corresponding member of the Academy of Arts of the USSR since 1966. Member of the CPSU since 1964.
He studied at the A. Azimzade Azerbaidzhan Art School in Baku from 1945 to 1950 and under P. D. Pokarzhevskii at the V. I. Surikov Moscow Art Institute from 1951 to 1957. He taught at the M. A. Aliev Azerbaidzhan Institute of the Arts in Baku between 1963 and 1974 and became a professor there in 1973. Since 1975 he has been a professor at the V. I. Surikov Moscow Art Institute.
His genre compositions, noted for their austere, bold imagery, are based on the life of Baku oil workers. He has also painted portraits, as well as subtly expressive still lifes and industrial landscapes. His works are characterized by monumental, graphically precise forms, strict, rhythmic composition, and tense, sometimes strongly contrasting colors. In 1968, he was awarded the State Prize of the USSR for three works: a portrait of the composer Kara Karaev (1960, Tret'iakov Gallery), *Repair Workers* (1961, R. Mustafaev Art Museum, Baku), and *By the Caspian Sea* (1967, property of the Artists' Union of the USSR, Moscow). His other works include *A New Sea* (1970, property of the Ministry of Culture of the Azerbaidzhan SSR, Baku; State Prize of the Azerbaidzhan SSR, 1970), a series of graphic works entitled *Around Apsheron* (1969, property of the Artists' Union of the USSR), and the stage set for the production of Shakespeare's *Antony and Cleopatra* at the M. Azizbekov Academic Dramatic Theater in Baku (1964; M. F. Akhundov Prize of the Azerbaidzhan SSR, 1965).
He was a deputy to the seventh and eighth convocations of the Supreme Soviet of the USSR and a delegate to the Twenty-third and Twenty-fourth Congresses of the CPSU. In 1973 he became first secretary of the board of the USSR Artists' Union. Decors.: Order of the Red Banner of Labour and various medals.

Salieva, Batysh
State official; b. 1922.
Kirghiz; mem. CPSU 1944–; grad. Party High School of Cen. Cttee. CPSU; pioneer leader, teacher, sec. of a raion cttee. of Kirghiz Komsomol 1938–44; dept. head 1944–51, head of a party bureau, sec. of a raion cttee. of Kirghiz CP 1944–51; instructor for a dept. of Cen. Cttee. CPSU 1952–53; Min. of Public Welfare Kirghiz SSR 1953–66; Dep. to Supr. Sov. Kirghiz SSR 1953–66; mem. Cen. Cttee. Kirghiz CP 1956–81; Chair. of Kirghiz Republican Trade Union Council 1966–73; Dep. to and Chair. of Comm. for Health and Public Welfare of Sov. of Nationalities USSR Supr. Sov. 1966–74; again Min. of Public Welfare Kirghiz SSR D. 1973–.

Salimov, Akil Umurzakovich
Party official.
Head of Dept. of Science and Educ. Institutions with Cen. Cttee. Uzbek CP 1968; Sec. and mem. Bureau 1970–, mem. Cen. Cttee. Uzbek CP 1971–.

Salishchev, Konstantin Alekseevich
Geographer and cartographer; b. N.7 (20), 1905, Tula.
Doctor of Technical Sciences (1940); Honored Scientist of the RSFSR (1966). Member of the CPSU since 1944. Professor at the Moscow Engineering Institute of Geodesy, Aerial Photographic Surveying and Cartography since 1942 and head of the subdepartment of geodesy and cartography of the department of geography at Moscow-State University since 1950. In 1926 and 1929–30, he was a geodesist in S. V. Obruchev's expeditions to the basins of the Indigirka and Kolyma rivers. His main works have dealt with the theory of cartography, map science, the his-

tory of cartography, and problems of comprehensive mapping. He has been vice-president of the Geographical Society of the USSR since 1964 and was president from 1968 to 1972 and vice-president from 1964 to 1968 and since 1972 of the International Cartographic Association. He was chairman of the commission on national and regional atlases of the International Geographical Union from 1956 to 1972. Decors.: honorary doctorate in natural sciences by the Humboldt University of Berlin, four Orders of the Red Banner of Labour, various medals. Works: *Osnovy kartovedeniia,* 3rd ed., vols. 1–2. Moscow 1959–62. *Sostavlenie i redaktirovanie kart,* part 1. Moscow 1947. *Kartografiia,* 2nd ed. Moscow 1971. *Atlas nationaux: Histoire, analyse, voies de perfectionnement et d'unification.* Moscow-Leningrad 1960. *Regional Atlases: Tendencies of Development. Subject Matters of the Maps on Natural Conditions and Resources.* Moscow-Leningrad 1964. *Einführung in die Kartographie,* vols. 1–2. Gotha-Leipzig 1967.

Salmanov, Grigorii Ivanovich
Military official; b. 1922.
Russian; mem. CPSU 1944–; grad. Higher Mil. Acad. 1949 and Gen.-Staff Mil. Acad. 1964; Army General; served in Sov. Army 1940–; platoon leader, company commander, chief-of-staff, dep. commander of a battalion, dep. dept. head of staff of a corps on Western, Cen., 1st and 2nd Belorussian Front, dep. head of dept. of staff of an army 1941–45; commander of a battalion, chief-of-staff of a regt., commander of various regts., chief-of-staff of a mechanized div., commander of a rifle div., 1st dep. chief-of-staff of a mil. district, commander of a unit 1949–69; Commander of Kiev Mil. District 1969–75, Transbaikal 1978–; Dep. Comm.-in-Chief USSR Ground Forces 1975–78; Dep. to USSR Supr. Sov. 1970–; mem. Cen. Cttee. Ukrainian CP 1971–76; cand. mem. Cen. Cttee. CPSU 1971–76; mem. 1981–. Decor.: Order of Red Star.

Saltykov, Aleksei Aleksandrovich
Motion picture director; b. My. 13, 1934, Moscow.
Honored Art Worker of the RSFSR (1964).
In 1959, he graduated from the department of direction of the All-Union State Institute of Cinematography. His first films included *Kids From Our Yard* (1959), *My Friend, Kol'ka!* (1961, with A. N. Mitta), and *Beat the Drum!* (1962). He then addressed himself to the fates of people and to real-life situations characterized by intense drama and openly expressed social conflicts. His major film is *Chairman* (1965), whose central character is a contemporary positive hero who is a Communist and a kolkhoz chairman.
The Russian countryside during the fascist occupation and the postwar restoration is depicted in his films *Women's Kingdom* (1968) and *A Point of No Return* (1974). These films, as well as the film *The Siberian Woman* (1972), which is about the secretary of a party raion committee, are distinguished by an interesting treatment of the female character. In 1970, he directed the film *Director,* which re-created the authentic biography of one of the builders of Soviet industry. In 1971, he directed *There Was Evening, and There Was Morning . . .* which is about the revolutionary seamen of the Baltic Fleet.

Salynskii, Afanasii Dmitrievich
Dramatist; b. S. 9, 1920, Smolensk.
Mem. CPSU 1942–; journalist 1939.
Publs.: "Bratia" (The Brothers) 1947, "Zabytyi drug" (Forgotten Friend) 1955, plays "Khleb i rozy" (Bread and Roses) 1957, "Kameshki na ladoni" (Pebbles in the Palm) 1965, "Muzhskie besedy" (Male Conversations) 1967, "Maria" 1969; film scripts "Chelovek s budushchim" (Man with a Future) 1961, "Vzorvannyi ad" (Blasted Hell) 1967, and others.

Samarskii, Aleksandr Andreevich
Mathematician; b. F. 19, 1919, Amvrosievka, now in Donetsk Oblast.
Corresponding member of the Academy of Sciences of the USSR (1966). Member of the CPSU since 1946.
He graduated from Moscow State University in 1945 and became a professor there in 1959. His principal works deal with mathematical physics and computer mathematics. Samarskii has developed a general theory of finite-difference schemes of operator and finite-difference-operator equations and a stability theory of finite-difference schemes. His work has been made use of in computational algorithms that have been successfully applied to the computer solution of important problems in various fields of science and engineering. Decors.: Lenin Prize in 1962 and the State Prize of the USSR in 1954. He has been awarded two Orders of Lenin, the Order of the October Revolution, the Order of the Red Banner of Labour, and various medals.
Works: *Uravneniia matematicheskoi fiziki,* 4th ed. Moscow 1972. (With A. N. Tikhonov.) *Vvedenie v teoriiu raznostnykh skhem.* Moscow 1972. *Ustoichivost' raznostnykh skhem.* Moscow 1973. (With A. V. Gulin.)

Samilyk, M. G.
Party official.
Sec. of Dnepropetrovsk Oblast Cttee. Ukrainian CP 1974–82; 1st Sec. of Kirovograd Oblast Cttee. Ukrainian CP F. 1982–.

Samodur, Semen Solomonovich
Puppet-theatre actor and director; b. Ja. 4 (17), 1911, Bui, now in Kostroma Oblast.
People's Artist of the RSFSR (1966).
In 1930, he began studying at the studio of P. Ia. Pavlinov. From 1931 to 1933 he studied at a representational arts technicum of the State Publishing House of Representational Arts. In 1933 he became an actor at the Central Puppet Theatre, and in 1947 the theatre's director.
He has worked with S. V. Obraztsov in staging Marshak's *Cat's House,* Tuzlukov's *Mine, Only Mine,* and Speranskii's *The Unusual Concert.* He has independently staged Matveev's *The Magic Galosh,* as well as *The Tale of Tsar Saltan,* based on Pushkin's work. His roles have included the Master of Ceremonies in *The Unusual Concert,* the Creator in Shtok's *Divine Comedy,* and the Village Master and the Devil in Speranskii's *Christmas Eve* (based on Gogol's work).

Samoilov, Boris Nikolaevich, Dr. of Phy.-Math. Sc.
Physicist; b. 1920.
Mem. CPSU 1952–; grad. Moscow Univ. 1947; Cand. of Phy.-Math. Sc. 1952, Dr. of Phy.-Math. Sc. 1964; full Prof. 1967–; mem. staff Inst. of Problems in Physics of USSR Acad. of Sciences 1947–55; mem. staff Kurchatov Inst. of Atomic Energy of USSR Acad. of Sciences 1955–, section head 1964–, concurrently dep. dept. head and Prof. at Physical-Tech. Inst. in Moscow 1969–.

Samoilov, David Samuilovich (real name: Kaufman)
Author and poet; b. Je. 1, 1920, Moscow.
Studied Moscow Inst. of Philosophy, Literature and History 1938–41; mil. service at front 1941–45; first works publ. 1941. Decors.: Order of Red Star, medals.
Publs.: collections "Blizhnie strany" (Close Countries) 1958, "Vtoroi pereval" (The Second Mountain Pass) 1963, "Dni" (Days) 1970, "Volna i kamen'" (Wave and Stone) 1974, and others.

Samoilov, Evgenii Valerianovich
Theatre and film actor; b. 1912.
Mem. CPSU 1951–; grad. Leningrad Polytech. Art School 1930; actor with V. E. Meierhold Theatre 1934–38, at Theatre of the Revolution (now V. Maiakovskii Theatre) 1940–67 and at Malyi Theatre in Moscow 1968–. Decors.: People's Artist of the USSR; 1974, State Prize of the USSR 1941, 1946, 1947.

Samoilova, Tat'iana Evgen'evna
Film actress; b. My. 4, 1934, Leningrad.
Honored Art Worker of the RSFSR (1963).
She graduated from the Shchukin Theatrical School in 1958. She began acting in motion pictures in 1956. She became famous for her roles as Veronika in the film *The Cranes Are Flying* (1957), for which she was awarded a prize at the 11th International Film Festival in Cannes. Another significant achievement was her portrayal of Anna Karenina in the film of the same name (1968).

Her other roles have included Tania in *The Letter That Was Never Sent* (1960), the Soviet Intelligence Agent in *Alba Regia* (1961, a Hungarian Film), Natasha in *Léon Garros Seeks a Friend* (1961, a Soviet-French film), Sonia in *They Went to the East* (1964, an Italian-Soviet film), Nastiura Shevtsova in *No Return,* and Masha in *Ocean.* Decor.: Order of the Badge of Honor.

Samsonov, Aleksandr Mikhailovich
Historian; b. D.26, 1907 (Ja.8, 1908), Nizhni Novgorod, now Gorkii.
Mem. CPSU 1943–; grad. Dept. of History at Leningrad State Univ. 1931; corresponding mem. USSR Acad. of Sciences 1964–81, mem. 1981–; after grad. lecturer at various colleges and mem. staff Museum of Religion and Atheism of USSR Acad. of Sciences 1948–61; Dir. of Publ. House of USSR Acad. of Sciences (now "Nauka") 1961–70; Sen. Scientist at Inst. of History of USSR Acad. of Sciences 1970–; Chief Ed. of "Istoricheskie zapiski"; wrote basic works on the history of the Great Patriotic War 1941–45. Decors.: 3 orders and various medals.

Samsonov, Fedor Aleksandrovich
Military official; b. N.8 (21), 1901, Zelenovka, now in Serdobsk Raion, Penza Oblast.
Colonel-general of the artillery (1944). Member of the CPSU since 1921.
He joined the Red Army in 1920 and graduated from the M. V. Frunze Military Academy in 1934 and from the Military Academy of the General Staff in 1941. He saw service in the Great Patriotic War as chief-of-staff of the artillery on the Briansk Front and chief of the artillery of the Twenty-eighth Army from August to December 1941. In December 1941 he became artillery chief-of-staff under the artillery commander of the Red Army.
He became chief of the artillery subdepartment of the K. E. Voroshilov Higher Military Academy in April 1946. In 1948 he was appointed academician secretary of the division of artillery warfare at the Academy of Artillery Science. He was a member of the academy's presidium from 1950 to 1953. From 1953 to 1964, he was head of the editorial board for the publication of the multivolume *Artillery and Artillery Supply in the Patriotic War.* He retired in 1964.
Decors.: Order of Lenin, three Orders of the Red Banner, the Order of Suvorov First and Second Class, the Order of Kutuzov First Class, the Order of the Patriotic War First Class, the Order of the Red Star, various medals, and an order of the Polish People's Republic.

Samsonov, Grigorii Valentinovich, Dr. of Tech. Sc.
Metallurgical scientist; b. F.15, 1918, Detskoe Selo, now Pushkin; s. of a clerk.
Mem. CPSU 1942–; grad. Inst. of Chemical Precision Technology in Moscow 1940; specializes in metallurgy, powder metallurgy, crystal chemistry and technology of rare metals and viscous compounds; corresponding mem. Acad. of Sciences Ukrainian SSR 1961–; Dr. of Tech. Sc. 1958; full Prof. 1959–; engineer in an inst. 1940–41; served in Sov. Army 1941–45; mem. staff Moscow Inst. of Nonferrous Metals and Gold 1947–56; Dep. Dir. of Inst. of Metal Ceramics and Special Alloys (now Inst. of Material Science) at Acad. of Sciences Ukrainian SSR 1956–61; Head of Dept. of Tech. Sciences of Acad. of Sciences Ukrainian SSR 1961–63; Head of Chair of Powder Metallurgy and Rare Metals at Kiev Polytech. Inst. 1962–; author of some 600 papers. Decors.: Order of Red Banner of Labour, various medals.
Publs.: "Tverdye soedineniia tugoplavkikh metallov" (co-author) 1957, "Silitsidy i ikh ispolzovanie v tekhnike" 1959, "Tugoplavkie soedineniia" 1963, "Berillidy" 1966, and others.

Samsonov, Rafael Mikhailovich
State official.
Cand. mem. Cen. Cttee. Armenian CP 1966–; Chair. of State Cttee. on Films Armenian SSR Jl. 1979–.

Samsonov, Semen Aleksandrovich
Author and journalist; b. S.1, 1931, Syrezshur, Udmurt ASSR.
Udmurt; mem. CPSU 1960–; grad. Udmurt Pedagogical Inst. 1956, concluded Advanced Courses in Literature in Moscow; with newspaper "Sovetskaia Udmurtiia" 1952–63; Ed. of journal "Molot" 1963–.
Publs.: "Doroga vpered" (The Way Ahead) 1959, "Cheremukha tsvetet" (Alderberry in Bloom) 1961, "Navstrechu vetru" (Facing the Wind) 1961, "Spasibo tebe, chelovek!" (Thank You, Man!) 1963, "Liubliu tebia" (I Love You) 1965, "Groza nad Kamoi" (Thunderstorm Over the Kama) 1969, and others.

Sanaev, Vsevolod Vasil'evich
Actor; b. 1912.
Mem. CPSU 1955–; grad. Moscow A. V. Lunacharski Inst. of Theatre Arts (FITIS) 1937; mem. Moscow Acad. M. Gorkii Arts Theatre of the Sov. Union (MHAT), Mossoviet Theatre and Film Actors' Theatre Studio. Decor.: People's Artist of the USSR 1969.

Sanakoev, Feliks Sergeevich
Party official; b. 1939.
Osset; mem. CPSU 1964–; grad. S. Ossetian Pedagogical Inst.; Raion Cttee. instructor, Oblast Cttee. dept. chief Georgian Komsomol, instructor at Cen. Cttee. Georgian Komsomol, 1st Oblast Cttee. Sec. Georgian Komsomol in S. Ossetia 1962–72; dept. supervisor Cen. Cttee. Georgian CP, 1st Dep. Chair., Chair. of Oblast Exec. Cttee. of People's Deps. in S. Ossetia 1972–73; 1st Oblast Cttee. Sec. Georgian CP in S. Ossetia 1973–; Dep. to USSR Supr. Sov. 1974–; mem. Cen. Cttee. 1976–, cand. mem. Bureau Cen. Cttee. Georgian CP 1979–.

Sandalov, Leonid Mikhailovich
Military official; b. Mr. 28 (Ap. 10), 1900, Vichuga, now in Ivanovo Oblast; s. of a worker.
Colonel-general (1944). Member of the CPSU since 1925.
He joined the Red Army in 1919. He graduated from training courses for the command staff in 1920. He was a platoon commander in the Civil War of 1918–20. He graduated from the Kiev Combined Military School in 1926, from the M. V. Frunze Military Academy in 1934, and from the Military Academy of the General Staff in 1938. He served in the Great Patriotic War as chief-of-staff of the Fourth Army of the Western Front (June–July 1941), chief of staff of the Central Front (August 1941), deputy chief of staff and chief of staff of the Briansk Front (September–November 1941), chief of staff of the Twentieth Army of the Western Front (December 1941–September 1942), chief-of-staff of the Briansk Front (September 1942–October 1943), chief-of-staff of the Second Baltic Front (October 1943–March 1945), and chief-of-staff of the Fourth Ukrainian Front (from April 1945).
After the war, he served as chief-of-staff of the Ciscarpathian Military District in 1945–46, deputy chief of the Main Headquarters of the Ground Forces in 1946–47, and simultaneously chief-of-staff and first deputy commander of the forces of the Moscow Military District from 1947 to 1953. He served at the disposal of the minister of defense of the USSR from 1953 to 1955 and went into the reserve in 1955 for reasons of health.
Decors.: three Orders of Lenin, four Orders of the Red Banner, the Order of Suvorov First Class, the Order of Kutuzov First Class, the Order of the Red Star, various medals, and three foreign orders.
Works: *Trudnye rubezhi.* Moscow 1965. *Perezhytoe.* Moscow 1966. *Na moskovskom napravlenii.* Moscow 1970.

Saneev, Viktor Danilovich
Track-and-field athlete and instructor; b. O. 3, 1945, Sukhumi.
Honored Master of Sports (1968).
He graduated from the Georgian Institute of Subtropical Farming in 1969 and the Georgian Institute of Physical Education in 1975. He was Olympic champion in the triple jump in 1968, 1972, and 1976. He was champion in the same category at the European track and field championships in 1969 and 1974 and at the European indoor track-and-field championship in 1970, 1971, 1972, and 1975. He was also triple-jump champion of the USSR seven times during the period 1968–75. Between 1968 and 1972 he broke the world's record three times (17 m 47 cm in 1972). Decors.: Order of

Lenin and the Order of the Red Banner of Labour.

Sangi, Vladimir Mikhailovich
Author; b. Mr. 18, 1935, Nabil, Sakhalin Oblast.
Mem. CPSU 1966–; grad. Leningrad Herzen Inst. of Pedagogics 1959 and Advanced Courses in Literature in Moscow 1965.
Publs.: "Solenye bryzgi" (Salty Spray) 1962, "Pervyi vystrel" (The First Shot) 1965, "Oktiabr'" (October 1965, "Druzhba narodov" (Friendship Between Nations) 1967, "Izgin" 1965, and others.

Sapozhnikov, Vladimir Konstantinovich
Author; b. My. 9, 1922, Klochki, Altai Krai; s. of a peasant.
Russian; studied Dept. of Linguistics and Literature at Tiumen Pedagogical Inst.; worked as journalist and teacher.
Publs.: "Belye koni" (White Horses) 1960, "Na fronte zatishe" (Quiet on the Front) 1967, and others.

Sarafanov, Gennadii Vasil'evich
Cosmonaut; b. 1942.
Mem. CPSU 1963–; Lieut.-Gen.; grad. Higher Mil. Aviation College, then served in Air Force in Balashov; cosmonaut 1965–; commander of spaceship "Soiuz 15" in which he (together with L. S. Demin) conducted a series of scientific-tech. experiments under various flight conditions and docking manoeuvre with spaceship "Saliut 3" during orbital flight Ag. 26–28, 1974. Decor.: Order of Lenin.

Sarbanov, Talgarbek
State official; b. 1927.
1st Sec. of Osh Oblast Cttee. Kirghiz Komsomol 1955; Head of Dept. of Propaganda and Agitation with Osh Oblast Cttee. Kirghiz CP 1965–73; Head of Dept. of Propaganda and Agitation with Cen. Cttee. Kirghiz CP 1973–80; mem. Cen. Cttee. Kirghiz CP 1976–; Chair. of State Cttee. of Publishing, Polygraphy and the Book Trade Kirghiz SSR Ja. 1980–.

Sarishvili, David Efremovich
State official.
Chair. of Georgian Republican Asscn. for Agricultural Technology "Gruzselkhoztekhnika" N. 1973–78; Chair. of State Cttee. for the Supply of Production Equipment for Agriculture Georgian SSR 1978–; mem. Auditing Comm. Georgian CP 1976–.

Sarkisian, A. G.
Komsomol official.
Sec. and mem. Bureau Armenian Komsomol Mr. 1978–; mem. Cen. Cttee. Armenian CP 1976–81.

Sarkisian, Fadei Tachatovich
State official; b. 1923.
Mem. CPSU 1945–; grad. Leningrad Mil. Acad. of Communications 1946, Dr. of Tech. Sc. 1975, Academician of Acad. of Sc. Armenian SSR 1977; held engineering posts in USSR Min. of Defense apparatus 1946–63; dir. of a scientific research inst. in Armenian SSR 1963–77; Chair. of Council of Mins. Armenian SSR Ja. 1977–; mem. Politbureau Cen. Cttee. Armenian CP Ap. 1977–; Dep. to USSR Supr. Sov. 1979–; mem. Cen. Auditing Comm. CPSU 1981–. Decor.: USSR State Prize 1971.

Sarkisov, Babken Esaevich
State official; b. 1913.
Armenian; mem. CPSU 1939–; grad. Azerbaidzhan Industrial Inst. 1938 and CPSU Cen. Cttee. Party Higher School 1944; locksmith apprentice, locksmith in various machine shops 1929–34; engineer, dep. Chief mechanic 1938–41 and Party organizer of the Cen. Cttee. CPSU in Baku petroleum processing plant 1941–43; instructor, branch chief Dept. of Propaganda and Agitation Cen. Cttee. CPSU 1944–55; mem. Cen. Cttee. 1956–, Sec. and mem. Bureau Cen. Cttee. Armenian CP 1956–61; Dep. to Supr. Sov. Armenian SSR 1959–; Chair. of State Cttee. on Science and Technology with Council of Mins. Armenian SSR 1961; Chair. of State Cttee. for the Coordination of Scientific Research with Council of Mins. Armenian SSR O. 1961–65; Min. of Motor Transport Armenian SSR 1966–70; Chair. of State Cttee. on Pricing with Council of Mins. Armenian SSR 1970–75; mem. Bureau Cen. Cttee. Armenian CP and Chair. Presidium of Supr. Sov. Armenian SSR Jl. 1975–; mem. Cen. Auditing Comm. CPSU 1976–81, Cand. mem. Cen. Cttee. CPSU 1981–; Dep. Chair. of the Presidium USSR Supr. Sov. O. 1977–.

Sarnov, Benedikt Mikhailovich
Literary historian and critic; b. Ja. 4, 1927, Moscow.
Grad. Gorkii Inst. of Literature 1951; first works publ. 1948; writes mainly on various trends of contemporary Soviet literature, aesthetic educ. of the general readership and children's literature.
Publs.: "Strana nashego detstva" (Land of Our Childhood) 1965, "Samuil Marshak. Ocherk poezii" (Samuil Marshak. Sketches on Poetry) 1968, and others.

Sarri, Gustav Eduardovich
State official.
Cand. mem. 1964–71, mem. Cen. Cttee. Esthonian CP 1971–; Head of Dept. of Science and Culture, resp. Dept. of Science and Educational Establishments with Cen. Cttee. Esthonian CP 1966–74; Min. of Public Welfare Esthonian SSR My. 1974–.

Sartakov, Sergei Venediktovich
Author; b. Mr. 13 (26), 1908, Omsk.
Russian; mem. CPSU 1951–; Sec. of USSR Writers' Union 1967; first works publ. 1934. Decors.: State Prize of the USSR 1970, Order of Lenin, 2 other orders.
Publs.: "Aleksei Khudonogov" 1945, "Sadik pod oknami" (Little Garden in Front of the Windows) 1959, "Khrebty Saianskie" (Saian Mountain Ridge) vols. 1–3, 1940–54, "Ledianoi klad" (The Ice Treasure) 1961, "Filosofskii kamen'" (The Stone of the Wise) vols. 1–2, 1966–71, "A ty gori, zvezda" (Shine, Star) vols. 1–2, 1974–75, trilogy "Gornyi veter" (Mountain Wind) 1957, "Ne otdavai korolevu" (Don't Give Away the Queen) 1960 and "Medlennyi gavot" (Slow Gavotte) 1967 (1st ed. publ. under the title "Kozia morda" - Goat's Mouth, 2nd ed. under the title "Barbinskie povesti" - Barbinsk Stories 1973).

Satarov, Kaldybai
State official.
Min. of the Light Ind. Kirghiz SSR O. 1965–; cand. mem. 1966–71, mem. Cen. Cttee. Kirghiz CP 1971–; Dep. to Supr. Sov. Kirghiz SSR 1967–.

Sats, Aleksander Johannesovitš
Actor and director; b. Ag. 21, 1914, Mariinsk, now in Kemerovo Oblast.
People's Artist of the Esthonian SSR (1964).
He graduated from the Tartu Theatrical Studio in 1937. He worked in the Esthonian Theatre of Drama in Tallinn from 1937 to 1941 and in the Vanemuine Theatre in Tartu from 1942 to 1944. In 1944 he worked as a stage director in the Ugala Theatre in Viljandi. He returned there in 1948 as principal stage director and occupied that position until 1971. He still directs in the Ugala Theatre. He has directed Sobko's *Behind the Second Front* (1951), Särgava's *The Educator* (1951), Priede's *Even Though It Is Autumn* (1957) and *Vika's First Ball* (1960), Heijermans' *The Good Hope* (1959), Liives' *Mistress Mürgi* (1967), Tammsaare's *Judith* (1969), and Gilroy's *Who'll Save the Plowboy?* (1972). Decor.: State Prize of the Esthonian SSR 1951.

Sats, Natalia Ilinichna
Theatre director; b. Ag. 14 (27), 1903, Irkutsk; d. of the composer I. A. Satz.
Grad. A. N. Skriabin Music Tech. College 1917 and Dept. of Theatre Arts of State Lunacharskii Inst. of Theatre Arts (GITIS) in Moscow 1953; Dir. and Artistic Manager of Moscow Children's Theatre 1920–36, of Cen. Children's Theatre 1936–; Dir. of Children's Music Theatre in Moscow 1964–. Decors.: People's Artist of the USSR 1975, State Prize of the USSR 1972, 3 orders and various medals.

Satybaldiev, Aben Sagimbekovich
Author and translator; b. D. 31, 1914, Aulie-Ata, now Dzhambul.
Mem. CPSU 1945–; grad. Dept. of Philology of Kazakh Univ. 1955; first works publ. 1939.
Publs.: "Akmaral", monograph "Istochnik dukhovnogo obogashcheniia" (The Source of Spiritual Enrichment) 1965, and others.

Saul, Bruno Eduardovich
State official.
Min. of Posts and Telecommunications Esthonian SSR 1971–75; cand. mem. 1971–76, mem. Cen. Cttee. Esthonian CP 1976–; Dep. Chair. of Council of Mins. Esthonian SSR Jl. 1975–.

Sauranbekov, T.
Party official.
Sec. and mem. Bureau Cen. Ctte. Kazakh Komsomol F. 1972–78; Sec. of Alma-Ata City Cttee. Kazakh CP 1978–.

Saushkin, Iulian Glebovich
Economic geographer; b. Je. 19 (Jl. 2), 1911, Moscow.
Doctor of geographic sciences (1947) and Honored Scientist of the RSFSR (1972). Member of the CPSU since 1940.
He graduated from Moscow State University in 1932. In 1948 he became a professor and head of the subdepartment of USSR economic geography at the university. His principal works have dealt with the economic geography of the USSR, general theoretical questions of economic geography, economic zoning, and the history of geography.
Works: *Geograficheskie ocherki prirody i sel'skokhoziaistvennoi deiatel'nosti naseleniia v razlichnykh raionakh Sovetskogo Soiuza*. Moscow 1947. *Moskva*, 4th ed. Moscow 1964. *Vvedenie v ekonomicheskuiu geografiiu*, 2nd ed. Moscow 1970. *Kompleksnoe razvitie narodnogo khoziaistva SSSR*. Moscow 1966. *Ekonomicheskaia geografiia Sovetskogo Soiuza*. parts 1–2. Moscow 1967–73. (With others.) *Ekonomicheskaia geografiia: istoriia, teoriia, metody, praktika*. Moscow 1973.

Savarenskii, Evgenii Fedorovich
Geophysicist and seismologist; b. Jl. 5 (18), 1911, Tula; s. of F. P. Savarenskii.
Corresponding member of the Academy of Sciences of the USSR (1966).
In 1932, he graduated from the Moscow Geological Prospecting Institute. From 1931 to 1933 he worked at the Geophysical Institute of the People's Commissariat of Agriculture of the USSR, and in 1934 he began working at the Institute of Seismology. Later, he worked at the O. Iu. Shmidt Institute of Physics of the Earth of the Academy of Sciences of the USSR. In 1950 he became a professor at Moscow University. In 1949 he was appointed head of a network of seismic stations in the USSR.
His research has dealth mainly with seismology, the interpretation of seismic waves, and the occurrence and propagation of seismic sea waves. The *Atlas of Earthquakes in the USSR* (1962) was compiled with him as co-author and executive editor. In 1956, Savarenskii became chairman of the Interdepartmental Council on Seismology and Earthquake-proof Construction of the Presidium of the Academy of Sciences of the USSR. In 1970 he became president of the European Seismological Commission and in 1971 president of the International Commission on Earthquake Prediction. Decors.: Order of Lenin, three other orders, and various medals.
Works: *Elementy seismologii i seismometrii*, 2nd ed. Moscow 1955. (With D. P. Kirnos.) *Seismicheskie volny*. Moscow 1972.

Savchenko, Nikolai Evseevich
State official.
Min. of Health Belorussian SSR Jl. 1966–; mem. Cen. Cttee. Belorussian CP 1971–.

Savenkov, Nikolai Efimovich
Trade union official.
Sec. of Mary Oblast Cttee. Turkmen CP, Head of Dept. for Party Organs with Cen. Cttee. Turkmen CP 1955–61; mem. Cen. Cttee. Turkmen CP 1956–61; Sec. and mem. Presidium of Turkmen Republican Trade Union Council Mr. 1962–.

Savin, Gurii Nikolaevich, Dr. of Phy.-Math. Sc.
Mechanics scientist; b. F. 1, 1907, Vesigonsk, now in Kalinin Oblast; s. of a teacher.
Mem. CPSU 1939–; grad. Inst. of Physics, Chemistry and Mathematics in Dnepropetrovsk 1932; mem. Acad. of Sciences Ukrainian SSR 1948–; Dr. of Phy.-Math. Sc. 1941; Prof. 1941–; Research Asst., Lecturer, Prof. at Inst. of Civil Engineering at Dnepropetrovsk Univ. 1932–41; Dir. of Inst. of Mining Mechanics (now Inst. of Mining Mechanics and Tech. Cybernetics) at Acad. of Sciences Ukrainian SSR in Lvov 1940–45; head of various research insts. in Lvov, Prof. at Lvov Univ. and Polytech. Inst. 1945–48; Rector of Lvov Univ. 1948–52; Vice-Pres. of Acad. of Sciences Ukrainian SSR and concurrently Chair. of Council on Research of Prod. Forces in the Ukrainian SSR of Acad. of Sciences Ukrainian SSR, head of a dept. at Inst. of Mathematics of Acad. of Sciences Ukrainian SSR, Prof. at Kiev Univ. 1952–57; head of a dept. 1957–58 and 1960–, Head of Inst. of Mechanics of Acad. of Sciences Ukrainian SSR 1958–59; author of some 180 papers, among them 10 monographs. Decors.: Merited Scientist of the Ukrainian SSR 1966, Order of Red Banner of Labour, 2 Orders "Badge of Honour", medals.
Publs.: "Dinamicheskaia teoriia rascheta shakhtnykh podemnykh kanatov" 1949, "Kontsentratsiia napriazhenii okolo otverstii" 1951, "Dinamika niti peremennoi dliny" 1962, "Pidiom z velykykh hlybyn" 1966, and others.

Savin, Sergei Nikolaevich
Laborer; b. 1924.
Mem. CPSU 1951–; concluded secondary school in Belozersk 1942; in Sov. Navy 1942–50; radio operator 1950–55; locksmith in a forestry enterprise in Belozersk 1950–55; electromechanic in Voronezh Radio Works 1955–; cand. mem. Cen. Cttee. CPSU 1971–81, mem. 1981–. Decor.: Hero of Socialist Labour.

Savinkin, Nikolai Ivanovich
Party official; b. 1913.
Russian; mem. CPSU 1937–; grad. Tech. College for the Mechanization and Electrification of Agriculture in Kashir (Moscow Oblast) 1932 and Mil.-Pol. Lenin Acad. 1950; Komsomol posts 1934–35; pol. activity in Sov. Army 1935–50; mem. Apparat Cen. Cttee. CPSU 1950–59; Dep. Head 1959–60 and 1st Dep. Head of Dept. of Admin. Organs with Cen. Cttee. CPSU 1960–68; Mem. Cen. Auditing Comm. CPSU 1966–71; Head of Dept. of Admin. Organs with Cen. Cttee. CPSU 1968–; Dep. to USSR Supr. Sov. 1970–; cand. mem. Cen. Cttee. CPSU 1971–81, mem. 1981–. Decor.: Order of Lenin.

Savitakhunov, A.
State official.
Mem. Cen. Cttee. Kirghiz CP 1971–; Head of Agricultural Dept. with Cen. Cttee. Kirghiz CP 1971–D. 1975; 1st Sec. of Naryn Oblast Cttee. Kirghiz CP D. 1975–80; Min. of Agriculture Kirghiz SSR Mr. 1980–.

Savitskaia, Svetlana Evgen'evna
Athlete, pilot, and engineer; b. Ag. 8, 1948, Moscow; d. of E. Ia. Savitskii.
Honored Master of Sports (1970). Member of the CPSU since 1975.
She graduated from the Sergo Ordzhonikidze Moscow Institute of Aviation in 1972. In 1970 she won the world championship in aerobatics in the piston aircraft category. She has set three world records in team sport parachuting from the stratosphere and nine world records for pilotage in jet airplanes. Decors.: gold medal and 14 diplomas by the International Aeronautical Federation and 12 gold medals by the USSR.

Savitskii, Evgenii Iakovl'evich
Military official; b. D. 11 (24), 1910, Novorossiisk.
Marshal of aviation (1961). Twice Hero of the Soviet Union (My. 11, 1944, and Je. 2, 1945). Honored Military Pilot of the USSR (1965). Member of the CPSU since 1931.
In the Soviet Army since 1929, he graduated from the Military Pilot's School in 1932 and the Military Academy of the General Staff in 1955. In the Great Patri-

otic War he was with the army in the field from January 1942, serving as commander of the air forces of the Twenty-fifth Army (March–April 1942), commander of the 205th Fighter Aviation Division (May–November 1942), and commander of an aviation group of the Seventeenth Air Army (November–December 1942). From December 1942 until the end of the war he commanded the III Fighter Aviation Corps on the Western, Voronezh, Southwestern, Stalingrad, Northern Caucasus, Southern, Fourth Ukrainian, and First and Third Belorussian fronts. He made 216 combat flights and shot down 22 enemy aircraft personally and two in group fighting.
After the war, he held responsible positions in the air force and commanded the aviation of the National Air Defense Forces (1948). He has been deputy commander-in-chief of the National Air Defense Forces since 1966. He was a candidate member of the Central Committee of the CPSU from 1961 to 1966 and a deputy to the Supreme Soviet of the USSR at the sixth convocation.
Decors.: three Orders of Lenin, five Orders of the Red Banner, the Order of Suvorov Second Class, the Order of Kutuzov Second Class, two Orders of the Red Star, and various medals, as well as several foreign orders and medals.

Savitskii, Evgenii Mikhailovich
Physical chemist and metallurgist; b. Ja. 17 (30), 1912, Moscow.
Mem. CPSU 1940–; grad. Moscow Inst. of Nonferrous Metals and Gold 1936; corresponding mem. USSR Acad. of Sciences 1966; mem. staff Kurshatov Inst. of Gen. and Inorganic Chemistry 1936–54; mem. staff Baibakov Inst. of Metallurgy of USSR Acad. of Sciences 1954–, Dir. of same inst. 1975–. Decors.: State Prize of the USSR 1953 and 1968, N. S. Kurnakov Prize of USSR Acad. of Sciences 1973, 3 orders and various medals.
Publs.: "Perspektivy razvitiia metallovedeniia" 1972, "Problema prognoza neorganicheskikh soedinenii s pomoshchiu EVM" (in "Vestnik Akademii Nauk SSSR" No. 1, 1975).

Savochko, Boris Nikolaevich
Party official; b. 1929; s. of a white-collar worker.
Ukrainian; mem. CPSU 1955–; grad. Polytechnical Inst. in L'vov; engineer, researcher in various enterprises and research institutions in Kazan, Leningrad and Bel'tsy 1952–65; 1st Dep. Chair. of Economic Council Moldavian SSR, head of Dept. for Ind. and Transport Cen. Cttee. Moldavian CP 1965–73; mem. Cen. Cttee. Moldavian CP 1966–; 1st Dep. Chair. of GOSPLAN Moldavian SSR 1973–76; Dep. Chair. of Council of Mins. and Chair. of GOSPLAN Moldavian SSR 1976–77; Sec. and mem. Bureau of Cen. Cttee. Moldavian CP O. 1976–.

Savva (Aleksandr Pavlovich Babinets), Cand. of Theol.
Russian Orthodox clergyman; b. Mr. 27, 1926, Transcarpathians.
Novice 1942; monk 1947; head clergyman in village parishes; studied Moscow Theological Seminary 1953–57 and Moscow Theological Acad. until 1961; pastor in Trostnik, Mukachevo Diocese; Main Clergyman and Provost of Uspenskii Parishes in Vinogradov, promoted to Abbot 1968; Bishop of Pereiaslav-Khmelnitskii, Substitute Bishop of Kiev Diocese 1969–72; Bishop of Chernovtsy and Bukovina Ja. 1972; Bishop of Mukachevo and Uzhgorod Mr. 1977–.

Savvin, Ivan Alekseevich
State official; b. 1914.
1st Dep. Min. of Finance Moldavian SSR 1964–79, Min. F. 1979–; mem. Cen. Cttee. Moldavian CP 1981–.

Sazandarian, Tatevik Tigranovna
Armenian mezzo-soprano; public figure; b. Ag. 20 (S. 2), 1916, Khudzoresk, now in Goris Raion, Armenian SSR.
People's Artist of the USSR (1956). Member of the CPSU since 1949.
In 1937, she became a soloist with the A. Spendiarov Armenian Theatre of Opera and Ballet in Yerevan. She was the first to perform the title role in Spendiarov's *Almast* and the roles of Parandzem in Chukhadzhian's *Arshak II,* Tamar in A. Tigranian's *David-bek,* Nazeli in Stepanian's *The Heroine* (State Prize of the USSR, 1951), and Anna in Arutiunian's *Saiat-Nova.* Her best roles have also included Liubasha in Rimsky-Korsakov's *The Tsar's Bride,* Konchakovna in Borodin's *Prince Igor,* Amneris in Verdi's *Aida,* and Dalila in Saint-Saëns' *Samson et Dalila.*
She started teaching at the Yerevan Conservatory in 1961, becoming a professor in 1970. In 1972 she also became head of the subdepartment of solo singing at the Yerevan Theatrical Institute. She was a deputy to the fifth convocation of the Supreme Soviet of the USSR. Decors.: Order of the Red Banner of Labour and various medals.

Sazhenev, Gennadii Ivanovich
Diplomat.
USSR Amb. to Grenada Jl. 1982–.

Sazhenin, Ivan Andreevich
State official.
Min. of Health Tadzhik SSR F. 1963–; Dep. to Supr. Sov. Tadzhik SSR 1963–; mem. Cen. Cttee. Tadzhik CP 1963–.

Sazhin, Petr Aleksandrovich
Author; b. Je. 17 (30), 1906, Bolshaia Lazovka, Gouvt. Tambov; s. of a peasant.
Mem. CPSU 1932–; worked as seaman and journalist, visited England, France, Sweden, Germany, India and other countries; first works publ. 1929.
Publs.: "Kapitan Kiribeev" (Captain Kiribeev) 1956, "Tramontana" 1959, "Siren'" (Lilac) 1962, "Tri chasa" (Three Hours) 1967, and others.

Sebris, Karl Karlovich
Actor; b. 1914.
Grad. Latvian Drama Courses (under the dir. of E. Zeltmanis) 1938; with Nat. Theatre (now Latvian A. Upta Drama Theatre) 1938–. Decor.: People's Artist of the USSR 1974.

Sedols, Vilnis Iorenovich
State official.
Mem. Cen. Cttee. Latvian CP 1971–81, cand. mem. 1981–; 1st Sec. of Leningradskii Raion Cttee. Latvian CP in Riga until O. 1976; Chair. of State Cttee. on Pricing Latvian SSR O. 1976–.

Sedov, Leonid Ivanovich
Scientist; b. N. 1 (14), 1907, Rostov on Don.
Grad. Moscow State Univ. 1930; specializes in mechanics and applied mathematics; mem. USSR Acad. of Sciences 1953–; mem. staff Central Aerodynamic Inst. 1930–47; Prof. at Moscow State Univ. 1937–; concurrently work at Inst. of Mathematics of USSR Acad. of Sciences 1945–; Chair. of Scientific Council, Div. of Mechanics of Fluids and Gases of USSR Acad. of Sciences 1965–. Decors.: Hero of Socialist Labour 1967, 4 Orders of Lenin, 2 (unspecified) orders and various medals, Order of the French Legion of Honour, State Prize 1952.
Publs.: "Vvedenie v mekhaniku sploshnoi sredy" 1962, "Metody podobiia i razmernostei v mekhanike" 1972, "Mekhanika sploshnoi sredy" 1973, and others.

Seidov, Gasan Niemat ogly
State official; b. 1932.
Mem. CPSU 1956–; grad. Azerbaidzhan Polytechnical Inst. 1956; asst. foreman, dep. shop superintendent, shop superintendent, dep. chief engineer at Lieutenant Shmidt Mechanical Engineering Plant 1956–65; Dir. of this plant 1965–71; Sec. of Cen. Cttee. and mem. Bureau of Cen. Cttee. Azerbaidzhan CP 1971–81; Chair. of Council of Mins. Azerbaidzhan SSR Ja. 1981–; cand. mem. Cen. Cttee. CPSU 1981–.

Seitov, Sagyngali
Author, critic and literary historian; b. N. 20, 1917.
Kazakh; mem. CPSU 1946–; grad. Alma-Ata Pedagogic Inst. 1941; has written mainly poetry (in Kazakh) on Komsomol topics, heroism of the Sov. soldier and friendship between nations.
Publs.: "Litsom k litsu" (Face to Face) 1959, "Ekho" (The Echo) 1967, "Ia vsegda s vami" (I Am Always With You) 1968, "Kartiny zhihni" (Pictures From Life) 1969, and others.

Semenenko, Evgenii Grigor'evich
Komsomol official.
Mem. Bureau 1978–, Sec. 1978–80, 2nd Sec. Cen. Cttee. Kirghiz Komsomol S. 1980–; mem. Cttee. of People's Control Kirghiz SSR 1980–; mem. Auditing Comm. Kirghiz CP 1981–.

Semenenko, Nikolai Panteleimonovich, Dr. of Geol.-Min. Sc.
Geologist; b. N. 16, 1905, Mariupol, now Zhdanov; s. of a worker.
Mem. CPSU 1932–; grad. Dnepropetrovsk Mining Inst. 1927; mem. Acad. of Sciences Ukrainian SSR 1948–; Prof. 1937–; Lecturer, Prof. at Dnepropetrovsk Mining Inst. 1927–41; Head of Chair at Inst. of Nonferrous Metals in Ordzhonikidze 1941–42; Head of Chair at Inst. of Metallurgy 1942–44; Head of Dept. of Ores at Geological Inst. of Acad. of Sciences Ukrainian SSR 1944; Dep. Head of Univ. Admin. Dept. with Council of Mins. Ukrainian SSR 1945–48; Acad. Sec. 1948–50, Vice-Pres. of Acad. of Sciences Ukrainian SSR 1950–70; author of numerous scientific works on petrography and geology of natural resources. Decors.: Merited Scientist of the Ukrainian SSR 1957, 2 Orders of Lenin, Order of Red Banner of Labour, Order "Badge of Honour".
Publs.: "Granitni permatyty Ukrainy" in "Geologichnyi zhurnal" No. 6, 1940, "Struktura rudnykh polei Krivorozhskikh zheleznorudnykh mestorozhdenii" 1946, "Rozvytok teorii metamorfizmu" in "Geologichnyi zhurnal" No. 18, 1958, "Metamorfizm podvizhnykh zon" 1966, and others.

Semenikhin, Vladimir Sergeevich
Scientist; b. Ja. 27 (F. 2), 1918, Sumy.
Mem. CPSU 1945–; grad. Moscow Inst. of Energetics 1941; specializes in automatics and telemechanics; mem. staff constructional engineering bureaus of various factories for mechanical appliances 1941–63; head of various scientific research orgs., Prof. of Moscow Inst. of Radio Technology, Electronics and Automatic Systems 1963–; Dep. USSR Min. of the Radio Ind. 1971–74; mem. USSR Acad. of Sciences 1972–; has publ. basic works on automated circuit systems and the construction of special appliances. Decors.: Order of Lenin, Order of the October Revolution, 3 other orders, and various medals.

Semenov, Bato Semenovich
State official.
Chair. of Presidium Supr. Sov. Buriat ASSR; Dep. to and Dep. Chair. of Presidium Supr. Sov. RSFSR D. 1970–.

Semenov, Georgii Vital'evich
Author; b. Ja. 12, 1931, Moscow.
Grad. Gorkii Inst. of Literature 1960; worked as sculptor; first works publ. 1961.
Publs.: "Sorok chetyre nochi" (Forty-four Nights) 1964, "Lebedi i sneg" (Swans and Snow) 1964, "Kto on i otkuda" (Who is he and Where does he come from?) 1968, "Vecherom, posle dozhdia" (At Evening, after the Rain) 1959, and others.

Semenov, Iulian Semenovich
Author; b. O. 8, 1931, Moscow.
Grad. Inst. of Oriental Studies 1953; first works publ. 1958.
Publs.: "Diplomaticheskii agent" (Diplomatic Agent) 1959, "49 chasov 25 minut" (49 Hours and 25 Minutes) 1960, "Ukhodiat, chtoby vernutsia" (They Leave in Order to Return) 1961, "Petrovka, 38" 1963, "Dunechka i Nikita" 1966, "Major Vikhr'" (Major Vikhr) 1967, "Semnadtsat' mgnovenii vesny" (Seventeen Spring Moments) 1969, "Vietnam. Laos" 1969, and others.

Semenov, Ivan Maksimovich
Graphic artist; b. 1906.
Grad. Rostov Tech. Art Inst. 1928; contributor to satirical magazine "Krokodil" 1935–; Chief Ed. of children's magazine "Veselye Kartinki" 1957–.

Semenov, Nikolai Nikolaevich
Physicist and physical chemist; b. Ap. 3 (15), 1896, Saratov.
Mem. CPSU 1947–; grad. Petrograd Univ. 1917; mem. USSR Acad. of Sciences 1932–; mem. staff Leningrad Inst. of Physical Technology 1920–31; concurrently Prof. at Leningrad Polytech. Inst. 1928–; Dir. of Inst. of Chemical Physics of USSR Acad. of Sciences 1931–; concurrently Prof. at Moscow State Univ. 1944–; Acad. Sec. of Dept. of Chemical Sciences of USSR Acad. of Sciences 1957–63; Chair. of Union Soc. "Znanie" 1960–63; Vice-Pres. of USSR Acad. of Sciences 1963–71; cand. mem. Cen. Cttee. CPSU 1961–66. Decors.: State Prize of the USSR 1941 and 1949, Nobel Prize 1956, 7 Orders of Lenin, Order of Red Banner of Labour, medals.
Publs.: "Tsepnye reaktsii" (Chain Reactions) 1934, "Nauka i obshchestvo" (Science and Society) 1973.

Semenov, Valentin Grigor'evich
Laborer; b. 1930.
Mem. CPSU 1958–; grad. Podolsk Industrial Technicum 1971; fitter in Podolsk Mechanical Plant 1947–50; served in Sov. Army 1950–53; brigade leader of fitter work brigade at Kalinin Mechanical Plant in Podolsk 1953–; mem. Cen. Cttee. CPSU. Decor.: Hero of Socialist Labour 1971.

Semenov, Viacheslav Mikhailovich
Diplomat; b. 1923.
Mem. Diplomatic Service 1948–; Emb. Counsellor at USSR Emb. in Ivory Coast Republic 1967–69; Emb. Counsellor-Envoy in Guinea 1969–72; mem. staff USSR Min. of Foreign Affairs 1972–75; USSR Amb. to Guinea-Bissau F. 1975–80; concurrently USSR Amb. to the Republic of the Cape Verde Islands D. 1975–80.

Semenov, Vladimir Semenovich
Diplomat; b. 1911.
Mem. CPSU 1938–; grad. Cherniashevskii Inst. of Philosophy, Literature and History 1937; Assoc. Prof., mem. Diplomatic Service 1939–; Counsellor at Sov. Mission in Lithuania 1939–40; Counsellor at USSR Mission in Germany 1940–41; Head of 3rd European Dept. of USSR People's Commissariat of Foreign Affairs 1941–42; Dep. Pol. Counsellor 1945–46 and Pol. Counsellor at Sov. Mil. Admin. in Germany 1946–49; Pol. Counsellor at Sov. Control Comm. in Germany 1949–53; Head of 3rd European Dept. of USSR Min. of Foreign Affairs 1953; mem. USSR High Comm. in Germany and USSR Amb. to the GDR 1953–54; mem. Collegium USSR Min. of Foreign Affairs and Head of its 3rd European Dept. 1954–55; USSR Dep. Min. of Foreign Affairs Mr. 1955–1978; USSR Amb. to the Federal Republic of Germany 1978–; mem. Cen. Auditing Comm. CPSU 1956–66; cand. mem. Cen. Cttee. CPSU 1966–. Decor.: Order of Lenin, Order of the October Revolution, 4 other orders and medals.

Semin, Vitalli Nikolaevich
Author; b. Je. 12, 1927, Rostov/Don.
Grad. Taganrog Pedagogical Inst. 1957; forced labour in Germany during World War II; after return studied at Rostov Pedagogical Inst., took part in construction work of Kuibyshev Hydro-Electric Power Station.
Publs.: "Shtorm na Tsimla" (Storm on the Tsimla) 1960, "Lastochka – zvezdochka" (Little Swallow, Little Star) 1963, "Sto dvadtsat' kilometrov do zheleznoi dorogi" (One Hundred and twenty Kilometres to the Railway) 1964, "Semero v odnom dome" (Seven in One House) 1965, "Asia Aleksandrovna" 1965, "Nashi starukhi" (Our Old Ones) 1966, "V sorok vtorom. Iz vospominanii" (In the Year Forty-Two. From the Memoirs) 1968, "Khoziain" (The Landlord) 1968, and others.

Semionov, Viktor Vladimirovich, Dr. of. Tech. Sc.
Cybernetics scientist; b. 1937.
Mem. CPSU 1964–; grad. N. E. Bauman Higher Tech. College (MVTU) in Moscow 1960; specialist in automatic control systems; Dr. of Tech. Sc. 1971; mem. staff MVTU 1960– (Research Asst., Lecturer 1966–). Decor.: State Prize of the USSR 1972.
Publs.: co-author of a series of engineering monographs "Tekhničeskaja kibernetika. Teorija avtomatičeskogo regulirovanija" (4 vols.) 1967–69.

Semioshkin, Aleksandr Semenovich
Diplomat; b. 1921.
Mem. Diplomatic Service 1947–; Emb. Counsellor at USSR Emb. in Iraq 1958–61; mem. staff USSR Min. of Foreign Affairs 1961–64 and 1968–74; Emb. Counsellor-Envoy at USSR Emb. in Egypt 1964–68; USSR Amb. to the People's Democratic Republic of Yemen Je. 1974–Ag. 1975.

Senkin, Ivan Ilich
Party official; b. 1915.
Karelian; mem. CPSU 1940–; grad. Timiriazev Agricultural Acad. 1939–; agronomist at a machine and tractor station, chief agronomist with a raion agricultural dept. in the Karelian-Finnish SSR 1939–41; served in Sov. Army 1941–46; sen. party and govt. posts in Sverdlovsk Oblast 1946–51 and 1953–54; Sec. of Sverdlovsk Oblast Cttee. CPSU 1951–52 and 1954–55; Dep. Chair. of Karelo-Finnish Council of Mins. 1955–56; 2nd Sec. of Cen. Cttee. of Karelo-Finnish CP 1956; 2nd Sec. Jl. 1956–58, 1st Sec. of Karelian Oblast Cttee. CPSU S. 1958–; mem. Cen. Auditing Comm. 1961–66, cand. mem. 1966–71, mem. Cen. Cttee. CPSU 1971–. Decors.: Order of Lenin, Order of the October Revolution.

Senko, Fedor Petrovich
State official.
Min. of Agriculture Belorussian SSR Mr. 1979–; mem. Cen. Cttee. Belorussian CP 1981–.

Sepelev, Iurii Faddeevich
Diplomat.
USSR Amb. to Mozambique D. 1982–.

Serafim (Vladimir Ivanovich Rodinov)
Russian Orthodox clergyman; b. Ap. 24, 1905, Moscow.
Studied Faculty of Physics and Mathematics of Moscow Univ.; after 1923 studied painting in Paris and Acad. of Fine Arts and Sorbonne; entered Orthodox Theological Sergios Inst. Paris 1934; novice with Patriarchate Representation in Paris 1935; studied philosophy, byzantology and patrology at Sorbonne; deacon and monk 1937; priest-monk 1939; Main Clergyman of Patriarchate Representation in Paris 1934–45; abbot 1945; Archimandrite 1952; Exarch of the Patriarch in W. Europe, Rep. of Russian Orthodox Church at World Council of Churches at time of its founding in Geneva 1946; Head of the Parish of the Resurrection in Zurich 1949; Bishop of Zurich 1971–.

Serapion (N. S. Fadeev), Cand. of Theol.
Soviet clergyman; b. My. 27, 1933, Moscow.
Grad. Theological Acad. in Moscow 1966; novice at Troitse-Sergieva Lavra (Monastery) 1951; served in Sov. Army 1952–55; Advisor with Dept. of Foreign Relations of Moscow Patriarchate 1969–71; Aide to Head of Troitse-Sergieva Lavra (Monastery) 1971–72; Bishop of Podolsk, Vicar of Moscow Diocese, Representative of the Patriarchate of Moscow at the See of the Patriarch of Antioch 1972–74; temporarily retired for health reasons 1974; appointed Bishop of Irkutsk and Chita Mr. 1975–80; Archbishop of Vladimir and Suzdal Ap. 1980–.

Serebrennikov, Boris Aleksandrovich
Linguist; b. F. 21 (Mr. 6), 1900, Kholmogory, Arkhangelsk Oblast.
Mem. CPSU 1950–; grad. Inst. of Philosophy, Literature and History 1940; corresponding mem. USSR Acad. of Sciences 1953–; Prof. 1969–; Dep. Dir. 1950–55, Dir. 1960–64, Head of Dept. of General Linguistics at Inst. of Linguistics of USSR Acad. of Sciences 1964–. Decors.: Order of Red Banner of Labour, other medals.
Publs.: "Istoricheskaia morfologiia permskikh iazykov" (Historical Morphology of the Perm Languages) 1963, "Istoricheskaia morfologiia mordovskikh iazykov" (Historical Morphology of the Mordovian Languages) 1967, "Obshchee iazykoznanie" (General Linguistics) 1970–73, and others.

Serdiuk, Aleksandr Ivanovich
Actor; b. Je. 14 (27), 1900, Bzov, Kiev Oblast.
Mem. CPSU 1943; grad. Lysenko Inst. of Music and Drama 1919; actor with 1st Ukrainian Shevchenko Theatre in Kiev 1919–22; with Shevchenko Theatre in Kharkov 1922; Dir. and Artistic Man. of same 1957–62. Decors.: People's Artist of the USSR 1951, Order of Lenin, Order of Red Banner of Labour, medals.

Serebriakova, Galina Iosifovna
Author; b. D. 7 (20), 1905, Kiev.
Mem. CPSU 1919–; grad. Dept. of Medicine Moscow Univ. 1925; first works publ. 1925. Decor.: Red Banner of Labour.
Publs.: trilogy "Prometei" ("Iunost' Marksa") vols. 1–2, 1933–34, "Pokhishchenie ognia" (Pillage of the Fire) 1961, "Vershiny zhizni" (Peaks of Life) 1962, "God kak zhizn'" (One Year Like One Life) 1966, "O drugikh i o sebe" (About Others and About Oneself) 1968, "Sobranie sochinenii" (Collected Works) vols. 1–5, 1967–69, "Marx i Engels" 1966, and others.

Serebrianyi, Iosif Aleksandrovich
Painter; b. Ap. 12 (25) 1907, Gorodnia, now in Chernigov Oblast.
People's Artist of the RSFSR (1965); corresponding member of the Academy of Arts of the USSR (1947). Member of the CPSU since 1943.
From 1927 to 1931, he studied at the Leningrad State Higher Arts and Technical Institute (reorganized in 1930 at the Institute of Proletarian Fine Arts); his instructors included V. E. Savinskii, A. A. Rylov, and M. P. Bobyshev. He directed the administrative board of the Leningrad department of the Union of Soviet Artists from 1948 to 1952 and again from 1954 to 1958. He teaches at the I. E. Repin Leningrad Institute of Painting, Sculpture, and Architecture (1948–51 and since 1954), where he has been a professor since 1949. He paints primarily historical revolutionary scenes and portraits. Using simple compositional elements, he creates realistic portraits of Soviet people (*The Students-Partisans from the P. F. Lesgaft Physical Training Institute,* 1942, Russian Museum, Leningrad: *Concert of the Leningrad Philharmonic Society.* 1957. Russian Museum: *Portrait of D. Shostakovich,* wax tempera, 1964. Tret'iakov Gallery).

Seregin, Nikolai Ivanovich
Diplomat.
USSR Amb. to Cape Verde Islands N. 1981–.

Sergeev, Aleksei Tikhonovich
Singer (bass); b. Ja. 24, 1919, Gerasimovka, Tambov Oblast.
People's Artist of the USSR (1967). Member of the CPSU since 1956.
In 1968, he graduated from the Gnesin's Music Pedagogic Institute, where he studied voice under E. V. Ivanov. In 1940 he became a member of the chorus of the Song and Dance Ensemble of the Soviet Army, and in 1950, a soloist. His repertoire includes folk songs as well as works by Russian, Western, and Soviet composers; among these compositions an A. V. Aleksandrov's "Poem About the Ukraine" and Eremka's song from Serov's opera *The Power of Evil.* He has toured many countries with the Song and Dance Ensemble of the Soviet Army. Decors.: two orders and several medals.

Sergeev, Evgenii Mikhailovich
Geologist; b. Mr. 10 (23), 1914, Moscow.
Mem. CPSU 1939–; grad. Moscow Univ. 1940; corresponding mem. USSR Acad. of Sciences 1966–; served in Sov. Army 1941–43; mem. staff Faculty of Geology of Moscow Univ. 1940–41 and 1943–; Prof. 1953–, Head of Dept. of Soil Science and Geological Engineering 1954–, Head of Dept. of Geology 1954–57, 1963–64 and Dep. Rector of Moscow Univ. 1964–; hon. Dr. Bratislava Univ. 1972, Warsaw Univ. 1974. Decors.: Order of Lenin, Order of the October Revolution, 4 other orders and various medals.

Sergeev, Nikolai Dmitrievich
Military official; b. S. 22 (O. 5), 1909, Starye Petrovtsy, now Kiev Oblast.
Mem. CPSU 1930–; grad. Higher Frunze Naval School 1931 and Naval Acad. 1941; Admiral of the Fleet 1970–; served in Sov. Navy 1928–; with Chief Staff of Sov. Navy 1941–43 and Jl. 1944–45;

Commander of Ship Brigade of Volga Flotilla My.–S. 1943, Chief-of-Staff of same S. 1943–Jl. 1944; head of a dept., Dep. Head of an Admin. Dept. of Chief Staff of Sov. Navy 1945–50, Head of Chief Admin. of Gen. Staff of Sov. Navy 1950–52; Commander of the White Sea Flotilla 1952–56; Dep. Head of Chief Staff of Sov. Navy 1956–64; Head of Chief Staff and 1st Dep. Chief Commander of Sov. Navy 1964–77. Decors.: 2 Orders of Lenin, 3 Orders of Red Banner, Order of Red Banner of Labour, Nakhimov Order 2nd Cl., Order of Patriotic War 1st Cl., 3 Orders of Red Star, various medals, 3 foreign orders.

Sergeev, Rostislav Aleksandrovich
Diplomat.
USSR Amb. to Mexico My. 1980–.

Sergienko, Raisa Mikhailovna
Singer; b. 1925.
Mem. CPSU 1956–; grad. Odessa Conservatory 1951–, soprano; soloist at Odessa Opera and Ballet Theatre 1951–. Decor.: People's Artist of the USSR 1973.

Sergii (S. V. Petrov), Cand. of Theol.
Russian Orthodox clergyman; b. O.5, 1924, Krasnodar.
Monk-deacon 1943; studied Moscow Theological Seminary and Acad. 1945–51; priest-monk 1951; teacher at Saratov Theological Seminary 1951; Head of Cathedral of the Ascension in Elets 1954; Asst. to the Supervisor, then Supervisor of Odessa Theological Seminary, Abbot 1958; Rector of Seminary, Archimandrite 1959; Bishop of Belgorod-Dnestrovskii, Substitute of Odessa Diocese 1960; Bishop of Voronezh 1961–63; Archbishop of Minsk and Belorussia 1963; Archbishop 1965 and Metropolitan of Kherson and Odessa 1971–.

Serkebaev, Ermek Bekmukhamedovich
Singer; b. Jl. 4, 1926, Petropavlovsk.
Mem. CPSU 1958–; studied violin at Alma-Ata Music School 1941; grad. in singing at Alma-Ata Conservatory 1951; lyric baritone; soloist at Kazakh Theatre for Opera and Ballet 1947–. Decor.: State Prize of the Kazakh SSR 1972–.

Sevastianov, Vitalii Ivanovich, Cand. of Tech. Sc.
Cosmonaut; b. Jl. 8, 1935, Krasnouralsk, Sverdlovsk Oblast.
Mem. CPSU 1963–; grad. Moscow Ordzhonikidze Aviation Inst. 1959; Cand. of Tech. Sc. 1965–; cosmonaut 1967–; space flight with "Soiuz 9" spacecraft Je. 1–19, 1970 (together with A. G. Nikolaev) and with "Soiuz 18" My. 24–Je. 26, 1975 (together with P. I. Klimuk). Decors.: twice Hero of the Sov. Union, 2 Orders of Lenin, various medals.

Severin, Sergei Evgen'evich
Biochemist; b. D. 8 (21), 1901, Moscow.
Grad. Moscow Univ. 1924; student of V. S. Gulevich; mem. USSR Acad. of Sciences 1968–; mem. USSR Acad. of Medicine 1948–; Prof. at 3rd Moscow Inst. of medicine (now Riazan Medical Inst.) 1932–48; Organizer of Laboratory 1935; Head of Dept. of Veterinary Biochemistry at Moscow State Univ. 1939–45; Dir. of Inst. of Nutrition of USSR Acad. of Medicine 1945–47; Dir. of Inst. of Biology and Medical Chemistry of USSR Acad. of Medical Sciences 1948–49; Acad. Sec. at Dept. of Medical-Biological Sciences of USSR Acad. of Medical Sciences 1949–57; Pres. of Biochemical Union Soc. 1969–. Decors.: 3 Orders of Lenin, Order of the October Revolution, Order of Red Banner of Labour.

Severnyi, Andrei Borisovich
Astrophysicist; b. Ap. 28 (My. 11), 1913, Tula.
Mem. CPSU 1941–; grad. Moscow Univ. 1935; mem. USSR Acad. of Sciences 1968–; mem. staff Shternberg Inst. of Astronomy 1938–46; mem. Staff 1946–52, Dir. of Astrophysical Observatory on Crimean Peninsula 1952–. Decors.: Hero of Socialist Labour 1973–, State Prize of the USSR 1952, Order of Lenin, 2 Orders of the October Revolution, 3 other orders and various medals.
Publs.: "Fizika solntsa" (Solar Physics) 1956, "Magnitnye polia solntsa i zvezd" (Magnetic Fields of the Sun and the Stars) in "Uspekhi fizicheskikh nauk" No. 88/1966.

Shabalin, Evgenii Pavlovich, Cand. of Tech. Sc.
Engineer-physicist; b. 1936.
Grad. Inst. of Physical Engineering Moscow 1959; mem. staff United Inst. of Nuclear Research in Dubna 1959–, head of a group at same 1966–. Decor.: State Prize of the USSR 1971.

Shabanov, Vitalii Mikhailovich
Military official; b. Ja. 1, 1923, Lobachi, Kostroma Oblast.
Mem. CPSU 1947–; grad. Leningrad Air Force Acad. 1945, Engineer Col.-Gen. 1981; practical coursework as part of study program with a fighter pilot regiment at the front 1941–45; engineer and test-pilot, asst. chief engineer for flight testing at scientific research insts. of the air force 1945–49; senior engineer, dep. chief of a laboratory, chief designer, dep. designer gen. in enterprises of USSR Min. of the Radio Ind. 1949–72; dir. of a scientific prod. asscn. 1972–74; Dep. USSR Min. of the Radio Ind. 1974–78; Dep. USSR Min. of Defense Ind. 1978–; cand. mem. Cen. Cttee. CPSU 1981–. Decors.: Lenin Prize 1963, USSR State Prize 1953, Order of Lenin, Order of the October Revolution, Order of Red Banner of Labour, Order of Red Star, Hero of Socialist Labour 1981, medals.

Shabasanov, Makhtumkuli Karaevich
State official; b. 1930.
Turkmen; mem. CPSU 1951–; grad. Azerbaidzhan Ind. Inst.; mem. Cen. Cttee. Turkmen CP 1960–; Chair. of Turkmen Republican Trade Union Council 1962–65; cand. mem. Bureau Cen. Cttee. Turkmen CP 1962–76; 1st Sec. of Ashkhabad City Cttee. Turkmen CP 1965–75; Head of Dept. of the Petroleum and Chemical Ind. with Cen. Cttee. Turkmen CP 1975–77; Head of Main Admin. for Gasification with Council of Mins. Turkmen CP F. 1977–78, Chair. of State Cttee. for Gasification Turkmen SSR D. 1978.

Shabashov, Sergei Mikhailovich
Party official; b. 1923, Draguny, Vitebsk Oblast.
Mem. CPSU 1945; grad. Mogilev Pedagogical Inst. 1956 and CPSU Cen. Cttee. Party Higher School 1961; Partisan in Belorussia 1941–44; Komsomol posts in Vitebsk 1944–56; party posts in Ezerishche and Kobrin 1956–62; Mem. Cen. Cttee. Belorussian CP 1961–66 and 1971–; Sec. of Brest Oblast Cttee. Belorussian CP 1963–71 (for Agriculture 1963–64); Head of Dept. of Propaganda and Agitation with Cen. Cttee. Belorussian CP 1971; 1st Sec. of Vitebsk Oblast Cttee. Belorussian CP Jl. 1971–. Decor.: Order of the October Revolution.

Shabliovskii, Evgenii Stepanovich, Dr. of Philol. Sc.
Literary historian; b. F. 25, 1908, Kamen-Kashirskii, Volhynia Oblast; s. of a teacher.
Mem. CPSU 1928–; grad. Inst. of Popular Educ. 1930 and Medical Inst. Kiev 1932; corresponding mem. Acad. of Sciences Ukrainian SSR 1934; Dr. of Philol. Sc. 1926–28; postgraduate studies at Shevchenko Inst. of People's Commissariat for Popular Educ. of the Ukrainian SSR 1931–32; Dir. of same Inst. 1933–35; Sen. Scientist at Shevchenko Inst. of Literature of Acad. of Sciences Ukrainian SSR 1967; author of some 200 scientific papers.
Publs.: "Internatsionalne i natsionalne v khudozhnii tvorchosti" 1963, "Gumanizm Shevchenka i nasha suchasnist" 1964, co-author of "T. G. Shevchenko. Biografia" 1964, "Shliakhamy ednannia. Ukrainska literatura v ii istorychnomu rozvytku" 1965, and others.

Shafarevich, Igor Rostislavovich
Mathematician; b. 1923.
Specializes in algebra; corresponding mem. USSR Acad. of Sciences 1958–; hon. mem. American Acad. of Sciences; mem. Cttee. for the Defense of Human Rights in the USSR. Decors.: Mathematics Prize of German Acad. of Sciences in Göttingen, Lenin Prize.

Shafran, Daniil Borisovich
Violoncellist; b. Ja. 13, 1923, Leningrad. Studied Leningrad Conservatory; numerous tours abroad. Decors.: 1st Prize at Union Competition for Cellists 1937, State Prize of the USSR, People's Artist of the RSFSR.

Shafranovskii, Ilarion Ilarionovich
Crystallographer; b. Mr. 11 (24), 1907, St. Petersburg.
Doctor of geological and mineralogical sciences (1943); professor (1945). Honored Scientist of the RSFSR (1973).
He graduated from Leningrad University in 1930. In 1931 he began working at the Leningrad Institute of Mines, where he became head of the subdepartment of crystallography in 1946. He also taught at Leningrad University from 1934 to 1948. His main works deal with crystal morphology, mineralogical crystallography, and the history of mineralogy and with the conditions. He linked the morphology of crystals with the conditions of crystal formation and the effects of environmental symmetry. He worked out the theory of distorted crystal forms and devised the goniometric technique of describing rough diamonds. Decors.: Order of the Red Banner of Labour and various medals.
Works: *Kristally mineralov,* parts 1–2. Leningrad-Moscow 1957–61. *Istoriia kristallografii v Rossii.* Moscow-Leningrad 1962. *Kristallografiia,* 5th ed. Moscow 1972. (With G. M. Popov.) *Ocherki po mineralogicheskoi kristallografii.* Leningrad 1974.

Shagazatov, Khabibulla Abdumazhitovich
State official.
Min. of Installation and Special Construction Work Uzbek SSR F. 1979–.

Shaginian, Grant Amazaspovich
Gymnast; b. Jl. 30, 1923, Giulasarak, Stepanavan Raion, Armenian SSR.
Honored Master of Sport (1951). Member of the CPSU since 1951.
He was Olympic champion in 1952, world champion in 1954, and Soviet champion six times between 1948 and 1955, taking part in individual competitions of combined events. He was all-around champion of the USSR in 1952. Decors.: two orders.

Shagonian, Robert Arutiunovich
State official.
Min. of Melioration and Water Econ. Armenian SSR D. 1977–; mem. Auditing Comm. Armenian CP 1981–.

Shaidurov, Sergei Afanasevich
State official; b. 1926.
Russian; mem. CPSU 1952–; grad. Irkutsk Inst. of Mining and Metallurgy 1948 and Party High School of Cen. Cttee. CPSU 1960; engineer, chief engineer, head of a mine of "Dal'stroi" Admin. 1948–58; party posts 1960–; 2nd Sec. 1961–68, 1st Sec. of Magadan Oblast Cttee. CPSU 1968–79; Dep. to USSR Supr. Sov. 1970–; cand. mem. Cen. Cttee. CPSU 1971–81; RSFSR Dep. Min. of Geology 1979–. Decor.: Order of Lenin 1976.

Shakarimov, Iuldash
Party official.
Cand. mem. Cen. Cttee. Tadzhik CP 1976–78, mem. 1978–; Head of Dept. of Ind. and Transport with Cen. Cttee. Tadzhik CP Ap.–D. 1978; 1st Sec. Dushanbe City Cttee. Tadzhik CP D. 1978–; cand. mem. Bureau Cen. Cttee. Tadzhik CP Mr. 1980–.

Shakhbazian Rafik Tatevosovich
State official.
Head of Main Admin. for Gasification Armenian SSR Jl. 1979–; mem. Auditing Comm. Armenian CP 1981–.

Shakirov, Midkhat Zakirovich
Party official; b. 1916.
Bashkir; mem. CPSU 1944–; grad. Bezhitsa Inst. Machine Construction 1942; engineer, chief engineer, dep. chief mechanic, chief mechanic, dep. prod. head of a plant 1942–47; Sec. of Leninskii Raion Cttee. CPSU in Ufa, Chief Engineer of Min. of Local Ind. Bashkir ASSR, head of a geophysical workshop-laboratory, dir. of a plant for geophysical appliances and apparatuses, head of a construction and assembly admin. 1947–54; Chief Enginner and Dep. Head of "Nefteprovodmontazh" Trust 1954–63; 1st Sec. of Ufa City Cttee. CPSU 1963–69; Dep. to 1966–, mem. Mandate Comm. of Sov. of Nationalities 1966–70, mem. Presidium USSR Supr. Sov. 1970–; 1st Oblast Sec. of CPSU in Bashkir ASSR 1969–; mem. Cen. Cttee. CPSU 1971–. Decor.: Order of Lenin 1976.

Shalaev, Stepan Alekseevich
Trade union official; b. 1929.
Mem. CPSU 1954–; grad. Moscow Technical Inst. of Forestry 1951; chief engineer of a forest production works in Udmurt ASSR 1951–53; dir. of various forest production works in Udmurt ASSR, Kalinin and Novgorod oblasts 1955–62; Chair. of Cen. Cttee. of the Timber, Paper, and Wood-Processing Industries Workers Union 1963–68; Sec. of All-Union Central Council of Trade Unions 1968–80; Dep. to USSR Supr. Sov. 1979–, mem. Presidium N. 1982–; USSR Min. of the Timber, Pulp and Paper, and Wood Processing Industries 1980–82; Chair. of All-Union Central Council of Trade Unions Mr. 1982–; cand. mem. Cen. Cttee. CPSU 1981–.

Shal'nikov, Aleksandr Iosifovich
Physicist; b. Ap. 27 (My. 10), 1905, St. Petersburg.
Corresponding member of the Academy of Sciences of the USSR (1946).
He graduated from the Leningrad Polytechnic Institute in 1928. In 1935 he began working at the Institute of Physical Problems of the Academy of Sciences of the USSR, and in 1938 he became a professor at Moscow State University. His main works are devoted to various aspects of experimental physics, for example, low-temperature physics and the study of properties of thin metal films and colloids.
Decors.: State Prize of the USSR in 1948, 1949, and 1953. He has been awarded the Order of Lenin and various medals.
Works: *Mekhanizm kondensatsii i obrazovanie kolloidov.* Leningrad-Moscow 1934. (Together with Iu. B. Khariton.) "Sverkhprovodiashchie svoistva tonkikh metallicheskikh sloev." *Zhurnal eksperimental'noi i teoreticheskoi fiziki,* 1940, vol. 10, issue 6. "Struktura sverkhprovodnikov v promezhutochnom sostoianii." *Ibid.,* 1946, vol. 16, issue 9.

Shamborskii, Viktor Korneevich
State official.
Chair. of State Cttee. on Pricing. Ukrainian SSR Ap. 1975–.

Shamiakin, Ivan Petrovich
Author; b. Ja. 30, 1921, Karma, Belorussia; s. of a peasant.
Belorussian; mem. CPSU 1943–; grad. Republican Party School of Cen. Cttee. Belorussian CP 1950; served in Sov. Army 1940–45; Chief Ed. of almanach "Sovetskaia Otchizna" 1952–53; Dep. Chair. 1954–66, Sec. 1966–68, 2nd Sec. 1968–71, 1st Dep. Chair. 1st Sec. of Belorussian Writers' Union 1971–81; Chair. of Supr. Sov. Belorussian SSR 1971–; mem. Auditing Comm. 1971–76, cand. mem. Cen. Cttee. Belorussian CP 1976–77, mem. 1977–. Decors.: 4 orders.
Publs.: "Eshelon u Germaniiu" (Troop Train to Germany) 1971, "Les maigo zemliaka" (The Forest of My Compatriot) 1970, and others.

Shamiev, Ingilab Shatagat ogly
State official.
Min. of Supplies Azerbaidzhan SSR 1969–75 and Mr. 1981–; Chair. of Agricultural-Ind. Asscn. for the Prod., Procurement, Processing and Sale of Vegetables and Fruit "Azplodovoshchprom" of Council of Mins. Azerbaidzhan SSR Ag. 1975–81; mem. Cen. Cttee. Azerbaidzhan CP 1976–.

Shamota, Nikolai Zakharovich, Dr. of Philol. Sc.
Literary historian; b. D. 17, 1916, Poltava; s. of a clerk.
Mem. CPSU 1945–; grad. Teachers' Training College 1937 and Gogol Pedagogical Inst. in Nezhin 1939; corresponding mem. Acad. of Sciences Ukrainian SSR 1959–; Dr. of Philol. Sc. 1961; Prof. 1962–; teacher 1939–41; served in Sov. Army 1941–45; party posts in Chernigov Oblast, attended Republican Party School 1945–48; Ed. of "Literaturnaia

gazeta", Dep. Head of Dept. of Science and Culture of Cen. Cttee. Ukrainian CP 1948–55; Dep. Chief Ed. of "Literaturnaia gazeta" 1955–56; Head of Dept. of Literary Theory 1957–59, Dep. Dir. 1959–62, Dir. of Shevchenko Inst. of Literature of Acad. of Sciences Ukrainian SSR 1962–; mem. Presidium Acad. of Sciences Ukrainian SSR, Acad. Sec. of Dept. of Literature, Language and Art 1966–. Decors.: Order of Patriotic War 2nd Cl., Order of Red Banner of Labour, medals.
Publs.: "Pro khudozhnist" 1957, "Talant i narod" 1958, "Kommunizm i sovoboda tvorchosti" 1959, "Za velinniam istorii" 1965, "O svobode tvorchestva" 1966, and others.

Shamov, Afzal Shigabutdinovich
Tartar writer; b. F.19 (Mr.4), 1901, Tatarskoe-Tanaevo, now Zelenodol'sk Raion.
Member of the CPSU since 1943.
A veteran of the Civil War of 1918–20, he studied in the literature department of Moscow State University from 1926 to 1930. He took part in the Great Patriotic War of 1941–45. In 1949 and 1950, Shamov was editor of the journal *Soviet tedebiyäti* (Soviet Literature).
He was first published in 1919. His main theme – the spiritual growth and political development of Soviet man – appears in such works as the novellas *Raufa* (1926) and *The Story of a Love* (1933) and the short-story collections *In Petroleum City* (1955) and *On One of the Days* (1959). Shamov's lyric works were published in the collection *On a Rainy Evening* (1965). He has translated into Tartar works by such writers as L. N. Tolstoy, N. V. Gogol, F. V. Gladkov, and N. A. Ostrovskii.
He was a member of the Presidium of the Supreme Soviet of the Tartar ASSR from 1959 to 1963. From 1958 to 1961 he was chairman of the administrative board of the Writers' Union of the Tartar ASSR.
Decors.: Order of the October Revolution, two other orders, and various medals.
Works: *Saylanma äsärlär.* Kazan 1954. *Ber mökhäbbät turïnda.* Kazan 1976. *In Russian translation: Rasskazy.* Moscow 1957. *Rasskazy.* Kazan 1960.

Shamshin, Vasilii Aleksandrovich
State official; b. 1926.
Mem. CPSU 1962–; grad. Moscow Electrotechnical Inst. for Communications 1949, Cand. of Tech. Sc. 1957, Asst. Prof. 1960; engineer, senior engineer, senior research associate, laboratory chief at a scientific research inst. of USSR Min. of Communications 1949–65; dep. chief for instruction at a scientific research inst. 1965–68; Dep. USSR Min. of Communications 1968–75, 1st Dep. Min. 1975–80, Min. O. 1980–; cand. mem. Cen. Cttee. CPSU 1981–.

Shapa, Petr Ivanovich
State official.
Cand. mem. Cen. Cttee. Moldavian CP 1976–81, mem. 1981–; Head of Dept. for Ind. and Transport with Cen. Cttee. Moldavian CP 1977–78; 1st Sec. of Tiraspol City Cttee. Moldavian CP until 1980; Dep. Chair. of Council of Mins. Moldavian SSR Ap. 1980–.

Shapiro, Lev Borisovich
Party official; b. 1927, in the Ukraine; s. of a bookkeeper.
Jew; mem. CPSU 1959–; grad. Moscow Steel and Alloys Inst.; engineer-metallurgist; master, steel caster, master of a blast furnace, head of an open-hearth furnace until 1966 and Sec. of party org. of "Azovstal'" Works in Komsomolsk-na-Amure 1966–70; 1st Oblast Sec. of CPSU in Jewish Autonomous Oblast 1970–; Dep. to and mem. Foreign Affairs Comm. of Sov. of Nationalities USSR Supr. Sov. 1974–; mem. Cen. Auditing Comm. CPSU 1976–81, Cand. mem. Cen. Cttee. CPSU 1981–. Decor.: Order of the October Revolution 1977.

Shaplyko, Vladimir Pavlovich
Komsomol official.
Mem. Bureau Cen. Cttee. Belorussian Komsomol F. 1976–; sec. 1976–80, 1st Sec. Ag. 1980–; mem. Cen. Cttee. Belorussian CP 1981–; cand. mem. Bureau Cen. Cttee. Belorussian CP 1981–.

Sharaev, Leonid Gavrilovich
Party official; b. 1935.
Mem. CPSU 1957–; grad. Nikolaev Shipbuilding Inst. 1958; Komsomol posts 1958–67; chief of a dept. of Nikolaev Oblast Cttee. Ukrainian CP 1967–68; 1st Sec. of Nikolaev City Cttee. Ukrainian CP 1968–73; posts in Cen. Cttee. apparatus Ukrainian CP 1973–75; Sec., 2nd Sec. of Voroshilovgrad Oblast Cttee. Ukrainian CP 1975–80; 1st Sec. of Nikolaev Oblast Cttee. Ukrainian CP O. 1980–; mem. Cen. Auditing Comm. CPSU 1981–.

Sharapov, Vasilii Ivanovich
State official; b. 1916.
Chair. of Exec. Cttee. of Minsk City Sov. of Workers' Deps. 1954–68; cand. mem. 1956–60, mem. Cen. Cttee. Belorussian CP 1960–76 and 1981–; 1st Sec. of Minsk City Cttee. Belorussian CP 1968–72; Head of Main Admin. for Public Highways of Council of Mins. Belorussian SSR 1972–73; Min. of Construction and Utilization of Main Public Highways Belorussian SSR D. 1973–.

Sharif, Aziz Alievich
Literary scholar and translator; b. Mr.16 (28), 1895, Nakhichevan.
Honored Scientist of the Azerbaidzhan SSR (1963).
He studied at the Moscow Commercial School from 1915 to 1917 and graduated from the M. Gorkii Institute of Literature in 1943. His doctoral dissertation, completed in 1956, was entitled "The Life and Work of Dzh. Mamedkulizade." He became a professor at Moscow State University in 1962.
He was first published in 1910. He is the author of many articles, pamphlets, critical-biographical studies, and monographs in Azerbaidzhani and in Russian dealing with classical and contemporary Azerbaidzhani literature, literary relations, and problems of translation. He has translated works by M. F. Akhundov, A. Akhverdov, N. Vezirov, Dzh. Dzhabarly, S. Ragimov, and M. Ibragimovo into Russian and works by N. V. Gogol, A. N. Ostrovskii, M. Gorkii, and A. A. Fadeev into Azerbaidzhani.
Works: *In Russian translation: Rozhdenie Molla Nasreddina.* Baku 1968. *Vagif – pevets liubvi i krasoty.* Moscow 1968.

Sharifov, Azad Aga Kerim ogly
State official.
Cand. mem. Cen. Cttee. Azerbaidzhan CP 1976–81; Chair. State Cttee. of Film Azerbaidzhan SSR Jl. 1979–.

Sharipov, Kholmurod
State official.
Mem. Cen. Cttee. Tadzhik CP 1956–60 and 1976–, cand. mem. 1960–76; Dep. Min. of Culture Tadzhik SSR until 1969; Head of Dept. of Culture with Cen. Cttee. Tadzhik CP 1969–79; Chair. of State Cttee. of Publishing, Polygraphy and the Book Trade Tadzhik SSR Ap. 1979–.

Sharipov, Takhir Iakhiaevich
State official.
Head of Main Admin. 1973–80, Chair. of Uzbek SSR State Cttee. for Material-Tech. Supply 1980–; Dep. to Supr. Sov. Uzbek SSR 1974–; mem. Cen. Cttee. Uzbek CP 1976–.

Shartava, Zhiuli Kalistratovich
Party official; b. 1944.
Georgian; grad. Georgian Kirov Polytech. Inst.; student, Dep. Sec., Sec. of Komsomol Cttee. of same inst., 1st Sec. of Ordzhonikidze Raion Cttee. Georgian Komsomol in Tbilisi 1961–73; 2nd Sec. Mr.–Je. 1973, 1st Sec. and mem. Bureau Cen. Cttee. Georgian Komsomol Je. 1973–80; cand. mem. Bureau Cen. Cttee. Georgian CP N. 1974–81; mem. Cen. Cttee. Georgian CP 1976–, Head of Dept. of Prop. and Agit. 1980–82, of Dept. of Org. Party Work with Cen. Cttee. Georgian CP Jl. 1982–.

Sharvin, Iurii Vasil'evich
Physicist; b. Je. 24, 1919, Moscow.
Corresponding member of the Academy of Sciences of the USSR (1970).
He graduated from Moscow State University in 1941. Beginning in 1943, he worked at the Institute of Physical Problems of the Academy of Sciences of the USSR. In 1965 he became a professor at

the Moscow Physicotechnical Institute. His main works deal with the experimental investigation of the properties of pure metals at low temperatures. Decors.: Order of the Red Banner of Labour and various medals.
Works: "Izmerenie poverkhnostnogo natiazheniia na granitse mezhdu sverkhprovodiashchei i normal'noi fazami." *Zhurnal eksperimental'noi i teoreticheskoi fiziki,* 1957, vol. 33, issue 6. "Nabliudenie fokusirovannykh puchkov elektronov v metalle." *Pis'ma v ZhETF,* 1965, vol. 1, issue 5. (With L. M. Fisher.) "Nabliudenie dinamicheskogo promezhutochnogo sostoianie sverkhprovodnikov s pomoshch'iu mikrokontaktov." *ibid.,* 1969, vol. 2, issue 6.

Shatalov, Aleksandr Stepanovich, Dr. of Tech. Sc.
Cybernetics scientist; b. 1914.
Mem. CPSU 1940–; grad. Novocherkassk Ind. Inst. 1938 and Artillery Mil. Acad. 1943; Dr. of Tech. Sc. 1951; Prof. 1952–; teaching work in Sov. Army 1941–; Scientific Consultant to F. E. Dzerzhinskii Acad. of Mil. Engineering 1970–. Decor.: State Prize of the USSR 1972 (for collaboration on engineering monograph "Tehničeskaija kibernetika. Teorija avtomatičeskogo regulirovanija" in 4 vols., 1967–69).

Shatalov, Vladimir Aleksandrovich
Cosmonaut; b. 1927.
Mem. CPSU 1953–; grad. from a special Air Force school 1945, a mil. aviation college 1949 and correspondence course of Air Force Acad. 1956; served in Sov. Air Force 1949–63; cosmonaut 1963–; commander of spaceship "Soiuz 4", took part in docking manoeuver with "Soiuz 5", landed 40 km from Karaganda Ja. 14–17, 1969; commander of spaceship "Soiuz 8" (with A. S. Eliseev), participated in group flight of 3 spaceships, landed 145 km north of Karaganda O. 13–18, 1969, "Soiuz 10" 1971. Decors.: twice Hero of the Sov. Union (Ja. 22 and O. 22, 1969), Order of Lenin and medals.

Shatilo, Voleslav Iosifovich
State official.
Dep. Min. of Finance 1962–65, Min. of Finance Belorussian SSR D. 1965–; mem. Cen. Cttee. Belorussian CP 1966–.

Shatirian, Mikhail Aleksandrovich
Armenian writer; b. F. 14 (27), 1916, Tbilisi.
Member of the CPSU since 1940.
He, who graduated from the Moscow Library Institute in 1939, took part in the Great Patriotic War of 1941–45. His work was first published in 1942. He wrote the novel *Soldiers* (1955; Russian translation, 1958), the historical novella on a revolutionary theme *The Spring Came Again* (1960), and the novel *The Tale of the Palm* (1967; Russian translation, 1971), which deals with the 26 Baku Commissars. Decors.: three orders and various medals.
Works: [Shat'iryan, M.] *P'ordzut'yun piesneri zhoghovatsu.* Yerevan 1948. *In Russian translation: Gnev zemli rodnoi: Roman.* Moscow 1963. *Muzykantskaia komanda.* Moscow 1970.

Shauro, Vasilii Filimonovich
Party official; b. 1912.
Belorussian; mem. CPSU 1940–; grad. Mogilev Teachers' Training Inst. 1936 and Party Higher School of Cen. Cttee. CPSU 1942; teacher, head of teaching dept., dir. of a secondary school in Vitebsk Oblast 1930–36 and 1937–40; served in Sov. Army 1936–37; worked for Cen. Cttee. CPSU 1942–47; worked for Cen. Cttee. Belorussian CP (B) 1947–48; Sec. 1948–56, 1st Sec. of Minsk Oblast Cttee. Belorussian CP 1956–60; Dep. to Supr. Sov. Belorussian SSR 1955–59 and 1963–67; Dep. to 1958–, mem. Budget Comm. Sov. of the Union 1962–66, mem. Comm. for Educ., Science and Culture Sov. of Nationalities USSR Supr. Sov. 1966–; mem. 1960–66, mem. Bureau, then Presidium and Sec. of Cen. Cttee. Belorussian CP 1960–D. 1965; Chair. of Supr. Sov. Belorussian SSR 1963–65; cand. mem. Cen. Cttee. CPSU 1966–; Head of Cultural Dept. of Cen. Cttee. CPSU 1966–. Decor.: Order of Red Banner of Labour.

Shavrov, Ivan Egorovich
Military official; b. 1916.
Belorussian; mem. CPSU 1940–; grad. Mil. Acad. of the Tank Troops 1941 and Acad. of the Gen. Staff 1948; Gen. of the Army; in Sov. Army 1935–; chief-of-staff of a tank army, chief-of-staff of a tank corps at Southwestern, Southern, Stalingrad, Don, 4th Ukrainian and 1st and 2nd Baltic Front 1941–45; head of staff dept. of an army, dept. head of staff of tank and mechanized troops, commander of a div., 1st dep. commander of a unit, Dep. Commander-in-Chief of Sov. Forces Group in Germany 1948–63; mem. Cen. Cttee. Esthonian CP 1966–71; 1st Dep. Commander of the Baltic Mil. District 1963–67; Commander of Leningrad Mil. District 1967–73; Dep. to USSR Supr. Sov. 1970–74, mem. Cen. Cttee. CPSU 1971–76; Head of Voroshilov Mil. Acad. 1973–78.

Shchedrin, Rodion Konstantinovich
Composer; b. D. 16, 1932, Moscow.
People's Artist of the RSFSR (1976).
In 1955, he graduated from the Moscow Conservatory, where he had studied composition with Iu. A. Shaporin and piano with Ia. V. Flier. He has created important works in a variety of genres, including the operas *Not For Love Alone* (1961) and *Dead Souls* (based on the novel by N. V. Gogol; libretto by the composer, 1977) and the ballets *The Humpback Horse* (1960), *The Carmen Suite* (based on the opera by G. Bizet, 1967), *Anna Karenina* (1972), and *The Seagull* (1980). All the above works were first staged at the Bolshoi Theatre in Moscow.
In his music he methodically combines elements of Russian folklore and innovative compositional devices. Among his works are the oratorio *Lenin in the Heart of the People* (1969), *Poetic Oratorio* (to words by A. A. Voznesenskii; 1968), and two symphonies (1958 and 1965; second version, 1976). Shchedrin has also composed two concerti for orchestra – *Mischievous Ditties* (1963) and *Chimes* (1968); three concerti for piano and orchestra (1954, 1966, and 1973); works for piano, including 24 preludes and fugues (1970); and the *Polyphonic Notebook* (1972).
He became chairman of the administrative board of the Composers' Union of the RSFSR in 1973. He was a deputy to the sixth, ninth, and tenth convocations of the Supreme Soviet of the RSFSR. Decors.: State Prize of the USSR in 1972, honorary corresponding member of the Bavarian Academy of Fine Arts in 1976, in 1962 member of the Soviet Committee for the Protection of Peace, two orders and various medals.

Shcheglov, Aleksandr Sergeevich
Mordovian writer; b. Mr. 20, 1916, Chindianovo, now Dubenki Raion, Mordovian ASSR; s. of a peasant.
Member of the CPSU since 1940.
He graduated from the Leningrad Institute of Journalism in 1939 and took part in the Great Patriotic War of 1941–45. He was first published in 1932. His works include the poetry collection *The Little Patriot* (1941), the narrative poem *Play, Bandura* (1944), the collections *Short Stories* (1951) and *Everybody Is Beautiful Here* (1963), the play *The Komsomol Card* (1950), and poems for children. His main themes are the heroism of the people during the Great Patriotic War and the contemporary life of the Mordovian village. He was editor-in-chief of the journal *Siatko* from 1967 to 1971. His works have been translated into several languages of the USSR. Decors.: Order of the Red Star and various medals.
Works: *Alkukson' vechkema.* Saransk 1960. *Vishka p'esat.* Saransk 1965. *Utsiaska,* 1968. *In Russia translation: Vernost': Povesti i rasskazy.* Saransk 1970. *Trudnaia liubov': Povesti i rasskazy.* Saransk 1976.

Shcheglov, Georgii Vasil'evich
State official.
Dep. Chair. of Econ. Council of Azerbaidzhan SSR 1963–65; Min. of the Timber, Cellulose, Paper and Wood-Processing Ind. Azerbaidzhan SSR 1965–70; mem. Auditing Comm. 1966–71, mem. Cen. Cttee. Azerbaidzhan CP 1971–; Dep. Chair. of Council of Mins. Azerbaidzhan SSR N. 1970–.

Shchegolev, Viacheslav Ivanovich
Checkers player (international checkers); b. D. 30, 1940, Moscow.
Honored Master of Sport (1964). International grand master (1960). Member of the CPSU since 1969.

He was world champion on several occasions during the years 1960–61 and 1964–65 and USSR champion in 1959, 1963, 1964, and 1976.
Work: *Ot novichka do chempiona.* [Moscow 1969].

Shchelokov, Nikolai Anisimovich
State official; b. 1910; s. of a worker.
Ukrainian; mem. CPSU 1931–; grad. Dnepropetrovsk Metallurgical Inst. 1933; mine worker 1926; engineer, dept. head at various metallurgical plants in the Ukraine 1933–34 and 1935–38; served in Sov. Army 1934–35 and 1941–46; Chair. of Dnepropetrovsk City Sov. 1939–41; Dep. Min. of Local Ind. Ukrainian SSR 1946–47; worked for Cen. Cttee. Ukrainian CP (B) 1947–51; 1st Dep. Chair. of Council of Mins. Moldavian SSR 1951–62; Dep. to 1954–, mem. Econ. Comm. 1962–66 and Chair. of Trade and Public Services Comm. of Sov. of Nationalities USSR Supr. Sov. 1966–68; Chair. of Econ. Council Moldavian SSR 1957–58; mem. and mem. Bureau 1956–66, 2nd Sec. of Cen. Cttee. Moldavian CP 1965–66; Dep. Chair. of Moldavian Council of Mins. 1962–63; mem. Ind. Bureau of Cen. Cttee. Moldavian CP 1962–64; Chair. of Econ. Council Moldavian SSR 1962–65; 1st Dep. Chair. of Council of Mins. Moldavian SSR F.–D. 1965; USSR Min. for the Maintenance of Public Order 1966–68; cand. mem. 1966–68, mem. Cen. Cttee. CPSU Ap. 1968–; USSR Mins. of the Interior N. 1968–D. 1982; promoted to Gen. of the Army S. 1976. Decors.: 2 Orders of Lenin.

Shchepetilnikov, Arkadii Nikolaevich
State official; b. 1930.
Mem. CPSU 1958–; construction engineer; laborer, foreman, chief of a technical prod. dept., chief engineer and chief of a trust, Chief of "Donetskzhytlobud" Combine 1946–66; Chief of "Ukrbud" Construction and Installation Trust during a construction project after a major earthquake in Tashkent 1966–68; Dep. Min. of Installation and Special Construction Work Ukrainian SSR 1968–75, 1st Dep. Min. 1975–79, Min. 1979–81; Min. of Industrial Construction Ukrainian SSR Mr. 1981–; mem. Cen. Cttee. Ukrainian CP 1981–.

Shchepotev, Fedor Lvovich, Dr. of Biol. Sc.
Botanist; b. Jl. 19, 1906, Alferovka, now Voronezh Oblast.
Mem. CPSU 1947–; grad. Voronezh Inst. of Forestry Technology 1932; corresponding mem. Acad. of Sciences Ukrainian SSR 1965–; Dr. of Biol. Sc. 1958; Prof. 1960–; mem. staff Ukrainian Research Inst. of Agro-Forestry Melioration in Kharkov 1933–41 and 1946–63; served in Sov. Army 1942–45; Head of Chair of Botany at Donetsk Univ. 1966–. Publs.: "Vplyv geberelinu na rist i rozvytok deiakykh derevnykh porid" in "Lisove hospodarstvo ta agrolisomelioratsiia" 1963.

Shcherbakov, A. I.
Komsomol official.
Sec. and mem. Bureau Cen. Cttee. Georgian Komsomol D. 1980–.

Shcherbakov, Ilia Sergeevich
Diplomat; b. 1912.
Mem. CPSU 1937–; concluded 4 terms at Cheliabinsk Pedagogic Inst. 1940; pedagogic work in elementary and secondary schools 1930–40; served in Sov. Army 1940–49; mem. staff Cen. Cttee. CPSU 1949–63; Emb. Counsellor-Envoy at USSR Emb. in China 1963–64; USSR Amb. to North Vietnam 1964–74; mem. Cen. Auditing Comm. CPSU 1966–, USSR Amb. to China Jl. 1978–. Decor.: Order of the October Revolution.

Shcherban, Aleksandr Nazarovich, Dr. of Tech. Sc.
Thermophysicist; b. Mr. 3, 1906, Dikanka, Poltava Oblast; s. of a railwayman.
Mem. CPSU 1927–; grad. Dnepropetrovsk Mining Inst. 1933; mem. Acad. of Sciences Ukrainian SSR 1957–; Dr. of Tech. Sc. 1953; Prof. 1954–; miner in the Don Basin 1924–29; worked for "Donshakhte proekt" Trust in Dnepropetrovsk 1933–41; served in World War II 1941–45; Dep. Dir. of Fedorov Inst. of Mining of Acad. of Sciences Ukrainian SSR 1946–53; Chief Acad. Sec. of Presidium Acad. of Sciences Ukrainian SSR 1953–57; Vice-Pres. of Acad. of Sciences Ukrainian SSR 1957–61; Dep. Chair. of Council of Mins. Ukrainian SSR, Chair. of State Cttee. for the Coordination of Scientific Research with Council of Mins. Ukrainian SSR 1961–65; concurrently Head of Dept. of Mining Thermotechnology of Inst. of Thermo-Energetics (now Dept. of Thermophysics of Inst. of Tech. Thermophysics) of Acad. of Sciences Ukrainian SSR 1958–; author of some 200 scientific papers. Decors.: Merited Scientist of the Ukrainian SSR 1966, Order of the Patriotic War 1st Cl., 2 Orders of Red Star, Order "Badge of Honour", medals.
Publs.: "Issledovanie atmosfernykh uslovii v shakhtakh Donbassa" 1950, "Obnaruzhenie i kontrol' metana v shakhtakh i promyshlennykh zdaniiakh" 1955, "Nauchnye osnovy rascheta i regulirovaniia teplovogo rezhima glubokikh shakht" vols. 1–2, 1959–60, and others.

Shcherbina, Boris Evdokimovich
State official; b. 1919.
Ukrainian; mem. CPSU 1939–; grad. Kharkov Inst. of Railroad Engineering 1942 and Party Higher School of Cen. Cttee. CPSU 1948; engineer at a railroad station 1942; Sec. of Kharkov Oblast Cttee. Ukrainian Komsomol 1942 and 1943–44; worked for Cen. Cttee. All-Union Komsomol 1942–43; leading party posts 1944–50; Sec. of Kharkov City Cttee. Ukrainian CP (B) 1950–51; Sec., 2nd Sec. of Irkutsk Oblast Cttee. CPSU 1951–61; Dep. to Supr. Sov. RSFSR 1959–63; cand. mem. 1961–76, mem. Cen. Cttee. CPSU 1976–; 1st Sec. of Tiumen Oblast Cttee. CPSU 1961–73 (for Agriculture 1963–64); Dep. to Supr. Sov. USSR 1962–; USSR Min. of Construction of Petroleum and Gas Ind. Enterprises D. 1973–. Decors.: 2 Orders of Lenin.

Shcherbina, Vladimir Rodionovich
Russian critic and literary scholar; b. Je. 26 (Jl. 7), 1908, Georgievsk, now Stavropol' Krai.
Corresponding member of the Academy of Sciences of the USSR (1976). Member of the CPSU since 1939.
He graduated from the literature department of the Krasnodar Pedagogical Institute in 1929. He was first published in 1933. He became deputy director of the A. M. Gorkii Institute of World Literature in 1953.
He is the author of books on A. S. Novikov-Priboi and A. N. Tolstoy. In his critical articles and scholarly works he propagandizes on behalf of the achievements of socialist realism and polemizes with bourgeois theorists of aesthetics. His works include *Our Contemporary: The Concept of the Human Being in 20th-century Literature* (1964), *Problems of Literary Scholarship in the Light of the Heritage of V. I. Lenin* (1971), and *The Writer in the Contemporary World: Literature, Ideology, Culture* (1973). Decor.: Order of the Red Banner of Labour 1968.

Shcherbitskii, Vladimir Vasil'evich
Party official; b. F. 17, 1918, Verkhnednepròvsk, now Dnepropetrovsk Oblast; s. of a worker.
Ukrainian; me. m. CPSU 1941–; grad. Dnepropetrovsk Inst. of Chemical Technology 1941; Komsomol work 1934–35; student 1936–41; command posts in Sov. Army 1941–45; engineer at a coke-chemical plant in Dneprodzerzhinsk 1945–46; sec. of party Cttee. at same plant 1946–48; Head of Dept. of Org. and Instruction with Dneprodzerzhinsk City Cttee. Ukrainian CP 1948; 2nd Sec. of Dneprodzerzhinsk City Cttee. Ukrainian CP (B) 1948–51; party organizer for Cen. Cttee. CPSU at a metal plant in Dneprodzerzhinsk 1951–52; 1st Sec. of Dneprodzerzhinsk City Cttee. Ukrainian CP 1952–54; mem. Auditing Comm. 1952–54; cand. mem. 1954–56, mem. Cen. Cttee. Ukrainian CP 1956–; 2nd Sec. 1954–55, 1st Sec. of Dnepropetrovsk Oblast Cttee. Ukrainian CP 1955–57; Dep. to Supr. Sov. Ukrainian SSR 1955–; mem. Cen. Auditing Comm. CPSU 1956–61; mem. Presidium, then mem. Politbureau 1957–, Sec. of Cen. Cttee. Ukrainian CP 1957–61; Dep. 1958–, mem. Presidium USSR Supr. Sov. 1974–; mem. 1961, cand. mem. 1961–63 and 1965–71, mem. Presidium, then Politbu-

reau Cen. Cttee. CPSU 1971–; Chair. of Ukrainian Council of Mins. 1961–63 and 1965–72; 1st Sec. of Dnepropetrovsk Oblast Cttee. Ukrainian CP 1963–65 (for Ind. 1963–64); 1st Sec. of Cen. Cttee. Ukrainian CP My. 1972–. Decors.: 3 Orders of Lenin, Hero of Socialist Labour.

Shchukin, Aleksandr Nikolaevich
Radio technologist; b. Jl. 9 (22), 1900.
Mem. CPSU 1944–; grad. Leningrad Inst. of Electrotechnology 1927; Maj.-Gen. in Engineering Service; teacher 1929–41, Prof. at Leningrad Inst. Electrotechnology 1939–; concurrently teacher at Naval Acad. and other scientific research institutes 1933–45; corresponding mem. 1946–53, mem. USSR Acad. of Sciences 1953–. Decors.: Order of Lenin, 2 other orders, various medals.

Shchukin, G. P.
Komsomol official.
Head of Dept. for Komsomol Organisations with Cen. Cttee. Latvian Komsomol 1978–82; Sec. and mem. Bureau Cen. Cttee. Latvian Komsomol O. 1980–.

Shefner, Vadim Sergeevich
Russian writer; b. D. 30, 1914 (Ja. 12, 1915), Petrograd.
Member of the CPSU since 1945.
He, the son of a military man, studied in the *rabfak* (workers' school) at the University of Leningrad and served in the Great Patriotic War (1941–45). His poetry collection *Defense* (1943), his narrative poem *Meeting in the Suburb* (1945), and his novella *Sister of Sorrow* (1970) were devoted to the heroic defenders of Leningrad. In his poetry collections *Signs of the Earth* (1961), *Vaults* (1967), and *Freeboard* (1970) he turned toward the traditions of Russian philosophical lyric poetry.
He has also written tales of fantasy, entertaining and ironic in tone, that raise moral questions. They include the novellas *Girl at the Precipice, or The Notes of Kovrigin* (1964), *The Man With Five "No's," or the Confession of a Simpleton* (1967), and *The Utter Mystery* (1970).
Works: *Izbr. proizv.*, vols. 1–2. Leningrad 1975. *Imia dlia ptitsy: Povesti.* Leningrad 1976.

Sheindlin, Aleksandr Efimovich
Specialist in heat physics and power engineering; b. Ag. 22 (S. 4), 1916, Kuibyshev.
Academician of the Academy of Sciences of the USSR (1974; corresponding member, 1964). Member of the CPSU since 1945.
After graduating from the Moscow Power Engineering Institute in 1937, he first worked as an engineer in a factory and then as an assistant at the institute. From 1941 to 1945 he was in the Red Army. From 1945 to 1967 he worked at the Moscow Power Engineering Institute, where he received the doctor of technical sciences degree in 1954 and became a professor in 1955. He became head of a laboratory at the Institute of High Temperatures of the Academy of Sciences of the USSR in 1960 and the institute's director in 1961. Since 1967 he has also been head of a subdepartment at the Moscow Physicotechnical Institute.
Of great importance have been his experimental and theoretical studies of the thermodynamic properties of water vapor at supercritical parameters. He directed the construction and successful operation of the world's first experimental model magnetohydrodynamic installation (U-02, 1964) for the direct conversion of heat energy into electricity and the first such industrial pilot installation (U-25, 1971). Decors.: Lenin Prize in 1959, the State Prize of the USSR in 1976, and the Polzunov Prize in 1963, Order of Lenin, two other orders, and various medals.
Works: *Issledovaniia termodinamicheskikh svoistv veshchestv.* Moscow-Leningrad 1963. (With V. A- Kirillin.) *Tekhnicheskaia termodinamika,* 2nd ed. Moscow 1974. (With V. A. Kirillin and V. V. Sychev.) "Issledovanie MGD-ustanovki U-02 pri ee rabote v dlitel'nykh rezhimakh." *Teplofizika vysokikh temperatur,* 1971, vol. 9, no. 5. (With others.) "Nekotorye itogi issledovaniia energeticheskoi MGD Ustanovki U-25." *Teplofizika vysokikh temperatur,* 1974, vol. 12, no. 2.

Shekinskii, Ekrem Mamed Eminovich
State official; b. 1929.
Head of Geology Admin. Azerbaidzhan SSR Ap. 1974–.

Shelaginov, E. S.
Trade union official.
Sec. and mem. Presidium Azerbaidzhan Republican Trade Union Council F. 1982–.

Shelar, Nikolai Nikanorovich
State official; b. 1929.
Moldavian; mem. CPSU 1956–; grad. Pedagogical Inst. in Kishinev and Party higher School with Cen. Cttee. CPSU; Teacher 1949–50; served in the Soviet Army 1950–52; Teacher, held Komsomol and party posts 1953–62; Chair. Oktiabrskii Raion Exec. Cttee. of People's Deps. 1962–67; mem. Auditing Comm. Moldavian CP 1963–76; mem. apparat Cen. Cttee. Moldavian CP 1967–68; Chair. of Bendery City Exec. Cttee. of People's Deps. 1968–73; 1st Sec. of Bendery City Cttee. Moldavian CP 1973–80; Chair. of State Cttee. on Vocational Training Moldavian SSR D. 1980–.

Shelenkov, Aleksandr Vladimirovich
Cameraman; b. D. 13 (26), 1903, Zamesh'e, now Smolensk Oblast.
Honored Art Worker of the RSFSR (1959).
He graduated from the department of camera operation of the State Technicum of Cinematography in 1929. Until 1945 he was a cameraman for the Soiuzdetfil'm motion picture studio, and after 1945 he worked for Mosfil'm. He worked with the cameraman I. E. Chen beginning in 1944.
His films have included *Dzhul'bars* (1936), *Salavat Iulaev* (1941), *Zoia* (1944), *Rainis* (1949), *Far From Moscow* (1950), *Admiral Ushakov* (1953), *The Ships Are Storming the Bastions* (1953), *Romeo and Juliet* (1955), the joint Soviet-Bulgarian film *A History Lesson* (1957), *The Communist* (1958), the joint film of the USSR and the German Democratic Republic (GDR) *Five Days, Five Nights* (1961), and the joint Soviet-Polish film *Jaroslaw Dąbrowski* (1974–76). He and Chen also shot a number of scenes for the film *War and Peace* (1966–67), including the battles of Schöngrabern and Austerlitz and the hunt. Decors.: State Prize of the USSR in 1946, 1947, 1950, and 1951, two orders and several medals.
Works: "Puti razvitiia izobrazitel'noi formy." In the collection *Mosfil'm,* no. 1. Moscow 1959. "Operator i fil'm." *Mosfil'm,* no. 2. Moscow 1961.

Shemiakin, Evgenii Ivanovich
Scientist specializing in the mechanics of rocks; b. D. 9, 1929, Novosibirsk.
Corresponding member of the Academy of Sciences of the USSR (1976). Member of the CPSU from 1963.
He graduated from Leningrad University in 1952. From 1955 to 1960 he worked as a research worker at the Institute of Chemical Physics of the Academy of Sciences of the USSR. From 1960 to 1971 he was head of a laboratory of the Institute of Theoretical and Applied Mechanics of the Siberian Division of the Academy of Sciences of the USSR. In 1971 he was named director of the Mining Institute of the Siberian Division of the Academy of Sciences of the USSR. He is also a professor at the University of Novosibirsk (since 1965).
His main works deal with the study of rocks, the effects of blasting and shock on rocks, and the deformation mechanics of solids. He developed a model of a solid medium with frictional properties and used this model to study the process of wave propagation in underground blasts. Decors.: 2 orders and a medal.
Work: *Dinamicheskie zadachi teorii uprugosti i plastichnosti.* Novosibirsk 1968.

Shendryk, Ludmila Karpovna
Trade union official; b. 1919.
Sec. and mem. of Presidium Ukrainian Republican Trade Union Council 1958–.

Shepelev, Iurii Faddeevich
Diplomat.
USSR Amb. to Mozambique N. 1982–.

Shepetis, Longinas Klemensovich
Party official.
Head of Dept. of Science and Culture of Cen. Cttee. Lithuanian CP 1964–67; mem. Cen. Cttee. Lithuanian CP 1966–; Min. of Culture Lithuanian SSR 1967–76; Sec. and mem. of Bureau Cen. Ctte. Lithuanian CP 1976–.

Sheremetev, Nikolai Vasil'evich
State official.
1st Dep. Chair. of City Exec. Cttee. of People's Deps. in Ashkhabad 1973–77; Min. for Construction Turkmen SSR D. 1977–; mem. Cen. Cttee. Turkmen CP 1981–.

Sheris, Iuozas Mikolovich
State official.
Cand. mem. 1964–66, mem. Cen. Cttee. Lithuanian CP 1966–; Dep. to Supr. Sov. Lithuanian SSR 1965–; Min. of Communal Econ. Lithuanian SSR Jl. 1971–.

Shermukhamedov, Said
State official.
Cand. mem. 1966–71, mem. Cen. Cttee. Uzbek CP 1971–; Head of Cultural Dept. of Cen. Cttee. Uzbek CP 1965–70; Dep. to 1967–, Chair. of Comm. for Science and Culture of Supr. Sov. Uzbek SSR 1967–70; Min. of Educ. Uzbek SSR Jl. 1970–.

Sheshial'gis, Kazis Kaze (Kazys Šešelgis)
Architect; b. Ja. 9 (22), 1915, Žeimiai, now Rokiškis Raion, Lithuanian SSR.
Honored Engineer of the Lithuanian SSR (1965). Doctor of architecture (1967).
He studied in the civil engineering department of the University of Kaunas from 1936 to 1943. He was a professor and head of a subdepartment at the polytechnical institute in Kaunas from 1951 to 1971 and has been head of a subdepartment at the civil engineering institute in Vilinius since 1971. He was chairman of the Kaunas branch of the Architects' Union of the Lithuanian SSR from 1954 to 1959.
His works include the master plans for Panevėžys (1947–53) and Kėdainiai (1962, with V. Miliukštis) and the reconstruction project for the old towns in Vilnius (1958, with S. Lasavickas and J. Šeibokas) and Kaunas (1962, with V. Jurkštas and S. Čerškutė).
Works: *Kolūkiu gyvenviečiy planavimas.* [Vilnius] 1951. [With M. Kleinas and V. Zubovas.] *Rajoninio planavimo ir urbanistikos pagrindai.* Vilnius 1975.

Sheshplaukis, Bronislovas Aleksovich
State official.
Min. of Construction Lithuanian SSR Je. 1978–; cand. mem. Cen. Cttee. Lithuanian CP 1981–.

Shestalov, Iuvan Nikolaevich (Ivan Nikolaevich Shestalov)
Mansi writer; b. Je. 22, 1937, Kamratka, Berezovo Raion, Khanty-Mansi National Okrug, Tiumen's Oblast.
Member of the CPSU since 1967.
He, the son of a kolkhoznik, graduated from the Tiumen' Pedagogical Institute in 1965. He writes in Mansi and Russian. An expert on the folklore and customs of the Mansi, he has made numerous trips through the Tiumen' north.
He is the author of collections of lyric and narrative poems, short stories, novellas, and essays, as well as children's books. The main themes of his works are the new life of the formerly deprived peoples of the Far North, the labour of the Soviet people, and his native countryside. His works have been translated into many languages of the peoples of the USSR and foreign languages.
Works: *Makem at.* Khanty-Mansiisk 1958. *Man'si pavyl niavramytn.* Leningrad 1960. *In Russian translation: Izbrannoe,* Introduction by Iu. Prokushev, Leningrad 1976. *Poite, moi zvesdy: Stikhi.* Leningrad 1959. *Ogon' na l'du.* Moscow 1961. *Sinii veter kaslaniia.* Moscow 1964. *Pesnia poslednego lebedia. Poemy.* Moscow 1969. *Iugorskaia kolybel',* 2nd ed. Moscow 1972. *Skazki taezhnogo igrishcha.* Moscow 1972. *Taina Sorni-nai.* Moscow 1976.

Shestopalov, Nikolai Fedorovich
Military official; b. D. 19 1919, Moscow. Mem. CPSU 1951–; grad. V. V. Kuibyshev Mil. Engineers Acad. 1941, Engineer Col.-Gen. 1978; entered Sov. Army 1937; chief engineer among various engineer troop forces on the Southern, Briansk, and Voronezh fronts 1941–43; brigade engineer, chief of engineer forces of 22nd Section of the guards of a motorized rifle brigade 1943–44; asst. chief of engineer troops staff of 3rd guard of the tank army on the 1st Ukrainian front 1944–45; dep. commander, commander of a construction battalion, dep. commander and chief engineer of an airfield construction regiment, chief engineer and chief of a dept. of a capital airfield construction district 1945–59; chief of a dept. of a construction admin. of a mil. district 1959–61, chief 1961–63; Dep. Commander of Troops of Belorussian Mil. District for Construction and Billeting of troops 1963–69; Chief of Main Admin. for Mil. Construction of USSR Min. of Defense 1969–72; 1st Dep. Chief for Construction and Billeting of Troops of USSR Min. of Defense 1972–78; Dep. USSR Min. of Defense for Construction and Billeting of Troops D. 1978–. Decors.: Order of the October Revolution, Order of the Great Fatherland War 1st and 2nd Class, Order of Red Banner of Labour (twice), Order of Red Star (twice), medals, foreign orders and medals.

Shevardnadze, Eduard Amvrosievich
Party official; b. 1928; s. of a teacher. Georgian; mem. CPSU 1948–, grad. Republican Party School Cen. Cttee. Georgian CP 1951 and Kutaisi Pedagogical Inst. 1959; Komsomol, sov., and party posts 1948–56; 2nd Sec. of Cen. Cttee. Georgian Komsomol 1956–57, 1st Sec. 1957–61; mem. Cen. Cttee. Georgian CP 1960–; cand. mem. Bureau of Cen. Cttee. Georgian CP 1960–61, mem. 1972–; 1st Sec. of Cen. Cttee. Georgian CP S. 1972–; 1st Sec. of Mtskheta Raion Cttee., 1st Sec. of Pervomaiskii Raion Cttee. in Tbilisi City Georgian CP, 1st Dep. Min. for Protection of Public Order Georgian SSR 1961–65; Min. of Internal Affairs Georgian SSR 1965–72; 1st Sec. of Tbilisi City Cttee. Georgian CP Jl.–S. 1972; mem. Cen. Cttee. CPSU 1976–; cand. mem. Politbureau Cen. Cttee. CPSU 1978–. Decor.: Hero of Socialist Labour 1981.

Shevchenko, Aleksandr Tikhonovich
State official; b. 1924.
After discharge from Sov. Army master, section head in various brickworks 1947–55; dir. of a plant for inflammable products and wall coverings, dir. of trust for construction materials, head of Main Admin. for Construction Materials Ind. Ukrainian SSR 1955–74; Dep. Min. 1974–79, Min. Construction Materials Ind. Ukrainian SSR N. 1979–; head of a dept. of State Planning Cttee. Ukrainian SSR 1977–79.

Shevchenko, Ivan Viktorovich
State official.
Dep. to Supr. Sov. Tadzhik SSR 1963–; Sec. of Party Cttee. of Prod. Admin. in Kairakkum 1963; mem. Cen. Cttee. Tadzhik CP 1963–81, cand. mem. 1981; Min. of the Construction Materials Ind. Tadzhik SSR S. 1969–.

Shevchenko, Valentina Semenovna
State official.
Sec. of Cen. Cttee. Ukrainian Komsomol 1962–69; Chair. of Ukrainian Soc. of Friendship and Cultural Relations with Foreign Countries 1972–75; Dep. to and Dep. Chair. of Presidium Supr. Sov. Ukrainian SSR 1975–; mem. Cen. Cttee. Ukrainian CP 1976–.

Shevelev, Firs Aleksandrovich
Hydraulics scientist; b. D. 1 (14), 1912, Moscow.
Doctor of technical sciences (1954); professor (1962). Member of the CPSU since 1940.
He graduated from the Moscow Construction Engineering Institute in 1937. From 1938 to 1941 and from 1946 to 1954 he worked at VODGEO (All-Union Scientific Research Institute of Water Supply, Sewage Hydrotechnical Construction, and Engineering Hydrogeology). From 1954 to 1957 he worked in the construction department of the Central

Committee of the CPSU. Between 1957 and 1964 he worked at the Academy of Construction and Architecture of the USSR. In 1965 he became director of the K. D. Pamfilov Academy of Municipal Services. His main works are devoted to engineering hydraulics, mainly the hydraulics of pipe systems. Decors.: eight orders and several medals.

Shibaev, Aleksei Ivanovich
State official; b. 1915.
Russian; mem. CPSU 1940–; grad. Dept. of Physics and Mathematics of Gorkii Univ. 1940; apprentice, tech. draftsman, designer at "Krasnoe Sormovo" Plant; after completing univ. educ. design engineer, dep. dept. head, dept. head at a plant in Gorkii; organizer for Cen. Cttee. All-Union CP(B) at a Novosibirsk plant 1943–47; dir. of various plants in Rostov and Saratov Oblast 1947–55; 2nd Sec. 1955–59, 1st Sec. of Saratov Oblast Cttee. CPSU (for Ind. 1963–64) until N. 1976; Dep. to RSFSR Supr. Sov. 1959–63; mem. Cen. Cttee. CPSU 1961–; Dep. to 1962–, mem. Budgetary Comm. 1962–66, mem. Draft Bills Comm. of Sov. of the Union USSR Supr. Sov. 1966–77, mem. Presidium of the USSR Supreme Soviet 1977–82; Chair. of Cen. Council of Sov. Trade Unions N. 1976–82; USSR Dep. Min. of Instrument-making, Automation Equipment and Control Systems Mr. 1982–. Decors.: 2 Orders of Lenin.

Shiian, Anatolii Ivanovich
Ukrainian writer; b. Mr. 23 (Ap. 5), 1906, Borisovka, now Belgorod Oblast.
Member of the CPSU since 1947.
He graduated from the Kiev Institute of Timber Technology in 1929. He was a member of General A. N. Saburov's partisan unit during the Great Patriotic War of 1941–45. He wrote the novella *Slops* (1930) and the novel *Main Line* (1934). He is also the author of the book of sketches *Partisan Country* (1946) and collections of plays, novellas, and short stories. His most important work, the novel *The Thunderstorm* (1936), deals with the life of a prerevolutionary village and the people's struggle for Soviet power; the film *Thunderstorm Over the Fields* (1959) was based on the novel. Decors.: three orders and various medals.
Works: *Tvory.* vols. 1–2. Kiev 1976. *In Russian translation: Groza.* Moscow 1959.

Shileikis, P.
Laborer.
Head of a painting crew at "Vilniusstroi" trust in Vilnius 1981; mem. Cen. Cttee. and cand. mem. Bureau of Cen. Cttee. Lithuanian CP 1981–.

Shilin, Afanasii Petrovich
Military official; b. S. 1, 1924, Petropavlovka, in present-day Novouzensk Raion, Saratov Oblast.
Twice Hero of the Soviet Union (F. 22, 1944, and Mr. 24, 1945). Lieutenant-general (1975). Member of the CPSU since 1944.
He was born into a worker family. He entered the Red Army in 1942 and graduated from the Second Tomsk Artillery School in 1943, the Higher Artillery School for Officers in 1946, the F. E. Dzerzhinskii Artillery Academy in 1952, and the Military Academy of the General Staff in 1966. During the Great Patriotic War (1941–45), he served from 1943 on the Southwestern, Third Ukrainian, and First Belorussian fronts. He commanded a platoon and was chief of reconnaissance for an artillery battalion in the 132nd Guards Artillery Regiment. He took part in the fighting for the Dnieper and the liberation of the Ukraine, Moldavia, and Poland. After the war, he held important command positions.
Decors.: Order of Lenin, the Order of the Red Banner, The Order of the Red Banner of Labor, the Order of the Patriotic War Second Class, and various medals. A bust of him has been erected in the city of Leninsk-Kuznetskii in Kemerovo Oblast.

Shil'nikovskii, Evstafii Pavlovich
Artist; b. F. 14 (26), 1890, Kuznetsovo, now Vologda Oblast.
Honored Art Worker of the RSFSR (1956).
He studied at the St. Petersburg Academy of Arts from 1913 to 1917 under V. V. Mate. He was a leading craftsman of Velikii Ustiug niello work in silver. From 1935 to 1962 he was artistic director of the Severnaia Chern' Artel (later reorganized as a factory) in Velikii Ustiug. His works include spoons with designs on themes from works by A. Pushkin (1937), a service on themes from P. P. Ershov's fairy tale "The Little Humpbacked Horse" (1947), and the box and cup "Moscow" (1956). Articles from these sets, all in niello, are housed in the Russian Museum in Leningrad, the Historical Museum and the Museum of Folk Art of the USSR in Moscow, and other museums of the USSR.

Shilo, Nikolai Alekseevich
Geologist; b. Mr. 25 (Ap. 7), 1913, Piatigorsk.
Academician of the Academy of Sciences of the USSR (1970); corresponding member (1964). Hero of Socialist Labor (1973). Member of the CPSU since 1941.
After graduating from the Leningrad Institute of Mines in 1937, he worked for various geological institutions in the northeastern part of the USSR. In 1960 he became director of the Northeastern Integrated Scientific Research Institute of the Far East Scientific Center of the Academy of Sciences of the USSR located in Magadan. He has been chairman of the presidium of the Far East Scientific Center since 1977.
His main works deal with the study of the geology and geochemistry of mineral deposits, including gold, silver, tin, mercury, tungsten, platinum, and copper. He developed a genetic classification of gold deposits and identified periglacial lithogenesis, revealing the mechanism of rock formation in periglacial zones. Decors.: two Orders of Lenin, three other orders, and medals.
Works: "Klassifikatsiia rossypei Iano-Kolymskogo zolotonosnogo poiasa." In *Geokhimiia, petrografiia i mineralogiia osadochnykii obrazovanii.* Moscow 1963. "Glavneishie cherty zolotogo i zoloto-serebrianogo orudeneniia vostochno-aziatskikh vulkanogennykh poiasov." In *Problemy metallogenii Sovetskogo Dal'nego Vostoka.* Moscow 1967.

Shimanskii, Vsevolod Pavlovich
State official.
Head of Section for Union Republics at Dept. of Party Organs with Cen. Cttee. CPSU 1962; Min. of Trade RSFSR O. 1972–.

Shimkus, Virginius-Kestutis Jono
Architect; b. 1937.
Grad. Dept. of Construction at Kaunas Polytech. Inst. 1961; civil engineer-architect; mem. staff Inst. of Design of Agricultural Construction Projects with "Gosstroi" Lithuanian SSR 1962–. Decor.: State Prize of the USSR 1971.

Shinkaruk, V.
Philosopher.
Corresponding mem. USSR Acad. of Sciences 1981–. Dir. of the Inst. of Philosophy Ukrainian Acad. of Sciences; specializes in inquiries into the historical origin of the problem of the unity of dialectic, logic, and the theory of cognition in modern philosophy, into classical German philosophy, and into the process of the evolution of Marxist philosophy; examines in particular the philosophy of Hegel, with research in the area of Russian philosophy; analyzed the philosophy of the outstanding Ukrainian scholar Grigori Skovoroda; author of more than 160 publications.

Shinkuba, Bagrat Vasil'evich
Abkhazian writer and state and public figure; b. Ap. 39 (My. 12), 1917, Chlou, now Ochamchira Raion, Abkhazian ASSR.
People's Poet of Abkhazia (1967). Member of the CPSU since 1946.
He graduated from the Sukhumi Pedagogical Institute in 1939. He first published his works in 1935. He is the author of two novels in verse: *My Countrymen* (1950), which is about the postwar life of kolkhozniks, and *Song of the Cliff* (1965; Russian translation 1967; film version entitled *The White Bashlyk*), which describes the revolutionary events in Abkhazia from 1905 to 1907. His other works include the novella *Chanta Has Come* (in Russian, 1969) and the historical novel *The Last to Leave* (1974; Russian translation, 1976).

He has been a member of the Auditing Commission of the Communist Party of Georgia since 1976. Since 1958 he has been a candidate member of the Central Committee of the Communist Party of Georgia and chairman of the Presidium of the Supreme Soviet of the Abkhazian ASSR. He was a deputy to the fifth convocation of the Supreme Soviet of the USSR. Decors.: Order of Lenin, three other orders, and several medals.
Works: *Ialkaau ifymtakua,* vols. 1–2. Akua 1967–68. *In Russian translation: Izbrannoe.* [Introduction by K. Simonov.] Moscow 1976.

Shiriev, Y.
Economist.
Corresponding mem. USSR Acad. of Sciences 1981–. Dir. of the International Inst. of Economic Problems of the World Socialist System of the Council for Mutual Economic Assistance; main areas of research: problems of the international socialist division of labor in the economic integration of CMEA countries, cooperation of socialist countries as an economic mechanism, improvement of planning and management of foreign economic ties; author of roughly 150 publications, mem. editorial advisory council of the periodical *Gesellschaftswissenschaften* (Social Sciences).

Shirkov, Dmitrii Vasil'evich
Physicist; b. Mr. 3, 1928, Moscow.
Corresponding member of the Academy of Sciences of the USSR (from 1960). Member of the CPSU from 1953.
He graduated from Moscow State University in 1949. From 1950 to 1960 he worked at the V. A. Steklov Institute of Mathematics of the Academy of Sciences of the USSR and at the Joint Institute for Nuclear Research in Dubna. From 1960 to 1969 he worked at the Institute of Mathematics of the Siberian Division of the Academy of Sciences of the USSR; at the same time, beginning in 1961, he was a professor at the University of Novosibirsk.
His main works deal with the theory of elementary particles, the theory of superconductivity, and the theory of neutron transport and moderation. In the theory of elementary particles, he developed the renormalization theory, created the method of the renormalization group, and constructed semiphenomenological schemes for the quantitative description of strong interactions at low energies. Decors.: Lenin Prize in 1958, three orders and various medals.
Works: *Novyi metod v teorii sverkhprovodimosti.* Moscow 1958. (With N. N. Bogoliubov and V. V. Tolmachev.) *Vvedenie v teoriiu kvantovannykh polei,* 3rd ed. Moscow 1976. (With N. N. Bogoliubov.) *Dispersionnye teorii sil'nykh vzaimodeistvii pri nizkikh energiiakh.* Moscow 1967. (With V. A. Meshcheriakov and V. V. Serebriakov.)

Shirobokov, Stepan Pavlovich
Udmurt writer; b. D. 5 (18), 1912, Malyi Kazess, now Sharkan Raion, Udmurt ASSR.
People's Poet of Udmurtia (1963).
He graduated from the Izhevsk Teachers Institute in 1939. He took part in the Great Patriotic War of 1941–45. He is the author of collections of poems and narrative poems; the drama *Every Wolf Has Its Own Path* (1956); the comedy *If There Is No Love* (1962); the first Udmurt comedy in verse, *Hey, Beautiful Girls* (1970); the novella *Songs Find the Road* (1967); books for children; literary sketches, and short stories. Shirobokov writes about modern youth and the people of kolkhozes.
Several of his poems have become folk songs. Decors.: two orders and various medals.
Works: *In Russian translation: Pust' oiut solov'i.* Izhevsk 1958. *Pesni polei.* Izhevsk 1968.

Shirokov, Aleksandr Zosimovich, Dr. of Geol.-Min. Sc.
Geologist; b. S. 5, 1905, Ivanovo; s. of a worker.
Grad. Dnepropetrovsk Mining Inst. 1929; corresponding mem. Acad. of Sciences Ukrainian SSR 1957–; Dr. of Geol.-Min. Sc. 1952; Prof. 1952–; Chief Geologist of "Donvugillia" Admin. 1929–32 and of "Donbasantratsit" 1932–33; Dep. Head of Geological Bureau "Golovvugillia", Specialist and Head of Geological Prospecting Projects of "Donbasvuglerozvidka" Trust 1933–41; Chief Engineer and Dep. Head of "Mosvuglerozvidka" Trust 1941–46; mem. staff Dnepropetrovsk Mining Inst. 1946–; Head of Chair of Historic Geology and Palaeontology at same inst. 1948–; author of more than 100 scientific papers. Decor.: Order of Red Banner of Labour.
Publs.: "Bolshoi Donbass" 1957, co-author of "Zakonomernosti uglenakopleniia na territorii Zapadnogo Donbassa" 1963, and others.

Shirshin, Grigorii Chooduevich
Party official; b. 1934.
Tuvin; mem. CPSU 1958–; grad. Kyzyl Pedagogical School 1957, CPSU Cen. Cttee. Party Higher School 1964, and CPSU Cen. Cttee. Acad. of Social Sciences 1972, Cand. of Hist. Sc. 1972; sec. of a people's court 1950–52; dept. chief of a Komsomol raion cttee. 1952–54; served in Sov. Army 1954–55; teacher 1955–57; 2nd sec., 1st sec. of a Komsomol raion cttee., dept. chief of a Komsomol oblast cttee., 1st Sec. of Kyzyl City Komsomol Cttee. 1957–60; Oblast Sec., 1st Oblast Sec. of Komosomol Cttee. Tuva ASSR 1960–67; dept. chief, Oblast Sec. of CPSU Cttee. Tuva ASSR 1967–73; Dep. to USSR Supr. Sov. 1970–; 1st Oblast Sec. of CPSU Cttee. Tuva ASSR Je. 1973–; cand. mem. Cen. Cttee. CPSU 1976–.

Shirshov, Anatolii Illarionovich
Mathematician; b. Ag. 8, 1921, Kolyvan', now Novosibirsk Oblast.
Corresponding member of the Academy of Sciences of the USSR (1964). Member of the CPSU since 1951.
He graduated from the Voroshilovgrad Pedagogical Institute in 1949. From 1950 to 1960 he was on the staff of Moscow State University, and in 1960 he became associated with the Institute of Mathematics of the Siberian Division of the Academy of Sciences of the USSR. His principal works deal with algebra, in particular, the general theory of nonassociative algebras, alternative rings, and Lie algebras. Decors.: two orders, medals.
Works: "Podalgebry svobodnykh hevykh algebr." *Mathematicheskii sbornik,* 1953, vol. 33, issue 2. "Nekotorye voprosy teorii kolets, blizkikh k assotsiativnym." *Uspekhi matematicheskikh nauk,* 1958, vol. 13, issue 6.

Shishigin, Firs Efimovich
Stage director; b. Ag. 17 (30), 1908, Borok-Gorodok, Arkhangel'sk Oblast.
People's Artist of the USSR (1964). Member of the CPSU since 1940.
He graduated from the Leningrad Technicum of Theater Arts in 1929. From 1928 to 1933 he worked at the Theater of Classical Miniatures and the Theater of Working Youth (TRAM), and from 1933 to 1946 he was artistic director of theaters in Ussuriisk, Spassk, and Vladivostok. In 1946–47 he was principal stage director of the Stavropol' Theater, and from 1947 to 1950 he worked at the Moscow Theater of Drama and Comedy. Shishigin was principal stage director of the Volgograd Theater from 1950 to 1956 and of the Voronez Theater from 1956 to 1960. In 1960 he became artistic director and principal stage director of the F. G. Volkov Iaroslavl Theater.
He, a master of the epic, monumental panorama, has been attracted mainly to heroic and romantic drama and folk tragedy. He has devoted particular attention to detailed mass scenes and to the class-related psychology of his heroes. His best productions include a dramatization of Fadeev's *The Rout* (1933), Kocherga's *The Watchmaker and the Hen* (1935), Simonov's *A Lad From Our Town* (1942), Gorkii's *The Old Man* (1946), Chepurin's *Conscience* (1950), Gorkii's *The Lower Depths* (1952), *Port Arthur* by Stepanov and Popov (1954), *Virineia* by Seifullina and Pravdukhin (1957), Korablinova's *Aleksei Kol'tsov* (1958), Pogodin's *Third Pathétique* (1959), Gorkii's *Children of the Sun* (1962), Nazarov's *The Panfilovites* (1965). Dangulov's *The Ambassador's Diary* (1970), Chepurin's *The Snows* (1971), and Ankilov's *The Soldier's Widow* (1972).
Decors.: State Prize of the USSR in 1951 and the State Prize of the RSFSR in 1971, Order of Lenin, the Order of the Red Banner of Labor, and various medals.

Shishkin, Aleksei Alekseevich
State official.
Head of Main Admin. 1959–66, Chair. of State Cttee. for Vocational Training Esthonian SSR 1966–; mem. Auditing Comm. 1960–61, mem. 1961–66 and 1976–, cand. mem. Cen. Cttee. Esthonian CP 1966–76.

Shishko, Sergei Fedorovich
Painter; b. Je. 12 (25), 1911, Nosovka, now Chernigov Oblast, Ukrainian SSR. People's Artist of the USSR (1974). Member of the CPSU since 1952.
He attended the Kiev Art Institute from 1929 to 1933 and the Leningrad Academy of Arts from 1936 to 1943. He paints lyrical landscapes, mostly devoted to Kiev. Notable examples are *Vladimir Hill* (1946, Kharkov Art Museum), *The Dnieper: View of the Podol* (1950, Museum of Ukrainian Fine Arts of the Ukrainian SSR, Kiev), *The Dnieper in the Morning* (1953, Museum of Ukrainian Fine Arts), *The New Kiev* (1960, Voroshilovgrad Artem Art Museum), and *The Vistas of Kiev* (1974).

Shishkov, Vladimir Petrovich
Veterinarian; b. My. 24, 1927, Odoev, Raion, Tula Oblast.
Academician of the V. I. Lenin All-Union Academy of Agricultural Sciences (1973). Member of the CPSU since 1951.
He graduated from the Moscow Veterinary Academy in 1951. Beginning in 1956, he worked at the academy, where he became rector in 1967. He has been academician-secretary of the veterinary division of the V. I. Lenin All-Union Academy of Agricultural Sciences since 1973. His work has dealt mainly with metabolic diseases of highly productive animals, diseases of young animals, leukosis, developmental physiology, geographic pathology, and veterinary aspects of stock raising. Decors.: Order of the Red Banner of Labor and several medals.

Shishlov, Fedor Vasil'evich
Laborer; b. 1931.
Mem. CPSU 1968–; shop driller in F. E. Dzerzhinskii Rail Car Manufacturing Plant in Uralsk 1950–54; served in Sov. Army 1954–57; fitter, boiler-maker in F. E. Dzerzhinskii Rail Car Manufacturing Plant in Uralsk 1957–81, fitter brigade leader 1981–; mem. Cen. Auditing Comm. CPSU 1981–. Decor.: Hero of Socialist Labor 1969.

Shishov, Lev Dmitrievich
Party official.
2nd Sec. and mem. Bureau Cen. Cttee. Esthonian Komsomol 1960–66; cand. mem. Cen. Cttee. 1961–66, mem. Auditing Comm. 1966–71, mem. Cen. Cttee. Esthonian CP 1971–; cand. mem. Bureau Cen. Cttee. Esthonian CP 1971–; Head of Dept. of Org. Party Work with Cen. Cttee. Esthonian CP 1971–.

Shitikov, Aleksei Pavlovich
State official; b. 1912.
Russian; mem. CPSU 1939–; grad. Gorkii Agricultural Inst. 1936 and Party Higher School of Cen. Cttee. CPSU 1941; veterinary technician with a raion agricultural dept., sec. of a raion Komsomol cttee. 1936–41; pol. work in Sov. Army 1941–45; worked for Khabarovsk Krai Cttee. All-Union CP(B) 1945–48; Sec. of Kamchatka Oblast Cttee. All-Union CP(B) 1948–50; 1st Sec. of Jewish Autonomous Oblast Cttee. CPSU 1952–55; Dep. to 1954–, mem. Foreign Affairs Comm. of Sov. of the Union USSR Supr. Sov. 1966–70; Sec. 2nd Sec. 1955–57, 1st Sec. of Khabarovsk Krai Cttee. CPSU Mr. 1957–Jl. 1970; mem. Cen. Cttee. CPSU 1961–; Chair. of Sov. of the Union USSR Supr. Sov. Jl. 1970–; Chair. of Parliamentarian Group of USSR Jl. 1970–; Chair. of Sov.-Polish Friendship Soc. D. 1970–; Chair. of Sov. Cttee. for European Security Je. 1971–. Decors.: 2 Orders of Lenin, Order of Red Banner of Labour, medals.

Shitov, Aleksandr Ivanovich
Party official; b. 1925.
Mem. CPSU 1955–; grad. Inst. of Water Transport Engineering in Gorkii 1952; milling machine operator, master, designer in various plants in Gorkii 1946–47; served in Sov. Army 1945–46; technologist, dept. head at a plant of USSR Min. of the Shipbuilding Ind. 1952–59; party and govt. work 1959–; 2nd Sec. of Volgograd City Cttee. CPSU 1962; 1st Dep. Chair. of Volgograd Oblast Exec. Cttee. of Workers' Deps. (for Agriculture) 1962–64; 1st Sec. of Volgograd City Cttee. CPSU 1964–67; mem. staff Cen. Cttee. CPSU 1967–70; 2nd Sec. and mem. Bureau Cen. Cttee. Tadzhik CP D. 1970–75; 1st Dep. Chair. USSR Cttee. for People's Control 1975–; Dep. to USSR Supr. Sov. 1970–; cand. mem. Cen. Cttee. CPSU 1971–.

Shivaza, Iasyr Dzhumazovich (pen name: Shchianma)
Dungan writer; b. My. 5 (18), 1906, Aleksandrovka, now Moscow Raion Kirghiz SSR.
People's Poet of Kirghizia. Member of the CPSU from 1940. Candidate member of the Central Committee of the Kirghiz Communist Party (1958–65).
He graduated from the G. Tukai Tartar Institute of Education in Tashkent in 1930 and later served in the Great Patriotic War (1941–45). From 1957 to 1964 he was editor of the Dungan-language newspaper *Shyiuedi Chi* (Banner of October), which is circulated in several republics of the USSR.
His works, which first appeared in print in 1929, include the poetry collections *Morning Star* (1931), *The Wave of Revolution* (1932), *Selected Verses* (1958), and *Hello, Spring* (1966). He is the author of novellas and the novel *Faithful Friends* (1958). A major theme in his works is friendship among peoples.
He was a secretary of the Kirghiz Writers' Union from 1938 to 1942 and from 1946 to 1954. Decors.: Five orders and several medals.
Works: *Shchin syvyn.* Frunze 1973. *In Russian translation: Moi novyi dom.* Frunze 1969. *Stikhi,* Moscow 1976.

Shkabardinia, Mikhail Sergeevich
State official; b. 1930.
Mem. CPSU 1960–; grad. Novocherkassk Polytechnical Inst. 1954, Cand. of Tech. Sc. 1971; engineer, technical laboratory chief, chief of a design bureau, chief engineer in an electric measuring instruments plant in Krasnodar 1954–68; chief of a main admin., chief of "Soiuzelektropribor" Asscn. of USSR Min. of Instrument-making, Automation Equipment, and Control Systems 1968–74, chief of an admin. in this min., Dep. Min. 1974–80, Min. S. 1980–; Dep. to USSR Supr. Sov. 1979–; cand. mem. Cen. Cttee. CPSU 1981–.

Shkadov, Ivan Nikolaevich
Military official; b. Ap. 19 (My. 2), 1913, Naumovo, now Spus-Demensk Raion, Kaluga Oblast; s. of a peasant.
General of the army (1975). Hero of the Soviet Union (F. 21, 1978). Member of the CPSU since 1938.
He joined the Red Army in 1935. He graduated from the Kharkov Armored School in 1938, advanced training courses for officers at the Military Academy for Armored and Mechanized Troops in 1953, and the Military Academy of the General Staff in 1959. In 1938, he commanded a tank platoon in fighting near Lake Khasan.
During the Great Patriotic War, he was assistant chief-of-staff of a tank regiment and commander of a tank battalion in 1941 and 1942 and commander of a separate tank regiment and a separate tank brigade between 1942 and 1945. He fought at Moscow and Stalingrad, in the Iasi-Kishinev Operation, and in the liberation of Rumania, Bulgaria, and Yugoslavia.
After the war, he continued to serve in command positions. He was named deputy commander of the Carpathian Military District in March 1961. From 1964 to 1967 he headed a group of Soviet military specialists in the Republic of Cuba. He was appointed commander of the Northern Group of Forces in May 1967, first deputy director of the Military Academy of the General Staff in December 1968, and chief of the Main Directorate of Military Educational Institutions of the Ministry of Defense in July 1969. He became chief of the Main Directorate of Personnel of the Ministry of Defense of the USSR in August 1972.
He was a deputy to the ninth convocation of the Supreme Soviet of the USSR. Decors.: Order of Lenin, the Order of the

October Revolution, five Orders of the Red Banner, the Order of Suvorov Third Class, two Orders of the Red Star, and various medals, also foreign orders and medals.

Shklovskii, Josif Samuilovich
Astrophysicist; b. Je. 18 (Jl. 1), 1916, Glukhov, now Sumy Oblast, Ukrainian SSR. Corresponding member of the Academy of Sciences of the USSR (1966).
After graduating from Moscow University in 1938, he began working at the Shternberg State Astronomical Institute. In 1968 he began working at the Institute for Space Research of the Academy of Sciences of the USSR. His main works deal with the theory of the solar corona, radio astronomy, and X-ray astronomy. In 1948 he investigated the presence of the 21-cm line in the radiation of neutral interstellar hydrogen. In 1953 he explained the properties of the Crab Nebula in terms of the radiation of relativistic electrons in a magnetic field. In 1965 he proposed a new concept of planetary nebulas, linking them to the end of the evolution of red giants and the birth of white dwarfs. In the 1970's, he turned to the study of the nature of galactic and extragalactic X-ray sources.
He is a foreign member of the US National Academy of Sciences and a member of the Royal Astronomical Society of Great Britain. Decors.: Lenin Prize in 1960, Order of the Badge of Honor and various medals.
Works: *Fizika solnechnoi korony,* 2nd ed. Moscow 1962. *Kosmicheskoe radioizluchenie.* Moscow 1956. *Vselennaia, zhizn', razum,* 4th ed. Moscow 1976. *Zvezdy: ikh rozhdenie, zhizn' i smert'.* Moscow 1975. *Sverkhnovye zvezdy.* Moscow 1976.

Shklovskii, Viktor Borisovich
Russian writer, literary scholar, and critic; b. Ja. 12 (24), 1893, St. Petersburg; s. of a teacher.
He studied in the philological faculty at the University of St. Petersburg. His first scholarly work on literature was published in 1914. He was close to the futurists and was a member of the Society for the Study of Poetic Language (OPOIAZ). He was a founder and a theorist of the formalist school of literary scholarship. His most important works from these years were collected in *On the Theory of Prose* (1925).
He became aware of the limitations of the formalist school in the 1930's. He began dealing with literature from a broad sociohistorical viewpoint in his article "Monument to a Scholarly Error" (1930). Beginning in the 1930's, he investigated the historical relations between literature and the social reality and dominant ideas of the age. His books and articles dealing with this problem include *Artistic Prose: Reflections and Analyses* (1959; 2nd ed., 1961), *Stories About Prose* (vols. 1–2, 1966), and *The Bowstring: On the Dissimilarity of the Similar* (1970).
He also wrote memoirs and lyric epistolary books of an autobiographical nature, such as *A Sentimental Journey* (parts 1–2, 1923), *Zoo: Letters Not About Love, or The Third Héloise* (1923), *Hamburg Account* (1928), and *Once Upon a Time* (1962). Much of Shklovskii's work deals with the Russian classics. In particular, he was the author of a series of works on L. N. Tolstoy, beginning with *Material and Style in Leo Tolstoy's Novel "War and Peace"* (1928) and ending with the detailed biography *Leo Tolstoy* (1963; 2nd ed., 1967; both editions part of the series *Zhizn' zamechatel'nykh liudei* [Lives of Outstanding People]). He also wrote works on film theory; literary portraits of S. M. Eisenstein, L. V. Kuleshov, Dziga Vertov, and A. P. Dovzhenko; and screenplays. Decors.: Order of the Red Banner of Labor.
Works: *Sobr. soch.,* vols. 1–3, Moscow 1973–74. *Poiski optimizma.* Moscow 1931. *O Maiakovskom.* Moscow 1940. *Zametki o proze russkikh klassikov,* 2nd ed. Moscow 1955. *Za i protiv: Zametki o Dostoevskom.* Moscow 1957. *Istoricheskie povesti i rasskazy.* Moscow 1958. *Za sorok let: Stat'i o kino.* [Introductory article by M. Bleiman.] Moscow 1965.

Shkolnikov, Aleksei Mikhailovich
State official; b. 1914; s. of a peasant.
Russian; mem. CPSU 1940–; concluded an ind. tech. school 1933, 3 terms at Industrial Acad. and Party Higher School of Cen. Cttee. CPSU 1952; sen. foreman, dept. head, chief power technician at a plant in Perm Oblast 1933–43; party work 1943–45; Rep. for Vladimir and later Voronezh Oblast Party Control Cttees. of Cen. Cttee. All-Union CP(B) 1945–47; 2nd Sec. of Kaluga Oblast Cttee. All-Union CP(B) 1947–49; 1st Sec. of Tambov Oblast Cttee. CPSU 1952–55; cand. mem. 1952–56, mem. Cen. Cttee. CPSU 1956–; Dep. to 1954–, mem. Foreign Affairs Comm. 1962–66, mem. Draft Bills Comm. of Sov. of the Union USSR Supr. Sov. 1966–74; 1st Sec. of Voronezh Oblast Cttee. CPSU 1955–60; 1st Sec. of Volgograd Oblast Cttee. CPSU 1960–65 (for Agriculture 1963–64); 1st Dep. Chair. of Council of Mins. RSFSR N. 1965–1974; Chair. of USSR Cttee. on People's Control Jl. 1974–. Decors.: 3 Orders of Lenin, medals.

Shkuratov, Ivan Fedorovich
Trade union official; b. 1912.
Mem. CPSU 1943–; grad. Moscow Zoo-Technol. Inst. 1935; zoo technologist in a sovkhoz in Orel Oblast, worked in USSR People's Commissariat for Agriculture and in a sovkhoz in Saratov Oblast 1935–42; served in Sov. Army 1942–46; administrative, economic and planning positions in a sovkhoz, at Union Inst. for Agricultural Machinery and in Min. USSR of Sovkhozes 1946–56; Dep. Chair. 1956–58, Sec. of Cen. Cttee. of Trade Union of Agricultural Workers and Workers of the Procurement Ind. 1958–62; Sec. of Cen. Union Council of USSR Trade Unions 1962–68; Chair. of Cen. Cttee. of Trade Union of Agricultural Workers Mr. 1968–; mem. Cen. Auditing Comm. CPSU 1971–.

Shliapnikov, German Evlampevich
Diplomat; b. 1929.
Hold diverse posts at USSR Emb. in China, Mongolia, New Zealand and Bolivia until 1975; USSR Amb. to Ecuador 1975–80, to Nicaragua F. 1980–.

Shlyk, Aleksandr Arkad'evich
Plant physiologist and biochemist; b. N. 1, 1928, Minsk.
Corresponding member of the Academy of Sciences of the USSR (1966). Member of the CPSU since 1948.
Upon graduation from the Belorussian University in 1950, he worked in the Institute of Biology of the Belorussian Academy of Sciences. In 1957 he became head of the academy's Laboratory of Biophysics and Isotopes. Since 1973 he has been director of the Institute of Photobiology, which evolved from the Laboratory of Biophysics and Isotopes.
His research has been directed at the development of the photosynthetic apparatus. He and his colleagues demonstrated the existence of a chlorophyll renewal process and elucidated its main mechanisms. He showed that chlorophyll b is synthesized from labile chlorophyll-a molecules and studied the metabolism of the precursors of chlorophyll-a molecules in green plants. Decors.: Order of the Red Banner of Labor and a medal.
Works: *Metod mechenykh atomov v izuchenii biosinteza khlorofilla.* Minsk 1956. *Metabolizm khlorofilla v zelenom rastenii.* Minsk 1956. "Biosintez khlorofilla i formirovanie fotosinteticheskikh sistem." In the collection *Teoreticheskie osnovy fotosinteticheskoi produktivnosti.* Moscow 1972. "Razvitie issledovanii metabolicheskoi geterogennosti fotosinteticheskikh membran." In the collection *Biosintez i sostoiante khlorofillov v rastenii.* Minsk 1975. "Tsentry biosinteza khlorofilla i regulirovanie protsessa formirovaniia pigmentnogo apparata fotosinteza." [With co-authors.] *Izv. AN SSSR: Ser. biologicheskaia.* 1976, no. 1.

Shmakov, Fedor Ivanovich
Belorussian actor; b. F. 16 (Mr. 1), 1917, Petrograd.
People's Artist of the USSR (1975). Member of the CPSU since 1948.
In 1939, he graduated from the Leningrad Theatrical Institute and joined the company of the Russian Theater of the Belorussian SSR in Mogilev. In 1941 he began working at the Belorussian Second State Theater in Vitebsk (now the Ia. Kolas Belorussian Drama Theater). He appeared in the role of V. I. Lenin in Popov's *The Family,* D. Zorin's *The Eternal Source, His The Sixth of July,* and Pogo-

din's *The Third Pathétique*. His other major parts have included Akim in Tolstoy's *The Power of Darkness*, Prisypkin in Mayakovsky's *The Bedbug*, Vaskov in Vasil'ev's *And Dawns Are Quiet Here*, and the title role in Vol'skii's *Nesterka*. He has also appeared in motion pictures, notably as Iusalov in *Immortal Song* (1958), Rubin in *Kochubei* (1958), and Andrei Razmetnov in *Virgin Soil Upturned* (1960–61). Decors.: State Prize of the Belorussian SSR in 1968, two orders and several medals.

Shmarinov, Dementii Alekseevich
Graphic artist; b. Ap. 29 (My. 12), 1907, Kazan.
Member of the Academy of Arts of the USSR (1953). People's Artist of the USSR (1967).
He studied at N. A. Prakhov's studio in Kiev from 1919 to 1922 and under D. N. Kardovskii in Moscow from 1923 to 1928. He was chairman of the administrative board of the Moscow organization of the Artist's Union of the RSFSR from 1959 to 1961, from 1966 to 1968, and in 1972. He became secretary of the administrative board of the Artists' Union of the USSR in 1968.
He works chiefly as an illustrator. He gives realistically precise graphic interpretations of literary works and conveys dramatic situations and the psychosocial characteristics of his heroes in a highly convincing manner. His works include illustrations for Dostoevsky's *Crime and Punishment* (charcoal, black watercolor, 1935–36), A. N. Tolstoy's *Peter the Great* (charcoal, black watercolor, 1940–45), and L. N. Tolstoy's *War and Peace* (black watercolor, 1953–55), all of which are housed at the Tret'iakov Gallery in Moscow. He also illustrated Lermontov's *A Hero of Our Time* (1941, Tret'iakov Gallery; Russian Museum, Leningrad) and Pushkin's *The Captain's Daughter* (1974) and *The Queen of Spades* (1976), all in charcoal and black watercolor.
During the Great Patriotic War (1941–45), he designed a number of political posters. He also executed a cycle of easel works entitled *We Won't Forget, We Won't Forgive!* (charcoal, black watercolor, 1942, Tret'iakov Gallery). This angry, impassioned series, which was awarded the State Prize of the USSR in 1943, vividly describes the suffering and courage of the Soviet people in their struggle against the enemy. He has also produced a number of paintings.
Decors.: Order of Lenin, two other orders, and several medals.

Shmatolian, Ivan Ivanovich
State official.
Min. of Sovkhozes Ukrainian SSR Ag. 1975–Ap. 1977; Min. of Supplies Ukrainian SSR Ap. 1977–; cand. mem. Cen. Cttee. Ukrainian CP 1976–77, mem. Ap. 1977–.

Shmidt, Evgenii Vladimirovich, Dr. of Med. Sc.
Neuropathologist; b. 1905.
Grad. Moscow Univ. 1949; Dr. of Med. Sc. 1952; Prof. 1953–; mem. USSR Acad. of Med. Sciences 1963–; Intern, Lecturer at 1st Med. Inst. Moscow 1930–49; mem. staff Inst. of Neurology of USSR Acad. of Med. Sciences 1949–63 and 1966–; Dir. of same inst. 1967–; mem. Presidium USSR Acad. of Med. Sciences 1964–66. Decor.: State Prize of the USSR 1971 (for his part in research on disturbances of blood circulation caused by pathological changes in the main arteries of the head).

Shmidt, Mikhail Herbertovich
State official.
Sec., 2nd Sec. of Chardzhou Oblast Cttee. Turkmen CP 1970–72; mem. Cen. Cttee. 1971–, cand. mem. Bureau 1972–76, mem. Bureau Cen. Cttee. Turkmen CP 1976–; Head of Dept. of Party Org. Work with Cen. Cttee. Turkmen CP O. 1972–79; Dep. Chair. Presidium Supr. Sov. Turkmen SSR Jl. 1979–.

Shmyga, Tat'iana Ivanovna
Russian operetta performer; b. D. 31, 1928, Moscow.
People's Artist of the USSR (1978).
In 1953, she graduated from the A. V. Lunacharskii State Institute of Theatrical Arts, where she studied voice under D. B. Beliavskaia and acting under I. M. Tumanov. She became a soloist at the Moscow Operetta Theater the same year. She masterfully combines the skills of a singer, actress, and dancer. Her roles include Violette and Ninon in Kálmán's *The Violet of Montmartre*, Chanita and Gloria Rosetti in Miliutin's *Chanita's Kiss* and *The Circus Show Is Up*, Tonia in Dunaevskii's *White Acacia*. Eliza Doolittle in Loewe's *My Fair Lady*, Verochka in Eshpai's *There's Nobody Happier Than I*, and Liubov' Iarovaia in Vl. Il'in's *Comrade Liubov'*.
She has toured abroad. Decors.: M. I. Glinka State Prize of the RSFSR in 1974, Order of the Badge of Honor and several medals.

Shokin, Aleksandr Ivanovich
State official; b. 1909; s. of a worker.
Mem. CPSU 1936–; grad Moscow Bauman Inst. of Mechanics and Machine Construction 1934; worker and fitter at an automobile workshop, student 1926–32; dept. head, head of design bureau at a shipbuilding plant 1932–38; Chief Engineer of Main Board of People's Commissariat of Shipbuilding 1938–43; leading engineering and tech. posts in the radio engineering ind. 1943–49; USSR Dep. Min. of Posts and Telecommunications 1949–53; USSR Dep. Min. 1954–55, 1st Dep. Min. of the Radio-Engineering Ind. 1955–57; 1st Dep. Chair. of State Cttee. for Radio and Electronics of USSR Council of Mins. 1958–61; Chair. (with rank of USSR Min.) of State Cttee. for Electronic Engineering of USSR Council of Mins. 1961–63; cand. mem. 1961–66, mem. Cen. Cttee. CPSU 1966–; Dep. to USSR Supr. Sov. 1962–; Chair. (with rank of USSR Min.) of State Cttee. for Electronic Engineering of USSR Supr. Econ. Council 1963–65; USSR Min. of the Electronics Ind. Mr. 1965–. Decors.: 2 Stalin Prizes, 2 Orders of Lenin, 2 medals.

Sholmov, Iurii Alekseevich
Diplomat.
USSR Amb. to Philippines Jl. 1981–.

Sholokhov, Mikhail Aleksandrovich
Author; b. 1905, on the farmstead Kruzhelinnyi, now Rostov Oblast.
Russian; mem. CPSU 1932–; attended secondary school until 1918; mem. of a mil. unit for the procurement and confiscation of concealed grain supplies 1918–22; house painter, loader and bookkeeper in Moscow 1922–26; literary work since 1923; has lived and worked in the Cossack village Veshenskaia since 1926; Dep. 1937–, mem. Foreign Affairs Comm. of Sov. of Nationalities USSR Supr. Sov. 1962–74; mem. Cen. Cttee. CPSU 1961–. Decors.: Stalin Prize, Lenin Prize, Nobel Prize for Literature 1965, Hero of Socialist Labour, 4 Orders of Lenin, medals.
Publs.: "Tikhii Don" (Quiet Flows the Don) 1934, "Podniataia tselina" (Virgin Soil Upturned), "Sobranie Sochinenii" vols. 1–8, 1956–60 and vols. 1–8, 1965–69, "Oni srazhalis' za rodinu" (They Fought for the Fatherland) 1971, and others.

Shomakhov, Amirkhan Kamizovich
Kabardin writer; b. S. 27 (O. 10), 1910, Nizhnii Kurp, now Terek Raion.
People's Poet of the Kabarda-Balkar ASSR (1977). Member of the CPSU since 1938.
He graduated from the Northern Caucasus Pedagogical Institute in 1933 and from the Higher Party School of the Central Committee of the CPSU in 1955. His works first appeared in print in 1930. He is the author of the novella *Kolkhoz Under Fire* (1958), about the events of the war years. His novel *Dawn Over the Terek* (1968; Russian translation, 1972) is about the Civil War and the first year of Soviet power in the republic. The novel *Horsemen on the Peaks* (1970) is about the formation of the working class in Kabarda-Balkaria. Decors.: two orders and several medals.
Works: *Unë nekhu*. Nal'chik 1953. *Zaranshu*. Nal'chik 1958. *In Russian translation: Zhambot i Kambot*. Moscow 1957. *Na beregu Psykhega*. Nal'chik 1960. *Talisman*. Moscow 1965.

Shonin, Georgii Stepanovich
Cosmonaut; b. 1935.
Mem. CPSU 1957–; grad. from a special aviation school in Odessa 1953, Navy

Aviation College 1957 and (without interrupting professional duties) Zhukovskii Acad of Mil. Aviation Engineering 1968; cosmonaut 1961–; commander of spaceship "Soiuz 6" participated in group flight of 3 spaceships (together with V. N. Kubasov) O. 11–16, 1969. Decor.: Hero of the Soviet Union 1969.

Shortanov, Askerbi Takhirovich
Kabardin writer; b. O. 3 (16), 1916, Anuzorovka, now the village of Lesken II. Urvan' Raion, Kabarda-Balkar ASSR.
Honored Art Worker of the Kabarda-Balkar ASSR (1939); Honored Art Worker of the RSFSR (1971). Member of the CPSU since 1941.
He studied at the State Institute of Theatrical Arts from 1934 to 1936. He was first published in 1939. His main theme is the history of the Kabardins and their life in the Soviet period. He ist the author of the historical novel *The Mountaineers* (1954; Russian translation, 1967); the chronicle-like novel *Betal Kalmykov* (1968; Russian translation under the title *Always in the Saddle*, 1970), which deals with the Civil War hero of that name; and the plays *Aul Batyr* (staged 1939) and *When The Light Breaks* (published 1947; Russian translation, 1950), which deals with life in a kolkhoz village. His play *The Party's Envoy* (published 1961) was staged in 1967.
He is a well-known folklorist. Decors.: five orders and various medals.
Works: *K"ëzënok"uë zhëbag"ï.* Nal'chik 1956. *Tkhig"ë k"ikhëkhakhë.* Nal'chik 1957. *In Russian translation: P'esy.* Nal'chik 1957.

Shostakovskii, Mikhail Fedorovich
Organic chemist; b. My. 24 (Je. 6), 1905, Novoselitsy, now Kirovograd Oblast.
Corresponding member of the Academy of Sciences of the USSR (1960). Member of the CPSU since 1946.
He graduated from the University of Irkutsk in 1929. From 1932 to 1962 he worked at the Institute of Organic Chemistry of the Academy of Sciences of the USSR. From 1957 to 1971 he was director of the Irkutsk Institute of Organic Chemistry of the Siberian Division of the Academy of Sciences of the USSR, and from 1971 to 1973, director of the Institute of Petrochemistry of the Siberian Division of the Academy of Sciences of the USSR. In 1973 he became head of a subdepartment of the All-Union Polytechnical Correspondence Institute.
His main works deal with the chemistry of vinyl ethers, organosilicon and organostannum compounds, and acetylene and its derivatives. He developed an antiphlogistic coating agent based on polyvinyl butyl ether (vinylene) known as Shostakovskii balsam, a blood substitute based on polyvinylpyrrolidone, and an agent to combat ringworm in animals. Decors.: State Prize of the USSR (1949), two orders and several medals. Works: *Prostye vinilovye efiry.* Moscow 1952. *Khimiia diatsetilena* $HC\equiv C-C\equiv CH$. Moscow 1971. (With A. V. Bogdanova.)

Shpak, Vladimir Stepanovich
Chemist; b. F. 7 (20), 1909, Pskov.
Corresponding member of the Academy of Sciences of the USSR (1968). Hero of Socialist Labor (1961). Member of the CPSU since 1939.
He graduated in 1931 from the Lensovet Leningrad Institute of Technology, where he worked from 1932 to 1975, becoming a professor in 1955. Beginning in 1948, he also worked at the Institute of Applied Chemistry, of which he was director from 1952 to 1977.
His main works deal with commercial organic chemistry. The synthesis and technological development of the manufacture of many nitro compounds, amines, ethers, esters, and other organic compounds, carried out by him, were introduced in industry. He was also involved in organizing the creation of new branches in the chemical industry, such as the manufacture of organofluorine compounds. Decors.: two Orders of Lenin, three other orders, and several medals.

Shped'ko, Ivan Fadeevich
Diplomat; b. 1918.
Grad. Kharkov Pedagogic Inst.; mem. Diplomatic Service 1941–; Emb. Counsellor at USSR Emb. in Afghanistan 1949–53; sen. posts in USSR Min. of Foreign Affairs 1953–56; USSR Amb. to Pakistan 1956–60; Dep. Head of South Asian resp. South-east-Asian Dept. of USSR Min. of Foreign Affairs 1960–63; USSR Amb. to Canada 1963–69; Dep. Head of II. European Dept. 1969–70, Head of II. Far East Dept. of USSR Min. of Foreign Affairs 1970–76; USSR Amb. to Indonesia Je. 1976–.

Shpinel', Iosif Aronovich
Motion-picture designer; b. S. 25 (O. 7), 1892, Belaia Tserkov', now Kiev Oblast.
Honored Art Worker of the RSFSR (1940).
He graduated from the architectural department of the Kiev Art School in 1914 and from the print-making department of the Vkhutemas. He began to work in the theater in 1921 and in the cinema in 1927. He created designs for the films *The Night Coachman* (1929), *Arsenal* (1929), *Ivan* (1932), *Boule de Suif* (1934), *Petersburg Night* (1934), *Parisian Dawns* (1937), *Alexander Nevsky* (1938), *Mashen'ka* (1942), *Ivan the Terrible* (1945, 1958), *The Story of a Real Man* (1948), *Sisters* (1957), *The Year 1918* (1958), *Gloomy Morning* (1959), and *A Year Like a Lifetime* (1965). He, a teacher since 1928, joined the faculty of the All-Union State Institute of Cinematography, becoming a professor in 1965. Decor.: State Prize of the USSR in 1951.

Shpogis, Kazimir Albinovich
State official.
Min. of Agriculture Latvian SSR Ap. 1980–; mem. Cen. Cttee. Latvian CP 1981–.

Shtein, Aleksandr Petrovich
Dramatist; b. S. 15 (28), 1906, Samarkand.
Mem. CPSU 1930–; studied Leningrad Univ.; Chief Ed. of periodical "Iskusstvo i zhizn'" 1935–40; Ed. of Navy paper on battleship "Oktiabrskaia Revoliutsiia" and contributor to newspaper "Krasnyi flot" in Leningrad 1941–45. Decor.: State Prize of the USSR 1951.
Publs.: "Vesna dvadtsat' pervogo" (Spring 1921) 1939, "Flag admirala" (The Admiral's Flag) 1950, "Gostinitsa 'Astoria'" (Hotel Astoria) 1956, "Piesy" (Theatre Plays) 1962, "Povest'o tom, kak voznikaiut siuzhety" (Report on How an Idea is Born) 1965, "Dramy" (Plays) 1966, and others.

Shternfeld, Ari Abromovich
Mechanical engineer; b. 1906, Sieradz, Poland.
Jew; grad. Nancy Univ. 1927; grad. mechanical engineer, specialist in theory of space flight; lectured at meeting of Warsaw Astronomical Observatory, later at Sorbonne in Paris and French Acad. of Sciences; has lived in the USSR 1935–; contributed significantly to Sov. space research; mem. Sov. Acad. of Sciences 1975. Decor.: Int. Prize for Astronautics.
Publs.: "Vvedenie v kosmonautiku" 1937 and 1974, "Iskustvennye sputniki Zemli" 1956 and 1958, "Na puti k Solntsu" 1959, "Osvoenie mezhplanetnykh prostranstv" 1965.

Shtogarenko, Andrei Iakovlevich
Composer; b. 1902.
Ukrainian; mem. CPSU 1944–; grad. Kharkov Conservatory 1936; Rector of Kiev Conservatory 1954–; composed 3 symphonies, the symphonic cantata "Ukraina moia", several choral and orchestral works; Sec. of Ukrainian Composers' Union. Decors.: People's Artist of the Ukrainian SSR 1972, State Prize of the USSR 1946, 1952.

Shtokalo, Iosip Zakharovich, Dr. of Phy.-Math. Sc.
Mathematician; b. N. 16, 1897, Skoromokhy, now Lvov Oblast; s. of a peasant.
Mem. CPSU 1941–; grad. Dnepropetrovsk Univ. 1931; corresponding mem. Acad. of Sciences Ukrainian SSR 1951–; Dr. of Phy.-Math. Sc. 1944; Prof. 1946–; corresponding mem. Int. Acad. of the History of Science in Paris 1965–; teacher in Dnepropetrovsk City and Raion and in Dneprodzerzhinsk 1915–31; mem. staff Inst. of Econ. Engineering, Aviation Inst. and Kharkov Univ. 1931–41; mem. staff Acad. of Sciences Ukrainian SSR 1941–; Sen. Scientist 1942–49 and Dep. Dir. of Inst. of Mathematics 1946–49;

mem. Presidium Acad. of Sciences Ukrainian SSR and Chair. of Presidium of Lvov branch of Acad. of Sciences Ukrainian SSR 1949–56; concurrently Prof. and Head of Chair at Kiev Univ. 1944–51 and 1956–; Head of Dept. of History of Mathematics of Inst. of Mathematics of Acad. of Sciences Ukrainian SSR 1956–63; Head of Section History of Natural Sciences and Technology of Inst. of the History of the Acad. of Sciences Ukrainian SSR 1963–; author of more than 70 scientific works. Decors.: Order of Lenin, medals.
Publs.: "Kriterii ustoichivosti i neustoichivosti reshenii lineinykh differentsialnykh uravnenii s kvaziperiodicheskimi koeffitsientami" in "Matematicheskii sbornik" No. 19, 1946, "Narys rozvytku matematyky na Ukraini za 40 rokiv Radianskoi vlady" 1958, "Operatsionnye metody i ikh razvitie v teorii lineinykh differentsialnykh uravnenii s peremennymi koeffitsientami" 1961, and others.

Shtokolov, Boris Timofeevich
Russian singer (bass); b. Mr. 19, 1930, Kuznetsk.
People's Artist of the USSR (1966).
He graduated from the Urals Conservatory in Sverdlovsk in 1954. He apprenticed at the Sverdlovsk Theater of Opera and Ballet from 1951 until 1954, when he became a soloist. In 1959 he became a soloist at the S. M. Kirov Theater of Opera and Ballet in Leningrad.
He, a leading Soviet vocalist, possesses a beautiful, expressive voice and portrays his characters vividly and convincingly. His roles have included the Miller in Dargomyzhskii's *The Mermaid,* Susanin and Ruslan in Glinka's *Ivan Susanin* and *Ruslan and Liudmila,* Godunov and Dositheos in Mussorgsky's *Boris Godunov* and *Khovanshchina,* the title role in Rubinstein's *The Demon,* Gremin in Tchaikovsky's *Eugene Onegin,* Mephistopheles in Gounod's *Faust,* Don Basilio in Rossini's *The Barber of Seville,* and Andrei Sokolov in Dzerzhinskii's *The Fate of a Man.* He also performs on the concert stage and has toured abroad. Decors.: 2 orders and several medals.

Shtyrbul, Kirill Antonovich (surname also spelled Shtirbul, Shtirbu)
Moldavian actor; b. F. 17 (Mr. 2), 1915, Brushteny, Rybnitsa Raion.
People's Artist of the USSR (1960). Member of the CPSU since 1947.
He graduated from the theatrical school in Odessa in 1937 and joined the First Moldavian Drama Theater in Tiraspol' (now the Moldavian Pushkin Academic Music and Drama Theater in Kishinev). He was one of the first actors to depict V. I. Lenin on the Moldavian stage, in Pogodin's *The Kremlin Chimes* and *Man With a Gun* and his *In the Name of Revolution.* His best roles were Shvandia in Trenev's *Liubov' Iarovaia,* Truffaldino in Goldoni's *The Servant of Two Masters,* the Mayor in Gogol's *The Inspector General,* and the title roles in Shakespeare's *King Lear* and Brecht's *Schweik in the Second World War.* His outstanding roles in musicals were Foma in Dunaevskii's *Free Wind* and Susik in Miliutin's *Trembita.*
He, a member of the Central Committee of the Communist Party of Moldavia from 1963 until 1976, was a deputy to the second through fourth convocations of the Supreme Soviet of the Moldavian SSR. Decors.: in 1970, State Prize of the Moldavian SSR, Order of Lenin, the Order of the October Revolution, the Order of the Red Banner of Labor, and several medals.

Shubenko-Shubin, Leonid Aleksandrovich, Dr. of Tech. Sc.
Mechanical engineer; b. Ag. 7, 1907, Turkey; s. of a physician.
Mem. CPSU 1943–; grad. Leningrad Univ. 1930 and All-Union Boiler-Turbine Inst. 1931; specializes in steam turbines; corresponding mem. Acad. of Sciences Ukrainian SSR 1951–; Dr. of Tech. Sc. 1961; teacher at Inst. of Road Construction Engineering 1934–41 and Leningrad Polytech. Inst. 1934–41; worked for Kirov Works 1941–44; Dir. of Cen. Boiler-Turbine Inst. 1944–50; Chief Designer at Kharkov Turbine Works 1950–; concurrently Head of Dept. of Internal Combustion Engines, Laboratory of Hydraulic Machines (now Kharkov branch of Inst. of Mechanics) of Acad. of Sciences Ukrainian SSR 1957–; author of approx. 90 works. Decors.: Order of Lenin, Order of Red Banner of Labour, medals.
Publs.: „Unikalnye turbiny" 1962, "Prochnost' elementov parovykh turbin" (co-author) 1962, and others.

Shubnikov, Nikolau Mikhailovich
Diplomat; b. 1924.
Mem. Diplomatic Service 1969–; Councellor Envoy, USSR Embassy in Korea 1969–82, Amb. D. 1982–.

Shukshin, Vasilii Makarovich
Film director and actor; b. 1929.
Mem. CPSU 1955–; grad. Directing Dept. of State Union Film Inst. (VGIK) 1960; with Mosfilm Studios 1960–62; Director with M. Gorkii Film Studio 1963–. Decor.: State Prize of the USSR 1971.

Shulgin, Nikolai Pavlovich
State official.
Cand. mem. Cen. Cttee. Ukrainian CP 1961–81; Dep. to 1963–, mem. Planning and Budget Comm. of Supr. Sov. Ukrainian SSR 1963–68; 1st Sec. of Donetsk City Cttee. Ukrainian CP 1963; Min. of Construction and Utilization of Public Highways Ukrainian SSR N. 1968–.

Shults, Mikhail Mikhailovich, Dr. of Chem. Sc.
Physical chemist; b. 1919.
Mem. CPSU 1946–; grad. Leningrad Univ. 1947; Dr. of Chem. Sc. and Prof. 1965–; corresponding mem. USSR Acad. of Sciences 1972–; served in Sov. Army 1941–45; postgraduate studies 1947–50; Research Asst. 1950–52, Head of Laboratory at Inst. for Research on Chemistry of Leningrad Univ. 1959–72; Lecturer at Leningrad Univ. 1953–59; Dir. of Inst. of Chemistry of Silicates of USSR Acad. of Sciences 1972–; concurrently Dean of Dept. of Chemistry 1967–72 and Prof. at Leningrad Univ. 1972–.

Shul'zhenko, Klavdiia Ivanovna
Russian singer; b. Mr. 11 (24), 1906, Kharkov.
People's Artist of the USSR (1971).
She was an actress at the Kharkov Drama Theater under the direction of N. N. Sinel'nikov from 1923 until 1928, when she began to perform on the variety stage (estrada). She was the first performer of many Soviet songs of everyday life, as well as many lyrical songs. Songs that she made famous included "The Blue Kerchief" by Petersburgskii, "Let's Light Up" by Tabachnikov, and "Pick Up the Guitar" by Lepin.
Decors.: Order of Lenin, the Order of the Red Banner of Labor, and several medals.

Shumakov, Valerii Ivanovich, Dr. of Med. Sc.
Medical scientist; b. 1931.
Mem. CPSU 1961–; grad. 1st Med. Inst. Moscow 1956; surgeon; Dr. of Med. Sc. 1966; Prof. 1969–; postgraduate studies 1956–59, Jun. Research Asst. at USSR Acad. of Medical Sciences 1959–63; mem. staff Inst. for Research on Clinical and Experimental Surgery 1963–, Head of Laboratory 1966–69, Head of Dept. at same inst. 1969–. Decor.: State Prize of the USSR 1971 (for his part in developing technique in kidney transplantation and its introduction into clinical practice).

Shumiatskaia, Elena Iakovlevna
Master of applied art; b. Mr. 16 (29), 1912, Turukhansk.
She attended the Moscow Textile Institute from 1930 to 1934. Her career has been closely linked with the Trekhgornaia Manufaktura Textile Combine, where she began working in 1934. She has made a thorough study of Russian folk art and has creatively employed traditional motifs in her decorative printed fabrics.

Shumskii, Petr Aleksandrovich, Dr. of Geog. Sc.
Geographer; b. 1915.
Mem. CPSU 1957–; grad. Leningrad Univ. 1939; specialist on glaziers; Dr. of Geog. Sc. 1948; Prof. 1962–; intermediate-school teacher 1939–40; geologist with a Pamir expedition 1941–45; mem. staff Arctics Research Inst. 1945–50; mem. staff 1950–63, Dir. of Inst. for Research on Permafrost of USSR Acad. of

Sciences 1958–; mem. Interdepartmental Comm. for the Exploration of the Antarctic of Presidium of USSR Acad. of Sciences, then Dep. Chair. of same comm. 1963–65; mem. staff Inst. of Geography of USSR Acad. of Sciences 1965–70; Sen. Scientist; head of a dept. at Inst. of Physical-Tech. Problems of the North at Yakut branch of Siberian div. of USSR Acad. of Sciences 1970; mem. staff and Sen. Scientist at Inst. of Mechanics at Moscow Univ. 1971–. Decor.: State Prize of the USSR 1971 (for his contribution to two-volume "Atlas Antarktiki" 1966 and 1969).

Shur, Iakov Shebselevich
Physicist; b. Ap. 7 (20), 1908, Dubrovno, now Vitebsk Oblast.
Corresponding member of the Academy of Sciences of the USSR (1970). Member of the CPSU since 1940.
He graduated from Leningrad State University in 1932. The same year, he joined the staff of the Institute of Physics of Metals of the Urals Scientific Center of the Academy of Sciences of the USSR.
His main works deal with the physics of magnetic effects. Shur discovered thermomagnetic hysteresis. He discovered and studied the anisotropy of the magnetic hysteresis of ferromagnetic crystals. He also discovered magnetoelastic hysteresis and explained the nature of magnetization and hysteresis processes in hard magnetic materials. He developed new magnetic and magnetostrictive materials, low-loss high-permeability electrical steels, and magnetic thin films. Decors.: State Prize of the USSR in 1967, four orders and various medals.
Works: *Ferromagnetizm.* Moscow-Leningrad 1948. (With S. V. Vonsovskii.) "Protsessy namagnichivaniia v magnitoodnoosnykh vysokoanizotropnykh ferromagnetikakh." In *Tr. mezhdunarodnoi konferentsii po magnetizmu: MKM-73,* vol. 5. Moscow 1974.

Shushanashvili, Aleksandr Amiranovich
State official.
Min. of Justice Georgian SSR Mr. 1976–; cand. mem. Cen. Cttee. Georgian CP 1981–.

Shustko, Lev Sergeevich
Military official.
Dep. of Supr. Sov. Armenian SSR 1980–; Mem. Cen. Cttee. and mem. Bureau Cen. Cttee. Armenian CP Ja. 1981–.

Shvedov, Aleksei Alekseevich
Diplomat; b. 1905.
Mem. Diplomatic Service 1944–; Adviser of I. European Dept. 1955–56; Dep. Head of Near East Dept. 1956–58, Head of African Dept. of USSR Min. of Foreign Affairs 1958–60; Head of I. African Dept. of USSR Min. of Foreign Affairs 1960–62; USSR Amb. to Morocco 1962–65; Head of I. African Dept. of USSR Min. of Foreign Affairs 1965–.

Shveikin, Gennadii Petrovich
Chemist; b. Ag. 29, 1926, Karabash, Cheliabinsk Oblast.
Corresponding member of the Academy of Sciences of the USSR (1976). Member of the CPSU since 1955.
He graduated from the S. M. Kirov Urals Polytechnical Institute in 1951. In the same year he joined the staff of the Institute of Chemistry of the Urals Branch of the Academy of Sciences of the USSR; in 1972 he was appointed director of the institute.
His principal works deal with the chemistry of refractory metals and their compounds. He proposed a method of preparing metallic niobium. He also developed methods of synthesizing the carbides and nitrides of the metals in groups IVa through VIa of the periodic table of elements.
Works: *Fiziko-khimicheskoe issledovanie okislov, karbidov i oksikarbidov vanadiia, niobiia i tantala.* Sverdlovsk 1970.

Shveitser, Mikhail Abramovich (Moisei Abramovich Shveitser)
Motion-picture director; b. Mr. 16, 1920, Penza.
People's Artist of the RSFSR (1977).
In 1944, he graduated from the department of film direction at the All-Union State Institute of Cinematography, where he studied in S. M. Eisenstein's workshop. His first work was *Path of Glory* (1949, in collaboration with B. A. Buneev and A. M. Rybakov).
His most significant motion pictures have been film versions of well-known works of classical and Soviet literature. They include *Someone Else's Relatives* (1956, based on V. F. Tendriakov's story "Not to the Yard"), *Resurrection* (1960–62, based on the L. N. Tolstoy novel), *Time, Forward!* (1966, in collaboration with S. A. Mil'kina; based on the novel by V. P. Kataev). *The Golden Calf* (1968, based on the novel by Il'f and Petrov), and *The Flight of Mister McKinley* (1976, based on a screenplay by L. M. Leonov). Decor.: State Prize of the USSR in 1977.

Shvetsov, Petr Filimonovich
Permafrost specialist and hydrogeologist; b. Ja. 14 (27), 1910, Kazanskaia, now Iglino Raion, Bashkir ASSR.
Corresponding member of the Academy of Sciences of the USSR (1953). Member of the CPSU since 1940.
After graduating from the Moscow Geological Prospecting Institute in 1935, he worked in the Main Administration of the Northern Sea Route. In 1941 he began working at the Institute of Geocryology, serving as its director from 1956 to 1958. In 1962 he became head of a laboratory at the All-Union Scientific Research Institute of Hydrogeology and Engineering Geology (VSEGINGEO).
His main works deal with groundwaters and the geothermal conditions of permafrost layers. He was scientific director and editor of the collective work *Fundamentals of Geocryology* (1959). Decors.: State Prize of the USSR in 1952, Order of the October Revolution, two other orders, and medals.
Works: *Zakonomernosti gidrogeotermicheskikh protsessov na Krainem Severe i Severo-Vostoke SSSR.* Moscow 1968. *Geotermicheskie usloviia mezozoisko-kainozoiskikh neftenosnykh basseinov.* Moscow 1974.

Sian-Belgin, Khasyr Bikinovich
Kalmyk writer; b. S. 12 (25), 1909, in the locality of Bulmukta, now in Iustin Raion, Kalmyk ASSR.
People's Poet of the Kalmyk ASSR (1966).
He began publishing in 1927. He wrote the collections of short stories and sketches *The Devils* (1933) and *Simple Steppe People* (1961), the verse collections *The Road* (1934), *Winged Youth* (1962), and *With the Sunrise* (1963), and the heroic dramatic narrative poem *The Fighting Orphan* (1934; play of the same name, 1936). His historical heroic drama *The Insurgent Steppe* (1968) dealt with the struggle for Soviet power in Kalmykia. He has translated folktales by Pushkin and works by Lermontov, Gorkii, T. G. Shevchenko, V. V. Mayakovsky, and R. Gamzatov. Decor.: Order of the Badge of Honor.
Works: *In Russian translation: Belyi bereg: Rasskazy, ocherki, skazki.* Elista 1970. *Dva brata: Stikhi i poemy.* Elista 1971. *Tridtsat' topolei: Stikhi i poemy.* [Foreword by B. Dzhimbinov.] Moscow 1974.

Sidorenko, Sergei Stepanovich, Cand. of Hist. Sc.
State official; b. 1916.
Mem. CPSU 1940–; grad. Kishinev Pedagogic Inst.; teacher, head of pedagogic dept. of a school 1937–39; served in Sov. Army 1939–46; instructor, dept. head with a regional cttee. of Moldavian CP, section head, Exec. Organizer for Cen. Cttee. Moldavian CP, dept. head with a regional cttee. of Moldavian CP, Sec. of Beltsy City Cttee. Moldavian CP 1946–58; mem. Cen. Cttee. Moldavian CP 1958–; 1st Sec. Beltsy City Cttee. Moldavian CP 1958–59; Chair. of Moldavian Republican Trade Union Council 1960–72; Dep. to 1959–, Chair. of Budget Comm. 1959–60, mem. Presidium Supr. Sov. Moldavian SSR 1960–67; cand. mem. Bureau 1960–61, mem. Bureau 1961–63, again cand. mem. Bureau Cen. Cttee. Moldavian CP 1963–72; Chair. of State Cttee. for Vocational Training with Council of Mins. Moldavian SSR F. 1972–80; Dep. Chair. Presidium Supr. Sov. Moldavian SSR N. 1980–. Decor.: Order of Red Banner of Labour.

Sidorov, Veniamin Aleksandrovich
Physicist; b. O. 19, 1930, Babarino, Suzdal' Raion, Vladimir Oblast.
Corresponding member of the Academy of Sciences of the USSR (1968).
After graduating from Moscow State University in 1953, he joined the staff of the Institute of Atomic Energy. In 1962 he became affiliated with the Institute of Nuclear Physics of the Siberian Division of the Academy of Sciences of the USSR. Together with other workers, he developed the colliding beam technique; he and the collective shared the Lenin Prize for this achievement in 1967. He has carried out experiments to test the applicability of the laws of quantum electrodynamics at short distances and to investigate double bremsstrahlung from colliding electron beams. He has also done work in fastneutron spectrometry. He has investigated vector mesons, multihadron annihilation, and the production of e^+e pairs by colliding beams of electrons and positrons.
Works: *Spektrometriia bystrykh neitronov.* Moscow 1968. (With B. V. Rybakov.) "Novosibirsk Colliding Beam Experiments." *Proceedings of the Symposium on Electron and Photon Interactions.* Ithaca, N. Y., Cornell University 1971.

Sidorov, Vladimir Vasil'evich
Military official; b. 1924.
Russian; mem. CPSU 1949–; grad. Pacific Ocean Naval College 1946, and Naval Acad. 1963, Admiral 1979; served in Sov. Army 1942–46; entered Sov. Navy 1946; Chief-of-Staff of Pacific Fleet, 1st Dep. Commander of Baltic Fleet 1946–78, Commander 1978–81; Dep. to USSR Supr. Sov. 1979–; cand. mem. Cen. Cttee. CPSU 1981–; Commander of Pacific Fleet 1981–.

Siforov, Vladimir Ivanovich
Scientist in the fields of radio engineering and electronics; b. My. 18 (31), 1904, Moscow.
Corresponding member of the Academy of Sciences of the USSR (1953). Member of the CPSU since 1941.
He graduated from the V. I. Lenin Leningrad Electrical Engineering Institute in 1929. From 1930 to 1941 and from 1946 to 1952, he taught at the institute, becoming a professor there in 1938. Between 1928 and 1941 he worked at the Institute of Radio Broadcasting Reception and Acoustics (until 1938, the Central Radio Laboratory), and from 1941 to 1953 he taught at the Leningrad Air Force Engineering Academy. In 1953 and 1954 and between 1955 and 1957, he worked at the Radio Research Institute, and in 1954 and 1955 he was deputy minister of the radio engineering industry in the USSR. In the years 1954–66 he was associated with the Institute of Radio Engineering and Electronics of the Academy of Sciences of the USSR. In 1966 he became director of the Institute of Information Transmission Problems of the Academy of Sciences of the USSR.
His main works have been devoted to theoretical questions on the stability of resonant amplifiers, the detection of signals and frequency conversion, radio reception at microwave frequencies, pulse radio communications, and radio navigation. He has also been concerned with the theory of information transmission and with methods of detecting radar signals and controlling radio interference.
In 1954, he became chairman of the Central Board of the A. S. Popov Scientific and Technical Society for Radio Engineering, Electronics, and Communications. Decors.: honorary member of the Hungarian Academy of Sciences (1973) and two Orders of Lenin, five other orders, and various medals.
Works: *Rezonansnye usiliteli.* Leningrad 1932. *Usiliteli vysokoi chastoty.* Moscow-Leningrad 1939. *Radiopriemniki sverkhvysokikh chastot,* 2nd ed. Moscow 1957. *Radiopriemnye ustroistva.* Moscow 1974. (With others.)

Sikachev, Nikolai Nikolaevich
Diplomat; b. 1920.
Mem. Diplomatic Service 1946–; mem. staff USSR Min. of Foreign Affairs 1955–65; Emb. Counsellor 1965–67 and Emb. Counsellor-Envoy at USSR Emb. in Hungary 1967–70; Head of V. European Dept. at USSR Min. of Foreign Affairs 1970–75; USSR Amb. to Kuwait Mr. 1975–.

Sikorskis, Romualdas Aleksandrovich
State official.
Min. of Finance Lithuanian SSR 1957–; cand. mem. 1958–60, mem. Cen. Cttee. Lithuanian CP 1960–; Dep. to Supr. Sov. Lithuanian SSR 1959–.

Silaev, Ivan Stepanovich
State official; b. 1930.
Mem. CPSU 1959–; grad. Kazan Aviation Inst. 1954; foreman, shop superintendent, dep. chief engineer, chief engineer, dir. of a plant in Gorkii 1954–74; Dep. USSR Min. of the Aviation Ind. 1974–77, 1st Dep. Min. 1977–80; USSR Min. of the Machine Tool and Tool Making Ind. 1980–81; USSR Min. of the Aviation Ind. F. 1981–; Dep. to USSR Supr. Sov. 1979–; mem. Cen. Cttee. CPSU 1981–. Decors.: Hero of Socialist Labour 1975, Lenin Prize 1972.

Silvet, Erki Petrovich
State official.
Min. of Justice Esthonian SSR Ja. 1976–; cand. mem. Cen. Cttee. Esthonian CP 1976–.

Simenenko, Grigorii Kalinovich
State official.
Cand. mem. Cen. Cttee. 1964–66 and 1981–, mem. Auditing Comm. Lithuanian CP 1976–81; Sec. of Vilna City Cttee. Lithuanian CP 1971–75; Min. of Local Ind. Lithuanian SSR Ap. 1975–.

Simeonov, Konstantin Arsenevich
Conductor; b. Je. 7 (20), 1910, Kaznakovo, Kalinin Oblast.
Grad. Leningrad Conservatory 1936; Principal Conductor of Ukrainian Theatre for Opera and Ballet in Kiev 1961–66; Principal Conductor of Leningrad Opera and Ballet Theatre 1967–; guest performances abroad, among them at La Scala, Milan. Decors.: People's Artist of the USSR 1962, 2 orders and various medals.

Simon (S. M. Novikov), Cand. of Theol.
Russian Orthodox clergyman; b. F. 5, 1928, Iaroslavl.
Studied Moscow Theological Seminary 1951–55; grad. Moscow Theological Acad., Cand. of Theology 1959; deacon, priest 1959; Head of the Cathedral of the Transfiguration in Lukino nr. Moscow 1964–65; Lecturer at Moscow Theological Acad. 1964; Supervisor of Moscow Theological Schools 1965–72; Bishop of Riazan and Kasinov 1972–.

Simonenko, P. M.
Komsomol official.
Sec. and mem. Bureau Cen. Cttee. Ukrainian Komsomol Ag. 1982–.

Simonov, Evgenii Rubenovich
Stage director; b. Je. 21, 1925, Moscow.
Grad. Shchukin Theatre School 1947; Asst. Stage Dir., Stage Dir. 1947–68, Principal Stage Dir. of Vakhtangov Theatre 1968–, of Malyi Theatre 1962–68. Decors.: Order of Red Banner of Labour, one medal.

Simonov, Kirill Stepanovich, Cand. of Tech. Sc.
Party official; b. 1917.
Mem. CPSU 1943–; grad. Leningrad Inst. of Railway Transport Engineering 1940 and Acad. of Railway Transport 1951; worked for Oktiabrskaia Railways 1940–47; Dept. Head, Chief Engineer with Admin. of USSR Min. of Transport 1947–49, 1951–52 and 1954–57; head of a dept. at Cen. Scientific Research Inst. of USSR Min. of Transport 1952–54; 1st Dep. Chair. of Scientific-Tech. Council of USSR Min. of Transport 1957–61; Head of Railway Admin. in Gorkii 1961–62; Head of Dept. of Transport, Posts and Telecommunications with Cen. Cttee. CPSU 1962–; mem. Cen. Auditing Comm. CPSU 1966–81, Cand. mem. Cen. Cttee. CPSU 1981–. Decor.: Order of Lenin 1977.

Simonov, Konstantin (Kirill) Mikhailovich
Author; b. N. 15 (28), Petrograd.
Russian; mem. CPSU 1942–; grad. Gorkii Inst. of Literature 1938; lathe-turner in Saratov until 1935; mem. Sov. Peace Cttee. 1949–; Dep. Chair. of USSR Writers' Union 1946–54; Sec. of USSR Writers' Union 1954–59 and 1967–; cand. mem. Cen. Cttee. 1952–56, mem. Cen. Auditing Comm. CPSU 1956–61 and 1976–; Chief Ed. of journal "Novyi mir"

1946–50 and 1954–58; Chief Ed. of "Literaturnaia gazeta" 1938 and 1950–54; first works publ. 1934. Decors.: State Prize of the USSR 1942, 1943, 1946, 1947, 1949, 1950, Hero of Socialist Labour 1974, Lenin Prize 1974 (for trilogy "The Living and the Dead") 1974.
Publs.: "Suvorov" 1939, "Paren' iz nashego goroda" (Boy From Our Town) 1941, "Tovarishchi po oruzhiiu" (Brothers in Arms) 1952, "Zhivye i mertvye" (The Living and the Dead) 1959, "Soldatami ne rozhdaiutsia" (Man Is Not Born a Soldier) 1963–64, "Sobranie sochinenii" (Collected Works), vols. 1–6, 1960–70, "Razgovor s tovarishchami" (Discussion with Friends) 1970, "Poslednee leto" (The Last Summer) 1970–71, and others.

Sinchenko, Georgii Zakharovich
State official; b. 1914.
Dep. Min. 1956–60, Min. of Posts and Telecommunications Ukrainian SSR 1960–; cand. mem. Cen. Cttee. 1961–66, Chair. 1960–71, mem. Auditing Comm. Ukrainian CP 1971–. Decors.: Order of Lenin, Order of Red Banner of Labour.

Siradze, Viktoria Moiseevna
State official; b. 1929.
Georgian; mem. CPSU 1951–; grad. Georgian Polytech. Inst.; Dep. Sec., Sec. of Komsomol Cttee. of Polytech. Inst. 1949–52; 1st Sec. of Tbilisi Raion Cttee., 2nd Sec., 1st Sec. of Tbilisi City Cttee. Georgian Komsomol, Sec. of Tbilisi Raion Cttee., Sec. of Tbilisi City Cttee. Georgian CP 1952–61; mem. Auditing Comm. 1956–58, cand. mem. 1958–60, mem. Cen. Cttee. Georgian CP 1960–; Sec. and mem. Bureau Cen. Cttee. Georgian CP 1961–62; Dep. Chair. of Council of Mins. Georgian SSR 1962–73; Sec. and mem. Bureau Cen. Cttee. Georgian CP F. 1973–81; Dep. Chair. of Presidium Supr. Sov. Georgian SSR D. 1978–.

Sirazhdinov, Sagdy Khasanovich
Scientist and state official.
Vice-Pres. of Acad. of Sciences Uzbek SSR Ap. 1970–; Dep. to Supr. Sov. Uzbek SSR 1971–; Chair. of Supr. Sov. Uzbek SSR Jl. 1971–80.

Sivolob, V. I.
Trade union official.
Sec. and mem. Presidium Ukrainian Republican Trade Union Council D. 1969–.

Sivtsev, Dmitrii Kononovich (pen name: Suorun Omollon)
Yakut writer; b. S. 1 (14), 1906, Third Zhekhsogonskii Nasleg (ancestra rural commune), now in Alekseevskii Raion, Yakut ASSR.
People's Writer of the Yakut ASSR (1967) and Honored Art Worker of the RSFSR and Yakut ASSR. Member of the CPSU since 1966.
He graduated from the Yakut Pedagogical Technicum in 1928; he had begun publishing in 1926. He wrote the short story collections *In the Flowering Greenery* (1933) and *Stories* (1937) and the dramas *The Blacksmith Kiukiur* (1932; Russian translation, 1960), about the collectivization of the Yakut village, *N'iurgun Bootur the Impetuous* (1940; Russian translation, 1963), *Saisary* (1943; 2nd version, 1958), and *Before the Dawn* (1970). He is also the author of the tragedy *Aiaal* (1940; 2nd version, 1958; Russian translation, 1966). His work has close ties with folklore. He has published several collections of his own versions of Yakut folktales. He also writes for children. Decors.: two orders and several medals.
Works: *Talyllybyt aiymn'ylar,* vols. 1–2, Yakutsk 1966–67. *In Russian translation: Izbrannoe.* Yakutsk 1958. "O chem pel khomus." *Poliarnaia Zvezda.* 1970, no. 2.

Sizenko, Evgenii Ivanovich
Party official; b. 1931.
Russian; mem. CPSU 1953–; grad. K. A. Timiriazev Agricultural Acad. in Moscow 1954, Cand. of Econ. Sc. 1975; agronomist in Riazan Oblast 1954; student 1954–57; teacher at K. A. Timiriazev Agricultural Acad. in Moscow 1957–60; party and sov. posts in Moscow Oblast 1961–70; Sec. of Moscow Oblast Cttee. CPSU 1970–78; 1st Sec. of Briansk Oblast Cttee. CPSU S. 1978–; Dep. to USSR Supr. Sov. 1979–; mem. Cen. Cttee. CPSU 1981–.

Sizov, Gennadii Fedorovich
Party official; b. 1903; s. of a peasant.
Russian; mem. CPSU 1926–; grad. Moscow Timiriazev Agricultural Acad. 1930; agricultural worker until 1922, then student at a workers' faculty; Dept. Head, Dir. of Inst. of Dairy Cattle Breeding 1930–33; dir. of a sovkhoz, of a sovkhoz trust, Head of Novosibirsk Oblast Agricultural Board 1933–41; served in Sov. Army 1941–47; dir. of various sovkhoz trusts, 2nd Sec. of Kurgan Oblast Cttee. CPSU, Chair. of Kurgan Oblast Sov. 1947–55; 1st Sec. of Kurgan Oblast Cttee. CPSU (1963–64 for Agriculture) 1955–66; cand. mem. 1956–64, mem. Cen. Cttee. CPSU 1964–66; Dep. to 1958–, mem. Credentials Comm. of Sov. of the Union USSR Supr. Sov. 1966–70; Chair. of Cen. Auditing Comm. CPSU Ap. 1966–. Decors.: Order of Lenin, Order of Red Banner of Labour.

Skachkov, Semen Andreevich
State official; b. 1907, Kharkov; s. of a worker.
Ukrainian; mem. CPSU 1936–; grad. Kharkov Inst. of Machine Construction 1930; joined Komsomol 1924; engineer, Dep. Dept. Head, Dept. Head, Chief Metallurgist at Kharkov Locomotive Construction Plant 1930–41; organizer for Cen. Cttee. All-Union CP (B) at a tank construction plant in the Urals 1941–45; dir. of a diesel plant in Leningrad, dir. of tractor plants in Nizhni Tagil and Cheliabinsk 1945–54; Dep. to USSR Supr. Sov. 1950–; USSR 1st Dep. Min. of Transport Machine Construction 1954–57; Chair. of Kharkov Econ. Council 1957–58; Chair. of USSR State Cttee. for Econ. Relations with Foreign Countries F. 1958–; cand. mem. 1961–71, mem. Cen. Cttee. CPSU 1971–. Decors.: 2 Orders of Lenin, 4 Orders of Red Banner of Labour, Order of Patriotic War 2nd Cl., 2 Orders of Red Star, Hero of Socialist Labour, Order of Lenin, Hammer and Sickle Medal in gold 1977.

Skatkin, Mikhail Nikolaevich
Educator; b. Jl. 29 (Aug. 11), 1900, Kozlovo, now in Konakovo Raion, Kalinin Oblast.
Corresponding member of the Academy of Pedagogical Sciences of the RSFSR (1950) and of the USSR (1968). Doctor of pedagogical sciences (1970) and professor (1970).
In the 1920's, he worked at the first experimental station of the People's Commissariat for Education of the RSFSR, where he also completed pedagogical courses in 1925. In the 1930's he worked at scientific research and teacher-training pedagogical institutes in Moscow. Beginning in 1945 he worked at institutes of the Academy of Pedagogical Sciences of the RSFSR, and beginning in 1968, at institutes of the Academy of Pedagogical Sciences of the USSR.
His chief works dealt with pedagogical methodology, didactics, and the content and methods of polytechnical education. He also wrote standard natural-science textbooks for the early grades of secondary school. Decors.: Order of Lenin, two other orders, and several medals, including the K. D. Ushinskii Medal (1954).
Works: *Metodika prepodavaniia estestvoznaniia v nachal'noi shkole,* 3rd ed. Moscow 1952. *Nekotorye voprosy didaktiki i svete ucheniia. I. P. Pavlova o vysshei nervnoi deiatel'nosti.* Moscow 1952. *Politekhnicheskoe obuchenie na sovremennom etape razvitiia shkoly.* Moscow 1965. *Sovershenstvovanie protsessa obucheniia.* Moscow 1971. *O shkole budushchego.* Moscow 1974.

Skiba, Ivan Ivanovich
Party official; b. 1937.
Ukrainian; mem. CPSU 1959–; grad. Odessa Agricultural Inst.; faculty asst. at Odessa Agricultural Inst., inst. of a dept. of Odessa Oblast Cttee. Komsomol, instructor, sector chief, dept. chief. of Cen. Cttee. Ukrainian Komsomol 1959–62; Sec. of Cen. Cttee. Ukrainian Komsomol 1962–68, 2nd Sec. 1968–70; inspector of a dept. of Cen. Cttee. Ukrainian CP 1970–72; 2nd Sec. of Transcarpathian Oblast Cttee. Ukrainian CP 1972–78; cand. mem. Cen. Cttee. Ukrainian CP 1976–80, mem. 1980–; 1st Sec. of Ivano-Frankovsk Oblast Cttee. Ukrainian CP O. 1978–.

Skliarevskii, Vadim Valentinovich, Cand. of Phy.-Math. Sc.
Physicist; b. 1922.
Mem. CPSU 1950–; grad. Leningrad Univ. 1946; Cand. of Phy.-Math. Sc. 1958–; served in Sov. Army 1941–42; mem. staff Kurchatov Inst. of Atomic Energy at USSR Acad. of Sciences 1946–; Head of Laboratory at same inst. 1969–. Decor.: State Prize of the USSR 1969.

Skliarov, Iurii Aleksandrovich
Journalist; b. 1925.
Mem. CPSU 1944–; grad. Kharkov Univ. 1951, Cand. of Hist. Sc. 1955, Asst. Prof. 1964; served in Sov. Army 1943–45; posts in Kharkov City and Oblast Cttee. apparatus Ukrainian CP 1955–63; teacher at Kharkov Univ. 1963–64; Sec. of Kharkov Oblast Cttee. Ukrainian CP 1964–69; Dep. Chief of Propaganda Dept. of Cen. Cttee. CPSU 1969–76; 1st Dep. Ed.-in-Chief of "Pravda" My. 1976–; cand. mem. Cen. Cttee. CPSU 1981–.

Skliarov, Vitalii Fedorovich
State official; b. 1935.
Mem. CPSU 1962–; higher education, electrical engineer; engineer-in-training, shift foreman, dep. dir. of a dept., chief engineer, dir. of an electric power station 1957–73; Chief Engineer 1973–74, Dir. 1974–76, Gen. Dir. of "Kievenergo" Engineering Production Assoc. 1976–77; 1st Dep. Min. 1977–82, Min. of Power and Electrification Ukrainian SSR My. 1982–.

Skoblikova, Lidiia Pavlovna
Speed skater; b. Mr. 8, 1939, Zlatoust.
Honored Master of Sports (1960) and teacher. Member of the CPSU since 1964. In 1974, she became head of the department of physical education at the Moscow Higher School of the Trade Union Movement of the All-Union Central Council of Trade Unions. She earned six gold medals at the winter Olympic games of 1960 and 1964 and was all-round world champion in 1963–64. She broke several records and was a world champion and USSR champion in speed skating over different distances, winning 40 gold medals, including 25 at world championships and 15 at USSR championships. Decors.: two orders and various medals.

Skomorokhov, Nikolai Mikhailovich
Military official; b. My. 19, 1920, Lapot, Saratov Oblast.
Mem. CPSU 1943–; grad. Bataisk Pilot School 1942, M. V. Frunze Mil. Acad. 1949, and Mil. Acad. of Gen. Staff 1958, Marshal of Sov. Air Forces 1981; pilot, commander of a squadron on the Caucasian, North Caucasian, Southern, Western, and 3rd Ukrainian fronts 1942–73; Chief of Iu. A. Gagarin Air Force Acad. 1973–. Decors.: Order of Lenin, Order of Red Banner of Labour (five), Order of Aleksandr Nevskii, Order of Great Fatherland War 1st Class, Order of Red Star, foreign orders, medals; Hero of the Soviet Union (F. 23, and Ag. 18, 1945).

Skrebnev, Vasilii Stepanovich
State official.
2nd Sec. Tashkent City Cttee. Uzbek CP 1963–66; Dep. Head of Main Admin. for Construction in Tashkent City with Council of Mins. Uzbek SSR 1966–78; mem. Cen. Cttee. Uzbek CP 1966–71 and 1981–; Chair. State Cttee. for Construction Uzbek SSR F. 1978–.

Skriabin, Georgii Konstantinovich
Microbiologist and biochemist; b. S. 4 (17), 1917, Leningrad.
Mem. CPSU 1947–; grad. Kazan Veterinary Inst. 1942; corresponding mem. USSR Acad. of Sciences 1968; served in Sov. Army 1943–46; mem. staff Inst. of Microbiology of USSR Acad. of Sciences 1949–, Head of Laboratory of same 1960–; Dep. Dir. 1962–67, Dir. of Inst. of Biochemistry and Physiology of Microorganisms of USSR Acad. of Sciences in Pushchino 1967; Dir. of Scientific Centre for Biological Research of USSR Acad. of Sciences in Pushchino 1967–74; Acting Scientific Chief Sec. of Presidium USSR Acad. of Sciences 1971–. Decors.: 2 Orders of Lenin, Order of Red Star, medals.

Skrigan, Ianka (Ivan) Alekseevich
Author; b. N. 3 (16), 1905, Trukhanovichi, now Minsk Oblast.
Belorussian; mem. CPSU 1964–; studied linguistics at Pedagogical Dept. of Belorussian State Univ. 1928–32; first works publ. 1924.
Publs.: "Vstrechi" (Encounters) 1935, "Natalia" 1957, "Krugi" (Circles) 1969, "Povorot u sosny" (Turning at the Pine) 1974.

Skrinskii, Aleksandr Nikolaevich
Nuclear physicist; b. Ja. 15, 1936, Orenburg.
Grad. Moscow State Univ. 1959; mem. USSR Acad. of Sciences 1970–; mem. staff Inst. of Nuclear Physics of Siberian Branch of USSR Acad. of Sciences 1959–; concurrently Prof. at Novosibirsk Univ. 1967–.
Publs.: "Ispolzovanie sinkhrotronnogo izlucheniia nakopitelia VEPP-3 dlia rentgenostrukturnykh issledovanii" in "Doklady Akademii Nauk SSSR" Nr. 4/1974.

Skripko, Nikolai Semenovich
Military official; b. N. 22 (D. 5), 1902, Bolderaia nr. Riga.
Mem. CPSU 1927–; grad. Pilots' School 1927, Higher Flight-Tactical School 1938; took part in advanced acad. courses at Voroshilov Higher Mil. Acad. 1950; Air Force Marshal 1944–; took part in Civil War 1918–20; Commander of 3rd Aviation Corps, Commander of Air Force 5th Army, Dep. Commander of Long-Range Air Force, took part in battle of Stalingrad, combat in Northern Caucasus, the Crimea, Kursk region, Belorussia, the Baltic States and East Prussia 1941–45; after the war 1st Dep. Commander of Long-Range Air Force 1946–49, Commander of Transport and Air Landing Forces 1950–55, Commander of Mil. Transport Air Force 1955–69; Mil. Inspector/Adviser of Group of Inspectors-Gen. of USSR Min. of Defence Ag. 1969–; mem. Cen. Auditing Comm. CPSU 1961–66. Decors.: 2 Orders of Lenin, Order of October Revolution, 5 Orders of Red Banner, Suvorov Order 1st and 2nd Cls., Kutuzov Order 1st Cl., medals.

Skulachev, Vladimir Petrovich
Biochemist; b. F. 21, 1935, Moscow.
Grad. Moscow State Univ. 1957; corresponding mem. USSR Acad. of Sciences 1974–; mem. staff Moscow State Univ. 1960– (Head of Dept. of Bioenergetics at Interfaculty Laboratory for Bio-organic Chemistry 1965–73, Head of same laboratory 1973–). Decors.: State Prize of the USSR 1975, Prize for Science and Technology of Lenin Komsomol 1967, Bakh Prize of USSR Acad. of Sciences 1972.

Sladkovskii, Mikhail Iosifovich, Dr. of Econ. Sc.
Economist; b. 1906.
Mem. CPSU 1928–; grad. Far Eastern Univ. 1930; corresponding mem. USSR Acad. of Sciences 1972–; Dr. of Econ. Sc. 1959; Prof. 1961–; mem. staff USSR Min. of Foreign Trade 1930–65; Dir. of Far Eastern Inst. of USSR Acad. of Sciences 1966–; main works deal with the economics of China and other Far Eastern countries and with economic relations between China, on one hand, and Russia and the USSR on the other. He is the co-author and editor of several works on the history of international relations, such as *International Relations in the Far East* (1973). Decors.: two Orders of Lenin, one Order of the October Revolution, five other orders, and various medals.
Works: *Ocherki razvitiia vneshneekonomicheskikh otnoshenii Kitaia*. Moscow 1953. *Ocherki ekonomicheskikh otnoshenii SSSR s Kitaem*. Moscow 1957. *Kitai i Iaponiia*. Moscow 1971. *Istoriia torgovo-ekonomicheskikh otnoshenii narodov Rossii s Kitaem (1917 g.)*. Moscow 1974.

Slavin, Lazar Moiseevich
Archaeologist; b. Je. 11, 1906, Vitebski; s. of a clerk.
Mem. CPSU 1940–; grad. Leningrad Univ. 1928; corresponding mem. Acad. of Sciences Ukrainian SSR 1939–; Dir. 1939–41 and 1944–45, Dep. Dir. of Inst. of Archaeology of Acad. of Sciences Ukrainian SSR 1946–49; concurrently Scientific Sec. of Dept. of Social Sciences of Acad. of Sciences Ukrainian SSR 1939–45; mem. staff Inst. of Archaeology of Acad. of Sciences Ukrainian SSR 1939–; Head of Chair of Archaeology

and Museum Science at Kiev Univ. 1945–; author of some 70 works. Decors.: Order "Badge of Honour", a medal.
Publs.: "Drevnii gorod Olvia" 1951, "Problemy istorii Severnogo Prichernomoria v antichnuiu epokhu" 1959, "Olviiski kvartaly tsentralnoi chastyny Verkhniogo mista" in "Arkheologichni pamiatky URSR" vol. 11, 1962, and others.

Slavin, Lev Isaevich
Author; b. O. 15 (27), 1896, Odessa.
Russian; served in World War I (1914–18), during Civil War (1918–20) and World War II (1941–45); correspondent for "Krasnaia zvezda" and "Izvestiia" 1941–45; first works publ. 1922. Decors.: 3 orders and various medals.
Publs.: "Naslednik" (The Heir) 1930, "Po tu storonu kholma" (On the Other Side of the Mountain) 1958, "Za nashu i vashu svobodu!" (For Our and Your Freedom!) 1968, "Izbrannoe" (Collected Works) 1970, and others.

Slavinskii, M. M.
Trade union official.
Head of Dept. of Cultural Mass Work until 1971; Sec. and mem. Presidium Ukrainian Republican Trade Union Council My. 1971–.

Slavskii, Efim Pavlovich
State official; b. 1898, Makeevka, Donets Basin; s. of a peasant.
Ukrainian; mem. CPSU 1918–; grad. Moscow Inst. of Non-Ferrous Metals and Gold 1933; Engineer, Dept. Head, Chief Engineer, Dir. of Ordzhonikidze "Elektrotsink" Plant 1933–40; dir. of an aluminium plant on the Dnepr 1940–41 and in the Urals 1941–45; USSR Dep. People's Commr. of Non-Ferrous Metallurgy 1945–46; dep. head of a main board with USSR Council of Mins. 1946–53; USSR Dep. Min. 1953–57, 1st Dep. Min. 1953–57, Min. of Medium Machine Construction 1957–63; Dep. to USSR Supr. Sov. 1958–; mem. Cen. Cttee. CPSU 1961–; Chair. of State Prod. Cttee. for Medium Machine Construction 1963–65; USSR Min. of Medium Machine Construction Mr. 1965–. Decors.: thrice Hero of Socialist Labour, State Prize, 6 Orders of Lenin, Order of Red Banner of Labour, Order of the October Revolution.

Sliede, Egon Edgarovich
State official.
Chair. of Riga City Exec. Cttee. of Workers' Deps. 1967–69; Min. of Motor Transport and Highways Latvian SSR Ag. 1969–; cand. mem. Cen. Cttee. Latvian CP 1971–.

Slinchenko, Vladimir Ivanovich
State official; b. 1931.
Cand. of Economic Sciences; Mem. CPSU 1951–; grad. Inst. for Economic Engineering in Kharkov; lathe operator, technician in a plant, various Komsomol and Party posts until 1972; in apparat of Cen. Cttee. Ukrainian CP 1972–73; 2nd Oblast Sec. Ukrainian CP in Chernovtsy 1973–77; 1st Dep. Min. 1979–80, Min. of Consumer Services Ukrainian SSR N. 1980–.

Slinko, Mikhail Gavrilovich
Chemical Scientist; b. S. 2 (15), 1914, Moscow.
Mem. CPSU 1943–; grad. Moscow State Univ. 1941; served in Sov. Army 1941–46; mem. staff Karpov Inst. of Physical Chemistry 1946–56; mem. Apparat Cen. Cttee. CPSU 1956–59; Dep. Dir. of Inst. of Catalysis at Siberian Branch of USSR Acad. of Sciences 1959–. Decors.: State Prize of the USSR 1946, Lenin Prize 1960, 2 Orders of Lenin, 4 other orders, medals.

Slipchenko, Sergei Aleksandrovich
Diplomat; b. 1912.
Mem. Diplomatic Service 1953–; mem. staff Min. of Foreign Affairs Ukrainian SSR 1953–54; Head of Pol. Dept. and mem. Collegium Min. of Foreign Affairs Ukrainian SSR 1954–55; Dep. Min. of Foreign Affairs Ukrainian SSR 1955–65; USSR Amb. to Sambia 1965–70; sen. posts at USSR Min. of Foreign Affairs 1972–Ap. 1980.

Sliunkov, Nikolai Nikitovich
Party official; b. 1929, in the village of Garadzets, Oblast Gomel.
Belorussian; mem. CPSU 1954–; grad. Belorussian Inst. of Mechanization of Agriculture 1962; After grad. from a mechanical-technical school assistant foreman, foreman, senior foreman, dep. shop superintendant, shop superintendant, dep. chair., chair. of the trade-union Cttee. of the Minsk Tractor Plant 1950–60; director of the Minsk Spare Parts Plant 1960–65; director of the Minsk Tractor Plant 1965–72; mem. Cen. Cttee. Belorussian CP 1966–76; 1st sec. of Minsk City Cttee. Belorussian CP 1972–74; dep. Chair. of the USSR GOSPLAN 1974–Ja. 1983; 1st sec. and mem. Bureau Cen. Cttee. Belorussian CP Ja. 1983–.

Sliusarenko, Zakhar Karpovich
Military official; b. S. 3 (16), 1907, Zmiev, Kharkov Oblast.
Twice Hero of the Soviet Union (S. 23, 1944, and May 31, 1945); lieutenant-general of the armored forces (1963). Member of the CPSU since 1929.
He graduated from the Orel Armored Troop School in 1934 and completed advanced academic courses for officer's training at the Military Academy for Armored and Mechanized Troops in 1949. He graduated from the Higher Academic Courses at the K. E. Voroshilov Higher Military Academy in 1957. In the Great Patriotic War of 1941–45 he commanded a tank battalion on the Southwest and Briansk fronts in 1941–42 and the 168th Tank Brigade on the Stalingrad Front and a separate tank regiment on the Leningrad Front from 1942 to 1944. In February 1944 he was put in command of the 56th Guard Tank Brigade on the First Ukrainian Front. After the war he held various command posts in the army. In September 1960 he became deputy commander of a group of troops in charge of combat training. He retired in 1965.
Decors.: two Orders of Lenin, the Order of the Red Banner, the Order of Suvorov Second Class, the Order of the Patriotic War Second Class, the Order of the Red Star, and medals, as well as two orders of socialist countries.

Slonimskii, Sergei Mikhailovich
Composer, pianist and musicologist; b. Ag. 12, 1932, Leningrad.
Grad. Leningrad Conservatory 1955; teacher at Leningrad Conservatory 1958–; wrote music for "Virineia" (opera) 1967, "The Master and Margarita" 1970–72, "Ikar" (ballet) and other works.

Slutskii, Boris Abramovich
Author; b. My. 7, 1919, Slaviansk, Donetsk Oblast.
Russian; mem. CPSU 1943–; grad. Gorkii Inst. of Literature 1941. Decors.: 3 orders and various medals.
Publs.: "Pamiat'" (Commemoration) 1957, "Vremia" (Time) 1959, "Segodnia i vchera" (Today and Yesterday) 1961, "Rabota" (Work) 1964, "Sovremennyo istorii" (Modern Stories) 1969, "Dobrota dnia" (Goodness of a Day) 1973, "Prodlennyi polden'" (Prolonged Noon) 1975, and others.

Slutskis, Mikolas (Sluckis, Mikolas)
Author; b. O. 20, 1928, Panevėžys.
Lithuanian; mem. CPSU 1950–; grad. Dept. of History and Philology Vilnius Univ.; first works publ. 1945. Decors.: State Prize of the Lithuanian SSR 1966, Order "Znak pocheta", various medals.
Publs.: "Pust' my luchshe ne vstretimsia" (We Had Better Not Meet Again) 1961, "Ulybki i sud'by" (Smiles and Fates) 1964, "Shagi" (Steps) 1965, "Lestnitsa v nebo" (Ladder to the Sky) 1963 (Russian translation 1965), "Adamovo iabloko" (Adam's Apple) 1966 (Russian translation 1969), "Chuzhie strasti" (Strange Passions) 1971, play "Ne beshenaia li tvoia sobaka" (Doesn't Your Dog Have Rabies?) 1974, collection of essays "Nachalo vsekh nachal" (The Beginning of all Beginnings) 1975, and others.

Smailov, Kamal Seitzhanovich
State official.
1st Dep. Chair. until 1971, Chair. of State Film Cttee. Kazakh SSR F. 1973–Ja. 1977; mem. Auditing Comm. Kazakh CP 1976–81; Head of Cultural Dept. of Council of Mins. Kazakh SSR 1977–80; Dep. Chair. of State Cttee. for Television and Radio Kazakh SSR Ja. 1980–.

Smerdov, Aleksandr Ivanovich
Author; b. Ag. 31 (S. 13), 1910, Teplaia Gora, Perm Oblast.
Mem. CPSU 1941–; studied Planning Inst. in Novosibirsk and at Gorkii Inst. of Literature.
Publs.: "Pushkinskie Gory" (Pushkin Hills) 1946–49, "Nash den'" (Our Day) 1953, "V stepi Kulundinskoi" (In the Kulundinskaia Steppe) 1947, "Kamen' na ladoni" (The Stone in the Palm) 1948–57, "Pis'ma s dorogi. Stikhi i poemy" (Letters from the Road. Verse and Poetry) 1961, "Stikhi" (Poems) 1966, "Vesniaia tiaga" (Forces of Spring) 1970, and others.

Smirnov, Adrian Anatolevich, Dr. of Phy.-Math. Sc.
Theoretical physicist; b. N. 16, 1908, Novgorod; s. of a teacher.
Mem. CPSU 1945–; grad. Leningrad Univ. 1932; corresponding mem. Acad. of Sciences Ukrainian SSR 1951–; Dr. of Phy.-Math. Sc. 1947; Prof. 1948–; mem. staff Uralsk Physical-Tech. Inst. 1932–39; Sen. Scientist at Uralsk Branch of USSR Acad. of Sciences 1939–41; served in Sov. Army 1941–; again with Uralsk Branch of USSR Acad. of Sciences 1941–49; mem. staff Inst. of Physics, Laboratory of Physics of Metals, at Inst. of Physics of Metals of Acad. of Sciences Ukrainian SSR 1949–; concurrently lecturer at various colleges in Sverdlovsk and Kiev 1932–; author of some 100 scientific papers. Decors.: Order "Badge of Honour", medals.
Publs.: co-author of "Teoriia uporiadochivaiushchikhsia splavov" 1958, "Teoriia elektrosoprotivleniia splavov" 1960, "Molekuliarnokineticheskaia teoriia metallov" 1966, and others.

Smirnov, Aleksandr Ivanovich
Diplomat; b. 1912; s. of a peasant.
Russian; mem. CPSU 1937–; completed agricultural tech. school 1931, 4 terms at Gorkii Agricultural Inst., and Party Higher School of Cen. Cttee. CPSU 1953; agronomist, chief veterinary technician with a raion agricultural dept. and a state cattlebreeding station 1931–37; Chief Veterinary Technician, Head of Cattle-Breeding Board and Dep. Head of Gorkii Oblast Agricultural Dept. 1937–43; Sec. of Gorkii Oblast All-Union CP (B) Cttee. 1943; worked for Cen. Cttee. CPSU 1943–47 and 1948–53; worked for Rostov Oblast All-Union CP (B) Cttee. 1947–48; 2nd Sec. of Chita Oblast CPSU Cttee. 1953–55; Chair. of Exec. Cttee. of Chita Oblast Sov. of Workers' Deps. 1955–61; Dep. to Supr. Sov. RSFSR 1955–63; 1st Sec. of Chita Oblast CPSU Cttee. 1961–73 (for Agriculture 1963–64); cand. mem. Cen. Cttee. CPSU 1961–; Dep. to USSR Supr. Sov. 1962–; USSR Amb. to Mongolia Jl. 1973–. Decor.: Order of Red Banner of Labour.

Smirnov, Aleksei Alekseevich
State official; b. 1921.
Russian; mem. CPSU 1947–; grad. Dnepropetrovsk Mining Inst.; student 1941; served in Sov. Army 1941–42; mining mechanic, engineer with a trust, postgraduate studies at Mining Inst. 1943–46; 1st Sec. of Dnepropetrovsk City and Oblast Cttee. Ukrainian Komsomol, dep. dept. head with Cen. Cttee. USSR Komsomol 1946–54; instructor for a dept. at Cen. Cttee. CPSU 1954–58; Head of Dept. of Party Organs with Cen. Cttee. Belorussian CP 1958–59; 1st Sec. of Brest Oblast Cttee. Belorussian CP 1959–63; mem. Cen. Cttee. Belorussian CP 1960–; Dep. to 1962–, mem. Mandate Comm. of Sov. of the Union 1962–66, Chair. of Comm. for Trade and Public Services 1966–70, Chair. of Comm. for Consumer Goods of Sov. of Nationalities USSR Supr. Sov. 1974–; Dep. Chair. of Bureau of Ind. with Cen. Cttee. Belorussian CP 1962–64; Chair. of Exec. Cttee. of Brest Oblast Sov. of Workers' Deps. 1963–65; cand. mem. 1965–66, mem. Bureau Cen. Cttee. Belorussian CP 1966–78; Sec. of Cen. Cttee. Belorussian CP Mr.–Ap. 1965 and Je. 1968–78; Dep. Chair. of Council of Mins. Belorussian SSR Ap.–D. 1965; Chair. of Cttee. on People's Control Belorussian SSR 1965–68; Cand. mem. Cen. Cttee. CPSU 1981–; Board Chair., USSR Cen. Union of Consumers Cooperatives 1978–.

Smirnov, Aleksei Semenovich
Military official; b. Ja. 25 (F. 7), 1917, Pal'tsevo, now in Rameshki Raion, Kalinin Oblast.
Colonel (1952). Twice Hero of the Soviet Union (S. 28, 1943, and F. 23, 1945). Member of the CPSU since 1941.
He graduated from a military aviation school for pilots in 1938 and in 1947 completed the Lipetsk Higher Officers' Tactical Flight Courses. During the Great Patriotic War (1941–45), he was a flight commander and a deputy squadron commander in the 153rd Fighter Aviation Regiment on the Leningrad, Volkhov, Voronezh, Northwestern, and Kalinin fronts from 1941 to 1943. In October 1943 he became a squadron commander in the 28th Guards Fighter Aviation Regiment on the First and Second Baltic fronts; in January 1945 he became deputy commander of the regiment on the Third Belorussian Front. He flew 457 missions, engaged in 72 air battles, and shot down 34 planes. After the war, he served in the Air Force; in 1954 he transferred to the reserves because of illness.
Decors.: two Orders of Lenin, five Orders of the Red Banner, the Order of the Patriotic War First Class, the Order of Alexander Nevsky, the Order of the Red Star, and various medals.

Smirnov, Anatolii Filippovich
Scientist specializing in structural mechanics and building design; b. N. 8 (21), 1909, in the *stanitsa* (large cossack village) of Racheika, Syzran' Raion, now in Kuibyshev Oblast.
Doctor of technical sciences (1949). Honored Worker in Sciences and Technology of the RSFSR (1967). Member of the CPSU since 1951.
In 1935, he graduated from the Moscow Institute of Railroad Transportation Engineering, where he taught structural mechanics beginning in 1936 and became a professor in 1947. In 1969 he became director of the V. A. Kucherenko Central Scientific Research Institute of Building Structures and editor-in-chief of the journal *Stroitel'naia mekhanika i raschet sooruzhenii* (Structural Mechanics and Design).
His principal works deal with research on the static and dynamic stability and vibrations of structures. He has proposed a matrix form of solution of various problems in structural mechanics. Decors.: Order of Lenin, two other orders, and various medals.
Works: *Staticheskaia i dinamicheskaia ustoichivost' sooruzhenii.* Moscow 1947. *Ustoichivost' i kolebaniia sooruzhenii.* Moscow 1958. *Raschet sooruzhenii s primeneniem vychislitel'nykh mashin.* Moscow 1964.

Smirnov, Boris Aleksandrovich
Artist; b. Mr. 1 (14), 1903, in St. Petersburg.
Honored Artist of the RSFSR (1968).
He studied under I. A. Fomin in the faculty of architecture of the Leningrad Vkhutein (Higher Art and Technical Institute) from 1920 to 1928. He taught at the I. E. Repin Leningrad Institute of Painting, Sculpture, and Architecture from 1946 to 1948 and at the V. I. Mukhina Leningrad Higher Industrial Arts School from 1952 to 1963. His students included G. A. Antonova, S. M. Beskinskaia, and A. M. Ostroumov.
In the 1920's and 1930's, he designed interiors and industrial and public buildings; he also did illustrations. Since 1948 he has specialized in decorative glassware and modern design in optical instruments. He has also arranged exhibitions. Later in his career he began working in ceramics and porcelain. Taking form and function into consideration, he incorporates pictorial, often topical, decorative elements into the articles he makes.
Work: *Khudozhnik o prirode veshchei.* Leningrad 1970.

Smirnov, Efim Ivanovich
Public-health official; b. O. 10 (23), 1904, Ozerki, now in Kovrov Raion. Vladimir Oblast; s. of a worker.
Colonel-general of the medical service (1944). Academician of the Academy of Medical Sciences of the USSR (1948).

Hero of Socialist Labor (1978). Member of the CPSU since 1929.
He began serving in the Soviet Army in 1928. He graduated from the Military Medical Academy in 1932 and from the M.V.Frunze Military Academy in 1938. He became director of the health department of the Leningrad Military District in April 1938 and head of the health directorate of the Workers' and Peasants' Red Army in May 1939. During the Great Patriotic War (1941–45), he was chief of the Main Health Directorate of the Red Army (August 1941 to April 1946). In 1946 and 1947 he served as chief of the Main Military Medical Directorate of Armed Forces of the USSR. He served as Minister of Health of the USSR from February 1947 to January 1953 and as director of the S.M.Kirov Military Medical Academy in 1953. He was chief of the Main Military Directorate of the Ministry of Defense of the USSR from 1955 to 1960 and has been chief of a directorate of the Ministry of Defense of the USSR since November 1960.
He has written works dealing mainly with the organization and tactics of the military medical service and history of military medicine. He developed and introduced into practice a system of specialized medical care that enabled 73 percent of the wounded and more than 90 percent of the sick to return to duty. The sanitary measures for epidemic control developed under his direction prevented epidemics among the field forces.
He is an honorary member of the Canadian Medical Association, the Royal Society of Medicine, and the Armed Forces Institute of Pathology of the USA. Decors.: six Orders of Lenin, three Orders of the Red Banner, the Order of Kutuzov First Class, two Orders of the Red Star, various medals, and foreign orders and medals.
Works: *Voprosy organizatsii i taktiki sanitarnoi sluzhby.* Moscow 1942. *Voennaia meditsina i Nikolai Ivanovich Pirogov.* Leningrad 1945. *Sovetskie voennye vrachi v Otechestvennuiu voinu.* Moscow 1945. *Nikolai Ivanovich Pirogov.* Moscow 1960 (co-author). *Voina i voennaia meditsina 1939–1945 gg. (Mysli i vospominaniia).* Moscow 1976.

Smirnov, Georgii Lukich
Party official; b. 1922.
Mem. CPSU 1943–; grad. Volgograd Pedagogical Inst. 1952 and CPSU Cen. Cttee. Acad. of Social Sciences 1957; Asst. Prof., Dr. of Philosophical Sc. 1971; Komsomol worker 1942–44; posts in econ. sphere 1944–47; various Party posts 1948–57; posts in Cen. Cttee. CPSU apparatus 1957–62; mem. of editorial staff of theoretical organ of CPSU "Kommunist" 1962–65; posts in Cen. Cttee. CPSU apparatus 1965–69; Dep. Chief of Propaganda Dept. of Cen. Cttee. CPSU 1969–74, 1st Dep. Chief 1974–; cand. mem. Cen. Cttee. CPSU 1976–.

Smirnov, Lev Nikolaevich
State official; b. Je.8 (21), 1911, St. Petersburg.
Mem. CPSU 1945–; worked for various sections of the Public Prosecutor's Office 1934–57; USSR Dep. Chief Prosecutor at Nuremberg War Crimes Trials 1945–46; USSR Dep. Prosecutor at Tokyo War Crimes Trials 1946–48; Dep. Chair. of USSR Supr. Court 1957–62; Chief Ed. of journal "Bulletin Verkhovnogo Suda SSSR" 1957–62; Chair. of RSFSR Supr. Court 1962–72; Dep. to USSR Supr. Sov. 1970–; Chair. of USSR Supr. Court S. 1972–; Pres. of Asscn. of Sov. Jurists 1964–; mem. Council of Int. Asscn. of Jurists-Democrats; mem. Cen. Cttee. CPSU 1976–; mem. Cen. Cttee. CPSU 1976–. Decors.: 2 Orders of Lenin, Order of the October Revolution, Order of the Red Banner of Labour, medals.

Smirnov, Leonid Vasil'evich
State official; b. 1916; s. of a worker.
Russian; mem. CPSU 1943–; grad. Novocherkassk Ind. Inst. 1939; electrical fitter, inspector for an electricity board 1930–39; engineer, dep. dept. head, dept. head, dep. chief power technician, head of a thermal electrical power plant, chief power technician at a plant of the USSR Min. of the Munitions Ind. 1939–48; dir. of a plant, dir. of a research inst. 1949–61; Dep. Chair. F.–Je. 1961, Chair. of State Cttee. for Defense Equipment with USSR Council of Mins. Je. 1961–Mr. 1963; mem. Cen. Cttee. CPSU 1961–; Dep. to USSR Supr. Sov. 1962; Dep. Chair. of USSR Council of Mins. Mr. 1963–. Decors.: Lenin Prize, Hero of Socialist Labour, Order of Lenin 1976.

Smirnov, Nikolai Ivanovich
Military official; b. S. 22 (O. 5), 1917, Rubtsovo, Kostroma Oblast.
Mem. CPSU 1942–; grad. Higher Naval School 1939 and Gen. Staff Mil. Acad. 1959; Fleet Admiral 1973–; in Sov. Army 1937–; dep. commander, commander of a submarine in the Pacific and Black Sea Fleet 1941–45; commander of a submarine, chief-of-staff of a div., div. commander, chief-of-staff of a group of submarine 1945–57; command posts with the Baltic and Black Sea Fleet, attached to staff of Sov. Navy 1959–69; commander with Pacific Ocean Fleet 1969–74; Dep. to Supr. Sov. USSR 1970–; cand. mem. Cen. Cttee. CPSU 1971–76; 1st Dep. Commander-in-Chief of Sov. Navy S. 1974–. Decors.: Order of Red Banner, Order of Red Banner of Labour, Order of Patriotic War 2nd Cl., 2 Orders of Red Star, other orders and medals.

Smirnov, Vasilii Aleksandrovich
Author; b. D.31, 1904 (Ja.13, 1905), Sinitsyno, Iaroslavl Gouvt.
Mem. CPSU 1925–; studied at external dept. of Gorkii Inst. of Literature 1939–40; wrote stories about life in the country; translated several books into languages of the various Sov. republics and into other languages.
Publs.: "Synovia" (Sons) 1940, "Otkrytel mira" (The Discovery of the World) vols. 1–3, 1947–67, and others.

Smirnov, Viktor Nikolaevich
Diplomat.
USSR Amb. to Malta O. 1981–.

Smirnov, Vitalii Stepanovich
Diplomat; b. 1930, Vitebsk Oblast.
Belorussian; grad. Inst. of Philosophy and Law in Minsk (1953); held Komsomol posts 1953–57; 2nd Sec. and mem. Bureau Komsomol Cen. Cttee. of Belorussian SSR 1957–61; Chair. Belorussian Society for Friendship and Cultural Ties with Foreign Countries 1961–69 and 1974–80; mem. Auditing Comm. Cen. Cttee. Belorussian CP 1966–71 and 1976–81; Permanent Representative of Belorussian SSR to United Nations in New York 1969–74; USSR Amb. to Pakistan Jl. 1980–.

Smirnov, Vladimir Ivanovich
Geologist; b. Ja.18 (31), 1910, Moscow.
Mem. CPSU 1940–; grad. Moscow Inst. of Geological Prospecting 1934; mem. USSR Acad. of Sciences 1962–; mem. Tadzhik Pamir Expedition 1934–37; concurrently mem. staff Moscow Inst. of Geological Prospecting 1934–41 and 1944–49; Chief Geologist of a Quicksilver Combine in Cen. Asia 1941–44; Chair. of Union Comm. for Natural Resources 1946–49; Dep. USSR Min. of Geology 1946–51; Prof. at Moscow Inst. of Nonferrous Metals and Gold 1949–52; Prof. 1951–, Head of Dept. of Natural Resources at Moscow State Univ. 1952–; Chief Ed. of journal "Geologiia rudnykh mestorozhdenii" 1968–; Acad. Sec. at Dept. of Geology, Geophysics and Geochemistry and mem. Presidium USSR Acad. of Sciences 1969–75; Vice-Pres. of Int. Asscn. for the Genesis of Ore Deposits 1964– and of Int. Union of Geological Sciences 1968–. Decors.: Lenin Prize 1972, 2 Orders of Lenin, Order of the October Revolution, Order of the Red Banner of Labour, medals.
Publs.: "Ocherki metallogenii" 1963, "Geologiia poleznykh iskopaemykh" 1969.

Smirnova, Lidia Nikolaevna
Film actress; b. 1915.
Mem. CPSU 1952–; grad. Municipal Drama College in Moscow 1938; worked with various Moscow theatres, at present with Cen. Film Actors' Theatre Studio. Decors.: People's Artist of the USSR 1974, State Prize of the USSR 1951.

Smirnova, Maria Nikolaevna
Film director; b. Ap.1 (14), 1905, Samaikono, Ulianovsk Oblast.
Mem. CPSU 1954–; grad. Dept. of Acting at State Film Inst. (GIK) Decors.: State Prize of the USSR 1948, Order of Red Banner of Labour, medals.

Smirnovskii, Mikhail Nikolaevich
Diplomat; b. 1921.
Mem. CPSU 1947–; grad. Ordzhonikidze Aircraft Construction Inst. in Moscow 1944; mem. Diplomatic Service 1948–; 1st Sec. at USSR Emb. in the USA 1952–55; mem. staff USSR Min. of Foreign Affairs 1955–58; Emb. Counsellor 1958–60 and Emb. Counsellor-Envoy at USSR Emb. in the USA 1960–62; Head of American Dept. of USSR Min. of Foreign Affairs 1962–66; USSR Amb. to Great Britain 1966–73 and concurrently USSR Amb. to Malta 1967–73; mem. Cen. Auditing Comm. CPSU 1966–76.

Smirtiukov, Mikhail Sergeevich
State official; b. 1909.
Russian; mem. CPSU 1940–; grad. Moscow State Univ.; sec. of a village sov., sec. of exec. cttee. of a village cooperative 1924–30; Dep. Head of Secr. of USSR Council of Mins. 1941–53; Dep. Dir. of Affairs 1953–64, Dir. of Affairs of USSR Council of Mins. 1964–; Dep. to USSR Supr. Sov. 1966–; mem. Cen. Auditing Comm. CPSU 1971–76; cand. mem. Cen. Cttee. CPSU 1976–81, mem. 1981–. Decors.: Order of Lenin, Hero of Socialist Labour and medals.

Smoktunovskii, Innokentii Mikhailovich
Actor; b. 1925.
Trained at Krasnoiarsk Pushkin Theatre 1945–46; served in Sov. Army 1943–45; with Gorkii Polar Circle Drama Theatre in Norilsk 1946–51, with Russian Gorkii Theatre in Makhachkala 1951–52, with Gorkii Theatre in Volgograd 1952–55, with Great Gorkii Drama Theatre in Leningrad 1957–60, with Theatre Studio of "Lenfilm" 1960–. Decors.: People's Artist of the USSR 1964, Lenin Prize 1965.

Smolenskii, Georgii Anatol'evich
Physicist; b. Je. 10 (23), 1910, Yalta.
Corresponding member of the Academy of Sciences of the USSR (1970). Member of the CPSU since 1943.
He graduated from the Leningrad Polytechnic Institute in 1938 and became a professor there in 1959. He was a staff member of the Institute of Silicate Chemistry of the Academy of Sciences of the USSR from 1951 to 1956 and the Institute of Semiconductors of the Academy of Sciences of the USSR from 1956 to 1972. He became affiliated with the Leningrad Physicotechnical Institute in 1972.
His principal works deal with ferroelectricity, ferrites, and the creation of a group of nonmetallic magnetic materials based on ferrites. He formulated an empirical criterion for the occurrence of ferroelectricity. Together with other workers, he discovered and investigated new ferroelectric materials, including ferroelectric-ferromagnetic materials. He developed concepts of the nature of magneto-optical phenomena, studied various types of interactions in crystals, and made advances in the frequency spectroscopic study of electrically and magnetically ordered substances. Decors.: State Prize of the USSR in 1952, Order of the Red Banner of Labour and several medals.
Works: *Segnetoelektriki i antisegnetoelektriki.* Leningrad, 1971. (With others.) *Fizika magnitnykh dielektrikov.* Leningrad 1974. (With others.)

Smolskii, Biakir Mikhailovich, Dr. of Tech. Sc.
Scientist; b. Je. 29 (Jl. 12), 1913, Minsk.
Belorussian; mem. CPSU 1942–; grad. Belorussian Polytech. Inst. 1935; specialist in thermal energy; Prof. 1961–; Dr. of Tech. Sc. 1958; corresponding mem. Acad. of Sciences Belorussian SSR 1967–. Decors.: 5 orders.

Smolskii, Nikolai Vladislavovich (Mikalai Uladzislavovich), Dr. of Biol. Sc.
Botanist; b. S. 24 (O. 7), 1905, Narovlia, Minsk Oblast.
Belorussian; mem. CPSU 1943–; grad. Siberian Inst. of Agriculture and Forestry 1927; Dr. of Biol. Sc. 1967; Prof. 1969–; mem. Acad. of Sciences Tadzhik SSR 1951–; mem. Acad. of Sciences Belorussian SSR 1969–.

Smorodintsev, Anatolii Aleksandrovich
Virologist and immunologist; b. Ap. 6 (19), 1901, Askino, now in the Bashkir ASSR.
Academician of the Academy of Medical Sciences of the USSR (1966; corresponding member, 1945). Member of the CPSU since 1967.
He graduated from the University of Tomsk in 1923. He worked at the L. Pasteur Institute of Epidemiology and Microbiology in Leningrad from 1933 to 1937, the All-Union Institute of Experimental Medicine in Moscow from 1938 to 1945, and the Institute of Experimental Medicine in Leningrad from 1946 to 1967. From 1967 to 1972 he was director of the All-Union Scientific Research Institute of Influenza, and beginning in 1972, he was head of one of the institute's departments.
His works have dealt primarily with the mechanisms of antiviral immunity and vaccinal protection against viral infections. He described hemorrhagic nephrosonephritis in 1944 and diphasic meningoencephalitis in 1953. He created vaccines against influenza, Russian spring-summer encephalitis, measles, and mumps. He verified the safety and effectiveness of the poliovirus vaccine against poliomyelitis. Decors.: State Prize of the USSR 1941 and the Lenin Prize 1963. He has been awarded the Order of Lenin, the Order of the Badge of Honor, and various medals.
Works: *Gripp.* Leningrad 1961. (With A. A. Korovin.) *Virusnye gemorragicheskie likhoradki.* Leningrad 1963. (With L. I. Kazbintsev and V. G. Chudakov.)

Smykov, Vladimir Il'ich
State official.
Min. of Construction and Utilization of Highways Turkmen SSR My. 1980–; mem. Auditing Comm. Turkmen CP 1981–.

Smyslov, Vasilii Vasil'evich
Chess player; b. Mr. 24, 1921, Moscow.
Grand Master of the USSR (1941) and Honored Master of Sports (1948); international grand master (1950).
He was champion of the USSR in 1949 and 1955 and world champion in the period 1957–58. He played against M. M. Botvinnik three times in world championship matches. In 1954, he and Botvinnik tied; he won in 1957, but in 1958 he lost his title to Botvinnik. He also played in a world championship tournament match in 1948, taking second place after Botvinnik. As part of the Soviet team, he played in nine chess olympiads between 1952 and 1972 and in the European championship between 1957 and 1973. Decor.: Order of Lenin.
Works: *Izbrannye partii.* Moscow 1952. *Teoriia ladeinykh okonchanii,* 2nd ed. Moscow 1959. (With G. Ia. Levenfish.)

Snegirev, Vladimir Vsevolodovich
Diplomat; b. 1923.
Mem. Diplomatic Service 1955–; Mem. Apparat USSR Min. of Foreign Affairs 1955–61 and 1968–78; Emb. Councellor, USSR Embassy in France 1961–63; Dep. Chief, 1st European Countries Dept., USSR Min. of Foreign Affairs 1963–64; USSR Amb. to Cameroon 1964–68, to Nigeria Je. 1978–.

Snezhkova, Nina Leonovna
State official.
Cand. mem. 1961–71, mem. Cen. Cttee. Belorussian CP 1971–; Sec. of Mogilev Oblast Cttee. Belorussian CP 1963–68 (for Agriculture 1963–64); Dep. Chair. of Council of Mins. Belorussian SSR O. 1970–; cand. mem. Bureau Cen. Cttee. Belorussian CP 1976–.

Snezhnevskii, Andrei Vladimirovich
Psychiatrist; b. My. 7 (20), 1904, Kostroma.
Academician of the Academy of Medical Sciences of the USSR (1962). Hero of Socialist Labour (1974). Member of the CPSU since 1945.
He graduated from the medical department of the University of Kazan in 1925. In 1950 and 1951 he was director of the V. P. Serbskii Institute of Forensic Psychiatry, and from 1951 to 1964 he was head of the subdepartment of psychiatry at the Central Institute for the Advanced Training of Physicians. Beginning in 1962, he was also director of the Institute of Psychiatry of the Academy of Medical Sciences of the USSR.
His major works deal with the comprehensive study of the nature and characteristics of schizophrenia, with lingering

somatogenic psychoses, and with senility. In 1952, he became editor of the *Zhurnal nevropatologii i psikhiatrii im. S. S. Korsakova* (S. S. Korsakov Journal of Neuropathology and Psychiatry).
He is an honorary member of the World Psychiatric Association (1972) and a foreign member of the Royal College of Psychiatrists in Great Britain, the American associations of psychiatrists and biological research in psychiatry, and societies of psychiatrists in the People's Republic of Bulgaria, the German Democratic Republic, and the Czechoslovak Socialist Republic. Decors.: Order of Lenin, four other orders, and various medals.
Works: "O Klinicheskikh zakonomernostiakh lecheniia psikhicheskikh boleznei." Vestnik AN SSSR 1962, no. 1. *Psikhiatriia,* 2nd ed. Moscow 1968 (Co-author). *Shizofreniia: Klinika i patogenez.* Moscow 1969. (Co-author.) *Shizofreniia.* Moscow 1972. (Editor.)

Sobchenko, Vladimir Fedorovich
Diplomat.
2nd Sec. of Vinnitsa Oblast Cttee. Ukrainian CP 1973–78; Dep. to Supr. Sov. Ukrainian SSR 1975–80; Cand. mem. Cen. Cttee. Ukrainian CP 1976–81; Student 1978–80; USSR Amb. to Laos O. 1980–.

Sobko, Vadim Nikolaevich
Author; b. My. 5 (18), 1912, Moscow.
Ukrainian; mem. CPSU 1940–; grad. Dept. of Philology Kiev Univ. 1939; contributor to various army newspapers; first works publs. 1930. Decors.: 7 orders and various medals.
Publs.: "Zvezdnye krylia" (Star Wings) 1937–50, "Put' zvezdy" (Path of a Star) 1943–47 (Russian translation 1959), "Dalekii front" (The Distant Front) 1948, "Zalog mira" (Pledge of Peace) 1950, "Madonna" 1973, "Pervye kapli dozhdia" (The First Raindrops) 1968, "Pochetnyi legion" (Legion of Honour) 1970, "Lykhobor" 1973, and others.

Sobolev, Feliks Mikhailovich
Theatre and film director; b. 1931.
Mem. CPSU 1956–; grad. Actors' Dept. 1953 and Directing Dept. of Kiev Theatre Inst. 1959; with Theatre for Young Spectators in Nikolaev 1953–56; Dir. of Film Studio for Popular Science Films in Kiev 1959–; directed more than 20 films. Decor.: State Prize of the USSR 1972.

Sobolev, Sergei L'vovich
Mathematician and mechanics scientist; b. S. 23 (O. 6), 1908, St. Petersburg.
Mem. CPSU 1940–; grad. Leningrad Univ. 1929; mem. USSR Acad. of Sciences 1939–; mem. staff Inst. of Seismology 1929–32, V. A. Steklov Inst. of Mathematics of USSR Acad. of Sciences 1932–43; Prof. at Moscow Univ. 1935–57; mem. staff Inst. of Atomic Energy of USSR Acad. of Sciences 1943–57; Dir. of Mathematics Inst. of Siberian Branch of USSR Acad. of Sciences and Prof. at Novosibirsk Univ. 1957–. Decors.: Hero of Socialist Labour 1968, State Prize of the USSR 1941, 1951, 1953, 7 Orders of Lenin, 2 other orders and various medals.
Publs.: "Uravnenie matematicheskoi fiziki", 4th ed. 1966, "Vvedenie v teoriiu kubaturnykh formul" 1974, and others.

Sobolev, Viktor Viktorovich
Astronomer; b. Ag. 20 (S. 2), 1915, Petrograd.
Corresponding member of the Academy of Sciences of the USSR (1958).
He graduated from Leningrad University in 1938. He became a professor at the university in 1948.
His principal works deal with theoretical astrophysics. He has developed a theory of the spectra of variable stars. He has introduced new methods into the theory of radiation transfer and has applied them to the solution of many astrophysical problems. He has investigated physical processes in gaseous nebulas and in the atmosphere of stars and planets. Decors.: two Orders of the Red Banner of Labour and several medals.

Sobolev, Vladimir Aleksandrovich
State official.
Cand. mem. Cen. Cttee. Georgian CP 1964–66, mem. 1976–; Dep. Chair. of Cttee. of Party and State Control of Cen. Cttee. Georgian CP and Council of Mins. Georgian SSR, Dep. Chair. of Cttee. of People's Control of Georgian SSR 1963–74; 1st Dep. Min. Mr.–Ag. 1974, Min. of Public Services Georgian SSR Ag. 1974–.

Sobolev, Vladimir Mikhailovich
Diplomat; b. 1924.
Mem. Diplomatic Service 1964–; mem. staff USSR Min. of Foreign Affairs 1964–65 and 1970–71; Emb. Counsellor 1965–68 and Emb. Counsellor-Envoy at USSR Emb. in Algiers 1968–70; USSR Amb. to Belgium Jl. 1971–Ap. 1975, to Finland Jn. 1979–. Decor.: Order of Friendship Between Nations.

Sobolev, Vladimir Stepanovich, Dr. of Geol.-Min. Sc.
Petrographer and mineralogist; b. My. 30, 1908, Lugansk; s. of a soldier.
Grad. Leningrad Mining Inst. 1930 mem. USSR Acad. of Sciences 1958–; corresponding mem. Acad. of Sciences Ukrainian SSR 1951–; Prof. 1938–; mem. staff Cen. Union Inst. for Geologist Research in Leningrad 1930–41; Lecturer at Leningrad Mining Inst. 1931–41 and 1943–45 and at Irkutsk Univ. 1941–43; Head of Chair of Petrography at Lvov Univ. 1945–58; concurrently mem. staff Inst. of Geology of Natural Resources of Acad. of Sciences Ukrainian SSR 1947–58; Dep. Dir. of Inst. of Geology and Geophysics of Siberian Branch of USSR Acad. of Sciences in Novosibirsk 1958–.
Publs.: "Vvedenie v mineralogiiu silikatov" 1949, "Do problemy metamorfizmu" in "Geologichnyi zhurnal" No. 11, 1951, "Sviaz svoistv silikatov so strukturoi" in "Geokhimiia" No. 6, 1956, and others.

Sofinskii, Vsevolod Nikolaevich
Diplomat; b. 1924.
Mem. Diplomatic Service 1963–; Emb. Counsellor at USSR Amb. in Great Britain 1963–69; sen. posts at USSR Min. of Foreign Affairs 1969–71; Head of Dept. of Cultural Relations With Foreign Countries 1971–73 and Head of Press Dept. of USSR Min. of Foreign Affairs 1973–79; USSR Amb. to New Zealand and Tonga Ja. 1979– and concurrently to Western Samoa Mr. 1979–.

Sofonov, G. P.
Association official.
Mem. Presidium Cen. Union Council of USSR Trade Unions 1959–68; Chair. of Union Soc. of Inventors and Rationalization Experts My. 1968–.

Sofronov, Anatolii Vladimirovich
Author; b. Ja. 6 (19), 1911, Minsk.
Russian; mem. CPSU 1940–; grad. Dept. of Literature of Rostov Pedagogical Inst. 1937; Chief Ed. of journal "Ogonok" 1953–; 1st Dep. Chair. of Sov. Cttee. on Relations with Writers of Asian and African Countries 1969–; wrote several popular songs, plays and novels. Decors.: State Prize of the USSR 1947 and 1949.
Publs.: "Sobranie sochinenii v piati tomakh, 1–2" (Collected Works in 5 vols., 1–2) 1971, "Izbrannyie proizvedeniia" (Selected Works) vols. 1–2, 1961, "Dramy i komedii" (Dramas and Comedies) 1964, "Godi zhizni. Stikhi, pesni, poemy" (Life Years. Verses, Songs, Poems) 1966, "3 komedii: Striapukha. Striapukha zamuzhem. Pavlina" (3 Comedies: The Cook. The Married Cook. Pauline) 1967, "Izbrannoe" (Selection) 1968, "Chest' roda. Rasskazy i ocherki" (Family Honour, Stories and Notes) 1970, play "Uragan" (Hurricane) 1972, and others.

Sokolov, Andrei Vasil'evich
Agricultural chemist; b. Jl. 25 (Ag. 6), 1898, Balashov, now in Saratov Oblast.
Corresponding member of the Academy of Sciences of the USSR (1964).
He graduated from the K. A. Timiriazev Moscow Agricultural Academy in 1922. He worked at agricultural experimental stations from 1923 to 1930, when he became director of the phosphorus laboratory at the Ia. V. Samoilov Scientific Research Institute of Fertilizers, Insecticides, and Fungicides. In 1942, maintaining his position at the laboratory, he became director of the department of agricultural chemistry at the V. V. Dokuchaev Soil Institute.
His research has focused on the agricultural chemistry of mineral fertilizers, the properties, from the standpoint of agri-

cultural chemistry, of soils, and the experimental techniques and analytical methods used in agricultural chemistry, in particular, isotope methods. He was one of the managing editors of the collection *Agrokhimicheskaia kharakteristika pochv SSSR* (Agricultural-Chemical Characteristics of Soils in the USSR, vols. 1–16, 1962–76) and of *Svodnye vypuski* (Summary Reports, vol. 1, 1974), which pertains to the above collection. Decors.: D. N. Prianishnikov Gold Medal of the Academy of Sciences of the USSR in 1971, Order of Lenin, four other orders, and various medals.
Works: *Raspredelenie pitatel'nykh veshchestv v pochve i urozhai rastenii.* Moscow-Leningrad 1947. *Agrokhimiia fosfora.* Moscow-Leningrad 1950. *Agrokhimicheskie metody issledovaniia pochv.* 4th ed. Moscow 1965. (With others.)

Sokolov, Arkadii Vasil'evich, Cand. of Tech. Sc.
Diplomat; b. 1923.
Mem. Diplomatic Service 1968–; Emb. Counsellor at USSR Emb. in Bulgaria 1968–71; USSR Consul-Gen. in Istanbul (Turkey) 1971–74; USSR Amb. to the Republic of Tchad S. 1974–N. 1977.

Sokolov, Boris Pavlovich, Dr. of Agr. Sc.
Plant breeder; b. 1897.
Grad. Kharkov Agricultural Inst. 1923; mem. V. I. Lenin Acad. of Agricultural Sciences (VASHNIL) 1956–; Dr. of Agr. Sc. 1953; Prof. 1954–; mem. staff All-Union Inst. of Maize Research (formerly Ukrainian Scientific Inst. for Research on Cereals Econ.) in Dnepropetrovsk 1924–; with inst. bred 14 hybrid maize and sorghum varieties in Saratov Oblast 1941–44; author of works on selection and breeding of maize seed. Decors.: Hero of Socialist Labour 1972, State Prize of the USSR 1951, Lenin Prize 1963.

Sokolov, Boris Sergeevich
Geologist and paleontologist; b. Mr. 27 (Ap. 9), 1914, Vyshnii Volochek, Kalinin Oblast.
Grad. Leningrad Univ. 1937; mem. staff Leningrad Univ. 1937–41 and 1945–58, mem. staff Novosibirsk Univ. 1960–; Full Prof. 1964–; mem. USSR Acad. of Sciences 1968–; Acad. Sec. for Dept. of Geology, Geochemistry and Geophysics and mem. Presidium of USSR Acad. of Sciences 1975–; Pres. of Union Asscn. of Paleontologists 1974–; Vice-Pres. of Int. Asscn. of Paleontologists 1972–; hon. mem. of Swedish Geological Soc. 1968–; mem. French Geological Soc. 1963–. Decors.: Lenin Prize 1967, Order of Lenin, 2 other orders and various medals.

Sokolov, Efrem Evseevich
Party official; b. 1926.
Belorussian; mem. CPSU 1955–; grad. Belorussian Agricultural Acad. 1956 and CPSU Cen. Cttee. Party Higher School 1961; served in Sov. Army 1944–50; laborer 1950–51; student 1951–56; chief agronomist, dir. of a machine and tractor station, chief of an inspection for agriculture of exec. cttee. of a people's sov. 1956–58; 2nd Sec. of Ivanovo Raion Cttee. Belorussian CP 1958–62; 1st Sec. of Urliutiub Raion Cttee. Kazakh CP, sec. of party cttee. of Zhelezno Collective and State Farm Prod. Asscn. Kazakh SSR 1962–65; Chair. of Brest Oblast Cttee. of Agriculture and Procurement Workers Trade Union 1965–67; 1st Sec. of Ivanovo Raion Cttee. Belorussian CP, Dep. Chief, Chief of Agriculture Dept. of Cen. Cttee. Belorussian CP 1967–77; cand. mem. Cen. Cttee. Belorussian CP 1971–76, mem. 1976–; 1st Sec. of Brest Oblast Cttee. Belorussian CP Mr. 1977–; Dep. to USSR Supr. Sov. 1979–; cand. mem. Cen. Cttee. CPSU 1981–.

Sokolov, Gennadii Dmitrievich
Diplomat; b. 1927.
Mem. Diplomatic Service 1950–; Emb. Counsellor at USSR Emb. in Senegal 1966–69; Emb. Counsellor-Envoy at USSR Emb. in Mali 1969–71; sen. posts at USSR Min. of Foreign Affairs 1971–72; USSR Amb. to the Republic of Niger Je. 1972–Ag. 1978.

Sokolov, Sergei Leonidovich
Military official; b. 1911; s. of a clerk.
Russian; mem. CPSU 1937–; grad. Mil. Acad. for Armoured and Mechanized Troops 1947 and Gen. Staff Mil. Acad. 1951; Marshal of the USSR 1968; in Sov. Army 1932–; chief-of-staff of a regt., then of a board, commander of armoured and mechanized troops on the Karelian front 1941–45; regimental commander, chief-of-staff of a div., commander of a div., chief-of-staff, then commander of an army, chief-of-staff, then 1st dep. commander of various mil. districts 1945–65; Commander of Leningrad Mil. District 1965–67; cand. mem. 1966–68, mem. Cen. Cttee. Ap. 1968–; Dep. to USSR Supr. Sov. 1966–; USSR 1st Dep. Min. of Defence Ap. 1967–. Decors.: Order of Lenin, 2 Orders of Red Banner, 2 Orders of Red Star, medals.

Sokolov, Tikhon Ivanovich
State official; b. 1913.
Russian; mem. CPSU 1940–; grad. Leningrad Agricultural Inst. 1936; teacher, dir. of a primary school 1930–41; agronomist with a machine and tractor station, chief agronomist, head of prod. board of an oblast agricultural dept. head of a kolkhoz experimental dept. 1936–39; dir. of an agricultural tech. school 1939–41; party and govt. posts in Perm Oblast 1941–45; Chair. of Exec. Cttee. of Novosibirsk Oblast Sov. of Workers' Deps. 1945–46; mem. Presidium and Head of Dept. of Kolkhoz Affairs with USSR Council of Mins. 1946–50; Representative of same council in the Ukrainian SSR 1950–53; Chair. of Exec. Cttee. of Smolensk Oblast Sov. of Workers' Deps. 1954–56; Dep. to RSFSR Supr. Sov. 1955–59; 1st Sec. of Novogorod Oblast Cttee. CPSU 1956–58; cand. mem. 1956–61, mem. Cen. Cttee. CPSU 1961–81; 1st Sec. of Perm Oblast Cttee. CPSU 1958–60; Dep. to 1958–, mem. Foreign Affairs Comm. 1962–66, mem. Agricultural Comm. Sov. of the Union USSR Supr. Sov. 1966–70; mem., mem. Bureau and Sec. of Cen. Cttee., 1st Sec. of Virgin Lands Krai Cttee. Kazakh CP 1960–63; relieved from all these posts for "gross errors in the management of the Krai Party Org. and unsatisfactory implementation of Party decisions in the development of agriculture in the Virgin Lands" Mr. 1963; 1st Dep. Chair. of Exec. Cttee. of Vladimir Oblast Sov. of Workers' Deps. Mr. 1963–64; Sec. of Vladimir Oblast Cttee. CPSU 1964–65; 1st Sec. of Orel Oblast Cttee. CPSU 1965–70; 1st Dep. Chair. of USSR GOSPLAN Ap. 1970–. Decor.: Order of Red Banner of Labour.

Sokolov, Viktor Viktorovich
Author; b. Ap. 17, 1919, Taburishche, now Kirovograd Oblast.
Ukrainian; mem. CPSU 1946–; works as journalist; Chair. of Donetsk branch of Ukrainian Writers' Union 1958–67; wrote several collections of poetry and novels in Ukrainian.
Publs.: "Chelovek khochet zit'" (A Man Wants to Live) 1963, "Sladkaia zemlia" (Sweet Earth) 1967, "Volodia" 1967, and others.

Sokolov, Vladimir Evgenevich, Dr. of Biol. Sc.
Zoologist; b. 1928.
Mem. CPSU 1956–; grad. Moscow Univ. 1950; mem. USSR Acad. of Sciences 1974–; Dr. of Biol. Sc. 1964; Prof. 1966–; postgraduate studies at Moscow Fur Inst. 1950–53; with Technological Inst. of the Fish Ind. and Econ. in Moscow 1953–56; Lecturer, then Prof. at Moscow Univ. 1956–; Dir. of Inst. of Evolutionary Morphology and Ecology of Animals of USSR Acad. of Sciences 1967–.

Soldatov, Aleksandr Alekseevich
Diplomat; b. 1915.
Mem. Diplomatic Service 1941–; Adviser at Sov. Mission in Australia 1944–46; various posts in USSR Min. of Foreign Affairs 1946–48; Adviser for Pol. Affairs at USSR Delegation at United Nations 1948–52; Head of Dept. of UN Affairs 1953–54, Head of American Dept. 1954–59, mem. Collegium USSR Min. of Foreign Affairs 1958–59; USSR Amb. to Great Britain 1960–66; mem. cen. Auditing Comm. CPSU 1961–71; Dep. USSR Min. of Foreign Affairs 1966–68; USSR Amb. to Cuba 1968–70; mem. Collegium USSR Min. of Foreign Affairs, Rector of Moscow Inst. for International Relations 1971–74; USSR Amb. to Lebanon O. 1974–.

Solntsev, Nikolai Adol'fovich (real surname: Solnnev-El'be)
Physical geographer; b. F. 8 (21), 1902, in the station of Jevje, now the city of Vevis, Trakai Raion, Lithuanian SSR.
One of the founders of the regional school of Soviet landscape science. Doctor of geographic sciences (1964).
He became a professor of the department of geography of Moscow University in 1965. He did most of his research in the arctic, including Kolguev Island, Novaia Zemlia, the Kara Sea, and the Kola Peninsula, and in the European part of the USSR – mainly the central regions. He studied the morphology and dynamics of landscapes and their constituent parts (for example, natural boundaries), the role that firn fields play in relief formation, the history of physical geography and landscape science, and problems of physicogeographical regionalization. An inlet on the eastern coast of Novaia Zemlia has been named in honor of him.

Solntseva, Iuliia Ippolitovna
Russian actress and film director; b. Jl. 25 (Ag. 7), 1901, Moscow.
People's Artist of the RSFSR (1964).
In 1922, she graduated from the State Institute of Musical Drama. Two years later she appeared in her first film; playing the title role in the film *Aelita*. Other early roles were Zina in *Cigarette Girl From the Mossel'prom* (1924) and Opanas' daughter in *Earth* (1930).
In 1930 she began working with her husband, the film director A. P. Dovzhenko, becoming a codirector in 1939 and then a director. Among the films she worked on were *Ivan* (1932), *Aerograd* (1935), *Shchors* (1939), and *Michurin* (1949) and the documentaries *Liberation* (1940), *The Battle for Our Soviet Ukraine* (1943), and *Victory in the Right-bank Ukraine* (1945). She made several films using Dovzhenko's scripts and themes, among them *Poem About the Sea* (1958), *Tale of Fiery Years* (1961), *The Bewitched Desna* (1963), *The Unforgettable* (1967), and *The Golden Gate* (1969), a film about Dovzhenko's artistic achievements. On the screen she re-created Dovzhenko's unique poetic vision, which was closely tied to Ukrainian folklore, his lofty civic spirit, and his lyricism. Decors.: three orders, including the Order of Lenin, as well as medals.

Solodovnikov, Vasilii Grigor'evich, Dr. of Econ. Sc.
Diplomat; b. 1918.
Mem. CPSU 1942–; grad. Kuibyshev Ind. Inst. 1942 and Union Acad. of Foreign Trade 1946; economist, corresponding mem. USSR Acad. of Sciences 1966–; Dr. of Econ. Sc. 1966; dir. of a dept. of UN Secretariat, Dep. Permanent USSR Representative at the UN 1961–64; Dir. of Africa Inst. of USSR Acad. of Sciences 1964–76; USSR Amb. to Sambia Jl. 1976–Jl. 1981.

Solodnikov, Vladimir Viktorovich, Dr. of Tech. Sc.
Cybernetics scientist; b. 1910.
Grad. Leningrad Inst. of Physical Mechanics 1934; specialist in automatic control systems; Dr. of Tech. Sc. 1950; Prof. 1951–; engineer at Physical-Tech. Inst. 1935–36; postgraduate studies, then mem. scientific staff Union Inst. of Electrotechnology 1936–41; served in Sov. Army 1942–45; Sen. Scientist, Head of Laboratory at Inst. of Automatics and Telemechanics of USSR Acad. of Sciences 1954–56; Dep. Dir. of Cen. Inst. of Research on Complex Automation 1956–59; head of chair at N. E. Bauman Higher College (MVTU) in Moscow 1959–.

Sologub, Vitalii Alekseevich, Cand. of Tech. Sc.
Trade union official; b. 1926.
Mem. CPSU 1947–; grad. Polytech. Union Inst. for Correspondence Studies 1952; lathe-turner at Kuibyshev Metal Works in Irkutsk 1941–44; lathe-turner, technologist, Chief Engineer-Designer, Head of Laboratory, Sec. of Party Cttee., Dir. of Ordzhonikidze Machine Works in Staro-Kramatorsk 1946–63; 2nd Sec. of Donetsk Oblast Cttee. Ukrainian CP 1963–71 (for Ind. 1963–64); Dep. to Supr. Sov. Ukrainian SSR 1965–; cand. mem. 1966–71, mem. Cen. Cttee. Ukrainian CP 1971–; cand. mem. 1971–73, mem. Politbureau Cen. Cttee. Ukrainian CP 1973–; Chair. of Ukrainian Republican Trade Union Council Mr. 1971–; cand. mem. Cen. Cttee. CPSU 1971–. Decor.: Order of Friendship Between Nations 1976.

Solomakha, Vitalii Konstantinovich
State official; b. 1931.
Mem. CPSU 1953–; Veterinary, Chair. of a Raion Exec. Cttee. of People's Deps., 1st Sec. of Kuibyshev Raion Cttee. Ukrainian CP, 1st Dep. Chair. of Zaporozhe Oblast Exec. Cttee. of People's Deps. until 1974; Dep. Min. of State Farms Ukrainian SSR 1974–75, 1st Dep. Min. 1975–78; Min. of Meat and Dairy Ind. Ukrainian SSR D. 1978–.

Solomentsev, Mikhail Sergeevich
State official; b. N. 7, 1913; s. of a peasant.
Russian; mem. CPSU 1940–; grad. Leningrad Polytech. Inst. 1940; kolkhoz worker, student 1930–40; foreman, dept. head, chief engineer, dir. of plants in Lipetsk, Cheliabinsk Oblast and Cheliabinsk 1940–54; Sec., 2nd Sec. of Cheliabinsk Oblast Cttee. CPSU 1954–57; Chair. of Cheliabinsk Econ. Council 1957–59; Dep. to USSR Supr. Sov. 1958–; 1st Sec. of Karaganda Oblast Cttee. Kazakh CP 1959–62; mem. Cen. Cttee. 1960–64, 2nd Sec. and mem. Presidium Cen. Cttee. Kazakh CP 1962–64; mem. Cen. Cttee. CPSU 1961–; Dep. to 1963–67, mem. Presidium Supr. Sov. Kazakh SSR 1963–66; 1st Sec. of Rostov Oblast Cttee. CPSU 1964–66; Head of Dept. of Heavy Ind. with Cen. Cttee. CPSU 1966–71; Sec. of Cen. Cttee. CPSU 1966–71; Chair. of Council of Mins. RSFSR Jl. 1971–; cand. mem. Politbureau Cen. Cttee. CPSU N. 1971–. Decors.: 2 Orders of Lenin, Order of Red Banner of Labour, medals, Hero of Socialist Labour.

Solomko, Frants Mikhailovich
State official; b. 1933; s. of a white-collar worker.
Ukrainian; mem. CPSU 1959–; grad. Moscow Inst. of Steel and Alloys and Cen. Cttee. Party Higher School CPSU; master, dispatcher in the shop of a foundry in Voroshilovgrad oblast 1955–56; engineer at the "Avtodetal'" factory in Kishinev, technologist, head of a dept., head of the technical dept., chief engineer at the engineering works in Rybnitsa 1957–63; Rybnitsa City Sec. CP Moldavia 1963–66; inspector of the Dept. for Party-Organizational Work Cen. Cttee. CP Moldavia 1966–68; Chair. of Exec. Cttee. of Raion Sov. of People's Deps. in Orgeev 1968–70; 2nd City Sec. 1970–72; 1st City Sec. CP Moldavia in Tiraspol' 1972–78; Dep. to Supr. Sov. Moldavian SSR 1975–; mem. Cen. Cttee. CP Moldavia 1976–; head of Dept. for Ind. Cen. Cttee. CP Moldavian 1978–82; Chair. of State Cttee. for Supervision of Safe Working Practices in Ind. and for Mine Supervision Moldavian SSR F. 1982–.

Soloukhin, Rem Ivanovich
Physicist; b. N. 19, 1930, Gus' Khrustal'nyi.
Corresponding member of the Academy of Sciences of the USSR (1968). Member of the CPSU since 1962.
He graduated from Moscow State University in 1953. From 1953 to 1958 he worked at the Power Engineering Institute of the Academy of Sciences of the USSR, and from 1959 to 1967, at the Institute of Hydrodynamics of the Siberian Division of the Academy of Sciences of the USSR. From 1967 to 1971 he was deputy director of the Institute of Nuclear Physics of the Siberian Division of the Academy of Sciences of the USSR, and in 1971 he became director of the Siberian Division's Institute of Theoretical and Applied Mechanics. He became a professor at the University of Novosibirsk in 1965.
He has done research on the combustion of gases and the kinetics of high-temperature chemical reactions in shock waves. He has investigated the structure of detonation waves in gases, processes by which combustion is transformed into detonation, and the propagation of shock waves in inhomogeneous media (in connection with the problem of shielding from shock waves). He has also developed methods for producing inverse media during rapid gas flow and has proposed techniques for measuring rapidly varying

pressures, densities, temperatures, and other parameters of processes in shock waves. Decors.: Lenin Prize (1965), two orders, and a number of medals.
Works: *Udarnye volny i detonatsiia v gazakh.* Moscow 1963. *Metody skorostnykh izmerenii v gazodinamike i fizike plasmy.* Moscow 1967. (With Iu. E. Nesterikhin.) *Opticheskie kharakteristiki vodorodnoi plazmy.* Novosibirsk 1973. (With Iu. A. Iakobi and A. V. Komin.)

Soloukhin, Vladimir Alekseevich
Author; b. Je. 14, 1924; Alepino, Vladimir Oblast.
Russian; mem. CPSU 1952–; grad. Gorkii Inst. of Literature 1951; publ. numerous poetry collections, novels and articles in journals.
Publs.: "Ruch'i na asfal'te" (Streams on Asphalt) 1958, "Stepnaia byl" (True Steppe Stories) 1959, "Svidanie v Viaznikakh" (Meeting Again in Viazniki) 1964, "Sorok zvonkikh kapel'. Osennie list'ia" (Forty Sonorous Drops. Autumn Leaves) 1968, "Zimnii den'" (Winter Day) 1969, "Chernye doski. Zapiski nachinaiushchego kollektsionera" (Black Boards. Notes of a Budding Collector) 1969, "Osennie list'ia (Autumn Leaves) 1970, "Serebrianyi kovshik" (Silver Vessel) 1971, and others.

Solovev, Gleb Mikhailovich
Surgeon; b. 1928.
Mem. CPSU 1950–; grad. 1st Medical Inst. in Moscow 1952; postgraduate studies at same inst. 1952–55; m. scientific staff USSR Acad. of Medical Sciences 1955–60; Head of Laboratory of Research Inst. of Clinical and Experimental Surgery 1960–69, Dep. Dir. of same 1963–69; Dir. of Inst. for Organ and Tissue Transplants of USSR Acad. of Sciences 1969–. Decor.: State Prize of the USSR 1971 (for his part in developing technique in kidney transplantation and its introduction into clinical practice).

Solovev, Iurii Filippovich
Party official; b. 1925.
Mem. CPSU 1955–; grad. Leningrad Engineering Inst. of Rail Transport 1951; served in Sov. Army 1943–44; shift supervisor, section chief, chief engineer, admin. chief at the construction of Leningrad subway 1951–73; Dep. Chair. of Exec. Cttee. of Leningrad City Sov. of People's Deps. 1973–74; Sec. of Leningrad Oblast Cttee. CPSU 1974–75, 2nd Sec. 1975–78; mem. Cen. Cttee. CPSU 1976–; 1st Sec. of Leningrad City Cttee. CPSU Ap. 1978–; Dep. to USSR Supr. Sov. 1979–.

Solovev, Iurii Vladimirovich
Ballet dancer; b. 1940.
Grad. Leningrad Choreographic School 1958; with Leningrad Opera and Ballet Theatre 1958–. Decor.: People's Artist of the USSR 1973.

Solovev, Sergei Leonidovich, Dr. of Phy.-Math. Sc.
Seismologist; b. 1930.
Grad. Leningrad Univ. 1953; corresponding mem. USSR Acad. of Sciences 1972–; Dr. of Phy.-Math. Sc. 1971; postgraduate studies at Geophysics Inst. of USSR Acad. of Sciences 1953–56; Jun. Scientist at Inst. of Physics of the Earth of USSR Acad. of Sciences 1956–59; Scientific Sec. of Council on Seismology of USSR Acad. of Sciences 1959–61; Dept. Head 1961–65, Dep. Dir. 1965–68, Sen. Scientist 1968–71, Dir. of Complex Research Inst. of Far Eastern centre of USSR Acad. of Sciences in Sakhalin 1971–.

Solovev-Sedoi, Vasilii Pavlovich (real surname: Solovev)
Composer and public figure; b. Ap. 12 (25), 1907, St. Petersburg.
People's Artist of the USSR (1967); Hero of Socialist Labour (1975).
He graduated in 1936 from the Leningrad Conservatory, where he studied composition under P. B. Riazanov. He is one of the outstanding masters of Soviet song. His most original compositions are his warm, sincere lyric songs, which glorify the patriotism of Soviet man, the battlefield camaraderie, the solidarity of peoples struggling for peace, and pure and true love. Examples include "Play, My Baian" (1941), "Evening on the Roadstead" (1941), "The Nightingales" (1944), "We've Been Gone a Long Time" (1945), "It's Time to Go" (1945), "Where Are You Now, My Comrades-in-Arms?" (1947), "Moscow Nights" ("Midnight in Moscow," 1956), "If the Lads of the Whole World" (1957), "Ballad of a Soldier" (1960), and "Song of Russia" (1971). Other compositions include the marches "March of the Nakhimov Cadets" (1949) and "Let Us Start" (1955) and the humorous songs "Beyond the Kama River" (1943), "In a Sunny Glade" (1943), "A Lad is Riding in a Cart" (1946), "An Accordion is Singing Beyond the Vologda" (1947), and "What a Soldier Needs" (1965).
His expressively national musical style is grounded in a synthesis of the intonations and rhythms of Russian peasant folk songs and urban lyric songs, *chiastushki* (folk ditties, often humorous), and instrumental folk music. Many of the composer's songs were written to poems by A.D. Churkin, A.I. Fat'ianov, S.B. Fogel'son, V.M. Gusev, and M. L. Matusovskii, as well as to poems by M. V. Isakovskii, V. I. Lebedev-Kumach, and A. A. Prokof'ev.
He composed music for the ballets *Taras Bulba* (1940, revised 1955) and *The "Rossiia" Sailed In* (1964), the musical comedies *A Faithful Friend* (1945), *The Dearest Thing* (1952), *Olympic Stars* (1962), *Eighteen Years* (1967), and *At the Home Berth* (1970), vocal cycles, and music for films and dramatic productions.
From 1948 to 1964, he was chairman of the administrative board of the Leningrad chapter of the Composers' Union. He was secretary of the Composers' Union of the USSR from 1957 to 1974; in 1960 he became secretary of the Composers' Union of the RSFSR. He was a deputy to the third through fifth convocations of the Supreme Soviet of the USSR. Decors.: Lenin Prize (1959) and the State Prize of the USSR (1943 and 1947), three Orders of Lenin, the Order of the Red Star, and various medals.
Work: *Stat'i, zametki, vystupleniia.* Leningrad-Moscow 1972.

Solov'ianenko, Anatolii Borisovich
Lyric-dramatic tenor; b. S. 25, 1932, Donetsk.
People's Artist of the USSR (1975).
He graduated from the Donetsk Polytechnic Institute in 1954. From 1952 to 1962 he studied voice with A. N. Korobeichenko. From 1963 to 1965 he apprenticed at La Scala in Milan. He became an apprentice at the Ukrainian Theatre of Opera and Ballet in 1962 and a soloist in 1965. He has performed the roles of Andrei in Gulak-Artemovskii's *Zaporozhian Cossack Beyond the Danube,* Alfredo, the Duke and Manrico in Verdi's *La Traviata, Rigoletto,* and *Il Trovatore,* Rodolfo in Puccini's *La Bohème,* Nadir in Bizet's *Les Pêcheurs de perles,* Edgar in Donizetti's *Lucia di Lammermoor,* and Des Grieux in Massenet's *Manon.*
He gives many concerts, performing romances by Russian composers, operatic arias, and Russian, Ukrainian, and Neapolitan songs. He has toured in Czechoslovakia, Bulgaria, the German Democratic Republic, Belgium, Mongolia, Italy, Canada, Japan, Australia, Cuba, and other countries.

Solozhenkin, Petr Mikhailovich
Scientist.
Vice-Pres. of Acad. of Sciences Tadzhik SSR S. 1975–.

Songaila, Ringaudas-Bronislovas Ignovich
State official; b. 1929.
Lithuanian; mem. CPSU 1953–; grad. Lithuanian Veterinary Acad. 1955; post at Lithuanian Veterinary Acad. 1955–56; posts in Cen. Cttee. Lithuanian CP and Council of Mins. Lithuanian apparatuses 1956–60; Dep. Min. of Agriculture Lithuanian SSR, Dep. Chair., 1st Dep. Chair. of Council of Mins. Lithuanian SSR and concurrently Min. of Prod. and Procurement of Agricultural Products Lithuanian SSR 1960–62; Sec. of Cen. Cttee. Lithuanian CP 1962–81; mem. Bureau of Cen. Cttee. Lithuanian CP 1962–; Chair. for agriculture of Bureau of Cen. Cttee. Lithuanian CP 1962–64; Dep. to USSR Supr. Sov. 1979–; Chair. of Council of Mins. Lithuanian SSR Ja. 1981–; mem. Cen. Auditing Comm. CPSU 1981–.

Somov, Vitalii Sergeevich
Journalist; b. 1913.
Univ. educ.; has worked for press organs 1946–; Dep. Head of Admin. for Polygraphical Ind., Publishing and Book Trade, Head of "Glavizdat'" of Min. of Culture Moldavian SSR 1946–53; Head of Dept. for the Introduction of New Technologies, Head of Dept. of Local Publishing Enterprises, Dep. Head of "Glavizdat'" of Min. of Culture Moldavian SSR 1953–64; Dep. Chief Ed. with Main Ed. Board for Art Literature of Printing and Publ. Cttee. of USSR Council of Mins. 1964–70; Dir. of "Khudozhestvennaia literatura" Publishing House Mr. 1971–.

Sonygina, Anastasia Petrovna
Party official; b. 1924.
Mem. CPSU 1946–; grad. Agricultural Tech. School in Usolsk 1942; agronomist with machine and tractor stations and kolkhozes in Kuibyshev Oblast 1942–67; Chief Agronomist of "Pobeda" Kolkhoz in Kuibyshev Oblast 1967–; mem. Cen. Auditing Comm. CPSU 1971–.

Sorokin, Mikhail Ivanovich
Military official; b. 1922.
Russian; mem. CPSU 1943–; grad. M.V. Frunze Mil. Acad. 1949 and Mil. Acad. of Gen. Staff 1964, Sov. Army General entered Sov. Army 1941; chief of a radio station, commander of a rifle company, battalion commander at war front 1941–45; command posts 1945–74; 1st Dep. Commander of Troops of Far Eastern Mil. District 1974–76; Commander of Troops of Leningrad Mil. District 1976–; Dep. to USSR Supr. Sov. 1979–; cand. mem. Cen. Cttee. CPSU 1981–.

Sorokin, Nerses Vladimirovich
State official.
Chair. of State Cttee. for the Supply of Petroleum Products Armenian SSR N. 1981–.

Sosnov, Ivan Dmitrievich
State official; b. 1908.
Russian; mem. CPSU 1940–; grad. Leningrad Inst. of Engineers for Rail Transport 1933; fitter 1922–33; railway division chief, construction section chief, chief engineer of a trust, chief of a construction office 1933–40; dep. chief of Riazan-Ural and Kalinin railroad 1940–45; posts in USSR Min. of Transport apparatus 1945–54; Chief of Main Admin. of USSR Min. of Transport 1954–58; Dep. USSR Min. of Transport Construction 1958–63, 1st Dep. Min. 1963–75, Min. Mr. 1975–; Dep. to USSR Supr. Sov. 1974–; cand. mem. Cen. Cttee. CPSU 1976–. Decors.: Order of Lenin, USSR State Prize 1976.

Spasskii, Boris Vasil'evich
Chess champion; b. Ja. 30, 1937, Leningrad.
Grad. Leningrad Univ. 1959; USSR and Int. Grand Master 1956; journalist; USSR Champion 1962 and 1973; World Champion 1969–72; winner of Championship Tournaments 1965 and 1968, of Int. Tournaments in Mar del Plata 1960, Belgrade 1964, Santa Monica 1966, Bevereik 1967, San Juan 1969, Leiden 1970; mem. of USSR team which participated in Chess Olympics 1960–70 and 1974. Decors.: Merited Master of Sports 1964, Order "Znak pocheta", various medals.

Speranskii, Evgenii Veniaminovich
Russian puppet-theatre actor, director, and playwright; b. Jl. 9 (22), 1903, Moscow.
People's Artist of the RSFSR (1966).
From 1925 to 1931, he worked at the Puppet Theatre of the Central Home for Children's Art Education. In 1931 he became an actor at the Central Puppet Theatre, which he had helped found. His roles have included Aladdin and the Wise Man in Gernet's *Aladdin's Magic Lamp,* the Master of Ceremonies in *An Unusual Concert,* and the Devil in Shtok's *Noah's Ark.* He has staged Poliakov's *Loves Me, Loves Me Not,* Borisova's *A Matter of Divorce, To the Rustle of Your Eyelashes, I-Ho-Ho!,* and *The Soldier and the Witch* (based on a story by H.C. Andersen). Decors.: Order of the Red Banner of Labour and several medals.
Works: *Akter teatra kukol.* Moscow 1965. *Povest' o strannom zhanre.* Moscow 1971.

Speranskii, Sergei Borisovich
Architect; b. 1914.
Mem. CPSU 1956–; grad. All-Russian Acad. of the Arts 1941; Sec. of the Board of USSR Union of Architects 1963–. Decor.: People's Architect of the USSR 1971.

Spirin, Aleksandr Sergeevich
Biochemist; b. S. 4, 1931, Kalinin Village, now Kaliningrad, Moscow Oblast.
Grad. Moscow State Univ. 1943; mem. USSR Acad. of Sciences 1970–; Head of Laboratory of A.N. Bakh Inst. of Biochemistry of USSR Acad. of Sciences 1962; Prof. 1964–73, Head of Chair of Biochemistry of Plants at Moscow State Univ. 1964–73; Head of (and founder of) Inst. of Proteins of USSR Acad. of Sciences in Pushchino nr. Moscow 1967–; mem. German Acad. of Natural Scientists "Leopoldina" 1974–. Decors.: Lenin Prize 1976, Order of Lenin, Hans Krebs Medal of Fed. of European Biochemicals Socs. 1969, hon. doctorate from Granada Univ. 1972.

Spirkin, Aleksandr Georgievich
Philospher; b. 1918.
Mem. CPSU 1960–; grad. Lenin Higher Pedagogical Inst. in Moscow 1941; corresponding mem. USSR Acad. of Sciences 1974–; mem. staff Inst. of Philosophy of USSR Acad. of Sciences, Sen. Scientist 1946–52 and 1962–; Dean of Pedagogical Inst. for Correspondence Studies Moscow 1952–54; Sen. Science Ed. 1954–60, Head of Ed. Dept. of "Sovetskaia entsiklopediia" 1960–62; Dep. Chief Ed. of "Filosofskaia entsiklopediia" 1960–70; Vice-Pres. of USSR Philosophical Soc. 1974–.

Spitskii, Ivan Stepanovich
Diplomat; b. 1920.
Mem. Diplomatic Service 1943–; Emb. Councellor at USSR Embassy in Afghanistan 1955–57, France 1963–64; mem. Apparat USSR Min. of Foreign Affairs 1957–63 and 1969–71; USSR Amb. to Congo (Brazzaville) 1964–69; Councellor Envoy, USSR Embassy in Iran 1971; USSR Amb. to Mauritania O. 1981–.

Spitsyn, Viktor Ivanovich
Chemical scientist; b. Ap. 12 (25), 1902, Moscow.
Mem. CPSU 1941–; grad. Moscow State Univ. 1922; Prof. 1942–, Dep. Rector of Moscow Univ. 1942–48; mem. staff 1949–, Dir. of Inst. of Physical Chemistry 1953–; mem. USSR Acad. of Science 1958–. Decors.: Hero of Socialist Labour 1969, 4 Orders of Lenin, Order of the October Revolution, Order of Red Banner of Labour, medals.

Sredin, Gennadii Vasil'evich
Military official; b. 1917.
Mem. CPSU 1940–; grad. Mil.-Pol. Acad. 1962; Col.-Gen.; mem. Mil. Council, Head of Pol. Admin. of Southern Forces Group 1964; mem. Mil. Council, Head of Pol. Admin. of Carpathian Forces Group 1968; Dep. Head 1972–74, 1st Dep. Head of Pol. Main Admin. of Sov. Army and Navy S. 1974–.

Stadniuk, Ivan Fotievich
Author; b. Mr. 8, 1920, Kordyshevka, Vinnitsa Oblast; s. of a peasant.
Russian; mem. CPSU 1940–; grad. Editors' Dept. of Moscow Polygraphic Inst. 1957; worked for various army newspapers 1941–45; first works publ. 1940. Decors.: 5 orders, various medals.
Publs.: "Maksim Perepelitsa" 1952 (film with same title 1956), "Liudi s oruzhiem" (People With Guns) 1956; "Serdtse pomnit" (The Heart Remembers) 1962, "Voennye povesti" (War Stories) 1967, "Liudi ne angely" (People Are Not Angels) vols. 1–2, 1962–65, "Voina" (War) vols. 1–2, 1970–74. "Gorkii khleb istiny" (The Bitter Bread of Truth) 1971, and others.

Stamo, Evgenii Nikolaevich
Architect; b. Ag. 17 (30), 1912, Kiev.
Honored Architect of the RSFSR (1969). He studied with V. N. Simbirtsev and L. V. Rudnev at the Moscow Architectural Institute from 1932 to 1936. His work in Moscow, done in collaboration with others, includes a residential building on Tchaikovsky Street (1952), the Kremlin Palace of Congresses (1959–61; Lenin Prize, 1962), 19-story residential buildings on Lenin Prospect (block numbers 32–33, 1965–67), the Matveevskoe residential district (construction began in 1966), blocks on Nesvizhskii and Khol'zunov lanes (1966–69), the embassy building of the Hungarian People's Republic (1967), and the Progress Publishing House (1976). Decor.: Order of Lenin.

Starinov, Il'ia Grigor'evich
Military official; b. Jl. 20 (Ag. 2), 1900, Voinovo, now in Bolkhov Raion, Orel Oblast.
A leader of the partisan movement during the Great Patriotic War of 1941–45; colonel. Member of the CPSU since 1924.
He served in the Red Army from June 1919 and fought in the Civil War of 1918–20. He graduated from the Military Transport Academy in 1935. He was an adviser attached to a special partisan unit in the Spanish Civil War in 1936 and 1937. He fought in the Soviet-Finnish War of 1939–40.
At the outset of the Great Patriotic War, he commanded an engineer-operational group on the Western and Southern fronts and headed a partisan school. In November 1941, on orders from the General Staff, he carried out an operation to strike at important military targets in occupied Kharkov through the use of radio-detonated mines. In 1942 he was commander of a special brigade on the Kalinin Front and assistant chief of the Central Headquarters of the Partisan Movement. In 1942 and 1943 he was director of the Central Partisan School and chief of the Headquarters of the Partisan Movement on the Southern Front. From April 1943 he was deputy chief of the Ukrainian Headquarters of the Partisan Movement, and from June 1944, deputy chief of the Polish Headquarters of the Partisan Movement. From September 1944 he was chief of staff of the Soviet military mission to Yugoslavia.
From 1945 to 1958, he taught at various institutions of higher military education.
Works: *Miny zhdut svoego chasa*. Moscow 1964. "Udary po vrazheskim kommunikatsiiam." In *Sovetskie partizany*. Moscow 1961. "Bitva na vrazheskikh kommunikatsiiakh." In *Voina v tylu vraga*, fasc. 1. Moscow 1974.

Starodubov, Kirill Fedorovich, Dr. of Tech. Sc.
Metallurgist; b. Ap. 19, 1904, Moscow; s. of a clerk.
Grad. Dnepropetrovsk Mining Inst. 1928; specialist on metal science and thermal metal processing; mem. Acad. of Sciences Ukrainian SSR 1957–; Dr. of Tech. Sc. and Prof. 1947–; worked at Petrovskii Metallurgical Works in Dnepropetrovsk 1925–37; Head of Chair of Thermal Metal Processing at Dnepropetrovsk Metallurgical Inst. 1928–; concurrently mem. staff Inst. for Research on Ferrous Metallurgy of USSR Min. of Ferrous Metallurgy (formerly Inst. of Ferrous Metallurgy of Acad. of Sciences Ukrainian SSR) and Head of Dept. of Thermal Steel Processing 1948–; author of some 250 papers. Decors.: Merited Scientist of the Ukrainian SSR 1964, Order of Lenin, 3 Orders of Red Banner of Labour, Order "Badge of Honour".
Publs.: "Vylyvnytsi" 1932, "Oborudovanie termicheskikh tsekhov metallurgicheskikh i mashinostroitelnykh zavodov" 1948, "Teoriia otpuska zakalennoi stali" in "Metallovedenie i termicheskaia obrabotka" 1961, and others.

Staros, Filipp Georgievich
Electronics engineer; b. 1917.
Mem. CPSU 1962–; grad. Toronto Univ. 1941; specialist in microelectronics; head of a laboratory 1956–60; chief designer in a design bureau 1960–. Decor.: State Prize of the USSR 1969.

Startsev, Vladimir Ivanovich
Diplomat; b. 1918.
Mem. Diplomatic Service 1947–; USSR Consul-Gen. in Aleppo (Syria) 1957–61; mem. staff USSR Min. of Foreign Affairs 1961–67 and 1972–75; USSR Consul-Gen. in Istanbul (Turkey) 1967–68; USSR Amb. to the Republic of Yemen 1968–72; USSR Amb. to Mauritania F. 1975–81; Chief Admin. for Servicing the Diplomatic Corps, USSR Min. of Foreign Affairs, 1981–.

Starunskii, Vladimir Gordeevich
State official; b. 1920.
Head of Dept. of Trade, Finance and Planning Organs resp. Dept. of Trade and Public Services of Cen. Cttee. Ukrainian CP 1962–70; Min. of Trade Ukrainian SSR Ag. 1970–; cand. mem. Cen. Cttee. Ukrainian CP 1971–.

Stasiunas, Ionas Antano (in Lithuanian: Jonas Antano Stasiūnas)
Baritone; b. S. 26, 1919, Ničiūnai, in present-day Kupiški Raion.
People's Artist of the USSR (1964). Member of the CPSU since 1957.
In 1948, he graduated from the Kaunas Conservatory, where he was a student of P. Oleka. He has been a soloist at the Lithuanian Theater of Opera and Ballet since 1941. His voice is distinguished by a wide range and a seamless transition between registers. He is a master of subtle expression and character portrayal. His principal roles have included Margiris in Klõva's *Pilėnai*, the title role in Borodin's *Prince Igor*, The title role in Mussorgsky's *Boris Godunov* (State Prize of the USSR, 1951), the title role in Tchaikovsky's *Mazepa*. Iago in Verdi's *Otello*, the title role in Verdi's *Rigoletto*, the King in Orff's *The Prudent Woman*, and Porgy in Gershwin's *Porgy and Bess*.
In 1962, he joined the teaching staff of the Vilnius Conservatory; he was made docent at the conservatory in 1974; deputy to the fourth convocation of the Supreme Soviet of the Lithuanian SSR. Decors.: Order of Lenin and various medals.

Stavrovskaia, Inna Arsenevna
State official.
Min. for Food Ind. Belorussian SSR F. 1980–; cand. mem. Cen. Cttee. Belorussian CP 1981–.

Stel'makh, Mikhail Afanasevich
Author; b. My. 11 (24), 1919, Diakovtsy, now Vinnitsa Oblast.
Ukrainian; grad. Vinnitsa Pedagogical Inst. 1933; mem. staff Inst. of History of Art, Folklore and Ethnography of Acad. of Sciences Ukrainian SSR (1945–53; has written numerous poems and novels which were translated into several languages. Decors.: State Prize of the USSR 1951, Lenin Prize 1961, Hero of Socialist Labour 1972.
Publs.: "Tvory" (Works), vols. 1–5, 1962–63, "Poezii" (Poetry) 1958, "Malen'ka Olenka" (Little Olenka) 1962, "Khleb i sol" (Bread and Salt) 1961, "Pravda i krivda" (Truth and Falsehood) 1963, "Zacharovannyi vetrak" (Enchanted Windmill) 1966, and others.

Stepakov, Vladimir Il'ich, Cand. of Econ. Sc.
Diplomat; b. 1912.
Russian; mem. CPSU 1937–; grad. Party Higher School of Cen. Cttee. CPSU 1935, Moscow Lenin Teachers' Training Inst. 1952 and Acad. of Social Sciences of Cen. Cttee. CPSU 1957; fitter and lumberjack 1927–32; served in Sov. Army 1935–37; worked for USSR People's Commissariat of the Heavy Ind., then USSR People's Commissariat of Transport 1937–40; dept. head with an ind. plant 1941–44; party work in Moscow 1944–71; worked for Moscow City Cttee. All-Union CP(B) 1948–50; 2nd Sec., 1st Sec. of Moscow Pervomaiskii Raion Cttee. All-Union CP(B) 1950–52; Dep. of Moscow Oblast Board of USSR Min. of State Security 1952–53; worked for Moscow City Cttee. CPSU 1957–59; Sec., 2nd Sec. of Moscow City Cttee. CPSU 1959–61; Head of Dept. of Propaganda and Agitation for the RSFSR of Cen. Cttee. CPSU, Head of Ideological Dept. (for Agriculture) for the RSFSR of Cen. Cttee. CPSU 1961–64; mem. Cen. Audit-

ing Comm. CPSU 1961–66; Dep. to and mem. Foreign Affairs Comm. Supr. Sov. RSFSR 1963–67; Chief Ed. of "Izvestiia" 1964–65; Head of Dept. of Propaganda of Cen. Cttee. CPSU 1965–71; mem. Cen. Cttee. CPSU 1966–81; Dep. to and mem. Comm. for Educ., Science and Culture of Sov. of the Union USSR Supr. Sov. 1966–70; USSR Amb. to Yugoslavia Ja. 1971–My. 1978. Decor.: Order of Red Banner of Labour.

Stepanenko, Igor Dmitrievich
State official; b. 1914, Globine, Poltava Oblast.
Mem. CPSU 1943–; served in Sov. Army 1942–45; worked in sugar ind. 1945–60; Dep. Chair. of Cherkassy Econ. Council 1960–61; Chair. of Cherkassy Exec. Cttee. of Workers' Deps. 1961–62; cand. mem. 1961–66, mem. Cen. Cttee. Ukrainian CP 1966–; Chair. of Econ. Council of Podolia 1962–65; Dep. to and mem. Draft Bills. Comm. Supr. Sov. Ukrainian SSR 1963–67; Min. of the Food Ind. Ukrainian SSR 1965–66; Dep. Chair. of Council of Mins. Ukrainian SSR O. 1966–. Decor.: Order of Red Banner of Labour.

Stepanenko, Ivan Nikiforovich
Military official; b. Ap. 13, 1920, Nekhaiki, now Drabov Raion, Cherkassy Oblast, Ukrainian SSR.
Major-general of aviation (1958). Honored Military Pilot of the USSR (1966); twice Hero of the Soviet Union (Ap. 4, 1944, and Ag. 18, 1945). Member of the CPSU since 1942.
The son of peasants, he joined the Soviet Army in 1940. He graduated from the Kacha A. F. Miasnikov Military Aviation School in 1941, the M. V. Frunze Military Academy in 1949, and the Military Academy of the General Staff in 1957. During the Great Patriotic War (1941–45), he was a pilot, flight commander, deputy squadron commander, and squadron commander in the Fourth Aviation Fighter Regiment on the Southern, Stalingrad, Northern Caucasus, Briansk, Second Baltic, and Leningrad fronts. He completed 414 combat missions and fought 118 air battles, in which he shot down 33 planes himself and eight more in group combat; he was wounded three times. After the war he held various command positions in the air force.
Decors.: Order of Lenin, three Orders of the Red Banner, the Order of the Patriotic War First and Second Class, the Order of Alexander Nevsky, five Orders of the Red Star, and various medals.

Stepanov, Boris Ivanovich,
Dr. of Phy.-Math. Sc.
Physicist; b. 1913.
Mem. CPSU 1939–; grad. Leningrad Univ. 1936; Prof.; mem. Acad. of Sciences Belorussian SSR 1953–; postgraduate studies at State Optical Inst. 1936–39; mem. staff at same inst. 1939–41 and 1943–53; served in Sov. Army 1941–43; head of a section at Inst. of Physical Technology of Acad. of Sciences Belorussian SSR 1953–55; mem. staff 1955–, Dir. of Inst. of Physics of Acad. of Sciences Belorussian SSR 1957–; concurrently Lecturer, Prof. at Inst. of Precision Mechanics and Optics in Leningrad 1946–51; Prof. at Leningrad Univ. 1951–52; head of a chair at Belorussian Univ. 1953–67. Decors.: State Prize of the USSR 1950 and 1972.

Stepanov, Georgii Afanasevich
State official; b. 1920; s. of a teacher.
Russian; mem. CPSU 1952–; grad. Frunze Inst. of Agriculture in Kishinev; served in Sov. Army 1941–47; driver, traffic supervisor of an auto crew 1947–51; technical zoologist, Chair. of "Biruintsa" collective farm in Strasheny 1951–53; Dir. of a machine and tractor station, chief technical zoologist at Min. of Agriculture Moldavian SSR 1953–59; held various posts in Council of Mins. Moldavian SSR, Chair. of Exec. Cttee. of Raion Sov. of Peoples' Deps. in Oknitsa, head of a territorial collective and state farm production asscn. in Leonovo 1959–63; Dep. Min. of Production and Supply of Agricultural Products Moldavian SSR, 1st Dep. Min. of Agriculture Moldavian SSR 1963–66; mem. Cen. Cttee. Moldavian CP 1966–; head of Agriculture Dept. Cen. Cttee. Moldavian CP 1966–69; Min. of Procurement Moldavian SSR 1969–77; Dep. Chair., 1st Dep. Chair. of Council of Mins. Moldavian SSR Jl. 1980–; mem. Bureau of Cen. Cttee. Moldavian CP Jl. 1980–. Decor.: Order of Red Banner of Labour.

Stepanov, Georgii Vladimirovich, Dr. of Philol. Sc.
Philologist; b. 1919.
Mem. CPSU 1956–; grad. Leningrad Univ. 1947; corresponding mem. USSR Acad. of Sciences 1974–81, mem. 1981–; Dr. of Philol. Sc. 1967; Prof. 1969–; served in Sov. Army 1941–45; postgraduate studies 1947–50, mem. staff, Lecturer, then Prof. at Leningrad Univ. 1950–71; head of a section at Inst. of Linguistic Research of USSR Acad. of Sciences 1971–.

Stepanov, Valentin Pavlovich
Diplomat; b. 1921.
Mem. Diplomatic Service 1949–; mem. Apparat USSR Min. of Foreign Affairs 1955–61, 1965–68 and 1970–76; Counsellor, USSR Embassy in Sri Lanka 1961–65; USSR Amb. to Sri Lanka and Maldive Islands 1968–70, to Bangladesh F. 1976–.

Stepanov, Vladimir Evgen'evich
Physicist; b. D. 1 (14), 1913, Shcheglovka, now Donetsk Oblast.
Corresponding member of the Academy of Sciences of the USSR (1968). Member of the CPSU since 1942.
From 1928 to 1931, he was a machinist and a worker on a cupola furnace in the Donbas. He graduated from Moscow State University in 1937 and from 1937 to 1941 worked at the Tashkent Observatory. From 1941 to 1946 he served in the Soviet Army. In the years 1946–53, he worked at the observatory of the University of L'vov, first as a senior research worker and then as director. From 1953 to 1955 he was at Moscow State University, and from 1955 to 1962, at the Crimean Astrophysical Observatory. In 1962 he became deputy director, and in 1964 director, of the Siberian Institute of Terrestrial Magnetism, the Ionosphere, and Radio-wave Propagation (Irkutsk). In 1972 he became chairman of the presidium of the East Siberian branch of the Academy of Sciences of the USSR.
He solved transport equations for a magnetic field with allowance for absorption and dispersion. He proved that the magnetic field of certain sunspots has a whirlpool structure, demonstrated the existence of anisotropy in the electrical conductivity of solar plasma, and proposed a mechanism for heating matter in the disturbed chromosphere. He proved the interconnection of magnetic regions on the sun's surface. He created a spectrograph having double reflection and a magnetograph capable of simultaneously recording the three components of a field.
He was a deputy to the ninth convocation of the Supreme Soviet of the USSR.
Decors.: five orders and various medals.
Works: "Solnechnyi magnitograf i registrator luchevykh skorostei." (With D. A. Kuznetsov and G. V. Kuklin.) In *Rezul'taty nabliudenii i issledovanii v period MGSS,* fasc. 1. Moscow 1966.

Stepanov, Vladimir Sevastianovich
Diplomat.
USSR Amb. to Finland D. 1973–Je. 1979.

Stepanov, V. T.
Party official.
Sec. N. Kazakhstan Oblast Cttee. Kazakh CP 1968–81, 1st Sec. Ap. 1981–.

Stepanova, Angelina Osipovna
Russian actress; b. N. 23, 1905, Nikolaevsk-na-Amure, now Khabarovsk Krai.
People's Artist of the USSR (1960); Hero of Socialist Labor (1975). Member of the CPSU since 1952.
She studied at the school of the Third Moscow Art Theater Studio and joined the company of the Moscow Art Theater in 1924. She was noted for her subtle and laconic characterizations, psychological depth, intelligence, and elegant restraint and irony. Her best roles included Zinaida in *Uncle's Dream,* after Dostoevsky, Irina in Chekhov's *The Three Sisters,* Betsy Tverskaia in *Anna Karenina,* after L. N. Tolstoy, Lida in Korneichuk's *Platon Krechet,* Queen Elizabeth in Schiller's *Maria Stuart,* Mrs. Patrick Campbell in Kilty's *Dear Liar,* Arkadina in Chekhov's

The Seagull, and Kol'tsova and Saburova in A. Tur and P. Tur's *Ambassador Extraordinary* and *The Only Witness*.
Decors.: State Prize of the USSR in 1952. Two Orders of Lenin, the Order of the October Revolution, two other orders, and various medals.

Stepanova, Evgeniia Akimovna
Historian; b. D.25, 1899 (Ja.6, 1900), Balashov, now Saratov Oblast.
Doctor of historical sciences (1959); Honored Scientist of the RSFSR (1971). Member of the CPSU since 1918; candidate member of the Central Committee of the CPSU from 1952 to 1956.
From 1918 to 1928, she worked in the party and taught. In 1931 she graduated from the Institute of Red Professors. Between 1931 and 1952 she often served as a senior research worker at the Institute of Marxism-Leninism of the Central Committee of the CPSU; she was deputy director from 1953 to 1958 and has been a consultant since 1958. She wrote the biography *Friedrich Engels* (published 1935; Lenin Prize of the Academy of Sciences of the USSR, 1936; 2nd ed. revised and expanded, 1956) and collaborated with a group of writers on the books *Karl Marx: A Biography* (1968; 2nd ed., 1973) and *Friedrich Engels: A Biography* (1970). She also compiled and edited the first and second editions of the works of Marx and Engels. Decors.: Order of the October Revolution, two other orders, and various medals.

Stoletov, Vsevolod Nikolaevich, Cand. of Biol. Sc.
Biologist.
Mem. CPSU 1940–; Prof.; Dir. of Timiriazev Acad. of Agricultural Sciences in Moscow until 1952; Dep. USSR Min. of Culture 1952–54; cand. mem. CPSU 1952–56; 1st Dep. USSR Min. of Higher Educ. 1954–59; Chair. of USSR-Norway Soc. 1958; Min. of Higher and Secondary Specialist Educ. RSFSR 1959–72; Pres. of USSR Pedagogical Acad. 1971–My. 1981; Chair. of Asscn. for Cultural Relations With Compatriots Abroad "Rodina" D. 1975–.

Stoliarov, Nikolai Georgievich
Military official; b. My. 22, 1922, Kazan. Twice Hero of the Soviet Union (Jl. 1, 1944 and Je. 27, 1945). Lieutenant-colonel (1953). Member of the CPSU since 1944.
In April 1941, he volunteered for the army. That same year he graduated from the Sverdlovsk Military Aviation School. Beginning in 1942, he served at the front in the Great Patriotic War (1941–45). He was a pilot in the 820th Attack Aviation Regiment, and from May 1943 he was a pilot and flight leader of the 667th Attack Aviation Regiment. Beginning in April 1944, he was deputy commander, squadron commander, and navigator of the 141st Guard Attack Aviation Regiment. He fought at the Kalinin, Voronezh, Steppe, and Second and First Ukrainian fronts and carried out more than 180 successful combat missions.
In 1946, he graduated from the Poltava Higher School of Navigation for Officers, and in 1954 he graduated from the Air Force Academy. A navigator and deputy commander of an air regiment, he has been in the reserves since 1956. Decors.: Order of Lenin, two Orders of the Red Banner, the Order of the Great Patriotic War First and Second Class, the Order of Alexander Nevsky, and various medals.

Stolper, Aleksandr Borisovich
Film director and screenwriter; b. Jl. 30 (Ag. 12), 1907, Dvinsk, present-day Daugavpils.
People's Artist of the USSR (1977).
From 1923 to 1927, he studied at L. V. Kuleshov's cinema studio and worked as an actor and director. In 1927 he began working for the Mezhrabpomfil'm studio's screenwriting department. Together with N. V. Ekk and R. V. Ianushkevich he wrote the screenplay for the film *A Start in Life* (1931). His first works as a director were *A Simple Story* (1930) and *Four Visits of Samuel Wolf* (1934). In 1938 he graduated from the department of directing at the All-Union State Institute of Cinematography, where he had studied under S. M. Eisenstein.
He has directed several films based on works of Soviet writers. His film adaptations of works by K. M. Simonov are *A Lad From Our Town* (1942) and *Wait for Me* (1943), both of which he directed in collaboration with B. G. Ivanov, *The Living and the Dead* (1964; first prize of the 14th International Film Festival in Karlovy Vary, 1964), and *Retribution* (1969), which is based on Simonov's novel *Nobody Is Born as Soldier*. He made films of B. N. Polevoi's *The Tale of a Real Man* (1948). V. N. Azhaev's *Far From Moscow* (1950), and Iu. M. Nagibin's *Difficult Happiness* (1958). He has also written several original screenplays for his films. He has taught at the All-Union State Institute of Cinematography since 1965. Decors.: State Prize of the USSR (1949, 1951), the Vasil'ev Brothers State Prize of the RSFSR (1966), three orders, and various medals.

Stomma, Zdislav Frantsevich
Belorussian actor; b. Ag. 4 (17), 1907, Minsk.
People's Artist of the USSR (1968). Member of the CPSU since 1963.
He graduated from thc drama studio of the Polish Theater of the Belorussian SSR in Minsk in 1929. In the same year he joined the theater's company; he later joined the company of the Polish Theater of the Ukrainian SSR in Kiev. In 1940 he became affiliated with the Ia. Kupala Belorussian Theater in Minsk.
He is a master of comedy. He has created memorable characters full of folk humor. His best roles have included Bykovskii in Kupala's *Pavlinka*. Levon Chmykh in Makaenok's *Levonikha in Orbit*, Glushak Khalimon in Melezh' *People in the Marshes*, Karavkin in Krapiva's *The Gates of Immortality*, Belogubov in Ostrovskii's *A Profitable Post*, Epikhodov in Chekhov's *The Cherry Orchard*, and Glapiex in Hugo's *A Thousand Francs Reward*. He has appeared in motion pictures since 1949.
He was a deputy to the seventh convocation of the Supreme Soviet of the Belorussian SSR. Decors.: State Prize of the Belorussian SSR in 1966, two orders and various medals.

Strautmanis, Petr Ekabovich
State official; b. 1919.
Latvian; mem. CPSU 1944–; grad. Party High School of Cen. Cttee. CPSU; farmhand 1937–40; dept. head with a regional Komsomol cttee. 1940–41; served in Sov. Army 1941–43; commander of a partisan detachment 1943–44; instructor with a regional cttee. of Latvian CP and ed. of a reg. newspaper 1944–50; 1st sec. of a raion cttee. Latvian CP, dept. head with Cen. Cttee. Latvian CP 1950–56; mem. Cen. Cttee. Latvian CP 1956–; Min. of Sovkhozes Latvian SSR 1956–58; 1st Sec. of Dobele Raion Cttee. Latvian CP 1958–60; Sec. of Cen. Cttee. 1960–65, mem. Bureau Cen. Cttee. Latvian CP 1960–; Chair. of Agricultural Bureau of Cen. Cttee. Latvian CP 1962–64; 1st Dep. Chair. of Council of Mins. Latvian SSR 1965–74; Dep. to 1966–, mem. Agricultural Comm. of Sov. of Nationalities 1966–74, Dep. Chair. of Presidium USSR Supr. Sov. D. 1974–; Chair. of Presidium Supr. Sov. Latvian SSR Ag. 1974–; mem. Cen. Auditing Comm. CPSU 1976–. Decors.: Order of Red Banner of Labour, Order of Lenin.

Strelkov, Viacheslav Sergeevich, Cand. of Phy.-Math. Sc.
Physicist; b. 1933.
Grad. Moscow Univ. 1956; Cand. of Phy.-Math. Sc. 1965–; mem. staff I. V. Kurchatov Inst. of Atomic Energy 1957–. Decor.: State Prize of the USSR 1971.

Streltsov, Mikhail Nikolaevich
Diplomat.
USSR Amb. to Iceland Jl. 1979–.

Striganov, Sergei Romanovich
Diplomat; b. 1916.
Mem. Diplomatic Service 1940–; Counsclor, USSR Embassy in Washington 1955–58; Dep. Head, Dept. of American Countries, USSR Min. of Foreign Affairs 1958–60; USSR Amb. to Uruguay 1960–65; mem. Apparat USSR Min. of Foreign Affairs 1965–78; USSR Amb. to Argentina S. 1978–.

Stroeva, Vera Pavlovna
Film director and screenwriter; b. S. 8 (21), 1903, Kiev.
People's Artist of the RSFSR (1973).
She studied in the acting division of the Kiev Theatrical Institute. From 1922 to 1925 she worked at the Studio of the Pedagogical Theater in Moscow. In 1925 she became a screenwriter, and in 1930 she directed her first film.
Her major films have revolutionary themes and include *Generation of Conquerors* (1936), *We Are the Russian People* (1966), and *The Heart of Russia* (1971). She directed a number of film adaptations, including *Petersburg Night* (1934), based on works by Dostoevsky, and *In Search of Joy* (1940), based on F. I. Panferov's novel *Bruski* (both with G. L. Roshal'). She also directed the film operas *Boris Godunov* (1955) and *Khovanshchina* (1959), both by Mussorgsky, and the films *The Grand Concert* (1951) and *Cheerful Stars* (1954). Together with cinematographers from the Union republics, she directed *Batyrs of the Steppes* (1942) in the Kazakh SSR and *Marite* (1947) in the Lithuanian SSR.
She has written many plays, as well as screenplays for her own films and for the films of other directors. Decors.: Order of the Badge of Honor and a medal.

Struchkova, Raisa Stepanovna
Ballerina; b. O. 5, 1925, Moscow.
People's Artist of the USSR (1959). Member of the CPSU since 1962.
A student of E. P. Gerdt, she graduated from a choreographic school in 1944 and was immediately engaged as a soloist in the ballet company of the Bolshoi Theater. Her best roles have included Lise in Hertel's *La Fille mal gardée,* Odette-Odile and Aurora in Tchaikovsky's *Swan Lake* and *The Sleeping Beauty,* Giselle in Adam's *Giselle,* Kitri in Minkus' *Don Quixote,* Cinderella, Juliet, and the Lady-in-waiting in Prokofiev's *Cinderella, Romeo and Juliet,* and *Lieutenant Kije,* Maria and Jeanne in Asaf'ev's *The Fountain of Bakhchisarai* and *The Flames of Paris,* and Parasha in Glière's *The Bronze Horseman.*
She became known for her virtuoso technique, precise, refined style, and accomplished character portrayals. She still gives solo performances and has frequently danced abroad. She has been teaching classical dance since 1966 (since 1968 at the A. V. Lunacharskii State Institute of Theatrical Arts). She has also performed in motion pictures, for example, *The Glass Slipper* and *Your Name*. Decors.: Order of Lenin, the Order of the Red Banner of Labor, and various medals.

Struev, Aleksandr Ivanovich
State official; b. 1906; s. of a worker.
Russian; mem. CPSU 1927–; miner 1917–25; sec. of a Komsomol cell, head of a raion All-Union CP(B) cttee., chair. of a settlement sov., dep. chair., chair. of exec. cttee. of a raion sov. of workers' deps., chair. of an oblast consumers' asscn. 1925–41; Dep. Chair., Chair. of Donetsk Oblast Exec. Cttee. of Workers' Deps. 1941–47; Dep. to USSR Supr. Sov. 1946–62 and 1966–; 1st Sec. of Donetsk Oblast Cttee. 1947–53, of Donetsk City Cttee. Ukrainian (CP(B) 1947–50; mem. Cen. Auditing Comm. CPSU 1952–56; 1st Sec. of Perm Oblast Cttee. CPSU 1954–58; mem. 1956–61, cand. mem. 1961–66, mem. Cen. Cttee. CPSU 1966–; Dep. Chair. of Council of Mins. RSFSR 1958–62; Chair. (with rank of USSR Min.) of State Trade Cttee. of USSR Council of Mins. 1962–65; USSR Min. of Trade S. 1965–. Decors.: 4 Orders of Lenin, Hero of Socialist Labour, Hammer and Sickle Medal in Gold.

Strugatskii, Arkadii Natanovich
Author; b. Ag. 28, 1925, Batumi.
Grad. Military Inst. of Foreign Languages in Moscow 1949; writes together with his brother Boris Natanovich Strugatskii; first works publ. 1957.
Publs.: co-author of "Strana bagrovykh tuch" (The Land of Purple Clouds) 1959, "Vozvrashchenie" (The Return) 1962, "Popytka k begstvu" (Escape Attempt) 1962, "Dalekaia Raduga" (The Far-Away Rainbow) 1964, "Khishchnye veshchi veka" (Rapacious Things of the Century) 1965, "Obitaemyi ostrov" (The Inhabited Island) 1971, "Trudno byt' bogom. Ponedel'nik nachinaetsia v subbotu" (It is Difficult to be God. Monday Begins on Saturday) 1966, "Polden', XXII vek. Malysh. Povesti" (Noon, 22nd Century. The Little One. Stories) 2nd ed. 1975, and others.

Strugatskii, Boris Natanovich
Author; b. Ap. 15, 1933, Leningrad.
Grad. Dept. of Mechanics and Mathematics Leningrad Univ. 1955; writes together with his brother Arkadii Natanovich Strugatskii; first works publ. 1957.
Publs.: co-author of "Strana bagrovykh tuch" (The Land of Purple Clouds) 1959, "Vozvrashchenie" (The Return) 1962, Popytka k begstvu" (Escape Attempt) 1962, "Dalekaia Raduga" (The Far-Away Rainbow) 1964, "Khishchnye veshchi veka" (Rapacious Things of the Century) 1965, "Obitaemyi ostrov" (The Inhabited Island) 1971, "Trudno byt' bogom. Ponedel'nik nachinaetsia v subbotu" (It is Difficult to be God. Monday Begins on Saturday) 1966, "Polden', XXII vek. Malysh. Povesti" (Noon, 22nd Century. The Little One. Stories) 2nd ed. 1975, and others.

Strukov, Anatolii Ivanovich, Dr. of Med. Sc.
Medical scientist; b. 1901.
Mem. CPSU 1942–; grad. Voronezh Univ. 1925; specialist in pathological anatomy; Dr. of Med. Sc. 1937; Prof.; mem. USSR Acad. of Med. Sciences 1966–; head of chair at 1st Med. Inst. in Kharkov 1938–45, in Orenburg 1941–44; Head of Laboratory at Inst. of Gen. and Pathological Anatomy 1945–48; mem. staff 1948–, head of chair at 1st Med. Inst. in Moscow 1952–; Scientific Consultant 1972–; concurrently ed. with Publishing House for Med. Literature 1948–59; Chief Ed. of journal "Arkhiv patologii" 1968; Head of Laboratory at Inst. of Human Morphology at USSR Acad. of Med. Sciences 1964–. Decor.: Lenin Prize 1974.

Struminskii, Vladimir Vasil'evich
Aerodynamics scientist; b. Ap. 16 (29), 1914, Orenburg.
Grad. Moscow Univ. 1938; mem. USSR Acad. of Sciences 1966; specializes in aerodynamics, flying devices and theoretical principles of airplane construction; mem. staff Cen. Zhukovskii Inst. of Aerodynamics 1941–66; Dir. of Inst. of Theoretical and Applied Mechanics at Siberian branch of USSR Acad. of Sciences 1966–71; Head of Dept. of Physical Aeromechanics of Inst. of Problems in Mechanics at USSR Acad. of Sciences 1971–. Decors.: State Prize of the USSR 1947 and 48, 1st Zhukovskii Prize and Zhukovskii Gold Medal 1947, Lenin Prize 1961, Order of Lenin, 2 other orders.

Struchkov, Viktor Ivanovich
Surgeon; b. Jl. 30 (Ag. 12), 1907, Riazhsk.
Academician of the Academy of Medical Sciences of the USSR (1965); Honored Scientist of the RSFSR (1963). Member of the CPSU since 1942.
He graduated from the Second Moscow Medical Institute in 1931. He served as chief surgeon in the Thirteenth Army during the Great Patriotic War (1941–45). In 1953 he became chairman of the subdepartment of general surgery at the First Moscow Medical Institute and from 1949 to 1965 served as chief surgeon of the Ministry of Public Health of the USSR. In 1965 he became editor of the journal *Grudnaia khirurgiia* (Thoracic Surgery) and in 1966 chief learned secretary of the Presidium of the Academy of Medical Sciences of the USSR.
His principal works deal with general surgery, lung surgery, surgery for suppurative conditions, traumatology, blood transfusion, anesthesiology, and clinical enzymology. He developed and introduced into practice radical surgical techniques for suppurative conditions and lung cancer, for which he was awarded the Lenin Prize in 1961. He was also awarded the S. I. Spasokukotskii Prize of the Academy of Medical Sciences of the USSR in 1965 for the monograph *Pulmonary Tumors* and the State Prize of the USSR in 1975 for the textbook *General Surgery.*
Decors.: Order of Lenin, the Order of the October Revolution, six other orders, and various medals.

Works: *Ocherki po obshchei i neotlozhnoi khirurgii*, 2nd ed. Moscow 1959. *Gnoinaia khirurgiia*, 2nd ed. Moscow 1967. *Gnoinye zabolevaniia legkikh i plevry*. Leningrad 1967. *Poroki razvitiia legkogo u cheloveka*. Moscow 1969. (Coauthor.) *Antibiotiki v khirurgii*. Moscow 1973. (Co-author.) *Gnoinaia rana*. Moscow 1975. (Co-author.)

Strzhel'chik, Vladislav Ignat'evich
Actor; b. Ja. 31, 1921, Petrograd.
People's Artist of the USSR (1974).
In 1947, he graduated from the studio of the Leningrad Bolshoi Drama Theater, where he had studied under B. A. Babochkin. He joined the theater's troupe in 1940, while he was still a student. From 1940 to 1946 he served in the Soviet Army.
His best roles have included Gania in an adaptation of Dostoevsky's *The Idiot*, Tsyganov in Gorkii's *The Barbarians*, He in Simonov's *The Fourth*, Kuligin in Chekhov's *The Three Sisters*, and the Prince in Tsagareli's *Khanuma*. He has appeared in films since 1941. His roles have included General Gottburg in *What Is Your Name Now?* (1965), Napoleon in *War and Peace* (1966), Antonov in *Liberation* (1971), and Nikolai Rubinshtein in *Tchaikovsky* (1972).
He taught at the Leningrad Ostrovskii Institute of Theater, Music and Cinematography from 1959 to 1966 and at the Institute of Culture from 1966 to 1969. In 1970 he joined the department of musical directing of the Leningrad Conservatory.
Decor.: State Prize of the RSFSR in 1971.

Stukalin, Boris Ivanovich
Party official; b. 1923.
Mem. CPSU 1943–; grad. Voronezh Pedagogical Inst. 1950; Journalist; Ed. of Voronezh Oblast Komsomol newspaper "Molodoi Kommunar" 1952–56; Ed. of Voronezh Oblast and City CPSU Cttee. newspaper "Kommuna" 1956–60; Head of RSFSR Section at Dept. of Propaganda and Agitation of Cen. Cttee. CPSU 1960–63; Chair. of State Cttee. on Printing and Publishing with Council of Mins. RSFSR 1963–65; Dep. Ed. of "Pravda" 1965–70; Chair. of USSR State Cttee. on Publishing, Polygraphy and the Book Trade 1970–82; Head of Propaganda Dept. of Cen. Cttee. CPSU D. 1982–; mem. Cen. Cttee. CPSU 1976–. Decor.: Order of Red Banner of Labour.

Stukalin, Viktor Fedorovich
Diplomat; b. 1927.
Mem. Diplomatic Service 1964–; Emb. Councellor at USSR Embassy in Pakistan 1964–66, Consul-general 1966–69; mem. Apparat, USSR Min. of Foreign Affairs 1969–80; USSR Dep. Min. of Foreign Affairs Ja. 1980–; Chair. USSR Cttee. for UNESCO Affairs Je. 1982–.

Stungrevits, Osvald Karlovich
State official.
Min. of Posts and Telecommunications of Latvian SSR Jl. 1973–; mem. Auditing Comm. 1976–81, cand. mem. Cen. Cttee. Latvian CP 1981–.

Styrikovich, Mikhail Adolfovich
Thermophysicist and thermo-energetics scientist; b. N. 3 (16) 1902, St. Petersburg.
Grad. Leningrad Inst. of Technology 1927; mem. staff Cen. Boiler and Turbine Inst. Leningrad 1928–46; teacher at Moscow Inst. of Energetics 1939–71; mem. Inst. of Energetics of USSR Acad. of Sciences 1939–60; mem. staff Inst. of High Temperatures of USSR Acad. of Sciences 1962–; mem. USSR Acad. of Sciences 1964–. Decors.: 2 Order of Red Banner of Labour, Hero of Socialist Labour 1972, various medals.

Subbotin, Aleksandr Mikhailovich
Trade union official; b. 1924.
Sec. Cen. Union Council of USSR Trade Unions Jl. 1980–. Decor.: Order of Red Banner of Labour.

Subbotin, Vasilii Efimovich
Author; b. F. 7, 1921, Subbotintsy, Kirov Oblast; s. of a peasant.
Grad. Advanced Courses in Literature of USSR Writers' Union 1956; first works publ. 1942.
Publs.: "Stikhi" (Poems) 1953, "Tanki v trave" (Tanks in the Grass) 1957, "Kniga moikh stikhov" (The Book of my Poems) 1964, "Kak konchaiutsia voiny" (How Wars End) 1965, "Malchik na delfine" (Boy on a Dolphin) 1965, "Zhizn' poeta" (The Life of a Poet) 1970, and others.

Sudarikov, Nikolai Georgievich, Cand. of Legal Sc.
Diplomat; b. 1913.
Mem. CPSU 1939–; grad. Moscow Juridical Inst. 1938; lecturer; sen. posts in various depts. of justice 1941–45 and 1946; Dean and Head of Pedagogic Dept. 1939–41, Dir. of Moscow Juridical Inst. 1945–46; Head of Legal Dept. of Council for Kolkhoz Affairs with USSR govt. 1946–48; teaching work 1948–56; Adviser, Emb. Counsellor at USSR Emb. in China 1956–62; mem. staff USSR Min. of Foreign Affairs 1962–67; USSR Amb. to North Korea 1967–74; mem. Cen. Auditing Comm. CPSU 1971–76; Head of Southern Asia Dept. and mem. Collegium USSR Min. of Foreign Affairs 1974–79; USSR Amb. to Australia O. 1979– and to Fiji Je. 1980–.

Sufiev, Kamol
State official.
Cand. mem. Cen. Cttee. Tadzhik CP 1976–78, mem. 1978–; Min. of Fruits and Vegetables Tadzhik SSR F. 1981–.

Suiunov, Nazar Toilievich
State official.
Head of Geological Admin. with Council of Mins. Turkmen SSR Ag. 1967–78; cand. mem. Cen. Cttee. Turkmen CP 1971–77, mem. D. 1977–; Dep. Chair. of Council of Mins. Jl. 1978–, and Min. of Foreign Affairs Turkmen SSR Ap. 1979–.

Suiumbaev, Akhmatbek Suttubaevich
State official; b. 1920.
Kirghiz; mem. CPSU 1942–; grad. Union Correspondence Inst. of Financial Econ. 1954; chief bookkeeper of a city financial dept. 1938–39; served in Sov. Army 1939–47; sec. of a raion sov., head of a raion, later of an oblast financial dept. 1947–55; Min. of Finance Kirghiz SSR 1955–60; Dep. to Supr. Sov. Kirghiz SSR 1959–; Chair. of Exec. Cttee. of Osh Oblast Sov. of Workers' Deps. 1960–62; mem. Cen. Cttee. 1960–, mem. Presidium; later Bureau Cen. Cttee. Kirghiz CP 1962–78; 1st Sec. of Osh Oblast Cttee. Kirghiz CP 1962–68 (for Agriculture 1963–64); Dep. to USSR Supr. Sov. 1962–; Chair. of Council of Mins. Kirghiz SSR Ja. 1968–78; cand. mem. Cen. Cttee. CPSU 1971–81; Min. of Housing and Communal Econ. Kirghiz SSR D. 1978–.

Sukhanov, Khodzhanazar
State official.
Cand. mem. Cen. Cttee. 1960–61, mem. 1961–63, Chair. of Auditing Comm. Turkmen CP D. 1963–; Head of Turkmen Republican Div. of USSR State Bank 1970–74; Min. of Finance Turkmen SSR Ja. 1974–.

Sulakauri, Archil Samsonovich
Georgian writer; b. D. 28, 1927, Tbilisi.
Member of the CPSU since 1971.
First published in 1945, he graduated from the department of languages and literature of the University of Tbilisi in 1951. He has written poetry, including the collection *Holiday on My Street* (1956) and the narrative poem *Lasha*, and has achieved success with his short stories and novellas, such as *Waves Strive for the Shore* (1960), *Pigeons, High Water* (1963), and *Return of Avel'* (1964). In the novel *The Golden Fish* (1966; Russian translation, 1970), he exposes the ugly vestiges of the past. His collection of children's tales *The Adventures of Salamura* (1968), was awarded the Sh. Rustaveli Prize in 1971. The inner world of today's young people forms the main theme of his works. His books have been translated into the various national languages of the USSR, and some have been adapted for the screen and stage.
Works: *In Russian translation: Maiskii dozhd': Stikhi*. Tbilisi 1957. *Rasskazy*. Tbilisi 1967. *Volny stremiatsia k beregu: Rasskazy, povest', roman, skazka*. Tbilisi 1973.

Sulakvelidze, Georgii Konstantinovich, Dr. of Geog. Sc.
Physicist; b. 1913.
Mem. CPSU 1948–; grad. Tbilisi Univ.; specialist in physics of the atmosphere; Cand. of Phy.-Math. Sc. 1948–; Dr. of Geog. Sc. 1957; Prof. 1958–; topographer with various geological expeditions in Transcaucasia 1930–34; postgraduate studies 1939–41 and 1946–48; served in Sov. Army 1941–46; Scientific Sec. of Dept. of Mathematics and Natural Sciences of Acad. of Georgian SSR 1948–51; Dep. Dir. of Inst. of Geophysics of Acad. of Sciences Georgian SSR 1951–55; dir. of a geophysical inst. 1955–.
Decor.: State Prize of the USSR 1969.

Suleimenov, Olzhas Omarovich
Author; b. My. 18, 1936, Alma-Ata.
Kazakh; grad. Dept. of Geology Kazakh Univ. 1959 and Gorkii Inst. of Literature 1961; first works publ. 1959. Decor.: Prize of the Lenin Komsomol 1967.
Publs.: "Zemlia, poklonis' cheloveku!" (Earth, Bow Before Man!) 1961, "Solnechnye nochi" (Sunny Nights) 1962, "Noch' – parizhanka" (The Night of a Parisian Woman) 1963, "Dobroe vremia voskhoda" (The Good Time of Rising) 1964, "God obeziany" (The Year of the Ape) 1967, "Nad belymi rekami" (On the White Rivers) 1970, film scripts "Zemlia ottsov" (Land of the Fathers) 1966, "Sinii marshrut" (The Blue March Route) 1968, "Kruglaia zvezda" (The Round Star) 1975, and others.

Suleimenov, Olzhas Omarovich
State official.
Chair. of State Cttee. on Films Kazakh SSR Mr. 1981–.

Suiiman, Rustam (Rustamnasade, Suleiman Ali-Abbas ogly)
Author; b. N. 14 (27), 1906, Baku.
Azerbaidzhani; mem. CPSU 1940–; grad. Azerbaidzhan State Univ. 1927; studied Moscow Univ. 1929; Dep. to 1955–, mem. Presidium 1955–, Chair. of Supr. Sov. Azerbaidzhan SSR 1971–; publ. numerous poetry collections dealing with social and political topics, and plays in Azerbaidzhan.
Publs.: "Izbrannoe" (Selected Works) 1948, "Ne mogu zabyt'" (I Cannot Forget) 1963, "Otkrytoe pis'mo" (Open Letter) 1963, "Kaspiiskie volny" (Caspian Waves) 1971, and others.

Sultanov, Faizulla Valeevich
State official; b. 1922.
Chair. of Supr. Sov. Bashkir ASSR 1968–; Dep. to 1968–, Dep. Chair. of Presidium Supr. Sov. RSFSR 1971–.

Sultanov, Sh.
Komsomol official.
1st Sec. and mem. Bureau Cen. Cttee. Tadzhik Komsomol S. 1981–; mem. Cen. Cttee. and cand. mem. Bureau Cen. Cttee. Tadzhik CP S. 1981–.

Sultanova, Saera Umarovna
State official.
Mem. Cen. Cttee. Uzbek CP 1966–; Dir. of "Krasnaia Zaria" Plant in Tashkent, 1st Sec. of Frunze Raion Cttee. Uzbek CP in Tashkent 1971–81; cand. mem. Bureau Uzbek CP 1971–81; Min. of Public Welfare Uzbek SSR Ja. 1981–.

Sumbatzade, Ali Soibat Sumbatovich
Historian and orientalist; b. Ja. 8 (21), 1907, Amiradzhan, in Surakhan (now Ordzhonikidze) Raion of the city of Baku.
Member of the Academy of Sciences of the Azerbaidzhan SSR (1958).
He graduated from the department of Oriental studies of Azerbaidzhan State University in 1929. In 1963 he became director of the Institute of the Peoples of the Near and Middle East of the Academy of Sciences of the Azerbaidzhan SSR. His main works are on the socioeconomic history of Azerbaidzhan in the 19th and 20th centuries.
Works: *Prisoedinenie Azerbaidzhana k Rossii i ego progressivnye posledstviia v oblasti ekonomiki i kul'tury (XIX-nach XX v.).* Baku 1955. *Kubinskoe vosstanie 1837.* Baku 1961. *Promyshlennost' Azerbaidzhana v XIX v.* Baku 1964. *Sotsial'no-ekonomicheskie predposylki pobedy Sovetskoi vlasti v Azerbaidzhane.* Moscow 1972.

Suprunenko, Dmitrii Alekseevich, Dr. of Phy.-Math. Sc.
Mathematician; b. O. 26 (N. 8), 1915, Maikop.
Grad. Rostov Univ. 1938; Dr. of Phy.-Math. Sc. 1956; Prof.; mem. staff Inst. of Mathematics of Acad. of Sciences Belorussian SSR 1963–. Decors.: Order "Badge of Honour", medals.

Surkov, Aleksei Aleksandrovich
Author; b. O. 1 (13), 1899, Serednevo, Iaroslavl Oblast.
Mem. CPSU 1925–; first works publ. 1918; grad. Dept. of Literature of Inst. of Red Professors 1934; took part in Civil War 1918–20 and World War II 1941–45; Chief Ed. of "Literaturnaia gazeta" 1944–46 and of journal "Ogonek" 1945–53; Chief Ed. of "Kratkaia Literaturnaia Entsiklopediia" (Short Literary Encyclopedia) 1962–; mem. Cen. Auditing Comm. 1952–56 and cand. mem. Cen. Cttee. CPSU 1956–66; mem. of World Peace Council 1953–; Sec. of USSR Writers' Union 1949– (1st Sec. of same 1953–59). Decors.: State Prize of the USSR 1946 and 1951, Inst. Boteev Prize 1976, 4 Orders of Lenin, 4 other orders, various medals and Bulgarian Kyrill and Method Order.
Publs.: "Dekabr' pod Moskvoi" (December near Moscow) 1942, "Nastuplenie" (The Attack) 1943, "Miru – Mir!" (Peace to the World!) 1950, "Vostok i Zapad" (East and West) 1957, "Chto takoe schaste?" (What is Happiness?) 1969, "Golosa Vremeni" (Voices of the Times) 1965, "Sobranie sochinenii" (Collected Works) 1965–66, "Posle voiny" (After the War) 1972, "Izbrannye stikhi" (Selected Poems) 1974.

Suseev, Aksen Iliumdzhinovich
Kalmyk poet; b. D. 12 (25), 1905, in the *khutor* (farmstead) of Atamanskii, now in Zimovniki Raion, Rostov Oblast.
People's Poet of the Kalmyk ASSR (1967). Member of the CPSU since 1929. He began publishing in 1927. He graduated from the Astrakhan Pedagogical Institute in 1938. He wrote the collections of poems and narrative poems *A Steel Heart* (1929), *Don't Forget!* (1932), and *The Native Steppe* (1941; second expanded edition, 1958). His major work is the three-part narrative poem *The Steppe, My Native Mother* (books 1–3, 1960–62; in Russian translation, books 1–3, 1962–66). The poem deals with the Civil War, the establishment of Soviet power in Kalmykia, and the heroism of the people during the Great Patriotic War (1941–45). He also wrote the play *In Search of Happiness* (1940; Russian translation, 1940) and the autobiographic novellas *The Beginning of Erdni's Path* (1973) and *In the Crucible of the People's Life* (1974). Decors.: In 1974, O. I. Gorodovikov State Prize of the Kalmyk ASSR for his narrative poem *Summoned by the Revolution* (1972), two orders and several medals.
Works: *In Russian translation. Syn stepei: Poema.* Moscow 1958. *Ocherki.* Elista 1965.

Suskin, V. I.
Party official.
Sec. Kashka-Daria Oblast Cttee. Uzbek CP 1971–74; Head of Dept. of Water Econ. of Cen. Cttee. Uzbek CP 1974–; cand. mem. Cen. Cttee. Uzbek CP 1981–.

Suslikov, Igor Pavlovich
Architect; b. 1929.
Decor.: State Prize of the USSR 1971.

Suslov, Vladimir Pavlovich
Diplomat; b. 1923.
Mem. Diplomatic Service 1948–; Adviser of Sov. Delegation at United Nations 1957–61; mem. staff USSR Min. of Foreign Affairs 1961–63; Dep. Sec.-Gen. for Pol. Questions and World Security Council Affairs of United Nations 1963–65; sen. posts at USSR Min. of Foreign Affairs 1965–73; Head of 2nd European Dept. of USSR Min. of Foreign Affairs 1973–; mem. Collegium USSR Min. of Foreign Affairs.

Suvorov, Georgii Dimitrevich, Dr. of Phy.-Math. Sc.
Mathematician; b. My. 19, 1919, Saratov; s. of a clerk.
Mem. CPSU 1943–; grad. Tomsk Univ. 1941; corresponding mem. Acad. of Sciences Ukrainian SSR 1965–; Dr. of Phy.-Math. Sc. and Prof. 1962–; served in Sov. Army 1941–45; postgraduate

studies, Research Asst., Prof., Head of Chair of Functional Theory at Tomsk Univ. 1946–66; Head of Dept. of Functional Theory of Donetsk Computing Centre of Acad. of Sciences Ukrainian SSR, concurrently Prof. and Head of Chair of Functional Theory and Mathematical Analysis at Donetsk Univ. 1966–; author of some 40 works.
Publs.: "Semeistva ploskikh topologicheskikh otobrazhenii" 1965, co-author of "Rasshirennoe poniatie kvazikonformnogo ploskogo otobrazheniia i lineinye sistemy differentsialnykh uravnenii smeshannogo tipa" in "Doklady AN SSSR" No. 2, 1966, and others.

Sverstiuk, Evgen
Author; b. 1928.
Ukrainian; former mem. staff Inst. of Psychology of Acad. of Sciences Ukrainian SSR; emphasized cultural autonomy of the Ukraine in numerous literary-historical Samizdat essays; interned in a strict-regime labour camp 1972–.

Sveshnikov, Aleksandr Vasil'evich
Choral director and music figure; b. Ag. 30 (S. 11), 1890, Kolomna.
People's Artist of the USSR (1956); Hero of Socialist Labour (1970). Member of the CPSU since 1950.
He attended the People's Conservatory in Moscow, where he studied music theory under B. L. Iavorskii. In 1913 he graduated from the Moscow Synod School. In 1928, he organized the vocal ensemble (subsequently, chorus) of the All-Union Radio; he was director of the chorus until 1936. Between 1937 and 1941 he served as music director of the Leningrad Academic Choir. In 1942 he became music director of the State Chorus of Russian Song of the USSR. In 1944 he founded the Moscow Choral School, where he also served as artistic director. In 1944, he began teaching at the Moscow Conservatory, assuming the position of professor in 1946; he was rector of the conservatory from 1948 to 1975. He also organized the All-Russian Choral Society, where he served as chairman until 1964.
An outstanding figure in Soviet choral art, he has created many adaptations of folk songs. He has toured abroad. Decors.: State Prize of the USSR (1946) and the Glinka State Prize of the RSFSR (1967) for his concert performances. He has also been awarded three Orders of Lenin, two other orders, and various medals.

Svetlanov, Evgenii Fedorovich
Composer, conductor, and pianist; b. S. 6, 1928, Moscow.
People's Artist of the USSR (1968).
In 1951, he graduated from the Gnesin Institute of Music and Pedagogy, where he studied composition under M. F. Gnesin and piano under M. A. Gurvich. In 1955 he graduated from the Moscow Conservatory, where he studied composition under Iu. A. Shaporin and conducting under A. V. Gauk. In 1954, while still a student, he became assistant conductor of the Full Symphony Orchestra of the All-Union Radio and Television. Beginning in 1955, he served as conductor and from 1963 to 1965 as principal conductor of the Bolshoi Theatre of the USSR. Here he conducted the operas *The Tsar's Bride* by Rimsky-Korsakov, *The Sorceress* by Tchaikovsky, *Not for Love Alone* by Shchedrin (premiere, 1961), and *October* by Muradeli (premiere, 1964); as well as premiere performance of the ballets *The Path of Thunder* by Karaev (1959), *The Pages of Life* by Balanchivadze (1960), *The Night City*, to music by Bartók (1962), and *Paganini*, to music by Rachmaninoff (1963). Since 1965 he has served as artistic director and principal conductor of the State Symphony Orchestra of the USSR.
He is an outstanding interpreter of Soviet Russian music. He was the first in the USSR to perform many works by Soviet and foreign composers, including *Joan of Arc at the Stake* by Honegger, *Turangalila-Symphonie* by Messiaen, and *A Survivor From Warsaw* by Schoenberg. He has also composed a symphony (1957), a cantata, symphonic poems, a string quartet, and vocal and instrumental works. He has toured abroad.
Decors.: Lenin Prize (1972) for his concert work, Grand Prix (France) for his recording of all of Tchaikovsky's symphonies, the Order of Lenin, the Order of the Red Banner of Labour, various medals, and the Bulgarian Order of Cyril and Methodius First Class.

Svetlova, Rema Khristoforovna
State official.
Mem. 1960–61 and 1976–, cand. mem. Cen. Cttee. Armenian CP 1966–76; 2nd Sec., Erevan City Cttee. Armenian CP until 1961; Head of Dept. of Light and Food Ind. at Cen. Cttee. Armenian CP 1967–75; Dep. Chair. of Council of Mins. Armenian SSR Jl. 1975–.

Svetovidov, Anatolii Nikolaevich
Zoologist-ichthyologist; b. O. 21 (N. 3), 1903, Moscow.
Corresponding member of the Academy of Sciences of the USSR (1953).
In 1925, he graduated from the K. A. Timiriazev Moscow Agricultural Academy. Beginning in 1932 he worked at the Zoological Institute of the Academy of Sciences of the USSR. His principal works dealt with intraspecific variation, as well as with fish taxonomy and phylogeny based on comparative and functional morphology. He also studied the geographic distribution, origin, and population dynamics of fishes, especially cod and herring. Decors.: six orders and various medals.
Works: *Treskoobraznye*. Moscow-Leningrad 1948 (*Fauna SSSR: Ryby*, vol. 9 issue 4.) *Sel'devye (Clupeidae)*. Moscow-Leningrad 1952. (*Fauna SSSR: Ryby*. vol. 2, issue 1.) *Ryby Chernogo moria*. Moscow-Leningrad 1964.

Sviridov, Georgii (Iurii) Vasil'evich
Composer; b. D. 3 (16), 1915, Fatezh, now Kursk Oblast.
Grad. Leningrad Conservatory 1941; pupil of P. R. Riazanov, D. D. Shostakovich; Sec. of the Board of USSR Composers' Union 1962–74; concurrently Sec. of the Board of RSFSR Composers' Union 1968–73. Decors.: State Prize 1946 and 1968, 2 Orders of Lenin, other orders and medals.
Works: "Kurskie pesni" (Kursk Songs) 1964, "Sneg idiet" (It is Snowing) 1965, "Vesenniaia kantata" (Spring Cantata) 1972, and others.

Sviridov, Nikolai Vasil'evich
State official.
Dep. Head of Dept. of Propaganda of Cen. Cttee. CPSU 1968; Chair. of Cttee. for the Printing and Publishing Ind. of Council of Mins. RSFSR 1969–72; Chair. of State Cttee. for Publishing, Polygraphy and the Book Trade RSFSR Ag. 1972–.

Svishchev, Georgii Petrovich
Scientist; b. D. 11 (24), 1912, Leningrad.
Specialist in aviation and mechanics. Corresponding member of the Academy of Sciences of the USSR (1966); Hero of Socialist Labour (1957). Member of the CPSU since 1945.
After graduating from the Moscow Institute of Dirigible Construction in 1935, he worked in a design bureau for dirigible construction. From 1940 to 1954 he worked in the N. E. Zhukovskii Central Aerohydrodynamics Institute, where he became deputy director in 1950. From 1954 to 1967 he was head of the P. I. Baranov Central Institute of Aircraft Engine Construction. In 1967 he became head of the Central Aerohydrodynamics Institute.
His principal works deal with the aerodynamics of aircraft and engine installations. Decors.: State Prize of the USSR (1946 and 1952); two Orders of Lenin, four other orders, and various medals.
Works: *Issledovanie profilia soprotivleniia s razlichnymi deformatsiiami nosika*. Moscow 1946. *Effektivnost' rulia i sharnirnye momenty*. Moscow 1948. *Sverkzhvukovye techeniia gaza v perforirovannykh granitsakh*. Moscow 1967. (Co-author.)

Svitlychnyi, Ivan Oleksiilovych
Literary historian and critic; b. 1929, Voroshilovgrad Oblast.
Ukrainian; grad. Kharkov State Univ. 1952; completed post-graduate studies at Inst. of Literature at Acad. of Sciences Ukrainian SSR, then mem. staff of same until 1964; played major part in reviving Sov.-Ukrainian literature in the Sixties; was dismissed as Ed. of Dept. of Linguistics and Dictionaries at "Naukova dumka" Publishing House and as contributor

to periodical "Dnipro" 1965; arrested for the first time in the beginning of September 1965, released Ap.30, 1966; private literary work 1966–72; interned in a strict-regime labour camp in Perm Oblast 1972–.

Sydykov, Maten
Party official.
Mem. Cen. Cttee. Kirghiz CP 1966–; Chair. of Naryn Oblast Exec. Cttee. of People's Deps. 1977–80; Dep. of Supr. Sov. Kirghiz SSR 1980–; 1st Sec. of Naryn Oblast Cttee. Kirghiz CP Mr. 1980–.

Syromiatnikov, Nikolai Aleksandrovich, Cand. of Phil. Sc.
Philologist; b. 1911.
Grad. Far Eastern Univ. 1936; Cand. of Philol. Sc. 1956–; concluded postgraduate studies at Leningrad Univ. 1941 and at Inst. of Oriental Studies at USSR Acad. of Sciences 1949; Lecturer at Far Eastern Univ. 1936–38; served in Sov. Navy 1942–44; mem. staff Moscow Univ. 1944–47; mem. staff Mil. Inst. of Foreign Languages 1946–50; Sen. Scientist at Inst. of Oriental Studies 1950–. Decor.: State Prize of the USSR 1972.
Publs.: co-author of 2-volume "Bol'shoi iaponsko-russkii slovar'" 1970.

Sysoev, Viktor Sergeevich
Military official; b. Ja.26 (F.8), 1915, Barakovo, Riazan Oblast.
Mem. CPSU 1942–; grad. Higher Frunze Naval College 1939 and Naval Acad. 1952; USSR Admiral 1970–; in USSR Navy 1937–; served on destroyers of the Black Sea and Northern Fleet 1941–45; commander of a destroyer after the war; chief-of-staff with Black Sea Fleet 1955–60; Head of Chair at USSR Navy Acad. 1960–65; 1st Dep. Commander 1965–68, Commander of the Black Sea Fleet 1968–74; Head of Naval Acad. Mr. 1974–. Decors.: Order of the October Revolution, 3 Orders of Red Banner, 2 Orders of Red Star, medals.

Tabeev, Fikhriat Akhmedzhanovich, Cand. of Econ. Sc.
Diplomat; b. 1928; s. of a peasant.
Tartar; mem. CPSU 1951–; grad. Kazan State Univ. 1951; kolkhoz worker, student 1945–51; Research Asst., Lecturer, Assoc. Prof. at Dept. of Pol. Econ. of Kazan State Univ. 1951–57; dept. head with Tartar Oblast CPSU Cttee. 1957–59; Dep. to 1958–, mem. Presidium USSR Supr. Sov. 1962–80; Sec. 1959–60, 1st Sec. of Tartar Oblast CPSU Cttee. O. 1960–79; mem. Cen. Cttee. CPSU 1961–; USSR Amb. to Afghanistan N. 1979–.

Tabyshaliev, Salmorbek
State official.
Cand. mem. 1956–63, mem. Cen. Cttee. Kirghiz CP 1963–; Rector of Kirghiz State Univ. 1963–; Dep. to 1963–, Chair. of Comm. for Educ. and Culture 1963–75, Chair. of Supr. Sov. Kirghiz SSR 1975–80; Decor.: Order of Red Banner of Labour.

Tadzhiev, Nasretdin
State official.
Chair. of State Cttee. on Pricing Uzbek SSR Mr. 1970–.

Tairov, Kalendar
Party official.
Diverse Party posts until 1971; cand. mem. Cen. Cttee. Uzbek CP 1966–76, mem. 1976–; Head of General Dept. Uzbek CP Mr. 1971–.

Tairova, Taira Akper kyzy
State official; b. 1913.
Mem. CPSU 1939–; grad. Azerbaizhan Petroleum Inst.; Dir. of Azerbaidzhan Petroleum Research Inst. in Kuibyshev 1937–54; Chair. of Azerbaidzhan Republican Trade Union Council 1954–57; mem. Cen. Cttee. 1956–61 and 1964–, cand. mem. Cen. Cttee. Azerbaidzhan CP 1961–64; Chair. of Scientific-Tech. Cttee. with Council of Mins. Azerbaidzhan SSR 1957–59; Min. of Foreign Affairs Azerbaidzhan SSR 1959–; concurrently Chair. of State Cttee. for the Coordination of Scientific Research with Council of Mins. Azerbaidzhan SSR 1961–63; Dep. Chair. of Council of Mins. Azerbaidzhan SSR 1963–68. Decors.: Order of Lenin, 3 Orders of Red Banner of Labour, medals.

Takaishvili, Sesil' Dmitrievna
Georgian actress; b. S.30 (O.13), 1906, Batumi.
People's Artist of the USSR (1966).
After graduating from A. N. Pagava's drama studio in 1926, she performed at the Georgian-language Rustaveli Theater. Between 1930 and 1967 she performed at the Georgian-language Mardzhanishvili Theater in Tbilisi. Her roles have included Anka in Pogodin's *Poem of the Ax,* Katharine in Shakespeare's *The Taming of the Shrew,* Ogudalova in Ostrovskii's *The Girl Without a Dowry,* Elizabeth in Schiller's *Maria Stuart,* and Grandmother in Dumbadze and Lordkipanidze's *Grandmother, Iliko, Illarion and I.* She has also appeared in motion pictures and variety shows. Her acting is marked by poignancy, originality, humor, and vividness. Decors.: Order of the Red Banner of Labour and several medals.

Takezhanov, Sauk Temirbaevich
State official.
Chief Engineer of Zinc Combine in Ust-Kamenogorsk 1967; Dep. Min. 1971–73, Min. of Non-Ferrous Metallurgy Kazakh SSR 1973–74 and 1980–; Dep. Chair. of Council of Mins. and Chair. of GOSPLAN Kazakh SSR O. 1974–80; mem. Cen. Cttee. Kazakh CP 1976–.

Takhtadzhian, Armen Leonovich
Botanist; b. My.28 (Je.10), 1910, Shusha.
Mem. CPSU 1946–; grad. Union Inst. of Subtropical Cultures in Tbilisi 1932; mem. USSR Acad. of Sciences 1972–, mem. Acad. of Sciences Armenian SSR 1971–; Head of Chair at Erevan Univ. 1938–48; Dir. of Botanical Inst. of Acad. of Sciences Armenian SSR 1944–48; Prof. at Leningrad Univ. 1949–61; Dept. Head at Botanical Inst. of USSR Acad. of Sciences 1954–; Pres. of Union Asscn. of Botanists 1973–; Pres. of Dept. of Botany of Int. Soc. of Biological Sciences 1975–; Pres. of Int. Asscn. of Plant Taxonomy 1975–; mem. US Acad. of Sciences 1971–; mem. Finnish Acad. of Sciences and Literature 1971– and German Acad. of Natural Scientists "Leopoldina" 1972–. Decors.: V. L. Komarov Prize of USSR Acad. of Sciences 1969, 2 Orders of Red Banner of Labour, medals.

Taktakishvili, Otar Vasil'evich
Composer and state official; b. 1924.
Mem. CPSU 1951–; grad. Tbilisi State Conservatory 1947; Head and Musical Dir. of State Ensemble of the Georgian SSR 1953; Chair. of Georgian Composers' Union 1956–62; cand. mem. 1960–64, mem. Cen. Cttee. Georgian CP 1964–; Min. of Culture Georgian SSR F. 1965–. Decors.: Order of Red Banner of Labour, 2 Stalin Prizes, State Prize, various medals.

Talyzin, Nikolai Vladimirovich
State official; b. 1929.
Mem. CPSU 1960–; grad. Moscow Electrical Engineering Inst. of Communications 1955, Dr. of Tech. Sc. 1970, Prof. 1975; electrical fitter 1942–50; posts at State Scientific Research Inst. of Radio Engineering 1955–65; Dep. USSR Min., 1st Dep. Min. of Communications 1965–75, Min. 1975–80; cand. mem. Cen. Cttee. CPSU 1976–81, mem. 1981–; Dep. to USSR Supr. Sov. 1979–; Dep. Chair of USSR Council of Ministers and USSR Permanent Representative to COMECON O. 1980–. Decors.: USSR State Prize (twice).

Tammur, Il'mar Augustovich (Il'mar Tammur)
Esthonian actor and stage director; b. My.11, 1921, Narva.
People's Artist of the Esthonian SSR (1964). Member of the CPSU since 1968. In 1941, he graduated from the School of Theater Arts in Tallinn. He worked as a stage director and actor at the Vanemuine Theater in Tartu from 1944 to 1946 and at the Tallinn Drama Theater from 1946 to 1948. He was principal director and manager of the Endla Theater in Pärnu from 1948 to 1952 and subsequently principal director of the V. Kingissepp Esthonian Drama Theater from 1952 to 1970 and of the Philharmonica of the Esthonian SSR (Tallinn) from 1970 to 1974. Since 1974 he has been a stage director and actor at the Rakvere Theater.

His roles have included the title role in Rostand's *Cyrano de Bergerac,* Petrucchio in Shakespeare's *The Taming of the Shrew,* Shvandia in Trenev's *Liubov Iarovaia,* and Wergan in Hauptmann's *The Beaver Coat.* Tammur has staged *Liubov' Iarovaia* (1952), Korneichuk's *The Destruction of the Squadron* (1953), Shakespeare's *Antony and Cleopatra* (1955), Brecht's *Mother Courage and Her Children* (1962), and Goethe's *Faust* (parts 1-2, 1968-69).
He taught at the Tallinn Conservatory from 1955 to 1960 and at the E. Vilde Pedagogical Institute from 1967 to 1970. Between 1953 and 1969 he was chairman of the Esthonian Theatrical Society. He has also appeared in motion pictures. Decors.: State Prize of the Esthonian SSR in 1958, two orders and several medals.

Tananaev, Ivan Vladimirovich
Chemical scientist; b. My. 22 (Je. 4), 1904, Serpovoe, Tambov Oblast.
Mem. CPSU 1942–; grad. Kiev Polytech. Inst. 1925; mem. USSR Acad. of Sciences 1958–; mem. staff Kiev Polytech. Inst. until 1934; mem. staff 1934–, Head of Laboratory 1939–, Dept. Head at N. S. Kurnakov Inst. of Gen. and Inorganic Chemistry of USSR Acad. of Sciences 1943–; concurrently (at various times) Head of Chair of Chemistry at Moscow Inst. of Chemical Engineering and Head of Chair of Inorganic Chemistry at Moscow Inst. of Chemical Precision Technology; mem. Section of Inorganic Chemistry of Int. Union of Theoretical and Applied Chemistry (IUPAC) 1959–; Chief Ed. of journal "Neorganicheskie materialy" 1965–. Decors.: State Prize of the USSR 1949, 1951 and 1971, D. I. Mendeleev Medal in Gold 1973, 2 Orders of Lenin, 4 other orders and various medals.

Tank, Maksim (real name: Skurko, Evgenii Ivanovich)
Author and poet; b. 1912, Pilkovshchina, Minsk Oblast.
Belorussian; mem. CPSU 1936–; worked in underground Komsomol org. in Poland 1927; Instructor of Cen. Cttee. Komsomol, arrested and sentenced to 6 years' imprisonment 1932; after Sov. occupation of Eastern Poland worked in press organs; contributor to front and partisan newspapers 1941–45; literary work 1945–; Dep. to Supr. Sov. Belorussian SSR 1947–59 and 1963–; Chair. of Belorussian section of Sov.-Polish Friendship Soc. 1958–; cand. mem. 1961–66, mem. Cen. Cttee. Belorussian CP 1966–; 1st Sec. of Belorussian Writers' Union and Sec. of USSR Writers' Union 1966–. Decors.: 2 Orders of Lenin, Order of Red Banner, 2 Orders of Red Banner of Labour, Stalin Prize, Hero of Socialist Labour 1974, Order of the October Revolution, Golden Peace Medal. Publs.: poetry "Na etapakh" (In Stages) 1936, "Pad machtay" (Before the Mast) 1938, "Vastrytse zbroiu" (Prepare to Fire) 1945, "Kab vedali" (That They May Know) 1948, "U daroze" (En Route) 1954, "Sled bliskavitsy" (Trace of Lightning) 1957, "Moi khleb nasushchny" (My Daily Bread) 1962, "Lirika" (Lyric) 1963, "Vybranye tvory" (Selected Works) 2 vols., 1952 and 1954, "Zbor tvorov" (Collected Works) 2 vols., 1958, "Tsvetnik shchastia" (Flower Garden of Happiness) 1965, "Navstrechu solntsu" (Toward the Sun) 1965; humorous verse "Seledka so stikhami" (Herring With Verse) 1966; "Zbor tvorau" (Collected Works), vols. 1–4, 1966–67, "Listki kalendara" (Calendar Leaves) 1970, "Khai budze sviatlo" (Let There be Light) 1972, and others.

Tankibaev, Zhansha Abilgalievich
State official.
Dep. Head of "Kazakhstanneft'" Asscn. 1967; Dep. Chair. of Council of Mins. Kazakh SSR 1968–70; Head of Kazakh SSR Main Admin. for Material-Tech. Supplies Ja. 1970–; mem. Cen. Cttee. Kazakh CP 1971–.

Tantsiura, Nikolai Dmitrievich
State official.
Cand. mem. Cen. Cttee. Kazakh CP 1981–; Min. of Food Ind. Kazakh SSR 1981–.

Tarakanov, Vladimir Samuilovich
State official.
Chair. RSFSR State Cttee. for the Supply of Petroleum Products S. 1981–.

Tarasenko, Nataliia Iuvenal'evna
Hygienist; b. Ag. 18 (31), 1911, Novozybkov, now Briansk Oblast.
Corresponding member of the Academy of Medical Sciences of the USSR (1967). Member of the CPSU since 1940.
She graduated from the Saratov Medical Institute in 1933. In 1965 she was appointed head of the subdepartment of occupational hygiene at the First Moscow Medical Institute. Her main works are concerned with radiation hygiene and safety in working with sources of ionizing radiation. Decors.: Lenin Prize in 1961, four orders and various medals.
Works: *Gigiena truda pri rabote na atomnykh elektrostantsiiakh.* Moscow 1960. (Co-author.) *Gigiena truda pri gammadefektoskopii metallicheskikh izdelii.* Moscow 1960. *Gigiena truda pri rabote s toriem.* Moscow 1963. *Zashchita i ochistka kozhnykh pokrovov ot radioaktivnykh zagriaznenii.* Moscow 1972.

Tarasov, Nikolai Konstantinovich
Diplomat; b. 1923.
Grad. Moscow State Univ.; concluded postgraduate studies; Consultant at Legal Dept., then Head of Sec. of Presidium USSR Supr. Sov. 1949–56; 1st Sec. Emb. Counsellor at USSR Emb. in Iran 1956–61; sen. posts in USSR Min. of Foreign Affairs 1961–68; Dep. Permanent USSR Rep. at United Nations 1968–72; USSR Amb. to Mexico Ag. 1972–Mr. 1976.

Tarasov, Nikolai Nikiforovich
State official; b. 1911; s. of a clerk.
Russian; mem. CPSU 1942–; grad. Moscow Textile Inst. 1935; foreman, Head of Mechanical and Repairs Dept., Head of Weaving Mill of the Cotton Trust (then Cotton Combine) in Orekhovo-Zuevo 1935–42; served in Sov. Army 1942–45; Chief Engineer of Drezna Spinning and Weaving Plant 1945–48; worked for the USSR Min. of the Textile and Light Ind. 1948–52; USSR Dep. Min. of the Light Ind. 1952–53; worked for USSR Council of Mins. 1952–55; Dep. Chair. of USSR GOSPLAN 1955–57; worked for USSR GOSPLAN 1957–58 and 1959–60; Dep. Chair. of Vladimir Econ. Council 1958–59; Dep. Chair. of Econ. Council RSFSR 1960–62; Chair. of State Cttee. for the Light Ind. of USSR Council of Mins. 1962–65 (of USSR GOSPLAN 1963–64); USSR Min. of the Light Ind. O. 1965–; cand. mem. 1966–76, mem. Cen. Cttee. CPSU 1976–; Dep. to USSR Supr. Sov. 1966–; Chair. of Sov.-Korean Friendship Soc. Je. 1966–. Decors.: Hero of Socialist Labour, Order of Lenin, medals.

Tarba, Ivan Konstantinovich
Abkhazian writer; b. Mr. 21, 1921, Beslakhuba, Ochamchire Raion.
Member of the CPSU since 1940.
He graduated from A. M. Gorkii Sukhumi Pedagogical Institute in 1947. He has published many collections of poems and narrative poems, as well as the novels *A Well-known Name* (1963; Russian translation, 1964) and *The Sun Rises Here* (1968; Russian translation, 1970), which deal with the new life of the contemporary Abkhazian village. His works have been translated into several national languages of the USSR. In 1958, he became chairman of the Abkhazian division of the Writers' Union of the USSR. Decors.: two orders and several medals.
Works: *Iälkaau i'ïmthakua,* vols. 1–3. Akua 1971–73. *In Russian translation: S pervykh dnei.* Moscow 1953. *O druz'iakh-tovarishchakh.* Moscow 1957. *Ispytanie.* Moscow 1968. *Stikhi i poemy.* Moscow 1959. *Vesennii sneg.* Sukhumi, 1970. *Kniga pesen.* Moscow 1973. *Ia vstrechaiu solntse: Stikhi.* Moscow 1973.

Tardzhumanian, Genrikh Vardanovich, Cand. of Tech. Sc.
State official; b. 1927.
Armenian; mem. CPSU 1955–; grad. Erevan Agricultural Inst.; controller-mechanic, chief engineer, dir. of a machine and tractor station, dir. of a tractor and repair station 1951–59; Instructor for Cen. Cttee. Armenian CP, Chief Engineer and Dep. Head of "Glavtorgmash" Admin. with Council of Mins. Ar-

menian SSR 1959–61; cand. mem. 1961–64, mem. Cen. Cttee. Armenian CP 1964–; Dep. Chair. of Armenian Republican "Armselkhoztekhnika" Asscn. for Agricultural Technology, 1st sec. of a raion cttee. Armenian CP, head of an agricultural prod. admin. 1961–65; Dep. to Supr. Sov. Armenian SSR 1963–; Chair. of Armenian Republican "Armselkhoztekhnika" Asscn. for Agricultural Technology 1965–67; 1st Sec. of Kirovakan City Cttee. Armenian CP 1967–69; Chair. of Armenian Republican Trade Union Council 1969–75; cand. mem. Bureau Cen. Cttee. Armenian CP 1969–76; Min. of Agriculture Armenian SSR Ja. 1975–.

Tarkovskii, Arsenii Aleksandrovich
Author and translator, b. Je. 12 (25), 1907, Elizavetgrad, now Kirovgrad.
Attended Advanced Courses in Literature 1925–29; worked for an army newspaper 1941–45; first works publ. 1926. Decors.: Makhtumkuli State Prize of the Turkman SSR 1971, Order of Red Star, various medals.
Publs.: "Pered snegom" (Before the Snow Falls) 1962, "Zemle – zemnoe" (The Earth – The Worldly) 1966, "Vestnik" (The Messenger) 1969, "Stikhotvoreniia" (Poems) 1974, and others.

Tashibekova, Dzhamal Chuibekovna
State official.
Dep. Chair. of Council of Mins. and Min. of Foreign Affairs Kirghiz SSR Mr. 1980–; mem. Cen. Cttee. and cand. mem. Bureau of Cen. Cttee. Kirghiz CP Ja. 1981–.

Tashmatov, A. Ia.
Trade union official.
Sec. of Uzbek Republican Trade Union Council F. 1982–.

Tatarinov, Leonid Petrovich
Biologist; b. 1926.
Mem. CPSU 1964–; grad. Moscow Univ. 1947; corresponding mem. USSR Acad. of Sciences 1974–; concluded postgraduate studies at Moscow Univ. 1952; mem. staff. Foreign Language Publishing House 1953–54; mem. staff Paleontological Inst. of USSR Acad. of Sciences 1955–73; sen. mem. scientific staff Inst. of Evolutionary Morphology and Ecology of Animals of USSR Acad. of Sciences 1973–.

Tatevosian, Garnik Anushavanovich
State official.
Mem. Cen. Cttee. Armenian CP 1961–; Head of Dept. of Construction and the Building Materials Ind. resp. Dept. of Construction and Communal Econ. with Cen. Cttee. Armenian CP 1961–72; mem. Bureau for Ind. with Cen. Cttee. Armenian CP 1962–64; Min. of Rural Construction Armenian SSR Mr. 1972–.

Tatliev, Suleiman Bairam ogly
State official.
Head of Main Admin. for Chemical Ind. Azerbaidzhan SSR 1966–70; cand. mem. Cen. Cttee. Azerbaidzhan CP 1971–76, mem. 1976–; 1st Dep. Chair. Council of Mins. Azerbaidzhan SSR Jl. 1978–; mem. Bureau Cen. Cttee. Azerbaidzhan CP O. 1978–.

Tauson, Lev Vladimirovich
Geochemist; b. O. 27, 1917, Kamyshlov, Sverdlovsk Oblast.
Corresponding member of the Academy of Sciences of the USSR (1966). Member of the CPSU since 1945. Son of V. O. Tauson.
He graduated from Moscow State University in 1947 and began working at the Institute of Geochemistry and Analytic Chemistry of the Academy of Sciences of the USSR that same year. In 1960 he became director of the Institute of Geochemistry of the Siberian Division of the Academy of Sciences of the USSR in Irkutsk. His main work has dealt with the geochemistry of rare elements in genetic series of igneous rocks and the geochemistry of ore zones confined to abyssal fractures.
He formulated the geochemical criteria for practical assessment of the ore content of magmas. Decor.: Order of the Badge of Honor.
Work: *Geokhimiia redkikh elementov v granitoidakh.* Moscow 1961.

Tazhibaev, Abdila
Author, poet and dramatist; b. 1909, Kzyl-Orda.
Kazakh; grad. Dep. of Literature Kazakh Abai Pedagogical Inst. 1929; has written numerous poems and dramas, the latter about the lives of the Kazakh writers Abaj and Dzhanbul and the Ukrainian writer Shevchenko.
Publs.: "Stikhi. Poemy. P'esy" (Verses. Poems. Dramas) 1957, "Vliublennye" (People in Love) 1963, and others.

Teder, Kheino Oskarovich
State official.
Head Cen. Admin. of Forestry and Nature Conservation with Council of Mins. Esthonian SSR 1963–66; Min. of Forestry and Nature Conservation Esthonian SSR 1966–; cand. mem. Cen. Cttee. Esthonian CP 1971–.

Tediashvili, Levan Kitoevich
Athlete and instructor; b. Mr. 15, 1948, Gegmoubani, Sagaredzho Raion, Georgian SSR.
Honored Master of Sports (1971). He was the Olympic champion in freestyle wrestling in 1972 and 1976. He was the world and USSR champion in freestyle wrestling in 1971, and again in 1973 and 1974; in 1975 he won the world championship in sambo in 1973. Decors.: Order of Lenin and the Order of the Badge of Honor.

Telzhanov, Kanafii Temir Bulatovich
Painter; b. My. 1, 1927, in the Aul Baitiuek, Omsk Oblast.
Mem. CPSU 1961–; studied at I. E. Repin Inst. of Painting, Sculpture and Architecture in Leningrad 1947–53 (pupil of M. I. Avilov); corresponding mem. USSR Acad. of the Arts 1967; mem. Cen. Cttee. Kazakh CP 1966–71; Chair. of Kazakh Painters' Union 1965–68. Decors.: Merited Artist of the Kazakh SSR 1963, Chokan-Valikhan State Prize of the Kazakh SSR 1967, 2 orders.

Tendriakov, Vladimir Fedorovich
Author; b. D. 5, 1923, Makarovskaia, Vologda Oblast.
Mem. CPSU 1948–; grad. Gorkii Inst. of Literature 1951; first works publ. 1947. Decors.: 2 orders.
Publs.: "Padenie Ivana Chuprova" (The Fall of Ivan Chuprov) 1953, "Nenastie" (Bad Weather) 1954, "Tugoi uzel" (Tied Knot) 1956, "Podenka – vek korotkii" (Day-Fly – A Short Life) 1965, "Konchina" (The End) 1968, "Tri meshka sornoi pshenitsy" (Three Sacks Full of Uncleaned Wheat) 1973, "Noch' posle vypuska" (The Night After Graduation) 1947, and others.

Tenishev, Ivan Ivanovich
Military official.
Maj.-Gen.; Commander of Cen. Forces Group (CSSR) 1973–.

Ter-Gazariants, Georgii Artashesovich
Diplomat; b. 1923.
Mem. CPSU 1942–; grad. Lenin Mil.-Pol. Acad. 1950; served in Sov. Army 1941–50; 1st Sec. of Cen. Cttee. Armenian Komsomol 1950–55; 1st Dep. Chair. of Cttee. for USSR Youth Orgs. 1955–57; mem. staff Cen. Cttee. USSR Komsomol 1957–58; various posts with Cen. Cttee. CPSU 1958–61; Sec. 1961–66, 2nd Sec. of Cen. Cttee. Armenian CP 1966–73; mem. Bureau Cen. Cttee. Armenian CP 1961–73; mem. Cen. Auditing Comm. CPSU 1966–76; USSR Amb. to Senegal and Gambia Ap. 1973–81, to Zimbabwe Jl. 1981–.

Terebilov, Vladimir Ivanovich, Cand. of Legal Sc.
State official; b. 1916.
Mem. CPSU 1940–; grad. Leningrad Inst. of Law 1939; worked in various depts. of the State Prosecutor's Office in Leningrad City and Oblast 1939–49; mem. scientific staff, section head at Inst. for Research on Criminology 1949–57; sen. posts with USSR State Prosecutor's Office 1957–62; Dep. Chair. of USSR Supr. Court 1962–70; USSR Min. of Justice S. 1970–; mem. 1971–73, Dep. Chair. of Cen. Auditing Comm. 1973–76, cand. mem. Cen. Cttee. CPSU 1976–. Decor.: Order of Lenin 1976.

Terekhov, Boris Pavlovich
State official.
Min. of the Furniture and Wood-Processing Ind. Moldavian SSR Mr. 1971–; cand. mem. Cen. Cttee. Moldavian CP 1971–76, mem. 1976–.

Tereshchenko, Nikolai Dmitrievich
Collective farm chairman; b. 1930.
Mem. CPSU 1955–; grad. North Ossetinian Agricultural Inst. 1960; served in Sov. Army 1950–53; veterinarian at a machine and tractor station in Stavropol Krai 1953–57; veterinary technician, chief zoo technologist of "Artezianskii" sheep-raising state farm in Stavropol Krai 1957–61; Chair. of "Put' k kommunizmu" Collective Farm in Stavropol Krai 1961–; cand. mem. Cen. Cttee. CPSU 1981–.

Teterin, Aleksandr Vasil'evich
Diplomat; b. 1927.
Mem. Diplomatic Service 1968–; Emb. Counsellor 1968–71, Emb. Counsellor-Envoy at USSR Emb. in Egypt 1971–74; USSR Amb. to Nigeria O. 1974–78; Chief, Consular Administration USSR Min. of Foreign Affairs 1978–.

Tiabut, Dmitrii Vasil'evich
Party official; b. 1914.
Belorussian; mem. CPSU 1939–; grad. Pedagogic Inst. and Party High School of Cen. Cttee. CPSU; teacher and head of a raion educ. dept. 1932–40; dep. dept. head of a raion cttee. Belorussian CP, chair. of an exec. cttee. of a raion sov. 1940–41; served in Sov. Army 1941–42; sec. of an underground raion party cttee., Commander of Partisan Brigade in Vitebsk Oblast 1942–44; sec. of a raion cttee. Belorussian CP, Sec. of Polotsk and Vitebsk Oblast Cttee. Belorussian CP 1944–60; Dep. to Supr. Sov. Belorussian SSR 1955–; cand. mem. 1960–61, mem. Cen. Cttee. Belorussian CP 1961–76; 2nd Sec. of Vitebsk Oblast Cttee. Belorussian CP 1960–61; 1st Dep. Min. of Supplies Belorussian SSR 1961–62; 1st Sec. of Minsk Oblast Cttee. Belorussian CP (for Agriculture 1963–64) 1962–64; Chair. of Minsk Exec. Cttee. of Oblast Sov. of Workers' Deps. D. 1964–76; Chair. of Auditing Comm. Belorussian CP 1976–; Dep. to USSR Supr. Sov. 1966–. Decors.: Order of Lenin, Order of Red Banner of Labour.

Tiazhelnikov, Evgenii Mikhailovich, Cand. of Hist. Sc.
Diplomat; b. Ja. 7, 1928, Verkhniaia Sanarka, Cheliabinsk Oblast; s. of a peasant.
Mem. CPSU 1951–; grad Cheliabinsk Pedagogical Inst. 1950; Lecturer; Sec. of Komsomol Cttee. at above inst., dep. dept. head with an oblast Komsomol cttee., after conclusion of postgraduate studies teacher and Sec. of Party Bureau of Cheliabinsk Pedagogical Inst. 1950–61; Rector of Cheliabinsk Pedagogical Inst. 1961–64; Sec. of Cheliabinsk Oblast Cttee. CPSU 1964–68; 1st Sec. and mem. Bureau Cen. Cttee. USSR Komsomol Je. 1968–77; mem. Cen. Cttee. CPSU 1971–; Head of Propaganda Dept. of Cen. Cttee. CPSU My. 1977–82; USSR Amb. to Rumania D. 1982–. Decors.: Order of Red Banner of Labour, Order of Lenin, medals.

Tibar, Marko Oskarovich
State official.
Min. of Interior Esthonian SSR Mr. 1979–; mem. Cen. Cttee. Esthonian CP 1981–.

Tikhomirov, Viktor Vasil'evich
Scientist in the fields of radio electronics and automation; b. D. 10 (23), 1912, Kineshma, now Ivanovo Oblast.
Corresponding member of the Academy of Sciences of the USSR (1953). Member of the CPSU from 1948.
He graduated from the Moscow Institute of Power Engineering in 1940. After being on the staff of a number of scientific research organizations, he became a department head at the Institute of Biophysics of the Academy of Sciences of the USSR in 1962. His major works deal with radio-engineering methods as applied to the development of aviation instruments, instruments for other engineering systems, and instruments for the automation of biological research. Decors.: State Prize of the USSR in 1943, 1945, and 1953, two Orders of Lenin, two other orders, and various medals.
Work: *Biotelemetricheskie sistemy*. Moscow 1974.

Tikhomirov, Vladimir Porfirevich
Laborer; b. 1934.
Mem. CPSU 1958–; finished school for working youth in Moscow 1969; lathe operator at Vladimir Ilich Electromechanical Plant in Moscow 1952–54 and 1957–; served in Sov. Army 1954–57; mem. Cen. Cttee. CPSU 1981–.

Tikhomirov, Vladimir Vladimirovich
Geologist; b. O. 12 (25), 1915, Petrograd.
Doctor of geological and mineralogical sciences (1949); professor (1955). Member of the CPSU from 1940.
He graduated from the Azerbaidzhan Industrial Institute in 1938. In 1949 he became a research worker at the geological history laboratory of the Geological Institute of the Academy of Sciences of the USSR; in 1956 he was made head of the laboratory. Since 1955 he has been chairman of the Committee on the Geological Study of the USSR.
His principal works deal with the history and methodology of geology, with regional geology, and with geotectonics. In 1958 he proposed a hypothesis concerning the metasomatic alterations of the material constituting the earth's crust.
Since 1966, he has been a member of the International Academy of the History of Science. He was the initiator of the International Committee on the History of the Geological Sciences and became its first president in 1968. Decors.: Order of the October Revolution, three other orders, and various medals.
Works: *Malyi Kavkaz v verkhnemelovoe vremia*. Moscow 1950. *Geologiia v Rossii 1-i poloviny 19 veka*, parts 1–2. Moscow 1960–63. "K voprosu o razvitii zemnoi kory i prirode granita." *Izv. AN SSSR: Seriia geologicheskaia*, 1958, no. 8.

Tikhonov, Aleksandr Ivanovich
Athlete in the biathlon; b. Ja. 1, 1947, Uiiskoe, Cheliabinsk Oblast.
Honored Master of Sport (1968). Military officer. Member of the CPSU since 1975. Competing in various biathlon events, he was a champion at the Olympics in 1968, 1972, and 1976, world champion seven times between 1969 and 1973, and champion of the USSR seven times between 1969 and 1972. Decors.: Order of the Red Banner of Labour, the Order of the Red Star, and various medals.

Tikhonov, Andrei Nikolaevich
Mathematician and geophysicist; b. O. 17 (30), 1906, Gzhatsk, now Gagarin, Smolensk Oblast.
Grad. Moscow Univ. 1927; mem. USSR Acad. of Sciences 1966–; Prof. 1936–, Head of Dept. of Computer Mathematics and Cybernetics at Moscow Univ. 1970–; Dep. Dir. of Inst. of Applied Mathematics of USSR Acad. of Sciences 1953–. Decors.: Hero of Socialist Labour 1953, State Prize of the USSR 1953 and 1976, Lenin Prize 1966, 5 Orders of Lenin, 4 other orders and various medals.
Publs.: "Uravneniia matematicheskoi fiziki", 4th ed. (with A. A. Samarski) 1972, "Teoriia funktsii kompleksnoi peremennoi", 3rd ed. (with A. G. Sveshnikov) 1974, "Metody resheniia nekorrektnykh zadach" (with V. Ia. Arsenin) 1974.

Tikhonov, Nikolai Aleksandrovich
State official; b. 1905; s. of an engineer.
Ukrainian; grad. Dnepropetrovsk Metallurgical Inst. 1930; asst. locomotive driver, technician 1924–30; engineer, dept. head chief engineer at various plants in Dnepropetrovsk and Pervouralsk 1930–47; dir. of a pipe factory in Nikopol 1947–50; Head of Main Board of USSR Min. of Ferrous Metallurgy 1950–55; USSR Dep. Min. of Ferrous Metallurgy 1955–57; Chair. of Dnepropetrovsk Econ. Council 1957–60; Dep. to USSR Supr. Sov. 1958–; mem. Cen. Cttee. Ukrainian CP 1960–61; Dep. Chair (with rank of USSR Min.) of State Econ. Council of USSR Council of Mins. 1960–63; cand. mem. 1961–66, mem. Cen. Cttee. CPSU 1966–; Dep. Chair. (with rank of USSR Min.) of USSR GOSPLAN 1963–65; Dep. Chair. of USSR Council of Mins. O. 1965–76; 1st Dep. Chair. of USSR Council of Mins. S. 1976–80, Chair. O. 1980–; cand. mem.

Politbureau Cen. Cttee. CPSU 1978–79, mem. N. 1979–. Decors.: 2 Orders of Lenin, Order of Red Banner of Labour, 2 Stalin Prizes, Hammer and Sickle Medal in Gold, Hero of Socialist Labour.

Tikhonov, Viacheslav Vasil'evich
film actor; b. F. 8, 1928, Pavlovskii Posad, Moscow Oblast.
People's Artist of the USSR (1974).
He graduated from the department of acting of the All-Union State Institute of Cinematography in 1950. He made his film debut in the role of Volodia Os'mukhin in *Young Guard* (1948), based on the novel by A. A. Fadeev. He won fame with his performance in the character parts of Matvei Morozov in *It Happened in Pen'kovo* (1958) and Viktor Raiskii in *An Extraordinary Incident* (1959). He has portrayed a number of lyric and heroic characters. The most important are Aleksei in *An Optimistic Tragedy* (1963), Prince Andrei in *War and Peace* (1966–67), and Mel'nikov in *Let's Live Till Monday* (1968). His greatest performance was in the role of the Soviet intelligence colonel Isaev (Shtirlits) in the television serial *Seventeen Moments of Spring* (1972). In his best work he enters into the inner world of the character he is portraying and reveals the true psychological, emotional and intellectual nature of the character. Decors.: State Prize of the USSR in 1970.

Timofeev, Leonid Ivanovich, Dr. of Philol. Sc.
Literary historian; b. D. 23, 1903 (Ja. 5, 1904).
Grad. V. Ia. Briusov Inst. of Belles Lettres 1925; Dr. of Philol. Sc. 1940; Prof., literary historian, mem. Acad. of Pedagogical Sciences 1947; Prof. at Moscow State Univ. and at other colleges; Head of Dept. of Soviet Literature at Inst. of World Literature of USSR Acad. of Sciences 1941–70; first works publ. 1926. Decors.: 4 orders, various medals.
Publs.: "Problemy teorii literatury" (Problems in Theory of Literature) 1955, "Sovetskaia literatura. Metod, stil, poetika" (Soviet Literature. Method, Style, Poetry) 1964, "Sovetskaia literatura i khudozhestvennyi progress" (Soviet Literature and Artistic Progress) in "Novyi Mir" No. 11, 1973.

Timofeev, Petr Petrovich, Dr. of Geol.-Min. Sc.
Geologist; b. 1918.
Mem. CPSU 1947–; grad. Moscow Univ. 1943; Dr. of Geol.-Min. Sc. 1968; Prof. 1971–; mem. staff Min. of the Coal Ind. 1943–47; postgraduate studies, Jun. Scientist 1947–51, Sen. Scientist 1951–60, Dep. Dir. of Geological Inst. of USSR Acad. of Sciences 1960–.
Publs.: "Geologiia i fatsii iurskoi uglenosnoi formatsii iuzhnoi Sibiri", "Iurskaia uglenosnaia formatsiia iuzhnoi Sibiri i usloviia ee obrazovaniia" 1969–70.

Timofeev, Petr Vasil'evich
Scientist in electronics and vacuum technology; b. Je. 12 (25), 1902, Moscow.
Corresponding member of the Academy of Sciences of the USSR (1953).
He graduated from Moscow State University in 1925. He taught at the Moscow Higher Technical School from 1925 to 1928, at Moscow State University from 1926 to 1941, at the Moscow Power Engineering Institute from 1933 to 1941, and at the N. E. Zhukovskii Air Force Engineering Academy from 1945 to 1948. He became a professor in 1935. In 1928 he joined the staff of the All-Union Electrotechnical Institute.
His main works deal with the photoelectric effect, secondary electron emission, gas discharges, and electron optics. He designed a number of photocells, electron multipliers, electron tubes, and image converters. His inventions from 1933 to 1936 included television camera tubes, among them the image iconoscope. Decors.: State Prize of the USSR in 1946 and 1951, five orders and various medals. Work: *Elektronnye i ionnye pribory.* Moscow-Leningrad 1940.

Timofeev, Timur Timofeevich
Historian and economist; b. N. 30, 1928; Ivanovo.
Mem. CPSU 1950–; corresponding mem. USSR Acad. of Sciences 1966–; Dep. Dir. of Inst. for World Trade and Int. Relations 1961–66, Dir. of Inst. of Int. Workers' Movts. at USSR Acad. of Sciences 1966–. Decors.: Red Banner of Labour, medals.
Publs.: "Proletariat protiv monopolii. Ocherki pro problemam klassovoi bor'by i obshchedemokraticheskikh dvizhenii v SShA" (The Proletariat Against Monopolies. Notes on Problems of the Class Struggle and General Democratic Movements in the USA) 1967, "Mezdunarodnoe rabochee dvizenie i voprosoy antiimperialisticheskoi bor'by" (The International Labour Movement and Problems of the Anti-Imperialist Struggle) 1968. "Filosofiia istoricheskogo optimizma" (The Philosophy of Historical Optimism) 1975; co-author and ed. of a number of collective works.

Timofeev-Resovskii, Nikolai Vladimirovich
Biologist; b. S. 7 (20), 1900, Moscow.
He graduated from Moscow State University in 1925. A student of S. S. Chetverikov and N. K. Kol'tsov, he was a research worker in the Institute of Experimental Biology from 1922 to 1925. He worked in Germany from 1925 to 1945. Having returned to the USSR, he headed various departments in the Institute of Biology of the Ural Branch of the Academy of Sciences of the USSR from 1955 to 1963 and in the Institute of Medical Radiology of the Academy of Medical Sciences of the USSR at Obninsk from 1964 to 1969. He has been a consultant to the Institute of Medical Biological Problems of the Ministry of Health of the USSR since 1969.
His principal works are in the fields of genetics, radiobiology, biogeocenology, and the theory of evolution. He is one of the founders of quantitative radiation genetics and radiation biogeocenology. He published many research reports on the genetic effects of radiation that led to some basic formulations in modern radiobiology – namely, the "hit principle" and the target theory. Together with M. Delbrück, he developed the first biophysical model of gene structure and proposed the possibility of genetic modification. His work includes studies on population genetics. Through his research on the initial stages of intraspecific differentiation, he formulated and developed the theory of microevolution.
He became a member of the presidium of the N. I. Vavilov All-Union Society of Geneticists and Selectionists in 1966, a member of the American Academy of Arts and Sciences in 1973, and a member of the Leopoldina Academy of the German Democratic Republic (GDR) in 1969. Decors.: GDR's Darwin Medal in 1959, the Czechoslovak Socialist Republic's Mendel Medal in 1965, the United States' Kimber Genetics Award and Gold Medal for an outstanding contribution to genetics in 1966, and the GDR's Mendel Medal in 1970.
Works: *Primenenie printsipa popadaniia v radiobiologii.* Moscow 1968 (With V. I. Ivanov and V. I. Korogodin.) *Kratkii ocherk teorii evoliutsii.* Moscow 1969. (With N. N. Vorontsov and A. V. Iablokov.) *Ocherk ucheniia o populiatsii.* Moscow 1973. (With A. V. Iablokov and N. V. Glotov.)

Timofeeva, Nina Vladimirovna
Ballerina; b. Je. 11, 1935, Leningrad.
People's Artist of the USSR (1969).
She graduated from the Leningrad Choreographic School, made her debut as a student in 1951, as Masha in Tchaikovsky's *The Nutcracker.* From 1953 to 1956 she was a soloist with the S. M. Kirov Leningrad Theater of Opera and Ballet, and since 1956 she has been a soloist with the Bolshoi Theater. A consummate technician, she masterfully embodies classical and modern choreography and plastique and dances with remarkable musicality. Her roles have included Odette-Odile in Tchaikovsky's *Swan Lake,* Kitri in Minkus' *Don Quixote,* the Mistress of the Copper Mountain in Prokofiev's *The Stone Flower,* Mekhmene-Banu in Melikov's *Legend of Love,* Leili in Balasanian's *Leili and Medzhnun,* and Aegina in Khachaturian's *Spartacus.*
She was a deputy to the seventh convocation of the Supreme Soviet of the USSR. Decors.: two orders and one medal.

Timonen, Antti Nikolaevich
Author; b. Ap. 18 (My. 1), 1915, Lusalma, now Karelian ASSR.
Karelian; mem. CPSU 1942–; grad. Advanced Courses in Literature of M. Gorkii Inst. of Literature 1956; has written novels and sketches in Finnish.
Publs.: "Ot Karelii do Karpat" (From Karelia to the Carpathians) 1948, "Pod grozoi i solntsem" (In Thunderstorm and in the Sun) 1964, "Zdes' moi dom" (Here Is My Home) 1966, "My – karely" (We Karelians) 1971, and others.

Tishchenko, Boris Ivanovich
Composer; b. Mr. 23, 1939, Leningrad.
In 1962, he graduated from the Leningrad Conservatory, where he studied composition with D. D. Shostakovich; in 1965 studied with Shostakovich as a graduate student. He is the composer of four symphonies (1961–74) and the *Sinfonia Robusta* (1970); the ballets *The Twelve,* based on the poem by A. A. Blok (1964, Leningrad Theater of Opera and Ballet), and *Iaroslavna,* based on *The Tale of Igor's Campaign* (1974, Leningrad Malyi Theater of Opera and Ballet), and three works on themes by K. I. Chukovskii – the ballet *The Chatterbox Fly,* the opera *Stolen Sun,* and the operetta *The Giant Cockroach* (1968). Other compositions include *Requiem* (1966, words by A. A. Akhmatova), two cello concerti (1963 and 1969), and concerti for piano (1962), violin (1958), and flute and piano (1972).

Titarenko, Aleksei Antonovich
Party official; b. 1915; s. of a worker.
Ukrainian; mem. CPSU 1940–; grad. Zhdanov Metallurgical Inst. 1937; fitter, Brigade-Leader, Foreman, Sen. Foreman at pipe-rolling shop of Ilich Plant in Zhdanov 1931–37; served in Sov. Army 1937–39; Foreman, Technologist, Dept. Head at Ilich Plant in Zhdanov and Nizhnii Tagil 1939–48; 1st sec. of a raion cttee., dept. head with an oblast cttee. Ukrainian CP 1948–51; Sec. of Donetsk Oblast Cttee. 1951–52, 1st Sec. of Donetsk City Cttee. Ukrainian CP 1952–60; cand. mem. 1954–65, mem. Cen. Cttee. Ukrainian DP 1965–; Dep. to Supr. Sov. Ukrainian SSR 1955–; 2nd Sed. of Donetsk Oblast Cttee. Ukrainian CP 1960–62; 1st Sec. of Zaporozhe Oblast Cttee. Ukrainian CP 1962–66 (for Ind. 1963–64); cand. mem. 1966–71, mem. Cen. Cttee. CPSU 1971–; mem. Politbureau 1966–, Sec. of Cen. Cttee. Ukrainian CP Mr. 1966–82, 2nd Sec. O. 1982–; Dep. to and mem. Comm. for Construction and the Building Materials Ind. of Sov. of the Union USSR Supr. Sov. 1966–. Decors.: 2 Orders of Lenin, Order of Red Banner of Labour.

Titov, German
Cosmonaut; b. 1935, Verkhnie Zhilino, Altai Krai.
Mem. CPSU 1961–; grad. Mil. Aviation College 1957; Maj.; first man ever to complete 17 orbits (travelled 700,000 km in 25 hours in Vostok-2) Ag. 6–7, 1961. Decors.: two Orders of Lenin, various medals, and many foreign decorations, including the titles of Hero of Socialist Labor of the People's Republic of Bulgaria, Hero of Labor of the Democratic Republic of Vietnam, and Hero of the Mongolian People's Republic. A crater on the far side of the moon has been named after Titov.
Works: *Semnadtsat' kosmicheskikh zor'.* Moscow 1963. *700,000 kilometrov v kosmose.* Moscow 1961. *Golubaia moia planeta.* Moscow 1973.

Titov, Iurij Evlampievich
Gymnast and gymnastics coach; b. N. 27, 1935, Omsk.
Honored Master of Sport (1956); international-class judge (1968). Member of the CPSU since 1969.
He was all-around champion of the USSR (1958, 1961), Europe (1959), and the world (1962). In 1956 he became an Olympic gymnastics champion in team competitions. Competing in individual sports in combined events, he was champion of the USSR five times between 1958 and 1962, European champion six times between 1957 and 1961, and world champion in 1958 and 1962. He won 13 gold medals and 20 silver and bronze medals at European and world championships and at the Olympic Games. He has been president of the World Gymnastic Federation since 1976. Decors.: three orders and various medals.
Work: *Summa ballov.* Moscow 1971.

Tiumebaev, Syrym Mukhamedzhanovich
State official.
Min. of Public Services Kazakh SSR Je. 1980–; mem. Auditing Comm. Kazakh CP 1981–.

Tiunov, Anatolii Ivanovich
State official.
Mem. Cen. Cttee. Moldavian CP 1960–76, cand. mem. 1981–; Dep. Chair. of Bureau of Agriculture of Cen. Cttee. Moldavian CP 1962–64; Head of Dept. of Agriculture at Cen. Cttee. Moldavian CP 1962–65; Dep. to Supr. Sov. Moldavian SSR 1963–; Min. of the Meat and Dairy Ind. Moldavian SSR O. 1965–.

Tiurebaev, Vil Nadzhimetdinovich
State official.
Min. of Posts and Telecommunications Kirghiz SSR Jl. 1975–.

Tiurenkov, Aleksandr Alekseevich
Sculptor; b. Ap. 24, 1924, Moscow.
Member of the CPSU since 1973.
Studied at the V. I. Surikov Moscow Art Institute from 1947 to 1953. His works include *Soldier With the Newspaper "Pravda"* (plaster of Paris, Museum of the Revolution of the USSR, Moscow) and *The Year 1919* (granite, 1958, Museum of the Armed Forces, Moscow). As part of a team headed by E. V. Vuchetich, he worked on the memorial ensemble to the heroes of the battle of Stalingrad on Mamai Hill in Volgograd (reinforced concrete, 1963–67; Lenin Prize, 1970) he also designed K. I. Skriabin's gravestone in Novodevicim Cemetery in Moscow (stone, 1974). Decor.: Order of the Red Banner of Labor.

Tkach, D. I.
Komsomol official.
Sec. and mem. Bureau Cen. Cttee. Ukrainian Komsomol Ap. 1982–.

Tkach, Dmitrii Vasil'evich
Author; b. Ag. 29 (S. 11), 1912, Orlik, Poltava Oblast.
Ukrainian; mem. CPSU 1943–; wrote numerous novels and sketches on patriotic subjects and on the life of metal workers, circus artists and gypsies.
Publs.: "Moriaki" (Sailors) 1948, "Arena" 1960, "Komendant moria" (Commander of the Sea) 1960, "U nas v obshchezhitii" (In Our Communal Living Quarters) 1966, "Suda ne budet" (The Trial Will Not Take Place) 1971, "Shtorm i shtil" (Storm and Calm) 1971, "Spokoinoe more" (The Calm Sea) 1974, "Genuezskaia bashnia" (The Tower of Genova) 1974.

Tkachenko, Anatolii Sergeevich
Author; b. Ap. 24, 1926, Gribskoe, Amur Oblast.
Russian; has written several novelettes and novels.
Publs.: "Solenyi bereg" (The Salty Shore) 1962, "Bereg dolgoi zimy" (The Shore of the Long Winter) 1966, "Nashe korotkoe leto" (Our Short Summer) 1971, and others.

Tkachenko, Ninel' Aleksandrovna
Soprano; b. N. 21, 1928, Kharkov.
People's Artist of the USSR (1964). Member of the CPSU since 1972.
Studied singing with T. Ia. Veske at the Kharkov Conservatory, graduating in 1958; in the same year she became an operatic soloist at the Kiev Conservatory, subsequently joining the L'vov Theater of Opera and Ballet in 1960, the Minsk Theater of Opera and Ballet in 1962, and the Odessa Theater of Opera and Ballet in 1968. Her artistry is characterized by virtuosity, and her performance by great sincerity. She has sung the title roles in Verdi's *Aïda* and Puccini's *Tosca,* as well as the parts of Liza and Tat'iana, respectively, in Tchaikovsky's *The Queen of Spades* and *Eugene Onegin,* Natasha in Dargomyzhskii's *The Mermaid,* Iaroslavna in Borodin's *Prince Igor,* and Leonora in Verdi's *Il Trovatore.*

Tkhapsaev, Vladimir Vasil'evich
Ossetian actor; b. Jl. 2 (15), 1910, Ardon, now in the Severnaia Ossetiia ASSR.
People's Artist of the USSR (1960). Member of the CPSU since 1944.

In 1933, began his studies at the studio of Russian theater in Ordzhonikidze, where he joined the Severnaia Osetiia Drama Theater in 1935. As an actor has asserted the heroic forces of life, combining dynamic passion with spiritual purity.
His best roles have included Sarmat in Salamov's *Sarmat and His Sons,* Dzambolat in Dzhimiev's *Black Fog,* the Governor in Gogol's *The Inspector-General,* Egor Bulychov in Gorkii's *Egor Bulychov and the Others,* General Murav'ev in Chirskov's *The Victors,* the title roles in Shakespeare's *Othello, King Lear,* and *Macbeth,* and Matthias Klausen in Hauptmann's *Before Sunset.* Decors.: K. Khetagurov Prize of the Severnaia Osetiia ASSR (1965), the K. S. Stanislavsky State Prize of the RSFSR (1968), the Order of Lenin, two other orders, and various medals.

Tliusten, Jusuf Ibragimovich
Adygeian poet; b. Ap. 25, 1913, in the aul (village) of Shakhancherikhabl', now Teuchezh Raion.
Member of the CPSU since 1941; studied at the Moscow Printing Institute and served in the Great Patriotic War (1941–45). He first published his works in 1938, wrote the collection *Adygeian Tales* (1939) and the novella *Light in the Mountains* (1949), in which he depicted the struggle of the Adygeian people against the fascist occupiers. He also wrote the novels *The Ozhbanokovs* (1962; translated into Russian as *It All Began in the Spring,* 1966) and *The Dawn of Maidens* (1971). He also translated L. N. Tolstoy's novella *Hadji Murad.* Decors.: four orders and various medals.
Works: *Shlum il "eg" okheshchkher.* Maikop 1975. *In Russian translation: Nemerknushchii svet.* Moscow 1952.

Todeson, Kuno Eduardovich
State official.
Min. of Trade Esthonian SSR D. 1963–; cand. mem. 1964–66, mem. Cen. Cttee. Esthonian CP 1966–.

Toichubekov, Iu. Zh.
Komsomol official.
Sec. and mem. Bureau Cen. Cttee. Kirghiz Komsomol S. 1978–.

Tokarev, Aleksandr Maksimovich
State official; b. 1921; s. of a worker.
Russian; grad. Kuibyshev Inst. of Civil Engineering 1949; served in Sov. Army 1940–45; Foreman, Works Manager at the construction of the Petroleum Refinery in Novokuibyshev 1949; Komsomol, party and govt. posts 1949–51; Sec. of Stavropol City Cttee. 1951–52, Sec. of Novokuibyshev City Cttee. CPSU 1952–55; dept. head, Sec. of Kuibyshev Oblast CPSU Cttee. 1955–59; Chair. of Exec. Cttee. of Kuibyshev Oblast Sov. of Workers' Deps. 1959–63; Dep. to 1962–, mem. Comm. for Construction and Construction Materials at Sov. of the Union USSR Supr. Sov. 1966–67; 1st Sec. of Kuibyshev Oblast Cttee. CPSU (for Ind. 1963–64) Ja. 1963–67; mem. Cen. Cttee. CPSU 1966–; USSR Min. of Ind. Construction F. 1967–. Decor.: Order of Lenin.

Tokhtaev, Timur Makhsudovich
State official.
Min. of Posts and Telecommunications Uzbek SSR Mr. 1971–; cand. mem. Cen. Cttee. Uzbek CP 1971–.

Tokombaev, Aaly (pseudonym: Balka)
Author; b. N. 7, 1904, Chon-Kemin, Kirghizia.
Mem. CPSU 1927–; grad. Cen. Asian Communist V. I. Lenin Univ. in Tashkent 1927; mem. Acad. of Sciences Kirghiz SSR 1954–; Chair. of Kirghiz Writers' Union 1934–49; first works publ. 1942. Decors.: People's Poet of the Kirghiz SSR 1945, State Prize of the Kirghiz SSR 1967.
Publs.: "Zerkalo zhenshchiny" (A Woman's Mirror) 1929, "Tsvety truda" (Flowers of Labour) 1932, "Dnestr vpadaet v glubokoe more" (The Dnestr Flows Into a Deep Sea) 1939, "Taina melodii" (The Secret of a Melody) 1940, "Vpered" (Forward) 1941, "Zemliak Manasa" (Mana's Fellow-Countryman) 1941, "Blagodarnost" (Gratitude) 1944, "Moia metrika" (My Birth Certificate) 1955, "Golos vremeni" (Voice of the Time) 1966, "Pered zarei" (Before Dawn) 1968, "Zhivaia istoriia" (A Lively Story) 1969, "Stikhotvoreniia" (Poems) 1974. Decors.: 2 Orders of Lenin, 4 other orders and various medals.

Tokombaev, Asanbek
State official.
Dep. to 1959–, Chair. of Supr. Sov. Kirghiz SSR 1963–67; dept. head at Cen. Cttee. Kirghiz CP 1959; Sec. and mem. Bureau Cen. Cttee. Kirghiz CP 1962–63; Chair. of State Cttee. for Television and Radio Kirghiz SSR 1964–; mem. Auditing Comm. 1971–76, cand. mem. Cen. Cttee. Kirghiz CP 1976–81, mem. 1981–.

Toktomushev, Abdrasul
Author and Poet; b. S. 2 (15), 1912, Chon-Kemin, Kirghiz SSR.
Kirghiz; mem. CPSU 1951–; grad. Frunze Pedagogical College; wrote poetry (mainly about Kirghiz life) in Kirghiz.
Publs.: "Syn Dzhakipa" (Dzhakip's Son) 1957, "Zolotaia gora" (The Golden Mountain), vol. 1, 1965, vol. 2, 1968, and others.

Toktonaliev, Aliaskar
State official.
Min. of Finance Kirghiz SSR Ap. 1960–; cand. mem. Cen. Cttee. 1960–63, mem. Auditing Comm. 1963–66, mem. Cen. Cttee. Kirghiz CP 1966–; Dep. to Supr. Sov. Kirghiz SSR 1963–.

Tolkunov, Lev Nikolaevich
Journalist; b. 1919.
Russian; mem. CPSU 1943–; studied 3 terms at Gorkii Inst. of Literature and completed Party Higher School of Cen. Cttee. CPSU 1946; mem. literary staff, Asst. Chief Ed., war correspondent for "Pravda" 1938–46; Special Correspondent for "Pravda", Dep. Exec. Sec., then dept. head with editorial board of newspaper "Za prochnyi mir, za narodnuiu demokratiiu" 1946–51; Dep. Ed. of and Dept. Ed. with "Pravda" 1951–57; worked for Cen. Cttee. CPSU 1957–65; Chief Ed. of "Izvestia" O. 1965–76; cand. mem. 1966–76, mem. Cen. Cttee. CPSU 1976–; Dep. to and mem. Foreign Affairs Comm. of Sov. of Nationalities USSR Supr. Sov. 1966–; Chair. of Sov. News Agency "Novosti" 1976–. Decors.: 2 Orders of Red Banner of Labour, Order of the October Revolution.

Tolpigo, Kirill Borisovich, Dr. of Phy.-Math. Sc.
Theoretical physicist; b. My. 3, 1916, Kiev; s. of a clerk.
Grad. Kiev Univ. 1939; corresponding mem. Acad. of Sciences Ukrainian SSR 1965–; Dr. of Phy.-Math. Sc. 1962; Prof. 1963–; served in Sov. Army 1939–45; postgraduate studies 1945–48, mem. staff Inst. of Physics of Acad. of Sciences Ukrainian SSR 1948–59; Lecturer, Prof., then Head of Chair of Theoretical Physics of Kiev Univ. 1959–65; Head of Theoretical Dept. of Physical Technology of Acad. of Sciences Ukrainian SSR in Donetsk Ja. 1966–; author resp., co-author of approx. 100 scientific papers.
Publs.: "Opticheskie, uprugie i piezoelektricheskie svoistva ionnykh i valentnykh kristallov s reshetkoi tipa ZnS" in "Fizika tverdogo tela" No. 10, 1960, "Termodinamika i statisticheskaia fizika" 1966, and others.

Tolstikov, Evgenii Ivanovich
Meteorologist; b. 1913.
Mem. CPSU 1943–; grad. Moscow Hydrometeorological Inst. 1937; took part in several polar expeditions, was Head of Drift Ice Station SP-4 1954–55 and Head of 3rd Soviet Antarctic Expedition 1957–59; dept. head at Antarctic Research Inst. 1946–50; Head of Polar Stations Admin. of Main Admin. for Northern Seaways at USSR Min. of Maritime Navigation (Glavsevmorput) 1951–54; Dep. Head of Admin. for Hydrometeorological Services with USSR Council of Mins. 1963–. Decor.: State Prize of the USSR 1971.

Tolubeev, Iurii Vladimirovich
Actor; b. Ap. 18 (My. 1), 1906, St. Petersburg.
People's Artist of the USSR (1956), Hero of Socialist Labour (1976).
He graduated from the Leningrad Institute of Stage Arts in 1929, made his debut while still a student in 1926. He acted in

various Leningrad theatres and, since 1942, has been with the A.S. Pushkin Drama Theatre. Tolubeev's interpretations of a wide range of characters are remarkable in their grasp of motivation and psychology. He is noted for his expressive powers, and his performances run the emotional gamut from the gently benevolent to the mercilessly satirical.
His finest roles were the *gorodnichii* (town chief) in Gogol's *The Inspector-General,* Polonius in Shakespeare's *Hamlet,* and Bubnov in Gorkii's *The Lower Depths.* He played the *gorodnichii* and Polonius in films as well. Tolubeev's greatest triumph, however, was the role of Vozhak in Vishnevskii's *An Optimistic Tragedy;* he was awarded the Lenin Prize for this performance in 1958. His film roles include Sancho Panza in *Don Quixote* (1957), Inspector Pilipenko in *The Accident* (1965), and Kuzmich in *Chronicle of a Dive-bomber* (1968). In 1967 a documentary film about his work, *Iurii Tolubeev,* was released.
Decors.: State Prize of the USSR in 1947 and the K.S. Stanislavsky State Prize of the RSFSR in 1970; Order of Lenin, two other orders, and various medals.

Tolubeev, Nikita Pavlovich
Diplomat; b. N. 11, 1922.
Russian; mem. CPSU 1947–; grad. Dneprodzerzhinsk Metallurgical Inst. 1951; controller for tech. control dept. of a metal plant 1941; served in Sov. Army 1941–45; technician, chief engineer, shift manager, dept. head at a metal plant 1946–54; leading party posts 1954–65; 2nd Sec. of Dnepropetrovsk Oblast Cttee. Ukrainian CP 1960–61; cand. mem. 1960–61, mem. Cen. Cttee. Ukrainian CP 1961–66; 1st Sec. 1961–63, 2nd Sec. of Dnepropetrovsk Oblast Cttee. Ukrainian CP 1963–65 (for Ind. 1963–64); mem. Cen. Cttee. CPSU 1961–66 and 1971–; Dep. to USSR Supr. Sov. 1962–66; USSR Amb. to Cyprus 1968–70, to Cuba D. 1970–79, to Bulgaria My. 1979–. Decors.: 2 orders, medals.

Tolubinskii, Vsevolod Ivanovich, Dr. of Tech. Sc.
Thermophysicist; b. Mr. 27, 1904, Tartak, now Vinnitsa Oblast; s. of a clerk.
Grad. Kiev Polytech. Inst. 1927; specializes in tech. thermophysics and thermal energy; mem. Acad. of Sciences Ukrainian SSR 1964–; Dr. of Tech. Sc. and Prof. 1951–; pedagogic and scientific work at Kiev Polytech. Inst. 1929–38; Head of Chair of Boiler Installations at Kiev Polytech. Inst. 1938–41; Head of Chair of Boiler Installations and Industrial Utilization of Thermal Energy at Cen. Asian Ind. Inst. 1941–44; Head of Chair of Boiler Installations 1944–64 and Head of Laboratory for Thermal Engineering Problems at Kiev Polytech. Inst. 1957–64; concurrently head of a dept. of Inst. of Thermal Energy (now Inst. of Tech. Thermal Physics) at Acad. of Sciences Ukrainian SSR 1939–41 and 1944–60; Dir. of Inst. of Tech. Thermophysics at Acad. of Sciences Ukrainian SSR 1963–; author of approx. 80 scientific papers. Decors.: Order of Red Banner of Labour, Order "Badge of Honour", medals.
Publs.: "Rabota i raschet kamernykh topok" 1939, "Kompleksne energotekhnologichne vykorystannia palyva" 1957, "Obshchaia teplotekhnika" 1963, and others.

Tolubko, Vladimir Fedorovich
Military official; b. N. 12 (25), 1914, Krasnograd, Kharkov Oblast.
Ukrainian; mem. CPSU 1939–; grad. Acad. of Tank and Armoured Troops 1941 and Acad. of the Gen. Staff 1951; Gen. of the Army 1970–; village teacher until 1932, then instructor of a raion Komsomol cttee.; in Sov. Army 1932–; Chief-of-Staff and Brigade Commander at Leningrad, Kalinin and 3rd Ukrainian Front, took part in battles of Iassy, Kishinev, Belgrade and Budapest 1941–45; chief-of-staff of a div., chief-of-staff, 1st dep. commander of a unit, Aide to Supr. Commander of Sov. Forces Group in Germany, commander of an army, 1st Dep. Supr. Commander of Rocket resp. Strategic Rocket Forces of Sov. Army 1945–68; Commander of Siberian and Far Eastern Mil. Districts 1968–72; Dep. to USSR Supr. Sov. 1970–; cand. mem. Cen. Cttee. CPSU 1971–76, mem. 1976–; USSR Dep. Min. of Defence, Commander-in-Chief of Strategic Rocket Forces Ap. 1972–. Decors.: 3 Orders of Lenin, 4 Order of Red Banner, Bogdan Khmelnitskii Order 2nd Cl., Order of Patriotic War 1st Cl., 2 Orders of Red Star, Order "For Service Rendered to the Fatherland in the USSR Armed Forces" 3rd Cl., medals and various foreign orders and medals.

Tomilin, Aleksandr Nikolaevich, Cand. of Math. Sc.
Mathematician; b. 1933.
Grad. Moscow Univ. 1956; Cand. of Math. Sc. 1969–; mem. staff Inst. of Precision Mechanics and Computer Technology of USSR Acad. of Sciences 1956–; Head of Engineering-Design Dept. of same inst. 1968–. Decor.: State Prize of the USSR 1969.

Tomskii, Nikolai Vasil'evich
Sculptor; b. D. 6 (19) 1900, Ramushevo, Novgorod Oblast.
Mem. CPSU 1950–; grad. Art Tech. College Leningrad 1927; mem. USSR Acad. of the Arts 1949–; Prof. 1949–; Pres. of USSR Acad. of the Arts 1968–; main field of work: monumental sculptures and busts. Decors.: Hero of Socialist Labour 1970, Stalin Prize 1941, 1948, 1949, 1950 and 1952, 2 Orders of Lenin, Order of Red Banner of Labour, medals.

Toome, Indrek
Party official.
Sec. Mr.–S. 1972, 1st Sec. and mem. Bureau Cen. Cttee. Esthonian Komsomol S. 1972–78; mem. Cen. Cttee. Esthonian CP 1976–; 1st Sec. of Tartu City Cttee. Esthonian CP N. 1978–; cand. mem. Bureau Cen. Cttee. Esthonian CP Ja. 1981–.

Toots, Erhard Iakobovich
State official.
Head of Cen. Admin. 1978–80, Chair. State Cttee. for Material and Tech. Supplies Esthonian SSR Mr. 1980–; cand. mem. Cen. Cttee. Esthonian CP 1981–.

Topcheev, Iurii Ivanovich, Dr. of Tech. Sc.
Cyberneticist; b. 1920.
Mem. CPSU 1948–; grad. Moscow Aviation Inst. 1943; specialist on automatic control systems; Dr. of Tech. Sc. 1966; Prof. 1968–; sen. scientist with a research inst. 1943–68; mem. staff Inst. of Physical Engineering in Moscow 1968–; head of a chair 1972–. Decor.: State Prize of the USSR 1972.
Publs.: co-author of engineering monograph series "Tekhnicheskaia kibernetika. Teoriia avtomaticheskogo regulirovaniia" 4 vols., 1967–69.

Topuzian, A.
Komsomol official.
Sec. 1975–77, 2nd Sec. O. 1977–, and mem. Bureau Cen. Cttee. Armenian Komsomol O. 1975–.

Torgov, Igor Vladimirovich, Dr. of Chem. Sc.
Chemical scientist; b. 1912.
Grad. Kazan Chemotechnological Inst. 1937; Prof. 1965–; corresponding mem. Acad. of Sciences 1972–; laboratory technician 1939–41, postgraduate student 1944–47, Jun. Scientist 1947–48, Sen. Scientist at Inst. of Organic Chemistry of USSR Acad. of Sciences 1948–59; served in Sov. Army 1941–44; Head of Laboratory at Inst. of Chemistry of Natural Compounds of USSR Acad. of Sciences 1959–.

Tovstonogov, Georgii Aleksandrovich
Stage director; b. S. 15 (28), 1915, Tbilisi.
People's Artist of the USSR (1957). Doctor of art studies (1968).
He graduated from the department of stage directing at the State Institute of Theatrical Arts in 1938. He was a director at the Tbilisi Griboedov Russian Theatre from 1938 to 1946 and worked at the Moscow Central Children's Theatre from 1946 to 1949. From 1950 to 1956 he was principal director of the Leningrad Lenin Komsomol Theatre, and in 1956 he became principal director of the M. Gorkii Bolshoi Drama Theatre in Leningrad.
A major aspect of his talent is his striving to express monumental generalized ideas, for example, in his stagings of *On the Road to Immortality* (1951, based on J. Fučik's *A Word Before the Execution*),

Korneichuk's *The Destruction of the Squadron* (1952), and his most important work, Vishnevskii's *An Optimistic Tragedy* (1955), which was awarded the Lenin Prize in 1958.
His best productions at the Bolshoi Drama Theatre have included a dramatization of Dostoevsky's *The Idiot* (1957 and 1966), Gorkii's *The Barbarians* (1959) and *Smug Citizens* (1966), Sholokhov's *Virgin Soil Upturned* (1964), Chekhov's *The Three Sisters* (1965), Rakhmanov's *Restless Old Age* (1970), and Tsagareli's *Khanuma* (1973).
He has trained a group of actors who share his theatrical views and has helped a number of playwrights begin successful careers in dramaturgy, including A.M. Volodin and V.S. Rozov. Tovstonogov is the author of books on the theory and practice of stage directing, including *On the Stage Director's Profession* (1965) and *Circle of Thoughts* (1972). From 1939 to 1946 he taught at the Sh. Rustaveli Georgian Theatrical Institute, and in 1962 he became head of the department of stage directing at the Leningrad Institute of Theater, Music, and Cinematography, where he had become a professor in 1960. Decors.: State Prize of the USSR (1950, 1952, 1968), two orders of Lenin, the Order of the Red Banner of Labour, and various medals.

Trapeznikov, Sergei Pavlovich, Dr. of Hist. Sc.
Party official; b. 1912; s. of a worker.
Russian; mem. CPSU 1931–; grad. Moscow Teachers' Training Inst. 1946, Party Higher School of Cen. Cttee. CPSU 1946 and Acad. of Social Sciences of Cen. Cttee. CPSU 1948; Assoc. Prof.; farmhand, worker in a sawmill 1925–29; Komsomol and party work in Penza Oblast 1929–34; Head of Dept. of Culture and Propaganda, 2nd Sec. of Vadinskii Raion Cttee., 1st Sec. of Narovchat and Serdobsk Raion Cttee. All-Union CP (B), Head of Dept. of Agriculture Penza Oblast Cttee. All-Union CP (B) 1935–48; Dir. of Party School of Cen. Cttee. Moldavian CP 1948–56; mem. Cen. Cttee. Moldavian CP 1956; worked for Cen. Cttee. CPSU 1956–60; Dep. Rector of Party Higher School of Cen. Cttee. CPSU 1960–65; Head of Dept. of Science and Educ. Institutions of Cen. Cttee. CPSU O. 1965–; mem. Cen. Cttee. CPSU 1966–; Chair. of Comm. for Educ., Science and Culture at Sov. of the Union 1966–70, mem. Comm. for Educ., Science and Culture at Sov. of Nationalities USSR Supr. Sov. 1970–. Decors.: Order of Lenin, Order of Red Banner of Labour, Order of the October Revolution.

Trapeznikov, Vadim Aleksandrovich
Mechanics scientist; b. N. 15 (28), 1905, Moscow.
Mem. CPSU 1951–; grad. Moscow Inst. of Technology 1928; specializes in automatic and electrical machine construction; mem. USSR Acad. of Sciences 1960–; mem. staff Union Inst. of Electrotechnology until 1933; Lecturer 1930–41, Prof. at Moscow Inst. of Energetics 1939–; mem. staff Inst. of Problems in Control Systems (Automatic and Telemechanical Devices) 1941–, Dir. of same insts. 1951–. Decors.: Hero of Socialist Labour 1965, State Prize of the USSR 1951, Order of Lenin, 3 Orders of Red Banner of Labour, various medals.

Tregubov, Anatolii Ivanovich
State official.
USSR Dep. Min. of Maritime Navigation until 1976; Dep. Chair. of Council of Mins. Estonian SSR S. 1978–; mem. Cen. Cttee. Estonian CP 1981–.

Treshnikov, Aleksei Fedorovich, Dr. of Geog. Sc.
Geographer and oceanologist; b. 1914.
Mem. CPSU 1944–; grad. Leningrad Univ. 1939; Dr. of Geog. Sc. 1963; Prof. 1967–; mem. staff Arctic and Antarctic Research Inst. 1939–, Dir. of same inst. 1960–; head of 22 polar expeditions. Decor.: State Prize of the USSR 1971.

Tretiak, Ivan Moiseevich
Military official; b. F. 1, 1923, Malaia Popovka, Poltava Oblast.
Ukrainian; mem. CPSU 1943–; grad. Frunze Mil. Acad. 1949 and Acad. of the Gen. Staff 1959; Army General; in Sov. Army 1939–; company commander, dep. commander of a battalion, of a regt., commander of a regt. at Western and 2nd Baltic Front 1941–45; commander of a regt., chief-of-staff, dep., 1st dep. commander of a unit, commander of a unit 1949–67; Dep. to USSR Supr. Sov. 1966–; mem. Cen. Cttee. Azerbaidzhan CP 1966–67; Commander of Belorussian Mil. District 1967–76; mem. Cen. Cttee. 1971–76, cand. mem. Bureau Cen. Cttee. 1971–76, mem. Bureau Cen. Cttee. Belorussian CP 1976; cand. mem. 1971–76, mem. Cen. Cttee. CPSU 1976–; Commander of Far Eastern Mil. District My. 1976–. Decors.: Hero of the Soviet Union 2 Orders of Lenin, 3 Orders of Red Banner, 2 Orders of Red Star, Aleksandr Nevskii Order, Kutuzov Order 3rd Cl., Order "For Services to the Fatherland in the USSR Armed Forces" 3rd Cl., foreign orders and medals.

Tretiakov, Dmitrii Nikolaevich
Physical chemist; b. 1935.
Grad. Leningrad Univ. 1957; Jun. Scientist at Physico-Tech. Inst. of USSR Acad. of Sciences 1957–. Decor.: Lenin Prize 1972.

Tretiakov, Petr Ivanovich
Party official; b. 1927.
Russian; mem. CPSU 1957–; grad. Uralsk Forestry Inst. 1951; engineer, chief of a timber station of a forest prod. works 1951–53; chair. of an exec. cttee. of a raion sov. of peoples' deps. 1961–70; 1st Sec. of Poronaisk City Cttee. CPSU, chief of a dept., Sec. of Sakhalin Oblast Cttee. CPSU 1970–78; 1st Sec. of Sakhalin Oblast Cttee. CPSU D. 1978–; Dep. to USSR Supr. Sov. 1979–; cand. mem. Cen. Cttee. CPSU 1981–.

Trifonov, Iurii Valentinovich
Author; b. Ag. 28, 1925, Moscow; s. of V. A. Trifonov.
Grad. Gorkii Inst. of Literature 1949; first works publ. 1947. Decors.: USSR State Prize 1951, Order "Badge of Honour", medals.
Publs.: "Studenty" (Students) 1950, "Utolenie zhazhdy" (Thirst Quenching) 1963 (Film with same title 1965), "Igry v sumerkakh" (Games at Dusk) 1970, "Dolgoe proshchanie" (A Long Farewell) 1971, "Drugaia zhizn" (Another Life) 1975, "Dom na naberezhnoi" (The House on the Pier) 1976, and others.

Troepolskii, Gavriil Nikolaevich
Author; b. N. 16 (29), 1905, Novospasovka, Voronezh Oblast; s. of a priest.
Grad. from an agricultural college 1924; worked as village teacher and agronomist; first works publ. 1953.
Publs.: "Prokhor semnadtsatyi i drugie" (Prokhor the Seventeenth and Others) 1954, "Kandidat nauk" (Candidate of Sciences) 1958, "Chernozem" (Black Earth) vols. 1–2, 1958–61, "V kamyshakh" (In the Reed) 1963, "Belyi Bim, Chernoe ukho" (White Bim, Black Ear) 1971, "Postoialtsy" (The Tenants) 1971, and others.

Trofimov, Iurii Nikolaevich
Party official.
Mem. Cen. Cttee. Kazakh CP 1976–; Sec. and mem. Bureau Cen. Cttee. Kazakh CP 1978–.

Trofimov, Vladimir Vasil'evich
State official; b. 1915.
Grad. 2nd Medical Inst. Moscow 1940; physician; served in Sov. Army 1940–46; assoc. of Main Mil. Medical Board at USSR Min. of Defense 1946–51; worked for Cen. Cttee. CPSU 1951–54; Dep. Min. of Health RSFSR 1954–62; Head of Foreign Relations Dept. and mem. Collegium of USSR Min. of Health 1962; Min. of Health RSFSR D. 1962–. Decors.: Order of Lenin, Order of Red Banner of Labour, Badge of Honour, various medals.

Trofimuk, Andrei Alekseevich
Petroleum geologist; b. Ag. 3 (16), 1911.
Mem. CPSU 1941–; grad. Kazan Univ. 1933; mem. USSR Acad. of Sciences 1958–; worked in petroleum ind. 1933–53; Dep. Dir. 1953–55, Dir. of All-Union Inst. for Petroleum and Gas Research 1955–57; Dir. of Inst. of Geology and Geophysics at Siberian branch of USSR Acad. of Sciences 1958–. Decors.: Hero of Socialist Labour 1944, State Prize of the USSR 1946, 1950, 2 Orders of Lenin, various medals.

Troian, Aleksandr Ivanovich
State official; b. 1930, Berezova Luka, Poltava Oblast.
Grad. Econo-Financial Inst. in Kiev; worked for financial orgs. and Ukrainian Econ. Council until 1967; dept. head, mem. Collegium of Main Statistical Admin. of Ukrainian SSR 1967–71; Dir. of Ukrainian branch of Union Inst. for Mechanization and Calculation. Work of USSR Cen. Statistical Admin. 1971–77; Head of Main Statistical Admin. Ukrainian SSR Jl. 1977–.

Troianovskii, Oleg Aleksandrovich
Diplomat; b. 1919.
Mem. CPSU 1947–; grad. Pedagogical Inst. of Foreign Languages in Moscow 1956; served in Soviet Army 1941–42; held posts in USSR Min. of Foreign Affairs 1942–53; Aide to USSR Min. of Foreign Affairs 1953–58; Aide to Chair. of USSR Council of Mins. 1958–67; USSR Amb. to Japan Ap. 1967–Ap. 1976; Permanent USSR Representative at UN and World Security Council D. 1976–; mem. CPSU Cen. Auditing Comm. 1981–.

Troitskii, Vsevolod Sergeevich
Astronomer; b. Mr. 12 (25), 1913, Mikhailovskoe, Bogoroditsk Raion, Tula Oblast.
Corresponding member of the Academy of Sciences of the USSR (1970). Member of the CPSU since 1944.
He graduated from the University of Gorkii in 1941 and joined the staff of the Gorkii Radiophysical Institute in 1948. His main works deal with radio astronomy and its applications. Decors.: A. S. Popov Prize of the Academy of Sciences of the USSR for his series of investigations called Radio Emissions of the Moon 1974, Order of the Red Banner of Labour and various medals.

Troshin, Afanasii Semenovich
Cytologist and cytophysiologist; b. D. 10 (23) 1912, Al'za, Chamzinka Raion, Mordovian ASSR.
Corresponding member of the Academy of Sciences of the USSR (1960). Member of the CPSU since 1944.
In 1936, he graduated from the department of biology at Leningrad State University, where he was a student of D. N. Nasonov. His works in the Institute of Cytology of the Academy of Sciences of the USSR. He became head of the institute's laboratory in 1957 and the institute's director in 1958. His main works deal with cell physiology. Decors.: Order of the October Revolution, three other orders, and several medals.
Works: *Problema kletochnoi pronitsaemosti*. Moscow-Leningrad 1956. *Transport sakharov cherez kletochnye membrany*. Leningrad 1973.

Trubachev, Oleg Nikolaevich
Philologist; b. 1930.
Grad. Dnepropetrovsk Univ. 1952; corresponding mem. USSR Acad. of Sciences 1972–; postgraduate student 1953–56, Jun. Scientist at Inst. of Slavic Studies of USSR Acad. of Sciences 1956–61; Sen. Scientist 1961–66, Dep. Dir. of Inst. of Russian Language at USSR Acad. of Sciences 1966–.

Trubitsyn, Evgenii Georgievich
State official.
Dir. of Kiev Railroads 1955–59; Dir. of Kalinin Railroads 1960–63; 1st Sec. for Ind. of Smolensk Oblast CPSU Cttee. 1963–64; Dir. of Privolzhsk Railroads 1964–67; Min. of Motor Transport and Public Highways RSFSR Ap. 1967–69; Min. of Motor Transport RSFSR Jl. 1969–. Decor.: Order of Lenin.

Trukhanovskii, Vladimir Grigorevich
Historian; b. Jl. 15, 1914.
Grad. Pokrovskii Pedagogical Inst. 1940 and Higher Foreign Service School of USSR Min. of Foreign Affairs 1941; mem. Diplomatic Service 1941–53; Head of Chair of Gen. History 1947–, then Head of Chair of History of Int. Relations and USSR Foreign Policy at Moscow Inst. of Int. Relations; Sen. Scientist 1953–57, Dep. Dir. of Inst. of History of USSR Acad. of Sciences 1957–60; Chief Ed. of journal "Voprosy istorii" (Questions of History) 1960–. Decors.: 3 orders.
Works: *Vneshniaia politika Anglii na pervom etape obshchego krizisa kapitalizma* (1918–1939). Moscow 1962. *Vneshniaia politika Anglii posle vtoroi mirovoi voiny*. Moscow 1951. *Noveishaia istoriia Anglii*. Moscow 1958. *Ocherki noveishei istorii Anglii 1917–1963*. Moscow 1964. *Vneshniaia politika Anglii v period vtoroi mirovoi voiny (1939)* Moscow 1965. *Uinston Cherchill': Politicheskaia biografiia*. Moscow 1968. *Leninskim vneshnepoliticheskim kursom*. Moscow 1971. *Antoni Iden*. Moscow 1976.

Trunov, Mikhail Petrovich
Party official; b. 1931.
Russian; mem. CPSU 1955–; grad. Moscow Agricultural Acad. 1954; Chief Agronomist of Kazinskaia Machine and Tractor Station 1954–56; Dir. of Veliko-Mikhailovskaia Machine and Tractor Station 1956–58; Sov. and party posts 1958–64; Sec. of Belgorod Oblast Cttee. CPSU 1964–69; Chair. of Belgorod Oblast Exec. Cttee. of Workers' Deps. 1969–71; 1st Sec. of Belgorod Oblast Cttee. CPSU Mr. 1971–; cand. mem. 1971–76, mem. Cen. Cttee. CPSU 1976–; Dep. to USSR Supr. Sov. 1974–. Decor.: Order of Lenin.

Tsagaraev, Maksim Nikolaevich
Ossetic writer; b. Ap. 22 (My. 5), 1916, Alagir, now the Severnaia Osetiia ASSR. Member of the CPSU since 1949.
He graduated from the Severnaia Osetiia Pedagogical Institute in 1941 and fought in the Great Patriotic War of 1941–45. He was minister of culture of the Severnaia Osetiia ASSR from 1958 to 1960 and editor of the journal *Makh dug* (Our Epoch) from 1960 to 1963. In 1947, he published the novella *On the Road of Happiness* (Russian translation under the title *The Story of the Kolkhoz Carpenter Sago*, 1952) and the book of wartime sketches *In the West*. Since the war his principal subject has been ordinary life in a kolkhoz village; this theme appears in such novellas as *The Shepherd of Black Mountain* (1961), *Campaign Song* (1965), and *When the Rocks Awake* (1970; Russian translation, 1972). He served as chairman of the administrative board of the Writer's Union of the Severnaia Osetiia ASSR from 1954 to 1958 and from 1963 to 1970. Decors.: four orders and various medals.
Works: *Auizbiyi Kästärtä*. Ordzhonikidze, 1973. *Razvärst uatsmistä*, vols. 1–2. Ordzhonikidze, 1975–76. *In Russian translation: Osetinskaia byl'*. Moscow, 1965.

Tsarapkin, Semen Konstantinovich
Diplomat; b. 1906.
Mem. Diplomatic Service 1937–; sen. posts in People's Commissariat of Foreign Affairs resp. USSR Min. of Foreign Affairs 1937–47; mem. Collegium USSR Min. of Foreign Affairs 1946 and 1954–66; Emb. Counsellor-Envoy at USSR Emb. in the USA 1947–49; USSR Dep. Permanent Rep. at UN World Security Council 1949–52 and 1953–54; Head of Dept. of Int. Orgs. at USSR Min. of Foreign Affairs 1954–64; Consultant to USSR Min. of Foreign Affairs 1964–66; USSR Amb. to the Fed. Republic of Germany 1966–71; Special Envoy of USSR Min. of Foreign Affairs 1971–. Decor.: Order of Red Banner of Labour.

Tsarev, Mikhail Ivanovich
Actor; b. 1903.
Mem. CPSU 1949–; grad. School of Russian Drama in Petrograd 1921; with Great Drama Theatre and Vasileostrovskii Theatre in Leningrad until 1924; with Comedy Theatre in Moscow, then theatres in Makhachkala, Kazan and Simferopol 1924–31; with Meierhold Theatre 1933–37, with Malyi Theatre 1937–; theatre dir. 1950–63; teaching work 1935–; Chair. of all-Russian Theatre Soc. 1964–. Decors.: People's Artist of the USSR 1949, State Prize of the USSR 1947 and 1969.

Tsarukaev, Aleksandr Ibragimovich
Author; b. D. 12, 1918, Karman-Sindzikau, now North Ossetian ASSR.
Osset; grad. Ordzhonikidze Pedagogical Inst. 1951 and M. Gorkii Inst. of Literature 1957; author of numerous poetry collections, novelettes and children's books; several translations from Ossetian into Russian.

Publs.: "Kolybol zhizni" (Cradle of Life) 1965, "Mezhdu zhizniu i smertiu" (Between Life and Death) 1972, and others.

Tselikov, Aleksandr Ivanovich
Metal engineering scientist; b. Ap. 7 (20), 1904.
Mem. CPSU 1945–; grad. Bauman Tech. College 1928; specialist in sheet rolling mills; mem. USSR Acad. of Sciences 1964–; design engineer with "Stahlprojekt", "Serp i Molot" Plant and a plant in Izhevsk 1925–35; teacher at various colleges 1935–; mem. staff Union Research Inst. and Inst. for Machine-Construction Design 1945–, Dir. of same inst. 1959–. Decors.: Hero of Socialist Labour 1964, State Prize of the USSR 1947, 1948 and 1951, Lenin Prize 1964.

Tsereteli, Zurab Konstantinovich
Artist; b. Ja. 4, 1934, Tsukulidze.
Honored Artist of the Georgian SSR (1967). Member of the CPSU since 1965.
He studied at the Academy of Arts in Tbilisi from 1952 to 1959. His decorative monumental works include panels, mosaics, stained-glass windows, and whimsically decorative sculpture. Artistically and compositionally integrated with the architecture and surrounding environment, they are distinguished for their richness of color and form and use of diverse materials. Tsereteli's designs for the ornamentation of health resorts in the Georgian SSR embody the unique beauty of the southern landscape. Tsereteli has also worked as a book and magazine illustrator and stage designer.
His principal works include mosaics executed in Tbilisi and Ul'ianovsk and designs for a children's centre in Adler (1973), the motor-coach terminal in Tbilisi (1973), the health resort complex in Titsunda (1967), and the Soviet embassy in Brazil (1975). Decors.: State Prize of the USSR (1970) and the Lenin Prize (1976).

Tsetsevich, Vladimir Platonovich, Dr. of Phy.-Math. Sc.
Astrophysicist; b. O. 11, 1907, Kiev; s. of an actor.
Mem. CPSU 1943–; grad. Leningrad Univ. 1927; corresponding mem. Acad. of Sciences Ukrainian SSR 1948–; Dr. of Phy.-Math. Sc. 1945; Prof. 1937–; mem. staff Astronomical Observatory of Leningrad Univ. 1927–33; Dir. of Tadzhik Astronomical Observatory in Dushanbe 1933–37; Prof. at Pedagogical Inst., Sen. Scientist at Astronomical Inst. of USSR Acad. of Sciences in Leningrad 1937–42; Prof., then Head of Chair. at Technological Inst. of the Canning Ind. in Dushanbe 1942–44; Dir. of Astronomical Observatory and Head of Chair. at Odessa Univ. 1944–; author of approx. 450 scientific and popular-science papers on astronomy.
Publs.: "Peremennye zvezdy i ikh znachenie dlia izucheniia Vselennoi" 1949, "Pidkorennia kosmosu" 1959.

Tsinev, Georgii Karpovich
State official; b. 1907.
Mem. CPSU 1932–; grad. Dnepropetrovsk Metallurgical Inst. 1934 and Mil. Acad. of the Gen. Staff 1953; Army General; Master, Engineer, Acting Sec. of Bureau, Acting Dept. Head at Liebknecht Works in Nizhnedneprovsk 1934–39; party posts in the Ukraine 1939–41; commr. of a regt., of the staff of an operative group, dep. head of the pol. admin. of a front, head of pol. dept. of an army at South-Western, Kalinin, 2nd and 3rd Ukrainian Front 1941–45; Dept. Head, Dep. High Commr., Chief-of-Staff of Sov. Section of Allied Comm. for Austria, Dep. High Commr. in Austria 1945–51; sen. posts in USSR Cttee. for State Security 1953–70; Dep. Chair. of same cttee. Jl. 1970–; mem. Cen. Auditing Comm. 1971–76, cand. mem. Cen. Cttee. CPSU 1976–81, mem. 1981–.

Tsintsadze, Sulkhan Fedorovich
Composer; b. Ag. 23, 1925, Gori.
People's Artist of the Georgian SSR (1961).
He graduated from the Moscow Conservatory, where he studied cello with S. M. Kozolupov in 1950 and composition with S. S. Bogatyrev in 1953. His compositions include the operas *The Golden Fleece* (1952) and *The Hermit* (staged 1972; based on the narrative poem by I. G. Chavchavadze) and the ballets *The Treasure of the Blue Mountain* (staged 1957) and *The Demon* (staged 1961; based on the narrative poem by Iu. M. Lermontov); the last three works received their premieres at the Georgian Theatre of Opera and Ballet.
He composed the ballet *Sketches of Antiquity* (1974) and the operettas *The Spiderweb* (staged 1963) and *Song in the Forest* (staged 1968), both of which received their premieres at the Tbilisi Theatre of Musical Comedy. He has written three symphonies, the second of which he later reworked as the ballet *Poem* (staged 1963 at the Stanislavsky and Nemirovich-Danchenko Moscow Music Theatre). Other notable works include piano and violin concerti; *Immortality* (1970), an oratorio written to honour the centennial of V. I. Lenin's birth; eight string quartets; and music for such motion pictures as *The Dragonfly, The Squabbler, Otar's Widow,* and *The Father of a Soldier.*
In 1965, Tsintsadze became rector of the Tbilisi Conservatory. Decors.: State Prize of the USSR in 1950 and the Z. P. Paliashvili Prize in 1974; two other orders and various medals.

Tsitsin, Nikolai Vasil'evich
Botanist, geneticist, and plant breeder; b. D. 6 (18), 1898, Saratov.
Academician of the Academy of Sciences of the USSR (1939) and of the V. I. Lenin All-Union Academy of Agricultural Sciences (1938; vice-president from 1938 to 1948). Twice Hero of Socialist Labour (1968, 1978). Member of the CPSU since 1938.
He graduated from the Saratov Institute of Agriculture and Land Reclamation in 1927. From 1931 to 1937 head of a laboratory of *Triticum-Agropyron* hybrids, which he had organized. From 1938 to 1948, chairman of the State Varietal Testing Commission of the Ministry of Agriculture of the USSR, and from 1940 to 1957 he headed the laboratory of remote hybridization of the Academy of Sciences of the USSR. In 1945 he became director of the Central Botanical Garden of the Academy of Sciences of the USSR.
His main works are devoted to distant hybridization of plants. He obtained a new species of wheat, *Triticum agropyrotriticum,* by crossing *Triticum* (wheat) with *Agropyron* (wheatgrass). He also developed varieties of *Triticum-Agropyron* hybrids.
Tsitsin is an honorary member of a number of academies in socialist countries. From 1958 to 1970 he was president, and since 1970 he has been vice-president, of the Soviet-Indian Society for Friendship and Cultural Ties. Decors.: State Prize of the USSR in 1943 and the Lenin Prize in 1978, five Orders of Lenin, and Order of the October Revolution, two other orders, and various medals.
Works: *Otdalennaia gibridizatsiia rastenii,* Moscow 1954. *Uspekhi sovetskoi nauki v oblasti otdallenoi gibridizatsii.* Moscow 1957. *Gibridy otdalennykh skreshchivanii i poliploidy.* [collection of articles]. Moscow 1963. *Rol'otdalennoi gibridizatsii v evoliutsii rastenii.* Moscow 1975.

Tsukanov, Georgii Emmanuilovich
Party official; b. 1919.
Russian; mem. CPSU 1941–; grad. Dneprodzerzhinsk Metallurgical Evening Inst. 1941; engineering-technological work in plants and combines of the Min. of Ferrous Metallurgy in Dnepropetrovsk and Cheliabinsk Oblasts 1937–58; mem. staff Cen. Cttee. CPSU 1958–60; mem. staff Presidium USSR Supr. Sov. 1960–63; sen. posts with Cen. Cttee. CPSU 1963–; mem. Cen. Auditing Comm. CPSU 1966–71; Dep. to USSR Supr. Sov. 1966–; Aide to Sec.-Gen. of Cen. Cttee. CPSU 1966–; mem. Cen. Cttee. CPSU 1971–.

Tsvetkov, Viktor Nikolaevich
Physical chemist; b. F. 3 (16), 1910, St. Petersburg.
Corresponding member of the Academy of Sciences of the USSR (1968).
He graduated from the A. I. Herzen Leningrad Pedagogical Institute in 1931. In 1934 he began working at Leningrad State University, where he became head of a subdepartment in 1945 and head of a special problems laboratory in 1958. At the same time, he worked at the Institute of Macromolecular Compounds of the Academy of Sciences of the USSR, where he became head of a laboratory in 1950.

His main works deal with the physics of polymers and with liquid crystals. Decors.: State Prize of the USSR in 1952. Two orders and various medals.
Work: *Struktura makromolekul v rastvorakh.* Moscow 1964. (With V. E. Eskin and S. Ia. Frenkel'.)

Tsvirko, Vitalii Konstantinovich
Painter; b. F. 1, (14), 1913, Radeevo, Gomel' Oblast.
People's Artist of the Belorussian SSR (1963). Member of the CPSU from 1953.
He studied unter G. G. Riazhskii at the Moscow Art Institute from 1935 to 1942. In 1952 he became an instructor at the Belorussian Institute of Theatrical Arts; he served as the institute's rector from 1958 to 1962. He was chairman of the governing board of the Artists' Union of the Belorussian SSR in 1962 and 1963.
His landscapes, executed in a rich picturesque manner, create a lyrical image of Belorussian nature. Examples are *March* (1947, Art Museum of the Belorussian SSR, Minsk), *Lake Naroch'* (1956, Art Museum of the Belorussian SSR), and the series *In the Belorussian Land* (1960's; State Prize of the Belorussian SSR, 1967). Tsvirko also painted a number of works devoted to the Great Patriotic War of 1941–45 and to revolutionary-historical themes. Decors.: Order of the Red Banner of Labor and various medals.

Tsybin, Vladimir Dmitrievich
Author; b. Mr. 11, 1932, Samsonovskaia, Frunze Oblast; s. of a peasant.
Russian; grad. Gorkii Inst. of Literature 1958; first works publ. 1952.
Publs.: "Puls" (The Pulse) 1963, "Au!" 1967, "Glagol" (The Verb) 1970, "Izby" (Cottages) 1971, "Sergei Vasilev", and others.

Tsybulko, Vladimir Mikhailovich
Party official; b. 1924.
Ukrainian; mem. CPSU 1944–; grad. Donetsk Pedagogic Inst. 1950, Party High School of Cen. Cttee. CPSU 1958 and Kiev Inst. of Econ. 1965; served in Sov. Army 1942–45; Sec. of Donetsk Oblast, 1st. Sec. of Donetsk Oblast Cttee. Ukrainian Komsomol 1949–56; party posts 1958–; 1st Sec. of Zhdanov City Cttee. Ukrainian CP 1966–68; cand. mem. 1966–69, mem. Cen. Cttee. Ukrainian CP 1969–; mem. staff Cen. Cttee. Ukrainian CP 1968–70; 1st Sec. of Kiev Oblast Cttee. Ukrainian CP Ap. 1970–; Dep. to USSR Supr. Sov. 1970–; mem. Cen. Cttee. CPSU 1971–; cand. mem. Politbureau Cen. Cttee. Ukrainian CP Jl. 1972–F. 1976.

Tsydendambaev, Chimit-Dorzhi
Buriat writer; b. F. 23, 1918, Tarbagatai, what is now Zaigraevo Raion, Buriat ASSR; s. of a peasant.
He first appeared in print in 1937 as a poet; he later turned to prose. His principal works include the novella *Banzarov's Inkwell* (1948) and the novels *Dorzhi, Son of Banzar* (1953; Russian translation, 1955) and *Far From the Native Steppes* (1957–58; Russian translation, 1962), all of which deal with the life of Dorzhi Banzarov, the first Buriat orientalist. Tsydendambaev's works also include short stories and the novellas *The Buriat Girl* (1961; Russian translation under the title *Bon Voyage, Zhargalma!,* 1963) and *The Year Begins With Spring* (1973). His works have been translated into various languages of the USSR. Decors.: Order of the Red Banner of Labour and various medals.
Works: *Zhelnai khabarhaa ekhildeg: Tuuzha,* Ulan-Ude 1973. *In Russian translation. Stikhotvoreniia.* Moscow 1953. *Buriatskie uzory.* Moscow 1970.

Tsydynzhapov, Gombozhap Tsydynzhapovich
Buriat actor, stage director, and playwright; b. Ap. 20 (My. 3), 1905, in the *ulus* (village) of Liubchikan, now Barguzin Raion, Buriat ASSR.
People's Artist of the USSR (1940). Member of the CPSU since 1939. One of the founders of professional theatre in Buriatia.
The son of peasants published his first works in 1940. In 1934 he graduated from the department of stage direction of the State Institute of Theatrical Arts. From 1934 to 1948 he served as artistic director of the Buriat Drama Theatre (since 1938 the Buriat Music and Drama Theatre). In 1948 he became principal stage director of the Kh. Namsaraev Buriat Theatre of Drama in Ulan-Ude, and from 1948 to 1960 he served as principal stage director of the Buriat Theatre of Opera and Ballet, also in Ulan-Ude.
He is known for his performances in the title role in Shakespeare's *Othello,* as Sharlai in *Erzhen* by Baldano and Edel', and in the title role in Korneichuk's *Platon Krechet.* In 1947 he played the role of V. I. Lenin in Pogodin's *The Kremlin Chimes.*
He has also staged operas, including Tchaikovsky's *Eugene Onegin* (1943), Knipper's *On Lake Baikal* (1948), and Dzerzhinskii's *The Quiet Don* (1953). He wrote the libretto of the musical drama *Bair* (together with A. Shadaev), the plays *The Sniper* (1943) and *Son of the People* (1943), and the screenplay for the film *House of Gold* (together with D. Batozhabai and V. I. Ezhov, 1960). Decors.: State Prize of the USSR (1949), the Order of Lenin, two other orders, and several medals.
Works: *Araday khübüün.* Bayär, Ulan-Ude 1958.

Tsypkin, Iakov Zalmanovich
Specialist in automatic control; b. S. 19, 1919, Dnepropetrovsk.
Corresponding member of the Academy of Sciences of the USSR (1974).
After his graduation from the Moscow Electrical Engineering Institute of Communications in 1941, worked at scientific research institutes. He served in the Soviet Army in 1942 and 1943. In 1950 he became a research worker and a laboratory head at the Institute of Problems of Control. Tsypkin taught at the Moscow Institute of Aviation in 1947 and 1948, at the All-Union Correspondence Electrical Engineering Institute from 1948 to 1965, and at the Moscow Power Engineering Institute from 1965 to 1972. In 1972, he assumed a teaching position at the Moscow Physiotechnical Institute. He was made a professor in 1949.
His principal works deal with the theory of time-delay systems, relay systems, pulse systems, and systems with a digital computer in the control circuit. He has also made important contributions to the study of adaptive and learning systems. Decors.: Lenin Prize in 1960, the Order of Lenin, the Order of the Red Banner of Labour, and various medals.
Works: *Adaptatsiia i obuchenie v avtomaticheskikh sistemakh.* Moscow 1968. *Osnovy teorii obuchaiushchikhsia sistem.* Moscow 1970. *Teoriia nelineinykh impul'snykh sistem.* Moscow 1973. (With Iu. S. Popkov.) *Releinye avtomaticheskie sistemy.* Moscow 1974.

Tsytovich, Nikolai Aleksandrovich
Specialist in soil mechanics, geomechanics, foundation engineering, and engineering geology; b. My. 13 (26), 1900, Mkhinichi, Cherikov Raion, Mogilev Oblast.
Corresponding member of the Academy of Sciences of the USSR (1943).
After graduation from the Leningrad Institute for Civil Engineers (now the Leningrad Construction Engineering Institute) in 1927, he took part in the building of large industrial and hydraulic engineering structures. Over a period beginning in 1930 he taught at a number of higher educational institutions in Leningrad. In 1943 he joined the staff of the Institute of Geocryology of the Academy of Sciences of the USSR. In 1952 he was appointed head of the subdepartment of soil mechanics, beddings, and foundations of the V. V. Kuibyshev Moscow Construction Engineering Institute. From 1947 to 1953, Tsytovich was chairman of the Presidium of the Yakutsk Branch of the Academy of Sciences of the USSR of the International Society for Soil Mechanics and Foundation Engineering.
His main works deal with the development of the scientific principle of the construction of buildings and other structures on permanently frozen ground. His textbook for higher educational institutions *Fundamentals of Soil Mechanics* (1934) has been translated into English, Czech, Hungarian, Polish, Rumanian, Korean, Vietnamese, and several other languages. Decors.: State Prize of the

USSR in 1950, two Orders of Lenin, four other orders, and various medals.
Works: *Osnovaniia i fundamenty*. Moscow 1970. (Co-author.) *Mekhanika merzlykh gruntov (obshchaia i prikladnaia)*. Moscow 1973. *Mekhanika gruntov*, 4th ed. Moscow-Leningrad 1963.

Tuchkevich, Vladimir Maksimovich
Physicist; b. D. 16 (29), 1904, Ianovtsy, Bukowina.
Mem. CPSU 1952–; grad. Kiev Univ. 1928; mem. staff Roentgen Inst. in Kiev 1928–31; Head of Laboratory of Ukrainian Roentgen Inst. 1931–35, of Leningrad Roentgen Inst. 1935–36; concurrently teacher at Kharkov Inst. of Electrotechnology 1931–35; Prof. at Leningrad Polytech. Inst. 1935–60; Head of Laboratory 1936–38, Dir. of Leningrad Ioffe Inst. of Physical Technology of USSR Acad. of Sciences 1968–; mem. USSR Acad. of Sciences 1970–. Decors.: State Prize of the USSR 1942, Lenin Prize 1966, Order of Lenin, Order of Red Banner of Labour, various medals.

Tufan, Khasan Fakhrievich
Tartar poet; b. N. 27 (D. 10), 1900, Karmat, Aksubaevo Raion, Tartar ASSR.
In 1924 and 1925 he studied at the eastern division of the Kazan Pedagogical Institute. He published his first works in 1924. His narrative poems *The Beginning of Beginnings, Between Two Epochs,* and *The Bibievs,* all written in 1927, gave rise to the working-class theme in Tartar poetry.
In the 1930's, he changed from tonic verse to classical Tartar meter and folk rhythms. In the 1940's and early 1950's wrote the optimistic poems "Karakums" (1947), "Solar" (1947), and "Floating Clouds" (1951). Later he primarily wrote philosophical lyric poetry, for example, the poems "O Life!" (1963), "Where Are You Going, Ursa Major?" (1965), and "You Are Old as Eternity" (1966). Decors.: G. Tukai Republican Prize in 1965. Two orders and various medals.
Works: *Saylanma äsärlär*. Kazan 1964. *Davilli ellar lirikasi*. Kazan 1970. *In Russian translation: Stikhotvoreniia*. [Afterword by L. Martynov.] Moscow 1958. *Stikhotvoreniia*. Moscow 1970.

Tulchinskaia, Vera Petrovna, Dr. of Vet. Sc.
Microbiologist; b. S. 13, 1907, Kazan; d. of a clerk.
Mem. CPSU 1942–; grad. Kazan Veterinary Inst. 1929; corresponding mem. Acad. of Sciences Ukrainian SSR 1948–; Dr. of Vet. Sc. 1941; Prof. 1941–; mem. staff Bacteriological Laboratory in Semipalatinsk 1930–32; mem. staff Inst. for Research on Veterinary Medicine of People's Commissariat of Agriculture of Tartar ASSR in Kazan 1932–36; Head of Chair of Microbiology at Kirghiz Inst. of Agriculture in Frunze 1936–45; Head of Chair of Microbiology at Agricultural Inst. in Belaia Tserkov 1945–47; in Odessa 1947–52 and at Odessa Univ. 1952–; author of approx. 110 scientific papers. Decor.: Order "Badge of Honour".
Publs.: "Antibiotiki" 1953, "Mikroflora morskikh vodoroslei i gubok kak istochnik biologicheski aktivnykh veshchestv (antibiotikov, vitaminov, fermentov i aminokislot)" in "IX. Mezhdunarodnyi kongress po mikrobiologii. Tezisy dokladov" 1966, and others.

Tulegenova, Bibigul' Akhmetovna
Coloratura soprano; b. D. 16, 1929, Semipalatinsk.
People's Artist of the USSR (1967).
She graduated from the conservatory in Alma-Ata in 1954, was a soloist with the Kazakh Theatre of Opera and Ballet from 1954 to 1956 and with the Kazakh Philharmonic Society from 1956 to 1971, at which time she returned to the Kazakh Theatre of Opera and Ballet, title role in Rimsky-Korsakov's *The Snow Maiden,* Violetta and Gilda in Verdi's *La Traviata* and *Rigoletto,* Rosina in Rossini's *The Barber of Seville,* Norina in Donizetti's *Don Pasquale,* the title role in Brusilovskii's *Kyz-Zhibek,* Gul'barshin in Rakhmadiev's *Alpamys,* and the title role in Zhubanova's *Enlik-Kebek*. She performs in concerts at home and has toured abroad. Decors.: Kuliash Baiseitova State Prize of the Kazakh SSR (1966), the State Prize of the USSR (1970), Order of Lenin, the Order of the Red Banner of Labour, and various medals.

Tulepbaev, B.
Historian.
Corresponding mem. USSR Acad. of Sciences 1981–. Vice-pres. Acad. of Sciences Kazakh SSR; works on problems of the history of agrarian reforms in Kazakhstan and the Central Asian republics; author of more than 70 publications.

Tuliaganov, Khabibulla Tuliaganovich
State official.
Head of Main Admin. for Geology and Min. Resources with Council of Mins. Uzbek SSR 1957–63; mem. Auditing Comm. 1960–71, mem. Cen. Cttee. Uzbek CP 1971–; Chair. of State Prod. Cttee. for Geology Uzbek SSR 1963–64; Min. of Geology Uzbek SSR O. 1965–.

Tulinov, Anatolii Filippovich, Dr. of Phy.-Math. Sc.
Physicist; b. 1924.
Mem. CPSU 1947–; grad. Moscow Univ. 1951; Dr. of Phy.-Math. Sc. 1967; Prof. 1969–; served in Sov. Army 1943–46; postgraduate student, Sen. Engineer, Sen. Scientist, Lecturer, Prof. at Moscow Univ. 1952–. Decor.: State Prize of the USSR 1972.

Tumanian, Sarkis Amaiakovich
State official.
Min. of Public Services Armenian SSR Ap. 1974–; mem. Auditing Comm. Armenian CP 1976–.

Tumanian, Stepan Levonovich
State official; b. 1921.
Dep. to Supr. Sov. Armenian SSR 1955–; Dir. of Kirov Works in Erevan 1955; cand. mem. 1956–60, mem. Cen. Cttee. Armenian CP 1960–; 1st Dep. Chair. of Econ. Council Armenian SSR 1959; Sec. and mem. Bureau Cen. Cttee. Armenian CP 1960–61; again Dir. of Kirov, Works in Erevan 1967–1973; Dep. Chair. of Council of Mins. and Chair. of GOSPLAN Armenian SSR F. 1973–Jl. 1977; Dep. Chair. USSR GOSPLAN 1977–79; Dep. Chair. Council of Mins. Armenian SSR O. 1979–.

Tumovs-Bekis, Iazep Donatovich
State official.
Mem. Cen. Cttee. Latvian CP 1971–; Min. of Public Services Mr. 1976–.

Tunkin, Grigorii Ivanovich
Association official; b. 1906.
Mem. Diplomatic Service 1939–; Emb. Counsellor at USSR Emb. in Canada 1942–44; Dep. Head of 2nd Europ. Dept. 1944–45 and Head of 1st Far East Dept. of USSR Min. of Foreign Affairs 1945–48; Emb. Counsellor at USSR Emb. in the People's Republic of Korea 1948–51; Head of 1st Far East Dept. 1951–52 and Head of Treaties and Legal Dept. of USSR Min. of Foreign Affairs 1952–65; Chair. of Sov. Asscn. of Int. Law 1957–; Head of Chair at Moscow Univ. 1965–.

Tupolev, Aleksei Andreevich, Dr. of Tech. Sc.
Airplane designer; b. My. 20, 1925, Moscow; s. of A. N. Tupolev.
Mem. CPSU 1959–; grad. Moscow Aviation Inst. 1949; Dr. of Tech. Sc. 1963; Prof. 1964; various supersonic jets were developed under his direction (among them the Tu-144, developed with his father in 1968). Decors.: USSR State Prize 1967, 3 Orders of Lenin, 2 other orders, medals.

Turaev, Aziz Appakovich
Party official.
Head of Cultural Dept. Cen. Cttee. Uzbek CP 1974–; mem. Cen. Cttee. Uzbek CP 1976–; Dep. to Supr. Sov. Uzbek SSR 1980–.

Turbin, Nikolai Vasil'evich
Geneticist and breeder; b. N. 20 (D. 3), 1912, Tuma. Klepiki Raion, Riazan' Oblast.
Academician of the Academy of Sciences of the Belorussian SSR (1953) and the V. I. Lenin All-Union Academy of Agricultural Sciences (1967). Member of the CPSU since 1950.
He graduated from Voronezh Agricultural Institute in 1935. Between 1945 and 1953 he was a professor at Leningrad State University; during this period, from 1948 to 1952, he was also director of the Biological Institute of Leningrad State

University. From 1953 to 1965 he was director of the Institute of Biology of the Academy of Sciences of the Belorussian SSR, and from 1965 to 1971, director of the Academy's Institute of Genetics and Cytology. In 1968 he became academician-secretary of the department of plant growing and breeding of the V. I. Lenin All-Union Academy of Agricultural Sciences. In 1974 he was appointed director of the All-Union Scientific Research Institute of Applied Molecular Biology and Genetics of the V. I. Lenin All-Union Academy of Agricultural Sciences.
His main research is concerned with the principles of plant breeding, for example, polyploidy of the sugar beet, the heterosis phenomenon, and the genetics of cytoplastic male sterility. He demonstrated the possibility of automixis in wheat in interspecific crossings (1968) and investigated the genetics of alkaloidness in lupines. He was president of the All-Union Society of Genetics and Selectionists from 1972 to 1976. Decors.: two orders and a medal.
Works: *Eksperimental'naia poliploidiia i geterozis u sakharnoi svekly.* Minsk 1972. (With V. E. Bormotov.) *Diallel'nyi analiz v selektsii rastenii.* Minsk 1974. (With L. V. Khotyleva and L. A. Tarutina.)

Turchin, Valentin, Dr. of Phy.-Math. Sc.
Mathematician; b. 1931.
Chair. of USSR Branch of Amnesty International.

Turgunbaev, Oirot
State official.
Min. of Health Kirghiz SSR F. 1979–; mem. Auditing Comm. Kirghiz CP 1981–.

Turian, Vladimir Agbanovich
Komsomol official.
1st Sec. Erevan City Cttee. Armenian Komsomol until 1975; mem. Bureau 1975–, 1st Sec. Cen. Cttee. Armenian Komsomol O. 1978–; mem. Cen. Cttee. and cand. mem. Bureau Cen. Cttee. Armenian CP 1981–.

Turishcheva, Liudmila Ivanovna
Athlete; b. O. 7, 1952, Groznyi.
Honored Master of Sport (1970).
She was all-around gymnastics champion of the USSR (1972, 1974), Europe (1971, 1973), the world (1970, 1974), and the Olympics (1972), champion of the USSR six times between 1970 and 1974 and European champion six times between 1971 and 1973. She was also world champion three times between 1970 and 1974 and Olympic champion in separate competitions of combined events in 1972. She won the World Cup in 1975 and the USSR Cup five times between 1967 and 1974 in combined events. As of S. 1, 1976, she had won 39 medals in official international competitions, including 24 gold medals, three of which were for team championships at the Olympics in 1968, 1972, and 1976. Decors.: Order of Lenin and the Order of the Red Banner of Labour.

Turkov, Andrei Mikhailovich
Literary historian and critic; b. Ag. 28, 1924, Mytishchi, Moscow Oblast.
Grad. Gorkii Inst. of Literature 1950. Publs.: "Aleksandr Tvardovskii" 1960, 2nd ed. 1970, "Nikolai Zabolotskii" 1966, "Saltykov-Shchedrin" 1964, "Po dorogram i stranitsam" (Over Roads and Pages) 1968, and others.

Turpaev, Tigran Melkumovich, Dr. of Biol. Sc.
Physiologist; b. O. 20, 1918, Astrakhan.
Grad. Moscow Univ. 1941; Dr. of Biol. Sc. 1962; Prof. 1972–; corresponding mem. USSR Acad. of Sciences 1972–; served in Sov. Army 1941–46; postgraduate studies 1946–48, Jun. Scientist 1948–55, Sen. Scientist 1955–61, Head of Laboratory at Inst. of Animal Morphology of USSR Acad. of Sciences 1961–67; Head of Laboratory and Dep. Dir. 1967–, Dir. of Inst. of Biology of Evolution at USSR Acad. of Sciences 1974–. Decors.: 2 orders, medals.

Tursunov, Askar
State official.
Min. of Educ. until 1979, Min. of Higher and Secondary Specialist Educ. Kirghiz SSR F. 1979–; cand. mem. Cen. Cttee. Kirghiz CP 1971–.

Tursunov, Minovar Tursunovich
State official; b. 1915.
Mem. Cen. Cttee. Uzbek CP 1956–; Chair. of Comm. for Sov. Control resp. Comm. for State Control with Council of Mins. Uzbek SSR 1959–62; Dep. Chair. of Cttee. on Party and State Control of Cen. Cttee. Uzbek CP and of Council of Mins. Uzbek SSR 1963–65; Chair. of Cttee. on People's Control Uzbek SSR 1965–69; cand. mem. Bureau Cen. Cttee. Uzbek CP 1966–71; Dep. Chair. of Council of Mins. My. 1969–, and Min. of Foreign Affairs Uzbek SSR My. 1969–80.

Tursunov, Sabirdin Tursunovich
State official.
Min. of the Meat and Dairy Ind. Kirghiz SSR 1965–70 and Mr. 1980–; cand. mem. Cen. Cttee. Kirghiz CP 1966–71, mem. 1971–; Permanent Rep. of the Kirghiz Council of Mins. at USSR Govt. 1971–76; Min. of Agriculture Kirghiz SSR My. 1976–80.

Turukalo, Nadezhda Prokop'evna
Underwater swimmer and teacher; b. Ap. 16, 1947, Ust'-Kamenogorsk.
International-class Master of Sport (1971).
She was on several occasions a champion and record holder of the USSR, Europe, and the world, won 53 medals, including 39 gold medals. Decor.: Order of the Badge of Honour.

Turysov, Karatai
Trade union official.
Sec. of Turgaiskii Oblast Cttee. Kazakh CP 1976–79; Chair. of State Cttee. for Supervision of Safety in Ind. and Mining Kazakh SSR 1980–81; Chair. of Kazakh Republican Trade Union Council D. 1981–; cand. mem. Cen. Cttee. Kazakh CP 1981–.

Tuzlov, Mikhail Ivanovich
State official.
Chair. of Komparskii Raion Exec. Cttee. of People's Deps, 1st Raion Sec. Moldavian CP in Soroki until 1977; cand. mem. Cen. Cttee. Moldavian CP 1960–66, mem. 1966–; Min. of Supplies Moldavian SSR Ja. 1977–.

Tverdokhlebov, Andrei
Scientist; b. 1940, Moscow; s. of former Dep. Min. of Culture Nikolai Tverdokhlebov.
Studied physics at Moscow State Univ.; postgraduate work at Dubno Inst. of Nuclear Research; publ. articles on elemental particles and electro-dynamics, worked for some time as an ed. of "Abstracts of Theoretical Physics" publ. by All-Union Inst. of Scientific and Tech. Information; because of his civil-rights activities his appointment with the All-Union Inst. was terminated F. 14, 1972; then worked at an experimental laboratory for concrete, working on problems of mechanical vibration; active in the Civil Rights Movt. in the USSR for the past 6 yrs., founding mem. of Moscow Human Rights Cttee.; with four others he founded Group 73 (an asscn. devoted to assisting pol. prisoners and their families) S. 1, 1973; Sec. of the first Amnesty Int. Group in the USSR which was officially recognized by Amnesty Int. in S. 1974; has defended in letters and appeals to USSR govt. and int. bodies many Sov. citizens prosecuted for their pol. convictions.

Tynspoeg, Gustav Augustovich
State official.
Mem. Auditing Comm. 1960–64, cand. mem. 1964–66, mem. Cen. Cttee. Esthonian CP 1966–; Min. of Meat and Dairy Ind. Esthonian SSR 1965–71; 1st Vice-Chair. of GOSPLAN of Esthonian SSR 1971–73; Vice-Chair. of Council of Mins. and Chair. of GOSPLAN of Esthonian SSR Mr. 1973–.

Tynybaev, Abubakir Alievich
Party official.
Min. of Melioration and Water Econ. Kazakh SSR F. 1974–81; cand. mem. Cen. Cttee. Kazakh CP 1976–; 1st Sec. of Taldy Kurgan Oblast Cttee. Kazakh CP Jl. 1981–.

Tyshler, Aleksandr Grigor'evich
Stage designer, painter, and graphic artist; b. Jl. 14 (26), 1898, Melitopol'.
Honored Art Worker of the Uzbek SSR (1943).

He studied at the Kiev Art School from 1912 to 1917 and became a member of the Society of Painters in 1925. Many of his works are noted for their whimsical fantasy, metaphorical imagery, and decorative beauty. As a stage designer, Tyshler often makes use of easily interchangeable sets based on the popular street theatre. Some of his other works, primarily his easel paintings, are characterized by traits of mystical symbolism and an expressionistic, highly subjective perception of reality. Tyshler's theatrical designs have included Shakespeare's *King Lear* (1935), Z. Schneer's *Freilechs* (1945; State Prize of the USSR, 1946), both of which were staged in the Moscow Jewish Theatre, and the series *The Makhno Movement* (1922–65) and *Fascism* (1966–68). Decor.: Order of the Badge of Honour.

Ubiriia, Aleksandr Evgenevich
State official; b. 1932.
Georgian; mem. CPSU 1952–; grad. Georgian Polytechnical Inst. 1957, Cand. of Tech. Sc., metallurgical engineer; asst. foreman, foreman, senior foreman in a tube-rolling shop at Rustavi Metallurgical Plant 1957–59; sec. of party organization of tube-rolling shop at Rustavi Metallurgical Plant 1959–70, chief 1970–74; Chair. of Rustavi City Exec. Cttee. of Sov. of Peoples' Deps. 1974–81; Chair. of State Cttee. for Supply of Petroleum Products Georgian SSR O. 1981–. Decors.: Badge of Honour, medals.

Uchaneishvili, Tengiz Yasonovich, Cand. of Econ. Sc.
State official; b. 1942.
Georgian; mem. CPSU 1969–; grad. Leningrad Tech. Inst. of the Cellulose and Paper Ind. 1965; engineer, chief engineer of a special planning, construction and technological office of Min. of Timber, Paper and Woodworking Ind. Georgian SSR 1965–59; asst. in a dept. of the Leningrad Tech. Inst. of the Cellulose and Paper Ind. 1969–73; lecturer at Tbilisi State Univ., research and teaching 1973–74; dep. head of Dept. for Ind. and Transport of Tbilisi City Cttee. CP Georgia 1974–75; dep. head of Dept. for Ind. and Transport Cen. Cttee. CP Georgia 1975–78; 1st Dep. Min. of the Timber and Woodworking Ind. Georgian SSR 1978–79, Min. Ag. 1979–; cand. mem. Cen. Cttee. CP Georgia 1981–.

Uchitel', Efim Iul'evich
Cameraman and motion-picture director who specializes in documentaries; b. O. 15 (28), 1913, Tiraspol'.
People's Artist of the USSR (1976). Member of the CPSU since 1940.
He studied in the department of camera operation at the Leningrad Institute of Motion-picture Engineers from 1930 to 1935. He was among the cameramen who shot the film *The Mannerheim Line* (1940). During the Great Patriotic War (1941–45) he was a cameraman at the front and helped shoot and direct several films, including *Struggling Leningrad* (1942) *Breaking the Blockade of Leningrad* (1944), *Berlin* (1945), and *Victory March* (1945).
More recently, Uchitel' has directed *The Russian Character* (1958), *Daughters of Russia* (1960), *Peace on Your House* (1961), *Workers' Stories* (1965), *Leningrad the Hero-City* (1974), and *Fidelity* (1974). The films *Struggling Leningrad* and *The Klooga Death Camp* (1944) were shown as evidence at the Nuremberg Trials.
In 1964, he became artistic director of the first creative association at the Leningrad Documentary Film Studio. Decors.: the State Prize of the USSR (1943), and the Vasil'ev Brothers State Prize of the RSFSR (1967), the Order of Lenin, two other orders, and various medals.

Udalaia, Raisa Silantevna
Laborer; b. 1931.
Party mem. CPSU 1974–, attended without completing secondary school to 1949; collective farmer 1949–50; riveter in a plant in Novosibirsk 1952–; mem. Cen. Auditing Comm. 1981–. Decor.: Heroine of Socialist Labour 1976.

Udumian, Kamo Babievich, Cand. of Hist. Sc.
Diplomat; b. 1927.
Lecturer; cand. mem. 1960–61 and 1964–66, mem. Cen. Cttee. Armenian CP 1961–64 and 1966–76; Min. of Culture 1967–72, Min. of Foreign Affairs Armenian SSR 1972–75; USSR Amb. to Nepal My. 1975–79, to Luxemburg S. 1979–.

Uglov, Fedor Grigorevich
Surgeon; b. S. 22 (O. 5), 1904, Chuguevo, Irkutsk Oblast.
Academician of the Academy of Medical Sciences of the USSR (1967). Member of the CPSU since 1931.
In 1929, he graduated from the medical department of the University of Saratov, where he studied under N. N. Petrov. In 1950 he became head of the subdepartment of hospital surgery of the First Leningrad Medical Institute and, while holding that position, also served as director of the All-Union Scientific Research Institute of Pulmonology from 1967 to 1972.
His work has been concerned primarily with esophageal surgery, portal hypertension, hypothermia in thoracic surgery, and related problems. In 1953, he was one of the first in the USSR to devise surgical methods for treating heart disease, developing several surgical techniques and instruments. In 1961 he was awarded a Lenin Prize for developing surgical methods for the treatment of lung diseases. He has been an editor of the journal *Vestnik khirurgii im. I. I. Grekov.* (I. I. Grekov Surgical Herald) since 1953. Decors.: two orders and various medals.
Works: *Rezektsiia legkikh,* 2nd ed. Leningrad 1954. *Rak legkogo,* 2nd ed. Leningrad 1962. *Kateterizatsiia serdtsa i selektivnaia angiokardiografiia.* Leningrad, 1974. (Co-author.) *Serdtse khirurga.* Moscow 1974.

Ugolev, Aleksandr Mikhailovich
Physiologist; b. Mr. 9, 1925, Dnepropetrovsk.
Corresponding member of the Academy of Sciences of the USSR (1966).
He graduated from the Dnepropetrovsk Medical Institute in 1948. He worked in Leningrad at the Stomatological Institute from 1949 to 1955 and in Moscow at the Institute of Normal and Pathological Physiology of the Academy of Medical Sciences of the USSR from 1955 to 1960. In 1960 he joined the I. P. Pavlov Institute of Physiology of the Academy of Sciences of the USSR, where he became director of the laboratory of nutrition physiology in 1963.
His work deals mainly with the physiology of digestion and nutrition. Ugolev discovered membrane-mediated (parietal) digestion (1959) and developed the theory of the threefold activity of the digestive system: cavitary digestion, membrane-mediated digestion, and absorption. He also formulated the excretory theory of the origin of external and internal secretions, the theory of the digestive-transport conveyer, and the metabolic theory of appetite regulation. Ugolev was the first to remove an entire duodenum under experimental conditions, a procedure that led to the discovery of the nondigestive hormonal function of the intestine. Decors.: Pavlov Prize of the Academy of Sciences of the USSR in 1963, Order of the Red Banner of Labour.
Works: *Pishchevarenie i ego prisposobitel'naia evoliutsiia.* Moscow 1961. *Pristenochnoe (kontaktnoe) pishchevarenie.* Moscow-Leningrad 1963. *Fiziologiia i patologiia pristenochnogo (kontaktnogo) pishchevareniia.* Leningrad 1967. *Membrannoe pishchevarenie: Polisubstratnye protsessy, organizm i reguliatsiia.* Leningrad 1972. "Fiziologiia membrannogo (pristenochnogo) pishchevareniia. (Co-author.) In *Fiziologiia pishchevarennia.* Leningrad 1974.

Uigun (real name: Atakuziev, Rakhmatulla)
Author and dramatist; b. My. 1 (14), 1905, Merke, Dzhambul Oblast.
Uzbek; mem. CPSU 1944–; grad. Samarkand Pedagogical Acad. 1930. Decors.: Merited Artist of the Uzbek SSR 1956, People's Poet of the Uzbek SSR 1965.
Publs.: "Radosti vesny" (The Joys of Spring) 1929, "Vetry Ukrainy" (The Winds of the Ukraine) 1935, "Mat'" (The Mother) 1942, "Riskovannaia shutka" (A Risky Joke) 1944, "Khurriiat" 1959, "Druzia" (Friends) 1961, "Dve epokhi" (Two Epochs) 1967 (Russian translation 1968), translated works by Pushkin, Chekhov, Tolstoi, Shevchenko and others into Uzbek.

Ukachin, Boris Ukachinovich
Author; b. F. 18, 1936, Kairylyk, Gorno-Altai Autonomous Krai.
Altai writer; grad. M. Gorkii Inst. of Literature 1967; has written several lyrical and prose works in Altai language (some of them were translated into other languages of the various Sov. republics); translation from Russian and Ukrainian. Publs.: "Est' takaia zemlia" (There Is Such a Country) 1962, "Vetka gornogo kedra" (Branch of a Mountain Cedar) 1966, "Gornye dukhi" (Mountain Spirits) 1971, and others.

Ukhsai, Iakov Gavrilovich
Chuvash poet; b. N. 13 (26), 1911, Slakbash, now Belebei Raion, Bashkir ASSR.
People's Poet of Chuvashia (1950). Member of the Communist Party since 1943.
He attended Moscow State University from 1930 to 1933 and served in the Great Patriotic War (1941–45). He first published his works in 1928. He is the author of the poetry anthologies *Verses* (1934), *Book of Joy* (1936), and *A Rural Chronicle* (1937) and the narrative poems *The Golden Book of Peoples* (1937), *A Tale of Happiness* (1940), *The Tale of a Regiment* (1943–47), and *On the Bank of the Vltava* (1945–47). Ukhsai has also written the lyrical epic poems *Grandfather Kel'buk* (published 1951; Russian translation, 1954), *The Mountain Pass* (1952), and *Earth* (1957–60). His trilogy of poems *The Star of My Childhood* was awarded the M. Gorkii State Prize of the RSFSR in 1972. Decors.: the Order of the October Revolution, three other orders, and various medals.
Works: *In Russian translation: Sel'biiskii rodnik: Poemy.* Moscow 1970. *Izbrannoe.* Moscow 1973.

Ukraintsev, Fedor Il'ich
Physicist; b. 1917.
Mem. CPSU 1946–; grad. Ordzhonikidze Electromechanical Tech. College 1937 and Leningrad Electrotech. Inst. 1949; served in Sov. Army 1941–46; mem. staff Inst. of Physical Energetics 1949–; Head of Laboratory 1956–70, Chief Engineer of a dept. at same inst. 1970–. Decor.: State Prize of the USSR 1971.

Ukueva, B.
Komsomol official.
Sec. and mem. Bureau Cen. Cttee. Kirghiz Komsomol Mr. 1978–.

Ulanov, Anatolii Andreevich
Diplomat.
Mem. CPSU 1950–; 1st Sec. of Dnepropetrovsk City Cttee. Ukrainian CP until 1970; Head of Dept. of Org-Party Wort with Cen. Cttee. Ukrainian CP 1970–72; mem. Cen. Cttee Ukrainian CP 1971–76; Sec. of Voroshilovgrad Oblast Cttee. Ukrainian CP 1972–76; USSR Amb. to Liberia N. 1977–.

Ulanov, Georgii Mikhailovich, Dr. of Tech. Sc.
Cybernetics scientist; b. 1917.
Grad. Bauman Higher Tech. College in Moscow 1941; specialist in automatic control systems; Dr. of Tech. Sc. 1958; Prof. 1961–; engineer in an experimental design bureau 1940–45; mem. staff Inst. of Control Problems (Automatics and Telemechanics) 1945–, Head of Laboratory of same inst. 1962–; concurrently Prof. at Bauman Higher Tech. College in Moscow. Decor.: State Prize of the USSR 1972 (for his contribution to engineering monograph series "Tekhnicheskaia kibernetika. Teoriia avtomaticheskogo regulirovaniia" in 4 vols., 1967–69).

Ulanova, Galina Sergeevna
Ballerina; b. D. 26, 1909 (Ja. 8, 1910); St. Petersburg.
People's Artist of the USSR (1951). Hero of Socialist Labour (1974).
The daughter of dancers, Ulanova graduated in 1928 from the Leningrad Choreographic School, where she studied with her mother, M. F. Romanova, and with A. Ia. Vaganova. That same year she was accepted into the ballet troupe of the Leningrad Theatre of Opera and Ballet (later renamed after S. M. Kirov). From 1944 to 1960 she was a soloist in the ballet troupe of the Bolshoi Theatre in Moscow. The characters Ulanova portrayed possessed a unique fragility, vulnerability, and femininity (for example, the title role in Adam's *Giselle*), yet, at the same time, were endowed with unyielding strength and heroic spirit (for example, Maria in Asaf'ev's *The Fountain of Bakhchisarai* and Juliet in Prokofiev's *Romeo and Juliet*). Ulanova was the leading exponent of the roles in P. I. Tchaikovsky's ballets, namely, Odette-Odile in *Swan Lake,* Aurora in *The Sleeping Beauty,* and Masha in *The Nutcracker.* Her dancing was flawless and harmonious, based on an ideal combination of the real and the conventional. It was characterized by a rare harmony of all means of expression and choreographic elements. Ulanova was able to transcend balletic conventions and express the truth of human emotion on the stage. Her dancing developed the principles and traditions of the Russian school of choreography. Among her other roles were the title roles in Glazunov's *Raymonda* and Prokofiev's *Cinderella,* Parasha in Glière's *The Bronze Horseman,* the waltz, nocturne, and mazurka in *Chopiniana* (also known as *Les Sylphides*), to music by Chopin, and the dance solo *The Dying Swan,* to music by Saint-Saëns. Ulanova, a consummate actress who embodied high tragedy on the ballet stage, created sweeping characterizations, conveying the most complex dramatic conflicts through dance. These qualities won her world acclaim.
Now retired from the stage, Ulanova works with young dancers. Her students include N. V. Timofeeva, E. S. Maksimova (Maximova), S. D. Adyrkhaeva, and L. I. Semenyaka. Over the years, she has toured extensively abroad. Decors.: Lenin Prize in 1957 and the State Prize of the USSR in 1941, 1946, 1947, and 1950, two Orders of Lenin, four other orders, and various medals.

Ulianov, Mikhail Aleksandrovich
Actor; b. 1927.
Mem. CPSU 1951–; actor, grad. B. V. Shchukin Theater School in Moscow 1950; studied at studio of Omsk Oblast Drama Theater 1945–46; at Vakhtangov Theater 1950–; teacher at Shchukin Theatre School 1954–; mem. Cen. Auditing Comm. CPSU 1976–. Decors.: Peoples' Artist of the RSFSR 1965 and USSR 1969, Lenin Prize 1966.

Ullas, Nikolai Nikolaevich
Architect and city planner; b. Ja. 14 (27), 1914, Smolensk.
People's Architect of the USSR (1975). Member of the CPSU from 1942.
He studied at the Moscow Architectural Institute under N. A. Ladovskii and N. Ia. Kolli from 1932 to 1936. During that same period he worked in the studio of I. A. Fomin. In 1951, Ullas joined the faculty of the institute; he was made a professor there in 1967.
Many of Ullas' major works are in Moscow. He, in collaboration with others, was responsible for the general plan for the southwestern section of Moscow (1948–54), the plan and construction of Frunze Esplanade and Komsomol'skii Prospect (1948–56), the reconstruction of the Moscow State University area (1950–51) and of the V. I. Lenin Central Stadium (1955–56; Lenin Prize, 1959), and the plan and construction of the Matveevskoe residential district (construction begun in 1966). Ullas also participated in drawing up the new master plan for Moscow (for the period 1966–85) and helped design a bridge for the Moscow Canal on the Leningrad Highway (1968–72). Decors.: five orders and various medals.

Ulmanis, Ivar Nikolaevich
State official.
Dep. Chair. of Riga City Exec. Cttee. of People's Deps. 1969–74 and Mr. 1982–; Min. of Construction Latvian SSR N. 1974–81.

Ulmasov, R.
Komsomol official.
Sec. and mem. Bureau Cen. Cttee. Tadzhik Komsomol Je. 1980–.

Umakhanov, Magomed-Salam Iliasovich
Party official; b. 1918.
Mem. CPSU 1939–; grad. Party High School of Cen. Cttee. CPSU 1952; served in Sov. Army 1939–48; dept. head with a raion CPSU cttee. 1948–52; chair. of exec. cttee. of a reg. sov., Min. of the Light and Food Ind., then Min. of the Consum-

er Goods Ind. Dagestan ASSR, sec. of a raion CPSU cttee. 1952–56; Chair. of Council of Mins. Dagestan ASSR 1956–67; Dep. to USSR Supr. Sov. 1954–; 1st Oblast Sec. of CPSU in Dagestan N. 1967–; mem. Cen. Cttee. CPSU 1971–. Decors.: 2 Orders of Lenin.

Umanets, Nikolai Vasil'evich
Party official; b. 1925.
Ukrainian; mem. CPSU 1953–; grad. Odessa Agricultural Inst.; served in Sov. Army 1944–45; agronomist, chair. of a kolkhoz 1950–53; Chair. of Novoselitsa Exec. Cttee. of Raion Sov. of Workers' Deps. 1953–58; dept. head with Chernovtsy Oblast Cttee. CPSU 1958–61; Inspector for Cen. Cttee. Ukrainian CP 1961; Sec., 2nd Sec. of Zhitomir Oblast Cttee. Ukrainian CP 1961–69 (for Agriculture 1963–64); Dep. Min. of Agriculture Ukrainian SSR 1969–70; 1st Sec. of Chernigov Oblast Cttee. Ukrainian CP Ap. 1970–; Dep. to and mem. Planning and Budget Comm. Sov. of Nationalities USSR Supr. Sov. 1970–; mem. Cen. Cttee. Ukrainian CP 1971–, mem. Cen. Auditing Comm. CPSU 1981–.

Umarov, Abdulla Khakimovich
State official.
Min. of the Furniture and Wood-Processing Ind. Uzbek SSR Je. 1977–.

Umarov, Khamdam
Party official; b. 1928.
Uzbek; mem. CPSU 1956–; grad. Gubkin Inst. of the Petrochemical and Gas Ind. in Moscow; held various posts at a petroleum enterprise 1962–1958; dir. of a petroleum processing plant, head administrator of the gas and petroleum ind. Uzbek Economic Council 1958–1962; Oblast Cttee Sec. CP Uzbekistan, Chair. of Exec. Cttee. of Sov. of Peoples' Deputies, manager of the Council of Mins. Uzbek SSR 1962–1978; mem. Cen. Cttee. CP Uzbekistan 1971–; First Sec. Oblast Cttee. CP Uzbekistan in Fergana 1978–.

Umarov, Tursun-Murat
Party official.
Mem. Auditing Comm. Uzbek CP 1960–61, cand. mem. Cen. Cttee. Uzbek CP 1961–66, mem. 1971–; Party posts until 1971; Administrator of Affairs Cen. Cttee. Uzbek CP 1971–.

Umarov, Uchkun
Party official; b. 1934.
Uzbek; mem. CPSU 1963–; grad. Moscow Inst. of Aviation; design engineer 1953–59; held various engineering posts in USSR Min. of the Electronics Ind. 1959–71; dir. of a plant 1971–78; head of a dept. of the Cen. Cttee. CP Uzbekistan 1978; First City Sec. CP Uzbekistan in Tashkent D. 1978–; mem. Cen. Cttee. 1981–, cand. mem. Bureau of Cen. Cttee. CP Uzbekistan F. 1981–.

Urbanovich, Galina Napoleonovna
Gymnast; b. S. 5, 1917, Baku.
Honored Master of Sport (1946). Honored Coach of the USSR (1963), international-class judge (1966).
She was an Olympic champion in a team championship (1952), seven-time all-around champion of the USSR (1943–48, 1950), and on several occasions champion of the USSR in individual competitions in combined events (more than 20 times between 1939 and 1952). Decor.: the Order of the Red Banner of Labour.

Upsi, Arthur-Bernhard Iuhannesovich
Party official; b. 1931.
Esthonian; mem. CPSU 1956–; grad. Party Higher School at Cen. Cttee. CPSU 1956; Sec., 1st Sec. of Rakvere Raion Cttee. Esthonian Komsomol, Dep. Chair. of Rakvere Raion Exec. Cttee. of Peoples Deps. 1950–59; 2nd Sec. of Rakvere Raion Cttee. Esthonian CP 1959–65; Inspector of Dept. of Org. Party Work with Cen. Cttee. Esthonian CP 1965–69; 1st Sec. of Rakvere Raion Cttee. Esthonian CP 1969–79; mem. Cen. Cttee. Esthonian CP 1971–; Dep. of Supr. Sov. Esthonian SSR 1971–; Sec. and mem. Bureau Cen. Cttee. Esthonian CP Ja. 1979–.

Urgalkin, Aleksei Konstantinovich
Chuvash actor; b. Mr. 17 (30), 1910, Elaur, now Sengilei Raion, Ul'ianovsk Oblast.
People's Artist of the USSR (1976).
He graduated from the studio of the Magnitogorsk Theatre of Young Workers in 1932. In 1936 he became an actor in the Chuvash Academic Theatre in Cheboksary. He has played leading roles in the Chuvash repertoire, for example, Setner and Mikheder in *Narspi* (after a work by Ivanov), Aidar and Pichura in Osipov's *Aidar,* Vavil in Terent'ev's *The Cuckoo Still Cuckoos,* Smolin and Entip in Rzhanov's *Entip,* and Sherkkei in Il'bek and Mikushkin's *Rye Bread.* In the Russian and Soviet classics he has interpreted the roles of Zhukhrai in *How the Steel Was Tempered* (after the novel by N. A. Ostrovskii), Vershinin in Vs. Ivanov's *Armored Train 14–69,* and Liubim Tortsov, Neschastlivtsev, and Knurov in A. N. Ostrovskii's *Poverty Is No Crime, The Forest,* and *The Girl Without a Dowry.* Decors.: the Order of the Badge of Honour and various medals.

Urin, Viktor Arkad'evich
Author; b. Je. 3, 1924, Kharkov.
Grad. Gorkii Inst. of Literature 1948.
Publs.: "Vesna pobeditelei" (Spring of the Victors) 1946, "Rekam sniatsia moria" (Rivers Dream of Oceans) 1956, "Bol'shaia Medveditsa" (The Great Bear) 1968, "Govori, Mamaev kurgan!" (Speak up, Mamaev Hill!) 1969, "Ballada o pervoi liubvi" (The Ballad of the First Love) 1970, "Izbrannaia lirika" (Selected Lyrics) 1970, "Izbrannye stikhi" (Selected Poems) 1972, and others.

Urkinbaev, Abdukarim Abdukadyrovich
State official.
Sec. of Kara-Kalpak Oblast Cttee. Uzbek CP 1961–71; Min. of Agriculture Uzbek SSR Ap. 1975–; mem. Cen. Cttee. Uzbek CP 1976–.

Urlanis, Boris Tsezar'evich
Demographer; b. Ag. 15 (28), 1906, Kiev.
Doctor of economic sciences (1940).
After graduating from Moscow State University in 1926, he engaged in scholarly and pedagogical work. From 1944 to 1949 he was a professor at Moscow State University, and from 1949 to 1956 he was a professor at the All-Union State Institute of Cinematography. In 1959 he became a staff member of the Institute of Economics of the Academy of Sciences of the USSR. Urlanis' most important works deal with general problems of population, the dynamics and structure of the population of the USSR, economic demography, and the general theory of statistics. Urlanis joined the International Union for the Scientific Study of Population in 1966 and became a member of the council of the World Population Society in 1973.
Works: *Istoriia amerikanskikh tsenzov.* Moscow 1938. *Rost naseleniia v Evrope.* Moscow 1941. *Voiny i narodonaselenie Evropy.* Moscow 1960. *Rozhdaemost' i prodolzhitel'nost' zhizni v SSSR.* Moscow 1963. *Dinamika i struktura naseleniia SSSR i SShA.* Moscow 1964. *Istoriia odnogo pokoleniia.* Moscow 1968. *Obshchaia teoriia statistiki,* 2nd ed. Moscow 1973. *Problemy dinamiki naseleniia SSSR.* Moscow 1974.

Usatyi, A. P.
Komsomol official.
Mem. Bureau 1974–, Sec. Cen. Cttee. Moldavian Komsomol 1978–.

Useinov, Mikael' Alekserovich
Architect and architectural historian; b. Ap. 6 (19), 1905, Baku.
People's Architect of the USSR (1970); member of the Academy of Sciences of the Azerbaidzhan SSR (1945).
He studied at the Azerbaidzhan Polytechnic Institute from 1922 to 1929 and taught there from 1930 to 1970, becoming a professor in 1942. He has been chairman of the administrative board of the Architects' Union of the Azerbaidzhan SSR since 1947 and director of the Institute of Architecture and Art of the Academy of Sciences of the Azerbaidzhan SSR since 1948.
Useinov worked with S. A. Dadashev until 1946. Works built by Useinov after 1946 include the M. F. Akhundov Public Library of the Azerbaidzhan SSR (1960), the complex of the Academy of Sciences of the Azerbaidzhan SSR (1960's), and the N. Narimanov subway station (1967), all in Baku. Decors.: the State Prize of the USSR in 1941. Two Orders of Lenin, two other orders, and various medals.

Works: *Pamiatniki azerbaidzhanskogo zodchestva.* Moscow 1951. *Istoriia arkhitektury Azerbaidzhana.* Moscow 1963. (With L. Bretanitskii and A.I. Salamzade.)

Ushakov, Andrei Petrovich
Trade union official.
Chair. of Cen. Cttee. Construction Workers' and Construction Ind. Workers' Trade Union until 1975; Sec. of Cen. Union Council of USSR Trade Unions Ja. 1975–.

Usikov, Aleksandr Iakovlevich, Dr. of Tech. Sc.
Physicist; b. Ja. 11, 1904, Iankovtsy, now Sumy Oblast.
Mem. CPSU 1944–; grad. Kharkov Inst. of Popular Educ. 1928; mem. Acad. of Sciences Ukrainian SSR 1964–; mem. staff Kharkov Polytech. Inst. 1928–33, Ukrainian Physicotech. Inst. (now Physicotech. Inst. of Acad. of Sciences Ukrainian SSR) in Kharkov 1932–55; Dir. of Inst. of Radiophysics and Electronics of Acad. of Sciences Ukrainian SSR 1966–; author of approx. 60 scientific works. Decors.: Order of Lenin, Order of Red Banner of Labour.
Publs.: "Impulsnyi generator upravliaemoi chastoty" in "Zhurnal tekhnicheskoi fiziki" No. 7, 1937, "Impulsnyi metod opredeleniia povrezhdenii v. kabeliakh" in "Elektrichestvo" No. 10, 1951, "Linzovyi effekt atmosfery Zemli" in "Geomagnetizm i aeronomiia" No. 2, 1962, and others.

Usmankhodzhaev, Inamzhon Buzrukovich
State official; b. 1930.
Uzbek; mem. CPSU 1958–; grad. Central Asian Polytechnical Inst. in Tashkent 1955; engineer, construction section chief of "Ferganavodstroi" Trust, chief architect of Margilan city 1955–60; instructor of an oblast cttee. Uzbek CP, Chair. of Syr-Daria City Exec. Cttee. of Sov. of People's Deps., Sec. of Syr-Daria Oblast Cttee. Uzbek CP 1960–69; instructor of a dept. of Cen. Cttee. CPSU 1969–72; Chair. of Namangan Oblast Exec. Cttee. of Sov. of Peoples' Deps. 1972–74; 1st Sec. of Andizhan Oblast Uzbek CP 1974–78; mem. Cen. Cttee. Uzbek CP 1976–; mem. Bureau of Cen. Cttee. Uzbek CP 1978–; Chair. of Presidium of Supr. Sov. Uzbek SSR D. 1978–; Dep. to and Dep. Chair. of Presidium of USSR Supr. Sov. 1979–; mem. Cen. Cttee. CPSU 1981–.

Usmanov, G. I.
Party official.
Until 1982 Chair. of Council of Mins. Tartar ASSR; 1st Sec. of Oblast Cttee. Tartar ASSR CPSU O. 1982–.

Usmanov, Inom Ikramovich
State official.
Min. of Construction and Utilization of Highways Tadzhik SSR D. 1979–.

Usmanov, Rashat Barievich
State official.
Min. of the Food Ind. Kirghiz SSR O. 1965–81; cand. mem. 1966–71, mem. Cen. Cttee. Kirghiz CP 1971–; Dep. to Supr. Sov. Kirghiz SSR 1967–; Dep. USSR Min. of Food Ind. Ap. 1981–.

Usmanov, Vakhobzhan
State official.
Min. of the Cotton-Cleaning Ind. Uzbek SSR Mr. 1973–; mem. Cen. Cttee. Uzbek CP 1976–.

Uspenskii, Viktor Konstantinovich
State official.
Min. of Local Ind. RSFSR Ap. 1969–.

Ustiian, Ivan Grigorevich
State official; b. 1939; s. of a peasant.
Moldavian; mem. CPSU 1961–; grad. Kishinev Agricultural Inst. 1965 and CPSU Cen. Cttee. Acad. of Social Sciences 1973, cand. of Econ. Sc. 1974; Komsomol posts 1958–65; Sec. of Cen. Cttee. Moldavian Komsomol 1965–70; cand. mem. Cen. Cttee. Moldavian CP 1966–71 and 1976–80, mem. 1980–; instructor of Organizational Party Work Dept. of Cen. Cttee. Moldavian CP 1970–73; 1st Sec. of Edintsy Raion Cttee. Moldavian CP 1973–77; Chief of Agriculture and Food Industry Dept. of Cen. Cttee. Moldavian CP 1977; Dep. Chair. of Council of Mins. and Chair. of GOSPLAN Moldavian SSR D. 1977–81; Chair. of Council of Mins. Ja. 1981–, Min. of Foreign Affairs Moldavian SSR Ja.–D. 1981; mem. Bureau of Cen. Cttee. Moldavian CP Ja. 1981–; mem. Cen. Auditing Comm. CPSU 1981–.

Ustinov, Dmitrii Fedorovich
State official; b. O. 30, 1908, Samara, now Kuibyshev; s. of a worker.
Russian; mem. CPSU 1927–; grad. Leningrad Inst. of Mil. Mechanical Engineering 1934; Marshal of the Soviet Union; completed Makarev Tech. Trade School, then worked as fitter and machine operator in various plants in Gorkii and Ivanovo Oblasts 1927–; after completing inst. Design Engineer, Dep. Chief Designer, Dir. of "Bol'shevik" Plant in Leningrad 1934–41; USSR People's Commr., then Min. of the Munitions Ind. 1941–53; Dep. to USSR Supr. Sov. 1946–50 and 1954–; mem. Cen. Cttee. CPSU 1952–; USSR Min. of the Defense Ind. 1953–57; Dep. Chair. 1957–Mr. 1963, 1st Dep. Chair. of USSR Council of Mins. Mr. 1963–Mr. 1965; Chair. of USSR Supr. Econ. Council Mr. 1963–Mr. 1965; Sec. Mr. 1965–76, cand. mem. 1965–76, mem. Politbureau Cen. Cttee. CPSU 1976–; USSR Min. of Defense Ap. 1976–. Decors.: twice Hero of Socialist Labour, Stalin Prize, 5 Orders of Lenin, Order of Suvorov 1st Cl., Kutuzov Order 1st Cl., various medals.

Ustinov, Viacheslav Aleksandrovich
Diplomat; b. 1925.
Mem. Diplomatic Service 1950–; sen. posts in USSR Min. of Foreign Affairs 1961–62; Emb. Counsellor at USSR Emb. in Tanganyika 1962–64 and in Tanzania 1964–66; mem. staff USSR Min. of Foreign Affairs 1966–69; USSR Amb. to Tanzania 1969–72; Head of 3rd African Dept. of USSR Min. of Foreign Affairs 1972.

Ustinova, Tat'iana Alekseevna
Choreographer; b. D. 19, 1908 (Ja. 1, 1909), Kalinin.
People's Artist of the USSR (1961). Member of the CPSU since 1928.
She graduated from the Moscow Choreographic School in 1931. Between 1933 and 1943 she directed the dance group of the Moscow Young People's Theatre, where she was also an actress. She became principal choreographer of the Piatnitskii Russian Folk Chorus in 1938. Ustinova's works develop the traditions of Russian folk dance. They include *Timonia, Moscow Round Dances, Northern Round Dances, Zhuravel'* (Crane Dance), *Gusachok* (Goose Dance), *The Mummers, The Voronezh Round Dance,* and *The Gold Chain.* Ustinova has also choreographed a number of works on contemporary themes, including *Celestial Round Dance, Hello, Volga,* and *Story of the Russian Land (Ivushka).* She is the author of books and articles on Russian choreography. Decors.: the State Prize of the USSR in 1949 and 1952, the Glinka State Prize of the RSFSR in 1971, two orders and several medals.
Works: *Russkie tantsy.* Moscow 1955. *Berech' krasotu russkogo tantsa.* Moscow 1959. *Zvezdnyi khorovod: Sovremennye siuzhetnye tantsy.* Moscow 1964. *Russkii narodnyi tanets.* Moscow 1976.

Usubaliev, Turdakun Usubalievich
Party official; b. 1919; s. of a peasant.
Kirghiz; mem. CPSU 1941–; grad. Party Higher School of Cen. Cttee. CPSU 1945 and Moscow Teachers' Training Inst. 1965; teacher at a primary school, after completing inst. head of the instruction dept. of a secondary school 1938–41; dep. dept. head with a raion cttee. of Kirghiz CP, Instructor for Cen. Cttee. Kirghiz CP, then studied at Party Higher School of Cen. Cttee. CPSU 1941–45; instructor with a dept. of Cen. Cttee. CPSU 1945–55; Ed. of newspaper "Sovettik Kirgizistan" 1955–56; Head of Dept. of Propaganda and Agitation Cen. Cttee. Kirghiz CP 1956–58; mem. Cen. Cttee. Kirghiz CP 1956–; Dep. to 1959–, Chair. 1959–61, mem. Presidium Supr. Sov. Kirghiz SSR 1961–; mem. Bureau 1960, 1st Sec. of Cen. Cttee. Kirghiz CP 1961–; mem. Cen. Cttee. CPSU 1961–; Dep. to 1962–, mem. Budget Comm., then Budget and Planning Comm. 1962–74, mem. Foreign Affairs Comm. Sov. of the Union USSR Supr. Sov. 1974–; mem. Bureau

for Cen. Asia Cen. Cttee. CPSU 1962–65. Decors.: 2 Orders of Lenin, Order of Red Banner of Labour.

Utevskii, Aron Mikhailovich, Dr. of Biol. Sc.
Biochemist; b. Jl. 3, 1904, Konotop, Sumy Oblast.
Grad. Dept. of Biology of Kharkov Inst. of Popular Educ. 1924; corresponding mem. Acad. of Sciences Ukrainian SSR 1939–; Dr. of Biol. Sc. 1935; Prof. 1932–; postgraduate studies 1925–28, mem. staff Ukrainian Inst. of Biochemistry 1928–30; Research Asst. at Chair of Biochemistry of Kharkov Med. Inst. 1930–32; Head of Chair of Biochemistry at 1st Med. Inst. in Kharkov and concurrently Head of Dept. at Ukrainian Inst. of Endocrinology resp. Ukrainian Inst. of Experimental Med. 1932–38; Head of Chair of Biochemistry at 2nd Med. Inst. in Kharkov 1938–41; Prof. of Biochemistry at Kirghiz Med. Inst. 1941–44; Head of Chair of Biochemistry at Kharkov Med. Inst. 1944–; author of approx. 130 scientific and popular-science publs.
Publs.: "Rol' biokhemicheskikh faktorov e patogenese gipertonii" in "Vrachebnoe delo" No. 5, 1949, "Obmen katekholaminov i ego funktsional'noe znachenie" in "Adrenalin i noradrenalin" 1964, and others.

Utkin, Vadim Ivanovich, Dr. of Tech. Sc.
Engineer; b. 1937.
Mem. CPSU 1966–; grad. Moscow Energetics Inst. 1960; specialist in electromechanics; Cand. of Tech. Sc. 1964–72, Dr. of Tech. Sc. 1972; Lecturer 1971–; mem. staff 1960–, Sen. Scientist at Inst. of Control Problems (Automatics and Telemechanics) 1967–. Decor.: Lenin Prize 1972.

Utkin, Vladimir Fedorovich
Mem. Cen. Cttee. CPSU; b. 1923.
Mem. CPSU 1945–; grad. Leningrad Mil. Mechanical Inst. 1952, Dr. of Tech. Sc. 1967, corresponding mem. Acad. of Sciences Ukrainian SSR 1972; served in Sov. Army 1941–45; worked as engineer 1952–; Dep. to USSR Supr. Sov. 1970–; mem. Cen. Cttee. CPSU 1976–. Decors: Hero of Socialist Labour 1969–1976 (twice), Lenin Prize 1964.

Uvachan, Vasilii Nikolaevich
Historian and party and state figure; b. D. 12 (25), 1917, in the nomad camp of Kresty, Katangskii Raion, Irkutsk Oblast. Doctor of Historical Sciences (1970). Professor (1974). Member of the CPSU since 1940.
The son of an Evenki hunter, graduated from the technicum of the Leningrad Institute of the Peoples of the North in 1937, from the Higher Party School Under the Central Committee of the ACP(B) in 1948, and from the Academy of Social Sciences of the Central Committee of the CPSU in 1954. In 1934 and 1935 he was involved in Komsomol work. From 1942 to 1946 he was a secretary of the Evenki Regional Committee of the CPSU, and from 1948 to 1951 and from 1961 to 1976 first secretary. From 1954 to 1961, Uvachan taught in Novosibirsk and Krasnoiarsk. In 1966 he became a member of the Central Auditing Commission of the CPSU, and in 1974 he became a committee member of the Parliamentary Group of the USSR. Since 1976 he has been with the Central Administration of the Council of Ministers of the RSFSR. His main works are devoted to the history of the peoples of the north and to problems connected with policies concerning the country's numerous nationalities and with relations between nationalities. Decors.: Order of Lenin, the Order of the October Revolution, two other orders, and various medals.
Works: "Oktiabr' i sud'by malykh narodov Severa." *Kommunist,* 1967, no. 14. *Put' narodov Severa k sotsializmu: Opyt sotsialisticheskogo stroitel'stva na Eniseiskom Severe (Istoricheskii ocherk).* Moscow 1971. "V. I. Lenin i narody Severa." In the book *K 70-letiiu 2-go s"ezda RSDRP,* fasc. 3, Krasnoiarsk 1974. *Narody Severa v usloviiakh razvitogo sotsializma.* Krasnoiarsk 1976.

Uvarov, A. T.
Komsomol official.
2nd Sec. and mem. Bureau Cen. Cttee. Tadzhik Komsomol O. 1979–; cand. mem. Cen. Cttee. Tadzhik CP 1981–.

Uzhvii, Natal'ia Mikhailovna
Ukrainian actress; b. Ag. 27 (S. 8), 1898, Liuboml'.
People's Artist of the USSR (1944). Hero of Socialist Labour (1973). Member of the CPSU since 1945.
She first appeared in amateur performances in 1918. In 1922 she began performing with the T. G. Shevchenko First State Dramatic Theatre of the Ukrainian SSR in Kiev. She performed in the Derzhdrama Theatre (now the October Revolution Odessa Theatre) in 1925, and in 1926 she joined the Berezil' Theatre in Kharkov. In 1936 she joined the I. Ia. Franko Ukrainian Dramatic Theatre in Kiev.
With great insight and realism, Uzhvii conveyed the innermost feelings of her heroines. Her stage characterizations were marked by an inspired poetic quality and were permeated with profound emotions and thoughts. Among her best roles were Anna in Franko's *Stolen Happiness,* Kruchinina and Tugina in Ostrovskii's *Guilty Though Guiltless* and *The Ultimate Sacrifice,* Ranevskaia in Chekhov's *The Cherry Orchard,* Beatrice in Shakespeare's *Much Ado About Nothing,* the title role in Trenev's *Liubov' Iarovaia,* Oksana and Natal'ia Kovshik in Korneichuk's *The Destruction of the Squadron* and *The Snowball Grove* and Tankabik in Karim's *On the Night of the Lunar Eclipse.* She began appearing in films in 1926; her film roles included Evdokiia in *The Vyborg Side* (1939) and Olena Kostiuk in *The Rainbow* (1944).
Became chairman of the Ukrainian Theatrical Society in 1954. Decors.: State Prize of the USSR in 1946, 1949, and 1951. Four Orders of Lenin, four other orders, and various medals.

Uzun, Nikolai Ivanovich
State official.
Sec. of Kishinev City Cttee. Moldavian CP 1973–76; Chair. of Kishinev City Exec. Cttee. of People's Deps. 1976–81; mem. Cen. Cttee. Moldavian CP 1976–; Dep. to 1980– and mem. of Presidium of Supr. Sov. Moldavian SSR 1980–81; Min. of Construction Moldavian SSR Ja. 1981–.

Vaarandi, Debora
Poet; b. 1915, Võru.
Esthonian; studied Tartu Univ. Decor.: Order of Lenin 1976.
Publs.: verse collection "Pod pylaiushchim nebom" (Under the Flaming Sky) 1945, "Pod shum priboia" (Under the Roar of the Surf) 1948, "V iasnoe utro" (On a Clear Morning) 1950, poem "Alla Mari" 1949, "Mechtatel' u okna" (A Dreamer at the Window) 1960, and others.

Vadachkoriia, Valerian Ivlianovich
State official.
Cand. mem. 1956–66, mem. Cen. Cttee. Georgian CP 1966–; Head of Admin. for the Machine Construction Ind. of Econ. Council Georgian SSR 1962; Head of Dept. of Ind. and Transport at Cen. Cttee. Georgian CP 1965–66; Min. of the Food Ind. Georgian SSR F. 1966–71; Dep. Chair. of Council of Mins. Georgian SSR F. 1971–.

Vagabzade, Bakhtiar Makhmud ogly
Author; b. 1925, Nukha.
Azerbaidzhani; concluded postgraduate studies at Azerbaidzhan State Univ. 1951; first works publ. 1944.
Publs.: "Druzhba" (Friendship, poem), "Moi druzia" (My Friends) 1949, "Vesna" (Spring) 1950, "Vechnyi pamiatnik" (The Eternal Monument) 1953, "Prostye liudi" (Simple People) 1956, "Konets stradanii" (The End of Suffering) 1956, "Noch' razluki" (The Night of Separation) 1959, and others.

Vagin, Andrei Grigor'evich
Painter and graphic artist; b. 1923, Ashmarino, W. Siberia.
Mem. CPSU 1947–; attended Gorkii Art School 1949–54, pupil of F. F. Kirillov, I. S. Miasnikov; exhibitions 1954–.

Vagin, Vladimir Vasil'evich
Graphic artist; b. 1937, Iarensk, Arkhangelsk Oblast.
Attended Moscow Polygraphic Inst. 1958–63, pupil of A. D. Goncharov, P. G. Zakharov; exhibitions 1962–.

Vagris, Jan Janovich
Party official; b. 1930.
Latvian; mem. CPSU 1958–; grad. Latvian State Univ. and Party Higher School of Cen. Cttee. CPSU; Engineer in a factory 1955–58; Dep. Chair. of City Exec. Cttee. of People's Deps. in Elgava, 2nd Sec. of City Cttee. Latvian CP in Elgave and Liepāja 1958–67; 1st Sec. of Liepāja City Cttee. Latvian CP 1967–73; mem. Cen. Cttee. Latvian CP 1971–; Head of Dept. of administrative Organs with Cen. Cttee. Latvian CP 1973–78; 1st Sec. of Riga City Cttee. Latvian CP F. 1978–; cand. mem. Bureau Cen. Cttee. Latvian CP Ap. 1978–81, mem. Ja. 1981–.

Vainauskene, O. I.
Trade union official.
Cand. mem. Cen. Cttee. Lithuanian CP 1971–76; Sec. of Lithuanian Republican Trade Union Council 1972–, mem. Auditing Comm. Lithuanian CP 1981–.

Vainberg, Moisei Samuilovich
Composer; b. D. 8, 1919, Warsaw.
Honored Art Worker of the RSFSR.
He was born into the family of a musician. In 1939 he graduated from the Warsaw Conservatory, where he studied piano with J. Turczyński, and in 1941 from the Belorussian State Conservatory, where he studied composition with V. A. Zolotarev. He has been living in Moscow since 1943.
The characteristic features of his style are fresh harmonic writing, brilliant imagery, and dance rhythms. He often uses Polish, Jewish, Moldavian, and Armenian folk songs (*Moldavian Rhapsody,* 1949; *Polish Melodies,* 1950; about 100 art songs to the words of J. Tuwim, I. Perets, O. Tumanian, and other poets; and the Sixth Symphony, 1963). The exposure of the inhuman nature of fascism, which is a major tragic and dramatic theme of his work, is combined with an expression of the joyful power of life and poetic love (the Eighth Symphony, called *The Flowers of Poland,* 1964: the cantata *Diary of Love,* written in 1965 and dedicated to the memory of the children who perished at Oświęcim). He has composed 11 symphonies, 12 string quartets, 19 sonatas for various instruments, music for the theatre and motion pictures, and other works.

Vainiunas, Stasis Andriaus
Composer, pianist, teacher, and public figure; b. Ap. 2, 1909, Riga.
People's Artist of the Lithuanian SSR (1959).
In 1933, he graduated from the Riga Conservatory in the piano and composition class. He studied composition under J. Vitol and completed his piano studies in Vienna under E. Sauer. He taught piano in Riga and at the Klaipėda and Šiauliai music schools, as well as at the Kaunas Conservatory. Since 1951, he has been head of the piano subdepartment at the Vilnius Conservatory (he has been a professor there since 1953). From 1954 to 1962 he served as chairman of the Board of the Union of Soviet Lithuanian Composers.
His work is characterized by optimism, vivid national color, and striking orchestration. He is the composer of a symphony (1957), the overture *For Peace* (1963), three concerti for piano and orchestra, a rhapsody for two pianos and orchestra, a concerto for organ and strings, a concerto for violin and orchestra, a rhapsody for violin and orchestra on Lithuanian themes (State Prize of the USSR, 1951), chamber instrumental works, including a piano quintet (State Prize of the Lithuanian SSR, 1967), songs, choral works, and other works. He has gone on tour as a pianist in many European countries. He won the International Piano Competition in Vienna in 1933. Decors.: three orders, medals.

Vaino, Karl Genrikhovich
Party official; b. 1923.
Esthonian; mem. CPSU 1947–; grad. Tomsk Electromechanical Inst. Railroad Engineering 1947 and CPSU Cen. Cttee. Party Higher School 1957; electrician 1947–48; instructor, dep. chief, chief of a dept. of Cen. Cttee. Esthonian CP 1948–57; mem. Cen. Cttee. Esthonian CP 1956–; Dep. Chair. of Sovnarkhoz Esthonian SSR, Min. Esthonian SSR 1957–60; mem. Bureau of Cen. Cttee. Esthonian CP 1960–; Sec. of Cen. Cttee. Esthonian CP 1960–78, 1st Sec. 1978–; Dep. to USSR Supr. Sov. 1979–; mem. Cen. Cttee. CPSU 1981–.

Vainshtein, Boris Konstantinovich
Crystallographer; b. Jl. 10, 1921, Moscow.
Corresponding member of the Academy of Sciences of the USSR. Member of the CPSU since 1957.
He graduated from Moscow State University (1945) and the Institute of Steel (1947). From 1949 he worked in the Institute of Crystallography of the AN SSSR; from 1958 he was chief of laboratories, and from 1962 director of the institute. His basic works are in the field of structural analysis of crystals, the theory of the diffraction of electrons and roentgen rays, and research on the structure of albumen molecules. He and Z. G. Pinsker and others created a method of structural electronography and worked out its theoretical foundations. He was the first to define the position of hydrogen atoms in a series of crystals and analyze the structures of many complex and organic compounds. Decor.: Order of the Red Banner of Labour.
Works: *Strukturnaia elektronografia.* Moscow 1956. *Difraktsiia rentgenovskikh luchei na tsepnykh molekulakh.* Moscow 1963.

Vainshtein, Lev Al'bertovich
Physicist; b. D. 6, 1920, Moscow.
Corresponding member of the Academy of Sciences of the USSR. Graduated from Moscow State University (1943). Since 1957, he has worked at the Institute of Physical Problems of the AN SSSR. His principal scientific works are in the area of radiophysics, especially the theory of diffusion of radio waves.
Works: *Elektromagnitnye volny.* Moscow 1957. *Difraktsiia elektromagnitnykh i zvukovykh voln na otkrytom kontse volnovoda.* Moscow 1953. *Vydelenie signalov na fone sluchainykh pomekh.* Moscow 1960. (With V. D. Zubakov.) *Teoriia difraktsii i metod faktorizatsii.* Moscow 1966. *Otkrytye rezonatory i otkrytye volnovody.* Moscow 1966.

Vaisfel'd, Il'ia Veniaminovich
Film critic and theoretician; b. Ag. 2 (15) 1909, Orenburg.
Honored Art Worker of the RSFSR (1969). Doctor of art studies (1966). Member of the CPSU since 1957.
In 1930, he graduated from the department of literature and arts of Moscow University, and in 1932 he finished a postgraduate course of study at the Scientific Research Institute of Film and Photography (division of film history and theory). His works began to appear in 1928. He was an editor at the Mosfil'm Motion Picture Studio. Since 1946 he has been a teacher (since 1967, a professor). The collections *Problems of Writing Film Scenarios* have been published since 1954 under his editorship.
Works: *G. Kozintsev i L. Trauberg: Tvorcheskii put'.* [Moscow] 1940. *Masterstvo kinodramaturga.* Moscow 1961. *Krushenie i sozidanie.* Moscow 1964. *Zavtra i segodnia.* Moscow 1968. *"Povest' i roman v kinematografe." Iskusstvo kino,* 1946, nos. 2–3. "Revoliutsiei mobilizovannoe i priznannoe. *Iskusstvo kino,* 1969, no. 2. "Gor'kii i voprosy kinodramaturgii." *Oktiabr',* 1953, no. 7.

Vaivadaite-Surgailiene, Aspaziia Antano
Graphic designer; b. 1928, Taurage, Lithuania.
Attended Art Inst. of the Lithuanian SSR 1949–58, pupil of I. M. Kuzminskis, A. A. Kuchas; exhibitions 1957–.

Vakht, Viktor Aleksandrovich
State official.
Mem. Auditing Comm. 1961–66, mem. Cen. Cttee. Esthonian CP 1966–; 1st Sec. of Kokhtla-Iarve City Cttee. Esthonian CP 1971; Dep. to 1971–, Sec. of Presidium Supr. Sov. Esthonian SSR Jl. 1972–.

Valamat-zade, Gafar Rustamovich
Ballet dancer, choreographer, and director; b. Ap. 26 (My. 9), 1916, Khodzhent, now Kairakkum, Tadzhik SSR.
People's Artist of the Tadzhik SSR (1945). Member of the CPSU since 1942.
In the 1930's and 1940's, he studied at a

music and theatrical technicum in Leninabad, at a choreographic school, and at the A. V. Lunacharskii Institute of Theatrical Arts in Moscow. From 1951 to 1963 he was the principal choreographer at the S. Aini Theater of Opera and Ballet in Dushanbe. He staged various dances in Lenskii's *Du Gul'* (Two Roses; 1941), the first Tadzhik ballet, as well as Balasanian's ballet *Leili and Medzhnun* (1947; State Prize of the USSR, 1949), Asaf'ev's ballet *The Fountain of Bakhchisarai* (1954), Vol'berg's ballet *The Blue Carpet,* (1958, scenario by Valamat-Zade), Balasanian's opera *The Vose Uprising* (1958), Iudakov's opera *Pranks of Maisara* (1962), and other works. In 1963 he became director and artistic manager of the Rudaki Tadzhik Philharmonic Society. He has been director of the dance ensemble Lola since 1965. Deputy to the second through fifth and the seventh convocations of the Supreme Soviet of the Tadzhik SSR. Decors.: Order of Lenin, three other orders, and various medals.

Valdmanis, Maldonis
Graphic artist; b. 1931, Riga.
Attended Art Acad. of Latvian SSR 1952–58; exhibitions 1958–.

Valeev, Akram Mukharramovich
Author; b. 1908, Ablai; s. of a peasant.
Bashkir; mem. CPSU 1944–; first works publ. 1928.
Publs.: "Sila" (Strength – poems) 1931, "Bashkiria, tebia poiu!" (Bashkiria, I'll Conquer You!) 1932, "Stoi, kto idet?" (Stop, Who Is It?) 1935, "Ia s Dal'nego Vostoka" (I'm From the Far East) 1940, novel "Pervye shagi" (The First Steps) 1952, "Maiskii dozhd" (May Rain) 1958, and others.

Validov, Batyr Khuziakhmedovich (pseudonym: Valid, Batyr)
Author; b. 1905, Tuktagulovo; s. of a miner.
Bashkir; worked on a farm as a child; after parents' death grew up in an orphanage, then worked as teacher and in the ed. div. of various regional newspapers.
Publs.: poems "Zlaia ulybka" (An Evil Smile) 1928, "Syn gor" (Son of the Mountains) 1929, "Pobeda" (Victory) 1932, "Moi podarok" (My Present) 1955, and others.

Valiev, Kamil Akhmetovich, Dr. of Phy.-Math. Sc.
Scientist; b. 1931.
Mem. CPSU 1954–; grad. Kazan Univ. 1954; Dr. of Phy.-Math. Sc. 1964; Prof. 1966–; corresponding mem. USSR Acad. of Sciences 1972–; postgraduate studies 1954–57; teacher, Lecturer, Head of Chair at Kazan Pedagogical Inst. 1957–64; dir. of a scientific research inst. 1965–.

Valiuškis, Gediminas Stanislovo
Architect; b. Ja. 24, 1927, Kaunas.
Honored builder of the Lithuanian SSR (1968). Member of the CPSU since 1963. He studied at the Vilnius Art Institute from 1946 to 1951. In 1963 he became chief architect of Vilnius. He is a specialist primarily in the field of housing construction and is responsible for the plans for standard housing units of the series 1-431 and 1-318 (1955–57, designed with co-planners), residential buildings in the area of Antakalnis in Vilnius (1961–62), and a 32-apartment building in Kėdainiai (1962, with E. Tamoševicius as co-planner). The buildings of him have a convenient layout, and their exteriors are characterized by a flicitous rhythmic alternation of balconies, gratings, eaves, and beautiful colour combinations.

Valnere, Rita Herbertovna
Painter; b. 1929, Bene, Latvia.
Attended Riga Middle School of Art 1944–49 and Art Acad. of the Latvian SSR 1949–56, pupil of A. A. Skride, G. Ia. Elias; exhibitions 1955–.

Valuevs, G. Ia.
Komsomol official.
Sec. and mem. Bureau Cen. Cttee. Latvian Komsomol D. 1980–.

Vanags, Iulii Petrovich
Author; b. 1903, Ekabspils Gouvt.; s. of an artisan.
Latvian; worked as railwayman, woodsman and librarian; first works publs. 1929.
Publs.: "Piatero" (The Five) 1942, "Boi u Liepny" (The Battle of Liepna) 1942, "Boitsy vtorogo vzvoda" (The Soldiers of the Second Section) 1943, libretto for the opera "Ruta" 1943, lyrics for Latvian national anthem 1944, play "Vstrecha na beregu" (Meeting on the Shore) 1947, "Korabli vykhodiat v more" (The Ships Go to Sea) 1955, and others.

Vanichev, Aleksandr Pavlovich
Heat engineer; b. Jl. 29 (Ag. 11), 1916, Kazan.
Corresponding member of the Academy of Sciences of the USSR (1962).
In 1938, he graduated from the Leningrad Polytechnic Institute. In 1947 he began teaching at the Moscow Physical-Technical Institute. (In 1953 he became a professor there.) His principal works are devoted to studies of energy plants and the development of methods of calculating the processes of heat conduction, combustion, and the flows of reacting gases, taking into account the kinetics of physical and chemical processes. Decors.: Lenin Prize (1957), the Order of Lenin, two other orders, and medals.

Vannakh, Sviatoslav Eduardovich
State official.
Mem. Cen. Cttee. Latvian CP 1961–; Dep. to Supr. Sov. Latvian SSR 1963–; dir. of a sovkhoz 1963; Min. of the Meat and Dairy Ind. Latvian SSR My. 1973.

Vannas, Meta Villemovna (née Iangolenko)
State official; b. 1924, Khiiumaa.
Esthonian; mem. CPSU 1949–; univ. educ.; interned in Stuthof Concentration Camp in Germany 1941–45; Komsomol work 1945–52; leading Sov. and party posts 1952–60; Chair. of Narva City Exec. Cttee. of Workers Deps. 1960–69; mem. Cen. Cttee. Esthonian CP 1961–; Dep. to 1963, Dep. Chair. of Presidium Supr. Sov. Esthonian SSR 1967–69 and 1975–; Min. of Public Services Esthonian SSR 1969–75.

Vanshenkin, Konstantin Iakovlevich
Author and poet; b. 1925, Moscow; s. of an engineer.
Russian; grad. Gorkii Inst. of Literature 1953.
Publs.: poems "Pesnia o chasovykh" (Song of the Sentinels) 1951, "Liricheskie stikhi" (Lyrical Verses) 1953, "Portret druga" (Portrait of a Friend) 1955, "Volny" (Waves) 1957, "Lirika" (Lyric) 1959, "Soldatskaia sud'ba" (Soldier's Fate) 1960, and others.

Varaksin, Vladimir Nikolaevich
Architect; b. Je. 22 (Jl. 5), 1901, Omsk.
Honored Builder of the Belorussian SSR (1961).
He studied at the Leningrad Academy of Art (1928–32) and the Leningrad Institute of Engineers of Communal Construction (1932–34). Since 1934 he has lived in the Belorussian SSR. His works include the Pioneers' Palace (1935–36) and the building of the Central Committee of the Belorussian CP (1940–41, completed in 1947) in Minsk, which were both done with A. P. Voinov; the Rodina Cinema Theatre in Mogilev (1939); the M. I. Kalinin Cinema Theatre in Gomel' (1944–48); the building of the oblast committee of the Party of Grodno (1948–50); the Institute of the National Economy in Minsk (1954); and a number of large apartment houses in Belorussian cities.

Vardzelashvili, Omar Georgievich
State official; b. 1933.
Mem. CPSU 1960–; grad. Agricultural Inst. in Tbilisi (1956) and Union Acad. of Foreign Trade (1963); head of a dept. at a state farm for citrus fruits, instructor of trust of state farms for tea and citrus fruits 1967–79; instructor of Georgian Komsomol Cen. Cttee. 1959–61; student 1961–63; held various posts in USSR Min. of Foreign Trade, Sov. trade orgs. abroad, Dep. Dir. of office of union asscn. "Soyuzplodoimport" 1963–73; dep. head of Trade, Planning and Finance Organs Dept. Cen. Cttee. CP Georgia, 1st Sec. of Mtskhetskii Raion Cttee. CP Georgia 1973–74; head of Dept. of Agriculture Cen. Cttee. CP

Georgia 1974–80; mem. Cen. Cttee. CP Georgia 1976–; Dep. Chair. of Council of Mins. Georgian SSR F. 1980–.

Varennikov, Valentin Ivanovich
Military official; b. 1923.
Russian; mem. CPSU 1944–; grad. M.V.Frunze Mil. Acad. and Acad. of Gen. Staff, Army Gen.; entered Sov. Army 1941–; served at the front 1941–45; command posts 1945–73; Commander of Troops of Carpathian Mil. District 1973–79; Dep. to USSR Supr. Sov. 1974–; mem. Cen. Cttee. Ukrainian CP 1976–79; 1st Dep. Chief of Gen. Staff of USSR Armed Forces 1979–.

Varfolomei (N. N. Gandorovskii), Cand. of Theol.
Russian Orthodox clergyman; b. O. 27, 1927, Nechaevka, Belgorod Oblast.
Monk 1954; priest; grad. Moscow Theological Acad., Cand. of Theol. 1959; teacher at Moscow Theological Seminary 1959–60; abbot, Dep. Head of Ecclesiastical Mission of Moscow Patriarchate in Jerusalem 1960; performed duties as Head of same Mission 1961; Head, Bishop of Uglich, Substitute of Iaroslavl Diocese, then Bishop of Saratov 1963; Bishop of Hungary and Austria 1964; Bishop of Tula 1968; Bishop of Kishinev and Moldavia 1969; Bishop 1972, Archbishop of Tashkent and Cen. Asia 1973–, mem. of Holy Synod 1976–.

Varlaam (Aleksei Timofeevich Iliushchenko)
Russian Orthodox clergyman; b. My. 13, 1929, Pruska, Klimovo district, Briansk Oblast.
Worker, choir-boy, lay-reader, chorister, sub-deacon; at Kiev Theological Seminary 1949; deacon, priest 1954; pastoral activity in Kiev and surroundings; archpriest, Head Clergyman in Church of the Resurrection in Demeevka, Head of Econ. Admin. in Ukrainian Exarchate 1968; Provost of the Kiev Churches 1969–; monk and Archimandrite, started external course at Moscow Theological Acad. 1970; Main Clergyman at Vladimir Cathedral Kiev, 1972; Bishop of Pereiaslav-Khmelnitskii, Substitute Bishop for the Diocese of Kiev O. 1972–77; Bishop of Chernovtsy and Bukovina Mr. 1977–.

Varnava (Kedrov, Vladimir Viktorovich)
Russian Orthodox clergyman; b. Ap. 21, 1931, Vysokoe, Riazan Oblast; s. of a priest.
Attended intermediate school; deacon in Riazan 1953–55; took vows as a monk 1954; in Sergieva Lavra (St. Sergius Monastery) 1956–70; Archimandrite 1965–; Bishop of Cheboksary and the Chuvash ASSR F. 1977–.

Varpakhovskii, Leonid Viktorovich
Stage director; b. Mr. 16 (29), 1908, Moscow.
People's Artist of the RSFSR (1966) and Honored Art Worker of the Georgian SSR (1964).
He graduated from the literature department of Moscow State University in 1931 and was a set and costume designer and assistant stage director at a Moscow theatre of working youth. From 1933 through 1935 he was a research assistant at the Meierkhol'd Theatre; in 1936–37, a director at the Alma-Ata Russian Theatre; from 1953 through 1955, a director at the Griboedov Theatre in Tbilisi; and in 1957, principal director of the Ermolova Theatre in Moscow. In 1962 he became director at the Malyi Theatre in Moscow and in 1970 at the Vakhtangov Theatre. He has staged Vishnevskii's *Optimistic Tragedy* in Alma-Ata in 1937, at the Franko Theatre in Kiev in 1961, and at the Malyi Theatre in 1967; Davurin's *Transport Train of the Second Class* at the Ermolova Theatre in Moscow in 1957; Lermontov's *Masquerade* (1962), De Filippo's *Christmas at the Cupiello House* (1965), Aleshin's *The Ward* (1962) and *The Diplomat* (1967), and A. N. Ostrovskii's *Easy Money* (1969) all at the Malyi Theatre; and Shatrov's *The Sixth of July* at the Moscow Art Theatre in 1966.

Vartanian, Stepan Amaiakovich
State official.
Dep. to Supr. Sov. Armenian SSR 1959–; Head of Agricultural Dept. of Cen. Cttee. Armenian CP 1958–61; mem. Cen. Cttee. Armenian CP 1958–64 and 1966–76; Min. of Supplies Armenian SSR 1961–62; mem. Agricultural Bureau of Cen. Cttee. Armenian CP 1962–64; Head of Ideological Dept., Head of Dept. of Propaganda and Agitation of Cen. Cttee. Armenian CP 1964–71; Min. of the Meat and Dairy Ind. Armenian SSR F. 1971–.

Varygin, Sergei Petrovich
Painter and graphic artist; b. 1923, Krasnoiarsk.
Attended Leningrad High School for Ind. Art 1947–50, pupil of A. I. Taran; exhibitions 1953–.

Vasadze, Akakii Alekseevich
Actor and director; b. Jl. 25 (Ag. 6), 1899, Kutaisi.
People's Artist of the USSR (1936). Member of the CPSU since 1939.
In 1917, he began to work on the stage. From 1920 he worked in the Sh. Rustaveli Georgian Theatre (Tbilisi), and from 1933 to 1955 he was the head director. His roles included Franz Moor (Schiller's *The Robbers*), Darispun (Kldiashvili's *Autumn Gentlefolk*), Kikvidze (Daraseli's *Brigade Commander Kikvidze*), Ananiia (Sumbatov's *Treason*), Abdul-Sadyk (Mosashvili's *Sunken Stones*), Iago (Shakespeare's *Othello*), and Shuiskii (Solov'ev's *The Great Tsar*). Among his productions were Shanshiashvili's *Arsen* (1936), *Brigade Commander Kikvidze* (1941), and *Sunken Stones* (1951). He is very much a character actor; his art is distinguished by mastery of social typification and variety of expressive means. In 1958 he became the head director of the L. Meskhishvili Georgian Theatre (in Kutaisi); he directed the plays *The Exile* by Vazha Pshavela (1958), *The Dawn of Kolkhida* by Lordkipanidze (1959), and others. Since 1969 he has been an actor of the Rustaveli Georgian Theatre. In 1923 he started acting in films. In 1924 he became involved in teaching, since 1947 as a professor at the Rustaveli Theater Institute in Tbilisi. He was a deputy to the first through fourth convocations of the Supreme Soviet of the Georgian SSR. Decors.: State Prize of the USSR (1942, 1946, 1951), two Orders of Lenin, two other orders, and medals.

Vasadze, Zurab Omerovich
State official.
Chair. of kolkhoz in Khala (Kubuleti Raion); Dep. to and Dep. Chair. of Supr. Sov. Georgian SSR 1975–.

Vasalattii, Grigorii Ivanovich
State official.
Mem. Cen. Cttee. Moldavian CP 1971–76, cand. mem. 1976–; Chair. of State Cttee. on Forestry Moldavian SSR Jl. 1975–78; Min. on Forestry Moldavian SSR Ja. 1978–.

Vashchenko, Grigorii Ivanovich
State official; b. 1920.
Ukrainian; mem. CPSU 1943–; grad. Kharkov College of Machine Construction 1938 and All-Union Polytech. Inst. for Correspondence Studies 1955; technician, technologist, chief technologist, head of a bureau, dept. head at various plants in Kharkov and Nizhnii Tagil 1938–58; Sec. of party cttee. of Kharkov Transport Machine Construction Plant 1958–59; Sec. 1959, 2nd Sec. 1959–63, 1st Sec. of Kharkov Oblast Cttee. Ukrainian CP 1963–72 (for Ind. 1963–64); Dep. to and mem. Trade and Public Catering Comm. of Supr. Sov. Ukrainian SSR 1963–67; mem. Cen. Cttee. Ukrainian CP Ja. 1965–; cand. mem. 1966–71, mem. Politbureau Cen. Cttee. Ukrainian CP 1971–; mem. Cen. Cttee. CPSU 1966–; Dep. to Supr. Sov. 1966–; 1st Dep. Chair. of Council of Mins. Ukrainian SSR Je. 1972–. Decors.: 2 Orders of Lenin.

Vashadze, Nodar Grigorevich
State official.
Min. of Housing and Communal Econ. Georgian SSR Jl. 1976–; mem. Cen. Cttee. Georgian CP 1976–.

Vashkiavichiene, S.
Komsomol official.
Sec. and mem. Bureau Cen. Cttee. Lithuanian Komsomol N. 1978–.

Vasiagin, Semen Petrovich
Military official; b. 1910.
Russian; mem. CPSU 1932–; intermediate-school educ.; Gen. of the Army; con-

struction worker 1924–32; in Sov. Army 1932–; head of pol. dept. of a div., then of a corps 1941–45; head of pol. admin. and mem. mil. council of various mil. districts 1948–58; Dep. to USSR Supr. Sov. 1954–; mem. Cen. Auditing Comm. CPSU 1966–81; Head of Pol. Admin., mem. Mil. Council of the Army N. 1967–. Decors.: Order of Lenin, 2 Orders of Red Banner, other orders and medals.

Vasikova, Mamliakat Sabirovna
State official.
Dep. Chair. of Supr. Court Uzbek SSR until 1961; mem. Auditing Comm. 1961–71, mem. Cen. Cttee. Uzbek CP 1971–; Min. of Justice Uzbek SSR N. 1970–.

Vasilenko, Vladimir Kharitonovich
Therapeutist; b. My. 25 (Je. 6), 1897, Kiev.
Academician of the Academy of Sciences of the USSR (1957). Hero of Socialist Labour (1967).
In 1922, he graduated from Kiev Medical Institute; he worked under the guidance of A. A. Bogomolets, N. D. Strazhesko, and F. G. Ianovskii. From 1935 he was head of a subdepartment at the Kiev Institute of Advanced Training of Physicians; during the Great Patriotic War he was chief therapeutist of the Northern Caucasus and First Ukrainian fronts. Since 1948 he has been head of the subdepartment of propaedeutics of internal diseases of the first Moscow Medical Institute and simultaneously (since 1967) director of the Moscow All-Union Scientific Research Institute of Gastroenterology of the Ministry of Public Health of the USSR. His principal works are devoted to the study of disturbances in metabolism with chronic circulatory insufficiency, kidney pathology, respiratory pathology, and digestive pathology. In collaboration with N. D. Strazhesko he proposed a new classification of insufficiency of the cardiovascular system. He created an important school of therapeutists. He is chairman of the All-Union Society of Gastroenterologists, editor of the journal *Klinicheskaia meditsina*, and editor-in-chief of the *Malaia meditsinskaia entsiklopediia*. Decors.: two Orders of Lenin and four other orders.
Works: *Krupoznoe vospalenie legkikh*. Kiev 1947. *Vnutrennie bolezni*. Moscow 1955.

Vasilenok, Evgenii Petrovich
Author; b. 1917, Orsha; s. of a railwayman.
Belorussian; studied Mogilev Pedagogic Inst.; literary work 1935–; worked for various oblast and republican newspapers; contributor to various front newspapers 1941–45.
Publs.: "Zelenye ogni" (Green Lights) 1954, "Prizvanie" (Calling) 1956, "Sluchainaia ostanovka" (Chance Stay) 1957, "Raznye dorogi" (Various Roads) 1957, plays "Korolevskii gambit" (King's Gambit) 1959, "Sed'moe okno" (The Seventh Window) 1958, and others.

Vasil'ev, Evgenii Alekseevich
Sculptor; b. 1930, Leningrad.
Attended Leningrad Art School 1947–52 and Repin Inst. for Painting, Sculpture and Architecture at USSR Acad. of Arts 1952–58; mem. CPSU 1952–; exhibitions 1958–.

Vasil'ev, Nikolai Fedorovich
State official.
Russian; mem. CPSU 1942–; grad. Tadzhik Agricultural Inst. 1940; bookkeeper with a raion finance dept., student 1932–40; served in Sov. Army 1940–46; dir. of a subsidiary farm, dir. of a machine and tractor station, chair. of a raion sov., 1st sec. of a raion cttee. Ukrainian CP 1946–58; 1st Sec. of Pavlograd City Cttee. Ukrainian CP 1958–60; dept. head with Dnepropetrovsk Oblast Cttee. Ukrainian CP 1960–61; Dep. to 1959–64, mem. Presidium Supr. Sov. Ukrainian SSR 1963–64; Chair. of Exec. Cttee. of Dnepropetrovsk Oblast Sov. of Workers' Deps. 1961–64; 1st Sec. of Belgorod Oblast Cttee. CPSU 1964–71 (for Agriculture My.–D. 1964); mem. Cen. Cttee. CPSU 1966–; Dep. to USSR Supr. Sov. 1966–; 1st Dep. Chair. of Council of Mins. RSFSR F. 1971–79; USSR Min. of Melioration and Water Econ. 1979–. Decor.: Order of Lenin 1976.

Vasil'ev, Oleg Borisovich
Painter and graphic artist; b. 1918, Ufa. Grad. Kiev Art Inst. 1949, pupil of V. I. Kasian, K. D. Trokhimenko; exhibitions 1945–.

Vasil'ev, Oleg Fedorovich
Scientist in the field of applied hydrodynamics and hydraulics; b. Ag. 1, 1925, Moscow.
Corresponding member of the Academy of Sciences of the USSR (1970). Became a member of the Communist Party of the Soviet Union in 1945.
He graduated from the Moscow Institute of Hydromelioration (1948) and taught in the Moscow Institute of Engineering and Construction until 1959. Since 1959 he has worked in the Institute of Hydrodynamics of the Siberian Division of the Academy of Sciences of the USSR. His research has focused mainly on the theory of nonstationary and eddy currents of liquids and gases, hydraulics of open river beds, pipe systems, and hydraulic works. Decors.: Order of the Badge of Honour and medals.
Works: *Osnovy mekhaniki vintovykh i tsirkuliatsionnykh potokov*. Moscow-Leningrad 1957. "O gazotermodinamicheskom raschete potokov v prostykh i slozhnykh truboprovodakh." Izv. *Sibirskogo otdeleniia AN SSSR: Seriia teknicheskikh nauk*, 1968, no. 13, issue 3. (Jointly with A. F. Voevodinyi.)

Vasil'ev, Vadim Petrovich, Dr. of Biol. Sc.
Zoologist-entomologist; b. D. 31, 1912, Odessa; s. of a teacher.
Mem. CPSU 1942–; grad. Dept. of Plant Protection of Kharkov Agricultural Inst. 1935; mem. Acad. of Sciences Ukrainian SSR 1964–; mem. former Ukrainian Acad. of Agricultural Sciences 1959–62; Dr. of Biol. Sc. 1952; Prof. 1955–; mem. staff Ukrainian (formerly Odessa) Inst. for Research on Viniculture 1935–37; sen. mem. scientific staff 1937–41, Head of Dept. of Plant Protection of Ukrainian Inst. for Research on Fruit Growing (now Ukrainian Inst. for Research on Horticulture) 1945–53; served during World War II 1941–45; Dir. of Inst. of Entomology and Phytopathology of Acad. of Sciences Ukrainian SSR (since 1956 Ukrainian Inst. for Research on Plant Protection 1953–; Chair. of Ukrainian Entomological Soc. 1959–; Vice-Pres. of Union Soc. of Entomologists 1960–. Decors.: 2 Orders of Patriotic War 2nd Cl., 2 Orders of Red Star, Order "Badge of Honour", several medals.
Publs.: "Vrediteli sadovykh nasazhdenii" 1955, "Vrediteli plodovykh kul'tur" 1958, "Dovidnyk po zakhystu roslyn" 1962, "Sveklovochnyi dolgonosik i borba s nim" 1956, "Pro suchasnu rol'i perspektyvy okremykh metodiv zakhystu silskohospodarskykh kul'tur vid shkidnykiv" 1961, "Sovremennye khimicheskie sredstva zashchity rastenii ot vreditelei" 1966, and others.

Vasil'ev, Vasili Ivanovich
Painter; b. 1927, Tumevaia, Mogilev, Oblast.
Attended Art Inst. of the Lithuanian SSR 1946–52, pupil of V. S. Matskiavichius; exhibitions 1952–; teacher at Minsk Art School 1961–.

Vasil'ev, Vladimir Viktorovich
Ballet dancer; b. 1940.
Grad. Moscow Choreographic School 1958; soloist with Moscow Bolshoi Theatre 1959–. Decors.: People's Artist of the RSFSR 1969, Lenin Prize 1970, People's Artist of the USSR 1973.

Vasil'ev, Vladimir Sergeevich
Graphic artist; b. 1926, Murmansk.
Attended Kazakh Art School 1951–56, pupil of N. I. Krutilnikov, A. M. Cherkasskii.

Vasil'ev-Borogonskii, Sergei Stepanovich
Yakut poet; b. S. 12 (25), 1907, now the Soto village soviet, Ust'-Aldansk Raion. His works were first published in 1929. The socialist construction of the Yakut countryside is reflected in his poems (*Roman's Artel*, 1929, and *A Flock of Cranes*, 1935). Between 1935 and 1939 the collections of poems *Megintsy on the Hill* and *On the Banks of the Niurba* were published. During the Great Patriotic War he wrote the poems *The Vow* (1941), *Poem About a Sabre* (1943), and *Glory* (1943).

Later he published the poem *Liberated Water* (1953) and the novel in verse *The Younger Son* (1960).
Works: *Talillibit ayimn'ilar,* vols. 1–2. Yakutsk 1966–67. *In Russian translation: Severnye poemy.* Moscow-Yakutsk 1953. *Sviashchennyi il'men'.* Yakutsk 1968.

Vasil'evich, Elena Semenovna
Author and poet; b. 1922, Lipniki, Minsk Oblast; d. of a peasant.
Belorussian; grad. Belorussian Univ.; first works publ. 1947.
Publs.: "Puti-dorogi" (Paths and Roads) 1950, "Blizkie znakomye" (Close Acquaintances) 1954, "Zavtra v shkolu" (To School tomorrow) 1956, "Druzia" (Friends) 1958, and others.

Vasilii (Krivoshein)
Russian Orthodox clergyman; b. Je. 30, 1900, St. Petersburg.
Studied Faculty of History and Philology of St. Petersburg, Moscow and Paris Univs.; monk, monastery sec. 1927; rep. of his monastery on Holy Mount Athos 1942–45; studied at Oxford 1951, attached to Moscow Patriarchate priest-monk, parson of a house church in Oxford answering the needs of Orthodox people at the Univ.; Principal of the Church of the Annunciation in Oxford 1959; took part in byzantological congresses; Archimandrite, Bishop of Volokolamsk, 2nd Substitute of the Exarch of the Moscow Patriarchate in Western Europe 1959; Bishop, then Archbishop of Brussels and Belgium 1960–.

Vasil'ko, Vasilii Stepanovich (pseudonym of V. S. Miliaev)
Ukrainian director and actor; b. Mr. 26 (Ap. 7), 1893, Burty, now in Cherkassy Oblast.
People's Artist of the USSR (1944).
He made his debut in the professional theatre in 1912. Subsequently, during various years, he worked in Kiev with N. K. Sadovskii's troupe, in the Molodoi Theatre, the Berezil' Theatre, and in the theatres of Kharkov, Chernovtsy, and other Ukrainian cities. In 1920 he began his directorial activities. From 1926 (with some interruptions) he worked at the October Revolution Ukrainian Drama Theatre in Odessa, becoming principal director in 1948. His roles have included Shpak and Shel'menko in Kvitka-Osnov'ianenko's *Shel'menko, the Orderly,* the Mayor in Gogol's *The Inspector General,* Globa in Simonov's *Russian People,* and Bublik in Korneichuk's *Platon Krechet.* Among his productions were Mamontov's *Republic on Wheels* (1927), Mikitenko's *Dictatorship* (1929), Gogol's *The Inspector General* (1934), Shakespeare's *Macbeth* (1938), Korneichuk's *Bogdan Khmelnitskii* (1939), Karpenko-Karyi's *The Woman Farm Laborer* (1955), and Kobylianskaia's *Early on Sundy She Gathered Herbs* (1956). Vasil'ko has been a theatrical teacher; he organized a number of theatrical studios and initiated the establishment of a theatrical museum in Kiev (1924). He retired from theatrical work in 1956. Decors.: two orders and various medals.
Works: *F. V. Levyts'kyi.* Kiev 1958. *Mykola Sadovs'kyi ta ioho teatr.* Kiev 1962. *Fragmenty rezhysury.* Kiev 1967.

Vassoevich, Nikolai Bronislavovich
Geologist; b. Mr. 17 (30), 1902, Rostov-on-Don.
Corresponding member of the Academy of Sciences of the USSR (1970).
In 1924, he graduated from the Leningrad Mining Institute. From 1924 to 1940 he studied the geological structure and the petroleum-gas content of the Northern Caucasus, Georgia, Azerbaidzhan, and the north of Siberia. From 1940 to 1963 he worked at the All-Union Petroleum Research Geological Prospecting Institute, where he studied basic problems of lithology and petroleum geography. He is head of the subdepartment of geology and geochemistry of inflammable minerals at the geological faculty of Moscow State University (1963). His principal works are on methods of studying flysch deposits and on the theory of the sediment-migration (organic) origin of oil. Decors.: Order of the Badge of Honour and medals.
Works: *Flish i metodika ego izucheniia.* Leningrad-Moscow 1948. *Usloviia obrazovaniia flisha.* Leningrad-Moscow 1951.

Vatchenko, Aleksei Fedoseevich
State official; b. 1914.
Ukrainian; mem. CPSU 1940–; grad. Dnepropetrovsk State Univ. 1938; teacher, school dir. 1938–41; served in Sov. Army 1941–45; Dep. Head, then Head of Dnepropetrovsk City Educ. Dept. 1945–48; party posts 1948–54; Sec., 2nd Sec. of Dnepropetrovsk Oblast Cttee. Ukrainian CP 1954–59; Dep. to 1958–63, mem. Credentials Comm. of Supr. Sov. Ukrainian SSR 1959–63; 1st Sec. of Khmelnitskii Oblast Cttee. Ukrainian CP 1959–63; mem. Cen. Cttee. Ukrainian CP 1960–; cand. mem. 1961–66, mem. Cen. Cttee. CPSU 1966–; Dep. to 1962–, mem. Econ. Comm. of Sov. of Nationalities 1962–66, mem. Comm. for Ind., Transport Posts and Telecommunications, then Ind. Comm. of Sov. of the Union USSR Supr. Sov. 1966–; 1st Sec. (for Agriculture) of Dnepropetrovsk Oblast Cttee. Ukrainian CP 1963–64; 1st Sec. of Cherkassy Oblast Cttee. Ukrainian CP 1964–65; 1st Sec. of Dnepropetrovsk Oblast Cttee. Ukrainian CP O. 1965–Je. 1976; mem. Presidium, then Politbureau Cen. Cttee. Ukrainian CP Mr. 1966–; Chair. of Presidium Supr. Sov. Ukrainian SSR Je. 1976–; Dep. Chair. of Presidium USSR Supr. Sov. O. 1976–. Decors.: 2 Orders of Lenin, medals.

Vatsietis, Oiar Ottovich
Author and poet; b. 1933.
Latvian; studied Latvian State Univ.; first works publ. 1950.
Publs.: "Veter dal'nikh dorog" (The Wind of Faraway Roads) 1956, "V ogne" (In the Fire) 1958, "Glazami tekh dnei" (With the Eyes of Those Days) 1959, "Porogi obchodit' nekogda" (No Time to Avoid the Threshold) 1960, and others.

Vauni, Suren (pseudonym of Suren Vaganovich Khachikian)
Armenian poet, publicist, and translator; b. D. 27, 1910, Pashalu, now the settlement of Azizbekov, Armenian SSR; s. of a teacher.
Honored Cultural Worker of the Armenian SSR.
He began to publish in 1926. His first collection was *Two Songs* (1930) and the second, *War Songs* (1938). *Ballad of a Partisan, His Wife, and His Horse* (1933) and *Ballad of a Blue-eyed Boy and the Sun* (1933) are devoted to the Civil War. In the poems *Song About Soldiers' Boots* (1943), *Volga,* and *Soviet Armenia* (1943) and the collection *To the West* (1943) the heroic days, thoughts, and aspirations of the Soviet people during the Great Patriotic War are portrayed. He wrote the poems *White House* (1948), *Sevan* (1949), and others. He translated *The Lay of the Host of Igor,* Pushkin's *The Fountain of Bakhchisarai,* and G. Boccaccio's *Decameron* into Armenian. Decors.: Order of the Badge of Honour.
Works: *Rhazmi erger.* Yerevan 1938. *Depi Arnmutk'.* Yerevan 1943. *Ergeri girk'.* Yerevan 1949. *In Russian translation: Kniga pesen.* Yerevan 1952.

Vaupshasov, Stanislav Alekseevich
Reconnaissance officer; b. Jl. 15 (27), 1899, Gruzdzhai, in present-day Šiauliai Raion.
Active participant in the partisan movement during the Great Patriotic War of 1941–45; Hero of the Soviet Union (N. 5, 1944), colonel. Member of the CPSU since 1940.
He volunteered for the Red Army in 1918. In 1920–24 he participated in the underground struggle in the western oblasts of Belorussia, which were occupied by bourgeois Poland. In later years he did reconnaissance work. In 1937–39 during the Civil War in Spain, he carried out combat assignments in the struggle against the Franco troops. Upon returning to the USSR he worked for organs of state security. In 1942 he was sent at the head of a group of cheka men to Belorussia, where he established contact with the partisan underground and participated in the partisan movement. For more than two years he led the large Mestnye Partisan Detachment, which operated successfully in Minsk Oblast. Decors.: four Orders of Lenin, the Order of the Red Banner, Orders of the Patriotic War First and Second Class, the Red Banner of Labour of the Belorussian SSR, and medals.

Vavilin, Ivan Mikhailovich
Author and poet; b. 1911, Boiarkeros, Vologda Oblast; s. of a peasant.
Komi; mem. CPSU 1944–; first works publ. 1939; his poems and features are publ. in literary journal of the Komi ASSR "Severnaia zvezda" (Northern Star).
Publs.: poetry "Ni shagu nazad" (Not One Step Backward) 1942, "Sovetskii voin" (Soviet Soldier) 1944, and others.

Vdovin, Valentin Petrovich
Diplomat; b. 1927.
Mem. Diplomatic Service 1959–; Emb. Counsellor at USSR Emb. in France 1959–65; USSR Amb. to the Republic of Chad 1965–69; mem. staff USSR Min. of Foreign Affairs 1969–72; USSR Amb. to Laos O. 1972–D. 1976, to Mozambique 1980–82, to Portugal D. 1982–.

Vechar, Aleksandr Stepanovich
Biochemist; b. Mr. 12 (25), 1905, Mashchytsy, Minsk Gouvt.
Belorussian; grad. Belorussian Agricultural Acad. 1929; mem. Acad. of Sciences Belorussian SSR 1966; mem. staff various scientific insts. of Acad. of Sciences Belorussian SSR and USSR; Prof. of Biochemistry at Belorussian State Univ. 1963–.

Vechkanov, Serafim Emelianovich
Author and poet; b. 1914, Kakino, Gorkii Oblast.
Mordvinian; first works publ. 1932.
Publs.: "Nachalo pesni" (The Beginning of the Song) 1933, "Moia zvezda" (My Star) 1940, "Venok" (The Wreath) 1946, "Stikhi" (Poems) 1952, "Solnyshko" (The Dear Sun) 1955, "Gde luchshe nam zhivetsia" (Where We Live Best), "Ded Bogdan" (Grandfather Bogdan) and others.

Veiian, Andris (real name: Kalnach, Donat Geronimovich)
Author and poet; b. 1927, Ludza.
Latvian; grad. Dept. of Philology of Riga Pedagogic Inst. 1950; first works publ. 1946.
Decor.: Order of Friendship Between Nations 1977.
Publs.: poetry "Molodost" (Youth) 1953, "Solntse podnimaetsia vyshe" (The Sun Is Climbing Higher) 1957, "Shkola u ozera" (The School at the Lake), "Pereklikaiutsia chaiki" (The Seagulls Screech), and others.

Veimarn, Boris Vladimirovich
Art historian and critic; b. O. 26 (N. 8), 1909, Sevastopol.
Mem. CPSU 1960–; corresponding mem. USSR Acad. of the Arts 1949–; mem. ed. staff and contributor to encyclopedia "Isskustvo stran i narodov mira" (Art of the Countries and Peoples of the World) 1962–.

Veinerman, Khanan Abramovich
Author and poet; b. 1902, Luginy, Zhitomir Oblast; s. of a cobbler.
Jew; grad. Odessa Inst. of Popular Educ. 1932; first works publ. 1925.
Publs.: "Zolotye vetki" (Golden Branches) 1935, "V zhizn' vliublen" (In Love With Life) 1940, "Obnovlennaia zemlia" (Land Renewed) 1948, and others.

Vekilova, Leila Makhat kyzy
Ballerina; b. Ja. 29, 1927, Baku.
People's Artist of the USSR (1967). Member of the CPSU since 1956.
She graduated from the Baku Choreographic School in 1943. From 1945 to 1946 she perfected her art at the Moscow Choreographic School. In 1943 she became a soloist at the M. F. Akhundov Theatre of Opera and Ballet in Baku. She has created expressive characters in national ballets, combining the technique of classical dance and features of native Azerbaidzhan choreography. These include Giul'ianak in Badalbeili's *The Maiden's Tower,* Aisha in Karaev's *Seven Beauties,* and Giul'shen in Gadzhibekov's work of the same name. Among her other roles are Tao Khoa in Gligere's *The Red Poppy,* Odette-Odile in Tchaikovsky's *Swan Lake,* Zarema in Asaf'ev's *The Fountain of Bakhchisarai,* and Shirin in Melikov's *Legend of Love.* She has been teaching at the Baku Choreographic School since 1953. Decor.: Order of Lenin.

Veldi, Heino Tynisovich
State official; b. 1936; s. of a farmer.
Esthonian; mem. CPSU 1964–; grad. Esthonian Acad. of Agrarian Sciences (1960); state technical inspector at Charyuskaya machine and tractor station, chief engineer of Paideski i Raion Production Admin. for Agriculture 1960–66; head of the Admin. for the Mechanization of Agriculture at Min. of Agriculture Esthonian SSR 1966–74; Dep. Min. of Agriculture Esthonian SSR 1974–79; Chair. of State Cttee. for Technical Production Supply to Agriculture Esthonian SSR 1979–81; Dep. Chair. of Council of Mins. Esthonian SSR 1981–.

Velichka, Ionas Ionovich
State official.
Chair. of State Cttee. for Melioration with Council of Mins. Lithuanian SSR 1964–65; Min. of Melioration and Water Econ. Lithuanian SSR O. 1965–; cand. mem. Cen. Cttee. Lithuanian CP 1966–; Dep. to Supr. Sov. Lithuanian SSR 1967–.

Veliev, Ali Kara ogly
Azerbaidzhan writer; b. 1901, Aguli, now Sisianskii Raion, Armenian SSR.
Member of the CPSU since 1925. A farm laborer, then a rabkor and a Party worker. First appeared in print in 1924.
He is the author of novels dedicated to the themes of the construction of socialism – *Gakhraman* (1941; Russian translation, *The Elapsed Years,* 1958), *In Our Chichekli* (1951; Russian translation, 1956), *Sincere Friends* (1956; Russian translation, *Campfires,* 1959), and *The Road to Turadzhly* (1958–59) – and the novellas *Giul'shen* (1949) and *Recollections of Budag* (books 1–2, 1963–64). He devotes attention to children's themes in the short stories "A Gift to the Children" (1962), "To My Dear Grandchildren" (1964), and "Little Star" (1967). His works contain a great deal of ethnographic information.
Works: *Sechilmish äsärläri,* vols. 1–2. Baku 1957–60. *Anagiz.* Baku 1965. *Äsärläri,* 6 vols.; vols. 1–4. Baku 1968–70.

Veliev, Dzhafar Dzhebrail ogly
State official.
Dep. Head of Dept. of Admin. Organs with Cen. Cttee. Azerbaidzhan CP 1962–76, Head 1976–78; can. mem. Cen. Cttee. Azerbaidzhan CP 1971–76, mem. 1976–; Min. of Interior Azerbaidzhan SSR N. 1978–.

Velikanov, Dmitrii Petrovich
Scientist in automobile technology and transportation; b. O. 12 (25) 1908, Odessa.
Corresponding member of the Academy of Sciences of the USSR (1968). Honored Worker in Science and Technology of the RSFSR (1968); professor (1958). Member of the CPSU since 1945.
In 1931, he graduated from Leningrad Polytechnical Institute. He worked in a number of scientific research institutes of automobile transportation, and in 1956 he joined the Institute for Problems of Integrated Transportation. His main work deals with the study of the operational qualities of the automobile, with the development of theoretical and experimental methods for determining (and evaluating) these qualities, and with research into the problems of motor vehicle transport within a unified system of transportation. He has conducted a number of government tests of lot-produced automobiles.
Works: *Ekspluatatsionnye kachestva avtomobilei.* Moscow 1962. *Effektivnost' avtomobilia.* Moscow 1969.

Velikhov, Evgenii Pavlovich, Dr. of Phy.-Math. Sc.
Physicist; b. 1935.
Mem. CPSU 1971–; grad. Moscow Univ. 1958; Dr. of Phy.-Math. Sc. 1964; mem. USSR Acad. of Sciences 1974–; mem. staff Inst. of Atomic Energy 1958–, Head of Laboratory 1962–70, Dep. Dir., Dir. of a branch of same inst. 1971–; concurrently mem. staff 1966–, Head of Chair at Moscow Univ. 1973–. His major works deal with the physics of low-temperature plasma and magnetic hydrodynamics.

Veliugin, Anatolii Stepanovich
Author and poet; b. 1923, Moshkany, Vitebsk Oblast; s. of a physician.
Belorussian; studied Pedagogic Insts. in Minsk and Sverdlovsk; first works publ. 1938.
Publs.: "Saliut u Minsku" (Salute of Honour in Minsk) 1947, "Zaria zanimaetsia" (Beginning Dawn) 1958, "Golub' na bashne" (A Dove on the Tower) 1957, and others.

Verbitskii, Nikolai Filippovich
State official.
Mem. Cen. Cttee. CP Moldavia 1963–66, cand. mem. 1981–; Chair. of the Agrarian-Ind. Asscn. for Tobacco Moldavian SSR D. 1977–.

Vereiskii, Orest Georgievich
Graphic artist; b. Jl. 7 (20), 1915, Anosovo, now in Smolensk Oblast; s. of G.S. Vereiskii.
People's Artist of the RSFSR (1970) and corresponding member of the Academy of Arts of the USSR (1958).
He, who audited courses at the Academy of Arts in Leningrad from 1936 to 1938, created illustrations and several series of easel paintings. His works are characterized by keen observation and a precise graphic style. They include illustrations for A.T. Tvardovskii's narrative poems *Vasilii Terkin* (1942–48) and *House by the Road* (1957), A.A. Fadeev's *The Defeat* (1949; published in 1967), and M.A. Sholokhov's *The Quiet Don* (1951–52), *Fate of a Man* (1958), and *Virgin Soil Upturned* (1967). He also did several series of drawings, watercolors, and lithographs, the result of his journey through Czechoslovakia, Syria, Lebanon, and Egypt (all in 1955), Finland (1957), Iceland (1958), and the USA (1960, 1963). Decors.: Order of the Red Star and various medals.

Vergelis, Aron Alterovich
Author and poet; b. 1918, Liubar, Ukraine.
Jew; grad. Dept. of Literature Lenin Pedagogic Inst. in Moscow 1940; first works publ. 1935.
Publs.: "U rodnika" (At the Source) 1940, "Birobidzhantsy" (Birobidzhan People) 1948, "Zhazhda" (Thirst) 1956, "Vtoraia vstrecha" (The Second Meeting) 1961, "Pesn' ob Josife Bumagine" (The Song About Josef Bumagin) 1947, "Poema o kosmose" (Poem About the Cosmos) 1962, and others.

Verich, Ian Georgievich
Sculptor; b. 1912, Fedorovka, Taganrog Oblast.
Grad. Repin Inst. for Painting, Sculpture and Architecture at the USSR Acad. of Arts 1947, pupil of V.A. Sinaiskii, G.A. Shults; exhibitions 1934–; teacher at Taganrog Pedagogic Inst. 1962–.

Verkhovets, Vladimir Lukich
State official.
Sec. Brest Oblast Cttee. Belorussian CP 1971–80; Chair. of State Cttee. for Vocational Training Belorussian SSR 1980–; cand. mem. Cen. Cttee. Belorussian CP 1981–.

Verkin, Boris Eremeevich, Dr. of Phy.-Math. Sc.
Physicist; b. Ag. 8, 1919, Kharkov.
Mem. CPSU 1945–; grad. Kharkov Univ. 1940; corresponding mem. Acad. of Sciences Ukrainian SSR 1961–; Dr. of Phy.-Math. Sc. 1958; Prof. 1963–; mem. staff Physiotech. Inst. of Acad. of Sciences Ukrainian SSR 1940–41 and 1946–60, concurrently at Kharkov Univ.; Dir. of Physicotech. Inst. of Low Temperatures of Acad. of Sciences Ukrainian SSR 1960–; author of approx. 60 scientific publs. Decors.: various medals.
Publs.: "Magnitnye svoistva metallov pri nizkikh temperaturakh" in "Zhurnal eksperimentalnoi i teoreticheskoi fiziki" No. 20/1950; "Issledovanie skhvatyvaniia aliuminiia i medi v glubokom vakuume" in "Avtomaticheskaia svarka" No. 2/1966.

Vernov, Sergei Nikolaevich
Physicist; b. Je. 28 (Jl. 11), 1910, Sestroretsk.
Academician of the Academy of Sciences of the USSR (1968; corresponding member, 1953).
Upon graduating from the Leningrad Polytechnical Institute in 1931, he worked at the Radium Institute of the Academy of Sciences of the USSR. In 1936 he began working at the Physics Institute of the Academy of Sciences of the USSR and in 1944, as a professor at Moscow State University. His principal works deal with the study of the nature and properties of cosmic rays in the upper layers of the atmosphere and beyond. He discovered and studied the latitudinal effect of cosmic rays in the stratosphere and determined the energy spectrum of primary radiation. He has investigated the transitional effects in cosmic rays and ascertained the origin of the electron-photon component of cosmic rays (State Prize, 1949). Together with A.E. Chudakov and others, he discovered the earth's external radiation belt (Lenin Prize, 1960). Decors.: Order of Lenin and three other orders.
Works: "Shirotnyi effekt kosmicheskikh luchei v stratosfere i proverka kaskadnoi teorii". *Trudy Fizicheskogo instituta im P.N. Lebedeva,* 1945, vol. 3, issue 1. "Izuchenie vzaimodeistviia pervichnoi komponenty kosmicheskikh luchei s veshchestvom v stratosfere." *Zhurnal eksperimental'noi i teoreticheskoi fiziki,* 1949, vol. 19, issue 7. "Radiatsionnye poiasa Zemli na vysotakh 180–250 km." *Iskusstvennye sputniki Zemli,* 1962, vol. 13. (Written in collaboration with others.) "Izmerenie intensivnosti pronikaiushchego izlucheniia na poverkhnosti Luny." *Dokl. AN SSSR,* 1966, vol. 169, no. 5. (Written in collaboration with others.) "Izmereniia pri polete stantsii 'Venera-3' i 'Zond-3' protonov solnechnogo proiskhozhdeniia s energiei 1–5 Mev." *Ibid.,* vol. 171, no. 4. (Written in collaboration with others.)

Verro, Rudolf Ottovich
State official; b. 1925.
Head of Agricultural Dept. of Cen. Cttee. Latvian CP 1961–65; mem. Cen. Cttee. Latvian CP 1961–; Dep. to Supr. Sov. Latvian SSR 1963–; mem. Agricultural Bureau of Cen. Cttee. Latvian CP 1963–64; Sec. of Cen. Cttee. 1965–74, mem. Bureau Cen. Cttee. Latvian CP 1965–; 1st dep. Chair. of Council of Mins. Latvian SSR Ag. 1974–.

Vershinin, Vladimir Viktorovich
State official.
Dep. Chair. of Council of Mins. Turkmen SSR Ap. 1979–; mem. Cen. Cttee. Turkmen CP 1981–.

Vertinskaia, Lidia Vladimirovna
Graphic artist and film actress; b. 1923, Harbin, China.
Attended Surikov Art Inst. Moscow 1949–55, pupil of M.I. Kurilko, A.M. Kuznetsov; exhibitions 1955–.

Vertsberg, Anna Ilinichna
Sculptor; b. 1927, Moscow.
Attended Moscow High School for Arts and Crafts 1945–54, pupil of G.I. Motovilov, G.A. Shults; exhibitions 1957–.

Veselov, Georgii Petrovich
State official.
1st Dep. Min. of Education RSFSR S. 1980–81, Min. Ja. 1981–.

Veskiviali, Viliar Hugovich
State official.
Min. of Local Ind. Esthonian SSR F. 1980–; cand. mem. Cen. Cttee. Esthonian CP 1981–.

Vezhev, Aleksandr Aleksandrovich
Literary historian and critic; b. Ag. 23 (S. 5), 1911, Ust-Sysolsk, now Syktyvkar.
Mem. CPSU 1943–. Decors.: Order of Patriotic War 1st Cl., medals.
Publs.: co-author of "Ocherki istorii komi literatury" 1958, and others.

Vezirov, Abdul Rakhman Khalil ogly
Diplomat.
Grad. Azerbaidzhan Ind. Institute; 2nd Sec. of Cen. Cttee. Azerbaidzhan Komsomol 1955–57, 1st Sec. 1957–59; Sec. and mem. Bureau of All-Union Komsomol Cen. Cttee. 1959–70; 1st Sec. of Kirovabad City Cttee. Azerbaidzhan CP 1970–74; mem. Cen. Cttee. Azerbaidzhan CP 1971–76; Head of Dept. of Industry and Transport with Cen. Cttee. Azerbaidzhan CP 1974–76; USSR Amb. to Nepal S. 1979–.

Vialias, Vaino Iosipovich
Diplomat; b. 1931; s. of a peasant.
Mem. CPSU 1952–; grad. Tartu State Univ. 1955; joined Komsomol 1948; Head of Dept. of Propaganda and Agitation of Khiiumaas Krai Cttee. Esthonian Komsomol 1949–50; Sec. of Komsomol Cttee. of Tartu State Univ. 1951–52; Sec. of Tartu City Cttee. Esthonian Komsomol 1952–55; 1st Sec. of Cen. Cttee. Esthonian Komsomol 1955–61; mem. Cen. Cttee. Esthonian CP 1956–81; cand. mem. 1960–61, mem. Bureau Cen. Cttee. Esthonian CP 1961–80; 1st Sec. of Tallinn City Cttee. Esthonian CP 1961–71; Sec. of Cen. Cttee. Esthonian CP F. 1971–80; USSR Amb. to Venezuela Jl. 1980– and concurrently to Trinidad and Tobago O. 1980–.

Vialov, Oleg Stepanovich, Dr. of Geol.-Min. Sc.
Geologist; b. Ja. 23, 1904, Tashkent; s. of a soldier.
Grad. Leningrad Univ. 1928– mem. Acad. of Sciences Ukrainian SSR 1948; Dr. of Geol.-Min. Sc. 1937; Prof. 1945–; mem. staff Geological Cttee. in Leningrad 1927–33; various posts with Union Petroleum Research Inst. 1933–49; Head of Dept. of Tectonics and Paleography of Inst. of Geology and Geochemistry of Mineral Resources of Acad. of Sciences Ukrainian SSR in Lvov 1949–; concurrently teaching work at Mining Inst. of Leningrad Univ. 1931–45; Head of Chair of Paleontology and Historical Geology at Lvov Univ. 1945–; author of approx. 330 scientific publs. on regional geology, tectonics, paleontology and hydrogeology. Decors.: State Prize of the USSR 1948, Order of Lenin, Order "Badge of Honour."
Publs.: "Mel i Paleogen Fergany" in book "Materialy Tadzhiksko-Pamirskoi Ekspeditsii" 1936, "Paleontologicheskaia seriia: Obshchee strukturnoe podrazdelenie zapadnykh oblastei Ukrainy" in "Izvestiia Akademii Nauk SSSR. Seriia geologischeskaia" 1953, "Paleogenovyi flish severnogo sklona Karpat" 1961, "Sledy zhiznedeiatelnosti organizmov i ikh paleontologicheskoe znachenie" 1966, and others.

Viatkin, Aleksandr Vasil'evich
Painter and graphic artist; b. 1922, Sarapul, Udmurt ASSR.
Attended Kharkov Art Inst. 1946–52, pupil of E. P. Svetlichnyi, G. A. Tomenko; exhibitions 1950–.

Viazankin, Nikolai Sergeevich, Dr. of Chem. Sc.
Chemical scientist; b. 1921.
Mem. CPSU 1967–; grad. Gorkii Univ. 1950; Cand. 1958–, Dr. of Chem. Sc. 1965; Prof. 1966–; served in Sov. Army 1943–45; worked in a tractor plant in the Altai 1950–53; postgraduate studies 1953–56; mem. staff Inst. for Research On Chemistry of Gorkii Univ. 1956–63; mem. staff, then Dep. Dir. and Head of Laboratory of Inst. of Chemistry of USSR Acad. of Sciences 1963–. Decor.: State Prize of the USSR 1971.

Videnskii, Mikhail Solomonovich
Graphic artist and painter; b. 1926, Berdichev.
Attended Union Inst. of Film-Making, Faculty of Painting, 1944–51, pupil of F. S. Bogorodskii, G. M. Shegal; exhibitions 1953–.

Vieru, Grigore Pavlovich
Moldavian poet; b. F. 14, 1935, Pereryta, Brichany Raion.
He graduated from the Kishinev Pedagogical Institute in 1958. His first work was published in 1954. He wrote several books of poetry for children. His book *Poems* was awarded the B. Glavan Prize in 1967 by the Central Committee of the Lenin Komsomol of Moldavia. The collection *Thy Name* was published in 1968. A love of life, the land, work, and song are the principal motifs of his creative work. His poems are characterized by a high level of craftmanship and a spontaneity of feeling.
Works: *Versur'.* Kishinev 1965. *In Russian translation: "Imia tvoe." Druzhba narodov,* 1967, no. 1.

Vieru, Igor Dmitrievich
Painter and graphic artist; b. 1923, Chernolevka, Moldavia.
Attended Kishinev Republican Art School 1946–49, pupil of R. A. Gabrikov; exhibitions 1955–; mem. CPSU 1962–. Decor.: Merited Artist of the Moldavian SSR 1963–.

Vigilianskaia, Natalia Valentinovna
Graphic artist; b. 1919, Krasnodar.
Attended Surikov Art Inst. Moscow 1939–45, pupil of A. A. Deineka, A. A. Osmerkin; exhibitions 1950–.

Viidalepp, Ants Feliksovich
Painter and graphic artist; b. 1921, Tallinn.
Attended Tartu Art Inst. 1944–48, pupil of I. Vyerahans; exhibitions 1948–.

Viidalepp, Helve Avgustovna
Painter; b. 1921, Paide, Esthonia.
Attended Tartu Art Inst. 1944–48, pupil of I. Vyerahans; exhibitions 1948–.

Viiding, Paul
Author and critic; b. 1904, Valga; s. of an engine driver.
Esthonian; grad. Tartu Univ. 1930; first works publ. 1932; very few of his poems and short stories were publ. before 1940.
Publs.: "Most" (The Bridge) 1946, "Prodvizhenie vpered" (Forward Movement) 1947, "Razmyshleniia i bespokoinoe eutro" (Meditations and a Restless Morning) 1958, and others.

Viies, Edgar Tinissovich
Sculptor; b. 1931, Zimititsy, Leningrad Oblast.
Attended Art Inst. of the Esthonian SSR 1951–52 and Repin Inst. for Paintin, Sculpture and Architecture at the USSR Acad. of Arts 1952–58, pupil of V. V. Lishev; exhibitions 1958–.

Viko, Renan Emelianovich
Painter; b. 1929, Buzluk.
Attended Tashkent Art Inst. 1946–50, pupil of A. P. Gan; exhibitions 1955–.

Viktorin (Vladimir Vasil'evich Bliaev)
Russian Orthodox clergyman; b. 1903, Bloshniki, Vitebsk Oblast.
Grad. Vilnius Theological Seminary 1924 and Theological Faculty of Warsaw Univ. 1931; completed external studies at Moscow Theological Acad. 1962; Bishop of Aleksin and Tula and of Perm and Solikamsk; Bishop of Vienna and Austria 1974–75; Substitute Bishop of Tula 1975–77; Bishop of Vilnius and Lithuania Ap. 1978–.

Viktorov, Aleksei Vasil'evich
Trade union official; b. 1917.
Mem. CPSU 1945–; concluded 2 terms at a machine construction college; apprentice as a machine fitter 1934–36; machine fitter at 1st State Ball-Bearing Plant 1936–38 and 1940–41; served in Sov. Army 1938–40; machine fitter at 4th State Ball-Bearing Plant 1941–47; Brigadier of machine-fitter brigade at 1st State Ball-Bearing Plant 1947–My. 1975; mem. Cen. Cttee. CPSU 1971–; Sec. and mem. Presidium Cen. Union Council of USSR Trade Unions My. 1975–. Decor.: Hero of Socialist Labour.

Vikulov, Sergei Vasil'evich
Author and poet; b. 1922, Emilianovskaia, Vologda Oblast.
Russian; mem. CPSU 1942–; grad. Dept. of Literature of Vologda Pedagogical Inst. 1951; first works publ. 1936.
Publs.: "Zavoevannoe schastie" (Fought for Happiness) 1949, "Khoziaeva zemli" (Masters of the Land) 1950, "Novyi bereg" (New Shore) 1952, "V Metel'" (In the Snowstorm) 1955, "Galinkino leto" (Galinka's summer) 1957, "Trudnoe schastie" (Difficult Happiness) 1958, and others.

Vikulov, Vladimir Ivanovich
Athlete; b. Jl. 20, 1946, Moscow.
Hockey player. Honored Master of Sport (1967). Officer in the Soviet Army. Member of the CPSU since 1971.
He has repeatedly been champion of the USSR (1966, 1968, and 1970), the world (1966–71), and Europe (1966–70). He was a champion at the Winter Olympic Game (1968) and holder of the Cup of the USSR (1966–69) and the Cup of European Champions (1969–70) for ice hockey. As one of the best forwards in modern hockey, he was included on a symbolic all-world team (1970).

Vilenkin, Vitalii Iakovlevich
Theatrical scholar; b. D.27, 1910 (Ja.9, 1911), Moscow.
Doctor of the Arts (1967).
Based on his studies of the work of the Moscow Art Theater, wrote several major works on the art of its leading figures. He was one of the organizers of the V.I. Nemirovich-Danchenko Studio School. He has been a professor since 1967.
Works: *Vl. I. Nemirovich-Danchenko: Ocherk tvorchestva.* Leningrad 1941. *I.M. Moskvin na stsene Moskovskogo Khudozhestvennogo teatra.* Moscow 1946. *Kachalov.* Moscow 1962. "Rabota Vl.I. Nemirovicha-Danchenko s akterom." In *Vl.I. Nemirovich-Danchenko vedet repetitsiiu: "Tri sestry" A.P. Chekhova v postanovke MKhAT 1940g.* [Moscow 1965.] *Amedeo Modil'iani.* Moscow 1970.

Vilenskii, Zinovii Moiseevich
Portrait sculptor; b. O.3 (15), 1899, now the city of Koriakovka, Chernigov Oblast.
People's Artist of the RSFSR (1969), corresponding member of the Academy of Arts of the USSR (1954).
He studied at the Moscow State Higher Arts and Technical Studios and State Higher Arts and Technical Institute from 1922 to 1928. He is characterized by meticulous execution and a thoughtful attitude toward the individual characteristics of the model. His works include portraits of K.N. Igumnov (plaster, 1939), S.A. Lavochkin (marble, 1947), and P.I. Tchaikovsky (marble, 1947) in the Tret'iakov Gallery; a monumental bust of S.E. Artemenko, twice Hero of the Soviet Union, in the village of Ratsulovo, Odessa Oblast, and I.Kh. Mikhailichenko, twice Hero of the Soviet Union, in the village of Almaznyi, Voroshilovgrad Oblast (both bronze, 1948–49); a portrait of S.A. Chaplygin (plaster, 1950, N.E. Zhukovskii Memorial Science Museum, Moscow); a monument to V.I. Lenin in Sochi (bronze, 1957); and a portrait of B. Russell (bronze, 1966, author's property).

Vilips, Pavel Martsevich
Author; b. 1901, Riga; s. of a worker.
Latvian; studied Dept. of Philosophy of Riga Univ.; first works publ. 1925.
Publs.: "Fabrika u Gaui" (Factory on the Gauiia) 1948, "V sadakh pustyni" (In the Gardens of the Desert) 1949, "Bol'shoi polet" (The Big Flight) 1951, "Solntsevorot" (Solstice) 1956, and others.

Vilks, Evalds Rudolfovich
Graphic artist; b. 1928, Annas, Latvia.
Attended Riga Middle School of Art 1947–50 and Riga School for Applied Art 1951–54, pupil of Friedrichsson; exhibitions 1956–.

Vinberg, Abram Il'ich, Dr. of Legal Sc.
Jurist and criminologist; b. Ag.29, 1908, St. Petersburg.
Mem. CPSU 1940–; Prof. 1949–; State Legal Consultant 3rd Cl. 1960–. Decor.: Merited Scientist of the RSFSR.

Vinogradov, Nikolai Ignat'evich
Military official; b. D.13 (26), 1905, Surikha, in present-day Ivanovo Raion, Kostroma Oblast.
Admiral since 1954. Member of the CPSU since 1930.
He graduated from the Frunze Higher Naval School in 1930 and from the Naval Academy in 1939. Until 1933 he served on submarines of the Black Sea Fleet; and from 1933 to 1936 he was commander of several submarines of the Pacific Fleet. From 1939 he was chief-of-staff and commander of submarine brigades of the Baltic and then of the Northern Fleet. During the Great Patriotic War he commanded a submarine brigade and then was commander of the submarine branch of the Northern Fleet and in 1943–45 deputy commander and commander of the submarine branch of the navy. From 1945 to 1948 he held high posts in the Baltic and Pacific fleets. In 1948–53 he was deputy chief of naval operations, in 1950–53 deputy minister of the navy, in 1953–58 deputy commander in chief of the navy, and in 1958–62 deputy chief of the general staff of the Armed Forces. Since 1962 he has been chief of naval educational institutions. Decors.: two Orders of Lenin, five Orders of the Red Banner, the Order of Ushakov First Class, the Order of the Red Star, and medals.

Vinogradov, Vladimir Alekseevich
Economics scientist; b. Jl.2, 1921, Kazan.
Mem. CPSU 1943–; grad. Inst. of Int. Relations in Moscow 1948; corresponding mem. USSR Acad. of Sciences 1966–; mem. staff Presidium of USSR Acad. of Sciences 1948–; concurrently Head of a Sector at Inst. of World Econ. and Int. Relations 1967–; Dep. Scientific Sec. of Presidium USSR Acad. of Sciences 1961–; Lecturer at Dept. of Pol. Econ. of Moscow Univ. 1954–60; Vice-Pres. of Int. Asscn. for Econ. History 1968–; mem. Sov. UNESCO Comm. 1965–; Dir. Inst. of scientic Information of Social Sciences 1979–. Decors.: Order of Red Banner of Labour, medals.

Vinogradov, Vladimir Mikhailovich
Diplomat; b. 1921.
Mem. CPSU 1942–; grad. Moscow Mendeleev Inst. of Chemistry and Technology 1945 and Union Acad. of Foreign Trade 1948; served in Sov. Army 1939–43; Dept. Head at USSR Trade Mission in Great Britain 1948–50; Dep. Head of USSR Trade Mission in Great Britain 1950–52; mem. staff USSR Min. of Foreign Trade 1952–62; USSR Amb. to Japan 1962–67; USSR Dep. Min. of Foreign Affairs 1967–70; USSR Amb. to the United Arab Republic 1970–74; cand. mem. Cen. Cttee. CPSU 1971–76; USSR Representative at Geneva Middle East Conference 1974–77; USSR Amb. to Iran Ja. 1977–82; Min. of Foreign Affairs RSFSR My. 1982–.

Vinokurov, Evgenii Mikhailovich
Author and poet; b. 1925.
Russian; mem. CPSU 1952–; grad. Gorkii Inst. of Literature 1951; first works publ. 1948.
Publs.: "Stikhi o dolge" (Poems on Duty) 1951, "Sineva" (Blue Sky) 1956, "Voennaia lirika" (Military Lyric) 1956, "Priznaniia" (Admissions) 1958, and others.

Vishnevskaia, Galina Pavlovna
Russian soprano; b. O.25, 1926, Leningrad.
People's Artist of the USSR (1966).
She studied singing with V.N. Garina in Leningrad. In 1944 she began performing with the Leningrad Oblast Operetta Theater and then worked with the Leningrad Philharmonic. In 1952 she became a soloist at the Bolshoi Theater of the USSR in Moscow. She is a performer possessing subtle musical and stage talent. Her performance is distinguished by fullness of sound and impeccable purity of intonation. Her roles include Tat'iana and Liza in Tchaikovsky's *Eugene Onegin* and *The Queen of Spades,* Kupava in Rimsky-Korsakov's *The Snow Maiden,* Aida and Violetta in Verdi's *Aida* and *La Traviata,* Cho-Cho-San in Puccini's *Madame Butterfly,* Natasha Rostova in Prokofiev's *War and Peace,* and Katherine in Shebalin's *The Taming of the Shrew* (first performer, 1957). As a concert singer, she is outstanding for her moving interpretations of art songs by M.P. Mussorgsky, P.I. Tchaikovsky, and D.D. Shostakovich. In 1966 she appeared in the film *Katerina Izmailova.* She has toured many countries of the world. Decor.: Order of Lenin.

Visnapuu, Donald Paulovich
Komsomol official.
Mem. Bureau 1976–, Sec. 1976–78, 1st Sec. Cen. Cttee. Esthonian Komsomol D. 1978–; mem. Cen. Cttee. and cand. mem. Bureau Cen. Cttee. Esthonian CP 1981–.

Vitka, Vasil (real name: Krysko, Timofei Vasil'evich)
Author, poet and dramatist; b. 1911, Evlichi, Minsk Oblast; s. of a peasant.
Belorussian; first works publ. 1928.
Publs.: "Zakalka" (Tempering) 1943, "Polden'" (Midday) 1946, "Luch budushchego" (Beam of the Future) 1948, "Vernost'" (Faithfulness) 1953, "Roza i shtyk" (The Rose and the Bayonet) 1958, "Schastie poeta" (Bliss of the Poet), and others.

Vitkiavichius, V.
Komsomol official.
Sec. and mem. Bureau Cen. Cttee. Lithuanian Komsomol Mr. 1978–.

Vitol, Leon Petrovich
State official.
Min. of Forestry and the Timber Ind. Latvian SSR D. 1971–; cand. mem. Cen. Cttee. Latvian CP 1976–.

Vivatenko, Leron Aleksandrovich
Painter; b. 1929, Kazan.
Attended Kazan Art School 1948–50, pupil of N. D. Kuznetsov, M. M. Timofeeva; worked in Ioshkar-Ole until 1965; exhibitions 1954–.

Vladimir (V. S. Kotliarov), Cand. of Theol.
Russian Orthodox clergyman; b. 1929, Aktiubinsk.
Studied Moscow Theological Seminary 1952; priest 1953; grad. Leningrad Theological Acad. 1958; Lecturer at Leningrad Theological Acad. 1959–62; at same time active abroad; monk, Dep. Head of Ecclesiastical Mission of Moscow Patriarchate at World Council of Churches in Geneva, Bishop of Zvenigorod 1962; Bishop of Voronezh 1964; Bishop of Podolsk, Rep. of Moscow Patriarchate at Patriarchate of Antioch in Damascus 1965; Bishop of Kirov 1966; Bishop of Berlin and Cen. Europe, Archbishop 1967; Archbishop of Rostov 1970; Archbishop of Irkutsk and Chita 1973–75; Archbishop of Vladimir and Suzdal 1975–80; Archbishop of Krasnodar and Suzdal Ap. 1980–.

Vladimir (Viktor Markianovich Sabodan), Cand. of Theol.
Russian Orthodox clergyman; b. N. 23, 1935, Khmelnitskii Oblast.
From 1954 attended Odessa Theological Seminary; grad. Leningrad Theological Acad., teacher at Odessa Theological Seminary 1962; deacon, priest and monk 1962; teacher and 1st Asst. to Supervisor at same Seminary, Sec. of Diocese of Odessa; concluded postgraduate studies at Moscow Theological Acad., Rector of Odessa Theological Seminary 1965; Dep. Head of Russian Ecclesiastical Mission in Jerusalem 1966; Bishop of Zvenigorod Jl. 1966–69; Bishop of Chernigov and Nezhiri 1969–73; Archbishop of Dmitrov, Substitute Bishop of Moscow Diocese and Rector of Moscow Theological Acad. Ap. 1973–82; Bishop of Rostov and Novocherkassk Jl. 1982–.

Vladimirov, Iurii Nikolaevich
Graphic artist; b. 1927, Naro-Fominsk, Moscow Oblast.
Attended Moscow School for Arts and Crafts 1946–49 and Moscow Polygraphic Inst. 1949–57, pupil of V. N. Baksheev; exhibitions 1958–.

Vladimirov, Vasilii Sergeevich
Mathematician; b. Ja. 9, 1923, Diaglevo, Leningrad Oblast.
Grad. Leningrad Univ. 1948; mem. USSR Acad. of Sciences 1970–; mem. staff V. A. Steklov Inst. of Mathematics 1948–; made several internationally recognized discoveries. Decors.: State Prize of the USSR 1953, Order of Red Banner of Labour and several medals.

Vladimirov, Vladimir Petrovich
Graphic artist; b. 1920, Moscow.
Attended Moscow Inst. for Applied and Decorative Art 1946–51, pupil of Ia. M. Sokolov, P. P. Sokolov-Skal; exhibitions 1959–.

Vladimirskii, Vasilii Vasil'evich
Physicist; b. Jl. 20 (Aug. 2), 1915, Zheleznovodsk.
Corresponding member of the Academy of Sciences of the USSR (1962).
He graduated from Moscow State University in 1938. He has been deputy director of the Institute of Theoretical and Experimental Physics since 1946. His principal works deal with problems in optics and the propagation of ultrasound and electromagnetic waves as well as with the theory of linear accelerators and neutron spectroscopy. He took part in the creation of the Serpukhov accelerator with an energy level of 70 giga-electronvolts. Decors.: State Prize of the USSR in 1953 and the Lenin Prize in 1970.

Vladimov, Georgii Nikolaevich (real name: Volosevich, G. N.)
Author; b. 1931, Kharkov.
Russian; grad. Dept. of Law of Leningrad Univ. 1953; first works publ. 1960.
Publ.: "Bol'shaia ruda" (The Big Ore) 1961, and others.

Vladislavskii, Valentin Vladimirovich
Graphic artist; b. 1925, Elan-Koleno, Voronezh Oblast.
Grad. Penza Art School 1947 and attended Kiev Art Inst. 1947–53, pupil of V. I. Kasian, I. N. Pleshchinskii; exhibitions 1952–.

Vladychenko, Ivan Maksimovich
State official; b. 1924.
Mem. CPSU 1943–; grad. Donetsk Ind. Inst. 1951; served in Sov. Army 1941–45; dep. head, section head at "Chistiakovantratsit" Trust Mine 1951–52; 2nd Sec. of Chistiakov (now Torez) City Cttee. Ukrainian CP, 1st Sec. of Snezhianskii Raion Cttee. Ukrainian CP in Donetsk Oblast 1952–59; Chair. of Cen. Cttee. of Coalworkers' Union 1959–64; mem. Cen. Auditing Comm. CPSU 1961–66; Sec. of All-Union Cen. Council of Trade Unions N. 1964–81; cand. mem. Cen. Cttee. CPSU 1966–; Chair. USSR State Cttee. for Supervision of Safe Working Practices in Industry and for Mine 1981–.

Vladychin, Iurii Nikolaevich
State official; b. 1917.
Cand. mem. 1956–60, mem. Cen. Cttee. Esthonian CP 1960–; Vice-Chair. of Econ. Council Esthonian SSR 1960–63; Head of Dept. of Ind. and Transport with Cen. Cttee. Esthonian CP 1963–65; Min. of Light Ind. Esthonian SSR O. 1965–77; Chair. of State Cttee. for Pricing Esthonian SSR D. 1977–. Decor.: Order of Red Banner of Labour.

Vlasiuk, Petr Antipovich
Plant physiologist and agrochemist; b. F. 3, 1905, Chemerinske, now Cherkassy Oblast; s. of a peasant.
Mem. CPSU 1930–; grad. Uman Agricultural Inst. 1926 and Leningrad Agricultural Inst. 1930; mem. Acad. of Sciences Ukrainian SSR 1948–; mem. Union Lenin Acad. of Agricultural Sciences 1948–; mem. and Pres. of former Ukrainian Acad. of Agricultural Sciences 1956–62; Dr. of Agr. Sc. 1946; Prof. 1931–; Head of Laboratory for Agrochemistry of Union Inst. for Sugar-Beet Research in Kiev 1931–41 and 1945–47; Dep. Dir. of Inst. of Botany of Acad. of Sciences Ukrainian SSR 1944–46; Dep. Dir. 1946–52, Dir. of Inst. of Plant Physiology and Agrochemistry of Acad. of Sciences Ukrainian SSR 1952–; concurrently Head of Chair. of Agrochemistry and Soil Science of Uman Agricultural Inst. 1931–41 and 1945–47; Head of Chair of Soil Science at Kiev Univ. 1947–56; Head of Dept. of Agricultural Sciences of Acad. of Sciences Ukrainian SSR 1948–56; Head of Chair of Agrochemistry at Ukrainian Acad. of Agricultural Sciences 1956–62; author of approx. 800 publs. on agrochemistry, plant physiology and soil science. Decors.: 2 Orders of Lenin, Order of Red Banner of Labour, Order "Badge of Honour", medals.
Publs.: "Novi margantsevi dobryva" 1941, "Agrofiziologicheskie osnovy pitaniia sakharnoi svekly" 1950, "Zhyvlennia ta udobrennia roslyn" 1955, "Metod mechenykh atomov v agrofiziologii" 1959, "Zimostoikost' ozimoi pshenitsy na Ukraine" 1959, "Margantseve zhyvelennia i udobrennia roslyn" 1962, "Primenenie radioaktivnykh izotopov v agrokhimii i biologii" 1965, and others.

Vlasov, Aleksandr Vladimirovich
Party official; b. 1932.
Mem. CPSU 1956–; grad. Irkutsk Metallurgical Mining Inst. 1954; Komsomol, sov., and party posts 1954–65; Oblast Sec., 2nd Oblast Sec. of CPSU Cttee. Yakut ASSR 1965–72; posts in Cen. Cttee. CPSU 1972–75; Dep. of USSR Supr. Sov. 1974–; 1st Oblast Sec. of CPSU Cttee. Chechen-Ingush ASSR D. 1975–; cand. mem. Cen. Cttee. CPSU 1976–81, mem. 1981–. Decors.: Order of Lenin.

Vlasov, Iurii Petrovich
Weight lifter; b. D. 5, 1935, Makeevka.
Honored Master of Sport of the USSR (1959). Member of the CPSU since 1960. He graduated from the N. E. Zhukovskii Air Force Engineering Academy in 1959. He was a five-time champion of the USSR, world champion in 1959 and 1961–63, and champion at the 17th Olympic Games (1960, Rome) in the heavyweight category. He holds 28 world records. His highest accomplishments are 580 kg in combined, 199 kg in the press, 172,5 kg in the snatch, and 215,5 kg in the clean and jerk. Decors.: Order of Lenin, the Order of the Badge of Honor, and medals.
Works: *Sebia preodolet': Rasskazy.* Moscow 1964.

Vlasova, Alevtina Nikolaevna
Painter and graphic artist; b. 1924, Konotop.
Attended Moscow Polygraphic Inst. 1943–48, pupil of G. T. Goroshchenko; exhibitions 1951–.

Vnodchenko, Iurii Fedorovich
Painter; b. 1927, Karabanovo, Odessa Oblast.
Grad. Kishinev Republican Art School 1949, attended Kiev Art Inst. 1950–56; exhibitions 1956–.

Vodop'ianov, Mikhail Vasil'evich
Pilot; b. N. 6 (18), 1899, Studenka, now a part of the city of Lipetsk.
One of the first Heroes of the USSR (Ap. 20, 1934); major-general of aviation (1943). Member of the CPSU since 1934. He was born into a peasant family. In February 1918 he joined the Red Army as a volunteer and served as a driver and mechanic. Later he served as an airplane mechanic and pilot. After the Civil War he piloted aircraft between Moscow and Irkutsk and between Moscow and Leningrad and developed the first air route to Sakhalin Island. In 1929 he graduated from a military school of aviation. In March and April 1934 he participated in the rescue of the crew of the icebreaker *Cheliuskin,* for which he was awarded the title of Hero of the Soviet Union. In 1937 he joined an air expedition to the north pole. During the Great Patriotic War, he commanded an air division. He retired in 1946.
He is the author of the novel *The Kireevs* (1956), tales, stories, the autobiographical *Polar Flier* (1952), and *Valerii Chkalov* (1954). Decors.: four Orders of Lenin, four Orders of the Red Banner, an Order of the Patriotic War First Class, and medals.

Voenushkin, Sergei Fedorovich
State official.
Dep. Min. of Building Materials Ind. RSFSR 1975–79, Min. D. 1979–.

Voitikov, Viktor Vladimirovich
Engineer and designer; b. 1923.
Grad. Ufa Aviation Inst. 1947; worker in a factory in Ufa 1941–43; worked in Minsk Tractor Works 1947–; Dep. Chief Designer-Engineer 1969–. Decor.: State Prize of the USSR 1971 (for his part in the development and introduction into mass prod. of unified 1.4–2-ton wheel, half-chain and chain tractor modelled on the "Belarus" MTZ-50 type for the complex mechanization of field work).

Voitovich, Nina Trofimovna, Dr. of Philol. Sc.
Philologist; b. 1913.
Belorussian; mem. CPSU 1940–; grad. Polotsk Pedagogical Tech. School 1930 and Minsk Pedagogical Inst. 1937; Cand. of Philol. Sc. 1958–; sen. mem. scientific staff 1958–; Dr. of Philol. Sc. 1967; village teacher 1930–33; postgraduate studies 1938–41 and 1945–46; mem. Orenburg Radio Cttee. 1942–43; sen. mem. scientific staff, then head of chair at Minsk Pedagogical Inst. 1947–56; sen. mem. scientific staff of Inst. of Linguistics of Acad. of Sciences Belorussian SSR 1956–. Decor.: State Prize of the USSR 1971.
Publ.: "Rodnaia literatura" (Native Literature), textbook on Belorussian literature.

Voitsekhovskii, Aleksandr Klimentevich
Painter; b. 1925, Dnepropetrovsk.
Grad. Dnepropetrovsk Art School 1949, attended Kiev Art Inst. 1949–54, pupil of M. N. Panin, I. N. Shtilmann; exhibitions 1955–; teacher at Dnepropetrovsk Art School 1958–.

Volchkov, Nikolai Nikolaevich
Actor; b. 1910.
Mem. CPSU 1946–; grad. Dikov-Studio in Moscow 1935; with Great Drama Theatre Leningrad 1935–37 and with V. Maiakovskii Russian Drama Theatre in Dushanbe 1937–. Decors.: People's Artist of the Tadzhik SSR 1960, People's Artist of the USSR 1974.

Volfkovich, Semen Isakovich
Inorganic chemist; b. O. 11 (23), 1896, Ananev, now Odessa Oblast.
Grad. Moscow Inst. of Econ. 1920; mem. USSR Acad. of Sciences 1946–; Prof. at Inst. of Econ. and at Moscow Inst. of Technology 1929– and at Moscow Univ. 1946–; mem. Presidium 1937–, Pres. of D. I. Mendeleev Union Soc. of Chemistry 1963–. Decors.: State Prize of the USSR 1941, 3 Orders of Lenin, Order of Red Banner of Labour, Golden Mendeleev Medal and several other medals.
Publs.: co-author of "Obščaja himičeskaja technologija" vols. 1–2, Moscow-Leningrad 1952–59; "Fiziko-himičeskij analiz i tehnologija mineral'nykh udobrenij" in "Žurnal prikladnoj himii" 34, 1961, No. 10, "Gidrotermičeskaja pererabotka fosfatov na udobrenija i kormovye sredstva" (co-author) Moscow-Leningrad 1964, "Problemy proizvodstva mineral'nykh udobrenij" Moscow 1965, "Problemy himii v sel'skom hozjajstve" Moscow 1969, and others.

Vol'kenshtein, Mikhail Vladimirovich
Physical chemist and biophysicist; b. O. 10 (23), 1912, St. Petersburg.
Corresponding member of the Academy of Sciences of the USSR (1966).
He graduated from the physics department of Moscow State University in 1935. He was head of the laboratory of the Institute of Macromolecular Compounds of the Academy of Sciences of the USSR from 1948 to 1967 and a professor at Leningrad University from 1945 to 1953 and from 1963 to 1968. In 1967 he became head of the laboratories of the Institute of Molecular Biology and the Institute of Biophysics of the Academy of Sciences of the USSR. His major works deal with the physics of molecules and with polymers and molecular biophysics. He is the originator of the theory of intensities in the vibrational spectra of molecules. He developed the statistical physics of macromolecules on the basis of the so-called rotational isomeric theory and has conducted theoretical and experimental studies on the structure and properties of molecules, polymers, and biopolymers. He has founded a school devoted to the theory of the physics of macromolecules. Decors.: State Prize of the USSR in 1950.
Works: *Kolebaniia molekul,* vols. 1–2. Moscow-Leningrad 1949. (With M. A. El'iashevich and B. I. Stepanov.) *Molekuliarnaia optika.* Moscow-Leningrad 1951. *Stroenie i fizicheskie svoistva molekul.* Moscow-Leningrad 1955. *Konfiguratsionnaia statistika polimernykh tsepei.* Moscow-Leningrad 1958. *Molekuly i zhizhn': Vvedenie v molekuliarnuiu biofiziku.* Moscow-Leningrad 1965. *Fizika fermentov.* Moscow 1967.

Volkotrubenko, Ivan Ivanovich
Military official; b. Je. 30 (Jl. 12), 1898, Cherntsy, in present-day Balta Raion, Odessa Oblast; s. of a peasant.
Became a colonel-general of the artillery in 1944. Joined the CPSU in 1924.
He fought in World War I as an enlisted man. He went into the Red Army in April 1918 and fought in the Civil War. He graduated from the First Moscow Artillery Courses in 1918, from the Higher Military Communications School in 1924, and from the advanced artillery training courses for commanders in 1931. From 1938 to 1941 he was chief of artillery supply for the Kiev Military District. During the Great Patriotic War he was appointed chief of artillery supply for the Southwestern Front in June 1941, and in February 1942 he was named deputy and first deputy chief of the Main Artillery Directorate. In 1950 he became head of the Main Artillery Directorate. From

1953 to 1967 he was chief of several military artillery educational institutions. He retired in 1967. Decors.: two Orders of Lenin, three Orders of the Red Banner, Orders of Kutuzov First and Second Class, the Order of Suvorov Second Class, the Order of the Red Banner of Labor, the Order of the Red Star, the Order of the Badge of Honor, and several foreign orders, as well as medals.

Volkov, Anatolii Andreevich
Literary scholar; b. My. 27 (Je. 9), 1909, Petriaevo, Moscow Province.
Doctor of philology (1947). Member of the CPSU since 1930.
He graduated from Moscow University in 1931. His works are devoted, for the most part, to the development of Russian literature at the turn of the 20th century. Works: *M. Gor'kii i literaturnoe dvizhenie kontsa XIX i nachala XX vekov.* Moscow 1954. *Ocherki russkoi literatury kontsa XIX i nachala XX vekov,* 2nd ed. Moscow 1955. *A. M. Gor'kii i literaturnoe dvizhenie sovetskoi epokhi.* Moscow 1958. *A. I. Kuprin,* Moscow 1959. *Proza Ivana Bunina.* Moscow 1969. *Put' khudozhnika: M. Gor'kii do Oktiabria.* Moscow 1969. *A. S. Serafimovich: Ocherk zhizni i tvorchestva.* Moscow 1969. *Russkaia literatura XX veka: Dooktiabr'skii period,* 5th ed. Moscow 1970.

Volkov, Anatolii Mikhailovich
Painter; b. 1927, Moscow.
Grad. Moscow School of Ind. Art 1949, attended Surikov Art Inst. Moscow 1949–55; exhibitions 1956–.

Volkov, Gavriil Moiseevich
State official.
Dep. Chair. of Cttee. of State Security Moldavian SSR 1963–79, Chair. Ja. 1979–; mem. Cen. Cttee. and cand. mem. Bureau Cen. Cttee. Moldavian CP Ja. 1981–.

Volkov, Valerii Aleksandrovich
Painter; b. 1928, Fergana; s. of A. N. Volkov.
Attended Republican Art School 1944–47 and Cen. Asian Univ., Faculty of Art, Tashkent 1947–52; exhibitions 1949–.

Volkovinskaia, Zinaida Vladimirovna
Painter and graphic artist; b. 1915, Odessa.
Attended Kiev Art Inst. 1934–40; pupil of P. G. Volokidin, S. A. Grigorev; exhibitions 1941–.

Volnov, Aleksandr Petrovich
Architect; b. N. 20 (D. 3), 1902, Kolpino, now Leningrad Oblast.
Mem. CPSU 1927–; grad. State Workshops for Arts and Crafts resp. Higher State Inst. for Arts and Crafts 1924–29; corresponding mem. Acad. of Sciences Belorussian SSR 1953–; Chair. of Belorussian Architects' Union 1941–49; Lecturer 1947–, Prof. at Minsk Polytech. Inst. 1951–. Decors.: State Prize of the Belorussian SSR 1968, 3 orders and 1 medal.

Volobuev, Pavel Vasil'evich
Historian; b. Ja. 1, 1923, Evgenovka, Kustanai Oblast.
Mem. CPSU 1944–; grad. Dept. of History of Moscow Univ. 1950; corresponding mem. USSR Acad. of Sciences 1970–; sen. mem. scientific staff Inst. of History of USSR Acad. of Sciences 1955–69; head of sector for the edition of multivolume "History of the USSR" 1966–; Dir. of Inst. of History of USSR Acad. of Sciences 1969–.

Volobuev, Vladimir Rodionovich
Soil scientist; b. Jl. 12 (25), 1909, Krasnodar.
Grad. Kuban Agricultural Inst. 1930; corresponding mem. Acad. of Sciences Azerbaidzhan SSR 1968–; Dir. of Inst. of Soil Science and Agrochemistry of USSR Acad. of Sciences 1952–57; Vice-Pres. 1957–59, Acad. Sec. of Dept. of Biol. Sciences of Acad. of Sciences Azerbaidzhan SSR 1969–. Decor.: State Prize of the USSR 1967.

Volodarskii, A. P.
Komsomol official.
2nd Sec. and mem. Bureau Cen. Cttee. Uzbek Komsomol Ag. 1978–.

Volodarskii, Lev Mordkovich
State official; b. 1911.
Mem. CPSU 1939–; grad. Leningrad Econ. Engineering Inst. 1934; Dr. of Econ. Sc. 1962, Prof. 1964; Dep. Chair. Planning Comm. Leningrad Oblast 1938–40; posts in USSR GOSPLAN apparatus; Dep. Chief of Cen. Statistical Admin. of USSR Council of Mins. 1948–53; Chief Ed. of State Publishing House for Statistics, dept. chief, admin. chief of Cen. Statistical Admin. of USSR Council of Mins. 1953–56; Dep. Chief of USSR Cen. Statistical Admin. 1956–67, 1st Dep. Chief 1967–75, Chief 1975–; cand. mem. Cen. Cttee. CPSU 1976–81, mem. 1981–; Dep. to USSR Supr. Sov. 1979–.

Volodin, Aleksandr Moiseevich (real name: Lifshits, A. M.)
Dramatist; b. F. 10, 1919, Minsk.
Grad. Dept. to Directing at Union Film Inst. 1949. Decors.: medals.
Publs.: "Fabrichnaia devochka" (Factory Girl) 1956, "Piat' vecherov" (Five Evenings) 1959, "V gostiakh i doma" (On Visit and at Home) 1960, "Starshaia sestra" (The Elder Sister) 1961, "Zvoniat, otkroite dver" (The Bell Is Ringing, Open the Door) 1965, "Pokhozhdeniia zubnogo vracha" (The Adventure of a Dentist) 1966, "Dulsineia Tobosskaia" 1971, and others.

Voloshin, Ivan Makarovich
Military official; b. 1923.
Ukrainian; mem. CPSU 1944–; grad. M. V. Frunze Mil. Acad. and Mil. Acad. of Gen. Staff, Col.-Gen.; entered Sov. Army 1941; company commander, regimental commander 1941–45; command posts 1945–74; Commander of Troops of Odessa Mil. District 1974–; Dep. to USSR Supr. Sov. 1974–; mem. Cen. Cttee. Ukrainian CP 1976–.

Volosiuk, Vasilii Mikhailovich
State official.
Cand. mem. 1960–61, mem. Cen. Cttee. Moldavian CP 1961–; Dep. to Supr. Sov. Moldavian SSR 1963–; Sec. of Party Cttee. of Kolkhoz-Sovkhoz Prod. Admin. in Dubossary 1963; Head of Dept. of Admin. Organs with Cen. Cttee. Moldavian CP 1963–70; Min. of Justice Moldavian SSR O. 1970–.

Volostnykh, A. I.
Trade union official.
Cand. mem. Cen. Cttee. Esthonian CP 1976–81; Sec. of Esthonian Republican Trade Union Council F. 1977–.

Volovik, Viktor Petrovich
Sculptor; b. 1924, Kharkov.
Attended Kharkov Art Inst. 1947–53, pupil of A. O. Strakhov; teacher at Kharkov Art Inst. 1953–.

Volskii, Iurii Ivanovich
Diplomat; b. 1922.
Mem. Diplomatic Service 1959–; Emb. Counsellor at USSR Emb. in the USA 1959–62; Dep. Min. of Foreign Affairs RSFSR 1962–66; USSR Amb. to Argentina 1966–72; mem. staff USSR Min. of Foreign Affairs 1972–76; USSR Amb. to Mexico Mr. 1976–80; concurrently to Jamaica Mr. 1977–78; Head, 2nd Latin American Countries Dept., USSR Min. of Foreign Affairs 1980–.

Vol'skii, Vitalii Fridrikhovich (pseudonym of Zeidel')
Belorussian writer and literary scholar; b. Ag. 23 (S. 5), 1901, St. Petersburg.
Member of the CPSU since 1925.
He took part in the Civil War, and during the years 1920–26 he served in the Red Army. He began publishing his work in 1926. His plays *The Miraculous Pipe* (1938), *Grangdad and the Crane* (1939), *Nesterka* (1940; film with the same title, 1955), and *Masheka* (1946), based on motifs of Belorussian national folklore, poeticize heroic folk characters. He is also the author of books of sketches and short stories about Belorussian nature: *Through Forest Paths* (1948), *In the Forests on the Bereza River* (1955), *Month After Month* (1956), *Sea Gulls Over the Naroch' River* (1959), *Native Land* (1961), *Three Journeys to the Land of the Belorussians* (1967), *Poles'e* (1970); also books of impressions gained while traveling abroad: *African Journey* (1963) and *El*

Maghreb (1965). He wrote *An Outline History of Belorussian Literature During the Epoch of Feudalism* (1958) and translated Goethe's *Reynard the Fox* into Belorussian. Decors.: 2 orders and medals.
Works: *In Russian translation: P'esy* [Foreword by P. N. Berkov.] Moscow 1954. *Pod nebom Afriki.* Minsk 1969.

Volynov, Boris Valentinovich
Cosmonaut; b. D.18, 1924, Irkutsk.
Mem. CPSU 1958–; grad. Volgograd Mil. Aviation College 1956 and Zhukovskii Acad. of Mil. Aircraft Engineering 1968; cosmonaut 1960–; as commander of spaceship "Soiuz 5" (together with A. S. Eliseev and E. V. Khrunov) undertook docking manoeuvre with "Soiuz 4" (commanded by V. A. Shatalov), travelled approx. 2 million km (50 orbits) in 72 hours 46 mins. on Ja. 15–18, 1969, "Soiuz 21" Jl.–Ag. 1976. Decor.: Order of Lenin.

Volzhenskii, Aleksandr Vasil'evich
Scientist in the field of technology of structural materials and products; b. Ag. 27 (S. 8), 1899, Alt-Kliuchi, Buguruslan District, in present-day Orenburg Oblast.
Became a doctor of technical sciences in 1946, and an Honored Scientist and Engineer of the RSFSR in 1960.
After graduating in 1925 from the Tomsk Technological Institute, he worked in enterprises of the structural building-materials industry, in higher educational institutions, and in scientific-research institutions. In 1949 he began teaching at the Moscow Civil Engineering Institute (having become a professor in 1948). He was the first to work out theoretical bases for the technology of autoclave materials; he did important work on the improvement of the properties of gypsum binders and the creation of gypsum-cement-pozzolana binders and also on the establishment of the technological principles of large-scale gypsum concrete and silicate concrete objects. Decors.: Lenin Prize (1962), the Order of Lenin, the Order of the Red Banner of Labor, and medals.
Works: *Vodotermicheskaia obrabotka stroitel'nykh materialov v avtoklavakh,* [Moscow] 1944. *Gipsotsementnye i gipsoshlakovye viazhushchie i izdeliia.* Moscow 1960. (Co-author.) *Betony i izdeliia iz shlakovykh i zol'nykh materialov* [2nd ed.]. Moscow 1969. (Co-author.)

Vonsovskii, Sergei Vasil'evich
Physicist; b. Ag. 20 (S. 2), 1910, Tashkent.
Grad. Leningrad Univ. 1932; mem. USSR Acad. of Sciences 1966–; Chair. of Presidium of Uralsk Scientific Centre of USSR Acad. of Sciences 1971–; mem. staff Uralsk Physicotech. Inst. in Sverdlovsk 1932–39; mem. staff Inst. of Metal Science, Metal Physics and Metallurgy of Uralsk Branch of USSR Acad. of Sciences 1939–47; mem. staff Inst. of Physics of Metals of USSR Acad. of Sciences 1947–. Decors.: Hero of Socialist Labour, Order of Lenin, 2 other orders and various medals.
Publs.: co-author of "Ferromagnetizm" Moscow-Leningrad 1948; "Sovremennoe uchenie o magnetizme" Moscow-Leningrad 1952, co-author of "Elektronnaia teoriia perekhodnykh metallov" in "Uspekhi fizicheskikh nauk" Nos. 77 and 78, 1962 p. 377 ff resp. 3 ff; co-author of "Austauschwechselwirkung in Isolatoren" in "Handbuch der Physik" ed. by S. Flügge, vol. 18/1, 1968.

Vorob'ev, Arkadii Nikitich
Sportsman (weight lifter); b. O. 3, 1924, Mordovo, Tambov Oblast.
Honored Master of Sport (1952) and Honored Coach of the USSR (1964).
He, a medical doctor, has been a member of the CPSU since 1954. Champion of the USSR ten times, he was the first Soviet light heavyweight weight lifter to win the world championship. He was world champion (1953–55, 1957–58) and champion of the sixteenth and seventeenth Olympics (1956, Melbourne; 1960, Rome) in the light heavyweight category. His highest total in the triathlon was 472,5 kg (1960). Decors.: Order of Lenin, two other orders, and medals.
Works: *Sovremennaia trenirovka tiazheloatleta.* Moscow 1964. *Na trekh Olimpiadakh: Khel'sinki – Mel'burn – Rim.* Sverdlovsk 1963.

Vorobev, Boris Sergeevich
Diplomat; b. 1923.
Sen. posts with Cen. Apparat of USSR Min. of Foreign Affairs 1952–62; 1st Sec., Emb. Councellor at USSR Emb. in Ghana 1962–68; mem. staff Cen. Apparat USSR Min. of Foreign Affairs 1968–70 and 1974–76; USSR Amb. to Nigeria 1970–74; USSR Amb. to Angola Mr. 1976–78, to Soudan Mr. 1982–.

Vorobev, Evgenii Zakharovich
Author; b. 1910, Riga.
Russian; grad. Leningrad Inst. of Journalism; worked for Komsomol press, then as correspondent for "Komsomolskaia pravda" 1930–41; correspondent for front newspaper "Krasnoarmeiskaia pravda" 1941–45; first works publ. 1946.
Publs.: novelettes "Pechotnaia gordost'" (The Pride of the Infantry) 1946, "Kaplia krovi" (Blood Drops) 1960, novel "Vysota" (Height) 1952, features "Litsom k solntsu" (With the Face to the Sun) 1956, "Sto chetyrnadtsat' flagov" (One Hundred and Fourteen Flags) 1956, "Polskie vstrechi" (Polish Encounters) 1956, "Na sed'mom nebe" (In Seventh Heaven) 1961, and others.

Vorobev, Fedor Dmitrievich
Painter and graphic artist; b. 1922, Poliany, Iaroslavl Oblast.
Attended Iaroslavl Art School 1937–41, pupil of P. S. Oparin, S. F. Shitov; exhibitions 1954–.

Vorobev, Georgii Ivanovich, Cand. of Econ. Sc.
State official; b. 1914.
Russian; mem. CPSU 1939–; grad. Timiriazev Acad. of Agriculture; agronomist, head of pol. dept. of a sovkhoz 1934–43; 2nd Sec. of Leningrad Oblast Cttee. CPSU, Chair. of Leningrad Exec. Cttee. of Oblast Sov. of Workers Deps., 1st Sec. of Udmurt ASSR Oblast Cttee. CPSU 1943–59; Dep. to USSR Supr. Sov. 1950–66; mem. Cen. Auditing Comm. 1956–61, mem. Cen. Cttee. CPSU 1961–66; Head of Agricultural Dept. of RSFSR with Cen. Cttee. CPSU 1959–60; 1st Sec. of Krasnodar Krai Cttee. CPSU 1960–66 (for Agriculture 1962–64); Dep. USSR Min. of Agriculture 1966–70; Chair. of USSR State Cttee. on Forestry My. 1970–; cand. mem. Cen. Cttee. CPSU 1976–. Decors.: 2 Orders of Lenin, Order of the October Revolution.

Vorob'ev, Ivan Alekseevich
Military official; b. Ag. 26, 1921, Gorbachevo, Odoev Raion, Tula Oblast; s. of a peasant.
Twice Hero of the Soviet Union (Ag. 19, 1944, and Je. 29, 1945), colonel (1956). Became a CPSU member in 1944.
He has been in the Soviet Army since 1940; he graduated from the Tambov Military Aviation School for Pilots in 1941 and from the Red Banner Air Force Academy in 1952. During the Great Patriotic War he advanced between August 1942 and May 1945 from pilot to squadron commander of the 76th Guards Storm Regiment and flew about 400 missions (over Stalingrad, Sevastopol', Belorussia, and East Prussia). After the war he held commander's posts and has been an instructor since 1958. Decors.: Order of Lenin, two Orders of the Red Banner, Orders of the Patriotic War First and Second Class, the Order of Alexander Nevsky, two Orders of the Red Star, and medals.

Voronin, Leonid Grigorevich, Dr. of Biol. Sc.
Physiologist; b. Jl. 22 (Ag. 4), 1908, Trituznoe, now Dnepropetrovsk Oblast.
Mem. CPSU 1927–; grad. Dept. of Biology of Dnepropetrovsk Pedagogical Inst. 1931; corresponding mem. USSR Acad. of Sciences and Acad. of Pedagogical Sciences 1968–; Head of Chair of Higher Nervous Activity at Moscow Univ. 1954– and concurrently Dir. of Inst. of Higher Nervous Activity of USSR Acad. of Sciences 1957–59. Decors.: 2 orders and various medals.
Publs.: "Analiz i sintez složnyh razdražitelej u vyših životnih" Leningrad 1952, "Lekcii po sravnitel'noj fiziologii vissej nervnoj dejatel'nosti" Moscow 1957, "Kurs lekcii po fiziologii vyssej nervnoj dejatel'nosti" Moscow 1965.

Voronin, Lev Alekseevich
State official; b. 1928.
Mem. CPSU 1953–; grad. Ural Polytechnical Inst. 1949; foreman, dept. chief, chief engineer in various plants in Sverdlovsk region 1949–63; chief of an admin. of the Agricultural Council 1963–65; Factory Dir. 1965–68; chief of an admin. of USSR Ministry of Defense Ind. 1968–72; Dep. Min., 1st Dep. Min. of USSR Defense Ind. 1972–80; 1st Dep. Chair. USSR GOSPLAN O. 1980–; Dep. to USSR Supr. Sov. 1979–; mem. Cen. Cttee. CPSU 1981–.

Voronin, Petr Vasil'evich, Cand. of Hist. Sc.
State official; b. 1918; s. of a white-collar worker.
Ukrainian; mem. CPSU 1943–; grad. Pedagogical Inst. in Poltava and Cen. Cttee. Party Higher School CPSU; teacher, served in Sov. Army 1940–46; instructor, dep. head of a dept. of Kishinev City Cttee. Moldavian CP 1946–50; Sec. of Party Bureau, senior teacher at Kishinev Univ. 1950–52; grad. student Kishinev Univ. 1952–55; head of Dept. for Propaganda and Agitation and Dept. for Party Organizational Work Kishinev City Cttee. Moldavian CP 1955–59; dep. head of Party Organs Dept. Cen. Cttee. Moldavian CP 1959–62; mem. Cen. Cttee. Moldavian CP 1961–; cand. mem. Bureau of Cen. Cttee. Moldavian CP 1961–62, mem. 1962–; Sec. of Cen. Cttee. Moldavian CP 1962–65; Dep. Chair. of Council of Mins. and Chair. of Cttee. for Party and State Control Cen. Cttee. Moldavian CP and of Council of Mins. Moldavian SSR 1962–65; Chair. of Cttee. for Peoples' Control Moldavian SSR D. 1965–.

Voronin, Sergei Alekseevich
Author; b. 1913; Liubim, Iaroslavl Oblast; s. of a clerk.
Russian; mem. CPSU 1947–; concluded factory school of Leningrad Metallurgical Works; has worked as lathe-turner, geologist and journalist.
Publs.: novelettes "Vstrechi" (Encounters) 1947, novel "Na svoei zemle" (In One's Own Country) vols. 1–2 1948–52, poem "Nenuzhnaia slava" (Superfluous Fame) 1955, novelletes "Dom naprotiv" (The House Opposite) 1954, "Vtoroi tsvet" (The Second Bloom) 1958, and others.

Voron'ko, Platon Nikitich
Author and poet; b. 1913, Chernetchina, Sumy Oblast; s. of a peasant.
Ukrainian; mem. CPSU 1943–; studied Gorkii Inst. of Literature. Decors.: Order of Red Banner, Order of Patriotic War 2nd Cl., medals.
Publs.: collection of poetry and songs "Karpatskii reid" (Carpathian Raid) 1944, "Poezii" (Poetry) 1950, "Moia Moskva" (My Moscow) 1953, "Dolg" (Duty), "Rozhdenie legendy" (Birth of a Legend) 1954, "Poedinok" (The Duel) 1958, poetry "Dobroe utro" (Good Morning) 1951, "Slaven mir" (Peace is Glorious) 1951, "Kazka pro Chugaistra" (The Tale of Chugaistra) 1957, and others.

Voronkova, Liubov' Fedorovna
Author; b. 1906, Moscow; d. of a peasant.
Russian; studied at a pedagogical inst.; worked for various village newspapers; first works publ. 1930.
Publs.: children's books "Devochka iz goroda" (A Girl From the City) 1943, "Selo Gorodishche" (The Village of Gorodishche) 1947, "Altaiskaia povest'" (Altai Poem), and others.

Voronkov, Mikhail Grigorevich, Dr. of Chem. Sc.
Chemical scientist; b. 1921.
Mem. CPSU 1964–; grad. Sverdlovsk Univ. 1942; Dr. of Chem. Sc. 1961; corresponding mem. Acad. of Sciences Latvian SSR 1966–, USSR Acad. of Sciences 1970–; Head of Laboratory of Inst. of Organic Synthesis of Acad. of Sciences Latvian SSR 1961–70; Dir. of Inst. of Organic Chemistry of Siberian Branch of USSR Acad. of Sciences in Irkutsk 1970–.

Voronov, Avenir Arkadevich, Dr. of Tech. Sc.
Cybernetics scientist; b. 1910.
Mem. CPSU 1943–; grad. Leningrad Polytech. Inst. 1938; Dr. of Tech. Sc. 1958; Prof. 1959–; mem. USSR Acad. of Sciences 1970–; mem. scientific staff Inst. of Automatics and Telemechanics of USSR Acad. of Sciences 1948–55; mem. staff Inst. of Electromechanics of USSR Acad. of Sciences 1955–64; Head of Laboratory, Dep. Dir. of Inst. of Control Processes 1964–70; mem. staff Far Eastern Scientific Centre of USSR Acad. of Sciences, Dep. Chair. of Presidium and concurrently Dir. of Inst. of Automation and Control Processes and Computer Centre 1971–.

Voronovich, Aleksandra Petrovna (married name: Kramova)
Russian actress; b. Jl. 7 (19), 1898, Kunino, Iaroslavl Oblast.
People's Artist of the USSR (1954).
She graduated from the theatrical studio of the Sukhodol'skiis Moscow Dramatic Theater in 1918 and was retained in the theater's troupe. Later she worked in the former Nezlobin Theater, the former Korsh Theater, and from 1933 to 1962 in the Pushkin Russian Drama Theater in Kharkov.
Her art is marked by a high degree of drama, profound lyricism, and emotional force. She created poetic images of Liza Kalitina in *A Nest of Gentlefolk,* based on Turgenev's novel, and of Anna Karenina in a stage version of L. N. Tolstoy's novel of the same name. Among her best roles were Negina in Ostrovskii's *Talents and Admirers,* Katiusha Maslova in the stage version of Tolstoy's novel *Resurrection,* and Mariia Aleksandrovna Ulianova in Popov's *The Family.* Decors.: two orders and various medals.

Vorontsov, Iulii Mikhailovich
Diplomat; b. 1929.
Mem. CPSU 1956–; grad. Moscow Inst. of Int. Relations 1952; held posts in USSR Min. of Foreign Affairs apparatus 1952–54; mem. of Sov. delegation at UN 1954–58 and 1963–65; councillor at USSR Emb. to USA 1966–70, ministerial councillor 1970–77; USSR Amb. to India 1977–; mem. Cen. Cttee. CPSU 1981–.

Voropaev, Mikhail Gavrilovich
Party official; b. 1919.
Russian; mem. CPSU 1945–; grad. Rostov Inst. of Railway Transport 1941 and Party High School of Cen. Cttee. CPSU 1960; engineer, chief technologist, engineer for locomotive repairs in Zlatoust (Cheliabinsk Oblast) 1941–44; Sec. of Komsomol Cttee. of Cheliabinsk Locomotive Depot, Raion Sec., City Sec. and 2nd Oblast Sec. of USSR Komsomol in Cheliabinsk 1944–49; engineer at Cheliabinsk Metallurgical Works, Sec. of Cheliabinsk Oblast Cttee. CPSU 1949–60; Head of Dept. of Ind. and Transport, 2nd, 1st Sec. of Cheliabinsk City Cttee., 2nd Sec. of Cheliabinsk Oblast Cttee. CPSU 1960–70, Dep. to USSR Supr. Sov. 1966–; 1st Sec. of Cheliabinsk Oblast Cttee. CPSU Jl. 1970–; mem. Cen. Cttee. CPSU 1971–. Decor.: Order of Lenin.

Vorotnikov, Vitalii Ivanovich
Party official; b. 1926.
Russian; mem. CPSU 1947–; grad. Kuibyshev Aviation Inst. 1954; controller, technologist, dept. head, sec. of party cttee., chief controller of a plant 1942–44 and 1947–60; party and sov. posts 1960–61; Sec. of Kuibyshev Oblast Cttee. 1961–63, 2nd Sec. (for Ind.) of Kuibyshev Oblast Cttee. 1963–64, 2nd Sec. of Kuibyshev Oblast Cttee. CPSU 1965–67; Chair. of Kuibyshev Exec. Cttee. of Oblast Sov. of Workers' Deps. 1967–71; Dep. to USSR Supr. Sov. 1970–; 1st Sec. of Voronezh Oblast Cttee. CPSU F. 1971–75; mem. Cen. Cttee. CPSU 1971–; 1st Dep. Chair. of Council of Mins. RSFSR Jl. 1975–79, Amb. USSR to Cuba 1979–82; 1st Sec. Krasnodar Krai Cttee. CPSU Jl. 1982–; Chair. of newly founded USSR-Angolan Friendship Soc. N. 1976–. Decor.: Order of the October Revolution.

Vorovich, Iosif Izrailevich
Mechanics scientist; b. 1920.
Mem. CPSU 1950–; studied 4 terms at Moscow Univ., grad. Zhukovskii Mil. Aviation Acad. 1944; corresponding mem. USSR Acad. of Sciences 1970–; served in Sov. Army 1941–50; mem. staff Rostov Univ. 1950–; Prof. and head of a chair 1959–.

Vorozheikin, Arsenii Vasil'evich
Military official; b. O.15 (28), 1912, Prokof'evo, in present-day Gorkii Oblast; s. of a peasant.
Twice Hero of the Soviet Union (F.4, 1944, and Ag.19, 1944), major-general of aviation (1954). Member of the CPSU beginning in 1932.
He was in the Soviet Army in 1931–33 and from 1934. He graduated from a military aviation pilot's school in 1937, from the Military Academy of Commanders and Navigators of the Air Force in 1942, and from the Military Academy of the General Staff in 1952. In 1939–41 he did political work in air force units and participated in battles on the Khalkhin-Gol River and in the Soviet-Finnish War (1939–40). During the Great Patriotic War he commanded a squadron of the 728th Aviation Fighter Regiment and was deputy commander of the 32nd Guards Aviation Fighter Regiment, making more than 240 combat flights and personally shooting down 52 enemy aircraft (on the Kalinin, Voronezh, and First Ukrainian fronts). After October 1944 he held high posts in the air force administration and then command and staff posts. Since 1957 he has been in the reserve for reasons of health. Decors.: Order of Lenin, four Orders of the Red Banner, the Order of Suvorov Third Class, the Order of Alexander Nevsky, two Orders of the Red Star, and medals.

Vorozhtsov, Aleksandr Petrovich
Diplomat.
USSR Amb. to Sierra Leone Ag. 1979–.

Vorozhtsov, Nikolai Nikolaevich
Organic chemist; b. My.24 (Je.6), 1907, Tomsk.
Mem. CPSU 1942–; grad. Dept. of Chemistry of Moscow Higher Tech. College 1929; mem. USSR Acad. of Sciences 1966–; teaching work at various institutes 1930–; Prof. 1939–45, head of a chair at D. I. Mendeleev Inst. of Chemotechnology in Moscow 1945–58; Dir. of Inst. of Organic Chemistry (founded under his supervision) of Siberian Branch of USSR Acad. of Sciences in Novosibirsk 1958–. Decors.: State Prize of the USSR 1952, Order of Lenin, Order of Red Banner of Labour and various medals.

Voskanian, Grant Mushegovich
Party official; b. 1924.
Mem. CPSU 1946–; grad. Erevan Pedagogical Inst., teacher, dir. of a boarding school, Head of Kirovakan City Dept. of Educ. until 1959; Sec. 1959–64, 1st Sec. of Kirovakan City Cttee. Armenian CP 1965–67; mem. Cen. Cttee. Armenian CP 1966–71 and 1976–; mem. Apparat Armenian CP 1967–73; Head of Dept. of Party Org. Work with Cen. Cttee. Armenian CP 1973–75; cand. mem. Bureau 1973–75, Sec. and mem. Bureau Cen. Cttee. Armenian CP F. 1975–.

Voskrekasenko, Sergei Illarionovich
Author and poet; b. 1906, Lazurtsy, Cherkassy Oblast; s. of a peasant.
Ukrainian; studied Kiev Pedagogical Inst.; first works publ. 1928.
Publs.: poetry collection "Shturm" (Storm) 1931, "Martovskaia noch'" (A Night in March) 1937, "Serzhant Ivan Rubakha" (Sergeant Ivan Rubakha), popular songs "Shakhterochka" (The Girl Miner), "Devushki iz Donbassa" (Girls from the Don Basin); translated Tvardovskii's "Vasilii Terkin" into Ukrainian.

Voss, August Eduardovich, Cand. of Econ. Sc.
Party official; b. 1916; s. of a peasant.
Latvian; grad. Party High School of Cen. Cttee. CPSU 1948 and Acad. of Social Sciences of Cen. Cttee. CPSU 1953; pedagogic work 1933–40; served in Sov. Army 1940–43; Dir. of an intermediate school 1943–45; instructor, section head with Cen. Cttee. Latvian CP 1943–45; mem. Cen. Cttee. Latvian CP 1954–; mem. Bureau 1960–, Sec. 1960–66, 1st Sec. of Cen. Cttee. Latvian CP 1966–; mem. Presidium Supr. Sov. Latvian SSR 1963–; Dep. to 1966–, mem. 1966–70, Chair. of Comm. for Popular Educ., Science and Culture of Sov. of the Union USSR Supr. Sov. 1970–; mem. Cen. Cttee. CPSU 1971–. Decors.: Order of Lenin, Order of Red Banner of Labour, Order of Lenin 1976.

Vostrukhov, Vladimir Ivanovich
Military official; b. My.31 (Je.12), 1895, at Novki station, in present-day Vladimir Oblast.
Colonel-general (1945). Member of the CPSU since 1944. Second lieutenant in World War I.
He joined the Soviet Army in January 1919. He fought in the Civil War. He graduated from Vystrel, the higher infantry school of the Soviet Army, in 1925 and from the M. V. Frunze Military Academy in 1933. In the Great Patriotic War of 1941–45, he was deputy commander of the army for logistics during August–October 1941, commander of the troops of the Twenty-second and Thirty-first armies from October 1941 to April 1942, and deputy commander of the troops for logistics of the Kalinin Front, of the Thirtieth Army, of the Steppe Military District, and of the Steppe, Second Ukrainian, and Transbaikal fronts. He was chief of the Military Academy of Logistics and Supplies during 1946–49 and chief of logistics of the Armed Forces from 1949 to 1953. In 1953 he retired for reasons of health. Decors.: Order of Lenin, two Orders of the Red Banner, the Order of Suvorov First Class, the Order of Kutuzov First Class, the Order of Bogdan Khmel'nitskii First Class, the Order of the Patriotic War First Class, and several foreign orders and medals.

Vovk, Aleksandr Ivanovich
Painter; b. 1922, Kiev.
Attended Kiev Art Inst. 1944–50, pupil of A. I. Sirotenko, A. I. Fomin, A. A. Shovkunenko; exhibitions 1949–.

Vovkushevskii, Rostislav Ivanovich
Painter and artist in monumental and decorative art; b. 1917, Polotsk.
Grad. Repin Inst. for Painting, Sculpture and Architecture at the USSR Acad. of Arts 1949; exhibitions 1952–.

Voznesenskii, Andrei Andreevich
Poet; b. My.12, 1933, Moscow; s. of a scientist.
Russian; first works publ. 1958.
Publs.: poem "Mastera" (Masters) 1959, poetry collection "40 liricheskikh otstuplenii iz poemy 'Treugol'naia grusha'" (40 Lyric Digressions on the Poem "The Three-Cornered Pear") 1962, poems "Lonzhiumo" 1963 and "Oza" 1964, "Antimiry" (Anti-Worlds) 1964, "Akhillesovo serdtse" (The Heart of Achilles) 1966, "Ten' zvuka" (The Shadow ov a Sound) 1970, and others.

Voznesenskii, Boris Mikhailovich
Graphic artist and painter; b. 1915, Shugurovo, Mordovian ASSR.
Mem. CPSU 1944–; grad. Sverdlovsk Art School 1947; exhibitions 1954–.

Vronskii, Vakhtang Ivanovich (pseudonym of V. I. Nadiradze)
Choreographer; b. Ag. 28 (S. 10), 1905, Tbilisi.
People's Artist of the USSR (1962). Member of the CPSU since 1940.
Upon graduating from the Tbilisi Ballet Studio in 1923, he became a soloist with opera and ballet theaters in Rostov-on-Don, Baku, Tashkent, and Saratov. He began his choreographic activities in 1932. From 1940 to 1954 he was a choreographer at the Odessa Theater of Opera and Ballet and from 1954 to 1959 at the Shevchenko Kiev Theater of Opera and Ballet. His productions include Rusinov's *Olesia* (1947), Dankevich's *Lileia* (1956), and Skorul'skii's *Forest Song* (1958). Since 1960 he has been directing a Ukrainian ice ballet company. Decors.: Order of Lenin, three other orders, and various medals.

Vrubel, Anatolii Vasil'evich
Sculptor; b. 1930, Moscow.
Grad. Moscow School of Arts and Crafts 1955, pupil of G. I. Motovilov, G. A. Shults; exhibitions 1956–; teacher at Moscow School of Arts and Crafts 1957–.

Vrublevskaia, V. B.
Komsomol official.
Sec. and mem. Bureau Cen. Cttee. Ukrainian Komsomol 1973–.

Vsevolozhskii, Mikhail Nikolaevich
Party official; b. 1917.
Russian; mem. CPSU 1944–; grad. Zaporozhe Aviation College 1936, Party School of Cen. Cttee. Ukrainian CP 1953 and Dnepropetrovsk Metallurgical Inst. 1965; technologist at various plants in Voronezh, Rybinsk and Ufa 1936–37 and 1941–43; Komsomol and party posts 1943–; organizer for Cen. Cttee. All-Union Komsomol in an ind. plant 1943–45; 2nd Sec. 1945–46, 1st Sec. of Zaporozhe City Cttee. Ukrainian Komsomol 1946–49; sec. of a raion cttee. Ukrainian CP, Head of Dept. of Party Organs of Zaporozhe Oblast Cttee. Ukrainian CP 1953–63; 2nd Sec. 1963–66, 1st Sec. of Zaporozhe Oblast Cttee. Ukrainian CP 1966– (for Ind. 1963–64); cand. mem. 1966–76, mem. Cen. Cttee. CPSU 1976–; cand. mem. 1966–67, mem. Cen. Cttee. Ukrainian CP N. 1967–; Dep. to and mem. Planning and Budget Comm. of Sov. of the Union USSR Supr. Sov. 1966–. Decors.: Hero of Socialist Labour, Order of Lenin, medals.

Vul, Betsion Moiseevich
Physicist and astronomer; b. My. 9 (22), 1903, Belaia Tserkov.
Mem. CPSU 1922–; grad. Kiev Polytech. Inst. 1928; mem. USSR Acad. of Sciences 1972–; mem. staff Physics Inst. of USSR Acad. of Sciences 1933–; Head of Laboratory for Physics of (Dielectric) Nonconductors (now Laboratory for Physics of Semiconductors) founded by him 1933–; mem. Chief Ed. Council of "Bol'shaia Sovetskaia Entsiklopediia" 1951–. Decors.: Hero of Socialist Labour 1969, State Prize of the USSR 1946, Lenin Prize 1964, 4 Orders of Lenin, 2 other orders and various medals.

Vyrgan, Ivan Anikievich
Author and poet; b. 1908, Matveevka, Poltava Oblast; s. of a peasant.
Ukrainian; grad. Kharkov Univ. 1940; worked as teacher; first works publ. 1930. Publs.: "Vooruzhennaia lirika" (Armed Lyrics) 1934, poetry collection "Sad druzhby" (Garden of Friendship) 1935, "Schastie-dolia" (Fortune and Fate) 1938, poem "Dzhigity" (The Dzhigits) 1937, "Povorot solntsa k letu" (The Sun's Turn Towards Summer) 1947, "Matviivka nad Suloiu" (Matviivka on the Sula River) 1949, "Tsvetushchie berega" (Flowering Banks) 1950, "Poezii" (Poetry 1954, "Izbrannoe" (Selected Works) 1956, and others.

Vyrk, Iurii Vasil'evich
State official.
Chair. of State Cttee. for the Supply of Petroleum Products Esthonian SSR Mr. 1980–.

Vyrodova-Gotie, Valentina Gavrilovna
Painter; b. 1934, Kiev.
Attended Kiev Art Inst. 1954–59, pupil M. I. Ivanov, V. G. Puzyrkov; exhibitions 1959–; teacher at Kiev Art Inst. 1959–.

Vyrzhikovskii, Edvard Iakovlevich
Painter; b. 1928, Irkutsk.
Attended Repin Inst. for Painting, Sculpture and Architecture at the USSR Acad. of Arts 1948–54, pupil of P. P. Belousov, A. D. Zaitsev; exhibitions 1952–.

Vysheslavskii, Leonid Nikolaevich
Author, poet and literary historian; b. 1914, Nikolaev.
Mem. CPSU 1943–; grad. Dept. of Philology of Kiev Univ. 1938; first works publ. 1931. Decors.: Badge of Honour, 2 other orders.
Publs.: "Vesna vdvoem" (Double Spring) 1936, "Zdravstvui, solntse" (Greetings, Sun) 1938, poem "Chaika" (The Seagull) 1946, "Podvig mechtatelei" (Feat of Dreamers) 1949, "Pesnia s Dnepra" (Song from the Dnepr) 1951, "Molodost' mira" (The Youth of the World) 1951, "Prostor" (Expanse) 1956, "Edinstvo zemli" (The Unity of Earth) 1958, and others.

Vyshku, Vasilii Konstantinovich
State official; b. 1940; s. of a farmer.
Moldavian; mem. CPSU 1964–; grad. Inst. of Trade and Economics in L'vov and Cen. Cttee. Party Higher School CPSU; chief accountant of Ataki raion consumers' cooperative 1961–62; held various posts in Cen. Cttee. Moldavian Komsomol 1962–63; attended Cen. Komsomol School of USSR Komsomol Cen. Cttee. 1963–64; 1st Sec. of Frunze Raion Cttee. Moldavian Komsomol in Kishinev 1965–66; 2nd, 1st City Sec. Moldavian Komsomol in Kishinev 1966–70; 2nd Sec. 1970–71, 1st Sec. of Cen. Cttee. Moldavian Komsomol 1971–74; cand. mem. Cen. Cttee. Moldavian CP 1971–73, mem. F. 1973–; 1st Sec. of Lenin Raion Cttee. Moldavian CP in Kishinev 1974–76; head of the Dept. for Propaganda and Agitation Cen. Cttee. Moldavian CP 1976–79; Dep. Chair. of Council of Mins. Moldavian SSR June 1981–.

Zababakhin, Evgenii Ivanovich
Physicist; b. Ja. 3 (16), 1917, Moscow.
Mem. CPSU 1949–; grad. Acad. of Mil. Aircraft Engineering 1944; mem. USSR Acad. of Sciences 1968–; specialist on gas dynamics.

Zabashta, Liubov' Vasil'evna
Author and poet; b. 1918, Priluki, Chernigov Oblast.
Ukrainian; mem. CPSU 1953–; worked as engineer; first works publ. 1935.
Publs.: "Chtoby molchali pushki" (To Keep the Cannons Silent) 1952, "Dorogi druzhby" (Roads of Friendship) 1953, "Oleksa Borkaniuk" 1954, "Malvy na kamne" (Mallows on the Stone) 1958, "Pesnia i khleb" (Song and Bread) 1963, and others.

Zabashta, Vasilii Ivanovich
Painter; b. 1918, Babenkovo, Kharkov Oblast.
Mem. CPSU 1945; attended Kharkov Art School 1937–39, pupil of S. Prokhorov; grad. Kharkov Art Inst. 1951, pupil of K. Trokhimenko, V. Kostetskii; teacher at Kharkov Art Inst. 1955–.

Zabelin, Igor Mikhailovich
Author; b. 1927, Leningrad.
Russian; grad. Dept. of Geography of Moscow State Univ. 1948; geographer; literary work 1952–.
Publs.: "Tam gde skhodiatsia tropy" (Where the Paths Meet) 1956, "Zona vzryvov" (Blasting Zone) 1960, novella "Pod severnym solntsem" (Under the Northern Sun) 1960, essays "Chokan Valikhanov" 1956, "Vstrechi, kotorykh ne bylo" (Meetings That Never Were) 1958, science-fiction novel "Poias zhizni" (Belt of Life) 1960, "Zagadki Khairkhana" (Khairkan's Riddles) 1961, African stories "Ochag zhizni" (Hearth of Life) 1959, and others.

Zabelina, Aleksandra Ivanovna
Fencer; b. Mr. 11, 1937, Moscow.
Honored Master of Sport since 1960.
She graduated from the Moscow Institute for History and Archives in 1970. In team competition she was an Olympic champion in 1960 (Rome) and in 1968 (Mexico City). A world champion seven times between 1956 and 1971, she was also a champion of the USSR in team competition 12 times between 1958 and 1970. In individual competition, she was a world champion in 1957 and 1967, and the champion of the USSR five times from 1957 to 1968.

Zabila, Nataliia L'vovna
Ukrainian writer; b. F. 20 (Mr. 5), 1903, St. Petersburg.
Became a member of the CPSU in 1940. Her work was first published in 1924. Her collections of poetry include *Distant Land* (1927), *Days of Anxiety* (1945), and *Poems* (1963). She is also the author of poetry collections for children, such as *Iasochka's Little Book* (1934), *The Swallows* (1937), *Our Homeland* (1947), and *To Little Ones About Big Deeds* (1958). Her prose works include *Relatives* (1934), *Katrusia Is Already Grown-up* (1955), and *Short Stories, Fairy Tales, and Novellas* (1962). She has adapted folk tales for children in her collection *Under the Green Oak Tree* (1952). She has also done translations. Her own works have been translated into many languages, particularly those of the people of the USSR. Decors.: two orders and various medals. Works: *In Russian translation: Solntse smeetsia: Stikhi i skazki.* Moscow 1963.

Zaborskii, Georgii Vladimirovich
Architect; b. O. 29 (N. 11), 1909, Minsk.
Honored Builder of the Belorussian SSR (1961).

He studied under S. S. Serfaimov at the Academy of Arts in Leningrad from 1933 to 1939 and began working in Belorussia in 1940. He drew the plan for the reconstruction of Lenin Square (1947–60, in collaboration with L. P. Matskevich) and plans for the construction of public buildings and apartment houses. His projects have included the headquarters of the Central Committee of the Komsomol of Belorussia (1952–53), the building complex of the pedagogical institute (1952–59), and apartment houses on Lenin Street (1952–56), as well as the Young Pioneer Movie Theater and the Puppet Theater (both built in 1964). In 1954, in collaboration with V. A. Korol', he designed the obelisk in Minsk commemorating the soldiers of the Soviet Army and the partisans who died in the Great Patriotic War. In the 1960's he directed the construction of several sovkhozes and kolkhozes in Belorussia. He received the State Prize of the USSR in 1971 for the planning and construction of the settlement of Vertelishki in Grodno Raion, Grodno Oblast. In 1970 he became the head architect of the Belorussian State Institute for the Design and Construction of Agricultural Buildings. From 1950 to 1964, he taught at the Belorussian Polytechnical Institute.

Zabulis, Genrikas Karolevich
State official.
Chair. of State Cttee. for Higher and Secondary Specialist Educ. with Council of Mins. Lithuanian SSR 1962–66; cand. mem. 1964–66, mem. Cen. Cttee. Lithuanian CP 1966–; Min. of Higher and Secondary Specialist Educ. Lithuanian SSR Mr. 1966–; Dep. to Supr. Sov. Lithuanian SSR 1967–.

Zadornov, Nikolai Pavlovich
Russian writer; b. N. 22 (D. 5), 1909, Penze.
Honored Cultural Worker of the Latvian SSR since 1969.
He is the author of two cycles of historical novels on the assimilation of the Far East by the Russian people in the 19th century and the heroic exploits of the explorers. The first cycle consists of four novels: *The Distant Land* (books 1–2, 1946–49), *The First Discovery* (1969, original title, *Toward the Ocean,* 1949), *Captain Nevel'skoi* (books 1–2, 1956–58), and *War for the Ocean* (books 1–2, 1960–62). The second cycle, which is about the development of the Far East by peasant settlers and is connected with the first in its theme, consists of two novels: *Amur the Father* (books 1–2, 1941–46) and *Gold Fever* (1969). In 1971 he published the novel *Tsunami,* which is devoted to the expedition of Admiral E. V. Putiatin to Japan in 1845–55. He is also the author of *Yellow, Green, Blue* ... (book 1, 1967), a novel about the contemporary period, and *The Blue Hour* (1968), travel essays. Decors.: State Prize of the USSR in 1952 for the novels *Amur the Father, The Distant Land,* and *Toward the Ocean,* three orders, as well as medals.

Zagladin, Vadim Valentinovich, Cand. of Hist. Sc.
Party official; b. 1927.
Mem. CPSU 1955–; grad. Moscow Inst. of Int. Relations 1949; concluded postgraduate studies 1952; Cand. of Hist. Sc.; pedagogic work at Moscow Inst. of Int. Relations until 1954; ed. work with journal "Novoe vremia" (Acting Sec. of Ed. Div. 1957–60) and "Problemy mira i sotsializma" (1961–64 in Prague) 1954–64; mem. Apparat Cen. Cttee. CPSU 1964–67; Dep. 1967–65, 1st Dep. Head of Int. Dept. of Cen. Cttee. CPSU 1975–; mem. Cen. Auditing Comm. 1971–76, cand. mem. Cen. Cttee. CPSU 1976–81, mem. 1981–.

Zagorodnikov, Nikolai Ivanovich
Scholar, specialist in criminal law; b. Mr. 22, 1918, Lamki, now Tambov Oblast.
Doctor of jurisprudence since 1962, professor since 1963. Honored Scholar of the RSFSR since 1971. Member of the CPSU since 1945.
His scholarly and teaching career began in 1948. He is the author of works on the theory of Soviet law, on socialist legality, and on some special problems of criminal law, including *Crimes Against Life According to Soviet Criminal Law* (1961) and *Crimes Against Health* (1969).

Zagrebelnyi, Pavel Arkhipovich
Author; b. 1924, Soloshino, Poltava Oblast; s. of a peasant.
Ukrainian; mem. CPSU 1960–; grad. Dnepropetrovsk State Univ., writer; served in Sov. Army 1941–45; collective farmer 1945–46; literary consultant of Ukrainian Union of Writers, literary ed. of newspaper "Dneprovskaia pravda" 1951–54; dept. chief, Dep. Ed.-in-Chief of journal "Otchizna" 1954–61; Ed.-in-Chief of newspaper "Literaturna Ukraina" 1961–64; Responsible Sec., Dep. Chair. of Ukrainian Union of Writers 1964–79; cand. mem. Cen. Cttee. Ukrainian CP 1976–81, mem. 1981–; 1st Sec. of Ukrainian Union of Writers 1979–.
Publs.: "Kakhovskie rasskazy" (Kakhovka stories) 1953, "Stepnye tsvety" (Steppe Flowers) 1955, "Uchitel'" (The Teacher) 1957, "Marevo" (Mirage) 1956, "Evropa, 45" (Europe 1945) 1959, "Evropa, zapad" (Europe, the West) 1961, and others.

Zagurskaia, Galina Nikolaevna
Actress; b. 1905.
Grad. Irkutsk Ballet School 1921; ballet dancer at Irkutsk Theatre 1921–32; actress with Iaroslavl Theatre 1932–; performances in Vladivostok, Irkutsk and other Sov. cities. Decors.: People's Artist of the Uzbek SSR 1944, People's Artist of the USSR 1969.

Zaichuk, Vladimir Ignatevich
State official.
Dep. Min. of Justice Ukrainian SSR 1955–62; Chair. of Supr. Court of Ukrainian SSR 1963–70; Min. of Justice Ukrainian SSR O. 1970–; mem. Auditing Comm. Ukrainian CP 1976–81.

Zaidov, Mirgiias Abbasovich
State official; b. 1930.
Uzbek; mem. CPSU 1953–; grad. Tashkent Pedagogical Inst.; teacher at an intermediate school 1951–53; Komsomol posts 1953–58; sec. of a raion cttee. Uzbek CP, dep. dept. head with Cen. Cttee. Uzbek CP, 2nd Sec. of Tashkent Oblast Cttee. Uzbek CP 1958–64; Chair. of Uzbek Republican Trade Union Council 1964–Mr. 1977; mem. Cen. Cttee. Uzbek CP 1966–81; Chair. State Cttee. on Labour Uzbek SSR S. 1982–; Dep. to USSR Supr. Sov. 1966–. Decor.: Order of Red Banner of Labour.

Zaikov, Lev Nikolaevich
City soviet official; b. 1923.
Russian; mem. CPSU 1957–; grad. Leningrad Econ. Engineering Inst. 1963; metal worker, group chief, foreman, dep. shop superintendent, shop superintendent, prod. chief at a plant 1941–61; dir. of a plant in Leningrad 1961–71; gen. dir. of prod., gen. dir. of a scientific prod. asscn. 1971–76; Chair. of Exec. Cttee. of Leningrad City Sov. of People's Deps. Ja. 1976–; Dep. to USSR Supr. Sov. 1979–; mem. Cen. Cttee. CPSU 1981–. Decors.: Hero of Socialist Labor 1971.

Zaionchkovskii, Petr Andreevich, Dr. of Hist. Sc.
Historian; b. S. 5 (18), 1904, Uralsk.
Mem. CPSU 1931–; grad. Inst. of History, Philosophy and Literature in Moscow 1937; Prof. at Moscow Univ. 1951–; his most important works are on the history of the Russian peasant, domestic policy and Russian war history of the 19th century; hon. mem. American Historians' Asscn. Decors.: 2 orders and various medals.
Publs.: "Voennye reformy 1860–1870 gg. v Rossii" Moscow 1952, "Provedenie v žizn' krest'janskoj reformy 1861 g." Moscow 1958, "Kirillo-Mefodievskoe obščestvo (1846–1847)" Moscow 1959, "Krizis samoderžavija na rubeže 1870–1880-ch gg." Moscow 1964, "Otmena krepostnogo prava v Rossii" Moscow 1968, "Rossijskoe samoderžavie v konce 19 stoletija" Moscow 1970.

Zairov, Mukhiddin
State official.
Head of Kolkhoz-Sovkhoz Prod. Admin. in Ordzhonikidzeabad 1963; cand. mem. 1966–71, mem. Cen. Cttee. Tadzhik CP 1971–; Head of Agricultural Dept. of Cen. Cttee. Tadzhik CP 1971–73; Min. of Agriculture Tadzhik SSR Ag. 1973–82; 1st Sec. of Gorno-Badakhshan Oblast Cttee. Tadzhik CP Je. 1982–.

Zaitsev, Anatolii Mikhailovich
State official.
Chair. of Kazakh SSR State Forestry Cttee. 1972–78; Min. of Forestry Kazakh SSR Ap. 1978–; Chair. Auditing Comm. Kazakh CP 1981–.

Zaitsev, Evgenii Alekseevich
Painter; b. D.21, 1907 (Ja.3, 1908), Nevel', present-day Pskov Oblast.
People's Artist of the Belorussian SSR from 1964. Became a member of the CPSU since 1939.
He studied at the Vitebsk Art Technicum from 1926 to 1930 and at the Academy of Arts in Leningrad under A. A. Osmerkin from 1930 to 1937. He began working in Belorussia in 1938. From 1948 to 1950 he was the chairman of the administration of the Union of Artists of the Belorussian SSR. In addition to his densely populated paintings dealing with the history of the struggle for Soviet power and the great Patriotic War, he has painted portraits and landscapes. His work *The Entry of the Red Army Into Minsk in 1930* (1940) has not been preserved. The paintings *A Hero's Funeral* (1946), *The Defense of Brest Stronghold in 1941* (1950), *Konstantin Zaslonov* (1957), and *Requiem, or Vow to Il'ka* (1970), are all in the Belorussian Art Museum in Minsk. Between 1963 and 1967, he painted a triptych entitled *My Republic in Patriotic Fervor.* Decors.: three orders and various medals.

Zaitsev, Mikhail Mitrofanovich
Military official; b. 1923.
Russian; mem. CPSU 1943–; grad. Mil. Acad. of Tank Troops 1954 and Mil. Acad. of Gen. Staff 1965, Army Gen.; entered Sov. Army 1941; held command and staff posts 1941–65; commander of various units 1965–72; mem. Cen. Cttee. Belorussian CP 1971–81; mem. Bureau Cen. Cttee. Belorussian CP 1976–81; 1st Dep. Commander of Troops of Belorussian Mil. District 1972–76, Commander 1976–80; Commander of Sov. Troops in Ger. 1980–; mem. Cen. Cttee. CPSU 1981–.

Zakharchenko, Vasilii Dmitrievich
Author; b. 1915, Petrograd.
Russian; mem. CPSU 1948–; grad. Moscow Energetics Inst. 1937 and Gorkii Inst. of Literature Moscow 1941.
Publs.: poetry "Kroviu etoi zemli" (With the Blood of This Earth) 1943, "Utro mira" (The Morning of Peace) 1950, "Puteshestvie v zavtra" (Journey Into Tomorrow) 1952, "Na poroge velikogo veka" (On the Threshold of a Great Century) 1959, "Vnimanie, govorit budushchee!" (Attention, the Future Is Speaking!) 1960, "Prislushaisia, eto zhizn' . . ." (Listen, This Is Life . . .) 1962, and others.

Zakharov, Aleksei Vasil'evich
Diplomat; b. 1913.
USSR Dep. Min. of Foreign Trade 1951–53; USSR Rep. 1953–54, Dep. USSR Rep. at Econ. Council of United Nations 1954–56; mem. Diplomatic Service 1956–; Dep. USSR Min. of Foreign Affairs 1956–59; USSR Amb. to Finland 1959–65; Head of Scandinavian Dept. at USSR Min. of Foreign Affairs 1965–67; Dep. Permanent USSR Rep. at United Nations 1967–72; USSR Amb. to Uganda Mr. 1972–N. 1976. Decor.: Order of Red Banner of Labour.

Zakharov, Gurii Filippovich
Graphic artist; b. N.27, 1926, Kimry, present-day Kalinin Oblast.
Honored Art Worker of the RSFSR (1968).
He studied under E. S. Teis at the Moscow Higher School of Arts and Industry from 1946 to 1954, receiving additional artistic counseling from V. A. Favorskii. He works in the media of linocutting and etching. His works are characterized by a contemporary perception of reality that is permeated by poetic feeling. Their strict architectonic and rhythmic composition is expressed by the contrasting juxtaposition of black and white tones. His linocuts include *The Mountain Ash* (1961), *Ten Minutes Past Midnight* (1964), and *Landscape With a Horseman* (1964). Examples of his etchings are *Rozhdestvenskii Boulevard* (1968) and *Solianka* (1969).

Zakharov, Rostislav Vladimirovich
Choreographer and director; b. 1907.
Mem. CPSU 1951–; grad. Leningrad Choreographic School 1926 and Dept. of Directing at Leningrad Theatre-Tech. Inst. 1932; ballet master and dir.; ballet master at Leningrad Kirov Opera and Ballet Theatre 1934–35; Ballet Master and Opera Dir. at Bolshoi Theatre 1936–56; Dir. and Artistic Man. of Moscow Choreographic School 1945–47; Head of Chair of Choreography 1946–, Prof. at Moscow State A. V. Lunacharskii Inst. of Theatre Arts (GITIS) 1951–. Decors.: State Prize of the USSR 1946 and 1949, People's Artist of the USSR 1969.
Publs.: "Iskusstvo baletmeistera" (The Art of the Ballet Master), "Sovetskii balet" (Soviet Ballet).

Zakharov, Semen Egorovich
Military official; b. Ja. 10 (23), 1906, Stolpovo, present-day Zaraisk Raion, Moscow Oblast.
Admiral (1966). Member of the CPSU since 1926.
He has been in the Soviet armed forces since 1932. He graduated from the Borisoglebsk School for Military Pilots and from the Lenin Military Political Academy in 1938. He was a member of the military council of the Pacific Fleet from March 1939 to November 1948. He graduated from the Voroshilov Higher Military Academy in 1950. He held a responsible position in the Ministry of the Navy of the USSR from 1950 to 1953 and managerial positions in the Main Political Directorate and the Main Directorate of Cadres of the Ministry of Defense of the USSR from 1953 to 1959. He was deputy chief of naval educational institutions from 1959 to 1969. He has been in the reserves since 1969. Decors.: Order of Lenin, the Order of the Red Banner, the Order of Ushakov First Class, three Orders of the Red Star, the Order of the Badge of Honor, and various medals.

Zakhidova, Liutfi
Ballerina; b. N.6, 1925, Kanibadam.
People's Artist of the USSR (1957). Member of the CPSU since 1957.
In 1938, after her appearance at the All-Union Competition in Moscow, she was accepted by the Song and Dance Ensemble of the All-Union Central Council of Trade Unions. In 1941 she joined the S. Aini Theater of Opera and Ballet in Dushanbe. She has also performed Tadzhik folk dances in concerts. She is the first ballet performer in the Tadzhik Republic to master the technique of classical ballet. In 1943 she made her debut in the role of Lise in Hertel's *La Fille mal gardée.* Her dancing is characterized by lyricism and gentleness as well as by a dramatic quality. Among her other roles are Maria in Asaf'ev's *The Fountain of Bakhchisarai,* Leili in Balsanian's *Leili and Medzhnun* (State Prize of the USSR, 1949), and Cinderella in Prokofiev's ballet of the same name. She has toured abroad as a performer of Tadzhik folk dances (in the German Democratic Republic, Rumania, India, and other countries). Decors.: two orders and various medals.

Zakhoder, Boris Vladimirovich
Author and translator; b. 1918, Kagul, Moldavia.
Russian; grad. Gorkii Inst. of Literature Moscow 1947; first works publ. 1938.
Publs.: children's poetry "Na zadnei parte" (On the Last Bench) 1955, "Martyshkino zavtra" (Martyshka's Morning) 1956, "Nikto i drugie" (Nobody and the Others) 1958, "Kto na kogo pokhozh" (Who Resembles Whom) 1960, and others.

Zakhritdinov, Guliam Zakhritdinovich
Party official.
Head of Dept. of Transport and Communications of Cen. Cttee. Uzbek CP Jl. 1978–; Dep. of Supr. Sov. Uzbek SSR 1980–; mem. Cen. Cttee. Uzbek CP 1981–.

Zakirov, Iakub Turgunovich
State official.
Dep. Head of Dept. of Party Organs of Cen. Cttee. Tadzhik CP 1961; cand. mem. Cen. Cttee. Uzbek CP 1961–71; Head of Dept. of Admin. Organs of Cen. Cttee. Uzbek CP 1961–67; Dep. to Supr. Sov. Uzbek SSR 1963–71; Chair. of State Cttee. of Labour Uzbek SSR Ap. 1967–F. 1977, 1st Dep. Chair. F. 1977–.

Zakrutkin, Vitalii Aleksandrovich
Author; b. Mr. 14 (27), 1908, Feodosia.
Grad. Kalinin Pedagogical Inst. in Blagoveshchensk 1932; post-graduate studies at Herzen Pedagogical Inst. in Leningrad 1936. Decors.: State Prize of the USSR 1951, 4 orders, various medals.
Publs.: "Akademik Pliushchov" (Academician Pliushchov) 1940, "Korichnevaia chuma" (The Brown Pest) 1941, "Sila" (Power) 1942, "Plavuchaia stanitsa" (Swimming Kossack Village) 1950, "Sotvorenie mira" (The Creation of the World) vol. 1, 1955–56, vol. 2, 1967, "Mat' chelovecheskaia" (Human Mother) 1969, "Mat' syra zemlia" (Mother, the Humid Earth) 1970, and others.

Zakusov, Vasilii Vasil'evich
Pharmacologist; b. Ap. 13 (26), 1903, St. Petersburg.
Academician of the Academy of Medical Sciences of the USSR (1952).
In 1926, he graduated from the Academy of Military Medicine, where he was a student of N. P. Kravkov. Between 1937 and 1954 he was the head of the pharmacology subdepartments in the Third Leningrad Medical Institute, Kuibyshev Academy of Military Medicine, and Leningrad Medical Institute. He became director of the Institute of Pharmacology of the Academy of Medical Sciences in 1954 and simultaneously (1956–65) head of the pharmacology subdepartment of the First Moscow Medical Institute. His main works deal with the pharmacology of the nervous, cardiovascular, and respiratory systems, general pharmacology, and industrial toxicology. He established a new school of thought in pharmacology. He was president of the All-Union Pharmacological Society. Decors.: Order of Lenin, two other orders, and medals.
Works: *Eksperimental'nye dannye po farmakologii tsentral'noi nervnoi sistemy.* Leningrad 1947. *Farmakologiia nervnoi sistemy.* Leningrad 1953. *Farmakologiia,* 2nd ed. Moscow 1966.

Zalmanov, Farkhad Rza ogly
State official.
Head of the Cen. Admin. 1980–81, Min. of the Cotton-Cleaning Ind. Azerbaidzhan SSR Ja. 1981–; mem. Auditing Comm. CP Azerbaidzhan 1981–.

Zalygin, Sergei Pavlovich
Author; b. N. 23 (D. 6), 1913, Durasovka, Bashkir ASSR.
Grad. Omsk Agricultural Inst. 1939; worked as engineer-hydro-technologist, hydrologist at Siberian Branch of USSR Acad. of Sciences; first works publ. 1936. Decors.: State Prize of the USSR 1968, Order of Red Banner of Labour, various medals.
Publs.: "Rasskazy" (Stories) 1941, "Vesnoi 1954 goda" (The Spring of 1954) 1955, "Krasnyi klever" (Red Clover) 1955, "Tropy Altaia" (Altai Paths) 1962, "Na Irtyshe" (On the Irtysh) 1964, "Solenaia Pad'" (The Salty Ravine) vols. 1–2, 1967–68, "Moi poet" (My Poet) 1969, "Cherty professii" (Traits of the Profession) 1970, and others.

Zamiatin, Leonid Mitrofanovich
Party official; b. 1922.
Mem. CPSU 1944–; grad. Ordzhonikidze Inst. of Aircraft Construction in Moscow 1944; mem. Diplomatic Service 1946–70; Counsellor with Sov. Del. to United Nations 1954–57; USSR Dep. Rep. at Preparatory Cttee. and later Council of Int. Atomic Energy Agency 1957–59; USSR Rep. at Int. Atomic Energy Agency 1959–60; Dep. Head 1960–63; Head of American Dept. 1962, Head of Press Dept. of USSR Min. of Foreign Affairs 1962–70; Dir.-Gen. of Sov. News Agency TASS Ap. 1970–78; mem. Cen. Auditing Comm. 1971–76, mem. Cen. Cttee. CPSU 1976–; Chair. of USSR Parliamentarian Group of Parliamentarian Relations with Federal Republic of Germany 1971–; Chair. of USSR-FRG Soc. 1972–; Chief of Cen. Cttee. CPSU Dept. for International Information 1978–.

Zankovich, Valentin Pavlovich
Architect; b. Je. 5, 1937, Priluki, Minsk Raion, Minsk Oblast, Belorussian SSR.
He studied in the Belorussian Polytechnic Institute in Minsk from 1953 to 1959. He was instrumental in designing the residential area of Minsk and helped design a series of memorials, including *To "Katiusha"* in Orsha (1966); the memorial park in Rossony, Vitebsk Oblast (1966; Lenin Komsomol Prize, 1967); and the memorial complex *Khatyn'* (1968–69; with the sculptor S. I. Selikhanov and architects Iu. M. Gradov and L. M. Levin; Lenin Prize, 1970).

Zapasko, Iakim Prokhorovich, Cand. of Art. Sc.
Art historian; b. 1923, Rosishki, Kiev Oblast.
Mem. CPSU 1949–; grad. Lemberg Polygraphic Inst. 1950, pupil of O. Kulchitska, V. Forostetskii; Cand. of Art Sc. 1958.

Zapolskii, M. A.
Trade union official.
Sec. of Ukrainian Republican Trade Union Council F. 1982–.

Zaporozhets, Aleksandr Vladimirovich
Psychologist; b. Ag. 30 (S. 12), 1905, Kiev.
Grad. Pedagogical Dept. of Moscow Univ. 1930; mem. USSR Acad. of Pedagogic Sciences 1968–; Dir. of Inst. of Preschool Educ. of USSR Acad. of Pedagogic Sciences 1960–.

Zardarian, Oganes Mkrtichevich
Painter; b. Ja. 7 (20), 1918, Kars, in present-day Turkey.
People's Artist of the Armenian SSR (1963). Corresponding member of the Academy of Arts of the USSR (1958).
He studied at the Leningrad Academy of Arts from 1937 to 1941. He became a professor at the Kh. Abovian Pedagogical Institute in Yerevan in 1969. His works include historical paintings (for example, *V. I. Lenin,* also known as *Eve of the Revolution,* 1952, Armenian State Picture Gallery, Yerevan), pictures devoted to the heroic spirit of labor, *Victory of the Builders of the Sevan Hydroelectric* (for example, *Power Plant,* 1947, Museum of Art of the Eastern Peoples, Moscow), landscapes that often include genre themes (for example, *Spring,* 1956, Tret'iakov Gallery), still lifes (for example, *Spring Flowers,* 1957, Tret'iakov Gallery), and portraits (*A. Gramsci's Comrade-in-arms – Eduardo d'Onofrio,* 1969). His works are characterized by a festive and decorative combination of sonorous, radiant planes of color. Decors.: Order of the Red Banner of Labor and various medals.

Zaretskii, Viktor Ivanovich
Painter; b. 1925, Bilipillia, Sumy Oblast.
Mem. CPSU 1944–; grad. Kiev Art Inst. 1953, pupil of G. Titov, S. Grigorev.

Zarian, Ruben Varosovich
Theatrical scholar; b. S. 1 (14), 1909, Aleksandropol', present-day Leninakan.
Doctor of art studies (1962). Honored Art Worker of the Armenian SSR (1961). Member of the CPSU since 1942.
He graduated from the philology department of the University of Yerevan in 1936. He began his literary activities in 1933. His works are devoted to Armenian theatrical art; they include *The Struggle for Russian Drama in the Armenian Theater* (1954), *Theatrical Portraits* (1956), *Siranuish, 1857–1932* (1957), *Arus Voskanian* (1957), *Adamian* (vols. 1–2, 1960–61), and *Adamian's Shakespeare* (1965). In 1948, he began to teach at the Yerevan Art and Theater Institute, becoming a professor there in 1963. He has been director of the Art Institute of the Academy of Sciences of the Armenian SSR in Yerevan since 1958 and director of the Armenian center for the study of Shakespeare since 1965. Decors.: Order of the Badge of Honor and medals.

Zarin', Margeris Ottovich (in Latvian: Margeris Zariņš)
Composer; b. My. 11 (24), 1910, Jaunpiebalga.
People's Artist of the USSR (1970).
He studied at the Latvian Conservatory in Riga from 1929 to 1933. From 1940 to 1950 he was musical director of the J. Rainis Art Theater in Riga. He wrote primarily vocal and vocal-instrumental works and music for plays. He was one of the founders of Latvian oratorio and Soviet Latvian opera. Among his compositions have been the operas *To a New Shore* (1955, Riga), *Green Mill* (1958, Riga), and *Beggars' Opera* (1965, Riga); the oratorios *Valmiera Heroes* (1950; State

Prize of the USSR, 1951), *Fighting Devil's Swamp* (1951), and *Black Butterfly* (1964); the cantata *Trumpeter from Talava* (1937); the vocal-instrumental suite *Ode to Swings* (1940); works for an acappella choir; and cycles for voice and piano. In 1951–52 and 1956–68, Zarin' was chairman of the board of the Composers' Union of the Latvian SSR. Decors.: Republic Prize of the Latvian SSR in 1957, Order of Lenin and various medals.

Zaritskii, Aleksei Aleksandrovich
Author; b. 1911, Khotimsk, Mogilev Oblast.
Belorussian; mem. CPSU 1943–; first works publ. 1927.
Publs.: "Povest' pro zolotoe dno" (Stories of the Golden Ground) 1954; has translated works of Russian, Ukrainian, Polish and German writers into Belorussian.

Zarkhi, Aleksandr Grigorevich
Film director and script writer; b. 1908.
Mem. CPSU 1948–; grad. Leningrad Tech. Film College 1927; work with "Sovkino" 1929–; with film studio "Belarusfilm" 1950–55 and "Mosfilm" (now Lenfilm) Leningrad 1957–; has written scripts for several films. Decors.: State Prize of the USSR 1941 and 1946, Merited Artist of the Belorussian SSR 1955, People's Artist of the USSR 1969.

Zarodov, Konstanin Ivanovich, Cand. of Hist. Sc.
Journalist; b. 1920.
Mem. CPSU 1940–; grad. Vologda Transport Tech. School 1939, Party Higher School of Cen. Cttee. CPSU 1951 and Acad. of Social Sciences of Cen. Cttee. CPSU 1954; served in Sov. Navy 1939–46; 1st Sec. of Vologda City Komsomol Cttee. 1946–47, of Vologda Oblast Komsomol Cttee. 1947–50; mem. Ed. Board. Dept. Ed. of "Komsomolskaia pravda" 1950–51; mem. ed. staff of newspaper "Za prochnyi mir, za narodnuiu demokratiiu" 1954–56; worked for Cen. Cttee. CPSU 1956–61; Chief Ed. of newspaper "Sovetskaia Rossiia" 1961–65; Dep. to and mem. Foreign Affairs Comm. of Supr. Sov. RSFSR 1963–67; 1st Dep. Chief Ed. of "Pravda" 1965–68; cand. mem. Cen. Cttee. CPSU 1966–; Chief Ed. of "Problemy mira i sotsializma" 1968–.

Zarubin, Ivan Ivanovich
State official.
Dep. Dir. of Affairs, RSFSR Council of Mins. 1975–82, Dir. Mr. 1982–.

Zarubin, Leonid Kirillovich
Military official.
Rear Admiral; Dep. to Supr. Sov. Esthonian SSR 1975–81; mem. Cen. Cttee. and cand. mem. Bureau Cen. Cttee. Esthonian CP 1976–81.

Zarudnyi, Nikolai Iakovlevich
Author and dramatist; b. 1921, Kiev.
Ukrainian; mem. CPSU 1948–. Decor.: Order of Red Banner of Labour.
Publs.: "Moi zemliaki" (My Compatriots) 1950, play "Vesna" (Spring) 1950, "Noch' i plamia" (The Night and the Fire) 1958, "Priletaiut zhuravli" (The Cranes Return) 1954, "Raduga" (Rainbow) 1958, "Esli ty liubish" (When You Love) 1959, "Mertvyi bog" (The Dead God) 1960, "Synovia idut dal'she" (Sons Go Further) 1961, and others.

Zaslavskaia, Sulamif' Aleksandrovna
Textile artist; b. Mr. 7, 1918, Cherkassy.
Studied in the Moscow Textile Institute (1936–41). From 1944 to 1971 she worked as an artist in the Krasnaia Roza Silkweaving Mill. She has created decorative panels (Women of Asia and Africa, 1968, cotton, batik) and designs for printed fabrics for dresses (Fair, 1957, Russian Museum, Leningrad), draperies (Little Horses, 1962, Russian Museum), souvenir kerchiefs (Peace, 1964), and handmade carpets (Romantics and March, both 1970). By treating floral and topical motifs as elements of linear flat ornaments. She organically blends her compositions with the fabric's structure and texture.

Zaslavskaia, Tat'iana Ivanovna
Economist; b. S. 9, 1927; Kiev.
Member of the Academy of Sciences of the USSR corresponding 1968–81, mem. 1981. Became a member of the CPSU in 1954.
She graduated from the economics department of Moscow State University in 1950. From 1950 to 1963 she was engaged in scholarly work in the Institute of Economics of the Academy of Sciences of the USSR; since 1963 she has worked in the Institute of Economics and Organization of Industrial Production in the Siberian division of the Academy of Sciences of the USSR. Her main studies have dealt with Kolkhoz workers' pay, social and economic problems of labor resources, and migration of the rural population. She is the co-author of the textbook *Political Economy* (4th ed., 1962).
Works: *Printsip material'noi zainteresovannosti i oplata truda v kolkhozakh.* Moscow 1958. *Sovremennaia economika kolkhozov.* Moscow 1960. *Raspredelenie po trudu v kolkhozakh.* Moscow 1966. *Migratsia sel'skogo naseleniia.* Moscow 1970. (Co-author.)

Zasukha, Mariia Iakovlevna
Laborer; b. 1939.
Mem. CPSU 1972–; finished a correspondence secondary school 1973; milkmaid of "Prapor Lenina" Collective Farm in Kiev Oblast 1955–65; milkmaid at a dairy products farm of an experimental station for livestock-breeding in Belaia Tserkov Raion in Kiev Oblast 1965–73, work brigade leader there 1973–74; mechanical milking operator at "Terezino" Experimental Works in Belaia Tserkov Raion in Kiev Oblast 1974–; cand. mem. Cen. Cttee. CPSU 1981–. Decor.: USSR State Prize 1977.

Zatsepin, Georgii Timofeevich
Physicist; b. My. 15 (28), 1917, Moscow. Corresponding member of the Academy of Sciences of the USSR (1968). Graduated from Moscow University in 1941. Since 1944 he has worked in the Physics Institute of the Academy of Sciences of the USSR, and since 1963 he has been the director of a laboratory. He has been a professor at Moscow University since 1958.
His main works are on the physics of cosmic rays and on neutrino astrophysics. In 1947–49 he developed a method for studying large cosmic-ray showers and, using experimental data, established that the basis for the development of such showers is a nuclear-cascade process, whereas electron-photon processes are secondary. From these data he found a number of characteristics of the elementary event of nuclear collisions at high particle energies and suggested an explanation of these processes (1949–62). He was the first to devote attention to the effects arising during the passage of ultrarelativistic particles through a photon gas. His works from 1960 to 1969 were devoted to the physics of muons and neutrinos, as well as to the methods and prospects of neutrino astrophysics. Decor.: State Prize of the USSR in 1951.
Works: "Iaderno-kaskadnyi protsess i ego rol' v razvitii shirokikh atmosfernykh livnei." *Dokl. AN SSSR: Seriia fizicheskaia,* 1949, vol. 67, fasc. 6. "O fotorasshcheplenii tiazhelykh chastits kosmicheskikh luchei, proiskhodiashchem pod vliianiem solnechnogo izlucheniia." *Dokl. AN SSSR: Seriia fizicheskaia,* 1951, vol. 80, fasc. 4.

Zatuashvili, Bogdan Antonovich
State official.
Dep. Min. of Construction Georgian SSR 1971–76; Head of Main Admin. for the Industrial Construction Georgian SSR F. 1976–.

Zatvornitskii, Vladimir Andreevich
Laborer; b. 1929.
Mem. CPSU 1958–; uncompleted intermediate-school educ.; bricklayer, construction worker, brigadier of complex brigade of "Mosstroi" Trust No. 1 1947–; mem. Cen. Auditing Comm. CPSU 1971–81, mem. Cen. Cttee. CPSU 1981–. Decor.: Hero of Socialist Labour.

Zavialov, Viktor Alekseevich
Komsomol official.
2nd Sec. and mem. Bureau Cen. Cttee. Lithuanian Komsomol S. 1979–; cand. mem. Cen. Cttee. Lithuanian CP 1981–.

Zbanatskii, Iurii Oliferovich
Author; b. 1914, Barsukovo, Chernigov Oblast.
Ukrainian; mem. CPSU 1939–; grad. Nezhin Pedagogic Inst. 1937; village teacher 1931–33, headmaster, correspondence studies 1933–37; Head of Dept. of Popular Educ. in Nezhin 1937–41; sent to work in German-occupied Chernigov by party to work and organize underground movt. and partisan orgs., arrested and interned in Chernigov prison, later in a concentration camp 1941–42; escaped from concentration camp and founded "Shchors" partisan org. in Oster Raion 1942–44; literary work 1944–; Dep. Chair. 1959–66, Sec. 1966–69, 1st Sec. 1969–71, Dep. Chair. 1971–73, Sec. of Ukrainian Writers' Union 1973–. Decors.: Order of Lenin, Order of Red Banner of Labour, Hero of the Sov. Union, Order of Friendship Between Nations.
Publs.: "Taemnytsia Sokolynogo Boru" (The Secret of Sokolinoe Pine Forest) 1947, "Lito v Sokolynomu" (Summer in Sokolinoe), "Lisova Krasunia" (Forest Beauty) 1953, "Mizh dobrymy liud'my" (Among Kind People), "Pryvitaite mene, druzi!" (Congratulate Me, Friends) 1956, "Kurylovi ostrova" (The Kurile Islands) 1963, and others.

Zdanovich, Aleksandr Aleksandrovich
State official; b. 1927, Orsha; s. of a worker.
Mem. CPSU 1954–; grad. Minsk Inst. of Law 1952; People's Judge in a raion in Vitebsk Oblast 1951–59; Instructor of Vitebsk Oblast Cttee. Belorussian CP 1959–61; Instructor of Dept. of Admin. Organs at Cen. Cttee. Belorussian CP 1961–66; Dep. Head of Dept. of Admin. Organs of Cen. Cttee. Belorussian CP 1966–70; Min. of Justice Belorussian SSR N. 1970–; cand. mem. Cen. Cttee. Belorussian CP 1976–.

Zedgenidze, Georgii Artemevich
Radiologist; b. F. 21 (Mr. 6), 1902, Tbilisi.
Mem. CPSU 1939–; Grad. Dept. of Medicine of Tbilisi Univ. 1927; mem. USSR Acad. of Medical Sciences 1960–; Head of Chair. at Navy Medical Acad. and S. M. Kirov Mil. Medical Acad. 1941–58; Dir. of Inst. for Research on Medical Radiology of USSR Acad. of Medical Sciences 1958–. Decors.: 3 orders and several medals.

Zefirov, Aleksei Petrovich
Specialist in the chemistry and metallurgy of rare and noble metals; b. Mr. 12 (25), 1907, Gomel'.
Corresponding member of the Academy of Sciences of the USSR (1968). Member of the CPSU since 1944. Graduated from the Moscow Institute of Nonferrous Metals and Gold in 1932.
He engaged in engineering, scientific research, and teaching activities. In 1958 he became a professor at the Moscow Physical Engineering Institute; he has been the head of a subdepartment at the Moscow D. I. Mendeleev Chemico-Technological Institute. Decors.: orders and medals.

Zeldovich, Iakov Borisovich
Physicist; b. Mr. 8, 1914, Minsk.
Mem. USSR Acad. of Sciences 1958–; mem. staff Inst. of Chemical Physics 1931–, Inst. of Applied Mathematics of USSR Acad. of Sciences 1964–; Prof. at Moscow Univ. 1966–. Decors.: Hero of Socialist Labour (3 times), Lenin Prize, 4 State Prizes of the USSR, 3 Orders of Lenin, 3 other orders and various medals.
Publs.: "Fizika udarnykh voln i vysokotemperaturnykh gidrodinamicheskikh iavlenii" 1963, 2nd ed. 1966, "Reliativistskaia astrofizika" 1967, "Teoriia tiagoteniia i evolutsiia zvezd" 1971, "Vysshaia matematika dlia nachinaiushchikh i ee prilozhenie k fizike" 1968, and others.

Zel'in, Konstantin Konstantinovich
Historian of classicial history; b. Mr. 13 (25), 1892, Moscow.
Doctor of historical sciences (1963).
He graduated from the history and philology department of Moscow University in 1916. He was a lecturer in the department of history of Moscow State University (MGU) from 1934 to 1955 and senior research associate at the State Academy of the History of Material Culture from 1932 to 1937. He was senior research associate at the Institute of History of the Academy of Sciences of the USSR from 1937 to 1941 and from 1950 to 1968. He became research consultant at the Institute of World History in 1968 and has been deputy editor-in-chief of the journal *Vestnik drevnei istorii* since 1966. His principal works are devoted to the problems of the economic structure and the social and political history of the classical world and to the problems of classical culture, historiography, and source study. One of the basic themes of his research is the history of Hellenism, which he examines not as a sociological category (a stage in the development of slave-owning social structures) but as a specific historical period in the history of the eastern Mediterranean.
Works: *Issledovaniia po istorii zemel'nykh otnoshenii v ellinisticheskom Egipte II–I vv. do n.e.* Moscow 1960. *Bor'ba politicheskikh gruppirovok v Attike v VI v. do n.e.* Moscow 1964. *Formy zavisimosti v Vostochnom Sredizemnomor'e v ellinisticheskii period.* Moscow 1969. (With M. K. Trofimova.)

Zemliannikova, L. A.
Trade union official.
Sec. and mem. Presidium Cen. Union Council of USSR Trade Union Ap. 1971–.

Zenkiavichius, Vitautas Mikolovich
State official.
Min. of Foreign Affairs Lithuanian SSR D. 1976–; cand. mem. Cen. Cttee. Lithuanian CP 1981–.

Zenkovich, Vsevolod Pavlovich
Oceanographer and geomorphologist; b. F. 4 (Ja. 22), 1910, Moscow.
Doctor of geographic sciences (1943), professor (1941), member of the CPSU (1953), director of the seacoast-zone laboratory of the institutes of oceanography (1944–71) and geography (1971–74) of the Academy of Sciences of the USSR.
He laid the foundations for the Soviet school of coastal research. He substantiated the principle of the morphological and dynamic unity between the seacoast and the underwater coastal slope, and he put a number of research methods into practice. He developed a classification of ocean coasts for the Marine Atlas (vol. 2, 1953). His most important works are *The Dynamics and Morphology of Seacoasts* (1946) and *Foundations of the Theory of the Development of Seacoasts* (1962; Lenin Prize, 1964). He has been a member of the International Commission on Coastal Geomorphology since 1960.

Zguridi, Aleksandr Mikhailovich
Director of popular science films; b. F. 10 (23), 1904, Saratov.
People's Artist of the USSR (1969).
He began his film work in 1930 in the scientific film laboratory of the State Institute of Microbiology and Epidemiology. He began working as a director at the Moscow Film Studio of Popular Science Film (presently Tsentrnauchfil'm) in 1932. He received prizes at international film festivals for his films *In the Sands of Central Asia* (1943), *The Forest Story* (1950), *In the Ocean Ice* (1953), *In the Pacific Ocean* (1957), *Along the Jungle Paths* (1959, with the Peking Film Studio), *Along the Ancestors' Way* (1962), *The Enchanted Islands* (1965), *Symphony of the Forest* (1967), and *The Black Mountain* (1971). He was vice-president of the International Association of Scientific Films from 1958 to 1962 and has been its president since 1971. He was vice-president of the Organizational Committee of the Union of Cinematographic Workers of the USSR from 1958 to 1965; in 1960 he became art director of Orbit, the creative association at Tsentrnauchfil'm.
He began teaching at the All-Union State Institute of Cinematography in 1947; as a professor from 1966 he directed the workshop for directors of popular science films. Decors.: State Prize of the USSR in 1941, 1946, and 1950 and the State Prize of the RSFSR in 1969, Order of Lenin, two other orders, and medals.
Works: "O nekotorykh voprosakh teorii i praktiki nauchno-populiarnoi kinematografii." In the collection *Nauchno-populiarnyi fil'm,* issue 1. Moscow 1959. *Nauchnoe kino v SSSR,* 2nd ed. Moscow

1959. (With I. Vasil'kov.) *V mire zhivotnykh.* Moscow 1961. *Tropoiu dzhunglei.* Leningrad, 1962. *Stranitsy avtobiografii.* Moscow 1968.

Zhabitskii, Gennadii Nikolaevich
State official; b. 1933.
Belorussian; mem. CPSU 1959–; grad. Belorussian State Univ.; held Komsomol posts 1955–59; Sec., 2nd Sec. 1959–64, 1st Sec. of the Cen. Cttee. Belorussian Komsomol 1964–70; mem. Cen. Cttee. Belorussian CP 1966–; cand. mem. Bureau of Cen. Cttee. Belorussian CP 1966–71; head of the Administrative Organs Dept. of the Cen. Cttee. CPSU 1971–76; Min. of the Interior Belorussian SSR D. 1978–.

Zhakov, Oleg Petrovich
Russian actor; b. Mr. 19 (Ap. 1), 1905, Sarapul, in present-day Udmurt ASSR.
People's Artist of the USSR (1960).
He graduated from the film division of the Leningrad Performing Arts Technicum in 1929. He began acting in films in 1925. His first important role was the radio operator Kurt Shefer in *The Bold Seven* (1936). Here and in subsequent roles Zhakov's acting was characterized by restraint and a profound psychological quality. His roles include Ian Draudin (*We Are From Kronstadt,* 1936), Toivo Antikainen (*For the Soviet Fatherland,* 1937), Rol'f (*Professor Mamlok,* 1938), Paul' (*The Swamp Soldiers,* 1938), Tomilin (*Courage,* 1939), Kostrov (*Submarine T-9,* 1943), Zhavoronkov (*March to April,* 1944), Ianis (*The Sons,* 1946), Dudyrev (*The Trial,* 1962), and Gastev (*The Road to the Moorage,* 1962). His portrayals of negative characters were multifaceted – for example, Docent Vorob'ev (*The Baltic Deputy,* 1937) and Borovskii (*The Great Citizen,* 1938–39), Zhakov's best role was Fedor Talanov in *The Invasion* (1945; State Prize of the USSR, 1946). In 1970 he played Barmin in *At the Lake* (State Prize of the USSR, 1971). Decors.: three orders.

Zhakupov, Anuar Kamzinovich
Party official.
Min. of Motor Transport Kazakh SSR Ag. 1969–78; cand. mem. 1971–76, mem. Cen. Cttee. Kazakh CP 1976–; Chair. of Alma-Ata City Exec. Cttee. of People's Deps. 1978–81; 1st Sec. of Alma-Ata City Cttee. Kazakh CP Ja. 1981–.

Zhanybekov, Shangerei
State official.
1st Sec. Kustanai City Cttee. Kazakh CP 1967; mem. Cen. Cttee. Kazakh CP 1971–; 2nd Sec. Kustanai Oblast Cttee. Kazakh CP 1975–76; Dep. Chair. Council of Mins. Kazakh SSR My. 1976–.

Zhapakov, Nauruz
Author; b. 1914, Takhta-Kupyrskii Raion, Karakalpak ASSR.
Karakalpak; mem. CPSU 1942–; literary work since mid-thirties.
Publs.: "Solntse" (The Sun) 1940, "Slovo poeta" (The Word of the Poet) 1949, "Na beregakh Amu" (On the Banks of the Amu) 1960, and others.

Zharikov, Vilen Andreevich, Dr. of Geol.-Min. Sc.
Geologist and mineralogist; b. 1926.
Mem. CPSU 1954–; grad. Moscow Inst. of Geological Prospecting 1950; Dr. of Geol.-Min. Sc. 1966; Prof. 1970–; corresponding mem. USSR Acad. of Sciences 1972–; sen. mem. scientific staff of Inst. of Geology of Ore Deposits, Petrography, Mineralogy and Geochemistry of USSR Acad. of Sciences 1960–69; Dep. Dir. of Inst. of Experimental Mineralogy of USSR Acad. of Sciences 1969–; concurrently Prof. at Moscow Univ. 1970–.

Zharov, Aleksandr Alekseevich
Author; b. Mr. 31 (Ap. 13), 1904, Semenovskaia, Moscow Oblast.
Mem. CPSU 1920–; studied Moscow State Univ.; first works publ. 1921. Decors.: 4 orders and various medals.
Publs.: "Kerim" 1942, "Boris Safonov" 1944, "Strana iunosti. 1921–1968" (The Land of Youth, 1921–1968) 1968, lyrics for songs "My za mir" (We Are for Peace), "Grustnye ivy" (Sad Willows), "Gde ty, utro rannee?" (Where Are You, Early Morning?) and others.

Zharov, Mikhail Ivanovich
Russian actor and director; b. O. 14 (27), 1900, Moscow.
People's Artist of the USSR (1949). Member of the CPSU since 1950.
He graduated from the drama studio attached to the Theater of the Artistic and Educational Union of Worker's Organizations in 1919. He was an actor with the mobile theater at the front, the Experimental Heroic Theater, the Meierkhol'd Theater, and the Baku Workers' Theater. Between 1931 and 1937 he acted at the Moscow Chamber Theater; his best roles there included his portrayal of Aleksei in Vishnevskii's *Optimistic Tragedy.* He also began to direct there, producing Sheinin's *Confrontation* in 1936. At first he was drawn to comic roles both on the stage and in film. He was especially effective in creating bright comic types, notable for their infectious folk humor, charm, color, and observation of the details of everyday life. In 1938 he began working with the Malyi Theater, where he played Kharitonov (*For Those at Sea,* by Lavrenev), Prokhor (*Vassa Zheleznova,* by Gorkii), Mitrich (*The Power of Darkness,* by L. N. Tolstoi), Lebedev (*Ivanov,* by Chekhov), Iagub (*The Poachers,* by Rannet, which he also directed), Dikii (*The Storm,* by Ostrovskii), and Vorontsov (*So Be It,* by Simonov). The breadth of his creative range was demonstrated particularly well in his work with the Malyi Theater. In addition to comic parts, he began to take sharp character and dramatic roles in both theater and film. He combined comic lightness with psychological depth, and his talent for acute satire found brilliant expression in his portrayal of negative characters.
He is one of the outstanding Soviet motion picture actors. His best parts in sound movies included Zhigan (*The Road to Life,* 1931), Kudriash (*The Storm,* 1934), Zaitsev (*Three Comrades,* 1935), the office worker Dymba (*Maksim's Return,* 1937, and *The Vyborg Region,* 1939), Smirnov (*The Bear,* 1938), Menshikov (*Peter I,* 1937, 1939), Deacon Gavrila (*Bogdan Khmel'nitskii,* 1941), the partisan Rusov (*The Raion Committee Secretary,* 1942), Eropkin (*The Twins,* 1945), Maliuta Skuratov (*Ivan the Terrible,* 1945, 1958), the Minister of War (*Cain XVIII,* 1963), and Aniskin (*The Country Detective,* 1969). Decors.: State Prize of the USSR in 1941, 1942, and 1947, Order of Lenin, three other orders, and various medals.
Work: *Zhizn', teatr, kino.* [Moscow 1967.]

Zhavoronkov, Nikolai Mikhailovich
Chemo-technologist; b. Jl. 25 (Ag. 7), 1907, Streletskie Vyselki, Riazan Oblast.
Mem. CPSU 1939–; grad. D. I. Mendeleev Inst. of Chemical Technology 1930; specializes in inorganic chemistry and chemical technology; Dir. of Karpov Physico-Chemical Inst. 1944–48; Rector of Moscow D. I. Mendeleev Inst. of Chemical Technology 1948–62; Dir. of N. S. Kurnakov Inst. of Gen. and Inorganic Chemistry of USSR Acad. of Sciences 1962–; mem. USSR Acad. of Sciences 1962–; Acad. Sec. of Dept. of Physical Chemistry and Technology of Inorganic Matter of USSR Acad. of Sciences 1963–. Decors.: Hero of Socialist Labour 1969, State Prize of the USSR 1953, D. I. Mendeleev Prize 1950, D. I. Mendeleev Medal in Gold 1969, 2 Orders of Lenin, 4 other orders, various medals.

Zhdanko, Grigorii Semenovich
State official.
Sec., 2nd Sec. of Gomel Oblast Cttee. Belorussian CP 1963–74; mem. Cen. Cttee. Belorussian CP 1971–76; Chair. of State Cttee. for Pricing Belorussian SSR Jl. 1975–.

Zhdanov, Iurii Andreevich
Chemical scientist; b. Ag. 20, 1919, Tver, now Kalinin.
Mem. CPSU 1944–; grad. Moscow Univ. 1941; corresponding mem. USSR Acad. of Sciences 1970–; Rector of Rostov Univ. 1957–70; Chair. of Council of Northern Caucasian Centre 1970–. Decors.: Order of Lenin, 3 other orders and various medals.

Zhdanov, Nikolai Gavrilovich
Author; b. 1909.
Grad. Dept. of Law of Leningrad Univ. 1931; first works publ. 1932.
Publs.: "Morskaia sol'" (Sea Salt) 1947, "Novoe more" (A New Sea) 1954, "Pe-

trogradskaia povest'" (Petrograd Story) 1960, "Veter veka" (The Wind of the Century) 1962, "Dzhiokonda" 1963, and others.

Zhdanov, Viktor Mikhailovich
Virologist; b. F.1 (13), 1914, Shtepino, now Donetsk Oblast.
Mem. CPSU 1941–; grad. Kharkov Medical Inst. 1936; mem. USSR Acad. of Sciences 1960–; Head of Laboratory and Dir. of Inst. of Microbiology and Epidemiology in Kharkov 1946–50; Head of Laboratory of D.I. Ivanovskii Inst. of Virology of USSR Acad. of Medical Sciences 1951– and Head of same Inst. 1961–. Decors.: D.I. Ivanovskii Prize 1969, 4 orders and various medals.

Zheleznikov, Vladimir Karpovich
Author; b. 1925, Vitebsk.
Russian; mem. CPSU 1956–; grad. Gorkii Inst. of Literature Moscow 1957.
Publs.: "Raznotsvetnaia istoriia" (Multi-Coloured Story) 1960, "Khoroshim liudam – dobroe utro" (A Good Morning to Good People) 1961, "Chudak iz shestogo 'B'-6" (The Strange One from 'B'-6) 1963, "Puteshestvennik s bagazhom" (Traveller With Luggage) 1963, "Belye parokhody" (White Ships) 1964, and others.

Zheltov, Aleksei Sergeevich
Military official; b. Ag. 15 (28), 1904, Kharkov; s. of a worker.
Colonel-general (1944). Member of the CPSU since 1929.
He joined the Soviet Army in 1924. He graduated from the Second Moscow Infantry School in 1927, from the Frunze Military Academy in 1937, and from the military political courses in 1938. In 1937 he began partypolitical work, serving as commissar of a division and as a member of a district military council. From February to August 1941 he was a member of the Military Council of the Far Eastern Front. During the Great Patriotic War (1941–45) he was a member of the Military Council of the Karelian Front from September 1941 to July 1942 and a member of the Military Council of the Sixty-third Army from July to September 1942. He was also a member of the military council of the Don Front in October 1942, of the Southwestern Front from October 1942 to October 1943, and of the Third Ukrainian Front from October 1943 to June 1945. He took part in the defense of the Soviet arctic, in the Battle of Stalingrad, and in the liberation of Ukraine, Moldavia, Rumania, Bulgaria, Hungary, Yugoslavia, and Austria. After the war he was the Soviet Union's deputy supreme commissar in Austria and a member of the Military Council of the Central Group of Troops (1945–50), a member of the Military Council of the Turkestan Military District (1950–51), head of the Main Directorate of the Cadres of the Soviet Army (1951–53), head of the Main Political Directorate of the Soviet Army and Navy (1953–57), and chief of a section of the Central Committee of the CPSU (1958–59). 1959–71 he has headed the V.I. Lenin Military Political Academy and was deputy at the second, fourth, and fifth convocations of the USSR Supreme Soviet. Decors.: four Orders of Lenin, four Orders of the Red Banner, an Order of Suvorov First Class, two Orders of Kutuzov First Class, Orders of the Red Star and of the Badge of Honor, nine foreign orders, and medals.

Zheltukhin, Nikolai Alekseevich
Scientist in mechanics and heat engineering; b. O.31 (N.13), 1915, Voronezh.
Corresponding member of the Academy of Sciences of the USSR (1968).
In 1950, he completed the advanced engineering courses at the N.E. Bauman Moscow Higher Technical School. From 1959 he worked at the Institute of Theoretical and Applied Mechanics of the Siberian division of the Academy of Sciences of the USSR. Decors.: Lenin Prize (in 1957), two orders, medals.

Zherdzinskii, Evgenii Fedorovich
Painter; b. 1928, Egorlitska, Rostov/Don Oblast.
Grad. Dnepropetrovsk Art School 1949 and Kharkov Art Inst. 1954, pupil of O. Kokel, G. Tomenko.

Zhigalin, Vladimir Fedorovich
State official; b. 1907; s. of a clerk.
Russian; mem. CPSU 1931–; grad. Leningrad Mechanics Inst. 1931; technician, asst. foreman, sen. foreman, dept. head at an assembly plant 1928–31; Dept. Head, Prod. Manager, Chief Engineer, Dir. of Leningrad "Krasnyi metallist" plant 1931–40; Dept. Head with USSR People's Commissariat of Heavy Machine Construction 1940–45; USSR Dep. People's Commr., Dep. Min., then 1st Dep. Min. of Heavy Machine Construction 1945–57; Dep. Chair. 1957–60, Chair. of Moscow City Econ. Council 1960–63; cand. mem. 1961–64, mem. Cen. Cttee. CPSU 1964–; Dep. to USSR Supr. Sov. 1962–; 1st Dep. Chair. (with rank of USSR Min.) of USSR Econ. Council 1963–65; USSR Min. of Heavy, Power and Transport Machine Building O. 1965–My. 1975; USSR Min. of Heavy Transport Machine Building My. 1975–. Decors.: Hero of Socialist Labour, Order of Lenin, Hammer and Sickle Medal in Gold 1977.

Zhiganov, Nazib Gaiazovich
Composer; b. Ja.2 (15), 1911, Ural'sk.
People's Artist of the USSR (1957). Member of the CPSU since 1944.
He graduated in 1938 from the course in composition taught by G.I. Litinskii at the Moscow Conservatory. In 1945 he began working as a rector and teacher at the Kazan Conservatory; he became a professor there in 1953. Heroism from the historical past is the main subject matter of his works. He wrote the first Tartar operas based on themes from contemporary life. He composed eight operas, three ballets, four symphonies, and other works. Among his best works are the operas *Kachkyn* (*The Fugitive,* produced in 1939), *Altyncha* (*The Golden-haired Girl,* 1941; State Prize of the USSR, 1948), and *Dzhalil'* (1957); a suite based on Tartar themes (1949; State Prize of the USSR, 1950); the overture *Nafisa* (1952); and Symphony No. 2, *Sabantui* (1968; State Prize of the USSR, 1970). He was a deputy to the seventh convocation of the Supreme Soviet of the USSR. Decors.: Order of Lenin and three other orders, medals.

Zhilin, Pavel Andreevich
Military historian; b. Mr. 5 (18), 1913, Vorob'evka, now Buturlinovka Raion, Voronezh Oblast.
Corresponding member of the Academy of Sciences of the USSR (1968); lieutenant-general (1968); member of the CPSU since 1942.
He graduated from the Frunze Military Academy in 1946. He was deputy editor-in-chief of *Voenno-istoricheskii zhurnal* from 1958 to 1964 and prorector of the Academy of Social Sciences of the Central Committee of the CPSU from 1964 to 1966. He has been head of the Institute of Military History of the Ministry of Defense of the USSR since 1966. His main works include *The Defeat of the Turkish Army in 1811* (1952), *The Counteroffensive of the Russian Army in 1812* (1953), *How Fascist Germany Prepared the Attack on the Soviet Union* (1965), and *The Defeat of the Napoleonic Army in Russia* (1968). Decors.: State Prize of the USSR in 1952, Order of the Red Banner of Labor, two Orders of the Red Star, and various medals.

Zhimerin, Dmitrii Georgevich
Energetics scientist and govt. official; b. O.12 (25), 1906, Dubki, now Tula Oblast.
Mem. CPSU 1928–; grad. Moscow Inst. of Energetics 1931; corresponding mem. USSR Acad. of Sciences 1970–; Dep. People's Commr. 1940–42, People's Commr. 1942–46, Min. for Power Plants in the USSR 1946–53; various sen. posts, among them with USSR GOSPLAN and RSFSR GOSPLAN 1955–58; Dir. of G.M. Krzhizhanovskii Inst. of Energetics 1964–71; 1st Dep. Chair. of State Cttee. of Science and Technology with USSR Council of Mins. 1971–. Decors.: 4 Orders of Lenin, Order of the October Revolution, 2 other orders and various medals.
Publs.: "Razvitie ėnergetikii SSSR" Moscow-Leningrad 1960, "Istorija ėlektrifikacii SSSR" Moscow 1962.

Zhitniuk, Galina Mikhailovna
State official.
Dep. to Supr. Sov. Moldavian SSR 1963–; dir. of a knitwear factory 1963;

mem. Cen. Cttee. Moldavian CP 1963–; Min. of the Light Ind. Moldavian SSR O. 1965–.

Zhnykrup, Oksana Leontievna
Sculptor; b. 1931, Chita.
Grad. Odessa Art School 1952.

Zhuchenko, Aleksandr Aleksandrovich
Scientist; b. 1935.
Mem. CPSU 1962–; corr. mem. USSR Acad. of Sciences 1979–; Pres. of Acad. of Sciences Moldavian SSR Je. 1977–.

Zhuganov, A. V.
Komsomol official.
Sec. and mem. Bureau, Cen. Cttee. All-Union Komsomol Ap. 1978–.

Zhuk, Stanislav Alekseevich
Athlete; b. Ja. 25, 1935, Ul'ianovsk.
Master of Sports (1957); Honored Coach of the USSR (1965); officer of the Soviet Army. Member of the CPSU since 1966. He was champion of the USSR in figure skating from 1957 to 1959 and in 1961 and, with N. A. Zhuk, European prize-winner from 1958 to 1960. He has been a coach at the Central Sports Club of the Army (TSSKA) since 1964. His trainees include I. K. Rodnina and A. N. Ulanov, Olympic champions and repeatedly champions of the USSR, and S. A. Chetverukhin, T. A. Zhuk, and A. Iu. Gorelik, prizewinners of European and world championships and at the Olympic Games. Decors.: Order of the Badge of Honor and medals.

Zhukov, Anatolii Borisovich
Forestry scientist; b. Jl. 24 (Ag. 6), 1901, Kharkov.
Grad. Kharkov Inst. of Agriculture and Forestry 1923; mem. USSR Acad. of Sciences 1966–; pedagogic and research work at various Ukrainian and Union insts. 1930–58; Dir. of Forestry and Timber Inst. of Siberian branch of USSR Acad. of Sciences 1958–. Decors.: 2 Orders of Lenin, Order of Red Banner of Labour, medals and other decorations. His main works dealt with oak forests and their renewal, the technology of creating state shelterbelts and oak groves of industrial importance, and methods of increasing the productivity of the forests of the USSR. He was the first to suggest the specialization of forestry and the development of a system of forest management determined by the level of development of forestry. Some of his studies are concerned with the technical properties of pine, oak, ash, and other kinds of wood in relation to the conditions under which they grow.
Works: *Dubravy UkrSSR i sposoby ikh vosstanovleniia.* Moscow-Leningrad 1949. *Sovremennye voprosy lesnogo khoziaistva i lesnoi promyshlennosti v zarubezhnykh stranakh.* Moscow 1962. (Co-author.) *Lesa SSSR,* vols. 1–5. Moscow 1966–70. (Co-author.)

Zhukov, Boris Petrovich, Dr. of Tech. Sc.
Technical chemist; b. 1912.
Mem. CPSU 1944–; grad. D. I. Mendeleev Chemotechnological Inst. Moscow 1937; Dr. of Tech. Sc. 1953; Prof. 1955–; mem. USSR Acad. of Sciences 1974–; scientific work 1937–. Decors.: State Prize of the USSR 1951 and 1967, Hero of Socialist Labour 1966.

Zhukov, Boris Vladimirovich
Painter; b. 1921, Levandalivka, Kharkov Oblast.
Mem. CPSU 1943–; attended Kharkov Art School 1934–39 and Art Inst. 1939–40 and 1946–50, pupil of M. Deregus, L. Chernov.

Zhukov, Georgii Aleksandrovich
Journalist; b. 1908.
Ukrainian; mem. CPSU 1943–; grad. Moscow Automobile and Tractor Inst. 1932; asst. machine operator, machine fitter, then journalist 1926–32; literary and ed. work for various newspapers and journals 1932–47; correspondent for "Pravda" in Paris 1947–52; Dep. Chief Foreign Ed. of "Pravda" 1952–57; mem. Cen. Auditing Comm. 1956–76, cand. mem. Cen. Cttee. CPSU 1976–; Chair. of State Cttee. for Cultural Relations With Foreign Countries with USSR Council of Mins. 1957–62; Dep. to USSR Supr. Sov. 1962–; Dep. Chair. of Sov. Peace Cttee., Pol. Commentator of "Pravda" 1962–. Decors.: Lenin Prize, Order of Red Banner of Labour.

Zhukov, Mikhail Fedorovich
Physicist; b. Ag. 28 (S. 6), 1917, Verkhove, Orel Oblast.
Mem. CPSU 1953–; grad. Mechanical-Mathematical Dept. of Moscow Univ. 1941; specialist on aerodynamics and plasmaphysics (low-temperature plasma); corresponding mem. USSR Acad. of Sciences 1968–; mem. staff Cen. N. E. Zhukovskii Inst. of Aerodynamics 1941–46; mem. staff Cen. Inst. of Aircraft Engine Construction 1946–59; Dep. Dir. of Inst. of Theoretical and Applied Mechanics of USSR Acad. of Sciences 1959–70; Dep. Dir. of Inst. of Thermophysics of Siberian Branch of USSR Acad. of Sciences 1970–. Decors.: Order of Red Banner of Labour and various medals.

Zhukov, Vladimir Semenovich
Author; b. 1920, Ivanovo; s. of a worker.
Russian; mem. CPSU 1944–; first works publ. 1937.
Publs.: "Soldatskaia slava" (Soldier's Glory) 1946, "Soldatskaia poema" (Soldier's Poem) 1947–57, "Nasledniki" (Heirs) 1954, "Pamiati Nikolaia Maiorova" (In Memory of Nikolai Majorov) 1959, "Ekho" (Echo) 1960, "Valia" 1959, "Utrenniaia smena" (Morning Shift) 1960, and others.

Zhukov, Vladislav Petrovich
Diplomat; USSR Amb. to Soudan 1979–82, to Arab Republic Yemen My. 1982–.

Zhukovskii, Igor Nazarievich
Diplomat; b. 1918.
Mem. Diplomatic Service 1946–; Emb. Counsellor at USSR Emb. in Turkey 1956–58; sen. posts in USSR Min. of Foreign Affairs 1958–62; USSR Consul-Gen. in Istanbul 1962–64; again in USSR Min. of Foreign Affairs 1964–66; Emb. Counsellor at USSR Emb. in Tunisia 1966–68; Emb. Counsellor-Envoy at USSR Emb. in Guinea 1968–69; USSR Amb. to Dahomey 1969–Ap. 1975.

Zhulenev, Viacheslav Fedorovich
State official.
Min. of Construction Turkmen SSR Ja. 1968–77; mem. Cen. Cttee. Turkmen CP 1971–; Dep. Chair. Council of Mins. Turkmen SSR 1977–79; Sec. and mem. Bureau Cen. Cttee. Turkmen CP Mr. 1979–.

Zhumasultanov, Tulebai Zhumasultanovich
State official.
Dep. Head 1970–75, 1st Dep. Head 1975–80, Head of Main Statistical Admin. Kazakh SSR O. 1980–; mem. Auditing Comm. Kazakh CP 1981–.

Zhurakovskii, Viktor Petrovich
Painter; b. 1928, Harbin, China.
Grad. Odessa Art School 1955.

Zhuravlev, Dmitrii Nikolaevich
Russian actor and artist of the variety stage; b. O. 11 (24), 1900, Alekseevka, present-day Kharkov Oblast.
People's Artist of the RSFSR (1960).
In 1927, he graduated from the school attached to the E. B. Vakhtangov Theater, where he acted until 1939. In 1931, after becoming acquainted with the dramatic reading of A. Ia. Zakushniak, he gave his first public performance, reading Pushkin's *Egyptian Nights* and poems by V. V. Mayakovsky. His repertoire included Pushkin's "Bronze Horseman" and *The Queen of Spades* (1940); Mayakovsky's "At the Top of My Voice," "Verses About a Soviet Passport." and "All Right"; A. P. Chekhov's *Lady With a Lapdog* (1939) and *The House With the Attic* (1953); and works by P. Merimée, A. A. Blok, M. Gorkii, and L. N. Tolstoy. His spirited renditions combine the lyrical with the heroic. Decors.: State Prize of the USSR in 1949, two orders and medals.

Zhuravlev, Tikhon Kononovich
Author; b. 1913, Zaluzhe nr. Voronezh; s. of a peasant.
Russian; mem. CPSU 1940–; his first stories and sketches were publ. in various front newspapers.
Publs.: "Zolotoi dozhd'" (Golden Rain)

1950, "Riadovoi Antipov, Zapiski ofitsera" (Private Antipov. An Officer's Diary) 1950, "Kombainery" (The Combine Operators) 1953, "Kurgan" (The Burial Mound) 1955, "Kombat" (The Battery Commander) 1957, "Ded Khariton" (Grandfather Khariton) 1959, "Kuznetsy" (The Blacksmiths) 1961, "Svezhii veter" (Fresh Wind) 1962, "Podvig" (Deed of Glory) 1963, and others.

Zhuravlev, Vasilii (Wilhelm) Andreevich
Author; b. 1914, Tulinovka nr. Tambov. Russian; mem. CPSU 1942–; grad. Gorkii Inst. of Literature in Moscow 1940; first works publ. 1934.
Publs.: "Pervaia beseda" (First Conversation) 1951, "Stikhi" (Poems) 1959, "Liubov' i vremia" (Love and Time) 1960, "Pritambovie" (The Tambov Region) 1962, and others.

Zhurkov, Serafim Nikolaevich
Physicist; b. My. 3 (16), 1905, Trubotkino, Tambov Oblast.
Mem. CPSU 1944–; grad. Voronezh Univ. 1929; mem. USSR Acad. of Sciences 1968–; mem. staff Leningrad Inst. of Physical Technology 1930–; Prof. at Leningrad Univ. 1947–.

Zidu, Dmitrii Georgievich
State official.
Cand. mem. Cen. Cttee. Moldavian CP 1966–68, mem. N. 1968–; Min. of Educ. Moldavian SSR D. 1978.

Ziedonis, Imants Ianovich
Latvian poet; b. My. 3, 1933, Ragatsiems, now Tukums Raion.
Member of the CPSU since 1962.
He graduated from the philology department of the University of Latvia in 1959 and from the Higher Literary Courses in 1964. He was first published in 1956. His first poetry collection, *The Earth and the Dream,* was published in 1961, followed by the collections *The Heart's Dynamite* (1963), *I Enter Into Myself* (1968), and *How It Burns as a Candle* (1971). His poetry is polemic and has a contemporary ring, including both romantic daring and philosophic meditation; his poetics combine classical tradition with a search for new means of expression and new rhythms. His prose essays, such as *The Poet's Diary* (1965; Russian translation, 1968) and *Kurzemite* (1970), deal with modern themes. In 1970 his collection of miniatures in prose, *The Epiphanies,* was published. He has translated into Latvian works of A.A. Blok, V.V. Mayakovsky, V.A. Lugovskoi, P. Sevak, and I. Drach. Decor.: State Prize of the Latvian SSR in 1967.
Works: *Motocikls,* Riga 1965. *Pa putu celu.* Riga 1967. *In Russian translation: Smola i iantar'.* Moscow 1965. *Izbr. lirika.* Moscow 1969. *U kazhdogo kolodtsa svoe ekho: Kniga puteshestvii.* Moscow 1971.

Zigmonte, Dagnia Andreevna
Latvian author; b. 1931, Riga.
Publs.: novels "Kholodnoe vremia" (Cold Times) 1957, "Deti i derevia tianutsia k solntsu" (Children and Trees Stretch Towards the Sun) 1958, "Morskie vorota" (Sea Gate) 1959, "Dai ruku utrennei zare" (Give Your Hand to Aurora) 1961, and others.

Zimianin, Mikhail Vasil'evich
Party official; b. N. 8 (21), 1914, Vitebsk; s. of a metal worker.
Belorussian; mem. CPSU 1939–; grad. Mogilev Teachers' Training Inst. 1939; worker at a depot, worked for newspaper "Kolkhoznik Belorussii", teacher, principal of a school 1929–36; served in Sov. Army 1936–38; sec. of an oblast cttee. Belorussian Komsomol 1939–40; 1st Sec. of Cen. Cttee. Belorussian Komsomol 1940–46; partisan in Belorussia 1941–45; 2nd Sec. of Gomel Oblast Cttee. Belorussian CP (B) 1946; Min. of Educ. Belorussian SSR 1946–47; Dep. to 1946–54 and 1966–, mem. Foreign Affairs Comm. of Sov. of Nationalities USSR Supr. Sov. 1966–; Sec., then 2nd Sec. of Cen. Cttee. Belorussian CP (B) 1947–53; mem. Cen. Cttee. CPSU 1952–56 and 1966–; Head of 4th European Dept. and mem. Collegium USSR Min. of Foreign Affairs 1953–56; USSR Amb. to North Vietnam 1956–57; mem. Cen. Auditing Comm. CPSU 1956–66; Head of Far East Dept. and mem. Collegium USSR Min. of Foreign Affairs 1957–60; USSR Amb. to the ČSSR 1960–65; USSR Dep. Min. of Foreign Affairs 1965; Chief Ed. of "Pravda" S. 1965–76; Chair. of USSR Journalists' Union O. 1966–76; Dep. to USSR Supr. Sov. 1970–; Sec. of Cen. Cttee. CPSU Mr. 1976–. Decors.: 2 Orders of Lenin, Order of Red Banner of Labour, 2 other orders and various medals.

Zimin, Aleksandr Aleksandrovich
Historian; b. F. 22, 1920, Moscow.
Doctor of historical sciences (1959), professor (1970).
He graduated from the history and philology department of the University of Central Asia in 1942 and became a senior research associate at the Institute of History of the Academy of Sciences of the USSR in 1951. He is a specialist on the history of Russia from the 11th to the 18th century, concentrating primarily on social and political history and social thought of the 16th century, as well as on historiography and textology. He has contributed, as author and editor, to a number of multivolume historical works, including *The History of Moscow* (vol. 1, 1952), *Essays on the History of the USSR: The Feudal Period, From the Late 15th to the Early 17th Century* (1955), *World History* (vol. 4, 1958), and *The History of the USSR: From the Earliest Times Until the Present* (vol. 2, 1966). He has also edited and compiled many collections of documents.
Works: *I. S. Peresvetov i ego sovremenniki.* Moscow 1958. *Metodika izdaniia drevnerusskikh aktov.* Moscow 1959. *Russkie letopisi i khronografy kontsa XV–XVI vv.* Moscow 1960. *Reformy Ivana Groznogo.* Moscow 1960. *Oprichnina Ivana Groznogo.* Moscow 1964. *Rossiia na poroge novogo vremeni.* Moscow 1972. *Kholopy na Rusi.* Moscow 1973.

Zinovev, Nikolai Mikhailovich
Painter; b. 1888.
Mem. CPSU 1941–; attended workshop for icon painting; specializes in miniatures; Lecturer at Palekh Art College 1928–68; Dir. of Palekh Museum 1948–56; one of the first painters to introduce Sov. themes into traditional lacquer painting. Decor.: People's Artist of the USSR 1974.

Zitmanis, August Karlovich
Trade union official; b. 1929.
Latvian; mem. CPSU 1949–; uncompleted college educ.; worker 1944–46; instructor, 1st sec. of a raion Komsomol cttee. 1946–49; served in Sov. Army 1952–56; dept. head, sec., 1st sec. of a raion cttee. Latvian Komsomol 1956–59; 1st Sec. and mem. Bureau Cen. Cttee. Latvian Komsomol 1959–62; mem. Cen. Cttee. 1960–, cand. mem. Bureau Cen. Cttee. Latvian CP 1961–; Head of Dept. of Party Organs resp. Party Org. Work of Cen. Cttee. Latvian CP 1963–66; 1st Sec. of Riga City Cttee. Latvian CP 1966–73; Chair. of Latvian Republican Trade Union Council My. 1973–.

Zivs, Samuil Lazarevich
Legal scholar; b. D. 24, 1921, Ventspils, Latvian SSR.
Doctor of jurisprudence (1962). Honored Scholar of the RSFSR (1972). Became a member of the CPSU in 1948.
He has been engaged in legal studies and teaching since 1948. He has written works on the general theory of the state and law and on comparative jurisprudence. Since 1972 he has been vice-president of the Institute of Soviet-American Relations, since 1965 of the Association of Soviet Jurists. He has been a member of the Soviet Peace Committee since 1967.

Zlobin, Anatolii Pavlovich
Author; b. 1923, Moscow.
Russian; mem. CPSU 1943–; grad. Gorkii Inst. of Literature Moscow 1950; first works publ. 1948.
Publs.: "Piat' chasov raznitsy" (Five Hours' Difference) 1959, "Geroi nashego vremeni" (Hero of Our Time) 1960, "Molodye serdtsa" (Young Hearts – together with Iu. Grachevskii) 1949, "Rozhdenie budushchego" (Birth of the Future) 1956, and others.

Zlobin, Gennadii Karpovich
State official.
Head of Dept. of Construction resp. Construction and Communal Econ. of Cen. Cttee. Ukrainian CP 1965–75; cand. mem. 1966–71, mem. Cen. Cttee. Ukrainian CP 1971–; Chair. of State Construction Cttee. Ukrainian SSR My. 1975–.

Znamenskii, Anatolii Dmitrievich
Russian author; b. 1923, Volgograd Oblast.
Publs.: "Ivan-chai" 1963, "Synovia Chistiakova" (Chistiakov's Sons) 1960, and others.

Znamenski, Iurii Stepanovich
Party official; b. 1929.
Russian; mem. CPSU 1959–; grad. Energy Inst. in Ivanovo and CPSU Cen. Cttee. Party Higher School; shift engineer, dep. dir., then dir. of a section of a plant, chief engineer of Boiler-Making Plant in Biisk 1953–73; 1st City Cttee. Sec. CPSU in Biisk 1973–78; 1st Oblast Cttee. Sec. CPSU in Gorno-Altaisk O. 1978–; Dep. to USSR Supr. Sov. 1979–.

Zolotarev, Nikolai Gavrilovich (pseudonym: N. Iakutskii)
Yakut writer; b. N.9 (22), 1908, Kharbalakh Nasleg (district), Verkhneviliuisk Raion.
Member of the CPSU since 1929.
He served in the frontier guards and was involved in party work. From 1948 to 1961, he served as chairman of the board of the Writers' Union of Yakutia. He began to appear in print in 1938. He is the author of short stories about the heroes of the Great Patriotic War, such as "He Who Has Vanquished Death, He Is Immortal" and "Kesha Berezkin." His other works include the novel *Destiny* (1947) about the difficult life and struggle of the Yakuts before the revolution; the novella *Golden Brook* (1948; Russian translation, 1962); the novel *Land* (1958) about collective farm life; and the novels *Dark Night* (1960) and *Before Dawn* (1962) about the heroic struggle to establish Soviet power in Yakutia. The novel *Surprise* (1967) and two later works make up the trilogy *Destiny* (Russian translation, 1968). Decors.: two orders, several medals.

Zolotov, Evgenii Vasil'evich
Mathematician; b. Ap.29, 1922, Tula.
Mem. CPSU 1955–; grad. Dzerzhinskii Artillery Acad. 1945; corresponding mem. USSR Acad. of Sciences 1970–; Prof. at Kalinin Polytech. Inst. 1968–; Dir. and Organizer of Computer Centre of Far Eastern Centre of USSR Acad. of Sciences 1971–.

Zolotukhin, Grigorii Sergeevich
State official; b. 1911; s. of a peasant.
Russian; mem. CPSU 1939–; grad. Chakinskii Agricultural Tech. School 1931 and Party Higher School of Cen. Cttee. CPSU 1949; econ. posts in Tambov Oblast 1931–38; chief agronomist with a machine and tractor station in Tambov Oblast 1938–39; Komsomol and party posts 1939–; Sec. 1944–46 and 1949–51, 2nd Sec. 1951–55, 1st Sec. of Tambov Oblast Cttee. CPSU 1955–66 (for Agriculture 1963–64); cand. mem. 1956–66, mem. Cen. Cttee. CPSU 1966–; Dep. to 1958–, mem., then Chair. of Agricultural Comm. of Sov. of the Union USSR Supr. Sov. 1966–73; 1st Sec. of Krasnodar Krai CPSU Cttee. 1966–73; USSR Min. of Supplies Ap. 1973–. Decor.: Order of Lenin.

Zorin, Karl Ernestovich
State official.
Cand. mem. 1960–61, mem. Cen. Cttee. Latvian CP 1961–; Dep. to 1963–, Dep. Chair. of Presidium 1963–73, Sec. of Presidium Supr. Sov. Latvian SSR 1973–; Head of Kolkhoz-Sovkhoz Prod. Admin. in Valmiera 1963.

Zorin, Leonid Genrikhovich
Dramatist; b. N.3, 1924, Baku.
Mem. CPSU 1952–; grad. Azerbaidzhan Kirov Univ. 1946 and Gorkii Inst. of Literature 1947.
Publs.: "Molodost'" (Youth) 1949, "Gosti" (Guests) 1954, "Svetlyi Mai" (Bright May) 1958, "Paluba" (The Deck) 1963, "Dion" 1965, "Dekabristy" (The Dekabrists) 1967, "Varshavskaia melodiia" (Warsaw Melody) 1967, "Serafim, ili tri glavy iz zhizni Kramolnikova" (Serafim, or Three Chapters From the Life of Kramolnikov) 1968, "Koronatsiia" (The Coronation) 1969, and others.

Zorskas, Albinas Teofilovich
State official.
Dep. to Supr. Sov. Lithuanian SSR 1959–; dir. of a tractor repair station 1959; Chair. of Lithuanian Republican Asscn. for Agricultural Technology "Litselkhoztekhnika" Mr. 1961–78; cand. mem. Cen. Cttee. Lithuanian CP 1961–81, mem. 1981–; mem. Agricultural Bureau of Cen. Cttee. Lithuanian CP 1962–64; Chair. of State Cttee. for the Supply of Production Equipment for Agriculture Lithuanian SSR Ap. 1978–.

Zubavin, Boris Mikhailovich
Author; b. 1915, Moscow.
Russian; mem. CPSU 1942; first works publ. 1932.
Publs.: "Khoziain" (Master of the House) 1947, "Samoe glavnoe" (The Main Thing) 1950, "Semeinaia khronika" (Family Chronicle) 1951, "Tretie pokolenie" (The Third Generation) 1952, "Malenkie istorii" (Little Stories) 1959, "Radost'" (Happiness) 1962, "Po prikazu Rodiny" (By Order of the Fatherland) 1963, and others.

Zubenko, Iu. S.
Trade union official.
Sec. of Ukrainian Republican Trade Union Council F. 1982–.

Zubkov, Iurii Aleksandrovich
Theater scholar; b. O.8 (21), 1914, Moscow.
Literary and drama critic. Honored Art Worker of the RSFSR (1969). Member of the CPSU since 1944.
He graduated from the A.V. Lunacharskii State Institute of Theatrical Arts (GITIS) in 1939 and in 1954 finished graduate studies at the Academy of Social Sciences of the CC CPSU. He began his literary work in 1931. During 1954–57 he was deputy editor-in-chief of the magazine *Oktiabr';* in 1958 he became editor-in-chief of *Teatral'naia zhizn'.* His articles on Soviet drama and the theater appear in newspapers and magazines, collections, and annuals. Decors.: two orders and medals.
Works: *Tvorchestvo N.K. Cherkasova: Problema perevoploshcheniia v iskusstve aktera.* Moscow 1964. *Sud'by akterskie.* Moscow 1969. *Vremia i teatr.* Moscow 1971.

Zuev, Valentin Petrovich
State official.
Head of Admin. of Geology Tadzhik SSR D. 1978–; mem. Auditing Comm. Tadzhik CP 1981–.

Zuev, Vladimir Evseevich
Physicist; b. Ja. 29, 1925, Malye Goly, Kachug Raion, Irkutsk Oblast.
Corresponding member of the Academy of Sciences of the USSR (1970). Member of the CPSU since 1945.
Graduated from the University of Tomsk in 1951. From 1955 to 1969 he worked in the Siberian Physicotechnical Institute, and in 1969 he became the director of the Institute of Atmospheric Optics of the Siberian Division of the Academy of Sciences of the USSR. Since 1951 he has also been teaching at the University of Tomsk (professor since 1964).
His main work has been in atmospheric optics and physics. He has developed a number of methods of estimating the energy loss of waves in the optical range in the atmosphere, particularly the absorption and scattering of laser radiation by atmospheric gases and aerosols. He was a deputy to the eighth convocation of the Supreme Soviet of the USSR. Decors.: Order of the Badge of Honor and medals.
Works: *Prozrachnost' atmosfery dlia vidimykh i infrakrasnykh luchei.* Moscow 1966. *Rasprostranenie vidimykh i infrakrasnykh voln v atmosfere.* Moscow 1970.

Zueva, Anastasiia Platonovna
Russian actress; b. D.5 (17), 1896, Spasskoe, in present-day Tula Oblast.
People's Artist of the USSR (1957).
She graduated from drama school in 1915 under the guidance of N.G. Alek-

sandrov, N.O. Massalitinov, and N.A. Podgornyi and entered the Second Studio of the Moscow Art Academic Theater. In 1916 she began acting with the Moscow Art Theater. A keen character actress, she has usually played elderly women, endowing them with the folksiness and spiritual warmth typical of the Russian national character.
Her major roles include the Mother (*Untilovsk* by Leonov), Matrena (*Resurrection,* based on L. N. Tolstoy), Korobochka (*Dead Souls,* based on Gogol), Domna Pantelevna (*Talents and Admirers* by Ostrovskii), Zobunova and Chugunova (*Egor Bulychev and the Others* and *Dostigaev and the Others* by Gorkii), Miss Wardle (*The Pickwick Club,* based on Dickens), Feklusha and Ulita (*The Storm* and *The Forest* by Ostrovskii), Marfa (*The Earth* by Virta), Glafira Firsovna (*The Last Victim* by Ostrovskii), and Anfisa (*The Three Sisters* by Chekhov). She also appears on the *estrada* (the variety stage) and on radio. Decors.: State Prize of the USSR in 1952, two Orders of Lenin, the Order of the Badge of Honor, and medals.

Zulfugarov, Akhmed Mamedovich
State official.
Chair. of State Pricing Cttee. Azerbaidzhan SSR Ag. 1970–.

Zurabashvili, Avlipii Davidovich
Psychiatrist; b. F. 2 (15), 1902, Bodbiskhevi, now Signakhi Raion, Georgian SSR.
Academician of the Academy of Sciences of the Georgian SSR (1955) and of the Academy of Medical Sciences of the USSR (1960); Honored Scientific Worker of the Georgian SSR (1946). Member of the CPSU since 1940.
In 1926, he graduated from the medical department of the University of Tbilisi; since 1937 he has been director of the M. M. Asatiani Scientific Research Institute of Psychiatry. His principal works have been devoted to the etiology, pathogenesis, clinical aspects, syndromes, and treatment of schizophrenia, epilepsy, oligophrenia, and certain other psychoses. He advanced the hypothesis of the structural-dynamic reversibility of the most subtle changes in the parenchymatous formations of nerve tissue. He created an original trend in neuromorphology – the experimental and clinicomorphological investigation of synapses. He is a member of the International Brain Research Organization and a member of the presidium of the Society of Neuropathologists and Psychiatrists of the Georgian SSR. Decors.: two Orders of Lenin, the Order of the Red Banner of Labor, and medals.
Works: *Nekotorye aktual'nye voprosy o psikhogennykh zabolevaniiakh.* Tbilisi 1940. *Sinapsy i obratimye izmeneniia nervnykh kletok,* 2nd ed. Tbilisi 1958. *Nekotorye klinikoteoreticheskie iskaniia v psikhiatrii.* Tbilisi 1961. *Aktual'nye problemy psikhiatrii.* Tbilisi 1964. *Problemy psikhologii i patopsikhologii lichnosti.* Tbilisi 1967.

Zurabian, Marlen Arshaluiskovich
State official.
Cand. mem. 1961–64, mem. Cen. Cttee. Armenian CP 1964–; Dep. to Supr. Sov. Armenian SSR 1963–; Head of Dept. of Admin. Organs of Cen. Cttee. Armenian CP 1965–70; Min. of Local Ind. Armenian SSR Je. 1970–.

Zverev, Aleksei Il'ich
State official; b. 1929.
Russian; mem. CPSU 1953–; grad. Belorussian Agricultural Acad.; chief agronomist, dir. of a machine and tractor station 1951–56; sec. of various raion party cttees. 1956–62; head of a kolkhoz-sovkhoz prod. admin., 2nd Sec. (for Agriculture) of Novosibirsk Oblast Cttee. CPSU 1962–64; Chair. of Novosibirsk Oblast Exec. Cttee. of Workers' Deps. 1964–73; Dep. to 1966–, mem. Agricultural Comm. of Sov. of the Union USSR Supr. Sov. 1966–74; Min. of Forestry RSFSR Je. 1973–.

Zverev, Il'ia (real name: Samdberg, Izold Iudovich)
Author; b. 1926, Aleksandria, Ukraine.
Russian; first works publ. 1947.
Publs.: "Na ugol'nom fronte" (At the Coal Front) 1948, "Tam, na shakhte ugol'noi . . ." (There, in the Coal Mine . . .) 1956, "Ne radi slavy'" (Not for Glory) 1958, "Dorogi vglub'" (Roads Into the Depth) 1957, "V dvukh kilometrakh ot schastia" (Two Kilometres to Happiness) 1960, and others.

Zverev, Mitrofan Stepanovich
Astronomer; b. Ap. 3 (16), 1903, Voronezh.
Corresponding member of the Academy of Sciences of the USSR (1953). Member of the CPSU since 1947.
He graduated from Moscow University (1931) and the Moscow Conservatory (1929). From 1931 to 1951 he worked at Moscow University (professor since 1948); in 1951 he became deputy director of the Pulkovo Observatory. He compiled a number of star catalogs. He is the initiator and director of international work on compiling the *Catalog of Faint Stars.* His principal works deal with time service, gravimetry, and variable stars. Decors.: Order of Lenin, two other orders, and medals.
Works: "Issledovanie rezul'tatov astronomicheskikh nabliudenii Sluzhby vremeni GAISH v. 1941–44 gg." *Tr. Gos. astronomicheskogo in-ta im P. K. Shternberga,* 1950, vol. 18, fasc. 2. "Predvaritel'nyi svodnyi Katalog fundamental'nykh slabykh zvezd so skloneniiami ot $+90°$ do $-20°$ (PFKSZ)." *Tr. Glavnoi astronomicheskoi observatorii AN SSSR,* 1958, vol. 72. (With D. D. Polozhentsev.) "Piat' let raboty astronomicheskoi ekspeditsii Pulkovskoi observatorii v Chili." *Izv. Glavnoi astronomicheskoi observatorii,* 1970, no. 185.

Zykina, Ludmila Georgievna
Singer; b. 1929.
Grad. Ippolitov-Ivanov Music School 1969; soloist with Piatnitskii Choir 1947–50; soloist with Choir for Russian Vocal Music of Union Radio and Television Station 1951–59; soloist with "Moskontsert" 1960–. Decors.: People's Artist of the RSFSR 1968, Lenin Prize 1970, People's Artist of the USSR 1973.

Zykov, Spartak Sergeevich
Diplomat; b. 1925.
Emb. Counsellor at USSR Embassy in Somalia 1960–63, in Ethiopia 1964–71; Counsellor Envoy, USSR Embassy in Mali 1971–74, in Algeria 1974–78; USSR Amb. to Chad O. 1979– and to Cameroon N. 1980–.

Index Section

Members of the Party and Government Leaders of the Komsomol and the Works Council as of January 1st 1983

USSR

The Communist Party of the Soviet Union

Politbureau of the Central Committee

Members
Aliev, A. G.
Andropov, Iu. V.*
Chernenko, K. U.
Gorbachev, M. S.
Grishin, V. V.
Gromyko, A. A.
Kunaev, D. A.
Pelshe, A. Ia.
Romanov, G. V.
Shcherbitskii, V. V.
Tikhonov, N. A.
Ustinov, D. F.

Candidate Members
Demichev, P. N.
Dolgikh, V. I.
Kiselev, T. Ia.
Kuznetsov, V. V.
Ponomarev, B. N.
Rashidov, Sh. R.
Shevardnadze, E. A.
Solomentsev, M. S.

Secretariat of the Central Committee

Secretary-General
– Andropov, Iu. V.**

Secretaries
– Dolgikh, V. I.
– Gorbachev, M. S.
– Kapitonov, I. V.
– Ponomarev, B. N.
– Rusakov, K. V.
– Ryzhkov, N. I.
– Zimianin, M. V.

Central Committee

Members
Abrasimov, P. A.
Afanasev, S. A.
Afanasev, V. G.
Aksenov, A. N.
Aksenov, N. F.
Aleksandrov, A. P.
Aleksandrov-Agentov, A. M.
Alekseev, P. F.
Aliev, G. A. R. o.
Alkhimov, V. S.
Altunin, A. T.
Andropov, Iu. V.*
Antonov, A. K.
Antonov, S. F.
Arbatov, G. A.
Aristov, B. I.
Arkhipov, I. V.
Ashimov, B. A.
Askarov, A. A.
Auelbekov, E. N.
Aukhadiev, K. M.
Avramenko, S. S.

Baibakov, N. K.
Bakhirev, V. V.
Bakin, B. V.
Balandin, A. N.
Balandin, Iu. N.
Bannikov, N. V.
Batitskii, P. F.
Bazovskii, V. N.
Bekturganov, Kh. Sh.
Beliak, K. N.
Berezin, A. I.
Bespalov, I. P.
Biriukova, A. P.
Bodiul, I. I.
Bogoliubov, K. M.
Bogomiakov, G. P.
Bondarenko, I. A.
Borisenkov, V. M.
Borodin, A. M.
Borodin, L. A.
Borodin, P. D.
Botvin, A. P.
Bratchenko, B. F.
Brekhov, K. I.
Brovikov, V. I.
Bugaev, B. P.
Bulgakov, A. A.

Chazov, E. I.
Chebrikov, V. M.
Chernenko, K. U.
Chernyi, A. K.
Chervonenko, S. V.
Chicherov, V. S.

Dementeva, R. F.
Demichev, P. N.
Demidenko, V. P.
Demirchian, K. S.
Dobrik, V. F.
Dobrynin, A. F.
Dolgikh, V. I.
Drygin, A. S.
Dymshits, V. E.

Egorov, A. G.
Egorov, M. V.
Elchenko, Iu. N.
Eliutin, V. P.
Eltsin, B. N.
Epishev, A. A.
Ermin, L. B.
Ezhevskii, A. A.

Fedirko, P. S.
Fedorov, V. S.
Fedoseev, P. N.
Filatov, A. P.
Finogenov, P. V.
Florentev, L. Ia.
Fomin, G. I.

Gapurov, M. G.
Garbuzov, V. F.
Glushko, V. P.
Goldin, N. V.
Golubeva, V. N.
Goncharenko, B. T.
Gorbachev, M. S.
Gorban, G. Ia.
Gorshkov, L. A.
Gorshkov, S. G.
Gostev, B. I.
Govorov, V. L.
Grekov, L. I.
Gribkov, A. I.
Grishin, V. V.
Grishkiavichius, P. P.
Gromyko, A. A.
Grossu, S. K.
Grushin, P. D.
Gudkov, A. F.
Gusev, V. K.
Gustov, I. S.
Guzhenko, T. B.

Isaev, V. Ia.
Iunak, I. Kh.
Ivanovskii, E. F.

Kachalovskii, E. V.
Kachin, D. I.
Kachura, B. V.
Kandrenkov, A. A.
Kapitonov, I. V.
Karavaev, G. A.
Karlov, V. A.
Karpova, E. F.
Katushev, K. F.

* Died February 9, 1984.
** Successor K. U. Chernenko.

Kavun, V. M.
Kazakov, L. D.
Kazanets, I. P.
Kebin, I. G.
Khitrov, S. D.
Khomiakov, A. A.
Khristoradnov, Iu. N.
Khudaiberdyev, N. D.
Kirichenko, N. K.
Kirilenko, A. P.
Kiselev, I. I.
Kiselev, T. Ia.
Klepikov, M. I.
Klimchenko, E. I.
Klimenko, I. E.
Kliuev, V. G.
Kniazev, F. K.
Kolbin, G. V.
Koldunov, A. I.
Kolesnikov, A. Ia.
Konoplev, B. V.
Konotop, V. I.
Konovalov, N. S.
Kornienko, G. M.
Korolev, A. M.
Kosolapov, R. I.
Kostandov, L. A.
Kostin, V. S.
Kovalenko, A. V.
Kozlov, N. T.
Krotov, V. V.
Kruchina, N. E.
Kruglova, Z. M.
Kudinov, I. P.
Kulichenko, L. S.
Kulikov, F. M.
Kulikov, V. G.
Kunaev, D. A.
Kurkotkin, S. K.
Kutakhov, P. S.
Kuznetsov, I. M.
Kuznetsov, V. V.

Lapin, S. G.
Lein, V. P.
Leonov, P. A.
Liashko, A. P.
Ligachev, E. K.
Lomakin, V. P.
Lomako, P. F.
Lomonosov, V. G.
Loshchenkov, F. I.
Lushev, P. G.
Lykova, L. P.

Makarenko, V. S.
Makeev, V. N.
Makeev, V. P.
Maltsev, V. F.
Maniakin, S. I.
Marchuk, G. I.
Marisov, V. K.
Markov, G. M.
Martynov, N. V.
Maslennikov, N. I.
Medunov, S. F.
Meshkov, F. S.
Mesiats, V. K.
Mikulich, V. A.
Miroshkin, O. S.
Modogoev, A. U.
Morgun, F. T.
Morozov, I. P.
Moskalenko, K. S.
Mozgovoi, I. A.
Murakhovskii, V. S.
Muravev, E. F.
Musakhanov, M. M.
Mysnichenko, V. P.

Neporozhnii, P. S.
Nikolaeva-Tereshkova, V. V.
Nikonov, V. P.
Novikov, I. T.
Nuriev, Z. N.

Ogarkov, N. V.
Orlov, V. P.

Pastukhov, B. N.
Patolichev, N. S.
Paton, B. E.
Pavlov, G. P.
Pavlov, G. S.
Pavlov, V. Ia.
Pavlovskii, I. G.
Pegov, N. M.
Pelshe, A. Ia.
Pereverzeva, N. V.
Petrov, V. A.
Petrov, V. I.
Petrovichev, N. A.
Pleshakov, P. S.
Poliakov, I. E.
Poliakov, V. N.
Ponomarev, B. N.
Ponomarev, M. A.
Popov, B. V.
Popov, N. S.
Popova, M. G.
Priezzhev, N. S.
Prokhorov, V. I.
Prokofev, M. A.
Prokopev, I. P.
Promyslov, V. F.
Ptitsyn, V. N.

Rakhmanin, O. B.
Rashidov, Sh. R.
Rekunkov, A. M.
Riabov, Ia. P.
Rodionov, N. N.
Romanov, G. V.
Ruben, V. P.
Rusakov, K. V.
Rybakov, A. Ia.
Rykov, V. N.
Ryzhkov, N. I.

Sakhniuk, I. I.
Salmanov, G. I.
Savin, S. N.
Savinkin, N. I.
Semenov, V. G.
Senkin, I. I.
Shakirov, M. Z.
Shalaev, S. A.
Shchelokov, N. A.
Shcherbina, B. E.
Shcherbitskii, V. V.
Shevardnadze, E. A.
Shibaev, A. I.
Shitikov, A. P.
Shkolnikov, A. M.
Shokin, A. I.
Sholokhov, M. A.
Silaev, I. S.
Sizenko, E. I.
Skachkov, S. A.
Slavskii, E. P.
Smirnov, L. N.
Smirnov, L. V.
Smirtiukov, M. S.
Sokolov, S. L.
Solomentsev, M. S.
Solovev, Iu. F.
Struev, A. I.
Stukalin, B. I.

Tabeev, F. A.
Talyzin, N. V.
Tarasov, N. N.
Taratuta, V. N.
Tiazhelnikov, E. M.
Tikhomirov, V. P.
Tikhonov, N. A.
Titarenko, A. A.
Tokarev, A. M.
Tolkunov, L. N.
Tolubeev, N. P.
Tolubko, V. F.
Trapeznikov, S. P.
Tretiak, I. M.
Trunov, M. P.
Tsinev, G. K.
Tsukanov, G. E.
Tsybulko, V. M.

Umakhanov, M.-S. I.
Usmankhodzhaev, I. B.
Ustinov, D. F.
Usubaliev, T. U.
Utkin, V. F.

Vaino, K. G.
Vashchenko, G. I.
Vasilev, N. F.
Vatchenko, A. F.
Viktorov, A. V.
Vlasov, A. V.
Volodarskii, L. M.
Voronin, L. A.
Vorontsov, Iu. M.
Voropaev, M. G.
Vorotnikov, V. I.
Voss, A. E.
Vsevolozhskii, M. N.

Zagladin, V. V.
Zaikov, L. N.
Zaitsev, M. M.
Zamiatin, L. M.
Zatvornitskii, V. A.
Zhigalin, V. F.
Zimianin, M. V.
Zolotukhin, G. S.

Candidate Members

Akhromeev, S. F.
Andreev, G. N.
Anisimov, P. P.
Antonov, N. A.

Balmont, B. V.
Barkauskas, A. S.

Bashilov, S.V.
Bashtaniuk, G.S.
Blatov, A.I.
Bobovikov, R.S.
Boitsov, V.V.
Brezhnev, Iu.L.
Burenkov, S.P.

Chakovskii, A.B.
Chaplin, B.N.
Cherdintsev, V.M.
Chernavin, V.N.
Cherniaev, A.S.
Churbanov, Iu.M.

Demchenko, V.A.
Demidenko, V.S.
Dmitriev, I.F.
Dmitriev, I.N.
Dmitriev, V.I.
Druzhinin, M.I.
Duisheev, A.D.
Dybenko, N.K.

Ermash, F.T.
Ershova, N.M.

Fomichenko, K.E.
Frolov, V.S.

Gerasimov, I.A.
Girenko, A.N.
Golikov, V.A.
Golubeva, M.A.
Gonchar, A.T.
Gorchakov, P.A.
Gordienko, A.F.
Gribachev, N.M.
Grintsov, I.G.

Iashin, A.I.
Iastrebov, I.P.
Iazkuliev, B.Ia.
Iazov, D.T.
Iksanov, M.B.
Ilichev, L.F.
Ivanova, T.G.

Kabaloev, B.E.
Kabkov, Ia.I.
Kalashnikov, A.M.
Kalin, I.P.
Kamalov, K.K.
Kamentsev, V.M.
Kapto, A.S.
Kazakov, V.I.
Khmara, I.G.
Kholov, M.Kh.
Khrennikov, T.N.
Klauson, V.I.
Kletskov, L.G.
Kochemasov, V.I.
Kolchina, O.P.
Kolomiets, Iu.A.
Komarov, N.D.
Korkin, A.G.
Kozlov, S.V.
Kozlovskii, E.A.
Krasnova, Z.I.
Kriulin, G.A.
Kuanyshev, O.S.
Kulidzhanov, L.A.
Kunaev, A.M.

Lebedev, K.V.
Lebedeva, V.S.
Listov, V.V.
Logunov, A.A.
Lunkov, N.M.

Maiorets, A.I.
Maksimov, Iu.P.
Malbakhov, T.K.
Malkov, N.I.
Maltsev, N.A.
Matafonov, M.I.
Merenishchev, N.V.
Mochalin, F.I.
Morozov, N.E.

Nenashev, M.F.
Nikulin, V.I.

Obraztsov, I.F.
Ovchinnikov, Iu.A.

Panov, K.N.
Parshina, V.R.
Paskar, P.A.
Pervyshin, E.K.
Pestov, V.S.
Petrishchev, A.G.
Piatkov, V.E.
Pogrebniak, Ia.P.
Pokryshkin, A.I.
Polukarov, Iu.I.
Popkov, M.D.
Poplavskii, A.M.
Posibeev, G.A.
Protozanov, A.K.
Pugachev, Iu.N.

Rachkov, A.I.
Radiukevich, L.V.
Rakhimov, B.R.
Razumov, E.Z.
Romanov, A.V.
Ruben, Iu.Ia.
Rustamova, Kh.
Rybakov, A.M.

Sarkisov, B.E.
Seidov, G.N.o.
Semenov, V.S.
Shabanov, V.M.
Shamshin, V.A.
Shapiro, L.B.
Shauro, V.F.
Shirshin, G.Ch.
Shitov, A.I.
Shkabardnia, M.S.
Sidorov, V.V.
Simonov, K.S.
Skliarov, Iu.A.
Smirnov, A.A.
Smirnov, A.I.
Smirnov, G.L.
Sokolov, E.E.
Sologub, V.A.
Sorokin, M.I.
Sosnov, I.D.

Terebilov, V.I.
Tereshchenko, N.D.
Tretiakov, P.I.

Vladychenko, I.M.
Vorobev, G.I.

Zasukha, M.Ia.
Zhukov, G.A.

Central Auditing Commission of the CPSU

Chairman
Sizov, G.F.

Members
Aleksankin, A.V.
Aslanova, Z.S.k.

Batsanov, B.T.
Bovin, A.V.
Bryzga, L.D.
Burakova, N.F.

Cherkashina, V.N.
Chudin, V.I.

Demidova, E.Ia.
Donenbaeva, K.B.
Dragunskii, D.A.
Dumachev, A.P.

Efremov, L.N.

Falin, V.M.
Fedulova, A.V.

Galkin, D.P.
Gilashvili, P.G.
Glushkov, N.T.
Golubev, V.N.
Golubtsova, T.V.
Grebeniuk, V.A.
Gromova, M.S.

Iakovlev, B.P.
Izrael, Iu.A.

Karryev, Ch.S.
Kasimova, V.A.
Kemashvili, V.B.
Khaldeev, M.I.
Khalilov, L.A.o.
Kharlamov, A.P.
Khitrun, L.I.
Komarova, D.P.
Konnov, V.F.
Konstantinov, A.U.
Koptiug, V.A.
Koshoev, T.Kh.
Kostiukov, I.I.
Kovalev, N.I.

Losev, S.A.
Lotsmanova, G.P.
Lukianov, A.I.

Malykhin, V.M.
Mamarasulov, S.
Mannapov, N.
Masol, V.A.
Medvedev, V.A.
Melnichenko, A.K.

Minovalova, N. F.
Mitskevich, V. F.
Motornyi, D. K.

Nabiev, R. N.
Naiashkov, I. S.
Nazarbaev, N. A.

Osetrov, T. N.

Pavlov, S. P.
Popov, V. I.
Postnikov, S. I.
Pudkov, I. I.

Rumiantsev, A. F.
Ryndina, A. N.
Ryzhov, N. S.

Sarkisian, F. T.
Sharaev, L. G.
Shcherbakov, I. S.
Shishlov, F. V.
Shkuratov, I. F.
Songaila, R.-B. I.
Strautmanis, P. Ia.

Troianovskii, O. A.

Udalaia, R. S.
Ulianov, M. A.
Umanets, N. V.
Ustian, I. G.

Party Control Committee

Chairman
– Pelshe, A. Ia.

City Committees Moscow and Leningrad and their First Secretaries

Moscow
– Grishin, V. V.
Leningrad
– Solovev, Iu. F.

Krai Committees and their First Secretaries

Altai
– Aksenov, N. F.
Khabarovsk
– Chernyi, A. K.
Krasnodar
– Vorotnikov, V. I.
Krasnoiarsk
– Fedirko, P. S.
Primore
– Lomakin, V. P.
Stavropol
– Murakhovskii, V. S.

Oblast Committees and their First Secretaries

Amur
– Avramenko, S. S.
Arkhangelsk
– Popov, B. V.
Astrakhan
– Borodin, L. A.
Belgorod
– Trunov, M. P.
Briansk
– Sizenko, E. I.
Cheliabinsk
– Voropaev, M. G.
Chita
– Matafonov, M. I.
Gorkii
– Khristoradnov, Iu. N.
Iaroslavl
– Loshchenkov, F. I.
Irkutsk
– Bannikov, N. V.
Ivanovo
– Kliuev, V. G.
Kalinin
– Leonov, P. A.
Kaliningrad
– Konovalov, N. S.
Kaluga
– Kandrenkov, A. A.
Kamchatka
– Kachin, D. I.
Kemerovo
– Gorshkov, L. A.
Kirov
– Bespalov, I. P.
Kostroma
– Balandin, Iu. N.
Kuibyshev
– Muravev, E. F.
Kurgan
– Kniazev, F. K.
Kursk
– Gudkov, A. F.
Leningrad
– Romanov, G. V.
Lipetsk
– Pavlov, G. P.
Magadan
– Malkov, N. I.
Moscow
– Konotop, V. I.
Murmansk
– Ptitsyn, V. N.
Novgorod
– Antonov, N. A.
Novosibirsk
– Filatov, A. P.
Omsk
– Maniakin, S. I.
Orel
– Meshkov, F. S.
Orenburg
– Balandin, A. N.
Penza
– Kulikov, F. M.
Perm
– Konoplev, B. V.
Pskov
– Rybakov, A. M.
Riazan
– Priezzhev, N. S.
Rostov
– Bondarenko, I. A.
Sakhalin
– Tretiakov, P. I.
Saratov
– Gusev, V. K.
Smolensk
– Klimenko, I. E.
Sverdlovsk
– Eltsin, E. N.
Tambov
– Khomiakov, A. A.
Tiumen
– Bogomiakov, G. P.
Tomsk
– Ligachev, E. K.
Tula
– Iunak, I. Kh.
Ulianovsk
– Kuznetsov, I. M.
Vladimir
– Ponomarev, M. A.
Volgograd
– Kulichenko, L. S.
Vologda
– Drygin, A. S.
Voronezh
– Ignatov, V. N.

ASSR Oblast Committees and their First Secretaries

Bashkir
– Shakirov, M. Z.
Buriat
– Modogoev, A. U.
Chechen-Ingush
– Vlasov, A. V.
Chuvash
– Prokopev, I. P.
Dagestan
– Umakhanov, M.-S. I.
Kabardino-Balkar
– Malbakhov, T. K.
Kalmyk
– Nikulin, V. I.
Karelian
– Senkin, I. I.
Komi
– Morozov, I. P.
Mari
– Posebeev, G. A.
Mordvinian
– Berezin, A. I.
North Ossetian
– Odintsov, V. E.
Tartar
– Usmanov, G. I.
Tuva
– Shirshin, G. Ch.

Udmurt
- Marisov, V. K.
Yakut
- Prokopev, Iu. N.

Autonomous Oblast Committees and their First Secretaries

Adygey
- Berzegov, N. A.
Gorno-Altai
- Znamenskii, Iu. S.
Jewish
- Shapiro, L. B.
Khakass
- Krylov, A. I.
Karachay-Circassian
- Inzhievskii, A. A.

The Government of the USSR

Supreme Soviet

Presidium

Chairman
- Chernenko, K. U.
First Deputy
- Kuznetsov, V. V.
Deputies
- Barkauskas, A. S.
- Gilashvili, P. G.
- Iasnov, M. A.
- Iazkuliev, B.
- Imashev, S. N.
- Kalin, I. P.
- Kebin, I. G.
- Khalilov, K. A.
- Kholov, M.
- Koshoev, T. Kh.
- Poliakov, I. E.
- Sarkisov, B. E.
- Strautmanis, P. Ia.
- Usmankhodzhaev, I. B.
- Vatchenko, A. F.
Secretary
- Menteshashvili, T. N.

Soviet of the Union

Chairman
- Shitikov, A. P.

Soviet of Nationalities

Chairman
- Ruben, V. P.

Council of Ministers

Chairman
- Tikhonov, N. A.
First Deputies
- Aliev, G. A. R. o.
- Arkhipov, I. V.
Deputies
- Antonov, A. K.
- Baibakov, N. K.
- Bodiul, I. I.
- Dymshits, V. E.
- Kostandov, L. A.
- Makeev, V. N.
- Marchuk, G. S.
- Martynov, N. V.
- Novikov, I. T.
- Nuriev, Z. N.
- Smirnov, L. V.
- Talyzin, N. V.

All-Union Ministries and their Ministers

Automotive Industry
- Poliakov, V. N.
Aviation Industry
- Silaev, I. S.
Chemical Industry
- Listov, V. V.
Chemical and Petroleum Machine Building
- Brekhov, K. I.
Civil Aviation
- Bugaev, B. P.
Communications Equipment Industry
- Pervyshin, E. K.
Construction for the Far East and Transbaikal Area
- Bashilov, S. V.
Construction of Petroleum and Gas Industry Enterprises
- Shcherbina, B. E.
Construction, Road and Municipal Machine Building
- Chudin, V. I.
Defense
- Ustinov, D. F.
Defense Industry
- Finogenov, P. V.
Electrical Equipment Industry
- Maiorets, A. I.
Electronics Industry
- Shokin, A. I.
Foreign Trade
- Patolichev, N. S.
Gas Industry
- Dinkov, V. A.
General Machine Building
- Afanasev, S. A.
Heavy and Transport Machine Building
- Zhigalin, V. F.
Instrument Making, Automation Equipment and Control Systems
- Shkabardnia, M. S.
Machine Building
- Bakhirev, V. V.
Machine Building for Animal Husbandry and Fodder Production
- Beliak, K. N.
Machine Building for Light and Food Industry and Household Appliances
- Pudkov, I. I.
Machine Tool and Tool Building Industry
- Balmont, B. V.
Maritime Fleet
- Guzhenko, T. B.
Medical Industry
- Melnichenko, A. K.
Medium Machine Building
- Slavskii, E. P.
Petroleum Industry
- Maltsev, N. A.
Power Machine Building
- Krotov, V. V.
Production of Mineral Fertilizers
- Pertishchev, A. G.
Radio Industry
- Pleshakov, P. S.
Railways
- Konarev, N. S.
Shipbuilding Industry
- Egorov, M. V.
Tractor and Agricultural Machine Building
- Ezhevskii, A. A.
Transport Construction
- Sosnov, I. D.

Union Republic Ministries and their Ministers

Agriculture
- Mesiats, V. K.
Coal Industry
- Bratchenko, B. F.
Communications
- Shamshin, V. A.
Construction
- Karavaev, G. A.
Construction of Heavy Industry Enterprises
- Goldin, N. V.
Construction Materials Industry
- Iashin, A. I.
Culture
- Demichev, P. N.
Education
- Prokofev, M. A.
Ferrous Metallurgy
- Kazanets, I. P.
Finance
- Garbuzov, V. F.
Fish Industry
- Kamentsev, V. M.
Food Industry
- Lein, V. P.
Foreign Affairs
- Gromyko, A. A.
Fruits and Vegetables
- Kozlov, N. T.
Geology
- Kozlovskii, E. A.
Health
- Burenkov, S. P.
Higher and Secondary Specialized Education
- Eliutin, V. P.
Industrial Construction
- Tokarev, A. M.

Installation and Special Construction Work
– Bakin, B. V.
Internal Affairs
– Fedorchuk, V. V.
Justice
– Terebilov, V. I.
Land Reclamation and Water Resources
– Vasilev, N. F.
Light Industry
– Tarasov, N. N.
Meat and Dairy Industry
– Antonov, S. F.
Nonferrous Metallurgy
– Lomako, P. F.
Petroleum Refining and Petrochemical Industry
– Fedorov, V. S.
Power and Electrification
– Neporozhnii, P. S.
Procurement
– Zolotukhin, G. S.
Rural Construction
– Danilenko, V. D.
Timber, Pulp and Paper, and Wood Processing Industry
– Busygin, M. I.
Trade
– Struev, A. I.

All-Union State Committees of the USSR and their Chairmen
Science and Technology
– Marchuk, G. I.
Inventions and Discoveries
– Naiashkov, I. S.
Standards
– Boitsov, V. V.
Foreign Economic Relations
– Skachkov, S. A.
Hydrometeorology and Environment Control
– Izrael, Iu. A.
Material Reserves
– Kovalenko, A. V.

Union Republic State Committees of the USSR and their Chairmen
State Planning Committee
– Baibakov, N. K.
Construction
– Novikov, I. T.
Material and Technical Supply
– Martynov, N. V.
Labor and Social Questions
– Lomonosov, V. G.
Prices
– Glushkov, N. T.
Vocational and Technical Education
– Bulgakov, A. A.
Television and Radio Broadcasting
– Lapin, S. G.
Cinematography
– Ermash, F. T.
Publishing Houses, Printing Plants and the Book Trade
– Pastukhov, B. N.
Forestry
– Vorobev, G. I.
State Security of the USSR
– Chebrikov, V. M.
Production-Technological Supply in Agriculture
– Khitrun, L. I.
Supervision of Labor Safety in Industry and Mining
– Vladychenko, I. M.
Supply of Petroleum Products
– Khuramshin, T. Z.
Committee for People's Control
– Shkolnikov, A. M.
USSR State Bank
– Alkhimov, V. S.
Central Statistical Administration
– Volodarskii, L. M.
Administrator of Affairs
– Smirtiukov, M. S.

USSR Ministry of Foreign Affairs

Minister
– Gromyko, A. A.
First Deputies
– Kornienko, G. M.
– Maltsev, V. F.
Deputies
– Firiubin, N. P.
– Ilichev, L. F.
– Kapitsa, M. S.
– Komplektov, V. G.
– Kovalev, A. G.
– Kozyrev, S. P.
– Ryzhov, N. S.
Secretary-General
– Fokin, Iu. E.

Apparatus of the Ministry
Administration of Affairs
– Duchkov, B. I.
Administration for Foreign Policy Planning
– Kovalev, A. G.
Administration for General International Problems
– Bogomolov, S. A.
Department for Cultural Relations with Foreign Countries
– Erofeev, V. Ia.
Protocol Department
– Nikiforov, D. S.
Press Department
– Cherniakov, Iu. N.
Information Department
– (not known)
International Organizations Department
– Petrovskii, V. F.
International Economic Organizations Department
– Makeev, E. N.
Treaty and Legal Department
– Rybakov, Iu. M.
Diplomatic Courier Service
– Lobanov, I. I.
Consular Administration
– Teterin, A. V.
Archives Administration
– Sevostianov, P. P.
Currency & Finance Administration
– Rybin, V. A.
Administration for Servicing the Diplomatic Corps
– Startsev, V. I.
Bureau of Translation
– Pastoev, V. V.
Moscow State Institute of International Relations
– Lebedev, N. I. (Rector)
Diplomatic Academy of the USSR Foreign Ministry
– Tikhvinskii, S. L.
Higher Courses in Foreign Languages
– (not known)
First African Department (Algeria, Central African Republic, Chad, Libya, Mali, Mauritania, Morocco, Senegal, Tunisia, Upper Volta)
– Shvedov, A. A.
Second African Department (all Countries south of the Sahara not belonging to the First or Third Department)
– Likhachev, V. A.
Third African Department (Burundi, Ethiopia, Kenya, Malawi, Mauritius, Rwanda, Zambia, Somalia, Tanzania, Uganda)
– Vasev, V. M.
First European Department (Belgium, France, Italy, Luxemburg, Netherlands, Portugal, Switzerland)
– Adamishin, A. I.
Second European Department (Australia, Canada, Great Britain, Ireland, New Zealand)
– Suslov, V. P.
Third European Department (Austria, Germany)
– Bondarenko, A. P.
Fourth European Department (Czechoslovakia, Poland)
– Deev, M. M.
Fifth European Department (Albania, Bulgaria, Cyprus, Greece, Hungary, Rumania, Yugoslavia)
– Grubiakov, V. F.
First Far Eastern Department (China, Korea, Mongolia)
– Kapitsa, M. S.
Second Far Eastern Department (Indonesia, Japan, Philippines)
– Solovev, N. N.

First Latin American Department (Caribbean, Central America, Guyana, Mexico, Suriname)
– Kazimirov, V. N.
Second Latin American Department (South America excluding Guyana and Suriname)
– Volskii, Iu. I.
Middle Eastern Department (Afghanistan, Iran, Turkey)
– Safronchuk, V. S.
Near Eastern Department (Egypt, Irak, Israel, Jordan, Kuwait, Lebanon, Saudi-Arabia, Soudan, Syria, Yemen Arab Republic, People's Democratic Republic of Yemen
– Grinevskii, O. A.
Scandinavian Department (Denmark, Finland, Iceland, Norway, Sweden)
– Farafonov, G. N.
South Asian Department (Bangladesh, Burma, India, Nepal, Pakistan, Sri Lanka)
– Valkov, A. I.
Southeast Asian Department (Kampuchea, Laos, Malaysia, Singapore, Thailand, Vietnam)
– Rogachev, I. A.
United States of America Department
– Komplektov, V. G.

Ambassadors
Afghanistan
– Tabeev, F. A.
Algeria
– Rykov, V. N.
Angola
– Loginov, V. P.
Argentina
– Striganov, S. R.
Australia
– Sudarikov, N. G.
Austria
– Efremov, M. T.
Bangladesh
– Stepanov, V. P.
Belgium
– Romanovskii, S. K.
Benin
– Agapov, V. I.
Bolivia
– Kovalev, S. I.
Botswana
– Petrov, M. N.
Brazil
– Chernyshev, V. I.
Bulgaria
– Tolubeev, N. P.
Burma
– Kuznetsov, V. N.
Burundi
– Levikov, V. Z.
Cameroon
– Zykov, S. S.*
Canada
– Iakovlev, A. N.
Cape Verde Islands
– Seregin, N. I.
Chad
– Zykov, S. S.*
China
– Shcherbakov, I. S.
Colombia
– Romanov, L. M.
Comoros
– Startsev, A. K.
Congo, People's Republic
– Lobachev, V. K.
Costa Rica
– Pavlov, Iu. I.
Cuba
– Katushev, K. F.
Cyprus
– Astavin, S. T.
Czechoslovakia
– Botvin, A. P.
Denmark
– Egorychev, N. G.
Djibouti
– Peryshkin, V. A.
Ecuador
– Kovalev, F. N.
Egypt, Arab Republic
– Poliakov, V. P.
Equatorial Guinea
– Krasnikov, B. A.
Ethiopia
– Fomichenko, K. E.
Fiji
– Sudarikov, N. G.
Finland
– Sobolev, V. M.
France
– Chervonenko, S. V.
Gabon
– Maksudov, L. M.
Gambia
– Belskii, Iu. I.
German Democratic Republic
– Abrasimov, P. A.
Germany, Federal Republic
– Semenov, V. S.
Ghana
– Ivantsov, A. I.
Greece
– Kaboshkin, V. F.
Grenada
– Sazhenev, G. I.
Guinea
– Kitaev, V. S.
Guinea-Bissau
– Krylov, L. V.
Guyana
– Kharchev, K. M.
Hungary
– Bazovskii, V. N.
Iceland
– Streltsov, M. N.

* Ambassador to Cameroon and Chad.

India
– Vorontsov, Iu. M.
Indonesia
– Shpedko, I. F.
Iran
– Boldyrev, V. K.
Iraq
– Minin, V. I.
Ireland
– Nesterenko, A. E.
Italy
– Lunkov, N. M.
Jamaica
– Musin, D. P.
Japan
– Pavlov, V. Ia.
Jordan
– Nishanov, R. N.
Kampuchea
– Bostorin, O. V.
Kenya
– Miroshnichenko, B. P.
Korea, Democratic People's Republic
– Shubnikov, N. M.
Kuwait
– Sikachev, N. N.
Laos
– Sobchenko, V. F.
Lebanon
– Soldatov, A. A.
Liberia
– Ulanov, A. A.
Libya
– Anisimov, A. V.
Luxembourg
– Udumian, K. B.
Madagascar
– Musatov, L. N.
Malaysia
– Kulik, B. T.
Maldives
– Kirnasovskii, B. E.**
Mali
– Fazylov, M. S.
Malta
– Smirnov, V. N.
Mauritania
– Spitskii, I. S.
Mauritius
– Pankov, N. A.
Mexico
– Sergeev, R. A.
Mongolia
– Smirnov, A. I.
Morocco
– Nersesov, E. V.
Mozambique
– Sepelev, Iu. F.
Nepal
– Vezirov, A. R. Kh. o.
Netherlands
– Beletskii, V. N.

** Ambassador to Maldives and Sri Lanka.

New Zealand
– Sofinskii, V. N.**
Nicaragua
– Shliapnikov, G. E.
Niger
– Kudashkin, V. N.
Nigeria
– Snegirev, V. V.
Norway
– Polianskii, D. S.
Pakistan
– Smirnov, V. S.
Peru
– Kuzmin, L. F.
Philippines
– Sholmov, Iu. A.
Poland
– Aristov, B. I.
Portugal
– Vdovin, V. P.
Rumania
– Tiazhelnikov, E. M.
Rwanda
– Rykov, G. V.
São Tomé and Principe
– Evsiukov, P. N.
Senegal
– Belskii, Iu. I.
Seychelles
– Orlov, M. G.
Sierra Leone
– Vorozhtsov, A. P.
Singapore
– Potapenko, F. I.
Somalia
– Aldoshin, V. V.
Soudan
– Vorobev, B. S.
Spain
– Dubinin, Iu. V.
Sri Lanka
– Kirnasovskii, B. E.*
Suriname
– Bubnov, I. D.
Sweden
– Pankin, B. D.
Switzerland
– Lavrov, V. S.
Syria
– Iukhin, V. I.
Tanzania
– Iukalov, Iu. A.
Thailand
– Kuznetsov, Iu. I.
Togo
– Iliukhin, I. A.
Tonga
– Sofinskii, V. N.**
Trinidad and Tobago
– Vialias, V. I.***
Tunisia
– Kizichenko, V. L.
Turkey
– Rodionov, A. A.
Uganda
– Bukin, S. A.
United Kingdom (London)
– Popov, V. I.
United States of America
– Dobranin, A. F.
Upper Volta
– Melnikov, E. N.
Uruguay
– Lebedev, Iu. V.
Venezuela
– Vialias, V. I.***
Vietnam
– Chaplin, B. N.
Western Samoa
– Sofinskii, V. N.**
Yemen Arab Republic
– Peresypkin, O. G.
Yemen, People's Democratic Republic
– Zhukov, V. P.
Yugoslavia
– Rodionov, N. N.
Zaire
– Marchuk, I. I.
Zambia
– Cherednik, V. I.
Zimbabwe
– Ter-Gazariants, G. A.

* Ambassador to Maldives and Sri Lanka.
** Ambassador to New Zealand, the Kingdom of Tonga and to Western Samoa.
*** Ambassador to Trinidad and Tobago, and Venezuela.

USSR Ministry of Defense

Minister
– Ustinov, D. F.
Marshal of the Soviet Union

First Deputy Ministers
– Kulikov, V. G.
Marshal of the Soviet Union
– Ogarkov, N. V.
Marshal of the Soviet Union
– Sokolov, L. S.
Marshal of the Soviet Union

Deputy Ministers
– Altunin, A. T.
Army General
– Gorshkov, S. G.
Naval Admiral of the Soviet Union
– Kurkotkin, S. K.
Army General
– Kutakhov, P. S.
Captain Marshal of the Air Force
– Moskalenko, K. S.
Marshal of the Soviet Union
– Petrov, V. I.
Army General
– Shabanov, V. M.
Colonel-General
– Shestopalov, N. F.
Colonel-General
– Tolubko, V. F.
Army General

Chief of General Staff of the Armed Forces
– Ogarkov, V. N.
Marshal of the Soviet Union

Chief of Main Political Directorate of the Soviet Army and Navy
– Epishev, A. A.
Army General

Commander-in-Chief of Air Forces
– Kutakhov, P. S.
Captain Marshal of the Air Force

Commander-in-Chief of Air Defense Forces
– Koldunov, A. I.
Marshal of the Air Force

Commander-in-Chief of Ground Forces
– Petrov, V. I.
Army General

Commander-in-Chief of Naval Forces
– Gorshkov, S. G.
Naval Admiral of the Soviet Union

Commander-in-Chief of Strategic Rocket Forces
– Tolubko, V. F.
Army General

Commander-in-Chief of Joint Armed Forces of Warsaw Pact Nations
– Kulikov, V. G.
Marshal of the Soviet Union

Commanders of Groups of Forces
Central Group of Forces (Czechoslovakia)
– Borisov, G. G.
Colonel-General

Group of Soviet Forces in Germany
– Zaitsev, M. M.
Army General

Northern Group of Forces (Poland)
– Zarudin, Iu. F.
Colonel-General

Southern Group of Forces (Hungary)
– Kochetov, K. A.
Lieutenant-General

Commanders of Military Districts
Baltic Military District
– Postnikov, S. I.
Colonel-General

Belorussian Military District
– Ivanovskii, E. F.
Army General

Carpathian Military District
– Belikov, V. A.
Colonel-General of Tank Troops

Central Asian Military District
– Iazov, D. T.
Colonel-General

Far East Military District
- Tretiak, I.M.
  Army General

Kiev Military District
- Gerasimov, I.A.
  Army General

Leningrad Military District
- Sorokin, M.I.
  Colonel-General

Moscow Military District
- Lushev, P.G.
  Army General

North Caucasus Military District
– –

Odessa Military District
- Voloshin, I.
  Colonel-General

Siberian Military District
- Popov, N.I.
  Colonel-General

Transbaikal Military District
- Salmanov, G.I.
  Army General

Transcaucasus Military District
- Kulishev, O.V.
  Colonel-General

Turkestan Military District
- Maksimov, Iu.P.
  Colonel-General

Ural Military District
- Tiagunov, M.A.
  Colonel-General

Volga Military District
- Riakhov, A.Ia.
  Colonel-General

Fleet Commanders

Baltic Fleet
- Sidorov, V.V.
  Admiral

Black Sea Fleet
- Khovrin, N.I.
  Admiral

Northern Fleet
- Chernavin, V.N.
  Vice-Admiral

Pacific Fleet
– –

All-Union Komsomol

First Secretary
- Mishin, V.M.

Secretaries
- Dergausov, Iu.A.
- Dolgushkin, N.K.
- Fedulova, A.V.
- Filippov, D.N.
- Koliakin, A.N.
- Okhromii, D.A.
- Shevtsova, L.I.
- Zhuganov, A.V.

All-Union Central Trade Union Council

Chairman
- Shalaev, S.A.

Deputy Chairman
- Prokhorov, V.I.

Secretaries
- Biriukova, A.P.
- Bogatikov, V.F.
- Gladkii, I.I.
- Kozlov, S.V.
- Matskevichius, K.Iu.
- Provotorov, V.P.
- Subbotin, A.M.
- Ushakov, A.P.
- Viktorov, A.V.
- Zemliannikova, L.A.

RSFSR

The Government of the RSFSR

Supreme Soviet

Presidium

Chairman
- Iasnov, M.A.

Deputies
- Kolchina, O.P.
- Sultanov, F.V.
- Semenov, B.S.
- Umalatov, A.D.
- Chabdarov, B.K.
- Namsinov, I.E.
- Mankin, I.P.
- Panev, Z.V.
- Romanov, V.I.
- Piksaev, A.O.
- Basiev, O.A.
- Batyev, S.G.
- Mendume, M.K.
- Tubylov, A.I.
- Bokov, Kh.Kh.
- Isliukov, S.M.
- Gorokhova, E.N.

Secretary
- Neshkov, Kh.P.

Chairman of the Supreme Soviet
Gribachev, N.M.

Council of Ministers

Chairman
- Solomentsev, M.S.

First Deputies
- Ermin, L.B.
- Orlov, V.P.

Deputies
- Aleksankin, A.V.
- Demchenko, V.A.
- Kalashnikov, A.M.
- Karpova, E.F.
- Kazakov, V.I.
- Kochemasov, V.I.
- Lykova, L.P.
- Maslennikov, N.I.

Union Republican Ministries and their Ministers

Agriculture
- Florentev, L.Ia.

Communications
- Baitsur, G.G.

Construction Materials Industry
- Voenushkin, S.F.

Culture
- Melentev, Iu.S.

Education
- Veselov, G.P.

Finance
- Bobrovnikov, A.A.

Fish Industry
- Peresypkin, I.P.

Food Industry
- Chistopliasov, S.I.

Foreign Affairs
- Vinogradov, V.M.

Forestry
- Zverev, A.I.

Fruits and Vegetables
- Naumov, V.I.

Geology
- Rovnin, L.I.

Health
- Trofimov, V.V.

Higher and Secondary Specialized Education
- Obraztsov, I.F.

Justice
- Blinov, V.M.

Land Reclamation and Water Resources
- Kalashnikov, V.I.

Light Industry
- Kondratkov, E.F.

Meat and Dairy Industry
- Konarygin, V.S.

Procurement
- Merkulov, P. I.

Rural Construction
- Maltsev, N. S.

Textile Industry
- Paramonov, A. M.

Trade
- Shimanskii, V. P.

Republican Ministries and their Ministers

Consumer Services
- Dudenkov, I. G.

Domestic Construction
- Bykhanov, A. G.

Fuel Industry
- Pankratov, Iu. A.

Highways
- Nikolaev, A. A.

Housing and Municipal Services
- Butusov, S. M.

Inland Water Transport
- Bagrov, L. V.

Local Industry
- Uspenskii, V. K.

Motor Transport
- Trubitsyn, E. G.

Social Security
- Komarova, D. P.

State Committees and their Chairmen

State Planning Committee
- Maslennikov, N. I.

Committee for People's Control
- Konnov, V. F.

State Committee for Construction Affairs
- Sabaneev, S. N.

State Committee for Labor
- Blokhin, A. P.

State Committee for Vocational and Technical Education
- Kamaev, G. L.

State Committee for Cinematography
- Filippov, A. G.

State Committee for Publishing Houses, Prinzing Plants and the Book Trade
- Sviridov, N. V.

State Committee for Prices
- Frolov, K. I.

State Committee for Production-Technological Supply in Agriculture
- Bosenko, N. V.

State Committee for the Supply of Petroleum Products
- Tarakanov, V. S.

State Committee for Material and Technical Supply
- Bolovinov, K. A.

State Committee on Viticulture and Wine-Making Industry
- Mishuk, A. S.

Central Statistical Administration
- Driuchin, A. P.

Ukrainian SSR

The Communist Party of the Ukraine

Politbureau of the Central Committee

Members

Elchenko, Iu. N.
Gerasimov, I. A.
Kachura, B. V.
Liashko, A. P.
Mozgovoi, I. A.
Shcherbitskii, V. V.
Sologub, V. A.
Titarenko, A. A.
Vashchenko, G. I.
Vatchenko, A. F.

Candidate Members

Dobrik, V. F.
Kachalovskii, E. V.
Kapto, A. S.
Kolomiets, Iu. P.
Pogrebniak, Ia. P.
Mukha, S. N.

Secretariat of the Central Committee

First Secretary
- Shcherbitskii, V. V.

Second Secretary
- Titarenko, A. A.

Secretaries
- Kachura, B. V.
- Kapto, A. S.
- Mozgovoi, I. A.
- Pogrebniak, Ia. P.

City Committee Kiev and its First Secretary

Rlchenko, Iu. N.

Oblast Committees and their First Secretaries

Cherkassy
- Lutak, I. K.

Chernigov
- Umanets, N. V.

Crimean
- Makarenko, V. S.

Chernovtsy
- Dikusarov, V. G.

Dnepropetrovsk
- Kachalovskii, E. V.

Donetsk
- Mironov, V. P.

Ivano-Frankovsk
- Skiba, I. I.

Kharkov
- Misnichenko, V. P.

Kherson
- Girenko, A. N.

Khmelnitskii
- Lisovyi, T. G.

Kiev
- Tsybulko, V. M.

Kirovograd
- Samilyk, M. G.

Lvov
- Dobrik, V. F.

Nikolaev
- Sharaev, L. G.

Odessa
- Kirichenko, N. K.

Poltava
- Morgun, P. T.

Rovno
- Panasenko, T. I.

Sumy
- Grintsov, I. G.

Ternopol
- Iarkovoi, I. M.

Transcarpathian
- Bandrovskii, G. I.

Vinnitsa
- Taratuta, V. N.

Volhynian
- Palazhchenko, L. I.

Voroshilovgrad
- Goncharenko, B. T.

Zaporozhe
- Vsevolozhskii, M. N.

Zhitomir
- Kavun, V. M.

The Government of the Ukrainian SSR

Supreme Soviet

Presidium

Chairman
- Vatchenko, A. F.

Deputies
- Bakhtin, Iu. G.
- Shcherbina, V. P.
- Shevchenko, V. S.

Secretary
- Kolotukha, Ia. Ia.

Chairman of the Supreme Soviet

Sytnik, K. M.

Council of Ministers

Chairman
- Liashko, A. P.

First Deputies
- Kolomiets, Iu. A.
- Vashchenko, G. I.

Deputies
- Esypenko, P. E.
- Gurenko, S. I.
- Masik, K. I.
- Masol, V. A.
- Nikolaev, N. F.
- Orlik, M. A.
- Stepanenko, I. D.

Union Republican Ministries and their Ministers

Agriculture
- Khorunzhii, M. V.

Coal Industry
- Grinko, N. K.

Communications
- Sinchenko, G. Z.

Construction of Heavy Industries Enterprises
- Lubenets, G. K.

Construction Materials Industry
- Shevchenko, A. T.

Culture
- Bezklubenko, S. D.

Education
- Fomenko, M. V.

Ferrous Metallurgy
- Galkin, D. P.

Finance
- Kozeruk, V. P.

Food Industry
- Kulinich, N. F.

Foreign Affairs
- Martinenko, V. N.

Forestry
- Baitala, V. D.

Fruits and Vegetables
- Lisitsyn, V. A.

Geology
- Gavrilenko, N. M.

Health
- Romanenko, A. E.

Higher and Secondary Specialized Education
- Efimenko, G. G.

Industrial Construction
- Shchepetilnikov, A. N.

Installation and Special Construction Work
- Borisovskii, B. S.

Internal Affairs
- Gladush, I. D.

Justice
- Zaichuk, V. I.

Land Reclamation and Water Resources
- Garkusha, N. A.

Light Industry
- Kasianenko, O. Ia.

Meat and Dairy Industry
- Solomakha, V. K.

Power and Electrification
- Skliarov, V. F.

Procurement
- Shmatolian, I. I.

Rural Construction
- Cherep, V. I.

State Farms
- Korzh, N. P.

Timber and Wood Processing Industry
- Grunianskii, I. I.

Trade
- Starunskii, V. G.

Republican Ministries and their Ministers

Construction and Utilization of Roads
- Shulgin, N. P.

Consumer Services
- Slinchenko, V. I.

Housing and Municipal Services
- Ploshchenko, V. D.

Local Industry
- Gaevskii, Iu. F.

Motor Transport
- Golovchenko, F. P.

Social Security
- Lukianenko, A. M.

State Committees and their Chairmen

State Planing Committee
- Masol, V. A.

State Committee for Construction Affairs
- Zlobin, G. K.

Committee for People's Control
- Kutsevol, V. S.

State Committee for Prices
- Shamborskii, V. K.

State Committee for Cinematography
- Olenenko, Iu. A.

State Committee for Vocational and Technical Education
- Kadatskii, E. M.

State Committee for Publishing Houses, Printing Plants and the Book Trade
- Pashchenko, A. Ia.

State Committee for Labor
- Panteleev, N. A.

State Committee for Television and Radio Broadcasting
- Okhmakevich, N. F.

State Committee for the Protecting of Nature
- Protesenko, D. I.

State Committee for the Supervision of Labor Safety in Industry and Mining
- Degtiarev, V. I.

State Committee for Material and Technical Supply
- Mostovoi, P. I.

Committee for State Security
- Mukha, S. N.

State Committee for Production-Technological Supply in Agriculture
- Filonenko, V. L.

Main Administration of Horticulture, Wine-growing and Wine Industry
- Lysenko, V. G.

Main Administration of the Petroleum Refining and Petrochemical Industry
- Lisnichii, G. E.

Central Statistical Administration
- Troian, A. I.

Komsomol of the Ukraine

First Secretary
- Kornienko, A. I.

Secretaries
- Borzov, V. F.
- Mironenko, V. I.
- Sakhan, I. Ia.
- Simonenko, P. M.
- Tkach, D. I.
- Vrublevskaia, V. B.

Ukrainian Republican Trade Union Council

Chairman
- Sologub, V. A.

Secretaries
- Efimenko, O. V.
- Poida, I. F.
- Sivolob, V. I.
- Slavinskii, M. M.
- Zapolskii, M. A.
- Zubenko, Iu. S.

Belorussian SSR

The Communist Party of Belorussia

Bureau of the Central Committee

Members
Aksenov, A. N.
Baluev, V. G.
Brovikov, V. I.
Dementei, N. I.
Firisanov, L. S.
Iakushev, I. F.
Ivanovskii, E. F.
Khusainov, Iu. M.
Kiselev, T. Ia.*
Kolokolov, Iu. B.
Kuzmin, A. T.
Lagir, M. I.
Mikulich, V. A.
Mitskevich, V. F.
Poliakov, I. E.
Polozov, N. N.

* Died January 11, 1983, sucessor N. N. Sliunkov.

Candidate Members
Lobanok, V. E.
Shaplyko, V. P.
Snezhkova, N. L.

Secretariat of the Central Committee

First Secretary
– Kiselev, T. Ia.*
Second Secretary
– Brovikov, V. I.
Secretaries
– Dementei, N. I.
– Firisanov, L. S.
– Kolokolov, Iu. B.
– Kuzmin, A. T.

City Committee Minsk and its First Secretary

Lepeshkin, V. A.

Oblast Committees and their First Secretaries

Brest
– Sokolov, E. E.
Gomel
– Malofeev, A. A.
Grodno
– Kletskov, L. G.
Minsk
– Mikulich, V. A.
Mogilev
– Prishchepchik, V. V.
Vitebsk
– Shabashov, S. M.

The Government of the Belorussian SSR

Supreme Soviet

Presidium
Chairman
– Poliakov, I. E.
Deputies
– Lobanok, V. E.
– Lopanik, G. Z.
Secretary
– Chagina, E. P.

Chairman of the Supreme Soviet
Shamiakin, I. P.

Council of Ministers

Chairman
– Aksenov, A. N.
First Deputies
– Khusainov, Iu. M.
– Mitskevich, V. F.
Deputies
– Danilov, D. A.
– Glazkov, I. M.
– Kovalev, M. V.
– Pilotovich, S. A.
– Snezhkova, N. L.

Union-Republican Ministries and their Ministers
Agriculture
– Senko, F. P.
Communications
– Gritsuk, I. M.
Construction Materials Industry
– Bildiukevich, V. L.
Culture
– Mikhnevich, Iu. M.
Education
– Minkevich, M. G.
Finance
– Shatilo, B. I.
Food Industry
– Stavrovskaia, I. A.
Foreign Affairs
– Gurinovich, A. E.
Forestry
– Moiseenko, S. T.
Fruits and Vegetables
– Iakushev, N. S.
Health
– Savchenko, N. E.
Higher and Secondary Education
– Meshkov, N. M.
Industrial Construction
– Mozoliako, I. M.
Installation and Special Construction Work
– Antonovich, I. A.
Internal Affairs
– Zhabitskii, G. M.
Justice
– Zhdanovich, A. A.
Land Reclamation and Water Resources
– Pavliuchuk, V. I.
Light Industry
– Nagibovich, L. N.
Meat and Dairy Industry
– Goncharov, A. A.
Procurement
– Kalitko, A. Ia.
Rural Construction
– Chura, L. M.
Timber and Wood Processing Industry
– Kiikov, A. Ia.
Trade
– Makaed, N. A.

Republican Ministries and their Ministers
Construction and Utilization of Highways
– Sharapov, V. I.
Consumer Services
– Grib, A. L.
Fuel Industry
– Filippov, G. A.
Housing and Municipal Services
– Bezliudov, A. I.
Local Industry
– Rusakov, L. V.
Motor Transport
– Andreev, A. E.
Social Security
– Kuzgin, V. I.

State Committees and their Chairmen
State Planning Committee
– –
State Committee for Construction Affairs
– Evtukh, V. G.
Committee for People's Control
– Lagir, M. I.
State Committee for Labor
– Romma, F. D.
State Committee for Television and Radio Broadcasting
– Buravkin, G. N.
State Committee for Cinematography
– Matveev, V. V.
State Committee for Prices
– Zhdanko, G. S.
State Committee for Publishing Houses, Printing Plants and the Book Trade
– Delets, M. I.

State Committee for Vocational and Technical Education
- Verkhovets, V. L.

State Committee for the Supply of Petroleum Products
- Chaukovskii, V. F.

State Committee for the Supervision of Labor Safety in Industry and Mining
- Bogdan, A. S.

State Committee for the Protection of Nature
- Kozlov, V. A.

State Committee for Material and Technical Supply
- Negerish, E. F.

Committee for State Security
- Baluev, V. G.

State Committee for Production-Technological Supply in Agriculture
- Pozharskii, B. M.

Central Statistical Administration
- Nechiporovich, V. N.

Main Administration for Gasification
- Dubovik, N. I.

Komsomol of Belorussia

First Secretary
- Shaplyko, V. P.

Secretaries
- Biriukova, M. M.
- Gurin, V. V.
- Kazakevich, V. V.
- Khandogin, A. P.
- Kurach, A. K.

Belorussian Republic Trade Union Council

Chairman
- Polozov, N. N.

Secretaries
- Arzhavkin, S. A.
- Barabanova, L. M.
- Bykovskii, V. I.
- Obukhovich, A. P.

Uzbek SSR

The Communist Party of Uzbekistan

Bureau of the Central Committee

Members
Aitmuratov, E.
Anisimkin, I. G.
Grekov, L. I.
Khodzhaev, A. A.
Khudaiberdyev, N. D.
Maksimov, Iu. P.
Musakhanov, M. M.
Osetrov, T. N.
Rashidov, Sh. R.
Salimov, A. U.
Usmankhodzaev, I. B.

Candidate Members
Kamalov, K.
Makhmudova, N. M.
Melkumov, L. N.
Umarov, U.

Secretariat of the Central Committee

First Secretary
- Rashidov, Sh. R.

Second Secretary
- Grekov, L. I.

Secretaries
- Aitmuratov, E.
- Anisimkin, I. G.
- Khodzhaev, A. A.
- Salimov, A. U.

City Committee Tashkent and its First Secretary

Umarov, U.

Oblast Committees and their First Secretaries

Andizhan
- Mamarasulov, S. M.

Bukhara
- Karimov, A.

Dzhizak
- Baimirov, T. B.

Fergana
- Umarov, Kh.

Kara-Kalpak (ASSR)
- Kamalov, K.

Kashka-Daria
- Gaipov, R. G.

Khorezm
- Khudaibergenov, M. Kh.

Namangan
- Kamalov, M. K.

Navoi
- Esin, V. P.

Samarkand
- Ashuraliev, R. S.

Surkhan-Daria
- Karimov, A.

Syr-Daria
- Khaidurov, V. A.

Tashkent
- Musakhanov, M. M.

The Government of the Uzbek SSR

Supreme Soviet

Presidium
Chairman
- Usmankhodzhaev, I. B.

Deputies
- Orlov, G. M.
- Rzaev, K.

Secretary
- Khaidarova, G. K.

Chairman of the Supreme Soviet
Khodzhaev, A. A.

Council of Ministers

Chairman
- Khudaiberdyev, N. D.

First Deputy
- Osetrov, T. N.

Deputies
- Abdullaeva, R. Kh.
- Akhmedov, K. A.
- Khodzhaev, A. R.
- Mikhailov, V. K.
- Rakhimov, B. R.
- Tursunov, M. T.

Union Republican Ministries and their Ministers

Agriculture
- Urkinbaev, A. A.

Communications
- Tokhtaev, T. M.

Construction
- Omerov, S. A.

Construction Materials Industry
- Isaev, G. G.

Cotton Cleaning Industry
- Usmanov, V.

Culture
– Rakhimbabaeva, Z.
Education
– Shermukhamedov, S.
Finance
– Muratkhodzhaev, V.
Food Industry
– Iuldashev, M.
Foreign Affairs
– Abdurazakov, B. A.
Forestry
– Iusupov, P. Iu.
Fruits and Vegetables
– Alimov, A. A.
Furniture and Wood Processing Industry
– Umarov, A. Kh.
Geology
– Tuliaganov, Kh. T.
Health
– Khudaibergenov, A. M.
Higher and Secondary Specialized Education
– Pulatov, S.
Installation and Special Construction Work
– Shagazatov, Kh. Sh.
Internal Affairs
– Ergashev, K.
Justice
– Vasikova, M. S.
Land Reclamation and Water Resources
– Dzhurabekov, I. Kh.
Light Industry
– Kurbanov, M. Kh.
Meat and Dairy Industry
– Mamadzhanov, Iu.
Power and Electrification
– Khamidov, A. Kh.
Procurement
– Kuibakarov, Kh. R.
Rural Construction
– Radzhanov, N. R.
Trade
– Khasanov, M. Kh.

Republican Ministries and their Ministers
Construction and Utilization of Roads
– Kaiumov, A. K.
Consumer Services
– Brodova, A. I.
Housing and Municipal Services
– Iliukhin, A. M.
Local Industry
– Nasreddinov, G.
Motor Transport
– Gavrilov, A. F.
Social Security
– Sultanova, S. U.

State Committees and their Chairmen
State Planning Committee
– Akhmedov, K. A.
State Committee for Construction Affairs
– Skrebnev, V. S.
Committee for People's Control
– Makhmudov, N.
State Committee for Labor
– Zaidov, M.
State Committee for Vocational and Technical Education
– Kaiumov, P.
State Committee for Cinematography
– Abdullaev, A.
State Committee for Fishery
– Makhmudov, A.
State Committee for Publishing Houses, Printing Plants and the Book Trade
– Esenbaev, Z. I.
State Committee for Supply of Petroleum Products
– Khamidov, V. A.
State Committee for Supervision of Labor Safety in Industry and Mining
– Akhmedbaev, A. G.
State Committee for Prices
– Tadzhiev, N.
State Committee for Television and Radio Broadcasting
– Ibragimov, U. Ia.
Committee for State Security
– Melkumov, L. N.
State Committee for Production-Technological Supply in Agriculture
– Igamberdyev, M.
Main Administration of Materials and Technical Supply
– Sharipov, T. Ia.
Main Administration for Construction in Tashkent City
– Nabiev, T.
State-Cooperative Administration for Construction on Collective Farms
– Ashurov, I.
Central Statistical Administration
– Makhmudov, S. M.

Komsomol of Uzbekistan

First Secretary
– Allamuradov, B. A.
Secretaries
– Daulanov, B. D.
– Ismailov, O. D.
– Khamidov, Kh. Kh.
– Makhmudova, Sh. N.
– Volodarskii, A. P.

Uzbek Republican Trade Union Council

Chairman
– Makhmudova, N. M.
Secretaries
– Adamanov, O. K.
– Kulibaev, M.
– Kuzmin, V. P.
– Nigmanov, Kh. V.
– Tashmatov, A. Ia.

Kazakh SSR

The Communist Party of Kazakhstan

Bureau of the Central Committee

Members

Ashimov, B. A.
Aukhadiev, K. M.
Bashmakov, E. F.
Grebeniuk, V. A.
Iazov, D. T.
Imashev, S. N.
Kamalidenov, Z. K.
Kazybaev, K. K.
Kunaev, D. A.
Miroshkhin, O. S.
Nazarbaev, N. A.
Trofimov, Iu. N.

Candidate Member
Dzhienbaev, S. S.

Secretariat of the Central Committee

First Secretary
– Kunaev, D. A.

Second Secretary
- Miroshkhin, O. S.

Secretaries
- Bashmakov, E. F.
- Kazybaev, K. K.
- Nazarbaev, N. A.
- Trofimov, Iu. N.

City Committee Alma-Ata and its First Secretary

Zhakupov, A. K.

Oblast Committees and their First Secretaries

Aktiubinsk
- Liventsov, V. A.

Alma-Ata
- Aukhadiev, K. M.

Chimkent
- Askarov, A. A.

Dzhambul
- Bekturganov, Kh. Sh.

Dzhezgazgan
- Davydov, N. G.

East Kazakhstan
- Protozanov, A. K.

Gurev
- Kushekov, U. K.

Karaganda
- Korkin, A. G.

Kokchetav
- Kuanyshev, O. S.

Kustanai
- Demidenko, V. P.

Kzyl-Orda
- Esetov, T.

Mangylashskii (Shevchenko)
- Mukashev, S. M.

North Kazakhstan
- Stepanov, V. T.

Pavlodar
- Erpilov, P. I.

Semipalatinsk
- Kubashev, S. K.

Taldy-Kurgan
- Tynybaev, A. A.

Tselinograd
- Morozov, N. E.

Turgaiskii (Arkalyk)
- Auelbekov, E. N.

Uralsk
- Iksanov, M. B.

The Government of the Kazakh SSR

Supreme Soviet

Presidium

Chairman
- Imashev, S. N.

Deputy
- Plotnikov, A. P.

Secretary
- Demesinov, Kh. Kh.

Chairman of the Supreme Soviet

Medeubekov, K. U.

Council of Ministers

Chairman
- Ashimov, B. A.

First Deputy
- Grebeniuk, V.-A.

Deputies
- Dzhienbaev, S. S.
- Gukasov, E. Kh.
- Kuppaev, T. B.
- Mukhamed-Rakhimov, T. G.
- Zhanybekov, Sh. Zh.

Union Republican Ministries and their Ministers

Agriculture
- Motoriko, M. G.

Communications
- Baizhanov, S. M.

Construction of Heavy Industry Enterprises
- Lomov, A. V.

Construction Materials Industry
- Parimbetov, B.

Culture
- Erkimbekov, Zh.

Education
- Balakhmetov, K. B.

Finance
- Baiseitov, R. S.

Fish Industry
- Sagdiev, M. R.

Food Industry
- Tantsiura, N. D.

Foreign Affairs
- Isinaliev, M. I.

Forestry
- Zaitsev, A. M.

Fruits and Vegetables
- Dzherembaev, E. I.

Geology
- Chakabaev, S. E.

Health
- Aliev, M.

Higher and Secondary Specialized Education
- Kataev, T.

Installation and Special Construction Work
- Erzhanov, B. M.

Internal Affairs
- Plataev, A. G.

Justice
- Dzhusupov, B.

Land Reclamation and Water Resources
- Kipshakbaev, N.

Light Industry
- Ibragimov, V. G.

Meat and Dairy Industry
- Alybaev, A.

Nonferrous Metallurgy
- Takezhanov, S. T.

Power and Electrification
- Ivanov, B. P.

Procurement
- Klevtsov, N. N.

Rural Construction
- Musin, K. N.

Timber and Wood Processing Industry
- Alderbaev, M.

Trade
- Ivanov, M. S.

Republican Ministries and their Ministers

Consumer Services
- Tiumebaev, S. M.

Local Industry
- Murzagaliev, G. M.

Motor Highways
- Bekbulatov, Sh. Kh.

Motor Transport
- Kadyrbaev, V. K.

Housing and Municipal Services
- Chernyshov, A. I.

Social Security
- Akhmetova, M. M.

State Committees and their Chairmen

State Planning Committee
- Mukhamed-Rakhimov, T. G.

State Committee for Construction Affairs
- Bektemisov, A. I.

Committee for People's Control
- Milkin, A. V.

State Committee for Publishing Houses, Printing Plants and the Book Trade
- Eleukenov, Sh. R.

State Committee for Cinematography
- Suleimenov, O. O.

State Committee for Labor
- Kasymkanov, A. K.

State Committee for the Supervision of Labor Safety in Industry and Mining
- Mukaev, T.

State Committee for Supply of Petroleum Products
- Bragin, A. K.

State Committee for Vocational and Technical Education
- Dzhandosov, S. U.

State Committee for Television and Radio Broadcasting
- Khasenov, Kh.

State Committee for Prices
- Nakipov, Sh. K.

Committee for State Security
- Kamalidenov, Z. K.

Main Administration for Material and Technical Supply
- Tankibaev, Zh. A.

State Committee for Production-Technological Supply in Agriculture
- Egorov, A. M.

Central Statistical Administration
- Zhumasultanov, T. Zh.

Komsomol of Kazakhstan

First Secretary
- Abdrakhmanov, S. A.

Secretaries
- Arystanbekov, A. Kh.
- Danenov, N. Zh.
- Diachenko, S. A.
- Eliseev, N. I.
- Nikitin, P. G.

Kazakh Republican Trade Union Council

Chairman
- Turysov, K. T.

Secretaries
- Abdrakhimova, D. E.
- Chulanov, B. G.
- Fomin, V. A.
- Krestiannikov, V. P.
- Plakhina, A. S.

Georgian SSR

The Communist Party of Georgia

Bureau of the Central Committee

Members
Andronikashvili, G. A.
Cherkeziia, O. E.
Chitanava, N. A.
Chkheidze, S. A.
Enukidze, G. N.
Gabuniia, G. D.
Gilashvili, P. G.
Inauri, A. N.
Kartvelishvili, D. L.
Khabeishvili, S. E.
Kolbin, G. V.
Kulishev, O. F.
Patiashvili, D. I.
Rostiashvili, T. B.
Shevardnadze, E. A.

Candidate Members
Adleiba, B. V.
Mosashvili, T. I.
Ordzhonikidze, I. N.
Papunidze, V. R.
Sanakoev, F. S.

Secretariat of the Central Committee

First Secretary
- Shevardnadze, E. A.

Second Secretary
- Kolbin, G. V.

Secretaries
- Andronikashvili, G. A.
- Enukidze, G. N.
- Khabeishvili, S. E.
- Patiashvili, D. I.

City Committee Tbilisi and its First Secretary

Gabuniia, D. G.

Oblast Committees and their First Secretaries

Abkhazian ASSR
- Adleiba, B. V.

Adzhar ASSR
- Papunidze, V. R.

South Ossetian Autonomous Oblast
- Sanakoev, F. S.

The Government of the Georgian SSR

Supreme Soviet

Presidium

Chairman
- Gilashvili, P. G.

Deputies
- Diasamidze, D. D.
- Kobakhiia, V. O.
- Siradze, V. M.

Secretary
- Lashkarashvili, T. V.

Chairman of the Supreme Soviet
Abashidze, I. V.

Council of Ministers

Chairman
- Kartvelishvili, D. L.

First Deputy
- Chitanava, N. A.

Deputies
- Chkheidze, Z. A.
- Cherkeziia, O. E.
- Kedishvili, Iu. I.
- Vadachkoriia, V. I.
- Vardzelashvili, O. G.

Union Republican Ministries and their Ministers

Agriculutre
- Mgeladze, G. D.

Communications
- Kobakhiia, V. I.

Construction
- Gogoladze, V. T.

Construction Materials Industry
- Lolashvili, O. I.

Culture
- Taktakishvili, O. V.

Education
- Kinkladze, O. D.

Finance
- Dvalishvili, D. N.

Food Industry
- Kontselidze, R. Kh.

Foreign Affairs
- –

Forestry
- Chalaganidze, Sh. I.

Fruits and Vegetables
- Dzhabadari, T. K.

Health
- Lezhava, G. G.

Higher and Secondary Specialized Education
- Chkhikvishvili, D. K.

Internal Affairs
- Gvetadze, G. I.

Justice
- Shuganashvili, A. A.

Land Reclamation and Water Resources
- Gadeliia, G. B.

Light Industry
- Gamtsemlidze, G. P.

Meat and Dairy Industry
- Movsesian, A. O.

Procurement
- Chakukvadze, Sh. I.

Rural Construction
- Kharatishvili, I. A.

Timber and Woodworking Industry
- Uchaneishvili, T. Ia.

Trade
- Koblianidze, G. L.

Republican Ministries and their Ministers

Consumer Services
- Sobolev, V. A.

Highways
- Robitashvili, G. V.

Housing and Municipal Services
– Vashadze, N. G.
Local Industry
– Kublashvili, D. D.
Motor Transport
– Davitashvili, T. N.
Social Security
– Gardabkhadze, K. K.

State Committees and their Chairmen
State Planning Committee
– Chkheidze, S. A.
State Committee for Construction Affairs
– Mirianashvili, G. Z.
Committee for People's Control
– Melkadze, O. V.
State Committee for Science and Technology
– Gverdtsiteli, I. G.
State Committee for Television and Radio Broadcasting
– Popkhadze, N. A.
State Committee for Vocational and Technical Education
– Rigvava, S. T.
State Committee for Publishing Houses, Printing Plants and the Book Trade
– Magradze, E. S.
State Committee for the Protection of Nature
– Kacharava, V. Ia.
State Committee for Cinematography
– Dvalishvili, A. A.
State Committee for Labor
– Dzhaparidze, R. A.
State Committee for Specialized Transport and Systems of its Automation
– Ordzhonikidze, Sh. M.
State Committee for Supply of Petroleum Products
– Ubiriia, A. E.
State Committee for the Supervision of Labor Safety in Industry and Mining
– Kvernadze, T. V.
State Committee for Prices
– Megrelishvili, M. A.
Committee for State Security
– Inauri, A. N.
State Committee for Production-Technological Supply in Agriculture
– Sarishvili, D. E.
State Committee for Material and Technical Supply
– Buadze, A. I.
State Committee for Gas Industry
– Lobzhanidze, B. F.
Main Administration for Installation and Specialized Construction Work
– Gamkrelidze, G. G.
Main Administration for the Industrial Construction
– Zatuashvili, B. A.
Central Statistical Administration
– Basariia, R. V.
Main Administration for Geology
– Gogitidze, I. K.

Georgian Komsomol

First Secretary
– Ordzhonikidze, I. N.
Secretaries
– Dzhikurashvili, I.
– Khachidze, M. I.
– Lordkipanidze, V. G.
– Shcherbakov, A. I.

Georgian Republican Trade Union Council

Chairman
– Mosashvili, T. I.
Secretaries
– Kharbediia, L. S.
– Khasaradze, I. G.
– Kvachadze, S. A.
– Labakhua, S. A.

Azerbaidzhan SSR

The Communist Party of Azerbaidzhan

Bureau of the Central Committee

Members
Bagirov, K. M.
Bagirov, O. A.
Gasanov, G. A.
Guseinov, V. A.
Iusif-zade, S. M.
Khalilov, K. A.
Kovtunov, A. V.
Mamedov, I. A.
Musaev, F. E.
Pugachev, Iu. N.
Seidov, G. N.
Tatliev, S. B.

Candidate Members
Efendiev, G. Sh.
Kevorkov, B. S.
Mekhtiev, R. E.
Muslim-zade, D. M.
Ragimov, K. N.
Rasulova, L. Kh.

Secretariat of the Central Committee

First Secretary
– Bagirov, K. M.
Second Secreatry
– Pugachev, Iu. N.
Secretaries
– Gasanov, G. A.
– Mamedov, I. A.
– Musaev, F. E.

City Committee Baku and its First Secretary

Guseinov, V. A.

Oblast Committees and their First Secretaries

Nakhichevan ASSR
– Ragimov, K. N.
Nagorno Karabakh Autonomous Oblast
– Kevorkov, B. S.

The Government of the Azerbaidzhan SSR

Supreme Soviet

Presidium

Chairman
– Khalilov, K. A. o.

Deputies
– Alieva, S. A. k.
– Aslanov, A. A.

Secretary
– Kazieva, R. S.

Chairman of the Supreme Soviet
Rustam-zade, S. A. o.

Council of Ministers

Chairman
– Seidov, G. N. o.

First Deputy
– Tatliev, S. B. o.

Deputies
– Akhmedov, F. G.
– Asanov, D. M. o.
– Lemberanskii, A. D.
– Mutalibov, A. N. o.
– Rasi-zade, Sh. A. o.
– Shcheglov, G. V.

Union Republican Ministries and their Ministers

Agriculture
– Askerov, M. G. o.

Communications
– Rasulbekov, G. D.

Construction Materials Industry
– Sadykov, R. D. o.

Cotton Cleaning Industry
– Zalmanov, F. R. o.

Culture
– Bagirov, Z. N. o.

Education
– Kafarova, E. M.

Finance
– Bakhshaliev, B. G.

Food Industry
– Mamedov, K. S. o.

Foreign Affairs
– Tairova, T. A.

Forestry
– Gasanov, G. R. o.

Fruits and Vegetables
– Orudzhev, T.

Health
– Kasumov, T. A. o.

Higher and Secondary Specialized Education
– Aliev, K. G. o.

Industrial Construction
– Safaraliev, T. Sh. o.

Installation and Specialized Construction Work
– Kiasimov, T. M. o.

Internal Affairs
– Veliev, D. D. o.

Justice
– Iusifov, N. G.

Land Reclamation and Water Resources
– Aliev, M. I. o.

Light Industry
– Ibragimov, S. M. o.

Meat and Dairy Industry
– Mamedov, A. G. K. o.

Petroleum Refining and Petrochemical Industry
– Aliev, N. K. M. o.

Procurement
– Shamiev, I. Sh.

Rural Construction
– Asanov, D. M. o.

Timber and Wood Processing Industry
– Airian, A. A.

Trade
– Kasumova, S. Ch. k.

Republican Ministries and their Ministers

Construction and Maintenance of Highways
– Khalafov, R. R. o.

Consumer Services
– Gasanova, Z. M. k.

Housing and Municipal Services
– Maiorov, V. S.

Local Industry
– Akhundov, S. S.

Motor Transport
– Kerimov, Sh. K. o.

Social Security
– Kaziev, M. Ia.

State Committees and their Chairmen

State Planning Committee
– Mutalibov, A. N. o.

State Committee for Construction Affairs
– Guseinov, D. T. o.

Committee for People's Control
– Efendiev, G. Sh. o.

State Committee for Vocational and Technical Education
– Allakhverdiev, T. A.

State Committee for Labor
– Kerimov, A. G. o.

State Committee for the Supervision of Labor Safety in Industry and Mining
– Aliev, N. M.

Committee for State Security
– Iusif-zade, Z. M. o.

State Committee for Viticulture and Winemaking
– Rzaev, Iu. K.

State Committee for the Protection of Nature
– Ibragimov, G. A. Kh. o.

State Committee for Television and Radio Broadcasting
– –

State Committee for Cinematography
– Sharifov, A. A. o.

State Committee for Publishing Houses, Printing Plants and the Book Trade
– Ibragimov, N. S. o.

State Committee for Prices
– Zulfugarov, A. M.

State Committee for Supply of Petroleum Products
– Mustafaev, A. I. o.

State Committee for Material and Technical Supply
– Guseinov, R. A. G. o.

Administration for Geology
– Shekinskii, E. M.

State Committee for Production-Technological Supply in Agriculture
– Kerimov, Kh. G. o.

Association for Nonferrous Metallurgy
– Rizaev, K. N. o.

Central Statistical Administration
– Mamedov, Kh. M.

Azerbaidzhan Komsomol

First Secretary
– Muslim-zade, D. M.

Secretaries
– Gadzhieva, S. R.
– Isaev, F. A.
– Movsumov, G. Sh.
– Petrov, V. V.

Azerbaidzhan Republican Trade Union Council

Chiarman
– Rasulova, L. Kh.

Secretaries
– Fataliev, N. S.
– Melkumian, G. A.
– Shelaginov, E. S.

Lithuanian SSR

The Communist Party of Lithuania

Bureau of the Central Committee

Members
Astrauskas, V. S.
Barkauskas, A.
Brazauskas, A.-M. K.
Dybenko, N. K.
Ferensas, A.
Grishkiavichius, P.
Kairialis, A.
Kairis, K.
Sakalauskas, V.
Shepetis, L.
Songaila, R.-B.

Candidate Members
Ignotas, P.-V. P.
Mikuchiauskas, V.
Petkiavichius, Iu.
Shileikis, P.

Secretariat of the Central Committee

First Secretary
- Grishkiavichius, P.

Second Secretary
- Dybenko, N. K.

Secretaries
- Astrauskas, V. S.
- Brazauskas, A.-M. K.
- Shepetis, L.

City Committee Vilnius and its First Secretary

Sakalauskas, V. V.

The Government of the Lithuanian SSR

Supreme Soviet

Presidium
Chairman
- Barkauskas, A. S.

Deputies
- Dirzhinskaite-Piliushenko, L. Iu.
- Mezhelaitis, E. B.

Secretary
- Nauialis, S. S.

Chairman of the Supreme Soviet
Shepetis, L. K.

Council of Ministers

Chairman
- Songaila, R.-B. I.

First Deputy
- Kairis, K. K.

Deputies
- Bernatavichius, Iu. I.
- Chesnavichius, A. Iu.
- Drobnis, A. A.
- Kazanavichius, V.-A.
- Rusenko, Iu. L.

Union Republican Ministries and their Ministers
Agriculture
- Grigaliunas, M. Iu.

Communications
- Onaitis, K. K.

Construction
- Sheshplaukis, B. A.

Construction Materials Industry
- Iasiunas, S. P.

Culture
- Belinis, Io. L.

Education
- –

Finance
- Sikorskis, R. A.

Food Industry
- Dulskas, S. A.

Foreign Affairs
- Zenkiavichius, V. M.

Forestry and Timber Industry
- Lukashevichius, V.-P.

Fruits and Vegetables
- Einoris, V. N.

Furniture and Woodworking Industry
- Minetas, K. B.

Health
- Platukis, Io. Io.

Higher and Secondary Specialized Education
- Zabulis, G. K.

Internal Affairs
- Mikalauskas, Iu. V.

Justice
- Kuris, P. M.

Land Reclamation and Water Resources
- Velichka, Io. Io.

Light Industry
- Ramanauskas, Io. K.

Meat and Dairy Industry
- Buklis, M. V.

Procurement
- Karetskas, L. Io.

Rural Construction
- Bagdonas, A. I.

Trade
- Mitskunas, P. P.

Republican Ministries and their Ministers
Consumer Services
- Plekhavichius, K. P.

Local Industry
- Simenenko, G. K.

Motor Transport and Roads
- Chernikov, I. S.

Municipal Services
- Sheris, Iu. M.

Social Security
- Patsevichene, Ia. B.

State Committees and their Chairmen
State Planning Committee
- Drobnis, A. A.

State Committee for Construction Affairs
- Sakalauskas, R. M.

Committee of People's Control
- Kairialis, A. K.

State Committee for Labor
- Gaigalas, B. P.

State Committee for Cinematography
- Brazaitis, A.-Io.

State Committee for Publishing Houses, Printing Plants and the Book Trade
- Nekroshius, Iu. P.

State Committee for Vocational and Technical Education
- Morkunas, V. A.

State Committee for Television and Radio Broadcasting
- Ianuitis, Io. Io.

State Committee for Prices
- Gruodis, M.

Committee for State Security
- Petkiavichius, Iu. Iu.

State Committee for Production-Technological Supply in Agriculture
- Zorskas, A. T.

State Committee for Gasification
- Krukauskas, Ch. S.

State Committee for the Protection of Nature
- Guniunas, K. P.

State Committee for the Supervision of Labor Safety in Industry and Mining
- Bernotas, Iu. M.

State Committee for Supply of Petroleum Products
- Laniauskas, Io. Iu.

Administration for Geology
- Mikalauskas, V. V.

Central Statistical Administration
- Zorskas, A. T.

State Committee for Material and Technical Supply
- Kiriushchenko, Ia. V.

Lithuanian Komsomol

First Secretary
- Ignotas, P.-V. P.

Secretaries
- Mikeliunas, V.
- Vashkiavichene, S.
- Vitkiavichius, V.
- Zavialov, V. A.

Lithuanian Republican Trade Union Council

Chairman
- Ferensas, A.

Secretaries
- Andrianov, I.
- Antanaitis, Iu.
- Baltrushaitis, R.
- Vainauskene, O.

Moldavian SSR

The Communist Party of Moldavia

Bureau of the Central Committee

Members
Grossu, S. K.
Eremei, G. I.
Kalenik, E. P.
Kalin, I. P.
Merenishchev, N. V.
Petrik, P. P.
Savochko, B. N.
Stepanov, G. A.
Ustiian, I. G.
Voronin, P. V.

Candidate Members
Dygai, G. G.
Kiktenko, V. K.
Volkov, G. M.

Secretariat of the Central Committee

First Secretary
- Grossu, S. K.

Second Secretary
- Merenishchev, N. V.

Secretaries
- Kalenik, E. P.
- Petrik, P. P.
- Savochko, B. N.

City Committee Kishinev and its First Secretary

Kiktenko, V. K.

The Government of the Moldavian SSR

Supreme Soviet

Presidium
Chairman
- Kalin, I. P.

Deputies
- Botezat, I. F.
- Sidorenko, S. S.

Secretary
- Melnik, A. V.

Chairman of the Supreme Soviet
Botsu, P. P.

Council of Ministers

Chairman
- Ustiian, I. G.

First Deputy
- Stepanov, G. A.

Deputies
- Kiriiak, N. P.
- Kutyrkin, V. G.
- Poliakov, N. D.
- Shapa, P. I.
- Vyshku, V. K.

Union Republican Ministries and their Ministers
Agriculture
- Lupashku, N. F.

Communications
- Russu, V. P.

Construction
- Uzun, N. I.

Construction Materials Industry
- Demchenko, I. I.

Culture
- Konstantinov, A. S.

Education
- Zidu, D. G.

Finance
- Savvin, I. A.

Food Industry
- Chekoi, A. I.

Foreign Affairs
- Komendant, P. V.

Forestry
- Vasalatii, G. I.

Fruits and Vegetables
- Bardash, V. N.

Furniture and Wood Processing Industry
- Terekhov, B. P.

Health
- Draganiuk, K. A.

Higher and Secondary Specialized Education
- Kerdivarenko, V. A.

Internal Affairs
- Bradulov, N. M.

Justice
- Volosiuk, V. M.

Land Reclamation and Water Resources
- Oleksich, V. N.

Light Industry
- Zhitniuk, G. M.

Meat and Dairy Industry
- Tiunov, A. I.

Procurement
- Tuzlov, M. I.

Rural Construction
- Iarutin, V. K.

Trade
- Cholak, M. I.

Republican Ministries and their Ministers
Construction and Utilization of Highways
- Bolbat, I. S.

Consumer Services
- Kuskevich, I. V.

Housing and Municipal Services
- Polozhenko, N. V.

Local Industry
- Nechaenko, A. V.

Motor Transport
- Fomin, V. M.

Social Security
- Bykova, O. V.

State Committees and their Chairmen
State Planning Committee
- Kutyrkin, V. G.

State Committee for Construction Affairs
- Grafov, S. S.

Committee for People's Control
- Voronin, P. V.

State Committee for Prices
- Chekati, G. Ia.

State Committee for Vocational and Technical Education
- Shelar, N. N.

State Committee for Labor
- Iakubovskii, P. I.

State Committee for Television and Radio Broadcasting
- Lozan, S. I.

State Committee for the Protection of Nature
- Kotiatsy, I. A.

State Committee for Cinematography
- Iordanov, I. E.

State Committee for Publishing Houses, Printing Plants and the Book Trade
- Khropotinskii, V. P.

Committee for State Security
- Volkov, G. M.

State Committee for the Supervision of Labor Safety in Industry and Mining
- Solomko, F. M.

State Committee for Supply of Petroleum Products
- Romanov, M. P.

Agrarian-Industrial Association for Viticulture and Winemaking
- Lukianov, N. N.

Agrarian-Industrial Association for Production of Volatile Oils, Perfumes and Cosmetics
- Khynku, M. S.

Cooperative State Scientific Production Association for Agro-Chemical Supply to Agriculture
- Atanovskii, L. N.

State Committee for Production-Technological Supply in Agriculture
- Bondarenko, M. V.

State Committee for Material and Technical Supply
- Parfenov, V. G.

Agrarian-Industrial Association for the Cultivation, Processing and Sale of Tobacco
- Verbitskii, N. F.

Central Statistical Administration
- Kozub, K. I.

Moldavian Komsomol

First Secretary
- Babii, V. V.

Secretaries
- Apostolova, V. G.
- Beiko, E. M.
- Kaziulin, V. A.
- Usatyi, A. P.

Moldavian Republican Trade Union Council

Chairman
- Eremei, G. S.

Secretaries
- Dmitriev, P. I.
- Iovits, V. D.
- Kokhanskii, V. I.
- Orlova, M. L.
- Russu, A. A.

Latvian SSR

The Communist Party of Latvia

Bureau of the Central Committee

Members
Anderson, I. A.
Aushkap, E. Ia.
Chemm, V. A.
Dmitriev, V. I.
Ikaunieks, A. E.
Postnikov, S. I.
Ruben, Iu. Ia.
Strautmanis, P. Ia.
Vagris, Ia. Ia.
Verro, R. O.
Voss, A. E.

Candidate Members
Ozols, E. M.
Pugo, B. K.
Zitmanis, A. K.

Secretariat of the Central Committee

First Secretary
- Voss, A. E.

Second Secretary
- Dmitriev, V. I.

Secretaries
- Anderson, I. A.
- Aushkap, E. Ia.
- Chemm, V. A.

City Committee Riga and its First Secretary

Vagris, Ia. Ia.

The Government of the Latvian SSR

Supreme Soviet

Presidium
Chairman
- Strautmanis, P. Ia.

Deputies
- Blum, V. A.
- Drizul, A. A.

Secretary
- Zorin, K. E.

Chairman of the Supreme Soviet
Klibik, V. S.

Council of Ministers

Chairman
- Ruben, Iu. Ia.

First Deputy
- Verro, R. O.

Deputies
- Bondaletov, I. V.
- Krumin, V. M.
- Peterson, E. K.
- Praude, R. V.
- Raman, M. L.

Union Republican Ministries and their Ministers
Agriculture
- Shpogis, K. A.

Communications
- Stungrevits, O. K.

Construction
- Kaftailov, N. I.

Construction Materials Industry
- Dorofeev, N. I.

Culture
- Kaupuzh, V. I.

Education
- Karklin, M. Ia.

Finance
- Blekte, Ia. O.

Food Industry
- Kuznetsova, I. D.

Foreign Affairs
- Krumin, V. M.

Forestry and Timber Industry
- Vitol, L. P.

Fruits and Vegetables
- Kurpnieks, V. E.

Health
- Kanep, V. V.

Higher and Secondary Specialized Education
- Linde, E. V.

Internal Affairs
- Drozd, M. F.

Justice
- Laivin, V. I.

Land Reclamation and Water Resources
- Bril, A. P.

Light Industry
- Iablonskii, E. Ia.

Meat and Dairy Industry
- Vannakh, S. E.

Procurement
- Girgenson, Z. V.

Trade
- Azan, V. D.

Wood-Working Industry
- Birkenfeld, V. Ia.

Republican Ministries and their Ministers

Consumer Services
- Tumovs-Bekis, Ia. D.

Local Industry
- Rubiks, A. P.

Motor Transport and Highways
- Sliede, E. E.

Municipal Services
- Berzin, I. P.

Social Security
- Pikhels, V. S.

State Committees and their Chairmen

State Planning Committee
- Raman, M. L.

State Committee for Construction Affairs
- Rubins, Ia. F.

Committee for People's Control
- Ozols, E. M.

State Committee for Labor
- Chevachin, V. N.

State Committee for Vocational and Technical Education
- Brodelis, Ia. Ia.

State Committee for Television and Radio Broadcasting
- Bartkevich, L. L.

State Committee for Cinematography
- Ivert, I. Ia.

State Committee for Publishing Houses, Printing Plants and the Book Trade
- Reimane, I. A.

State Committee for Prices
- Sedols, V. I.

State Committee for Supervision of Labor Safety in Industry and Mining
- Laskin, E. N.

State Committee for Gasification
- Greizha-Lisovskii, V. S.

State Committee for Supply of Petroleum Products
- Diachkov, E. P.

Committee for State Security
- Pugo, B. K.

State Committee for Production-Technological Supply in Agriculture
- Babkin, N. A.

State Committee for Material and Technical Supply
- Lubotskii, L. D.

Central Statistical Administration
- Baltin, G. A.

Latvian Komsomol

First Secretary
- Rukmane, M. K.

Secretaries
- Bogomolov, A. I.
- Brenchev, S. A.
- Shchukin, G. P.
- Valuevs, G. Ia.

Latvian Republican Trade Union Council

Chairman
- Zitmanis, A. K.

Secretaries
- Anchupan, A. P.
- Kachurovskii, A. P.
- Karklish, N. N.

Kirghiz SSR

The Communist Party of Kirghizia

Bureau of the Central Committee

Members

Duisheev, A. D.
Dzhumagulov, A. D.
Khodos, P. M.
Koshoev, T. Kh.
Kulmatov, K. N.
Lomov, N. P.
Makarenko, V. A.
Masaliev, A. M.
Minich, N. G.
Moldobaev, K. M.
Naumov, P. I.
Usubaliev, T. U.

Candidate Members

Abakirov, E. A.
Dolmatov, V. K.
Tashibekova, D. Ch.

Secretariat of the Central Committee

First Secretary
- Usubaliev, T. U.

Secretaries
- Dzhumagulov, A. D.
- Kulmatov, K. K.
- Makarenko, V. A.
- Naumov, P. I.

City Committee Frunze and its First Secretary

Moldobaev, K. M.

Oblast Committees and their First Secretaries

Issyk-Kul
- Masaliev, A. M.

Naryn
- Sydykov, M.

Osh
- Kulmatov, R. S.

The Government of the Kirghiz SSR

Supreme Soviet

Presidium

Chairman
- Koshoev, T. Kh.

Deputies
- Kabanov, M. F.
- Saginbaeva, K.

Secretary
- Omurkulova, S.

Chairman of the Supreme Soviet

Askarov, T.

Council of Ministers

Chairman
- Duisheev, A.

First Deputy
- Khodos, P. M.

Deputies
- Atashev, K. K.
- Begaliev, S.
- Moiseev, S. G.
- Ponomarev, O. B.
- Tashibekova, D. Kh.

Union Republican Ministries and their Ministers

Agriculture
- Savitakhunov, A.

Communications
- Tiurebaev, V. N.

Construction
- Ablesov, S. A.

Construction Materials Industry
- Bessmertnyi, I. Sh.

Culture
- Nusupova, D.

Education
- Bazarkulov, M.

Finance
- Toktonaliev, A.

Food Industry
- Ryspaev, B.

Foreign Affairs
- Tashibekova, D. Ch.

Health
- Turgunbaev, O.

Higher and Secondary Specialized Education
- Tursunov, A.

Internal Affairs
- Akmatov, D.

Justice
- Abakirov, D. A.

Land Reclamation and Water Resources
- Batyrkanov, K. M.

Light Industry
- Satarov, K.

Meat and Dairy Industry
- Tursunov, S. T.

Procurement
- Abdrakmanov, S.

Rural Construction
- Kondrashov, I. P.

Trade
- Chilebaev, T. B.

Republican Ministries and their Ministers

Consumer Services
- Aldasheva, Sh.

Housing and Municipal Services
- Suiumbaev, A. S.

Local Industry
- Konurbaev, M. O.

Motor Transport and Highways
- Orozaliev, K. A.

Social Security
- Salieva, B.

State Committees and their Chairmen

State Planning Committee
- Begaliev, S.

State Committee for Construction Affairs
- Khokhalachev, M. E.

Committee for People's Control
- Minich, N. G.

State Committee for Labor
- Endovitskii, M. V.

State Committee for Forestry
- Bekbaev, D. B.

State Committee for Vocational and Technical Education
- Kasendeev, I.

State Committee for Television and Radio Broadcasting
- Tokombaev, A.

State Committee for Cinematography
- Baialinov, M. K.

State Committee for Publishing Houses, Printing Plants and the Book Trade
- Sarbanov, T.

State Committee for Prices
- Chonoev, A.

Committee for State Security
- Lomov, N. P.

State Committee for Production-Technological Supply in Agriculture
- Boiko, A. G.

State Committee for Material and Technical Supply
- Dadabaev, H. D.

State Committee for Supervision of Labor Safety in Industry and Mining
- Karmanov, I. V.

State Committee for Supply of Petroleum Products
- Sadykov, E. A.

Administration of Geology
- Igemberdyev, S. A.

Central Statistical Administration
- Almaev, T. M.

Kirghiz Komsomol

First Secretary
- Abdykalykov, O.

Secretaries
- Matiukhin, V. I.
- Semenenko, E. G.
- Toichubekov, Iu. Zh.
- Ukueva, B.

Kirghiz Republican Trade Union Council

Chairman
- Abakirov, E. A.

Secretaries
- Briukhov, A. O.
- Dyikanbaev, K. D.
- Omurova, A.

Tadzhik SSR

The Communist Party of Tadzhikistan

Bureau of the Central Committee

Members
Babaev, M. B.
Bobosadykova, G. B.
Dadabaev, A.
Dedov, I. F.
Karmyshev, L. K.
Khaidarov, A. Kh.
Kholov, M.
Makhkamov, K. M.
Nabiev, R.
Perventsev, E. I.
Polukarov, Iu. I.

Candidate Members
Novichkov, V. E.
Shakarimov, Iu.
Sultanov, Sh.

Secretariat of the Central Committee

First Secretary
– Nabiev, R. N.
Second Secretary
– Polukarov, Iu. I.
Secretaries
– Babaev, M. B.
– Bobosadykova, G. B.
– Dadabaev, A. D.

City Committee Dushanbe and its First Secretary

Shakarimov, Iu. A.

Oblast Committees and their First Secretaries

Gorno-Badakhshan
– Zairov, M.
Kuliab
– Khisamutdinov, A. B.
Kurgan-Tiube
– Pallaev, G.
Leninabad
– Khodzhiev, R.

The Government of the Tadzhik SSR

Supreme Soviet

Presidium
Chairman
– Kholov, M.
Deputies
– Oplanchuk, V. Ia.
– Zaripova, N.
Secretary
– Rakhimova, I.

Chairman of the Supreme Soviet
Khasanov, U.

Council of Ministers

Chairman
– Makhkamov, K.
First Deputy
– Novichkov, V. E.
Deputies
– Gafurov, R.
– Grishina, R. M.
– Iusufbekov, R.
– Koshlakov, G. V.
– Maksumov, A. N.
– Mirkhalikov, T.

Union Republican Ministries and their Ministers
Agriculture
– Babaev, A.
Communications
– Popov, D. I.
Construction
– Mikirtychev, E. N.
Construction Materials Industry
– Shevchenko, I. V.
Cotton Cleaning Industry
– Kobilov, U.
Culture
– Mirzoshoev, S.
Education
– Dadaboev, R.
Finance
– Lafizov, D.
Food Industry
– Kurbanov, I. I.
Foreign Affairs
– Iusufbekov, R.
Fruits and Vegetables
– Sufiev, K.
Health
– Sazhenin, I. A.
Higher and Secondary Specialized Education
– Davliatkadanov, Kh.
Internal Affairs
– Kurbanov, I.
Justice
– Makhmudov, Sh.
Land Reclamation and Water Resources
– Nasreddinov, Kh.
Light Industry
– Kalandarov, Sh.
Meat and Dairy Industry
– Khaeev, I.
Procurement
– Prokopenko, S. K.
Rural Construction
– Khalimov, A.
Trade
– Kataev, O. K.

Republican Ministeries and their Ministers
Construction and Utilization of Highways
– Usmanov, I. I.
Consumer Services
– Kasymova, A. T.
Local Industry
– Baimatov, A.
Motor Transport
– Iakubov, N. Kh.
Municipal Services
– Kurbanov, R. M.
Social Security
– Mirzoalieva, K.

State Committees and their Chairmen
State Planning Committee
– Mirkhalikov, T.
State Committee for Construction Affairs
– Iakubov, V. A.
State Committee for Forestry
– –
Committee for People's Control
– Karmyshev, L.
State Committee for Publishing Houses, Printing Plants and the Book Trade
– Sharipov, Kh.
State Committee for Cinematography
– Saidov, Sh.
State Committee for Labor
– Narzibekov, M.
State Committee for Vocational and Technical Education
– Khasanov, K. G.
State Committee for Television and Radio Broadcasting
– Kalandarov, G.
State Committee for Prices
– Rafiev, M. M.
Committee for State Security
– Perventsev, E. I.
State Committee for Material and Technical Supply
– Merzaev, B. A.
State Committee for Supply of Petroleum Products
– Rakutsev, Iu. I.
State Committee for Gasification
– Kostarev, Iu. B.
State Committee for the Supervision of Labor Safety in Industry and Mining
– Malygin, V. I.
State Committee for Production-Technological Supply in Agriculture
– Babaev, A. M.
Administration of Geology
– Zuev, V. P.
Central Statistical Administration
– Kapustin, E. D.

Tadzhik Komsomol

First Secretary
- Sultanov, Sh. M.

Secretaries
- Eftakova, S. Kh.
- Musaeva, R. G.
- Ulmasov, R. U.
- Uvarov, A. T.

Tadzhik Republican Trade Union Council

Chairman
- Khaidarov, A. Kh.

Secretaries
- Babadzhanov, A. A.
- Danilchenko, V. I.
- Khakimov, M.
- Rechkin, N. V.

Armenian SSR

The Communist Party of Armenia

Bureau of the Central Committee

Members
Andreev, G. N.
Dallakian, K. L.
Demirchian, K. S.
Galumian, V. B.
Gambarian, K. A.
Nersesian, L. N.
Petrosian, S. M.
Sarkisian, F. T.
Sarkisov, B. E.
Shustko, L. S.
Voskanian, G. M.

Candidate Members
Arutiunian, D. A.
Arzumanian, M. B.
Iuzbakhshian, M. A.
Martirosian, G. A.
Muradian, M. O.
Saakian, L. G.
Turian, V. A.

Secretariat of the Central Committee

First Secretary
- Demirchian, K. S.

Second Secretary
- Andreev, G. N.

Secretaries
- Dallakian, K. L.
- Galumian, V. B.
- Gambarian, K. A.
- Voskanian, G. M.

City Committee Erevan and its First Secretary

Nersesian, L. N.

The Government of the Armenian SSR

Supreme Soviet

Presidium
Chairman
- Sarkisov, B. E.

Deputy
- Bagdasarian, O. M.

Secretary
- Bakhchinian, M. M.

Chairman of the Supreme Soviet
Tonoian, S. Kh.

Council of Ministers

Chairman
- Sarkisian, F. T.

First Deputy
- Kirakosian, A. M.

Deputies
- Airapetian, G. A.
- Movsesian, V. M.
- Petrosian, F. G.
- Sagoian, G. S.
- Svetlova, R. Kh.
- Tumanian, Sh.

Union Republican Ministries and their Ministers
Agriculture
- Tardzhumanian, G. V.

Communications
- –

Construction Materials Industry
- Nazarian, L. V.

Culture
- Arakelian, G. A.

Education
- Akhumian, S. T.

Finance
- Dzhanoian, D. A.

Food Industry
- Danielian, A. A.

Foreign Affairs
- Kirakosian, A. M.

Fruits and Vegetables
- Arakelian, S. V.

Health
- Gabrielian, E. S.

Higher and Secondary Specialized Education
- Garibdzhanian, L. P.

Industrial Construction
- Avakian, E. P.

Internal Affairs
- Patalov, E. G.

Justice
- Gevorkian, A. A.

Land Reclamation and Water Resources
- Shagoian, R. A.

Light Industry
- Gevorkian, A. A.

Meat and Dairy Industry
- Vartanian, S. A.

Procurement
- Ovakimian, O. T.

Rural Construction
- Tatevosian, G. A.

Timber and Wood Processing Industry
- Oganian, S. Kh.

Trade
- Safarian, S. R.

Republican Ministries and their Ministers
Construction and Utilization of Highways
- Melkumian, G. A.

Consumer Services
- Tumanian, S. A.

Housing and Municipal Services
- Minasian, R. A.

Local Industry
- Zurabian, M. A.

Motor Transport
- Drampian, Kh. A.

Social Security
- Galstian, R. S.

State Committees and their Chairmen
State Planning Committee
- Sagoian, G. S.

State Committee for Construction Affairs
- Marutian, K. O.

Committee for People's Control
- Martirosian, G. A.

State Committee for Utilization and Conservation of Surface and Underground Water Resources
- Movsesian, K. M.

State Committee for Labor
- Oganesian, O. G.

State Committee for Cinematography
- Samsonov, R. M.

State Committee for Forestry
- Avakian, G. A.

State Committee for Publishing Houses, Printing Plants and the Book Trade
- Mugnetsian, G. M.

State Committee for Vocational and Technical Education
- Mikaelian, S. A.

State Committee for Television and Radio Broadcasting
- Pogosian, S. K.

State Committee for Prices
- -

Committee for State Security
- Iuzbakhshian, M. A.

State Committee for the Supervision of Labor Safety in Industry and Mining
- Akopian, A. A.

State Committee for Supply of Petroleum Products
- Sorokin, N. V.

State Committee for Pruduction-Technological Supply in Agriculture
- Oganesian, S. S.

State Committee for Material and Technical Supply
- Astsatrian, E. T.

Main Administration for Installation and Special Construction Work
- Gambarian, G. O.

Administration of Geology
- Gulian, E. Kh.

Administration of Nonferrous Metallurgy
- Petrosian, F. A.

Main Administration for Gasification
- Shakhbazian, R. T.

Central Statistical Administration
- Mutafian, S. G.

Armenian Komsomol

First Secretary
- Turian, V. A.

Secretaries
- Akopian, G.
- Oganesian, N. S.
- Sarkisian, A. G.
- Topuzian, A. O.

Armenian Republican Trade Union Council

Chairman
- Saakian, L. G.

Secretaries
- Akopian, G. S.
- Arutiunian, V. F.
- Balaian, N. M.

Turkmenian SSR

The Communist Party of Turkmenistan

Bureau of the Central Committee

Members

Annaorazov, R.
Boiko, A. S.
Charyev, M. A.
Gapurov, M. G.
Iazkuliev, B.
Karryev, Ch. S.
Makarkin, N. V.
Mishchenko, G. S.
Mollaeva, M.
Rachkov, A. I.
Shmidt, M. G.
Zhulenev, V. F.
Kharkov, V. A.
Niiazov, S. A.

Candidate Members

Ishankulieva, O. I.

Secretariat of the Central Committee

First Secretary
- Gapurov, M.

Second Secretary
- Rachkov, A. I.

Secretaries
- Charyev, M. A.
- Mollaeva, M.
- Zhulenev, V. F.

City Committee Ashkhabad and its First Secretary

Niiazov, S. A.

Oblast Committees and their First Secretaries

Ashkhabad
- Annaorazov, P. A.

Chardzhou
- Khudaiberdyev, R.

Krasnovodsk
- Mitrin, E. T.

Mary
- Akgaev, A.

Tashauz
- Ataev, B.

The Government of the Turkmenian SSR

Supreme Soviet

Presidium

Chairman
- Iazkuliev, B.

Deputies
- Karryeva, R. M.
- Shmidt, M. G.

Secretary
- Nazarova, O.

Chairman of the Supreme Soviet

Ovezgeldyev, O.

Council of Ministers

Chairman
- Karryev, Ch. S.

First Deputy
- Mishchenko, G. S.

Deputies
- Abramov, V. E.
- Bazarova, R. A.
- Orazmukhamedov, N.
- Suiunov, N. T.
- Vershinin, V. V.

Union Republican Ministries and their Ministers

Agriculture
- Gurbanov, G. G.

Communications
- Abdykerimov, A.

Construction Affairs
- Sheremetev, N. V.

Construction Materials Industry
- Gladkii, V. I.

Cotton Cleaning Industry
- Izotov, V. D.

Culture
- Mamiliev, A.

Education
- Alieva, M.

Finance
- Sukhanov, Kh.

Food Industry
- Dzhumaev, Ia. K.

Foreign Affairs
- Suiunov, N. T.

Forestry
- Khudaikuliev, M.

Fruits and Vegetables
- -

Health
- Kiiatkhanov, B. A.

Higher and Secondary Specialized Education
- Muradov, S. N.

Internal Affairs
- Berdyev, R. N.

Justice
- Khaitov, Iu.

Land Reclamation and Water Resources
- Khodzhamuradov, A.

Light Industry
- Baimakhanova, A.

Meat and Dairy Industry
- Gukasov, S. B.

Procurement
- Babaev, S.

Rural Construction
- Gulmanov, B.

Trade
- Rybalov, E. G.

Republican Ministries and their Ministers

Construction and Utilization of Highways
- Smykov, V. I.

Consumer Services
- Khodzhamamedov, K.

Local Industry
- Kafurov, O.

Motor Transport
- Gurbangeldyev, Kh.

Municipal Services
- Orazmamedov, S.

Social Security
- Kurbanov, S.

State Committees and their Chairmen

State Planning Committee
- Abramov, V. E.

State Committee for Construction Affairs
- Kushpel, V. P.

Committee for People's Control
- Makarkin, N. V.

State Committee for Television and Radio Broadcasting
- Annakurbanov, Ch.

State Committee for Cinematography
- Redzhepov, P.

State Committee for Publishing Houses, Printing Plants and the Book Trade
- Khodzhaev, E.

State Committee for Vocational and Technical Education
- Bairamsakhatov, N.

State Committee for Labor
- Alovov, N.

State Committee for Prices
- Nazarov, E. K.

Committee for State Security
- Boiko, A. S.

State Committee for Production-Technological Supply in Agriculture
- Krugliakov, L. A.

State Committee for Supervision of Labor Safety in Industry and Mining
- Klychev, N.-D.

State Committee for Material and Technical Supply
- Zubrilin, A. F.

State Committee for Supply of Petroleum Products
- Karadurdyev, Ch.

State Committee for Gasification
- Shabasanov, M. K.

Administration for Geology
- Karryev, A.

Central Statistical Administration
- Pirkuliev, A.

Turkmen Komsomol

First Secretary
- Charyeva, Zh. K.

Secretaries
- Chagylov, A.
- Dovodov, Kh. N.
- Ovezova, M. K.
- Ovsiannikov, A. A.

Turkmen Republican Trade Union Council

Chairman
- Ishankulieva, O. I.

Secretaries
- Ashirov, T.
- Nazarova, E.
- Poltaeva, E. V.
- Savenkov, N. E.

Esthonian SSR

The Communist Party of Esthonia

Bureau of the Central Committee

Members

Iuhanson, N. O.
Kebin, I. G.
Kiao, V.
Klauson, V. I.
Kudriavtsev, A. I.
Lentsman, L. N.
Merimaa, O. O.
Ristlaan, R.
Riuitel, A.
Upsi, A.-B. Io.
Vaino, K. G.

Candidate Members

Pedak, M. A.
Shishov, L. D.
Toome, I. Kh.
Visnapuu, D. P.

Secretariat of the Central Committee

First Secretary
- Vaino, K. G.

Second Secretary
- Kudriavtsev, A. I.

Secretaries
- Kiao, V.
- Ristlaan, R.
- Upsi, A.-B. Io.

City Committee Tallinn and its First Secretary

Pedak, M. A.

The Government of the Esthonian SSR

Supreme Soviet

Presidium
Chairman
- Kebin, I. G.

Deputies
- Taal, A. E.
- Vannas, M. V.

Secretary
- Vakht, V. A.

Chairman of the Supreme Soviet
Pedak, M. A.

Council of Ministers

Chairman
- Klauson, V. I.

First Deputy
- Riuitel, A. F.

Deputies
- Gren, A. K.
- Saul, B. R.
- Tregubov, A. I.
- Tynspoeg, G. A.
- Veldi, Kh. T.

Union Republican Ministries and their Ministers
Agriculture
- Lind, V. N.

Communications
- Kaldma, A. U.

Construction
- Palu, P. K.

Construction Materials Industry
- Klauson, V. R.

Culture
- Lott, Io. A.

Education
- Grechkina, E. R.

Finance
- Miandmaa, E. Io.

Food Industry
- Laos, A. Iu.

Foreign Affairs
- Gren, A. K.

Forestry and Protection of Nature
- Teder, H. O.

Fruits and Vegetables
- Miannik, H. A.

Health
- Riatsep, V. Io.

Higher and Secondary Specialized Education
- Nuut, I. Iu.

Internal Affairs
- Tibar, M. O.

Justice
- Silvet, E. P.

Light Industry
- Kraft, Iu. A.

Meat and Dairy Industry
- Essenson, A. R.

Procurement
- Raevskii, V. I.

Timber and Wood Processing Industry
- Chernyshev, V. V.

Trade
- Todeson, K. E.

Republican Ministries and their Ministers
Consumer Services
- Khalmiagi, V. A.

Housing and Municipal Services
- Eks, V. P.

Local Industry
- Veskiviali, V. H.

Motor Transport and Highways
- Kruger, H. O.

Social Security
- Sarri, G. E.

State Committees and their Chairmen
State Planning Committee
- Tynspoeg, G. A.

State Committee for Construction Affairs
- Kveli, R. V.

Committee for People's Control
- Merimaa, O. O.

State Committee for Prices
- Vladychin, Iu. N.

State Committee for Cinematography
- Penu, R. A.

State Committee for Publishing Houses, Printing Plants and the Book Trade
- Kaik, L. Iu.

State Committee for Supply of Petroleum Products
- Vyrk, Iu. V.

State Committee for Supervision of Labor Safety in Industry and Mining
- Kask, E. B.

State Committee for Vocational and Technical Education
- Cherevashko, E. I.

State Committee for Land Reclamation and Water Resources
- Pyldroo, A. Io.

State Committee for Television and Radio Broadcasting
- Kullaste, A. A.

State Committee for Labor
- Konstantinov, V. N.

Committee for State Security
- Kortelainen, K. E.

State Committee for Production-Technological Supply in Agriculture
- Nakhkur, A. Ia.

Main Administration of Material and Technical Supply
- Toots, E. Ia.

Central Statistical Administration
- Kimask, G. K.

Esthonian Komsomol

First Secretary
- Visnapuu, D.

Secretaries
- Raa, A.
- Rei, K.
- Saarniit, Ia.
- Savvin, L.

Esthonian Republican Trade Union Council

Chairman
- Juhanson, N. O.

Secretaries
- Butel, R.
- Murtasina, L.
- Volostnykh, A.

Scientists

A

Abaev, V.I., **philologist**
Abdullaev, G.M.B. o.
Abdullaev, S.N., **philologist**
Abov, I.G., **physicist**
Abramov, N.N.
Abrikosov, A.A., **physicist**
Abzianidze, G.N., **literary historian**
Afanasev, G. D., **geologist and petrographer**
Agababian, S.B., **literary scholar and critic**
Agaian, T.P., **historian**
Aganbegian, A.G., **economist**
Ageev, N.V., **chemical and metallurgical scientist**
Agoshkov, M.I.
Akhiezer, A.I., **theoretical physicist**
Akhiezer, N.I., **mathematician**
Aksenenok, G.A., **lawyer**
Alekin, O.A., **hydrochemist**
Aleksandrov, A.D., **mathematician**
Aleksandrov, A.P., **nuclear physicist**
Aleksandrov, K.S., **physicist**
Alekseev, A.E., **electrical engineer**
Alekseev, V., **historian**
Alekseevskii, N.E., **physicist**
Alferov, Z.I., **physicist**
Ali-zade, A.A. o., **orientalist**
Alikhanian, A.I., **physicist**
Alikhanov, A.I., **nuclear physicist**
Alikhanov, E.N.
Alimarin, I.P., **chemical scientist**
Alimov, A.N., **economist**
Alsnik, P.I.
Ambartsumian, R.S.
Ambartsumian, V.A., **astrophysicist**
Amiranashvili, Sh.I., **historian of Georgian and Iranian art**
Amosov, N.M., **medical scientist**
Ananev, B.G., **psychologist**
Andreev, N.N., **acoustical physicist**
Andreev, V.M., **electronic engineer**
Andrianov, K.A., **chemist**
Andriiashev, A.P., **ichthyologist and zoogeographer**
Anikst, A.A., **literary and theater critic**
Anokhin, P.K., **physiologist**
Antonov, O.K., **aircraft designer**
Antroshchenko, V.I., **chemical technologist**
Arakelian, B.N., **archaeologist and historian**
Arbatov, G.A.
Arbuzov, B.A., **organic chemist**
Arif (Dadash-zade, M.A.M. o.), **literary scholar and critic**
Ariste, P.A., **linguist**
Arkhabov, V.I., **physicist**
Artamonov, M.I., **archaeologist and historian**
Artisikhovskii, A.V., **archaeologist**
Artsimovich, L.A., **nuclear physicist**
Arutiunov, A.I., **neurosurgeon**
Asatiani, T.L., **physicist**
Asimov, M.S., **philosopher**
Asratian, E.A., **physiologist**
Astaurov, B.L., **biologist**
Atlas, Z.V., **economist**
Atroshchenko, V.I., **chemical technologist**
Avenesov, R.I., **linguist**
Avdeev, V.N., **specialist in electronics**
Avrorin, V.A., **philologist, specialist in Tunguso-Manchurian language**
Avsiuk, G.A., **glaciologist**
Azimov, P.A., **linguist**
Azizbekov, Sh.A., **geologist**

B

Babaev, A.G., **geographer**
Babaev, J.N., **physicist**
Babinets, A.E., **hydrogeologist**
Babko, A.K., **chemical scientist**
Babskii, E.B., **physiologist**
Bader, O.N., **archaeologist**
Baev, A.A., **biochemist**
Bagdasarian, G.St., **physical chemist**
Baishev, S.B., **economist and historian**
Baramidze, A.G., **literary critic**
Baranov, S.N., **chemist**
Baratov, R.B.
Barmin, V.P.
Barsanov, G.P., **mineralogist**
Bashkirov, A.N., **organic chemist and technological engineer**
Basov, N.G., **physicist**
Bazanov, V.G., **literary scholar**
Bei-Bienko, G.I., **entomologist**
Bekhtereva, N.P., **medical scientist**
Beletskaia, I.P., **chemist**
Belevtsev, I.N., **geologist**
Beliaev, D.K., **biologist**
Beliaev, S.T., **physicist**
Beliakov, R.A.
Belitser, V.A., **biochemist**
Belotserkovskii, O.M., **mathematician**
Belousov, V.V., **geologist**
Belov, A.F., **metallurgist**
Belza, I.F., **musicologist**
Berezanskii, I.M., **mathematician**
Berezhnoi, A.S., **chemical technologist**
Berezin, I.V., **chemical scientist**
Bergelson, L.D., **chemical scientist**
Beriev, G.M., **aircraft designer**
Bezrukov, P.L., **geologist**
Bialik, B.A., **literary historian**
Bibikov, S.N., **archaeologist**
Bilai, V.I., **botanist**
Bilyi, M.U.
Bitsadze, A.V., **mathematician**
Biushgens, G.S., **technologist**
Blinova, E.N.
Blokhin, N.N., **medical scientist**
Blokhintsev, D.I., **physicist**
Bogoliubov, M.N., **philologist**
Bogoliubov, N.N., **mathematician, mechanic, physicist**
Bogomolets, O.A., **pathophysiologist**
Bogomolov, A.F., **radiation technologist**
Bogomolov, O.T., **economist**
Bogorodskii, V.V., **oceanographer**
Bogorov, V.G., **oceanographer, hydrobiologist**
Bogush, L.K., **medical scientist**
Boiarskii, A.I., **economist and statistician**
Bokii, G.B., **crystallographer, chemist**
Bolotin, V.V.
Bolshakov, K.A., **inorganic chemist**
Bolshev, L.N., **mathematician**
Bondarchuk, V.G., **geologist**
Bongard-Levin, G., **orientalist**
Boreskov, G.K., **physical chemist**
Borisenko (Barysenka), V.V.
Borisevich, N.A.
Boriskovskii, P.I., **archaeologist**
Borkovskii, V.I., **linguist**
Borovik-Romanov, V.-A.St., **physicist and astronomer**
Borovkov, A.A., **mathematician**
Bozhenov, P.I.
Braginskii, I.S.
Braichevskyi, N., **historian and archaeologist**
Bratukhin, I.P., **designer and scientist in the field of helicopter construction**
Braude, S.I., **physicist**
Braunshtein, A.E., **biochemist**
Brekhovskikh, L.M., **physicist**
Brezhnev, D.D., **grower and breeder of plants**
Bromlei, I.V., **historian**
Brovman, G.A., **literary critic**
Budagov, R.A., **philologist and linguist**
Budyko, M.I., **geophysicist**
Bulanzhe, I.D., **geophysicist**
Bulashevich, I.P., **geophysicist**
Buniatian, M.R., **mathematician**
Bunin, A.V., **art historian**
Bunin, K.P., **specialist in physical metallurgy**
Bunkin, B.V., **physicist**
Bursov, B.I., **literary historian**
Bushmin, A.S., **literary historian**
Buslaev, I.A., **chemical scientist**
Butenko, R.G., **biologist**
Bykhovskii, B.E., **philosopher**

C

Chagin, B.A., **philosopher**
Chagovets, R.V., **biochemist**
Chailakhian, M.K., **plant physiologist**
Chalmaev, V.A., **literary historian and critic**
Chaplin, B. N., **dipl. and Cand. of Tech. Sciences**
Chazov, E.I., **medical scientist**
Chekaniuk, A.T., **historian and party official**
Chelomei, V.N., **mechanical engineer**
Chelyshev, A., **linguist**
Cherenkov, P.A., **physicist**
Chernigovskii, V.N., **physiologist**
Chernyi, G.G., **scientist in mechanics**
Cherskii, N.V., **specialist in mechanics and devel. of petroleum and gas deposits**

Chibisov, K.V., **physical chemist**
Chichikanov, V., **expert in the field of economics**
Chikobava, A.S., **linguist**
Chistov, K., **ethnographist**
Chudakov, A.E., **physicist**
Chukhanov, Z.F., **specialist in thermal power engineering**
Chukhrov, F.V., **mineralogist and geochemist**
Chumakov, M.P., **microbiologist and virologist**
Chunikhin, S.A., **mathematician**
Chutkikh, A.S., **innovator in light industry**

D

Dadenkov, I.N.
Danchenko, A.S., **art historian**
Daniliuk, I.I., **mathematician**
Davidov, A.S., **physicist**
Davitaia, F.F., **agrometeorologist and climatologist**
Delimarskii, I.K., **chemical scientist**
Demidov, A.M., **physicist**
Demin, V.G.
Denisiuk, I.N.
Denisov, A.I., **jurist**
Deriagin, B.V., **chemical scientist**
Desnitskaia, A.V., **linguist**
Deviatkov, N.D., **electrical engineer**
Deviatykh, G.G., **chemist**
Diakonov, I.M., **historian, orientalist**
Dmitriev, I.A.
Dmitrochenko, A.P., **agronomist**
Dolenko, G.N., **geologist**
Dolgoplosk, B.A., **chemical scientist**
Dollezhal, N.A., **engineer**
Dorfman, I.G., **physicist**
Dorodnitsyn, A.A., **mathematician, geophysicist and specialist in mechanics**
Drozdov, O.A., **climatologist**
Druskin, M.S., **musicologist and teacher**
Druzhinina, Y., **historian**
Dubinin, M.M., **physical chemist**
Dubinin, N.P., **geneticist**
Dubovskoi, B.V., **astronomer**
Dzhamgerchinov, B.D., **historian**
Dzhelepov, B.S., **physicist**
Dzhelepov, V.P., **physicist**
Dzialoshinskii, I.E., **physicist**

E

Efimov, A.N., **economist**
Egorov, A.G., **philosopher**
Eliakov, G.B., **chemist and biochemist**
Emanuel, N.M., **physical chemist**
Emelianov, I.I., **specialist in energetics**
Emelianov, S.V., **specialist in automatic control systems**
Emelianov, V.S., **metallurgist**
Enikolopov (Enikolopian), N.S., **physical chemist**
Eremenko, V.N., **physicist and chemist**
Eremian, S.T., **historian**
Eroshin, V.K., **biologist**
Ershov, A.P., **mathematician**
Ershov, I.L., **mathematician**
Ershov, V.S., **parasitologist**
Evstratov, V.F., **organic chemist**

F

Fabrikant, L.B., **engineer**
Faddeev, D.K., **mathematician**
Faddeev, L.D., **mathematician and physicist**
Fedorchenko, I.M., **specialist in powder metallurgy and metallurgy**
Fedorenko, N.P., **specialist in chemical technology**
Fedorenko, N.T., **philologist and orientalist**
Fedorov, A.A., **botanist**
Fedorov, A.A., **botanist (brother of Fedorov, A.A.)**
Fedorov, E.P., **astronomer**
Fedorov (Fedarau), F.I., **physicist**
Fedorovich, B.A., **physical geographer and geomorphologist**
Fedoseev, P.N.
Fedotov, S.A., **geophysicist**
Feinberg, I.B., **physicist**
Feofilov, P.P., **physicist**
Filchakov, P.F., **mathematician**
Filin, F.P., **linguist**
Firsov, O.B., **physicist**
Flerov, G.N., **physicist**
Florensov, N.A., **geologist**
Fomin, V.V., **chemist**
Fradkin, E.S., **physicist**
Frank, I.M., **physicist**
Frantsevich, I.N.
Frolov, I.T., **philosopher**
Frolov, K.V., **authority on machine vibrations**
Frolov, N.A., **poet and mathematician**

G

Gainullin, M.K., **philologist**
Gakhov, F.D., **mathematician**
Galazii, G.I., **botanist and hydrobiologist**
Galin, L.A., **scientist in the field of mechanics**
Galkin, A.A., **physicist**
Gaponov-Grekov, A.V., **physicist**
Gatovskii, L.M., **economist**
Gavrilov, M.A., **specialist in the theory of automatic control systems**
Gazenko, O.G., **physiologist**
Geld, P.V., **physical chemist**
Gelfand, I.M., **mathematician**
Georgiev, G.P., **biologist**
Gerasimov, I.I., **chemist**
Gerasimov, I.P., **geographer, geomorphologist and soil scientist**
Gershuni, G.V., **medical scientist**
Gikhman, I.I., **mathematician**
Giliarov, M.S., **zoologist**
Ginzburg, V.L., **physicist**
Gippius, E.V., **music critic and musical ethnographer**
Glezerman, G.E., **philosopher**
Glushko, V.P.
Gnedenko, B.V., **mathematician**
Goian, G.I., **theatrical scholar**
Goldanskii, V.I., **physical chemist**
Golobutskii, V.A., **historian**
Gonchar, A.A., **mathematician**
Gorbunov, G.I., **geologist**
Gorizontov, P.D., **pathophysiologist**
Gorkov, L.P., **physicist**
Gorshkov, G.S., **volcanologist**
Govorun, N.N., **mathematician**
Grabin, V.G., **designer of artillery weapons**
Gradova, N.B., **soil scientist, agro-chemist**
Gribanov, I.E., **engineer**
Gribov, V.N., **physicist**
Grigorianis, A.N.
Grinberg, G.A., **physicist**
Gromyko, A.A., **political scientist**
Grushin, P.D., **scientist in aviation technology**
Grutso, A.P., **philologist**
Gulii, M.F., **biochemist**
Gurevich, I.I., **physicist**
Gurfinkel, V.S., **biophysicist**
Gutyria, V.S., **specialist in petroleum and gas technology**
Gvishiani, D.M., **philosopher and sociologist**
Gvozdetskii, N.A., **physical geographer**

I

Iakovlev, A.S., **aircraft designer**
Ianenko, N.N., **mechanics scientist**
Ianshin, A.L., **geologist**
Iatsimirskii, K.B., **chemical scientist**
Ibragimov, I.A., **mathematician**
Ievlev, V.M., **physicist**
Iliushin, A.A., **scholar in continuum mechanics**
Imshenetskii, A.A., **microbiologist**
Ioannisian, A.R., **historian**
Iovchuk, M.T.
Isachenko, A.G., **geographer**
Isanin, N.N., **engineer**
Ishlinskii, A.I., **specialist in mechanics**
Islambekov, R.K., **endocrinologist**
Ivakhnenko, A.G., **specialist in automation and technical cybernetics**
Ivanenko, D.D., **physicist**
Ivanov, A.V., **zoologist**
Ivanov, S.N., **geologist**
Ivanov, V.E., **physicist**
Ivanov, V.K., **mathematician**
Izrael, I.A., **geophysicist**

K

Kabachnik, M.I., **organic chemist**
Kabaev, L.N., **geologist and geophysicist**
Kabanov, V.A., **chemical scientist**
Kadomtsev, B.B., **physicist**
Kafarov, V.V., **chemical engineer**
Kagan, G.I., **microbiologist**
Kagan, I.M., **physicist**
Kalashnikov, M.T., **engineer**
Kantorovich, L.V., **mathematician and economist**
Kapitsa, A.P., **geographer and geomorphologist**
Karaganov, A.V., **literary and theatrical historian and film critic**
Karakeev, K.-G., **historian**
Karamian, A.I., **neurophysiologist**
Karaulov, Y., **linguist**
Kargapolov, M.I.
Karger, M.K., **archaeologist and art historian**
Karpachev, S.V., **chemical scientist**
Karryev, B.A., **literary scholar**

Kasianenko, V.G., **veterinary anatomist and morphologist**
Kataev, S.I., **television scientist**
Katanian, V.A., **literary historian**
Kavalerov, K.V., **mining engineer and geophysicist**
Kazanskii, V.B., **physical chemist**
Kedrov, B.M., **philosopher**
Keldysh, I. (G.) V., **musicologist**
Keldysh, L.V., **physicist**
Kemenov, V.S., **art historian and critic**
Kerimov, D.A.-A., **jurist**
Khachaturov, T.S., **economist**
Khalatnikov, I.M., **physicist**
Khariton, I.B., **physical chemist**
Kharlamov, P.V., **specialist in mechanics**
Khechinashvili, S.N., **otorhinolaryngologist**
Khesin-Lur'e, R.B., **biologist**
Khimich, G.L., **specialist in heavy machine building**
Khitarov, N.I., **geochemist**
Khodorovich, T.S.
Khokhlov, A.S., **chemist**
Khomentovskii, A.S., **geologist**
Khoroshilov, I.I., **agronomist**
Khrapchenko, M.B., **literary historian**
Khristianovich, S.A., **mathematician and mechanics scientist**
Khromov, P.A., **economist**
Kiiak, G.S., **specialist in plant breeding**
Kikoin, I.K., **physicist**
Kilchevskii, N.A.
Kim, M.P., **historian**
Kimiagarov, B.A., **film director**
Kirichenko, F.G., **specialist in selection and plant breeding**
Kirpotin, V.I., **literary scholar and critic**
Kirsanov, A.V., **organic chemist**
Kishkin, S.T., **metallurgist**
Kisunk'ko, G.V., **scientist in radioelectronics**
Knipovich, E.F., **literary scholar and critic**
Knorozov, I.V., **ethnologist and linguist**
Knuniants, I.L., **organic chemist**
Kobzarev, I.B., **radio physicist**
Kochetkov, N.K., **organic chemist**
Kochina (Polubarinova-Kochina), P.I., **hydrodynamics scientist**
Kolesnikov, A.G., **specialist in marine physics**
Kolesnikov, B.P., **biologist**
Kolesnikov, I.S., **medical scientist**
Kolmogorov, A.N., **mathematician**
Kolomiichenko, A.I., **otolaryngologist**
Kolosov, M.N., **organic chemist**
Kolotyrkin, I.M., **physical chemist**
Komar, A.P., **physicist**
Komar, I.V., **economic geographer**
Komisarenko, V.P., **pathophysiologist and endocrinologist**
Kondratev, K.I., **geophysicist**
Konen, V.D., **musicologist**
Kononenko, V.O., **mechanical engineer**
Kononov, A.N., **orientalist**
Konovalov, E.G. (Kanavalau, I.R.), **engineer**
Konstantinov, F.V., **philosopher**
Kontrimavichus, V.L., **biologist**
Koptiug, V.A., **organic chemist**
Koroid, A.S., **economist**
Koroliuk, V.D., **historian**
Korostovtsev, M.A., **historian**
Korshak, V.V., **chemical scientist**
Korzhinskii, D.S., **geologist, petrographer and mineralogist**
Kosmodamianskii, A.S., **mechanical engineer**
Kostenko, M.V., **specialist in energetics**
Kostiuk, G.S., **psychologist**
Kostiuk, P.G., **physiologist**
Kosygin, I.A., **geologist**
Kotelnikov, V.A., **radio and electronics engineer**
Koton, M.M., **organic chemist**
Kotov, V.T., **epizootologist**
Kovalchenko, I.D., **historian**
Kovalenko, A.D.
Kovalenko, I.R., **veterinary microbiologist**
Kovalev, F.L., **process engineer**
Kovalev, N.N.
Kovalev, S.A., **biologist and geneticist**
Kovalskii, A.A., **physical chemist**
Kovanov, V.V., **medical scientist**
Kovda, V.A., **soil scientist**
Kozhevnikov, F.I., **jurist**
Kozlov, G.A., **economist**
Kozlov, N.S. (Kazlou, M.S.), **organic chemist**
Kozlov, V.I., **mathematician**
Kozyrev, B.M., **physicist**
Krasnovskii, A.A., **biochemist and biophysicist**
Krasnyi, L.I., **geologist**
Krasovskii, A.A., **specialist in automatic control system**
Krasovskii, N.N., **specialist in mechanics and control processes**
Kratts, K.O., **geologist**
Krauchenko (Krauchanka), I.S., **historian**
Kraulin', K.I., **literary critic**
Kreps, E.M., **physiologist and biochemist**
Kretovich, V.L., **biochemist**
Kropotkin, P.N., **geologist**
Krushanov, A.I., **historian**
Krutikova, N.E., **literary historian**
Kruus, K.K., **historian**
Kruzhilin, G.N., **specialist in thermotechnology and atomic energy**
Kruzhkov, V.S., **philosopher**
Krylov, A.P., **scientist specializing in the exploitation of petroleum deposits**
Kucher, R.V., **chemical scientist**
Kukhtenko, A.I., **automation specialist**
Kukin, D.M., **historian**
Kunaev, A.M., **metallurgist**
Kundzin', K.E., **historian of literature and theater**
Kurbatov, L.N., **physicist**
Kurdiumov, G.V., **metal physicist**
Kursanov, A.L., **physiologist and biochemist**
Kursanov, D.N., **organic chemist**
Kutateladze, S.S., **thermophysicist**
Kuzin, A.M., **biochemist and radiobiologist**
Kuznetsov, I.A., **geologist**
Kuznetsov, N.D., **specializing in jet aircraft engines**
Kuznetsov, S.I., **microbiologist**
Kuznetsov, V.A., **geologist and geographer**
Kuznetsov, V.I., **mechanics scientist**
Kvasnikov, E.I., **microbiologist**
Kvasnitskii, A.V., **animal physiologist**

L

Ladyzhenskaia, O.A., **mathematician**
Lakshin, V.I., **literary historian**
Lapcheva, V.F., **geophysicist and meteorologist**
Lashkarev, V.E., **physicist**
Lavrenko, E.M., **geobotanist and botanic geographer**
Lavrent'ev, M.M., **mathematician**
Lavrov, S.S., **mathematician**
Lazarenko, E.K., **geologist**
Lazarev, B.G., **physicist**
Lebedev, A.A., **literary historian**
Leontev, A.F., **mathematician**
Leontev, O.K., **geomorphologist**
Leontovich, M.A., **physicist**
Levich, V.G., **(theoretical) physicist**
Lifshits, E.M., **physicist**
Lifshits, I.M., **physicist**
Likhachev, D.S., **literary and cultural historian**
Likhachev, N.V., **veterinarian, virologist and immunologist**
Lisovyi, V.
Litvinenko, L.M., **chemical scientist**
Liubareva, E.P., **literary historian and critic**
Liulka, A.M., **aircraft engine designer**
Liusternik, L.A., **mathematician**
Livanov, M.N., **medical scientist**
Lobashev, V.M., **physicist**
Logunov, A.A., **theoretical physicist**
Lomidze, G.I., **literary historian**
Luchitskii, I.V., **geologist**
Lupanov, O.B., **mathematician**
Lur'e, A.I., **scientist in theoretical and applied mechanics**
Lvovich, M.I., **hydrologist**

M

Maergoiz, I.M., **economic geographer**
Magak'ian, I.G., **geologist**
Magnitskii, V.A., **geophysicist**
Makarevskii, A.I., **aircraft engineer**
Maliusov, V.A.
Malmeister, A.K.
Marchenko, V.A., **mathematician**
Mardzhanishvili, K.K., **mathematician**
Margulova, **specialist in nuclear energy**
Markevich, A.P., **zoologist**
Markish, S.P., **literary historian and translator**
Markov, A.A., **mathematician**
Markov, D.F., **Slavic scholar and literary historian**
Markov, K.K., **geographer and geomorphologist**
Markov, M.A., **physicist**
Masumi, N.A., **literary historian**
Matora, I.M., **physicist**
Matskevich, I.F., **philologist**
Matulis, I.I., **physical chemist**
Meilakh, B.S., **literary historian**
Meisel', M.N., **microbiologist and cytologist**
Melentev, L.A.
Melikishvili, G.A., **historian**
Meliukhin, S., **philosopher**
Melnichuk, A., **linguist**
Mel'nikov, N.V., **mining scientist**

Melnikov, O.A., **astronomer**
Melnikov, P.I., **geologist**
Menner, V.V., **geologist and palaeontologist**
Mergelian, S.N., **mathematician**
Meshcheriankov, M.G., **physicist**
Migdal, A.B., **physicist**
Migulin, V.V., **radio physicist**
Mikhailov, B.M., **organic chemist**
Mikhailov, N.I., **physical geographer**
Mikhailovskii, V.N., **specialist in automation, telemechanics and surveying technology**
Mikulinskii, S.R., **philosopher and historian of science**
Miliakh, A.N., **engineer**
Mil'kov, F.N., **physical geographer**
Minachev, K.M., **organic chemist**
Mints, I.I., **historian**
Mirnov, S.V., **physicist**
Mirzo-zade, K., **literary historian and critic**
Mishin, V.P., **aircraft engineer**
Mishustin, E.N., **microbiologist**
Mitin, M.B., **philosopher**
Mitropolskii, I.A., **mathematician**
Moiseev, N.N., **scientist in general mechanics and applied mathematics**
Molchanov, A.A., **forestry scientist**
Monin, A.S., **oceanologist**
Morozov, A.A., **tank designer**
Motyleva, T.L., **literary historian and critic**
Movchan, B.A., **metallurgist**
Mozgov, I.E., **veterinary pharmacologist**
Mukhamedov, T.M., **hydrotechnologist and state official**
Mukhovatov, V.S., **physicist**
Murashko, A.G., **philologist**
Muratov, M.V., **geologist**
Murzaev, E.M., **physical geographer and toponymist**
Musabaev, I.K., **specialist in infectious diseases**
Mustel', E.R., **astronomer**

N

Nabiev, I.A. o., **cybernetic scientist**
Nalivkin, V.D., **geologist**
Nametkin, N.S., **organic chemist**
Narochnitskii, A.L., **historian**
Nasonov, V.N., **scientist in structural design**
Naumov, A.A., **physicist**
Naumov, E.I., **literary historian**
Nechkina, M.V., **historian**
Nekrasov, B.V., **chemical scientist**
Nekrasov, N.N., **economist**
Nekrasov, Z.I., **metallurgist**
Nersisian, M.G., **historian**
Nesmeianov, A.N., **radiochemist**
Nesterikhin, I.E., **physicist**
Nesturkh, M.F., **anthropologist and primatologist**
Neunylov, B.A., **soil scientist, agricultural chemist and specialist in rice growing**
Nezhevenko, G.S.
Nikitin, S.A., **historian**
Nikolaev, G.A., **engineer**
Nikolskii, B.P., **physical and radiochemical scientist**
Nikolskii, S.M., **mathematician**
Novichenko, L.N., **literary historian and critic**
Novikov, I.I., **physicist**
Novikov, S.P., **mathematician**
Novoselova, A.V., **chemical scientist**
Nurdzhanov, N.K., **theater scholar and doctor of the arts**

O

Obraztsov, I.F., **scientist in structural mechanics and the theory of strength of aircraft**
Obreimov, I.V., **physicist**
Obukhov, A.M., **geophysicist**
Odintsov, M.M., **geologist**
Ognev, V.F., **literary historian and critic**
Oizerman, T.I., **philosopher**
Okhotsimskii, D.E., **scientist in mechanics**
Okladnikov, A.P., **archaeologist**
Okun', L.B., **theoretical physicist**
Omel'ianovskii, M.E., **philosopher**
Oreshnonkova, A.V., **philologist**
Orlov, I.F., **physicist**
Orlov, I.V., **helminthologist**
Orlov, V.N., **literary scholar**
Ostrianin, D.F., **philosopher**
Ovanesova, A.A., **film director**
Ovcharenko, E.I., **literary historian and critic**
Ovcharenko, F.D., **chemical scientist**
Ovchinnikov, I.A., **biochemist**
Ovchinnikov, L.N., **geologist**
Ovsiannikov, L.V., **mathematician**
Ozerov, V.M., **literary historian and critic**

P

Pankin, V.E., **cybernetics and computer scientist**
Pankov, V.K., **literary historian**
Papanin, I.D., **geographer and arctic explorer**
Papernyi, Z.S., **literary historian and critic**
Parasiuk, O.S., **mathematician**
Pariiskii, N.N., **geophysicist and astronomer**
Parkhomenko, M.N., **literary historian**
Pashkov, A.I., **economist**
Pasichnik, M.V., **physicist**
Paton, B.E., **engineer**
Pavlov, I.M., **metallurgist**
Pavlovskii, E.V., **geologist**
Peive, A.V., **geologist**
Pekar, S. I., **theoretical physicist**
Penkov, A.M., **mechanics scientist**
Perepelkin, I.I., **egyptologist**
Persianinov, L.S., **gynaecologist and obstetrician**
Petrianov-Sokolov, I.V., **physical chemist**
Petrov, A.A., **organic chemist**
Petrov, A.P., **railroad transportation engineer**
Petrov, G.I., **hydro-aeromechanics and gas-dynamics engineer**
Petrov, M.P., **geographer**
Pidoplichko, I.G., **zoologist and palaeontologist**
Pidoplichko, N.M., **botanist-mycologist**
Piliugin, N.A., **scientist in automation and remote control**
Piotrovskii, B.B., **historian, archaeologist and oriental scholar**
Pisarenko, G.S., **specialist in mechanical engineering**
Plate, N.A., **chemist**
Plotnikov, K.N., **economist**
Pogorelov, A.V., **mathematician**
Pokrovskii, G.I., **physicist and engineer**
Pokshishevskii, V.V., **economic geographer**
Poliakov, A.A., **scientist in veterinary sanitation and disinfection**
Poliakov, I.M., **biologist**
Poliakov, N.S., **scientist in the field of mining engineering**
Polianskii, I. (G.) I., **protozoologist**
Polikanov, S.M., **atomic physicist**
Pomerants, G.S., **philosopher and Indian scholar**
Pontekorvo, B.M., **physicist**
Pontriagin, L.S., **mathematician**
Popkov, V.I., **energetics scientist**
Popov, E.P., **automation specialist**
Popov, K.M., **geographer and oriental scholar**
Porai-Koshits, M.A., **crystallographer**
Porfirev, V.B., **geologist**
Pospelov, G.N., **literary historian**
Pospelov, G.S., **automation specialist**
Postovskii, I.I., **organic chemist**
Potapov, L.P., **ethnographer and historian**
Povkh, I.L., **physical-thechnical scientist**
Preobrazhenskii, N.A., **otorhinolaryngologist**
Prikhotko, A.F., **physicist**
Primakov, E.M., **economist and historian**
Prokhorov, A.M., **radiophysicist**
Prokhorov, I.V., **mathematician**
Prokoshkin, I.D., **physicist**
Prokushev, I.L., **journalist**
Puchkovskaia, N.A., **ophthalmologist**
Pudovik, A.N., **chemist**
Pugachov, V.S., **engineer**
Pukhov, G.E., **computer scientist**
Pyshnov, V.S., **scientist specializing in the aerodynamics of airplanes**

R

Rabotnov, I.N., **physicist**
Rafikov, S.R., **chemist**
Rakitin, I.V., **plant physiologist**
Rakitskii, P.F., **biologist and geneticist**
Razin, N.V., **hydraulic engineer**
Razuvaev, G.A., **physical chemist**
Rebane, K.K., **physicist**
Reformatskii, A.A., **linguist**
Reimers, F.E., **plant physiologist**
Reizov, B.G., **literary critic**
Remeslo, V.N., **biologist**
Reshetnikov, F.G., **physical chemist**
Reutov, O.A., **chemical scientist**
Riabushkin, T.V., **economist**
Riazanskii, M.S., **radio engineer**
Riznichenko, I.V., **seismologist**
Roginskii, I.I., **anthrophologist**
Roitbak, A.I., **physiologist**
Roiter, V.A., **physical chemist**
Romankov, P.G., **chemical scientist-technologist**
Romanovich, E.M., **linguist**
Ronov, A.B., **geologist and geochemist**

Rostotskii, B.N.I., **theater historian and critic**
Rozanov, B.S., **surgeon**
Rozin, M.S., **economic geographer**
Rubinov, A.N., **physicist**
Rudenko, V.T., **engineer and physicist**
Rumiantsev, A.M., **economist**
Rumiantsev, V.V., **mechanics scientist**
Rutkevich, M.N., **philosopher**
Rybakov, B.A., **archaeologist and historian**
Rykalin, N.N., **metallurgical scientist**
Rytov, S.M., **physicist**
Ryzhikov, K.M., **helminthologist**
Rzhanov, A.V., **physicist**
Rzhevskii, V.V., **mining scientist**

S

Sadovskii, M.A., **physicist**
Sadovskii, V.D., **metallurgist**
Sadykov, A.S., **organic chemist**
Sagdeev, R.Z., **physicist**
Sakharov, A.D., **physicist**
Saks, V.N., **geologist and palaeontologist**
Salishchev, K.A., **geographer and cartographer**
Samarskii, A.A., **mathematician**
Samoilov, B.N., **physicist**
Samsonov, A.M., **historian**
Samsonov, G.V., **metallurgical scientist**
Sarnov, B.M., **literary historian and critic**
Saushkin, I.G., **economic geographer**
Savarenskii, E.F., **geophysicist and seismologist**
Savin, G.N., **mechanics scientist**
Savitskii, E.M., **physical chemist and metallurgist**
Sedov, L.I., **specialist in mechanics and applied mathematics**
Semenenko, N.P., **geologist**
Semenikhin, V.S., **specialist in automatics and telemechanics**
Semenov, N.N., **physicist and physical chemist**
Semionov, V.V., **cybernetics scientist**
Serebrennikov, B.A., **linguist**
Sergeev, E.M., **geologist**
Severin, S.E., **biochemist**
Severnyi, A.B., **astrophysicist**
Shabalin, E.P., **engineer-physicist**
Shabliovskii, E.S., **literary historian**
Shafarevich, I.R., **mathematician**
Shafranovskii, I.I., **crystallographer**
Shal'nikov, A.I., **physicist**
Shamota, N.Z., **literary historian**
Sharif, A.A., **literary scholar and translator**
Sharvin, I.V., **physicist**
Shatalov, A.S., **cybernetics scientist**
Shchepotev, F.L., **botanist**
Shcherban, A.N., **thermophysicist**
Shcherbina, V.R., **critic and literary scholar**
Shchukin, A.N., **radio technologist**
Sheindlin, A.E., **specialist in heat physics and power engineering**
Shemiakin, E.I., **scientist specializing in the mechanics of rocks**
Shevelev, F.A., **hydraulics scientist**
Shilo, N.A., **geologist**
Shinkaruk, V., **philosopher**
Shiriev, Y., **economist**
Shirkov, D.V., **physicist**
Shirokov, A.Z., **geologist**
Shirshov, A.I., **methematician**
Shishkov, V.P., **veterinarian**
Shklovskii, I.S., **astrophysicist**
Shlyk, A.A., **plant physiologist and biochemist**
Shmidt, E.V., **neuropathologist**
Shostakovskii, M.F., **organic chemist**
Shpak, **chemist**
Shternfeld, A.A., **mechanical engineer**
Shtokalo, J.Z., **mathematician**
Shubenko-Shubin, L.A., **mechanical engineer**
Shults, M.M., **physical chemist**
Shumakov, V.I., **medical scientist**
Shumskii, P.A., **geographer**
Shur, **physicist**
Shveikin, G.P., **chemist**
Shvetsov, P.F., **permafrost specialist and hydrogeologist**
Sidorov, V.A., **physicist**
Siforov, V.I., **scientist in the fields of radio engineering and electronics**
Sirazhdinov, S.K.
Skatkin, M.N., **educator**
Skliarevskii, V.V., **physicist**
Skriabin, G.K., **microbiologist and biochemist**
Skrinskii, A.N., **nuclear physicist**
Skulachev, V.P., **biochemist**
Sladkovskii, M.I., **economist**
Slavin, L.M., **archaeologist**
Slinko, M.G., **chemical scientist**
Smirnov, A.A., **theoretical physicist**
Smirnov, A.F., **scientist in structural mechanics and building design**
Smirnov, V.I., **geologist**
Smolenskii, G.A., **physicist**
Smolskii, B.M.
Smolskii, N.V. (M.U.), **botanist**
Smorodintsev, A.A., **virologist and immunologist**
Snezhnevskii, A.V., **psychiatrist**
Sobolev, **petrographer and mineralogist**
Sobolev, S.L., **mathematician and mechanics scientist**
Sobolev, V.V., **astronomer**
Sokolov, A.V., **agricultural chemist**
Sokolov, B.P., **plant breeder**
Sokolov, B.S., **geologist and palaeontologist**
Sokolov, V.E., **zoologist**
Solntsev, N.A., **physical geographer**
Solodovnikov, V.G., **economist**
Solodovnikov, V.V., **cybernetics scientist**
Soloukhin, R., **physicist**
Solovev, G.M., **surgeon**
Solovev, S.L., **seismologist**
Solozhenkin, P.M.
Sonygina, A.P., **agronomist**
Spirin, A.S., **biochemist**
Spirkin, A.G., **philosopher**
Spitsyn, V.I., **chemical scientist**
Starodubov, K.F., **metallurgist**
Staros, F.G., **electronics engineer**
Stepanov, B.I., **physicist**
Stepanov, G.V., **philologist**
Stepanov, V.E., **physicist**
Stepanova, E.A., **historian**
Stoletov, V.N., **biologist**
Strelkov, V.S., **physicist**
Struchkov, V.I., **surgeon**
Strukov, A.I., **medical scientist**
Struminskii, V.V., **aerodynamics scientist**
Styrikovich, M.A., **thermophysicist and thermo-energetics scientist**
Sulakvelidze, G.K., **physicist**
Sumbatzade, A.S.S., **historian and orientalist**
Suprunenko, D.A., **mathematician**
Suvorov, G.D., **mathematician**
Svetovidov, A.N., **zoologist-ichthyologist**
Svishchev, G.P., **specialist in aviation and mechanics**
Svitlychnyi, I.O., **literary historian and critic**
Syromiatnikov, N.A., **philologist**

T

Tabyshaliev, S.
Takhtadzhian, A.L., **botanist**
Tananaev, I.V., **chemical scientist**
Tarasenko, N.I., **hygienist**
Tatarinov, L.P., **biologist**
Tauson, L.V., **geochemist**
Tikhomirov, V.V., **scientist in the fields of radio electronics and automation**
Tikhomirov, V.V., **geologist**
Tikhonov, A.N., **mathematician and geophysicist**
Timofeev, L.I., **literary historian**
Timofeev, P.P., **geologist**
Timofeev, P.V., **scientist in electronics and vacuum technology**
Timofeev, T.T., **historian and economist**
Timofeev-Resovskii, N. V., **biologist**
Tolpigo, K.B., **theoretical physicist**
Tolstikov, E.I., **meteorologist**
Tolubinskii, V.I., **thermophysicist**
Tomilin, A.N., **mathematician**
Topcheev, I.I., **cyberneticist**
Torgov, I.V., **chemical scientist**
Trapeznikov, V.A., **mechanics scientist**
Treshnikov, A.F., **geographer and oceanologist**
Tretiakov, D.N., **physical chemist**
Trofimuk, A.A., **petroleum geologist**
Troitskii, V.S., **astronomer**
Troshin, A.S., **cytologist and cytophysiologist**
Trubachev, O.N., **philologist**
Trukhanovskii, V.G., **historian**
Tselikov, A.I., **metal engineering scientist**
Tsetsevich, V.P., **astrophysicist**
Tsitsin, N.V., **botanist, geneticist and plant breeder**
Tsvetkov, V.N., **physical chemist**
Tsypkin, I.Z., **specialist in automatic control**
Tsytovich, N.A., **specialist in soil mechanics, geomechanics**
Tuchkevich, V.M., **physicist**
Tulchinskaia, V.P., **microbiologist**
Tulepbaev, B., **historian**
Tulinov, A.F., **physicist**
Tupolev, A.A., **airplane designer**
Turbin, N.V., **geneticist and breeder**
Turchin, V., **mathematician**
Turkov, A.M., **literary historian and critic**
Turpaev, T.M., **physiologist**
Tverdokhlebov, A.

U

Uglov, F.G., **surgeon**
Ugolev, A.M., **physiologist**
Ukraintsev, F.I., **physicist**
Ulanov, G.M., **cybernetics scientist**
Urlanis, B.T., **demographer**
Usikov, A.I., **physicist**
Utevskii, A.M., **biochemist**
Utkin, V.I., **engineer**
Uvachan, V.N., **historian, and party and state figure**

V

Vainshtein, B.K., **crystallographer**
Vainshtein, L.A., **physicist**
Vaisfel'd, I.V., **film critic and theoretician**
Valiev, K.A.
Vanichev, A. P., **heat engineer**
Vasilenko, V.K., **therapeutist**
Vasil'ev O.F., **scientist in the field of applied hydrodynamics and hydraulics**
Vasilev, V.P., **zoologist-entomologist**
Vassoevich, N.B., **geologist**
Vechar, A.S., **biochemist**
Veimarn, B.V., **art historian and critic**
Velikanov, D.P., **scientist in automobile technology**
Velikhov, E.P., **physicist**
Verkin, B.E., **physicist**
Vernov, S.N., **physicist**
Vezhev, A.A., **literary historian and critic**
Vialov, O.S., **geologist**
Viazankin, N.S., **chemical scientist**
Vilenkin, V.I., **theatrical scholar**
Vinberg, A.I., **jurist and criminologist**
Vinogradov, V.A., **economics scientist**
Vladimirov, V.S., **mathematician**
Vladimirskii, V.V., **physicist**
Vlasiuk, P.A., **plant physiologist and agrochemist**
Voitikov, V.V., **engineer and designer**
Voitovich, N.T., **philologist**
Volfkovich, S.I., **inorganic chemist**
Vol'kenshtein, M.V., **physical chemist and biophysicist**
Volkov, A.A., **literary scholar**
Volobuev, P.V., **historian**
Volobuev, V.R., **soil scientist**
Volzhenskii, A.V., **scientist in the field of technology of structural materials and products**
Vonsovskii, S.V., **physicist**
Voronin, L.G., **physiologist**
Voronkov, M.G., **chemical scientist**
Voronov, A.A., **cybernetics scientist**
Vorovich, I.I., **mechanics scientist**
Vorozhtsov, N.N., **organic chemist**
Vul, B.M., **physicist and astronomer**

Z

Zababakhin, E.I., **physicist**
Zaionchkovskii, P.A., **historian**
Zakusov, V.V., **pharmacologist**
Zapasko, I.P., **art historian**
Zaporozhets, A.V., **psychologist**
Zarian, R.V., **theater historian**
Zaslavskaia, T.I., **economist**
Zatsepin, G.T., **physicist**
Zedgenidze, G.A., **radiologist**
Zefirov, A.P., **specialist in the chemistry and metallurgy of rare and noble metals**
Zeldovich, I.B., **physicist**
Zel'in, K.K., **historian of classical history**
Zenkovich, V.P., **oceanographer and geomorphologist**
Zguridi, A.M., **director of popular science films**
Zharikov, V.A., **geologist and mineralogist**
Zhavoronkov, N.M., **chemo-technologist**
Zhdanov, I.A., **chemical scientist**
Zhdanov, V.M., **virologist**
Zheltukhin, N.A., **scientist in mechanics and heat engineering**
Zhilin, P.A., **military historian**
Zhimerin, D.G., **scientist in the field of energy**
Zhuchenko, A.A.
Zhukov, A.B., **forestry scientist**
Zhukov, B.P., **technical chemist**
Zhukov, M.F., **physicist**
Zhurkov, S.N., **physicist**
Zimin, A.A., **historian**
Zolotov, E.V., **mathematician**
Zubkov, I.A., **theater scholar**
Zuev, V.E., **physicist**
Zurabashvili, A.D., **psychiatrist**
Zverev, M.S., **astronomer**

Authors

A

Abashidze, G. G., **poet**
Abashidze, I. V., **poet**
Abdukadyrov, K. A., **author**
Abdullin, I. A., **author and dramatist**
Abdulzhalilov, F. A., **author and poet**
Abdumomunov, T., **author**
Abilev, D., **poet**
Abishev, A., **author**
Abramov, F. A., **author**
Abramov, K. G., **author**
Abulgasan (Alekper-zade, A. A. B. o.), **author**
Adamian, N. (Adamova, E. G.), **author**
Adamovich, A. M., **author**
Adzhie, D. (Dzhauari, A. D.), **author**
Adzhiev, A. A., **poet, author, translator**
Agakov, L. I., **author and dramatist**
Agashina, M. K., **author and poet**
Agishev, R. K., **author**
Agranovich, L. D., **writer, actor and film producer**
Aimurzaev, Z. M., **author**
Aitmatov, Ch., **author**
Akhavni (Grigorian, A. A.), **author**
Akhmadulina, B. (I.) A., **poet**
Akhmatova, R. S., **poet**
Aksenov, V. P., **author**
Aladzhadzhian, S. E., **writer**
Aleksandrova, Z. N., **poet**
Alekseev, M. N., **author**
Aleksin, A. G., **author**
Aleshin, S. I. (Kotliar), **dramatist**
Alga, A. E., **poet**
Aliev, A. (Rukhi), **author and literary scholar**
Alieva, F., **poet**
Aliger, M. I., **poet**
Alimdzhan, Kh., **poet, journalist and critic**
Alimzhanov, A., **author**
Altaiskii, K. N., **author**
Alver, B., **writer and translator**
Amin Avdal, **author**
Amin-zade, M., **poet**
Amiri, G. G., **author**
Amirov, M. M. (pseudonym: Mirsai Amir), **author and dramatist**
Ananev, A. A., **author**
Ananian, V. S., **author**
Andreev, A. (Andreev, V. D.), **author**
Andronikov, I. L. (Andronikashvili), **author**
Angervaks, H. (Piall, E. N.), **author**
Antonenko-Davidovich, B. D., **author**
Antonov, S. P., **author**
Arkhipov, T. A., **writer**
Armen, M. (Arutiunian, M. G.), **author**
Arslan, T. (Arslanov, T. G.), **author**
Asadov, E. A., **poet**
Asaev, R. N., **author and poet**
Ashirov, Ch., **poet**
Asipenko, A. K., **writer**
Astafev, V. P., **author**
Ata Salikh, **poet**
Avdeenko, A. O., **author**
Avdeev, V. F., **author**
Avizhius (Avyžius), J., **author**
Avramenko, I. K., **poet**
Avshalumov, K. D., **writer**
Azarov, V. B., **poet**
Azeroglu (Abizade, B.), **poet**

B

Baal Khabyryys (Veshnikov, G. G.), **poet**
Babadzhan, R. (Babadzhanov, R. N.), **author**
Babaev, M., **author**
Babaev, N. K. (N. A. O.), **poet**
Babaevskii, S. P., **author**
Bagmut, I. A., **author**
Baialinov, K., **writer**
Bairamov, B. S. o., **author**
Bairamukova, K., **poet**
Baitemirov, N., **poet and author**
Baklanov, G. I., **author**
Baliasnaia, R. N., **poet**
Balodis, A. M., **poet**
Baltakis, A., **poet**
Baltrunas, A., **writer**
Baltushis, I. (Iuozenas, A.), **author**
Banykin, V. I., **author**
Baratashvili, M. G., **poet**
Barskii, L. M., **author**
Barto, A. L., **poet**
Baruzdin, S. A., **author**
Bash, I. V. (Bashmak), **author**
Bashirov, G. B., **author**
Battal, S., **writer**
Battalov, S. (Batalov, S. V.), **author**
Baukov, I. P., **poet**
Bazhan, M. P., **author and poet**
Bazorkin, I. M., **author**
Beban, M. A., **poet**
Bednyi, B. V., **author**
Bedzyk, D. I., **author**
Bedzyk, I. D., **author**
Bekkhozhin, K. N., **poet**
Bekoev, E. D., **author**
Belevich, A. P., **poet**
Beliaev, V. P., **writer**
Beliauskas, A., **writer**
Belinskii, I. L., **poet**
Belous, D. G., **poet and satirist**
Belov, V. I., **author**
Berestov, V. D., **poet**
Berezko, G. S., **autor**
Bershadskii, R. I., **author**
Bertse, V. A., **author**
Berulava, K. M., **poet**
Besaev, T. U., **author and poet**
Bgazhba, K. S., **author and linguist**
Biisheva, Z. A., **author**
Bikbai, B. G., **author**
Bikchentaev, A. G., **author**
Bikkulov, S. S., **poet**
Birze, M. I. (Berzin), **author**
Bogdanov, N. V., **author**
Bokov, V. F., **poet**
Bondarev, I. V., **author and state official**
Bondarin, S. A., **writer**
Borodulin, G. I., **poet**
Borovikov, G. F., **author**
Borshchagovskii, A. M., **author**
Borshosh-Kumiatskii, J. V., **poet**
Botsu, P. P., **author and poet**
Bragin, M. G., **author**
Bratun, R. A., **author**
Braun, N. L., **author**
Brodele, A. I., **author**
Brutane, V. I., **author**
Bryl, I. (I. A.), **author**
Bubennov, M. S., **author**
Bukov, E. N., **author**
Buriak, B. S., **author**
Busakov, D., **writer**
Bychko, V. V., **author**
Bykov, V. V., **author**

C

Chakovskii, A. B., **author**
Chelidze, O. S., **author and poet**
Chepurov, A. N., **author and poet**
Chigranov, I. G., **writer**
Chikovani, G. S., **writer**
Chiladze, T. I., **writer**
Chilaia, S. E., **writer, critic and literary scholar**
Chivilikhin, V. A., **author**

D

Daian, K. (Daianov, K. K.), **author**
Damdinov, N. G., **author**
Dar, D. I., **author**
Darchiev, D. G., **author**
Davtian, V. A., **author**
Davydov, I. V., **author**
Davydova, N. (M. M.), **author**
Davydychev, L. I., **author**
Dementev, N. S., **author**
Demin, M. (Trifonov, G. E.), **author**
Devin, I. M., **author**
Diagilev, V. I., **author**
Dinere, T. R., **author**
Dmiterko, L. D., **writer**
Dmitrevskii, V. I., **author**
Dneprov, A. P. (Mitskevich), **author**
Dobin, G. I., **author**
Dobrovolskii, V. A., **author**
Dolidze, S. V., **film director and screenwriter**
Dolmatovskii, E. A., **author**
Donskoi, M. S., **film director and screenwriter**
Donzhashvili, T. G., **author**
Dordzhiev, B. B., **author**
Dorizo, N. K., **author**
Dorokhov, A. A., **author**
Doroshko, P. O., **author**
Dostian, R. M., **author**
Dovidaitis, I., **author**
Drach, I. F., **author and poet**
Drunina, I. V., **author**
Dubov, N. I., **author**

Dudin, M. A., **author and poet**
Dudintsev, V. D., **author**
Dumbadze, N. V., **author**
Durdyev, K., **author**
Durova, N. I., **author**
Dvoretskii, I. M., **author**
Dzesov, K. G., **writer**
Dzhamil', A., **lyric poet**
Dzhanelidze, D. S., **theatrical scholar**
Dzhaparidze, R. A., **author**
Dzhasme, D. O., **poet**
Dzhatiev, T. I., **author**
Dzhimbinov, B. O., **author**
Dzhioev, D. A., **author**
Dzhonua, A. N., **author and poet**
Dzhonua, C. M., **poet**
Dzhusoev, N. G., **poet and literary historian**
Dzhusoity, N. G., **writer and literary scholar**
Dziuba, I. M., **literary critic**
Dzugaev, G. K., **poet**

E

Efimov, M. D., **poet**
Egorov, B. A., **author**
Eibek (Tashmukhamedov, M.), **author**
Emin, G. (Muradian, K. G.), **author and poet**
Enikeev, A. N. (Eniki, A.), **author**
Ergaliev, K., **author**
Evdokimov, N. S., **author**
Evseeva, S. G., **author**
Evtushenko, E. A., **author**
Evtykh, A. K., **author**
Ezhov, V. I., **author**

F

Fedorov, G. A., **writer**
Fedorov, V. D., **poet**
Firsov, V. I., **author, poet and translator**
Fokina, O. A., **poet**
Fomenko, L. N., **author and critic**
Fomenko, V. D., **writer**

G

Gabdrakhmanov, R. B., **author**
Gabdullin, M., **author and literary historian**
Gabrilovich, E. I., **author and script writer**
Gabyshev, N. A., **author**
Gaidaenko, I. P., **author**
Gairbekova, M. G., **author**
Galeev, G. G. (Gali, M.), **poet**
Galeev, G. Z., **author**
Galsanov, T. G., **poet**
Gamzatov, R. G., **poet**
Ganina, M. A., **author**
Gareev, I. I., **author**
Gashokov, K. K., **poet**
Gatov, A. B., **poet, translator and literary scholar**
Gatserelia, A. K., **writer and literary critic**
Gavrusev, S. Z., **poet**
Georgievskaia, S. M., **writer**
Gerasimov, E. N., **author**
Gerasimov, S. A., **film director, actor, screenwriter and theorist**
Geraskina, L. B., **dramatist**
Gernet, N. V., **author**
Giliazhev, L.-K. G., **poet**
Gintsburg, E. S., **author**
Ginzburg, L. V., **writer and translator**
Glazov, G. S., **author and poet**
Godenko, M. M., **author**
Gogolev, I. M., **poet**
Gogua, A. N., **author**
Goliavkin, V. V., **author**
Golovanivskii, S. E., **poet and dramatist**
Gomiashvili, A. K., **poet**
Gonchar, O., **writer**
Goncharenko, I. I., **poet**
Goncharov, I. D., **author**
Gontar, A. I., **author**
Gorbov, E. K., **author**
Gorbovskii, G. I., **poet**
Gordeichev, V. G., **poet**
Gordienko, I. P., **poet and translator**
Gordon, I. Z., **writer**
Gordon, S. V., **author**
Gorshman, S., **author**
Goryshin, G. A., **author**
Gotua, L. P., **writer**
Grachev, A. M., **author**
Granin, D. A. (German), **author**
Grebenev, N. I., **poet and translator**
Gribachev, N. M., **author and public figure**
Grigulis, A. P., **author and literary historian**
Grin, E. (Iakimov, A. V.), **author**
Grinberg, I. L., **author and critic**
Gromovich, I. I., **author**
Gross, V., **author**
Grubian, M. M., **author and poet**
Gubai, G. (Gubaidullin, G. S.), **author**
Gubarevich, K. L., **dramatist**
Gubnitskii, A. M., **author**
Gulia, G. D., **author**
Gureev, A. I., **author**
Gurtuev, B. I., **author**
Guseinov, I. M. o., **author**
Guseinzade, G. D. o., **author**
Guzhel, A. M., **author**
Gzhitskii, V. Z., **author**

I

Iashen, K. (Nugmanov), **author**
Ibragimov, M. A. o., **author**
Ikhsanova, L. F., **author**
Ikrami, D., **author**
Il'chenko, A. E., **writer**
Ilina, N. I., **author**
Imermanis, A. A., **author**
Indzhiev, L. O., **author**
Infantev, V. N., **author**
Iroshnikova, I. I., **author**
Isaev, E. (G.) A., **author**
Isanbet, N. S., **poet, dramatist and folklorist**
Isangulov, F. A., **author**
Iskander, F. A., **author**
Iskhak, A. A., **poet**
Istru, B. S., **author**
Ivanenko, O. D., **author**
Ivanov, A. G., **film director and screenwriter**
Ivanov, A. S., **author**
Ivanov, L. I., **author**
Iziumskii, B. V., **author**
Iziurov, I. V., **author**

K

Kaal, A. (Houn, A.), **author**
Kaalep, A., **author and translator**
Kaitukov, G. K., **author**
Kalachinskii, M. I., **author and poet**
Kaladze, K. R., **poet and playwright**
Kalgan, A. D., **author**
Kaliaev, S. K., **writer**
Kalinin, A. V., **author**
Kamianskii, V. K., **author**
Kantorovich, V. I., **author**
Kapitsa, P. I., **author**
Kapler, A. I., **actor and filmscript writer**
Kaputikian, S. B., **author and poet**
Karataev, M., **author, critic and translator**
Karavaeva, A. A., **author**
Kardin, V. (E. V.), **literary and theater critic**
Karelin, L. V., **author**
Karim, M. (Karimov, M. S.), **author and poet**
Karimov, K. K., **author and poet**
Karpeko, V. K., **author and poet**
Karpiuk, A. N., **author**
Karpov, V. B., **author**
Kasumov, I. A. o., **author**
Kataev, V. P., **author**
Katsiev, K. K., **writer**
Katsnelson, A. I., **author and poet**
Kaverin, V. A., **author**
Kazakov, I. P., **author**
Kazakov, N. (M.) I., **poet**
Kazakova, R., **author and poet**
Kazantsev, A. P., **author**
Kazin, V. V., **poet**
Kedrina, Z. S., **literary critic**
Kekilov, A., **author, poet and literary historian**
Kerashev, T. M., **writer**
Kesamaa, M. A., **author and poet**
Keshokov, A. P., **author and poet**
Kezhun, B. A., **author and poet**
Khakim, S. T., **poet and public figure**
Khanzadian, S. N., **writer**
Khappalaev, I. R., **poet**
Kheifits, I. E., **film director and screenwriter**
Khodzher, G. G., **writer**
Khubiev, O. A., **writer**
Kilchichakov, M. E., **poet**
Kimiagarov, B. A., **film director and screenwriter**
Kireenko, K. T., **author and poet**
Kireev, A. N. (Kirei Mergen), **author and literary historian**
Kiselev, V. L., **author**
Kishniakov, I. P., **author**
Kislik, N. Z., **author and poet**
Kldiashvili, S. D., **writer**
Kleshchenko, A. D., **author**
Kobzev, I. I., **author and poet**
Kochin, N. I., **writer**
Kochnev, M. K., **author**
Kolesnik, P. I., **author and literary historian**
Kolomasov, V. M., **author**
Kolosov, M. B., **author**

Koluntsev, F. (Barkhudarian, T. A.), **author**
Kondratev, N. D., **author**
Konetskii, V. V., **author**
Konovalov, G. I., **author**
Koptelov, A. L., **writer**
Koptiaeva, A. D., **writer**
Korban, V. I., **author**
Kornilov, V. N., **author and poet**
Korotkevich, V. S., **author**
Korotych, V. A., **author and poet**
Korsakas, K. P., **author, poet and literary historian**
Korsakiene, G. P., **author**
Korzhavin, N. (Mandel, N. M.), **author and poet**
Kosarik, D. M. (Kovalenko), **author and literary historian**
Kostenko, L. V., **author and poet**
Kostenko, N. F., **writer**
Kostiukovskii, B. A., **author**
Kostiukovskii, I. A., **author**
Kostrov, V. A., **author and poet**
Kotov, V. P., **author and poet**
Kovaldzhi, K. V., **author and poet**
Kovalenkov, A. A., **author**
Kovalev, D. M., **author and poet**
Kovalev, P. N., **author**
Kozachenko, V. P., **author**
Kozhevnikov, A. V., **writer**
Kozhevnikov, V. M., **author**
Kozlovskii, I. A., **author, poet and translator**
Kramov, I. N., **author and literary critic**
Krapiva, K. (Atrakhovich, K. K.), **author and playwright**
Krasilnikov, G. D., **author**
Krasnov-Asli, V. I., **writer**
Krasynskyj, S. M., **poet**
Kron, A. A., **author and dramatist**
Krucheniuk, P. A., **author and poet**
Krusten, E., **writer**
Krutilin, S. A., author
Kublitskii, G. I., **author**
Kudashev, S. F., **poet**
Kugul'tinov, D. N., **poet**
Kuklin, L. V., **poet**
Kulakovskii, A. N., **author**
Kulibaev, S. G., (Kulibai, S.), **poet**
Kuliev, K. S., **author and poet**
Kuniaev, S. I., **poet and author**
Kuprianov, I. P., **dramatist**
Kurbansakhatov, K., **author**
Kustov, P. P., **poet**
Kutateli, A. N., **writer**
Kutov, N. N., **poet**
Kuusberg, P., **author and journalist**
Kuznetsov, A. V., **author**
Kuznetsov, V. N., **poet**
Kuznetsova, S. A., **poet**

L

Lagin, L. I., **writer**
Lamutskii, P. (Stepanov, P. A.), **poet**
Lankauskas, R., **author**
Lanskoi, M. Z., **author**
Lapin, K. K., **author**
Laptsui, L. V., **author**
Laskin, B. S., **author**
Lavrentev, V. V., **dramatist**
Lavrov, I. M., **author**
Lazarev, L. (Shindel, L. I.), **literary critic**
Lebedev, K. V., **author**
Lekhtmets, T., **author**
Leonov, L. M., **author**
Levada, A. (Kosiak, A. S.), **author**
Levakovskaia, E. V., **author**
Levashov, M. A., **author**
Levchaev, P. I., **author**
Levchenko, I. N., **author**
Levidova, V. I., **author**
Levik, V. V., **literary historian, critic, translator and poet**
Levina, H. M., **poet**
Levitanskii, I. D., **poet and translator**
Lialenkov, V. D., **author**
Liamin, M. A., **author**
Lidin, V. G., **author**
Lifshits, V. A., **author**
Likhodeev, L. I. (Lides), **author**
Lipatov, V. V., **writer**
Lipkin, S. I., **poet and translator**
Lisianskii, M. S., **poet**
Liushnin, G. I., **poet**
Loban, (N. P.) M., **author**
Loginov, V. N., **author**
Lipatov, V. V., **author**
Lisnianskaia, I. L., **author**
Lordkipanidze, K. A., **author**
Loriia, P. O., **writer**
Los', E. I., **poet**
Lukin, I. B., **critic and film script writer**
Lupan, A. P., **author**
Lupan, A. P., **author and poet**
Lupsiakov, M. R., **author**
Lur'e, N. M., **writer**
Luzhanin, M., **author**
Lvov, M. D., **poet**
Lvov, S. L., **author, journalist and literary critic**
Lyiurov, A. A., **author**

M

Maasik, E., **author**
Maksimov, M. D. (Lipovich), **poet and author**
Maksimov, V. E., **author**
Maldonis, A. M., **author and poet**
Malikov, K. I., **poet and dramatist**
Malkin, A. S., **poet**
Maltinskii, K. I., **poet**
Maltsev, E. I. (Pupko), **author**
Mamedkhanly, E. G. o., **writer**
Mamlin, G. S., **author**
Mandelshtam, N. I., **author**
Mannur, S. F., **poet**
Margiani, R. A., **poet**
Margiev, K. I., **poet**
Markarian, M. E., **author**
Markov, A. I., **poet**
Markov, G. M., **writer and public figure**
Martianov, S. N., **author**
Martsinkiavichius, I. M., **author**
Martynov, L. N., **author**
Marukas, K. (Krasauskas, M. I.), **author**
Marynov, A. K., **poet**
Matevosian, G. I., **author**
Matsiavichius, I., **poet**
Matusovskii, M. L., **poet**
Matveeva, N. N., **poet**
Maulenov, S., **poet**
Mdivani, G. D., **playwright and screenwriter**
Medynskii, G. A., **writer**
Melesh, I. P., **author**
Meniuk, G. N., **author**
Merilaas, K. (Sang, K.), **author**
Mezhelaitis, E. B., **author**
Mezhirov, A. P., **poet**
Mikelinskas, I., **author**
Mikhailov, A. A., **literary critic**
Mikhailov, I. L., **poet**
Mikhailov, N. N., **writer**
Mikhalkov, S. V., **author**
Miller, B. I. (Miller, B., Riter, D., etc.), **author**
Minchkovskii, A. M. (Minchikovskii), **author**
Minko, V. P., **writer**
Mirmukhsin (Mirsaidov, M.), **author**
Mirzaev, A. T., **author**
Mirzagitov, A. M., **dramatist**
Mirzoev, G., **author**
Mordinov, N. E., **writer**
Moskovkin, V. F., **author**
Mozhaev, B. A., **author**
Mrevlishvili, M. N., **writer**
Mukhamedov, M., **author**
Mukhammadiev, F., **author**
Mukhtar, A., **author**
Muratov, I. L., **author**
Mushketik, I. M., **author**
Muslimov, M. N. (Marat, M.), **author**
Musrepov, G. M., **author**
Mustafin, G., **author**
Muzaev, N. D., **poet**

N

Nadiradze, K. (N.) G., **poet**
Nadzhmi, N. (Nazmutdinov, N. M.), **author**
Nagibin, I. M., **author**
Nagnibeda, M. (N. L.), **author**
Narmaev, M. B., **author**
Narovchatov, S. S., **author**
Natroshvili, G. K., **author and literary historian**
Naumenko, I. I., **author**
Navolochkin, N. D., **author**
Nekrasov, V. P., **author**
Niedre, I. I., **writer and literary scholar**
Niiazi, F., **author and journalist**
Niit, E. (Kross), **author and translator**
Nikitin, S. K., **author**
Nikolaev, S. N., **playwright**
Nikolaevskaia, E. M., **poet and translator**
Nikulkov, A. V., **author**
Nilin, P. F., **author**
Nishnianidze, S. G., **poet**
Noneshvili, I. E., **author**
Nosov, E. I., **author**
Novikov, I. G., **author**
Novikov, V. M., **poet**
Nurberdyev, P., **author**
Nuretdinov, Z. S. (Nuri Zaki), **author**
Nurme, M., **author**
Nurpeisov, A. K., **author**

O

Obukhova, L. A., **author**
Ocheretin, V. K., **author**

Odintsov, A. B. (Oder), **author**
Ognetsvet, E. (Kagan, E. S.), **author, poet and translator**
Okudzhava, B. S., **poet**
Oleinik, B. I., **author**
Olshanskii, I. G., **author and film script writer**
Omarov, S., **author**
Orlov, V. I., **author**
Ormanov, G., **poet**
Orovetskii, P. A., **author**
Osetrov, E. I., **author and literary critic**
Oshanin, L. I., **author and lyricist**
Osherovich, G. S., **author and poet**
Osin, D. D., **author**
Osipov, P. N., **writer**
Ostrovoi, S. G., **author and poet**
Ovalov, L. S. (Shapovalov), **author**
Ovanesian, R. K., **author and poet**
Ozerov, L. A. (Goldberg), **author, critic and translator**

P

Paderin, I. G., **author**
Pagirev, G. V., **author**
Pakalnis, A., **author**
Palman, V. I., **author**
Panchenko, P. E., **author**
Pankratov, I. I., **author**
Panov, N. N., **author**
Papaskiri, I. G., **author**
Parnov, E. I., **author**
Parve, R., **author**
Passar, A. A., **poet**
Paushkin, G. A., **author**
Pavlichko, D. V., **author and poet**
Pavlova, D. G. (Gilevich), **author**
Pavlova, M. K., **author and daramatist**
Perevalov, N. I. (Kolesnikov), **author and poet**
Permiak, E. A., **author**
Perventsev, A. A., **writer**
Peskov, V. M., **author and journalist**
Pidsukha, A. N., **author**
Piliar, I. E., **author**
Pliev, G. D., **poet and dramatist**
Pliev, I. K., **author**
Pochivalin, N. M., **author**
Podelkov, S. A., **author and poet**
Pogodin, R. P., **author**
Polevoi, B. N. (Kampov), **author**
Poliakova, N. M., **author and poet**
Polianker, G. I., **author**
Poltoratskii, V. V., **writer**
Pomozov, I. F., **author**
Ponomar, F. M., **author and poet**
Popov, L. A., **author**
Popov, S. A., **poet and author**
Popovskii, M. A., **author**
Porfirev, B. A., **author**
Pospelov, G. N., **literary critic**
Potseluev-Snegin, D. F., **author and state official**
Poznanskaia, M. A., **author and poet**
Priede, G. R., **dramatist**
Prigara, M. A., **author**
Prilezhaeva, M. P., **author**
Pristavkin, A. I., **author**
Privalov, B. A., **author and poet**
Promet, L. A., **author**
Proskurin, P. L., **author**
Prut, I. L., **dramatist**
Ptashnikov, I. N., **author**
Purs, L. (Strazdinsh, L. I.), **author**
Pysin, A. V., **author and poet**

R

Radzinskii, E. S., **dramatist**
Rafikov, S. K., **author**
Ragimov, S. G. o., **writer**
Rakhmanov, L. N., **author**
Ramazanov, G. Z., **author and poet**
Ramz, B. (Babadzhanov, R.), **poet and dramatist**
Rashidov, R. M., **author and poet**
Rashidov, S. R., **writer**
Rasputin, V. G., **author**
Rassadin, S. B., **literary critic**
Rasul, R. (Rzaev, R. I. o.), **author**
Ratner, E. I., **author**
Rebrin, P. N., **author**
Rekemchuk, A. E., **author**
Reshetnikov, L. V., **author**
Reshetov, A. E., **author and poet**
Revich, A. M., **poet and translator**
Riabokliach, I. A., **author**
Rinkule-Zemzare, D., **author**
Rochev, I. M., **author**
Rosliakov, V. P., **author**
Rozen, A. G., **author**
Rozhdestvenskii, R. I., **author and poet**
Rozov, V. S., **dramatist**
Rudenko, M. (N.) D., **author**
Rugoev, I. V., **author**
Rusetskii, A. S. (Burdelev), **author**
Rustam-zade, S. A.-A. o. (see Suliman, R.)
Rybakov, A. N., **writer**
Rytkheau, I. S., **author**
Ryvina, E. I., **author**
Rzhevskaia, E. M., **author**

S

Sachenko, B. I., **author**
Sadri, M. (Sadretdinov, M. K.), **author and poet**
Sadybekov, T., **author**
Safarov, N., **author**
Sakhnin, A. I., **author**
Sakse, A., **author**
Salynskii, A. D., **dramatist**
Samoilov, D. S., **poet**
Samsonov, S. A., **author and journalist**
Sangi, V. M., **author**
Sapozhnikov, V. K., **author**
Sartakov, S. V., **author**
Satybaldiev, A. S., **author and translator**
Sazhin, P. A., **author**
Seitov, S., **author, critic and literary historian**
Semenov, G. V., **author**
Semenov, I. S., **author**
Semin, V. N., **author**
Serebriakova, G. I., **author**
Shamiakin, I. P., **writer and state figure**
Shamov, A. S., **writer**
Shatirian, M. A., **writer**
Shcheglov, A. S., **writer**
Shefner, V. S., **writer**
Shestalov, I. N., **writer**
Shiian, A. I., **writer**
Shinkuba, B. V., **writer and state and public figure**
Shirobokov, S. P., **writer**
Shivaza, I. D., **writer**
Shklovskii, V. B., **writer, literary scholar and critic**
Sholokhov, M. A., **author**
Shomakhov, A. K., **writer**
Shortanov, A. T., **writer**
Shtein, A. P., **dramatist**
Sian-Belgin, K. B., **writer**
Simonov, K. M., **author**
Sivtsev, D. K., **writer**
Skrigan, J. (I.) A., **author**
Slavin, L. I., **author**
Slutskii, B. A., **author**
Slutskis, M., (Sluckis), **author**
Smerdov, A. I., **author**
Smirnov, V. A., **author**
Smirnova, M. N., **film director**
Sobko, V. N., **author**
Sofronov, A. V., **author**
Sokolov, V. V., **author**
Soloukhin, V. A., **author**
Stadiuk, I. F., **author**
Stel'makh, M. A., **author**
Stolper, A. B., **film director and screenwriter**
Stroeva, V. P., **film director and screenwriter**
Strugatskii, A. N., **author**
Strugatskii, B. N., **author**
Subbotin, V. E., **author**
Sulakauri, A. S., **writer**
Suleimenov, O. O., **author**
Suliman, R. (Rustamnasade, S. A.-A. o.), **author**
Surkov, A. A., **author**
Suseev, A. I., **poet**
Sverstiuk, E., **author**

T

Tank, M. (Skurko, E. I.), **author and poet**
Tarba, I. K., **writer**
Tarkovskii, A. A., **author and translator**
Tazhibaev, A., **author, poet and dramatist**
Tendriakov, V. F., **author**
Timonen, A. N., **author**
Tkach, D. V., **author**
Tkachenko, A. S., **author**
Tliusten, I. I., **poet**
Tokombaev, A. (Balka), **author**
Toktomushev, A., **author and poet**
Trifonov, I. V., **author**
Troepolskii, G. N., **author**
Tsagaraev, M. N., **writer**
Tsarukaev, A. I., **author**
Tsybin, V. D., **author**
Tsydendambaev, Ch.-D., **writer**
Tufan, K. F., **poet**

U

Uigun (Atakuziev, R.), **author and dramatist**
Ukachin, B. U., **author**
Ukhsai, I. G., **poet**
Urin, V. A., **author**

V

Vaarandi, D., **poet**
Vagabzade, B. M. o., **author**
Valeev, A. M., **author**
Validov, B. K. (Valid, B.), **author**
Vanags, I. P., **author**
Vanshenkin, K. I., **author and poet**
Vasilenok, E. P., **author**
Vasil'ev-Borogonskii, S. S., **poet**
Vasilevich, E. S., **author and poet**
Vatsietis, O. O., **author and poet**
Vauni, S. (Khachikian, S. V.), **poet, publicist and translator**
Vavilin, I. M., **author and poet**
Vechkanov, S. E., **author and poet**
Veiian, A. (Kalnach, D. G.), **author and poet**
Veinerman, K. A., **author and poet**
Veliev, A. K. o., **writer**
Veliugin, A. S., **author and poet**
Vergelis, A. A., **author and poet**
Vieru, G. P., **poet**
Viiding, P., **author and critic**
Vikulov, S. V., **author and poet**
Vilips, P. M., **author**
Vinokurov, E. M., **author and poet**
Vitka, V. (Krysko, T. V.), **author, poet and dramatist**
Vladimov, G. N. (Volosevich), **author**
Volodin, A. M. (Lifshits), **dramatist**
Vol'skii, V. F., **writer and literary scholar**
Vorobev, E. Z., **author**
Voronin, S. A., **writer**
Voron'ko, P. N., **author and poet**
Voronkova, L. F., **author**
Voskrekasenko, S. I., **author and poet**
Voznesenskii, A. A., **poet**
Vyrgan, I. A. (Vergun), **author and poet**
Vysheslavskii, L. N., **author, poet and literary historian**

Z

Zabashta, L. V., **author and poet**
Zabelin, I. M., **author**
Zabila, N. L., **writer**
Zadornov, N. P., **writer**
Zagrebelnyi, P. A., **author**
Zakharchenko, V. D., **author**
Zakhoder, B. V., **author and translator**
Zakrutkin, V. A., **author**
Zalygin, S. P., **author**
Zaritskii, A. A., **author**
Zarudnyi, N. I., **author and dramatist**
Zbanatskii, I. O., **author**
Zhapakov, N., **author**
Zharov, A. A., **author**
Zhdanov, N. G., **author**
Zheleznikov, V. K., **author**
Zhukov, V. S., **author**
Zhuravlev, T. K., **author**
Zhuravlev, V. A., **author**
Ziedonis, I. I., **poet**
Zigmonte, D. A., **author**
Zlobin, A. P., **author**
Znamenskii, A. D., **author**
Zolotarev, N. G., **writer**
Zorin, L. G., **dramatist**
Zubavin, B. M., **author**
Zverev, I. (Samdberg, I. I.), **author**

Artists

A

Aarela (Roos), R.I., **graphic artist**
Aavik, P.E., **painter**
Abakumtsev, N.I., **sculptor**
Abdrakhimov (Abdurakhimov), A.K., **sculptor**
Abdulkhalyk, A.A. o., **painter**
Abdullaev, A.A., **painter**
Abdullaev, A.B. o., **painter**
Abdullaev, L.A., **painter and designer**
Abdullaev, M.G. o., **painter**
Abdullaev, S.A., **painter**
Abdullaev, S.F., **painter**
Abdullaeva, G.G., **sculptor**
Abdulov, B.N., **painter**
Abdurakhmanov, I.A., **painter**
Abdurakhmanov, N.G. o., **painter**
Abegian, M.M., **painter and graphic artist**
Abelite, O.M., **graphic artist**
Abesadze, D.P., **sculptor**
Aboev, Z.-B.P., **graphic artist**
Abol, O.G., **painter and graphic artist**
Abolin (Abolinsh), M.A., **stage painter**
Abraamian (Abramian) A.S., **painter**
Abraamian (Abramian), E.L., **painter**
Abramavichius, K.V.A., **graphic artist**
Abramenkov, I.V., **painter**
Abramov, A.A., **sculptor**
Abramov, A.A., **sculptor**
Abramov, A.F., **painter**
Abramov, G.(I.)D., **painter**
Abramov, N.A., **painter**
Abramova, S.A., **sculptor**
Abseliamov, M., **painter**
Abugov, M.E., **sculptor**
Abutkova, N.V., **painter**
Adalian, R.M., **painter and graphic artist**
Adamkevich, V.V., **painter and graphic artist**
Adamov, E.B., **graphic artist**
Adamova, E.M., **painter**
Adamova, G.V., **artist in applied art**
Adamovich, S.F., **graphic artist and painter**
Adamovskaia-Andreeva, B.M., **stage designer**
Adashkevich, S.A., **sculptor**
Adomonis, I.A., **ceramic artist**
Advadze, V.N., **painter**
Adylov, N.I., **sculptor**
Adylov, V., **painter**
Adzhiashvili, I.I., **artist in applied art**
Aerova, B.M., **artist in applied art**
Afanasev, V.V., **sculptor**
Agababaev, O.I., **painter**
Agababaev, T.M.A. o., **artist in applied art**
Agadzhanian, N.G., **artist in applied art**
Agaev, I.I. o., **graphic artist**
Agamalova, L.N., **artist in applied art**
Agapov, I.A., **painter**
Agapova, N.N., **graphic artist**
Agasian, G.K., **painter and graphic artist**
Ageev, M.S., **painter**
Agriian, L.S., **painter and graphic artist**
Aide, B.A., **graphic artist**
Airapetian, R.A., **sculptor**
Airiev, A.A., **sculptor**
Aitiev, G.A., **painter and state official**
Aiusheev, E.E., **painter**
Aivazian, V.T., **graphic artist**
Aizman, N.S., **dramatist**
Akazeev, L.L., **graphic artist**
Akhloian, V.K., **sculptor**
Akhmedov, F.M., **painter and graphic artist**
Akhmedov, R., **painter**
Akhobadze, M.A., **painter**
Akhundov, N.S., **painter and graphic artist**
Akhverdov (Akhverdiev), S.I. o., **painter**
Akhvlediani, V.G., **painter**
Akimenko, V.I., **painter**
Akimushkin, N.N., **graphic artist**
Akimushkina, V.M., **sculptor**
Aknuni, P.A., **painter**
Akopian, G.A., **sculptor**
Akopian, G.S., **sculptor**
Akopian, K.V.(B.), **sculptor**
Akopian, N.S., **painter**
Akopian, S.A., **painter**
Akopian, S.O., **stage designer**
Akritas, A.G., **painter and graphic artist**
Aksenov, N.S., **painter**
Aksirov, M.A., **painter**
Akulov, G.V., **graphic artist**
Alberg, V.K., **sculptor**
Aleev, S.M., **painter**
Aleeva, T.A., **artist of monumental and decorative art**
Aleinikov, A.A., **graphic artist**
Alekhin, N.A., **painter**
Alekperov, K.D. o., **sculptor**
Aleksandr, N.M., **painter**
Aleksandrin, M.S., **stage designer**
Aleksandrov, A.A., **graphic artist**
Aleksandrov, B.A., **composer, conductor**
Aleksandrov, G.V., **film director**
Aleksandrov, I.V., **sculptor**
Aleksandrov, K.I., **sculptor**
Aleksandrov, N.P., **painter and graphic artist**
Aleksandrov, S.L., **painter**
Aleksandrov-Sedelnikov, S.K., **painter**
Aleksandrova, E.N., **graphic artist**
Aleksandrova, I.D., **sculptor**
Aleksandrova-Roslavleva, V.V., **sculptor**
Aleksandrova-Sanamiants, A.A., **sculptor**
Aleksanian-Shakhaziz, N.N., **graphic artist**
Alekseev, A.E., **painter and graphic artist**
Alekseev, A.I., **painter**
Alekseev, B.A., **actor**
Alekseev, B.V., **painter**
Alekseev, G.L., **graphic artist**
Alekseev, M.I., **graphic artist**
Alekseev, V.A., **stage designer and painter**
Alekseev, V.M., **graphic artist**
Alekseev, V.P., **painter**
Aleksidze, N.A., **sculptor**
Aleskerov, G.A.R. o., **painter and artist in applied art**
Alferov, A.P., **painter**
Alferov, S.P., **graphic artist and painter**
Aliamovskii, A.D., **graphic artist**
Aliev, S.K. o., **painter**
Alikov, S.V., **graphic artist**
Alimov, B.A., **graphic artist**
Alimov, G.V., **graphic artist**
Aliverdieva (Aliverdova), S.S. k., **sculptor**
Allendorf, M.A., **graphic artist**
Altaev, I.K., **painter and graphic artist**
Altaev, S.P., **painter**
Altshuller, M.S., **sculptor**
Alygulov, B., **designer**
Amangeldyev, A., **painter**
Amangeldyev, Ch., **painter**
Amankozhoev, K., **painter**
Amashukeli, E.D., **sculptor**
Amasian, P.G., **painter**
Ambardnishvili, O.A., **painter and sculptor**
Ambartsumian, R.E., **sculptor**
Ambitskii, I.I., **artist in applied art**
Ambitskii, M.I., **artist in applied art**
Ambokadze, E.M., **graphic artist and painter**
Ambraziunas, A.-V.I., **sculptor**
Ambur, L.M., **artist in applied art**
Amiian, I.A., **painter and graphic artist**
Aminov, A.U., **artist in applied art**
Amirkhanova, M.S., **painter**
Amurskii, S.I., **painter**
Ananev, I.A., **painter**
Ananev, M.A., **painter**
Ananikian, A.A., **painter**
Ananin, A.M., **painter and graphic artist**
Ananin, I.A., **sculptor**
Ananin, K.M., **painter**
Ananko, V.I., **sculptor**
Anatolev, N.A., **painter**
Anatskii, S.D., **sculptor**
Anderson, E.A., **graphic artist**
Andreenkov, V.E., **graphic artist**
Andreev, I.I., **painter**
Andreev, K.G., **painter and graphic artist**
Andreev, N.A., **painter**
Andreev, V.A., **painter**
Andreeva, K.G., **painter**
Andreeva (Rumiantseva), G.A., **painter**
Andreichenko, S.S., **sculptor**
Andreichuk, I.I., **sculptor**
Andrianov, G.I., **graphic artist**
Andrianov, P.N., **painter**
Andrienko, V.T., **painter**
Andriiuk, P.S., **painter and graphic designer**
Andrikanis, E.N., **cameraman and director**
Andriushchenko, T.T., **painter**
Andronikashvili, E.N., **graphic artist**
Andronov, N.I., **painter**
Andrushkevich, V.I., **graphic artist**
Andzhaparidze, V.I., **actress**
Andzhaparidze, Z.I., **singer (lyrical-dramatic tenor)**
Anfimov, L.V., **graphic artist**
Anikeenok, A.A., **painter**

Anikeev, M.K., **painter and graphic artist**
Anikeichik, A.A., **sculptor**
Anikushin, M.K., **painter**
Anisiforova, N.N., **artist**
Anisimov, A.S., **sculptor**
Anisimov, L.P., **graphic artist**
Aniskin, P.S., **painter**
Annakulieva, A., **singer (soprano)**
Annenkov, N.A., **actor**
Anni, A.A., **painter**
Anokhin, I.N., **painter**
Anosov, A.I., **graphic artist**
Anosov, V.M., **painter**
Ansulis, V.F., **artist in applied art**
Antadze, I.(D.)K., **stage director**
Antinite-Povilaitiene, D.R., **graphic designer**
Antipov, N.G., **painter**
Antipova, E.P., **painter**
Antonchik, M.V., **painter**
Antonenko, P.K., **painter and graphic artist**
Antonian, L.G., **painter**
Antonov, E.I., **painter**
Antonov, E.N., **painter and graphic artist**
Antonov, I.I., **graphic artist**
Antonov, S.G., **painter and graphic artist**
Antonova, G.A., **artist in applied art**
Antonovskii, V.A., **painter**
Anufriev, E.I., **painter**
Apanovich, V.A., **painter**
Apine-Zhiglevitsa, B.K., **painter**
Apollonov, I.G., **artist in applied art**
April, A.I., **painter and graphic artist**
Arabidze, V.V., **painter**
Arakcheev, B.V., **painter and graphic artist**
Arakelian, A.S., **painter**
Arakelian, G.G., **painter**
Aramian, G.L., **graphic artist**
Arapov, V.M., **painter**
Arbuzov, A.N., **dramatist**
Arbuzov, A.V., **painter and graphic artist**
Aref'ev, A.V., **designer of stage sets**
Aristakesova, S.L., **graphic artist**
Aristov, I.N., **painter**
Aristova, I.F., **painter**
Arkhipov, A.A., **sculptor**
Arkhipov, A.P., **painter**
Arkhipov, M.V., **painter**
Arkhipova, I.K., **singer (mezzo-soprano)**
Arkhipova (Arkhi), L.F., **painter**
Arkhipova (Surenkova), I.N., **sculptor**
Armalas, V.I., **graphic artist**
Arnautov, A.V., **painter**
Arndt, I.V., **architect, graphic artist and stage designer**
Arnshtam, L.O., **film director and screenwriter**
Arrak, E.I., **painter**
Arrak (until 1958: Soosuu), M.E., **painter**
Ars, I.L., **painter**
Arsenin, D.D., **graphic artist**
Arsentev, A.I., **graphic artist**
Arshakuni, Z.P., **painter**
Artamonov, V.E., **painter**
Artiukhin, M.F., **painter**
Artiukhov, D.F., **painter**
Artmane, V.(A.F.), **actress**
Arustamian, R.S., **painter and graphic artist**
Arutiunian, A.A., **sculptor**
Arutiunian, A.G., **composer**
Arutiunian, A.N., **painter**
Arutiunian, M.(A.)R., **painter**
Arutiunian, M.G., **painter**
Arutiunian, Z.(E.)A., **painter**
Arutiuniants (Arutiunian), A.S., **graphic artist**
Arzhanov, M.N., **painter**
Arzhekaev, B.A., **graphic artist**
Arzumanov, L.S., **painter and graphic artist**
Asatiani (Astiani), T.G., **sculptor**
Aseev, P.P., **designer**
Asetskii, L.P., **graphic artist**
Ashkenazi, I.M., **painter**
Ashrafmurad (Muradov), A.M. o., **painter**
Asitashvili (Asiatishvili), T.G., **graphic artist**
Aslamazian, M.A., **painter**
Aslanian, V.S., **painter**
Aslian, A.A., **graphic artist**
Asmanov, E.A.(M.A.), **graphic artist**
Asmanova, T.I., **graphic artist**
Astafev, L.E., **painter**
Astakhin, V.I., **graphic artist**
Astakhov, E.P., **painter**
Astapov, V.P., **sculptor**
Astashev, E.I., **graphic artist and painter**
Astaulov, S.P., **sculptor**
Astrakhantsev, A.A., **graphic artist**
Ataian, A.A., **painter**
Atamanov, L.K., **producer of animated cartoon films**
Atlanov, I.M., **graphic artist**
Atlantov, V.N., **painter**
Atoian, R.G., **painter**
Atsmanchuk, A.P., **painter**
Aunapuu, H., **painter and graphic artist**
Auza, L.K., **painter**
Avagimov, Z.A., **painter**
Avakian, G.S., **graphic artist**
Avakian, V.A., **sculptor**
Avakimian, A.B., **painter**
Avalishvili, B.G., **sculptor**
Avdeev, D.I., **painter**
Avdeev, I.K., **painter**
Avdeev, I.S., **graphic artist**
Avdeeva, T.T., **sculptor**
Avdyshev (Avedyshev) A.I., **graphic artist**
Avdysheva (Avedysheva) V.M., **painter**
Avedikov, N.V., **sculptor**
Averbakh, A.M., **stage designer**
Averbukh, I.A., **graphic artist**
Averbukh, L.V., **sculptor**
Averianov, B.I., **graphic artist**
Avetisian, A.A., **painter and graphic artist**
Avetisian, M.K., **painter and stage designer**
Avetisian, R.G., **painter**
Avgustovich, A.I., **painter and graphic artist**
Avidzba, K.T., **painter**
Avramchenko, O.N., **sculptor**
Avrutis, K.A., **graphic artist and painter**
Avsiian, O.(I.)A., **graphic artist and painter**
Avtokratova, K.M., **artist in applied art**
Avvakumov, B.P., **stage designer and graphic artist**
Azanov, N.F., **painter of miniatures**
Azarova, L.P., **artist in applied art**
Azeev, I.P., **sculptor**
Azgur, G.G., **painter**
Azgur, Z.I., **sculptor**
Azhgikhin, B.S., **painter and graphic artist**
Azhiev, U., **painter**
Azizian, G.M., **painter**
Azizian, S.K., **painter**
Azizian, T.L., **painter**

B

Babadzhanian, A.A., **composer and pianist**
Babaev, R.G. o., **graphic artist and painter**
Babaian, R.S., **graphic artist**
Babak, P.I., **painter**
Babakulov, A., **singer**
Babanova, M.I., **actress**
Babentsov (Babintsev), V.V., **painter and graphic artist**
Babian, R.A., **painter and stage designer**
Babikov, S.G., **painter**
Babitsin, V.A., **painter and graphic artist**
Babkov, S.F., **painter and graphic artist**
Babloian, G.T., **graphic artist**
Bachinin, N.I., **painter**
Bachurin, E.V., **graphic artist**
Badakva, V.P., **painter**
Badalian, G.B., **sculptor**
Badova, R.G., **sculptor**
Baenskaia, I.V., **artist in applied art**
Bagautdinov, R.M., **graphic artist**
Bagdasarian, G.G., **painter**
Bagdasarian, L.I., **painter**
Bagdasarian, S.I., **sculptor**
Bagdavadze, E.N., **painter**
Baglanova, R.T., **singer (soprano)**
Bagrov, G.K., **painter and graphic artist**
Baidashvili, M.A., **artist in applied art**
Bakaev, S.I., **painter**
Bakhlulzade, S.B. o., **painter**
Bakhtin, I.A., **artist in applied art**
Bakradze, D.K., **painter**
Bakuradze, K.N., **sculptor and artist in monumental and decorative art**
Balabanov, P.N., **painter**
Balabin, A.F., **painter**
Balabin, B.N., **painter**
Balakai, V.L., **painter**
Balanchivadze, A.M., **composer**
Balandiuk, M.V., **graphic artist**
Balanova, K.E., **painter**
Balasanian, S.A., **composer, teacher, musical and public figure**
Balashka, I.I., **painter**
Balavadze, I.I., **painter**
Baldano, N.G., **playwright, actor, director**
Baldina, I.M., **painter**
Balikov, I.E., **painter**
Balsy, E., **composer and public figure**
Baltrunas, I.F., **painter**
Balykin, A.P., **painter and graphic artist**
Balzhak (Bolzhak), R.I., **graphic artist**
Bandzeladze, A.M., **painter and graphic artist**
Banis, T.K., **painter and graphic artist**
Baraisov, M.S., **painter and graphic artist**
Barancheev, M.S., **painter and graphic artist**
Barannikov, N.A., **graphic artist**
Baranov, A.P., **painter and graphic artist**
Baranov, G.N., **sculptor**
Baranovskaia, O.P., **sculptor**

Barchenkov, N.I., **painter**
Bardadze, T.D., **painter**
Bardina (Shpirko), R.N., **graphic artist**
Barilkis, G.L., **painter**
Barinova, M.N., **sculptor**
Barisaite, L.I., **graphic artist**
Barkovskii, M.A., **painter**
Barpal, T.V., **artist in applied art**
Barshch, G.A., **painter**
Barskii, V.I., **painter**
Bartulis, P.K., **sculptor**
Barvenko, V.P., **painter and artist in monumental and decorative art**
Baryshnikov, R.V., **painter and graphic artist**
Baskakov, N.V., **painter**
Basov, I.M., **painter**
Basov, L.S., **sculptor**
Bassus, O.S., **sculptor**
Batalov, A.V., **film actor and director**
Batechko, I.V., **graphic artist**
Bauma, D.K., **painter and graphic artist**
Bauman (Baumanis), K.E., **sculptor and graphic artist**
Baumane, B.F., **painter**
Bazelian, I.F., **painter**
Bazelian, N.M., **painter**
Bazhbeuk-Melikian, L.A., **painter**
Bazhenov, G.I., **painter**
Bazhenova, E.I., **painter**
Bazilev, N.I., **painter and graphic artist**
Bazilevich, A.D., **graphic artist**
Bazilevskii, I.I., **painter**
Beda, D.M., **painter**
Bednoshei, D.P., **painter**
Bednov, V.A., **painter**
Beibutov, R.M. o., **singer (lyric tenor)**
Beliaev, E.M., **singer (lyric tenor)**
Beliaev, G.V., **sculptor**
Beliaev, S.S., **graphic artist and painter**
Belinnik, P.S., **singer (lyric tenor)**
Belostotskii, A.E., **sculptor**
Belov, I.V., **painter**
Belskaia, V.N., **sculptor**
Beltsov, G.I., **painter**
Belyi, B.M., **painter**
Belykh, A.P., **painter**
Bembel', A.O., **sculptor**
Bendinger, V.A., **graphic artist**
Berezin, V.S., **sculptor**
Berzin, B.A., **painter**
Berzins, L., **actress**
Beschastnov, I.V., **sculptor**
Beskakotov, V.V., **graphic artist**
Bessarabova, N.I., **ceramic artist**
Bevzenko, I.P., **painter and graphic artist**
Beziuk, L.L., **sculptor**
Bieshu, M.L., **singer**
Blanter, M.I., **composer**
Bogatyrev, A.V., **composer**
Bogatyrev, M.G., **painter**
Bogdanas, K.A., **sculptor**
Bogdanov, G.V., **painter**
Bogdanov, M.A., **film artist**
Bogdesko, I.T., **graphic artist**
Bogoliubov, N.I., **actor**
Boiko, V.E., **sculptor**
Bolduman, M.P., **actor**
Borisov, A.F., **actor**
Borisova, I.K., **actress**
Borodai, V.Z., **sculptor**
Bozhko, N.G., **graphic artist**
Briede, A.I., **sculptor**
Brusilovskii, E.G., **composer**
Brusilovskii, M.S., **graphic artist**
Bublev, V.P., **sculptor**
Bugrimova, I.N., **circus artist and animal trainer**
Bukeeva, K., **actress**
Bukovskii, L.V., **sculptor**
Burkhanov, A.B., **actor and director**
Burkhanov, M.M., **composer and public figure**
Burkhanov, S., **actor**
Burmistrova, N.M., **actress**
Butiai, V.D., **sculptor**
Bychkov, S.A., **painter and graphic artist**
Bykova, E.I., **chess player**
Bystritskaia, E.A., **actress**

C

Chaikovskii, B.A., **composer**
Chavdar, E.I., **singer (coloratura soprano) and teacher**
Cheban, T.S., **singer (soprano)**
Chervoniuk, E.I., **singer (bass)**
Chirkov, B.P., **actor**
Chkheidze, R.D., **motion-picture director**
Chkoniia, L.G., **singer (lyric-coloratura soprano)**
Chukhral, G.N., **motion-picture director**

D

Dal'skii, V.M., **actor**
Danchenko, A.G., **graphic artist**
Daneliia, G.N., **film director**
Dan'kevich, K.F., **composer and conductor**
Darienko, D.T., **actress**
Davydova, V.A., **singer (mezzo-soprano) and teacher**
Davydova-Medene, L.M., **portrait sculptor**
Deregus, M.G., **painter and graphic artist**
Deregus, M.M., **painter**
Deregus, N.M., **sculptor**
D'iakonov, N.M., **playwright and director**
Dimitriadi, O.A., **conductor**
Diordiev, E.I., **actor**
Dobrzhanskaia, L.I., **actress**
Doronina, T.V., **actress**
Dotsenko, N.P., **actress**
Dovboshinskii, D.D., **painter**
Dovgan, B.S., **sculptor**
Dubinovskii, L.I., **sculptor**
Dudinskaia, N.M., **ballerina**
Dveirin, A.B., **graphic artist**
Dzhamanova, R.U., **singer (soprano)**
Dzhanibekian, G.D., **actor**
Dzhaparidze, U.M., **painter and graphic artist**
Dzigan, E.L., **film director**
Dziuban, I.F., **graphic artist**

E

Efimov, B.E., **graphic artist**
Efremov, O.N., **actor and stage director**
Egorov, E.P., **graphic artist and painter**
Egorov, I.P., **motion-picture director**

F

Fadeechev, N.B., **ballet dancer**
Faintsimmer, A.M., **film director**
Faiziev, L.A., **film director**
Fazylova, T.F., **actress and singer (soprano)**
Freindlikh, B.A., **actor**
Furnike, I.D., **ballet dancer**

G

Gabashvili, D.A., **graphic artist, painter and sculptor**
Gabe, M.R., **graphic artist**
Gabitashvili, D.D., **painter**
Gabrielian, V.T., **graphic artist**
Gabyshev, L.M., **painter**
Gadaskina, D.G., **artist in applied art**
Gadzhiev, A.D.I. o., **composer**
Gadzhieva, B.A.B. k., **graphic artist**
Gaev, G.P., **painter**
Gagarin, I.P., **painter**
Gaidai, I.I., **artist in monumental and decorative art**
Gaiduk, V.G., **graphic artist**
Gaivanenko, A.P., **painter**
Galaganov, I.A., **painter**
Galakhov, N.N., **painter**
Galents, A.S., **painter**
Galitskii, R.N., **painter**
Galkov, R.A., **artist in monumental and decorative art**
Galkus, I.I., **graphic artist**
Galochkin, V.A., **sculptor**
Galperin, I.A., **graphic artist**
Ganiev, A.A., **sculptor**
Gaponenko, T.G., **painter**
Garanko, E.V., **sculptor**
Garin, E.P., **actor and stage manager**
Garokalns, V.E., **painter**
Garusova, V.L., **sculptor**
Gashinskii, A.E., **actor**
Gasparian, G.M., **singer (soprano)**
Gatin, S.Z., **painter**
Gaumigs, T.I., **sculptor**
Gavdzinskii, A.S., **painter**
Gavrilenko, G.I., **graphic artist**
Gavriliuk, N.(M.)S., **painter**
Gavrilov, E.G., **graphic artist**
Gazhenko-Zlotnik, A.T., **sculptor**
Gazizova, M.G., **painter**
Gedmantaite-Galkuvene, V.A., **graphic artist**
Genere, M.G., **artist in applied art**
Georgievskaia, A.P., **actress**
Giatsintova, S.V., **actress and director**
Gilels, E.G., **pianist**
Gladkyi, V.I., **painter**
Glikman, G.N., **graphic artist and sculptor**
Glushchuk, F.T., **graphic artist**
Gnatiuk, D.M., **singer (baritone)**
Gnatiuk, O.R., **graphic artist**
Gogoleva, E.N., **actress**
Golovkina, S.N., **ballet dancer**
Golozubov, V.V., **graphic artist**
Goncharov, A.D., **graphic artist, primarily for books**
Gordnaov, V.V., **cameraman**
Grachov, O.V., **painter**
Grebennik, V.S., **painter**

Grechanyk, V.V., **sculptor**
Greeku, M.G., **painter**
Grigor'ev, S.A., **painter**
Grigorovich, I.N., **ballet dancer and choreographer**
Gritsai, A.M., **painter**
Gritsenko, N.O., **actor**
Gruzberg, S.B., **graphic artist**
Gudiashvili, L.D., **painter and graphic artist**
Guliaev, I.A., **singer (lyric baritone)**
Gutman, G.P., **sculptor**
Gvelesiani, S.F., **painter**

I

Iaunushan, A.I., **actor and film director**
Ignatovich, B.V., **photography expert**
Il'in, B.F., **actor**
Ilina, L.A., **graphic artist**
Il'inskii, I.V., **actor**
Ilko, I.I., **painter**
Il'skaia, F.S., **actress**
Iltner, E.K., **painter**
Iokubonis, G.A., **sculptor**
Ird, K.K., **director, theater and public figure**
Isakova, N.S., **singer (mezzo-soprano)**
Ishanturaeva, S.A., **actress**
Ishmurat, R.F., **playwright and public figure**
Iskenderov, A. o., **director and actor**
Ivanov, A.P., **singer (baritone)**
Ivanov, I., **composer**
Ivanov, K.K., **conductor**
Ivanov-Vano, I.P., **film director and artist**
Izmailova, G.B., **ballerina**

K

Kabalevskii, D.B., **composer, teacher and public figure**
Kadochnikov, P.P., **film actor**
Kaiumov, M.K., **cameraman and director of documentary films**
Kalinovskaia, V.F., **ballerina**
Kalnroze, V., **landscape painter**
Kalnyn', E.F., **painter**
Kaluga, H.M., **ceramist**
Kaluga, V.I., **graphic artist**
Kaminnoi, N.I., **graphic artist**
Kandelaki, V.A., **singer (bass-baritone)**
Kapitan, A.A., **graphic artist**
Kapitan, L.A., **graphic artist**
Kaplan, S.S., **graphic artist**
Kaplanian, R.N., **theater director**
Kapp, E., **composer and public figure**
Karaffa-Korbut, S.P., **graphic artist**
Karamian, E.A., **motion-picture director**
Karandash, **circus artist**
Karas, I.I., **painter**
Karieva, B.R., **ballet dancer**
Karm, K.I., **actor**
Karmaliuk, P.P., **singer (baritone)**
Katrushenko, I.G., **graphic artist**
Kazarinova, N.N., **actress**
Kazimirova, E.G., **actress**
Kedainis, J.J., **sculptor**
Kerbel', L.E., **sculptor**
Kevorkov, S.A., **film director**
Khabibullaev, Z.N., **artist**
Khaikin, B.E., **conductor**
Khalitov, F.I., **actor**
Khanov, A.A., **actor**
Kharlamova, V.B., **actress**
Khodulov, D.F., **actor**
Khokhriakov, V.I., **actor and film director**
Khrennikov, T.N., **composer**
Khushvakhtov, K.D., **painter**
Khutsiev, M.M., **motion-picture director**
Kibal'nikov, A.P., **sculptor**
Kimantaite, K.M., **actress and film director**
Kirichenko, O.S., **painter**
Kirichenko, S.A., **painter**
Kiselitsia, O.A., **painter**
Klimova, A.I., **actress**
Klokov, V.M., **sculptor**
Klova, V.J., **composer**
Klychev, I.N., **painter**
Knebel', M.O., **director and teacher**
Kniazeva, L.N., **actress**
Kniazik, A.B., **sculptor**
Kobizeva, K.S., **sculptor**
Kocharian, S.A., **variety performer**
Kogan, L.B., **violinist**
Kokin, M.A., **painter**
Kolmanovskii, E.S., **composer**
Kolpakova, I.A., **ballerina**
Kondrashin, K.P., **conductor**
Koretskii, V.B., **graphic and poster artist**
Kormashov, N.I., **painter**
Korzhev, G.M., **painter**
Kovalenko, V.K., **painter**
Kovalev, A.A., **sculptor**
Kozelkova, E.G., **actress**
Kozhamkulov, S., **actor**
Kozhukhov, A.D., **graphic artist**
Kozlovskii, I.S., **singer (lyric tenor)**
Krainiev, V., **pianist**
Kremer, G., **violinist**
Kriuchkov, N.A., **actor**
Kruglikova, E.D., **singer (soprano)**
Kruglov, G.G., **ceramic artist**
Krupenina, L.I., **ballet dancer**
Krylovich, V.N., **actor**
Ktorov, A.P., **actor**
Kuchas, A.A., **graphic artist**
Kudriashova, K.K., **singer**
Kugach, I.P., **painter**
Kuiukova, D., **actress**
Kulidzhanov, L.A., **motion-picture director**
Kulieva, M., **singer (lyric soprano)**
Kulmamedov, A., **actor and theater director**
Kumanchenko, P.V., **actress**
Kurbanova, O.A., **actress**
Kurilich, M.V., **painter**
Kusenko, O.I., **actress**
Kuuzik, T.(D.I.), **singer**
Kuz'mina, E.A., **motion-picture actress**
Kuzminskis, I.M., **graphic artist**
Kuznetsov, M.A., **actor**
Kuznetsov, M.P., **actor**
Kuznetsov, V.V., **painter**
Kydykeeva, B., **actress**

L

Labenok, L.V., **painter**
Ladynina, M.A., **actress**
Lapin, V.I., **painter and graphic artist**
Laptev, K.A., **singer**
Lavrovskii, M.L., **ballet dancer**
Lebedev, E.A., **actor**
Leonov, P.V., **ceramist**
Lepeshinskaia, O.V., **ballet dancer**
Levchuk, T.A., **director**
Levitan, I.B.
Levitska, I.G., **painter**
Line, V.M., **actress**
Lingis, I.I., **ballet dancer and choreographer**
Linkhovoin, L.L., **singer (bass)**
Lishchiner, M.I., **painter**
Lisitsian, P.G., **singer (baritone)**
Listov, K.I., **composer**
Liubeznov, I.A., **actor**
Liubimov, I.P., **actor and theater director**
Livanova, T.N., **musicologist**
Livshits, S.N., **painter**
Logvin, I.D., **painter**
Lopukhov, A.M., **painter**
Luchenko, E.D., **painter**
Lutskevich, I.P., **painter**

M

Machavariani, A.D., **composer**
Mackevičius, V.S., **painter**
Maiboroda, G.I., **composer**
Maikanova, S., **actress**
Maisenberg, O., **pianist**
Makarev, L.F., **actor, stage director and dramatist**
Makarova, I.V., **actress**
Makarova, T.F., **actress**
Makaseev, B.K., **cameraman**
Makatukha, V.I., **painter**
Maksimova, E.S., **ballerina**
Mambeev, S.A., **painter**
Manastyrskii, V.A., **painter**
Manuilova, O.M., **sculptor**
Martinson, S.A., **actor**
Masal'skaia, N.N., **actress**
Matrosov, V.E., **sculptor**
Mavlianova, K.M., **singer (soprano)**
Medvedkin, A.I., **film director**
Meitus, I.S., **composer**
Melik-Avakian, G.G., **film director**
Melikhov, G.S., **painter**
Melnichenko, V.V., **painter**
Menovshchikova, N.I., **ballerina**
Messerer, A.M., **ballet dancer**
Miasnikov, G.A., **actor**
Miasnikova, L.V., **singer (mezzo-soprano)**
Mikhailenok, L.L., **sculptor**
Milashkina, T.A., **singer**
Miltinis, I.I., **director and actor**
Min'kovskii, A.Z., **conductor**
Miroshnichenko, E.S., **singer**
Mirzoian, E.M., **composer**
Mochal'skii, D.K., **painter and graphic artist**
Moiseenko, E.E., **painter**
Moiseev, I.A., **ballet dancer and choreographer**
Monakhov, V.V., **cameraman and director**
Mordiukova, N.V., **film actress**
Mravinskii, E.A., **conductor**
Mshvelidze, S.M., **composer**
Mubariakov, A.K., **actor and stage director**

Mukhin, V.I., **sculptor**
Muradova, B.S.M., **actress**
Myl'nikov, A.A., **painter**

N

Nadezhdina, N.S., **ballet dancer and choreographer**
Nalbandian, D.A., **painter**
Narimanbekov, T.F., **painter and graphic artist**
Naruzetskii, M.I., **sculptor**
Nasretdinova, Z.A., **ballet dancer**
Nasyrova, K., **singer (soprano)**
Naumov, V.N., **film director**
Naurzbaev, K.I., **sculptor**
Nebylitskii, B.R., **cinematographer**
Nechiporenko, M.S., **painter**
Nefed, V.I., **theater scholar, educator and playwright**
Nenado, V.S., **graphic artist**
Neprintsev, I.M., **painter and graphic artist**
Neroda, G.V., **sculptor**
Nikulin, I.V., **circus artist und carpet clown**
Nirod, F.F., **stage designer**
Nisskii, G.G., **painter**
Noreika, V.L., **singer (tenor)**
Novikov, A.G., **composer**
Novikov, A.S., **sculptor**
Nurmukhammedov, N.-B.D.-E., **painter and art scholar**

O

Obraztsov, S.V., **puppet artist**
Ochiauri, I.A., **sculptor, master of chased metalwork**
Oistrakh, I.D., **violinist**
Oreshnikov, V.M., **painter**
Orlov, P.P., **graphic artist**
Oshurkov, M.F., **cinematographer**
Osipov, A.N., **painter**
Ossovskii, P.P., **painter**
Ovanesova, A.A., **film director**
Ovanisian, N.M., **singer**
Ovcharenko, I.P., **sculptor**
Ozerov, I.N., **film director**

P

Pakhmutova, A.N., **composer**
Pakul', E.I., **singer (coloratura soprano)**
Papanov, A.D., **actor**
Pavlovskaia, A.G., **graphic artist**
Petrenko, V.V., **sculptor**
Petrov, I.I., **singer (bass)**
Pinchuk, V.B., **sculptor**
Plisetskaia, M.M., **ballet dancer**
Pluchek, V.N., **stage director**
Pokrovskii, B.A., **operatic director and teacher**
Polkovnikov, V.I., **film director**
Polonyk, V.P., **sculptor**
Polosin, E.M., **actor**
Ponomarenko, E.P., **actor**
Ponomarev, N.A., **graphic artist**
Popov, I.A., **painter**
Popov, O.K., **circus artist**
Potapova, E.M., **ballet dancer**
Primachenko, M.A., **master of decorative graphics**
Prisiazhniuk, A.A., **actor**
Prudkin, M.I., **actor**

R

Raikin, A.I., **estrada (variety stage) entertainer**
Raizman, I.I., **film director**
Rakhlenko, L.G., **actor and director**
Rakhlin, N.G., **conductor**
Randviir, T.A., **ballet dancer**
Rapoport, V.A., **cinematographer**
Ravenskikh, B.I., **stage director**
Redlikh, V.P., **stage director**
Reize, M.O., **singer (bass)**
Reshetnikov, F.P., **painter and graphic artist**
Rikhter, S.T., **pianist**
Rodionov, L.A., **sculptor**
Romadin, N.M., **painter**
Romanitskii, B.V., **actor and director**
Roshal', G.L., **teacher and film director**
Rozhdestvenskii, G.N., **conductor**
Rozhdestvenskii, K.I., **designer**
Rudenko, B.A., **singer**
Rudenko, L.A., **singer (mezzo-soprano)**
Rudenko, L.V., **chess player and economic planner**
Rybnov, A.V., **choral conductor**
Ryzhov, N.I., **actor**

S

Sabirova, M.A., **ballet dancer**
Saifiddinov, S.S., **composer**
Sakh'ianova, L.P., **ballerina**
Salakhov, T.T., **painter and graphic artist**
Saltykov, A.A., **motion picture director**
Salynskii, A.D., **playwright**
Samodur, S.S., **puppet-theater actor and director**
Samoilov, E.V., **theater and film actor**
Samoilova, T.E., **film actress**
Sanaev, V.V., **actor**
Sats, A.J., **actor and director**
Sats, N.I., **theater director**
Sazandarian, T.T., **singer (mezzo-soprano)**
Sebris, K.K., **actor**
Semenov, I.M., **graphic artist**
Serdiuk, A.I., actor
Serebrianyi, I.A., **painter**
Sergeev, A.T., **singer (bass)**
Sergienko, R.M., **singer**
Serkebaev, E.B., **singer**
Shafran, D.B., **violoncellist**
Shchedrin, R.K., **composer**
Shelenkov, A.V., **cameraman**
Shil'nikovskii, E.P., **artist**
Shishigin, F.E., **stage director**
Shishko, S.F., **painter**
Shmakov, F.I., **actor**
Shmarinov, D.A., **graphic artist**
Shmyga, T.I., **operetta performer**
Shpinel', I.A., **motion-picture designer**
Shtogarenko, A.I., **composer**
Shtokolov, B.T., **singer (bass)**
Shtyrbul, K.A., **actor**
Shukshin, V.M., **film director and actor**
Shul'zhenko, K.I., **singer**
Shumiatskaia, E.I., **master of applied art**
Shveitser, M.A., **motion-picture director**
Simeonov, K.A., **conductor**
Simonov, E.R., **stage director**
Slonimskii, S.M., **composer, pianist and musicologist**
Smirnov, B.A., **artist**
Smirnova, L.N., **film actress**
Smoktunovskii, I.M., **actor**
Sobolev, F.M., **theater and film director**
Solntseva, I.I., **actress and film director**
Solov'ev-Sedoi, V.P., **composer and public figure**
Solov'ianenko, A.B., **singer (lyric-dramatic tenor)**
Soloviov, I.V., **ballet dancer**
Speranskii, E.V., **puppet-theater actor**
Stasiunas, I.A., **singer (baritone)**
Stepanova, A.O., **actress**
Stomma, Z.F., **actor**
Struchkova, R.S., **ballerina**
Strzhel'chik, V.I., **actor**
Sveshnikov, A.V., **choral director**
Svetlanov, E.F., **composer, conductor and pianist**
Sviridov, G.(I.)V., **composer**

T

Takishvili, S.D., **actress**
Tammur, I.A., **actor and stage director**
Telzhanov, K.T.B., **painter**
Tikhonov, V.V., **film actor**
Timofeeva, N.V., **ballerina**
Tishchenko, B.I., **composer**
Tiurenkov, A.A., **sculptor**
Tkachenko, N.A., **singer (soprano)**
Tkhapsaev, V.V., **actor**
Tolubeev, I.V., **actor**
Tomskii, N.V., **sculptor**
Tovstonogov, G.A., **stage director**
Tsarev, M.I., **actor**
Tsereteli, Z.K., **artist**
Tsintsadze, S.F., **composer**
Tsvirko, V.K., **painter**
Tsydynzhapov, G.T., **actor, stage director and playwright**
Tulegenova, B.A., **singer (coloratura soprano)**
Tyshler, A.G., **stage designer, painter and graphic artist**

U

Uchitel', E.I., **cameraman and motion-picture director**
Ulanova, G.S., **ballerina**
Ulianov, M.A., **actor**
Urgalkin, A.K., **actor**
Ustinova, T.A., **choreographer**
Uzhvii, N.M., **actress**

V

Vagin, A.G., **painter and graphic artist**
Vagin, V.V., **graphic artist**
Vainberg, M.S., **composer**
Vainius, S.A., **composer, pianist and teacher**
Vaivadaite-Surgailiene, A.A., **graphic designer**
Valamat-zade, G.R., **ballet dancer, choreographer and director**

Valdmanis, M., **graphic artist**
Valnere, R.H., **painter**
Varpakhovskii, L.V., **director**
Varygin, S.P., **painter and graphic artist**
Vasadze, A.A., **actor and director**
Vasilev, E.A., **sculptor**
Vasilev, O.B., **painter and graphic artist**
Vasilev, V.I., **painter**
Vasilev, V.S., **graphic artist**
Vasilev, V.V., **ballet dancer**
Vasil'ko, V.S., **director and actor**
Vekilova, L.M., **ballerina**
Vereiskii, O.G., **graphic artist**
Verich, I.G., **sculptor**
Vertinskaia, L.V., **graphic artist and film actress**
Vertsberg, A.I., **sculptor**
Viatkin, A.V., **painter and graphic artist**
Videnskii, M.S., **graphic artist and painter**
Vieru, I.D., **painter and graphic artist**
Vigilianskaia, N.V., **graphic artist**
Viidalepp, A.F., **painter and graphic artist**
Viidalepp, H.A., **painter**
Viies, E.T., **sculptor**
Viko, R.E., **painter**
Vilenskii, Z.M., **portrait sculptor**
Vilks, E.R., **graphic artist**
Vishnevskaia, G.P., **singer (soprano)**
Vivatenko, L.A., **painter**
Vladimirov, I.N., **graphic artist**
Vladimirov, V.P., **graphic artist**
Vladislavskii, V.V., **graphic artist**
Vlasova, A.N., **painter and graphic artist**
Vnodchenko, I.F., **painter**
Voitsekhovskii, A.K., **painter**
Volchkov, N.N., **actor**
Volkov, A.M., **painter**
Volkov, V.A., **painter**
Volkovinskaia, Z.V., **painter and graphic artist**
Volovik, V.P., **sculptor**
Vorobev, F.D., **painter and graphic artist**
Voronovich, A.P., **actress**
Vovk, A.I., **painter**
Vovkushevskii, R.I., **painter and artist in monumental and decorative art**
Voznesenskii, B.M., **graphic artist and painter**
Vronskii, V.I., **choreographer**
Vrubel, A.V., **sculptor**
Vyrodova-Gotie, V.G., **painter**
Vyrzhikovskii, E.I., **painter**

Z

Zabashta, V.I., **painter**
Zagurskaia, G.N., **actress**
Zaitsev, E.A., **painter**
Zakharov, G.F., **graphic artist**
Zakharov, R.V., **choreographer and director**
Zakhidova, L., **ballerina**
Zardarian, O.M., **painter**
Zaretskii, V.I., **painter**
Zarian, R.V., **theatrical scholar**
Zarin', M.O., **composer**
Zarkhi, A.G., **film director and script writer**
Zaslavskaia, S.A., **textile artist**
Zhakov, O.P., **actor**
Zharov, M.I., **actor and director**
Zherdzinskii, E.F., **painter**
Zhiganov, N.G., **composer**
Zhnykrup, O.L., **sculptor**
Zhukov, B.V., **painter**
Zhurakovskii, V.P., **painter**
Zinovev, N.M., **painter**
Zueva, A.P., **actress**
Zykina, L.G., **singer**

Military Leaders

Agaltsov, F.A.
Akhromeev, S.F.
Andrianov, V.I.
Arkhipov, V.S.
Asmolov, A.N.
Baidukov, G.F.
Batov, P.I.
Begel'dinov, T.I.
Beliakov, A.V.
Belov, A.I.
Bobylev, S.A.
Brandys, A.I.
Chistiakov, M.N.
Diadkin, I.A.
Druzhinin, M.I.
Efimov, A.N.
Efremov, V.S.
Egorov, G.M.
Evstigneev, K.A.
Fedorov, A.F.
Fedorov, E.P.
Gareev, M.G.
Getman, A.L.
Golovachev, P.I.
Gorchakov, P.A.
Govorov, V.L.
Gribkov, A.I.
Grizodubova, V.S.
Gromov, M.M.
Gulaev, N.D.
Gusakovskii, I.I.
Iazov, D.T.
Ivanov, S.P.
Ivashutin, P.I.
Kamanin, N.P.
Kamozin, P.M.
Karatsiupa, N.F.
Karitskii, K.D.
Kasatonov, V.A.
Kazakov, K.P.
Kazakov, M.I.
Kharlamov, N.M.
Khomulo, M.G.
Khrenov, A.F.
Klimov, I.D.
Kokkinaki, V.K.
Konstantinov, A.U.
Kovtunov, A.V.
Kozhedub, I.N.
Kozlov, M.M.
Krasovskii, S.A.
Krivda, F.
Kuleshov, P.N.
Kulishev, O.F.
Kungurtsev, E.M.
Kurochkin, P.A.
Kuznetsov, M.V.
Lashchenko, P.N.
Lavrinenkov, V.D.
Liashchenko, N.G.
Litovtsev, D.I.
Losik, O.A.
Luchinskii, A.A.
Luganskii, S.D.
Maiorov, A.M.
Managarov, I.M.
Mares'ev, A.P.
Maslov, V.P.
Mazurenko, A.E.
Mednikov, I.S.
Mikhailichenko, I.K.
Mikhailin, V.V.
Mikmailin, V.V.
Molokov, V.S.
Nagornyi, N.N.
Naumenko, I.A.
Nedbailo, A.K.
Odintsov, M.P.
Okunev, V.V.
Orel, A.E.
Osipov, V.N.
Panteleev, I.A.
Pavlov, I.F.
Pavlovskii, I.G.
Peredelskii, G.E.
Plotnikov, P.A.
Pokryshkin, A.I.
Poluboiarov, P.P.
Popkov, M.D.
Popkov, V.I.
Prokhorov, A.N.
Radzievskii, A.I.
Rakov, V.I.
Rechkalov, G.A.
Riazanov, A.K.
Rudenko, S.I.
Rybalchenko, S.D.
Samsonov, F.A.
Sandalov, L.M.
Savitskii, E.I.
Sergeev, N.D.
Shavrov, I.E.
Shilin, A.P.
Shkadov, I.N.
Shustko, L.S.
Silchenko, N.K.
Sivtorov, V.V.
Skomorokhov, N.M.
Skripko, N.S.
Sliusarenko, Z.K.
Smirnov, A.S.
Smirnov, E.I.
Smirnov, N.I.
Sredin, G.V.
Starinov, I.G.
Stepanenko, I.N.
Stoliarov, N.G.
Sysoev, V.S.
Tenishchev, I.I.
Varennikov, V.I.
Vasiagin, S.P.
Vaupshasov, S.A.
Vinogradov, N.I.
Vodopianov, M.V.
Volkotrubenko, I.I.
Vorobev, I.A.
Vorozheikin, A.V.
Vostrukhov, V.I.
Zakharov, S.E.
Zarubin, L.K.
Zheltov, A.S.

Ecclesiastical Dignitaries

Aleksii (V.A. Konoplev)
Aleksii (A.M. Ridiger)
Anatolii (E.V. Kuznetsov)
Antonii (O.I. Vakarik)
Antonii (A.S. Melnikov)
Antonii (A.M. Savgorodnii)
Antonii (A. Blum)
Damaskin (A.I. Bordry)
Damian (D.G. Marchuk)
Feodosii (M.N. Dikun)
Feodosii (I.I. Protsiuk)
Filaret (K.V. Vakhromeev)
Filaret (M.A. Denisenko)
Gedeon (A.N. Dokukin)
German (G.E. Timofeev)
Gleb (I.I. Smirnov)
Hiob (D.I. Tyvoniuk)
Iakov (Akkersdaik)
Ioann (D.A. Razumov)
Ioann (I.M. Synchev)
Ioann (K.N. Vendland)
Ioasaf (V.S. Ovsiannikov)
Ionafan (I.M. Kopalovich)
Iosif (I.M. Savrash)
Irenei
Irenei (I.V. Zuzemil)
Irenei (I.P. Seredni)
Iuvenalii (S.A. Tarasov)
Iuvenalii (V.K. Poiarkov)
Kassian (S.N. Iaroslaviskii)
Khrizostom (G.F. Martishkin)
Kliment (A.A. Perestiuk)
Leonid (L.K. Poliakov)
Leontii (L.F. Bondar)
Leontii (I.A. Gudimov)
Makarii (L.N. Svistun)
Maksim (B. Krokha)
Mark, Bishop of Ladoga
Meliton (M.D. Solovev)
Melkhisedek (V.M. Lebedev)
Mikhail (M.A. Chub)
Mikhail (M.N. Mudiugin)
Nikodim (B.G. Rotov)
Nikodim (N.S. Rusnak)
Nikolai (E.N. Iurik)
Nikolai (N.V. Kutepov)
Nikolai (P. Sajama)
Nikon (N.V. Fomichev)
Panteleimon (S.A. Mitrukovskii)
Pimen (S.M. Izvekov)
Pimen (P.V. Khmelevskii)
Piotr (P. l'Huillier)
Pitrim (K.V. Nechaev)
Platon (V.P. Udovenko)
Savva (A.P. Babinets)
Serafim (V.I. Rodinov)
Serapion (N.S. Fadeev)
Sergii (S.V. Petrov)
Simon (S.M. Novikov)
Varfolomei (N.N. Gandorovskii)
Varlaam (A.T. Iliushchenko)
Varnava (V.V. Kedrov)
Vasilii (Krivoshein)
Viktorin (V.V. Beliaev)
Vladimir (V.S. Kotliarov)
Vladimir (V.M. Sabodan)

Jurists

Aksenenok, G.A.
Bannikov, S.G.
Bondar, A.G.
Bratus', S.N.
Chekharin, E.M.
Chkhikvadze, V.M.
Farberov, N.P.
Gorkin, A.F.
Iakimenko, A.N.
Kerimov, D.A.-A.
Koretskii, V.M.
Kozhevnikov, F.I.
Luts, L.A.
Orlov, A.K.
Tunkin, G.I.
Zagorodnikov, N.K.
Zivs, S.L.

Journalists

Lukovets, A.I.
Medvedev, I.N.
Prokushev, I.L.
Somov, V.S.

Cosmonauts

Beregovoi, G.T.
Bykovskii, V.F.
Demin, L.S.
Egorov, B.B.
Eliseev, A.S.
Feoktistov, K.P.
Filipchenko, A.V.
Gorbatko, V.V.
Khrunov, E.V.
Klimuk, P.I.
Kubasov, V.N.
Lazarev, V.G.
Lebedev, V.V.
Leonov, A.A.
Makarov, O.G.
Nikolaev, A.G.
Popovich, P.R.
Rukavishnikov, N.N.
Sarafanov, G.V.
Sevstianov, V.I.
Shatalov, V.A.
Shonin, G.S.
Titov, G.
Volynov, B.V.

Architects

Asaris, G.K.
Balchunas, V.-K.M.
Baranov, N.V.
Barshch, M.O.
Belopolskii, I.B.
Chechulin, D.N.
Chekanauskas, V.A.
Chkhenkeli, I.N.
Demin, V.A.
Dobrovol'skii, A.V.
Driba, D.I.
Elizarov, V.D.
Fomin, I.I.
Isakovich, G.G.
Israelian, R.S.
Ivanov, V.M.
Konstantinov, M.P.
Korol', V.A.
Kurdiani, A.G.
Levin, L.M.
Loveiko, I.I.
Nikolaev, I.S.
Orlov, G.M.
Pokrovskii, I.A.
Polianskii, A.T.
Posokhin, M.V.
Priimak, B.I.
Rozhin, I.E.
Rubanenko, B.R.
Sheshial'gis, K.K.
Shimkus, V.-K.J.
Speranskii, S.B.
Stamo, E.N.
Suslikov, I.P.
Ullas, N.N.
Useinov, M.A.
Valiuškis, G.S.
Varaksin, V.N.
Voinov, A.P.
Zaborskii, G.V.
Zankovich, V.P.

Athletes

Bobrov, V.M. **(soccer and hockey)**
Bogdanov, A.I. **(sports marksman)**
Botvinnik, M.M. **(chess player)**
Brumel', V.N. **(field athlete)**
Chaikovskaia, E.A. **(figure skater)**
Chernyshev, A.I.
Chukarin, V.I. **(gymnast)**
Davydov, V.S. **(ice hockey player)**
Dumbadze, N.I. **(track athlete)**
Goncharenko, O.G. **(coach)**
Granatkin, V.A. **(public figure)**
Grishin, E.R. **(ice skater)**
Isakova, M.G.
Ivanov, V.N. **(oarsman of single shells)**
Kadyrov, G.F. **(motor racer)**
Kapitanov, V.A.
Katulin, A.Z. **(coach and teacher)**
Kharlamov, V.B. **(ice hockey player)**
Kizimov, I.M. **(equestrian sportsman)**
Kolchin, P.K.
Kolchina, A.P.
Konovalenko, V.S. **(hockey player)**
Korkiia, O.M.
Korolev, N.F. **(boxer)**
Kulakova, G.A. **(skier)**
Kuzin, V.S.
Kuz'kin, V.G. **(hockey player)**
Kuperman, I.I. **(checker)**
Kurentsov, V.G.
Loktev, K.B. **(ice hockey)**

Retired

Abdulchakov, N. **(Dec 1979)**
Afanasenko, E.I. **(Sept 1978)**
Afanasev, P.V. **(Jan 1980)**
Afteniuk, G.T.
Aimamedov, A. **(Nov 1979)**
Aksomitas, A.I. **(Aug 1982)**
Akulintsev, V.K. **(Aug 1979)**
Alekseev, A.I. **(June 1980)**
Allakhverdiev, T.A.G. o. **(Dec 1982)**
Andreev, A.N.
Andrianov, S.N. **(April 1975)**
Ani, V.F. **(March 1979)**
Areshkovich, V.D. **(March 1981)**
Arpentev, V.A. **(Feb 1979)**
Arutiunian, G.T. **(Oct 1981)**
Arutiunian, N.K. **(July 1975)**
Asadov, I.M. **(Dec 1977)**
Ashimbaev, T. **(Dec 1981)**
Aslanov, A.I. **(Dec 1979)**
Ataev, K. **(Dec 1979)**
Avdiukevich, L.I. **(Nov 1980)**
Azizov, S.G. o. **(March 1980)**
Badaev, M.B. **(Nov 1979)**
Bagramian, G.A. **(Dec 1977)**
Bagratuni, G.R. **(Dec 1979)**
Bandura, N.I. **(Jan 1976)**
Baranov, N.V. **(1981)**
Baranovskii, A.M. **(Aug 1979)**
Barauskas, A.B. **(Nov 1982)**
Basilov, D.P. **(July 1979)**
Bastin, I.G. **(Oct 1976)**
Baturov, T.I. **(Oct 1980)**
Bavrin, V.I. **(Jan 1981)**
Begmatova, S. **(Feb 1980)**
Belokolos, D.Z. **(July 1976)**
Belous, N.A. **(Sept 1980)**
Belukha, N.A. **(Jan 1978)**
Blum, A.I. **(Nov 1979)**
Bobo-zade, I. **(Dec 1975)**
Borodin, A.M. **(April 1981)**
Brolish, I.V. **(Feb 1978)**
Burka, M.I. **(May 1975)**
Buss, A.A. **(1979)**
Chemodurov, T.N. **(Dec 1979)**
Chikhladze, M.F. **(Feb 1976)**
Dairov, M.D. **(Sept 1982)**
Demidov, N.V. **(Aug 1978)**
Diukarev, S.P. **(Sept 1978)**
Dolgov, P.S. **(May 1979)**
Dzhaparidze, E.M. **(April 1979)**
Dzhavakhishvili, G.D. **(Dec 1975)**
Eisen, F.M. **(July 1980)**
Elibaev, A.A. **(Jan 1981)**
Erei, A.I. **(Feb 1980)**
Galimzhanova, L.G. **(April 1981)**
Gamkharashvili, A.G. **(June 1979)**
Geldiashvili, Z.V. **(Feb 1980)**
Gerasimov, K.M. **(May 1974)**
Gladyrevskii, A.V. **(Jan 1980)**
Golovchenko, I.K. **(June 1982)**
Goncharov, L.B. **(Jan 1981)**
Gorelik, L.I. **(Feb 1977)**
Goriachev, F.S. **(Dec 1978)**
Goriunov, D.P. **(Oct 1978)**
Gorodovikov, B.B. **(Dec 1978)**
Grigorii **(March 1977)**
Grushetskii, I.S. **(Oct 1976)**
Guseinov, K.A. o. **(March 1980)**
Iakovlev, M.D. **(Sept 1982)**
Ibragimov, A.I. o. **(Jan 1981)**
Ilin, I.S. **(April 1979)**
Ilnitskii, I.V. **(Dec 1980)**
Ishchenko, D.T. **(Jan 1976)**
Ishkov, A.A. **(Feb 1979)**
Iurgens, F.E. **(Feb 1980)**
Iusupov, G.G. **(March 1979)**
Izmailov, I.A.A. o. **(Nov 1978)**
Kakharov, A. **(July 1973)**
Kalchenko, N.T. **(Feb 1976)**
Kantseliaristov, P.S. **(Aug 1981)**
Kashcheev, N.V.
Kazanskii, A.N. **(Aug 1981)**
Kazantsev, B.A. **(April 1981)**
Ketiladze, K.E. **(May 1979)**
Khitrov, S.D. **(Dec 1982)**
Kiknadze, Sh.D.
Kleiza, V.-A.I. **(Nov 1980)**
Klemenchuk, A.P. **(March 1979)**
Klimov, A.I. **(March 1980)**
Klychev, A.-M. **(Dec 1978)**
Kochinian, A.E. **(Nov 1974)**
Kokarev, A.A. **(Dec 1980)**
Kokhonov, F.L. **(Sept 1978)**
Kolotukha, I.I. **(Dec 1982)**
Konakbaev, K.D. **(June 1980)**
Kornev, K.S. **(Sept 1982)**
Korytkov, N.G. **(Dec 1978)**
Kospanov, S.K.
Kostousov, A.I. **(Dec 1980)**
Kozyr', P.P. **(July 1977)**
Kron, V.A. **(Dec 1977)**
Kuchava, M.I. **(Aug 1976)**
Kuchkin, S.A. **(July 1978)**
Kulatov, T.
Kulikov, I.P. **(Nov 1981)**
Kurbanov, M.K. o. **(July 1979)**
Kuris, P.M. **(Nov 1980)**
Kurmakaev, Z.A. **(Dec 1980)**
Kusainov, S.K.
Leonova, V.V. **(Dec 1981)**
Lesechko, M.A. **(Oct 1980)**
Liivik, F.I. **(March 1982)**
Losev, K.S. **(Nov 1982)**
Luzgin, V.I. **(Jan 1983)**
Magradze, M.K. **(July 1982)**
Maksarev, I.E. **(Aug 1978)**
Mamedov, A.A. o. **(Jan 1980)**
Mamedova, D. **(Oct 1979)**
Maniushis, I.A. **(Jan 1981)**
Marakov, G.N. **(Dec 1979)**
Martinaitis, V.P. **(Aug 1976)**
Martsiniuk, I.V. **(Dec 1976)**
Matulionis, A.A. **(Aug 1978)**
Mazurov, K.T. **(Dec 1978)**
Mette, A.A. **(July 1979)**
Mikhaelian, G.A. **(June 1978)**
Mikhail (Voskresenskii, D.A.) **(July 1975)**
Molochko, N.P. **(Aug 1980)**
Mukhitdinova, K.K. **(June 1982)**
Muzhitskii, A.M. **(Jan 1973)**
Mzhavanadze, V.P. **(Dec 1972)**
Nasriddinova, I.S. **(July 1979)**
Niiazbekov, S.B. **(Dec 1978)**
Nikulkin, I.P. **(Aug 1980)**
Novikov, V.N. **(Dec 1980)**
Novoselov, E.S. **(Dec 1980)**
Nyrtsev, M.P. **(Dec 1975)**
Omarova, Z.S. **(Nov 1980)**
Paatashvili, V.A. **(July 1976)**
Pakhlevanian, G.A. **(May 1981)**
Palvanova, B. **(Dec 1979)**
Petrosian, G.S. **(April 1978)**
Pork, A.P. **(June 1982)**
Pozhidaev, D.P. **(Jan 1980)**
Putsilev, S.M. **(Sept 1978)**
Rakhimova, A.M. **(Oct 1978)**
Raudsalu, V.J. **(Nov 1976)**
Romanov, N.N.
Rozenko, P.A. **(Jan 1979)**
Rustamov, N.G. o. **(March 1980)**
Sadykova, V.S. **(Jan 1981)**
Safarmamedov, A. **(March 1980)**
Safonov, D.F. **(Nov 1977)**
Saiko, G.I. **(Dec 1976)**
Samolevskii, V.N. **(Jan 1980)**
Samsonov, G.E. **(Dec 1978)**
Sanov, N.M. **(Feb 1981)**
Selianinov, O.P. **(Jan 1979)**
Semenikhin, G.V. **(July 1975)**
Shamsudinov, F.Sh. **(Dec 1978)**
Shelest, P.E. **(May 1973)**
Shevel, G.G. **(Nov 1980)**
Shinkuba, B.V. **(Oct 1978)**
Shishkin, A.A. **(Aug 1982)**
Shkorupeev, I.S. **(March 1977)**
Shpakovskii, L.K. **(Nov 1980)**
Sibul, R.I. **(Oct 1979)**
Sinitsyn, I.F. **(Oct 1980)**
Sitchikhin, V.K. **(Oct 1976)**
Sizov, P.K. **(Sept 1977)**
Skachko, N.A. **(Feb 1978)**
Sliusarenko, P.K. **(Sept 1978)**
Smirnov, S.A. **(June 1979)**
Stafiichuk, I.I.
Starovskij, V.N.
Startsev, A.K. **(Aug 1981)**
Suetov, I.G. **(Jan 1980)**
Suntsov, N.K.
Sveshnikov, M.N. **(Oct 1976)**
Tavamaishvili, G.D. **(March 1977)**
Tepandi, I.I. **(Dec 1979)**
Terekhov, K.P. **(May 1978)**
Titov, F.E. **(1982)**
Tokarev, S.P. **(Dec 1977)**
Tolstikov, V.S. **(June 1982)**
Topchiev, A.R. **(March 1980)**
Trebukhin, F.V. **(Oct 1979)**
Trotsenko, Z.P. **(Oct 1980)**
Udaltsov, I.I. **(Sept 1979)**
Umuraliev, M. **(Sept 1982)**
Vaniaev, N.A. **(June 1982)**
Vazalinskas, V.M. **(Nov 1980)**
Veselov, I.A. **(May 1978)**
Vildzhiunas, I.I. **(June 1976)**

Viniukov, I.K. **(April 1975)**
Voltovskii, B.I. **(April 1978)**
Voronov, G.I. **(May 1973)**
Vorontsov, A.I. **(March 1979)**
Zakharov, V.A. **(Jan 1975)**
Zakhvatov, V.E. **(Nov 1982)**
Zhiliakov, G.V. **(May 1978)**
Zhurin, N.I.

Deceased

Abbasaliev, S.K. o. **(23/5/1982)**
Abdalin, A.S. **(23/10/1978)**
Abdulla Sabir (Abdullaev, S.) **(1972)**
Abdurakhmanov, F.G. o. **(1971)**
Absaliamov, A.S. **(1979)**
Agababian, G.G. **(9/10/1977)**
Agapov, B.N. **(1973)**
Agish, S. (Agishev, S.I.) **(1973)**
Akhmed Dzhamil (Dzhemilzade, A.S. o.) **(24/9/1977)**
Akhmetov, K.A. **(26/11/1976)**
Aleksandrov, A.N.
Aleksandrov, P.S. **(16/11/1982)**
Alekseev, M.P. **(Sept 1981)**
Alekseevskii, E.E. **(1/1/1979)**
Andrianov, K.A. **(13/3/1978)**
Andropov, Iu.V. **(9/2/1984)**
Antokolskii, P.G. **(1978)**
Antosiak, G.F. **(12/2/1977)**
Apryatkin, S.S. **(12/2/1977)**
Artobolevskii, I.I. **(21/9/1977)**
Arzumanian, G.A. **(28/11/1976)**
Asadullaev, A.-B.B. o. **(6/3/1981)**
Astreika, A. (P.), **(23/8/1978)**
Atarov, N.S. **(1978)**
Babadzhanian, A.K. **(1/11/1977)**
Bagramian, I.K. **(21/9/1982)**
Baialinov, K. **(3/9/1979)**
Bairamov, M.B. o. **(6/1/1981)**
Baklanov, G.M. **(20/10/1979)**
Bardzimov, G.N. **(27/6/1977)**
Barto, A.L. **(3/4/1981)**
Bauzha, A. **(20/12/1977)**
Belevich, A.P. **(11/4/1978)**
Belik, P.A. **(12/6/1980)**
Belonozhko, S.E. **(12/12/1978)**
Belov, N.V. **(March 1982)**
Berg, A.I. **(9/7/1979)**
Berggolts, O.F. **(1975)**
Beshchev, B.P. **(27/5/1981)**
Bilialov, K. **(6/12/1977)**
Bilodid, I.K. **(Sept 1981)**
Biriukov, N.Z. **(1966)**
Blokhintsev, D.I. **(27/1/1979)**
Bobokalonov, N.K. **(21/3/1976)**
Boltinskii, V.N. **(1977)**
Borisenko, N.M. **(8/5/1980)**
Brezhnev, L.I. **(10/11/1982)**
Brodskii, A.I. **(1969)**
Brovka, P.U. **(24/3/1980)**
Budennyi, S.M.
Budker, G.I. **(4/7/1977)**
Bukhrashvili, Sh.E. **(22/3/1979)**
Burashnikov, B.F. **(8/9/1980)**
Burdin, D.I. **(1978)**
Butoma, B.E. **(11/7/1976)**
Chekmarev, A.P. **(1975)**
Cherepnin, L.V. **(12/6/1977)**
Chiriaev, G.I. **(9/5/1982)**
Chuev, A.V. **(20/11/1976)**
Chuikov, V.I. **(18/3/1982)**
Churaev, V.M. **(March 1982)**
Danilov, A.I. **(27/11/1980)**
Danilov, S.P. **(1978)**
Darienko, P.S. **(5/7/1976)**
Davitaia, F.F. **(29/6/1979)**
Dediushko, V.I. **(1973)**
Dementev, P.V. **(14/5/1977)**
Dikushin, V.I. **(12/1/1979)**
Doenin, V.N. **(23/2/1977)**
Dolidze, Sh.A. **(23/1/1976)**
Drozdenko, V.I. **(30/11/1982)**
Dzhumabaev, M.A. **(15/8/1980)**
Dzotsenidze, G.S. **(5/5/1976)**
Eger, K.O. **(20/1/1981)**
Engibarov, L.G. **(1972)**
Eshtokin, A.F. **(19/8/1974)**
Ezhov, I.M. **(22/4/1980)**
Faiko, A.M. **(1978)**
Fedin, K.A. **(15/7/1977)**
Fedorov, E.K. **(30/12/1981)**
Fomichev, G.F. **(24/1/1978)**
Frank, G.M. **(10/10/1976)**
Furtseva, E.A. **(24/10/1974)**
Gadoev, D. **(8/11/1978)**
Gadzhibekov, S.I. o. **(1974)**
Gafurov, B.G. **(12/7/1977)**
Gamsakhurdia, K.S. **(1975)**
Gavrilov, I.G. **(1973)**
Geidarov, A.N. o. **(29/6/1978)**
Georgadze, M.P. **(23/11/1982)**
Georgiev, A.V. **(9/4/1976)**
Gerasimova, V.A. **(1970)**
Germogen (G.V.O.) **(27/1/1980)**
Glushkov, V.M. **(Jan 1982)**
Gorchakov, A.K. **(11/9/1974)**
Gorinov, A.V. **(1975)**
Gorlach, M.V. **(11/4/1978)**
Grechko, A.A. **(26/4/1976)**
Grishmanov, I.A. **(4/1/1979)**
Griva, Z. **(June 1982)**
Gross, E.F. **(1972)**
Grosul, I.S. **(29/9/1976)**
Grushevoi, K.S. **(10/2/1982)**
Gudaitis-Guziavichius, A. **(1969)**
Gurzhii, I.A. **(1971)**
Gusenkov, P.V. **(26/1/1975)**
Iakubovskii, F.B. **(27/3/1975)**
Iakubovskii, I.I. **(30/11/1976)**
Ibraimov, S. **(4/12/1980)**
Iliashenko, K.F. **(21/4/1980)**
Iliushin, S.V. **(9/2/1977)**
Inozemtsev, N.N. **(12/8/1982)**
Isbakh, A.A. (Isbakh-Bakhrakh) **(1977)**
Iusifzade, K.I. o. **(June 1982)**
Ivanchenko, M.L. **(22/10/1978)**
Ivanov, V.E. **(24/12/1980)**
Ivchenko, A.G. **(1968)**
Ivnev, R. (Kovalev, M.A.)
Kakhkhar, A. **(1968)**
Kalmykov, V.D. **(22/3/1974)**
Kamenskii, V.A. **(1975)**
Karryev, A.K. **(27/12/1976)**
Kasian, V.I. **(26/6/1976)**
Kalesnik, S.V. **(13/9/1977)**
Kaplin, A.S. **(3/8/1979)**
Karpov, V.B. **(Aug 1977)**
Kasianenko, V.G. **(18/1/1981)**
Kasumov, I.A. o. **(20/4/1981)**
Kazakov, V.A. **(17/2/1981)**
Keldysh, M.V. **(24/6/1978)**
Kempe, M.I. **(1974)**
Khachaturian, A.I. **(1/5/1978)**
Khaikin, B.E. **(1978)**
Khalilov, M.R. o. **(9/10/1976)**
Khodzhaev, F. **(27/4/1975)**
Khokhlov, R.V. **(8/8/1977)**
Kibrik, E.A. **(20/8/1978)**
Kilchevskii, N.A. **(14/6/1979)**
Kiselev, T.I. **(11/1/1983)**
Klimov, A.P. **(7/9/1979)**
Klimovskoi, A.A. **(9/10/1978)**
Knipper, L.K. **(1974)**
Kochetov, V.A.
Kondratev, V.N. **(Feb 1979)**
Kononova, M.S. **(11/6/1977)**
Konstantinov, P.A. **(1973)**
Kopelian, E.Z. **(1975)**
Korol', V.A. **(28/5/1980)**
Korostylev, V.N.
Kosmatov, L.V. **(1977)**
Kosygin, A.N. **(18/12/1980)**
Kovalchuk, N.M. **(3/9/1981)**
Kozhevnikov, E.F. **(29/4/1979)**
Kozlov, G.A. **(Sept 1981)**
Krakhmalev, M.K. **(17/9/1977)**
Krylov, A.P. **(11/5/1981)**
Kulakov, F.D. **(17/7/1978)**
Kuleshov, A.A. **(4/2/1978)**
Kunin, V.N. **(1976)**
Kurbanov, Iu.R. **(31/1/1981)**
Lavrentev, M.A. **(15/10/1980)**
Lazarenko, E.K. **(Feb 1979)**
Le, I. (Moisia, I.L.) **(9/10/1978)**
Lebedev, E.I. **(18/2/1974)**
Lebedev, S.A. **(1974)**
Lebedev, V.D. **(3/12/1978)**
Lelashvili, M.M. **(14/4/1976)**
Leontev, A.N. **(1979)**
Leontovich, M.A. **(April 1981)**
Lipatov, V.V. **(May 1979)**
Lobov, S.M. **(12/7/1977)**
Los', E.I. **(July 1977)**
Lukianenko, P.P. **(1973)**
Lukianov, B.N. **(30/12/1979)**
Lukonin, M.K. **(4/8/1976)**
Lukovets, A.I. **(8/11/1977)**
Luriia, A.R. **(14/8/1977)**
Lysenko, T.D. **(20/11/1976)**
Malik, I.A. **(11/2/1980)**
Malyshko, A.S. **(1970)**
Mamakaev, M.A. **(1973)**
Mardashev, S.R. **(1974)**
Mardzhanishvili, K.K. **(13/2/1981)**
Markov, K.K. **(Sept 1980)**
Masherov, P.M. **(4/10/1980)**
Melezh, I.P. **(9/8/1976)**
Melnikov, N.V. **(7/12/1980)**
Merkurev, V.V. **(1978)**
Mikoian, A.I. **(21/10/1978)**
Minasiants, T.S. **(14/12/1982)**
Mirchink, M.F. **(1976)**
Momotenko, N.P. **(13/1/1981)**
Morozov, A.A. **(14/6/1979)**
Mstislav (D.I. Volonsevich) **(2/4/1978)**
Mukhamedov, A.M. **(19/12/1977)**
Muratov, M.V. **(Sept 1982)**
Murtazaev, K. **(25/5/1982)**

Musatov, A.I.
Musin, R.M. **(2/10/1982)**
Muskhelishvili, N.I. **(15/7/1976)**
Mustafaev, M.G. o. **(7/10/1979)**
Nabiev, I.M. **(1/12/1980)**
Nadezhdina, N.S. **(Oct 1979)**
Nasrullaev, N.I. o. **(26/4/1978)**
Neiman, L.R. **(1975)**
Nemchina, S.S. **(15/5/1978)**
Nesmeianov, A.N. **(17/1/1980)**
Nesterov, A.I. **(1979)**
Nikolaev, A.V. **(13/2/1977)**
Nikolai (O.N. Sychkowskii) **(3/6/1981)**
Noneshvili, I.E. **(6/10/1980)**
Nosov, N.N.
Novikov, P.S. **(1975)**
Obreimov, I.V. **(Dec 1981)**
Odintsov, M.M. **(March 1980)**
Okladnikov, A.P. **(Nov 1981)**
Oleinik, S.I. **(Jan 1982)**
Omelianovskii, M.E. **(1/12/1979)**
Oparin, A.I. **(21/4/1980)**
Orlov, S.S. **(1977)**
Ormanov, G. **(3/9/1978)**
Orozbekov, T. **(24/10/1978)**
Orudzhev, S.A. **(20/4/1981)**
Otarov, K.S. **(1974)**
Ots, G.K. **(1975)**
Pachomov, A.F. **(1973)**
Palladii (G.M. Kaminskii) **(6/6/1978)**
Palladii (Sherstiannikov) **(23/4/1976)**
Panch, P. (Panchenko, P.I.) **(1/12/1978)**
Panova, V.F. **(3/3/1975)**
Paputin, V.S. **(28/12/1979)**
Paranuk, M.S. **(1970)**
Pataridze, Z.A. **(5/6/1982)**
Peive, I.V. **(13/9/1976)**
Peresypkin, I.T. **(12/10/1978)**
Pervomaiskii, L.S. (Gurevich, I.S.) **(9/12/1973)**
Petrov, B.N. **(23/8/1980)**
Petrov, G.N. **(1977)**
Petrushevskii, I.P. **(1977)**
Pigalev, P.F. **(15/4/1975)**
Piliugin, N.A. **(2/8/1982)**
Pimenov, P.T. **(27/5/1980)**
Platon (P.E. Lobankov) **(27/10/1975)**
Pogrebniak, P.L. **(1/2/1980)**
Polevai, B.N. (Kampov) **(12/7/1981)**
Poltoratskii, A.I. **(16/3/1977)**
Popliovkin, T.T. **(25/2/1977)**
Popov, P.I. **(16/9/1977)**
Pospelov, P.N. **(23/4/1979)**
Postovskii, I.I. **(Nov 1980)**
Prokkonen, P.S. **(18/7/1979)**
Psurtsev, N.D. **(9/3/1980)**
Radzievskii, A.I. **(Sept 1979)**
Randakiavichius, A.B. **(10/12/1978)**
Rasulov, D. **(4/4/1982)**
Revutskii, L.N. **(30/3/1977)**
Riabikov, V.M.
Rimkus, A.S. **(20/12/1982)**
Rotmistrov, P.A. **(April 1982)**
Rozhdestvenskii, V.A. **(1977)**
Rtveliashvili, J.G. **(29/8/1976)**
Rudenko, R.A. **(23/1/1981)**
Rudnev, K.N. **(13/8/1980)**
Sakse, A. **(2/3/1981)**
Seitakov, B. **(1979)**
Serafim (V.M. Nikitin) **(22/4/1979)**
Serbin, I.D. **(15/2/1981)**
Sevriukov, V.K. **(27/2/1977)**
Shashin, V.D. **(23/3/1977)**
Shchetinin, S.N. **(6/12/1975)**
Shevardnadze, I.A. **(8/11/1978)**
Shtemenko, S.M. **(23/4/1976)**
Shuleikin, V.V. **(25/4/1979)**
Shumauskas, M.I. **(28/5/1982)**
Sidorenko, A.V. **(23/3/1982)**
Simonov, K.M. **(28/8/1979)**
Skobeltsyn, D.V.
Skochilov, A.A. **(6/7/1977)**
Slazhnev, I.G. **(9/9/1978)**
Smirnov, A.A. **(14/2/1982)**
Smirnov, B.A. **(Dec 1982)**
Smolych, I.K. **(26/8/1976)**
Snechkus, A.I.
Sochava, V.B. **(1978)**
Sokolov, I.Z. **(1/10/1982)**
Sozykin, A.G. **(18/4/1982)**
Surganov, F.A. **(26/12/1976)**
Stepanian, A.A. **(20/11/1979)**
Strakhov, N.M. **(July 1978)**
Sudets, V.A. **(6/5/1981)**
Sultankhodzhaev, A.S. **(22/6/1979)**
Suslov, M.A. **(25/1/1982)**
Suslov, N.I. **(5/9/1982)**
Tamm, I.E.
Tarasov, A.M. **(27/6/1975)**
Telkov, N.T. **(4/2/1979)**
Tikhonov, N.S. **(8/2/1979)**
Tikunov, V.S. **(16/7/1980)**
Timakov, V.D. **(21/6/1977)**
Timofeevskii, A.D.
Titov, G.A. **(19/10/1980)**
Titov, V.N. **(9/9/1980)**
Toigombaev, D. **(22/11/1976)**
Tolmadzhev, K.G. **(14/1/1977)**
Tolstov, S.P. **(1976)**
Trifonov, I.V. **(April 1981)**
Tronov, M.V.
Tsvigun, S.K. **(19/1/1982)**
Tukhikian, N.M. **(17/8/1978)**
Tursun-zade, M. **(24/9/1977)**
Tychinin, V.S. **(2/7/1977)**
Ushanev, F.S. **(30/4/1981)**
Vader, A.P. **(25/5/1978)**
Vasliaev, V.A. **(4/10/1980)**
Veimer, A.T. **(3/3/1977)**
Veinshtein, O.L.
Veksler, V.I. **(1966)**
Vekua, I.N. **(2/12/1977)**
Veniamin, N. **(14/10/1976)**
Vereshchagin, L.F. **(20/2/1977)**
Vernov, S.N. **(Sept 1982)**
Vigdorova, F.A. **(17/8/1965)**
Vilde, I. (Drobiazko, D.D.) **(Nov 1982)**
Vilks, E. (Latsis, E.M.) **(1976)**
Vinogradov, A.P.
Vinogradov, I.M.
Vishnevskii, A.A. **(1975)**
Vlasiuk, P.A. **(18/3/1980)**
Vlasov, A.A. **(1975)**
Voronovskii, N.A. **(28/11/1973)**
Zakolupin, N.A. **(27/6/1975)**
Zavoiskii, E.K. **(1976)**
Zavriev, K.S. **(1978)**
Zelenov, D.A. **(24/4/1979)**
Zemskov, I.N. **(5/4/1982)**
Zhukov, A.B. **(Sept 1979)**
Zhukov, D.A. **(24/3/1981)**
Zhukov, E.M. **(11/3/1980)**
Ziemelis-Kadikis, G.K. **(19/2/1978)**
Zinovev, N.M. **(July 1979)**
Zolotov, I.M. **(10/8/1975)**
Zorev, N.N. **(1978)**
Zverev, I. (Samdberg, I.I.)
Zverev, S.A. **(17/12/1978)**

Relieved and Dismissed

Abaeva, N.B. **(relieved 13/6/81)**
Abakirov, A. **(relieved 29/9/79)**
Abdukarimov, I. **(relieved 13/12/79)**
Abdullaev, G.M. o. **(dismissed 17/6/79)**
Abdullaeva, Sh.A. **(relieved 12/3/78)**
Abdurakhmanov, G. **(relieved 26/7/78)**
Abilova, G.A. k. **(relieved 28/3/80)**
Adigesalov, B.M.S. o. **(relieved 24/1/81)**
Airian, G.A. **(relieved 3/7/79)**
Aitbaev, T.S. **(relieved 1/12/82)**
Akhmedov, I.S. **(relieved 4/4/76)**
Akimov, K.A. o. **(relieved 22/1/80)**
Akinfiev, V.I. **(relieved 17/1/80)**
Akopian, K.A. **(dismissed 28/1/79)**
Aliev, F.G. **(relieved 10/7/77)**
Almakaev, P.A. **(relieved 1980)**
Aloviddinov, U.A. **(relieved 26/3/80)**
Altukhov, N.G. **(relieved 10/3/82)**
Amerkhanova, Zh.A. **(relieved 25/3/82)**
Amosova, E.G. **(relieved 1980)**
Ananiashvili, P.A. **(dismissed 8/5/81)**
Annaeva, Sh. **(relieved 25/3/80)**
Anspok, K.S. **(relieved 23/4/80)**
Arakelian, A.G. **(relieved 21/3/80)**
Arutiunian, S.G. **(relieved April 1978)**
Arutiunian, U. **(relieved 7/10/77)**
Arzumanian, R.A. **(relieved 24/1/81)**
Asamov, S. **(relieved 28/6/75)**
Asankulov, D. **(relieved 29/4/78)**
Astaikin, I.P. **(relieved 1980)**
Atadzhanov, A.R. **(relieved 4/7/79)**
Avanian, S.O. **(relieved 13/12/75)**
Avgitov, L.G. **(relieved 19/7/81)**
Babaev, M.A. o. **(dismissed 9/2/79)**
Bagdasarian, S.B. **(relieved 3/7/81)**
Baigushev, F.K. **(relieved 1/12/76)**
Bakradze, V. **(relieved 18/2/78)**
Balpanov, S. **(relieved 20/2/82)**
Barkar, A.A. **(relieved 8/1/77)**
Bartsits, P.Kh. **(relieved 10/1/76)**
Bazarbaev, M. **(relieved 31/7/81)**
Bazarov, A.U. **(relieved 1981)**
Beloblotskii, M.P. **(relieved 11/1/79)**
Beman, E.K. **(relieved 20/3/80)**
Berdyev, K. **(relieved 11/3/77)**
Berezhnoi, I.N. **(relieved 8/4/80)**
Beschastnov, A.D. **(relieved 14/7/76)**
Bezruk, P.F. **(relieved 5/10/78)**
Bleskov, A.A. **(relieved)**
Boloshko, L. **(relieved 1980)**
Bolshak, V.G. **(relieved 10/12/79)**
Borodawko, I.K. **(relieved 14/2/79)**
Bratus', V.D. **(relieved 25/4/75)**
Briushkov, I.V. **(relieved 9/10/79)**
Bukharbaev, T.K. **(relieved 28/7/77)**
Burmistrov, A.A. **(relieved 27/3/80)**
Burmistrov, F.P. **(relieved 26/7/75)**
Bushuev, V.M. **(relieved 31/10/77)**
Bychkovskaia, Z.M. **(relieved March 1980)**
Chaika, E.M. **(relieved April 1982)**
Charyev, B.Ch. **(dismissed 1/7/77)**
Cherednichenko, E.T. **(relieved 1976)**
Chernyi, V.I. **(dismissed March 1978)**
Chervanev, D.L. **(relieved March 1980)**
Cherviakova, R.I. **(relieved 11/1/79)**
Chikvaidze, A.D. **(relieved 5/11/78)**
Churkin, A.N. **(dismissed 15/4/75)**
Chvertko, P.V. **(relieved 19/12/75)**
Davidchik, A.L. **(relieved 18/4/79)**
Davydov, N.I. **(relieved 6/3/76)**
Denisenko, A.I. **(relieved 13/10/76)**
Denisova, R.N. **(relieved 6/3/76)**
Diakov, V.A. **(relieved 14/2/76)**
Dobynde, I.G. **(relieved 2/4/80)**
Dodor, A.P. **(relieved 4/3/81)**
Dolchanmaa, B.-K.S. **(relieved 1980)**
Dolgoshei, G.A. **(relieved 12/1/77)**
Duchidze, G.A. **(relieved 8/7/76)**
Durdyev, A. **(relieved 1976)**
Durdyev, A.A. **(relieved 19/12/80)**
Durdyev, K. **(relieved 22/4/77)**
Dzenitis, I.E. **(relieved 22/3/80)**
Dzhergalbaev, U. **(relieved 17/9/76)**
Dzhibladze, G.N. **(relieved 20/1/80)**
Dzhumaev, A. **(relieved 18/4/79)**
Dzhumanaliev, K. **(relieved 24/12/78)**
Dzhuraev, K.Sh. **(relieved 26/3/80)**
Efimov, S.V. **(relieved 1980)**
Ekhala, E.S. **(relieved 19/3/80)**
Erme, R. **(relieved 9/9/77)**
Ermolaev, S.A. **(relieved 4/3/81)**
Esbulatov, M. **(relieved 19/1/80)**
Esenov, R.M. **(relieved 15/11/79)**
Eshimbetov, D. **(relieved 15/3/80)**
Evdokimenko, G.S. **(relieved 5/11/75)**
Fakhrutdinov, F. **(relieved 9/2/78)**
Fedan, V.I. **(relieved 6/7/75)**
Fedorov, A.F. **(relieved 19/3/79)**
Filimonov, V.P. **(relieved 4/3/81)**
Filippov, I.F. **(relieved Aug 1979)**
Fominykh, A.M. **(relieved 4/3/81)**
Frolov, M.I. **(relieved 13/12/79)**
Furs, I.L. **(relieved 6/3/76)**
Gabidulin, A.K. **(relieved 13/7/82)**
Gabrielliants, G.A. **(relieved 11/1/77)**
Gafurzhanov, E.G. **(relieved 22/3/82)**
Galanshin, K.I. **(relieved)**
Galimzhanov, K.G. **(relieved 3/7/80)**
Garipov, R.K. **(dismissed 13/2/76)**
Gasparian, G.A. **(relieved 21/3/80)**
Geldyev, B. **(relieved 25/8/76)**
Gettuev, M.I. **(relieved 1980)**
Golubkov, V. **(relieved 8/9/79)**
Goncharov, V.I. **(relieved 6/3/76)**
Gorbach, F.S. **(relieved 6/3/76)**
Goregliad, A.A. **(relieved 1976)**
Gorovoi, N.I. **(relieved 1980)**
Gorozhaeva, R.F. **(relieved 4/3/81)**
Greichiuvene (Greičuviene), I.P. **(relieved 10/4/80)**
Gusein-zade, A.M.A.K. o. **(relieved 3/11/78)**
Guseinova, Z.I. k. **(relieved 6/2/82)**
Iakhshibaev, U. **(relieved 20/10/79)**
Iaroshovets, V.S. **(relieved April 1978)**
Ibragimov, I.A. o. **(relieved July 1978)**
Ildyrym-Zade, D.S. o. **(relieved 25/1/76)**
Isaev, B.V. **(relieved 16/1/82)**
Isaev, N.A.K. o. **(relieved 17/6/76)**
Ishchenko, A.I. **(relieved 24/4/75)**
Ismailov, A. **(relieved 30/12/77)**
Israilova, Z. **(relieved 15/3/80)**
Iurisoo, U.B. **(relieved 27/5/76)**
Iurna, J.-K.M. **(relieved 23/11/78)**
Iurnaste, Iu.K. **(relieved 4/2/77)**
Iushkis, E.I. **(relieved 14/6/79)**
Iusupova, I.T. **(relieved 23/6/76)**
Ivannikova, M.S.
Ivashchuk, D.I. **(relieved 16/5/81)**
Izotov, V.D. **(relieved 6/1/81)**
Kaarma, Ia. **(relieved 8/10/76)**
Kabaloev, B.E. **(dismissed 16/1/82)**
Kabulova, T.S. **(relieved 1981)**
Kambarov, V.I. **(relieved 24/1/82)**
Kanimetov, A. **(relieved 14/1/77)**
Karba, O.N. **(relieved 28/3/80)**
Kasimov, A. **(relieved 30/1/80)**
Kavtarov, P.I. **(relieved 23/2/78)**
Kazakbaev, K. **(relieved 6/2/82)**
Ketebaev, K. **(relieved 21/1/77)**
Khakimov, A.K. **(relieved 2/7/75)**
Khaknazarov, S. **(relieved 13/2/81)**
Khalmatov, I. **(relieved 15/3/80)**
Khalmuradov, P. **(relieved 15/3/80)**
Khanaev, K. **(relieved 11/10/79)**
Khertsa, K.G. **(relieved 29/1/78)**
Khilimonchik, V.P. **(relieved 18/12/79)**
Khintba, V.M. **(relieved 20/4/78)**
Khodzhakuliev, G. **(relieved 23/3/80)**
Khodzhaniiazov, M.N. **(relieved 6/7/76)**
Khodzhibaev, A. **(relieved 11/2/77)**
Khramtsov, A.I. **(relieved 4/3/81)**
Kim, N.L. **(relieved 31/7/79)**
Kirilenko, A.P. **(relieved 22/11/82)**
Kirillin, V.A. **(relieved)**
Kiselev, I.P. **(relieved 5/1/79)**
Kliger, E.I. **(relieved 30/1/82)**
Kobuliia, G.S. **(relieved 31/3/77)**
Kobylchak, M.M. **(relieved 28/2/82)**
Kochetkov, N.G. **(relieved 4/3/81)**
Kodua, M.G. **(dismissed 10/7/76)**
Kolesov, O.A. **(relieved 14/12/78)**
Konduchalova, K. **(relieved 21/3/80)**
Konstantinov, V.Z. **(relieved 7/2/82)**
Korotkov, B.F. **(relieved 6/3/76)**
Kortunov, V.V. **(relieved 4/3/81)**
Kostin, I.I. **(relieved 12/1/78)**
Kotandzhian, G.S. **(relieved 1981)**
Kozhomkulov, A. **(relieved 31/7/79)**
Kramarev, V.G. **(relieved 6/12/80)**
Krasilnikov, V.S. **(relieved 18/6/80)**
Kudratov, S. **(relieved 6/3/76)**
Kudriavtsev, S.V. **(relieved Aug 1978)**
Kuznetsov, A.N. **(relieved 5/6/82)**
Kuznetsov, Iu.N. **(relieved 19/1/80)**
Kuznetsov, V. **(relieved 5/9/81)**
Labakhua, A.M. **(relieved 19/9/76)**
Lapshin, A.F. **(relieved 11/5/77)**
Lazarev, A.M. **(relieved 12/4/80)**
Lazebnyi, N.S. **(relieved 7/10/78)**
Lebedeva, V.S. **(relieved 4/3/81)**
Lekhepuu, E.P. **(relieved 7/9/76)**
Lenshin, A.T. **(relieved 1/3/78)**
Leonov, N.N. **(relieved 26/3/82)**
Leosk, M. **(relieved May 1978)**
Levin, G.M. **(relieved 1980)**
Liashchenko, N.G. **(relieved)**

Lopato, B.I. **(relieved 6/2/82)**
Maisuradze, V.I. **(relieved 22/3/76)**
Maksimov, L.G. **(relieved 22/1/80)**
Malanchuk, V.I. **(relieved 27/4/79)**
Mamaeva, T.N. **(relieved 1980)**
Mamedzade, R.G. **(dismissed 20/7/78)**
Marinich, A.M. **(relieved 4/7/79)**
Matchanov, N.M. **(relieved)**
Matskevich, V.V. **(relieved Feb 1980)**
Mekhti-zade, M.M. o. **(relieved 29/3/80)**
Melkov, J.D. **(relieved 20/12/73)**
Melnikov, O.S. **(relieved 21/9/77)**
Melnikov, P.V. **(relieved 26/2/78)**
Melnikova, Z.V. **(relieved 6/3/76)**
Meshchaninov, B.B. **(relieved 13/10/79)**
Mikhailov, G.L. **(relieved)**
Moldasanov, Z. **(relieved 4/3/81)**
Molotievskii, L.A. **(dismissed 4/4/79)**
Movlamov, S. **(relieved 11/4/78)**
Mozhaiskii, V.T. **(relieved 2/3/78)**
Mstoian, M.A. **(relieved 21/3/80)**
Muidinov, M. **(relieved 4/3/81)**
Mukasheva, K. **(relieved 29/3/80)**
Mukhortov, M.D. **(relieved 8/12/79)**
Muradian, B.A. **(relieved 6/3/76)**
Musabekov, F.S. o. **(relieved 17/6/76)**
Nadzhimov, G. **(relieved 29/4/81)**
Nasreddinov, S.K. **(relieved 26/2/78)**
Nasyrov, K.M. **(relieved 21/12/78)**
Nauialis, S.S. **(relieved 7/12/82)**
Nazarov, M.N. **(relieved 17/11/79)**
Negrutsa, V.B. **(relieved 5/9/76)**
Norak, A.A. **(relieved 15/7/79)**
Nordmann, E.B. **(relieved 19/12/78)**
Novoselov, V.D. **(relieved 15/3/80)**
Novozhilova, Z.G. **(relieved Feb 1981)**
Nurmukhamedov, R. **(relieved 27/1/81)**
Nuzhnyi, V.I. **(relieved 21/3/80)**
Oganesian, O.G. **(relieved 10/6/82)**
Oganesian, V.M. **(relieved 22/5/78)**
Ol'kov, N.P. **(relieved 16/10/79)**
Orazmukhamedov, O.N. **(dismissed 19/12/75)**
Orazov, N. **(relieved 14/12/79)**
Osmokesku, R.G. **(relieved 29/1/78)**
Ovchinnikova, A.I. **(relieved 1980)**
Paalmann, E.R. **(relieved 29/3/80)**
Pachuashvili, M. **(relieved 11/12/80)**
Panchenko, G.P. **(relieved 12/4/80)**
Pasamanova, E.N. **(relieved 14/3/80)**
Pavlovskii, I.G. **(relieved 29/11/82)**
Peller, V.I. **(relieved 6/3/76)**
Peregudova, V.A. **(relieved 10/4/80)**
Pereudin, V.M. **(relieved 21/12/80)**
Petrosiants, V.A. **(relieved 10/2/79)**
Petrova, L.A. **(relieved 6/3/76)**
Petrovskii, B.V. **(relieved 23/12/80)**
Piotrovskii, Iu.M. **(relieved 19/2/76)**
Pitenkov, N.A. **(relieved 23/9/82)**
Piuskiulian, Z.E. **(dismissed 12/5/76)**
Plakhov, V.B. **(relieved 17/3/76)**
Platonov, K.M. **(relieved Nov 1980)**
Plaude, A.K. **(relieved 30/12/80)**
Podgornyi, N.V. **(disgraced May 1977)**
Podoinikova, N.I. **(relieved 29/3/80)**
Podrez, V.I. **(relieved 7/2/76)**
Ponomarev, V.A. **(relieved 24/6/76)**
Popova, N.V. **(relieved 6/3/76)**
Posokhin, V.S. **(relieved March 1980)**
Pulatov, E.P. **(relieved 6/3/76)**
Pulatov, S. **(relieved 7/4/79)**
Pulatova, K.M. **(relieved 15/3/80)**
Purga, A.P. **(relieved 4/1/76)**
Radomskii, V.I. **(relieved 3/2/76)**
Radzhabov, S. **(relieved 29/10/78)**
Ragozin, A.P. **(relieved 20/1/79)**
Ramazanov, A.G. **(relieved 10/10/82)**
Reikhmanis, G.O. **(relieved 30/6/77)**
Romanovskii, A.K. **(relieved 8/6/77)**
Rozmanas, A.V. **(relieved 30/12/77)**
Rudnev, O.A. **(relieved 11/4/82)**
Rybakova, G.I. **(relieved 4/3/81)**
Sadykov, V. **(relieved 5/12/79)**
Safarmamedov, A. **(relieved 20/3/80)**
Safarov, R.R. **(relieved 23/12/81)**
Safonov, D.F. **(relieved Nov 1977)**
Saidmamatov, U.A. **(relieved 1981)**
Sakhatliev, T. **(relieved 21/7/78)**
Sakhatmuradov, K. **(relieved 23/11/80)**
Salnikova, E.A. **(relieved 6/3/76)**
Sarkisov, G.M. **(relieved 3/2/77)**
Sarsenova, D.Z. **(relieved 29/3/80)**
Satorov, A. **(relieved 12/9/81)**
Savchenko, B. **(relieved July 1977)**
Savchenkov, N.G. **(relieved 25/3/80)**
Sbarazskii, V.V. **(relieved 24/11/76)**
Semichastnyi, V.E. **(relieved 15/5/81)**
Sennikov, A.A. **(relieved 9/12/78)**
Serdiukov, P.T. **(relieved 5/3/77)**
Sergeev, M.I. **(relieved 4/3/81)**
Serova, A.S. **(relieved 4/3/81)**
Severnev, M.M. **(relieved 16/1/76)**
Shakhsuvarov, G.S. o. **(relieved 17/2/77)**
Shakhzadeian, M. **(relieved 7/10/77)**
Shamkhalov, Sh.M. **(relieved 1980)**
Sharipov, M. **(relieved 6/8/82)**
Sharmanov, T.S. **(relieved 17/8/82)**
Shatovskii, A.F. **(relieved 10/1/78)**
Shcherbachevich, A.F. **(relieved July 1975)**
Shelepin, A.N. **(relieved 23/5/75)**
Shengeliia, G.S. **(relieved 21/2/76)**
Sheremetov, A.S. **(relieved 6/3/76)**
Shevchenko, A.F. **(relieved 4/3/81)**
Shevchenko, V.T. **(relieved 31/3/ 82)**
Shevchenko, V.V. **(dismissed 15/12/73)**
Shevchuk, G.I. **(relieved 26/6/76)**
Shevelukha, V.S. **(relieved 7/4/79)**
Shibalov, A.N. **(relieved 4/3/81)**
Shodyeva, Z. **(relieved 29/3/80)**
Shpak, P.F. **(dismissed 14/7/82)**
Sirazhdinov, S.Kh. **(relieved 15/3/80)**
Sitnin, V.K. **(relieved 27/7/74)**
Smirnov, V.A. **(relieved 6/3/76)**
Starovoitov, V.K. **(relieved 14/3/80)**
Stepakov, V.I. **(relieved)**
Steshov, B.A. **(relieved 28/10/77)**
Stetsenko, S.O. **(relieved 1975)**
Struzhuk, E.F. **(relieved 10/4/80)**
Sychev, G.A. **(relieved 10/4/80)**
Sysoev, P.P. **(relieved 1980)**
Tairov, S.M. **(relieved 9/2/78)**
Tarasov, M.P. **(relieved 15/3/80)**
Tartyshev, N.N. **(relieved 26/4/75)**
Tarushkin, A.P. **(relieved 24/11/76)**
Tashiev, K. **(relieved 4/2/76)**
Tashpulatova, D. **(relieved 18/4/79)**
Tatarchuk, P.M. **(relieved 14/9/76)**
Telkov, N.T. **(relieved 25/3/80)**
Terlemesian, E.O. **(relieved 29/1/82)**
Tikk, V.A. **(relieved 29/3/80)**
Timofeev, N.V. **(relieved 31/10/80)**
Titberidze, Sh.A. **(relieved 5/4/79)**
Troian, N.V. **(relieved 14/5/76)**
Tron'ko, P.T. **(relieved 3/4/78)**
Truusa, K.A. **(relieved 4/2/77)**
Tsagelnik, A.K. **(relieved 14/3/80)**
Tsetsegov, S.S. **(relieved 18/4/79)**
Tsimakuridze, K.S. **(relieved 28/3/80)**
Tsitsishvili, I.N. **(relieved 3/2/78)**
Tumenbaeva, D. **(relieved 6/8/81)**
Tupitsyn, N.K. **(relieved March 1975)**
Tur, V.Z. **(relieved 4/3/81)**
Tynel, L.G. **(relieved 18/4/79)**
Tynurist, E.G. **(relieved 18/1/79)**
Ubilava, I.D. **(relieved 22/9/78)**
Udalov, A.P. **(relieved 4/3/81)**
Uguzhakov, V.A. **(relieved 1980)**
Ulfsak, I.K. **(relieved 6/2/82)**
Ulitina, L.V. **(relieved 4/3/81)**
Umarova, M. **(relieved 4/3/81)**
Usmanov, S.N. **(relieved 6/3/76)**
Usmanov, U.G. **(relieved 25/3/80)**
Utegaliev, I.M. **(relieved 17/7/80)**
Uvachan, V.N. **(relieved 4/3/81)**
Vakhidova, M. **(relieved 13/2/77)**
Valing, O.I. **(relieved 12/9/79)**
Vartanian, G.O. **(dismissed 17/1/80)**
Velimamedov, S.M. **(relieved 1981)**
Vertikov, D.A. **(relieved 13/7/77)**
Vikhvelin, L.A. **(relieved 23/1/80)**
Vladimirov, V.I. **(relieved 15/10/75)**
Volkov, A.P. **(relieved 27/7/74)**
Volodkevich, I.Kh. **(relieved 26/2/82)**
Volovchenko, I.P. **(relieved Dec 1977)**
Voronin, A.L. **(relieved of position as Ambassador 7/6/78)**
Zagorskii, V.N. **(relieved 20/2/82)**
Zaichenko, N.M. **(relieved 5/7/73)**
Zairov, K.S. **(relieved 9/1/79)**
Zakharov, M.E. **(relieved 6/3/76)**
Zaripova, N. **(relieved 26/3/80)**
Zaritskii, E.E. **(relieved 18/1/79)**
Zbaraszkii, V.V. **(relieved 9/1/81)**
Zedginidze, I.Sh. **(relieved 20/1/78)**
Zeinalov, G.I. **(dismissed 21/6/79)**
Zhakeev, A.E. **(relieved 26/11/80)**
Zhaksybekov, S.Sh. **(relieved 18/4/79)**
Zhitenev, V.A. **(relieved April 1978)**
Zhorzhikashvili, G.D. **(relieved 25/7/76)**
Zhvirble (Žvirble), S.Ia. **(relieved 11/2/78)**
Zubrilin, A.F. **(relieved 20/3/82)**
Zuev, G.G. **(relieved 25/3/82)**